OBJECTIVE NCERT PUNCH 4.0

MCQs Extracted from NCERT Line by Line

AIIMS PYQs along with NEET to level up

CHEMISTRY

UPDATED WITH HIGH ORDER TIME INTENSIVE MCQs
AS PER 2025 NEET PAPER

Pankaj Sijairya

EDITION: Fourth

Published By: Physicswallah Limited

Physics Wallah Publication

ISBN: 978-93-6897-850-3

MRP: 749/-

Edited by : Pankaj Sijairya

Mobile App: Physics Wallah (Available on Play Store)

Website: www.pw.live

Youtube Channel: Physics Wallah - Alakh Pandey
Competition Wallah
NCERT Wallah

Email: publication@pw.live

SKU Code: fd26d78b-63fb-4c05-8550-76d12acd411c

Printed By : Taj Press

Preface

A dedicated team at Physics Wallah (PW) works diligently to provide students with quality content for the NEET exam.

From the beginning, the whole content team comprising Associate faculties, DTP operators, Proofreaders and others are involved in shaping the material to their best knowledge and experience to produce powerful content for the students.

Faculties have adopted a new style of presenting the content in easy-to-understand language and have provided the team with their guidance and supervision throughout the creation of this Study Material.

Physics Wallah (PW) strongly believes in conceptual and fun-based learning. PW provides exam-focused content designed to offer clarity and value to students.

A plethora of NEET Study Material is available in the market but PW professionals are continuously Working to provide comprehensive study material for our NEET students.

This Study Material adopts a multi-faceted approach to master and understanding the concepts by having a rich diversity of questions asked in the examination and equip the students with the knowledge for the competitive exam.

The main objective of the study material is to provide a large number of quality problems with varying cognitive levels to facilitate the teaching-learning of concepts that are presented through the book.

It is widely used by aspirants because of its easy-to-understand language and clear illustrations with diagrams, mnemonics, flow charts and tables.

Students can benefit themselves by attempting the exercise given in this Study Material for the self-assessment and also mastering the basic techniques of problem-solving.

The questions are strictly designed in accordance with the exam relevant topics that helps to develop examination temperament within the students.

By mastering the Physics Wallah (PW) study material curated by the PW team, students can improve their chances of qualifying for the NEET exam and achieving a strong rank.

In each chapter, for better understanding, questions have been classified according to the latest syllabus of NEET

- The nature and diversity of the questions help students to ace the examination
- Quality questions to strengthen the concept of the topic at the zenith level

This Book Consists of

The following features are included in each chapter for systematic preparation:

- **NEET PYQs Analysis:** Year wise number of questions and topicwise number of questions
- **NCERT-Picks:** Provides crisp theory, formulas and statements asked in NEET & JEE-Mains exams.
- **NCERT Topic-wise MCQs:** Aligned with the latest NCERT and NEET syllabus.
- **High-Order Time Intensive MCQs:** Assists in practicing high order time consuming questions with detailed explanations.
- **Match the Columns, Correct & Incorrect, Statement-based, Assertion & Reason MCQs:** Provides a variety of question types involving conceptual understanding for practice.
- **Multi-Correct MCQs:** Includes Questions as per latest NEET 2025 exam pattern.
- **NCERT Exemplar MCQs:** Includes questions from the NCERT Exemplar with explanations.
- **NEET & AIIMS Past Year MCQs and NEET 2025 Paper:** Provides Past Years Questions for Practice.
- **Kattar NEET:** Preparing you to tackle unpredictable patterns and evolving question formats

About the Editor

Sir **Pankaj Sijairya,** from **jhansi**, is a renowned Chemistry Professor at PW with **15 yrs** of experience who received his **B.Tech** in Electronic Engineering from **HBTI Kanpur** where also he got excited about teaching chemistry to students, his way of teaching chemistry and grasp on the subject made the subject more accessible to masses. Till now he has taught more than **1.5 million** students studied and still Counting. Mentor of AIR 3, 6, 29 in JEE Advanced and 24, 28, 39, 42, 59, 67 in NEET and many more

He is recognized nationally for his teaching style and has great influence as a professor of chemistry. Pankaj Sijairya has also been invited for TEDxTALK for the difference he is making in the Education Field.

Pankaj Sir & Team together have curated the book which will help students to understand the concepts of chemistry easily and effectively. The book is designed for both **JEE and NEET** aspirants covering every topic. The book is consist of many tips and tricks as well for aspirants that they can use and save time in the examination.

Topper's Testimonials

Reading and understanding NCERT text books is a must do for NEET aspirants. Many students fail to realise the importance of NCERT. Once they realise, they find it challenging. To retain the information given in NCERT, one needs to practice a lot of questions. NCERT punch offers a variety of questions which helps in buliding grip on core concepts for NEET, boards and JEE main

Samuel Harshith
NEET AIR-24

NEET PAST YEAR WEIGHTAGE ANALYSIS

Chemistry-XI

Chapters Name	Year Wise Number of Questions															
	2025	2024 Re	2024	2023	2023 Manipur	2022	2022 Re	2021	2020	2020 Covid	2019	2018	2017	2016 II	2016 I	2015
Some Basic Concepts of Chemistry	2	1	3	1	1	1	1	1	1	1	1	2	0	1	0	0
Structure of Atom	2	2	2	2	2	2	2	1	1	1	2	2	1	2	1	1
Classification of Elements and Periodicity in Properties	2	1	2	1	3	1	2	1	1	0	1	0	1	0	1	2
Chemical Bonding and Molecular Structure	2	4	3	2	2	2	2	3	2	3	2	1	2	5	2	5
Thermodynamics	1	3	3	1	1	1	1	2	2	2	2	1	2	1	2	1
Equilibrium	3	3	3	1	2	2	2	1	3	2	3	3	3	3	2	3
Redox Reactions	1	1	1	1	1	2	1	1	1	1	2	1	0	0	1	0
The p-Block Elements	0	0	0	2	1	2	2	0	3	1	2	2	1	2	0	0
Organic Chemistry-Some Basic Principles and Techniques	3	4	4	2	2	2	3	2	2	1	0	3	3	1	0	8
Hydrocarbons	2	2	2	2	1	3	2	4	2	2	4	2	3	6	4	2

Chemistry-XII

Chapters Name	Year Wise Number of Questions															
	2025	2024 Re	2024	2023	2023 Manipur	2022	2022 Re	2021	2020	2020 Covid	2019	2018	2017	2016 II	2016 I	2015
Solutions	3	2	2	1	1	2	1	2	2	2	2	0	2	2	2	3
Electrochemistry	1	2	2	2	3	2	2	2	2	2	2	1	0	5	1	1
Chemical Kinetics	3	3	3	2	2	2	2	2	2	2	2	2	2	1	2	2
The p-Block Elements (Group 15 to 18)	2	3	2	2	0	2	2	2	2	4	3	4	3	1	5	1
The d and f-Block Elements	2	4	3	2	2	2	2	2	2	3	1	2	3	1	2	3
Coordination Compounds	3	3	3	2	2	2	2	2	1	1	1	3	3	2	1	2
Haloalkanes and Haloarenes	2	2	2	2	3	1	1	2	1	1	0	0	1	1	1	0
Alcohols, Phenols and Ethers	1	1	2	2	2	2	2	2	2	2	2	3	3	0	1	1
Aldehydes, Ketones and Carboxylic Acids	3	2	3	4	4	4	4	2	2	2	1	2	2	1	3	1
Amines	4	3	2	2	2	2	2	2	1	1	1	1	1	2	1	1
Biomolecules	2	1	1	1	1	1	1	1	2	2	1	2	0	2	3	0
Principles Related to Practical Chemistry	1	3	2	–	–	–	–	–	–	–	–	–	–	–	–	–

Contents

CLASS-XI

CLASS-XII

Class-XI

CHAPTER 1

Some Basic Concepts of Chemistry

Topicwise Number of Questions Analysis (2025-2016)

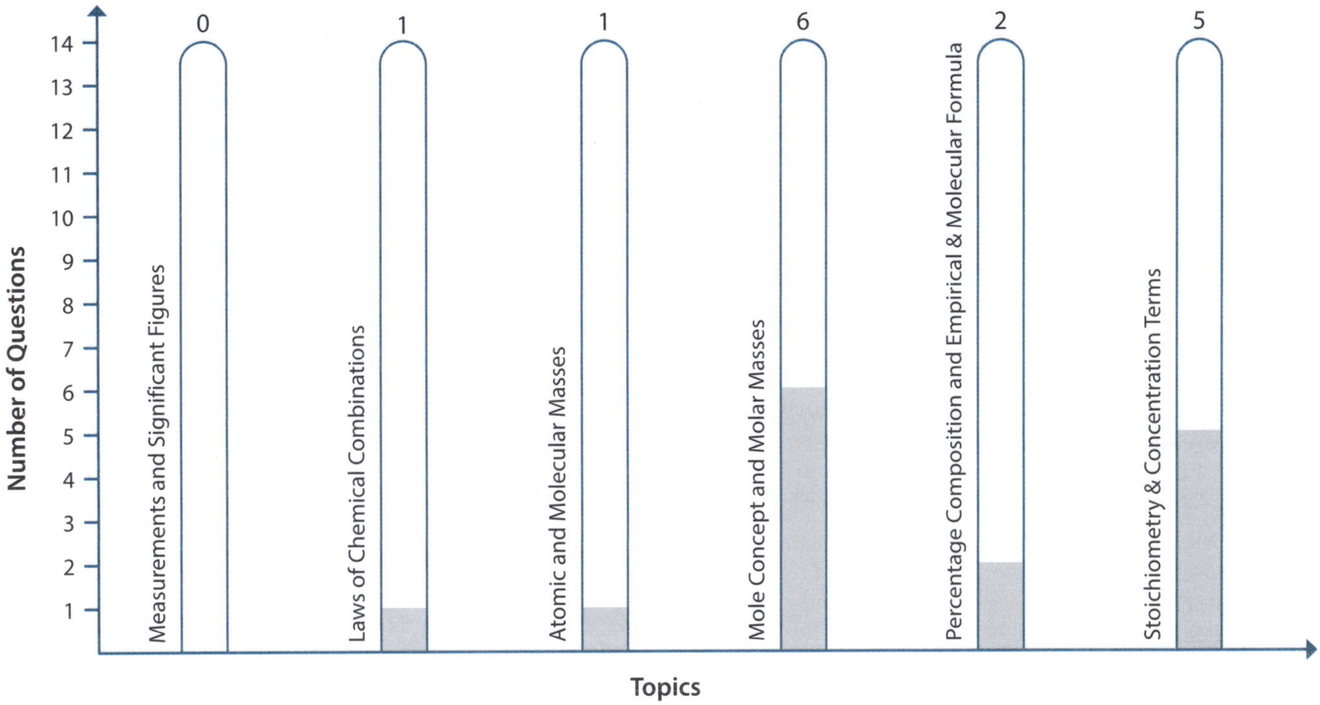

NCERT-Picks

General Introduction

Chemistry plays a central role in science and is often intertwined with other branches of science. Principles of chemistry are applicable in diverse areas, such as weather patterns, functioning of brain and operation of a computer, production in chemical industries etc. Chemistry provides methods for the isolation of lifesaving drugs from natural sources and makes possible synthesis of such drugs. Some of these drugs are cisplatin and taxol, which are effective in cancer therapy. The drug AZT (Azidothymidine) is used for helping AIDS patients. To be a good chemist and to accept new challenges, one needs to understand the basic concepts of chemistry, which begin with the concept of matter.

Matter & Its Classification

Anything which has mass and occupies space is called matter. Everything around us, for example: book, pen, pencil, water, air, all living beings, etc., is composed of matter.

States of Matter: Matter exist in three physical states viz. solid, liquid and gas. Because of different arrangement of particles, states of matter exhibit the following characteristics:

1. Solids have definite volume and definite shape.
2. Liquids have definite volume but do not have definite shape. They take the shape of the container in which they are placed.
3. Gases have neither definite volume nor definite shape. They completely occupy the space in the container in which they are placed.

These three states of matter are interconvertible by changing the conditions of temperature and pressure.

$$\text{Solid} \underset{\text{cool}}{\overset{\text{heat}}{\rightleftharpoons}} \text{liquid} \underset{\text{cool}}{\overset{\text{heat}}{\rightleftharpoons}} \text{Gas}$$

Usually a solid changes to a liquid, and the liquid on further heating changes to gas (or vapour). But when solid directly changes to the vapour state, the process is called **Sublimation**.

Classification of Matter: At the macroscopic level matter is classified as, mixture and pure substances. These are further classified as,

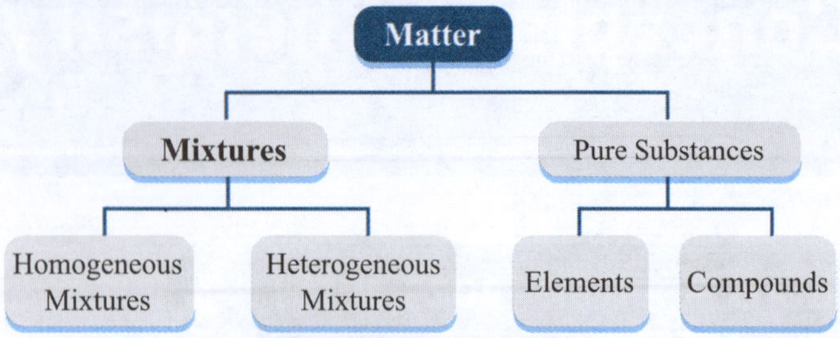

Classification of matter

In **a homogeneous mixture**, the components completely mix with each other. Sugar solution and air are the examples of homogeneous mixtures.

In contrast to this, **in a heterogeneous mixture**, the composition is not uniform throughout and sometimes different components are visible. For example, mixtures of salt and sugar, grains and pulses along with some dirt.

In **pure substances** constituent particles have fixed composition. Copper, silver, gold, water and glucose are some examples of pure substances. Pure substances can further be classified into **elements** and **compounds**. Particles of an **element** consist of **only one type of particles**. These particles may exist as **atoms or molecules**.

Sodium (Na), copper (Cu), silver (Ag), hydrogen (H), oxygen (O), etc. are some examples of elements. Whereas, Hydrogen (H_2), nitrogen (N_2) and oxygen (O_2) gases consist of molecules, in which two atoms combine to give their respective molecules.

When two or more atoms of different elements combine together in a definite ratio, the molecule of a **compound** is obtained. Examples of some compounds are water (H_2O), ammonia (NH_3), carbon dioxide (CO_2), sugar etc.

Measurement of Properties of Matter

The SI system has **seven** base units and they are listed in the given Table.

Base Physical Quantity	Symbol for Quantity	Name of SI Unit	Symbol for SI Unit
Length	l	metre	m
Mass	m	kilogram	kg
Time	t	second	s
Electric current	I	ampere	A
Thermodynamic temperature	T	kelvin	K
Amount of substance	n	mole	mol
Luminous intensity	I_v	candela	cd

The SI system allows the use of prefixes to indicate the multiples or submultiples of a unit. These prefixes are listed in following Table.

Multiple	Prefix	Symbol
10^{-15}	femto	f
10^{-12}	pico	p
10^{-9}	nano	n
10^{-6}	micro	m
10^{-3}	milli	m
10^{-2}	centi	c
10^{-1}	deci	d
10	deca	da
10^{2}	hecto	h
10^{3}	kilo	k
10^{6}	mega	M
10^{9}	giga	G
10^{12}	tera	T
10^{15}	peta	P

The SI unit of mass is kilogram. However, its fraction named as gram (1 kg = 1000 g), is used in laboratories due to the smaller amount of chemicals used in chemical reactions.

In SI system, volume has units of m^3. But again, in chemistry laboratories, smaller volumes are used. Hence, volume is often denoted in cm^3 or dm^3 units. A common unit, litre (L) which is not an SI unit, is used for measurement of volume of liquids.

$1\ L = 1000\ mL,\ 1000\ cm^3 = 1\ dm^3$

There are three common scales to measure temperature — °C (degree celsius), °F (degree fahrenheit) and K (kelvin). Here, K is the SI unit. The temperatures on two scales are related to each other by the following relationship:

$°F = 9/5\ (°C) + 32.$

The kelvin scale is related to celsius scale as follows:

$K = °C + 273.15$

Uncertainty in Measurement

Significant Figures: The uncertainty in the experimental or the calculated values is indicated by mentioning the number of significant figures. Significant figures are meaningful digits which are known with certainty plus one which is estimated or uncertain.

Rules for determining the number of significant figures:

1. All non-zero numbers are significant. For example in 421 cm, there are three significant figures.
2. Zeros preceding to first non-zero digit are not significant. Such zero indicates the position of decimal point. For example in 0.04 has only one significant figure and 0.0052 has only two significant figures.
3. Zeros between two non-zero digits are significant. Thus, 5.005 has four significant figures.
4. Zeros at the end or right of a number are significant, provided they are on the right side of the decimal point. For example, 0.200 g has three significant figures. Terminal zeros are not significant if there is no decimal point. For example, 400 has only one significant figure, but 400. has three significant figures and 400.0 has four significant figures. Such numbers are better represented in scientific notation. We can express the number 400 as 4×10^2 for one significant figure, 4.0×10^2 for two significant figures and 4.00×10^2 for three significant figures.
5. Counting the numbers of object, For example, 2 balls or 20 eggs, have infinite significant figures. As these are exact numbers and can be represented by writing infinite number of zeros after placing a decimal i.e., 2 = 2.000000 or 20 = 20.000000

Laws of Chemical Combinations

Law of Conservation of Mass: In all physical and chemical changes, there is no net change in mass during the process. In other words it states that matter can neither be created nor Destroyed.

Law of Definite Proportions: Irrespective of the source, a given compound always contains same elements combined together in the same proportion by mass. This law is also known as Law of definite Composition.

Law of Multiple Proportions: When two elements combine to form two or more compounds then the masses of one of the elements which combine with a fixed mass of the other element, are in the ratio of small whole numbers. For example, water and hydrogen peroxide are two compounds which are obtained by the combination of hydrogen and oxygen but in both the compounds mass of hydrogen is fixed i.e. 2g whereas, mass of oxygen is 16g and 32g respectively. Means mass of oxygen bears a simple ratio i.e. 1:2.

Gay Lussac's Law of Gaseous Volumes: Gay Lussac in 1808 observed that when gases combine or are produced in a chemical reaction they do so in simple ratio by volume (at same T and P). This law was explained properly by the work of Avogadro in 1811.

Avogadro's Law: In 1811, Avogadro proposed that equal volumes of all gases at the same temperature (T) and pressure (P) should contain equal number of molecules. Avogadro made a distinction between atoms and molecules.

Dalton's Atomic Theory

In 1808, Dalton published 'A New System of Chemical Philosophy', in which he proposed the following :

1. Matter consists of indivisible atoms.
2. All atoms of a given element have identical properties, including identical mass. Atoms of different elements differ in mass.
3. Compounds are formed when atoms of different elements combine in a fixed ratio.
4. Chemical reactions involve reorganisation of atoms. These are neither created nor destroyed in a chemical reaction.

Note: Dalton's Atomic theory could explain law of conservation of mass, law of multiple proportion and law of constant proportion. However, it could not explain law of gaseous volumes

Atomic & Molecular Masses

Atomic mass is actually very-very small because atoms are very small. In the present system atomic mass is based on the carbon-12 isotope as the standard. Carbon-12 is assigned a mass of exactly 12 atomic mass unit (amu) and masses of other atoms are given relative to this standard.

One **atomic mass unit** is defined as a mass exactly equal to one-tweleth of mass of one carbon-12 atom.

And 1 amu = 1.66056×10^{-24} g

Mass of an atom of hydrogen = 1.6736×10^{-24} g

Thus, in terms of amu, the mass

$$\text{of hydrogen atom} = \frac{1.6736\times10^{-24}\,g}{1.66056\times10^{-24}\,g}$$

$$= 1.0078 \text{ amu} = 1.0080 \text{ amu}$$

Similarly, the mass of oxygen –16(^{16}O) atom would be 15.995 amu.

At present '**amu**' has been replaced by '**u**', which is known as unified mass.

Average atomic mass: It is the sum of the products of fractional abundances (f_i) of the isotopes and their corresponding mass number (A_i), i.e..

$A = \Sigma f_1 A_1 = A_1 \times f_1 + f_2 \times A_2 + ...$

Carbon has the following three isotopes with relative abundance and masses as shown against each of them.

Isotope	Relative Abundance (%)	Atomic Mass (amu)
^{12}C	98.892	12
^{13}C	1.108	13.00335
^{14}C	2×10^{-10}	14.00317

From the above data, the average atomic mass of carbon will come out to be:

$(0.98892)(12\text{ u}) + (0.01108)(13.00335\text{ u}) + (2 \times 10^{-12})(14.00317\text{ u})$
$= 12.011\text{ u}$

Molecular Mass: It is the sum of atomic masses of the elements present in a molecule. It is obtained by multiplying the atomic mass of each element by the number of its atoms and by adding them together. For example:

Molecular mass of methane,

$(CH_4) = (12.011\text{ u}) + 4\,(1.008\text{ u}) = 16.043\text{ u}$

relation between molecular mass and vapour density:

Molecular mass = 2 × vapour density

Molecular mass = Mass of 22.4 L of vapour of STP

$$V.D = \frac{\text{Density of a gas at certain temp. and pressure}}{\text{Density of hydrogen at same temp. and pressure}}$$

Mole Concept and Molar Mass

The mole, **(symbol mol)**, is the SI unit of amount of substance. One mole contains exactly $6.02214076 \times 10^{23}$ elementary entities. This number is the fixed numerical value of the Avogadro constant, N_A, when expressed in the unit mol^{-1} and is called the Avogadro number.

- 1 mole of atoms : Gram atomic mass (or 1g atom)
  : 6.022×10^{23} atoms.
- 1 mole of molecules : Gram molecular mass (or 1g molecule)
  : 6.022×10^{23} molecules
  : 22.4 L at STP
- 1 mole of ionic compound : Gram formula mass
  : 6.022×10^{23} formula units
- Molar mass of H_2O = 18.02 g/mol

Important Formulas

- $\text{Number of moles} = \dfrac{\text{Given Mass}}{\text{Atomic or Molar mass}}$
- $\text{Eq. Wt. of an ion/radical} = \dfrac{\text{Formula Weight}}{\text{total charge}}$
- $\text{Eq. Wt. of acids} = \dfrac{\text{Molar Mass}}{\text{Basicity}}$

- Eq. Wt. of bases $= \frac{\text{Molar Mass}}{\text{Acidity}}$
- Eq. Wt. of Salts $= \frac{\text{Molar Mass}}{\text{Net Positive or Negative charge}}$

Empirical & Molecular Formula

An **empirical formula** represents the simplest whole number ratio of various atoms present in a compound, whereas, the **molecular formula** shows the exact number of different types of atoms present in a molecule of a compound. The steps to find molecular formula from empirical formula is given as,

Step 1: Conversion of mass per cent to grams

Step 2: Convert into number of moles of each element

Step 3: Divide each of the mole values obtained above by the smallest number amongst them

Step 4: Write down the empirical formula by mentioning the numbers after writing the symbols of respective elements

Step 5: Writing molecular formula

In final step divide molar mass or (molecular mass) by empirical formula mass as,

$$\frac{\text{Molecular Mass}}{\text{Empirical Formula Mass}} = n$$

Molecular formula = n × Empirical formula

or Molecular formula = $(\text{Empirical formula})_n$

Stoichiometric Calculations

Limiting Reagent: The reactant that limits the amount of product is called the limiting reactant or limiting reagent.

Reactants not completely consumed are called excess reactants.

It can be explained by the following example,

In the preparation of ammonia:

$N_2(g) + 3H_2(g) \rightarrow 2NH_3\,(g)$

When one mole of N_2 reacts with three moles of H_2, two moles of NH_3 are produced. What would happen if two moles of N_2 reacted with three moles of H_2? Would more than two moles of NH_3 be formed? Answer is given below.

Before the reaction takes place, nitrogen and hydrogen are present in a 1:3 molar ratio. The reaction takes place according to the balanced equation. One mole of N_2 reacts with three moles of H_2 to produce two moles of NH_3. At this point, all the hydrogen has been used up, and the reaction stops. One mole of unreacted nitrogen is left in addition of the two moles of NH_3 that have been produced by the reaction.

In this reaction, only the hydrogen is completely used up. It is the limiting reagent, or the reagent that determines the amount of product that can be formed by a reaction. The reaction occurs only until the limiting reagent is used up. By contrast, the reactant that is not completely used up excess reagent because some nitrogen will remain unreacted.

Sometimes in stoichiometric problems, the given quantities of reactants are expressed in units other than moles. In such cases, the first step in the solution is to convert each reactant to moles. Then the limiting reagent can be identified. The amount of product formed in a reaction can be determined from the given amount of limiting reagent.

Methods of Expressing the Concentration of a Solution

The concentration of a solution can be expressed in a number of ways. The important methods are:

1. **Mass percentage or percent by mass:** It is defined as the amount of solute in grams present in 100 grams of the solution.

$$= \frac{\text{mass of solute}}{\text{mass of solute + mass of solvent}} \times 100$$

2. **Mole fraction:** This method is used when the solution is constituted by mixing two or more components.

Mole Fraction = The ratio of number of moles of one component to the total number of moles of the solution (i.e all the components)

If a substance 'A' dissolves in substance B and their number of moles are n_A and n_B respectively then the mole fraction of $A(x_A)$ and $B(x_B)$ given as-

$$x_A = \frac{n_A}{n_A + n_B},\ x_B = \frac{n_B}{n_A + n_B}$$

$x_A + x_B = 1$

In a binary solution,

Mole fraction of solute + Mole fraction of solvent = 1

Mole fraction is independent of temperature.

3. **Molality:** It is defined as number of moles of the solute dissolved in 1 kg of the solvent. It is denoted by 'm'. Its unit is mol kg^{-1}. It is also called molal.

$$\text{Molality (m)} = \frac{\text{moles of solute}}{\text{weight of solvent (in kg)}}$$

4. **Molarity (Molar concentration):** It is defined as the number of moles of the solute per liter or per dm^3 of the solution. The unit of molarity is mol $liter^{-1}$ or mol dm^{-3}.

$$\text{Molarity (M)} = \frac{\text{moles of solute}}{\text{volume of solution (L)}}$$

Representation	Molarity	Common name
1M	1	Molar solution
0.5 M or M/2	1/2	Semimolar solution
0.1 M or M/10	1/10	Decimolar solution
0.01M or M/100	1/100	Centimolar solution
0.001 M or M/1000	1/1000	Millimolar solution

Molarity is dependent on volume and temperature. It decreases with increase of temperature, due to increase in temperature the water expands that is why the volume increases hence molarity decreases.

5. **Normality:** It is defined as the number of gram equivalents of solute present per litre of solution. It is denoted by 'N'.

Or Normality × Volume of solution in litres

= Number of gram equivalents of the solute.

Relationship between normality and molarity:

Normality = n-factor (acidity or basicity) × Molarity

- Molality of a solution does not change with temperature as it involves masses of the solute and solvent. Normality and molarity change with temperature because they involve volumes which change with temperature.

MCQs

NCERT Topic-wise MCQs

Uncertainty in Measurement

1. The number of significant figures in value 5.041 is
   1) 5 2) 2
   3) 3 4) 4

2. Express the result of (0.582 + 324.65) to the appropriate number of significant figures:
   1) 325.24 2) 325.23
   3) 325.2 4) 325.232

3. Two students X and Y report the mass of the same substance as 7.0 g and 7.00 g respectively, which of the following statement is correct?
   1) Both are equally accurate
   2) X is more accurate than Y
   3) Y is more accurate than X
   4) Both are inaccurate scientifically

4. The number of significant figures in value of π are:
   1) 1 2) 2
   3) 3 4) ∞

5. The correctly reported answer of the addition of 29.4406, 3.2 and 2.25 will have significant figures:
   1) 3 2) 4
   3) 2 4) 5

6. The correctly reported answer of the area of rectangle which is 12.34 cm long and 1.23 cm wide is:
   1) 15.2 m^2 2) 15.2 cm^2
   3) 15.1 cm^2 4) 15.17 cm^2

7. If an object has a mass of 0.2876 g, then find the mass of nine such objects:
   1) 2.5884 g 2) 2.5886 g
   3) 2.588 g 4) 2.5 g

Laws of Chemical Combinations

8. x L of N_2 at S.T.P. contains 3×10^{22} molecules. The number of molecules in x/2 L of ozone at S.T.P. will be:
   1) 3×10^{22} 2) 1.5×10^{22}
   3) 1.5×10^{21} 4) 1.5×10^{11}

9. Which of the following data illustrates the law of conservation of mass?
   1) 56 g of C reacts with 32 g of Oxygen to produce 44 g of CO_2
   2) 1.70 g of $AgNO_3$ reacts with 100 mL of 0.1M HCl to produce 1.435 g of AgCl and 0.63 g of HNO_3
   3) 12 g of C is heated in vacuum and on cooling, there is no change in mass
   4) 36 g of S reacts with 16 g of O_2 to produce 48 g of SO_2

10. Two elements X and Y combine in gaseous state to form XY in the ratio 1:35.5 by mass. The mass of Y that will be required to react with 2 g of X is:
    1) 7.1 g 2) 3.55 g
    3) 71 g 4) 35.5 g

11. 4.4 g of an oxide of nitrogen gives 2.24 L of nitrogen and 60 g of another oxide of nitrogen gives 22.4 L of nitrogen at S.T.P. The data illustrates:
    1) Law of conservation of mass
    2) Law of constant proportions
    3) Law of multiple proportions
    4) Law of reciprocal proportions

12. If law of conservation of mass was to hold true, then 20.8 g of $BaCl_2$ on reaction with 9.8 g of H_2SO_4 will produce 7.3 g of HCl and $BaSO_4$ equal to:
    1) 11.65 g 2) 23.3 g
    3) 25.5 g 4) 30.6 g

13. The law of conservation of mass holds good for all of the following except.
    1) All chemical reactions
    2) Nuclear reaction
    3) Endothermic reactions
    4) Exothermic reactions

14. Hydrogen and oxygen combine to form H_2O_2 and H_2O containing 5.93% and 11.2% hydrogen respectively, the data illustrates:
    1) Law of conservation of mass
    2) Law of Constant proportions
    3) Law of reciprocal proportions
    4) Law of multiple proportions

15. Two elements X (of mass 16) and Y (of mass 14) combine to form compounds A, B and C. The ratio of different masses of Y which combine with a fixed mass of X in A, B and C is 1 : 3 : 5, if 32 parts by mass of X combines with 84 parts by mass of Y in B, then in C, 16 parts by mass of X will combine with;

1) 14 parts by mass of Y 2) 42 parts by mass of Y
3) 70 parts by mass of Y 4) 84 parts by mass of Y

Atomic and Molecular Masses

16. Boron has two isotopes B^{10} and B^{11} whose relative abundances are 20% and 80% respectively. Atomic weight of Boron is

1) 10 2) 11
3) 10.5 4) 10.8

17. Insulin contains 3.4% sulphur by mass. What will be the minimum molecular weight of insulin?

1) 94.117 u 2) 1884 u
3) 941 u 4) 976 u

18. The molecular mass of glucose ($C_6H_{12}O_6$)

1) 160.8 u 2) 180.162 u
3) 216.4 u 4) 134.821 u

Mole Concept and Molar Masses

19. 1 g-atom of nitrogen represents:

1) 6.02×10^{23} N_2 molecules
2) 22.4 L of N_2 at S.T.P
3) 11.2 L of N_2 at S.T.P
4) 28 g of nitrogen

20. Which is correct for 10 g of $CaCO_3$?

1) It contains 1 g atom of carbon
2) It contains 0.3 g atoms of oxygen
3) It contains 12 g of calcium
4) It refers to 0.1 g equivalent of $CaCO_3$

21. The number of oxygen atoms present in 14.6 g of magnesium bicarbonate is:

1) 6 N_A 2) 0.6 N_A
3) N_A 4) $\frac{N_A}{2}$

22. Which of the following has the highest mass?

1) 20 g of sulphur
2) 4 mol of carbon dioxide
3) 12×10^{24} atoms of hydrogen
4) 11.2 L of helium at N.T.P.

23. If we assume 1/24 th part of mass of carbon instead of 1/12 th part of it as 1 amu., mass of 1 mole of a substance will

1) Remain unchanged 2) get doubled
3) Get halved 4) can't be predicted

24. Which among the given gases contain the equal number of molecules as that of 16 grams of oxygen?

1) 16 gm of SO_2 2) 32 grams of SO_2
3) 12 gm of O_3 4) All

25. One atom of an element weighs 1.8×10^{-22} g, its atomic mass is:

1) 29.9 u 2) 18 u
3) 108.36 u 4) 154 u

26. If H_2SO_4 ionises as $H_2SO_4 + 2H_2O \rightarrow 2H_3O^+ + SO_4^{2-}$. Then total number of ions produced by 0.1 mol H_2SO_4 will be:

1) 9.03×10^{21} 2) 3.01×10^{22}
3) 6.02×10^{22} 4) 1.8×10^{23}

27. Which of the following will not have a mass of 10 g?

1) 0.1 mol $CaCO_3$.
2) 1.51×10^{23} Ca^{2+} ions
3) 0.16 mol of CO_3^{2-} ions
4) 7.525×10^{22} Br atom

28. 2 g of oxygen contain number of atoms equal to that contained by:

1) 2 g of calcium 2) 4 g sulphur
3) 9 g nitrogen 4) 4 g sodium.

29. A person adds 1.71 gram of sugar ($C_{12}H_{22}O_{11}$) in order to sweeten his tea. The number of carbon atoms added are: (mol mass of sugar = 342).

1) 3.6×10^{22} 2) 7.2×10^{21}
3) 0.05 4) 6.6×10^{22}

30. The number of atoms present in 0.1 mole of P_4 (at. mass = 31) are:

1) 2.4×10^{24} atoms
2) Same as in 0.05 mol of S_8
3) 6.02×10^{22} atoms
4) Same as in 3.1g of phosphorus

31. Which one contains maximum number of molecules?

1) 2.5 g molecule of N_2
2) 4 g atom of nitrogen
3) 3.01×10^{24} atoms of nitrogen
4) 82 g of dinitrogen

32. Out of 1.0 g dioxygen, 1.0 g (atomic) oxygen and 1.0 g ozone, the maximum number of oxygen atoms are contained in:

1) 1.0 g of atomic oxygen
2) 1.0 g of ozone
3) 1.0 g of oxygen gas
4) All contain same number of atoms

33. What is the total number of atoms present in 25.0 mg of camphor, $C_{10}H_{16}O$?

1) 9.89×10^{19} 2) 6.02×10^{20}
3) 9.89×10^{20} 4) 2.67×10^{21}

34. If N_A is Avogadro's number, then the number of oxygen atoms in one g-equivalent of oxygen is:

1) N_A 2) $N_A/2$
3) $N_A/4$ 4) $2N_A$

35. If 224 ml. of a triatomic gas has a mass of 1g at 273 K and 1 atm pressure, then the mass of one atom is:

1) 8.30×10^{-23} g 2) 6.24×10^{-23}
3) 2.08×10^{-23} g 4) 5.54×10^{-23} g

36. The rest mass of an electron is 9.11×10^{-31} kg. Molar mass of the electron is:

1) 1.5×10^{-31} kg mol^{-1} 2) 9.11×10^{-31} kg mol^{-1}
3) 5.5×10^{-7} kg mol^{-1} 4) 6.02×10^{23} kg mol^{-1}

37. A sample of ammonium phosphate, $(NH_4)_3 PO_4$, contains 3.18 moles of hydrogen atoms. The number of moles of oxygen atoms in the sample is:

1) 0.265 2) 0.795
3) 1.06 4) 3.18

38. 7.5 grams of a gas occupy 5.8 litres of volume at STP, the gas is:

1) NO 2) N_2O
3) CO 4) CO_2

39. Which of the following samples Contain the largest number of atoms?

1) 1 g of CO_2 2) 1 g of C_8H_8
3) 1 g of C_2H_6 (g) 4) 1 g of LiF (s)

40. 4.0 g of caustic soda (NaOH) (mol mass 40) contains same number of sodium ions as are present in-

1) 10.6 g of Na_2CO_3 (mol. mass 106)
2) 58.5 g of NaCl (Formula mass 58.5)
3) 100 ml of 0.5 M Na_2SO_4 (Formula mass 142)
4) 1mol of $NaNO_3$ (mol. mass 85)

41. Total number of atoms present in 64 gm of SO_2 is-

1) $2 \times 6.02 \times 10^{23}$ 2) 6.02×10^{23}
3) $4 \times 6.02 \times 10^{23}$ 4) $3 \times 6.02 \times 10^{23}$

42. The total number of protons, electrons and neutrons in 12gm of ${}_6C^{12}$ is-

1) 1.084×10^{25} 2) 6.022×10^{23}
3) 6.022×10^{22} 4) 18

43. Number of Ca^{+2} and Cl^- ion in 111 g of anhydrous $CaCl_2$ are-

1) N_A, $2N_A$ 2) $2N_A$, N_A
3) N_A, N_A 4) None

44. The maximum volume at N.T.P. is occupied by-

1) 12.8 gm of SO_2
2) 6.02×10^{22} molecules of CH_4
3) 0.5 mol of NO_2
4) 1 gm-molecule of CO_2

45. Number of moles of water in 488 gm of $BaCl_2.2H_2O$ are- (At. mass of Ba = 137)

1) 2 moles 2) 4 moles
3) 3 moles 4) 5 moles

46. 23g of sodium will react with ethyl alcohol to give

1) 1 mole of H_2 2) 1/2 mole of H_2
3) 1 mole of O 4) 1 mole of NaOH

47. One mole of nitrogen gas has volume equal to

1) 1 litre of nitrogen at S.T.P.
2) 32 litre of nitrogen at S.T.P.
3) 22.4 litre of nitrogen at S.T.P.
4) 11.2 litre of nitrogen at S.T.P.

48. 10 grams of each O_2, N_2 and Cl_2 are kept in three bottles. The correct order of arrangement of bottles containing decreasing number of Molecules.

1) O_2, N_2, Cl_2 2) Cl_2, N_2, O_2
3) Cl_2, O_2, N_2 4) N_2, O_2, Cl_2

Percentage Composition and Empirical & Molecular Formula

49. An element A (at wt = 75) and another element B (at. wt. = 25) combine to form a compound. The compound contains 75% A by weight. The formula of the compound will be-

1) A_2B 2) A_3B
3) AB_3 4) AB

50. 60 g of a compound on analysis gave 24 g C, 4 g H and 32 g O. The empirical formula of the compound is:

1) $C_2H_4O_2$ 2) $C_2H_2O_2$
3) CH_2O_2 4) CH_2O

51. A compound made of two elements A and B are found to contain 25% A (at mass 12.5) and 75% B (at mass 37.5). The empirical formula of the compound is:

1) AB 2) AB_2
3) AB_3 4) A_3B

52. An oxide of a metal (M) contains 40% by mass of oxygen. Metal (M) has atomic mass of 24. The empirical formula of the oxide is-

1) M_2O 2) MO
3) M_2O_3 4) M_3O_4

53. 400 mg of capsule contains 100 mg of ferrous fumarate. The percentage of Fe present in the capsule is approximately: (formula of ferrous fumarate is $(CHCOO)_2Fe$).

1) 8.2% 2) 25%
3) 16% 4) Unpredictable

54. The percentage of oxygen in NaOH is:

1) 40 2) 60
3) 8 4) 10

55. A sample of pure compound contains 1.15 g of sodium, 3.01×10^{22} atoms of carbon and 0.1 mol of oxygen atom. Its empirical formula is

1) Na_2CO_3 2) $NaCO_2$
3) Na_2CO 4) Na_2CO_2

56. A hydrocarbon is composed of 75% carbon. The empirical formula of the compound is

1) CH_2 2) CH_3
3) C_2H_3 4) CH_4

57. A compound has 20% of nitrogen by weight. If one molecule of the compound contains two nitrogen atoms, the molecular weight of the compound is

1) 35 2) 70
3) 140 4) 280

58. An alkaloid contains 17.28% of nitrogen and its molecular mass is 162. The number of nitrogen atoms present in one molecule of alkaloid is

1) Five 2) Four
3) Three 4) Two

59. The molecular mass of a compound having the empirical formula (C_3H_4O) is 170 ± 5. Find the molecular formula of the compound.

1) $C_4H_3O_2$ 2) $C_6H_8O_2$
3) $C_6H_{12}O_3$ 4) $C_9H_{12}O_3$

60. A gaseous hydrocarbon on complete combustion gives 3.38 g of CO_2 and 0.690 g of H_2O and no other products. The empirical formula of hydrocarbon is:

1) CH 2) CH_2
3) CH_3 4) The data is not complete

61. Empirical formula of a compound is CH_2O and its molecular mass is 90. The molecular formula of the compound is

1) $C_3H_6O_3$ 2) $C_2H_4O_2$
3) $C_6H_{12}O_6$ 4) CH_2O

62. On analysis a certain compound was found to contain iodine and oxygen in the ratio of 254 g of iodine (at mass 127) and 80 g oxygen (at mass 16). What is the formula of compound?

1) IO 2) I_2O
3) I_5O_3 4) I_2O_5

63. 14 g of element X combine with 15g of oxygen. On the basis of this information, which of the following is a correct statement

1) The element X could have an atomic weight of 7 and its oxide formula XO
2) The element X could have an atomic weight of 14 and its oxide formula X_2O
3) The element X could have an atomic weight of 7 and its oxide is X_2O
4) The element X could have an atomic weight of 14 and its oxide is XO_2

64. Caffeine has a molecular weight of 194. If it contains 28.9% by mass of nitrogen, number of atoms of nitrogen in one molecule of caffeine is:

1) 4 2) 6
3) 2 4) 3

Stoichiometric Calculations

65. The moles of O_2 required for reacting with 6.8 g ammonia. (.....NH_3+....O_2→.....NO +.....H_2O) is:

1) 5 2) 2.5
3) 1 4) 0.5

66. What mass of $CaCl_2$ in grams would be enough to produce 14.35 gm of AgCl?

1) 5.55 g 2) 8.29 g
3) 16.59 g 4) 10 g

67. The mass of 70% H_2SO_4 by mass solution which is required for neutralisation of 1 mole of NaOH is:

1) 65 2) 98
3) 70 4) 54

68. If potassium chlorate is 80% pure then 48 g of oxygen would be produced from:

1) 153.12 g of $KClO_3$ 2) 120 g of $KClO_3$
3) 20 g of $KClO_3$ 4) 90 g of $KClO_3$

69. A solution containing 0.1 mole of a metal chloride MCl_x required 500 mL of 0.6 M $AgNO_3$ solution for complete precipitation. The value of x is:

1) 5 2) 4
3) 3 4) 1

70. If 20 g of $CaCO_3$ is treated with 100 mL 20% HCl solution. The amount of CO_2 produced is:

1) 22.4 l 2) 8.8 g
3) 2.2 g 4) 8 l

71. The mass of $CaCO_3$ required to react with 25 mL of 0.75 molar HCl is:

1) 0.94 g 2) 0.68 g
3) 0.76 g 4) 0.52 g

72. 2 moles of H_2S and 11.2 L of SO_2 at N.T.P. reacts to form x moles of sulphur. The value of x is:

1) 1.5 2) 3.5
3) 7.8 4) 12.7

73. What weight of HNO_3 is needed to convert 5 g of iodine into iodic acid according to the reaction,

$I_2 + HNO_3 \rightarrow HIO_3 + NO_2 + H_2O$

1) 12.4 g 2) 24.8 g
3) 0.248 g 4) 49.6 g

74. The molarity of pure water is:

1) 100 M 2) 55.6 M
3) 50 M 4) 18 M

75. If 0.30 mol of zinc are added to 0.52 mol of HCl, the moles of H_2 formed are:

1) 0.52 2) 0.30
3) 0.26 4) 0.60

High Order Time Intensive MCQs

1. Light travels with a speed of 3×10^8 m/sec. The distance travelled by light in 1 Femto sec is:
   1) 0.03 mm
   2) 0.003 mm
   3) 3 mm
   4) 0.0003 mm

2. The molecular formula of a commercial resin used for exchanging ions in water softening is $C_8H_7SO_3^-Na^+$ (mol. wt 206). What would be the maximum uptake of Ca^{2+} ions by the resin when expressed in mole per gram resin?
   1) $\frac{1}{103}$
   2) $\frac{1}{206}$
   3) $\frac{2}{309}$
   4) $\frac{1}{412}$

3. If 10^{21} molecules are removed from 200 mg CO_2, then the number of moles of CO_2 left are:
   1) 2.88×10^{-3}
   2) 1.66×10^{-3}
   3) 1.66×10^{-2}
   4) 4.54×10^{-3}

4. If 1 ml of water contains 20 drops, then the number of molecules in a drop of water is:
   1) 6×10^{23}
   2) 1.376×10^{26}
   3) 1.673×10^{21}
   4) 4.34×10^{20}

5. If x g of A (atomic mass 50) contains n atoms, how many atoms are there in $20x$ g of elements B: (at wt. 100)
   1) n
   2) 10n
   3) 20n
   4) n/10

6. Diborane (B_2H_6) can be prepared by the following reaction-
   $$3NaBH_4 + 4BF_3 \longrightarrow 3NaBF_4 + 2B_2H_6$$
   If the reaction has a 70% yield, how many moles of $NaBH_4$ should be used with excess BF_3 in order to obtain 0.200 mol of B_2H_6?
   1) 0.21 moles
   2) 0.429 moles
   3) 0.300 mol
   4) 0.175 moles

7. An ore contains 1.24% of mineral argentate, Ag_2S by mass. How many grams of this ore would have to be processed in order to obtain 1g of pure solid silver?
   1) 92.6 g
   2) 88.1 g
   3) 101.11 g
   4) 107.25 g

8. An aqueous solution of ethanol has density 1.025 g/mL and molarity 8 M. Calculate molality of this solution:
   1) 20.27 m
   2) 17.12 m
   3) 12.18 m
   4) 1.117 m

9. The molar ratio of Cr^{2+} to Cr^{3+} in a mixture of $CrSO_4$ and $Cr_2(SO_4)_3$ having equal number of sulphate ions in both sulphates is
   1) 1 : 3
   2) 3 : 2
   3) 2 : 3
   4) 3 : 4

10. $A + 2B + 3C \rightleftharpoons AB_2C_3$
    Reaction of 6 g of A, 6×10^{23} atoms of B & 0.036 mole of C yields 4.8 g of compound AB_2C_3. If the atomic masses of A & C are 60 & 80 amu respectively, the atomic mass of B is:
    1) 60 amu
    2) 50 amu
    3) 90 amu
    4) 120 amu

11. A 100 g of a sample of haemoglobin on analysis was found to contain 0.34% Fe by mass. If each haemoglobin molecule has four Fe^{2+} ions, the molecular mass of haemoglobin is- (Fe = 56 amu)
    1) 77099.9 g
    2) 12735 g
    3) 65882 g
    4) 96359.9 g

12. What volume of 5 M Na_2SO_4 must be added to 250 mL of 1 M $BaCl_2$ to produce 10 g of $BaSO_4$?
    1) 8.58 mL
    2) 6.4 mL
    3) 18 mL
    4) 22 mL

13. A compound is composed of O and Mn in equal weight ratio. Find the empirical formula of the compound.
    1) Mn_3O_4
    2) MnO_2
    3) Mn_2O_3
    4) Mn_2O_7

14. If isotopic distribution of C-12 and C-14 is 98% and 2% respectively, then the number of C-14 atoms in 12 g of carbon is:
    1) 1.032×10^{22}
    2) 3.01×10^{22}
    3) 5.88×10^{23}
    4) 6.02×10^{23}

15. Sulphuric acid reacts with sodium hydroxide as follows:
    $$H_2SO_4 + 2NaOH \rightarrow Na_2SO_4 + 2H_2O$$
    When 1 L of 0.1 M sulphuric acid solution is allowed to react with 1 L of 0.1 M sodium hydroxide solution, the amount of sodium sulphate formed is:
    1) 0.15 g
    2) 7.10 g
    3) 0.025 g
    4) 3.55 g

16. X gm of Ag was dissolved in HNO_3 and the solution was treated with excess of NaCl. When 2.87 g of AgCl was precipitated, the value of x is
    $$Ag + 2HNO_3 \rightarrow AgNO_3 + NO_2 + H_2O$$
    $$AgNO_3 + NaCl \rightarrow AgCl + NaNO_3$$
    1) 1.08 g
    2) 2.16 g
    3) 2.70 g
    4) 1.62 g

17. The % loss in weight after heating a pure sample of potassium chlorate (M. wt. 122.5) will be-
    $$2KClO_3 \xrightarrow{\Delta} 2KCl + 3O_2$$
    1) 12.25
    2) 24.50
    3) 39.18
    4) 49.00

18. The candela is the luminous intensity, in a given direction, of a source that emits monochromatic radiation of frequency 'A' × 10^{12} hertz and that has a radiant intensity in that direction of $\frac{1}{'B'}$ watt per steradian. 'A' and 'B' are respectively.

1) 540 and $\frac{1}{683}$ 2) 540 and 683
3) 450 and $\frac{1}{683}$ 4) 450 and 683

19. In an organic compound of molar mass 108 g mol^{-1} C, H and N atoms are present in 9 : 1 : 3.5 by weight. Molecular formula can be-

1) $C_6H_8N_2$ 2) $C_7H_{10}N$
3) $C_5H_6N_3$ 4) $C_4H_{18}N_3$

20. 6.02 × 10^{20} molecules of urea are present in 100 mL of its solution. The molarity of urea solution is-

1) 0.001 M 2) 0.01 M
3) 0.02 M 4) 0.1 M

(Avogadro constant, $N_A = 6.02 \times 10^{23}$ mol^{-1})

21. In the reaction,

$2Al(s) + 6HCl(aq) \rightarrow 2Al^{3+}(aq) + 6Cl^-(aq) + 3H_2(g)$,

1) 6L HCl(aq) is consumed for every 3L $H_2(g)$ produced
2) 33.6 L $H_2(g)$ is produced regardless of temperature and pressure for every mole Al that reacts
3) 67.2 L $H_2(g)$ at STP is produced for every mole Al that reacts
4) 11.2 L $H_2(g)$ at STP is produced for every mole HCl(aq) consumed

22. 4.4 g of CO_2 and 2.24 litre of H_2 at STP are mixed in a container. The total number of molecules present in the container will be -

1) 6.022 × 10^{23} 2) 1.2044 × 10^{23}
3) 2 moles 4) 6.023 × 10^{24}

23. 1 g sample of alkaline earth metal react completely with 4.08 g H_2SO_4 and yields an ionic product MSO_4. Then find out the atomic mass of alkaline earth metal (M)?

1) 9 2) 24
3) 40 4) 87

24. A compound contains 36% C by mass. If each molecule contains two C atoms the number of moles of the compound in its 10 g is:

1) 0.15 2) 1.5
3) 150 4) 1500

25. When a certain amount of octane is burnt completely, 7.04 g of CO_2 is formed. What mass of H_2O is formed simultaneously?

1) 3.24 g 2) 6.68 g
3) 6.48 g 4) 6.16 g

26. A 1.50 g sample of $KHCO_3$ having 80% purity is strongly heated. Assuming the impurity to be thermally stable, the loss in weight of the sample, on heating is:

1) 26.4 g 2) 2.64 g
3) 0.264 g 4) 0.0264 g

27. A metal nitride M_3N_2 contains 28% of nitrogen. The atomic mass of metal M is:

1) 24 2) 54
3) 9 4) 87.62

28. Chlorine can be prepared by reacting HCl with MnO_2. The reaction is represented by the equation

$MnO_2(g) + 4HCl(aq.) \rightarrow MnCl_2(aq.) + Cl_2(g) + 2H_2O(l)$

Assuming that the reaction goes to completion. What mass of conc. HCl solution (36% by mass) is needed to produce 2.5 g Cl_2?

1) 14.27 g 2) 25 g
3) 25.21 g 4) 30 g

29. 29.2% (w/w) HCl stock solution has a density of 1.25 g mL^{-1}. The molecular weight of HCl is 36.5 g mol^{-1}. The volume (in mL) of stock solution required to prepare a 200 mL solution of 0.4 M HCl is-

1) 16 mL 2) 61 mL
3) 80 mL 4) 8 mL

30. The molality of 1 L solution with x% H_2SO_4 (w/v) is equal to 9. The weight of solvent present in the solution is 910 g. The value of x is-

1) 90 2) 80.3
3) 100 4) 95.5

31. The composition of residual mixture will be, if 30 g of Mg combines with 30 g of O_2-

1) 40 g MgO + 20 g O_2 2) 45 g MgO + 15 g O_2
3) 50 g MgO + 10 g O_2 4) 60 g MgO only

32. How many moles of magnesium phosphate, $Mg_3(PO_4)_2$ will contain 0.25 moles of oxygen atoms?

1) 0.02 2) 3.125 × 10^{-2}
3) 1.25 × 10^{-2} 4) 4.25 × 10^{-2}

33. The atomic weights of two elements A and B are 40 u and 80 u respectively. If x g of A contains y atoms, how many atoms are present in 2x g of B:

1) $\frac{y}{2}$ 2) $\frac{y}{4}$
3) y 4) 2y

34. The haemoglobin from red blood corpuscles of most mammals contain approximately 0.33% of iron by mass. The molecular mass of haemoglobin is 67200 g mol^{-1}. The number of iron atoms in each molecule of haemoglobin is:

1) 3 2) 4
3) 2 4) 6

35. 0.8 mole of a mixture of CO and CO_2 requires exactly 40 gram of NaOH in solution for complete conversion of all the CO_2 into Na_2CO_3, if the mixture (0.8 mole) is completely oxidised to CO_2, find further required moles of NaOH.

1) 0.2 2) 0.6
3) 1 4) 1.5

36. 3.68 g of a mixture of $CaCO_3$ and $MgCO_3$ is heated to liberate 0.04 mole of CO_2. The mole % of $CaCO_3$ and $MgCO_3$ in the mixture is respectively:

1) 50%, 50%
2) 60%, 40%
3) 40%, 60%
4) 30%, 70%

37. An element is found in nature in two isotopic forms with mass numbers (A–1) and (A + 3). If the average atomic mass of the element is found to be A, then the relative abundance of the heavier isotope in the nature will be.

1) 66.6%
2) 75%
3) 25%
4) 33.3%

38. 0.5 mol of potassium ferrocyanide contains carbon equal to: (Formula of potassium ferrocyanide is $K_4[Fe(CN)_6]$.

1) 1.5 mol
2) 36 g
3) 18 g
4) 3.6 g

39. Mass of 0.224 litre of hydrogen gas at 273 K and 1 atm (assuming ideal behaviour) is

1) 1 g
2) 0.2 g
3) 0.02 g
4) 0.002 g

40. 0.078 grams of a hydrocarbon occupy 22.4 mL of volume at STP. The empirical formula of the hydrocarbon is CH. The molecular formula of hydrocarbon is

1) C_2H_2
2) C_6H_6
3) C_8H_8
4) C_4H_4

41. 0.16 g of dibasic acid required 25 ml of N/10 NaOH for complete neutralization. Molecular weight of acid is:

1) 32
2) 64
3) 128
4) 256

42. Which of the following molarity values of ions in a aqueous solution of 5.85% w/v NaCl, 5.55% w/v $CaCl_2$ and 6% w/v NaOH are correct [Na = 23, Cl = 35.5, Ca = 40,O = 16]

1) $[Cl^{-1}] = 2M$
2) $[OH^-] = 1.5$ M
3) $[Ca^{2+}] = 0.5$ M
4) All of these

43. Match the column:

	Column-I		Column-II
A.	1 mole $(NH_4)_2CO_3$	p.	Moles of carbon atom is less than 1
B.	0.5 mole NH_2CONH_2	q.	Mass of oxygen is greater than 16 g in sample
C.	1 mole C_6H_5N	r.	At least two elements have same number of atoms in given sample
D.	1 mole C_9N_6	s.	Number of neutrons and protons are same
		t.	Moles of nitrogen atom is greater than and equal to 1

1) A → q, t; B → p, r, t; C → t; D → s, t
2) A → s, r; B → q, t, p; C → t; D → q, r
3) A → q, p; B → p, t; C → p; D → p, r, t
4) A → t; B → s, p, t; C → s; D → q, t

44. How many moles of ferric alum, $(NH_4)_2SO_4 \cdot Fe_2(SO_4)_3.24H_2O$ can be made from the sample of Fe containing 0.0056 g of it?

1) 10^{-4} mol
2) 0.5×10^{-4} mol
3) 0.33×10^{-4} mol
4) 2×10^{-4} mol

45. 1 gram of a carbonate (M_2CO_3) on treatment with excess HCl produces 0.01186 mole of CO_2. The molar mass of M_2CO_3 (in g mol^{-1}) is

1) 118.6
2) 11.86
3) 1186
4) 84.3

46. Suppose the elements X and Y combine to form two compounds XY_2 and X_3Y_2. When 0.1 mole of XY_2 weighs 10 g and 0.05 mole of X_3Y_2 weighs 9 g, the atomic weights of X and Y are

1) 40, 30
2) 60, 40
3) 20, 30
4) 30, 20

47. Concentrated aqueous sulphuric acid is 98% H_2SO_4 by mass and has a density of 1.80 g mL^{-1}. Volume of acid required to make one litre of 0.1 M H_2SO_4 solution is

1) 11.10 mL
2) 16.65 mL
3) 22.20 mL
4) 5.55 mL

48. The following substances are present in different containers

(i) One gram atom of nitrogen
(ii) One mole of calcium
(iii) One atom of silver
(iv) One mole of oxygen molecules
(v) 10^{23} atoms of carbon
(vi) One gram of iron.

The correct order of increasing masses (in grams) is/are

1) (iii) < (iv) < (i) < (v)
2) (iii) < (vi) < (iv) < (ii)
3) (vi) < (v) < (i) < (iv)
4) (2) and (3) both are correct

49. Consider the following statements:

(i) A balanced chemical reaction should follow law of conservation of mass on either side.
(ii) 2 moles of $H_2(g)$ and 3 moles of $O_2(g)$ produce 3 moles of water.
(iii) Equal wt. of carbon and oxygen are taken to produce CO_2, then O_2 is limiting reagent.

The above statements (i), (ii), (iii) respectively are (T = True, F = False)

1) TTT
2) FTF
3) FFF
4) TFT

50. How many molecules of HCl gas will be produced by reacting 112 L of H_2 (0 °C, 1 atm) with 213 g of Cl_2?

1) 3.61×10^{24} 2) 6.13×10^{23}
3) 6.13×10^{24} 4) 1.63×10^{24}

51. 20 mL of 0.4 M $AgNO_3$ (molar mass = 170 g) is reacted with 15 mL of 0.6 M $BaCl_2$ (molar mass = 208.4 g). The mass of AgCl (molar mass = 143.5 g) produced is

1) 11.48 g 2) 18.14 g
3) 14.18 g 4) 1.148 g

52. 85 g $CaCO_3$ (limestone sample), on heating produces exactly the same amount of CO_2 which converts 30 g of MgO to $MgCO_3$. The percentage purity of limestone sample is

1) 80% 2) 82.4%
3) 88.24% 4) 84.8%

53. Mole fraction of acetic acid in an aqueous sample is 0.1. The molality of the solution is

1) 7.16 mol kg^{-1} 2) 1.67 mol kg^{-1}
3) 6.17 mol kg^{-1} 4) 5.25 mol kg^{-1}

54. Molarity of H_2SO_4 (density 1.8 g/mL) is 18 M. The molality of this H_2SO_4 is:

1) 36 2) 200
3) 500 4) 18

55. 1.5 moles of each of XY_2 and XY_3 if weigh 96 g and 120 g respectively. The atomic masses of X and Y respectively are

1) 4, 8 2) 8, 16
3) 32, 16 4) 32, 64

Match the Columns MCQs

1. Match the following:

	Column-I		Column-II
(a)	88 g of CO_2	(i)	0.25 mol
(b)	6.022×10^{23} molecules of H_2O	(ii)	2 mol
(c)	5.6 litres of O_2 at STP	(iii)	1 mol
(d)	96 g of O_2	(iv)	6.022×10^{23} molecules
(e)	1 mol of any gas	(v)	3 mol

1) (a)-(i) (b)-(iv) (c)-(iii) (d)-(ii) (e)-(v)
2) (a)-(ii) (b)-(iii) (c)-(i) (d)-(v) (e)-(iv)
3) (a)-(v) (b)-(ii) (c)-(iii) (d)-(iv) (e)-(i)
4) (a)-(iii) (b)-(v) (c)-(ii) (d)-(i) (e)-(iv)

2. Match the following physical quantities with units

	Physical quantity		Unit
(a)	Molarity	(i)	mol kg^{-1}
(b)	Molality	(ii)	mol L^{-1}
(c)	Mol fraction	(iii)	Unit less
(d)	Mole	(iv)	Mol

1) (a)-(i) (b)-(ii) (c)-(iii) (d)-(iv)
2) (a)-(ii) (b)-(i) (c)-(iv) (d)-(iii)
3) (a)-(i) (b)-(ii) (c)-(iv) (d)-(iii)
4) (a)-(ii) (b)-(i) (c)-(iii) (d)-(iv)

3. Match the following:

	Column-I		Column-II
(a)	Element	(i)	20 carat gold
(b)	Compound	(ii)	Iodized common salt
(c)	Homogeneous mixture	(iii)	Silica
(d)	Heterogeneous mixture	(iv)	Radium

1) (a)-(i) (b)-(ii) (c)-(iii) (d)-(iv)
2) (a)-(iv) (b)-(ii) (c)-(iii) (d)-(i)
3) (a)-(iv) (b)-(iii) (c)-(i) (d)-(ii)
4) (a)-(ii) (b)-(i) (c)-(iii) (d)-(iv)

4. Match the following

	Column-I		Column-II
(a)	mega	(i)	10^9
(b)	peta	(ii)	10^{-12}
(c)	pico	(iii)	10^{15}
(d)	giga	(iv)	10^6

1) (a)-(i) (b)-(iii) (c)-(ii) (d)-(iv)
2) (a)-(iv) (b)-(ii) (c)-(iii) (d)-(i)
3) (a)-(iv) (b)-(iii) (c)-(ii) (d)-(i)
4) (a)-(iii) (b)-(i) (c)-(ii) (d)-(iv)

5. Match the following

	Column-I (Symbol)		Column-II (Unit)
(a)	Newton	(i)	Kg m^2 sec^{-2}
(b)	Joule	(ii)	Kg m sec^{-2}
(c)	Acceleration	(iii)	m sec^{-2}
(d)	Mass	(iv)	kg

1) (a)-(i) (b)-(ii) (c)-(iii) (d)-(iv)
2) (a)-(iii) (b)-(i) (c)-(ii) (d)-(iv)
3) (a)-(ii) (b)-(i) (c)-(iii) (d)-(iv)
4) (a)-(ii) (b)-(iii) (c)-(i) (d)-(iv)

Correct & Incorrect MCQs

1. Identify the incorrect statement from the following
   1) Glucose is not a pure substance.
   2) Water, ammonia, sugar, CO_2 are compounds.
   3) H_2 and O_2 are gases whereas the compound formed by their combination i.e., water is a liquid, used as fire extinguisher.
   4) Sodium, copper, silver, hydrogen are elements.

2. Identify the incorrect statement from the following.
   1) The multiple of prefix femto is 10^{-15}
   2) The multiple of prefix pico is 10^{-12}
   3) The multiple of prefix nano is 10^{-18}
   4) The multiple of prefix micro is 10^{-6}

3. Which statement regarding mass and weight is not correct ?
   1) Mass of a substance is the amount of matter present in it.
   2) Weight is the force exerted by gravity on an object.
   3) The SI unit of mass is kilogram.
   4) Mass of a substance may vary from one place to another while weight of a substance is constant.

4. The incorrect statement regarding significant figures is:
   1) There are three significant figures in 285 cm.
   2) There are two significant figures in 0.25 mL.
   3) There are three significant figures in 0.03
   4) There are four significant figures in 2.005.

5. Select the incorrect statement regarding laws of chemical combination:
   1) According to law of conservation of mass, matter can neither be created nor be destroyed
   2) According to law of definite proportions, a given compound always contains exactly the same proportion of elements by weight.
   3) According to law of multiple proportions, if two elements can combine to form more than one compound, the masses of one element that combine with a fixed mass of the other element, are in the ratio of small whole numbers.
   4) According to Gay Lussac's law, equal volumes of all gases at the same temperature and pressure should contain equal number of molecules

6. Choose the correct statement:
   1) Molecular mass of methane is 16.043 u.
   2) Molecular mass of water is 24.0 u.
   3) Molecular mass of glucose is 324.68 u.
   4) Molecular mass of ammonia is 36.0 u.

7. Which statement regarding moles is not correct?
   1) The mole is the SI unit of amount of substance.
   2) One mole of a substance contains exactly 12.044×10^{23} elementary entities.
   3) 1 mol of hydrogen atoms contain 6.022×10^{23} atoms.
   4) The mass of one mole of a substance in grams is called its molar mass.

8. Which amongst the following statements is not correct about the given reaction?

   $CH_4\,(g) + 2O_2\,(g) \rightarrow CO_2\,(g) + 2\,H_2O\,(g)$
   1) One mole of CH_4(g) reacts with two moles of O_2(g) to give one mole of CO_2(g) and two moles of H_2O(g).
   2) One molecule of CH_4(g) reacts with 2 molecules of O_2(g) to give one molecule of CO_2(g) and 2 molecules of H_2O(g).
   3) 22.7 L of CH_4 (g) reacts with 45.4 L of O_2 (g) to give 22.7 L of CO_2 (g) and 45.4 L of H_2O(g)
   4) 16 g of CH_4 (g) reacts with 32 g of O_2 (g) to give 44 g of CO_2 (g) and 18 g of H_2O (g).

9. Choose the incorrect statement about molarity:
   1) Molarity is defined as the number of moles of the solute present in 1 litre of the solution.
   2) Molarity is denoted by symbol M.
   3) 1 M NaOH means 1 mol of NaOH present in 1 litre of the solution.
   4) Molarity is independent on temperature.

10. Choose the correct statement/s:

    A. 0.22 g of CO_2 contains 0.005 moles

    B. 320 mg of SO_2 contains 0.005 moles

    C. 560 mg of SO_2 contains 0.005 moles

    D. 0.085 g of NH_3 contains 0.005 moles

    1) A, B and C
    2) Only C
    3) A, B and D
    4) B, C and D

Statement Based MCQs

Directions: These questions consist of two statements each, printed as Statement-I and Statement-II. While answering these questions, you are required to choose any one of the following four responses.

1) **Both Statement-I and Statement-II are correct.**
2) **Both Statement-I and Statement-II are incorrect.**
3) **Statement-I is correct & Statement-II is incorrect.**
4) **Statement-I is incorrect & Statement-II is correct.**

1. **Statement-I:** The mathematical approach to obtain the amount of the products or reactants is called stoichiometry.

   Statement-II: The stoichiometric calculations are derived from chemical equations.

2. **Statement-I:** A certain element X, forms three binary compounds with Chlorine containing 59.68%, 68.95% and 74.75% Chlorine respectively. These data illustrate the law of multiple proportions.

   Statement-II: According to law of multiple proportions, the relative amounts of an element combining with some fixed amount of a second element in a series of compounds are the ratios of small whole numbers.

3. **Statement-I:** 32 g of O_2 and 48 g of O_3 have same number of atoms.

   Statement-II: Number of moles in 32 g of O_2 and 48 g of O_3 are different.

4. **Statement-I:** Equal moles of different substances contain same number of constituent particles.

   Statement-II: Equal weights of different substances contain the same number of constituent particles.

5. **Statement-I:** Molality, mole fraction and mass fraction change with temperature.

   Statement-II: Molarity and Normality do not change with temperature.

6. **Statement-I:** Air is a homogeneous mixture.

   Statement II: The components of air completely mix with each other and its composition is uniform throughout.

7. **Statement-I:** Mass of a substance is constant where as its weight may vary from one place to another.

   Statement-II: Weight changes due to change in earth's gravity.

8. **Statement-I:** Dalton's atomic theory could explain all the laws of chemical combination including laws of gaseous volumes.

   Statement-II: In all physical and chemical changes, the total mass of the reactants is never equal to the total mass of the products.

9. **Statement-I:** Molality of a solution depends upon temperature.

   Statement-II: Molarity of a solution depends upon temperature.

10. **Statement-I:** One mole is the amount of a substance that contains as many particles or entities as there are atoms in exactly 12 g of the 12C isotope.

    Statement-II: One mole of different substances contain different number of entities.

Assertion & Reason MCQs

Directions: These questions consist of two statements each, printed as Assertion and Reason. While answering these questions, you are required to choose any one of the following four responses.

1) **Both Assertion and Reason are True and the Reason is a correct explanation of the Assertion.**
2) **Both Assertion and Reason are True but Reason is not a correct explanation of the Assertion.**
3) **Assertion is True but the Reason is False.**
4) **Assertion is False but Reason is True.**

1. **Assertion:** 1 mole of any gas occupies 22.4 lit at NTP.

   Reason: Volume of gas depends on temperature, pressure and nature of gas.

2. **Assertion:** The no. of Oxygen atoms present in 16 g of oxygen and 16 g of O_3 is same.

   Reason: Both these species represents 1 g atom of oxygen.

3. **Assertion:** The number of significant figures in 1502 cm is two.

   Reason: All non-zero digits are significant and zeroes lie between any two non-zeroes will be significant.

4. **Assertion:** The ratio by volume of NH_3 : HCl : NH_4Cl is 1 : 1 : 1 in the reaction

   $NH_3(g) + HCl(g) \rightarrow NH_4Cl(s)$

   Reason: Gay Lussac's law deals with gaseous reactants and products only.

5. **Assertion:** 64 g of SO_2 and 80g of SO_3 will occupy same volume at NTP.

   Reason: 1 mole of any gas occupies 22.4 L at NTP.

6. **Assertion:** The empirical mass of ethene is half of its molecular mass.

   Reason: The empirical formula represents the simplest whole number ratio of various atoms present in a compound.

7. **Assertion:** One atomic mass unit is defined as one tweleth of the mass of one carbon-12 atom.

   Reason: Carbon-12 isotope is the most abundant isotope of carbon and has been chosen as standard.

8. **Assertion:** Significant figures for 0.200 is 3 whereas for 200 it is 1.

   Reason: Zero at the end or right of a number are significant provided they are not on the right side of the decimal point.

9. **Assertion:** Combustion of 16 g of methane gives 18 g of water.

   Reason: In the combustion of methane, water is one of the products.

10. **Assertion:** 1 amu represents a mass equal to $1{\cdot}66 \times 10^{-27}$ kg.

    Reason: 1 amu represents the actual mass of one atom of carbon–12.

Multi Correct MCQs

1. Which of the following have same number of significant figures?

   A. 0.00253 B. 1.0003
   C. 15.0 D. 163

   Choose the correct answer from the options given below

   1) (A), (B) and (C) only 2) (C) and (D) only
   3) (A), (C) and (D) only 4) (B) and (C) only

2. The incorrect postulates of the Dalton's atomic theory are:

   A. Atoms of different elements differ in mass.
   B. Matter consists of divisible atoms.
   C. Compounds are formed when atoms of different element combine in a fixed ratio.
   D. All the atoms of given element have different properties including mass.
   E. Chemical reactions involve reorganisation of atoms.

   Choose the correct answer from the options given below:

   1) (B), (D), (E) only 2) (A), (B), (D) only
   3) (C), (D), (E) only 4) (B), (D) only

3. 116 g of a substance upon dissociation reaction yields 7.5 g of hydrogen, 60g of oxygen and 48.5 g of carbon. Given that the atomic masses of H, O and C are 1,16 and 12 g/mol respectively. The data agrees with which of the following formulae?

   A. CH_3COOH B. HCHO
   C. CH_3OOCH_3 D. CH_3CHO

   1) (A) and (B) 2) (B), (C) and (D)
   3) (C) and (D) 4) (B) and (C)

4. Which of the following methods of expressing the concentration of a solution are not temperature dependent?

   A. Molarity B. Mole Fraction
   C. Molality D. Weight percent

   1) (A), (B) and (C) 2) (B), (C) and (D)
   3) (A) and (C) 4) (A) and (D)

5. One of the statements of Dalton's atomic theory is given below: "Compounds are formed when atoms of different elements combine in a fixed ratio"

   Which of the following laws is not related to this statement?

   A. Law of conservation of mass
   B. Law of definite proportions
   C. Law of multiple proportions
   D. Avogadro law

   1) Only (A) 2) Both (B) and (C)
   3) Both (A) and (D) 4) Only (D)

NCERT Exemplar MCQs

1. Two students performed the same experiment separately and each one of them recorded two readings of mass which are given below. Correct reading of mass is 3.0 g. On the basis of given data, mark the correct option out of the following statements:

Students	Readings	
	(i)	(ii)
A	3.01	2.99
B	3.05	2.95

   1) Results of both the students are neither accurate nor precise
   2) Results of student A are both precise and accurate
   3) Results of student B are neither precise nor accurate
   4) Results of student B are both precise and accurate

2. What will be the molarity of a solution, which contains 5.85 g of NaCl (s) per 500 mL?

   1) 4 mol L^{-1} 2) 20 mol L^{-1}
   3) 0.2 mol L^{-1} 4) 2 mol L^{-1}

3. If 500 mL of a 5 M solution is diluted to 1500 mL, what will be the molarity of the solution obtained?

   1) 1.5 M 2) 1.66 M
   3) 0.017 M 4) 1.59 M

4. The number of atoms present in one mole of an element is equal to Avogadro number. Which of the following element contains the greatest number of atoms?

   1) 4 g He 2) 46 g Na
   3) 0.40 g Ca 4) 12 g He

5. If the concentration of glucose ($C_6H_{12}O_6$) in blood is 0.9 g L^{-1}, what will be the molarity of glucose in blood?

1) 5 M 2) 50 M 3) 0.005 M 4) 0.5 M

6. What will be the molality of the solution containing 18.25 g of HCl gas in 500 g of water?

1) 0.1 m 2) 0.05 m 3) 0.5 m 4) 1 m

7. One mole of any substance contains 6.022×10^{23} atoms/molecules. Number of molecules of H_2SO_4 present in 100 mL of 0.02 M H_2SO_4 solution is:

1) 12.044×10^{20} molecules 2) 6.022×10^{23} molecules
3) 1×10^{23} molecules 4) 12.044×10^{23} molecules

8. What is the mass per cent of carbon in carbon dioxide?

1) 0.034% 2) 27.27% 3) 3.4% 4) 28.7%

9. The empirical formula and molecular mass of a compound are CH_2O and 180 g respectively. What will be the molecular formula of the compound?

1) $C_9H_{18}O_9$ 2) CH_2O 3) $C_6H_{12}O_6$ 4) $C_2H_4O_2$

10. If the density of a solution is 3.12 g mL^{-1}, the mass of 1.5 mL solution in significant figures is:

1) 4.7 g 2) 4680×10^{-3} g
3) 4 .680 g 4) 46.80 g

11. Which of the following statements about a compound is incorrect?

1) A molecule of a compound has atoms of different elements
2) A compound cannot be separated into its constituent elements by physical methods of separation
3) A compound retains the physical properties of its constituent elements
4) The ratio of atoms of different elements in a compound is fixed

12. Which of the following statements is correct about the reaction given below?

$$4Fe(s) + 3O_2(g) \rightarrow 2Fe_2O_3(g)$$

1) Total mass of iron and oxygen in reactants = total mass of iron and oxygen in product therefore it follows law of conservation of mass
2) Total mass of reactants = total mass of product, therefore, law of multiple proportions is followed
3) Amount of Fe_2O_3 can be increased by taking any one of the reactants (iron or oxygen) in excess
4) Amount of Fe_2O_3 produced will decrease if the amount of any one of the reactants (iron or oxygen) is taken in excess

13. Which of the following statements indicates that law of multiple proportion is being followed?

1) Sample of carbon dioxide taken from any source will always have carbon and oxygen in the ratio 1 : 2
2) Carbon forms two oxides namely CO_2 and CO, where masses of oxygen which combine with fixed mass of carbon are in the simple ratio 2 :1
3) When magnesium burns in oxygen, the amount of magnesium taken for the reaction is equal to the amount of magnesium in magnesium oxide formed
4) At constant temperature and pressure, 200 mL of hydrogen will combine with 100 mL oxygen to produce 200 mL of water vapour

Kattar NEET

1. 15 gm $Ba(MnO_4)_2$ sample containing inert impurity is completely reacting with 100 ml of '11.2 V' H_2O_2, then what will be the purity of $Ba(MnO_4)_2$ in the sample?

1) 5% 2) 10% 3) 50% 4) none

2. SO_2Cl_2, sulphuryl chloride reacts with water to give a mixture of H_2SO_4 and HCl. What volume of 0.2 M $Ba(OH)_2$ is needed to completely neutralize 50 ml of 0.2 M SO_2Cl_2 solution?

1) 100 ml 2) 200 ml 3) 300 ml 4) 400 ml

3. Two acids H_2SO_4 and H_3PO_4 are neutralised separately by the same amount of a base, producing sulphate and dihydrogen phosphates respectively. The ratio of mass of H_2SO_4 and H_3PO_4 reacted with base respectively is

1) 1 : 1 2) 2 : 1
3) 1 : 2 4) Data insufficient

4. A clay sample contains 50% silica and 15% water. Now, the sample is heated until it loses 10 gm water. If the partially dried clay now contains 52% silica, what should be the percentage of water in it? All the percentages are given by mass. Assume that only moisture is lost from clay, on heating.

1) 5.56% 2) 11.6% 3) 5.0% 4) 15.6%

5. A gaseous oxide contains 30.4% of nitrogen, one molecule of which contains one nitrogen atom. The vapour density of the oxide relative to dioxygen is

1) 0.94 2) 1.44 3) 1.50 4) 3.0

6. Dieldrin, an insecticide, contains C, H, Cl and O. Combustion of 29.72 mg of Dieldrin gave 41.21 mg CO_2 and 5.63 mg of H_2O. In a separate analysis Dieldrin was converted into 67.16 mg AgCl. What is the empirical formula of Dieldrin?

1) $C_6H_4Cl_3O$ 2) C_8H_8ClO
3) $C_{12}H_8Cl_6O$ 4) $C_6H_4Cl_3O_2$

7. Suppose two elements X and Y combine to form two compounds XY_2 and X_2Y_3. 0.05 mole of XY_2 weighs 5 g while 3.011×10^{23} molecules of X_2Y_3 weigh 85 g. The atomic masses of X and Y are respectively

1) 20, 30 2) 30, 40 3) 40, 30 4) 80, 60

8. Which of the following statements indicates that law of multiple proportion is being followed.
   1) Sample of carbon dioxide taken from any source will always have carbon and oxygen in the ratio 1:2
   2) Carbon forms two oxides namely CO_2 and CO, where masses of oxygen which combine with fixed mass of carbon are in the simple ratio 2:1
   3) When magnesium burns in oxygen, the amount of magnesium taken for the reaction is equal to the amount of magnesium in magnesium oxide formed
   4) At constant temperature and pressure, 200 mL of hydrogen will combine with 100 mL oxygen to produce 200 mL of water vapour

9. Match the following:

	List-I		List-II
I.	4.5 m solution of $CaCO_3$ of density 1.45 gm/ml	P.	Mole fraction of solute is 0.2
II.	3 M 100 ml H_2SO_4 mixed with 1 M 300 ml H_2SO_4 solution	Q.	Mass of the solute is 360 g
III.	14.5 m solution of $Ca(NO_3)_2$	R.	Molarity = 4.5
IV.	In 2 litre solution of 4 M NaOH, 40 g NaOH is added	S.	Molarity = 1.5

   1) I-P; II-S; III- R; IV-Q, R
   2) I-R; II-S; III- P; IV-Q, R
   3) I-R; II-P; III- S; IV-Q, R
   4) I-R; II-Q; III- P; IV-S

10. Match the following:

	List-I (Atomic Masses)				List-II (% Composition of Lighter Isotope)
	Isotope-I	Isotope-II	Avg.		
I.	(a+ 4)	(a – 1)	a	P.	66.67% by moles
II.	a	5a	2a	Q.	50% by moles
III.	(a+ 3)	(a + 1)	(a +2)	R.	% by mass independent of 'a'
IV.	(a + 2)	(a – 1)	a	S.	80% by moles

   1) I-R; II-S; III-Q; IV-P   2) I-S; II-Q; III-R; IV-P
   3) I-S; II-R; III-P; IV-Q   4) I-S; II-R; III-Q; IV-P

11. Match the following:

	Column-I		Column-II
I.	1.7 g of NH_3	P.	0.4 N_0 atom
II.	3.2 g oxygen	Q.	2.24 L at NTP
III.	2.6 g C_2H_2	R.	N_0 no. of electrons
IV.	6.4 g of SO_2	S.	0.2 N_0 atoms
		T.	0.1 N_0 molecule

   1) I-P,Q,R,T; II-Q,S,T; III-P,Q,T; IV-Q, T
   2) I-Q,S,T; II-P,Q,R,T; III-P,Q,T; IV-R, T
   3) I-S,Q,R,T; II-P,Q,T; III-Q,S,T; IV-Q, S
   4) I-P,Q,S,T; II-Q,S,T; III-Q, T IV-P,Q,T;

12. A compound contains 4% oxygen, $\frac{14}{3}$% nitrogen, 4% sulphur. Then empirical formula of compound contains ________ number of nitrogen atoms.
   1) 6   2) 8
   3) 3   4) Can not be determined

13. 56 gm of N_2 and 9 gm of H_2 are made to react completely to produce a mixture of NH_3 and N_2H_4. The ratio of moles of NH_3 and N_2H_4 is:
   1) 1 : 1   2) 3 : 2
   3) 2 : 3   4) None of these

14. Iodobenzene is prepared from aniline ($C_6H_5NH_2$) in a two-step process as shown here.

$$C_6H_5NH_2 + HNO_2 + HCl \rightarrow C_6H_5N_2^+Cl^- + 2H_2O$$

$$C_6H_5N_2^+Cl^- \xrightarrow{+KI} C_6H_5I + N_2 + KCl$$

In an actual preparation, 9.30 g of aniline was converted to 16.32 g of iodobenzene. The percentage yield of iodobenzene is (I = 127)
   1) 8%   2) 50%   3) 75%   4) 80%

15. A protein isolated from a bovine preparation, was subjected to amino acid analysis. The amino acid present in the smallest amount was lysine, $C_6H_{14}N_2O_2$ and the amount of lysine was found to be 365 mg per 100 g protein. What is the minimum molecular mass (in g/mol) of the protein?
   1) 40,000,000   2) 40,000
   3) 40   4) 4,00,000

16. Elements X and Y can combine to form two different compounds. If 1.6 g of X reacts with exactly 1.6 g of Y the compound produced has formula XY_2. However under different conditions, 2.4 g of X will react with 1.6 g of Y to form a second compound, whose empirical formula is ____.
   1) X_3Y_4 2) XY   3) XY_3 4) X_2Y

17. A graph is plotted for an element by putting its mass on X-axis and the corresponding number of atoms on Y-axis. What is the atomic mass of the element for which the graph is plotted? ($N_A = 6.0 \times 10^{23}$)

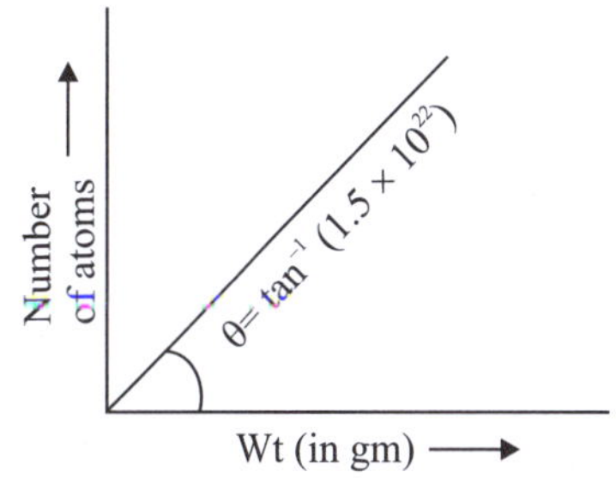

   1) 80   2) 40   3) 0.025   4) 20

18. A drug, marijuana, owes its activity to tetrahydrocannabinol, which contains 70 percent as many carbon atoms as hydrogen atoms and 15 times as many hydrogen atoms as oxygen atoms. The number of moles in a gram of terahydrocannabinol is 0.00318. Determine its molecular formula.

1) CH_3O_2 2) $C_{21}H_{30}O_2$ 3) $C_{15}H_{30}O_2$ 4) $C_{70}H_{15}O$

19. From the following reactions,

$2CoF_2 + F_2 \rightarrow 2CoF_3$

$(CH_2)_n + 4n\ CoF_3 \rightarrow (CF_2)_n + 2n\ HF + 4n\ CoF_2$

Calculate how much F_2 will be consumed to produce 1 kg of $(CF_2)_n$. (F = 19)?

1) 1.52 kg 2) 2.04 kg
3) 0.76 kg 4) 4.56 kg

20. A person needs on average of 2.0 mg of riboflavin (vitamin B_2) per day. How many gm of butter should be taken by the person per day if it is the only source of riboflavin? (Butter contains 5.5 microgram riboflavin per gm.)

1) 363.6 gm 2) 2.75 mg
3) 11 gm 4) 19.8 gm

21. What volume (in ml) of 0.2 M H_2SO_4 solution should be mixed with the 40 ml of 0.1 M NaOH solution such that the resulting solution has the concentration of H_2SO_4 as $\frac{6}{55}$ M.

1) 70 2) 45
3) 30 4) 58

NEET & AIIMS Past Year MCQs

1. 1.0 g of H_2 has same number of molecules as in: **(2024 Re)**

1) 14 g of N_2 2) 18 g of H_2O
3) 16 g of CO 4) 28 g of N_2

2. The highest number of helium atoms is in **(2024)**

1) 4 g of helium
2) 2.271098 L of helium at STP
3) 4 mol of helium
4) 4 u of helium

3. A compound X contains 32% of A, 20% of B and remaining percentage of C. Then, the empirical formula of X is:

(Given atomic masses of A = 64; B = 40; C = 32u) **(2024)**

1) AB_2C_2 2) ABC_4
3) A_2BC_2 4) ABC_3

4. 1 gram of sodium hydroxide was treated with 25 mL of 0.75 M HCl solution, the mass of sodium hydroxide left unreacted is equal to **(2024)**

1) Zero mg 2) 200 mg
3) 750 mg 4) 250 mg

5. The right option for the mass of CO_2 produced by heating 20 g of 20% pure limestone is (Atomic mass of Ca = 40)

$\left[CaCO_3 \xrightarrow{1200\text{K}} CaO + CO_2\right]$ **(2023)**

1) 1.32 g 2) 1.12 g
3) 1.76 g 4) 2.64 g

6. What mass of 95% pure $CaCO_3$ will be required to neutralise 50 mL of 0.5 M HCl solution according to the following reaction? **(2022)**

$CaCO_{3(s)} + 2HCl_{(aq)} \rightarrow CaCl_{2(aq)} + CO_{2(g)} + 2H_2O_{(l)}$

[Calculate upto second place of decimal point]

1) 9.50 g 2) 1.25 g
3) 1.32 g 4) 3.66 g

7. An organic compound contains 78% (by wt.) carbon and remaining percentage of hydrogen. The right option for the empirical formula of this compound is: [Atomic wt. of C is 12, H is 1] **(2021)**

1) CH_2 2) CH_3
3) CH_4 4) CH

8. If a compound consist of 4% H, 24% Carbon, Cl=72% with molar mass = 99 g: **(AIIMS 2016)**

Find molecular formula of compound.

1) C_2H_5Cl 2) $C_2H_2Cl_2$
3) $C_2H_4Cl_2$ 4) C_4H_2Cl

9. The weight of a metal of equivalent weight 12, which will give 0.475 g of its chloride, is: **(AIIMS 1994)**

1) 0.18 g 2) 0.12 g
3) 0.24 g 4) 0.16 g

Answer Key

NCERT Topic-wise MCQs

1	2	3	4	5	6	7	8	9	10	11	12	13	14	15	16	17
4	2	3	4	1	2	3	2	2	3	3	2	2	4	3	4	3
18	19	20	21	22	23	24	25	26	27	28	29	30	31	32	33	34
2	3	2	2	2	1	2	3	4	3	2	1	2	4	4	4	2
35	36	37	38	39	40	41	42	43	44	45	46	47	48	49	50	51
4	3	3	1	3	3	4	1	1	4	2	2	3	4	4	4	1
52	53	54	55	56	57	58	59	60	61	62	63	64	65	66	67	68
2	1	1	2	4	3	4	4	1	1	4	3	1	4	1	3	1
69	70	71	72	73	74	75										
3	2	1	1	1	2	3										

High Order Time Intensive MCQs

1	2	3	4	5	6	7	8	9	10	11	12	13	14	15	16	17
4	4	1	3	2	2	1	3	2	2	3	1	4	1	2	2	3
18	19	20	21	22	23	24	25	26	27	28	29	30	31	32	33	34
2	1	2	4	2	2	1	1	3	1	1	4	2	3	2	3	2
35	36	37	38	39	40	41	42	43	44	45	46	47	48	49	50	51
2	1	3	2	3	2	3	4	1	2	4	1	4	4	4	1	4
52	53	54	55													
3	3	3	3													

Match the Columns MCQs

1	2	3	4	5
2	4	3	3	3

Correct & Incorrect MCQs

1	2	3	4	5	6	7	8	9	10
1	3	4	3	4	1	2	4	4	3

Statement Based MCQs

1	2	3	4	5	6	7	8	9	10
1	1	2	3	2	1	1	2	4	3

Assertion & Reason MCQs

1	2	3	4	5	6	7	8	9	10
3	1	4	4	1	1	2	2	4	3

Multi Correct MCQs

1	2	3	4	5
3	4	1	2	3

NCERT Exemplar MCQs

1	2	3	4	5	6	7	8	9	10	11	12	13
2	3	2	4	3	4	1	2	3	1	3	1	2

Kattar NEET

1	2	3	4	5	6	7	8	9	10	11	12	13	14	15	16	17
3	1	3	2	2	3	3	2	2	4	1	2	3	4	2	1	2
18	19	20	21													
2	1	1	1													

NEET & AIIMS Past Year MCQs

1	2	3	4	5	6	7	8	9
1	3	4	4	3	3	2	3	2

CHAPTER

2 Structure of Atom

Topicwise Number of Questions Analysis (2025-2016)

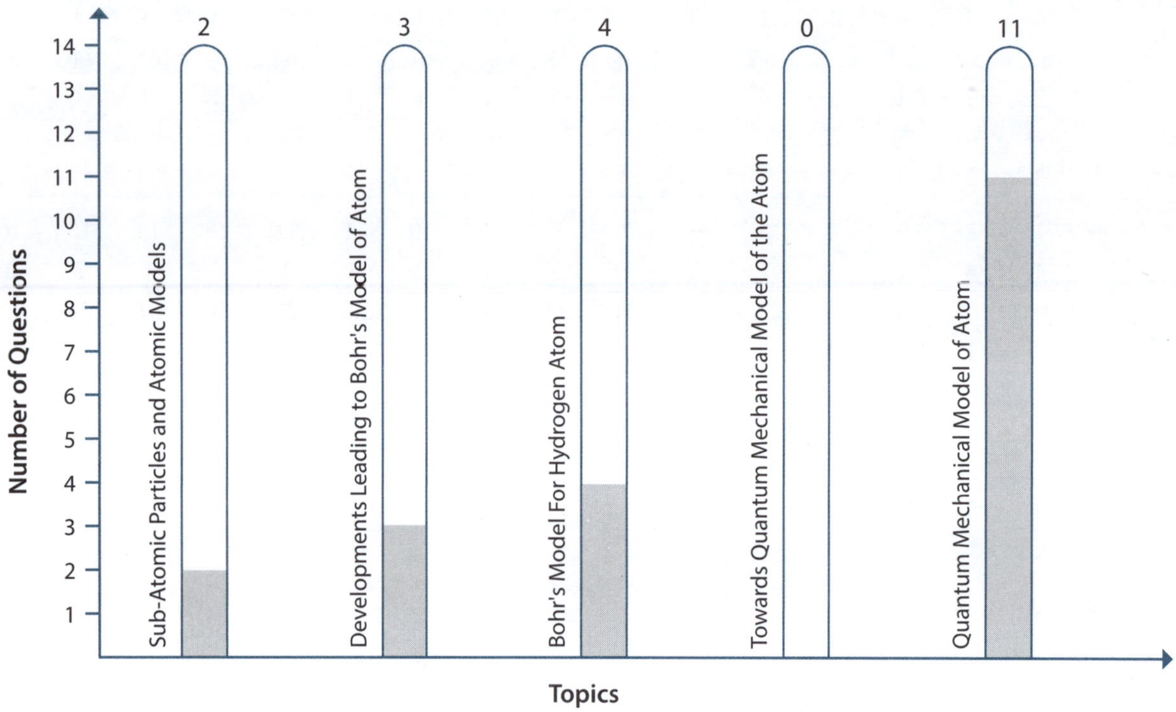

NCERT-Picks

Atomic Models

Thomson Model

J.J. Thomson proposed the first atomic model known as the Plum Pudding model.

An important feature of this model is that the **mass** of the **atom** is assumed to be **uniformly distributed over the atom**.

Drawback of Thomson's model

Although this model was able to explain the overall neutrality of the atom, but was not consistent with the results of later experiments.

Rutherford's atomic model

Rutherford proposed the nuclear model of atom. According to this model,

- The positive charge and most of the mass of the atom was densely concentrated in extremely small region called nucleus.
- The nucleus is surrounded by electrons that move around the nucleus in circular paths called orbits.
- Electrons and nucleus are held together by electrostatic force of attraction.
- **Drawback of Rutherford's Model:** A serious drawback of the Rutherford model is that it says nothing about distribution of the electrons around the nucleus and the energies of these electrons. Also, it was unable to explain the stability of an atom.

Atomic number and Mass number

- **Atomic Number:** The number of protons present in the nucleus of an atom is known as Atomic Number. It is represented by "Z".
- **Mass Number:** The mass of the nucleus is due to the presence of protons and neutrons, and collectively they are called as nucleons. It is represented by "A".
- No. of neutrons = A – Z

Planck's Quantum and Photoelectric Effect

The energy (E) of a quantum of radiation is proportional to its frequency (ν) and is expressed by equation:

$E = h\nu$

The proportionality constant, 'h' is known as Planck's constant and has the value 6.626×10^{-34} J s.

The Photoelectric effect is the phenomenon in which metals emit electrons when light is incident on it.

Kinetic energy of the ejected electron is proportional to the frequency of the electromagnetic radiation. Since the striking photon has energy equal to hν and the minimum energy required to eject the electron is $h\nu_0$ (also called work function, W_0, then the difference in energy ($h\nu - h\nu_0$) is transferred as the kinetic energy of the photoelectron. Following the conservation of energy principle, the kinetic energy of the ejected electron is given by the equation where m_e is the mass of the electron and v is the velocity associated with the ejected electron.

$$h\nu = h\nu_0 + \frac{1}{2}m_e v^2$$

or

$$E = W_0 + \frac{1}{2}mv^2$$

where; E = Energy of photon

W_0 = Work function

- **Planck's equation:** $E = h\nu = \frac{hc}{\lambda}$

Bohr's Model

Bohr gave the model of stationary energy orbits within the atom. He also gave emission spectrum of hydrogen atom.

$$\nu = \frac{1}{\lambda} = R_H Z^2\left[\frac{1}{n_1^2} - \frac{1}{n_2^2}\right]$$

$$R_H = 1.097 \times 10^7 m^{-1}$$

For Lyman series; $n_1 = 1$; $n_2 > 1$

For Balmer series; $n_1 = 2$; $n_2 > 2$

For Paschen series; $n_1 = 3$; $n_2 > 3$

For Brackett series; $n_1 = 4$; $n_2 > 4$

For Pfund series; $n_1 = 5$; $n_2 > 5$

- **Angular Momentum:** $mvr = \frac{nh}{2\pi}$

  n = 1, 2, 3.....

- **Bohr's Radius:** $r_n = \frac{r_0 n^2}{Z} = 0.529 \times \frac{n^2}{Z}\overset{0}{A}$

- **Energy of Electron in n^th^ Bohr's orbit:**

  $$E_n = \frac{-13.6Z^2}{n^2} eV/atom$$

- **Potential Energy:** $(E_P) = \frac{-Ze^2}{4\pi r\varepsilon_0}$

- **Speed of electron in n^th^ orbit (V_n):** $V_n = \frac{2.18 \times 10^6 Z}{n}$ m/s

- **Frequency:** $c = h\nu, \nu = \frac{\Delta E}{h} = \frac{R_H}{h}\left[\frac{1}{n_1^2} - \frac{1}{n_2^2}\right]$

- Energy of electron in its stationary state

  $$E_n = -R_H\left(\frac{1}{n^2}\right)$$

  n = 1, 2, 3.....

- **Number of Spectral Line:** $\frac{(n_2 - n_1)(n_2 - n_1 - 1)}{2}$

Limitations of bohr's model

- **Zeeman Effect:** Splitting of spectral line by magnetic field.
- **Stark Effect:** Splitting of spectral line by electric field.
- Unable to explain the ability of atoms to form molecules by chemical bonds.

Towards Quantum Mechanical Model of Atom

De-Broglie Relationship

Matter which exhibit both particle as well as wave nature is called as de-broglie relationship.

$$\lambda = \frac{h}{p} = \frac{h}{mv}$$

Heisenberg's uncertainty principle

- According to this principle, it is impossible to determine simultaneously the exact position and exact momentum (or velocity) of an electron.
- **Heisenberg's Uncertainty Principle:** $\Delta x \times \Delta p \geq \frac{h}{4\pi}$

Quantum Mechanical Model of Atom

Quantum Number

There are 4 types of quantum numbers:

1. **Principal Quantum Number (n):** It represents the **name, size** and **energy** of the shell to which electron belongs
2. **Angular/Azimuthal Quantum Number(*l*):**

   It represents the shape of the **subshell and orbital**.

   It is used for calculating **orbital angular momentum**.

   Value of *l* is between 0 to (n – 1)

   $l = 0$ = s-orbital $l = 1$= p-orbital

   $l = 2$ = d-orbital $l = 3$ = f-orbital
3. **Magnetic Quantum Number (m):** Shows subshell (orbital) preferred orientation in 3-D shape.

Value of m = All integral values from $-l$ to $+l$ including zero.

Total number of $m_l = 2l + 1$

4. **Spin Quantum Number (s):** Shows the direction of spin of electrons.

$$s = \pm\frac{1}{2}$$

Shapes of Atomic Orbitals

The probability of locating the electron at different distances from the nucleus can be plotted graphically by plotting probability (ψ^2) against the distance (r) from the nucleus of the atom.

- The plane on which the probability of finding the electron is zero is known as nodal plane.
- For 1s, the probability density is maximum at the nucleus and it decreases sharply as we move away from it.

s-orbital

- The shape of s orbital ($l = 0$) **is spherical.**
- Size of s orbital increases with increase in n, that is 4s > 3s > 2s > ls

p-orbital

- In a p-sub shell, the three orbitals are represented as p_x, p_y, and p_z. These are degenerate orbitals.
- The shape of a p-orbital ($l = 1$) **is dumbbell.**
- With the increase of principle quantum number, size, and energy of 'p' orbitals increases, that is, **4p > 3p > 2p**

d-orbital

- In a d-sub shell, the five orbitals are represented as d_{xy}, d_{yz}, d_{zx}, $d_{(x^2-y^2)}$ and d_{z^2}.
- The shape of a d – orbital ($l = 2$) is **double dumb bell.**
- d_{xy}, d_{yz} d_{zx} orbitals are oriented in between the **axes.** $d_{(x^2-y^2)}$ and $d_{(z^2)}$ orbitals are oriented **along the axes.**

Nodes

- Number of nodal planes (or) angular nodes for an orbital = l.
- Number of radial nodes = n – l – 1 where n = principal quantum number, l = Azimuthal quantum number.
- Total numbers of nodes = (n–1) i.e., sum of angular nodes and radial nodes.

Energy of orbitals

The energy of an electron in a hydrogen atom is determined by the principal quantum number. Thus, the energy of the orbitals increases as follows:

1s < 2s = 2p < 3s = 3p = 3d < 4s = 4p = 4d = 4f

Rules For Filling of Orbitals

Aufbau principle

According to this principle, electron enters in the subshells of an atom in the increasing order of energy.

Exceptions of aufbau's principle

- In some cases, it is observed that the actual electronic configuration of an element is slightly different from the arrangement given by aufbau principle. Reason is that half filled & fully filled subshell have got extra stability due to symmetry and exchange energy.

(n + l) rule

- The subshell with lowest (n + l) value is filled up first, but when two or more subshells have same (n + l) value, then the subshell with lowest value of n is filled up first.

Pauli's exclusion principle

It states that "no two electrons in the same atom can have the same set of four quantum numbers".

Total number of nth orbital = n^2

Total number of electron in n^{th} orbital = $2n^2$

Hund's rule

According to this rule, pairing of electrons in the orbitals belonging to the same subshell (p, d, or f) does not take place until each orbital belonging to that subshell is singly occupied.

MCQs

NCERT Topic-wise MCQs

Sub-Atomic Particles and Atomic Models

1. The charge of an electron was discovered by:
   1) J.J. Thomson 2) Neil Bohr
   3) Chadwick 4) Millikan
2. In lithium nitride, the nitride ion is composed of:
   1) 7 – protons + 10 electrons
   2) 12 – protons + 5 electrons
   3) 12 – protons + 8 electrons
   4) 10 – protons + 7 electrons
3. The nucleus of helium contains
   1) Four protons
   2) Four neutrons
   3) Two neutrons and two protons
   4) Four protons and two electrons
4. The element used by Rutherford in his famous scattering experiment was?
   1) Gold 2) Tin
   3) Silver 4) Lead
5. Be^{2+} is isoelectronic with which of the following ions?
   1) H^+ 2) Li^+
   3) Na^+ 4) Mg^{2+}
6. ($_{32}Ge^{76}$, $_{34}Se^{76}$) and ($_{14}Si^{30}$, $_{16}S^{32}$) are the examples of-
   1) Isotopes and isobars 2) Isobars and isotones
   3) Isotones and isotopes 4) Isobars and isotopes
7. Neutrons are present in all atoms except:
   1) He 2) C
   3) H 4) Ca
8. The ratio of charge and mass would be greater for:
   1) Proton 2) Electron
   3) Neutron 4) α-particle
9. An isotope of $^{112}_{50}Sn$ contains 68 neutrons, its mass number is:
   1) 118 2) 90
   2) 50 4) 112
10. An element X has the following composition ^{200}X : 90%, ^{199}X : 8.0%, ^{202}X : 2.0%. The average atomic mass of the naturally occurring element X is closest to:
    1) 200 u 2) 201 u
    3) 202 u 4) 199 u
11. Which pair is of isoelectronic species?
    1) K^+, Cl^-, Na^+ 2) K^+, Cl^-, Ca^{2+}
    3) Fe^-, Al^{3+}, K^+ 4) Fe^{2+}, Cu^{2+}, V^{3+}
12. The nitride ion in lithium nitride is composed of:
    1) 7-protons + 10 electrons
    2) 10-protons + 10 electrons
    3) 7-protons + 7 electrons
    4) 10-protons + 7 electrons
13. In which one of the following pairs, the two species are both isoelectronic and isotopic? (Atomic number: Cr = 24, Ar = 18, K = 19, Mg = 12, Fe = 26, Na = 11)
    1) $^{24}Mg^{2+}$ and ^{40}Ar 2) $^{39}K^+$ and $^{40}K^+$
    3) $^{24}Mg^{2+}$ and ^{25}Mg 4) ^{23}Na and $^{24}Na^+$
14. The ratio of neutrons in C and Si with respective atomic masses 12 and 28 is:
    1) 2:3 2) 3:2
    3) 3:7 4) 7:3
15. If a species has 16 protons, 18 electrons and 16 neutrons. Find the species and its charge
    1) S^- 2) Si^{2-}
    3) S^{3-} 4) S^{2-}
16. In which one of the following, the number of protons is greater than the number of neutrons but number of protons is less than the number of electrons?
    1) D_3O^+ 2) SO_2
    3) H_2O 4) OH^-

Developments Leading to The Bohr's Model of Atom

17. The scientist who proposed the atomic model based on the quantisation of energy for the first time is:
    1) Max Planck 2) Neil Bohr
    3) De-Broglie 4) Heisenberg
18. The wavelength (in Å) of an emission line obtained for Li^{2+} during a electronic transition from $n_2 = 2$ to $n_1 = 1$ is: (R = Rydberg constant)
    1) 3R/4 2) 27R/4
    3) 4/3R 4) 4/27R
19. The threshold frequency ν_0 for a metal is 6×10^{14} s^{-1}. Calculate the kinetic energy of an electron emitted when radiation of frequency = 1.1×10^{15} s^{-1} hits the metal.
    1) 3.313×10^{-19}J 2) 3.6×10^{-19}J
    3) 0.3313×10^{-17} 4) 0.3313×10^{-20}J

20. The value of Rydberg constant is:

1) 109, 677 cm^{-1} 2) 19876 cm^{-1}
3) 108769 cm^{-1} 4) 108976 cm^{-1}

21. What is the lowest energy of the spectral line emitted by the hydrogen atom in the Lyman series?

1) $\frac{5hR_Hc}{36}$ 2) $\frac{4hR_Hc}{3}$
3) $\frac{3hR_Hc}{4}$ 4) $\frac{7hR_Hc}{144}$

22. A metal surface is exposed to solar radiations

1) The emitted electrons have energy less than a maximum value of energy depending upon the frequency of the incident radiation.
2) The emitted electrons have energy less than the maximum value of energy depending upon intensity of incident radiations.
3) The emitted electrons have zero energy.
4) The emitted electrons have energy equal to energy of photons of incident light.

23. A 100 watt bulb emits monochromatic light of wavelength 400nm. Calculate the number of photons emitted per second by the bulb.

1) 20.12×10^{20} s^{-1} 2) 2.012×10^{20} s^{-1}
3) 4.969×10^{-19} s^{-1} 4) 49.69×10^{-19} s^{-1}

24. The quantum of light energy is called:

1) Photon 2) Neutron
3) Electron 4) Proton

Bohr's Model For Hydrogen Atom

25. The emission spectrum of hydrogen discovered first and the region of the electromagnetic spectrum to which it belongs, respectively are-

1) Lyman, Ultraviolet 2) Lyman, Visible
3) Balmer, Ultraviolet 4) Balmer, Visible

26. The velocity of electron in second shell of hydrogen atom is:

1) 10.94×10^6 m/sec 2) 18.88×10^6 m/sec
3) 1.88×10^6 m/sec 4) 1.094×10^6 m/sec

27. For which of the following species, Bohr's theory is not applicable?

1) Be^{3+} 2) Li^{2+}
3) He^{2+} 4) H

28. If the first ionisation energy of H- atom is 13.6 eV, then the second ionisation energy of He- atom is:

1) 27.2 eV 2) 40.8 eV
3) 54.4 eV 4) 108.8 eV

29. When the electrons of hydrogen atom return to L-shell from shell of higher energy, we get a series of lines in the spectrum. This series is called:

1) Balmer series 2) Lyman series
3) Brackett series 4) Paschen series

30. When the electron of a hydrogen atom jump from n = 4 to n = 1 state, the number of different spectral line emitted are:

1) 15 2) 9
3) 6 4) 3

31. The wave number of the spectral line in the emission spectrum of hydrogen will be equal to 8/9 times the rydberg constant if the electron jumps from:

1) n = 3 to n = 1 2) n = 10 to n = 1
3) n = 9 to n = 1 4) n = 2 to n = 1

32. The radius of the first Bohr orbit of the H-atom is r. Then, the radius of the first orbit of Li^{2+} will be:

1) r/9 2) r/3
3) 3r 4) 9r

33. The energy ratio of a photon of wavelength 3000Å and 6000Å is:

1) 1 : 1 2) 2 : 1
3) 1 : 2 4) 1 : 4

34. Bohr's model can explain:

1) The solar spectrum
2) Spectrum of hydrogen molecule
3) Spectrum of any atom or ion containing one electron only
4) Spectrum of hydrogen atom only

35. The first line emission of hydrogen atom spectrum in the Balmer species appears at:

1) 5R/36 cm^{-1} 2) 3R/4 cm^{-1}
3) 7R/144 cm^{-1} 4) 9R/400 cm^{-1}

36. The radius of 2nd Bohr's orbit of hydrogen atom is:

1) 0.053 nm 2) 0.106 nm
3) 0.2116 nm 4) 0.4256 nm

37. The maximum energy possessed by an electron is when it is present:

1) In nucleus
2) In ground energy state
3) In 1st excited state
4) At infinite distance from the nucleus

38. Which one of the following transition have minimum wavelength?

1) $n_4 \rightarrow n_1$ 2) $n_2 \rightarrow n_1$
3) $n_4 \rightarrow n_2$ 4) $n_3 \rightarrow n_1$

39. Select the pair where both species have same radius?

1) r_4He^+ and r_1H 2) r_2Be^{3+} and r_1H
3) r_2He and r_1H 4) r_3Li^{2+} and r_1H

40. The radius of the first orbit of hydrogen atom is 0.52×10^{-8} cm. The radius of the first orbit of He^+ ion is

1) 0.26×10^{-8} cm 2) 0.52×10^{-8} cm
3) 1.04×10^{-8} cm 4) 2.08×10^{-8} cm

41. The ratio of ionization energy of H and Be^{+3} is
1) 1 : 1
2) 1 : 3
3) 1 : 9
4) 1 : 16

42. As we move away from nucleus, the energy of orbit
1) Decreases
2) Increases
3) Remains unchanged
4) None of these

43. The ratio of the energy of the electron in ground state of hydrogen to the electron in Ist excited state of Be^{3+} is:
1) 1 : 4
2) 3 : 2
3) 18 : 1
4) 1 : 18

44. In which transition, one quantum of energy is emitted?
1) $n_2 = 4 \rightarrow n_1 = 2$
2) $n_2 = 3 \rightarrow n_1 = 1$
3) $n_2 = 2 \rightarrow n_1 = 1$
4) All of these

45. The wavelength of first line of Balmer spectrum of hydrogen will be :
1) 4340 Å
2) 4101 Å
3) 6569 Å
4) 4861 Å

46. If the radius of 2nd Bohr orbit of hydrogen atom is r_2. The radius of third Bohr orbit will be:-
1) $\frac{4}{9}r_2$
2) $4r_2$
3) $\frac{9}{4}r_2$
4) $9r_2$

47. The ratio of highest possible wavelength to lowest possible wavelength of Lyman series is
1) 4/3
2) 9/8
3) 27/5
4) 16/5

48. Magnitude of kinetic energy in an orbit is equal to:
1) Half of the P.E.
2) Twice of the P.E.
3) 1/4th of the P.E.
4) None of the above

49. According to Bohr's theory, the angular momentum of an electron in the 4th orbit is:
1) $\frac{h}{2\pi}$
2) $\frac{2h}{\pi}$
3) $\frac{3h}{2\pi}$
4) $\frac{3h}{\pi}$

50. If the radius of 1st Bohr orbit be a_0, then radius of 3rd Bohr orbit would be:
1) $3a_0$
2) $6a_0$
3) $9a_0$
4) $\frac{1}{9}a_0$

Towards Quantum Mechanical Model of The Atom

51. The uncertainty in the velocity of particle of mass 6.626×10^{-28} kg is 10^{-6} m/sec. What is the uncertainty in its position in nm?
1) $1/2\pi$
2) $2.5/\pi$
3) $4/\pi$
4) $1/4\pi$

52. The ratio of de-broglie wavelength of a deuterium atom to that of α-particle when the velocity of former is five times greater than that of latter is:
1) 4
2) 10.2
3) 2
4) 0.4

53. As per de-broglie formula, a microscopic particle of mass 100 g and moving at a velocity of 100 cm/sec will have a wavelength of:
1) 6.6×10^{-29} cm/s
2) 6.6×10^{-30} cm/s
3) 6.6×10^{-31} cm/s
4) 6.6×10^{-32} cm/s

54. The de-broglie wavelength associated with particle of mass 10^{-6} kg moving with a velocity of 10 m/sec is:
1) 6.63×10^{-7} m
2) 6.63×10^{-16} m
3) 6.63×10^{-21} m
4) 6.63×10^{-29} m

55. If E_e, E_α, and E_p represents the kinetic energy of an electron, α-particle and proton respectively and each moving with same de-broglie wavelength then:
1) $E_e = E_\alpha = E_p$
2) $E_e > E_\alpha > E_p$
3) $E_\alpha > E_p > E_\alpha$
4) $E_e > E_p > E_\alpha$

56. It is given that, 0.1 m/sec and 0.05 m/sec are the velocities of particles "X" and "Y" respectively. If the mass of particle Y is 5 times that of particle X, then find the ratio of de-broglie wavelength (λ) associated with particle X and Y.
1) 1 : 2
2) 4 : 3
3) 2 : 3
4) 5 : 2

57. Uncertainty principle was given by-
1) Einstein
2) Heisenberg
3) Rutherford
4) Thomson

58. A particle having a mass of 1 mg has a velocity 1 km/sec Calculate the wavelength of the particle:
1) 6.626×10^{-28} cm
2) 6.626×10^{-29} cm
3) 6.626×10^{-30} cm
4) 6.626×10^{-31} cm

59. The uncertainty in the position of an electron moving with a velocity of 3×10^4 cm/sec accurate upto 0.011% will be:
1) 1.92 cm
2) 7.66 cm
3) 0.175 cm
4) 3.84 cm

60. If uncertainty in position and velocity are equal, then uncertainty in momentum will be:
1) $\frac{1}{2}\sqrt{\frac{mh}{\pi}}$
2) $\frac{1}{2}\sqrt{\frac{h}{\pi m}}$
3) $\frac{h}{4\pi m}$
4) $\frac{mh}{4\pi}$

61. The velocity of particle "A" is 0.1m/sec and that of particle "B" is 0.05m/sec. If the mass of particle B is 5 times that of particle A, then the ratio of de-broglie wavelength associated with particle A and B:
1) 2 : 5
2) 3 : 4
3) 6 : 4
4) 5 : 2

62. If de-broglie wavelength of mass 'm' is 100 times of its velocity then its value in term of its mass 'm' and planck's constant "h" is:

1) $\frac{1}{10}\times\sqrt{\frac{m}{h}}$ 2) $10\sqrt{\frac{h}{m}}$

3) $\frac{1}{10}\sqrt{\frac{h}{m}}$ 4) $10\sqrt{\frac{m}{h}}$

63. The wavelength of electron waves in two orbit is in ratio 3 : 5. The ratio of K.E. of electrons will be:

1) 25 : 9 2) 15 : 3

3) 9 : 25 4) 3 : 5

64. If uncertainty in position and momentum are equal, then uncertainty in velocity is:

1) $\frac{1}{2m}\sqrt{\frac{h}{\pi}}$ 2) $\sqrt{\frac{h}{2\pi}}$

3) $\frac{1}{m}\sqrt{\frac{h}{\pi}}$ 4) $\sqrt{\frac{h}{\pi}}$

65. The de-broglie wavelength associated with a mass of 1kg having K.E. 0.5J is:

1) 6.626×10^{-34} m 2) 13.20×10^{-34} m

3) 10.38×10^{-31} m 4) 6.626×10^{-37} m

66. A tennis ball weighing 0.91 kg and an electron travel with the same velocity of 10^2 m/sec. The ratio of their de-Broglie wavelengths would be:

1) $10^{-16} : 4$ 2) $10^{-28} : 1$

3) $4 : 10^{-16}$ 4) $10^{-30} : 1$

67. For an electron, if the uncertainty in velocity is Δv, the uncertainty in position Δx is given by:

1) $\frac{hm}{4\pi\Delta v}$ 2) $\frac{4\pi}{hm\Delta v}$

3) $\frac{h}{4\pi m\Delta v}$ 4) $\frac{4\pi m}{h\Delta v}$

68. If uncertainty in the position of an electron is zero, the uncertainty in its momentum will be:

1) $<\frac{h}{2\pi}$ 2) $>\frac{h}{2\pi}$

3) $<\frac{h}{4\pi}$ 4) infinite

69. de-Broglie wavelength associated with a material particle is

1) Inversely proportional to momentum
2) Inversely proportional to its energy
3) Directly proportional to momentum
4) Directly proportional to its energy

70. The uncertainty in the position of an electron and proton is equal, the ratio of the uncertainties in the velocity of an electron and proton is:

1) $10^3 : 1$ 2) 1 : 1838

3) 3672 : 1 4) 1836 : 1

Quantum Mechanical Model of Atom

71. Which among the following species have the same number of electron in its outermost as well as penultimate shell?

1) Mg^{2+} 2) O^{2-}

3) F^- 4) Ca^{2+}

72. Which have the same number of s-electrons as the d-electrons in Fe^{2+}?

1) Li 2) Na

3) N 4) P

73. The number of waves made by an electron moving in an orbit having maximum magnetic quantum number + 3 is-

1) 4 2) 3

3) 5 4) 6

74. The set of quantum no. not applicable for an electron-

1) $2,0,0,-\frac{1}{2}$ 2) $3,1,-2,+\frac{1}{2}$

3) $4,2,-2,-\frac{1}{2}$ 4) $6,0,0,+\frac{1}{2}$

75. What is the orbital angular momentum of an electron in f-orbital?

1) $\frac{1.5h}{\pi}$ 2) $\frac{\sqrt{6}h}{\pi}$

3) $\frac{\sqrt{3}h}{\pi}$ 4) $\frac{\sqrt{3}h}{\pi}$

76. Given K L M N The number of electrons present
2 8 11 2
in $l = 2$ is

1) 3 2) 6

3) 5 4) 4

77. The maximum number of electrons which can be held by subshell with azimuthal quantum number "ℓ" in an atom is given by:

1) $2\ell + 1$ 2) $2\ell + 2$

3) $2(2\ell + 1)$ 4) $2(2\ell + 2)$

78. What is the maximum number of electrons that can be associated with a following set of quantum number n = 3, l = 1 and m = –1?

1) 10 2) 6

3) 4 4) 2

79. The quantum number "m" of a free gaseous atom is associated with:

1) Effect volume of orbital
2) Shape of orbital
3) Spatial orientation of orbital
4) Energy of orbital in the absence of magnetic field

80. Which of the following element is represented by electronic configuration $1s^2\ 2s^2\ 2p_x^1\ 2p_y^1\ 2p_z^1$?

1) N 2) O

3) P 4) S

81. The atomic number of element is 17. The number of orbital containing electron pair in its valence shell is:
    1) 3 2) 4
    3) 6 4) 8

82. The total number of electrons present in all the p–orbital of bromine are (Given: Atomic no. of Br = 35)
    1) 5 2) 15
    3) 17 4) 35

83. The number of unpaired electron in Fe^{3+}(Z = 26) are:
    1) 5 2) 6
    3) 3 4) 4

84. The orbital angular momentum of p- electron is given as:
    1) $\sqrt{2}\pi$ 2) $\frac{h}{\sqrt{2}\pi}$
    3) $\frac{\sqrt{3}h}{2\pi}$ 4) $\frac{h}{\pi}$

85. Which of the following statement in relation to the hydrogen atom is correct?
    1) 3s, 3p and 3d orbital all have the same energy
    2) 3s and 3p orbital are of lower energy than 3d orbital
    3) Both of these
    4) None of these

86. The total number of orbitals in a shell with principle quantum number n is:
    1) 2n 2) $2n^2$
    3) n^2 4) n + 1

87. Which of the following is not possible?
    1) n = 3, ℓ = 0, m = 0
    2) n = 3, ℓ = 1, m = –1
    3) n = 2, ℓ = 0, m = –1
    4) n = 2, ℓ = 1, m = –1

88. Which of the following represents the correct set of quantum numbers of a 4d electron?
    1) 4, 3, 2, +1/2 2) 4, 2, 1, 0
    3) 4, 3, –2, +1/2 4) 4, 2, 1, –1/2

89. Any f-orbital can accommodate upto
    1) 2 electrons with parallel spin
    2) 6 electrons
    3) 2 electrons with opposite spin
    4) 14 electrons

90. The atomic numbers of elements X,Y and Z are 19, 21 and 25 respectively. The number of electrons present in the M shells of these elements follow the order.
    1) Z > X > Y 2) X > Y > Z
    3) Z > Y > X 4) Y > Z > X

91. This electronic configuration shows element of:

↑↓	↑↓	↑↓ ↑↓ ↑
1s	2s	2p

    1) Oxygen 2) Neon
    3) Fluorine 4) Nitrogen

High Order Time Intensive MCQs

1. If the binding energy of 2nd excited state of H like sample is 24 eV approximately, then the ionisation energy is approximately:
    1) 54.4 eV 2) 24 eV
    3) 122.4 eV 4) 216 eV

2. The de-broglie wavelength of an electron moving in a circular orbit is λ.
   The minimum radius of the orbit is given by:
    1) $\frac{\lambda}{\pi}$ 2) $\frac{\lambda}{2\pi}$
    3) $\frac{\lambda}{4\pi}$ 4) $\frac{\lambda}{3\pi}$

3. A sodium street light gives off yellow light that has a wavelength of 600 nm. Then:
   $$\left(\text{Energy of photon} = \frac{12400\text{eV}\,\text{Å}}{\lambda(\text{Å})}\right)$$
    1) Frequency of this light is $7 \times 10^{14}\ s^{-1}$
    2) Frequency of this light is $5 \times 10^{14}\ s^{-1}$
    3) Energy of photon is approximately 2.07 eV
    4) Both (2) and (3)

4. Photon having wavelength 310 nm is used to break the bond of A_2 molecule having bond energy 288 kJ/mol, then the percentage of energy of photon converted to KE is (1 eV = 96 kJ/mol):
    1) 25 2) 50
    3) 75 4) 80

5. The radii of two of first six Bohr's orbits of H atom are in the ratio 4 : 9. The energy difference between them may be:
    1) Either 12.09 eV or 3.4 eV
    2) Either 2.55 eV or 10.2 eV
    3) Either 1.89 eV or 0.49 eV
    4) 1.89 eV only

6. Last line of Lyman series for H-atom has wavelength λ_1 Å. The 2nd line of Balmer series has wavelength λ_2 Å, then:
    1) $\frac{16}{\lambda_1} = \frac{9}{\lambda_2}$ 2) $\frac{16}{\lambda_2} = \frac{3}{\lambda_1}$
    3) $\frac{4}{\lambda_1} = \frac{1}{\lambda_2}$ 4) $\frac{16}{\lambda_1} = \frac{3}{\lambda_2}$

7. Which of the following statements is/are correct for an electron having quantum numbers n = 4 & m = 2?
   1) The value of ℓ may be 2
   2) The value of ℓ may be 3
   3) The value of s may be +1/2
   4) All of these

8. Change in orbit angular momentum when an electron makes a transition corresponding to 3rd line of Balmer series in Li^{2+} ion is:
   1) $\frac{h}{2\pi}$ 2) $\frac{2h}{2\pi}$
   3) $\frac{3h}{2\pi}$ 4) $\frac{4h}{2\pi}$

9. The ionisation energy of H atom is x J/atom. The wavelength of first Balmer line for He^+ ion is-
   1) $\frac{5x}{36}$ 2) $\frac{5x}{9}$
   3) $\frac{5x}{36hc}$ 4) $\frac{9hc}{5x}$

10. An energy of 24.6 eV is required to remove one of the electrons from a He atom. The energy required to remove both the electrons from He atom is-
    1) 38.2 eV 2) 49. 2 eV
    3) 51.8 eV 4) 79 eV

11. Photons of frequency 3.2×10^{16} Hz is used to irradiate a metal surface, the maximum kinetic energy of the emitted photo-electron is $\frac{3}{4}^{th}$ of the energy of the irradiating photon? What is the threshold frequency of the metal?
    1) 2.4×10^{25} Hz 2) 2.4×10^{16} Hz
    3) 1.6×10^{15} Hz 4) 8×10^{15} Hz

12. The total number of orbitals in the principal shell of He^+ that has energy equal to $\frac{-Rhc}{4}$ (where R is Rydberg's constant) is.
    1) 4 2) 8
    3) 16 4) 32

13. Not considering the electron spin, the degeneracy of second excited state is 9, while the degeneracy of second excited state of H^- is:
    1) 1 2) 2
    3) 3 4) 4

14. Match the column-I with column-II.

	Column-I		Column-II
(a)	Radius of 2nd orbit of H-atom	(i)	2 a_0
(b)	Radius of 1st orbit of He^+ ion	(ii)	4 a_0
(c)	Radius of 2nd orbit of Be^{3+} ion	(iii)	0.5 a_0
(d)	Radius of the nodal surface of 2s orbital	(iv)	a_0

    1) (a)-(ii); (b)-(iii); (c)-(iv); (d)-(i)
    2) (a)-(ii); (b)-(iii); (c)-(i); (d)-(iv)
    3) (a)-(iii); (b)-(ii); (c)-(iv); (d)-(i)
    4) (a)-(i); (b)-(iii); (c)-(iv); (d)-(ii)

15. If ΔE is the energy emitted in eV when an electronic transition occurs from higher energy level to lower energy level in H–atom, the λ of the line produced is approximately equal to-
    1) $\frac{12375}{\Delta E}$Å 2) $\frac{19800}{\Delta E}$Å
    3) $\frac{13600}{\Delta E}$Å 4) $\frac{21800}{\Delta E}$Å

16. Which of the following is the energy of a possible excited state of hydrogen?
    1) + 13.6 eV 2) – 3.4 eV
    3) + 6.8 eV 4) – 2.8 eV

17. If an electron travels with a velocity of $\frac{1}{100}$th speed of light in Bohr's first orbit, what is its velocity (relative to the speed of light) in the 5th Bohr orbit?
    1) 0.002 2) 0.1
    3) 0.5 4) 0.7

18. The ionisation energy of He^+ is 19.6×10^{-18} J $atom^{-1}$. The energy of the first stationary of Li^{2+} is-
    1) -2.2×10^{-15} J $atom^{-1}$ 2) 8.82×10^{-17} J $atom^{-1}$
    3) 4.41×10^{-16} J $atom^{-1}$ 4) -4.41×10^{-17} J $atom^{-1}$

19. Photoelectric emission is observed from a metal surface with incident frequencies ν_1 & ν_2 where $\nu_1 > \nu_2$. If the kinetic energies of the photoelectrons emitted in the two cases are in the ratio 2:1, then the threshold frequency ν_o of the metal is-
    1) $2\nu_2 - \nu_1$ 2) $2\nu_1 - \nu_2$
    3) $\frac{\nu_1 - \nu_2}{2}$ 4) $\nu_2 - \nu_1$

20. The ratio of area covered in 2nd orbit to first orbit is-
    1) 1 : 1 2) 16 : 1
    3) 1 : 16 4) 4 : 1

21. An electron in an atom undergoes transition in such a way that its kinetic energy changes from x to $\frac{x}{4}$, the change in potential energy will be :
    1) $+\frac{3}{2}x$ 2) $-\frac{3}{8}x$
    3) $+\frac{3}{4}x$ 4) $-\frac{3}{4}x$

22. What transition in He^+ ion shall have the same wave number as the first line in Balmer series of hydrogen atom ?
    1) $3 \rightarrow 2$ 2) $6 \rightarrow 4$
    3) $5 \rightarrow 3$ 4) $7 \rightarrow 5$

23. In an atom, an electron is moving with a speed of 600 m/s with an accuracy upto 0.005%. What is the uncertainty in position?
    1) 1.52×10^{-4} m 2) 5.10×10^{-3} m
    3) 1.92×10^{-3} m 4) 3.84×10^{-3} m

24. The energy absorbed by each molecule A_2 of a substance is 4.4×10^{-19} J and bond energy per molecule is 4×10^{-19} J. The K.E. of the molecule per atom is-

1) 2.2×10^{-19}J
2) 2×10^{-19}J
3) 4×10^{-20}J
4) 2×10^{-20}J

25. What is the maximum number of electrons which can be accomodated in principal quantum number 4?

1) 16
2) 18
3) 22
4) 32

26. What is the maximum number of electrons which can be accommodated in an atom in which the highest principal quantum number is 4?

1) 10
2) 18
3) 36
4) 54

27. How many elements would be in the IInd period of the Periodic Table if the spin quantum numbers could have the value $+\frac{1}{2}, 0, -\frac{1}{2}$?

1) 18
2) 12
3) 10
4) 8

28. Suppose that a hypothetical H-like atom gives a red, green, blue and violet line in spectrum. Which jump, according to figure, would give off the violet spectral line?

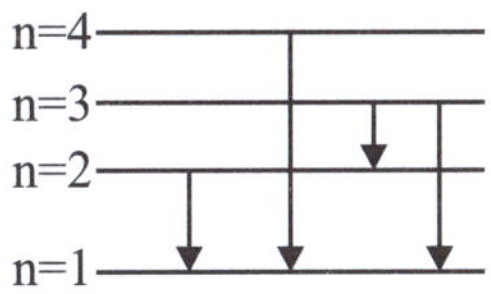

1) $3 \rightarrow 1$
2) $2 \rightarrow 1$
3) $4 \rightarrow 1$
4) $3 \rightarrow 2$

29. Which of the following set of quantum number is possible?

1) n = 4, l = 2, m = –2, s = –2
2) $n = 4, l = 4, m = 0, s = \frac{1}{2}$
3) $n = 4, \ell = 3, m = -3, s = \frac{1}{2}$
4) n = 4, l = 0, m = 0, s = 0

30. In an atom, which has 2K, 8L, 18M and 2N electrons in the ground state. The total number of electrons having magnetic quantum number, m = 0 is

1) 6
2) 10
3) 7
4) 14

31. The probability density curve for 2s electron appears like

1) 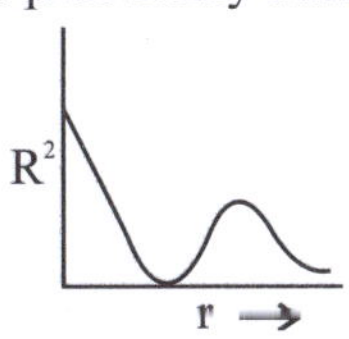

2) 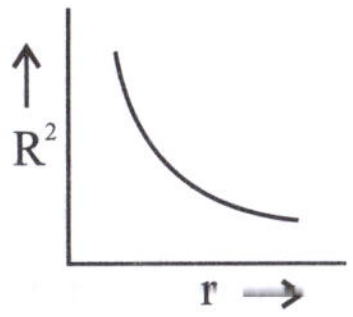

3) 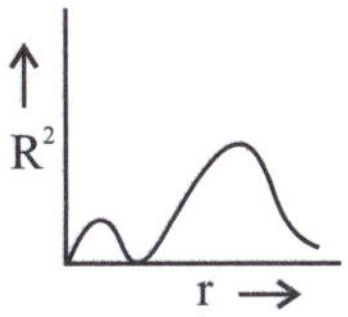

4) 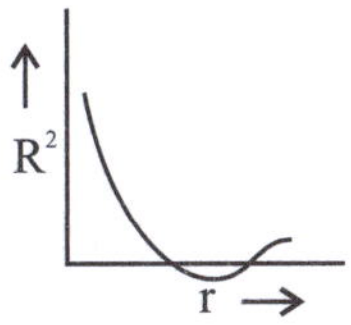

32. Which of the following statement concerning the four quantum number is incorrect?

1) n gives the size of an orbital
2) *l* gives the shape of an orbital
3) m gives the energy of the electron in orbital
4) Both (1) & (2)

33. Consider the following sets of quantum number

	n	l	m	s
(i)	3	0	0	+1/2
(ii)	2	2	1	+1/2
(iii)	4	3	–2	–1/2
(iv)	1	0	–1	–1/2
(v)	3	2	3	+1/2

Which of the following sets of quantum number is not possible?

1) (i), (ii), (iii) and (iv)
2) (ii), (iv) and (v)
3) (i) and (iii)
4) (ii), (iii) and (iv)

34. Which of the following sets of quantum number is/are incorrect ?

1) $n = 3, \ell = 3, m = 0, s = \frac{1}{2}$
2) $n = 3, \ell = 2, m = 2, s = -\frac{1}{2}$
3) $n = 3, \ell = 1, m = 2, s = -\frac{1}{2}$
4) $n = 3, \ell = 0, m = 0, s = +\frac{1}{2}$

35. If an electron in H atom has an energy of –78.4 kcal/mol. The orbit in which the electron is present is:-

1) 1st
2) 2nd
3) 3rd
4) 4th

36. Difference between nth and (n + 1)th Bohr's radius of H–atom is equal to its (n – 1)th Bohr's radius. The value of n is:-

1) 1
2) 2
3) 3
4) 4

37. The dissociation energy of H_2 is 430.53 kJ mol^{-1}. If H_2 is dissociated by illumination with radiation of wavelength 253.7 nm. The fraction of the radiant energy which will be converted into kinetic energy is given by -

1) 8.86%
2) 2.33%
3) 1.3%
4) 90%

38. Light of wavelength λ shines on a metal surface with intensity x and the metal emits Y electrons per second of average energy, Z. What will happen to Y and Z if x is doubled?

1) Y will be double and Z will become half
2) Y will remain same and Z will be doubled
3) Both Y and Z will be doubled
4) Y will be doubled but Z will remain same

39. The momentum of radiation of wavelength 0.33 nm is ……. kg m sec^{-1}

1) 2×10^{-24} 2) 2×10^{-12}
3) 2×10^{-26} 4) 2×10^{-48}

40. Which one is the wrong statement?

1) The uncertainty principle is $\Delta E \times \Delta t \geq \frac{h}{4\pi}$
2) Half filled and fully filled orbitals have greater stability due to greater exchange energy, greater symmetry and more balanced arrangement.
3) The energy of 2s-orbital is less than the energy of 2p-orbital in case of hydrogen like atoms.
4) de-Broglie's wavelength is given by $\lambda = \frac{h}{mv}$, where m = mass of the particle, v = group velocity of the particle.

41. Arrange the electrons represented by the following set of quantum numbers in the decreasing order of energy

(i) n = 4, l = 0, m = 0, s = + 1/2
(ii) n = 3, l = 1, m = 1, s = – 1/2
(iii) n = 3, l = 2, m = 0, s = + 1/2
(iv) n = 3, l = 0, m = 0, s = – 1/2

1) (i) > (ii) > (iii) > (iv) 2) (iv) > (iii) > (ii) > (i)
3) (iii) > (i) > (ii) > (iv) 4) (i) > (iii) > (ii) > (iv)

42. Select the correct statement(s):

1) The value of spin only magnetic moment of Co^{3+} ion (in BM) = $\sqrt{24}$
2) The number of radial nodes in a 3p-orbital = 1
3) The number of electrons with (m = 0) in Mn^{2+} ion = 11
4) All are correct

43. Which among the following is an incorrect order against the property mentioned?

1) No. of Radial nodes : 3s > 2s > 1s
2) Order of energies of orbitals : $E_{2s(H)} > E_{2s(Li)} > E_{2s(K)}$
3) Order of de Broglies wavelength if their velocities are same : He > H_2 > O_2 > N_2
4) Order of distance between successive orbits: $(r_2 - r_1) < (r_3 - r_2)$

44. An isotone of $^{76}_{32}Ge$ is

(i) $^{77}_{32}Ge$ (ii) $^{77}_{33}As$
(iii) $^{77}_{34}Se$ (iv) $^{78}_{34}Se$

1) Only (i) and (ii) 2) Only (ii) and (iii)
3) Only (ii) and (iv) 4) Only (ii), (iii) and (iv)

45. Splitting of spectral lines under the influence of electric field is called:

1) Stark effect 2) Zeeman effect
3) Photoelectric effect 4) None of these

46. The energy associated with first orbit of He^+ is

1) 0 J 2) -8.72×10^{-18} J
3) -4.58×10^{-18} J 4) -0.545×10^{-18} J

Match the Columns MCQs

1. Match the following species with their corresponding ground state electronic configuration.

	Atom/Ion		Electronic configuration
(a)	Cu	(i)	$1s^2\ 2s^2\ 2p^6\ 3s^2\ 3p^6\ 3d^{10}$
(b)	Cu^{2+}	(ii)	$1s^2\ 2s^2\ 2p^6\ 3s^2\ 3p^6\ 3d^{10}\ 4s^2$
(c)	Zn^{2+}	(iii)	$1s^2\ 2s^2\ 2p^6\ 3s^2\ 3p^6\ 3d^{10}\ 4s^1$
(d)	Cr^{3+}	(iv)	$1s^2\ 2s^2\ 2p^6\ 3s^2\ 3p^6\ 3d^9$
		(v)	$1s^2\ 2s^2\ 2p^6\ 3s^2\ 3p^6\ 3d^3$

1) (a)-(ii) (b)-(iv) (c)-(i) (d)-(iii)
2) (a)-(iv) (b)-(v) (c)-(ii) (d)-(i)
3) (a)-(iii) (b)-(iv) (c)-(i) (d)-(v)
4) (a)-(v) (b)-(i) (c)-(iii) (d)-(iv)

2. Match the quantum numbers with the information provided by these.

	Quantum number		Information provided
(a)	Principal quantum number	(i)	orientation of the orbital
(b)	Azimuthal quantum number	(ii)	energy and size of orbital
(c)	Magnetic quantum number	(iii)	spin of electron
(d)	Spin quantum number	(iv)	shape of the orbital

1) (a)-(ii) (b)-(iv) (c)-(i) (d)-(iii)
2) (a)-(iv) (b)-(i) (c)-(ii) (d)-(v)
3) (a)-(iii) (b)-(ii) (c)-(iv) (d)-(i)
4) (a)-(iv) (b)-(iii) (c)-(i) (d)-(ii)

3. Match the following rules with their statements:

	Rules		Statements
(a)	Hund's Rule	(i)	No two electrons in an atom can have the same set of four quantum numbers.
(b)	Aufbau Principle	(ii)	Half-filled and completely filled orbitals have extra stability
(c)	Pauli Exclusion Principle	(iii)	Pairing of electrons in the orbitals belonging to the same subshell does not take place until each orbital is singly occupied.
(d)	Heisenberg's Uncertainty principle	(iv)	It is impossible to determine the exact position and exact momentum of a subatomic particle simultaneously.
		(v)	In the ground state of atoms, orbitals are filled in the order of their increasing energies.

1) (a)-(iv) (b)-(iii) (c)-(i) (d)-(ii)
2) (a)-(iv) (b)-(i) (c)-(ii) (d)-(v)
3) (a)-(iii) (b)-(ii) (c)-(iv) (d)-(i)
4) (a)-(iii) (b)-(v) (c)-(i) (d)-(iv)

4. Match the following

Column-I		Column-II	
(a)	X-rays	(i)	$v = 10^0 - 10^4$ Hz
(b)	UV	(ii)	$v = 10^{10}$ Hz
(c)	Long radio waves	(iii)	$v = 10^{16}$ Hz
(d)	Microwave	(iv)	$v = 10^{18}$ Hz

1) (a)-(iv) (b)-(iii) (c)-(i) (d)-(ii)
2) (a)-(ii) (b)-(iv) (c)-(i) (d)-(iii)
3) (a)-(iii) (b)-(ii) (c)-(iv) (d)-(i)
4) (a)-(iv) (b)-(i) (c)-(ii) (d)-(iii)

5. Match the following

Column-I		Column-II	
(a)	Photon	(i)	Value is 4 for N shell
(b)	Electron	(ii)	Probability density
(c)	Ψ^2	(iii)	Always positive value
(d)	Principal quantum number n	(iv)	Exhibits both momentum and wavelength

1) (a)-(iii) (b)-(ii) (c)-(iv) (d)-(i)
2) (a)-(ii) (b)-(iv) (c)-(i) (d)-(iii)
3) (a)-(iv) (b)-(iv) (c)-(ii)(iii) (d)-(i)(ii)
4) (a)-(iv) (b)-(i) (c)-(iii) (d)-(ii)

6. Match species given in Column I with the electronic configuration given in Column II

Column-I		Column-II	
(a)	Cr	(i)	$[Ar]3d^84s^0$
(b)	Fe^{2+}	(ii)	$[Ar]3d^{10}4s^1$
(c)	Ni^{2+}	(iii)	$[Ar]3d^64s^0$
(d)	Cu	(iv)	$[Ar]\ 3d^54s^1$
		(v)	$[Ar]3d^64s^2$

1) (a)-(iv) (b)-(i) (c)-(ii) (d)-(iii)
2) (a)-(iv) (b)-(iii) (c)-(i) (d)-(ii)
3) (a)-(iii) (b)-(ii) (c)-(iv) (d)-(i)
4) (a)-(ii) (b)-(iv) (c)-(i) (d)-(iii)

Correct & Incorrect MCQs

1. Which is an incorrect statement?
   1) Zeeman effect explains the splitting of spectral lines in a magnetic field.
   2) Cu^+ (z = 29) is paramagnetic
   3) Number of orbitals present in N-shell = 16
   4) Number of subshells present in M-shell – 3
2. The incorrect statement regarding Thomson's model of atom is:
   1) An atom possesses a spherical shape in which the positive charge is uniformly distributed.
   2) The electrons are embedded into the positively charged sphere.
   3) Atom as a whole is electrically neutral.
   4) Number of positive charges are more than number of negative charges in an atom.
3. Which statement regarding Rutherford's scattering experiment is not correct ?
   1) Alpha particle had enough energy to pass from gold foil in the Rutherford's experiment.
   2) Most of the α–particles passed through the gold foil undeflected.
   3) A small fraction of the α–particles was deflected by small angles.
   4) Nearly all α–particles bounced back, i.e., were deflected by nearly 180°.
4. Choose the correct statement/s:
   A. Atomic number is equal to the number of protons in the nucleus of an atom.
   B. Mass number is equal to the number of protons in the nucleus of an atom.
   C. Atomic number is equal to the number of electrons in a neutral atom.
   D. Mass number is equal to the number of neutrons in the nucleus of an atom.
   1) A, B and C 2) A and C
   3) A, C and D 4) B and D
5. Choose incorrect statement among the following:
   1) In an electromagnetic wave, the electric and magnetic field are mutually perpendicular.
   2) Electric field & magnetic field are perpendicular to the direction of propagation of the wave.
   3) Electromagnetic wave require medium, and cannot move in vacuum.
   4) Microwave region around 10^{10} Hz is used for radar.

6. Which statement regarding photoelectric effect is not correct ?
   1) In photoelectric effect, there are ejection of electrons from metal surface when radiation strikes it.
   2) The number of electrons ejected is proportional to the intensity or brightness of light.
   3) Photoelectric effect is observed below threshold frequency.
   4) At a frequency $\nu > \nu_0$, the ejected electrons come out with certain kinetic energy.

7. Select the incorrect statement :
   1) The five d-orbitals are designated as d_{xy}, d_{yz}, d_{xz}, $d_{x^2-y^2}$ and d_{z^2}.
   2) The shapes of d_{xy}, d_{yz}, d_{xz}, $d_{x^2-y^2}$ are similar to each other.
   3) The shape of d_{z^2} is different from rest d-orbitals.
   4) All five 3d orbitals have different energy.

8. Which of the following statements regarding Pauli Exclusion Principle is correct ?
   1) Two electrons in an atom can have the same set of four quantum numbers.
   2) Only two electrons may exist in the same orbital and these electrons must have same spin.
   3) The maximum number of electrons present in s, p and d subshell can be 8, 6 and 14.
   4) The maximum number of electrons in the shell with principal quantum number n is equal to $2n^2$.

9. Choose incorrect statement among the following:
   1) For n = 3, the possible l values are 0, 1 and 2.
   2) For $l = 2$, m_l can be –1, 0 and + 1
   3) For n = 2, the possible value of l can be 0 and 1.
   4) For a given value of l, $(2l + 1)$ values of m_l are possible.

10. Correct statement is/are :
   A. Number of nodes in 3p orbital is 2.
   B. The number of nodes is given by n-1.
   C. Number of radial and angular nodes for 3p-orbital are 1, 1 respectively.
   D. Number of radial and angular nodes depends only on principal quantum number.

   1) A, B and C   2) B, C and D
   3) A and D   4) B and C

Statement Based MCQs

Directions: These questions consist of two statements each, printed as Statement-I and Statement-II. While answering these questions, you are required to choose any one of the following four responses.

1) **Both Statement-I and Statement-II are correct.**
2) **Both Statement-I and Statement-II are incorrect.**
3) **Statement-I is correct & Statement-II is incorrect.**
4) **Statement-I is incorrect & Statement-II is correct.**

1. **Statement-I:** Orbital angular momentum of 1s, 2s, 3s are zero.

   Statement-II: 1s, 2s, 3s all have spherical shape.

2. **Statement-I:** Fe^{+2} has 24 electrons, hence, its electronic configuration is similar to that of Cr(24) = [Ar]$3d^5\,4s^1$.

   Statement-II: Half filled sub-shell give stability to the ion.

3. **Statement-I:** In third energy level, there is no f subshell.

   Statement-II: For n = 3, the possible value of ℓ are 0,1 and 2.

4. **Statement-I:** The orbital angular momentum of 2s-electron is equal to that of 3s-electron.

   Statement-II: The orbital angular momentum is given by the relation $\frac{h}{2\pi}\sqrt{\ell(\ell+1)}$ and the value of it is same for 2s-electron and 3s-electron.

5. **Statement-I:** The p-orbital is dumb-bell shaped.

   Statement-II: Electrons present in p-orbital can have any one of the three values of magnetic quantum number i.e. +1, 0, –1.

6. **Statement-I:** All the isotopes of a given element show different chemical behaviour.

   Statement-II: Chemical properties of atoms are controlled by the number of neutrons.

7. **Statement-I:** Rutherford's model of an atom resembles the solar system in which the nucleus plays the role of a sun and the electron is that of revolving planet.

   Statement-II: According to Rutherford's model, electrons and the nucleus are held together by electrostatic forces of attraction.

8. **Statement-I:** X-rays are used to study the interior structure of the object.

   Statement-II: X-rays are highly deflected by the electric & magnetic field, and have very less penetrating power through the matter.

9. **Statement-I:** The amount of light emitted (intensity of radiation) from a black body and its spectral distribution depends only on its temperature.

   Statement-II: At a given temperature, intensity of radiation emitted increases with the increase of wavelength, reaches a maximum value at a given wavelength and then starts decreasing with further increase of wavelength.

10. **Statement-I:** In absorption spectrum $n_i > n_f$ while in emission spectrum $n_f > n_i$.

    Statement-II: Energy is emitted in absorption spectrum while energy is absorbed in the emission spectrum.

Assertion & Reason MCQs

Directions: These questions consist of two statements each, printed as Assertion and Reason. While answering these questions, you are required to choose any one of the following four responses.

1) **Both Assertion and Reason are True and the Reason is a correct explanation of the Assertion.**
2) **Both Assertion and Reason are True but Reason is not a correct explanation of the Assertion.**
3) **Assertion is True but the Reason is False.**
4) **Assertion is False but Reason is True.**

1. **Assertion:** The 19th electron in K-atom enters into 3p and not the 4s-orbital.
   Reason: $(n + l)$ rule is followed for determining the orbital of the lowest energy state.
2. **Assertion:** Black body is an ideal body that emits and absorbs radiations of all frequencies.
   Reason: The frequency of radiation emitted by a body goes from a lower frequency to higher frequency with an increase in temperature.
3. **Assertion:** It is impossible to determine the exact position and exact momentum of an electron simultaneously.
   Reason: The path of an electron in an atom is clearly defined.
4. **Assertion:** Hydrogen has only one electron in its orbit but produces several spectral lines.
   Reason: There are many excited energy levels available.
5. **Assertion:** The two lobes of a p-orbital have opposite sign.
   Reason: The two lobes of a p-orbital lie on the opposite side of the nucleus.
6. **Assertion:** 2p-orbitals don't have any spherical node.
   Reason: In p orbitals, the number of spherical nodes is given by (n-2), where n is the principal quantum number.
7. **Assertion:** In $3d_{z^2}$, the number of radial nodes is zero.
   Reason: No. of radial nodes of atomic orbitals is equal to value of $(n - \ell - 1)$.
8. **Assertion:** For multi electron species, energy of 3d-orbital is more than that of 4s-orbital.
   Reason: Orbital energy is calculated by $(n + \ell)$ rule.
9. **Assertion:** $1s^2\ 2s^2\ 2p^6\ 3s^2\ 3p^6\ 4s^1$ is the electronic configuration of K(19).
   Reason: Energy of 4s < 3d, therefore, 4s is filled before 3d as decided by Aufbau principle.
10. **Assertion:** Most of the alpha particles striking gold foil were deflected in Rutherford's experiment.
    Reason: Nucleus present inside the atom is heavy.

Multi Correct MCQs

1. Consider the following pairs of electrons

   A. (*a*) $n = 3, l = 1, m_l = 1, m_s = +\frac{1}{2}$

   (*b*) $n = 3, l = 2, m_l = 1, m_s = +\frac{1}{2}$

   B. (*a*) $n = 3, l = 2, m_l = -2, m_s = -\frac{1}{2}$

   (*b*) $n = 3, l = 2, m_l = -1, m_s = -\frac{1}{2}$

   C. (*a*) $n = 4, l = 2, m_l = 2, m_s = +\frac{1}{2}$

   (*b*) $n = 3, l = 2, m_l = -2, m_s = +\frac{1}{2}$

   The pairs of electrons present in degenerate orbitals is/are:

   1) Only (A) 2) Only (B)
   3) Only (C) 4) (B) and (C)

2. Given below are the quantum numbers for 4 electrons.

   I. $n = 3, l = 2, m_l = 1, m_s = +1/2$
   II. $n = 4, l = 1, m_l = 0, m_s = +1/2$
   III. $n = 4, l = 2, m_l = -2, m_s = -1/2$
   IV. $n = 3, l = 1, m_l = -1, m_s = +1/2$

   Which of the following represents the correct order of increasing energy for the above given electrons?

   A. (II) < (I) and (III) < (IV)
   B. (I) < (II)
   C. (IV) < (I) and (II) < (III)
   D. (IV) < (II)

   1) Only (A) 2) (A), (B) and (C)
   3) (B), (C) and (D) 4) Only (D)

3. Number of radial nodes are same for which of the following orbitals?

   A. 4d B. 2s
   C. 3s D. 5f
   E. 2p

   Choose the most appropriate option from the following.

   1) (C), (D) and (E) 2) (A) and (B)
   3) (B), (C), (D) and (E) 4) (A), (B) and (D)

4. Consider the following set of quantum numbers.

	n	l	m_l
A.	3	3	–3
B.	3	2	–2
C.	2	1	+1
D.	2	2	+2

   Choose the correct sets of quantum numbers.

   1) (A), (B) and (C) 2) (A) and (B)
   3) (B) and (C) 4) (C) and (D)

NCERT Exemplar MCQs

1. Which of the following conclusions could not be derived from Rutherford's α-particle scattering experiment?
   1) Most of the space in the atom is empty
   2) The radius of the atom is about 10^{-10} m, while that of nucleus is 10^{-15} m.
   3) Electrons move in a circular path of fixed energy called orbits.
   4) Electrons and the nucleus are held together by electrostatic forces of attraction.
2. Which of the following options does not represent the ground state electronic configuration of an atom?
   1) $1s^2 2s^2 2p^6 3s^2 3p^6 3d^8 4s^2$ 2) $1s^2 2s^2 2p^6 3s^2 3p^6 3d^9 4s^2$
   3) $1s^2 2s^2 2p^6 3s^2 3p^6 3d^{10} 4s^1$ 4) $1s^2 2s^2 2p^6 3s^2 3p^6 3d^5 4s^1$
3. The probability density plots of 1s and 2s orbitals are given in figure:

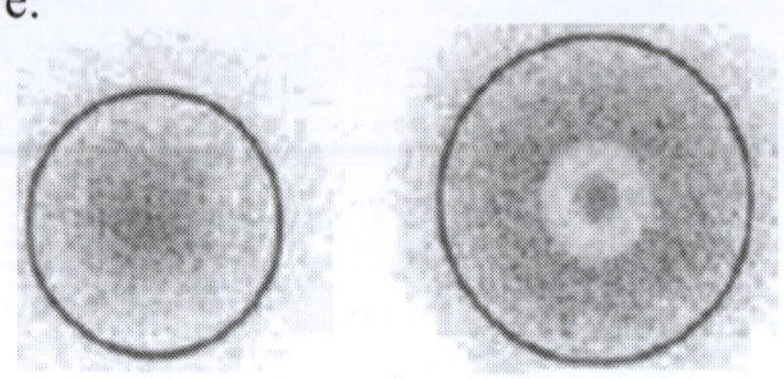

1*s* 2*s*

   The density of dots in a region represents the probability density of finding electrons in the region.
   On the basis of above diagram which of the following statements is incorrect?
   1) 1s and 2s orbitals are spherical in shape
   2) The probability of finding the electron is maximum near the nucleus
   3) The probability of finding the electron at a given distance is equal in all directions
   4) The probability density of electrons for 2s orbital decreases uniformly as distance from the nucleus increases
4. Which of the following statement is not correct about the characteristics of cathode rays?
   1) They start from the cathode and move towards the anode
   2) They travel in straight line in the absence of an external electrical or magnetic field
   3) Characteristics of cathode rays do not depend upon the material of electrodes in cathode ray tube
   4) Characteristics of cathode rays depend upon the nature of gas present in the cathode ray tube
5. Which of the following statements about the electron is incorrect?
   1) It is a negatively charged particle
   2) The mass of electron is equal to the mass of neutron
   3) It is a basic constituent of all atoms
   4) It is a constituent of cathode rays
6. Which of the following properties of atom could be explained correctly by Thomson model of atom?
   1) Overall neutrality of atom
   2) Spectra of hydrogen atom
   3) Position of electrons, protons and neutrons in atom
   4) Stability of atom
7. Two atoms are said to be isobars if:
   1) They have same atomic number but different mass number
   2) They have same number of electrons but different number of neutrons
   3) They have same number of neutrons but different number of electrons
   4) Sum of the number of protons and neutrons is same but the number of protons is different
8. The number of radial nodes for 3p orbital is:
   1) 3 2) 4
   3) 2 4) 1
9. Number of angular nodes for 4d orbital:
   1) 4 2) 3
   3) 2 4) 1
10. Which of the following is responsible to rule out the existence of definite paths or trajectories of electrons?
    1) Pauli's exclusion principle
    2) Heisenberg's uncertainty principle
    3) Hund's rule of maximum multiplicity
    4) Aufbau principle
11. Total number of orbitals associated with third shell will be:
    1) 2 2) 4
    3) 9 4) 3
12. Orbital angular momentum depends on:
    1) l 2) n and l
    3) n and m 4) m and s
13. Chlorine exists in two isotopic forms. Cl-37 and Cl-35 but its atomic mass is 35.5. This indicates the ratio of Cl-37 and Cl-35 is approximately:
    1) 1 : 2 2) 1 : 1
    3) 1 : 3 4) 3 : 1
14. The pair of ions having same electronic configuration is:
    1) Cr^{3+}, Fe^{3+} 2) Fe^{3+}, Mn^{2+}
    3) Fe^{3+}, Co^{3+} 4) Sc^{3+}, Cr^{3+}
15. For the electrons of oxygen atom, which of the following statements is correct?
    1) Z_{eff} for an electron in a 2s orbital is the same as Z_{eff} for an electron in a 2p orbital
    2) An electron in the 2s orbital has the same energy as an electron in the 2p orbital
    3) Z_{eff} for an electron in 1s orbital is the same as Z_{eff} for an electron in a 2s orbital
    4) The two electrons present in the 2s orbital have spin quantum numbers, m_s but of opposite sign
16. If waves travelling at same speeds, which of the following matter waves have the shortest wavelength?
    1) Electron 2) Alpha particle (He^{2+})
    3) Neutron 4) Proton

Kattar NEET

1. For a hypothetical H like atom which follows Bohr's model, some spectral lines were observed as shown. If it is known that line 'E' belongs to the visible region, then the lines possibly belonging to ultra violet region will be (n_1 is not necessarily ground state)

[Assume for this atom, no spectral series shown overlap with other series in the emission spectrum]

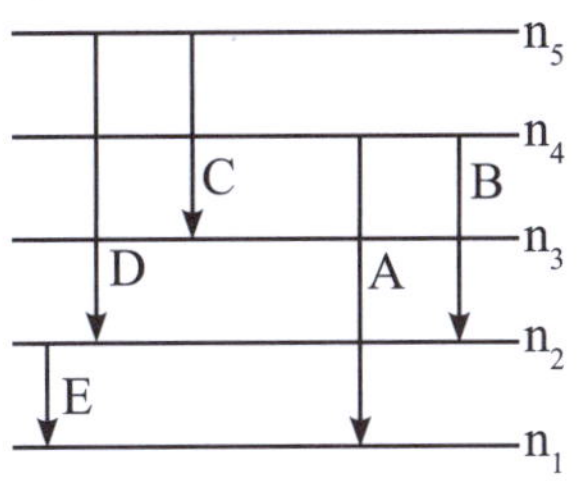

1) B and D only   2) D only
3) C only   4) A only

2. The photon emitted due to electronic transition from 5th excited state to 2nd excited state in Li^{2+}, is used to excite He^+ already in first excited state. He^+ ion after absorbing the photon reaches in an orbit having total energy equal to:

1) –3.4 eV   2) –13.6 eV
3) –6.8 eV   4) –27.2 eV

3. A proton accelerated from rest through a potential difference of 'V' volts has a wavelength λ associated with it. An alpha particle in order to have the same wavelength must be accelerated from rest through a potential difference of

1) V volt   2) 4V volt
3) 2V volt   4) $\frac{V}{8}$ volt

4. Hydrogen atoms in a particular excited state 'n' when all return to ground state six different photons are emitted. Which of the following is correct?

1) Total number of angular nodes in all the orbitals in $(n-1)^{th}$ shell is 23
2) Total number of radial nodes in all the orbitals of nth shell is 24
3) If highest energy photon emitted from the above sample is incident on metal plate having work function 8 eV then KE of liberated photoelectron may be equal to or less than 4.75 eV
4) Out of 6 different photons only two photons have speed equal to that of visible light

5. Calculate the wavelength of spectral line obtained in spectrum of Li^{+2} when the transition takes place between two levels whose sum is 4 and difference is 2.

1) 11.4×10^{-6} cm   2) 1.14×10^{-6} cm
3) 0.114×10^{-6} cm   4) 1.14×10^{-8} cm

6. The charge on the electron and proton are reduced to $\frac{2}{3}$rd of their original values. Let the present value of Rydberg constant of H-atom be R. What will be the new value of Rydberg constant?

1) $\frac{2R}{3}$ 2) $\frac{4R}{9}$
3) $\frac{8R}{27}$ 4) $\frac{16R}{81}$

7. Consider a large number of hydrogen atoms with electrons randomly distributed in the n = 1, 2, 3 and 4 orbits. How many different wavelengths of light are emitted by these atoms as the electrons fall into lower energy orbits.

1) 1   2) 3
3) 6   4) 9

8. $\sqrt{V}$ of two particles A and B are plotted against de-Broglie wavelengths. Where V is potential difference applied on the particles. Which of the following relation is correct about the mass of particles? (Assume equal charge on the two particles)

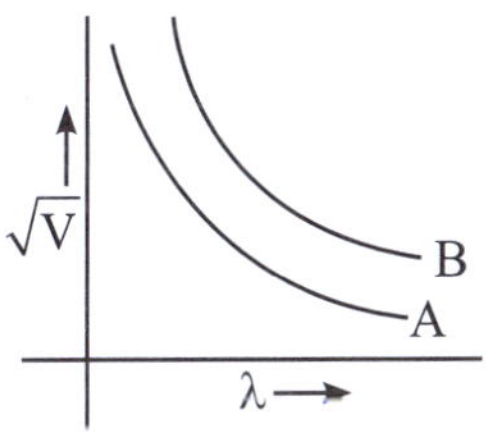

1) $m_A = m_B$ 2) $m_A > m_B$
3) $m_A < m_B$ 4) $m_A \le m_B$

9. The Schrodinger wave equation for hydrogen atom is

$$\psi(\text{radial}) = \frac{1}{16\sqrt{4}}\left(\frac{Z}{a_0}\right)^{3/2}\left[(\sigma-1)(\sigma^2-8\sigma+12)\right]e^{-\sigma/2}$$

where a_0 and Z are the constant in which answer can be expressed and $\sigma = \frac{2Zr}{a_0}$ minimum and maximum position of radial modes from nucleus are ...... respectively.

1) $\frac{a_0}{z}, \frac{3a_0}{z}$ 2) $\frac{a_0}{2z}, \frac{a_0}{z}$
3) $\frac{a_0}{2z}, \frac{3a_0}{z}$ 4) $\frac{a_0}{2z}, \frac{4a_0}{z}$

10. In a sample of hydrogen atoms, electrons jump from 10th excited state to ground state. If x are the number of different ultraviolet radiations, y are number of different visible

radiations and z are number of different infrared radiations. The value of z– (x + y) is. (Assume all the Balmer lines lie within visible region)

1) 17 2) 18
3) 19 4) 36

11. If spin quantum number has 4 values instead of two then identify the incorrect statement.

1) First period will be shortest with 4 elements
2) Fourth period can have maximum 36 elements
3) Each orbital can have maximum 4 electrons.
4) For a given value of 'n' (principal quantum number) number of electrons will be $2n^2$.

12. The orbital angular momentum of an electron is $\sqrt{3}\frac{h}{\pi}$.

Which of the following may be the permissible value of angular momentum of this electron revolving in an unknown Bohr orbit?

1) $\frac{h}{2\pi}$ 2) $\frac{h}{\pi}$
3) $3\cdot\frac{h}{2\pi}$ 4) $2\cdot\frac{h}{\pi}$

13. For an unielectronic species, compare the energy of the following orbitals (represented by S)

S_1 = a spherically symmetrical orbital having two spherical nodes

S_2 = an orbital which is double - dumb bell and has no radial nodes.

S_3 = an orbital with orbital angular momentum zero and three radial nodes

S_4 = an orbital having one planar and one radial node.

1) $S_1 = S_2 = S_3 = S_4$ 2) $S_1 = S_2 = S_4 < S_3$
3) $S_1 > S_2 > S_3 > S_4$ 4) $S_1 < S_2 < S_3 < S_4$

14. The angular momentum of an electron of H-atom is proportional to

1) r^2 2) $\frac{1}{r}$
3) $\sqrt{r}$ 4) $\frac{1}{\sqrt{r}}$

15. The uncertainty in the location of circulating electron is equal to its de-Broglie wavelength. The minimum percent error in its measurement of velocity under this circumstance will be approximately

1) 4 2) 8
3) 18 4) 22

16. Consider the following six electronic configurations (remaining inner orbitals are completely filled) and mark the incorrect option.

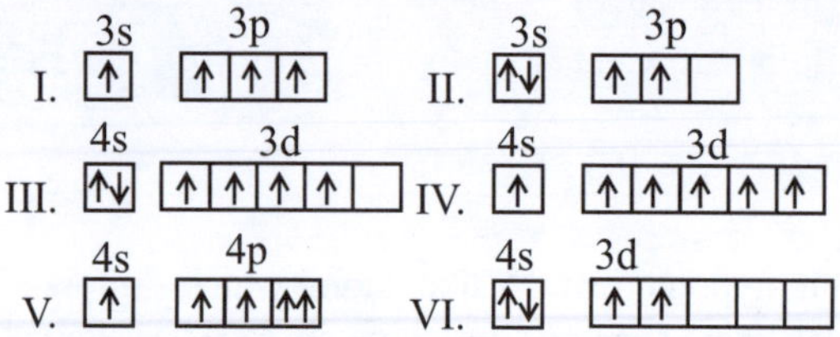

1) Stability order : IV > III
2) Order of spin multiplicity : IV > III = I > II
3) V does not violate all the three rules of electronic configuration
4) If VI represents A then A^+ when kept near a magnet, acts as diamagnetic substance.

17. In the hydrogen spectrum, the longest wavelength in the Lyman series is x Å and the shortest wavelength in the Balmer series is y Å. Hence, the longest wavelength of light needed to ionise a hydrogen atom from its ground state should be

1) (x + y) Å 2) (y – x) Å
3) $\frac{xy}{(x+y)}$ Å 4) $\frac{(x+y)}{xy}$ Å

18. Calculate the minimum and maximum number of electrons which may have magnetic quantum number, m = +1 and spin quantum number $s = -\frac{1}{2}$ in chromium (Cr):

1) 0, 1 2) 1, 2
3) 4, 6 4) 2, 3

19. For H-atom, as the orbit number increases, the distance between two successive orbits (r_1 = radius of first orbit)

1) Increases by $2r_1$
2) increases by $(2n + 1)r_1$, where n is higher orbit
3) increases by $(2n + 1)r_1$, where n is lower orbit
4) remains constant

20. In hydrogen atom during de-excitation of electron from an excited state to ground state 15 spectral lines are emitted. Find the longest wavelength of U.V. region during de-excitation.

1) $\frac{4}{3R_H}$ 2) $\frac{36}{35R_H}$
3) $\frac{25}{24R_H}$ 4) $\frac{16}{15R_H}$

NEET & AIIMS Past Year MCQs

1. Given below are two statements:

Statement - I: The Balmer spectral line for H atom with lowest energy is located at $\frac{5}{36}R_H\,cm^{-1}$.

(R_H = Rydberg constant)

Statement - II: When the temperature of blackbody increases, the maxima of the curve (intensity and wavelength) shifts to shorter wavelength.

In the light of the above statements, choose the correct answer from the options given below: **(2024 Re)**

1) Statement - I is true but Statement - II is false
2) Statement - I is false but Statement - II is true
3) Both Statement - I and Statement - II are true
4) Both Statement - I and Statement - II are false

2. The quantum numbers of four electrons are given below: **(2024 Re)**

I. $n = 4; l = 2; m_l = -2; s = -\frac{1}{2}$

II. $n = 3; l = 2; m_l = 1; s = +\frac{1}{2}$

III. $n = 4; l = m_l = 0; s = +\frac{1}{2}$

IV. $n = 3; l = 1; m_l = -1; s = +\frac{1}{2}$

The correct decreasing order of energy of these electrons is

1) IV > II > III > I
2) I > III > II > IV
3) III > I > II > IV
4) I > II > III > IV

3. The energy of an electron in the ground state (n = 1) for He^+ ion is –xJ, then that for an electron in n = 2 state for Be^{3+} ion in J is: **(2024)**

1) $-4x$
2) $-\frac{4}{9}x$
3) $-x$
4) $-\frac{x}{9}$

4. Match List I with List II. **(2024)**

List I Quantum Number		List II Information provided	
A.	m_l	I	shape of orbital
B.	m_s	II	size of orbital
C.	l	III	orientation of orbital
D.	n	IV	orientation of spin of electron

Choose the correct answer from the options given below:

1) A-III, B-IV, C-II, D-I
2) A-II, B-I, C-IV, D-III
3) A-I, B-III, C-II, D-IV
4) A-III, B-IV, C-I, D-II

5. Select the correct statements from the following: **(2023)**

A. Atoms of all elements are composed of two fundamental particles.
B. The mass of the electron is 9.10939×10^{-31} kg.
C. All the isotopes of a given element show same chemical properties.
D. Protons and electrons are collectively known as nucleons.
E. Dalton's atomic theory, regarded the atom as an ultimate particle of matter.

Choose the correct answer from the options given below:

1) B, C and E only
2) A, B and C only
3) C, D and E only
4) A and E only

6. The relation between n_m,(n_m = the number of permissible values of magnetic quantum number (m)) for a given value of azimuthal quantum number (l), is **(2023)**

1) $n_m = \ell + 2$
2) $\ell = \frac{n_m - 1}{2}$
3) $\ell = 2n_m + 1$
4) $n_m = 2\ell^2 + 1$

7. Identify the incorrect statements from the following. **(2022)**

1) The shapes of d_{xy}, d_{yz}, and d_{zx} orbitals are similar to each other; and $d_{x^2-y^2}$ and d_{z^2} are similar to each other.
2) All the five 5d orbitals are different in size when compared to the respective 4d orbitals.
3) All the five 4d orbitals have shapes similar to the respective 3d orbitals.
4) In an atom, all the five 3d orbitals are equal in energy in free state.

8. If radius of second Bohr orbit of the He^+ ion is 105.8 pm, what is the radius of third Bohr orbit of Li^{2+} ion? **(2022)**

1) 158.7 Å
2) 158.7 pm
3) 15.87 pm
4) 1.587 pm

9. A particular station of All India Radio, New Delhi, broadcasts on a frequency of 1,368 kHz (kilohertz). The wavelength of the electromagnetic radiation emitted by the transmitter is:[speed of light, c = 3.0×10^8 ms^{-1}] **(2021)**

1) 219.2 m
2) 2192 m
3) 21.92 cm
4) 219.3 m

10. Wave length of particular transition for H atom is 400 nm. What can be wavelength of He^+ for same transition? **(AIIMS 2018)**

1) 400 nm
2) 100 nm
3) 1600 nm
4) 200 nm

11. When on metal sheet fall λ_1 light will eject electron with v_1 velocity and with λ_2 light eject electron of v_2 velocity, what is $v_2^2 - v_1^2$ value? **(AIIMS 2018)**

1) $\frac{2hc}{m}\left(\frac{1}{\lambda_2} - \frac{1}{\lambda_1}\right)$
2) $\frac{hc}{m}\left(\frac{1}{\lambda_2} - \frac{1}{\lambda_1}\right)$
3) $\frac{2hc}{m}\left(\frac{1}{\lambda_1} - \frac{1}{\lambda_2}\right)$
4) $\frac{m}{2hc}\left(\frac{1}{\lambda_2} - \frac{1}{\lambda_1}\right)$

12. The wavelength of a spectral line emitted by hydrogen atom in the Lyman series is $\frac{16}{15R}$ cm. What is the value of n_2? (R = Rydberg constant) **(AIIMS 2011)**

1) 2
2) 3
3) 4
4) 1

13. The most probable radius (in pm) for finding the electron in He^+ is: **(AIIMS 2005)**

1) 0.0
2) 52.9
3) 26.5
4) 105.8

14. For principle quantum number n = 4 the total number of orbitals having $l = 3$ is: **(AIIMS 2004)**

1) 3
2) 7
3) 5
4) 9

Answer Key

NCERT Topic-wise MCQs

1	2	3	4	5	6	7	8	9	10	11	12	13	14	15	16	17
4	1	3	1	2	2	3	2	1	1	2	1	2	3	4	4	2
18	19	20	21	22	23	24	25	26	27	28	29	30	31	32	33	34
4	1	1	3	1	2	1	4	4	3	3	1	3	1	2	2	3
35	36	37	38	39	40	41	42	43	44	45	46	47	48	49	50	51
1	3	4	1	2	1	4	2	1	4	3	3	1	1	2	3	4
52	53	54	55	56	57	58	59	60	61	62	63	64	65	66	67	68
4	3	4	4	4	2	2	3	1	4	2	1	1	1	4	3	4
69	70	71	72	73	74	75	76	77	78	79	80	81	82	83	84	85
1	4	4	4	1	2	4	1	3	4	3	1	1	3	1	2	1
86	87	88	89	90	91											
3	3	4	3	3	3											

High Order Time Intensive MCQs

1	2	3	4	5	6	7	8	9	10	11	12	13	14	15	16	17
4	2	4	1	3	2	4	3	4	4	4	1	3	1	1	2	1
18	19	20	21	22	23	24	25	26	27	28	29	30	31	32	33	34
4	1	2	1	2	3	4	4	3	2	3	3	4	1	3	2	1, 3
35	36	37	38	39	40	41	42	43	44	45	46					
2	4	1	4	1	3	3	4	3	3	1	2					

Match the Columns MCQs

1	2	3	4	5	6
3	1	4	1	3	2

Correct & Incorrect MCQs

1	2	3	4	5	6	7	8	9	10
2	4	4	2	3	3	4	4	2	1

Statement Based MCQs

1	2	3	4	5	6	7	8	9	10
1	4	1	1	1	2	1	3	1	2

Assertion & Reason MCQs

1	2	3	4	5	6	7	8	9	10
4	2	3	1	2	1	1	1	1	4

Multi Correct MCQs

1	2	3	4
2	3	4	3

NCERT Exemplar MCQs

1	2	3	4	5	6	7	8	9	10	11	12	13	14	15	16
3	2	4	4	2	1	4	4	3	2	3	1	3	2	4	2

Kattar NEET

1	2	3	4	5	6	7	8	9	10	11	12	13	14	15	16	17
4	1	4	3	2	4	3	2	3	1	4	4	2	3	2	4	3
18	19	20														
4	3	1														

NEET & AIIMS Past Year MCQs

1	2	3	4	5	6	7	8	9	10	11	12	13	14
3	2	3	4	1	2	1	2	4	2	1	3	3	2

CHAPTER 3

Classification of Elements and Periodicity in Properties

Topicwise Number of Questions Analysis (2025-2016)

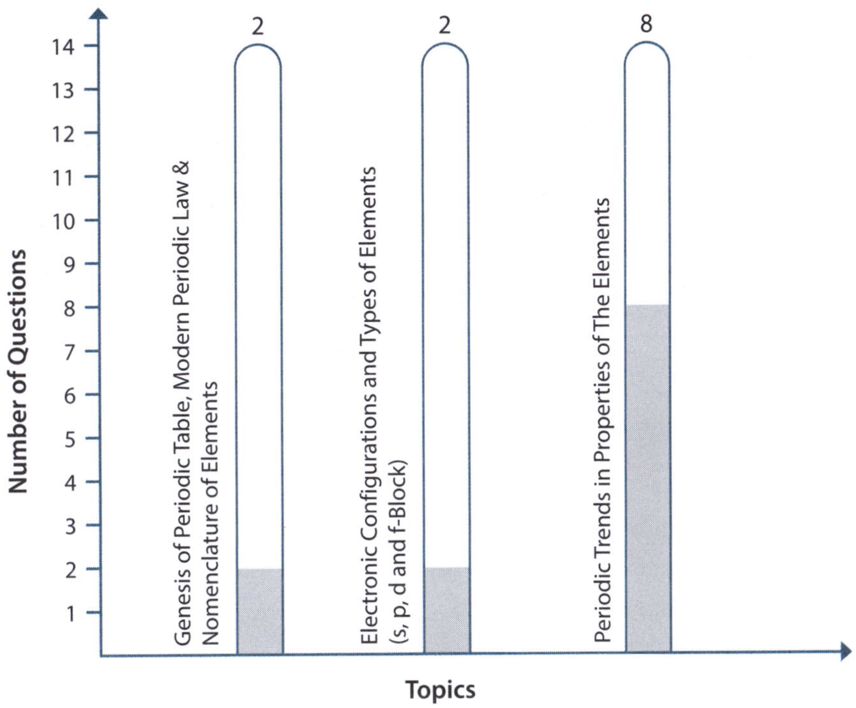

NCERT-Picks

Classification

Dobereiner's Law of Triads

He arranged the elements in groups of three,called triads such that atomic weight of middle element is average of first and third element.

These elements resembled in their physical and chemical properties

Newlands Law of Octaves

He arranged the elements in increasing order of their atomic weights and found that any given element has similar properties to the eighth element that followed it.

Lothar Meyer

- Lothar Meyer plotted a graph between atomic weight and atomic volume (i.e., atomic weight in solid state/density). Elements with similar properties of the element occupied the similar positions on the graph.

Mendeleev's Periodic Law

Properties of elements are periodic function of their atomic weights

- First period is known as shortest period and contain only two elements. Second and third periods are called short periods containing eight elements each.
- Second period elements are called bridge elements.
- Seventh period is an incomplete period containing nineteen elements. Numbers 2,8,8,18,18,32 are called magic numbers.

Long form of the Periodic Table/Modern Periodic Law

- Moseley (1909) studied the frequency of X-rays by the bombardment of a strong beam of electrons on metal target. He found that the square root of the frequency of X-rays is directly proportional to the atomic number (Z) of metal.
- **Modern periodic table** was based on atomic number rather than atomic mass. Acc. to Henry Moseley, square root of the frequency emitted by metal is directly proportional to atomic no. & not the atomic weight.

- The long form of periodic table is also called as Bohr's periodic table.

Bohr's Classification of Elements

On the basis of electronic configuration of the incomplete shells, the elements are classified into five main categories:

1. Inert gases
2. Representative elements
3. Transition elements
4. Inner transition elements
5. Transuranium elements

- **s-block:** Elements of this group are never found pure in nature. *s*-block elements have high reactivity & low ionisation enthalpy.
- **p-block:** The group III B to VII B and Zero groups i.e., groups 13 to 18 of the periodic table makes p-block together.

Prediction of Period, Group and Block

- Period of an element corresponds to the principal quantum number of the valence shell.
- Block of an element corresponds to the type of subshell which receives the last electron.
- Group No. for s-block elements = No. of valence electron.
- Group No. for p-block elements = 10 + No. of valence electrons.
- Group No. for d-block elements = no. of electrons in (n-1)d subshell + no. of electrons in ns shell.

Periodic Trends in Properties of Elements

Atomic Radius

Covalent Radius

- In a heteronuclear diatomic AB molecule, if both atoms are linked by a single covalent bond having nearly same electronegativity, then bond length d_{A-B}.

$$d_{A-B} = r_A + r_B$$

If the covalent bond is formed between two elements of different electronegativity, then we use the following relation: $d_{A-B} = r_A + r_B - 0.09\ (x_A - x_B)$

x_A and x_B are electronegativity of A and B respectively

(Formula not given in NCERT)

Vander Waal's Radii

- It is half of the distance between the nuclei of two non-bonded neighboring atoms of two adjacent molecules.

$r_{covalent} < r_{crystal} < r_{vander\ waals}$

- Atomic radius decreases across a period, due to increased nuclear charge.
- Atomic radius increases down the group due to presence of extra shell & shielding of valence electrons.

Ionic Radius

- Ionic radii also increase in a group from top to bottom.
- The sizes of ions increases as we go down a group (considering the ions of same charge).

For example: Li^+ (0.76) < Na^+ (1.02) < K^+ (1.38) < Rb^+ → (in Å)

$Be^{2+} < Mg^{2+} < Ca^{2+} < Sr^{2+}$

$F^- < Cl^- < Br^- < I^-$

- Among isoelectronic species, the cation with the greater positive charge will have a smaller radius. Greater is the attraction of the electrons to the nucleus, smaller is the size of atom/ion.

$\therefore Na^+ > Mg^{2+} > Si^{4+} > Cl^{+7}$

- For isoelectronic species, more is the negative charge, greater will be the size of atom/ion.

$\therefore O^{2-} > F^-$

Ionization Enthalpy (ΔH)

- This is the energy required to remove an outermost electron from an isolated gaseous atom.
- I.E decreases with increase in size of atom or ion.
- Higher the nuclear charge, higher is the I.E.
- More the screening effect, less is the I.E.
- I.E decreases down the group and increases along the period.
- Noble gases have largest ionization enthalpy due to fully filled stable valence electronic configuration.

Electron Gain Enthalpy

It is the amount of energy released when an extra electron is added to an isolated gaseous atom.

First $\Delta_{eg}H$ is exothermic, whereas the second $\Delta_{eg}H$ is endothermic.

- **Smaller the size of atom and more** the nuclear charge, more **negative is the value of $\Delta_{eg}H$.**
- $\Delta_{eg}H$ becomes more negative **along a period and becomes less negative down the group.**
- Cl has the most negative electron gain enthalpy.
- Electron affinity ∝ Effective nuclear charge

Electronegativity

- Electronegativity is a qualitative measure of the ability of an atom in a chemical compound to attract shared electrons to itself.
- Smaller the size of atom, higher is the value of E.N.
- EN increases along a period and decreases down the group.
- Electronegativity trend:

F > O > N

Cl > Br > I

S > C > H

Nature of Oxides and Hydroxides

- The acidic nature of oxides increases in a period from left to right but basic nature increases in a group from top to bottom.

- E. g., acidic nature in a period: $Na_2O < MgO < Al_2O_3 < SiO_2 < P_4O_{10} < SO_3 < Cl_2O_7$
- Basic nature in a group: $Li_2O < Na_2O < K_2O < Rb_2O < Cs_2O$
- The basic **nature of hydroxides** of elements of group 1st increases on descending the group with increase in size of cation.

Chemical Reactivity

- Metallic character decreases from left to right in a period as there is increase in effective nuclear charge & decrease in tendency to lose valence electrons.
- On moving down the group, metallic character of element increases due to shielding of valence e^-, as they are being farther from the nucleus as a result, the tendency to lose valence electrons increases.

The pictorial representation of periodic trends of elements in the periodic table is given below-

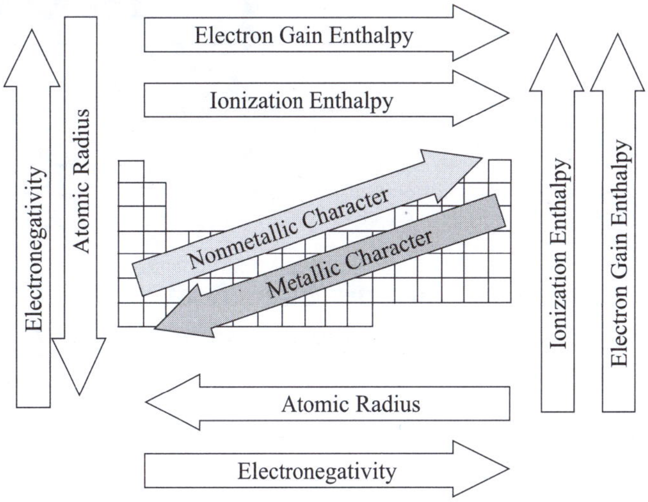

MCQs

NCERT Topic-wise MCQs

Genesis of Periodic Classification, Modern Periodic Law & Nomenclature of Elements

1. The third period of periodic table contains
   1) 8 elements 2) 32 elements
   3) 3 elements 4) 18 elements

2. Dobereiner triads is
   1) Na, K, Rb 2) Mg, S, As
   3) Cl, Br, I 4) P, S, As

3. The tenth element in the periodic table resembles the element with atomic number
   1) 2 as well as 30 2) 2 as well as 54
   3) 8 as well as 18 4) 8 only

4. The number of elements in the 5th period of the periodic table are
   1) 8 2) 10
   3) 18 4) 32

5. The plot of square root of frequency of X-rays emitted against atomic number led to the suggestion of which law(rule)?
   1) Periodic law 2) Modern Periodic law
   3) Hund's rule 4) Newland's law

6. In the modern periodic table, elements are arranged in
   1) Increasing mass
   2) Increasing volume
   3) Increasing atomic number
   4) Alphabetically

7. Which element was named as Eka-silicon in Mendeleev's classification of elements?
   1) Germanium 2) Gallium
   3) Indium 4) Thallium

8. Number of periods in the long form of periodic table are:
   1) 3 2) 9
   3) 5 4) 7

9. The long form of periodic table was based on
   1) Electron gain enthalpy 2) Atomic mass
   3) Atomic number 4) Effective nuclear charge

10. What is the name of element with atomic number 105?
    1) Kurchatovium 2) Dubnium
    3) Nobelium 4) Holmium

11. Which of the following is the name of element having atomic number 103?
    1) Lawrencium 2) Dubnium
    3) Kurchatovium 4) Ytterbium

Electronic Configuration and Types of Elements

12. The electronic configuration of elements of group 18 (except He), can be represented by.
    1) ns^2np^5 2) ns^1
    3) ns^2np^6 4) $(n-1)\ d^8\ ns^2$

13. In transition element , the incoming electron occupies (n-1)d subshell, in preference to:
    1) np-level 2) ns-level
    3) (n-1) p-level 4) (n+1) s-level

14. Without looking at the periodic table, select from each of the following list, the elements belonging to the same group.
    1) Z = 12, 38, 4, 88 2) Z = 9, 16, 3, 35
    3) Z = 5, 11, 27, 19 4) Z = 24, 47, 42, 55

15. Electronic configuration of the 4th transitional element is:
    1) $1s^2\ 2s^2\ 2p^6\ 3s^2\ 3p^6\ 4s^2$ 2) $1s^2\ 2s^2\ 2p^6\ 3s^2\ 3p^6\ 3d^4\ 4s^2$
    3) $1s^2\ 2s^2\ 2p^6\ 3s^2\ 3p^6\ 3d^4$ 4) $1s^2\ 2s^2\ 2p^6\ 3s^2\ 3p^6\ 3d^5\ 4s^1$

16. Which of the following ions does not have the configuration of Argon?
    1) Cl^- 2) K^+
    3) Ca^{2+} 4) I^-

17. Which pair of atomic numbers represents s-block elements?
    1) 7, 15 2) 6, 12
    3) 9, 17 4) 3, 12

18. Name an element of p-block of the periodic table in which last electron goes to the s-orbital of valence shell instead of p-orbital:
    1) As 2) Ga
    3) No such element is there 4) He

19. Find the total no. of d-block elements in the following atomic numbers?
    48, 28, 70, 100, 55, 45, 34, 36
    1) 5 2) 3
    3) 6 4) 2

20. Which of the following electronic configuration is of transition elements?
1) $1s^2\ 2s^2\ 2p^6\ 3s^2\ 3p^6\ 4s^2$
2) $1s^2\ 2s^2\ 2p^6\ 3s^2\ 3p^6 3d^{10}\ 4s^2 4p^1$
3) $1s^2\ 2s^2\ 2p^6\ 3s^2\ 3p^6\ 3d^{10}\ 4s^2 4p^6$
4) $1s^2\ 2s^2\ 2p^6\ 3s^2\ 3p^6 3d^2\ 4s^2$

21. Which block of the periodic table contains the element with configuration $1s^2\ 2s^2\ 2p^6\ 3s^2\ 3p^6\ 3d^{10} 4s^1$?
1) s-block 2) p-block
3) d-block 4) f-block

22. The electronic configuration of gadolinium (atomic number 64) is:
1) $[Xe]\ 4f^8\ 5d^9\ 6s^2$ 2) $[Xe]\ 4f^7\ 5d^1\ 6s^2$
3) $[Xe]\ 4f^3\ 5d^5\ 6s^2$ 4) $[Xe]\ 4f^6\ 5d^2\ 6s^2$

23. Which of the following will have total number of d-electrons equal to the difference in the number of total p and s electrons?
1) He 2) Ne
3) Ar 4) Kr

24. An element of atomic number 29 belongs to
1) s-block 2) p-block
3) d-block 4) f-block

25. Which metal has 2 electrons in the outermost orbit?
1) Na 2) Cu
3) Au 4) Be

26. Pd has exceptional electronic configuration of $4d^{10}\ 5s^0$. It belongs to
1) 4th period, group 11 2) 5th period, group 10
3) 6th period, group 9 4) 3rd period, group 16

27. Atomic number of Ag is 47. In the same group, the atomic numbers of elements placed above and below Ag in long form of periodic table will be
1) 29, 65 2) 39, 79
3) 29, 79 4) 39, 65

28. Which of the following sequence contain atomic number of only representative elements?
1) 3, 33, 53, 87 2) 13, 39, 54, 83
3) 22, 23, 66, 54 4) 3, 13, 48, 53

29. Which is mismatched regarding the position of the element as given below?
1) X(Z = 89) - f block, 6th period
2) Y(Z = 100) - f block, 7th period
3) Z(Z = 115) - d block, 7th period
4) Both (1) & (3)

Periodic Trends and Properties of The Elements

30. The radii of the F, F^-, O and O^{2-} are in the order:
1) $O^{2-} > O > F^- > F$ 2) $F^- > O^{2-} > F > O$
3) $O^{2-} > F^- > F > O$ 4) $O^{2-} > F^- > O > F$

31. Which of the following is not correct for iso-electronic ions?
1) They have the same number of electrons around their nuclei
2) Higher the atomic number, higher will be positive charge in a series of isoelectronic ions of same period
3) Isoelectronic ions have same electric charge
4) An isoelectronic series may have both positively and negatively charged ions

32. Which of the following is correct?
1) $r_{ionic} \propto Z$ 2) $r_{ionic} \propto Z_{eff}$
3) $r_{ionic} \propto \frac{1}{Z_{eff}}$ 4) $r_{ionic} \propto Z_{eff}^2$

33. Which of the following isoelectronic ions has lowest ionisation enthalpy?
1) K^+ 2) Ca^{2+}
3) Cl^- 4) S^{2-}

34. The elements which occupy the peaks of ionisation enthalpy curve are
1) Na, K, Rb, Cs 2) Na, Mg, Cl, I
3) Cl, Br, I, F 4) He, Ne, Ar, Kr

35. Which out of the following has the largest ionisation enthalpy?
1) $_{11}Na$ 2) $_{19}K$
3) $_{12}Mg$ 4) $_{37}Rb$

36. Which of the process requires largest energy among them?
1) $Al(g) \rightarrow Al^+(g) + e^-$
2) $Al^{2+}(g) \rightarrow Al^{3+}(g) + e^-$
3) $Al^+(g) \rightarrow Al^{2+}(g) + e^-$
4) All the processes require same amount of energy

37. Which of the following process refers to the ionisation enthalpy?
1) $X(s) \rightarrow X^+(g) + e^-$ 2) $X(g) + 2e^- \rightarrow X^{2-}(g)$
3) $X(g) \rightarrow X^+(g) + e^-$ 4) $X(g) + e^- \rightarrow X^-(g)$

38. Which one of the following has the largest size?
1) Rb^+ 2) Mg^{2+}
3) Li^+ 4) Na^+

39. Which of the following units are used frequently for atomic radii?
1) Meter 2) Picometres
3) Kilometers 4) Centimeters

40. An atom of an element has electronic configuration 2, 8, 1. Which of the following statement is correct?
1) The valency of element is 7
2) The element exists as a triatomic molecule
3) The element is metalloid
4) The element forms basic oxide

41. The correct order of ionic radii for the ions S^{2-}, Cl^-, P^{3-}, Ca^{2+} is
1) $Ca^{2+} > Cl^- > S^{2-} > P^{3-}$ 2) $S^{2-} > P^{3-} > Cl^- > Ca^{2+}$
3) $P^{3-} > S^{2-} < Cl^- > Ca^{2+}$ 4) $Ca^{2+} < Cl^- < S^{2-} < P^{3-}$

42. The correct order of electron affinity of halogens is:
1) F > Cl > Br > I
2) I > Br > Cl > F
3) Cl > F > Br > I
4) Cl < F > Br > I

43. The electron affinity of Be is also similar to that of:
1) Li
2) B
3) Na
4) Ne

44. Which of the following pairs show diagonal relationship?
1) Li, N
2) Li, K
3) Li, Mg
4) None of these

45. Which of the following element is expected to have highest electron affinity?
1) $1s^2 2s^2 2p^6 3s^2 3p^5$
2) $1s^2 2s^2 2p^3$
3) $1s^2 2s^2 2p^4$
4) $1s^2 2s^2 2p^5$

46. Which of the following oxide is amphoteric?
1) Na_2O
2) Al_2O_3
3) SO_3
4) P_2O_5

47. The correct order of electronegativity of N, O, F and P is.
1) F > N > P > O
2) F > O > P > N
3) F > O > N > P
4) N > O > F > P

48. Which set of elements have strongest tendency to form anions?
1) Na, Cl, Al
2) Cu, Ag, Au
3) Be, F, N
4) F, Cl, Br

49. The ions which are arranged in correct order of increasing radii are:
1) K^+, Ca^{2+}, S^{2-}
2) Be^{2+}, Mg^{2+}, Na^+
3) O^{2-}, F^-, N^{3-}
4) S^{2-}, O^{2-}, As^{3-}

50. Compared to the first ionisation enthalpy, the value of second ionisation enthalpy of an element is.
1) Negligible
2) Smaller
3) Greater
4) Double

51. Outermost configuration of most electronegative element of the periodic table is.
1) $3s^2 3p^6$
2) $2s^2 2p^5$
3) $4s^2 4p^5$
4) $2s^2 2p^4$

52. In which of the following pairs, the ionisation enthalpy of the first species is less than that of the second:
1) N, P
2) Be^{2+}, Be
3) N, N^-
4) S, P

53. Isoelectronic ions are those which have:
1) Same size
2) Same ionisation energy
3) Same electronic configuration
4) Same nuclear charge

54. Which of the following is the correct order for the size of C, N, P, S?
1) N < C < P < S
2) C < N < S < P
3) C < N < P < S
4) N < C < S < P

55. Which of the following has smallest size?
1) Al^{3+}
2) Al^{2+}
3) Al^+
4) Al

56. Which one of the following is an incorrect statement?
1) The ionisation enthalpy of nitrogen is greater than that of oxygen
2) The electron affinity of fluorine is greater than that of chlorine
3) The ionization enthalpy of beryllium is greater than that of boron
4) The electronegativity of fluorine is greater than that of chlorine

57. Which of the following has the greatest electron affinity?
1) I
2) Br
3) F
4) Cl

58. Identify the correct order of the size of the following
1) $Ca^{2+} < K^+ < Ar < Cl^- < S^{2-}$
2) $Ar < Ca^{2+} < K^+ < Cl- < S^{2-}$
3) $Ca^{2+} < Ar < K^+ < Cl^- < S^{2-}$
4) $Ca^{2+} < K^+ < Ar < S^{2-} < Cl^-$

59. Which of the following configuration represent atoms of element having the highest second ionisation enthalpy?
1) $1s^2 2s^2 2p^4$
2) $2s^2 2s^2 2p^6$
3) $1s^2 2s^2 2p^6 3s^1$
4) $1s^2 2s^2 2p^6 3s^2$

60. Correct order of I[st] ionization enthalpy among elements Be, B, C, N, O is
1) B < Be < C < O < N
2) B < Be < C < N < O
3) Be < B < C < N < O
4) Be < B < C < O < N

61. The size of ionic species is correctly given in the order
1) $Na^+ > Mg^{+2} > Cl^{+7} > Si^{4+}$
2) $Na^+ > Mg^{+2} > Si^{4+} > Cl^{+7}$
3) $Cl^{+7} > Si^{+4} > Mg^{+2} > Na^+$
4) $Cl^{+7} > Na^+ > Mg^{+2} > Si^{+4}$

62. A neutral atom (A) is converted to (A^{3+}) by the following process:

$$A \xrightarrow[-e^-]{E_1} A^+ \xrightarrow[-e^-]{E_2} A^{2+} \xrightarrow[-e^-]{E_3} A^{3+}$$

The correct order of E_1, E_2 and E_3 energies is
1) $E_1 < E_2 < E_3$
2) $E_1 > E_2 > E_3$
3) $E_1 = E_2 = E_3$
4) $E_1 > E_2 < E_3$

63. Which of the following is a favourable factor for cation formation?
1) High electronegativity
2) High electron affinity
3) Low ionisation enthalpy
4) Smaller atomic size

64. Which one of the following is correct order of the size of aluminium species?
1) $Al > Al^+ > Al^{2+}$
2) $Al^{2+} > Al^+ > Al$
3) $Al^{2+} = Al^+ = Al$
4) All of these

65. Generally, the first ionisation enthalpy increases along a period, but there are some exceptions. The one which is not an exception is
1) Na and Mg
2) Be and B
3) N and O
4) Mg and Al

High Order Time Intensive MCQs

1. The element californium belongs to the family of
   1) Actinoid series 2) Alkali metals
   3) Lanthanoid series 4) Alkali Earth metals

2. In 4th period of periodic table, how many elements have one or more 4d electrons?
   1) 2 2) 0
   3) 13 4) 6

3. Few reactions are given below-
   a. $O_{(g)} + e^- \longrightarrow O^-_{(g)}, \Delta_1H$
   b. $F_{(g)} + e^- \longrightarrow F^-_{(g)}, \Delta_2H$
   c. $Cl_{(g)} + e^- \longrightarrow Cl^-_{(g)}, \Delta_3H$
   d. $O^-_{(g)} + e^- \longrightarrow O^{2-}_{(g)}, \Delta_4H$

   Which of the following statement(s) is/are correct with respect to the reactions mentioned above?
   1) Δ_3H is more –ve than Δ_1H and Δ_2H
   2) Δ_1H is less negative than Δ_2H
   3) Δ_1H, Δ_2H and Δ_3H are –ve whereas Δ_4H is +ve
   4) All are correct

4. Which of the following is not a periodic property?
   1) Atomic mass 2) Atomic volume
   3) Covalent radius 4) Electronegativity

5. Which of the following is/are incorrect?
   1) I.E of $O_{(g)}$ is less than I.E of $O^-_{(g)}$
   2) I.E of $Ne_{(g)}$ is greater than I.E of $Ne^+_{(g)}$
   3) I.E of $N_{(g)}$ is greater than I.E of $N^+_{(g)}$
   4) All are incorrect

6. Energy required for the ionisation of 0.02 gram atom of magnesium is x kJ. The amount of energy required to ionise 1 atom of magnesium is
   1) x kJ 2) $\frac{x}{0.02N_A}$ J
   3) $\frac{x \times 10^3}{0.02N_A}$ J 4) xN_A kJ

7. Out of N, O, Ne, Na and Na^+, which of these species will have the maximum and minimum I.E respectively:
   1) Na^+, O 2) Na^+, Na
   3) N, Ne 4) Ne, Na^+

8. The species Ar, K^+ and Ca^{2+} contain the same number of electrons. In which order do their radii increase?
   1) $Ca^{2+} < K^+ < Ar$ 2) $K^+ < Ar < Ca^{2+}$
   3) $Ar < K^+ < Ca^{2+}$ 4) $Ca^{2+} < Ar < K^+$

9. Which of the following statements is/are incorrect?
   1) Helium has the highest first ionisation enthalpy in the periodic table.
   2) Nitrogen has more electron affinity than that of phosphorus
   3) Mercury and bromine are liquids at room temperature.
   4) In any period, atomic radius of alkali metal is the highest

10. Which one of the following represents a d-block element?
    1) [Rn] $6d^{10}\,7s^2\,7p^2$ 2) [Xe] $4f^7\,5d^1\,6s^2$
    3) [Xe] $4f^{14}\,5d^1\,6s^2$ 4) [Kr] $4d^{10}\,5s^1$

11. Which of the following sets contain only isoelectronic ions?
    1) Zn^{2+}, Ca^{2+}, Ga^{3+}, Al^{3+} 2) K^+, Ca^{2+}, Sc^{3+}, Cl^-
    3) O^{2-}, S^{2-}, Cl^-, Mg^{2+} 4) Ti^{4+}, Ar, Cr^{3+}, V^{5+}

12. The electronic configuration having maximum difference in first and second ionization enthalpies is
    1) $1s^2\,2s^2\,2p^6\,3s^2$ 2) $1s^2\,2s^2\,2p^6\,3s^2\,3p^1$
    3) $1s^2\,2s^2\,2p^6\,3s^2\,3p^2$ 4) $1s^2\,2s^2\,2p^6\,3s^1$

13. Which one of the following oxides is neutral?
    1) CO 2) SnO_2
    3) ZnO 4) SiO_2

14. If you are given Avogadro's number of atoms of a gas 'X'. If half of the atoms are converted into $X^+_{(g)}$ by energy ΔH. The IE of X is
    1) $\frac{2\Delta H}{N_A}$ 2) $\frac{2N_A}{\Delta H}$
    3) $\frac{\Delta H}{2N_A}$ 4) $\frac{N_A}{\Delta H}$

15. The energy required to convert all atoms present in 1.2 g magnesium to Mg^{2+} ions if IE_1, and IE_2 of magnesium are 120 kJ mol^{-1} and 240 kJ mol^{-1} respectively
    1) 18 kJ 2) 36 kJ
    3) 360 kJ 4) 24 kJ

16. Identify the wrong statement in the following.
    1) Atomic radius of the elements increases as one moves down the first group of the periodic table
    2) Atomic radius of the elements decreases as one moves across from left to right in the 2nd period of the periodic table
    3) Amongst isoelectronic species, smaller the positive charge on the cation, smaller is the ionic radius
    4) Amongst isoelectronic species, greater the negative charge on the anion, larger is the ionic radius

17. What is the value of electron gain enthalpy of Na^+ if IE_1 of Na = 5.1 eV?

1) +2.55 eV 2) +10.2 eV
3) –5.1 eV 4) –10.2 eV

18. Among the elements Ca, Mg, P and Cl, the order of increasing atomic radii is

1) Mg < Ca < Cl < P
2) Cl < P < Mg < Ca
3) P < Cl < Ca < Mg
4) Ca < Mg < P < Cl

19. The electronic configuration of an element is $1s^2\ 2s^2\ 2p^6\ 3s^2\ 3p^6\ 4s^2\ 3d^{10}\ 4p^2$. What is the atomic number of the element, which is just below the above element in the periodic table?

1) 36 2) 49
3) 50 4) 54

20. An element X with Z = 112 has been recently discovered. To which group and period will it belong, respectively-

1) 12, 7 2) 10, 6
3) 8, 4 4) 6, 6

21. Which of the following has lowest Ist ionisation enthalpy?

1) Li 2) C
3) O 4) F

22. Which of the following order best depicts the oxides' progressively less acidic nature?

1) $Li_2O > BeO > B_2O_3 > CO_2 > N_2O_3$
2) $N_2O_3 > CO_2 > B_2O_3 > BeO > Li_2O$
3) $BeO > CO_2 > N_2O_3 > Li_2O > B_2O_3$
4) $CO_2 > N_2O_3 > B_2O_3 > BeO > Li_2O$

23. The correct order of atomic size/ionic size is

1) $H^- > Li^+ > Be^{2+} > H^+$
2) $C^{4-} > O^{2-} > N^{3-} > F^-$
3) $Na^+ > Mg^{2+} > Al^{3+} > F^-$
4) All of these

24. On going from right to left in a period in the periodic table, the electronegativity of the elements

1) Increases
2) Decreases
3) Remain unchanged
4) Decreases first then increases

25. The diagram below is a part of the skeleton of the periodic tables in which elements are indicated by letter which are not their usual symbols.

	D	
A	C	F
52	B	E

The correct option is:-

1) 'D' has maximum electron affinity
2) 'B' exist in nature in a liquid state
3) 'A' is a 5th period element in a periodic table
4) (2) & (3)

26. Which of the following statement can help a chemistry student to predict chemical properties of an element?

(i) Position of element in the periodic table
(ii) Atomic number of the element
(iii) Number of shells in the atom
(iv) Number of electron in the outer most shell

1) (i), (ii) and (iii) 2) (i), (ii) and (iv)
3) (i), (iii) and (iv) 4) (ii), (iii) and (iv)

27. Which one of the following sets of ions represents the collection of isoelectronic species?

1) $K^+, Cl^-, Mg^{2+}, Sc^{3+}$ 2) $Na^+, Ca^{2+}, Sc^{3+}, F^-$
3) $K^+, Ca^{2+}, Sc^{3+}, Cl^-$ 4) $Na^+, Mg^{2+}, Al^{3+}, Cl^-$

28. The correct option regarding the following electronic configuration:-

1) $[Rn]\ 7s^2\ 6d^2$ – d block
2) $[Xe]\ 6s^2\ 4f^{14}\ 5d^1$ – f block
3) $[Xe]\ 6s^2\ 4f^{14}\ 5d^{10}\ 6p^1$ – d block
4) $[Xe]\ 6s^2\ 4f^1\ 5d^1$ – d block

29. In which of the following element, 2nd I.P. < Ist IP individually.

1) Mg 2) Ne
3) K 4) None of these

30. From the following given electronic configuration, identify the correct order of electron affinity.

(I) $[He]\ 2s^2\ 2p^5$ (II) $[He]\ 2s^2\ 2p^3$
(III) $[Ne]\ 3s^2\ 3p^5$ (IV) $[Ne]\ 3s^2\ 3p^3$

1) I > II > III > IV 2) III > I > IV > II
3) I < II < III < IV 4) II > III > IV > I

31. Consider the following as a portion of the periodic table from Group No. 13 to 17. Which of the following statements is/are true about the elements shown in it?

(i) V, W, Y and Z are less electropositive than X
(ii) V, W, X and Y are more electronegative than Z
(iii) Atomic size of Y is greater than that of W.
(iv) Atomic size of W is smaller than that of X.

			V	Z
				Y
W				
X				

1) (i), (ii) and (iii) 2) (ii) and (iii)
3) (i) and (iv) 4) (iii) and (iv)

32. Chloride ion and potassium ion are isoelectronic. Then

1) Potassium ion is relatively bigger
2) They contain different number of electrons
3) Their size are same
4) Chloride ion is bigger than potassium ion

33. In the given graph, a periodic property (R) is plotted against atomic numbers (Z) of the elements. Which property is shown in the graph and how is it correlated with reactivity of the elements?

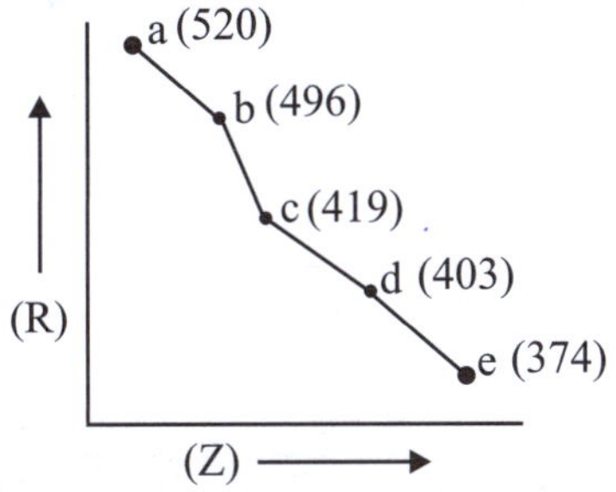

1) Ionisation enthalpy in a group decreases and reactivity decreases from a → e.
2) Ionisation enthalpy in a group decreases and reactivity increases from a → e.
3) Atomic radius in a group increases and reactivity decreases from a → e.
4) Metallic character in a group decreases and reactivity increases from a → e.

34. Match the entries of List I with appropriate entries of List II and select the correct answer using the codes given below the lists :

List I		List II	
P.	Rutherfordium (At. No. = 104)	(i)	Period number = 7
Q.	Rontgenium (At. No. = 111)	(ii)	Group number = 4
R.	Thorium (At. No. = 90)	(iii)	d-block elements
S.	Neptunium (At. No. = 93)	(iv)	f-block elements

	P	Q	R	S
1)	(iii)	(ii), (iv)	(ii),(iii)	(i),(ii),(iii)
2)	(i),(ii),(iii)	(i),(iii)	(i),(iv)	(i),(iv)
3)	(i),(ii)	(ii),(i)	(iv),(iii)	(iii),(i)
4)	(iii),(i)	(i),(iii)	(ii),(iv)	(iv),(i),(ii)

35. The correct sequence which shows decreasing order of the ionic radii of the element is:
1) $Na^+ > F^- > Mg^{2+} > O^{2-} > Al^{3+}$
2) $O^{2-} > F^- > Na^+ > Mg^{2+} > Al^{3+}$
3) $Al^{3+} > Mg^{2+} > Na^+ > F^- > O^{2-}$
4) $Na^+ > Mg^{2+} > Al^{3+} > O^{2-} > F^-$

36. In which of the following arrangements, the order is not correct according to the property indicated against it?
1) $Al^{3+} < Mg^{2+} < Na^+ < F^-$
   Increasing ionic size
2) $B < C < N < O$
   Increasing first ionization enthalpy
3) $I < Br < F < Cl$
   Increasing electron gain enthalpy (with negative sign)
4) $Li < Na < K < Rb$
   Increasing metallic radius

37. In a periodic table, the basic character of oxides
1) Increases from left to right and decreases from top to bottom
2) Decreases from right to left and increases from top to bottom
3) Decreases from left to right and increases from top to bottom
4) Decreases from left to right and increases from bottom to top.

38. Identify the incorrect statement.
1) The first ionisation enthalpy of Al is less than the first ionisation enthalpy of Mg.
2) The second ionisation enthalpy of Mg is greater than the second ionisation enthalpy of Na.
3) The first ionisation enthalpy of Na is less than the first ionisation enthalpy of Mg.
4) The third ionisation enthalpy of Mg is greater than the third ionisation enthalpy of Al.

39. Elements X, Y and Z have atomic numbers 19, 37 and 55 respectively. Which of the following statements is true about them?
1) Their ionization potential would increase with increasing atomic number.
2) Y would have an ionization potential between those of X and Z.
3) Z would have the highest ionization potential.
4) Y would have the highest ionization potential.

40. The element Z = 114 has been discovered recently. It will belong to which of the following family/group and will possess which of following electronic configurations?
1) Carbon family, $[Rn]\ 5f^{14}\ 6d^{10}\ 7s^2\ 7p^2$
2) Oxygen family, $[Rn]\ 5f^{14}\ 6d^{10}\ 7s^2\ 7p^4$
3) Nitrogen family, $[Rn]\ 5f^{14}\ 6d^{10}\ 7s^2\ 7p^6$
4) Halogen family, $[Rn]\ 5f^{14}\ 6d^{10}\ 7s^2\ 7p^5$

41. Which of the following is/are correct statement(s)?
1) The electron affinity of Si is smaller than that of P.
2) Penetrating power of p-orbital is more than s-orbital.
3) The numerical value of electronegativity of an atom depends on its ionisation potential and electron affinity.
4) All of the above.

42. Fluorine has low electron affinity than chlorine because of
1) Smaller radius of fluorine, high electron density
2) Smaller radius of chlorine, high electron density
3) Bigger radius of fluorine, less electron density
4) Smaller radius of chlorine, less electron density

Match the Columns MCQs

1. Match the correct atomic radius with the element

Element		Atomic radius (pm)	
(a)	Be	(i)	74
(b)	C	(ii)	88
(c)	O	(iii)	111
(d)	B	(iv)	77
(e)	N	(v)	66

1) (a)-(iv) (b)-(i) (c)-(ii) (d)-(iii) (e)-(v)
2) (a)-(i) (b)-(v) (c)-(ii) (d)-(iii) (e)-(iv)
3) (a)-(iv) (b)-(iii) (c)-(ii) (d)-(v) (e)-(i)
4) (a)-(iii) (b)-(iv) (c)-(v) (d)-(ii) (e)-(i)

2. Match the correct ionisation enthalpies and electron gain enthalpies of the following elements.

Elements			ΔH_1	ΔH_2	$\Delta_{eg}H$
(a)	Most reactive non metal	(i)	419	3051	– 48
(b)	Most reactive metal	(ii)	1681	3374	– 328
(c)	Least reactive element	(iii)	738	1451	– 40
(d)	Metal forming binary halide	(iv)	2372	5251	+ 48

1) (a)-(iv) (b)-(i) (c)-(ii) (d)-(iii)
2) (a)-(ii) (b)-(i) (c)-(iv) (d)-(iii)
3) (a)-(i) (b)-(iii) (c)-(ii) (d)-(iv)
4) (a)-(iv) (b)-(i) (c)-(iii) (d)-(ii)

3. Electronic configuration of some elements is given in Column-I and their electron gain enthalpies are given in Column-II. Match the electronic configuration with electron gain enthalpy.

Column-I Electronic configuration		Column-II Electron gain enthalpy/ kJ mol^{-1}	
(a)	$1s^2 2s^2 2p^6$	(i)	–53
(b)	$1s^2 2s^2 2p^6 3s^1$	(ii)	–328
(c)	$1s^2 2s^2 2p^5$	(iii)	–141
(d)	$1s^2 2s^2 2p^4$	(iv)	+ 48

1) (a)-(ii) (b)-(i) (c)-(iv) (d)-(iii)
2) (a)-(i) (b)-(iv) (c)-(iii) (d)-(ii)
3) (a)-(iv) (b)-(i) (c)-(ii) (d)-(iii)
4) (a)-(ii) (b)-(iv) (c)-(iii) (d)-(i)

4. Match atomic number of the element given in Column-I with their position in the periodic table given in Column-II.

Column-I		Column-II	
(a)	60	(i)	s-Block
(b)	47	(ii)	p-Block
(c)	56	(iii)	d-Block
(d)	52	(iv)	f-Block

1) (a)-(iii) (b)-(iv) (c)-(ii) (d)-(i)
2) (a)-(iv) (b)-(iii) (c)-(i) (d)-(ii)
3) (a)-(i) (b)-(ii) (c)-(iv) (d)-(iii)
4) (a)-(ii) (b)-(i) (c)-(iii) (d)-(iv)

5. Match the type of element given in Column-I with their general electronic configuration given in Column-II.

Column-I (Type of element)		Column-II (General Elect. Config.)	
(a)	Noble gases	(i)	ns^1 to $ns^2\ np^5$
(b)	Normal elements	(ii)	ns^2, $ns^2\ np^6$
(c)	Transition elements	(iii)	$(n–2)f^{1-14}$ $(n\ –1)$ d^{0-1} ns^2
(d)	Inner-transition elements	(iv)	$(n – 1)\ d^{1-10}\ ns^{1-2}$

1) (a)-(ii) (b)-(iii) (c)-(i) (d)-(iv)
2) (a)-(i) (b)-(iv) (c)-(ii) (d)-(iii)
3) (a)-(iii) (b)-(i) (c)-(iv) (d)-(ii)
4) (a)-(ii) (b)-(i) (c)-(iv) (d)-(iii)

Correct & Incorrect MCQs

1. Which of the following statement (s) is/are correct?

A. Elements Li, Na & K form a Dobereiner's Triad.

B. Elements Ca, Sr & Ba form a Dobereiner's Triad.

C. Elements Be, B & S form a Dobereiner's Triad.

D. Elements Cl, Br & I form a Dobereiner's Triad.

1) A, B and D 2) A and C
3) A, C and D 4) Only C

2. Incorrect statement(s) is/are

A. According to Henry Moseley square root of the frequency emitted by metal is directly proportional to atomic number & not the atomic weight.

B. Modern periodic table is based on atomic mass rather than atomic number.

C. In long form of periodic table, elements having similar outer electronic configurations in their atoms are arranged in horizontal rows, referred to as groups.

D. The period number corresponds to the highest principal quantum number (n) of the elements in the period.

1) A and B
2) B, C and D
3) Only B
4) B and C

3. Identify the incorrect statement from the following.

1) The IUPAC name of an element with atomic number 101 is Unnilunium.
2) The IUPAC name of an element with atomic number 102 is Ununseptium.
3) The IUPAC name of an element with atomic number 103 is Unniltrium
4) The IUPAC name of an element with atomic number 104 is Unnilquadium.

4. The incorrect statement regarding s-block elements is:

1) They are reactive metals with low ionization enthalpies.
2) They lose the outermost electron(s) readily to form unipositive ion (in the case of alkali metals) or divalent positive ion (in the case of alkaline earth metals).
3) The metallic character and the reactivity decreases while moving down the group.
4) The s-block elements are never found pure in nature.

5. Choose the correct statement/s:

A. The atomic size generally increases across a period.

B. The first ionization enthalpy generally increases as we go across a period and decreases as we descend in a group.

C. Effective nuclear charge decreases from left to right in a period.

D. Electronegativity generally increases across a period from left to right and decrease down a group in the periodic table.

1) A only
2) A and C
3) A and B
4) B and D

6. Which statement regarding isoelectronic species is not correct?

1) The cation with the greater positive charge will have a smaller radius.
2) The anion with the greater negative charge will have th larger radius.
3) The cation Mg^{2+} will have larger radius than that of cation Na^+.
4) The anion F^- will have smaller radius than that of anion O^{2-}.

7. Which of the following statement about p-block elements is incorrect?

1) It comprises of elements belonging to group 13 to 18.
2) These together with the s-block elements are called the representative or main group elements.
3) The outermost electronic configuration varies from ns^1 to ns^2 in each period.
4) At the end of each period, there is a noble gas element with a closed valence shell ns^2np^6 configuration.

8. Which amongst the following is incorrect statement ?

1) The ionisation enthalpy of elements are always positive.
2) Ionisation enthalpy for 'X' is less than that of X^+ .
3) First ionization enthalpy of oxygen is smaller than that of nitrogen.
4) First ionization enthalpy of beryllium is less than that of boron.

9. Identify the incorrect statement from the following:

1) Electron gain enthalpy provides a measure of the ease with which an atom adds an electron to form anion.
2) Noble gases have large negative electron gain enthalpies.
3) Cl has the most negative electron gain enthalpy.
4) In general, electron gain enthalpy becomes more negative with increase in the atomic number across a period.

10. Which amongst the following is incorrect statement ?

1) Na_2O is an acidic oxide.
2) Al_2O_3 is an amphoteric oxide.
3) Cl_2O_7 is an acidic oxide
4) CO is a neutral oxide.

Statement Based MCQs

Directions: These questions consist of two statements each, printed as Statement-I and Statement-II. While answering these questions, you are required to choose any one of the following four responses.

1) **Both Statement-I and Statement-II are correct.**
2) **Both Statement-I and Statement-II are incorrect.**
3) **Statement-I is correct & Statement-II is incorrect.**
4) **Statement-I is incorrect & Statement-II is correct.**

1. **Statement-I:** Boron has a smaller first ionization enthalpy than beryllium.

Statement-II: The first ionization enthalpy of nitrogen is smaller as compared to oxygen.

2. **Statement-I:** Ionization enthalpy is the energy released to remove an electron from an isolated gaseous atom in its ground state.

Statement-II: Every element has a tendency to lose the electron to attain the stable configuration.

3. **Statement-I:** Helium and beryllium have similar outer electronic configuration of the type ns^2.

   Statement-II: Both helium and beryllium are chemically inert.

4. **Statement-I:** The first ionisation enthalpy of Al is lower than magnesium.

   Statement-II: Atomic radius of Al is smaller than magnesium.

5. **Statement-I:** Ionization enthalpy decreases on moving down the group.

   Statement-II: Force of attraction between nucleus and electrons decreases on moving down the group.

6. **Statement-I:** Modern periodic table is based on atomic number rather than atomic mass.

   Statement-II: Elements in the same vertical column of the periodic table exhibit similar properties.

7. **Statement-I:** In general, the metallic character increases from left to right in a period.

   Statement-II: In general, while moving down the group, metallic character of element decreases.

8. **Statement-I:** Atomic radius of S is larger than that of Mg.

   Statement-II: Atomic radius decreases across a period, due to increased nuclear charge.

9. **Statement-I:** Noble gases have largest ionization enthalpy.

   Statement-II: Noble gases have stable fully filled valence electron configuration.

10. **Statement-I:** The electronegativity of any given element is not constant, it varies depending on the element to which it is bound.

    Statement-II: The electronegativity values of N and Cl are not the same on Pauling scale.

Assertion & Reason MCQs

Directions: These questions consist of two statements each, printed as Assertion and Reason. While answering these questions, you are required to choose any one of the following four responses.

1) **Both Assertion and Reason are True and the Reason is a correct explanation of the Assertion.**
2) **Both Assertion and Reason are True but Reason is not a correct explanation of the Assertion.**
3) **Assertion is True but the Reason is False.**
4) **Assertion is False but Reason is True.**

1. **Assertion:** Generally, ionisation enthalpy increases from left to right in a period.

   Reason: When successive electrons are added to the orbitals in the same principal quantum level, the shielding effect of inner core of electrons does not increase very much to compensate for the increased attraction of the electron to the nucleus.

2. **Assertion:** Electron gain enthalpy always becomes less negative as we go down a group.

   Reason: Size of the atom increases on going down the group and the added electron would be farther from the nucleus.

3. **Assertion:** Electron gain enthalpy of oxygen is less –ve than that of fluorine but more –ve than that of nitrogen.

   Reason: Ionization enthalpy is as follows: N > O > F.

4. **Assertion:** A qualitative measure of the ability of an atom in a chemical compound to attract shared electrons to itself is called electronegativity.

   Reason: Electronegativity is a measurable quantity.

5. **Assertion:** The electron gain enthalpy of chlorine is more negative than that of fluorine.

   Reason: The fluorine atom is much larger in size than that of chlorine.

6. **Assertion:** Cl_2O_7 is an amphoteric oxide

   Reason: Amphoteric oxides react with both acids and bases.

7. **Assertion:** Na^+ and Al^{3+} are isoelectronic but the magnitude of ionic radius of Al^{3+} is less than that of Na^+.

   Reason: The magnitude of effective nuclear charge of the outer shell electrons in Al^{3+} is less than that in Na^+.

8. **Assertion:** Cations are always greater than the neutral atom.

   Reason: Electronic configuration of atom and its corresponding ions is different.

9. **Assertion:** He is a p-block element.

   Reason: Last filling e^- goes into p-subshell.

10. **Assertion:** The period number corresponds to the highest principal quantum number (n) of the elements in the period.

    Reason: The first period in the long form of the periodic table contains 8 elements.

Multi Correct MCQs

1. Consider the following elements.

Group ↓ A'B' ⟶ Period
C'D'

Which of the following is/are true about A', B',C' and D' ?

A. Order of atomic radii: B' < A' < D' < C'

B. Order of metallic character: B' < A' < D' < C'

C. Size of the element: D' < C' < B' < A'

D. Order of ionic radii: $B'^{+} < A'^{+} < D'^{+} < C'^{+}$

Choose the correct answer from the options given below:

1) (A) only 2) (A), (B) and (D) only
3) (A) and (B) only 4) (B), (C) and (D) only

2. The electron affinity value are negative for:

A. $Be \rightarrow Be^{-}$ B. $N \rightarrow N^{-}$
C. $O \rightarrow O^{2-}$ D. $Na \rightarrow Na^{-}$
E. $Al \rightarrow Al^{-}$

Choose the most appropriate answer from the options given below:

1) (D) and (E) only 2) (A), (B), (D) and (E) only
3) (A) and (D) only 4) (A), (B) and (C) only

3. The statement(s) that are correct about the species O^{2-}, F^{-}, Na^{+} and Mg^{2+}.

A. All are isoelectronic

B. All have the same nuclear charge

C. O^{2-} has the largest ionic radii

D. Mg^{2+} has the smallest ionic radii

Choose the most appropriate answer from the options given below:

1) (B), (C) and (D) only 2) (A), (B), (C) and (D)
3) (C) and (D) only 4) (A), (C) and (D) only

4. For the given species, choose the set with isoelectronic species

I. O^{2-} II. F^{-}
III. Al IV. Mg^{2+}
V. Na^{+} VI. O^{+}
VII. Mg VIII. Al^{3+}
IX. F

1) (I), (II), (IV) and (V)
2) (III), (I), (IV), (IX), (VIII) and (V)
3) (II), (III), (IV), (VI) and (IX)
4) (I), (II), (IV), (V) and (VIII)

5. Which of the following electronic configurations corresponds to a main group element?

I. [Ar] $3d^{10}$ $4s^{2}$ $4p^{5}$ II. [Ar] $3d^{6}$ $4s^{2}$
III. [Ar] $4s^{1}$ IV. [Kr] $4d^{10}$ $5s^{1}$

1) Both II and IV 2) Only I
3) Only III 4) Both I and III

NCERT Exemplar MCQs

1. Consider the isoelectronic species, Na^{+}, Mg^{2+}, F^{-} and O^{2-}. The correct order of increasing length of their radii is:

1) $F^{-} < O^{2-} < Mg^{2+} < Na^{+}$ 2) $Mg^{2+} < Na^{+} < F^{-} < O^{2-}$
3) $O^{2-} < F^{-} < Na^{+} < Mg^{2+}$ 4) $O^{2-} < F^{-} < Mg^{2+} < Na^{+}$

2. Which of the following is not an actinoid?

1) Curium (Z = 96) 2) Californium (Z = 98)
3) Uranium (Z = 92) 4) Terbium (Z = 65)

3. The order of screening effect of electrons of s, p, d and f orbitals of a given shell of an atom on its outer shell electrons is:

1) s > p > d > f 2) f > d > p > s
3) p < d < s > f 4) f > p > s > d

4. The first ionisation enthalpies of Na, Mg, Al and Si are in the order:

1) Na < Mg > Al < Si 2) Na > Mg > Al > Si
3) Na < Mg < Al < Si 4) Na > Mg > Al < Si

5. The electronic configuration of Gadolinium (Atomic number 64) is:

1) [Xe] $4f^{3}$ $5d^{5}$ $6s^{2}$ 2) [Xe] $4f^{7}$ $5d^{2}$ $6s^{1}$
3) [Xe] $4f^{7}$ $5d^{1}$ $6s^{2}$ 4) [Xe] $4f^{8}$ $5d^{6}$ $6s^{2}$

6. The statement that is not correct for periodic classification of elements is:

1) The properties of elements are periodic function of their atomic numbers
2) Non-metallic elements are less in number than metallic elements
3) For transition elements, the 3d-orbitals are filled with electrons after 3p-orbital and before 4s-orbitals
4) The first ionisation enthalpies of elements generally increase with increase in atomic number as we go along a period

7. Among halogens, the correct order of amount of energy released (electron gain enthalpy) is:

1) F > Cl > Br > I 2) F < Cl < Br < I
3) F < Cl > Br > I 4) F < Cl < Br < I

8. The period number in the long form of the periodic table is equal to:

1) Magnetic quantum number of any element of the period
2) Atomic number of any element of the period
3) Maximum principal quantum number of any element of the period
4) Maximum azimuthal quantum number of any element of the period

9. The elements in which electrons are progressively filled in 4f-orbital are called:

1) Actinoids 2) Transition elements
3) Lanthanoids 4) Halogens

10. Which of the following is the correct order of size of the given species:

1) $I > I^{-} > I^{+}$ 2) $I^{+} > I^{-} > I$
3) $I > I^{+} > I$ 4) $I^{-} > I > I^{+}$

11. The formation of oxide ion $O^{2-}(g)$, from oxygen atom requires first an exothermic and then an endothermic step as shown below:

$$O(g)+e^- \rightarrow O^-(g); \Delta H^\ominus = -141 \text{ kJ mol}^{-1}$$

$$O^-(g)+e^- \rightarrow O^{2-}(g); \Delta H^\ominus = +780 \text{ kJ mol}^{-1}$$

Thus, process of formation of O^{2-} in gas phase is unfavourable even though O^{2-} is isoelectronic with neon. It is due to the fact that:
1) Oxygen is more electronegative
2) Addition of electron in oxygen results in larger size of the ion
3) Electron repulsion outweighs the stability gained by achieving noble gas configuration
4) O^- ion has comparatively smaller size than oxygen atom

Comprehension given below is followed by some multiple choice questions. Each question has one correct option. Choose the correct option.

In the modern periodic table, elements are arranged in order of increasing atomic numbers. Depending upon the type of orbitals receiving the last electron, the elements in the periodic table have been divided into four blocks, viz s, p, d and f.

The modern periodic table consists of 7 periods and 18 groups. Each period begins with the filling of a new energy shell. In accordance with the Aufbau principle, the seven periods (1 to 7) have 2, 8, 8, 18, 18, 32 and 32 elements respectively.

The seventh period is still incomplete. To avoid the periodic table being too Long, the two series of f-block elements, called lanthanoids and actinoids are placed at the bottom of the main body of the periodic table

12. The element with atomic number 57 belongs to:
1) s - block 2) p - block
3) d - block 4) f - block

13. The last element of the p-block in 6th period is represented by the outermost electronic configuration:
1) $7s^2\,7p^6$ 2) $5f^{14}\,6d^{10}\,7s^2\,7p^0$
3) $4f^{14}\,5d^{10}\,6s^2\,6p^6$ 4) $4f^{14}\,5d^{10}\,6s^2\,6p^4$

14. Which of the elements whose atomic numbers are given below, cannot be accommodated in the present set up of the long form of the periodic table?
1) 107 2) 118
3) 126 4) 102

15. The electronic configuration of the element which is just above the element with atomic number 43 in the same group is:
1) $1s^2\,2s^2\,2p^6\,3s^2\,3p^6\,3d^5\,4s^2$
2) $1s^2\,2s^2\,2p^6\,3s^2\,3p^6\,3d^5\,4s^3\,4p^6$
3) $1s^2\,2s^2\,2p^6\,3s^2\,3p^6\,3d^6\,4s^2$
4) $1s^2\,2s^2\,2p^6\,3s^2\,3p^6\,3d^7\,4s^2$

16. The elements with atomic numbers 35, 53 and 85 are all:
1) Noble gases 2) Halogens
3) Heavy metal 4) Light metals

17. Electronic configuration of four elements A, B, C, and D are given below:
A. $1s^2\,2s^2\,2p^6$ B. $1s^2\,2s^2\,2p^4$
C. $1s^2\,2s^2\,2p^6\,3s^1$ D. $1s^2\,2s^2\,2p^5$

Arrange them in increasing order of electron gain enthalpies.
1) $A<C<B<D$ 2) $A<B<C<D$
3) $D>B<C<A$ 4) $A>B<C>D$

Kattar NEET

Scan Me for Detailed Explanations

1. Which of the following order is incorrect?
1) Electronegativity of central atom: $CF_4 > CH_4 > SiH_4$
2) Hydration energy: $Al^{3+} > Be^{2+} > Mg^{2+} > Na^+$
3) Electrical conductance : $F^-_{(aq.)} > Cl^-_{(aq.)} > S^{2-}_{(aq.)}$
4) Magnetic moment: $Ni^{4+} > V^{3+} > Sr^{2+}$

2. The ground state electronic configurations of the elements, U, V, W, X and Y (these symbols do not have any chemical significance) are as follows:

U $1s^22s^22p^3$

V $1s^22s^22p^63s^1$

W $1s^22s^22p^63s^23p^2$

X $1s^22s^22p^63s^23p^63d^54s^2$

Y $1s^22s^22p^63s^23p^63d^{10}4s^24p^6$

Determine which sequence of elements satisfy the following statement:

(*i*) Element forms a carbonate which is not decomposed by heating

(*ii*) Element is most likely to form coloured ionic compounds

(*iii*) Element has larger atomic radius

(*iv*) Element forms only acidic oxide
1) V W Y U 2) V X Y W
3) V W Y X 4) V X W U

3. A, B and C are hydroxy-compounds of the elements X, Y and Z respectively. X, Y and Z are in the same period of the periodic table. A gives an aqueous solution of pH less than seven. B reacts with both strong acids and strong alkalis. C gives an aqueous solution which is strongly alkaline.

Which of the following statements is/are true?

(*i*) The three elements are metals.

(*ii*) The electronegativities decrease from X to Y to Z.

(*iii*) The atomic radius decreases in the order X, Y and Z.

(*iv*) X, Y and Z could be phosphorus, aluminium and sodium respectively.

1) Only (*i*), (*ii*), (*iii*) correct
2) Only (*i*), (*iii*) correct
3) Only (*ii*), (*iv*) correct
4) Only (*ii*), (*iii*), (*iv*) correct

4. Consider the following ionisation reactions:

I. E. (kJ mol^{-1})

$A_{(g)} \longrightarrow A^{+}_{(g)} + e^{-}, A_1$ $\quad$ $B_{(g)} \longrightarrow B^{2+}_{(g)} + e^{-}, B_2$

$B^{+}_{(g)} \longrightarrow A^{+}_{(g)} + e^{-}, A_1$ $\quad$ $C_{(g)} \longrightarrow C^{+}_{(g)} + e^{-}, C_1$

$C^{+}_{(g)} \longrightarrow C^{2+}_{(g)} + e^{-}, C_2$ $\quad$ $C^{2+}_{(g)} \longrightarrow C^{3+}_{(g)} + e^{-}, C_3$

If monovalent positive ion of A, divalent positive ion of B and trivalent positive ion of C have zero electron. Then incorrect order of corresponding I.E. is:

1) $C_3 > B_2 > A_1$ $\quad$ 2) $B_1 > C_1 > A_1$
3) $C_3 > C_2 > B_2$ $\quad$ 4) $B_2 > C_3 > A_1$

5. Which one of the following arrangements represents the correct order of least negative to most negative electron gain enthalpy for C, Ca, Al, F and O?

1) Al < Ca < O < C < F $\quad$ 2) Al < O < C < Ca < F
3) C < F < O < Al < Ca $\quad$ 4) Ca < Al < C < O < F

6. There are four elements 'p'. 'q', 'r' and 's' having atomic numbers Z – 1, Z, Z + 1 and Z + 2 respectively. If the element 'q' is an inert gas, select the correct answers from the following statements.

(*i*) 'p' has most negative electron gain enthalpy in the respective period.
(*ii*) 'r' is an alkali metal
(*iii*) 's' exists in +2 oxidation state.

1) (*i*) and (*ii*) only $\quad$ 2) (*ii*) and (*iii*) only
3) (*i*) and (*iii*) only $\quad$ 4) (*i*), (*ii*) and (*iii*)

7. Consider the following statements and arrange in the order of true/false as given in the codes.

S_1: In modern periodic table each block contains a number of columns equal to the number of electrons that can occupy that sub-shell.

S_2: The greatest increase in ionization enthalpy is experienced on removal of electron from core noble gas configuration.

S_3: The size of the isoelectronic species is affected by electron-electron interaction in the outer orbitals.

S_4: Any thing that Influences the valence electrons will affect the chemistry of element and the valence shell is not affected by nuclear mass.

1) FFTT $\quad$ 2) TTFT
3) FTFT $\quad$ 4) TTTT

8. Match the following:

Column-I (Consider Elements of 4th Period)		Column-II	
I.	Number of elements which have ns^1 configuration	P.	4
II.	(n + *l*) for last electron of last element	Q.	5
III.	Maximum unpaired electron in any of the element	R.	6
IV.	(n + *l*) for outermost electron of 1st element	S.	3

1) I-S; II-Q; III-R; IV-P $\quad$ 2) I-P; II-Q; III-R; IV-S
3) I-Q; II-P; III-R; IV-S $\quad$ 4) I-R; II-Q; III-S; IV-P

9. Match the following:

Column-I (Elements (Electrons in K, L, M, N...)		Column-II (Statements)	
I.	W(2, 8, 7)	P.	Paramagnetic
II.	X(2, 8, 18, 8)	Q.	3rd group element
III.	Y(2, 8, 14, 2)	R.	Last electron does not enter to valence shell
IV.	Z(2, 8, 18, 25, 8, 2)	S.	Reactive non-metal
		T.	diamagnetic

1) I–Q, R; II–P; III–S, T; IV–S
2) I-P, S; II-T; III-P, R; IV-P, Q, R
3) I–T; II–Q, R; III–Q; IV–T, S
4) I–R; II–S; III–Q, T; IV–Q

10. Match the values of ionization energy and electron gain enthalpy listed in Column-I with characteristic(s) of elements listed in Column-II.

Column-I				Column-II	
	$\Delta_{IE}H_1, \Delta_{IE}H_2, \Delta_{eg}H$ (in kJ mol^{-1})				
I.	2372	5251	+48	P.	Element which acts as a strong reducing agent
II.	419	3051	–48	Q.	Element which exists as a monoatomic molecule.
III.	1681	3374	–333	R.	Less reactive than halogen
IV.	1008	1846	–295	S.	Element which acts as a, strong oxidising agent.
				T.	Element which oxide is a stronger basic in nature.

1) I–P, S; II–Q; III–P, R; IV–T
2) I–S; II–R; III–Q, T; IV–P
3) I-Q,R; II-P,T; III-S; IV-R
4) I–T; II–S; III–P, R; IV–Q

11. Which of the following orders is incorrect for size?

1) Al ≈ Ga $\quad$ 2) $Te^{2-} > I^{-} > Cs^{+} > Ba^{2+}$
3) $Cr^{3+} < Cr^{6+}$ $\quad$ 4) Pd ≈ Pt

12. Which of the following statements is incorrect?

1) The second ionization enthalpy of oxygen element is greater than that of fluorine element.
2) The third ionization enthalpy of phosphorus is greater than that of Aluminium.
3) The first ionization enthalpy of Aluminium is slightly greater than that of Gallium.
4) The second ionization enthalpy of copper is greater than that of zinc.

13. Which of the following match is incorrect regarding B, Al, C and S elements?
    1) The highest first ionisation enthalpy : C
    2) The largest atomic size : Al
    3) The most negative electron gain enthalpy : C
    4) The most metallic character : Al

14. The following qualitative plots depict the first, second and third ionization energies (I.E.) of Mg, Al and K. Among the following, the correct match of I.E. and the metal is

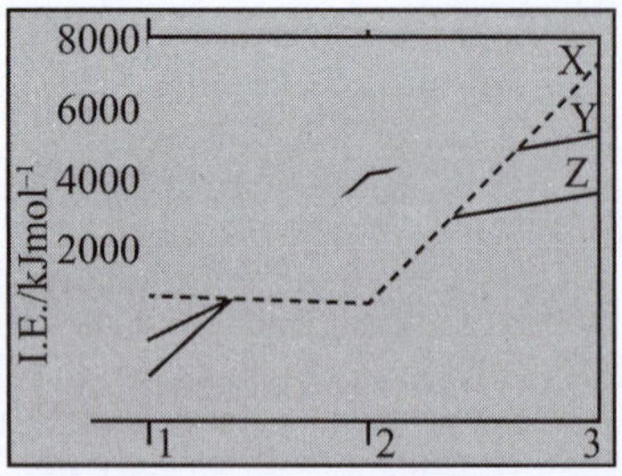

    1) X-Al; Y-Mg; Z-K
    2) X-Mg; Y-Al; Z-K
    3) X-Mg; Y-K; Z-Al
    4) X-Al; Y-K ; Z-Mg

15. The ground state electronic configurations of the elements, U, V, W, X and Y (these symbols do not have any chemical significance) are as follows:
    U. $1s^2\ 2s^2\ 2p^3$
    V. $1s^2\ 2s^2\ 2p^6\ 3s^1$
    W. $1s^2\ 2s^2\ 2p^6\ 3s^2\ 3p^2$
    X. $1s^2\ 2s^2 2p^6\ 3s^2\ 3p^6\ 3d^5\ 4s^2$
    Y. $1s^2\ 2s^2\ 2p^6\ 3s^2\ 3p^6\ 3d^{10}\ 4s^2\ 4p^6$

    Determine which sequence of elements satisfy the following statements:
    (*i*) Element forms a carbonate which is not decomposed by heating
    (*ii*) Element is most likely to form coloured ionic compounds
    (*iii*) Element has largest atomic radius
    (*iv*) Element forms only acidic oxide
    1) V W Y U
    2) V X Y W
    3) V W Y X
    4) V X W U

16. Which of the following set has highest sum of I.E. indicated as compared to all other sets mentioned?
    1) $I.E._1$ of He, $I.E_2$ of Li, $I.E._2$ of Ca
    2) $I.E._1$ of He, $I.E._2$ of Li, $I.E._3$ of Mg
    3) $I.E._1$ of Li, He and K
    4) $I.E._2$ of Na, Ba and Sr

17. If X(OH) > Y(OH) (acidic strength) where 'Y' and 'X' both belong to same group then select the correct statement
    1) electron gain enthalpy of 'Y' must be greater than 'X'
    2) atomic size of 'X' must be greater than 'Y'
    3) O–H bond will be easily broken in case of hydroxide of 'X'
    4) O–H bond will be easily broken in case of hydroxide of 'Y'

18. Variation of ionization energy Be < B < C < N < O < F is CORRECT for order of
    1) $I.E._1$ of given elements
    2) $I.E._2$ of given elements
    3) $I.E._3$ of given elements
    4) I.E. of their cations which have $2s^1$ valence electronic configuration

19. Match column-I with column-II and choose the correct answer from the code given below:

	Column-I		Column-II
A.	Strongest reductant	p.	Aurum/Au^{3+}
B.	Half filled d-orbital	q.	Cerium/Ce^{3+}
C.	Coinage metal	r.	Chromium
D.	Lanthanide	s.	Iodide ion

    1) A – s, B – r, C – p, D – q
    2) A – q, B – r, C – p, D – q
    3) A – q, B – s, C – r, D – p
    4) A – p, B – q, C – r, D – s

20. If there were more periods in the modern long form of the periodic table, then find the maximum number of elements in the 10th period.
    1) 25
    2) 27
    3) 72
    4) 29

21. If there were 9 periods in the periodic table & each orbital can have maximum 5 electrons, then find the maximum number of elements will be present in period 9?
    1) 125
    2) 127
    3) 172
    4) 129

NEET & AIIMS Past Year MCQs

1. The correct decreasing order of atomic radii (pm) of Li, Be, B and C is **(2024 Re)**
   1) Be > Li > B > C
   2) Li > Be > B > C
   3) C > B > Be > Li
   4) Li > C > Be > B

2. Arrange the following elements in increasing order of electronegativity: N, O, F, C, Si **(2024)**
   Choose the correct answer from the options given below:
   1) O < F < N < C < Si
   2) F < O < N < C < Si
   3) Si < C < N < O < F
   4) Si < C < O < N < F

3. Arrange the following elements in increasing order of first ionization enthalpy: Li, Be, B, C, N **(2024)**

Choose the correct answer from the options given below:

1) Li < Be < C < B < N 2) Li < Be < N < B < C

3) Li < Be < B < C < N 4) Li < B < Be < C < N

4. The element expected to form largest ion to achieve the nearest noble gas configuration is: **(2023)**

1) Na 2) O 3) F 4) N

5. The IUPAC name of an element with atomic number 119 is. **(2022)**

1) Ununcotium 2) Ununennium

3) Unnilennium 4) Unununnium

6. From the following pairs of ions which one is not an iso-electronic pair? **(2021)**

1) Na^{+}, Mg^{2+} 2) Mn^{2+}, Fe^{3+}

3) Fe^{2+}, Mn^{2+} 4) O^{2-}, F^{-}

7. **Assertion:** First ionization energy for nitrogen is lower than oxygen. **(AIIMS 2005)**

Reason: Across a period effective nuclear charge decreases.

1) If both Assertion & Reason are True & the Reason is a correct explanation of the Assertion.

2) If both Assertion & Reason are True but Reason is not a correct explanation of the Assertion.

3) If Assertion is True but the Reason is False.

4) If both Assertion & Reason are False.

8. If a neutral atom is converted into a cation, then its **(AIIMS 1996)**

1) Atomic weight increases

2) Size increases

3) Atomic weight decreases

4) Size decreases

9. In the P^{3-}, S^{2-} and Cl^{-} ions, the increasing order of size is **(AIIMS 1996)**

1) $S^{2-} < Cl^{-} < P^{3-}$ 2) $Cl^{-} < S^{2-} < P^{3-}$

3) $S^{2-} < P^{3-} < Cl^{-}$ 4) $P^{3-} < S^{2-} < Cl^{-}$

10. The element, with atomic number 118, will be **(AIIMS 1995)**

1) A transition element

2) An alkali metal

3) An alkaline earth metal

4) A noble gas

11. Which of the following elements will have the highest electron affinity? **(AIIMS 1994)**

1) Chlorine 2) Nitrogen

3) Phosphorus 4) Flourine

Answer Key

NCERT Topic-wise MCQs

1	2	3	4	5	6	7	8	9	10	11	12	13	14	15	16	17
1	3	2	3	2	3	1	4	3	2	1	3	1	1	4	4	4
18	**19**	**20**	**21**	**22**	**23**	**24**	**25**	**26**	**27**	**28**	**29**	**30**	**31**	**32**	**33**	**34**
4	2	4	3	2	4	3	4	2	3	1	4	4	3	3	4	4
35	**36**	**37**	**38**	**39**	**40**	**41**	**42**	**43**	**44**	**45**	**46**	**47**	**48**	**49**	**50**	**51**
3	2	3	1	2	4	4	3	4	3	1	2	3	4	2	3	2
52	**53**	**54**	**55**	**56**	**57**	**58**	**59**	**60**	**61**	**62**	**63**	**64**	**65**			
4	3	4	1	2	4	1	3	1	2	1	3	1	1			

High Order Time Intensive MCQs

1	2	3	4	5	6	7	8	9	10	11	12	13	14	15	16	17
1	2	4	1	4	3	2	1	2	4	2	4	1	1	1	3	3
18	**19**	**20**	**21**	**22**	**23**	**24**	**25**	**26**	**27**	**28**	**29**	**30**	**31**	**32**	**33**	**34**
2	3	1	1	2	1	2	1	2	3	2	4	2	3	4	2	2
35	**36**	**37**	**38**	**39**	**40**	**41**	**42**									
2	2	3	2	2	1	3	1									

Match the Columns MCQs

1	2	3	4	5
4	2	3	2	4

Correct & Incorrect MCQs

1	2	3	4	5	6	7	8	9	10
1	4	2	3	4	3	3	4	2	1

Statement Based MCQs

1	2	3	4	5	6	7	8	9	10
3	2	3	1	1	1	2	4	1	3

Assertion & Reason MCQs

1	2	3	4	5	6	7	8	9	10
1	4	3	3	3	4	3	4	3	3

Multi Correct MCQs

1	2	3	4	5
2	4	4	4	4

NCERT Exemplar MCQs

1	2	3	4	5	6	7	8	9	10	11	12	13	14	15	16	17
2	4	1	1	3	3	3	3	3	4	3	3	3	3	1	2	1

Kattar NEET

1	2	3	4	5	6	7	8	9	10	11	12	13	14	15	16	17
3	2	3	4	4	4	2	1	2	3	3	3	3	3	2	2	3
18	**19**	**20**	**21**													
4	2	3	1													

NEET & AIIMS Past Year MCQs

1	2	3	4	5	6	7	8	9	10	11
2	3	4	4	2	3	4	4	2	4	1

CHAPTER 4

Chemical Bonding and Molecular Structure

Topicwise Number of Questions Analysis (2025-2016)

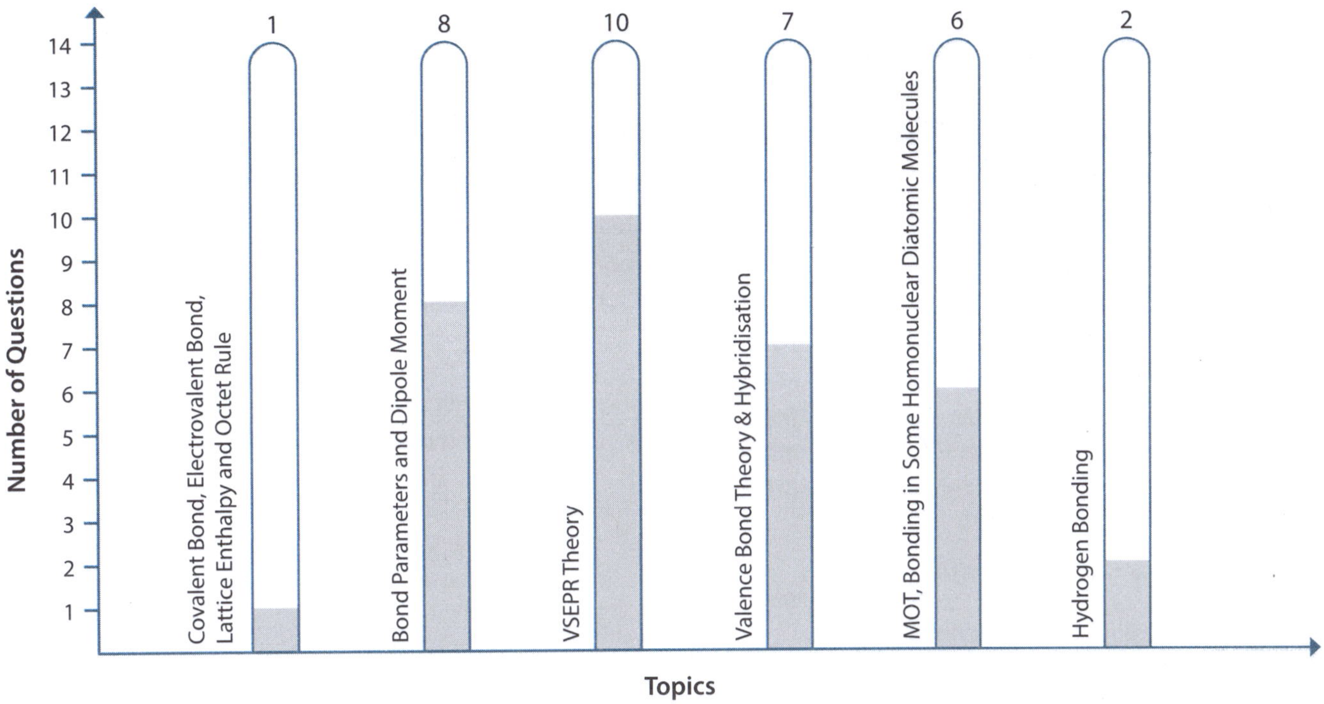

NCERT-PICKS

Kossel -Lewis Approach to Chemical Bonding

Octet rule

- According to "Electronic theory of chemical bonding", the valency of an element is the number of electrons that its atoms can gain, lose or share to acquire stable nearest noble gas configuration.

Covalent bond

- Covalent bonding is present between two non- metals.
- The compound $CaCO_3$ contains both ionic & covalent bonds. Calcium ion & carbonate ion are bonded by ionic bond while covalent bond is present between C & O in carbonate ion.

Lewis Structures

The Lewis dot structures basically represent a picture of bonding in molecules and ions in terms of electrons pair that are being shared.

Formal charge

The formal charge of an atom can be calculated by the difference between the number of valence electrons that a neutral atom would have and the number of electrons that belong to it in the Lewis structure.

The formal charge of any atom in molecule can be calculated as:

$$F.C = V - N - \frac{1}{2}(B)$$

where, V is total no. of valence electron in its ground state

- N is no. of non bonding valence electrons on this atom
- B is total no. of e–s shared in bonds with other atoms in molecules

Limitations of the octet rule

1. Electron deficient central atom: Number of electron < 8
   BeH_2, BF_3, BCl_3, BBr_3, BI_3, $AlCl_3$, $AlBr_3$, AlI_3.
2. Electron rich central atom: Number of electron > 8
   PCl_5, IF_7, SF_6, XeF_2.

3. Odd electron species central atom: has odd electron
   NO, NO_2, ClO_2, ClO_2.

Ionic Bond

- Ionic bond formed b/w metallic and non-metallic element. Metallic elements have low ionization enthalpy & non-metallic element have large –ve ΔegH. So, transfer of electrons done easily. Ionic bond is also known as electrovalent bond.

Condition for formation of ionic bond

- Electropositive atoms have low ionisation energy.
- Electronegative atom have high electron affinity.
- The ionic crystal possess high lattice energy.

Factors affecting L.E.

(i) Lattice energy (L.E.) $\propto 1/r$

(ii) L.E. $\propto Z^+, Z^-$

$Z^+ \Rightarrow$ charge on cation in terms of electronic charge

$Z^- \Rightarrow$ charge on anion in terms of electronic charge

(iii) Size of cation ↑ C.N. ↑ L.E. ↑

- Ionic compounds have high melting and boiling points because a large amount of thermal energy is required to separate the ions which are bound by strong electrical forces.
- Most ionic compounds are hard.
- Lattice enthalpy $\propto$ charge on its ions $\propto \dfrac{1}{\text{size of ions}}$
- % of covalent character in ionic bond $\propto$

  charge on ions $\propto$ size of anion $\propto \dfrac{1}{\text{size of cation}}$

Bond Parameters

- Bond Angle: It is defined as the angle between the orbitals containing bonding e– pairs around the central atom in a molecule.
- Bond angle helps in determining the shape of molecule. It gives idea regarding distribution of orbital around the central atom in a molecule / complex ion.

Factors affecting bond angle

(i) Hybridisation

Bond angle $\propto$ % of s-character

(ii) No. of lp/bp

Bond angle $\propto$ 1/ lone pair

(iii) Type of Central atom : Applicable when:

Hybridization No. of lp/bp are same

Bond angle $\propto$ Electronegativity of central atom

Bond Length: It is defined as the equilibrium distance between the nuclei of two bonded atoms in a molecule. Each atom of the bonded pair contributes to the bond length. In the case of a covalent bond, the contribution from each atom is called the covalent radius of that atom.

Bond Angle: It is defined as the angle between the orbitals containing bonding electron pairs around the central atom in a molecule/complex ion. It is expressed in degree which can be experimentally determined by spectroscopic methods.

For example: H–O–H bond angle in water can be represented as:

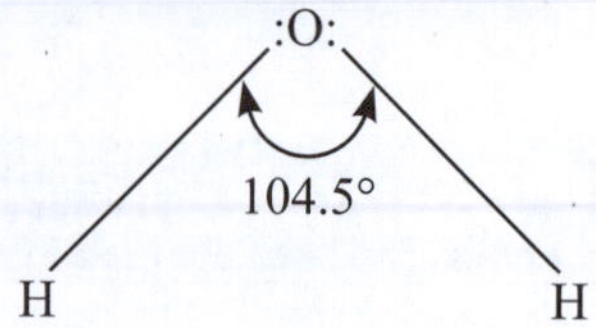

Bond Enthalpy: It is defined as the amount of energy required to break one mole of bonds of a particular type between two atoms in a gaseous state. The unit of bond enthalpy is kJ mol^{-1}. For example, the H – H bond enthalpy in hydrogen molecule is 435.8 kJ mol^{-1}.

$H_2(g) \rightarrow H(g) + H(g)$; $\Delta_a H^\ominus = 435.8$ kJ mol^{-1}

Dipole Moment (μ)

- It is equal to the product of magnitude of the charge and the distance between the centres of + ve and - ve charges of a dipole.

  μ = Charge (Q) × distance (d)
- μ increases with increase in electronegativity difference
- μ predicts the shape of molecule
- The net dipole moment of BeF_2 is zero as equal dipole point present in opposite direction & cancel the effect of each other.
- The dipole moment of NH_3 is greater than that of NF_3 because in case of NH_3, the orbital dipole due to lone pair is in same direction as the resultant dipole moment of N-H bonds, whereas in case of NF_3, the orbital dipole due to lone pair of the nitrogen is in the direction opposite to the resultant dipole moment of three N-F bonds.

Valence Shell Electron Pair Repulsion (VSEPR) Theory

Main postulates

- The electron pair surrounding the central atom repel one another to the extent that there are no further repulsions between them. Thus, the molecule will have minimum energy and maximum stability.
- If a central atom is bonded to similar atoms and contains no lone pair of electrons, the shape of molecule is symmetrical or regular.
- e. g. BF_3 – 120° – triangular

  CO_2 – 180° – linear

  CH_4 – 109°28' – tetrahedral
- If central atom is bonded with different atoms or is surrounded by lone pairs of electrons also along with bond pairs, then the shape of molecule is unsymmetrical or it has irregular geometry.
- The order of repulsion between electron pairs is as: lone pair - lone pair > lone pair - bond pair > bond pair - bond pair.

Valence Bond Theory

- Valence bond theory was introduced by Heitler and London (1927) and developed further by Pauling and others. This theory is based upon quantum mechanical principles.

Postulates of VBT

- A covalent bond is formed by the overlapping between two half filled atomic orbitals having electrons with opposite spin.
- Sigma bond (σ - bond) - The following overlappings result in the formation of sigma bond

  s - s overlapping

  s - p overlapping

  p - p head on overlapping

  pi - bond (π - bond) - It is formed by the sidewise or lateral overlapping between unhybridised p - atomic orbitals

  p - p side wise or lateral overlapping.

  π-bond is a weaker bond than σ - bond.
- The strength of bond depends upon the extent of overlapping between atomic orbitals. Greater the overlapping, stronger the σ - bond is. It follows the following order: $s - s > s - p > p - p$
- Single bond is always σ bond.
- In multiple bonds, only one is σ bond, others are π - bonds.
- π - bond is not formed by hybrid orbitals.

Compounds with the type of hybridization, geometry, number of bp-lp of electrons

Type of Molecule	Total No. of electron pairs	Bond pairs	Lone pairs	Type of hybridization	Geometry and shapes of molecules	Examples
AB_3	3	3	0	sp^2	Trigonal planar	BF_3, $AlCl_3$
AB_2L	3	2	1	sp^2	V-shaped	$SnCl_2$, $PbCl_2$
AB_2	2	2	0	sp	Linear	BeF_2, $[Ag(NH_3)_2]^+$
AB_4	4	4	0	sp^3	Tetrahedral	CH_4, SiF_4, CCl_4
AB_2L_2	4	2	2	sp^3	V-shaped	H_2O, OF_2, SCl_2
AB_3L	4	3	1	sp^3	Trigonal pyramidal	NH_3, PX_3 (X = F, Cl, Br, I)
AB_5	5	5	0	sp^3d	Trigonal bipyramidal	PF_5, PCl_5, $SbCl_5$
AB_4L	5	4	1	sp^3d	See-saw	SF_5, $TeBr_4$
AB_3L_2	5	3	2	sp^3d	T-shaped	ClF_3, $XeOF_2$
AB_2L_3	5	2	3	sp^3d	Linear	XeF_2, ICl_2^-, I_3^-
AB_5L	6	5	1	sp^3d^2	Square	IF_5, ClF_5,$[SbF_5]^{2-}$
AB_6	6	6	0	sp^3d^2	Octahedral	SF_6, $[SbF_6]^-$
AB_4L_2	6	4	2	sp^3d^2	Square planar	SF_4, XeF_4, ICl_4^-
AB_7	7	7	0	sp^3d^3	Pentagonal bipyramidal	IF_7, XeF_6

Hybridization

Hybridization involves th intermixing of orbitals having slightly different energies which results in the formation of new set of orbitals having equivalent energies and shape. Only valence shell orbitals of the atom will get hybridized.

Calculation of hybridization

$$H = \frac{1}{2}(V + M - C + A)$$

V = No of e^-s in valence shell of central atom

M = No of monovalent atom

C = Charge on cation

A = Charge on anion.

Molecular Orbital Theory

- The atomic orbitals of comparable energies and proper symmetry of two atoms are mixed up to form molecular orbitals.
- Total number of MO = Total number of A.O (antibonding orbital)
- Two A.O. combine to form two M.Os. One is anti bonding M.O and the other is bonding M.O. The energy of bonding MO is less than anti bonding MO.
- The filling up of electrons in M.Os must follow Aufbau's principle, Pauli exclusion principle and Hund's rule.
- The energy order of molecular orbitals for number of electrons ≤ 14.

  $\sigma 1s, \sigma^* 1s, \sigma 2s, \sigma^* 2s, \pi 2p_x \approx \pi 2p_y, \sigma 2p_z, \pi^* 2p_x \approx \pi^* 2p_y, {}^*\sigma 2p_z$

  For number of electrons > 14

  $\sigma 1s, \sigma^* 1s, \sigma 2s, \sigma^* 2s, \sigma 2p_z, \pi 2p_x \approx \pi 2p_y, \pi^* 2p_x \approx \pi^* 2p_y, \sigma^* 2p_z$

- Bond order (BO) $= \frac{N_b - N_a}{2}$
- Bond order $\propto$ Bond energy $\propto$ Stability of molecule $\propto \frac{1}{\text{bond length}}$
- As bond order increases, bond strength increases consequently bond enthalpy increases.
- Eg: Increasing bond enthalpy order is $H_2 < O_2 < N_2$.
- If a molecule contains unpaired electrons, then it is paramagnetic and if it contains paired electrons, then nature will be diamagnetic.

Stability of Molecules

(i) If $N_b > N_a$; then molecule is stable

(ii) If $N_b < N_a$; then molecule is unstable

Hydrogen Bond

Nitrogen, oxygen and fluorine are the highly electronegative elements. When they are attached to a hydrogen atom to form covalent bond, the electrons of the covalent bond are shifted towards the more electronegative atom. This partially positively charged hydrogen atom forms a bond with the other more electronegative atom. This bond is known as hydrogen bond and is weaker than the covalent bond.

Types of hydrogen bonds

(A) Intramolecular H-Bonding: It is formed when hydrogen atom is in between the two highly electronegative (F, O, N) atoms present within the same molecule. For example, in o - nitrophenol, the hydrogen is in between the two oxygen atoms.

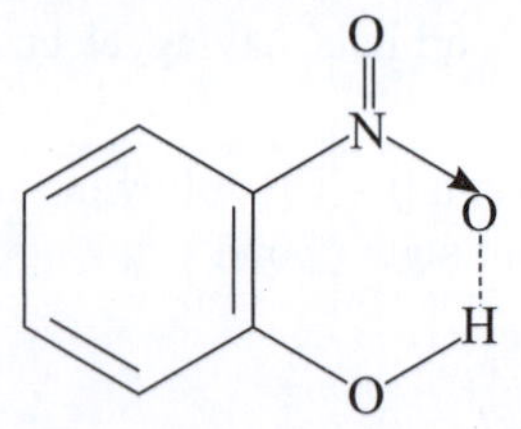

Intramolecular hydrogen bonding in o-nitrophenol molecule

(B) Intermolecular H-Bonding: It is formed between two different molecules of the same or different compounds. For example, H-bond in case of HF molecule, alcohol or water molecules, etc.

$$---H^{\delta+} - F^{\delta-} --- H^{\delta+} - F^{\delta-} --- H^{\delta+} - F^{\delta-}$$

MCQs ?

NCERT Topic-wise MCQs

Covalent Bond and Ionic or Electrovalent Bond

1. In an ionic compound A^+X^-, the degree of covalent bonding is greater when
   1) A^+ and X^- ions are small
   2) A^+ is small and X^- is large
   3) A^+ and X^- ions are approximately of the same size
   4) X^- is small and A^+ is large
2. Which of the following has highest ionic character?
   1) $MgCl_2$ 2) $CaCl_2$
   3) $BaCl_2$ 4) $BeCl_2$
3. The number of π-bonds in the structure given below are: $(NC)_2\,C = C\,(CN)_2$
   1) 9 2) 7
   3) 5 4) 2
4. The molecule that deviates from octet rule is
   1) NaCl 2) $BeCl_2$
   3) MgO 4) NH_3
5. What is the formal charge on O-atom in carbonate ion?
   1) –1 2) 0
   3) +1 4) Both (1) & (2)
6. Which among the following compounds has a maximum value of lattice energy?
   1) NaF 2) NaCl
   3) NaBr 4) NaI
7. The compound which does not contain ionic bond is:
   1) NaOH 2) HCl
   3) K_2S 4) LiH
8. Which of the following chloride has considerable covalent character?
   1) LiCl 2) NaCl
   3) KCl 4) CsCl
9. An ionic bond A^+B^- is most likely to be formed when
   1) The ionization energy of A is high and the electron affinity of B is low
   2) The ionization energy of A is low and the electron affinity of B is high
   3) The ionization energy of A and the electron affinity of B is high
   4) The ionization energy of A and the electron affinity of B is low
10. Which compound has ionic, covalent, coordinate and hydrogen bond ?
    1) $CuSO_4{\cdot}5H_2O$ 2) $ZnSO_4{\cdot}7H_2O$
    3) $FeSO_4{\cdot}7H_2O$ 4) $FeCl_3{\cdot}6H_2O$
11. Molecule having maximum number of covalent bonds is:
    1) NH_4OH 2) NH_4Cl
    3) $CO(NH_2)_2$ 4) CH_3OH
12. Number of bonded electrons in ethane molecule are:
    1) 7 2) 12
    3) 10 4) 14
13. The number of electron pairs involved in the formation of hydrogen cyanide molecule are:
    1) Two 2) Eight
    3) Three 4) Four
14. The electronegativities of two elements are 0.7 and 3.0. The bond formed between them would be:
    1) Ionic 2) Covalent
    3) Co-ordinate covalent 4) Metallic
15. Highly ionic compound is formed by the combination of elements belonging to:
    1) I A group and VII A group
    2) II A group VI A group
    3) III A group and V A group
    4) 'O' group and VII A group
16. Potassium forms a highly ionic compound when it combines with
    1) Chlorine 2) Fluorine
    3) Bromine 4) Iodine
17. Most ionic compound among the following is:
    1) Sodium fluoride 2) Sodium Chloride
    3) Sodium bromide 4) Sodium iodide
18. Which contains both polar and non-polar bonds ?
    1) NH_4Cl 2) HCN
    3) H_2O_2 4) CH_4
19. Which of the following is non-polar?
    1) H_2S 2) NaCl
    3) Cl_2 4) H_2SO_4
20. Covalent molecules are usually held in a crystal structure by
    1) Dipole-dipole attraction
    2) Electrostatic attraction
    3) Hydrogen bond
    4) Van-der Waal's attraction

21. What is the total number of electrons in the correct Lewis dot formula of the sulphite ion?
1) 8 2) 24
3) 26 4) 30

22. Indicate the nature of bonding in diamond.
1) Ionic 2) Covalent
3) Molecular 4) Metallic

Bond Parameters

23. In which of the following species, the angle around the central atom is exactly equal to 109°28′?
1) SF_4 2) NH_3
3) NH_4^+ 4) None of these

24. The bond angle in H_2S is:
1) $> NH_3$ 2) Same as in $BeCl_2$
3) $> H_2Se, < H_2O$ 4) Same as in CH_4

25. The largest bond angle around the central atom will be present in
1) H_2O 2) BeF_2
3) CH_4 4) NH_3

26. Which of the following molecule has shortest C-H bond?
1) Ethene 2) Ethane
3) Ethyne 4) Methane

27. The correct sequence of decrease in the bond angles of the following hydrides is.
1) $NH_3 > PH_3 > AsH_3 > SbH_3$
2) $NH_3 > AsH_3 > PH_3 > SbH_3$
3) $SbH_3 > AsH_3 > PH_3 > NH_3$
4) $PH_3 > NH_3 > AsH_3 > SbH_3$

28. In O_3, there are:
1) 1σ, 1π bond 2) 1σ, 2π bonds
3) 2σ, 2π bonds 4) 2σ, 1π bond

29. Which one of the following has the smallest bond angle?
1) NH_3 2) BeF_2
3) H_2O 4) CH_4

30. Bond angle in water is.
1) 120° 2) 109.5°
3) 107° 4) 104.5°

31. The correct order in which the O–O bond length increases in the following:
1) $H_2O_2 < O_2 < O_3$ 2) $O_3 < H_2O_2 < O_2$
3) $O_2 < H_2O_2 < O_3$ 4) $O_2 < O_3 < H_2O_2$

32. The ONO angle is maximum in:
1) HNO_3 2) NO_2^+
3) HNO_2 4) NO_2

33. Which of the following is correct order of bond length?
1) $BF_4^- < BF_3$ 2) $NO_2^+ < NO_2^-$
3) $CCl_4 < CF_4$ 4) $^+CH_3 > CH_4$

34. The correct order towards bond angle is:
1) Bond angle does not depend on hybridisation
2) $sp < sp^2 < sp^3$
3) $sp^2 < sp < sp^3$
4) $sp^3 < sp^2 < sp$

35. Resonance is not shown by:
1) C_6H_6 2) CO_2
3) CO_3^{2-} 4) SiO_2

36. H_2O is dipolar, whereas BeF_2 is not. It is because
1) H_2O involves hydrogen bonding whereas BeF_2 is a discrete molecule
2) H_2O is linear and BeF_2 is angular
3) H_2O is angular and BeF_2 is linear
4) The electronegativity of F is greater than that of O

37. The bond length of the S–O bond is maximum in which of the following compound ?
$SOBr_2$, $SOCl_2$, SOF_2
1) $SOCl_2$ 2) $SOBr_2$
3) SOF_2 4) All have same length

38. The correct order of dipole moment is
1) $CH_4 < NF_3 < NH_3 < H_2O$
2) $NF_3 < CH_4 < NH_3 < H_2O$
3) $NH_3 < NF_3 < CH_4 < H_2O$
4) $H_2O < NH_3 < NF_3 < CH_4$

39. Find the maximum number of identical angle in CH_2F_2 molecule.
1) 2 2) 4
3) 6 4) 8

40. In which of the following, bond angle is maximum?
1) NH_3 2) NH_4^+
3) PCl_3 4) SCl_2

41. Which one of the following compounds has the smallest bond angle in its molecule?
1) OH_2 2) SH_2
3) NH_3 4) SO_2

42. The correct order of bond angles (smallest first) in, H_2S, NH_3, BF_3 and SiH_4 is:
1) $H_2S < NH_3 < SiH_4 < BF_3$
2) $NH_3 < H_2S < SiH_4 < BF_3$
3) $H_2S < SiH_4 < NH_3 < BF_3$
4) $H_2S < NH_3 < BF_3 < SiH_4$

43. Which one of the following bonds has the highest average bond energy?
1) $S = O$ 2) $C \equiv C$
3) $C \equiv N$ 4) $N \equiv N$

44. The correct decreasing order of bond angles is:
1) $ClF_3 > PF_3 > NF_3 > BF_3$
2) $BF_3 > PF_3 > NF_3 > ClF_3$
3) $BF_3 > NF_3 > PF_3 > ClF_3$
4) $BF_3 > ClF_3 > PF_3 > NF_3$

45. Which of the following shows the correct order of bond lengths ?
   1) C–C > C–H > O–H   2) C–H > C–C > O–H
   3) O–H > C–H > C–C   4) C–H > O–H > C–C

VSEPR Theory

46. The shape of covalent molecule AX_3 is.
   1) Triangular   2) T-shape
   3) Pyramidal   4) Any of the above

47. Of the three molecules XeF_4, SF_4, SiF_4, one which have tetrahedral structure is:
   1) All the three   2) Only SiF_4
   3) Both SF_4 & XeF_4 4) Only SF_4 & XeF_4

48. The atomic number of Sn is 50. The shape of gaseous $SnCl_2$ molecule is:
   1) Cl – Sn – Cl   2) 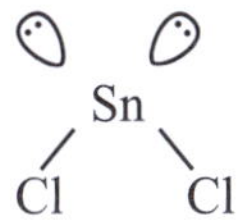
   3) 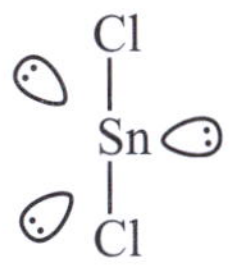   4) 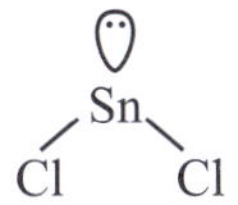

49. The shape of SO_4^{2-} ion is:
   1) Square planar   2) Square Pyramidal
   3) Tetrahedral   4) None of these

50. Which of the following will be trigonal planar?
   1) PCl_3 2) NH_3
   3) ClF_3 4) $AlCl_3$

51. Which of the angle is most influenced by the presence of lone pair in SF_4 ?
   1) Equatorial angle
   2) Axial angle
   3) Angle between axial and equatorial
   4) None of these

52. The shape of CO_2 molecule is similar to:
   1) H_2O 2) BeF_2
   3) SO_2 4) None of these

53. Out of $CHCl_3$, CH_4 and SF_4, the molecules having regular geometry are:
   1) $CHCl_3$ only   2) $CHCl_3$ and SF_4
   3) CH_4 only   4) CH_4 and SF_4

54. Which one of the following has the regular tetrahedral structure?
   1) XeF_4 2) $[Ni(CN)_4]^{2-}$
   3) BF^-_4 4) SF_4

55. The ion which is not tetrahedral shape is:
   1) BF_4^- 2) NH_4^+
   3) NH_3 4) $NiCl_4^{2-}$

56. The shape of formaldehyde molecule as per the VSEPR theory is
   1) Linear   2) Trigonal Planar
   3) Pyramid   4) Tetrahedron

57. Which of the following pair contains iso-structural species?
   1) CH_3^+ and CH_3^+ 2) NH_4^+ and NH_3
   3) SO_4^{2-} and BF_4^- 4) NH_2^- and BeF_2

58. The interhalogen compound that cannot exist is
   1) IBr_3 2) ICl_7
   3) IF_4 4) BrF_5

59. The molecule ML_x is planar with six pair of electrons around M in the valence shell. The value of x is
   1) 6   2) 2
   3) 4   4) 3

60. Cyanogen, $(CN)_2$, has a ___ shape/structure:
   1) Linear   2) Zig-zag
   3) V-shape   4) Cyclic

61. The geometry of electron pairs around I in IF_5 is:
   1) Octahedral   2) Trigonal Bipyramidal
   3) Square Pyramidal   4) Pentagonal Planar

62. Which of the following tetra atomic species is non-polar and planar?
   1) BrF_4^- 2) NH_4^+
   3) ClF_3 4) BF_3

63. Which of the possible molecule/species is having maximum values for dipole moment. (where "A" is the central atom)?
   1) AX_3 (having one lone pair on central atom)
   2) AX_4 (Tetrahedral)
   3) AX_4Y (having no lone pair on central atom)
   4) Can't be predicted

64. Which of the following species given below have shape similar to $XeOF_4$?
   1) XeO_3 2) IOF_4^+
   3) PCl_5 4) XeF_5^+

65. The total number of lone pairs of electrons in NO_3^- :
   1) 1   2) 2
   3) 6   4) 8

66. According to VSEPR theory,
   1) The shape of the molecule depends upon the bonded electron pair
   2) The pair of electrons tend to occupy such positions that maximize repulsions
   3) The pair of electrons tend to occupy such positions that minimise repulsions
   4) The pairs of electrons tend to occupy such positions that minimise distance from each other.

67. In PO_4^{3-} ion, number of bond pairs and lone pairs of electrons on Phosphorus atom are:
   1) 2,2   2) 3,1
   3) 1,3   4) 5,0

68. In OF_2 molecule, the total number of bond pairs and lone pairs of electrons are respectively:
   1) 2, 6   2) 2, 8
   3) 2, 10   4) 2, 9

69. The shape of $XeOF_2$ on the basis of VSEPR theory is:
1) See-saw 2) V-shaped
3) Trigonal planar 4) T-shaped

70. Which of the following is a pyramidal species?
1) CO_2 2) H_3O^+
3) $SnCl_2$ 4) NH_4^+

71. In CCl_4, the four valencies of carbon are directed towards the corners of a
1) Cube 2) Hexagon
3) Prism 4) Tetrahedron

Valence Bond Theory & Hybridisation

72. A hybrid orbital formed from s and p-orbital can contribute to
1) σ- bond only 2) π-bond only
3) Either σ or π-bond 4) Cannot be predicted

73. The hybrid state of C in CS_2 should be:
1) sp^2 2) sp
3) sp^3 4) Not specific

74. In which of the following, the central atom does not use sp^3 hybrid orbitals in its bonding?
1) BeF_3^- 2) OH_3^+
3) NH_2^- 4) NF_3

75. The compound 1, 2-butadiene has
1) Only sp hybridized carbon atoms.
2) Only sp^2 hybridized carbon atoms.
3) Both sp & sp^2 hybridized carbon atoms.
4) sp, sp^2 and sp^3 hybridized carbon atoms.

76. The hybridisation of $\overset{+}{ICl_2}$ is
1) sp 2) sp^3
3) sp^2 4) dsp^2

77. The hybridisation scheme for the central atom includes a d-orbital contribution in:
1) I_3^- 2) PCl_3
3) NO_3^- 4) H_2Se

78. The species/molecule is having different shape and different hybridisation:
1) XeF_2, CO_2 2) I_3^-, H_3O^+
3) OCl^-, CO 4) SO_2, OCl_2

79. All the four sigma bonds in perchlorate ion are formed by which orbital?
1) sp^3d^3 2) sp^3d^2
3) sp^3d 4) sp^3

80. Which one of the following conversions involve change in both hybridisation and shape?
1) $CH_4 \rightarrow C_2H_6$ 2) $NH_3 \rightarrow \overset{+}{NH_4}$
3) $BF_3 \rightarrow BF_4^-$ 4) $H_2O \rightarrow H_3O^+$

81. The bond angle and hybridization in ether (CH_3OCH_3) is:
1) 106°51', sp^3 2) 104°31', sp^3
3) > 109° 28' sp^3 4) None of these

82. Hybrid orbital having maximum p - character is:
1) sp^3d 2) sp^3
3) sp 4) sp^3d^2

83. In which of the following central atom does not have sp^3 hybridisation?
1) SF_4 2) BF_4^-
3) NH_4^+ 4) CH_4

84. Hybridisation of C in CO_3^{2-}:
1) sp 2) sp^2
3) sp^3 4) sp^3d

85. The hybridisation of Nitrogen in Nitrate ion is:
1) sp 2) sp^2
3) sp^3 4) sp^3d

86. Hybridisation of iodine in iodine hepta fluoride molecule is:
1) sp^3d^3 2) sp^3d
3) sp^3d^2 4) dsp^2

87. The molecule which contains sp^3d^2 and p orbitals overlapping in it is
1) PCl_5 2) BrF_5
3) ClF_3 4) IF_7

88. Hybridisation of central atom is sp^2 in:
1) H_2S 2) CS_2
3) N_2O 4) NO_3^-

89. The type of hybrid orbitals used by the oxygen atom in Cl_2O molecule is:
1) sp^3 2) sp^2
3) sp 4) None of these

90. Which type of overlapping results in the formation of a π bond?
1) Axial overlapping of s–s orbitals
2) Lateral overlapping of p–p orbitals
3) Axial overlapping of p–p orbitals
4) Axial overlapping of s–p orbitals

91. Indicate the wrong statement according to Valence bond theory:
1) A sigma bond is stronger than π - bond
2) p-orbitals always have only sidewise overlapping
3) s-orbitals never form π - bonds
4) There can be only one sigma bond between two atoms

92. The strongest P–O bond is found in the molecule
1) F_3PO 2) Cl_3PO
3) Br_3PO 4) $(CH_3)_3PO$

93. The correct order of hybridization of the central atom in the following species:
NH_3, XeO_2F_2, SeF_4, NO_2^+
1) sp^3, sp^3, sp^3d, sp 2) sp^3, sp^3d, sp^3d, sp
3) sp^3, sp^3d^2, sp^3d, sp^2 4) sp^2, sp^3d, sp^3d^2, sp

94. What is the hybridization of Te in $TeCl_6$?

1) sp
2) sp^2
3) sp^3d
4) sp^3d^2

95. The hybridisation of the central atom will change when

1) NH_3 combines with H^+
2) H_3BO_3 combines with OH^-
3) NH_3 forms NH_2^-
4) H_2O combines with H^+

96. What is the state of hybridisation of anionic part of solid N_2O_5?

1) sp
2) sp^2
3) sp^3
4) Not applicable

97. $BF_3 + F^- \rightarrow BF_4^-$

What is the hybridisation state of B in BF_3 and BF_4^- ?

1) sp^2, sp^3
2) sp^3, sp^3
3) sp^2, sp^2
4) sp^3, sp^3d

98. Propyne molecule contains

1) 6 sigma and 2 pi bonds
2) 8 pi bonds
3) 5 pi bonds and 1 sigma bond
4) 2 sigma and 3 pi bonds

99. sp^3d hybridization is considered to be a combination of two hybridization. They are

1) p^3 + sd
2) sp^2 + pd
3) spd + p^2
4) None of these

100. Considering the X-axis as the internuclear axis, a sigma bond will not be formed in:

1) 1s and 2s
2) 1 and $2p_x$
3) $2p_y$ and $2p_y$
4) 1s and 1s

101. On hybridisation of one 's' and one 'p' orbitals we get

1) Two mutually perpendicular orbitals
2) Two orbitals at 180°
3) Four orbitals directed tetrahedrally
4) Three orbitals in a plane

102. s - p overlapping is present in

1) Br_2
2) H_2
3) O_2
4) HF

103. The hybridization of atomic orbitals of nitrogen in NO_2^+, NO_3^- and NH_4^+ are

1) sp^2, sp^3 and sp^2 respectively
2) sp, sp^2 and sp^3 respectively
3) sp^2, sp and sp^3 respectively
4) sp^2, sp^3 and sp respectively

104. The 'S' atom among the following that does not have sp^3 hybridization is

1) SO_4^{2-}
2) SF_4
3) SF_2
4) S_8

105. Which one of the following is a correct set?

1) H_2O,sp^3,angular
2) H_2O,sp^2, linear
3) NH_4^+,dsp^2, square planar
4) CH_4,dsp^2, tetrahedral

106. The ground state electronic configuration of sulphur is $3p^2$, $3p^4$. How does it form the compound SF_6?

1) Due to octahedral shape of sulphur atoms
2) Due to presence of vacant 3d-orbitals which provide 6 unpaired electrons in excited state.
3) Due to d^2sp^3 hybridisation of sulphur atom which provides 8 electrons to 4F atoms.
4) Due to presence of 3 sigma and 3 pi bonds between sulphur and fluorine

Molecular Orbital Theory

107. The species having bond order different from that in CO is

1) NO^-
2) NO^+
3) CN^-
4) N_2

108. Which of the following molecules/ions does not contain unpaired electrons?

1) C_2
2) O_2
3) He_2
4) Both (1) and (3)

109. Which one of the following species does not exist under normal conditions?

1) Li_2
2) Be_2^+
3) Be_2
4) B_2

110. Which of the following is/are paramagnetic?

1) NO
2) NO^+
3) NO^-
4) N_2

111. Which one of the following has the strong O-O bond?

1) O_2^+
2) O_2
3) O_2^-
4) O_2^{2-}

112. In which of the following pairs, the two species have identical bond order?

1) N_2^- , O_2^{2-}
2) N_2^- , O_2^-
3) N_2^- , O_2^+
4) O_2^+ , N_2^{2-}

113. How many bonds does B_2 have?

1) 0
2) 1
3) 2
4) 3

114. Out of the following, which has smallest bond length?

1) O_2
2) O_2^+
3) O_2^-
4) O_2^{2-}

115. Which of the following is diamagnetic?

1) H_2^+
2) O_2
3) Li_2
4) He_2^+

116. Molecular orbital theory was given by:

1) Kossel
2) Moseley
3) Mulliken
4) Werner

117. Molecular orbital electronic configuration for 'X' anion is- KK σ^*2s^2 σ^*2s^2 $\pi\, 2p_x^2$ $\pi\, 2p_y^2$ $\sigma 2p_z^2$ $\pi^*\, 2p_x^1$, the anion 'X' is.

1) N_2^-
2) O_2^-
3) N_2^{2-}
4) O_2^{2-}

118. The number of anti-bonding electrons pairs in O_2^{2-} molecular ion on the basis of molecular orbital theory is (atomic no. of O is 8)
1) 3 2) 2
3) 5 4) 4

119. Which of the following is paramagnetic with bond order 0.5?
1) F_2 2) H_2^+
3) N_2 4) O_2^-

120. Which of the following pairs have identical bond order?
1) N_2, O_2^{2+} 2) N_2, O_2^-
3) N_2^-, O_2 4) O^{2+}, N_2

121. The bond order of O_2^- is:
1) 0.5 2) 1.5
3) 3.5 4) 2.5

122. The molecule having bond order 3 is
1) H_2 2) N_2
3) O_2 4) He^{2+}

123. In which of the following set, the values of bond orders will be 2.5?
1) O_2^+, NO, NO^{2+}, CN
2) CN, NO_2^+, CN^-, F_2
3) O_2^+, NO_2^+, O_2^{2+}, CN^-
4) O_2^{2-}, O_2^-, O_2^+, O_2

124. Which of the following species is paramagnetic?
1) CO 2) NO
3) O_2^{2-} 4) CN^-

125. In which of the following ionization processes, the bond order has increased and the magnetic behaviour has changed?
1) $C_2 \rightarrow C_2^+$ 2) $NO \rightarrow NO^+$
3) $O_2 \rightarrow O_2^+$ 4) $N_2 \rightarrow N_2^+$

126. The molecular electronic configuration of H_2^- ion is
1) $(\sigma 1s)^2$ 2) $(\sigma 1s)^2(\sigma^* 1s)^2$
3) $(\sigma 1s)^2(\sigma^* 1s)^1$ 4) $(\sigma 1s)^3$

127. Combination of two atomic orbitals results in the formation of two molecular orbitals namely_________
1) One bonding and one anti-bonding orbital
2) Two bonding orbitals
3) Two anti-bonding orbitals
4) Two bonding and anti-bonding orbitals

Hydrogen Bonding

128. Maximum number of H-bonds in one molecule of water is:
1) 1 2) 2
3) 3 4) 4

129. H-bond is not present in:
1) Water 2) Glycerol
3) Hydrogen fluoride 4) Hydrogen sulphide

130. Which of the following has strongest hydrogen bonding?
1) Ethyl amine 2) Ammonia
3) Ethyl alcohol 4) Diethyl Ether

131. The high boiling point of water is due to-
1) Co-ordinate bonding
2) Covalent bonding
3) Electrostatic force of attraction
4) Hydrogen Bonding

132. When two ice cubes are pressed over each other, they unite to form one cube. Which of the following forces is responsible to hold them together?
1) Dipole forces 2) Vander waal's forces
3) Covalent forces 4) Hydrogen bond forces

133. Hydrogen bonding is maximum in
1) Ethyl chloride 2) Triethyl amine
3) Ethanol 4) Diethyl ether

134. What is the dominant intermolecular force or bond that must be overcome in converting liquid CH_3OH to a gas?
1) Hydrogen bonding
2) Dipole-dipole force
3) Covalent bond
4) London dispersion force

135. Which of the following is least volatile?
1) H_2O 2) H_2S
3) H_2Se 4) H_2Te

136. The hydrogen bonding is maximum strong in:
1) S-H-----S 2) N-H-----O
3) S-H-----O 4) F-H------O

137. The comparatively high boiling point of HF is due to
1) High reactivity of fluorine
2) Small size of hydrogen atom
3) Formation of hydrogen bonds and consequent association
4) High IE of fluorine

138. Intramolecular H-bonding is present in:
1) Meta- nitrophenol 2) Salicyaldehyde
3) HCl 4) Benzophenone

139. Which of the following species has intramolecular H-bonds?
1) Phenol 2) o-Nitrophenol
3) p-Nitrophenol 4) Nitroethane

140. The interparticle forces in liquid hydrogen are.
1) H-bonds 2) Vander waal's forces
3) Covalent bonds 4) None of these

141. Which of the following H-bonds is expected to have maximum strength?
1) H-O----H 2) H-N----H
3) H-S-----H 4) All have same strength

142. Which of the following is the correct order of strength of H-bonding in the given compound ?
1) HF < NH_3 2) $H_2O > H_2O_2$
3) $H_2O_2 > H_2O$ 4) $NH_3 > H_2O$

143. Among the following, the boiling point is high for

1) Ethyl alcohol 2) Dimethyl ether
3) Acetone 4) Chloroform

144. Water molecules contain __________ hydrogen bonds.

1) Intermolecular
2) Intramolecular
3) Both intermolecular and intramolecular
4) Neither intermolecular nor intramolecular

145. Which of the following is miscible with water?

1) CS_2 2) C_2H_5OH
3) CCl_4 4) $CHCl_3$

146. Which of the following form hydrogen bond?

1) CH_3F 2) CH_3Cl
3) CH_3Br 4) None of these

147. In which of the following, hydrogen bond is present?

1) H_2 2) Ice
3) Sulphur 4) Hydrocarbon

148. Which concept best explains that o-nitrophenol is more volatile than p-nitrophenol?

1) Resonance
2) Hyperconjugation
3) Hydrogen bonding
4) Steric hindrance

High Order Time Intensive MCQs

1. If equatorial plane in PCl_5 molecule is the X–Y plane, the orbital hybridising to produce axial bonds are:

1) p_z , d_{z^2} 2) p_x , d_{xy}
3) p_y , p_z 4) $d_{xy}, d_{x^2-y^2}$

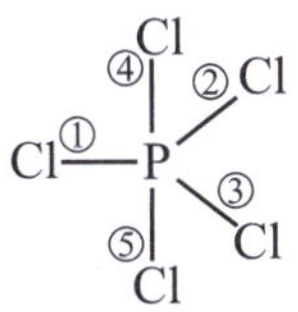

2. Which one of the following statement is incorrect?

1) In SF_6, all the S – F bond lengths are equal
2) SF_6 has regular geometry
3) In SF_6, hybridisation is $sp^3 d^2$.
4) None of these

3. Which of the following pair of molecules have the same shape but difference in polarity (polar or non-polar)?

1) H_2O and NH_3
2) $SnCl_2$ and SO_2
3) CO_2 and N_2O
4) SO_2 and SO_3

4. Among the following pairs, which two species are not isostructural?

1) PF_6^- and SF_6 2) SiF_4 and SF_4
3) IO_3^- and XeO_3 4) BH_4^- , NH_4^+

5. Assuming 2s–2p mixing is not operative, the paramagnetic species among the following is:

1) Be_2 2) B_2
3) C_2 4) N_2

6. Among KO_2, AlO_2^- and BaO_2, the unpaired electron is present in:

1) KO_2 only
2) KO_2, BaO_2
3) Only AlO_2^-
4) Only BaO_2

7. Which of the following is correct w.r.t the property mentioned?

1) $C_2H_2 > BF_3 > CH_4 > NH_3 > H_2O$ (Bond angle)
2) $NH_4^+ > NH_3 > NH_2^-$ (Bond angle)
3) $NH_3 > NF_3$ (Dipole moment)
4) All are correct

8. The decreasing order of O – O bonds length in the following molecules is:

1) $Na_2O_2 > KO_2 > O_2 > O_2\ (AsF_6)$
2) $KO_2 = Na_2O_2 > O_2 > O_2\ (AsF_6)$
3) $KO_2 > Na_2O_2 > O_2 > O_2\ (AsF_6)$
4) $Na_2O_2 > O_2 > KO_2 > O_2\ (AsF_6)$

9. The predicted geometry of TeF_4 by VSEPR theory is:

1) Octahedral 2) Square planar
3) Trigonal bipyramidal 4) Tetrahedral

10. The fluoride whose value of dipole moment is not equal to zero, is:

1) XeF_4 2) CF_4
3) SF_4 4) PF_5

11. $\ddot{N}=N=\ddot{N}$ (I) (II) (III) Calculate the formal charge of each I, II and III nitrogen atom respectively.

1) +1, –1, 0 2) –1, –1, +1
3) –1, +1, –1 4) 0, –1, –1

12. An element (X) can form different compounds of the formula XCl_3, X_2O_5 & Ca_3X_2 but does not form XCl_5. What is X?

1) B 2) Al
3) N 4) P

13. BrO_3^- is iso-structural with which of the following-

1) XeO_3 2) XeF_3
3) XeF_4 4) XeO_2

14. The bond order of individual carbon-carbon bonds in benzene is:

1) One
2) Two
3) Between 1 and 2
4) One and two alternately

15. Which compound exhibits maximum dipole moment among the following?

1) 
2) 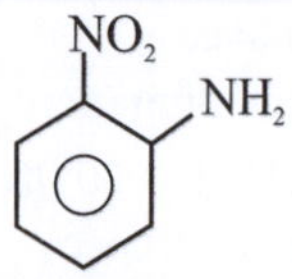
3) 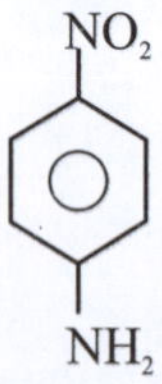
4) 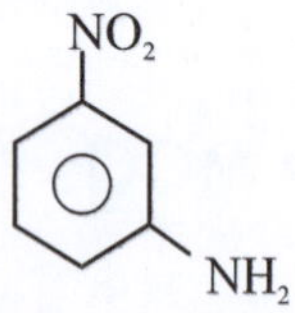

16. Which of the following is a wrong order with respect to the property mentioned against each

1) $O_2^{2-} > O_2 > O_2^+$ [Paramagnetic moment]
2) Bond Angle : $H_2O > H_2S > H_2Se$
3) $H_2 > H_2^+ > He_2^+$ [bond energy]
4) $NO_2^+ > NO_2 > NO_2^-$ [bond angle]

17. Point out incorrect statement about resonance:

1) Resonance structures should have equal energy
2) In resonance structures, the constituent atoms should be in the same position
3) In resonance structures, there should not be the same number of electron pairs
4) Resonance structures should differ only in the location of electrons around the constituent atoms

18. In which of the following compounds B–F length is shortest?

1) BF_4^-
2) $BF_3 \rightarrow NH_3$
3) BF_3
4) $BF_3 \rightarrow N(CH_3)_3$

19. Which of the following statement is incorrect

1) O_2 is paramagnetic, O_3 is also paramagnetic
2) Different observation is found in their bond length when $NO \rightarrow NO^+$ and $CO \rightarrow CO^+$
3) B_2 is paramagnetic, C_2 is also paramagnetic
4) All are incorrect

20. O_2F_2 is an unstable yellow orange solid and H_2O_2 is a colourless liquid, both have O–O bond and O–O bond length in H_2O_2 and O_2F_2 respectively is-

1) 1.22 Å, 1.48 Å
2) 1.48 Å, 1.22 Å
3) 1.22 Å, 1.22 Å
4) 1.48 Å, 1.48 Å

21. Choose the polar molecule from the following:

1) CCl_4
2) CO_2
3) $CH_2 = CH_2$
4) $CHCl_3$

22. Among the following select the correct the statements:-

I. PH_6 does not exist
II. pπ-dπ bond is present in SO_2
III. SeF_4 and CH_4 have the same shape
IV. I_3^+ has a bent shape

1) II, III
2) I, II
3) I, IV
4) I, II, IV

23. The p – p orbital overlapping is present in the following molecule

1) Hydrogen molecule
2) Hydrogen bromide
3) Hydrogen chloride
4) Chlorine molecule

24. The boiling point of a compound is raised by:

1) Intramolecular hydrogen bonding
2) Intermolecular hydrogen bonding
3) Covalent bonding
4) Ionic covalent

25. Which of the following pair of molecules contain odd electron molecule and an expanded octet molecule?

1) BCl_3 and SF_6
2) NO and H_2SO_4
3) SF_6 and H_2SO_4
4) BCl_3 and NO

26. Which of the following hydrocarbons has the lowest dipole moment?

1) $\overset{CH_3}{\underset{H}{>}}C=C\overset{CH_3}{\underset{H}{<}}$
2) $CH_3C \equiv CCH_3$
3) $CH_3CH_2C \equiv CH$
4) $CH_2 = CH - CH_3$

27. Which of the following sets of species does not follow octet rule?

1) CO, PCl_5, PCl_3, $AlCl_3$
2) CO, B_2H_6, NH_3, H_2O
3) $AlCl_3$, BF_3, PCl_5, SF_6
4) H_2O, NH_3, CO_2, $AlCl_3$

28. Which one of the following is an incorrect statement?

1) A pi bond is formed when a sigma bond already exists.
2) A pi bond may be formed by the overlapping of 'p' or 'd' orbitals.
3) A pi bond is formed by the overlapping of hybrid orbitals.
4) A pi bond is formed by the lateral overlapping of atomic orbitals.

29. The molecule which contains $\sigma_{sp^3-sp^3}$ and σ_{sp^3-p} bonds in it is

1) CH_3CH_3
2) CH_3CHO
3) CH_3CH_2Cl
4) $CHCl_3$

30. Bond angle between two hybrid orbitals is 107°, s-orbital character of hybrid orbital is nearly

1) 50%
2) 33.33%
3) 16.6%
4) 25%

31. The difference in energy between the actual structure and the lowest energy resonance structure for the given compound is
1) Electromeric energy
2) Resonance energy
3) Ionization energy
4) Hyperconjugation energy

32. Which of the following has fractional bond order
1) O_2^{2+} 2) O_2^{2-}
3) F_2^{2-} 4) H_2^-

33. In O_2^-, O_2 and O_2^{2-} molecular species, the total number of anti bonding electrons respectively are
1) 7, 6, 8 2) 1, 0, 2
3) 6, 6, 6 4) 8, 6, 8

34. Match List-I (Molecules) with List-II (Bond order) and select the correct answer using the codes.

List-I		List-II	
I.	Li_2	A.	3
II.	N_2	B.	1.5
III.	Be_2	C.	1
IV.	O_2	D.	0
		E.	2

Codes
1) I - B, II - C, III - A, IV - E
2) I - C, II - A, III - D, IV - E
3) I - D, II - A, III - E, IV - C
4) I - C, II - B, III - E, IV - A

35. What is common between the following molecules SO_3, CO_3^{2-}, NO_3^-?
1) All molecules have octahedral shape
2) All molecules have trigonal planar shape
3) All molecules have tetrahedral shape
4) All molecules have trigonal pyramidal shape

36. The atomic numbers of three elements A,B and C are a, a + 1 and a + 2. C is an alkali metal. In a compound of A and C, the nature of bonding is
1) Coordinate
2) Covalent
3) Ionic
4) Metallic

37. If Na^+ ion is larger than Mg^{2+} ion, and S^{2-} ion is larger than Cl^- ion, which of the following will be less soluble in water
1) NaCl 2) Na_2S
3) $MgCl_2$ 4) MgS

38. The compound which contains ionic as well as covalent bonds is
1) $C_2H_4Cl_2$ 2) CH_3I
3) KCN 4) H_2O_2

39. Which type of overlapping is shown by p(p_x, p_y and p_z)-orbitals?
1) Four end to end and three sidewise overlap
2) Two sidewise and one end to end overlap
3) Three sidewise overlaps
4) Two end to end and one sidewise overlap

40. The electronegativity difference between N and F is greater than N and H, yet the dipole moment of NH_3 (1.5D) is greater than that of NF_3 (0.2D). This is because:
1) In NH_3 as well as NF_3, the orbital dipole and bond dipole are in opposite direction
2) In NH_3, the orbital dipole and bond dipole are in the opposite direction, where as in NF_3 these are in same direction
3) In NH_3, as well as in NF_3 the orbital dipole and bond dipole are in same direction.
4) In NH_3, the orbital dipole and bond dipole are in same direction where as in NF_3 these are in opposite direction.

41. Which one of the following pairs of molecules will have permanent dipole moment for both members?
1) SiF_4 and NO_2 2) NO_2 and CO_2
3) NO_2 and O_3 4) SiF_4 and CO_2

42. In a molecule having the shape of an octahedron, the s-character in the hybrid orbital of the central atom is:
1) 45 % 2) 40 %
3) 65 % 4) 16.66 %

43. Which one of the following is the correct set with reference to molecular formula, hybridisation of central atom and shape of the molecule?
1) CO_2, sp^2, bent 2) H_2O, sp^2, bent
3) $BeCl_2$, sp, linear 4) H_2O, sp^3, linear

44. Which statement is correct about O_2^+?
1) Paramagnetic and bond order < O_2
2) Paramagnetic and bond order > O_2
3) Diamagnetic and bond order < O_2
4) Diamagnetic and bond order > O_2

45. Select the incorrect statement regarding molecular orbital theory?
1) The atomic orbitals of comparable energies and proper symmetry combine to form molecular orbitals.
2) An atomic orbital is monocentric while a molecular orbital is polycentric.
3) Bonding molecular orbital has higher energy than antibonding molecular orbital.
4) The molecular orbitals like atomic orbitals are filled in accordance with the aufbau principle obeying the Pauli's exclusion principle and the Hund's rule.

46. The force responsible for the union of two ice blocks as a single block is
1) Vander waals force
2) Hydrogen bonds
3) Dipole interaction
4) Vander waals repulsion

47. Which of the following geometry is not possible when the central atom is having sp^3d-hybridization?
1) Trigonal bipyramidal 2) Trigonal planar
3) Linear 4) T-shaped

48. Which of the following statements are correct?
I. The hybridisation found in cation of solid PCl_5 is sp^3.
II. In AB_2L_2 type, the BAB bond angle is always greater than the normal tetrahedral bond angle.
III. In ClO_3^-, NH_3 and XeO_3, the hybridisation and the number of lone pairs on the central atoms are same.
IV. In discrete P_4 molecule, there are six P—P bonds and four lone pairs of electrons.
1) I, II and III only 2) I, III and IV only
3) III and IV only 4) All are correct

49. Which of the following pairs will have same bond order?
1) F_2 and O_2^{2-} 2) N_2 and CO_2
3) O_2 and O_2^- 4) N_2 and N_2^+

50. Which of the following statements are correct? As the s-character of a hybrid orbital decreases
I. The bond angle decreases
II. The bond strength increases
III. The bond length increases.
1) I and III 2) II and III
3) I and II 4) All are correct

51. The correct order of decreasing polarisability of the ions is
1) Cl^-, Br^-, I^-, F^- 2) F^-, I^-, Br^-, Cl^-
3) F^-, Cl^-, Br^-, I^- 4) I^-, Br^-, Cl^-, F^-

52. 2s and 2p-atomic orbitals combine to give how many molecular orbitals?
1) 1 2) 5
3) 8 4) 7

53. The electronic configuration of the four elements are P : $2s^2$; Q : $1s^2, 2s^2, 2p^2$; R : $1s^2, 2s^2, 2p^5$; S : $1s^2, 2s^2, 2p^6$. The tendency to form electrovalent bond is maximum in
1) P 2) Q
3) R 4) S

54. Which among the following is an incorrect combination?

	Compound	Hybridisation	Shape
1)	$XeOF_4$	sp^3d^2	Square Pyramidal
2)	XeO_4	sp^3d	Tetrahedral
3)	XeO_2F_2	sp^3d (Xe in 3rd excited state)	Distorted tetrahedral
4)	XeF_4	sp^3d^2 (Xe in 2nd excited state)	Square Planar

55. Which among the following is a Non-linear molecule?
1) $BeCl_2$ 2) ICl^-_2
3) SO_2 4) XeF_2

56. Which of the following is a non-polar molecule?
1) NH_3 2) BF_3
3) CHF_3 4) CH_3Cl

57. Select the correct statement about $IO_2F_2^-$:
1) Its structure is analogous to SF_4.
2) It has sp^2 hybridisation and is bent in shape.
3) The electrons are located at the corners of a trigonal bipyramidal but one of the equatorial pairs is unshared.
4) Both (1) and (3)

58. A molecule AB_2 contains two σ, two π bonds and one lone pair of electrons surrounding a. The arrangement of electron pairs and hybridisation state of central atom are, respectively
1) Square pyramidal, sp^2
2) Trigonal planar, sp^2
3) Linear, sp
4) Can't be predicted

59. The hybridisation which leads to a regular geometry with none of the bond angles equal to 90°, is–
1) sp^3d 2) sp^3
3) sp^3d^2 4) sp^3d^3

60. The C–H bond length in C_2H_6, C_2H_4 and C_2H_2 decreases in the order
1) $C_2H_6 > C_2H_4 > C_2H_2$ 2) $C_2H_2 > C_2H_4 > C_2H_6$
3) $C_2H_4 > C_2H_6 > C_2H_2$ 4) $C_2H_4 = C_2H_6 = C_2H_2$

61. XeF_4 has
1) Square planar shape
2) Two lone pairs resting On Xe
3) Five atoms in the same plane
4) All are correct

62. Axial d-orbital(s) is/are used in the hybridisation of-
1) SF_6 2) PCl_5
3) XeF_5^+ 4) All of these

63. In which change, bond order & number of unpaired electron(s), both are increased?
1) $N_2 \rightarrow N_2^+$ 2) $O_2 \rightarrow O_2^+$
3) $O_2^- \rightarrow O_2^{2-}$ 4) $O_2^- \rightarrow O_2$

64. Which of the following cannot exist? (according to MOT)
1) H_2^{2+} 2) He_2
3) H_2^+ 4) Both (1) & (2)

65. If Hund's rule does not hold good, then which of the following pairs is diamagnetic?
1) B_2, O_2^+ 2) B_2, O_2
3) O_2^-, N_2 4) N_2, NO

66. In the molecule of the type AX_2L_n (where L represents lone pairs & n its number), there exists a bond b/w element A & X. The ∠XAX
1) Always decreases if n increases
2) Always increases if n increases
3) Will be max. for n = 3
4) Generally decreases if n decreases

67. Identify the incorrect statement.
   1) If pentagonal bipyramidal is the electronic geometry, 'pentagonal planar' is possible molecular shape
   2) If octahedral is the electronic geometry, 'square pyramidal' is the possible molecular shape.
   3) If tetrahedral is the electronic geometry, 'bent' is the possible molecular shape.
   4) If trigonal bipyramidal is the electronic geometry, triangular planar is the possible molecular shape.

68. The correct order of increasing s-character (in % age) in the hybrid orbitals of

I. CO_3^{2-} II. XeF_4
III. I_3^- IV. NCl_3
V. $BeCl_2$
   1) V > I > IV > III > II
   2) I > V > III > II > IV
   3) V > IV > III > II > I
   4) None of these

69. Select the correct match.

	Molecules	Properties (Hybridisation & Planarity)
(1)	PCl_5	sp^3d, Non-planar
(2)	ICl_2^-	sp^3d, Planar
(3)	PCl_4^+	sp^3, Non-planar
(4)	BrF_3	sp^3d, Planar
(5)	NO_3^-	sp^2, Planar
(6)	$CH_3^{\oplus}$	sp^2, planar

   1) Only 1, 2 & 3 are correct
   2) Only 3, 4, 5 & 6 are correct
   3) Only 1, 2 & 5 are correct
   4) All are correct

70. Correct order of bond angle in CH_2F_2:
   1) $H\hat{C}H > H\hat{C}F > F\hat{C}F$ 2) $F\hat{C}F > H\hat{C}F > H\hat{C}H$
   3) $H\hat{C}H > F\hat{C}F > H\hat{C}F$ 4) $H\hat{C}F > F\hat{C}F > H\hat{C}H$

71. Which of the following compounds have the same no. of lone pairs with their central atom ?

[I] XeF_5^- [II] BrF_3 [III] XeF_2 [IV] H_3O^+
   1) IV and III 2) I and III
   3) I and II 4) II, and IV

72. On catalytic hydrogenation, ethylene gives ethane during this reaction
   1) Hybridization of carbon atoms changes from sp^2 to sp^3
   2) bond angle decreases from120^0 to 109.5^0
   3) C-C bond length increases from$1.34\ A^0$ to $1.54A^0$
   4) All of these.

73. Anti- bonding molecular orbital is formed by
   1) Addition of wave functions of atomic orbitals
   2) Substraction of wave functions of atomic orbitals
   3) Multiplication of wave functions of atomic orbitals
   4) Finding the arithmetic mean

74. How many molecule which has sp^3d hybridisation?

$[ICl_2]^-$, BrF_5, SOF_4, SF_6, $XeOF_4$, ClF_3, $[I(CN)_2]^-$, PCl_5
   1) 2 2) 4
   3) 6 4) 8

75. In a compound

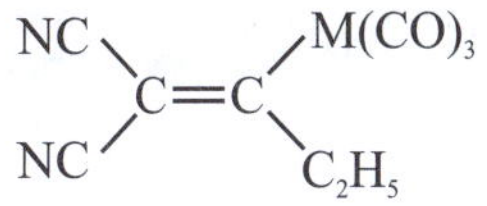

The number of sigma and pi bonds respectively are
   1) 19, 11 2) 19, 5
   3) 13, 11 4) 7, 3

76. Which of the following is incorrect match ?
   1) SiF_4 : Can act as lewis acid.
   2) Benzyne : All C-atoms are sp^2 hybridised
   3) PBr_3 : Non polar
   4) CHF = C = CHF:Nodal planes of π-bonds are not lying in same plane

77. Choose the correct set from the following options regarding the hybridisation of central atom and shape:
   1) $SnCl_2$, sp^2 , linear 2) $AlCl_4^-$, sp^3, square planar
   3) $SOCl_2$, sp^2 , planar 4) COF_2, sp^2 , planar

78. Choose the correct statement:
   1) NH_3 is having bond angle of 109°28'.
   2) The direction of the dipole moment of NF_3 is as shown in the diagram

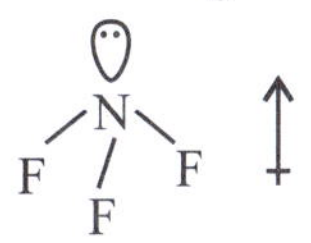

   3) In the Lewis structure of SO_3^{2-}, there is no π-bond.
   4) sp^2 hybrid orbital is consisting of 33.33 % 'p' character.

79. The geometry of ammonia molecule can be best described as:
   1) Nitrogen at one vertex of a regular tetrahedron, the other three vertices being occupied by the three hydrogens
   2) Nitrogen at the centre of the tetrahedron, three of the vertices being occupied by three hydrogens
   3) Nitrogen at the centre of an equilateral triangle, three corners being occupied by three hydrogens
   4) Nitrogen at the junction of a T, three open ends being occupied by three hydrogens

80. In which of the following sets, all the species are diamagnetic in nature?
   1) $[O_2^{2-}, O_2^{2+}, N_2^{2-}]$ 2) $[C_2, O_2^{2-}, N_2^{2+}]$
   3) $[O_2^{2-}, O_2^{+}, O_2^{2-}]$ 4) $[N_2^{+}, O_2^{2+}, B_2]$

81. Stability of the species Li_2, Li_2^- and Li_2^+ increases in the order of
   1) $Li_2^- < Li_2 < Li_2^+$
   2) $Li_2 < Li_2^+ < Li_2^-$
   3) $Li_2^- < Li_2^+ < Li_2$
   4) $Li_2 < Li_2^- < Li_2^+$

Match the Columns MCQs

1. Match the species in Column-I with the type of hybrid orbitals in Column-II.

Column-I	Column-II
(a) SF_4	(i) sp^3d^2
(b) IF_5	(ii) d^2sp^3
(c) NO_2^+	(iii) sp^3d
(d) NH_4^+	(iv) sp^3
	(v) sp

1) (a)-(i) (b)-(ii) (c)-(v) (d)-(iii)
2) (a)-(ii) (b)-(i) (c)-(iv) (d)-(v)
3) (a)-(iv) (b)-(iii) (c)-(v) (d)-(i)
4) (a)-(iii) (b)-(i) (c)-(v) (d)-(iv)

2. Match the species in Column-I with the geometry/shape in Column-II.

Column-I	Column-II
(a) H_3O^+	(i) Linear
(b) $HC \equiv CH$	(ii) Angular
(c) ClO_2^-	(iii) Tetrahedral
(d) NH_4^+	(iv) Trigonal bipyramidal
	(v) Pyramidal

1) (a)-(i) (b)-(ii) (c)-(v) (d)-(iii)
2) (a)-(v) (b)-(i) (c)-(ii) (d)-(iii)
3) (a)-(iv) (b)-(iii) (c)-(v) (d)-(i)
4) (a)-(iii) (b)-(i) (c)-(v) (d)-(iv)

3. Match the species in Column-I with the bond order in Column-II.

Column-I	Column-II
(a) NO	(i) 1.5
(b) CO	(ii) 2.0
(c) O_2^-	(iii) 2.5
(d) O_2	(iv) 3.0

1) (a)-(iii) (b)-(iv) (c)-(i) (d)-(ii)
2) (a)-(v) (b)-(i) (c)-(ii) (d)-(iii)
3) (a)-(iv) (b)-(iii) (c)-(v) (d)-(i)
4) (a)-(iii) (b)-(i) (c)-(v) (d)-(iv)

4. Match the items given in Column-I with examples given in Column-II.

Column-I	Column-II
(a) Hydrogen bond	(i) C
(b) Resonance	(ii) LiF
(c) Ionic Solid	(iii) H_2
(d) Covalent solid	(iv) HF
	(v) O_3

1) (a)-(iii) (b)-(iv) (c)-(i) (d)-(ii)
2) (a)-(v) (b)-(i) (c)-(ii) (d)-(iii)
3) (a)-(iv) (b)-(v) (c)-(ii) (d)-(i)
4) (a)-(iii) (b)-(i) (c)-(v) (d)-(iv)

5. Match the shape of molecules in Column-I with the type of hybridisation in Column-II.

Column-I	Column-II
(a) Tetrahedral	(i) sp^2
(b) Trigonal	(ii) sp
(c) Linear	(iii) sp^3

1) (a)-(iii) (b)-(ii) (c)-(i)
2) (a)-(i) (b)-(ii) (c)-(iii)
3) (a)-(iii) (b)-(i) (c)-(ii)
4) (a)-(ii) (b)-(iii) (c)-(i)

Correct & Incorrect MCQs

1. Identify the incorrect statement from the following :
   1) Only valence electrons takes part in chemical combination.
   2) The inner shell electrons are well protected and are generally not involved in the combination process.
   3) The number of valence electrons helps to calculate the common or group valence of the element.
   4) The group valence of the elements is generally equal to 2.
2. Incorrect statement is/are
   A. Ionic bonds will be formed more easily between elements with comparatively high ionization enthalpies.
   B. Ionic bonds will be formed more easily between elements with comparatively low negative value of electron gain enthalpy.
   C. Most ionic compounds have cations derived from metallic elements and anions from non-metallic elements.
   D. Ionic compounds in the crystalline state consist of orderly 3-D arrangements of cations and anions held together by coulombic interaction energies.

   1) A and B 2) B, C and D
   3) A and D 4) B and C

3. Identify the incorrect statement/s from the following:

A. The bond length is defined as the angle between the orbitals containing bonding electron pairs around the central atom in a molecule/complex ion.

B. Each atom of the bonded pair contributes to the bond length.

C. Bond angle helps in determining the shape of molecule.

D. Bond angle gives idea regarding distribution of orbital around the central atom in a molecule / complex ion.

1) Only C 2) B, C and D

3) Only A 4) B and C

4. Which of the following statement(s) about bond enthalpy is/are correct?

A. It is defined as the amount of energy required to break one mole of bonds of a particular type between two atoms in a gaseous state.

B. The unit of bond enthalpy is kJ mol^{-1}.

C. Increasing bond enthalpy order is $N_2 < O_2 < H_2$.

D. Larger the bond dissociation enthalpy, stronger will be the bond in the molecule.

1) A, B and D 2) A and C

3) Only C 4) B, C and D

5. Which amongst the following is incorrect statement ?

1) Bond order is given by the number of bonds between the two atoms in a molecule.

2) The bond order of H_2 is 1.

3) The bond order of F_2 is 1 and the bond order of O_2^{2-} is 4.

4) The bond enthalpy increases with increase in bond order.

6. The incorrect statement regarding dipole moment is:

1) The net. dipole moment of BeF_2 is zero.

2) The dipole moment of H_2O is less than that of H_2S.

3) The net dipole moment of BF_3 is zero.

4) The dipole moment of NH_3 is greater than that of NF_3.

7. Which of the following statements is incorrect ?

1) According to VSEPR theory, the shape of a molecule depends upon the number of valence shell electron pairs (bonded or nonbonded) around the central atom.

2) The repulsive interaction of electron pairs decrease in the order: lp-lp > lp-bp > bp-bp.

3) The VSEPR Theory is able to predict geometry of a large number of molecules.

4) The molecular geometry of BF_3 molecule is linear according to VSEPR theory.

8. Select the incorrect statement :

1) Order of decreasing bond length is : HI > HBr > HCl > HF.

2) Smaller the bond dissociation enthalpy, stronger will be the bond in the molecule.

3) Triple bond enthalpy are in the order: $C \equiv O > N \equiv N > C \equiv N > C \equiv C$.

4) The bond length decreases with increase in bond order.

9. Identify the incorrect statement from the following:

1) Be_2 molecule does not exist.

2) C_2 molecule is diamagnetic.

3) Bond order of H_2 molecule is 1.

4) O_2 molecule is diamagnetic.

10. Select the incorrect statement:

1) Molecular geometry of BF_3 molecule is trigonal planar.

2) dsp^2 hybridisation is present in square planar molecules.

3) PF_5 is square pyramidal in shape.

4) Molecular geometry of SF_6 molecule is octahedral.

Statement Based MCQs

Directions: These questions consist of two statements each, printed as Statement-I and Statement-II. While answering these questions, you are required to choose any one of the following four responses.

1) Both Statement-I and Statement-II are correct.

2) Both Statement-I and Statement-II are incorrect.

3) Statement-I is correct & Statement-II is incorrect.

4) Statement-I is incorrect & Statement-II is correct.

1. **Statement-I:** C_2 is paramagnetic.

Statement-II: The highest occupied molecular orbital is of s(sigma)-type.

2. **Statement-I:** CO_2 is a non-polar molecule.

Statement-II: In the molecule C-O bond is non polar.

3. **Statement-I:** NO_2 is bent molecule.

Statement-II: It is sp^2 hybridised, having repulsion in between an unpaired e^- and bond pairs.

4. **Statement-I:** In a change from $PCl_3 \rightarrow PCl_5$, the hybrid state of P remain same.

Statement-II: d-orbitals are not available for P to expand its octet.

5. **Statement-I:** H_2O_2 is more acidic than H_2O.

Statement-II: H_2O_2 is planar but H_2O is non-planar.

6. **Statement-I:** The octet rule does not hold good for H_2SO_4.

Statement-II: In H_2SO_4, central atom has 12 valence electrons around it instead of having '8' according to octet rule.

7. **Statement-I:** Molecules N_2 , CO & NO^+ have equal bond order.

Statement-II: Resonance destabilizes the structure of molecules.

8. **Statement-I:** A bonding molecular orbital is less stable than antibonding molecular orbital.

Statement-II: In bonding orbital repulsion between nuclei is very less and tends to hold the nuclei together while in antibonding orbital greater repulsion of electrons takes place.

9. **Statement-I:** Bond energy of O-H is greater than S-H.

Statement-II: S-H has more size difference than O-H.

10. **Statement-I:** SF_4 molecule has see-saw geometry.

Statement-II: In SF_4 molecule, sulphur has four bond pairs and two lone pairs.

Assertion & Reason MCQs

Directions: These questions consist of two statements each, printed as Assertion and Reason. While answering these questions, you are required to choose any one of the following four responses.

1) **Both Assertion and Reason are True and the Reason is a correct explanation of the Assertion.**
2) **Both Assertion and Reason are True but Reason is not a correct explanation of the Assertion.**
3) **Assertion is True but the Reason is False.**
4) **Assertion is False but Reason is True.**

1. **Assertion:** Sodium chloride formed by the action of chlorine gas on sodium metal is a stable compound.

Reason: This is because sodium and chloride ions acquire octet in sodium chloride formation.

2. **Assertion:** Though the central atom of both NH_3 and H_2O molecules are sp^3 hybridised, yet H – N – H bond angle is greater than that of H– O – H.

Reason: This is because nitrogen atom has one lone pair and oxygen atom has two lone pairs.

3. **Assertion:** Nitrogen is unreactive at room temperature but becomes reactive at elevated temperatures (on heating or in the presence of catalysts).

Reason: In nitrogen molecule, there is extensive delocalisation of electrons.

4. **Assertion:** NO_3^- is planar while NH_3 is pyramidal.

Reason: N in NO^-_3 is sp^2 and in NH_3, it is sp^3 hybridised with one lone pair.

5. **Assertion:** N_2 and NO^+ both are diamagnetic substance.

Reason: NO^+ is isoelectronic with N_2.

6. **Assertion:** The value of dipole moment helps to predict whether a molecule is polar or nonpolar.

Reason: The dipole moment also helps to predict the geometry of the molecules.

7. **Assertion:** The bond angle in SO_2 is 119° whereas the bond angle in SO_3 is 120°

Reason: Sulphur atom in both SO_2 and SO_3 is sp^2 hybridized.

8. **Assertion:** σ-bond is a strong bond, while π–bond is weak bond.

Reason: Atoms rotate freely about π–bond.

9. **Assertion:** In the O_3 molecule, the oxygen-oxygen bonds are intermediate between a single and double bond.

Reason: Resonance averages the bond characteristic as whole.

10. **Assertion:** Geometry of NH_3 molecule is tetrahedral.

Reason: Nitrogen is sp^3 hybridized in NH_3.

Multi Correct MCQs

1. Order of Covalent bond;

A. KF > KI; LiF > KF

B. KF < KI; LiF > KF

C. $SnCl_4 > SnCl_2$; CuCl > NaCl

D. LiF > KF; CuCl < NaCl

E. KF < KI; CuCl > NaCl

1) C, E only 2) B, C only

3) B, C, E only 4) A, B only

2. Bonding in which of the following diatomic molecule(s) become(s) stronger, on the basis of MO Theory, by removal of an electron?

A. NO B. N_2 C. O_2

D. C_2 E. B_2

Choose the most appropriate answer from the options given below:

1) (A), (B), (C) only 2) (B), (C), (E) only

3) (A), (C) only 4) (D) only

3. Choose the species from the following in which the central atom uses sp^3 hybrid orbitals in its bonding.

A. NH_3 B. SO_2 C. SiO_2

D. $BeCl_2$ E. CO_2 F. H_2O

G. CH_4 H. BF_3

1) (A), (B), (C) and (E) 2) (D), (E), (F), (G) and (H)

3) (A), (C), (F) and (G) 4) (C), (E), (A) and (G)

4. Which of the following contains the bent-shaped molecule/s from the following ?

A. N_3^-, B. NO_2^-, C. I_3^-,

D. O_3, E. SO_2

1) (A), (B) and (D) 2) (B), (C) and (D)

3) (A), (C), (D) and (E) 4) (B), (D) and (E)

NCERT Exemplar MCQs

1. Isostructural species are those which have the same shape and hybridisation. Among the given species, identify the isostructural pairs:
   1) [NF_3 and BF_3]
   2) [BF_4^- and NH_4^+]
   3) [BCl_3 and $BrCl_3$]
   4) [NH_3 and NO_3^-]

2. Polarity in a molecule and hence the dipole moment depends primarily on electronegativity of the constituent atoms and shape of a molecule. Which of the following has the highest dipole moment?
   1) CO_2 2) HI
   3) H_2O 4) SO_2

3. The types of hybrid orbitals of nitrogen in NO_2^+, NO_3^- and NH_4^+ respectively are expected to be:
   1) sp, sp^3 and sp^2 2) sp, sp^2 and sp^3
   3) sp^2, sp and sp^3 4) sp^2, sp^3 and sp

4. Hydrogen bonds are formed in many compounds e.g., H_2O, HF, NH_3. The boiling point of such compounds depends to a extent on the strength of hydrogen bond and the number of hydrogen bonds. The correct decreasing order of the boiling points above compounds is:
   1) HF > H_2O > NH_3 2) H_2O > HF > NH_3
   3) NH_3 > HF > H_2O 4) NH_3 > H_2O > HF

5. In PO_4^{3-} ion, the formal charge on the oxygen atom of P–O bond is:
   1) + 1 2) –1
   3) – 0.75 4) + 0.75

6. In NO_3^- ion, the number of bond pairs and lone pairs of electrons on nitrogen atom are:
   1) 2, 2 2) 3, 1
   3) 1, 3 4) 4, 0

7. Which of the following species has tetrahedral geometry?
   1) BH_4^- 2) NH_2^-
   3) CO_3^{2-} 4) H_3O^+

8. Number of π bonds and σ bonds in the following structure is:

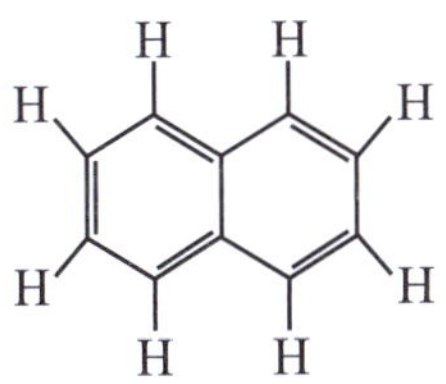

   1) 6, 19 2) 4, 20
   3) 5, 19 4) 5, 20

9. Which molecule/ion out of the following does not contain unpaired electrons?
   1) N_2^+ 2) O_2
   3) O_2^{2-} 4) B_2

10. In which of the following molecule/ion, all the bonds are not equal?
   1) XeF_4 2) BF_4^-
   3) C_2H_4 4) SiF_4

11. If the electronic configuration of an element is $1s^2\ 2s^2\ 2p^6\ 3s^2\ 3p^6\ 3d^2\ 4s^2$, the four electrons involved in chemical bond formation will be:
   1) $3p^6$ 2) $3p^6, 4s^2$
   3) $3p^6, 3d^2$ 4) $3d^2, 4s^2$

12. Which of the following angle corresponds to sp^2 hybridization?
   1) 90° 2) 120°
   3) 180° 4) 109°

Direction: *The electronic configurations of the elements A, B and C are given below. Answer the questions from 13 to 16 on the basis of these configurations.*

A. $1s^2\ 2s^2\ 2p^6$
B. $1s^2\ 2s^2\ 2p^6\ 3s^2\ 3p^3$
C. $1s^2\ 2s^2\ 2p^6\ 3s^2\ 3p^5$

13. Stable form of A may be represented by the formula:
   1) A 2) A_2
   3) A_3 4) A_4

14. Stable form of C may be represented by the formula:
   1) C 2) C_2
   3) C_3 4) C_4

15. The molecular formula of the compound formed from B and C will be:
   1) BC 2) B_2C
   3) BC_2 4) BC_3

16. The bond between B and C will be:
   1) Ionic 2) Covalent
   3) Hydrogen 4) Coordinate

17. Which of the following order of energies of molecular orbitals of N_2 is correct?
   1) $(\pi 2p_y) < (\sigma 2p_z) < (\pi^* 2p_x) \approx (\pi^* 2p_y)$
   2) $(\pi 2p_y) > (\sigma 2p_z) > (\pi^* 2p_x) \approx (\pi^* 2p_y)$
   3) $(\pi 2p_y) < (\sigma 2p_z) > (\pi^* 2p_x) \approx (\pi^* 2p_y)$
   4) $(\pi 2p_y) > (\sigma 2p_z) < (\pi^* 2p_x) \approx (\pi^* 2p_y)$
   $< \sigma 2p_z < (\pi^* 2p_x \approx \pi^* 2p_y) < \sigma^* 2p_z$

18. Which of the following statement is not correct from the view point molecular orbital theory?

1) Be_2 is not a stable molecule
2) He_2 is not stable but He_2^+ is expected to exist.
3) Bond strength of N_2 is maximum amongst the homonuclear diatomic molecules belonging to the second period.
4) The order of energies of molecular orbitals in N_2 molecule is $\sigma 1s^2 \sigma 1s^2, \sigma 2s^2, \sigma 2s^*, \pi 2p_x^2 = \pi 2p_y^2, \sigma 2p_x^2$

19. Which of the following options represent the correct bond order?

1) $O_2^- > O_2 > O_2^+$
2) $O_2^- < O_2 < O_2^+$
3) $O_2^- > O_2 < O_2^+$
4) $O_2^- < O_2 > O_2^+$

20. The electronic configuration of the outer most shell of the most electronegative element is:

1) $2s^2 2p^5$
2) $3s^2 3p^5$
3) $4s^2 4p^3$
4) $5s^2 5p^5$

21. Amongst the following elements whose electronic configuration are given below, the one having the highest ionization enthalpy is:

1) [Ne] $3s^2 3p^1$
2) [Ne] $3s^2 3p^3$
3) [Ne] $3s^2 3p^2$
4) [Ar] $3d^{10} 4s^2 4p^3$

Kattar NEET

Scan Me for Detailed Explanations

1. Which of the molecule have zero dipole moment?

1) Cl_2BBCl_2
2) NF_3
3) $PbCl_2$
4) ClNO

2. If $F_2C_1 = C_2$ part of $F_2C_1 = C_2 = C_3 = C_4F_2$ lies in yz-plane, then incorrect statement is

1) Nodal plane of π-bond between C_1 and C_2 lies in yz-plane, formed by sideways overlapping of p_x -orbitals
2) Nodal plane of π-bond between C_2 and C_3 lies in xz-plane, formed by sideways overlapping of p_y -orbitals
3) Nodal plane of π-bond between C_3 and C_4 lies in yz-plane, formed by sideways overlapping of p_y -orbitals
4) Nodal plane of π-bond between C_2 and C_3 lies in xy-plane, formed by sideways overlapping of p_z-orbitals

3. O_2F_2 is an unstable yellow orange solid and H_2O_2 is a colourless liquid. Both have O—O bond and O—O bond length in H_2O_2 and O_2F_2 respectively is:

1) 1.22 Å, 1.48 Å
2) 1.48 Å, 1.22 Å
3) 1.22 Å, 1.22 Å
4) 1.48 Å, 1.48 Å

4. The bond length of the S–O bond is maximum in which of the following compound.

$SOBr_2$, $SOCl_2$, SOF_2

1) $SOCl_2$
2) $SOBr_2$
3) SOF_2
4) All have same length

5. Hybridisation of central atom does not always change due to back bonding. This statement is valid for which of the following compounds?

(*i*) CCl_3^+
(*ii*) BF_3
(*iii*) $(SiH_3)_2O$
(*iv*) $N(SiH_3)_3$

1) (*ii*) only
2) (*i*), (*iii*) only
3) (*ii*), (*iii*) only
4) All

6. Which of the following statement is not correct from the view point of molecular orbital theory ?

1) Be_2 is not a stable molecule.
2) He_2 is not stable but He_2^+ is expected to exist.
3) Bond strength of N_2 is maximum amongst the homonuclear diatomic molecules belonging to the second period.
4) The order of energies of molecular orbitals in N_2 molecule is $\sigma 2s < \sigma^* 2s < \sigma 2p_z < (\pi 2p_x = \pi 2p_y) < (\pi^* 2p_x = \pi^* 2p_y) < \sigma^* 2p_z$

7. In P_4S_3 how many P–P bonds are presents

1) 3
2) 4
3) 5
4) 2

8. Which of the following statement(s) is(are) NOT correct?

(*i*) Bond length order is $H_2^- = H_2^+ > H_2$
(*ii*) O_2^+, NO, NO_2^- have bond order 2.5
(*iii*) Bond order varies from 0 to 3.
(*iv*) In NO_3^- and BO_3^{3-}, the bond order is same between central atom and side atom.

1) (*i*), (*ii*) only
2) (*ii*), (*iii*) only
3) (*i*), (*iv*) only
4) (*i*), (*iii*) only

9. List-I contains properties and List-II contains moles or ions.

	List-I		List-II
I	2 lone pair on central atom	P.	XeF_5^+
II	Zero dipole moment (excluding any moment due to cationic or an ionic charge)	Q.	NF_3
III	Planar	R.	ICl_3
IV	All adjacent bond angles are equal (Neglect any bond angles ≥ 180°)	S.	XeF_4
		T.	PCl_3F_2

1) I-P,Q; II-P,Q; III-R,S; IV-S,T
2) I-P,Q,R; II-P,S,T; III-P,R,S; IV-P,S
3) I-P,R,S; II-P,S,T; III-Q,R; IV-Q,R,S
4) I-P,R,S; II-P,S,T; III-P,R,S; IV-P,Q,R,S

10. Identify the correct match.

A. XeF_2	p. Central atom has sp^3 hybridisation and bent geometry.
B. N_3^-	q. Central atom has sp^3d^2 hybridisation and octahedral.
C. PCl_5 (s) anion	r. Central atom has sp hybridisation and linear geometry.
D. I_2Cl_6 (l) cation	s. Central atom has sp^3d hybridisation and linear geometry.

1) $A \to p$; $B \to q$; $C \to r$; $D \to s$
2) $A \to s$; $B \to q$; $C \to s$; $D \to r$
3) $A \to q$; $B \to r$; $C \to p$; $D \to s$
4) $A \to s$; $B \to r$; $C \to q$; $D \to p$

11. Carbon atoms in $C_2(CN)_4$ are:
1) sp-hybridized
2) sp^2-hybridized
3) sp-and sp^2 hybridized
4) sp, sp^2 and sp^3-hybridized

12. In compounds type ECl_3, where E = B, P, As or Bi, the angles Cl - E - Cl for different E are in the order
1) $B > P = As = Bi$
2) $B > P > As > Bi$
3) $B < P = As = Bi$
4) $B < P < As < Bi$

13. The most likely representation of resonance structure of p-nitrophenoxide is:

1) 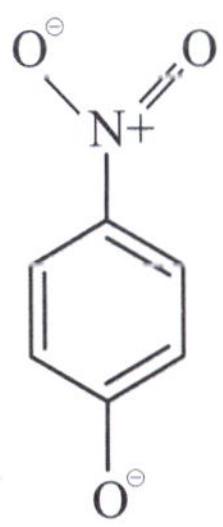

2) 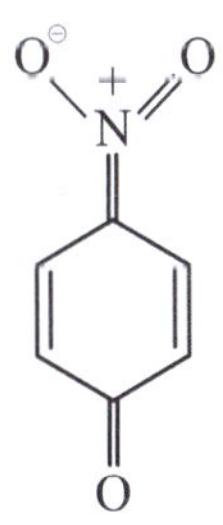

3) 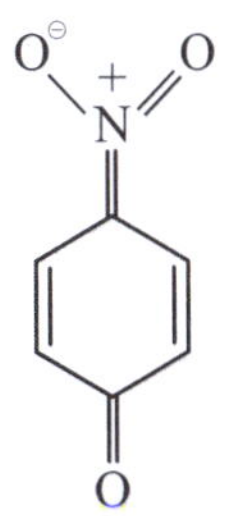

4) All of these

14. Which has the smallest bond angle (X – S – X) in the given molecules?
1) OSF_2
2) $OSCl_2$
3) $OSBr_2$
4) OSI_2

15. Which of the following has been arranged in order of decreasing dipole moment?
1) $CH_3Cl > CH_3F > CH_3Br > CH_3I$
2) $CH_3F > CH_3Cl > CH_3Br > CH_3I$
3) $CH_3Cl > CH_3Br > CH_3I > CH_3F$
4) $CH_3F > CH_3Cl > CH_3I > CH_3Br$

16. Which of the following molecules are expected to exhibit intermolecular H-bonding?
(I) Acetic acid
(II) o-nitrophenol
(III) m-nitrophenol
(IV) o-boric acid
1) (I), (II), (III)
2) (I), (II), (IV)
3) (I), (III), (IV)
4) (II), (III), (IV)

17. Geometry (i.e. arrangement of electron pairs around central atom) of $ClOF_3$ is similar to the:
1) XeF_4
2) $SOCl_2$
3) I_3^-
4) ClO_4^-

18. Which of the following statement(s) is/are incorrect?
1) Ethyne gas is more soluble in acetone than in water.
2) CH_3F is more polar than CD_3F due to deuterium (D) being less electronegative than hydrogen (H).
3) Silyl isocyanate (SiH_3NCO) is linear in shape around N while methyl isocyanate (CH_3NCO) is bent in shape around N.
4) In $CH_2{=}C{=}C{=}C{=}CH_2$, the 2H–atoms on one C–atom lie in a plane perpendicular to the plane in which 2H–atoms on other C–atom lie.

19. Which of the following compounds contain 3-centered 2-electron bonding?
(i) $[BeF_2]_n$
(ii) $[Be(CH_3)_2]_n$
(iii) $[BeCl_2]_n$
(iv) $[BeH_2]_n$
1) (i) and (ii)
2) (ii) and (iii)
3) (ii) and (iv)
4) (iii) and (iv)

20. Which of the following has the minimum heat of dissociation of $N \to B$ bond?
1) $[(CH_3)_3N \to BF_3]$
2) $[(CH_3)_3N \to B(CH_3)F_2]$
3) $[(CH_3)_3N \to B(CH_3)_2F]$
4) $[(CH_3)_3N \to B(CH_3)_3]$

21. In the structure H_2CSF_4, to decide the plane in which C = S is present the following bond angle value are given
Axial FSF angle (idealised $= 180°) \Rightarrow 170°$
Equatorial FSF angle (idealised $= 120°) \Rightarrow 97°$
After deciding the plane of double bond, which of the following statements is correct?
1) Two C-H bonds are in the same plane of axial S-F bonds
2) Two C-H bonds are in the same plane of equatorial S-F bonds.
3) Maximum five atoms are in the same plane.
4) Equatorial S-F bonds are perpendicular to plane of π bond.

22. For F_2 and OF molecules consider the following statements:
    I. Bond order for both is one
    II. OF is paramagnetic but F_2 is diamagnetic
    III. F_2 is more likely to dissociate into atoms than OF
    IV. Both have greater number of electrons in BMO than that in AMBO

    Select the correct set of codes using 'T' for true statement and 'F' for incorrect statement in the given sequence.
    1) F T T T 2) T F T F
    3) T T T F 4) T T F F

23. The correct order of stability for the following species is
    1) $Li_2 < He_2^+ < O_2^+ < C_2$
    2) $C_2 < O_2^+ < Li_2 < He_2^+$
    3) $He_2^+ < Li_2 < C_2 < O_2^+$
    4) $O_2^+ < C_2 < Li_2 < He_2^+$

24. The order strength of $p\pi$-$d\pi$ interaction in the compounds containing bond between Si/P/S/Cl and oxygen is in the order
    1) $P > Si > Cl > S$ 2) $Si < P < S < Cl$
    3) $S < Cl < P < Si$ 4) $Si > P > S > Cl$

NEET & AIIMS Past Year MCQs

1. Which of the following molecules has "NON ZERO" dipole moment value? **(2024 Re)**
   1) CCl_4 2) HI
   3) CO_2 4) BF_3

2. Arrange the following compounds in increasing order of their solubilities in chloroform:
   NaCl, CH_3OH, cyclohexane, CH_3CN **(2024 Re)**
   1) NaCl < CH_3CN < CH_3OH < Cyclohexane
   2) CH_3OH < CH_3CN < NaCl < Cyclohexane
   3) NaCl < CH_3OH < CH_3CN < Cyclohexane
   4) Cyclohexane < CH_3CN < CH_3OH < NaCl

3. Match List-I with List-II: **(2024 Re)**

	List-I Molecule		List-II Bond enthalpy ($kJmol^{-1}$)
A.	HCl	I.	435.8
B.	N_2	II.	498
C.	H_2	III.	946.0
D.	O_2	IV.	431.0

   Choose the correct answer from the options given below:
   1) A-III, B-IV, C-I, D-II 2) A-IV, B-I, C-III, D-II
   3) A-IV, B-III, C-II, D-I 4) A-IV, B-III, C-I, D-II

4. Identify the incorrect statement about PCl_5. **(2024 Re)**
   1) PCl_5 possesses two different Cl – P – Cl bond angles
   2) All five P – Cl bonds are identical in length
   3) PCl_5 exhibits sp^3d hybridisation
   4) PCl_5 consists of five P – Cl (sigma) bonds

5. Identify the correct answer. **(2024)**
   1) Dipole moment of NF_3 is greater than that of NH_3.
   2) Three canonical forms can be drawn for CO_3^{2-} ion
   3) Three resonance structures can be drawn for ozone.
   4) BF_3 has non-zero dipole moment.

6. Match List I with List II. **(2024)**

List I (Compound)		List II (Shape/geometry)	
A.	NH_3	I.	Trigonal Pyramidal
B.	BrF_5	II.	Square Planar
C.	XeF_4	III.	Octahedral
D.	SF_6	IV.	Square Pyramidal

   Choose the correct answer from the options given below:
   1) A-III, B-IV, C-I, D-II 2) A-II, B-III, C-IV, D-I
   3) A-I, B-IV, C-II, D-III 4) A-II, B-IV, C-III, D-I

7. Intramolecular hydrogen bonding is present in **(2024)**
   1) 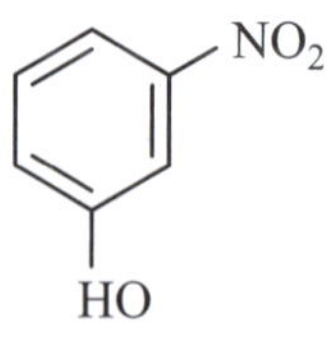
   2) HF
   3) 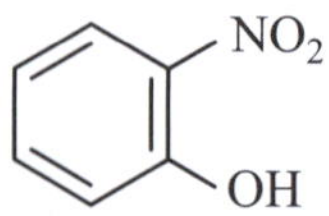
   4) 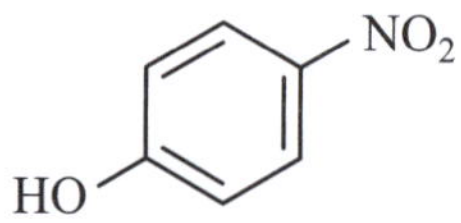

8. Amongst the following, the total number of species NOT having eight electrons around central atom in its outer most shell, is: NH_3, $AlCl_3$, $BeCl_2$, CCl_4, PCl_5 **(2023)**
   1) 1 2) 3
   3) 2 4) 4

9. The correct order of energies of molecular orbitals of N_2 molecule is: **(2023)**
   1) $\sigma 1s < \sigma^* 1s < \sigma 2s < \sigma^* 2s < (\pi 2p_x = \pi 2p_y) < (\pi^* 2p_x = \pi^* 2p_y) < \sigma 2p_z < \sigma^* 2p_z$
   2) $\sigma 1s < \sigma^* 1s < \sigma 2s < \sigma^* 2s < (\pi 2p_x = \pi 2p_y) < \sigma 2p_z < (\pi^* 2p_x = \pi^* 2p_y) < \sigma^* 2p_z$
   3) $\sigma 1s < \sigma^* 1s < \sigma 2s < \sigma^* 2s < \sigma 2p_z < (\pi 2p_x = \pi 2p_y) < (\pi^* 2p_x = \pi^* 2p_y) < \sigma^* 2p_z$
   4) $\sigma 1s < \sigma^* 1s < \sigma 2s < \sigma^* 2s < \sigma 2p_z < \sigma^* 2p_z < (\pi 2p_x = \pi 2p_y) < (\pi^* 2p_x = \pi^* 2p_y)$

10. Amongst the following which one will have maximum 'lone pair-lone pair' electron repulsions? **(2022)**
    1) XeF_2 2) ClF_3
    3) IF_5 4) SF_4

11. Which amongst the following is incorrect statement? **(2022)**
    1) O^+_2 ion is diamagnetic
    2) The bond orders of O_2^+, O_2, O_2^- and O_2^{2-} are 2.5, 2, 1.5 and 1, respectively
    3) C_2 molecule has four electrons in its two degenerate π molecular orbitals
    4) H_2^+ ion has one electron

12. BF_3 is planar and electron deficient compound. Hybridization and number of electrons around the central atom, respectively are: **(2021)**
    1) sp^3 and 6 2) sp^2 and 6
    3) sp^2 and 8 4) sp^3 and 4

13. Match List-I with List-II. **(2021)**

List-I		List-II	
(A)	PCl_5	(i)	Square pyramidal
(B)	SF_6	(ii)	Trigonal planar
(C)	BrF_5	(iii)	Octahedral
(D)	BF_3	(iv)	Trigonal bipyramidal

Choose the correct answer from the options given below.

1) A-ii B-iii C-iv D-i
2) A-iii B-i C-iv D-ii
3) A-iv B-iii C-ii D-i
4) A-iv B-iii C-i D-ii

14. Which of the following molecules is non-polar in nature? **(2021)**
    1) CH_2O 2) $SbCl_5$
    3) NO_2 4) $POCl_3$

15. Which is incorrect regarding s and p mixing (along Z axis)? **(AIIMS 2018)**
    1) Nodal plane(s) present in ABMO
    2) Nodal plane is absent in BMO
    3) MO formed may have higher energy than parent AO
    4) MO formed are asymmetric

16. The AsF_5 molecule is trigonal bipyramidal. The hybrid orbitals used by the As atoms for bonding are: **(AIIMS 2017)**
    1) $d_{x^2-y^2}, d_{z^2}, s, p_x, p_y$ 2) d_{xy}, s, p_x, p_y, p_z
    3) $d_{x^2-y^2}, s, p_x, p_y$ 4) $s, p_x, p_y, p_z, d_{z^2}$

17. **Assertion:** All F-S-F angle in SF_4 is greater than 90° but less than 180°.
    Reason: The lone pair-bond pair repulsion is weaker than bond pair – bond pair repulsion. **(AIIMS 2017)**
    1) If both Assertion & Reason are True & the Reason is a correct explanation of the Assertion.
    2) If both Assertion & Reason are True but Reason is not a correct explanation of the Assertion.
    3) If Assertion is True but the Reason is False.
    4) If both Assertion & Reason are False.

18. In O_3 molecule, the formal charge on the central O-atom is: **(AIIMS 2014)**
    1) 0 2) –1
    3) –2 4) +1

19. The molecules having the same hybridization, shape and number of lone pairs of electrons are **(AIIMS 2009)**
    1) SeF_4, XeO_2F_2 2) SF_4, XeF_2
    3) $XeOF_4$, TeF_4 4) $SeCl_4$, XeF_4

20. Which of the following option w.r.t. increasing bond order is correct? **(AIIMS 2001)**
    1) $NO < C_2 < O_2^- < He_2^+$
    2) $C_2 < NO < He_2^+ < O_2^-$
    3) $He_2^+ < O_2^- < NO < C_2$
    4) $He_2^+ < O_2^- < C_2 < NO$

21. **Assertion:** B_2H_6, Si_2H_6 are said to have similar structure.
    Reason: They have same number of σ and π bonds. **(AIIMS 2001)**
    1) If both Assertion & Reason are True & the Reason is a correct explanation of the Assertion.
    2) If both Assertion & Reason are True but Reason is not a correct explanation of the Assertion.
    3) If Assertion is True but the Reason is False.
    4) If both Assertion & Reason are False.

22. The compound containing co-ordinate bond is **(AIIMS 1994)**
    1) SO_3 2) O_3
    3) H_2SO_4 4) All of these

Answer Key

NCERT Topic-wise MCQs

1	2	3	4	5	6	7	8	9	10	11	12	13	14	15	16	17
2	3	1	2	3	1	2	1	2	1	3	4	4	1	1	2	1

18	19	20	21	22	23	24	25	26	27	28	29	30	31	32	33	34
3	3	4	3	2	3	3	2	3	1	4	3	4	4	2	2	4

35	36	37	38	39	40	41	42	43	44	45	46	47	48	49	50	51
4	3	2	1	2	2	2	1	4	3	1	4	2	4	3	4	1

52	53	54	55	56	57	58	59	60	61	62	63	64	65	66	67	68
2	3	3	3	2	3	3	3	1	3	4	4	4	4	3	4	2

69	70	71	72	73	74	75	76	77	78	79	80	81	82	83	84	85
4	2	4	1	2	1	4	2	1	2	4	3	3	2	1	2	2

86	87	88	89	90	91	92	93	94	95	96	97	98	99	100	101	102
1	2	4	1	2	2	1	2	4	2	2	1	1	2	3	2	4

103	104	105	106	107	108	109	110	111	112	113	114	115	116	117	118	119
2	2	1	2	1	4	3	1,3	1	3	2	2	3	3	1	4	2

120	121	122	123	124	125	126	127	128	129	130	131	132	133	134	135	136
1	2	2	1	2	2	3	1	4	4	3	4	4	3	1	1	4

137	138	139	140	141	142	143	144	145	146	147	148
3	2	2	2	1	2	1	1	2	4	2	3

High Order Time Intensive MCQs

1	2	3	4	5	6	7	8	9	10	11	12	13	14	15	16	17
1	4	3	2	3	1	4	1	3	3	3	3	1	3	3	1	3

18	19	20	21	22	23	24	25	26	27	28	29	30	31	32	33	34
3	4	2	4	4	4	2	2	2	1	3	3	4	2	4	1	2

35	36	37	38	39	40	41	42	43	44	45	46	47	48	49	50	51
2	3	4	3	2	4	3	4	3	2	3	2	2	2	1	1	4

52	53	54	55	56	57	58	59	60	61	62	63	64	65	66	67	68
3	3	2	3	2	4	2	2	1	4	4	4	4	2	3	4	1

69	70	71	72	73	74	75	76	77	78	79	80	81
4	1	3	4	2	2	1	3	4	3	2	2	3

Match the Columns MCQs

1	2	3	4	5
4	2	1	3	3

Correct & Incorrect MCQs

1	2	3	4	5	6	7	8	9	10
4	1	3	1	3	2	4	2	4	3

Statement Based MCQs

1	2	3	4	5	6	7	8	9	10
2	3	1	2	3	1	3	4	1	3

Assertion & Reason MCQs

1	2	3	4	5	6	7	8	9	10
1	1	3	1	2	1	2	3	1	1

Multi Correct MCQs

1	2	3	4
3	3	3	4

NCERT Exemplar MCQs

1	2	3	4	5	6	7	8	9	10	11	12	13	14	15	16	17
2	3	2	2	3	4	1	3	3	3	4	2	1	2	4	2	1
18	**19**	**20**	**21**													
4	2	1	2													

Kattar NEET

1	2	3	4	5	6	7	8	9	10	11	12	13	14	15	16	17
1	3	2	2	1	4	1	3	4	4	3	2	1	1	1	3	3
18	**19**	**20**	**21**	**22**	**23**	**24**										
2	3	4	1	1	3	2										

NEET & AIIMS Past Year MCQs

1	2	3	4	5	6	7	8	9	10	11	12	13	14	15	16	17
2	3	4	2	2	3	3	2	2	1	1	2	4	2	2	4	4
18	**19**	**20**	**21**	**22**												
4	1	4	4	4												

CHAPTER 5 Thermodynamics

Topicwise Number of Questions Analysis (2025-2016)

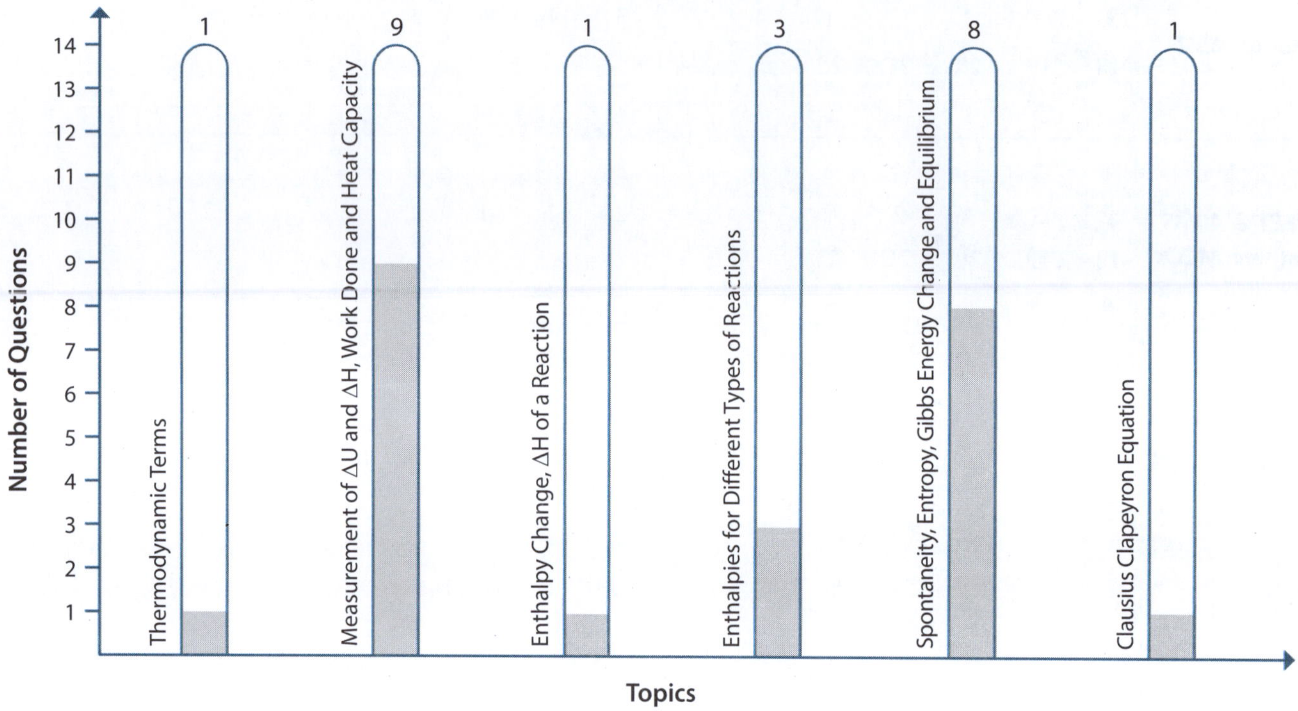

NCERT-PICKS

Introduction

Chemical energy is used to do lots of works like mechanical works, electrical works etc. Various forms of energy are interrelated and under certain conditions, these may be transformed from one form to another. The study of these energy transformations forms the subject matter of thermodynamics. The laws of thermodynamics deal with energy changes of macroscopic systems involving large number of molecules.

Laws of thermodynamics apply only when a system is in equilibrium or moves from an equilibrium state to another equilibrium state.

Thermodynamic Terms

- **System:** Part of the universe which is under study for energy changes.
- **Surrounding:** Rest of the universe.
- **Universe:** Universe = The system + The surroundings

Ex. Air in a room, water in a bottle, any living body.

- **Boundary:** Anything which separates system & surroundings is called boundary.

  Boundary can be real or imaginary.

  Boundary can be adiabatic(non-conducting) or diathermic (conducting).

Types of system

1. **Open System:** The system where matter and energy are exchanged with surroundings.

   Boundary is neither sealed nor insulated

   Eg. All living beings, A cup containing water.

2. **Closed system:** The system where only the energy but not the matter is exchanged with the surroundings.

   Boundary is sealed but not insulated

   Eg. A closed steel container having hot water.

3. **Isolated system:** The system which does not exchange either the matter or energy with the surroundings.

   Boundary is sealed and insulated

   Eg. A perfectly insulated, closed flask containing water.

Types of properties

Extensive properties:

- Properties of the system which are dependent on mass or on size of the system are called extensive properties.
- Extensive functions are additive in nature (The addition of the volumes of the two parts equals the volume of the whole of the room.)

  e.g. Volume, mass, heat capacity, internal energy (E), enthalpy(H), Gibbs free energy(G), entropy(S).

Intensive properties:

- Functions or properties which are not mass dependent or size dependent are called intensive function.
- Intensive properties are not additive in nature.

  e.g. Temperature, pressure, molar heat capacity, specific heat capacity, density, vapour pressure.

Thermodynamic process: Any method/process by which system can change its state from one state of thermodynamic equilibrium to another state of thermodynamic equilibrium.

There can be infinite type of thermodynamic processes, out of these the following are important ones:

1. **Isothermal process:** $T = \text{constant}$, $dT = 0$, $\Delta T = 0$
2. **Isochoric process:** $V = \text{constant}$, $dV = 0$, $\Delta V = 0$
3. **Isobaric process:** $P = \text{constant}$, $dP = 0$, $\Delta P = 0$
4. **Adiabatic process :** $q = \text{constant}$

   or heat exchange with the surrounding = 0(zero)

First Law of Thermodynamics

Law of energy conservation

Energy of total universe is always conserved.

or

Total energy of an isolated system is always conserved.

or

Hence, absolute value of E can never be calculated. Only change in value of E can be calculated for a particular process.

Mathematical form of First Law of thermodynamics.

If a system is initially in a particular state in which its total internal energy is E_1. Now q amount of heat is given to it and w amount of work is done on it so that in new state its total internal energy becomes E_2. Then according to 1st Law of thermodynamics.

$$E_2 = E_1 + q + w$$

$$\text{so } \Delta E = (E_2 - E_1) = q + w$$

$$\Delta U = \Delta q + \Delta w \qquad \text{Since} \quad (\Delta w = -P\Delta V)$$

$$\therefore \Delta U = \Delta q - P\Delta V$$

Note: To represent internal energy either 'E' or 'U' symbols can be used.

(1) Constant volume process (Isochoric)

$$dU = dq + dw$$

$$dU = dq - pdV$$

$$\therefore dU = (dq)_V$$

Heat given at constant volume = change in internal energy

$$dU = (dq)_V$$

$$\therefore dU = (nCdT)_v$$

$$dU = nC_v dT$$

C_v is specific molar heat capacity at constant volume.

(2) Constant pressure process (Isobaric):

$$dU = dq + dw$$

$$dU = dq - PdV \Rightarrow dq = dU + PdV \qquad \text{...(i)}$$

Defining a new thermodynamic function

H – Enthalpy

- It is a state function and extensive property.
- It is mathematically defined as:

$$H = U + PV$$

$$\text{as } dH = dU + d(PV)$$

$$\text{as } P = \text{constant}$$

$$dH = dU + PdV \qquad \text{...(ii)}$$

from equation (i) & (ii)

$\therefore dH = (dq)$ only at constant pressure.

Heat given at constant pressure = Change in enthalpy

$$dH = (nCdT)_p$$

$$dH = nC_p\, dT$$

Heat absorbed by an ideal gas during isothermal expansion in vacuum is zero.

As $q = \Delta U + P\Delta V$

In isothermal expansion $\Delta T = 0$; $\Delta U = 0$, $P = 0$

Internal energy & pressure is zero so heat is zero.

Work Done

Reversible expansion of ideal gas

$$w_{rev} = -2.303nRT \log\frac{V_2}{V_1}$$

$$w_{rev} = -2.303nRT \log\frac{P_1}{P_2}$$

Work done in isothermal

Irreversible Expansion:

$$w_{irr} = -P_{ext}\Delta V$$

In a cyclic process, net work done is equal to (-) ve amount of heat energy transferred. In cyclic process, system returns to its initial state.

$\Delta U = q + w \therefore \Delta U = 0 \Rightarrow w = -q$

Enthalpy Change

Enthalpy is expressed as the sum of the internal energy and the product of pressure and volume of a system.

$$H = U + PV$$

Relation between : ΔH & ΔE

ΔH of a process can be calculated as

$$\Delta H = \Delta E + W; \quad \Delta H = \Delta U + P\Delta V$$

For finite changes at constant pressure, we can write equation

$$\Delta H = \Delta U + P\Delta V$$

Since P is constant, we can write

$$\Delta H = \Delta U + P\Delta V$$

- For any chemical reaction, at any constant temperature $\Delta H = \Delta U + \Delta n_g RT$
- Δn_g = no. of gaseous moles of products- no. of gaseous moles of reactants.
- For any process which does not involve gases.

$$\Delta H = \Delta U$$

Heat Capacity

- **Specific heat capacity** of a gas is **infinite**, during isothermal process.

$$C = \frac{q}{m\Delta T}$$

$$\therefore \frac{q}{0} = \infty$$

- Specific heat capacity of a gas is **zero** during adiabatic process.

As $C = \frac{q}{m\Delta T}$

for adiabatic process q = 0

$$\therefore C = \frac{0}{m\Delta T}$$

- **Relation between C_p ad C_v for an ideal gas**

At constant volume $q_V = C_V\Delta T = \Delta U$

At constant pressure $q_P = C_P\Delta T = \Delta H$

For a mole of an ideal gas

$$\Delta H = \Delta U + \Delta(PV)$$
$$= \Delta U + \Delta(RT)$$
$$= \Delta U + R\Delta T$$

On putting the values of ΔH and ΔU. We have,

$$C_P\Delta T = C_V\Delta T + R\Delta T$$

$$C_P = C_V + R \rightarrow C_P - C_V = R$$

$\Rightarrow$ **$C_P - C_V = R$; only for ideal gas (Mayer's Relationship)**

Hess's Law of Constant Heat Summation

- The heat energy released or absorbed in a process is same whether the process occurs in one step or in several steps. This is known as Hess law of constant heat summation.
- According to Hess law, the heat energy released or absorbed in a process depends only on the initial state and final state but not on the path, in which the process occurs.

Spontaneous/Natural/ Irreversible Processes

A spontaneous process is a process that can actually occur in a finite time period under the existing conditions. Any change over time in the state of a system that we observe experimentally is a spontaneous process.

A spontaneous process is sometimes called a natural process, feasible process, possible process, allowed process, or real process.

Limitation

First law cannot predict the direction of any process. But in nature a process is found to have a specific direction at certain conditions.

1. **Energy can be the factor :** To achieve stability

* All exothermic processes are spontaneous.

* All endothermic processes are non spontaneous.

Enthalpy diagram for Exothermic Reactions

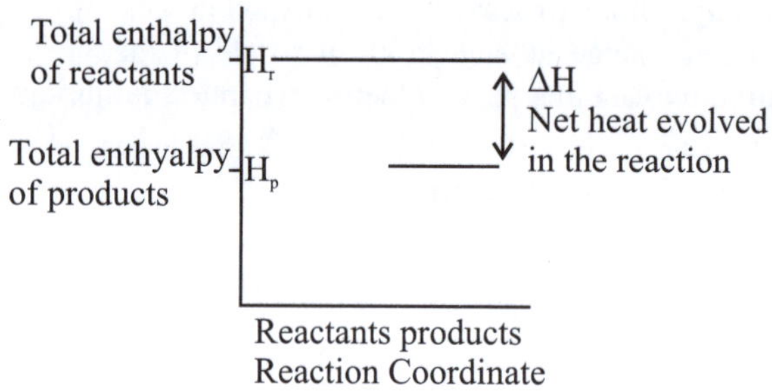

Enthalpy diagram for Endothermic Reactions

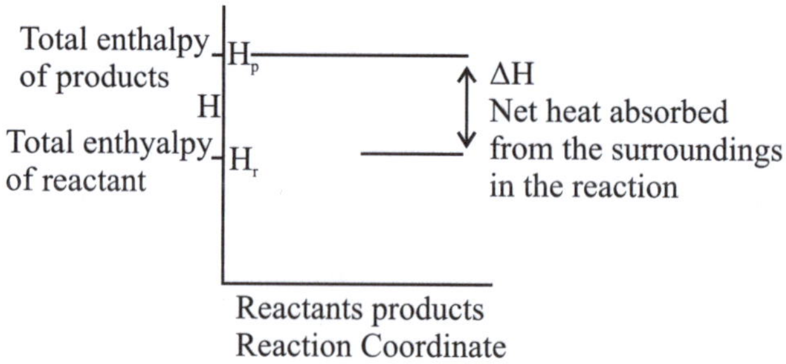

Standard enthalpy of a reaction,

- $\Delta H^0_r = \Sigma\, \Delta H^0_f(\text{Products}) - \Sigma\, \Delta H^o_f(\text{reactants})$
- ΔH reaction = Σ bond energies of reactants – Σ bond energies of products
- Hess's law of constant heat of summation

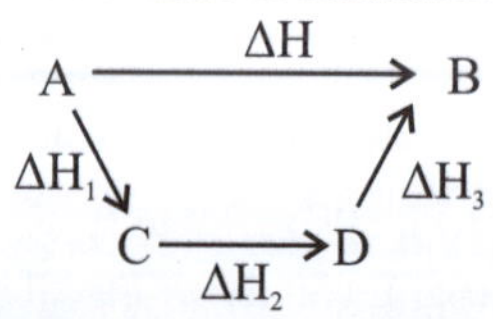

Then; according to Hess's $\Delta H = \Delta H_1 + \Delta H_2 + \Delta H_3$

Second law of thermodynamics :

Entropy of the universe is constantly increase.

$\Delta S_{universe} = \Delta S_{system} + \Delta S_{surrounding} > 0$ for a spontaneous process.

Where **(S)** = entropy is the measure of randomness or disorder. It is a state function and extensive property.

Mathematically.

$$dS = \frac{dq_{rev}}{T} \text{ or } \Delta S = \int \frac{dq_{rev}}{T}$$

dq is state function, measure this value considering reversible process.

$$dS \propto dq_{rev}, dS \propto \frac{1}{T}$$

At Higher T - entropy is already high. Therefore, heat addition will not introduce much change in the entropy.

- **Clausius statement of second law of thermodynamics:**

  "The energy of the universe remains constant, the entropy of the universe tends towards a maximum."

- Entropy change of an ideal gas is given by

$$\Delta S = nC_p \ell n\left(\frac{T_2}{T_1}\right) - nR\,\ell n\left(\frac{P_2}{P_1}\right)$$

- Entropy of mixing is defined as the difference between the entropy of the mixture of gases and the sum of entropies of the separate gases, each at a pressure P.
- Entropy change of an ideal gas (for 1 mole),

$$\Delta S = C_V \ell n\frac{T_2}{T_1} + R\ell n\frac{V_2}{V_1} = C_p \ell n\frac{T_2}{T_1} + R\ell n\frac{P_1}{P_2}$$

- **For chemical reaction**

  aA + bB → cC + dD

$$\Delta S^0_{system} = \left(\sum n_1 S^0_m\right)_{product} - \left(\sum n_1 S^0_m\right)_{reactant}$$

  where S_m^0 = standard molar entropy. It can be calculated by using third law of thermodynamics.

$$\Delta S_{surr} = \frac{-q_{system}}{T} = -\frac{-\Delta H_{system}}{T}$$

- * For a perfectly crystalline substance at 0 K, entropy = 0
- **For phase transformations**

$$\Delta S_{fusion} = \int \frac{dp}{T} = \frac{1}{T}\int dp = \frac{q}{T}$$

For constant pressure $\Delta S_{fus.} = \frac{\Delta H_{fus}}{T}$

For constant volume $\Delta S_{fus.} = \frac{\Delta U_{fusion}}{T}$

$$\Delta S_{vap} = \int \frac{dq}{T} = \frac{1}{T}\int dq = \frac{q}{T}$$

For constant Pressure, $\Delta S_{vap} = \frac{\Delta H_{vap}}{T}$

For constant Volume, $\Delta S_{vap} = \frac{\Delta U_{vap}}{T}$

The entropy of a system increases as heat is added to it. Entropy ∝ disorderness

∴ when heat is added, disorderness of molecule increases.

Entropy decreases as a liquid crystallise into solid. After freezing, the molecule attains an ordered state.

Gibb's Free Energy (G)

A system parameter to produce the Spontaneity of chemical reaction was introduced by Gibb's so that entropy calculation for the surrounding need not be carried out.

for a spontaneous change

$$\Delta S_{universe} = \Delta S_{system} + \Delta S_{surrounding}$$

$$\Delta S_{surr} = \frac{-q_{system}}{T} = -\frac{-\Delta H_{system}}{T}$$

$$T\Delta S_{univ} = T\Delta S_{system} - \Delta H_{system} \quad ...(i)$$

Introducing a new thermodynamic function G = Gibb's free energy (State function and an extensive property)

$$G_{system} = H_{system} - TS_{system}$$

$$\therefore \quad \Delta G = \Delta H - T\Delta S$$

$$\Delta G_{system} = \Delta H_{system} - T\Delta S_{system} \quad ...(ii)$$

Comparing equation (i) and (ii)

$$\Delta G_{system} = -T\Delta S_{universe}$$

A reaction with +ve enthalpy change & entropy change can be spontaneous. As ΔG = ΔH – TΔS for ΔG < 0, TΔS should be large enough to outweigh the enthalpy.

New criteria of Spontaneity :

(i) If ΔG_{system} is (–ve) < 0 ⇒ process is spontaneous

(ii) If ΔG_{system} is > 0 ⇒ process is non spontaneous

(iii) If ΔG_{system} = 0 ⇒ system is at equilibrium

Criteria for Spontaneity of Reaction

ΔH	ΔS	ΔG	Reaction characteristics	Example
–	+	Always negative	Reaction is spontaneous at all temperatures	$2O_3(g) \rightarrow 3O_2(g)$
+	–	Always positive	Reaction is non-spontaneous at all temperature	$3O_2(g) \rightarrow 2O_3(g)$
–	–	Negative at low temperature and positive at high temperatures	Reaction is spontaneous at low temperatures Reaction is non spontaneous at high temperatures	$CaO(g) + CO_2(g) \rightarrow CaCO_3(g)$
+	+	Positive at low temperature but negative at high temperature	Reaction is non-spontaneous at low temperatures but becomes spontaneous at high temperature	$CaCO_3(g) \rightarrow CaO(g) + CO_2(g)$

Third Law Of Thermodynamics: The entropy of perfect crystals of all pure elements & compounds is zero at the absolute zero of temperature.

Thus absolute value of entropy (unlike the absolute value of enthalpy) for any pure substance can be calculated at any given temperature . In standard state (298 K, 1 atm), it is standard absolute entropy S^o .

These values are tabulated : $\Delta S^o = \Sigma S^o_{products} - \Sigma S^o_{reactants}$.

Important Points

- The properties associated with a macroscopic system (i.e., consisting of large number of particles) are called macroscopic properties. The properties are pressure, volume, temperature, density, etc.
- Heat required to vaporise 1 mol of acetone is less than that of 1 mol of water. Strong H-bonding present in water while weak force dipole-dipole interaction are present in acetone.
- Enthalpy of fusion is always positive. Fusion is an endothermic reaction, so heat is absorbed in this reaction.
- Melting is an isothermal process. Temperature of substance remains unchanged during melting.
- The presence of reactants in a closed vessel made of conducting material is an example of a closed system.
- Oxygen, carbon, mercury and sulphur and oxygen gas, graphite, liquid mercury and rhombic sulphur at 1 atm pressure at 298 K is standard state.
- For an isothermal process involving in an ideal gas, ΔT, ΔU and ΔH are zero.
- Inversion temperature is the temperature below which a gas becomes cooler on expansion.
- When a gas expands freely, i.e., when it expands against vacuum such as $P_{ext} = 0$, no work is done by the system i.e., $P_{ext} = 0$, $w = 0$
- To determine the enthalpies of transition of allotropic modifications such as graphite to diamond, rhombic to monoclonic sulphur, yellow to red phosphorus etc, Hess's law is useful.
- Measure of randomness or disorder of the system is entropy.
- For elementary substances; $\Delta G^0{}_f = 0$

MCQs

NCERT Topic-wise MCQs

Thermodynamic Terms and Applications, Measurement of DU and DH Calorimetry

1. Which of the following is not a state function?
   1) Internal energy 2) Free energy
   3) Work 4) Enthalpy

2. Which among the following state function is an extensive property of system?
   1) Temperature 2) Volume
   3) Refractive index 4) Viscosity

3. A cup of tea placed in the room eventually acquires room temperature by losing heat. The process may be considered close to:
   1) Cyclic process 2) Reversible process
   3) Isothermal process 4) None of these

4. A thermally isolated gaseous system can exchange energy with the surroundings. The mode of transference of energy can be:
   1) Heat 2) Work
   3) Heat and Radiation 4) Only Radiations

5. Which of the following is an intensive property?
   1) Volume 2) Mass
   3) Density 4) Energy

6. What is correct about isothermal expansion of ideal gas?
   1) $W_{rev} = W_{irr}$ 2) $W_{rev} + W_{irr} = 0$
   3) $W_{rev} > W_{irr}$ 4) $q_{rev} = q_{irr}$

7. A well stoppered thermos flask contains some ice cubes. This is an example of:
   1) Closed system
   2) Open system
   3) Isolated system
   4) Non-thermodynamics system

8. A gas expands isothermally and reversibly, the work done by a gas is:
   1) Zero 2) Maximum
   3) Minimum 4) Cannot be determined

9. 4L - atm is equal to:
   1) 40.50 cal 2) 78.75 cal
   3) 95.23 cal 4) 96 cal

10. When ammonium chloride is dissolved in water, the solution becomes cold. The change is:
    1) Endothermic 2) Exothermic
    3) Supercooling 4) None of above

11. The internal energy of one mole of an ideal gas is
    1) $\frac{3}{2}RT$ 2) $\frac{1}{2}KT$
    3) $\frac{1}{2}RT$ 4) $\frac{3}{2}KT$

12. Internal energy does not include:
    1) Vibration energy
    2) Rotational energy
    3) Energy arising by gravitational pull
    4) Nuclear energy

13. This graph express the various steps of the system containing 1 mole of gas. Which type of process system has, when it moves from C to A?

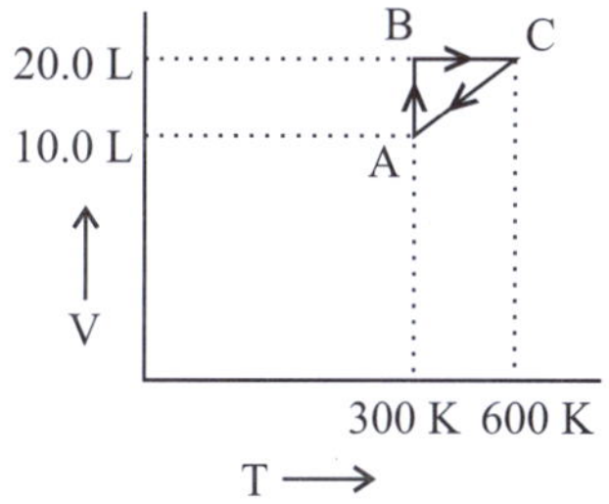

    1) Isothermal 2) Adiabatic
    3) Isobaric 4) Isochoric

14. Which of the following is a path function?
    1) Internal energy 2) Enthalpy
    3) Work 4) Entropy

15. Solubility curve of a hydrated salt in water with temperature is given. The curve indicates that the solution process is:

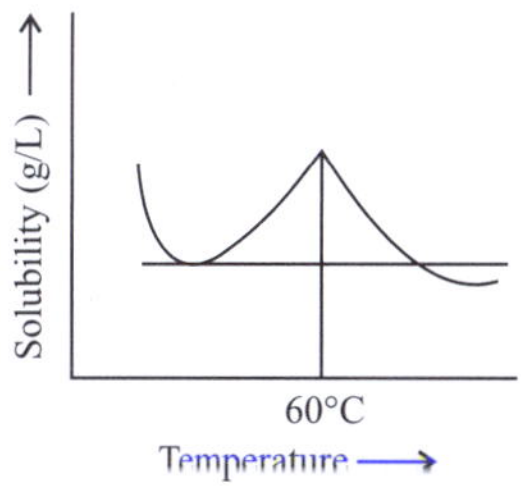

    1) Exothermic
    2) Endothermic
    3) Endothermic till 60°C and endothermic after 60°C
    4) Endothermic till 60°C and exothermic thereafter

16. Which of the following is not an endothermic reaction?
1) Dehydrogenation of ethane
2) Decomposition of water
3) Combustion of propane
4) Change of chlorine molecule into chlorine atoms

17. For the gaseous reaction ($N_2O_4 \rightarrow 2NO_2$):
1) $\Delta H > \Delta U$
2) $\Delta U > \Delta H$
3) $\Delta H = \Delta U$
4) $\Delta H = 0$, but ΔU may have any value

18. A monoatomic neon possesses:
1) Only potential energy
2) Potential as well as vibrational energy
3) Vibrational as well as translational energy
4) Translational as well as potential energy

19. In a change from state A to state B:
1) "q" depends only on the initial and final state
2) "w" depends only on the initial and final state
3) ΔU depends only on the initial and final state
4) ΔU depends upon the path adopted by "A" to change into "B"

20. The heat of combustion of solid benzoic acid at constant volume is –321.30 kJ at 27°C. The heat of combustion at constant pressure is:
1) – 321.30 – 300 R 2) – 321.30 + 300 R
3) – 321.30 – 150 R 4) –321.30 + 900 R

21. The heat absorbed in a reaction at constant temperature and constant volume is:
1) ΔU 2) ΔH
3) $-\Delta A$ 4) $-\Delta G$

22. For hypothetical reaction

$A_{(g)} + B_{(g)} \rightarrow C_{(g)} + D_{(g)}$

Which of the following statements is correct?
1) $\Delta H = \Delta U$ 2) $\Delta H > \Delta U$
3) $\Delta H < \Delta U$ 4) Unpredictable

23. The value of ΔH for cooling 2 moles of an ideal monoatomic gas from 225°C to 125°C at constant pressure will be: (Given, $C_p = 5R/2$)
1) 250 R 2) – 500 R
3) 500 R 4) – 250 R

24. The heat absorbed at constant volume is equal to the system's change in:
1) Enthalpy
2) Entropy
3) Entropy × temperature
4) Internal energy

Work Done & Enthalpy Change, ΔH of a Reaction

25. An ideal gas expands in volume from 1×10^{-3} m^3 to 1×10^{-2} m^3 at 300 K against a constant pressure 1×10^5 Nm^{-2}. The work done is:
1) – 900 J 2) – 900 kJ
3) 900 kJ 4) 900 kJ

26. Change in internal energy, when 4 kJ of work is done on the system and 1 kJ heat is given out by the system, is:
1) + 1 kJ 2) – 5 kJ
3) + 5 kJ 4) + 3 kJ

27. 16 g of oxygen of gas expands isothermally and reversibly at 300 K from 10 dm^3 to 100 dm^3. The work done is: (in J)
1) Zero 2) –2875 J
3) + 2875 J 4) Infinite

28. The work done when two moles of an ideal gas is compressed from a volume of 5 m^3 to 1 dm^3 at 300 K, under a pressure of 100k Pa is:
1) 499.9 kJ 2) – 499.9 kJ
3) – 99.5 kJ 4) 42495 kJ

29. Under which of the following conditions is the relation ($\Delta H = \Delta U + P\Delta V$) valid for a closed system?
1) Constant pressure
2) Constant temperature
3) Constant temperature and pressure
4) Constant temperature, pressure and composition

30. The reaction given below

$BaCl_2(g) + 2H_2O(\ell) \rightarrow BaCl_2 . 2H_2O(s)$; $\Delta H = -x$ kJ

The value of ΔH represents:
1) Enthalpy of solution
2) Enthalpy of dilution
3) Enthalpy of combustion
4) Enthalpy of hydration

31. The heat of combustion of benzene determined in a bomb calorimeter is – 870 k cal mol^{-1} at 298 K. The value of ΔU for the reaction is:
1) – 1740 k cal mol^{-1} 2) + 870 k cal mol^{-1}
3) – 870 k cal mol^{-1} 4) + 1740 k cal mol^{-1}

32. The work done during expansion of a gas from a volume of 4 dm^3 to 6 dm^3 against a constant external pressure of 3 atm is:
1) – 6 J 2) – 608 J
3) + 304 J 4) – 304 J

33. Calculate the work done when 1 mole of an ideal gas is compressed reversibly from 1 bar to 4 bar at a constant temperature of 300 K:
1) 4.01 kJ 2) – 8.02 kJ
3) 18.02 kJ 4) 3.458 kJ

34. Based on the first law of thermodynamics, which of the following is correct?
1) For an isothermal process , q = +w
2) For an isochoric process , $\Delta U = -q$
3) For an adiabatic process , $\Delta U = -w$
4) For a cyclic process , q = –w

35. The volume of a gas decreases from 500 cc to 300 cc when a sample of gas is compressed by an average pressure of 0.6 atm. During this process, 10 J of heat is liberated. The change in internal energy is:
1) – 2.16 J
2) 12.156 J
3) 2.16 J
4) 101.3 J

Enthalpies of Different Types of Reactions

36. Which of the following statements / relationship is not correct?
1) In an exothermic reaction, the enthalpy of products is less than that of the reactants.
2) $\Delta H_{fusion} = \Delta H_{sublimation} - \Delta H_{vapourisation}$
3) A reaction for which $\Delta H^0 < 0$ and $\Delta S > 0$, is possible at all temperatures
4) ΔH is less than ΔE for combustion of carbon to carbon dioxide

37. "X" gm of ethanal was subjected to combustion in a bomb calorimeter and the heat produced is "Y" Joules. Then:
1) $\Delta U_{(combustion)} = -XJ$
2) $\Delta U_{(combustion)} = -YJ$
3) $\Delta U_{(combustion)} = -44Y/X\ J\ mol^{-1}$
4) $\Delta H_{(combustion)} = -44Y/X\ J\ mol^{-1}$

38. Enthalpy of combustion of carbon to CO_2 is –393.5 kJ mol^{-1}. What amount of heat will be released upon formation of 35.2 g of CO_2 from carbon and oxygen gas?
1) 214.8 kJ
2) – 314.8 kJ
3) 414.8 kJ
4) 514.8 kJ

39. In an order to decompose 9 g water, 142.5 kJ heat is required. Hence; enthalpy of formation of water is:
1) – 142.5 kJ
2) + 142.5 kJ
3) – 285 kJ
4) + 285 kJ

40. Given:

$NH_3(g) + 3Cl_2(g) \rightleftharpoons NCl_3(g) + 3HCl(g), -\Delta H_1$

$N_2(g) + 3H_2(g) \rightleftharpoons 2NH_3(g); -\Delta H_2$

$H_2(g) + Cl_2(g) \rightleftharpoons 2HCl(g); \Delta H_3$

The heat of formation of $NCl_3(g)$ in terms of ΔH_1, ΔH_2 and ΔH_3 is:
1) $\Delta H_f = -\Delta H_1 + \frac{\Delta H_2}{2} - \frac{3}{2}\Delta H_3$
2) $\Delta H_f = \Delta H_1 + \frac{\Delta H_2}{2} - \frac{3}{2}\Delta H_3$
3) $\Delta H_f = \Delta H_1 + \frac{\Delta H_2}{2} + \frac{3}{2}\Delta H_3$
4) None of the above

41. 4 g of graphite is burnt in a bomb calorimeter of heat capacity 30 kJ K^{-1}. In excess of oxygen at 1 atm pressure the temperature rises from 300 K to 304 K. What is the enthalpy of combustion of graphite (in kJ mol^{-1})?
1) 360
2) 1440
3) –360
4) –1440

42. The enthalpy of formation of $CH_{4(g)}$, $H_2O_{(cl)}$ and $CO_{2(g)}$ are respectively – 74.8 kJ mol^{-1}, –285.8 kJ mol^{-1} and –393.5 kJ mol^{-1}. Then, the standard enthalpy of combustion of $CH_{4(g)}$ is:
1) + 890.3 kJ mol^{-1}
2) – 604.5 kJ mol^{-1}
3) + 604.5 kJ mol^{-1}
4) – 890.3 kJ mol^{-1}

43. If bond dissociation energies of XY, X_2 and Y_2 (all diatomic molecules) are in the ratio of 1 : 1 : 0.5 and ΔH_f for the formation of XY is –200 kJ mol^{-1} , the bond dissociation energy of X_2 will be:
1) 400 kJ/mol
2) 300 kJ/mol
3) 200 kJ/mol
4) 800 kJ/mol

44. Enthalpy of vaporisation, $\Delta_{vap}H$ for water at 100°C is 40.66 kJ/mol. The internal energy of vaporisation of water at 100°C (in kJ/mol) is:

(Assume vapour behave as an ideal gas)
1) +37.56
2) –43.76
3) +43.76
4) +40.66

45. The heat of neutralisation of four acids A, B, C and D when neutralised against a common base are 13.7, 9.4, 11.2 and 12.4 kcal respectively. The weakest among these acids is:
1) A
2) B
3) C
4) D

46. For an endothermic reaction, energy of activation is E_a and enthalpy of the reaction is ΔH (both of these in kJ/mol) minimum value of E_a will be:
1) Equal to zero
2) Less than ΔH
3) Equal to ΔH
4) More than ΔH

Spontaneity, Entropy, Gibbs Energy Change and Equilibrium

47. For the reaction, $X_2O_{4(l)} \rightarrow 2XO_{2(g)}$

$\Delta U = 2.1$ kcal, $\Delta S = 20$ cal K^{-1} at 300 K

Hence: ΔG is:
1) 2.7 kcal
2) – 2.7 kcal
3) 9.3 kcal
4) – 9.3 kcal

48. Which reaction, with the following value of ΔH, ΔS at 400 K is spontaneous and endothermic?
1) $\Delta H = -48$ kJ , $\Delta S = +135$ J/K
2) $\Delta H = -48$ kJ , $\Delta S = -135$ J/K
3) $\Delta H = +48$ kJ , $\Delta S = +135$ J/K
4) $\Delta H = +48$ kJ , $\Delta S = -135$ J/K

49. If the enthalpy change for the transition of liquid water to steam is 300 kJ mol^{-1} at 27°C, the entropy change for the process would be:

1) 1000 JK^{-1} mol^{-1}
2) 10 JK^{-1} mol^{-1}
3) 1 JK^{-1} mol^{-1}
4) 0.1 JK^{-1} mol^{-1}

50. In which of the following process, max. increase in entropy is observed?

1) Melting of ice
2) Sublimation of naphthalene
3) Condensation of water
4) Dissolution of salt in water

51. If the heat of neutralisation for a strong acid-base reaction is –57.1 kJ. What would be the heat released when 350 cm^3 of 0.20M H_2SO_4 is mixed with 650 cm^3 of 0.10M NaOH?

1) 37.1 kJ
2) 3.71 kJ
3) 3.17 kJ
4) 0.317 kJ

52. A reaction can occur spontaneously if:

1) $T\Delta S < \Delta H$ and both ΔH and ΔS are positive
2) $T\Delta S > \Delta H$ and both ΔH and ΔS are positive
3) $T\Delta S = \Delta H$ and both ΔH and ΔS are positive
4) $T\Delta S < \Delta H$ and ΔH is positive

53. The enthalpies of combustion of carbon and carbon monoxide are –393.5 and –283 kJ mol^{-1} respectively. The enthalpy of formation of carbon monoxide per mole is:

1) 110.5 kJ
2) – 110.5 kJ
3) – 676.5 kJ
4) 676.5 kJ

54. When 50 cm^3 of 0.2 N H_2SO_4 is mixed with 50 cm^3 of 1 N KOH, the heat liberated is:

1) 11.46 kJ
2) 57.3 kJ
3) 573 J
4) 53 J

55. The correct relationship between standard free energy change in a reaction and the corresponding equilibrium constant K_C is:

1) $\Delta G = RT\ln K_C$
2) $-\Delta G = RT\ln K_C$
3) $\Delta G° = RT\ln K$
4) $-\Delta G° = RT\ln K_C$

56. Which one of the following has $\Delta S°$ greater than zero?

1) $CaO(s) + CO_2(g) \rightleftharpoons CaCO_3(s)$
2) $NaCl(aq) \rightleftharpoons NaCl(s)$
3) $NaNO_3(s) \rightleftharpoons Na^+(aq) + NO_3^-(aq)$
4) $N_2(g) + 3H_2(g) \rightleftharpoons 2NH_3(g)$

57. The intensive property among these quantities is:

1) Mass
2) Volume
3) Enthalpy
4) Mass / Volumc

58. According to the second law of thermodynamics, a process (reaction) is spontaneous, if during the process:

1) $\Delta S_{universe} > 0$
2) $\Delta S_{universe} = 0$
3) $\Delta H_{system} > 0$
4) $\Delta S_{universe} = \Delta S$ system

59. Which one of the following is correct?

1) $-\Delta G = \Delta H - T\Delta S$
2) $\Delta H = \Delta G - T\Delta S$
3) $\Delta S = 1/T(\Delta G - \Delta H)$
4) $\Delta S = 1/T(\Delta H - \Delta G)$

60. Which thermodynamics property provides a measures of randomness in the system?

1) Enthalpy
2) PV-work
3) Free-energy
4) Entropy

61. According to third law of thermodynamics, the entropy at 0 K is zero for:

1) Elements in their stable form
2) Perfectly crystalline solids
3) Substance as 1 atm and 25°C
4) N_2O

62. Which of the following has highest entropy?

1) Water
2) Graphite
3) Mercury
4) Hydrogen

63. An ideal gas is allowed to expand under adiabatic conditions, what is zero for such a process?

1) $\Delta G = 0$
2) $\Delta T = 0$
3) $\Delta S = 0$
4) None of these

64. In any natural process:

1) The entropy of universe remains constant
2) The entropy of universe tends towards maximum
3) The entropy of universe tends towards minimum
4) Any of the above can happen

65. The unit of entropy is:

1) JK^{-1} mol^{-1}
2) J g^{-1}
3) J mol^{-1}
4) K^{-1} mol

66. Entropy of system depends upon:

1) Volume only
2) Temperature only
3) Pressure only
4) Temperature, pressure and volume

67. For the gaseous reaction, involving the complete combustion of isobutane:

1) $\Delta H = \Delta U$
2) $\Delta H > \Delta U$
3) $\Delta H = \Delta U = 0$
4) $\Delta H < \Delta U$

68. For a hypothetical reaction:

$A + B \rightarrow C$ if $\Delta G° > 0$

1) The reaction tends to proceed slowly towards the formation of C.
2) The increase in temperature increases the yield of C
3) A and B predominate in the reaction
4) C predominates in the reaction mixture

69. In which of the following conditions, the reaction will be always spontaneous?

1) $\Delta H < 0$ and $\Delta S < 0$
2) $\Delta H > 0$ and $\Delta S > 0$
3) $\Delta H > 0$ and $\Delta S < 0$
4) $\Delta H < 0$ and $\Delta S > 0$

High Order Time Intensive MCQs

1. Which of the following is correct for adiabatic expansion of ideal gas?
   1) PV = constant   2) PV^{γ} = constant
   3) PT^{γ} = constant   4) TV^{γ} = constant

2. A process will be spontaneous at all temperatures if:
   1) $\Delta H > 0$ and $\Delta S < 0$
   2) $\Delta H < 0$ and $\Delta S > 0$
   3) $\Delta H > 0$ and $\Delta S > 0$
   4) $\Delta H < 0$ and $\Delta S < 0$

3. The molar enthalpy of vaporization of C_6H_6 at its boiling point (353K) is 7.4 kcal/mol. The molar internal energy change of vaporization is:
   1) 7.4 kcal/mol   2) 8.106 kcal/mol
   3) 62.47 kcal/mol   4) 6.694 kcal/mol

4. Given the bond dissociation enthalpy of $CH_3 - H$ bond as 103 kcal/mol and the enthalpy of formation of $CH_4(g)$ is – 18 kcal/mol, find the enthalpy of formation of methyl radical. The dissociation energy of H_2 is 103 kcal/mol:
   1) 18 kcal/mol   2) 33.5 kcal/mol
   3) –33.5 kcal/mol   4) 21 kcal/mol

5. Calculate the enthalpy change (in kcal) for the reaction-

   $XeF_4 \longrightarrow Xe^+F^- + F_2 + F$

   The average Xe – F bond enthalpy is 34 kcal/mol, first I.E. of Xe is 279 kcal/mol, electron gain enthalpy of F is –85 kcal/mol and the bond dissociation enthalpy of F_2 is 38 kcal/mol:
   1) 292 kcal mol^{-1}
   2) 383 kcal mol^{-1}
   3) 521 kcal mol^{-1}
   4) 528 kcal mol^{-1}

6. The bond dissociation energies for Cl_2, I_2 and ICl are 242.3, 151, 211.3 kJ/mol respectively. The enthalpy of sublimation of iodine is 62.8 kJ/mol. What is the standard enthalpy of formation of ICl (g) nearly equal to?
   1) 16.8 kJ/mole   2) 33.5 kJ/mole
   3) – 14.6 kJ/mole   4) –211.3 kJ/mole

7. The enthalpy change for the reaction, $4H(g) \rightarrow 2H_2(g)$ is –869.6 kJ. The dissociation energy of H–H bond is:
   1) 217.4 kJ   2) – 434.8 kJ
   3) –869.6 kJ   4) + 434.8 kJ

8. Standard enthalpy of vaporization, $\Delta H°$ for water at 100°C is 40.66 kJ/mol. The internal energy change of vaporization of water at 100°C (in kJ/mol) is:
   1) 37.56   2) – 43.16
   3) + 43.76   4) + 40.66

9. The standard enthalpies of formation of $CO_2(g)$, $H_2O(l)$ and glucose (s) at 25°C are – 400 kJ/mole, – 300 kJ/mole, –1300 kJ/mole respectively. The standard enthalpy of combustion per gram of glucose at 25°C is:
   1) – 2900 kJ   2) + 2900 kJ
   3) –16.11 kJ   4) + 16.11 kJ

10. Standard entropies of X_2, Y_2 and XY_3 are 60, 40 and 50 JK^{-1} mol^{-1} respectively. For the reaction

    $\frac{1}{2}X_2 + \frac{3}{2}Y_2 \rightleftharpoons XY_3$, ΔH = –30 kJ/mole, to be at equilibrium, the temperature should be:
    1) 500 K   2) 750 K
    3) 1000 K   4) 1250 K

11. For a sample of perfect gas when its P is changed isothermally from P_i to P_f, the entropy change is given by:
    1) $\Delta S = nR\ln\left(\frac{P_f}{P_i}\right)$ 2) $\Delta S = nR\ln\left(\frac{P_i}{P_f}\right)$
    3) $\Delta S = nRT\ln\left(\frac{P_f}{P_i}\right)$ 4) $\Delta S = RT\ln\left(\frac{P_i}{P_f}\right)$

12. For an isothermal free expansion of an ideal gas into vacuum, which one of the following set correct?
    1) $\Delta U = 0, q > 0, w < 0$
    2) $\Delta U = 0, q > 0, w = 0$
    3) $\Delta U = 0, q = 0, w = 0$
    4) $\Delta U < 0, q = 0, w < 0$

13. The ΔH_f^0 of $CaCO_3(s)$, CaO(s) and $CO_2(g)$ are –1206.9, –635.1 and –393.5 kJ/mol respectively. Their S° values are 92.9, 38.2 and 213.7 J/K respectively at 1000°C and 1 atm. The partial pressure of CO_2 in the reaction

    $CaCO_3(s) \rightleftharpoons CaO(s) + CO_2(g)$ is approximately:
    1) 1.25 atm   2) 9.75 atm
    3) 0.03 atm   4) 0.98 atm

14. The difference between ΔH and ΔU (ΔH – ΔU), when the combustion of one mole of heptane (*l*) is carried out at a temperature T, is equal to
    1) 3RT   2) –3RT
    3) – 4RT   4) 4RT

15. The enthalpy change for the hydration

    $CuSO_4(s) + 5H_2O(g) \rightarrow CuSO_4 . 5H_2O(s)$ is –71.50 kcal/mol. If the enthalpy of vaporisation of water is 10.5 kcal / mol at 25°C, the enthalpy change for the hydration:

    $CuSO_4(s) + 5H_2O(\ell) \rightarrow CuSO_4.5H_2O(s)$ is:
    1) 61 kcal/mol   2) –61 kcal/mol
    3) 19 kcal/mol   4) –19 kcal/mol

16. The enthalpy of formation of liquid $C_{10}H_{20}$ (alkene) is equal to:
Data given is:
$\Delta H_{atomisation}$ of $C(s) = x$ kJ mol^{-1}
$\Delta H_{vap.}$ of $C_{10}H_{20}(\ell) = y$ kJ mol^{-1}
B.E of $C = C \Rightarrow w$ kJ mol^{-1}
B.E of $C - C \Rightarrow z$ kJ mol^{-1}
B.E of $C - H \Rightarrow a$ kJ mol^{-1}
B.E of $H - H \Rightarrow b$ kJ mol^{-1}

1) $4z + w + 20a + 10x - 5b - 4$
2) $10b + 10x - 8z + w - 20a - y$
3) $8z + w + 20a - 10x - 10b + y$
4) $10b + 10x - 8z - w - 20a - y$

17. If the reaction achieves equilibrium at 298 K & one bar pressure i.e., standard conditions, then equilibrium constant is:

1) 1 2) 10
3) Zero 4) 6

18. The standard enthalpy of formation of NH_3 is –46 kJ/mol. If the enthalpy of formation of H_2 from its atoms is –436 kJ mol^{-1} & that of N_2 is –712 kJ mol^{-1}, the average bond enthalpy of N–H bond in NH_3 is:

1) 964 kJ mol^{-1} 2) 352 kJ mol^{-1}
3) –1102 kJ mol^{-1} 4) +1056 kJ mol^{-1}

19. Consider the reactions at 300K
1. $4NO_2(g) + O_2(g) \rightarrow 2N_2O_5(g)$; $\Delta H = -111$ kJ
2. $4NO_2(g) + O_2(g) \rightarrow 2N_2O_5(s)$; $\Delta H = -219$ kJ
Thus entropy change for the sublimation of one mole $N_2O_5(s)$ is:

1) 180 J mol^{-1} K^{-1} 2) 360 J mol^{-1} K^{-1}
3) 108 J mol^{-1} K^{-1} 4) 54 J mol^{-1} K^{-1}

20. If x_1, x_2 & x_3 are the enthalpies of H–H, O–O & O–H bonds respectively & x_4 is the enthalpy of vaporisation of water, estimate the standard enthalpy of combustion of hydrogen:

1) $x_1 + \frac{x_2}{2} - 2x_3 + x_4$ 2) $x_1 + \frac{x_2}{2} - 2x_3 - x_4$
3) $x_1 + \frac{x_2}{2} - x_3 + x_4$ 4) $2x_3 - x_1 - \frac{x_2}{2} - x_4$

21. The heat of vaporisation & heat of fusion of H_2O are 540 cal/g & 80 cal/g. The ratio of $\frac{\Delta S_{vap}}{\Delta S_{fusion}}$ for H_2O is:

1) 6.75 2) 9.23
3) 4.94 4) 0.202

22. Which of the following relation is not correct?

1) $\Delta H = \Delta U - P\Delta V$ 2) $\Delta U = q + W$
3) $\Delta S_{sys} + \Delta S_{surr} \geq 0$ 4) $\Delta G = \Delta H - T\Delta S$

23. One mole of NaCl(s) on melting absorbed 30.5 kJ of heat & its entropy increased by 28.8 JK^{-1}. What is the melting point of NaCl?

1) 1059 K 2) 1260 K
3) 1179 K 4) 1333 K

24. A solution of 500 ml of 0.2 M KOH and 500 ml of 0.2 M HCl is mixed and stirred; the rise in temperature is T_1. The experiment is repeated using 250 ml of each solution, the temperature raised is T_2. Which of the following is true -

1) $T_1 = T_2$ 2) $T_1 = 2T_2$
3) $T_1 = 4T_2$ 4) $T_2 = 9T_1$

25. In a system, a piston caused an expansion against an external pressure of 1.25 bar giving a change in volume of 32 L for which $\Delta E = -51$ kJ. What was the value of heat involved:

1) –55 kJ 2) –11 kJ
3) –47 kJ 4) –91 kJ

26. Which of the following processes is not accompanied by increase of entropy ?

1) Dissolution of NH_4Cl in water
2) Burning of rocket fuel
3) Sublimation of dry ice
4) Condensing steam

27. According to the diagram given below, the value of ΔH for conversion of A to B is-

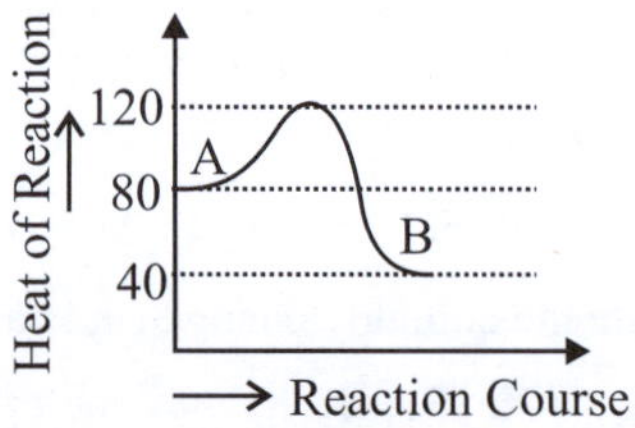

1) – 40 2) + 40
3) –120 4) + 120

28. The work done on the system when one mole of an ideal gas at 500 K is compressed isothermally and reversibly to 1/10th of its original volume (R = 2 cal)-

1) 500 kcal 2) 1.51 kcal
3) –23.03 kcal 4) 2.303 kcal

29. ΔH for $CaCO_3(s) \rightarrow CaO(s) + CO_2(g)$ is 176 kJ mol^{-1} at 1240 K. The ΔU for the change is equal to :

1) 160 kJ 2) 165.6 kJ
3) 186.3 kJ 4) 180.0 kJ

30. For a certain reaction, the change in enthalpy and change in entropy are 40.63 kJ mol^{-1} and 100 JK^{-1}. What is the value of ΔG at 27°C and indicate whether the reaction is spontaneous or not-

1) +10630 J mol^{-1}; spontaneous
2) + 10630 J mol^{-1}; non spontaneous
3) – 7990 J mol^{-1}; spontaneous
4) + 7900 J mol^{-1}; spontaneous

31. The solubility of NaCl in water at 25°C is about 6 moles per litre. Suppose you add 1 mole of NaCl to a litre of water. For the reaction, NaCl $+ H_2O \rightarrow$ Salt solution-

1) $\Delta G > 0, \Delta S > 0$ 2) $\Delta G < 0, \Delta S > 0$
3) $\Delta G > 0, \Delta S < 0$ 4) $\Delta G < 0, \Delta S < 0$

32. The enthalpy of formation of ammonia is –46.0 kJ mol^{-1}. The enthalpy for the reaction $2N_2(g) + 6H_2(g) \rightarrow 4\ NH_3(g)$ is equal to -

1) –46.0 kJ 2) 46.0 kJ
3) 184.0 kJ 4) –184.0 kJ

33. (i) $H_2(g) + Cl_2(g) \rightarrow 2HCl(g)$; $\Delta H = -x$ kJ
(ii) $NaCl + H_2SO_4 \rightarrow NaHSO_4 + HCl$; $\Delta H = -y$ kJ
(iii) $2H_2O + 2Cl_2 \rightarrow 4HCl + O_2$; $\Delta H = -z$ kJ
From the above equations, the value of ΔH_f of HCl is -

1) –x kJ 2) –y kJ
3) –z kJ 4) –x/2 kJ

34. On the basis of the following thermochemical data $\Delta_f G°(H^+)_{aq} = 0$

$H_2O(l) \rightarrow H^+(aq) + OH^-(aq)$; $\Delta H = 57.32$ kJ

$H_2(g) + \frac{1}{2}O_2(g) \rightarrow H_2O(l)$; $\Delta H = -286.20$ kJ

The value of enthalpy of formation of OH^- ion at 25°C is -

1) – 228.88 kJ 2) + 228.88 kJ
3) – 343.52 kJ 4) – 22.88 kJ

35. The P-T graph, as given below, was observed for a process on an ideal gas. Which of the following statement is true:

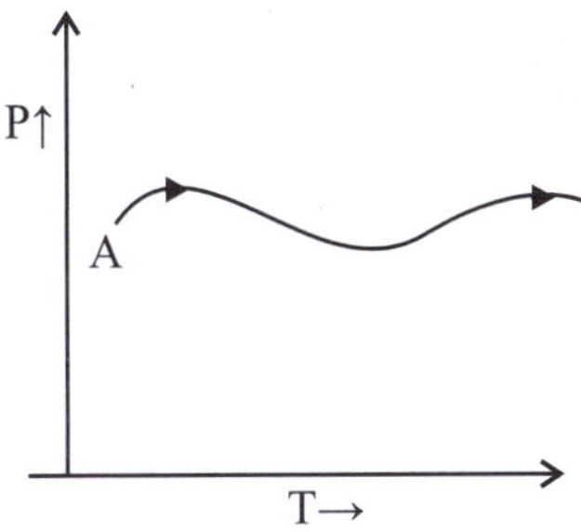

1) W = +ve, ΔH = +ve
2) W – ve, ΔH – ve
3) W = –ve, ΔH = +ve
4) W = +ve, ΔH = –ve

36. For a particular reversible reaction at temperature T, ΔH and ΔS were found to be both +ve. If T_e is the temperature at equilibrium, the reaction would be spontaneous when

1) $T = T_e$ 2) $T_e > T$
3) $T > T_e$ 4) T_e is 5 times T

37. An ideal gas is allowed to expand from 1 L to 10 L against a constant external pressure of 1 bar. The work done in kJ is

1) –9.0 2) +10.0
3) –0.9 4) –2.0

38. Calculate the lattice energy of LiF(s), given that:
(i) ΔH_{sub} for Lithium is 155.2 kJ mol^{-1}
(ii) ΔH_{diss} of 1/2 mole of F_2 (g) is 75.3 kJ
(iii) I.E of Lithium = 520 kJ mol^{-1}
(iv) E.A. of Fluorine = –333 kJ mol^{-1}
(v) ΔH_f for LiF(s) = –594.1 kJ mol^{-1}

1) –1011.6 kJ mol^{-1} 2) +2056.6 kJ mol^{-1}
3) –906.6 kJ mol^{-1} 4) +573.6 kJ mol^{-1}

39. Which of the following is correct option for free expansion of an ideal gas under adiabatic condition?

1) $q = 0, \Delta T \neq 0, w = 0$ 2) $q \neq 0, \Delta T = 0. w = 0$
3) $q = 0, \Delta T = 0, w = 0$ 4) $q = 0, \Delta T < 0, w \neq 0$

40. Equal volumes of two monoatomic gases, A and B, at same temperature and pressure are mixed. The ratio of specific heats (C_P/C_V) of the mixture will be:

1) 0.83 2) 1.50
3) 3.3 4) 1.67

41. 20 g Ar gas is allowed to expand reversibly and isothermally at 300 K from 5 L to 10 L. Calculate the approximate value of work done. (Take R = 25/3 J/K/mole, at. wt of Ar = 40 u)

1) –862.5 J 2) –375 J
3) –1437.5 J 4) –1250 J

42. Which of the following reaction can be used to calculate the value of ΔH_f^0 of the product?

1) $\frac{1}{2}H_2(g) + \frac{1}{2}Br_2(g) \longrightarrow HBr(g)$
2) $C_{diamond} + O_2(g) \longrightarrow CO_2(g)$
3) $P_4(white) + 3O_2(g) \longrightarrow P_4O_6(s)$
4) $CO(g) + \frac{1}{2}O_2(g) \longrightarrow CO_2(g)$

43. An ideal gas with $C_V = 3R$ expands adiabatically into a vacuum thus doubling its volume. The final temperature is given by:

1) $T_2 = T_1[2^{-1/3}]$ 2) $T_2 = T_1$
3) $T_2 = 2T_1$ 4) $T_2 = \frac{T_1}{2}$

44. 4.4 g of an ideal gas (mol. wt = 44) at 300K and 10 atm is allowed to expand isothermally against a external pressure of 1 atm. Taking 1 L-atm = 100 J, heat absorbed by the gas is:

1) 200.5 J 2) 212.5 J
3) 221.4 J 4) 242.4 J

45. One mole of an ideal gas at 450K is expanded isothermally & reversibly from an initial volume of 5L to 20L . Then, ΔU for the process is–

1) 263.7 cal 2) 538.1 cal
3) Zero 4) 92.5 cal

46. The standard enthalpy of formation of Mg^{2+} (aq) is calculated using the following data:
$\Delta H_f^\circ[Mg(OH)_2(s)] = -924$ kJ/mol
$\Delta H_f^\circ[OH^-(aq)] = -230$ kJ/mol
Enthalpy of crystallization of $Mg(OH)_2$ (s) = –8 kJ/mol
What is the standard enthalpy of formation (ΔH_f°) of $Mg^{2+}(aq)$?

1) –456 kJ/mol 2) –472 kJ/mol
3) –462 kJ/mol 4) –484 kJ/mol

Match the Columns MCQs

1. Match the following :

	A		B
(a)	Adiabatic process	(i)	Heat
(b)	Isolated system	(ii)	At constant volume
(c)	Isothermal change	(iii)	No exchange of energy and matter
(d)	Path function	(iv)	No transfer of heat
(e)	$\Delta U = q$	(v)	Constant temperature

1) (a)-(iv) (b)-(iii) (c)-(v) (d)-(i) (e)-(ii)
2) (a)-(i) (b)-(iii) (c)-(iv) (d)-(v) (e)-(ii)
3) (a)-(i) (b)-(ii) (c)-(iv) (d)-(v) (e)-(iii)
4) (a)-(ii) (b)-(iii) (c)-(iv) (d)-(i) (e)-(v)

2. Match the following processes with entropy change:

	Reaction		Entropy change
(a)	A liquid vapourises	(i)	$\Delta S = 0$
(b)	Reaction is non-spontaneous at all temperatures and ΔH is positive	(ii)	ΔS = positive
(c)	Reversible expansion of an ideal gas	(iii)	ΔS = negative

1) (a)-(i) (b)-(iii) (c)-(ii)
2) (a)-(ii) (b)-(iii) (c)-(i)
3) (a)-(iii) (b)-(ii) (c)-(i)
4) (a)-(ii) (b)-(i) (c)-(iii)

3. Match the following parameters with description for spontaneity :

Δ (Parameters)				Description	
	$\Delta_r H^\ominus$	$\Delta_r S^\ominus$	$\Delta_r G^\ominus$		
(a)	+	-	+	(i)	Non-spontaneous at high temperature.
(b)	-	-	+ at high T	(ii)	Spontaneous at all temperatures
(c)	-	+	-	(iii)	Non-spontaneous at all temperatures

1) (a)-(i) (b)-(iii) (c)-(ii)
2) (a)-(ii) (b)-(iii) (c)-(i)
3) (a)-(iii) (b)-(ii) (c)-(i)
4) (a)-(iii) (b)-(i) (c)-(ii)

4. Match the following :

	Column-I		Column-II
(a)	Entropy of vapourisation	(i)	decreases
(b)	K for spontaneous process	(ii)	is always positive
(c)	Crystalline solid state	(iii)	lowest entropy
(d)	ΔU in adiabatic expansion of ideal gas	(iv)	$\frac{\Delta H_{vap}}{T_b}$

1) (a)-(ii)(iv) (b)-(ii) (c)-(iii) (d)-(i)
2) (a)-(iii) (b)-(i) (c)-(ii) (d)-(iv)
3) (a)-(iv) (b)-(iii) (c)-(ii) (d)-(i)
4) (a)-(ii) (b)-(i) (c)-(iv) (d)-(i)

5. Match the following :

	A		B
(a)	Law of Conservation of energy	(i)	First law of thermodynamics
(b)	Reversible process	(ii)	Infinitely slow process which proceeds through a series of equilibrium states.
(c)	Free expansion	(iii)	$P_{ext} = 0$
(d)	$\Delta H = q$	(iv)	At constant pressure

1) (a)-(iv) (b)-(i) (c)-(ii) (d)-(iii)
2) (a)-(iii) (b)-(i) (c)-(iv) (d)-(ii)
3) (a)-(i) (b)-(ii) (c)-(iii) (d)-(iv)
4) (a)-(iv) (b)-(i) (c)-(iii) (d)-(ii)

Correct & Incorrect MCQs

1. Which of the following statement (s) is/are correct?
   A. Temperature is a path function.
   B. Volume is a state function.
   C. The internal energy is a state function.
   D. Pressure is a state function.
   1) A, B and C 2) A and D
   3) Only A 4) B, C and D

2. Which of the following statement is incorrect?
   1) Heat absorbed at constant volume by a system is equal to the work done by the system.
   2) $\Delta U = q + w$ is mathematical statement of the first law of thermodynamics.
   3) Heat absorbed by an ideal gas during isothermal expansion in vacuum is zero.
   4) Both (2) and (3)

3. Incorrect statement is/are
   A. Density is an intensive property.
   B. Heat capacity is an intensive property.
   C. Internal energy is an extensive property.
   D. Temperature is an extensive property.
   1) A and B 2) B, C and D
   3) Only C 4) B and D

4. Identify the incorrect statement from the following :
   1) Water has a large heat capacity.
   2) Heat capacity is an extensive property.
   3) Specific heat capacity of a gas is zero, during isothermal process.
   4) Specific heat capacity of a gas is zero during adiabatic process.

5. Select the incorrect statement/s from the following :
   1) Enthalpy of fusion is always negative.
   2) The magnitude of the enthalpy change depends on the strength of the intermolecular interactions in the substance undergoing the phase transfomations.
   3) Heat required to vaporise 1 mol of acetone is less than that of 1 mol of water.
   4) Both (1) and (2)

6. Which of the following statements is/are incorrect?
   A. For most of the ionic compounds, $\Delta_{sol}H^{-}$ is positive and the dissociation process is endothermic.
   B. $\Delta_f H°$ of F_2 is taken as zero.
   C. The enthalpy of formation of $H_2O(g)$ is greater than that of H_2O (l).
   D. Combustion of N_2 to give NO is endothermic.
   1) A and B 2) B, C and D
   3) Only C 4) B and C

7. Which of the following statements is not correct about entropy ?
   1) Entropy is denoted by symbol S.
   2) The entropy of a system decreases as heat is added to it.
   3) Entropy is directly proportional to disorderness.
   4) Entropy decreases as a liquid crystallise into solid.

8. Select the correct statement/s :
   A. Entropy of system increases for a spontaneous reaction.
   B. Enthalpy of reaction always decreases for spontaneous reaction.
   C. ΔG is positive for spontaneous process.
   D. ΔG is negative for non-spontaneous process.
   1) A, B and C 2) A and B
   3) A, C and D 4) B, C and D

9. Which amongst the following is incorrect statement ?
   1) A process is called adiabatic if the system does not exchange heat with the surroundings.
   2) Adiabatic process does not involve increase or decrease in temperature of the system.
   3) There is no exchange of energy or matter between the system and the surroundings in an isolated system.
   4) Both (2) and (3)

10. Select the incorrect statement :
   1) Enthalpy change is positive for the reaction $H_2O\ (g) \rightarrow H_2O(l)$.
   2) Dissolution of CaO is disfavoured at high temperature.
   3) The solubility of most salts in water increases with rise of temperature.
   4) All enthalpies of fusion are positive.

Statement Based MCQs

Directions: These questions consist of two statements each, printed as Statement-I and Statement-II. While answering these questions, you are required to choose any one of the following four responses.

1) **Both Statement-I and Statement-II are correct.**
2) **Both Statement-I and Statement-II are incorrect.**
3) **Statement-I is correct & Statement-II is incorrect.**
4) **Statement-I is incorrect & Statement-II is correct.**

1. **Statement-I:** An exothermic process, non-spontaneous at high temperature, may become spontaneous at low temperature.
   Statement-II: With decrease in temperature, randomness (entropy) decrease.

2. **Statement-I:** Internal energy change in a cyclic process is zero.
   Statement-II: Internal energy is a state function.

3. **Statement-I:** For the combustion reactions, the value of ΔH is always negative.
   Statement-II: The combustion reactions are always endothermic.

4. **Statement-I:** Decrease in free energy causes spontaneous reaction.
   Statement-II: Spontaneous reactions are invariably exothermic.

5. **Statement-I:** Hess's law is an extension of I[st] law of thermodynamics.
   Statement-II: Enthalpy is a state function.

6. **Statement-I:** Heat required to vaporise 1 mol of acetone is more than that of 1 mol of water
   Statement-II: Strong H-bonding is present in water while weak dipole-dipole interactions are present in acetone.

7. **Statement-I:** The gaseous state of a given substance has more entropy than that in the solid state.
   Statement-II: In solid state, particles are already fixed at their position while in gaseous state particles are much disordered.

8. **Statement-I:** Melting of a solid is exothermic.
   Statement-II: Temperature of substance remains unchanged during melting.

9. **Statement-I:** In a adiabatic system, $\Delta U = w_{ad'}$
Statement-II: In adiabatic system, transfer of heat takes place.

10. **Statement-I:** Addition of q and w give ΔU.
Statement-II: Addition of two path function can not give state function.

Assertion & Reason MCQs

Directions: These questions consist of two statements each, printed as Assertion and Reason. While answering these questions, you are required to choose any one of the following four responses.

1) **Both Assertion and Reason are True and the Reason is a correct explanation of the Assertion.**
2) **Both Assertion and Reason are True but Reason is not a correct explanation of the Assertion.**
3) **Assertion is True but the Reason is False.**
4) **Assertion is False but Reason is True.**

1. **Assertion:** Combustion of all organic compounds is an exothermic reaction.
Reason: The enthalpies of all elements in their standard state are zero.

2. **Assertion:** Spontaneous process is an irreversible process and may be reversed by some external agency.
Reason: Decrease in enthalpy is a contributory factor for spontaneity

3. **Assertion:** A liquid crystallises into a solid and is accompanied by decrease in entropy.
Reason: In crystals, molecules organise in an ordered manner.

4. **Assertion:** For the combustion of methane, $\Delta U > \Delta H$.
Reason: ΔH is related to ΔU by the expression, $\Delta U = \Delta E + \Delta n_g RT$.

5. **Assertion:** The heat absorbed during isothermal expansion of an ideal gas against vacuum is zero.
Reason: The volume occupied by the molecules is zero.

6. **Assertion:** The enthalpy of formation of $H_2O(l)$ is greater than that of H_2O (g).
Reason: Enthalpy change is positive for the vaporization reaction $H_2O\ (l) \rightarrow H_2O(g)$

7. **Assertion:** In the following combustion reaction.
$C(s) + O_2(g) \rightarrow CO_2(g);$
The value of ethalpy change is calculated as:
$\Delta H = \Delta U - RT$
Reason: ΔH is related to ΔU by equation.
$$\boxed{\Delta H = \Delta U + \Delta n_g RT}$$

8. **Assertion:** The formation of NH_3 from nitrogen and hydrogen is an exothermic reaction.
Reason: The process passes into equilibrium state when ΔG becomes zero.

9. **Assertion:** An adiabatic process is a process in which the system does not exchange energy with its surroundings.
Reason: It does not involve any change in temperature of the system.

10. **Assertion:** The combustion reaction of N_2 to produce NO is an endothermic reaction.
Reason: N-N bond is very strong and therefore has high bond energy.

Multi Correct MCQs

1. Among the following the set of parameters that represents path function is:
A. q + w B. q
C. w D. H – TS
1) (B) and (C) 2) (B), (C) and (D)
3) (A), (B) and (C) 4) (A) and (D)

2. Which of the following relations are correct?
A. $\Delta U = q + p\Delta V$ B. $\Delta G = \Delta H - T\Delta S$
C. $\Delta S = \frac{q_{rev}}{T}$ D. $\Delta H = \Delta U - \Delta nRT$
Choose the most appropriate answer from the options given below:
1) (C) and (D) only 2) (B) and (C) only
3) (A) and (B) only 4) (B) and (D) only

3. During which of the following processes, does entropy decrease?
A. Freezing of water to ice at 0°C
B. Freezing of water to ice at –10°C
C. $N_2(g) + 3H_2(g) \rightarrow 2NH_3(g)$
D. Adsorption of CO(g) on lead surface
E. Dissolution of NaCl in water
Choose the correct answer from the options given below:
1) (A) and (E) only
2) (A), (B), (C) and (D) only
3) (A), (C) and (E) only
4) (B) and (C) only

4. Identify the intensive properties from the following list:
A. Volume B. Molar heat capacity
C. Molarity D. E^{θ} cell
E. Gibbs free energy change
F. Molar mass
G. Mole
1) A, B and C 2) B, C and D
3) A, B, C and D 4) B, C, D and F

NCERT Exemplar MCQs

1. Thermodynamics is not concerned about:
   1) energy changes involved in a chemical reaction
   2) the extent to which a chemical reaction proceeds
   3) the rate at which a reaction proceeds
   4) the feasibility of a chemical reaction

2. Which of the following statement is correct?
   1) The presence of reacting species in a covered beaker is an example of open system
   2) There is an exchange of energy as well as matter between the system and the surroundings in a closed system
   3) The presence of reactants in a closed vessel made up of copper is an example of a closed system
   4) The presence of reactants in a thermos flask or any other closed insulated vessel is an example of a closed system

3. The state of a gas can be described by quoting the relationship between:
   1) Pressure, volume, temperature
   2) Temperature, amount, pressure
   3) Amount, volume, temperature
   4) Pressure, volume, temperature, amount

4. The volume of gas is reduced to half from its original volume. The specific heat will be:
   1) reduce to half
   2) be doubled
   3) remain constant
   4) increase four times

5. During complete combustion of one mole of butane, 2658 kJ of heat is released. The thermo-chemical reaction for above change is:
   1) $2C_4H_{10}(g) + 13O_2(g) \rightarrow 8CO_2(g) + 10H_2O(l)$; $\Delta_c H = -2658.0 \text{ kJ mol}^{-1}$
   2) $C_4H_{10}(g) + \frac{13}{2}O_2(g) \rightarrow 4CO_2(g) + 5H_2O(l)$, $\Delta_c H = -1329.0 \text{ kJ mol}^{-1}$
   3) $C_4H_{10}(g) + \frac{13}{2}O_2(g) \rightarrow 4CO_2(g) + 5H_2O(l)$; $\Delta_c H = -2658.0 \text{ kJ mol}^{-1}$
   4) $C_4H_{10}(g) + \frac{13}{2}O_2(g) \rightarrow 4CO_2(g) + 5H_2O(l)$; $\Delta_c H = +2658.0 \text{ kJ mol}^{-1}$

6. $\Delta_f U^0$ of formation of CH_4 (g) at certain temperature is -393 kJ mol^{-1}. The value of $\Delta_f H$ is:
   1) Zero
   2) $< \Delta_f U^0$
   3) $> \Delta_f U^0$
   4) Equal to $\Delta_f U^0$

7. In an adiabatic process, no transfer of heat takes place between system and surroundings. Choose the correct option for free expansion of an ideal gas under adiabatic condition from the following:
   1) $q = 0, \Delta T \neq 0, W = 0$
   2) $q \neq 0, \Delta T = 0, W = 0$
   3) $q = 0, \Delta T = 0, W = 0$
   4) $q = 0, \Delta T < 0, W \neq 0$

8. The pressure-volume work for an ideal gas can be calculated by using the expression $W = -\int_{V_i}^{V_f} p_{ex} dV$. The work can also be calculated from the pV - plot by using the area under the curve within the specified limits. When an ideal gas is compressed (a) reversibly or (b) irreversibly from volume V_i to V_f. Choose the correct option:
   1) W (reversible) = W (irreversible)
   2) W (reversible) < W (irreversible)
   3) W (reversible) = W (irreversible)
   4) W (reversible) = W (irreversible + $p_{ex}.\Delta V$)

9. The entropy change can be calculated by using the expression $\Delta S = \frac{q_{rev}}{T}$.

   When water freezes in a glass beaker, choose the correct statement amongst the following:
   1) ΔS (system) decreases but ΔS (surroundings) remains the same
   2) ΔS (system) increases but ΔS (surroundings) decreases
   3) ΔS (system) decreases but ΔS (surroundings) increases
   4) ΔS (system) decreases but ΔS (surroundings) also decreases

10. On the basis of thermochemical equations (a), (b) and (c), find out which of the algebraic relationships given in options (1) to (4) is correct:

    a. $C(\text{graphite}) + O_2(g) \rightarrow CO_2(g); \Delta_r H = x \text{ kJ mol}^{-1}$

    b. $C(\text{graphite}) + \frac{1}{2}O_2(g) \rightarrow CO(g); \Delta_r H = y \text{ kJ mol}^{-1}$

    c. $CO(g) + \frac{1}{2}O_2(g) \rightarrow CO_2(g); \Delta_r H = z \text{ kJ mol}^{-1}$

    1) $z = x + y$
    2) $x = y - z$
    3) $x = y + z$
    4) $y = 2z - x$

11. Consider the reactions given below. On the basis of these reactions find out which of the algebraic relationship given in options (1) to (4) is correct?

    a. $C(g) + 4H(g) \rightarrow CH_4(g); \Delta_r H = x \text{ kJ mol}^{-1}$

    b. $C(\text{graphite}) + 2H_2(g) \rightarrow CH_4(g); \Delta_r H = y \text{ kJ mol}^{-1}$

    1) $x = y$
    2) $x = 2y$
    3) $x > y$
    4) $x < y$

12. The enthalpies of elements in their standard states are taken as zero. The enthalpy of formation of a compound:
    1) is always negative
    2) is always positive
    3) may be positive or negative
    4) is never negative

13. Enthalpy of sublimation of a substance is equal to:
   1) enthalpy of fusion + enthalpy of vaporisation
   2) enthalpy of fusion
   3) enthalpy of vapourisation
   4) twice the enthalpy of vapourisation

14. Which of the following is not correct?
   1) ΔG is zero for a reversible reaction.
   2) ΔG is positive for a spontaneous reaction.
   3) ΔG is negative for a spontaneous reaction.
   4) ΔG is positive for a non-spontaneous reaction.

Kattar NEET

Scan Me for Detailed Explanations

1. The following curves represent adiabatic expansion of gases He, O_2 and O_3 not necessarily in order. Which curve represents for O_3?

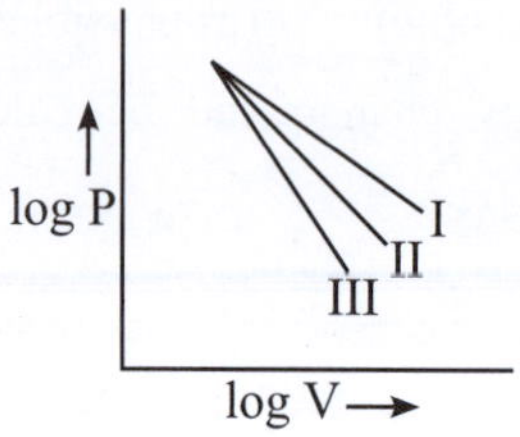

   1) I
   2) II
   3) III
   4) Both I and II

2. A cyclic process ABCD is shown in P–V diagram for an ideal gas. Which of the following diagram represents the same process?

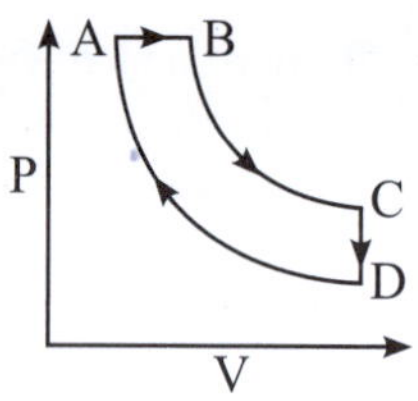

1)
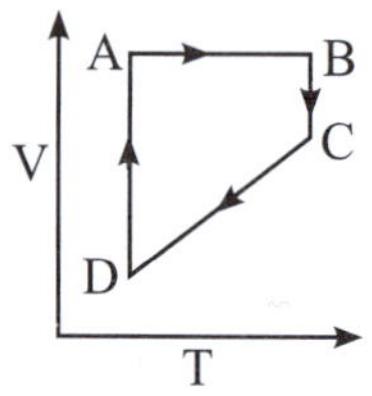

2)
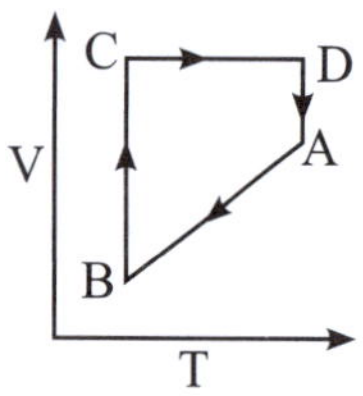

3)
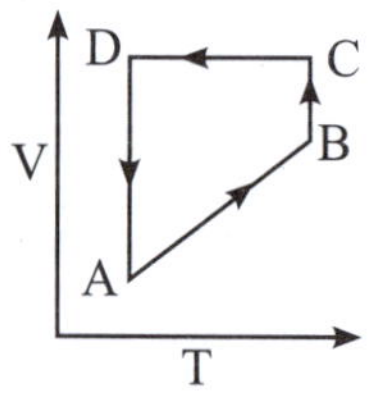

4)
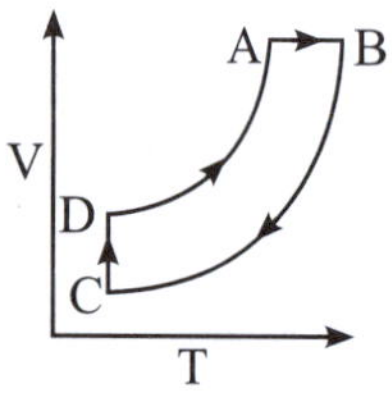

3. Boron can undergo the following reactions with the given enthalpy changes:

$$2B(s) + \frac{3}{2}O_2(g) \longrightarrow B_2O_3(s);\ \Delta H = -1260 \text{ kJ}$$

$$2B(s) + 3H_2(g) \longrightarrow B_2H_6(g);\ \Delta H = 30 \text{ kJ}$$

If in a container (operating at constant pressure) which is isolated from the surrounding, mixture of H_2 (gas) and O_2 (gas) are passed over excess of B(s), then calculate the molar ratio ($O_2 : H_2$) so that temperature of the container do not change.
   1) 15 : 3
   2) 42 : 1
   3) 1 : 42
   4) 1 : 84

4. One mole of an ideal gas is carried through the reversible cyclic process as shown in figure. The max. temperature attained by the gas during the cycle

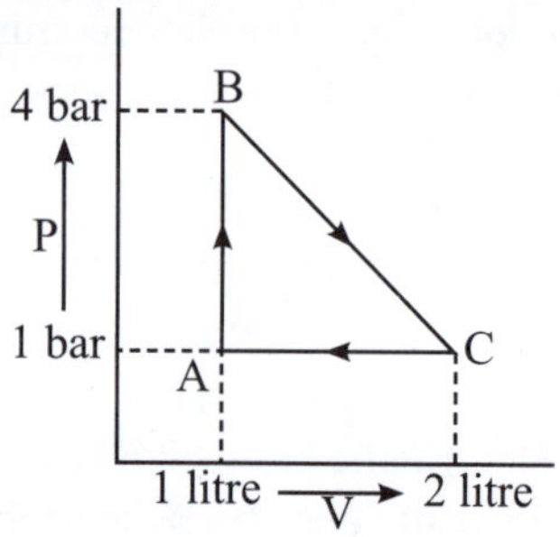

   1) $\frac{7}{6R}$
   2) $\frac{12}{49R}$
   3) $\frac{49}{12R}$
   4) $\frac{72}{56R}$

5. $LiCl(s) + H_2O(l) \rightleftharpoons Li^+(aq) + Cl^-(aq)$
   $\Delta H = -37.1 \text{ kJ mol}^{-1}$
   Variation of solubility S with temperature T is given by:

1)
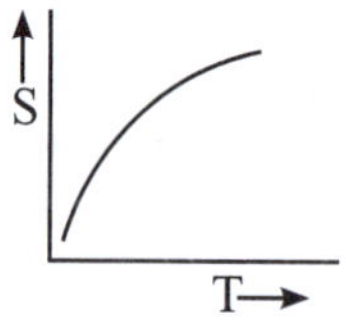

2)
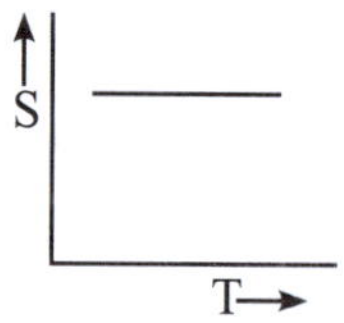

3)
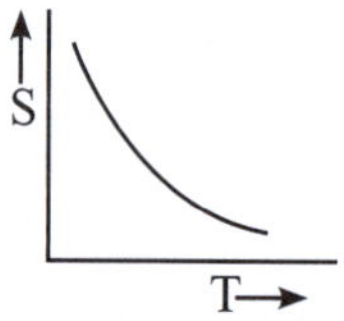

4)
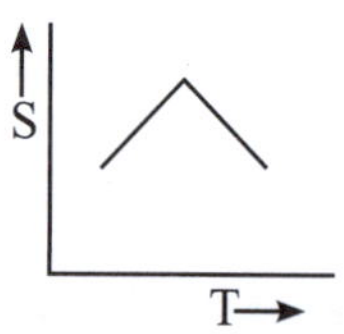

6. One mole of ideal gas ($C_v = 5/2\ R$) beginning at 1 atm and 300 K is put through the following cycle

Step–I: Heating to twice its initial pressure at constant volume.

Step–II: Adiabatic reversible expansion to its initial temperature.

Step–III: Isothermal compression back to 1 atm.

What is the volume at point 3?

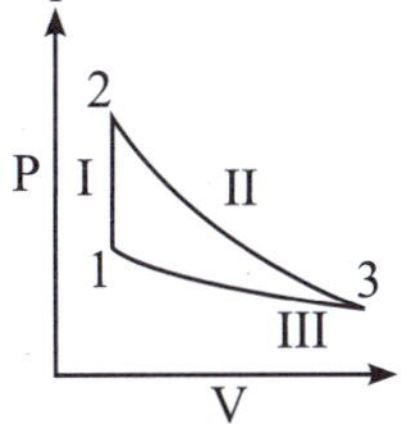

1) 40.4 L
2) 65.0 L
3) 139 L
4) 279 L

7. A diatomic ideal gas changes its state from A to B as shown in the figure.

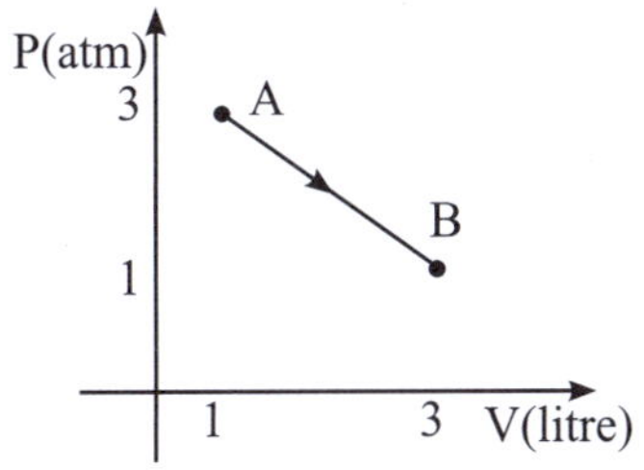

Choose the correct U versus V graph: (where U is internal energy of the gas)

1) 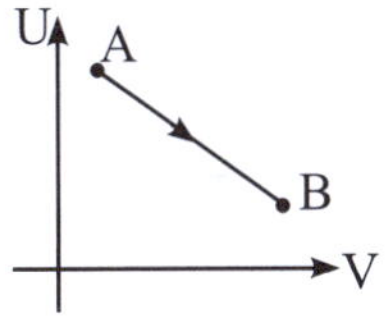

2) 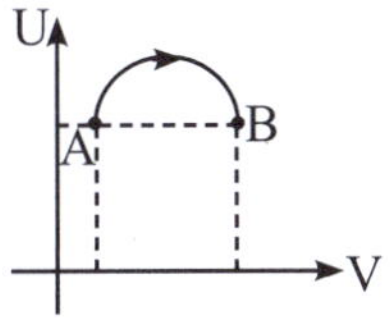

3) 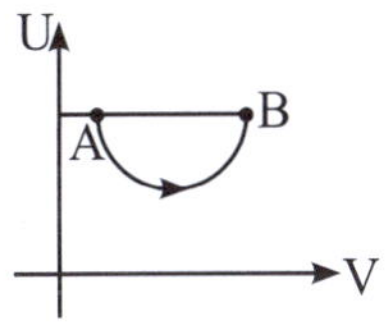

4) 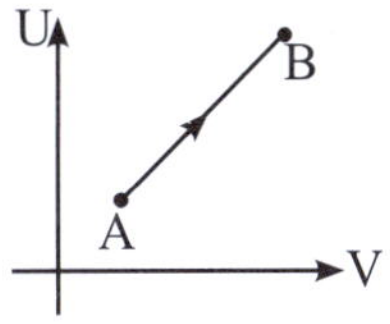

8. For a perfectly crystalline solid, $C_{P,m} = aT^3$, where a is constant. If $C_{P,m}$ is 0.84 J $K^{-1}mol^{-1}$ at 10 K, molar entropy at 10 K is:

1) 0.12 J K^{-1} mol^{-1}
2) 0.28 J K^{-1} mol^{-1}
3) 4.2 J K^{-1} mol^{-1}
4) 2.8 J K^{-1} mol^{-1}

9. A diatomic ideal gas undergoes a process in which the ratio of P to V^2 at any instant is constant and equals to 1. What is molar heat capacity of gas [excluding vibrational degree of freedom]

1) 2R
2) $\frac{5R}{2}$
3) $\frac{17R}{6}$
4) $\frac{4R}{3}$

10. Which of the following options contain correct set of cases where entropy of system is increasing?

Case-I: Dissociation of ammonia gas into N_2 and H_2 gases.

Case-II: Polymerization process of ethene to form polythene.

Case-III: Adiabatic irreversible compression of an ideal gas occurring spontaneously.

Case-IV: Condensation of $H_2O_{(vap)}$ to give $H_2O(l)$

Case-V: Isothermal free expansion of Ideal gas.

1) All the cases
2) Case-I, III and V only
3) Case-I, II and V only
4) Case-I, II, III and V only

11. The heat capacity of a monoatomic ideal gas subjected to a process such that absolute temperature is proportional to cube of its volume is given by:

1) $\frac{15R}{2}$
2) $\frac{17R}{2}$
3) $\frac{11R}{6}$
4) $\frac{3R}{2}$

12. The enthalpy of tetramerization of X in gas phase ($4X(g) \longrightarrow X_4(g)$) is – 100 kJ/mol at 300 K.

The enthalpy of vaporisation for liquid X and X_4 are respectively 30 kJ/mol and 72 kJ/mol respectively.

ΔS for tetramerization of X in liquid phase is – 125 J/K mol at 300 K.

What is the ΔG at 300 K for tetramerization of X in liquid phase?

1) –52 kJ/mol
2) –89.5 kJ/mol
3) –14.5 kJ/mol
4) –26.8 kJ/mol

13. Which of the following reaction is expected to be spontaneous at low temperatures and non-spontaneous at high temperatures.

1) Dissociation of PCl_5 to give PCl_3 and Cl_2
2) Formation of H_2O from $H_2(g)$ and $O_2(g)$
3) Melting of ice
4) Vaporisation of any volatile liquid

14. Calculate ΔG(kJ) for the reaction at 300K.

$N_2(g) + O_2(g) \rightleftharpoons 2NO(g)$

at an instant where partial pressure of N_2, O_2 and NO are 10^{-1} bar, 10^{-3} bar, 10^{-3} bar. If $\Delta H_f^\circ(NO(g))$ at 300K = 90.5 kJ/mole and $\Delta S_f^\circ(NO(g))$ at 300K = 12.5J/Kmole [2.303 × R × 300 = 5750J/mole]

1) 173.5 kJ/mole
2) 185 kJ/mole
3) 162 kJ/mole
4) 84.25 kJ/mole

15. A cooking gas cylinder is assumed to contain 11.2 kg isobutane. The heat of combustion of isobutane is 2658 kJ/mol. If a family needs 15×10^3 kJ of energy per day for cooking, then how long would the cylinder last? Assume that 30% of the gas goes waste due to incomplete combustion.

1) 34 days
2) 24 days
3) 40 days
4) 12 days

16. The average energy required to break a P – P bond in $P_4(s)$ into gaseous atoms is 53.2 kcal mol^{-1}. The bond dissociation energy of $H_2(g)$ is 104.2 kcal mol^{-1}; of $PH_3(g)$ from $P_4(s)$ is 5.5 kcal mol^{-1}. The P-H bond energy in kcal mol^{-1} is [Neglect presence of Van der Waals forces in $P_4(s)$]

1) 85.2
2) 57.6
3) 76.9
4) 63.3

17. 1 mol of a liquid of molar volume 100 mL is kept in an adiabatic container under a pressure of 1 bar. The pressure is steeply increased to 100 bar. Under this constant pressure of 100 bar, the volume of the liquid decreases by 1 mL. Calculate magnitude of ΔU of the process. (in J)

1) 10
2) 15
3) 20
4) 25

NEET & AIIMS Past Year MCQs

1. Choose the correct statement for the work done in the expansion and heat absorbed or released when 5 litres of an ideal gas at 10 atmospheric pressure isothermally expands into vacuum until volume is 15 litres: **(2024 Re)**
   1) Both the heat and work done will be greater than zero
   2) Heat absorbed will be less than zero and work done will be positive
   3) Work done will be zero and heat will also be zero
   4) Work done will be greater than zero and heat will remain zero

2. For an endothermic reaction: **(2024 Re)**
   A. q_p is negative. B. Δ_rH is positive.
   C Δ_rH is negative. D. q_p is positive.
   Choose the correct answer from the options given below:
   1) B and D 2) C and D
   3) A and B 4) A and C

3. For the following reaction at 300 K
   $A_2(g) + 3B_2(g) \rightarrow 2AB_3(g)$
   the enthalpy change is +15 kJ, then the internal energy change is : **(2024 Re)**
   1) 19988.4 J 2) 200 J
   3) 1999 J 4) 1.9988 kJ

4. The work done during reversible isothermal expansion of one mole of hydrogen gas at 25 °C from pressure of 20 atmosphere to 10 atmosphere is: **(2024)**
   (Given R = 2.0 cal K^{-1} mol^{-1})
   1) 413.14 calories 2) 100 calories
   3) 0 calorie 4) – 413.14 calories

5. Match List I with List II. **(2024)**

List I (Process)		List II (Conditions)	
A.	Isothermal process	I.	No heat exchange
B.	Isochoric process	II.	Carried out at constant temperature
C.	Isobaric process	III.	Carried out at constant volume
D.	Adiabatic process	IV.	Carried out at constant pressure

   Choose the correct answer from the options given below:
   1) A-I, B-II, C-III, D-IV 2) A-II, B-III, C-IV, D-I
   3) A-IV, B-III, C-II, D-I 4) A-IV, B-II, C-III, D-I

6. In which of the following processes entropy increases?
   A. A liquid evaporates to vapour **(2024)**
   B. Temperature of a crystalline solid lowered from 130K to 0 K.
   C. $2\,NaHCO_{3(s)} \rightarrow Na_2CO_{3(s)} + CO_{2(g)} + H_2O_{(g)}$
   D. $Cl_{2(g)} \rightarrow 2\,Cl_{(g)}$
   Choose the correct answer from the options given below:
   1) A, C and D 2) C and D
   3) A and C 4) A, B and D

7. Which amongst the following options is the correct relation between change in enthalpy and change in internal Energy? **(2023)**
   1) $\Delta H + \Delta U = \Delta nR$ 2) $\Delta H = \Delta U - \Delta n_gRT$
   3) $\Delta H = \Delta U + \Delta n_gRT$ 4) $\Delta H - \Delta U = -\Delta nRT$

8. Which of the following p-V curve represents maximum work done? **(2022)**

1) 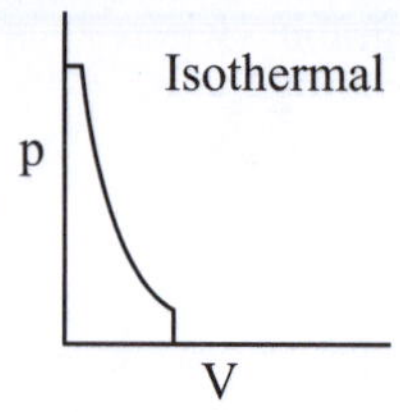

2) 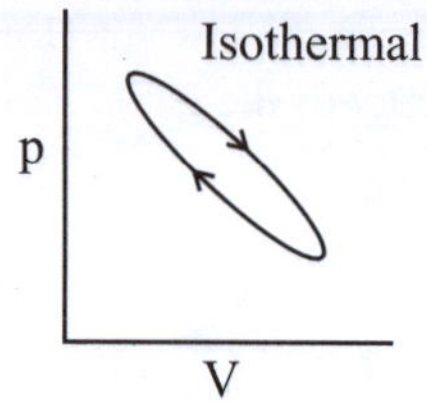

3) 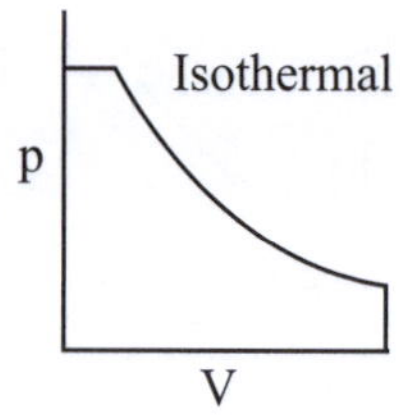

4) 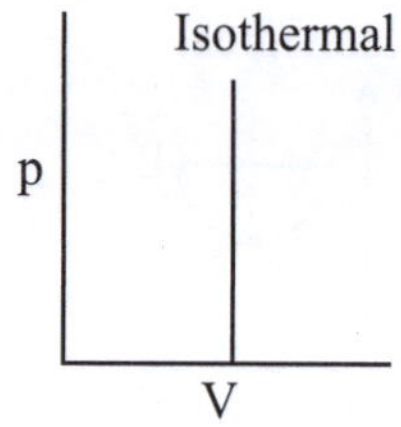

9. Which one among the following is the correct option for right relationship between C_P and C_V for one mole of ideal gas? **(2021)**
   1) $C_P - C_V = R$ 2) $C_P = RC_V$
   3) $C_V = RC_P$ 4) $C_P + C_V = R$

10. For irreversible expansion of an ideal gas under isothermal condition, the correct option is: **(2021)**
    1) $\Delta U \neq 0, \Delta S_{total} \neq 0$ 2) $\Delta U = 0, \Delta S_{total} \neq 0$
    3) $\Delta U \neq 0, \Delta S_{total} = 0$ 4) $\Delta U = 0, \Delta S_{total} = 0$

11. One monoatomic gas is expanded adibatically from 2 L to 10 L at 1 atm external pressure find ΔU (in atm L)? **(AIIMS 2018)**
    1) –8 2) 0
    3) –66.7 4) 58.2

12. A diatomic gas at pressure P, compressed adiabatically to half of its volume, what is the final pressure? **(AIIMS 2014)**
    1) $(2)^{1.4}P$ 2) $P/(2)^{1.4}$
    3) $(2)^{5/3}P$ 4) $P/(2)^{5/3}$

13. For adiabatic process, which is correct? **(AIIMS 2011)**
    1) $\Delta T = 0$ 2) $\Delta S = 0$
    3) $q = 0$ 4) $q_p = 0$

14. The heat liberated when 1.89 g of benzoic acid is burnt in a bomb calorimeter at 25°C and it increases the temperature of 18.94 kg of water by 0.632°C. If the specific heat of water at 25°C and it increases the temperature of 18.94 kg of water by 0.632°C. If the specific heat of water at 25°C is 0.998 cal/g-deg, the value of the heat of combustion of benzoic acid is: **(AIIMS 1994)**

1) 881.1 kcal
2) 771.124 kcal
3) 981.1 kcal
4) 871.2 kcal

Answer Key

NCERT Topic-wise MCQs

1	2	3	4	5	6	7	8	9	10	11	12	13	14	15	16	17
3	2	2	2	3	3	3	2	4	1	1	3	3	3	4	3	1
18	19	20	21	22	23	24	25	26	27	28	29	30	31	32	33	34
1	3	3	1	1	2	4	1	4	2	1	3	4	3	2	4	4
35	36	37	38	39	40	41	42	43	44	45	46	47	48	49	50	51
3	4	3	2	3	4	1	4	4	1	2	4	2	3	1	2	2
52	53	54	55	56	57	58	59	60	61	62	63	64	65	66	67	68
2	2	3	4	3	4	1	4	4	2	4	3	2	1	4	2	3
69																
4																

High Order Time Intensive MCQs

1	2	3	4	5	6	7	8	9	10	11	12	13	14	15	16	17
2	2	4	2	1	1	4	1	3	2	2	3	2	3	4	2	1
18	19	20	21	22	23	24	25	26	27	28	29	30	31	32	33	34
2	1	2	3	1	1	1	3	4	1	4	2	2	2	4	4	1
35	36	37	38	39	40	41	42	43	44	45	46					
3	3	3	1	3	4	1	3	2	3	3	1					

Match the Columns MCQs

1	2	3	4	5
1	2	4	1	3

Correct & Incorrect MCQs

1	2	3	4	5	6	7	8	9	10
4	1	4	3	1	3	2	2	2	1

Statement Based MCQs

1	2	3	4	5	6	7	8	9	10
1	1	3	3	1	4	1	4	3	3

Assertion & Reason MCQs

1	2	3	4	5	6	7	8	9	10
2	2	1	1	3	1	4	2	3	1

Multi Correct MCQs

1	2	3	4
1	2	2	4

NCERT Exemplar MCQs

1	2	3	4	5	6	7	8	9	10	11	12	13	14
3	3	4	3	3	2	3	2	3	3	3	3	1	2

Kattar NEET

1	2	3	4	5	6	7	8	9	10	11	12	13	14	15	16	17
1	3	4	3	3	3	2	2	3	2	3	3	2	3	2	3	1

NEET & AIIMS Past Year MCQs

1	2	3	4	5	6	7	8	9	10	11	12	13	14
3	1	1	4	2	1	3	3	1	2	1	1	3	2

CHAPTER 6

Equilibrium

Topicwise Number of Questions Analysis (2025-2016)

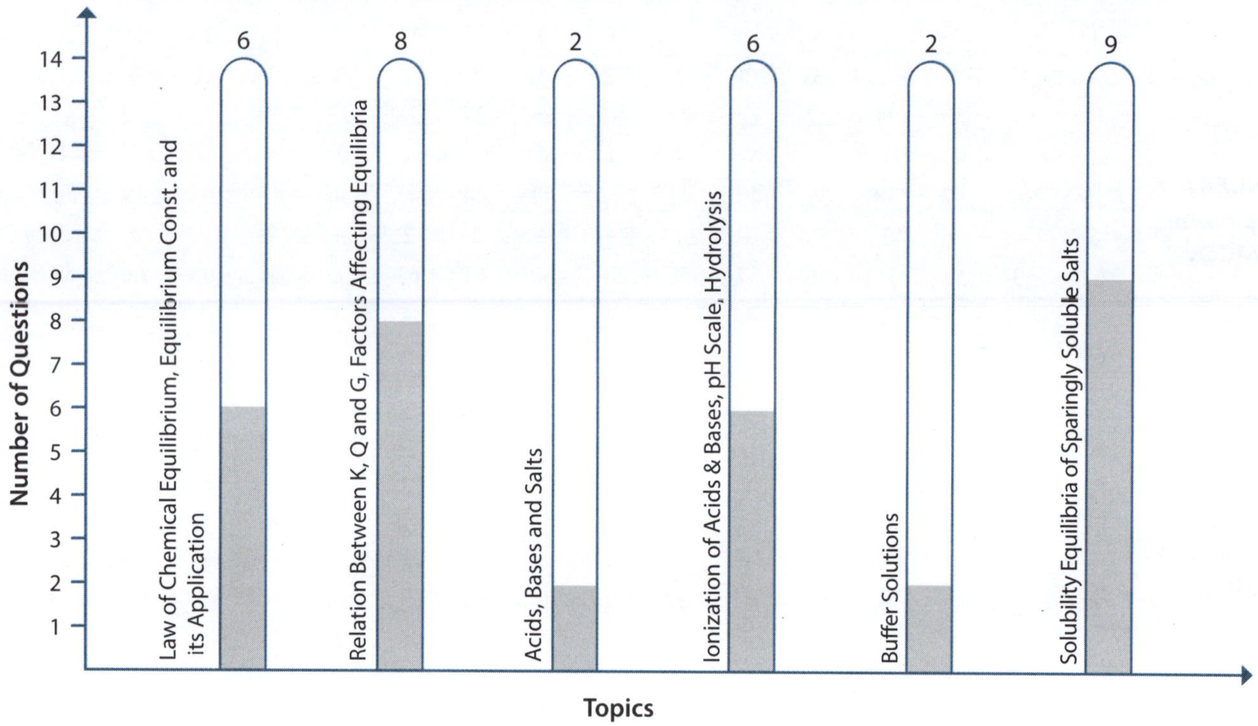

NCERT-Picks

General Introduction

State of equilibrium means the balance of driving forces i.e. the factors taking the reaction in forward direction and the backward direction are balancing each other

It is not possible to reach chemical equilibrium in an open system. Rate of condensation is much less than evaporation rate.

Chemical Equilibrium

In a reversible reaction like

reactants $\rightleftharpoons$ products

Ultimately a stage is reached when both the rates become equal. Such a state is known as **"Chemical Equilibrium"** or "state of Equilibrium".

Law of Chemical Equilibrium

The rate of reaction is directly proportional to product of Active masses of substances raised to the power of their respective stoichiometric coefficient."

Rate $\propto$ Active mass of A $\times$ Active mass of B

Rate = k [A] [B]

and if any reaction is of the type

aA + bB $\rightarrow$ product

Rate = $k[A]^a [B]^b$

Equilibrium Constant(K)

For a general reaction

aA + bB $\rightleftharpoons$ cC + dD

Rate of forward Reaction (R_f) = $k_f [A]^a [B]^b$

Rate of backward Reaction (R_b) = $k_b[C]^c [D]^d$

At equilibrium

$R_f = R_b$

$k_f[A]^a[B]^b = k_b[C]^c[D]^d \Rightarrow \frac{k_f}{k_b} = \frac{[C]^c[D]^d}{[A]^a[B]^b}$

$K_c = \frac{k_f}{k_b}$

$K_c \rightarrow$ equilibrium constant in terms of concentration

$K_p = \frac{p_C^c \times p_D^d}{p_A^a \times p_B^b}$ (K_p = eq. constant in terms of partial pressure)

- Consider the reaction

  $N_2(g) + 3H_2(g) \rightleftharpoons 2NH_3(g)$

  then $K_C = \frac{[NH_3]^2}{[N_2][H_2]^3}$

  $K_p = \frac{p_{NH_3}^2}{p_{N_2} \times p_{H_2}^3}$

Relation Between K_p and K_c

$K_p = K_c(RT)^{\Delta n_g}$

Δn_g = no. of moles of gaseous Reactants – no. of moles of gaseous products

if $\Delta n_g = 0 \Rightarrow K_p = K_c$

Δn_g = +ve, $K_p > K_c$

Δn_g = –ve, $K_p < K_c$

Equilibrium constant in terms of mole-fraction is written as K_x

$PCl_5(g) \rightleftharpoons PCl_3(g) + Cl_2(g)$

$K_x = \frac{(x_{PCl_3})(x_{Cl_2})}{(x_{PCl_5})}$

Where x represents mole-fraction of each component

e.g. $x_{PCl_3} = \frac{n_{PCl_3}}{n_{total}}$

Also $K_p = K_x \times (P)^{\Delta n}$

Unit of Equilibrium constants:

- Unit of K_p **is (atm)$^{\Delta n}$**
- Unit of K_c **is (mole/Lit)$^{\Delta n}$** = (conc.)$^{\Delta n}$
- Concentration of **pure solids and pure liquids** are **taken to be unity** because their change in conc. is very much less compared to gases.

Applications of Equilibrium Constant

1. **Predicting the direction of the reaction**

   Reaction Quotient (Q)

   At each point in a reaction, we can write a ratio of concentration terms having the same form as the equilibrium constant expression. This ratio is called the reaction quotient denoted by symbol Q.

   It helps in predicting the direction of a reaction.

   The expression $Q = \frac{[C]^c[D]^d}{[A]^a[B]^b}$ at any time during reaction is called reaction quotient. The concentrations [C] , [D] , [A], [B] are not necessarily at equilibrium.

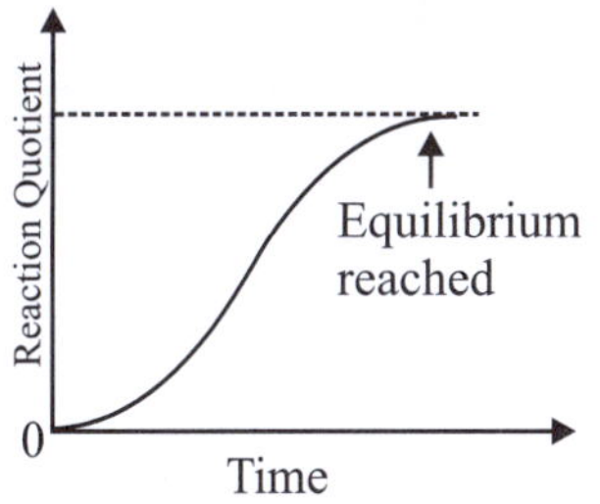

- The reaction quotient is a variable quantity with time.
- It helps in predicting the direction of a reaction.

  if $Q_c > K_c$ reaction will proceed in **backward direction** until equilibrium in reached.

  if $Q_c < K_c$ reaction will proceed in **forward direction** until equilibrium is established.

  if $Q_c = K_c$ Reaction is at **equilibrium**.

  e.g. $2A(g) + B(g) \rightleftharpoons C(g) + D(g)$

  Q_C = Reaction quotient in terms of concentration

  $Q_C = \frac{[C][D]}{[A]^2[B]}$

  $K_C = \frac{[C]_{eq}[D]_{eq}}{[A]_{eq}^2[B]_{eq}}$ (Here all the conc. are at equilibrium)

2. **Predicting the extent of the reaction**

   $K = \frac{[Product]}{[Reactant]}$

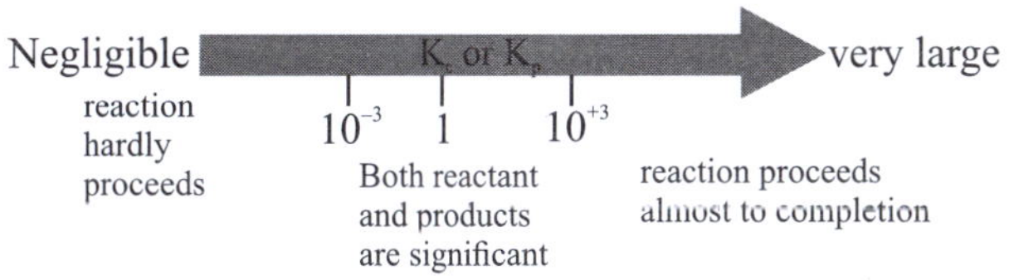

   Case-I If K is large ($K > 10^3$)

   Then product concentration is very larger than the reactant ([Product] >>[Reactant]) Hence, concentration of reactant can be neglected with respect to the product. In this case, the reaction is product favourable and equilibrium will be more in forward direction than in backward direction.

   Case-II If K is very small ($K < 10^{-3}$)

   [Product] << [Reactant]

   Hence, concentration of Product can be neglected as compared to the reactant.

   In this case, the reaction is reactant favourable.

3. **Calculating equilibrium concentrations**

   Equilibrium constant

- $A(g) \rightleftharpoons B(g)$ K_1

  $B(g) \rightleftharpoons C(g)$ K_2

  On adding

  $A(g) \rightleftharpoons C(g)$

  $\mathbf{K = K_1 \cdot K_2}$

- $A(g) + B(g) \rightleftharpoons C(g) + D(g)$

  On reversing,

  $C(g) + D(g) \rightleftharpoons A(g) + B(g)$ $\mathbf{K = 1/K_1}$
- If a chemical reaction having equilibrium constant K_1 is multiplied by a factor n then the resulting equation has equilibrium constant $\mathbf{K = (K_1)^n}$, n can be fractional
- **Relation between equilibrium constant & standard free energy change.**
- $\Delta G^o = -2.303RT \log K_c$

  or $\ln K_c = \dfrac{\Delta G^0}{RT}$
- $K_c = e^{-\Delta G/RT}$
- For gaseous reaction ; $\Delta G^o = -2.303\ RT \log K$ or $K = e^{-\Delta G^0/RT}$

  Where ΔG° = standard free energy change

  T = Absolute temperature,

  R = universal gas constant

Factors Affecting Equilibrium

Le Chatelier's Principle

- If a chemical reaction at equilibrium is subjected to a change in temperature, pressure (or) concentration of the reactants (or) products that govern the chemical equilibrium, the position shifts to the side where the change is reduced (or) nullified.

1. Effect of Concentration

- Increase in concentration of reactants favours the shift of equilibrium towards products side (forward reaction is favoured).
- Increase in the concentration of the products favours the shift of equilibrium towards the reactant side. (backward reaction is favoured).
- In the reaction:

  $Fe^{3+}(aq) + SCN^-(aq) \rightleftharpoons Fe(SCN)^{2+}(aq)$

  yellow colourless deep red

  Addition of aq. $HgCl_2$ decreases red colour. Hg^{2+} reacts with SCN^- ions to form stable complex ion $[Hg(SCN)_4]^{2-}$. Removal of free SCN^- (aq) shift the equilibrium from right to left to replenish SCN^- ions.

2. Effect of Pressure:

When pressure increases, equilibrium shifts in a direction of decrease of volume (or) towards less number of moles (or) molecules.

- Decrease of external pressure on the reaction at equilibrium favours the reaction in the direction in which the volume (or) number of moles (or) molecules increase.
- Pressure has no effect on equilibrium with same mole number ($n_p = n_r$) i.e $\Delta n = 0$

  eg : $H_2(g) + I_2(g) \rightleftharpoons 2HI(g)$
- Pressure does not show any marked effect on equilibrium reactions taking place in the solution phase (or) in the solid phase.

3. Effect of Temperature:

- **Equilibrium constant is dependent only on the temperature.**

  It means K_p and K_c will remain constant at constant temperature no matter how much changes are made in pressure, concentration, volume or catalyst.
- However if temperature is changed,

  $$\log\frac{K_2}{K_1} = \frac{\Delta H}{2.303R}\left[\frac{1}{T_1} - \frac{1}{T_2}\right];\ \Delta H = \text{Enthalpy of reaction}$$

  If $T_2 > T_1$ then $K_2 > K_1$ provided ΔH = +ve (endothermic reaction)

  $K_2 < K_1$ if ΔH = –ve (exothermic reaction)

  In the above equation, the unit of R and $\Delta H/T$ should be same.
- Increase of temperature of the equilibrium system favours endothermic reaction.
- Decrease of temperature of the equilibrium system favours exothermic reaction.

4. Effect of Addition of Inert Gas

- Addition of inert gas at constant volume: The total pressure of the system is increased, but the partial pressure of each reactant and product remains the same. Hence, no effect on the state of equilibrium.
- Addition of inert gas at constant pressure. The total volume is increased, the number of moles per unit volume of each reactant and product is decreased. Hence, equilibrium will shift to the side where number of moles are increased.

  e.g: $PCl_5(g) \rightleftharpoons PCl_3(g) + Cl_2(g)$

5. Effect of a Catalyst

- A catalyst increases the rate of the chemical reaction by making available a new low energy pathway for the conversion of reactants to products. It increases the rate of forward and reverse reactions that pass through the same transition state and does not effect equilibrium.
- Catalyst lowers the activation energy for the forward and reverse reactions by exactly the same amount. Catalyst does not effect the equilibrium composition of a reaction mixture.

Ionic Equilibrium

Degree of Dissociation (α)

It is the fraction of one mole dissociated into the products.

(Defined for one mole of substance)

So, α = no. of moles dissociated / initial no. of moles taken

Ostwald's Dilution Law (for weak electrolyte's)

- For a weak electrolyte A^+B^- dissolved in water, if α is the degree of dissociation then

	AB $\rightleftharpoons$	A^+ +	B^-
initial conc	C	0	0
conc-at eq.	$C(1-\alpha)$	$C\alpha$	$C\alpha$

Then according to law of mass action,

$$K_{eq} = \frac{[A^+][B^-]}{[AB]} = \frac{C\alpha.C\alpha}{C(1-\alpha)} = \frac{C\alpha^2}{(1-\alpha)} = \text{dissociation constant}$$

of the weak electrolyte

$[C = \frac{1}{V}$, then V = 1/C(volume of Solution in which 1 mole is present) is called **dilution**, so $K_{eq} = \frac{\alpha^2}{(1-\alpha)V}$]

If α is negligible in comparison to unity then, $1 - \alpha \simeq 1$. So

$$K_{eq} = \alpha^2 C \Rightarrow \alpha = \sqrt{\frac{K_{eq}}{C}} = \sqrt{K_{eq}.V} \ .$$

Acids and Bases

Arrhenius Concept

- **Arrhenius Acid:** Substance which gives H^+ ion on dissolving in water (H^+ donor). E.g., HNO_3, $HClO_4$, HCl, HI, HBr, H_2SO_4, H_3PO_4 etc
- **Arrhenius base:** Any substance which releases OH^- (hydroxyl) ion in water (OH^- ion donor).

Bronsted - Lowry Concept

- **A Bronsted-Lowry acid** is any species that is capable of donating a proton H^+.
- **A Bronsted-Lowry base** is any species that is capable of accepting a proton, which requires a lone pair of electrons to bond to the H^+
- **Amphoteric (amphiprotic):** Substances which can act as acid as well as base are known as amphoteric.
- The **conjugate base** of a Bronsted-Lowry acid is the species formed after an acid donates a proton. The **conjugate acid** of a Bronsted-Lowry base is the species formed after a base accepts a proton.

Conjugate Acid - Base Pairs

In a typical acid base reaction

$HX + B \rightleftharpoons X^- + HB^+$

Acid		Base		Conjugate Acid		Conjugate Base
HCl	+	H_2O	$\rightleftharpoons$	H_3O^+	+	Cl^-
HSO_4^-	+	NH_3	$\rightleftharpoons$	NH_4^+	+	SO_4^{-2}
$[Fe(H_2O)_6]^{3+}$	+	H_2O	$\rightleftharpoons$	H_3O^+	+	$[Fe(H_2O_5)(OH)]^{2+}$

Lewis Concept (Electronic Concept)

An acid is a molecule/ion which can accept an electron pair.

E.g. Electron deficient molecules : BF_3, $AlCl_3$

Cations : H^+, Fe^{2+}, Na^+

Molecules with vacant orbitals : SF_4, PF_3

A base is any molecule/ion which has a lone pair of electrons which can be donated.

Base $\longrightarrow$ (One electron pair donate)

E.g. Molecules with lone pairs : NH_3, PH_3, H_2O, CH_3OH

Ionization Constants of Acids and Base

Dissociation Constant for Weak Acid

Consider an acid which, when dissolved in water ionizes as,

$HA \rightleftharpoons H^+ + A^-$

Applying the law of mass action,

$K_a = [H^+][A^-]/[HA]$

Dissociation Constant for Weak Base

The equilibrium of NH_4OH (a weak base) can be represented as,

$NH_4OH \rightleftharpoons NH_4^+ + OH^-$

Applying the law of mass action,

$K_b = [NH_4^+][OH^-]/[NH_4OH]$

Ionic Product of Water (K_W)

- The product of the concentration of hydrogen ions (or) hydronium ions and hydroxide ions at any temperature in pure water or in any aqueous solution is known as the ionic product of water. It is represented by K_w.

  $K_w = [H^+][OH^-]$

  $K_w = [H_3O^+][OH^-]$

  $\boxed{pK_w = -\log K_w}$

- The value of K_w depends only on temperature.

pH Scale

pH is defined as negative logarithm of hydrogen ion concentration.

i. $pH = -\log [H^+]$

ii. $pH = \log \frac{1}{[H^+]}$

iii. $[H^+] = 10^{-pH}$

- $pOH = -\log [OH^-]$
- $pK_w = -\log K_w$
- $pK_a = -\log K_a$
- $pK_b = -\log K_b$
- For any aqueous solution, the sum of pH and pOH depends on the ionic product of water.
- $pH + pOH = pK_w$ at any temperature
- pH of neutral water decreases as its temperature is raised. Ionization of H_2O increases $\therefore$ K_w also increases.
- At 25° C, for any aqueous solution. pH + pOH = 14
- If two strong acids are mixed

  $$[H^+] = \frac{N_1V_1 + N_2V_2}{V_1 + V_2}$$

- If two strong bases are mixed

  $$= [OH^-] = \frac{N_1V_1 + N_2V_2}{V_1 + V_2}$$

- If excess of strong acid is added to limited quantity of strong base.

  $$= [H^+] = \frac{N_aV_a - N_bV_b}{V_a + V_b}$$

- If excess of strong base is added to limited quantity of strong acid

$$\left[OH^-\right]=\frac{N_bV_b-N_aV_a}{V_a+V_b}$$

	$[H^+]$	$[OH^-]$	pH	pOH
Acidic solution	$>10^{-7}$	$<10^{-7}$	<7	>7
Neutral solution	10^{-7}	10^{-7}	7	7
Basic solution	$<10^{-7}$	$>10^{-7}$	>7	<7

Relation between K_a and K_b

$K_a \times K_b = K_w$

- K_a and K_b represent the strength of an acid and a base, respectively. In case of a conjugate acid-base pair, they are related in a simple manner so that if one is known, the other can be deduced.

Ionisation Constant of some Common Polyprotic Acids

- Poly basic acids ionise stepwise as for example ortho phosphoric acid ionises in three steps and each step has its own ionisation constant

$H_3PO_4 \rightleftharpoons H^+ + H_2PO_4^-$ I step

$H_2PO_4^- \rightleftharpoons H^+ + HPO_4^{2-}$ II step

$HPO_4^{2-} \rightleftharpoons H^+ + PO_4^{3-}$ III step

Let K_1, K_2 and K_3 be the ionisation constants of first, second and third steps respectively, thus

$$K_1=\frac{[H^+][H_2PO_4^-]}{[H_3PO_4]}$$

$$K_2=\frac{[H^+][HPO_4^{2-}]}{[H_2PO_4^-]}$$

$$K_3=\frac{[H^+][PO_4^{3-}]}{[HPO_4^{2-}]}$$

is general $K_1 > K_2 > K_3$.

- The overall dissociation constant (k) is given by the relation

$K = K_1 \times K_2 \times K_3$

Relative Strengths of Acids

For weak acid, $K_a = C\alpha^2$. For two acids with dissociation constants K_{a1} and K_{a2} at the same concentration C_1

$$\frac{\alpha_1}{\alpha_2}=\sqrt{\frac{K_{a_1}}{K_{a_2}}}=\frac{\text{Strength of acid } HA_1}{\text{Strength of acid } HA_2}$$

Relative Strengths of Bases

For weak base, $K_b = C\alpha^2$. For two bases with dissociation constants K_{b1} and K_{b2} at the same concentration C.

$$\frac{\alpha_1}{\alpha_2}=\sqrt{\frac{K_{b_1}}{K_{b_2}}}=\frac{\text{Strength of base } BOH_1}{\text{Strength of base } BOH_2}$$

Common on I Effect

The degree of dissociation of an electrolyte (weak) is suppressed by the addition of another electrolyte (strong) containing a common ion, this is termed as common ion effect.

Acetic acid is a weak electrolyte and its ionisation is suppressed in presence of a strong acid (H^+ ion as common ion) or a strong salt like sodium acetate (acetate ion is a common ion).

Buffer Solution

- $pH = pK_a + \log\frac{(\text{Conc. of common ion from salt})}{(\text{Conc. of common ion from acid})}$ is known as Henderson's equation and enables the calculations of pH values of buffer solutions made by mixing known concentrations of a weak acid and its salt.
- For a **buffer made up of weak base** and **its salt with a strong acid**, the Henderson's equation looks like this :

  $pOH = pK_b + \log(\text{salt form})/(\text{Base form})$
- pH of the buffer solution is not affected by dilution.

Salt Hydrolysis

- Salt hydrolysis is the phenomenon of the interaction of anions and cations of the salt with the H^+ and OH^- ions furnished by water yielding acidic or alkaline or sometimes even neutral solution.

Salt of	Nature of solution	Hydrolysis Constant (K_h)	Degree of hydrolysis (h)	$[H^+]$ in solution	pH of solution
Weak acid and strong base e.g., CH_3COONa	Basic	$\frac{K_w}{K_a}$	$\sqrt{\frac{K_b}{C}}$ or $\sqrt{\frac{K_w}{K_aC}}$	$\left(\frac{K_wK_a}{C}\right)^{1/2}$	$-\log\sqrt{\frac{K_w.K_a}{C}}$ or $\frac{1}{2}[pK_w+pK_a+\log C]$
Strong acid and weak base e.g., NH_4Cl	Acidic	$\frac{K_w}{K_b}$	$\sqrt{\frac{K_b}{C}}$ or $\sqrt{\frac{K_w}{K_bC}}$	$\left(\frac{K_wC}{K_b}\right)^{1/2}$	$-\log\sqrt{\frac{K_w.C}{K_b}}$ or $\frac{1}{2}[pK_w-pK_b-\log C]$

Weak acid and weak base e.g., CH_3COONH_4	Acidic/Basic/Neutral depending on K_a and K_b of weak acid and weak base	$\frac{K_w}{K_a.K_b}$	$\sqrt{K_h}$ or $\sqrt{\frac{K_w}{K_a.K_b}}$	$\left(\frac{K_w K_a}{K_b}\right)^{1/2}$	$-\log\sqrt{\frac{K_a.K_w}{K_b}}$ or $\frac{1}{2}[pK_w+pK_a-pK_b]$
Strong acid and strong base e.g., NaCl	Neutral	Does not undergo Hydrolysis			

Solubility and Solubility Product

- A substance's **solubility product** is the mathematical product of its dissolved ion concentrations raised to the power of their stoichiometric coefficients.
- The solubility product is a heterogeneous equilibrium constant, a specific form of the equilibrium constant. It is relevant in saturated solutions in which an ionic compound has not fully dissolved.
- The smaller the solubility product of a substance, the lower is its solubility.
- For the dissolution of a sparingly soluble salt

  $A_xB_y \rightleftharpoons xA^{y+} + yB^{x-}$

  $K_{sp} = [A^{y+}]^x\ [B^{x-}]^y = (xS)^x\ (yS)^y = x^x\ .\ y^y\ .\ S^{x+y}$

  Where, S = Solubility of sparingly soluble salt

  K_{sp} = Solubility
- If ionic product > K_{sp} , precipitation occurs.
- If ionic products < K_{sp} , no precipitation occurs
- If ionic product = K_{sp} , the solution is just saturated and precipitation just begins.

Some Important Points

- Molecules or ions that can behave both as Bronsted acid and base are amphiprotic substances.
- Some common acids have been arranged in the following order of their acid strengths $HClO_4 > HBr > H_2SO_4 > HCl > HNO_3$
- An acid - base indicator is an organic dye that signals the end point by a visual change in colour.
- A suitable indicator for a given titration may defined as one which has a narrow pH range as possible that lies entirely on the upright part of the titration curve.
  - Suitable indicators for strong acids and strong base titrations are both methyl orange and phenolphthalein.
  - The pH range of phenolphthalein is 8.3 – 10.0 and methyl orange is 3.1 – 4.4
  - For a weak - acid base titrations, phenolphthalein is a suitable indicator while methyl orange is not.
  - Silver iodide has lower solubility product than silver chloride, and so the former gets precipitated in preference to the later.

MCQs ?

NCERT Topic-wise MCQs

Equilibrium in Physical Processes

1. Which of the following reactions can be followed by its colour change?
   1) $SO_2(g) + 1/2\, O_2(g) \rightleftharpoons SO_3(g)$
   2) $N_2(g) + 3H_2(g) \rightleftharpoons 2NH_3(g)$
   3) $2HI(g) \rightleftharpoons H_2(g) + I_2(g)$
   4) $C(graphite) + CO_2(g) \rightleftharpoons 2CO(g)$

2. For the physical equilibrium $H_2O(s) \rightleftharpoons H_2O(l)$ which of the following is true ?
   1) The pressure changes do not affect the equilibrium
   2) More of ice melts if pressure on the system is increased.
   3) More of liquid freezes if pressure on the system is increased.
   4) At low pressure, the nature of equilibrium changes to forward direction

3. A reaction reaches a state of equilibrium only when:
   1) The reactants and products stop reacting
   2) The concentration of reactants and products become equal
   3) The products react together at the same rate at which they are formed
   4) All the reactants and products are in the same state of matter

4. Attainment of equilibrium in a coloured gaseous reversible reaction is detected by the constancy of
   1) Colour
   2) Density
   3) Pressure
   4) All the above properties of the mixture

5. Which of the following is not a general characteristic of equilibrium involving physical processes?
   1) Equilibrium is possible only in a closed system at a given temperature
   2) The equilibrium is dynamic in nature
   3) Measurable properties of the system keep changing
   4) Equilibrium can be attained from both sides of the reaction.

6. Which of the following is a characteristic of a reversible reaction ?
   1) Number of moles of reactants and products are equal
   2) It can be influenced by a catalyst
   3) It can never proceed to completion
   4) It can be attained in open vessel

7. A reaction is said to be in equilibrium when
   1) The rate of transformation of reactants to products is equal to the rate of transformation of products to the reactants
   2) 50% of the reactants are converted to products
   3) The reaction is near completion and all the reactants are converted to products
   4) The volume of reactants is just equal to the volume of the products

8. Which of the following is irreversible reaction?
   1) $PCl_5 \rightarrow PCl_3 + Cl_2$
   2) $2SO_2 + O_2 \rightarrow 2SO_3$
   3) $N_2 + 3H_2 \rightarrow 2NH_3$
   4) $2KClO_3 \rightarrow 2KCl + 3O_2$

9. Which of the following is not true about a reversible reaction?
   1) The reaction does not proceed to completion
   2) It cannot be influenced by a catalyst
   3) Number of moles of reactants and products is always equal
   4) It can be attained only in a closed container

10. The following is a reversible reaction:
    1) $KClO_3$ heated in a sealed tube
    2) Na_2CO_3 heated in a closed vessel
    3) $CaCO_3$ heated in a closed vessel
    4) CH_4 heated with excess of O_2 in a closed vessel

Law of Chemical Equilibrium and Equilibrium Constant (K)

11. In the dissociation of PCl_5 as:-
    $PCl_5(g) \rightarrow PCl_3(g) + Cl_2(g)$, If the degree of dissociation is α at equilibrium pressure "P", then the equilibrium constant for the reaction is:
    1) $K_p = \dfrac{\alpha^2}{1+\alpha^2 p}$
    2) $K_p = \dfrac{\alpha^2 p^2}{1-\alpha^2}$
    3) $K_p = \dfrac{\alpha p^2}{1-\alpha^2}$
    4) $K_p = \dfrac{\alpha^2 p}{1-\alpha^2}$

12. Consider the equilibrium reactions:

$H_3PO_4 \xrightleftharpoons{K_1} H^+ + H_2PO_4^-$

$H_2PO_4^- \xrightleftharpoons{K_2} H^+ + HPO_4^{2-}$

$HPO_4^{2-} \xrightleftharpoons{K_3} H^+ + PO_4^{3-}$

The equilibrium constant 'K' for the following dissociation $H_3PO_4 \rightleftharpoons 3H^+ + PO_4^{3-}$ is:

1) K_1/K_2K_3 2) $K_1K_2K_3$
3) K_2/K_1K_3 4) $K_1 + K_2 + K_3$

13. For the equilibrium.

$2NO(g) + Cl_2(g) \rightleftharpoons 2NOCl(g)$

K_p is related to K_c by the reaction:

1) $K_p = K_c(RT)$ 2) $K_p = K_c(RT)^2$
3) $K_p = \frac{K_c}{RT}$ 4) $K_p = \frac{K_c}{(RT)^2}$

14. For the reaction

$2A(g) + B(g) \rightleftharpoons 3C(g) + 4D(g)$

Two moles each of A and B were taken into a 1L flask. The following must always be true when the system attained equilibrium:

1) $[A] = [B]$ 2) $[A] < [B]$
3) $[B] = [C]$ 4) $[A] + [B] < [C] + [D]$

15. For the reaction,

$N_2 + O_2(g) \rightleftharpoons 2NO(g)$

the equilibrium constant is K_1.

The equilibrium constant is K_2 for the reaction

$2NO(g) + O_2(g) \rightleftharpoons 2NO_2(g)$

What is "K" for the reaction:

$NO_2(g) \rightleftharpoons \frac{1}{2}N_2(g) + O_2(g)$?

1) $\frac{1}{(K_1K_2)}$ 2) $\frac{1}{(2K_1K_2)}$
3) $\frac{1}{(4K_1K_2)}$ 4) $\left[\frac{1}{(K_1K_2)}\right]^{\frac{1}{2}}$

16. Calculate K_c for the reversible process given below if $K_p = 167$, and $T = 800°C$

$CaCO_3(s) \rightleftharpoons CaO(s) + CO_2(g)$

1) 1.95 2) 1.85
3) 1.89 4) 1.60

17. Two moles of each reactants A and B are taken in the reaction flask. They react in the following manner,

$A(g) + B(g) \rightleftharpoons C(g) + D(g)$

At equilibrium, it was found that the concentration of "C" is triple that of B. The equilibrium constant for the reaction is:

1) 4.5 2) 6
3) 9 4) 1/6

18. For the reaction,

$AB(g) \rightleftharpoons A(g) + B(g)$

AB is 33% dissociated at a total pressure "P". Therefore, P is related to K_p by which of following options:

1) $P = K_p$ 2) $P = 3K_p$
3) $P = 4K_p$ 4) $P = 8K_p$

19. $2HI(g) \rightleftharpoons H_2(g) + I_2(g)$

The equilibrium constant of the above reaction is 6.4 at 300 K. If 0.25 mole each of H_2 and I_2 are added to the system, the equilibrium constant will be:

1) 0.8 2) 3.2
3) 1.6 4) 6.4

20. The dissociation equilibrium of a gas AB_2 can be represented as

$2AB_2(g) \rightleftharpoons 2AB(g) + B_2(g)$

The degree of dissociation is "X" and is small compared to the expression relating the degree of dissociation (x) with equilibrium constant K_p and total pressure "P" is:

1) $\frac{2K_p}{P}$ 2) $\left(\frac{2K_p}{P}\right)^{\frac{1}{3}}$
3) $\left(\frac{2K_p}{P}\right)^{\frac{1}{2}}$ 4) $\frac{K_p}{P}$

21. At 600°C, K_p for the following reaction is 1 atm,

$X(g) \rightleftharpoons Y(g) + Z(g)$

at equilibrium, 50% of X(g) is dissociated. The total pressure of the equilibrium system is P atm. What is the partial pressure (in atm) of X(g) at equilibrium?

1) 1 2) 4
3) 2 4) 0.5

22. The reaction $CaCO_3(s) \rightleftharpoons CaO(s) + CO_2(g)$ goes to completion in lime kiln because of the

1) High temperature
2) CaO is more stable than the $CaCO_3$
3) CaO is not dissociated
4) CO_2 escapes continuously

23. For reactions involving gaseous reactants and products the equilibrium constant K_p is written in terms of

1) The pressure of the gases
2) The molar volumes of the gases
3) The partial pressures of the gases
4) The mole fraction of the gases

24. At 500 K, the equilibrium constant for the reaction *cis* – $C_2H_2Cl_2 \rightleftharpoons$ *trans* $C_2H_2Cl_2$ is 0.6. At the same temperature, the equilibrium constant for the reaction

trans – $C_2H_2Cl_2 \rightleftharpoons$ *cis* – $C_2H_2Cl_2$

1) 0.6 2) 1.66
3) 1 4) 16.6

25. The value of K_c for the reaction N_2 (g)+ $3H_2$ (g) $\rightleftharpoons$ $2NH_3$ (g) depends on

1) Temperature 2) Pressure
3) Collision 4) Atomic number

26. The following statement is incorrect about chemical equilibrium:

1) There is no change in the concentrations of reactants and products with time
2) Equilibrium can be attained by starting with either reactants or products
3) Equilibrium is dynamic
4) Position of equilibrium cannot be disturbed by changing the concentrations of reactants or products

27. K_c for the reaction $N_2 + 3H_2 \rightleftharpoons 2NH_3$ is K then the value of K_c for the reaction $2NH_3 \rightleftharpoons N_2 + 3H_2$ will be

1) K 2) 2K
3) 1/ K 4) K^2

28. In the reaction $H_2(g) + I_2(g) \rightarrow 2HI(g)$

1) $K_p = K_c$ 2) $K_p \neq K_c$
3) $K_p > K_c$ 4) $K_p < K_c$

29. K_p for the reaction

$CaCO_3(s) \rightleftharpoons CaO(s) + CO_2(g)$ is correctly expressed as

1) $K_P = \frac{P_{CaO} \times P_{CO_2}}{P_{CaCO_3}}$ 2) $K_P = \frac{P_{CaCO_3}}{P_{CaO} . P_{CO_2}}$

3) $K_p = P_{CO_2}$ 4) $K_P = \frac{P_{CO_2}}{P_{CaCO_3}}$

30. The equilibrium constant for the reaction $A \rightleftharpoons B$ is K. The equilibrium constant for the reaction $mA \rightleftharpoons mB$ is

1) mK 2) K/m
3) K^m 4) K^{-m}

31. For the homogeneous reaction $4NH_3(g) + 5O_2(g) \rightleftharpoons 4NO(g) + 6H_2O(g)$ the equilibrium constant K_c has the unit of

1) $(Conc)^{-1}$ 2) Conc
3) $(Conc)^{+10}$ 4) it is dimensionless

32. What is the unit of K_p for the reaction?

$CS_2(g) + 4H_2(g) \rightleftharpoons CH_4(g) + 2H_2S(g)$

1) atm 2) atm^{-2}
3) atm^2 4) atm^{-1}

33. At 527°C, the reaction given below has $K_c = 4$

$NH_3(g) \rightleftharpoons \frac{1}{2}N_2(g) + \frac{3}{2}H_2(g)$

What is the K_p for the reaction?

$N_2(g) + 3H_2(g) \rightleftharpoons 2NH_3(g)$

1) $16 \times (800\,R)^2$ 2) $\left(\frac{800\,R}{4}\right)^{-2}$

3) $\left(\frac{1}{4 \times 800\,R}\right)^2$ 4) None of these

34. The degree of dissociation of PCl_5

1) Increases with increasing pressure
2) Decreases with increasing pressure
3) No effect on change in pressure
4) Decreases with decreasing pressure

35. In which of the following gaseous reactions K_p is less than K_c?

1) $N_2O_4 \rightleftharpoons 2NO_2$ 2) $2HI \rightleftharpoons H_2 + I_2$

3) $2SO_2 + O_2 \rightleftharpoons 2SO_3$ 4) $N_2 + O_2 \rightleftharpoons 2NO$

36. The unit of equilibrium constant (K_c) in general is

1) (Mole / lit) 2) (Lits / Mole)
3) $(Mole / lit)^{\Delta n}$ 4) $(Lit / Mol)^{\Delta n}$

37. For the reaction

$CO(g) + Cl_2(g) \rightleftharpoons COCl_2$ (g), K_p/K_c is equal to:

1) 1/RT 2) RT
3) $\sqrt{RT}$ 4) $(RT)^2$

38. For the reaction equilibrium,

$N_2O_4(g) \rightleftharpoons 2NO_{2(g)}$

the concentrations of N_2O_4 and NO_2 at equilibrium are 4.8 × 10^{2-} and 1.2 × 10^{-2} $molL^{-1}$ respectively. The value of K_c for the reaction is:

1) 3.3 × 10^2 mol L^{-1} 2) 3.3 × 10^{-1} mol L^{-1}
3) 3.3 × 10^{-3} mol L^{-1} 4) 3.3 × 10^3 mol L^{-1}

39. N_2O_4 is dissociated to 33% and 40% at total pressure P_1 and P_2 atm respectively. Then the ratio P_1/P_2 is:

1) 7/4 2) 7/3
3) 8/3 4) 8/5

40. $A(g) + B(g) \rightleftharpoons AB(g)$ is a reversible reaction. At equilibrium 0.4 mole of AB is formed when each A and B are taken one mole. How much of A changes into AB?

1) 20% 2) 40%
3) 60% 4) 4%

41. Eight mole of a gas AB_3 attain equilibrium in a closed container of volume 1dm³ as, $2AB_3 \rightleftharpoons A_2(g) + 3B_2(g)$. If at equilibrium 2 mole of A_2 are present then, equilibrium constant is:

1) 72 mol^2L^{-2} 2) 36 mol^2L^{-2}
3) 3 mol^2L^{-2} 4) 27 mol^2L^{-2}

42. $NH_2COONH_4(s) \rightleftharpoons 2NH_3(g) + CO_2(g)$ If equilibrium pressure is 3 atm for the above reaction, K_p will be

1) 4 2) 20
3) 25 4) 15

43. The K_p of the reaction is $NH_4HS(s) \rightleftharpoons NH_3(g) + H_2S(g)$, if the total pressure at equilibrium is 30 atm

1) 15 atm^2 2) 225atm^2
3) 30 atm^2 4) 15 atm

Application of Equilibrium Constant

44. Predict the effect of increased pressure on the following reaction's equilibrium.

$2SO_2(g) + O_2(g) \rightleftharpoons 2SO_3(g)$

1) Equilibrium shift to the right
2) Equilibrium shift to the left
3) No effect on equilibrium
4) Reaction stops

45. Given exothermic reaction

$CoCl_4^{2-} + 6H_2O_{(l)} \rightleftharpoons [Co(H_2O)_6]^{2+} + 4Cl^-$

Which one of the following will decrease the equilibrium concentration of $CoCl_4^{2-}$?

1) Addition of HCl
2) Addition of $Co(NH_3)_2$
3) The solution is diluted with water
4) The temperature is increased

46. Inert gas has been added to the following equilibrium system at constant volume

$SO_2(g) + 1/2\ O_2(g) \rightleftharpoons SO_3(g)$

To which direction will the equilibrium shift?

1) Forward 2) Backward
3) No effect 4) Unpredictable

47. Which of the following reaction is favoured by increase of temperature? ΔH values are given in every reaction.

1) $N_2 + 3H_2 \rightleftharpoons 2NH_3\ (\Delta H_1 = +22.9$ kcal)
2) $N_2(g) + O_2(g) \rightleftharpoons 2NO(g)\ (\Delta H_2 = -42.8$ kcal)
3) $2SO_2(g) + O_2(g) \rightleftharpoons SO_3(g)\ (\Delta H_3 = +45.3$ kcal)
4) None of these

48. $H_2 + I_2 \rightleftharpoons 2HI$

(g) (g) (g)

Which one of the following will not affect the rate of this reaction?

1) Pressure change 2) Temperature change
3) Concentration change 4) Catalyst

49. Reaction, $2BaO_2(g) \rightleftharpoons 2BaO(s) + O_2(g)$; ΔH = +ve

In equilibrium condition, pressure of O_2 depends upon:

1) Increase mass of BaO_2
2) Increase mass of BaO
3) Increased temperature of equilibrium
4) Increased mass of BaO_2 and BaO both

50. The reaction favored at low pressure is:

1) $H_2(g) + I_2(g) \rightleftharpoons 2HI(g)$
2) $PCl_5(g) \rightleftharpoons PCl_3(g) + Cl_2(g)$
3) $N_2(g) + O_2(g) \rightleftharpoons 2NO(g)$
4) $N_2(g) + 3H_2(g) \rightleftharpoons 2NH_3(g)$

51. For a reversible reaction, if the concentrations of the reactants are doubled, then the equilibrium constant value

1) Gets doubled 2) Gets halved
3) Remains the same 4) Increases four times

52. An increase in pressure would favour the forward reaction

1) $N_2(g) + O_2(g) \rightleftharpoons 2NO(g)$
2) $2SO_2(g) + O_2(g) \rightleftharpoons 2SO_3(g)$
3) $PCl_5(g) \rightleftharpoons PCl_3\ (g) + Cl_2(g)$
4) $H_2(g) + I_2(g) \rightleftharpoons 2HI(g)$

53. In which of the equilibria, the position of the equilibrium shifts towards products, if the total pressure is increased?

1) $N_2(g) + 3H_2(g) \rightleftharpoons 2NH_3(g)$
2) $I_2(g) + H_2(g) \rightleftharpoons 2HI(g)$
3) $N_2(g) + O_2(g) \rightleftharpoons 2NO(g)$
4) $N_2O_4(g) \rightleftharpoons 2NO_2(g)$

54. In the reaction $NH_4SH(s) \rightleftharpoons NH_3(g) + H_2S(g)$ on doubling the concentration of ammonia the equilibrium concentration of H_2S is

1) Reduced to half to its initial value
2) Increases by two times
3) Remains unchanged
4) Increases by four times

55. When the following reaction come to equilibrium,

$N_2(g) + 2H_2(g) \rightleftharpoons N_2H_4(g)$

$K_c = 7.4 \times 10^{-26}$

the equilibrium mixture contains.

1) Mostly products
2) Mostly reactants
3) It can not be predicted
4) Some times reactants and sometimes products

56. With increase in temperature, the value of equilibrium constant

1) Increases
2) Decreases
3) May increase or decrease
4) Remains constant

57. The equilibrium constants of a reaction at 298 K and 308 K are 1.0×10^{-2} and 2×10^{-2} respectively, the reaction is

1) Exothermic
2) Endothermic
3) May be endothermic or exothermic
4) Cannot be predicted

58. In which of the following, reaction goes to more completion?

1) $K_c = 1$ 2) $K_c = 10^{10}$
3) $K_c = 10^{-10}$ 4) $K_c = 10$

59. For a system to be in equilibrium, ΔG = 0 under conditions of constant.

1) Temperature and pressure
2) Temperature and volume
3) Energy and volume
4) Pressure and volume

60. The following reactions occurs at 500 K. Arrange them in order of increasing tendency to proceed to completion.

(I) $2NOCl(g) \rightleftharpoons 2NO_2(g); K_p = 1.7 \times 10^{-2}$

(II) $N_2O_4(g) \rightleftharpoons 2NO_2(g); K_p = 1.5 \times 10^{-5}$

(III) $2SO_3(g) \rightleftharpoons SO_2(g) + O_2(g); K_p = 1.3 \times 10^{-5}$

(IV) $2NO_2(g) \rightleftharpoons 2NO(g) + O_2(g); K_p = 5.9 \times 10^{-5}$

1) II < I < IV < III   2) III < II < IV < I
3) I < III < IV < II   4) III < IV < II < I

61. Law of mass action cannot be applied to
1) Decomposition of gaseous HI
2) Decomposition of gaseous PCl_5
3) Transition of Rhombic Sulphur to Monoclinic sulphur
4) Decomposition of Calcium Carbonate

62. Consider the following equilibrium in a closed container:

$N_2O_4(g) \rightleftharpoons 2NO_2(g)$

At a fixed temperature, the volume of the reaction container is halved. For this change, which of the given statements holds true regarding the equilibrium constant (K_p) and degree of dissociation (α)?
1) Neither K_p nor α changes
2) Both K_p and α changes
3) K_p changes, but α does not change
4) K_p does not changes, but α changes

63. At constant temperature, the equilibrium constant (K_p) for the decomposition reaction. $N_2O_4(g) \rightleftharpoons 2NO_2$ is expressed by,

$$K_p = \frac{(4x^2P)}{(1-x^2)}$$

where P = pressure, x = extent of decomposition. Which of the following statements is true?
1) K_p increases with increase of P
2) K_p increases with increase of x
3) K_p increase with decrease of x
4) K_p remains constant with change in P or x

64. In a system: $A(s) \rightleftharpoons 2B(g) + 3C(g)$. If the concentration of C at equilibrium is increased by a factor 2, it will cause the equilibrium concentration of B to change to:
1) Two times of its original value
2) One half of its original value
3) $2\sqrt{2}$ times of its original value
4) $\frac{1}{2\sqrt{2}}$ times of its original value

65. When a catalyst is introduced into a reversible reaction
1) Forward reaction rate increases
2) Backward reaction rate increases
3) Equilibrium is not changed
4) Attains equilibrium quickly

66. The equilibrium constant of a reaction is 300, if the volume of the reaction flask is tripled, the equilibrium constant will be
1) 100   2) 300
3) 250   4) 150

Acids, Bases and Salts

67. Which is strongest acid in the following?
1) H_2SO_4 2) $HClO_3$
3) $HClO_4$ 4) H_2SO_3

68. The correct order of basic strength:
1) $H_2O < OH^- < CH_3OH < CH_3O^-$
2) $CH_3OH < H_2O < CH_3O^- < OH^-$
3) $H_2O < CH_3OH < OH^- < CH_3O^-$
4) $OH^- < H_2O < CH_3O^- < CH_3OH$

69. Which one of the following can be classified as a Bronsted base?
1) NO_3^- 2) H_3O^+
3) CH_3COOH 4) NH_4^+

70. Which of the following is not lewis acid?
1) $AlCl_3 . 6H_2O$ 2) $AlCl_3$
3) $SnCl_4$ 4) $FeCl_3$

71. $H_3O^+ + OH^- \rightarrow 2H_2O$ is
1) Arrhenius neutralisation
2) Bronsted neutralisation
3) Lewis neutralisation
4) Both Lewis neutralization and Bronsted neutralisation.

72. The no. of conjugate acid-base pairs present in the aqueous solution of H_3PO_3 is
1) 2   2) 3
3) 4   4) 5

73. The conjugate base of $H_2PO_4^-$ is:
1) PO_4^{3-} 2) P_2O_5
3) H_3PO_4 4) HPO_4^{2-}

74. Which of the following has no conjugate base?
1) Oxide ion
2) Hydroxide ion
3) Hydro peroxide ion
4) Hydrogen sulphate ion

75. The strongest base among the following is
1) Cl^- 2) CH_3COO^-
3) HSO_4^- 4) NO_3^-

76. In the reaction $NH_3 + H_2O \rightleftharpoons NH_4^+ + OH^-$, the conjugate acid-base pair is
1) NH_3 and H_2O 2) NH_3 and OH^-
3) H_2O and NH_4^+ 4) NH_4^+ and NH_3

77. The conjugate base of hydrazoic acid is
1) N_2H_4 2) $N_2H_5^+$
3) N_3^- 4) NH_2OH

78. Conjugate base of $[Al(H_2O)_6]^{3+}$ is
1) $[Al(H_2O)_5OH]^{3+}$ 2) $[Al(H_2O)_6]^{2+}$
3) $[Al(H_2O)_5OH]^{2+}$ 4) $[Al(H_2O)_5OH]^+$

79. The Bronsted acid which has the weakest conjugate base is:
1) HCl 2) HF
3) H_2S 4) H_2O

80. Which of the following acts as Lewis acid?
1) Cu^{2+} 2) $AlCl_3$
3) CO_2 4) All the above

81. Which of the following is a strong electrolyte?
1) NH_3 2) Glucose
3) $BaCl_2$ 4) H_3PO_4

Ionization of Acids and Bases

82. A mono-basic weak acid solution has a molarity of 0.005 M and pH of 5. What this percentage ionisation in this solution?
1) 2.0 2) 0.2
3) 0.5 4) 20.0

83. Degree of dissociation of 0.1N CH_3COOH is:
($K_{acid} = 1 \times 10^{-5}$)
1) 10^{-5} 2) 10^{-4}
3) 10^{-3} 4) 10^{-2}

84. The dissociation constant value of four acids are as follows. Which of the values stand for strongest acid?
1) 2×10^{-4} 2) 2×10^{-2}
3) 3×10^{-4} 4) 0.02×10^{-2}

85. Extent of ionisation depends upon:
1) Pressure 2) Volume
3) Dilution 4) None of these

86. If pK_b for CN^- at 25°C is 4.7, the pH of 0.5 M aqueous NaCN solution is:
1) 12 2) 10
3) 11.5 4) 11

87. Liquid ammonia ionises to a slight extent. At –50°C, its self ionisation constant, $K_{NH_3} = [NH_4^+][NH_2^-] = 10^{-30}$
How many amide ions, are present per cm^3 of pure liquid ammonia? (Assume $N_A = 6.0 \times 10^{23}$)
1) 6×10^6 ions 2) 6×10^5 ions
3) 6×10^{-5} ions 4) 6×10^{-6} ions

88. If pH of the solution is one, what weight of HCl present in one litre of solution?
1) 3.65 g 2) 36.5 g
3) 0.365 g 4) 0.0365 g

89. How many gram of NaOH must be present in one litre of the solution to give it a pH = 12?
1) 0.20 g 2) 0.4 g
3) 4.0 g 4) 0.10 g

90. Number of H^+ ions present in 10 mL of a solution of pH = 3 is:
1) 10^{13} 2) 6.02×10^{18}
3) 6.02×10^{13} 4) 6.02×10^{10}

91. Which of the following is not true for acidic solutions at room temperature?
1) $[H^+] > [OH^-]$ 2) $[H^+] > 10^{-7}M$
3) $[H^+] < 10^{-7}M$ 4) $[OH^-] < 10^{-7}$

92. If NaOH is added to a solution of acetic acid
1) H^+ ions increases
2) pH decreases
3) $[CH_3COO^-]$ increases
4) $[CH_3COOH]$ increases

93. 100 ml of HCl + 35 ml of NaOH, colour of methyl orange in the solution will be:
1) Red
2) Yellow
3) Can't be predicted
4) Methyl orange is not suitable indicator

94. What is the pH at equivalence point in the titration of 0.1M CH_3COOH and 0.1 M NaOH?
(K_a for acetic acid is 1.8×10^{-5})
1) 7 2) Between 7 and 8
3) Between 8 and 9 4) Between 6 and 7

95. The pH of a dilute solution of acetic acid was found to be 4.3. The addition of a small crystal of sodium acetate will cause pH to:
1) Become less than 4.3 2) Become more than 4.3
3) Remains equal to 4.3 4) Unpredictable

96. For anionic hydrolysis, pH is given by:
1) $\frac{1}{2}pK_w - \frac{1}{2}pK_b - \frac{1}{2}\log c$
2) $\frac{1}{2}pK_w + \frac{1}{2}pK_a - \frac{1}{2}pk_b$
3) $\frac{1}{2}pK_w + \frac{1}{2}pK_a + \frac{1}{2}\log c$
4) None

97. At certain temperature dissociation constant of formic acid and acetic acid are 1.8×10^{-4} and 1.8×10^{-5} respectively. At what concentrations of acetic acid solution, the H_3O^+ ion concentration is same as that in 0.001 M formic acid solution?
1) 0.01 M 2) 1×10^{-3} M
3) 1×10^{-1}M 4) 1×10^{-4} M

98. Which of the following salts will not change the pH of pure water on dissociation?

1) KCl 2) $AlCl_3$
3) Na_2CO_3 4) $Al_2(SO_4)_3$

99. What happens to ionic product of H_2O as temperature increases?

1) Increases
2) Decreases
3) Remains same
4) Increase then decrease

100. A 100 mL, 0.1 M solution of ammonium acetate is diluted by adding 100 mL of water. The pH of the resulting solution will be: (pKa of acetic acid is nearly equal to pK_b of NH_4OH)

1) 4.9 2) 5.0
3) 7.0 4) 10.0

101. The K_a for formic acid and acetic acid are 2×10^{-4} and 2×10^{-5} respectively. Calculate the relative strength of acids with the same molar concentration.

1) $\sqrt{10}$ 2) $\sqrt{7}$
3) $\sqrt{8}$ 4) $\sqrt{5}$

102. The aqueous solution of HCOONa, $C_6H_5NH^+{}_3Cl^-$ and KCN are respectively:

1) Acidic, acidic, basic 2) Acidic, basic, neutral
3) Basic, neutral, neutral 4) Basic, acidic, basic

103. The ionisation constant of ammonium hydroxide is 1.77×10^{-5} at 298 K. Hydrolysis constant of ammonium chloride is:

1) 5.65×10^{-10} 2) 6.50×10^{-12}
3) 5.65×10^{-13} 4) 5.65×10^{-12}

104. At 25ºC, the $[H^+]$ of a solution is 2×10^{-9}M. The nature of the solution is

1) Neutral 2) Acidic
3) Basic 4) Can not be predicted

105. NH_4OH is a weak base but it becomes still weaker in the aqueous solution of:

1) 0.1 M HCl 2) 0.1 M NH_4Cl
3) 0.1 M H_2SO_4 4) 0.1 M CH_3COOH

106. Which pair will show common ion effect?

1) $BaCl_2 + Ba(NO_3)_2$ 2) NaCl + HCl
3) $NH_4OH + NH_4Cl$ 4) AgCN + KCN

107. 30 cc of $\frac{M}{3}$ HCl, 20 cc of $\frac{M}{2}$ HNO_3 & 40 cc $\frac{M}{4}$ NaOH solution are mixed and the volume was made upto 1 dm^3. The pH of the resulting solution is:

1) 2 2) 1
3) 3 4) 8

108. The product of molar concentrations of hydrogen ions and hydroxide ions in a 0.01 M aqueous solution of sodium chloride is known as:

1) Hydrolysis constant of salt
2) Dissociation constant of acid
3) Dissociation constant of base
4) Ionic product of water

109. A solution of pH 9.0 is 1000 times as basic as a solution. The pH of this solution is:

1) 9 2) 8
3) 7 4) 6

110. pH of HCl (conc. 0.1 mol/L) is one. Compute the pH of 0.05 mol/L solution of H_2SO_4.

1) 0.05 2) 0.5
3) 1 4) 2

Buffer Solutions

111. Which of the following pairs constitutes a buffer?

1) HNO_2 and $NaNO_2$ 2) NaOH and NaCl
3) HNO_3 and NH_4NO_3 4) HCl and KCl

112. A buffer solution is prepared in which the concentration of NH_3 is 0.30 M and the concentration of NH_4^+ is 0.20 M. If K_b for NH_3 equals 1.8×10^{-5}. What is the pH of this solution?

1) 8.73 2) 9.08
3) 9.44 4) 11.72

113. The pK_b value of NH_3 is 5. Calculate the pH of the buffer solution, 1 L of which contains 0.01 M NH_4Cl and 0.10 M NH_4OH:

1) 4 2) 6
3) 8 4) 10

114. Which of the following combinations will constitute buffer solution?

1) $CuSO_4/H_2SO_4$
2) CH_3COOH/CH_3COONH_4
3) CH_3COOH/CH_3COONa
4) NaCl/NaOH

115. Which of the following is not a buffer solution?

1) CH_3COOH/CH_3COONa 2) HCl/NaCl
3) HCOOH/HCOONa 4) NH_4OH / NH_4Cl

116. A buffer solution is used in:

1) Preparation of potash alum
2) The removal of PO_4^{3-} ions
3) Increasing the pH value of a solution
4) Precipitation of $Cr(OH)_3$ from $CrCl_3$

117. A certain buffer solution contains equal concentration of X^- and HX. The K_a for HX is 10^{-8}. The pH of buffer is:

1) 3 2) 8
3) 11 4) 14

118. A buffer solution contains 100 mL of 0.01 M CH_3COOH and 200 mL of 0.02 M CH_3COONa. 700 ml of water is added, pH before and after dilution are: ($pK_a = 4.74$)

1) 5.04, 5.04 2) 5.04, 0.504
3) 5.04, 1.54 4) 5.34, 5.34

119. Which can act as buffer?

1) $NH_4Cl + HCl$
2) $CH_3COOH + H_2CO_3$
3) 40ml of 0.1 M NaCN+20ml of 0.1M HCN
4) NaCl+NaOH

120. The mixed solution of phthalic acid and potassium hydrogen phthalate is

1) Basic buffer 2) Acid buffer
3) Not buffer 4) An acid

121. Which of the following mixture in aqueous solution of equimolar concentration acts as a buffer solution?

1) HNO_3 + NaOH
2) H_2SO_4 + KOH
3) NH_4OH (excess) + HCl
4) CH_3COOH +NaOH (large excess)

122. For an acid buffer solution, the pH is 3. The pH can be increased by:

1) Increasing the concentration of salt
2) Increasing the concentration of acid
3) Decreasing the concentration of salt
4) Independent of concentration of acid & salt

Solubility Equilibria of Sparingly Soluble Salts

123. Which of the following metal sulphide has maximum solubility in water?

1) CdS ($K_{sp} = 36 \times 10^{-30}$)
2) FeS ($K_{sp} = 11 \times 10^{-20}$)
3) HgS ($K_{sp} = 32 \times 10^{-54}$)
4) ZnS ($K_{sp} = 11 \times 10^{-22}$)

124. If the solubility of Ag_2CrO_4 is S mole / litre. Its solubility product is:

1) S^2 2) S^3
3) $4S^3$ 4) $2S^3$

125. Solubility product constants (K_{sp}) of salt MX, MX_2, M_3X at temperature "T" are 4.0×10^{-8}, 3.2×10^{-14}, and 2.7×10^{-15} respectively. Solubility (mol dm^{-3}) of the salts at temperature "T" are in the order:

1) $MX > MX_2 > M_3X$
2) $M_3X > MX_2 > MX$
3) $MX_2 > M_3X > MX$
4) $MX > M_3X > MX_2$

126. Solubility product of a sparingly soluble salt AX_2 is 3.2×10^{-11}. Its solubility in mol/litre is:

1) 5.6×10^{-6} 2) 3.1×10^{-4}
3) 2×10^{-4} 4) 4×10^{-4}

127. K_{sp} of salt A_3B_2, for solubility x mol L^{-1}, is:

1) 36x 2) $72x^6$
3) $108x^5$ 4) $108x^6$

128. For the electrolyte of type, A_2B, K_{sp} is given then its solubility is calculated by

1) $K_{sp}/4$ 2) $\sqrt[3]{\frac{K_{sp}}{4}}$
3) $\sqrt[3]{K_{sp}}$ 4) $\sqrt{\frac{K_{sp}}{4}}$

129. The solubility of CaF_2 is 2×10^{-4} mole/litre. Its solubility product is

1) $2\times8\times10^{-12}$
2) 4.0×10^{-8}
3) $4\times8.0\times10^{-12}$
4) 3.2×10^{-4}

130. The solubility of AgCl in 0.1 M NaCl is (K_{sp} of AgCl $= 1.2\times10^{-10}$)

1) 0.05 2) 1.2×10^{-6}
3) 2×10^{-5} 4) 1.2×10^{-9}

High Order Time Intensive MCQs

1. For a reversible reaction

$A \underset{K_2}{\overset{K_1}{\rightleftharpoons}} B$, the initial molar concentration of A & B are aM and bM, respectively. If the equilibrium concentration are (a – x) & (b + x) respectively, express x in terms of K_1, K_2, a, b.

1) $\frac{K_1a - K_2b}{K_1 + K_2}$ 2) $\frac{K_1a - K_2b}{K_1 - K_2}$
3) $\frac{K_1a - K_2b}{K_1K_2}$ 4) $\frac{K_1a + K_2b}{K_1 + K_2}$

2. 40% of a mixture of 0.2 moles of N_2 & 0.6 moles of H_2 react to give NH_3 according to the equation:

$$N_2(g) + 3H_2(g) \rightleftharpoons 2NH_3(g)$$

at constant T & P. Then, ratio of final volume to initial volume of gases is:

1) 4 : 5
2) 5 : 4
3) 7 : 10
4) 8 : 5

3. For the reaction,

$$\underset{(g)}{2NOBr} \rightleftharpoons \underset{(g)}{2NO} + \underset{(g)}{Br_2}$$

The ratio $\frac{K_P}{P}$, where P is the total pressure of gases at equilibrium and $P_{Br_2} = P/9$ at a certain temperature is:

1) 1/9 2) 1/81
3) 1/27 4) 1/3

4. In the reaction, $N_2(g) + 3H_2(g) \rightleftharpoons 2NH_3(g)$. One mole of N_2 reacts with 3 moles of H_2. If at equilibrium 'x' moles of N_2 combined, then the value of n in terms of K_p and total equilibrium pressure P, is:

1) $\frac{3P\sqrt{3K_p}}{8}$ 2) $\frac{8P}{3\sqrt{3K_p}}$
3) $\frac{3\sqrt{3K_p}}{8P}$ 4) $\frac{8}{P}\sqrt{\frac{K_p}{3}}$

5. The reaction, A(g) + B(g) $\rightleftharpoons$ C(g) + D(g) occurs in a single step. The rate constant of forward reaction is 2×10^{-3} mol^{-1} L sec^{-1}. When the reaction is started with equimolar amounts of A & B, it is found that the concentration of A is twice that of C at equilibrium. The rate constant of backward reaction is:

1) 5×10^{-4} mol^{-1} L sec^{-1} 2) 8×10^{-3} mol^{-1} L sec^{-1}
3) 1.25×10^{2} mol^{-1} L sec^{-1} 4) 2×10^{3} mol^{-1} L sec^{-1}

6. In the equilibrium mixture : $\underset{(g)}{H_2} + \underset{(g)}{I_2} \rightleftharpoons \underset{(g)}{2HI}$

the mole ratio of gases are $\sqrt{2} : \sqrt{2} : 10$, respectively, what would be the effect on the mole ratio on adding 5 mole of He gas at constant V?

1) No change
2) New molar ratio is 1 : 1 : 5
3) New molar ratio is 2 : 2 : 5
4) None of these

7. In a reaction,

$AB(g) \rightleftharpoons A(g) + B(g)$

the value of K_p is numerically equal to 1/3rd of total equilibrium pressure. What is the percentage dissociation of AB?

1) 25% 2) 30%
3) 45% 4) 50%

8. If pressure of N_2/H_2 mixture in closed apparatus is 100 atm, 20% of the mixture then reacts, the total pressure at the same temperature is

1) 60 atm 2) 30 atm
3) 90 atm 4) 45 atm

9. Log $\frac{K_p}{K_c}$ + log RT = 0, This relationship is valid for which reaction?

1) $\underset{(g)}{PCl_5} \rightleftharpoons \underset{(g)}{PCl_3} + \underset{(g)}{Cl_2}$
2) $\underset{(g)}{2SO_2} + \underset{(g)}{O_2} \rightleftharpoons \underset{(g)}{2SO_3}$
3) $\underset{(g)}{H_2} + \underset{(g)}{I_2} \rightleftharpoons \underset{(g)}{2HI}$
4) $\underset{(g)}{N_2} + \underset{(g)}{3H_2} \rightleftharpoons \underset{(g)}{2NH_3}$

10. Variation of equilibrium constant K_c with temperature is given by:

1) Arrhenius equation
2) Clausius clapeyron equation
3) Van't Hoff equation
4) Gibbs equation

11. At 600 K & 200 bar, a 1 : 3 (molar ratio) mixture of A_2 & B_2 react to form an equilibrium mixture containing $\chi_{AB_3} = 0.60$. Assuming ideal gas behaviour, calculate K_p for the reaction:

$A_2(g) + 3B_2(g) \rightleftharpoons 2AB_3(g)$

1) 3.33×10^{-3} 2) 3.33×10^{-11}
3) 5.61×10^{-7} 4) 9.87×10^{-12}

12. The partial pressure of $CH_3OH(g)$, CO(g) & $H_2(g)$ in equilibrium mixture for the reaction, $CO(g) + 2H_2(g) \rightleftharpoons CH_3OH(g)$ are 2, 1 & 0.1 atm respectively at 427° C. The value of K_p for the decomposition of CH_3OH to CO & H_2 is-

1) 10^2 atm 2) 2×10^2 atm^{-1}
3) 5×10^{-3} atm^2 4) 5×10^{-3} atm^4

13. Following reaction is given

$CH_3COCH_3(g) \rightleftharpoons CH_3 - CH_3 + CO(g)$, initial pressure of CH_3COCH_3 is 100 mm of Hg. When equilibrium is set up, the mole fraction of $CO_{(g)}$ is 1/3, hence K_p is-

1) 10 mm 2) 50 mm
3) 25 mm 4) 150 mm

14. For the reaction $CaCO_3(s) \rightleftharpoons CaO(s) + CO_2(g)$ K_p = 1.16 atm at 800° C. If 20 g of $CaCO_3$ were kept in a 10 litre container & heated upto 800°C, what % age by weight of $CaCO_3$ would remain unreacted at equilibrium?

1) 35% 2) 6.8%
3) 10.2% 4) None of those

15. Two moles of HI were heated in a sealed tube at 440°C till the given equilibrium was reached. HI was found to be 20% decomposed. The equilibrium constant for dissociation is:

2HI (g) $\rightleftharpoons$ H_2 (g) + I_2 (g)

1) $\frac{1}{16}$ 2) $\frac{1}{32}$
3) $\frac{1}{64}$ 4) $\frac{1}{128}$

16. K_p for the reaction $2BaO_2(s) \rightleftharpoons 2BaO(s) + O_2(g)$ is 1.6×10^{-4} atm at 400°C. Heat of reaction is –25.14 kcal. What will be the number of moles of O_2 gas produced at 500°C temperature. The reaction is carried in 2 litre vessel.
   1) 0.022×10^{-10}
   2) 10^{-8}
   3) 10^{-24}
   4) 0.046×10^{-5}

17. Consider the following reactions;
   (i) $PCl_5(g) \rightleftharpoons PCl_3(g) + Cl_2(g)$
   (ii) $2HI(g) \rightleftharpoons H_2(g) + I_2(g)$
   The addition of an inert gas at constant volume:
   1) Will increase the dissociation of PCl_5 as well as HI.
   2) Will reduce the dissociation of PCl_5 and not affect the dissociation of HI.
   3) Will increase the dissociation of PCl_5 and not affect the dissociation of HI.
   4) Will not disturb the equilibrium of the reactions.

18. The thermal dissociation equilibrium of $NH_4HS(s)$ is studied under different conditions.
   $NH_4HS(s) \rightleftharpoons NH_3(g) + H_2S(g)$
   For this equilibrium, the correct statement(s) is (are) :
   1) K depends on temperature
   2) K is independent of the initial amount of $NH_4HS(s)$
   3) If pressure of $NH_3(g)$ is increased at equilibrium then pressure of $H_2S(g)$ decreases at given temperature.
   4) All option are correct

19. 200 mL of M/5 dibasic acid is mixed with 150 mL of M/2 monoacid base. The pH of the resulting solution is likely to be
   1) More than 7
   2) Less than 7
   3) Equal to 7
   4) Uncertain

20. The value of equilibrium constant of a reaction changes with change of temperature and the change is given by van't Hoff equation, $\frac{d\ln K_p}{dT} = \frac{\Delta H^\circ}{dT} = \frac{\Delta H^\circ}{RT^2}$ where enthalpy change, ΔH° is taken as constant in the small temperature range.
   If for reaction, $A_{(g)} + 3B_{(g)} \rightleftharpoons 2C_{(g)}$, a plot of ln K_{eq} versus for 1/T a reaction is shown, then which of the following condition will be favourable for formation of product C?

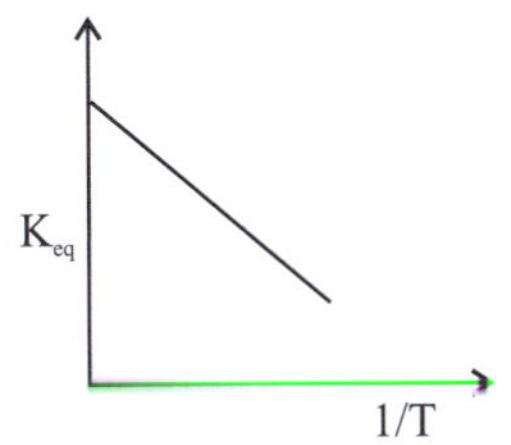

   1) Low temperature and high pressure
   2) High temperature and high pressure
   3) High temperature and low pressure
   4) Low temperature and low pressure.

21. The value of ΔG° for the phosphorylation of glucose in glycolysis is 13.8 kJ/mol. What would be the value of K_C at 298 K for the above process?
   1) $e^{-55.69}$
   2) $e^{-5.569}$
   3) $e^{-0.5569}$
   4) $e^{-556.9}$

22. What is the expression of K_p for the following equilibrium $2NH_3(g) \rightleftharpoons N_2(g) + 3H_2(g)$ if initially 'a' mole of NH_3 is present and at equilibrium 'x' mole of it decomposed and total equilibrium pressure is p?
   1) $\frac{27P^2x^4}{16(a-x)^2(a+x)^2}$
   2) $\frac{9Px^2}{8(a-x)(a+x)}$
   3) $\frac{3Px^2}{4(a-x)^2}$
   4) $\frac{3P^2x^2}{8(a^2-x^2)^2}$

23. A saturated solution of $Ca(RCOO)_2$ contains 0.6 mole in 2 litre solution pOH of the solution is 5.60. If 90% dissociation of the salt takes place then what is the pK_a of the carboxylic acid?
   1) 2.8 – log (0.54)
   2) 2.8 + log (0.54)
   3) 2.8 – log (0.27)
   4) None of these

24. What is the degree of dissociation of weak acid HA having concentration 0.1 M in the presence of strong acid HB having concentration 0.1 M?
   Given K_a for weak acid = 10^{-6}
   1) 10^{-5}
   2) 10^{-6}
   3) 10^{-4}
   4) 10^{-3}

25. The ratio of pH of a solution containing 1 mole of CH_3COONa + 1 mole of HCl per litre and other solution containing 1 mole of CH_3COONa + 1 mole of CH_3COOH per litre will be:
   1) 1 : 2
   2) 2 : 1
   3) 1 : 3
   4) 3 : 1

26. To 1 litre solution containing 0.1 mole each of NH_3 & NH_4Cl, 0.05 moles of NaOH are added, ΔpH will be:
   (pK_b for NH_3 = 4.74)
   1) 0.92
   2) 0.96
   3) 1.2
   4) 0.47

27. If pH of a saturated solution of $Ba(OH)_2$ is 12, the value of its K_{sp} is:
   1) 5×10^{-7} M^3
   2) 4×10^{-6} M^3
   3) 4×10^{-7} M^3
   4) 5×10^{-6} M^3

28. How many litres of water must be added to 1 litre of an aqueous solution of HCl with pH = 1 to create an aqueous solution with pH = 2?
   1) 9 L
   2) 0.1 L
   3) 0.9 L
   4) 2.0 L

29. The pH of solution produced by mixing equal volumes of 2×10^{-3} M $HClO_4$ and 1×10^{-2} M $KClO_4$ is:
   1) 2.7
   2) 2.3
   3) 3.0
   4) 1.0

30. The solubility of $Pb(OH)_2$ in water is 6×10^{-6} M. The solubility of $Pb(OH)_2$ in a buffer solution of pH = 8 is:
   1) 8.64 M
   2) 2.16×10^{-16} M
   3) 8.64×10^{-16} M
   4) 864×10^{-6} M

31. The solubility of sparingly soluble salt A_3B_2 (having molar mass = Mg/mol) in water is X g/L. The ratio of molar concentration of B^{3-} to the K_{sp} of the salt is:

1) $\frac{108X^5}{M^5}$ 2) $\frac{X^4}{108M^4}$ 3) $\frac{X^4}{54M^4}$ 4) $\frac{M^4}{54X^4}$

32. Which of the following will have the highest pH in water?

1) KCl 2) NaCl
3) Na_2CO_3 4) $CuSO_4$

33. The minimum concentration of silver ions that is required to start the precipitation of Ag_2S ($K_{sp} = 10^{-51}$) in a 0.1 M solution of S^{2-} is:

1) 10^{-49} M 2) 10^{-50} M 3) 10^{-26} M 4) 10^{-25} M

34. Select the correct statement –

1) If $[H^+] = y \times 10^{-x}$ then pH = x – log y
2) If $[H^+] = \frac{1}{y} \times 10^{-x}$ then pH = x + log y
3) pH of a solution = 14 + log $[OH^-]$
4) All of the above

35. In a saturated solution of the sparingly soluble strong electrolyte $AgIO_3$ (molecular mass = 283), the equilibrium which sets in is -

$AgIO_3(s) \rightleftharpoons Ag^+(aq) + IO_3^-(aq)$

If the solubility product K_{sp} of $AgIO_3$ at a given temperature is 1.0×10^{-8}, what is the mass of $AgIO_3$ contained in 100 mL of its saturated solution ?

1) 28.3×10^{-2} g 2) 2.83×10^{-3} g
3) 1.0×10^{-7} g 4) 1.0×10^{-4} g

36. Which of the following salts has maximum solubility –

1) HgS, $K_{sp} = 1.6 \times 10^{-54}$
2) $PbSO_4$, $K_{sp} = 1.3 \times 10^{-8}$
3) ZnS, $K_{sp} = 7.0 \times 10^{-26}$
4) AgCl, $K_{sp} = 1.7 \times 10^{-10}$

37. The extent of dissociation of PCl_5 at a certain temperature is 20 % at one atm pressure. Calculate the pressure at which this substance is half dissociated at the same temperature.

1) 0.123 2) 0.246 3) 0.826 4) 0.111

38. At 25°C, what will be the solubility of silver carbonate in 0.1 M Na_2CO_3 solution. At this temperature, K_{sp} of silver carbonate is 4×10^{-13}.

1) 2×10^{-7} 2) 2×10^{-6}
3) 10^{-6} 4) 10^{-7}

39. Solid $Ba(NO_3)_2$ is gradually dissolved in a 1.0×10^{-4} M Na_2CO_3 solution. At what concentration of Ba^{2+} will a precipitate begin to form?
(K_{sp} for $BaCO_3 = 5.1 \times 10^{-9}$)

1) 5.1×10^{-5} M 2) 8.1×10^{-8} M
3) 8.1×10^{-7} M 4) 4.1×10^{-5} M

40. At 25°C, the solubility product of $Mg(OH)_2$ is 1.0×10^{-11}. At which pH, will Mg^{2+} ions start precipitating in the form of solution of 0.001 M Mg^{2+} ions?

1) 8 2) 9 3) 10 4) 11

41. When equal volumes of the following solutions are mixed, precipitation of CaF_2 ($K_{sp} = 1.7 \times 10^{-10}$) will occur only with –

1) 10^{-4} M Ca^{2+} and 10^{-4} M F^-
2) 10^{-2} M Ca^{2+} and 10^{-3} M F^-
3) 10^{-5} M Ca^{2+} and 10^{-3} M F^-
4) 10^{-3} M Ca^{2+} and 10^{-5} M F^-

42. For the given reaction, choose the correct expression of K_C from the following:

$Fe^{3+}_{(aq)} + SCN^-_{(aq)} \rightleftharpoons (FeSCN)^{2+}_{(aq)}$

1) $K_C = \frac{[(FeSCN)^{2+}]}{[Fe^{3+}][SCN^-]}$ 2) $K_C = \frac{[Fe^{3+}][SCN^-]}{[FeSCN^{2+}]}$

3) $K_C = \frac{[FeSCN^{2+}]}{[Fe^{3+}]^2[SCN^-]^2}$ 4) $K_C = \frac{[FeSCN^{2+}]^2}{[Fe^{3+}][SCN^-]}$

43. Two buffers (X) & (Y) of pH 4 & 6 respectively are prepared from acid HA & salt NaA. Both the buffers are 0.5 M in HA. What would be the pH of the solution if equal volumes (1 L each) of both are mixed? (K_a of HA = 10^{-4})

1) 5.703 2) 7.50 3) 3.50 4) 2.77

44. pOH = 7 – 0.5 pK_a + 0.5 pK_b is true for aqueous solution containing which pair of cation and anion?

1) NH_4^+, F^- 2) $C_6H_5NH_3^+$, CH_3COO^-
3) Both (1) and (2) 4) None of these

45. To 100 mL of an aqueous solution of 0.1M CH_3COOH ($K_a = 2 \times 10^{-5}$), 0.01 mol of HCl(g) is passed. Select correct options regarding the resulting solution.

1) Degree of dissociation of acetic acid in resulting solution is 10^{-4}
2) pH of resulting solution is nearly 3
3) Degree of dissociation of water in resulting solution is 1.8×10^{-15}
4) Concentration of OH^- ions contributed by water in resulting solution is 10^{-7} M.

46. If K_w for H_2O at 90°C is 10^{-12} M^2, then what would be the dissociation constant of water at the same temperature? [Assume d_{H_2O} = 1 g/mL]

1) 10^{-14} 2) 1.8×10^{-16}
3) 1.8×10^{-14} 4) 10^{-12}

47. Which of the following should have maximum pH value?

1) 1 M HCl solution 2) Human blood
3) Milk of magnesia 4) Human saliva

48. In which of the following equilibrium reactions, the equilibrium would shift to the right side, if total pressure is decreased:

1) $N_2 + 3H_2 \rightleftharpoons 2NH_3$ 2) $H_2 + I_2 \rightleftharpoons 2HI$
3) $N_2O_4 \rightleftharpoons 2NO_2$ 4) $H_2 + Cl_2 \rightleftharpoons 2HCl$

49. Manufacture of ammonia by Haber's process involves the reaction $N_2 + 3H_2 \rightleftharpoons 2NH_3$; ΔH = –22.4 kcals. The effect of increase of temperature on the equilibrium is

1) Equilibrium is shifted to the right
2) Equilibrium is unaffected
3) Equilibrium is shifted to the left
4) Equilibrium is shifted first to right then to left

50. What should be the ionic product of $CaSO_4$, whose $K_{sp} = 2.4 \times 10^{-5}$, when 150 mL of 0.0008 M ammonium sulphate solution is mixed with 50 mL of 0.04 M calcium nitrate solution?

1) $< K_{sp}$ 2) $> K_{sp}$
3) $\approx K_{sp}$ 4) None of the above

51. Which may be added to one litre of water to act as a buffer?

1) One mole of CH_3COOH and one mole of HCl
2) One mole of NH_4OH and one mole of NaOH
3) One mole of NH_4Cl and one mole of HCl
4) One mole of CH_3COOH and 0.5 mole of NaOH

52. (I) $H_2O(l) \rightleftharpoons H_2O(g)$ (II) $I_2(s) \rightleftharpoons I_2(vap)$
(III) $H_2O(l) \rightleftharpoons H_2O(s)$ (IV) $CO_2(g) \rightleftharpoons CO_2(l)$
Rise of Temperature shifts equilibrium towards right in the case of

1) I & IV   2) II, III & IV
3) I & II   4) I, II & III

53. For a dibasic acid,

$H_2A \rightleftharpoons HA^- + H^+ \ (K_1)$
$HA^- \rightleftharpoons A^{2-} + H^+ \ (K_2)$
$H_2A \rightleftharpoons 2H^+ + A^{2-} \ (K)$

then

1) $K = K_1 + K_2$ 2) $K = K_1 - K_2$
3) $K = K_1/K_2$ 4) $K = K_1.K_2$

54. Among the following which one is wrong?

1) Degree of dissociation of a weak electrolyte increases with dilution.
2) Increase in temperature increases the ionisation
3) Strong electrolytes are ionised completely even at moderate concentrations.
4) Addition of NH_4Cl to NH_4OH increases the ionisation of the latter.

55. The correct statements

A) pH of water decreases with increase in temperature
B) pH of water remains same with increase in temperature
C) pH of water decreases with the addition of acid
D) Degree of dissociation of water is independent of temperature

1) All are correct   2) A and C are correct
3) B, C and D are correct   4) D is correct

56. (A) Value of K_w is independent of temperature
(B) pH of the solution cannot be zero.
(C) pOH = 14–pH is always correct at 90° C

1) All are correct   2) B only is correct
3) Only C wrong   4) All are wrong

57. In the third group of qualitative analysis, the precipitating reagent is $NH_4Cl + NH_4OH$. The function of NH_4Cl is to

1) Increase the ionization of NH_4OH
2) Supress the ionization of NH_4OH
3) Stabilise the hydroxides of group cations
4) Convert the ions of group third into their respective chlorides.

58. The dissociation of $CaCO_3$ takes place as per the equation
$CaCO_3(s) \rightarrow CaO(s) + CO_2(g); = 110$ kJ.
If the reaction is carried out in a closed vessel, the pressure of CO_2 increases and reaches a constant value when

1) Temperature is increased
2) Temperature is decreased
3) Volume of vessel is increased
4) Amount of $CaCO_3$ is decreased

59. Le chatelier's principle is not applicable to which of the following ?

1) $2SO_2(g) + O_2(g) \rightleftharpoons 2SO_3(g)$
2) $Fe(s) + S(s) \rightleftharpoons FeS(s)$
3) $N_2(g) + 3H_2(g) \rightleftharpoons 2NH_3(g)$
4) $N_2(g) + O_2(g) \rightleftharpoons 2NO(g)$

60. If K_1 and K_2 are the respective equilibrium constants for the two reactions

$XeF_6(g) + H_2O(g) \rightleftharpoons XeOF_4(g) + 2HF(g)$
$XeO_4(g) + XeF_6(g) \rightleftharpoons XeOF_4(g) + XeO_3F_2(g)$

the equilibrium constant of the reaction
$XeO_4(g) + 2HF(g) \rightleftharpoons XeO_3F_2(g) + H_2O(g)$ will be

1) $K_1 K_2$ 2) K_2 / K_1
3) $K_1 / (K_2)^2$ 4) K_1 / K_2

61. If the equilibrium constant for the reaction $2AB \rightleftharpoons A_2 + B_2$ is 49. What is the value of equilibrium constant for $AB \rightleftharpoons \frac{1}{2}A_2 + \frac{1}{2}B_2$?

1) 49   2) 4   3) 7   4) 0.02

62. Solubility product of $Ba(OH)_2$ and $Al(OH)_3$ are 1.8×10^{-10} and 2.4×10^{-20} respectively. If both Al^{+3} and Ba^{+2} ions are present in a solution, which one will be precipitated on addition of Ammonium hydroxide solution?

1) $Ba(OH)_2$
2) $Al(OH)_3$
3) Both are precipitated at same time
4) Both are not precipitated

63. The correct order of increasing solubility of AgCl in (A) water (B) 0.1 M NaCl (C) 0.1 $BaCl_2$(D) 0.1M NH_3 is

1) D > A > B > C   2) D > C > B > A
3) B > A > D > C   4) A > D > B > C

64. 2 g of NaOH per 250 ml of solution is added to a buffer solution of buffer capacity 0.2. Then the change in pH is

1) 0.5   2) 1   3) 1.5   4) 2.0

65. The pH of 0.1 M K_2SO_4 is 7. This solution is diluted by 10 times. Then the pH of resulting solution is

1) 7   2) 6   3) 8   4) 14

66. Sulphuric acid ionizes in two steps with their ionization constant values K_{a_1} and K_{a_2}, respectively, while K is the overall ionization constant. Which of the following statements are true?

A. H_2SO_4 is a weaker acid than HSO_4^-

B. $\log(K) = \log K_{a_1} + \log K_{a_2}$

C. $K_{a_1} = K_{a_2} / 2$

D. K_{a_1} is greater than K_{a_2}

Choose the correct answer from the options given below:

1) A and B only 2) B and D only

3) B, C and D only 4) A, B and C only

67. Consider the equilibrium reaction: $C(g) \rightleftharpoons 3D(g)$

At a temperature of 1200 K, the K_b = is $5000 \times K_f$. [Given: R = 0.0831 L atm mol^{-1} K^{-1}] K_p for the reaction at 1200K is

1) 0.003 2) 1.5

3) 2.4 4) 1.98

Match the Columns MCQs

1. Match the following equilibria with the corresponding condition

	Column-I		Column-II
(a)	Liquid $\rightleftharpoons$ Vapour	(i)	Saturated solution
(b)	Solid $\rightleftharpoons$ Liquid	(ii)	Boiling point
(c)	Solid $\rightleftharpoons$ Vapour	(iii)	Sublimation point
(d)	Solute(s) $\rightleftharpoons$ Solute (solution)	(iv)	Melting point
		(v)	Unsaturated Solution

1) (a)-(iv) (b)-(i) (c)-(iii) (d)-(ii)

2) (a)-(ii) (b)-(iii) (c)-(i) (d)-(iv)

3) (a)-(ii) (b)-(iv) (c)-(iii) (d)-(i)

4) (a)-(i) (b)-(iv) (c)-(ii) (d)-(iii)

2. For the reaction : $N_2(g) + 3H_2(g) \rightleftharpoons 2NH_3(g)$

Equilibrium constant $K_c = \dfrac{[NH_3]^2}{[N_2][H_2]^3}$

Some reactions are written below in Column-I and their equilibrium constants in terms of K_c are written in Column-II. Match the following reactions with the corresponding equilibrium constant

	Column-I (Reaction)		Column-II (Equilibrium constant)
(a)	$2N_2(g) + 6H_2(g) \rightleftharpoons 4NH_3(g)$	(i)	$2K_c$
(b)	$2NH_3(g) \rightleftharpoons N_2(g) + 3H_2(g)$	(ii)	$K_c^{\frac{1}{2}}$
(c)	$\frac{1}{2}N_2(g) + \frac{3}{2}H_2(g) \rightleftharpoons NH_3(g)$	(iii)	$\frac{1}{K_c}$
		(iv)	K_c^2

1) (a)-(i) (b)-(ii) (c)-(iv)

2) (a)-(iv) (b)-(iii) (c)-(ii)

3) (a)-(ii) (b)-(i) (c)-(iii)

4) (a)-(iv) (b)-(ii) (c)-(iii)

3. Match standard free energy of the reaction with the corresponding equilibrium constant

	Column-I		Column-II
(a)	$\Delta G^\ominus > 0$	(i)	$K > 1$
(b)	$\Delta G^\ominus < 0$	(ii)	$K = 1$
(c)	$\Delta G^\ominus = 0$	(iii)	$K = 0$
		(iv)	$K < 1$

1) (a)-(i) (b)-(ii) (c)-(iv)

2) (a)-(iv) (b)-(iii) (c)-(ii)

3) (a)-(ii) (b)-(i) (c)-(iii)

4) (a)-(iv) (b)-(i) (c)-(ii)

4. Match the following species with the corresponding conjugate acid

	Species		Conjugate acid
(a)	NH_3	(i)	CO_3^{2-}
(b)	HCO_3^-	(ii)	NH_4^+
(c)	H_2O	(iii)	H_3O^+
(d)	HSO_4^-	(iv)	H_2SO_4
		(v)	H_2CO_3

1) (a)-(v) (b)-(iii) (c)-(ii) (d)-(i)

2) (a)-(ii) (b)-(v) (c)-(iii) (d)-(iv)

3) (a)-(ii) (b)-(iv) (c)-(v) (d)-(i)

4) (a)-(i) (b)-(v) (c)-(ii) (d)-(iii)

5. Match the following graphical variation with their description

	Column-I		Column-II
(a)	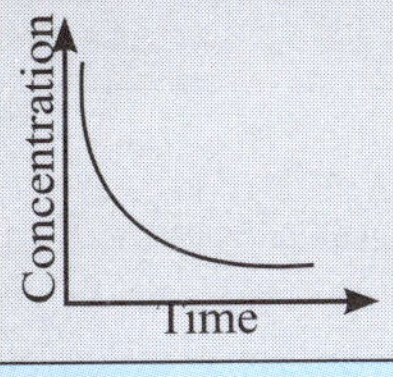	(i)	Variation in product concentration with time
(b)	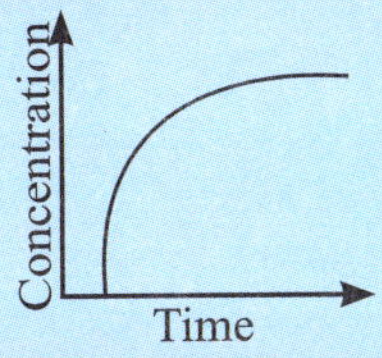	(ii)	Reaction at equilibrium
(c)	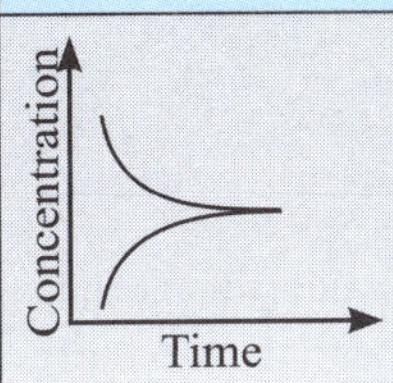	(iii)	Variation in reactant concentration with time

1) (a)-(iii) (b)-(i) (c)-(ii)
2) (a)-(i) (b)-(iii) (c)-(ii)
3) (a)-(ii) (b)-(i) (c)-(iii)
4) (a)-(ii) (b)-(iii) (c)-(i)

6. Match the Column-I with Column-II.

	Column-I		Column-II
(a)	$N_2(g) + 3H_2(g) \rightleftharpoons 2NH_3(g)$	(i)	$\Delta n_g > 0$
(b)	$PCl_5(g) \rightleftharpoons PCl_3(g) + Cl_2(g)$	(ii)	$K_p < K_c$
(c)	$2SO_3(g) \rightleftharpoons 2SO_2(g) + O_2(g)$	(iii)	K_p not defined
(d)	$CH_3COOC_2H_5(l) + H_2O(l) \rightleftharpoons CH_3COOH(l) + C_2H_5OH(l)$ (assume temperature = 25°C)	(iv)	$K_p > K_c$

1) (a)-(ii); (b)-(i,iv); (c)-(i,iv); (d)-(iii)
2) (a)-(i,iv); (b)-(ii); (c)-(i,iv); (d)-(iii)
3) (a)-(i,iv); (b)-(iii); (c)-(i); (d)-(i,iv)
4) (a)-(iii); (b)-(i,iv); (c)-(ii); (d)-(i,iv)

Correct & Incorrect MCQs

1. Which of the following statement (s) is/are correct?
   A. At equilibrium the pressure exerted by the water molecules at a given temperature remains constant and is called the equilibrium vapour pressure of water.
   B. Vapour pressure of water increases with increase in temperature.
   C. Volatile liquids have very low vapour pressure than other liquid at the same temperature.
   D. More volatile liquid has lower boiling point.
   1) A, B and D 2) A and D
   3) A, B and C 4) Only C

2. Which of the following statement is incorrect?
   1) When a soda water bottle is opened some of the carbon dioxide gas dissolved in it fizzes out rapidly.
   2) Solubility of carbon dioxide in water is same at different pressures.
   3) The soda water bottle is sealed under pressure of gas when its solubility in water is high.
   4) Both (1) and (2)

3. Correct statement(s) is/are :
   A. The equilibrium equation is also known as the law of mass action.
   B. Equilibrium constant for the reverse reaction is the inverse of the equilibrium constant for the reaction in the forward direction.
   C. All the reactants and products are in the same phase in a homogeneous system.
   D. The equilibrium between water vapour and liquid water in a closed container is an example of homogeneous equilibrium.
   1) A, B and C 2) B, C and D
   3) A and D 4) Only D

4. Which of the following statement is not correct about equilibrium constant?
   1) Expression for equilibrium constant is applicable only when concentrations of the reactants and products have attained constant value at equilibrium state.
   2) The value of equilibrium constant is independent of initial concentrations of the reactants and products.
   3) Equilibrium constant helps to predict the extent of a reaction on the basis of its magnitude.
   4) Equilibrium constant is independent on temperature.

5. Incorrect statement is/are
   A. If $K_c > 10^3$, products predominate over reactants
   B. If $K_c < 10^{-3}$, reactants predominate over products.
   C. If $K_c < 10^{-3}$, products predominate over reactants.
   D. If K_c is in the range of 10^{-3} to 10^3, appreciable concentrations of both reactants and products are present.
   1) A and C 2) B, C and D
   3) Only C 4) B and C

6. Which amongst the following is incorrect statement ?
   A. Net reaction goes from left to right if $Q_c < K_c$.
   B. Net reaction goes from right to left if $Q_c > K_c$.
   C. Net reaction goes from left to right if $Q_c > K_c$.
   D. Net reaction goes from right to left if $Q_c < K_c$.
   1) A, B and C 2) C and D
   3) A, C and D 4) B, C and D

7. Select the incorrect statement :
   1) The conjugate acid of $HCOO^-$ is HCOOH.
   2) The conjugate base of HCO_3^- is CO_3^{2-}.
   3) The conjugate base of H_2O is H_3O^+.
   4) The conjugate acid of NH_3 is NH_4^+.

8. Identify the incorrect statement :
   1) At constant volume when argon (Ar) is added to the reaction mixture, the equilibrium remains undisturbed.
   2) Addition of an inert gas at constant volume does not change the partial pressure or the molar concentration of substance involved in the reaction.
   3) The equilibrium constant for an exothermic reaction increases as the temperature increases.
   4) The equilibrium constant for an endothermic reaction increases as the temperature increases.

9. Which of the following statement is not correct ?
   1) BCl_3 acts as a lewis acid.
   2) $AlCl_3$ acts as a lewis acid.
   3) H_2O acts as a lewis base.
   4) NH_3 acts as a Lewis acid.

10. Select the correct statement :
   1) K_p is always greater than K_c.
   2) $K_p = K_c (RT)^{\Delta n}$ depends on the number of moles of gaseous reactant & products.
   3) Equilibrium constant does not change on changing the temperature.
   4) Both (1) and (3)

Statement Based MCQs

Directions: These questions consist of two statements each, printed as Statement-I and Statement-II. While answering these questions, you are required to choose any one of the following four responses.

1) **Both Statement-I and Statement-II are correct.**
2) **Both Statement-I and Statement-II are incorrect.**
3) **Statement-I is correct & Statement-II is incorrect.**
4) **Statement-I is incorrect & Statement-II is correct.**

1. **Statement-I:** K_p is related to K_c by the relation. $K_p = K_c[RT]^{\Delta n_g}$
   Statement-II: K_p has same dimensions as K_c
2. **Statement-I:** If Q_c (reaction quotient) < K_c (Equilibrium constant) reaction moves in the direction of reactants.
   Statement-II: Reaction quotient is defined in the same way as equilibrium constant at any stage of the reaction.
3. **Statement-I:** pH of hydrochloric acid solution is less than that of acetic acid solution of the same concentration.
   Statement-II: In equimolar solutions, the number of titrable protons present in hydrochloric acid is less than that present in acetic acid.
4. **Statement-I:** NaCl solution in water will have higher pH than aqueous solution of Na_2CO_3.
   Statement-II: NaCl hydrolysis to give HCl.
5. **Statement-I:** Value of K_p and K_c for the reaction $H_2(g) + I_2(g) \rightleftharpoons 2HI(g)$ is same.
   Statement-II: Because Δn is zero for this reaction.
6. **Statement-I:** The concentration of partial pressure of pure solid or liquid also appear in the expression of the equilibrium constant.
   Statement-II: Molar concentration of pure solid or liquid is constant.
7. **Statement-I:** While production of ammonia, if temperature is raised then the equilibrium shift towards left.
   Statement-II: Production of ammonia is an endothermic reaction.
8. **Statement-I:** Acidic strength increases as HF < HCl < HBr < HI.
   Statement-II: As size of X(X = F, Cl, Br, I) increases, H – X bond strength increases.
9. **Statement-I:** The conjugate acid should have one extra proton.
   Statement-II: BCl_3 act as a lewis acid.
10. **Statement-I:** The pH of NH_4Cl solution in water is more than 7 and pH of CH_3COONa solution is less than 7.
    Statement-II: NH_4Cl is a salt of weak base NH_4OH and strong acid HCl whereas CH_3COONa is salt of a weak acid CH_3COOH and strong base NaOH.

Assertion & Reason MCQs

Directions: These questions consist of two statements each, printed as Assertion and Reason. While answering these questions, you are required to choose any one of the following four responses.

1) **Both Assertion and Reason are True and the Reason is a correct explanation of the Assertion.**
2) **Both Assertion and Reason are True but Reason is not a correct explanation of the Assertion.**
3) **Assertion is True but the Reason is False.**
4) **Assertion is False but Reason is True.**

1. **Assertion:** Increasing order of acidity of hydrogen halides is HF < HCl < HBr < HI
   Reason: While comparing acids formed by the elements belonging to the same group of periodic table, H–A bond strength is a more important factor in determining acidity of an acid than the polar nature of the bond.
2. **Assertion:** The ionisation of hydrogen sulphide in water is low in the presence of hydrochloric acid.
   Reason: Hydrogen sulphide is a weak acid
3. **Assertion:** For any chemical reaction at a particular temperature, the equilibrium constant is fixed and is a characteristic property.
   Reason: Equilibrium constant is independent of temperature.

4. **Assertion:** Aqueous solution of ammonium carbonate is basic.
Reason: Acidic/basic nature of a salt solution of a salt of weak acid and weak base depends on K_a and K_b value of the acid and the base forming it.

5. **Assertion:** For the reaction, $N_2 + O_2 \rightleftharpoons 2NO$, an increase in pressure at equilibrium does not affect the reaction.
Reason: Σ Moles of gaseous product – Σ moles of gaseous reactant = 0.

6. **Assertion:** K_p is related to K_c by the relation. $K_p = \frac{K_c[RT]}{\Delta n}$
Reason: K_p has different dimensions as that of K_c

7. **Assertion:** H_2SO_4 acts as an acid in presence of $HClO_4$
Reason: Perchloric acid is stronger acid than H_2SO_4

8. **Assertion:** pH of 10^{-3} M HCl is equal to 3.
Reason: HCl being a strong base is completely ionized.

9. **Assertion:** On diluting the solution of CH_3COONH_4 by 100 times its degree of hydrolysis doesn't change.
Reason: Degree of hydrolysis for this type of salt is $\sqrt{\frac{K_W \times K_a}{K_b}}$.

10. **Assertion:** An aqueous solution of NH_4Br is acidic in character.
Reason: NH_4Br in an aqueous solution undergoes cationic hydrolysis.

Multi Correct MCQs

1. Consider the following equation:
$2SO_2(g) + O_2(g) \rightleftharpoons 2SO_3(g), \Delta H = -190kJ$
Which of the following factors will increase the yield of SO_3 at equilibrium?
A. Increasing temperature B. Increasing pressure
C. Adding more SO_2 D. Adding more O_2
E. Addition of catalyst
1) (A), (C), (D) and (E) 2) (B), (C) and (D)
3) (C), (D) and (E) 4) Only (B)

2. The following reaction occurs in the Blast furnace where iron ore is reduced to iron metal
$Fe_2O_3(s) + 3CO(g) \rightleftharpoons Fe(s) + 3CO_2(g)$
Using the Le-Chatelier's principle, predict which of the following will disturb the equilibrium.
A. Addition of Fe_2O_3 B. Addition of CO_2
C. Removal of CO D. Removal of CO_2
1) Only (B) 2) (B), (C) and (D)
3) (A) and (D) 4) Only (A)

3. Arrange the following solutions in the decreasing order of pOH.
I. 0.01 M HCl II. 0.01 M NaOH
III. 0.01 M CH_3COONa IV. 0.01 M NaCl
A. (I) < (III)
B. (I) > (IV) and (III) > (II)
C. (II) < (III) and (I) > (III)
D. (II) > (IV) and (III) = (IV)
Choose the most appropriate option from the following:
1) (B) and (C) 2) (A), (B) and (D)
3) Only (C) 4) (A), (C) and (D)

4. Which of the following compounds acts as a Lewis base?
A. ClF_3 B. PCl_5
C. NF_3 D. SF_4
Choose the most appropriate option from the following.
1) (A), (C) and (D) 2) Only (B)
3) (C) and (D) 4) (A) and (C)

NCERT Exemplar MCQs

1. We know that the relationship between K_c and K_p is:
$K_p = K_c (RT)^{\Delta n}$
What would be the value of Δn_g for the reaction?
$NH_4Cl(s) \rightleftharpoons NH_3(g) + HCl(g)$
1) 1 2) 0.5
3) 1.5 4) 2

2. For the reaction,
$H_2(g) + I_2(g) \rightleftharpoons 2HI(g)$, the standard free energy is $\Delta G^{\circ} > 0$. The equilibrium constant (K) would be:
1) K = 0 2) K > 1
3) K = 1 4) K < 1

3. Which of the following is not a general characteristic of equilibrium involving physical processes?
1) Equilibrium is possible only in a closed system at a given temperature
2) All measurable properties of the system remain constant
3) All the physical processes stop at equilibrium
4) The opposing processes occur at the same rate and there is dynamic but stable condition

4. PCl_5, PCl_3, and Cl_2 are at equilibrium at 500 K in a closed container and their concentrations are 0.8×10^{-3} mol L^{-1}, 1.2×10^{-3} mol L^{-1} and 1.2×10^{-3} mol L^{-1}, respectively. The value of K_c for the reaction $PCl_5(g) \rightleftharpoons PCl_3(g) + Cl(g)$ will be:
1) 1.8×10^3 mol L^{-1} 2) 1.8×10^{-3}
3) 1.8×10^{-3} L mol^{-1} 4) 0.55×10^4

5. Which of the following statements is incorrect?
   1) In equilibrium mixture of ice and water kept in perfectly insulated flask, mass of ice and water does not change with time
   2) The intensity of red colour increases when oxalic acid is added to a solution containing iron (III) nitrate and potassium thiocyanate
   3) On addition of catalyst the equilibrium constant value is not affected
   4) Equilibrium constant for a reaction with negative ΔH value decreases as the temperature increases

6. When hydrochloric acid is added to cobalt nitrate solution at room temperature, the following reaction takes place and the reaction mixture becomes blue. On cooling the mixture it becomes pink. On the basis of this information mark the correct answer:

$$\underset{\text{(pink)}}{[Co(H_2O_6)]^{3+}}(aq) + 4Cl^-(aq) \rightleftharpoons \underset{\text{(Blue)}}{[CoCl_4]^{2-}}(aq) + 6H_2O(l)$$

   1) $\Delta H > 0$ for the reaction
   2) $\Delta H < 0$ for the reaction
   3) $\Delta H = 0$ for the reaction
   4) The sign of ΔH cannot be predicted on the basis of this information

7. The pH of neutral water at 25°C is 7.0. As the temperature increases, ionisation of water increases, however, the concentration of H^+ ions and OH^- ions are equal. What will be the pH of pure water at 60°C?
   1) Equal to 7.0
   2) Greater than 7.0
   3) Less than 7.0
   4) Equal to zero

8. The ionisation constant of an acid, K, is the measure of strength of an acid. The K_a values of acetic acid, hypochlorous acid and formic acid are 1.74×10^{-5}, 3.0×10^{-8} and 1.8×10^{-4} respectively. Which of the following orders of pH of 0.1 mol dm^{-3} solutions of these acids is correct?
   1) Acetic acid > hypochlorous acid > formic acid
   2) Hypochlorous acid > acetic acid > formic acid
   3) Formic acid > hypochlorous acid > acetic acid
   4) Formic acid > acetic acid > hypochlorous acid

9. K_{a_1}, K_{a_2} and K_{a_3} are the respective ionisation constants for the following reactions.

$$H_2S \rightleftharpoons H^+ + HS^-$$

$$HS^- \rightleftharpoons H^+ + HS^{2-}$$

$$H_2S \rightleftharpoons 2H^+ + S^{2-}$$

The correct relationship between $K_{a_1}, K_{a_2}, K_{a_3}$ is:
   1) $K_{a_3} = K_{a_1} \times K_{a_2}$
   2) $K_{a_3} = K_{a_1} + K_{a_2}$
   3) $K_{a_3} = K_{a_1} - K_{a_2}$
   4) $K_{a_3} = K_{a_1} / K_{a_2}$

10. Acidity of BF_3 can be explained on the basis of which of the following concepts?
   1) Arrhenius concept
   2) Bronsted Lowry concept
   3) Lewis concept
   4) Bronsted Lowry as well as Lewis concept

11. Which of the following will produce a buffer solution when mixed in equal volumes?
   1) 0.1 mol dm^{-3} NH_4OH and 0.1 mol dm^{-3} HCl
   2) 0.05 mol dm^{-3} NH_4OH and 0.1 mol dm^{-3} HCl
   3) 0.1 mol dm^{-3} NH_4OH and 0.05 mol dm^{-3} HCl
   4) 0.1 mol dm^{-3} CH_4COONa and 0.1 mol dm^{-3} NaOH

12. In which of the following solvents is silver chloride most soluble?
   1) 0.1 mol dm^{-3} $AgNO_3$ solution
   2) 0.1 mol dm^{-3} HCl solution
   3) H_2O
   4) Aqueous ammonia

13. What will be the value of pH of 0.01 mol dm^{-3} CH_3COOH ($K_a = 1.74 \times 10^{-5}$)?
   1) 3.4
   2) 3.6
   3) 3.9
   4) 3.0

14. K_a for CH_3COOH is 1.8×10^{-5} and K_b for NH_4OH is 1.8×10^{-5}. The pH of ammonium acetate will be:
   1) 7.005
   2) 4.75
   3) 7.0
   4) Between 6 and 7

15. Which of the following options will be correct for the stage of half completion of the reaction $A \rightleftharpoons B$?
   1) $\Delta G^\ominus = 0$
   2) $\Delta G^\ominus > 0$
   3) $\Delta G^\ominus < 0$
   4) $\Delta G^\ominus = -RT \ln K$

16. On increasing the pressure, in which direction will the gas phase reaction proceed to re-establish equilibrium, is predicted by applying the Le-Chatelier's principle. Consider the reaction,

$$N_2(g) + 3H_2(g) \rightleftharpoons 2NH_3(g)$$

Which of the following is correct, if the total pressure at which equilibrium is established, is increased without changing the temperature?
   1) K will remain same
   2) K will decrease
   3) K will increase
   4) K will increase initially and decrease when pressure is very high

17. What will be the correct order of vapour pressure of water, acetone and ether at 30°C? Given that among these compounds, water has maximum boiling point and ether has minimum boiling point?
   1) Water < ether < acetone
   2) Water < acetone < ether
   3) Ether < acetone < water
   4) Acetone < ether < water

18. At 500 K, equilibrium constant, K_c, for the following reaction is 5.

$$\frac{1}{2}H_2(g)+\frac{1}{2}I_2(g)\rightleftharpoons HI(g)$$

What would be the equilibrium constant K_c for the reaction?

$$2HI(g)\rightleftharpoons H_2(g)+I_2(g)$$

1) 0.04 2) 0.4
3) 25 4) 2.5

19. In which of the following reactions, the equilibrium remains unaffected on addition of small amount of argon at constant volume?

1) $H_2(g)+I_2(g)\rightleftharpoons 2HI(g)$
2) $PCl_5(g)\rightleftharpoons PCl_3(g)+Cl_2(g)$
3) $N_2(g)+3H_2(g)\rightleftharpoons 2NH_3(g)$
4) The equilibrium will remain unaffected in all the three cases

Scan Me for Detailed Explanations

Kattar NEET

1. For the equilibrium:

$LiCl.3NH_3 \rightleftharpoons LiCl.NH_3 + 2NH_3$, $K_P = 9\ atm^2$

At 27°C, a 8.2 L vessel contains 0.1 mol of $LiCl.NH_3(s)$. How many moles of NH_3 should be added to the flask at this temperature to derive the backward reaction for completion? (use R = 0.082 atm L/mol-K)

1) 20.5 mol 2) 3 mol
3) 1.2 mol 4) 0.21 mol

2. The $CaCO_3$ is heated to 600 K in a closed vessel of volume 1 litre to form CaO and CO_2. The minimum weight of $CaCO_3$ required to establish the equilibrium $CaCO_3(s) \rightleftharpoons CaO(s) + CO_2(g)$ is (K_P = 2.25 atm)

1) 2 g 2) 4.57 g
3) 10 g 4) 100 g

3. The equilibrium constant for the reaction H_3BO_3 + glycerine $\rightleftharpoons$ (H_3BO_3-glycerine) is 0.9. How much glycerine should be added per litre of 0.1 M H_3BO_3 solution so that 60% of the H_3BO_3 is converted to the boric acid-glycerine complex?

1) 0.7 M 2) 1.73 M
3) 2.27 M 4) 3.7 M

4. The vapour density of N_2O_4 at a certain temperature was 30. What is the percentage of dissociation of N_2O_4 at this temperature?

1) 53% 2) 58%
3) 68% 4) 62%

5. Attainment of the equilibrium $A(g) \rightleftharpoons 2C(g) + B(g)$ gave the following graph. Find the correct option. (% dissociation = fraction dissociated × 100)

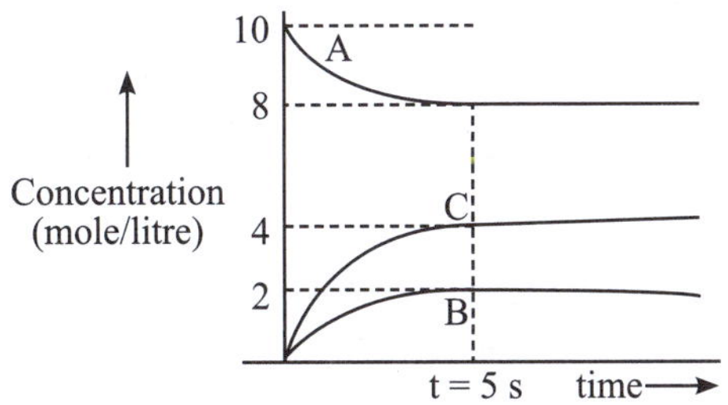

1) At t = 5 s equilibrium has been reached and K_C = 40 $(mol/litre)^2$
2) At t = 5 s equilibrium has been reached and % dissociation of A is 20 %
3) At t = 5 s equilibrium has been reached and % dissociation of A is 30 %
4) At t = 5 s equilibrium has been reached and K_C = 20 $(mol/litre)^2$

6. The graph which represents all the equilibrium concentrations for the reaction $N_2O_4(g) \rightleftharpoons 2NO_2(g)$. Then select the correct graph for concentrations of NO_2 against concentrations of N_2O_4.

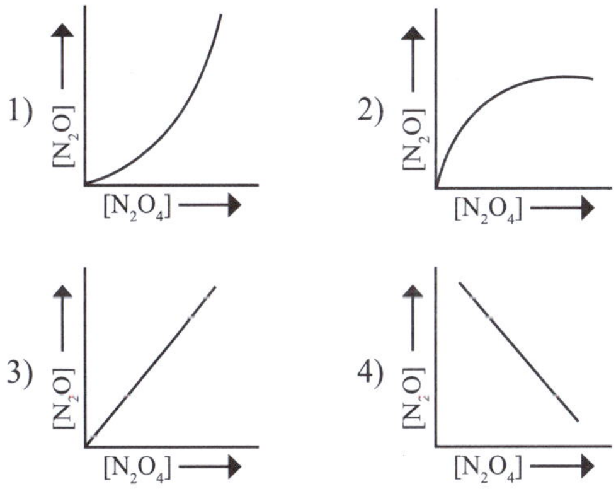

7. A vessel of 250 litre was filled with 0.01 mole of Sb_2S_3 and 0.01 mole of H_2 to attain the equilibrium at 440°C as $Sb_2S_3(s) + 3H_2(g) \rightleftharpoons 2Sb(s) + 3H_2S(g)$. After equilibrium, the H_2S formed was analysed by dissolving it in water and treating with excess of Pb^{2+} to give 1.19 g of PbS as precipitate. What is the value of K_C, at 440°C?

1) 1 2) 2
3) 4 4) 8

8. For the reaction $N_2O_4 \rightleftharpoons 2NO_2(g)$, if percentage dissociation of N_2O_4 are 20%, 45%, 65%, 80% then the sequence of observed vapour densities will be:

1) $d_{20} > d_{45} > d_{65} > d_{80}$ 2) $d_{80} > d_{65} > d_{45} > d_{20}$
3) $d_{20} = d_{45} = d_{65} = d_{80}$ 4) $d_{20} = d_{45} = d_{65} = d_{80}$

9. 0.1 mole each of CH_3COOH and C_2H_5OH are allowed to react and at equilibrium the acid was exactly neutralised by 100 ml of 0.85N NaOH. If no hydrolysis of ester is supposed to have undergone, find K_C?

1) 0.31 2) 0.031
3) 0.25 4) 2.5

10. For the equilibrium starting with 2 moles of SO_2 and 1 mole O_2 in 1 litre flask; $2SO_2(g) + O_2(g) \rightleftharpoons 2SO_3(g)$. Equilibrium mixture required 0.4 moles MnO_4^- in acidic medium, K_c for the given equilibrium is

1) 2 2) 0.4
3) 1.0 4) 2.8

11. A 1.458 g of Mg reacts with 80.0 ml of a HCl solution whose pH is –0.477. The change in pH when all Mg has reacted (Assume constant volume. Mg = 24.3 g/mol.) (log 3 = 0.477, log 2 = 0.301)

1) –0.176 2) +0.477
3) –0.2385 4) 0.3

12. Ionisation constant of each HA (weak acid) and BOH (weak base) are 3.0×10^{-7} each at 298 K. The percentage degree of hydrolysis of BA at the dilution of 10L is:

1) 25 2) 50
3) 75 4) 40

13. A sample of water has a hardness expressed as 80 ppm of Ca^{2+}. This sample is passed through an ion exchange column and the Ca^{2+} is replaced by H^+. What is the pH of the water after it has been so treated? [Atomic mass of Ca = 40]

1) 3 2) 2.7
3) 5.4 4) 2.4

14. An aqueous solution of a metal chloride MCl_2 (0.05 M) is saturated with H_2S (0.1 M). The minimum pH at which metal sulphide will be precipitated is given $K_{sp}(MS) = 5 \times 10^{-21}$, $K_1(H_2S) = 1 \times 10^{-7}$, $K_2(H_2S) = 1 \times 10^{-14}$

1) 1.25 2) 1.50
3) 2.50 4) 3.25

15. Calcium lactate is a salt of weak organic acid and strong base represented as $Ca(LaC)_2$. A saturated solution of $Ca(LaC)_2$ contains 0.6 mole in 2 litre solution. pOH of solution is 5.60. If 90% dissociation of the salt takes place then what is pK_a of lactic acid

1) 2.8 – log(0.54) 2) 2.8 + log(0.54)
3) 2.8 + log(0.27) 4) Done

16. The pH of blood is 7.4. What is the ratio of $\frac{[HPO_4^{2-}]}{[H_2PO_4^-]}$ in the blood? Given: K_{a_1}, K_{a_2} and K_{a_3} for H_3PO_4 are 10^{-3}, 8×10^{-8} and 10^{-12} respectively.

1) 2 : 1 2) 1 : 2
3) 3 : 1 4) 1 : 3

17. Given : Enthalpy of ionization of two acids :

$\Delta H°(HCN) = 45.2$ kJ mol^{-1}

$\Delta H°(CH_3COOH) = 2.1$ kJ mol^{-1}

Which relationship for the two acids is true?

1) $pK_a(HCN) = pK_a(CH_3COOH)$
2) $pK_a(HCN) > pK_a(CH_3COOH)$
3) $pK_a(HCN) < pK_a(CH_3COOH)$
4) $pK_a(HCN) = \frac{45.2}{2.1} pK_a(CH_3COOH)$

18. Match the effect of addition of 0.1 M KOH to 0.1 M, 50 ml H_3PO_4. K_{a_1}, K_{a_2}, K_{a_3} are the I, II, III ionisation constant of H_3PO_4:

	List-I		List-II
I.	75 ml of KOH	P.	$pH = p^{K_{a_1}}$
II.	25 ml of KOH	Q.	$pH = p^{K_{a_2}}$
III.	150 ml of KOH	R.	$pH = \frac{p^{K_{a_2}} + p^{K_{a_3}}}{2}$
IV.	100 ml of KOH	S.	$pH = 7 + \frac{1}{2}[p^{K_{a_3}} + \log C]$
		T.	$pOH = 7 - \frac{1}{2}[p^{K_{a_3}} + \log C]$

1) I-Q; II-P; III-S, T; IV-R
2) I-R, S; II-P, S; III-Q; IV-S, T
3) I-P; II-Q; III-R, T; IV-S
4) I-P; II-R; III-S; IV-Q

19. Match List-I having components with List-II having pH or pOH of solution.

	List-I	List-II	
I.	Saturated solution of $A(OH)_3$ having $K_{sp} = 5 \times 10^{-31}$	P.	7
II.	Mixing equal volumes of 0.1 M CH_3COONa and 0.1 M XNO_3	Q.	10
III.	Solution obtained at $\frac{10}{11}$ of the equivalence point when 0.1 M XOH is titrated with HCl 0.1 M	R.	5
IV.	Mixing equal volume of 0.1 M CH_3COOH and 0.1 M $CH_3COO^-Na^+$	S.	4

[Given data: $K_h(X^+) - 10^{-5}$, K_a, $CH_3COOH - 10^{5}$]

1) I-S; II-P; III-Q; IV-P
2) I-P; II-Q; III-R; IV-S
3) I-Q; II-P; III-R; IV-S
4) I-P; II-R III-Q,S; IV-R

20. Match the following:

	List-I		List-II
I.	1 litre solution having 0.1 M CH_3COOH ($pK_a = 5$) and 0.1 M CH_3COONa	P.	pH < 7 at 25°C
II.	1 litre solution having 0.1 M CH_3COONa and 0.1 M HCl (pK_a of $CH_3COOH = 5$)	Q.	pH > 7 at 25°C
III.	1 litre solution having 0.1 M CH_3COOH and 0.1 M HCl (K_a of $CH_3COOH = 10^{-5}$)	R.	pH changes with dilution
IV.	1 litre solution having 0.1 M CH_3COOH and 0.1 M NaOH(K_a of $CH_3COOH = 10^{-5}$)	S.	pH do not change with dilution

1) I-P, S; II-P, R; III-Q, R, IV-Q, S
2) I-P, R; II-Q, R; III-Q, S; IV-P, S
3) I-P, S; II-P, R; III-P, R; IV-Q, R
4) I-Q, S; II-P, S; III-P, R; IV-Q, R

NEET & AIIMS Past Year MCQs

1. At a given temperature and pressure, the equilibrium constant values for the equilibria are given below: **(2024 Re)**

$3A_2 + B_2 \rightleftharpoons 2A_3B, K_1$

$A_3B \rightleftharpoons \frac{3}{2}A_2 + \frac{1}{2}B_2, K_2$

The relation between K_1 and K_2 is:

1) $K_1^2 = 2K_2$ 2) $K_2 = \frac{K_1}{2}$

3) $K_1 = \frac{1}{\sqrt{K_2}}$ 4) $K_2 = \frac{1}{\sqrt{K_1}}$

2. For the reaction in equilibrium **(2024 Re)**

$N_2(g) + 3H_2(g) \rightleftharpoons 2NH_3(g), \Delta H = -Q$

Reaction is favoured in forward direction by:

1) use of catalyst
2) decreasing concentration of N_2
3) low pressure, high temperature and high concentration of ammonia
4) high pressure, low temperature and higher concentration of H_2

3. The ratio of solubility of AgCl in 0.1 M KCl solution to the solubility of AgCl in water is:

(Given : Solubility product of AgCl = 10^{-10}) **(2024 Re)**

1) 10^{-4} 2) 10^{-6}
3) 10^{-9} 4) 10^{-5}

4. For the reaction $2A \rightleftharpoons B + C$, $K_c = 4 \times 10^{-3}$. At a given time, the composition of reaction mixture is: **(2024)**

$[A] = [B] = [C] = 2 \times 10^{-3}$ M.

Then, which of the following is correct?

1) Reaction has a tendency to go in backward direction.
2) Reaction has gone to completion in forward direction.
3) Reaction is at equilibrium.
4) Reaction has a tendency to go in forward direction.

5. In which of the following equilibria, K_p and K_c are **NOT** equal? **(2024)**

1) $CO_{(g)} + H_2O_{(g)} \rightleftharpoons CO_{2(g)} + H_{2(g)}$
2) $2BrCl_{(g)} \rightleftharpoons Br_{2(g)} + Cl_{2(g)}$
3) $PCl_{5(g)} \rightleftharpoons PCl_{3(g)} + Cl_{2(g)}$
4) $H_{2(g)} + I_{2(g)} \rightleftharpoons 2HI_{(g)}$

6. Consider the following reaction in a sealed vessel at equilibrium with concentrations of $N_2 = 3.0 \times 10^{-3}$ M, $O_2 = 4.2 \times 10^{-3}$ M and $NO = 2.8 \times 10^{-3}$ M. **(2024)**

$2NO_{(g)} \rightleftharpoons N_{2(g)} + O_{2(g)}$

If 0.1 mol L^{-1} of $NO_{(g)}$ is taken in a closed vessel, what will be degree of dissociation (α) of $NO_{(g)}$ at equilibrium ?

1) 0.8889 2) 0.717
3) 0.00889 4) 0.0889

7. The equilibrium concentrations of the species in the reaction $A + B \rightleftharpoons C + D$ are 2, 3, 10 and 6 mol L^{-1} respectively at 300 K. ΔG^0 for the reaction is **(2023)**

(R = 2 cal/mol K)

1) –13.73 cal 2) 1372.60 cal
3) –137.26 cal 4) –1381.80 cal

8. The pH of the solution containing 50 mL each of 0.10 M sodium acetate and 0.01 M acetic acid is

[Given pK_a of $CH_3COOH = 4.57$] **(2022)**

1) 2.57 2) 5.57 3) 3.57 4) 4.57

9. $3O_2(g) \rightleftharpoons 2O_3(g)$ for the given reaction at 298 K, K_c is found to be 3.0×10^{-59}. If the concentration of O_2 at equilibrium is 0.040 M then concentration of O_3 in M is **(2022)**

1) 1.2×10^{21} 2) 4.38×10^{-32}
3) 1.9×10^{-63} 4) 2.4×10^{31}

10. The pK_b of dimethylamine and pK_a of acetic acid are 3.27 and 4.77 respectively at T(K). The correct option for the pH of dimethylammonium acetate solution is: **(2021)**

1) 5.50 2) 7.75 3) 6.25 4) 8.50

11. $2ICl \rightarrow I_2 + Cl_2$ $K_C = 0.14$

Initial concentration of ICl is 0.6 M then equilibrium concentration of I_2 is : **(AIIMS 2018)**

1) 0.37 M 2) 0.128 M
3) 0.224 M 4) 0.748 M

12. K_a for HCN is 5×10^{-10} at 25°C. For maintaining a constant pH = 9, the volume of 5 M KCN solution required to be added to 10 mL of 2 M HCN solution is: **(AIIMS 2017)**

1) 4 mL
2) 2.5 mL
3) 2 mL
4) 6.4 mL

13. The equilibrium constant for the reaction $\frac{1}{2} H_2(g) + \frac{1}{2} I_2(g) \rightleftharpoons HI(g)$ is K_c

Equilibrium constant for the reaction,

$2HI(g) \rightleftharpoons H_2(g) + I_2(g)$ is? **(AIIMS 2014)**

1) $1/K_c$
2) $1/(K_c)^2$
3) $2/K_c$
4) $2/(K_c)^2$

14. The following equilibria are given:

I. $N_2 + 3H_2 \rightleftharpoons 2NH_3;\ K_1$

II. $N_2 + O_2 \rightleftharpoons 2NO;\ K_2$

III. $H_2 + \frac{1}{2}O_2 \rightleftharpoons H_2O;\ K_3$

The equilibrium constant for the reaction,

$2NH_3 + \frac{5}{2}O_2 \rightleftharpoons 2NO + 3H_2O$

In terms of K_1, K_2 and K_3 will be **(AIIMS 2010)**

1) $K_1K_2K_3$
2) $\frac{K_1K_2}{K_3}$
3) $\frac{K_1K_3^2}{K_2}$
4) $\frac{K_2K_3^3}{K_1}$

15. What is the pH of 0.01 M glycine solution?

For glycine $K_{a_1} = 4.5 \times 10^{-3}$ and $K_{a_2} = 1.7 \times 10^{-10}$ at 298 K. **(AIIMS 2010)**

1) 3.0
2) 10.0
3) 7.06
4) 8.2

16. Of the following which change will shift the reaction towards the product: **(AIIMS 2004)**

$I(g) \rightleftharpoons 2I(g)$, $H°_r$ (298K) = +150kJ

1) Increase in concentration of I
2) Decrease in concentration of I_2
3) Increase in temperature
4) Increase in total pressure

17. The pH value of decinormal solution of NH_4OH which is 20% ionised is: **(AIIMS 2001)**

1) 12.95
2) 12.30
3) 14.70
4) 13.30

18. **Assertion:** Relative strength of acids can be known by knowing the value of dissociation constant. **(AIIMS 2001)**

Reason: It gives the value of H^+ dissolved in solution.

1) If both Assertion & Reason are True & the Reason is a correct explanation of the Assertion.
2) If both Assertion & Reason are True but Reason is not a correct explanation of the Assertion.
3) If Assertion is True but the Reason is False.
4) If both Assertion & Reason are False.

Answer Key

NCERT Topic-wise MCQs

1	2	3	4	5	6	7	8	9	10	11	12	13	14	15	16	17
3	2	3	1	3	3	1	4	3	3	4	2	3	2	4	3	3
18	19	20	21	22	23	24	25	26	27	28	29	30	31	32	33	34
4	4	2	1	4	3	2	1	4	3	1	3	3	2	2	3	2
35	36	37	38	39	40	41	42	43	44	45	46	47	48	49	50	51
3	3	1	3	4	2	4	1	2	1	3	3	2	1	3	2	3
52	53	54	55	56	57	58	59	60	61	62	63	64	65	66	67	68
2	1	1	2	3	2	2	1	2	3	4	4	4	4	2	3	3
69	70	71	72	73	74	75	76	77	78	79	80	81	82	83	84	85
1	1	2	2	4	1	2	4	3	3	1	4	3	2	4	2	3
86	87	88	89	90	91	92	93	94	95	96	97	98	99	100	101	102
3	2	1	2	2	3	3	1	3	2	3	1	1	1	3	1	4
103	104	105	106	107	108	109	110	111	112	113	114	115	116	117	118	119
1	3	2	3	1	4	4	3	1	3	4	3	2	2	2	4	3
120	121	122	123	124	125	126	127	128	129	130						
2	3	1	2	3	4	3	3	2	3	4						

High Order Time Intensive MCQs	1	2	3	4	5	6	7	8	9	10	11	12	13	14	15	16	17
	1	1	2	1	2	1	4	3	2	3	1	3	2	1	3	4	4
	18	19	20	21	22	23	24	25	26	27	28	29	30	31	32	33	34
	4	2	2	2	1	1	1	1	4	1	1	3	4	4	3	4	4
	35	36	37	38	39	40	41	42	43	44	45	46	47	48	49	50	51
	2	2	1	3	1	3	2	1	1	3	3	3	3	3	3	1	4
	52	53	54	55	56	57	58	59	60	61	62	63	64	65	66	67	
	3	4	4	2	4	2	1	2	2	3	2	1	2	1	2	4	

Match the Columns MCQs	1	2	3	4	5	6
	3	2	4	2	1	1

Correct & Incorrect MCQs	1	2	3	4	5	6	7	8	9	10
	1	2	1	4	3	2	3	3	4	2

Statement Based MCQs	1	2	3	4	5	6	7	8	9	10
	3	4	3	2	1	4	3	3	1	4

Assertion & Reason MCQs	1	2	3	4	5	6	7	8	9	10
	1	2	3	1	1	4	4	1	1	1

Multi Correct MCQs	1	2	3	4
	2	2	1	1

NCERT Exemplar MCQs	1	2	3	4	5	6	7	8	9	10	11	12	13	14	15	16	17
	4	4	3	2	2	1	3	2	1	3	3	4	1	3	1	1	2
	18	19															
	1	4															

Kattar NEET	1	2	3	4	5	6	7	8	9	10	11	12	13	14	15	16	17
	3	2	2	1	2	2	1	1	2	1	4	1	4	2	1	1	2
	18	19	20														
	1	4	3														

NEET & AIIMS Past Year MCQs	1	2	3	4	5	6	7	8	9	10	11	12	13	14	15	16	17
	4	4	1	1	3	2	4	2	2	2	2	3	2	4	3	3	2
	18																
	1																

CHAPTER

7 Redox Reactions

Topicwise Number of Questions Analysis (2025-2016)

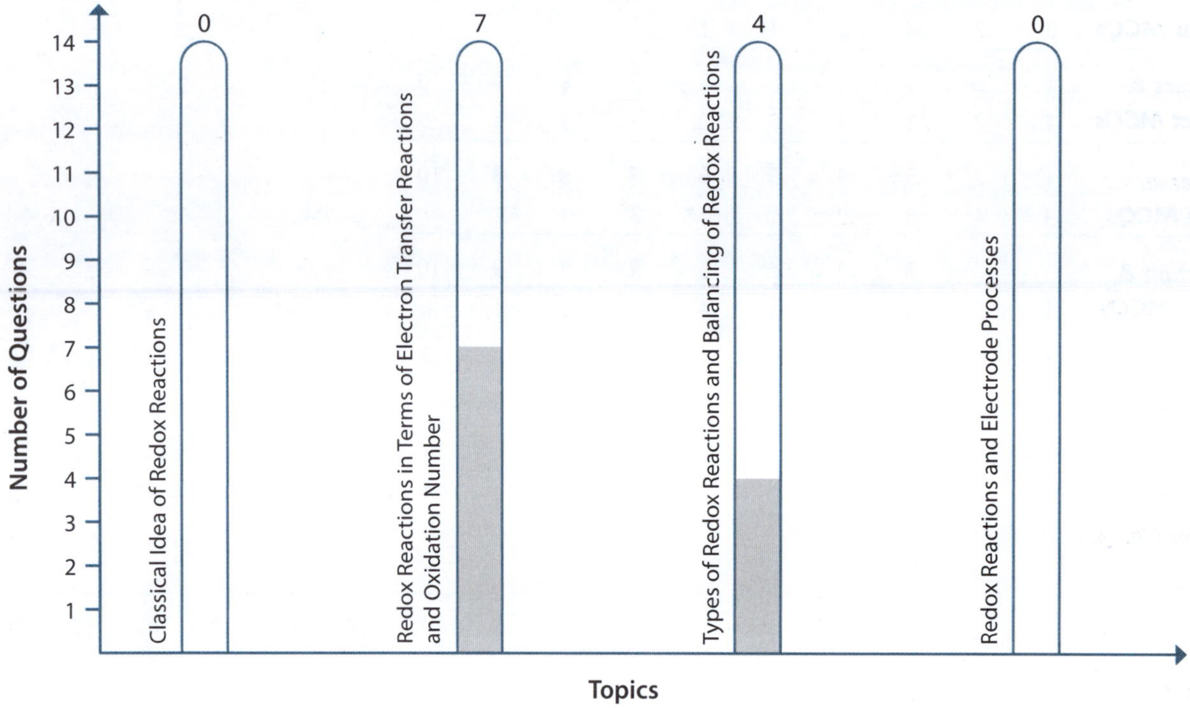

NCERT-Picks

Classical Idea of Redox Reactions

- **Oxidation** is the addition of oxygen/electronegative element to a substance or removal of hydrogen/electropositive element from a substance.
- **Reduction** is the addition of hydrogen/electropositive element to a substance or removal of oxygen/electronegative element from a substance.

Redox Reaction in Terms of Electron Transfer Reactions

- In terms of electron transfer reactions, the redox reactions are explained by the following examples. The formation of sodium chloride is represented as-

  $2Na(s) \rightarrow 2Na^{+}(g) + 2e^{-}$, $Cl_2(g) + 2e^{-} \rightarrow 2Cl^{-}(g)$

  Each step is called a half reaction, and shows involvement of electrons. Sum of half reactions gives the overall reaction:

  $2Na(s) + Cl_2(g) \rightarrow 2Na^{+}Cl^{-}(s)$ or $2NaCl(s)$

  Hence, the half reactions that involves loss of electrons are called oxidation reaction and that involves gain of electrons are called reduction reactions.
- The species which gain electrons in redox reactions is called **oxidising agent**. Oxidising agent itself undergo reduction, and its oxidation number decreases in redox reactions.
- The species which lose electrons in redox reaction is called **reducing agent**. Reducing agent itself undergoes oxidation and its oxidation number increases in redox reactions.
- Common oxidising agents are

  $KMnO_4$, $K_2Cr_2O_7$, $KClO_3$, $KClO_4$, MnO_2, Cl_2, $CaOCl_2$.
- Common reducing agents are:

  H_2, CO, Coke, metals, H_2S, $FeSO_4$, $SnCl_2$, etc.
- Substances which can behave both as oxidising agents and reducing agents are:

  O_3, HNO_2, SO_2, I_2 etc.

Oxidation Number

It is an imaginary or apparent charge developed over the element when it goes from its elemental free state to combined state in molecules.

- The oxidation number of oxygen is –2 in most of its compounds (an exception is O in H_2O_2 and other peroxides, where oxidation number is –1 and in oxygen fluoride (OF_2), it is +2).

Group	Common oxidation numbers (States) excepts zero in free state
I A	+1
II A	+2
III A	+ 3, +1
IV A	+4, +3, +2, +1, –1, –2, –3, –4
V A	+ 5, + 3, + 1, –1, –3
VI A	+ 6, +4, + 2, –2
VII A	+ 7, + 5, + 3, + 1, –1

Special Examples of Oxidation State

In order to determine the exact or individual oxidation number, we need to take help from the structures of the molecules. Some special cases are discussed as follows:

- The structure of CrO_5 is 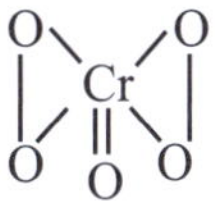

From the structure, it is evident that in CrO_5 there are two peroxide linkages and one double bond. The contribution of each peroxide linkage is –2. Let the oxidation number of Cr is x.

$\therefore$ x + (–2)2 + (–2) = 0 or x = 6

$\therefore$ Oxidation number of Cr = + 6

- The structure of H_2SO_5 is 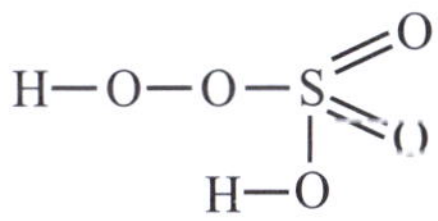

From the structure, it is evident that in H_2SO_5, there is one peroxide linkage, two sulphur-oxygen double bonds and one OH group. Let the oxidation number of S = x.

$\therefore$ (+ 1) + (– 2) + x + (–2) 2+ (–2) + 1 = 0

or x + 2 8 = 0 or x – 6 = 0 or x = 6

$\therefore$ Oxidation number of S in H_2SO_5 is + 6

- Oxidation state in carbon suboxide

$$O=\overset{+2}{C}=\overset{0}{C^*}=\overset{+2}{C}=O$$

Structure of C_3O_2 (Carbon suboxide)

- Oxidation state in Tetrathionate ion

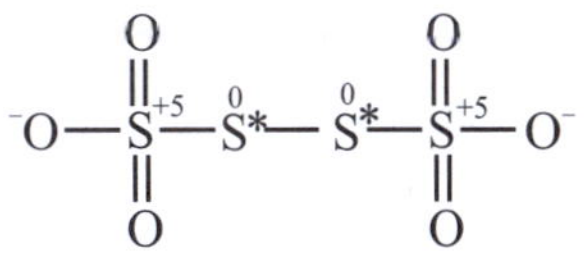

Structure of $S_4O_6^{2-}$ (tetrathionate ion)

Types of Redox Reactions

(A) Combination Reaction:

It is denoted as A + B → C

Either or both A and B must be in the elemental form for such a reaction to be redox reaction. All combustion reactions which make use of elemental dioxygen as well as other electronegative elements are redox reactions

$$\overset{0}{(C)} + \overset{0}{O_2}(g) \xrightarrow{\Delta} \overset{+4\,-2}{CO_2}(g)$$

$$3\,\overset{0}{Mg}(s) + \overset{0}{N_2}(g) \xrightarrow{\Delta} \overset{+2\,-3}{Mg_3N_2}(g)$$

(B) Decomposition Reaction:

These are opposite of combination reactions. It leads to the breakdown of a compound into two or more components, at least one of which must be in the elemental state.

$$2\overset{+1\,-2}{H_2O}(\ell) \xrightarrow{\Delta} 2\overset{0}{H_2}(g) + \overset{0}{O_2}(g),\ 2\overset{+1\,-1}{NaH}(s) \xrightarrow{\Delta} 2\overset{0}{Na}(s) + \overset{0}{H_2}(g)$$

(C) Disproportionation Reaction:

Disproportionation reaction is when reduction and oxidation take place on same element of a compound.

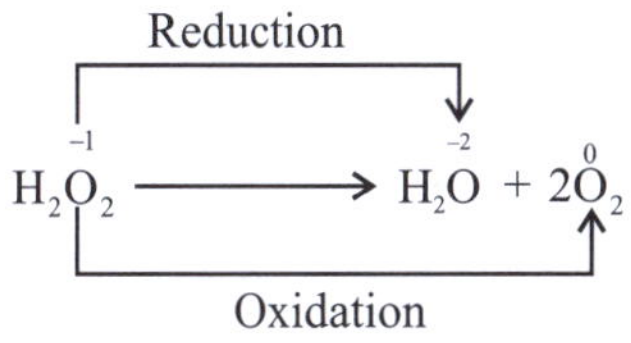

(D) Displacement Reaction:

These are of two types:

(i) Metal displacement reaction:

Eg: $\overset{+2\,+6\,-2}{CuSO_4} + \overset{0}{Zn} \xrightarrow{\Delta} \overset{0}{Cu} + \overset{+2\,+6\,-2}{ZnSO_4}$

(ii) Non-metal displacement reaction :

Eg: $2\overset{0}{Na} + 2\overset{+1\,-2}{H_2O} \xrightarrow{\Delta} 2\overset{+1\,-2\,+1}{NaOH} + \overset{0}{H_2}$

Balancing of Redox Reaction

- To balance a redox reaction, several methods are used.

  These are:

  (i) Oxidation number method

  (ii) Half equation or ion-electron method

- **Spectator ions:**

  Species that are present in the solution but do not take part in the reaction that occurs and are omitted in writing the net ionic reaction.

  $Zn + 2H^+ + 2Cl^- \rightarrow Zn^{2+} + 2Cl^- + H_2$

  Cl^- ions are omitted. These omitted ions are called as spectator ions or bystander ions, in order to indicate that they do not take part in the reaction. The spectator ions appear on the reactant as well as on the product side.

MCQs

NCERT Topic-wise MCQs

Classical Idea of Redox Reactions

1. In which of the following reactions, the underlined substance has been oxidised?
1) $\underline{Br_2} + H_2S \rightarrow 2HBr + S$
2) $\underline{2HgCl_2} + SnCl_2 \rightarrow Hg_2Cl_2 + SnCl_4$
3) $Cl_2 + \underline{2KI} \rightarrow 2KCl + I_2$
4) $\underline{2Cu^{2+}} + 4I^- \rightarrow Cu_2I_2 + I_2$

2. In the reaction $2Na(s) + H_2(g) \longrightarrow 2NaH(s)$, sodium is-
1) Oxidised
2) Reduced
3) Neither oxidised nor reduced
4) Oxidised as well as reduced

3. In the reaction, $2\ Mg\ (s) + O_2\ (g) \rightarrow 2\ MgO\ (s)$
1) Element Mg gets oxidized
2) Element Mg gets reduced
3) Element Mg gets oxidized as well as reduced
4) None of these

4. What is reduction according to classical concept ?
1) Addition of hydrogen
2) Addition of electropositive element
3) Removal of either oxygen or some electronegative element
4) All of these

Redox Reactions in Terms of Electron Transfer Reactions and Oxidation Number

5. Which of the following examples does not represent disproportionation ?
1) $MnO_2 + 4HCl \rightarrow MnCl_2 + Cl_2 + 2H_2O$
2) $2H_2O_2 \rightarrow 2H_2O + O_2$
3) $4KClO_3 \rightarrow 3KClO_4 + KCl$
4) $3Cl_2 + 6NaOH \rightarrow 5NaCl + NaClO_3 + 3H_2O$

6. In oxidation process:
1) Oxidaton number decreases
2) Number of electron decrease
3) Oxygen content decreases
4) Number of ions decreases

7. In which of the following reactions, H_2O_2 acts as a reducing agent?
1) $HOCl + H_2O_2 \xrightarrow{H^+} H_3O^+ + Cl^- + O_2$
2) $Mn^{2+} + H_2O_2 \xrightarrow{OH^+} Mn^{4+} + 2OH^-$
3) $2Fe^{2+} + H_2O_2 \xrightarrow{OH^-} 2Fe^{3+} + 2OH^-$
4) $PbS(s) + 4H_2O_2(aq) \xrightarrow{H^+} PbSO_4(s) + 4H_2O(l)$

8. In the reaction,
$Ag_2O + H_2O_2 \rightarrow 2Ag + H_2O + O_2$
H_2O_2 act as:
1) Reducing agent
2) Oxidising agent
3) Bleaching agent
4) None of these

9. In $Cu^{2+} + Ag \rightarrow Cu + Ag^+$, oxidation half-reaction is:
1) $Cu^{2+} \rightarrow Cu$
2) $Ag \rightarrow Ag^+$
3) $Cu \rightarrow Cu^{2+}$
4) All of these

10. In one of the following reactions, HNO_3 does not behave as an oxidising agent, identify it:
1) $I_2 + 10\ HNO_3 \rightarrow 2\ HIO_3 + 10NO_2 + 4H_2O$
2) $3Cu + 8\ HNO_3 \rightarrow 3Cu\ (NO_3)_2 + 2NO + 4H_2O$
3) $4Zn + 10HNO_3 \rightarrow 4Zn\ (NO_3)_2 + NH_4NO_3 + 3H_2O$
4) $2HNO_3 + P_2O_5 \rightarrow 2HPO_3 + N_2O_5$

11. Which changes requires an oxidising agent?
1) $2S_2O_3^{2-} \rightleftharpoons S_4O_6^{2-}$
2) $Zn^{2+} \rightleftharpoons Zn$
3) $ClO^- \rightleftharpoons Cl^-$
4) $SO_3 \rightleftharpoons SO_4^{2-}$

12. Which of the following molecules can act as an oxidising as well as reducing agent?
1) H_2S
2) SO_3
3) H_2O_2
4) F_2

13. Which of the following is a redox reaction?
1) $2CuSO_4 + 4KI \rightarrow Cu_2I_2 + 2K_2SO_4 + I_2$
2) $SO_2 + H_2O \rightarrow H_2SO_3$
3) $Na_2SO_4 + BaCl_2 \rightarrow BaSO_4 + 2NaCl$
4) $CuSO_4 + 4NH_3 \rightarrow [Cu(NH_3)_4]\ SO_4$

14. Oxidation takes place with-
1) Gain of electrons
2) Loss of electrons
3) Increase in the valency of negative part
4) Decrease in the valency of positive part

15. The oxidation number of hydrogen in CaH_2 is

1) 2 2) 0
3) –1 4) 3

16. The oxidation number and covalency of sulphur in sulphur molecules (S_8) are:

1) 0 and 2 2) + 6 and 8
3) 0 and 8 4) + 6 and 2

17. Oxidation number of N in HNO_3 is:

1) –3.5 2) +3.5
3) –3, +5 4) +5

18. In which of the following compounds, iron has an oxidation number of +3?

1) $Fe(NO_3)_2$ 2) FeC_2O_4
3) $[Fe(H_2O)_6]Cl_3$ 4) None of these

19. Oxidation states of P in $H_4P_2O_5$, $H_4P_2O_6$ $H_4P_2O_7$, respectively are:

1) + 3, + 4, + 5 2) + 3, + 5, + 4
3) + 5, + 3, + 4 4) + 5, + 4. + 3

20. State the oxidation number of carbonyl carbon in methanal and methanoic acid respectively:

1) 0 and 0 2) 0 and + 2
3) + 1 and + 2 4) + 1 and + 3

21. When copper turnings are added to silver nitrate solution, a blue coloured solution is formed after some time. It is because,

1) Cu is more easily reduced than Ag
2) Ag is more easily reduced than Cu
3) Nitrate ion acts as an oxidising agent
4) Nitrate ion acts as a reducing agent

22. Oxidation number of iodine in IO_3^-, IO_4^-, KI and I_2 respectively are:

1) –1, –1, 0 + 1 2) + 3, + 5, + 7, 0
3) + 5, + 7, –1, 0 4) –1, –5, 1, 0

23. When sulphur dioxide is passed in an acidified $K_2Cr_2O_7$ solution, the oxidation state of sulphur is changed from:

1) + 4 to 0 2) + 4 to + 2
3) + 4 to + 6 4) + 6 to + 4

24. Oxidation number of C in HNC is:

1) +2 2) –3
3) +3 4) +4

25. What is the oxidation state of P in Ba $(H_2PO_2)_2$?

1) +1 2) +2
3) +3 4) –1

26. When a manganous salt is fused with the mixture of KNO_3 and the solid NaOH, the oxidation number of Mn changes from +2 to:

1) +4 2) +3
3) +6 4) +7

27. In acidic medium, H_2O_2 changes $Cr_2O_7^{2-}$ to CrO_5 which has two (–O–O–) bonds. Oxidation state of Cr in CrO_5 is:

1) +5 2) +3
3) +6 4) –10

28. A mixture of potassium chlorate, oxalic acid and sulphuric acid is heated. During the reaction which element undergoes maximum change in oxidation number?

1) S 2) H
3) Cl 4) C

29. Which of the following have been arranged in the decreasing order of oxidation number of sulphur?

1) $Na_2S_4O_6 > H_2S_2O_7 > Na_2S_2O_3 > S_8$
2) $H_2SO_4 > SO_2 > H_2S > H_2S_2O_8$
3) $SO_2^{2+} > SO_4^{2-} > SO_3^{2-} > HSO_4^-$
4) $H_2SO_4 > H_2SO_3 > SCl_2 > H_2S$

30. The oxidation states of S-atoms in $S_4O_6^{2-}$ from left to right respectively are:

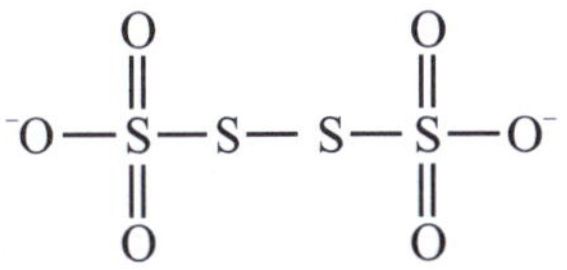

1) + 6, 0, 0, + 6 2) + 3, 1, + 1, + 3
3) + 5, 0, 0, +5 4) +4, +1, + 1, +4

31. In the conversion of Br_2 to BrO_3^-, the oxidation number of Br changes from:

1) Zero to +5 2) + 1 to +5
3) Zero to - 3 4) + 2 to + 5

32. The oxidation number of the sulphur atoms in peroxomonosulphuric acid (H_2SO_5) and peroxodisulphuric acid ($H_2S_2O_8$) respectively, are:

1) + 8 and + 7
2) + 3 and +3
3) + 6 and +6
4) + 4 and +6

33. The average oxidation number of Fe in Fe_3O_4 is:

1) + 2 2) +3
3) $\frac{8}{3}$ 4) $\frac{2}{3}$

34. The reaction of aqueous $KMnO_4$ with H_2O_2 in acidic condition gives:

1) Mn^{+4} and O_2 2) Mn^{2+} and O_2
3) Mn^{2+} and O_3 4) Mn^{4+} and MnO_2

35. Carbon is in lowest oxidation state in:

1) CO_2 2) CF_4
3) CCl_4 4) CH_4

36. Oxidation state of phosphorus in pyrophosphoric acid is:

1) + 5 2) + 3
3) +4 4) + 1

37. Oxidation state of nitrogen is correctly given for:

Compound	Oxidation state
1) Mg_3N_2	– 3
2) NH_2OH	+ 1
3) $(N_2H_5)_2SO_4$	+ 2
4) $[Co(NH_3)_5Cl]Cl_2$	0

38. Oxidation number of 'N' in N_3H (hydrazoic acid) is:

1) –1/3 2) + 3
3) 0 4) –3

39. The oxidation number of cobalt in $K[Co(CO)_4]$ is:

1) + 1 2) +3
3) –1 4) –3

40. The oxidation number of P is + 3 in-

1) H_3PO_3 2) H_3PO_4
3) HPO_3^- 4) $H_4P_2O_7$

41. In the reaction,

$3Br_2 + 6CO_3^{2-} + 3H_2O \rightarrow 5Br^- + BrO_3^- + 6HCO_3^-$

1) Bromine is oxidised and carbonate is reduced
2) Bromine is oxidised as well as reduced
3) Bromine is reduced and water is oxidised
4) Br_2 is neither oxidised nor reduced

42. Which one of the following is not a redox reaction?

1) $CaCO_3 \rightarrow CaO + CO_2$
2) $2H_2 + O_2 \rightarrow 2H_2O$
3) $Na + H_2O \rightarrow NaOH + \frac{1}{2}H_2$
4) $MnCl_3 \rightarrow MnCl_2 + \frac{1}{2}Cl_2$

43. Which of following is not a redox change ?

1) $2H_2S + SO_2 \rightarrow 2H_2O + 3S$
2) $2BaO + O_2 \rightarrow 2BaO_2$
3) $BaO_2 + H_2SO_4 \rightarrow BaSO_4 + H_2O_2$
4) $2KClO_3 \rightarrow 2KCl + 3O_2$

44. Which of the following halogens always shows only one oxidation state in compounds?

1) Cl 2) F
3) Br 4) I

45. In which reaction, there is change in oxidation number of N

1) $2NO_2 \rightarrow N_2O_4$
2) $NH_4OH \rightarrow NH_4^+ + OH^-$
3) $N_2O_5 + H_2O \rightarrow 2HNO_3$
4) $2NO_2 + H_2O \rightarrow HNO_3 + HNO_2$

46. In a reaction,

$H_2O + C \rightarrow CO + H_2$

1) H_2O is the reducing agent
2) H_2O is the oxidising agent
3) Carbon is the oxidising agent
4) Oxidation-reduction does not occur

47. The oxidation number of chlorine in HOCl is-

1) – 1 2) 0
3) + 1 4) 2

48. Which one can act as oxidising & reducing agent both?

1) HNO_2 2) H_2O_2
3) H_2SO_3 4) All

49. Which compound can not be used as oxidising agent?

1) O_3 2) HNO_3
3) $KMnO_4$ 4) NH_3

50. Oxidation number of S in H_2SO_5 is 6. This is observed because –

1) There are five oxygen atoms in the molecule
2) The hydrogen atom is directly linked with non-metal
3) There is peroxide linkage in the molecule
4) The sulphur atom shows co-ordinate linkage

51. In which of the following compound oxidation number of Cl is + 3?

1) ICl 2) ClO_3^-
3) ClF_3 4) $HClO_4$

52. Reduction is

1) Loss of electrons by any species
2) Gain of electrons by any species
3) Both of these
4) None of these

53. An oxidising agent is a substance which can

1) Accept electron
2) Donate electrons
3) Accept protons
4) Donate protons

54. In acid solution, the reaction $MnO_4^- \rightarrow Mn^{2+}$ involves

1) Oxidation by 3 electrons
2) Reduction by 3 electrons
3) Oxidation by 5 electrons
4) Reduction by 5 electrons

55. In given reaction $Cu^{2+} + Ag \rightarrow Cu + Ag^+$, what is the reduction half-reaction?

1) $Cu \rightarrow Cu^{2+}$ 2) $Ag \rightarrow Ag^+$
3) $Cu^{2+} \rightarrow Cu$ 4) All of these

56. Which of the following is the correct representation of $HAuCl_4$, as per stock notation?

1) $HAu(III)Cl_4$ 2) $HAu(II)Cl_4$
3) $HAu(IV)Cl_4$ 4) $HAu(I)Cl_4$

57. The oxidation number of Cl in Cl_2O_7 is

1) + 7 2) + 5
3) + 3 4) – 7

Balancing of Redox Reactions

58. How many moles of iodine are liberated when 1 mole of potassium dichromate reacts with potassium iodide?
1) 1
2) 2
3) 3
4) 4

59. In the redox reaction,
$xKMnO_4 + NH_3 \rightarrow KNO_3 + yMnO_2 + KOH + H_2O$
x and y are:
1) x = 4, y = 6
2) x = 3, y = 8
3) x = 8, y = 6
4) x = 8, y = 8

60. The number of electrons involved in the reduction of one nitrate ion to hydrazine is:
1) 8
2) 5
3) 3
4) 7

61. In an ionic equation,
$BiO_3^- + 6H^+ + xe^- \rightarrow Bi^{3+} + 3H_2O$
the value of x is:
1) 6
2) 2
3) 4
4) 3

62. In the following redox reaction,
$xUO^{2+} + Cr_2O_7^{2-} + yH^+ \rightarrow xUO_2^{2+} + zCr^{3+} + 4H_2O$
the value of coefficients x, y and z respectively are:
1) 3, 8, 2
2) 3, 8, 7
3) 3,2,4
4) 3,1,8

63. In the reaction,
$IO_3^- + SO_2 + 4H_2O \rightarrow I_2 + SO_4^{2-} + 8H^+$
the coefficient of SO_2 is:
1) 3
2) 4
3) 5
4) 6

64. The coefficients of I^-, IO_3^- and H^+ in the redox reaction,
$I^- + IO_3^- + H^+ \rightarrow I_2 + H_2O$
in the balanced form respectively are:
1) 5,1,6
2) 1, 5, 6
3) 6, 1, 5
4) 5, 6, 1

65. In alkaline medium ClO_2 oxidises H_2O_2 to O_2 and reduced itself to Cl^-, then how many moles of H_2O_2 will oxidise by one mole of ClO_2?
1) 1.0
2) 1.5
3) 2.5
4) 3.5

66. For the redox reaction
$Zn + NO_3^- + H_2O \rightarrow Zn^{2+} + NH_4^+ + OH^-$
In the basic medium, coefficient of Zn, NO_3^- and OH^- in the balanced reaction respectively, are:
1) 4, 1, 7
2) 7, 4, 1
3) 4, 1, 10
4) 1, 4, 10

67. For the redox reaction
$MnO_4^- + Fe^{2+} + H^+ \rightarrow Mn^{2+} + Fe^{3+} + H_2O$
in the balanced equation, correct coefficient are

	MnO_4^-	Fe^{2+}	H^+
1)	1	5	8
2)	16	5	2
3)	5	16	2
4)	2	16	5

Redox Reactions and Electrode Processes

68. Which of the following is the most powerful oxidising agent?
1) F_2
2) O_2
3) Br_2
4) I_2

69. Given : $Zn^{2+}/Zn = -0.76$ V, $Mg^{2+}/Mg = -2.37$ V, then the correct statement about the reaction $Zn(s) + MgCl_2(aq) \rightarrow$
1) Solid zinc dissolves
2) Zinc chloride precipitates
3) Magnesium chloride precipitates
4) No reaction takes place.

70. A negative E^0 means that the redox couple is
1) A stronger reducing agent
2) A stronger oxidizing agent
3) A weaker reducing agent
4) None of these

High Order Time Intensive MCQs

1. The equivalent weight of $MnSO_4$ is equal to its molecular weight when it is converted to:
1) Mn_2O_3
2) MnO_2
3) MnO_4^-
4) MnO_4^{2-}

2. The equivalent mass of $KMnO_4$ in the following reaction is:
$MnO_4^- + 5Fe^{2+} + 8H^+ \rightarrow Mn^{2+} + 5Fe^{3+} + 4H_2O$
(M = Molecular mass)
1) $\frac{M}{2}$
2) $\frac{M}{3}$
3) $\frac{M}{4}$
4) $\frac{M}{5}$

3. Equivalent weight of Mn^{3+} in the following reaction is (Mn = 55)
$Mn^{3+} \rightarrow Mn^{2+} + MnO_2$
1) 27.5
2) 55
3) 110
4) 165

4. In the following change,
$3Fe + 4H_2O \rightarrow Fe_3O_4 + 4H_2$
If the atomic weight of iron is 56, then its equivalent weight will be:
1) 42
2) 21
3) 63
4) 84

5. The equivalent mass of potassium permanganate in strong alkaline medium is:

1) $\frac{\text{Molar mass}}{5}$ 2) $\frac{\text{Molar mass}}{3}$

3) $\frac{\text{Molar mass}}{2}$ 4) Molar mass itself

6. Equivalent weight of FeC_2O_4 in the reaction,

$FeC_2O_4 \rightarrow Fe^{3+} + CO_2$ is

1) M/3 2) M/6

3) M/2 4) M/1

7. For the reaction, $N_2 + 3H_2 \rightarrow 2NH_3$; if E_1 & E_2 are the equivalent masses of NH_3 & N_2; then $E_1 - E_2$ is:

1) 1 2) 2

3) 3 4) 4

8. What mass of HNO_3 is required to make 1 ltr. of 2N solution to be used as oxidising agent in the reaction?

$3Cu + 8HNO_3 \rightarrow 3Cu(NO_3)_2 + 2NO + 4H_2O$

1) 63 gm 2) 42 gm

3) 84 gm 4) 126 gm

9. If equal volumes of 1 M $KMnO_4$ & 1 M $K_2Cr_2O_7$ solutions are allowed to oxidise Fe^{2+} to Fe^{3+} in acidic medium, then Fe^{2+} oxidised will be:

1) More by $KMnO_4$ 2) More by $K_2Cr_2O_7$

3) Equal in both cases 4) Can't be determined

10. Calculate the equivalent weight of KHC_2O_4 when it is neutralised by KOH.

1) 128 2) 64

3) 42 4) None of these

11. The number of moles of MnO_4^- required by 1 mole of ferric oxalate in acidic medium for oxidation is/are-

1) 0.2 2) 1.2

3) 2.6 4) 5.2

12. The oxidation states of S in SO_4^{2-}, $S_2O_8^{2-}$ & H_2SO_5 are respectively:

1) +6, +6, +7 2) +6, +7, +6

3) +7, +7, +7 4) +6, +6, +6

13. In basic medium, CrO_4^{2-} oxidises $S_2O_3^{2-}$ to form SO_4^{2-} and itself changes to $Cr(OH)_4^-$. How many mL of 0.154 M CrO_4^{2-} are required to react with 40 mL of 0.246 M $S_2O_3^{2-}$?

1) 200 mL 2) 156.4 mL

3) 170.4 mL 4) 190.4 mL

14. Which of the following is/are redox reaction (s)?

1) $SO_2 + 2H_2S \rightarrow 2H_2O + 3S$

2) $2BaO + O_2 \rightarrow 2BaO_2$

3) $2KClO_3 \rightarrow 2KCl + 3O_2$

4) All are correct

15. When $Cr(s) + OH^-(aq) \rightarrow Cr(OH)_4^- (aq) + H_2(g)$ (basic solution) is balanced, the sum of the coefficients of all the reactants and products is:

1) 14 2) 15

3) 17 4) 9

16. A 0.1097 g sample of As_2O_3 required 36.10 mL of $KMnO_4$ solution for its titration. The molarity of $KMnO_4$ solution is.

1) 0.02 M 2) 0.04 M

3) 0.0122 M 4) 0.3 M

17. Standard electrode potentials of redox couples A^{2+}/A, B^{2+}/B, C^{2+}/C and D^{2+}/D are 0.3 V, –0.5 V, –0.75 V and 0.9 V respectively. Which of these is best oxidising agent and reducing agent respectively?

1) D^{2+}/D and B^{2+}/B 2) B^{2+}/B and D^{2+}/D

3) D^{2+}/D and C^{2+}/C 4) C^{2+}/C and D^{2+}/D.

18. The weight of sodium bromate required to prepare 55.5 mL of 0.672 N solution for cell reaction,

$BrO_3^- + 6H^+ + 6e^- \rightarrow Br^- + 3H_2O$, is

1) 1.56 g 2) 0.9386 g

3) 1.23 g 4) 1.32 g

19. What volume of 0.1 M $KMnO_4$ is needed to oxidize 100 mg of FeC_2O_4 in acid solution ?

1) 4.1 mL 2) 8.2 mL

3) 10.2 mL 4) 4.6 mL

20. When one mole NO_3^- is converted into 1 mole NO_2, 0.5 mole N_2 and 0.5 mole N_2O respectively, it accepts x, y and z mole of electrons. Then x, y and z are respectively:

1) 1, 5, 4 2) 1, 2, 3

3) 2, 1, 3 4) 2, 3, 4

21. A, B and C are three elements forming a part of compound in oxidation states of +2, +5 and –2 respectively. What could be the compound if molecule is neutral?

1) $A_2(BC)_2$ 2) $A_2(BC_4)_3$

3) $A_3(BC_4)_2$ 4) ABC

22. The oxidation number of S in $Na_2S_4O_6$ is-

1) + 3.5 for each S atom

2) + 2 and + 3 (two S have + 2 and other two have + 3)

3) + 2 and + 3 (three S have + 2 and one S has + 3)

4) + 5 and 0 (two S have + 5 and the other two have 0)

23. Which one is correctly matched?

	Substance	O.N. of S
1)	H_2S	+2
2)	H_2SO_5	+4
3)	H_2SO_4	+4
4)	H_2SO_3	+4

24. In the following reaction
$O_3 + 6I^- + 6H^+ \rightarrow 3I_2 + 3H_2O$
equivalent weight of O_3 (with molecular weight M) is -

1) $\frac{M}{2}$ 2) $\frac{M}{4}$

3) $\frac{M}{24}$ 4) $\frac{M}{6}$

25. How many electrons should X_2H_4 liberate so that in the new compound, X shows oxidation number of –1/2? [E.N. of X_4 > H]

1) 10 2) 4

3) 3 4) 2

26. The oxidation state of oxygen in O_2PtF_6 is -

1) Zero 2) + 1/2

3) + 1 4) –1/2

27. $(NH_4)_2Cr_2O_7 \xrightarrow{\text{heat}} Cr_2O_3 + N_2 + 4H_2O$
In the above reaction the oxidation state of N and chromium changes respectively as -

1) – 3 to 0 and + 6 to + 3

2) + 3 to 0 and + 3 to + 6

3) – 4 to 0 and + 6 to + 2

4) Unpredictable

28. When beryllium carbide is reacted with water, gas X of characteristic smell is given out. The oxidation state of heaviest constituent element of gas X should be -

1) – 1 2) + 1

3) – 4 4) –2

29. In the reaction,

$S_2O_3^{2-}(aq) + X_2(s) \text{ or } (l) \longrightarrow S_4O_6^{2-}(aq) + 2X^-(aq)$

X in the above reaction will be

1) Both Br and I 2) Only I

3) Only Br 4) Only S

30. Identify the disproportionation reaction:

1) $H_2O_2(aq) \rightarrow 2H_2O(l) + O_2(g)$

2) $S_8(s) + 12OH^-(aq) \rightarrow 4S^{2-}(aq) + 2S_2O_3^{2-}(aq)$

3) $Cl_2(g) + 2OH^- \rightarrow ClO^-(aq) + Cl^-(aq) + H_2O(l)$

4) All of these

31. One mole of N_2H_4 loses 10 moles of electrons to form a compound Y. Assuming that all nitrogen appears in the new compound, if there is no change in O.N. of hydrogen, the O.N. of nitrogen in Y is

1) + 3 2) – 3

3) + 1 4) + 5

32. Equivalent weight of oxidising agent will be-
$2H_2 + O_2 \rightarrow 2H_2O$

1) 1 2) 32

3) 2 4) 8

33. In the following reaction (unbalanced), equivalent weight of As_2S_3 is related to molecular weight M by :
$As_2S_3 + H^+ + NO_3^- \rightarrow NO + H_2O + AsO_4^{3-} + SO_4^{2-}$

1) $\frac{M}{2}$ 2) $\frac{M}{4}$

3) $\frac{M}{28}$ 4) $\frac{M}{24}$

34. Consider the following statement(s)

A. Iron reacts with dilute H_2SO_4 to evolve $H_2(g)$ but Ag does not.

B. Copper dissolves in dilute HNO_3 but not in dilute HCl.

C. Amongst halogens, fluorine is the best reductant.

Choose the correct statement(s)

1) A & B 2) B & C

3) A & C 4) A, B & C

35. Which of the following element does not show disproportionation tendency?

1) O_2 2) F_2

3) I_2 4) N_2

36. An element A in a compound ABD has oxidation number A^{n-}. It is oxidised by $Cr_2O_7^{2-}$ in acidic medium. In the experiment, 1.68×10^{-3} mole of $K_2Cr_2O_7$ were used for 3.26×10^{-3} mole of ABD. The new oxidation number of A after oxidation is:

1) 3 2) 3 – n

3) n – 3 4) +n

37. Which of the following reactions is an example of redox reaction?

1) $CH_2{=}CH_2(g) + H_2(g) \rightarrow H_3C - CH_3(g)$

2) $NaOH(aq) + HCl(aq) \rightarrow NaCl(aq) + H_2O(l)$

3) $XeF_6(g) + 3H_2O(l) \rightarrow XeO_3 + 6HF$

4) All of these

38. Equivalent weight of Br_2 is 96 in the following disproportionation reaction:
$Br_2 + OH^- \rightarrow Br^- + H_2O + X$
The oxidised product is: (Br = 80)

1) BrO^- 2) BrO_2^-

3) BrO_3^- 4) BrO_4^-

39. Which of the following reactions represent reduction process?

1) $S(s) + O_2(g) \rightarrow SO_2(g)$

2) $2Mg(s) + O_2(g) \rightarrow 2MgO(s)$

3) $2HgO(s) \rightarrow 2Hg(l) + O_2(g)$

4) None of these

40. Among NH_3, HNO_3, NaN_3 and Mg_3N_2 the number of molecules having nitrogen in negative oxidation state is

1) 1 2) 2

3) 3 4) 4

41. Among the species, ClO^-, ClO_2^-, ClO_3^- and ClO_4^-, which one does not show disproportionation reaction?

1) ClO^- 2) ClO_2^-
3) ClO_3^- 4) ClO_4^-

42. Which of the following represents the correct decreasing order of oxidation number of oxygen in compounds BaO_2, O_3, KO_2 and OF_2?

1) $BaO_2 > KO_2 > O_3 > OF_2$
2) $OF_2 > O_3 > KO_2 > BaO_2$
3) $O_3 > OF_2 > KO_2 > BaO_2$
4) $BaO_2 > O_3 > OF_2 > KO_2$

43. In the balanced chemical reaction,

$Cr_2O_7^{2-}(aq) + aSO_3^{2-}(aq) + bH^+(aq) \rightarrow xCr^{3+}(aq) + ySO_4^{2-}(aq) + zH_2O(l)$

a, b, x, y and z in the above reaction correspond to

1) 2,3,4,5,6 2) 3,8,2,3,4
3) 1,2 ,5, 9,4 4) 2,8,5,3,5

44. The oxidation numbers of phosphorus in PO_4^{3-}, of sulphur in SO_4^{2-} and that of chromium in $Cr_2O_7^{2-}$ are respectively

1) + 3, + 6 and + 5 2) + 5, + 3 and + 6
3) − 3, + 6 and + 6 4) + 5, + 6 and + 6

Match the Columns MCQs

1. Match Column-I with Column-II for the oxidation states of the central atoms

Column-I		Column-II	
(a)	$Cr_2O_7^{2-}$	(i)	+3
(b)	MnO_4^-	(ii)	+4
(c)	VO_3^-	(iii)	+5
(d)	FeF_6^{3-}	(iv)	+6
		(v)	+7

1) (a)-(iv) (b)-(i) (c)-(iii) (d)-(ii)
2) (a)-(iv) (b)-(v) (c)-(iii) (d)-(i)
3) (a)-(ii) (b)-(iv) (c)-(iii) (d)-(i)
4) (a)-(i) (b)-(iv) (c)-(ii) (d)-(iii)

2. Match the items in Column-I with relevant items in Column-II.

Column-I		Column-II	
(a)	Ions having positive charge	(i)	+7
(b)	The sum of oxidation number of all atoms in a neutral molecule	(ii)	–1
(c)	Oxidation number of hydrogen ion (H^+)	(iii)	+1
(d)	Oxidation number of fluorine in NaF	(iv)	0
(e)	Ions having negative charge	(v)	Cation
		(vi)	Anion

1) (a)-(iv) (b)-(i) (c)-(ii) (d)-(ii) (e)-(v)
2) (a)-(vi) (b)-(iii) (c)-(ii) (d)-(iv) (e)-(v)
3) (a)-(i) (b)-(iv) (c)-(vi) (d)-(ii) (e)-(iii)
4) (a)-(v) (b)-(iv) (c)-(iii) (d)-(ii) (e)-(vi)

3. Match the Column-I with column-II.

Column-I		Column-II	
(a)	$\text{Eq. wt.} = \frac{\text{Molecular weight}}{33}$	(i)	When CrI_3 oxidise into $Cr_2O_7^{2-}$ and IO_4^-
(b)	$\text{Eq. wt.} = \frac{\text{Molecular weight}}{27}$	(ii)	When Fe $(SCN)_2$ oxidise into Fe^{3+}, SO_4^{2-},
(c)	$\text{Eq. wt.} = \frac{\text{Molecular weight}}{28}$	(iii)	When NH_4SCN oxidise into SO_4^{2-}, CO_3^{2-}
(d)	$\text{Eq. wt.} = \frac{\text{Molecular weight}}{24}$	(iv)	When As_2S_3 oxidise into AsO_3^- and SO_4^{2-}

1) (a)-(ii) (b)-(i) (c)-(iii) (d)-(iv)
2) (a)-(iv) (b)-(ii) (c)-(i) (d)-(iii)
3) (a)-(iv) (b)-(i) (c)-(iii) (d)-(ii)
4) (a)-(ii) (b)-(i) (c)-(iv) (d)-(iii)

4. Match the items in Column-I with relevant items in Column-II.

Column-I		Column-II	
(a)	$P_2H_4 \rightarrow PH_3 + P_4H_2$	(i)	$E = \frac{3M}{4}$
(b)	$I_2 \rightarrow I^- + IO_3^-$	(ii)	$E = \frac{3M}{5}$
(c)	$MnO_4^- + Mn^{2+} + H_2O \rightarrow Mn_3O_4 + H^+$	(iii)	$E = \frac{15M}{26}$
(d)	$H_3PO_2 \rightarrow PH_3 + H_3PO_3$	(iv)	$E = \frac{5M}{6}$

1) (a)-(iv) (b)-(ii) (c)-(iii), (d)-(i)
2) (a)-(iv) (b)-(ii) (c)-(i) (d)-(iii)
3) (a)-(iv) (b)-(ii) (c)-(iii) (d)-(ii)
4) (a)-(i) (b)-(iii) (c)-(iv) (d)-(i)

5. Match the column-I with column-II.

	Column-I		Column-II
(a)	$V_2O_5(s) + 5Ca(s) \rightarrow 2V(s) + 5CaO(s)$	(i)	Disproportionation reaction
(b)	$CH_4(g) + 2O_2(g) \xrightarrow{\Delta} CO_2(g) + 2H_2O(l)$	(ii)	Decomposition reaction
(c)	$P_4(s) + 3OH^-(aq) + 3H_2O(l) \rightarrow PH_3(g) + 3H_2PO_2^-(aq)$	(iii)	Combustion reaction
(d)	$2KClO_3(s) \xrightarrow{\Delta} 2KCl(s) + 3O_2(g)$	(iv)	Displacement reaction

1) (a)-(iv) (b)-(iii) (c)-(i) (d)-(ii)
2) (a)-(iv) (b)-(ii) (c)-(i) (d)-(iii)
3) (a)-(iii) (b)-(iv) (c)-(i) (d)-(ii)
4) (a)-(ii) (b)-(iv) (c)-(i) (d)-(iii)

Correct & Incorrect MCQs

1. Identify the incorrect statement :
   1) The oxidation number of oxygen in most compounds is –2.
   2) The oxidation number of oxygen in OF_2 is +2.
   3) The oxidation number of oxygen in O_2F_2 is +1.
   4) Each oxygen atom is assigned an oxidation number of –1/2 in peroxides like H_2O_2 and Na_2O_2.

2. Which of the following statement is not correct about oxidation number ?
   1) In elements, in the free or the uncombined state, each atom bears an oxidation number of zero.
   2) For ions composed of only one atom, the oxidation number is equal to the charge on the ion.
   3) The algebraic sum of the oxidation number of all the atoms in a compound must be 1.
   4) In polyatomic ion, the algebraic sum of all the oxidation numbers of atoms of the ion must equal the charge on the ion.

3. Which of the following statement (s) is/are correct?
   A. Displacement reaction of Cl, Br, I using fluorine are not generally carried out in the aqueous solution.
   B. Fluorine is so reactive that it can replace chloride, bromide & iodide ions in the solution.
   C. There is no way to convert F^- to F_2 by the chemical reaction.
   D. Fluorine is the strongest reducing agent.
   1) A, B and C 2) A and D
   3) A, C and D 4) B, C and D

4. Which of the following statement is incorrect?
   1) In a disproportionation reaction, an element in one oxidation state is simultaneously oxidised and reduced.
   2) ClO_4^- undergoes disproportionation reaction.
   3) Cl, Br, & I show disproportionation but fluorine doesn't show.
   4) Phosphorous, sulphur and chlorine undergo disproportionation in the alkaline medium.

5. Correct statement(s) is/are
   A. Redox reaction is also known as electron transfer reaction.
   B. In redox reactions, electrons are transferred from the oxidant to the reductant.
   C. $CaCO_3 \xrightarrow{\Delta} CaO + CO_2$ is not a redox reaction.
   D. $Cu(s) + 2Ag^+(aq) \rightarrow Cu^{2+}(aq) + 2Ag(s)$ is not a redox reaction.
   1) Only C 2) B, C and D
   3) A and C 4) B and D

6. Which of the following statements is incorrect ?
   1) In redox systems, the titration method can be adopted to determine the strength of a reductant/oxidant using a redox sensitive indicator.
   2) MnO_4^- acts as a self indicator in the titration.
   3) $Cr_2O_7^{2-}$ acts as a self indicator in the titration.
   4) Iodine itself gives an intense blue colour with starch and has a very specific reaction with thiosulphate ions ($S_2O_3^{2-}$).

7. Select the incorrect statement :
   1) Fluorine is the strongest oxidizing agent.
   2) Sulphur dioxide and hydrogen peroxide can act as oxidising as well as reducing agents in their reactions.
   3) HNO_2 can neither act as an oxidising nor as a reducing agent.
   4) Ozone and nitric acid act only as oxidants.

8. Incorrect statements are:
   A. Oxidation is the increase in the oxidation number of the element in the given substance.
   B. $AgCN + KCN \rightarrow K[Ag(CN)_2]$, Ag undergoes oxidation.
   C. $2Cu_2O(s) + Cu_2S(s) \rightarrow 6Cu(s) + SO_2(g)$ is a redox reaction.
   D. In $MnO_2 + HCl \rightarrow MnCl_2 + Cl_2 + H_2O$, HCl behave as an oxidising agent.
   1) A and B 2) B, C and D
   3) A and D 4) B and D

9. Which of the following statements is/are correct ?

A. In a redox reaction, reductant gets reduced and oxidant gets oxidised.

B. In a reaction $Zn(s) + CuSO_4(aq) \rightarrow ZnSO_4(aq) + Cu(s)$, Zn is a reductant and itself get oxidized.

C. In a reaction, $Cu(s) + AgNO_3(aq) \rightarrow Cu(NO_3)_2(aq) + Ag(s)$, $AgNO_3$ behaves as an oxidising agent.

D. Redox reaction occurs in photosynthesis.

1) A only
2) A and C
3) B, C and D
4) A and D

10. Identify the incorrect statement from the following :

1) Conversion of potassium ferrocyanide to potassium ferricyanide is an oxidation process.
2) The oxidation number of carbon in C_3O_2 is 4/3.
3) Oxidation no. of Mg^{2+} ion is '+2', Cl^- ion is '-1', O^{2-} ion is '–2'.
4) Aluminium has most common +2 oxidation number in all its compounds.

Statement Based MCQs

Directions: These questions consist of two statements each, printed as Statement-I and Statement-II. While answering these questions, you are required to choose any one of the following four responses.

1) **Both Statement-I and Statement-II are correct.**
2) **Both Statement-I and Statement-II are incorrect.**
3) **Statement-I is correct & Statement-II is incorrect.**
4) **Statement-I is incorrect & Statement-II is correct.**

1. **Statement-I:** Cu forms hydrogen gas with dil. HCl.

Statement-II: Position of Cu is above H_2 in electrochemical series.

2. **Statement-I:** In the redox reaction $8\,H^+(aq) + 4\,NO_3^- + 6\,Cl^- + Sn(s) \rightarrow SnCl_6^{2-} + 4\,NO_2 + 4\,H_2O$, the reducing agent is Sn(s).

Statement-II: In balancing half-reaction, $S_2O_3^{2-} \rightarrow S(s)$, the number of electrons added on the left is 4.

3. **Statement-I:** The oxidation number of Cr in CrO_5 is +6.

Statement-II: In CrO_5, four oxygen atoms are involved in peroxide linkage.

4. **Statement-I:** $Fe_2O_3 + 2Al \rightarrow Al_2O_3 + 2Fe$

The above chemical equation is an example of a displacement reaction.

Statement-II: Aluminium is more reactive than iron, displacing Fe from its oxide.

5. **Statement-I:** Equivalent weight of NH_3 in the reaction, $N_2 \rightarrow NH_3$ is 17/3 while that of N_2 is 28/6.

Statement-II: Equivalent weight

$$= \frac{\text{Molecular weight}}{\text{number of } e^- \text{ lost or gained}}$$

6. **Statement-I:** Chlorine, bromine and iodine when combined with oxygen, for example in oxoacids and oxoanions, have positive oxidation numbers.

Statement-II: Fluorine has an oxidation number of +1 in all its compounds.

7. **Statement-I:** In a disproportionation reaction, an element in one oxidation state is only oxidized.

Statement-II: Decomposition of hydrogen peroxide is an example of disproportionation reaction.

8. **Statement-I:** Oxidation number of carbon in CH_2O is +2.

Statement-II: Oxidation number of hydrogen in CaH_2 is -1.

9. **Statement-I:** Cu is stronger reducing agent than H_2.

Statement-II: E° of Cu^{2+}/Cu is negative.

10. **Statement-I:** Oxidation number exhibited by an atom generally increases across the period in periodic table.

Statement-II: ClO_4^- does not disproportionate.

Assertion & Reason MCQs

Directions: These questions consist of two statements each, printed as Assertion and Reason. While answering these questions, you are required to choose any one of the following four responses.

1) **Both Assertion and Reason are True and the Reason is a correct explanation of the Assertion.**
2) **Both Assertion and Reason are True but Reason is not a correct explanation of the Assertion.**
3) **Assertion is True but the Reason is False.**
4) **Assertion is False but Reason is True.**

1. **Assertion:** Among halogens, fluorine is the best oxidant.

Reason: Fluorine is the most electronegative atom.

2. **Assertion:** In the reaction between potassium permanganate and potassium iodide, permanganate ions act as oxidising agent.

Reason: Oxidation state of manganese changes from +2 to +7 during the reaction.

3. **Assertion:** The decomposition of hydrogen peroxide to form water and oxygen is an example of disproportionation reaction.

Reason: The oxygen of peroxide is in –1 oxidation state and it is converted to zero oxidation state in O_2 and –2 oxidation state in H_2O.

4. **Assertion:** Redox couple is the combination of oxidised and reduced form of a substance involved in an oxidation or reduction half cell.

Reason: In the representation $E^0_{Fe^{3+}/Fe^{2+}}$ and $E^0_{Cu^{2+}/Cu}$, Fe^{3+}/Fe^{2+} and Cu^{2+} / Cu are redox couples.

5. **Assertion:** Silver can be precipitated by adding zinc powder to $AgNO_3$ powder.
Reason: $ZnSO_4$ can be stirred with silver spoon.

6. **Assertion:** Oxidation state of N in N_3H is –1/3.
Reason: Nitrogen is less electronegative than hydrogen.

7. **Assertion:** Bromide ion acts as a reducing agent in a reaction.
$$2MnO_4^-(aq) + Br^-(aq) + H_2O(l) \rightarrow 2MnO_2(aq) + 3\ BrO_3^-(aq) + 2OH(aq)$$
Reason: Oxidation number of bromine increases from, –1 to 5.

8. **Assertion:** $KMnO_4$ acts as oxidant only.
Reason: Mn in $KMnO_4$ is in its highest oxidation state.

9. **Assertion:** $AgCN + KCN \rightarrow K[(Ag(CN)_2]$
Ag undergoes reduction in this reaction.
Reason: Decrease in oxidation number is known as reduction.

10. **Assertion:** HNO_3 acts as a reductant.
Reason: In HNO_3 molecule, central atom is in its maximum oxidation state.

Multi Correct MCQs

1. Iron (III) catalyses the reaction between iodide and persulphate ions, in which
A. Fe^{3+} oxidises the iodide ion
B. Fe^{3+} oxidises the persulphate ion
C. Fe^{2+} reduces the iodide ion
D. Fe^{2+} reduces the persulphate ion
Choose the most appropriate answer from the options given below:
1) B and C only
2) B only
3) A only
4) A and D only

2. Which of the following options are correct for the reaction
$2[Au(CN)_2]^-_{(aq)} + Zn(s) \rightarrow 2Au(s) + [Zn(CN)_4]^{-2}_{(aq)}$?
A. Redox reaction
B. Displacement reaction
C. Decomposition reaction
D. Combination reaction
Choose the correct answer from the options given below:
1) A and B only
2) A only
3) C and D only
4) A and D only

3. From the given list, choose the set of compounds with + 4 oxidation state of Sulphur.
A. SO_3
B. H_2SO_3
C. $SOCl_2$
D. SF_4
E. $BaSO_4$
F. $H_2S_2O_7$
1) (A), (B) (C) and (D)
2) (A), (C) and (D)
3) (B), (C), (D), (E) and (F)
4) (B), (C) and (D)

4. Choose the set from the following which can undergo disproportionation reaction.
A. H_2O_2
B. ClO_3^-
C. P_4
D. Cl_2
E. Ag
F. Cu^{+1}
G. F_2
H. NO_2
I. K^+
1) (C), (D), (E), (F) and (G)
2) (B), (A), (C), (D), (F), (H)
3) (C), (D), (E), (F), (H)
4) (A), (B), (G) and (I)

NCERT Exemplar MCQs

1. Which of the following is not an example of redox reaction?
1) $CuO + H_2 \rightarrow Cu + H_2O$
2) $Fe_2O_3 + 3CO \rightarrow 2Fe + 3CO_2$
3) $2K + F_2 \rightarrow 2KF$
4) $BaCl_2 + H_2SO_4 \rightarrow BaSO_4 + 2HCl$

2. $E^\ominus$ value of some redox couples are given below. On the basis of these values choose the correct option:
$E^\ominus$ values : $Br_2/Br^- = +1.90$
$Ag^+/Ag(s) = +0.80$
$Cu^{2+}/Cu(s) = +0.34$; $I_2(s)/I^- = +0.54$
1) Cu will reduce Br^-
2) Cu will reduce Ag
3) Cu will reduce I^-
4) Cu will reduce Br_2

3. Using the standard electrode potential, find out the pair between which the reaction is not feasible.
$E^\ominus$ values: $Fe^{3+}/Fe^{2+} = +0.77$; $I_2/I^- = +0.54$;
$Cu^{2+}/Cu = +0.34$; $Ag^+/Ag = +0.80$ V
1) Fe^{3+} and I^-
2) Ag^+ and Cu
3) Fe^{3+} and Cu
4) Ag and Fe^{3+}

4. Thiosulphate reacts differently with iodine and bromine in the reactions given below:
$$2S_2O_3^{2-} + I_2 \rightarrow S_4O_6^{2-} + 2I^-$$
$$S_2O_3^{2-} + 2Br_2 + 5H_2O \rightarrow 2SO_4^{2-} + 4Br^- + 10H^+$$
Which of the following statements justifies the above dual behaviour of thiosulphate?
1) Bromine is a stronger oxidant than iodine
2) Bromine is a weaker oxidant than iodine
3) Thiosulphate undergoes oxidation by bromine and reduction by iodine in these reactions
4) Bromine undergoes oxidation and iodine undergoes reduction in these reactions

5. The oxidation number of an element in a compound is evaluated on the basis of certain rules. Which of the following is incorrect in this respect?
   1) The oxidation number of hydrogen is always +1
   2) The algebraic sum of all the oxidation numbers in a compound is zero
   3) An element in the free or the uncombined state bears oxidation number zero
   4) In all its compounds, the oxidation number of fluorine is –1

6. In which of the following compounds, an element exhibits two different oxidation states?
   1) NH_2OH
   2) NH_4NO_3
   3) N_2H_4
   4) N_3H

7. Which of the following arrangements represent increasing oxidation number of the central atom?
   1) $CrO_2^-, ClO_3^-, CrO_4^{2-}, MnO_4^-$
   2) $ClO_3^-, CrO_4^{2-}, MnO_4^-, CrO_2^-$
   3) $CrO_2^-, ClO_3^-, MnO_4^-, CrO_4^{2-}$
   4) $CrO_4^{2-}, MnO_4^-, CrO_2^-, ClO_3^-$

8. The largest oxidation number exhibited by an element depends on its outer electronic configuration. With which of the following outer electronic configurations, the element will exhibit largest oxidation number?
   1) $3d^1 4s^2$
   2) $3d^3 4s^2$
   3) $3d^5 4s^1$
   4) $3d^5 4s^2$

9. Identify disproportionation reaction:
   1) $CH_4 + 2O_2 \rightarrow CO_2 + 2H_2O$
   2) $CH_4 + 4Cl_2 \rightarrow CCl_4 + 4HCl$
   3) $2F_2 + 2OH^- \rightarrow 2F^- + OF_2 + H_2O$
   4) $2NO_2 + 2OH^- \rightarrow NO_2^- + NO_3^- + H_2O$

10. Which of the following elements does not show disproportionation tendency?
    1) Cl
    2) Br
    3) F
    4) I

Kattar NEET

Scan Me for Detailed Explanations

1. X gram of pure As_2S_3 is completely oxidised to respective highest oxidation states by 50 ml of 0.1 M hot acidified $KMnO_4$ then X, mass of As_2S_3 taken is: (Molar mass of $As_2S_3 = 246$)
   1) 22.4 g
   2) 0.22 g
   3) 64.23 g
   4) 1.06 g

2. Volume V_1 mL of 0.1 M $K_2Cr_2O_7$ is needed for complete oxidation of 0.678 g N_2H_4 in acidic medium. The volume of 0.3 M $KMnO_4$ needed for the same oxidation in acidic medium will be:
   1) $\frac{2}{5}V_1$
   2) $\frac{5}{2}V_1$
   3) 113 V_1
   4) $\frac{7}{5}V_1$

3. 0.4 g of a polybasic acid H_nA (all the hydrogens are acidic) requires 0.5g of NaOH for complete neutralization. The number of replaceable hydrogen atoms in the acid and the molecular weight of 'A' would be: (Molecular weight of the acid is 96 gms/mole.)
   1) 1, 95
   2) 2, 94
   3) 3, 93
   4) 4, 92

4. In an iodometric estimation, the following reaction occur
   $2Cu^{2+} + 4I^- \longrightarrow Cu_2I_2 + I_2;$
   $I_2 + 2Na_2S_2O_3 \longrightarrow 2NaI + Na_2S_4O_6$
   0.12 mole of $CuSO_4$ was added to excess of KI solution and the liberated iodine required 120 mL of hypo. The molarity of hypo solution was:
   1) 2
   2) 0.20
   3) 0.1
   4) 1.0

5. In the redox reaction:
   $x\,MnO + y\,PbO_2 + z\,HNO_3 \rightarrow a\,HMnO_4 + b\,Pb(NO_3)_2 + c\,H_2O$
   1) x = 2, y = 5, z = 10
   2) x = 2, y = 7, z = 8
   3) x = 2, y = 5, z = 8
   4) x = 2, y = 5, z = 5

6. Which of the following is not correct for the reaction?
   $(CN)_2 + 2OH^- \rightarrow CNO^- + CN^- + H_2O$
   1) It is a disproportionation reaction
   2) N atom disproportionate and oxidation number of N are –3 in $(CN)_2$, –2 in CN^- and –5 in CNO^-
   3) C atom disproportionate and oxidation number of carbon are +3 in $(CN)_2$, + 4 in CNO^- and +2 in CN^-
   4) $(CN)_2$ undergoes auto redox

7. Lead dioxide(PbO_2) and potassium permanganate are used as oxidising agents in acidic medium for oxidation of ferrous ammonium sulphate to ferric sulphate. The ratio of number of moles of PbO_2 required per mole of ferrous ammonium sulphate to the number of moles of $KMnO_4$ required per mole of ferrous ammonium sulphate, is
   1) 2.5
   2) 0.2
   3) 0.4
   4) 2.0

8. Titration of I_2 produced from 0.1045 g of primary standard KIO_3 required 30.72 mL of sodium thiosulphate as shown below:
$IO_3^- + 5I^- + 6H^+ \rightarrow 3I_2 + 3H_2O$
$I_2 + 2S_2O_3^{2-} \rightarrow 2I^- + S_4O_6^{2-}$
The molarity of sodium thiosulphate ion is :
1) 0.095 2) 0.079
3) 0.084 4) 0.064

9. 2 mole, equimolar mixture of $Na_2C_2O_4$ and $H_2C_2O_4$ required V_1 L of 0.1 M $KMnO_4$ in acidic medium for complete oxidation. The same amount of the mixture required V_2 L of 0.1 M NaOH. The ratio of V_1 to V_2 is :
1) 2 : 1 2) 4 : 5
3) 5 : 4 4) 2 : 5

10. On passing 1.8 L of oxygen through ozonizer gives 1.6 L of a mixture of ozone and oxygen. What is the mole fraction of ozone in the resultant mixture?
1) 0.15 2) 0.25
3) 0.40 4) 0.50

11. A metal oxide Z_2O_3 can be reduced by hydrogen to give free metal and water. 0.1596 gm of metal oxide requires 6 mg of hydrogen for complete reduction. The atomic weight of the metal is
1) 27.9 2) 159.6
3) 79.8 4) 55.8

12. $K_2Cr_2O_7 + HCl \rightarrow KCl + CrCl_3 + Cl_2 + H_2O$. One mole of HCl reacts completely with $K_2Cr_2O_7$ to give Cl_2. How many grams of MnO_2 will be required for the production of same amount of chlorine from HCl?
1) 18.642 g 2) 1.19717 g
3) 14.034 g 4) 43.5 g

13. The equivalent weight of HNO_3 in following reaction $3Cu + 8HNO_3 \rightarrow 3Cu(NO_3)_2 + 2NO + 4H_2O$ is
1) $4\times\frac{63}{3}$ 2) $\frac{63}{5}$
3) $\frac{63}{3}$ 4) 63

14. Match the following:

List-I		List-II	
I.	$\underset{3.5\text{ mole}}{Sn^{+2}} + \underset{1.2\text{ mole}}{MnO_4^-}$ (acidic)	P.	Amount of oxidant available decides the number of electrons transfer.
II.	$\underset{8.4\text{ mole}}{H_2C_2O_4} + \underset{3.6\text{ mole}}{MnO_4^-}$ (acidic)	Q.	Amount of reductant available decides the number of electrons transfer
III.	$\underset{7.2\text{ mole}}{S_2O_3^{2-}} + \underset{3.6\text{ mole}}{I_2}$	R.	Number of electrons involved per mole of oxidant > Number of electrons involved per mole of reductant.
IV.	$\underset{9.2\text{ mole}}{Fe^{+2}} + \underset{1.6\text{ mole}}{Cr_2O_7^{-2}}$ (acidic)	S.	Number of electrons involved per mole of oxidant < Number of electrons involved per mole of reductant.

1) I-P, R; II-Q, R; III-P, Q, R; IV-Q, R
2) I-Q, R; II-P, R; III-P, Q, R; IV-S, R
3) I-R; II-Q, R; III-P, S, R; IV-Q, R
4) I-S, R; II-Q, R; III-P, Q, R; IV-Q, R

15. Match the following:

Column-I		Column-II (Equivalent Mass)	
I.	60 % metal in metal oxide	P.	12
II.	64.4 % metal in metal oxide	Q.	8
III.	29 % metal in metal chloride	R.	14.5
IV.	Mg	S.	35.5
		T.	Eq. mass $= \frac{\text{Molar mass}}{\text{n - factor}}$

1) I-P,Q,T; II-Q,R,T; III-R,S,T; IV-P,T
2) I–R, S; II–P, S; III–P, Q; IV–Q, R
3) I–S; II–P,Q,T; III–Q; IV–R, S
4) I–R; II–S; III–P, Q; IV–Q

16. 1 litre solution of KIO_3 of unknown molarity is given to titrate with KI in strong acid medium. 50 mL solution of KIO_3 requires 10 mL of 0.1 M KI for complete reduction to I_2. The molarity of KIO_3 solution is:
1) 4×10^{-3} M 2) 4×10^{-4} M
3) 4×10^{-5} M 4) 4×10^{-2} M

17. $Hg_5(IO_6)_2$ oxidizes KI to I_2 in acid medium and the other product containing iodine is K_2HgI_4. If the I_2 liberated in the reaction requires 0.004 mole of $Na_2S_2O_3$, the number of moles of $Hg_5(IO_6)_2$ that have reacted is:
1) 10^{-3} 2) 10^{-4}
3) 2.5×10^{-4} 4) 2.5×10^{-2}

18. Cerric ammonium sulphate and potassium permanganate are used as oxidising agents in acidic medium for the oxidation of ferrous ammonium sulphate to ferric sulphate. The ratio of number of moles of cerric ammonium sulphate required per mole of ferrous ammonium sulphate to the number of moles of $KMnO_4$ required per mole of ferrous ammonium sulphate, is
1) 5.0 2) 0.2
3) 0.6 4) 2.0

19. Compounds in which oxidation state of at least one oxygen is (–1):
(1) H_2O_2 (2) O_2F_2
(3) H_2SO_5 (4) CrO_5
(5) $H_2S_2O_8$ (6) $[Na_2B_2H_4O_8]$
1) 1,2,3,5 2) 1,3,4,5,6
3) 1,3,4 4) 1,2,3,4,5,6

20. Equivalent weight of underlined substances in the given reactions are respectively

$\underline{K_4[Fe(CN)_6]} + MnO_4^- + H^+ \rightarrow Fe^{3+} + CO_2 + NO_3^- + Mn^{2+} + H_2O + K^+$

$Cu(s) + \underline{HNO_3}(aq) \rightarrow Cu(NO_3)_2(aq) + NO(g) + H_2O(\ell)$

{M_1 = Mol.wt. of $K_4[Fe(CN)_6]$

M_2 = Mol.wt. of HNO_3 }

1) $\frac{M_1}{61}, \frac{M_2}{3}$ 2) $\frac{M_1}{1}, \frac{M_2}{3}$

3) $\frac{M_1}{61}, \frac{4M_2}{3}$ 4) $\frac{M_1}{48}, \frac{M_2}{4}$

NEET & AIIMS Past Year MCQs

1. The oxidation states not shown by Mn in given reaction is:

$$3MnO_4^{2-} + 4H^+ \longrightarrow 2MnO_4^- + MnO_2 + 2H_2O$$

A. +6 B. +2 C. +4 D. +7 E. +3

Choose the most appropriate answer from the options given below : **(2024 Re)**

1) D and E only 2) B and D only
3) A and B only 4) B and E only

2. Which reaction is **NOT** a redox reaction? **(2024)**

1) $H_2 + Cl_2 \rightarrow 2\,HCl$
2) $BaCl_2 + Na_2SO_4 \rightarrow BaSO_4 + 2\,NaCl$
3) $Zn + CuSO_4 \rightarrow ZnSO_4 + Cu$
4) $2\,KClO_3 + I_2 \rightarrow 2\,KIO_3 + Cl_2$

3. On balancing the given redox reaction, **(2023)**

$aCr_2O_7^{(2-)} + bSO_3^{(2-)}(aq) + cH^+(aq) \rightarrow 2aCr^{3+}(aq) + bSO_4^{2-}(aq) + c/2\,H_2O\,(l)$

the coefficients a, b and c are found to be, respectively:

1) 8, 1, 3 2) 1, 3, 8
3) 3, 8, 1 4) 1, 8, 3

4. At 298 K, the standard electrode potentials of Cu^{2+}/Cu, Zn^{2+}/Zn, Fe^{2+}/Fe and Ag^+/Ag are 0.34 V, – 0.76 V, – 0.44 V and 0.80 V, respectively. **(2022)**

On the basis of standard electrode potential, predict which of the following reaction can not occur?

1) $2CuSO_4(aq) + 2Ag(s) \rightarrow 2Cu(s) + Ag_2SO_4(aq)$
2) $CuSO_4(aq) + Zn(s) \rightarrow ZnSO_4(aq) + Cu(s)$
3) $CuSO_4(aq) + Fe(s) \rightarrow FeSO_4(aq) + Cu(s)$
4) $FeSO_4(aq) + Zn(s) \rightarrow ZnSO_4(aq) + Fe(s)$

5. Which of the following reactions is the metal displacement reaction? Choose the right option. **(2021)**

1) $Cr_2O_3 + 2Al \xrightarrow{\Delta} Al_2O_3 + 2Cr$
2) $Fe + 2HCl \longrightarrow FeCl_2 + H_2\uparrow$
3) $2Pb(NO_3)_2 \longrightarrow 2PbO + 4NO_2 + O_2\uparrow$
4) $2KClO_3 \xrightarrow{\Delta} 2KCl + 3O_2$

6. MnO_4^{2-} (1 mol) in neutral aqueous medium is disproportionate to **(AIIMS 2017)**

1) 2/3 mole of MnO_4^- and 1/3 mol of MnO_2
2) 1/3 mole of MnO_4^- and 2/3 mol of MnO_2
3) 1/3 mole of Mn_2O_7 and 1/3 mol of MnO_2
4) 2/3 mole Mn_2O_7 and 1/3 mol of MnO_2

7. Assertion: In some cases oxygen shows positive oxidation number though it is an electronegative element.

Reason: Fluorine is more electronegative than oxygen. **(AIIMS 2001)**

1) If both Assertion & Reason are True & the Reason is a correct explanation of the Assertion.
2) If both Assertion & Reason are True but Reason is not a correct explanation of the Assertion.
3) If Assertion is True but the Reason is False.
4) If both Assertion & Reason are False.

8. Oxidation state of sulphur in S_8, H_2S and S_2F_2 is: **(AIIMS 2000)**

1) 0, –2, –1 2) 0, +2, –1
3) 1, –2, +1 4) 0, –2, +1

9. What is the oxidation number of sulphur in $Na_2S_4O_6$? **(AIIMS 1998)**

1) $\frac{3}{5}$ 2) $\frac{3}{2}$
3) $\frac{2}{5}$ 4) $\frac{5}{2}$

10. In the reaction: $4Fe + 6O_2 \rightarrow 4Fe^{3+} + 6O_2^{2-}$ which of the following statements is incorrect? **(AIIMS 1998)**

1) Metallic iron is a reducing agent
2) Fe^{3+} is an oxidising agent
3) It is a redox reaction
4) Metallic iron is reduced to Fe^{3+}

Answer Key

NCERT Topic-wise MCQs

1	2	3	4	5	6	7	8	9	10	11	12	13	14	15	16	17
3	1	1	4	1	2	1	1	2	4	1	3	1	2	3	1	4
18	19	20	21	22	23	24	25	26	27	28	29	30	31	32	33	34
3	1	2	2	3	3	1	1	3	3	3	4	3	1	3	3	2
35	36	37	38	39	40	41	42	43	44	45	46	47	48	49	50	51
4	1	1	1	3	1	2	1	3	2	4	2	3	4	4	3	3
52	53	54	55	56	57	58	59	60	61	62	63	64	65	66	67	68
2	1	4	3	1	1	3	4	4	2	1	3	1	3	3	1	1
69	70															
4	1															

High Order Time Intensive MCQs

1	2	3	4	5	6	7	8	9	10	11	12	13	14	15	16	17
1	4	2	2	4	1	1	2	2	1	2	4	3	4	2	3	3
18	19	20	21	22	23	24	25	26	27	28	29	30	31	32	33	34
2	1	1	3	4	4	4	3	2	1	3	2	4	1	4	3	1
35	36	37	38	39	40	41	42	43	44							
2	2	1	3	3	3	4	2	2	4							

Match the Columns MCQs

1	2	3	4	5
2	4	4	1	1

Correct & Incorrect MCQs

1	2	3	4	5	6	7	8	9	10
4	3	1	2	3	3	3	4	3	4

Statement Based MCQs

1	2	3	4	5	6	7	8	9	10
2	1	1	1	1	3	4	4	2	1

Assertion & Reason MCQs

1	2	3	4	5	6	7	8	9	10
2	3	1	2	2	3	1	1	4	4

Multi Correct MCQs

1	2	3	4
4	1	4	2

NCERT Exemplar MCQs

1	2	3	4	5	6	7	8	9	10
4	4	4	1	1	2	1	4	4	3

Kattar NEET

1	2	3	4	5	6	7	8	9	10	11	12	13	14	15	16	17
2	1	3	4	1	2	1	1	4	2	4	1	1	1	1	1	3
18	19	20														
1	2	3														

NEET & AIIMS Past Year MCQs

1	2	3	4	5	6	7	8	9	10
4	2	2	1	1	1	1	4	4	4

CHAPTER

8

The p-Block Elements

Topicwise Number of Questions Analysis (2025-2016)

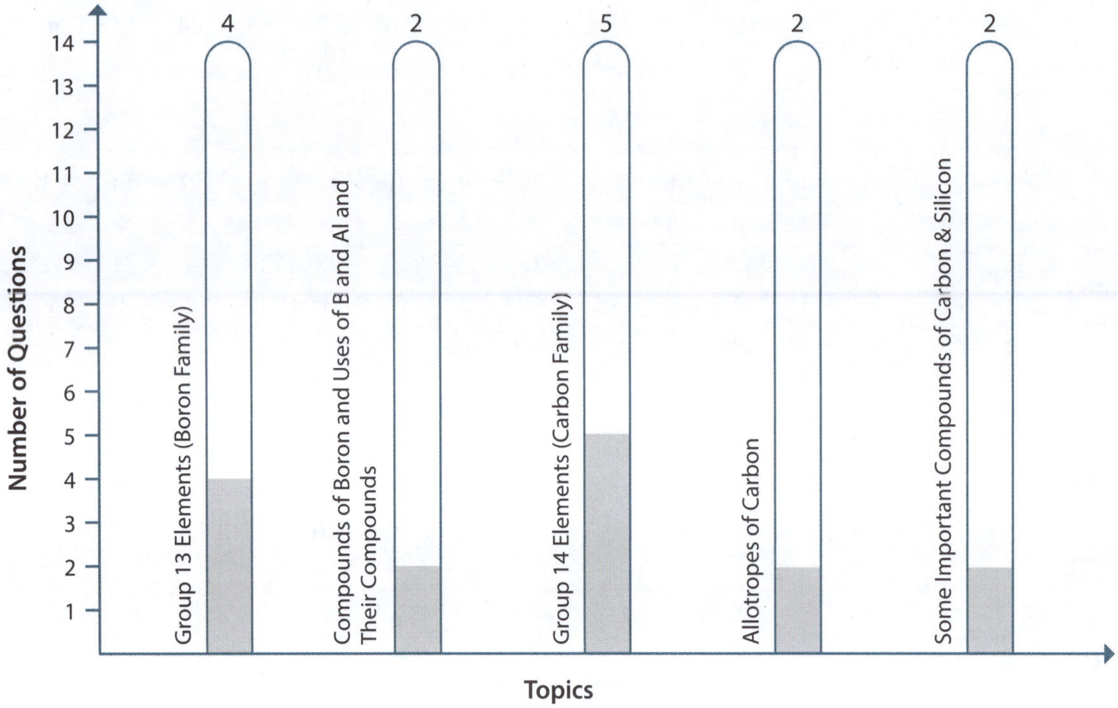

NCERT-Picks

General Introduction

- In p-block elements, the last electron enters the outermost p-orbital as the number of p-orbitals is 3 and therefore, the maximum number of electrons that can be accommodated is six. Consequently, six groups of p-block is observed which are from group 13 to group 18. In this unit, we will study the chemistry of group 13 and group 14 elements.

Group 13 Elements: The Boron family

- The elements of this group are boron (B), aluminum (Al), gallium (Ga), indium (In), thallium (Tl),
- The outer configuration of elements is ns^2np^1.

Atomic and ionic radii

- The atomic and ionic radii down the group increases due to the addition of a new shell. There is a deviation, however, on moving from Aluminium (143 pm) to Gallium (135 pm). This arises due to the poor shielding of the intervening d-orbitals in Gallium, making the size smaller than Aluminium.
- Boron $<$ Aluminium $>$ Gallium $<$ Indium $<$ Thallium

Ionization energy

- Down the group, the values of Ionization Enthalpy do not decrease smoothly. From Boron to Aluminium, the Ionization Enthalpy increase as expected. But from Aluminium to Gallium, the Ionization Enthalpy increase slightly. The first Ionization Enthalpy of Thallium is greater than Aluminium.

Electronegativity

- The electronegativity first decreases from B to Al, then increases slightly from Aluminium to Tl. This can be attributed to the poor shielding of the intervening d and f orbitals.

Chemical properties

- B does not form B^{3+} ions while Al is able to form Al^{3+} ions. The sum of first three ionization enthalpies is very high for B, while it considerably decreases for Al.
- Only $AlCl_3$ sublimes on heating, out of all the chlorides of group 13 elements.

Reactivity with air

- All of the elements of group 13 react at high temperature forming trioxides, M_2O_3.
- Amorphous boron and aluminium metal on heating in air form **B_2O_3 and Al_2O_3 respectively**. With dinitrogen at high temperature, they form nitrides.

  $2E(s) + 3O_2(g) \xrightarrow{\Delta} 2E_2O_3(s)$

  $2E(s) + N_2(g) \xrightarrow{\Delta} 2EN(s)$

Reactivity with acids and alkali

- Boron does not react with non-oxidising acids like HCl, but at higher temperatures, it reacts with strong oxidizing acids like a mixture of hot concentrated H_2SO_4 and HNO_3 to give boric acid.
- All the other elements of group 13 react both with non-oxidising and oxidising acids liberating hydrogen gas.

Complex formation

- Group 13 elements have more tendency to form complexes than s-block elements due to their smaller size and more polarising power.
- Boron can form many complexes like $[BF_4]^-$. Other elements also form complex compounds like $Li[AlH_4]$, $[GaCl_6]^{3-}$.
- Boron is a powerful reducing agent which reduces CO_2 and SiO_2 to C and Si respectively.
- $Al(NO_3)_3$, $Al(SO_4)_3$, AlF_3 are ionic compounds while $AlCl_3$, $AlBr_3$ are covalent.

Anomalous properties of boron

- Other members decompose by steam but Boron does not.
- Concentrated nitric acid oxides boron to boric acid but no such action is noticed by other group members.

  $B + 3HNO_3 \rightarrow H_3BO_3 + 3NO_2$
- Except for boron, the compounds of the elements of the boron family like tetrahedral $[M(OH)_4]^-$ and octahedral $[M(H_2O)_6]^{3+}$ (where M denotes the member of boron family) exists in an aqueous medium.
- The maximum covalency of boron is 4 due to the absence of d orbital.

Some Important Compounds of Boron

Diborane: It is simplest Boron hydride and prepared as-

- $4BF_3 + 3LiAlH_4 \xrightarrow{\text{Ether}} 2B_2H_6 + 3Li[AlF_4]$
- **Laboratory Preparation**: Oxidation of Sodium borohydride with iodine gives diborane.

  $2NaBH_4 + I_2 \rightarrow B_2H_6 + 2NaI + H_2$
- **Industrial Preparation** - Reaction of BF_3 with Sodium hydride produces B_2H_6.

  $2BF_3 + 6NaH \xrightarrow{450\text{ K}} B_2H_6 + 6NaF$
- Diborane catches fire spontaneously in air and explodes with O_2. Reaction with oxygen is extremely exothermic.

  $B_2H_6 + 3O_2 \rightarrow B_2O_3 + 3H_2O; \quad \Delta H^{\ominus} = -1976$ kJ/mol
- Borazole or inorganic benzene is $B_3N_3H_6$ when boron trichloride and hydrogen or diborane and NH_3 are subject to electric discharge at 200°C, borazine is formed. It has three isomers.

Orthoboric acid

Preparation

$Na_2B_4O_7 + 2HCl / H_2SO_4 + 5H_2O \rightarrow 2NaCl / Na_2SO_4 + 4H_3BO_3 \downarrow$

- Boric acid is considered as a weak acid. It is not able to release H^+ ions on its own.
- When heated, it first forms metaboric acid (HBO_2) and then boron trioxide.

Borax ($Na_2B_4O_7.10H_2O$)

Preparation

$4H_3BO_3 + Na_2CO_3 \rightarrow Na_2B_4O_7 + 6H_2O + CO_2 \uparrow$

sodium meta borate

Action of heat

- When borax powder is heated, it first swells due to loss of water in the form of steam but at 740°C, it gets converted into colourless transparent borax bead

Uses of group 13 elements

- Alums are water softeners. They are also used in leather tanning.
- For blood coagulation, potash alums $K_2SO_4.Al_2(SO_4)_3.2H_2O$ is commonly used because Al^{3+} ions given by it neutralize the negatively charge albuminoid of blood.
- Fire proof clothes is made up of $Al(SO_4)_3$.

Group 14 Elements: The Carbon Family

The group 14 elements are the second group in the p-block of the periodic table. It is also called the carbon group.

The elements of Group 14 are carbon (C), silicon (Si), germanium (Ge), tin (Sn), lead (Pb),

The valence shell electronic configuration of these elements is $ns^2\ np^2$

Covalent radii

The radii of group 14 elements are lesser than that of group 13 elements. This can be explained by the increase in the effective nuclear charge.

Ionization enthalpy

The ionization energy of group 14 elements is greater than that of group 13 elements. This can be attributed to size.

Down the group, the Ionization Enthalpy decreases. There is a sharp decrease from C to Si, after which the decrease is nominal.

The order is as follows:

C > Si > Ge > Pb >Sn

Metallic character

Down the group, the metallic character increases. C and Si are non-metals, Ge a metalloid, and Sn and Pb are soft metals with low melting points.

Chemical properties

- In group–14, the tendency of heavier elements to show +2 oxidation state increases as Ge < Sn < Pb due to inability of ns^2 electrons of valence shell to participate in the bonding.

Reactivity with oxygen

- Group 14 elements form oxides of the type MO and MO_2. Lead also form an oxide Pb_3O_4 which is a mixed oxide of PbO and PbO_2. Among the monoxides, CO is neutral, GeO is basic while SnO and PbO are amphoteric
- Pb_3O_4 is red lead (sindoor) and $2PbCO_3$. $Pb(OH)_2$ is white lead.
- Litharge is chemically PbO.

Reactivity with halogens

- They form tetrahalides of the form MX_4. The central atom is sp^3 hybridized and adopts a tetrahedral shape.
- Carbon is not attacked by sodium hydroxide solution. CCl_4 is not hydrolyzed easily.
- SiF_6^{2-} is formed due to small size of fluorine.

Anomalous behaviour of carbon

- Heavier elements of group – 14 do not form $p\pi - p\pi$ bonds. In group-14, atomic orbitals are too large and diffuse to have effective overlapping.
- Carbon cannot exceeds its covalence more than 4 due to absence of d–orbitals in carbon.

Allotropes of carbon

- Carbon exhibit many allotropic forms both crystalline as well as amorphous. Diamond and graphite are two well known crystalline forms.
- Diamond is covalent, yet it has high melting point. It has a 3–dimensional network involving strong C – C bonds.
- Graphite has layered structure and layers are held together by vander waals forces.
- Graphite is used as a lubricant where as oil cannot be used. Graphite cleaves easily between the layers, so it's very soft and slippery.
- In graphite, electrons are mobile and delocalised over the whole sheet therefore, it conducts electricity along the sheet.

Uses of carbon

- Graphite fibres embedded in plastic material form high strength, light weight composites.

Carbon Compounds

(a) Carbon monoxide

Pure CO is prepared by dehydration of formic acid with conc. H_2SO_4 at 373 K.

$$HCOOH \xrightarrow[\text{Conc.}H_2SO_4]{373} H_2O + CO$$

On commercial scale, it is prepared by passing steam over coke.

$$C(s) + H_2O(g) \xrightarrow[1273K]{473} \underbrace{CO(g) + H_2(g)}_{\text{Water gas}}$$

Instead of using steam, a mixture of CO and N_2 is produced.

$$2C(s) + O_2(g) + 4N_2(g) \xrightarrow{1273K} \underbrace{2CO(g) + 4N_2(g)}_{\text{Producer gas}}$$

(b) Carbon dioxide

CO_2 is prepared by complete combustion of carbon and carbon containing fuels in excess of air.

$$C(s) + O_2(g) \xrightarrow{\Delta} CO_2(g)$$

$$CH_4(g) + 2O_2(g) \xrightarrow{\Delta} CO_2(g) + 2H_2O(g)$$

In lab, it is prepared by the action of dil. HCl on $CaCO_3$.

$$CaCO_3(s) + 2HCl(aq) \rightarrow CaCl_2(aq) + CO_2(g) + H_2O(l)$$

Silicon Compounds

Silicon dioxide-SiO_2

- Silicon dioxide commonly known as silica.
- Silica in its normal form is almost non-reactive because of very high Si–O bond enthalpy.
- It resists the attack by halogens, dihydrogen and most of the acids and metals even at elevated temperatures. However, it is attacked by HF and NaOH.
- Monosilane on contact with air burns forming vortex rings of SiO_2.
- Acid in which silica is soluble is HF due to the formation of silicon tetrafluoride.

Silicones

- These are the group of organosilicon polymers which have $\left(R_2SiO \right)$ group as a repeating unit. The starting materials for the manufacturing of silicons are alkyl or aryl substituted silicon chlorides $RnSiCl_{(4-n)}$ where R is alkyl or aryl group.

Silicates

- They are the minerals containing silicon and oxygen in tetrahedral SiO_4^{4-} units, which are linked together in several patterns.
- Zeolites are alumina silicates.
- Asbestos can withstand red hot flame without any damage. Soft water contains CO_3^{2-} ions, which form a coating of lead carbonate (Which is insoluble) on lead pipes.

MCQs

NCERT Topic-wise MCQs

Group 13 Elements : The Boron Family

1. Which of the following is least stable compound?
   1) BCl_3 2) $GaCl_3$
   3) $InCl_3$ 4) $TlCl_3$
2. Which of the following pairs has almost same radii?
   1) Al,Ga 2) Be,Mg
   3) Mg,Al 4) B, Be
3. The tendency of BF_3 , BCl_3 and BBr_3 to behave as lewis acid decreases in the sequence:
   1) $BF_3 > BCl_3 > BBr_3$ 2) $BCl_3 > BF_3 > BBr_3$
   3) $BBr_3 > BCl_3 > BF_3$ 4) $BBr_3 > BF_3 > BCl_3$
4. Which one of the following has highest Lewis acid strength?
   1) BI_3 2) BBr_3
   3) BF_3 4) BCl_3
5. Bauxite has the composition:
   1) Al_2O_3 2) $Al_2O_3 . H_2O$
   3) $Al_2O_3 . 2H_2O$ 4) None of these
6. The stability of +1 oxidation state increase in the sequence:
   1) Ga < In < Al < Tl 2) Al < Ga < In < Tl
   3) Tl < In < Ga < Al 4) In < Tl < Ga < Al
7. The ion (s) that act as an oxidizing agent in solution is/are:
   1) Tl^+ and Al^{3+} 2) B^{3+} and Al^{3+}
   3) Tl^{3+} only 4) B^{3+} only
8. Conc. HNO_3 is stored in container made up of:
   1) Cu 2) Zn
   3) Al 4) Sn
9. Among the following, the maximum covalent character is shown by the compound:
   1) $SnCl_2$ 2) $AlCl_3$
   3) $MgCl_2$ 4) $FeCl_2$
10. Which of the following compound(s) undergoes disproportionation in aqueous solution?
    1) $TlCl_3$ 2) GaCl
    3) InCl 4) TlCl
11. BF_3 is:
    1) Electron deficient compound
    2) Lewis base
    3) Used as rocket fuel
    4) Ionic compound
12. Which element of Group 13 is the strongest reducing agent?
    1) B 2) Al
    3) Fe 4) Ga
13. Which of the following does not exhibit inert pair effect?
    1) Bi 2) Pb
    3) B 4) Tl
14. The outer electronic configuration of group-13 elements is:
    1) $ns^2 np^2$ 2) $ns^2 np^6$
    3) $ns^2 np^1$ 4) $ns^2 np^4$
15. The maximum oxidation state shown by a p-block element is equal to the
    1) Sum of the s-and p-electrons
    2) Sum of the p-and d-electrons
    3) Sum of the s-and d-electrons
    4) Sum of the d-and f-electrons
16. The electronegativity of group-13 elements
    1) First decreases from B to Al and then increases marginally
    2) First increases from B to Al and then decreases marginally
    3) Decreases from B to Tl
    4) None of these
17. Which of the following statements is incorrect about boron?
    1) It is metallic in nature
    2) It is extremely hard and black coloured solid
    3) It has unusually high melting point
    4) It exists in many allotropic forms
18. The element that exhibits negative oxidation state in IIIA group elements is:
    1) B 2) Al
    3) Ga 4) Ti
19. Aluminium chloride in acidified aqueous solution forms
    1) Octahedral ion 2) Tetrahedral ion
    3) Icosahedral ion 4) None of these
20. Amorphous boron and aluminium metal on heating in air form:
    1) BO_4 and $Al(OH)_3$ 2) B_2H_6 and $AlCl_3$
    3) $B(OH)_2$ and $Al(OH)_3$ 4) B_2O_3 and Al_2O_3
21. Aluminium and gallium oxides are
    1) Acidic 2) Basic
    3) Amphoteric 4) None of these

22. The maximum covalence of B is 4 due to
   1) The presence of d orbitals
   2) The absence of d orbitals
   3) The presence of p orbitals
   4) The absence of p orbitals

Important Compounds of Boron & Uses of Boron and Aluminium and their Compounds

23. Which of the following is electron deficient molecule?
   1) B_2H_6 2) C_2H_6
   3) PH_3 4) SiH_4
24. Inorganic benzene is:
   1) $B_3H_3N_3$ 2) BH_3NH_3
   3) $B_3H_6N_3$ 4) $H_3B_3N_6$
25. The number of isomers possible for disubstituted borazine, $B_3N_3H_4X_2$, is:
   1) 3 2) 4
   3) 5 4) 2
26. Which is correct statement about diborane structure?
   1) All BHB bond angles are equal
   2) All H-B bond lengths are equal
   3) It has three - centre - 2 electron bonds
   4) All hydrogens and boron atoms are in one plane
27. The type of hybridization of boron in diborane is:
   1) sp- hybridization 2) sp^2 - hybridization
   3) sp^3 - hybridization 4) sp^3d^2 - hybridization
28. A bond that does not exist in diborane is
   1) B–H 2) B–B
   3) B–H–B 4) H–B–H
29. The maximum number of atoms present in the same plane in diborane molecule is
   1) 2 2) 6
   3) 4 4) 3
30. What happens when borax is heated in a Bunsen burner flame with CoO on a loop of platinum wire?
   1) A blue coloured $Co(BO_2)_2$ bead is formed
   2) A red coloured $Co(BCl)_2$ bead is formed
   3) A green coloured $CoCl_2$ bead is formed
   4) None of these
31. Which of the following is formed when orthoboric acid is heated above 370K ?
   1) Borax 2) Metaboric acid
   3) Paraboric acid 4) Diborane
32. HBO_2 (obtained on heating orthoboric acid above 370 K) on further heating yields
   1) Orthoboric acid 2) Diborane
   3) Boric oxide 4) Borax
33. Diborane is produced on an industrial scale by
   1) The reaction of BF_3 with sodium hydride
   2) Acidifying an aqueous solution of borax
   3) The reaction of HBO_2 with air
   4) None of these
34. Which among the following is not a borane?
   1) B_2H_6 2) B_3H_6
   3) B_4H_{10} 4) None of these
35. Which of the following is used in making bullet-proof vest?
   1) Aluminium wire 2) Boron fibre
   3) Gallium 4) Thallium
36. Borax and boric acid are used in the manufacture of
   1) Glass-wool 2) Heat resistant glasses
   3) Fibre glass 4) All of these
37. Which of the following is used as a flux for soldering metals ?
   1) Diborane 2) Borax
   3) Metaboric acid 4) Orthoboric acid
38. An aqueous solution of which of the following is generally used as a mild antiseptic?
   1) Borax 2) Diborane
   3) Orthoboric acid 4) None of these
39. Which of the following statements about aluminium is/are correct?
   1) It is a bright silvery-white metal
   2) It has high tensile strength
   3) It has a high electrical and thermal conductivity
   4) All of these

Group 14 Elements : The Carbon Family

40. Which element is used in making transistor?
   1) Sn 2) Sb
   3) Si 4) Mg
41. The stability of dihalides of Si, Ge, Sn and Pb increases steadily in the sequence:
   1) $PbX_2 << SnX_2 << GeX_2 << SiX_2$
   2) $GeX_2 << SiX_2 << SnX_2 << PbX_2$
   3) $SiX_2 << GeX_2 << PbX_2 << SnX_2$
   4) $SiX_2 << GeX_2 << SnX_2 << PbX_2$
42. Which of the following oxides is not expected to react with sodium hydroxide?
   1) CaO 2) SiO_2
   3) BeO 4) B_2O_3
43. Which of the following oxidation states are the most characteristic for lead and tin respectively?
   1) + 2, +2 2) + 4, +2
   3) + 2, +4 4) + 4, +4

44. Which of the following of group 14 has the highest melting and boiling point?
1) Pb 2) Si
3) Sn 4) C

45. The reducing character (M^{2+} species) in dihalides decreases as:
1) $PbCl_2 > SnCl_2 > GeCl_2$
2) $PbCl_2 > GeCl_2 > SnCl_2$
3) $SnCl_2 > PbCl_2 > GeCl_2$
4) $GeCl_2 > SnCl_2 > PbCl_2$

46. Which of the following oxides is acidic?
1) SiO_2 2) CO
3) SnO_2 4) PbO_2

47. Which of the following is not hydrolysed?
1) CCl_4 2) $SiCl_4$
3) $GeCl_4$ 4) $SnCl_4$

48. The strongest reductant among the following is:
1) $GeCl_2$ 2) $SnCl_2$
3) $PbCl_2$ 4) $SnCl_4$

49. The outer shell electronic configuration of group 14 elements is
1) ns^2np^1 2) ns^2np^2
3) ns^2np^8 4) ns^2np^6

50. Which of the following represents the common oxidation states exhibited by group 14 elements?
1) +4 and +2 2) +6 and +5
3) +1 and +5 4) +3 and +6

51. Lead compounds in +4 oxidation state are
1) Strong reducing agents
2) Strong oxidising agents
3) Strong reductant
4) None of these

52. Which of the following is/are correct about monoxides of group 14 elements ?
1) GeO is acidic
2) CO is neutral
3) SnO and PbO are amphoteric
4) All of these

53. Which of the following decomposes steam to form dioxide and dihydrogen gas ?
1) Carbon 2) Silicon
3) Germanium 4) Tin

54. Select the incorrect statement.
1) SnF_4 and PbF_4 are covalent in nature
2) PbI_4 does not exist
3) PbX_2 is more stable than PbX_4
4) GeX_4 is more stable than GeX_2

55. Orthoboric acid (H_3BO_3) and metaboric acid (HBO_2) differ in respect of:
1) Basicity 2) Structure
3) Melting point 4) All of these

Allotropes of Carbon

56. The geometry in which carbon atoms in diamond are bonded to each other is:
1) Linear 2) Tetrahedral
3) Square planar 4) Octahedral

57. The hybridization state of carbon in fullerene is:
1) sp 2) sp^2
3) sp^3 4) sp^3d

58. With respect to graphite and diamond, which of the given statement is correct?
1) Graphite is harder than diamond
2) Graphite has higher thermal conductivity than diamond
3) Graphite has higher C-C bond order than diamond
4) None of these

59. Graphite has:
1) Tetrahedral structure
2) Hexagonal sheet like structures
3) Linear structure
4) Three dimensional structure

Some Important Compounds of Carbon & Silicon

60. The repeating structural unit of silicone is:
1) SiO_2
2) 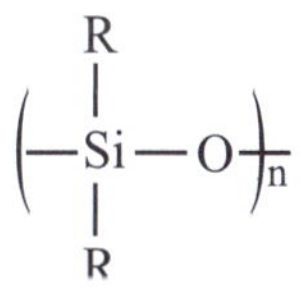
3) 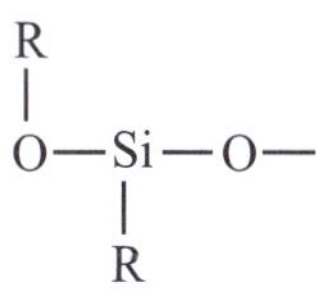
4) 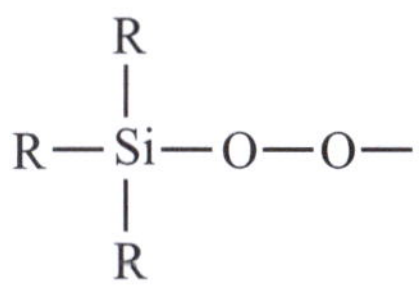

61. Which of these is not a monomer for a high molecular mass silicone polymer?
1) $MeSiCl_3$ 2) Me_2SiCl_2
3) Me_3SiCl 4) $PbSiCl_3$

62. The ratio of "Si" and "O" atoms in silica is:
1) 1 : 2 2) 2 : 1
3) 1 : 4 4) 4 : 1

63. Silica dissolves in NaOH solution to form:
1) Quartz 2) Sodium silicate
3) Carborundum 4) Jaspar

64. In silicones, silicon is strongly linked to

1) Oxygen 2) Nitrogen

3) Sulphur 4) Carbon

65. Producer gas is a mixture of:

1) $CO + H_2$ 2) $CO + N_2$

3) $C + O_2$ 4) $N + O_2$

66. Select the correct statement about carbon monoxide:

1) In CO molecule, there are four sigma and one π bonds between carbon and oxygen

2) It is a violet coloured gas

3) It is highly poisonous in nature

4) It does not form complex with haemoglobin

67. Which of the following statements is incorrect?

1) CO_2 can be obtained as a solid in the form of dry ice

2) Dry ice is used as a refrigerant for ice-cream

3) Gaseous CO_2 is extensively used to carbonate soft drinks

4) CO_2 is a supporter of combustion

68. Silicones are used as

1) Sealant 2) Electrical insulators

3) Greases 4) All of these

69. ZSM-5 is used to convert

1) Alcohols directly into gasoline

2) Phenols directly into ether

3) Methane directly into alcohols

4) None of these

High Order Time Intensive MCQs

1. In which of the following compounds, does the vacant orbital can participate in hybridisation?

1) B_2H_6 2) Al_2Cl_6

3) NH_3 4) Both (1) and (2)

2. Choose the correct statements:

1) Borazine is also known as inorganic benzene

2) Borazine is more reactive than organic benzene

3) Borazine contains 3 B atoms

4) All are correct

3. The incorrect statement from the following for borazine is:

1) It has electronic delocalization

2) It contains banana bonds

3) It can react with water

4) It is a cyclic compound

4. Choose the correct stability order of group 13 elements in their +1 oxidation state.

1) Al < Ga < In < TI 2) Tl < In < Ga < Al

3) Al < Ga < TI < In 4) Al < TI < Ga < In

5. The bond dissociation energy of M–M bond decreases in the order:

1) Sn–Sn > Ge–Ge > Si–Si > C–C

2) C–C > Si–Si > Ge–Ge > Sn–Sn

3) C–C > Ge–Ge > Si–Si > Sn–Sn

4) C–C > Si–Si > Sn–Sn > Ge–Ge

6. $Na_2B_4O_7.10H_2O \xrightarrow{\Delta} X + NaBO_2 + H_2O$

$X + Cr_2O_3 \xrightarrow{\Delta} Y$ (green colored)

The X & Y are:

1) Na_3BO_3 & $Cr(BO_2)_3$ 2) $Na_2B_4O_7$ & $Cr(BO_2)_3$

3) B_2O_3 & $Cr(BO_2)_3$ 4) B_2O_3 & $CrBO_3$

7. A π bond between which of the following species is not successful.

1) C & O (in CO_2) 2) S & O (in SO_2)

3) Si & O (in SiO_2) 4) Both (1) & (2)

8. Which property is common in diamond and graphite?

1) Electrical conductivity 2) Crystal structure

3) Atomic weight 4) Density

9. Silica is soluble in

1) HCl 2) HNO_3

3) H_2SO_4 4) HF

10. $BCl_3 + LiAlH_4 \rightarrow A + LiCl + AlCl_3$

$A + H_2O \rightarrow B + H_2$

$B \xrightarrow{\text{Red heat}} C$. In this reaction sequence, A, B, and C compounds respectively are:

1) B_2H_6, B_2O_3, B 2) B_2H_6, H_3BO_3, B_2O_3

3) B_2H_6, H_3BO_3, B 4) HBF_4, H_3BO_3, B_2O_3

11. Silica reacts with hydride of superhalogen to form 'X'. On hydrolysis of 'X', another compound 'Y' is formed. 'Y' on heating at 1000°C loses water to form 'Z'. The 'Z' can also be prepared in the following reaction.

1) $SiO_2 + 4HF \rightarrow SiF_4 + 2H_2O$

2) $Si + O_2 \rightarrow SiO_2$

3) $K_2SiF_6 + 4K \rightarrow 6KF + Si$

4) $Si + 2NaOH + H_2O \rightarrow Na_2SiO_3 + 2H_2$

12. Which is not true about SiO_2?

1) It is a network solid

2) It is not attacked by molten NaOH

3) It is attacked by HF

4) All of the above

13. The first ionisation energies of the elements of group 14 follow the order.
1) C > Si > Ge > Pb > Sn
2) C > Si > Ge > Sn > Pb
3) C < Si > Ge > Sn > Pb
4) C > Si > Ge > Sn > Pb

14. Aqueous solution of borax is __ in nature
1) Neutral 2) Acidic
3) Alkaline 4) None

15. Borax is the sodium salt of
1) Ortho boric acid 2) Meta boric acid
3) Tetra boric acid 4) Pyro boric acid

16. Thallous chloride is more stable than thallic chloride because of
1) More ionic character
2) Larger size of Tl^+ ion
3) High hydration energy of Tl^+ ion
4) Inert pair effect

17. The metal that does not give the borax bead test
1) Chromium 2) Nickel
3) Lead 4) Manganese

18. Which of the following is the correct formula of borax?
1) $Na_2[B_4O_5(OH)_4]$. $2H_2O$
2) $Na_2[B_4O_5(OH)_4]$. $4H_2O$
3) $Na_2[B_4O_5(OH)_4]$. $6H_2O$
4) $Na_2[B_4O_5(OH)_4]$. $8H_2O$

19. Orthoboric acid contains
1) Triangular BO_3^{3-} units 2) Linear BO_3^{3-} units
3) T-shaped BO_3^{3-} units 4) Pyramidal BO_3^{3-} units

20. H_3BO_3 is
1) Monobasic acid and weak lewis acid
2) Monobasic acid and weak bronsted acid
3) Monobasic acid and strong lewis acid
4) Tribasic acid and weak bronsted acid

21. Which of the following is not an ionic trihalide?
1) AlF_3 2) BF_3
3) In F_3 4) GaF_3

22. Two elements X and Y react separately with highly electropositive metal to form binary compounds, which upon hydrolysis yield mixture of boranes and silanes. X and Y respectively are:
1) B, Si 2) Si, B
3) Al, B 4) B, Al

23. Aluminate ion in aqueous solution actually exists as:
1) $[Al(OH)_6]^{3-}$ 2) $[Al(OH)_4(H_2O)_2]^-$
3) $[Al(OH)_3(H_2O)_3]$ 4) $[Al(OH)_5H_2O]^{-2}$

24. The inactiveness of aluminium in the presence of conc. HNO_3 is due to the formation of a layer of
1) AlN 2) Al_2S_3
3) Al_4C_3 4) Al_2O_3

25. Al and Ga have the same covalent radius because of
1) Greater shielding of s-electrons of Ga atoms
2) Poor shielding of s-electrons of Ga atoms
3) Poor shielding of d-electrons of Ga atoms
4) Greater shielding of d-electrons of Ga atoms

26. Which species does not exist?
1) $[BF_6]^{3-}$ 2) $[AlF_6]^{3-}$
3) $[GaF_6]^{3-}$ 4) $[InF_6]^{3-}$

27. The element that does NOT show catenation is:
1) Ge 2) Si
3) Sn 4) Pb

28. Boron halides are
1) Electron deficient compounds
2) Ionic compounds
3) Lewis bases
4) Used as refractory compounds

29. An alkali metal hydride (NaH) reacts with diborane in Y to give a tetrahedral compound 'Z' which is extensively used as reducing agent in organic synthesis. The 'Y' and 'Z' in the above reaction are:
1) C_2H_6, $C_2H_5NH_2$
2) $C_2H_5OC_2H_5$, $NaBH_4$
3) NH_3, $B_3N_3H_6$
4) C_3H_8, $B_3N_3H_6$

30. Beryllium and aluminium exhibit many properties which are similar. But, the two elements differ in
1) Exhibiting maximum covalency in compounds
2) Exhibiting amphoteric nature in their oxides
3) Forming polymeric hydrides
4) Forming covalent halides.

31. When a nitrate is boiled with aluminium turnings and conc. NaOH solution, the gas liberated is:
1) NO_2 2) NH_3
3) NO 4) N_2O

32. Which of the following occurs as a consequence of inert pair effect?
A. $SnCl_2$ acts as a reducing agent
B. $SnCl_4$ acts as an oxidising agent
C. SnO_2 is amphoteric
D. PbO_2 is an oxidant
E. CCl_2 is unstable but $PbCl_2$ is stable
1) A, D, E 2) D, E
3) A, B, C, D, E 4) A, B, C

33. For the properties mentioned, the correct trend for the different species is in
1) Strength as Lewis acids – $BCl_3 > AlCl_3 > GaCl_3$
2) Inert pair effect – Al > Ga > In
3) Oxidising property – $Al^{3+} > In^{3+} > Tl^{3+}$
4) First ionization enthalpy – B > Al > Tl

34. In diborane, the number of electrons that account for bonding in the bridges is:
1) Six 2) Two 3) Eight 4) Four

35. Which one of the following is the correct statement?
1) $B_2H_6.2NH_3$ is known as 'inorganic benzene'.
2) Boric acid is a protonic acid
3) Beryllium exhibits coordination number of six
4) Chlorides of both beryllium and aluminium have bridged chlorine structures in solid phase

36. The hybridization of the central atom will change when-
1) NH_3 combines with H^+
2) H_3BO_3 combines with OH^-
3) NH_3 forms NH_2^-
4) H_2O combines with H^+

37. Graphite is a good conductor of electricity due to
1) Its giant tetrahedral polymer structure
2) Its high refractive index
3) Presence of free and mobile electrons
4) High IP value of carbon

38. Boron cannot form which one of the following anions?
1) BF_6^{3-} 2) BH_4^- 3) $B(OH)_4^-$ 4) BO_2^- BO_2^-

39. In the preparation of amorphous silicon, HF acid is used to remove:
1) Mg 2) SiO_2
3) Si 4) None of these

40. Compound that combines with haemoglobin of blood to form carboxyhaemoglobin is:
1) N_2O 2) OF_2 3) CO_2 4) CO

41. Which of the following is used in making fire proof clothes?
1) $Al_2(SO_4)_3$ 2) Fe_2O_3
3) Cr_2O_3 4) K_2SO_4

42. Which of the following is a chief component of silvery paints?
1) B 2) Al 3) C 4) Si

43. Organosilicon polymers containing which type of linkages?
1) Si–S–Si 2) Si–O–Si
3) Si–N–Si 4) Si–C–Si

44. The basic structure unit of silicates is:
1) SiO^- 2) SiO_4^{4-} 3) SiO_3^{2-} 4) SiO_4^{2-}

45. A gas other than HCl is obtained in:
1) $BCl_3 + H_2 \xrightarrow[\text{Discharge}]{\text{Silent electric}}$
2) $LiAlH_4 + BF_3 \rightarrow$
3) Both (1) and (2)
4) None of these

46. Name the type of the structure of silicate in which one oxygen atom of $[SiO_4]^{4-}$ is shared?
1) Linear chain silicate 2) Sheet silicate
3) Pyrosilicate 4) Three dimensional

47. Among the following substituted silanes, the one which will give rise to cross linked silicone polymer on hydrolysis:
1) R_3SiCl 2) R_4Si 3) $RSiCl_3$ 4) R_2SiCl_2

48. Which of the following anions is present in the chain structure of silicates?
1) SiO_4^{-4} 2) $Si_2O_7^{6-}$ 3) $(Si_2O_5^{2-})_n$ 4) $(SiO_3^{2-})_n$

49. Which of the following group elements represent p-block elements?
1) Group 10 to group 16 2) Group 3 to group 9
3) Group 13 to group 18 4) Group 1 to group 6

50. Which of the following statements is correct about boron and carbon families?
1) The group oxidation state is the least stable state for the lighter elements in the group
2) The group oxidation state is the most stable state for the lighter elements in the group
3) The group oxidation state is the most stable state for the heavier elements in the group
4) The oxidation state two unit less than the group oxidation state is more stable for the lighter elements in the group

51. The non-metals and metalloids exist in
1) Only p-block elements
2) Only s-block elements
3) Both s and p-block elements
4) None of these

52. The hybridisation state of Al in the complex ion, $[Al(H_2O)_6]^{3+}$ is:
1) sp^3 2) sp^2 3) sp^3d^2 4) sp

53. Borohydrides, are prepared by the reaction of
1) Metal hydrides with H_3BO_3 in ethanol
2) Metal hydrides with B_2H_6 in diethyl ether
3) Metal hydrides with HBO_2 in methyl ether
4) None of these

54. Select the correct statements regarding the physical properties of Group 13 elements from boron to thallium:
1) Boron like caesium is a metallic element that is soft and has a high melting point.
2) Gallium has an unusually low melting point, allowing it to exist in a liquid state during high summer temperatures.
3) The members of Group 13, with the exception of boron, are characterised by high melting points and low electrical conductivity.
4) Aluminium (Al) has a higher density than Gallium (Ga).

Match the Columns MCQs

1. Match the species given in Column-I with the properties mentioned in Column-II.

Column-I		Column-II	
(a)	BF_4^-	(i)	Oxidation state of central atom is +4
(b)	$AlCl_3$	(ii)	Strong oxidising agent
(c)	SnO	(iii)	Lewis acid
(d)	PbO_2	(iv)	Can be further oxidised
		(v)	Tetrahedral shape

1) (a)-(iv) (b)-(iv) (c)-(i) (d)-(ii) (e)-(v)
2) (a)-(v) (b)-(iii) (c)-(iv) (d)-(i)(ii) (e)-(iv)
3) (a)-(i) (b)-(ii) (c)-(v) (d)-(iii) (e)-(i)
4) (a)-(ii) (b)-(i) (c)-(iv) (d)-(ii) (e)-(iv)

2. Match the species given in Column-I with properties given in Column-II.

Column-I		Column-II	
(a)	Diborane	(i)	Used as a flux for soldering metals
(b)	Gallium	(ii)	Crystalline form of silica
(c)	Borax	(iii)	Banana bonds
(d)	Aluminosilicate	(iv)	Low melting, high boiling, useful for measuring high temperatures
(e)	Quartz	(v)	Used as catalyst in petrochemical industries

1) (a)-(ii) (b)-(i) (c)-(iii) (d)-(iv) (e)-(v)
2) (a)-(v) (b)-(ii) (c)-(iii) (d)-(i) (e)-(iv)
3) (a)-(ii) (b)-(i) (c)-(iv) (d)-(v) (e)-(iii)
4) (a)-(iii) (b)-(iv) (c)-(i) (d)-(v) (e)-(ii)

3. Match the species given in Column-I with the hybridisation given in Column-II.

Column-I		Column-II	
(a)	Boron in $[B(OH)_4]^-$	(i)	sp^2
(b)	Aluminium in $[Al(H_2O)_6]^{3+}$	(ii)	sp^3
(c)	Boron in B_2H_6	(iii)	sp^3d^2
(d)	Carbon in Buckminsterfullerene		
(e)	Silicon in SiO_4^{4-}		
(f)	Germanium in $[GeCl_6]^{2-}$		

1) (a)-(ii) (b)-(iii) (c)-(ii) (d)-(i) (e)-(ii) (f)-(iii)
2) (a)-(iii) (b)-(i) (c)-(i) (d)-(ii) (e)-(iii) (f)-(i)
3) (a)-(i) (b)-(ii) (c)-(iii) (d)-(iii) (e)-(ii) (f)-(i)
4) (a)-(iii) (b)-(iii) (c)-(i) (d)-(i) (e)-(ii) (f)-(ii)

4. Match the species given in Column-I with the chemical formula given in Column-II.

Column-I		Column-II	
(a)	Most abundant element	(i)	B
(b)	Lowest m. pt.	(ii)	Tl
(c)	Inert Pair Effect	(iii)	Ga
(d)	No d-orbitals	(iv)	Al

1) (a)-(ii) (b)-(i) (c)-(iv) (d)-(iii)
2) (a)-(iv) (b)-(ii) (c)-(i) (d)-(iii)
3) (a)-(iv) (b)-(i) (c)-(iii) (d)-(ii)
4) (a)-(iv) (b)-(iii) (c)-(ii) (d)-(i)

5. Match the elements given in Column-I with their electronic configuration given in Column-II.

Column-I (Symbol)		Column-II (Unit)	
(a)	Ga	(i)	$[He]2s^22p^2$
(b)	In	(ii)	$[Ne]3s^23p^2$
(c)	C	(iii)	$[Ar]3d^{10}4s^24p^1$
(d)	Si	(iv)	$[Kr]4d^{10}5s^25p^1$

1) (a)-(iii) (b)-(iv) (c)-(i) (d)-(ii)
2) (a)-(ii) (b)-(i) (c)-(iv) (d)-(iii)
3) (a)-(i) (b)-(ii) (c)-(iii) (d)-(iv)
4) (a)-(iii) (b)-(ii) (c)-(i) (d)-(i)

Correct & Incorrect MCQs

1. Which of the following statement (s) are correct?
   A. The valence shell electronic configuration of p-block elements is $ns^2 np^{1-6}$ (except for He).
   B. The maximum oxidation state shown by a p-block element is equal to the total number of valence electrons.
   C. The number possible oxidation states of p-block elements decreases towards right across a period.
   D. The oxidation state two unit less than the group oxidation state becomes progressively more stable for the heavier elements in boron and carbon families.
   1) A, B and D 2) A and C
   3) A, C and D 4) Only A and D
2. Which of the following statement is incorrect?
   1) The heaviest element in each p-block group is the most metallic in nature.
   2) The third period p-block elements can expand their covalence above four.
   3) Atomic radius of Ga is less than that of Al.
   4) The ionization enthalpy of Tl is smaller than that of In.
3. Which of the following statements is incorrect about boron?
   1) It is metallic in nature.
   2) It exists in many allotropic forms.
   3) It has high melting point
   4) It has very strong crystalline lattice.
4. Correct statement(s) is/are
   A. B is able to form B^{3+} ions while Al does not form Al^{3+} ions.
   B. BF_3 molecule is acidic in nature.
   C. Aluminium oxides are basic in nature.
   D. Aluminium dissolves in mineral acids and aqueous alkalies.
   1) A and B 2) B, C and D
   3) Only A 4) B and D
5. Incorrect statement(s) is/are
   A. There is a small increase in radius from Si to Pb.
   B. The ionization enthalpy of Sn is slightly greater than that of Pb.
   C. The electronegativity values for elements from Si to Pb are almost the same.
   D. Carbon can exceeds its covalence more than 4.
   1) Only A 2) A, C and D
   3) B and D 4) A and C
6. Identify the incorrect statement from the following:
   1) Heavier elements of group – 14 do not form $p\pi - p\pi$ bonds.
   2) The order of catenation is Sn > Ge > Si > C.
   3) Carbon is able to show allotropic forms.
   4) Both (1) and (3)
7. Which amongst the following is incorrect statement ?
   1) In diamond, each carbon atom undergoes sp^3 hybridisation.
   2) Each layer in graphite is composed of planar hexagonal rings of carbon atoms.
   3) Layers of graphite are held together by van der Waals forces of attraction.
   4) Fullerenes contain twelve six- membered rings and twenty five-membered rings.
8. Select the incorrect statement :
   1) Boric acid is a weak monobasic acid.
   2) Diborane catches fire spontaneously upon exposure to air.
   3) The four terminal hydrogen atoms and the two boron atoms in diborane lie in different plane.
   4) Boric acid is not a protonic acid but acts as a Lewis acid by accepting electrons from a hydroxyl ion.
9. Which of the following statements is correct ?
   1) The mixture of CO and H_2 is known as producer gas.
   2) In CO molecule, there are two sigma and one π bonds between carbon and oxygen.
   3) Dry ice is used as a refrigerant for ice-cream and frozen food.
   4) In CO_2 molecule, carbon atom undergoes sp^3 hybridisation.
10. Identify the incorrect statement :
    1) Silicon dioxide is a covalent, three-dimensional network solid in which each silicon atom is covalently bonded in a tetrahedral manner to four oxygen atoms.
    2) Silicones are a group of organosilicon polymers, which have (R_2SiO) as a repeating unit.
    3) Feldspar, zeolites, mica and asbestos are examples of silicates.
    4) The basic structural unit of silicates is $Si_2O_7^{2-}$.

Statement Based MCQs

Directions: These questions consist of two statements each, printed as Statement-I and Statement-II. While answering these questions, you are required to choose any one of the following four responses.

1) **Both Statement-I and Statement-II are correct.**
2) **Both Statement-I and Statement-II are incorrect.**
3) **Statement-I is correct & Statement-II is incorrect.**
4) **Statement-I is incorrect & Statement-II is correct.**

1. **Statement-I:** PbI_4 is not formed.

   Statement-II: Pb^{2+} is more stable than that of Pb^{4+} due to inert pair effect.

2. **Statement-I:** NaOH cannot be stored in the vessel made up of Al or Zn.

   Statement-II: A protective layer of oxide is formed on the surface of the metal.

3. **Statement-I:** BF_3 behaves as Lewis acid.

   Statement-II: Carbon cannot expand its coordination number beyond 4 because of the absence of d-orbitals.

4. **Statement-I:** Al forms $[AlF_6]^{3-}$ but B does not form $[BF_6]^{3-}$.

   Statement-II: In carbon dioxide, the carbon is sp^3 hybridized.

5. **Statement-I:** Boric acid is a weak monobasic acid.

   Statement-II: Boranes are readily hydrolysed by water to give boric acid.

6. **Statement-I:** Boron-10 isotope has very low ability to absorb neutrons.

   Statement-II: Metal borides are used in nuclear industry as protective shields and control rods.

7. **Statement-I:** $[SiCl_6]^{2-}$ is known whereas $[SiF_6]^{2-}$ not.

   Statement-II : Boron forms only $[BF_4]^-$ while aluminium gives $[AlF_6]^{3-}$ ion.

8. **Statement-I:** Diamond is covalent, yet it has high melting point.

   Statement-II: Diamond has a 3–dimensional network involving strong C – C bonds.

9. **Statement-I:** Graphite has layered structure.

   Statement-II: Graphite do not conduct electricity.

10. **Statement-I:** Carbon monoxide is not a poisonous gas.

    Statement-II: CO forms a complex with haemoglobin and prevents it from carrying oxygen around the body.

Assertion & Reason MCQs

Directions: These questions consist of two statements each, printed as Assertion and Reason. While answering these questions, you are required to choose any one of the following four responses.

1) **Both Assertion and Reason are True and the Reason is a correct explanation of the Assertion.**
2) **Both Assertion and Reason are True but Reason is not a correct explanation of the Assertion.**
3) **Assertion is True but the Reason is False.**
4) **Assertion is False but Reason is True.**

1. **Assertion:** If aluminium atoms replace a few silicon atoms in three dimensional network of silicon dioxide, the overall structure acquires a negative charge.

   Reason: Aluminium is trivalent while silicon is tetravalent.

2. **Assertion:** Silicons are water repelling in nature.

   Reason: Silicons are organosilicon polymers, which have ($-R_2SiO-$) as repeating unit.

3. **Assertion:** The compound called inorganic benzene is borazine.

   Reason: The chemical formula of borazine is $B_3N_3H_6$.

4. **Assertion:** Diamond is the hardest possible substance and is a network covalent solid.

   Reason: All the C-atoms in diamond are sp^2-hybridized.

5. **Assertion:** Boron is non-metallic in nature.

   Reason: Boron exists in many allotropic forms.

6. **Assertion:** Tl^{+3} is less stable than Tl^+.

   Reason: Tl shows inert pair effect.

7. **Assertion:** Aluminium oxides are amphoteric in nature.

   Reason: Gallium oxides are basic in nature.

8. **Assertion:** Si–Si bonds are much weaker than Si–O bonds.

   Reason: Silicon forms double bonds with itself.

9. **Assertion:** The first ionization energy of Al is lower than that of Mg.

   Reason: There is one e^- in outermost shell in Al whereas in Mg all electrons are paired.

10. **Assertion:** Pb^{4+} can be reduced easily to Pb^{2+}.

    Reason: The stability of lower oxidation state is due to inert pair effect.

Multi Correct MCQs

1. Consider the oxides of group 14 elements:.

A. SiO_2 B. GeO_2
C. SnO_2 D. PbO_2
E. CO F. GeO.

Choose the most appropriate option from the following containing oxides which are amphoteric in nature.

1) (A), (B), (F)
2) (B), (C), and (D)
3) (C) and (D)
4) (A), (B) and (C)

2. Which of the following materials are semiconductors?

A. Germanium B. Graphite
C. Silicon D. Copper oxide

1) (A), (B) and (D) 2) (B), (C) and (D)
3) (A), (C) and (D) 4) (B), (C) and (D)

3. Which of the following ions are expected to behave as an oxidising agents?

A. Sn^{2+} B. Pb^{2+} C. Tl^{3+}
D. Pb^{4+} E. Tl^{+}

Choose the most appropriate answer from the options given below.

1) (A), (B), (D) and (E) 2) (B), (C), (D) and (E)
3) (A), (B), and (C) 4) (C) and (D)

4. Consider the following molecules:

A. PH_3 B. B_2H_6
C. CCl_4 D. NH_3
E. LiH F. BCl_3

Choose the option from the following with the electron deficient molecules.

1) (A), (B), (E) and (F) 2) (B) and (F)
3) (A), (B) and (C) 4) (B), (C), (D) and (F)

NCERT Exemplar MCQs

1. The element which exists in liquid state for a wide range of temperature and can be used for measuring high temperature is:

1) B 2) Al
3) Ga 4) In

2. Which of the following is a Lewis acid?

1) $AlCl_3$ 2) $MgCl_2$
3) $CaCl_2$ 4) $BaCl_2$

3. The geometry of a complex species can be understood from the knowledge of type of hybridisation of orbitals of central atom. The hybridisation of orbitals of central atom in $[B(OH_4)]^-$ and the geometry of the complex are respectively:

1) sp^3, tetrahedral 2) sp^3, square planar
3) sp^3d^2, octahedral 4) dsp^2, square planar

4. Which of the following oxides is acidic in nature?

1) B_2O_3 2) Al_2O_3
3) Ga_2O_3 4) In_2O_3

5. The exhibition of highest coordination number depends on the availability of vacant orbitals in the central atom. Which of the following elements is not likely to act as central atom in MF_6^{3-} ?

1) B 2) Al
3) Ga 4) In

6. Boric acid is an acid because its molecule:

1) Contains replaceable H^+ ion
2) Gives up a proton
3) Accepts OH^- from water releasing proton
4) Combines with proton from water molecule

7. Catenation i.e., linking of similar atoms depends on size and electronic configuration of atoms. The tendency of catenation in group 14 elements follows the order:

1) $C > Si > Ge > Sn$ 2) $C >> Si > Ge \approx Sn$
3) $Si > C > Sn > Ge$ 4) $Ge > Sn > Si > C$

8. Silicon has a strong tendency to form polymers like silicones. The chain length of silicone polymer can be controlled by adding:

1) $MeSiCl_3$ 2) Me_2SiCl_2
3) Me_3SiCl 4) Me_4Si

9. Ionisation enthalpy ($\Delta_i H$ kJ mol^{-1}) for the elements of group 13 follows the order:

1) $B > Al > Ga > In > Tl$ 2) $B < Al < Ga < In < Tl$
3) $B < Al > Ga < In > Tl$ 4) $B > Al < Ga > In < Tl$

10. In the structure of diborane:

1) All hydrogen atoms lie in one plane and boron atoms lie in a plane perpendicular to this plane
2) 2 boron atoms and 4 terminal hydrogen atoms lie in the same plane and 2 bridging hydrogen atoms lie in the perpendicular plane
3) 4 bridging hydrogen atoms and boron atoms lie in one plane and two terminal hydrogen atoms lie in a plane perpendicular to this plane
4) All the atoms are in the same plane

11. A compound X, of boron reacts with NH_3 on heating to give another compound Y which is called inorganic benzene. The compound X can be prepared by treating BF_3 with lithium aluminium hydride. The compounds X and Y are represented by the formulas:

1) B_2H_6, $B_3N_3H_6$ 2) B_2O_3, $B_3N_3H_6$
3) BF_3, $B_3N_3H_6$ 4) $B_3N_3H_6$, B_2H_6

12. Quartz is extensively used as a piezoelectric material, it contains:
    1) Pb
    2) Si
    3) Ti
    4) Sn

13. The most commonly used reducing agent is:
    1) $AlCl_3$
    2) $PbCl_2$
    3) $SnCl_4$
    4) $SnCl_2$

14. Dry ice is:
    1) Solid NH_3
    2) Solid SO_2
    3) Solid CO_2
    4) Solid N_2

15. Cement, the important building material is a mixture of oxides of several elements. Besides calcium, iron and sulphur, oxides of elements of which of the group (s) are present in the mixture?
    1) Group 2
    2) Groups 2, 13 and 14
    3) Groups 2 and 13
    4) Groups 2 and 14

Scan Me for Detailed Explanations

Kattar NEET

1. Which is correct statement about diborane structure?
   1) All HBH bond angles are equal
   2) All H-B bond lengths are equal
   3) It has two three-centre-2 electron bonds
   4) All hydrogens and boron atoms are in one plane

2. What is the hybridisation of boron atoms in compound $Mg[B_2O(OH)_6]$?
   1) Both sp^3
   2) One sp^2 and other sp^3
   3) Both sp^2
   4) One sp^3 and other sp^3d

3. Choose the incorrect statement of the following
   1) There is a change in hybridization of P when PCl_5 vapours undergoes solidification.
   2) There is a change in hybridization of Si when SiF_4 vapours are passed in HF (aq.).
   3) The hybridization of Be in $BeCl_2(s)$ is sp^3.
   4) The species $[SiCl_6]^{2-}$ and $[SiF_6]^{2-}$ both do not exist.

4. Which is not true about borax?
   1) Contains 4B – O – B linkage
   2) Bicyclo compound
   3) Aqueous solution may act as buffer solution
   4) Contains sp^3 hybridised as well as sp^2 hybridised B

5. H_3BO_3 when react with ${}_1^3T_2O$ gives
   1) ${}_1^3T_3O^+$ ions
   2) ${}_1^2D_2O^+$ ions
   3) ${}_1^3TH_2O^+$ ions
   4) ${}_1^1H_3O^+$ ions

6. The silicate anion in the mineral kinoite is a chain of three SiO_4 tetrahedra, that share corners with adjacent tetrahedra. The charge of the silicate anion is:
   1) – 4
   2) – 8
   3) – 6
   4) – 2

7. A red coloured mixed oxide (X) on treatment with conc. HNO_3 gives a compound (Y). (Y) with HCl produces a chloride (Z) which is insoluble in cold water but soluble in hot water. (Z) can also be formed by treating (X) with conc. HCl. Compounds (X), (Y) and (Z) are
   1) Pb_3O_4, PbO_2, $PbCl_2$
   2) Mn_3O_4, MnO_2, $MnCl_2$
   3) Fe_3O_4, Fe_2O_3, $FeCl_3$
   4) Fe_3O_4, FeO, $FeCl_2$

8. $B_3N_3H_6$ + solution of hydrochloric acid →?

   Select the correct statement about above reaction.
   1) No reaction
   2) $B_3N_3H_6$ shows substitution reaction and produces $B_3N_3Cl_6$
   3) $B_3N_3H_6$ shows addition reaction and produces $B_3N_3H_9Cl_3$ in which Cl is bonded to boron
   4) $B_3N_3H_6$ shows addition reaction and produces $B_3N_3H_9Cl$ in which Cl is bonded to nitrogen

9. Aqueous solution of borax reacts with two mole of acids. This leads to:
   1) Formation of 2 mol of $B(OH)_3$ only
   2) Formation of 2 mol of $[B(OH)_4]^-$ only
   3) Formation of 1mol each of $B(OH)_3$ and $[B(OH)_4]^-$
   4) Formation of 4 mol of $B(OII)_3$

10. Match the following:

	Column-I		Column-II
I	Negative charge on the anion is equal to the number of terminal oxygen atoms	P	$Si_4O_{13}^{10-}$
II	Three shared corners and ten unshared corners	Q	SiO_4^{4-}
III	Silicon atom(s) is/are present at the center of geometry and every oxygen atom is present at each corner of the geometry	R	$Si_4O_{12}^{8-}$
IV	Non-planar geometry	S	$Si_2O_7^{6-}$

   1) I-P, Q, R, S; II-P; III-P, Q, R, S; IV-P, Q, R, S
   2) I–T; II–Q, R; III–T; IV–T,,P,Q
   3) I–P,Q,R,T; II–S, T; III–T; IV–T
   4) I–T; II–Q; III–T; IV–Q, T

11. Borax is actually made of two tetrahedra and two triangular units joined together and should be written as: $Na_2[B_4O_5(OH)_4]\cdot 8H_2O$

Consider the following statements about borax:

(P) Each boron atom has four B—O bonds.

(Q) Each boron atom has three B—O bonds.

(R) Two boron atoms have four B—O bonds while other two have three B—O bonds.

(S) Each boron atom has one —OH groups.

Select correct statements:

1) P and Q
2) Q and R
3) R and S
4) P and R

12. Consider the following statements for diborane:

1. Boron is approximately sp^3 hybridized
2. B-H-B angle is 180°
3. There are two terminal B-H bonds for each boron atom
4. There are only 12 bonding electrons available of these statements:

1) 1, 3 and 4 are correct
2) 1, 2 and 3 are correct
3) 2, 3 and 4 are correct
4) 1, 2 and 4 are correct

13. Consider the following statements and which of the following are correct?

S_1: $B_4O_7^{2-}$ on hydrolysis with acid / water yields $B(OH)_3$

S_2: SiO_4^{4-} on hydrolysis with acid / water yields $Si_2O_7^{6-}$

S_3: $MeSiCl_3$ on hydrolysis and then condensations gives a complex cross-linked polymer of silicones.

S_4: Among CO_2, CuO, CaO and H_2O, CO_2 is most acidic oxide while CaO is most basic oxide.

1) S_1, S_2, S_3 only
2) S_1, S_3, S_4 only
3) S_1, S_2, S_3 and S_4
4) S_2, S_3, S_4 only

14. A complex cross-linked polymer (silicone) is formed by

1) Hydrolysis of $(CH_3)_3SiCl$.
2) Hydrolysis of a mixture of $(CH_3)_3SiCl$ and $(CH_3)_2SiCl_2$
3) Hydrolysis of CH_3SiCl_3
4) Hydrolysis of $SiCl_4$.

15. Which of the following statements is correct regard the C–O bond of a CO molecule upon the bond formation between BH_3 and CO?

1) It gets longer because there will be π-back donation from BH_3 to CO.
2) It gets longer because CO donates π-bonding electrons to BH_3.
3) No or slight change on it because CO donates its mainly non-bonding electrons to BH_3.
4) It gets shorter because CO donates π* anti-bonding electrons to BH_3.

16. Reaction of ammonia with diborane gives an ionic product ($B_2H_6.2NH_3$). The hybridization of boron in the cation and anion of this product are respectively

1) sp^3 in both
2) sp^3 and sp^2
3) sp^2 & sp^3
4) sp^2 in both

17. The correct order of Lewis acid strengths of BF_2Cl, BFClBr, BF_2Br and $BFBr_2$ is:

1) $BF_2Cl > BFClBr > BF_2Br > BFBr_2$
2) $BFBr_2 > BFClBr > BF_2Br > BF_2Cl$
3) $BF_2Cl > BF_2Br > BFClBr > BFBr_2$
4) $BFClBr > BFBr_2 > BF_2Cl > BF_2Br$

18. Anion $Si_3O_9^{x-}$ is present in benitoite (a barium titanium silicate mineral). Thus value of x is:

1) 5
2) 6
3) 7
4) 12

NEET & AIIMS Past Year MCQs

1. Taking stability as the factor, which one of the following represents correct relationship? **(2023)**
   1) $TlI > TlI_3$
   2) $TlCl_3 > TlCl$
   3) $InI_3 > InI$
   4) $AlCl > AlCl_3$

2. Match List-I with List-II **(2023)**

List-I		List-II	
(a)	Coke	(i)	Carbon atoms are sp^3 hybridised
(b)	Diamond	(ii)	Used as a dry lubricant
(c)	Fullerene	(iii)	Used as a reducing agent
(d)	Graphite	(iv)	Cage like molecules

Choose the correct answer from the options given below:
   1) (a)-(iii) (b)-(iv) (c)-(i) (d)-(ii)
   2) (a)-(ii) (b)-(iv) (c)-(i) (d)-(iii)
   3) (a)-(iv) (b)-(i) (c)-(ii) (d)-(iii)
   4) (a)-(iii) (b)-(i) (c)-(iv) (d)-(i)

3. Choose the correct statement: **(2022)**
   1) Both diamond and graphite are used as dry lubricants.
   2) Diamond and graphite have two dimensional network
   3) Diamond is covalent and graphite is ionic
   4) Diamond is sp^3 hyridised and graphite is sp^2 hybridized.

4. Which of the following statement is not correct about diborane? **(2022)**
   1) Both the Boron atoms are sp^2 hybridised.
   2) There are two 3-centre-2-electron bonds.
   3) The four terminal B-H bonds are two centre two electron bonds
   4) The four terminal Hydrogen atoms and the two Boron atoms lie in one plane.

5. Hydride of boron occurs as B_2H_6 but B_2Cl_2 does not exist. This is because **(AIIMS 2017)**
   1) $p\pi$–$d\pi$ back bonding is possible in B_2Cl_2
   2) Boron and hydrogen have almost equal values of electronegativity
   3) Boron and chlorine have almost equal atomic sizes
   4) Small hydrogen atoms can easily fit in between boron atoms but large chlorine atoms do not

6. Chain length of silicon is controlled by: **(AIIMS 2015)**
   1) CH_3SiCl_3
   2) $(CH_3)_2SiCl_2$
   3) $(CH_3)_3SiCl$
   4) $(CH_4)S$

7. Iodine oxidises sodium borohydride to gives: **(AIIMS 2011)**
   1) B_2H_6
   2) Sodium hydride
   3) HI
   4) I_3^-

8. Which of the following is an example of chain silicates? **(AIIMS 2009)**
   1) Kaolinite
   2) Zircon
   3) Benitonite
   4) Diopside

9. Which of the following is stable (inert) to fire? **(AIIMS 2001)**
   1) CCl_4
   2) C_2H_5OH
   3) CH_4
   4) C_4H_{10}

10. Silicon is an important constituent of **(AIIMS 1996)**
   1) Alloys
   2) Rocks
   3) Vegetables
   4) Animals

11. Which of the following is the least stable and its existence is doubtful? **(AIIMS 1995)**
   1) SnI_4
   2) CI_4
   3) PbI_4
   4) GeI_4

12. Which of the following bonds is shown in silicone? **(AIIMS 1994)**
   1) Si–C–Si–C–Si
   2) Si–Si–Si–Si
   3) –Si–O–Si–O–Si–
   4) Si–C–Si–O–Si

Answer Key

NCERT Topic-wise MCQs

1	2	3	4	5	6	7	8	9	10	11	12	13	14	15	16	17
4	1	3	1	3	2	3	3	2	2,3	1	2	3	3	1	1	1
18	**19**	**20**	**21**	**22**	**23**	**24**	**25**	**26**	**27**	**28**	**29**	**30**	**31**	**32**	**33**	**34**
1	1	4	3	2	1	3	2	3	3	2	2	1	2	3	1	2
35	**36**	**37**	**38**	**39**	**40**	**41**	**42**	**43**	**44**	**45**	**46**	**47**	**48**	**49**	**50**	**51**
2	4	2	3	4	3	4	1	3	4	4	1	1	1	2	1	2
52	**53**	**54**	**55**	**56**	**57**	**58**	**59**	**60**	**61**	**62**	**63**	**64**	**65**	**66**	**67**	**68**
4	4	1	4	2	2	3	2	2	3	1	2	1	2	3	4	4
69																
1																

High Order Time Intensive MCQs

1	2	3	4	5	6	7	8	9	10	11	12	13	14	15	16	17
4	4	2	1	2	3	3	3	4	2	2	2	1	3	3	4	3
18	**19**	**20**	**21**	**22**	**23**	**24**	**25**	**26**	**27**	**28**	**29**	**30**	**31**	**32**	**33**	**34**
4	1	1	2	1	2	4	3	1	4	1	2	1	2	2	1	4
35	**36**	**37**	**38**	**39**	**40**	**41**	**42**	**43**	**44**	**45**	**46**	**47**	**48**	**49**	**50**	**51**
4	2	3	1	2	4	1	2	2	2	3	3	3	4	3	2	1
52	**53**	**54**														
3	2	2														

Match the Columns MCQs

1	2	3	4	5
2	4	1	4	1

Correct & Incorrect MCQs

1	2	3	4	5	6	7	8	9	10
1	4	1	4	3	2	4	3	3	4

Statement Based MCQs

1	2	3	4	5	6	7	8	9	10
1	3	1	3	1	4	4	1	3	4

Assertion & Reason MCQs

1	2	3	4	5	6	7	8	9	10
1	2	2	3	2	1	3	3	1	1

Multi Correct MCQs

1	2	3	4
3	3	4	2

NCERT Exemplar MCQs

1	2	3	4	5	6	7	8	9	10	11	12	13	14	15
3	1	1	1	1	3	2	3	4	2	1	2	4	3	2

Kattar NEET

1	2	3	4	5	6	7	8	9	10	11	12	13	14	15	16	17
3	1	4	1	1	2	1	3	4	1	3	1	3	3	2	1	2
18																
2																

NEET & AIIMS Past Year MCQs

1	2	3	4	5	6	7	8	9	10	11	12
1	4	4	1	4	3	1	4	1	2	3	3

CHAPTER

9

Organic Chemistry-Some Basic Principles and Techniques

Topicwise Number of Questions Analysis (2025-2016)

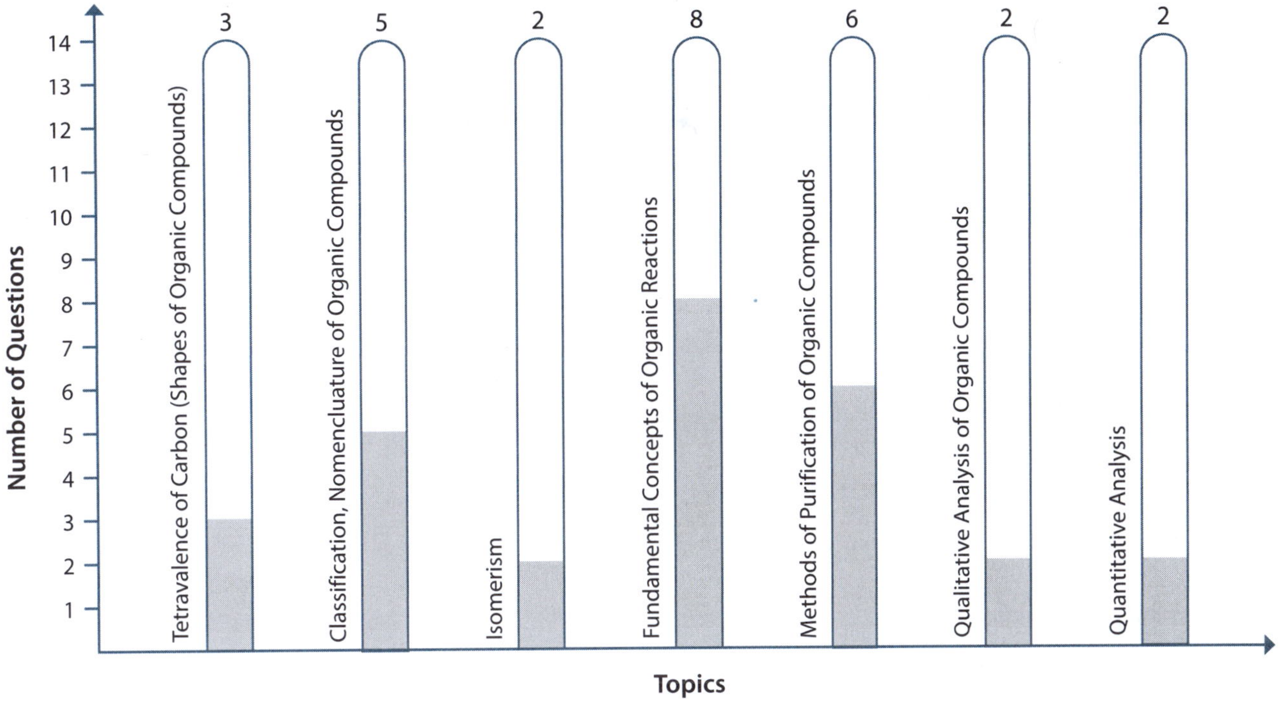

NCERT-Picks

General Introduction

- Organic compounds are vital for sustaining life on earth and these include the complex molecules like genetic information bearing DNA and proteins that constitute essential compounds of our blood, muscles and skin. Organic chemistry is about 200 year old. In 1928, F. Wohler synthesized an organic compound urea from an inorganic compound NH_4CNO

Tetravalence of Carbon

- The tetravalence of carbon and the formation of covalent bonds by it are explained in terms of its **electronic configuration** and the hybridisation of s and p orbitals
- **Hybridisation** influences the bond length and bond enthalpy in organic compounds. More the s character of hybrid orbital, shorter and stronger will be the bond.
- The C — H bond length decreases as $C(sp^3) — H > C(sp^2) — H > C(sp) — H$. The size of the orbital of carbon decreases.
- The order of size of different orbitals is $p > sp^3 > sp^2 > sp > s$
- $CH_3{-}C{\equiv}CH$ is more polar than $CH_3{-}CH{=}CH_2$. sp carbon is more electro-negative than sp^2 carbon.

Some characteristic feature of pi bond

- In pi bond formation, parallel orientation of the two 'p' orbitals on adjacent atoms is necessary for a proper sidewise overlap.
- In $H_2C = CH_2$ molecule, The 'p' orbitals are mutually parallel and both the p orbitals are perpendicular to the plane of the molecule. Rotation about C-C double bond in C = C is restricted.

Nomenclature

- **Trivial names:** Initially organic compounds were named after the source from which they were obtained or from their characteristic properties.
- **IUPAC system**

Prefix	+ Root word	+ Suffix
Tells us about the branches or substitutes except in the case of multiple bonds	Tells us about the length of longest possible chain.	Tell us about the principle functional groups and multiple bonds mostly except in case of halogens.

- Formula for writing name of organic compound is secondary prefix + primary prefix + word root + primary suffix + secondary suffix.

Nomenclature of branched chain alkanes

1. First of all, the **longest carbon chain in the molecule is identified.**
2. The numbering is done in such a way that the branched carbon atoms get the lowest possible numbers
3. The names of alkyl groups attached as a branch are then prefixed to the name of the parent alkane and position of the substituents is indicated by the appropriate numbers. If different alkyl groups are present, they are listed in alphabetical order

   3 $\rightarrow$ ethyl and 4 $\rightarrow$ methyl not 4 $\rightarrow$ ethyl and 3 $\rightarrow$ methyl
4. If two or more identical substituent groups are present then the numbers are separated by commas. The names of identical substituents are not repeated, instead prefixes such as di (for 2), tri (for 3), tetra (for 4), penta (for 5), hexa (for 6) etc. are used
5. If the two substituents are found in equivalent positions, the **lower number is given to the one coming first in the alphabetical listing**
6. The branched alkyl groups can be named by following the above mentioned procedures. However, the **carbon atom of the branch that attaches to the root alkane** is numbered
7. In multi-substituted compounds, the following rules may be remembered:

   If there happens to be two chains of equal size, then that chain is to be selected which contains more number of side chains.

   After selection of the chain, numbering is to be done from the end closer to the substituent.

Nomenclature of organic compounds having functional group(s)

The longest chain of carbon atoms containing the functional group is numbered in such a way that the functional group is attached at the carbon atom possessing lowest possible number in the chain.

The order of decreasing priority for some functional groups is:

$-COOH > -SO_3H > -COOR$ (R = alkyl group) $> COCl > -CONH_2 > -CN > -HC = O, > C = O > -OH > -NH_2 > C = C > -C \equiv C- > X$ (halogen)

Isomerism

(The phenomenon shown by 2 or more organic compounds having same molecular formula but different properties is known as isomerism)

Isomerism

- **Structural isomerism**
  Shown by compounds having same molecular formula but different structural formula.
  - **Chain isomerism**
    Shown by compounds having same molecular formula but different carbon chains.
  - **Position isomerism**
    In this, the position of substituent or functional groups are different.
  - **Functional isomerism**
    It is due to the difference in nature or functional groups are different.
  - **Metamerism**
    Due to the different nature of alkyl group around a polyvalent functional group.
  - **Tautomerism**
    The phenomenon in which a single compound exists in 2 readily inter convertable structures that differ in relative position.
- **Stereo isomerism**
  Shown by compounds having same molecular formula but different spatial arrangment.
  - **Geometrical isomerism**
    Shown by compounds possesses same structural formula but differ in spatial arrangement of the groups around a doubly bonded carbon atoms.
    - **Cis**
      Like atoms or groups at the same side of the double bond.
    - **trans**
      Like atoms or groups across the double bond.
  - **Optical isomerism**
    Arises from different arrangement of atoms or groups in 3-D space resulting in 2 isomers which are mirror images of each other.

Fundamental Concept of Organic Reaction

Fission of a covalent bond

- Depending upon the nature of the given compound,the nature of attacking reagent & reaction conditions bond cleavage occurs in two different ways
  1. Homolytic Cleavage (Equal cleavage of covalent bond)
  2. Heterolytic Fission (Unequal cleavage of covalent bond)

Reaction intermediates

- The important reaction intermediates are free radicals, carbocations, carbanions, carbenes, nitrenes, benzyne
- If positive charge is present on the carbon then cation formed is **termed as Carbocation**, if negative charge is present on the carbon then the anion is called **Carbanion**.
- Carbene (R_2C:) are reactive species containing a formally divalent carbon atom.
- Allyl free radicals are more stable than alkyl free radicals, Allyl free radicals are more stabilized due to resonance.
- The stability of alkyl carbanions decreases as methyl > primary > secondary > tertiary. As the number of alkyl groups increase the intensity of (-)ve charge on the central carbon increase.

  Stability of carbanions

  $$(C_6H_5)_3C^{\ominus} > (C_6H_5)_2CH^{\ominus} > C_6H_5CH_2^{\ominus} >$$

  $$CH_2=CH-CH_2^{\ominus} > CH_2=CH^{\ominus} > CH_3^{\ominus} >$$

  $$CH_3-CH_2^{\ominus} > CH_3-\underset{CH_3}{\overset{\ominus}{C}}-H > CH_3-\underset{CH_3}{\overset{CH_3}{C^{\ominus}}}$$

Electron displacement in organic compounds

Types of electronic displacement are:

Electron displacement in covalent bond

1. Inductive effect
2. Hyperconjugation
3. Resonance
4. Mesomeric effect
5. Electromeric effect (temporary effect)

Inductive effects are of two types

- Decreasing order of +I effect is

  $-O > -COO- > (CH_3)_3C- > (CH_3)_2CH- > CH_3CH_2- > -CH_3$
- Decreasing order of - I effect is

  $NR_3 > NO_2 > CN > F > Cl > Br > I > OH > OCH_3 > C_6H_6 > -CH=CH_2$
- A resonance hybrid is always more stable than any of its canonical structures. In resonance hybrid, there is delocalisation of electrons.
- These are electron donating groups, –OH, –OR, –NHCOR which donate electrons to conjugated system.
- Nitrobenzene does not undergo friedel-crafts reaction. Nitro group in nitro benzene deactivates the benzene ring.

Methods of Purification of Organic Compound

1. **Crystallisation:** is based on the difference in solubilities of compound and the impurities in a suitable solvent.
2. **Distillation:** used to separate
   (i) Volatile liquid from non-volatile impurities.
   (ii) Difference in their boiling points
3. **Fractional distillation:** If the difference in boiling point of 2 liquids is less than 25°.
4. **Steam distillation:** To separate compounds which are steam volatile and immiscible with water.
5. **Differential extraction:** Organic compound is present in an aqueous medium. It is separated by shaking it with an organic solvent in which it is more soluble than water.
6. **Sublimation:** Naphthalene, camphor, $NH_4^+ Cl^-$, benzoic acid. These are not going in liquid directly first go to vapour state.

MCQs ?

NCERT Topic-wise MCQs

Classification and Nomenclature

1. The correct IUPAC name of the organic compound

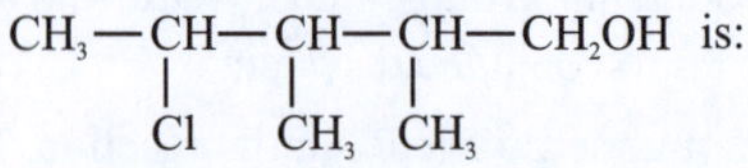

   1) 4-Chloro-2,3-dimethyl pentan-1-ol
   2) 2-chloro-3-4-dimethyl pentan-5-ol
   3) 2,3-dimethyl-4-chlorpentan-1-ol
   4) 2-chloro-3,4-dimethyl n-pentyl alcohol

2. The IUPAC name of:

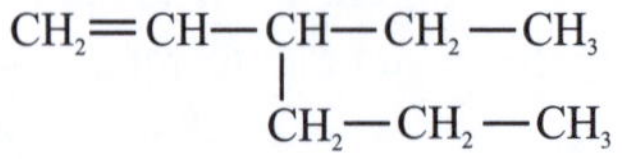

   1) 3-Propyl pent-l-ene
   2) 3-Ethyl pent-l-ene
   3) 4-Ethyl hex-1-ene
   4) 3-Ethyl hex-1-ene

3. For the given alkane,

   The IUPAC name is:
   1) 1,1-Dimethyl-5-ethyl octane
   2) 6-Ethyl-2-methyl nonane
   3) 4-Ethyl-8- methyl nonane
   4) 2-Methyl-6-propyl octane

4. The IUPAC name of:

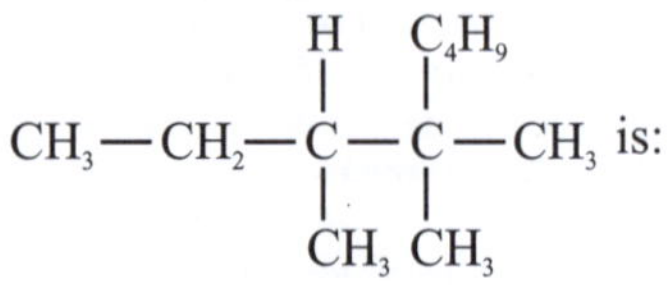

   1) 3,4,4-Trimethyl heptane
   2) 3,4,4-Trimethyl octane
   3) 2-Butyl-2-methyl-3-ethyl butane
   4) 2-Ethyl -3, 3- dimethyl heptane

5. The IUPAC name of the following compound is:
   $CH_2 = C = CH - CH_3$
   1) But -2, 3 diene 2) Buta-1,2-diene
   3) But-1-ene-2-yne 4) Conjugated butadiene

6. IUPAC name is:

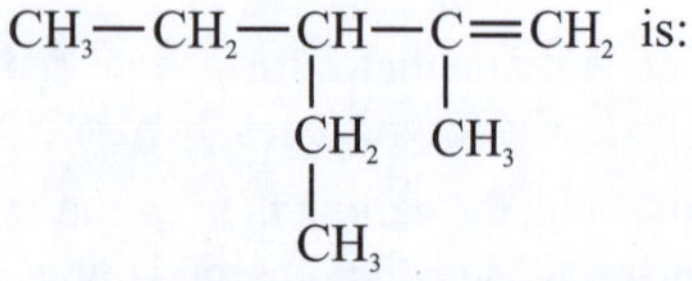

   1) 2-methyl-3-ethyl pent-1-ene
   2) 3-ethyl-4-methyl-4-pent -4-ene
   3) 3-ethyl-2 -methyl-1-pentene
   4) 3-methyl-2-ethyl-1-pentene

7. The IUPAC name of CH_3COCH_3 is:
   1) Acetone 2) 2-Propanone
   3) DI-methyl ketone 4) Propanal

8. IUPAC name of the given compond is:

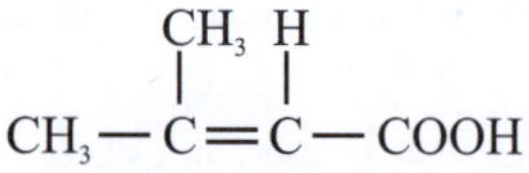

   1) 2-Methyl but-2-enoic acid
   2) 3-Methyl but-2-enoic acid
   3) 3-Methyl but-3-enoic acid
   4) 2-Methyl but-3-enoic acid

9. IUPAC name of $(CH_3)_3$ CCl is:
   1) 3-Chloro butane
   2) 2-Chloro-2-methyl propane
   3) t-Butyl chloride
   4) n-Butyl chloride

10. The correct structure of 4-Bromo-3-methyl but-1-ene is:
    1) $Br - CH = C(CH_3)_2$
    2) $CH_2 = CH - CH\ (CH_3) - CH_2Br$
    3) $CH_2 = C(CH_3)\ CH_2\ CH_2\ Br$
    4) $CH_3 - C(CH_3) = CH\ CH_2{-}Br$

11. Which is the incorrect name of CH_3NC?
    1) Aceto isonitrile 2) Methyl isocyanide
    3) Methyl carbylamine 4) Methyl isonitrile

12. The IUPAC name of the molecule:

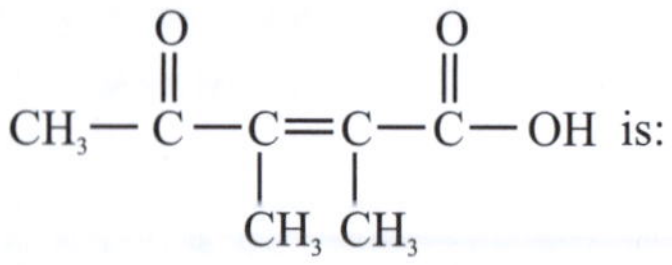

    1) 4-Oxo-2,3-dimethylpent-2-en-1 oic acid
    2) 2-Carboxy-3-methylpent-2-en-3-one
    3) 4-Carboxy-3-methylpent-3 en-2-one
    4) 2,3-Dimethyl-4 -oxo pent-2-en-1-oic acid

13. The IUPAC name of the compound:

$C_2H_5-\underset{\underset{CH_2}{\|}}{C}-CH_2OH$ is:

1) 2-ethyl prop-2-en-1-ol
2) 2-hydroxymethyl butan-1-ol
3) 2-methylenebutan-1-ol
4) 2-ethyl-3-hydroxy prop-1-ene

14. The IUPAC name of:

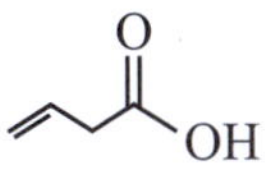

is:

1) But-3-enoic acid
2) But-1-enoic acid
3) Pent-4-enoic acid
4) Prop-2-enoic acid

15. The IUPAC name of tertiary butyl iodide is:

1) 4-Iodobutane
2) 2-Iodo butane
3) 1-Iodo-3-methyl propane
4) 2-Iodo-2- methylpropane

16. Correct IUPAC name is:

$H_3C-CH=\underset{\underset{CH_3-CH_2-CH_2}{|}}{C}-CH_2-CH_3$

1) 3-Ethyl hex-2-ene
2) 3-Ethyl pent-2-ene
3) 3-Ethylpent-3-ene
4) 3-Propyl pent-2-ene

17. The IUPAC name is:

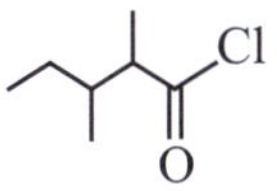

is:

1) 3,4-Dimethyl pentanoyl chloride
2) 1-Chloro-1-oxo-2,3-dimethyl pentane
3) 2-Ethyl-3-methyl butanoyl chloride
4) 2, 3, Dimethyl pentanoyl chloride

18. The IUPAC name for the hydrocarbon represented by  swastik sign is:

1) Neo-nonane
2) Tetraethylcarbon
3) 2-Ethyl pentane
4) 3, 3-Diethyl pentane

19. I.U.P.A.C name of

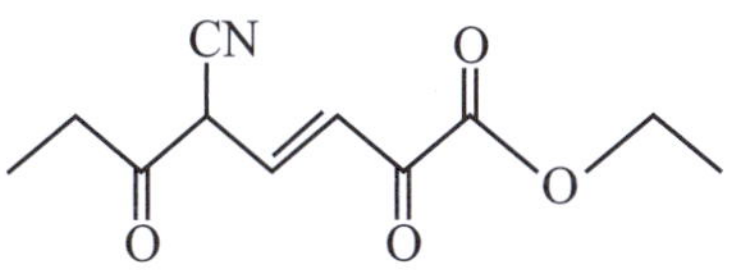

1) 8-ethoxy-3, 7, 8-trioxo-4-cyano oct-5-ene
2) 2 ethyl carbonyl, 5-ethoxy carbonyl 5-oxo, pent-3 ene nitrile
3) 5-cyano-1-epoxy-2, 6-dioxo-oct-3-en-1-al
4) Ethyl 5-cyano-2-6-dioxo-oct-3-en-1-oate

20. According to Huckel's rule a compound is said to be aromatic if it contains:

1) 4n bonds
2) 4n atoms
3) (4n+2) atoms
4) (4n+2) π electrons

21. Which of the following is an aromatic compound?

1) Phenol
2) Naphthalene
3) Pyridine
4) All of these

22. In IUPAC system of naming of organic compounds,which of the following functional group has more preference than others (in a poly functional compound)?

1) – OH
2) – CHO
3) – CO
4) $-CONH_2$

23. The IUPAC name of $CH_3-\underset{OH}{\overset{|}{CH}}-CH_2-\underset{CH_3}{\overset{|}{CH}}-CHO$ is,

1) 4-Hydroxy-2-methylpentanal
2) 2-Hydroxy-4-methylpentanal
3) 4-Hydroxy-2-methylpentanol
4) 2-Hydroxy-4-methylpentanol

24. Formula of phenyl carbinol and chloral are respectively:

1) $C_6H_5CH_2CH_2OH$ and $CHCl_2\ CHO$
2) $C_6H_5CH_2OH$ and $CCl_3\ CHO$
3) $C_6H_5CH_2OH$ and $CH_2Cl.CHO$
4) C_6H_5CHO and $CHCl_2CHO$

25. In the following molecule how many carbon atoms arrange linearly:

$CH_2=CH-C\equiv C-CH_2-\underset{\underset{CH_3}{|}}{CH}-CH_3$

1) 2
2) 3
3) 4
4) 5

26. In which of the following molecule the bond strength of carbon-carbon bond is maximum?

1) CH_3-CH_3
2) $HC\equiv CH$
3) $CH_2=CH_2$
4) $CH_3-CH_2-CH_3$

27. Select structure of iso-propylidene radical:

1) $CH_3-CH_2-\dot{C}H_2$
2) $CH_3-\dot{C}H-\underset{\underset{CH_3}{|}}{CH_2}$
3) $CH_3-\underset{\underset{CH_3}{|}}{\ddot{C}}$
4) $CH_3-\underset{\underset{CH_3}{|}}{\dot{C}}-CH_3$

28. Propargyl alcohol has sum of no of σ, π bonds & lp:

$CH\equiv C-CH_2-\ddot{O}H$

1) 10
2) 12
3) 11
4) 13

29. $C_6H_5-\underset{\bullet}{\overset{\overset{C_6H_5}{|}}{C}}-C_6H_5$ is:–

1) Triphenyl radical
2) Ter-phenyl radical
3) Trityl radical
4) None of these

30. Match the given general formula with their respective functional groups:-

A. $C_nH_{2n}O$	P – Hydroxy Carbonyl
B. C_nH_{2n-2}	Q – Alkenyne
C. C_nH_{2n-4}	R – Cyclic ether
D. $C_nH_{2n}O_2$	S – Alkadiene

1) A-Q, B-R, C-S, D-P
2) A-R, B-Q, C-S, D-P
3) A-P, B-S, C-Q, D-R
4) A-R, B-S, C-Q, D-P

31. Structure of Anthranilic & sulfanilic acid are represented by:

1) 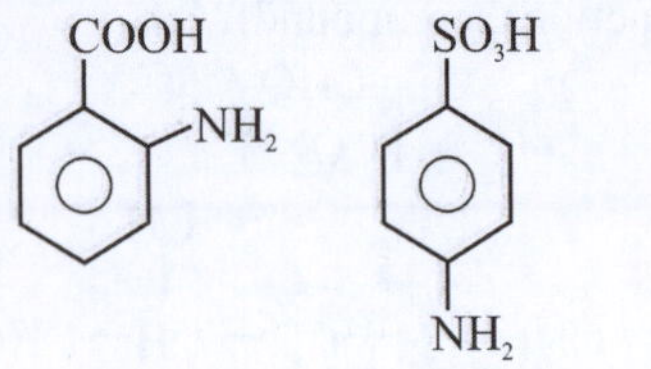

2) 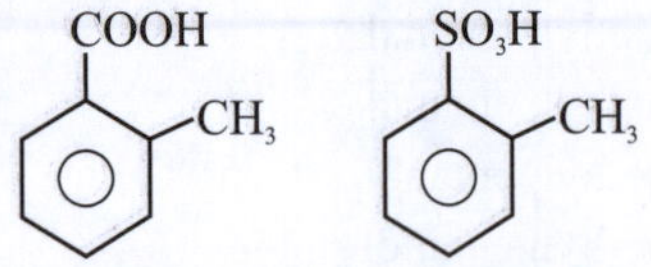

3) 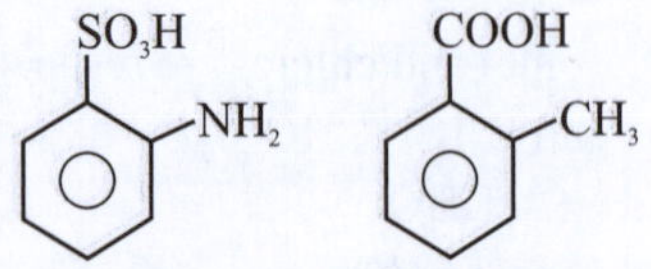

4) 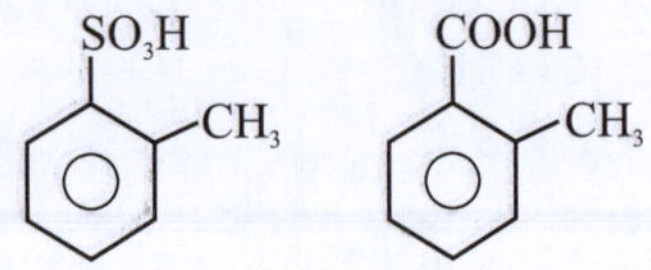

32. The structure of isobutyl isonitrile is:

1) $CH_3-\underset{CH_3}{\underset{|}{CH}}-CH_2-CN$ 2) $CH_3-\underset{CH_3}{\underset{|}{CH}}-CH_2-NC$

3) $CH_3-\underset{CH_3}{\underset{|}{CH}}-C\equiv N$ 4) $CH_3-\underset{CH_3}{\underset{|}{CH}}-N\equiv C$

Isomerism

33. Which of following compounds can not show Geometrical Isomerism?

1) 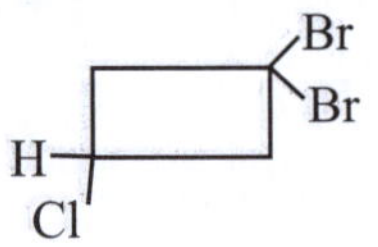 2) 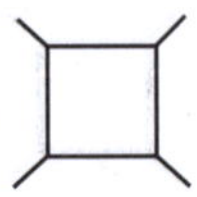

3) 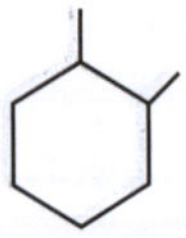 4) 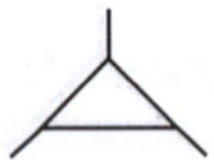

34. Methoxy methane and ethanol are:

1) Position isomers 2) Chain isomers

3) Functional isomers 4) Optical isomers.

35. In which of the following reactions, the product obtained is chiral?

1) $CH_3COCH_3 \xrightarrow{NaBH_4}$

2) $CH_3COCl \xrightarrow[\text{reaction}]{\text{Rosemund's}}$

3) $CH_3CH_2COCH_2CH_3 \xrightarrow{Sn,HCl}$

4) $CH_3CH_2COCH_3 \xrightarrow{LiAlH_4}$

36. Which will exhibit geometrical isomerism?

1) 3-phenyl-1-butene

2) 1-phenyl-2-butene

3) 2-phenyl-1-butene

4) 1, 1-diphenyl-1-propene

37. Different structures generated due to rotation about, C–C axis, of an organic molecule , are the examples of:

1) Geometrical isomerism

2) Conformational isomerism

3) Optical isomerism

4) Structural isomerism

38. Out of the following, which compound exhibits geometrical isomerism?

1) $CH_3CH_2\ CH_2\ CH_3$

2) $CH_3 - CH\ (CH_3) - CH(CH_3) - CH_3$

3) $CH_3 - CH\ (CH_3) - CH_3$

4) $CH_3\ CH = CH - CH_3$

39. Which of the following molecules is achiral?

1) 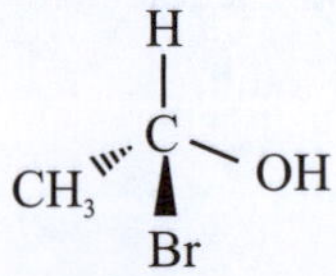 2) 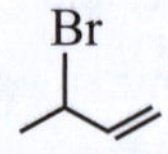

3) 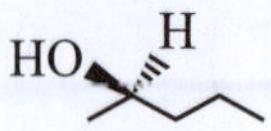 4) 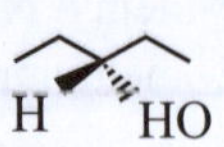

40. Optical isomerism is shown by:

1) Propan-2-ol 2) butan-2-ol

3) Ethanol 4) Methanol

41. The type of isomerism observed in urea molecule is:

1) Chain 2) Position

3) Geometrical 4) Tautomerism

42. Optical isomerism arises from the presence of:

1) A centre of symmetry

2) A line of symmetry

3) An asymmetric carbon atom

4) All of the above

43. The number of isomeric ethers with molecular formula $C_4H_{10}O$ are:

1) One 2) Two

3) Three 4) Four

44. Which of the following shows optical isomerism?

1) Butan-1-ol 2) Butan-2-ol

3) Butene 4) But-2-enol

45. Which one of the following alkenes does not exhibit geometric isomerism?

1) $C_4\ H_8$ 2) C_3H_6

3) C_5H_{10} 4) C_8H_{16}

46. The number of optical isomers of 2, 3-dihydroxy butanal is/are:

1) Two 2) Three

3) Four 4) One

47. Which of the following is chiral?

1) $Cl - CH_2 - CH_2 - CH_2 - CH_2 - CH_3$

2) $CH_3-CH_2-\underset{Cl}{\underset{|}{CH}}-CH_2-CH_3$

3) $Cl - CH = CH - CH_2 - CH_2 - CH_3$

4) $H-\underset{Cl}{\underset{|}{C}}=C=CH-CH_2-CH_3$

48. Number of Geometrical isomers in given compound:

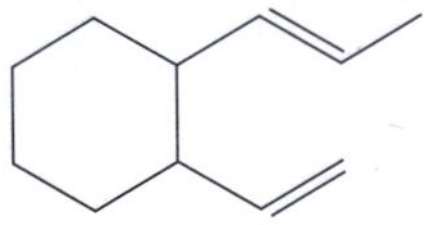

1) 2 2) 4
3) 8 4) 6

49. α-hydroxy propanoic acid has asymmetric carbon atoms:
1) 4 2) 3
3) 2 4) 1

50. How many isomeric vicinal-dihalides are possible for the compound having molecular formula $C_3H_6Cl_2$?
1) 1 2) 2
3) 3 4) 4

51. The most stable conformation of n - butane is:
1) Skew 2) Eclipsed
3) Staggered 4) None of the above

52. Number of possible position isomers for dichlorobenzene is:
1) 2 2) 3
3) 4 4) 5

53. Alkanols and Alkoxyalkanes are:
1) Functional isomers
2) Keto-enol tautomers
3) Geometrical isomers
4) Not isomers at all

54. n-Propyl alcohol and isopropyl alcohol are examples of:
1) Position isomerism
2) Chain isomerism
3) Tautomerism
4) Geometrical isomerism

55. The number of structural alcoholic isomers for $C_4H_{10}O$ is:
1) 2 2) 3
3) 4 4) 5

56. What is the configuration of the following molecule?

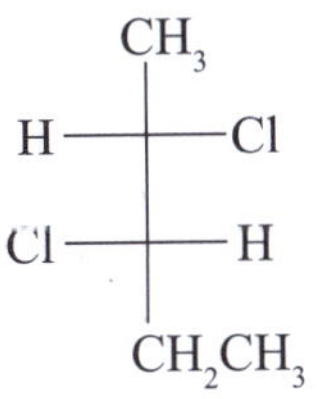

1) 2S, 3R 2) 3R, 3S
3) 2S, 3S 4) 2R, 3R

57. Number of possible structures of monochlorobenzene is:
1) 1 2) 3
3) 5 4) 6

58. The total number of structural isomers for the compound of the formula $C_4H_{10}O$ is
1) 7 2) 6
3) 4 4) 3

59. The compound which is not isomeric with diethyl ether is:
1) n-Propyl methyl ether 2) Butan-1-ol
3) 2- Methylpropan-2-ol 4) Butanone

60. $CH_3-CH_2-CH_2-Br$, $CH_3-CH(Cl)-CH_3$

these show which type of isomerism?
1) Positional isomerism (P.I.)
2) Chain isomerism (C.I.)
3) Functional group isomerism (F.G.I.)
4) None of these

61. $CH_2=CH-CH=CH_2$ (I) $CH_3-CH_2-C\equiv CH$ (II)
$CH_3-C\equiv C-CH_3$ (III)

Which among these are positional isomers?
1) I, II 2) I, III
3) II, III 4) None of these

62. 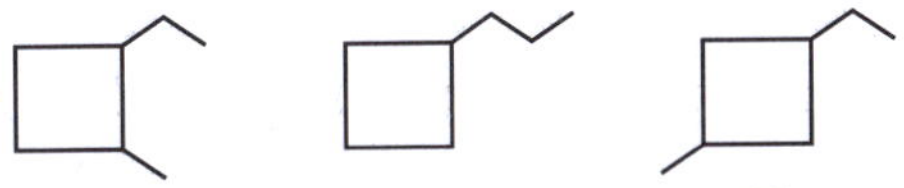

I II III

Among these, positional isomers are:–
1) I, II 2) II, III
3) I, III 4) All of these

63. The type of isomerism exhibited by the compound with formula $C_5H_{12}O$ is:
1) Chain & position
2) Functional & position
3) Metamerism
4) Chain, Position, Functional & Metamerism

64. The total no. of structural isomers of $C_3H_6Br_2$:
1) 2 2) 4
3) 6 4) 3

65. The total no. of structural isomers of dichloro-cyclohexane:
1) 2 2) 3
3) 4 4) 5

66. The total number of structural isomers of C_3H_6Cl Br:
1) 3 2) 4
3) 6 4) 5

67. 

I II III

Among these, ring-chain isomers are:-
1) I, II 2) II, III
3) I, III 4) All of these

68. 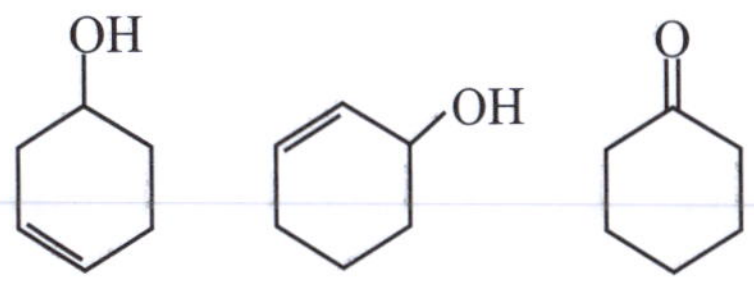

I II III

II is the functional isomer of:-
1) I only 2) III only
3) Both I & III 4) None of these

69. Which of the following is correct order of physical properties of Geometrical Isomers:

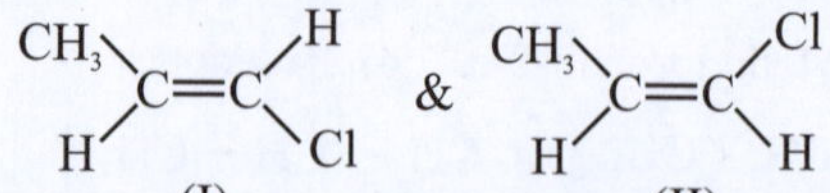

(I) & (II)

	Dipole moment	B.P.	M.P.	Stability
1)	I > II	I > II	II > I	I > II
2)	II > I	II > I	II > I	II > I
3)	I > II	I > II	I > II	I > II
4)	II > I	II > I	I > II	I > II

Fundamental Concepts of Organic Reactions

70. Heterolysis of propane gives:
1) Methyl and ethyl free radicals
2) Methyl Carbocation and ethyl free radicals
3) Methyl anion and ethyl carbocation
4) Methyl free radical and ethyl carbocation

71. $CH_4 + Cl_2 \xrightarrow{hv} CH_3Cl + HCl$ is an example for
1) Free radical substitution reaction
2) Electrophilic substitution reaction
3) Nucleophilic substitution reaction
4) Nucleophilic addition reaction

72. Which of the following is an electrophilic reagent?
1) H_2O 2) OH^-
3) NO_2^+ 4) All of these

73. Decreasing order of -I effect of given groups is:
a. CN^- b. NO_2
c. NH_3^+ d. F
1) b > a > d > c   2) b > c > d > a
3) c > b > d > a   4) c > b > a > d

74. The reaction intermediate carbenes are produced from:
1) Diazomethane   2) Ketene
3) $CHCl_3/C_2H_5ONa$ 4) All of these

75. Which hybridised state of carbon exerts strongest -I effect having how much percentage of 's' character?
1) 25%   2) 33%
3) 50%   4) Equal in all

76. 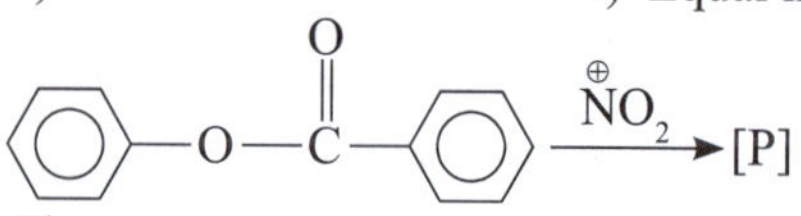

The product [P] is:

1) 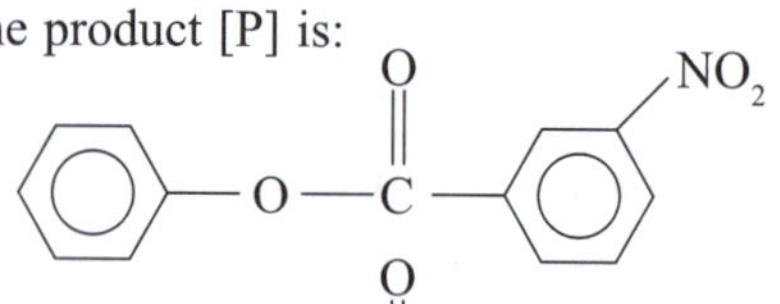

2) 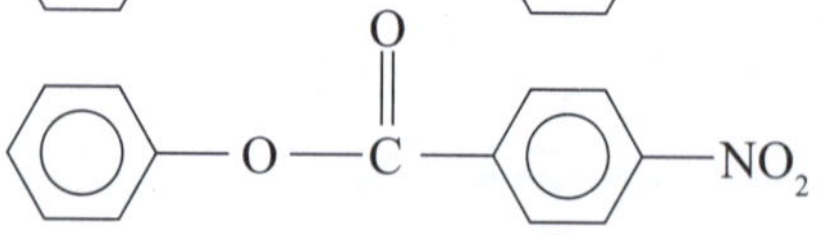

3) 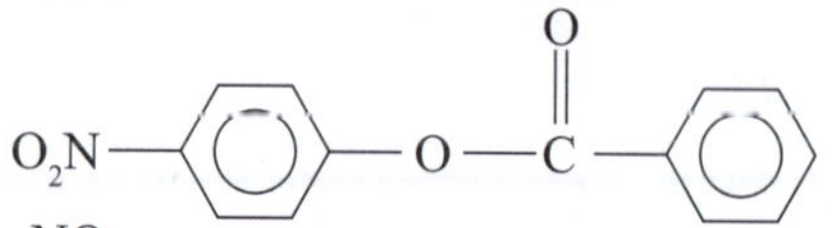

4) 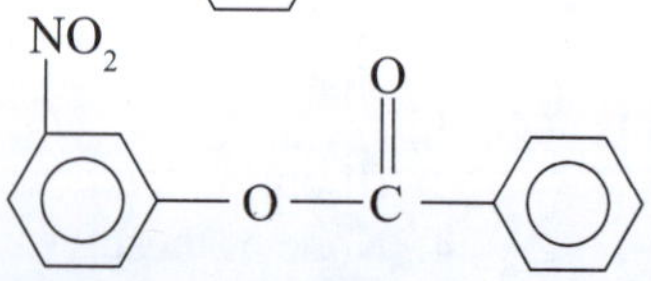

77. Arrange in order of decreasing pK_b:
A. $F-CH_2COOH$
B. $Cl-CH(Cl)-CH_2-COOH$
C. $F-CH_2-CH_2-COOH$
D. $Br-CH_2-CH_2-COOH$
1) A > B > C > D   2) B > A > C > D
3) D > C > B > A   4) D > C > A > B

78. Find correct order of stability of carbocation

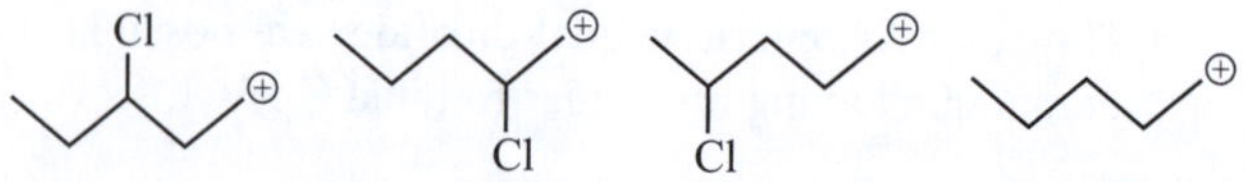

I   II   III   IV

1) IV > III > II > I   2) IV > I > II > III
3) IV > III > I > II   4) IV > II > I > III

79. The presence of Chlorine atom on benzene ring makes the second substituent enter at position.
1) Ortho   2) Meta
3) Para   4) Ortho & para

80. Arrange following in order of K_a value
A. C_6H_5OH B. CH_3OH
C. CH_3COOH D. CH_3SO_3H
1) A > C > D > B   2) D > C > A > B
3) A > C > A > B   4) B > A > C > D

81. Give correct order of increasing acidity of following compounds:

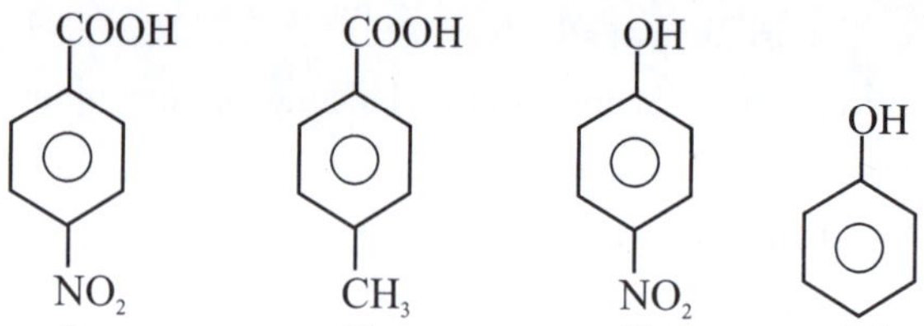

I   II   III   IV

1) I > III > II > IV   2) I > II > III > IV
3) I > III > IV > II   4) IV > I > II > III

82. Select correct order of stability of alkene in following organic compound:

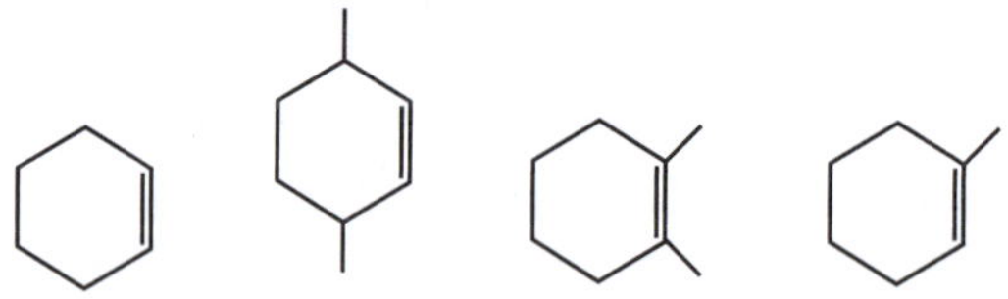

A   B   C   D

1) C > D > B > A   2) C > D > A > B
3) B > A > C > D   4) A > B > C > D

83. Which of following has most negative heat of hydrogenation?

1)    2) 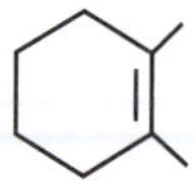

3) 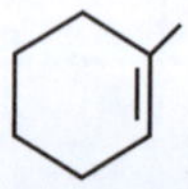   4) 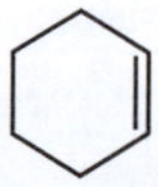

84. The correct orientation of dipoles in pyrrole & pyridine is:

1) 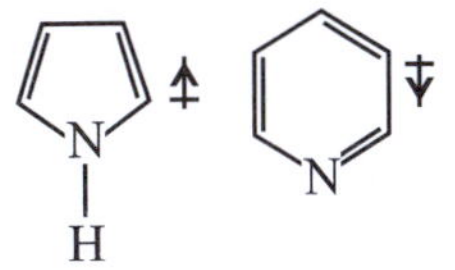 2) 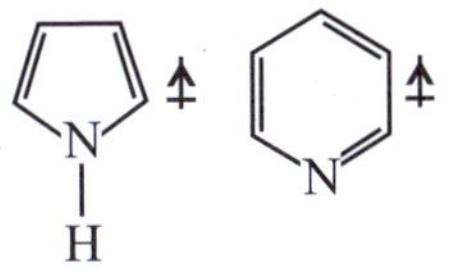

3) 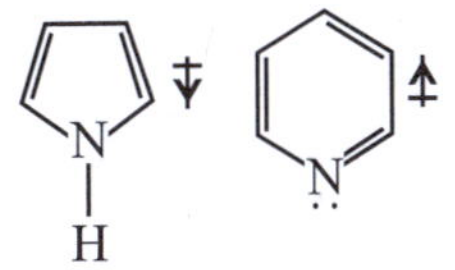 4) None of these

85. The correct order of electron density in aromatic ring of following compound is:

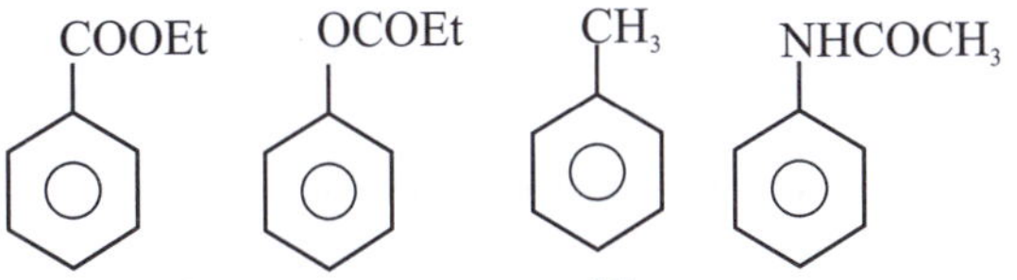

I II III IV

1) IV > III > II > I 2) I > II > III > IV
3) IV > II > I > III 4) IV > II > III > I

Methods of Purification of Organic Compounds

86. How will you separate a solution (miscible) of benzene and $CHCl_3$?
1) Sublimation 2) Filtration
3) Distillation 4) Crystallisation

87. Which one of the following methods can be used to separate a mixture of ortho and para-nitrophenols?
1) Crystallisation 2) Solubility
3) Sublimation 4) Steam distillation

88. Which one of the following can be purified by sublimation?
1) F_2 2) Cl_2
3) Br_2 4) I_2

89. Which is purified by steam distillation?
1) Aniline 2) Benzoic acid
3) Petroleum 4) Naphthalene

90. Sublimation is a process in which a solid:
1) Changes into vapour form
2) Changes into another allotropic from
3) Changes into liquid form
4) None of the above

91. A mixture of camphor and benzoic acid can be separated by:
1) Sublimation 2) Extraction with a solvent
3) Chemical method 4) Fractional distillation

92. The best method for the separation of naphthalene and benzoic acid from their mixture is:
1) Chromatography 2) Crystallisation
3) Distillation 4) Sublimation

93. For the purification, isolation and separation of organic compounds, the latest technique followed is:
1) Chromatography
2) Steam distillation
3) Fractional crystallization
4) Sublimation

94. Which of the following reagents will be fruitful for separation of a mixture of nitrobenzene and aniline?
1) Aqueous $NaHCO_3$ 2) H_2O
3) Aqueous HCl 4) Aqueous NaOH

95. A liquid which decomposes at its boiling point can be purified by:
1) Distillation at atmospheric pressure
2) Distillation under reduced pressure
3) Fractional distillation
4) Steam distillation

96. Turpentine oil can be purified by:
1) Steam distillation 2) Vacuum distillation
3) Fractional distillation 4) Sublimation

97. Paper chromatography is:
1) Adsorption chromatography
2) Partition chromatography
3) Ion exchange chromatography
4) All of these

98. Vacuum distillation is used to purify liquids which:
1) Are highly volatile
2) Are explosive in nature
3) Soluble in water
4) Decomposes below their B.P`s.

99. Impure Naphthalene is purified by:
1) Fractional crystallisation
2) Fractional distillation
3) Solvent extraction
4) Sublimation

100. A mixture of camphor and KCl can be separated by:
1) Evaporation 2) Sublimation
3) Filtration 4) Decantation

101. In steam distillation, the sum of the vapour pressure of the volatile compound and that of water is:
1) Equal to atmospheric pressure
2) Less than atmospheric pressure
3) More than atmospheric pressure
4) Exactly half of the atmospheric pressure

Qualitative Analysis of Organic Compounds

102. Lassaigne's test in qualitative analysis is used to detect:
1) Nitrogen 2) Sulphur
3) Chlorine 4) All of these

103. The compound formed in the positive test for nitrogen with the Lassaigne solution of an organic compound is:
1) $Fe_4[Fe(CN)_6]_3$ 2) $Na_3[Fe(CN)_6]$
3) $Fe(CN)_3$ 4) $Na_4[Fe(CN)_5COS]$

104. Lassaigne's test is not used for the detection of:
1) Carbon 2) Halogens
3) Nitrogen 4) Sulphur

105. During Lassaigne's test, N and S present in an organic compound change into:
1) Na_2S and NaCN 2) NaCNS
3) Na_2SO_4 and NaCN 4) H_2S and NO_2

106. Presence of halogen in organic compounds can be detected by using:
1) Leibig's test 2) Duma's test
3) Kjeldahl's test 4) Beilstein test

107. The prussian blue colour obtained in the Lassaigne's test for nitrogen is due to formation of:
1) Iron (II) hexa cyano ferrate (II)
2) Iron (III) hexa cyano ferrate (II)
3) Iron (III) hexa cyano ferrate (III)
4) Iron (II) hexa cyano ferrate (III)

108. In an organic compound, phosphorus is estimated as:
1) $Mg_2P_2O_7$ 2) $Mg_3(PO_4)_2$
3) H_3PO_4 4) P_2O_5

109. During the fusion of an organic compound with sodium metal, nitrogen of the compound is converted into:
1) $NaNO_2$ 2) $NaNH_2$
3) NaCN 4) NaNC

110. Presence of nitrogen in which among the following compounds can not be detected by Lassaigne's method?
1) Hydrazine 2) Aniline
3) p-toluidine 4) Picric acid

111. Violet coloured complex obtained in the detection of sulphur is:
1) $Na_2[Fe(NO)(CN)_5]$ 2) $Na_3[Fe(ONS)(CN_5)]$
3) $Na_4[Fe(CN)_5NOS]$ 4) Both (2) and (3)

112. In Lassaigne's method organic compound is fused with:
1) Sodium metal
2) Zinc dust
3) Sodium carbonate and Zinc dust
4) Calcium metal

113. Medium of sodium extract is:
1) Neutral 2) Basic
3) Acidic 4) Depends on organic compound

114. In the Lassaigne's test the sulphur present in the organic compound first changes into
1) Na_2SO_3 2) CS_2
3) Na_2SO_4 4) Na_2S

115. A mixture of calcium sulphate and camphor can be separated by-
1) Filtration 2) Evaporation
3) Sublimation 4) Chromatography

116. A compound which does not give a positive result in the Lassaigne's test for nitrogen is:
1) Urea 2) $NH_2OH.HCl$
3) Glycine 4) Phenylhydrazine

117. Sodium extract gives blood red colour when treated with $FeCl_3$. Formation of blood red colour confirms the presence of:
1) Only nitrogen 2) Only sulphur
3) Only halogens 4) Both Nitrogen and Sulphur

118. Which of the following compounds will give Lassaigne's test for nitrogen?
1) NH_2NH_2 2) NH_2OH
3) NaCN 4) $NaNO_3$

Quantitative Analysis

119. In Duma's method for estimation of nitrogen 0.3 g of an organic compound gave 50 mL of nitrogen collected at 300 K temperature and 715 mm pressure. Calculate the % composition of nitrogen in the compound: (Aqueous tension at 300 k = 15 mm)
1) 11.46% 2) 17.46%
3) 22.06% 4) 39.65%

120. In Carius method of estimation of halogen, 0.15 g of an organic compound gave 0.12 g of AgBr. Find out the % of bromine in the compound:
1) 34.04% 2) 44.04%
3) 96.05% 4) 88%

121. Which of the following methods is not used for estimation of nitro and azo compounds?
1) Duma's method 2) Kjeldahl's method
3) Carius method 4) All of the above

122. 4g of hydrocarbon (C_xH_y) on complete combustion gave 12 g of CO_2. What is the empirical formula of the hydrocarbon?
1) CH_3 2) C_4H_9
3) CH 4) C_3H_8

123. In Kjeldahl's method, ammonia from 5 g of food neutralises 30 cm^3 of 0.1 N acid. The percentage of nitrogen in the food is:
1) 0.84 2) 8.4
3) 16.8 4) 1.68

124. An unknown compound has molecular mass 240g/mol. If empirical formula is CH_2O then its molecular formula will be:
1) $C_4H_8O_4$ 2) $C_{12}H_{12}O_{12}$
3) $C_6H_{12}O_6$ 4) $C_8H_{16}O_8$

125. On complete combustion, 0.246 g of an organic compound gave 0.198g of carbon dioxide and 0.101g of water. Determine the percentage composition of carbon and hydrogen in the compound:
1) 8.9% (H) 4.58% (C)
2) 6.8% (C) 4.58% (H)
3) 4.58% (H) 21.95% (C)
4) 3.2% (H) 21.95% (C)

126. A sample of 0.5 g of an organic compound was treated according to Kjeldahl's method. The ammonia evolved was absorbed in 50 ml of 0.5 M H_2SO_4. The remaining acid after neutralilsation by ammonia consumed 80 ml of 0.5 M NaOH. The percentage of nitrogen in the organic compound is:
1) 14 2) 28
3) 42 4) 56

127. In the estimation of sulphur by Carius method, 0.480 g of an organic compound gives 0.699 g of barium sulphate. The percentage of sulphur in this compound is: (atomic masses : Ba = 137, S = 32 and O = 16)

1) 20 % 2) 15%
3) 35% 4) 30%

128. Following is estimated by Liebig method:

1) C and H 2) Nitrogen
3) Chlorine 4) Bromine

129. In Carius method halogens are estimated as:

1) X_2 2) BaX_2
3) PbX_2 4) AgX

130. In Duma's method nitrogen in organic compound is estimated in the form of:

1) N_2 2) NO
3) NH_3 4) N_2O_5

131. In organic compounds, sulphur is estimated as:

1) $BaSO_4$ 2) $BaCl_2$
3) $Ba_3(PO_4)_2$ 4) H_2SO_4

132. Kjeldahl's method cannot be used for the estimation of nitrogen in:

1) $C_6H_5-N=N-C_6H_5$ 2) 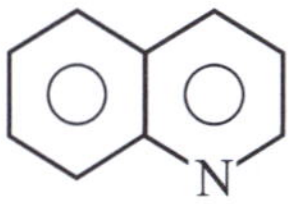

3) 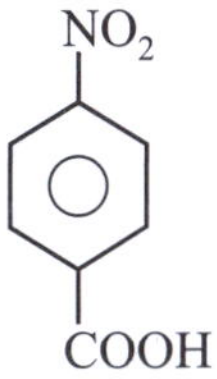 4) All of these

133. In Kjeldahl's method, nitrogen present in the organic compound is first converted into:

1) NH_3 2) $(NH_4)_2SO_4$
3) N_2 4) NO

134. Which gas is introduced into the combustion tube in Liebig's method?

1) Pure and dry CO_2 2) Pure and dry N_2
3) Pure and dry O_2 4) Pure and dry He

135. If two compounds have the same empirical formula but different molecular formula they must have:

1) Different percentage composition
2) Different molecular weight
3) Same viscosity
4) Same vapour density

136. In Dumas method 0.5 g of an organic compound containing nitrogen gave 112 ml of nitrogen at S.T.P. The percentage of nitrogen in the given compound is:

1) 28 2) 38
3) 18 4) 48

137. 0.73 g of organic compound on oxidation gave 1.32 g of carbon dioxide. The percentage of carbon in the given compound will be:

1) 49.32 2) 59.32
3) 29.32 4) 98.64

138. In an estimation of S by Carius method 0.2175g of the compound gave 0.5825g of $BaSO_4$ Percentage of S is:

1) 36.78% 2) 45.50%
3) 39.48% 4) 30.69%

High Order Time Intensive MCQs

1. Which of the following orders is correct for the stability of carbocations given below?

I. 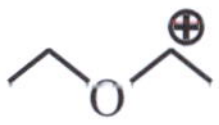 II. 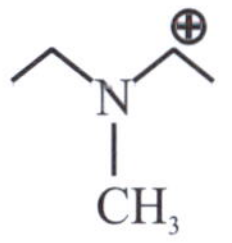 III. 

1) I > II > III 2) III > II > I
3) II > I > III 4) II > III > I

2. The increasing order of basicity for the following intermediates is (from weak to strong)

$(CH_3)_3C^{\ominus}$ (i) $H_2C=CH-\overset{\ominus}{C}H_2$ (ii) $HC\equiv\overset{\ominus}{C}$ (iii) $\overset{\ominus}{C}H_3$ (iv) $\overset{\ominus}{C}N$ (v)

1) (v) < (iii) < (ii) < (iv) < (i)
2) (iii) < (i) < (ii) < (iv) < (v)
3) (v) < (i) < (iv) < (ii) < (iii)
4) (iii) < (iv) < (ii) < (i) < (v)

3. Which of the following orders is correct for the energy required for heterolytic cleavage of the indicated C–Cl bonds producing carbocations:

(I) 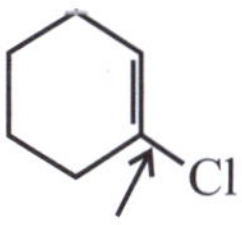 (II) 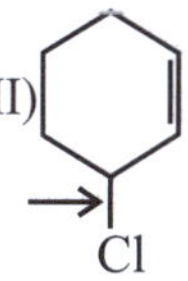 (III) 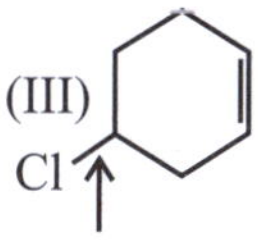

1) I > II > III 2) III > II > I
3) I > III > II 4) II > III > I

4. Which of the following carbanion is resonance stabilised?

1) 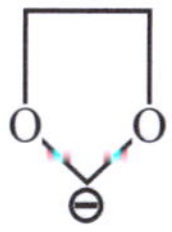 2) 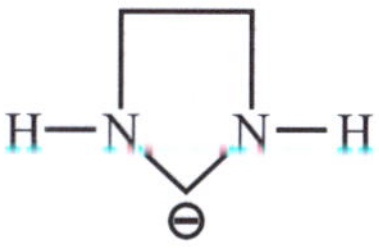

3)  4) 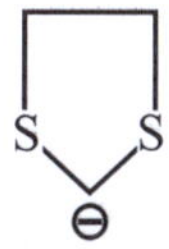

5. In which of the following, extra bond will be formed in its resonating structures?

1) $H_2C{=}CH{-}\overset{\oplus}{C}H_2$
2) 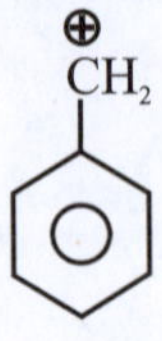
3) $H_3C{-}\overset{\oplus}{C}H{-}OCH_3$
4) 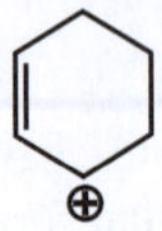

6. Which of the following carbocation is not resonance stabilised?

1) 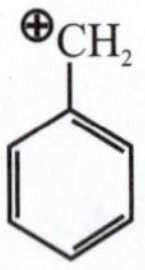
2) 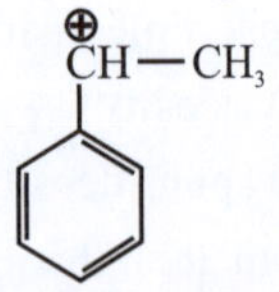
3) 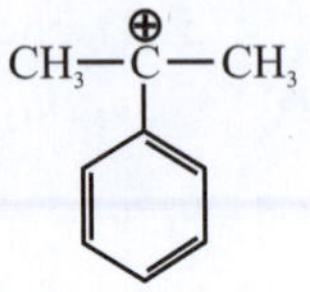
4) 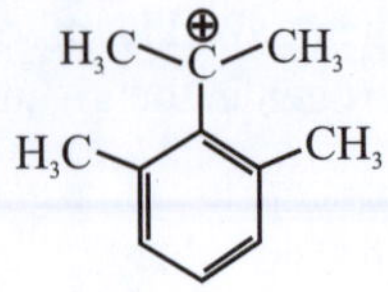

7. Which carbanion is maximum stable?

1) 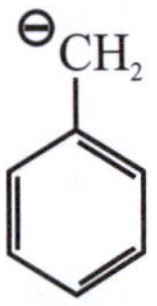
2) 
3) 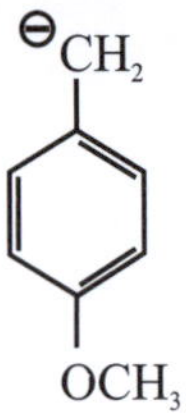
4) 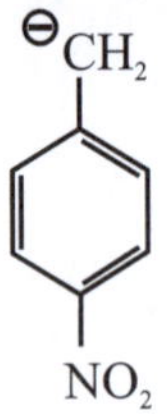

8. In which of the following compound, lone pair of N is not involved in resonance?

1) 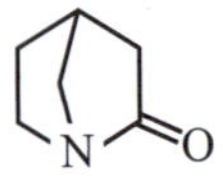
2) 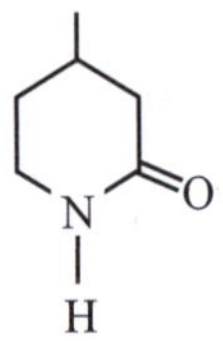
3) 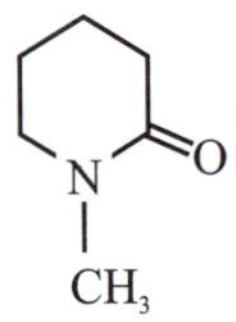
4) 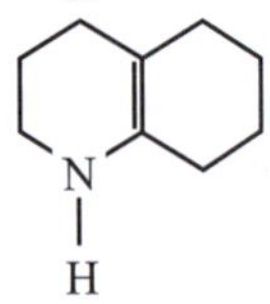

9. Order of K_a in following compounds

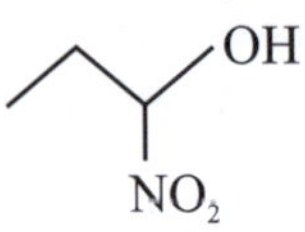
I

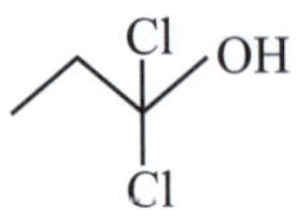
II

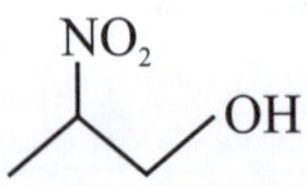
III

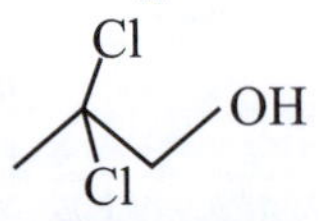
IV

1) IV > I > II > III
2) III > IV > I > II
3) II > I > IV > III
4) IV > III > II > I

10. The most stable resonating structure among the above given structures is:

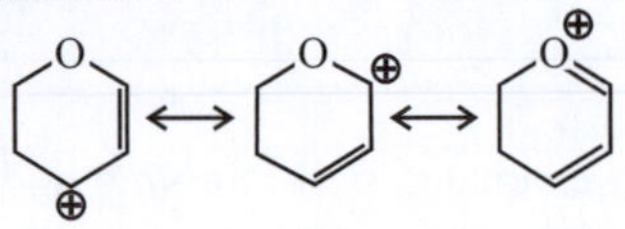
(I) (II) (III)

1) I
2) II
3) III
4) All of these

11. In pyridine; 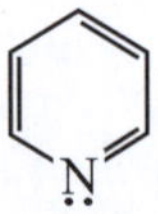 Number of conjugated electrons are:

1) 6
2) 8
3) Zero
4) 5

12. Which of the following molecules has longest C = C length?

1) $CH_2{=}C{=}CH_2$
2) $CH_3{-}CH{=}CH_2$
3) $CH_3{-}C(CH_3)_2{-}CH{=}CH_2$
4) $CH_3{-}C(CH_3){=}CH_2$

13. 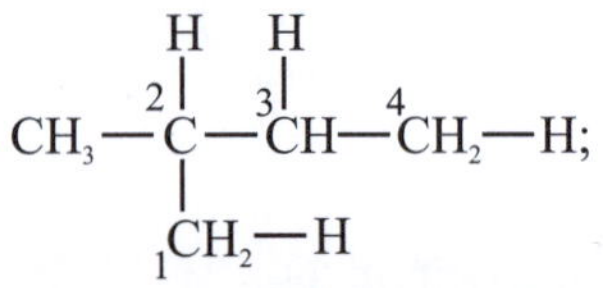

$C_1 - H$, $C_2 - H$, $C_3 - H$ and $C_4 - H$ the homolytic bond dissociation energy is in the order :

1) $C_2 - H > C_3 - H > C_4 - H > C_1 - H$
2) $C_1 - H > C_4 - H > C_2 - H > C_3 - H$
3) $C_2 - H > C_3 - H > C_1 - H > C_4 - H$
4) $C_1 - H > C_4 - H > C_3 - H > C_2 - H$

14. The decreasing order of stability of following anions is-

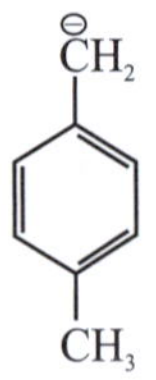
(P)

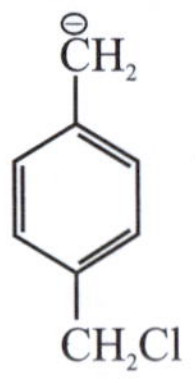
(Q)

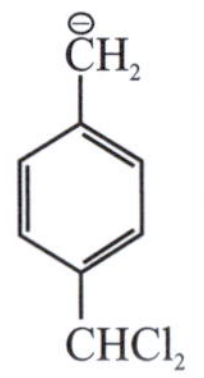
(R)

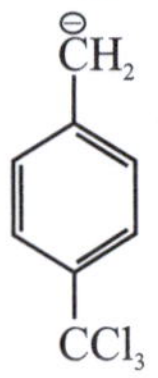
(S)

1) P > Q > R > S
2) S > R > Q > P
3) Q > S > R > P
4) Q > R > S > P

15. The decreasing order of stability of following anions is-

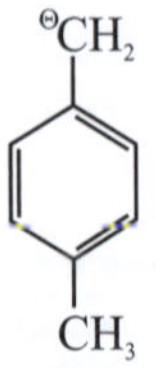
(P)

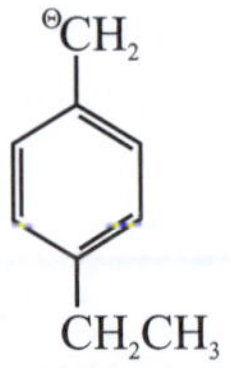
(Q)

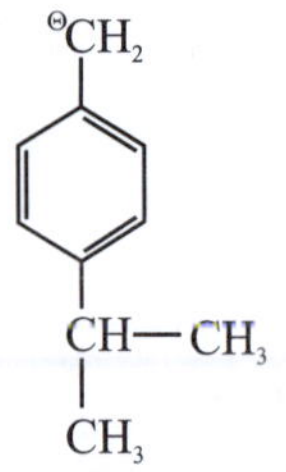
(R)

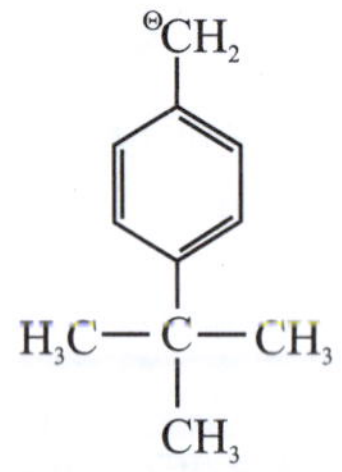
(S)

1) R > S > Q > P
2) P > S > R > Q
3) Q > P > S > R
4) P > Q > R > S

16. The correct order of nucleophilicity is
1) $F^- > OH^-$
2) $H_2O > OH^-$
3) $R-\ddot{O}H > RO^-$
4) $NH_2^- > NH_3$

17. Which of the following orders is correct for hyperconjugation of these radicals ?

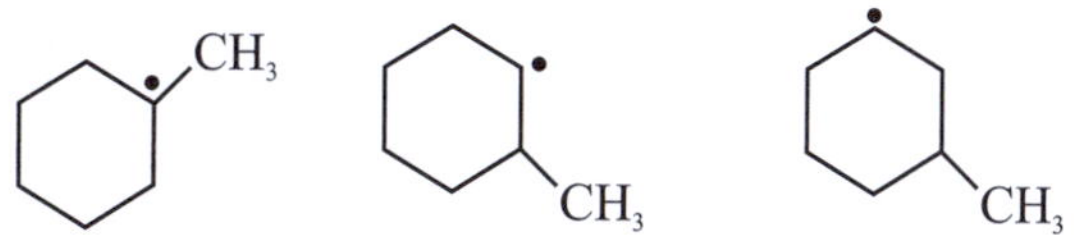

(P) (Q) (R)

1) P > Q > R
2) R > Q > P
3) Q > P > R
4) P > R > Q

18. Hyperconjugation occurs in :

1) 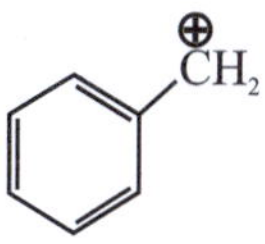

2) 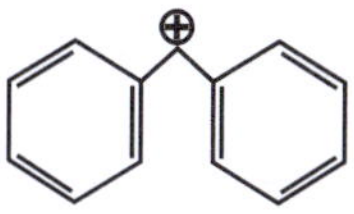

3) 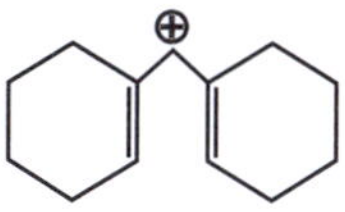

4) None of these

19. Which is most stable species?

1) 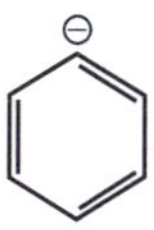

2) $(H_3C)_3-C-\overset{\ominus}{C}H_2$

3) $(H_3C)_2-CH-\overset{\ominus}{C}H_2$

4) $HC\equiv\overset{\ominus}{C}$

20. Which is an aromatic compound?

1) 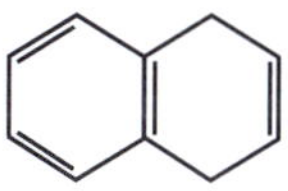

2) 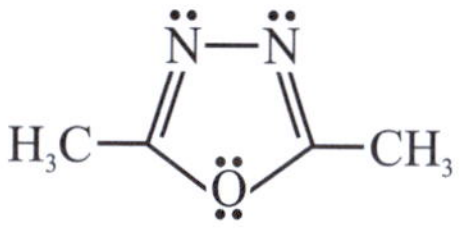

3) 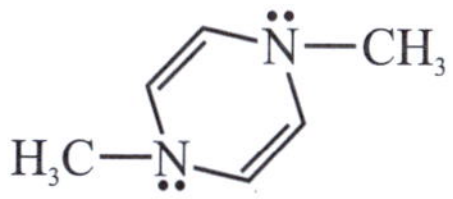

4) 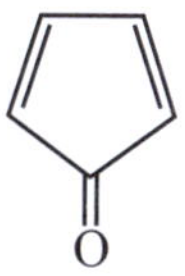

21. Which is incorrect statement?

1) 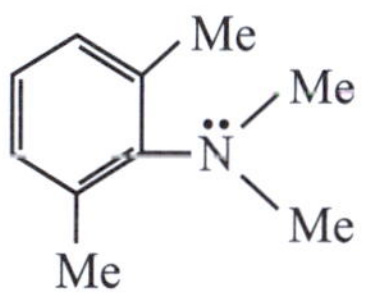 is more basic than 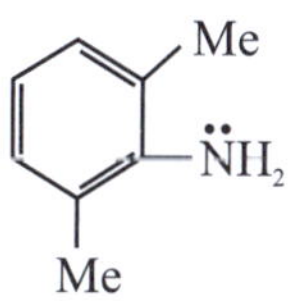

2) 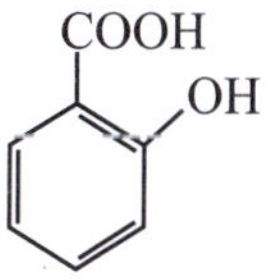 is more acidic than 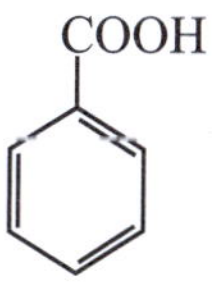

3) 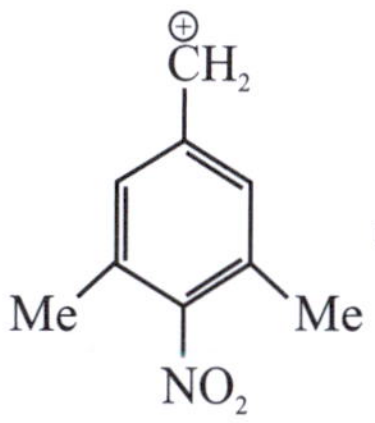 is less stable than 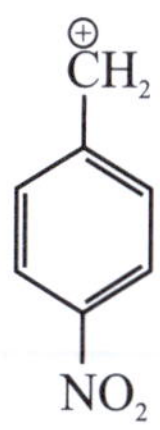

4) 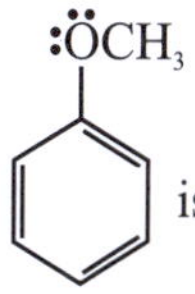 is less reactive than 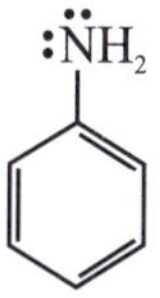 towards ESR.

22. Which carbocation is the most stabilized ?

1) 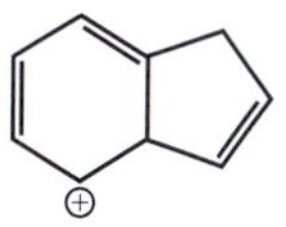

2) 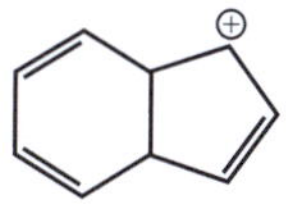

3) 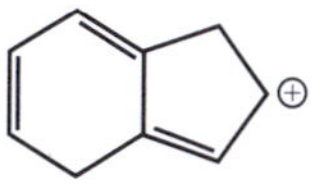

4) 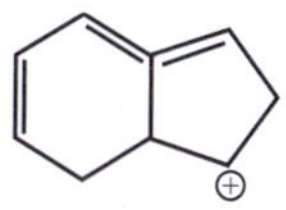

23. The compound which gives the most stable carbonium ion on dehydration

1) $CH_3-CH(CH_3)-CH_2-OH$

2) $H_3C-C(CH_3)_2-OH$

3) $CH_3-CH_2-CH_2-CH_2OH$

4) $CH_3-CH(OH)-CH_2-CH_3$

24. What is the decreasing order of stability (most stable ® least stable) of the following carbocations?

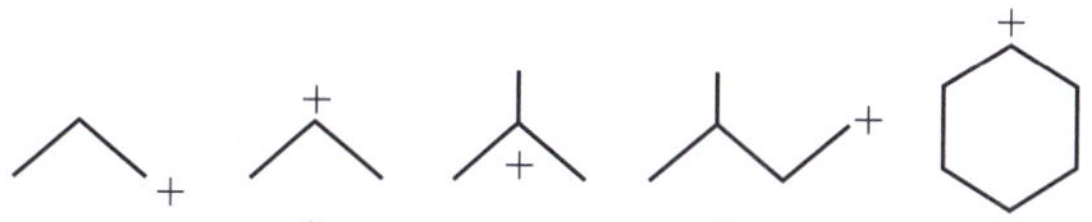

a b c d e

1) c > b > a > d > e
2) c > b > e > d > a
3) a ≈ d > b ≈ e > c
4) c > a ≈ d > b ≈ e

25. Arrange in order of stability:

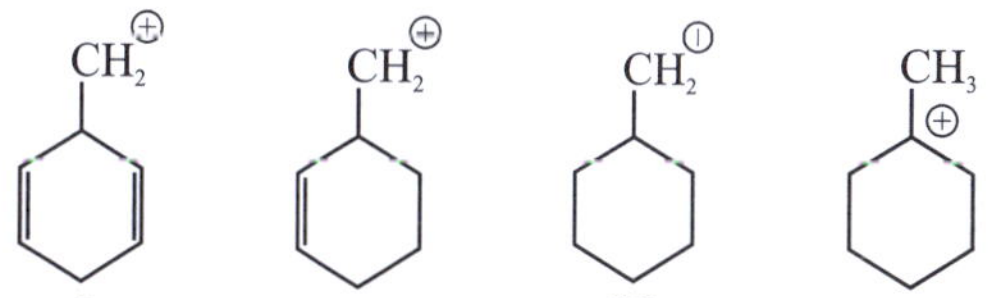

I II III IV

1) I > II > III > IV
2) I > II > IV > III
3) III > II > I > IV
4) IV > III > II > I

26. Choose the correct comparison of stability:

1) 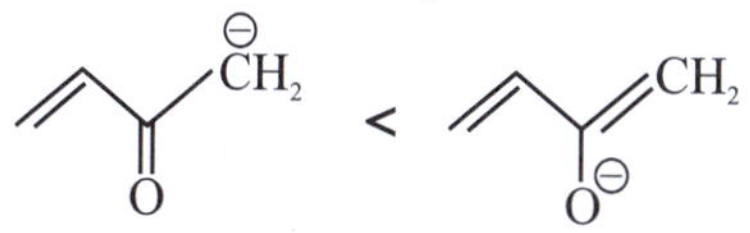

2) 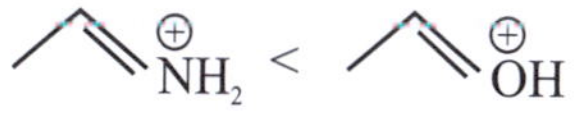

3) 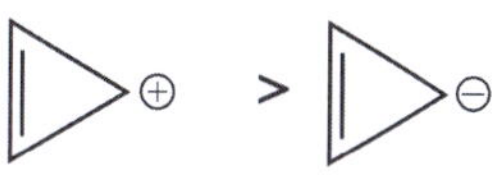

4) Both (1) & (3)

27. The correct stability order of following is:

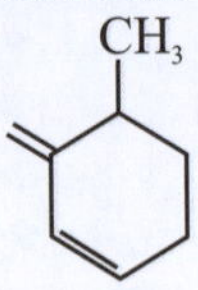 (I) 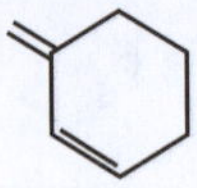 (II) 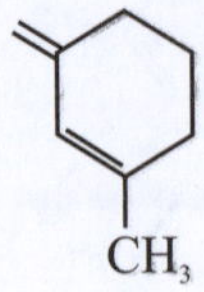 (III) 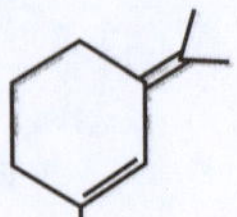 (IV)

1) I > II > III > IV
2) III > IV > II > I
3) III > IV > II > I
4) IV > III > II > I

28. Arrange following compounds in decreasing order of their dipole moment.

$CH_3-CH_2-NO_2$ (I) $CH_3-CH_2-NH_2$ (II)

CH_3-CH_2-F (III) $CH_3-CH_2-C\equiv N$ (IV)

1) IV > III > I > II
2) IV > I > III > II
3) I > III > IV > II
4) I > IV > III > II

29. Identify which of the following does not show + M effect?

1) $-NH_2$
2) $-O^{\ominus}$
3) $-NH-\overset{O}{\overset{\|}{C}}-CH_3$
4) $-\overset{O}{\overset{\|}{C}}-NH-CH_3$

30. The correct IUPAC name of $CH_3-\underset{CH_3}{\underset{|}{C}}=\underset{C_2H_5}{\underset{|}{C}}-CH_3$ is:

1) 1, 2-diethyl butene
2) 2-ethyl-3- methyl pentene
3) 3, 4-dimethyl hex-3-ene
4) 2, 3-dimethyl pent-2-ene

31. IUPAC name of $CH\equiv C-CH=CH_2$ is

1) But-3-ene-1-yne
2) But-1-ene-3-yne
3) But-1-yne-3-ene
4) But-3-yne-1-ene

32. The IUPAC name of the given compound 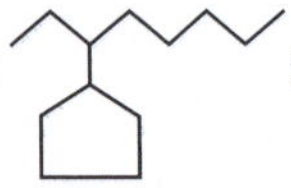 is

1) Octyl cyclopentane
2) 3- cyclopentyl octane
3) Cyclopentane octane
4) 6- cyclopentyl octane

33. Structure of 4-Methylpent-2-enal is

1) $H_2C=\underset{H}{\underset{|}{C}}-\overset{CH_3}{\overset{|}{\underset{H}{\underset{|}{C}}}}-CH_2-\overset{O}{\overset{\|}{C}}-H$
2) $CH_3-CH_2-\underset{CH_3}{\underset{|}{C}}=CH-\overset{O}{\overset{\|}{C}}-H$
3) $CH_3-CH_2-CH=\underset{CH_3}{\underset{|}{C}}-\overset{O}{\overset{\|}{C}}-H$
4) $CH_3-\underset{CH_3}{\underset{|}{CH}}-CH=CH-\overset{O}{\overset{\|}{C}}-H$

34. IUPAC name of the compound is:

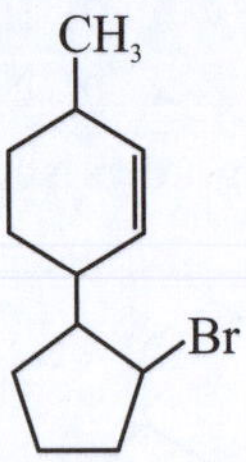

1) 3-(2-bromo cyclopentyl)-6-methyl cyclohex-1-ene
2) 3-methyl-6-(2-bromo cyclopentyl) cyclohex-1-ene
3) 1-methyl-4-(2-bromo cyclopentyl) cyclohex-2-ene
4) 4-methyl-1-(2-bromo cyclopentyl) cyclohex-2-ene

35. Which of the following represent correct pair(s) of homologous?

1) 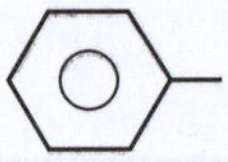 & 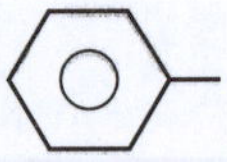
2) $Me-OH$ & $MeCH_2CH_2OH$
3) $H-\overset{O}{\overset{\|}{C}}-O-CH_3$ & $CH_3-\overset{O}{\overset{\|}{C}}-O-CH_3$
4) $CH_3-CH_2-NH_2$ & $CH_3-CH_2-N(H)(CH_3)$

36. 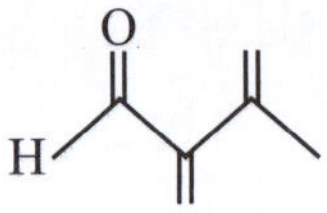 is named as:

1) 2, 3-Dimethylenebutanal
2) 3-Methyl-2-methylenebut-3-enone
3) 3-Methyl-2-methylenebut-3-enal
4) 2, 3-Dimethylenebutanal

37. Which of the following pairs of compounds are not isomers?

1) 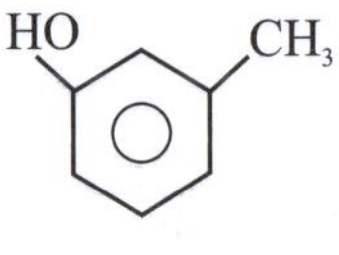 and 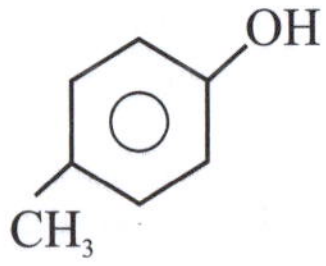
2) 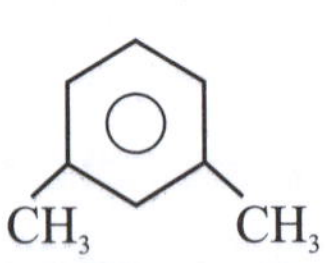 and 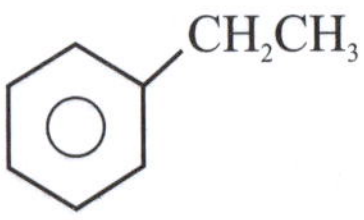
3) 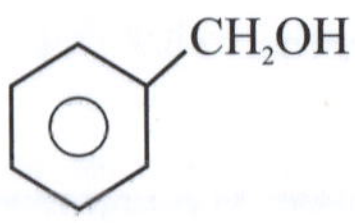 and 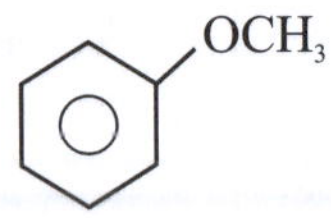
4) 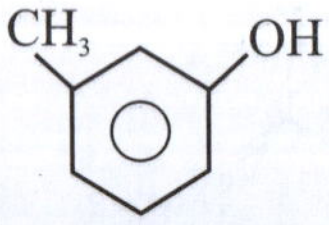 and 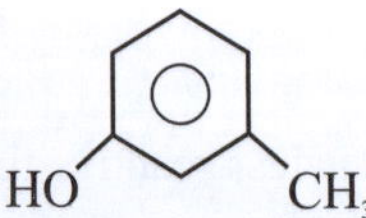

38. Identify the position isomer:

1) 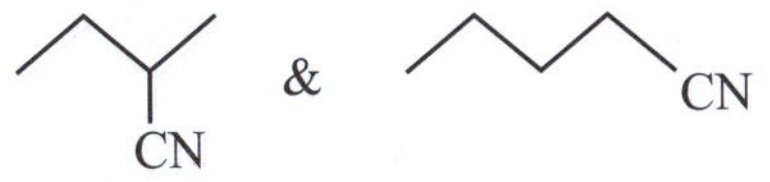

2) 

3) 

4) 

39. Which of the following is not correctly matched?

1) $CH_3-\overset{O}{\overset{||}{C}}-OH$ and $H-\overset{O}{\overset{||}{C}}-OCH_3$
Metamers

2) $CH_3-CH_2-C\equiv CH$ and $CH_3-C\equiv C-CH_3$
Position isomers

3) CH_3CH_2OH and $(CH_3)_2O$
Functional isomers

4) All the the above

40. Match the following :

	Column I		Column II
(a)	$CH_3-\overset{O}{\overset{||}{C}}-O-C_3H_7$ and $Et-\overset{O}{\overset{||}{C}}-O-Et$	(i)	Chain isomers
(b)	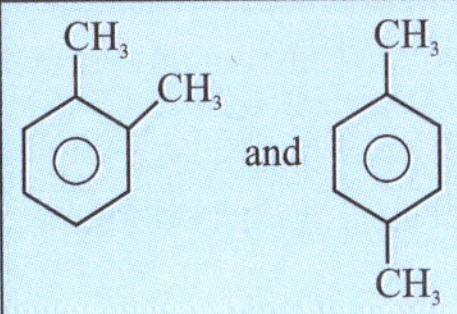	(ii)	Position isomers
(c)	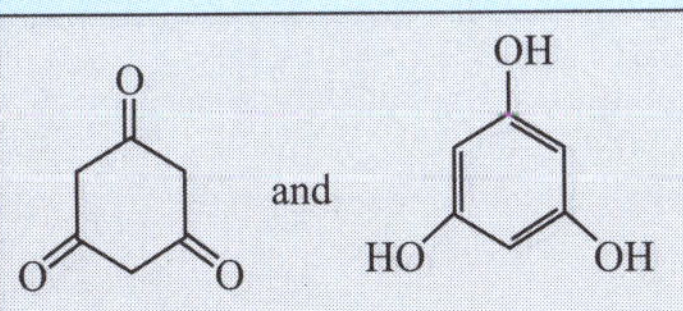	(iii)	Metamers
(d)	$CH_3-\underset{CH_3}{\underset{\vert}{CH}}-OH$ and $CH_3-CH_2-CH_2-OH$	(iv)	Tautomers

1) (a) → (iii); (b) → (ii); (c) → (iv); (d) → (i)
2) (a) → (ii); (b) → (i); (c) → (iv); (d) → (ii)
3) (a) → (iii); (b) → (ii); (c) → (i); (d) → (ii)
4) (a) → (i); (b) → (ii); (c) → (iv); (d) → (ii)

41. The total number of benzene derivatives having the molecular formula C_7H_7Cl are-

1) 2
2) 3
3) 4
4) 5

42. In the given functional groups, the correct arrangement of priority is-

1) $-OH > -CH_2OH > -CHO > -COOH$
2) $-OH > -COOH > -CHO > -CH_2OH$
3) $-COOH > -OH -CHO > -CH_2OH$
4) $-COOH > -CHO > R-CO-R > -OH$

43. The correct decreasing order of priority of functional groups in naming an organic compound as per IUPAC system of nomenclature is

1) $-COOH > -CONH_2 > -COCl > -CHO$
2) $-SO_3H > -COCl > -CONH_2 > -CN$
3) $-COOR > -COCl > -NH_2 > C=O$
4) $-COOH > -COOR > -CONH_2 > -COCl$

44. Which of the following pairs of compounds are identical?

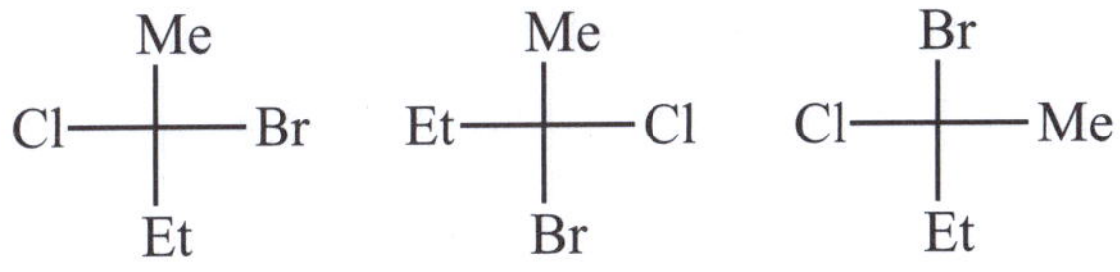

(I) (II) (III)

1) I & II
2) II & III
3) I & III
4) I, II & III

45. Number of structurally isomeric carbonyl compounds possible with molecular formula $C_5H_{10}O$ are:

1) 5
2) 6
3) 7
4) 8

46. Which of the compounds will show G.I.?

1) 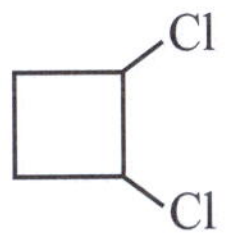
2) $C_6H_5-CH=NOH$
3) $CH_3-CH=CH-CH_3$
4) All of these

47.

	Column I		Column II
(a)	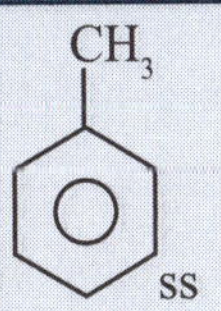	(i)	Aniline
(b)	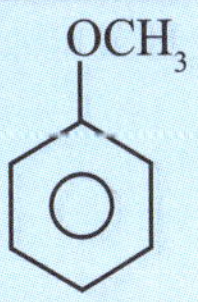	(ii)	Toluene
(c)	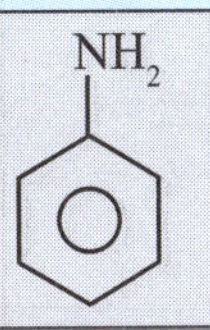	(iii)	Anisole

1) (a) → (i), (iv); (b) → (ii); (c) → (iii)
2) (a) → (ii); (b) → (i); (c) → (iii)
3) (a) → (iii); (b) → (iii); (c) → (ii)
4) (a) → (ii); (b) → (iii); (c) → (i)

48. 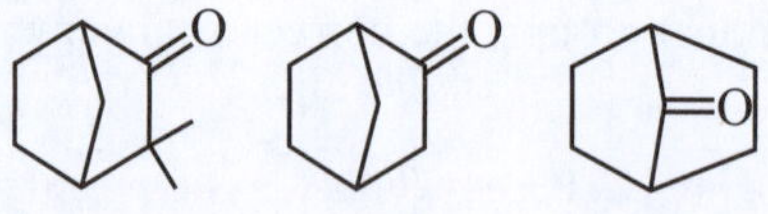

I II III

Which among these can exhibit tautomerism?

1) II only
2) I only
3) III only
4) II & III

49. A species having carbon with sextet of electrons and can act as electrophile is called

1) carbon free radical 2) carbanion
3) carbocation 4) pentavalent carbon

50. The correct stability order of the following resonance structures of CH_3–CH = CH–CHO is

$\overset{\ominus}{C}H_2$...

$CH_3-\overset{\ominus}{C}H-CH=\underset{|}{C}(\ddot{\underset{..}{O}}^{\oplus})-H \leftrightarrow CH_3-\overset{\oplus}{C}H-CH=C(\ddot{\underset{..}{O}}^{\ominus})-H$

I II

$\leftrightarrow CH_3-CH=CH-C(=\ddot{\underset{..}{O}})-H$

III

1) II > III > I 2) III > II > I
3) I > II > III 4) II > I > III

51. Which will not show Geometrical Isomerism?

1) 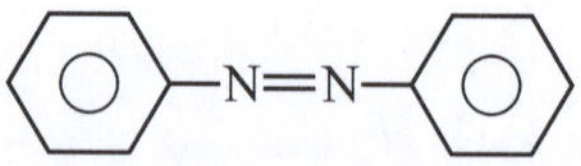

2) 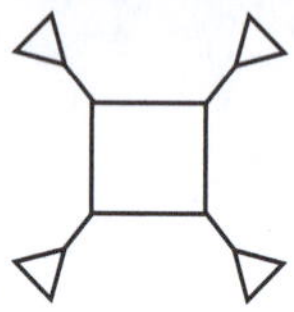

3) 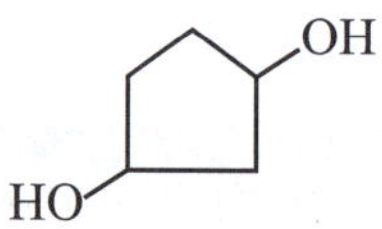

4) $CH_3-\underset{CH_3}{\underset{|}{CH}}-CH=\underset{CH_2D}{\underset{|}{C}}-CH_2D$

52. Which of following compounds has a zero dipole moment?

1)  2) 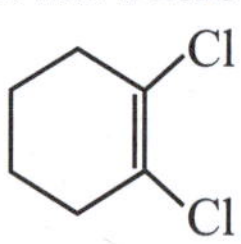

3) 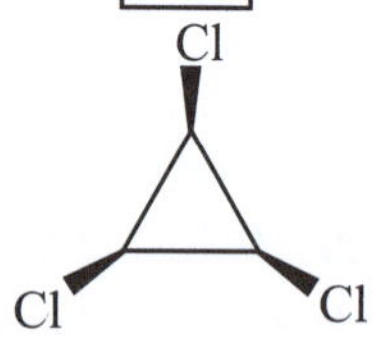 4) 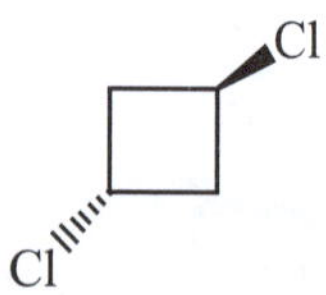

53. Resonance stabilized cation is:

1) 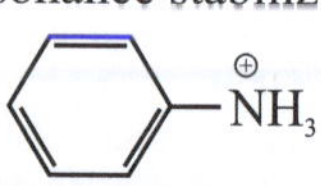 2) 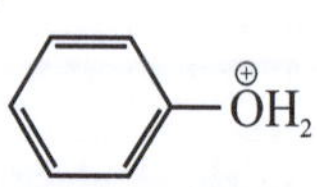

3) 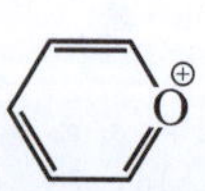 4) 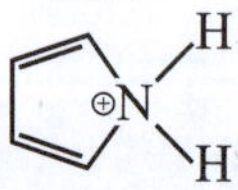

54. Decreasing order of resonance energy for the given molecules is:

(i) 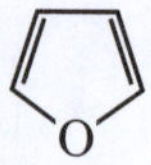 (ii) 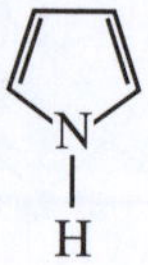 (iii) 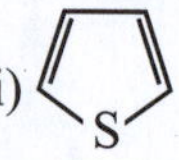 (iv) 

1) i > ii > iii > iv 2) iii > ii > i > iv
3) ii > i > iii > iv 4) Iv > ii > i > iii

55. Which of the following is/are aromatic compounds?

(i) 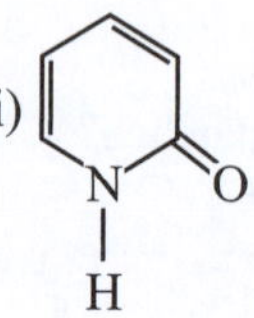 (ii) 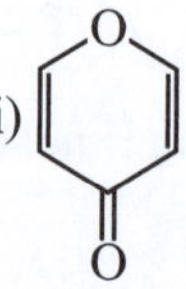 (iii) 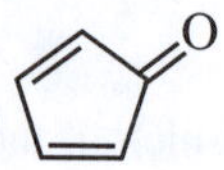

(iv) 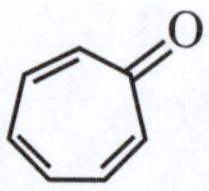 (v) 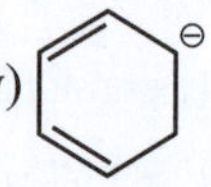

1) i, ii & iii only 2) ii, iii & iv only
3) i, ii & iv only 4) All of them

56. What is the relation between the following compounds?

(i) 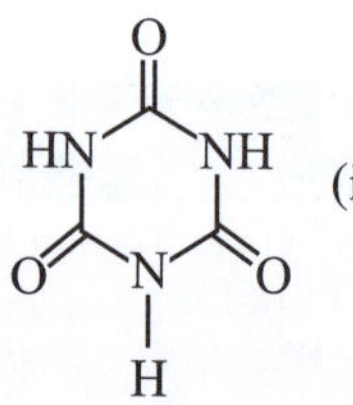 (ii) 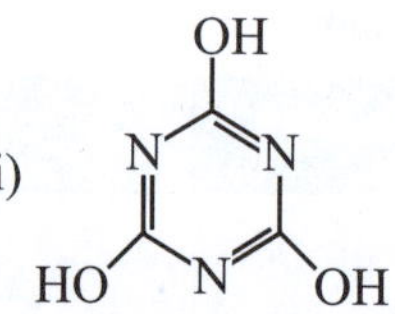 (iii) 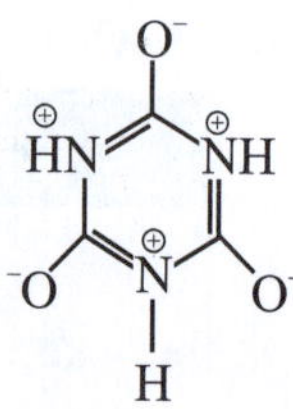

1) i and ii are tautomers
2) i and iii are resonating structures
3) i and ii are resonating structures
4) Both (1) & (2)

57. Order of K_{a_1} & K_{a_2} in the following carboxylic acids:

I: CH_2-CH_2-COOH / CH_2-CH_2-COOH (the two CH_2 groups bonded)

II: CH_2-COOH / CH_2-COOH (the two CH_2 groups bonded)

III: $CH_2(COOH)_2$

IV: COOH–COOH

1) K_{a_1} = I > II > III > IV, K_{a_2} = IV > III > II > I
2) K_{a_1} = IV > III > II > I, K_{a_2} = I > II > III > IV
3) K_{a_1} = II > III > IV > I, K_{a_2} = I > IV > III > II
4) K_{a_1} = II > IV > I > III, K_{a_2} = I > III > IV > II

58. Order of acidic strength of following alcohols:

A. $CH \equiv C-CH_2OH$ B. $CH_2=CH-CH_2OH$

C. 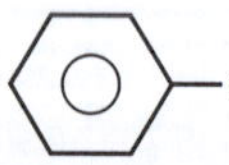 $-CH_2-OH$ D. $CH_3-CH_2-CH_2-OH$

1) A > B > C > D 2) D > C > B > A
3) A > C > B > D 4) B > C > D > A

59. Order of stability of carbocations:-

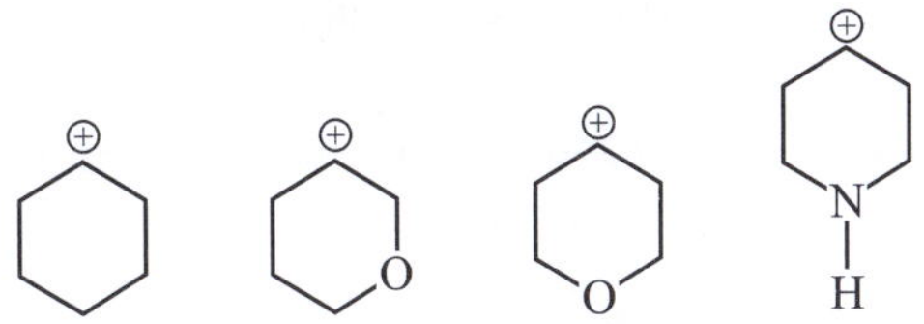

I II III IV

1) I > IV > III > II
2) I > II > III > IV
3) I > III > IV > II
4) II > III > IV > I

60. Match correct common name of compounds:-

	Column I		Column II
(a)	Ph—C(OH)(H)—COOH	(i)	Tartaric acid
(b)	COOH—CHOH—CHOH—COOH	(ii)	Mandelic acid
(c)	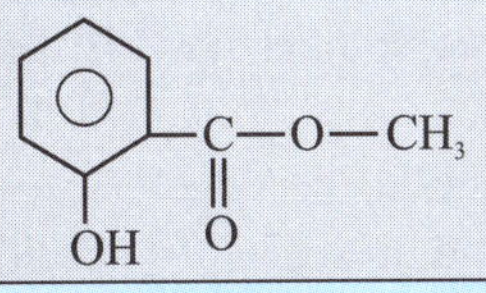	(iii)	Aspirin
(d)	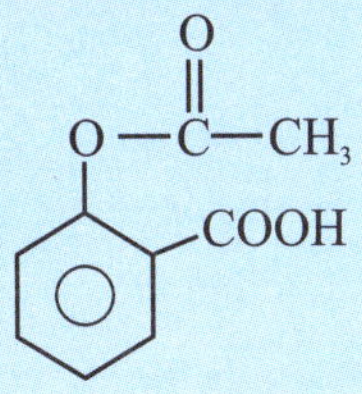	(iv)	Oil of winter green (Methyl salicylate)

1) (a)-(i), (b)-(ii), (c)-(iv), (d)-(i)
2) (a)-(ii), (b)-(i), (c)-(iv), (d)-(iii)
3) (a)-(i), (b)-(iv), (c)-(ii), (d)-(iii)
4) (a)-(iii), (b)-(iv), (c)-(ii), (d)-(i)

61. The correct structure of 2-Ethyl 4-Methyl hexa, 1, 4-diene:–

1) 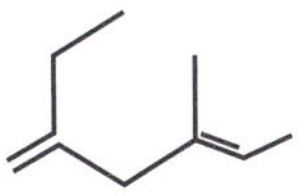
2) 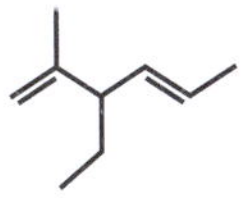
3) 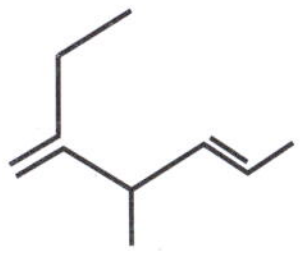
4) 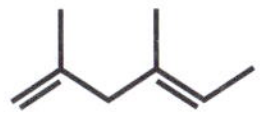

62. Which one of following compounds does not show tautomerism?

1) $CH_3—C(CH_3)_2—NO_2$
2) $CH_3—NH—C(=O)—CH_3$
3) $CH_3—CH(CH_3)—NO_2$
4) 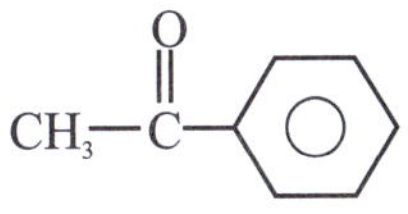

63. Which compound would exhibit optical isomerism ?

1) 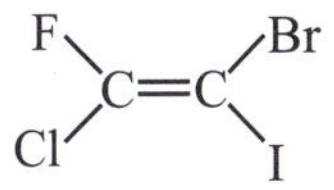
2) $HOOC—C(NH_2)(H)—CH_2C_6H_5$
3) 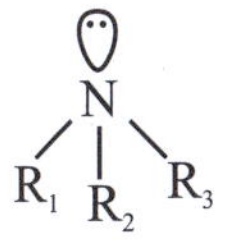
4) 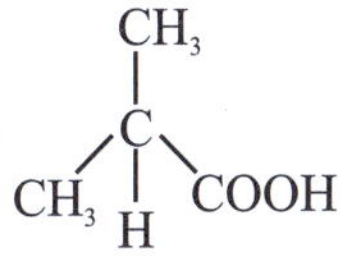

64. Select Aromatic, Antiaromatic in the following molecules:

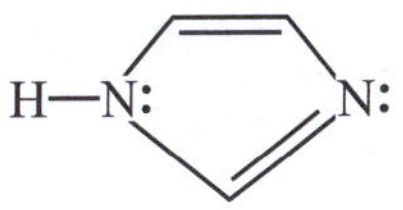

1

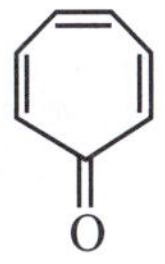

2

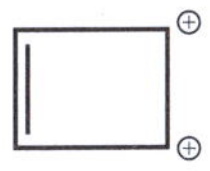

3

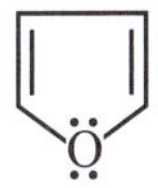

4

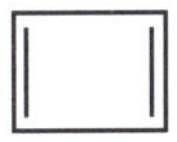

5

1) 1, 2, 3, 4 = Aromatic
   5 = Antiaromatic
2) 2, 3, 4 Aromatic
   5, 5 Antiaromatic
3) 1, 3, 4 = Aromatic
   2, 5 = Antiaromatic
4) All aromatic

65. In which of following option I[st] is more stable than IInd ?

1) 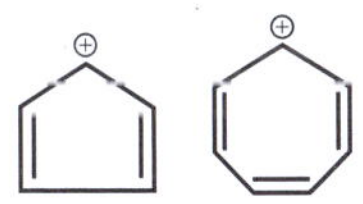
2) 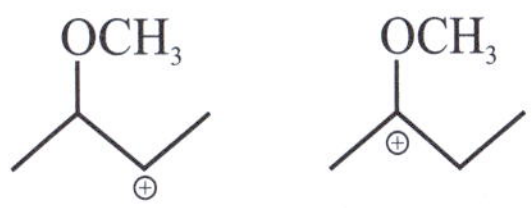
3) $CH_3—O—\overset{\oplus}{C}H—CH_3$, $CH_3—\overset{\oplus}{C}(CH_3)—CH_2—CH_3$
4) 

66. In which of following, o/p directing groups are arranged in correct order?

1) $–NH_2 > NO_2 > CH_3 > CHO$
2) $–NH_2 > –OH > –OCH_3 > CH_3$
3) $–\overset{\ominus}{O} > –\overset{\ominus}{N}H > –NHCOCH_3 > NO_2$
4) $–CH_3 > –NH_2 > OH$

67. Which of following compound have maximum bond length between C = C?

1) 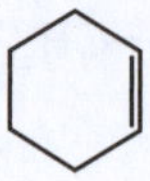 2) $CH_2 = CH_2$

3) 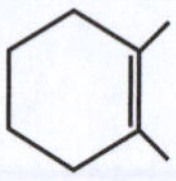 4) 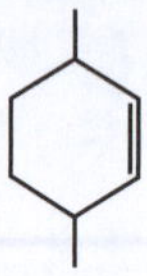

68. Correct matching is:–

	Column I		Column II
(a)	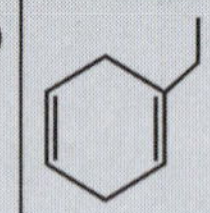	(i)	phenylethene
(b)	C≡CH 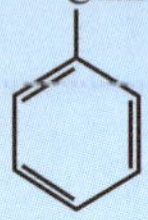	(ii)	6-ethyl-3, 3-dimethyl cyclohexene
(c)	$CH=CH_2$ 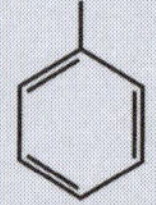	(iii)	1-ethylcyclohexa-1, 4-diene
(d)	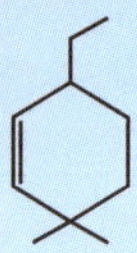	(iv)	phenylethyne

1) (a)-(ii); (b)-(i); (c)-(iv); (d)-(iii)
2) (a)-(iii); (b)-(iv); (c)-(ii); (d)-(i)
3) (a)-(iii); (b)-(iv); (c)-(i); (d)-(ii)
4) (a)-(iii); (b)-(i); (c)-(iv); (d)-(ii)

69. Which of the following molecule has highest dipole moment?

1) 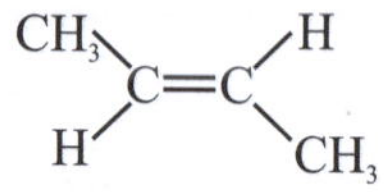 2) 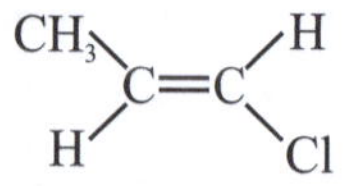

3) 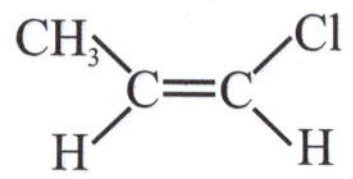 4) 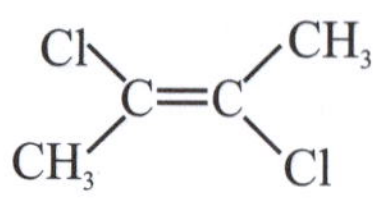

70. Select correct statement?

1) Anti aromatic species have negligible tendency to form dimer.
2) Compounds having more no. of resonating structures are always more stable than compound having lesser no. of resonating structures.
3) The one which has higher resonance energy is always more stable than the other which has lower resonance energy
4) –Cl have lesser negative inductive effect as well as lesser mesomeric effect than –F group.

71. The ascending acidity order of the following H atoms is

A. HC ≡ C –Ⓗ B. $H_2C = CH$–Ⓗ

C. $(H_3C)_3C$—Ⓗ D. $H_3C - CH_2$– Ⓗ

1) C < D < B < A 2) A < B < C < D
3) A < B < D < C 4) D < C < B < A

72. Methods used for purification of organic compounds are based on:

1) neither on nature of compound nor on the impurity present.
2) nature of compound only.
3) nature of compound and presence of impurity.
4) presence of impurity only.

73. Primary, secondary and tertiary amines are:

1) Chain isomers 2) Position isomers
3) Functional isomers 4) Tautomers

74. Order of Acidic strength in given molecule

CHF_3	$CHCl_3$	$CHBr_3$	CHI_3
I	II	III	IV

1) I > II > III > IV 2) II > I > III > IV
3) I > III > IV > II 4) III > IV > I > II

75. Select the correct order of basicity

1) $CH_3\overset{\ominus}{C}H_2 > CH_2{=}\overset{\ominus}{C}H > CH{\equiv}\overset{\ominus}{C} > \overset{\ominus}{O}H > \overset{\ominus}{N}H_2$
2) $CH_3{-}\overset{\ominus}{C}H_2 > CH_2{=}\overset{\ominus}{C}H > \overset{\ominus}{N}H_2 > CH{\equiv}\overset{\ominus}{C} > \overset{\ominus}{O}H$
3) $CH_3{-}\overset{\ominus}{C}H_2 > \overset{\ominus}{O}H > CH{\equiv}\overset{\ominus}{C} > \overset{\ominus}{N}H_2 > CH_2{=}\overset{\ominus}{C}H$
4) $CH_3{-}\overset{\ominus}{C}H_2 > CH_2{=}\overset{\ominus}{C}H > CH{\equiv}\overset{\ominus}{C} > \overset{\ominus}{N}H_2 > \overset{\ominus}{O}H$

Match the Columns MCQs

1. Match the type of mixture of compounds in Column-I with the technique of separation/purification given in Column-II.

	Column-I		Column-II
(a)	Two solids which have different solubilities in a solvent and which do not undergo reaction when dissolved in it.	(i)	Steam distillation
(b)	Liquid that decomposes at its boiling point	(ii)	Fractional distillation
(c)	Steam volatile liquid	(iii)	Simple distillation
(d)	Two liquids which have boiling points close to each other	(iv)	Distillation under reduced pressure
(e)	Two liquids with large difference in boiling points.	(v)	Crystallisation

1) (a)-(v) (b)-(iv) (c)-(i) (c)-(i) (c)-(i)
2) (a)-(ii) (b)-(iii) (c)-(i) (c)-(i) (c)-(i)
3) (a)-(v) (b)-(iv) (c)-(iii) (c)-(iii) (c)-(iii)
4) (a)-(iii) (b)-(iv) (c)-(v) (c)-(v) (c)-(v)

2. Match the terms mentioned in Column-I with the terms in Column-II.

	Column-I		Column-II
(a)	Carbocation	(i)	Cyclohexane and 1-hexene
(b)	Nucleophile	(ii)	Conjugation of electrons of C–H σ bond with empty p-orbital present at adjacent positively charged carbon
(c)	Hyperconjugation	(iii)	sp^2 hybridised carbon with empty p-orbital
(d)	Isomers	(iv)	Ethyne
(e)	sp hybridisation	(v)	Species that can receive a pair of electrons
(f)	Electrophile	(vi)	Species that can supply a pair of electrons

1) (a)-(i) (b)-(iii) (c)-(ii) (d)-(v) (e)-(vi) (f)-(iv)
2) (a)-(iv) (b)-(i) (c)-(iii) (d)-(v) (e)-(ii) (f)-(vi)
3) (a)-(iii) (b)-(vi) (c)-(ii) (d)-(i) (e)-(iv) (f)-(v)
4) (a)-(v) (b)-(vi) (c)-(i) (d)-(iv) (e)-(iii) (f)-(ii)

3. Select correct I.U.P.A.C name of following Alkyl halides:

	Column-I		Column-II
(a)	$CH_3-CH(Cl)-CH(Br)-CH_3$	(i)	2-Bromo-3-Chlorobutane
(b)	$CHF_2-C(Br)(Cl)-F$	(ii)	1-Bromo-1-Chloro-1,2,2-Trifluoroethane
(c)	$ClCH_2-C\equiv C-CH_2-Br$	(iii)	1-bromo-4-chlorobut-2-yne
(d)	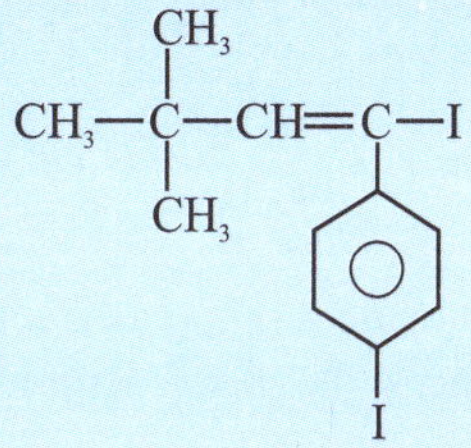	(iv)	1-Iodo-1-(4 iodophenyl)-3,3-dimethylbutene

1) (a)-(i) (b)-(iii) (c)-(ii) (d)-(iv)
2) (a)-(i) (b)-(ii) (c)-(iii) (d)-(iv)
3) (a)-(iii) (b)-(iv) (c)-(ii) (d)-(i)
4) (a)-(iv) (b)-(i) (c)-(ii) (d)-(iii)

4. Correct matching is:

	Column-I		Column-II
(a)	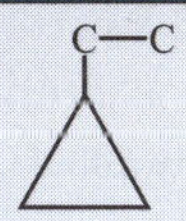	(i)	n-propyl cyclopropane
(b)	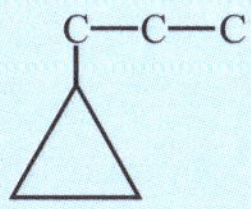	(ii)	isopropylcyclobutane
(c)	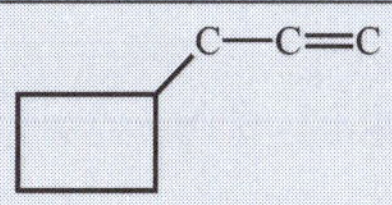	(iii)	ethylcyclopropane
(d)	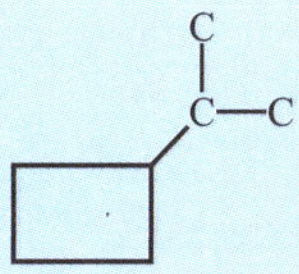	(iv)	3-cyclobutylpropene

1) (a)-(i) (b)-(ii) (c)-(iv) (d)-(iii)
2) (a)-(iii) (b)-(i) (c)-(iv) (d)-(ii)
3) (a)-(iii) (b)-(ii) (c)-(iii) (d)-(iv)
4) (a)-(iii) (b)-(i) (c)-(ii) (d)-(iv)

5. Match the ions given in Column-I with their nature given in Column-II.

	Column-I		Column-II
(a)	$CH_3-\ddot{\underset{\cdot\cdot}{O}}-\overset{\oplus}{C}H-CH_3$	(i)	Stable due to resonance
(b)	$F_3-C^{\oplus}$	(ii)	Destabilised due to inductive effect
(c)	$CH_3-\overset{CH_3}{\underset{CH_3}{C^{\ominus}}}$	(iii)	Stabilised by hyperconjugation
(d)	$CH_3-\overset{\oplus}{C}H-CH_3$	(iv)	A secondary carbocation

1) (a)-(iii) (b)-(i) (c)-(ii) (d)-(iv)
2) (a)-(i)(ii) (b)-(ii) (c)-(ii) (d)-(iii)(iv)
3) (a)-(iii) (b)-(ii) (c)-(iv) (d)-(i)
4) (a)-(iv) (b)-(i) (c)-(iii) (d)-(ii)

Correct & Incorrect MCQs

1. Which of the following statement (s) are correct?
   A. Hybridisation influences the bond length and bond enthalpy (strength) in compounds.
   B. sp hybrid orbitals form stronger bonds than sp^3 hybrid orbitals.
   C. A carbon atom having an sp^3 hybrid orbital is more electronegative than that possessing sp or sp^2 hybridised orbitals.
   D. The greater the s character of the hybrid orbitals, the greater is the electronegativity.
   1) A, B and D 2) A and C
   3) A, C and D 4) B, C and D

2. Which of the following statement is incorrect?
   1) In a sigma (σ) bond formation, parallel orientation of the two p orbitals on adjacent atoms is necessary for a proper sideways overlap.
   2) In ethene, rotation about carbon-carbon double bond is restricted.
   3) The electron charge cloud of the π bond is located above and below the plane of bonding atoms.
   4) In general, π bonds provide the most reactive centres in the molecules containing multiple bonds.

3. Identify the incorrect statement :
   1) Acyclic compounds consist of straight or branched chain compounds.
   2) Alicyclic (aliphatic cyclic) compounds contain carbon atoms joined in the form of a ring.
   3) Heterocyclic compounds consist of straight chain containing only carbon atom.
   4) Both (1) and (2).

4. Which of the following statement (s) is/are incorrect?
   A. Acetaldehyde is an example of alicyclic compound.
   B. Ethane is an example of acyclic compound.
   C. Cyclopropane is an example of heterocyclic compound.
   D. Tetrahydrofuran is an example of homocyclic compound.
   1) A, B and C 2) Only A
   3) A, C and D 4) B, C and D

5. Which of the following statement is incorrect ?
   1) Nitrogen, sulphur, halogens and phosphorus present in an organic compound are detected by "Lassaigne's test".
   2) In Lassaigne's test ,the elements present in the compound are converted from ionic form into the covalent form by fusing the compound with sodium metal.
   3) Freshly prepared $FeSO_4$ solution is used in the lassaigne's test for nitrogen.
   4) Both (1) and (2)

6. Select the correct statement (s) :
   A. In heterolytic cleavage, the bond breaks in such a fashion that the shared pair of electrons remains with one of the fragments.
   B. Heterolytic cleavage of bromomethane will give ethyl carbocation and bromide ion.
   C. A species having a carbon atom possessing sextet of electrons and a positive charge is called a carboanion.
   D. The organic reactions which proceed through heterolytic bond cleavage are called ionic or heteropolar reactions.
   1) A, B and D 2) A and C
   3) A, C and D 4) Only C

7. Which of the following statements is incorrect ?
   1) Carbocations are highly unstable and reactive species.
   2) Alkyl groups directly attached to the positively charged carbon stabilise the carbocations due to inductive and hyperconjugation effects.
   3) Carbon in carbanion is generally sp hybridised and its structure is distorted trigonal planar shape.
   4) Carbanions are unstable and reactive species.

8. Which of the following statement is/are not correct about nucleophile and electrophile ?

A. Hydroxide ion acts as nucleophile.

B. Cyanide ion acts as an electrophile.

C. Neutral molecule such as $H_2\ddot{O}$: act as nucleophiles.

D. Carbocations $\overset{+}{\underset{\cdot\cdot}{C}}H_3$ acts as an electrophile.

1) A, B and C
2) Only B
3) A, C and D
4) B, C and D

9. Which of the following statements is incorrect?

1) Chloroform (b.p 334 K) and aniline (b.p. 457 K) are easily separated by the technique of distillation
2) Glycerol can be separated from spent-lye in soap industry by distillation under reduced pressure.
3) A mixture of camphor and benzoic acid can be separated by sublimation.
4) Aniline is separated by steam distillation method from aniline – water mixture.

10. Choose the correct Statement:

1) The inductive effect is related to the ability of substituent(s) to either withdraw or donate electron density to the attached carbon atom.
2) Methyl and ethyl groups are considered as electron withdrawing groups.
3) Electromeric effect is a permanent effect.
4) Hyperconjugation is a temporary effect.

Statement Based MCQs

Directions: These questions consist of two statements each, printed as Statement-I and Statement-II. While answering these questions, you are required to choose any one of the following four responses.

1) **Both Statement-I and Statement-II are correct.**
2) **Both Statement-I and Statement-II are incorrect.**
3) **Statement-I is correct & Statement-II is incorrect.**
4) **Statement-I is incorrect & Statement-II is correct.**

1. **Statement-I:** Although F has more electro negativity than Cl; yet p-fluoro benzoic acid is a weaker acid than p-chloro benzoic acid.

Statement-II: Due to matching size of 2p orbital of F and C, F has a stronger +R effect than Cl.

2. **Statement-I:** Pyrrole is more basic than pyridine.

Statement-II: In pyrrole, the nitrogen atom is sp^3 hybridized.

3. **Statement-I:** Glycerol is purified by distillation under reduced pressure.

Statement-II: Organic compounds in liquid state are purified by distillation.

4. **Statement-I:** Polarisation of one σ bond caused by the polarisation of adjacents bond is referred to as the inductive effect.

Statement-II: The substituents can be classified as electron withdrawing or electron donating groups relative to hydrogen.

5. **Statement-I:** Structural isomers have different physical and chemical properties.

Statement-II: Structural isomers have different skeleton of covalent bonds so have different stability and polarity which affect the chemical and physical properties of the compounds.

6. **Statement-I:** C_3H_6O shows functional group isomerism.

Statement-II: Propanone and propanal have molecular formula C_3H_6O.

7. **Statement-I:** Electrophiles behave as lewis base.

Statement-II: Electrophiles have a strong tendency to attract electrons from other sources.

8. **Statement-I:** The energy of actual structure of the molecule (the resonance hybrid) is more than that of any of the canonical structures.

Statement-II: The difference in energy between the actual structure and the lowest energy resonance structure is called the resonance stabilisation energy.

9. **Statement-I:** The observed order of carbocation stability is $\overset{+}{C}H_3 < CH_3\overset{+}{C}H_2 <. (\overset{+}{C}H_3)_2\ CH < (CH_3)_3\overset{+}{C}$

Statement-II: The observed order of alkyl radical stability is $\dot{C}H_3 > \dot{C}H_2CH_3 >. \dot{C}H\ (CH_3)_2 < \dot{C}\ (CH_3)_3$,

10. **Statement-I:** The greater the s character of the hybrid orbitals, the greater is the electronegativity.

Statement-II: The change in hybridization is associated with a change in electronegativity.

Assertion & Reason MCQs

Directions: These questions consist of two statements each, printed as Assertion and Reason. While answering these questions, you are required to choose any one of the following four responses.

1) **Both Assertion and Reason are True and the Reason is a correct explanation of the Assertion.**
2) **Both Assertion and Reason are True but Reason is not a correct explanation of the Assertion.**
3) **Assertion is True but the Reason is False.**
4) **Assertion is False but Reason is True.**

1. **Assertion:** Simple distillation can help in separating a mixture of propan-1-ol (boiling point 97°C) and propanone (boiling point 56°C).
   Reason: Liquids with a difference of more than 20°C in their boiling points can be separated by simple distillation
2. **Assertion:** Pent-1-ene and pent-2-ene are position isomers.
   Reason: Position isomers differ in the position of functional group or a substituent.
3. **Assertion:** Sulphur present in an organic compound can be estimated quantitatively by Carius method in which it can be oxidised to H_2SO_4.
   Reason: Sulphur is separated easily from other atoms in the molecule and gets precipitated as light yellow solid.
4. **Assertion:** Components of a mixture of red and blue inks can be separated by distributing the components between stationary and mobile phases in paper chromatography.
   Reason: The coloured components of inks migrate at different rates because paper selectively retains different components according to the difference in their partition between the two phases.
5. **Assertion:** Glycerol can be purified by distillation under reduced pressure.
   Reason: Organic compounds in liquid state are purified by distillation.
6. **Assertion:** Alkanes which contain more than three carbon atoms exhibit chain isomerism.
   Reason: All the carbon atoms are sp^2 hybridised in alkanes.
7. **Assertion:** Anisole undergoes electrophilic substitution at ortho and para positions.
   Reason: Anisole is an aromatic compound.
8. **Assertion:** In heterolytic cleavage, the bond breaks in such a way that both the electrons of the shared pair are carried away by one of the atoms.
   Reason: Heterolytic fission occurs readily in polar covalent bonds.
9. **Assertion:** In carbonate ion, carbon-oxygen bonds are of equal length.
   Reason: Bond length decreases with the multiplicity of bond between two atoms.
10. **Assertion:** $CH_3–C≡CH$ is more polar than that of $CH_3–CH=CH_2$.
    Reason: sp carbon is more electro-negative than sp^2 carbon.

Multi Correct MCQs

1. The adsorbent used in adsorption chromatography is/are
   A. Silica gel B. Alumina
   C. Quick lime D. Magnesia
   Choose the most appropriate answer from the options given below:
   1) B only
   2) C and D only
   3) A and B only
   4) A only
2. Ionic reactions with organic compounds proceed through:
   A. Homolytic bond cleavage
   B. Heterolytic bond cleavage
   C. Free radical formation
   D. Primary free radical
   E. Secondary free radical
   Choose the correct answer from the options given below:
   1) (A) only 2) (C) only
   3) (B) only 4) (D) and (E) only
3. Among the given species the Resonance stabilized carbocations are:

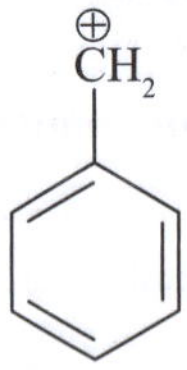
(A)
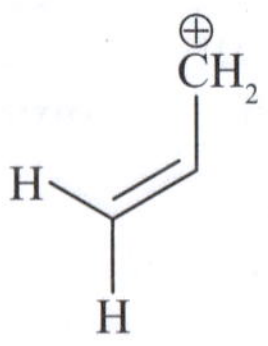
(B)
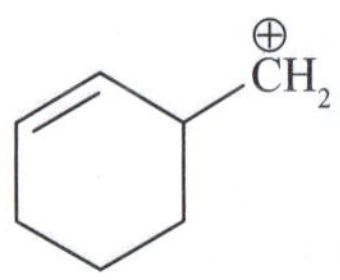
(C)
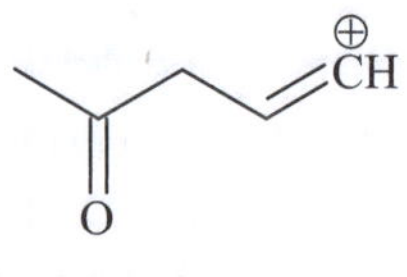
(D)

   1) (A), (B) and (C) only 2) (C) and (D) only
   3) (A), (B) and (D) only 4) (A) and (B) only
4. Which pair shows Geometrical isomerism?

(I) 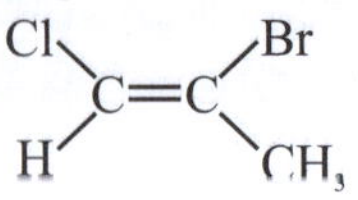
(II) 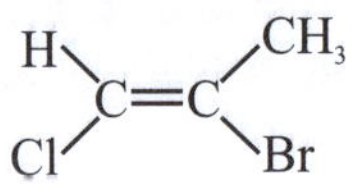
(III) 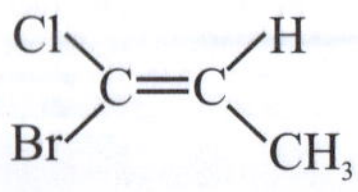
(IV) 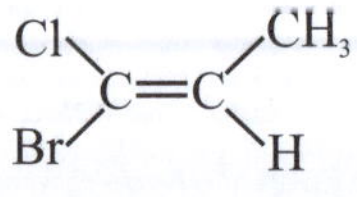

   1) I & II 2) I & III
   3) II & IV 4) III & IV

NCERT Exemplar MCQs

1. Which of the following is the correct IUPAC name?
   1) 3-Ethyl-4, 4-dimethylheptane
   2) 4, 4-Dimethyl-3-ethylheptane
   3) 5-Ethyl-4,4-dimethylhepane
   4) 4, 4-Bis(methyl)-3-ethylheptane

2. The IUPAC name for

$$CH_3-\overset{\overset{\displaystyle O}{\|}}{C}-CH_2-CH_2-\overset{\overset{\displaystyle O}{\|}}{C}-OH \text{ is:}$$

   1) 1-Hydroxypentane-1, 4-dione
   2) 1, 4-Dioxopentanol
   3) 1-Carboxybutan-3-one
   4) 4-Oxopentanoic acid

3. The IUPAC name for:

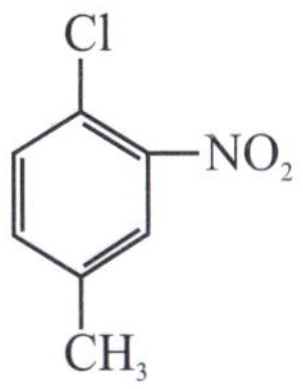

   1) 1-Chloro-2-nitro-4-methylbenzene
   2) 1-Chloro-4-methyl-2-nitrobenzene
   3) 2-Chloro-1-nitro-5-methylbenzene
   4) m-Nitro-p-chlorotoluene

4. Electronegativity of carbon atoms depends upon their state of hybridization. In which of following compounds, the carbon marked with asterisk is most electronegative?
   1) $CH_3-CH_2-\overset{*}{C}H_2-CH_3$
   2) $CH_3-\overset{*}{C}H=CH-CH_3$
   3) $CH_3-CH_2-C\equiv\overset{*}{C}H$
   4) $CH_3-CH_2-CH=\overset{*}{C}H_2$

5. In which of the following, functional group isomerism is not possible?
   1) Alcohols
   2) Aldehydes
   3) Alkyl halides
   4) Cyanides

6. The fragrance of flowers is due to the presence of some steam volatile organic compounds called essential oils. These are generally insoluble in water at room temperature but are miscible with water vapour in vapour phase. A suitable method for the extraction of these oils from the flowers is:
   1) Distillation
   2) Crystallization
   3) Distillation under reduced pressure
   4) Steam distillation

7. During hearing of a court case, the judge suspected that some changes in the documents had been carried out. He asked the forensic department to check the ink used at two different places. According to you which technique can give the best results?
   1) Column chromatography
   2) Solvent extraction
   3) Distillation
   4) Thin layer chromatography

8. The principle involved in paper chromatography is:
   1) Adsorption
   2) Partition
   3) Solubility
   4) Volatility

9. What is the correct order of decreasing stability of the following cations?

   I. $CH_3-\overset{\oplus}{C}H-CH_3$

   II. $CH_3-\overset{\oplus}{C}H-OCH_3$

   III. $CH_3-\overset{\oplus}{C}H-CH_2-OCH_3$

   1) II > I > III
   2) II > III > I
   3) III > I > III
   4) I > II > III

10. Correct I.U.P.A.C name for:

$$H_3C-\underset{\displaystyle C_2H_5}{\underset{|}{C}H}-\underset{\displaystyle C_2H_5}{\underset{|}{C}H}-CH_3 \text{ is:}$$

   1) 2-ethyl-3-methylpentane
   2) 3, 4-dimcthylhcxanc
   3) 2-sec-butylbutane
   4) 2, 3-dimethylbutane

11. In which of the following compounds the carbon marked with asterisk is expected to have greatest positive charge?
   1) $\overset{*}{C}H_3-CH_2-Cl$
   2) $\overset{*}{C}H_3-CH_2-Mg^+Cl^-$
   3) $\overset{*}{C}H_3-CH_2-Br$
   4) $\overset{*}{C}H_3-CH_2-CH_3$

Kattar NEET

Scan Me for Detailed Explanations

1. The correct IUPAC name of the compound:

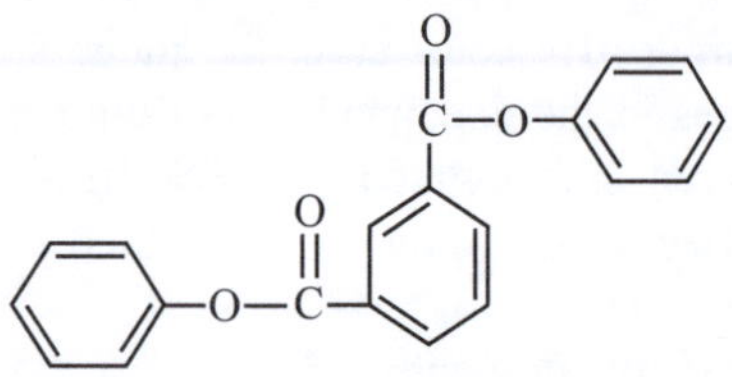

   1) Phenyl-3-phenoxycarbonylbenzoate
   2) Phenyl-3-phenoxycarbonylbenzenecarboxylate
   3) Diphenylbenzene-1, 3-dicarboxylate
   4) 1, 3-Diphenoxycarbonylbenzene

2. Professor I.L. Finar instructed his student to convert the α-Amino acid (X) into a new compound (Y) in which the $-NH_2$ group is replaced by –COOH group. The student was confused in writing the correct IUPAC names of these two compounds. The correct IUPAC names of X and Y will be respectively.

(X) = 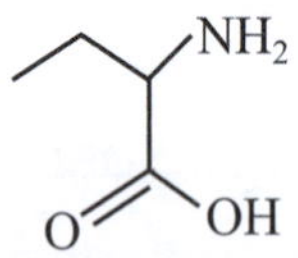

   1) 2-Ethyl-2-amino ethanoic acid, 2-Ethyl-2-carboxy ethanoic acid
   2) 2-Aminobutanoic acid, Ethylpropanedioic acid
   3) 2-Aminobutanoic acid, 2-Carboxybutanoic acid
   4) 2-Carboxypropan-1-amine, Propane-1, 1-dicarboxylic acid

3. IUPAC name of 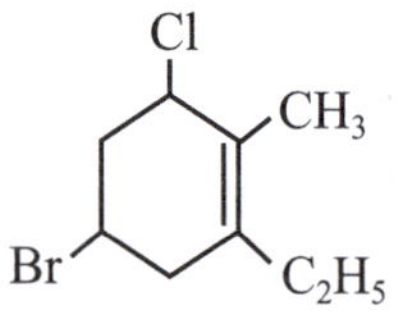 is

   1) 4-Bromo-6-chloro-2-ethyl-1-methylcyclohex-1-ene
   2) 5-Bromo-1-chloro-3-ethyl-2-methylcyclohex-2-ene
   3) 5-Bromo-3-chloro-1-ethyl-2-methylcyclohex-1-ene
   4) 1-Bromo-5-chloro-3-ethyl-4-methylcyclohex-3-ene

4. A hydrocarbon (R) has six membered ring in which there is no unsaturation. Two alkyl groups are attached to the ring adjacent to each other. One group has 3 carbon atoms with branching at 1st carbon and another has 4 carbon atoms. The larger alkyl group has main chain of three carbon atoms of which second carbon is substituted. Correct IUPAC name of compound (R) is

   1) 1-(1-Methylethyl)-2-(1-methylpropyl)cyclohexane
   2) 1-(2-Methylethyl)-2-(1-methylpropyl)cyclohexane
   3) 1-(1-Methylethyl)-2-(2-methylpropyl)cyclohexane
   4) 1-(1-Methylethyl)-2-butylcyclohexane

5. Correct order of magnitude of resonance energy of following compounds:

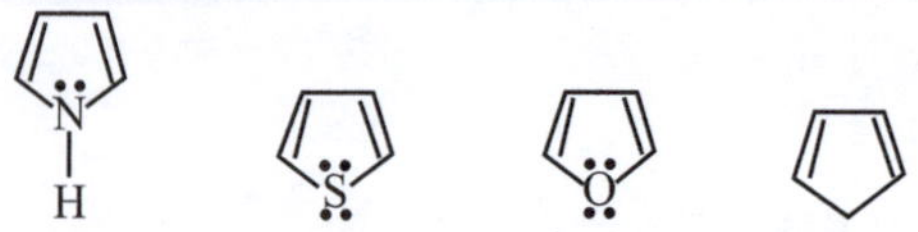

(I) (II) (III) (IV)

   1) II > I > IV > III
   2) II > IV > I > III
   3) I > II > III > IV
   4) II > I > III > IV

6. The order of heat of hydrogenation in following compound is:

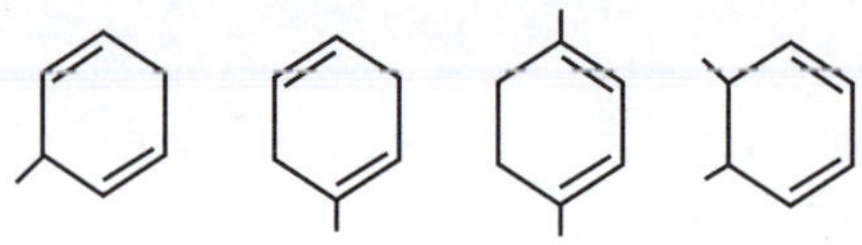

(I) (II) (III) (IV)

   1) I < II < IV < III
   2) III < IV< II < I
   3) II < III < I < IV
   4) II < IV < I < III

7. The acid strength order is

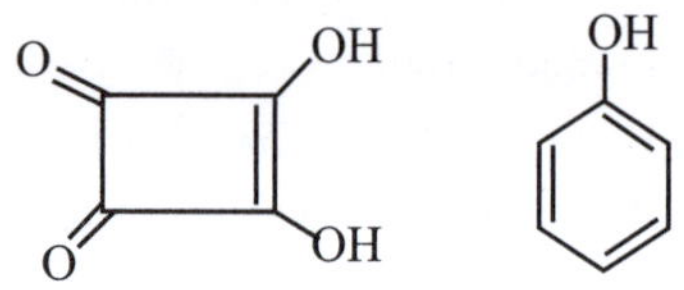

I II

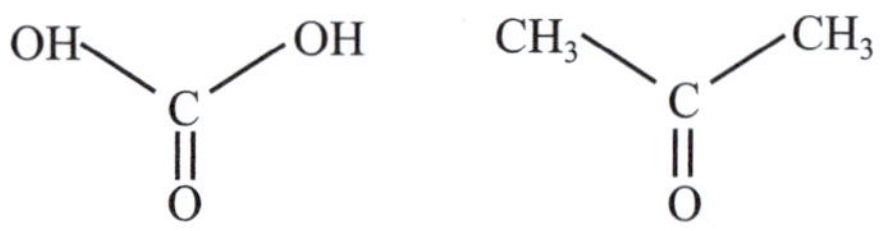

III IV

   1) I > IV > II > III
   2) III > I > II > IV
   3) II > III > I > IV
   4) I > III > II > IV

8. The acid dissociation constants of the following acids are given as under:

Compound	K_a
$ClCH_2COOH$	136×10^{-5}
$CH_3CH_2-CH(Cl)COOH$	139×10^{-5}
$CH_3CH(Cl)CH_2COOH$	8.9×10^{-5}
$CH_2(Cl)CH_2CH_2COOH$	2.96×10^{-5}
$CH_3CH_2CH_2COOH$	1.52×10^{-5}

From this data, the following observations can be made. Select the correct statements for above mentioned compounds.

(i) The above variation in acidities of the above acids are due to inductive effect only.

(ii) The above variation in acidities is due to are both due to inductive and resonance effects.

(iii) Inductive effect varies sharply with distance.

(iv) –I effect of chlorine is not much.

1) i & ii only 2) i & iii only

3) ii & iii only 4) iii & iv only

9. The correct order of acidic strength is:

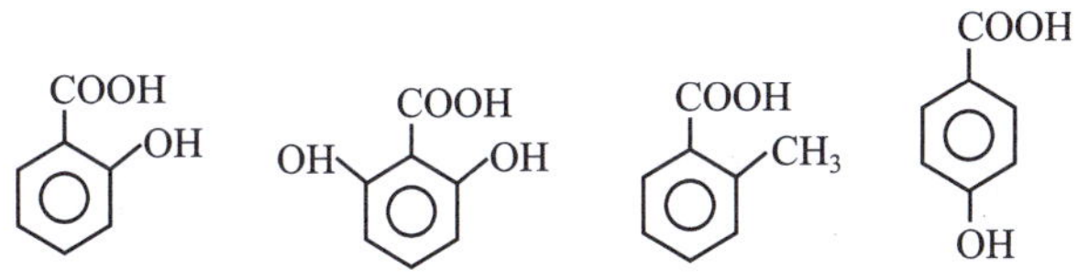

(I) (II) (III) (IV)

1) III > IV > I > II 2) II > I > III > IV

3) II > III > I > IV 4) II > I > IV > III

10. The correct order of the stability of the following carbocations is

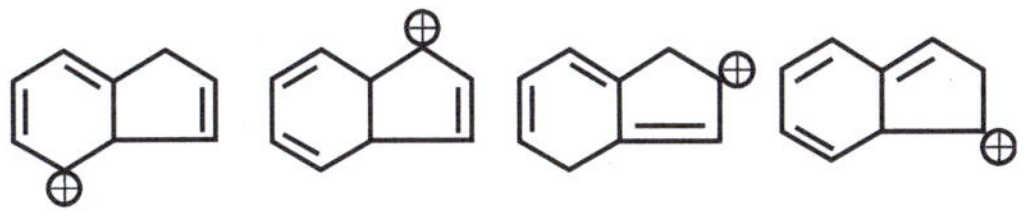

(i) (ii) (iii) (iv)

1) (i) > (iv) > (iii) > (i) 2) (i) > (iv) > (ii) > (iii)

3) (i) > (ii) > (iii) > (iv) 4) (iii) > (i) > (ii) > (iv)

11. Lassaigne's test is used to detect nitrogen in organic compounds. Which of the following compounds respond to this test?

1) $PhNH_2$ 2) PhN_2Cl

3) NH_2NH_2 4) NH_2OH

12. The incorrect statement about the given below two compounds 'A' and 'B' is

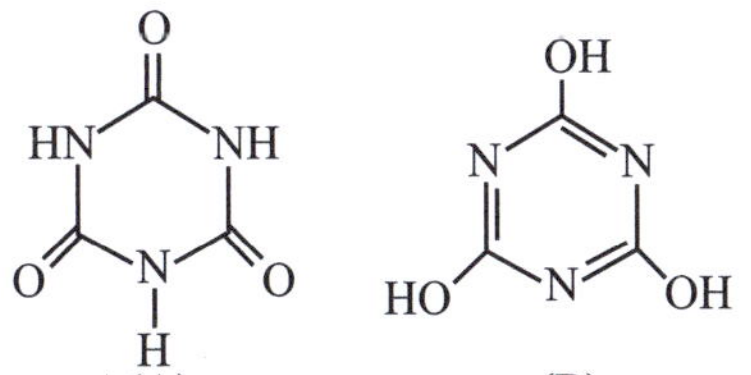

(A) (B)

1) They are structural isomers of each other

2) 'B' is more stable than 'A'

3) Double bond equivalent of both is 4

4) 'A' is more acidic than 'B'

13. Which of the following will have least pK_b value?

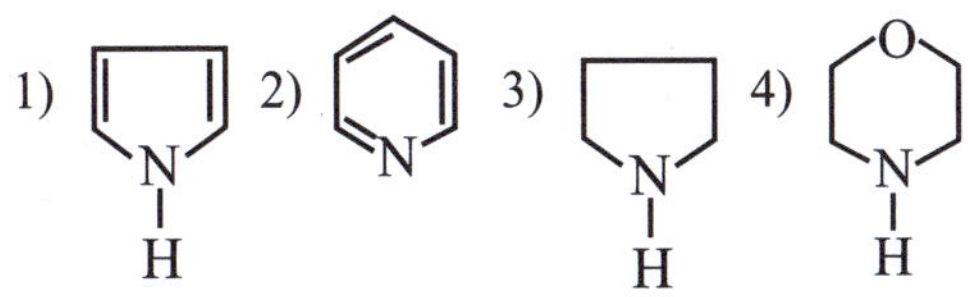

14. Correct order of basic strength of given amines is:

1) $Me_2NH > MeNH_2 > Me_3N > NH_3$ (Protic solvent)

2) $Et_2NH > Et_3H > EtNH_2 > NH_3$ (Protic solvent)

3) $Me_3N > Me_2NH > Me - NH_2 > NH_3$ (Gas phase)

4) All are correct

15. Compare the physical properties (μ, b.p., m.p., & stability) in the geometrical isomers of

$CH_3 - C = CH - CN$. Which of the following option is correct

	μ	b.p.	m.p.	stability
1)	cis > trans	cis > trans	tans > cis	tans > cis
2)	cis > trans	trans > cis	trans > cis	trans > cis
3)	cis > trans	cis > trans	trans > cis	cis > trans
4)	trans > cis	trans > cis	trans > cis	trans > cis

16. Which of the following is pair of resonance structure?

1) 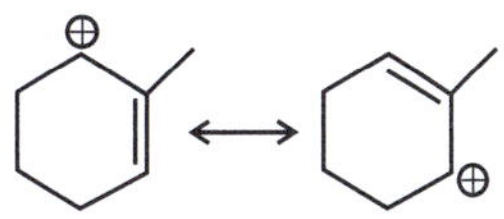

2) 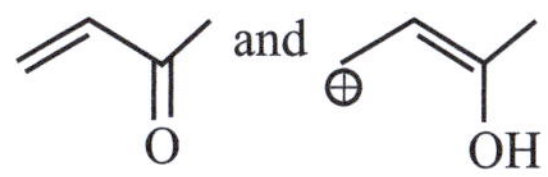

3) 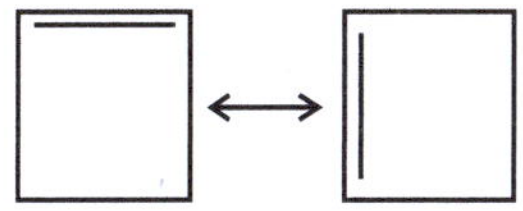

4) 

17. Match the contents of List-I with the contents of List-II.

	List-I		List-II
I.	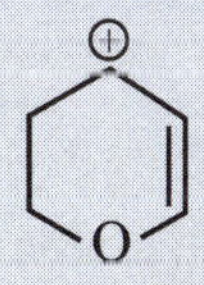	P.	Stabilised by Hyper-conjugation
II.	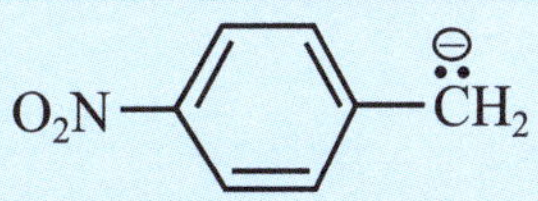	Q.	Stabilised by - I effect
III	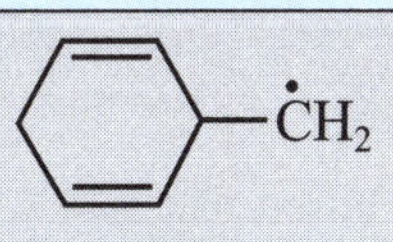	R	Stabilised by + m effect
IV	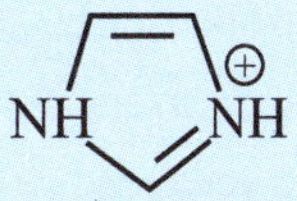	S	Stabilised by – m effect

1) I-P,R; II-Q,S; III-P; IV-R

2) I-Q,S; II-P; III-Q; IV-R,S

3) I-Q,R; II-Q; III-P; IV-R

4) I-Q; II-P; III-R; IV-S

18. Match the pK_a values with the given compounds:

List-I (Acid)		List-II (pK_a)	
I	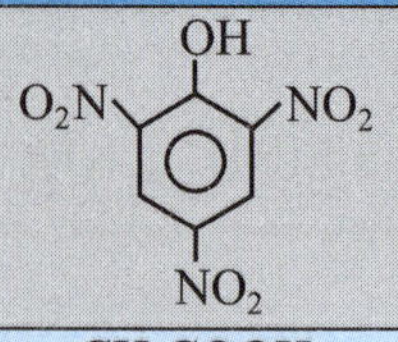	P	4
II	CH_3COOH	Q	9
III	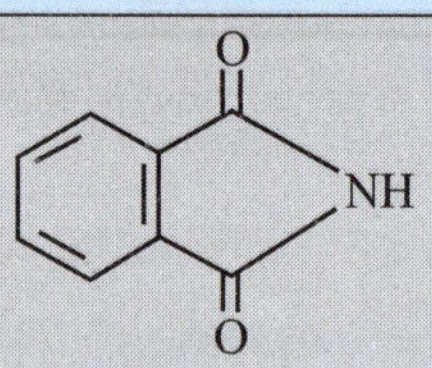 Phthalimide	R	5
IV	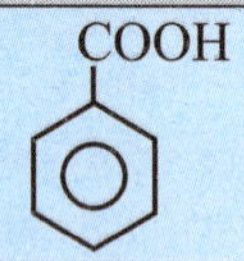	S	0.74

1) I-Q; II-R; III-P; IV-S
2) I-S; II-R; III-Q; IV-P
3) I-P; II-Q; III-R; IV-S
4) I-P; II-R; III-S; IV-Q

19. Match the following:

List-I (Reactive Intermediate)		List-II (Effects Operating)	
I	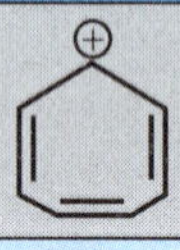	P	+I effect
II	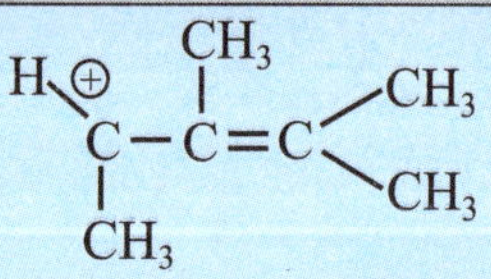	Q	Resonance
III	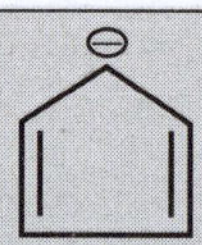	R	Aromaticity
IV	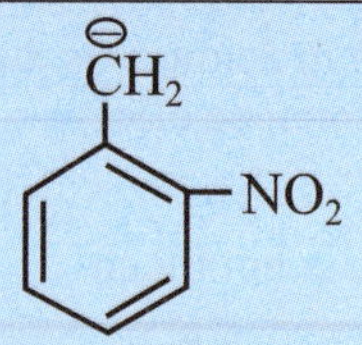	S	–I effect
		T	Hyper-conjugation

1) I-R,S; II-P,Q; III-Q,S; IV-Q
2) I-Q,R; II-P,Q,T; III-Q,R; IV-Q,R,S
3) I-S,T; II-R,S; III-R,S; IV-P,Q
4) I-P,Q; II-R,S; III-S,T; IV-P,Q

20. Identify correct I.U.P.A.C. name of the following compound.

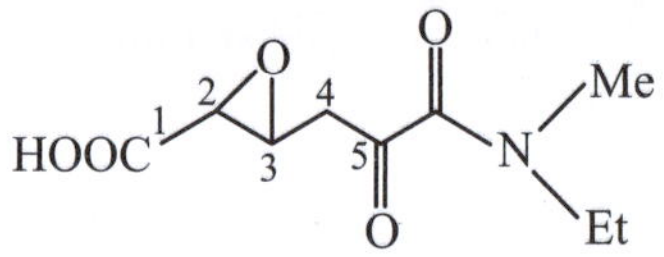

1) 5,6-Dioxo-6-(N-ethyl-N-methyl-amino)-2,3-epoxy hexanoic acid.
2) 6-(N-Ethyl-N-methylamino)-2,3-epoxy-5,6-dioxohexanoic acid.
3) 5-(N-Ethyl-N-methylcarbamoyl)-2,3-epoxy-5-oxopentanoic acid
4) 5-(N-Ethyl-N-methylcarbamoyl)-1,2-epoxy-4-oxobutanoic carboxylic acid.

NEET & AIIMS Past Year MCQs

1. The correct IUPAC name of the compound **(2024 Re)**

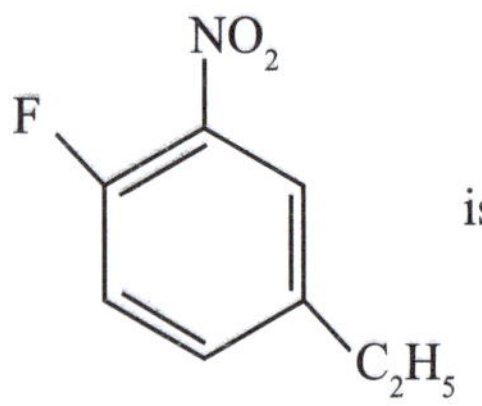

is:

1) 4-ethyl-1-fluoro-2-nitrobenzene
2) 4-ethyl-1-fluoro-6-nitrobenzene
3) 3-ethyl-6-fluoro-1-nitrobenzene
4) 1-ethyl-4-fluoro-3-nitrobenzene

2. Methyl group attached to a positively charged carbon atom stabilizes the carbocation due to **(2024 Re)**

1) -I inductive effect
2) electromeric effect
3) hyperconjugation
4) mesomeric effect

3. A steam volatile organic compound which is immiscible with water has a boiling point of 250°C. During steam distillation, a mixture of this organic compound and water will boil: **(2024 Re)**

1) above 100°C but below 250°C
2) above 250°C
3) at 250°C
4) close to but below 100°C

4. On complete combustion, 0.3 g of an organic compound gave 0.2 g of CO_2 and 0.1 g of H_2O. The percentage composition of carbon and hydrogen in the compound, respectively is: **(2024 Re)**

1) 4.07% and 15.02%
2) 18.18% and 3.70%
3) 15.02% and 4.07%
4) 3.70% and 18.18%

5. On heating some solid substances change from solid to vapour state without passing through liquid state. The technique used for the purification of such solid substances based on the above principle is known as **(2024)**

1) Distillation 2) Chromatography
3) Crystallization 4) Sublimation

6. A compound with a molecular formula of C_6H_{14} has two tertiary carbons. Its IUPAC name is: **(2024)**

1) 2,3-dimethylbutane
2) 2,2-dimethylbutane
3) n-hexane
4) 2-methylpentane

7. Match List I with List II. **(2024)**

List I (Molecule)		List II (Number and types of bonds between two carbon atoms)	
A.	ethane	I.	one σ-bond and two π-bonds
B.	ethene	II.	two π-bonds
C.	carbon molecule, C_2	III.	one σ-bond
D.	ethyne	IV.	one σ-bond and one π-bond

Choose the correct answer from the options given below:

1) A-III, B-IV, C-II, D-I 2) A-III, B-IV, C-I, D-II
3) A-I, B-IV, C-II, D-III 4) A-IV, B-III, C-II, D-I

8. The most stable carbocation among the following is: **(2024)**

1) 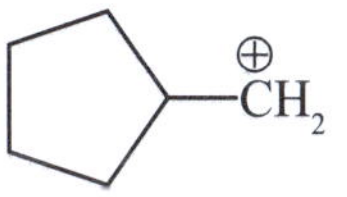 2) 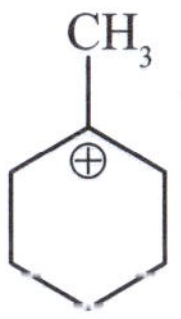

3) 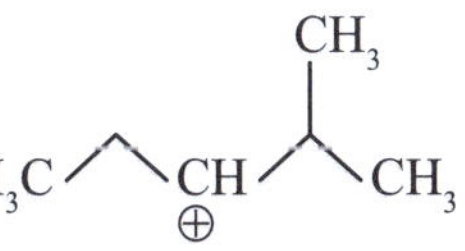 4) 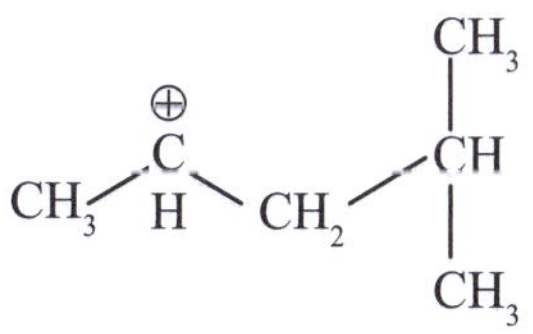

9. In Lassaigne's extract of an organic compound, both nitrogen and sulphur are present, which gives blood red colour with Fe^{3+} due to the formation of **(2023)**

1) $[Fe(SCN)]^{2+}$ 2) $Fe_4[Fe(CN)_6]_3.xH_2O$
3) NaSCN 4) $[Fe(CN)_5NOS]^{4-}$

10. The number of σ bonds, π bonds and lone pair of electrons in pyridine, respectively are **(2023)**

1) 12, 2, 1 2) 11, 2, 0
3) 12, 3, 0 4) 11, 3, 1

11. The Kjeldahl's method for the estimation of nitrogen can be used to estimate the amount of nitrogen in which one of the following compounds? **(2022)**

1) 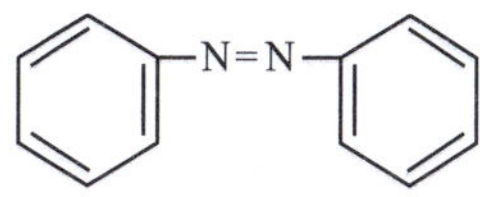 2) 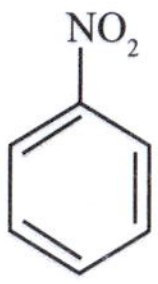

3) 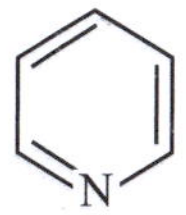 4) 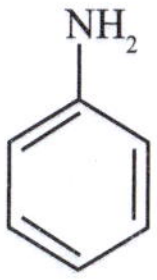

12. The correct IUPAC name of the following compound is: **(2022)**

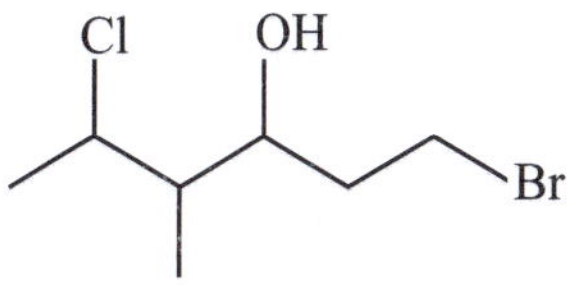

1) 6-bromo-4-methyl-2-chlorohexan-4-ol
2) 1-bromo-5-chloro-4-methylhexan-3-ol
3) 6-bromo-2-chloro-4-methylhexan-4-ol
4) 1-bromo-4-methyl-5-chlorohexan-3-ol

13. The compound which shows metamerism is: **(2021)**

1) C_3H_8O 2) C_3H_6O
3) $C_4H_{10}O$ 4) C_5H_{12}

14. The correct structure of 2,6-Dimethyl-dec-4-ene is: **(2021)**

1) 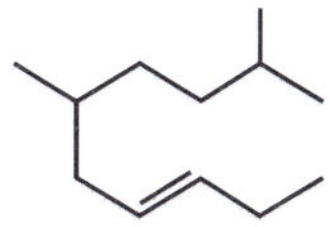 2) 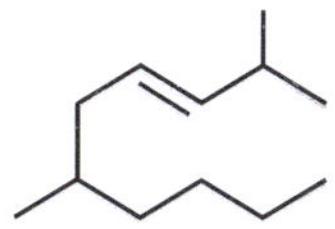

3) 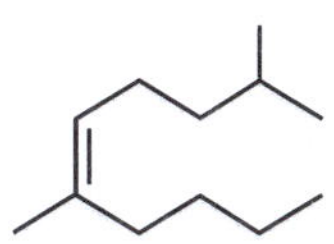 4) 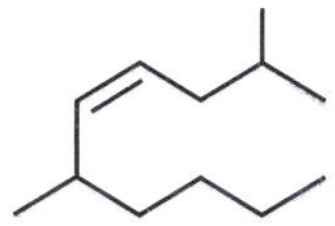

15. Order of increasing acidic strength **(AIIMS 2018)**

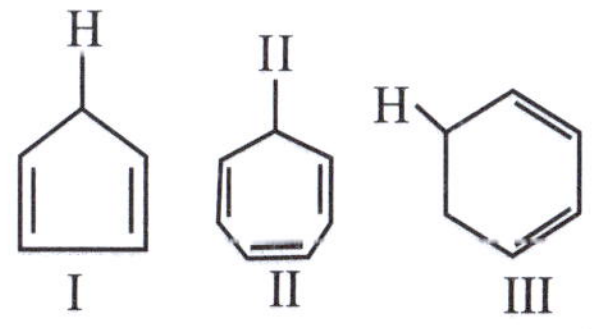

1) I > II > III 2) II > III > I
3) I > III > II 4) III > II > I

16. **Assertion:** trans-pent-2-ene is polar but trans-but-2-ene is non-polar.

Reason: The polarity of cis-isomer is more than trans-isomer, which are either non-polar or less polar. **(AIIMS 2017)**

1) If both Assertion & Reason are True & the Reason is a correct explanation of the Assertion.
2) If both Assertion & Reason are True but Reason is not a correct explanation of the Assertion.
3) If Assertion is True but the Reason is False.
4) If both Assertion & Reason are False.

17. The IUPAC name of **(AIIMS 2017)**

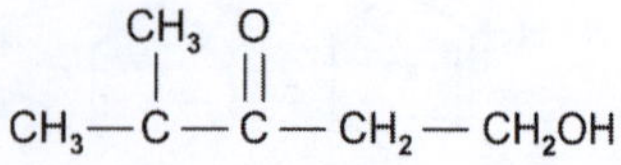

1) 1-hydroxy-4-methyl pentan-3-one
2) 5- methyl-5-hydroxyl pentane-3-one
3) 2- methyl-5-hydroxyl-pentane-3-one
4) 4-methyl-3-oxopentane-1-ol

18. Give saw – horse projection for **(AIIMS 2016)**

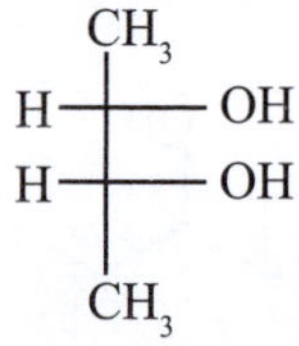

1) 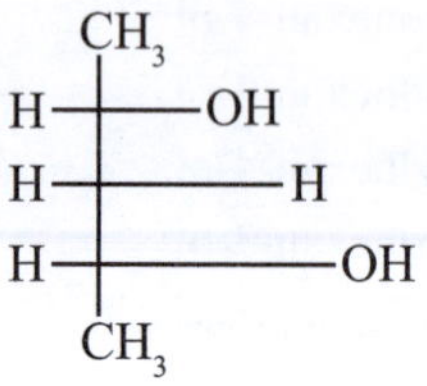

2) 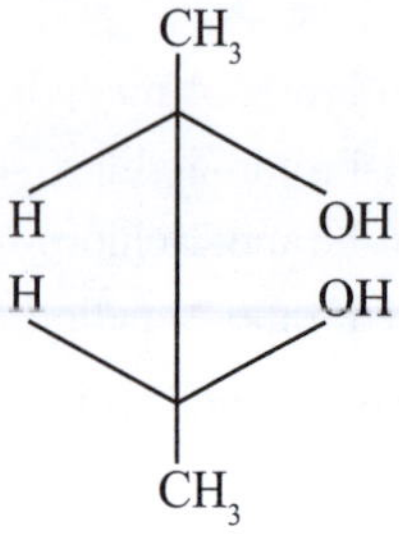

3) 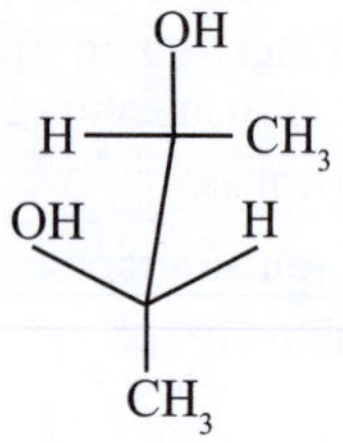

4) 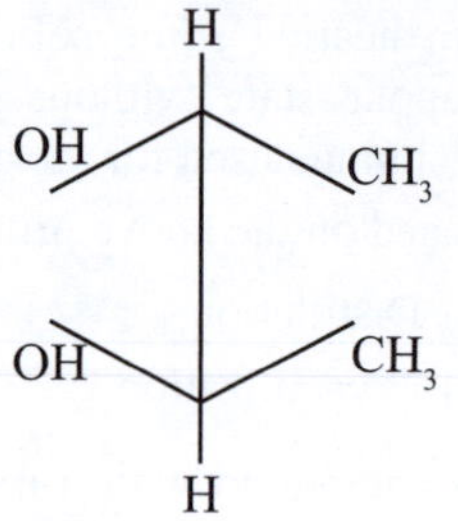

19. **Assertion:** Electromeric effect is brought into play only at the requirement of the reagent.

Reason: It is a temporary effect in which bond pair is shifted to one of the constituent atoms. **(AIIMS 2009)**

1) If both Assertion & Reason are True & the Reason is a correct explanation of the Assertion.
2) If both Assertion & Reason are True but Reason is not a correct explanation of the Assertion.
3) If Assertion is True but the Reason is False.
4) If both Assertion & Reason are False.

20. Beilstein test is used for the detection of: **(AIIMS 1999)**

1) Cl^- 2) Na
3) N_2 4) CO_2

Answer Key

NCERT Topic-wise MCQs

1	2	3	4	5	6	7	8	9	10	11	12	13	14	15	16	17
1	4	2	2	2	3	2	2	2	2	1	4	1	1	4	1	4
18	19	20	21	22	23	24	25	26	27	28	29	30	31	32	33	34
4	4	4	4	4	1	2	3	2	3	3	3	4	1	2	1	3
35	36	37	38	39	40	41	42	43	44	45	46	47	48	49	50	51
4	2	2	4	4	2	4	3	3	2	2	3	4	2	4	1	3
52	53	54	55	56	57	58	59	60	61	62	63	64	65	66	67	68
2	1	1	3	3	1	1	4	4	3	3	4	2	3	4	4	2
69	70	71	72	73	74	75	76	77	78	79	80	81	82	83	84	85
4	3	1	3	4	4	3	3	1	3	4	2	2	2	1	1	4
86	87	88	89	90	91	92	93	94	95	96	97	98	99	100	101	102
3	4	4	1	1	1	4	1	3	2	1	2	4	4	2	1	4
103	104	105	106	107	108	109	110	111	112	113	114	115	116	117	118	119
1	1	2	4	2	1	3	1	3	1	2	4	3	2	4	3	2
120	121	122	123	124	125	126	127	128	129	130	131	132	133	134	135	136
1	2	4	1	4	3	2	1	1	4	1	1	4	2	3	2	1
137	138															
1	1															

High Order Time Intensive MCQs

1	2	3	4	5	6	7	8	9	10	11	12	13	14	15	16	17
3	1	3	4	3	4	4	1	3	1	1	4	4	2	4	4	4
18	**19**	**20**	**21**	**22**	**23**	**24**	**25**	**26**	**27**	**28**	**29**	**30**	**31**	**32**	**33**	**34**
3	4	2	3	3	2	2	4	4	4	4	4	4	2	2	4	1
35	**36**	**37**	**38**	**39**	**40**	**41**	**42**	**43**	**44**	**45**	**46**	**47**	**48**	**49**	**50**	**51**
2	3	4	3	1	1	3	4	2	1	3	4	4	1	3	2	4
52	**53**	**54**	**55**	**56**	**57**	**58**	**59**	**60**	**61**	**62**	**63**	**64**	**65**	**66**	**67**	**68**
3	3	2	3	4	2	3	1	2	1	1	2	1	3	3	3	3
69	**70**	**71**	**72**	**73**	**74**	**75**										
2	4	1	3	3	2	2										

Match the Columns MCQs

1	2	3	4	5
1	3	2	2	2

Correct & Incorrect MCQs

1	2	3	4	5	6	7	8	9	10
1	1	3	3	2	1	3	2	3	1

Statement Based MCQs

1	2	3	4	5	6	7	8	9	10
1	2	1	1	1	1	4	4	3	1

Assertion & Reason MCQs

1	2	3	4	5	6	7	8	9	10
1	1	3	1	2	3	2	2	2	1

Multi Correct MCQs

1	2	3	4
3	3	4	4

NCERT Exemplar MCQs

1	2	3	4	5	6	7	8	9	10	11
1	4	2	3	3	4	4	2	1	2	1

Kattar NEET

1	2	3	4	5	6	7	8	9	10	11	12	13	14	15	16	17
3	2	3	3	4	2	4	2	2	4	1	4	3	4	4	1	1
18	**19**	**20**														
2	2	3														

NEET & AIIMS Past Year MCQs

1	2	3	4	5	6	7	8	9	10	11	12	13	14	15	16	17
1	3	4	2	4	1	1	2	1	4	4	2	3	4	3	2	1
18	**19**	**20**														
3	1	1														

CHAPTER

10 Hydrocarbons

Topicwise Number of Questions Analysis (2025-2016)

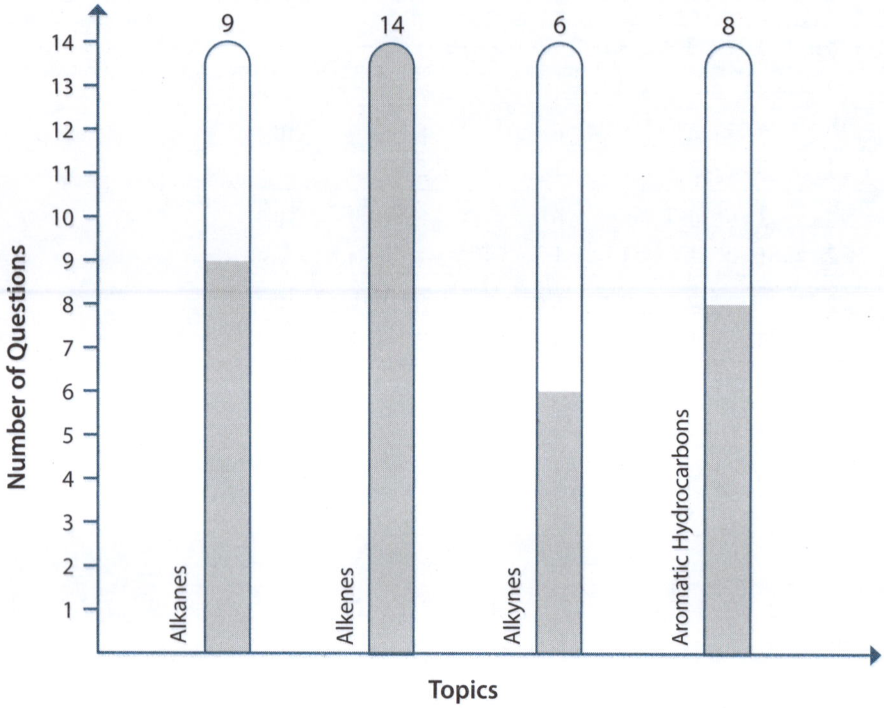

NCERT-Picks

- The major source of aliphatic hydrocarbons is petroleum and are also known as paraffins because of their low .reactivity towards chemicals.

Alkanes

Preparation

- By reduction of alcohols, aldehyde, ketones and carboxylic acids hydrocarbons are formed.

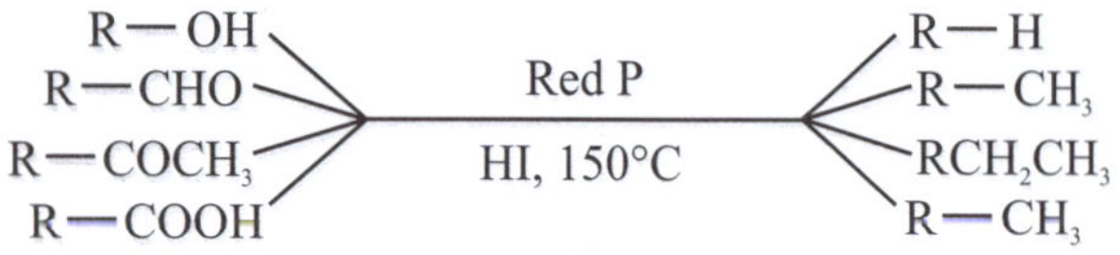

- From grignard reagent:

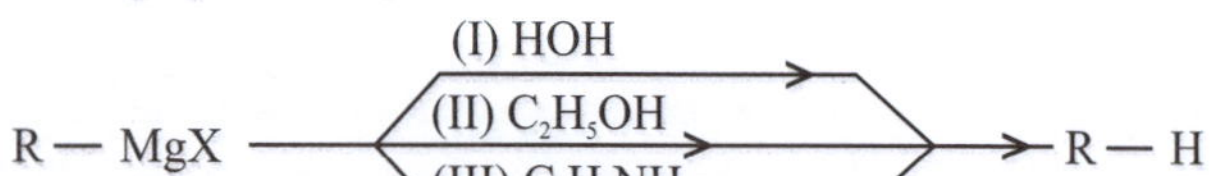

- By wurtz reaction

  Alkyl halides on treatment with sodium metal in dry ether (free from moisture) solution give higher alkanes. This reaction is known as Wurtz reaction and is used for the preparation of higher alkanes containing even number of carbon atoms.

$$2R-X + 2Na \xrightarrow{\text{dry ether}} R-R + 2NaX$$

$$R-X + R'-X \xrightarrow[\text{ether (dry)}]{Na} R-R, R-R', R'-R'$$

Mechanism: (a) Ionic mechanism: $2Na \rightarrow 2Na^{\oplus} + 2e^{\ominus}$;

$$R-X + 2e^- \rightarrow R^{\ominus} + X^{\ominus}$$

(1°, 2°)

$$\overset{\ominus}{R} + R-X \xrightarrow{S_N2} R-R; \quad \overset{\oplus}{Na} + \overset{\ominus}{X} \longrightarrow NaX$$

(1° or 2°)

(b) Free radical mechanism :

$$\overset{\oplus}{Na} \rightarrow \overset{\ominus}{e}; R-X + \overset{\ominus}{e} \rightarrow R^{\bullet} + \overset{\ominus}{X}; R^{\bullet} + R^{\bullet} \rightarrow R-R$$

By Kolbe's electrolysis:

An aqueous solution of sodium or potassium salt of a carboxylic acid on electrolysis gives alkane containing **even number of carbon atoms at the anode**.

$$2RCOOK + 2HOH \xrightarrow{\text{Electrolysis}} R-R + 2CO_2 + H_2 + 2KOH$$

Mechanism : $RCO_2K \rightleftharpoons RCO_2^- + K^+$

At Anode:

$RCO_2^- \rightarrow RCO_2^{\cdot} + e^-$; $RCO_2^{\cdot} \rightarrow R^{\cdot} + CO_2$; $R^{\cdot} + R^{\cdot} \rightarrow R-R$

At Cathode:

$2H_2O + 2e^- \rightarrow 2OH^- + 2H^{\cdot} + 2K^+ \rightarrow H_2(g) + 2KOH$

- Methane cannot be prepared by kolbe's electrolytic method. Alkanes formed by this method contain twice the number of carbon atoms present in the sodium / potassium salt of carboxylic acid.

Physical properties

The compound C_4H_{10} i.e.butane can exist in two isomeric forms.

- Alkanes with even number of carbon atoms have higher melting points than alkanes with odd number of carbon atoms due to symmetry.
- Except for the very small alkanes, the boiling point rises 20 to 30 degrees for each carbon that is added to the chain.
- Branched chain alkanes have lower boiling point. As molecular size increases, boiling point increases.
- Isomerisation:

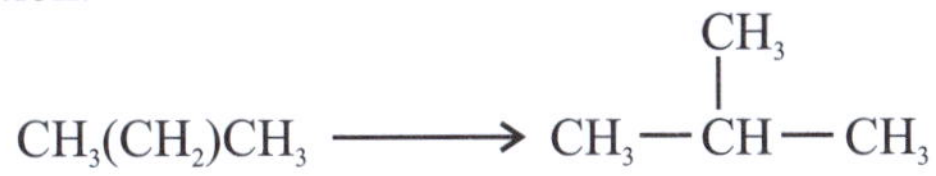

n-Butane iso-Butane

Chemical properties

- At room temperature methane and chlorine reacts in presence of light.
- reaction is given as- $CH_4 + Cl_2 \xrightarrow{h\nu} CH_3Cl + HCl$

  $CH_3Cl + Cl_2 \xrightarrow{h\nu} CH_2Cl_2 + HCl$

  $CH_2Cl_2 + Cl_2 \xrightarrow{h\nu} CHCl_3 + HCl$

  $CHCl_3 + Cl_2 \xrightarrow{h\nu} CCl_4 + HCl$

  Fluorination is too violent to be controlled. Iodination is very slow and reversible. it can be carried out in presence of oxidising agent like HNO_3.
- The rate of reactions of alkanes with halogens decreases as $F > Cl_2 > Br_2 > I_2$. Electro negativity of halogens decreases as $F > Cl_2 > Br_2 > I_2$.
- Complete Combustion of alkanes:

$$C_nH_{2n+2} + \left(\frac{3n+1}{2}\right)O_2 \rightarrow nCO_2 + (n+1)H_2O$$

- Controlled Oxidation: $2CH_4 + O_2 \xrightarrow[\text{100 atm,}]{\text{Cu, tube}} 2CH_3OH$ (9 : 1) methanol

$$CH_4 + O_2 \xrightarrow[\Delta]{Mo_2O_3} HCHO + H_2O$$

$$2CH_3CH_3 + 3O_2 \xrightarrow{(CH_3COO)_2Mn} 2CH_3COOH + 2H_2O$$

ethanoic acid

$(CH_3)_3CH \xrightarrow[\Delta]{KMnO_4} (CH_3)_3COH$ [For alkanes having 3^0 H atoms]

Pyrolysis-Decomposition of higher alkanes into smaller fragments on strong heating in the absence of air is called pyrolysis or cracking. Pyrolysis is a free radical reaction

$$C_{12}H_{26} \xrightarrow[\text{973 K}]{\text{Pd/Pt/Ni}} C_7H_{16} + C_5H_{10} + \text{Other Products}$$

Heptane Pentene

Conformational isomers

- The potential energy difference among these conformations of ethane is about 3 k cal/mol called torsional strain (or energy barrier). Due to small difference in energy, the two conformations are readily inter-convertible with ethane.
- Staggered form of ethane is the most stable and eclipsed form is the least stable. In staggered form, the electron clouds of C – H bonds are as far as possible and it comes closer to each other in eclipsed form.

Alkenes

Preparation

1. By partial hydrogenation of alkyne

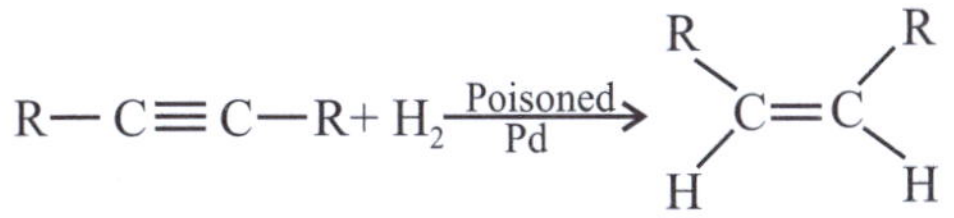

(cis alkene)

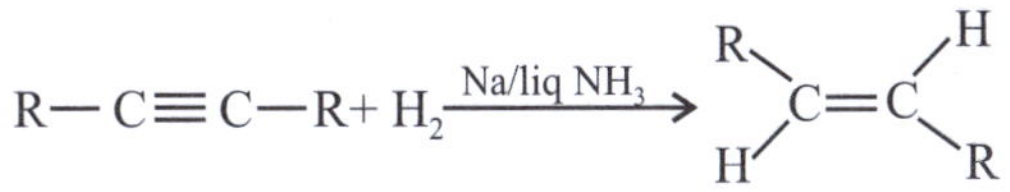

(Trans alkene)

2. By Dehydrohalogenation of alkyl halides

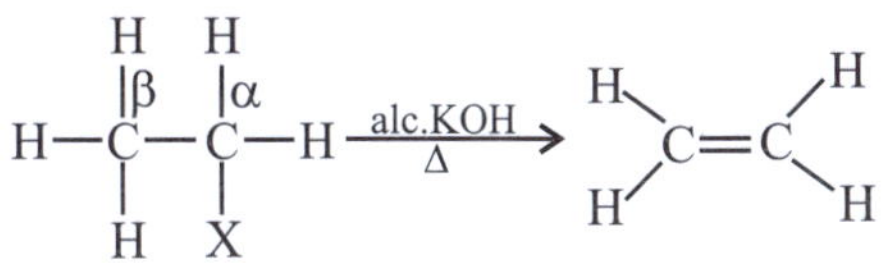

(X=Cl, Br, I)

Chemical Properties

- Alkenes react with cold concentrated sulphuric acid to form compounds of the general formula $ROSO_3H$ known as alkyl hydrogen sulphates.

Reaction with halogen:

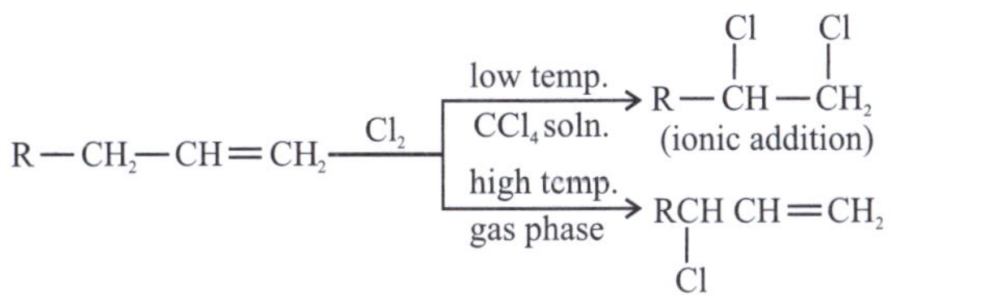

(free - radical substitution)

Halogens like bromine or chlorine add up to alkene to form vicinal dihalides.Addition of halogen to alkene is an example of electrophilic addition reaction. Iodine does not show addition reaction under normal condition.

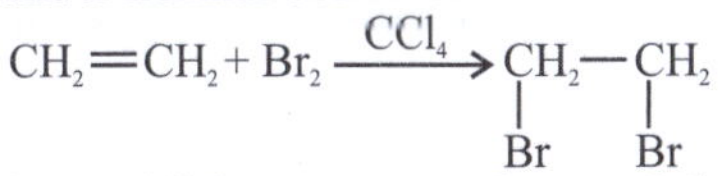

Note: The reddish orange colour of bromine solution in carbon tetrachloride is discharged when bromine adds up to an unsaturation site. This reaction is used as a test for unsaturation.

Mechanism

Step 1 :

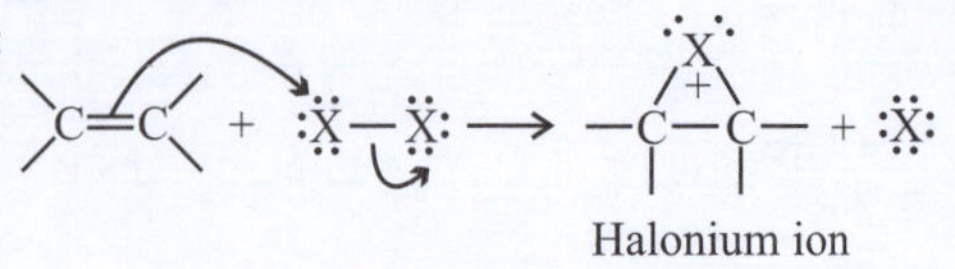

Step 2 :

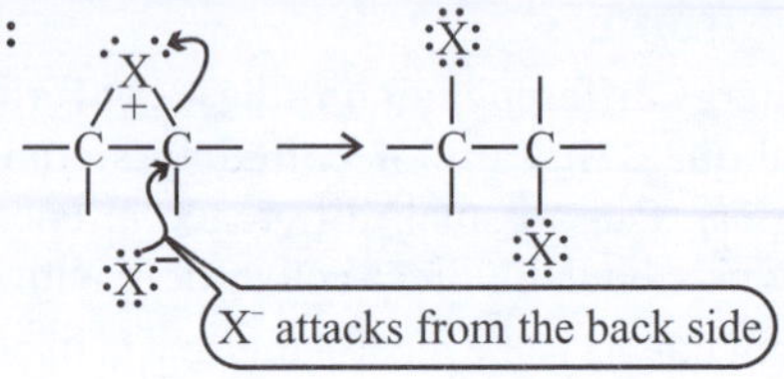

Addition of hydrogen halides

Addition of HX (HCl, HBr, HI) to unsymmetrical alkene place according to Markovnikov rule.

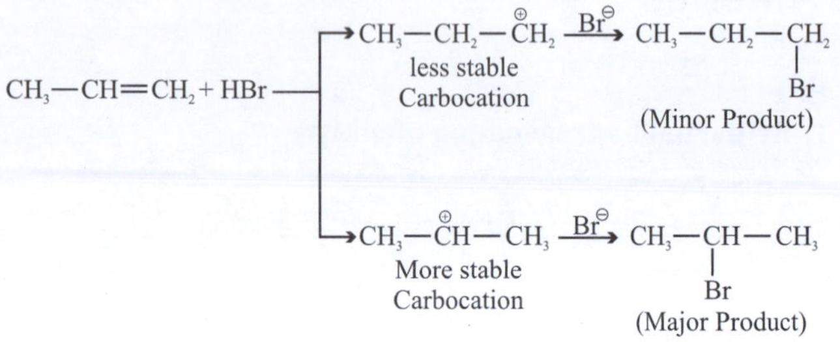

- Peroxide effect is observed in addition of HBr but not in addition of HCl. H — Cl is stronger than H — Br bond and is not cleaved by the free radical.
- Propene reacts with HBr in presence of peroxide to give 1 – Bromopropane. In presence of peroxide, propene reacts with HBr by free radical mechanism according to antimarkovnikov rule.
- **Ozonolysis:** This reaction is used to detect the position of unsaturation i.e. double and triple bond.
- **Hydroxylation:**

  Test for unsaturation, called Baeyer test and alkaline $KMnO_4$ (MnO_4 + OH) is called baeyer's reagent in this reaction. The alkaline $KMnO_4$ oxidises alkene to glycol and pink colour fades.

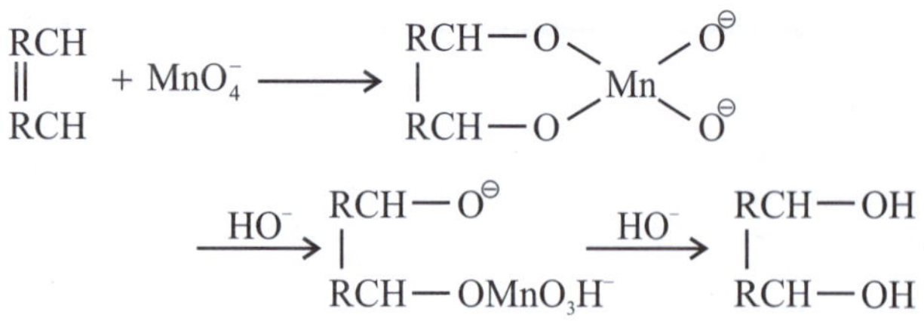

- Propene undergoes allylic bromination when it is treated with N-bromo succinimide (NBS) in CCl_4 in presence of peroxides or light.
- cis-form of alkene is found to be more polar than the trans-form. The groups are in opposite directions in the trans-form, the dipole moments of bonds cancel each other making trans-form almost non-polar.
- Oxidation of cycloalkenes leads to the ring openings and gives dicarboxylic acids or keto acids.

Alkynes

- Ethyne is acidic in nature in comparison to ethene and ethane. sp hybridised orbitals of carbon atoms in ethyne have highest electro negativity hence can attract shared e– pair of C — H bond to a greater extent.

Preparations of alkynes

(1) Hydrolysis: From calcium carbide (Industrial Method)

$$\underset{\text{Calcium carbide}}{CaC_2} + H_2O \xrightarrow{\text{hydrolysis}} \underset{\text{Acetylene}}{H-C\equiv C-H} + Ca(OH)_2$$

$$Mg_2C_3 + 4H_2O \rightarrow CH_3-C\equiv CH + 2Mg(OH)_2$$

(2) From vicinal dihalides

Ex: $Br-H_2C-CH_2-Br \xrightarrow[-KBr,-H_2O]{\text{Alcoholic KOH}}$

$$H_2C=CH-Br \xrightarrow{NaNH_2} CH\equiv CH + NaBr + NH_3$$

- Acetylene can be prepared by the controlled high temperature partial oxidation of methane.

Chemical properties

- Alkynes are less reactive than alkenes towards electrophilic addition reaction because the electrons in alkynes are more tightly held than alkenes.
- Reaction with sodium in liquid ammonia: Terminal alkynes react with sodium in liquid ammonia or sodamide to form sodium alkynide
- Addition reactions:

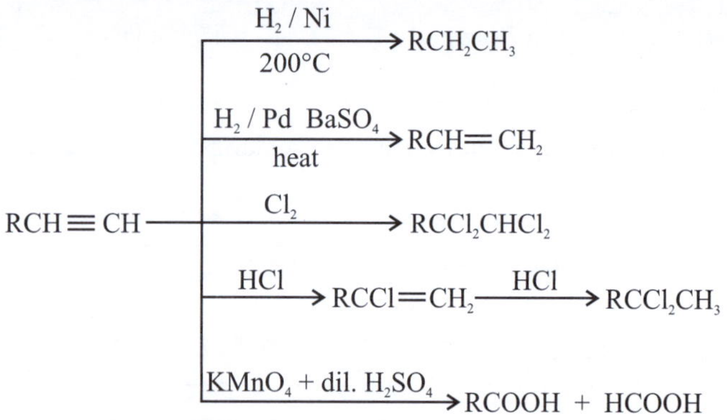

- Substitution reactions:

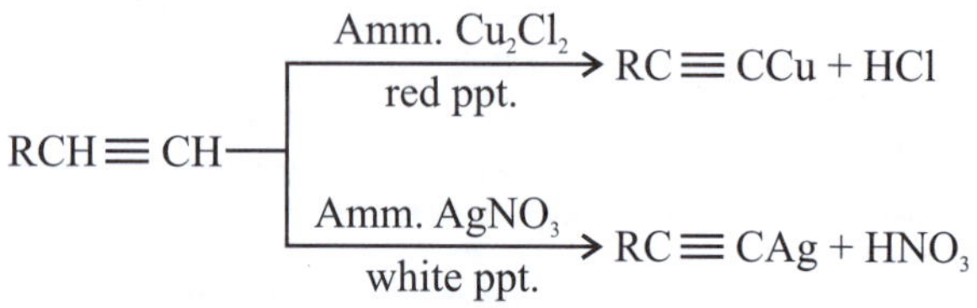

Aromatic Compounds

- Huckel rule of aromaticity: is applied to all the ring systems whether have benzene ring or not and possess the following characteristics ;
- Planarity: Complete delocalisation of π - electrons in the ring.
- Presence of $(4n+2)$ π - electrons in the ring. (n = 0, 1, 2,...)

- Benzene is isolated from coal tar by fractional distillation
- Under U-V light three chlorine molecules add to benzene to produce benzene Hexachloride C6H6Cl6 which is also called gammaxane.

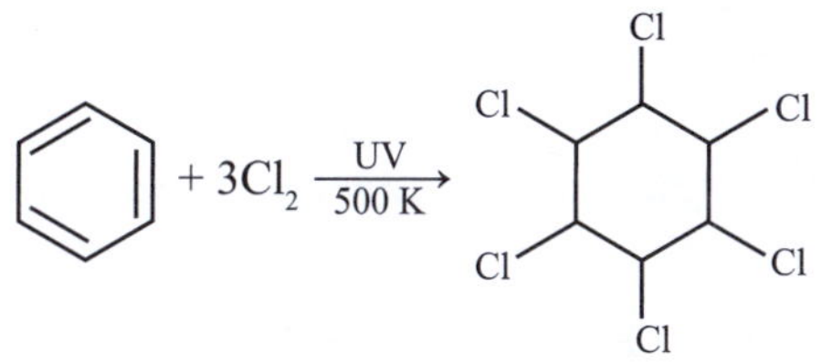

(BHC)

- Direct the incoming group to ortho and para positions of benzene ring, for example : CH_3 , $-OCH_3$, $-NH_2$, –NHR, $-NHCOCH_3$ etc is also known as ortho and para directing group is called activating group.
- Halogens are O - and P directing groups but are moderately deactivating due to their strong - I effect.
- Meta directing groups also called deactivating groups, direct the incoming group to meta position of benzene ring for example: $-NO_2$, –CN, –CHO, –COR, –COOH, –COOR, $-SO_3H$, etc.
- Toluene has the highest electron density in the benzene ring & hence can be most readily sulphonated due to +I effect of $-CH_3$ group.

MCQs

NCERT Topic-wise MCQs

Alkanes

1. Which of the following compounds will react with Na to form 4,5 - diethyloctane?
   1) $CH_3\,CH_2\,CH_2\,CH_2\,Br$
   2) $CH_3CH_2CH_2-CH(CH_3)-CH_2-CH_2Br$
   3) $CH_3CH_2CH_2CH_2-CH(Br)-CH_3$
   4) $CH_3CH_2CH_2-CH(Br)-CH_2-CH_3$

2. Which of the following is the most stable free radical?
   1) $CH_3\,\dot{C}H_2$
   2) $CH_3\,\dot{C}HCH_3$
   3) $CH_3\,\dot{C}HC_6H_5$
   4) $C_6H_5CH_2\,\dot{C}H_2$

3. Hydrocarbon which is liquid at room temperature is:
   1) Pentane
   2) Butane
   3) Propane
   4) Ethane

4. The major constituent of natural gas is:
   1) Methane
   2) Propane
   3) Butane
   4) Hexane

5. Which of the following products is formed when n-heptane is passed over $(Al_2O_3 + Cr_2O_3)$ catalyst at 773 K?
   1) Benzene
   2) Toluene
   3) Polyheptane
   4) cycloheptane

6. Which compound cannot be formed from Wurtz reaction?
   1) Propane
   2) Butane
   3) Ethane
   4) Methane

7. Sodium salt of X is heated with soda lime and gives ethane, X is:
   1) Ethanoic acid
   2) Methanoic acid
   3) Propanoic acid
   4) Either (1) or (3)

8. Which of the following is called marsh gas?
   1) C_2H_4
   2) C_2H_6
   3) C_2H_2
   4) CH_4

9. Among the following, select the alkane that is expected to have lowest boiling point?
   1) Hexane
   2) 2-methylpentane
   3) 3-methylpentane
   4) 2,2-dimethyl propane

10. The compound with highest boiling point is:
    1) n-hexane
    2) n-pentane
    3) n-butane
    4) 2,2-dimethylpropane

11. An alkyl halide by formation of its Grignard reagent and heating with water gives propane. What is the original alkyl halide?
    1) Methyl iodide
    2) Ethyl iodide
    3) Ethyl bromide
    4) Propyl bromide

12. For preparing an alkane a saturated solution of sodium or potassium salt of a carboxylic acid is subjected to:
    1) Hydrolysis
    2) Oxidation
    3) Hydration
    4) Electrolysis

13. Of the five isomeric hexanes, the isomer which can give two monochlorinated compounds is:
    1) n-hexane
    2) 2, 3-dimethyl butane
    3) 2, 2-dimethylbutane
    4) 2-methyl pentane

14. Which of the following reactions is not correct for preparation of alkanes?
    1) $(CH_3)_2\,CHCl \xrightarrow{LiAlH_4} (CH_3)_2\,CH_2$
    2) $(CH_3)_3\,CCl \xrightarrow{LiAlH_4} (CH_3)_3\,CH$
    3) $(CH_3)_2\,CHCl \xrightarrow{NaBH_4} (CH_3)_2\,CH_2$
    4) $(CH_3)_3\,CCl \xrightarrow{NaBH_4} (CH_3)_3\,CH$

15. Pick out the alkane which differs from the other member of the group:
    1) 2,2-Dimethyl propane
    2) Pentane
    3) 3-Methyl butane
    4) 2,2-Dimethylbutane

16. Which of the following isomeric heptanes can yield seven different monochlorinated products upon free radical chlorination?
    1) 3-methyl hexane
    2) 2-2, dimethyl pentane
    3) 2-methyl hexane
    4) 2, 3-dimethyl pentane

17. What happens when methane reacts with conc. HNO_3 at high temperature?
    1) Nitro methane is formed
    2) Methanol is formed
    3) CO_2 and H_2O are formed
    4) CO and H_2O are formed

18. The dihedral angle between the hydrogen atoms of 2 methyl groups in staggered conformation of ethane is

1) 0° 2) 60°
3) 120° 4) 240°

19. The most stable conformation of ethane is

1) Eclipsed
2) Skew
3) Staggered
4) All are equally stable

20. Which of the following alkanes cannot be produced by Kolbe electrolysis of sodium or potassium salts of carboxylic acids?

1) Methane 2) Ethane
3) Butane 4) Hexane

21. The I.U.P.A.C. name of neopentane is:

1) 2-Methyl butane
2) 2,2-Dimethyl propane
3) 2-Methyl propane
4) 2,2-Dimethyl butane

22. Methyl bromide is converted into ethane by heating it in ether medium with:

1) Al 2) Mg
3) Na 4) Cu

23. Which of following catalyst is best for single step conversion of CH_4 into formaldehyde?

1) $Cu/300°C$ 2) Mo_2O_3/Δ
3) $(CH_3COO)_2Mn/\Delta$ 4) $KMnO_4/\Delta$

24. Pyrolysis of methane and ethane respectively are:

1) Exothermic & Endothermic
2) Endothermic & Exothermic
3) Both are endothermic
4) Both are exothermic

25. Wrong statement regarding alkanes:

1) All alkanes burn in air to give CO_2 and H_2O
2) At room temperature, some alkanes are gases, some are liquids and some are solids
3) What ever the molecular weight of alkane, its density is less than water
4) Most of the alkanes are soluble in water.

26. The reaction represents:

$C_{12}H_{26} \longrightarrow C_6H_{12} + C_6H_{14}$

1) Substitution 2) Synthesis
3) Cracking 4) Polymerization

27. Bromination of alkane is initiated by a process of:

1) Pyrolysis 2) Substitution
3) Homolysis 4) Peroxidation

28. Final products of complete oxidation of hydrocarbon is:

1) Acid 2) Dihydric alcohol
3) Aldehyde 4) $H_2O + CO_2$

29. Isomerisation in alkane can be brought by using:

1) Al_2O_3 2) Fe_2O_3
3) $Anh.AlCl_3/HCl$ 4) $Conc.H_2SO_4$

30. The number of sigma bonds formed in ethane by the overlapping of sp^3-sp^3 orbitals:

1) 7 2) 5
3) 1 4) 4

31. Ethylene is converted to ethane in the presence of Ni at . 300ºC In this reaction the hybridisation of carbon changes from:

1) sp to sp^2 2) sp^2 to sp^3
3) sp^3 to sp 4) sp to sp^3

32. Which conformer of butane molecule has lowest energy?

1) Ellipsed
2) Skew
3) Staggered
4) Anti staggered

33. What is difference in monomer unit of polyisoprene & Neoprene polymer respectively?

1) $-CH_3$ group in Isoprene & –Cl group in chloroprene at 2nd carbon
2) $-CH_3$ group at 2nd carbon while Cl group at 3rd carbon
3) Both unit have no different
4) $-CH_3$ group at 3rd carbon while Cl group at 2nd carbon

34. The molecule which is free from angular strain is:

1) Cyclopropane
2) Cyclobutane
3) Cyclopentane
4) Cyclohexane

35. Which cycloalkane has the lowest heat of combustion per CH_2 group?

1) Cyclopropane
2) Cyclobutane
3) Cyclopentane
4) Cyclohexane

36. In the following reaction, the major product is:

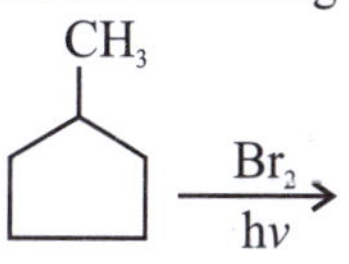

1) 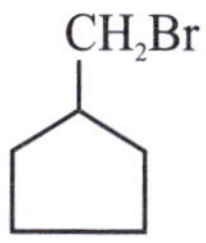

2) 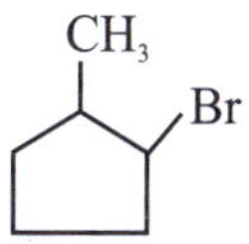

3) 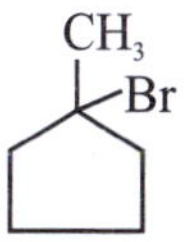

4) 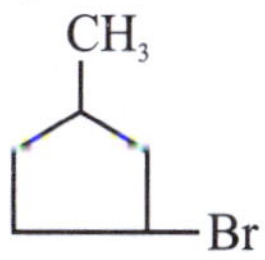

37. Cycloalkanes are isomeric with:

1) Alkane 2) Alkenes
3) Alkynes 4) Arenes

Alkenes

38. Which of the following reactions does not show the correct products of the reaction?
    1) $CH_3-CH=CH_2 \xrightarrow[\text{Peroxide}]{\text{HBr}} CH_3-CH_2-CH_2-Br$
    2) $CH_3-CH=CH_2 \xrightarrow[\text{Peroxide}]{\text{HCl}} CH_3-CH_2-CH_2-Cl$
    3) $CH_3-CH=CH_2 \xrightarrow{\text{HBr}} CH_3-\underset{|}{\overset{Br}{CH}}-CH_3$
    4) $CH_3-CH=CH_2 \xrightarrow{\text{HCl}} CH_3-\underset{|}{\overset{Cl}{CH}}-CH_3$

39. Which of the following compounds will show cis-trans isomerism?
    1) $(CH_3)_2C=CH\text{-}C_2H_5$
    2) $H_2C=CCl_2$
    3) $CH_3HC=CCl\,CH_3$
    4) $HClC=CH_2$

40. The ease of dehydrohalogenation for different Halogens is in the order:
    1) Iodide > Bromide > Chloride
    2) Bromide > Iodide > Chloride
    3) Chloride > Bromide > Iodide
    4) Iodide > Chloride > Bromide

41. Ozonolysis of 2,3- dimethyl but - 1 - ene followed by reduction with zinc and water gives:
    1) Methanal and hexanoic acid
    2) Methanoic acid and butanone
    3) Methanal and 3-methylbutane - 2- one
    4) Butanoic acid and 2,3 - dimethyl butanoic acid

42. Arrange the following alkyl halides in decreasing order of the rate of β elimination reaction with alcoholic KOH:

    (i) $CH_3-\underset{\underset{CH_3}{|}}{\overset{\overset{H}{|}}{C}}-CH_2Br$

    (ii) CH_3-CH_2-Br

    (iii) $CH_3-CH_2-CH_2-Br$

    1) (i) > (ii) > (iii) 2) (iii) > (ii) > (i)
    3) (ii) > (iii) > (i) 4) (i) > (iii) > (ii)

43. For the given reaction, how many products (mono substituted) will be obtained including all isomers?

 $\xrightarrow{Br_2 / h\nu}$ Products

    1) 6 2) 5
    3) 4 4) 3

44. Which of the following is the predominant product in the reaction of HOBr with propene?
    1) 2- bromo-1- propanol
    2) 3-bromo-1-propanol
    3) 2-bromo-2-propanol
    4) 1-bromo-2-propanol

45. For synthesis of 1-butene, CH_3MgI should be treated with:
    1) Propene 2) 2- chloropropene
    3) Allyl chloride 4) Ethyl chloride

46. An alkene on vigorous oxidation with hot $KMnO_4$ gives only acetic acid, the alkene is:
    1) $CH_3CH_2CH=CH_2$
    2) $CH_3CH=CH\,CH_3$
    3) $(CH_3)_2C=CH_2$
    4) $CH_3CH=CH_2$

47. When an alkene on ozonloysis gives formaldehyde and acetaldehyde, the alkene is:
    1) Ethene 2) Propene
    3) But-1-ene 4) but-2-ene

48. Ethylene can be separated from acetylene by passing the mixture through:
    1) Fuming H_2SO_4 2) Pyrogallol
    3) Ammonical Cu_2Cl_2 4) Charcoal power

49. Aqueous solution of sodium succinate on electrolysis gives:
    1) CH_3-CH_3 2) $CH_2=CH_2$
    3) $CH\equiv CH$ 4) CO_2

50. 1,3,5,7-Octatetraene contains X σ bonds and Y π bonds. "X" and "Y" are:
    1) 23, 4 2) 17,4
    3) 18,5 4) 33,2

51. The alkene that exhibits geometrical isomerism is:
    1) Propene 2) 2-methyl propene
    3) 2-butene 4) 2-methyl-2-butene

52. The compound formed when alcoholic solution of ethylene dibromide is heated with granulated zinc is:
    1) Ethene 2) Ethyne
    3) Ethane 4) bromoethene

53. Addition of cold conc. H_2SO_4 with alkene is an example of
    1) Electrophilic addition reaction
    2) Nucleophilic substitution reaction
    3) Electrophilic substitution reaction
    4) Nucleophilic addition reaction

54. $CH_3-\underset{\underset{CH_3}{|}}{\overset{\overset{CH_3}{|}}{C}}-CH=CH_2 \longrightarrow CH_3-\underset{\underset{CH_3}{|}}{\overset{\overset{CH_3}{|}}{C}}-\overset{\overset{OH}{|}}{CH}-CH_3$

    above conversion is possible in presence of following reagents-
    1) B_2H_6/THF & NaOH
    2) Hg $(OCOCH_3)_2$/THF & $NaBH_4 + H_2O$
    3) dil H_2SO_4
    4) Al_2O_3/Δ

55. Vinyl group among the following is
    1) $(CH_3)_2CH-$ 2) $HC\equiv C-$
    3) $H_2C=CH-CH_2-$ 4) $CH_2=CH-$

56. 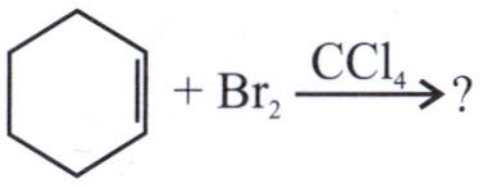

1) 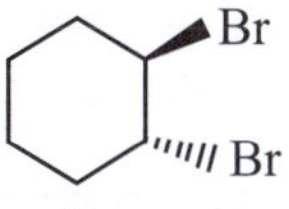 2) 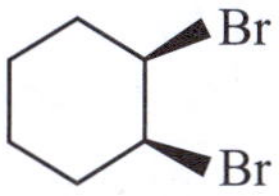

3) 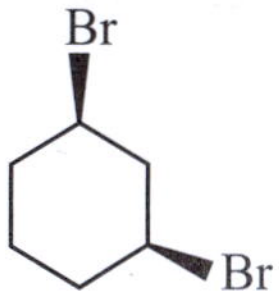 4) 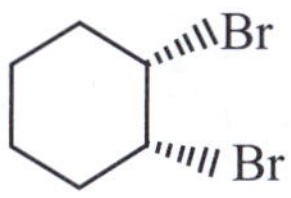

57. IUPAC name of the following alkene is,

$(CH_3)_2CH—CH=CH—CH_2—CH=CH—\underset{\displaystyle C_2H_5}{\underset{|}{CH}}—CH_3$

1) 2, 8 Dimethyldeca-3,6-diene
2) 4,8 Dimethyl deca- 3, 6-diene
3) 2,4 Dimethyl deca- 3, 6-diene
4) 3,9 Dimethyldeca-4,7-diene

58. IUPAC name of the following alkene is,

$H_2C=CH—HC=CH_2—\underset{\displaystyle CH_3}{\underset{|}{C}}=CH—HC=CH_2$

1) 5-Methylocta-1,3,5,7- tetraene
2) 4-Methylocta-1,3,5,7- tetraene
3) 4-Butenylocta-1,3- diene
4) Octa-1, 5- diene

59. 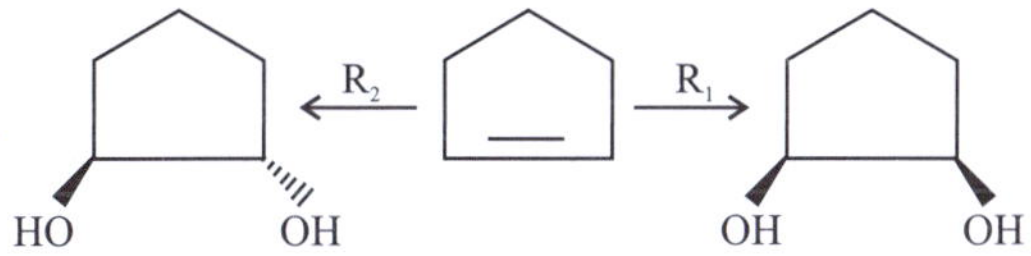

R_1 & R_2 are respectively

1) Cold alkaline $KMnO_4$, OsO_4 / H_2O_2
2) Cold alkaline $KMnO_4$, HCO_3H, $H_3O^{\oplus}$
3) Cold alkaline $KMnO_4$, $Pb(NO_3)_2$
4) CH_3CO_3H/H_2O, Cold alkaline $KMnO_4$

60. Lindlar's catalyst is used for controlled.
1) Oxidation
2) Hydrogenation
3) Halogenation
4) Dehydration

61. Ethylene readily undergoes the following type of reaction:
1) Elimination 2) Addition
3) Rearrangement 4) Substitution

62. On reductive ozonolysis of ethylene gives:
1) Aldehyde 2) Ketone
3) Carboxylic acid 4) Ether

63. When ethanol vapours are passed over alumina heated at 350°C, the main product obtained is:
1) C_2H_6 2) C_2H_4
3) C_2H_2 4) $C_2H_5OC_2H_5$

64. In which of the following will Kharasch effect operate?
1) $CH_3-CH_2-CH=CH_2+HCl$
2) $CH_3-CH_2-CH=CH_2+HBr$
3) $CH_3-CH=CH-CH_3+HBr$
4) $CH_3-CH_2-CH=CH_2+HI$

65. The olefin which on ozonolysis gives CH_3CH_2CHO and CH_3CHO is:
1) But-1-ene 2) But-2-ene
3) Pent-1-ene 4) Pent-2-ene

66. The oxidation product of ethylene by air at 200–400°Cin presence of silver catalyst is:
1) Ethylene glycol
2) Ethylene oxide
3) 1,2-Ethanediol
4) Ethylene chlorohydrin

67. What are X and Y in the reaction,

$$C_2H_4+H_2SO_4\xrightarrow{80^0C}X\xrightarrow[\Delta]{H_2O}y$$

1) C_2H_6, C_2H_5OH
2) C_2H_4, C_2H_5SH
3) $C_2H_5OSO_3H$, C_2H_5OH
4) C_2H_4, C_2H_5OH

68. In the following sequence of reactions the compound A is,

$$A\xrightarrow{HBr}B\xrightarrow{alc\ KOH}C\xrightarrow{O_3,\ Zn/H_2O}CH_3CHO+HCHO$$

1) Ethylene 2) Acetic acid
3) Propene 4) 1-Butene

69. IUPAC names of the products obtained by addition of HBr to hex - 1 - ene in the absence of peroxide and in the presence of peroxide respectively are:
1) 2-Bromohexane, 1-Bromohexane
2) 2-Bromohexane, 3-Bromohexane
3) 1-Bromohexane, 3-Bromohexane
4) 1-bromo hexane, 4-bromo hexane

70. $Z-\text{isomer}\xleftarrow{y}2-\text{butyne}\xrightarrow{x}E-\text{isomer}$

x,y respectively are:
1) Na /$NH_{3(liq)}$ and Pd / $BaSO_4+H_2$
2) Ni / 140°C and Pd / $BaSO_4+H_2$
3) Ni / 140°C and Na /$NH_{3(liq)}$
4) $BaSO_4+H_2$ and Na /$NH_{3(aq)}$

71. The final product in the given reaction:

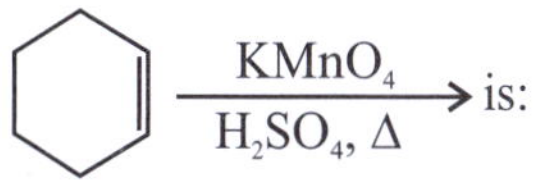

1) 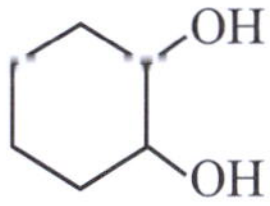 2) 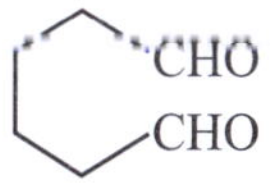

3) 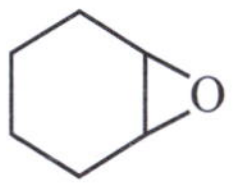 4) 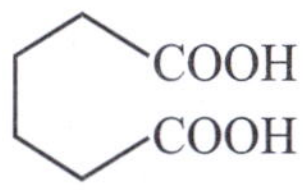

72. Which of the following has the maximum heat of hydrogenation?

1) 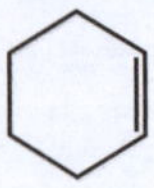 2) 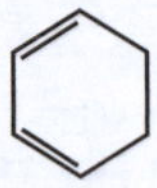

3)  4) 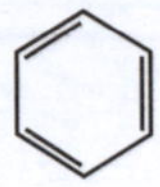

Alkynes

73. Acetylene when treated with dilute HCl at 60°C (333K) in presence of $HgCl_2$ produces –

1) Methyl chloride
2) Vinyl chloride
3) Acetaldehyde
4) Formaldehyde

74. $CH_3CH_2C{\equiv}CH \xrightarrow{NaNH_2\,/\,NH_3\,(l)} \xrightarrow{CH_3CH_2Br} \xrightarrow{Li,\,NH_3\,(l)}$

1) $CH_3CH_2CH{=}CH{-}CH_3$
2) $CH_3CH_2CH{=}CH_2$
3) 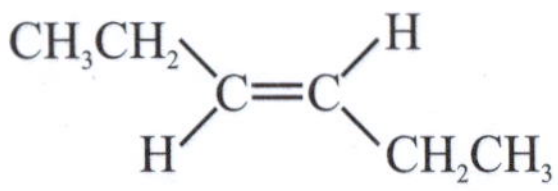
4) 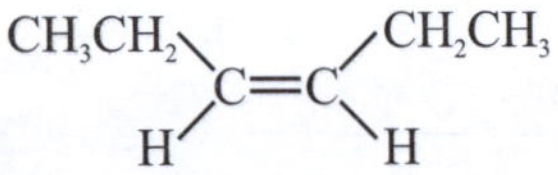

75. Which of the following alkynes is most acidic?

1) $CH_3\ C \equiv CH$
2) $CH_3\ C \equiv C\ CH_3$
3) $CH_3 - CH_2 - C \equiv CH$
4) $CH \equiv CH$

76. What happens when calcium carbide is treated with water?

1) Ethane is formed
2) Methane and ethane is formed
3) Ethyne is formed
4) Ethene and ethyne are formed

77. Ozonolysis products of 2-pentyne after decomposition of ozonide with water and subsequent oxidation are:

1) Ethanoic acid and propanoic acid
2) Ethanoic acid and propanone
3) Ethanoic acid
4) Formic acid and glyoxal

78. Which of the following will give 2,2- dibromopropane on reaction with HBr?

1) $CH_3\ CH = CH_2$ 2) $CH_3\ C \equiv CH$
3) $CH_3\ CH = CHBr$ 4) $CH \equiv CH$

79. In the reaction,

$B \xleftarrow{\text{lindlar catalyst}+H_2} RC \equiv CR \xrightarrow{Na/NH_3} A$ .A and B are geometrical isomers:

1) A is cis and B is trans 2) A is trans and B is cis
3) A and B are cis 4) A and B are trans

80. The reagent used to convert an alkyne to alkene is:

1) Zn/HCl
2) Sn/HCl
3) Zn-Hg/HCl
4) Lindlar's catalyst

81. What is the product formed when acetylene reacts with hypochlorous acid?

1) $CH_3\ COCl$
2) $Cl - CH_2 - CH_2\ CHO$
3) Cl_2CHCHO
4) $ClCH_2\ COOH$

82. Ethylidene dichloride is obtained by the reaction of excess of HCl with:

1) Ethylene 2) Acetylene
3) Propene 4) Methane

83. 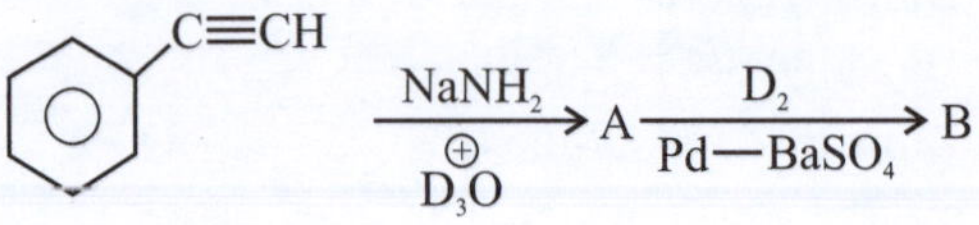

End product B is -

1) 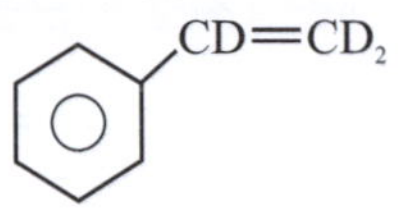 2) 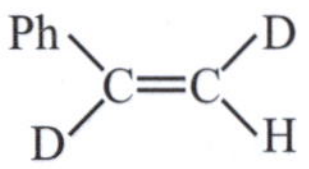

3) 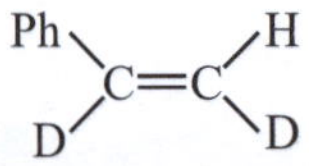 4) $Ph{-}C{\equiv}CD$

84. 1- butyne on hydration gives:

1) Butane-1, 2-diol 2) Butan - 1- ol
3) Butan -2-ol 4) Butan - 2-one

85. Which of the following organic compounds has same hybridisation as its combustion product - (CO_2)?

1) Ethane 2) Ethyne
3) Ethene 4) Ethanol

86. Which of the following reacts with $KMnO_4$ but does not react with tollen's reagent?

1) $C_2\ H_6$ 2) CH_4
3) $C_2\ H_4$ 4) C_2H_2

87. Which of the following gives propyne on hydrolysis?

1) $Al_4\ C_3$ 2) $Mg_2\ C_3$
3) $B_4\ C$ 4) $Ca_4\ C_3$

88. Which of the following compounds reacts with an aqueous solution of $[Ag(NH_3)_2]^+OH$?

1) Ethane 2) Ethene
3) 1- butyne 4) 2- butyne

89. How many structures are possible for $C_5\ H_8$ with one triple bond?

1) 4 2) 3
3) 2 4) 1

90. The most acidic hydrogen atoms are present in:

1) Ethane 2) Ethene
3) Ethyne 4) Benzene

91. Excess of CH_3COOH is reacted with CH≡CH in presence of Hg^{2+}, the product is –
1) $CH_3CH(OOCCH_3)_2$
2) $CH_2{=}CH(OOCCH_3)$
3) $(CH_3COO)CH_2{-}CH_2(OOCCH_3)$
4) None of these

92. $C_4H_6 \xrightarrow{NaNH_2} X \xrightarrow{CH_3CH_2Br} Y \xrightarrow{Pd/C,\ BaSO_4} Z \xrightarrow[Zn/H_2O]{O_3} W$
1) W is acetone 2) W is propanal
3) W is acetaldehyde 4) W is 2 methyl propanal

93. Alkynes exhibit:
1) Chain isomerism 2) Position isomerism
3) Functional isomerism 4) All of these

94. The IUPAC name of the compound having the formula $CH \equiv C - CH = CH_2$
1) Butene-2-yne 2) But-2-yne-3-ene
3) 3-butane-1-ene 4) But-1-ene-3-yne

95. Acetylene can be obtained by the electrolysis of the following compound:
1) Potassium fumerate 2) Potassium succinate
3) Potassium acetate 4) Potassium formate

96. The gas obtained when ethylene chloride reacts with alcoholic potash and sodamide is:
1) C_2H_4 2) C_2H_6
3) C_2H_2 4) C_2H_5Cl

97. Iodoform on heating with silver powder gives:
1) CH_4 2) C_2H_2
3) C_2H_4 4) C_6H_6

98. The reagent used for converting Acetylene tetrabromide into ethyne is:
1) Alcoholic KOH 2) Conc. H_2SO_4 at 170°C
3) Aq. Potash 4) Zinc dust

99. Pure acetylene has sweet smell, whereas impure gives garlic odour due to presence of:
1) NH_3 2) PH_3
3) SbH_3 4) HCl

100. 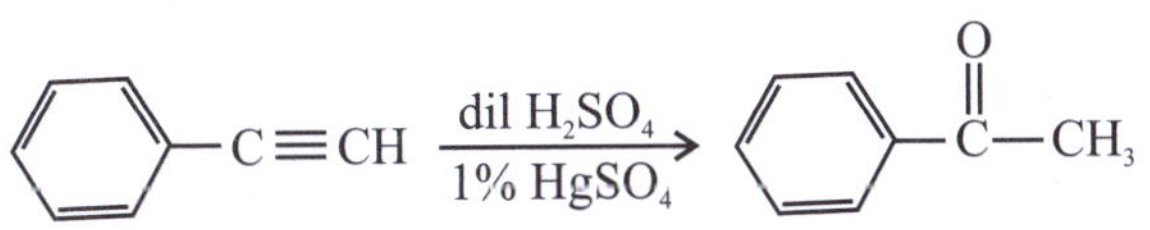

Intermediate of this conversion is

1) 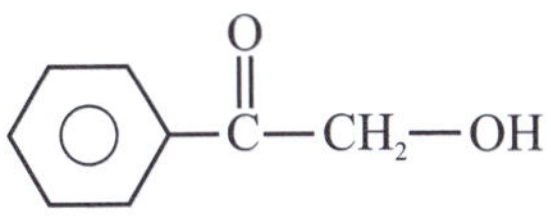

2) 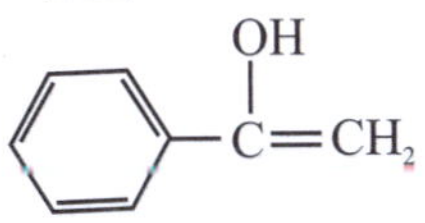

3) 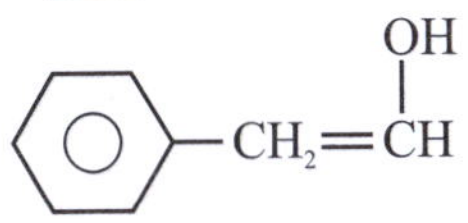

4) None of these

101. The reagent used for obtaining trans alkene from alkyl substituted acetylene with hydrogen is:
1) Na in liq.NH_3 2) $LiAlH_4$
3) Zn + HCl 4) H_2 in presence of Ni

102. Gem dihalides on treatment with alcoholic KOH give:
1) Alkyne 2) Alkene
3) Alkane 4) Cyclo alkanes

103. $CH \equiv CH \xrightarrow{HOCl} X \xrightarrow{HOCl} Y \xrightarrow[KOH]{aq} Z$

Z is:
1) CH_2-Cl | CH_2-Cl 2) $CH_2{=}O$ | $CHCl_2$
3) CHO | CHO 4) CH_2-OH | CH_2-OH

104. 2-Butyne when treated with lithium in presence of liquid ammonia gives:
1) Cis-2-butene 2) Trans-2-butene
3) n-Butane 4) I-Butyne

105. Which one of the following possess the minimum boiling point?
1) 1-Pentyne 2) 1-Butyne
3) n-Butane 4) Isobutane

106. 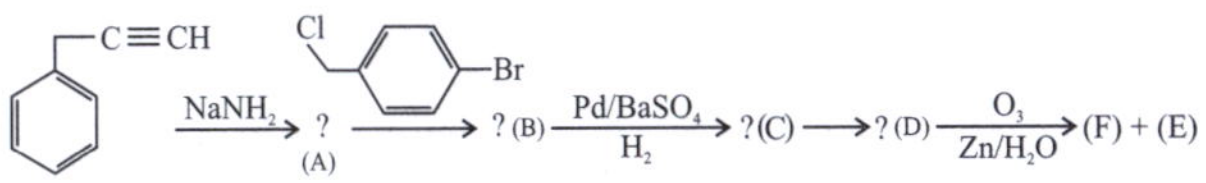

E & F are

1) 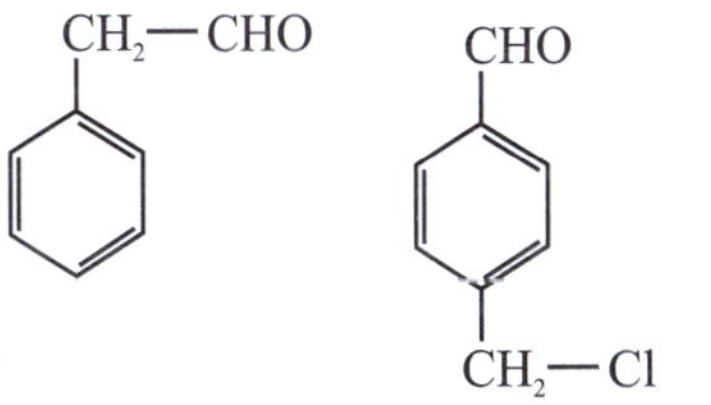

2) 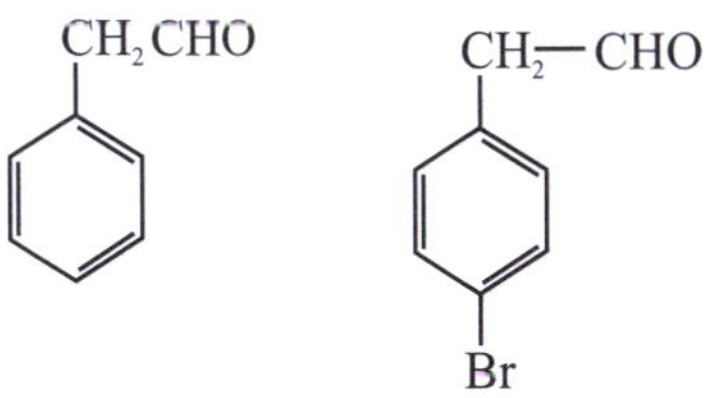

3) 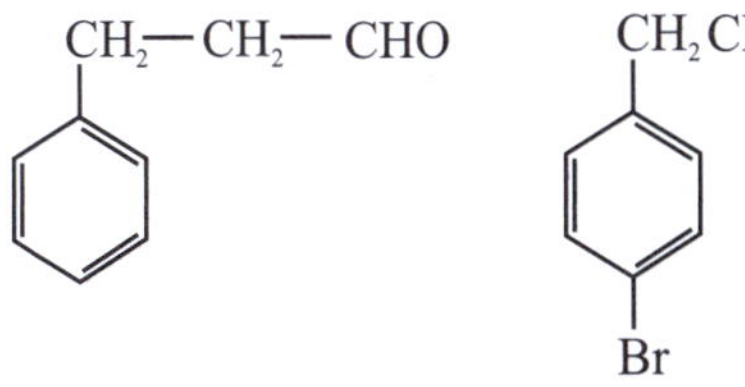

4) 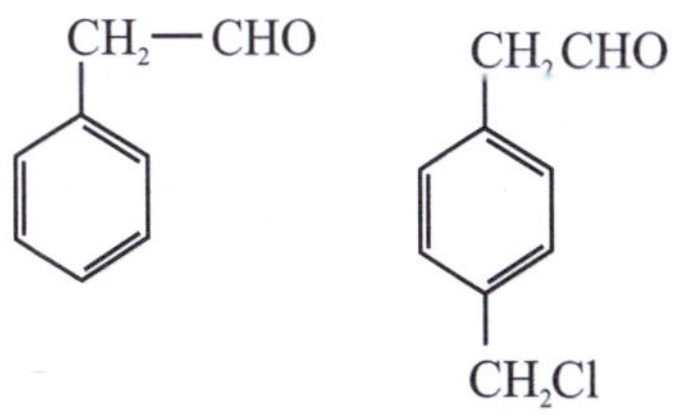

107. An alkyne (C_7H_{12}) on ozonolysis gives isobutyric and & propionic acid, find structure of hydrocarbon
1) 3 hexyne 2) 2 methyl 3 - hexyne
3) 2 methyl 2 - hexyne 4) 2 methyle hexyne

108. Which of these will not react with acetylene?
1) NaOH 2) Na
3) Ammonical $AgNO_3$ 4) HCl

109. $CH_3-C\equiv CH \xrightarrow[NaOH/H_2O_2]{B_2H_6} A \Rightarrow$ relation b/w A & B is
$CH_3-C\equiv CH \xrightarrow[HgSO_4]{dil\ H_2SO_4} B$
1) Position isomer
2) Identical
3) Functional isomer
4) No relation

110. What is the product when acetylene reacts with HCN:
1) CH_3COCl 2) $CH_2=CH-CN$
3) Cl_2CHCHO 4) $ClCH_2COOH$

111. 'x' on ozonolysis gives a dial while 'y' reacts with Baeyer's reagent to give a diol. Then 'x' and 'y' respectively are
1) C_2H_2 & C_6H_6 2) C_2H_4 & C_2H_2
3) C_2H_2 & C_2H_4 4) C_2H_4 & C_6H_6

Aromatic Hydrocarbons

112. Identify correct product of the following reaction:

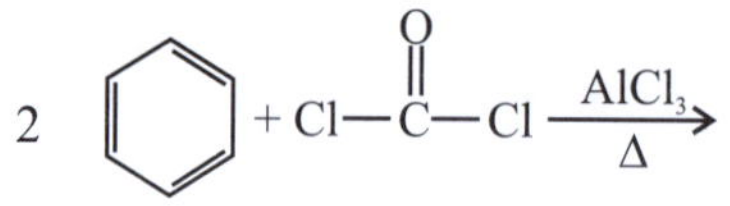

1) 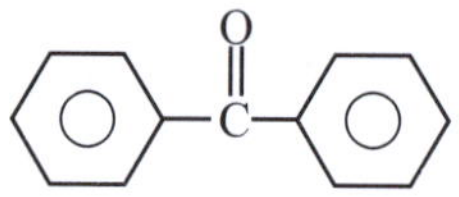 2) 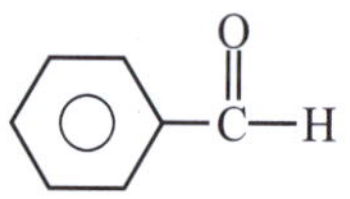

3) 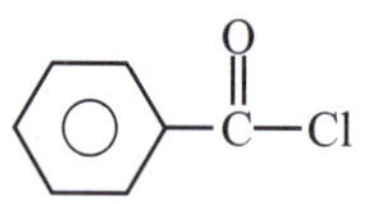 4) None of these

113. The treatment of benzene with iso - butene in the presence of sulphuric acid gives:
1) iso - butyl benzene 2) Tert - butyl benzene
3) n- butyl benzene 4) No reaction

114. Which one is the most reactive towards electrophilic reagent?

1) 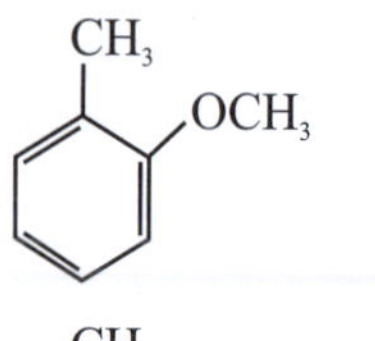 2) 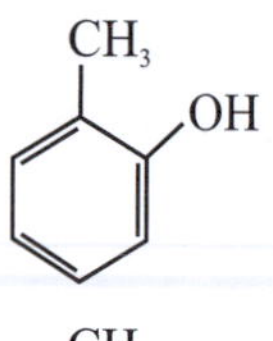

3) 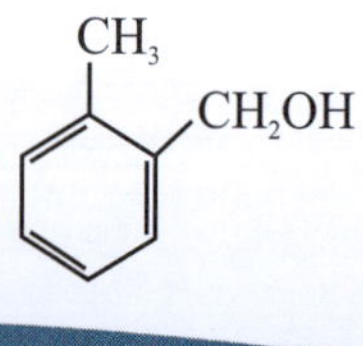 4) 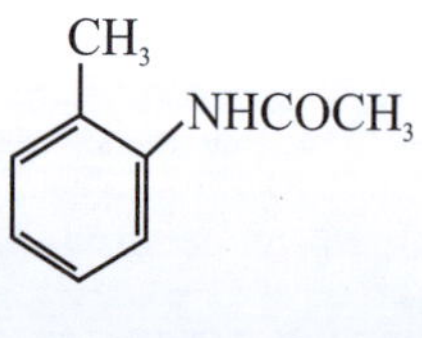

115. A compound that undergoes bromination easily:
1) Toluene 2) Benzoic acid
3) Phenol 4) Benzene

116. Which of the following will not be soluble in sodium hydrogen carbonate?
1) 2,4,6 - trinitrophenol 2) Benzoic acid
3) o - nitrophenol 4) Benzene sulphonic acid

117. Nitration and chlorination of benzene are:
1) Nucleophilic and electrophilic substitution respectively
2) Electrophilic and nucleophilic substitution respectively
3) Electrophilic substitution in both the reactions
4) Nucleophilic substitution in both the reactions

118. Similar to alkenes and alkynes benzene also undergoes ozonolysis. In the sequence of the given reaction identify X and Y:

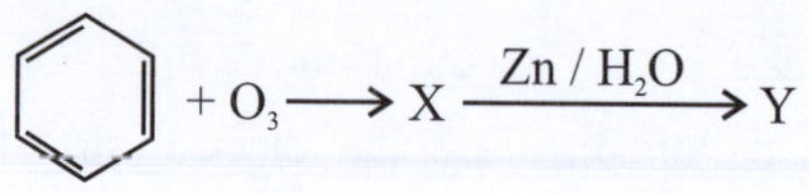

1) X = Triozonide , Y = Glyoxal
2) X = Diozonide , Y = succinic acid
3) X = monoozonide , Y = Benzoic acid
4) X = Triozonide , Y = Benzaldehyde

119. The following reaction is known as:
$C_6H_6 + CH_3Cl \xrightarrow[(Anhyd.)]{AlCl_3} C_6H_5CH_3 + HCl$
1) Wurtz - Fittig's reaction
2) Friedel - Craft's reaction
3) Rosenmund reaction
4) Sandmeyer reaction

120. Mark the incorrect statement from the following:
1) Benzene has a planar structure
2) Benzene is an unsaturated hydrocarbon and shows addition reactions like alkenes
3) In benzene, carbon uses two p- orbitals for hybridisation
4) Aromatic hydrocarbons contain high percentage of carbon hence burn with sooty flame

121. Benzene easily shows:
1) Ring fission reactions since it is unstable
2) Addition reactions since it is unsaturated
3) Electrophilic substitution reaction due to stable ring and high π e density
4) Nucleophilic substitution reactions due to stable ring and minimum electron density.

122. The most common reactions shown by benzene are electrophilic substitution reactions, which of the following undergoes electrophilic substitution reactions most readily?
1) Benzene
2) Benzoic acid
3) Nitrobenzene
4) Toluene

123. Identify the product obtained in the following reaction:

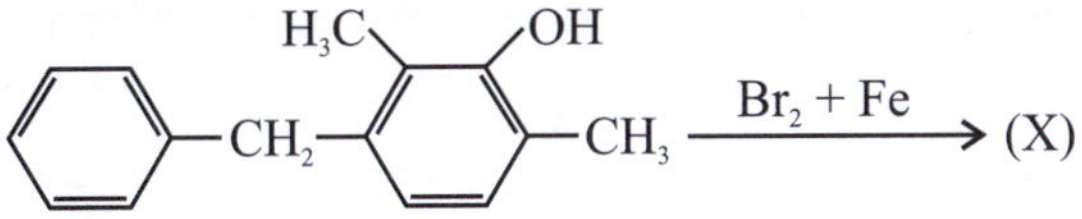

1) 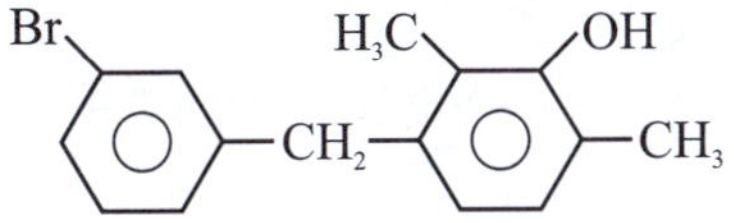

2) 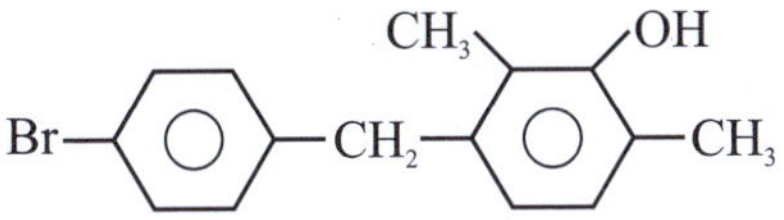

3) 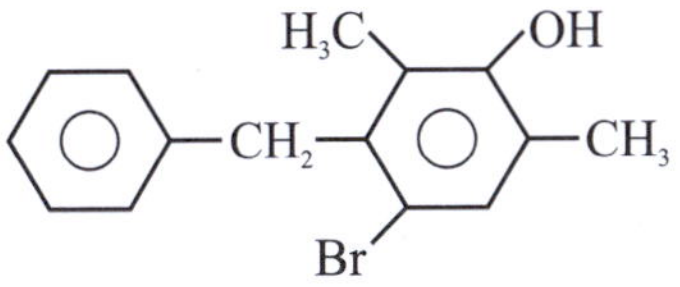

4) 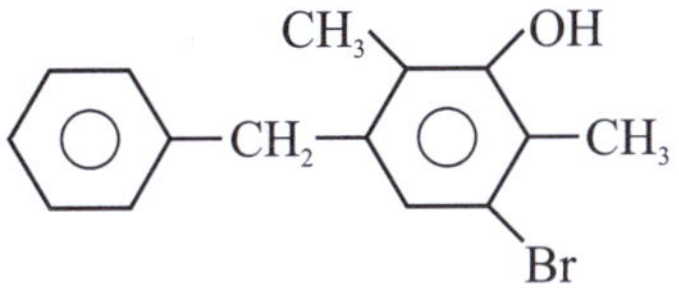

124. 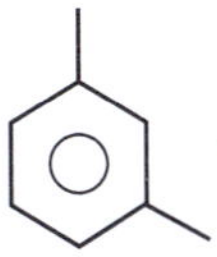 $+ CH_3—Cl \xrightarrow{AlCl_3} x$

1) 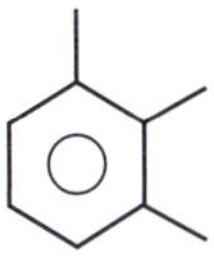
2) 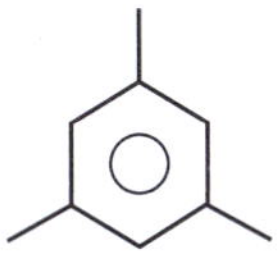
3) 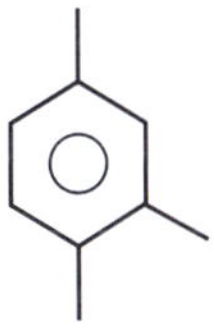
4) 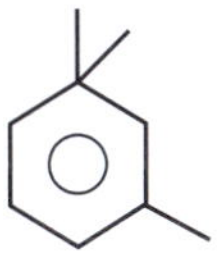

125. Formation of which of the following compound confirms the unsaturation character of benzene?

1) Cyclohexane
2) Gammexane
3) Triozonide
4) All of the above

126. Which of the following compounds reacts slower in electrophilic substitution?

1) $C_6H_5NO_2$
2) C_6H_5OH
3) $C_6H_5CH_3$
4) $C_6H_5NH_2$

127. Aromatic compounds give smoky flame because:

1) Hydrogen percentage is more
2) Carbon percentage is more
3) Delocalisation
4) Saturation

128. Bond length of C – C in benzene:

1) 1.34 A°
2) 1.39 A°
3) 1.54 A°
4) 1.20 A°

129. 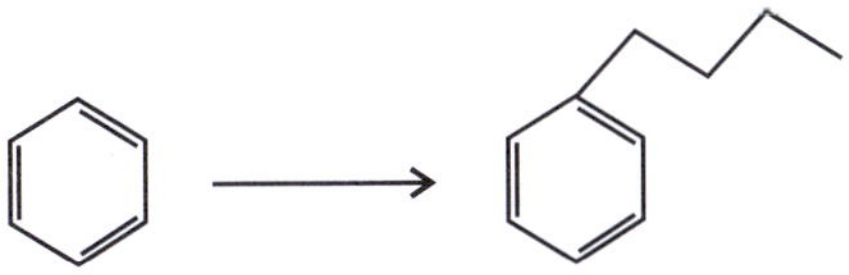

This conversion is possible by which method

1)  Br + $AlCl_3$ (friedel craft alkylation)
2) F. C Acylation $CH_3CH_2CH_2—\overset{O}{\overset{\|}{C}}—Cl + AlCl_3$ followed by clemmensen reduction
3) F. C alkylation  Cl + $AlCl_3$
4) All of the above

130. Chlorination of toluene in presence of light & heat followed by treatment with aq KOH & subsequently with dil HCl gives

1) 
2) 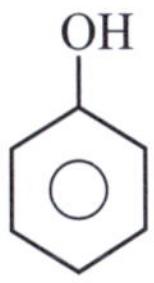
3) 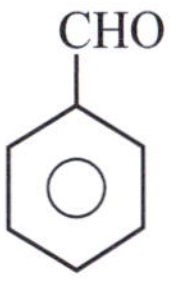
4) 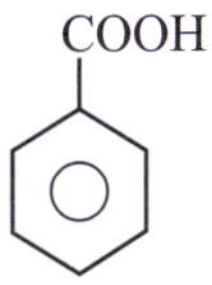

131. Chemical name of the insecticide gammaxene is

1) DDT
2) Benzene hexachloride
3) Chloral
4) Hexachlor ethane

132. Preparation of benzene from phenol is:

1) Reduction
2) Oxidation
3) Addition
4) Dehydrogenation

133. Benzene is resonance hybrid of mainly two Kekule structures. Hence,

1) Half the molecules correspond to one structure and half to the second structure
2) At low temperature benzene can be separated into two structures
3) Two structures make equal contribution to resonance hybrid
4) An individual benzene molecule changes back and forth between two structures

134. Cyclobutadiene is said to be:

1) Aromatic
2) Aliphatic
3) Antiaromatic
4) Heterocyclic

135. Rate determining step in nitration of benzene is

1) Formation of $NO_2^{\oplus}$
2) Formation of carbocation
3) Replacement of H atm
4) None of these

136. Which one among the following gives a dicarbonyl compound with O_3 followed by reduction with zinc and water?

1) C_2H_6
2) C_6H_6
3) C_2H_2
4) Both (2) & (3)

137. With which one of the following reagents benzene does not undergo substitution reaction?

1) Fuming sulphuric acid
2) Nitration mixture
3) Chlorine in presence of sunlight
4) Acyl halide in presence of $AlCl_3$

138. 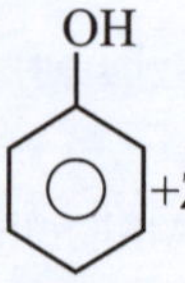 +Zn → X Here the product X is used as a:

1) Insecticide
2) For welding purpose
3) For dry cleaning
4) Artificial ripening of fruits

139. The conversion

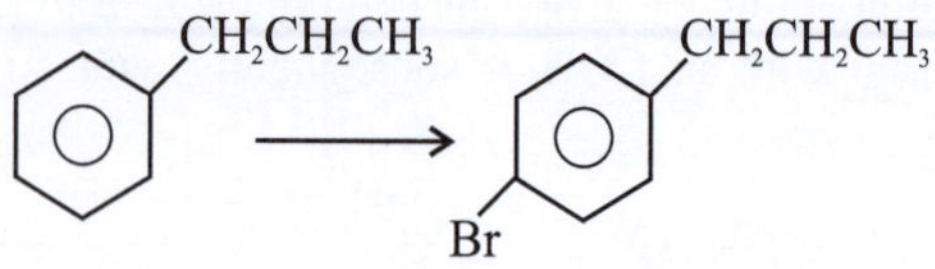

can be carried out using-

1) Br_2 / CCl_4
2) Br_2 / H_2O
3) Br_2 / Fe
4) Br_2 / benzoyl peroxide

140. The order of reactivity towards S_E reactions of – $\phi - CH_3$ (I), $\phi - CH_2 - CH_3$ (II), $\phi - CH(CH_3)_2$ (III) and $\phi - C(CH_3)_3$ (IV)

Where $\phi = C_6H_5$

1) I > II > III > IV
2) IV > III > II > I
3) II > I > III > IV
4) III > II > I > IV

High Order Time Intensive MCQs

1. 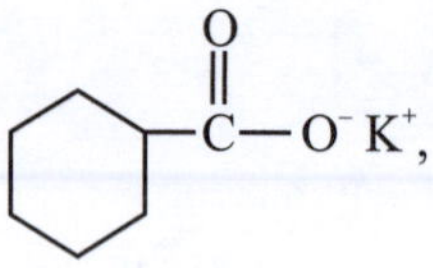, major product A:

1) 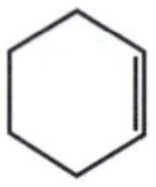
2) 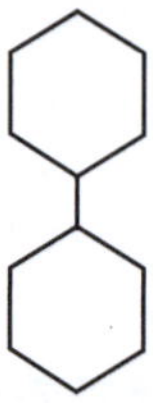
3) 
4) 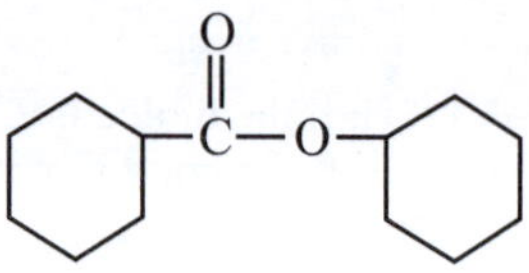

2.  $\xrightarrow[hv]{Br_2}$ (A) $\xrightarrow[\text{dry ether}]{Na}$ (B). Product B is:

1) 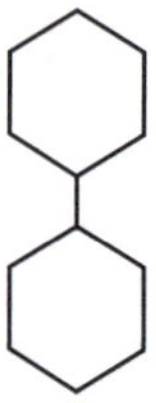
2) 
3) 
4) 

3. 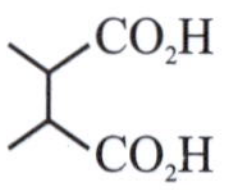 $\xrightarrow{2\ KOH}$ A $\xrightarrow{\text{electrolysis}}$ B + C.

Products B & C are:

1) Position isomers
2) Enantiomer
3) Geometric isomers
4) None of these

4. How many distinct monochlorinated products, including stereo-isomers, can result when the alkane below is heated in the presence of Cl_2?

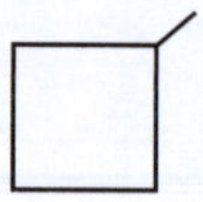

1) 2
2) 4
3) 5
4) 8

5. Which of the following is/are the major product, of the following reaction?

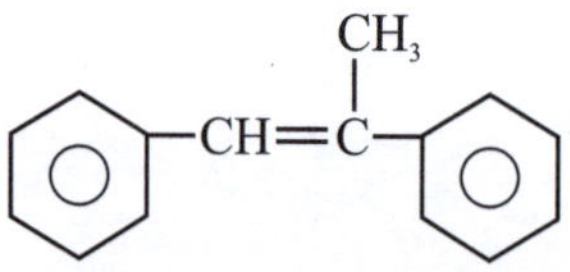 $\xrightarrow{HBr}$

1) 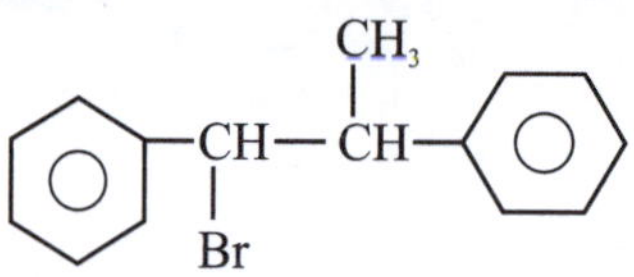
2) 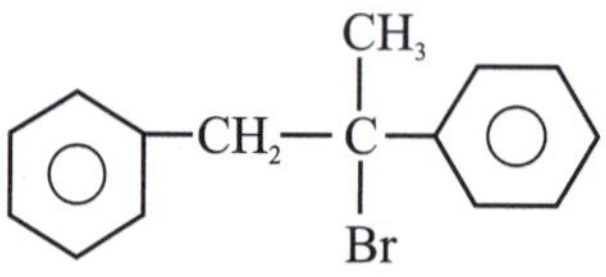
3) 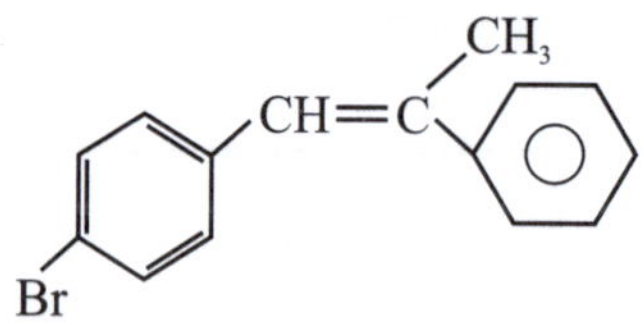
4) None of these

6. The final major product of the following reaction is:

$CH_3—CH_2—C\equiv C—H \xrightarrow[(2)\ CH_3CH_2Br]{(1)\ NaNH_2}$

1) $CH_3 - CH - C \equiv C - CH_2 - CH_3$
2) $CH_3—CH(CH_2CH_3)—C\equiv C—H$
3) $CH_3—CH_2—CH(NH_2)—CH_3$
4) $H_3C - CH_2 - C \equiv C - NH_2$

7. Which of the following reactions result in the formation of cumene 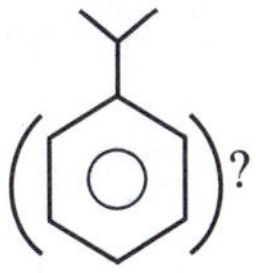?

1) 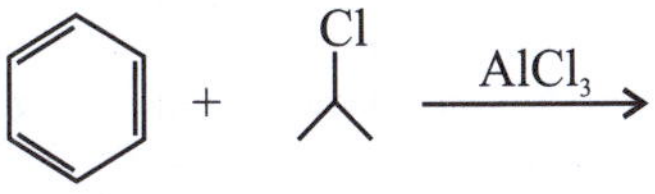

2) 

3) 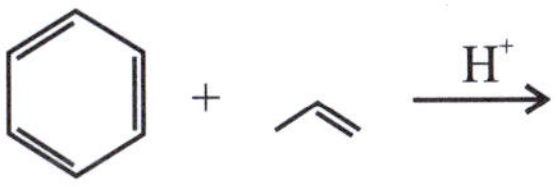

4) All of these

8. In the following sequence of reactions, the P is:

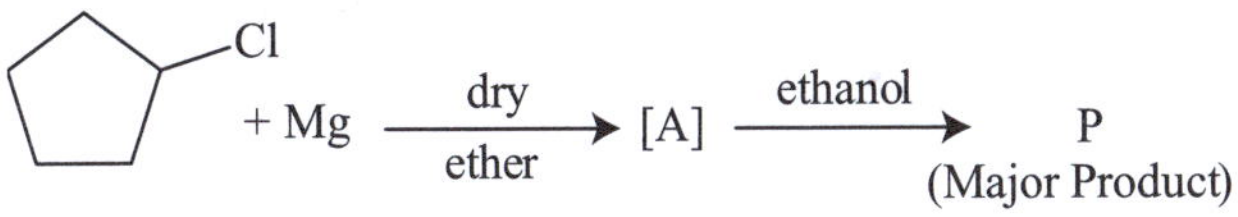

1) 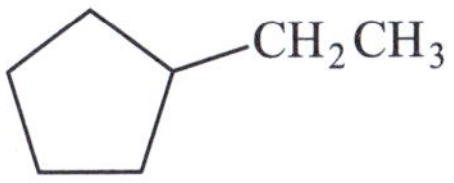 2) 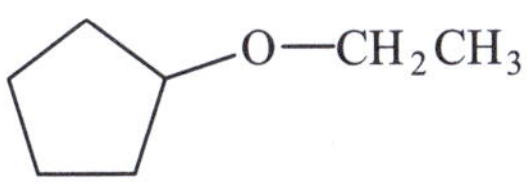

3)  4) 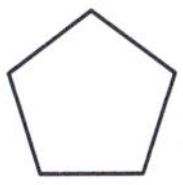

9. Identify A in the given chemical reaction.

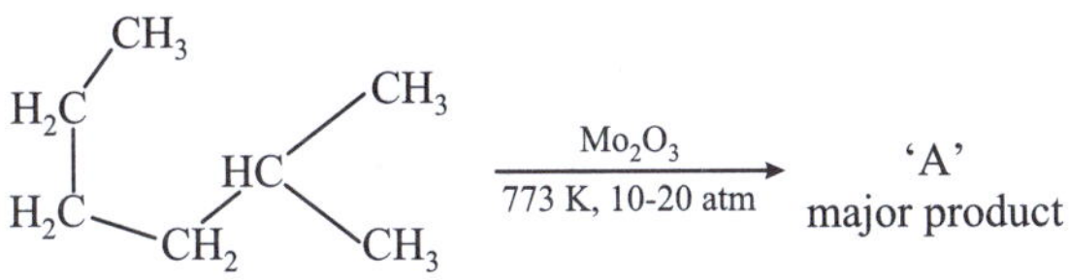

1) 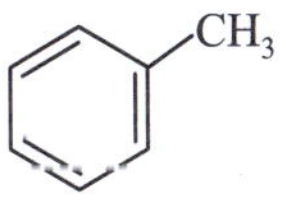 2) 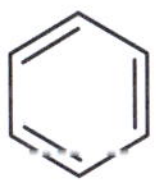

3)  4) 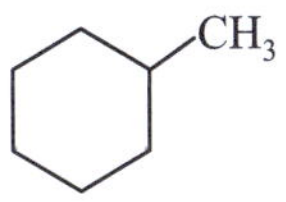

10. Which of the following species is an intermediate in the reaction shown below?

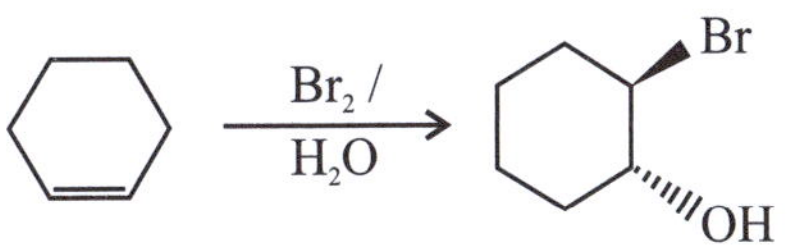

1) 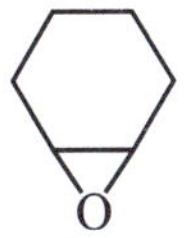 2) 

3) 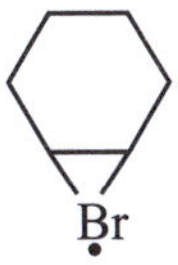 4) 

11. 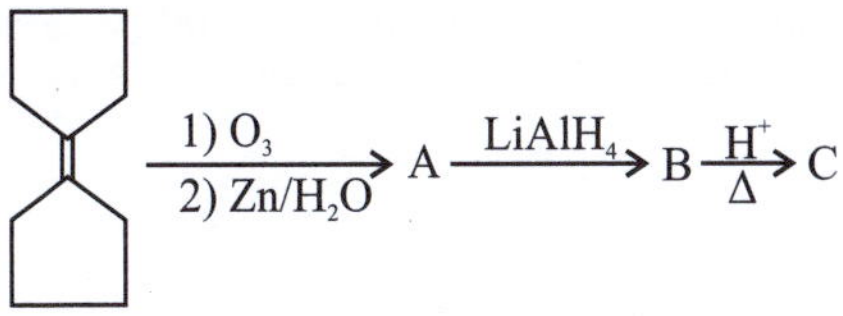

identify the product C-

1)  2) 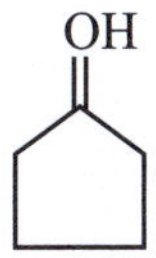

3)  4) 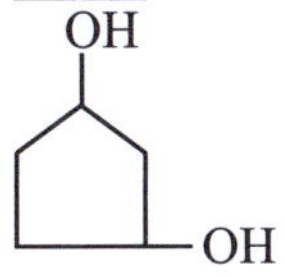

12. Which of the following molecule is not aromatic?

1)  2) 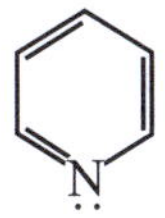

3) 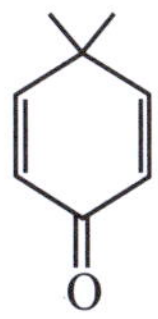 4) 

13. Ethylene is treated with Br_2, in the presence of C_2H_5OH. The products formed are:

1) $\underset{Br}{CH_2}-\underset{Br}{CH_2}$

2) $\underset{OC_2H_5}{CH_2}-\underset{OC_2H_5}{CH_2}$

3) $\underset{Br}{CH_2}-\underset{OC_2H_5}{CH_2}$

4) Both (1) & (3)

14. But-2-ene is treated with HCl. The product formed is:

1) (+) sec butyl chloride

2) (–) sec butyl chloride

3) (±) sec butyl chloride

4) None of these

15. Number of σ and π bonds present in ethylene molecule is respectively:

1) 3 and 1 2) 5 and 2

3) 4 and 1 4) 5 and 1

16. Which of the following reactivity orders is/are correct?

1) $CH_3-CH=CH_2 > CH_2=CH_2$ (towards electrophilic addition reaction)

2) $CH_2=CH_2 > CH_3-CH=CH_2$ (towards catalytic hydrogenation)

3) $CH_2=CH_2 > H-C\equiv C-H$ (towards electrophilic addition)

4) All of these are correct

17. Which of the following methods yield saturated hydrocarbon?

(A) $R-CH=CH_2 \xrightarrow[CH_3COOH]{BH_3}$

(B) $R-CH=CH_2 \xrightarrow[\Delta]{CH_2N_2}$

(C) 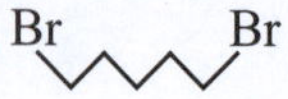 $\xrightarrow[\Delta]{Na / ether}$

(D) 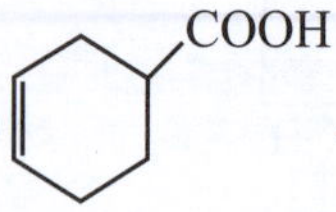 $\xrightarrow[\Delta]{NaOH + CaO}$

The correct code is -

1) A 2) A, B

3) A, B, C 4) A, B, C, D

18. Which of the following compounds is most reactive for Wurtz reaction?

1) $CH_3-CH(Br)-CH_3$ 2) 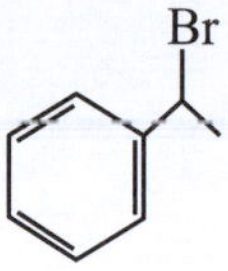

3) 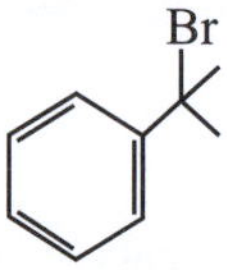 4) $CH_3-C(CH_3)_2-Br$

19. The major product of the following reaction is:

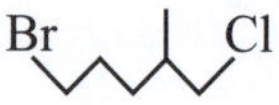 $\xrightarrow{Na / Ether} \xrightarrow[\Delta]{Br_2} \xrightarrow{alc. KOH}$

1)  2) 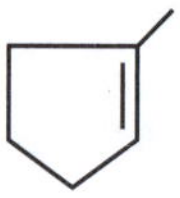

3) 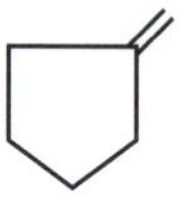 4) 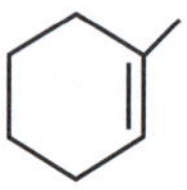

20. Which compound is a possible product from addition of Br_2 to 1-butene ?

1) 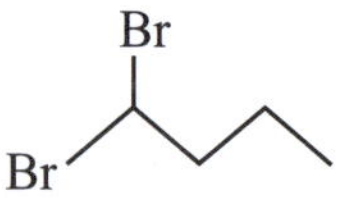 2) 

3) 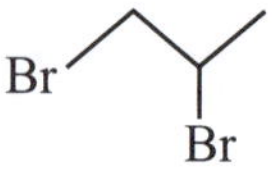 4) 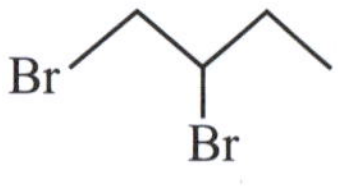

21. In the reaction 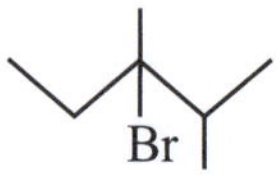 $\xrightarrow[C_2H_5OH]{C_2H_5O^{\ominus}}$

The major product obtained is:

1) 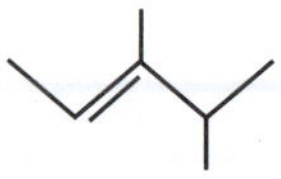 2) 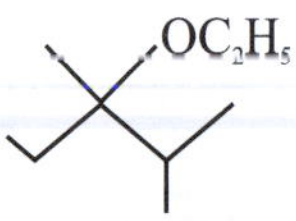

3) 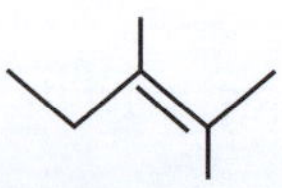 4) 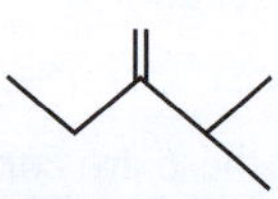

22. Which of the following is true?

1) Acetylene is more reactive than ethylene to an electrophilic attack
2) Acetylene is less reactive than ethylene towards electrophilic attack
3) Acetylene may show more reactivity or less reactivity towards electrophilic reagent.
4) Acetylene and ethylene show identical reactivities towards an electrophilic attack

23. A compound 'X' on ozonolysis followed by reduction gives an aldehyde, C_2H_4O and 2-Butanone. Compound 'X' is :

1) 3-methyl pent-2-ene 2) 3-methyl pent-3-ene

3) 3-methyl hex-3-ene 4) 3-ethyl pent-3-ene

24. A $\xleftarrow{Br_2}$  $\xrightarrow{KMnO_4}$ B

Compound A and B respectively are -

1) O-bromostyrene, benzoic acid
2) p-bromostyrene, benzaldehyde
3) m-bromostyrene, benzaldehyde
4) styrenedibromide, benzoic acid

25. 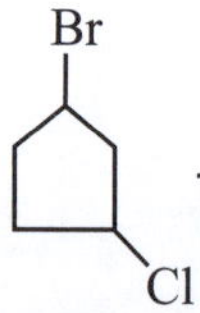 $\xrightarrow{Mg / ether} \xrightarrow{D_2O} \xrightarrow{Na / ether}$ (X)

The compound (X) is :

1) 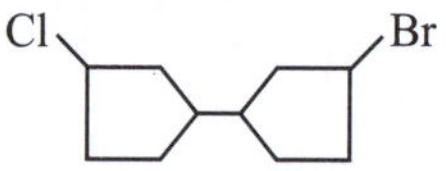 2) 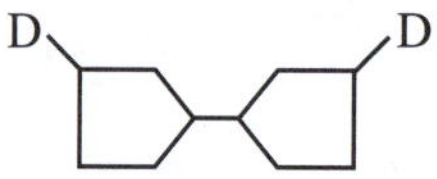

3) 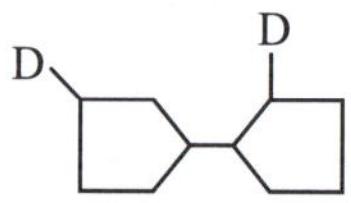 4) 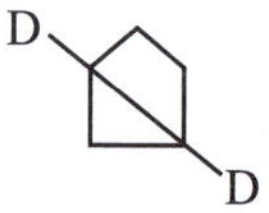

26. The major products 'A' and 'B', respectively, are

$$'A' \xleftarrow[H_2SO_4]{Cold} H_3C-\overset{CH_3}{\overset{|}{C}}=CH_2 \xrightarrow[80°C]{H_2SO_4} 'B'$$

1) $H_3C-\underset{OSO_3H}{\underset{|}{\overset{CH_3}{\overset{|}{C}}}}-CH_3$ & $CH_3-\overset{CH_3}{\overset{|}{C}}=CH-\underset{CH_3}{\underset{|}{\overset{CH_3}{\overset{|}{C}}}}-CH_3$

2) $CH_3-\overset{CH_3}{\overset{|}{C}}=CH-\underset{CH_3}{\underset{|}{\overset{CH_3}{\overset{|}{C}}}}-CH_3$ & $CH_3-\underset{OSO_3H}{\underset{|}{\overset{CH_3}{\overset{|}{C}}}}-CH_3$

3) $CH_3-\overset{CH_3}{\overset{|}{CH}}-CH_2CH_2\overset{CH_3}{\overset{|}{CH}}CH_3$ & $H_3C-\underset{OSO_3H}{\underset{|}{\overset{CH_3}{\overset{|}{C}}}}-CH_3$

4) $H_3C-\underset{OSO_3H}{\underset{|}{\overset{CH_3}{\overset{|}{C}}}}-CH_3$ & $CH_3-\overset{CH_3}{\overset{|}{CH}}-CH_2CH_2-\overset{CH_3}{\overset{|}{HC}}-CH_3$

27. Which of the following triene gives following products on ozonolysis?

Products = 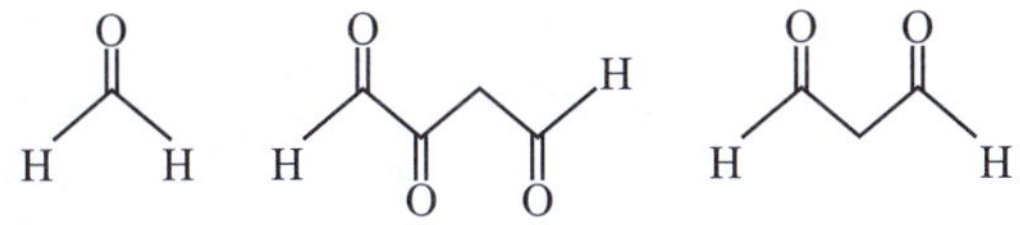

1) 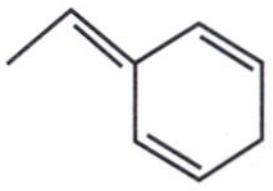 2) 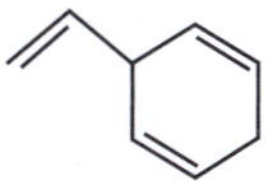

3) 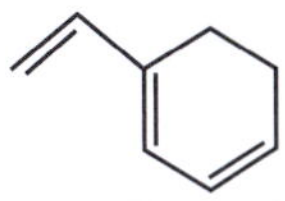 4) 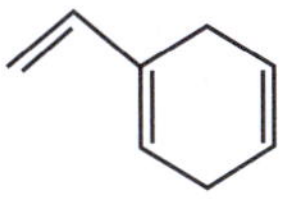

28.  $+ CH_3-\underset{CH_3}{\underset{|}{C}}=CH_2 \xrightarrow{H_3PO_4} X;$

The compound 'X' is :

1) 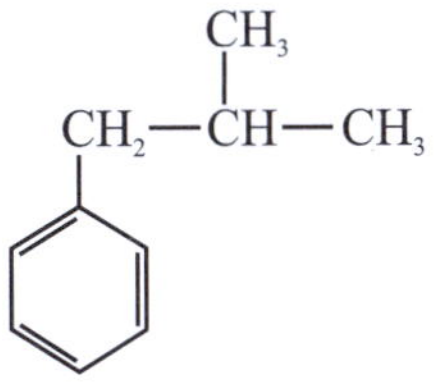

2) 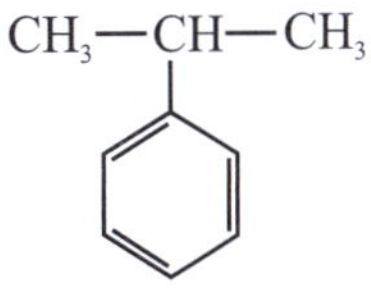

3) 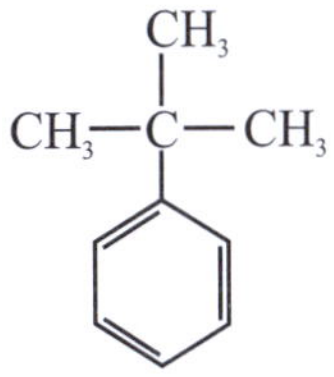

4) None of these

29. Which of the following procedure would be best for achieving the following reaction ?

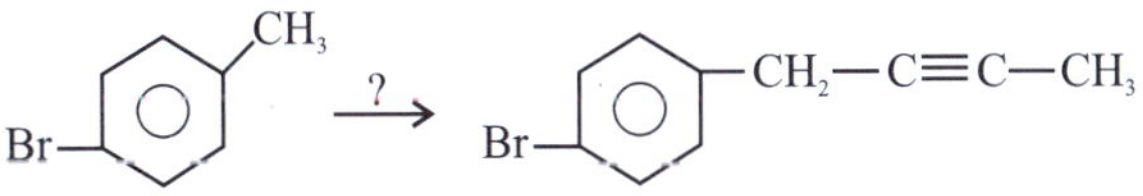

1) (i) KOH and heat (ii) $CH_3-C\equiv C-Br$

2) (i) $KMnO_4$ and heat (ii) $CH_3-C\equiv C^-Na^+$ (iii) Excess of H_2O

3) (i) NBS in CCl_4 and hcat (ii) CH_3 $C\equiv C^-Na^+$

4) (i) Mg in ether (ii) $CH_3-C\equiv C^-Na^+$ (iii) excess of H_3PO_4

30. In the given reaction

i. $CH_3-C\equiv C-CH_2CH_3 + H_2 \xrightarrow[BaSO_4]{Pd/C} A$

ii. $CH_3-C\equiv C-CH_2CH_3 + H_2 \xrightarrow{Na/liq.NH_3} B;$

A & B respectivcly arc

1) Cis & trans alkene
2) Trans & Cis alkene
3) Alkane & alkene
4) Alkane & alkane

31. cis-2-Butene on treatment with bromine gives:

1) (+)-2, 3-dibromobutane
2) (–)-2, 3-dibromobutane
3) (±)-2, 3-dibromobutane
4) meso-2, 3-dibromobutane

32. 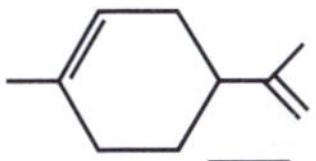 $\xrightarrow[Cl_2 + H_2O]{\text{Chlorine water}}$ Major product is:

1) 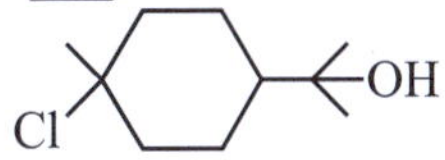 2) 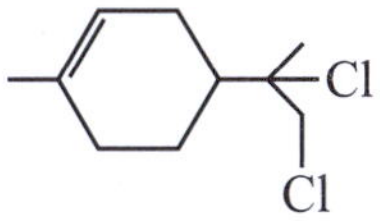

3) 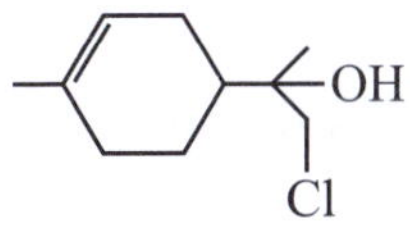 4) 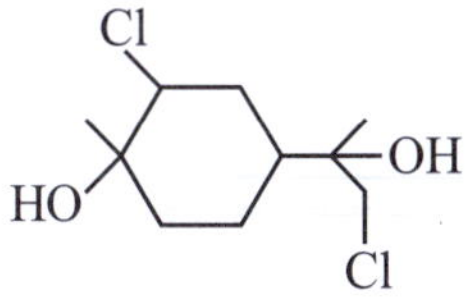

33. Which of the following alkenes can be used to prepare 1-Bromo-1-methylcyclopentane by the addition of HBr?

1) 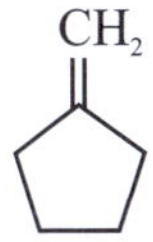 2) 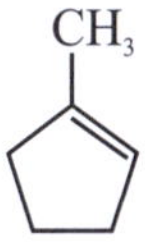

3) 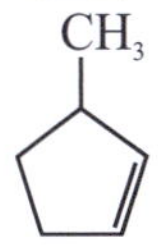 4) All of these

34. The gases liberated at anode in the electrolysis of sodium acetate are

1) CO_2 & H_2 2) C_2H_6 & CO_2
3) H_2 & C_2H_6 4) H_2 & O_2

35. The following substance reacts with water to give ethane

1) CH_4 2) C_2H_5MgBr
3) C_2H_4OH 4) $C_2H_5OC_2H_5$

36. 2 – hexene $\xrightarrow[\text{(ii) } H_2O]{\text{(i) } O_3}$ Products

The two products formed in above reaction are

1) Butanoic acid and acetic acid
2) Butanal and acctic acid
3) Butanal and acetaldehyde
4) Butanoic acid and acetaldehyde

37. The hydrocarbon that cannot be prepared effectively by Wurtz reactions:

1) 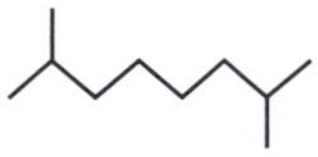

2) 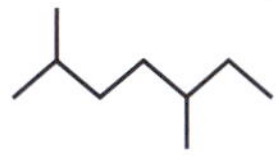

3) 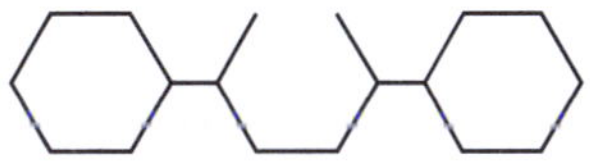

4) 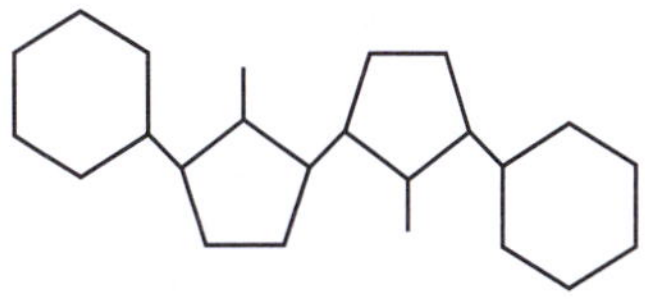

38. Amount of oxygen consumed for complete combustion of 29 g butane?

1) 104 g 2) 52 g
3) 29 g 4) 208 g

39. In the following reaction, A and B respectively are,

$$A \xrightarrow{HBr} C_2H_5Br \xrightarrow{B} A$$

1) C_2H_4 and alcoholic KOH/Δ
2) C_2H_5Cl and aqueous KOH/Δ
3) C_2H_5OH and aq KOH/Δ
4) C_2H_2 and Br_2

40. Planar molecule among the following is

1) $CH_2 = CH_2$ 2) $CH_3 - CH_3$
3) $CH \equiv C - CH_3$ 4) Cyclohexane

41. The relative reactivity of the following towards Br_2 in chloroform is:

(i) $CH_2 = CH - COCH_3$
(ii) $CH_2 = CH - CH_3$
(iii) $CH_2 = CH - O - CH_3$
(iv) $CH_2 = CH_2$

1) i > ii > iii > iv 2) iii > ii > iv > i
3) i > iii > ii > iv 4) iii > i > ii > iv

42. The compound that would produce Heptane-2, 6-dione on ozonolysis is:

1) 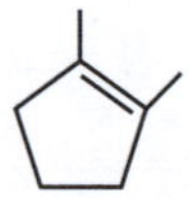 2) 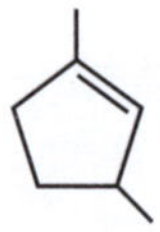

3) 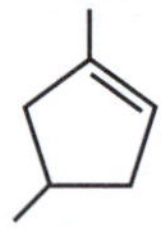 4) 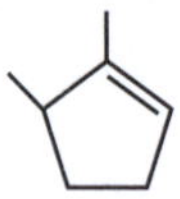

43. An organic compound C_6H_{10} on treatment with dil H_2SO_4 & $HgSO_4$ give X which on reduction with $LiAlH_4/H_2O$ give Y, Y on reaction with conc H_2SO_4 give Z. On heating Z, 2, 3 dimethyl but - 2 ene is formed. Find the structure of A.

1) $CH_3-CH(CH_3)-C\equiv C-CH_3$
2) $CH_3-CH_2-C\equiv C-CH_2-CH_3$
3) $CH_3-C(CH_3)_2-C\equiv CH$
4) None of these

44. 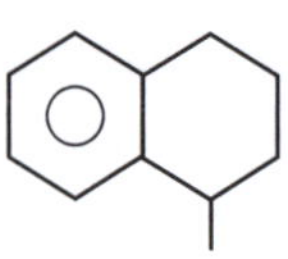 $\xrightarrow[\text{1 eq.}]{\text{NBS}}$ X (major product) ; X is:

1) 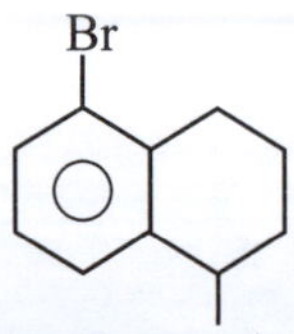 2) 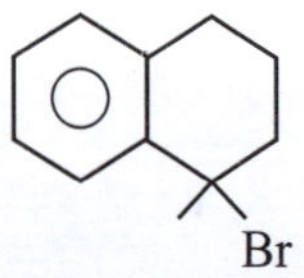

3) 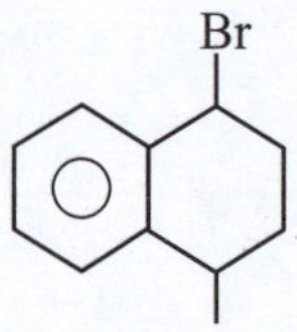 4) 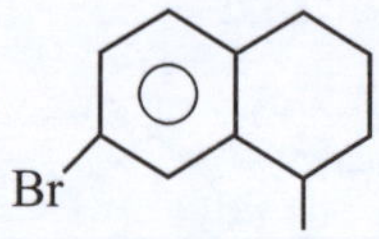

45. $CH_3 - C \equiv CH \xrightarrow[H_2SO_4]{HgSO_4} Y \xrightarrow{NaBH_4} Z$ is :

1) Acetone 2) Propan-2-ol
3) Propan-1-ol 4) Propanal

46. Select the correct match

	Compound		Ozonolysis products
A.	Acetylene	1.	HCHO & CH_3CHO
B.	Ethylene	2.	CH_3CHO
C.	Benzene	3.	One mole of $(CHO)_2$
D.	2-Butene	4.	3 moles of $(CHO)_2$
		5.	CH_2O

1) A-3 B-2 C-5 D-4
2) A-4 B-2 C-1 D-3
3) A-3 B-5 C-4 D-2
4) A-5 B-1 C-4 D-4

47. In the given reactions identify A and B.

$H_2 + A \xrightarrow{Pd/C}$ 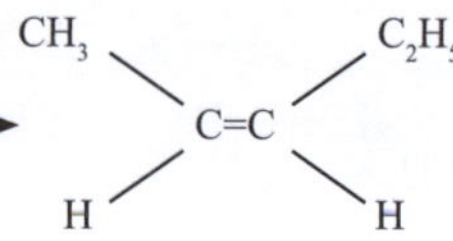

$$CH_3 - C \equiv C - CH_3 + H_2 \xrightarrow{Na/Liquid NH_3} \text{"B"}$$

1) A : 2 – Pentyne B : trans – 2 – butene
2) A : n – Pentane B : trans – 2 – butene
3) A : 2 – Pentyne B : Cis – 2 – butene
4) A : n – Pentane B : Cis – 2 – butene

48. Cold and dil.Alk.$KMnO_4$ will oxidise acetylene to

1) Ethylene glycol 2) Ethyl alcohol
3) Oxalic acid 4) Acetic acid

49. Acetylene gives white precipitate with ammonical silver nitrate but ethylene cannot give because

1) Acetylene possess sp^2 carbon
2) Acetylene posses acidic hydrogen
3) Acetylene possess low electronegative carbon
4) Acetylene possess triple bond

50. The monosodium salt of acetylene on treating with methyl chloride forms

1) $CH \equiv C.COOH$ 2) $CH \equiv C - CH_3$
3) $CH_3C \equiv CCH_3$ 4) $H - C \equiv C - CH_2 - CH_3$

51. 1- pentyne and 2-pentyne can be distinguished by

1) Silver mirror test 2) Iodoform test
3) Addition of H_2 4) Baeyer's test

52. Consider the following reaction sequence–

$$CH_3-C\equiv C-CH_3 \xrightarrow[H_2]{Pd-BaSO_4} A \xrightarrow{Br_2} B$$

A & B, respectively, are–

1) 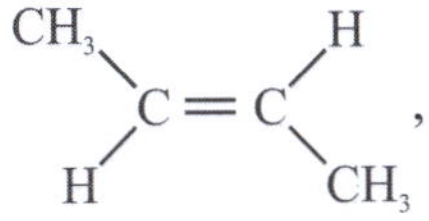 , (meso) CH_3 CH(Br) CH(Br)CH_3

2) 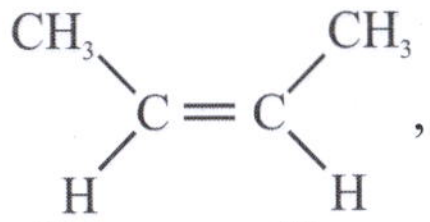 , (±) CH_3 CH(Br) CH(Br)CH_3

3) 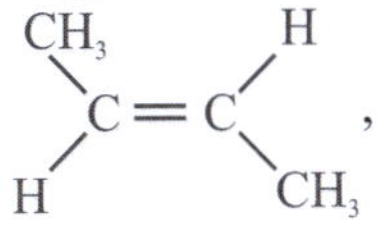 , (±) CH_3 CH(Br) CH(Br)CH_3

4) None of these

53. In nitrating mixture HNO_3 acts as a

1) Base
2) Acid
3) Reducing agent
4) Catalyst

54. Which of the following cannot form ozonide

1) Benzene
2) Ethene
3) Ethyne
4) Ethane

55. The order of activities of the various Ortho and Para director is

1) $-O^- > -OH > -OCH_3 > -OCOCH_3$
2) $-OH > -O^- > -OCOCH_3 > -OCH_3$
3) $-OH > -O^- > -COCH_3 > -OCOCH_3$
4) $-O^- > -OCH_3 > -OCOCH_3 > -OH$

56. Which of the following is not meta directing group ?

1) $-SO_3H$
2) $-NO_2$
3) $-CN$
4) $-NH_2$

57. In the nitration mixture concentrated sulphuric acid is used

1) As a sulphonating agent
2) As dehydrating agent
3) For the formation of nucleophile
4) As a solvent

58. Which one of the following compounds is not aromatic?

1) 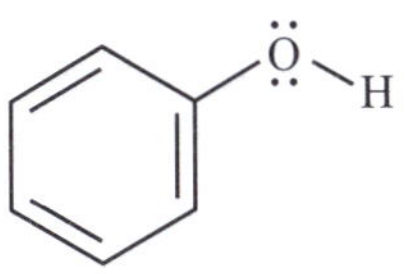
2) 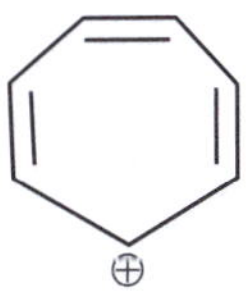
3) 
4) 

59. Which of following carboxylic acid undergo fastest decarboxylation when reaction with sodalime?

1) 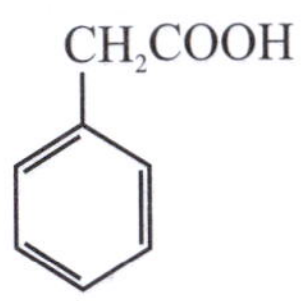
2) 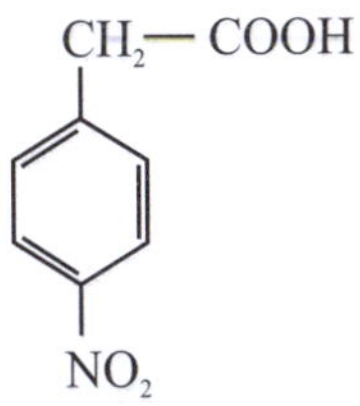
3) 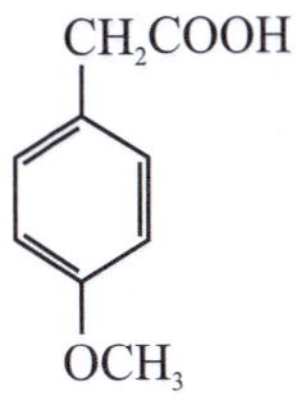
4) 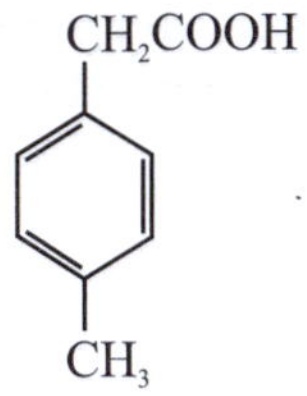

60. 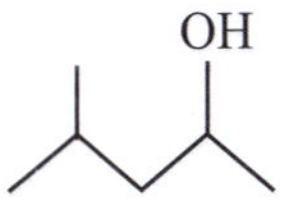 $\xrightarrow[\Delta]{Conc.H_2SO_4}$ Major

1) 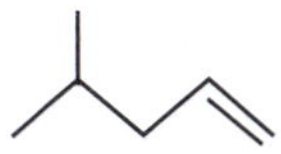
2) 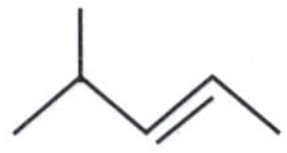
3) 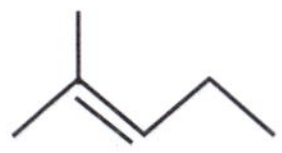
4) 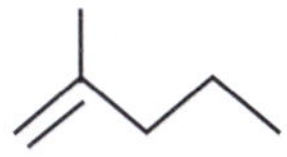

61. $R-Br + Mg \xrightarrow{dry\ ether} A \xrightarrow{D_2O} CH_3-\underset{}{\overset{D}{\overset{|}{C}}}H-CH_3$

$R-Br \xrightarrow[dry\ ether]{Na} B$; B is

1) $CH_3-CH_2-CH_2-CH_2-CH_2-CH_3$
2) $CH_3-CH_2-CH-CH_2-\overset{CH_3}{\overset{|}{C}}H-CH_3$
3) $CH_3-\overset{CH_3}{\overset{|}{C}}H-\overset{CH_3}{\overset{|}{C}}H-CH_3$
4) $CH_3-\underset{CH_3}{\underset{|}{C}}H-CH-\underset{CH_3}{\underset{|}{C}}H-CH_3$

62. $\underset{(A)}{CH_3-C\equiv CH} + C_6H_5MgBr \longrightarrow X + W \xrightarrow{D_2O} (Y)$

Which of following statements are correct regarding A & Y?

1) A & Y both have same molecular mas
2) A & Y both show acidic behaviuor
3) A & Y are same compound
4) Y have lesser molecular mass than A

63. $CH_3-C\equiv C-CH_3 + \xrightarrow[(l)]{Na/NH_3} x \xrightarrow{Br_2/CCl_4} y$

product y is

1) Racemic Mixture
2) Optically active
3) Meso compound
4) Form of product

64. $CH_3-CH_2-CH_2-CH_2-OH \xrightarrow[H_2SO_4/\Delta]{Conc.} X$

$CH_3-CH_2-CH_2-CH_2-Cl \xrightarrow[\Delta]{alc.\ KOH} Y$

Relation between X & Y are-

1) Same product
2) Position isomers
3) Both can show Geometrical isomer
4) Chain isomer

65. Find total possible isomers formed when n - octane undergo catalytic reforming by treatment of Al_2O_3 - Cr_2O_3 at 500°C - 600°C
1) 3 2) 4
3) 5 4) 6

66. Match the following reaction with their mechanism involved

(P)	Cracking	a.	Free radical substitution
(Q)	Catalytic reforming	b.	Free radical elimination
(R)	Halogenation of alkane	c.	Free radical addition
(S)	Wurtz Reaction	d.	Dehydrogenation

1) P – b, Q – a, R – d, S – c
2) P – b, Q – d, R – a, S – c
3) P – a, Q – b, R – c, S – b
4) P – d, Q – b, R – a, S – c

67. How much volume of air will be needed for complete combustion of 5 lit of propane?
1) 25 lit 2) 125 lit
3) 100 lit 4) 135 lit

68. An organic compound 2, 2 dibromo propane on treatment with Zn dust give X which on treatment with HI give y; y is:
1) $H_3C-CH(CH_3)-C(I)(CH_3)-CH_3$
2) $H_3C-CH_2CH(I)-CH_2-CH_2-CH_3$
3) $H_3C-CH(I)-CH_3$
4) $H_3C-HC(CH_3)-CH(I)-H_2C-CH_3$

69. Which of the following compounds will exhibit geometrical isomerism?
1) 1-Phenyl-but-2-ene 2) 3-Phenyl-but-1-ene
3) 2-Phenyl-but-1-ene 4) 1, 1-Diphenyl-prop-1-ene

70. What is/are end product of following?

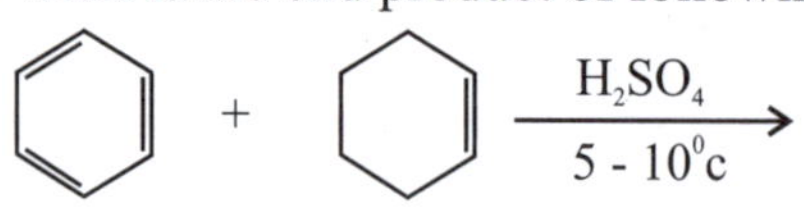

1) 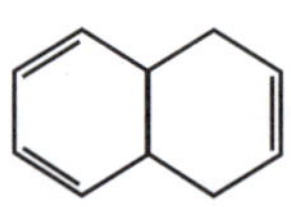 2) 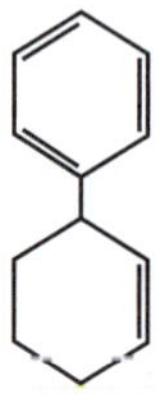

3) 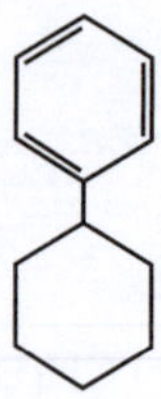 4) 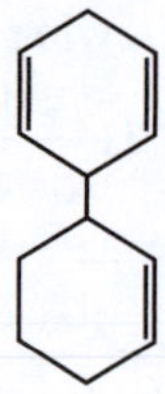

71. Which of the following compound has the lowest dipole moment?
1) 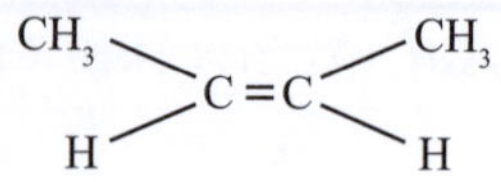

2) $CH_3C \equiv CCH_3$
3) $CH_3CH_2C \equiv CH$
4) $CH_2 = CH - C \equiv CH$

72. 2-Methylbutane on reacting with bromine in the presence of sunlight gives mainly
1) 1–bromo–2-methylbutane
2) 2–bromo–2-methylbutane
3) 2–bromo–3-methylbutane
4) 1–bromo–3-methylbutane

73. Phenyl magnesium bromide reacts with methanol to give
1) a mixture of benzene and Mg (OMe)Br
2) a mixture of toluene and Mg(OH)Br
3) a mixture of phenol and Mg(Me)Br
4) a mixture of anisole and Mg(OH)Br

74. The IUPAC name of 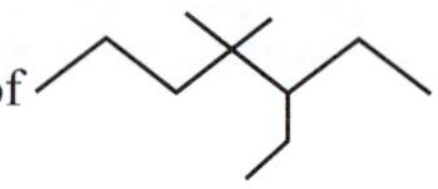 is
1) 4,4-dimethyl-5,5-diethylpentane
2) 5,5-diethyl-4,4-diethylpentane
3) 3-ethyl-4,4-dimethylheptane
4) 1,1-diethyl-2,2-dimethylpentane

75. Of the isomeric hexanes, the isomers that give the minimum and maximum number of monochloro derivatives are respectively
1) 3-methylpentane and 2, 3-dimethylbutane
2) 2,3-dimethylbutane and n-hexane
3) 2, 2-dimethylbutane and 2-methylpentane
4) 2-methylpentane and 2, 2-dimethylbutane

76. An organic compound with molecular formula C_6H_{12} upon ozonolysis gave only acetone as the product. The compound is
1) 2,3-dimethyl-1butene
2) 3-hexene
3) 2-hexene
4) 2,3-dimethyl-2-butene

Match the Columns MCQs

1. Match the reagent from Column-I which on reaction with $CH_3 — CH = CH_2$ gives some product given in Column-II as per the codes given below :

	Column-I		Column-II
(a)	$O_3/Zn + H_2O$	(i)	Acetic acid and CO_2
(b)	$KMnO_4/H^+$	(ii)	Propan-1-ol
(c)	$KMnO_4/OH^-$	(iii)	Propan-2-ol
(d)	H_2O/H^+	(iv)	Acetaldehyde and formaldehyde
(e)	B_2H_6/NaOH and H_2O_2	(v)	Propane-1,2-diol

1) (a)-(i) (b)-(iv) (c)-(v) (d)-(ii) (e)-(iii)
2) (a)-(iv) (b)-(i) (c)-(v) (d)-(iii) (e)-(ii)
3) (a)-(iii) (b)-(ii) (c)-(iv) (d)-(v) (e)-(i)
4) (a)-(v) (b)-(iv) (c)-(i) (d)-(iii) (e)-(ii)

2. Match the hydrocarbons in Column-I with the boiling points given in Column-II.

	Column-I		Column-II
(a)	n–Pentane	(i)	282.5 K
(b)	iso-Pentane	(ii)	309 K
(c)	neo-Pentane	(iii)	301 K

1) (a)-(ii) (b)-(iii) (c)-(i)
2) (a)-(i) (b)-(ii) (c)-(iii)
3) (a)-(iii) (b)-(i) (c)-(ii)
4) (a)-(iii) (b)-(ii) (c)-(i)

3. Match the following reactants in Column-I with the corresponding reaction products in Column-II.

	Column-I		Column-II
(a)	Benzene + $Cl_2 \xrightarrow{AlCl_3}$	(i)	Benzoic acid
(b)	Benzene + $CH_3Cl \xrightarrow{AlCl_3}$	(ii)	Methyl phenyl ketone
(c)	Benzene + $CH_3COCl \xrightarrow{AlCl_3}$	(iii)	Toluene
(d)	Toluene $\xrightarrow{KMnO_4/NaOH}$	(iv)	Chlorobenzene
		(v)	Benzene hexachloride

1) (a)-(ii) (b)-(i) (c)-(iii) (d)-(iv)
2) (a)-(i) (b)-(ii) (c)-(iv) (d)-(iii)
3) (a)-(iii) (b)-(iv) (c)-(i) (d)-(ii)
4) (a)-(iv) (b)-(iii) (c)-(ii) (d)-(i)

4. Match the reactions given in Column-I with the reaction types in Column-II.

	Column-I		Column-II
(a)	$CH_2 = CH_2 + H_2O \xrightarrow{H^+} CH_3CH_2OH$	(i)	Hydrogenation
(b)	$CH_2 = CH_2 + H_2 \xrightarrow{Pd} CH_3 — CH_3$	(ii)	Halogenation
(c)	$CH_2 = CH_2 + Cl_2 \longrightarrow Cl — CH_2 — CH_2 — Cl$	(iii)	Polymerisation
(d)	$3\ CH \equiv CH \xrightarrow[\text{Heat}]{\text{Cu tube}} C_6H_6$	(iv)	Hydration
		(v)	Condensation

1) (a)-(iv) (b)-(i) (c)-(ii) (d)-(iii)
2) (a)-(i) (b)-(ii) (c)-(iv) (d)-(iii)
3) (a)-(iii) (b)-(iv) (c)-(i) (d)-(ii)
4) (a)-(iv) (b)-(iii) (c)-(ii) (d)-(i)

5. Match the following

	Column-I		Column-II
(a)	Benzene	(i)	Wurtz reaction
(b)	Ethene	(ii)	Evolves H_2 when heated with Na metal
(c)	Ethyne	(iii)	Dehydration of ethanol
(d)	Butane	(iv)	Electrophilic substitution

1) (a)-(iii) (b)-(iv) (c)-(i) (d)-(ii)
2) (a)-(iv) (b)-(iii) (c)-(ii) (d)-(i)
3) (a)-(i) (b)-(iv) (c)-(ii) (d)-(iii)
4) (a)-(ii) (b)-(iii) (c)-(i) (d)-(iv)

Correct & Incorrect MCQs

1. Which of the following statements is incorrect?
   1) Alkanes formed in wurtz reaction contain odd number of carbon atoms.
   2) Sodium salt of butanoic acid can be used to prepare propane.
   3) Methane cannot be prepared by kolbe's electrolytic method.
   4) Both (2) and (3).

2. Which of the following statement is not correct about alkanes ?
   1) The first four alkanes are gases.
   2) Alkanes possess weak van der Waals forces.
   3) Alkanes are almost non-polar molecules.
   4) There is a steady decrease in boiling point of alkanes with increase in molecular mass.

3. Which of the following statement (s) are correct?
   A. n-pentane has higher boiling point than neopentane.
   B. The rate of reaction of alkanes with halogens is $F_2 > Cl_2 > Br_2 > I_2$.
   C. Rate of replacement of hydrogens of alkanes is : $3° > 2° > 1°$.
   D. Boiling point increases with increase in branching.
   1) A, B and C
   2) A and D
   3) A, C and D
   4) B, C and D

4. Select the incorrect statement/s :
   A. Alkanes contain carbon-carbon sigma (σ) bonds.
   B. Magnitude of torsional strain depends upon the angle of rotation about C–C bond.
   C. Staggered form of ethane has the maximum torsional strain and the eclipsed form has the least torsional strain.
   D. Staggered conformation of ethane is less stable than the eclipsed conformation.
   1) A and B
   2) B, C and D
   3) C and D
   4) B and C

5. Which of the following statements is incorrect about alkenes ?
   1) Carbon-carbon double bond in alkenes consists of one strong sigma (σ) bond and one weak pi (π) bond.
   2) Alkenes are not attacked by electrophilic reagents.
   3) The presence of weaker π-bond makes alkenes unstable molecules in comparison to alkanes.
   4) Strength of the double bond is greater than that of a carbon-carbon single bond in ethane.

6. Which amongst the following is incorrect statement ?
   1) The boiling points of the cis – isomers are higher than those of the corresponding trans-isomers.
   2) Cis form of alkene is found to be less polar than the trans form.
   3) The cis – isomers have higher dipole moments than the corresponding trans – isomers.
   4) In the case of solids, it is observed that the trans isomer has higher melting point than the cis form.

7. Select the correct statement:
   1) Partially deactivated palladised charcoal is known as Ziegler catalyst.
   2) Alkynes on reduction with sodium in liquid ammonia form cis alkenes.
   3) Vicinal dihalides on treatment with zinc metal lose a molecule of ZnX_2 to form an alkene. This reaction is known as dehalogenation.
   4) Both (1) and (2).

8. Which amongst the following is incorrect statement ?
   1) Hydrogen halides (HCl, HBr,HI) add up to alkenes to form alkyl halides. The order of reactivity of the hydrogen halides is HI > HBr > HCl.
   2) According to Markovnikov's rule, the negative part of the addendum gets attached to that carbon atom which possesses lesser number of hydrogen atoms.
   3) Peroxide effect is also observed in addition of HCl.
   4) Peroxide effect is not observed in addition of HI though H — I bond is weaker than H — Br bond.

9. Select the incorrect statement/s about alkynes ?
   A. First three members of alkynes are gases.
   B. They are weakly polar in nature.
   C. They are heavier than water and miscible with water but soluble in organic solvents.
   D. Their melting point, boiling point and density increase with increase in molar mass.
   1) A, B and C
   2) A and D
   3) A, C and D
   4) Only C

10. Which amongst the following is incorrect statement ?
    1) –OH group activates the benzene ring for the attack of an electrophile at o and p- positions.
    2) Halogens are moderately deactivating group in the case of aryl halides.
    3) Chlorobenzene is more reactive than benzene towards the electrophilic substitution reaction.
    4) $-NO_2$ is a meta directing group.

Statement Based MCQs

Directions: These questions consist of two statements each, printed as Statement-I and Statement-II. While answering these questions, you are required to choose any one of the following four responses.

1) **Both Statement-I and Statement-II are correct.**
2) **Both Statement-I and Statement-II are incorrect.**
3) **Statement-I is correct & Statement-II is incorrect.**
4) **Statement-I is incorrect & Statement-II is correct.**

1. **Statement-I:** Acetylene reacts with sodamide to evolve NH_3 gas.
   Statement-II: Acetylene is a weaker acid than ammonia.
2. **Statement-I:** Benzene readily undergoes electrophilic substitution reaction.
   Statement-II: Benzene is an unsaturated compound.
3. **Statement-I:** Nitration of nitrobenzene using nitrating mixture yields m-dinitrobenzene.
   Statement-II: Nitro group at benzene ring lower the electron density at the m-position.
4. **Statement-I:** Sod. Acetate on kolbe electrolysis gives ethane
   Statement-II: Hydrocarbon is obtained at cathode.
5. **Statement-I:** But-2-ene shows geometrical isomerism but But-1-ene does not show.
   Statement-II: Geometrical isomerism arises due to the restricted rotation of the double bond.
6. **Statement-I:** Iodination of alkanes is very fast.
   Statement-II: Iodination of alkanes is carried out in the presence of an oxidizing agent.
7. **Statement-I:** Alkanes are used as fuels.
   Statement-II: A larger amount of heat is evolved during combustion of alkanes.
8. **Statement-I:** In staggered form, the electron clouds of C – H bonds are as far as possible and it comes closer to each other in eclipsed form.
   Statement-II: Staggered form of ethane is the least stable and eclipsed form is the most stable.
9. **Statement-I:** Tertiary carbon atom is attached to three carbon atoms.
   Statement-II: 2,2-Dimethylbutane does not have any tertiary carbon atom.
10. **Statement-I:** The second substituent may enter the monosubstituted benzene ring at either ortho, para or at meta position.
    Statement-II: The position of the incoming group depends on the nature of the entering group.

Assertion & Reason MCQs

Directions: These questions consist of two statements each, printed as Assertion and Reason. While answering these questions, you are required to choose any one of the following four responses.

1) **Both Assertion and Reason are True and the Reason is a correct explanation of the Assertion.**
2) **Both Assertion and Reason are True but Reason is not a correct explanation of the Assertion.**
3) **Assertion is True but the Reason is False.**
4) **Assertion is False but Reason is True.**

1. **Assertion:** The compound cyclooctane has the following structural formula:

   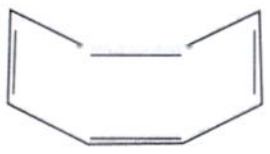

   It is cyclic and has conjugated 8π-electron system but it is not an aromatic compound.
   Reason: (4n + 2) π electrons rule does not hold good and ring is not planar.
2. **Assertion:** Toluene on Friedel Crafts methylation gives o– and p–xylene.
   Reason: CH_3-group bonded to benzene ring increases electron density at o– and p– position.
3. **Assertion:** Nitration of benzene with nitric acid requires the use of concentrated sulphuric acid.
   Reason: The mixture of concentrated sulphuric acid and concentrated nitric acid produces the electrophile, NO_2^+.
4. **Assertion:** Tropylium cation is aromatic in nature

   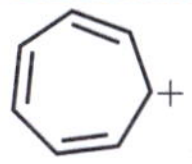

   Reason: The only property that determines the aromatic behavior is it's planar structure.
5. **Assertion:** Addition of HBr to 2-butene gives two isomeric products.
   Reason: Addition of HBr to 2-butene follows Markovnikov's rule.
6. **Assertion:** Benzene on treatment with excess of chlorine in the presence of light give BHC.
   Reason: BHC is also known as gammexane or 666.
7. **Assertion:** The nitrating reagent for carrying out nitration of benzene contains conc. H_2SO_4 and conc. HNO_3.
   Reason: HNO_3 acts as a base and produces $2NO_2^+$ ions in the presence of conc. H_2SO_4.
8. **Assertion:** Propene reacts with HBr in presence of benzoyl peroxide to give 2-bromopropane.
   Reason: In presence of peroxide, the addition of HBr to propene follows free radical mechanism.
9. **Assertion:** Boiling point of tert-butyl alcohol is more than that of n-butyl alcohol.
   Reason: Boiling point decreases with branching.
10. **Assertion:** Both cyclopropane and propene give addition reactions readily.
    Reason: Cyclopropane and propene are isomers of each other.

Multi Correct MCQs

1. Consider the following reactions:

Which of these reactions are possible?

A. 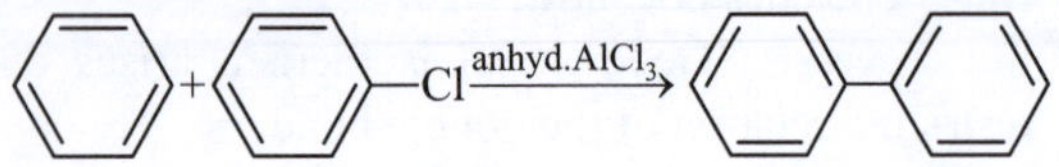

B. 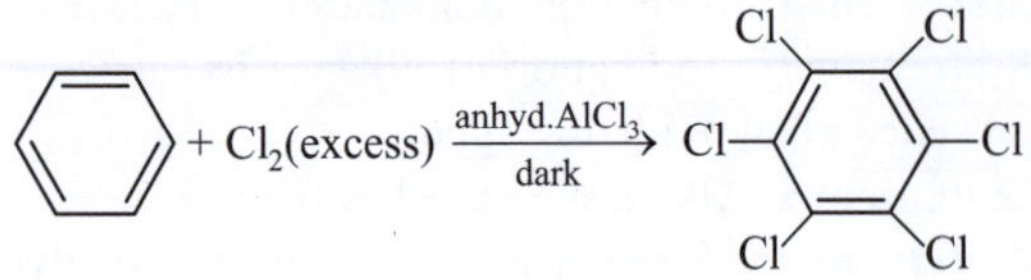

C. 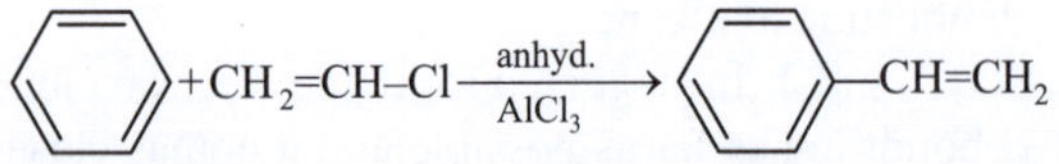

D. 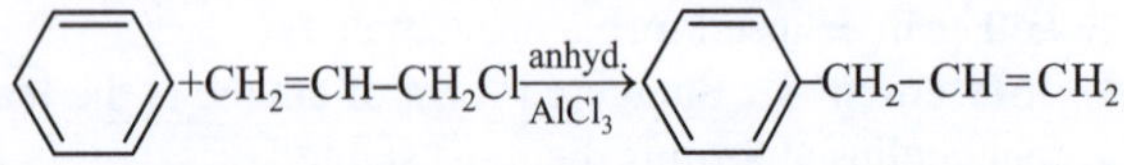

1) (B) and (D) 2) (A) and (D)
3) (A) and (B) 4) (B), (C) and (D)

2. Consider the carbanions:

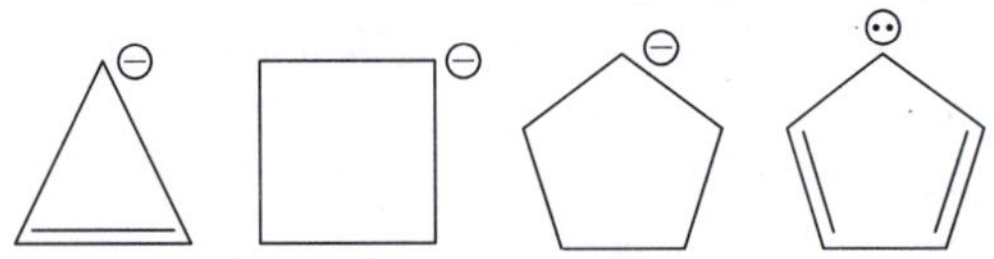

(a) (b) (c) (d)

Which of the following represents the correct order of stability of above given carbanions?

A. (a) > (d) and (c) < (b)
B. (b) < (a)
C. (c) > (b) and (a) < (d)
D. (b) < (d)

1) Both (A) and (B) 2) Only (B)
3) (A), (C) and (D) 4) Both (C) and (D)

3. Which of the following are aromatic?

A. 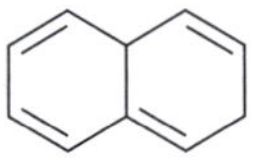

B. 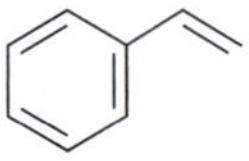

C. 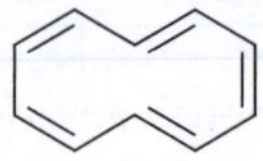

D. 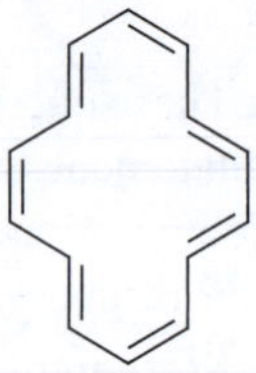

1) B and D only 2) A and C only
3) A and B only 4) C and D only

4. Among the following, the aromatic compounds are:

(A) 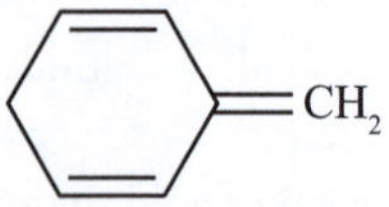

(B) 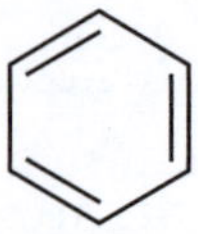

(C) 

(D) 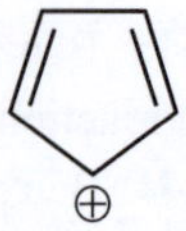

Choose the correct answer from the following options:

1) (B), (C) and (D) only 2) (A) and (B) only
3) (A), (B) and (C) only 4) (B) and (C) only

5. Which of the following compounds can decolourise bromine water?

A. 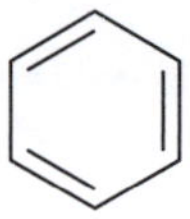

B. 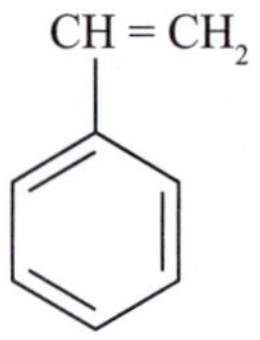

C. 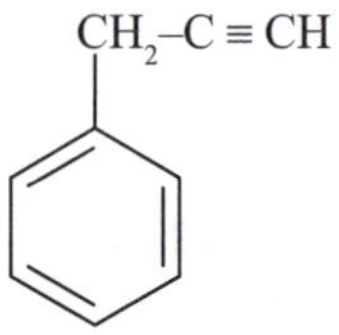

D. 

1) only (A) 2) only (D)
3) (A), (B) and (C) 4) Both (B) and (C)

NCERT Exemplar MCQs

1. Arrange the following in decreasing order of their boiling points:

A. n -butane B. 2-methylbutane
C. n -pentane D. 2, 2-dimethylpropane

1) A > B > C > D 2) B > C > D > A
3) D > C > B > A 4) C > B > D > A

2. Arrange the halogens F_2, Cl_2, Br_2, I_2, in order of their increasing reactivity with alkanes:

1) $I_2 < Br_2 < Cl_2 < F_2$ 2) $Br_2 < Cl_2 < F_2 < I_2$
3) $F_2 < Cl_2 < Br_2 < I_2$ 4) $Br_2 < I_2 < Cl_2 < F_2$

3. The increasing order of reduction of alkyl. halides with zinc and dilute HCl is:

1) R – Cl < R – I < R – Br 2) R– Cl < R – Br < R – I
3) R – I < R – Br < R – Cl 4) R – Br < R – I < R – Cl

4. The correct IUPAC name of the following alkane is:

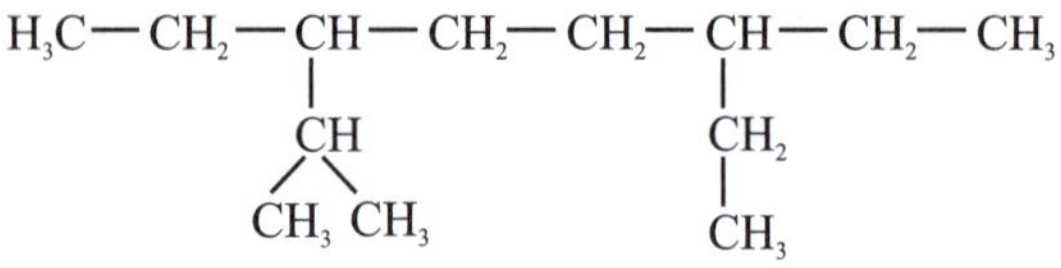

1) 3,6-diethyl-2-methyloctane
2) 5-isopropyl-3-ethyloctane
3) 3-ethyl-5-isopropyloctane
4) 3-isopropyl-6-ethyloctane

5. The addition of HBr to 1-butene gives a mixture of products A, B and C:

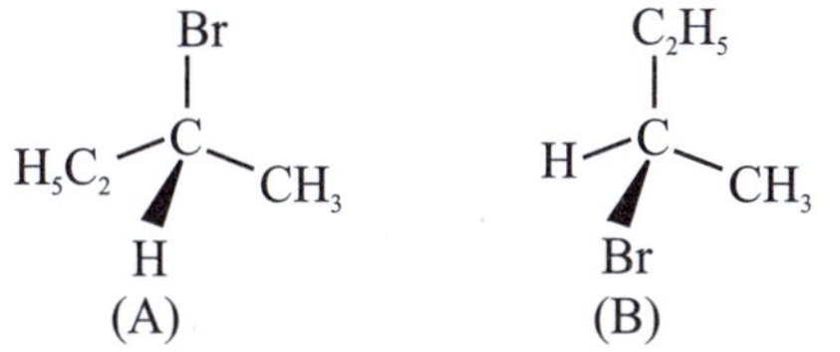

$CH_3-CH_2-CH_2-CH_2-Br$

(C)

The mixture consists of:

1) A and B as major and C as minor products
2) B as major, A and C as minor products
3) B as minor, A and C as major products
4) A and B as minor and C as major products

6. Which of the following will not show geometrical isomerism?

1) 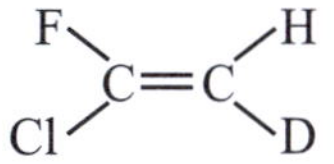

2) 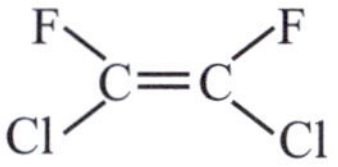

3) 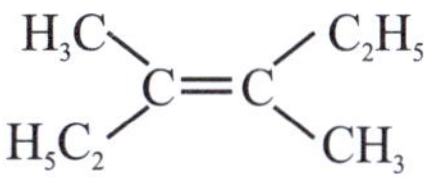

4) 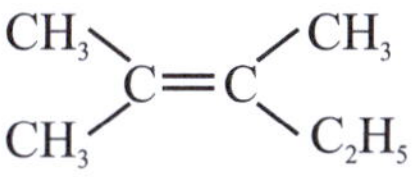

7. Arrange the following hydrogen halides in order of their decreasing reactivity with propene:

1) HCl > HBr > HI
2) HBr > HI > HCl
3) HI > HBr > HCl
4) HCl > HI > HBr

8. Arrange the following carbanions in order of their decreasing stability:

A. $H_3C-C\equiv C^-$

B. $H-C\equiv C^-$

C. $H_3C-\bar{C}H_2$

1) A > B > C
2) B > A > C
3) C > B > A
4) C > A > B

9. Arrange the following alkyl halides in decreasing order of the rate of β-elimination reaction with alcoholic KOH:

A. $CH_3-\underset{CH_3}{\overset{H}{\underset{|}{\overset{|}{C}}}}-CH_2Br$

B. CH_3-CH_2-Br

C. $CH_3-CH_2-CH_2-Br$

1) A > B > C
2) C > B > A
3) B > C > A
4) A > C > B

10. Which of the following reactions of methane is incomplete combustion?

1) $2CH_4+O_2\xrightarrow{Cu/523\ K/100\ atm}2CH_3OH$

2) $CH_4+O_2\xrightarrow{Mo_2O_3}HCHO+H_2O$

3) $CH_4+O_2\longrightarrow C(s)+2H_2O(l)$

4) $CH_4+2O_2\longrightarrow CO_2(g)+2H_2O(l)$

Kattar NEET

Scan Me for Detailed Explanations

1. Which of the following mechanism best describes the reaction between propene and BrCl?

1) 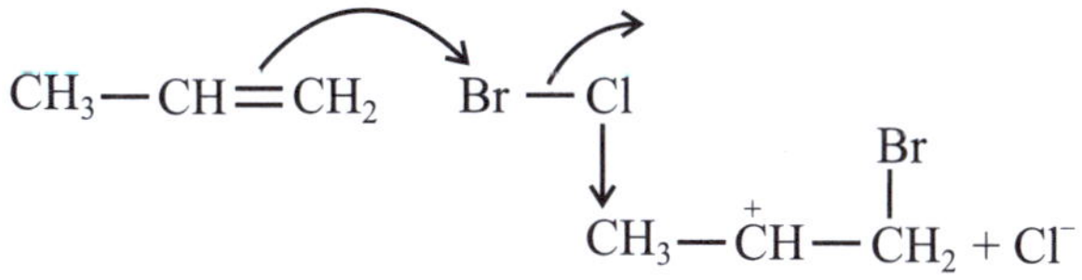

2) 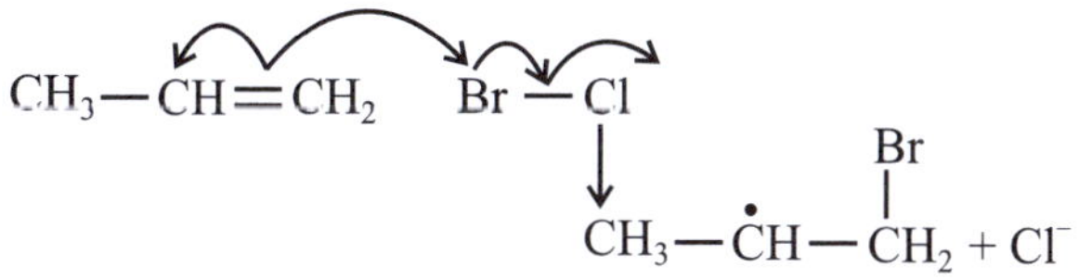

3) 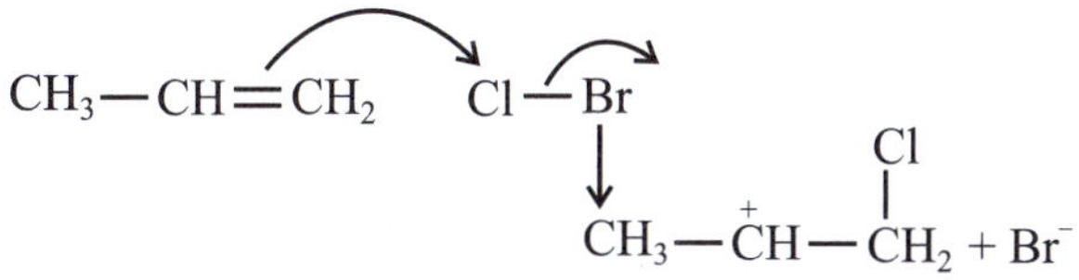

4) 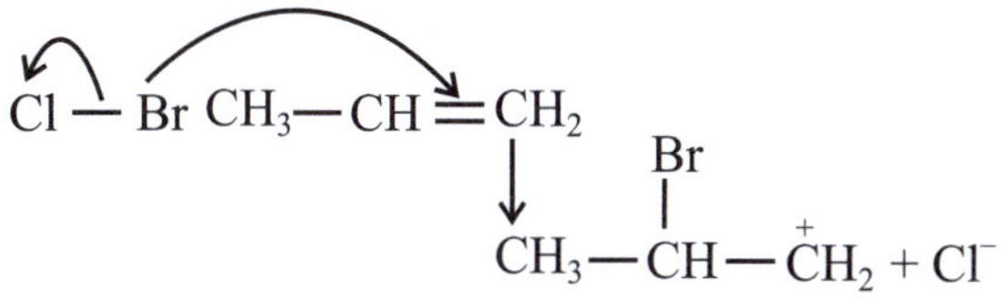

2. The reactions which follow free radical mechanism is/are

I. $CH_2=CH_2+HBr\xrightarrow{H_2O_2}CH_3CH_2Br$

II. $C_6H_5CH_3\xrightarrow{Cl_2/hv}C_6H_5CH_2Cl$

III. $C_6H_5CH=CH_2\xrightarrow{Br_2/CCl_4}C_6H_5-\underset{}{\overset{Br}{\overset{|}{C}H}}-CH_2Br$

IV. $C_6H_6\xrightarrow{Fe/Br_2}C_6H_5Br$

1) I & II only
2) II only
3) I, III & IV only
4) II & IV only

3. Major product (C) in the following sequence is

$CaO+C\xrightarrow{Heat}(A)\xrightarrow{H_2O}(B)\xrightarrow[dil.\ H_2SO_4]{Hg^{2+}}(C)$

1) Ethanol
2) Ethyl hydrogen sulphate
3) Ethanal
4) Ethylene glycol

4. What is the major product in the following reaction?

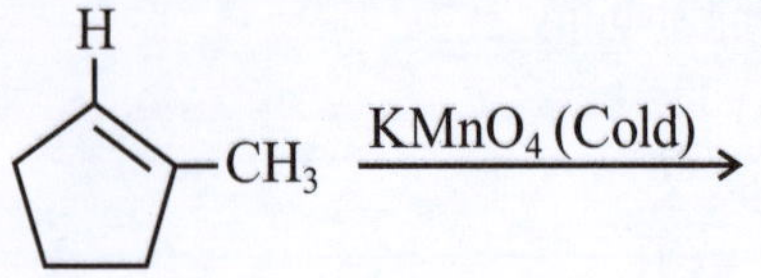

1) $H_3C-\overset{\overset{O}{\|}}{C}-CH_2-CH_2-CH_2-\overset{\overset{O}{\|}}{C}-H$

2) $CH_3-\overset{\overset{O}{\|}}{C}-CH_2-CH_2-CH_2-\overset{\overset{O}{\|}}{C}-OH$

3) 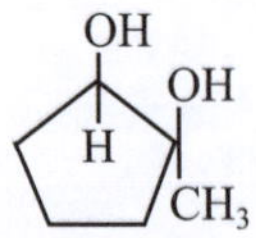

4) 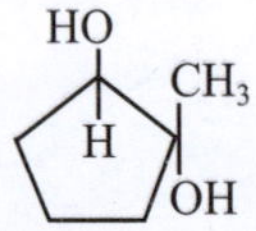

5. 2-Methyl propene is treated with catalytic amount of conc. H_2SO_4 for a short reaction time, C_8H_{16} product can be isolated. What is the structure of the compound?

1) $CH_3-\overset{\overset{CH_3}{|}}{C}=\overset{\overset{H}{|}}{C}-\overset{\overset{CH_3}{|}}{CH}-CH_2CH_3$

2) $CH_3-\overset{\overset{CH_3}{|}}{C}=\overset{\overset{H}{|}}{C}-\underset{\underset{CH_3}{|}}{\overset{\overset{CH_3}{|}}{C}}-CH_3$

3) 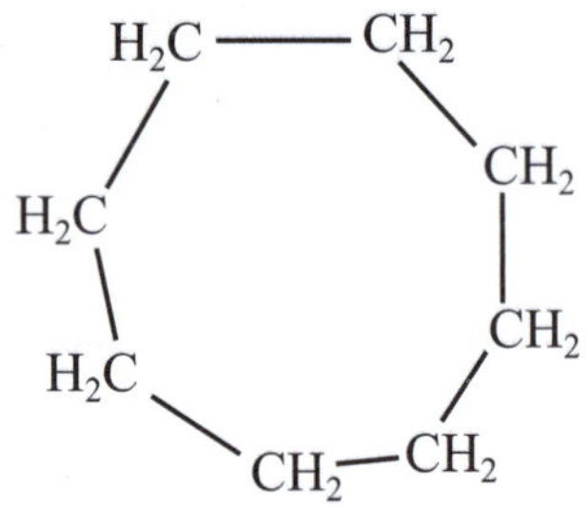

4) $CH_3-CH_2-\overset{\overset{CH_3}{|}}{C}=\overset{\overset{CH_3}{|}}{C}-CH_2-CH_3$

6. $CH_2=CH-CH=CH_2 \xrightarrow{\text{1 equi } H_2/Pt} (A) \xrightarrow{O_3/H_2O} (B)$

Structure of (A) and (B) respectively are

1) $CH_3-CH_2-CH=CH_2, (CH_3CH_2COOH + CO_2)$
2) $CH_3-CH=CH-CH_3, CH_3COOH$ (2 mol)
3) $CH_3-CH=CH-CH_3, CH_3CHO$ (2 mol)
4) $CH_3-CH_2-CH=CH_2, (CH_3CH_2CHO + HCHO)$

7.  $\xrightarrow[\text{Lindlar's catalyst}]{H_2}$ A (Major product)

Identify the major product (A)

1) 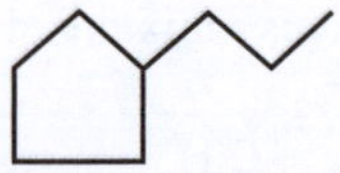 2) 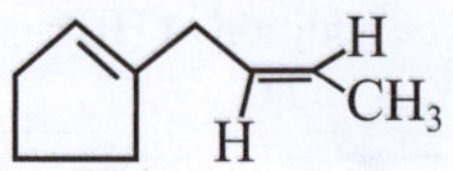

3) 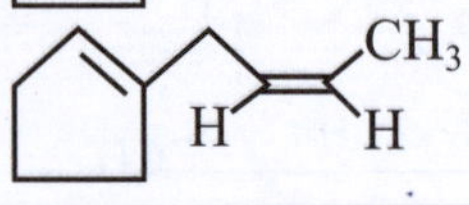 4) 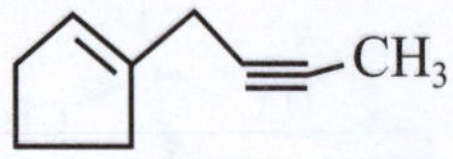

8. $CH\equiv CH \xrightarrow[NH_4Cl]{Cu_2Cl_2} A \xrightarrow{H_2+Pd/BaSO_4} B \xrightarrow[\Delta]{HC\equiv CH} C$

Product C will be

1) 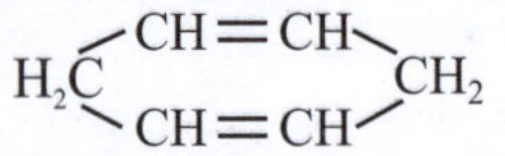
2) $CH_2=CH-CH=CH_2$
3) $CH_2=CH-C\equiv C-CH-CH_2$
4) 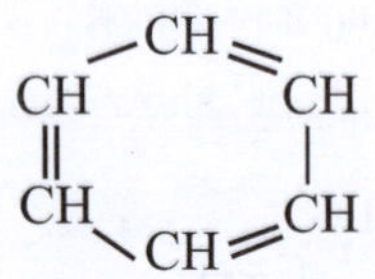

9. The best sequence of reactions to prepare 2 heptanone is

1) propyne $\xrightarrow{NaNH_2} X \xrightarrow{n-C_4H_9Br} Y \xrightarrow[H_2SO_4]{H_2O,\ Hg^{2+}}$
2) ethyne $\xrightarrow{NaNH_2} X \xrightarrow{n-C_5H_{11}Br} Y \xrightarrow[H_2SO_4]{H_2O,\ Hg^{2+}}$
3) 1 − hexyne $\xrightarrow{NaNH_2} X \xrightarrow{CH_3Br} Y \xrightarrow[H_2SO_4]{H_2O,\ Hg^{2+}}$
4) 1 − pentyne $\xrightarrow{NaNH_2} X \xrightarrow{C_2H_5Br} Y \xrightarrow[H_2SO_4]{H_2O,\ Hg^{2+}}$

10. Rank the following reactions A, B and C in order of increasing rate,

A) 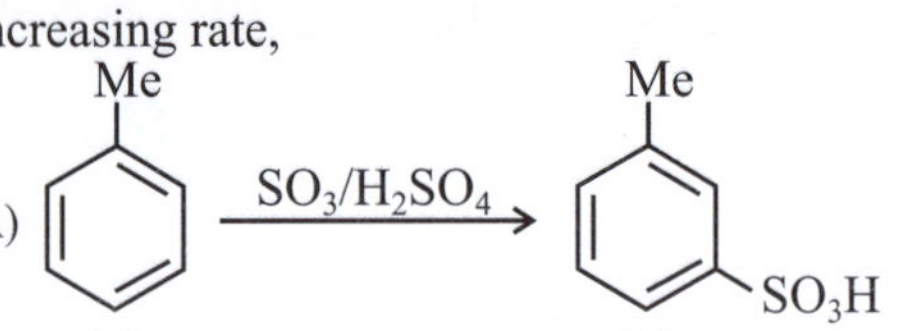

B) 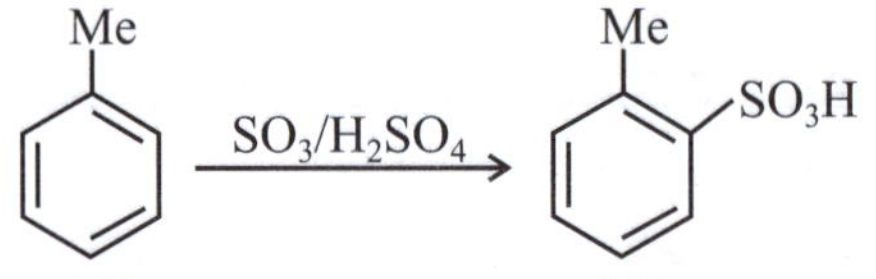

C) 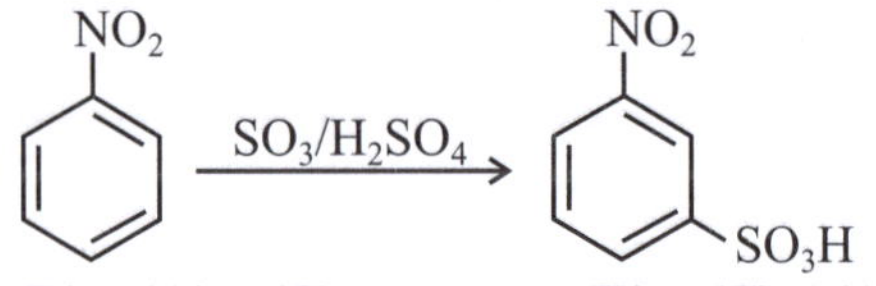

1) (B) > (A) > (C) 2) (B) > (C) > (A)
3) (A) > (B) > (C) 4) (A) > (C) > (B)

11. Arrange the following compounds in increasing order of heat evolved when one mole of each of them is hydrogenated using Pt-catalyst.

I. 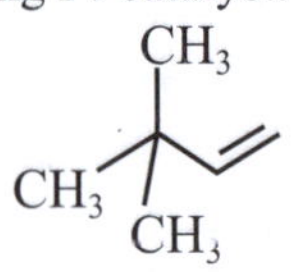 II. 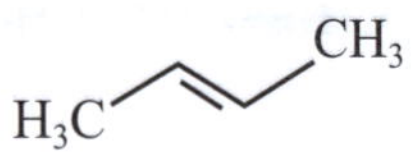

III. 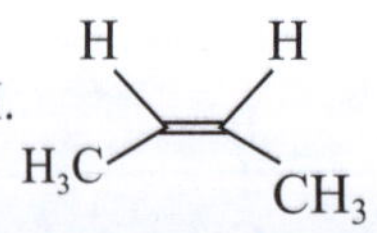 IV. 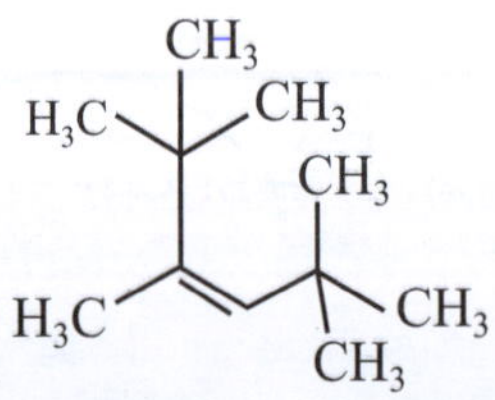

1) I < II < III < IV 2) III < I < II < IV
3) IV < II < III < I 4) II < III < IV < I

12. Which of the following compound will lose optical activity after the reaction?

1) 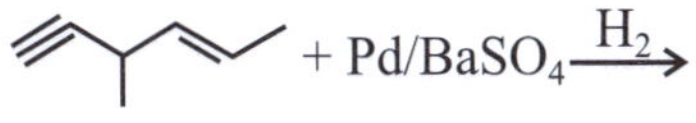

2) 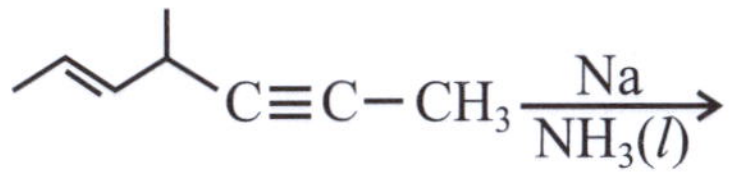

3) 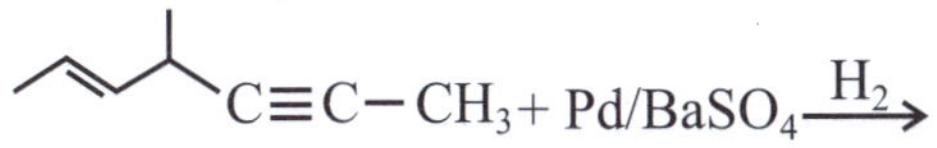

4) 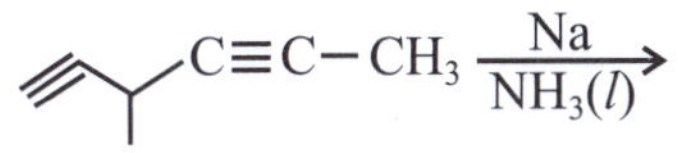

13. Observe the following reaction sequence

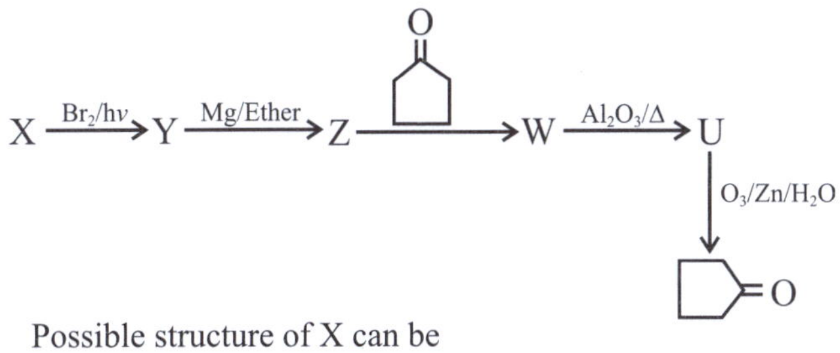

Possible structure of X can be

1) 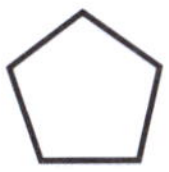 2) 

3)  4) 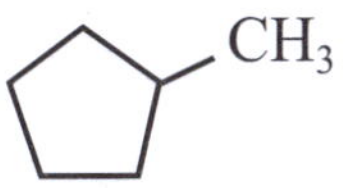

14. The final major product of the following reaction is:

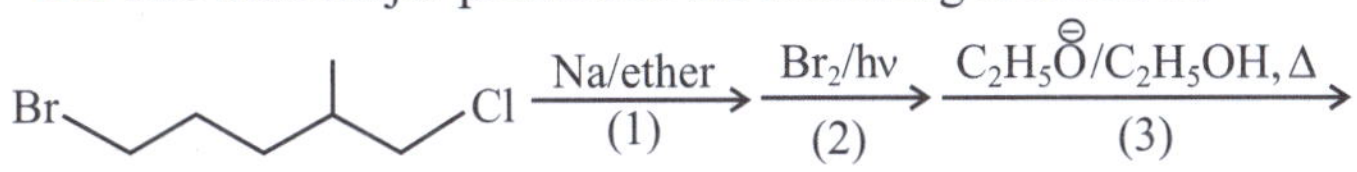

1)  2) 

3) 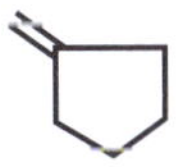 4) 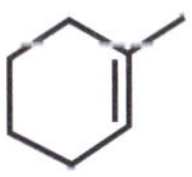

15. 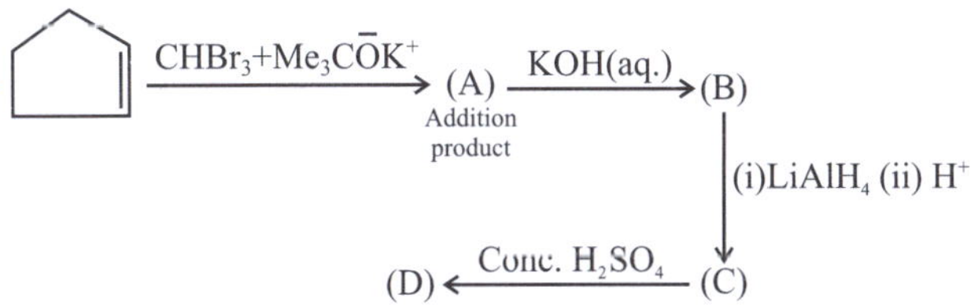

Identify structure of compound (*d*) is the above reaction sequence

1) 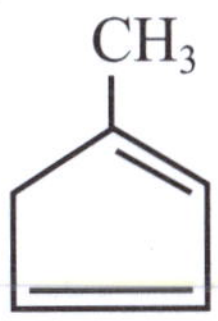 2) 

3) 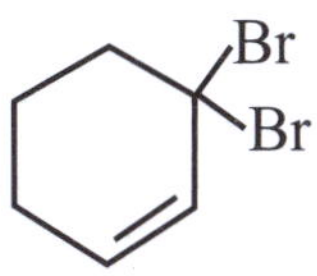 4) 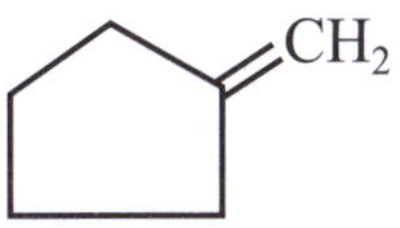

16.

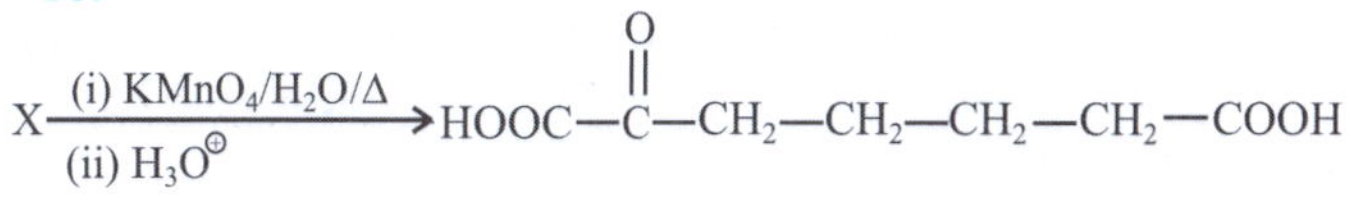

1) 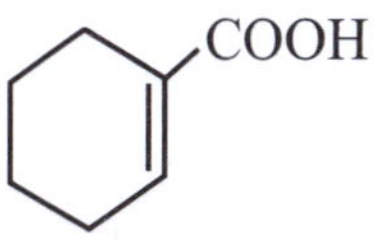 2) 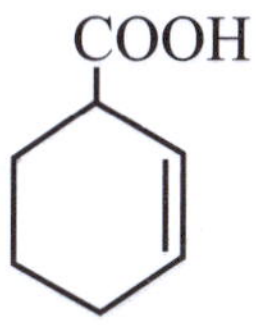

3) 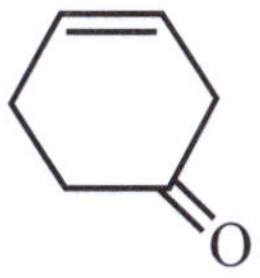 4) 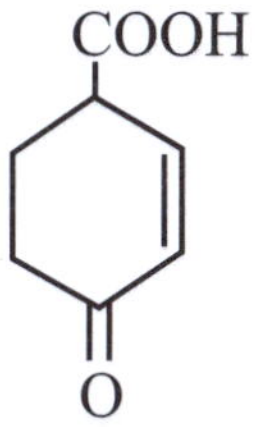

17. Match the following:

List-I (Reaction)		List-II (Intermediate)	
I.	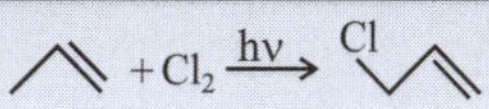	P.	Cyclic transition state
II.	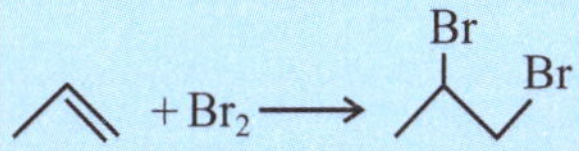	Q.	Carbocation
III.	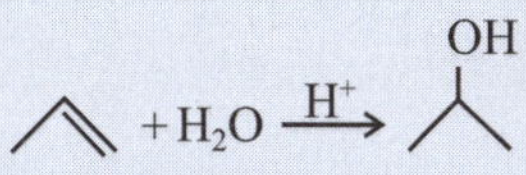	R.	Free radicals
IV.	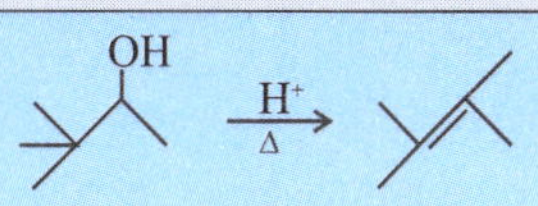	S.	Rearrangement

1) I-R; II-Q; III-P; IV-Q, S 2) I-R; II-P; III-Q; IV-P, S
3) I-R; II-P; III-S; IV-Q, S 4) I-R; II-P; III-Q; IV-Q, S

18. Probability factor $3^\circ H : 2^\circ H : 1^\circ H$ = $5 : 3.8 : 1$

$$CH_3-\underset{}{CH}(CH_3)CH_2CH_2CH_3 + Cl_2 \xrightarrow[h\nu]{1\text{ eq}} \text{Products.}$$

Match the following:

	List-I (Product formed)		List-II (% of Product)
I	$ClCH_2CH(CH_3)-CH_2-CH_2CH_3$	P	17.1%
II	$CH_3-C(Cl)(CH_3)-CH_2-CH_2CH_3$	Q	10.3%
III	$CH_3-C(H)(CH_3)-C(H)(Cl)-CH_2-CH_3$	R	26.0%
IV	$CH_3CH(CH_3)-CH_2-CH_2CH_2Cl$	S	20.6%

1) I-P; II-Q; III-R; IV-S 2) I-R; II-S; III-P; IV-Q
3) I-S; II-P; III-R; IV-Q 4) I-Q; II-P; III-R; IV-S

19. Match the following:

	List-I (Product Formed)		List-II
I	Reaction of alkanes with halogens	P	2-methyl pentane
II	Corey House reaction can form	Q	Butane
III	Wurtz reaction is not useful in formation of	R	Acidic strength
IV	$CH_3-CH_3 < CH_2=CH_2 < CH\equiv CH$	S	$F_2 > Cl_2 > Br_2 > I_2$ (reactivity)
		T	Methane cannot be prepared

1) I-Q; II-P, S, T; III-P, R; IV-T
2) I-R; II-T, Q, P; III-P, T; IV-R
3) I-T; II-S, Q, T; III-R, T; IV-P
4) I-S; II-P, Q, T; III-P, T; IV-R

20. Which one of the following reactions would be the best for the formation of 2-bromobutane?

1) $CH_3CH = CHCH_2CH_3 \xrightarrow{HBr}$
2) $CH_3CH_2CH = CH_2 \xrightarrow{HBr}$
3) $CH_3CH = CHCH_3 \xrightarrow{Br_2}$
4) $CH_3CH_2CH = CH_2 \xrightarrow[\text{Peroxide}]{HBr}$

NEET & AIIMS Past Year MCQs

1. The alkane that can be oxidized to the corresponding alcohol by $KMnO_4$ as per the equation **(2024 Re)**

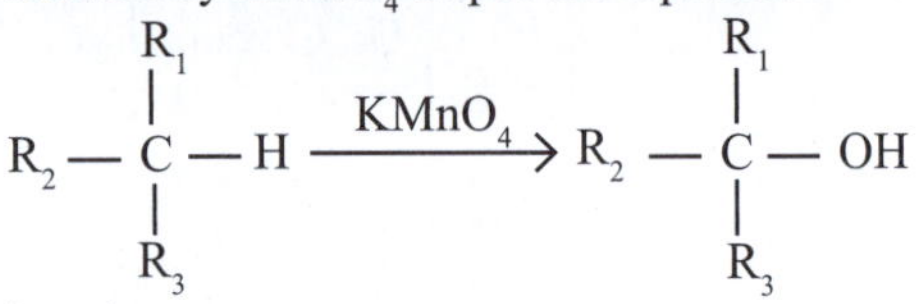

is, when:
1) $R_1 = H; R_2 = H; R_3 = H$
2) $R_1 = CH_3; R_2 = CH_3, R_3 = CH_3$
3) $R_1 = CH_3; R_2 = H; R_3 = H$
4) $R_1 = CH_3; R_2 = CH_3; R_3 = H$

2. Baeyer's reagent is : **(2024 Re)**
1) Acidic potassium permanganate solution
2) Acidic potassium dichromate solution
3) Cold, dilute, aqueous solution of potassium permanganate
4) Hot, concentrated solution of potassium permanganate

3. Given below are two statements: **(2024)**
Statement - I: The boiling point of three isomeric pentanes follows the order
n-pentane > isopentane > neopentane
Statement - II: When branching increases, the molecule attains a shape of sphere. This results in smaller surface area for contact, due to which the intermolecular forces between the spherical molecules are weak, thereby lowering the boiling point.
In the light of the above statements, choose the most appropriate answer from the options given below:
1) Statement - I is correct but Statement - II is incorrect.
2) Statement - I is incorrect but Statement - II is correct.
3) Both Statement - I and Statement - II are correct.
4) Both Statement - I and Statement - II are incorrect.

4. For the given reaction **(2024)**

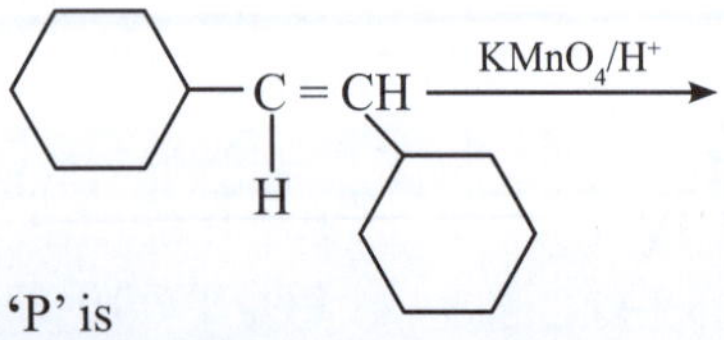

'P' (major product)

'P' is

1) 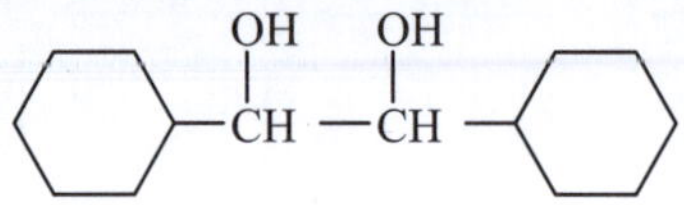

2) 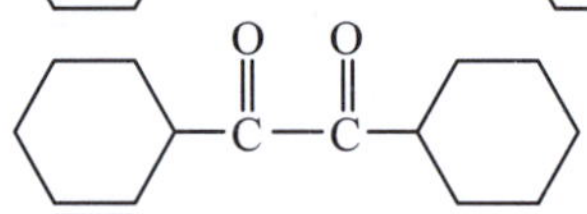

3) 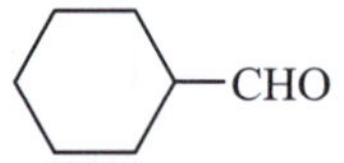

4) 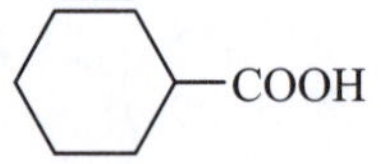

5. Weight (g) of two moles of the organic compound, which is obtained by heating sodium ethanoate with sodium hydroxide in presence of calcium oxide is: **(2023)**
1) 18 2) 16 3) 32 4) 30

6. Consider the following compounds/species : **(2023)**

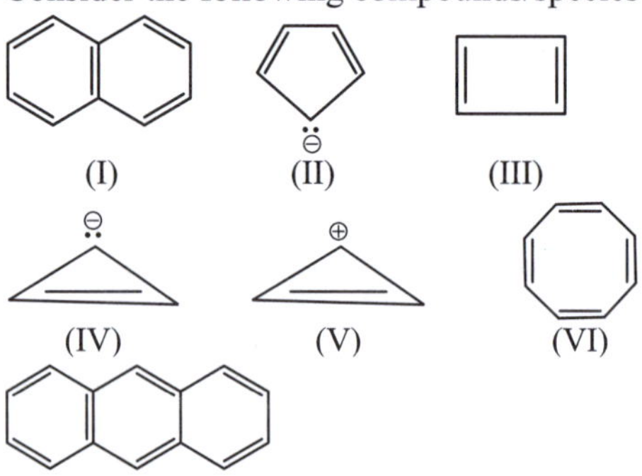

(I) (II) (III) (IV) (V) (VI) (VII)

The number of compounds/species which obey Huckel's rule is ________.
1) 5 2) 4 3) 6 4) 2

7. Which compound amongst the following is not an aromatic compound? **(2022)**

1) 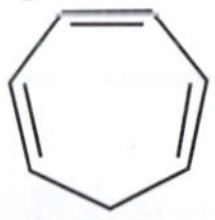

2) 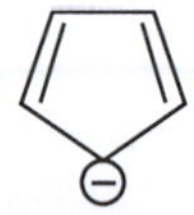

3) 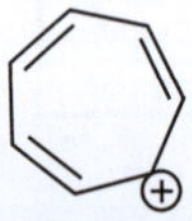

4) 

8. Which of the following is **suitable** to synthesize chlorobenzene? **(2022)**

1) 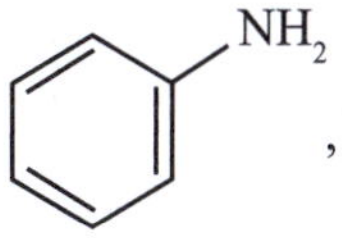 , HCl Heating
2) Benzene, Cl_2, anhydrous $FeCl_3$
3) Phenol, $NaNO_2$, HCl, CuCl
4) 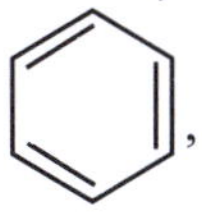 , HCl

9. Compound X on reaction with O_3 followed by Zn/H_2O gives formaldehyde and 2-methyl propanal as products. The compound X is: **(2022)**

1) Pent-2-en 2) 3-Methylbut-1-ene
3) 2-Methylbut-1-ene 4) 2-Methylbut-2-ene

10. The correct structure of 2,6-Dimethyl-dec-4-ene. **(2021)**

1) 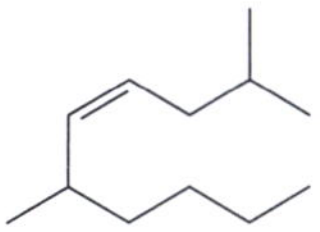
2) 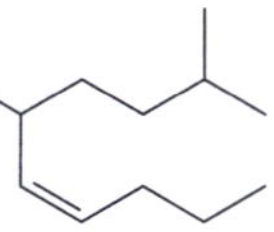
3) 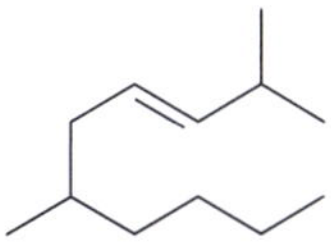
4) 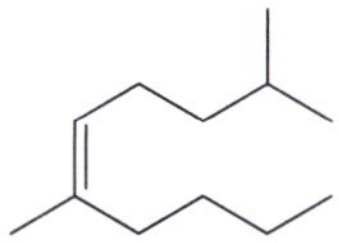

11. Dihedral angle (D.A.) of least stable conformer of ethane is **(2021)**

1) 120° 2) 180° 3) 60° 4) 0°

12. The major product of the following chemical reaction is **(2021)**

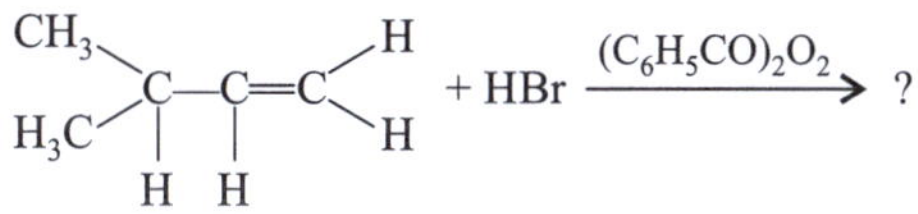

1) 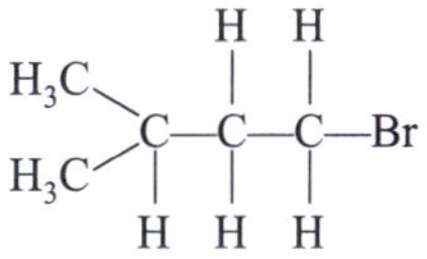
2) 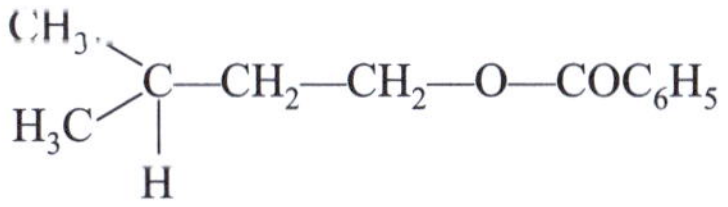
3) 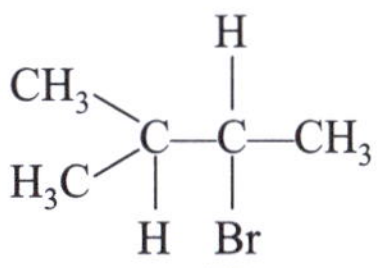
4) 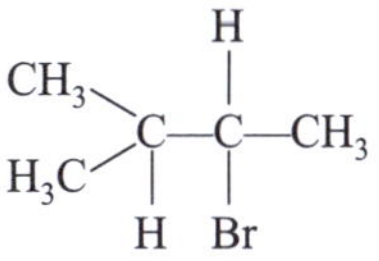

13. $CH_3CH_2COO^-Na^+ \xrightarrow[\text{Heat}]{\text{NaOH} + ?} CH_3{-}CH_3 + Na_2CO_3$

Consider the above reaction and Identify the missing reagent/chemical. **(2021)**

1) B_2H_6 2) Red. phosphorus
3) CaO 4) DIBAL–H

14. **Assertion:** All C—C—C bonds angles in Isobutene ($CH_3{-}C(CH_3){=}CH_2$) are different.

Reason: CH_3 (Methyl group) show steric crowding. **(AIIMS 2018)**

1) If both Assertion & Reason are True & the Reason is a correct explanation of the Assertion.
2) If both Assertion & Reason are True but Reason is not a correct explanation of the Assertion.
3) If Assertion is True but the Reason is False.
4) If both Assertion & Reason are False.

15. **Assertion:** $HC{\equiv}C^-$ is more stale than. $H_2C = CH^-$

Reason: $HC{\equiv}C^-$ has more s-character than $H_2C = CH^-$. **(AIIMS 2012)**

1) If both Assertion & Reason are True & the Reason is a correct explanation of the Assertion.
2) If both Assertion & Reason are True but Reason is not a correct explanation of the Assertion.
3) If Assertion is True but the Reason is False.
4) If both Assertion & Reason are False.

16. How many geometrical isomers are possible in the following two alkenes? **(AIIMS 2009)**

(i) $CH_3 - CH = CH - CH = CH - CH_3$
(ii) $CH_3 - CH = CH - CH = CH - Cl$

1) 4 and 4 2) 4 and 3 3) 3 and 3 4) 3 and 4

17. The function of $AlCl_3$ in Friedel-Craft's reaction is to: **(AIIMS 2008)**

1) Absorb HCl 2) Absorb water
3) Produce nucleophiled 4) Produce electrophile

18. $C_6H_6 \xrightarrow[H_2SO_4]{HNO_3} X \xrightarrow[FeCl_3]{Cl_2} Y$. In the above sequence Y can be: **(AIIMS 2008)**

1) 3-nitrochlorobenzene 2) 1-nitrochlorobenzene
3) 4-nitrochlorobenzene 4) None of these

19. Which of the following is aromatic? **(AIIMS 2008)**

1) 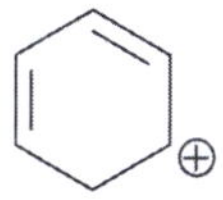
2) 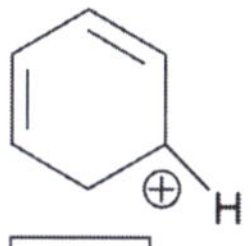
3) 
4) 

20. 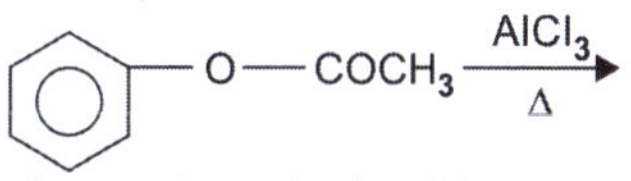

The product obtained is/are: **(AIIMS 2007)**

1) o-product 2) m-product
3) o- and p-products 4) o-, m- and p-products.

21. The treatment of benzene with isobutene in the presence of sulphuric acid gives: **(AIIMS 2003)**

1) Isobutyl benzene 2) t-butyl benzene
3) n-Butyl benzene 4) No reaction

22. **Assertion:** The carbon atoms of the benzene ring may be numbered for identification of substituent groups, just as a continuous chains of carbon atoms are numbered.

Reason: Smallest set of numbers designating the substitutents is the preferred set. **(AIIMS 2000)**

1) If both Assertion & Reason are True & the Reason is a correct explanation of the Assertion.
2) If both Assertion & Reason are True but Reason is not a correct explanation of the Assertion.
3) If Assertion is True but the Reason is False.
4) If both Assertion & Reason are False.

Answer Key

NCERT Topic-wise MCQs

1	2	3	4	5	6	7	8	9	10	11	12	13	14	15	16	17
4	3	1	1	2	4	3	4	4	1	4	4	2	2	4	1	1
18	19	20	21	22	23	24	25	26	27	28	29	30	31	32	33	34
2	3	1	2	3	2	3	4	3	3	4	3	3	2	4	1	4
35	36	37	38	39	40	41	42	43	44	45	46	47	48	49	50	51
4	3	2	2	3	1	3	4	1	4	3	2	2	3	2	2	3
52	53	54	55	56	57	58	59	60	61	62	63	64	65	66	67	68
1	1	2	4	1	1	2	2	2	2	1	2	2	4	2	3	3
69	70	71	72	73	74	75	76	77	78	79	80	81	82	83	84	85
1	1	4	3	2	3	4	3	1	2	2	4	3	2	1	4	2
86	87	88	89	90	91	92	93	94	95	96	97	98	99	100	101	102
3	2	3	2	3	1	2	4	4	1	3	2	4	2	2	1	1
103	104	105	106	107	108	109	110	111	112	113	114	115	116	117	118	119
3	2	4	2	2	1	3	2	3	1	2	2	3	3	3	1	2
120	121	122	123	124	125	126	127	128	129	130	131	132	133	134	135	136
2	3	4	3	3	4	1	2	2	2	4	2	1	3	3	2	4
137	138	139	140													
3	3	3	1													

High Order Time Intensive MCQs

1	2	3	4	5	6	7	8	9	10	11	12	13	14	15	16	17
2	1	3	4	2	1	4	4	1	4	3	3	4	3	4	4	3
18	19	20	21	22	23	24	25	26	27	28	29	30	31	32	33	34
1	2	4	3	2	1	4	2	1	4	3	3	1	3	4	4	2
35	36	37	38	39	40	41	42	43	44	45	46	47	48	49	50	51
2	1	2	1	1	1	2	1	3	2	2	3	1	3	2	2	1
52	53	54	55	56	57	58	59	60	61	62	63	64	65	66	67	68
2	1	4	1	4	2	3	2	2	3	2	3	2	2	2	2	1
69	70	71	72	73	74	75	76									
1	3	2	2	1	3	4	4									

Match the Columns MCQs

1	2	3	4	5
2	1	4	1	2

Correct & Incorrect MCQs

1	2	3	4	5	6	7	8	9	10
1	4	1	3	2	2	3	3	4	3

Statement Based MCQs

1	2	3	4	5	6	7	8	9	10
3	1	3	4	1	4	1	3	1	3

Assertion & Reason MCQs

1	2	3	4	5	6	7	8	9	10
1	1	1	3	3	2	1	4	4	2

Multi Correct MCQs

1	2	3	4	5
1	4	1	4	4

NCERT Exemplar MCQs

1	2	3	4	5	6	7	8	9	10
4	1	2	1	1	4	3	2	4	3

Kattar NEET

1	2	3	4	5	6	7	8	9	10	11	12	13	14	15	16	17
1	1	3	3	2	2	3	1	2	1	4	2	1	2	2	1	4
18	19	20														
3	4	2														

NEET & AIIMS Past Year MCQs

1	2	3	4	5	6	7	8	9	10	11	12	13	14	15	16	17
2	3	3	4	3	2	1	2	2	1	4	1	3	1	1	4	4
18	19	20	21	22												
1	3	3	2	2												

Class-XII

CHAPTER

1 Solutions

Topicwise Number of Questions Analysis (2025-2016)

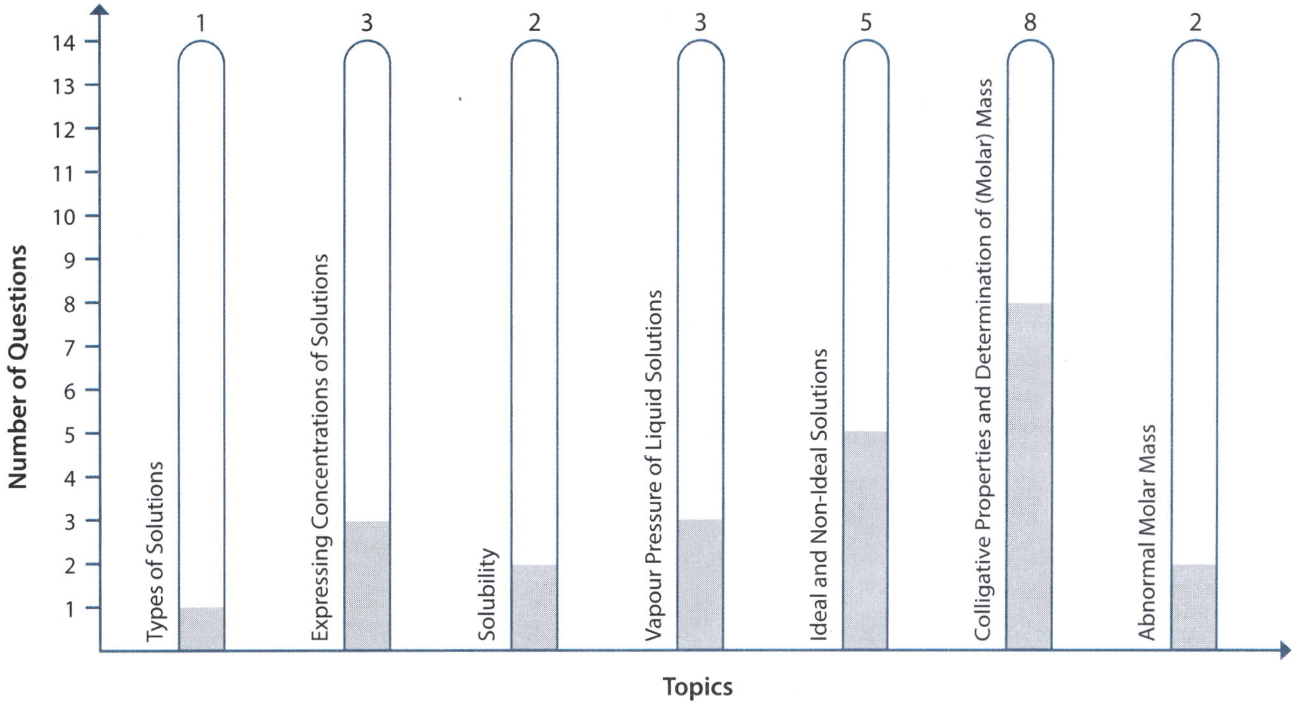

NCERT-PICKS

General Introduction

- Solutions are homogeneous mixtures of two or more than two components. By homogenous mixture we mean that its composition and properties are uniform throughout the mixture. Generally, the component that is present in the largest quantity is known as solvent. Solvent determines the physical state in which solution exists. One or more components present in the solution other than solvent are called solutes. In this Unit we shall consider only binary solutions (i.e., consisting of two components). Here each component may be solid, liquid or in gaseous state and are summarised in the table given below.

Types of solutions

Type of Solution	Solute	Solvent	Common Examples
Gaseous Solutions	Gas	Gas	Mixture of oxygen and nitrogen gases
	Liquid	Gas	Chloroform mixed with nitrogen gas
	Solid	Gas	Camphor in nitrogen gas
Liquid Solutions	Gas	Liquid	Oxygen dissolved in water
	Liquid	Liquid	Ethanol dissolved in water
	Solid	Liquid	Glucose dissolved in water
Solid Solutions	Gas	Solid	Solution of hydrogen in palladium
	Liquid	Solid	Amalgam of mercury with sodium
	Solid	Solid	Copper dissolved in gold

Methods of Expressing the Concentration of a Solution

The concentration of a solution can be expressed in a number of ways. The important methods are:

1. **Mass percentage or percent by mass:** It is defined as the amount of solute in grams present in 100 grams of the solution.

$$\text{Mass \% age of solute} = \frac{\text{mass of solute}}{\text{mass of solution}} \times 100$$

$$= \frac{\text{mass of solute}}{\text{mass of solute + mass of solvent}} \times 100$$

2. **Percent by volume:** It is defined as the volume of solute in mL present in 100 mL solution.

$$\text{\% of solute by volume} = \frac{\text{volume of solute}}{\text{volume of solution}} \times 100$$

3. **Mass by volume percent:** it is defined as the mass of solute present in 100 mL of solution.

$$\text{\% of solute (mass by volume)} = \frac{\text{mass of solute}}{\text{volume of solution}} \times 100$$

4. **Parts per million (ppm):** When the solute is present in trace quantities, it is convenient to express the concentration in parts per million (ppm). It is defined as the quantity of the solute in grams present in 10^6 grams of the solution.

$$\text{ppm} = \frac{\text{mass of solute}}{\text{mass of solution}} \times 10^6$$

5. **Mole fraction:** The ratio of number of moles of one component to the total number of moles of the solution (i.e., all the components).

   In a binary solution,

   Mole fraction of solute + Mole fraction of solvent = 1

   Mole fraction is independent of temperature.

6. **Molality:** It is defined as number of moles of the solute dissolved in 1 kg of the solvent. It is denoted by 'm'. Its unit is mol kg^{-1}.

$$\text{Molality (m)} = \frac{\text{moles of solute}}{\text{mass of solvent (in kg)}}$$

$$= \frac{\text{moles of solute}}{\text{mass of solvent (in gm)}} \times 1000$$

   If w_B grams of solute of molecular mass 'm_B' are present in w_A grams of the solvent, then

$$m = \frac{n_B \times 1000}{w_A} = \frac{w_B \times 1000}{m_B \times w_A}$$

   where, n_B is number of moles of solute

7. **Molarity (Molar concentration):** It is defined as the number of moles of the solute per litre or per dm^3 of the solution. The unit of molarity is mol $litre^{-1}$ or mol dm^{-3}.

$$\text{Molarity (M)} = \frac{\text{moles of solute}}{\text{volume of solution (L)}}$$

   Molarity × volume of solution(L) = moles of solute

   If w_B g of the solute of molecular mass m_B are dissolved in V litre of solution, then

$$M = \frac{w_B}{m_B \times V}$$

$$\text{Molarity} \times m_B = \frac{w_B}{V} = \text{Strength of solution (g/L)}$$

$$M = \frac{w_B}{m_B \times V(cm^3)} \times 1000$$

$$M = \frac{w_B \times 1000}{m_B \times V}$$

   Molarity is dependent on volume and temperature. It decreases with increase of temperature, as volume increases with increase in temperature.

8. **Normality:** It is defined as the number of gram equivalents of solute present per litre of solution. It is denoted by 'N'.

$$\text{Normality(N)} = \frac{\text{Number of gram equivalent of solute}}{\text{Volume of solution in litres}}$$

   Or Normality × Volume of solution in litres

   = Number of gram equivalents of the solute.

Before dilution		After dilution
M_1V_1	=	M_2V_2
N_1V_1	=	N_2V_2

- Molality of a solution does not change with temperature as it involves masses of the solute and solvent. Normality and molarity change with temperature because they involve volumes which change with temperature.
- Molarity and Normality are related to each other as follows;

  Normality = Molarity × n

$$\text{Normality} = \text{Molarity} \times \frac{\text{Molecular mass of solute}}{\text{equivalent mass of solute}}$$

Solubility

- Solubility of a substance is its maximum amount that can be dissolved in a specified amount of solvent at a specified temperature.
- It is dependent on the nature of solute and solvent along with temperature and pressure.

Solubility of a gas in a liquid

- The solubility of all gases in water decrease with increasing temperature. That is why, carbonated drinks that has **CO_2 gas dissolved in them** will become **"flat" tasting when heated**. The sparkle of the drink will have disappeared along with the CO_2 gas.

- **Henry's law:** Henry was the first to give a quantitative relation between pressure and solubility of a gas in a solvent which is known as Henry's law. The law states that at a constant temperature, the solubility of a gas in a liquid is directly proportional to the partial pressure of the gas present above the surface of liquid or solution. The most commonly used form of Henry's law states that "the partial pressure of the gas in vapour phase (p) is proportional to the mole fraction of the gas (x) in the solution" and is expressed as: $p \propto x$

  $p = K_H x$ (K_H = Henry's law constant.)

  - K_H is dependent upon nature of gas. Generally, higher the value of K_H at a particular temperature, lower is the solubility of gas in the liquid.
  - Different gases have different K_H values at the same temperature.
  - Henry's law finds several applications in industry and explains some biological phenomena. One of this is given as, to increase the solubility of CO_2 in soft drinks and soda water, the bottle is sealed under high pressure.

Vapour Pressure of Liquid Solutions

- **Raoult's Law:** It states that in a solution, the partial pressure of a component at a given temperature is equal to the mole fraction of that component in the solution multiplied by the vapour pressure of that component in the pure state.
- $P_{total} = P_A + P_B$

$$= p_A^0 x_A + p_B^0 x_B = p_A^0(1 - x_B) + p_B^0 x_B$$
$$= p_A^0 - p_A^0 x_B + p_B^0 x_B = p_A^0 + x_B(p_B^0 - p_A^0)$$

Ideal and Non-ideal Solutions

The solution in which the interactions between A and B are of the same magnitudes as in the pure components, or a solution which obeys Raoult's law at all temperatures and concentrations is known as **ideal solution**. An ideal solution will show no change in volume or enthalpy on mixing, i.e,

$\Delta V_{mixing} = 0, \ \Delta H_{mixing} = 0$

- Benzene + toluene is an example of ideal solution.

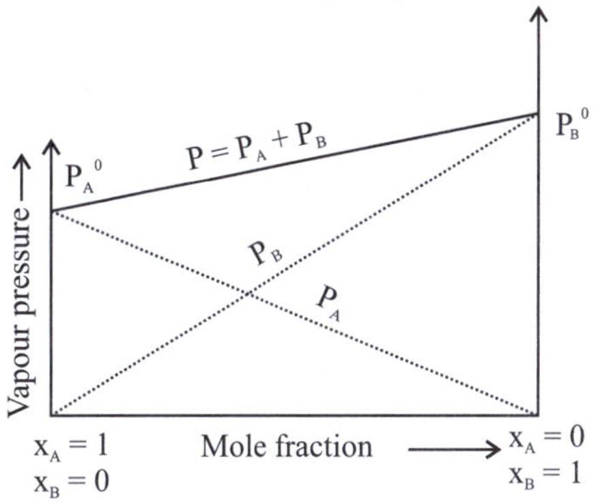

Non-ideal solutions

When a solution does not obey Raoult's law over the entire range of concentration, then it is called non-ideal solution. The vapour pressure of such a solution is either higher or lower than that predicted by Raoult's law. If it is higher, the solution exhibits positive deviation and if it is lower, it exhibits negative deviation from Raoult's law.

Non-ideal solutions are the solutions in which A–B interactions are of different magnitudes than in pure components or

$\Delta V_{mixing} \neq 0, \Delta H_{mixing} \neq 0.$

There are two types of deviation in non-ideal solutions:

- C_2H_5OH + cyclohexane shows **positive deviations** in non-ideal solution. The H-bonding present in pure C_2H_5OH are cut off on adding cyclohexane. For such solutions, ΔV and ΔH are +ve.
- Chloroform + Acetone shows **negative deviations** in non-ideal solution. The new interactions are stronger than in the pure components. For such solutions, ΔV and ΔH are –ve.

Positive deviation	Negative deviation
The total vapour pressure determined experimentally is higher than the calculated from Raoult's law. This is due to weak interactions than in pure components	The total vapour pressure determined experimentally is lower than the calculated from Raoult's law. This is due to strong interactions than in pure components.
$\Delta V_{mixing} = +ve, \Delta H_{mixing} = +ve$ 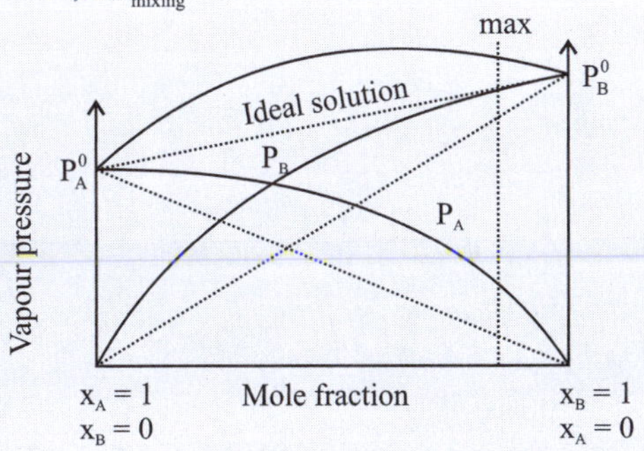	$\Delta V_{mixing} = -ve, \Delta H_{mixing} = -ve$ 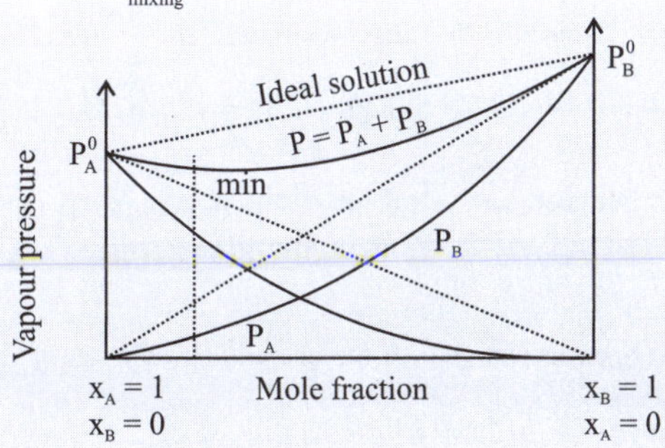

Colligative Properties

Relative lowering of vapour pressure

- When a non-volatile solute is added to a solvent, the vapour pressure is lowered, this proccess is known as lowering in vapour pressure.
- Relative lowering of vapour pressure:

$$\frac{p_1^0 - p_1}{p_1^0} = x_2 = \frac{n_2}{n_1 + n_2} = \frac{n_2}{n_1} \ (\because \text{ For dilute solution, } n_2 << n_1)$$

$$\frac{p_1^0 - p_1}{p_1^0} = \frac{w_2 \times M_1}{M_2 \times w_1}$$

Elevation of boiling point

- The increase in boiling point when a non volatile solute is added to solvent, addition of the solute lowers the vapour pressure of solvent, hence more heat is required to increase the vapour pressure upto the atmospheric pressure. It is called as elevation in boiling point.
- For dilute solution : $\Delta T_b \propto m$ or $\Delta T_b = K_b m = K_b \left(\frac{w_B \times 1000}{M_B \times w_A}\right)$

$$\Rightarrow M_B = \left(\frac{1000 \times w_B \times K_b}{\Delta T_b \times w_A}\right)$$

Where, $\Delta T_b = T_b - T_b^0$

m = molality

K_b = boiling point elevation constant or molal elevation constant or Ebullioscopic constant with unit : K kg mol^{-1}

Depression of freezing point

- Addition of a non-volatile solute lowers the vapour pressure of the solvent, hence the solid form separates out at a lower temperature known as depression in freezing point.
- It is the decrease in freezing point when a non-volatile solute is added to the solvent.
- Depression of freezing point:

For dilute solution : $\Delta T_f \propto m$ or $\Delta T_f = K_f \times m$

$$= K_f \left(\frac{w_B \times 1000}{M_B \times w_A}\right)$$

$$M_B = \left(\frac{1000 \times w_B \times K_f}{\Delta T_f \times w_A}\right)$$

Where, $\Delta T_f = T_f^0 - T_f$

m = molality

K_f = freezing point depression constant or cryoscopic constant or molal depression constant with unit : K kg mol^{-1}

Osmosis

- The equilibrium hydrostatic pressure of the column set up as a result of osmosis known as osmotic pressure and also defined as the minimum external pressure applied to stop osmosis.
- Two solution of different substances having same osmotic pressure at same temperature are known as **isotonic solutions**.
- A solution having osmotic pressure more than the other solution is known as **hypertonic solution**.
- A solution having osmotic pressure less than the other solution is known as hypotonic solution.
- Osmosis and osmotic pressure :

$$\pi = CRT = \left(\frac{n_2}{V}\right)RT$$

$$\pi = \frac{w_2 RT}{M_2 V} \Rightarrow M_2 = \frac{w_2 RT}{\pi V}$$

- Abnormal molecular weights and colligative properties are observed in some cases where the experimental and theoretical values differ considerably.

Since; colligative property $\propto \dfrac{1}{\text{molecular weight}}$

Van't Hoff factor

The ratio of the experimental value of the colligative property to the calculated value of the colligative property is defined by Van't Hoff factor.

$$i = \frac{\text{Observed value of colligative property}}{\text{Calculated value of colligative property}}$$

$$= \frac{\text{Normal molar mass}}{\text{Abnormal molar mass of solute}}$$

$$i = \frac{\text{Total number of moles of particles after association / dissociation}}{\text{Total number of moles of particles before association / dissociation}}$$

For association, $i < 1$

For dissociation, $i > 1$

- Modified equations for colligative properties:-

$$\frac{p_1^0 - p_1}{p_1^0} = i \cdot \frac{n_2}{n_1}, \ \Delta T_b = iK_b m, \ \Delta T_f = iK_f m, \ \Pi = in_2 RT/V$$

- Relation between Van't Hoff factor and degree of dissociation,

$$\alpha = \frac{i - 1}{n - 1}$$

- Relation between Van't Hoff factor and degree of association,

$$\alpha = \frac{1 - i}{1 - 1/n}$$

MCQs

NCERT Topic-wise MCQs

Expressing Concentrations of Solutions

1. The molality of the solution prepared by dissolving 125 mL of pure methanol ($d = 0.8 gmL^{-1}$) in 375 g of ethanol is:
   1) 10.8 2) 11.2
   3) 8.33 4) 10.4
2. When the volume of the solution is doubled, the following becomes exactly half
   1) Molality 2) Mole-fraction
   3) Molarity 4) Weight percent
3. What is the mole fraction of glucose in 10% w/w glucose solution?
   1) 0.01 2) 0.02
   3) 0.03 4) 0.04
4. 138 g ethyl alcohol is mixed with 72 g of water. The ratio of mole fraction of alcohol to water is:
   1) 3 : 4 2) 1 : 2
   3) 1 : 4 4) 1 : 1
5. 2.5 cm^3 of 0.2 M H_2SO_4 solution is diluted to 0.5 dm^3. Find normality of the diluted solution:
   1) 0.2 N 2) 0.02 N
   3) 0.002 N 4) 0.04 N
6. What is the molarity of 0.2 N Na_2CO_3 solution?
   1) 0.1 M 2) 0 M
   3) 0.4 M 4) 0.2 M
7. The normality of orthophosphoric acid having purity of 70% by weight and specific gravity 1.54 gm/mL is:
   1) 11 N 2) 22 N
   3) 33 N 4) 44 N
8. 0.126 g of an acid requires 20 mL of 0.1 N NaOH for complete neutralization. The equivalent weight of the acid is:
   1) 63 g 2) 50 g
   3) 53 g 4) 23 g
9. Mole fraction of a solute in benzene is 0.2, then find molality of solution:
   1) 3.2 2) 2
   3) 4 4) 3.6
10. How much volume of 1 M H_2SO_4 is required to neutralise 20 mL of 1 M NaOH?
    1) 10 mL 2) 20 mL
    3) 5 mL 4) 15 mL
11. A solution which is described as 10 % glucose in water by mass means
    1) 10 g of glucose is dissolved in 90 g of water resulting in a 100 g solution
    2) 90 g of glucose is dissolved in 10 g of water resulting in a 100 g solution
    3) 100 g glucose is dissolved in 100 g of water resulting in a 100 g solution
    4) 10 g of glucose is dissolved in 110 g of water resulting in a 100 g solution
12. In a solution of 7.8 gm benzene (C_6H_6) and 46.0 gm toluene ($C_6H_5CH_3$), the mole fraction of benzene in this solution is
    1) 1/6 2) 1/5
    3) 1/2 4) 1/3
13. Which of the following is temperature dependent?
    1) Parts per million 2) Molality
    3) Mole fraction 4) Molarity
14. 171 g of cane sugar ($C_{12}H_{22}O_{11}$) is dissolved in 1 litre of water. The molarity of the solution is
    1) 2.0 M 2) 1.0 M
    3) 0.5 M 4) 0.25 M

Solubility

15. H_2S is a toxic gas used in qualitative analysis. If solubility of H_2S in water at STP is 0.195 m, what is the value of K_H?
    1) 0.0263 bar 2) 69.16 bar
    3) 192 bar 4) 282 bar
16. The law which indicates the relationship between solubility of a gas in liquid and pressure is:
    1) Roult's law
    2) Henry's law
    3) Lowering of Vapour pressure
    4) Van't Hoff's law
17. Which one of the following gases has the lowest value of Henry's law constant?
    1) N_2 2) He
    3) CO_2 4) O_2

18. Which of the following equations represents Henry's law?
1) $p = K_H x$ 2) $p_{total} = p_1 + p_2$
3) $p = K_H + p_1$ 4) $p = K_H \div p_2$

19. Solubility of gases in liquids
1) Increases with decrease in temperature
2) Increases with rise in temperature
3) Remain unaffected with rise in temperature
4) None of these

Vapour Pressure of Liquid Solutions

20. Which of the following gas mixture is used by the divers inside the sea?
1) $O_2 + He$ 2) $O_2 + Xe$
3) $O_2 + Ar$ 4) $O_2 + CO_2$

21. If two components A and B have $p_A^0 : p_B^0 = 1 : 2$ and have mole fraction in solution as 1 : 2 then mole fraction of A in vapour is:
1) 0.33 2) 0.25
3) 0.52 4) 0.2

22. Vapour pressure of chloroform ($CHCl_3$) and dichloromethane (CH_2Cl_2) at 25°C are 200 mm Hg and 41.5 mm Hg respectively. Vapour pressure of the solution obtained by mixing 25.5 g of $CHCl_3$ and 40 g of CH_2Cl_2 at the same temperature will be (Molecular mass of $CHCl_3$ = 119.5 u and molecular mass of CH_2Cl_2 = 85 u):
1) 173.9 mmHg 2) 615.0 mmHg
3) 347.9 mmHg 4) 90.63 mmHg

23. Among the following substance the lowest vapour pressure is exerted by:
1) Water 2) Alcohol
3) Ether 4) Mercury

24. Partial pressure of a solution component is directly proportional to its mole fraction. This is known as:
1) Henry's law 2) Raoult's law
3) Distribution law 4) Ostwald's dilution law

25. 3 moles of 'P' and 2 moles of "Q" are mixed, what will be their total vapour pressure in the solution if their partial vapour pressures in pure state are 80 and 60 torr respectively:
1) 80 torr 2) 140 torr
3) 72 torr 4) 70 torr

26. In which case Raoult's law is not applicable?
1) 1 m NaCl 2) 1 m urea
3) 1 m glucose 4) 1 m sucrose

27. Two liquids X and Y form an ideal solution. The mixture has a vapour pressure of 400 mm at 300 K when mixed in the molar ratio of 1 : 1 and a vapour pressure of 350 mm when mixed in the molar ratio of 1 : 2 at the same temperature. The vapour pressures of the two pure liquids X and Y respectively are:
1) 250 mm, 550 mm 2) 350 mm, 450 mm
3) 350 mm, 700 mm 4) 550 mm, 250 mm

28. Two liquids X and Y form an ideal solution. At 300 K, vapour pressure of the solution containing 1 mol of X and 3 mol of Y is 550 mm Hg. At the same temperature if 1 mol of Y is further added to this solution, vapour pressure of the solution increases by 10 mm Hg. Vapour pressure (in mm Hg) of X and Y in their pure states will be respectively
1) 200 and 300 2) 300 and 400
3) 400 and 600 4) 500 and 600

29. The vapour pressure of two liquids P and Q are 80 and 60 torr, respectively. The total vapour pressure of solution obtained by mixing 3 mole of P and 2 mole of Q would be
1) 140 torr 2) 20 torr
3) 68 torr 4) 72 torr

Ideal and Non-Ideal Solutions

30. The system that forms maximum boiling azeotrope is:
1) Carbon disulphide - acetone
2) Benzene - toluene
3) Acetone - chloroform
4) n - hexane - n heptane

31. A solution of acetone in ethanol:
1) Shows a negative deviation from Raoult's law
2) Shows a positive deviation from Raoult's law
3) Behaves like a near ideal solution
4) Obey Raoult's law

32. Which of the following will show a negative deviation from Raoult's law?
1) n-Hexane and n-Heptane
2) Acetone and ethanol
3) Benzene and toluene
4) Acetone and chloroform

33. A solution containing components A and B follows Raoult's law:
1) A – B attraction is greater than A – A and B – B
2) A – B attraction forces is less than A – A and B – B
3) A – B attraction forces remains same as A – A and B – B
4) Volume of solution is different from sum of volumes of solute and solvent

34. What are the conditions for an ideal solution which obeys Raoult's law over the entire range of concentration?
1) $\Delta H_{mix} = 0, \Delta V_{mix} = 0, P_{total} = p_A^0 x_A + p_B^0 x_B$
2) $\Delta H_{mix} = +ve, \Delta V_{mix} = 0, P_{total} = p_A^0 x_A + p_B^0 x_B$
3) $\Delta H_{mix} = 0, \Delta V_{mix} = +ve, P_{total} = p_A^0 x_A + p_B^0 x_B$
4) $\Delta H_{mix} = 0, \Delta V_{mix} = 0, P_{total} = p_B^0 x_B$

35. An ideal solution was obtained by mixing methanol and ethanol. If the partial vapour pressure of methanol and ethanol in solution are 2.619 kPa and 4.556 kPa respectively, the composition of vapour (in terms of mole fraction) will be :

1) 0.635 MeOH, 0.365 EtOH
2) 0.365 MeOH, 0.635 EtOH
3) 0.574 MeOH, 0.326 EtOH
4) 0.173 MeOH, 0.827 EtOH

36. Which of the following azeotropes is not correctly matched?

1) HNO_3 (68%) + H_2O (32%): Maximum boiling azeotrope, B.P = 393.5 K
2) H_2O (43%) + HI (56.7%): Minimum boiling azeotrope, B.P = 290 K
3) C_2H_5OH(95.5%) + H_2O(4.5%): Minimum boiling azeotrope, B.P. = 351.15 K
4) Chloroform (93.2%) + C_2H_5OH (6.8%): Minimum boiling azeotrope, B.P. = 332.3 K

37. Study the figures given below and mark the correct statement:

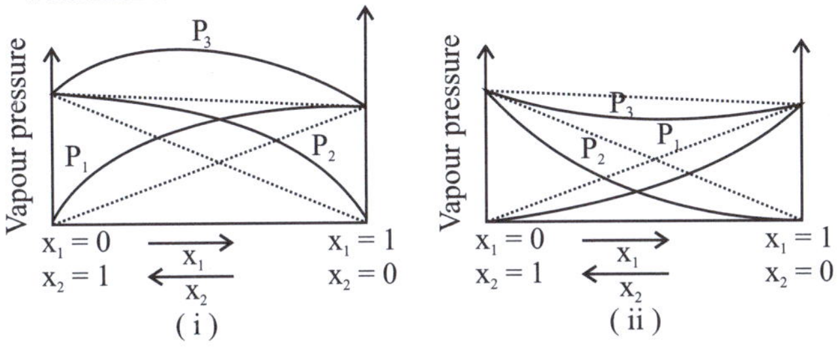

1) (i) Nitric acid + water (ii) Acetone + ethyl alcohol
2) (i) Water + ethyl alcohol (ii) Acetone + Benzene
3) (i) Acetone + ethyl alcohol (ii) Acetone + chloroform
4) (i) Benzene + chloroform (ii) Acetone + chloroform

38. A solution of two liquids boils at a temperature more than the boiling point of either of them. Hence, the binary solution shows:

1) Negative deviation from Raoult's law
2) Positive deviation from Raoult's law
3) No deviation from Raoult's law
4) Positive or negative deviation from Raoult's law depending upon the composition.

39. Out of the compounds given below, the vapour pressure of (B) at a particular temperature is:

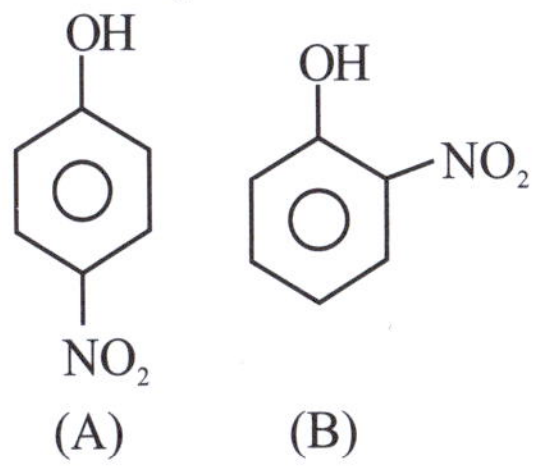

1) Higher than that of A
2) Lower than that of B
3) Higher of lower than (A), depending on the size of the vessel
4) Same as that of (A)

Colligative Properties & Determination of (Molar) Mass

40. The vapour pressure of pure solvent is 0.8 mm of Hg at a particular temperature. On addition of a non - volatile solute 'A' the vapour pressure of solution becomes 0.6 mm of Hg. The mole fraction of component 'A' is:

1) 0.25
2) 0.75
3) 0.5
4) 0.35

41. Lowering of vapour pressure of an aqueous solution of a non-volatile, non-electrolyte 1 molal aqueous solution at 100°C is:

1) 14.12 torr
2) 312 torr
3) 13.45 torr
4) 352 torr

42. The amount of solute (molar mass 60 g mol^{-1}) that must be added to 180 g of water so that the vapour pressure of water is lowered by 10% is:

1) 30 g
2) 60 g
3) 120 g
4) 12 g

43. A solution containing 12.5 g of non-electrolyte substance in 185 g of water shows boiling point elevation of 0.80 K. Calculate the molar mass of the substance: (K_b = 0.52 K kg mol^{-1})

1) 53.06 g mol^{-1}
2) 25.3 g mol^{-1}
3) 16.08 g mol^{-1}
4) 43.92 g mol^{-1}

44. 10% solution of urea is isotonic with 6% solution of a non-volatile solute X, what is the molecular mass of solute X?

1) 6 g mol^{-1}
2) 60 g mol^{-1}
3) 36 g mol^{-1}
4) 32 g mol^{-1}

45. Osmotic pressure of a solution containing 2 g dissolved protein per 300 cm^3 of solution is 20 mm of Hg at 27°C. The molecular mass of protein is:

1) 6239.6 g mol^{-1}
2) 12315.5 g mol^{-1}
3) 3692.1 g mol^{-1}
4) 7368.4 g mol^{-1}

46. A solution is made by dissolving 20 g of a substance in 500 mL of water. Its osmotic pressure was found to be 600 mm of Hg at 15°C. Find the molecular weight of the substance:

1) 1198 g mol^{-1}
2) 500 g mol^{-1}
3) 1200 g mol^{-1}
4) 1000 g mol^{-1}

47. An aqueous solution of 2% non-volatile solute exerts a pressure of 1.004 bar at a normal boiling point of the solvent. What is the molecular mass of the solute?

1) 23.4 g mol^{-1}
2) 41.35 g mol^{-1}
3) 10 g mol^{-1}
4) 20.8 g mol^{-1}

48. Which of the following statement is correct?

1) A saturated solution will remain saturated at all temperatures
2) A plant cell swells when placed in hypertonic solution
3) The depression in freezing point is directly proportional to molality of the solution
4) None of these

49. A solution containing 10.2 g glycerine per litre is isotonic with a 2% (w/v)solution of glucose. What is the molecular mass of glycerine?

1) 91.8 g 2) 1198 g
3) 83.9 g 4) 890.3 g

50. What weight of glycerol should be added to 600 g of water in order to lower its freezing point by 10°C? (K_f = 1.86 °C kg mol^{-1})

1) 496 g 2) 297 g
3) 310 g 4) 426 g

51. If 1 g of solute (molar mass = 50 g mol^{-1}) is dissolved in 50 g of solvent and the elevation in boiling point is 1 K. The molal elevation constant of the solvent is:

1) 2 2) 3
3) 2.5 4) 5

52. The colligative property is not represented by:

1) Elevation in boiling point
2) Osmotic pressure
3) Optical activity
4) Relative lowering of vapour pressure

53. If the elevation in boiling point of a solution of 10 g of solute (molecular weight = 100) in 100 g of water is ΔT_b, the ebullioscopic constant of water is:

1) 10 2) 100 T_b
3) ΔT_b 4) $\frac{\Delta T_b}{10}$

54. The osmotic pressure is expressed in the unit of:

1) MeV 2) Calorie
3) cm/sec 4) atm

55. Dissolution of 1.5 g of a non-volatile solute (molecular weight = 60) in 250 g of a solvent reduces its freezing point by 0.01°C. Find the molal depression constant of the solvent.

1) 0.01 2) 0.001
3) 0.0001 4) 0.1

56. A solution has higher osmotic pressure than its standard solution. Which of the following term will be used for this solution?

1) Isotonic
2) Hypertonic
3) Dilute
4) Hypotonic

57. 20 g of a non-volatile solute is added to 500 g of solvent, freezing point of pure solvent = 5.48°C and that of solution is 4.47°C, K_f = 1.93 °C kg mol^{-1}, molecular mass of solute is:

1) 77.2 2) 76.4
3) 73.2 4) 70.6

58. The boiling point of $CHCl_3$ was raised by 0.323 K when 0.5143 g of anthracene was dissolved in 35 g $CHCl_3$. Calculate the molecular mass of anthracene. K_b' for $CHCl_3$ = 3.9 K mol^{-1} kg.

1) 177.42 g mol^{-1} 2) 246.82 g mol^{-1}
3) 132.51 g mol^{-1} 4) 54.92 g mol^{-1}

59. The mass of glucose that should be dissolved in 100 g of water in order to produce same lowering of vapour pressure as is produced by dissolving 1 g of urea (mol. Mass = 60) in 50 g of water is : (Assume dilute solution in both cases)

1) 1 g 2) 2 g
3) 6 g 4) 12 g

60. If all the four compounds were sold at the same price, which would be cheapest for preparing an antifreeze solution for a car radiator?

1) CH_3OH 2) C_2H_5OH
3) $C_2H_4(OH)_2$ 4) $C_3H_5(OH)_3$

61. The vapour pressure of a solvent decreases by 10 mm of Hg when a non-volatile solute was added to the solvent. The mole fraction of solute is 0.2, what should be the mole fraction of the solvent if the decrease in vapour pressure is to be 20 mm of Hg.

1) 0.6 2) 0.4
3) 0.2 4) 1

62. The unit of K_f is

1) Mol L^{-1} 2) K kg mol^{-1}
3) K g L^{-1} 4) Mol kg^{-1}

63. The vapour pressure lowering caused by the addition of 100 g of sucrose(molecular mass = 342) to 1000 g of water if the vapour pressure of pure water at 25°C is 23.8 mm Hg is:

1) 1.25 mm Hg 2) 0.125 mm Hg
3) 1.15 mm Hg 4) 00.12 mm Hg

64. The rise in the boiling point of a solution containing 1.8 gram of glucose in 100g of a solvent is 0.1°C. The molal elevation constant of the liquid is

1) 0.01 K/m 2) 0.1K/m
3) 1K/m 4) 10K/m

65. 1.00 gm of a non-electrolyte solute dissolved in 50 gm of benzene lowered the freezing point of benzene by 0.40 K. K_f for benzene is 5.12 K kg mol^{-1}. Molecular mass of the solute will be

1) 256 g mol^{-1} 2) 2.56 g mol^{-1}
3) 512×10^3 g mol^{-1} 4) 2.56×10^4 g mol^{-1}

Abnormal Molar Mass

66. A compound MX_2 has observed and normal molar masses 65.6 and 164 respectively. Calculate the apparent degree of ionization of MX_2 :

1) 75% 2) 85%
3) 65% 4) 25%

67. Which has the least freezing point?
   1) 1% sucrose 2) 1% NaCl
   3) 1% $CaCl_2$ 4) 1% glucose

68. The freezing point of one molal NaCl solution, assuming NaCl to be 100% dissociated in water is: (molal depression constant is 1.86)
   1) – 2.72°C 2) – 3.72°C
   3) 2.72°C 4) 3.72°C

69. Which of following representations of i (van't hoff factor) is not correct?
   1) $i = \dfrac{\text{Observed colligative property}}{\text{Expected colligative property}}$
   2) $i = \dfrac{\text{Normal molecular mass of solute}}{\text{Observed molecular mass of solute}}$
   3) $i = \dfrac{\text{No. of molecules actually present}}{\text{No. of molecules expected to be present}}$
   4) $i = \dfrac{\text{Total particles before association / dissociation}}{\text{Total particles after association / dissociation}}$

70. Which of the following will have same value of Van't Hoff factor as that of $K_4[Fe(CN)_6]$?
   1) $Al_2(SO_4)_3$ 2) $AlCl_3$
   3) $Al(NO_3)_3$ 4) $Al(OH)_3$

71. What will be the degree of dissociation of 0.1 M $Mg(NO_3)_2$ solution if Van't Hoff factor is 2.74?
   1) 75% 2) 87%
   3) 100% 4) 92%

72. Which of the following has the highest freezing point?
   1) 1m NaCl solution
   2) 1m KCl solution
   3) 1m $AlCl_3$ solution
   4) 1m $C_6H_{12}O_6$ solution

73. If α is the degree of dissociation of Na_2SO_4, the van't Hoff factor (i) used for calculating the molecular mass is:
   1) 1 + α 2) 1 - α
   3) 1 + 2α 4) 1 - 2α

74. The Van't Hoff factor of 0.005 M aqueous solution of KCl is 1.95. The degree of ionisation of KCl is:
   1) 0.95 2) 0.97
   3) 0.94 4) 0.96

75. A 0.2 molal aqueous solution of weak acid (HA) is 20% ionised. What would be the freezing point of this solution? (Given, K_f = 1.86 °C m^{-1} for water)
   1) –0.45°C 2) –0.50°C
   3) –0.31°C 4) –0.53°C

76. Van't Hoff factor more than unity indicates that the solute in solution is:
   1) Dissociated
   2) Associated
   3) Both (1) and (2)
   4) Cannot say anything

77. KBr is 80% dissociated in aqueous solution of 0.5 m concentration (given, K_f for water = 1.86 K kg mol^{-1}). The solution freezes at:
   1) 271.326 K 2) 272 K
   3) 270.5 K 4) 268.5 K

78. Van't Hoff factor of $Ca(NO_3)_2$ is:
   1) One 2) Two
   3) Three 4) Four

79. Benzoic acid is dissolved in benzene, van't Hoff factor will be:
   1) 1 2) 0.5
   3) 1.5 4) 2

80. The freezing point depression constant for water is 1.86 K kg mol^{-1}. If 5.00 g Na_2SO_4 is dissolved in 45.0 g H_2O, the freezing point is changed by –3.82°C. Calculate the van't Hoff factor for Na_2SO_4:
   1) 0.381 2) 2.05
   3) 2.63 4) 3.11

81. The van't Hoff factor (i) for a compound which undergoes dissociation in one solvent and association in other solvent is respectively:
   1) Greater than one and greater than one
   2) Less than one and less greater than one
   3) Less than one and less than one
   4) Greater than one and less than one

82. What is the freezing point of a solution containing 8.1 g of HBr in 100 g water, assuming the acid to be 90% ionised? (K_f for water = 1.86 K kg mol^{-1})
   1) 0.85°C 2) –3.53°C
   3) 0°C 4) –0.35°C

83. Maximum lowering of vapour pressure is observed in the case of:
   1) 0.1 M glucose 2) 0.1 M $BaCl_2$
   3) 0.1 M $MgSO_4$ 4) 0.1 M NaCl

84. Van't Hoff factor of Hg_2Cl_2 in its aqueous solution will be : (Hg_2Cl_2 is 80% ionized in the solution)
   1) 1.6 2) 2.6
   3) 3.6 4) 4.6

High Order Time Intensive MCQs

1. A non-ideal solution was prepared by mixing 30 mL chloroform and 50 mL acetone. The volume of mixture will be:

1) > 80 mL 2) < 80 mL
3) = 80 mL 4) ≥ 80 mL

2. Equal weight of a solute are dissolved in equal weight of two solvents A & B and formed a very dilute solution. The relative lowering of vapour pressure for solution B has twice the relative lowering of vapour pressure for solution A. If M_A and M_B are molecular weights of solvents A & B respectively, then:

1) $M_B = 2M_A$ 2) $M_A = 4M_B$
3) $M_A = 2M_B$ 4) $M_A = M_B$

3. If M_{normal} is the normal molecular mass and α is the degree of ionization of $K_3[Fe(CN)_6]$, then the abnormal molecular mass of the complex is:

1) $M_{normal}(1+3\alpha)^{-1}$
2) $M_{normal}(1+2\alpha)^{-1}$
3) $M_{normal}(1+\alpha)^{-1}$
4) None of these

4. A living cell contains a solution which is isotonic with 0.2 M glucose solution. What osmotic pressure develops when the cell is placed in 0.05 M $BaCl_2$ solution at 300 K?

1) 1.23 atm 2) 6.15 atm
3) 3.69 atm 4) None of these

5. A quantity of 10 gm of solute 'A' and 20 g of solute 'B' is dissolved in 500 mL water. The solution is isotonic with the solution obtained by dissolving 6.67 g of 'A' and 30g of 'B' in 500 mL water at the same temperature. The ratio of molar masses $M_A : M_B$ is:

1) 1 : 1 2) 3 : 1
3) 1 : 3 4) 2 : 3

6. The vapour pressure of water is 12.3 kPa at 300 K. What is the vapour pressure of 1 molal aqueous solution of a non-volatile solute in it at 300 K?

1) 1.208 kPa 2) 12.08 kPa
3) 1208 kPa 4) 2.08 kPa

7. Which one of the following aqueous solutions will show the highest boiling point?

1) 0.015 M glucose
2) 0.10 M Na_2SO_4
3) 0.01 M KNO_3
4) 0.015 M urea

8. For an ideal binary liquid solution with $p_A^0 > p_B^0$, which relation between X_A (mole fraction of A in liquid phase) and Y_A (mole fraction of A in vapour phase) is correct?

1) $Y_A = Y_B$ 2) $\frac{Y_A}{Y_B} = \frac{X_A}{X_B}$
3) $\frac{Y_A}{Y_B} > \frac{X_A}{X_B}$ 4) $\frac{Y_A}{Y_B} < \frac{X_A}{X_B}$

9. 18 g glucose ($C_6H_{12}O_6$) is added to 178.2 g water. The vapour pressure of water (in torr) for this aqueous solution is:

1) 76 2) 752.4
3) 759 4) 760.4

10. How many grams of sucrose (M. mass = 342) should be dissolved in 100 grams H_2O in order to produce a solution with 105°C difference between the freezing point & boiling point temperature?

(K_f = 1.86 °C kg/mole) & (K_b = 0.51 °C kg/mole)

1) 34.2 gm 2) 72 gm
3) 342 gm 4) 460 gm

11. To 500 cm³ of water, 3×10^{-3} kg of CH_3COOH is added. If 23% of CH_3COOH is dissociated, what will be the depression in freezing point? (K_f for water is 1.86 K kg mol^{-1})

1) 0.372 K 2) 0.228 K
3) 0.328 K 4) 0.556 K

12. An aqueous solution of a compound AB freezes at –0.48°C. At this temperature, AB behaves as a non-electrolyte. The same solution boils at 100.26°C at a P of 1 atm (for water, $\frac{K_f}{K_b} = 3.7$) we can conclude that compound AB at the boiling point of solution,

1) Behaves as a non-electrolyte
2) Ionises to the extent of 50%
3) Dimerises completely in solution
4) Dissociates in solution

13. The freezing point of a solution containing 50 cm³ of ethylene glycol in 50 g water is found to be –34°C. Assuming ideal behaviour, the density of ethylene glycol is-

(K_f of water is 1.86 K kg mol^{-1})

1) 1.13 g/cm³ 2) 1.99 g/cm³
3) 2 g/cm³ 4) 6 g/cm³

14. A solution is prepared by dissolving 0.6 g of urea (molar mass = 60 g mol^{-1}) and 1.8 g of glucose (molar mass = 180 g mol^{-1}) in 100 mL of water at 27ºC. The osmotic pressure of the solution is:
($R = 0.08206$ L atm K^{-1} mol^{-1})
1) 8.2 atm 2) 1.64 atm
3) 4.92 atm 4) 2.46 atm

15. The quantity which changes with temperature is:
1) Molarity 2) Mass percentage
3) Molality 4) Mole fraction

16. If we have 10 molal urea solution, calculate mole fraction of urea in this solution and also calculate % w/w of urea. (MW = 60)
1) 0.152 and 37.5% 2) 0.352 and 37.5%
3) 0.152 and 58.6% 4) 0.521 and 48.6%

17. The vapour pressure of ethyl alcohol at 25°C is 59.2 torr. The vapour pressure of a solution of urea in ethyl alcohol is 51.3 torr. What is the molality of the solution?
1) 1.1 m 2) 3.34 m
3) 2.9 m 4) 4.86 m

18. Depression in freezing point of 0.01 molal aqueous acetic acid solution is found to be 0.02046 °C. One molal aq. urea solution freezes at –1.86°C. Assuming molarity = molality, what is the pH of acetic acid solution?
1) 3 2) 2
3) 3.89 4) 5.2

19. The pH of 1 M solution of a weak monobasic acid (HA) is 2. Then, what is its van't hoff factor?
1) 1.01 2) 2.3
3) 3.5 4) 7

20. Elevation in boiling point of 1 M aqueous glucose solution ($\rho = 1.2$ g/mL) is-
1) K_b 2) 1.2 K_b
3) 1.02 K_b 4) 0.98 K_b

21. A solution is prepared by mixing 100 mL each of methanol (d = 0.791 g/mL) and ethanol (d = 0.998 g/mL). If M_1 is molarity of solution described as MeOH solute in EtOH solvent and M_2 is molarity of solution described as EtOH solute in MeOH solvent. Then:
1) $M_1 < M_2$ 2) $M_1 > M_2$
3) $M_1 = M_2$ 4) $M_1 + M_2 > 50$

22. Total vapour pressure of mixture of 1 mol B (p^o_B = 150 torr) and 2 mol A (p^o_A = 240 torr) is 200 torr. In this case:
1) There is positive deviation from Raoult's law
2) There is negative deviation from Raoult's law
3) There is no deviation from Roult's law
4) Molecular masses of A and B are also required

23. Molality (m) of 3M aqueous solution of NaCl is: (Given: Density of solution = 1.25 g mL^{-1}, Molar mass in g mol^{-1} : Na-23,Cl-35.5)
1) 2.90 m 2) 2.79 m
3) 1.90 m 4) 3.85 m

24. A mixture of two immiscible liquids at a constant pressure of 1 atm boils at a temperature
1) Equal to the normal boiling point of more volatile liquid
2) Equal to the mean of the normal boiling points of the two liquids
3) Greater than the normal boiling point of either of the liquid
4) Smaller than the normal boiling point of either of the liquid

25. $PtCl_4 . 6H_2O$ can exist as a hydrated complex, 1 molal aqueous solution of the salt has value of ΔT_f of 3.72° C. Assuming 100% ionisation of the salt and K_f (H_2O) = 1.86° C mol^{-1} kg, the correct representation of the complex is
1) $[Pt(H_2O)_6]Cl_4$
2) $[Pt(H_2O)_4Cl_2].2H_2O$
3) $[Pt(H_2O)_3Cl_3]Cl.3H_2O$
4) $[Pt(H_2O)_2Cl_4].4H_2O$

26. At 300 K, the two solutions of glucose A and B in water with respective concentration 0.01 M and 0.001 M are separated by semipermeable membrane. How much external pressure is needed to be applied and on which solution so as to prevent osmosis?
1) 0.2463 atm pressure is applied on solution B
2) 0.2217 atm pressure is applied on solution A
3) 0.0246 atm pressure is applied on solution B
4) 0.0217 atm pressure is applied on solution A

27. At a given temperature, total vapour pressure in torr of a mixture of volatile components A and B is given by $P = 120 - 75\ x_B$, hence vapour pressure of pure components A and B respectively (in torr) are -
1) 120, 75 2) 120, 195
3) 120, 45 4) 75, 45

28. Mole fraction of the component A in vapour phase is x_1 and mole fraction of component A in liquid phase is x_2. (If p^0_A = vapour pressure of pure A; p^0_B = vapour pressure of pure B), then total vapour pressure of the liquid mixture is:
1) $\frac{p^0_A x_2}{x_1}$ 2) $\frac{p^0_A x_1}{x_2}$
3) $\frac{p^0_B x_1}{x_2}$ 4) $\frac{p^0_B x_2}{x_1}$

29. The boiling point of an aqueous solution of a non volatile solute is 100.15°C. What is the freezing point of an aqueous solution obtained by diluting the above solution with an equal volume of water? The values of K_b and K_f for water are 0.512 and 1.86 °C molality^{-1}

1) –0.544°C
2) –0.512°C
3) –0.272°C
4) –1.86°C

30. 'V_1' mL of an aqueous urea solution having osmotic pressure equal to 2.4 atm is mixed with 'V_2' mL of another aqueous urea solution having osmotic pressure equal to 4.6 atm at same temperature. What will be value of $\frac{V_1}{V_2}$ if resulting solution after mixing has osmotic pressure equal to 2.95 atm?

1) 1
2) 1.5
3) 2
4) 3

31. In which of the following pairs of solutions, will the values of the van't Hoff factor be the same?

1) 0.05 M $K_4[Fe(CN)_6]$ and 0.10 M $FeSO_4$
2) 0.10 M $K_4[Fe(CN)_6]$ and 0.05 M $FeSO_4.(NH_4)_2SO_4.6H_2O$
3) 0.20 M NaCl and 0.10 M $BaCl_2$
4) 0.05 M $FeSO_4(NH_4)_2SO_4.6H_2O$ and 0.02 M KCl. $MgCl_2.6H_2O$

32. The vapour pressure of the solution of two liquids A(p^0 = 80 mm) and B(p^0 = 120 mm) is found to be 100 mm of Hg when x_A = 0.4. The result shows that

1) Solution exhibits ideal behaviour
2) $\Delta_{mix}H < 0$
3) Solution shows positive deviation
4) Solution will show positive deviation for lower concentration and negative deviation for higher concentration.

33. Dissolving 120 g urea (mol.wt = 60) in 1000 g of water results in a solution of density 1.15 g/mL. The molarity of the solution is:

1) 1.78 M
2) 3 M
3) 2.05 M
4) 2.22 M

Match the Columns MCQs

1. Match the items given in Column-I and Column-II.

	Column-I		Column-II
(a)	Saturated solution	(i)	Solution having same osmotic pressure at a given temperature as that of given solution
(b)	Binary solution	(ii)	A solution whose osmotic pressure is less than that of another.
(c)	Isotonic solution	(iii)	Solution with two components
(d)	Hypotonic solution	(iv)	A solution which contains maximum amount of solute that can be dissolved in a given amount of solvent at a given temperature
(e)	Solid solution	(v)	A solution whose osmotic pressure is more than that of another
(f)	Hypertonic solution	(vi)	A solution in solid phase

1) (a)-(i) (b)-(iii) (c)-(ii) (d)-(i) (e)-(v) (f)-(vi)
2) (a)-(ii) (b)-(iii) (c)-(vi) (d)-(v) (e)-(i) (f)-(vi)
3) (a)-(iv) (b)-(iii) (c)-(i) (d)-(ii) (e)-(vi) (f)-(v)
4) (a)-(iv) (b)-(iii) (c)-(ii) (d)-(i) (e)-(v) (f)-(vi)

2. Match the items given in Column-I with the type of solutions given in Column-II.

	Column-I		Column-II
(a)	Soda water	(i)	A solution of gas in solid
(b)	Sugar solution	(ii)	A solution of gas in gas
(c)	German silver	(iii)	A solution of solid in liquid
(d)	Air	(iv)	A solution of solid in solid
(e)	Hydrogen gas in palladium	(v)	A solution of gas in liquid
		(vi)	A solution of liquid in solid

1) (a)-(v) (b)-(iii) (c)-(iv) (d)-(ii) (e)-(i)
2) (a)-(ii) (b)-(i) (c)-(iii) (d)-(v) (e)-(iv)
3) (a)-(v) (b)-(iv) (c)-(iii) (d)-(ii) (e)-(i)
4) (a)-(i) (b)-(v) (c)-(vi) (d)-(ii) (e)-(iii)

3. Match the laws given in Column-I with expressions given in Column-II.

	Column-I		Column-II
(a)	Raoult's law	(i)	$\Delta T_f = K_f m$
(b)	Henry's law	(ii)	$\Pi = CRT$
(c)	Elevation of boiling point	(iii)	$p = x_1 p_1^0 + x_2 p_2^0$
(d)	Depression in freezing point	(iv)	$\Delta T_b = K_b m$
(e)	Osmotic pressure	(v)	$p = K_H.x$

1) (a)-(ii) (b)-(i) (c)-(iii) (d)-(v) (e)-(iv)
2) (a)-(iv) (b)-(ii) (c)-(i) (d)-(iii) (e)-(v)
3) (a)-(iv) (b)-(iii) (c)-(ii) (d)-(i) (e)-(v)
4) (a)-(iii) (b)-(v) (c)-(iv) (d)-(i) (e)-(ii)

4. Match the terms given in Column-I with expressions given in Column-II.

	Column-I		Column-II
(a)	Mass percentage	(i)	$\frac{\text{Number of mole of the solute component}}{\text{Volume of solution in litres}}$
(b)	Volume percentage	(ii)	$\frac{\text{Number of moles of a component}}{\text{Total number of moles of all the components}}$
(c)	Mole fraction	(iii)	$\frac{\text{Volume of the solute component in solution}}{\text{Total volume of solution}} \times 100$
(d)	Molality	(iv)	$\frac{\text{Mass of the solute component in solution}}{\text{Total mass of the solution}} \times 100$
(e)	Molarity	(v)	$\frac{\text{Number of moles of the solute components}}{\text{Mass of solvent in kilograms}}$

1) (a)-(v) (b)-(ii) (c)-(i) (d)-(iv) (e)-(iii)
2) (a)-(iv) (b)-(iii) (c)-(ii) (d)-(v) (e)-(i)
3) (a)-(i) (b)-(ii) (c)-(iv) (d)-(v) (e)-(iii)
4) (a)-(ii) (b)-(iii) (c)-(i) (d)-(v) (e)-(iv)

5. Match the terms given in Column-I with expressions given in Column-II.

	Column-I (van't Hoff factor)		Column-II (Behaviour of compound)
(a)	$i < 1$	(i)	Association in the solution
(b)	$i > 1$	(ii)	Dissociation in the solution
(c)	$i = 1$	(iii)	Impossible
(d)	$i < 0$	(iv)	No association or dissociation in the solution

1) (a)-(i) (b)-(ii) (c)-(iii) (d)-(iv)
2) (a)-(ii) (b)-(i) (c)-(iii) (d)-(iv)
3) (a)-(i) (b)-(ii) (c)-(iv) (d)-(iii)
4) (a)-(ii) (b)-(iii) (c)-(i) (d)-(iv)

Correct & Incorrect MCQs

1. Which of the following statement (s) are correct?
A. Polar solutes dissolve in polar solvents.
B. Sodium chloride and sugar dissolve readily in benzene.
C. Non polar solutes dissolve in non-polar solvents.
D. Naphthalene and anthracene dissolve readily in water.
1) A, B and C
2) A and C
3) A, C and D
4) B, C and D

2. Which of the following statements is correct, if the intermolecular forces in liquids A, B and C are in the order $A < B < C$?
1) B evaporates more readily than A
2) B evaporates less readily than C
3) A and B evaporate at the same rate
4) A evaporates more readily than C

3. If 0.1 M solution of glucose and 0.1 M urea solution are placed on two sides of a semipermeable membrane to equal heights, then it will be correct to say that -
1) There will be no net movement across the membrane
2) Glucose will flow towards urea solution
3) Urea will flow towards glucose solution
4) Water will flow from urea solution towards glucose solution

4. Select the correct statement:
1) Mole fraction has no units.
2) Unit of molarity is mol Kg^{-1}.
3) Mass percentage is dependent on temperature.
4) Molarity is independent of temperature.

5. Correct statement(s) is/are:
A. The solubility of gases in liquid increases with increase of temperature.
B. Aquatic species are more comfortable in warm water rather than in cold water
C. At high altitudes, the partial pressure of oxygen is less than that at the ground level.
D. There is low concentration of O_2 in the blood and tissue of the people living at high altitude.
1) A and B
2) B, C and D
3) C and D
4) Only A

6. Incorrect statement(s) is/are
A. A mixture of benzene and toluene is an ideal solution.
B. For ideal solution, $\Delta_{mix}H = 0$ and $\Delta_{mix}V = 0$.
C. A mixture of ethanol and acetone shows negative deviation from Raoult's law.
D. A mixture of chloroform and acetone forms a solution with positive deviation from Raoult's law
1) A and B
2) C and D
3) B, C and D
4) B and C

7. Which of the following statement about osmosis is not correct?
   1) The flow of solvent from concentrated solution to the dilute solution across a semipermeable membrane is due to osmosis.
   2) A raw mango placed in concentrated salt solution loses water via osmosis and shrivel into pickle.
   3) People taking a lot of salt or salty food experience water retention in tissue cells and intercellular spaces because of osmosis.
   4) Water movement from soil into plant roots and subsequently into upper portion of the plant is partly due to osmosis.

8. Which of the following statements is incorrect?
   1) In a solution containing several non-volatile solutes, the lowering of the vapour pressure depends on the sum of the mole fraction of different solutes.
   2) For dilute solutions the elevation of boiling point is directly proportional to the molal concentration of the solute in a solution.
   3) The value of K_f depends upon the nature of the solvent.
   4) Osmosis takes place in two isotonic solutions separated by semipermeable membrane.

9. Select the incorrect statement:
   1) The observed molar mass of acetic acid in benzene is more than the normal molar mass of acetic acid.
   2) The extent to which a substance associates or dissociates in a solution is described by the van't hoff factor (i).
   3) In case of association, value of i is greater than unity.
   4) In case of dissociation, value of i is greater than unity.

10. Select the correct statement:
   1) Higher the value of K_H at a given pressure, the higher is the solubility of the gas in the liquid.
   2) Higher the value of K_H at a given pressure, the lower is the solubility of the gas in the liquid.
   3) K_H is not a function of the nature of the gas.
   4) All statements are incorrect.

Statement Based MCQs

Directions: These questions consist of two statements each, printed as Statement-I and Statement-II. While answering these questions, you are required to choose any one of the following four responses.

1) **Both Statement-I and Statement-II are correct.**
2) **Both Statement-I and Statement-II are incorrect.**
3) **Statement-I is correct & Statement-II is incorrect.**
4) **Statement-I is incorrect & Statement-II is correct.**

1. **Statement-I:** Relative lowering of vapour pressure is equal to mole fraction of the solvent.
   Statement-II: Relative lowering of vapour pressure is not a colligative property.

2. **Statement-I:** In case of positive deviation from Raoult's law, A-B interactions are weaker than those between A-A or B-B.
   Statement-II: In case of negative deviations from Raoult's law, the intermolecular attractive forces between A-A and B-B are weaker than those between A-B.

3. **Statement-I:** The molality of the solution does not change with change in temperature.
   Statement-II: The molality is defined as the number of moles of solute per kilogram (kg) of the solvent.

4. **Statement-I:** NaCl in water and organic acids in benzene show abnormal molecular mass.
   Statement-II: Abnormal molecular mass is obtained when the substance in the solution undergoes dissociation or association.

5. **Statement-I:** If a liquid solute more volatile than the solvent, is added to the solvent,the vapour pressure of the solution may increases i.e. $P_S > P^0$.
   Statement-II: In the presence of a more volatile liquid solute, only the solute will form the vapours and solvent will not.

6. **Statement-I:** At equilibrium, vapour phase will be always rich in component which is more volatile.
   Statement-II: The composition of vapour phase in equilibrium with the solution is determined by the partial pressures of the components.

7. **Statement-I:** A solution of phenol and aniline will show negative deviations from Raoult's law.
   Statement-II: In case of negative deviations from Raoult's law, A – B forces are weaker than A – A and B – B forces.

8. **Statement-I:** 1 M solution of glucose has greater osmotic pressure than 1 M solution of KCl at the same temperature.
   Statement-II: In solution, KCl dissociates to produce more number of particles.

9. **Statement-I:** Azeotropes are binary mixtures having the same composition in liquid and vapour phase and boil at a constant temperature.
   Statement-II: Nitric acid and water form minimum boiling azeotrope.

10. **Statement-I:** An ideal solution does not obey Raoult's law over the entire range of concentration.
   Statement-II: In an ideal solution, solute-solute as well as solvent-solvent interactions are similar to solute-solvent interactions.

Assertion & Reason MCQs

Directions: These questions consist of two statements each, printed as Assertion and Reason. While answering these questions, you are required to choose any one of the following four responses.

1) **Both Assertion and Reason are True and the Reason is a correct explanation of the Assertion.**
2) **Both Assertion and Reason are True but Reason is not a correct explanation of the Assertion.**
3) **Assertion is True but the Reason is False.**
4) **Assertion is False but Reason is True.**

1. **Assertion:** Molarity of a solution in liquid state changes with temperature.
   Reason: The volume of a solution changes with change in temperature.
2. **Assertion:** When methyl alcohol is added to water, boiling point of water decreases.
   Reason: When a volatile solute is added to a volatile solvent, elevation in boiling point is observed.
3. **Assertion:** When NaCl is added to water, a depression in freezing point is observed.
   Reason: The lowering of vapour pressure of a solution causes depression in the freezing point.
4. **Assertion:** When a solution is separated from the pure solvent by a semipermeable membrane, the solvent molecules pass through it from pure solvent side to the solution side.
   Reason: Diffusion of solvent occurs from a region of high concentration solution to a region of low concentration solution.
5. **Assertion:** Mole fraction has no units.
   Reason: Mole fraction is independent of temperature.
6. **Assertion:** Solubility of a substance is its maximum amount that can be dissolved in a specified amount of solvent at a specified temperature.
   Reason: Solubility does not depend upon the nature of solute and solvent.
7. **Assertion:** 0.02 m solutions of urea and sucrose will freeze at same temperature.
   Reason: Freezing point of solution is inversely related to the concentration of solution.
8. **Assertion:** A solution of table salt in a glass of water is homogeneous.
   Reason: Solution having same composition throughout is heterogeneous.
9. **Assertion:** The freezing point is the temperature at which a solid dissolves in the solution.
   Reason: Depression of freezing point for dilute solution is directly proportional to molality, m of the solution.
10. **Assertion:** Osmotic pressure is a colligative property.
    Reason: Two solutions having same osmotic pressure at a given temperature are called isotonic solutions.

Multi Correct MCQs

1. A graph of vapour pressure and temperature for three different liquids X, Y and Z is shown below.

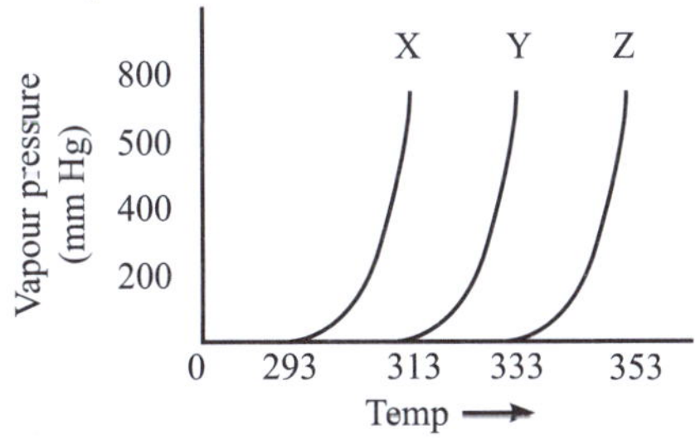

The following inferences are made
A. X has higher intermolecular interactions as compared to Y.
B. X has lower intermolecular interactions as compared to Y.
C. Z has lower intermolecular interactions as compared to Y.
The correct inferences is/are
1) (B) 2) (C)
3) (A) and (C) 4) (A)

2. Among the given units, which of the following set is used to express concentration of solutions?
A. Mass percent B. Mole
C. Mole fraction D. Molarity
E. ppm F. Molality.
1) (A), (C), (D), (E) and (F)
2) (D) and (F) only
3) Only (B)
4) (A), (B), (C) and (D)

3. Among the given pairs of the solution, choose the one with the same value of the osmotic pressure from the following. (Assume 100% ionization)
A. 0.500 M C_2H_5OH and 0.25 M KBr (aq)
B. 0.100 M $K_4[Fe(CN)_6]$ (aq) and 0.100 M $FeSO_4(NH_4)_2SO_4$(aq)
C. 0.05 M $K_4[Fe(CN)_6]$ (aq) and 0.25 M NaCl (aq)
D. 0.15 M NaCl (aq) and 0.1 M $BaCl_2$ (aq)
E. 0.02 M $KCl.MgCl_2.6H_2O$ (aq) and 0.05 M KCl (aq)
1) (A), (B), (C) and (D) 2) (C) and (D)
3) (B), (C), (D) and (E) 4) (A), (B), (D) and (E)

4. Of the following four aqueous solutions, which solutions have freezing point lower than that of 0.10 M C_2H_5OH.
A. 0.10 M $Ba_3(PO_4)_2$ B. 0.10 M Na_2SO_4
C. 0.10 M KCl D. 0.10 M Li_3PO_4
1) (A), (B), (C) and (D) 2) (B), (C) and (D)
3) (C) and (D) 4) (A), (C), (D)

NCERT Exemplar MCQs

1. Which of the following units is useful in relating concentration of solution with its vapour pressure?
   1) Mole fraction 2) Parts per million
   3) Mass percentage 4) Molality

2. On dissolving sugar in water at room temperature solution feels cool to touch. Under which of the following cases dissolution of sugar will be most rapid?
   1) Sugar crystals in cold water
   2) Powdered sugar in cold water
   3) Sugar crystals in hot water
   4) Powdered sugar in hot water

3. At equilibrium the rate of dissolution of a solid solute in a volatile liquid solvent is:
   1) less than the rate of crystallisation
   2) greater than the rate of crystallisation
   3) equal to the rate of crystallisation
   4) zero

4. A beaker contains a solution of substance A. Precipitation of substance A takes place when small amount of A is added to the solution The solution is:
   1) saturated 2) supersaturated
   3) unsaturated 4) concentrated

5. The maximum amount of a solid solute that can be dissolved in a specified amount of a given liquid solvent does not depend upon:
   1) Temperature 2) Nature of solute
   3) Pressure 4) Nature of solvent

6. Low concentration of oxygen in the blood and tissues of people living at high altitude is due to:
   1) low temperature
   2) low atmospheric pressure
   3) high atmospheric pressure
   4) both low temperature and high atmospheric pressure

7. Considering the formation, breaking and strength of hydrogen bond, predict which of the following mixtures will show a positive deviation from Raoult's law?
   1) Methanol and acetone 2) Chloroform and acetone
   3) Nitric acid and water 4) Phenol and aniline

8. Colligative properties depend on:
   1) the nature of the solute particles dissolved in solution
   2) the number of solute particles in solution
   3) the physical properties of the solute particles dissolved in solution
   4) the nature of solvent particles

9. The unit of ebullioscopic constant is:
   1) K kg mol^{-1} or K $(molality)^{-1}$
   2) mol kg K^{-1} or K^{-1} (molality)
   3) kg mol^{-1} K^{-1} or K^{-1} $(molality)^{-1}$
   4) K mol kg^{-1} or K (molality)

10. In comparison to a 0.01 M solution of glucose, the depression in freezing point of a 0.01 M $MgCl_2$ solution is:
    1) the same 2) about twice
    3) about three times 4) about six times

11. An unripe mango placed in a concentrated salt solution to prepare pickle shrivels because:
    1) It gains water due to osmosis
    2) It loses water due to reverse osmosis
    3) It gains water due to reverse osmosis
    4) It loses water due to osmosis

12. At a given temperature, osmotic pressure of a concentrated solution of a substance:
    1) is higher than that of a dilute solution
    2) is lower than that of a dilute solution
    3) is same as that of a dilute solution
    4) cannot be compared with osmotic pressure of dilute solution

13. Which of the following statements is false?
    1) Two different solutions of sucrose of same molality prepared in different solvents will have the same depression in freezing point.
    2) The osmotic pressure of a solution is given by the equation π = CRT (where, C is the molarity of the solution.
    3) Decreasing order of osmotic pressure for 0.01 M aqueous solutions of barium chloride, potassium chloride, acetic acid and sucrose is $BaCl_2 > KCl > CH_3COOH >$ sucrose
    4) According to Raoult's law, the vapour pressure exerted by a volatile component of a solution is directly proportional to its mole fraction in the solution

14. The values of Van't Hoff factors for KCl, NaCl and K_2SO_4 respectively are:
    1) 2, 2 and 2 2) 2, 2 and 3
    3) 1, 1 and 2 4) 1, 1 and 1

15. Which of the following statements is false?
    1) Units of atmospheric pressure and osmotic pressure are the same
    2) In reverse osmosis, solvent molecules move through a semipermeable membrane from a region of lower concentration of solute to a region of higher concentration
    3) The value of molal depression constant depends on the nature of the solvent
    4) Relative lowering of vapour pressure is a dimensionless quantity

16. Value of Henry's constant K_H:
    1) increases with increase in temperature
    2) decreases with increase in temperature
    3) remains constant
    4) first increases then decreases

17. The value of Henry's constant, K_H is:
    1) greater for gases with higher solubility
    2) greater for gases with lower solubility
    3) constant for all gases
    4) not related to the solubility of gases

18. Consider the figure and mark the correct option:

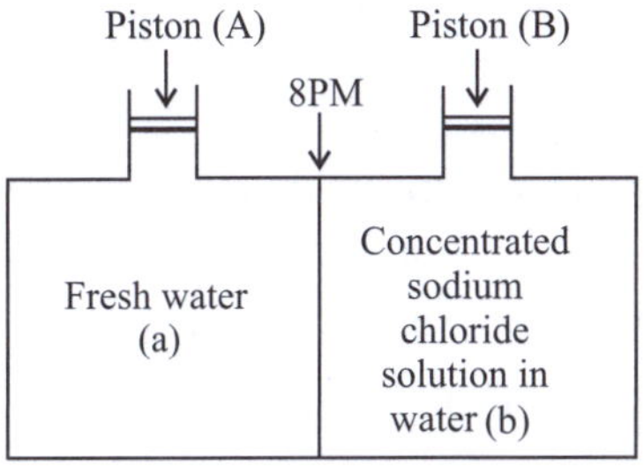

    1) Water will move from side (a) to side (b) if a pressure lower than osmotic pressure is applied on piston (b)
    2) Water will move from side (b) to side (a) if a pressure greater than osmotic pressure is applied on piston (b)
    3) Water will move from side (b) to side (a) if a pressure equal to osmotic pressure is applied on piston (b)
    4) Water will move from side (a) to side (b) if pressure equal to osmotic pressure is applied on piston (a)

19. We have three aqueous solutions of NaCl labelled as 'A', 'B' and 'C with concentrations 0.1 M, 0.01 M and 0.001 M, respectively. The value of van't Hoff factor for these solutions will be in the order:
    1) $i_A < i_B < i_C$
    2) $i_A > i_B > i_C$
    3) $i_A = i_B = i_C$
    4) $i_A < i_B > i_C$

20. On the basis of information given below mark the correct option: Information

    (i) In bromoethane and chloroethane mixture intermolecular interactions of A—A and B—B type are nearly the same as A—B type interactions.

    (ii) In ethanol and acetone mixture A—A or B—B type intermolecular interactions are stronger than A—B type interactions.

    (iii) In chloroform and acetone mixture A—A or B—B type intermolecular interactions are weaker than A—B type interactions:
    1) Solution (i) will follow Raoult's law
    2) Solution (ii) will show negative deviation from Raoult's law
    3) Solution (ii) and (iii) will follow Raoult's law.
    4) Solution (iii) will show positive deviation from Raoult's law

21. Two beakers of capacity 500 mL were taken. One of these beakers, labelled as "A", was filled with 400 mL water whereas the beaker labelled "B' was filled with 400 mL of 2 M solution of NaCl. At the same temperature, both the beakers were placed in closed containers of same material and same capacity as shown in figure:

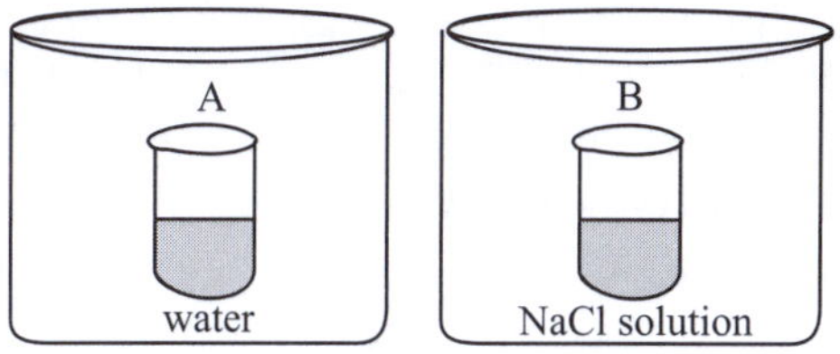

At a given temperature, which of the following statement is correct about the vapour pressure of pure water and that of NaCl solution?
    1) Vapour pressure in container (A) is more than that in container (B)
    2) Vapour pressure in container (A) is less than that in container (B)
    3) Vapour pressure is equal in both the containers
    4) Vapour pressure in container (B) is twice the vapour pressure in container (A)

22. In isotonic solutions:
    1) Solute and solvent both are same
    2) Osmotic pressure is same
    3) Solute and solvent may or may not be same
    4) Solute is always the same solvent may be different

23. For a binary ideal liquid solution, the variation in total vapour pressure versus composition of solution is given by which of the curves?

1) 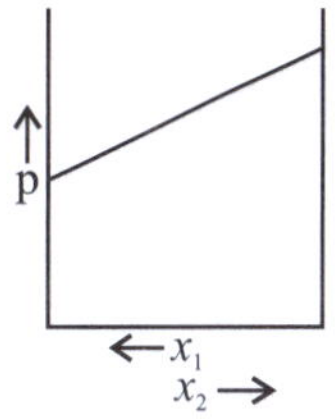

2) 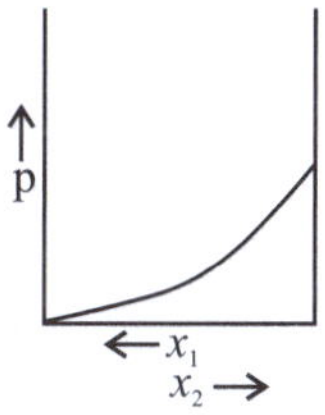

3) 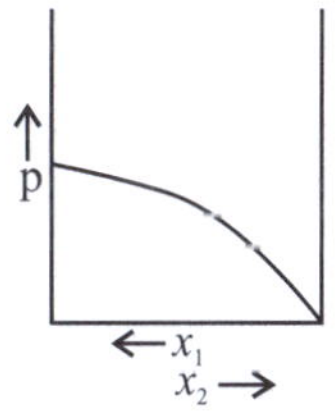

4) 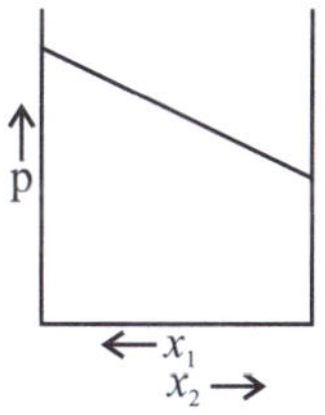

24. On the basis of information given below, mark the correct option.

    Information : On adding acetone to methanol, some of the hydrogen bonds between methanol molecules break:
    1) At specific composition, methanol-acetone mixture will form minimum boiling azeotrope and will show positive deviation from Raoult's law
    2) At specific composition, methanol-acetone mixture will form maximum boiling azeotrope and will show positive deviation from Raoult's law
    3) At specific composition, methanol-acetone mixture will form minimum boiling azeotrope and will show negative deviation from Raoult's law
    4) At specific composition, methanol-acetone mixture will form maximum boiling azeotrope and will show negative deviation from Raoult's law

25. K_H value for Ar(g), CO_2 (g), HCHO (g) and CH_4 (g) are 40.39, 1.67, 1.83×10^{-5} and 0.413 respectively. Arrange these gases in the order of their increasing solubility:
1) HCHO < CH_4 < CO_2 < Ar
2) HCHO < CO_2 < CH_4 < Ar
3) Ar < CO_2 < CH_4 < HCHO
4) Ar < CH_4 < CO_2 < HCHO

26. Which of the following factor(s) affect the solubility of a gaseous solute in the fixed volume of liquid solvent?
(i) Nature of solute
(ii) Temperature
(iii) Pressure
1) (i) and (iii) at constant T
2) (i) and (ii) at constant P
3) (ii) and (iii) only
4) Only (iii)

27. Intermolecular forces between two benzene molecules are nearly of same strength as those between two toluene molecules. For a mixture of benzene and toluene, which of the following are not true?
1) $\Delta_{mix} H$ = zero
2) $\Delta_{mix} V$ = zero
3) These will form minimum boiling azeotrope
4) These will not form ideal solution

28. Relative lowering of vapour pressure is a colligative property because:
1) It depends on the concentration of a non-electrolyte solute in solution and does not depend on the nature of the solute molecules
2) It depends on number of particles of electrolyte solute in solution and does not depend on the nature of the solute particles
3) It depends on the concentration of a non-electrolyte solute in solution as well as on the nature of the solute molecules
4) It depends on the concentration of an electrolyte or non-electrolyte solute in solution as well as on the nature of solute molecules

29. Van't Hoff factor (i) is given by the expression:
1) $i = \dfrac{\text{normal molar mass}}{\text{abnormal molar mass}}$
2) $i = \dfrac{\text{abnormal molar mass}}{\text{normal molar mass}}$
3) $i = \dfrac{\text{observed colligative property}}{\text{calculated colligative property}}$
4) $i = \dfrac{\text{calculated colligative property}}{\text{observed colligative property}}$

30. Isotonic solutions must have the same:
1) Solute
2) Density
3) Elevation in boiling point
4) Depression in freezing point

31. Which of the following binary mixtures will have same composition in liquid and vapour phase?
1) Benzene-toluene
2) Water-nitric acid
3) Water-ethanol
4) *n*-hexane – *n*-heptane

Kattar NEET

Scan Me for Detailed Explanations

1. Osmotic pressure [in atm] of a 0.1 M solution of $K_4[Fe(CN)_6]$, which undergoes 50% dissociation, will be _____ at 27°C:
1) 7.38
2) 3.69
3) 405.9
4) none of these

2. A solution of x moles of sucrose in 100 grams of water freezes at –0.2°C. As ice separates, the freezing point goes down to –0.25°C. How many grams of ice would have separated?
1) 18 grams
2) 20 grams
3) 25 grams
4) 23 grams

3. If freezing point of an aqueous urea solution is 271.14 K at 1 atmospheric pressure (if K_f of water = 1.86 K/m) then what should be mole fraction of urea in this solution?
1) $\dfrac{1}{56.5}$
2) 1
3) $\dfrac{55.5}{56.5}$
4) $\dfrac{1.86}{56.5}$

4. 1.22 g of benzoic acid is dissolved in 100 g of acetone and 100 g of benzene separately. Boiling point of the solution in acetone increases by 0.17°C while in benzene increases by 0.13°C. K_b for acetone and benzene is 1.7 K kg mol^{-1} and 2.6 K kg mol^{-1}. Molecular weight of benzoic acid in two cases are respectively are
1) 122 and 244
2) 244 and 122
3) 122 and 122
4) 224 and 122

5. Based on the given diagram, which of the following statements regarding the homogeneous solutions of two volatile liquids are correct?

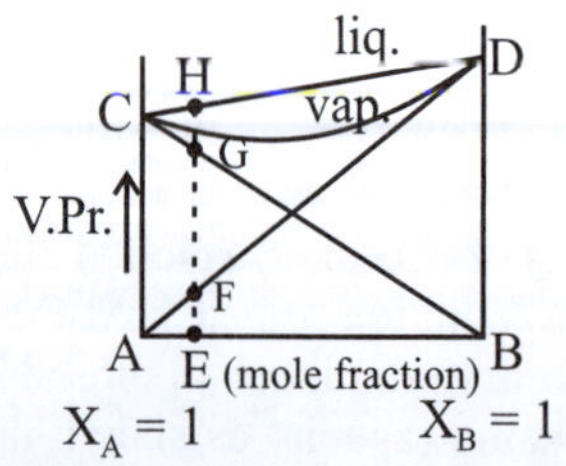

(A) Plots AD and BC show that Raoult's law is obeyed for the solution in which B is a solvent and A is the solute and as well as for that in which A is solvent and B is solute.

(B) Plot CD (vap.) shows that Dalton's law of partial pressures is obeyed by the binary solution of components A and B.

(C) EF + EG = EH; and AC and BD correspond to the vapour pressures of the pure solvents A and B respectively.

Select the correct answer using the options given below

1) Only A 2) B and C only

3) A and C only 4) A, B and C

6. Insulin is dissolved in a suitable solvent and the osmotic pressure (π) (in atm) of solutions of various concentrations C(g/cm^3) is measured at 20°C. The slope of a plot of π against C is found to be 4.65×10^{-3}. The molecular weight of the insulin (g/mol) is

1) 3×10^5 2) 9×10^5

3) 4.5×10^5 4) 5.16×10^6

7. A solute 'S' undergoes a reversible trimerization when dissolved in a certain solvent. The boiling point elevation of its 0.1 molal solution was found to be identical to the boiling point elevation in case of a 0.08 molal solution of a solute which neither undergoes association nor dissociation. To what percent had the solute 'S' undergone trimerization?

1) 30% 2) 40%

3) 50% 4) 60%

8. The boiling point of 0.2 mol kg^{-1} solution of X in water is greater than equimolal solution of Y in water. Which one of the following statement could be true in this case ?

1) Molecular mass of X is less than the molecular mass of Y

2) Y is undergoing dissociation in water while X undergoes no change

3) X is undergoing dissociation in water, while Y undergoing no change

4) Molecular mass of X is greater than the molecular mass of Y

9. Match the following:

	Column-I		Column-II (Osmotic Pressure)
I.	0.5 M NaCl, 0.2 M $K_4[Fe(CN)_6]$, 1 M glucose	P.	2 : 1 : 2
II.	0.4 M CaF_2, 2 M sugar, 1 M KCl	Q.	1.2 : 2 : 2
III.	1 M glucose, 1 M sucrose, 1 M urea	R.	1 : 1 : 1
IV.	1 M KCl, 0.25 M $AlCl_3$, 2 M urea	S.	1 : 1 : 2
		T.	Osmotic pressures are equal

1) I-R, T; II-Q, III-R, T; IV-P

2) I-S, R; II-T, III-R, T; IV-P

3) I-R; II-Q, III-S, T; IV-P

4) I-T; II-Q, III-R, T; IV-Q, P

10. Match the following:

	Column-I		Column-II
I.	Acetone + $CHCl_3$	P.	$\Delta S_{mix.} > 0$
II.	Ethanol + Water	Q.	$\Delta V_{mix.} > 0$
III.	$C_2H_5Br + C_2H_5I$	R.	$\Delta H_{mix.} < 0$
IV.	Acetone + Benzene	S.	Maximum boiling azeotropes
		T.	Minimum boiling azeotropes

1) I-S; II-P, Q, T; III-Q; IV-T

2) I-P, S, R; II-P, Q, T; III-P; IV- P, Q, T

3) I-R; II-Q, T; III-P; IV-Q

4) I-R; II-P, T; III-Q, S; IV-Q

11. $FeCl_3$ on reaction with $K_4[Fe(CN_6)]$ in aqueous solution gives blue colour. These two solutions are separated by a semipermeable membrane AB as shown. Due to osmosis, there is

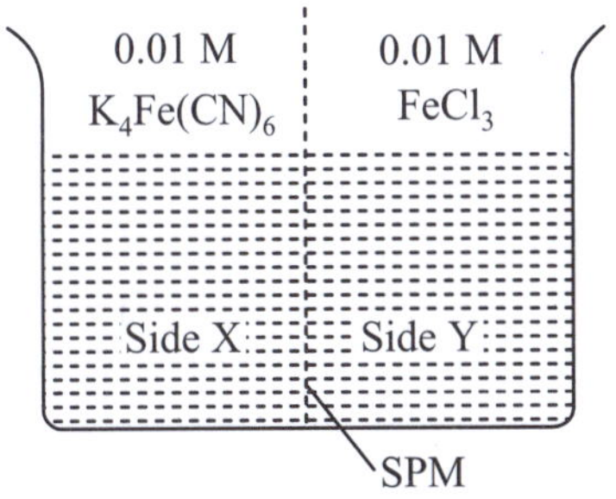

1) Blue colour formation on side X

2) Blue colour formation on side Y

3) Blue colour formation on both sides of X and Y

4) No blue colour formation

12. A very small amount of non-volatile solute (that does not dissociate) is dissolved is 56.8 cm^3 of benzene (density 0.889 g cm^{-3}). At room temperature, vapour pressure of this solution is 100 mm Hg while that of benzene is 102 mm Hg. If the freezing temperature of this solution is 1.3 degree lower than that of benzene, then what is the value of molal freezing point depression constant of benzene?

1) 5.07 deg/molal 2) 1.3 deg/molal

3) 3.9 deg/molal 4) 4.97 deg/molal

13. Pressure over ideal binary liquid mixture containing 10 moles each of liquid A and B is gradually decreased isothermally. If $P^\circ_A = 200$ mm Hg and $P^\circ_B = 100$ mm Hg, find the pressure at which half of the liquid is converted into vapour.

1) 150 mm Hg 2) 166.5 mm Hg

3) 133 mm Hg 4) 141.4 mm Hg

14. The amino acid alanine has two isomers, such as α-alanine and β-alanine. When equal masses of these two compounds are dissolved in equal mass of a solvent, the solution of α-alanine freezes at relatively lower temperature. Which forms of α-alanine or β-alanine, has the larger equilibrium constant for ionization?

1) α-alanine 2) β -alanine

3) Same for both 4) Unpredictable

15. Assuming all the solutes are non-volatile and all solutions are ideal and neglect the hydrolysis of cation and anion.

	Column-I		Column-II
A.	10 mL of 0.1 M NaOH aqueous solution is mixed with 10 mL of 0.1 M HCl aqueous solution	p.	Osmotic pressure of solution decreases
B.	10 mL of 0.1 M NaOH aqueous solution is added to 10 mL of 0.1 M CH_3COOH aqueous solution	q.	Vapour pressure of solution increases
C.	10 mL of 0.1 M HCl aqueous solution is added to 10 mL of 0.1 M NH_3 aqueous solution	r.	Boiling point of solution decreases
D.	10 mL of 0.1 M HCl aqueous solution is added to 10 mL of 0.1 M KOH aqueous solution	s.	Freezing point of solution increases

1) A→ q, r, s; B→ p, q, r, s; C→ p, q, r; D→ p, r, s
2) A→ p, q; B→ q, r, s; C→ p, r, s; D→ p, q, r, s
3) A→ p, q, r, s; B→ p, q, r, s; C→ p, q, r, s; D→ p, q, r, s
4) A→ p, q, r; B→ p, q, s; C→ p, q, r, s; D→ p, q,

16. A solution containing compound 'X' in water and a solution containing glucose in water were put in a closed system. By doing this, some water vapour was removed from one solution and got condensed in the other. It is found that when both the solutions were at equilibrium vapour pressure, one solution contains 10% (w/w) of 'X' and the other 5% (w/w) glucose. The molar mass (in g/mol) of 'X' is

1) 340 2) 360
3) 300 4) 380

17. Safrole is contained in oil of sassafras and was once used to flavour root beer. A 2.4 mg sample of safrole was dissolved in 100.0 mg of diphenyl ether. The solution has a freezing of 25.64°C. The freezing point of pure diphenyl ether is 26.84°C and the freezing-point-depression constant K_f, is 8.00°C/m. The molecular mass (in g/mol) of safrole is

1) 160 2) 180
3) 200 4) 150

18. 1.22 g of benzoic acid is dissolved in 100 g of acetone and 100 g of benzene separately. Boiling point of the solution in acetone increases by 0.17 °C while in benzene increases by 0.13°C. K_b for acetone and benzene are 1.7 K kg mol^{-1} and 2.6 K kg mol^{-1}. If x and y are the molecular weight of benzoic acid in acetone and benzene respectively, then the value of (x + y) is:

1) 300 2) 250
3) 366 4) 400

19. Vapour pressure of an equimolar mixture of benzene and toluene at a given temperature was found to be 80 mm Hg. If vapour above the liquid phase is condensed in a beaker, vapour pressure of this condensate at the same temperature was found to be 100 mm Hg. If the pure sate vapour pressure of benzene and toluene is respectively x and y. Then determine the value of $\frac{x+2y}{50}$.

1) 2 2) 1
3) 4 4) 5

20. In 10^3 Litre sample of hard water $CaSO_4$ and $MgSO_4$ is present. If elevation in Boiling point is 0.000052°C. Calculate the degree of Hardness (in ppm) of hard water. (K_b for H_2O = 0.52 K.kg/mol)

1) 4 2) 5
3) 6 4) 7

NEET & AIIMS Past Year MCQs

1. The amount of glucose required to prepare 250 mL of $\frac{M}{20}$ aqueous solution is:
(Molar mass of glucose: 180 g mol^{-1}) **(2024 Re)**

1) 2.25 g 2) 4.5 g 3) 0.44 g 4) 1.125 g

2. Mass of glucose ($C_6H_{12}O_6$) required to be dissolved to prepare one litre of its solution which is isotonic with 15 g L^{-1} solution of urea (NH_2CONH_2) is (Given: Molar mass in g mol^{-1} C : 12, H : 1, O : 16, N : 14) **(2024 Re)**

1) 55 g 2) 15 g 3) 30 g 4) 45 g

3. The Henry's law constant (K_H) values of three gases (A, B, C) in water are 145, 2×10^{-5} and 35 kbar, respectively. The solubility of these gases in water follow the order: **(2024)**

1) A > C > B 2) A > B > C
3) B > A > C 4) B > C > A

4. The plot of osmotic pressure (Π) vs concentration (mol L^{-1}) for a solution gives a straight line with slope 25.73 L bar mol^{-1}. The temperature at which the osmotic pressure measurement is done is: **(2024)**
(Use R = 0.083 L bar mol^{-1} K^{-1})

1) 25.73°C 2) 12.05°C
3) 37°C 4) 310°C

5. Given below are two statements: one is labelled as Assertion A and the other is labelled as Reason R:

Assertion (A): Helium is used to dilute oxygen in diving apparatus.

Reasons (R): Helium has high solubility in O_2

In the light of the above statements, choose the correct answer from the options given below: **(2023)**

1) A is false but R is true.
2) Both A and R are true and R is the correct explanation of A.

3) Both A and R are true and R is NOT the correct explanation of A.
4) A is true but R is false.

6. In one molal solution that contains 0.5 mole of a solute, there is **(2022)**

1) 1000 g of solvent
2) 500 mL of solvent
3) 500 g of solvent
4) 100 mL of solvent

7. Which one is not correct mathematical equation for Dalton's Law of partial pressure? Here p = total pressure of gaseous mixture **(2022)**

1) $p_i = x_i p_i^o$, where x_i = mole fraction of ith gas in gaseous mixture
p_i^o = pressure of ith gas in pure state
2) $p = p_1 + p_2 + p_3$
3) $p = n_1\frac{RT}{V} + n_2\frac{RT}{V} + n_3\frac{RT}{V}$
4) $p_i = x_i p$, where p_i = partial pressure of ith gas
x_i = mole fraction of ith gas in gaseous mixture

8. The following solutions were prepared by dissolving 10 g of glucose ($C_6H_{12}O_6$) in 250 ml of water (P_1), 10 g of urea (CH_4N_2O) in 250 ml of water (P_2) and 10 g of sucrose ($C_{12}H_{22}O_{11}$) in 250 ml of water (P_3). The right option for the decreasing order of osmotic pressure of these solutions is: **(2021)**

1) $P_1 > P_2 > P_3$
2) $P_2 > P_3 > P_1$
3) $P_3 > P_1 > P_2$
4) $P_2 > P_1 > P_3$

9. The correct option for the value of vapour pressure of a solution at 45°C with benzene to octane in molar ratio 3 : 2 is:

[At 45°C vapour pressure of benzene is 280 mm Hg and that of octane is 420 mm Hg. Assume Ideal gas] **(2021)**

1) 168 mm of Hg
2) 336 mm of Hg
3) 350 mm of Hg
4) 160 mm of Hg

10. **Assertion:** A non volatile solute added in solvent liquid then freezing point of mixture decreases.

Reason: Vapour pressure decrease by addition of non volatile solute, so equilibrium point where vapour pressure of solid and vapour pressure of liquid are equal can reach at lower temp. **(AIIMS 2018)**

1) If both Assertion & Reason are True & the Reason is a correct explanation of the Assertion.
2) If both Assertion & Reason are True but Reason is not a correct explanation of the Assertion.
3) If Assertion is True but the Reason is False.
4) If both Assertion & Reason are False.

11. **Assertion:** A non volatile solute is mixed in a solution then elevation in boiling point and depression in freezing point both are 2K. **(AIIMS 2011)**

Reason: Elevation in boiling point and depression in freezing point both depend on melting point of nonvolatile solute.

1) If both Assertion & Reason are True & the Reason is a correct explanation of the Assertion.
2) If both Assertion & Reason are True but Reason is not a correct explanation of the Assertion.
3) If Assertion is True but the Reason is False.
4) If both Assertion & Reason are False

12. For a dilute solution, Raoult's law states that: **(AIIMS 2009)**

1) The relative lowering of vapour pressure is proportional to the amount of solute in solution
2) The relative lowering of vapour pressure is equal to the mole fraction of solute
3) The lowering of vapour pressure is equal to the mole fraction of the solute
4) The vapour pressure of the solution is equal to the mole fraction of the solvent

13. **Assertion:** If red blood cells were removed from the body and placed in pure water, pressure inside the cells increases. **(AIIMS 2006)**

Reason: The concentration of salt content in the cells increases.

1) If both Assertion & Reason are True & the Reason is a correct explanation of the Assertion.
2) If both Assertion & Reason are True but Reason is not a correct explanation of the Assertion.
3) If Assertion is True but the Reason is False.
4) If both Assertion & Reason are False

14. The vapour pressure of benzene at a certain temperature is 640 mm of Hg. When a non-volatile and non-electrolyte solid weighing 2. 175 g is added to 39.08 g of benzene, the vapour pressure of the solution becomes 600 mm of Hg. What is the molecular weight of solid substance? **(AIIMS 1999)**

1) 69.60
2) 49.59
3) 59.60
4) 108.30

15. Which of the following salt has the same value of vant's Hoff factor as that of $K_3[Fe(CN)_6]$? **(AIIMS 1998)**

1) Na_2SO_4
2) $Al(NO_3)_3$
3) $Al_2(SO_4)_3$
4) Fe_3O_4

16. If active mass of a 6% solution of a compound is 2, its molecular weight will be: **(AIIMS 1996)**

1) 30
2) 15
3) 60
4) 22

Answer Key

NCERT Topic-wise MCQs

1	2	3	4	5	6	7	8	9	10	11	12	13	14	15	16	17
3	3	1	1	3	1	3	1	1	1	1	1	4	3	4	2	3
18	19	20	21	22	23	24	25	26	27	28	29	30	31	32	33	34
1	1	1	4	4	4	2	3	1	4	3	4	3	2	4	3	1
35	36	37	38	39	40	41	42	43	44	45	46	47	48	49	50	51
2	2	3	1	1	1	3	2	4	3	1	1	2	3	1	2	3
52	53	54	55	56	57	58	59	60	61	62	63	64	65	66	67	68
3	3	4	4	2	2	1	3	1	1	2	2	3	1	1	3	2
69	70	71	72	73	74	75	76	77	78	79	80	81	82	83	84	
4	1	2	4	3	1	1	1	1	3	2	3	4	2	2	2	

High Order Time Intensive MCQs

1	2	3	4	5	6	7	8	9	10	11	12	13	14	15	16	17
2	1	1	1	3	2	2	3	2	2	2	4	1	3	1	1	2
18	19	20	21	22	23	24	25	26	27	28	29	30	31	32	33	
1	1	4	2	2	2	4	3	2	3	1	3	4	2,4	2	3	

Match the Columns MCQs

1	2	3	4	5
3	1	4	2	3

Correct & Incorrect MCQs

1	2	3	4	5	6	7	8	9	10
2	4	1	1	3	2	1	4	3	2

Statement Based MCQs

1	2	3	4	5	6	7	8	9	10
2	1	1	1	3	1	3	4	3	4

Assertion & Reason MCQs

1	2	3	4	5	6	7	8	9	10
1	3	1	2	2	3	2	3	4	2

Multi Correct MCQs

1	2	3	4
1	1	4	1

NCERT Exemplar MCQs

1	2	3	4	5	6	7	8	9	10	11	12	13	14	15	16	17
1	4	3	2	3	2	1	2	1	3	4	1	1	2	2	1	2
18	19	20	21	22	23	24	25	26	27	28	29	30	31			
2	3	1	1	2, 3	1, 4	1	3	1, 2	3, 4	1, 2	1, 3	3, 4	2, 3			

Kattar NEET

1	2	3	4	5	6	7	8	9	10	11	12	13	14	15	16	17
1	2	1	1	4	4	1	3	1	2	4	1	4	1	3	2	1
18	19	20														
3	3	2														

NEET & AIIMS Past Year MCQs

1	2	3	4	5	6	7	8	9	10	11	12	13	14	15	16
1	4	4	3	4	3	1	4	2	1	4	2	1	1	2	1

CHAPTER

2 Electrochemistry

Topicwise Number of Questions Analysis (2025-2016)

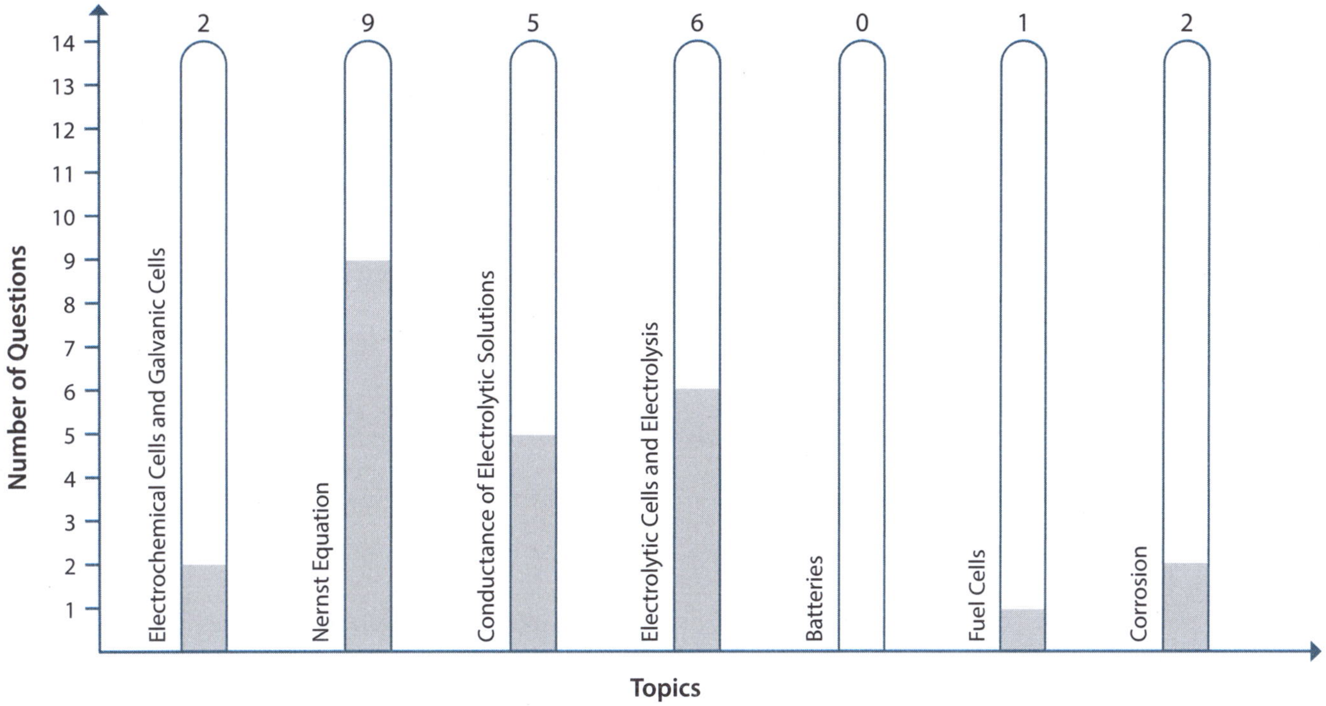

NCERT-PICKS

- Electrochemistry is the study of production of electricity from energy released during spontaneous chemical reactions and the use of electrical energy to bring about non-spontaneous chemical transformations.

Electrochemical Cell

- Device in which the free energy of physical or chemical process is converted into electrical energy is known as electrochemical cell. Following are the types of electrochemical cells:

Electrolytic cell	Galvanic cell
1. Electrical energy is converted into chemical energy.	1. Chemical energy is converted into electrical energy.
2. Anode = +ve electrode Cathode = –ve electrode	2. Anode = –ve electrode Cathode = +ve electrode
3. Ions are discharged on both the electrodes.	3. Ions are discharged only on the cathode.
4. Both the electrodes can be fitted in the same compartments.	4. The electrodes are fitted in different compartments. But it is not always necessary, as in case of lead storage battery.

	Electrolytic cell		Voltaic cell or Galvanic cell	
	Anode	Cathode	Anode	Cathode
Sign Electron flow Half - Reaction	+ in Oxidation	– out reduction	– out Oxidation	+ in reduction

In galvanic cell,

$Cu(s) | Cu^{2+} || Ag^{+} (aq) | Ag(s)$, silver electrode act as cathode & copper electrode act as anode.

Reduction takes place at cathode & Ag deposited at silver electrode while copper dissolved at copper electrode.

- The Daniell cell becomes dead after sometimes as oxidation potential of zinc anode decreases & that of copper increases.

Electrode potential

- Electrode potential in electrochemical series is, the tendency of an electrode to lose or gain electron when it is in contact with its own ion in solution.
- Electrode potential is of two types:
  1. Reduction potential (tendency to get reduced)
  2. Oxidation potential (tendency to get oxidised)
- The reduction potential of an electrode is exactly equal in magnitude but opposite in sign to its oxidation potential

Emf of the cell/cell potential

- The difference in electrode potentials of the two half cell reactions (oxidation half cell and reduction half cell) is known as emf of the cell or cell potential.
- Ecell = Eocathode - Eoanode = Eoredn. - Eooxdn.
- When reduction potential of both electrodes are taken into account:

  $$E_{cell} = E^0_{cathode} - E^0_{anode}$$

- When oxidation potential of both electrodes are taken into account;

  $$E_{cell} = E^0_{anode} - E^0_{cathode}$$

Nernst Equation

The relationship of electrolyte concentration of electrode potential of the cell is shown by this equation.

If we write the electrode reaction in general,

as oxidised state + ne^- $\rightleftharpoons$ reduced state

Then electrode potential:

$$E_{cell} = E^0_{cell} - \frac{RT}{nF} \ln \frac{[\text{Reduced state}]}{[\text{Oxidised state}]}$$

$$E_{cell} = E^0_{cell} + \frac{RT}{nF} \ln \frac{[\text{Oxidised state}]}{[\text{Reduced state}]}$$

$$E_{cell} = E^0_{cell} - \frac{2.303RT}{nF} \log Q$$

Equilibrium constant and Nernst equation

If the standard EMF (E^0) is known, then as the equilibrium constant of the cell reaction can be calculated by the use of equation as:

$$-\Delta G^0 = RT\ln K = nFE^0$$

$$\ln K = \frac{nFE^\circ}{RT}$$

$$\log K = \frac{nFE^0}{2.303RT} = \frac{nE^0}{0.0591} \text{ at } 25^0C$$

EMF and gibbs energy of the reaction

- To determine standard free energy change ($\Delta G0$) of the reaction.

  $$-\Delta G^0 = nFE^0$$

  $$E^0_{cell} = \frac{0.0591}{n} \log K_c \text{ [at 298 K]}$$

- Electromotive force and equilibrium constant of a cell reaction are related as:

  $-\Delta G^0 = nFE^0$

  we know that that, $\Delta G^0 = -RT \ln K = -2.303\ RT \log K$

Conductance of Electrolytic Solutions

- Conductivity or conductance is the power of electrolytes to conduct electrical current.
- The electrical resistance of any object is directly proportional to its length l and inversely proportional to its area of cross section A. i.e.-

  $$R \propto \frac{\ell}{A} \Rightarrow R = \rho\frac{\ell}{A}$$

- Conductance (G): $G = \frac{1}{R}$

  This term implies the ease with which current flows through the conductor

- Specific conductance (κ): Specific conductance of a conductor is the conductance of ions shown by unit volume of an electrolyte.

  $$C = \frac{1}{R} \quad R = \rho\frac{\ell}{A} \text{ and } \kappa = \frac{1}{\rho}$$

  $$R = \frac{1}{\kappa} \times \frac{\ell}{A}$$

  $\frac{\ell}{A}$ = Cell constant, depends upon the dimensions of the cell.

- Equivalent conductance (Λ_{eq}) : The conducting power of all the ions produced by one gram equivalent of an electrolyte in the solution is defined by equivalent conductance.

- If 1 cm^3 solution contains 1gm equivalent weight of electrolyte, the conductance of the solution = specific conductance = equivalent conductance
- If 1 gm equivalent weight is present in Vcc of solution. then, Λ = specific conductance × V = κV
- If conc C, given in gm equivalent per litre, then

$$V=\frac{1000}{C} \text{ or, } \Lambda=\frac{\kappa\times1000}{C};\ C=\text{gm equivalent/litre}=\text{normality}$$

- Molar Conductance (Λm): It is defined as the conducting power of all the ions produced by one mole of the electrolyte in solution.

$\Lambda_m = \kappa \times V_m$

V_m = volume of solution containing 1 mole of electrolyte.

$$\Lambda_m = \frac{\kappa\times1000}{C_m};\ C_m = \text{Conc. in moles litre}^{-1} \text{ or molarity}$$

$\Lambda_m = ohm^{-1}\ m^2\ mol^{-1}$

Variation of conductivity with concentration

- When electrolytes are dissolved in water, conductivity of water increases. Very pure water has small amount of hydrogen & OH–(~ 10^{-7} M) , when electrolytes added, they furnish their own ions in the solution.
- Conductivity always decreases with decrease in concentration both, for weak & strong electrolytes. This is due to number of ions per unit volume that carry the current in a solution decrease on dilution.
- Molar conductivity increase with decrease in concentration. This is because the total vol. (V) of sol^n containing one mole of electrolyte also increases.

Debye Huckel Onsager's equation (for strong electrolytes)

$$\Lambda_m = \Lambda_m^\infty - A\sqrt{c}$$

- Kohlrausch's law: $\Lambda_m^\infty = \lambda_+^\infty + \lambda_-^\infty$
- Kohlrausch law helps to find the molar conductivity of weak electrolyte at infinite dilution. Molar conductivity of a weak electrolyte at infinite dilution cannot be determined experimentally.
- Degree of dissociation (α)

$$=\frac{\text{Molar conductivity at Conc.c}}{\text{Molar conductivity at infinite dilution}}=\frac{\Lambda_m}{\Lambda_m^O}$$

Electrolysis

- A chemical reaction brought about the passage of electrical current through an electrolyte is electrolysis.

For example: When current is passed through molten NaCl, Na^+ and Cl^- ions moves towards the oppositely charged electrodes.

The electrochemical reactions may takes place as:

$Na^+ + e^- \rightarrow Na$ (reduction at cathode)

$Cl^- \rightarrow \frac{1}{2}Cl_2 + e^-$ (Oxidation at anode)

Faraday's law of electrolysis

- **First' law:** The amount of any substance that is deposited or liberated at an electrode is directly proportional to the quantity of electricity passed through the electrolyte such as:

$m \propto Q$ or $m = ZQ$

$m = z I t$

$$(Q=It)\begin{bmatrix} Q = \text{quantiy of electricity} & \\ I = \text{current}, & m = \text{amount of substance} \\ t = \text{time} & z = \text{electrochemical equivalent}\end{bmatrix}$$

- **Second law:** When the same quantity of electricity flows through different electrolytes, the amount of different substances produced at the same electrodes is directly proportional to their equivalent mass such as:

$\frac{m_1}{m_2}=\frac{E_1}{E_2}$: E_1 and E_2 are equivalent masses.

- **Electrochemical equivalent (Z):** is the weight of ion deposited by the passage of one ampere current in one second.

1 Faraday is defined as charge on 1 mole of electrons.

$F = N_A \times e^- = 6.023 \times 10^{23} \times 1.60 \times 10^{-19} C = 96485\ C\ mol^{-1}$

Cells and Batteries

- Primary Cells are those in which the redox reaction occurs only once and the cell becomes dead after sometime. Example: dry cell, mercury cell.
- Secondary Cells are those which can be recharged by passing an electric current through them and hence can be used over and again. Example lead storage and Ni-Cd cell.
- Fuel Cells are those in which the energy produced from the combustion of fuels like H_2, CO, CH_4 etc. is converted into electric current.Ex H_2-O_2 cell

Corrosion

- Corrosion slowly coats the surfaces of metallic objects with oxides or other salts of the metal. The rusting of iron, tarnishing of silver, development of green coating on copper and bronze are some of the examples of corrosion. It causes enormous damage to buildings, bridges, ships and to all objects made of metals especially that of iron.
- Corrosion of Iron (Rusting): Is an electrochemical phenomenon and occurs in the presence of moisture and oxygen.

  At anode : $2Fe(s) \rightarrow Fe^{2+}(aq) + 4e^-$

  At cathode : $O_2(g) + 4H^+(aq) + 4e^- \rightarrow 2H_2O(l)$

  Overall reaction : $2Fe(s) + O_2(g) + 4H^+(aq) \rightarrow 2Fe^{2+}(aq) + 2H_2O(l)$

One of the simplest methods of preventing corrosion is to prevent the surface of the metallic object to come in contact with atmosphere. This can be done by covering the surface with paint or by some chemicals (e.g. bisphenol). Another simple method is to cover the surface by other metals (Sn, Zn, etc.) that are inert or react to save the object. An electrochemical method is to provide a sacrificial electrode of another metal (like Mg, Zn, etc.) which corrodes itself but saves the object.

MCQs

NCERT Topic-wise MCQs

Electrochemical Cells

1. In a Daniel cell,
   1) The chemical energy liberated during the redox reaction is converted to electrical energy
   2) The electrical energy of the cell is converted to chemical energy
   3) The energy of the cell is utilised in conduction of the redox reaction
   4) The potential energy of the cell is converted into electrical energy
2. The electro-chemical cell stops working after some time because
   1) Electrode potentials of both electrodes become zero
   2) Electrode potentials of both electrodes become equal
   3) Temperature of the cell increases
   4) The reaction starts proceeding in opposite direction
3. A cell from the following which converts electrical energy into chemical energy
   1) Dry cell 2) Electrochemical cell
   3) Electrolytic cell 4) None of these

Galvanic Cell

4. What will be standard cell potential of galvanic cell with the following reaction?
   $2Cr(s) + 3Cd^{2+}(aq) \rightarrow 2Cr^{3+}(aq) + 3Cd(s)$
   [Given : $E^0_{Cr^{3+}/Cr} = -0.74V$ and $E^0_{Cd^{2+}/Cd} = -0.4V$]
   1) 0.70 V 2) 1.14 V
   3) 0.34 V 4) –0.34 V
5. The emf of a galvanic cell, with electrode potential of $Zn^{2+} = -0.76$ V and that of $Cu^{2+} = 0.34$ V , is
   1) 0.76 V 2) 1.1 V
   3) –1.1 V 4) 1.6 V
6. If the half - cell reactions are given as
   I. $Fe^{2+}(aq) + 2e^- \rightarrow Fe(s)$, $E^0 = -0.44$ V
   II. $2H^+(aq) + \frac{1}{2}O_2(g) + 2e^- \rightarrow H_2O(\ell)$, $E^0 = +1.23$ V
   The E^0 for the reaction,
   $Fe(s) + 2H^+ + 1/2\ O_2(g) \rightarrow Fe^{2+}(aq) + H_2O(l)$ is
   1) + 1.67 V 2) –1.67 V
   3) + 0.79 V 4) –0.79 V
7. For cell reaction, $Zn + Cu^{2+} \rightarrow Zn^{2+} + Cu$, cell representation is:
   1) $Zn / Zn^{2+} || Cu^{2+} / Cu$
   2) $Cu / Cu^{2+} || Zn^{2+} / Zn$
   3) $Cu / Zn^{2+} || Zn / Cu^{2+}$
   4) $Cu^{2+} / Zn || Zn^{2+} / Cu$
8. In a galvanic cell, the reactions taking place in the anodic half cell and the cathodic half cell will be
   1) Reduction
   2) Oxidation
   3) Oxidation and reduction
   4) Reduction and oxidation
9. Which of the following is not true for a galvanic cell represented in IUPAC system?
   1) Right hand electrode is a +ve terminal
   2) Right hand electrode acts as cathode
   3) Electrons are given out in the external circuit from the anode
   4) Electrons are given out in the external circuit from the cathode.
10. The standard reduction potentials at 298 K for the following half-reactions are given against each.
    $Zn^{2+}(aq) + 2e^- = Zn(s); -0.762$
    $Cr^{3+}(aq) + 3e^- = Cr(s); -0.74$ V
    $2H^+ + 2e^- = H_2(g); = 0.0V$
    $Fe^{3+}(aq) + e^- = Fe^{2+}(aq); + 0.77V]$
    Which is the strongest oxidising agent?
    1) Zn(s) 2) Cr(s)
    3) $H_2(g)$ 4) $Fe^{3+}(aq)$
11. Which of the following statements about galvanic cell is incorrect?
    1) Anode is positive
    2) Oxidation occurs at the electrode with lower reduction potential
    3) Cathode is positive
    4) Reduction occurs at cathode
12. A potential difference developed between the electrode and the electrolyte is known as
    1) Electrode potential
    2) Electromotive force
    3) Strong electrolyte
    4) Weak electrolyte

13. Select the incorrect statement:
1) The half-cell in which oxidation takes place is called anode
2) The half-cell in which reduction takes place is called cathode
3) The direction of current flow is same to that of electron flow
4) The potential difference between the two electrodes of a galvanic cell is called the cell potential.

Nernst Equation

14. The relation between standard reduction potential of a cell and equilibrium constant is shown by
1) $E^o_{cell} = \frac{n}{0.059}\log K_c$ 2) $E^o_{cell} = \frac{0.059}{n}\log K_c$
3) $E^o_{cell} = 0.059\,n\log K_c$ 4) $E^o_{cell} = \frac{\log K_c}{n}$

15. For a certain redox reaction, E^0 is positive. This means that:
1) $\Delta G°$ is positive, K_{eqm} is greater than 1
2) $\Delta G°$ is positive, K_{eqm} is less than 1
3) $\Delta G°$ is negative, K_{eqm} is greater than 1
4) $\Delta G°$ is negative, K_{eqm} is less than 1

16. $\Delta G°$ for the reaction, $Cu^{2+} + Fe \rightarrow Fe^{2+} + Cu$ is:
[given: $E^o_{Cu^{2+}/Cu} = +0.34V$, $E^o_{Fe^{2+}/Fe} = -0.44V$]
1) 11.44 KJ 2) 180.8 KJ
3) –150.5 KJ 4) 28.5 KJ

17. $E^o_{c\,ell}$ for the reaction, $2H_2O \rightarrow H_3O^+ + OH^-$ at 25°C is –0.8277 V. The equilibrium constant for the reaction is:
1) 10^{-14} 2) 10^{-23}
3) 10^{-7} 4) 10^{-21}

18. $\Delta_r G°$ for the cell with cell reaction:
$Zn(s) + Ag_2O(s) + H_2O(l) \rightarrow Zn^{2+}(aq) + 2Ag(s) + 2OH^-(aq)$, is
$[E^o_{Ag_2O/Ag} = +0.344V, E^o_{Zn^{2+}/Zn} = -0.76\,V]$
1) 2.13×10^5 J mol^{-1} 2) -2.13×10^5 J mol^{-1}
3) 1.06×10^5 J mol^{-1} 4) -1.06×10^5 J mol^{-1}

19. Mark the correct Nernst equation for the given cell:
$Fe(s)/Fe^{2+}(0.001M)//H^+(1M)/H_2(g)(1bar)/Pt(s)$
1) $E_{cell} = E^o_{cell} - \frac{0.0591}{2}\log\frac{[Fe^{2+}][H^+]^2}{[Fe][H_2]}$
2) $E_{cell} = E^o_{cell} - \frac{0.0591}{2}\log\frac{[Fe][H^+]^2}{[Fe^{2+}][H_2]}$
3) $E_{cell} = E^o_{cell} - \frac{0.0591}{2}\log\frac{[Fe^{2+}][H_2]}{[Fe][H^+]^2}$
4) $E_{cell} = E^o_{cell} - \frac{0.0591}{2}\log\frac{[Fe][H_2]}{[Fe^{2+}][H^+]^2}$

20. The oxidation potential of hydrogen half-cell will be negative if -
1) $p(H_2) = 1$ atm and $[H^+] = 1$ M
2) $p(H_2) = 1$atm and $[H^+] = 2$M
3) $p(H_2) = 0.2$ atm and $[H^+] = 1$M
4) Both (2) & (3)

21. If the E^o_{cell} for a given reaction has a negative value, then which of the following gives the correct relationships for the values of $\Delta G°$ and K_{eq}?
1) $\Delta G° > 0$; $K_{eq} < 1$ 2) $\Delta G° > 0$; $K_{eq} > 1$
3) $\Delta G° < 0$; $K_{eq} > 1$ 4) $\Delta G° < 0$; $K_{eq} < 1$

22. Calculate the emf of half cell:
$$Pt, \begin{matrix} Cl_2(g) \\ (10\text{ atm}) \end{matrix} \Bigg| \begin{matrix} HCl \\ (0.1)M \end{matrix} \quad E^o_{Cl_2|Cl^-} = 1.36V$$
1) 1.45 V 2) 1.27 V
3) –1.45 V 4) –1.27 V

23. During electrochemical process:
1) Gibbs free energy increases
2) Gibbs free energy remains constant
3) No prediction can be made about Gibbs free energy
4) Gibbs free energy decreases

24. Alkali metals have high oxidation potential, hence, they behave as good:
1) Oxidising agent 2) Lewis bases
3) Reducing agents 4) Electrolytes

25. Zn gives H_2 gas with H_2SO_4 and HCl but not with HNO_3 because:
1) Zn act as oxidising agent when react with HNO_3
2) HNO_3 is weaker acid than H_2SO_4 and HCl
3) In electrochemical series, Zn is above hydrogen
4) NO_3^- is reduced in preference to hydronium ion

26. What is the electrode potential (in volt) of the following electrode at 25°C?
$Ni^{2+}(0.1M)/Ni(s)$
[standard reduction potential of
$[Ni^{2+}/Ni]$ is $-0.25V, \frac{2.303RT}{F} = 0.06)$
1) –0.28 V 2) –0.34 V
3) –0.82 V 4) –0.22 V

27. Emf of hydrogen electrode in terms of pH is: [at 1atm pressure]
1) $E_{H_2} = \frac{RT}{F}pH$ 2) $E_{H_2} = \frac{RT}{F}\frac{1}{pH}$
3) $E_{H_2} = \frac{2.303RT}{F}pH$ 4) $E_{H_2} = -0.0591pH$

28. For $Cr_2O_7^{2-} + 14H^+ + 6e^- \rightarrow 2Cr^{3+} + 7H_2O$,
$E° = +1.33V$ at $[Cr_2O_7^{2-}] = 4.5$ millimole per ltr, $[Cr^{3+}] = 15$ millimole per ltr, E is 1.06 V. The pH of the solution is nearly equal to:
1) 2 2) 3
3) 5 4) 4

29. Given that $E_{H_2O|H_2|Pt} = 0$ at 298K. The pressure of H_2 gas would be

1) 10^{-7} atm 2) 10^{-14}atm
3) 10^{-10}atm 4) 10^{-12}atm

30. Following two half cells form a complete cell which has $\Delta G°$ (in kJ) value

$2H^+ + 1/2\ O_2 + 2e^- \rightarrow H_2O$; $E° = +1.23$ V
$Fe^{2+} + 2e^- \rightarrow Fe(s)$; $E° = -0.44$ V

1) –122 kJ 2) –222 kJ
3) –322 kJ 4) –422 kJ

31. E° for the cell Zn | Zn^{2+} (aq) || Cu^{2+} (aq) | Cu is 1.10 V at 25 °C, the equilibrium constant for the reaction, Zn + Cu^{2+} (aq) $\rightleftharpoons$ Cu + Zn^{2+} (aq) is of the order of

1) 10^{-28} 2) 10^{+37}
3) 10^{+18} 4) 10^{+17}

32. Which of the following statements is correct ?

1) A positive E° means that the redox couple is a weaker reducing agent than the H^+/H_2 couple
2) A positive E° means that the redox couple is a stronger reducing agent than the H^+/H_2 couple
3) A positive E° means that the redox couple can neither behave as oxidizing nor as reducing agent
4) None of these

33. If $E°_{Ni^{2+}|Ni} = 0.25V, E°_{Cu^{2+}|Cu} = 0.34V, E°_{Ag^+|Ag} = 0.8V$ and $E°_{Zn^{2+}|Zn} = -0.76V$, then which of the following reactions under standard condition will not take place in the specified direction spontaneously?

1) $Cu(s) + Ni^{2+}(aq.) \rightarrow Cu^{2+}(aq.) + Ni(s)$
2) $Cu(s) + 2Ag^+(aq.) \rightarrow Cu^{2+}(aq.) + 2Ag(s)$
3) $Cu(s) + Zn^{+2}(aq.) \rightarrow Cu^{2+}(aq.) + Zn(s)$
4) Both (1) & (3)

34. Fluorine is the best oxidising agent because it has:

1) Highest electron affinity
2) Highest reduction potential
3) Highest oxidising potential
4) Lowest electron affinity

35. Which of the following is the cell reaction that occurs when the following half – cells are combined?

$I_2 + 2e^- \rightarrow 2I^-$ (1M) ; E = + 0.54 V
$Br_2 + 2e^- \rightarrow 2\ Br^-$ (1M) ; E = + 1.09 V

1) $2Br^- + I_2 \rightarrow Br_2 + 2I^-$
2) $I_2 + Br_2 \rightarrow 2I^- + 2Br^-$
3) $2I^- + Br_2 \rightarrow I_2 + 2Br^-$
4) $2I^- + 2Br^- \rightarrow I_2 + Br_2$

36. Which one of the following metal will not reduce H_2O?

1) Ca 2) Fe
3) Cu 4) Li

37. Arrange the following in the order of their decreasing electrode potentials: Mg, K, Ba,Ca

1) K > Ba > Ca > Mg
2) Ba > Ca > K > Mg
3) Ca > Mg > K > Ba
4) Mg > Ca > Ba > K

38. When Zn metal is added to $CuSO_4$ solution, Cu is precipitated. It is due to

1) Oxidation of Cu^{2+} 2) Reduction of Cu^{2+}
3) Hydrolysis of $CuSO_4$ 4) Ionization of $CuSO_4$

Conductance of Electrolytic Solutions

39. The ionic conductances of H^+ and OH^- at 298 K are 349.8 and 198.5 $ohm^{-1}\ cm^2\ eq^{-1}$ respectively. The equivalent conductance of H_2O at infinite dilution is:

1) 548.3 2) 151.3
3) 699.6 4) 54.83

40. The specific conductivity of N/10 KCl solution at 20°C is 0.0212 $ohm^{-1}\ cm^{-1}$ and the resistance of the cell containing this solution at 20°C is 55 Ω. The cell constant is:

1) 3.324 cm^{-1} 2) 1.166 cm^{-1}
3) 2.372 cm^{-1} 4) 3.682 cm^{-1}

41. NaCl, $MgCl_2$ and $CaSO_4$ are known as:

1) 1-1, 2-1, 2-2 types electrolyte respectively
2) Strong, weak and strong electrolytes respectively
3) Electrolytes with different value of A
4) Electrolytes with same molar conductivity

42. Calculate molar conductivity of 0.15 M solution of KCl at 298 K if its conductivity is 0.0152 S cm^{-1}.

1) 124 $\Omega^{-1}\ cm^2\ mol^{-1}$ 2) 204 $\Omega^{-1}\ cm^2\ mol^{-1}$
3) 101 $\Omega^{-1}\ cm^2\ mol^{-1}$ 4) 300 $\Omega^{-1}\ cm^2\ mol^{-1}$

43. Specific conductance of 0.1 M NaCl solution is $1.01 \times 10^{-2}\ \Omega^{-1}\ cm^{-1}$. Its molar conductance in $\Omega^{-1}\ cm^2\ mol^{-1}$ is:

1) 1.01×10^2 2) 1.01×10^3
3) 1.01×10^4 4) 1.01

44. Limiting molar conductivity of NaBr is:

1) Λ^o_m NaBr = Λ^o_m NaCl + Λ^o_m KBr
2) Λ^o_m NaBr = Λ^o_m NaCl + Λ^o_m KBr – Λ^o_m KCl
3) Λ^o_m NaBr = Λ^o_m NaOH + Λ^o_m NaBr – Λ^o_m NaCl
4) Λ^o_m NaBr = Λ^o_m NaCl – Λ^o_m NaBr

45. Molar conductivity of 0.025 mol L^{-1} methanoic acid is 46.1 Scm^2mol^{-1}, the degree of dissociation and dissociation constant will be:

(Given: $\lambda^o_{H^+}$ = 349.6 $Scm^2\ mol^{-1}$ and $\lambda^o_{HCOO^-}$ = 54.6 $Scm^2\ mol^{-1}$)

1) 0.114, 3.67×10^{-4} mol L^{-1}
2) 0.228, 1.83×10^{-4} mol L^{-1}
3) 0.522, 4.25×10^{-4} mol L^{-1}
4) 0.011, 3.67×10^{-4} mol L^{-1}

46. Molar conductivity of NH_4OH can be calculated by the equation:

1) $\Lambda^o_{NH_4OH} = \Lambda^o_{Ba(OH)_2} + \Lambda^o_{NH_4Cl} - \Lambda^o_{BaCl_2}$

2) $\Lambda^o_{NH_4OH} = \Lambda^o_{BaCl_2} + \Lambda^o_{NH_4Cl} - \Lambda^o_{Ba(OH)_2}$

3) $\Lambda^o_{NH_4OH} = \dfrac{\Lambda^o_{Ba(OH)_2} + 2\Lambda^o_{NH_4Cl} - \Lambda^o_{BaCl_2}}{2}$

4) $\Lambda^0_{NH_4OH} = \dfrac{2\Lambda^o_{NH_4Cl} - \Lambda^o_{Ba(OH)_2}}{2}$

47. Units of the properties measured are given below, which of the properties has not been matched correctly?

1) Molar conductance = $Sm^2\ mol^{-1}$
2) Cell constant = m^{-1}
3) Specific conductance = Sm^2
4) Equivalent conductance = $Sm^2\ (geq)^{-1}$

48. When water is added to an aqueous solution of an electrolyte, what is the change in specific conductivity of the electrolyte?

1) Conductivity increases
2) Conductivity decreases
3) Conductivity remains same
4) Conductivity does not depends on number of ions

49. What would be the equivalent conductivity of a cell in which 0.5 N salt solution offers a resistance of 40 Ω whose electrodes are 2 cm apart and 5 cm^2 in area?

1) $10\Omega^{-1}\ cm^2\ eq^{-1}$
2) $20\Omega^{-1}\ cm^2\ eq^{-1}$
3) $30\Omega^{-1}\ cm^2\ eq^{-1}$
4) $25\Omega^{-1}\ cm^2\ eq^{-1}$

50. The equivalent conductance of Ba^{2+} and Cl^- respectively are 127 and 76 $\Omega^{-1}\ cm^2\ eq^{-1}$ at infinite dilution. What will be the equivalent conductance of $BaCl_2$ at infinite dilution?

1) 139.5 $\Omega^{-1}\ cm^2\ eq^{-1}$
2) 203 $\Omega^{-1}\ cm^2\ eq^{-1}$
3) 279 $\Omega^{-1}\ cm^2\ eq^{-1}$
4) 101.5 $\Omega^{-1}\ cm^2\ eq^{-1}$

51. Electrical conductance through metals is called metallic or electronic conductance and is due to the movement of electrons. The electronic conductance depends on:

1) The nature and structure of the metal
2) The nature of valence electrons per atom
3) Change in temperature
4) All of these

52. What will be the molar conductivity of Al^{3+} ions at infinite dilution if molar conductivity of $Al_2(SO_4)_3$ is 858 $Scm^2\ mol^{-1}$ and molar ionic conductance of SO_4^{2-} is 160 $Scm^2\ mol^{-1}$ at infinite dilution?

1) 189 $Scm^2\ mol^{-1}$
2) 698 $Scm^2\ mol^{-1}$
3) 1018 $Scm^2\ mol^{-1}$
4) 429 $Scm^2\ mol^{-1}$

53. Mark the correct choice of electrolytes represented in the graph:

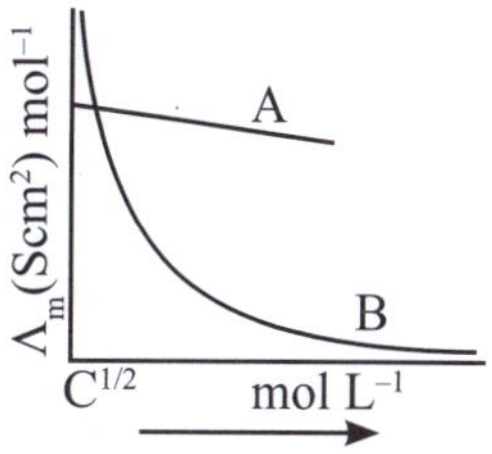

1) A → NH_4OH , B → NaCl
2) A → NH_4OH , B → NH_4Cl
3) A → CH_3COOH , B → CH_3COONa
4) A → KCl , B → NH_4OH

54. Match the column-I with column-II and mark the appropriate choice.

	Column-I		Column-II
A.	Kohlrausch's law	(i)	$\Lambda^o_{eq} = \Lambda^o_c + \Lambda^o_a$
B.	Molar conductivity	(ii)	$\Lambda_m = \dfrac{\kappa}{Molarity}$
C.	Degree of dissociation	(iii)	$\alpha = \dfrac{\Lambda_m}{\Lambda^o_m}$
D.	Dissociation constant	(iv)	$K_a = \dfrac{c\alpha^2}{1-\alpha}$

1) A → (iii), (B) → (iv), (C) → (i), (D) → (ii)
2) A → (i), (B) → (ii), (C) → (iii), (D) → (iv)
3) A → (iv), (B) → (i), (C) → (ii), (D) → (iii)
4) A → (ii), (B) → (iii), (C) → (iv), (D) → (i)

55. Conductivity of a saturated solution of a sparingly soluble salt AB at 298 K is $1.85 \times 10^{-5}\ Sm^{-1}$. Solubility product of the salt AB at 298 K is:

[Given; Λ^o_m (AB) = $140 \times 10^{-4}\ Sm\ mol^{-1}$]

1) $5.7 \times 10^{+12}$
2) 1.32×10^{-12}
3) 7.5×10^{12}
4) 1.74×10^{12}

56. At 18°C, the conductance of H^+ and CH_3COO^- at infinite dilution are 315 and 35 mho $cm^2\ eq^{-1}$ respectively. The equivalent conductivity of CH_3COOH at infinite dilution is---- (mho $cm^2\ eq^{-1}$):

1) 350
2) 280
3) 30
4) 315

57. The specific conductance (κ) of an electrolyte of 0.1 N concentration is related to equivalent conductance Λ by the following formula:

1) $\Lambda = \kappa$
2) $\Lambda = 10\ \kappa$
3) $\Lambda = 100\ \kappa$
4) $\Lambda = 10000\ \kappa$

58. The variation of equivalent conductance of strong electrolyte with $\sqrt{concentration}$ is represented by

1) 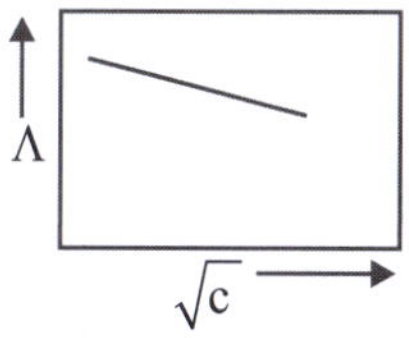

2) 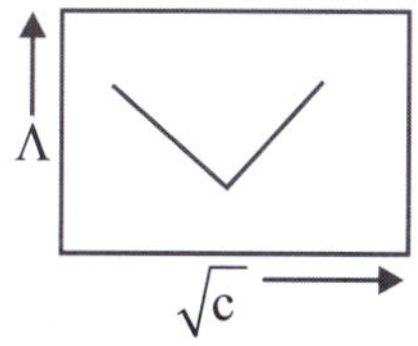

3) 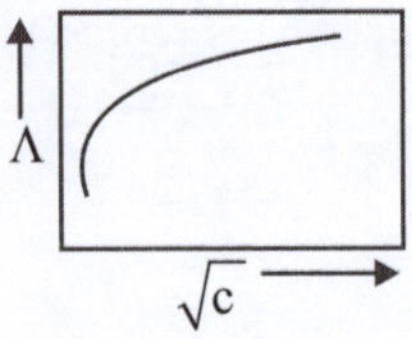

4) 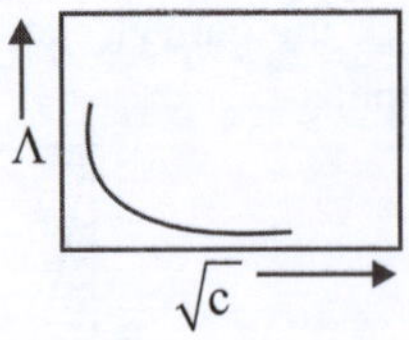

59. Pure water does not conduct electricity because it is
1) Neutral
2) Readily decomposed
3) Almost unionised
4) Completely ionised

60. In aqueous solution, strong electrolytes ionize and yield
1) Ions
2) Electrons
3) Acids
4) Oxides

61. Which of the following is a poor conductor of electricity?
1) CH_3COONa
2) C_2H_5OH
3) NaCl
4) KOH

62. What happens at infinite dilution in a given solution?
1) The degree of dissociation is unity for the weak electrolytes
2) The electrolyte is 100% ionised
3) All interionic attractions disappear
4) All the above

63. During electric conduction, the composition of which of the following is changed ?
1) Graphite
2) Zinc wire
3) Copper wire
4) H_2SO_4

64. Conductivity of a solution is directly proportional to
1) Dilution
2) Number of ions
3) Current density
4) Volume of the solution

65. On which of the following factors, the degree of dissociation of an electrolyte does not depend?
1) Temperature
2) Catalytic action
3) Dilution
4) Nature of electrolyte

66. Conductivity is represented by:
1) κ
2) G^*
3) Λ_m
4) $\frac{l}{A}$

67. Which of the following is the SI unit of conductance?
1) Mho
2) Siemens
3) Ohm
4) Metre

68. Which of the following statements about cell constant is incorrect?
1) The quantity l/A is known as cell constant
2) It is denoted by the symbol, G*
3) It depends on the distance between the electrodes and their area of cross-section
4) It has the dimension of $mass^{-1}$

69. Resistance of a conductivity cell filled with 0.1 mol L^{-1} KCl solution is 100 Ω. If the resistance of the same cell when filled with 0.02 mol L^{-1} KCl solution is 520 Ω, calculate the conductivity and molar conductivity of 0.02 mol L^{-1} KCl solution. The conductivity of 0.1 mol L^{-1} KCl solution is 1.29 S/m.
1) 0.164 S m^{-1} and 198×10^{-4} S m^2mol^{-1}
2) 0.248 S m^{-1} and 124×10^{-4} S m^2mol^{-1}
3) 0.591 S m^{-1} and 332×10^{-4} S m^2mol^{-1}
4) 0.361 S m^{-1} and 339×10^{-4} S m^2mol^{-1}

70. Which one is not a conductor of electricity?
1) NaCl (aqueous)
2) NaCl (solid)
3) NaCl (molten)
4) Ag metal

71. When concentration approaches zero, the molar conductivity is known as
1) Conductance
2) Equivalent conductivity
3) Molar conductance
4) Limiting molar conductivity

72. Which of the following is a 2 – 1 type of electrolyte?
1) NaCl
2) $MgSO_4$
3) $CaCl_2$
4) None of these

73. Λ°_m for NaCl, HCl and NaAc are 126.4 , 425.9 and 91.0 S $cm^2 mol^{-1}$ respectively. Calculate Λ° for HAc.
1) 135.8 S cm^2 mol^{-1}
2) 148.6 S cm^2 mol^{-1}
3) 224. 8 S cm^2 mol^{-1}
4) 390.5 S cm^2 mol^{-1}

74. The resistance of 0.5 M solution of an electrolyte in a cell was found to be 50 Ω. If the electrodes in the cell are 2.2 cm apart & having an area of 4.4 cm^2 then the molar conductivity (in Sm^2 mol^{-1}) of solution is:
1) 0.2
2) 0.02
3) 0.002
4) None of these

Electrolytic Cell And Electrolysis

75. Electrolytic conduction differs from metallic conduction. In case of metallic conduction:
1) The resistance increases with increasing temperature
2) The resistance decreases with increasing temperature
3) The flow of current does not generate heat
4) The resistance is independent of the length of electrolytic conductor

76. If 54 g of silver is deposited during an electrolysis reaction, how much aluminium will be deposited by the same amount of electric current?
1) 2.7 g
2) 4.5 g
3) 27 g
4) 5.4 g

77. The quantity of electricity needed to separately electrolyse 1 M solution of $ZnSO_4$, $AlCl_3$, and $AgNO_3$ completely is in the ratio of:
1) 2 : 3 : 1
2) 2 : 1 : 1
3) 2 : 1 : 3
4) 2 : 2 : 1

78. What is the time required (in seconds) for depositing all the silver present in 125 mL of 1 M $AgNO_3$ solution by passing a current of 241.25 A [1F = 96500 C]?
1) 10 sec
2) 50 sec
3) 1000 sec
4) 100 sec

79. When 9.65 coulomb of electricity is passed through a solution of silver nitrate (atomic mass of Ag 108 g mol^{-1}). The amount of silver deposited is:
1) 16.2 mg
2) 21.2 mg
3) 10.8 mg
4) 6.4 mg

80. When 0.1 mol MnO_4^{2-} is oxidised, the quantity of electricity required to completely oxidise MnO_4^{2-} to MnO_4^- is:
1) 96500 C
2) 2 × 96500 C
3) 9650 C
4) 96.50 C

81. Products of electrolysis depend on
1) The nature of material being electrolyzed
2) The type of electrodes being used
3) Both (1) and (2)
4) None of these

Batteries and Fuel Cells

82. When lead storages battery discharges:
1) SO_2 is evolved
2) $PbSO_4$ is consumed
3) Lead is formed
4) H_2SO_4 is consumed

83. Select the correct statements about fuel cells.
1) They run till reactants are active
2) They are free from pollution
3) They are more efficient
4) All of these

84. The overall reaction of a hydrogen - oxygen fuel cell is:
1) $2H_2(g) + O_2(g) \rightarrow 2H_2O(l)$
2) $2H_2(g) + 4OH^-(aq) \rightarrow 4H_2O(\ell) + 4e^-$
3) $O_2(g) + 2H_2O(l) + 4e^- \rightarrow 4OH^-(eq)$
4) $4OH^-(aq) + 4e^- \rightarrow 2H_2O(l)$

85. Which of the following is an example of primary battery ?
1) Lead storage battery
2) Leclanche cell
3) Nickel-cadmium cell
4) None of these

86. Hydrogen–oxygen fuel cells are used in space–craft to supply
1) Oxygen
2) Power
3) Water
4) Both (2) and (3)

Corrosion

87. When iron is rusted, it is:
1) Reduced
2) Decomposed
3) Oxidised
4) Changed in the fine power

88. Chemical formula of rust is:
1) $Fe_2O_3 . H_2O$
2) $Fe_2O_3 . 5H_2O$
3) $Fe_2O_3 . xH_2O$
4) None of these

89. Which of the following is/are example/s of corrosion ?
1) Tarnishing of silver
2) Rusting of iron
3) Development of green coating on copper
4) All of these

High Order Time Intensive MCQs

1. For the electrode Cu^{2+}/Cu. log (Cu^{2+}) (along x-axis) is plotted against E_{red} (along y-axis). The plot obtained is shown in the figure. The electrode oxidation potential of half cell Cu/Cu^{2+} (0.1 M) will be:

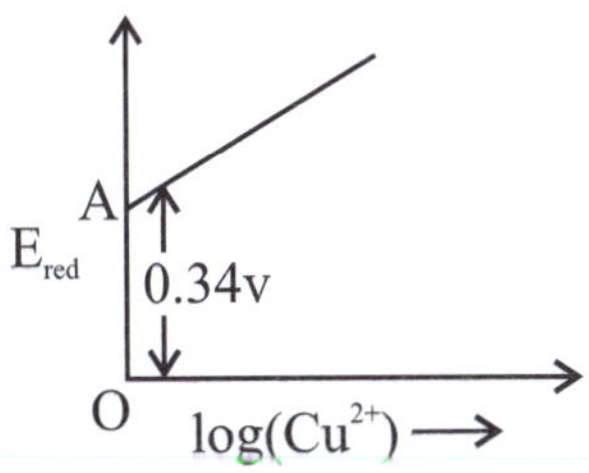

1) $-0.34 + \frac{0.0591}{2}$ V
2) 0.34 V
3) $0.34 + \frac{0.0591}{2}$ V
4) None of these

2. Degree of dissociation of pure H_2O is 1.9×10^{-9}. Molar ionic conductances of H^+ & OH^- ions at infinite dilution are 200 S cm^2 mol^{-1} & 350 S cm^2 mol^{-1} respectively. Molar conductance of water is:
1) 3.8×10^{-7} S cm^2 mol^{-1}
2) 5.7×10^{-7} S cm^2 mol^{-1}
3) 9.5×10^{-7} S cm^2 mol^{-1}
4) 1.045×10^{-6} S cm^2 mol^{-1}

3. A cell constructed by coupling a standard copper electrode and a standard magnesium electrode has emf of 2.7 volts. If the standard reduction potential of copper electrode is + 0.34 volt that of magnesium electrode is
1) + 3.04 volts
2) +2.36 volts
3) + 3.04 volts
4) – 2.36 volts

4. The cell, $Zn|Zn^{2+}(1M)||Cu^{2+}(1M)|Cu$ (E^0_{cell} = 1.10V) was allowed to be completely discharged at 298 K. The relative concentration of Zn^{2+} to Cu^{2+} is

1) 37.3 2) $10^{37.3}$
3) 27.8 4) $10^{24.04}$

5. The emf of the cell $M | M^{n+}$ (0.02M) || H^+ (1M), H_2(g) 1 atm Pt at 25°C is 0.81 V. The valency of the metal if the standard oxidation potential of the metal is 0.76 V?

1) 5 2) 2
3) 4 4) 3

6. $E^o_{Cu^{2+}/Cu}$ = +0.34V, what will be the reduction potential at pH = 14 for same couple. Given K_{sp} of $Cu(OH)_2 = 10^{-19}$:

1) –0.22 V 2) –0.71 V
3) +0.22 V 4) +0.71 V

7. When 6×10^{22} electrons are used in electrolysis of metallic salt, 1.9 g of metal is deposited at cathode. The atomic weight of that metal is 57. What is the oxidation state of the metal?

1) +1 2) +2
3) +3 4) +4

8. An electric current is passed through copper voltmeter & water voltameter connected in series. If Cu of Copper voltameter now weighs 16 mg less, H_2 liberated at cathode of water voltameter measured at STP is about:

1) 22.4 mL 2) 3.8 mL
3) 5.6 mL 4) 7.2 mL

9. In an experiment, 0.04 F was passed through 400 mL. of 1 M solution of NaCl. What would be the pH of solution after electrolysis?

1) 13 2) 11
3) 13.6 4) 12

10. During the electrolysis of aqueous solution of copper sulphate by using Pt electrodes, what will happen?

1) Copper will deposit at cathode
2) Copper and oxygen both will deposit at anode.
3) Oxygen will be released at anode
4) Copper will deposit at anode

11. The equivalent conductance of monobasic acid at infinite dilution is 348 ohm^{-1} cm^2 eq^{-1}. If the resistivity of the solution containing 15 g acid (mol. wt. 49) in 1 litre is 18.5 ohm cm, what is the degree of dissociation of acid?

1) 0.459 2) 0.402
3) 0.604 4) 0.507

12. At what concentration of $ZnSO_4$, the reduction potential of Zn electrode is –0.82 V (SRP of Zn electrode is –0.76 V)

1) 1 M 2) 0.1 M
3) 0.05 M 4) 0.01 M

13. 2 litres of 1 M $CuSO_4$ solution is electrolysed. The molarity of $CuSO_4$ solution after passing 2F of electrical charge is (ignore the change in volume of solution) is:

1) 1 M 2) 0.5 M
3) 2 M 4) 1.5 M

14. The resistance of a solution 'A' is 40 ohm and that of solution 'B' is 70 ohm, both solutions being taken in the same conductivity cell. If equal volumes of solution 'A' and 'B' are mixed, what will be the resistance of the mixture using the same cell? (Assume that there is no increase in the degree of dissociation of 'A' and 'B' on mixing.)

1) 50.9 ohm 2) 101.8 ohm
3) 110 ohm 4) 55 ohm

15. $Zn|ZnSO_4 || CuSO_4 | Cu$
(C_1) (C_2)

For the above cell, when the concentration of Zn^{2+} is 10 times the concentration of Cu^{2+}, the expression for ΔG is $\left(E^\circ_{cell} = 1.1V\right)$

1) 2.303 RT + 1.1F 2) 2.303 RT – 2.2 F
3) 2.2 F 4) 2.303 RT – 1.1 F

16. The molar conductance of acetic acid at infinite dilution is 390.7 and for 0.1 M acetic acid solution is 5.2 mho cm^2 mol^{-1}. The degree of dissociation of 0.1 M CH_3COOH solution is:

1) 13.3% 2) 0.0133%
3) 1.33% 4) 133%

17. The position of some metals in the electrochemical series in decreasing electropositive character is given as Mg > Al > Zn > Cu > Ag. What will happen if a copper spoon is used to stir a solution of aluminium nitrate :

1) The spoon will get coated with aluminium
2) An alloy of copper and aluminium is formed
3) The solution becomes blue
4) There is no reaction

18. A solution containing one mol per litre each of $Cu(NO_3)_2$ $AgNO_3$, $Hg_2(NO_3)_2$ and $Mg(NO_3)_2$ is being electrolysed by using inert electrodes. The values of standard electrode potentials in volt (reduction potentials) are,

$Ag^+|Ag = 0.80$, $Hg_2^{+2} | 2\,Hg = 0.79$
$Cu^{+2} | Cu = +0.34$ and $Mg^{+2} | Mg = -2.37$

With increasing voltage, the sequence of deposition of metals on the cathode will be -

1) Mg, Hg, Cu, Ag 2) Mg, Cu, Hg, Ag
3) Ag, Hg, Cu 4) Cu, Hg, AgS

19. $I_2(s)|I^-(0.1M)$ half cell is connected to a $H^+(aq)|H_2$ (1 bar)|Pt half cell and e.m.f. is found to be 0.7714V. If $E_{I_2|I^-} = 0.535V$, find the pH of $H^+|H_2$ half-cell.

1) 1 2) 3
3) 5 4) 7

20. Total charge required for the oxidation of two moles Mn_3O_4 into MnO_4^{2-} in presence of alkaline medium is-

1) 5 F 2) 10 F
3) 20 F 4) None of these

21. Zinc is used to protect iron from rusting is because
1) E°_{red} of Zn^{2+} is greater than that of Fe^{2+}
2) E°_{ox} of Zn is greater than that of Fe
3) E°_{red} of Zn^{2+} is nearly equal to that of Fe^{2+}
4) Zn is cheap

22. For a cell reaction involving two electron change, the standard EMF of the cell is 0.295 V at 25°C. The equilibrium constant of the reaction at 25 °C will be
1) 29.5×10^{-2} 2) 10
3) 1×10^{10} 4) 2.95×10^{-10}

23. The electromotive force for the following cell Cu(s) | Cu^{2+} (0.01M) || Cu^{+2} (0.04 M) | Cu(s) is:-
1) $+\frac{RT}{F}\ln(0.5)$ 2) $+\frac{RT}{F}\ln(0.25)$
3) $-\frac{RT}{F}\ln(0.5)$ 4) $-\frac{RT}{2F}\ln(0.5)$

24. Which of the following is not correct for dry cell?
1) Zinc is used as anode
2) Manganese is reduced from + 4 to + 3 state
3) It is a primary cell
4) NH_3 gas is liberated out

25. The electrode potentials for
$Cu^{2+}(aq) + e^- \rightarrow Cu^+(aq)$
and $Cu^+(aq) + e^- \rightarrow Cu(s)$ are 0.15 V
and + 0.50 V respectively. The value of $E^0_{Cu^{2+}/Cu}$ will be
1) 0.150 V
2) 0.500 V
3) 0.325 V
4) 0.650 V

26. The equilibrium constant of the reaction,
$Cu(s) + 2Ag^+(aq) \rightarrow Cu^{2+}(aq) + 2Ag(s)$
$E^\circ = 0.46$ V at 298 K is:
1) 2.4×10^{10} 2) 2.0×10^{10}
3) 4.0×10^{10} 4) 4.0×10^{15}

27. Fused NaCl has less electrical conductance than NaCl in the aqueous solution. This is due to
1) Fused NaCl has less number of ions
2) Incomplete ionization occurs in the fused state
3) Na^+, Cl^- ions do not move freely in the fused salt
4) Fused NaCl has no ions

28. On passing 3A of electricity for 50 min, 1.8 g metal deposits, the equivalent mass of metal is:
1) 9.3 2) 19.3
3) 38.3 4) 39.9

29. In electrolysis of dil. H_2SO_4 using platinum electrodes:
1) H_2 is evolved at cathode
2) SO_2 is produced at anode
3) O_2 is obtained at cathode
4) SO_2 is produced at cathode

30. Which of the following statements is true for the electrochemical Daniel cell?
1) Electrons flows from copper electrode to zinc electrode
2) Currents flows from zinc electrode to copper electrode
3) Cations move towards copper electrode
4) Cations move towards zinc electrode

31. The reaction which is taking place in nickel-cadmium battery can be represented by which of the following equation?
1) $Cd + NiO_2 + 2H_2O \rightarrow Cd(OH)_2 + Ni(OH)_2$
2) $Cd + NiO_2 + 2OH \rightarrow Ni + Cd(OH)_2$
3) $Ni + Cd(OH)_2 \rightarrow Cd + Ni(OH)_2$
4) $Ni(OH)_2 + Cd(OH)_2 \rightarrow Ni + Cd + 2H_2O$

32. Which of the following reactions does not take place during rusting?
1) $H_2CO_3 \rightleftharpoons 2H^+ + CO_3^{2-}$
2) $4Fe^{2+} + O_2(dry) \rightarrow Fe_2O_3$
3) $4Fe^{2+} + O_2 + 4H_2O \rightarrow 2Fe_2O_3 + 8H^+$
4) $Fe_2O_3 + xH_2O \rightarrow Fe_2O_3xH_2O$

33. The conductivity of 0.001028 mol L^{-1} acetic acid is 4.95×10^{-5} S cm^{-1} Calculate its dissociation constant if Λ°_m for acetic acid is 390.5 S cm^2 mol^{-1}.
1) 1.78×10^{-5} mol L^{-1}
2) 2.14×10^{-3} mol L^{-1}
3) 3.21×10^{-2} mol L^{-1}
4) 1.18×10^{-6} mol L^{-1}

34. Given standard electrode potentials
$Fe^{++} + 2e^- \rightarrow Fe$; $E^\circ = -0.440$ V
$Fe^{+++} + 3e^- \rightarrow Fe$; $E^\circ = -0.036$ V
The standard electrode potential (E°) for $Fe^{+++} + e^- \rightarrow Fe^{++}$ is
1) –0.476 V 2) –0.404 V
3) +0.404 V 4) +0.771 V

35. The specific conductance of a solution is 0.2 $ohm^{-1}cm^{-1}$ and conductivity is 0.04 ohm^{-1}. The cell constant would be
1) 4 cm^{-1} 2) 2 cm^{-1}
3) 5 cm^{-1} 4) 10 cm^{-1}

Match the Columns MCQs

1. Match the terms given in Column-I with the units given in Column-II

	Column-I		Column-II
(a)	$\wedge_m$	(i)	$S\ cm^{-1}$
(b)	E_{cell}	(ii)	m^{-1}
(c)	κ	(iii)	$S\ cm^2mol^{-1}$
(d)	G^*	(iv)	V

1) (a)-(ii) (b)-(iii) (c)-(iv) (d)-(i)
2) (a)-(ii) (b)-(iv) (c)-(iii) (d)-(i)
3) (a)-(iii) (b)-(iv) (c)-(i) (d)-(ii)
4) (a)-(ii) (b)-(i) (c)-(iii) (d)-(iv)

2. Match the terms given in Column-I with the items given in Column-II.

	Column-I		Column-II
(a)	Λ_m	(i)	intensive property
(b)	$E^{\ominus}_{cell}$	(ii)	depends on number of ions/ volume
(c)	κ	(iii)	extensive property
(d)	Δ_rG_{cell}	(iv)	increases with dilution

1) (a)-(i) (b)-(ii) (c)-(iii) (d)-(iv)
2) (a)-(iv) (b)-(i) (c)-(ii) (d)-(iii)
3) (a)-(i) (b)-(iv) (c)-(iii) (d)-(ii)
4) (a)-(iii) (b)-(ii) (c)-(i) (d)-(iv)

3. Match the items of Column-I and Column-II.

	Column-I		Column-II
(a)	Lead storage battery	(i)	maximum efficiency
(b)	Mercury cell	(ii)	prevented by galvanisation
(c)	Fuel cell	(iii)	gives steady potential
(d)	Rusting	(iv)	Pb is anode, PbO_2 is cathode

1) (a)-(i) (b)-(iv) (c)-(iii) (d)-(ii)
2) (a)-(ii) (b)-(iii) (c)-(i) (d)-(iv)
3) (a)-(iii) (b)-(ii) (c)-(iv) (d)-(i)
4) (a)-(iv) (b)-(iii) (c)-(i) (d)-(ii)

4. Match the items of Column-I and Column-II.

	Column-I		Column-II
(a)	κ	(i)	$I \times t$
(b)	$\wedge_m$	(ii)	$\Lambda_m / \Lambda^{\circ}_m$
(c)	α	(iii)	$\frac{\kappa}{c}$
(d)	Q	(iv)	$\frac{G^*}{R}$

1) (a)-(iv) (b)-(iii) (c)-(ii) (d)-(i)
2) (a)-(i) (b)-(ii) (c)-(iv) (d)-(iii)
3) (a)-(iii) (b)-(iv) (c)-(ii) (d)-(i)
4) (a)-(ii) (b)-(i) (c)-(iv) (d)-(iii)

5. Match the items of Column-I and Column-II.

	Column-I		Column-II
(a)	Leclanche cell	(i)	cell reaction $2H_2 + O_2 \longrightarrow 2H_2O$
(b)	Ni–Cd cell	(ii)	does not involve any ion in solution and is used in hearing aids.
(c)	Fuel cell	(iii)	rechargeable
(d)	Mercury cell	(iv)	reaction at anode, $Zn \longrightarrow Zn^{2+} + 2e^-$
		(v)	converts energy of combustion into electrical energy

1) (a)-(v) (b)-(i) (c)-(ii) (d)-(iii)(iv)
2) (a)-(iv)(v) (b)-(i) (c)-(ii) (d)-(iii)
3) (a)-(iv) (b)-(iii) (c)-(i)(v) (d)-(ii)
4) (a)-(ii) (b)-(i)(iii) (c)-(iv) (d)-(v)

Correct & Incorrect MCQs

1. Which of the following statement (s) are correct?
   A. In Daniell cell when $E_{ext} = 1.1$ V, electricity is produced.
   B. The Daniell cell becomes dead after sometimes.
   C. In galvanic cell, the Gibbs energy of the spontaneous redox reaction is converted into electrical work.
   D. Galvanic cells are used for electrical gadgets like heater, fan, geyser etc.
   1) A, B and C 2) A and D
   3) A, C and D 4) B, C and D

2. Which of the following statement is incorrect?
   1) If the standard electrode potential of an electrode is greater than zero then its reduced form is more stable as compared to hydrogen gas.
   2) If the standard electrode potential is negative, then hydrogen gas is more stable than the reduced form of the species.
   3) Fluoride ion is the strongest reducing agent.
   4) Lithium ion is the weakest oxidising agent.

3. Choose incorrect statement among the following:
   1) Three metallic cations X , Y & Z has standard reduction potential value 0.52, -3.03 & -1.18V respectively, its order of reducing power is Y > Z > X.
   2) Greater the reduction potential, less is the reducing power.
   3) Fluorine gas is the weakest reducing agent.
   4) Lithium has the highest electrode potential.

4. Select the incorrect statement:
   1) If length of any object is increased, its electrical resistance remain unaffected.
   2) Electric resistance is inversely proportional to the area of cross section.
   3) The SI unit of resistivity is ohm metre.
   4) The inverse of resistance, R, is called conductance.

5. Correct statement(s) is/are :
   A. Electrical conductance through metals is called metallic or electronic conductance.
   B. The electronic conductance do not depend on temperature.
   C. The conductance of electricity by ions present in the solutions is called electrolytic or ionic conductance.
   D. The conductivity of electrolytic (ionic) solutions depends on the nature of the electrolyte added, the nature of the solvent and its viscosity, temperature etc.
   1) A and B 2) A, C and D
   3) Only A and D 4) Only B

6. Which of the following statements is incorrect ?
   1) Conductivity always decreases with the decrease in concentration both for weak and strong electrolytes.
   2) Molar conductivity decreases with decrease in concentration.
   3) Limiting molar conductivity is represented by the symbol Δ°_{M}
   4) All are incorrect

7. Select the correct statement:
   1) When electrolytes are dissolved in water, they do not furnish their own ions in the solution and hence, its conductivity does not get affected.
   2) The conductance of electricity by ions present in the solutions is called limiting conductance.
   3) Passage of direct current through ionic solution over a prolonged period can lead to change in its composition due to electrochemical reactions
   4) Both (1) and (2)

8. Which amongst the following is incorrect statement ?
   1) Products of electrolysis depend on the nature of material being electrolysed and the type of electrodes being used.
   2) The products of electrolysis may be different for reactive & inert electrodes.
   3) If the electrode is reactive, it participates in the electrode reaction.
   4) Metals like platinum or gold are reactive electrodes.

9. Which of the following statements is incorrect ?
   1) The mercury cell is a type of primary cell.
   2) Primary cells are rechargeable and can be used again and again.
   3) Lead storage battery is an example of secondary cell.
   4) Nickel-cadmium cell has longer life than the lead storage cell.

10. Which of the following statement is not correct about corrosion ?
    1) Corrosion causes enormous damage to buildings, bridges, ships and to all objects made of metals especially that of iron.
    2) In corrosion, a metal is oxidised by loss of electrons to oxygen and formation of oxides.
    3) The chemical formula of rust is $FeCl_2.5\ H_2O$.
    4) Galvanised iron does not rust.

Statement Based MCQs

Directions: These questions consist of two statements each, printed as Statement-I and Statement-II. While answering these questions, you are required to choose any one of the following four responses.

1) **Both Statement-I and Statement-II are correct.**
2) **Both Statement-I and Statement-II are incorrect.**
3) **Statement-I is correct & Statement-II is incorrect.**
4) **Statement-I is incorrect & Statement-II is correct.**

1. **Statement-I:** $CuSO_4$ solution cannot be stored in iron vessel.
   Statement-II: Cu atoms are oxidised by iron ions.
2. **Statement-I:** Absolute value of E°_{red} of an electrode cannot be determined.
   Statement-II: In redox reaction, neither oxidation nor reduction can take place alone.
3. **Statement-I:** Salt Bridge is used generally in the electrochemical cells.
   Statement-II: The ions of the electrolyte used in salt bridge should have almost same ionic mobility.
4. **Statement-I:** The cell constant of a conductivity cell do not depend on the distance between the electrodes and their area of cross-section.
   Statement-II: Unit of cell constant is cm^{-1}.
5. **Statement-I:** In electrolysis the quantity of electricity needed for depositing 1mole of silver is different from that required for 1 mole of copper.
   Statement-II: The molecular weights of silver and copper are different.
6. **Statement-I:** In standard hydrogen electrode, the pressure of H_2 gas is 100 bar and concentration of hydrogen ion is 2 molar.
   Statement-II: In standard hydrogen electrode, the concentration of both the reduced and oxidised forms of hydrogen is maintained at unity.
7. **Statement-I:** We generally use KCl solutions to determine cell constant.
   Statement-II: KCl is a solution whose conductivity is known accurately at various concentrations.
8. **Statement-I:** The products of electrolysis may be different for reactive and inert electrodes.
   Statement-II: Inert electrodes participate while reactive electrodes do not participate in the electrode reaction.
9. **Statement-I:** Cu^{2+} ions get reduced more easily than H^+ ions.
   Statement-II: Standard electrode potential of copper is –1.81 V.
10. **Statement-I:** Kohlrausch law helps to find the molar conductivity of weak electrolyte at infinite dilution.
    Statement-II: Conductivity decreases but molar conductivity increases with decrease in concentration.

Assertion & Reason MCQs

Directions: These questions consist of two statements each, printed as Assertion and Reason. While answering these questions, you are required to choose any one of the following four responses.

1) **Both Assertion and Reason are True and the Reason is a correct explanation of the Assertion.**
2) **Both Assertion and Reason are True but Reason is not a correct explanation of the Assertion.**
3) **Assertion is True but the Reason is False.**
4) **Assertion is False but Reason is True.**

1. **Assertion:** Cu is less reactive than hydrogen.
   Reason: $E^{\ominus}_{Cu^{2+}/Cu}$ is negative.
2. **Assertion:** E_{Cell} should have a positive value for the cell to function.
   Reason: $E_{cathode} < E_{anode}$
3. **Assertion:** Conductivity of all electrolytes decreases on dilution.
   Reason: On dilution, number of ions per unit volume decreases.
4. **Assertion:** Λ_m for weak electrolytes shows a sharp increase when the electrolytic solution is diluted.
   Reason: For weak electrolytes degree of dissociation increases with dilution of solution.
5. **Assertion:** Mercury cell does not give steady potential.
   Reason: In the cell reaction, ions are not involved in solution.
6. **Assertion:** Electrolysis of NaCl solution gives chlorine at anode instead of O_2.
   Reason: Formation of oxygen at anode requires overvoltage.
7. **Assertion:** Current stops flowing when $E_{Cell} = 0$.
   Reason: Equilibrium of the cell reaction is attained.
8. **Assertion:** Tin can displace lead from lead bromide solution.
   Reason: The standard reduction potential of tin is less than that of lead.
9. **Assertion:** If an aqueous solution of NaCl is electrolysed, the product obtained at the cathode is H_2 gas and not Na.
   Reason: Gases are liberated faster than the metals.
10. **Assertion:** According to Kohlrausch's law, the molar, conductance of a strong electrolyte at infinite dilution is sum of molar conductance of its ions.
    Reason: The current carried by cation and anion is always equal.

Multi Correct MCQs

1. The metals that are employed in the battery industries are
   A. Fe B. Mn
   C. Ni D. Cr
   E. Cd
   Choose the correct answer from the options given below:
   1) (B), (C) and (E) only
   2) (A), (B), (C), (D) and (E)
   3) (A), (B), (C) and (D) only
   3) (B), (D) and (E) only
2. The decreasing order of electrical conductivity of the following aqueous solution is:
   0.1 M Formic acid (I), 0.1 M Acetic acid (II),
   0.1 M Benzoic acid (III),
   A. (I) > (II)
   B. (I) > (III) and (II) < (III)
   C. (I) > (III) and (II) > (III)
   D. (II) > (I)
   1) Both (A) and (B)
   2) (A), (B) and (C)
   3) (B) and (D)
   4) Only (C)
3. Fuel cell, using hydrogen and oxygen as fuels,
   A. has been used in spaceship
   B. has an efficiency of 40% to produce electricity
   C. uses aluminium as catalysts
   D. is eco-friendly
   E. is actually a type of Galvanic cell only
   1) (A), (B), (C) only 2) (A), (B), (D) only
   3) (A), (B), (D), (E) only 4) (A), (D), (E) only
4. Identify the factor/s from the following that affects electrolytic conductance of a solution.
   A. The nature of the electrolyte added.
   B. The nature of the electrode used.
   C. Concentration of the electrolyte.
   D. The nature of solvent used.
   1) Only (A) 2) Both (A) and (B)
   3) (B), (C) and (D) 4) (A), (C) and (D)

NCERT Exemplar MCQs

1. Which cell will measure standard electrode potential of copper electrode?
   1) $Pt(s)|H_2(g,0.1\,bar)|H^+(aq.,1M)||Cu^{2+}(aq.,1M)|Cu$
   2) $Pt(s)|H_2(g,1\,bar)|H^+(aq,1M)||Cu^{2+}(aq.,2M)|Cu$
   3) $Pt(s)|H_2(g,1\,bar)|H^+(aq,1M)||Cu^{2+}(aq.,1M)|Cu$
   4) $Pt(s)|H_2(g,0.1\,bar)|H^+(aq,0.1M)||Cu^{2+}(aq.,1M)|Cu$
2. Which of the following statement is correct?
   1) E_{cell} and Δ_rG of cell reaction both are extensive properties
   2) E_{cell} and Δ_rG of cell reaction both are intensive properties
   3) E_{cell} is an intensive property while Δ_rG of cell reaction is an extensive property
   4) E_{cell} is an extensive property while Δ_rG of cell reaction is an intensive property
3. The difference between the electrode potentials of two electrodes when no current is drawn through the cell is called:
   1) Cell potential 2) Cell emf
   3) Potential difference 4) Cell voltage
4. Which of the following statement is not correct about an inert electrode in a cell?
   1) It does not participate in the cell reaction
   2) It provides surface either for oxidation or for reduction reaction
   3) It provides surface for conduction of electrons
   4) It provides surface for redox reaction
5. An electrochemical cell can behave like an electrolytic cell when:
   1) $E_{cell} = 0$ 2) $E_{cell} > E_{ext}$
   3) $E_{ext} > E_{cell}$ 4) $E_{cell} = E_{ext}$
6. Which of the following statements about solutions of electrolytes is not correct?
   1) Conductivity of solution depends upon size of ions
   2) Conductivity depends upon viscosity of solution
   3) Conductivity does not depend upon solvation of ions present in solution
   4) Conductivity of solution increases with temperature
7. Using the data given below, find out the strongest reducing agent:
   $E^{\oplus}Cr_2O_7^{2-}/Cr^{3+} = 1.33V; E^{\oplus}Cl_2/Cl^- = 1.36V$
   $E^{\oplus}MnO_4^-/Mn^{2+} = 1.51V; E^{\oplus}Cr^{3+}/Cr = -0.74\ V$
   1) Cl^- 2) Cr
   3) Cr^{3+} 4) Mn^{2+}
8. Use the data given in Q. 7 and find out which of the following is the strongest oxidising agent?
   1) Cl^- 2) Mn^{2+}
   3) MnO_4^- 4) Cr^{3+}

9. Using the data given in Q. 7 find out in which option the order of reducing power is correct:
   1) $Cr^{3+} < Cl^- < Mn^{2+} < Cr$
   2) $Mn^{2+} < Cl^- < Cr^{3+} < Cr$
   3) $Cr^{3+} < Cl^- < Cr_2O_7^{-2} < MnO_4^-$
   4) $Mn^{2+} < Cr^{3+} < Cl^- < Cr$

10. Use the data given in Q. 7 and find out the most stable ion in its reduced form:
    1) Cl^- 2) Cr^{3+}
    3) Cr^{2+} 4) Mn^{2+}

11. Use the data of Q. 7 and find out the most stable oxidised species:
    1) Cr^{3+} 2) MnO_4^-
    3) $Cr_2O_7^{2-}$ 4) Mn^{2+}

12. The quantity of charge required to obtain one mole of aluminium from Al_2O_3 is:
    1) 1 F 2) 6 F
    3) 3 F 4) 2 F

13. The cell constant of a conductivity cell:
    1) Changes with change of electrolyte
    2) Changes with change of concentration of electrolyte
    3) Changes with temperature of electrolyte
    4) Remains constant for a cell

14. While charging the lead storage battery:
    1) $PbSO_4$ anode is reduced to Pb
    2) $PbSO_4$ cathode is reduced to Pb
    3) $PbSO_4$ anode is oxidised to Pb
    4) $PbSO_4$ anode is oxidised to PbO_2

15. In the electrolysis of aqueous sodium chloride solution, which of the half cell reaction will occur at anode?
    1) $Na^+(aq) + e^- \rightarrow Na(s); E^{\oplus}_{cell} = -2.71\,V$
    2) $2H_2O(l) \rightarrow O_2(g) + 4H^+(aq) + 4e^-; E^{\oplus}_{cell} = -1.23\,V$
    3) $H^+(aq) + e^- \rightarrow \frac{1}{2}H_2(g); E^-_{cell} = 0.00\,V$
    4) $Cl^-(aq) \rightarrow \frac{1}{2}Cl_2(g) + e^-; E^-_{cell} = 1.36\,V$

16. The positive value of the standard electrode potential of Cu^{2+}/Cu indicates that:
    1) This redox couple is a stronger reducing agent than the H^+/H_2 couple
    2) This redox couple is a stronger oxidising agent than H^+/H_2
    3) Cu can displace H_2 from acid
    4) Cu cannot displace H_2 from acid

17. For the given cell, $|Mg|Mg^{2+}||Cu^{2+}|Cu$:
    1) Mg is cathode
    2) Cu is cathode
    3) The cell reaction is $Mg + Cu^{2+} \rightarrow Mg^{2+} + Cu$
    4) Cu is the oxidising agent

18. Conductivity of an electrolytic solution depends on:
    1) Nature of electrolyte
    2) Concentration of electrolyte
    3) Power of AC source
    4) Distance between the electrodes

19. What will happen during the electrolysis of aqueous solution of $CuSO_4$ by using platinum electrodes?
    1) Copper will deposit at cathode
    2) Copper will deposit at anode
    3) Oxygen will be released at anode
    4) Copper will dissolve at anode

20. Iron can be protected by coating with zinc or tin. If coating is broken
    1) Iron will corrode faster if coated with zinc
    2) Iron will corrode faster if coated with tin
    3) Iron will corrode faster in both cases
    4) Iron will not undergo any corrosion in both cases

21. Conductivity κ, is equal to:
    1) $\frac{1}{R}\frac{l}{A}$ 2) $\frac{G^*}{R}$
    3) A_m 4) $\frac{l}{A}$

22. Molar conductivity of ionic solution depends on:
    1) Temperature
    2) Distance between electrodes
    3) Concentration of electrolytes in solution
    4) Surface area of electrodes

23. The more positive the value of E°, the greater is the tendency of the species to get reduced. Using the standard electrode potential of redox couples given below, find out which of the following is the strongest oxidizing agent?
    E° values : $Fe^{3+}/Fe^{2+} = +0.77$
    $I_2(s)/I^- = +0.54$;
    $Cu^{2+}/Cu = +0.34$; $Ag^+/Ag = 0.80\,V$
    1) Fe^{3+} 2) $I_2(s)$
    3) Cu^{2+} 4) Ag^+

Kattar NEET

Scan Me for Detailed Explanations

1. Adding powdered lead and iron to a solution that is 1.0 molar each in Pb^{2+} and Fe^{2+} ions, would result to a reaction in which [Assume Pb and Fe are not present initially]

$E^\circ_{Fe^{2+}/Fe} = -0.44$ V,

$E_{Pb^{2+}/Pb} = -0.13$ V

1) Conc. of both Pb^{2+} and Fe^{2+} are increased
2) More lead and iron are formed
3) More of iron and Pb^{2+} ions are formed
4) More of lead and Fe^{2+} ions are formed

2. If $E^0_{XO_4^-/XO_3^-} = -0.36\text{ V}$ and $E^0_{XO_3^-/XO_2^-} = -0.342\text{ V}$ at 27°C, the concentration of XO_4^- ion at equilibrium, which was initially 1.0 M in XO_3^- when reaction started to attain the equilibrium,

$2XO_3^- \rightleftharpoons XO_2^- + XO_4^-$

$\left[\frac{2.303RT}{F} = 0.06 \text{ and } \log 5 = 0.7\right]$

1) 0.40 M
2) 0.2 M
3) 0.122 M
4) 0.02 M

3. For which of these electrodes will the reduction potential vary with pH?

I. AmO_2^{2+} / AmO_2^{+}

II. AmO_2^{2+} / Am^{4+}

III. Am^{4+} / Am^{2+}

1) I only
2) II only
3) I and II only
4) I, II and III only

4. A graph is plotted between electrode potential of quinhydrone half cell and pH of solution. Intercept is 0.699 V. At what pH, electrode potential will become 0.492 V?

1) 3.5
2) 4.2
3) 7
4) 10

5. The pH of 0.5 L of 1.0 M NaCl after the electrolysis for 965 seconds using 5.0 A current (100% efficiency) is

1) 1.02
2) 13.0
3) 12.7
4) 1.3

6. Find the value of log K_{eq} at 25°C for; $HNO_2 \rightleftharpoons NO + NO_3^-$, if stoichiometric coefficient of HNO_2 is 3 and $E^\circ_{HNO_2/NO} = 0.99\text{ V}, E^\circ_{NO_3^-/HNO_2} = 0.94\text{ V}$.

1) 0.85
2) 1.7
3) 2.5
4) 5.1

7. Equivalent conductance of saturated $BaSO_4$ solution is 400 ohm^{-1} cm^2 $equivalent^{-1}$ and its specific conductance is 8×10^{-5} ohm^{-1} cm^{-1}; hence solubility product K_{sp} of $BaSO_4$ is

1) 4×10^{-8} M^2
2) 1×10^{-8} M^2
3) 2×10^{-4} M^2
4) 1×10^{-4} M^2

8. Match the following:

	List-I		List-II
I.	Very dilute solution of HCl	P.	O_2 evolved at anode
II.	Very dilute solution of NaCl	Q.	H_2 evolved at cathode
III.	Concentrated solution of NaCl	R.	Cl_2 evolved at anode
IV.	Fairly concentrated solution of $AgNO_3$ with inert electrodes	S.	Ag deposition at cathode

1) I-Q, R; II-P, Q; III-Q, R; IV-P, S
2) I-P, Q; II-P, Q; III-Q, R; IV-P, S
3) I-P, S; II-P, R; III-Q, R; IV-P, S
4) I-P, Q; II-P, Q; III-P, S; IV-Q, R

9. For H_2O at 298 K, $\lambda^\circ (H^+) = 3.5 \times 10^{-2}$ S m^2 mol^{-1}, $\lambda^\circ (OH^-) = 2.0 \times 10^{-2}$ S m^2 mol^{-1}

Specific conductance (k) = 5.7×10^{-6} S m^{-1},

Density of water = 1.0 g cm^{-3}

Match the parameters in **List-I** (in terms of –log (value) i.e. p value) with their values in **List-II**.

	List-I		List-II
I	px = –log x (x = degree of dissociation)	P	10.0
II	$p\Lambda_c$ (Λ_c = molar conductance)	Q	8.73
III	pH	R	14.0
IV	pK_w(K_w = ionic product)	S	7.00

1) I-P; II-Q; III-R; IV-S
2) I-S; II-Q; III-R; IV-P
3) I-Q; II-P; III-S; IV-R
4) I-Q; II-S; III-P; IV-R

10. Λ^o_m (CH_3COOH) = 390.5 S cm^2 mol^{-1}, K_a(CH_3COOH) = 1.8 × 10^{-5}, [CH_3COOH] = 0.040 M

Cell constant of the cell used = 0.206 cm^{-1}

Match the parameter given in List-I with the values in List-II and select answer from the codes given at the end.

	List-I		List-II
I	Λ_m (molar conductance at given concentration in S m^2 mol^{-1})	P	8.28×10^{-4}
II	κ (conductivity in S m^{-1})	Q	1.61×10^{-3}
III	Ω (resistance in kilo ohm)	R	3.312×10^{-2}
IV	S (conductance in ohm^{-1})	S	0.622

1) I-P; II-R; III-S; IV-Q
2) I-P; II-Q; III-R; IV-S
3) I-Q; II-R; III-P; IV-S
4) I-P; II-Q; III-S; IV-R

11. The standard reduction potential for the process : $[Co(H_2O)_6]^{3+} + e^- \rightarrow [Co(H_2O)_6]^{2+}$ is 1.8 V. The standard reduction potential for the process : $[Co(NH_3)_6]^{3+} + e^- \rightarrow [Co(NH_3)_6]^{2+}$ is 0.1 V. Which of the complex ion, $[Co(H_2O)_6]^{2+}$ or $[Co(NH_3)_6]^{2+}$ can be oxidized to the corresponding cobalt (III) complex, by oxygen, in basic medium, under standard condition? [Given: $E^o_{O_2|OH^-} = 0.4$ V]

1) $[Co(H_2O)_6]^{2+}$ 2) $[Co(NH_3)_6]^{2+}$
3) both 4) none of these

12. A hydrogen electrode placed in a buffer solution of CH_3COONa and CH_3COOH in the molar ratios x : y and y : x has oxidation potentials E_1 and E_2 volts respectively at 298 K. $(p_{H_2} = 1\,atm)$. The pK_a value of acetic acid will be given by.

1) $\frac{E_1 - E_2}{0.118}$ 2) $\frac{E_2 - E_1}{0.118}$
3) $\frac{E_1 + E_2}{0.118}$ 4) None of these

13. 100 mL of 0.3 M Fe^{3+}(aq) ions were electrolyzed by a charge of 0.072 F. In electrolysis, metal was deposited and O_2 (g) was evolved. At the end of electrolysis, it is desired to oxidize the unelectrolyzed metal ion.

$Fe^{3+} + e^- \rightarrow Fe^{2+}$

$Fe^{2+} + 2e^- \rightarrow Fe$

The moles of Fe^{2+} ions left unelectrolyzed in the solution is

1) 0.009 2) 0.021
3) 0.072 4) 0.042

14. Marshall's acid ($H_2S_2O_8$) or peroxo disulphuric acid is prepared by the electrolytic oxidation of H_2SO_4 as

$2H_2SO_4 \rightarrow H_2S_2O_8 + 2H^+ + 2e^-$

O_2(g) and H_2(g) are obtained as by products. In such electrolysis, 4.48 L of H_2(g) and 1.12 L of O_2(g) were produced at STP. The weight of $H_2S_2O_8$ formed is

1) 9.7 g 2) 19.4 g
3) 14.5 g 4) 29.1 g

15. In a 500 mL of 0.5 M $CuSO_4$ solution during electrolysis, 1.5×10^{23} electrons were passed using copper electrodes. Assume the volume of solution remains unchanged during electrolysis, which of the following statements is incorrect?

1) At the end of electrolysis, the concentration of the solution is 0.5 M
2) 7.9 g of Cu is deposited on the cathode
3) 4 g of Cu is dissolved from the anode
4) 7.9 g of Cu^{2+} ions are discharged

16. The conductivity of 0.001 M Na_2SO_4 solution is 2.6×10^{-4} S cm^{-1} and increases to 7.0×10^{-4} S cm^{-1} when the solution is saturated with $CaSO_4$. The molar conductivities of Na^+ and Ca^{2+} are 50 and 120 S cm^2 mol^{-1}, respectively. Neglecting conductivity of used water, what is the solubility product of $CaSO_4$?

1) 4×10^{-6} 2) 1.57×10^{-3}
3) 4×10^{-4} 4) 2.46×10^{-6}

17. The standard reduction potential of a silver chloride electrode is 0.2 V and that of a silver electrode is 0.79 V. The maximum amount of AgCl that can dissolve in 10^6 L of a 0.1 M $AgNO_3$ solution is

1) 0.5 m mol 2) 1.0 m mol
3) 2.0 m mol 4) 2.5 m mol

18. Zn-amalgam is prepared by electrolysis of aqueous $ZnCl_2$ using Hg cathode (9 gm). How much current is to be passed through $ZnCl_2$ solution for 1000 seconds to prepare a Zn-amalgam with 25% Zn by weight. (Atomic mass of Zn = 65.4)?

1) 5.6 amp 2) 7.2 amp
3) 8.85 amp 4) 11.2 amp

19. The following electrochemical cell has been set up:

Pt(s) | Fe^{3+}, Fe^{2+} (a = 1) || Ce^{4+}, Ce^{3+} (a = 1) | Pt(s);

Given: $E^o(Fe^{3+} | Fe^{2+}) = 0.77$ V; $E^o(Ce^{4+} | Ce^{3+}) = 1.61$ V.

If an ammeter is connected between the two platinum electrodes, predict the direction of flow of current. Will the current increase or decrease with time?

1) Ce electrode to Fe electrode, decrease
2) Ce electrode to Fe electrode, increase
3) Fe electrode to Ce electrode, decrease
4) Fe electrode to Ce electrode, increase

20. The ionization constant of a weak electrolyte is 25×10^{-6} while the equivalent conductance of its 0.01 M solution is 19.6 S cm^2 eq^{-1}. The equivalent conductance of the electrolyte at infinite dilution (in S cm^2 eq^{-1}) will be

1) 250 2) 196
3) 392 4) 384

NEET & AIIMS Past Year MCQs

1. The standard cell potential of the following cell Zn|Zn^{2+}(aq)|Fe^{2+}(aq)|Fe is 0.32 V. Calculate the standard Gibbs energy change for the reaction : **(2024 Re)**

$$Zn(s) + Fe^{2+}(aq) \rightarrow Zn^{2+}(aq) + Fe(s)$$

(Given: 1 F = 96487 C)

1) –61.75 kJ mol^{-1} 2) +5.006 kJ mol^{-1}
3) –5.006 kJ mol^{-1} 4) +61.75 kJ mol^{-1}

2. From the following select the one which is not an example of corrosion. **(2024 Re)**

1) Rusting of iron object
2) Production of hydrogen by electrolysis of water
3) Tarnishing of silver
4) Development of green coating on copper and bronze ornaments

3. Match List-I with List-II. **(2024)**

	List-I (Conversion)		List-II (Number of Faraday required)
A.	1 mol of H_2O to O_2	I.	3F
B.	1 mol of MnO_4^- to Mn^{2+}	II.	2F
C.	1.5 mol of Ca from molten $CaCl_2$	III.	1F
D.	1 mol of FeO to Fe_2O_3	IV.	5F

Choose the correct answer from the options given below:

1) A-II, B-III, C-I, D-IV 2) A-III, B-IV, C-II, D-I
3) A-II, B-IV, C-I, D-III 4) A-III, B-IV, C-I, D-II

4. Mass in grams of copper deposited by passing 9.6487 A current through a voltmeter containing copper sulphate solution for 100 seconds is: **(2024)**

(Given : Molar mass of Cu: 63 g mol^{-1}, 1 F = 96487 C)

1) 31.5 g 2) 0.0315 g 3) 3.15 g 4) 0.315 g

5. The conductivity of centimolar solution of KCl at 25°C is 0.0210 ohm^{-1} cm^{-1} and the resistance of the cell containing the solution at 25 °C is 60 ohm. The value of cell constant is: **(2023)**

1) 3.34 cm^{-1} 2) 1.34 cm^{-1}
3) 3.28 cm^{-1} 4) 1.26 cm^{-1}

6. Given below are two statements : one is labelled as Assertion A and the other is labelled as Reason R:

Assertion (A): In equation $\Delta_rG = -nFE_{cell}$, value of Δ_rG depends on n.

Reasons (R): E_{cell} is an intensive property and Δ_rG is an extensive property.

In the light of the above statements, choose the correct answer from the options given below: **(2023)**

1) A is false but R is true.
2) Both A and R are true and R is the correct explanation of A.
3) Both A and R are true but R is NOT the correct explanation of A.
4) A is true but R is false

7. Given below are half cell reactions: **(2022)**

$$MnO_4^- + 8H^+ + 5e^- \rightarrow Mn^{2+} + 4H_2O$$

$$E^o_{Mn^{2+}/MnO_4^-} = -1.510\,V$$

$$\frac{1}{2}O_2 + 2H^+ + 2e^- \rightarrow H_2O$$

$$E_{O_2/H_2O} = +1.223\,V$$

Will the permanganate ion, MnO_4^- **liberate** O_2 from water in the presence of an acid?

1) No because $E^o_{cell} = -2.733\,V$
2) Yes, because $E^o_{cell} = +0.287\,V$
3) No, because $E^o_{cell} = -0.287\,V$
4) Yes, because $E^o_{cell} = +2.733\,V$

8. Find the emf of the cell in which the following reaction takes place at 298 K **(2022)**

$Ni(s) + 2Ag^+ (0.001\ M) \rightarrow Ni^{2+} (0.001\ M) + 2Ag(s)$

(Given that $E^\circ_{cell} = 10.5$ V, $\frac{2.303\,RT}{F} = 0.059$ at 298 K)

1) 1.05 V 2) 1.0385 V
3) 1.385 V 4) 0.9615 V

9. The molar conductance of NaCl, HCl and CH_3COONa at infinite dilution are 126.45, 426.16 and 91.0 S cm^2 mol^{-1} respectively. The molar conductance of CH_3COOH at infinite dilution is. Choose the right option for your answer. **(2021)**

1) 390.71 S cm^2 mol^{-1} 2) 698.28 S cm^2 mol^{-1}
3) 540.48 S cm^2 mol^{-1} 4) 201.28 S cm^2 mol^{-1}

10. The molar conductivity of 0.007 M acetic acid is 20 S cm^2 mol^{-1}. What is the dissociation constant of acetic acid? Choose the correct option. **(2021)**

$$\begin{bmatrix} \Lambda^\circ_{H^+} = 350\ S\ cm^2 mol^{-1} \\ \Lambda^\circ_{CH_3COO^-} = 50\ S\ cm^2 mol^{-1} \end{bmatrix}$$

1) 2.50×10^{-4} mol L^{-1} 2) 1.75×10^{-5} mol L^{-1}
3) 2.50×10^{-5} mol L^{-1} 4) 1.75×10^{-4} mol L^{-1}

11. Cell equation : $A + 2B^+ \rightarrow A^{2+} + 2B$ **(AIIMS 2018)**

$A^{2+} + 2e \rightarrow A\ E^o = +0.34$ V

and $\log_{10} K = 15.6$ at 300 K for cell reactions

Find E° for $B^+ + e \rightarrow B$ Given $\left[\frac{2.303RT}{nF} = 0.059\right]_{at\,300\,K}$

1) 0.80 V 2) 1.26 V
3) 0.54 V 4) +0.94 V

12. **Assertion:** As the lead storage battery get discharged, density of electrolyte, present in it, decreases.

Reason: Lead and lead dioxide both react with sulphuric acid to form lead sulphate. **(AIIMS 2017)**

1) If both Assertion & Reason are True & the Reason is a correct explanation of the Assertion.
2) If both Assertion & Reason are True but Reason is not a correct explanation of the Assertion.
3) If Assertion is True but the Reason is False.
4) If both Assertion & Reason are False.

13. The time period of coat a metal surface of 80 cm^2 with 5×10^{-3} cm thick layer of silver (density 1.05 g cm^{-3}) with the passage of 3 A current through is silver nitrate solution is: **(AIIMS 2008)**

1) 115 sec 2) 125 sec
3) 135 sec 4) 145 sec

Answer Key

NCERT Topic-wise MCQs

1	2	3	4	5	6	7	8	9	10	11	12	13	14	15	16	17
1	2	3	3	2	1	1	3	4	4	1	1	3	2	3	3	1
18	**19**	**20**	**21**	**22**	**23**	**24**	**25**	**26**	**27**	**28**	**29**	**30**	**31**	**32**	**33**	**34**
2	3	4	1	1	4	3	4	1	4	1	2	3	2	1	4	2
35	**36**	**37**	**38**	**39**	**40**	**41**	**42**	**43**	**44**	**45**	**46**	**47**	**48**	**49**	**50**	**51**
3	3	4	2	1	2	1	3	1	2	1	3	3	2	2	1	4
52	**53**	**54**	**55**	**56**	**57**	**58**	**59**	**60**	**61**	**62**	**63**	**64**	**65**	**66**	**67**	**68**
1	4	2	4	1	4	1	3	1	2	4	4	2	2	1	2	4
69	**70**	**71**	**72**	**73**	**74**	**75**	**76**	**77**	**78**	**79**	**80**	**81**	**82**	**83**	**84**	**85**
2	2	4	3	4	3	1	2	1	2	3	3	3	4	4	1	2
86	**87**	**88**	**89**													
4	3	3	4													

High Order Time Intensive MCQs

1	2	3	4	5	6	7	8	9	10	11	12	13	14	15	16	17
1	4	4	2	2	1	3	3	1	1,3	4	4	2	1	2	3	4
18	**19**	**20**	**21**	**22**	**23**	**24**	**25**	**26**	**27**	**28**	**29**	**30**	**31**	**32**	**33**	**34**
3	2	3	2	3	3	4	3	4	3	2	1	3	1	2	1	4
35																
3																

Match the Columns MCQs

1	2	3	4	5
3	2	4	1	3

Correct & Incorrect MCQs

1	2	3	4	5	6	7	8	9	10
4	3	4	1	2	2	3	4	2	3

Statement Based MCQs

1	2	3	4	5	6	7	8	9	10
3	1	1	4	1	4	1	3	3	1

Assertion & Reason MCQs

1	2	3	4	5	6	7	8	9	10
3	3	1	1	4	1	1	1	3	3

Multi Correct MCQs

1	2	3	4
1	1	4	4

NCERT Exemplar MCQs

1	2	3	4	5	6	7	8	9	10	11	12	13	14	15	16	17
3	3	2	4	3	3	2	3	2	4	1	3	4	1	4	2,4	2,3
18	**19**	**20**	**21**	**22**	**23**											
1, 2	1, 3	2	1, 2	1, 3	4											

Kattar NEET

1	2	3	4	5	6	7	8	9	10	11	12	13	14	15	16	17
4	1	2	1	2	2	2	2	3	1	2	3	1	2	3	1	2
18	**19**	**20**														
3	1	3														

NEET & AIIMS Past Year MCQs

1	2	3	4	5	6	7	8	9	10	11	12	13
1	2	3	4	4	2	2	None	1	2	1	1	2

CHAPTER 3

Chemical Kinetics

Topicwise Number of Questions Analysis (2025-2016)

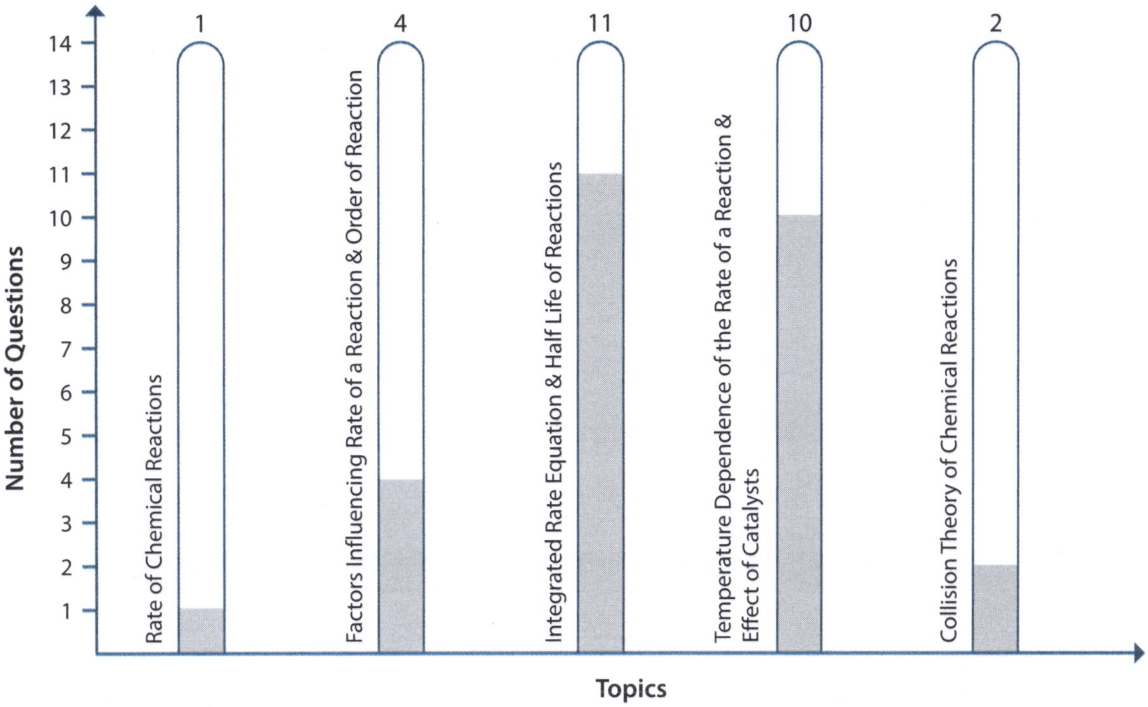

NCERT-Picks

The branch of chemistry which deals with rate and mechanism of chemical reaction is known as chemical kinetics.

Rate of a Chemical Reaction

- Some of the chemical reactions occur very fast, for example neutralization reactions between acids and bases, some are very slow, for example rusting of iron, whereas, some reactions occur at moderate speed for example hydrolysis of starch etc.
- To find out the speed or rate of the reaction, change in concentration of the reactants or products is required.
- The rate of the reaction may be expressed in either of the two ways:
  - (a) Disappearance or decrease in the rate of concentration of reactant with time.
  - (b) Appearance or increase in the rate of concentration of product with time.
- For a reaction, $A \rightarrow B$
- $$\text{Rate of reaction} = \frac{\text{Total change in concentration of reactants/products}}{\text{Change in time (in sec)}}$$

  i.e., $\text{Rate of reaction} = \frac{-d[A]}{dt} = \frac{+d[B]}{dt}$

  Units of rate of reaction : mol L^{-1} sec^{-1} (in general)

  atm sec^{-1} (in gaseous reactions)

Factors Influencing Rate of a Reaction

There are many factors which influence the rate of the reaction but mainly five factors affects the rate of the reaction. These are,

1. Concentration of the reactants
2. Surface area of the reactants
3. Presence of catalyst
4. Effect of sunlight
5. Effect of temperature

1. **Concentration of the reactants**

   Since we have been observed that the rate of a chemical reaction is directly proportional to the concentration of the reactants means rate of reaction decreases with decreasing in concentration and vice-versa.

2. **Surface area of the reactants**

   Larger the surface area of reactant, the probability of collisions on the surface of the reactant particles by the surrounding molecules increases and thus rate of reaction increases.

3. **Presence of catalyst**

   A catalyst increases the rate of the reaction, but does not affect the state of equilibrium. This is because the activation energy for the forward and backward reaction decreases to the same extent.

4. **Effect of sunlight**

   There are many chemical reactions whose rates are influenced by radiations particularly by ultraviolet and visible light. Such reactions are called photochemical reactions.

   Example:

   $H_2 + Cl_2 \xrightarrow{\text{sunlight}} 2HCl$

   The radiant energy initiates the chemical reaction by supplying the necessary activation energy required for the reactions.

5. **Effect of temperature**

   Temperature dependence of the rate of reaction is explained by Arrhenius equation. Generally rate of reaction increases with an increase in temperature.

Rate Expression and Rate Constant

- Rate law or rate equation is a mathematical expression which expresses the observed rate of the reaction in terms of the concentration of the reaction species.

  eg: $aA + bB \rightarrow$ product

  The rate expression for this reaction is Rate $\propto [A]^a [B]^b$

  Rate = $k [A]^a [B]^b$

  Where, k is rate constant.
- The rate constant of a reaction not affected by concentration of reactants but increases with increase of temperature.
- The units of rate constant depend on the order of reaction.

Order of a Reaction

- The sum of exponents of the concentration terms in the rate law is the order of reaction.
- The order of a reaction can have a fractional value. Order of reaction depends upon the power to which the concentration is raised to express the rate of reaction as found experimentally.
- Rate constant of a **zero order reaction** has same unit as the rate of reaction. k = conc/time = mol L^{-1} s^{-1} , i.e. it depends upon unit of the concentration.
- The specific rate constant of a first order reaction is only **temperature dependent**.
- For the first order reaction $A \rightarrow$ products, plot of log [A] vs time is **linear with negative slope**.

  The units of rate constants for different order reactions are given in the following table.

Units of rate constant

- Unit of rate constant (k) = $\dfrac{\text{Rate}}{[A]^x[B]^y} = \dfrac{\text{Conc}}{\text{time}} \times \dfrac{1}{(\text{Conc.})^n}$

Reaction	Order	Rate Law	Rate constant	Units
$R \rightarrow P$	0	Rate = $k[R]^0$	$k = \dfrac{[R_0]-[R]}{t}$	mol L^{-1} s^{-1}
$R \rightarrow P$	1	Rate = $k [R]^1$	$k = \dfrac{2.303}{t} \log \dfrac{[R_0]}{[R]}$	s^{-1}
$2R \rightarrow P$	2	Rate = $k [R]^2$	$k = \dfrac{1}{t}\left[\dfrac{1}{(R)} - \dfrac{1}{(R_0)}\right]$	mol^{-1} L s^{-1}

Molecularity

- The number of reacting species (atoms, ions or molecules) taking part in an elementary reaction, which must collide simultaneously in order to bring about a chemical reaction is called molecularity of a reaction. The reaction can be unimolecular when one reacting species is involved, for example, decomposition of ammonium nitrite.

  $NH_4NO_2 \rightarrow N_2 + 2H_2O$
- Bimolecular reactions involve simultaneous collision between two species, for example, dissociation of hydrogen iodide.

  $2HI \rightarrow H_2 + I_2$
- Trimolecular or termolecular reactions involve simultaneous collision between three reacting species, for example,

  $2NO + O_2 \rightarrow 2NO_2$
- The probability that more than three molecules can collide and react simultaneously is very small. Hence, **reactions with the molecularity three are very rare and slow to proceed.**
- It is, therefore, evident that complex reactions involving more than three molecules in the stoichiometric equation must take place in more than one step.

  For example:

  $KClO_3 + 6FeSO_4 + 3H_2SO_4 \rightarrow KCl + 3Fe_2(SO_4)_3 + 3H_2O$

  This reaction which apparently seems to be of tenth order is actually a second order reaction. This shows that this reaction takes place in several steps. The slowest step is the rate determining step.

Integrated Rate Equations

Differential and Integrated Rate Laws for Zero and First Order reactions						
Order	**Reaction type**	**Differential rate law**	**Integrated rate law**	**Straight line plot**	**Half-life**	**Units of k**
0	$R \rightarrow P$	$d[R]/dt = -k$	$kt = [R]_0 - [R]$	$[R]$ vs t	$[R]_0/2k$	conc time^{-1} or mol $L^{-1}s^{-1}$
1	$R \rightarrow P$	$d[R]/dt = -k[R]$	$[R] = [R]_0 e^{-kt}$ or $kt = \ln\{[R]_0/[R]\}$	$\ln[R]$ vs t	$\ln 2/k$	time^{-1} or s^{-1}

- **For a First-Order Reaction:** Relationship between $t_{99.9\%}$ and $t_{½}$

$$t_{99.9\%} = \frac{\ln(1000)}{k} = \frac{6.91}{k}$$

$$t_{1/2} = \frac{\ln(2)}{k} = \frac{0.693}{k}$$

$$\frac{t_{99.9\%}}{t_{1/2}} = \frac{6.91}{0.693} \approx 10$$

$$t_{99.9\%} \approx 10 \times t_{1/2}$$

- **Pseudo first order reactions:** Those reactions which are not truly first order but under certain conditions behaves as of the first order.

  eg : Acid hydrolysis of ethyl acetate:

  Rate = $k'[CH_3COOC_2H_5][H_2O] = k[CH_3COOC_2H_5]$

  (concentration of water does not get altered much during the course of reaction)

- Acid catalysed inversion of cane sugar:

$$\underset{\text{Cane sugar}}{C_{12}H_{22}O_{11}} + H_2O \xrightarrow{H^+} \underset{\text{Glucose}}{C_6H_{12}O_6} + \underset{\text{Fructose}}{C_6H_{12}O_6}$$

  Rate = $k[C_{12}H_{22}O_{11}]$

Difference Between Order and Molecularity

Order	Molecularity
1. It is an experimental quantity.	1. It is a theoretical quantity.
2. It is the sum of powers of the concentration terms in rate law	2. It is the number of species which gives effective collisions.
3. It may have fractional values.	3. It has only whole number values.
4. It can be zero.	4. It cannot be zero.

Temperature Dependence of Rate of Reaction

- For a chemical reaction with rise in temperature by 10° C, the rate constant is nearly doubled. At T + 10° C, the fraction of molecules having energy equal to or greater than activation energy gets doubled.

- The temperature dependence of the rate of chemical reaction can be explained by Arrhenius equation:

  $k = Ae^{-E_a/RT}$ (1)

  A is Arrhenius factor or frequency factor or pre-exponential factor

  E_a is Activation energy

  T is temperature in Kelvin

  From reaction (1)

$$\log k = \log A - \frac{E_a}{RT \times 2.303}$$

  (i) Reaction is slow, when E_a is high

  (ii) Reaction is fast, when E_a is low

- $$\log\frac{k_2}{k_1} = \frac{E_a}{2.303R}\left[\frac{1}{T_1} - \frac{1}{T_2}\right]$$

- The minimum amount of energy which the colliding molecules must possess in known as **threshold energy**.

- The excess energy (above the average energy of the reactant) required by the reactant to undergo chemical reaction is called **activation energy**.

- **Activation energy = threshold energy-average K.E. of reactants.**

- When the temperature of the substance increases the fraction of molecule also increases ,that collide with energies greater than E_a.

  Then the fraction of molecules get doubled leading to doubling the rate of reaction.

- A reaction can have zero activation energy. Free radical reactions have zero activation energy.

  For Ex: $\dot{C}H_3 + \dot{C}H_3 \rightarrow H_3C - CH_3 (E_a = 0)$

Collision Theory of Chemical Reactions

- **Collision frequency (Z)** is the number of collision that takes place per second per unit volume of a reaction mixture.

- For a bimolecular elementary reaction, A + B ⟶ Products Rate can be expressed as Rate = $Z_{AB}e^{(-E_a/RT)}$

- The collisions which actually produces the products and result in the chemical reaction are **effective collision.**

  Rate = $PZ_{(AB)}e^{(-E_a/RT)}$

  P is steric factor

  $Z_{(AB)}$ is collision frequency of reactant A and B.

MCQs

NCERT Topic-wise MCQs

Rate of a Chemical Reaction

1. When a chemical reaction takes place, during the course of the reaction the rate of reaction:(excluding zero order)
   1) Keeps on increasing with time
   2) Remains constant with time
   3) Keeps on decreasing with time
   4) Shows irregular trend with time

2. For the reaction $2NH_3 \rightarrow N_2 + 3H_2$,

   If $\frac{-d[NH_3]}{dt} = K_1[NH_3]$, $\frac{d[N_2]}{dt} = K_2[NH_3]$,

   $\frac{d[H_2]}{dt} = K_3[NH_3]$

   then the relation between K_1, K_2 and K_3 is:
   1) $K_1 = K_2 = K_2$
   2) $K_1 = 3\,K_2 = 2\,K_3$
   3) $1.5\,K_1 = 3\,K_2 = K_3$
   4) $2\,K_1 = K_2 = 3\,K_3$

3. In a reaction $2X \rightarrow Y$, the concentration of X decreases from 3.0 moles/litre to 2.0 moles/litre in 5 minutes. The rate of reaction is:
   1) 0.1 mol L^{-1} min^{-1} 2) 5 mol L^{-1} min^{-1}
   3) 1 mol L^{-1} min^{-1} 4) 0.5 mol L^{-1} min^{-1}

4. Consider the reaction, $2N_2O_5 \rightarrow 4NO_2 + O_2$

   In the reaction NO_2 is being formed at the rate of 0.0125 mol L^{-1} s^{-1}. What is the rate of reaction at this time?
   1) 0.0018 mol^{-1} L s^{-1} 2) 0.0031 mol^{-1} L^{-1} s^{-1}
   3) 0.0041 mol^{-1} L s^{-1} 4) 0.050 mol L^{-1} s^{-1}

5. On increasing the pressure three fold, the rate of reaction of $2H_2S + O_2 \rightarrow$ products, would increase
   1) 3 times 2) 9 times
   3) 12 times 4) 27 times

Factors Influencing Rate of Reaction

6. Units of rate constant for second order reaction:
   1) s^{-1} 2) mol L^{-1} s^{-1}
   3) L mol^{-2} s^{-1} 4) L mol^{-1} s^{-1}

7. The rate of a first order reaction is 1.8×10^{-3} mol L^{-1} min^{-1} when the initial concentration is 0.3 mol L^{-1}. The rate constant is
   1) 1×10^{-2} s^{-1} 2) 1×10^{-4} s^{-1}
   3) 6×10^{-2} s^{-1} 4) 4×10^{-4} s^{-1}

8. For a reaction, $I^- + OCl^- \rightarrow IO^- + Cl^-$ in an aqueous medium, the rate of reaction is given by

   $$\frac{d[IO^-]}{dt} = \frac{k[I^-][OCl^-]}{[OH^-]}$$

   The overall order of reaction is:
   1) 1 2) 0
   3) 3 4) 2

9. The order of reaction is decided by:
   1) Temperature
   2) Molecularity
   3) Pressure
   4) Mechanism of reaction as well as relative concentration of reactants

10. The unit of rate and rate constant are same for a:
    1) Zero order reaction
    2) First order reaction
    3) Second order reaction
    4) Third order reaction

11. The number of molecules of the reactants taking part in a single step of the reaction is indicative of:
    1) Order of a reaction
    2) Molecularity of a reaction
    3) Fast step of the mechanism of a reaction
    4) Half-life of the reaction

12. The rate law for the reaction, $xA + yB \rightarrow mP + nQ$ is rate = k $[A]^c$ $[B]^d$, what is the total order of the reaction?
    1) x + y 2) m + n
    3) c + d 4) Two

13. Rate constant of a reaction (K) is 300 L^2 mol^{-2} s^{-1}, what is the order of reaction?
    1) First 2) Second
    3) Third 4) Zero

14. Which of the following statement is not correct?
   1) In zero order reaction, the rate of the reaction remains constant throughout
   2) A second order reaction would becomes a pseudo first order reaction, when one of the reactants is taken in large excess
   3) The units of first order rate constant depends on the units of the concentration terms used
   4) In the first order reaction, the plot of log (a-x) vs time follows straight line equation

15. Which one of the following statements for the order of a reaction is incorrect?
   1) Order of reaction is always whole number
   2) Order can be determined only experimentally
   3) Order is not influenced by stoichiometric coefficient of the reactants
   4) Order of reaction is sum of power to the concentration terms of reactant to express the rate of reaction while writing the rate law.

16. The reaction:

$$N_2O_5(\text{in } CCl_4) \longrightarrow 2NO_2 + \frac{1}{2}O_2(g) \text{ is first order in}$$

N_2O_5 with rate constant $6.2 \times 10^{-4}\ s^{-1}$. What is the value of rate of reaction when $[N_2O_5] = 1.25$ mol L^{-1}

   1) 7.75×10^{-4} mol $L^{-1} s^{-1}$
   2) 6.35×10^{-3} mol $L^{-1} s^{-1}$
   3) 5.15×10^{-5} mol $L^{-1} s^{-1}$
   4) 3.85×10^{-4} mol $L^{-1} s^{-1}$

Integrated Rate Equations

17. For a first order reaction, the time taken to reduce the initial concentration by a factor of 1/4 is 20 min. The time required to reduce initial concentration by a factor of 1/16 is:
   1) 20 min
   2) 10 min
   3) 80 min
   4) 40 min

18. The rate equation for a reaction, $A \rightarrow B$ is $r = K[A]^0$. If the initial concentration of the reactant is a mol dm^{-3}, the half period of the reaction is:
   1) $\frac{k}{a}$
   2) $\frac{a}{k}$
   3) $\frac{2a}{k}$
   4) $\frac{a}{2k}$

19. Which of the following does not effect half-life of first order reaction?
   1) Initial concentration
   2) Catalyst
   3) Temperature
   4) Pressure

20. The value of rate constant for a first order is $2.303 \times 10^{-2} sec^{1}$ What will be the time required to reduce the concentration to $\frac{1}{10}$th of the initial concentration?
   1) 100 sec
   2) 10 sec
   3) 2303 sec
   4) 23.03 sec

21. For which order half-life period is independent of initial concentration?
   1) Zero
   2) First
   3) Second
   4) Third

22. The half-life period for a zero order reaction is:
   1) Proportional to initial concentration
   2) Independent of initial concentration
   3) Inversely proportional to initial concentration
   4) Inversely proportional to the square of the initial concentration

23. 75% of a first order reaction is completed in 30 minutes, what is the time required for 93.75% of the reaction in minutes?
   1) 45
   2) 120
   3) 90
   4) 60

24. If 60% of first order reaction was completed in 60min, 50% of the same reaction would be completed in approximately:
   1) 45 min
   2) 60 min
   3) 40 min
   4) 50 min

25. For a second order reaction, rate at a particular time is x. If the initial concentration is tripled, the rate will become:
   1) 3x
   2) $9x^2$
   3) 9x
   4) 27zx

26. The expression to calculate time required for completion of zero order reaction is:
   1) $t = \frac{[R_o]}{k}$
   2) $t = [R] - [R_o]$
   3) $t = \frac{k}{[R_o]}$
   4) $t = \frac{[R_o] - [R]}{[R_o]}$

27. A first order reaction is 20% complete in 10 minutes. What is the specific rate constant for the reaction?
   1) 0.0970 min^{-1}
   2) 0.009 min^{-1}
   3) 0.0233 min^{-1}
   4) 2.223 min^{-1}

28. In a first order reaction, the concentration of reactants decreases from 400mol L^{-1} to 25 mol L^{-1} in 200 seconds. The rate constant for the reaction is:
   1) 1.01386 s^{-1}
   2) $2 \times 10\ s^{-1}$
   3) $1.386 \times 10^{-2}\ s^{-1}$
   4) $3.4 \times 10\ s^{-1}$

29. For a first order reaction, the rate constant is 6.909 min^{-1}. The time taken for 75% conversion in minutes is:
   1) $\frac{3}{2}\log 2$
   2) $\frac{2}{3}\log 3$
   3) $\frac{2}{3}\log 2$
   4) $\frac{3}{2}\log\frac{3}{4}$

30. In the first order reaction, the relation between the velocity constant (k_1) and half-life period ($t_{1/2}$) is:
   1) $t_{1/2} = \frac{0.693}{k_1}$
   2) $k_1 = \frac{t_{1/2}}{0.693}$
   3) $t_{1/2} = 0.693 + k_1$
   4) $t_{1/2} = \frac{k_1}{0.693}$

31. The first order integrated rate equation is:

1) $k = \frac{x}{t}$
2) $k = \frac{-2.303}{t}\log\frac{a}{a-x}$
3) $k = \frac{1}{t}\ln\frac{a}{a-x}$
4) $k = \frac{1}{t}\cdot\frac{x}{a(a-x)}$

32. For the first order reaction, half-life is 14 sec, the time required for the initial concentration to reduce to 1/8 of its value is:

1) $(14)^3$ sec
2) 28 sec
3) 42 sec
4) $(14)^2$sec

33. $A \rightarrow B$ is a first order reaction. The initial concentration of A is 0.2 mol L^{-1}. After 10 min, the concentration of B is found to be 0.18 mol L^{-1}. The rate constant (in min^{-1}) for the reaction is:

1) 0.2303
2) 2.303
3) 0.693
4) 0.01

34. The ratio $t_{7/8} : t_{1/2}$ for the first order reaction is -

1) 3
2) 5
3) 2
4) 7

35. Half life ($t_{1/2}$) and completion time (T) of the zero order reaction are- (k = 0.001 mol/litre / sec and a = 1 M.)

1) 500 min, 750 min
2) 500 sec, 750 sec
3) 500 sec, 1000 sec
4) None of these

Pseudo First Order Reaction

36. The hydrolysis of ethyl acetate

$$CH_3OOC_2H_5 + H_2O \xrightarrow{H^+} CH_3COOH + C_2H_5OH$$

is the reaction of:

1) Zero order
2) Pseudo first order
3) Second order
4) Third order

37. Which one of the following is wrongly matched?

1) Saponification of $CH_3COOC_2H_5$ – second order reaction
2) Hydrolysis of CH_3COOCH_3 – Pseudo unimolecular reaction
3) Decomposition of H_2O_2 – first order reaction
4) Combination of H_2 and Br_2 to give HBr – zero order reaction

38. Under what conditions a bimolecular reaction may be kinetically of first order?

1) When both reactants have same concentration.
2) When one of the reacting species is in large excess
3) When the reaction is at equilibrium
4) When the activation energy of reaction is less

39. The order of a reaction for an esterification process is :

1) Zero
2) First
3) Second
4) Pseudo first order

40. For the reaction, $H_2(g) + Br_2(g) = 2HBr(g)$, the reaction rate = $k[H_2][Br_2]^{1/2}$. Which statement is true about this reaction?

1) The reaction is of second order
2) Molecularity of the reaction is 3/2
3) The unit of k is sec^{-1}
4) Molecularity of the reaction is 2

Temperature Dependence of The Rate of a Reaction

41. The activation energy in a chemical reaction is defined as:

1) The difference in energy of reactants and products
2) The sum of energies of reactant and products
3) The difference in energy of intermediate complex with the average energy of reactants and products
4) The difference in energy of intermediate complex and the average energy of reactants

42. The temperature dependence of the rate of a chemical reaction can be explained by Arrhenius equation which is:

1) $k = Ae^{-E_a/RT}$
2) $k = Ae^{E_a/RT}$
3) $k = A \times \frac{E_a}{RT}$
4) $k = A \times \frac{RT}{E_a}$

43. When a catalyst is used in an equilibrium process:

1) It increases the rate of forward reaction
2) It decreases the rate of backward reaction
3) It decreases activation energy of forward process and decreases activation energy of backward process
4) It fastens the attainment of equilibrium by lowering activation energy

44. The rate constant for a first order reaction at 300°C for which E_a is 35 kcal mol^{-1} and frequency constant 'A' is 1.45×10^{11} sec^{-1} is:

1) 10×10^{-2} s^{-1}
2) 5.37×10^{10} s^{-1}
3) 5×10^{-4} s^{-1}
4) 7.94×10^{-3} s^{-1}

45. Which of the following statements about the catalyst is true?

1) A catalyst makes the reaction feasible by making ΔG more negative
2) A catalyst makes equilibrium constant more favorable for forward reaction
3) A catalyst accelerate rate of reaction by bringing down the activation energy
4) A catalyst always increases the rate of reaction

46. In a zero order reaction for every 10°C rise of temperature, the rate is doubled. If the temperature is increased from 10°C to 100°C, rate of reaction will becomes:

1) 256 times
2) 512 times
3) 64 times
4) 128 times

47. The rate constant K_1 and K_2 for two different reactions are $10^{16}.e^{-2000/T}$ and $10^{15}.\ e^{-1000/T}$, respectively. The temperature at which $K_1 = K_2$ is:

1) 1000K
2) $\frac{2000}{2.303}$ K
3) 2000K
4) $\frac{1000}{2.303}$ K

48. The potential energy diagram for a reaction R → P is given below.

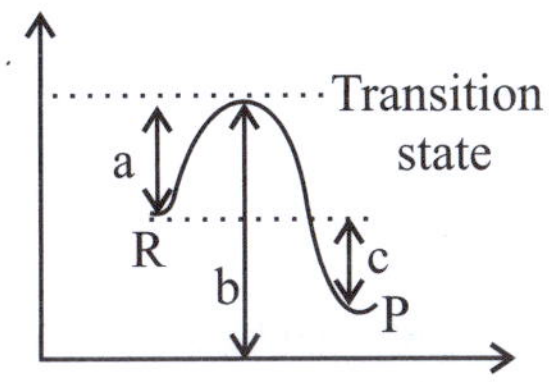

ΔH^0 of the reaction corresponding to the energy:

1) a
2) b
3) c
4) a + b

49. According to Arrhenius equation, the slope of log k vs $\frac{1}{T}$ plot is:

1) $\frac{-E_a}{2.303R}$
2) $\frac{-E_a}{2.303}$
3) $\frac{-E_a}{2.303RT}$
4) $\frac{E_a}{2.303RT}$

50. According to Arrhenius hypothesis, rate of reaction increases with:

1) Rise in temperature
2) Decrease in temperature
3) Rise in pressure
4) Decrease in pressure

51. Activation energy (E_a) and rate constant (K_1 and K_2) of a chemical reaction at two different temperatures (T_1 and T_2) are related by: (where T_1 and T_2 are is final temp.)

1) $\ln\frac{k_2}{k_1} = \frac{-E_a}{R}\left(\frac{1}{T_1} - \frac{1}{T_2}\right)$
2) $\ln\frac{k_2}{k_1} = \frac{-E_a}{R}\left(\frac{1}{T_2} - \frac{1}{T_1}\right)$
3) $\ln\frac{k_2}{k_1} = \frac{-E_a}{R}\left(\frac{1}{T_2} + \frac{1}{T_1}\right)$
4) $\ln\frac{k_2}{k_1} = \frac{E_a}{R}\left(\frac{1}{T_1} + \frac{1}{T_2}\right)$

52. A reaction having equal energies of activation for forward and reverse reaction has:

1) $\Delta S = 0$
2) $\Delta G = 0$
3) $\Delta H = 0$
4) $\Delta H = \Delta G = \Delta S = 0$

53. The effect of temperature on reaction rate is shown by:

1) Kirchoff's equation
2) Arrhenius equation
3) Gibb's- Helmholtz equation
4) Clausius- Clapeyron equation

54. The rate constant of a first order reaction at 27°C is 10^{-3} min^{-1}. The temperature coefficient of this reaction is 2. The rate constant (in min^{-1}) at 17°C for this reaction?

1) 10^{-3}
2) 5×10^{-4}
3) 2×10^{-3}
4) 10^{2}

55. $N_2(g) + 3H_2(g) \rightleftharpoons 2NH_3(g) + 22\,kcal$

the activation energy for forward reaction is 50kcal. What is the activation energy for the backward reaction?

1) –72 kcal
2) –28 kcal
3) + 28 kcal
4) +72 kcal

56. Chemical reaction with high E_a values are generally:

1) Very fast
2) Very slow
3) Moderately fast
4) Spontaneous

57. A reactant (A) forms two products:

$A \xrightarrow{k_1} B$, Activation energy, E_{α_1}

$A \xrightarrow{k_2} B$, Activation energy, E_{α_2}

If $E_{\alpha_2} = 2E_{\alpha_1}$, then k_1 and k_2 are related as

1) $k_2 = k_1 e^{E_{\alpha_1}/RT}$
2) $k_2 = k_1 e^{E_{\alpha_2}/RT}$
3) $k_1 = Ak_2 e^{E_{\alpha_1}/RT}$
4) $k_1 = 2k_2 e^{E_{\alpha_2}/RT}$

58. The activation energy of a reaction at a give temperature is found to be 2.303 RT $Jmol^{-1}$. The ratio of rate constant (k) to the Arrhenius factor (A) is:

1) 0.01
2) 0.1
3) 0.02
4) 0.001

59. The Arrhenius equation $k = Ae^{-\frac{E_a}{RT}}$ shows the dependence of rate constant on temperature. In which condition the rate constant k is smallest for a reaction?

1) Small E_a and low T
2) Small E_a and high T
3) Large E_a and low T
4) Large E_a and high T

60. Activation energy is ________ to rate of reaction

1) Directly proportional
2) Inversely proportional
3) Equal
4) Not related

61. The rate of a reaction can be increased in general by all the factors except

1) Using a catalyst
2) Increasing the temperature
3) Increasing the activation energy
4) Increasing the concentration of reactants

62. The energy of activation of positive catalyzed reaction as compared to that of an uncatalyzed reaction is

1) More
2) Less
3) Same
4) May be more or less

63. Chemical reactions which have high E_a values are generally:

1) Moderately fast
2) Very slow
3) Very fast
4) Explosive

64. For the decomposition of hydrogen iodide at 127°C, the fraction of molecules having energy equal to or greater than activation energy can be given as

1) $e^{\frac{-E_a}{2.303R}}$ 2) $e^{\frac{-E_a}{920R}}$

3) $10^{\frac{-E_a}{2.303R}}$ 4) $10^{\frac{-E_a}{921.2R}}$

65. The decomposition of a hydrocarbon follows the equation $k = (4.5 \times 10^{11}\ s^{-1})\ e^{-28000\ K/T}$. What will the value of activation energy ?

1) 669 $kJ\ mol^{-1}$ 2) 232.79 $kJ\ mol^{-1}$

3) 4.5×10^{11} $kJ\ mol^{-1}$ 4) 28000 $kJ\ mol^{-1}$

66. The rate constant, activation energy and Arrhenius parameter of a chemical reaction at 25 °C are $3.0 \times 10^{-4}\ s^{-1}$, 104.4 $kJ\ mol^{-1}$ and $6.0 \times 10^{14}\ s^{-1}$ respectively. The value of the rate constant as $T \to \infty$ is

1) $2.0 \times 10^{18}\ s^{-1}$ 2) $6.0 \times 10^{14}\ s^{-1}$

3) Infinity 4) $3.6 \times 10^{30}\ s^{-1}$

Collision Theory of Chemical Reactions

67. The threshold energy of a chemical reaction depends upon

1) Nature of reacting species
2) Temperature
3) Concentration of species
4) Number of collisions

68. The increase in concentration of the reactants lead to change in:

1) ΔH 2) Collision frequency

3) Activation energy 4) Equilibrium constant

69. The activation energy of a reaction is zero. The rate constant of the reaction.

1) Increases with increase of temperature.
2) Decreases with increase of temperature.
3) Decreases with decrease of temperature.
4) Is nearly independent of temperature.

70. Threshold energy is equal to:

1) Activation energy
2) Activation energy – energy of molecules
3) Activation energy + energy of molecules
4) None of these

71. The rate constant is given by the equation $K = PZe^{-Ea/RT}$ which factor should register a decrease for the reaction to proceed more rapidly?

1) T 2) Z

3) E_a 4) P

72. For a certain reaction a large fraction of molecules has energy more than the threshold energy, still the rate of reaction is very slow. The possible Statement-II for this could be that:

1) The colliding molecules could be large in size
2) The colliding molecules must not be properly oriented for effective collisions.
3) The rate of reaction could be independent of the energy
4) One of the reactant could be in excess

73. Rate of a general reaction A + B → Products can be expressed as follows on the basis of collision theory.

Rate = $Z_{AB}e^{-Ea/RT}$

Which of the following statement is not correct for the above expression?

1) Z is collision frequency and is equal to number of collisions per second per unit volume of the reaction mixture
2) $e^{-Ea/RT}$ is the fraction of molecules with kinetic energy equal to or greater than E_a.
3) E_a is activation energy of the reaction
4) All the molecules which colloid with one other are effective collisions

High Order Time Intensive MCQs

1. The reaction : $A_{(g)} + 2B_{(g)} \rightarrow C_{(g)} + D_{(g)}$ is an elementary process. In an experiment, the initial partial pressure of A and B are $P_A = 0.60$ atm and $P_B = 0.80$ atm. When $P_C = 0.20$ atm, the rate of reaction relative to the initial rate is:

1) $\frac{1}{16}$ 2) $\frac{1}{24}$

3) $\frac{1}{6}$ 4) $\frac{1}{48}$

2. For a reaction $A \xrightarrow{K_1} B \xrightarrow{K_2} C$

If the rate of formation of B is set to be zero then the concentration of B is given by:

1) $K_1K_2[A]$ 2) $(K_1 - K_2)[A]$

3) $(K_1 + K_2)[A]$ 4) $(K_1/K_2)[A]$

3. Following is the graph between $(a-x)^{-1}$ and time t for 2nd order reaction:

Given , $\theta = \tan^{-1}(0.5)$

OA = 2

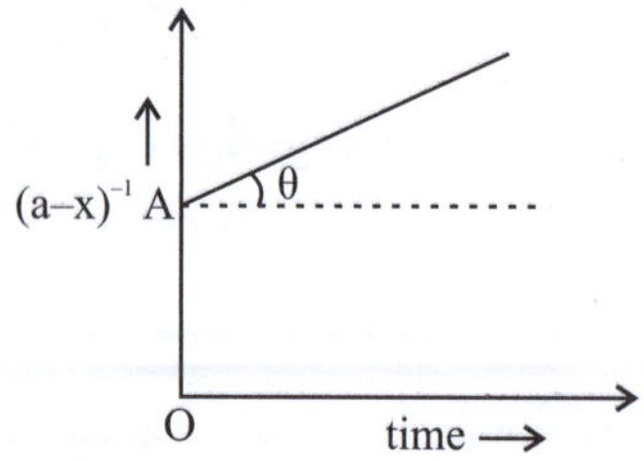

What is the initial rate of reaction?

1) 0.250 $mol\ ltr^{-1}\ min^{-1}$ 2) 0.125 $mol\ ltr^{-1}\ min^{-1}$

3) 0.375 $mol\ ltr^{-1}\ min^{-1}$ 4) None of these

4. For a first order reaction, the time required for completion of 90% reaction is ‘x’ times the half life of the reaction. The value of ‘x’ is

(Given : ln 10 = 2.303 and log 2 = 0.3010)

1) 1.12 2) 2.43
3) 3.32 4) 33.31

5. The rate of a I[st] order reaction is 0.04 mol ltr^{-1} sec^{-1} at 10 seconds and 0.03 mol ltr^{-1} sec^{-1} at 20 seconds after initiation of reaction. What is $t_{1/2}$ of the reaction?

1) 48.2 sec 2) 24.1 sec
3) 37 sec 4) 96 sec

6. Decomposition of H_2O_2 follows a first order reaction. In fifty minutes, the concentration of H_2O_2 decreases from 0.5 to 0.125 M in one such decomposition. When the concentration of H_2O_2 reaches 0.05 M, the rate of formation of O_2 will be:

1) 2.66 L min^{-1} at STP 2) 1.34×10^{-2} mol min^{-1}
3) 6.93×10^{-2} mol min^{-1} 4) 6.93×10^{-4} mol min^{-1}

7. Two reactions R_1 & R_2 have identical pre-exponential factors. Activation energy of R_1 exceeds that of R_2 by 10 kJ mol^{-1}. If k_1 & k_2 are the rate constants for reactions R_1 & R_2 respectively at 300 K, then ln (k_2/k_1) is equal to:

(R = 8.314 J mol^{-1} K^{-1})

1) 12 2) 6
3) 4 4) 8

8. For the equilibrium, $A_{(g)} \rightleftharpoons B_{(g)}$; $\Delta H = -40$kJ/mol.

It the ratio of activation energies of the forward (E_f) & the backward reaction (E_b) is 2 : 3, then:

1) E_f = 30kJ/mol, E_b = 70 kJ/mol
2) E_f = 80kJ/mol, E_b = 120 kJ/mol
3) E_f = 60kJ/mol, E_b = 100 kJ/mol
4) None of these

9. Which of the following statement are correct?

1) In an Arrehenius equation $K = Ae^{-Ea/RT}$, if $T \rightarrow \infty$ K = A
2) Alkaline hydrolysis of ester is irreversible reaction
3) Rate of reaction increase with increase in temperature
4) All of these

10. A reaction obeys zero order. The time required to decompose 50 g out of 100 g of the reactant A is 10 minute. Calculate the time required when half of the reactant A is decomposed if its initial mass is 200 g.

1) 5 min 2) 10 min
3) 15 min 4) 20 min

11. With respect to the figure given below, which of the following statement is correct?

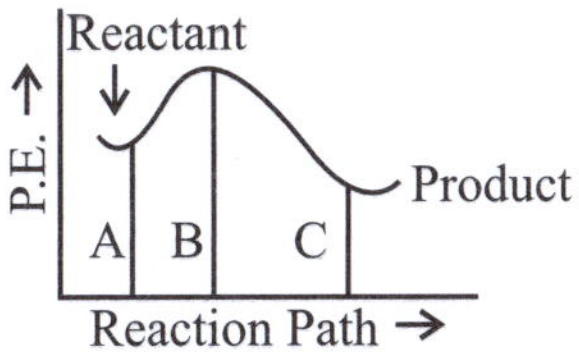

1) E_a for the forward reaction is C – B
2) E_a for the forward reaction is B – A
3) $E_{a(forward)} > E_{a(backward)}$
4) E_a(for reverse reaction) = C – A

12. Mark the correct statement about given graph :

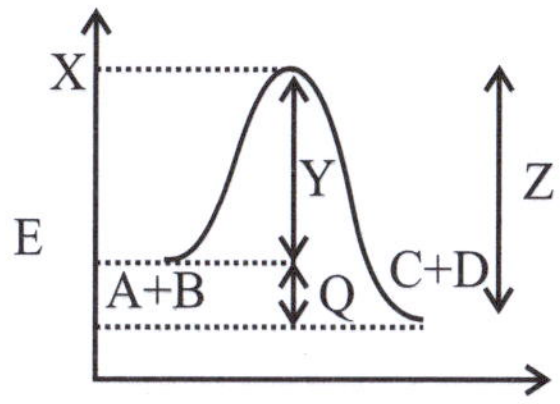

$A+B \rightarrow C+D+Q$ (reaction of the above graph)

1) X is threshold energy level
2) Y and Z are energy of activation for forward and backward reaction respectively
3) Q is heat of reaction and reaction is exothermic
4) All of these

13. The activation energy for a reaction is 9.0 kcal/mol. The increase in the rate constant when its temperature is increased from 298 K to 308 K is -

1) 10 % 2) 100 %
3) 50 % 4) 63 %

14. For a first order reaction, the correct option is

1) The degree of dissociation is equal to $(1-e^{-kt})$
2) The time taken for the completion of 75% reaction is thrice the $t_{1/2}$ of the reaction
3) The pre-exponential factor in the Arrhenius equation has the dimension of time, (T^{-1})
4) Both (1) & (3)

15. The rate constant is numerically the same for three reactions of first, second and third order respectively. Which of the following is correct ?

1) If [A] = 1 then $r_1 = r_2 = r_3$
2) If [A] < 1 then $r_1 > r_2 > r_3$
3) If [A] > 1 then $r_3 > r_2 > r_1$
4) All of these

16. The inversion of cane sugar proceeds with half-life of 500 minute at pH 5 for any concentration of sugar. However if pH = 6, the half-life changes to 50 minute. The rate law expression for the sugar inversion can be written as

1) $r = k[sugar]^2[H^+]^6$ 2) $r = k[sugar]^1[H^+]^0$
3) $r = k[sugar]^0[H^+]^6$ 4) $r = k[sugar]^0[H^+]^1$

17. For the reaction $2A + 3B + \frac{3}{2}C \rightarrow 3P$, which statement is correct?

1) $\frac{dn_A}{dt} = \frac{2}{3}\frac{dn_B}{dt} = \frac{3}{4}\frac{dn_C}{dt}$ 2) $\frac{dn_A}{dt} = \frac{dn_B}{dt} = \frac{dn_C}{dt}$

3) $\frac{dn_A}{dt} = \frac{3}{2}\frac{dn_B}{dt} = \frac{3}{4}\frac{dn_C}{dt}$ 4) $\frac{dn_A}{dt} = \frac{2}{3}\frac{dn_B}{dt} = \frac{4}{3}\frac{dn_C}{dt}$

18. In presence of a catalyst, the activation energy is lowered by 2 kcal at 127°C. The rate as compared to old rate is-

1) e^2 times 2) e^{-2} times
3) $e^{2.5}$ times 4) $e^{-4.5}$ times

19. Consider the reaction 2A + B → Product. When concentration of B alone was doubled, the half life did not change. When concentration of A alone was doubled, the rate increased by two times. The unit of rate constant for this reaction is-

1) s^{-1} 2) $L\ mol^{-1}s^{-1}$
3) $L\ mol^2s^{-1}$ 4) $mol\ L^{-1}\ s^{-1}$

20. The time required for 10% completion of a first order reaction at 298 K is equal to that required for its 25% completion at 308. If the value of A is $4 \times 10^{10}\ s^{-1}$. The value of E_a is (log 2.728 = 0.436)

1) 76.623 kJ/mol 2) 100 kJ/mol
3) 94 kJ/mol 4) 52.261 kJ/mol

21. The order of a reaction A → product in which half the reagent is reacted in half an hour, three quarters in one hour and seven - eighth in one and half hour is

1) 2 2) 1
3) Zero 4) –1

22. Assign T/F to the following statements.

A : The frequency factor has the same unit as the rate constant k

B : A plot ℓn C for the nth order reaction gives a straight line with slope -n and intercept k_n

C : The order of a reaction A → product in which half the reagent is reacted in half an hour, three quarters in one hour and seven - eighth in one and half hours must be 1 (unity).

D : The unit of rate constant for a second order reaction will be $M^{-1}s^{-1}$. (M is representing the molarity of solution)

1) T F T T 2) T T T T
3) F F T T 4) T F F T

23. If temperature of a reaction is increased from 30°C to 70°C and the value of rate increases by 16 times, then the value of temperature coefficient will be

1) 2.15 2) 2.25
3) 2 4) 2.5

24. The rate law for a reaction between the substances A and B is given by rate = $k[A]^n[B]^m$. On doubling the concentration of A and halving the concentration of B, the ratio of the new rate to the earlier rate of reaction will be

1) $\frac{1}{2^{(m+n)}}$ 2) m + n
3) n – m 4) $2^{(n-m)}$

25. Reactant (A) forms two products:

$A \xrightarrow{k_2} B$ Activation energy, E_{a_1}

$A \xrightarrow{k_1} B$ Activation energy, E_{a_2} If $E_{a_2} = 2E_{a_1}$, then k_1 and k_2 are related as

1) $k_2 = k_1e^{E_{a_1}/RT}$ 2) $k_2 = Ak_1e^{E_{a_2}/RT}$
3) $k_1 = k_2e^{E_{a_1}/RT}$ 4) $k_1 = 2k_2e^{E_{a_2}/RT}$

26. For two reactions, A → Products & B → Products have rate constants K_A & K_B at temperature T & activation energies E_A & E_B respectively. If $K_A > K_B$ & $E_A < E_B$ & assuming that A for both the reactions is same, then-

1) At higher temperatures, K_A will be greater than K_B.
2) At lower temperature, K_A & K_B will differ more & $K_A > K_B$.
3) As temperature rises, K_A & K_B will be close to each other in magnitude.
4) All of the above are correct

27. Half-life period of a first order reaction is 10min. What percentage of the reaction will be completed in 100 min?

1) 25% 2) 50%
3) 99.9% 4) 75%

28. $t_{1/2}$ for a first order reaction is 10 min. Starting with 10 M, the rate after 20 min is:

1) 0.0693 M min^{-1} 2) 0.0693 × 5 M min^{-1}
3) 0.0693 × 2.5 M min^{-1} 4) 0.0693 × 10 M min^{-1}

29. Consider the following statements.

The rate law for the acid catalysed hydrolysis of an ester being given is-

rate = $k[H^+]$[ester] = k' [ester], if the acid concentration is doubled at constant ester concentration.

I. The second order rate constant, k is doubled

II. The pseudo first order rate constant, k' is doubled

III. The rate of the reaction is doubled

Which of the above statement are correct?

1) I and II 2) II and III
3) I and III 4) I, II and III

30. In the presence of a catalyst, activation energy of a reaction is lowered by 2 kcal at 27° C. Hence, rate will be:

1) 20 times 2) 28 times
3) 14 times 4) Remain same

31. A reaction was found to be second order with respect to concentration of carbon monoxide. If the concentration of carbon monoxide is doubled, with everything else kept the same, the rate of reaction will

1) triple 2) increase by a factor of 4
3) double 4) remain unchanged

32. For a reaction : $\frac{1}{2}A \longrightarrow 2B$ rate of disappearance of 'A' is related to the rate of appearance of B by the expression:

1) $-\frac{d[A]}{dt} = \frac{d[B]}{dt}$ 2) $-\frac{d[A]}{dt} = \frac{4d[B]}{dt}$
3) $-\frac{d[A]}{dt} = \frac{1}{2}\frac{d[B]}{dt}$ 4) $-\frac{d[A]}{dt} = \frac{1}{4}\frac{d[B]}{dt}$

33. For a first order reaction, (A) → Products, the concentration of A changes from 0.1 M to 0.025 M in 40 minutes. The rate of reaction when the concentration of A is 0.01 M is

1) 3.47×10^{-4}M/min 2) 3.47×10^{-5} M/min
3) 1.73×10^{-4}M/min 4) 1.73×10^{-5}M/min

34. At 500 K, the half life period of a gaseous reaction at an initial pressure of 80k/Pa is 350 sec. When the pressure is 40k/Pa, the half life period is 175 sec. The order of the reaction is

1) zero 2) one
3) two 4) three

35. The activation energy for a reaction at the temperature T K was found to be 2.303 RT J mol^{-1}. The ratio of the rate constant to Arrhenius factor is

1) 10^{-1} 2) 10^{-2}
3) 2×10^{-3} 4) 2×10^{-2}

36. For the reaction 2A + B → C, the values of initial rate at different reactant concentrations are given in the table below: The rate law for the reaction is:

[A]($molL^{-1}$)	[B]($molL^{-1}$)	Initial Rate (mol $L^{-1}s^{-1}$)
0.05	0.05	0.045
0.10	0.05	0.090
0.20	0.10	0.72

1) Rate = $k[A]^2[B]^2$ 2) Rate = $k[A][B]^2$
3) Rate = k[A][B] 4) Rate = $k[A]^2[B]$

37. Plots showing the variation of the rate constant (k) with temperature (T) are given below. The plot that follow Arrhenius equation is

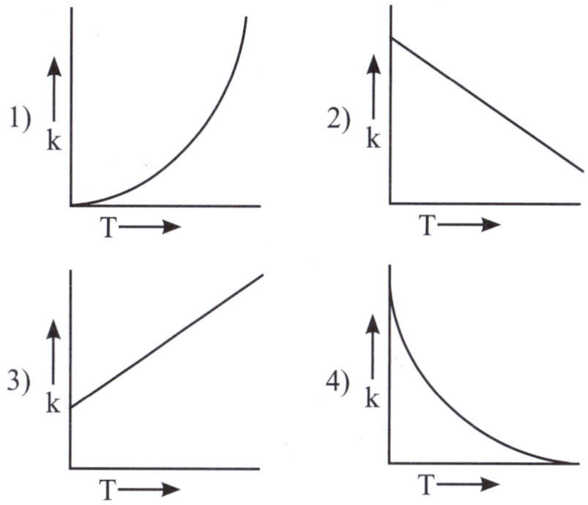

38. For a reaction of order n, the unit of the rate constant is:

1) $mol^{1-n} L^{n-1} s^{-1}$
2) $mol^{1-n} L^{1-n} s^{-1}$
3) $mol^{1-n} L^{2n} s^{-1}$
4) $mol^{1-n} L^{1-n} s$

Match the Columns MCQs

1. Match the graph given in Column-I with the order of reaction given in Column-II. More than one item in Column-I may link to the same item of Column-II.

	Column-I		Column-II
(a)	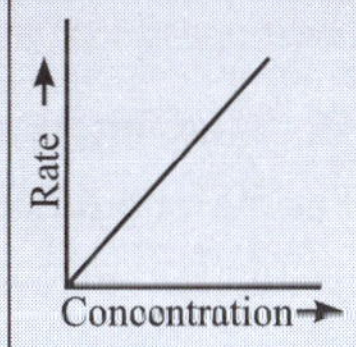	(i)	I st order
(b)	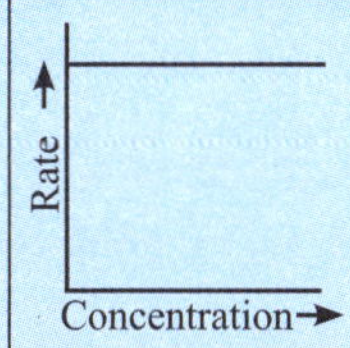	(ii)	Zero order
(c)	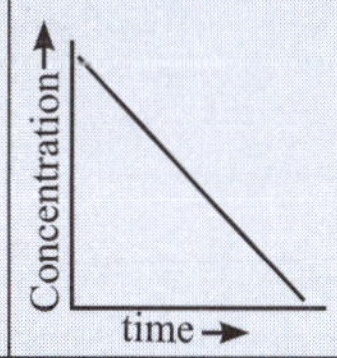		
(d)	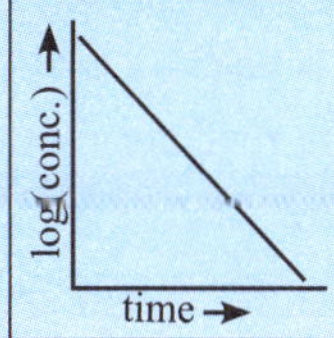		

1) (a)-(i) (b)-(ii) (c)-(ii) (d)-(i)
2) (a)-(i) (b)-(i) (c)-(ii) (d)-(ii)
3) (a)-(ii) (b)-(i) (c)-(i) (d)-(ii)
4) (a)-(ii) (b)-(ii) (c)-(i) (d)-(i)

2. Match the statements given in Column-I and Column-II

	Column-I		Column-II
(a)	Catalyst alters the rate of reaction	(i)	cannot be fraction or zero
(b)	Molecularity	(ii)	proper orientation is not there always
(c)	Second half life of first order reaction	(iii)	by lowering the activation energy
(d)	$e^{-E_a/RT}$	(iv)	is same as the first
(e)	Energetically favourable reactions are sometimes slow	(v)	total probability is one
(f)	Area under the Maxwell Boltzman curve is constant	(vi)	refers to the fraction of molecules with energy equal to or greater than activation energy

1) (a)-(ii) (b)-(i) (c)-(iii) (d)-(v) (e)-(vi) (f)-(iv)
2) (a)-(iv) (b)-(v) (c)-(i) (d)-(iii) (e)-(ii) (f)-(vi)
3) (a)-(iii) (b)-(i) (c)-(iv) (d)-(vi) (e)-(ii) (f)-(v)
4) (a)-(ii) (b)-(iii) (c)-(i) (d)-(iv) (e)-(vi) (f)-(v)

3. Match the items of Column-I and Column-II.

	Column-I		Column-II
(a)	Diamond	(i)	short interval of time
(b)	Instantaneous rate	(ii)	ordinarily rate of conversion is imperceptible
(c)	Average rate	(iii)	long duration of time

1) (a)-(ii) (b)-(i) (c)-(iii)
2) (a)-(iii) (b)-(i) (c)-(ii)
3) (a)-(ii) (b)-(iii) (c)-(i)
4) (a)-(i) (b)-(ii) (c)-(iii)

4. Match the items of Column-I and Column-II.

	Column-I		Column-II
(a)	Mathematical expression for rate of reaction	(i)	rate constant
(b)	Rate of reaction for zero order reaction is equal to	(ii)	rate law
(c)	Units of rate constant for zero order reaction is same as that of	(iii)	order of slowest step
(d)	Order of a complex reaction is determined by	(iv)	rate of a reaction

1) (a)-(ii) (b)-(i) (c)-(iii) (d)-(iv)
2) (a)-(iii) (b)-(iv) (c)-(i) (d)-(ii)
3) (a)-(ii) (b)-(i) (c)-(iv) (d)-(iii)
4) (a)-(i) (b)-(ii) (c)-(iv) (d)-(iii)

5. Match the items of Column-I and Column-II.

	Column-I		Column-II
(a)	Half life of zero order reaction	(i)	$[A]_0/2k$
(b)	Total life of zero order reaction	(ii)	$1/k$
(c)	Half- life of 2nd order reaction	(iii)	$[A]_0/k$
(d)	Average life of a 1st order reaction	(iv)	$1/(k[A]_0)$

1) (a)-(i) (b)-(iii) (c)-(ii) (d)-(iv)
2) (a)-(iv) (b)-(iii) (c)-(ii) (d)-(i)
3) (a)-(i) (b)-(iii) (c)-(iv) (d)-(ii)
4) (a)-(ii) (b)-(iii) (c)-(iv) (d)-(i)

Correct & Incorrect MCQs

1. Which of the following statement (s) are correct?

A. Average rate cannot be used to predict the rate of a reaction at a particular instant.

B. Average rate depends upon the change in concentration of reactants or products and the time taken for that change to occur.

C. Unit of rate of reaction is concentration time^{-1}.

D. Rate of reaction does not depend upon the experimental conditions such as concentration of reactants, temperature and catalyst.

1) A, B and C
2) A and D
3) A, C and D
4) B, C and D

2. Which of the following statement is not correct about order of a reaction ?

1) It is an experimental quantity.
2) It is applicable to elementary as well as complex reactions
3) For complex reaction, order is given by the fastest step
4) It can be zero and even can have a fractional value.

3. Identify the incorrect statement from the following :

1) Molecularity of a reaction cannot be zero or a non integer.
2) The molecularity of the reaction $H_2 + Br_2 \rightarrow 2HBr$ is 2.
3) Reactions with the molecularity three are very rare and slow to proceed.
4) Molecularity is applicable to elementary as well as complex reactions.

4. Identify the incorrect statement from the following :

1) Zero order reaction means that the rate of the reaction is proportional to zero power of the concentration of reactants.
2) The decomposition of gaseous ammonia on a hot platinum surface is a zero order reaction at high pressure.
3) The thermal decomposition of HI on gold surface is an example of zero order reaction.
4) Rate constant of a zero order reaction has unit s^{-1}.

5. Correct statement(s) is/are :

A. Half life of a reaction is represented as $t_{1/2}$.

B. If concentration of the reactant is doubled, half life remain same in the zero order reaction.

C. For a zero order reaction, $t_{1/2}$ is directly proportional to the initial concentration of the reactants and inversely proportional to the rate constant.

D. For first order reaction, $t_{1/2}$ is independent of initial concentration of the reacting species.

1) Only A and B
2) A, C and D
3) Only B
4) B and C

6. Which of the following statements is correct ?
   1) Activation energy is directly proportional to the rate of the reaction
   2) Activation energy is inversely proportional to the rate of the reaction
   3) Activation energy has no relation with the rate of the reaction.
   4) None of these
7. Which of the following statements is incorrect regarding catalysts ?
   1) It catalyses the spontaneous reactions.
   2) It does not change the equilibrium constant of a reaction.
   3) It alters Gibbs energy of a reaction.
   4) It catalyses the forward as well as the backward reactions to the same extent
8. Which of the following statement is correct regarding threshold energy ?
   1) It is the minimum amount of energy required by the colliding molecules to yield the products
   2) It is the maximum quantity of external energy required for the conversion of reactant into product
   3) It is the number of collisions per second per unit volume of the molecules in a chemical reaction
   4) All are incorrect
9. Select the incorrect statement :
   1) For chemical reaction with rise in temperature by 10°C, the rate constant remains unaffected.
   2) According to Arrhenius concept, when temperature is increased, rate of reaction is also increased.
   3) A in Arrhenius equation is the Arrhenius factor or the frequency factor.
   4) E_a in Arrhenius equation is activation energy which is measured in joules/mole.
10. Select the incorrect statement :
   1) In collision theory, activation energy and proper orientation of the molecules together determine the criteria for an effective collision.
   2) $k = Ae^{-\frac{E_a}{RT}}$ represents Arrhenius equation.
   3) The plot of ln k vs 1/T gives a parabola.
   4) Increasing the temperature or decreasing the activation energy will result in an increase in the rate of the reaction and an exponential increase in the rate constant.

Statement Based MCQs

Directions: These questions consist of two statements each, printed as Statement-I and Statement-II. While answering these questions, you are required to choose any one of the following four responses.

1) **Both Statement-I and Statement-II are correct.**
2) **Both Statement-I and Statement-II are incorrect.**
3) **Statement-I is correct & Statement-II is incorrect.**
4) **Statement-I is incorrect & Statement-II is correct.**

1. **Statement-I:** For zero order reaction, the concentration vs time graph is a straight line.

   Statement-II: The rate of change of concentration per unit time in zero order reaction remains constant.

2. **Statement-I:** For the first order reaction half-life period is expressed as

   $$t_{1/2} = \frac{2.303}{k}\log_{10} 2$$

   Statement-II: The half-life time of a first order reaction is not always constant and it depends upon the initial concentration of reactants.

3. **Statement-I:** In effective collision, the energy of molecules can be greater than threshold energy.

   Statement-II: In effective collisions, molecules do not have proper orientation.

4. **Statement-I:** Spontaneous reaction may be slow or fast.

   Statement-II: Spontaneous nature deals with the feasibility of the reaction but not the rate.

5. **Statement-I:** In a multi step reaction, the molecularity of overall reaction has no significance.

   Statement-II: Molecularity refers to the order of rate determining step.

6. **Statement-I:** A plot of $t_{1/2}$ versus $[A]°$ for a zero order reaction will be a straight line passing through the origin and slope $= \frac{1}{2k}$.

   Statement-II: Half – life period of a zero order reaction is inversely proportional to initial concentration.

7. **Statement-I:** The order of a reaction cannot have a fractional value.

   Statement-II: Order of reaction depends upon the power to which the concentration is raised to express the rate of reaction as found experimentally.

8. **Statement-I:**

   $$2CO_{(g)} + \frac{1}{2}O_{2(g)} \rightarrow 2CO_{2(g)} \ ...(i)$$

   $$2NO_{(g)} + \frac{1}{2}O_{2(g)} \rightarrow 2NO_{2(g)} \ ...(i)$$

   Reaction (i) is slower than (ii)

   Statement-II: Activation energy of reaction (i) is greater than that of (ii).

9. **Statement-I:** Reaction can have zero activation energy.

   Statement-II: Catalyst changes the equilibrium constant of a reaction.

10. **Statement-I:** Rate of a reaction depends upon the concentration of reactant.

    Statement-II: Rate generally increase when the reactant concentration increase

Assertion & Reason MCQs

Directions: These questions consist of two statements each, printed as Assertion and Reason. While answering these questions, you are required to choose any one of the following four responses.

1) **Both Assertion and Reason are True and the Reason is a correct explanation of the Assertion.**
2) **Both Assertion and Reason are True but Reason is not a correct explanation of the Assertion.**
3) **Assertion is True but the Reason is False.**
4) **Assertion is False but Reason is True.**

1. **Assertion:** Order of the reaction can be zero or fractional.
   Reason: We cannot determine order from balanced chemical equation.
2. **Assertion:** Order and molecularity are same.
   Reason: Order is determined experimentally and molecularity is the sum of the stoichiometric coefficient of rate determining elementary step.
3. **Assertion:** The enthalpy of reaction remains constant in the presence of a catalyst.
   Reason: A catalyst participating in the reaction, forms different activated complex and lowers down the activation energy but the difference in energy of reactant and product remains the same.
4. **Assertion:** All collision of reactant molecules lead to product formation.
   Reason: Only those collisions in which molecules have correct orientation and sufficient kinetic energy lead to compound formation.
5. **Assertion:** Rate constants determined from Arrhenius equation are fairly accurate for simple as well as complex molecules.
   Reason: Reactant molecules undergo chemical change irrespective of their orientation during collision.
6. **Assertion:** If in a zero order reaction, the concentration of the reactant is doubled, the half-life period is also doubled.
   Reason: For a zero order reaction, the rate of reaction is independent of initial concentration.
7. **Assertion:** Lower the activation energy, faster is the reaction.
   Reason: Catalyst does not affect activation energy of the reaction.
8. **Assertion:** When of ethyl acetate is hydrolysed in acidic medium a pseudo first order reaction occurs.
   Reason: $CH_3COOC_2H_5 + H_2O \xrightarrow{H^{\oplus}} CH_3COOH + C_2H_5OH$, here water does not take part in this reaction
9. **Assertion:** Fraction of total molecules having energy equal to or greater than activation energy $\frac{K}{A} = e^{-E_a/RT}$
   Reason: Activation energy of a chemical reaction is always constant at a given temperature.
10. **Assertion:** Temperature coefficient of different reactions lie between 2 to 3.
    Reason: The rate of a reaction is increases between 2 to 3 times by 10°C rise in temperature.

Multi Correct MCQs

1. Mark the correct statements:
   A. Larger the activation energy, smaller is the value of the rate constant.
   B. The higher is the activation energy, higher is the value of the temperature coefficient.
   C. At lower temperatures, increase in temperature causes more change in the value of k than at higher temperature.
   D. A plot of ln k vs $\frac{1}{T}$ is a straight line with slope equal to $-\frac{E_a}{R}$

   Options:
   1) A, B and C
   2) B, C, and D
   3) A, B and D
   4) All of these

2. For the following graphs,

A. 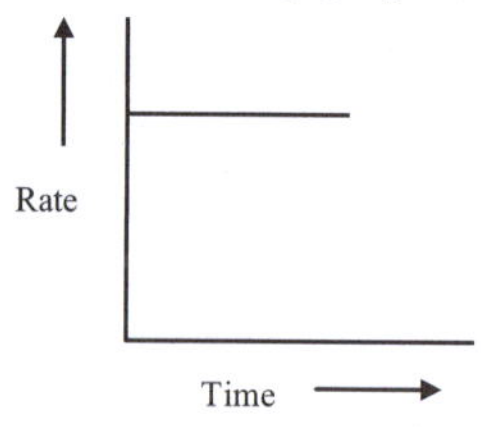

B. 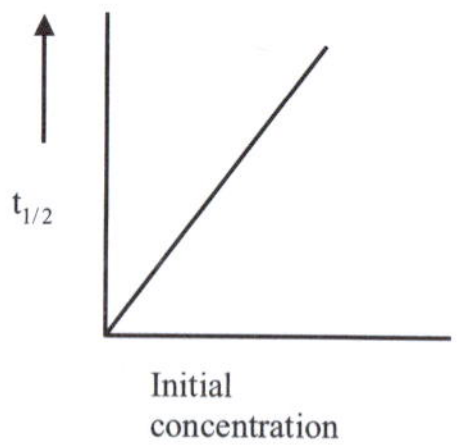

C. 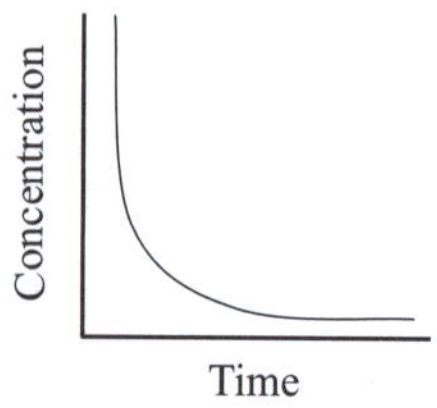

D. 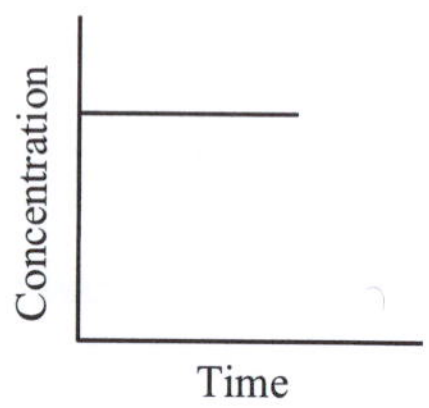

E. 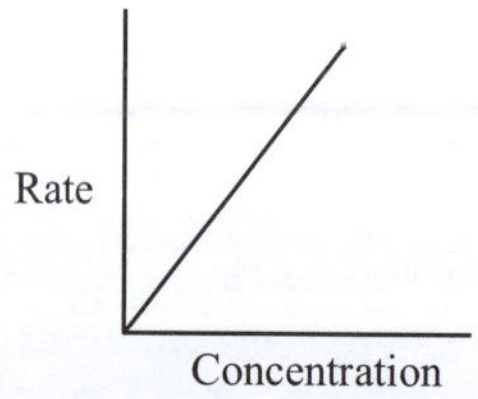

Choose from the options given below, the correct one regarding order of reaction is:

1) (A) and (B) zero order (E) First order
2) (B) and (D) zero order (E) First order
3) (A) and (B) zero order (C) and (E) First order
4) (B) zero order (C) and (E) First order

3. Mark the incorrect statement

A. Catalyst provides an alternative pathway to reaction mechanism.
B. Catalyst raises the activation energy.
C. Catalyst lowers the activation energy.
D. Catalyst alters enthalpy change of the reaction.

Options:

1) Both A and B
2) Both B and D
3) A, C and D
4) A, B and D

4. Identify which of the following is not a unit of the rate constant in chemical kinetics:

A. Mol/L/sec
B. Mol/L-sec
C. Sec^{-1}
D. $Mol^{-1}L\ sec^{-1}$
E. Min^{-1}
F. $Mol^{-1}L^{-1}min$

Options:

1) A and F
2) B and C
3) B, C and D
4) A, D and F

NCERT Exemplar MCQs

1. The role of a catalyst is to change:
   1) Gibbs energy of reaction
   2) Enthalpy of reaction
   3) Activation energy of reaction
   4) Equilibrium constant

2. In the presence of a catalyst, the heat evolved or absorbed during the reaction:
   1) Increases
   2) Decreases
   3) Remains unchanged
   4) May increase or decrease

3. Activation energy of a chemical reaction can be determined by:
   1) Determining the rate constant at standard temperature
   2) Determining the rate constant at two temperatures
   3) Determining probability of collision
   4) Using catalyst

4. Consider figure and mark the correct option:

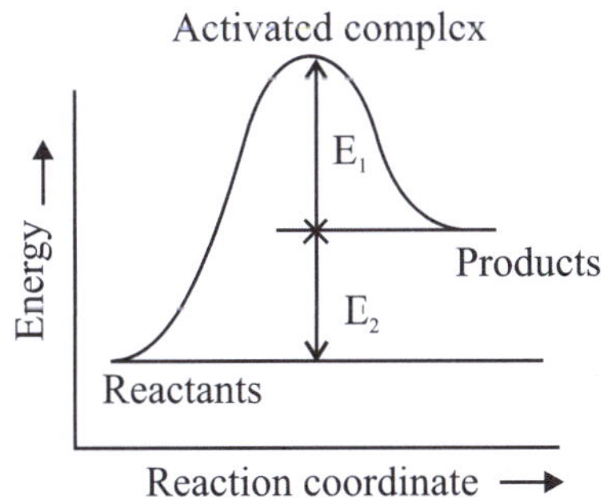

   1) Activation energy of forward reaction is $E_1 + E_2$ and product is less stable than reactant.
   2) Activation energy of forward reaction is $E_1 + E_2$ and product is more stable than reactant.
   3) Activation energy of both forward and backward reaction is $E_1 + E_2$ and reactant is more stable than product.
   4) Activation energy of backward reaction is E_1 and product is more stable than reactant.

5. Consider a first order gas phase decomposition reaction given below:

$$A(g) \rightarrow B(g) + C(g)$$

The initial pressure of the system before decomposition of A was p_i. After lapse of time 't' total pressure of the system increased by x units and became 'p_t'. The rate constant k for the reaction is given as:

1) $k = \frac{2.303}{t}\log\frac{p_i}{p_i - x}$
2) $k = \frac{2.303}{t}\log\frac{p_i}{2p_i - p_t}$
3) $k = \frac{2.303}{t}\log\frac{p_i}{2p_i + p_t}$
4) $k = \frac{2.303}{t}\log\frac{p_i}{p_i + x}$

6. According to Arrhenius equation, rate constant k is equal to $Ae^{-Ea/RT}$.

Which of the following options represents the graph of $\ln k$ vs $\frac{1}{T}$?

1) 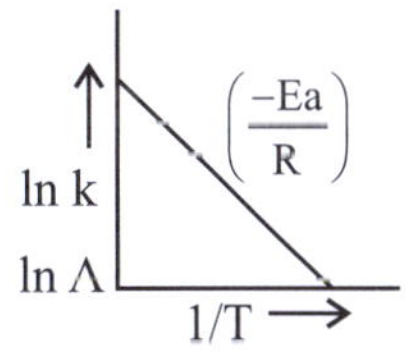

2) 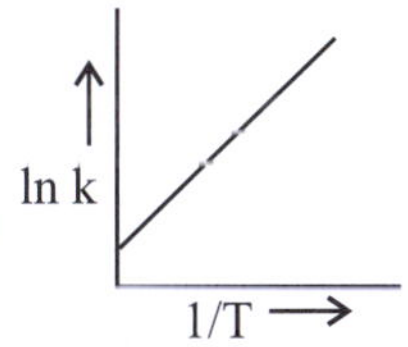

3) 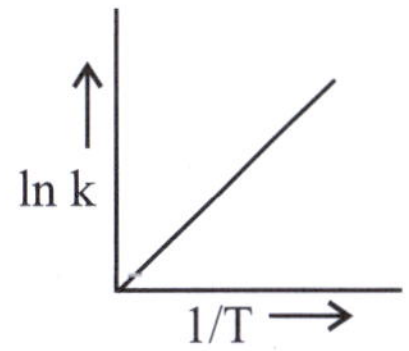

4) 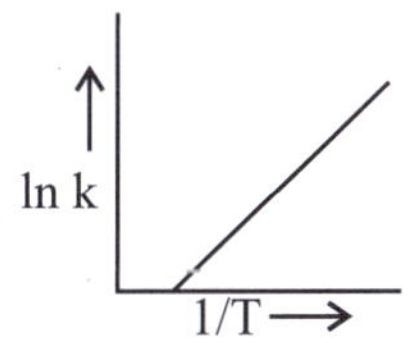

7. Consider the Arrhenius equation given below and mark the correct option:

$$k = Ae^{\frac{-Ea}{RT}}$$

1) Rate constant increases exponentially with increasing activation energy and decreasing temperature
2) Rate constant decreases exponentially with increasing activation energy and increasing temperature
3) Rate constant increases exponentially with decreasing activation energy and decreasing temperature
4) Rate constant increases exponentially with decreasing activation energy and increasing temperature

8. A graph of volume of hydrogen released vs time for the reaction between zinc and dil. HCl is given in figure. On the basis of this mark the correct option:

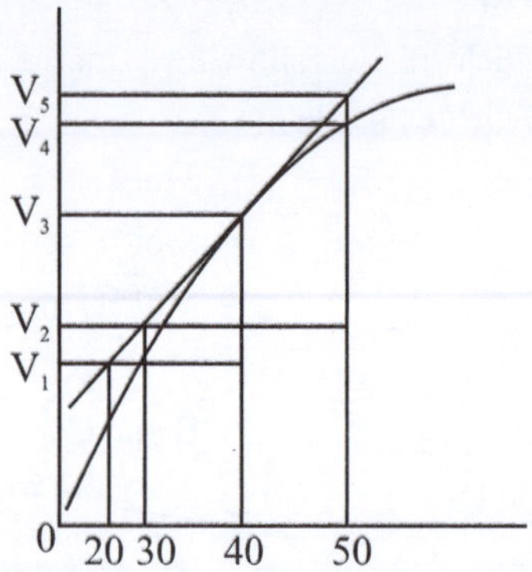

1) Average rate upto 40 s is $\frac{V_3 - V_2}{40}$
2) Average rate upto 40 s is $\frac{V_3 - V_2}{40 - 30}$
3) Average rate upto 40 s is $\frac{V_3}{40}$
4) Average rate upto 40 s is $\frac{V_3 - V_1}{40 - 20}$

9. Which of the following statements is not correct about order of a reaction?
   1) The order of a reaction can be a fractional number
   2) Order of a reaction is experimentally determined quantity
   3) The order of a reaction is always equal to the sum of the stoichiometric coefficients of reactants in the balanced chemical equation for a reaction
   4) The order of a reaction is the sum of the powers of molar concentration of the reactants in rate law expression

10. Which of the following statements is correct?
   1) The rate of a reaction decreases with passage of time as the concentration of reactants decreases
   2) The rate of a reaction is same at any time during the reaction
   3) The rate of a reaction is independent of temperature change
   4) The rate of a reaction decreases with increase in concentration of reactant (s)

11. Which of the following graphs represents exothermic reaction?

(I) 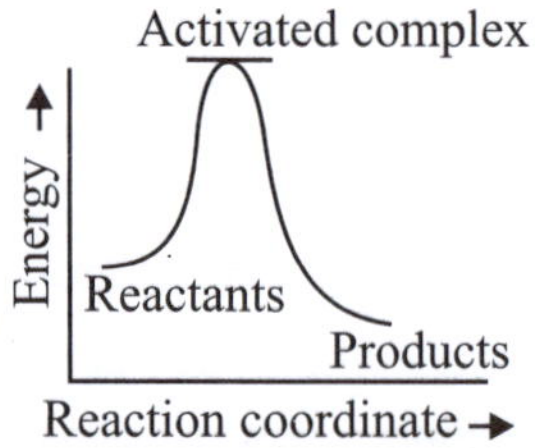

(II) 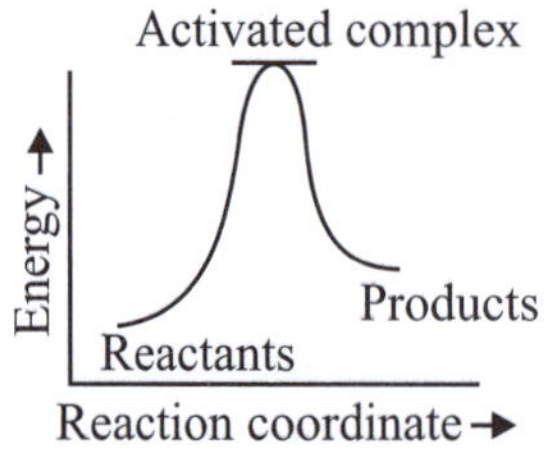

(III) 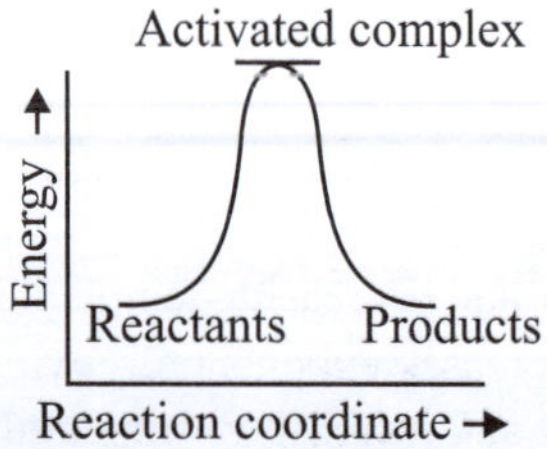

1) Only (I) 2) Only (II)
3) Only (III) 4) Both (I) and (II)

12. Rate law for the reaction $A + 2B \rightarrow C$ is found to be
Rate = k[A][B]
Concentration of reactant 'B' is doubled, keeping the concentration of 'A' constant, the value of rate constant will be:
   1) The same 2) Doubled
   3) Quadrupled 4) Halved

13. Which of the following statements is incorrect about the collision theory of chemical reaction?
   1) It considers reacting molecules or atoms to be hard spheres and ignores their structural features.
   2) Number of effective collisions determines the rate of reaction.
   3) Collision of atoms or molecules possessing sufficient threshold energy results into the product formation.
   4) Molecules should collide with sufficient threshold energy and proper orientation for the collision to be effective.

14. A first order reaction is 50% completed in 1.26×10^{14} s. How much time would it take for 100% completion?
   1) 1.26×10^{15} s 2) 2.52×10^{14} s
   3) 2.52×10^{28} s 4) Infinite

15. Which of the following statement is not correct for the catalyst?
   1) It catalyses the forward and backward reactions to the same extent
   2) It alters ΔG of the reaction
   3) It is a substance that does not change the equilibrium constant of a reaction
   4) It provides an alternate mechanism by reducing activation energy between reactants and products

16. The value of rate constant of a pseudo first order reaction:
   1) Depends on the concentration of reactants presents in small amount
   2) Depends on the concentration of reactants present in excess
   3) Is independent of the concentration of reactants
   4) Depends only on temperature

17. Consider the reaction $A \rightarrow B$. The concentration of both the reactants and the products varies exponentially with time. Which of the following figures correctly describes the change in concentration of reactants and products with time:

1) 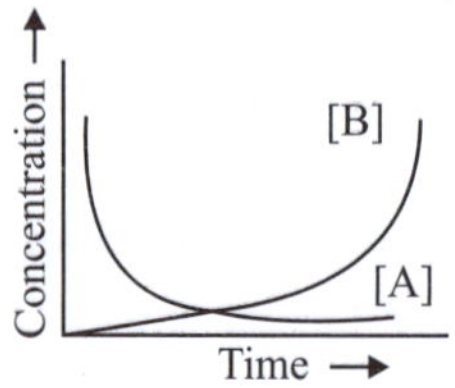

2) 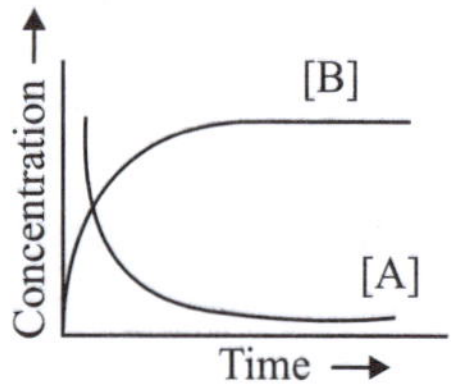

3) 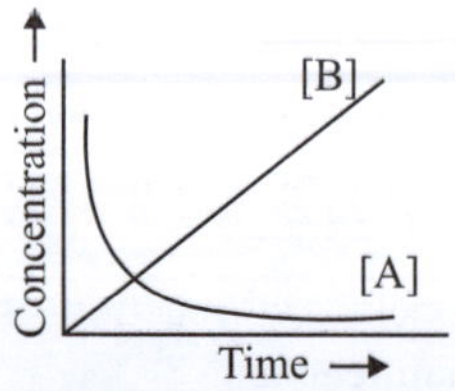

4) 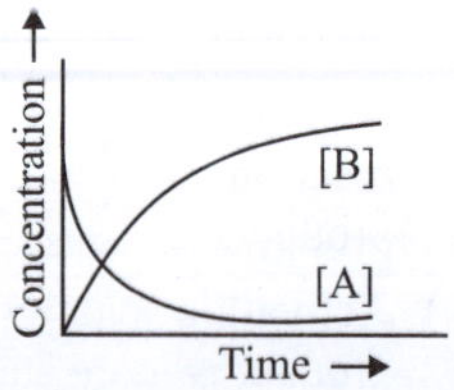

18. Rate law cannot be determined from balanced chemical equation if:
    1) Reverse reaction is involved.
    2) It is an elementary reaction.
    3) It is a sequence of elementary reactions.
    4) Any of the reactants is in excess.

19. Which of the following statements are applicable to a balanced chemical equation of an elementary reaction?
    1) Order is same as molecularity.
    2) Order is less than the molecularity.
    3) Order is greater than the molecularity.
    4) Molecularity can never be zero.

20. In any unimolecular reaction:
    1) Only one reacting species is involved in the rate determining step.
    2) The order and the molecularity of slowest step are equal to one.
    3) The molecularity of the reaction is one and order is zero.
    4) Both molecularity and order of the reaction are one.

21. For a complex reaction:
    1) Order of overall complex reaction is same as molecularity of the slowest step.
    2) Order of overall reaction is less than the molecularity of the slowest step.
    3) Order of overall reaction is greater than molecularity of the slowest step.
    4) Molecularity of the slowest step is never zero or non-integer.

22. At high pressure the following reaction is zero order.

$$2NH_3(g) \xrightarrow[\text{Platinum catalyst}]{1130\text{ K}} N_2(g) + 3H_2(g)$$

Which of the following options are correct for this reaction?
    1) Rate of reaction = Rate constant.
    2) Rate of the reaction depends on concentration of ammonia.
    3) Rate of decomposition of ammonia will remain constant until ammonia disappears completely.
    4) Further increase in pressure will change the rate of reaction.

23. During decomposition of an activated complex:
    1) Energy is always released.
    2) Energy is always absorbed.
    3) Energy does not change.
    4) Reactants may be formed.

24. According to Maxwell Boltzmann distribution of energy:
    1) The fraction of molecules with most probable kinetic energy decreases at higher temperatures.
    2) The fraction of molecules with most probable kinetic energy increase at higher temperatures.
    3) Most probable kinetic energy increases at higher temperatures.
    4) Most probable kinetic energy decreases at higher temperatures.

25. In the graph showing Maxwell Boltzmann distribution of energy:
    1) Area under the curve must not change with increase in temperature.
    2) Area under the curve increases with increase in temperature.
    3) Area under the curve decreases with increase in temperature.
    4) With increase in temperature curve broadens and shifts to the right hand side.

26. Which of the following statements are in accordance with the Arrhenius equation?
    1) Rate of a reaction increases with increase in temperature.
    2) Rate of a reaction increases with decrease in activation energy.
    3) Rate constant decreases exponentially with increase in temperature.
    4) Rate of reaction decreases with decrease in activation energy.

27. Mark the incorrect statements:
    1) Catalyst provides an alternative pathway to reaction mechanism.
    2) Catalyst raises the activation energy.
    3) Catalyst lowers the activation energy.
    4) Catalyst alters enthalpy change of the reaction.

28. Which of the following graphs is correct for a zero order reaction?

1) 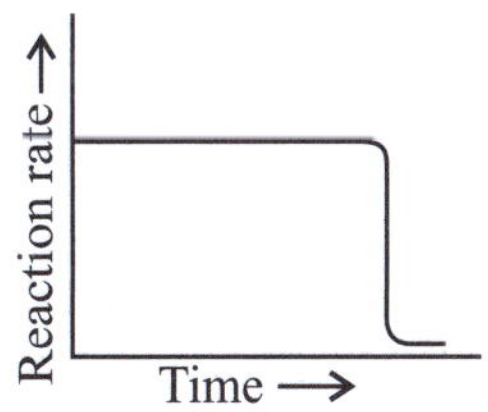

2) 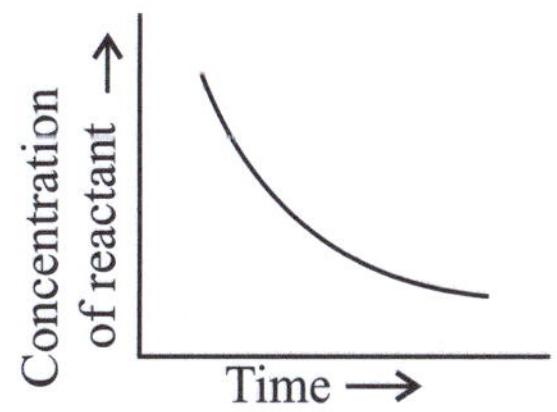

3) 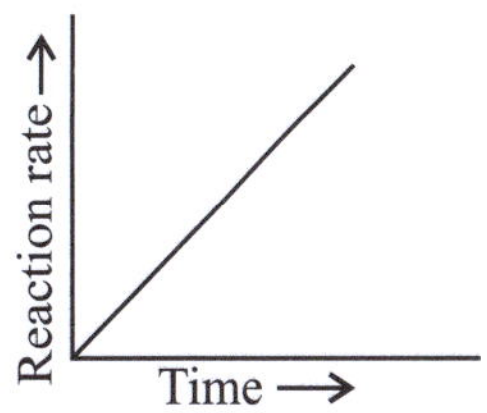

4) 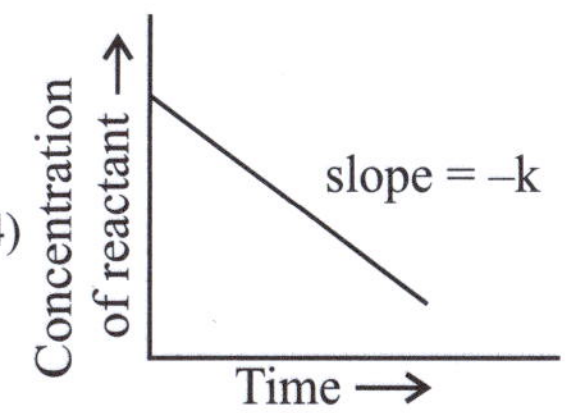

Kattar NEET

Scan Me for Detailed Explanations

1. For a first order reaction, nA→ B whose concentration vs time curve is as shown in the figure. If half-life for this reaction is 24 minutes. Find out the value of n.

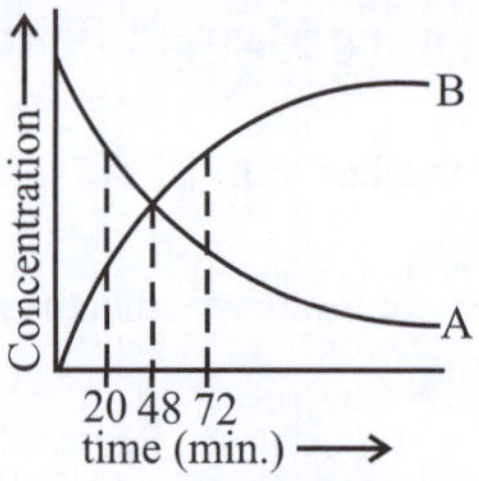

1) 1 2) 2
3) 3 4) 4

2. A graph between log $t_{1/2}$ and log a (abscissa), a being the initial concentration of A in the reaction is given For reaction A → Product, the rate law is:

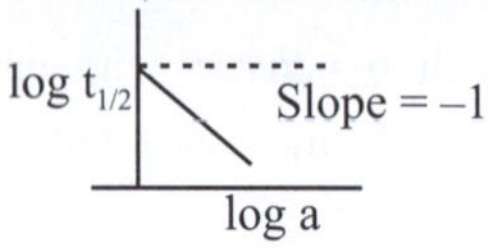

1) $\frac{-d[A]}{dt} = K$ 2) $\frac{-d[A]}{dt} = K[A]$
3) $\frac{-d[A]}{dt} = K[A]^2$ 4) $\frac{-d[A]}{dt} = K[A]^3$

3. In a gaseous phase reaction, $A_2(g) \longrightarrow B(g) + \frac{1}{2}C(g)$. The increase in pressure from 100 mm to 110 mm is observed in 5 minute. The ratio of disappearance of A_2 in mm min^{-1} is:

1) 4 2) 8
3) 16 4) 2

4. For a 1st order reaction (gaseous) (constant V, T):
$aA \longrightarrow (b-1)B + C$ (with b > a), the pressure of the system rose by $50\left(\frac{b}{a}-1\right)\%$ in a time of 10 min. The half life of the reaction is therefore.

1) 10 min 2) 20 min
3) 30 min 4) 40 min

5. For a certain reaction, the variation of the rate constant with temperature is given by the equation

$$\ln k_t = \ln k_0 + \frac{(\ln 3)t}{10} \quad (t \geq 0°C)$$

The value of the temperature coefficient of the reaction is:

1) 0.1 2) 1.0
3) 10 4) 3

6. A reaction takes place in three steps; the respective rate constants are k_1, k_2 and k_3. The overall rate constant $k = \frac{k_1 k_3}{k_2}$. If energies of activation are 40, 30 and 20 kJ, the overall energy of activation is (in kJ assuming 'A' to be constant for all):

1) 10 2) 15
3) 30 4) 60

7. In a certain reaction, 10% of the reactant decomposes in one hour, 20% in two hours, 30% in three hours and so on. The dimensions of the rate constant is:

1) $hour^{-1}$ 2) mole $litre^{-1}$ sec^{-1}
3) litre $mole^{-1}$ sec^{-1} 4) mole sec^{-1}

8. Compounds A and B react with a common reagent with first order kinetics in both cases. If 99% of A must react before 1% of B has reacted, what is the minimum ratio of their respective rate constants?

1) 916 2) 229
3) 500 4) 458

9. A reaction with respect to X is zero order till the concentration is reduced to half of initial concentration. Then the reaction become first order with respect to it. If the value of rate constant for the zero order and first order are equal to k (in magnitude), then find the time taken to reduce the concentration of X to $\frac{1}{16}$th of its original concentration X_0.

1) $\frac{1}{2k} + \frac{3\ell n2}{k}$ 2) $\frac{x_0}{2k} + \frac{3\ell n2}{k}$
3) $\frac{x_0}{2k} + \frac{4\ell n2}{k}$ 4) $\frac{1}{2k} + \frac{4\ell n2}{k}$

10. Consider the following statements and rearrange in the order of true/false as given in the codes.

S_1 : The rate of the reaction A → B having the rate law $-\frac{d[A]}{dt} = k[A][B]$ when plotted against time will exhibit a maximum at some time.

S_2 : A catalyst in a chemical reaction increases the forward E_a and decreases the backward E_a

S_3 : A catalyst in a chemical reaction decreases both forward and backward E_a

S_4 : For a first-order reaction, the time required to reduce successively the concentration of reactant by a constant fraction is always same.

1) TTTT 2) FFFF
3) FTFT 4) TFTT

11. Which of the following is correct w.r.t. increasing temperature?

1) Rate constant increases for exothermic reaction and decreases for endothermic reaction
2) Rate constant increases for both exothermic and endothermic reaction
3) Rate constant decreases for both exothermic and endothermic reaction
4) Rate constant increases for endothermic and decreases for exothermic reaction

12. For the reaction, $2SO_2(g) + O_2(g) \rightarrow 2SO_3(g)$, the rate of consumption of O_2 is 6.4 kg/min., then rate of formation of SO_3 in kg/s is

1) 12.8 2) 0.53
3) 33.3 4) 32

13. For a gas phase first order reaction,

$A(g) \longrightarrow B(g) + 2C(g) + D(g)$

The initial pressure was P_o while pressure after 't' time becomes $P_t(P_t > P_o)$. The expression for the rate constant (K) would be

1) $\frac{2.303}{t}\log\frac{P_o}{P_t}$ 2) $\frac{2.303}{t}\log\frac{4P_o}{(3P_o - P_t)}$

3) $\frac{2.303}{t}\log\frac{3P_o}{(4P_o - P_t)}$ 4) $\frac{2.303}{t}\log\frac{5P_o}{(3P_o - P_t)}$

14. If $t_{1/2}$ of a first order reaction is 2 h, then time required to complete 93.75% of reaction is

1) 16 h 2) 8 h

3) 4 h 4) 6 h

15. For the reaction $2A + B \rightarrow P$, when concentration of 'B' alone was doubled, the half life did not change but when conc. of 'A' alone was doubled, the rate of reaction increases 4 times. The unit of rate constant is

1) s^{-1} 2) $L\ mol^{-1}\ s^{-1}$

3) $L^2\ mol^{-2}\ s^{-1}$ 4) No unit

16. The decomposition of azo methane, at certain temperature according to the equation

$(CH_3)_2N_2 \rightarrow C_2H_6 + N_2$ is a first order reaction.

After 40 minutes from the start, the total pressure developed is found to be 350 mm Hg in place of initial pressure 200 mm Hg of azo methane. The value of rate constant k is:

1) $2.88 \times 10^{-4}\ s^{-1}$ 2) $1.25 \times 10^{-4}\ s^{-1}$

3) $5.77 \times 10^{-4}\ s^{-1}$ 4) None of these

17. $A \rightarrow B$ first order reaction, A is optically active and B is optically inactive. A series of experiment were conducted on a solution of A

Time	0	60 min	∞
Optical rotation	82°	77°	2°

Assume some impurity is present, calculate the optical rotation (in degrees) after 5 hours.

(Given, ln 1.066 = 0.064 $e^{0.16} = 1.17$)

1) 60 2) 30

3) 20 4) 120

18. Rate law of nth order is given by,

$$\frac{dx}{dt} = k[A]^n$$

Match different properties in Column-I with their related values of half-life in Column-II.

	Column-I		Column-II
I	Graph between $\log\left(\frac{dx}{dt}\right)$ and log [A] is a straight line with a slope 45°	P	$t_{1/2} \propto [A]$
II	Graph between $\log\left(\frac{dx}{dt}\right)$ and log [A] is a straight line parallel to log[A] axis	Q	$t_{1/2}$ is independent of concentration
III	If $[A] = [H^+]$ rate becomes 100 times on decreasing pH by 1 unit	R	$t_{1/2}$ is 100 min at 1 M and 25 at 2 M
IV	On Doubling [A], rate becomes 8 times	S	Half-life $\propto \frac{1}{[A]}$

1) I-Q, II-P, III-S, IV-R 2) I-R, II-S, III-P, IV-R

3) I-S, II-P, III-R, IV-Q 4) I-R, II-P, III-R, IV-S

19. Match the following:

	Column-I		Column-II
I	$2H_2O_2 \rightarrow 2H_2O + O_2$ Rate = $K[H_2O_2]$	P	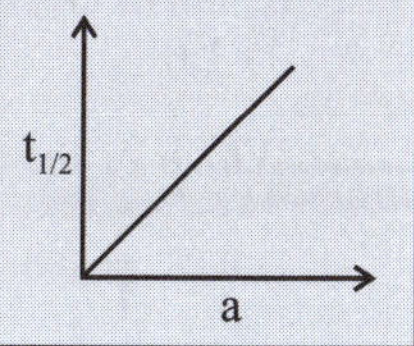
II	$2NH_3 \longrightarrow N_2 + 3H_2$ Rate = $\frac{K_1[NH_3]}{1+K_2[NH_3]}$ High concentration of NH_3	Q	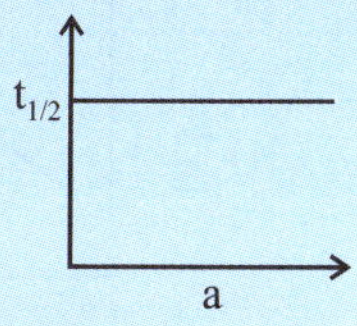
III	$2NH_3 \longrightarrow N_2 + 3H_2$ Rate = $\frac{K_1[NH_3]}{1+K_2[NH_3]}$ Very low concentration of NH_3	R	Second order reaction
IV	$2CH_3CHO \rightarrow 2CH_4 + CO$ Rate = $K[CH_3CHO]^2$	S	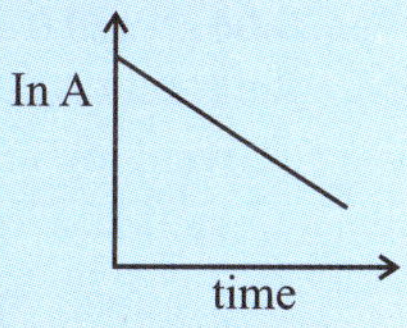
		T	First order reaction

1) I-Q, S, T; II-P; III-Q, S, T; IV-R

2) I- T; II-P; III-Q, T; IV-R

3) I-S, T; II-P; III- T; IV-Q

4) I-Q, T; II-R; III-Q, P, T; IV-S

20. Match the following:

	Column-I		Column-II
I	Hydrolysis of ester in acidic medium	P	Pseudo-unimolecular reaction
II	Inversion of cane sugar	Q	Second order with none of the reagents in excessive amount
III	Decomposition of H_2O_2	R	First order
IV	Hydrolysis of ester in alkaline medium	S	Zero order
		T	Rate = k[A][B]

1) I-P, R; II-R; III-R, S; IV- T

2) I-P, R; II-P, R; III-R; IV-Q, T

3) I-P, R; II-P, Q; III-S; IV-Q

4) I-S,R; II-Q; III-P; IV-Q, T

NEET & AIIMS Past Year MCQs

1. Following data is for a reaction between reactants A and B:

Rate mol L^{-1} s^{-1}	[A]	[B]
2×10^{-3}	0.1 M	0.1 M
4×10^{-3}	0.2 M	0.1 M
1.6×10^{-2}	0.2 M	0.2 M

The order of the reaction with respect to A and B, respectively, are **(2024 Re)**

1) 1,0 2) 0,1 3) 1,2 4) 2,1

2. Which of the following plot represents the variation of ln k versus $\frac{1}{T}$ in accordance with Arrhenius equation? **(2024 Re)**

1) 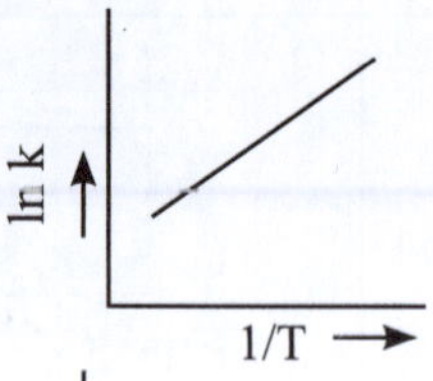

2) 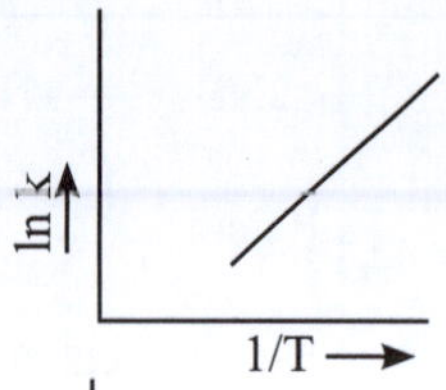

3) 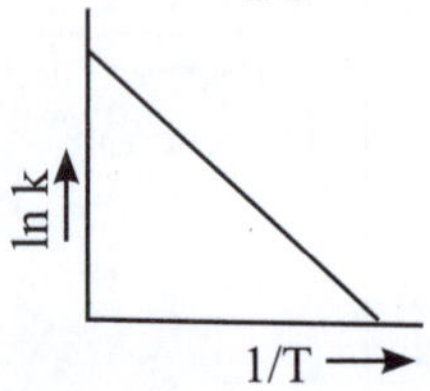

4) 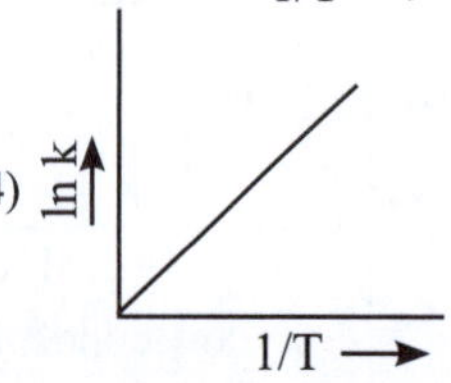

3. Rate constants of a reaction at 500 K and 700 K are 0.04 s^{-1} and 0.14 s^{-1}, respectively; then, activation energy of the reaction is: **(2024 Re)**

(Given: log3.5 = 0.5441, R = 8.31 JK^{-1} mol^{-1})

1) 182310 J 2) 18500 J 3) 18219 J 4) 18030 J

4. Activation energy of any chemical reaction can be calculated if one knows the value of **(2024)**
   1) orientation of reactant molecules during collision
   2) rate constant at two different temperatures
   3) rate constant at standard temperature
   4) probability of collision

5. Which plot of ln k vs $\frac{1}{T}$ is consistent with Arrhenius equation? **(2024)**

1) 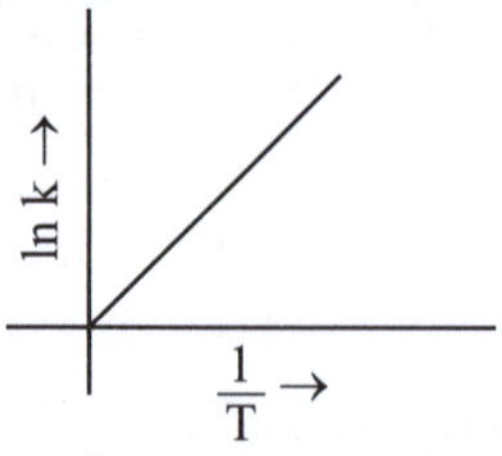

2) 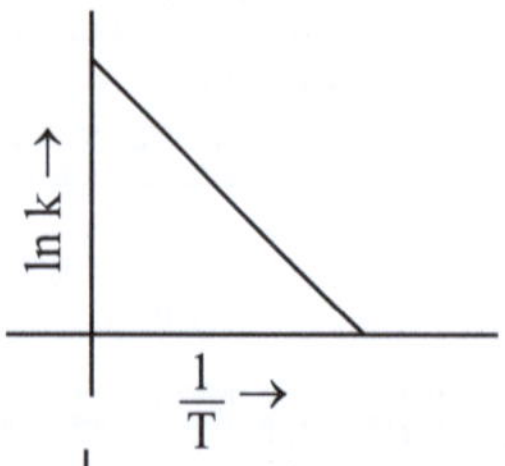

3) 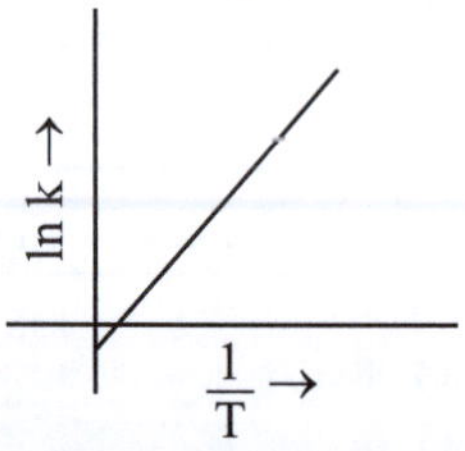

4) 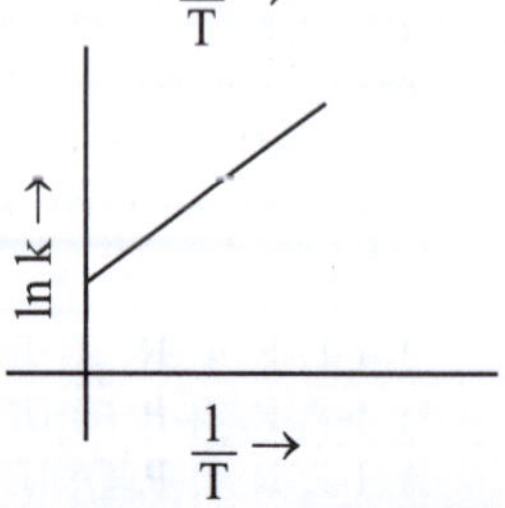

6. The rate of a reaction quadruples when temperature changes from 27°C to 57°C. Calculate the energy of activation. **(2024)**

Given R = 8.314J K^{-1} mol^{-1}, log 4 = 0.6021

1) 3.80 kJ/mol 2) 3804 kJ/mol
3) 38.04 kJ/mol 4) 380.4 kJ/mol

7. For a certain reaction, the rate = $k[A]^2[B]$. When the initial concentration of A is tripled keeping concentration of B constant, the initial rate would **(2023)**
   1) increase by a factor of three.
   2) decrease by a factor of nine.
   3) increase by a factor of six.
   4) increase by a factor of nine

8. Given below are two statements: one is labelled as Assertion A and the other is labelled as Reason R:

Assertion A: A reaction can have zero activation energy.

Reason R: The minimum extra amount of energy absorbed by reactant molecules so that their energy becomes equal to threshold value, is called activation energy.

In the light of the above statements, choose the correct answer from the options given below: **(2023)**
   1) A is false but R is true.
   2) Both A and R are true and R is the correct explanation of A.
   3) Both A and R are true and R is NOT the correct explanation of A.
   4) A is true but R is false.

9. The given graph is a representation of kinetics of a reaction. **(2022)**

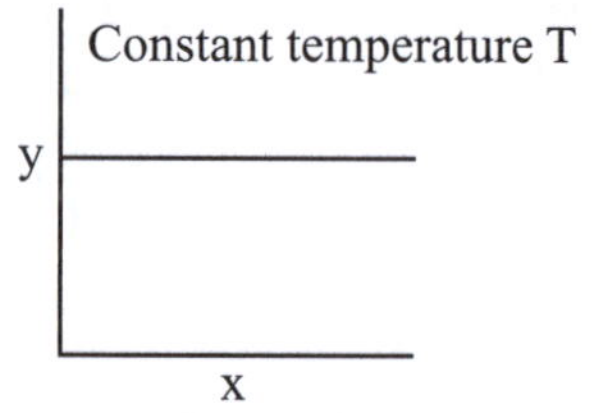

The y and x axes for zero and first order reactions, respectively are
   1) zero order (y = rate and x = concentration), first order (y = rate and x = $t_{1/2}$)
   2) zero order (y = concentration and x = time), first order (y = $t_{1/2}$ and x = c concentration)
   3) zero order (y = concentration and x = time), first order (y = rate constant and x = concentration)
   4) zero order (y = rate and x = concentration), first order (y = $t_{1/2}$ and x = concentration)

10. For a first order reaction A→Products, initial concentration of A is 0.1 M, which becomes 0.001 M after 5 minutes. Rate constant for the reaction in min^{-1} is **(2022)**

1) 0.2303
2) 1.3818
3) 0.9212
4) 0.4606

11. For a reaction A → B, enthalpy of reaction is –4.2 kJ mol^{-1} and enthalpy of activation is 9.6 kJ mol^{-1}. The correct potential energy profile for the reaction is shown in option. **(2021)**

1) 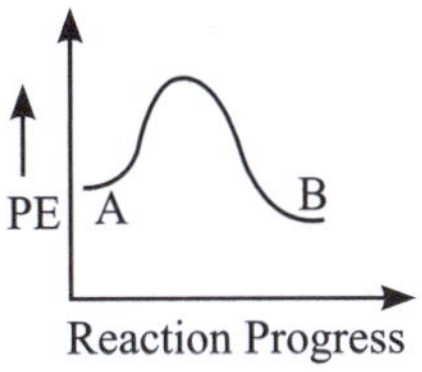

2) 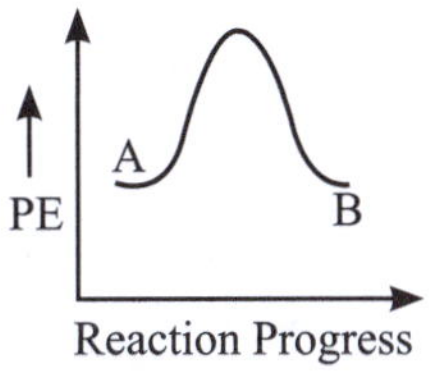

3) 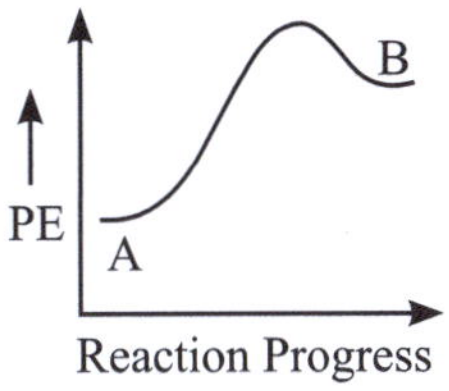

4) 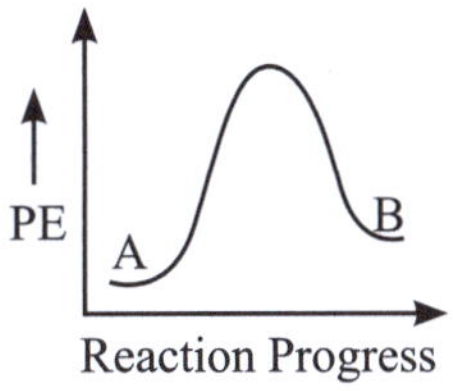

12. The slope of Arrhenius Plot (ln K v/s 1/T) of first order reaction is -5×10^3 K. The value of E_a of the reaction is. Choose the correct option for your answer. **(2021)**

[Given R = 8.314 JK^{-1} mol^{-1}]

1) 83.0 kJ mol^{-1}
2) 166 kJ mol^{-1}
3) –83 kJ mol^{-1}
4) 41.5 kJ mol^{-1}

13. Which option is valid for zero order reaction. **(AIIMS 2018)**

1) $t_{1/2} = \frac{3}{2} t_{1/4}$
2) $t_{1/2} = \frac{4}{3} t_{1/4}$
3) $t_{1/2} = 2t_{1/4}$
4) $t_{1/4} = (t_{1/2})^2$

14. Half life of I^{131} is 8.1 days. What is the activity of 0.8 mg of I^{131}? **(AIIMS 2015)**

1) 3.6×10^9 Bq
2) 4.98×10^{10} Bq
3) 9.85×10^{10} Bq
4) 9.85×10^9 Bq

15. If $t_{1/2}$ vs $\frac{1}{a^2}$ is a straight line graph then determine the order of reaction: **(AIIMS 2012)**

1) Zero order
2) First order
3) Second order
4) Third order

16. Which curve corresponds to the temperature dependance of the rate *R* of a simple one step reaction? **(AIIMS 2009)**

1) 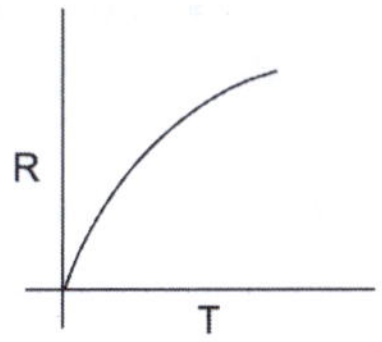

2) 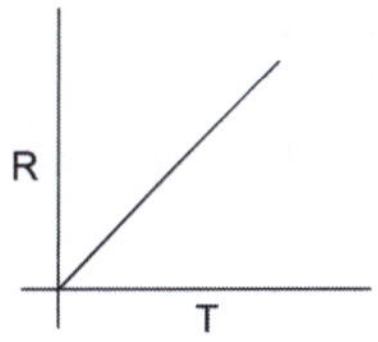

3) 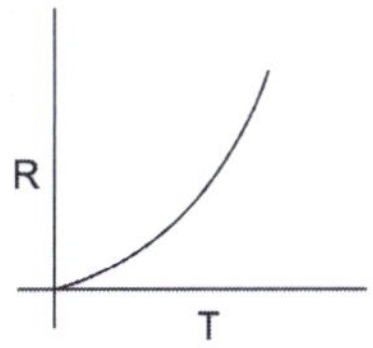

4) 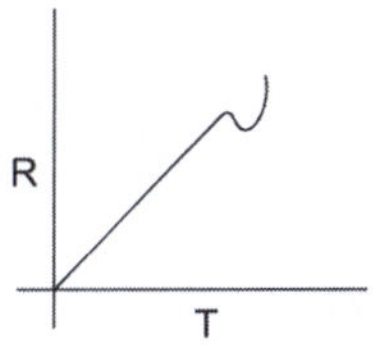

17. Assertion: According to transition state theory, for the formation of an activated complex, one of the vibrational degrees of freedom is converted into a translation degree of freedom. **(AIIMS 2006)**

Reason: Energy of the activated complex is higher than the energy of reactant molecules.

1) If both Assertion & Reason are True & the Reason is a correct explanation of the Assertion.
2) If both Assertion & Reason are True but Reason is not a correct explanation of the Assertion.
3) If Assertion is True but the Reason is False.
4) If both Assertion & Reason are False

18. An endothermic reaction with high activation energy for the forward reaction is given by the diagram: **(AIIMS 2005)**

1) 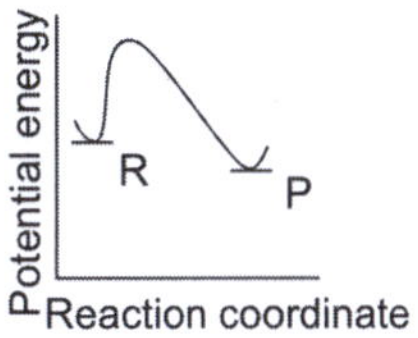

2) 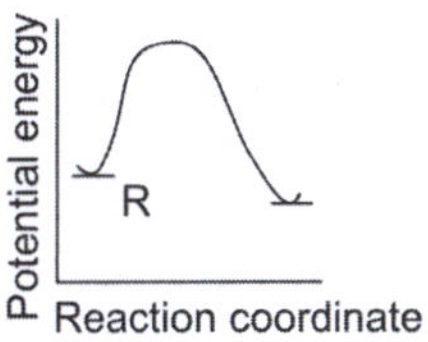

3) 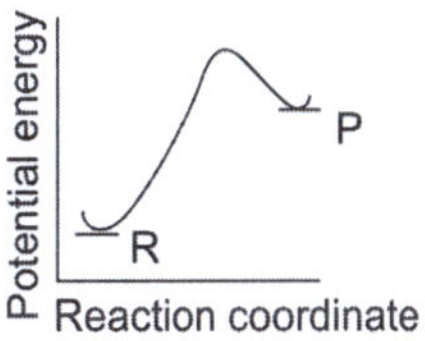

4) 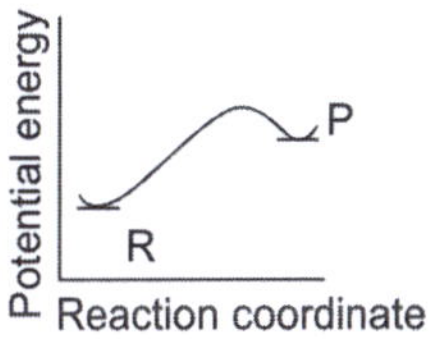

19. For the reaction: **(AIIMS 2002)**

$H_2 + Cl_2 \xrightarrow{\text{sunlight}} 2HCl$

the order of reaction is

1) 0
2) 2
3) 1
4) 3

20. The rate, at which a substance reacts, depends upon its: **(AIIMS 1996)**

1) Equivalent mass
2) Molecular mass
3) Active mass
4) Atomic mass

Answer Key

NCERT Topic-wise MCQs

1	2	3	4	5	6	7	8	9	10	11	12	13	14	15	16	17
3	3	1	2	4	4	2	1	4	1	2	3	3	3	1	1	4
18	**19**	**20**	**21**	**22**	**23**	**24**	**25**	**26**	**27**	**28**	**29**	**30**	**31**	**32**	**33**	**34**
4	1	1	2	1	4	1	3	1	3	3	3	1	3	3	1	1
35	**36**	**37**	**38**	**39**	**40**	**41**	**42**	**43**	**44**	**45**	**46**	**47**	**48**	**49**	**50**	**51**
3	2	4	2	3	4	4	1	4	4	3	2	4	3	1	1	2
52	**53**	**54**	**55**	**56**	**57**	**58**	**59**	**60**	**61**	**62**	**63**	**64**	**65**	**66**	**67**	**68**
3	2	2	4	2	3	2	3	2	3	2	2	4	2	2	1	2
69	**70**	**71**	**72**	**73**												
4	3	3	2	4												

High Order Time Intensive MCQs

1	2	3	4	5	6	7	8	9	10	11	12	13	14	15	16	17
3	4	2	3	2	4	3	2	4	4	2	4	4	4	4	2	4
18	**19**	**20**	**21**	**22**	**23**	**24**	**25**	**26**	**27**	**28**	**29**	**30**	**31**	**32**	**33**	**34**
3	2	1	2	1	3	4	3	4	3	3	2	2	2	4	1	1
35	**36**	**37**	**38**													
1	2	1	1													

Match the Columns MCQs

1	2	3	4	5
1	3	1	3	3

Correct & Incorrect MCQs

1	2	3	4	5	6	7	8	9	10
1	3	4	4	2	2	3	1	1	3

Statement Based MCQs

1	2	3	4	5	6	7	8	9	10
1	3	3	1	3	3	4	1	3	1

Assertion & Reason MCQs

1	2	3	4	5	6	7	8	9	10
2	4	1	4	3	2	3	3	3	1

Multi Correct MCQs

1	2	3	4
3	3	2	1

NCERT Exemplar MCQs

1	2	3	4	5	6	7	8	9	10	11	12	13	14	15	16	17
3	3	2	1	2	1	4	3	3	1	1	2	3	4	2	2	2
18	**19**	**20**	**21**	**22**	**23**	**24**	**25**	**26**	**27**	**28**						
1, 3, 4	1, 4	1, 2	1, 4	1, 3, 4	1, 4	1, 3	1, 4	1, 2	1, 4	1, 4						

Kattar NEET

1	2	3	4	5	6	7	8	9	10	11	12	13	14	15	16	17
3	3	1	1	4	3	2	4	2	4	2	2	3	2	3	3	1
18	**19**	**20**														
1	1	2														

NEET & AIIMS Past Year MCQs

1	2	3	4	5	6	7	8	9	10	11	12	13	14	15	16	17
3	3	3	2	2	3	1	3	1	3	1	1	3	1	1	2	1
18	**19**	**20**														
3	1	3														

CHAPTER 4

The p-Block Elements (Group 15 to 18)

Topicwise Number of Questions Analysis (2025-2016)

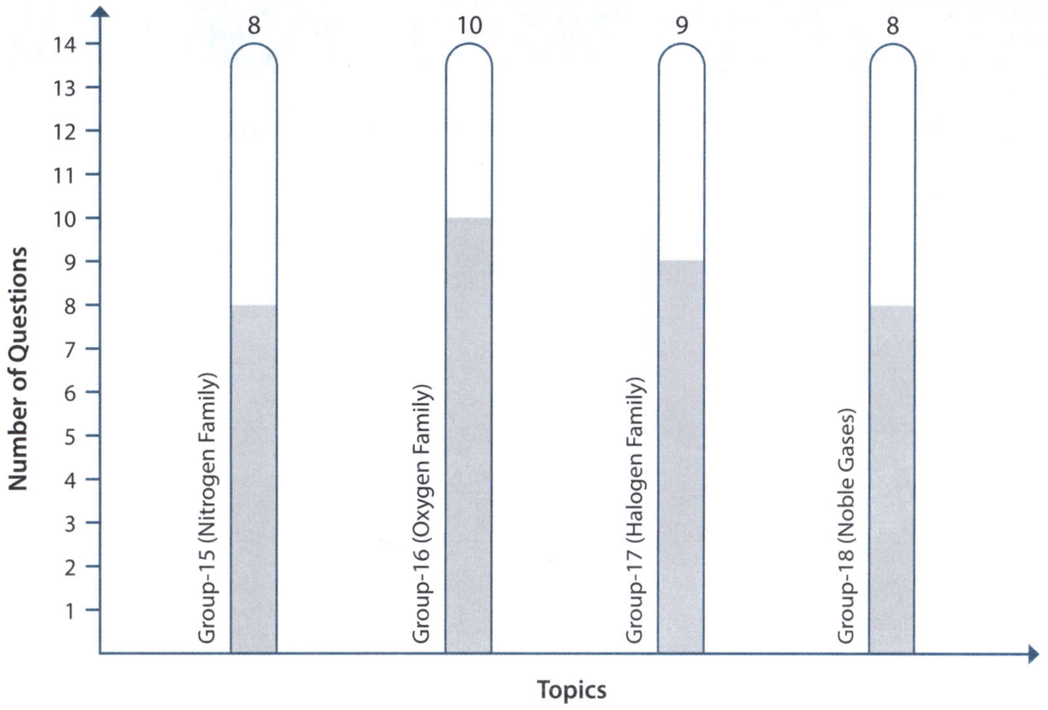

NCERT-Picks

The p-block elements are place in group 13 to 18 in the periodic table. Here we will study from group 15 to group 18 elements.

Group 15 Elements

- Nitrogen (N), phosphorus (P), Arsenic (As), Antimony (Sb), and Bismuth (Bi) are group 15 elements, and are collectively called pnictogens.
- The general valence shell electronic configuration of these elements is $ns^2\ np^3$.

  Molecular formula, physical state and metallic character: N_2 (unreactive gas), P_4 (Solid non-metal), As_4 and Sb_4 (Solid metalloids), Bi (metal).
- Due to the tendency of nitrogen to form multiple bond ($p\pi - p\pi$), nitrogen exists as N_2 and is unreactive due to high dissociation energy of the triple bond.
- N – N bond is weaker than that of P – P bond because of high inter electronic repulsion of non – bonding electrons owing to the small bond length.
- Except N and Bi, all elements exhibit allotropism.
- On moving down the group, reducing nature of these halides increases. The bond angles and basic strength of the hydrides decrease on moving down the group.
- Covalent & ionic radii increase down the group but small increase from As to Bi. This is due to the presence of completely filled d & f orbitals.
- Ionisation enthalpy of group 15 element is much greater than that of group 14 elements, in corresponding period. This is due to extra stable half – filled p orbitals electronic configuration & smaller size.

Atomic and physical properties

- **Ionisation enthalpy:** N > P > As > Sb > Bi (IA. of 15 group >>> 14group)
- **Electronegativity:** N > P > As > Sb = Bi
- **Covalent radius:** Bi > Sb > As > P > N
- **Melting point:** As > Sb > Bi > P > N
- **Boiling Point:** Sb > Bi > As > P > N

- **Density:** Bi > Sb > As > P > N
- **Basicity:** $NH_3 > PH_3 > AsH_3 > SbH_3 > BiH_3$.
- **Boiling point:** $BiH_3 > SbH_3 > NH_3 > AsH_3 > PH_3$

Dinitrogen

Preparation

- **By heating ammonium dichromate:** $(NH_4)_2Cr_2O_7 \xrightarrow{\Delta} N_2 + 4H_2O + Cr_2O_3$.
- **By thermal decomposition of sodium or barium azide** $Ba(N_3)_2 \xrightarrow{\Delta} Ba + 3N_2$ **(Very pure nitrogen)**

Name	Formula	Oxidation state of nitrogen	Common methods of preparation	Physical appearance and chemical nature
Dinitrogen oxide [Nitrogen(I) oxide]	N_2O	+ 1	$NH_4NO_3 \xrightarrow{Heat} N_2O + 2H_2O$	colourless gas, neutral
Nitrogen monoxide [Nitrogen(II) oxide]	NO	+ 2	$2NaNO_2 + 2FeSO_4 + 3H_2SO_4 \rightarrow Fe_2(SO_4)_3 + 2NaHSO_4 + 2H_2O + 2NO$	colourless gas, neutral
Dinitrogen trioxide [Nitrogen(III) oxide]	N_2O_3	+ 3	$2NO + N_2O_4 \xrightarrow{250\ K} 2N_2O_3$	blue solid, acidic
Nitrogen dioxide [Nitrogen(IV) oxide]	NO_2	+ 4	$2Pb(NO_3)_2 \xrightarrow{673\ K} 4NO_2 + 2PbO + O_2$	brown gas, acidic
Dinitrogen tetroxide [Nitrogen(IV) oxide]	N_2O_4	+ 4	$2NO_2 \underset{Heat}{\overset{Cool}{\rightleftharpoons}} N_2O_4$	colourless solid/ liquid, acidic
Dinitrogen pentoxide [Nitrogen(V) oxide]	N_2O_5	+ 5	$4HNO_3 + P_4O_{10} \rightarrow 4HPO_3 + 2N_2O_5$	colourless solid, acidic

- **Brown Ring test:** Used to indicate the presence of nitrate ion in the solution.

$$NO_3^- + 3Fe^{2+} + 4H^+ \rightarrow NO + 3Fe^{3+} + 2H_2O$$

$$[Fe(H_2O)_6]^{2+} + NO \rightarrow \underset{Brown}{[Fe(H_2O)_5(NO)]^{2+}} + H_2O$$

Phosphorous

- **Allotropes of phosphorus:**

White Phosphorus	Red Phosphorus	Black Phosphorus	
• Poisonous and insoluble in water but soluble in CS_2. • Glows in dark (Chemiluminescence) 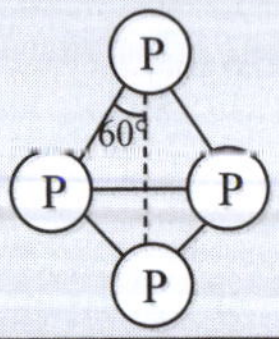	Non-Poisonous and insoluble in water as well as in CS_2. It is less reactive than white phosphorus and does not glow in dark. 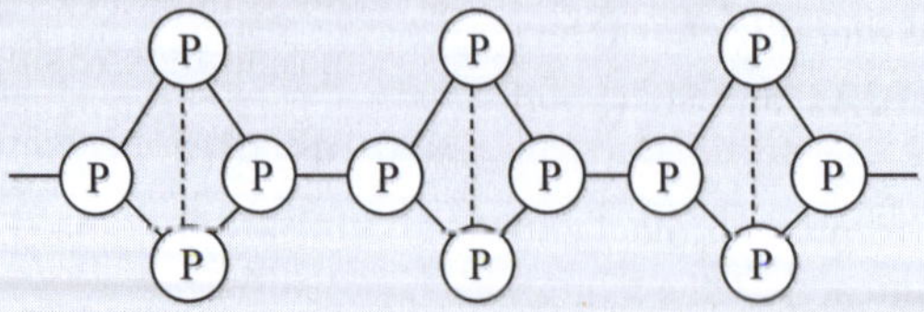	It occurs in 2 forms • α-Black: Phosphorus has opaque monoclinic or rhombohedral crystal and can be sublimed in air	• β-Black: Phosphorus has layered structure.

- Phosphorus forms two types of halides, PX_3 and PX_5 [X = F, Cl, Br].
- As electronegativity of X in PX_3 increases, bond angle decreases. $PF_3 < PCl_3 < PBr_3 < PI_3$.

Phosphorus pentachloride

Preparation:

$P_4 + 10Cl_2 \rightarrow 4PCl_5$

$P_4 + 10SO_2Cl_2 \rightarrow 4PCl_5 + 10SO_2$

Properties:

- It hydrolysis to $POCl_3$, and finally gets converted to phosphoric acid.

  $PCl_5 + H_2O \rightarrow POCl_3 + 2HCl$

  $POCl_3 + 3H_2O \rightarrow H_3PO_4 + 3HCl$

- When heated it sublimes but decomposes on stronger heating

  $PCl_5 \xrightarrow{\Delta} PCl_3 + Cl_2$

Oxoacids of Phosphorus

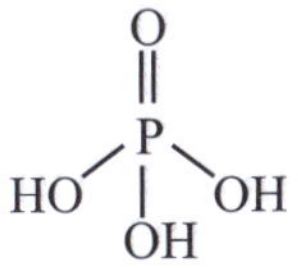

H_3PO_4
Orthophosphoric acid

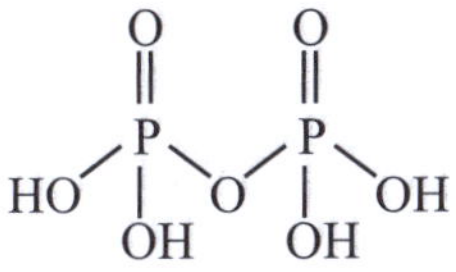

$H_4P_2O_7$
pyrophosphoric acid

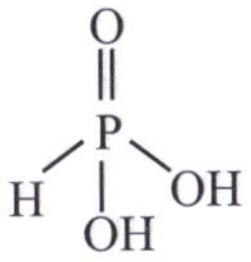

H_3PO_3
Orthophosphorous acid

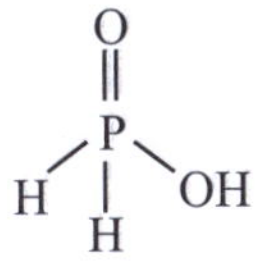

H_3PO_2
Hypophosphorous acid

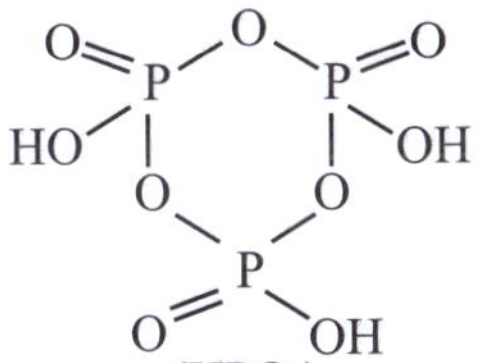

$(HPO_3)_3$
cyclictrimetaphosphoric acid

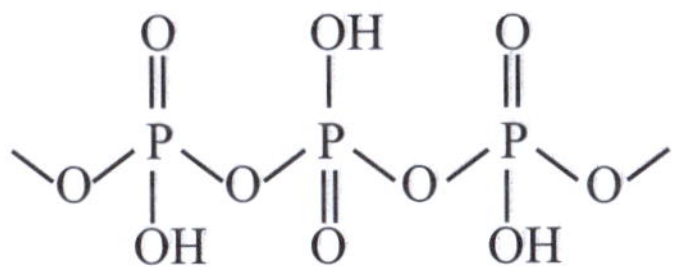

$(HPO_3)_n$
Polymetaphosphoric acid

- H_3PO_4 is a tribasic acid as it contains three ionisable H-atoms.
- Hypophosphorous acid is a good reducing agent, as it contains two P – H bonds.

Group 16 Elements

- O, S, Se, Te, Po are group 16 elements and are collectively called as 'chalcogens' which means ore forming.

Properties

- All elements of group 16 exhibit allotropism.
- $p\pi$-$p\pi$ bond is formed by O_2, while other form $d\pi$-$p\pi$ bonds.
- Sulphuric acid, oil of vitriol, is prepared by lead-chamber process or by contact process.
- Po is used in photography to remove undecomposed silver-halide.
- 'S' has many allotropic forms such as rhombic sulphur, monoclinic sulphur, cyclo-S_6 form etc.
- O_2 has less negative electron gain enthalpy than sulphur due to compact nature of oxygen atom.
- The thermal stability of hydrides decreases from H_2O to H_2PO due to the decrease in bond (H – E) dissociation enthalpy down the group.

  Order: $H_2O > H_2S > H_2Se > H_2Te > H_2PO$

- Ozone is a powerful oxidising agent in comparison to O_2. Ozone on easy decomposition, it liberates atoms of nascent oxygen $[O_3 \rightarrow O_2 + [O]]$ which is more reactive than O_2.

Oxoacids of sulphur

- Sulphur forms a number of oxoacid such as H_2SO_3, $H_2S_2O_3$, $H_2S_2O_4$, $H_2S_xO_6$ (x = 2 to 5) H_2SO_4, $H_2S_2O_7$, $H_2S_2O_8$, Structures of some important oxoacids are shown in figure below:

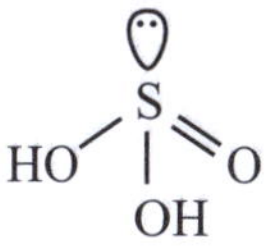

H_2SO_3
Sulphurous acid

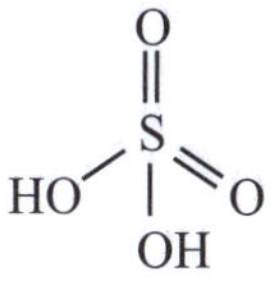

H_2SO_4
Sulphuric acid

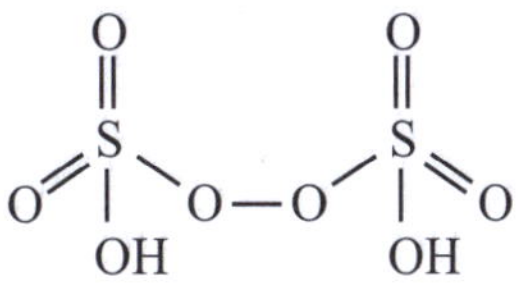

$H_2S_2O_8$
Peroxodisulphuric acid

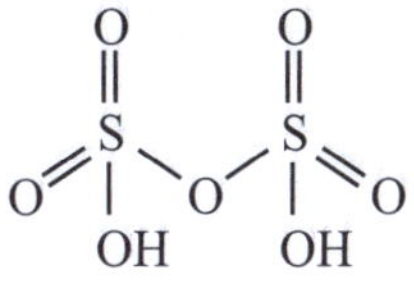

$H_2S_2O_7$
Pyrosulphuric acid (oleum)

- In vapour state, S_2 molecule shows paramagnetic behaviour. S_2 molecule has two unpaired electron in the anti bonding π^* orbitals like O_2.
- SO_2 is a reducing agent while SO_3 is a oxidising agent.

Group 17 Elements

- The group **17 elements** F, Cl, Br, I, At are collectively called as **halogens**.
- F, is most electronegative of all the halogens.

Properties

- Order of oxidising power of halogens is

  $F_2 > Cl_2 > Br_2 > I_2$

- **Oxoacids of halogens** are monobasic. As the number of oxygen atoms increases, oxidising power decreases, thermal stability and acidic strength also increases.
- Fluorine forms two oxides OF_2 and O_2F_2 called oxygen fluorides. Other halogen form oxides in which oxidation states of these halogens range from +1 to +7.
- Chlorine, Bromine and Iodine show -1, +1, +3, +5 and +7 oxidation states. Higher oxidation states are due to the presence of vacant d-orbitals.

- Aqua regia (3 part of Conc. HCl and 1 part of Conc. HNO_3) is used for dissolving noble metals. E.g: gold, platinum.
- Fluorine forms only one oxo acid (HOF), while all other halogens form oxo acid of type HOX, HXO_2, HXO_3 and HXO_4.
- The negative electron gain enthalpy of fluorine is less than that of chlorine. Due to small size of fluorine, there are strong interelectronic repulsion in relatively small 2p orbital which cause incoming electron does not experience much attraction.
- HF is not keep in glass vessels as it reacts with SiO_2 of glass.
- Inter halogen compounds are more reactive than halogens (except fluorine). This is because X – X′ bond in inter halogens is weaker than X – X bond in halogens except F – F

Chlorine

- Chlorine is a powerful bleaching agent. Bleaching action is due to oxidation. $Cl_2 + H_2O \rightarrow 2HCl + O$
  Coloured substance + O → Colourless substance
- Chlorine water on standing loses its yellow colour due to the formation of HOCl & HCl.
- Bleaching powder $Ca(OCl)_2$ is prepared by the action of Cl_2 on dry slaked lime $Ca(OH)_2$.

Group 18 Elements

- He, Ne, Ar, Kr, Xe, Rn are **18th group/zero group elements**. These are known as rare gas or inert gases. or noble gases. All these gases are monoatomic.

Properties

- The atomic radius of noble gases generally correspond to the vander waals's radii.
- Due to low ionisation energy, Xe forms several compounds with F and O.
- Noble gases have large positive electron gain enthalpy. Noble gas have stable electronic configuration, they have no tendency to accept the electrons.
- **Structure of compounds of Xenon:**

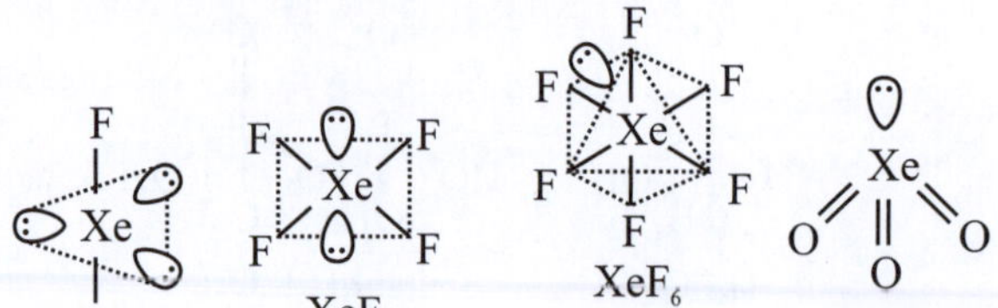

XeF_2 linear, sp^3d | XeF_4 square planar sp^3d^2 | XeF_6 distorted ↑ octahedral sp^3d^3 | XeO_3 pyramidal sp^3

- Radon is used in radio therapy for the treatment of cancer and the neon lamps are used in botanical garden.

MCQs ?

NCERT Topic-wise MCQs

Group-15 (Nitrogen Family)

1. The decreasing order of the boiling points of the following hydrides is:
   1) $H_2O > SbH_3 > AsH_3 > PH_3 > NH_3$
   2) $H_2O > NH_3 > SbH_3 > AsH_3 > PH_3$
   3) $H_2O > SbH_3 > NH_3 > AsH_3 > PH_3$
   4) $H_2O > PH_3 > AsH_3 > SbH_3 > NH_3$
2. For H_3PO_3, the correct choice is:
   1) Dibasic and non-reducing
   2) Dibasic and reducing
   3) Tribasic and reducing
   4) Tribasic and non-reducing
3. Gaseous product obtained on thermal decomposition of $(NH_4)_2Cr_2O_7$ is:
   1) NH_3 2) N_2
   3) O_2 4) NO
4. Which one of the following is obtained when excess of NH_4OH is added to $CuSO_4$ solution?
   1) A blue ppt. at the end
   2) A black ppt. Initially
   3) A deep blue solution at the end
   4) A colourless solution at the end
5. Among the following, the oxidation state of N is lowest in:
   1) NH_3 2) HN_3
   3) N_2H_4 4) NO_2
6. One mole of calcium phosphide on reaction with excess of water gives:
   1) One mole of phosphine
   2) Two mole of phosphoric acid
   3) Two mole of phosphine
   4) One mole of phosphorus acid
7. Among the following, the dipole moment is maximum in case of:
   1) NH_3 2) SbH_3
   3) PH_3 4) AsH_3
8. In the reaction of NaOH with white P, PH_3 and NaH_2PO_2 are obtained. The reaction is an example of:
   1) Reduction 2) Oxidation
   3) Neutralisation 4) Disproportionation
9. The number of steps in which orthophosphoric acid ionizes is/are:
   1) 3 2) 1
   3) 4 4) 2
10. The acid which forms two series of salt is:
    1) H_3PO_3 2) H_3PO_2
    3) H_3PO_4 4) HPO_3
11. Which of the following is kept in water?
    1) White phosphorous 2) Sodium
    3) Potassium 4) Calcium
12. The oxidation state(s) of nitrogen in ammonium nitrate is/are:
    1) – 3 only 2) + 5 only
    3) – 3, – 5 4) – 3, + 5
13. Which of the following is an anhydride of nitrous acid?
    1) N_2O_3 2) NO
    3) NO_2 4) N_2O_5
14. The number of bond pairs and lone pairs of electrons, on nitrogen atom in nitrate ions, are respectively:
    1) 2, 2 2) 3, 1
    3) 1, 3 4) 4, 0
15. The two stable isotopes of nitrogen atom are
    1) ^{16}N and ^{17}N 2) ^{14}N and ^{15}N
    3) ^{12}N and ^{18}N 4) ^{14}N and ^{20}N
16. Which one of the following order of compounds is according to the decreasing order of the oxidation state of nitrogen?
    1) HNO_3, NO, NH_4Cl, N_2 2) HNO_3, NO, N_2, NH_4Cl
    3) HNO_3, NH_4Cl, NO, N_2 4) NO, HNO_3, NH_4Cl, N_2
17. Nitrogen can form only one chloride with chlorine which is NCl_3 whereas P can form PCl_3 and PCl_5. This is:
    1) Due to absence of vacant d-orbitals in nitrogen
    2) Due to difference in size of N and P
    3) Due to high reactivity of P towards Cl than N
    4) Due to presence of multiple bonding in nitrogen
18. Which oxide of nitrogen is obtained on heating ammonium nitrate at 250°C?
    1) Nitric oxide, NO 2) Nitrous oxide, N_2O
    3) Nitrogen dioxide 4) Dinitrogen tetraoxide

19. The structure of white phosphorus is:
1) Square planar 2) Pyramidal
3) Tetrahedral 4) Trigonal planar

20. PCl_3 on hydrolysis gives:
1) H_3PO_3 2) HPO_3
3) H_3PO_4 4) $POCl_3$

21. Which of the following is tetrabasic acid?
1) Hypo-phosphorous acid
2) Metaphosphoric acid
3) Pyrophosphoric acid
4) Orthophosphoric acid

22. Nitrous oxide is:
1) Acidic 2) Basic
3) Amphoteric 4) Neutral

23. A brown ring is formed in the ring test for NO_3^- ion. It is due to formation of:
1) $[Fe(H_2O)_5(NO)]^{2+}$ 2) $Fe SO_4 NO_2$
3) $[Fe(H_2O)_4(NO)_2]^{2+}$ 4) $FeSO_4. HNO_3$

24. The element of 15th group which cannot exhibit allotropy
1) P 2) As
3) Sb 4) Bi

25. Nitrogen shows allotropy in -------- state
1) gaseous 2) liquid
3) solid 4) Liquid and Solid

26. Dinitrogen is inert at room temperature because
1) It has high bond enthalpy of N≡N bond
2) It has low bond enthalpy of N≡N bond
3) It has low bond enthalpy of N≡N bond
4) It is solid at room temperature

27. The trend in the hydrides from Bi to N is
1) Bond length increases
2) Bond length decreases
3) Acidic nature increases
4) Bond energy decreases

28. The largest bond angle is in
1) AsH_3 2) NH_3
3) H_2O 4) PH_3

29. Which of the following is a Hydride of Nitrogen?
1) NH_3 2) N_2H_4
3) HN_3 4) All

30. Oxide of nitrogen formed in the atmosphere during the lightenings is
1) NO 2) N_2O
3) NO_2 4) None

31. How many bond pairs and lone pair of electrons respectively are present in ammonia molecule?
1) 2 and 4 2) 3 and 1
3) 1 and 2 4) 4 and 2

32. Which of the following has maximum complex forming ability with a given metal ion?
1) PH_3 2) BiH_3
3) NH_3 4) SbH_3

33. Ammonia molecule acts as a
1) Lewis base 2) Lewis acid
3) Bronsted acid 4) None of these

Group-16 (Oxygen Family)

34. Which of the following compound is a peroxide?
1) NO_2 2) KO_2
3) BaO_2 4) MnO_2

35. α and β forms of sulphur are at equilibrium at a temperature known as:
1) Critical temperature 2) Transition temperature
3) Boyle's temperature 4) Inversion temperature

36. Select the correct statement about α-sulphur :
1) It is stable above 369 K and transforms into β-sulphur below this.
2) It is stable below 369 K and transforms into β-sulphur above this.
3) It is stable above 569 K and transforms into β-sulphur below this.
4) It is stable below -286 K and transforms into β-sulphur above this.

37. Ozone is:
1) An isobar of oxygen 2) An isotope of oxygen
3) A polymer of oxygen 4) An allotrope of oxygen

38. Which of the following is used as fluorinating agent?
1) SF_6 2) SF_4
3) S_2Cl_2 4) SF_2

39. Hydridisation of S in SF_6 is:
1) sp^3 2) sp^3d^3
3) sp^3d^2 4) sp^3d

40. In which of the following reaction, H_2SO_4 behaves as an oxidizing agent?
1) $2\ PCl_5 + H_2SO_4 \rightarrow 2\ POCl_3 + 2HCl + SO_2Cl_2$
2) $2HI + H_2SO_4 \rightarrow I_2 + SO_2 + 2H_2O$
3) $S + 2H_2SO_4 \rightarrow 3SO_2 + 2H_2O$
4) Both (2) and (3)

41. Which of the following possesses pπ-dπ bonding?
1) NO_3^- 2) BO_3^{3-}
3) SO_3^{2} 4) CO_3^{2-}

42. The number of σ and π bonds present in a molecule of H_2SO_4 are respectively:
1) 2, 6 2) 4,4
3) 4, 2 4) 6, 2

43. Which one of the following is not oxidized by ozone?
1) $FeSO_4$ 2) $KMnO_4$
3) KI 4) K_2MnO_4

44. Conc H_2SO_4 displaces hydrogen chloride from chlorides because:
1) Chlorides are the salt of acid HCl which is more volatile than H_2SO_4
2) Sulphates are more soluble than chlorides
3) H_2SO_4 is stronger acid than HCl
4) Sulphates are less soluble that chlorides

45. The type of bonds present in sulphuric anhydride is:
1) 3σ and 3 $p\pi - d\pi$ 2) 3σ, 1 $p\pi - p\pi$ and $2p\pi - d\pi$
3) 2σ and 3 $p\pi - d\pi$ 4) 2σ and 2 $p\pi - d\pi$

46. Which of the following is the wrong statement?
1) Ozone is paramagnetic gas
2) S^{2-} and O^{2-} are not isoelectronic
3) O_3 molecule is bent
4) Ozone is violet-black in solid state

47. Dissolution of sulphuric acid in water is:
1) An endothermic process
2) An exothermic process
3) A dehydrating process
4) A displacement process

48. What is the name of the compound $H_2S_2O_8$?
1) Peroxodisulphuric acid.
2) Peroxonitric acid.
3) Sulphuric acid.
4) Pyrosulphuric acid.

49. Ozonisation of water is carried to remove
1) Bacterial impurities
2) Bad taste
3) Excess of chlorine present
4) Calcium and magnesium salts present in it

50. Catalyst used in making H_2SO_4 in Contact process is:
1) V_2O_5 2) Fe_2O_3
3) Cr_2O_3 4) CrO_3

51. BaO_2 and Ozone reacts to produce:
1) Ba 2) Ba_2O_3
3) BaO 4) $Ba(OH)_3$

52. Sulphur molecule is:
1) Diatomic 2) Triatomic
3) Tetratomic 4) Octa-atomic

53. In which of the following, sulphur is present in + 5 oxidation state?
1) Dithionic acid 2) Sulphurous acid
3) Sulphuric acid 4) Pyrosulfuric acid

54. Metal reacts with sulphur to give
1) Sulphide 2) Sulphite
3) Sulphate 4) Thiosulphate

55. Which of the following is a true Chalcogen?
1) Nitrogen 2) Oxygen
3) Phosphorus 4) Chlorine

56. Caro's acid is
1) H_2SO_5 2) $H_2S_2O_8$
3) H_2SO_3 4) $H_2S_2O_3$

57. Which of the following does not contain S–S bond ?
1) $H_2S_2O_4$ 2) $H_2S_2O_5$
3) $H_2S_2O_6$ 4) $H_2S_2O_7$

58. The number of unpaired electrons in the p-subshell of oxygen atom is
1) 1 2) 2
3) 3 4) 4

59. Strongest reducing agent of the following is
1) H_2O 2) H_2S
3) H_2Se 4) H_2Te

60. The maximum covalency of oxygen is 4 due to
1) Absence of d orbitals 2) Presence of d orbitals
3) Presence of p orbitals 4) None of these

61. Which of the following is an anhydride of H_2SO_4?
1) H_2S 2) H_2S_2
3) SO_2 4) SO_3

62. O^{2-}_2 is the symbol of .... ion
1) Oxide 2) Super oxide
3) Peroxide 4) Monoxide

63. Which one of the gas dissolves in H_2SO_4 to give oleum?
1) SO_2 2) H_2S
3) S_2O 4) SO_3

64. What happens when H_2SO_4 is added to CaF_2?
1) Gives $CaSO_4$ 2) Gives HF
3) Gives (1) & (2) both 4) Gives $CaHF_2$

65. A and B in the given reaction respectively are
$2KClO_3 \xrightarrow[MnO_2]{Heat} A + B$
1) HNO_3 and O_2 2) KCl and O_2
3) NaCl and N_2 4) KCl and Cl_2

Group-17 (Halogen Family)

66. Which of the following statements about halogens is incorrect ?
1) They contain seven electrons in their outermost shell
2) They have the smallest atomic radii in their respective periods
3) They have very low ionization enthalpy
4) They have very high electronegativity

67. The molecular geometry of the interhalogen AB_5 is:
1) T-Shaped 2) Tetrahedral
3) Square pyramidal 4) See-saw

68. Iodine has lowest oxidation state in:

1) KI_3 2) KIO_3
3) KI 4) IF_5

69. Cl_2O is an anhydride of:

1) Hypochlorous acid 2) Chloric acid
3) Chlorous acid 4) Perchloric acid

70. Cl_2O_6 is an anhydride of:

1) Chlorous acid
2) Chloric acid
3) Perchloric acid
4) Both (2) and (3) together

71. Among the known interhalogens, the maximum number of halogen atoms in a molecule is/are-

1) 8 2) 7
3) 5 4) 6

72. Which one of the following will further combine with F_2?

1) SF_6 2) CaF_2
3) IF_5 4) None of these

73. In oxoacids of chlorine, Cl=O bond is formed by the overlapping which is:

1) dπ-dπ 2) dπ-pπ
3) pπ-pπ 4) Any of these

74. The ion that cannot undergo disproportionation is:

1) ClO_4^- 2) ClO_3^-
3) ClO^- 4) ClO_2^-

75. The halogen which gives two gases, on treatment with water is:

1) Cl_2 2) I_2
3) F_2 4) Br_2

76. Least volatile hydrogen halide is:

1) HF 2) HCl
3) HI 4) HBr

77. The bond dissociation enthalpy of F_2 is

1) Smaller than the bond dissociation enthalpy of Cl_2
2) Larger than the bond dissociation enthalpy of Cl_2
3) Equal to the bond dissociation enthalpy of Cl_2
4) None of these

78. Bleaching action of $CaOCl_2$ is due to:

1) Nascent oxygen 2) Chlorine
3) HClO 4) HCl

79. The correct order of increasing hydration energy of the following conjugate bases of oxyacids of chlorine is:

1) $ClO^- < ClO_2^- < ClO_3^- < ClO_4^-$
2) $ClO_4^- < ClO_3^- < ClO_2^- < ClO^-$
3) $ClO_4^- < ClO_3^- < ClO^- < ClO_2^-$
4) $ClO_3^- < ClO_4^- < ClO_2^- < ClO^-$

80. Which of the following represents the correct order of acidic strength ?

1) HI < HCl < HBr < HF 2) HBr < HI < HCl < HF
3) HF < HCl < HBr < HI 4) HF < HI < HCl < HBr

81. Which of the following is an isoelectronic pair?

1) ICl_2, ClO_2 2) BrO_2^-, BrF_2^+
3) ClO_2, BrF 4) CN^-, O_3

82. The halogen which is solid at room temperature is

1) I_2 2) Br_2
3) Cl_2 4) F_2

83. Iodine is more soluble in aqueous solution of Kl because it forms soluble

1) KI_2 2) KI_3
3) KI_4 4) K_2I_2

84. Which of the following halogen sublimes?

1) Cl_2 2) Br_2
3) I_2 4) F_2

85. The halogen which can form both cations and anions also

1) F 2) Cl
3) Br 4) I

86. Fluorine exhibits

1) +5 oxidation state only
2) –1 oxidation state only
3) +3 and +5 oxidation state
4) +2 and +7 oxidation state

87. The correct sequence of arrangements of the following compounds in order of decreasing oxidation number as iodine is

1) ICl, HIO_4, I_2, HI 2) HIO_4, HI, I_2, ICl_5
3) HIO_4, ICl_5, I_2, HI 4) HIO_4, ICl_5, I_2, ICl

88. The bond dissociation energy of F_2 is very low due to

1) Low density
2) Repulsions between non bonding electrons
3) Its low atomic number
4) Attractions between non bonding electrons

89. The property that correspond to the order HI > HBr > HCl > HF is

1) Thermal stability 2) Reducing power
3) Ionic character 4) Bond strength

90. The set with correct order of acidity is

1) $HClO < HClO_2 < HClO_3 < HClO_4$
2) $HClO_4 < HClO_3 < HClO_2 < HClO$
3) $HClO < HClO_4 < HClO_3 < HClO_2$
4) $HClO_4 < HClO_2 < HClO_3 < HClO$

91. The number of lone pairs on chlorine atom in ClO^-, ClO_2^-, ClO_3^-, ClO_4^- ions are

1) 0, 1, 2, 3 2) 1, 2, 3, 4
3) 4, 3, 2, 1 4) 3, 2, 1, 0

92. Which one of the following oxyacids of chlorine is the least reducing agent?
1) HOCl 2) $HClO_2$
3) $HClO_4$ 4) $HClO_3$

Group-18 (Noble Gases)

93. In Xenon tetrafluoride, the hybridised state of Xe is:
1) sp^3 2) sp^3d^3
3) dsp^2 4) sp^3d^2

94. Among the following inert gas elements, the element that shows highest chemical reactivity is:
1) Ne 2) Ar
3) He 4) Xe

95. Which of the following rare gas is the most abundant in air?
1) He 2) Ne
3) Ar 4) Kr

96. The noble gas compound prepared by Bartlett was:
1) XeO_3 2) $XePtF_6$
3) KrF_2 4) XeF_2

97. Which of the following statements about noble gases is incorrect ?
1) They are monoatomic
2) They are colourless, odourless and tasteless gases
3) They are sparingly soluble in water
4) They have extremely high melting and boiling points

98. Which is monoatomic?
1) Sulphur 2) Helium
3) Phosphorus 4) Chlorine

99. In XeF_2, XeF_4 and XeF_6 the number of lone pair on Xe is / are respectively:
1) 2,3,1 2) 1,2,3
3) 4,1,2 4) 3,2,1

100. 'A' in the given reaction is $XeF_4 + O_2F_2 \rightarrow A + O_2$
1) XeF_6 2) XeF_2
3) XeO_4 4) None of these

101. Cationic part of solid XeF_6 having the shape is:-
1) Linear 2) Angular
3) Square pyramidal 4) Tetrahedral

102. Which one of the following configuration represents Ar.?
1) $1s^2 2s^2 2p^6 3s^2$ 2) $1s^2 2s^2 2p^6 3s^2 3p^5$
3) $1s^2 2s^2 2p^6 3s^2 3p^6$ 4) $1s^2 2s^2 2p^6 3s^2 3p^2$

103. Which one of the following noble gases is not found in atmosphere?
1) Rn 2) Kr
3) Ne 4) Ar

104. Number of unpaired electrons in inert gas is
1) Zero 2) 8
3) 4 4) 18

105. Noble gases form compounds very easily with
1) Sulphur 2) Nitrogen
3) Lithium 4) Fluorine

106. The lightest gas which is non-inflammable is
1) H_2 2) He
3) N_2 4) Ar

107. XeF_2 molecule is
1) Trigonal planar 2) Square planar
3) Linear 4) Pyramidal

108. Which of the following is non-existing?
1) H_2 2) O_2
3) N_2 4) He_2

109. XeF_6 on complete hydrolysis gives.
1) Xe 2) XeO_2
3) XeO_3 4) XeO_4

110. The number of lone pairs of electrons on xenon atom in $XeOF_4$ molecule is
1) 4 2) 3
3) 1 4) zero

111. Which of the following noble gases is used in the treatment of cancer?
1) Xe 2) Ar
3) Rn 4) Kr

High Order Time Intensive MCQs

1. Which of the following does not give oxygen on heating?
1) $KClO_3$ 2) $Zn(ClO_3)_2$
3) $K_2Cr_2O_7$ 4) $(NH_4)_2Cr_2O_7$

2. Select the incorrect statement:
1) Xe is the most reacting among rare gases
2) Ar is an inert gas
3) Rn is obtained by the decay of Ra
4) The most abundant rare gas in the atmosphere is He

3. Thermal and electrical conductivity is highest in:
1) N 2) P
3) As 4) Bi

4. Which is most acidic in nature?
1) Bi_2O_3 2) Bi_2O_4
3) Sb_2O_3 4) Bi_2O_5

5. Maximum covalency of Sb will be:
1) 3 2) 4
3) 5 4) 6

6. Incorrect reaction among the following is/are:
   1) $NH_4NO_3 \xrightarrow{\Delta} N_2O + H_2O$
   2) $(NH_4)_2Cr_2O_7 \xrightarrow{\Delta} Cr_2O_7 + O_2$
   3) $NO + NO_2 \xrightarrow{<250K} N_2O_3$
   4) $NO_2 + H_2O \longrightarrow HNO_2 + HNO_3$
7. Sodium nitrate decomposes above 800°C to give:
   1) N_2 2) O_2
   3) Na_2O 4) All of the above
8. Which of the following on thermal decomposition can produce very pure nitrogen?
   1) Potassium chloride 2) Ammonium chloride
   3) Barium azide 4) Sodium hydroxide
9. Which of the following compound(s) have peroxy linkage?
   1) NH_4NO_3 2) H_2SO_5
   3) $H_2S_2O_7$ 4) $H_2S_2O_8$
10. The nitrogen oxide (s) that contain (N – N) bonds is (are):
    1) N_2O 2) N_2O_3
    3) N_2O_4 4) All of these
11. Which of the following statement(s) are correct for SO_2 gas?
    1) It acts as bleaching agent in moist conditions
    2) It's molecule has linear geometry
    3) It's dilute solution is used as disinfectant
    4) It can be prepared by the reaction of dilute H_2SO_4 with metal sulphide
12. Among halogens, the one which can oxidise water to oxygen is:
    1) Iodine 2) Chlorine
    4) Bromine 3) Fluorine
13. $A + H_2O \rightarrow B + HCl$
    $B + H_2O \rightarrow C + HCl$
    Compound (A), (B) and (C) will be respectively:
    1) PCl_5, $POCl_3$, H_3PO_3 2) PCl_5, $POCl_3$, H_3PO_4
    3) $SOCl_2$, $POCl_3$, H_3PO_3 4) PCl_3, $POCl_3$, H_3PO_4
14. Aqua regia is a mixture of?
    1) 3 : 1 ratio of conc HCl and conc HNO_3
    2) 3 : 1 ratio of conc HNO_3 and conc HCl
    3) 1 : 3 ratio of conc HCl and conc HNO_3
    4) 1 : 3 ratio of conc HCl and conc H_2SO_4
15. During bleaching action, SO_2 converts into
    1) SO_3 2) H_2SO_3
    3) H_2SO_4 4) $H_2S_2O_7$
16. Regarding N_2O, the wrong statement is
    1) It is called laughing gas
    2) It is called nitrous oxide
    3) It is a linear molecule
    4) It is a more reactive oxide
17. IBr_7 cannot exist but IF_7 exist. This fact can be explained on the basis of
    1) Electronegativities 2) Electron affinities
    3) Ratio of radii of atoms 4) Reducing abilities
18. Iodine reacts with concentrated HNO_3 to yield Y along with other products. The oxidation state of iodine in Y, is:
    1) 5 2) 7
    3) 1 4) 3
19. Which of the following is correct about V group Hydrides (from Ammonia to Bismuthine)
    1) Their thermal stability gradually increases
    2) Their ease of preparation gradually increases
    3) The electron pair donating nature gradually decreases
    4) The bond energies gradually increases
20. $FeSO_4$ forms brown ring with
    1) NO^-_3 2) NO
    3) NO_2 4) N_2O_3
21. Both allotropic forms of sulphur i.e., α- and β-sulphur are stable at temperature 369 K, this temperature is called as
    1) Curie temperature
    2) Kraft temperature
    3) Transition temperature
    4) Absolute temperature
22. Among the following, which one is more stable?
    1) PH_3 2) NH_3
    3) AsH_3 4) SbH_3
23. The oxidation state of nitrogen is a fractional value in
    1) Hydroxyl amine 2) Hydrazoic acid
    3) Nitrate ion 4) Hydrazine
24. Aqua regia is used for dissolving noble metals (Au, Pt, etc.). The gas evolved in this process is
    1) NO 2) N_2
    3) N_2O_5 4) N_2O_3
25. Which of the following trihalides is not hydrolysed?
    1) NF_3 2) PCl_3
    3) $AsCl_3$ 4) $SbCl_3$
26. Which of the following undergoes hydrolysis very easily?
    1) NCl_3 2) PCl_3
    3) $AsCl_3$ 4) $BiCl_3$
27. Ozone is obtained from oxygen
    1) By oxidation at high temperature
    2) By oxidation using a catalyst
    3) By silent electric discharge
    4) By conversion at high pressure
28. Amongst H_2O, H_2S, H_2Se and H_2Te the one with the highest boiling point is
    1) H_2O because of hydrogen bonding
    2) H_2 Te because of higher molecular weight
    3) H_2S because of hydrogen bonding
    4) H_2Se because of lower molecular weight
29. Which of the following contains pπ – dπ bonds?
    1) NO_2 2) SO_3
    3) SO_2 4) Both (2) and (3)

30. Which of the following dissociates to give H^+ most easily?
1) H_2O 2) H_2S 3) H_2Te 4) H_2Se

31. Chemically, inert halide of sulphur is
1) S_2Cl_2 2) SF_4 3) SCl_2 4) SF_6

32. Commercial name of peroxymonosulphuric acid is
1) Marshall's acid 2) Oleum
3) Fuming sulphuric acid 4) Caro's acid

33. Ozone is a
1) Mild oxidizing agent
2) Powerful oxidizing agent
3) Powerful reducing agent
4) Mild reducing agent

34. In the preparation of sulphuric acid, V_2O_5 is used in the reaction, which is
1) $S + O_2 \rightarrow SO_2$ 2) $2SO_2 + O_2 \rightarrow 2SO_3$
3) $SO_2 + H_2O \rightarrow H_2SO_4$ 4) $N_2 + 3H_2 \rightarrow 2NH_3$

35. Although electron gain enthalpy of fluorine is less negative as compared to chlorine, Fluorine is stronger oxidizing agent than chlorine because of
(i) Low enthalpy of dissociation of F-F bond
(ii) High Hydration Enthalpy of F^-
(iii) High Electron affinity of fluorine than chlorine
1) i, ii 2) i, iii 3) i, ii, iii 4) ii, iii

36. Among C-X bond (where X = Cl, Br, I), the correct decreasing bond energy order is
1) $C-Cl > C-Br > C-I$
2) $C-I > C-Br > C-Cl$
3) $C-Br > C-Cl > C-I$
4) $C-I < C-Br < C-Cl$

37. Find the correct statements.
A. Electron affinity of F is less than that of Cl
B. Number of lone pairs at central chlorine atom of ClF_3 is 2
C. Iodine absorbs radiation of yellow colour and appear in violet colour.

Find the correct answer.
1) A 2) B
3) Only C 4) All are correct

38. Anomalous behaviour of Nitrogen is due to
1) Small size and high electronegativity
2) Non availability of d-orbitals in valency shell
3) Ease of multiple bond formation
4) All are correct

39. Conc. H_2SO_4 is diluted
1) By adding water in H_2SO_4
2) By adding H_2SO_4 in water
3) By adding glacial acetic acid in H_2SO_4
4) None of the above

40. Which of the following statements are true?
1) ClO_2 in alkaline solution undergoes disproportionation.
2) Ionisation enthalpy of molecular oxygen is very close to that of xenon.
3) Both P_4O_6 and P_4O_{10} contain 12 P – O bonds
4) All are correct

41. Sulphur reacts with chlorine in 1 : 2 ratio and forms X. Hydrolysis of X gives a sulphur compound Y. What is the hybridisation state of central atom in the compound Y?
1) sp 2) sp^3 3) sp^2 4) dsp^2

42. Which of the following allotropes of phosphorus is thermodynamically most stable?
1) Red 2) White
3) Yellow 4) Black

43. Which of the following statement is not correct for a noble gas?
1) Argon is used to fill the incandescent bulbs
2) Krypton is obtained in nuclear fission.
3) Radon is present in the atmosphere
4) Xenon cannot form XeF_3

44. Ozone oxidises lead sulphide to
1) Lead sulphate 2) Lead chloride
3) Lead nitrate 4) None of these

45. Sulphur trioxide is dissolved in heavy water to form a compound X. The hybridisation state of sulphur in X is
1) sp^2 2) sp^3 3) sp 4) dsp^2

46. Which of the following is a false statement?
1) Radon is obtained by the decay of radium
2) Helium is an inert gas
3) Xenon is the most reacting among rare gases
4) The most abundant rare gas in the atmosphere is helium

47. Which of the following pairs is not correctly matched
1) A halogen which is liquid at room temperature—Bromine
2) The most electronegative element—Fluorine
3) The most reactive halogen—Fluorine
4) The strongest oxidizing halogen—Iodine

48. When chlorine is passed over dry slaked lime at room temperature, the main reaction product is
1) $Ca(ClO_2)_2$ 2) $CaCl_2$
3) $Ca(OCl)_2$ 4) $Ca(OCl_2)_2$

49. Perchloric acid is prepared by the action of concentrated sulphuric acid on
1) $NaClO_4$ 2) $KClO_3$
3) $Ca(ClO_3)$ 4) $Ba(ClO_3)_2$

50. Shape of $XeOF_4$ is
1) Octahedral 2) Square pyramidal
3) Pyramidal 4) T-Shaped

51. Which of the following reactions is quite violent in nature?
1) Between hydrogen and fluorine
2) Between hydrogen and chlorine
3) Between hydrogen and bromine
4) Between hydrogen and iodine

52. Which one of the following represents the reaction between fluorine and cold dilute NaOH?
1) $2F_2 + 4NaOH \rightarrow 4NaF + 2H_2O$
2) $3F_2 + NaOH \rightarrow 5NaF + NaFO_3$
3) $F_2 + 2NaOH \rightarrow NaF + NaOF + H_2O$
4) $2F_2 + 2NaOH \rightarrow 2NaF + OF_2 + H_2O$

53. In cold water, bleaching powder ionises to form
1) Ca^{2+}, Cl^- and ClO^- 2) CaO, Cl^-
3) Ca^{2+}, Cl^-, and ClO_3^- 4) Ca^{2+}, Cl^- and ClO_2^-

54. In hot water, Bleaching powder does not give
1) Ca^{2+} ions 2) ClO^- ions
3) Cl^- ions 4) ClO_3^- ions

55. The reaction of NO with N_2O_4 at 250 K gives
1) N_2O_3 2) N_2O_5 3) N_2O 4) NO_2

56. Given below are the oxides:
Na_2O, As_2O_3, N_2O, NO and Cl_2O,
Number of amphoteric oxides is:
1) 0 2) 1 3) 2 4) 3

57. Identify the mismatch

	Element		Oxidation states
1)	Cl	(i)	−1, 0, +1, +3, +5, +7
2)	N	(ii)	−3, 0, +1, +3, +5, +7
3)	F	(iii)	0, −1
4)	Both (1) and (2)	(iv)	6.022×10^{23} molecules

58. Which of the following reactions is an example of a redox reaction?
1) $XeF_2 + PF_5 \rightarrow [XeF]^+ [PF_6^-]$
2) $XeF_6 + H_2O \rightarrow XeF_4 + 2HF$
3) $XeF_6 + 2H_2O \rightarrow XeO_2F_2 + 4HF$
4) $XeF_4 + O_2F_2 \rightarrow XeF_6 + O_2$

59. Among the given oxides of nitrogen ; N_2O, N_2O_3, N_2O_4 and N_2O_5, the number of compound/(s) having N – N bond is :
1) 1 2) 2 3) 3 4) 4

60. Which of the following statements about group 15 hydrides and oxides is incorrect?
1) Nitrogen forms ammonia (NH_3), which is a basic and stable hydride.
2) Arsenic forms arsine (AsH_3), which is less stable and less basic than ammonia.
3) Antimony forms antimony pentoxide (Sb_2O_5), showing +5 oxidation state.
4) Bismuth forms a stable hydride and shows a +5 oxidation state commonly.

Match the Columns MCQs

1. Match the compounds given in Column-I with the hybridisation and shape given in Column-II and mark the correct option

	Column-I		Column-II
(a)	XeF_6	(i)	sp^3d^3-distorted octahedral
(b)	XeO_3	(ii)	sp^3d^2-square planar
(c)	$XeOF_4$	(iii)	sp^3-pyramidal
(d)	XeF_4	(iv)	sp^3d^2-square pyramidal

1) (a)-(i) (b)-(iii) (c)-(iv) (d)-(ii)
2) (a)-(i) (b)-(ii) (c)-(iv) (d)-(iii)
3) (a)-(iv) (b)-(iii) (c)-(i) (d)-(ii)
4) (a)-(iv) (b)-(i) (c)-(ii) (d)-(iii)

2. Match the formulas of oxides given in Column-I with the type of oxide given in Column-II, and mark the correct option.

	Column-I		Column-II
(a)	Pb_3O_4	(i)	Neutral oxide
(b)	N_2O	(ii)	Acidic oxide
(c)	Mn_2O_7	(iii)	Basic oxide
(d)	Bi_2O_3	(iv)	Mixed oxide

1) (a)-(i) (b)-(ii) (c)-(iii) (d)-(iv)
2) (a)-(iv) (b)-(i) (c)-(ii) (d)-(iii)
3) (a)-(iii) (b)-(ii) (c)-(iv) (d)-(i)
4) (a)-(iv) (b)-(iii) (c)-(i) (d)-(ii)

3. Match the items of Columns-I and II and mark the correct option.

	Column-I		Column-II
(a)	H_2SO_4	(i)	Highest electron gain enthalpy
(b)	CCl_3NO_2	(ii)	Chalcogen
(c)	Cl	(iii)	Tear gas
(d)	Sulphur	(iv)	Storage batteries

1) (a)-(iv) (b)-(iii) (c)-(i) (d)-(ii)
2) (a)-(iii) (b)-(iv) (c)-(i) (d)-(ii)
3) (a)-(iv) (b)-(i) (c)-(ii) (d)-(iii)
4) (a)-(ii) (b)-(i) (c)-(iii) (d)-(iv)

4. Match the species given in Column-I with the shape given in Column-II and mark the correct option.

Column-I	Column-II
(a) SF_4	(i) Tetrahedral
(b) BrF_3	(ii) Pyramidal
(c) BrO_3^-	(iii) Sea-saw shaped
(d) NH_4^+	(iv) Bent T-shaped

1) (a)-(iii) (b)-(ii) (c)-(i) (d)-(iv)
2) (a)-(iii) (b)-(iv) (c)-(ii) (d)-(i)
3) (a)-(i) (b)-(ii) (c)-(iii) (d)-(iv)
4) (a)-(i) (b)-(iv) (c)-(iii) (d)-(ii)

5. Match the items of Columns-I and II and mark the correct option.

Column-I	Column-II
(a) Its partial hydrolysis does not change oxidation state of central atom	(i) He
(b) It is used in modern diving apparatus	(ii) XeF_6
(c) It is used to provide inert atmosphere for filling electrical bulbs	(iii) XeF_4
(d) Its central atom is in sp^3d^2 hybridisation	(iv) Ar

1) (a)-(i) (b)-(iv) (c)-(ii) (d)-(iii)
2) (a)-(i) (b)-(ii) (c)-(iii) (d)-(iv)
3) (a)-(ii) (b)-(i) (c)-(iv) (d)-(iii)
4) (a)-(i) (b)-(iii) (c)-(ii) (d)-(iv)

Correct & Incorrect MCQs

1. Which of the following statement (s) are correct?
A. In group 15 elements, covalent and ionic radii increases down the group but there is small increase from As to Bi.
B. Ionisation enthalpy of group 15 element is less than that of 14 elements, in corresponding period.
C. In group 15 elements, electronegativity decreases down the group with increasing atomic size.
D. In group 15 elements, metallic character increases down the group.
1) A, B and C 2) Only A and D
3) A, C and D 4) B, C and D

2. Which of the following statement about nitrogen is incorrect?
1) Nitrogen has higher electronegativity than that of Bismuth.
2) Nitrogen is restricted to a maximum covalency of 4.
3) Bond enthalpy of nitrogen is very less.
4) Nitrogen has unique ability to form pπ-pπ multiple bonds with itself and with other elements having small size and high electronegativity.

3. Choose incorrect statement among the following:
1) NH_3 is only a mild reducing agent.
2) BiH_3 is the strongest reducing agent.
3) NH_3 exhibits hydrogen bonding in solid as well as liquid state.
4) NH_3 has lower melting and boiling points than that of PH_3.

4. Select the correct statement:
1) Nitrogen forms pentahalide.
2) Among trihalides of nitrogen, only NF_3 is known to be stable.
3) Trihalides of group 15 elements except BiF_3 are predominantly covalent in nature.
4) Both (2) and (3).

5. Correct statement(s) are
A. H_3PO_4 is a tribasic acid.
B. H_3PO_4 contains three P – OH groups.
C. Hypophosphorous acid is a good oxidising agent.
D. Hypophosphorous acid contains two P – H bonds .
1) Only A and B 2) A, B and D
3) A and C 4) B and C

6. Incorrect statement(s) is/are
A. Sulphur has less negative electron gain enthalpy than oxygen atom..
B. Metallic character increases from oxygen to polonium.
C. The thermal stability of hydrides increases from H_2O to H_2Po.
D. The acidic character increases from H_2O to H_2Te.
1) A and B 2) B, C and D
3) A and C 4) Only B

7. Select the incorrect statement :
1) H_2Te has higher boiling point than H_2Se.
2) Ozone (O_3) and sulphur dioxide (SO_2) are gases while selenium dioxide (SeO_2) is solid.
3) SO_2 is an oxidising while TeO_2 is a reducing agent.
4) H_2O is most stable while H_2Te is least stable.

8. Which of the following statement is not correct about fluorine?
1) Fluorine is the most electronegative element in the periodic table.
2) F_2 has smaller enthalpy of dissociation as compared to that of Cl_2.
3) The negative electron gain enthalpy of chlorine is less than that of fluorine.
4) F_2 is the strongest oxidising halogen.

9. Which of the following statement is correct about chlorine?
   1) With cold and dilute alkalis, it gives chloride and chlorate.
   2) With hot and concentrated alkalis, it produces a mixture of chloride and hypochlorite.
   3) With dry slaked lime, it gives bleaching powder.
   4) Both (1) and (2)

10. Which of the following statement is incorrect ?
   1) Radon is obtained as a decay product of ^{226}Ra.
   2) All noble gases have general electronic configuration $ns^2 np^6$, except helium.
   3) Noble gases have very high melting and boiling points
   4) Noble gases have large positive electron gain enthalpy.

Statement Based MCQs

Directions: These questions consist of two statements each, printed as Statement-I and Statement-II. While answering these questions, you are required to choose any one of the following four responses.

1) **Both Statement-I and Statement-II are correct.**
2) **Both Statement-I and Statement-II are incorrect.**
3) **Statement-I is correct & Statement-II is incorrect.**
4) **Statement-I is incorrect & Statement-II is correct.**

1. **Statement-I:** Liquid NH_3 is used in refrigeration.
   Statement-II: NH_3 has lower dielectric constant than H_2O.
2. **Statement-I:** Both Cl_2 and SO_2 are bleaching agents.
   Statement-II: The bleaching action of Cl_2 and SO_2 is by oxidation process.
3. **Statement-I:** PH_3 is more basic than NH_3.
   Statement-II: Electronegativity of 'P' is more than 'N'.
4. **Statement-I:** In $HClO_4$, chlorine has the oxidation number of +4.
   Statement-II: $HClO_4$ (perchloric) acid has two peroxide linkages.
5. **Statement-I:** Among nitrogen halides of type NX_3, the dipole moment is highest for NI_3, and lowest for NF_3.
   Statement-II: Nitrogen halides NX_3, have pyramidal structure.
6. **Statement-I:** The single N – N bond is weaker than the single P – P bond.
   Statement-II: Bismuth has higher electronegativity than that of Nitrogen.
7. **Statement-I:** Ozone is a powerful oxidising agent.
   Statement-II: Ozone on decomposition liberates atoms of nascent oxygen $[O_3 \rightarrow O_2 + [O]]$.
8. **Statement-I:** The halogens have the smallest atomic radii in their respective periods.
   Statement-II: Halogens have very less effective nuclear charge.
9. **Statement-I:** Interhalogen compounds are less reactive than halogens (except fluorine).
   Statement-II: X–X′ bond in interhalogens is stronger than X–X bond in halogen (except fluorine).
10. **Statement-I:** Hydrolysis of XeF_4 and XeF_6 with water gives oxyfluorides, $XeOF_4$ and XeO_2F_2.
   Statement-II: Partial hydrolysis of XeF_6 gives XeO_3 .

Assertion & Reason MCQs

Directions: These questions consist of two statements each, printed as Assertion and Reason. While answering these questions, you are required to choose any one of the following four responses.

1) **Both Assertion and Reason are True and the Reason is a correct explanation of the Assertion.**
2) **Both Assertion and Reason are True but Reason is not a correct explanation of the Assertion.**
3) **Assertion is True but the Reason is False.**
4) **Assertion is False but Reason is True.**

1. **Assertion:** N_2 is less reactive than P_4.
   Reason: Nitrogen has more electron gain enthalpy than phosphorus.
2. **Assertion:** HNO_3 makes iron passive.
   Reason: HNO_3 forms a protective layer of ferric nitrate on the surface of iron.
3. **Assertion:** HI cannot be prepared by the reaction of KI with concentrated H_2SO_4
   Reason: HI has lowest H – X bond strength among halogen acids.
4. **Assertion:** Both rhombic and monoclinic sulphur exist as S_8 but oxygen exists as O_2.
   Reason: Oxygen forms $p\pi - p\pi$ multiple bond due to small size and small bond length but $p\pi - p\pi$ bonding is not possible in sulphur.
5. **Assertion:** NaCl reacts with concentrated H_2SO_4 to give colourless fumes with pungent smell. But on adding MnO_2 the fumes become greenish yellow.
   Reason: MnO_2 oxidises HCl to chlorine gas which is greenish yellow.
6. **Assertion:** SF_6 cannot be hydrolysed but SF_4 can be.
   Reason: Six F atoms in SF_6 prevent the attack of H_2O on sulphur atom of SF_6.
7. **Assertion:** Nitrogen does not form pentahalide.
   Reason: Nitrogen has the absence of vacant 'd' orbital.
8. **Assertion:** FCl_3 does not exist.
   Reason: Fluorine contains vacant 'd' orbitals.

9. **Assertion:** Dry green leaf can be bleached by Cl_2 gas.
Reason: Cl_2 is a bleaching agent.

10. **Assertion:** Water is liquid but H_2S is a gas.
Reason: Molecular weight of H_2O is higher than H_2S.

Multi Correct MCQs

1. Given below are some of the interhalogen compounds:
A. ClF_3 B. IF_7 C. BrF_5
D. BrF_3 E. I_2Cl_6 F. IF_5
G. ClF H. ClF_5
Choose from the following the set of compounds with square pyramidal structure.
1) (A), (B), (D), (E) and (G)
2) (C), (F) and (H)
3) (A), (B), (D), (E) and (F)
4) (E), (F) and (H)

2. Bond dissociation energy of E–H bond of the "H_2E" hydrides of group 16 elements (given below), follows order.
I. O II. S
III. Se IV. Te
A. (I) > (II) and (III) < (IV)
B. (I) > (IV) and (I) > (II)
C. (II) > (I)
D. (II) > (III) and (III) > (IV)
1) Only (A) 2) Both (B) and (D)
3) (A), (B) and (C) 4) (C) and (D)

3. Consider the following sulphur based oxoacids.
A. H_2SO_3 B. H_2SO_4
C. $H_2S_2O_8$ D. $H_2S_2O_7$
Amongst these oxoacids, choose the set of oxoacids with peroxo (O-O) bond.
1) Only (C) 2) Both (C) and (D)
3) (B), (C) and (D) 4) Only (D)

4. Choose the species from the following carrying a single lone pair on central atom Xenon.
A. XeF_5^+ B. XeO_3 C. XeO_2F_2
D. XeF_5^- E. XeO_3F_2 F. $XeOF_4$
G. XeF_4
1) (B), (C) and (E)
2) (E), (F) and (G)
3) (A, (B), (C) and (F)
4) (A), (B), (D), (E) and (F)

NCERT Exemplar MCQs

1. On addition of conc H_2SO_4 to a chloride salt, colourless fumes are evolved but in case of iodide salt, violet fumes come out. This is because:
1) H_2SO_4 reduces HI to I_2 2) HI is of violet colour
3) HI gets oxidised to I_2 4) HI changes to HIO_3

2. In qualitative analysis when H_2S is passed through an aqueous solution of salt acidified with dil. HCl, a black precipitate is obtained. On boiling the precipitate with dil. HNO_3, it forms a solution of blue colour. Addition of excess of aqueous solution of ammonia to this solution gives:
1) Deep blue precipitate of $Cu(OH)_2$
2) Deep blue solution of $[Cu(NH_3)_4]^{2+}$
3) Deep blue solution of $Cu(NO_3)_2$
4) Deep blue solution of $Cu(OH)_2 \cdot Cu(NO_3)_2$

3. In a cyclotrimetaphosphoric acid molecule, how many single and double bonds are present?
1) 3 double bonds; 9 single bonds
2) 6 double bonds; 6 single bonds
3) 3 double bonds; 12 single bonds
4) Zero double bond; 12 single bonds

4. Which of the following elements can be involved in pπ–dπ bonding?
1) Carbon 2) Nitrogen
3) Phosphorus 4) Boron

5. Which of the following pairs of ions are isoelectronic and iso-structural?
1) CO_3^{2-}, NO_3^- 2) ClO_3^-, CO_3^{2-}
3) SO_3^{2-}, NO_3^- 4) ClO_3^-, SO_3^{2-}

6. Affinity for hydrogen decreases in the group from fluorine to iodine. Which of the halogen acids should have highest bond dissociation enthalpy?
1) HF 2) HCl
3) HBr 4) HI

7. Bond dissociation enthalpy of E–H (E = element) bonds are given below. Which of the compounds will act as strongest reducing agent?

Compound	NH_3	PH_3	AsH_3	SbH_3
Δ_{diss} (E—H)/kJ mol^{-1}	389	322	297	255

1) NH_3 2) AsH_3
3) PH_3 4) SbH_3

8. On heating with concentrated NaOH solution in an inert atmosphere of CO_2, white phosphorus gives a gas. Which of the following statement is incorrect about the gas?
1) It is highly poisonous and has smell like rotten fish
2) It's solution in water decomposes in the presence of light
3) It is more basic than NH_3
4) It is less basic than NH_3

9. Which of the following acids forms three series of salts?
1) H_3PO_2 2) H_3BO_3
3) H_3PO_4 4) H_3PO_3

10. On heating lead nitrate, it forms oxides of nitrogen and lead. The oxides formed are:
1) N_2O, PbO 2) NO_2, PbO
3) NO, PbO 4) NO, PbO_2

11. Which of the following elements does not show allotropy?
1) Nitrogen 2) Bismuth
3) Antimony 4) Arsenic

12. Maximum covalency of nitrogen is:
1) 3 2) 5
3) 4 4) 6

13. Which of the following statements is wrong?
1) Single N–N bond is stronger than the single P–P bond
2) PH_3 can act as a ligand in the formation of coordination compound with transition elements
3) NO_2 is paramagnetic in nature
4) Covalency of nitrogen in N_2O_5 is four

14. Elements of group –15 form compounds in +5 oxidation state. However, bismuth forms only one well characterised compound in + 5 oxidation state. The compound is:
1) Bi_2O_5 2) BiF_5
3) $BiCl_5$ 4) Bi_2S_5

15. On heating ammonium dichromate and barium azide separately we get:
1) N_2 in both cases
2) N_2 with ammonium dichromate and NO with barium azide
3) N_2O with ammonium dichromate and N_2 with barium azide
4) N_2O with ammonium dichromate and NO_2 with barium azide

16. In the preparation of HNO_3, we get NO gas by catalytic oxidation of ammonia. The moles of NO produced by the oxidation of two moles of NH_3 will be:
1) 2 2) 3
3) 4 4) 6

17. The oxidation state of central atom in the anion of compound NaH_2PO_2 will be:
1) + 3 2) + 5
3) + 1 4) – 3

18. Which of the following is not tetrahedral in shape?
1) NH_4^+ 2) $SiCl_4$
3) SF_4 4) SO_4^{2-}

19. Which of the following are peroxoacids of sulphur?
1) H_2SO_5 and $H_2S_2O_8$ 2) H_2SO_3 and $H_2S_2O_7$
3) $H_2S_2O_7$ and $H_2S_2O_8$ 4) $H_2S_2O_6$ and $H_2S_2O_7$

20. Hot conc. H_2SO_4 acts as moderately strong oxidising agent. It oxidises both metals and non-metals. Which of the following element is oxidised by conc. H_2SO_4 into two gaseous products?
1) Cu 2) S
3) C 4) Zn

21. A black compound of manganese reacts with a halogen acid to give greenish yellow gas. When excess of this gas reacts with NH_3, an unstable trihalide is formed. In this process the oxidation state of nitrogen changes from:
1) – 3 to + 3 2) – 3 to 0
3) – 3 to + 5 4) 0 to – 3

22. In the preparation of compounds of Xe, Bartlett had taken $O_2^+PtF_6^-$ as a base compound. This is because:
1) Both O_2 and Xe have same size.
2) Both O_2 and Xe have same electron gain enthalpy.
3) Both O_2 and Xe have almost same ionisation enthalpy.
4) Both Xe and O_2 are gases.

23. In solid state, PCl_5 is a:
1) Covalent solid
2) Octahedral structure
3) Ionic solid with $[PCl_6]^+$ octahedral and $[PCl_4]^-$ tetrahedral
4) Ionic solid with $[PCl_4]^+$ tetrahedral and $[PCl_6]^-$ octahedral

24. Reduction potentials of some ions are given below. Arrange them in decreasing order of oxidizing power:

Ion	ClO_4^-	IO_4^-	BrO_4^-
Reduction potential E°/V	E° = 1.19 V	E° = 1.65 V	E° = 1.74 V

1) $ClO_4^- > IO_4^- > BrO_4^-$ 2) $IO_4^- > BrO_4^- > ClO_4^-$
3) $BrO_4^- > IO_4^- > ClO_4^-$ 4) $BrO_4^- > ClO_4^- > IO_4^-$

25. If chlorine gas is passed through hot NaOH solution, two changes are observed in the oxidation number of chlorine during the reaction. These are ....and:
1) 0 to + 5 2) 0 to + 3
3) 0 to – 1 4) 0 to + 1

26. Which of the following options are not in accordance with the property mentioned against them?
1) $F_2 > Cl_2 > Br_2 > I_2$ (Oxidising power)
2) MI > MBr > MCl > MF (Ionic character of metal halide)
3) $F_2 > Cl_2 > Br_2 > I_2$ (Bond dissociation enthalpy)
4) HI < HBr < HCl < HF (Hydrogen-halogen bond strength)

27. Which of the following is correct for P_4 molecule of white phosphorus?
1) It has 6 lone pairs of electron
2) It has six P – P single bonds
3) It has three P – P single bonds
4) It has four lone pairs of electrons

28. Which of the following statements are correct?
1) Among halogens, radius ratio between iodine and fluorine is maximum
2) Leaving F – F bond, all halogens have weaker X – X bond than X – X' in inter-halogens.
3) Among inter-halogen compounds, the maximum number of atoms are present in iodine fluoride.
4) Inter-halogen compounds are more reactive than halogen compounds

29. Which of the following statements are correct?
1) All the three N – O bond lengths in HNO_3 are equal
2) All P – Cl bond lengths in PCl_5 molecule in gaseous state are equal
3) P_4 molecule in white phosphorus have angular strain therefore white phosphorus is very reactive
4) PCl_5 is ionic in solid state in which cation is tetrahedral and anion is octahedral

30. Which of the following orders are correct as per the properties mentioned against each

1) $As_2O_3 < SiO_2 < P_2O_3 < SO_2$ (Acid strength)
2) $AsH_3 < PH_3 < NH_3$ (Enthalpy of vaporization)
3) $S < O < Cl < F$ (More negative electron gain enthalpy)
4) $H_2O > H_2S > H_2Se > H_2Te$ (Thermal stability)

31. Which of the following statements are correct?

1) S–S bond is present in $H_2S_2O_6$
2) In peroxosulphuric acid (H_2SO_5), sulphur is in + 6 oxidation state.
3) Iron powder along with Al_2O_3 and K_2O is used as a catalyst in the preparation of NH_3 by Haber's process
4) Change in enthalpy is positive for the preparation of SO_3 by catalytic oxidation of SO_2

32. In which of the following reactions, conc. H_2SO_4 is used as an oxidizing reagent?

1) $CaF_2 + H_2SO_4 \rightarrow CaSO_4 + 2HF$
2) $2HI + H_2SO_4 \rightarrow I_2 + SO_2 + 2H_2O$
3) $Cu + 2H_2SO_4 \rightarrow CuSO_4 + SO_2 + 2H_2O$
4) $NaCl + H_2SO_4 \rightarrow NaHSO_4 + HCl$

33. Which of the following statements are true?

1) Only type of interactions between particles of noble gases are due to weak dispersion forces
2) Ionisation enthalpy of molecular oxygen is very close to that of xenon
3) Hydrolysis of XeF_6 is a redox reaction
4) Xenon fluorides are not reactive

34. Strong reducing behaviour of H_3PO_2 is due to:

1) Low oxidation state of phosphours
2) Presence of two –OH groups and one P–H bond
3) Presence of one –OH group and two P–H bond
4) High electron gain enthalpy of phosphorus

35. In which of the following molecules, all the bonds are not equal :-

1) ClF_3 2) BF_3
3) AlF_3 4) NF_3

36. Basicity of phosphinic acid is

1) 1 2) 2
3) 3 4) 4

37. Match the Column-I with Column-II

	Column-I (Oxoacid)		Column-II (Oxidation state of central element)
(a)	Phosphoric acid	(i)	+6
(b)	Marshall's acid	(ii)	+5
(c)	Perchloric acid	(iii)	+4
(d)	Sulphurous acid	(iv)	+7

1) (a)-(ii), (b)-(i), (c)-(iv), (d)-(iii)
2) (a)-(iv), (b)-(iii), (c)-(ii), (d)-(i)
3) (a)-(iv), (b)-(ii), (c)-(iii), (d)-(i)
4) (a)-(i), (b)-(iii), (c)-(ii),(d)-(iv)

Kattar NEET

Scan Me for Detailed Explanations

1. During hydrolysis of hydrolith (CaH_2) a compound 'A' and hydrogen gas is obtained. When chlorine gas is bubbled in solution of 'A' it produces 'B' which has permanent bleaching action. Find oxidation state of chlorine in 'B'

1) +1 2) +3
3) +5 4) +7

2. Which of the following statement is correct?

1) Allotropy of sulphur arises partly because of the great variety of molecular forms that can be achieved by —S—S— catenation and partly because of the numerous ways in which the molecules so formed can be arranged within the crystal
2) Both rhombic and monoclinic sulphur have S_8 molecules and, therefore, possess same value of specific gravities.
3) Ozone on reaction with KI solution buffered with a borate buffer (pH 9.2) yields potassium iodate.
4) The ozone layer protects the earth's surface from an excessive concentration of mainly infrared radiations.

3. Which of the following statements is false?

1) H_3PO_2, H_3PO_3 and H_3PO_4 all are tribasic and reducing in nature.
2) Among anions, NO_3^-, SO_3^{2-}, CO_3^{2-} and BO_3^{2-}, only SO_3^{2-} have $p\pi$-$d\pi$ bonding
3) Among anions SO_3^{2-}, SO_4^{2-} and HSO_4^-, SO_3^{2-} is basic and reducing in nature.
4) Number of lone pair(s) of electrons on Xe atoms in XeF_2, XeF_4 and XeF_6 are 3, 2 and 1 respectively.

4. Consider the following statements and arrange in the order of true/false as given in the codes:

S_1: Ammonia on heating with concentrated solution of sodium hypochlorite gives nitrogen gas.

S_2: The phosphine gas dissolves in water in presence of sunlight and produces phosphonium hydroxide like ammonium hydroxide.

S_3: Barium azide on heating gives pure nitrogen gas.

S_4: The oxo-acids of phosphorus in which phosphorus has lower oxidation state less than +5 contain either P—P or P—H bonds but not both in addition to P=O and P—OH bonds.

1) TFTT 2) FFTT
3) TTTF 4) TFTF

5. Consider the following statements and arrange in the order of true/false as given in the codes:

S_1: Argon is used in arc welding of metals or alloys to provide an inert atmosphere.

S_2: XeF_2, XeF_4 and XeF_6 are colourless crystalline solids and sublime readily at 298 K.

S_3: XeF_2, XeF_4 and XeF_6 are readily hydrolyzed.

S_4: Xenon fluorides react with fluoride ion acceptor to form cationic species and fluoride ion donors to form fluoro anions.

1) FFFF 2) TFTF
3) TTTT 4) FTFT

6. One of the products of the following reaction $KCNO + (NH_4)_2SO_4 \xrightarrow{\Delta}$ is:

1) NH_4NO_2 2) NH_4NO_3
3) NH_2—CO—NH_2 4) Both N_2 and NO_2

7. Which of the following option is correctly matched?

1) S_3O_9; contains no S—S linkage
2) $S_2O_6^{2-}$; contains —O—O— linkage
3) $(HPO_3)_3$; contains P—P linkage
4) $S_2O_8^{2-}$; contains S—S linkage

8. Which one of the following statements regarding helium is incorrect?

1) It is used to produce and sustain powerful superconducting magnets.
2) It is used as a cryogenic agent for carrying out experiments at low temperatures.
3) It is used in filling balloons for meteorological observations because it is lighter and non-inflammable.
4) It is used as diluent for oxygen in modern diving apparatus because of its high solubility in blood.

9. Which of the following properties of halogens increase with increasing atomic number?

(I) Ionization energy
(II) Ionic radius
(III) Bond energy of the X_2 molecule
(IV) Enthalpy of vaporisation

1) I, II and III only 2) I and III only
3) II and IV 4) IV only

10. The three elements X, Y and Z with electronic configurations shown below all form hydrides

Element	Electronic configuration
X	$1s^22s^22p^2$
Y	$1s^22s^22p^63s^1$
Z	$1s^22s^22p^63s^23p^63d^{10}4s^24p^5$

Which set of properties match correctly with properties of the hydrides of these elements.

	Hydride of X	Hydride of Y	Hydride of Z
1)	Colourless gas insoluble in H_2O	Silver/grey solid reacts with H_2O to form an alkali	Colourless gas forms a strong acid in H_2O
2)	Colourless liquid, no reaction with H_2O	Silver/grey solid forms H_2 and H_2O	Ionic solid with formula ZH
3)	Non-polar compound reacts with Cl_2 in light	Silver/grey ionic solid with formula YH_2	Forms when water is added to phosphorus and elemental Z
4)	Colourless gas which burns with air	Silver/grey solid which reacts violently with acids	Colourless, corrosive liquid at STP

11. Among the following, cyclic species are

(I) $H_5P_3O_{10}$
(II) $[B_3O_3(OH)_5]^{2-}$
(III) Trimer of SO_3 (S_3O_9)
(IV) $P_3N_3Cl_6$

1) I and III only 2) I, III and IV only
3) II, III and IV only 4) I, II, III and IV only

12. Bromine is commercially prepared from sea water by displacement reaction

$$Cl_2 + 2Br^-(aq.) \longrightarrow 2Cl^-(aq.) + Br_2$$

Br_2 gas thus formed is dissolved into solution of Na_2CO_3 and then pure Br_2 is obtained by treatment of the solution with

1) $Ca(OH)_2$ 2) NaOH
3) H_2SO_4 4) HI

13. When sulphur is boiled with Na_2SO_3, a compound (A) is produced. (A) with excess of $AgNO_3$ solution gives a compound (B) which is water soluble and produces a black coloured sulphide (C). Compounds (A), (B) and (C) will be respectively.

1) $Na_2S_2O_3$, $Ag_2S_2O_3$, Ag_2S 2) Na_2SO_4, Ag_2SO_4, Ag_2S
3) $Na_2S_2O_7$, Ag_2SO_4, Ag_2S 4) Na_2SO_5, Ag_2SO_5, Ag_2SO_4

14. On addition of conc. H_2SO_4 to a chloride salt, colourless fumes are evolved but in case of iodide salt, violet fumes come out. This is because:

1) H_2SO_4 reduces HI to I_2 2) HI is of violet colour
3) HI gets oxidised to I_2 4) HI changes to HIO_3

15. In a cyclotrimetaphosphoric acid molecule, how many single and double bonds are present?

1) 3 double bonds; 9 single bonds
2) 6 double bonds; 6 single bonds
3) 3 double bonds; 12 single bonds
4) Zero double bonds; 12 single bonds

16. Hot conc. H_2SO_4 acts as moderately strong oxidising agent. It oxidises both metals and non-metals. Which of the following element is oxidised by conc. H_2SO_4 into two gaseous products?

1) Cu 2) S
3) C 4) Zn

17. Reduction potentials of some ions are given below. Arrange them in decreasing order of oxidising power.

Ion	ClO_4^-	IO_4^-	BrO_4^-
Reduction potential	$E° = 1.19$ V	$E° = 1.65$ V	$E° = 1.74$ V

1) $ClO_4^- > IO_4^- > BrO_4^-$ 2) $IO_4^- > BrO_4^- > ClO_4^-$
3) $BrO_4^- > IO_4^- > ClO_4^-$ 4) $BrO_4^- > ClO_4^- > IO_4^-$

18. Which of the following statements is not valid for oxoacids of phosphorus?
   1) Orthophosphoric acid is used in the manufacture of triple superphosphate
   2) Hypophosphorous acid is a diprotic acid
   3) All oxoacids contain tetrahedral four coordinated phosphorus
   4) All oxoacids contain atleast one P = O unit and one P—OH group

19. Consider the following statements(s):
   (I) Hypophosphorous acid shows reducing properties
   (II) Pyrophosphoric acid can show basicities from one to four because it has 4-OH groups attached to the phosphorus.
   (III) Orthophosphoric acid is obtained during disproportionation of phosphorous acid.
   (IV) Metaphosphoric acid on heating gives P_4O_{10}.
   Which of the following is/are correct?
   1) I, II and III only 2) I, II and IV only
   3) I, III and IV only 4) All are correct

20. Match the reactions listed in List-I with characteristic(s)/ type of reactions listed in List-II:

	List-I		List-II
I.	$PCl_5 \xrightarrow[\text{Air}]{\text{Moist}}$	P.	Hydrolysis
II.	$P_4 + NaOH(conc.) + H_2O \xrightarrow{\text{Warm}}$	Q.	At least one of the products has tetrahedral hybridisation
III.	$H_3PO_3 \xrightarrow{200°C}$	R.	Disproportionation
IV.	$P_4O_6 + H_2O \xrightarrow{200°C}$	S.	At least one of the products has $p\pi$–$d\pi$ bonding.

   1) I-P, Q, S; II-P, Q, S; III-Q, P, R; IV-P, Q, R, S
   2) I-P, Q, S; II-P, Q, R, S; III-Q, R, S; IV- P, Q, R, S
   3) I-P, R, S; II-P, Q, R, S; III-Q, R, S; IV- P, Q, S
   4) I-P, R, S; II-P, Q, R, S; III-Q, R, S; IV-P, Q, R, S

NEET & AIIMS Past Year MCQs

1. Match List-I with List-II : **(2024 Re)**

List-I (Atom/Molecule)		List-II (Property)	
A.	Nitrogen atom	I.	Paramagnetic
B.	Fluorine molecule	II.	Most reactive element in group 18
C.	Oxygen molecule	III.	Element with highest ionisation enthalpy in group 15
D.	Xenon atom	IV.	Strongest oxidising agent

   Identify the correct answer from the options given below:
   1) A-III, B-I, C-IV, D-II 2) A-I, B-IV, C-III, D-II
   3) A-II, B-IV, C-I, D-III 4) A-III, B-IV, C-I, D-II

2. Identify the incorrect statement. **(2024 Re)**
   1) PEt_3 and $AsPh_3$ as ligands can form $d\pi$– $d\pi$ bond with transition metals
   2) The N – N single bond is as strong as the P – P single bond
   3) Nitrogen has unique ability to form $p\pi$– $p\pi$ multiple bonds with nitrogen, carbon and oxygen
   4) Nitrogen cannot form $d\pi$ – $p\pi$ bond as other heavier elements of its group

3. Identify the incorrect statement from the following: **(2024 Re)**
   1) The acidic strength of HX(X = F, Cl, Br and I) follows the order: HF > HCl > HBr > HI.
   2) Fluorine exhibits -1 oxidation state whereas other halogens exhibit +1, +3, +5 and +7 oxidation states also.
   3) The enthalpy of dissociation of F_2 is smaller than that of Cl_2.
   4) Fluorine is stronger oxidising agent than chlorine.

4. Given below are two statements: **(2024)**
   Statement-I: The boiling point of hydrides of Group 16 elements follow the order
   $H_2O > H_2Te > H_2Se > H_2S$.
   Statement-II: On the basis of molecular mass, H_2O is expected to have lower boiling point than the other members of the group but due to the presence of extensive H-bonding in H_2O, it has higher boiling point.
   In the light of the above statements, choose the correct answer from the options given below:
   1) Statement-I is true but Statement-II is false.
   2) Statement-I is false but Statement-II is true.
   3) Both Statement-I and Statement-II are true.
   4) Both Statement-I and Statement-II are false.

5. Among Group 16 elements, which one does **NOT** show –2 oxidation state? **(2024)**
   1) Te 2) Po 3) O 4) Se

6. Amongst the given options, which of the following molecules/ion acts as a Lewis acid? **(2023)**
   1) OH^- 2) NH_3 3) H_2O 4) BF_3

7. Match List-I with List-II: **(2023)**

List-I (Oxoacids of Sulphur)		List-II (Bonds of Sulphur)	
(a)	Peroxodisulphuric acid	(i)	Two S-OH, Four S=O, One S-O-S
(b)	Sulphuric acid	(ii)	Two S-OH, One S=O
(c)	Pyrosulphuric acid	(iii)	Two S-OH, Four S=O,One S-O-O-S
(d)	Sulphurous acid	(iv)	Two S-OH, Two S=O

   Choose the correct answer from the options given below:
   1) (a)-(iii); (b)-(iv); (c)-(ii); (d)-(i)
   2) (a)-(i); (b)-(iii); (c)-(ii); (d)-(iv)
   3) (a)-(iii); (b)-(iv); (c)-(i); (d)-(ii)
   4) (a)-(i); (b)-(iii); (c)-(iv); (d)-(ii)

8. Given below are two statements: one is labelled as Assertion (A): and the other is labelled as Reason (R) **(2022)**

Assertion (A): ICl is more reactive than I_2.

Reason (R): I-Cl bond is weaker than I-I bond

In the light of the above statements, choose the most appropriate answer from the options given below:

1) (A) is not correct but (R) is correct.
2) Both (A) and (R) are correct and (R) is the correct explanation of (A).
3) Both (A) and (R) are correct but (R) is not the correct explanation of (A).
4) (A) is correct but (R) is not correct.

9. Given below are two statements: **(2022)**

Statement-I: The boiling points of the following hydrides of group 16 elements increases in the order:

$H_2O < H_2S < H_2Se < H_2Te$.

Statement-II: The boiling points of these hydrides increase with increase in molar mass.

In the light of the above statements, choose the most appropriate answer from the options given below:

1) Statements I is incorrect but Statements II is correct.
2) Both Statements I and Statements II are correct
3) Both Statements I and Statements II are incorrect
4) Statements I is correct but Statements II is incorrect

10. Noble gases are named because of their inertness towards reactivity. Identify an incorrect statement about them. **(2021)**

1) Noble gases have very high melting and boiling points
2) Noble gases have weak dispersion forces
3) Noble gases have large positive values of electron gain enthalpy
4) Noble gases are sparingly soluble in water

11. In which one of the following arrangements the given sequence is not strictly according to the properties indicated against it? **(2021)**

1) $H_2O < H_2S < H_2Se < H_2Te$: Increasing pK_a values
2) $NH_3 < PH_3 < AsH_3 < SbH_3$: Increasing acidic character
3) $CO_2 < SiO_2 < SnO_2 < PbO_2$: Increasing oxidizing power
4) $HF < HCl < HBr < HI$: Increasing acidic strength

12. **Assertion:** Pure N_2 is formed from $Ba(N_3)_2$

Reason: Mass of Barium is high **(AIIMS 2018)**

1) If both Assertion & Reason are True & the Reason is a correct explanation of the Assertion.
2) If both Assertion & Reason are True but Reason is not a correct explanation of the Assertion.
3) If Assertion is True but the Reason is False.
4) If both Assertion & Reason are False.

13. **Assertion:** HOF bond angle is higher than HOCl bond angle in HOX. **(AIIMS 2017)**

Reason: Oxygen is more electronegative than halogen.

1) If both Assertion & Reason are True & the Reason is a correct explanation of the Assertion.
2) If both Assertion & Reason are True but Reason is not a correct explanation of the Assertion.
3) If Assertion is True but the Reason is False.
4) If both Assertion & Reason are False.

14. Paramagnetic gas among the following: **(AIIMS 2016)**

1) S_2^- 2) S_8
3) S^{2-} 4) S_6

15. **Assertion:** Charcoal is used in Separation of noble gases. **(AIIMS 2016)**

Reason: Charcoal has porous structure.

1) If both Assertion & Reason are True & the Reason is a correct explanation of the Assertion.
2) If both Assertion & Reason are True but Reason is not a correct explanation of the Assertion.
3) If Assertion is True but the Reason is False.
4) If both Assertion & Reason are False.

16. When chlorine reacts with cold and dilute solution of sodium hydroxide, the products obtained are: **(AIIMS 2015)**

1) $Cl^- + ClO^-$ 2) $Cl^- + ClO_2^-$
3) $Cl^- + ClO_3^-$ 4) $Cl^- + ClO_4^-$

17. Which of the following product is not formed by the reaction of PH_4I and KOH: **(AIIMS 2010)**

1) KI 2) PH_3
3) H_2O 4) PH_4OH

Answer Key

NCERT Topic-wise MCQs

1	2	3	4	5	6	7	8	9	10	11	12	13	14	15	16	17
3	2	2	3	1	3	1	4	1	1	1	4	1	4	2	2	1
18	19	20	21	22	23	24	25	26	27	28	29	30	31	32	33	34
2	3	1	3	4	1	4	3	1	2	2	4	1	2	3	1	3
35	36	37	38	39	40	41	42	43	44	45	46	47	48	49	50	51
2	2	4	2	3	4	3	4	2	1	2	1	2	1	1	1	3
52	53	54	55	56	57	58	59	60	61	62	63	64	65	66	67	68
4	1	1	2	1	4	2	4	1	4	3	4	3	2	3	3	1
69	70	71	72	73	74	75	76	77	78	79	80	81	82	83	84	85
1	4	1	3	2	1	3	1	1	1	1	3	2	1	2	3	4
86	87	88	89	90	91	92	93	94	95	96	97	98	99	100	101	102
2	3	2	2	1	4	3	4	4	3	2	4	2	4	1	3	3
103	104	105	106	107	108	109	110	111								
1	1	4	2	3	4	3	3	3								

High Order Time Intensive MCQs

1	2	3	4	5	6	7	8	9	10	11	12	13	14	15	16	17
4	4	4	4	3	2	4	3	2,4	4	1,3	4	2	1	3	4	3
18	19	20	21	22	23	24	25	26	27	28	29	30	31	32	33	34
1	3	2	3	2	2	1	1	1	3	1	4	3	4	4	2	2
35	36	37	38	39	40	41	42	43	44	45	46	47	48	49	50	51
1	1	4	4	2	4	2	4	3	1	2	4	4	3	1	2	1
52	53	54	55	56	57	58	59	60								
4	1	2	1	2	2	4	3	4								

Match the Columns MCQs

1	2	3	4	5
1	2	1	2	3

Correct & Incorrect MCQs

1	2	3	4	5	6	7	8	9	10
3	3	4	4	2	3	3	3	3	3

Statement Based MCQs

1	2	3	4	5	6	7	8	9	10
1	3	2	2	1	3	1	3	2	2

Assertion & Reason MCQs

1	2	3	4	5	6	7	8	9	10
3	3	2	1	1	1	1	3	2	3

Multi Correct MCQs

1	2	3	4
2	2	1	3

NCERT Exemplar MCQs

1	2	3	4	5	6	7	8	9	10	11	12	13	14	15	16	17
3	2	3	3	1	1	4	3	3	2	1	3	1	2	1	1	3
18	19	20	21	22	23	24	25	26	27	28	29	30	31	32	33	34
3	1	3	1	3	4	3	1,3	2,3	2,4	1,3,4	3,4	1,4	1,2	2,3	1,2	3
35	36	37														
1	1	1														

Kattar NEET

1	2	3	4	5	6	7	8	9	10	11	12	13	14	15	16	17
1	1	1	1	3	3	1	4	3	1	3	3	1	3	3	3	3
18	19	20														
2	4	2														

NEET & AIIMS Past Year MCQs

1	2	3	4	5	6	7	8	9	10	11	12	13	14	15	16	17
4	2	1	3	2	4	3	2	3	1	1	2	4	3	1	1	4

CHAPTER

5 The d-and f-Block Elements

Topicwise Number of Questions Analysis (2025-2016)

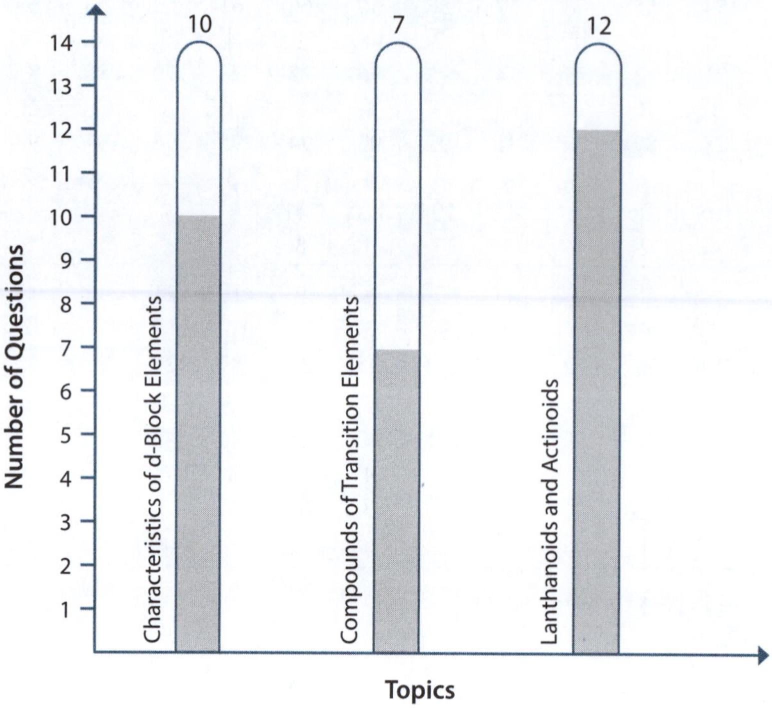

NCERT-PICKS

General Introduction

- The d-block of the periodic table contains the elements of the groups 3-12 in which the d orbitals are progressively filled in each of the four long periods. The f-block consists of elements in which 4 f and 5 f orbitals are progressively filled. They are placed in a separate panel at the bottom of the periodic table. The names transition metals and inner transition metals are often used to refer to the elements of d-and f-blocks respectively.
- In general, electronic configuration of outer orbitals of these elements is $(n-1)\,d^{1-10}\,ns^{1-2}$. The (n–1) stands for the inner d-orbitals which may have one to ten electrons and the outermost 'ns' orbitals may have one or two electrons.

Electronic configuration of outer orbitals of first series of transition elements

Element	Sc	Ti	V	Cr	Mn	Fe	Co	Ni	Cu	Zn
Atomic number	21	22	23	24	25	26	27	28	29	30
Electronic Configuration										
M	$3d^14s^2$	$3d^24s^2$	$3d^34s^2$	$3d^54s^1$	$3d^54s^2$	$3d^64s^2$	$3d^74s^2$	$3d^84s^2$	$3d^{10}4s^1$	$3d^{10}4s^2$
M^+	$3d^14s^1$	$3d^24s^1$	$3d^34s^1$	$3d^5$	$3d^54s^1$	$3d^64s^1$	$3d^74s^1$	$3d^84s^1$	$3d^{10}$	$3d^{10}4s^1$
M^{2+}	$3d^1$	$3d^2$	$3d^3$	$3d^4$	$3d^5$	$3d^6$	$3d^7$	$3d^8$	$3d^9$	$3d^{10}$
M^{3+}	[Ar]	$3d^1$	$3d^2$	$3d^3$	$3d^4$	$3d^5$	$3d^6$	$3d^7$	–	–

General properties of d-block elements

- The transition elements exhibit typical characteristic properties. This is due to their small atomic sizes, large nuclear charges and the presence of unpaired d - electrons.
- Nearly all the transition elements display typical metallic properties such as high tensile strength, ductility, malleability, high thermal and electrical conductivity and metallic lustre. With the exceptions of Zn, Cd, Hg and Mn, they have one or more typical metallic structures at normal temperatures.

Atomic radii

- **Atomic radii** of these elements first decreases and then increases, with the increase in atomic number.
- Due to lanthanoid contraction, atomic radii increase down the group but radii of third series are virtually the same as those of the corresponding members of second series.

Metallic/ionic radii/pm	M	164	147	135	129	137	126	125	125	128	137
	M^{2+}	-	-	79	82	82	77	74	70	73	75
	M^{3+}	73	67	64	62	65	65	61	60	-	-

Ionisation enthalpy

- **Ionisation energy** increases with increase in nuclear charge along each series. The first ionisation enthalpies of Zn, Cd, Hg are very high because of fully filled configurations.

$\Delta_i H^\ominus$	I	631	656	650	653	717	762	758	736	745	906
$\Delta_i H^\ominus$	II	1235	1309	1414	1592	1509	1561	1644	1752	1958	1734
$\Delta_i H^\ominus$	III	2393	2657	2833	2990	3260	2962	3243	3402	3556	3837

- Ionization enthalpy of 5d – series are much higher than 3d & 4d – compounds. Weak shielding of nucleus by 4f electrons in 5d series by which greater effective nuclear charge act on valence electron.

Oxidation state

- Transition elements have different O.S., this is due to the participation of inner (n – 1) d-electrons in addition to outer ns-electrons. The energies of the ns and (n – 1) d-sub-shells are nearly same.
- The compound in +2 and +3 oxidation state are mostly ionic in d-block elements.

Sc	**Ti**	**V**	**Cr**	**Mn**	**Fe**	**Co**	**Ni**	**Cu**	**Zn**
	+2	+2	+2	**+2**	**+2**	**+2**	**+2**	+1	**+2**
+3	+3	+3	**+3**	+3	+3	+3	+3	**+2**	
	+4	+4	+4	+4	+4	+4	+4		
		+5	+5	+5					
			+6	+6	+6				
				+7					

Electrode potential

- Mn^{+3} and Co^{+3} are strong oxidising agents due to high Standard Reduction Potential of M^{+3}/M^{+2} values.
- Due to its low hydration enthalpy, Cu has positive E^0 value.
- Ni^{2+} has exceptional E° value high negative enthalpy of hydration.
- All the elements of 3d – series are good reducing agent except copper. All elements of 3d – series has negative $E°(M^{2+}/M)$ but copper has high positive electrode potential.

Magnetic properties

When a magnetic field is applied to substances, mainly two types of magnetic behaviour are observed: diamagnetism and paramagnetism.

Diamagnetic substances are repelled by the applied field while the paramagnetic substances are attracted. Substances which are attracted very strongly are said to be ferromagnetic. In fact, ferromagnetism is an extreme form of paramagnetism.

Many of the transition metal ions are paramagnetic Paramagnetism arises from the presence of unpaired electrons, each such electron having a magnetic moment associated with its spin angular momentum and orbital angular momentum. For the compounds of the first series of transition metals, the contribution of the orbital angular momentum is effectively quenched and hence is of no significance. For these, the magnetic moment is determined by the number of unpaired electrons and is calculated by using the 'spin-only' formula, i.e.,

$$\mu = \sqrt{n(n+2)}$$

where n is the number of unpaired electrons and μ is the magnetic moment in units of Bohr magneton (BM). A single unpaired electron has a magnetic moment of 1.73 Bohr magnetons (BM).

Colours of transition metal ions

- The colour of the transition metal ion depends on the number of electrons undergoing the d-d transition and the energy difference between these d-orbitals
- Cu^+, Ag^+, Au^+ are colourless. They have no vacant d – orbital for promotion of electrons from ground state to excited state.

Catalytic properties of transition metals

- Transition metals and their compounds act as good catalysts. This is due to the presence of free valencies and also variable oxidation states.

Formation of interstitial compounds

- Transition metals also have a tendency to form interstitial compounds with H, C, B, or N-atoms.

Some Important d-block Metal Compounds

Potassium dichromate ($K_2Cr_2O_7$)

Preparation

$4\,FeCr_2O_4 + 8Na_2CO_3 + 7O_2 \rightarrow 8\,Na_2CrO_4 + 2Fe_2O_3 + 8CO_2$

$2Na_2CrO_4 + H_2SO_4 \rightarrow Na_2Cr_2O_7 + Na_2SO_4 + H_2O$

$Na_2Cr_2O_7 + 2KCl \rightarrow K_2Cr_2O_7 + 2NaCl$

Properties

- In $Cr_2O_7^{2-}$, two CrO_3 units are bonded to sp^3 hybridised oxygen atom.
- Cr atom is surrounded by four oxygen atoms tetrahedrally.
- $K_2Cr_2O_7$ can be used as a primary standard in volumetric analysis.
- Potassium dichromate are orange crystals and are moderately soluble in cold water but readily soluble in hot water.
- Cr(VI) in the form of dichromate in acidic medium is a strong oxidising agent where MoO_3 and WO_3 are not. In d-block elements higher, oxidation states are favoured by heavier members.

Chemical properties

- On heating $K_2Cr_2O_7$ with alkalies, it is converted to chromate, i.e., the colour changes from orange to yellow. On acidifying, $Cr_2O_4^{2-}$ yellow colour again changes to orange.

  $K_2Cr_2O_7 + 2KOH \rightarrow 2K_2CrO_4 + H_2O$

  Orange Yellow

  $2CrO_4^{2-} + 2H^+ \rightarrow Cr_2O_7^{2-} + H_2O$

  Yellow Orange

  The structures of CrO_4^{2-} and $Cr_2O_7^{2-}$ are-

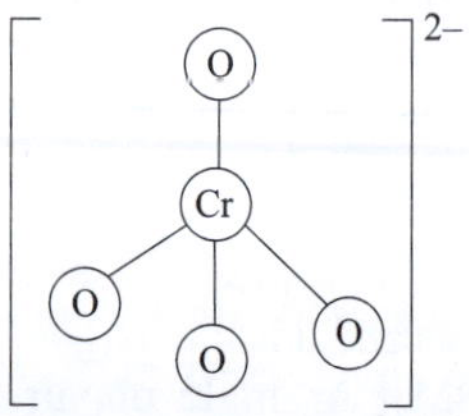

Chromate ion

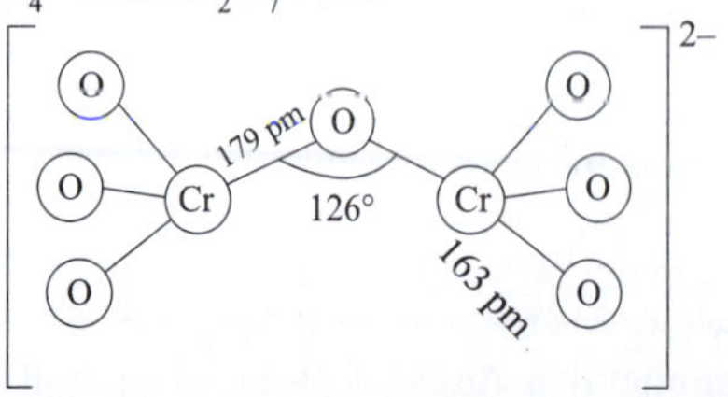

Dichromate ion

Chromyl chloride test: A metal chloride on heating with solid $K_2Cr_2O_7$ and conc.H_2SO_4 gives red vapours of chromyl chloride (CrO_2Cl_2).

$K_2Cr_2O_7 + 4KCl + 6H_2SO_4 \rightarrow 2CrO_2Cl_2 + 6KHSO_4 + 3H_2O$

(red vapours)

- Dichromate ions forms deep blue colour with H_2O_2 due to the formation of peroxo compound [CrO_5] which is soluble in ether.

Potassium permanganate ($KMnO_4$)

Preparation

$2\,MnO_2 + 4KOH + O_2 \rightarrow 2K_2MnO_4 + 2H_2O$

Disproportionation reaction takes place as-

$3MnO_4^{2-} + 4H^+ \rightarrow 2MnO_4^- + MnO_2 + 2H_2O$

Commercially it is prepared by the alkaline oxidative fusion of MnO_2 followed by the electrolytic oxidation of manganate (VI).

$MnO_2 \xrightarrow[\text{KOH}]{\text{Fused with}} MnO_4^{2-}$ (manganate ion)

$MnO_4^{2-} \xrightarrow[\text{alkaline solution}]{\text{electrolytic oxidation in}} MnO_4^-$

- It is also shown as-

  At anode: $MnO_4^{2-} \rightarrow MnO_4^- + e^-$

  Green (manganate ion) → purple (per manganate)

  At cathode : $2H^+ + e^- \rightarrow H_2$

  Structures of both manganate and permanganate ions are given as:

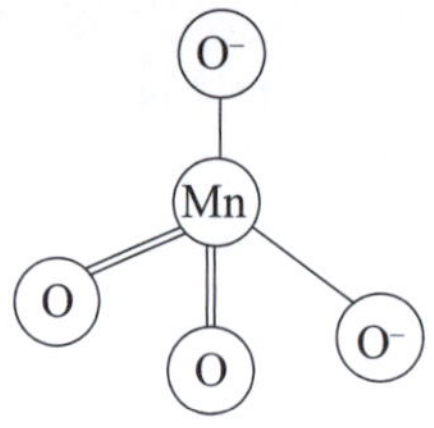

Tetrahedral manganate ion (green)

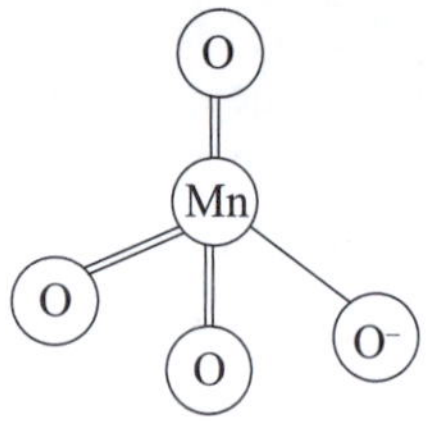

Tetrahedral permanganate ion (purple)

Properties

Potassium permanganate is sparingly soluble in water and dark purple (almost black) crystals.

- The manganate and permanganate ions are tetrahedral; the green manganate is paramagnetic with one unpaired electron but the permanganate is diamagnetic.
- The π-bonding takes place by overlap of p orbitals of oxygen with d orbitals of manganese.

The Inner Transition Elements (f-Block)

- **f - block:** It is also called inner - transition elements
- **General electronic configuration:** $(n–2)\,f^{1-14}$, $(n–1)\,d^{0-10}$, ns^2
- They have 2 series.

  4f – elements = lanthanoids, 5f elements = actinoids.

  Lanthoids

  General electronic configuration of Lanthanoides is-

  $4f^{1-14}$, $5d^{0-1}$, $6s^2$

Atomic size

- The radii of members of 5d-series are similar to those of corresponding members of 4d-series due to lanthanoid contraction. eg: Zr and Hf have almost same radii and due to same radii, they have similar physical and chemical properties. That's why they are also known as chemical twins of the periodic table.
- There is fairly regular decrease in the sizes with increasing atomic number.

Ionisation energy

- Lanthanoids have fairly low ionisation energies.
- The IE_1 & IE_2 values are quite comparable to those of alkaline earth metals particularly calcium. (IE_1 600 kJ/mole, IE_2 1200 kJ/mole).
- La, Gd, Lu have low IE_3 values due to empty, half filled and completely filled f orbitals respectively

Magnetic behaviour

- Lanthanide ions (M^{3+}) generally show paramagnetism due to the unpaired electrons in f-orbitals.

Color

- The colour is attributed to f-f transitions since they have partly filled f-orbitals. (Absorption bands are narrow probably because of the excitation within f-level)

Oxidation state

- The common oxidation state exhibited by Lanthanoids is +3. This oxidation state in Lanthanum, Gadolinium and Lutetium is stable because +3 ions of these elements have an empty (f^0), a half filled (f^7) and completely filled (f^{14}) configurations.
- The formation of Ce(IV) is favoured by noble gas configuration
- Ce(IV) is a good analytical reagent. The reaction rate is very slow.
- $La(OH)_3$ is most basic while $Lu(OH)_3$ is least basic. As the size of lanthanoids decreases covalent character of hydroxide increase & basic character decreases.

Actinoids

- General outer electronic configuration is $5f^{1-14}$, $6d^{0-1}$, $7s^2$.
- **Ionic Sizes:** The general trend in lanthanoids is observable in the actinoids as well. There is a gradual decrease in the size of atoms or M^{3+} ions across the series. This may be referred to as the **actinoids contraction** (like lanthanoid contraction). The contraction is, however, greater from elements to element in this series resulting from poor shielding by 5f electrons.

Oxidation states

- Actinoids exhibit a larger number of oxidation state because of the comparable energies of 5f, 6d and 7s energy levels.
- The actinoids show in general +3 oxidation state. The elements, in the first half of the series frequently exhibit higher oxidation states.

Comparison between lanthanoids and actinoids

Lathanoids	Actinoids
▪ Lanthanum - 57 though a d - block element is also included in the lanthanoid series because it closely resembles Lanthanoids	▪ Actinoid series also includes Actinium 89.
▪ Due to poor shielding of one 4f-electrons by another, they show lanthanoid contraction	▪ Due to poor shielding effect of the 5f-electrons, they show actinoid contraction.
▪ Trivalent lanthanoid ions (La^{3+}) are coloured both in solid and in aqueous solutions.	▪ Actinoids are generally radioactive and shows in general +3 oxidation state.
▪ The compounds of lanthanoids are less basic.	▪ Actinoids compounds are radioactive in nature.

MCQs ?

NCERT Topic-wise MCQs

Electronic Configurations & Characteristics of d-block Elements

1. Which of the following statements is correct?
   1) Iron belongs to third transition series of the periodic table
   2) Iron belongs to f-block of the periodic table
   3) Iron belongs to first transition series
   4) Iron belongs to s-block of the periodic table

2. Which one of the following transition metals ions is colourless in aqueous solution?
   1) Ti^{4+} 2) V^{4+}
   3) Mn^{2+} 4) Fe^{3+}

3. Which one of the following is a d-block element?
   1) Gd 2) Zn
   3) Es 4) Cs

4. Which of the following has maximum unpaired d-electrons?
   1) Zn^{2+} 2) Fe^{2+}
   3) Ni^{2+} 4) Cu^{+}

5. The spin only magnetic moment of Fe^{3+} ion (in BM) is approximately:
   1) 4 2) 7
   3) 5 4) 6

6. The highest oxidation state exhibited by transition metals is:
   1) +7 2) +8
   3) +6 4) +5

7. The magnetic moment μ, of transition metals is related to the number of unpaired electron, n as:
   1) $\mu = n(n+2)$ 2) $\mu = n^2(n+2)$
   3) $\mu = \frac{n}{n+2}$ 4) $\mu = \sqrt{n(n+2)}$

8. Transition metal show paramagnetic behaviour. This is because of their:
   1) High lattice energy
   2) Variable oxidation state
   3) Characteristic configuration
   4) Unpaired electrons

9. Which of the following pairs of ions have the same electronic configuration?
   1) Cu^{2+} , Cr^{2+} 2) Fe^{3+} , Mn^{2+}
   3) Co^{3+} , Ni^{3+} 4) Sc^{3+} , Cr^{3+}

10. The magnetic moment of a divalent ion in aqueous solution with atomic number 25 is:
    1) 5.9 BM 2) 2.9 BM
    3) 6.9 BM 4) 9.9 BM

11. The correct order of number of unpaired electron is:
    1) $Cu^{2+} > Ni^{2+} > Cr^{3+} > Fe^{3+}$
    2) $Ni^{2+} > Cu^{2+} > Fe^{3+} > Cr^{3+}$
    3) $Fe^{3+} > Cr^{3+} > Ni^{2+} > Cu^{2+}$
    4) $Cr^{3+} > Fe^{3+} > Ni^{2+} > Cu^{2+}$

12. In which one of the following pairs, metals have same number of unpaired electrons?
    1) Cr, Cu 2) Mn, Zn
    3) Sc, Zn 4) Sc, Cu

13. The variable oxidation state are exhibited by the elements of 3d-series because:
    1) They contain unpaired electrons in 3d
    2) 4s-electrons are readily involved in bonding
    3) Energy difference between 3d and 4s orbitals is small
    4) 3d- electrons only are involved in bonding

14. Only one oxidation state is possible with:
    1) Cu 2) Zn
    3) Fe 4) Mn

15. The oxidation state of +1 is possible with:
    1) Cu and Cr 2) Cu and Zn
    3) Cr and Zn 4) Mn and Zn

16. In which one of the following, the first is more stable than the second?
    1) Ti^{3+}, Ti^{4+} 2) Mn^{3+}, Mn^{2+}
    3) Na, Na^{+} 4) Fe^{3+}, Fe^{2+}

17. The magnetic moment of Ni^{2+} ion (atomic number of Ni = 28) in BM unit is:
    1) 1.73 2) 4.81
    3) 5.96 4) 2.84

18. Which of the following transition metal is not coloured?
    1) Cu^{+} 2) V^{3+}
    3) Co^{2+} 4) Ni^{2+}

19. The ion of least magnetic moment among the following is:
    1) Ti^{3+} 2) Ni^{2+}
    3) Co^{2+} 4) Mn^{3+}

20. Which of the following pairs of elements cannot form an alloy?
    1) Zn , Cu 2) Fe, Hg
    3) Fe, C 4) Hg, Na

21. Essential component of amalgam is:

1) Fe 2) Pb 3) Hg 4) Cr

22. For which of the following pairs magnetic moment is same?

1) $MnCl_2$, $CuSO_4$ 2) $CuCl_2$, $TiCl_3$ 3) TiO_2 , $CuSO_4$ 4) $TiCl_3$, $NiCl_2$

23. The stability of ferric ion is due to:

1) Half-filled f-orbitals
2) Half-filled d-orbitals
3) Completely filled f-orbitals
4) Completely filled d-orbitals

24. Which of the following compounds is not coloured?

1) $Na_2[CuCl_4]$ 2) $Na_2[CdCl_4]$ 3) $K_4[Fe(CN)_6]$ 4) $K_3[Fe(CN)_6]$

25. Electronic configuration of a transition element X in +3 oxidation state is [Ar] $3d^5$. What is its atomic number?

1) 25 2) 26 3) 27 4) 24

26. Which of the following is not an amphoteric ion?

1) Al^{3+} 2) Cr^{3+} 3) Fe^{3+} 4) Zn^{2+}

27. Which of the following d-block elements have same configuration of (n-1) d subshell?

1) Cu , Zn 2) Cr, Cu 3) Cr, Zn 4) Co, Ni

28. The correct order of decreasing second ionization enthalpy of Ti (22), V(23), Cr(24), and Mn (25) is:

1) Ti > V > Cr > Mn 2) Cr > Mn > V > Ti 3) V > Mn > Cr > Ti 4) Mn > Cr > Ti > V

29. The maximum number of unpaired electrons are present in:

1) Fe 2) Fe^{2+} 3) Fe^{4+} 4) Fe^{3+}

30. Among the following, the most stable ion is:

1) Fe^{2+} 2) Sc^{4+} 3) Mn^{3+} 4) Mn^{2+}

31. Which one of the following compounds is expected to be coloured?

1) $TiCl_3$ 2) Cu_2Cl_2 3) $ZnSO_4$ 4) $ScCl_3$

32. In the reaction, $Cu \rightarrow Cu^+ + e^-$, the electron is lost from the subshell:

1) 3d 2) 4s 3) 3p 4) Either 3d or 4s

33. The ion, Cu^+ is isoelectronic with:

1) Cu^{2+} 2) Mn^{2+} 3) Zn^{2+} 4) Sc^{3+}

34. Which one of the following species will impart colour in its aqueous solution?

1) Cr^{3+} 2) Cu^+ 3) Zn^{2+} 4) Ti^{4+}

35. In which one of the following pairs, both elements exhibit same highest oxidation state?

1) Ti, Ni 2) Ti, Cu 3) Cr, Cu 4) Sc, Zn

36. Which one of the following ions is coloured in its aqueous solution?

1) V^{3+} 2) Zn^{2+} 3) Cu^+ 4) Ti^{4+}

37. Which one of the following has the highest number of unpaired electrons?

1) Co^{2+} 2) Ni^{2+} 3) Fe^{2+} 4) Mn^{2+}

38. Which one of the following configurations represents the ion which would be colourless?

1) $3d^4, 4s^0$ 2) $3d^3, 4s^0$ 3) $3d^5, 4s^0$ 4) $3d^{10}, 4s^0$

39. The elements of 3d-series will give coloured compounds of its ions containing:

1) Vacant 3d-orbitals
2) Unpaired electrons in 3d
3) Unpaired electrons in 4s
4) Both (1) and (2)

40. In which one of the following pairs of ions, both are colorless?

1) Sc^{3+}, Ti^{3+} 2) Cu^+, Fe^{2+} 3) Ti^{4+}, Cu^+ 4) V^{3+}, Zn^{2+}

41. Among the following, the lightest transition element is:

1) Fe 2) Os 3) Sc 4) Co

42. The d-block elements include:

1) Metals and non-metals
2) Only non-metals
3) Only metals
4) Metals, non-metals metalloids

43. The actual configuration of Cr is:

1) [Ar] $3d^5 4s^1$ 2) [Ar] $3d^4 4s^2$ 3) [Kr] $3d^5 4s^1$ 4) [Kr] $3d^4 4s^2$

44. Transition metal with low oxidation number will act as

1) A base 2) An acid 3) Both (1) and (2) 4) None of these

45. Which of the following pair will have effective magnetic moment equal?

1) Cr^{+3} and Mn^{+2} 2) Cr^{+2} and Fe^{+2} 3) V^{+2} and Sc^{+3} 4) Ti^{+2} and V^{+2}

46. The number of d-electrons in Fe^{2+} (at no. of Fe = 26) is not equal to that of the
1) p – electrons in Ne (at. no. = 10)
2) s – electrons in Mg (at. no. = 12)
3) d – electrons in Fe
4) p – electrons in Cl^- (at. no. of Cl = 17)

47. The atomic number of vanadium (V), chromium (Cr), manganese (Mn) and iron (Fe) are respectively 23, 24, 25 and 26 respectively. Which one of these may be expected to have the highest second ionization enthalpy
1) V 2) Cr
3) Mn 4) Fe

48. The correct order of magnetic moments (spin only values in B.M.) among is
1) $[Fe(CN)_6]^{4-} > [MnCl_4]^{2-} > [CoCl_4]^{2-}$
2) $[MnCl_4]^{2-} > [Fe(CN)_6]^{4-} > [CoCl_4]^{2-}$
3) $[MnCl_4]^{2-} > [CoCl_4]^{2-} > [Fe(CN)_6]^{4-}$
4) $[Fe(CN)_6]^{4-} > [CoCl_4]^{2-} > [MnCl_4]^{2-}$

49. What is the magnetic moment of $[FeF_6]^{3-}$?
1) 5.92 2) 5.49
3) 2.32 4) 4

50. Which of the following has diamagnetic character?
1) $[NiCl_4]^{2-}$ 2) $[CoF_6]^{3-}$
3) $[Fe(H_2O_6)]^{2+}$ 4) $[Ni(CN)_4]^{2-}$

51. Which metal is used to make alloy steel for armour plates, safes and helmets?
1) Al 2) Mn
2) Cr 4) Pb

52. Solder is an alloy of
1) 70% lead, 30% tin 2) 33% lead, 67% tin
3) 80% lead, 20% tin 4) 90% Cu, 10% tin

53. Which of the following is the stable electronic configuration of Fe^{+3} ion?
1) $3d^6 4s^0$ 2) $3d^5 4s^0$
3) $3d^6 4s^2$ 4) $3d^4 4s^2$

54. Which of the following elements does not belong to the first transition series (3d series)?
1) Fe 2) V
3) Pd 4) Cu

55. In the first transition series, the highest boiling and melting point is of
1) Cr 2) V
3) Ni 4) Fe

56. The correct order of the following elements with respect to their density is:
1) Cr < Zn < Co < Cu < Fe
2) Cu < Fe < Co < Cr < Zn
3) Zn < Cr < Fe < Co < Cu
4) Cr < Fe < Co < Cu < Zn

57. Electronic configurations of Cr and Cu respectively are
1) $[Ar]\ 3d^4 4s^2$, $[Ar]\ 3d^9 4s^2$
2) $[Ar]\ 3d^5 4s^1$, $[Ar]\ 3d^{10} 4s^1$
3) $[Ar]\ 3d^2 4s^2$, $[Ar]\ 3d^8 4s^2$
4) $[Ar]\ 3d^6 4s^2$, $[Ar]\ 3d^4 4s^1$

58. Select the correct statement :
1) When d-block elements form ions, (n – 1) d electrons are lost before ns electrons.
2) When d-block elements form ions, ns electrons are lost before (n – 1) d electrons.
3) When d-block elements form ions, (n – 1) f electrons are lost before ns electrons
4) None of these

59. The number of oxidation states of manganese is
1) 3 2) 6
3) 2 4) 8

60. The number of unpaired electrons in cobalt atom is (atomic number of Co = 27)
1) 2 2) 3
3) 4 4) 1

61. If 22 is the atomic number of an element, then the highest oxidation state exhibited by it in its compounds will be :
1) 1 2) 2
3) 3 4) 4

62. What would be the charge on Fe^{x+} if the magnetic moment is $\sqrt{24}$?
1) + 2 2) + 3
3) Zero 4) None of these

63. The ability of d- block elements to form complexes is due to -
1) Small and highly charged ions
2) Vacant low energy orbital to accept lone pair of electrons from ligands
3) Both (1) & (2) are correct
4) None is correct

64. Mn show oxidation state from + 2 to + 7. The most oxidizing state known in aqueous solution is
1) + 7 2) + 1
3) + 3 4) + 4

65. The number of unpaired electrons present in Cr^+ would be
1) 3 2) 4
3) 5 4) 6

66. The highest oxidation state exhibited by Cr would be
1) 2 2) 3
3) 4 4) 6

67. Which of the following is not true for transition elements?
1) They are all metals
2) They show variable valency
3) They form coloured ions
4) They do not form co-ordinate compounds

68. Of the ions Zn^{2+}, Ni^{2+} and Cr^{3+} [atomic number of Zn = 30, Ni = 28, Cr = 24]

1) Only Zn^{2+}, is colourless and Ni^{2+} and Cr^{3+} are coloured
2) All three are colourless
3) All three are coloured
4) Only Ni^{2+} is coloured and Zn^{2+} and Cr^{3+} are colourless

69. Zinc and mercury do not show variable valency like d-block elements because

1) They are soft
2) Their d- shells are complete
3) They have only two electrons in the outermost sub shell
4) Their d-shells are incomplete

70. On applying magnetic field, the paramagnetic substances get

1) Repelled 2) Attracted
3) Both (1) and (2) 4) None of these

71. Ti^{3+} transition metal ion is purple in colour while Ti^{4+} metal ion is colourless due to

1) No unpaired electrons in d-orbitals
2) Unpaired electrons in d-orbitals
3) Completely filled d orbitals
4) None of the above

72. Which of the following statements is not true?

1) Colourless compounds of transition elements are paramagnetic.
2) Coloured compounds of transition elements are paramagnetic.
3) Colourless compounds of transition elements are diamagnetic.
4) Transition elements form the complex compounds.

Compounds of Transition Elements

73. Which of the following reactions is not correct?

1) $MnO_4^- + 8H^+ + 5Fe^{2+} \rightarrow 5\,Fe^{3+} + Mn^{2+} + 4H_2O$
2) $2MnO_2 + 4KOH + O_2 \rightarrow 4KMnO_4 + 2H_2O$
3) $2Na_2CrO_4 + 2H^+ \rightarrow Na_2Cr_2O_7 + 2Na^+ + H_2O$
4) $K_2Cr_2O_7 + 7H_2SO_4 + 6KI \rightarrow 4K_2SO_4 + Cr_2(SO_4)_3 + 3I_2 + 7H_2O$

74. The basic character of the transition metal monoxides follows the order:

1) VO > CrO > TiO > FeO
2) CrO > VO > FeO > TiO
3) TiO > FeO > VO > CrO
4) TiO > VO > CrO > FeO

75. One mole of acidified $K_2Cr_2O_7$ on reaction with excess KI will liberate .... mole (s) of I_2:

1) 6 2) 3
3) 7 4) 2

76. Potassium permanganate acts as an oxidant in neutral, alkaline as well as acidic media. The final products obtained from it in the three conditions are, respectively

1) MnO_2, MnO_2, Mn^{2+} 2) MnO_4^{2-}, Mn^{3+}, Mn^{2+}
3) MnO_2, MnO_4^{2-} 4) MnO, MnO_4, Mn^{2+}

77. The oxidation state of chromium is + 6 in:

1) Potassium chromate 2) Potassium dichromate
3) Both (1) & (2) 4) None of these

78. The oxidation state of Mn in potassium manganate is:

1) +7 2) +6
3) +4 4) +5

79. Identify the correct structure of dichromate ion:

1) 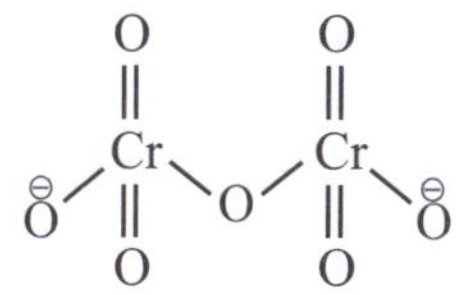
2) 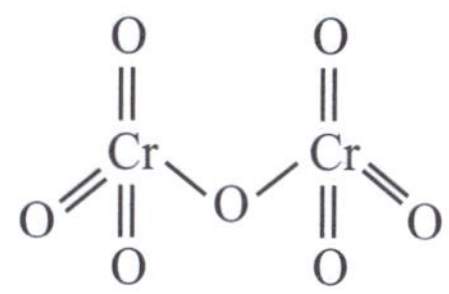
3) 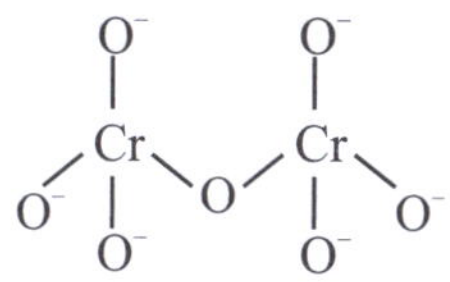
4) 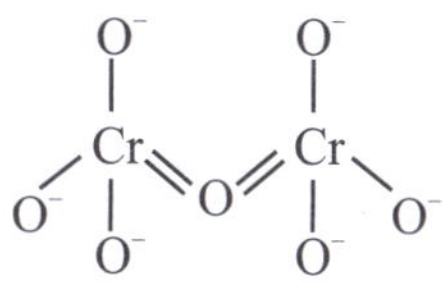

80. In which of the following ionic radii of chromium would be smallest

1) K_2CrO_4 2) CrO_2
3) $CrCl_3$ 4) CrF_2

81. $KMnO_4$ in basic medium is reduced to

1) K_2MnO_4 2) MnO_2
3) $Mn(OH)_2$ 4) Mn^{2+}

82. A group of acidic oxide is

1) CrO_3, Mn_2O_7 2) ZnO, Al_2O_3
3) CaO, ZnO 4) Na_2O, Al_2O_3

83. The yellow colored solution of Na_2CrO_4 changes to orange red on passing CO_2 gas due to the formation of:

1) CrO_5 2) CrO_3
3) $Na_2Cr_2O_7$ 4) Na_3CrO_8

84. $CrO_4^{2-} \underset{pH=Y}{\overset{pH=X}{\rightleftharpoons}} Cr_2O_7^{2-}$

The value of X and Y respectively are

1) 4, 8 2) 8, 4
3) 8, 10 4) 10, 2

85. What is the oxidation number of Mn in $KMnO_4$?

1) 2 2) 6
3) 4 4) 7

86. Which of the following statements about $KMnO_4$ is incorrect ?

1) $KMnO_4$ forms green crystals which are isostructural with those of $KClO_4$
2) The salt $KMnO_4$ is not very soluble in water
3) When $KMnO_4$ is heated, it decomposes at 513 K
4) All of these

87. In acidic solution, oxidising action of potassium dichromate can be represented as

1) $Cr_2O_7^{2-} + 3H^+ + Mn^{2+} \longrightarrow 2Cr^{2+} + NH_3$

2) $Cr_2O_7^{2-} + 14H^+ + 6e^- \longrightarrow 2Cr^{3+} + 7H_2O$

3) $Cr_2O_7^{2-} + 18H^+ + 6Fe^{2+} \longrightarrow 4Cr^{3+} + Cr_2O_5$

4) $CrO_4^{2-} + 14H^+ + 5Cl_2 \longrightarrow 2Cr^{3+} + Mn^{2+} + 7H_2O$

88. Select the correct statement about $K_2Cr_2O_7$:

1) $K_2Cr_2O_7$ is used as a primary standard in volumetric analysis
2) $K_2Cr_2O_7$ is used as a secondary standard in volumetric analysis
3) $K_2Cr_2O_7$ are green coloured crystals.
4) None of these

89. The manganate ion is

1) Paramagnetic
2) Diamagnetic
3) Neither paramagnetic nor diamagnetic
4) None of these

Lanthanoids and Actinoids

90. The common oxidation state shown by lanthanides in their compounds is:

1) +1 2) +3
3) +5 4) +16

91. Zr and Hf have almost equal atomic and ionic radii because of:

1) Diagonal relationship 2) Lanthanoid contraction
3) Actinoid contraction 4) Belong to the same group

92. The correct configuration of f-block elements is:

1) $(n-2)\ f^{1-14}\ (n-1)\ d^{0-1}\ ns^2$
2) $(n-1)\ f^{1-14}\ (n-1)\ d^{0-1}\ ns^2$
3) $(n-3)\ f^{1-14}\ (n-2)\ d^{0-1}\ (n-1)\ s^2$
4) $(n-2)\ f^{0-1}\ (n-1)\ d^{0-1}\ ns^2$

93. In lanthanoids, the last electron enters (n–2)f subshell where n is equal to:

1) 4 2) 5
3) 6 4) 7

94. The number of unpaired electrons present, as per accepted electronic configuration in Yb (70) is / are:

1) 0 2) 1
3) 2 4) 7

95. The total number of unpaired electrons present in an atom of Gd (64), are:

1) 7 2) 8
3) 1 4) 10

96. Gd (64) gives an oxidation state of only +3 because:

1) It contains 3 unpaired electrons
2) In +3 oxidation state, it has unstable configuration
3) In +3 oxidation state, it has stable configuration
4) There are 3 electrons present in 4f in +3 oxidation state

97. Which of the following is the non-lanthanide element?

1) La 2) Lu
3) Pr 4) Pm

98. The incorrect statement among the following is:

1) d-block elements show irregular and erratic chemical properties
2) La and Cu have partially filled d-orbitals and no other partially filled orbitals
3) The chemistry of various lanthanoids is very similar
4) 4f and 5f orbitals are equally shielded

99. The atomic number of 4f-series range from:

1) 57 to 71 2) 58 to 71
3) 58 to 72 4) 57 to 72

100. Which of the following ion is paramagnetic?

1) La^{3+} (Z = 57) 2) Lu^{3+} (Z = 71)
3) Yb^{2+} (Z = 70) 4) Sm^{3+} (Z = 62)

101. What is the total number of inner transition elements in the periodic table?

1) 10 2) 14
3) 30 4) 28

102. The electronic configuration of cerium is

1) $[Xe]4f^0 5d^1 6s^2$ 2) $[Xe]4f^1 5d^1 6s^2$
3) $[Xe]4f^2 5d^0 6s^2$ 4) Both (2) and (3)

103. Which of the following is the most stable oxidation state of actinoids?

1) +7 2) +3
3) +4 4) +5

104. Lanthanum is grouped with f-block elements because

1) It has partially filled f -orbitals
2) It is just before Ce in the periodic table
3) It has both partially filled f and d-orbitals
4) The properties of Lanthanum are very similar to the elements of 4f block

105. Which of the following is not an actinide?

1) Curium 2) Californium
3) Uranium 4) Terbium

106. Cerium (Z = 58) is an important member of the lanthanoids. Which of the following statements about cerium is incorrect

1) The +4 oxidation state of cerium is not known in solutions
2) The +3 oxidation state of cerium is more stable than the +4 oxidation state
3) The common oxidation states of cerium are +3 and +4
4) Cerium (IV) acts as an oxidizing agent

107. The electronic configuration of Thorium is
1) [Rn] $6d^5\ 7s^2$
2) [Rn] $6d^2\ 7s^2$
3) [Rn] $5f^{10}\ 7s^2$
4) [Rn] $5f^6\ 7s^2$

Some Applications of d-and f-Block Elements

108. The oxidation of ethyne to ethanal is catalysed by which of the following in the Wacker process ?
1) SO_2 2) NH_3
3) V_2O_5 4) $PdCl_2$

109. Which of the following compounds is used in the pigment industry ?
1) TiO 2) MnO_2
3) V_2O_5 4) None of these

110. Which of the given compound is used in dry battery cells?
1) TiO 2) $KMnO_4$
3) V_2O_5 4) MnO_2

111. Which of the following is correct?
1) Wacker process - $PtCl_4$ catalyst
2) Ziegler catalyst - $TiCl_4 + Al(CH_3)_3$
3) Haber process - SO_2 catalyst
4) None of these

High Order Time Intensive MCQs

1. Which one of the following reactions is incorrect?
1) $FeCr_2O_4 + Na_2CO_3 + O_2 \rightarrow Na_2CrO_4 + Fe_2O_3 + CO_2$
2) $Na_2Cr_2O_7 + KCl \rightarrow K_2Cr_2O_7 + NaCl$
3) $K_2Cr_2O_7 + NaCl + H_2SO_4 \rightarrow KHSO_4 + NaHSO_4 + CrO_3 + H_2O$
4) None of these

2. Maximum oxidation state is shown by:
1) Os 2) Mn
3) Cr 4) Co

3. E° values for the couples Cr^{3+}/Cr^{2+} & Mn^{3+}/Mn^{2+} are –0.41 & + 1.51 V respectively. Considering these values, select the correct option:
1) Cr^{2+} acts as a reducing agent and Mn^{3+} acts as an oxidising agent in their aqueous solutions
2) $Cr^{2+}_{(aq)}$ is more stable than $Cr^{3+}_{(aq)}$
3) $Mn^{3+}_{(aq)}$ is more stable than $Mn^{2+}_{(aq)}$
4) None of these

4. Select the correct statement:
1) CrO_5 is stable in pyridine because it forms an adduct
2) $KMnO_4$ on being treated with concentration H_2SO_4 gives Mn_2O_7
3) Potassium dichromate is used as a primary standard in volumetric analysis.
4) All are correct

5. In the dichromate dianion:
1) 4 Cr-O bonds are equivalent
2) 6 Cr-O bonds are equivalent
3) All Cr-O bonds are equivalent
4) All Cr-O bonds are non-equivalent

6. The stability of a particular oxidation state of a metal in aqueous solutions is determined by:
1) Enthalpy of sublimation
2) Ionisation enthalpy
3) Enthalpy of hydration
4) All of these

7. Which of the following factor may be regarded as the main cause of Lanthanide contraction?
1) Poor shielding of one of the 4f-electrons by another in the sub-shell
2) Effective shielding of one of the 4f-electrons by another in the sub-shell
3) Greater shielding of 5d electron by 4f electrons
4) None of these

8. Pyrolusite is MnO_2 used to prepare $KMnO_4$. Steps are,
$MnO_2 \xrightarrow{I} MnO_4^{2-} \xrightarrow{II} MnO_4^-$.
Steps I and II are respectively :
1) Fuse with KOH / air and electrolytic oxidation.
2) Fuse with KOH / KNO_3 and electrolytic oxidation.
3) Both (1) and (2).
4) Dissolve in H_2O and oxidation.

9. Which of the following metals react with chlorine to form their respective chlorides?
1) Cu 2) Ag
3) Au 4) All

10. Lanthanoid contraction occurs because
1) The 4f electrons, which are gradually added, create a strong shielding effect
2) The 4f orbitals are greater in size than the 3d and 3f orbitals
3) The 5f orbitals strongly penetrate into the 4f orbitals
4) The poor shielding effect of 4f electrons is coupled with increased attraction between the nucleus and the added electrons

11. The element with the electronic configuration $[Xe]4f^{14}5d^{1}6s^{2}$ is a
1) Representative element
2) Transition element
3) Actinide element
4) Lanthanide element

12. Which oxide of Mn is acidic in nature?
1) MnO 2) Mn_2O_7
3) Mn_2O_3 4) MnO_2

13. Which of the following ion is not diamagnetic?
1) La^{3+} (Z = 57) 2) Lu^{3+} (Z = 71)
3) Yb^{2+} (Z = 70) 4) Sm^{3+} (Z = 62)

14. Which of the following is not the configuration of lanthanoid?
1) $[Xe]\ 4f^{10}6s^{2}$ 2) $[Xe]\ 4f^{1}5d^{1}6s^{2}$
3) $[Xe]\ 4f^{14}5d^{10}6s^{1}$ 4) $[Xe]\ 4f^{14}5d^{1}6s^{2}$

15. The Lanthanoids contraction is responsible for the fact that
1) Zr and Y have about the same radius
2) Zr and Nb have similar oxidation state
3) Zr and Hf have about the same radius
4) Zr and Zn have the same oxidation state

16. Which of the following is the strongest base?
1) $Sc(OH)_3$ 2) $La(OH)_3$
3) $Lu(OH)_3$ 4) $Yb(OH)_3$

17. Which of the following set of elements does not belong to transitional elements?
1) Fe, Co, Ni 2) Cu, Ag, Au
3) Ti, Zr, Hf 4) Ga, In, Tl

18. Which of the following oxides of chromium is amphoteric in nature?
1) CrO 2) Cr_2O_3
3) CrO_3 4) CrO_5

19. The inner transition elements are the elements in which the added electrons go to
1) (n-1)d-orbitals
2) (n-2)f-orbitals
3) (n-1) d-orbitals and (n-1) f-orbitals
4) (n-1)d-orbitals and ns orbitals

20. The ground state electronic configuration of chromium is against
1) Hund's rule 2) Pauli's principle
3) Aufbau principle 4) Boyle principle

21. What factor make the separation of Lanthanides a formidable task
1) Similarity in ionic size
2) Constant charge of +3
3) Small charge radius ratio
4) All of these

22. Lanthanides are characterized by the filling of the
1) Penultimate 4f energy level
2) Antipenultimate 4f energy level
3) Penultimate 5f energy level
4) Antipenultimate 5f energy level

23. Which of the following is not an amphoteric oxide?
1) HgO 2) PbO_2
3) ZnO 4) SnO_2

24. Transition metal present in the alloy Gun metal is
1) Ni 2) Zn 3) Sn 4) Cu

25. The following is not a typical transition element
1) Cu 2) Ag 3) Au 4) Hg

26. Which of the following is not an element of first transition series?
1) Fe 2) Co 3) Ni 4) Ag

27. When $KMnO_4$ reacts with acidified $FeSO_4$
1) Only $FeSO_4$ is oxidised
2) Only $KMnO_4$ is oxidised
3) $FeSO_4$ is oxidized and $KMnO_4$ is reduced
4) $KMnO_4$ and $FeSO_4$ oxidised

28. The group numbers of d-block elements
1) 1 to 10 2) 1 to 9
3) 3 to 11 4) 3 to 12

29. General electronic configuration of d-block elements is
1) $ns^2np^6nd^{1-10}$ 2) $(n-1)d^{1-10}\ ns^{0-2}\ np^{0-6}$
3) $(n-1)d^{1-10}\ ns^{1-2}$ 4) $nd^{1-9}ns^{0-2}$

30. Which of the following is an alloy of a metal and a non-metal?
1) Bronze 2) Electron
3) Nichrome 4) Steel

31. In which of the following elements, the electronic configuration is against Aufbau rule ?
1) Ni, Pd, Pt 2) Sc, Ti, Zr
3) Pd, Pt, Cu 4) Fe, Cr, Mn

32. An element M has the electronic configuration $[Ar]3d^{5}4s^{2}$. Which one of its oxide is unlikely to exist?
1) MO_2 2) M_2O_3 3) MO_4 4) M_2O_7

33. Which of the following elements exhibits maximum oxidation state?
1) Mn 2) Co 3) Fe 4) Zn

34. The following belongs to d-block but it is not a transition element?
1) Mn 2) Fe 3) Zn 4) Cr

35. Which of the following metal gives more than one chloride?
1) Cu 2) Al 3) Ag 4) Na

36. Which of the following elements form interstitial compounds?
1) Alkali metals 2) Transition metals
3) Halogens 4) Noble gases

37. When $KMnO_4$ acts as oxidising agent in acidic medium, the oxidation number of Mn decreases by
 1) 1 2) 2
 3) 3 4) 5

38. Chromyl chloride when dissolves in NaOH solution gives yellow solution. The yellow solution contains
 1) $Cr_2O_7^{2-}$ 2) CrO_4^{2-}
 3) CrO_5 4) Cr_2O_3

39. An inorganic compound 'y' on heating gives a green colouration & evolves O_2 gas. Then, y is-
 1) $(NH_4)_2Cr_2O_7$ 2) $K_2Cr_2O_7$
 3) CrO_2Cl_2 4) Both (1) & (3)

40. Select the correct statement(s).
 1) Cerium (Z = 58) forms a tetra positive ion in aq. solution
 2) Lanthanum, Gadolinium & Lutetium are stable in + 3 oxidation states
 3) $La(OH)_3$ is more basic than $Lu(OH)_3$
 4) All are correct

41. Which of the following arrangements does not represent the correct order of the property stated against it?
 1) $V^{2+} < Cr^{2+} < Mn^{2+} < Fe^{2+}$; Paramagnetic behaviour
 2) $Ni^{2+} < Co^{2+} < Fe^{2+} < Mn^{2+}$; ionic size
 3) $Co^{3+} < Fe^{3+} < Cr^{3+} < Sc^{3+}$, stability in aq. solution
 4) Sc < Ti < Cr < Mn ; number of oxidation states

42. Acidified potassium dichromate is treated with hydrogen sulphide. In the reaction, the oxidation number of chromium
 1) Increases from + 3 to + 6
 2) Decreases from +6 to +3
 3) Remains unchanged
 4) Decreases from +6 to +2

43. $4K_2Cr_2O_7 \xrightarrow{\text{heat}} 4K_2CrO_4 + 3O_2 + X$. In the above reaction, X is
 1) CrO_3 2) Cr_2O_7 3) Cr_2O_3 4) CrO_5

44. Arrange Ce^{+3}, La^{+3}, Pm^{+3} and Yb^{+3} in increasing order of their ionic radii
 1) $Yb^{+3} < Pm^{+3} < Ce^{+3} < La^{+3}$
 2) $Ce^{+3} < Yb^{+3} < Pm^{+3} < La^{+3}$
 3) $Yb^{+3} < Pm^{+3} < La^{+3} < Ce^{+3}$
 4) $Pm^{+3} < La^{+3} < Ce^{+3} < Yb^{+3}$

45. The colour of $K_2Cr_2O_7$ changes from red orange to lemon yellow on treatment with aqueous KOH because of
 1) The reduction of Cr^{VI} to Cr^{III}
 2) The formation of chromium hydroxide
 3) The conversion of dichromate to chromate
 4) The oxidation of potassium hydroxide to potassium peroxide

46. Stainless steel contains:
 1) Fe + Cr + Cu 2) Fe + C + Ni
 3) Fe + Cr + Ni 4) Fe + Ni + Cu

47. Stainless steel does not rust because
 1) Chromium and nickel combine with iron
 2) Chromium forms an oxide layer and protects iron from rusting
 3) Nickel present in it, does not rust
 4) Iron forms a hard chemical compound with chromium present in it

48. The elements of transition series are those which have:
 1) Incomplete outermost shell
 2) Completely filled outermost shells
 3) Incomplete inner orbitals
 4) None of these

49. Mn^{2+} can be converted into Mn^{7+} by reacting with
 1) SO_2 2) Cl_2 3) PbO_2 4) $SnCl_2$

50. Acidified $K_2Cr_2O_7$ solution turns green when Na_2SO_3 is added to it. This is due to the formation of:
 1) $CrSO_4$ 2) $Cr_2(SO_4)_3$
 3) CrO_4^{2-} 4) $Cr(SO_3)_3$

51. The electronic configuration for Neodymium is:
 [Atomic Number for Neodymium 60]
 1) $[Xe]\ 4f^4\ 6s^2$ 2) $[Xe]\ 5f^4 7s^2$
 3) $[Xe]\ 4f^6\ 6s^2$ 4) $[Xe]\ 4f^1 5d^1 6s^2$

52. The correct order of basicity of oxides of vanadium is
 1) $V_2O_3 > V_2O_4 > V_2O_5$ 2) $V_2O_3 > V_2O_5 > V_2O_4$
 3) $V_2O_5 > V_2O_4 > V_2O_3$ 4) $V_2O_4 > V_2O_3 > V_2O_5$

53. Metal oxides which decomposes on heating is
 1) ZnO 2) Al_2O_3 3) HgO 4) Na_2O

54. Which is mild oxidising agent?
 1) Ag_2O 2) $KMnO_4$
 3) $K_2Cr_2O_7$ 4) Cl_2

55. Amalgams are
 1) Highly coloured alloys
 2) Always solid
 3) Alloys which contain mercury as one of the contents
 4) Alloys which have great resistance to abrasion

56. Pure conc. HNO_3 makes iron passive as the surface is covered with protective layer of
 1) Fe_2O_3 2) FeO
 3) Fe_3O_4 4) $Fe(NO_3)_3$

57. A nitrate when mixed with common salt gives a white precipitate which is soluble in dilute NH_4OH. It is the nitrate of
 1) Copper 2) Mercury
 3) Silver 4) Gold

58. Zinc when reacted with excess of NaOH gives
 1) Zinc hydroxide 2) Zinc oxide
 3) Di sodium zincate 4) Sodium zincate

59. Reaction of zinc with cold and very dilute nitric acid yields
 1) $Zn(NO_3)_2 + N_2O$ 2) $Zn(NO_3) + NO$
 3) $Zn(NO_3)_2 + NH_4NO_3$ 4) $Zn(NO_3) + NO_2$

60. The pair that is referred as 'chemical twins' is:
 1) Ac, Cf 2) Hf, Ta
 3) Tc, Re 4) La,Ac

61. Across the lanthanide series, the basicity of the lanthanide hydroxides:
1) Increases
2) Decreases
3) First increases and then decreases
4) First decreases and then increases

62. The 4f-electrons do not affect the properties of lanthanoids because they:
1) Are inactive in nature
2) Are all paired
3) lie well in the interior of the atom
4) lie well in the interior of the atom and hence not exposed of surroundings

63. Which of the following statements is/are correct?

A. The number of unpaired electrons in Sc^{3+} is more than that in La^{2+}

B. Ferromagnetism is not considered as an extreme form of paramagnetism.

C. Atomic size of Ce is more than that of Lu.

D. Ce in +3 oxidation state is more stable than in +4.

1) Both (A) and (B)
2) Both (C) and (D)
3) All the statements are correct
4) (A), (B) and (C) only

Match the Columns MCQs

1. Match the catalysts given in Column-I with the processes given in Column-II.

	Column-I		Column-II
(a)	Ni in the presence of hydrogen	(i)	Zieglar Natta catalyst
(b)	Cu_2Cl_2	(ii)	Contact process
(c)	V_2O_5	(iii)	Vegetable oil to ghee
(d)	Finely divided iron	(iv)	Sandmeyer reaction
(e)	$TiCl_4 + Al(CH_3)_3$	(v)	Haber's Process
		(vi)	Decomposition of $KClO_3$

1) (a)-(iv) (b)-(i) (c)-(ii) (d)-(iii) (e)-(v)
2) (a)-(i) (b)-(iii) (c)-(iv) (d)-(v) (e)-(ii)
3) (a)-(iii) (b)-(iv) (c)-(ii) (d)-(v) (e)-(i)
4) (a)-(iv) (b)-(ii) (c)-(v) (d)-(iii) (e)-(i)

2. Match the properties given in Column-I with the metals given in Column-II.

	Column-I		Column-II
(a)	An element which can show +8 oxidation state	(i)	Mn
(b)	3d block element that can show upto +7 oxidation state	(ii)	Cr
(c)	3d block element with highest melting point	(iii)	Os
		(iv)	Fe

1) (a)-(iii) (b)-(ii) (c)-(i)
2) (a)-(ii) (b)-(i) (c)-(iii)
3) (a)-(iii) (b)-(i) (c)-(ii)
4) (a)-(i) (b)-(iii) (c)-(ii)

3. Match the statements given in Column-I with the oxidation states given in Column-II.

	Column-I		Column-II
(a)	Oxidation state of Mn in MnO_2 is	(i)	+2
(b)	Most stable oxidation state of Mn is	(ii)	+3
(c)	Most stable oxidation state of Mn in oxides is	(iii)	+4
(d)	Characteristic oxidation state of lanthanoids is	(iv)	+5
		(v)	+7

1) (a)-(ii) (b)-(i) (c)-(iv) (d)-(iii)
2) (a)-(ii) (b)-(i) (c)-(v) (d)-(iv)
3) (a)-(iv) (b)-(iii) (c)-(i) (d)-(ii)
4) (a)-(iii) (b)-(i) (c)-(v) (d)-(ii)

4. Match the solutions given in Column-I and the colours given in Column-II.

	Column-I (Aqueous solution of salt)		Column-II (Colour)
(a)	$FeSO_4.7H_2O$	(i)	Green
(b)	$NiCl_2.4H_2O$	(ii)	Light pink
(c)	$MnCl_2.4H_2O$	(iii)	Blue
(d)	$CoCl_2.6H_2O$	(iv)	Pale green
(e)	Cu_2Cl_2	(v)	Pink
		(vi)	colourless

1) (a)-(i) (b)-(iii) (c)-(ii) (d)-(iv) (e)-(v)
2) (a)-(i) (b)-(iii) (c)-(i) (d)-(v) (e)-(vi)
3) (a)-(iv) (b)-(i) (c)-(ii) (d)-(v) (e)-(vi)
4) (a)-(iii) (b)-(ii) (c)-(iv) (d)-(vi) (e)-(v)

5. Match the property given in Column-I with the element given in Column-II.

	Column I (Property)		Column II (Element)
(a)	Lanthanoid which shows +4 oxidation state	(i)	Pm
(b)	Lanthanoid which can show +2 oxidation state	(ii)	Ce
(c)	Radioactive lanthanoid	(iii)	Lu
(d)	Lanthanoid which has $4f^7$ electronic configuration in +3 oxidation state	(iv)	Eu
(e)	Lanthanoid which has $4f^{14}$ electronic configuration in +3 oxidation state	(v)	Gd
		(vi)	Dy

1) (a)-(ii) (b)-(i) (c)-(iv) (d)-(v) (e)-(vi)
2) (a)-(i) (b)-(v) (c)-(ii) (d)-(iii) (e)-(iv)
3) (a)-(iii) (b)-(ii) (c)-(iv) (d)-(v) (e)-(i)
4) (a)-(ii) (b)-(iv) (c)-(i) (d)-(v) (e)-(iii)

6. Match the statements given in Column-I with the oxidation states given in Column-II.

	Column I (Property)		Column II (Metal)
(a)	Element with highest second ionisation enthalpy	(i)	Co
(b)	Element with highest third ionisation enthalpy	(ii)	Cr
(c)	M in $M(CO)_6$ is	(iii)	Cu
(d)	Element with highest heat of atomisation	(iv)	Zn
		(v)	Ni

1) (a)-(v) (b)-(ii) (c)-(iii) (d)-(i)
2) (a)-(iii) (b)-(iv) (c)-(ii) (d)-(i)
3) (a)-(iii) (b)-(ii) (c)-(i) (d)-(v)
4) (a)-(iv) (b)-(i) (c)-(ii) (d)-(iii)

Correct & Incorrect MCQs

1. Which of the following statement (s) is/are correct?
   A. The d-block of the periodic table contains the elements of the groups 3-12.
   B. Transition metals are defined as metals which have incomplete d subshell either in neutral atom or in their ions.
   C. Zinc, cadmium and mercury are regarded as transition metals.
   D. The d–orbitals of the penultimate energy level of atoms receive electrons giving rise to four rows of the transition metals, i.e., 3d, 4d, 5d and 6d.
   1) A, B and D 2) Only A and D
   3) Only C 4) B, C and D

2. Which of the following statement regarding electronic configuration is incorrect?
   1) The electronic configuration of Scandium is $3d^1 4s^2$.
   2) The electronic configuration of Iron is $3d^6 4s^2$.
   3) The electronic configuration of Chromium is $3d^4 4s^2$.
   4) The electronic configuration of Zinc is $3d^{10} 4s^2$.

3. Choose incorrect statement among the following:
   1) The transition metals (with the exception of Zn, Cd and Hg) are very hard and have low volatility
   2) The melting and boiling points of transition metals are very low.
   3) The metals of the second and third series have greater enthalpies of atomisation than the corresponding elements of the first series.
   4) Both (1) and (3).

4. Correct statements are :
   A. Manganese, exhibits all the oxidation states from +2 to +7.
   B. Titanium (III) is more stable than Ti(IV).
   C. The only oxidation state of zinc is +2.
   D. Cr(VI) is found to be more stable than Mo(VI) and W(VI).
   1) A and B 2) B, C and D
   3) A and C 4) B and C

5. Select the correct statement/s:
   1) The $E^\ominus$ value for copper is negative.
   2) Only oxidising acids (nitric and hot concentrated sulphuric) react with Cu, the acids being reduced.
   3) The high energy to transform Cu(s) to Cu^{2+}(aq) is not balanced by its hydration enthalpy.
   4) Both (2) and (3)

6. Incorrect statement(s) is/are:
   A. The oxidation state of Mn in Mn_2O_7 is +7.
   B. The oxidation state of Mn in MnO_2 is +6.
   C. The oxidation state of Mn in Mn_2O_3 is +5.
   D. The oxidation state of Mn in MnO is +2.
   1) A and B
   2) A, C and D
   3) Only A
   4) B and C

7. Identify the incorrect statement from the following:
   1) Diamagnetic substances get repelled by the applied magnetic field.
   2) Paramagnetic substances get attracted by the applied magnetic field.
   3) Paramagnetism arises from the presence of unpaired electrons.
   4) A single unpaired electron has a magnetic moment of 2.54 Bohr magnetons (BM).

8. Pick out the incorrect statement:
   1) MnO_2 dissolves in dilute HCl, but does not form Mn^{4+} ions
   2) MnO_2 oxidizes hot concentrated H_2SO_4 liberating oxygen
   3) K_2MnO_4 is formed when MnO_2 is fused with KOH in presence of KNO_3
   4) Decomposition of $KMnO_4$ is not catalysed by sunlight.

9. Select the incorrect statement:
   1) Potassium dichromates are orange coloured crystals.
   2) Sodium and potassium dichromates are strong oxidising agents.
   3) Sodium dichromate is less soluble than potassium dichromate.
   4) Potassium dichromate is used as a primary standard in volumetric analysis.

10. Select the correct statement:
   1) The earlier members of lanthanoid series resemble calcium in their chemical properties.
   2) The extent of actinoid contraction is almost the same as lanthanoid contraction.
   3) In general, lanthanoid and actinoids do not show variable oxidation states.
   4) Ce^{4+} in aqueous solution is not known.

Statement Based MCQs

Directions: These questions consist of two statements each, printed as Statement-I and Statement-II. While answering these questions, you are required to choose any one of the following four responses.

1) **Both Statement-I and Statement-II are correct.**
2) **Both Statement-I and Statement-II are incorrect.**
3) **Statement-I is correct & Statement-II is incorrect.**
4) **Statement-I is incorrect & Statement-II is correct.**

1. **Statement-I:** Mn^{2+} is a colourless ion

   Statement-II: Ti shows +2, +3, and +4 oxidation states due to $4s^2$ and $3d^2$ electrons.

2. **Statement-I:** Most of the trivalent lanthanide ions are coloured both in the solid state and in aqueous solution.

   Statement-II: The elements with xf electrons have a similar colour to those of (14-x)f electrons.

3. **Statement-I:** $K_2Cr_2O_7$ is a strong oxidising agent.

   Statement-II: There is no electron present in d-orbitals of manganese in MnO_4^-.

4. **Statement-I:** $K_2Cr_2O_7$ is used as primary standard in volumetric analysis.

   Statement-II: Potassium dichromate is more soluble than sodium dichromate.

5. **Statement-I:** Fe^{3+} is more stable than Fe^{2+}.

   Statement-II: Half filled and completely filled subshells are more stable.

6. **Statement-I:** Sc^{3+} and Ti^{4+} are coloured ions.

   Statement-II: Ti^{3+} and V^{4+} are colourless ions.

7. **Statement-I:** Fe^{2+} is paramagnetic in nature.

   Statement-II: Fe^{2+} has 4 unpaired electron.

8. **Statement-I:** Ce(IV) is a good analytical reagent.

   Statement-II: Tb^{IV} has half-filled f-orbitals and is an oxidant.

9. **Statement-I:** Cr_2O_3 is acidic in nature.

   Statement-II: CrO is amphoteric in nature.

10. **Statement-I:** The transition metals and their compounds are known for their catalytic activity.

    Statement-II: Transition metals have ability to adopt multiple oxidation states.

Assertion & Reason MCQs

Directions: These questions consist of two statements each, printed as Assertion and Reason. While answering these questions, you are required to choose any one of the following four responses.

1) **Both Assertion and Reason are True and the Reason is a correct explanation of the Assertion.**
2) **Both Assertion and Reason are True but Reason is not a correct explanation of the Assertion.**
3) **Assertion is True but the Reason is False.**
4) **Assertion is False but Reason is True.**

1. **Assertion:** Cu^{2+} iodide is not known.

   Reason: Cu^{2+} oxidises I^- to iodine.

2. **Assertion:** Separation of Zr and Hf is difficult.

   Reason: Because Zr and Hf lie in the same group of the periodic table.

3. **Assertion:** Actinoids form relatively less stable complexes as compared to lanthanoids.

   Reason: Actinoids can utilise their 5f orbitals along with 6d orbitals in bonding but lanthanoids do not use their 4f orbital for bonding.

4. **Assertion:** Cu cannot liberate hydrogen from acids.

   Reason: Because it has positive electrode potential.

5. **Assertion:** The highest oxidation state of osmium is +8.

   Reason: Osmium is a 5d-block element.

6. **Assertion:** The actinoids include the fourteen elements from Th to Lr.

   Reason: All the actinoids are believed to have the electronic configuration of $7s^2$ and variable occupancy of the 5f and 6d subshells.

7. **Assertion:** The electronic configuration of Mn is $3d^{10}\,4s^1$.

   Reason: Manganese exhibits all the oxidation states from +2 to +7.

8. **Assertion:** Solution of Na_2CrO_4 in water is intensely coloured.

   Reason: Oxidation state of Cr in Na_2CrO_4 is +VI.

9. **Assertion:** Cuprous ion(Cu^+) is colourless whereas cupric ion (Cu^{++}) is blue in the aqueous solution.

   Reason: Cuprous ion (Cu^+) has unpaired electrons while cupric ion(Cu^{++}) does not.

10. **Assertion:** ZnO is an amphoteric.

    Reason: Zn react with acid as well as base.

Multi Correct MCQs

1. Which of the following elements have half-filled f-orbitals in their ground state?

   (Given : atomic number)

   Sm = 62; Eu = 63; Tb = 65;Gd = 64, Pm = 61

   A. Sm B. Eu C. Tb

   D. Gd E. Pm

   Choose the correct answer from the options given below:

   1) B and D only 2) A and E only

   3) A and B only 4) C and D only

2. Which one amongst the following are good oxidizing agents?

   A. Sm^{2+} B. Ce^{2+}

   C. Ce^{4+} D. Tb^{4+}

   Choose the most appropriate answer from the options given below:

   1) C only 2) D only

   3) A and B only 4) C and D only

3. Which of the following reactions are disproportionation reactions?

   A. $Cu^+ \rightarrow Cu^{2+} + Cu$

   B. $3MnO_4^{2-} + 4H^+ \rightarrow 2MnO_4^- + MnO_2 + 2H_2O$

   C. $3KMnO_4 \rightarrow K_2MnO_4 + MnO_2 + O_2$

   D. $2MnO_4^- + 3Mn^{2+} + 2H_2O \rightarrow 5MnO_2 + 4H^+$

   Choose the correct answer from the options given below:

   1) (A) and (B)
   2) (B), (C) and (D)
   3) (A), (B) and (C)
   4) (A) and (D)

4. Choose the set of ions from the following with the noble gas configuration.

   A. Sr^{2+} (Z = 38) B. Cs^+ (Z = 55)

   C. La^{2+} (Z = 57) D. Pb^{2+} (Z = 82)

   E. Yb^{2+} (Z = 70) F. Fe^{2+} (Z = 26)

   1) (A) and (B)
   2) (C), (D), (E) and (F)
   3) (B), (C) and (D)
   4) (D), (E) and (F)

5. Consider the elements given below:

   A. Sc B. Cr C. V

   D. Ti E. Mn

   Choose the elements from the following having unpaired electrons greater than or equal to 5.

   1) Only (A)
   2) Both (B) and (C)
   3) (B), (C) and (D)
   4) Both (B) and (E)

NCERT Exemplar MCQs

1. Which of the following characteristic property of $KMnO_4$ is/are correct?

1) $KMnO_4$ forms dark purple crystals
2) SO_2 is oxidised to SO_4^{2-} with $KMnO_4$ in acidic medium
3) Purple colour of ion is due to charge transfer
4) All of these

2. The electronic configuration of Cu(II) is $3d^9$ where as that of Cu(I) is $3d^{10}$. Which of the following is correct?

1) Cu(II) is more stable
2) Cu(II) is less stable
3) Cu(I) and Cu(II) are equally stable
4) Stability of Cu(I) and Cu(II) depends on nature of copper salts

3. Metallic radii of some transition elements are given below. Which of these elements will have highest density:

Element	Fe	Co	Ni	Cu
Metallic radii/pm	126	125	125	128

1) Fe 2) Ni
3) Co 4) Cu

4. Generally, transition elements form coloured salts due to the presence of unpaired electrons. Which of the following compounds will be coloured in solid state?

1) Ag_2SO_4 2) CuF_2
3) ZnF_2 4) Cu_2Cl_2

5. On addition of small amount of $KMnO_4$ to concentrated H_2SO_4, a green oily compound is obtained which is highly explosive in nature. Identify the compound from the following:

1) Mn_2O_7 2) MnO_2
3) $MnSO_4$ 4) Mn_2O_3

6. The magnetic nature of elements depends on the presence of unpaired electrons. Identify the configuration of transition element, which shows highest magnetic moment:

1) $3d^7$ 2) $3d^5$
3) $3d^8$ 4) $3d^2$

7. Which of the following oxidation state is common for all lanthanoids?

1) + 2 2) + 3
3) + 4 4) + 5

8. Which of the following reactions are disproportionation reactions?

(i) $Cu^+ \rightarrow Cu^{2+} + Cu$
(ii) $3MnO_4^{2-} + 4H^+ \rightarrow 2MnO_4^- + MnO_2 + 2H_2O$
(iii) $2KMnO_4 \rightarrow K_2MnO_4 + MnO_2 + O_2$
(iv) $2MnO_4^- + 3Mn^{2+} + 2H_2O \rightarrow 5MnO_2 + 4H^+$

1) (i) and (ii) 2) (i), (ii) and (iii)
3) (ii), (iii) and (iv) 4) (i) and (iv)

9. When $KMnO_4$ solution is added to oxalic acid solution, the decolourisation is slow in the beginning but becomes instantaneous after some time because:

1) CO_2 is formed as the product
2) Reaction is exothermic
3) MnO_4^- catalyses the reaction
4) Mn^{2+} acts as autocatalyst

10. There are 14 elements in actinoid series. Which of the following elements does not belong to this series?

1) U 2) Np
3) Tm 4) Fm

11. $KMnO_4$ acts as an oxidising agent in acidic medium. The number of moles of $KMnO_4$ that will be needed to react with one mole of sulphide ions in acidic solution is:

1) $\frac{2}{5}$ 2) $\frac{3}{5}$
4) $\frac{4}{5}$ 4) $\frac{1}{5}$

12. Which of the following is amphoteric oxide?

Mn_2O_7, CrO_3, Cr_2O_3, CrO, V_2O_5, V_2O_4

1) V_2O_5, Cr_2O_3 2) Mn_2O_7, CrO_3
3) CrO, V_2O_5 4) V_2O_5, V_2O_4

13. Gadolinium belongs to 4f series. Its atomic number is 64. Which of the following is the correct electronic configuration of gadolinium?

1) $[Xe]4f^7 5d^1 6s^2$ 2) $[Xe]4f^6\ 5d^2\ 6s^2$
3) $[Xe]4f^8\ 6d^2$ 4) $[Xe]4f^9\ 5s^1$

14. Interstitial compounds are formed when small atoms are trapped inside the crystal lattice of metals. Which of the following is not the characteristic property of interstitial compounds?

1) They have high melting points in comparison to pure metals
2) They are very hard
3) They retain metallic conductivity
4) The are chemically very reactive

15. The magnetic moment is associated with its spin angular momentum and orbital angular momentum. Spin only magnetic moment value of Cr^{3+} ion is:

1) 2.87 BM 2) 3.87 BM
3) 3.47 BM 4) 3.57 BM

16. $KMnO_4$ acts as an oxidising agent in alkaline medium. When alkaline $KMnO_4$ is treated with KI, iodide ion is oxidised to:

1) I_2 2) IO^-
3) IO_3^- 4) IO_4^-

17. Which of the following statements is not correct?
    1) Copper liberates hydrogen from acids
    2) In its higher oxidation states, manganese forms stable compounds with oxygen and fluorine
    3) Mn^{3+} and Co^{3+} are oxidising agents in aqueous solution are reducing agents in aqueous solution
    4) Ti^{2+} and Cr^{2+} are reducing agents in aqueous solution
18. When acidified $K_2Cr_2O_7$ solution is added to Sn^{2+} salt then Sn^{2+} changes to:
    1) Sn 2) Sn^{3+}
    3) Sn^{4+} 4) Sn^{+}
19. Highest oxidation state of manganese in fluoride is +4 (MnF_4) but highest oxidation state in oxides is +7(Mn_2O_7) because:
    1) Fluorine is more electronegative than oxygen
    2) Fluorine does not possess d orbitals
    3) Fluorine stabilises lower oxidation state
    4) In covalent compounds, fluorine can form single bond only while oxygen forms double bond
20. Although Zirconium belongs to 4d transition series and Hafniun to 5d transition series even then they show similar physical and chemical properties because:
    1) Both belong to d-block
    2) Both have same number of electrons
    3) Both have similar atomic radius
    4) Both belong to the same group of the Periodic Table
21. Why is HCl not used to make the medium acidic in oxidation reactions of $KMnO_4$ in acidic medium?
    1) Both HCl and $KMnO_4$ act as oxidising agents
    2) $KMnO_4$ oxidises HCl into Cl_2 which is also an oxidising agent
    3) $KMnO_4$ is a weaker oxidising agent than HCl
    4) $KMnO_4$ acts as a reducing agent in the presence of HCl
22. Generally transition elements and their salts are coloured due to the presence of unpaired electrons in metal ions. Which of the following compounds are coloured?
    1) $KMnO_4$ 2) $Ce(SO_4)_2$
    3) $TiCl_4$ 4) Cu_2Cl_2
23. Transition elements show magnetic moment due to spin and orbital motion of electrons. Which of the following metallic ions have almost same spin only magnetic moment?
    1) Co^{2+} 2) Cr^{2+}
    3) Mn^{2+} 4) Cr^{3+}
24. In the form of dichromate, Cr(VI) is a strong oxidising agent in acidic medium but Mo(VI) in MoO_3 and W(VI) in WO_3 are not because:
    1) Cr (VI) is more stable than Mo(VI) and W(VI)
    2) Mo (VI) and W (VI) are more stable than Cr(VI)
    3) Higher oxidation states of heavier members of group 6 of transition series are more stable
    4) Lower oxidation states of heavier members of group-6 of transition series are more stable
25. Which of the following actinoids show oxidation states upto +7?
    1) Am 2) Pu
    3) U 4) Np
26. General electronic configuration of actinoids is $(n-2)\ f^{1-14}\ (n-1)\ d^{0-2}\ ns^2$. Which of the following actinoids have one electron in 6d orbital?
    1) U (Atomic number. 92)
    2) Np (Atomic number. 93)
    3) Pu (Atomic number. 94)
    4) Am (Atomic number. 95)
27. Which of the following lanthanoids show +2 oxidation state besides the characteristic oxidation state + 3 of lanthanoids?
    1) Ce 2) Eu
    3) Yb 4) Ho
28. Which of the following ions show higher spin only magnetic moment value?
    1) Ti^{3+} 2) Mn^{2+}
    3) Fe^{2+} 4) Co^{3+}
29. Transition elements form binary compounds with halogens. Which of the following elements will form MF_3 type compounds?
    1) Cr 2) Co
    3) Cu 4) Ni
30. Which of the following will not act as oxidising agents?
    1) CrO_3 2) MoO_3
    3) WO_3 4) CrO_4^{2-}
31. Although +3 is the characteristic oxidation state for lanthanoids but cerium also shows +4 oxidation state because:
    1) It has variable ionisation enthalpy
    2) It has a tendency to attain noble gas configuration
    3) It has a tendency to attain f^0 configuration
    4) It resembles Pb^{4+}

Scan Me for Detailed Explanations

Kattar NEET

1. Consider the following statements and arrange in the order of true/false as given in the codes.

S_1: Mn^{2+} compounds more stable than Fe^{2+} towards oxidation to their + 3 states.

S_2 : Titanium and copper both in the first series of transition metals exhibit +1 oxidation state most frequently.

S_3 : Cu^+ ion is stable in aqueous solutions.

1) T T F 2) T F F
3) T F T 4) F F T

2. Consider the following statements and arrange in the order of true/false as given in the codes.

S_1 : Interstitial compounds have high melting points, higher than those of pure metals.

S_2 : Permangnate titrations in presence of hydrochloric acid are unsatisfactory.

S_3 : $KMnO_4$ does not act as an oxidising agent in strong alkaline medium.

S_4 : For 1st transition series, the value for M^{2+}/M indicate a decreasing tendency to form cation across the series.

1) T T F T 2) T F F T
3) T F T T 4) F F T F

3. A sulphate of metal 'A' on heating evolves two gases B and C and an oxide D. Gas 'B' turns $K_2Cr_2O_7$ paper green while gas 'C' forms a trimer in which there is no S – S bond. Compound 'D' with HCl forms a Lewis acid E which exists as a dimer. Compounds A, B, C, D and E are respectively

1) $FeSO_4$, SO_2, SO_3, Fe_2O_3, $FeCl_3$
2) $Al_2(SO_4)_3$, SO_2, SO_3, Al_2O_3, $FeCl_3$
3) FeS, SO_2, SO_3, $FeSO_4$, $FeCl_3$
4) FeS, SO_2, $Fe_2(PO_4)_3$, $FeCl_3$, SO_3

4. When $Mn(OH)_2$ is made by adding an alkali to a solution containing Mn^{2+} ions, the precipitate quickly darkens, and eventually goes black. What might be the chemical giving black colour?

1) MnO 2) MnO_2
3) Mn_2O_7 4) MnO_4^-

5. Which of the following statements about cerium (Z = 58) is incorrect?

1) The + 3 oxidation state of cerium is more stable than the +4 oxidation state
2) The common oxidation states of cerium are +3 and +4
3) Cerium (IV) acts as an oxidising agent
4) The + 4 oxidation state of cerium is not known in solution

6. When aqueous sodium hydroxide is added to an aqueous solution of chromium (III) ions, a green blue precipitate is first formed which re-dissolves to give a green solution. This green colour is due to

1) $[Cr(H_2O)_6]^{3+}$ 2) CrO_4^{2-}
3) $[Cr(OH)_4]^-$ 4) $[Cr(OH)_3(H_2O)_3]$

7. A metal gives two chlorides 'A' and 'B'. 'A' gives black precipitate with NH_4OH and 'B' gives white precipitate. With KI 'B' gives a red precipitate soluble in excess of KI. 'A' and 'B' are respectively.

1) $HgCl_2$ and Hg_2Cl_2 2) Hg_2Cl_2, $HgCl_2$
3) $HgCl_2$ and HgCl 4) None of these

8. The maximum and minimum melting points of first and second transition series respectively are observed with

1) Cr and Zn 2) Cr and Hg
3) V and Cd 4) Cr and Cd

9. Unknown salt 'A' + $K_2Cr_2O_7$ + conc. H_2SO_4 Reddish brown fumes. Which is the correct statement regarding the above observation

1) It confirms the presence of Cl^- ion
2) It confirms the presence of Br^- ion
3) It confirms the presence of both
4) It neither confirms Cl^- nor Br^- unless it is passed through NaOH solution

10. Which of the following statements is incorrect?

1) Mercurous ion exist as Hg^+
2) Mercurous ion is diamagnetic and exist as dimer
3) Mercurous ion is colourless
4) There is a covalent bond between two Hg^+ ions

11. Match the compounds given in Column-I with characteristic(s) of products given in Column-II.

List-I		List-II	
I.	K_2MnO_4	P.	Transition element in +6 state.
II.	$KMnO_4$	Q.	Colour due to charge transfer.
III.	$K_2Cr_2O_7$	R.	Manufactured from pyrolusite ore.
IV.	K_2CrO_4	S.	Manufactured from chromite ore.

1) I-Q, S; II-P,R; III-R,S; IV-P,Q
2) I-P, Q, R; II-Q, R; III-P, Q, S; IV-P, Q, S
3) I-P, R; II-P, R; III-Q, S; IV-P, R, S
4) I-Q, R; II-Q, R; III-P, Q; IV-Q, S

12. A black compound of manganese react with a halogen acid to give greenish yellow gas. When excess of this gas reacts with NH_3 an unstable trihalide is formed. In this process the oxidation state of nitrogen changes from____________.
   1) – 3 to +3 2) – 3 to 0
   3) – 3 to +5 4) 0 to – 3

13. The atomic volumes of the transition elements are low compared with elements in neighbouring groups 1 and 2, because
   1) The nuclear charge is poorly screened and so attracts all the electrons more strongly.
   2) The extra electrons added occupy inner orbitals.
   3) (1) and (2) both
   4) None of these

14. First IE of 5d series elements are higher than those of 3d and 4d series elements. This is due to
   1) Bigger size of atoms of 5d-series elements than 3d-series elements.
   2) Greater effective nuclear charge is experienced by Valence electrons because of the weak shielding of the nucleus by 4*f*-electrons in 5d series.
   3) (1) and (2) both
   4) None of these

15. Which of the following statement is correct?
   1) The lesser number of oxidation states in 3d-series in the beginning of the series is due to the presence of too few electrons to loose or share.
   2) The lesser number of oxidation states in 3d-series towards the end of the series is due to the presence of too many electrons and thus fewer empty orbitals to share electrons with the ligands.
   3) (1) and (2) both
   4) None is correct

16. Magnetic moment of Cr^{2+} (Z = 24), Mn^{2+} (Z = 25) and Fe^{2+}(Z = 26) are x, y, z. They are in the order
   1) $x < y < z$ 2) $x > y > z$
   3) $z < x = y$ 4) $x = z < y$

17. Which of the following group of ions is paramagnetic in nature?
   1) Cu^{+}, Zn^{2+}, Sc^{3+} 2) Mn^{2+}, Fe^{3+}, Ni^{2+}
   3) Cr^{2+}, Mn^{3+}, Sc^{3+} 4) Cu^{2+}, Ni^{2+}, Ti^{4+}

18. The number of moles of acidified $KMnO_4$ required to convert one mole of sulphite ion into sulphate ion is
   1) 2/5 2) 3/5
   3) 4/5 4) 1

19. When MnO_2 is fused with KOH, a coloured compound is formed, the product and its colour is
   1) K_2MnO_4, green 2) $KMnO_4$, purple
   3) Mn_2O_3, brown 4) Mn_3O_4, black

20. Which of the following compound has colour but no unpaired electrons?
   1) $KMnO_4$ 2) K_2MnO_4
   3) $MnSO_4$ 4) $MnCl_2$

NEET & AIIMS Past Year MCQs

1. Which of the following set of ions act as oxidising agents? **(2024 Re)**
   1) Ce^{4+} and Tb^{4+} 2) La^{3+} and Lu^{3+}
   3) Eu^{2+} and Yb^{2+} 4) Eu^{2+} and Tb^{4+}

2. The UV-visible absorption bands in the spectra of lanthanoid ions are 'X', probably because of the excitation of electrons involving 'Y'. The 'X' and 'Y', respectively, are: **(2024 Re)**
   1) Broad and f orbitals
   2) Narrow and f orbitals
   3) Broad and d and f orbitals
   4) Narrow and d and forbitals

3. Match List-I with List-II: **(2024 Re)**

List-I (Block/group in periodic table)		List-II (Element)	
A.	Lanthanoid	I.	Ce
B.	d-block element	II.	As
C.	p-block element	III.	Cs
D.	s-block element	IV.	Mn

   Choose the correct answer from the options given below:
   1) A-I, B-II, C-IV, D-III 2) A-I, B-IV, C-III, D-II
   3) A-I, B-IV, C-II, D-III 4) A-IV, B-I, C-II, D-III

4. Which of the following pairs of ions will have same spin only magnetic moment values within the pair? **(2024 Re)**
   A. Zn^{2+}, Ti^{2+} B. Cr^{2+}, Fe^{2+}
   C. Ti^{3+}, Cu^{2+} D. V^{2+}, Cu^{+}
   Choose the correct answer from the options given below:
   1) C and D only 2) A and D only
   3) A and B only 4) B and C only

5. The pair of lanthanoid ions which are diamagnetic is **(2024)**
   1) Gd^{3+} and Eu^{3+} 2) Pm^{3+} and Sm^{3+}
   3) Ce^{4+} and Yb^{2+} 4) Ce^{3+} and Eu^{2+}

6. The E° value for the Mn^{3+}/Mn^{2+} couple is more positive than that of Cr^{3+}/Cr^{2+} or Fe^{3+}/Fe^{2+} due to change of **(2024)**
   1) d^4 to d^5 configuration 2) d^3 to d^5 configuration
   3) d^5 to d^4 configuration 4) d^5 to d^2 configuration

7. 'Spin only' magnetic moment is same for which of the following ions? **(2024)**
   A. Ti^{3+} B. Cr^{2+}
   C. Mn^{2+} D. Fe^{2+}
   E. Sc^{3+}
   Choose the most appropriate answer from the options given below:
   1) B and C only 2) A and D only
   3) B and D only 4) A and E only

8. The stability of Cu^{2+} is more than Cu^{+} salts in aqueous solutions due to: **(2023)**
   1) second ionisation enthalpy
   2) first ionisation enthalpy
   3) enthalpy of atomisation
   4) hydration energy

9. Which of the following statements are INCORRECT? **(2023)**
   A. All the transition metals except scandium form MO oxides which are ionic.
   B. The highest oxidation number corresponding to the group number in transition metal oxides is attained in Sc_2O_3 to Mn_2O_7.
   C. Basic character increases from V_2O_3 to V_2O_4 to V_2O_5.
   D. V_2O_4 dissolves in acids to give VO_4^{3-} salts.
   E. CrO is basic but Cr_2O_3 is amphoteric.

   Choose the correct answer from the options given below:
   1) B and C only 2) A and E only
   3) B and D only 4) C and D only

10. Gadolinium has a low value of third ionisation enthalpy because of **(2022)**
    1) High basic character
    2) Small size
    3) High exchange enthalpy
    4) Hight electronegativity

11. In the neutral or faintly alkaline medium, $KMnO_4$ oxidises iodide into iodate. The change in oxidation state of manganese in this reaction is from **(2022)**
    1) + 6 to + 5 2) + 7 to + 4
    3) + 6 to + 4 4) + 7 to + 3

12. Zr (Z = 40) and Hf (Z = 72) have similar atomic and ionic radii because of **(2021)**
    1) Diagonal relationship
    2) Lanthanoid contraction
    3) Having similar chemical properties
    4) Belonging to same group

13. The incorrect statement among the following is **(2021)**
    1) Most of the trivalent Lanthanoid ions are colorless in the solid state
    2) Lanthanoids are good conductors of heat and electricity
    3) Actinoids are highly reactive metals, especially when finely divided
    4) Actinoid contraction is greater for element to element that Lanthanoid contraction

14. The colourless species is: **(AIIMS 2017)**
    1) VCl_3
    2) $VOSO_4$
    3) Na_3VO_4
    4) $[V(H_2O)_6SO_4.H_2O$

15. The one rupee coin is made of: **(AIIMS 2015)**
    1) Steel 2) Cu and Ni
    3) Cu 4) Ni

16. When calomel is treated with ammonium hydroxide a black substance is formed. The black substance is: **(AIIMS 2013)**
    1) Hg + HgO 2) $HgO.HgCl_2$
    3) $H_2N - Hg - Cl + Hg$ 4) $Hg(NH_2)_2 + HgO$

17. **Assertion:** Aq. solution of $CoCl_2$ is pink in colour. It turns blue in presence of conc. HCl. **(AIIMS 2013)**

    Reason: It is due to the formation of $[CoCl_4]^{2-}$
    1) If both Assertion & Reason are True & the Reason is a correct explanation of the Assertion.
    2) If both Assertion & Reason are True but Reason is not a correct explanation of the Assertion.
    3) If Assertion is True but the Reason is False.
    4) If both Assertion & Reason are False.

18. MnO_3 in an acidic medium dissociates into: **(AIIMS 2012)**
    1) MnO_2 and MnO^{-}_4 2) MnO and $Mn\ O^{-}_4$
    3) MnO_2 and MnO_4 4) MnO_2 and MnO_3

19. Philosopher's wool when heated with BaO at 1100°C gives a compound identify the compound: **(AIIMS 2000)**
    1) $BaZnO_2$ 2) $Ba + ZnO_2$
    3) $BaCdO_2$ 4) $BaO_2 + Zn$

20. **Assertion:** Tailing of mercury occurs on passing ozone through it. **(AIIMS 1999)**

    Reason: Due to oxidation of mercury.
    1) If both Assertion & Reason are True & the Reason is a correct explanation of the Assertion.
    2) If both Assertion & Reason are True but Reason is not a correct explanation of the Assertion.
    3) If Assertion is True but the Reason is False.
    4) If both Assertion & Reason are False.

Answer Key

NCERT Topic-wise MCQs

1	2	3	4	5	6	7	8	9	10	11	12	13	14	15	16	17
3	1	2	2	4	2	4	4	2	1	3	4	3	2	1	4	4
18	19	20	21	22	23	24	25	26	27	28	29	30	31	32	33	34
1	1	2	3	2	2	2	2	3	1	2	4	4	1	2	3	1
35	36	37	38	39	40	41	42	43	44	45	46	47	48	49	50	51
1	1	4	4	4	3	3	3	1	2	2	4	2	3	1	4	2
52	53	54	55	56	57	58	59	60	61	62	63	64	65	66	67	68
2	2	3	1	3	2	2	2	2	4	1	3	1	3	4	4	1
69	70	71	72	73	74	75	76	77	78	79	80	81	82	83	84	85
2	2	1	1	2	4	2	1	3	2	1	1	2	1	3	1	4
86	87	88	89	90	91	92	93	94	95	96	97	98	99	100	101	102
1	2	1	1	2	2	1	3	1	2	3	1	4	2	4	4	2
103	104	105	106	107	108	109	110	111								
2	4	4	1	2	4	1	4	2								

High Order Time Intensive MCQs

1	2	3	4	5	6	7	8	9	10	11	12	13	14	15	16	17
3	1	1	4	2	4	1	3	4	4	4	2	4	3	3	2	4
18	19	20	21	22	23	24	25	26	27	28	29	30	31	32	33	34
2	2	3	4	2	1	4	4	4	3	4	3	4	3	3	1	3
35	36	37	38	39	40	41	42	43	44	45	46	47	48	49	50	51
1	2	4	2	2	4	1	2	3	1	3	3	2	3	3	2	1
52	53	54	55	56	57	58	59	60	61	62	63					
1	3.	1	3	3	3	4	3	3	2	4	2					

Match the Columns MCQs

1	2	3	4	5	6
3	3	4	3	4	2

Correct & Incorrect MCQs

1	2	3	4	5	6	7	8	9	10
1	3	2	3	4	4	4	4	3	1

Statement Based MCQs

1	2	3	4	5	6	7	8	9	10
4	1	1	3	1	2	1	1	2	1

Assertion & Reason MCQs

1	2	3	4	5	6	7	8	9	10
1	2	4	1	2	2	4	2	3	1

Multi Correct MCQs

1	2	3	4	5
1	4	1	1	4

NCERT Exemplar MCQs

1	2	3	4	5	6	7	8	9	10	11	12	13	14	15	16	17
4	1	4	2	1	2	2	1	4	3	1	1	1	4	2	3	1
18	19	20	21	22	23	24	25	26	27	28	29	30	31			
3	4	3	2	1, 2	1, 4	2, 3	2, 4	1, 2	2, 3	2, 3	1, 2	2, 3	2, 3			

Kattar NEET

1	2	3	4	5	6	7	8	9	10	11	12	13	14	15	16	17
2	1	1	2	4	3	2	4	4	1	2	1	3	2	3	4	2
18	19	20														
1	1	1														

NEET & AIIMS Past Year MCQs

1	2	3	4	5	6	7	8	9	10	11	12	13	14	15	16	17
1	2	3	4	3	3	3	4	4	3	2	2	1	3	2	3	1
18	19	20														
1	1	1														

CHAPTER 6

Coordination Compounds

Topicwise Number of Questions Analysis (2025-2016)

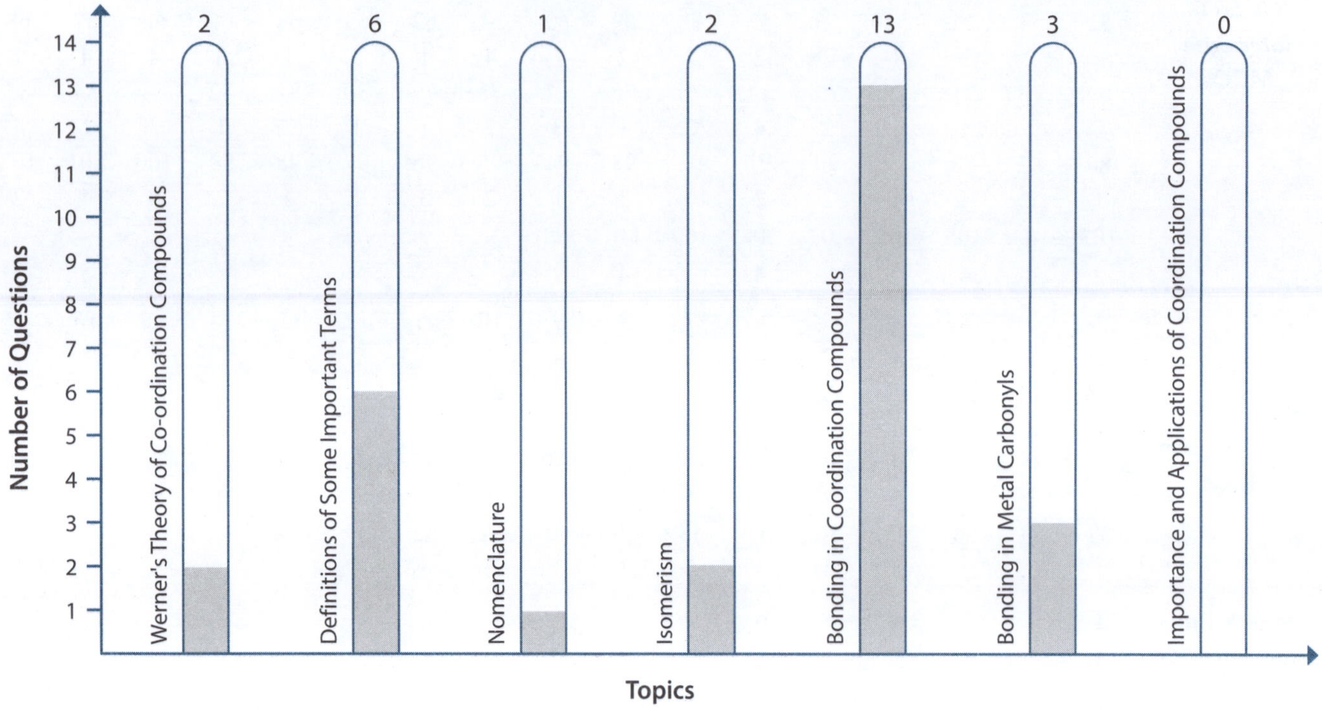

NCERT-Picks

- The compounds in which the central metal atom or ion is linked to a number of anions or neutral molecules by sharing of electrons in a co-ordination sphere are called coordination compounds.

Werner's Theory

According to this theory, metal possess two types of valencies: Primary valency and secondary valency.

1. **Primary valency:** These are normally ionisable and are satisfied by negative ions.
2. **Secondary valency:** The atoms within the square brackets form a single entity which does not dissociate under the reaction conditions. Werner proposed the term secondary valence for the number of groups bound directly to the metal ion. These are non-ionisable. These are satisfied by anions or neutral molecules. The groups satisfying secondary valencies are called **ligands**. The number of secondary valencies is fixed for a metal and is called coordination number.

The number of ions given by some cobalt complexes

By adding excess of $AgNO_3$ solution to one mole each of the given complexes, the number of moles of AgCl formed is 3, 2, 1, 0 respectively.

$CoCl_3.6NH_3$ gives $[Co(NH_3)_6]^{3+}$ & 3 Cl^-

$CoCl_3.5NH_3$ gives $[Co\ (NH_3)_5Cl]^{+2}$ & 2 Cl^-

$CoCl_3.4NH_3$ gives $[Co\ (NH_3)_4Cl_2]^+$ & 1 Cl^-

$CoCl_3.3NH_3$ is not ionised

Hence, all the given complexes have different primary valencies that is 3, 2, 1 & 0 respectively and fixed secondary valencies that is six.

Difference Between Double Salt and Coordination Compound

Double salts	Coordination Compounds
1. Double salts exist only in the solid state.	1. Coordination compounds exist in both the form i.e. solid as well as aqueous.
2. In aqueous solution, they dissociate completely into ions.	2. They will remain as a compound in aqueous state and does not dissociates into simpler ions.

Some Important Terms Related to Coordination Compound

Ligand

Species, which is linked to the central metal atom / ion and which donate electron pair is called as ligand.

Unidentate ligand (in which metal is bound to the single donor atom) : Cl^-, H_2O, $\ddot{N}H_3$

Bidentate ligand (in which metal is bound to the two donor atom) : $H_2\ddot{N}CH_2CH_2\ddot{N}H_2$, $C_2O_4^{2-}$

Chelate ligand (when ligand is bound to the metal with two or more donor atoms) : EDTA, it is a hexadentate ligand and forms a ring structure with the metal.

Ambidentate ligand (ligand which has two donor sites but either of the two is ligands): NO_2^-, SCN^-

Coordination Number

The coordination number (C.N.) of a metal ion in a complex can be defined as the number of ligand donor atoms to which the metal is directly bonded.

- Coordination number of Fe and Co in $[Fe(C_2O_4)_3]^{3-}$ and $[Co(en)_3]^{3+}$ respectively is 6. Both $C_2O_4^{2-}$ (oxalate) and en (ethane-1, 2-diamine) are bidentate ligands.

Oxidation Number

The oxidation number of the central atom in a complex is defined as the charge it would carry if all the ligands are removed along with the electron pairs that are shared with the central atom.

- **Coordination polyhedron:** It is the spatial arrangement of the ligands around the central metal atom/ion.
- **Homoleptic Complexes:** The complex in which the central metal atom (or) ion bound with only one kind of ligands. Ex: $[Co(NH_3)_6]^{3+}$
- **Heteroleptic complexes:** The complex in which the central metal atom (or) ion bound with more than one kind of ligands. Ex: $[Co(NH_3)_4Cl_2]^+$
- **Effective Atomic Number** (EAN): Predicting stability of a coordination compound.

  EAN = Atomic number of central metal – Electron lost by central atom + Electron gained by ligands.

Nomenclature of Coordination Compounds

Rules For Writing The Formula of Coordination Compounds

- Central atom listed first then ligands are listed in alphabetical order.
- Polydentate ligands are also listed alphabetically. In case of abbreviated ligand, first letter of abbreviation is used to determine the alphabetical order.
- Formula of the cation whether simple or complex must be written first followed by anion.
- The formula of the entire coordination entity, whether charged or not, is enclosed in square brackets. When ligands are polyatomic, then their formulas are enclosed in parentheses.
- The number of cations or anions to be written in the formula is calculated on the basis of charges present. The total positive charge must be equal to the total negative charge, as the complex as a whole is electrically neutral.
- Within the coordination sphere i.e. in the square bracket first the metal atom is written, followed by anionic ligand then finally cationic ligand.

Rules For Naming Coordination Compounds

- The anion is named after the cation.
- The complex part is written as one word and name of coordination compounds is started with a small letter.
- In naming of the coordination sphere, ligands are named first in alphabetical order followed by metal atom and then oxidation state of metal in by Roman numeral (in parenthesis).
- Name of the anionic ligand ends with -"O".
- Name of the cationic and neutral ligands are the same except aqua for H_2O, ammine for NH_3, carbonyl for CO and nitrosyl for NO.
- Name of cationic ligand ends in - ium.
- The prefixes di- , tri-, tetra- , penta- and hexa- are used to indicate the number of each ligand. If the ligand name includes a prefix, then the ligand name should placed in parentheses and preceded by bis-(2), tris (3), tetrakis (4), penta kis(5), hexakis(6) Ex: $[C(NH_3)_3(H_2O)_3]Cl_3$– Triamminetria quachromimum(III) chloride.
- If the complex ion is a cation, the metal is named same as the element. For ex., Co in complex cation is called cobalt and Pt is called platinum. If complex ion is an anion, the name of the metal ends with the suffix-ate. For ex., Co in a complex anion is called cobaltate. For same metals, the latin names are used in the complex anions ex. ferrate for Fe.

Isomerism in Coordination Compound

In co-ordination compounds mainly two types of isomerism is observed:

a. Structural isomerism

b. Stereoisomerism

These are further divided into two sub-categories as given below:

	(a) Structural Isomerism	(b) Stereoisomerism		
(i)	Ionisation $[Pt(NH_3)_4(OH_2)](SO_4)$ $[Pt(NH_3)_4(SO_4)]\ (OH_2)$	**Geometrical Isomerism**		**Optical Isomerism**
(ii)	Coordination $[Cr(NH_3)_6][Co(CN)_6]$ $[Cr(CN)_6][Co(NH_3)_6]$	(i) cis-trans Ex. 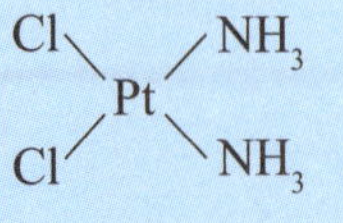 cis 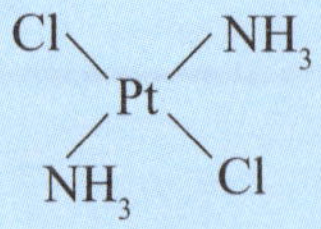 trans		(i) dextro (d) Rotate plane polarised light to the right Ex. 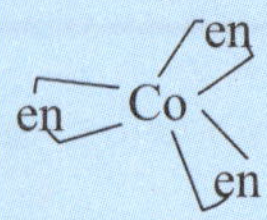
(iii)	Hydrate or solvate $[Cr(H_2O)_6]Cl_3$: violet-colored $[Cr(H_2O)_5Cl]Cl_2.H_2O$: blue green in color $[Cr(H_2O)_4Cl_2]Cl.2H_2O$: dark green in color	(ii) Facial (fac) Ex. 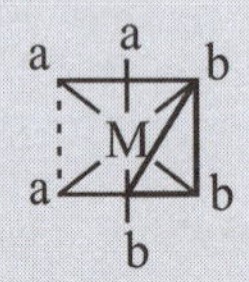	Meridional (mer) Ex. 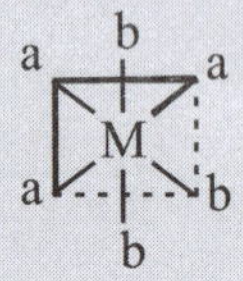	(ii) Laevo (l) Rotate plane polarised light to the left. 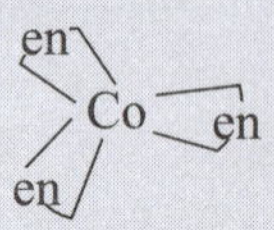
(iv)	Linkage $[CO(NH_3)_2][(NO_2)_2(Py)_2]NO_3$ $[CO(NH_3)_2][(ONO)_2(Py)_2]NO_3$			

Some Important Points of Stereoisomerism

- Tetrahedral complexes do not show geometrical isomerism. The relative position of atoms with respect to each other will be the same.
- Geometrical isomerism arises in heteroleptic complexes. Due to the difference in geometrical arrangement of the ligands around the central atom.
- Optical isomerism is not shown by square planar complex because square planar complexes have horizontal plane of symmetry. Square planar complexes do not possess chiral structures.
- Optical isomers are also generally provided by the octahedral coordination compounds having bidentate (or) polydentate ligands.

  Ex: $[CoCl_2(en)_2]^+$, $[Cr(C_2O_4)_3]^{2-}$

  $[Cr(NH_3)_2Cl_2en]^+$; $[PtCl_2(en)_2]^{2+}$

  Here 'en' represents $H_2NCH_2CH_2NH_2$
- Optical isomerism is shown by the complexes having chiral structures.

Isomerism in Coordination Compound

- **Valence bond theory:** According to this theory, the metal atom or ion under the influence of ligands can use its (n-1)d, ns, np or ns, np, nd orbitals for hybridisation to yield a set of equivalent orbitals of definite geometry such as octahedral, tetrahedral, square planar and so on. These hybridised orbitals are allowed to overlap with ligand orbitals that can donate electron pairs for bonding. This is illustrated by the following examples

Table: Number of Orbitals and Types of Hybridisations

Coordination number	Types of hybridisation	Distribution of hybrid orbitals in space
4	sp^3	Tetrahedral
4	dsp^2	Square planar
5	sp^3d	Trigonal bipyramidal
6	sp^3d^2	Octahedral
6	d^2sp^3	Octahedral

It is usually possible to predict the geometry of a complex from the knowledge of its magnetic behaviour on the basis of the valence bond theory.

In the diamagnetic octahedral complex, $[Co(NH_3)_6]^{3+}$, the cobalt ion is in +3 oxidation state and has the electronic configuration $3d^6$. The hybridisation scheme is as shown in diagram.

Orbitals of Co^{3+}ion

d^2sp^3 hybridised orbitals of Co^{3+}

$[Co(NH_3)_6]^{3+}$ (inner orbital or low spin complex)

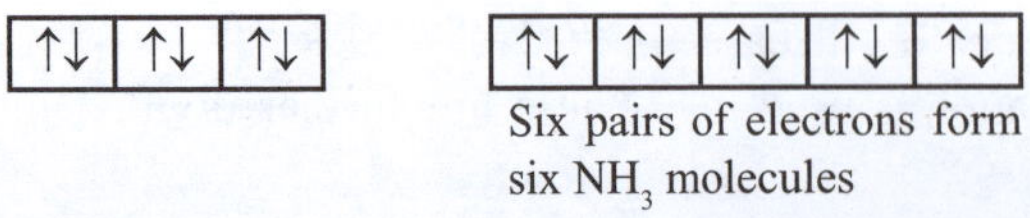

Six pairs of electrons, one from each NH_3 molecule, occupy the six hybrid orbitals. Thus, the complex has octahedral geometry and is diamagnetic because of the absence of unpaired electron. In the formation of this complex, since the inner d orbital (3d) is used in hybridisation, the complex, $[Co(NH_3)_6]^{3+}$ is called **an inner orbital or low spin or spin paired complex**. The paramagnetic octahedral complex, $[CoF_6]^{3-}$ uses outer orbital (4d) in hybridisation (sp^3d^2). It is thus called **outer orbital or high spin or spin free complex**. Thus:

	3d		
Orbitals of Co ion	↑↓ ↑ ↑ ↑ ↑ (3d)	☐ (4s) ☐☐☐ (4p)	☐☐☐☐☐ (4d)
sp^3d^2 hybridised orbitals of Co^{3+}	↑↓ ↑ ↑ ↑ ↑ (3d)	☐☐☐☐☐ (sp^3d^3 hybrid)	☐☐☐ (4d)
$[CoF_6]^{3-}$ (outer orbital or high spin complex)	↑↓ ↑ ↑ ↑ ↑ (3d)	↑↓ ↑↓ ↑↓ ↑↓ ↑↓ (Six pairs of electrons form six F^- ions)	☐☐☐ (4d)

In tetrahedral complexes one s and three p orbitals are hybridised to form four equivalent orbitals oriented tetrahedrally. This is illustrated below for $[NiCl_4]^{2-}$. Here nickel is in +2 oxidation state and the ion has the electronic configuration $3d^8$. The hybridisation scheme is as shown below,

Orbitals of Ni^{2+} ion	↑↓ ↑↓ ↑↓ ↑ ↑ (3d)	☐ (4s) ☐☐☐ (4p)
sp^3 hybridised orbitals of Ni^{2+}	↑↓ ↑ ↑ ↑ ↑ (3d)	☐☐☐☐ (sp^3 hybrid)
$[NiCl_4]^{2-}$ (high spin complex)	↑↓ ↑↓ ↑↓ ↑ ↑	↑↓ ↑↓ ↑↓ ↑↓ (Four pairs of electrons from 4 Cl^-)

Each Cl^- ion donates a pair of electrons. The compound is paramagnetic since it contains two unpaired electrons. Similarly, $[Ni(CO)_4]$ has tetrahedral geometry but is diamagnetic since nickel is in zero oxidation state and contains no unpaired electron. In the square planar complexes, the hybridisation involved is dsp^2. An example is $[Ni(CN)_4]^{2-}$. Here nickel is in +2 oxidation state and has the electronic configuration $3d^8$. The hybridisation scheme is as shown in diagram:

Orbitals of Ni^{2+} ion	↑↓ ↑↓ ↑↓ ↑ ↑ (3d)	☐ (4s)	☐☐☐ (4p)
dsp^2 hybridised orbitals of N^{2+}	↑↓ ↑↓ ↑↓ ↑ ↑ (3d)	☐☐☐☐ (sp^3 hybrid)	☐ (4p)
$[Ni(CN)_4]^2$ (low spin complex)	↑↓ ↑↓ ↑↓ ↑↓ (3d)	↑↓ ↑↓ ↑↓ ↑↓ (Four pairs of electrons from 4 CN^- groups)	☐ (4p)

Each of the hybridised orbitals receives a pair of electrons from a cyanide ion. The compound is diamagnetic as evident from the absence of unpaired electron. It is important to note that the hybrid orbitals do not actually exist. In fact, hybridisation is a mathematical manipulation of wave equation for the atomic orbitals involved.

Difference between inner and outer orbital complexes:

Inner orbital complex	Outer orbital complex
Complex involved inner (n-1)d orbitals	Complex involved outer (n-1) orbitals
Hybridisation: d^2sp^3	Hybridisation: sp^3d^2
Low spin or spin paired complex	High spin or spin free complex
Have less number of unpaired electrons	Have high number of unpaired electrons.
Eg: $[Co(NH_3)_6]^{3+}$, $[Co(CN)_6]^{4-}$	Eg: $[MnF_6]^{3-}$, $[CoF_6]^{3-}$

Magnetic Properties of Coordination Compounds

- The magnetic moment of the complexes can be easily calculated using the formula

$$\mu = \sqrt{n(n+2)}\text{B.M} \quad \text{where n = no. of unpaired electrons.}$$

Limitation of VBT

- Fails to explain the strength of magnetic field, relative energies of different shapes, colour and spectral properties of complex.

Crystal Field Theory (CFT)

- In crystal field theory, the interaction between ligands and metal ion is purely electrostatic.
- In the influence of ligand field, degeneracy of the d-orbital is destroyed and it splits into two or more energy levels and split the d-orbital into two group e_g and t_{2g} and the energy difference between these two is known as field splitting parameters. (Δ_0)

Crystal Field Splitting in Octahedral Complexes

- The energy of the two e_g orbitals will increase by $(3/5)\Delta_o$ and three t_2g will decrease by$(2/5)\ \Delta_o$. Consequently. In the **e_g set of orbitals has higher energy than t_{2g} orbitals.**
- In general, ligands can be arranged in a series in the order of increasing field strength as given below:

  $I^- < Br^- < SCN^- < Cl^- < S^{2-} < F^- < OH^- < C_2O_4^{2-} < H_2O < NCS < EDTA^{4-} < NH_3 <$ en $< CN^- < CO$

  This is termed on spectrochemical series.
- If $\Delta_0 > P$ (Pairing energy) (i.e., Δ_0 large, strong field ligand) $\rightarrow$ electron pair-up in t_{2g} orbital $\rightarrow$ forms low spin complex.
- If $\Delta_0 < P$ (Δ_0 small, weak field ligand) $\rightarrow$ electron spread out $\rightarrow$ form high spin complex.

Crystal Field Splitting in Tetrahedral Complexes

- **In this situation, the t_{2g} set of orbital** lie relatively nearer to the approaching ligands and therefore t_{2g} set of d-orbitals **have higher energy than e_g set of orbitals.**
- The relation between Δt and Δ_O for the same metal complex is given as,
- In tetrahedral complexes low spin configurations are rarely observed. $\Delta_t = \left(\frac{4}{9}\right)\Delta_O$

Stability

- More is the stability greater the charge on central metal ion.
- More is the stability more the basic strength of ligand.
- Increases the stability of complex on formation of chelate ring,
- More is the stability, smaller the size of metal ion.
- More is the electronegative and polarising power of the metal ion, more stable is the complex,

Limitation of CFT

- CFT treat the metal ligand bond as purely ionic and do not consider the covalent character of bonding between ligand and central atom.

Bonding in Metal Carbonyls

- The homoleptic carbonyls are formed by most of the transition metals. These carbonyls have simple, well defined structures.
- The metal-carbon bond in metal carbonyls possess both σ and π-characters.
- The M–C σ bond is formed by the donation of lone pair of electrons on the carbonyl carbon into a vacant orbital of the metal.
- The M–C π bond is formed by the donation of a pair of electrons from a filled d-orbital of metal into the vacant anti-bonding π^* orbital of carbon monoxide
- C-O bond in metal carbonyl is long. There is a delocalization of electrons from filled d orbitals into the empty orbitals on the CO ligand.
- The metal to ligand bonding creates a **synergic effect** which strengthens the bond between CO and the metal

Structure of Metal Carbonyls

- Tetracarbonyl nickel (0) is tetrahedral. Some carbonyls have metal-metal bonds. Ex: $[Mn_2(CO)_{10}]$
- The IUPAC nomenclature of $[Mn_2(CO)_{10}]$ is decacarbonyl dimanganese (0) and it is made up of two square pyramidal $Mn(CO)_5$ units joined by Mn-Mn bond.

Application of Complexes

- Fe(III) ion in group III is identified as $[Fe(NCS)_6]^{3-}$ ion which has blood red colour.
- In the electroplating of Ag, large amounts of silver are held as the complex, $K[Ag(CN)_2]$.
- Haemoglobin is a complex which contains Fe^{+2} ions.
- Chlorophyll is a complex which contains Mg^{+2} ions.
- Vitamin - B_{12} is a complex which contains Co^{3+} ions.

MCQs ?

NCERT Topic-wise MCQs

Werner's Theory of Co-ordination Compounds

1. The number of ions given by $[Pt(NH_3)_6]Cl_4$ in a aqueous solution will be:
   1) Two 2) Three
   3) Five 4) Eleven

2. According to Werner's theory of coordination compounds:
   1) Primary valancy is ionisable
   2) Secondary valancy is ionisable
   3) Primary and secondary valency are ionisable
   4) Primary and secondary valency are non-ionisable

3. According to Werner's theory, the stereochemistry of a complex is determined by:
   1) Only primary valencies
   2) Only by secondary valencies
   3) Both of these
   4) None of these

4. The coordination number of the central metal atom is 6 in:
   1) $[Ni(NH_3)_6]^{2+}$ 2) $[Co(en)_3]^{2+}$
   3) $[Fe(C_2O_4)_3]^{3-}$ 4) All of these

5. The central metal ion iron is present in ferric ion is:
   1) $K_4[Fe(CN)_6]$ 2) $K_3[Fe(CN)_6]$
   3) Both of these 4) None of these

6. Fill in the blanks:-
   According to Werner's theory, there are ____ kinds of valency, ____ and ____. The primary valency, of a central metal ion is satisfied with _____.
   1) Three, negative, positive, cations.
   2) Different, negative, positive, anions
   3) Two, primary, secondary, anions.
   4) Two, saturated, unsaturated, cations.

7. The oxidation number of Co in $[Co(NH_3)_3(NO_2)_3]$ is:
   1) +3 2) 0
   3) –3 4) +6

8. According to Werner's theory, the secondary valencies of a central metal atom correspond to its:
   1) Oxidation state
   2) Co-ordination number
   3) Any of the two above
   4) Neither of the (1) and (2)

9. The molecular formula of various hexacoordinate complexes are given as below.
   (a) $CrCl_3.6NH_3$ (b) $CrCl_3.5NH_3$
   (c) $CrCl_3.4NH_3$
   If the number of NH_3 ligands attached to central metal ion respectively are 6, 5 and 4, then the primary valencies in (a), (b) and (c) respectively are :
   1) 3, 3, 3 2) 0, 1, 2
   3) 3, 2, 1 4) 6, 5, 4

10. When $[Co(NH_3)_6]Cl_3$ ionises in water, the number of ions produced are :
   1) 4 2) 6
   3) 3 4) 2

11. In which of the following transition metal complexes, does the metal exhibit zero oxidation state?
   1) $[Co(NH_3)_6]Cl_3$ 2) $[Fe(H_2O)_6]SO_4$
   3) $[Ni(CO)_4]$ 4) $[Fe(H_2O)_6]Cl_3$

12. The number of moles of ions produced when one mole of Potassium Ferricyanide is dissolved in water is
   1) 2 2) 4
   3) 5 4) 6

13. Silver Chloride dissolves in excess ammonia due to the formation of a soluble complex whose formula is:
   1) $[Ag(NH_3)]Cl$ 2) $[Ag(NH_3)_2]Cl$
   3) $[Ag(NH_3)_3]Cl$ 4) $[Ag(NH_3)_4]Cl$

14. Copper sulphate solution forms blue coloured complex with excess of ammonia. Its formula is:
   1) $[Cu(NH_3)_4]^{+3}$ 2) $[Cu(NH_3)_4]^{+2}$
   3) $[Cu(NH_4)_3]^{+2}$ 4) $[Cu(NH_4)_3]^{+3}$

15. Which of the following cannot act as a ligand?
   1) BF_3 2) NH_3
   3) NO^+ 4) CN^-

16. Total number of moles of ions that can be obtained from each mole of $[Co(NH_3)_3Cl_3]$ is:
   1) 1 2) 0
   3) 3 4) 4

17. The number of ions formed in aqueous solution by the compound $[Co(NH_3)_4Cl_2]Cl$ is:
   1) 2 2) 3
   3) 4 4) 5

18. Which of the following does not give a precipitate either with $AgNO_3$ or $BaCl_2$?
1) $[Co(NH_3)_5Cl]SO_4$ 2) $[Co(NH_3)_3Cl_3]$
3) $[Co(NH_3)_4Cl_2]Cl$ 4) $[Co(NH_3)_2Cl_4]Cl$

19. The no. of moles of AgCl obtained when excess $AgNO_3$ is added to one mole of $[Cr(NH_3)_4Cl_2]Cl$:
1) 1 2) 2
3) 3 4) 4

20. An example of double salt is :
1) Bleaching powder 2) $K_4[Fe(CN)_6]$
3) Hypo 4) Potash alum

21. Which of the following is not a double salt but is a complex salt ?
1) $KCl.MgCl_2.6H_2O$
2) $FeSO_4.(NH_4)_2SO_4.6H_2O$
3) $K_2SO_4.Al_2(SO_4)_3.2H_2O$
4) $4KCN.Fe(CN)_2$

22. The groups satisfying the secondary valencies of a cation in a complex are called:
1) Radicals
2) Electron deficient Molecules
3) Primary valencies
4) Ligands

23. The primary valency of the metal ion is satisfied by:
1) Neutral molecules 2) Positive ions
3) Negative ions 4) All of the above

24. No of ionizable & non-ionizable Cl^- ions in $CoCl_3.5NH_3$ respectively are:
1) 3, 0 2) 2, 1
3) 1, 2 4) 0, 3

Definitions of Some Important Terms Related to Coordination Compound

25. Which of the following ligands form a chelate?
1) Acetate 2) Oxalate
3) Cyanide 4) Ammonia

26. The complex ion which has no d-electrons in the central metal atom is:
1) $[MnO_4]^-$ 2) $[Co(NH_3)_6]^{3+}$
3) $[Fe(CN)_6]^{3-}$ 4) $[Cr(H_2O)_6]^{3+}$

27. Which of the following is not a neutral ligand?
1) H_2O 2) NH_3
3) NO_2 4) CO

28. The ligand $N(CH_2CH_2NH_2)_3$ is:
1) Bidentate 2) Tridendate
3) Tetradentate 4) Pentadentate

29. Metal attached with EDTA in an octahedral complex, has coordination number:
1) 3 2) 4
3) 5 4) 6

30. Among the following the ambidentate ligand is / are:
1) NO_2^- 2) CN^-
3) Cl^- 4) Both (1) & (2)

31. Ligand act as:
1) Bronsted acids 2) Bronsted bases
3) Lewis acids 4) Lewis bases

32. The ligand, sulphite is:
1) SO_4^{2-} 2) S^{2-}
3) SO_3^{2-} 4) SO_2

33. Among the following, the bidentate ligand(s) is/are:
1) Pyridine 2) Ethane-1, 2-diamine
3) Dimethyl glyoximato 4) Both (2) and (3)

34. The hexadentate ligand is:
1) PPh_3 2) DMG
3) EDTA 4) en

35. The heteroleptic complex (es) is/are:
1) $[Co(NH_3)_6]Cl_3$ 2) $[Co(NH_3)_4(ONO)_2]$
3) $[Cr(NH_3)_4(Cl)_2]^+$ 4) Both (2) and (3)

36. Which of the following ligands will not show chelation?
1) EDTA 2) DMG
3) Ethane-1, 2-diamine 4) SCN^-

37. Transition elements form complexes readily because of having:
(i) Small size of cation (ii) Large ionic charge
(iii) Vacant d orbitals
The correct statement is/are:
1) (i) only 2) (ii) only
3) (i) & (ii) 4) (i), (ii) & (iii)

38. Diethylenetriamine is :
1) Chelating agent 2) Tridentate neutral molecule
3) Tridentatemonoanion 4) (1) & (2) both

39. Which of the following are bidentate monoanion ligands ?
(1) Acetylacetonato
(2) Oxalato ion
(3) Dimethylglyoximato
Select the correct answer using the codes given below :
1) 1 only 2) 1 and 3 only
3) 3 only 4) 2 and 3 only

40. Match the following

	Column-I		Column-II
(a)	en	(i)	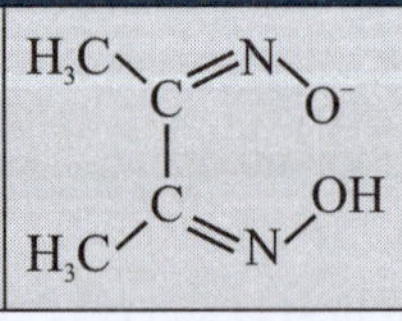

(b)	dmg	(ii)	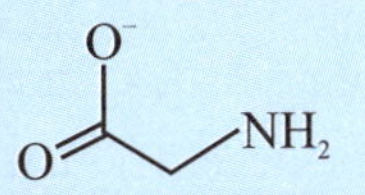
(c)	EDTA	(iii)	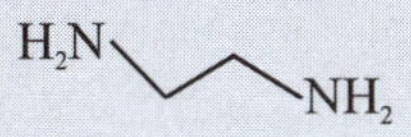
(d)	gly	(iv)	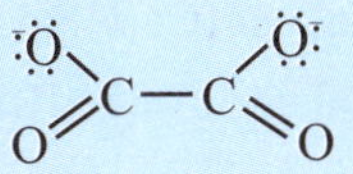
(e)	ox	(v)	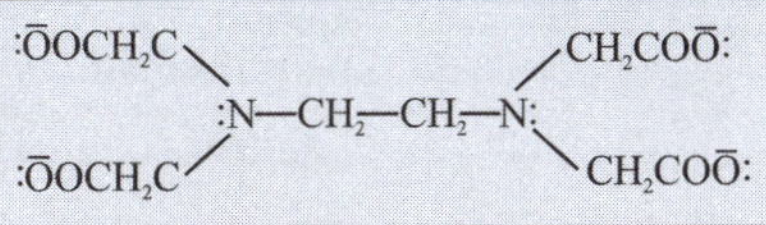

	(a)	(b)	(c)	(d)	(e)
1)	(iii)	(i)	(v)	(ii)	(iv)
2)	(iii)	(i)	(v)	(ii)	(iv)
3)	(i)	(iv)	(ii)	(iii)	(v)
4)	(iv)	(ii)	(v)	(i)	(iii)

41. Among the following which are ambidentate ligands?

A. NO_2^- B. NO_3^-
C. $EDTA^{4-}$ D. $C_2O_4^{2-}$
E. SCN^- F. $H_2NCH_2CH_2NH$

1) A and B 2) C and E
3) A and F 4) A and E

42. Which one of the following is a tridentate ligand?

1) NO_2^- 2) oxalate ion
3) glycinate ion 4) diene

43. The oxidation state of Ni in $K_2[Ni(CN)_4]$:

1) +1 2) +2
3) +3 4) +4

44. Which of the following is cationic complex?

1) $K_4[Fe(CN)_6]$ 2) $[Ni(CO)_4]$
3) $[Co(NH_3)_3Cl_3]$ 4) $[Cu(NH_3)_4]SO_4$

45. Which of the following ligands form a chelate?

1) Acetate 2) Oxalate
3) Cyanide 4) Ammonia

46. The number of ligands in $K_4[Fe(CN)_6]$:

1) 4 2) 6
3) 8 4) 5

47. Ligand in a metal carbonyl complex is:

1) NH_3 2) CO
3) CN^- 4) SCN^-

48. Which is a coordination compound?

1) Ferrous ammonium sulphate
2) Carnallite
3) Potassium ferrocyanide
4) Gypsum

49. The coordination number and oxidation number of Co in $[Co(NH_3)_5SO_4]Cl$ will be:

1) 6 and 3 2) 2 and 6
3) 10 and 3 4) 6 and 4

50. 2 , 4 Dinitrophenylhydrazine is an example for:

1) Tridentate ligand
2) Monodentate ligand
3) Polydentate ligand
4) Didentate ligand

Nomenclature of Coordination Compounds

51. Which among the following will be named as dibromidobis (ethylene diamine) chromium (III) bromide?

1) $[Cr(en)_2Br_2]Br$ 2) $[Cr(en)Br_4]^-$
3) $[Cr(en)Br_2]Br$ 4) $[Cr(en)_3]Br_3$

52. The correct IUPAC name of the following compound is: $[Cr(NH_3)_5(NCS)][ZnCl_4]$

1) Pentaammineisothiocyanato chromium (III) tetra-chlorozincate (II)
2) Pentammineisothiocyanate zinc chlorido chromate (III)
3) Pentaammine isothiocyanato chromate (II)
4) Isothiocyanato pentaamine chromium (II) zinc chloride (III)

53. The IUPAC name of the complex $[Pt(NH_3)_3Br(NO_2)Cl]Cl$ is:

1) Triamminechlorobromonitroplatinum (IV) chloride
2) Triamminebromonitrochloroplatinum (IV) chloride
3) Triamminebromochloronitroplatinum (IV) chloride
4) Triamminenitrochlorobromoplatinum (IV) chloride

54. The IUPAC name of $K_2[Ni(CN)_4]$ is:

1) Potassium tetracyanonickelate (II)
2) Potassium tetracyanatonickelate (III)
3) Potassium tetracyanatonickel (II)
4) Potassium tetracyanonickel (III)

55. The IUPAC name of the complex ion formed when gold dissolves in aqua - regia is:

1) Tetra chloridoaurate (III)
2) Tetra chloridoaurate (I)
3) Tetra chloridoaurate (II)
4) Dichloridoaurate (III)

56. The IUPAC name of $[Cr(H_2O)_4Cl_2]Cl$ is:

1) Tetraaquadichlorochromate (III) chloride
2) Tetraaquadichlorochromium (III) chloride
3) Tetraaquadichlorochromium (I) chloride
4) None of these

57. The IUPAC name of the compound $[Cu(NH_3)_4(NO_3)_2]$ is:
1) Cuprammonium nitrate
2) Tetraamminedinitrato copper (II)
3) Tetrramminecopper (II) dinitrate
4) Tetraamminecopper (II) nitrate

58. The IUPAC name of potassium ferrocyanide is:
1) Potassium hexacyano iron (II)
2) Potassiu iron (II) hexacyanide
3) Potassium hexacyanoferrate (II)
4) Potassium hexacyanoferrate (III)

59. Correct formula of tetraaminechloronitro platinum (IV) sulphate can be written as:
1) $[Pt(NH_3)_4(ONO)Cl]SO_4$
2) $[Pt(NH_3)_4Cl_2NO_2]_2\ SO_4$
3) $[Pt(NH_3)_4(NO_2)Cl]SO_4$
4) $[PtCl(ONO)NH_3(SO_4)]$

60. The IUPAC name for $[Co(NCS)(NH_3)_5]Cl_2$ is:
1) Pentaamminethiocyanato-N-cobalt(III) chloride
2) Pentaamminethiocyanato-S-cobalt(III) chloride
3) Pentaamineisothiocyanato-N,S-cobalt(III) chloride
4) Pentaammine (mercapto-N) cobalt(III) chloride

61. The IUPAC name of the complex $[Pt(NH_3)_2Cl_2]$ is:
1) Diamminedichloridoplatinum (IV)
2) Diamminedichloridoplatinum (II)
3) Diamminedichloridoplatinum (0)
4) Dichloridodiammineplatinum (IV)

62. IUPAC name for the complex $[Cu(NH_3)_4]SO_4$ is:
1) Cuprammonium sulphate
2) Copper sulphate tetraammonia
3) Tetraamminecopper (II) sulphate
4) Copper ammonium (IV) sulphate

63. Hexaaquatitanium (III) Chloride is represented as:
1) $[Ti(H_2O)_6]Cl_3$ 2) $[TiCl_3]6H_2O$
3) $[Ti(H_2O)_5Cl]Cl_2$ 4) $[Ti(H_2O)_3Cl_3]$

64. The correct IUPAC name of $KAl\ (SO_4)_2.12H_2O$ is:
1) Aluminium potassium sulphate –12 water
2) Potassium aluminium (III) sulphate –12 water
3) Potassium aluminium (III) sulphate hydrate
4) Aluminium (III) potassium sulphate hydrate 12.

65. The IUPAC name for the coordination compound $Ba[BrF_4]_2$ is:
1) Barium tetrafluorobromate (V)
2) Bariumtetrafluorobromate (III)
3) Bariumbis (tetrafluorobromate) (III)
4) None of these

66. The IUPAC name of $[Ni(NH_3)_4]\ [NiCl_4]$ is
1) Tetrachloronickel(II) - tetraamminenickel(II)
2) Tetraamminenickel(II) - tetrachloronickel(II)
3) Tetraamminenickel(II) - tetrachloronickelate(II)
4) Tetrachloronickel(II) - tetraamminenickelate(0)

67. The IUPAC name of $[Co(NH_3)_5ONO]^{2+}$ ion is
1) Pentaamminenitritocobal(III) ion
2) Pentaamminenitrocobalt(III) ion
3) Pentaamminenitrocobalt(IV) ion
4) Pentaamminenitritocobalt(IV) ion

68. IUPAC name of $Na_3[CoCl(NO_2)_5]$ is:
1) Sodium chloropentanitrocobaltate(III)
2) Sodium cobaltnitrate
3) Trisodium chloropentanitro cobalt
4) Pentanitrocobalt (III) trisodium complex

69. The IUPAC name of the Wilkinsons catalyst $[RhCl(P\,Ph_3)_3]$ is:
1) Chlorotris(triphenylphosphine) rhodium (I)
2) Chlorotris(triphenylphosphine) rhodium (IV)
3) Chlorotris(triphenylphosphine) rhodium (0)
4) Chlorotris (triphenylphosphine) rhodium (VI)

Isomerism in Coordination Compounds

70. $[Pt(NH_3)_4][CuCl_4]$ and $[Cu(NH_3)_4]\ [PtCl_4]$ are known as:
1) Ionisation isomers 2) Coordination isomers
3) Linkage isomers 4) Polymerisation isomers

71. Two isomers of a compound $[Co\ (NH_3)_3Cl_3]$ (MA_3B_3type) are shown in figures

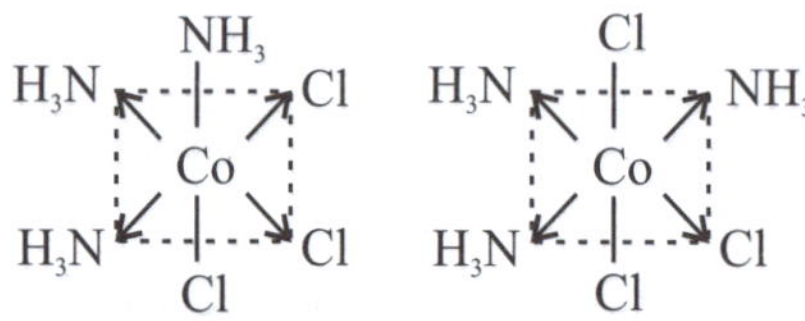

(i) (ii)

The isomers can be classified as:
1) (i) fac-isomers (ii) mer-isomers
2) (i) optical-isomers (ii) trans-isomers
3) (i) mer-isomers (ii) fac-isomer
4) (i) trans-isomers (ii) cis-isomer

72. Which of the following shows maximum number of isomers?
1) $[Co(NH_3)_4Cl_2]$ 2) $[Ni(en)(NH_3)_4]^{2+}$
3) $[Ni(C_2O_4)(en)_2]^{2-}$ 4) $[Cr(SCN)_2(NH_3)_4]^+$

73. Compounds $[PtCl_2(NH_3)_4]Br_2$ and $[PtBr_2(NH_3)_4]Cl_2$ shows the following type of isomerism:
1) Hydrate isomerism
2) Ionisation isomerism
3) Linkage isomerism
4) Coordination position isomerism

74. The complex $[Co(NH_3)_6][Cr(CN_6)]$ and $[Cr(NH_3)_6][Co(CN)_6]$ are the examples of which types of isomerism:
1) Geometrical isomerism
2) Linkage isomerism
3) Ionisation isomerism
4) Coordination isomerism

75. The existence of two different coloured complexes with the composition $[Co(NH_3)_4Cl_2]^+$ is due to:
1) Ionisation Isomerism
2) Linkage Isomerism
3) Geometrical Isomerism
4) Coordination Isomerism

76. The isomerism which is possible with complex, $[Co(NH_3)_5Br]SO_4$, is:
1) Linkage isomerism
2) Hydrate isomerism
3) Coordination isomerism
4) Ionisation isomerism

77. Linkage isomerism is possible when ligand is:
1) Neutral 2) Negative
3) Ambidentate 4) Positive

78. In the octahedral complex $[Ma_3b_3]$, If three identical ligands occupy positions at the corners of the same face of octahedron, it is called:
1) cis-Isomer 2) trans-Isomer
3) Fac-isomer 4) mer-isomer

79. The complex $[Co(NH_3)_4(SCN)_2]Cl$ exhibits the isomerism which are:
1) Linkage, ionization and geometrical
2) Linkage, ionization and hydrate
3) Ionization, geometrical and coordination
4) Linkage, geometrical and polymerization

80. $[Co(NH_3)_4Cl_2]NO_2$ and $[Co(NH_3)_4Cl(NO_2)]Cl$ exhibit which type of isomerism?
1) Geometrical 2) Optical
3) Linkage 4) Ionisation

81. Three arrangement are shown for the complex $[Co(en)(NH_3)_2Cl_2]^+$. Pick up the wrong statement.

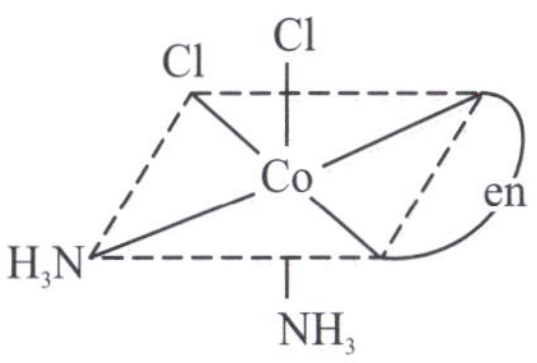

(I)

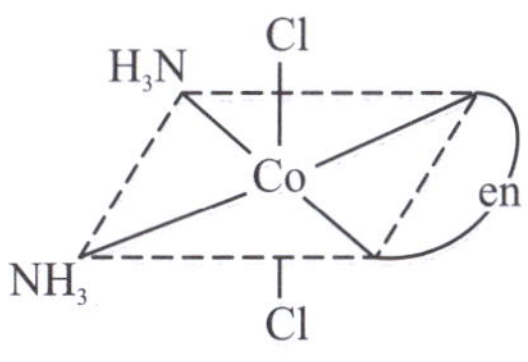

(II)

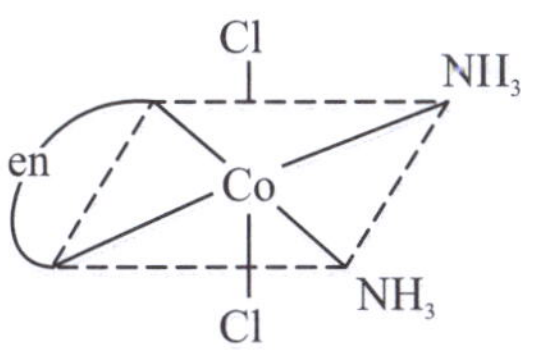

(III)

1) I and II are geometrical isomers
2) II and III are optical isomers
3) I and III are optical isomers
4) II and III are geometrical isomers

82. Which of the following complex species is not expected to exhibit optical isomerism?
1) $[Co(en)(NH_3)_2Cl_2]^+$ 2) $[Co(en)_3]^{3+}$
3) $[Co(en)_2Cl_2]^+$ 4) $[Co(NH_3)_3Cl_3]$

83. $CrCl_3.6H_2O$ exists in different isomeric forms which show different colours like violet and green. This is due to:
1) Ionisation isomerism
2) Coordination isomerism
3) Optical isomerism
4) Hydrate isomerism

84. Which of the following does/do not exhibit optical isomerism
1) Tetrahedral complexes 2) Square planar complexes
3) Octahedral complexes 4) Polynuclear complexes

85. The complex, $[Pt(Py)(NH_3)BrCl]$ will have how many geometrical isomers?
1) 2 2) 3
3) 4 4) 5

86. How many geometrical isomers are possible for the complex ion $[Cr(C_2O_4)_3]^{3-}$?
1) 0 2) 1
3) 2 4) 3

87. The complexes $[Co(NH_3)_6]$ $[Cr(C_2O_4)_3]$ and $[Cr(NH_3)_6]$ $[Co(C_2O_4)_3]$ are:
1) Geometrical isomerism
2) Ionization isomerism
3) Co-ordination isomerism
4) Linkage isomerism

88. Which of the following complex or the complex ion will show geometrical isomerism?
1) $[Pt(NH_3)_2Cl_2]$ 2) $[Pt(NH_3)Cl_5]^-$
3) $[Pt(NH_3)_5Cl]^{3+}$ 4) $[Co(NH_3)_6]Cl_3$

89. Which of the following is not optically active?
1) $[Co(en)_3]^{3+}$ 2) $[Cr(ox)_3]^{3-}$
3) cis – $[CoCl_2(en)_2]^+$ 4) trans – $[CoCl_2(en)_2]^+$

90. Cis-trans isomerism is found in square planar complexes of the molecular formula (a and b are monodentate ligands):
1) Ma_4 2) Ma_3b
3) Ma_2b_2 4) Mab_3

91. Which of the following complexes does not show geometrical isomerism ?
1) $[Co(en)(SCN)_4]^-$ 2) $[Pt(en)_2(NH_3)(H_2O)]^{4+}$
3) $[Cr(gly)_3]$ 4) $[Pt (gly)_2Cl_2]^{2+}$

Bonding In Coordination Compounds

92. The magnitude of magnetic moment of $[NiCl_4]^{2-}$ will be:

1) 2.82 B.M 2) 3.25 B.M
3) 1.23 B.M 4) 5.64 B.M

93. Which of the following energy level diagram for $[FeF_6]^{3-}$ is correct on the basis of crystal field theory?

1)
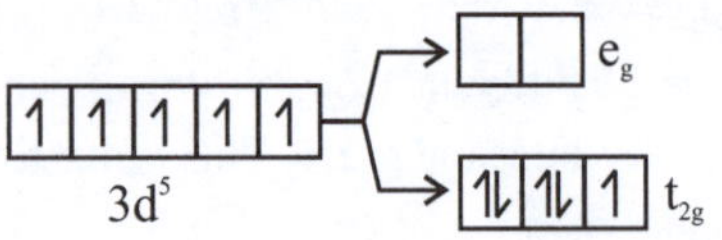

2)
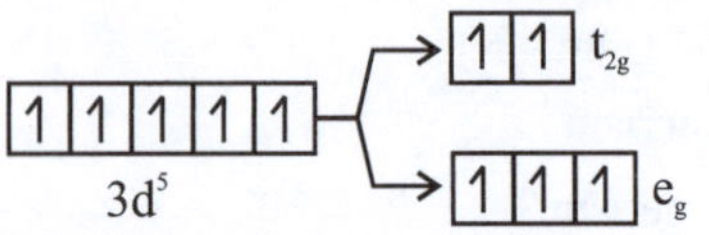

3)
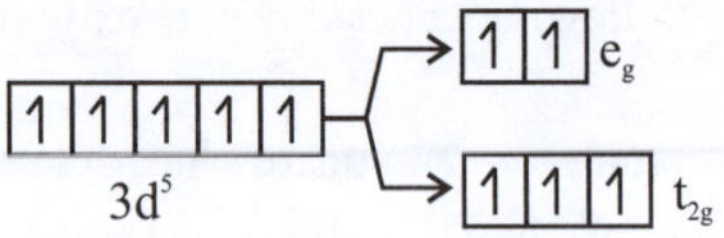

4)
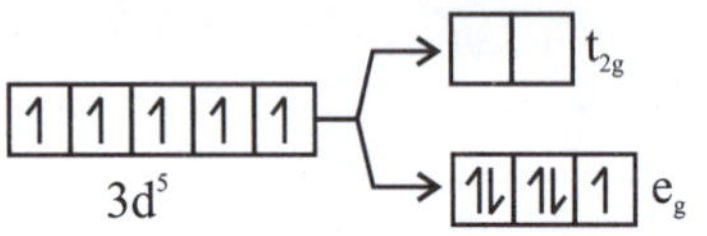

94. $CuSO_4 . 5H_2O$ is blue in colour while $CuSO_4$ is colourless due to:

1) Presence of strong field ligand in $CuSO_4 . 5H_2O$
2) Due to absence of water (ligand), d-d transitions are not possible in $CuSO_4$
3) Anhydrous $CuSO_4$ undergoes d-d transitions due to crystal field splitting
4) Colour is lost due to loss of unpaired electrons.

95. Which one of the following is wrongly match?

1) $[Cu(NH_3)_4]^{2+}$ = Square planar
2) $[Ni(CO)_4]$ = Neutral ligand
3) $[Fe(CN)_6]^{3-} = sp^3d^2$
4) $[Co(en)_3]^{3+}$ = Follows EAN rule

96. The value of the 'spin only' magnetic moment for one of the following configurations is 2.84 BM. the correct one

1) d^4 (in strong ligand field)
2) d^4 (in weak ligand field)
3) d^3 (in weak as well as in strong fields)
4) d^5 (in strong ligand field)

97. The most stable complex among the following is:

1) $[Pd(CN)_4]^{4-}$ 2) $[Fe(CO)_5]$
3) $[Ni(CN)_4]^{4-}$ 4) $[Ni(CN)_4]^{3-}$

98. The complex ion which has only three d-electrons in the central metal ion is:

1) $[Co(NH_3)_6]^{3+}$ 2) $[MnO_4]^-$
3) $[Fe(CN)_6]^{3-}$ 4) $[Cr(CN)_6]^{3-}$

99. According to CFT, Δ_o is called:

1) Pairing energy
2) Crystal field excitation energy
3) Crystal field splitting energy
4) Resonance energy

100. In crystal field splitting, in octahedral complexes, the energy order of d-orbitals is:

1) $e_g > t_{2g}$ 2) $t_{2g} > e_g$
3) $d_{xy} > d_{x^2-y^2}$ 4) $d_{y^3} > d_{zx}$

101. The complex, $[Ni(CN)_4]^{2-}$, is diamagnetic in nature, its geometry is:

1) Tetrahedral 2) Square planar
3) Octahedral 4) Unpredictable

102. Which of the following complex ion is expected to absorb visible light?

1) $[Sc(H_2O)_3(NH_3)_3]^{3+}$
2) $[Ti(en)_2(NH_3)_2]^{4+}$
3) $[Cr(NH_3)_6]^{3+}$
4) $[Zn (NH_3)_6]^{2+}$

103. What will be the theoretical value of magnetic momentum (μ) when CN^- ligands join Fe^{3+} ion to yield complex?

1) 3.87 BM 2) 2.83 BM
3) 1.73 BM 4) 5.92 BM

104. The increasing order of crystal field spitting strength of the given ligands is:

1) $NH_3 < Cl^- < CN^- < F^- < CO < H_2O$
2) $F^- < Cl^- < NH_3 < CN^- < H_2O < CO$
3) $Cl^- < F^- < H_2O < NH_3 < CN^- < CO$
4) $CO < CN^- < NH_3 < H_2O < F^- < Cl^-$

105. The spin only magnetic moment value of $Cr(CO)_6$ is:

1) 2.84 BM 2) 4.90 BM
3) 5.92 BM 4) 0 BM

106. The hybridisation involved in $[Co(C_2O_4)_3]^{3-}$ is:

1) sp^3d^2 2) sp^3d^3
3) dsp^3 4) d^2sp^3

107. Which of the following complexes will show maximum paramagnetism?

1) $3d^4$ 2) $3d^5$
3) $3d^6$ 4) $3d^7$

108. Which of the following complex shows zero crystal field stabilisation energy?

1) $[Co(H_2O)_6]^{3+}$ 2) $[Cr(H_2O)_6]2^+$
3) $[Co(H_2O)_6]^{2+}$ 4) $[Fe(H_2O)_6]^{3+}$

109. Bonds present in $K_4 [Fe (CN)_6]$ are:

1) Only ionic
2) Only covalent
3) Ionic and Covalent
4) Ionic, covalent and coordinate covalent

110. The hybridisation of the complex $[NiCl_4]^{-2}$ is:
1) sp^3 2) dsp
3) sp^3d 4) sp^3d^2

111. The correct order for the CFSE (numerical value) for the following complexes is

Complex	P	Q	R	S
Formula	$[CoF_6]^{3-}$	$[Co(CN)_6]^{3-}$	$[Co(NH_3)_6]^{3+}$	$[Co(H_2O)_6]^{3-}$

1) P > Q > R > S 2) Q > R > S > P
3) S > R > P > Q 4) R > Q > P > S

112. The shape of the complex species will be square planar if its coordination number is:
1) 2 2) 6
3) 5 4) 4

113. Which of the following is outer orbital complex?
1) $[CoF_6]^{-3}$
2) $[Cu(H_2O)_6]^{+2}$
3) $[Co(NH_3)_6]^{+2}$
4) Both (1) and (2)

114. Hybridisation of $[Ni(NH_3)_6]^{2+}$ and $[Zn(NH_3)_6]^{2+}$ are:
1) Both d^2sp^3
2) Both sp^3d^2
3) d^2sp^3 and sp^3d^2 respectively
4) sp^3d^2 and d^2sp^3 respectively

115. $[Cu(NH_3)_4]^{2+}$ has hybridisation and magnetic moment?
1) sp^3, 1.73 B.M. 2) sp^3d, 1.73 B.M.
3) dsp^2, 2.83 B.M. 4) dsp^2, 1.73 B.M.

116. Among the following compounds that is both paramagnetic and coloured is:
1) $K_2Cr_2O_7$ 2) $(NH_4)_2[TiCl_6]$
3) $VOSO_4$ 4) $K_3[Cu(CN)_4]$

Bonding in Metal Carbonyls

117. What is the oxidation number of Fe in Fe $(CO)_5$?
1) +3 2) Zero
3) +2 4) +5

118. The charge on the central metal ion in the complex $[Ni(CO)_4]$ is:
1) +2 2) +4
3) 0 4) +1

119. Oxidation state of cobalt in $Na[Co(CO)_4]$ is
1) + 1 2) 0
3) –2 4) –1

Importance and Applications of Coordination Compounds

120. Zeigler - Natta catalyst is:
1) $ZnCl_2$ 2) $Et_3Al + TiCl_4$
3) V_2O_5 4) Pt

121. Which of the following is used as antidote in lead poisoning?
1) EDTA 2) $CoCl_3$
3) DMG 4) Cis-platin

122. Which of the following is organometallic compound?
1) CH_3MgBr 2) $(CH_3COO)_2Pb$
3) CH_3COONa 4) All of these

123. The central metal ion in heme is:
1) Fe^{+2} 2) Co^{+3}
3) Mn^{+2} 4) Mg^{+2}

124. The coordination number of Fe(II) in oxyhaemoglobin is
1) 6 2) 4
3) 8 4) 10

High Order Time Intensive MCQs

1. Which of the following complexes is non - conducting?
1) $CoCl_3{\cdot}3NH_3$ 2) $CoCl_3{\cdot}6NH_3$
3) $CoCl_3{\cdot}4NH_3$ 4) $CoCl_3{\cdot}5NH_3$

2. Which of the following statement is not correct ?
1) $[Ni(CN)_4]^{2-}$ involves dsp^2 hybridisation
2) $[Ni(CN)_4]^{2-}$ is tetrahedral
3) $[Fe(H_2O)_6]^{3+}$ is inner orbital complex
4) Both $[Co(ox)_3]^{3-}$ and $[COF_6]^{3-}$ are outer orbital complexes.

3. In which of the following octahedral complexes of Co(At no. 27) will the magnitude of Δ_0 be the highest ?
1) $[Co(NH_3)_6]^{3+}$ 2) $[Co(CN)_6]^{3-}$
3) $[Co(C_2O_4)_3]^{3-}$ 4) $[Co(H_2O)_6]^{3+}$

4. Which of the following pairs represents linkage isomers?
1) $[Cu(NH_3)_4][Pt(NH_3)]$ and $[Pt(NH_3)_4]$ $[CuCl_4]$
2) $[Pd(PPh_3)_2(NCS)_2]$ and $[Pd\ (PPh_3)_2(SCN)_2]$
3) $[Co(NH_3)_5(NO)_3]SO_4$ and $[Co(NH_3)_5(SO_4)]NO_3$
4) $[PtCl_2(NH_3)_4]Br_2$ and $[PtBr_2(NH_3)_4]Cl$

5. Which one of the following has an optical isomer ?
1) $[Co(H_2O)_4(en)]^{3+}$ 2) $[Zn(en)_2]^{2+}$
3) $[Zn(en)(NH_3)_2]^{2+}$ 4) $[Co(en)_3]^{3+}$

6. Which of the following facts about the complex $[Cr(NH_3)_6]Cl_3$ is wrong ?
1) The complex involves d^2sp^3 hybridisation and is octahedral in shape.

2) The complex is paramagnetic.
3) The complex is an outer orbital complex.
4) The complex gives white precipitate with silver nitrate solution.

7. In the brown ring complex $[Fe(H_2O)_5(NO)]SO_4$, nitric oxide behaves as

1) NO^+ 2) NO
3) NO^- 4) NO^{2-}

8. which of the following is a negatively charged bidentate ligand?

1) dimethyl glyoximato 2) Cyano
3) Ethylene diamine 4) Acetato

9. The effective atomic number for $[Rh(H_2O)_6]^{3+}$ (atomic number for Rh is 45) is

1) 42 2) 45
3) 48 4) 54

10. Amongst $[NiCl_4]^{2-}$, $[Ni(H_2O)_6]^{2+}$ $[Ni(PPh_3)_2Cl_2]$, $[Ni(CO)_4]$ and $[Ni(CN)_4]^{2-}$ the paramagnetic species are

1) $[NiCl_4]^{2-}$,$[Ni(H_2O)_6]^{2+}$, $[Ni(PPh_3)_2]$
2) $[Ni(CO)_4]$, $[Ni(PPh_3)_2Cl_2]$, $[NiCl_4]^{2-}$
3) $[Ni(CN)_4]^{2-}$ $[Ni(H_2O)_6]^{2+}$,$[NiCl_4]^{2-}$
4) $[Ni(PPh_3)_2Cl_2]$, $[Ni(CO)_4]$,$[Ni(CN)_4]^{2-}$

11. Amongst the following ions which one has the highest magnetic moment value? (At. no. :Co = 27, Ni = 28)

1) $[Co(NH_3)_6]^{3+}$ 2) $[CoF_6]^{3-}$
3) $[NiCl_4]^{2-}$ 4) $[Ni(CN)_4]^{2-}$

12. The ionization isomer of $[Cr(H_2O)_4Cl(NO_2)]Cl$ is

1) $[Cr(H_2O)_4(O_2N)]Cl_2$ 2) $[Cr(H_2O)_4Cl_2]NO_2$
3) $[Cr(H_2O)_4Cl(ONO)]Cl$ 4) $[Cr(H_2O)_3Cl_2(NO_2)]H_2O$

13. Consider the following complex ions, P, Q and R.

P = $[FeF_6]^{3-}$, Q = $[V(H_2O)_6]^{2+}$and R = $[Ni(H_2O)_6]^{2+}$. The correct order of the complex ions, according to their spin only magnetic moment values (in B.M.) is

1) R< Q <P 2) Q < R <P
3) R < P < Q 4) Q < P < R

14. $[Mn(CO)_6]$ and $[V(CO)_6]$ can act as respectively:

1) Oxidising agent, oxidising agent
2) Reducing agent, reducing agent
3) Reducing agent, oxidising agent
4) Oxidising agent, reducing agent

15. Only cis Pt (NH_3) Cl_2 reacts with oxalic acid $(H_2C_2O_4)$ to form $[Pt\ Cl_2(ox)]^{2-}$ and not the trans isomer because:

1) Oxalate ion is a monodentate ligand
2) Oxalate ion is an unsymmetrical bidentate ligand
3) The oxalate ion is a bidentate ligand which occupies adjacent positions only
4) None of these

16. Which of the following is/are correct?

1) After splitting of d-orbitals in octahedral field, Hund's rule is not violated anywhere
2) NO is a $3e^-$ donor
3) Dithioxalate is a bidentate ambidentate ligand
4) All are correct

17. Mabcd (tetrahedral geometry) can show which one of following isomerism?

1) Geometrical 2) Optical
3) Co-ordination position 4) Both (1) and (3)

18. The magnetic moment of octahedral complex having CFSE $= -0.8\ \Delta_0$ and surrounded by weak field ligands can be:

1) $\sqrt{15}$ B.M. 2) $\sqrt{8}$ B.M.
3) $\sqrt{22}$ B.M. 4) Both (1) and (2)

19. Which of the following can show optical isomerism?

1) MA_4B_2 2) MA_3B_3
3) MA_4BC 4) None of these

(Here A, B, and C are monodentate ligands)

20. Which of the compounds show ionization isomerism?

1) $[Cr(NH_3)_6]Cl_3$ 2) $[Cr(en)_3]Cl_3$
3) $[Co(NH_3)_5Br]SO_4$ 4) Both (1) and (2)

21. Which of following is d^2sp^3 hybridised?

1) $[NiF_6]^{2-}$ 2) $[IrCl_6]^{3-}$
3) $[Co(H_2O)_6]^{3+}$ 4) All of these

22. A compound is made by mixing cobalt(III) nitrite and potassium nitrite solutions in the ratio of 1 : 3. The aqueous solution of the compound showed 4 particles per molecule whereas molar conductivity reveals the presence of six electrical charges. The formula of the compound is :

1) $Co(NO_2)_3 . 2KNO_2$ 2) $Co(NO_2)_3 . 3KNO_2$
3) $K_3[Co(NO_2)_6]$ 4) $K[Co(NO_2)_4]$

23. Match the geometry given in column A with the complex given in column B using the codes given below :

Column (A)		Column (B)	
(a)	Tetrahedral	(i)	$[Cu(NH_3)_4]^{2+}$
(b)	Octahedral	(ii)	$[Ag(NH_3)_2]^+$
(c)	Square planar	(iii)	$Fe(CO)_5$
(d)	Trigonal bipyramidal	(iv)	$[Cr(H_2O)_6]^{3+}$
(e)	Linear	(v)	$[NiCl_4]^{2-}$

1) (a)-(v), (b)-(iv), (c)-(i), (d)-(iii) ,(e)-(ii)
2) (a)-(iv), (b)-(v), (c)-(i), (d)-(iii) ,(e)-(ii)
3) (a)-(iv), (b)-(v), (c)-(ii), (d)-(i) ,(e)-(iii)
4) (a)-(iii), (b)-(v), (c)-(ii), (d)-(i) ,(e)-(iv)

24. Which of the following complex species is not expected to show optical isomerism?

1) $[Co(NH_3)Cl_3]$ 2) $[Co(en)(NH_3)_2Cl_2]^+$
3) $[Co(en)_3]^{3+}$ 4) $[Co(en)_2Cl_2]^+$

25. Which of the following complexes is example of strongest reducing agent?

1) $[Co(CN)_6]^{4-}$ 2) $[Co(CN)_6]^{3-}$
3) $[Ag(CN)_2]^-$ 4) $[Cu(CN)_4]^{3-}$

26. For the octahedral complexes of Fe^{3+} in SCN^- (thiocyanate – S) and in CN^- ligand environments, the difference between the spin only magnetic moments (in B.M) is:

1) 4 2) 3
3) 7 4) 8

27. Match the following using the codes given below :

1.	Double salt	a.	$[Co(NH_3)_3Cl_3]$
2.	Zeise's salt	b.	Hexadentate
3.	Neutral molecule	c.	Bidentate
4.	EDTA	d.	Paramagnetic
5.	$Ni(CO)_4$	e.	$FeSO_4.(NH_4)_2SO_4.\ 6H_2O$
6.	$[Cr(NH_3)_6]^{3+}$	f.	$K_4Fe(CN)_6$
7.	Low spin complex	g.	Diamagnetic
8.	Glycine	h.	An organometallic compound

1) (1.e); (2.h); (3.a); (4.b); (5.g); (6.d); (7.f); (8.c).
2) (1.h); (2.e); (3.a); (4.b); (5.g); (6.d); (7.f); (8.c).
3) (1.h); (2.e); (3.a); (4.b); (5.d); (6.g); (7.c); (8.f).
4) (1.h); (2.a); (3.e); (4.b); (5.d); (6.g); (7.f); (8.c)

28. Which of the following complexes has a geometry different from others ?

1) $[Ni\ Cl_4]^{2-}$ 2) $[Ni\ (CO)_4]$
3) $[Ni(CN)_4]^{2-}$ 4) $[Zn(NH_3)_4]^{2+}$

29. Which of the following statements is incorrect ?

1) The stability constant of $[Co(NH_3)_6]^{3+}$ is greater than that of $[Co(NH_3)_6]^{2+}$.
2) Among F^-, Cl^-, Br^- and I^-, F^- forms strongest complexes due to small size.
3) $[Cu(NH_3)_4]^{2+}$ is thermodynamically more stable than $[Zn(NH_3)_4]^{2+}$.
4) Among $[Fe(CN)_6]^{3-}$, $[Fe(H_2O)_6]^{2+}$ and $[Fe(en)_3]^{3+}$, $[Fe(CN)_6]^{3-}$ is most stable.

30. Which of the following complexes show optical isomerism?

1) Cis $[Co(en)_2Cl_2]$ Cl 2) Trans $[Co(en)_2Cl_2]Cl$
3) $[Co(NH_3)_4Cl_2]Cl$ 4) $[Co(NH_3)_3Cl_3]$

31. The maximum number of atoms which are in same plane in $[Ni(CO)_4]$ are:

1) 3 2) 2
3) 4 4) 5

32. The complexes $[Pt(NH_3)_4]$ $[PtCl_6]$ and $[Pt(NH_3)_4Cl_2]$ $[PtCl_4]$ are :

1) Linkage isomers 2) Optical isomers
3) Co-ordination isomers 4) Ionisation isomers

33. Which of the following complexes show geometrical as well as optical isomerism ?

(1) $[Cr(ox)_3]^{3-}$ (2) $[Rh(en)_2Cl_2]^+$
(3) $[Co(NH_3)_2(Cl)_2(en)]^+$

1) 1 only 2) 1 and 2 only
3) 2 and 3 only 4) All 1, 2, 3

34. Which of the following is correct for the Zeise's salt ?

1) The complex ion is square planar.
2) The central metal ion, platinum is in +2 oxidation state.
3) $H_2C = CH_2$ molecules is perpendicular to the $PtCl_3$ plane
4) All of these.

35. Which of the following is not an ambidentate ligand?

1) CN^- 2) SCN^-
3) NH_3 4) NO_2^-

36. The compound(s) that exhibit(s) geometrical isomerism or optical isomerism is/are:

1) $[Pt(en)_3]Cl_4$ 2) $[Pt(NH_3)_2(CN)_2]$
3) $K_2[NiCl_2Br_2]$ 4) Both (1) and (2)

37. A d-block element forms octahedral complex but its spin magnetic moment remains same either in strong field or in weak field ligand. Which of the following is /are correct?

1) Element always forms colourless compound.
2) Number of electrons in t_{2g} orbitals are higher than in e_g orbitals.
3) It can have either d^3 or d^8 configuration.
4) Both (2) and (3).

38. In the crystal field of the complex $[Fe(Cl)\ (CN)_4(O_2)]^{4-}$ the electronic configuration of metal is found to be t_{2g}^6, e_g^0 then which of the following is true about this complex ion :

1) It is a paramagnetic complex
2) Its IUPAC name will be chlorotetracyanosuperoxido-ferrate (II) ion
3) Both (1) and (2)
4) It is a diamagnetic complex

39. If $\Delta_0 < P$, the correct electronic configuration for d^4 system will be

1) $t_{2g}^4 e_g^0$ 2) $t_{2g}^3 e_g^1$
3) $t_{2g}^0 e_g^4$ 4) $t_{2g}^2 e_g^2$

40. The non-existent metal carbonyl among the following is–

1) $[Cr(CO)_6]$ 2) $[Mn(CO)_5]$
3) $[Ni(CO)_4]$ 4) $[Fe(CO)_5]$

41. The nature of Mn_2O_7, V_2O_5 and CrO are, respectively:
1) Acidic, acidic & basic
2) Basic, amphoteric & acidic
3) Acidic, amphoteric & basic
4) Acidic, basic & amphoteric.

42. Which of the following is not correctly matched ?
1) NO_2^- – Bidentate ligand
2) Ethylenediamine–Bidentate ligand
3) SCN^- – Monodentate ligand
4) CO – Monodentate ligand

43. The solution of the complex $[Cu(NH_3)_4]SO_4$ in water :
1) Will give the tests of Cu^{2+} ions.
2) Will give the tests of NH_3 .
3) Will give the tests of SO_4^{2-} ions.
4) Will not give the tests of any of the above species.

44. Dipole moment will be zero in the complexes
1) $[Ni(CN)_4]^{2-}$
2) Cis – $[Pt(NH_3)_2Cl_2]$
3) Trans – $[Pt(NH_3)_2Cl_2]$
4) Both (1) and (3)

45. Which of the following solutions require three moles of $AgNO_3$ for the complete precipitation of all the chloride ions present in it:
1) One litre of 1M $[Co(NH_3)_6]Cl_3$
2) Three litres of 1M $[Co(NH_3)_4Cl_2]Cl$
3) One litre of 1.5M $[Co(NH_3)_5Cl]Cl_2$
4) All the above

46. The IUPAC name of $K_2[Cr(CN)_2O_2(O_2)(NH_3)]$ is:
1) Potassium amminecyanoperoxodioxochromatic (VI).
2) Potassium amminedicyanoperoxodioxochromium (VI).
3) Potassium amminecyanoperoxodioxochromium (VI).
4) Potassium amminedicyanodioxoperoxochromate (VI).

47. Which of the following is true about the complex $[PtCl_2(NH_3)(OH_2)]$? [Atomic number of Pt = 78]
(i) It will have two geometrical isomeric forms, cis and trans
(ii) The hybridisation state of Pt(II) is sp^3
(iii) It is a square planar complex
(iv) It is a diamagnetic complex
(v) It can show hydrate isomerism
(vi) It is a tetrahedral complex
1) (i), (iii), (iv)
2) (ii), (iv), (v)
3) (ii), (v), (vi)
4) (i), (v), (vi)

48. $[Co(NH_3)_4(NO_2)_2]Cl$ exhibits:
1) Linkage isomerism, ionization isomerism and geometrical isomerism
2) Ionisation isomerism, geometrical isomerism and optical isomerism
3) Linkage isomerism, geometrical isomerism and optical isomerism
4) Linkage isomerism, ionization isomerism and optical isomerism.

49. Which is incorrect statement?
1) $[Co(en)_3]$ $[Cr(CN)_6]$ will display coordination isomerism.
2) $[Mn(H_2O)_5(SCN)]$ will display linkage isomerism.
3) $[Co(NH_3)_5(NO_2)]SO_4$ will display ionization isomerism
4) None of these.

50. Which of the following statements are correct?
(A) $[Ni(H_2O)_6]^{2+}$ and $[Ni(NH_3)_6]^{2+}$ have same value of CFSE
(B) $[Ni(H_2O)_6]^{2+}$ and $[Ni(NH_3)_6]^{2+}$ have same value of magnetic moment
(C) $[Ni(H_2O)_6]^{2+}$ and $[Ni(NH_3)_6]^{2+}$ have same hybridisation of nickel
1) (A) and (C) only
2) (B) and (C) only
3) (A) and (B) only
4) All of these

51. Which of the following statements is incorrect?
1) $Ni(CO)_4$ _____Tetrahedral, diamagnetic
2) $[Ni(CN)_4]^{2-}$ _____Square planar, diamagnetic
3) $[Ni(PPh_3)_4]$ _____ Square planar, diamagnetic
4) $[NiCl_4]^{2-}$ _____Tetrahedral, paramagnetic

52. The effective atomic number of central metal ion is wrongly calculated in the following complex?
1) In $[Ni(CO)_4]$, the EAN of Ni is 36
2) In $K_2[Ni(CN)_4]$, the EAN of Ni is 36
3) In $K_3[Fe(CN)_6]$, the EAN of Fe 35
4) In $[Cr(NH_3)_6]Cl_3$, the EAN of Cr is 33

53. Arrange the following in order of decreasing number of unpaired electrons?
I. $[Fe(H_2O)_6]^{2+}$
II. $[Fe(CN)_6]^{3-}$
III. $[Fe(CN)_6]^{4-}$
IV. $[Fe(H_2O)_6]^{3+}$
1) IV, I, II, III
2) I, II, III, IV
3) III, II, I, IV
4) II, III, I, IV

54. AgCl dissolved in excess of NH_3, KCN and $Na_2S_2O_3$ solutions, the complex produces ions:
1) $[Ag(NH_3)_2]^+$, $[Ag(CN)_2]^-$ & $[Ag(S_2O_3)_2]^{3-}$
2) $[Ag(NH_3)_2]^{2+}$, $[Ag(CN)_2]^{3-}$ & $[Ag_4(S_2O_3)_5]^{2+}$
3) $[Ag(NH_3)_2]^{2+}$, $[Ag(CN)_2]^+$ & $[Ag_2(S_2O_3)_2]^{2-}$
4) $[Ag(NH_3)_4]^+$, $[Ag(CN)_4]^{3-}$ & $[Ag_2(S_2O_3)_2]^{2-}$

55. The compound that inhibits the growth of tumors is:

1) cis-$[Pd(Cl_2)(NH_3)_2]$ 2) cis-$[Pt(Cl)_2(NH_3)_2]$

3) trans-$[Pt(Cl)_2(NH_3)_2]$ 4) trans-$[Pd(Cl)_2(NH_3)_2]$

56. Coordination compounds have great importance in biological systems. In this context which of the following statement is incorrect?

1) Chlorophylls are green pigments in plants and contain calcium.

2) Haemoglobin is the red pigment of blood and contains iron.

3) Cyanocobalamin is B_{12} and contains cobalt.

4) Carboxypeptidase-A is an enzyme and contains zinc.

57. Consider the following complex, $[Co(NH_3)_5CO_3]ClO_4$. The coordination number, oxidation number, number of d-electrons and number of unpaired d-electrons on the metal are respectively:

1) 6, 3, 6, 0 2) 7, 2, 7, 1

3) 7, 1, 6, 4 4) 6, 2, 7, 3

58. Which of the following pairs will show the same magnetic moment (spin only)?

1) $[Cr(H_2O)_6]^{3+}$ and $[Fe(H_2O)_5NO]^{2+}$

2) $[Mn(CN)_6]^{4-}$ and $[Fe(CN)_6]^{3-}$

3) $[Ni(CO)_4]$ and $[Zn(NH_3)_4]^{2+}$

4) All of these

59. In the given complex compounds the stable complex (based on EAN rule) is:

1) $K_4[Fe(CN)_6]$

2) $[Co(NH_3)_5Cl]Cl_2$

3) $[Ni(CO)_4]$

4) All the above

60. The effective atomic number of Cr in $[Cr(NH_3)_6]Cl_3$ is:

1) 35 2) 27

3) 33 4) 36

61. The most stable complex among the following is:

1) $[Pd(CN)_4]^{4-}$

2) $[Fe(CO)_5]$

3) $[Ni(CN)_4]^{4-}$

4) $[Ni(CN)_4]^{3-}$

Match the Columns MCQs

1. Match the complex ions given in Column-I with the colours given in Column-II and assign the correct code:

	Column-I (Complex ion)		Column-II (Colour)
(a)	$[Co(NH_3)_6]^{3+}$	(i)	Violet
(b)	$[Ti(H_2O)_6]^{3+}$	(ii)	Green
(c)	$[Ni(H_2O)_6]^{2+}$	(iii)	Pale blue
(d)	$[Ni(H_2O)_4(en)]^{2+}$ (aq)	(iv)	Yellowish orange
		(v)	Blue

1) (a)-(i) (b)-(ii) (c)-(iv) (d)-(v)

2) (a)-(iv) (b)-(iii) (c)-(ii) (d)-(i)

3) (a)-(iii) (b)-(ii) (c)-(iv) (d)-(i)

4) (a)-(iv) (b)-(i) (c)-(ii) (d)-(iii)

2. Match the coordination compounds given in Column-I with the central metal atoms given in Column-II and assign the correct code:

	Column-I (Coordination Compound)		Column-II (Central metal atom)
(a)	Chlorophyll	(i)	rhodium
(b)	Blood pigment	(ii)	cobalt
(c)	Wilkinson catalyst	(iii)	calcium
(d)	Vitamin B_{12}	(iv)	iron
		(v)	magnesium

1) (a)-(v) (b)-(iv) (c)-(i) (d)-(ii)

2) (a)-(iii) (b)-(iv) (c)-(v) (d)-(i)

3) (a)-(iv) (b)-(iii) (c)-(ii) (d)-(i)

4) (a)-(iii) (b)-(iv) (c)-(i) (d)-(ii)

3. Match the complex ions given in Column-I with the hybridisation and number of unpaired electrons given in Column-II and assign the correct code:

	Column-I (Complex ion)		Column-II (Hybridisation, number of unpaired electrons)
(a)	$[Cr(H_2O)_6]^{3+}$	(i)	dsp^2, 1
(b)	$[Co(CN)_4]^{2-}$	(ii)	sp^3d^2, 5
(c)	$[Ni(NH_3)_6]^{2+}$	(iii)	d^2sp^3, 3
(d)	$[MnF_6]^{4-}$	(iv)	sp^3, 4
		(v)	sp^3d^2, 2

1) (a)-(iii) (b)-(i) (c)-(v) (d)-(ii)

2) (a)-(iv) (b)-(iii) (c)-(ii) (d)-(i)

3) (a)-(iii) (b)-(ii) (c)-(iv) (d)-(i)

4) (a)-(iv) (b)-(i) (c)-(ii) (d)-(iii)

4. Match the complex species given in Column-I with the possible isomerism given in Column-II and assign the correct code:

	Column-I (Complex species)		Column-II (Isomerism)
(a)	$[Co(NH_3)_4Cl_2]^+$	(i)	optical
(b)	cis-$[Co(en)_2Cl_2]^+$	(ii)	ionisation
(c)	$[Co(NH_3)_5(NO_2)]Cl_2$	(iii)	coordination
(d)	$[Co(NH_3)_6][Cr(CN)_6]$	(iv)	geometrical
		(v)	linkage

1) (a)-(i) (b)-(ii) (c)-(iv) (d)-(v)
2) (a)-(iv) (b)-(iii) (c)-(ii) (d)-(i)
3) (a)-(iv) (b)-(i) (c)-(v) (d)-(iii)
4) (a)-(iv) (b)-(i) (c)-(ii) (d)-(iii)

5. Match the compounds given in Column I with the oxidation state of cobalt present in it (given in Column II) and assign the correct code.

	Column I (Compound)		Column II (Oxidation state of Co)
(a)	$[Co(NCS)(NH_3)_5](SO_3)$	(i)	+4
(b)	$[Co(NH_3)_4Cl_2]SO_4$	(ii)	0
(c)	$Na_4[Co(S_2O_3)_3]$	(iii)	+1
(d)	$[Co_2(CO)_8]$	(iv)	+2
		(v)	+3

1) (a)-(i) (b)-(ii) (c)-(iv) (d)-(v)
2) (a)-(iv) (b)-(iii) (c)-(ii) (d)-(i)
3) (a)-(v) (b)-(i) (c)-(iv) (d)-(ii)
4) (a)-(iv) (b)-(i) (c)-(ii) (d)-(iii)

Correct & Incorrect MCQs

1. Which of the following statement (s) are correct?

a. The primary valences are non ionisable.
b. The secondary valences are normally ionisable.
c. $CrCl_3$, $CoCl_2$ or $PdCl_2$ have primary valence of 3, 2 and 2 respectively.
d. The secondary valence is equal to the coordination number and is fixed for a metal.

1) (a), (b) and (c) 2) (c) and (d)
3) (a), (c) and (d) 4) (b), (c) and (d)

2. Which of the following statement is incorrect?

1) 1 mol $CoCl_3.6NH_3$ gives 3 mol AgCl on adding excess of silver nitrate solution.
2) 1 mol $CoCl_3.5NH_3$ gives 5 mol AgCl on adding excess of silver nitrate solution.
3) 1 mol $CoCl_3.4NH_3$ gives 1 mol AgCl on adding excess of silver nitrate solution.
4) None of these

3. Which of the following statement regarding double salt and complex salt is not correct ?

a. Double salts dissociate into simple ions completely when dissolved in water.
b. Carnallite is an example of complex salt.
c. Complex salt do not dissociate into simple ions when dissolved in water.
d. Potassium ferrocyanide is an example of double salt.

1) (a) and (b) 2) (b), (c) and (d)
3) (b) and (d) 4) (a) and (c)

4. Which of the following statement regarding ligands is not correct ?

1) The ions or molecules bound to the central atom/ion in the coordination entity are called ligands.
2) Cl^- is an example of unidentate ligand.
3) Ethylenediaminetetraacetate ion is a bidentate ligand.
4) NO_2^- ion can coordinate either through nitrogen or through oxygen to a central metal atom/ion.

5. Choose the incorrect statement among the following:

1) The central atom/ion in the coordination entity $[NiCl_2(H_2O)_4]$ is Ni^{2+}.
2) The coordination number of Pt in the complex ion $[PtCl_6]^{2-}$ is 6.
3) In the complex $K_4[Fe(CN)_6]$, the coordination sphere is $[Fe(CN)_6]^{4-}$ and the counter ion is K^+.
4) $[Co(NH_3)_4Cl_2]^+$ is an example of homoleptic complex.

6. Select the incorrect statement :

1) Geometrical isomerism arises in heteroleptic complexes.
2) Geometrical isomerism is possible for a tetrahedral geometry.
3) Optical isomers are mirror images that cannot be superimposed on one another.
4) Optical isomerism is common in octahedral complexes involving didentate ligands.

7. Which of the following statements is correct ?

1) In a coordination entity of the type $[PtCl_2(en)_2]^{2+}$, only the trans-isomer shows optical activity
2) In a coordination entity of the type $[PtCl_2(en)_2]^{2+}$, only the cis-isomer shows optical activity
3) In a coordination entity of the type $[PtCl_2(en)_2]^{2+}$, both cis and trans-isomer shows optical activity
4) None of these

8. Identify the incorrect statement from the following :
   1) The complex $[Ti(H_2O)_6]^{3+}$is violet in colour.
   2) Removal of water from $[Ti(H_2O)_6]Cl_3$ on heating renders it colourless.
   3) In the absence of ligand, crystal field splitting does not occur.
   4) Anhydrous $CuSO_4$ is blue in colour.

9. Select the correct statement :
   1) Tetrahedral complexes are sp^3 hybridised.
   2) Square planar complexes are sp^3d^2 hybridised.
   3) Trigonal bipyramidal complexes are d^2sp^3 hybridised.
   4) Octahedral complexes are sp^3d hybridised.

10. Which of the following statements is incorrect ?
    1) In tetrahedral complexes low spin configurations are rarely observed.
    2) Inner orbital complexes are low spin complexes.
    3) $[Co(NH_3)_6]^{3+}$ is an outer orbital complex.
    4) $[CoF_6]^{3-}$ is an outer orbital complex.

Statement Based MCQs

Directions: These questions consist of two statements each, printed as Statement-I and Statement-II. While answering these questions, you are required to choose any one of the following four responses.

1) **Both Statement-I and Statement-II are correct.**
2) **Both Statement-I and Statement-II are incorrect.**
3) **Statement-I is correct & Statement-II is incorrect.**
4) **Statement-I is incorrect & Statement-II is correct.**

1. **Statement-I:** Optical isomerism is not shown by square planar complex.
   Statement-II: Square planar complexes do not possess chiral structures.

2. **Statement-I:** $K_4[Fe(CN)_6]$ and $K_3[Fe(CN)_6]$ have same magnetic moment.
   Statement-II: Magnetic moment is controlled by number of paired electrons.

3. **Statement-I:** Chelates are more stable than non chelated complex.
   Statement-II: Complexes containing ligands which can be easily replaced by other ligands are called labile complexes.

4. **Statement-I:** Potassium ferrocyanide is diamagnetic whereas potassium ferricyanide is paramagnetic.
   Statement-II: Crystal field splitting in ferrocyanide ion is greater than that of ferricyanide ion.

5. **Statement-I:** The $[Ni(en)_3]Cl_2$ (en = ethylene diamine) has lower stability than $[Ni(NH_3)_6]Cl_2$
   Statement-II: In $[Ni(en)_3]Cl_2$ the geometry of Ni is trigonal bi-pyramidal.

6. **Statement-I:** The metal-carbon bond in metal carbonyls possess σ character.
   Statement-II: In metal carbonyls, the M – C σ bond is formed by the donation of a pair of electrons from a filled d orbital of metal into the vacant antibonding π^* orbital of carbon monoxide.

7. **Statement-I:** NO_2^- and SCN^- ion are ambidentate ligand.
   Statement-II: SCN^- and NO_2^- ion can coordinate only through 'N' to a central atom or ion.

8. **Statement-I:** $[CO(NH_3)_6]^{3+}$ is a heteroleptic complex.
   Statement-II: $[Co(NH_3)_4Cl_2]^+$ is a homoleptic complex.

9. **Statement-I:** Inner orbital complexes are high spin complexes.
   Statement-II: Outer orbital complexes are low spin complexes.

10. **Statement-I:** Linkage isomerism arises in coordination compounds containing ambidentate ligand.
    Statement-II: Ambidentate ligand has two different donor atoms.

Assertion & Reason MCQs

Directions: These questions consist of two statements each, printed as Assertion and Reason. While answering these questions, you are required to choose any one of the following four responses.

1) **Both Assertion and Reason are True and the Reason is a correct explanation of the Assertion.**
2) **Both Assertion and Reason are True but Reason is not a correct explanation of the Assertion.**
3) **Assertion is True but the Reason is False.**
4) **Assertion is False but Reason is True.**

1. **Assertion:** Toxic metal ions are removed by the chelating ligands.
   Reason: Chelate complexes tend to be more stable.

2. **Assertion:** $[Cr(H_2O)_6]Cl_2$ and $[Fe(H_2O)_6]Cl_2$ are reducing in nature.
   Reason: Unpaired electrons are present in their d-orbitals.

3. **Assertion:** Linkage isomerism arises in coordination compounds containing ambidentate ligand.
   Reason: Ambidentate ligand has two different donor atoms.

4. **Assertion:** Complexes of MX_6 and MX_5L type (X and L are unidentate) do not show geometrical isomerism.
   Reason: Geometrical isomerism can be shown by complexes of coordination number 6.

5. **Assertion:** $[Fe(CN)_6]^{3-}$ ion shows magnetic moment corresponding to two unpaired electrons.
   Reason: Because it has d^2sp^3 type hybridisation.

6. **Assertion:** Aqueous solution of Mohr's salt exhibits the test for NH^+_4, Fe^{+2} and SO_4^{2-} ions.
   Reason: Mohr's salt is a double salt.
7. **Assertion:** The ligands nitro and nitrito are called ambidentate ligands.
   Reason: These ligands give linkage isomers.
8. **Assertion:** [Mabcd] has three geometrical isomers.
   Reason: It is optically inactive.
9. **Assertion:** All the octahedral complexes of Ni^{2+} must be outer orbital complexes.
   Reason: Outer orbital octahedral complexes are given by weak ligands.
10. **Assertion:** $CrCl_3.4H_2O$ gives white ppt with $AgNO_3$.
   Reason: It contains free chloride ion.

Multi Correct MCQs

1. If Ni^{2+} is replaced by Pt^{2+} in the complex $[NiCl_2Br_2]^{2-}$, which of the following properties are expected to get changed?
   A. Geometry B. Geometrical isomerism
   C. Optical isomerism D. Magnetic properties
   1) A, B and C 2) A, B and D
   3) A and D 4) B and C
2. Consider the following complex ions
   $P = [FeF_6]^{3-}$
   $Q = [V(H_2O)_6]^{2+}$
   $R = [Fe(H_2O)_6]^{2+}$
   Choose the appropriate option according to their spin only magnetic moment values (in B.M.) is:
   A. $P < Q < R$ B. $Q = R$
   C. $Q < R$ and $R < P$ D. $P > R$
   1) Both A and B 2) Only C
   3) Both C and D 4) A, C and D
3. Number of complexes from the following with even number of unpaired "d" electrons is____.
   I. $[V(H_2O)_6]^{3+}$ II. $[Cr(H_2O)_6]^{2+}$
   III. $[Fe(H_2O)_6]^{3+}$ IV. $[Ni(H_2O)_6]^{2+}$
   V. $[Cu(H_2O)_6]^{2+}$
   [Given atomic numbers : V = 23, Cr = 24, Fe = 26, Ni = 28, Cu = 29]
   1) I and II 2) III and V
   3) I, II and IV 4) I and IV
4. Choose the complexes from the following having unpaired electrons greater than or equal to 3.
   A. $[Fe(CN)_6]^{3-}$ B) $[FeF_6]^{3-}$
   C. $[CoF_6]^{3-}$ D) $[Cr(oxalate)_3]^{3-}$
   E. $[Ni(CO)_4]$
   1) A and E 2) B, C and D
   3) Only B 4) A, B and E

NCERT Exemplar MCQs

1. Which of the following complexes formed by Cu^{2+} ions is most stable?
   1) $Cu^{2+} + 4NH_3 \rightarrow [Cu(NH_3)_4]^{2+}$, log K = 11.6
   2) $Cu^{2+} + 4CN^- \rightarrow [Cu(CN)_4]^{2-}$, log K = 27.3
   3) $Cu^{2+} + 2en \rightarrow [Cu(en)_2]^{2+}$, log K = 15.4
   4) $Cu^{2+} + 4H_2O \rightarrow [Cu(H_2O)_4]^{2+}$, log K = 8.9
2. The colour of the coordination compounds depends on the crystal field splitting. What will be the correct order of absorption of wavelength of light in the visible region, for the complexes, $[Co(NH_3)_6]^{3+}$, $[Co(CN)_6]^{3-}$, $[Co(H_2O)_6]^{3+}$:
   1) $[Co(CN)_6]^{3-} > [Co(NH_3)_6]^{3+} > [Co(H_2O)_6]^{3+}$
   2) $[Co(NH_3)_6]^{3+} > [Co(H_2O)_6]^{3+} > [Co(CN)_6]^{3-}$
   3) $[Co(H_2O)_6]^{3+} > [Co(NH_3)_6]^{3+} > [Co(CN)_6]^{3-}$
   4) $[Co(CN)_6]^{3-} > [Co(NH_3)_6]^{3+} > [Co(H_2O)_6]^{3+}$
3. When 0.1 mol $CoCl_3(NH_3)_5$ is treated with excess of $AgNO_3$, 0.2 mol of AgCl are obtained. The conductivity of solution will correspond to:
   1) 1 : 3 electrolyte 2) 1 : 2 electrolyte
   3) 1 : 1 electrolyte 4) 3 : 1 electrolyte
4. When 1 mole of $CrCl_3.6H_2O$ is treated with excess of $AgNO_3$, 3 moles of AgCl are obtained. The formula of the complex is:
   1) $[CrCl_3(H_2O)_3].3H_2O$ 2) $[CrCl_2(H_2O)_4].2H_2O$
   3) $[CrCl(H_2O)_5]Cl_2.H_2O$ 4) $[Cr(H_2O)_6]Cl_3$
5. The correct IUPAC name of $[Pt(NH_3)_2Cl_2]$ is:
   1) Diamminedichloridoplatinum (II)
   2) Diamminedichloridoplatinum (IV)
   3) Diamminedichloridoplatinum (0)
   4) Dichloridodiammineplatinum (IV)
6. The stabilization of coordination compounds due to chelation is called the chelate effect. Which of the following is the most stable complex species?
   1) $[Fe(CO)_5]$ 2) $[Fe(CN)_6]^{3-}$
   3) $[Fe(C_2O_4)_3]^{3-}$ 4) $[Fe(H_2O)_6]^{3+}$
7. Indicate the complex ion which shows geometrical isomerism:
   1) $[Cr(H_2O)_4Cl_2]^+$ 2) $[Pt(NH_3)_3Cl]$
   3) $[Co(NH_3)_6]^{3+}$ 4) $[Co(CN)_5(NC)]^{3-}$

8. The CFSE for octahedral $[COCl_6]^{4-}$ is 18,000 cm^{-1}. The CFSE for tetrahedral $[CoCl_4]^{2-}$ will be:

1) 18,000 cm^{-1} 2) 16,000 cm^{-1}
3) 8,000 cm^{-1} 4) 20,000 cm^{-1}

9. Due to the presence of ambidentate ligands, coordination compounds show isomerism. Palladium complexes of the type $[Pd(C_6H_5)_2(SCN)_2]$ and $[Pd(C_6H_5)_2(NCS)_2]$ are:

1) Linkage isomers 2) Coordination isomers
3) Ionisation isomers 4) Geometrical isomers

10. The compounds $[Co(SO_4)(NH_3)_5]Br$ and $[Co(SO_4)(NH_3)_5]Cl$ represent:

1) Linkage isomerism
2) Ionisation isomerism
3) Coordination isomerism
4) No isomerism

11. A chelating agent has two or more than two donor atoms to bind to a single metal ion. Which of the following is not a chelating agent?

1) Thiosulphato 2) Oxalato
3) Glycinato 4) Ethane-1, 2-diamine

12. Which of the following species is not expected to be a ligand?

1) NO 2) NH_4^+
3) $NH_2CH_2CH_2NH_2$ 4) CO

13. What kind of isomerism exists between $[Cr(H_2O)_6]Cl_3$ (violet) and $[Cr(H_2O)_5Cl]Cl_2 . H_2O$ (grayish-green)?

1) Linkage isomerism 2) Solvate isomerism
3) Ionisation isomerism 4) Coordination isomerism

14. IUPAC name of $[Pt(NH_3)_2Cl . (NO_2)]$ is:

1) Platinum diaminechloronitrite
2) Chloronitrito-N-animineplatinum (II)
3) Diamminechloridonitrito-N-platinum (II)
4) Diamminechloronitrito-N-platinate (II)

Kattar NEET

Scan Me for Detailed Explanations

1. Select the correct statement for $[M(AB)b_2cd]$

1) All geometrical isomers are optically active
2) It has four trans form with respect to b
3) It has 7 geometrical isomers
4) It has four cis isomers with respect to b

2. A complex $[CoA_6]^{3+}$ is red coloured while $[CoB_6]^{3+}$ is green coloured, which of the following is correct statement?

1) Ligand A is producing larger crystal field splitting
2) $[CoB_6]^{3+}$ is easily oxidized
3) Both are equally oxidized
4) Ligand B is producing larger crystal field splitting

3. Which of the following statement is correct?

(i) The ligand thiosulphato $S_2O_3^{2-}$ can give rise to linkage isomers
(ii) In metal carbonyls the ligand CO molecule acts both as donor and acceptor
(iii) The complex $K[Cu(CN)_2]$ has coordination number 2
(iv) The complex ferricyanide does not follow effective atomic number (EAN) rule

1) (i), (ii) and (iii) only 2) (i), (ii) and (iv) only
3) (i), (iii) and (iv) only 4) (ii), (iii) and (iv) only

4. Which of the following statements is incorrect?

1) Complex $[Co(NH_3)_4(H_2O)Cl]Br_2$ can show both hydrate as well as ionization isomerism
2) Complex $[Co(NH_3)_5(H_2O)](NO_3)_3$ can show hydrate isomerism
3) Complex $[Pt(NH_3)_4][PtCl_5]$ cannot show coordination isomerism
4) $[Co(NH_3)_4(NO_2)Cl]Cl$ can show both ionization as well as linkage isomerism

5. In the isoelectronic series of metal carbonyl, the CO bond strength is expected to increase in the order:

1) $[Mn(CO)_6]^+ < [Cr(CO)_6] < [V(CO)_6]^-$
2) $[V(CO)_6]^- < [Cr(CO)_6] < [Mn(CO)_6]^+$
3) $[V(CO)_6]^- < [Mn(CO)_6]^+ < [Cr(CO)_6]$
4) $[Cr(CO)_6] < [Mn(CO)_6]^+ < [V(CO)_6]^-$

6. Which of the following order of CFSE is incorrect?

1) $[Co(en)_3]^{3+} > [Co(NH_3)_6]^{3+} > [Co(H_2O)_6]^{3+}$
2) $[PtCl_4]^{2-} > [PdCl_4]^{2} > [NiCl_4]^{2}$
3) $[Ni(DMG)_2] < [Ni(en)_2]^{2+}$
4) $[Co(ox)_3]^{3-} < [Co(en)_3]^{3+}$

7. Complex compound(s) having even number of space (stereo) isomers is/are: (where AA-symmetrical bidentate ligand and a, b, c, d, e-monodentate ligands)

1) $[M(AA)_2b_2]^{n\pm}$ 2) $[Ma_3b_3]^{n+}$
3) $[Ma_3bcd]^{n\pm}$ 4) $[Ma_2bcde]^{n\pm}$

8. Find out the wrong statement for an octahedral complex.

1) An ion with d^5 configuration has one unpaired electron both in weak and strong fields.
2) A central metal ion with d^8configuration has two unpaired electrons.
3) An ion with d^6 configuration is diamagnetic in a strong field.
4) In d^4, d^5, d^6 and d^7 configurations, weak and strong field complexes have different number of unpaired electrons.

9. A solution contains 2.675 g of $CoCl_3.6NH_3$(molar mass = 267.5 g mol^{-1}) is passed through a cation exchanger. The chloride ions obtained in solution were treated with excess of $AgNO_3$ to give 4.305g of AgCl. The formula of the complex is

1) $[CoCl(NH_3)_5]Cl_2$
2) $[Co(NH_3)_6]Cl_3$
3) $[Co(NH_3)_4Cl_2]Cl$
4) $[Co(NH_3)_3Cl_3]$

10. In the following complexes of manganese what is the electronic configuration in d-orbitals of manganese?

(i) $[Mn(H_2O)_6]^{2+}$ (ii) $[Mn(CN)_6]^{4-}$

1) $t_{2g}^3\ e_g^2$ in both
2) $t_{2g}^5\ e_g^0$ in both
3) $t_{2g}^3\ e_g^2$ in (i) and $t_{2g}^5\ e_g^0$ in (ii)
4) $t_{2g}^5\ e_g^0$ in (i) and $t_{2g}^3\ e_g^2$ in (ii)

11. Which of the following has five donor (coordinating) sites?

1) Triethylene tetramine
2) Ethylenediamine tetracetate ion
3) Ethylenediamine triacetate ion
4) Diethylene triamine

12. Consider the following spatial arrangements of the octahedral complex ion $[Co(NH_3)_4Cl_2]^+$

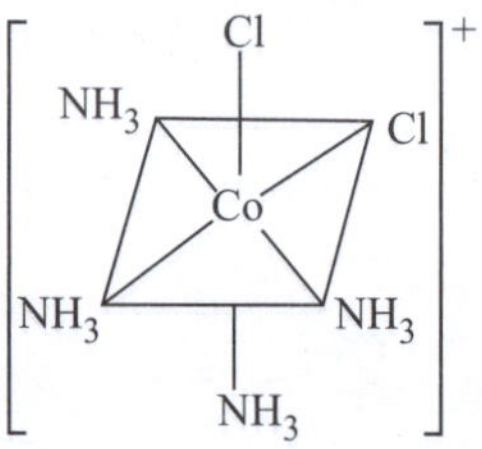

(I)

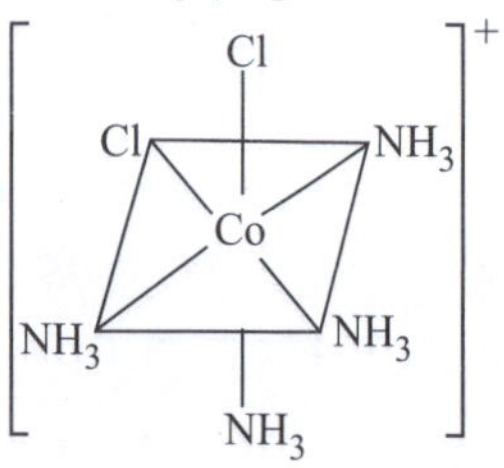

(II)

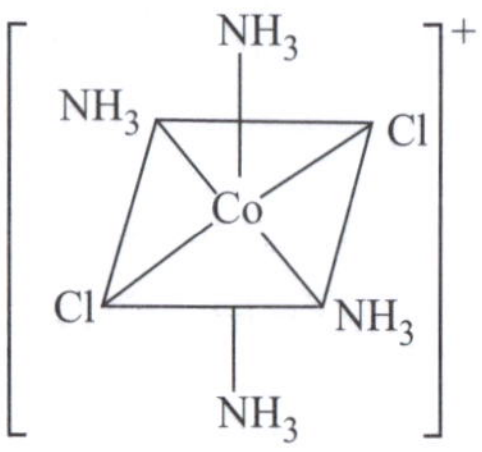

(III)

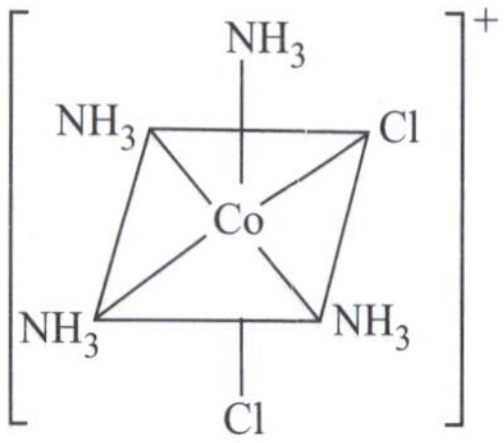

(IV)

Which of the following statements is incorrect regarding these structures ?

1) I and II are enantiomers
2) II and III are cis and trans isomers respectively
3) III and IV are trans and cis isomers respectively
4) II and IV have identical structures

13. In nitroprusside ion, the iron and NO exist as Fe^{II} and NO^+ rather than Fe^{III} and NO. These forms can be differentiated by:

1) Estimating the concentration of iron
2) Measuring the concentration of CN^-
3) Measuring the solid state magnetic moment
4) Thermally decomposing the compound

14. Which among the following will be named as dibromidobis(ethylenediamine) chromium (III) bromide?

1) $[Cr(en)_3]Br_3$
2) $[Cr(en)_2Br_2]Br$
3) $[Cr(en)Br_4]^-$
4) $[Cr(en)Br_2]Br$

15. The solutions of two compounds listed in List-I when taken in molecular proportions give coordination compounds on evaporation. Match these compounds with their properties listed in List-II:

	List-I		List-II
I.	$Fe(CN)_2.4KCN$	P.	Coordination number of central metal ion is six
II.	$CuSO_4.4NH_3$	Q.	Coordination number of central metal ion is four
III.	$3NaF.AlF_3$	R.	Coordination number of central metal ion is two
IV.	AgCN.KCN	S.	Oxidation state of central metal ion is +2
		T.	Overall charge on complex ion is +2

1) I-P, S; II-Q, S, T; III-P; IV-R
2) I-Q, S, T; II-P, S; III-P; IV-R
3) I-P, S, T; II-P, S; III-R; IV-P
4) I-Q, S, T; II-P, T; III-P, S; IV-R

16. Match the pair of complexes given in Column-I and the characteristic(s) given in Column-I:

	Column-I		Column-II
I.	$(NH_4)_2[NiCl_4]$ and $(NH_4)_2[Ni(CN)_4]$	P.	Both show same or nearly same electrical conductance
II.	$[Co(NH_3)_6]Cl_3$ and $[Pt(NH_3)_5Cl]Cl_3$	Q.	Both show same effective atomic number
III.	$[Pt(NH_3)_2Cl_2]$ and $(NH4)_2[PtCl_4]$	R.	Both show same primary valencies
IV.	$K_2[Fe(H_2O)_6]$ and $K_4[FeCl_6]$	S.	Both gives white precipitate with $AgNO_3$ solution

1) I-P, S; II–Q, S, T; III–P; IV–R
2) I–Q, S, T; II–P, S; III–P; IV–R
3) I–P, Q, R; II–P, S; III–Q, R; IV–Q, R
4) I–Q, S, T; II–P, T; III–P, S; IV–R

17. Match the column:

	Column-I		Column-II
I.	$[CoCl_3(NH_3)_3]$	P.	Show facial isomer form
II.	$[Cr(Ox)_3]^{3-}$	Q.	cis form is optically active
III.	$[CrCl_2(Ox)_2]$	R.	Trans form is optically inactive
IV.	$[RhCl_3(Py)_3]$	S.	Show meridional form
		T.	Two optically active isomer

1) I–P, S; II–Q, S, T; III–P; IV–R
2) I–Q, S, T; II–P, S; III–P; IV–R
3) I–P, S, T; II–P, S; III–R; IV–P
4) I–P, Q, R, S, T; II–T; III–Q, R, T; IV–P, Q, R, S, T

18. Co-ordination number of Cr in $CrCl_3{\cdot}5H_2O$ is six. The maximum volume of 0.1N $AgNO_3$ needed to precipitate the chlorine in outer sphere in 200 ml of 0.01 M solution of the complex (among all possible isomers) is/are:

1) 140 ml
2) 40 ml
3) 80 ml
4) 20 ml

19. Arrange the following cyano complexes in decreasing order of their magnetic moment.

1) $[Cr(CN)_6]^{3-} > [Mn(CN)_6]^{3-} > [Fe(CN)_6]^{3-} > [Co(CN)_6]^{3-}$
2) $[Mn(CN)_6]^{3-} > [Cr(CN)_6]^{3-} > [Fe(CN)_6]^{3-} > [Co(CN)_6]^{3-}$
3) $[Fe(CN)_6]^{3-} > [Cr(CN)_6]^{3-} > [Mn(CN)_6]^{3-} > [Co(CN)_6]^{3-}$
4) $[Co(CN)_6]^{3-} > [Cr(CN)_6]^{3-} > [Mn(CN)_6]^{3-} > [Fe(CN)_6]^{3-}$

20. Select complex which has all the following characteristic

(*i*) Paramagnetic
(*ii*) No unpaired electron in t_{2g} orbitals
(*iii*) Exist in geometrical isomeric form

1) $K_4[Fe(NO_2)_2(CN)_4]$ 2) $[Ni(NH_3)_4(en)]Cl_2$
3) $[Co(en)_3]Cl_3$ 4) $[Ni(NH_3)_2(en)_2]SO_4$

NEET & AIIMS Past Year MCQs

1. Ethylene diaminetetraacetate ion is a/an: **(2024 Re)**

1) hexadentate ligand 2) ambidentate ligand
3) monodentate ligand 4) bidentate ligand

2. Which of the following is not an ambidentate ligand? **(2024 Re)**

1) $C_2O_4^{2-}$ 2) SCN^- 3) NO_2^- 4) CN^-

3. $[Mn_2(CO)_{10}]$ and $[Co_2(CO)_8]$ structures have **(2024 Re)**

A. Metal-Metal linkage B. Terminal CO groups
C. Bridging CO groups D. Metal in zero oxidation state

Choose the correct answer from the options given below

1) Only A, B, C 2) Only B, C, D
3) Only A, C, D 4) Only A, B, D

4. Match List-I with List-II. **(2024)**

	List-I (Complex)		List-II (Type of isomerism)
A.	$[Co(NH_3)_5(NO_2)]Cl_2$	I.	Solvate isomerism
B.	$[Co(NH_3)_5(SO_4)]Br$	II.	Linkage isomerism
C.	$[Co(NH_3)_6][Cr(CN)_6]$	III.	Ionization isomerism
D.	$[Co(H_2O)_6]Cl_3$	IV.	Coordination isomerism

Choose the correct answer from the options given below:

1) A-I, B-IV, C-III, D-II 2) A-II, B-IV, C-III, D-I
3) A-II, B-III C-IV, D-I 4) A-I, B-III, C-IV, D-II

5. Given below are two statements: **(2024)**

Statement-I: Both $[Co(NH_3)_6]^{3+}$ and $[CoF_6]^{3-}$ complexes are octahedral but differ in their magnetic behaviour.

Statement-II: $[Co(NH_3)_6]^{3+}$ is diamagnetic whereas $[CoF_6]^{3-}$ is paramagnetic.

In the light of the above statements, choose the correct answer from the options given below:

1) Statement-I is true but Statement-II is false.
2) Statement-I is false but Statement-II is true.
3) Both Statement-I and Statement-II are true.
4) Both Statement-I and Statement-II are false.

6. Given below are two statements: **(2024)**

Statement-I: $[Co(NH_3)_6]^{3+}$ is a homoleptic complex whereas $[Co(NH_3)_4Cl_2]^+$ is a heteroleptic complex.

Statement-II: Complex $[Co(NH_3)_6]^{3+}$ has only one kind of ligands but $[Co(NH_3)_4Cl_2]^+$ has more than one kind of ligands.

In the light of the above statements, choose the correct answer from the options given below:

1) Statement-I is true but Statement-II is false.
2) Statement-I is false but Statement-II is true.
3) Both Statement-I and Statement-II are true.
4) Both Statement-I and Statement-II are false.

7. Homoleptic complex from the following complexes is: **(2023)**

1) Triamminetriaquachromium (III) chloride
2) Potassium trioxalatoaluminate (III)
3) Diamminechloridonitrito – N- platinum(II)
4) Pentaamminecarbonatocobalt (III) chloride

8. Which complex compound is most stable? **(2023)**

1) $[Co(NH_3)_6]_2(SO_4)_3$ 2) $[Co(NH_3)_4(H_2O)Br](NO_3)_2$
3) $[Co(NH_3)_3(NO_3)_3]$ 4) $[CoCl_2(en)_2]NO_3$

9. The IUPAC name of the complex: **(2022)**
$[Ag(H_2O)_2][Ag(CN)_2]$ is

1) Diaquasilver(I) dicyanidoargentate(I)
2) Dicyanidosilver(II) diaquaargentate(II)
3) Diaquasilver(II) dicyanidoargentate(II)
4) Dicyanidosilver(I) diaquaargentate(I)

10. The order of energy absorbed which is responsible for the color of complexes **(2022)**

A. $[Ni(H_2O)_2(en)_2]^{2+}$
B. $[Ni(H_2O)_4(en)]^{2+}$ and
C. $[Ni(en)_3]^{2+}$ is

1) B > A > C 2) A > B > C
3) C > B > A 4) C > A > B

11. Ethylene diaminetetraacetate (EDTA) ion is: **(2021)**

1) Unidentate ligand
2) Bidentate ligand with two "N" donor atoms
3) Tridentate ligand with three "N" donor atoms
4) Hexadentate ligand with four "O" and two "N" donor atoms

12. Match the column-I with column-II. **(2021)**

	Column-I		Column-II
(a)	$[Fe(CN)_6]^{3-}$	(i)	5.92 BM
(b)	$[Fe(H_2O)_6]^{3+}$	(ii)	0 BM
(c)	$[Fe(CN)_6]^{4-}$	(iii)	4.90 BM
(d)	$[Fe(H_2O)_6]^{2+}$	(iv)	1.73 BM

Choose the correct answer from the options given below.
1) (a)-(ii); (b)-(iv); (c)-(iii); (d)-(i)
2) (a)-(i); (b)-(iii); (c)-(iv); (d)-(ii)
3) (a)-(iv), (b)-(i); (c)-(ii); (d)-(iii)
4) (a)-(iv), (b)-(ii); (c)-(i); (d)-(iii)

13. Which of the following is diamagnetic complex **(AIIMS 2018)**
1) $[Co(ox)_3]^{3-}$, $[Fe(CN)_6]^{4-}$
2) $[Co(ox)_3]^{3-}$, $[FeF_6]^{3-}$
3) $[Fe(ox)_3]^{3-}$, $[FeF_6]^{3-}$
4) $[Fe(CN)_6]^{3-}$, $[CoF_6]^{3-}$

14. The wavelength of light absorbed is highest in: **(AIIMS 2011)**
1) $[Co(NH_3)_5Cl]^{2+}$
2) $[Co(NH_3)_5H_2O]^{3+}$
3) $[Co(NH_3)_6]^{3+}$
4) $[Co(en)_3]^{3+}$

15. $[CoCl_2(NH_3)_4]^+ + Cl^- \rightarrow [CoCl_3(NH_3)_3] + NH_3]$
In the reaction only one isomer of complex product is obtained. The initial complex is: **(AIIMS 2010)**
1) Cis isomer
2) Trans isomer
3) Not having stereoisomers
4) Either cis or trans

16. Which of the following is considered as an anticancer species? **(AIIMS 2008)**

1) 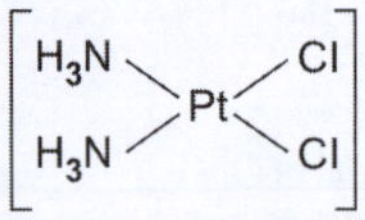

2) 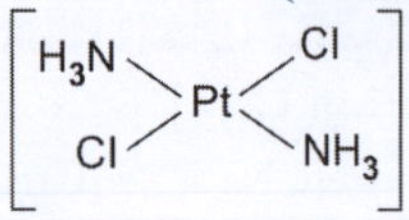

3) 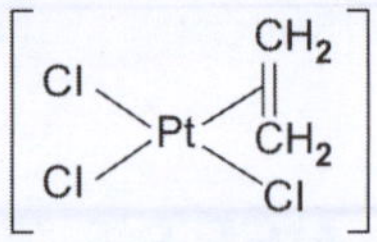

4) 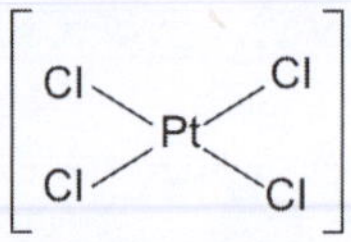

17. The correct order for the wavelength of absorption in the visible region is: **(AIIMS 2005)**
1) $[Ni(NO_2)_6]^{4-} < [Ni(NH_3)_6]^{2+} < [Ni(H_2O)_6]^{2+}$
2) $[Ni(NO_2)_6]^{4-} < [Ni(H_2O)_6]^{2+} < [Ni(NH_3)_6]^{2+}$
3) $[Ni(NO_2)_6]^{4+} < [Ni(H_2O)_6]^{2+} < [Ni(NH_3)_6]^{2+}$
4) $[Ni(NH_3)]^{2+} < [Ni(H_2O)_+ < Ni(NO_a\}_6]^{4-}$

18. Which one of the following shows maximum paramagnetic character? **(AIIMS 1998)**
1) $[Fe(CN)_6]^{3-}$
2) $[Fe(CN)_6]^{4-}$
3) $[Cr(H_2O)_6]^{3+}$
4) $[Cu(H_2O_6]^{2+}$

19. In the compound lithium tetrahydroaluminate, the ligand is: **(AIIMS 1997)**
1) H
2) H^+
3) H^-
4) F^-

20. Which of the following compound is not coloured? **(AIIMS 1997)**
1) $Na_2[CuCl_4]$
2) $Na_2[CdCl_4]$
3) $K_4[Fe(CN)_6]$
4) $K_3[Fe(CN)_6]$

21. Which of the following complex has square planar structure? **(AIIMS 1995)**
1) $[Ni(CN)_4]^{2-}$
2) $Ni(CO)_4$
3) $[Zn(NH_3)_4]^{2+}$
4) $[NiCl_4]^{2-}$

Answer Key

NCERT Topic-wise MCQs

1	2	3	4	5	6	7	8	9	10	11	12	13	14	15	16	17
3	1	2	4	2	3	1	2	3	1	3	2	2	2	1	2	1
18	19	20	21	22	23	24	25	26	27	28	29	30	31	32	33	34
2	1	4	4	4	3	2	2	1	3	3	4	4	4	3	4	3
35	36	37	38	39	40	41	42	43	44	45	46	47	48	49	50	51
4	4	4	4	2	1	4	4	2	4	2	2	2	3	1	2	1
52	53	54	55	56	57	58	59	60	61	62	63	64	65	66	67	68
1	3	1	1	2	2	3	3	1	2	3	1	2	2	3	1	1
69	70	71	72	73	74	75	76	77	78	79	80	81	82	83	84	85
1	2	1	4	2	4	3	4	3	3	1	4	2	4	4	2	2
86	87	88	89	90	91	92	93	94	95	96	97	98	99	100	101	102
1	3	1	4	3	1	1	3	2	3	1	2	4	3	1	2	3
103	104	105	106	107	108	109	110	111	112	113	114	115	116	117	118	119
3	3	4	4	2	4	4	1	2	4	4	2	4	3	2	3	4
120	121	122	123	124												
2	1	1	1	1												

High Order Time Intensive MCQs																	
	1	2	3	4	5	6	7	8	9	10	11	12	13	14	15	16	17
	1	3	2	2	4	3	1	1	4	1	2	2	1	3	3	4	2
	18	19	20	21	22	23	24	25	26	27	28	29	30	31	32	33	34
	4	4	3	4	3	1	1	1	1	1	3	4	1	4	3	3	4
	35	36	37	38	39	40	41	42	43	44	45	46	47	48	49	50	51
	3	4	4	3	2	2	3	1	3	4	4	4	1	1	4	2	3
	52	53	54	55	56	57	58	59	60	61							
	2	1	1	2	1	1	4	4	3	2							

Match the Columns MCQs	1	2	3	4	5
	2	1	2	3	1

Correct & Incorrect MCQs	1	2	3	4	5	6	7	8	9	10
	2	2	3	3	4	2	2	4	1	3

Statement Based MCQs	1	2	3	4	5	6	7	8	9	10
	1	2	1	3	2	2	3	2	2	1

Assertion & Reason MCQs	1	2	3	4	5	6	7	8	9	10
	1	2	1	2	4	1	2	3	2	1

Multi Correct MCQs	1	2	3	4
	2	3	3	2

NCERT Exemplar MCQs	1	2	3	4	5	6	7	8	9	10	11	12	13	14
	2	3	2	4	1	3	1	3	1	4	1	2	2	3

Kattar NEET																	
	1	2	3	4	5	6	7	8	9	10	11	12	13	14	15	16	17
	3	1	2	3	2	3	2	1	2	3	3	1	3	2	1	3	4
	18	19	20														
	2	1	4														

NEET & AIIMS Past Year MCQs																	
	1	2	3	4	5	6	7	8	9	10	11	12	13	14	15	16	17
	1	1	4	3	3	1	2	4	1	4	4	3	1	1	2	1	1
	18	19	20	21													
	3	3	2	1													

CHAPTER 7

Haloalkanes and Haloarenes

Topicwise Number of Questions Analysis (2025-2016)

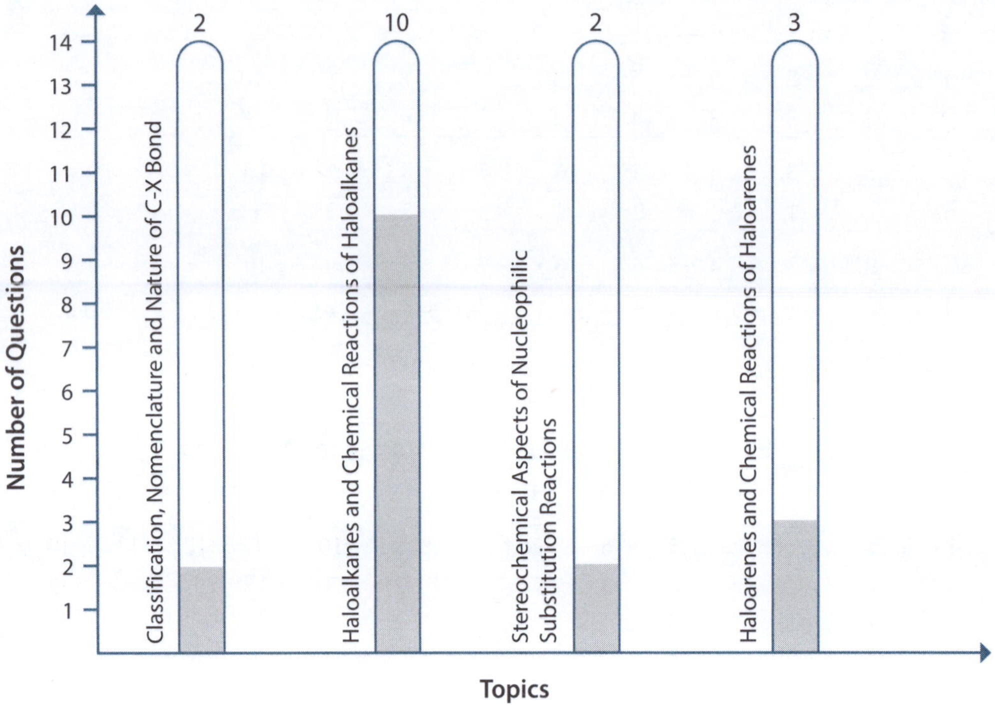

NCERT-Picks

Introduction

- The compounds which are obtained by the replacement of one or more hydrogen atoms in hydrocarbon, aliphatic and aromatic by halogen atoms, are called **haloalkanes** and **haloarenes**.
- In haloalkane, the carbon atom attached to halogen is **sp^3**-hybridised whereas is case of haloarene, the carbon atom attached to halogen is **sp^2**-hybridised.

Haloalkanes

Nature of C - X bond

- Since halogen atoms are more electronegative than carbon, therefore the carbon-halogen bond of alkyl halide is polarised and as a result carbon atom bears a partial positive charge whereas, halogen atom attains partial negative charge.

$$\overset{\delta+}{\text{C}}-\overset{\delta-}{\text{X}}$$

- The size of halogen atom increases as we go down the group in the periodic table, consequently, the C – X bond length increases from C – F to C – I.
- Reactivity of alkyl halides: R – I > R – Br > R – Cl >> R – F

Preparation of haloalkane

(i) From alkanes:

$$R-H \xrightarrow[\text{By halogenation}]{Cl_2/\text{uv light or heat}} R-Cl + HCl$$

(ii) From alkenes:

$$R-CH=CH_2 \xrightarrow{HX} R-\underset{}{\overset{X}{\overset{|}{C}}}H-CH_3$$

$$R-CH=CH_2 \xrightarrow{HX,\ \text{Peroxide}} R-CH_2-CH_2X$$

$$H_2C=CH_2 \xrightarrow[CCl_4]{Br_2} Br\,CH_2-CH_2Br \text{ (vic-Dibromide)}$$

(iii) From alcohols:

$$(R-OH) \xrightarrow[\text{conc. HCl + anhy. ZnCl}_2]{\text{Lucas Reagent}} RCl$$

$$(R-OH) \xrightarrow[PI_3]{\text{Red P} + I_2} R-I$$

$$(R-OH) \xrightarrow{PCl_5 / PCl_3} R-Cl$$

$$(R-OH) \xrightarrow{PBr_3} R\text{–}Br$$

$$(R-OH) \xrightarrow[\text{(Darzen's method)}]{SOCl_2\text{, Pyridine}} R-Cl$$

$$(R-OH) \xrightarrow[PBr_3]{\text{Red P} + Br_2} R-Br$$

- The reactivity of the alcohol is 3° > 2° > 1°. The stability of carbocation decreases in the order : 3° > 2° > 1°
- Reactivity of the halogen acids decrease in the order HI > HBr > HCl. Reactivity of the halogen acid is directly proportional to nucleophilicity.

(iv) Halogen Exchange Reactions.

Two types of halogen reaction are observed given as:

Swartz reaction: RCl/RBr + AgF → RF + AgBr/AgCl

AgF can be replaced by other metallic fluorides like Hg_2F_2, CoF_2, SbF_3 and so on.

Finkelstein reaction:

$$RCl/Br + NaI \xrightarrow[\Delta]{\text{Acetone}} RI + NaCl / NaBr$$

Physical properties

- Haloalkanes are only very slightly soluble in water. In order for a haloalkane to dissolve in water, energy is required to overcome the attractions between the haloalkane molecules. Less energy is released when new attractions are set up between the haloalkane and the water molecules.
- The solubility decreases with increase in the number of carbon atoms in their molecules.
- In alkyl halides B.P. increases with increase of atomic weight of halogen.

  R – F < R – Cl < R – Br < R – I
- Among isomeric alkyl halides B.P. decreases with increase in branching.
- With increase in the number of carbon atoms in their molecules, density of alkyl halides goes on decreasing.

Chemical Reactions

The reactions of haloalkanes may be divided into the following categories:

(i) Nucleophilic substitution reactions

$$RX \xrightarrow{\text{aq. KOH}} R-OH$$

$$RX \xrightarrow{\text{alc. KOH}} \text{Alkene}$$

$$RX \xrightarrow{\text{alc. KCN}} R-CN$$

$$RX \xrightarrow{AgCN} R-NC$$

$$RX \xrightarrow{Na} R-R + Na\,X$$

$$RX \xrightarrow{\text{Mg/dry ether}} R\,Mg\,X$$

$$RX \xrightarrow[R'ONa]{\text{Williamson's synthesis}} R-O-R'$$

$$RX \xrightarrow{KNO_2} R-O-NO$$

$$RX \xrightarrow{AgNO_2} R-NO_2$$

$$RX \xrightarrow{\text{Dry } Ag_2O} R-O-R$$

$$RX \xrightarrow{NH_3} R-NH_2$$

- Treatment of chloroethane with a saturated solution of AgCN gives ethyl isocyanide as the major product. Due to covalent nature of AgCN compound it act as a nucleophile through lone pair of electrons present on 'N' to give ethyl isocyanide.
- Alkyl halides undergoes **nucleophilic substitution reaction by two mechanism S_N1 or S_N2.** S_N1 forms carbocation.
- In S_N2 inversion of configuration occurs.

	S_N2 (Biomolecular)		S_N1 (Unimolecular)
(1)	It is second order reaction	(1)	It is first order reaction
(2)	Rate of reaction: 1° > 2° > 3°	(2)	Rate of reaction: 3° > 2° > 1°
(3)	It is one step reaction	(3)	It is two step reaction
(4)	Inversion of configuration occurs	(4)	Racemisation process takes place.

- Benzyl bromide when kept in acetone – water produces benzyl alcohol. Benzyl halide under go hydrolysis by S_N1 mechanism.
- Primary halide react most rapidly in S_N2 while tertiary halide is least. Bulky groups hinder the approaching nucleophile so tertiary halide is least reactive.

Stereochemistry

- A mixture containing two **enantiomers** in equal proportion will have zero optical rotation. The rotation due to one isomer will cancelled by the rotation due to other isomer.
- Propan – 2 – ol does not contain an asymmetric carbon. All the four groups attached to the tetrahedral carbon are not different.
- BrClCHI & $H_3C - CHBr - COOH$ are chiral molecule. Both have 4- different groups attached the tetrahedral carbon.

Stereochemical Aspects of Nucleophilic Substitution Reactions

1. Retention

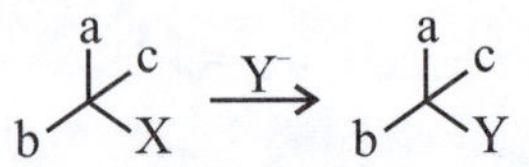

2. Inversion, retention and racemisation

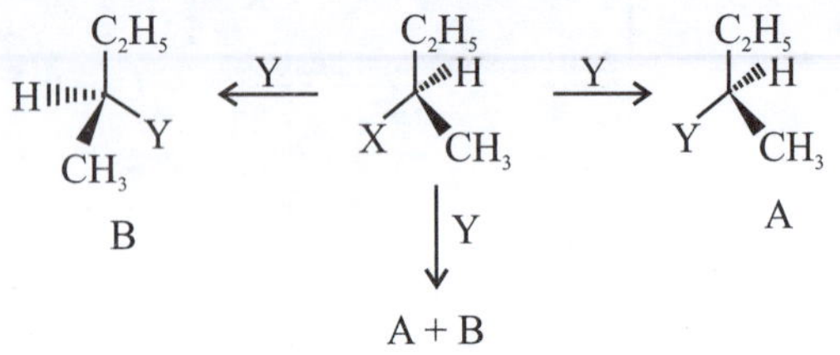

(ii) Elimination reaction

When a haloalkane with β-H atom is heated with concentrated alcoholic solution of KOH, a molecule of hydrogen halide is eliminated and alkene is formed.

$$R\,CH_2-\underset{\displaystyle X}{\underset{|}{CH}}-CH_3 + KOH\,(alc.) \longrightarrow R\,CH{=}CHCH_3 + KX + H_2O$$

Saytzeff rule:

- In dehydro halogenation reaction, the preferred product is that alkene which has greater number of alkyl group attached to the double bonded carbon atom.

$$CH_3-\overset{\displaystyle Br}{\overset{|}{CH}}-CH_2-CH_3 \xrightarrow[\text{alc. KOH}]{\text{hot and}} \underset{\substack{\text{1-Butene}\\\text{(minor)}}}{CH_2{=}CH-CH_2-CH_3} + \underset{\substack{\text{2-Butene}\\\text{(major)}}}{CH_3-CH{=}CH-CH_3}$$

- In E_2 and E_1 mechanism, reactivity of halogen follows the sequence for alkyl halide.

  tertiary > secondary > primary

- Unsymmetrical alkenes can be obtained from alkyl halides by Corey - House reaction, the reaction is carried by lithium dialkyl copper with an alkyl halide.
- In the presence of finely divided nickel, palladium or platinum (catalytic reduction) haloalkanes can be reduced to alkanes by H_2.

(iii) Reactions with metals

Wurtz Reaction

$$2RX + 2Na \xrightarrow{\text{dry ether}} R-R + 2NaX$$

Haloarenes

Preparation of haloarene

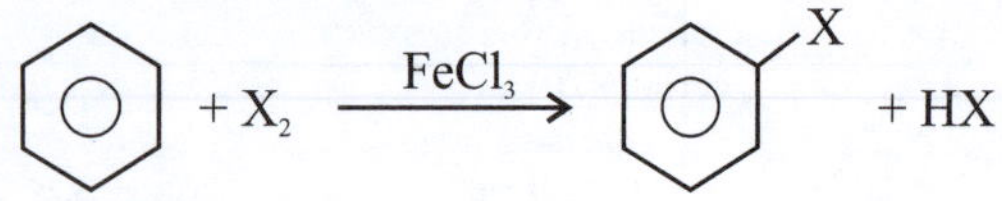

(X = Cl, Br)

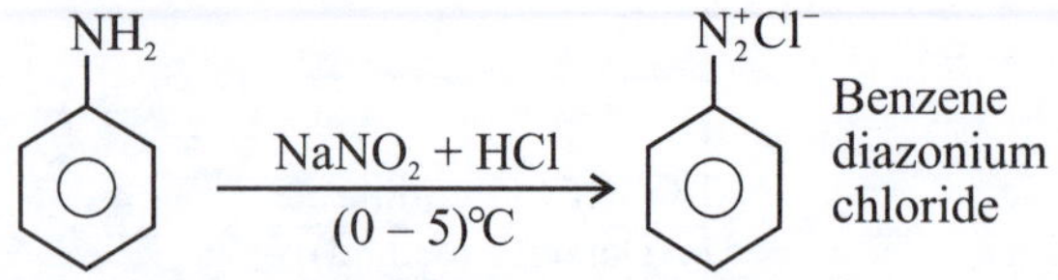

- Aryl halide cannot be prepared by the action of halogen acids, $SOCl_2$ on phenol. Due to resonance, C – O bond in phenols is much stronger than C – O bond in alcohols & hence cannot be easily broken.

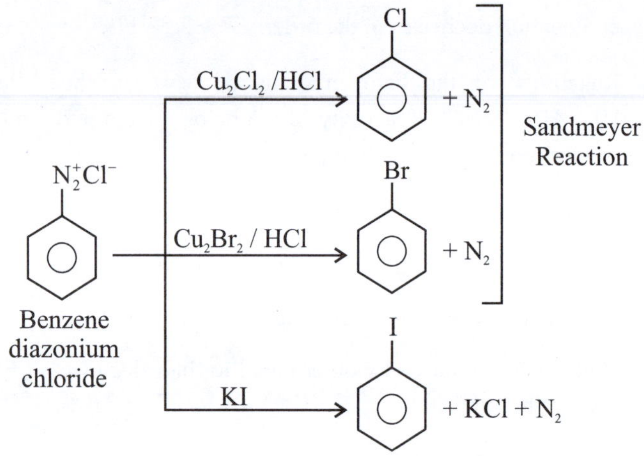

Properties

- Decreasing order of boiling point in a same aryl group is

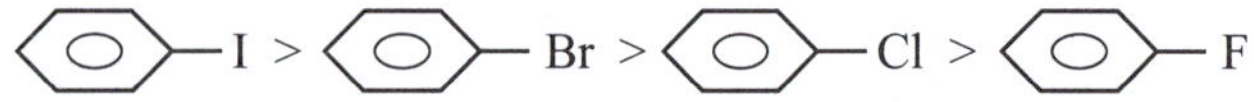

  As molecular mass of aryl group is increases, Van der Waals forces of attraction increases & hence melting / Boiling point increases.
- Aryl halides are extremely less reactive towards nucleophilic substitution reactions. The bond cleavage in haloarene is difficult than haloalkane.
- Para – isomers are high melting than ortho & meta – isomers in halobenzene. This is due to symmetry of para – isomers that fits in crystal lattice better as compared to ortho & meta isomers.

Reactions of Haloarenes

Reaction with metals

- **Fittig reaction:**

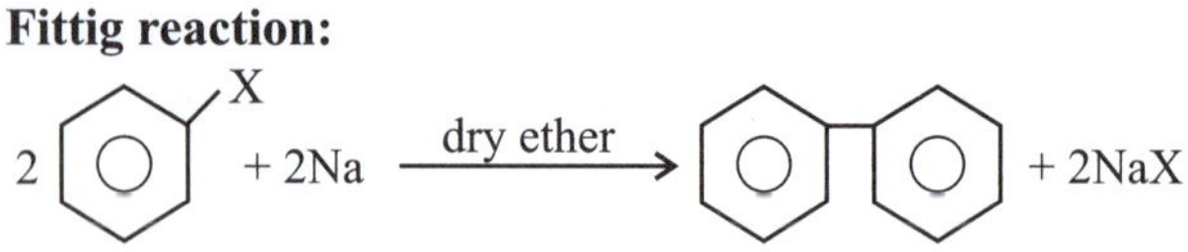

Diphenyl

- **Wurtz-Fittig reaction:**

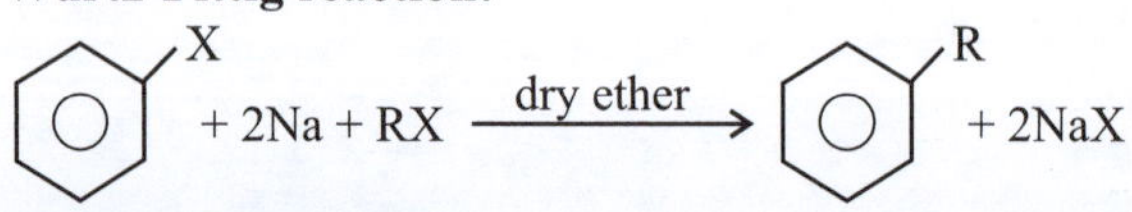

Electrophilic and nucleophilic substitution

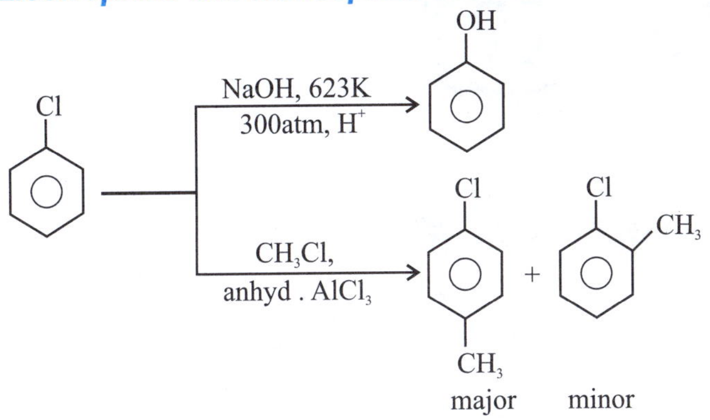

Polyhalogen Compounds

Tetrachloromethane (CCl_4): It is used in the production of refrigerants, propellants for aerosol cans

Dichlorodifluoro methane: CCl_2F_2, is called freon. It is used in air conditioning, refrigeration purposes

Benzene hexachloride (BHC)

- It is an organochlorine chemical compound which is an isomer of gammaxene with the formula $C_6H_6Cl_6$.
- Benzene hexachloride is used as an insecticide and pesticide in agriculture, under trade name gammaxane or lindane or 666.

Chloroform

- Stored in closed dark coloured bottles completely filled so that air is kept out. Chloroform is slowly oxidised by air in the presence of light to an extremely poisonous gas **phosgene**.
- Inhalation of chloroform vapours produces loss of consciousness and is used as a general anesthetic in surgery.

DDT

- It is first chlorinated organic insecticide.
- Chloral reacts with phenyl chloride to form DDT. Due to electrophilic substitution reaction.

MCQs

NCERT Topic-wise MCQs

Classification

1. Which of the following is a primary halide?

1) Iso - propyliodide 2) sec - Butyliodide
3) tert - Butylbromide 4) Neo - hexylchloride

2. Which one of the following is not an allylic halide?

1) 4-Bromopent-2-ene 2) 3-Bromo-2-methylbut-1-ene
3) 1-Bromobut-2-ene 4) 4-Bromobut-1-ene

3. When two halogen atoms are attached to same carbon atom then it is:

1) vic-dihalide
2) gem-dihalide
3) α, ω-dihalide
4) α, β-dihalide

4. C_4H_9X is a:

1) Pri - Alkyl halide
2) sec - Alkyl halilde
3) ter - Alkyl halide
4) Primary, sec and tertiary alkyl halide

Nomenclature

5. The IUPAC name of tertiary Butyl chloride is:

1) 2 - Chloro - 2 - methyl propane
2) 3 - Chlorobutane
3) 4 - chlorobutane
4) 1, 2 - chloro - 3 - methyl propane

6. The common name for the molecule given below is-

$CH_2 = CH- CH_2 - Cl$

1) Vinyl chloride 2) Aryl chloride
3) Allyl chloride 4) None of these

7. The structure of 2 - Bromopentane:

1) 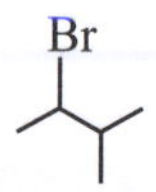
2) 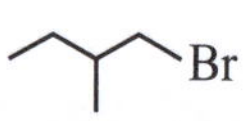
3) 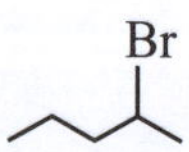
4) 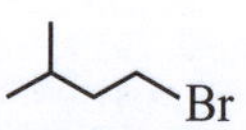

8. The structure of 2 - Bromo - 2 - methyl propane:

1) $CH_3-\overset{\overset{Br}{|}}{CH}-\overset{\overset{CH_3}{|}}{CH_2}$ 2) $CH_3 - CH_3 - CH_2 - Br$

3) $CH_3-\overset{\overset{Br}{|}}{\underset{\underset{CH_3}{|}}{C}}-CH_3$ 4) None of these

9. Identify the incorrect IUPAC name-

1) $CH_3-\overset{\overset{H}{|}}{\underset{\underset{Cl}{|}}{C}}-CH_2COOH \longrightarrow$ 3-Chlorobutanoic acid

2) $CH_2=CH-\underset{\underset{Cl}{|}}{CH}-CH_3 \longrightarrow$ 3-Chlorobut-1-ene

3) 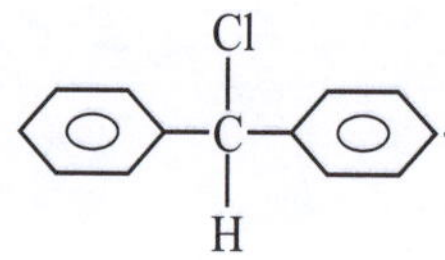 ⟶ 1-Chloro-1,1-diphenylmethane

4) None of these

Methods of Preparation

10. In the reaction $ROH \xrightarrow{Red\,P+I_2} RI + A$; A is :

1) H_3PO_2 2) H_3PO_3
3) H_3PO_4 4) HPO_3

11. 1-Butene when treated with HBr gives:

1) sec. - Butyl bromide 2) 3 - Butyl bromide
3) 1 - Bromo butane 4) None of these

12. What is the final product of the given reaction :

$$CH_3 - C \equiv CH \xrightarrow{Na} \xrightarrow{CH_3-CH_2-I} \text{Product}$$

1) $CH_2{=}CH{-}CH_2{-}CH_3$
2) $CH_3{-}CH_2{-}C{\equiv}C{-}CH_2{-}CH_3$
3) $CH \equiv C - CH_2 - CH_2 - CH_3$
4) $CH_3 - C \equiv C - CH_2 - CH_3$

13. Which of the following reaction follows Markovnikov's rule?

1) $C_2H_4 + HBr$ 2) $C_3H_6 + Cl_2$
3) $C_3H_6 + HBr$ 4) $C_3H_6 + Br_2$

14. 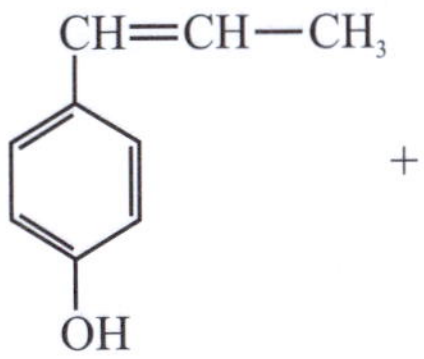 + HBr → X, here X in the reaction is:

1) 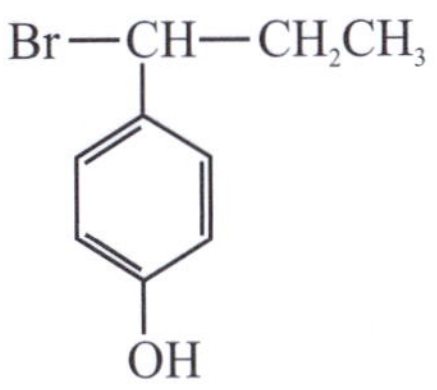
2) 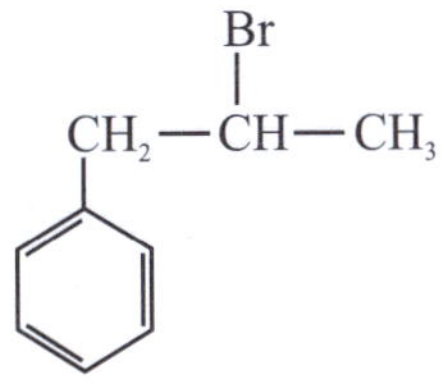
3) 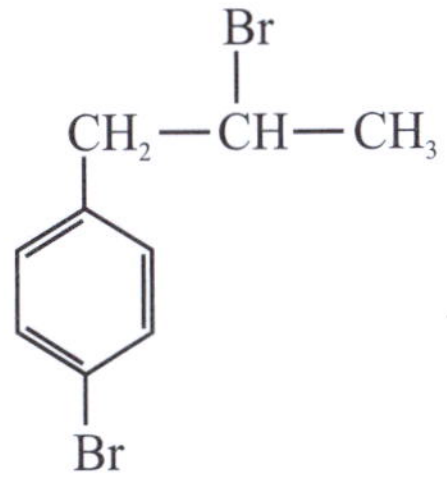
4) 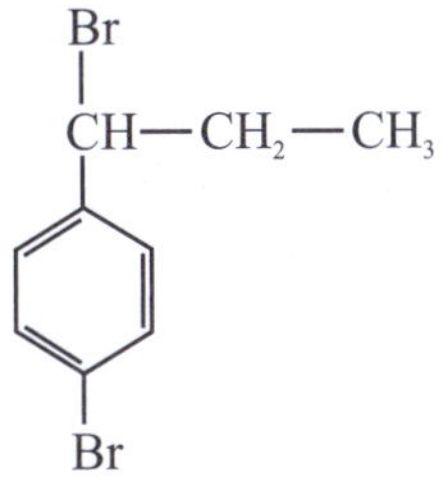

15. Which one of the following methods cannot yield alkyl halide?

1) $C_2H_5OH + HCl \rightarrow$
2) $CH_3COOAg + Br_2 \xrightarrow{CCl_4}$
3) $CH_2 = CH_2 \xrightarrow{SOCl_2}$
4) $CH_3CH_2OH + HX \xrightarrow{ZnCl_2}$

16. Identify B in the following scheme:

$CH_3CH_2CH_2OH \xrightarrow{PCl_5} A \xrightarrow{alc.KOH} B$

1) Propyne
2) Propene
3) Propanol
4) Propanone

17. In the following reactions:

I. $CH_3-CH(CH_3)-CH(OH)CH_3 \xrightarrow{H^+ / Heat} \underset{(Major)}{A} + \underset{(Minor)}{B}$

II. $A \xrightarrow[\text{in the absence of peroxide}]{\text{HBr, dark}} \underset{(Major)}{C} + \underset{(Minor)}{D}$

The major product (A) and (C) are respectively

1) $CH_3=C(CH_3)-CH_2CH_3$ and $CH_3-C(CH_3)(Br)-CH_2CH_3$
2) $CH_3=C(CH_3)-CH_2-CH_3$ and $CH_2(Br)-C(CH_3)-CH_2CH_3$
3) $CH_3=C(CH_3)-CH_2-CH_3$ and $CH_2(Br)-C(CH_3)-CH_2CH_3$
4) $CH_3-C(CH_3)=CHCH_3$ and $CH_3-C(CH_3)(Br)-CH_2-CH_3$

18. Which of the following compounds can yield only one monochlorinated product upon free radical chlorination?

1) 2, 2 - Dimethyl propane
2) 2 - Methyl propane
3) 2 - Methyl butane
4) n - Butane

19. In the following reaction, A is:

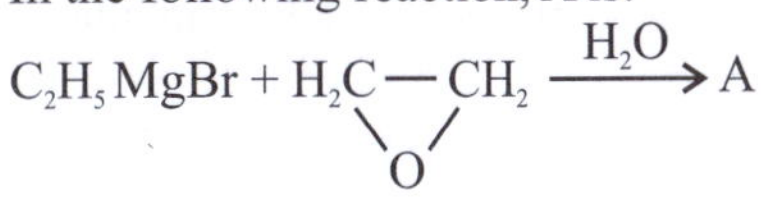

1) $C_2H_5CH_2CHO$
2) $C_2H_5CH_2CH_2OH$
3) $C_2H_5CH_2OH$
4) $C_2H_5HSO_4$

20. 1 - Chloropropane $\xrightarrow{alc.KOH} B \xrightarrow{HX} C$. In this reaction C is:

1) 1 - Chloropropene
2) Alkyl halide
3) 2 - Chloropropane
4) sec. Alkyl halide

21. The reaction of toluene with chlorine in the presence of $FeCl_3$ gives predominantly:

1) A mixture of o - and p - Chlorotoluene
2) Benzyl chloride
3) m - Chlorotoluene
4) Benzoyl chloride

22. What is 'A' in the following reaction?

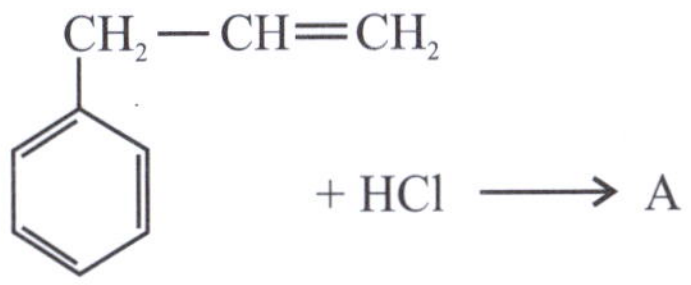 + HCl ⟶ A

1) 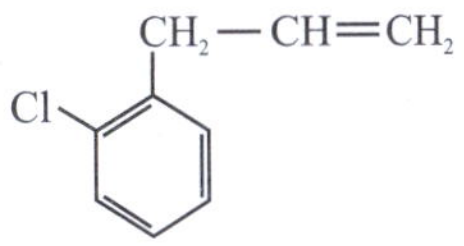
2) 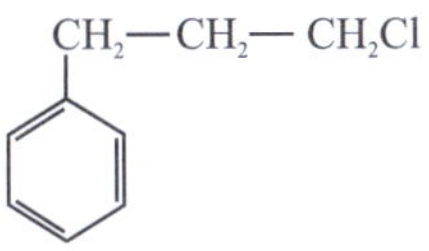
3) 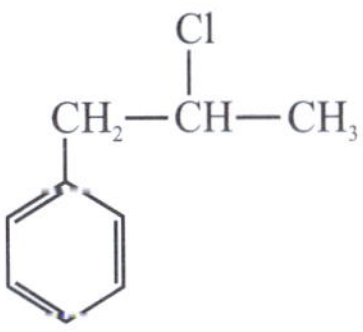
4) 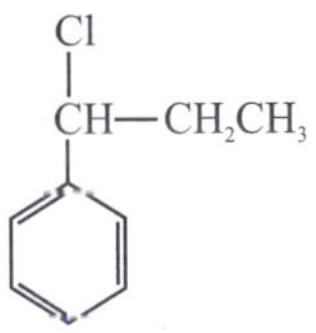

Physical Properties

23. Which of the following molecules has highest dipole moment?

1) CH_3Cl
2) CH_2Cl_2
3) $CHCl_3$
4) CCl_4

24. Decreasing order of reactivity of hydrogen halide acid in the conversion of ROH → RX is:

1) HCl > HBr > HI > HF
2) HI > HBr > HCl > HF
3) HF > HCl > HBr > HI
4) HF > HBr > HI > HCl

25. The density of glycerol is higher than propanol due to:

1) Van der waals' attraction
2) Hydrogen bonding
3) Ionic bonding
4) More no. of covalent bonds

26. The correct order of boiling point for isomeric primary (1°), secondary (2°) and tertiary (3°) alcohols is:
1) 1° > 2° > 3° 2) 3° > 2° > 1°
3) 2° > 1° > 3° 4) 2° > 3° > 1°

27. Which is the correct increasing order of boiling points of the following compounds?
1 - Iodobutane, 1 - Bromobutane, 1 - Chlorobutane, Butane.
1) Butane < 1 - Chlorobutane < 1 - Bromobutane < 1 - Iodobutane
2) 1 - Iodobutane < 1 - Bromobutane < 1 - Chlorobutane < Butane
3) Butane < 1 - Iodobutane < 1 - Bromobutane < 1 - Chlorobutane
4) Butane < 1 - Cholorobutane < 1 - Iodobutane < 1 - Bromo butane.

Chemical Reactions

28. A primary alkyl halide would prefer to undergo:
1) S_N1 reaction 2) S_N2 reaction
3) α - elimination 4) Racemisation

29. The halogen atom attached to C - atom in the benzene ring is:
1) o - directing and activating
2) p - directing and activating
3) o, p - directing and deactivating
4) m - directing and deactivating

30. A S_N2 reaction takes place with:
1) Retention of configuration
2) Inversion of configuration
3) Racemisation
4) Formation of meso form

31. Consider two reactions.
1. $(CH_3)_2CH-CH_2Br \xrightarrow{C_2H_5OH} (CH_3)_2CH-CH_2-O-C_2H_5 + HBr$
2. $(CH_3)_2-CH-CH_2Br \xrightarrow{C_2H_5O^-} (CH_3)_2CH-CH_2-O-C_2H_5 + Br^-$
The mechanism of reactions (1) and (2) are respectively:
1) S_N2, S_N2 2) S_N2, S_N1
3) S_N1, S_N2 4) S_N1, S_N1

32. Identify the products (A) and (B) in the reactions:
$RX + AgCN \rightarrow (A) + AgX$
$RX + KCN \rightarrow (B) + KX$
1) $(A) \rightarrow R-CN, (B) \rightarrow R-CN$
2) $(A) \rightarrow R-CN, (B) \rightarrow R-NC$
3) $(A) \rightarrow R-NC, (B) \rightarrow RCN$
4) $(A) \rightarrow R-NC, (B) \rightarrow R-NC$

33. The alkyl halide is converted into an alcohol by:
1) Elimination 2) Dehydrogenation
3) Addition 4) Substitution

34. Grignard reagents are formed by the reaction of alkyl halides by warming with:
1) With alcoholic solution
2) With $MgCl_2$
3) Mg in presence of dry ether
4) With $MgCO_3$

35. Which of the following alkyl halides is hydrolysed by S_N1 mechanism?
1) CH_3Cl 2) CH_3CH_2Cl
3) $CH_3CH_2CH_2Cl$ 4) $(CH_3)_3C\ Cl$

36. Among the choices of alkyl bromide, the least reactive bromide in S_N2 reaction is:
1) 1 - Bromopentane
2) 2 - Bromo - 2 - methyl butane
3) 1 - Bromo - 3 methyl butane
4) 1 - Bromo - 2 - methyl butane

37. 2 - Chloro - 2 - methyl propane on reaction with alcoholic KOH gives X as the product. X is:
1) But - 2 - ene
2) 2 - Methyl but - 1 -ene
3) 2 - Methyl prop - 1 - ene
4) 2 - Methyl butan - 2 - ol

38. Which of the following haloalkanes reacts with aqueous KOH most easily (via S_N1)?
1) 1 - Bromobutane
2) 2 - Bromobutane
3) 2 - Bromo - 2 - methylpropane
4) 2 - Chlorobutane

39. In the following pairs of halogen compounds, which compound undergoes faster S_N1 reaction?
(i)  and 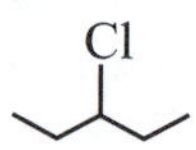
(ii) 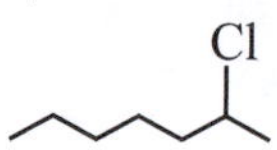 and 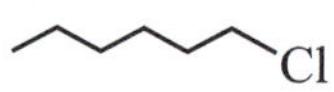
1) (i) 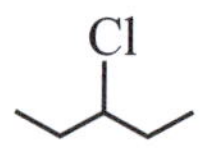 (ii) 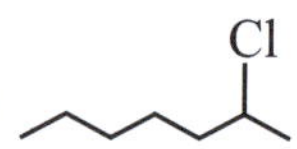
2) (i)  (ii) 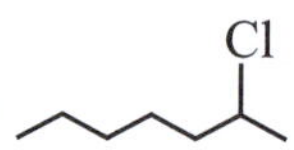
3) (i) 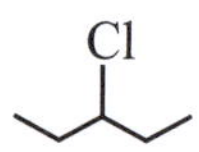 (ii) 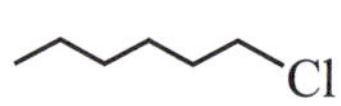
4) (i)  (ii) 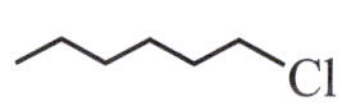

40. Which of the following haloalkane is most reactive towards S_N1?
1) 1 - Chloropropane 2) 1 - Bromopropane
3) 2 - Chloropropane 4) 2 - Bromopropane

41. Reaction of ethyl halide with alkoxide is called:
1) Wurtz reaction 2) Williamson's synthesis
3) Elimination reaction 4) Carbylamine reaction

42. Friedel craft's acetylation of benzene ring involves the use of:
1) CH_3Cl 2) CH_3COCl
3) $(CH_3CO)_2O$ 4) Either (2) or (3)

43. In the reaction $CH_3-CH_2-CH=CH_2 \xrightarrow{HX} B \xrightarrow[Ag_2O]{moist} C$; C is:
1) n - propyl alcohol 2) Isopropyl alcohol
3) Primary alcohol 4) Secondary alcohol

44. Alc. KOH is used for:
   1) Dehydrogenation 2) Dehydrohalogenation
   3) Dehalogenation 4) Dehydration

45. In the reaction. $C_2H_5I \xrightarrow{alc.KOH} X \xrightarrow{Br_2} Y \xrightarrow{KCN} Z$;
   X, Y and Z are respectively:
   1) C_2H_4, C_2H_5Br, C_2H_5CN
   2) C_2H_5OH, C_2H_5Br, C_2H_5CN
   3) C_2H_4, CH_2BrCH_2Br, $CNCH_2CH_2CN$
   4) C_2H_4, $C_2H_4Br_2$, C_2H_5CN

46. Wurtz reaction is not possible with:
   1) CH_3I 2) $(CH_3)_3C-Cl$
   3) $(CH_3)_2CH.I$ 4) C_2H_5I

47. 2 - Bromobutane on treatment with alc . KOH gives:
   1) But - 1 - ene 2) Butan - 2 - ol
   3) But - 2 - ene 4) Both (1) and (3)

48. The reaction, $C_2H_5Br + NaI \xrightarrow[\Delta]{Acetone} C_2H_5I + NaBr$ is called:
   1) Wurtz reaction 2) Finkelstein reaction
   3) Hunsdiecker reaction 4) Swartz reaction

49. In the following sequence of reactions.
   $CH_3CH_2CH_2OH \xrightarrow{HBr} A \xrightarrow{alc.KOH} B \xrightarrow[Peroxide]{HBr} C \xrightarrow{aq.KOH} D$
   The original compound and D are:
   1) Same 2) Isomers
   3) Metamers 4) Homologues

50. In the following sequence of reaction,
   $CH_3-Br \xrightarrow{KCN} (A) \xrightarrow{H_3O^+} B \xrightarrow[ether]{LiAlH_4} C$. C is:
   1) Acetone 2) Methane
   3) Acetaldehyde 4) Ethyl alcohol

51. Following is the substitution in which - CN replace - Cl
   $R-Cl + KCN \xrightarrow{\Delta} R-CN + KCl$
   (alcoholic)
   To obtain propanenitrile, R - Cl should be:
   1) Chloroethane 2) 1 - Chloropropane
   3) Chloromethane 4) 2 - Chloropropane

52. The correct leaving group ability order is :
   1) $\overset{\ominus}{O}H > H_2O$ 2) $\overset{\ominus}{O}H > \overset{\ominus}{S}H$
   3) 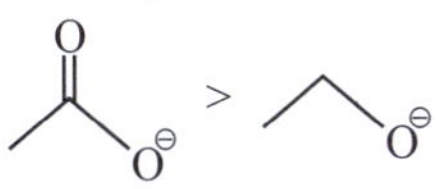 4) $\overset{\ominus}{Cl} > \overset{\ominus}{I}$

53. The compound that reacts the fastest with sodium methoxide is:
   1) 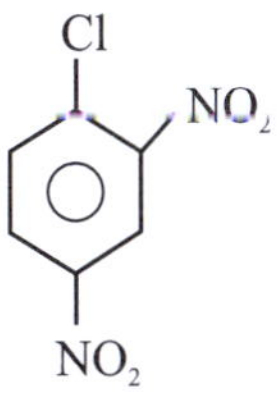 2) 
   3) 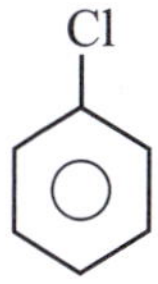 4) 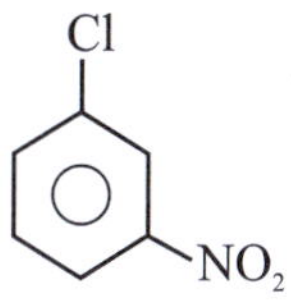

54. In the given reaction, 'Y' is:

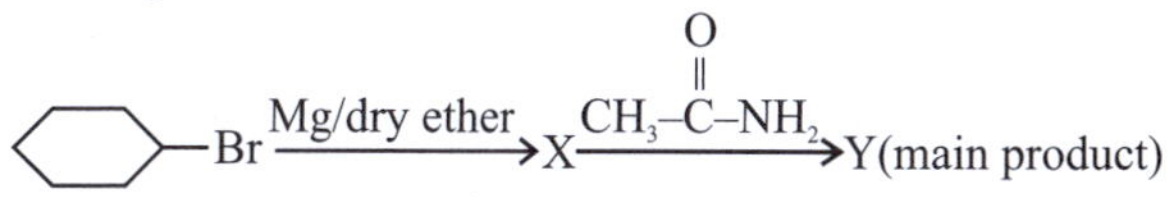

   1) Hexane
   2) Cyclohexane
   3) Cyclohexyl cyclohexane
   4) Cyclohexyl ether.

55. What is the name of the following reaction?
   $CH_3-CH_2-CH_2Cl \xrightarrow[Dry\ aetone]{NaI} CH_3CH_2CH_2I$
   1) Sandmeyer reaction 2) Gattermann reaction
   3) Finkelstein reaction 4) Swarts reaction

56. $CH_3CH_2CH_2Br \xrightarrow[alcohol]{KOH} CH_3CH=CH_2$
   The above reaction is an example of:
   1) Substitution 2) Elimination
   3) Addition 4) Rearrangment

57. Which one of the following species will be most reactive in S_N2 reaction?
   1) 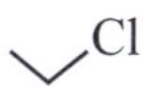 2) 
   3) 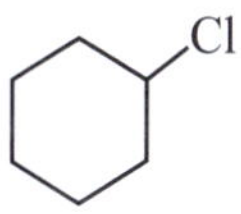 4) 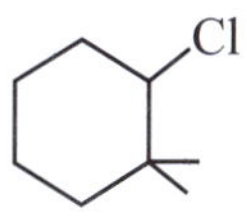

58. Iso - propyl bromide on Wurtz reaction gives:
   1) Hexane 2) Propane
   3) 2,3 - dimethyl butane 4) Neo - hexane

Polyhalogen Compounds

59. An organic halogen compound which is used as refrigerant in refrigerators and air conditioners is:
   1) BHC 2) CCl_4
   3) Freon 4) $CHCl_3$

60. Gammexane is chemically known as:
   1) Benzene hexachloride 2) Hexa chlorobenzene
   3) Benzene hexabromide 4) Hexa bromobenzene

61. Freon used as refrigerant is:
   1) $CF_2=CF_2$ 2) CH_2F_2
   3) CCl_2F_2 4) CF_4

62. Which one of the following has highest chlorine content?
   1) Pyrene 2) DDT
   3) Chloral 4) BHC

High Order Time Intensive MCQs

1. Alkyl halides can be obtained by all methods except:

1) $CH_3-CH_2-CH_2-CH_3+Cl_2 \xrightarrow[light]{UV}$
2) $CH_3CH_2OH+HCl/ZnCl_2 \xrightarrow{\Delta}$
3) $C_2H_2OH + NaCl \rightarrow$
4) $CH_3COOAg+Br_2 \xrightarrow{CCl_4}$

2. $CH_3-\underset{Br}{\overset{Br}{C}}-\underset{Br}{\overset{Br}{C}}-CH_3 \xrightarrow{(a)\ NaNH_2} x \xrightarrow{(b)\ CH_3I} y$

$CH_3-C\equiv C-C\equiv C-CH_3$

In the above reaction what is the value of a + b?

1) 10 2) 8
3) 4 4) 7

3. Identify correct reactivity order for S_N1 reaction.

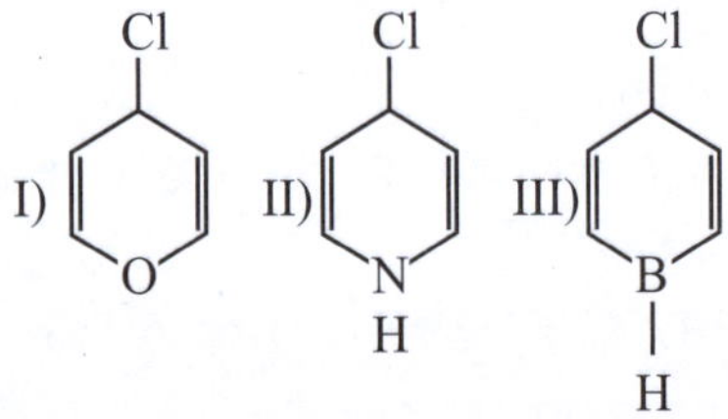

1) I > II > III 2) II > I > III
3) I > III > II 4) I = III = II

4. Identify the correct reactivity order for addition elimination reaction (ArSN):

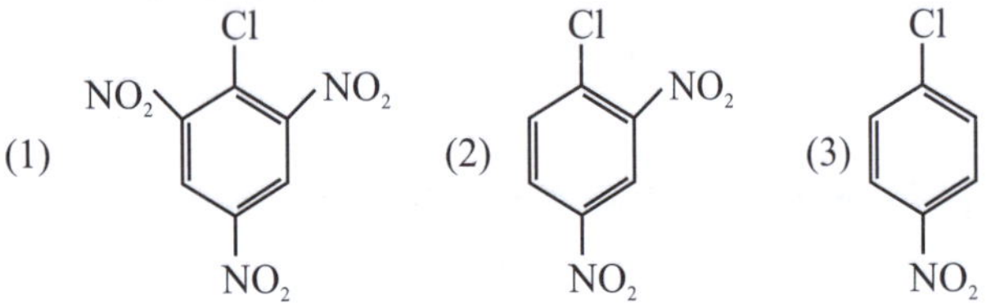

1) 1 > 2 > 3 2) 2 > 1 > 3
3) 3 > 2 > 1 4) 3 = 2 > 1

5. Which of the following is most reactive towards S_N2 reaction?

1) $CH_2=CH-CH_2-Cl$ 2) $Ph-CH_2-Cl$
3) $MeO-CH_2-Cl$ 4) $C_6H_5-\overset{O}{\overset{\|}{C}}-CH_2-Cl$

6. Which one of the following can neither undergo S_N1 nor S_N2 ?

1) 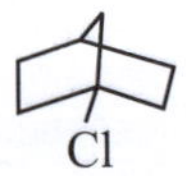 2) $CH_3-\underset{CH_3}{\overset{CH_3}{C}}-CH_2-Cl$
3) CH_3-Cl 4) None of these

7. In the reactions given below

$R-Cl \xrightarrow{(i)\ KCN,(ii)\ LiAlH_4}$ Product A

$R-Cl \xrightarrow{(i)\ AgCN,(ii)\ LiAlH_4}$ Product B

the compounds A and B are :

1) Chain isomers 2) Position isomers
3) Functional isomers 4) Metamers

8. What is the major product of the following reaction

$H_2C=CH-CH_2-OH \xrightarrow[excess]{HBr}$ Product

1) $CH_3-\overset{Br}{\overset{|}{CH}}-CH_2-Br$
2) $H_2C=CH-CH_2-Br$
3) $CH_3-\overset{Br}{\overset{|}{CH}}-CH_2-OH$
4) $CH_3-\overset{OH}{\overset{|}{CH}}-CH_2-OH$

9. Increasing the concentration of a nucleophile in a typical S_N2 reaction by a factor of 10 will cause the reaction rate to:

1) Increase by a factor of 10
2) Increase by a factor of 10^2
3) Decrease by a factor of 10
4) Remain about the same

10. Treatment of 1-phenyl-2-chloropropane with alcoholic KOH gives mainly

1) 3-phenylpropene
2) 1-phenylpropene
3) 1-phenylpropan-3-o1
4) 1-phenylpropan-2-o1

11. The final product obtained in the reaction

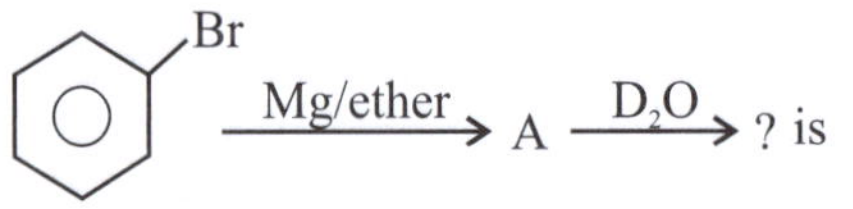

1)  2) 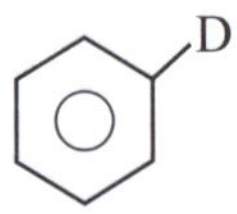
3) 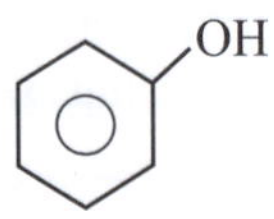 4) 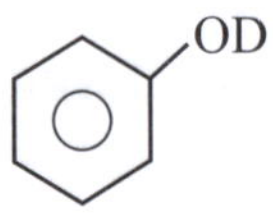

12. Of the following, which is an S_N1 reaction ?

1) $(CH_3)_3CBr+H_2O \rightarrow$
2) $CH_3CH_2CH_2Cl+I^- \rightarrow$
3) $(CH_3)_3CBr+CN^- \rightarrow$
4) $CH_3CHBrCH_3+\bar{O}H\ (alc.) \rightarrow$

13. The major product obtained in the reaction

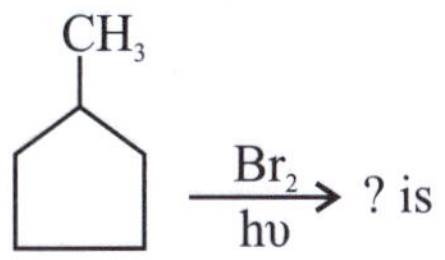

1) 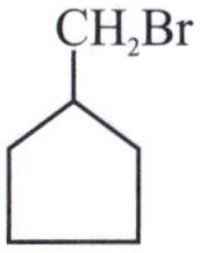
2) 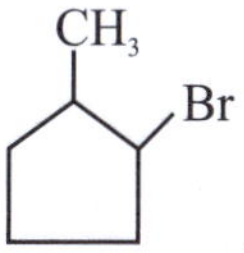
3) 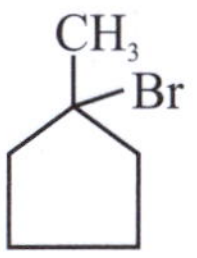
4) 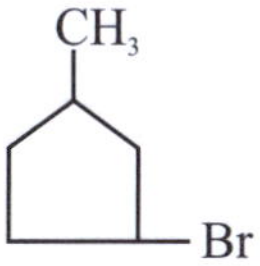

14. The ease of dehydrohalogenation with alcoholic KOH in case of chloroethane (I), 2-chloropropane (II) and 2-chloro-2-methylpropane (III) is of the order

1) III > II > I
2) I > II > III
3) II > I > III
4) I > III > II

15. An unknown alkyl halide (A) reacts with alcoholic KOH to produce a hydrocarbon (C_4H_8). Ozonolysis of the hydrocarbon affords one mole of propionaldehyde and one mole of formaldehyde. Suggest which organic structure among the following is the correct structure of the above alkyl halide (A)?

1) $CH_3(CH_2)_3Br$
2) $CH_3CH(Br)CH(Br)CH_3$
3) $CH_3CH_2CH(Br)CH_3$
4) $Br(CH_2)_4Br$

16. Example of vinylic halide is

1) 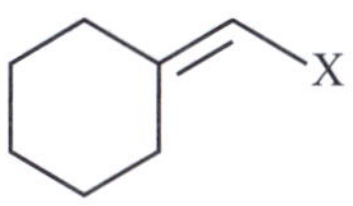
2) 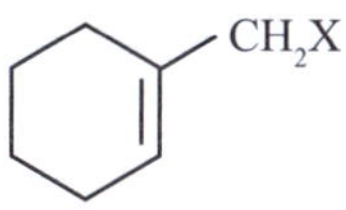
3) 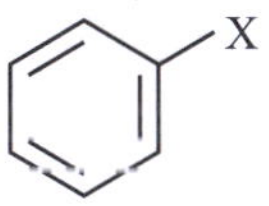
4) 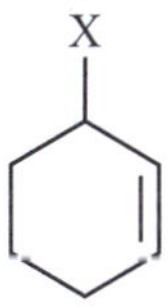

17. The product obtained in the reaction

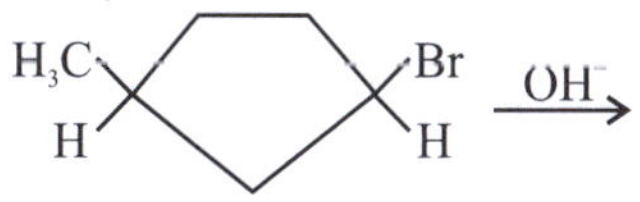

1) 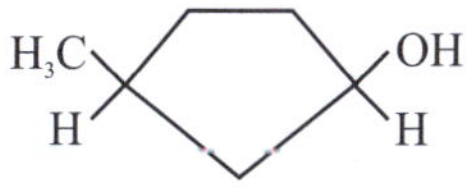
2) 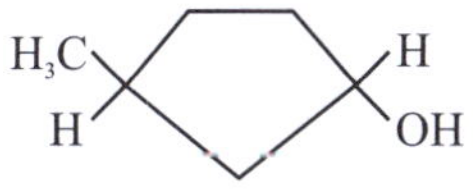
3) 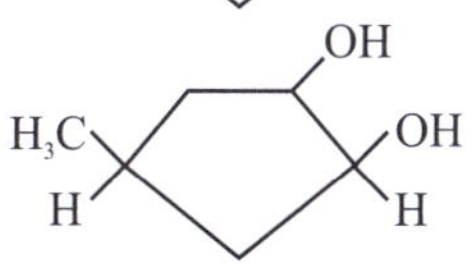
4) None of these

18. Which of the following alkyl halide can not be solvolysed easily with aqueous $AgNO_3$ solution?

1) $CH_2{=}CH{-}Cl$
2) $CH_3{-}CH(CH_3){-}Cl$
3) $CH_3{-}C(CH_3)_2{-}Cl$
4) $Ph{-}CH_2{-}Cl$

19. Which one of the following undergoes nucleophilic aromatic substitution at the fastest rate?

1) 
2) 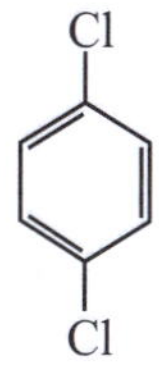
3) 
4) 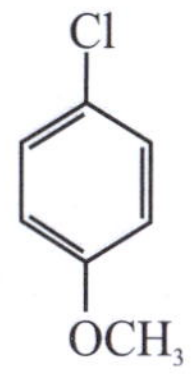

20. Identify the principal organic product of the following reaction,

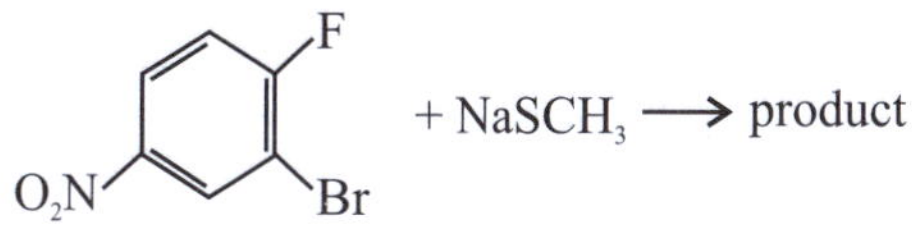

1) 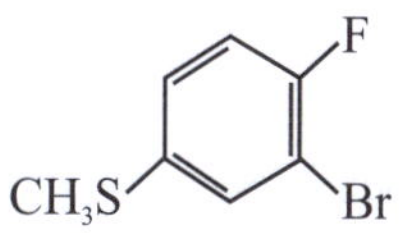
2) 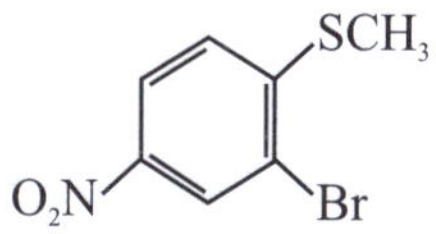
3) 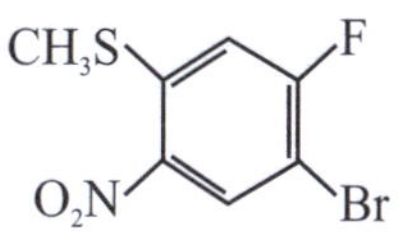
4) 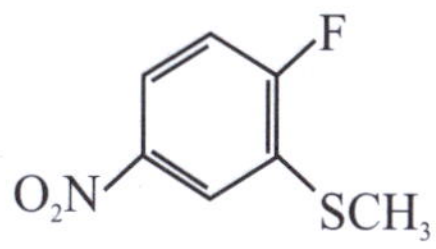

21. 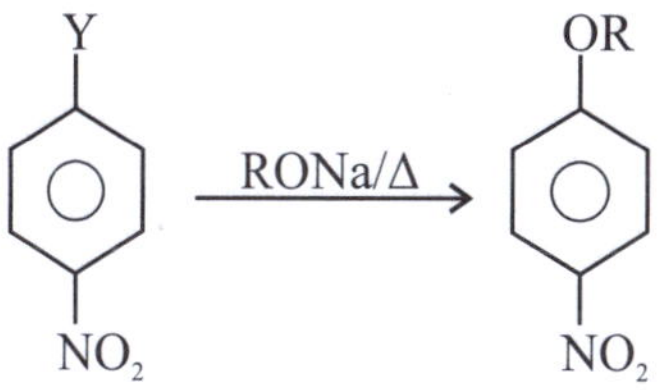

Above reaction has maximum rate when:

1) Y = –I
2) Y = –Br
3) Y = –Cl
4) Y = –F

22. Identify product A and product B:

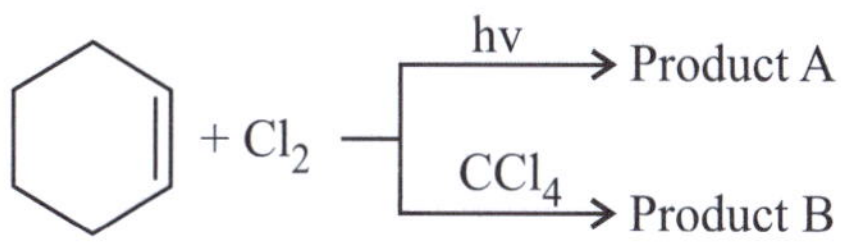

1) A: 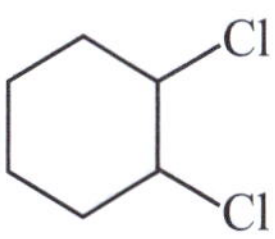 B: 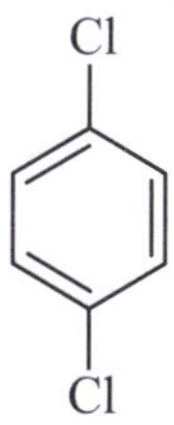

2) A:  B: 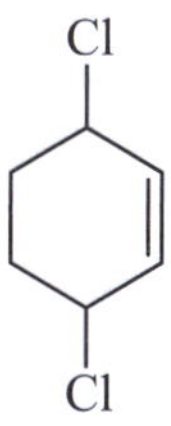

3) A: 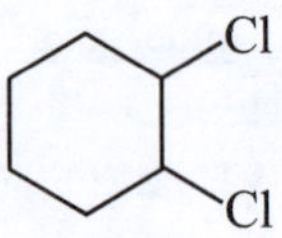 B: 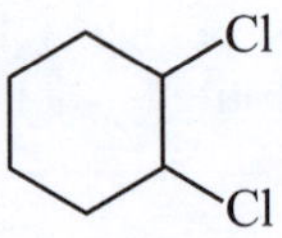

4) A: 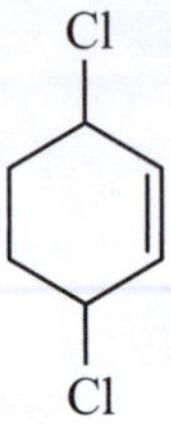 B: 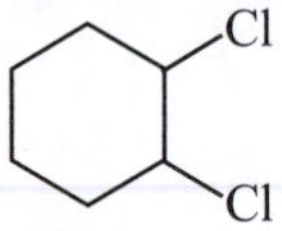

23. For the following reaction, rate is expressed as,

$CH_3-CH(Br)-CH_3 + NaOH \longrightarrow CH_3-CH(OH)-CH_3$

rate = 4.7×10^{-5} [RX] [OH^-] + 0.24×10^{-5} [RX]

1) The reaction proceeds by S_N1 route only
2) The reaction proceeds by S_N2 route only
3) The reaction proceeds by S_N1 as S_N2 routes (follows mixed mechanism pathway)
4) If concentration of RX is doubled then rate will increase four times

24. Alkyl halide is converted into alkyl isocyanide by reaction with

1) NaCN 2) NH_4CN
3) KCN 4) AgCN

25. For the following the increasing order of nucleophilicity would be: (in protic solvent)

(i) I^- (ii) Cl^- (iii) Br^-

1) $I^- < Cl^- < Br^-$ 2) $Br^- < Cl^- < I^-$
3) $I^- < Br^- < Cl^-$ 4) $Cl^- < Br^- < I^-$

26. Which among the following species is an ambident nucleophile?

1) Ethene 2) Benzene
3) Cyanide ion 4) Acetone

27. Which one of the following has maximum nucleophilicity?

1) $CH_3S^{\ominus}$ 2) $C_6H_5-\overset{\ominus}{O}$
3) Et_3N 4) $F^{\ominus}$

28. The product (C) in the below mentioned reaction is:

$CH_3-CH_2-CH_2-Br \xrightarrow[\Delta]{KOH(alc)} A \xrightarrow{HBr} B \xrightarrow[KOH(aq)]{\Delta} C$

1) Propan-1-ol 2) Propene
3) Propyne 4) Propan-2-ol

29. Which configuration will be adopted by the product at carbon atoms marked (I) and (II) respectively in the given reaction.

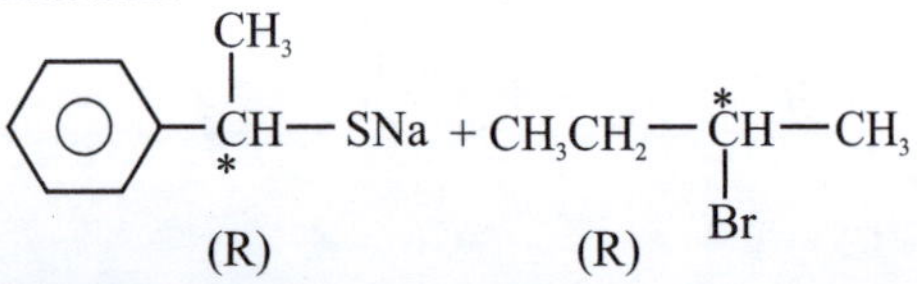

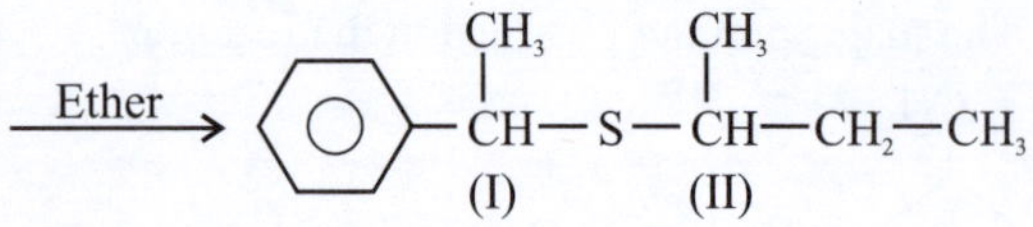

1) R, R 2) R, S
3) S, S 4) S, R

30. What is the final product of the given reaction?

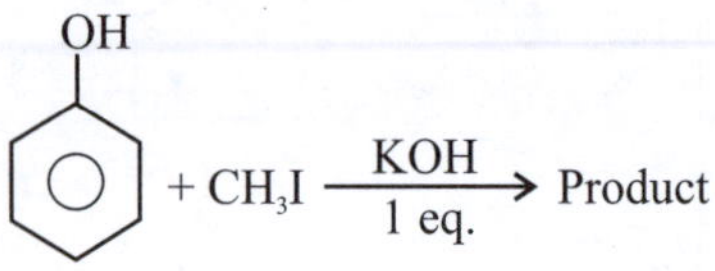

1) 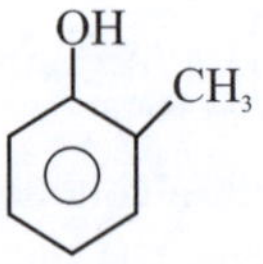 2) 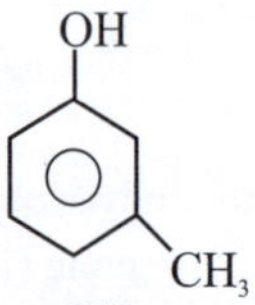
3) 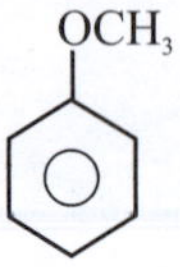 4) 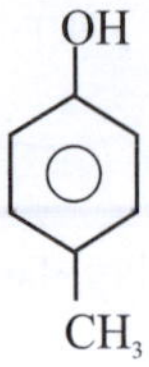

31. Which one of the following compound is most reactive for E_1 reaction ?

1) $CH_3-CH_2-C(Br)(CH_3)-CH_3$ 2) $CH_3-CH_2-C(Br)(CH_3)-C_6H_5$
3) $CH_3-CH_2-C(Br)(CH_3)-C_2H_5$ 4) $CH_3-CH_2-CH(Br)-CH_3$

32. The major product in the reaction,

$CH_3-C(CH_3)_2-Cl + \overset{+}{K}\overset{-}{O}-C(CH_3)_2-CH_3 \longrightarrow ?$ is

1) t-Butyl ethyl ether 2) 2, 2 -Dimethylbutane
3) 2-Methylpent-1-ene 4) 2-Methylprop-1-ene

33. The correct order of S_N2 reaction mechanism for the product of the following halide is :

(P) $CH_3-CH_2-C(Ph)(I)-CH_3$

(Q) $CH_3-CH(Ph)-CH(I)-CH_3$

(R) CH_3-CH_2-I

(S) $CH_3-CH_2-CH(I)-CH_3$

1) R > S > Q > P 2) R > Q > S > P
3) P > R > S > Q 4) Q > P > R > S

34. By passing excess of $Cl_2(g)$ in boiling toluene, which one of the following compounds is exclusively formed?

1) 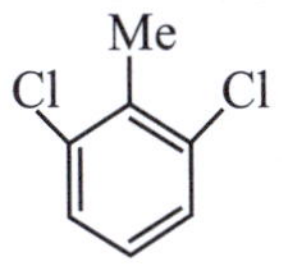
2) 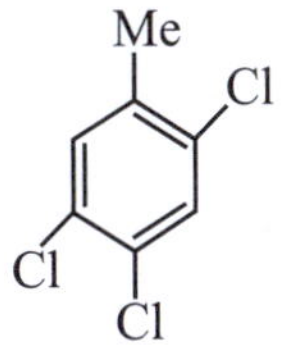
3) 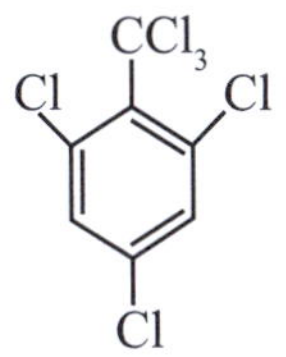
4) 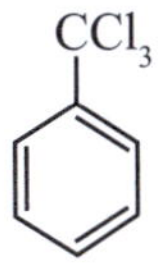

35. The major product of the following reaction is:

$$\underset{\text{Br}\quad\ \text{Br}}{CH_3CH_2CH{-}CH_2} \xrightarrow[\text{(ii) NaNH}_2 \text{ in liq NH}_3]{\text{(i) KOH alc.}}$$

1) $CH_3CH{=}C{=}CH_2$
2) $CH_3CH_2CH(NH_2){-}CH_2NH_2$
3) $CH_3CH{=}CHCH_2NH_2$
4) $CH_3CH_2C{\equiv}CH$

36. Thionyl chloride method is preferred over phosphorus pentachloride method for the preparation of alkyl chloride because :
1) The reaction goes to completion.
2) The by-products being gases escape into the atmosphere leaving behind almost pure alkyl chloride.
3) Thionyl chloride is cheap while phosphorus pentachloride is costly.
4) None of the above

37. S_N1 reaction is most feasible in :
1)  Cl + KOH (aq.) ⟶
2)  + KOH (aq.) ⟶
3) 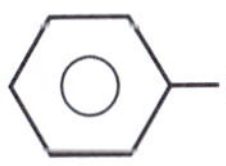 Cl + KOH (aq.) ⟶
4) 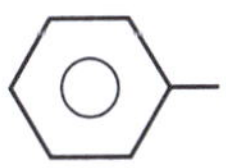 CH_2Cl + KOH (aq.) ⟶

38. Which one of the following has maximum nucleophilicity?
1) $\overset{\ominus}{C}H_3$
2) $\overset{\ominus}{N}H_2$
3) $CH_3O^{\ominus}$
4) $(CH_3)_3C{-}O^{\ominus}$

39. The decreasing order of rate of S_N2 reaction is :

$CH_3{-}Cl$ (I)

$CH_3{-}C(=O){-}CH_2{-}Cl$ (II)

$CH_3{-}CH(CH_3){-}CH_2{-}Cl$ (III)

$CH_3{-}CH_2{-}Cl$ (IV)

1) IV > III > II > I
2) II > III > I > IV
3) II > I > IV > III
4) None

40. The correct sequence of reagents for following reaction is:

$$CH_3-CH_3 \rightarrow CH_3-CH_2-CN$$

1) (i) HCl (ii) KCN
2) KCN
3) (i) Cl_2/hv (monochlorination) (ii) KCN
4) (i) NaCl (ii) KCN

41. Which of them is correct order for solvolysis rate in aqueous acetone? (via S_N1)

P. $CH_3O{-}CH{=}CH{-}CH_2{-}Cl$

Q. $CH_3{-}O{-}C(=CH_2){-}CH_2{-}Cl$

R. $CH_3O{-}CH_2{-}CH{=}CH{-}Cl$

S. $CH_3O{-}CH_2{-}CH_2{-}CH_2{-}Cl$

1) R > P > Q > S
2) P > Q > R > S
3) Q > P > S > R
4) P > Q > S > R

42. Which of the following can work as dehydrating agent for alcohols ?
1) H_2SO_4
2) Al_2O_3
3) H_3PO_4
4) All of these

43. In the reaction:

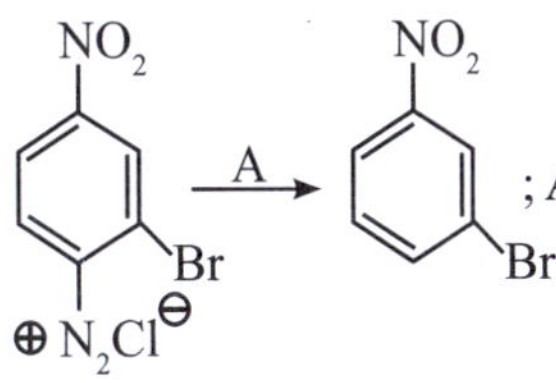

; A is-

1) $HgSO_4/H_2SO_4$
2) Cu_2Cl_2
3) H_3PO_2 and H_2O
4) H^+/H_2O

44. In the given reaction, CH_3 $CH_2X + CH_3SNa \rightarrow$

The fastest reaction occurs when 'X' is:
1) –OH
2) –F
3) $-OCOCF_3$
4) $-OCOCH_3$

45. Which of the following reactions cannot occur?

1) 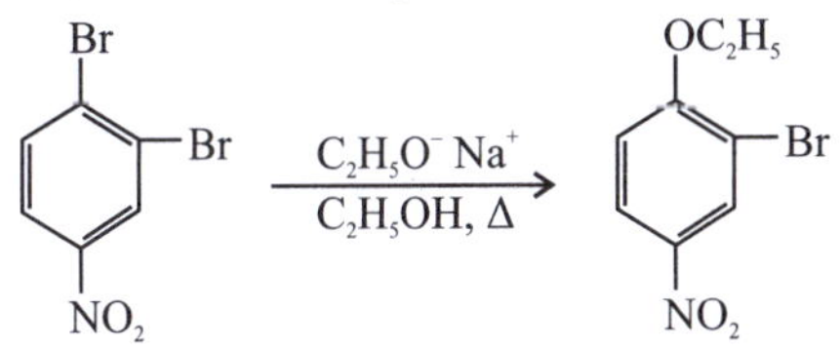
2) 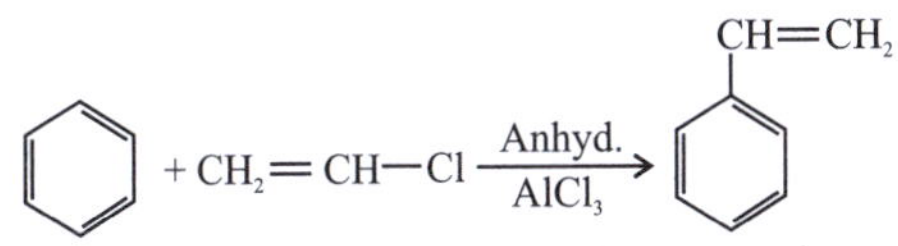
3) 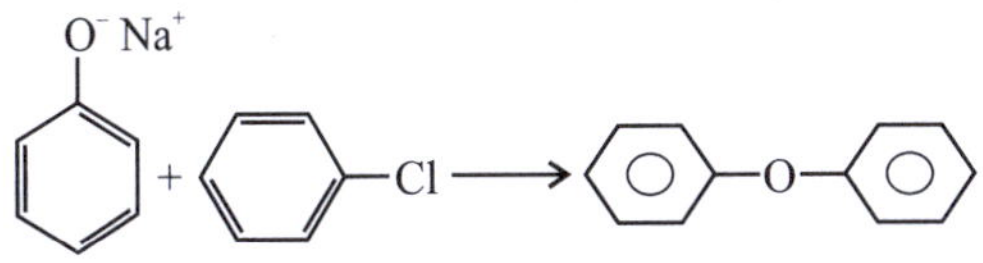
4) Both (2) & (3)

46. Identify the mismatch:

Reaction	Product	Mechanism followed
1) 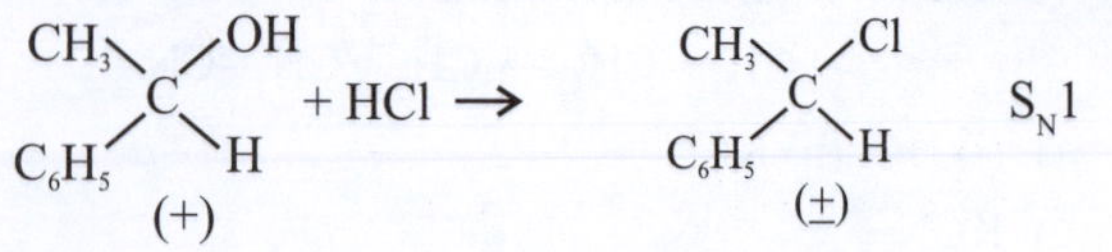		S_N1
2) 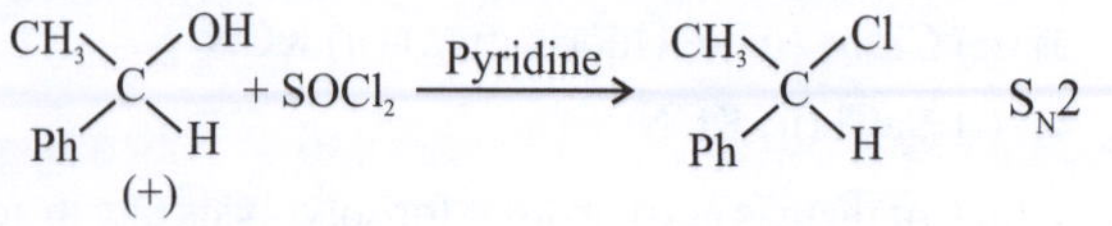		S_N2
3) 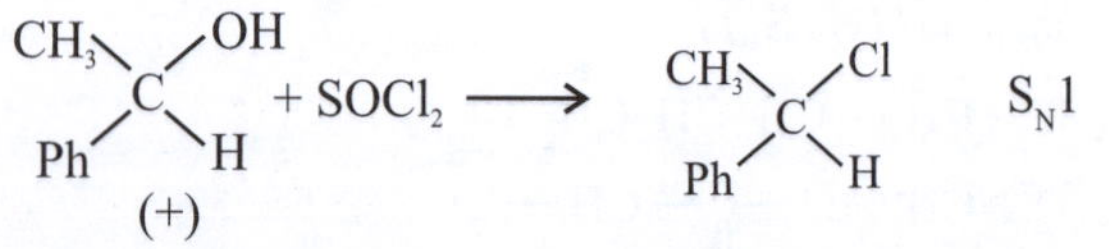		S_N1

4) None of these

47. 3-methylpent-2-ene on reaction with HBr in the presence of peroxide forms an addition product. The number of possible stereoisomers for the product is:

1) Zero
2) Two
3) Four
4) Six

48. Which of the following reactions take place by S_N2 mechanism?

1) $CH_3CH_2O^-Na^+ + CH_3 - CH_2 - Cl \rightarrow$

2) 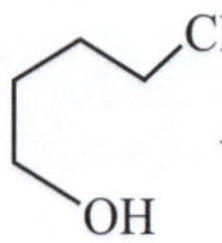 $+ NaOH \longrightarrow$

3) $CH_3 - CH_2 - OH + HBr \rightarrow$

4) All of these

49. Which of the following statements are correct?

i. Benzyl halides are more reactive than vinyl and aryl halides

ii. Vinyl halides are more reactive than alkyl halides

iii. Aryl halides are less reactive than alkyl halides

iv. Aryl halides are more reactive than benzyl halides

1) i & ii
2) i & iii
3) i , ii & iii
4) iii & iv

50. Identify the set of reagents/reaction conditions 'X' and 'Y' in the following set of transformation:

$$CH_3-CH_2-CH_2Br \xrightarrow{X} \text{Product} \xrightarrow{Y} CH_3-\underset{\substack{|\\Br}}{CH}-CH_3$$

1) X = dilute aqueous solution, 20°C;
Y = HBr/ acetic acid at 20°C

2) X = dilute aqueous NaOH, 20°C;
Y = HBr/acetic acid at 20°C

3) dilute aqueous NaOH, 20°C;
Y = $Br_2/CHCl_3$, 0°C

4) X=concentrated alcoholic NaOH, 80°C;
Y=Br/$CHCl_3$ 0°C

51. What would be the product formed when 1-bromo-3chloro cyclobutane reacts with two equivalents of metallic sodium in ether?

1) 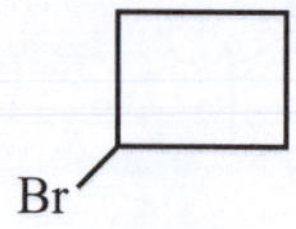
2) 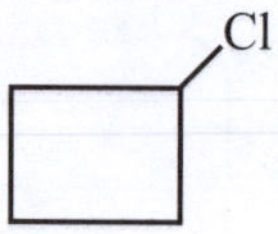
3) 
4) 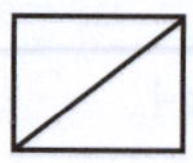

52. The structure of the major product formed in the formed in the following reaction is:

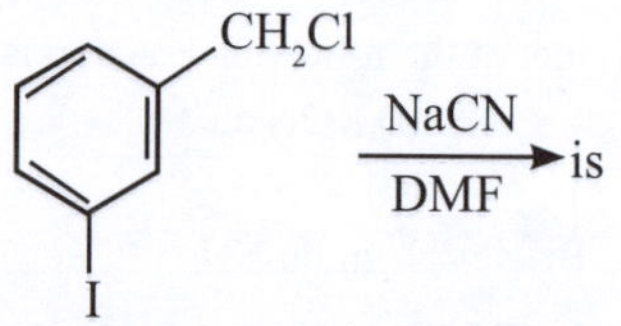

1) 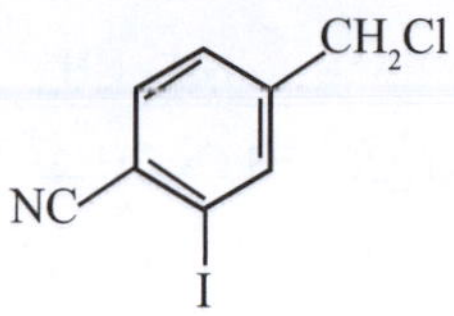
2) 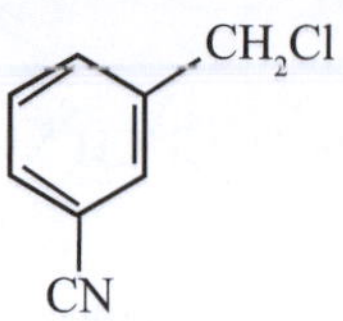
3) 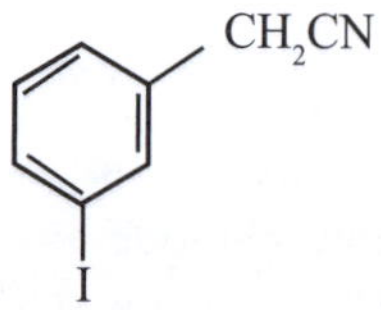
4) 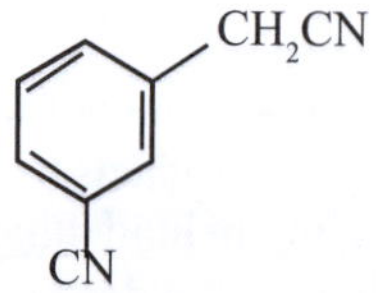

53. Which of the following shows S_N1 reaction most readily?

1) 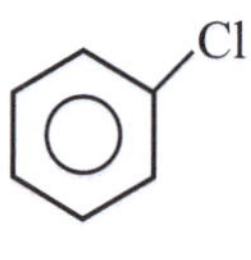
2) 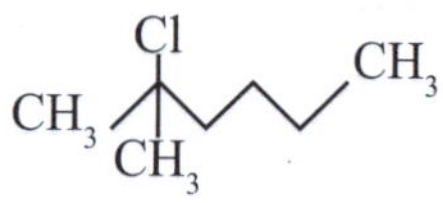
3) 
4) 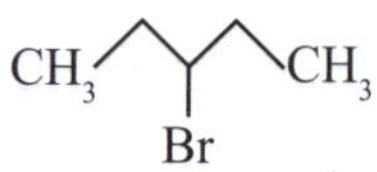

54. Find the major product of the following reaction:

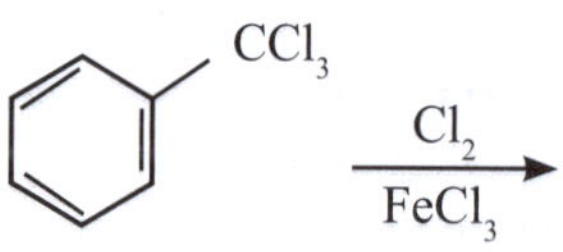

1) 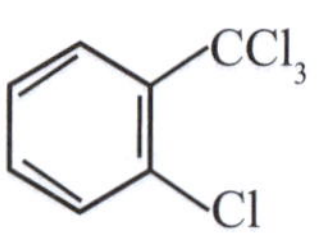
2) 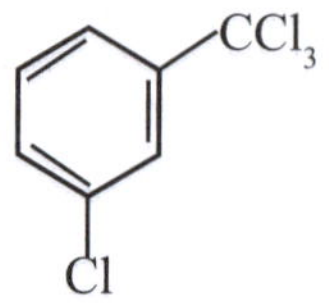
3) 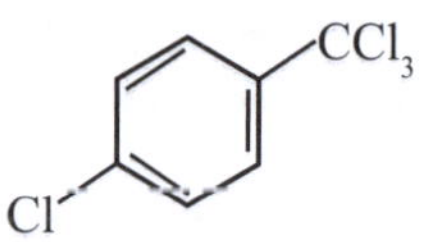
4) 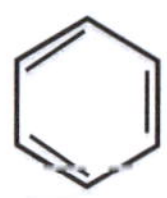

55. The alkyl halide that undergoes S_N1 reaction more readily

1) ethyl bromide
2) isopropyl bromide
3) vinyl bromide
4) t-butyl bromide

Match the Columns MCQs

1. Match the compounds given in Column-I with the effects given in Column-II.

	Column-I		Column-II
(a)	Chloramphenicol	(i)	Malaria
(b)	Thyroxine	(ii)	Anaesthetic
(c)	Chloroquine	(iii)	Typhoid fever
(d)	Chloroform	(iv)	Goitre
		(v)	Blood substituent

1) (a)-(i) (b)-(iii) (c)-(iv) (d)-(ii)
2) (a)-(iii) (b)-(iv) (c)-(i) (d)-(ii)
3) (a)-(ii) (b)-(iii) (c)-(iv) (d)-(i)
4) (a)-(iii) (b)-(i) (c)-(ii) (d)-(iv)

2. Match the items of Column-I and Column-II.

	Column-I		Column-II
(a)	S_N1 reaction	(i)	vic-dibromides
(b)	Chemicals in fire extinguisher	(ii)	gem-dihalides
(c)	Bromination of alkenes	(iii)	Racemisation
(d)	Alkylidene halides	(iv)	Saytzeff rule
(e)	Elimination of HX from alkylhalide	(v)	Chlorobromocarbons

1) (a)-(iv) (b)-(ii) (c)-(i) (d)-(iii) (e)-(v)
2) (a)-(v) (b)-(iv) (c)-(ii) (d)-(i) (e)-(iii)
3) (a)-(iii) (b)-(iv) (c)-(v) (d)-(ii) (e)-(i)
4) (a)-(iii) (b)-(v) (c)-(i) (d)-(ii) (e)-(iv)

3. Match the structures of compounds given in Column-I with the classes of compounds given in Column-II.

	Column-I		Column-II
(a)	$CH_3-\underset{X}{CH}-CH_3$	(i)	Aryl halide
(b)	$CH_2=CH-CH_2-X$	(ii)	Alkyl halide
(c)	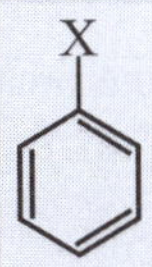	(iii)	Vinyl halide
(d)	$CH_2=CH-X$	(iv)	Allyl halide

1) (a)-(ii) (b)-(iv) (c)-(i) (d)-(iii)
2) (a)-(i) (b)-(iii) (c)-(iv) (d)-(ii)
3) (a)-(iv) (b)-(ii) (c)-(iii) (d)-(i)
4) (a)-(iv) (b)-(iii) (c)-(i) (d)-(ii)

4. Match the reactions given in Column-I with the types of reactions given in Column-II.

	Column-I		Column-II
(a)	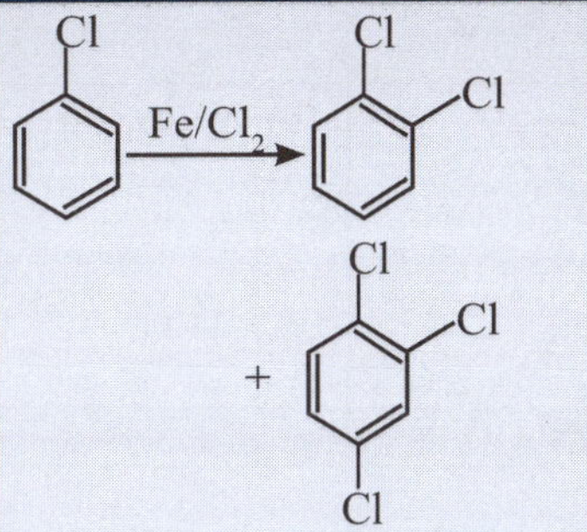	(i)	Nucleophilic aromatic substitution
(b)	$CH_3-CH=CH_2+HBr \longrightarrow CH_3-\underset{Br}{CH}-CH_3$	(ii)	Electrophilic aromatic substitution
(c)	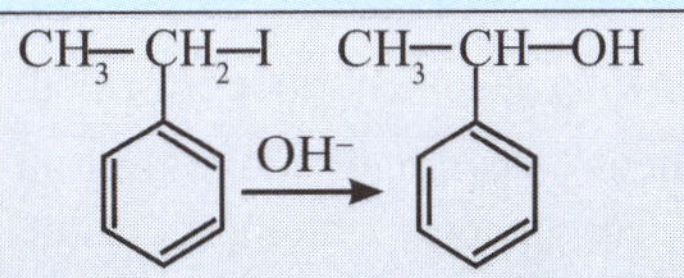	(iii)	Saytzeff elimination
(d)	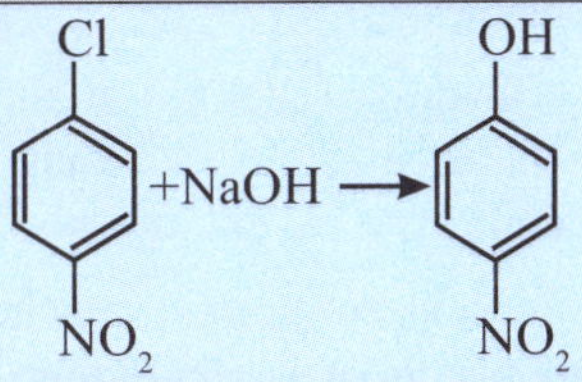	(iv)	Electrophilic addition
(e)	$CH_3CH_2\underset{Br}{CH}CH_3 \xrightarrow{alc.KOH} CH_3CH=CHCH_3$	(v)	Nucleophilic substitution (S_N1)

1) (a)-(ii) (b)-(iv) (c)-(v) (d)-(i) (e)-(iii)
2) (a)-(iv) (b)-(i) (c)-(ii) (d)-(iii) (e)-(v)
3) (a)-(v) (b)-(iv) (c)-(ii) (d)-(i) (e)-(iii)
4) (a)-(v) (b)-(iii) (c)-(i) (d)-(ii) (e)-(iv)

5. Match the structures given in Column-I with the names in Column-II

	Column-I		Column-II
(a)	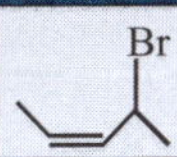	(i)	4-Bromopent-2-ene
(b)	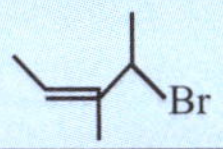	(ii)	4-Bromo-2-methylpent-2-ene
(c)	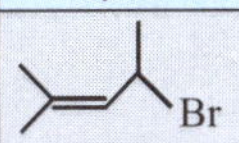	(iii)	4-Bromo-3-methylpent-2-ene
(d)	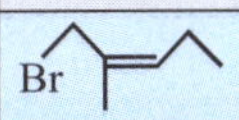	(iv)	1-Bromo-2-methylpent-2-ene

1) (a)-(ii) (b)-(i) (c)-(iii) (d)-(iv)
2) (a)-(i) (b)-(iii) (c)-(ii) (d)-(iv)
3) (a)-(ii) (b)-(iii) (c)-(i) (d)-(iv)
4) (a)-(iv) (b)-(i) (c)-(ii) (d)-(iii)

6. Match the reactions given in Column-I with the names given in Column-II.

	Column-I		Column-II
(a)	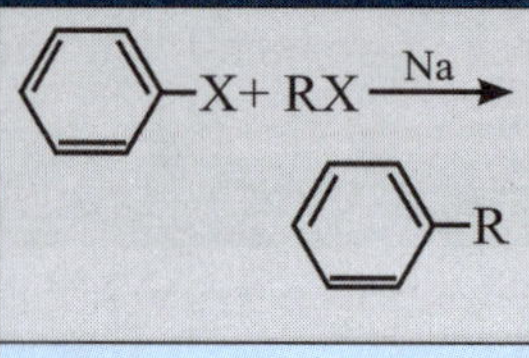	(i)	Fittig reaction
(b)	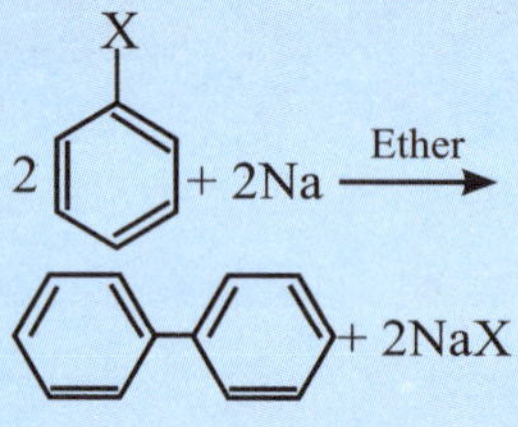	(ii)	Wurtz Fittig reaction
(c)	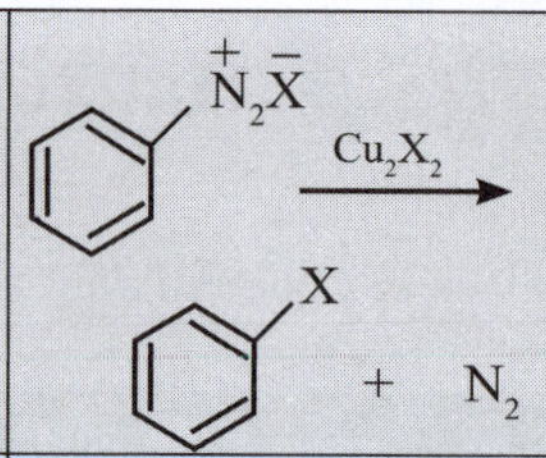	(iii)	Finkelstein reaction
(d)	$C_2H_5Cl + NaI \xrightarrow[\text{acetone}]{\text{dry}} C_2H_5I + NaCl$	(iv)	Sandmeyer reaction

1) (a)-(i) (b)-(iii) (c)-(ii) (d)-(iv)
2) (a)-(iv) (b)-(ii) (c)-(i) (d)-(iii)
3) (a)-(ii) (b)-(i) (c)-(iv) (d)-(iii)
4) (a)-(iv) (b)-(i) (c)-(iii) (d)-(ii)

Correct & Incorrect MCQs

1. Which of the following statement (s) are correct?
   (a) The replacement of hydrogen atom(s) in an aliphatic hydrocarbon by halogen atom(s) results in the formation of haloalkane.
   (b) The replacement of hydrogen atom(s) in an aromatic hydrocarbon by halogen atom(s) results in the formation of haloarene.
   (c) Haloalkanes contain halogen atom(s) attached to the sp^2 hybridised carbon atom of an alkyl group.
   (d) Haloarenes contain halogen atom(s) attached to sp^3 hybridised carbon atom(s) of an aryl group.
   1) (a), (b) and (c) 2) (a) and (b)
   3) (a), (c) and (d) 4) (c) and (d)
2. Which of the following statement is incorrect?
   1) In allylic halides, the halogen atom is bonded to an sp^3-hybridised carbon atom adjacent to carbon-carbon double bond.
   2) In benzylic halides, the halogen atom is bonded to an sp^3-hybridised carbon atom attached to an aromatic ring.
   3) In allylic halides, the halogen atom is bonded to a sp^2-hybridised carbon atom of a carbon-carbon double bond.
   4) In aryl halides, the halogen atom is directly bonded to the sp^2-hybridised carbon atom of an aromatic ring.
3. Choose incorrect statement among the following:
   1) The IUPAC name of n-propyl bromide is 1-Bromopropane.
   2) The IUPAC name of isopropyl chloride is 1-Chloro-2-methylpropane.
   3) The IUPAC name of bromobenzene is bromobenzene.
   4) The IUPAC name of m-dibromobenzene is 1,3-Dibromobenzene.
4. Select the correct statement :
   1) In vic-dihalides, both the halogen atoms are present on the same carbon atom.
   2) In gem-dihalides, halogen atoms are present on adjacent carbon atoms.
   3) 1,1-Dichloroethane is an example of gem-dihalide.
   4) 1,2-Dichloroethane is an example of gem-dihalide.
5. Which of the following statement is not correct ?
   1) Halogen atoms are more electronegative than carbon
   2) Carbon-halogen bond of alkyl halide is polarized.
   3) The carbon-halogen bond length decreases from C — F to C — I.
   4) Alkyl halides are best prepared from alcohols, which are easily accessible.
6. Which of the following statement(s) regarding preparation of haloalkanes is/are correct ?
   (a) The reactions of primary and secondary alcohols with HCl require the presence of a catalyst, $ZnCl_2$.
   (b) The order of reactivity of alcohols with a given haloacid is 3° > 2 ° > 1°.
   (c) In Finkelstein reaction, alkyl chlorides are often prepared by the reaction of alkyl iodides/ bromides with $ZnCl_2$ in dry acetone.
   (d) The synthesis of alkyl fluorides is best accomplished by heating an alkyl chloride/bromide in the presence of a metallic fluoride such as AgF, Hg_2F_2, CoF_2 or SbF_3.
   1) (a) and (b) 2) (b), (c) and (d)
   3) (a), (b) and (d) 4) (b) and (c)
7. Select the correct statement :
   1) Methyl chloride is solid at room temperature
   2) Methyl bromide is gas at room temperature
   3) Ethyl chloride is liquid at room temperature
   4) None of these

8. Incorrect statement(s) is/are :
1) Iodoalkanes formed by Finkelstein reaction
2) Aryl halides cannot be prepared by replacement of hydroxyl group of phenol by halogen atom.
3) Treatment of chloroethane with a saturated solution of AgCN gives ethyl cyanide as the major product.
4) Both (1) and (3)

9. Which amongst the following is incorrect statement ?
1) Primary halide react most rapidly in S_N2 reactions.
2) 1° alkyl halides undergo S_N1 reaction very fast.
3) A mixture containing two enantiomers in equal proportion will have zero optical rotation.
4) Propan – 2 – ol does not contain an asymmetric carbon

10. Which of the following statement about tetrachloromethane is not correct ?
1) It is used as feedstock in the synthesis of chlorofluorocarbons and other chemicals.
2) Exposure to tetrachloromethane can make the heart beat irregularly or stop.
3) The chemical may irritate the eyes on contact
4) When carbon tetrachloride is released into the air, it rises to the atmosphere and forms the ozone layer

Statement Based MCQs

Directions: These questions consist of two statements each, printed as Statement-I and Statement-II. While answering these questions, you are required to choose any one of the following four responses.

1) Both Statement-I and Statement-II are correct.
2) Both Statement-I and Statement-II are incorrect.
3) Statement-I is correct & Statement-II is incorrect.
4) Statement-I is incorrect & Statement-II is correct.

1. **Statement-I:** Molecules that are not superimposable on their mirror images are chiral.
Statement-II: All chiral molecules have chiral centres.

2. **Statement-I:** Aryl halides undergo nucleophilic substitution reaction.
Statement-II: In aryl halide, partial double bond character is observed.

3. **Statement-I:** S_N2 reactions lead to Walden inversion.
Statement-II: During transition state formation, nucleophile attacks the side opposite to that where halide is attached.

4. **Statement-I:** The main product of reaction of silver nitrite with ethyl bromide is nitroethane.
Statement-II: Silver nitrite is a complex compound.

5. **Statement-I:** Tert-butyl chloride preferably undergo hydrolysis by S_N1 mechanism.
Statement-II: S_N1 mechanism completed in two steps.

6. **Statement-I:** C_2H_5I is the least reactive alkyl halide.
Statement-II: C – I bond is the weakest and I^- is a better leaving group than Br^-, Cl^- & F^- ions.

7. **Statement-I:** For the same alkyl group, the boiling points of alkyl halides decreases in the order: RI < RBr < RCl < RF.
Statement-II: The boiling points of isomeric haloalkanes decrease with increase in branching.

8. **Statement-I:** The dipole moment of chlorobenzene is lower than that of cyclohexyl chloride.
Statement-II: The sp^2 hybrid of C–Cl bond in chlorobenzene has less tendency to release electrons to Cl than a sp^3 hybrid carbon of cyclohexyl chloride.

9. **Statement-I:** S_N2 reactions proceed with retention of configuration.
Statement-II: S_N1 reactions are accompanied by racemisation.

10. **Statement-I:** Aryl halides are extremely less reactive towards nucleophilic substitution reactions.
Statement-II: The bond cleavage in haloarene is difficult than haloalkane

Assertion & Reason MCQs

Directions: These questions consist of two statements each, printed as Assertion and Reason. While answering these questions, you are required to choose any one of the following four responses.

1) Both Assertion and Reason are True and the Reason is a correct explanation of the Assertion.
2) Both Assertion and Reason are True but Reason is not a correct explanation of the Assertion.
3) Assertion is True but the Reason is False.
4) Assertion is False but Reason is True.

1. **Assertion:** The boiling points of alkyl halides decrease in the order: RI > RBr > RCl > RF
Reason: The boiling points of alkyl chlorides, bromides and iodides are considerably higher than that of the hydrocarbon of comparable molecular mass.

2. **Assertion:** KCN reacts with methyl chloride to give methyl isocyanide
Reason: CN^- is an ambident nucleophile.

3. **Assertion:** tert-Butyl bromide undergoes Wurtz reaction to give 2, 2, 3, 3-tetramethylbutane.

   Reason: In Wurtz reaction, alkyl halides react with sodium in dry ether to give hydrocarbon containing double the number of carbon atoms present in the halide.

4. **Assertion:** Presence of a nitro group at ortho or para position increases the reactivity of haloarenes towards nucleophilic substitution.

   Reason: Nitro group, being an electron withdrawing group decreases the electron density over the benzene ring.

5. **Assertion:** In monohaloarenes, further electrophilic substitution occurs at ortho and para positions.

   Reason: Halogen atom is a ring deactivator.

6. **Assertion:** Aryl iodides can be prepared by reaction of arenes with iodine in the presence of an oxidising agent.

   Reason: Oxidising agent oxidises I_2 into HI.

7. **Assertion:** It is difficult to replace chlorine by –OH in chlorobenzene in comparison to that in chloroethane.

   Reason: Chlorine-carbon (C—Cl) bond in chlorobenzene has a partial double bond character due to resonance.

8. **Assertion:** Hydrolysis of (–)-2-bromooctane proceeds with inversion of configuration.

   Reason: This reaction proceeds through the formation of a carbocation.

9. **Assertion:** Nitration of chlorobenzene leads to the formation of m-nitrochlorobenzene

   Reason: —NO_2 group is a m-directing group.

10. **Assertion:** Alkyl halides form alkenes when heated above 300°C

    Reason: CH_3CH_2I react slowly with strong base as compared to CD_3CH_2I.

Multi Correct MCQs

1. Which among the following halide/s will not show S_N1 reaction:

   A. $H_2C=CH-CH_2Cl$ B. $CH_3-CH=CH-Cl$

   C. 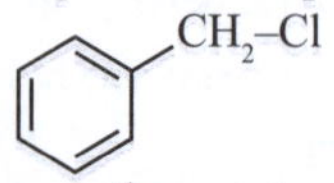 D. 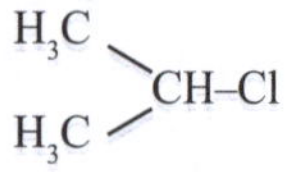

   Choose the most appropriate answer from the options given below:

   1) (A), (B) and (D) only 2) (A) and (B) only
   3) (B) and (C) only 4) (B) only

2. Which of the following compounds contain achiral carbon atoms.

   I. 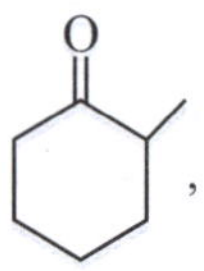 , II. 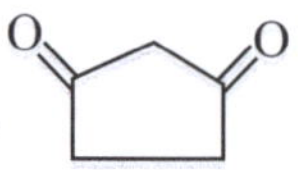

   III. $CH_3-CH_2-CH(NO_2)-COOH$
   IV. $CH_3-CH_2-CHBr-CH_2-CH_3$
   V. $CH_3-CH(I)-CH_2-NO_2$
   VI. $CH_3-CH_2-CH(OH)-CH_2OH$
   VII. $CH_3-CH(I)-CH(I)-C_2H_5$

   1) Both I & III 2) Both II and IV
   3) Both III and V 4) Both VI and VII

3. Which of the following that can be prepared by Sandmeyer's reaction are:

   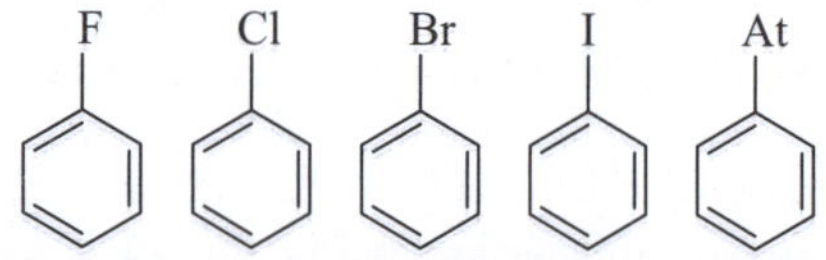

   I II III IV V

   1) I and II 2) II and III
   3) IV and V 4) I and IV

4. The mechanism of S_N1 reaction is given as:

   $$R-X \rightarrow \underset{\text{Ion pair}}{\overset{\oplus}{R}\overset{\ominus}{X}} \rightarrow \underset{\text{Solvent Separated ion pair}}{\overset{\oplus}{R} \parallel \overset{\ominus}{X}} \xrightarrow{\overset{\ominus}{Y}} R-Y + \overset{\ominus}{X}$$

   A student writes general characteristics based on the given mechanism as: **[3 Sept, 2020 (Shift-I)]**

   (A) The reaction is favoured by weak nucleophiles.
   (B) $R^{\oplus}$ would be easily formed if the substituents are bulky.
   (C) The reaction is accompanied by racemization.
   (D) The reaction is favoured by non-polar solvents.

   Which observations are correct?

   1) (B) and (D) 2) (A), (B) and (C)
   3) (A) and (C) 4) (A) and (B)

NCERT Exemplar MCQs

1. The order of reactivity of following alcohols with halogen acids is:

a. $CH_3CH_2-CH_2-OH$

b. $CH_3CH_2-CH(CH_3)-OH$

c. $CH_3CH_2-C(CH_3)_2-OH$

1) a > b > c 2) c > b > a
3) b > a > c 4) a > c > b

2. Which of the following alcohols will yield the corresponding alkyl chloride on reaction with concentrated HCl at room temperature?

1) $CH_3CH_2-CH_2-OH$
2) $CH_3CH_2-CH(CH_3)-OH$
3) $CH_3CH_2-CH(CH_3)-CH_2OH$
4) $CH_3CH_2-C(CH_3)_2-OH$

3. Identify the compound Y in the following reaction:

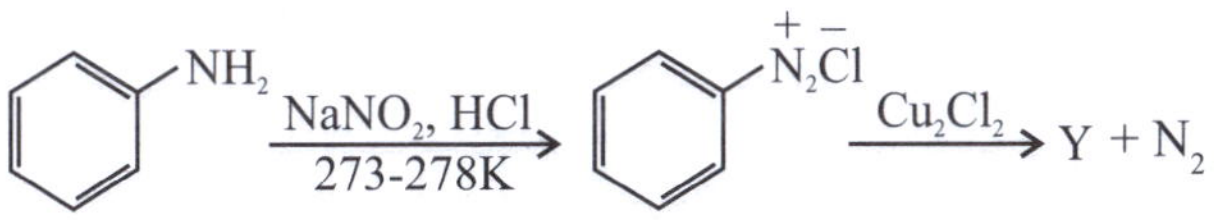

1) 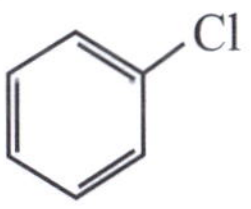 2) 

3) 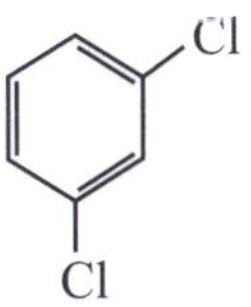 4) 

4. Toluene reacts with a halogen in the presence of iron (III) chloride giving ortho and para halo compounds. The reaction is:

1) Electrophilic elimination reaction
2) Electrophilic substitution reaction
3) Free radical addition reaction
4) Nucleophilic substitution reaction

5. Which of the following is halogen exchange reaction?

1) $RX + NaI \rightarrow RI + NaX$

2) 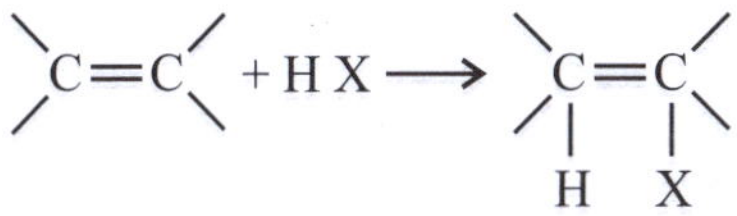

3) $R-OH + HX \xrightarrow{ZnCl_2} R-X + H_2O$

4) 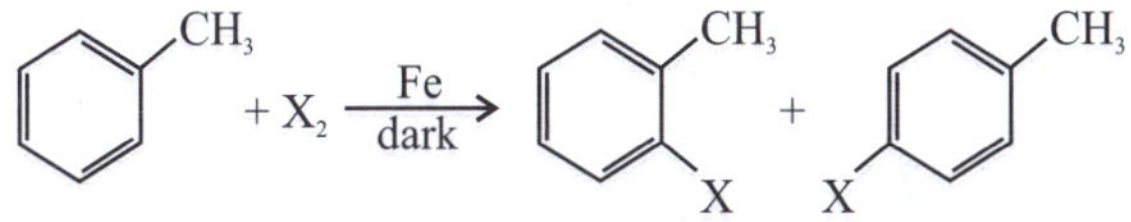

6. Which reagent will you use for the following reaction?

$CH_3CH_2CH_2CH_3 \rightarrow CH_3CH_2CH_2CH_2Cl + CH_3CH_2CHClCH_3$

1) Cl_2/UV light
2) $NaCl + H_2SO_4$
3) Cl_2 gas in dark
4) Cl_2 gas in the presence of iron in dark

7. Arrange the following compounds in the increasing order of their densities:

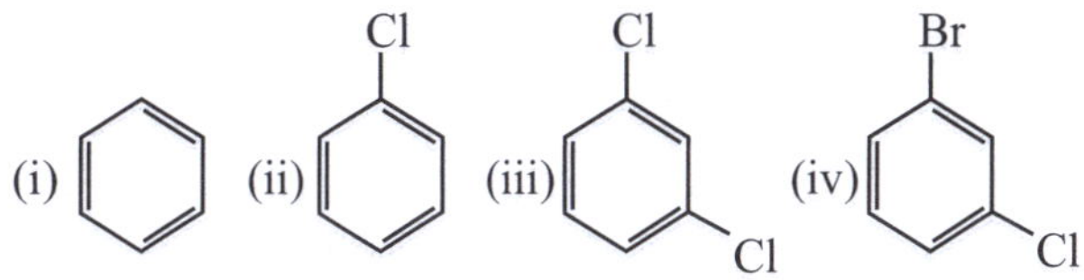

1) (i) < (ii) < (iii) < (iv) 2) (i) < (iii) < (iv) < (ii)
3) (iv) < (iii) < (ii) < (i) 4) (ii) < (iv) < (iii) < (i)

8. Arrange the following compounds in increasing order of their boiling points:

(i) $(CH_3)_2CH-CH_2Br$ (ii) $CH_3CH_2CH_2CH_2Br$ (iii) $H_3C-C(CH_3)(Br)-CH_3$

1) (ii) < (i) < (iii) 2) (i) < (ii) < (iii)
3) (iii) < (i) < (ii) 4) (iii) < (ii) < (i)

9. In which of the following molecules carbon atom marked with asterisk (*) is asymmetric?

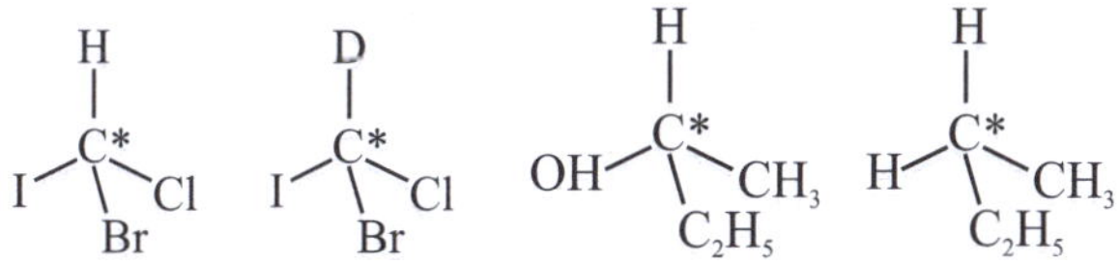

(i) (ii) (iii) (iv)

1) (i), (ii), (iii) and (iv) 2) (i), (ii) and (iii)
3) (ii), (iii) and (iv) 4) (i), (iii) and (iv)

10. Which of the following structures is enantiomeric with the molecule (A) given below?

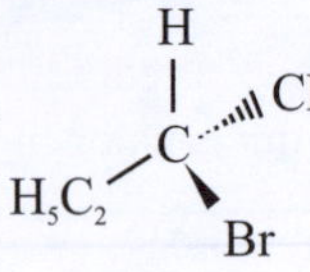

(A)

1) 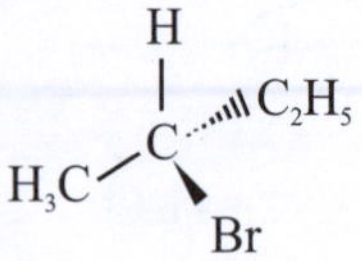

2) 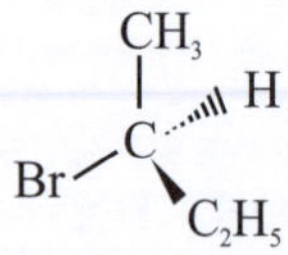

3) 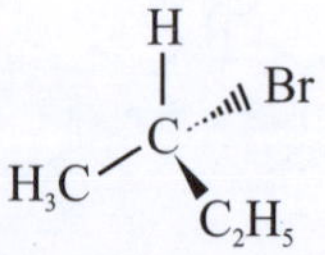

4) 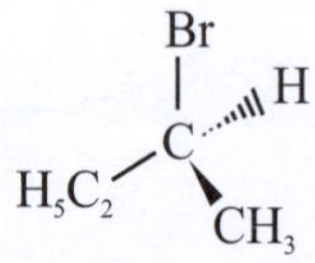

11. Which of the following is an example of vic-dihalide?

1) Dichloromethane
2) 1, 2-Dichloroethane
3) Ethylidene chloride
4) Allyl chloride

12. The position of Br in the compound in $CH_3CH=CHC(Br)(CH_3)_2$ can be classified as:

1) Allyl
2) Aryl
3) Vinyl
4) Secondary

13. Chlorobenzene is formed by reaction of chlorine with benzene in the presence of $AlCl_3$. Which of the following species attacks the benzene ring in this reaction?

1) Cl^-
2) Cl^+
3) $AlCl_3$
4) $[AlCl_4]^-$

14. Ethylidene chloride is a/an:

1) vic-dihalide
2) gem-dihalide
3) allylic halide
4) vinylic halide

15. What is 'A' in the following reaction?

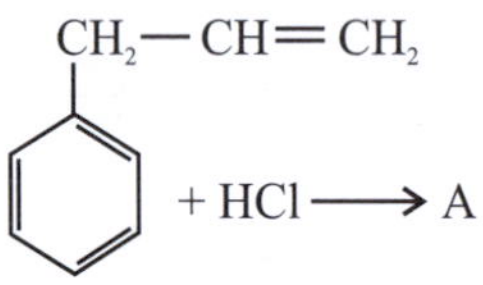

1) 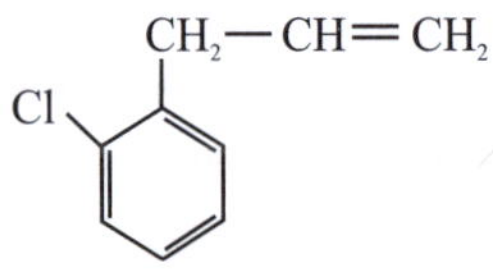

2) 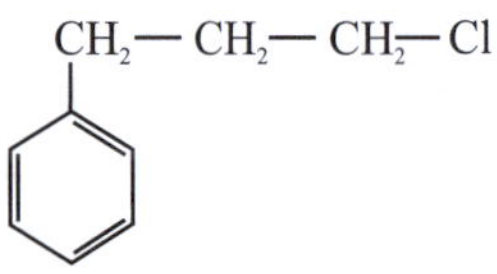

3) 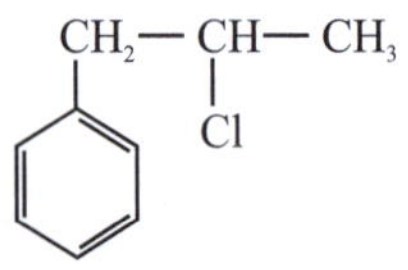

4) 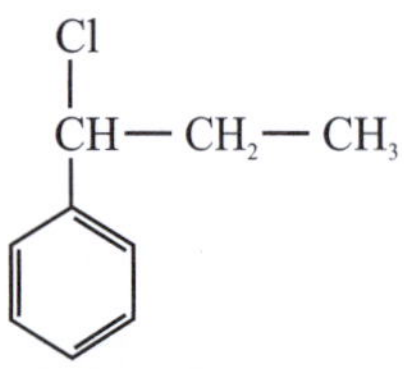

16. A primary alkyl halide would prefer to undergo:

1) S_N1 reaction
2) S_N2 reaction
3) Alpha-elimination
4) Racemisation

17. Which of the following alkyl halides will undergo S_N1 reaction most readily?

1) $(CH_3)_3C-F$
2) $(CH_3)_3C-Cl$
3) $(CH_3)_3C-Br$
4) $(CH_3)_3C-I$

18. Which is the correct IUPAC name for:

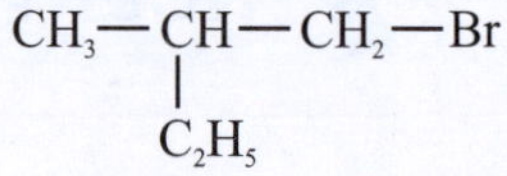

1) 1-Bromo-2-ethylpropane
2) 1-Bromo-2-ethyl-2-methylethane
3) 1-Bromo-2-methylbutane
4) 2-Methyl-1-bromobutane

19. What should be the correct IUPAC name for diethylbromomethane?

1) 1-Bromo-1, 1-diethylmethane
2) 3-Bromopentane
3) 1-Bromo-1, 1-ethylmethane
4) 1-Bromopentane

20. The reaction of toluene with chlorine in the presence of iron and in the absence of light yields:

1) 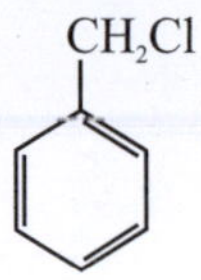

2) 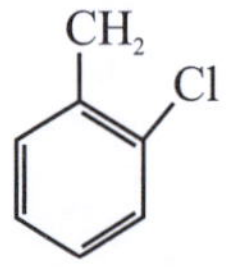

3) 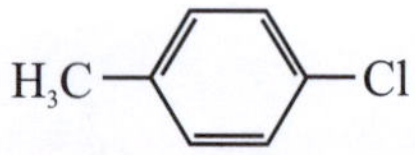

4) Mixture of (2) and (3)

21. Reactivity towards nucleophilic substitution:

(i) 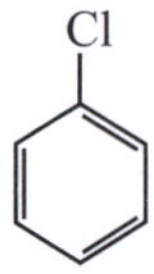 (ii) 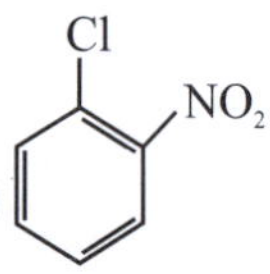 (iii) 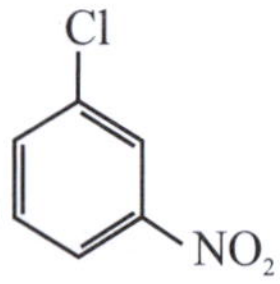

1) (i) < (ii) < (iii)
2) (iii) < (ii) < (i)
3) (i) < (iii) < (ii)
4) (iii) < (i) < (ii)

22. Increasing order towards nucleophilic substitution:

(i) 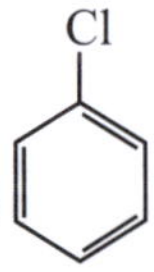 (ii) 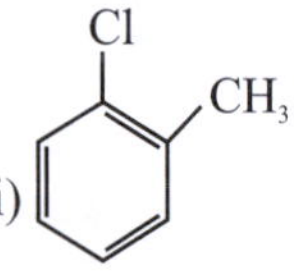 (iii) 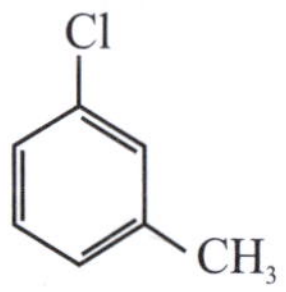

1) (i) < (ii) < (iii)
2) (i) < (iii) < (ii)
3) (iii) < (ii) < (i)
4) (ii) < (iii) < (i)

23. Increasing order towards nucleophilic substitution:

(i) 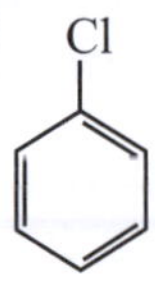 (ii) 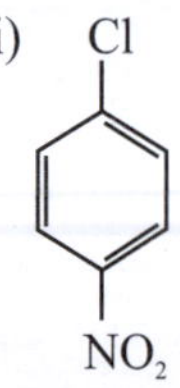 (iii) 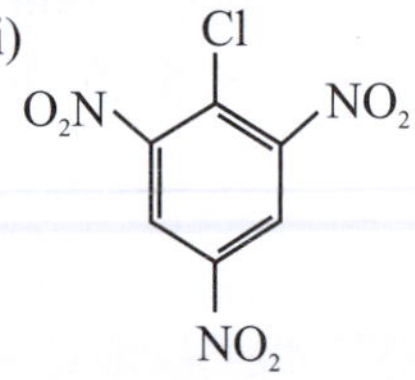

1) (iii) < (ii) < (i)
2) (ii) < (iii) < (i)
3) (i) < (iii) < (ii)
4) (i) < (ii) < (iii)

24. Rate of nucleophilic substitution in the following is of the order:

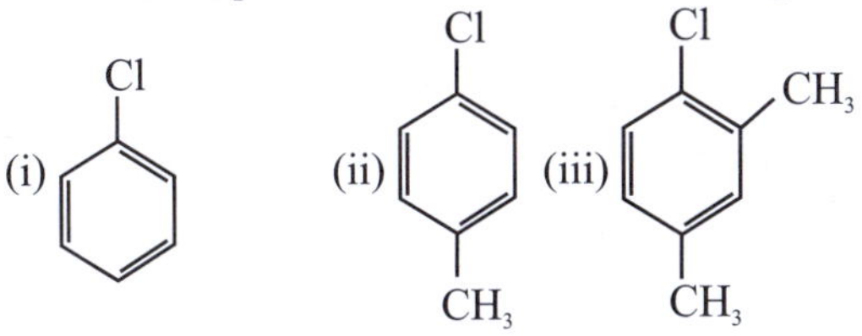

1) (i) < (ii) < (iii) 2) (ii) < (i) < (iii)
3) (iii) < (ii) < (i) 4) (i) < (iii) < (ii)

25. Which of following compounds is not chiral?
1) 3-Chloro-2-methyl pentane
2) 1-Chloro 2 methyl pentane
3) 2-Chloro pentane
4) 1-Chloropentane

26. Which is the correct increasing order of boiling points of the following compounds?

1-Bromoethane, 1-Bromopropane, 1-Bromobutane, Bromobenzene

1) Bromobenzene < 1-Bromobutane < 1-Bromopropane < 1-bromoethane
2) Bromobenzene < 1-Bromoethane < 1-Bromopropane < 1-Bromobutane
3) 1-Bromopropane < 1-Bromobutane < 1-Bromoethane < Bromobenzene
4) 1-Bromoethane < 1-Bromopropane < 1-Bromobutane < Bromobenzene

Kattar NEET

Scan Me for Detailed Explanations

1. A compound A has the molecular formula C_5H_9Cl. It does not react with bromine in carbon tetrachloride. On treatment with strong base it produces a single structural isomer B. B has a molecular formula C_5H_8 and reacts with bromine in carbon tetrachloride. Ozonolysis of B produces a compound C which has a molecular formula $C_5H_8O_2$. Which of the following structure is that of A?

1) 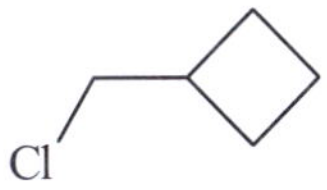

2) 

3) 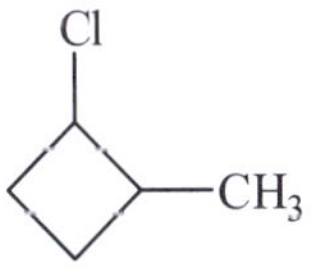

4) 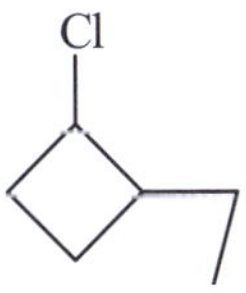

2. S_N1 reaction undergoes through a carbocation intermediate as follows:

$$R-X(aq.) \underset{}{\overset{Slow}{\rightleftharpoons}} R^+(aq.) + X^-(aq.) \xrightarrow[fast]{H_2O} ROH(aq.) + H^+(aq.)$$

[R = t-Bu, iso-Pr, Et, Me] (X = Cl, Br, I)

The correct statements are:

(I) The decreasing order of rate of S_N1 reaction is t-BuX > iso-PrX > EtX > MeX

(II) The decreasing order of C – X bond energy is MeX > EtX > iso-PrX > t-BuX

(III) The decreasing order of energy of activation is t-BuX > iso-ProX > EtX > MeX

1) I and II only 2) I and III only
3) II and III only 4) I, II and III only

3. The product in the given reaction is:

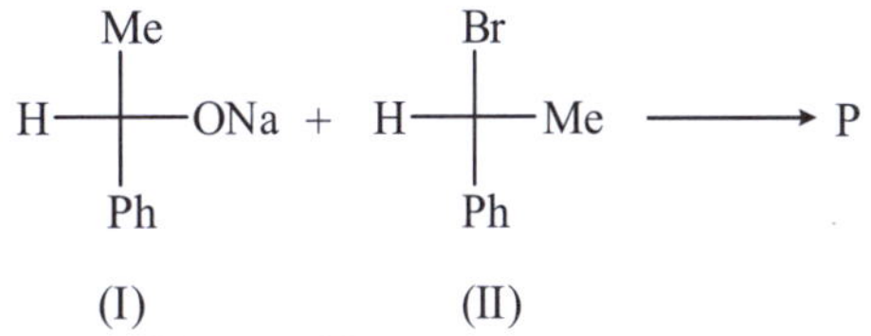

1) 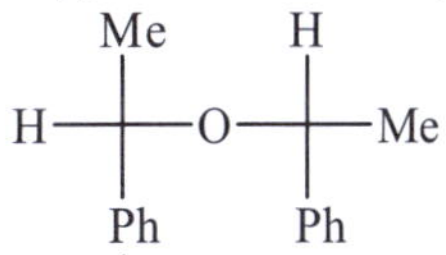

2) 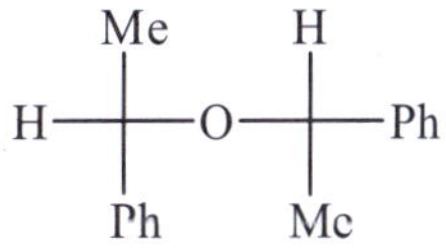

3) 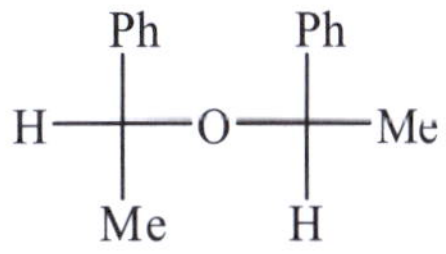

4) 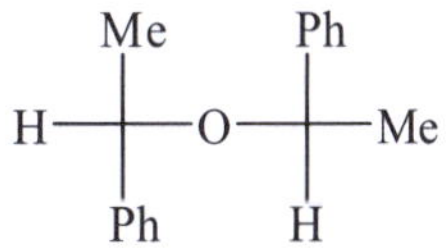

4. The product (P) of the reaction is

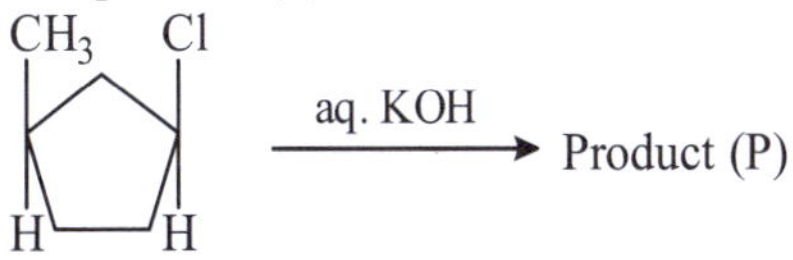

1) 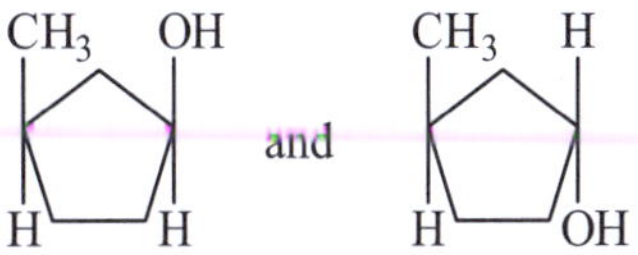

2) 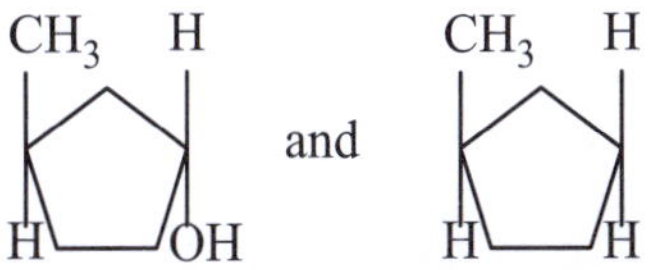

3) 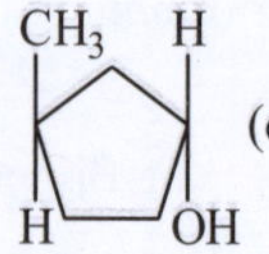 (only)

4) 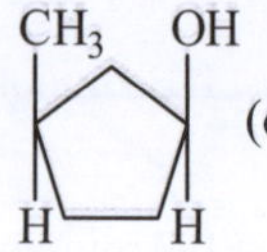 (only)

5. 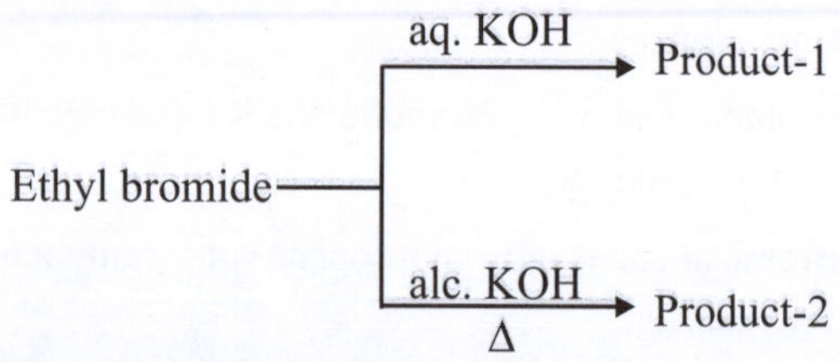

Which of the following statement is correct regarding product-1 and product-2?

1) Product-1 is obtained by the elimination reaction
2) Product-2 is obtained by the substitution reaction
3) The molecular formula of product 1 is C_2H_4, while the molecular formula of Product-2 is C_2H_6O
4) Product-1 is the isomer of dimethylether, while Product-2 is the dehydrated compound of Product-1

6. In which of the following S_N2Ar reaction is feasible at room temperature?

1) 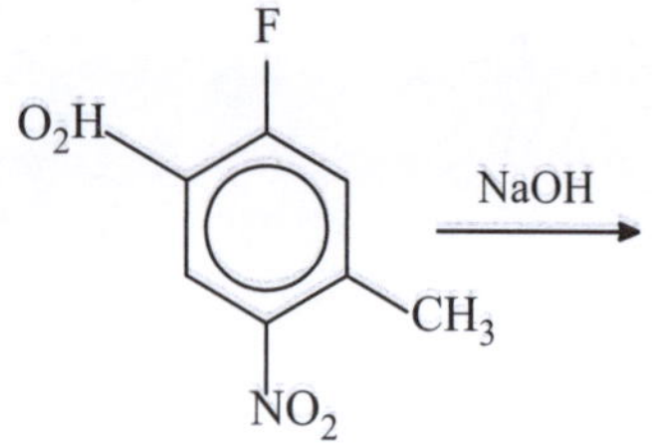

2) 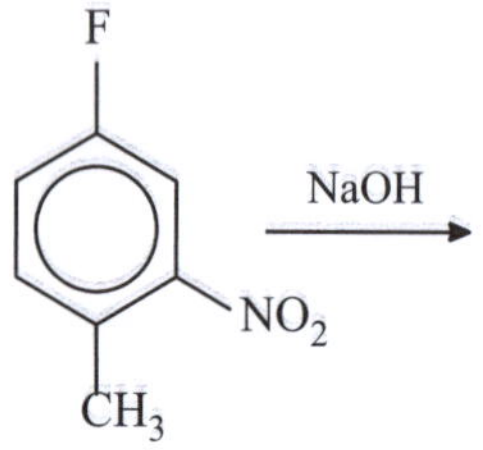

3) 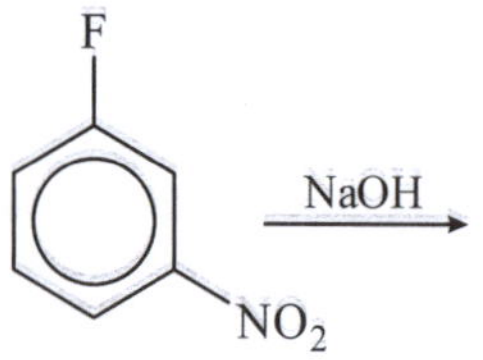

4) All of these.

7. In two separate experiments equal quantity of C_4H_9Cl were treated with equal volume of 0.2 molar and 0.5 molar solution of aq. NaOH. If found reaction intermediate in major possibilities was identical in both the experiments. The most likely structure of halide is

1) 1-chlorobutane
2) 2-chlorobutane
3) 2-chloro-2-methylpropane
4) 1-chloro-2-methylpropane

8. In the following photochemical reaction a mixture of products are formed

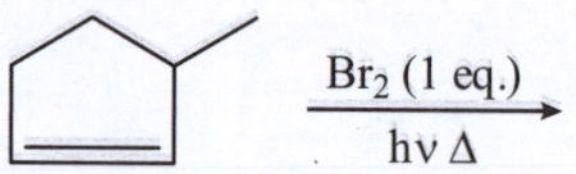

isomeric monobrominated products. The least probable product is

1) 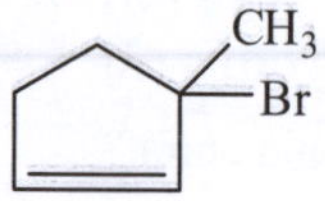

2) 

3) 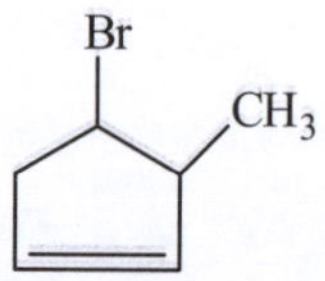

4) 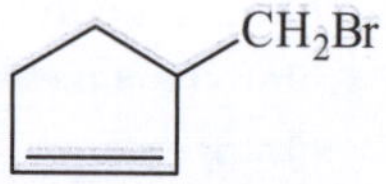

9. Total number of mono-chlorinated products are

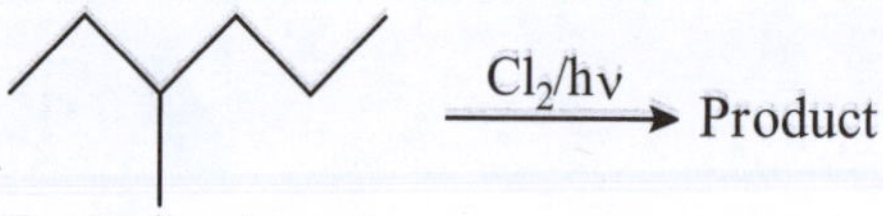

(Racemic mixture)

1) 7
2) 6
3) 20
4) 18

10. How many dichloroisomers (consider only structural isomers) can be obtained from the following reaction? (Do not consider rearranged products)

$$CH_3-\underset{}{\overset{CH_3}{\overset{|}{C}}}=CH-CH_3 \xrightarrow[h\nu]{Cl_2}$$

1) 1
2) 2
3) 3
4) 4

11. Which of the following reactions results in the formation of a pair of diastereomers?

1) 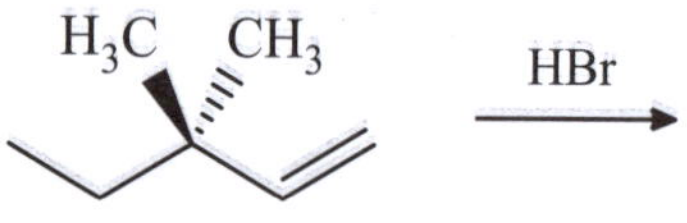

2) 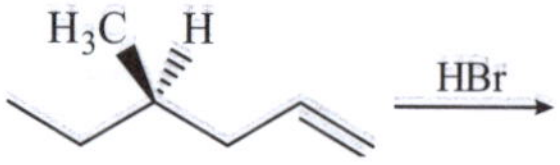

3) 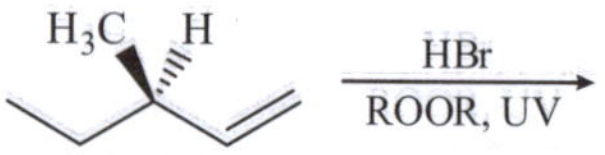

4) 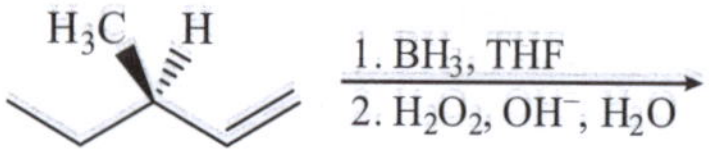

12. Rank the following in the increasing order of rate of reaction with HBr.

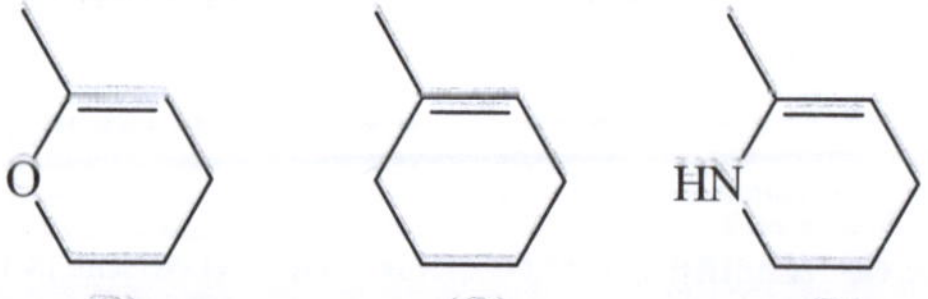

(P) (Q) (R)

1) R > P > Q
2) R > Q > P
3) P > R > Q
4) P > Q > R

13. Which of the following compound give diastereomers when treated with Br_2 in CCl_4?

1) 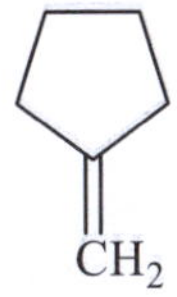
Methylcyclopentane

2) 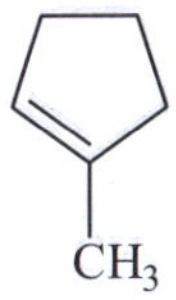
1-Methylcyclopentane

3) 
3-Methylcyclopentene

4) 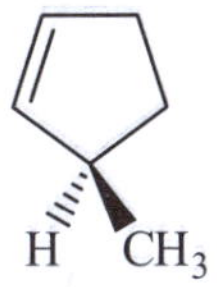
4-Methylcyclopentene

14. Select the correct statement for the given reaction.

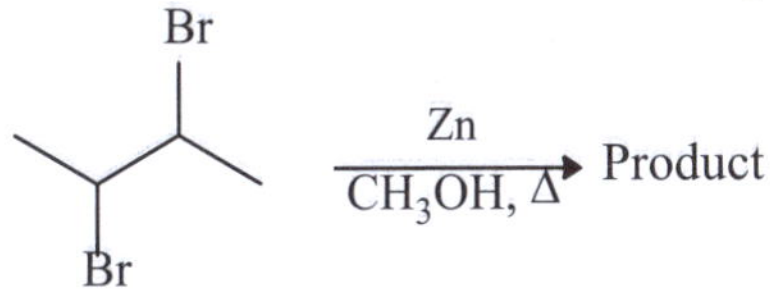

1) The above reaction proceeds via α-elimination.
2) It proceeds via β-elimination and the product is but-2-yne.
3) It proceeds via β-elimination and the product is but-2-ene.
4) It proceeds by γ-elimination.

15. Which of the following reaction gives product via both S_N1 and E1 mechanism?

1) $(CH_3)_3CBr + {}^{-}OH \longrightarrow$
2) $(CH_3)_3CBr + \bar{O}C_2H_5 \longrightarrow$
3) $CH_3CH = CHBr + KNH_2 \longrightarrow$
4) $(CH_3)_3CBr + C_2H_5OH \longrightarrow$

16. Which is the correct reaction coordinate diagram for the following solvolysis reaction?

1) 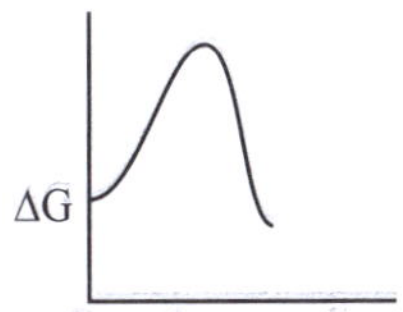

2) 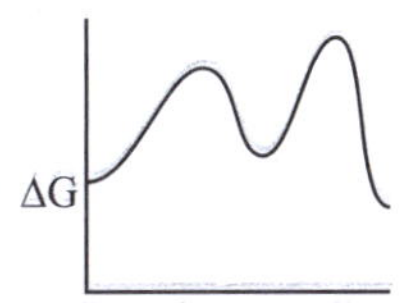

3) 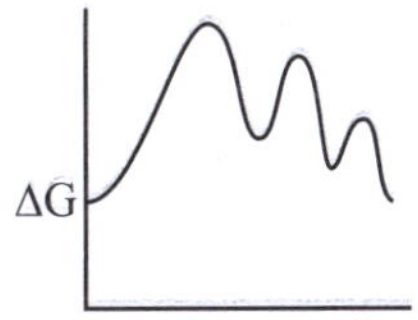

4) 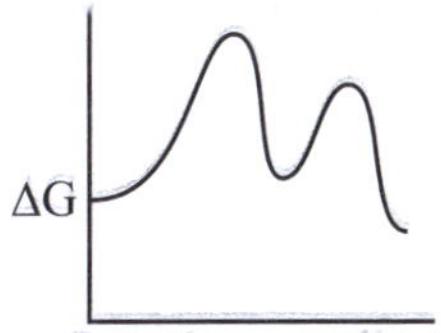

17. Select incorrect statement amongst the following.

1) S_N1 reaction occurs in polar solvent

2) 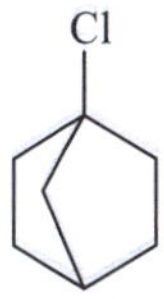 (I) will react more readily than 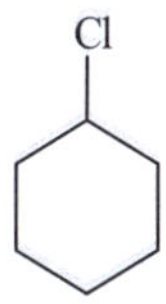 II) for S_N1 reaction.

3) 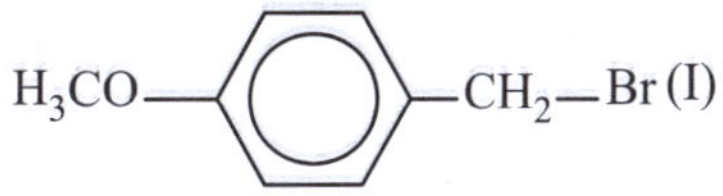 (I) will react more

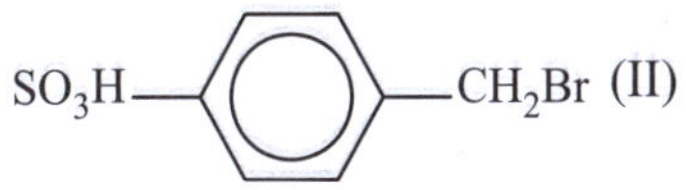 (II) for S_N1 reaction.

4) 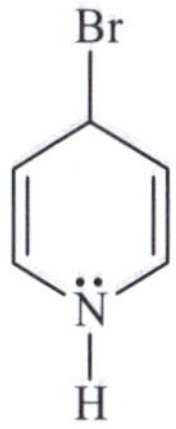 (I) will react more readily than

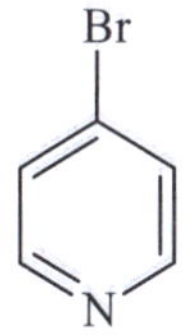 (II) for S_N1 reaction.

18. Match the following:

	List-I		List-II
I.	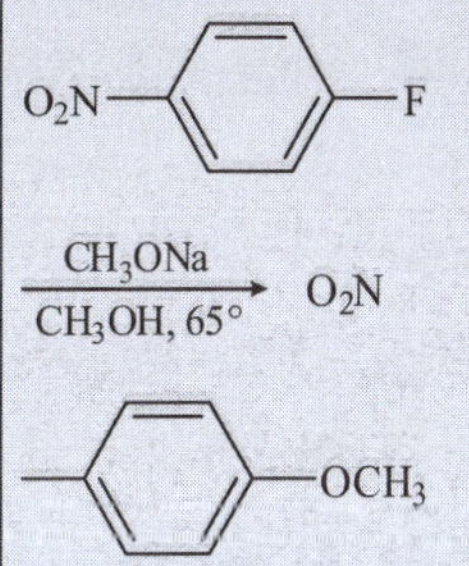	P.	Addition-Elimination
II.	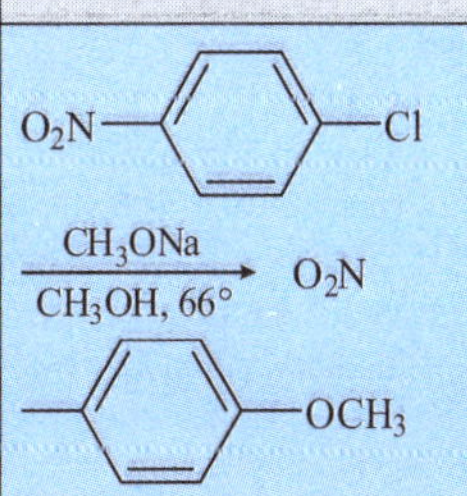	Q.	Elimination-Addition
III.	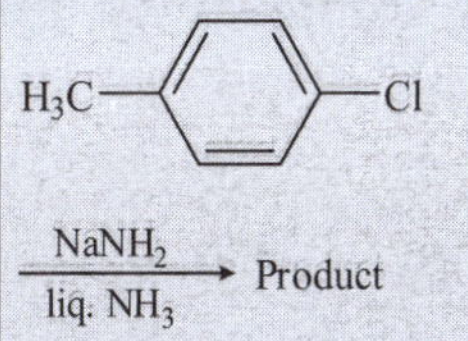	R.	ArS_N2
IV	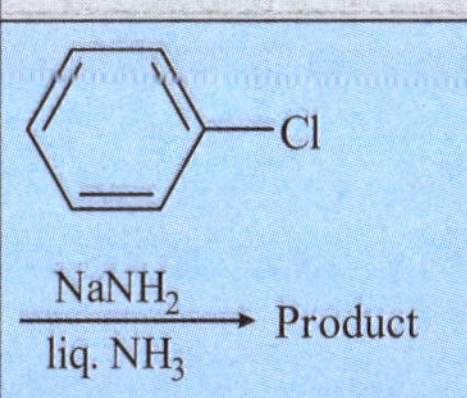	S.	Benzyne

1) I-P, R; II-P, R; III-Q, S; IV-Q, S
2) I-P; II-R; III-P, Q; IV-Q, S
3) I-P, R; II-P, R; III-P, Q; IV-Q, S
4) I-Q; II-P, R; III-Q, S; IV-P, R

19. Match the following:

List-I		List-II	
I.	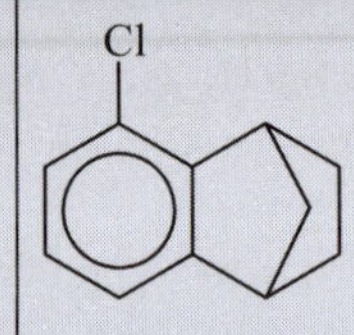	P.	S_N1 reaction can take place
II.	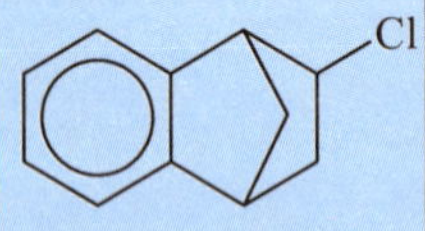	Q.	S_N2 reaction can take place
III.	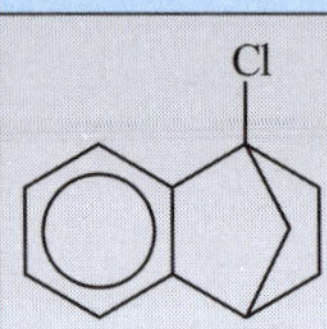	R.	S_N1 is not possible
IV.	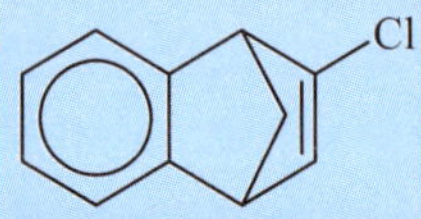	S.	S_N2 is not possible

1) I-R, S; II-P, Q; III-R, S; IV-R, S
2) I-R, S; II-R; III-R, S; IV-R, S
3) I-R, S; II-P, Q; III-Q; IV-R, S
4) I-P, Q; II-R, S; III-P, Q; IV-R, S

20. Match the following:

List-I		List-II	
I.	$CHCl_2-CF_3 \xrightarrow[\Delta]{\text{alc. KOH}}$	P.	Carbanion
II.	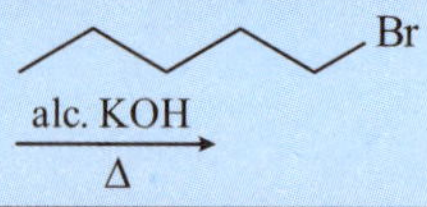	Q.	Two step process
III.	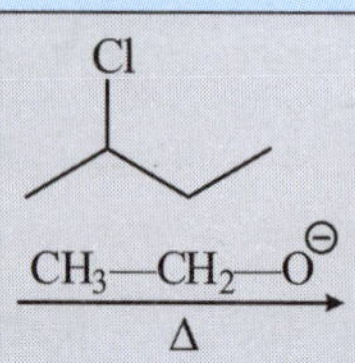	R.	Carbocation
IV.	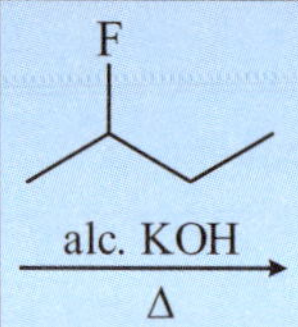	S.	Transition state

1) I-P, Q; II-S; III-S; IV-S
2) I-P,Q,S; II-S; III-S; IV-P,Q,S
3) I-P, Q; II-S; III-S; IV-Q, R
4) I-P, Q; II-S; III-S; IV-R

NEET & AIIMS Past Year MCQs

1. The major product C in the below mentioned reaction is: **(2024 Re)**

$$CH_3CH_2CH_2Br \xrightarrow[\Delta]{\text{alc.KOH}} A \xrightarrow{\text{HBr}} B \xrightarrow[\Delta]{\text{aq.KOH}} C$$

1) Propan-1-ol
2) Propan-2-ol
3) Propane
4) Propyne

2. The following reaction method

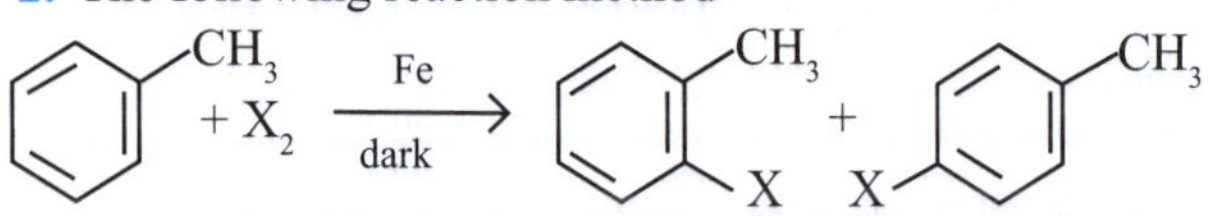

is not suitable for the preparation of the corresponding haloarene products, due to high reactivity of halogen, when X is: **(2024 Re)**

1) F
2) I
3) Cl
4) Br

3. Major products A and B formed in the following reaction sequence, are **(2024)**

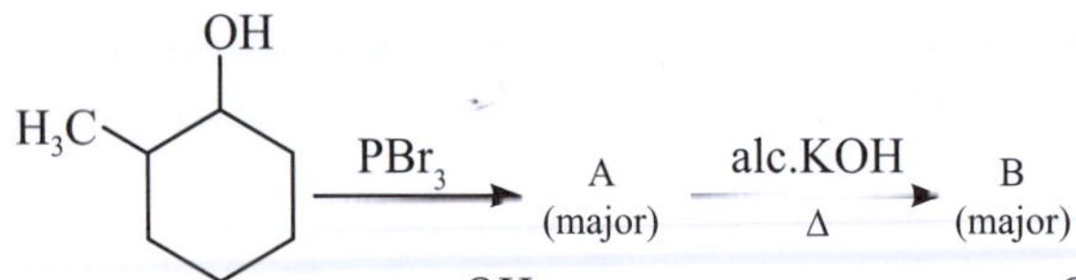

1) A = 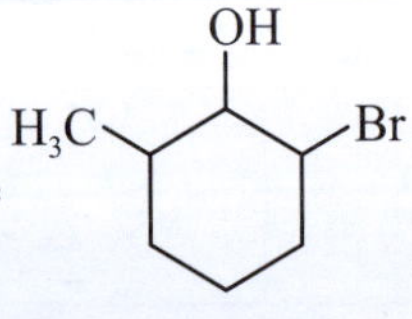 ; B = 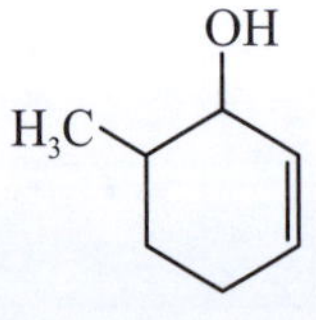

2) A = 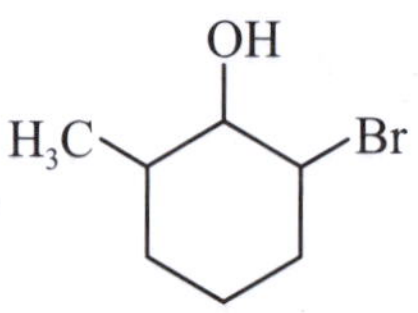 ; B = 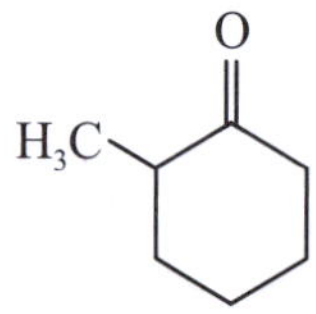

3) A = 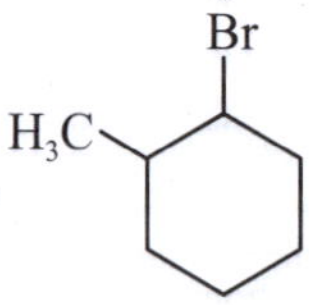 ; B = 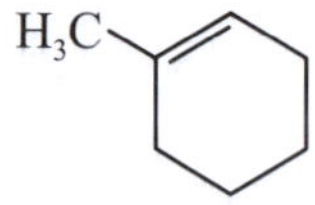

4) A = 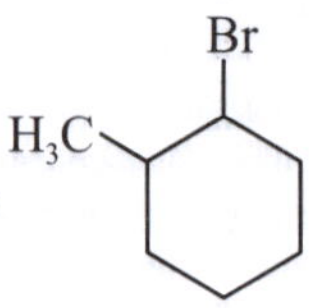 ; B = 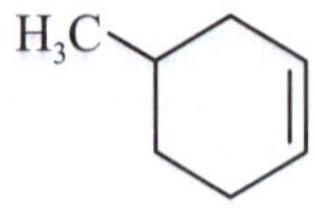

4. The compound that will undergo S_N1 reaction with the fastest rate is **(2024)**

1) 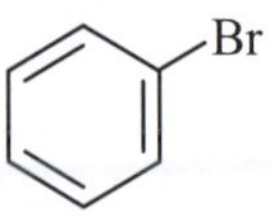

2) 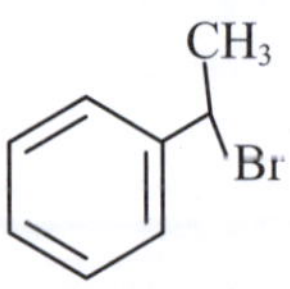

3) 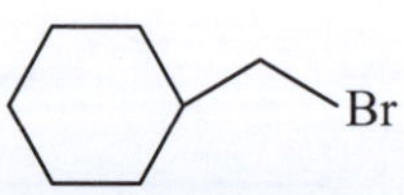

4) 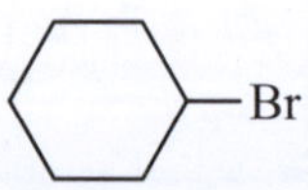

5. Consider the following reaction and identify the product (P).

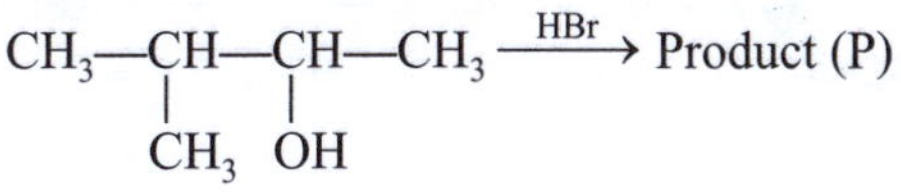

3–Methylbutan –2–o1 (2023)

1) 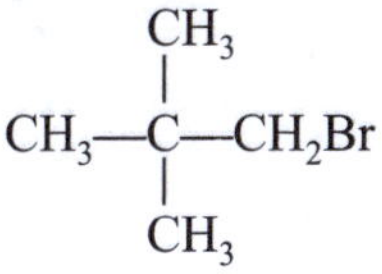
2) 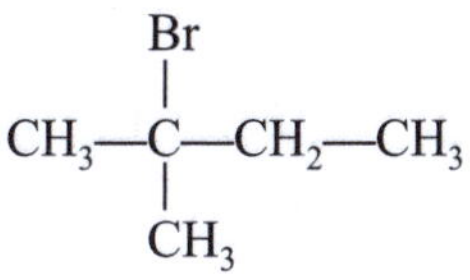
3) $CH_3CH = CH—CH_3$
4) 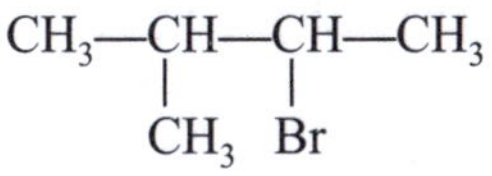

6. The given compound

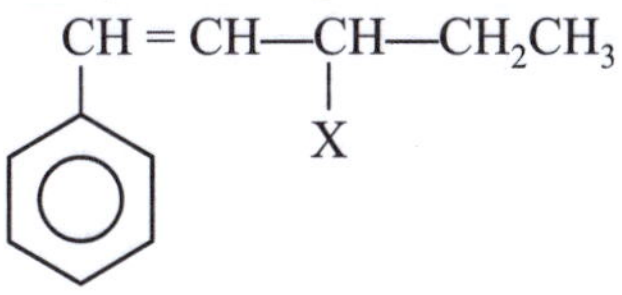

is an example of ________. (2023)

1) vinylic halide
2) benzylic halide
3) aryl halide
4) allylic halide

7. The incorrect statement regarding chirality is: (2022)

1) A racemic mixture shows zero optical rotation.
2) S_N1 reaction yields 1 : 1 mixture of both enantiomers.
3) The product obtained by S_N2 reaction of haloalkane having chirality at the reactive site shows inversion of configuration.
4) Enantiomers are superimposable mirror images on each other.

8. The correct sequence of bond enthalpy of 'C–X' bond is: (2021)

1) $CH_3 – F > CH_3 – Cl > CH_3 – Br > CH_3 – I$
2) $CH_3 – F < CH_3 – Cl > CH_3 – Br > CH_3 – I$
3) $CH_3 – Cl > CH_3 – F > CH_3 – Br > CH_3 – I$
4) $CH_3 – F < CH_3 – Cl < CH_3 – Br < CH_3 – I$

9. The major product formed in dehydrohalogenation reaction of 2-Bromo pentane is Pent-2-ene. This product formation is based on? (2021)

1) Hund's rule
2) Hofmann rule
3) Huckel's rule
4) Saytzeff's rule

10. 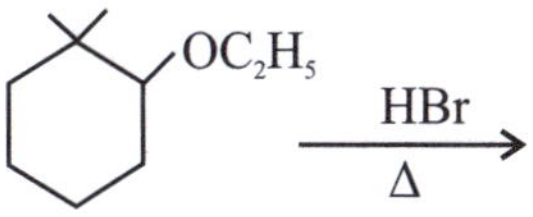 (AIIMS 2018)

1) 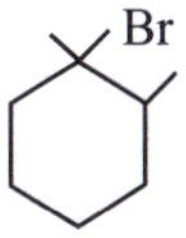
2) 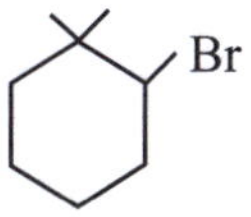
3) 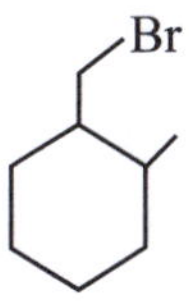
4) 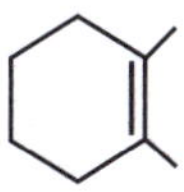

11. The major product formed in the following reaction is: (AIIMS 2017)

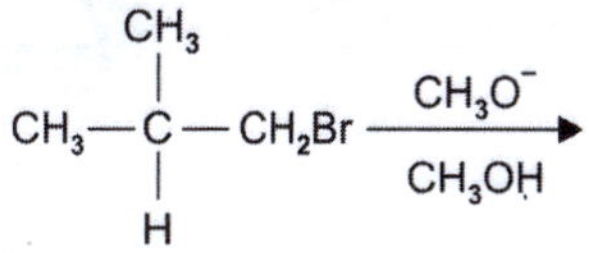

1) 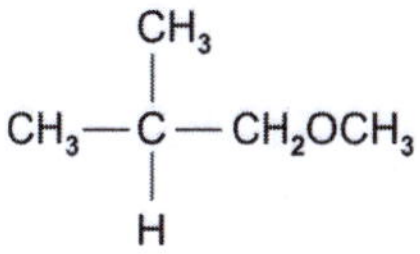
2) 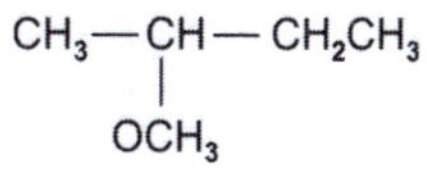
3) 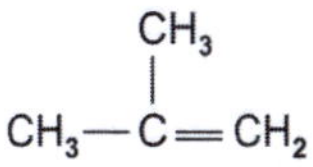
4) 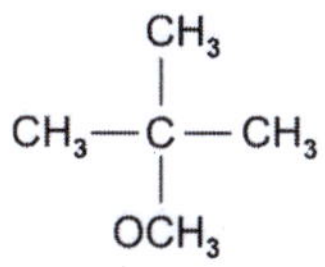

12. Nucleophilicity decreasing order for: (AIIMS 2016)

a. $C_6H_5O^-$
b. $C_6H_5^-$
c. $C_6H_5CH_2^-$
d. $C_6H_5COCH_2^-$

1) a > b > c > d
2) d > a > b > c
3) b > c > d > a
4) d > c > b > a

13. Which of the following compounds are optically active: (AIIMS 2012)

a. 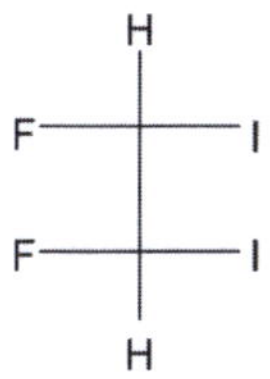
b. 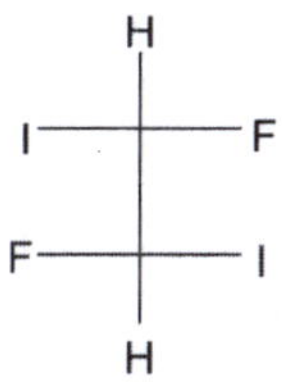
c. 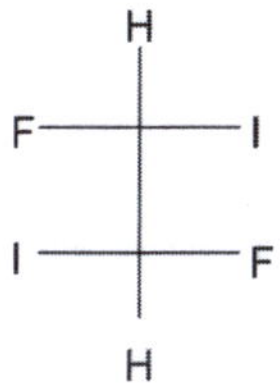
d. 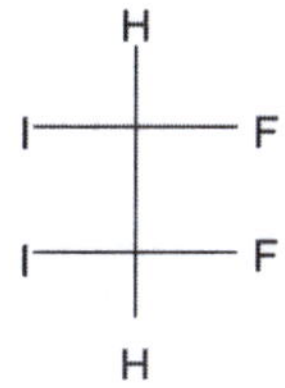

1) a and b
2) b and c
3) c and d
4) a and d

14. Which of the following molecules will not show optical activity? (AIIMS 2008)

1) 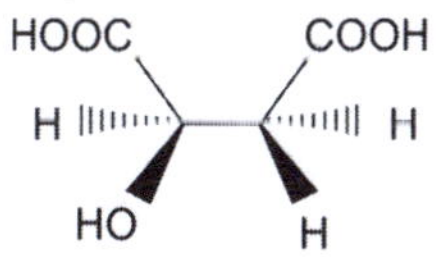
2) 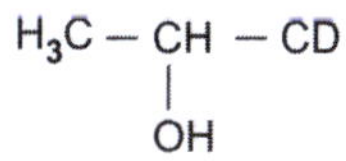
3) 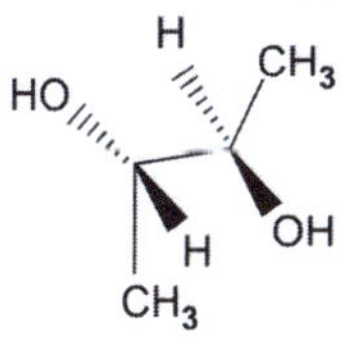
4) 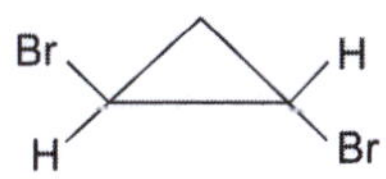

15. Aromatic nitriles (ArCN) are not prepared by reaction: (AIIMS 2004)

1) ArX +KCN
2) ArN_2^+ + CuCN
3) $ArCONH_2 + P_2O_5$
4) $ArCONH_2 + SOCl_2$

16. An acyl halide is formed when PCl_5 reacts with: (AIIMS 1998)

1) Amide
2) Alcohol
3) Acid
4) Ester

Answer Key

NCERT Topic-wise MCQs

1	2	3	4	5	6	7	8	9	10	11	12	13	14	15	16	17
4	4	2	4	1	3	3	3	4	2	1	4	3	1	3	2	4
18	19	20	21	22	23	24	25	26	27	28	29	30	31	32	33	34
1	2	4	1	3	1	2	2	1	1	2	3	2	1	3	4	3
35	36	37	38	39	40	41	42	43	44	45	46	47	48	49	50	51
4	2	3	3	2	4	2	4	4	2	3	2	4	2	1	4	1
52	53	54	55	56	57	58	59	60	61	62						
3	1	2	3	2	1	3	3	1	3	1						

High Order Time Intensive MCQs

1	2	3	4	5	6	7	8	9	10	11	12	13	14	15	16	17
3	2	2	1	4	1	3	1	1	2	2	1	3	1	1	1	2
18	19	20	21	22	23	24	25	26	27	28	29	30	31	32	33	34
1	1	2	4	4	3	4	4	3	1	4	2	3	2	4	1	4
35	36	37	38	39	40	41	42	43	44	45	46	47	48	49	50	51
4	2	1	1	3	3	4	4	3	3	4	4	3	4	2	2	4
52	53	54	55													
3	2	2	4													

Match the Columns MCQs

1	2	3	4	5	6
2	4	1	1	2	3

Correct & Incorrect MCQs

1	2	3	4	5	6	7	8	9	10
2	3	2	3	3	3	2	3	2	4

Statement Based MCQs

1	2	3	4	5	6	7	8	9	10
3	4	1	3	1	4	4	1	4	1

Assertion & Reason MCQs

1	2	3	4	5	6	7	8	9	10
2	4	1	1	2	3	1	3	4	3

Multi Correct MCQs

1	2	3	4
4	2	2	2

NCERT Exemplar MCQs

1	2	3	4	5	6	7	8	9	10	11	12	13	14	15	16	17
2	4	1	2	1	1	1	3	2	1	2	1	2	2	3	2	4
18	19	20	21	22	23	24	25	26								
3	2	4	3	4	4	3	4	4								

Kattar NEET

1	2	3	4	5	6	7	8	9	10	11	12	13	14	15	16	17
2	1	1	1	4	1	3	4	3	4	2	1	4	3	4	3	2
18	19	20														
1	1	2														

NEET & AIIMS Past Year MCQs

1	2	3	4	5	6	7	8	9	10	11	12	13	14	15	16
2	1	3	2	2	4	4	1	4	1	3	3	2	3	1	3

CHAPTER 8

Alcohols, Phenols and Ethers

Topicwise Number of Questions Analysis (2025-2016)

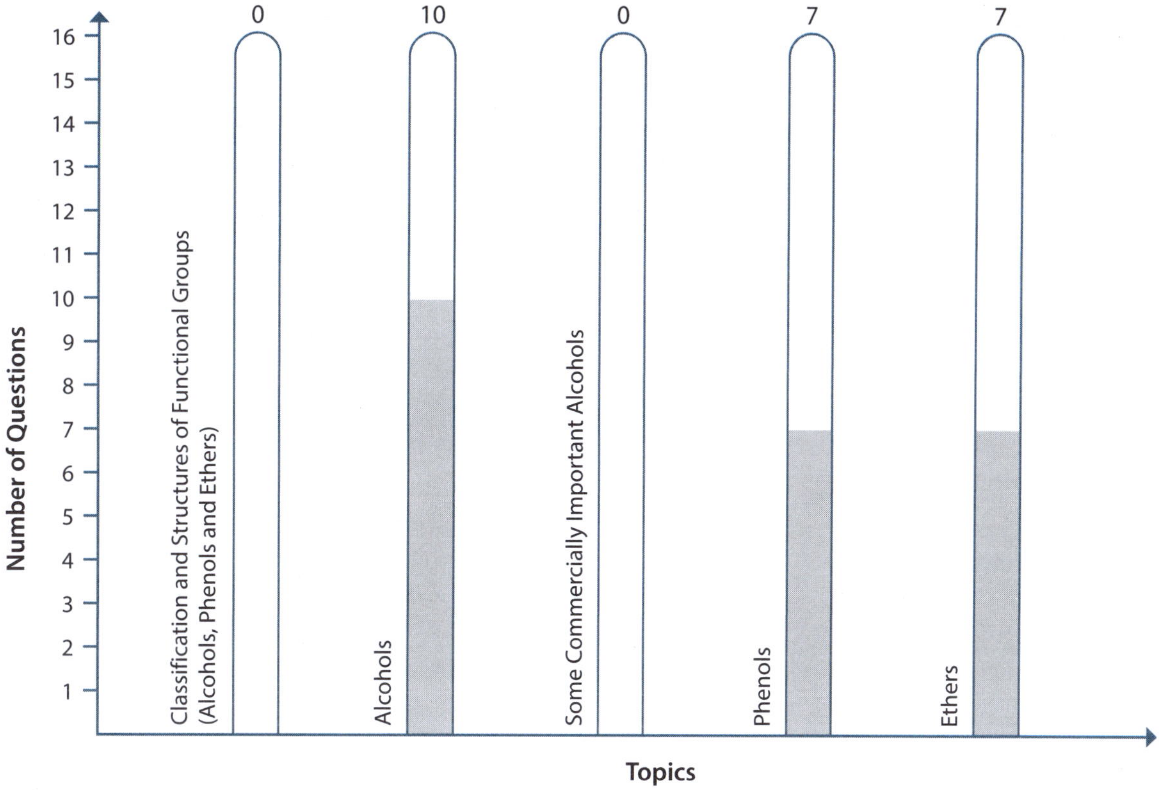

NCERT-Picks

Alcohols

- Alcohols are the compounds with the general formula ROH where R is an alkyl or substituted alkyl group.
- $(CH_2)_n(OH)_2$ where n = 2, 3, 4 .....etc is the general formula of dihydric alcohols.
- Another name of dihydric alcohols are glycols because of their sweet taste.
- The bond angle in alcohols is slightly less than the tetrahedral angle. Due to the repulsion between the unshared electron pairs of oxygen.
- When haloalkanes (alkyl halides) are heated with aqueous sodium or potassium hydroxide or moist silver oxide, alcohols are produced.

Preparation of Alcohols

From Alkenes

1. By Acid catalysed hydration

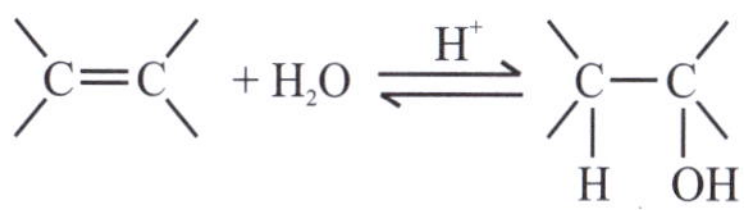

$$CH_3CH{=}CH_2 + H_2O \xrightleftharpoons{H^+} CH_3-\underset{\substack{|\\ OH}}{CH}-CH_3$$

2. By hydroboration-oxidation

$$CH_3-CH=CH_2 + (H-BH_2)_2 \longrightarrow CH_3-\underset{H}{CH}-\underset{BH_2}{CH_2}$$

$$\downarrow CH_3-CH=CH_2$$

$$(CH_3-CH_2-CH_2)_3B \xleftarrow{CH_3-CH=CH_2} (CH_3-CH-CH_2)_2BH$$

$$H_2O \downarrow 3H_2O_2, \bar{O}H$$

$$3CH_3-CH_2-CH_2-OH + B(OH)_3$$

Propan–1–ol

From Carbonyl Compounds

1. By reduction of aldehydes and ketones

$$RCHO + H_2 \xrightarrow{Pd} RCH_2OH$$

$$RCOR' \xrightarrow{NaBH_4} R-\underset{OH}{CH}-R'$$

2. By reduction of carboxylic acids and esters

$$RCOOH \xrightarrow[(II)\ H_2O]{(I)\ LiAlH_4} RCH_2OH$$

$$RCOOH \xrightarrow[H^+]{R'OH} RCOOR' \xrightarrow[Catalyst]{H_2} RCH_2OH + R'OH$$

From Grignard Reagents

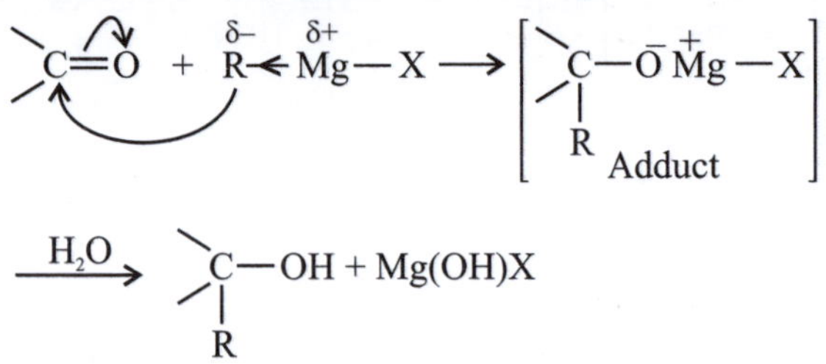

Physical Properties

- Alcohols:

1. B.P $\propto$ No. of carbon atoms $\propto \dfrac{1}{\text{Branching in C– chain}}$

$\downarrow$ B.P $\propto \uparrow$ branching $\propto \downarrow$ surface area

2. Solubility $\propto \dfrac{1}{\text{Size of alkyl group}}$

$\downarrow$ solubility $\uparrow$ size of alkyl group

Chemical Reactions

- The reactions in which O - H bond breaks, the reactivity order of alcohols is **1° > 2° > 3°**
- The reactions in which C - O bond breaks, the reactivity order of alcohols is **3° > 2° > 1°**

- **Reactions involved cleavage of C - O bond**

R—OH:

- $\xrightarrow[\Delta]{HX / Anhy.\ ZnCl_2} R-X + H_2O$
- $\xrightarrow{PX_3} R-X + H_3PO_3$
- $\xrightarrow{PX_5} R-X + POX_3 + HX$
- $\xrightarrow[\Delta]{Red\ P/ X_2} R-X + H_3PO_3$
- $\xrightarrow{SOCl_2,\ Pyridine} RCl + SO_2\uparrow + HCl\uparrow$
- $\xrightarrow[conc.\ H_2SO_4 / \Delta]{RCOOH} RCOOR + H_2O$
- $\xrightarrow[or\ conc.\ H_3PO_4]{conc.\ H_2SO_4,\ 443K} >C=C< + H_2O$
- $\xrightarrow[or\ K_2Cr_2O_7 / H^+]{KMnO_4 / H^+\ or\ OH^-} RCOOH$
- $\xrightarrow{[O]} RCHO \xrightarrow{[O]} RCOOH$
- $\xrightarrow[or\ CrO_3\ .\ C_5H_5N\ .\ HCl]{PCC} RCHO$
- $\xrightarrow{Cu / 573\ K}$ RCHO (1° alcohol) or RCOR (2° alcohol) or alkene (3° alcohol)

- Tertiary alcohols do not undergo oxidation reaction. They undergo dehydration.
- **Dehydration of Alcohols:**
  (i) Protonation of alcohol
  (ii) Formation of carbocation: It is slowest step and hence rate determining step.
  (iii) Elimination of a proton from the C—H bond adjacent to the carbocation to form alkene.
- **Alcohol can be distinguished by the following methods:**

With Lucas reagent (anhyd. $ZnCl_2$ + Conc. HCl)

(i) Tertiary alcohols $\rightarrow$ immediate turbidity
(ii) Secondary alcohols $\rightarrow$ turbidity after 5 min
(iii) Primary alcohols $\rightarrow$ no turbidity at room temperature.

Uses

- Glycol is used as an antifreeze in automobile radiators. Glycol has high boiling point & miscibility with H_2O.
- The commercial alcohol is made unfit for drinking by mixing in it some $CuSO_4$. Commercial alcohol gets denatured after adding $CuSO_4$.

Phenols

Structure

- The C – O bond length in phenol is slightly less than that in methanol. (i) Phenol has partial double bond character (ii) sp^2 hybridised state of carbon to which oxygen is attached.
- In a resonating structure of phenol; oxygen of –OH group depicted as positive. The hydroxyl group, in a phenol is directly attached with sp^2 hybridised carbon of benzene ring which act as electron withdrawing group.

- The hydroxyl group in phenol is directly attached to highly electronegative sp^2 hybridised C - atom of benzene ring which decrease the electron density on O atom. This increases the polarity of O - H bond and results in an increase in ionisation of phenols.
- Phenoxide ion is more stable than alkoxide ion. Due to delocalisation of negative charge in phenols makes phenol more stable.

Preparation of Phenols

1. From benzenesulphonic acid:

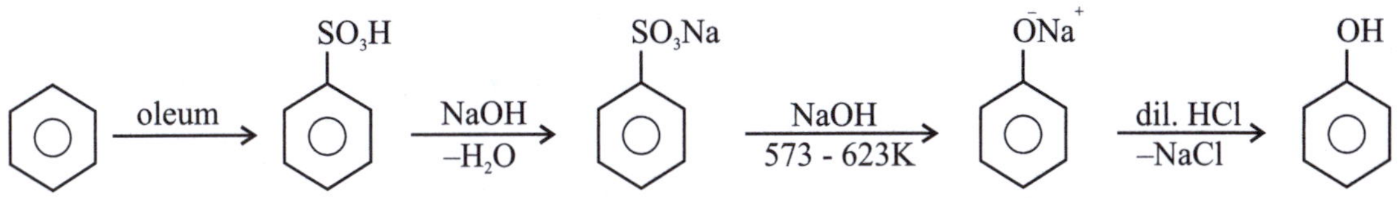

2. From dizonium salt:

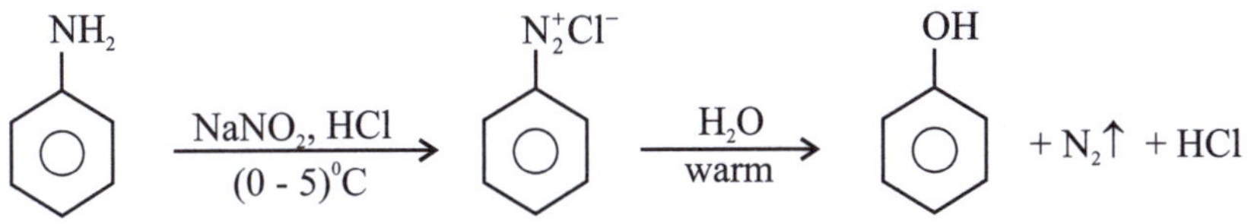

Benzene diazonium chloride

3. From haloarenes (Dow's process):

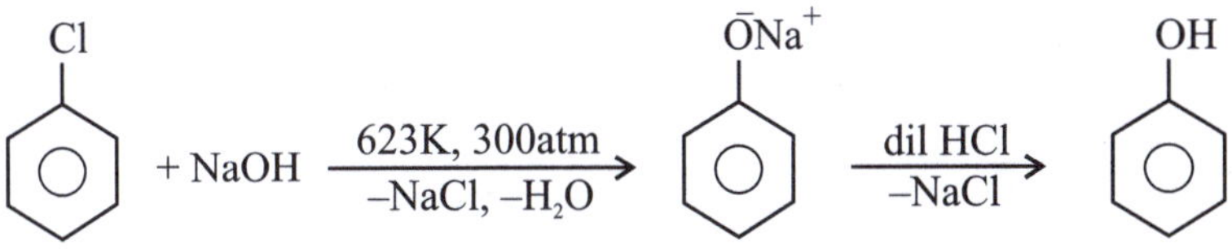

4. From cumene:

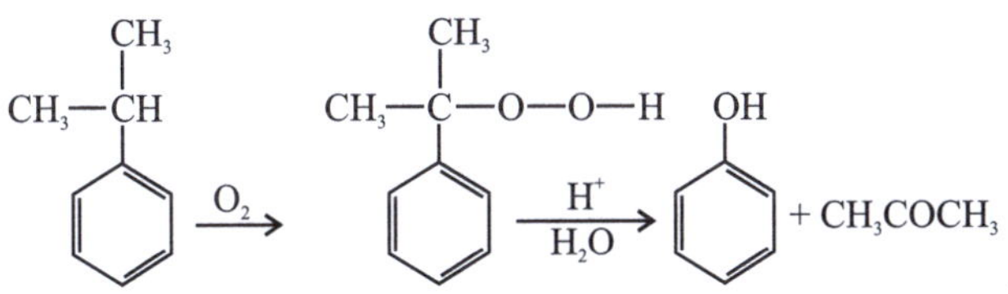

Cumene

Cumene hydroperoxide

Physical Properties

- Phenols are colorless crystalline solid or liquid having characteristic odour.
- Due to hydrogen bonding phenols has higher boiling point.
- The solubility of phenols in water is decreases with increase in size of aryl groups. Hydrophobic character increases as number of aryl groups increases.

Reactions Involving Cleavage of O–H Bond

1. Acidity of alcohols

   The acid strength of alcohols decrease in the following order. $R \rightarrow CH_2OH > R_2CHOH >> R_3C—OH$. An electron – releasing group increases electron density on oxygen tending to decrease the polarity of O – H bond.

2. Acidity of phenols

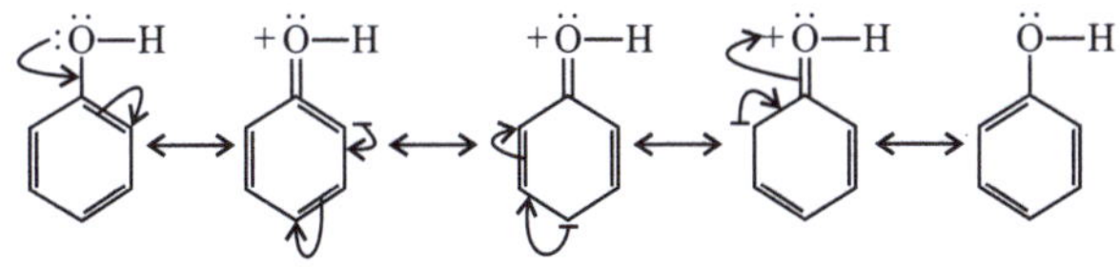

Nitro group enhances the acidic strength of phenol. Due to effective delocalisation of negative charge in phenoxide ion. Cresols are less acidic than phenols. R : Alkyl group ($-CH_3$) present on the cresols do not favour the formation of phenoxide ion.

Esterification

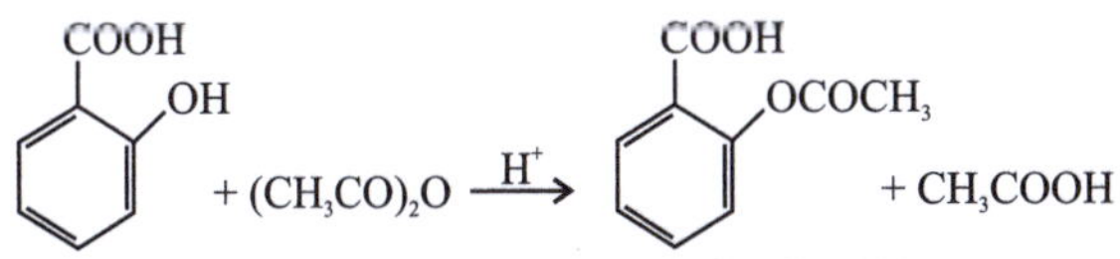

Salicylic acid

Acetylsalicylic acid (Aspirin)

Kolbe's Reaction

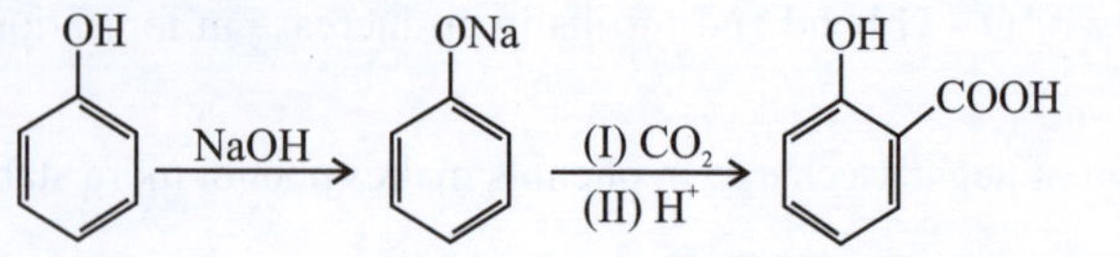

2-Hydroxybenzoic acid (Salicylic acid)

Reimer-Tiemann Reaction

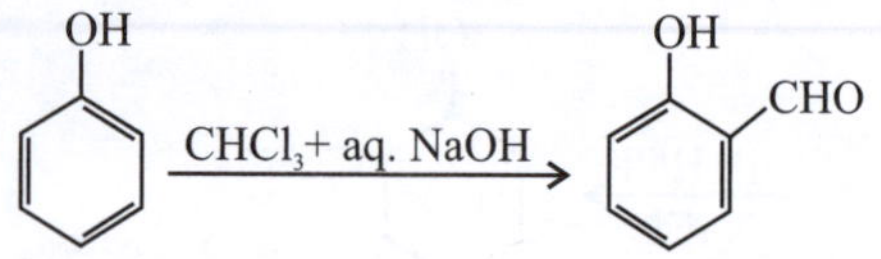

Salicylaldehyde

Reaction of Phenol with Zinc Dust

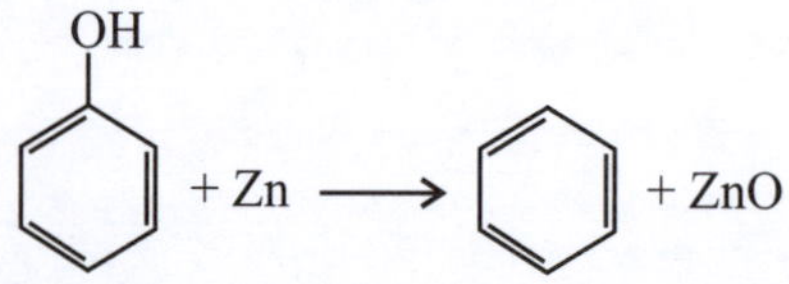

Oxidation

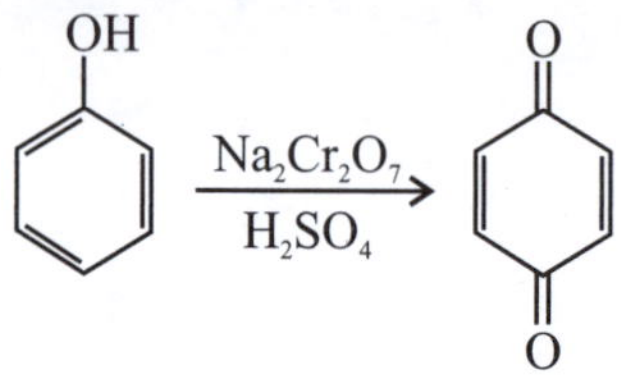

Benzoquinone

Reactions of Phenol

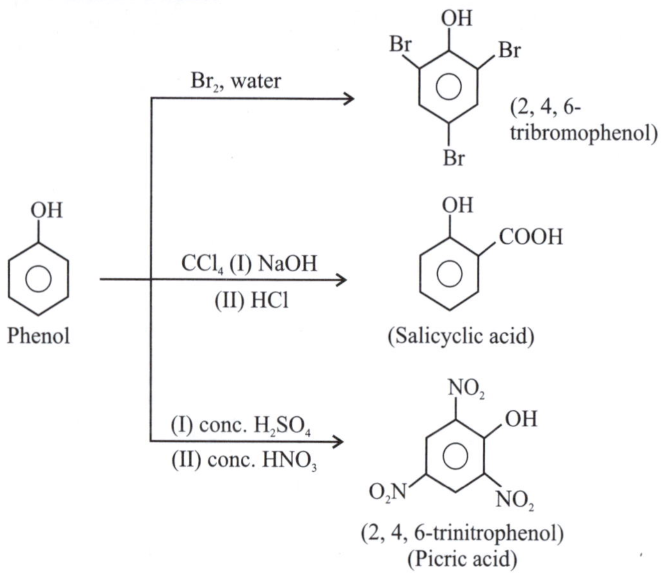

Some Commercially Important Alcohols

1. Methanol

$$CO + 2H_2 \xrightarrow[\substack{200\text{-}300\ atm \\ 573\text{-}673\ K}]{ZnO-Cr_2O_3} CH_3OH$$

2. Ethanol

$$C_{12}H_{22}O_{11} + H_2O \xrightarrow{Invertase} \underset{Glucose}{C_6H_{12}O_6} + \underset{Fructose}{C_6H_{12}O_6}$$

$$C_6H_{12}O_6 \xrightarrow{Zymase} 2C_2H_5OH + 2CO_2 \uparrow$$

Ethers

- The substitution of a hydrogen atom in a hydrocarbon by an alkoxy (OR) aryloxy (OAr) group yield ethers.
- The bond angle of ether is slightly greater than the tetrahedral angle due to the repulsive interaction between the two bulky (-R) groups in ether.

Preparation of Ethers

1. $CH_3CH_2OH \longrightarrow$

$$\xrightarrow[443\ K]{H_2SO_4} CH_2{=}CH_2$$

$$\xrightarrow[413\ K]{H_2SO_4} C_2H_5OC_2H_5$$

2. Williamson Synthesis

$$R{-}X + R'{-}\ddot{\underset{..}{O}}{}^{-}\ \overset{+}{Na} \longrightarrow R{-}\ddot{\underset{..}{O}}{-}R' + Na\,X$$

Chemical Reactions

Cleavage of C–O bond in ethers

- Alkyl-aryl ether are cleaved at the alkyl – oxygen bond, due to more stable aryl – oxygen bond.
- Ether are less reactive as they do not contain any active functional group.

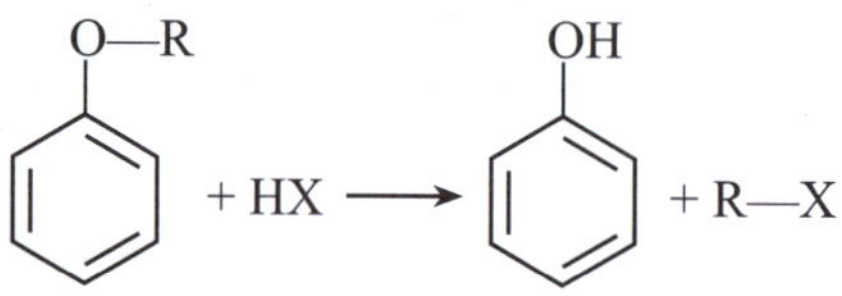

Reactivity order of HX:

HI > HBr >> HCl

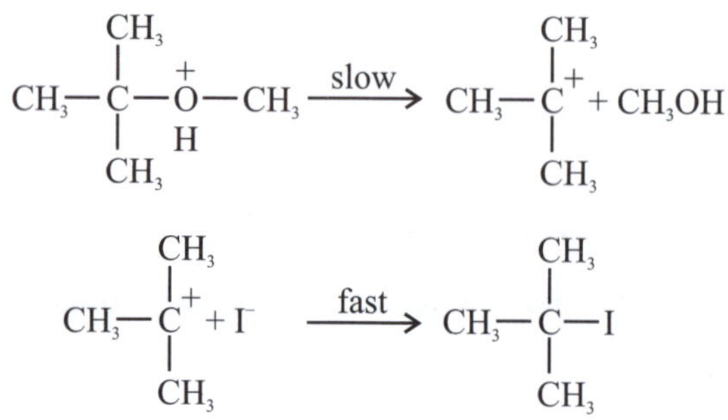

Electrophilic Substitution

1. Halogenation

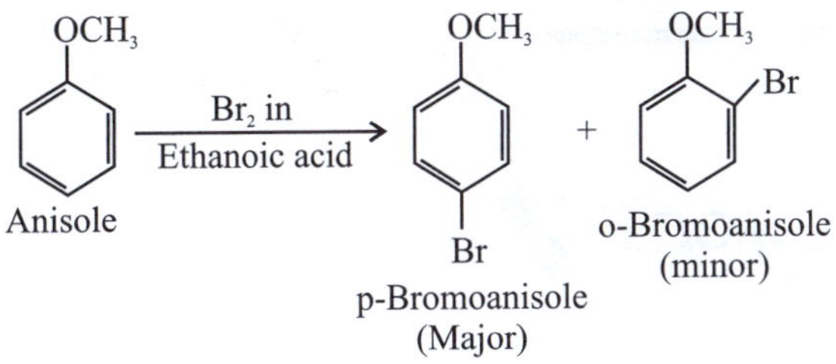

2. Friedel-Crafts Reaction

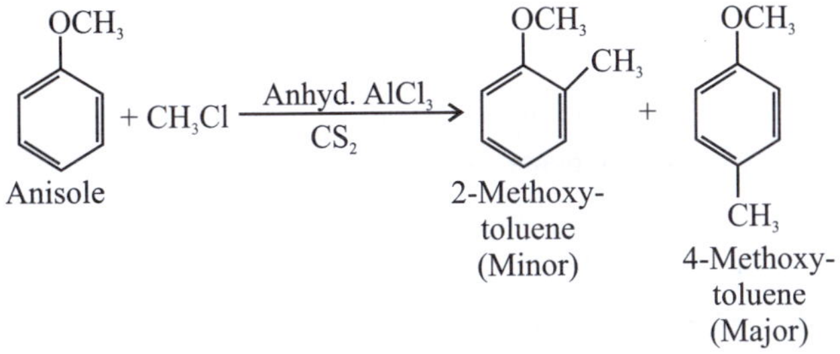

3. Friedel-Crafts Acylation

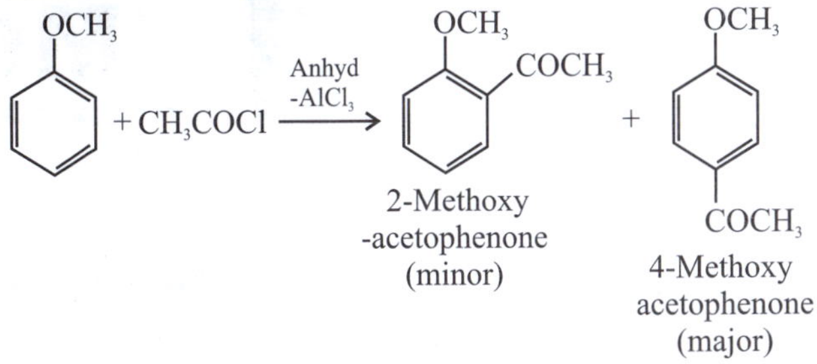

4. Nitration

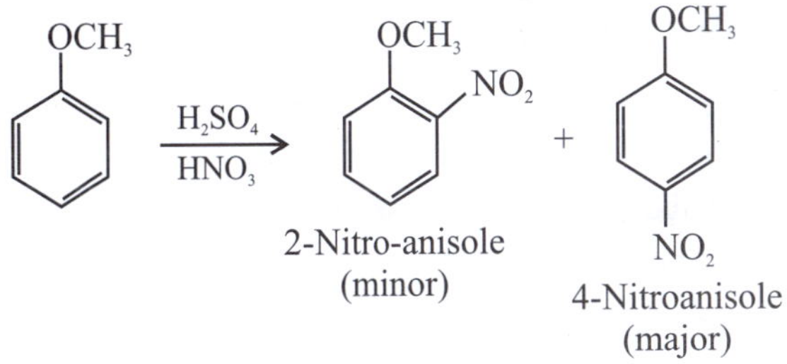

MCQs

NCERT Topic-wise MCQs

Introduction

1. Which class of compound is obtained upon substitution of hydrogen by alkoxy/aryloxy group in hydrocarbon?
   1) Methoxy 2) Ether
   3) Alcohol 4) Phenol

2. Which one of the following is used for Polishing wooden furniture?
   1) Ethanol 2) Methanol
   3) Propanol 4) Butanol

Classification

3. Butan - 2 - ol is a/an:
   1) Aldehyde 2) Primary alcohol
   3) Secondary alcohol 4) Ketone

4. Which of the following is an example of allylic alcohol?
   1) $CH_3CH_2—CH_2—OH$ 2) CH_3CH_2OH
   3) $CH_2=CH—CH_2—OH$ 4) $CH_2=CH_2—OH$

5. o-cresol is a/an:
   1) Monohydric 2) Dihydric
   3) Trihydric 4) Polyhydric

6. The formula C_3H_8O represents:
   1) An ether only 2) An alcohol only
   3) Both ether and alcohols 4) Neither ether and alcohols

7. Which of the following is a simple ether?
   1) $C_2H_5OCH_3$ 2) CH_3OCH_3
   3) $C_6H_5OCH_3$ 4) $C_6H_5OC_2H$

Nomenclature

8. The structure of 3 - Methylbutan -2 - ol is:
   1) $CH_3—CH(OH)—CH_2—CH_3$
   2) $CH_3—CH(OH)—CH(CH_3)—CH_3$
   3) $CH_3—C(OH)(CH_3)—CH_2—CH_3$
   4) None of these

9. The IUPAC name of $CH_3CH_2CH_2CH(CH_2OH)–CH_2CH_3$ is:
   1) 3 - Propyl butan - 1 - ol
   2) 2 - ethylpentan - 1 - ol
   3) 3 - methyl hydroxyhexane
   4) 2 - ethyl - 2 - propyl ethanol

10. Which of the following is the correct common name of the given compound?
    $CH(OH)—CH(OH)—CH_2(OH)$
    1) Glycol 2) Ethylene glycol
    3) Glycerol 4) Glycerin

11. Which of the following is the correct IUPAC name of the given compound?
    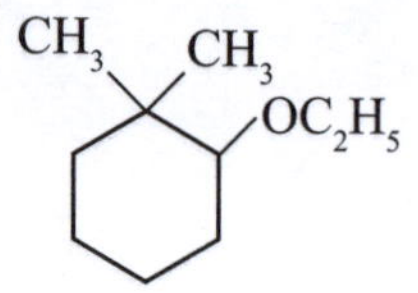
    1) 2-ethoxy-1, 1-dimethylcyclohexane
    2) 1,2-dimethyl cyclohexane
    3) 3-ethoxy-2, 2-dimethyl cyclohexane
    4) 1-Ethoxy-2, 2-dimethyl cyclohexane

Structures of Functional Groups

12. Bond length of phenol is less than the alcohol why?
    1) Steric hindrance 2) +R effect of –Ph ring
    3) sp^3 hybridisation 4) –R effect of –Ph ring

13. The C – O – C angle in ether is about:
    1) 180° 2) 190°28
    3) 110° 4) 105

Alcohols and Phenols

14. In the reaction
    $CH_3–CH=CH_2 \xrightarrow{H_2SO_4} B \xrightarrow{H_2O} C$
    B and C are respectively:
    1) Propyl hydrogen sulphate and propyl alcohol
    2) Iso - propyl hydrogen sulphate and propanol
    3) n - propyl hydrogen sulphate and propan - 2 - ol
    4) iso - propyl hydrogen sulphate and propan - 2 - ol

15. In the reaction, acetone $\xrightarrow{Red^n} B \xrightarrow{dil.H_2SO_4} C$, C is:
    1) Propene 2) But - 1 - ene
    3) Iso - butyl alcohol 4) n - butyl alcohol

16. Lucas reagent is:
    1) Conc. HCl and anhydrous $ZnCl_2$
    2) Conc. HNO_3 and hydrous $ZnCl_2$
    3) Conc. HCl and hydrous $ZnCl_2$
    4) Conc. HNO_3 and anhydrous $ZnCl_2$

17. Which reagent converts propene to 1-propanol?
    1) H_2O, H_2SO_4
    2) B_2H_6, H_2O_2, OH^-
    3) $Hg(OAc)_2, NaBH_4/H_2O$
    4) Aq. KOH

18. In the reaction given below. Identify A is

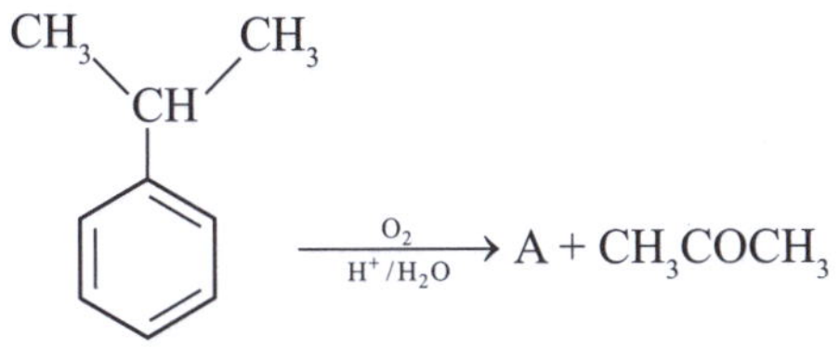

    1) Phenol
    2) Benzoic acid
    3) Cumene hydroperoxide
    4) Benzaldehyde

19. Which of the following product is formed when benzene diazonium salt is treated with H_2O?
    1) Phenol 2) Chlorobenzene
    3) Benzene 4) o-chloro phenol

20. Lower alcohol are:
    1) Soluble in water
    2) Insoluble in organic solvents
    3) Soluble in water only on heating
    4) Insoluble in all solvents

21. Methanol and ethanol are miscible in water due to:
    1) Covalent character
    2) Hydrogen bonding character
    3) Oxygen bonding character
    4) None of these

22. The molecule with maximum boiling point is:
    1) $CH_3-CH(OH)-CH_3$
    2) $CH_3CH_2CH_2CH_2Cl$
    3) $CH_3-CH(OH)-CH_2-CH_2-OH$
    4) $CH_3-CH(Cl)-CH_3$

23. The best reagent to convert pent - 3 - en - 2 - ol into pent - 3 - en - 2 - one is:
    1) Acidic permanganate
    2) Acidic dichromate
    3) Chromic anhydride in glacial acetic acid
    4) Pyridinium chlorochromate

24. Benzoquinone is prepared by reaction of phenol with:
    1) $Na_2Cr_2O_7, H_2SO_4$ 2) $KMnO_4, H_2SO_4$
    3) Na_2CrO_4, HCl 4) K_2MnO_4, H_2SO_4

25. The major product obtained on reaction of phenol with sodium hydroxide and carbon dioxide is:
    1) Benzoic acid 2) Salicylaldehyde
    3) Salicylic acid 4) Phthalic acid

26. Tertiary butyl alcohol gives tertiary butyl chloride on treatment with:
    1) Conc. HCl/ anhydrous $ZnCl_2$
    2) KCN
    3) NaOCl
    4) Cl_2

27. $R-OH + HX \rightarrow RX + H_2O$

    In the above reaction, the reactivity of alcohols is:
    1) Tertiary > Secondary > Primary
    2) Tertiary < Secondary < Primary
    3) Tertiary < Primary < Secondary
    4) Secondary > Primary > tertiary

28. Lucas reagent reacts fastest with:
    1) $CH_3-CH(OH)-CH_3$ 2) $C_2H_5CH_2OH$
    3) CH_3OH 4) $CH_3-C(OH)(CH_3)-CH_3$

29. In the reaction, $C_2H_5OH \xrightarrow[ZnCl_2]{HX} B \xrightarrow{Na/ether} C$, C is
    1) Butane 2) Ethane
    3) Iso butane 4) 2 - methyl butane

30. Which one of the following is obtained when phenol is shaken with excess of conc. HNO_3?

    1) 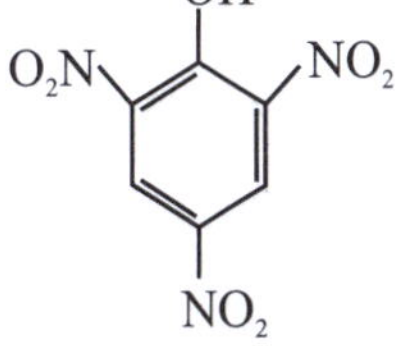

    2) 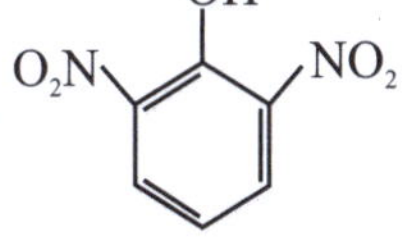

    3) 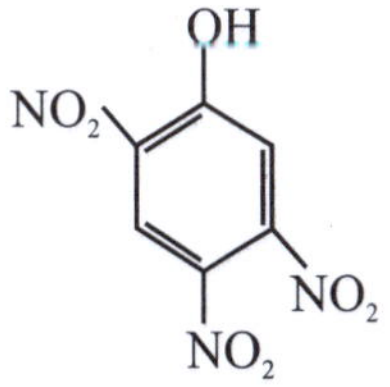

    4) None of these

31. Dehydration of n - butyl alcohol with Al_2O_3 gives:
1) But - 2 - ene 2) But - 1 - ene
3) But - 1 - yne 4) n - butane

32. $CH_2OH - CH_2OH$ on heating with periodic acid gives:
1) $2CO_2$ 2) 2HCOOH
3) CHO–CHO (CHO | CHO) 4) 2 $H_2C=O$

33. Identify B in the following sequence:

$$CH_3CH_2CH_2OH \xrightarrow{PCl_5} A \xrightarrow{alc.KOH} B$$

1) Propyne 2) Propene
3) Propanol 4) Propanone

34. Which one can differentiate between C_2H_5OH and CH_3OH?
1) H_2O 2) $Na_2CO_3 + I_2$
3) NH_3 4) HCl

35. Chlorobenzene $\xrightarrow{\text{reaction X}}$ Phenol $\xrightarrow{\text{reaction Y}}$ salicylaldehyde

X and Y reactions are respectively:
1) Fries rearrangement and kolbe - schmitt
2) Cumene and Reimer - Tiemann
3) Dow and Reimer - Tiemann
4) Dow and Friedel craft

36. Phenol is heated with $CHCl_3$ and alcoholic KOH when salicylaldehyde is produced. This reaction is known as:
1) Rosenmund's reaction
2) Reimer - Tiemann reaction
3) Friedel - crafts reactions
4) Sommelet reaction

37. Phenol is treated with bromine water and shaken well. The white precipitate formed during the process is:
1) m - bromophenol
2) 2, 4, 6 - tribromophenol
3) 2, 4 - dibromophenol
4) a mixture of o – and p – bromophenol

38. Hydrogenation of phenol gives:
1) Salicylic acid 2) p - benzoquinone
3) Cyclohexanol 4) None of these

39. Phenol $\xrightarrow{X}$ forms a tribromo derivative. X is:
1) Bromine in benzene
2) Bromine in water
3) Potassium bromide solution
4) Bromine in carbon tetrachloride at 0°C

40. At low temperature, phenol reacts with Br_2 in CS_2 to form:
1) m – Bromophenol
2) o and p – Bromophenol
3) p - Bromophenol
4) 2, 4, 6 – Tribromo phenol

41. Aspirin is an acetylation product of:
1) p - Dihydroxy benzene
2) o - Hydroxy benzoic acid
3) o - Dihydroxy benzene
4) m - Hydroxy benzoic acid

42. Which one of the following on oxidation gives a ketone?
1) Primary alcohol 2) Secondary alcohol
3) Tertiary alcohol 4) All of these

43. The correct order of acidic strength of the following is:

(I) 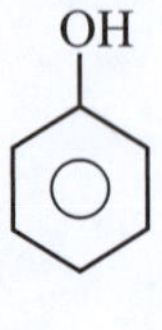 (II) 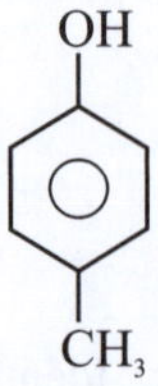 (III) 

1) I > II > III 2) III > I > II
3) II > III > I 4) I > III > II

44. In the given reaction, the compound X is:

'X' $\xrightarrow[H_2SO_4]{Conc.}$ 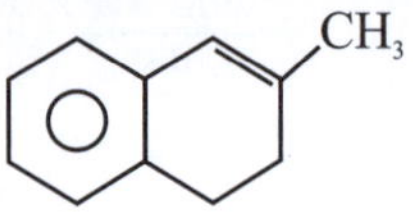

1) 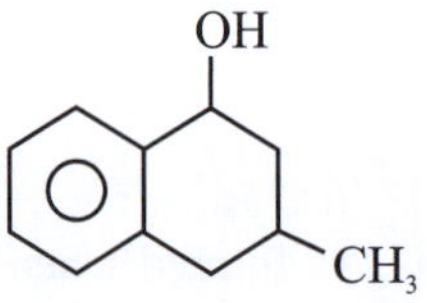
2) 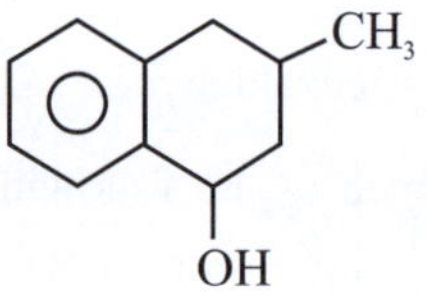
3) 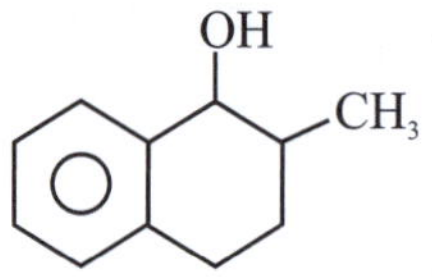
4) 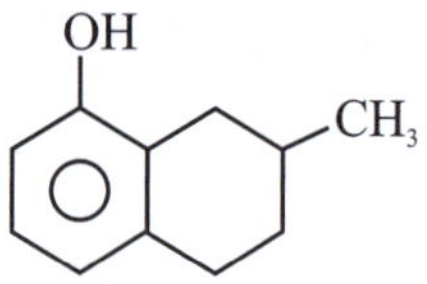

45. When 3, 3-dimethyl-2-butanol is heated with H_2SO_4, the major product obtained is
1) 2, 3-dimethyl-2-butene
2) cis and trans isomers of 2, 3-dimethyl-2-butene
3) 2, 3-dimethyl-1-butene
4) 3, 3-dimethyl-1-butene

46. The groups which when present in para position tend to decrease the acidity of phenol are-
1) $-NO_2$ 2) –CN
3) $-OCH_3$ 4) –F

47. The strongest acid among the following aromatic compounds is
1) p-chlorophenol 2) p-nitrophenol
3) m-nitrophenol 4) o-nitrophenol

48. An alkene obtained by the dehydration of an alcohol (A) on ozonolysis gives two molecules of acetaldehyde for each molecule of the alkene. The alcohol (A) is
1) $CH_3CH_2CH_2OH$ 2) CH_3CH_2OH
3) $CH_3 - CH = CHCH_2OH$ 4) $CH_3CH_2CHOHCH_3$

49. Phenol is used as a starting material for the manufacture of a drug is known as:

1) Phenyl acetate 2) Bakelite
3) Aspirin 4) Dettol

50. Which one of the following has higher pk_a value?

1) o-Nitrophenol 2) m-Nitrophenol
3) p-Nitrophenol 4) Phenol

51. Which of the following will have less acidity?

a) 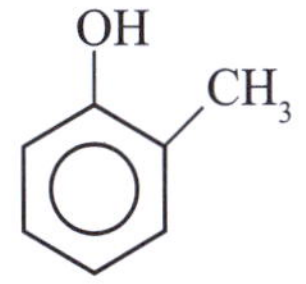

o-cresol

b) 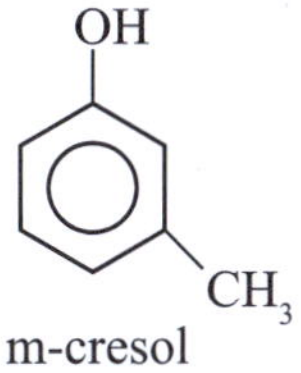

m-cresol

c) 

p-cresol

d) 

phenol

1) a > b > c > d 2) d > a > b > c
3) d > b > a ≈ c 4) c > a > b > d

Some Commercially Important Alcohols

52. $CO + H_2 \xrightarrow[X(\text{catalyst})]{300^\circ C/300atm} CH_3OH$

The catalyst X is:

1) Fe 2) Cr_2O_3/ZnO
3) V_2O_5 4) Al_2O_3

53. Denatured alcohol is:

1) Ethanol + methanol
2) Rectified spirit + methanol + naphthalene
3) Undistilled ethanol
4) Rectified spirit

54. The enzyme which can catalyse the conversion of glucose to ethanol is:

1) Invertase 2) Zymase
3) Maltase 4) Diastase

55. Denatured alcohol is commonly known as:

1) Methanol 2) Methylated spirit
3) Ethylated spirit 4) Cordite

56. Starch is converted to ethanol by fermentation. The sequence of enzyme used are:

1) Amylase, maltase, zymase
2) Diastase, Maltase, zymase
3) Amylase, invertase, zymase
4) Amylase, zymase, maltase

57. In wash or wort, the percentage of ethyl alcohol present is

1) 30-40% 2) 11-20%
3) 20-30% 4) 6-10%

58. Which of the following is known as wood spirit?

1) Ethanol 2) Ethanal
3) Methanol 4) Methanal

59. Drinking of methyl alcohol mainly cause:

1) Fatigue 2) Headache
3) Heart failure 4) Blindness

Ethers

60. Which of the following alcohols gives the best yield of dialkyl ether on being heated with a trace of sulphuric acid?

1) 2-pentanol 2) 2-methyl-2 butanol
3) 1-pentanol 4) 2-propanol

61. Willamson's synthesis of preparing dimethyl ether is a/an:

1) Electrophilic substitution
2) S_N1 reaction
3) S_N2 reaction
4) Electrophilic addition

62. Anisole can be prepared by the action of methyl iodide on sodium phenoxide. The reaction is called:

1) Wurtz's reaction 2) Williamson's reaction
3) Fittig's reaction 4) Etard's reaction

63. Ether is obtained from ethyl alcohol in presence of H_2SO_4 at:

1) 113 K 2) 443 K
3) 413 K 4) 213 K

64. In continuous etherification process, the intermediate obtained is:

1) Alkyl hydrogen sulphite
2) Ethyl hydrogen sulphite
3) Ethyl sulphate
4) Ester of sulphuric acid

65. The major product of the reaction between tert - butyl chloride and sodium ethoxide is:

1) 2 - methyl prop - 1 - ene 2) 1 - butane
3) 2 - butane 4) Ethene

66. Dehydration of methyl alcohol with conc. H_2SO_4 yields:

1) Methane 2) Ethane
3) Dimethyl ether 4) Acetone

67. Which of the following is the best way to prepare phenyl-N propyl ether?

1) 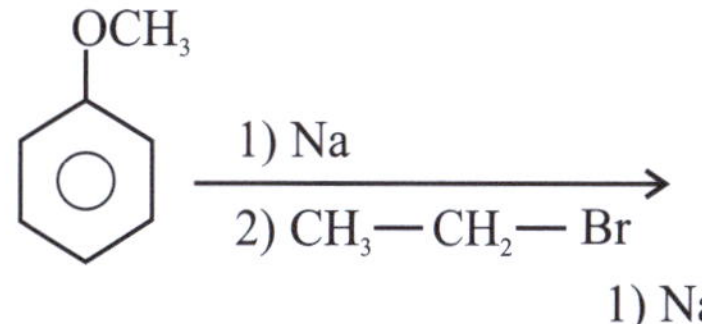

2) $CH_3—CH_2—CH_2—OH$ 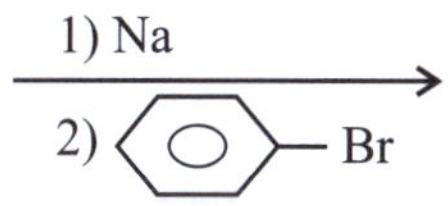

3) 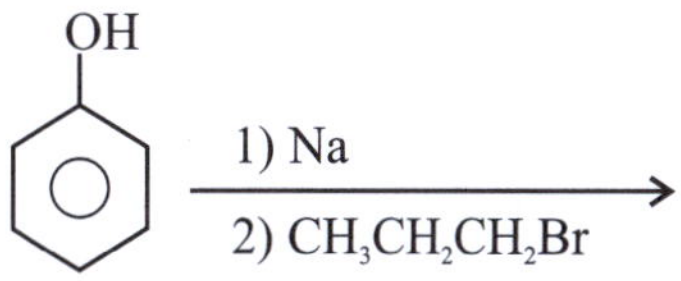

4) Both (1) & (2)

68. Identity Z in the sequence of reactions $CH_3CH_2CH = CH_2 \xrightarrow{HBr/H_2O_2} Y \xrightarrow{C_2H_5ONa} Z$
   1) $CH_3-(CH_2)_3-O-CH_2CH_3$
   2) $(CH_3)_2CH_2-O-CH_2CH_3$
   3) $CH_3(CH_2)_4-O-CH_3$
   4) $CH_3CH_2-CH(CH_3)-O-CH_2CH_3$

69. The correct order of boiling point is:
   1) $C_2H_5 - O - C_2H_5 > C_4H_9OH > CH_3 - O - CH_3$
   2) $C_2H_5 - O - C_2H_5 < C_4H_9OH < CH_3 - O - CH_3$
   3) $CH_3 - O - CH_3 < C_2H_5 - O - C_2H_5 < C_4H_9OH$
   4) $CH_3 - O - CH_3 > C_2H_5 - O - C_2H_5 > C_4H_9OH$

70. When $CH_2 = CH - O - CH_2 - CH_3$ reacts with one mole of HI, one of the products formed is:
   1) Ethane 2) Ethanol
   3) Iodoethene 4) Ethanal

71. Which of the following are the products shown by the reaction of methoxyethane with HI?
   1) $C_2H_5I + CH_3OH$ 2) $CH_3I + H_2O$
   3) $C_2H_5OH + H_2O$ 4) $C_2H_5OH + CH_3I$

72. The ether that undergoes electrophilic substitution reaction is:
   1) $CH_3OC_2H_5$ 2) $C_6H_5OCH_3$
   3) CH_3OCH_3 4) $C_2H_5OC_2H_5$

73. In the reaction 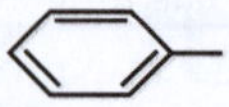 $-OCH_3 \xrightarrow{HBr}$ the products are:
   1) Br– 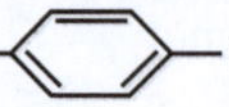 $-OCH_3$ and H_2
   2) 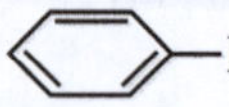 –Br and CH_3Br
   3) 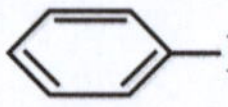 –Br and CH_3OH
   4) 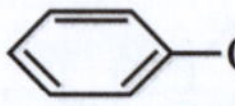 –OH and CH_3Br

74. An ether on hydrolysis with dil H_2SO_4 gives two products which are successive members of a homologous series. The ethers may be:
   1) Dimethyl ether
   2) Diethyl ether
   3) Methyl propyl ether
   4) Ethyl methyl ether

High Order Time Intensive MCQs

1. Arrange the following compounds according to decreasing boiling points.

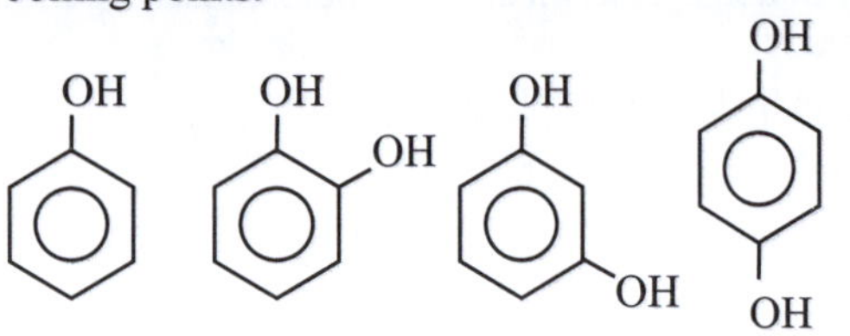

(I) (II) (III) (IV)

   1) IV > III > II > I 2) III > IV > II > I
   3) I > II > III > IV 4) III > II > I > IV

2. In the following sequence of reaction :

$$H_3C\,CH_2OH \xrightarrow{P,\,I_2} A \xrightarrow[\text{ether}]{Mg} B \xrightarrow{HCHO} C \xrightarrow{H_2O} D$$

   The compound D is -
   1) n - Butyl compound 2) n- Propyl alcohol
   3) Propanal 4) Butanal

3. An alcohol of molecular formula $C_5H_{11}OH$ on dehydration gives an alkene, which on oxidation yields a mixture of ketone & an acid. The alcohol is -
   1) $CH_3CH_2CH(OH)\,CH_2CH_3$
   2) $CH_3\,CH(OH)\,CH_2CH_2CH_3$
   3) $(CH_3)_2CHCH\,(OH)\,CH_3$
   4) $(CH_3)_3\,(CH_2OH)$

4. The order of reactivity of the following alcohols towards

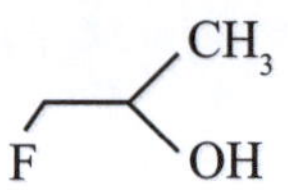

(I)

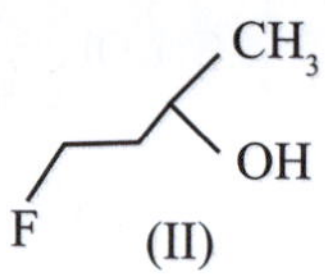

(II)

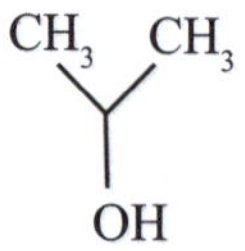

(III)

(IV)

   options are -
   1) I > II > III > IV 2) I > III > II > IV
   3) IV > III > II > I 4) IV > III > I > II

5. Common name of Benzene-1, 2-diol is
   1) quinol 3) resorcinol
   3) catechol 4) o-cresol

6. Chlorination of toluene in the presence of light and heat followed by treatment with aqueous KOH gives
   1) o-cresol 2) m-cresol
   3) p-cresol 4) benzyl alcohol

7. How many different stereoisomers are possible for the given molecule?

$$CH_3-\underset{\underset{OH}{|}}{CH}-CH=CH-CH_3$$

   1) 1 2) 2 3) 3 4) 4

8. Predict the product - 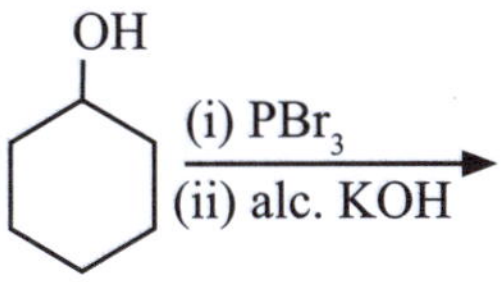

1) 

2) 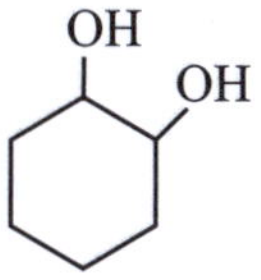

3) 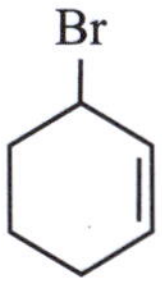

4) 

9. 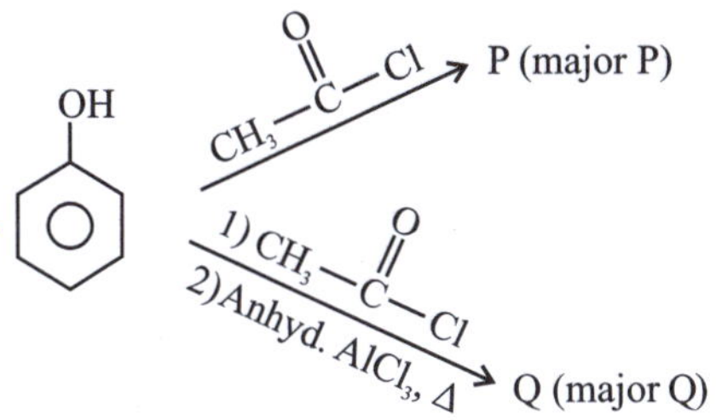

The correct statement for P & Q is/are

1) They can be distinguished by neutral $FeCl_3$

2) Both are functional isomers

3) Both are soluble in hot aq. NaOH

4) All are correct

10. Most acidic among the following is

1) 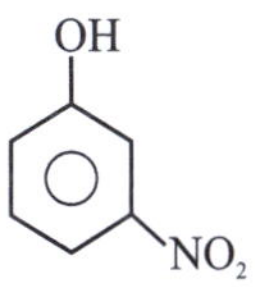

2) 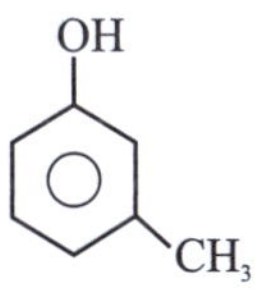

3) 

4) 

11. Molecule which does not oxidise using PCC is

1) 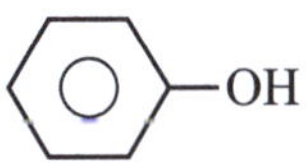

2) 

3) 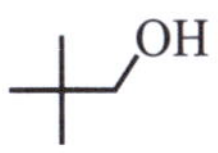

4) 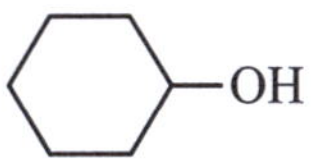

12. $CH_3-CH(CH_3)-CH=CH_2 \xrightarrow[\text{(ii) } H_2O_2/OH^-]{\text{(i) } B_2H_6} X \xrightarrow[140°]{H_2SO_4} Y$

What is Y?

1) $CH_3-CH(CH_3)-CH_2-CH_2-O-CH_2-CH_2-CH(CH_3)-CH_3$

2) $CH_3-CH(CH_3)-CH=CH_2$

3) $CH_3-CH(CH_3)-CH(CH_3)-O-CH(CH_3)-CH(CH_3)-CH_3$

4) $CH_3-C(CH_3)(C_2H_5)-O-C(CH_3)(C_2H_5)-CH_3$

13. The main product of the following reaction

$C_6H_5CH_2CH(OH)CH(CH_3)_2 \xrightarrow{\text{Conc. } H_2SO_4}$ is

1) 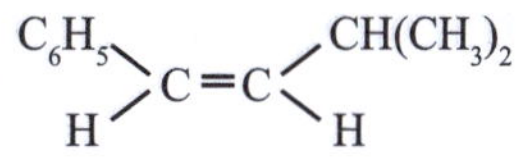

2) 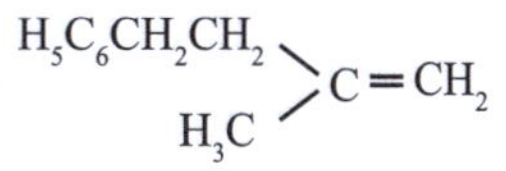

3) 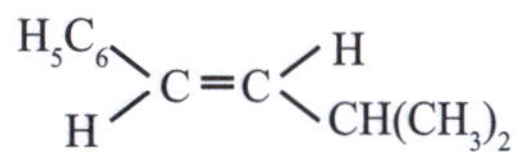

4) 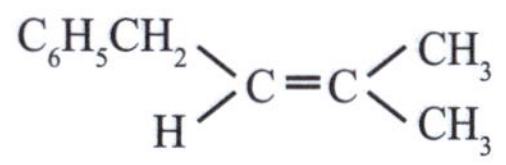

14. Iso-propyl alcohol $\xrightarrow[\Delta]{Al_2O_3} B \xrightarrow[H_2SO_4]{\text{Conc.}} C \xrightarrow[\text{boil}]{\text{water}} D$.

In this reaction D is:

1) Iso - propyl alcohol
2) Propene
3) n - butyl alcohol
4) n - propyl alcohol

15. Alcoholic group can be identified by a positive:

1) Phthalein dye test
2) Lucas test
3) Tollen's test
4) Carbylamine test

16. Which of the following reagents would distinguish cis-cyclopenta-1, 2-diol from the trans-isomer?

1) Aluminium isopropoxide

2) Acetone

3) Ozone

4) MnO

17. Correct acidic order of acidity is

1) 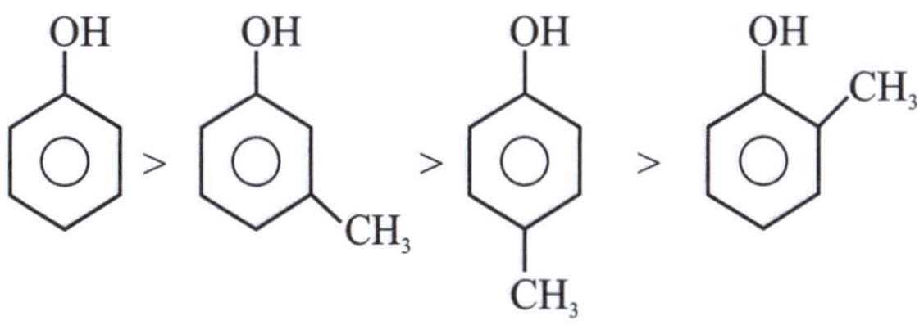

2) 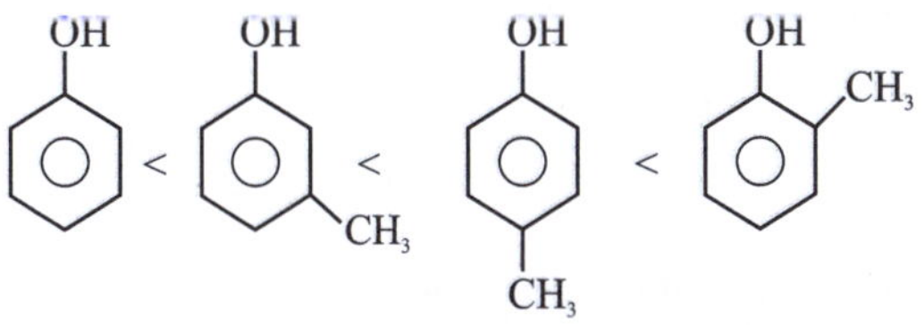

3) 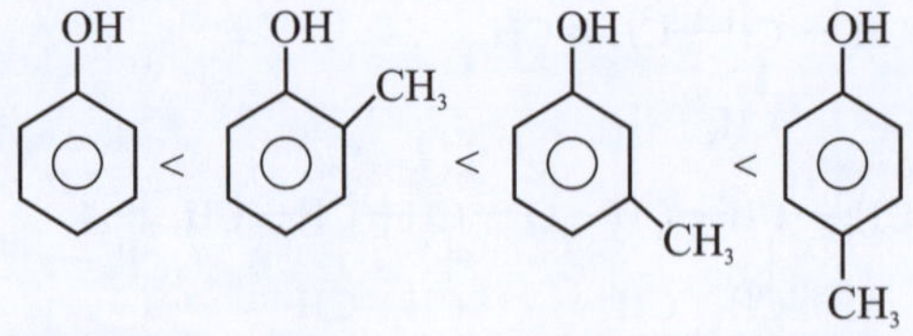

4) 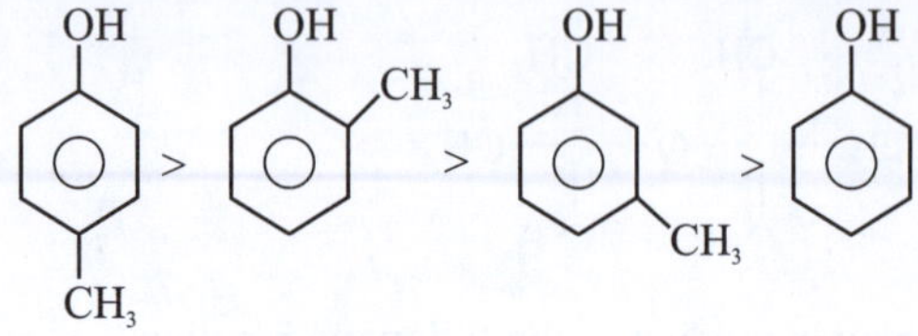

18. Which of the following is not the product of acid catalysed dehydration of 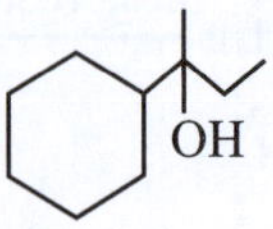 ?

1) 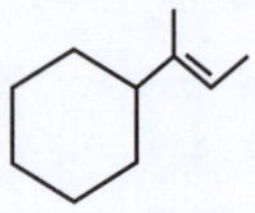
2) 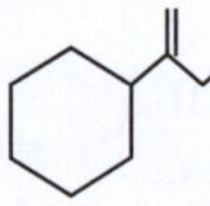
3) 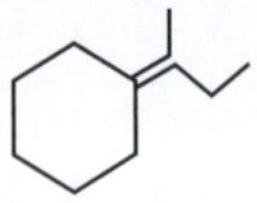
4) 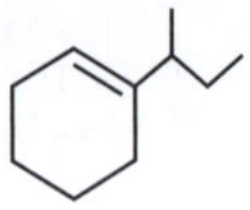

19. 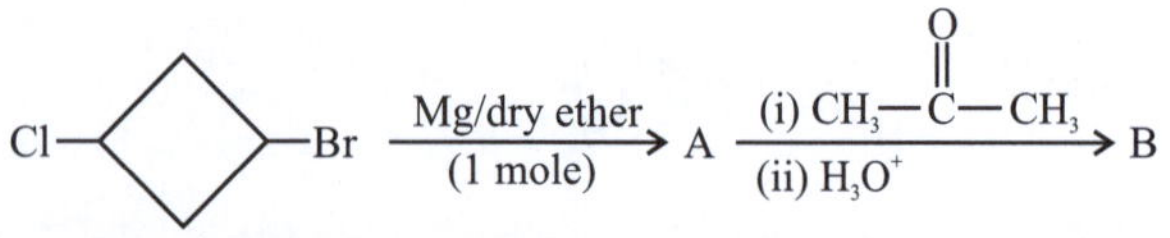

What is B?

1) 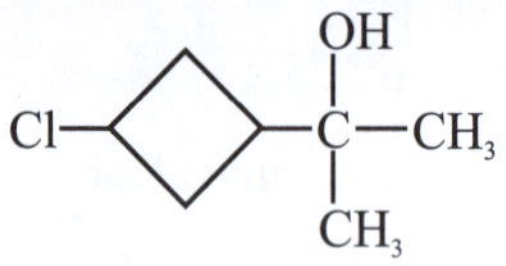

2) 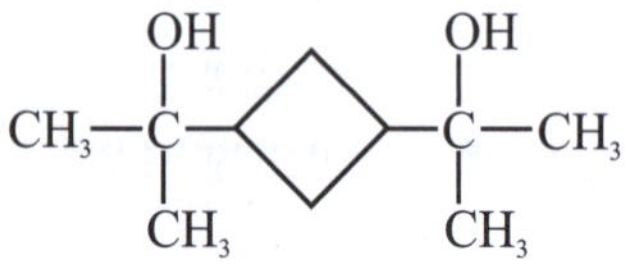

3) 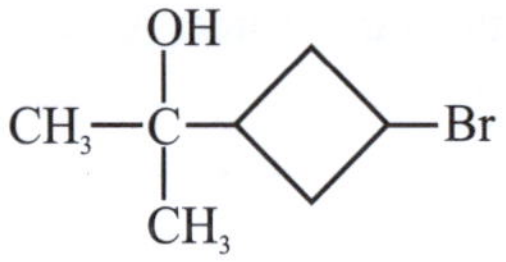

4) 

20. In the following reactions,

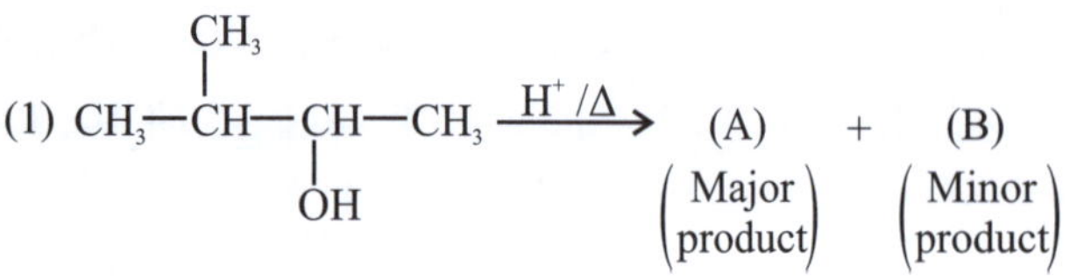

(2) A $\xrightarrow[\text{in the absence of } H_2O_2]{\text{HBr, dark}}$ (C) (Major product) + (D) (Minor product)

The major products (A) and (C) are respectively

1) $CH_3-\underset{}{C}(CH_3)=CH-CH_3$ and $CH_3-C(CH_3)(Br)-CH_2-CH_3$

2) $CH_2=C(CH_3)-CH_2-CH_3$ and $CH_3-CH(CH_3)-CH(Br)-CH_3$

3) $CH_3-C(CH_3)=CH-CH_3$ and $CH_3-CH_2-CH(Br)-CH_3$

4) $CH_2=C(CH_3)-CH_2-CH_3$ and $CH_2(Br)-CH_2-CH_2-CH_3$

21. Which one of the following compounds will be most readily dehydrated?

1) 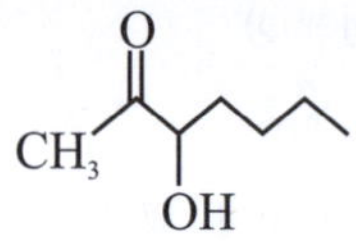
2) 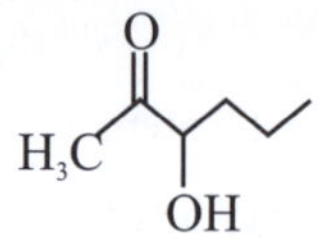
3) 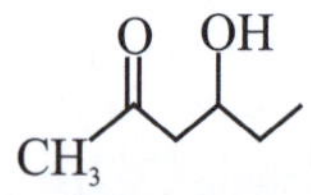
4) 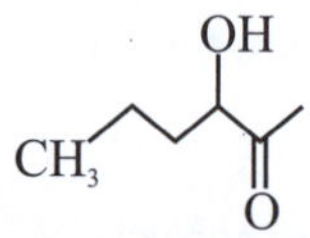

22. Salicylaldehyde is synthesized from phenol, when reacted with

1) 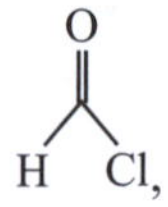, NaOH
2) CO_2, NaOH
3) CCl_4, NaOH
4) $HCCl_3$, NaOH

23. The compound 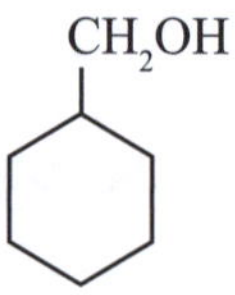 is prepared from two substrate A & B. these are -

1) A = CH_3-OH B = 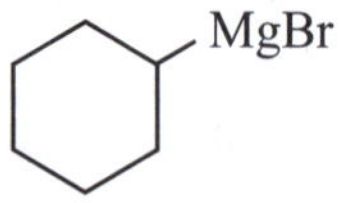

2) A = CH_2O B = 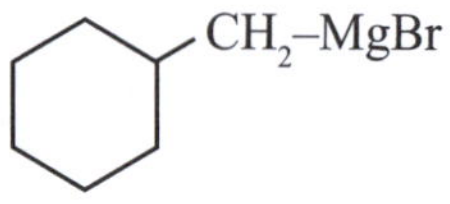

3) A = CH_2O B = 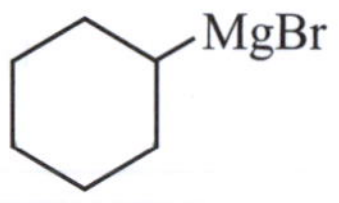

4) A = CH_3-OH B = 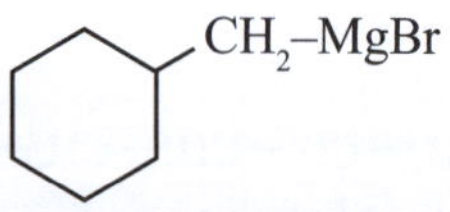

24. 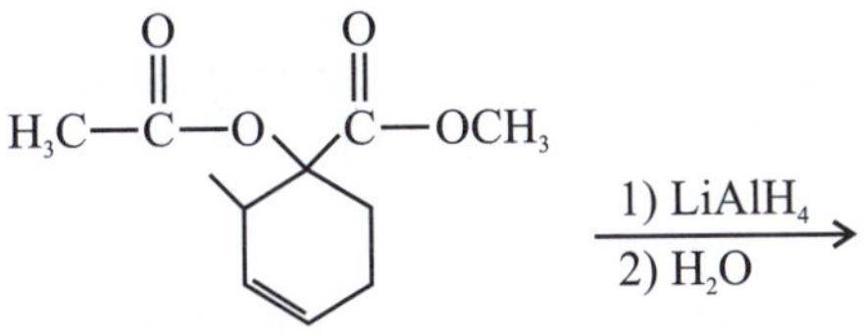

The product of above reaction is:

1) 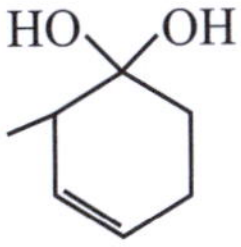

2) 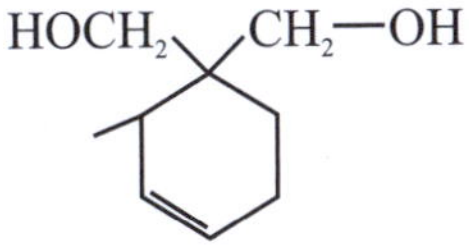

3) 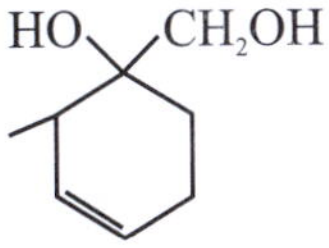

4) 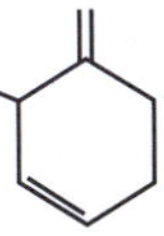

25. Which one the following compounds will readily react with dilute NaOH?

1) $C_6H_5CH_2OH$
2) C_2H_5OH
3) $(CH_3)_3COH$
4) C_6H_5OH

26. 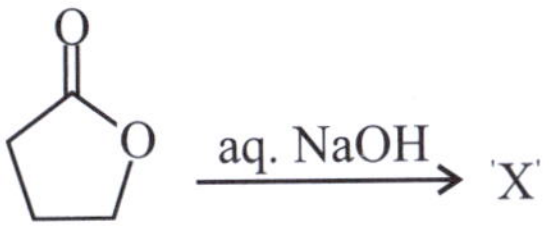

The product 'X' is

1) 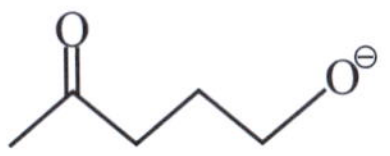

2) 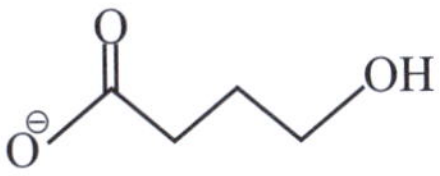

3) 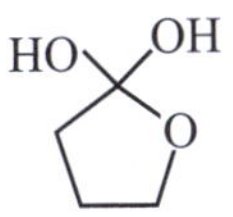

4) 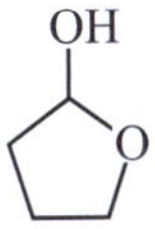

27. 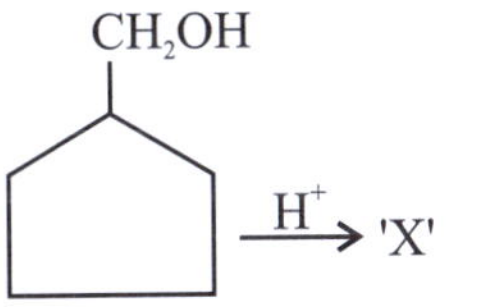

The major product 'X' in the above reaction is-

1) 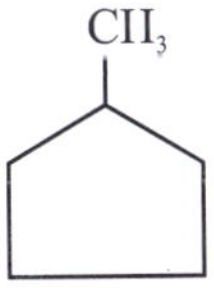

2) 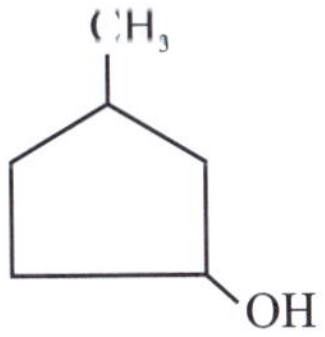

3) 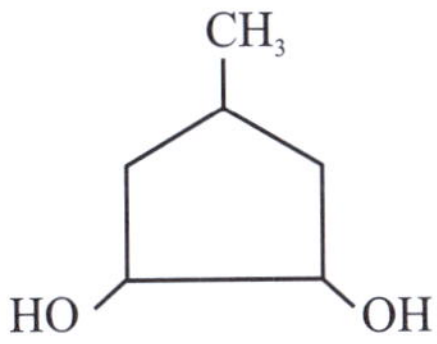

4) 

28. 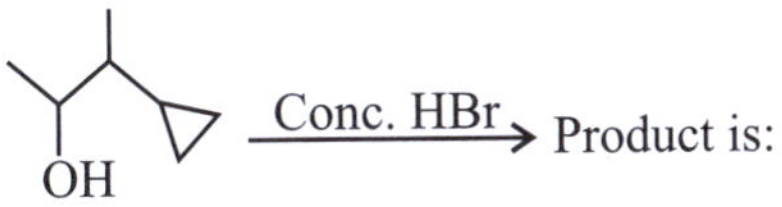

Product is:

1) 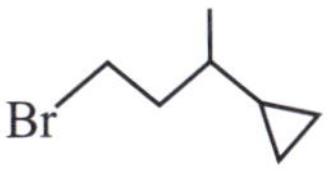

2) 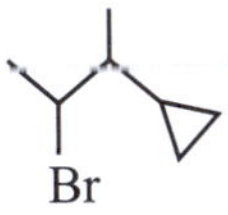

3) 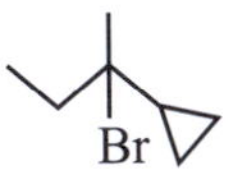

4) 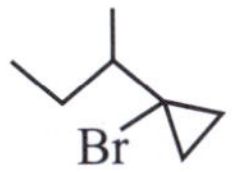

29. The correct options for the products A and B of the following reactions are

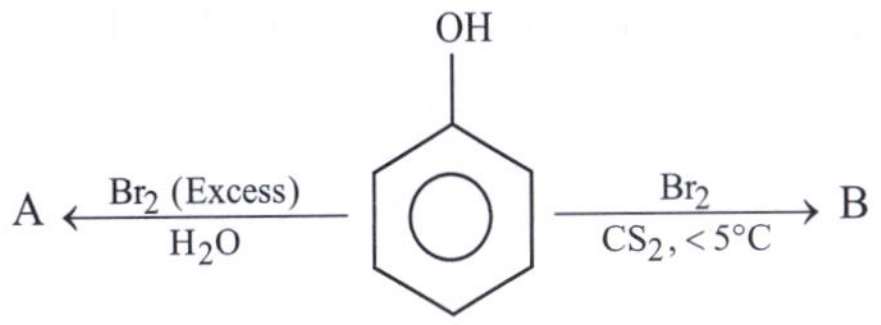

1) A = , B = 

2) A = 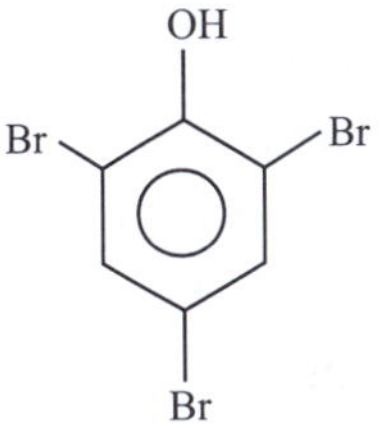, B = 

3) A = 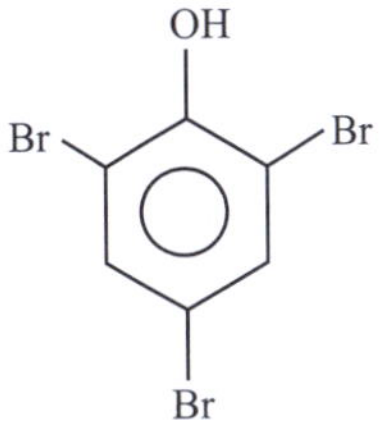, B = 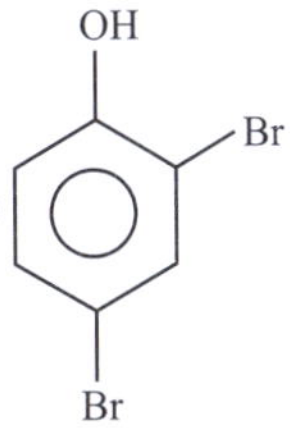

4) A = 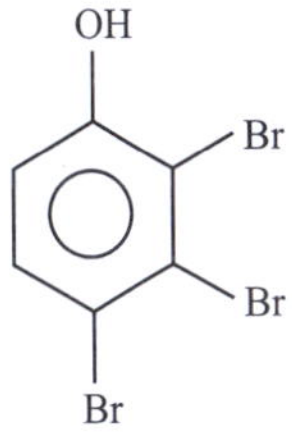, B = 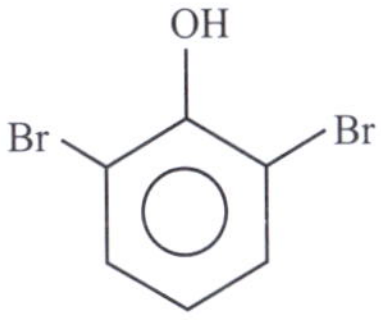

30. (p)  (q)  (r) 

Order of reduction with $NaBH_4$ is:

1) q > p > r
2) r > p > q
3) r > q > p
4) p > q > r

31. Identify the product [Z] in the given sequence of reaction

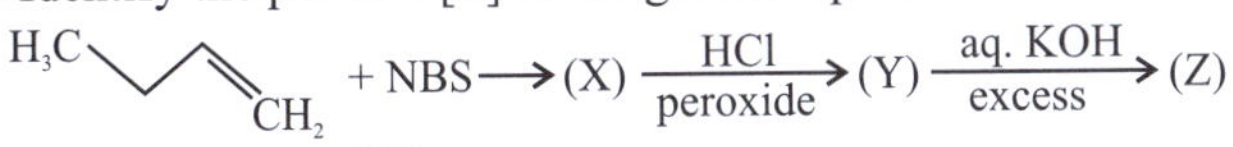

1) 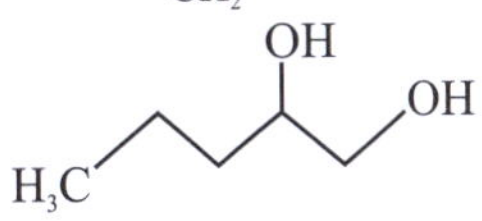

2) 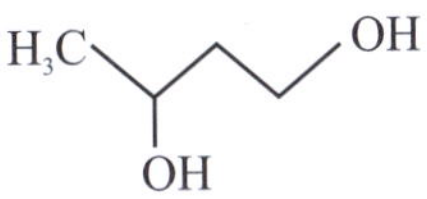

3) 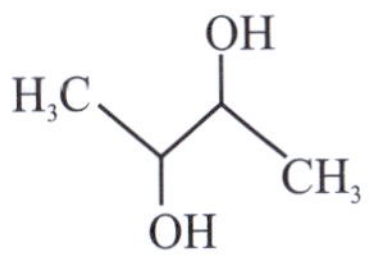

4) None of these

32. 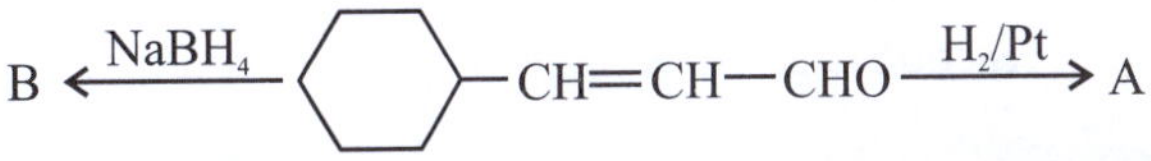

A and B are, respectively-

1) 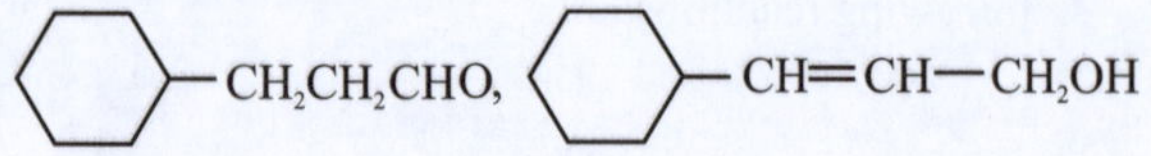

2) 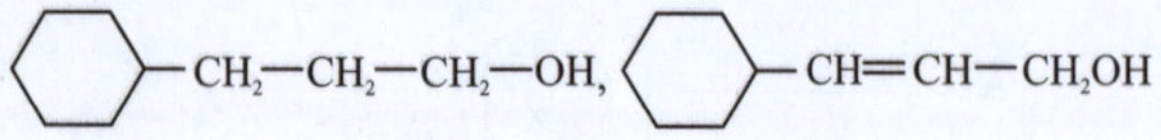

3) 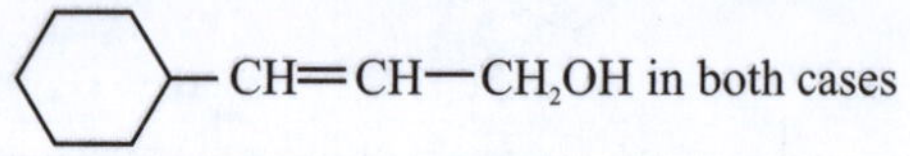

4) 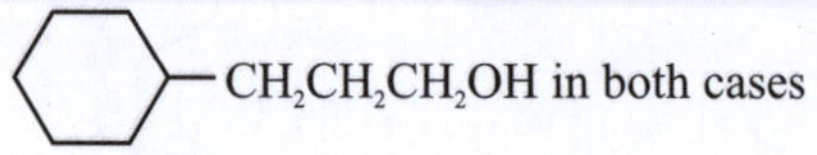

33. Which one of the following is most acidic?

1) p-nitrophenol 2) Phenol

3) m-nitrophenol 4) o-nitrophenol

34. Increasing order of acid strength among p -methoxy phenol, p-methyl phenol and p-nitrophenol is-

1) p- nitrophenol, p-methoxyphenol, p-methylphenol

2) p-methylphenol, p-methoxyphenol, p-nitrophenol

3) p- nitrophcnol, p-mcthylphenol, p-methoxyphenol

4) p-methoxyphenol, p-methylphenol, p- nitrophenol

35. The major product of dehydration of the following

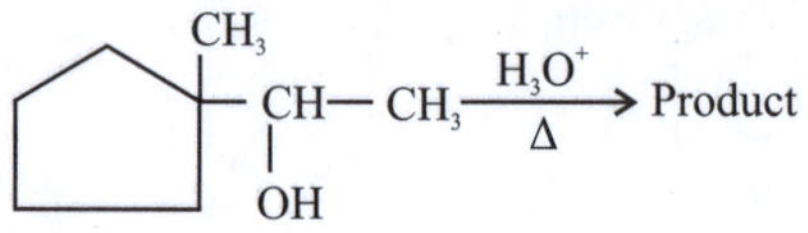

1) 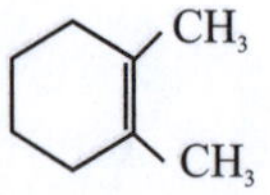

2) 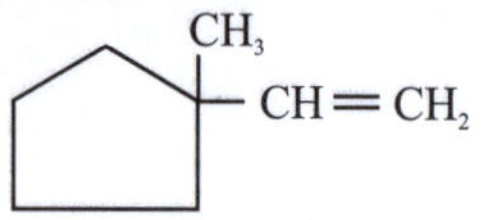

3) 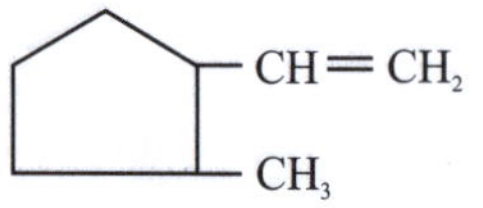

4) 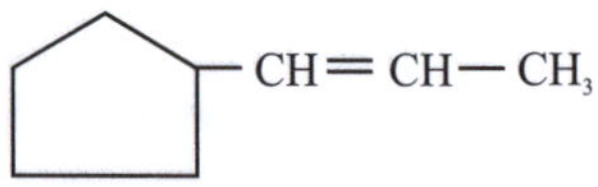

36. The major product (B) in the following reactions is:

$CH_3-CH_2-CH(CH_3)-CH_2-O-CH_2-CH_3$

$\xrightarrow[\text{Heat}]{\text{HI}}$ [A] alcohol $\xrightarrow[\Delta]{H_2SO_4}$ [B]

1) $CH_3-CH_2-CH=CH-CH_3$

2) $CH_3-CH_2-C(CH_3)=CH_2$

3) $CH_3-CH=C(CH_3)-CH_3$

4) $CH_2=CH_2$

37. A reaction is given as-

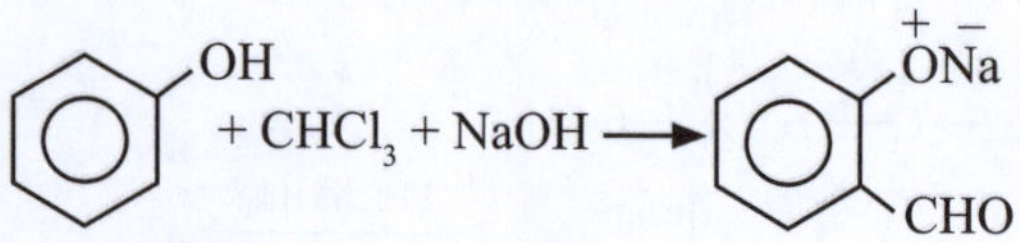

The electrophile involved in the above reaction is -

1) trichlromethyl anion ($\overset{\ominus}{C}Cl_3$)

2) formyl cation ($\overset{\oplus}{C}HO$)

3) dichloromethyl cation ($\overset{\oplus}{C}HCl_2$)

4) dichlorocarbene (: CCl_2)

38. Which of the following reaction does not form ether as major product?

1) 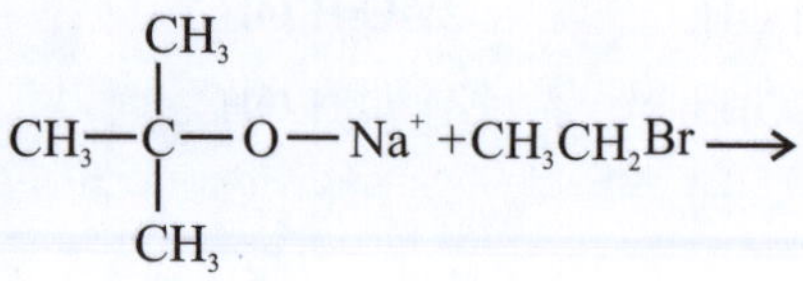

2) 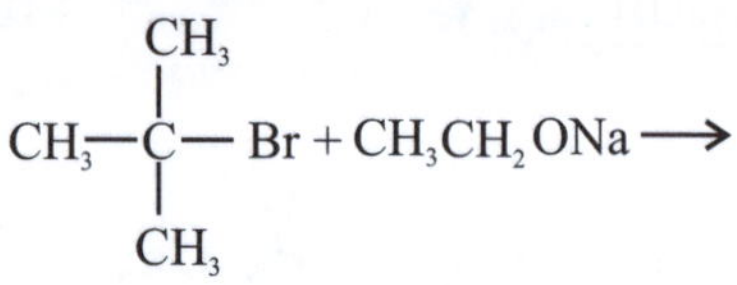

3) 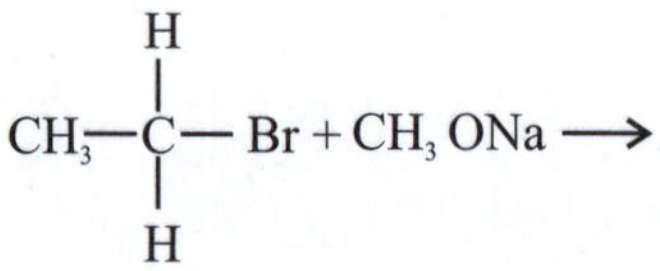

4) 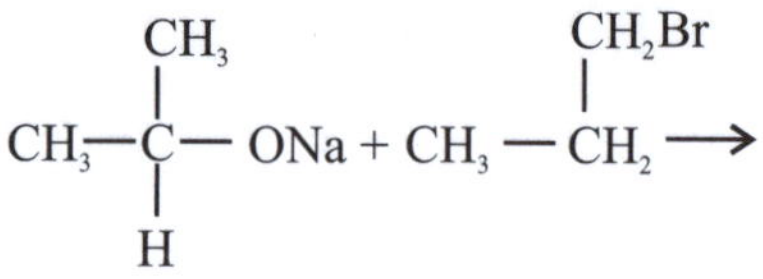

39. 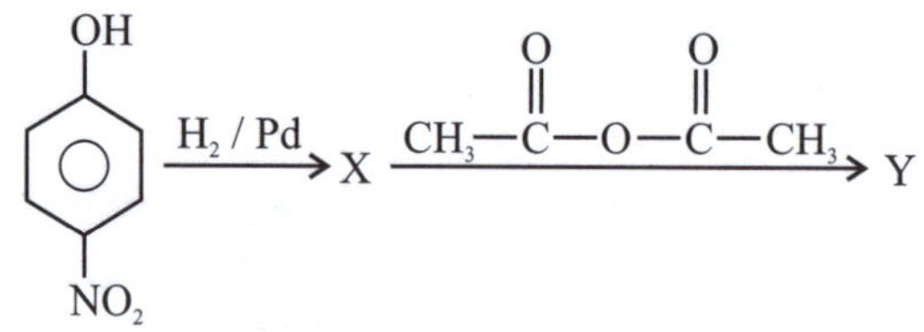

The final product (Y) is:

1) 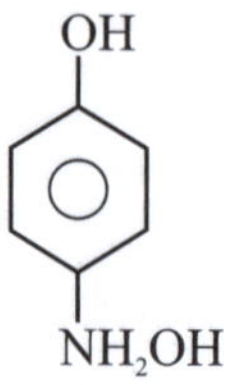

2) 

3) 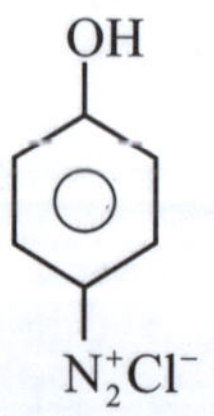

4) 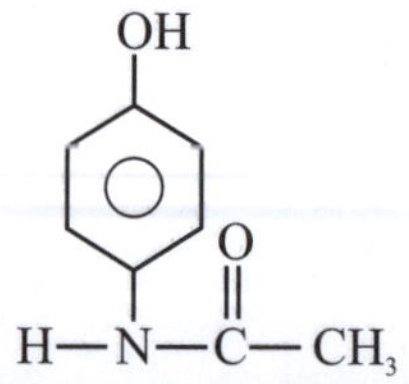

40. Major product formed in the following reaction is a mixture

of:

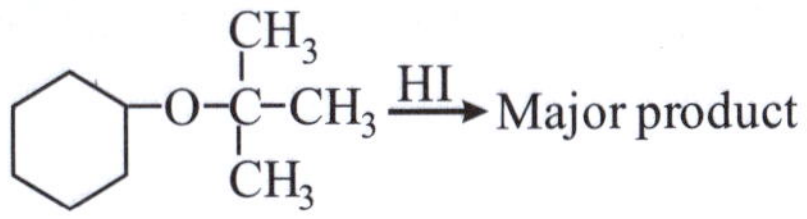

1) 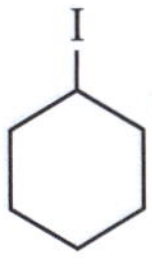 and $(CH_3)_3Cl$ 2) 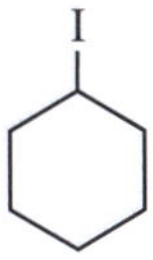 and $(CH_3)_3COH$ 3) 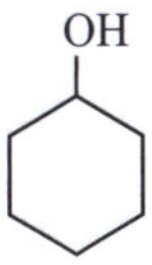 and $(CH_3)_3COH$ 4) 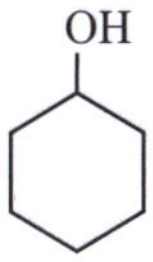 and $CH_3-C(CH_3)_2-I$

Match the Columns MCQs

1. Match the structures of the compounds given in Column-I with the name of the compounds given in Column-II.

Column-I	Column-II
(a) 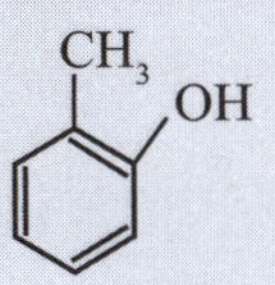	(i) Hydroquinone
(b) 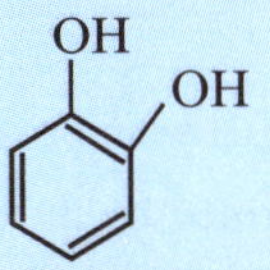	(ii) Phenetole
(c) 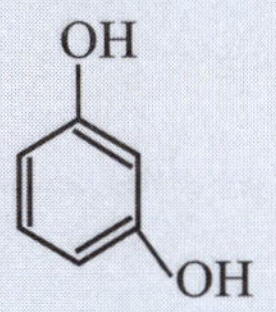	(iii) Catechol
(d) 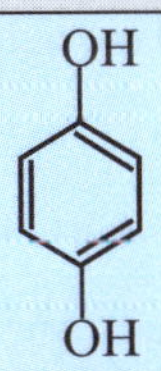	(iv) o-Cresol
(e) 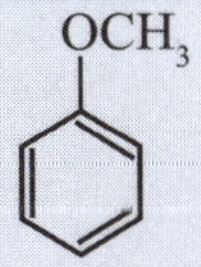	(v) Quinone
(f) 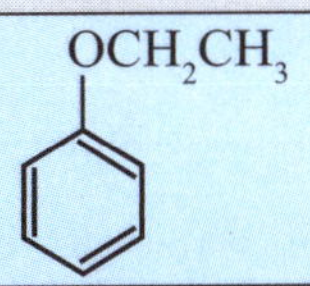	(vi) Resorcinol
	(vii) Anisole

1) (a)-(iv) (b)-(iii) (c)-(vi) (d)-(i) (e)-(vii) (f)-(ii)
2) (a)-(vii) (b)-(iii) (c)-(i) (d)-(iv) (e)-(v) (f)-(ii)
3) (a)-(i) (b)-(iii) (c)-(iv) (d)-(v) (e)-(vi) (f)-(ii)
4) (a)-(vi) (b)-(i) (c)-(ii) (d)-(iv) (e)-(v) (f)-(iii)

2. Match the starting materials given in Column-I with the products formed by these Column-II in the reaction with HI.

Column-I	Column-II
(a) CH_3-O-CH_3	(i) 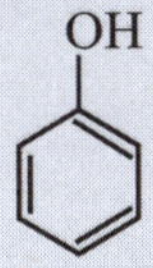 $+ CH_3I$
(b) $(CH_3)_2CH-O-CH_3$	(ii) $CH_3-C(CH_3)_2-I + CH_3OH$
(c) $H_3C-C(CH_3)_2-O-CH_3$	(iii) 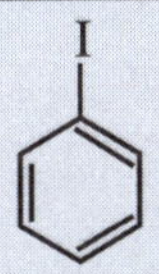 $+ CH_3OH$
(d) 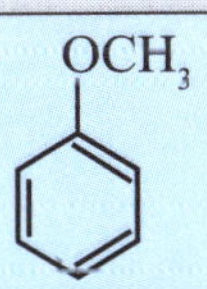	(iv) $CII_3-OII + CII_3-I$
	(v) $(CH_3)_2CH-OH + CH_3I$
	(vi) $(CH_3)_2CH-I + CH_3OH$
	(vii) $CH_3-C(CH_3)_2-OH + CH_3I$

1) (a)-(iv) (b)-(v) (c)-(i) (d)-(ii)
2) (a)-(ii) (b)-(iii) (c)-(vii) (d)-(vi)
3) (a)-(iv) (b)-(v) (c)-(ii) (d)-(i)
4) (a)-(ii) (b)-(iii) (c)-(iv) (d)-(i)

3. Match the items of Column-I with items of Column-II.

Column-I	Column-II
(a) Antifreeze used in car engine	(i) Neutral ferric chloride
(b) Solvent used in perfumes	(ii) Glycerol
(c) Starting material for picric acid	(iii) Methanol
(d) Wood spirit	(iv) Phenol
(e) Reagent used for detection of phenolic group	(v) Ethyleneglycol
(f) By product of soap industry used in cosmetics	(vi) Ethanol

1) (a)-(i) (b)-(iii) (c)-(iv) (d)-(ii) (e)-(v) (f)-(vi)
2) (a)-(vi) (b)-(v) (c)-(iii) (d)-(i) (e)-(ii) (f)-(iv)
3) (a)-(iv) (b)-(v) (c)-(iii) (d)-(ii) (e)-(i) (f)-(vi)
4) (a)-(v) (b)-(vi) (c)-(iv) (d)-(iii) (e)-(i) (f)-(ii)

4. Match the items of Column-I with items of Column-II.

Column-I	Column-II
(a) Methanol	(i) Conversion of phenol to o-hydroxysalicylic acid
(b) Kolbe's reaction	(ii) Ethyl alcohol
(c) Williamson's synthesis	(iii) Conversion of phenol to salicylaldehyde
(d) Conversion of 2° alcohol to ketone	(iv) Wood spirit
(e) Reimer-Tiemann reaction	(v) Heated copper at 573K
(f) Fermentation	(vi) Reaction of alkyl halide with sodium alkoxide

1) (a)-(v) (b)-(ii) (c)-(vi) (d)-(iv) (e)-(i) (f)-(iii)
2) (a)-(iv) (b)-(i) (c)-(vi) (d)-(v) (e)-(iii) (f)-(ii)
3) (a)-(ii) (b)-(i) (c)-(iii) (d)-(iv) (e)-(v) (f)-(vi)
4) (a)-(vi) (b)-(v) (c)-(iii) (d)-(i) (e)-(ii) (f)-(iv)

Correct & Incorrect MCQs

1. Which of the following statement (s) are correct?
   (a) In symmetrical ether, the alkyl or aryl groups attached to the oxygen atom are different.
   (b) In unsymmetrical ether, the alkyl or aryl groups attached to the oxygen atom are the same.
   (c) $C_2H_5OC_2H_5$, is a symmetrical ether.
   (d) $C_2H_5OCH_3$ is an unsymmetrical ethers.
   1) (a), (b) and (c)
   2) (c) and (d)
   3) (a), (c) and (d)
   4) (b), (c) and (d)

2. Which of the following statement regarding IUPAC nomenclature is incorrect ?
   1) The IUPAC name of methyl alcohol is methanol.
   2) The IUPAC name of n-propyl alcohol is Propan-1-ol.
   3) The IUPAC name of isobutyl alcohol 2-Methylpropan-2-ol.
   4) The IUPAC name of n-butyl alcohol is Butan-1-ol.

3. Identify the correct statement :
   1) The IUPAC name of dimethyl ether is Ethoxyethane.
   2) The IUPAC name of Diethyl ether is Methoxymethane.
   3) The IUPAC name of methyl n-propyl ether is 1-Methoxypropane.
   4) The IUPAC name of ethyl phenyl ether is 1-Phenoxyheptane.

4. Correct statement is/are
   (a) In alcohols, the oxygen of the –OH group is attached to carbon by a sigma (σ) bond formed by the overlap of a sp^3 hybridised orbital of carbon with a sp^3 hybridised orbital of oxygen.
   (b) The bond angle in alcohols is slightly less than the tetrahedral angle (109° 28').
   (c) In phenols, the –OH group is attached to sp^2 hybridised carbon of an aromatic ring.
   (d) The carbon– oxygen bond length (136 pm) in phenol is more than that in methanol.
   1) (a), (b) and (c)
   2) (b), (c) and (d)
   3) (a) and (d)
   4) (b) and (c)

5. Which of the following statement about alcohols is correct ?
   1) The solubility of alcohols in water decrease with increase in size of alkyl/aryl groups.
   2) The solubility of alcohols in water increase with increase in size of alkyl/aryl groups.
   3) The solubility of alcohols in water remain unaffected with increase in size of alkyl/aryl groups.
   4) All of these

6. Which of the following statement is incorrect?
   1) The boiling points of alcohols and phenols increase with increase in the number of carbon atoms.
   2) In alcohols, the boiling points increase with increase of branching in carbon chain.
   3) The – OH group in alcohols and phenols is involved in intermolecular hydrogen bonding.
   4) Alcohols and phenols have higher boiling point than the ethers and hydrocarbons of comparable molecular masses.

7. Select the incorrect statement :
   1) The acid strength of alcohols decreases in the order : primary > secondary > tertiary.
   2) An alkoxide ion is a better proton acceptor than hydroxide ion.
   3) Alcohols act as Bronsted bases.
   4) Phenoxide ion is less stable than alkoxide ion.

8. Identify the incorrect statement :
   1) Nitro group enhances the acidic strength of phenol.
   2) Cresols are more acidic than phenols.
   3) The relative ease of dehydration of alcohols follows the order: Tertiary > Secondary > Primary
   4) Tertiary alcohols do not undergo oxidation reaction.

9. Which of the following statement regarding Williamson synthesis is correct ?
   1) Better results are obtained if the alkyl halide used is primary.
   2) Better results are obtained if the alkyl halide used is secondary.
   3) Better results are obtained if the alkyl halide used is tertiary.
   4) None of these

10. Which of the following statement is correct regarding chemical reactions of ethers?
   1) The alkoxy group is only ortho directing and deactivates the aromatic ring towards electrophilic substitution reactions
   2) The alkoxy group is meta directing and deactivates the aromatic ring towards electrophilic substitution reactions
   3) The alkoxy group is ortho, para directing and activates the aromatic ring towards electrophilic substitution reactions
   4) The alkoxy group is only para directing and deactivates the aromatic ring towards electrophilic substitution reactions

Statement Based MCQs

Directions: These questions consist of two statements each, printed as Statement-I and Statement-II. While answering these questions, you are required to choose any one of the following four responses.

1) Both Statement-I and Statement-II are correct.
2) Both Statement-I and Statement-II are incorrect.
3) Statement-I is correct & Statement-II is incorrect.
4) Statement-I is incorrect & Statement-II is correct.

1. **Statement-I:** Ether has lower boiling points than isomeric alcohols.
   Statement-II: Alcohols form H-bond but there is no H- bonding among ether molecules.

2. **Statement-I:** Phenol does not show protonation reaction.
   Statement-II: Lone pair of electrons in phenol is involved in resonance.

3. **Statement-I:** p-Hydroxybenzoic acid has a lower boiling point than o-hydroxybenzoic acid.
   Statement-II: o-Hydroxybenzoic acid has intra molecular hydrogen bonding.

4. **Statement-I:** Alcohols react with acid anhydride and acid chloride to form esters.
   Statement-II: Dissociation constant of alcohols are higher than the dissociation constant of water.

5. **Statement-I:** Phenol does not react with $NaHCO_3$.
   Statement-II: Phenol is less acidic than H_2CO_3.

6. **Statement-I:** The bond angle of ether is less than the tetrahedral angle.
   Statement-II: There is repulsive interaction b/w the two bulky (-R) groups in ether.

7. **Statement-I:** An alkoxide ion, is a better proton acceptor than hydroxide ion.
   Statement-II: The acid strength of alcohols decrease in the following order.

$$R_3C-OH > R_2CHOH > R-CH_2OH$$

8. **Statement-I:** In Lucas test, turbidity is not produced with tertiary alcohols.
   Statement-II: Tertiary alcohol easily form halides which is immiscible in water.

9. **Statement-I:** Nitro group reduces the acidic strength of phenol.
   Statement-II: Cresol is less acidic than phenol.

10. **Statement-I:** The commercial alcohol is made unfit for drinking by mixing in it some $CuSO_4$.
   Statement-II: Commercial alcohol gets denatured after adding $CuSO_4$.

Assertion & Reason MCQs

Directions: These questions consist of two statements each, printed as Assertion and Reason. While answering these questions, you are required to choose any one of the following four responses.

1) **Both Assertion and Reason are True and the Reason is a correct explanation of the Assertion.**
2) **Both Assertion and Reason are True but Reason is not a correct explanation of the Assertion.**
3) **Assertion is True but the Reason is False.**
4) **Assertion is False but Reason is True.**

1. **Assertion:** p-nitrophenol is more acidic than phenol.

   Reason: Nitro group helps in the stabilisation of the phenoxide ion by dispersal of negative charge due to resonance.

2. **Assertion:** IUPAC name of the compound is

   $CH_3-\underset{\underset{CH_3}{|}}{CH}-O-CH_2-CH_2-CH_3$

   is 2-Ethoxy-2-methylethane.

   Reason: In IUPAC nomenclature, ether is regarded as hydrocarbon derivative in which a hydrogen atom is replaced by —OR or —OAr group [where R = alkyl group and Ar = aryl group]

3. **Assertion:** Bond angle in ethers is slightly less than the tetrahedral angle.

   Reason: There is a repulsion between the two bulky (—R) groups.

4. **Assertion:** Like bromination of benzene, bromination of phenol is also carried out in the presence of Lewis acid.

   Reason: Lewis acid polarises the bromine molecule.

5. **Assertion:** o-Nitrophenol is less soluble in water than the m- and p-isomers.

   Reason: m- and p- Nitrophenols exist as associated molecules.

6. **Assertion:** Ethanol is a weaker acid than phenol.

   Reason: Sodium ethoxide may be prepared by the reaction of ethanol with aqueous NaOH.

7. **Assertion:** Phenols give o- and p-nitrophenol on nitration with conc. HNO_3 and H_2SO_4 mixture.

   Reason: —OH group in phenol is o-, p- directing.

8. **Assertion:** Ether has lower boiling points than isomeric alcohols.

   Reason: o–Hydroxybenzoic acid has intramolecular hydrogen bonding.

9. **Assertion:** Tert-butyl alcohol undergoes acid catalysed dehydration readily than that of propanol.

   Reason: Tertiary alcohols do not give victor-meyers test.

10. **Assertion:** $(CH_3)_3CONa$ and CH_3CH_2Br react to form $(CH_3)_3C\text{-}O\text{-}CH_2CH_3$.

    Reason: Good yield of ethers are obtained when tert alkyl halides are treated with alkoxides.

Multi Correct MCQs

1. Select the correct order of acidity

   A. Bu – OH

   B. 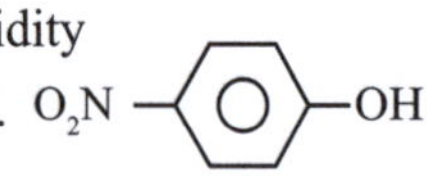

   C. 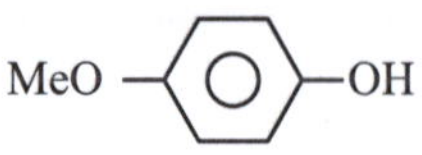

   D. 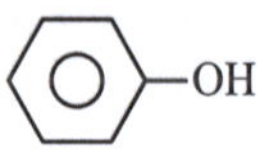

   E. 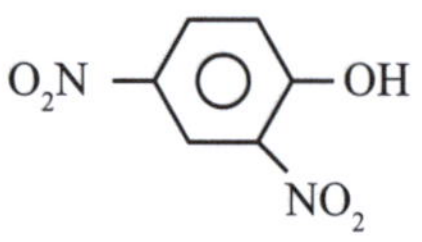

   I. (B) > (E) and C = D
   II. (D) < (C)
   III. (A) = (B) < (C)
   IV. (B) > (C) and (E) > (D)

   Choose the correct answer from the options given below:

   1) (II) only
   2) (II) and (III)
   3) (III) and (IV)
   4) (IV) only

2. The compound/s which will show significant intermolecular H–bonding is/are:

   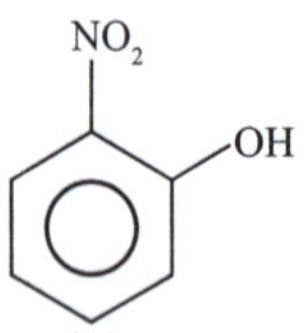

   (A)

   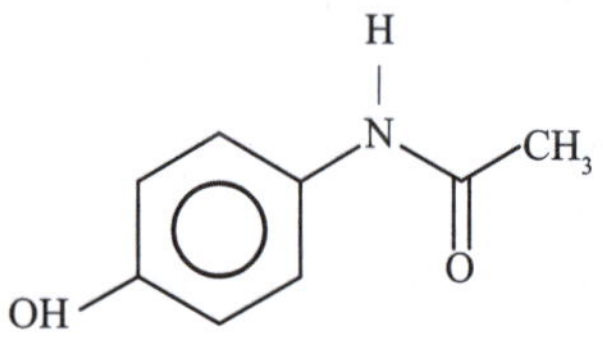

   (B)

   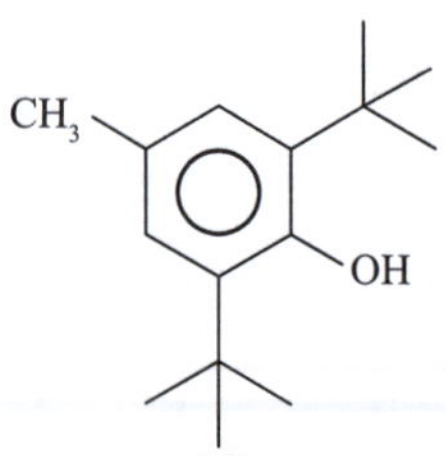

   (C)

   1) (C) only
   2) (B) only
   3) (A), (B) and (C)
   4) (A) and (B) only

3. Which one of the following reactions is NOT possible?

A. 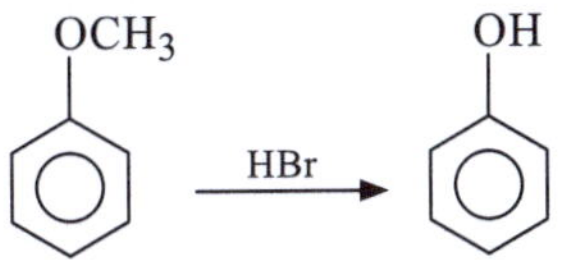

B. 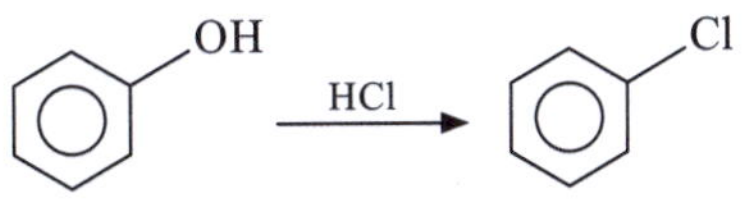

C. 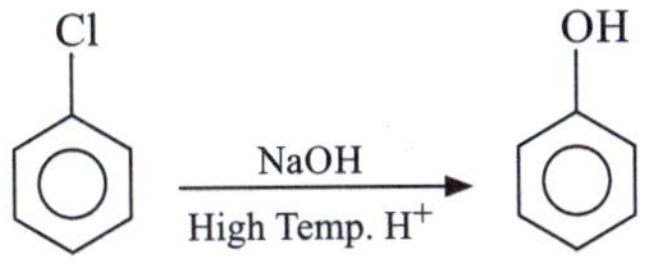

D. 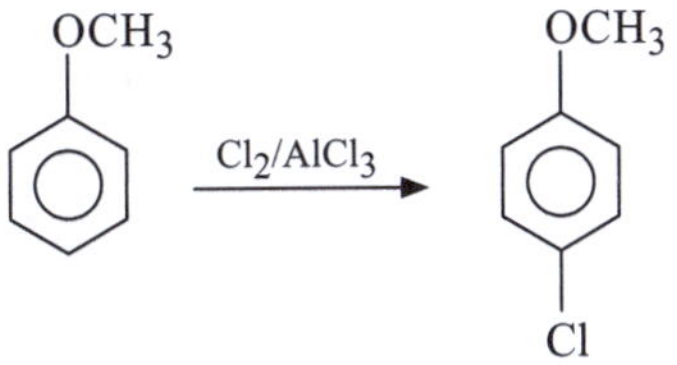

1) (A) and (B)　　2) (B) and (C)
3) (C) and (D) only　　4) (A), (C) and (D)

4. 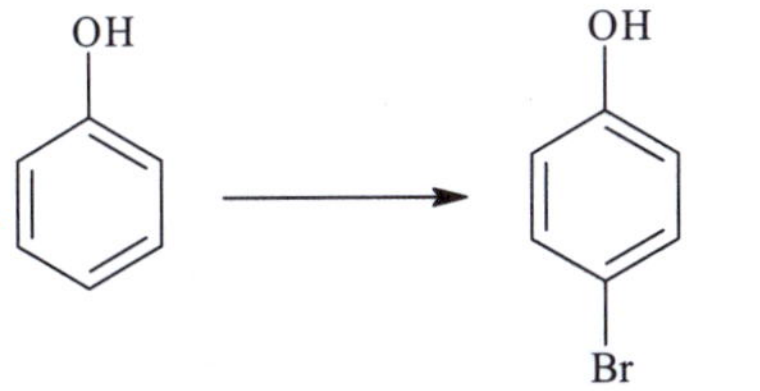

(Major Product)

The given reaction can occur in the presence of:

A. Bromine water　　B. Br_2 in CS_2, 273 K
C. Br_2 / $FeBr_3$　　D. Br_2 in $CHCl_3$, 273 K

Choose the correct answer from the options given below:

1) (A) and (C) only
2) (B), (C) and (D) only
3) (B) and (D) only
4) (A), (B) and (D) only

NCERT Exemplar MCQs

1. Mono-chlorination of toluene in sunlight followed by hydrolysis with aq. NaOH yields:
   1) o-cresol　　2) m-cresol
   3) 2, 4-dihydroxytoluene　　4) benzyl alcohol

2. How many alcohols with molecular formula $C_4H_{10}O$ are chiral in nature?
   1) 1　　2) 2
   3) 3　　4) 4

3. What is the correct order of reactivity of alcohols in the following reaction?

   $$R-OH + HCl \xrightarrow{ZnCl_2} R-Cl + H_2O$$

   1) $1° > 2° > 3°$　　2) $1° < 2° > 3°$
   3) $3° > 2° > 1°$　　4) $3° > 1° > 2°$

4. CH_3CH_2OH can be converted into CH_3CHO by:
   1) Catalytic hydrogenation
   2) Treatment with $LiAlH_4$
   3) Treatment with pyridinium chlorochromate
   4) Treatment with $KMnO_4$

5. The process of converting alkyl halides into alcohols involves:
   1) Addition reaction
   2) Substitution reaction
   3) Dehydrohalogenation reaction
   4) Rearrangement reaction

6. Which of the following compounds is are aromatic alcohol?

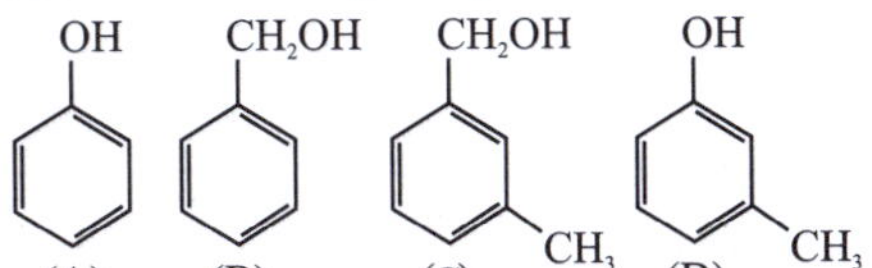

(A)　(B)　(C)　(D)

   1) A, B, C, D　　2) A, D
   3) B, C　　4) A

7. Give IUPAC name of the compound given below:

   $$CH_3-\underset{\underset{Cl}{|}}{CH}-CH_2-CH_2-\underset{\underset{OH}{|}}{CH}-CH_3$$

   1) 2-chloro-5-hydroxyhexane
   2) 2-hydroxy-5-chlorohexane
   3) 5-chlorohexan-2-ol
   4) 2-chlorohexan-5-ol

8. IUPAC name of m-cresol is:
   1) 3-methylphenol　　2) 3-chlorophenol
   3) 3-methoxyphenol　　4) benzene-1, 3-diol

9. IUPAC name of the compound

   $$CH_3-\underset{\underset{CH_3}{|}}{CH}-OCH_3$$ is:

   1) 1-methoxy-1-methylethane
   2) 2-methoxy-2-methylethane
   3) 2-methoxypropane
   4) isopropylmethyl ether

10. Which of the following species can act as the strongest base?
    1) $^{\ominus}OH$　　2) $RO^{\ominus}$
    3) $C_6H_5O^{\ominus}$　　4) 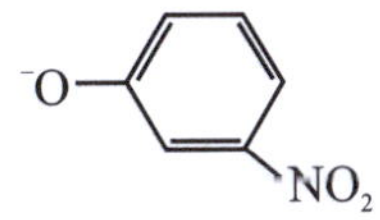

11. Which of the following compounds will react with sodium hydroxide solution in water?
    1) C_6H_5OH　　2) $C_6H_5CH_2OH$
    3) $(CH_3)_3COH$　　4) C_2H_5OH

12. Phenol is less acidic than:
   1) Ethanol 2) o-nitrophenol
   3) o-methylphenol 4) o-methoxyphenol

13. Which of the following is most acidic?
   1) Benzyl alcohol 2) Cyclohexanol
   3) Phenol 4) m-chlorophenol

14. Mark the correct order of decreasing acid strength of the following compounds:

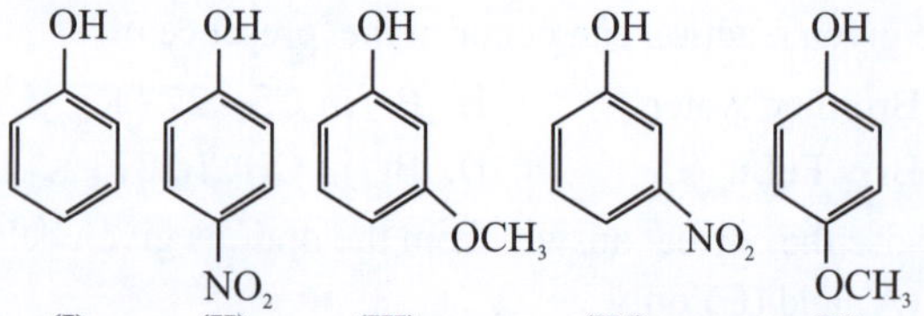

(I) (II) (III) (IV) (V)

   1) V > IV > II > I > III 2) II > IV > I > III > V
   3) IV >V > III > II > I 4) V > IV > III > II > I

15. Mark the correct increasing order of reactivity of the following compounds with HBr /HCl:

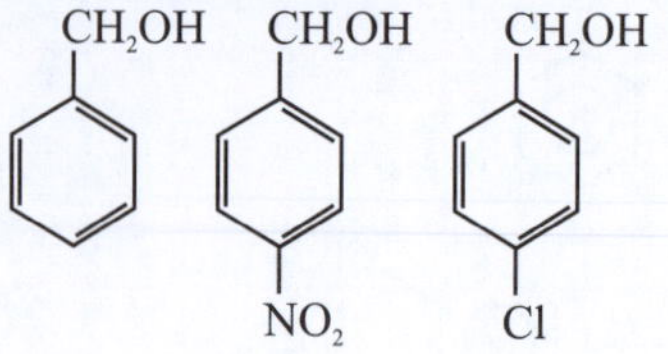

(I) (II) (III)

   1) I < II < III 2) II < I < III
   3) II < III < I 4) III < II < I

16. Arrange the following compounds in increasing order of boiling point:
   Propan-1-ol, butan-1-ol, butan-2-ol, pentan-1-ol
   1) Propan-l-ol, butan-2-ol, butan-1-ol, pentan-1-ol
   2) Pentan-1-ol, butan-1-ol, butan-2-ol, propan-l-ol
   3) Propan-1-ol, butan-2-ol, pent-1-o1, butan-l-ol
   4) Propan-1-ol, butan-1-ol, butan-2-ol, pentan-1-ol

Scan Me for Detailed Explanations

Kattar NEET

1. Hydroboration of 1-methylcyclopentene using B_2D_6, followed by treatment with alkaline hydrogen peroxide, yields (major product)

1) 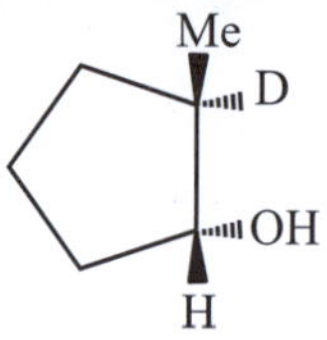

2) 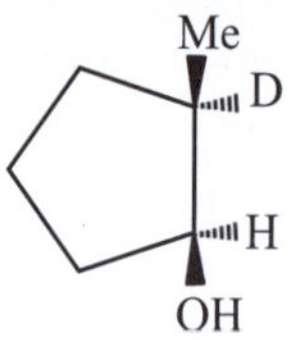

3) 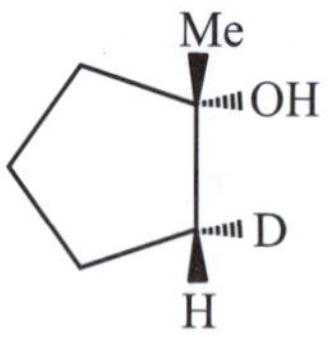

4) 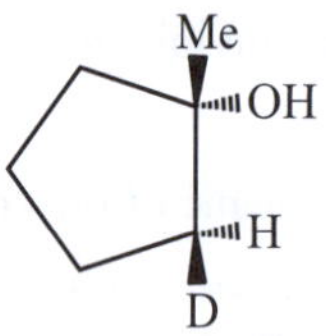

2. An optically active compound X with molecular formula C_8H_{14} undergoes catalytic hydrogenation to yields meso compound, the structure of (X) is:

1) 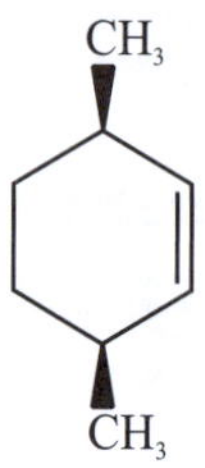

2) 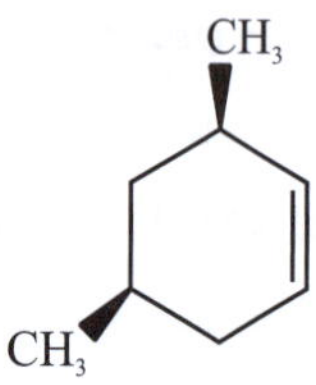

3) 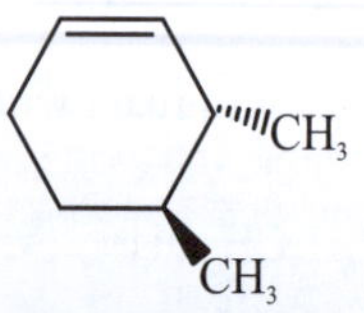

4) 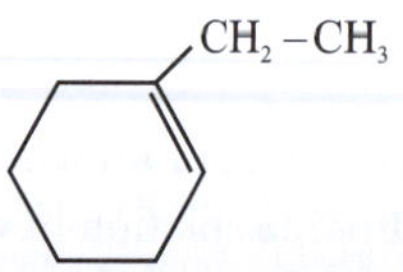

3. 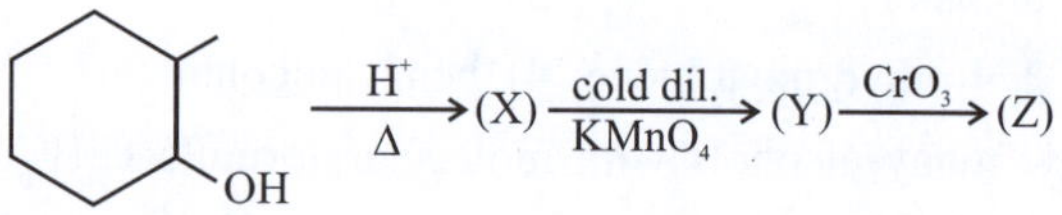

Product (Z) of the reaction is:

1) 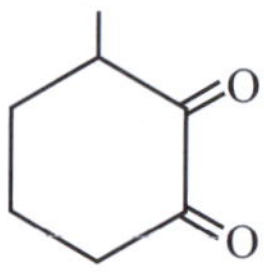

2) 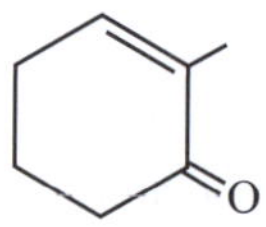

3) 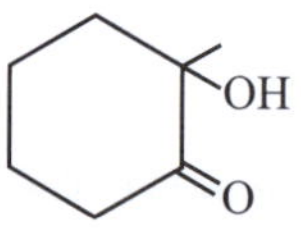

4) 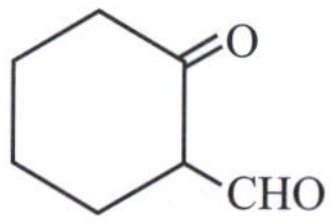

4. 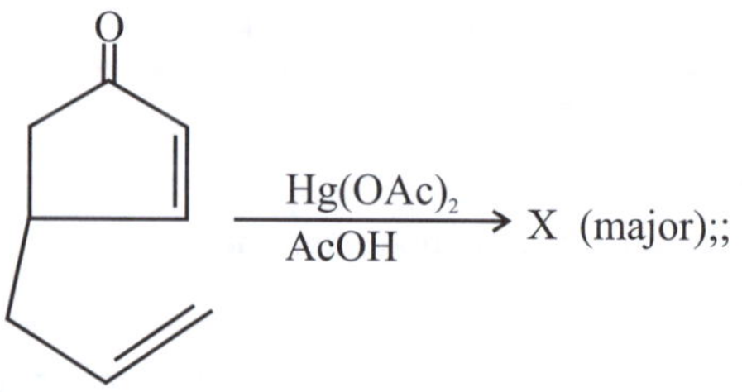

The structure of product (X) is:

1) 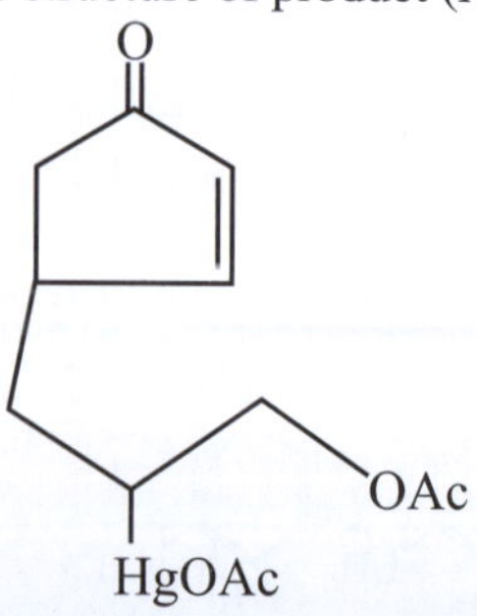

2) 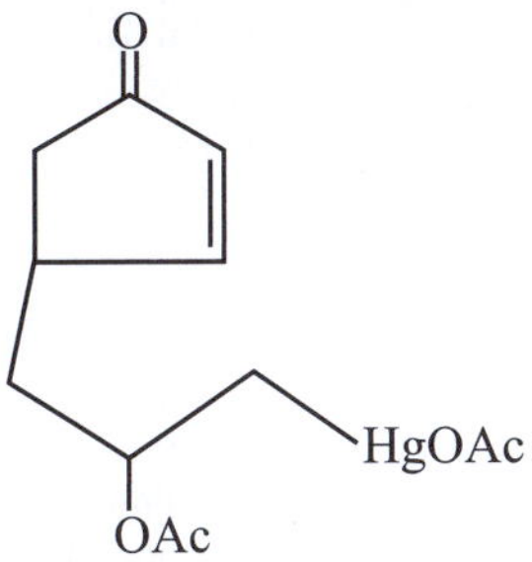

3) 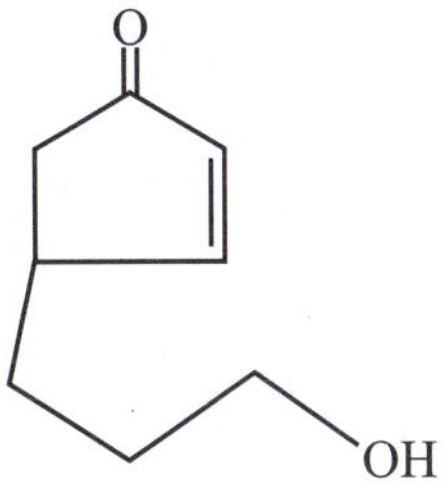

4) 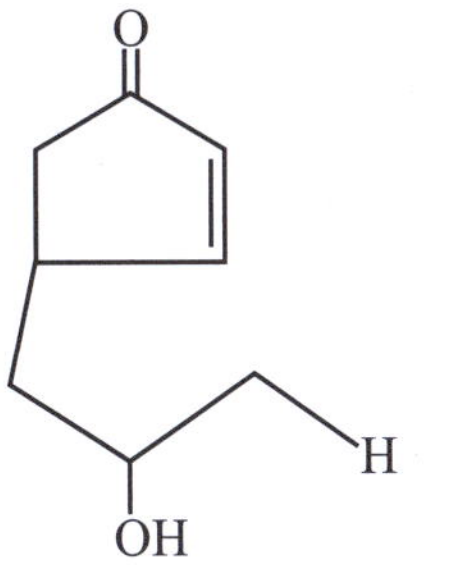

5. 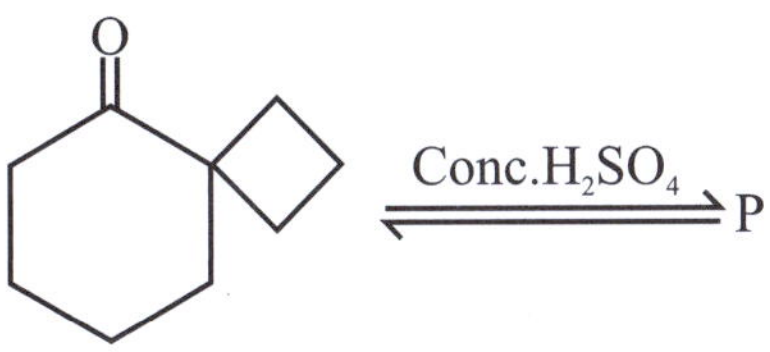

Identify product (P).

1) 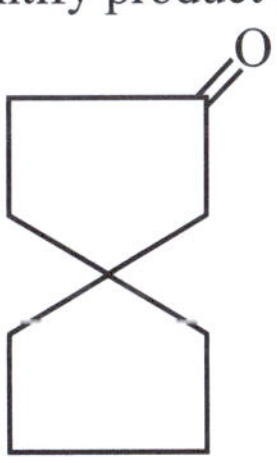

2) 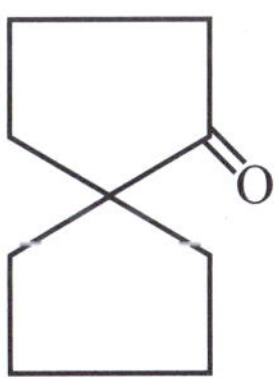

3) 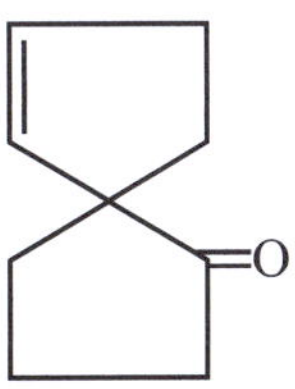

4) 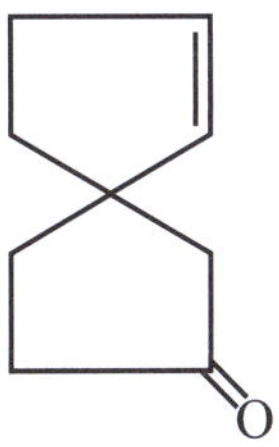

6. Energy profile diagram for dehydration of 2-butanol using conc. H_2SO_4 is given below:

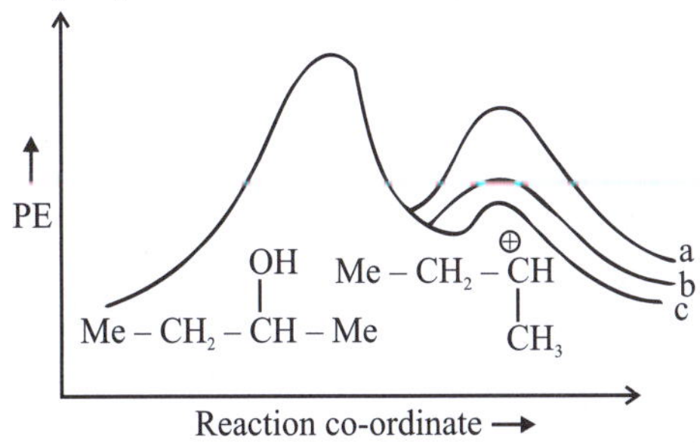

Product (b) of above reaction is:

1) 1-butene
2) cis-2-butene
3) trans-2-butene
4) iso-butene

7. 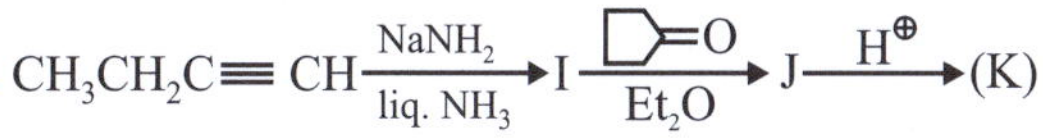

Product (K) of the above reaction is

1) 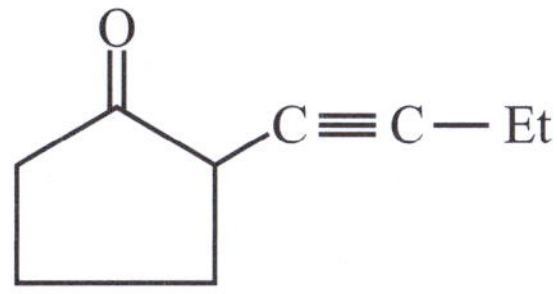

2) 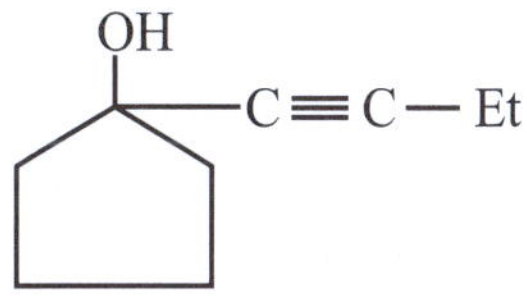

3) 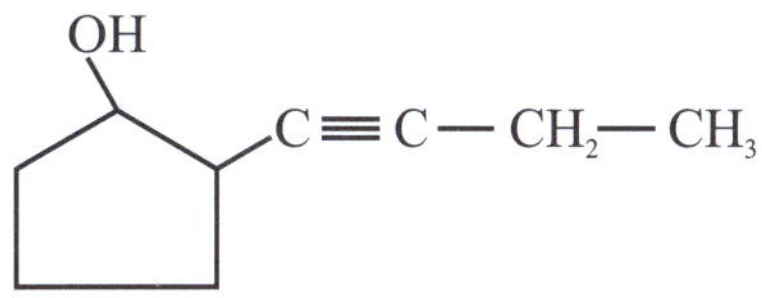

4) 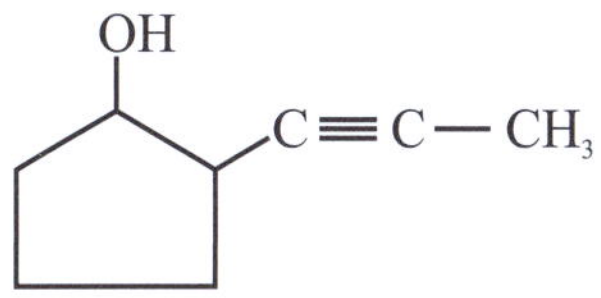

8. 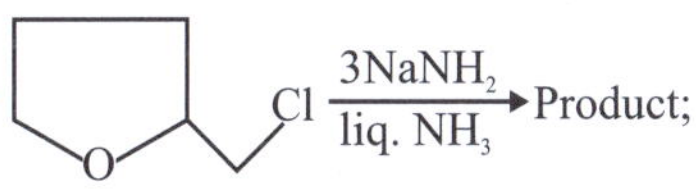

The product is

1) 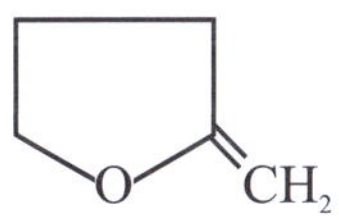

2) $HC\equiv C-(CH_2)_3ONa$

3) $NaC\equiv C-(CH_2)_3ONa$

4) $H-C\equiv C-(CH_2)_3OH$

9. In the given pair of alcohols, in which pair first alcohol is less reactive than second towards hydrogen bromide?

1) 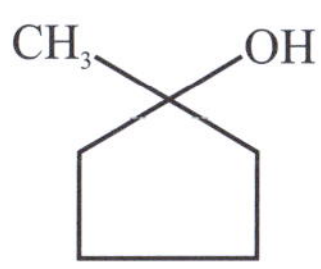 and 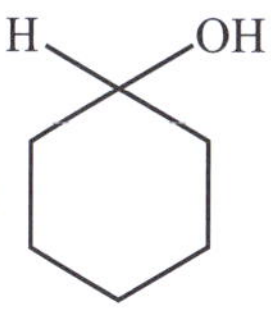

2) 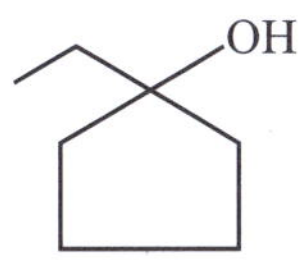 and 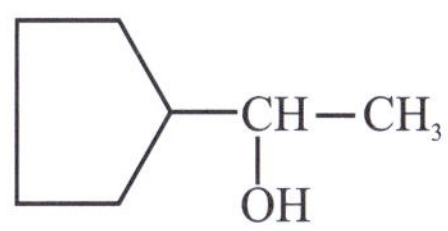

3) $CH_3CH(OH)CH_2CH_3$ and $CH_3CH_2CH(CH_3)CH_2OH$

4) $CH_3CH(OH)CH_2CH_3$ and $(CH_3)_2C(OH)CH_2CH_3$

10. 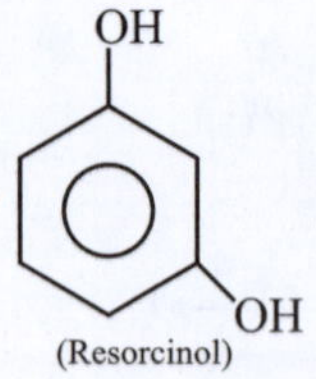 $\xrightarrow{I_2 + NaOH} CHI_3 + (A)$; (Iodoform)

(Resorcinol)

Compound (A) is

1) 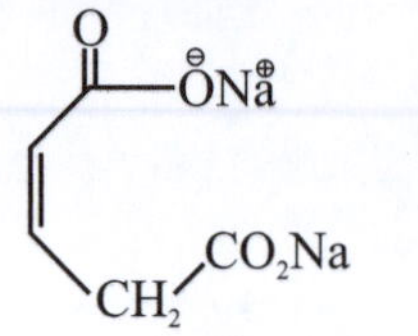

2) $H-O-\overset{O}{\overset{\|}{C}}-(CH_2)_2-\overset{O}{\overset{\|}{C}}-OH$

3) $CH_3-CH_2-CH_2-CH_2-\overset{O}{\overset{\|}{C}}-OH$

4) $CH_3-(CH_2)_3-\overset{O}{\overset{\|}{C}}-ONa$

11. Which of the following reduction methods is not suitable for preparing an alcohol?

1) $CH_3COOC_2H_5 + NaBH_4 \rightarrow$
2) $CH_3COOC_2H_5 + Na/EtOH \rightarrow$
3) $CH_3CH_2COCl + LiAlH_4 \rightarrow$
4) $CH_3COOH + H_2 \xrightarrow{Ni}$

12. Mark the correct increasing order of reactivity of the following compounds with HBr/HCl. (Visualise S_N1 reaction)

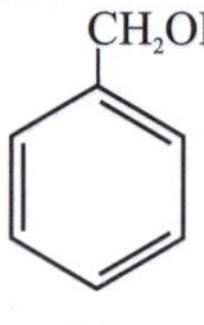
(a)

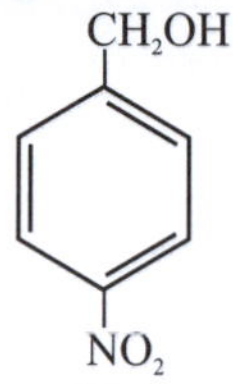
(b)

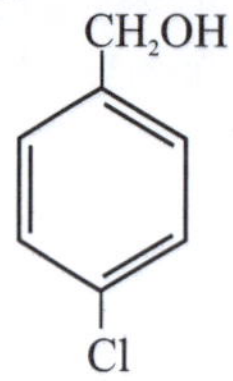
(c)

1) a < b < c　　2) b < a < c
3) b < c < a　　4) c < b < a

13. Arrange the following compounds in increasing order of boiling point. Propan-1-ol, butan-1-ol, butan-2-ol, pentan-1-ol.

1) Propan-1-ol, butan-2-ol, butan-1-ol, pentan-1-ol
2) Propan-1-ol, butan-1-ol, butan-2-ol, pentan-1-ol
3) Pentan-1-ol, butan-2-ol, butan-1-ol, propan-1-ol
4) Pentan-1-ol, butan-1-ol, butan-2-ol, propan-1-ol

14. The structure of the compound that gives tribromo derivative on treatment with bromine water is:

1) 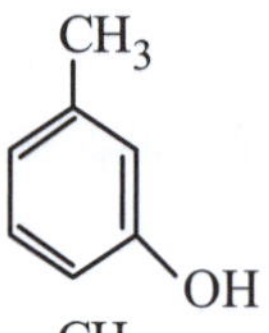

2) 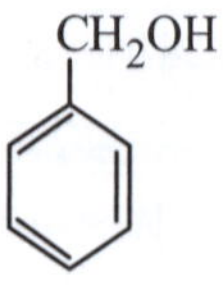

3) 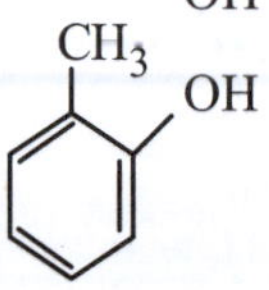

4) 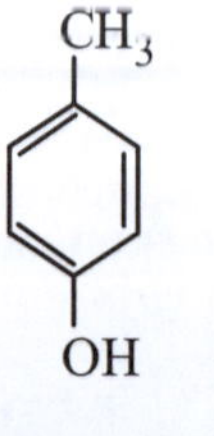

15. Number of isomeric alcohols of molecular formula which give positive iodoform test is (only structural isomers).

1) three　　2) four
3) five　　4) two

16. In the following reactions,

I. $CH_3\underset{}{\overset{CH_3}{\overset{|}{C}}}HCHCH_3 \xrightarrow{H^+/heat} \underset{\text{[Major product]}}{A} + \underset{\text{[Minor product]}}{B}$

II. $A \xrightarrow[\text{in absence of peroxide}]{\text{HBr, dark}} \underset{\text{[Major product]}}{C} + \underset{\text{[Minor product]}}{D}$

the major product (A) and (C) are respectively:

1) $CH_2=\overset{CH_3}{\overset{|}{C}}CH_2CH_3$ and $\overset{CH_3}{\overset{|}{C}H_2}\underset{Br}{\underset{|}{C}}HCH_2CH_3$

2) $CH_3\overset{CH_3}{\overset{|}{C}}=CHCH_3$ and $CH_3\underset{Br}{\underset{|}{\overset{CH_3}{\overset{|}{C}}}}CH_2CH_3$

3) $CH_2\overset{CH_3}{\overset{|}{C}}=CHCH_3$ and $CH_3\overset{CH_3}{\overset{|}{C}}H\underset{Br}{\underset{|}{C}}HCH_3$

4) $CH_3\overset{CH_3}{\overset{|}{C}}=CHCH_3$ and $CH_3\overset{CH_3}{\overset{|}{C}}HCHCH_2Br$

17. An optically active alcohol of formula produced the following compound when refluxed with

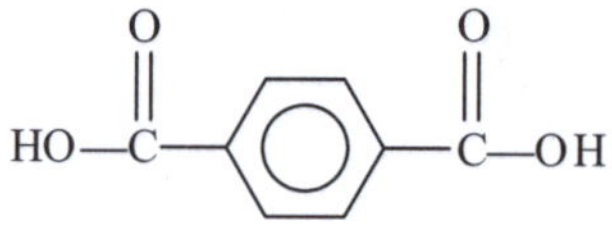

The original compound showed these properties also

$C_9H_{12}O_2$ (A) $\xrightarrow{Na}$ H_2 liberated
$C_9H_{12}O_2$ (A) $\xrightarrow{Br_2}$ No rapid reaction
$C_9H_{12}O_2$ (A) $\xrightarrow[\text{cool}]{CrO_3/H^+}$ $C_9H_8O_3$

The structure of compound (A) is?

1) 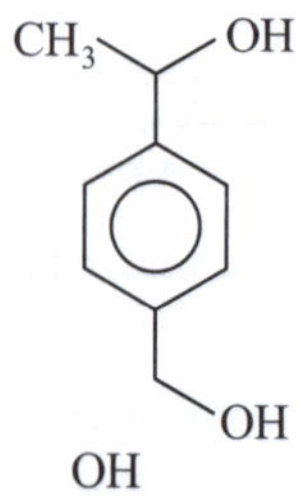

2) 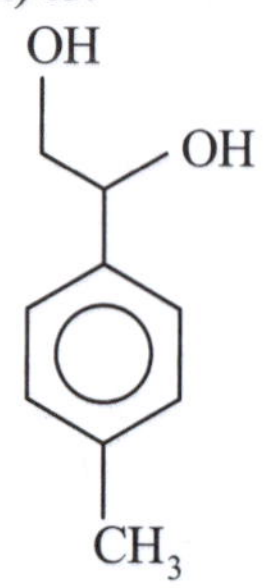

3) 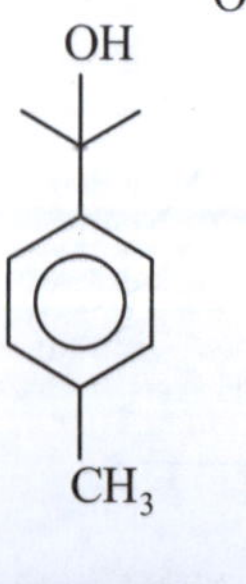

4) Both (a) and (b)

18. 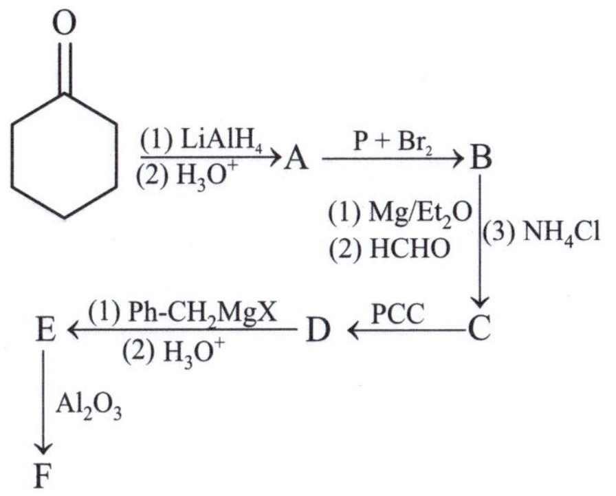

The product (F) is

1) 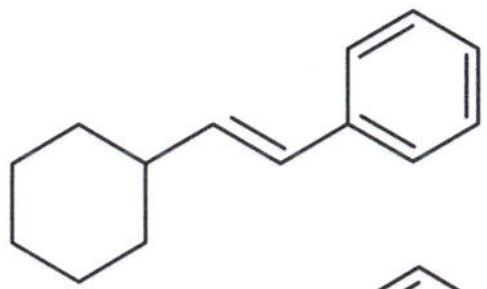

2) 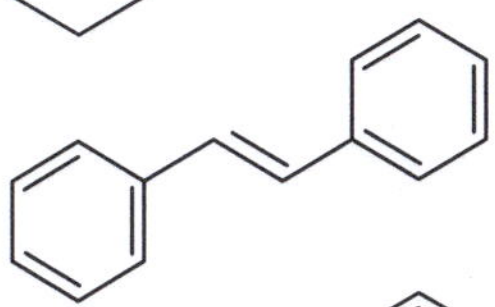

3) 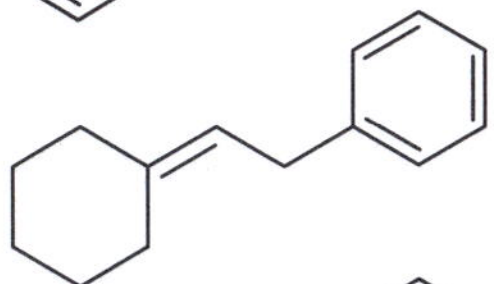

4) 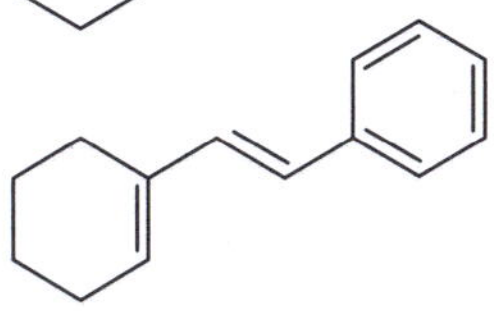

19. A chiral $C_5H_{10}O$ alcohol is reduced by catalytic hydrogenation to an achiral alcohol $C_5H_{12}O$. The original alcohol is oxidized by MnO_2. activated to an achiral carbonyl compound (C_5H_8O) . Which of the following might be the chiral alcohol?

1) 1-penten-3-ol 2) 4-penten-2-ol
3) 3-methyl-2-penten-l-ol 4) 2-methyl-2-buten-l-ol

20. Match the following:

	List-I		List-II
I.	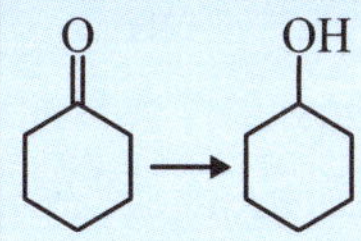	P.	H_2/Ni
II.	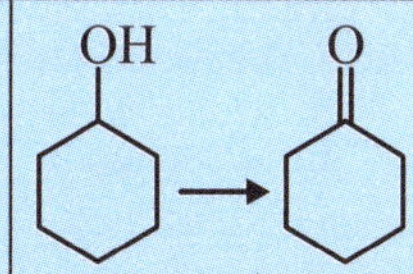	Q.	PCC
III.	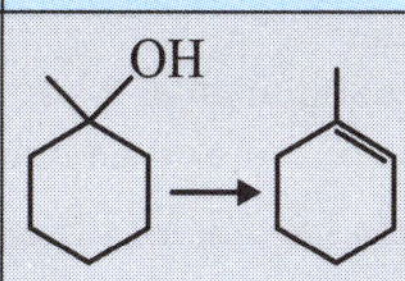	R.	H_3PO_4/Δ
IV.	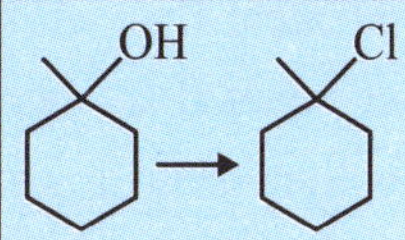	S.	Conc. HCl

1) I-R; II-P; III-Q; IV-S 2) I-P; II-Q; III-R; IV-S
3) I-P; II-R; III-S; IV-Q 4) I-Q; II-P; III-R; IV-S

NEET & AIIMS Past Year MCQs

1. Given below are two statements:

Statement-I: Propene on treatment with diborane gives an addition product with the formula $((CH_3)_2-CH)_3B$.

Statement-II: Oxidation of $((CH_3)_2-CH)_3B$ with hydrogen peroxide in presence of NaOH gives propan-2ol.

In the light of the above statements, choose the most appropriate answer from the options given below: **(2024 Re)**

1) Statement-I is correct but Statement-II is incorrect
2) Statement-I is incorrect but Statement-II is correct
3) Both Statement-I and Statement-II are correct
4) Both Statement-I and Statement-II are incorrect

2. Which one of the following alcohols reacts instantaneously with Lucas reagent? **(2024)**

1) $CH_3-CH(CH_3)-CH_2OH$ 2) $CH_3-C(CH_3)_2-OH$

3) $CH_3-CH_2-CH_2-CH_2O$ 4) $CH_3-CH_2-CH(CH_3)-OH$

3. The products A and B obtained in the following reactions, respectively, are **(2024)**

$3ROH + PCl_3 \rightarrow 3RCl + A$

$ROH + PCl_5 \rightarrow RCl + HCl + B$

1) H_3PO_4 and $POCl_3$ 2) H_3PO_3 and $POCl_3$
3) $POCl_3$ and H_3PO_3 4) $POCl_3$ and H_3PO_4

4. Consider the following reaction

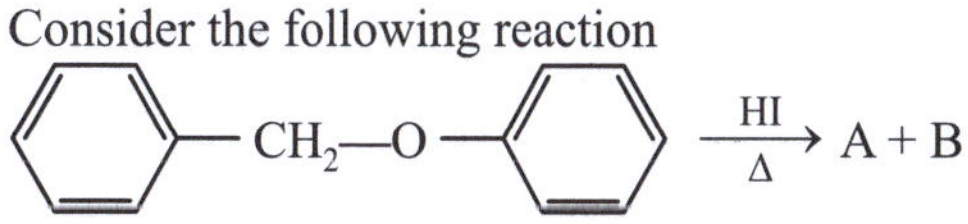

Identify products A and B **(2023)**

1) A = 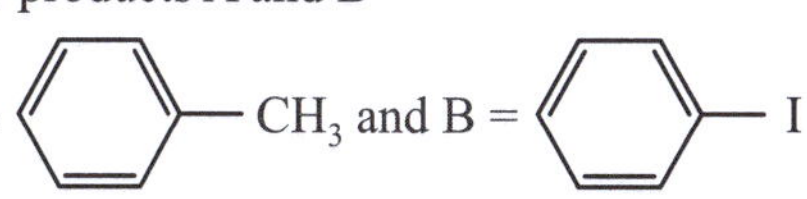

2) A = 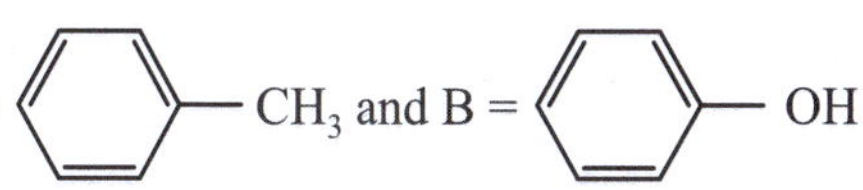

3) A = 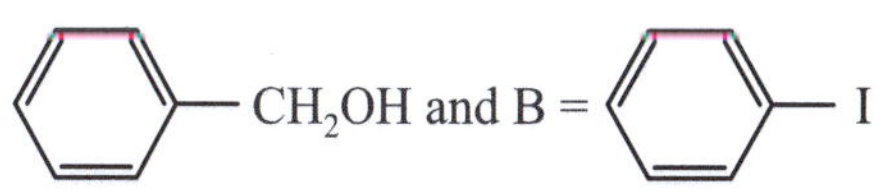

4) A = 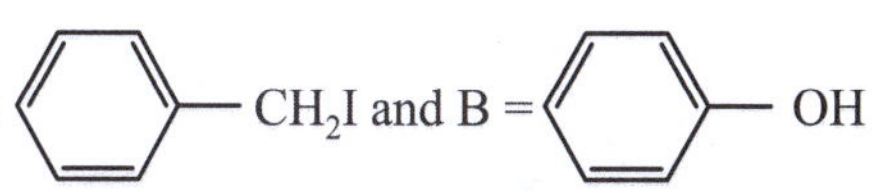

5. Which amongst the following will be most readily dehydrated under acidic conditions? **(2023)**

1) 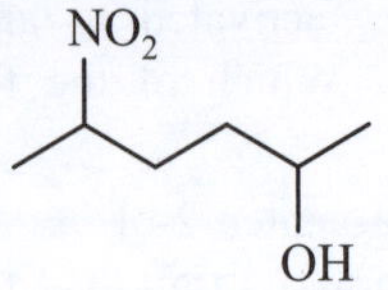

2) 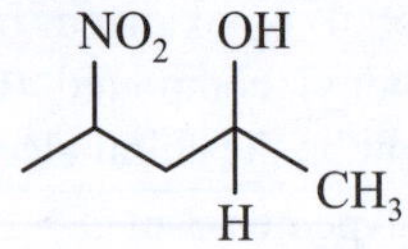

3) 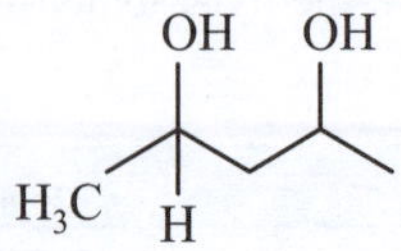

4) 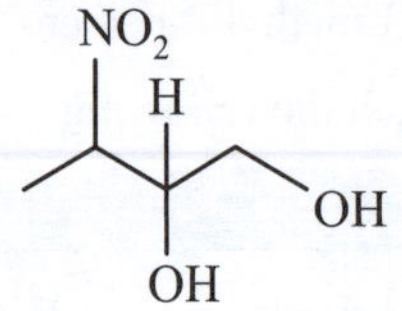

6. Given below are two statements: **(2022)**

Statement-I: The acidic strength of monosubstituted nitrophenol is higher than phenol because of electron withdrawing nitro group.

Statement-II: o-nitrophenol, m-nitrophenol and p-nitrophenol will have same acidic strength as they have one nitro group attached to the phenolic ring.

In the light of the above statements, choose the most appropriate answer from the options given below:

1) Statement-I is incorrect but Statement-II is correct.
2) Both Statement-I and Statement-II are correct.
3) Both Statement-I and Statement-II are incorrect.
4) Statement-I is correct but Statement-II is incorrect.

7. Given below are two statements: **(2022)**

Statement-I: In Lucas test, primary, secondary and tertiary alcohols are distinguished on the basis of their reactivity with conc. $HCl + ZnCl_2$, known as Lucas Reagent.

Statement-II: Primary alcohols are most reactive and immediately produce turbidity at room temperature on reaction with Lucas Reagent.

In the light of the above statements, choose the most appropriate answer from the options given below:

1) Statement-I is incorrect but Statement-II is correct.
2) Both Statement-I and Statement-II are correct.
3) Both Statement-I and Statement-II are incorrect.
4) Statement-I is correct but Statement-II is incorrect.

8. What is the IUPAC name of the organic compound formed in the following chemical reaction? **(2021)**

$$\text{Acetone} \xrightarrow[\text{(ii)} H_2O,\ H^+]{\text{(i)} C_2H_5MgBr,\ \text{dry ether}} \text{Product}$$

1) Pentan-2-ol
2) Pentan-3-ol
3) 2-methyl butan-2-ol
4) 2-methyl propan-2-ol

9. The product formed in the following chemical reaction is: **(2021)**

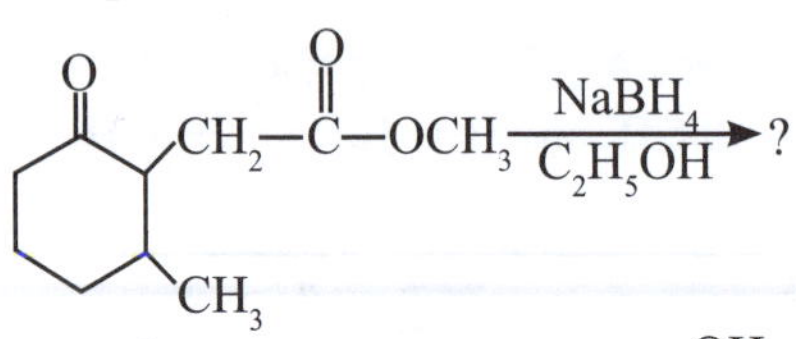

1) 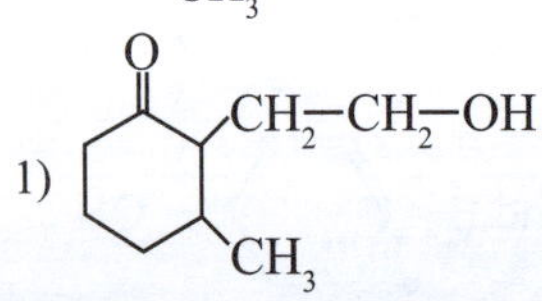

2) 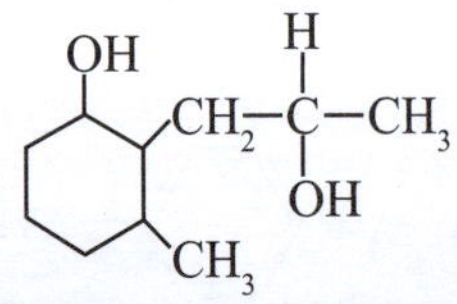

3) 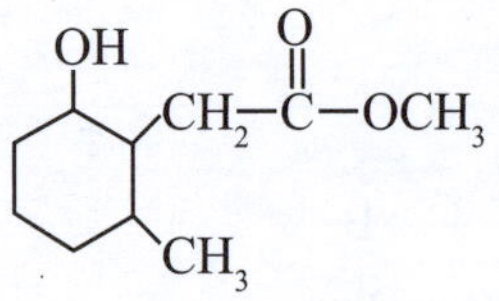

4) 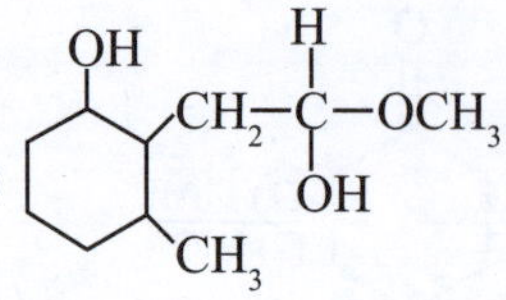

10. What are the suitable rectant for the following ether synthesis $CH_3-C(CH_3)_2-O-CH_2-Ph$ **(AIIMS 2018)**

1) $CH_3-C(CH_3)_2-\bar{O}\overset{\oplus}{Na} + Ph-CH_2-Cl$

2) $Ph-CH(Cl)-CH_2-CH_3 + CH_3-CH(\bar{O}\overset{\oplus}{Na})-CH_3$

3) $CH_3-C(CH_3)_2-Cl + PhCH_2\overset{\ominus}{O}\overset{\oplus}{Na}$

4) $CH_3-C(CH_3)_2-\bar{O}\overset{\oplus}{Na} + CH_3-Cl$

11. Phenol $\xrightarrow[H^+]{CHCl_3/NaOH}$ Salicylaldehyde

The above reaction is known as: **(AIIMS 2017)**

1) Reimer– Tiemann reaction
2) Bucherer reaction
3) Gattermann synthesis
4) Perkin reaction

12. **Assertion:** Phenol forms 2, 4, 6-tribromophenol on treatment with Br_2– water at 273 K.

Reason: Phenol is o, p-directing group. **(AIIMS 2017)**

1) If both Assertion & Reason are True & the Reason is a correct explanation of the Assertion.
2) If both Assertion & Reason are True but Reason is not a correct explanation of the Assertion.
3) If Assertion is True but the Reason is False.
4) If both Assertion & Reason are False.

13. 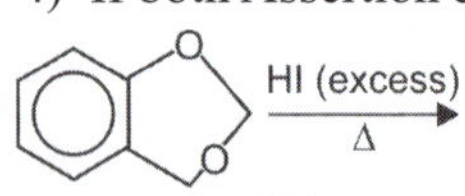 **(AIIMS 2013)**

1) 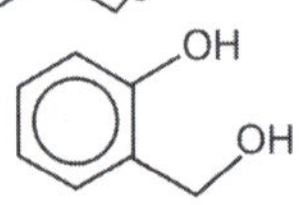 + CH_2I_2

2) 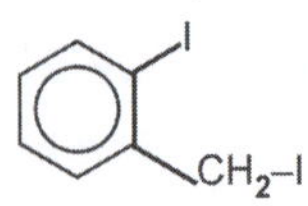 + HCHO

3) 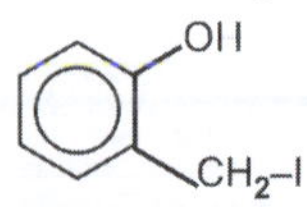 + $I-CH_2-OH$

4) 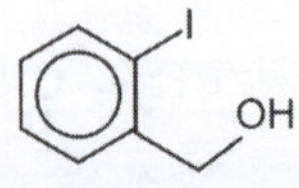 + $I_2 + CH_3OH$

Answer Key

NCERT Topic-wise MCQs

1	2	3	4	5	6	7	8	9	10	11	12	13	14	15	16	17
2	1	3	3	1	3	2	2	2	3	1	4	3	4	1	1	2
18	19	20	21	22	23	24	25	26	27	28	29	30	31	32	33	34
1	1	1	2	3	4	1	3	1	1	4	1	1	2	4	2	2
35	36	37	38	39	40	41	42	43	44	45	46	47	48	49	50	51
3	2	2	3	2	2	2	2	2	3	1	3	2	4	3	4	3
52	53	54	55	56	57	58	59	60	61	62	63	64	65	66	67	68
2	1	2	2	2	4	3	4	3	3	2	3	4	1	3	3	1
69	70	71	72	73	74											
3	4	4	2	4	4											

High Order Time Intensive MCQs

1	2	3	4	5	6	7	8	9	10	11	12	13	14	15	16	17
1	2	3	3	3	4	4	1	4	4	1	1	3	1	2	2	1
18	19	20	21	22	23	24	25	26	27	28	29	30	31	32	33	34
4	1	1	3	4	3	3	4	2	4	3	2	1	3	2	1	4
35	36	37	38	39	40											
1	3	4	2	4	4											

Match the Columns MCQs

1	2	3	4
1	3	4	2

Correct & Incorrect MCQs

1	2	3	4	5	6	7	8	9	10
2	3	3	1	1	2	4	2	1	3

Statement Based MCQs

1	2	3	4	5	6	7	8	9	10
1	1	4	3	1	4	3	4	4	1

Assertion & Reason MCQs

1	2	3	4	5	6	7	8	9	10
1	4	4	4	2	3	4	2	2	3

Multi Correct MCQs

1	2	3	4
4	2	2	2

NCERT Exemplar MCQs

1	2	3	4	5	6	7	8	9	10	11	12	13	14	15	16
4	1	3	3	2	3	3	1	3	2	1	2	4	2	3	1

Kattar NEET

1	2	3	4	5	6	7	8	9	10	11	12	13	14	15	16	17
1	2	3	2	2	2	2	3	4	1	1	3	1	1	2	2	4
18	19	20														
1	1	2														

NEET & AIIMS Past Year MCQs

1	2	3	4	5	6	7	8	9	10	11	12	13
2	2	2	4	3	4	4	3	3	1	1	1	2

CHAPTER 9

Aldehydes, Ketones and Carboxylic Acids

Topicwise Number of Questions Analysis (2025-2016)

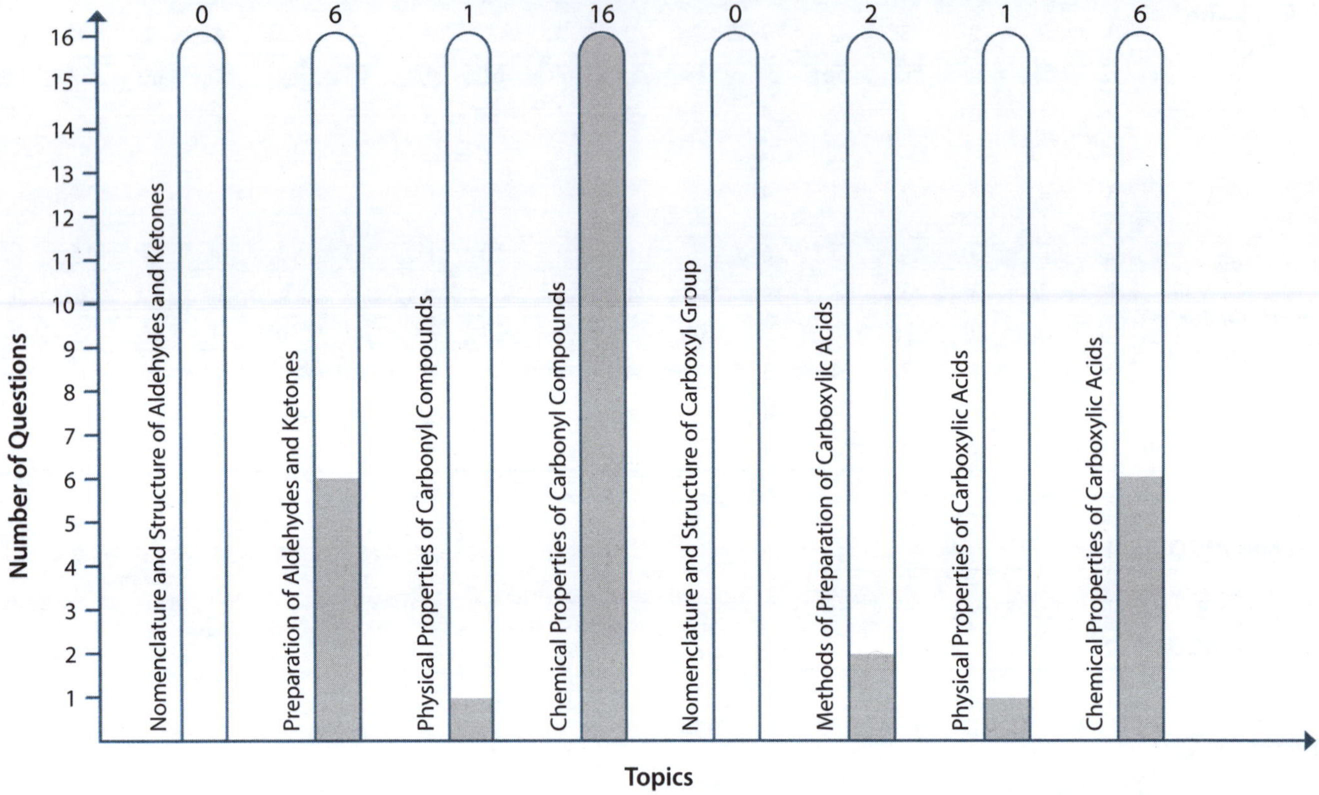

NCERT-Picks

- $C_nH_{2n}O$ is the general formula for aldehyde and ketone.
- By the controlled oxidation of lo alcohols using acidified potassium permanganate or acidified potassium dichromate, aldehydes are prepared.

Preparation of Aldehydes

1. **From acyl chloride (acid chloride) (Rosenmund reduction)**

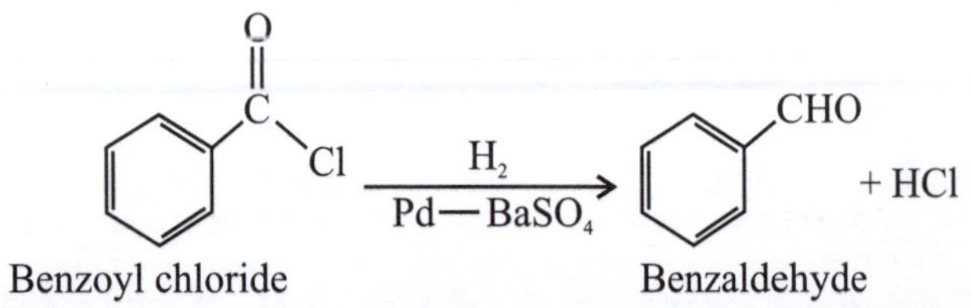

2. **From nitriles and esters (Stephen Reduction)**

$$RCN + SnCl_2 + HCl \longrightarrow RCH{=}NH \xrightarrow{H_3O^+} RCHO$$

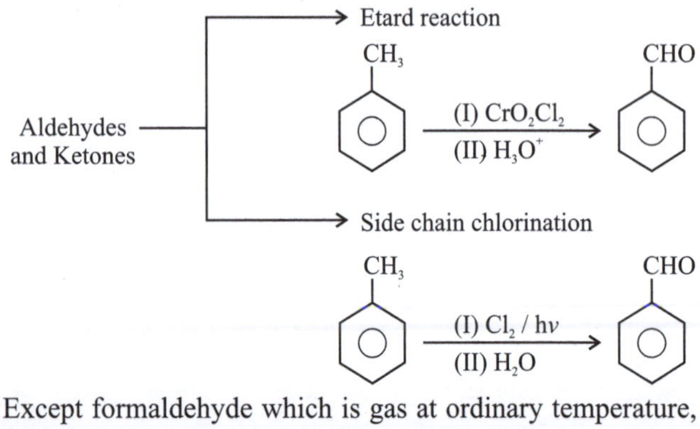

- Except formaldehyde which is gas at ordinary temperature, other lower members of aldehydes and ketone (upto C10) are colourless, volatile liquid.

Preparation of Ketones

1. From acyl chloride

$$2\,R-Mg-X + CdCl_2 \longrightarrow R_2Cd + 2Mg(X)Cl$$

$$2\,R'-\underset{\underset{O}{\|}}{C}-Cl + R_2Cd \longrightarrow 2\,R'-\underset{\underset{O}{\|}}{C}-R + CdCl_2$$

2. From nitriles

$$CH_3-CH_2-C\equiv N + C_6H_5MgBr \xrightarrow{\text{ether}} CH_3CH_2-C(=NMgBr)-C_6H_5 \xrightarrow{H_3O^+} C_2H_5-C(=O)-C_6H_5$$

3. From benzene or substituted benzenes

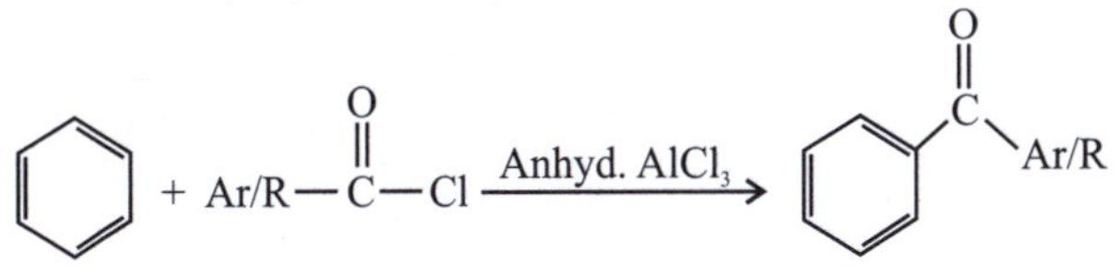

Preparation of Aldehydes and Ketones

Aldehydes and Ketones

- $RCH_2OH + [O] \xrightarrow[H_2SO_4]{K_2Cr_2O_7+} RCHO + H_2O$
- $RCH_2OH \xrightarrow[573K]{Cu} RCHO + H_2\uparrow$
- $R-\overset{\overset{R}{|}}{C}H-OH \xrightarrow[573K]{Cu} R-\overset{\overset{O}{\|}}{C}-R + H_2\uparrow$
- $RCH{=}CHR + O_3 \xrightarrow[196K]{CCl_4}$ 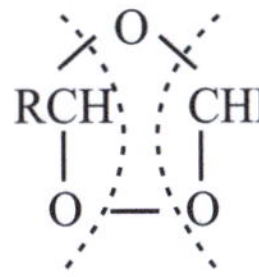 $\xrightarrow[H_2O]{Zn,} 2\,RCHO + ZnO$
- **Rosenmund reduction**

  $RCOCl + H_2 \xrightarrow[\text{boiling xylene}]{Pd-BaSO_4} RCHO + HCl$
- $RCN \xrightarrow[\text{DIBAL–H}]{AlH(i\text{ - }Bu)_2} RCHO$
- $RCN \xrightarrow[\text{(II) ether}]{\text{(I) }SnCl_2 + HCl} RCHO + NH_4Cl$ **(stephen's reaction)**
- $RCOOR \xrightarrow[\text{(II) }H_2O]{\text{(I) DIBAL - H}} RCHO$
- **Gatterman - Koch reaction**

  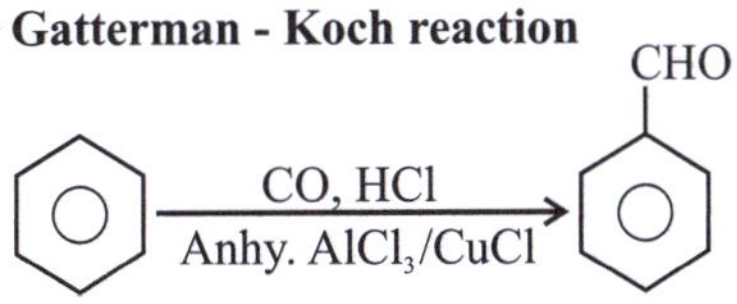
- **Fridel - Craft acylation reaction**

  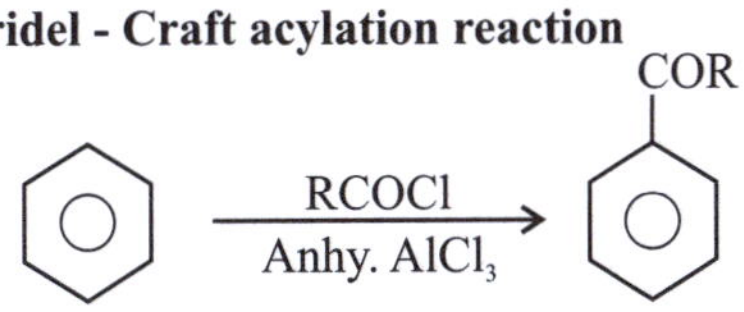
- $-C\equiv C- \xrightarrow{\text{dil. }H_2SO_4} RCHO \text{ or } RCO-R$
- $-C\equiv C- \xrightarrow[H_2O_2\,/\,OH^-]{B_2H_6,\ THF} RCHO \text{ or } RCOR$ (hydroboration of alkynes)

Chemical Reactions of Aldehydes and Ketones

Addition of ammonia and its derivatives

$$>C=O + H_2N-Z \rightleftharpoons \left[>C\begin{matrix} OH \\ NHZ \end{matrix}\right] \rightarrow >C=N-Z+H_2O$$

Some N-substituted derivatives of aldehydes and ketones (>C = N – Z)

Z	Reagent Name	Carbonyl derivative	Product Name
–H	Ammonia	$>C=NH$	Imine
–R	Amine	$>C=NR$	Substituted imine (Schiff's base)
–OH	Hydroxylamine	$>C=N-OH$	Oxime
$-NH_2$	Hydrazine	$>C=N-NH_2$	Hydrazone
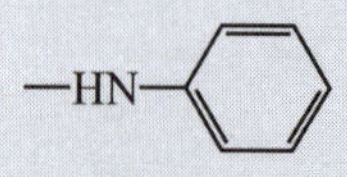	Phenylhydrazine	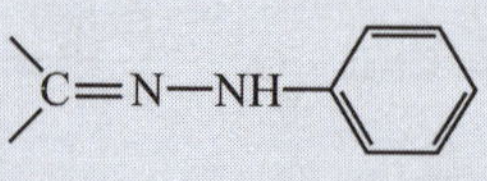	Phenylhydrazone
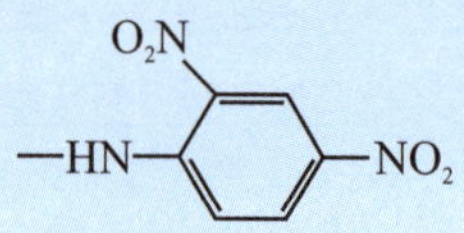	2,4-Dinitrophenylhydrazine	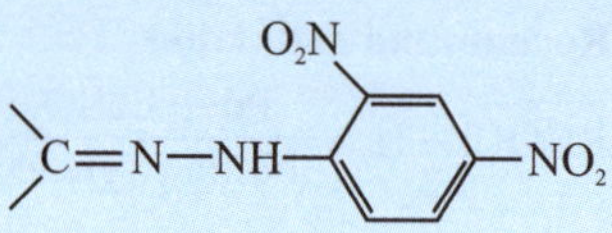	2,4-Dinitrophenylhydrazone
$-NH-\overset{O}{\overset{\|}{C}}-NH_2$	Semicarbazide	$>C=N-NH-\overset{O}{\overset{\|}{C}}-NH_2$	Semicarbazone

- Clemmensen Reaction: $R-\overset{O}{\overset{\|}{C}}-R' \xrightarrow{Zn-Hg/HCl} R-CH_2-R'$
- Wolff-kishner reaction: $R-\overset{O}{\overset{\|}{C}}-R' \xrightarrow{NH_2-NH_2/KOH} R-CH_2-R'$
- Fehling's solution test: Given only by aldehyde: $R-CHO \xrightarrow[+\text{Rochelle Salt}]{2Cu^{2+}, 5OH^-} RCOO^- + Cu_2O\downarrow$

  Red-brown ppt.
- Silver mirror test: given only by aldehyde: $R-CHO \xrightarrow[\text{Tollen test}]{2[Ag(NH_3)_2]^+, OH^-} RCOO^- + 2Ag\downarrow$
- Benedict's solution test: only given by aldehydes. $RCHO \xrightarrow[+\text{Sodium citrate}]{CuO \text{ or } Cu(OH)_2} RCOO^- + Cu_2O$
- Haloform reaction: Given by the compounds having $CH_3-\overset{O}{\overset{\|}{C}}-$ group or group $CH_3-\overset{OH}{\overset{|}{C}H}-$

  $2NaOH + I_2 \rightarrow NaI + NaOI + H_2O$

  $R-\overset{O}{\overset{\|}{C}}-CH_3 + 3NaOI \rightarrow RCOONa + CHI_3\downarrow + 2NaOH$

  (Yellow ppt.)

- Aldol condensation:

  At least one α - hydrogen atoms should be present in aldehydes and ketones.

$$2R-\overset{\alpha}{CH_2}-\overset{O}{\overset{\|}{C}}-H \xrightarrow{\text{dil. NaOH}} R-CH_2-\overset{OH}{\overset{|}{C}H}-\overset{R}{\overset{|}{C}H}-\overset{O}{\overset{\|}{C}}-H \xrightarrow[\Delta]{-H_2O} R-CH=\overset{R}{\overset{|}{C}}-\overset{O}{\overset{\|}{C}}-H$$

- Canizzaro reaction: Only those aldehydes which do not have an α - hydrogen atom.

$$\underset{\text{formaldehyde}}{HCHO + HCHO} \xrightarrow{\text{Conc.KOH}} \underset{\text{methanol}}{CH_3OH} + \underset{\text{potassium formate}}{HCOOK}$$

- Relative basic strength of leaving group:

  $NH_2^- > R'O^- > R'COO^- > Cl^-$

- Reactivity: $\underset{\text{Acid chloride}}{R-\overset{O}{\overset{\|}{C}}-Cl} > \underset{\text{Anhydride}}{R-\overset{O}{\overset{\|}{C}}-O-\overset{O}{\overset{\|}{C}}-R'} > \underset{\text{Ester}}{R-\overset{O}{\overset{\|}{C}}-OR'} > \underset{\text{Amide}}{R-\overset{O}{\overset{\|}{C}}-NH_2}$

Methods of Preparation of Carboxylic Acids

RCOOH —

- $RCH_2OH \xrightarrow[(2)\ H_3O^+]{(1)\ \text{Alk. } KMnO_4} RCOOH$
- $RCOCl \xrightarrow{H_2O} RCOOH + Cl^-$
- $RCOOR \xrightarrow{H_3O^+} RCOOH + ROH$

- From alkylbenzenes

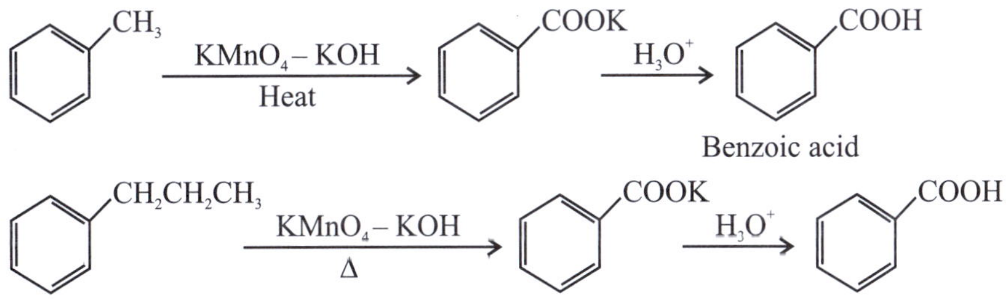

Benzoic acid

Benzoic acid

- From nitriles and amides

$$R-CN \xrightarrow[H_2O]{\overset{+}{H} \text{ or } \overset{-}{OH}} R-\overset{O}{\overset{\|}{C}}-NH_2 \xrightarrow[\Delta]{\overset{+}{H} \text{ or } \overset{-}{OH}} RCOOH$$

$$\underset{\text{Ethanamide}}{CH_3CONH_2} \xrightarrow[\Delta]{H_3O^+} \underset{\text{Ethanoic acid}}{CH_3COOH} + NH_3$$

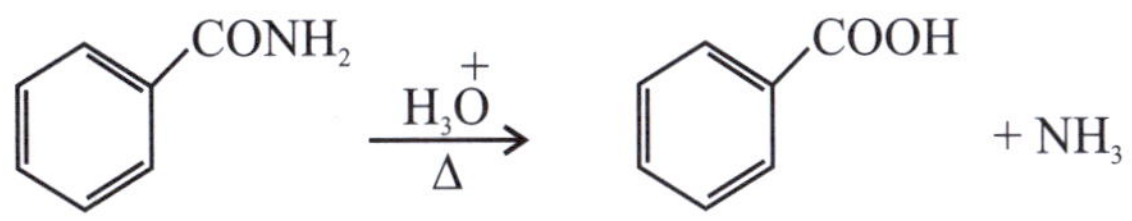

Benzamide   Benzoic acid

- From Grignard reagents

$$R-Mg-X + O=C=O \xrightarrow{\text{Dry ether}} R-C\begin{matrix}\nearrow O \\ \searrow O^-MgX^+\end{matrix} \xrightarrow{H_3O^+} RCOOH$$

Chemical Properties of Carboxylic Acids

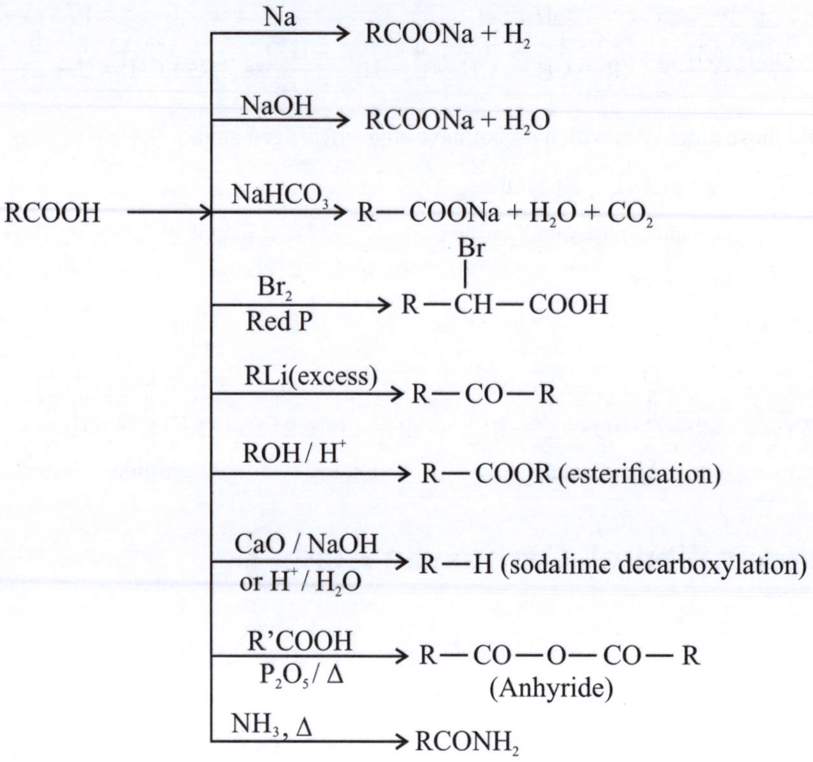

Important Mechanisms

Basic-catalyzed hydrolysis of a nitrile

Attack by hydroxide ion and reprotonation

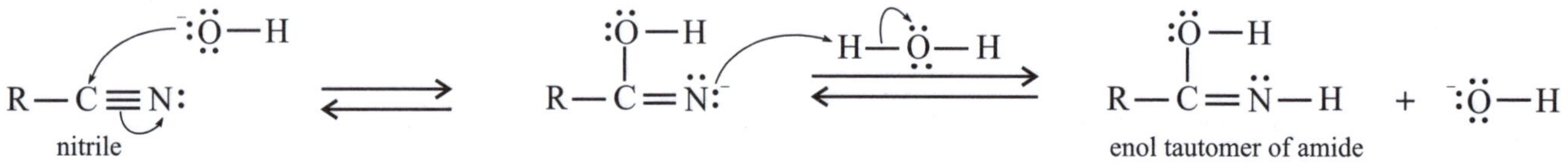

Removal and replacement of a proton (tautomerization)

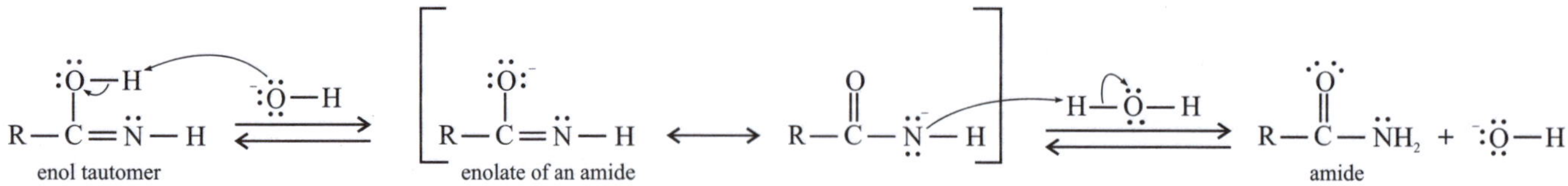

Basic hydrolysis of an amide

Step 1: Additon of the nucleophile.

Step 2: Elimination of the leaving group

Step 3: Proton transfer.

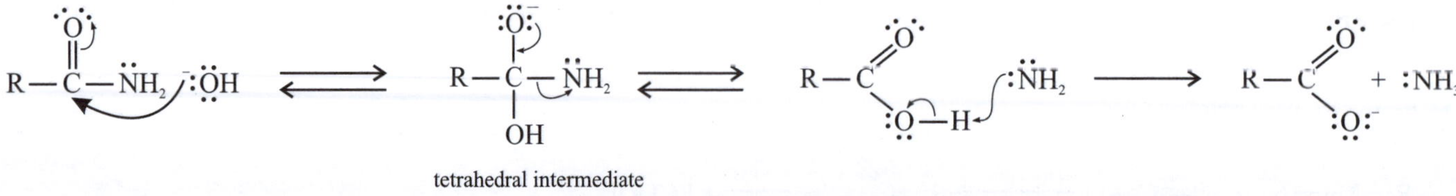

Base-catalyzed aldol condensation

Step 1: Additon of the enolate. *Step 2*: Protonation.

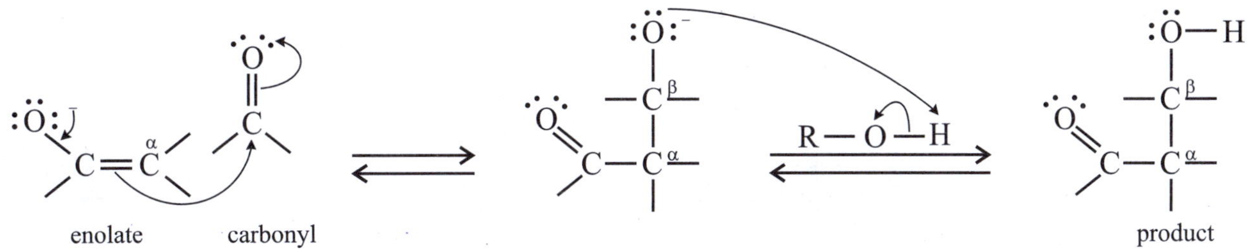

Acid-catalyzed aldol condensation

Step 1: Addition of the enol to the protonated carbonyl. *Step 2*: Deprotonation.

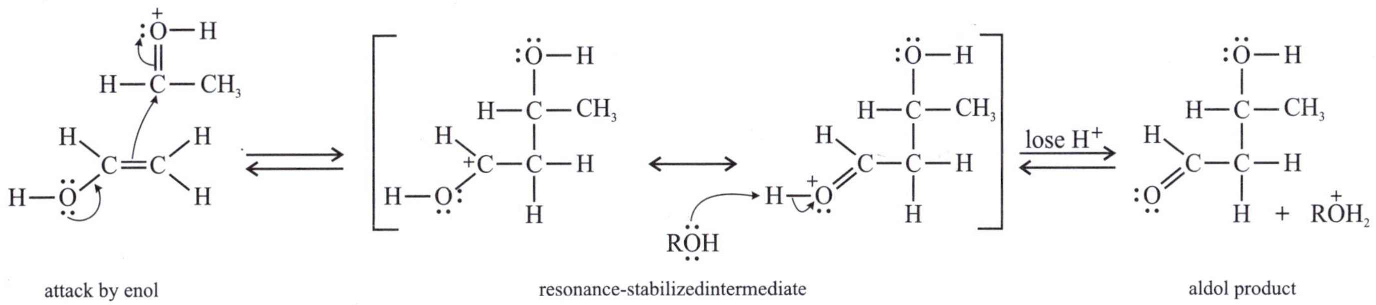

Final steps of the haloform reaction

Step 1: Additon of the nucleophile. *Step 2*: Elimination of the leaving group. *Step 3*: Proton transfer.

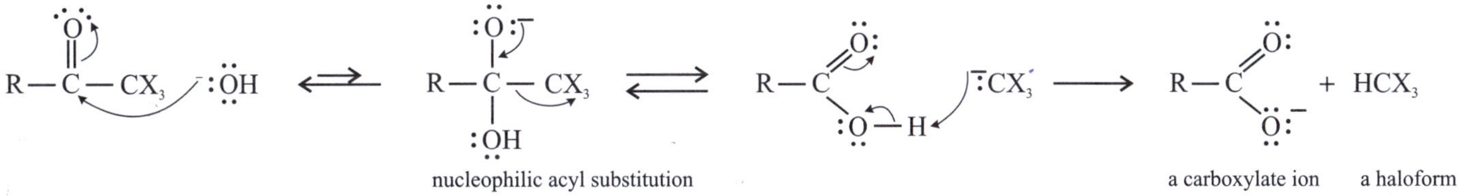

Conversion of an acid chloride to an ester

Step 1: Additon of the nucleophile. *Step 2*: Elimination of the leaving group *Step 3*: Loss of a proton.

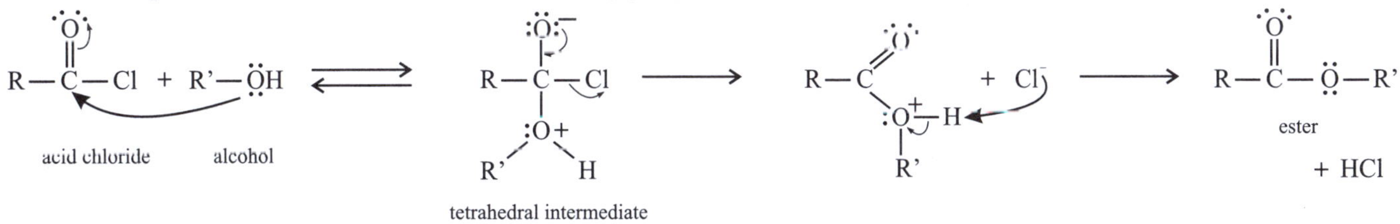

Conversion of an acid anhydride to an ester

Step 1: Additon of the nucleophile. *Step 2*: Elimination of the leaving group *Step 3*: Loss of a proton.

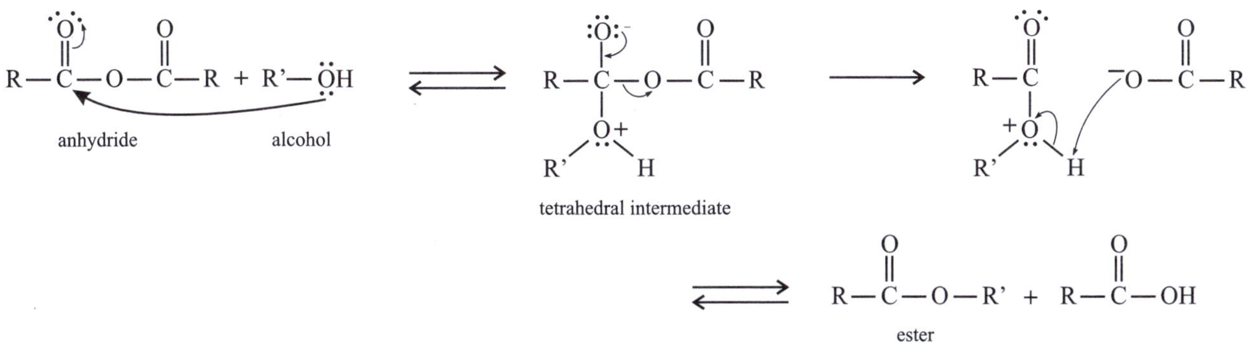

Reaction of an ester with 2 moles of a grignard reagent

Step 1: Additon of the Grignard. *Step 2*: Elimination of alkoxide *Step 3*: Additon of another Grignard.

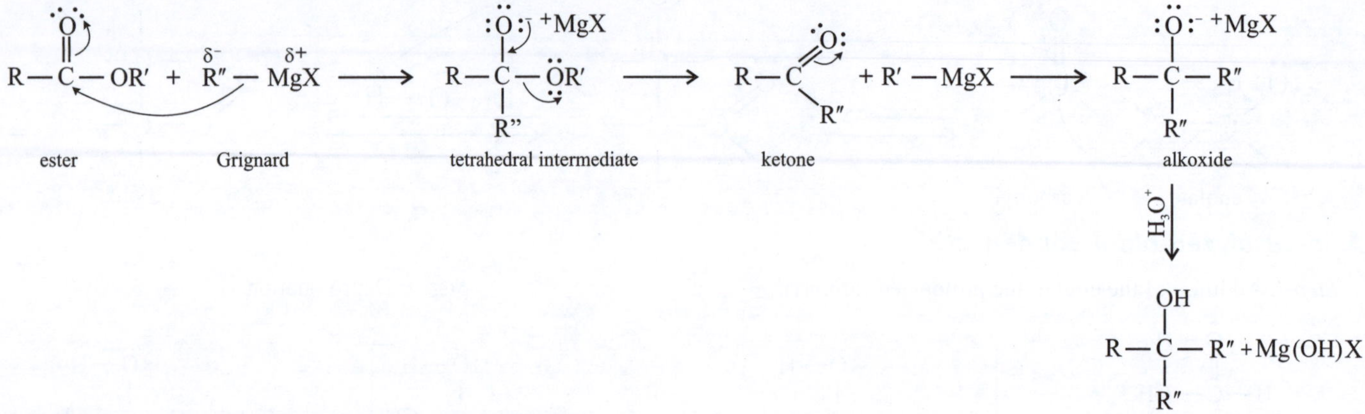

Reduction of esters by lithium aluminium hydride

Step 1: Additon of the nucleophile (hydride). *Step 2*: Elimination of alkoxide

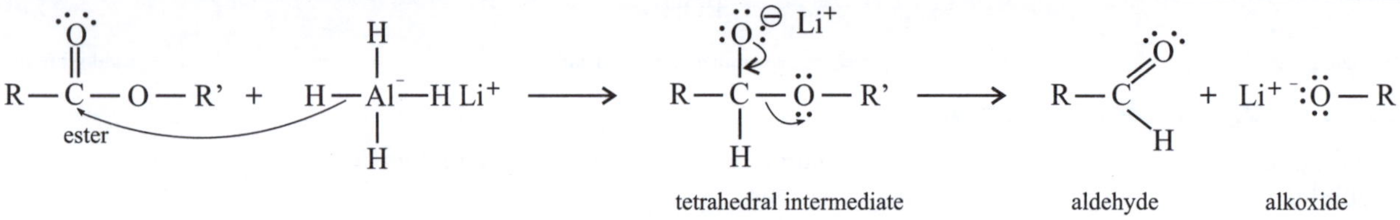

Step 3: Addition of a second hydride ion.

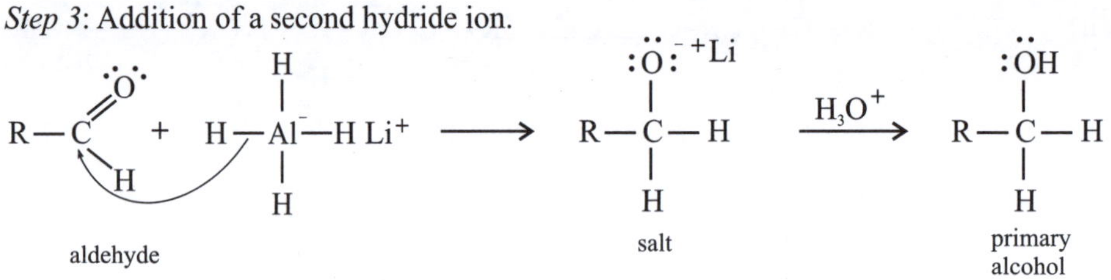

MCQs

NCERT Topic-wise MCQs

Nomenclature and Structure of Carbonyl Group

1. IUPAC name of α-hydroxybutyraldehyde is:
   1) 1-Hydroxy butanal 2) 2-Hydroxy butanol
   3) 2-Hydroxy butanal 4) 2-Hydroxy butyraldehyde

2. Which of the following is not a monovalent functional group?
   1) Aldehydic 2) Ketonic
   3) Carboxylic 4) Hydroxy

3. The formula $C_5H_{10}O$ represents:
   1) An aldehyde and a symmetrical ketone only
   2) An aldehyde and an unsymmetrical ketone only
   3) A symmetrical and an unsymmetrical ketone only
   4) An aldehyde and an symmetrical as well as an unsymmetrical ketone

Preparation of Aldehydes and Ketones

4. $A \xrightarrow[(ii)H_3O^+]{(i)CH_3MgBr} R \xrightarrow[(ii)\Delta]{(i)Dil.NaOH}$ 4 - methylpent - 3 -en - 2 - one. What is 'A'?
   1) Propanone 2) Ethanamine
   3) Ethane nitrile 4) Ethanal

5. Reaction by which benzaldehyde cannot be prepared is:
   1) 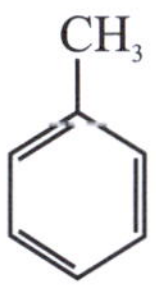 $+ CrO_2Cl_2$ in CS_2 followed by H_3O^+
   2) 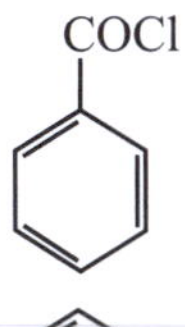 $+ H_2$ in the presence of Pd - $BaSO_4$
   3)  + CO + HCl in the presence of anhydrous $AlCl_3$
   4) 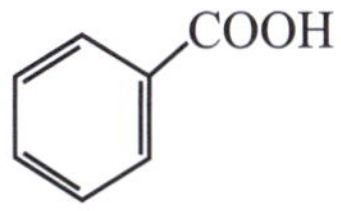 + Zn/Hg and Conc. HCl

6. Consider the following reaction:

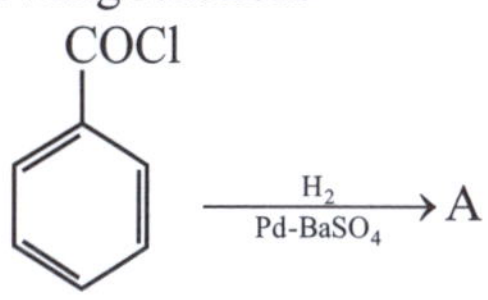

   The product A is:
   1) C_6H_5CHO 2) C_6H_5OH
   3) $C_6H_5COCH_3$ 4) C_6H_5Cl

7. The reaction of formaldehyde with magnesium methyl bromide followed by hydrolysis gives
   1) secondary alcohol 2) primary alcohol
   3) ketone 4) tertiary alcohol

8. The reagent used in Gatterman Koch aldehyde synthesis is:
   1) $Pb/BaSO_4$ 2) Alkaline $KMnO_4$
   3) Acidic $KMnO_4$ 4) CO + HCl

9. In the reaction,

   $HCOOC_2H_5 \xrightarrow[ether]{CH_3MgI} X \xrightarrow{H_3O^+} y$, **y is:**
   1) Acetaldehyde 2) Acetone
   3) Propanal 4) Propanone

10. In the reaction,

    Acetonitrile $\xrightarrow{C_2H_5MgI} A \xrightarrow{HOH} B$; B is :
    1) Acetone 2) Acetaldehyde
    3) Butan-2-one 4) Butanal

11. The oxidation of Toluene to Benzaldehyde by Chromyl Chloride is called:
    1) Etard reaction 2) Riemer - Tiemann reaction
    3) Wurtz reaction 4) Cannizzaro's reaction

12. In the following reaction, product (P) is:

    $$R-\overset{\overset{\displaystyle O}{\|}}{C}-Cl \xrightarrow[Pd\,/\,BaSO_4]{H_2} P$$
    1) RCHO 2) RCH_3
    3) RCOOH 4) RCH_2OH

13. Propanone can be prepared from ethyne by:
    1) Passing a mixture of ethyne and steam over a catalyst, magnesium at 420°C
    2) Passing a mixture of ethyne and ethanol over a catalyst zinc chromite
    3) Boiling ethyne with water and H_2SO_4
    4) Treating ethyne with Iodine and NaOH

Physical Properties & Chemical Properties

14. Ketone is much less soluble than aldehyde because of:
1) Steric hindrance 2) Hydrogen bonding
3) Resonance 4) Both (1) and (2)

15. There is a large difference in boiling points of butanal and butan - 1 - ol due to:
1) Intermolecular hydrogen bonding in butan - 1 - ol
2) Intramolecular hydrogen bonding in butanal
3) High molecular mass of butan - 1 - ol
4) Resonance shown by butanal

16. Acetaldehyde and Propyne can be distinguished by
1) $NaHCO_3$ 2) $I_2/NaOH$
3) Lucas reagent 4) Neutral $FeCl_3$

17. $CH_3-CH_2C\equiv N \xrightarrow{X} CH_3CH_2CHO$

The compound X is:
1) $SnCl_2$ / HCl / H_2O, boil
2) H_2 / Pd – $BaSO_4$
3) $LiAlH_4$ / ether
4) $NaBH_4$ / ether / H_3O^+

18. Which of the following compound on treatment with $LiAlH_4$ will give a product that will give positive iodoform test?
1) CH_3CH_2CHO 2) $CH_3CH_2CO_2CH_3$
3) $CH_3CH_2OCH_2CH_3$ 4) CH_3COCH_3

19. The most reactive compound towards formation of cyanohydrin on treatment with KCN followed by acidification is:
1) Benzaldehyde 2) p-nitrobenzaldehyde
3) Phenylacetaldehyde 4) p-hydroxybenzaldehyde

20. 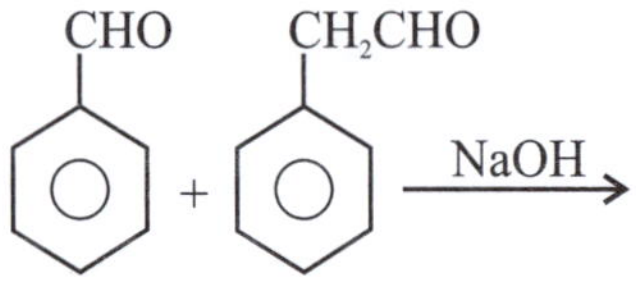

Product can be

1) 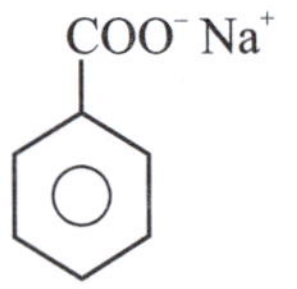

2) 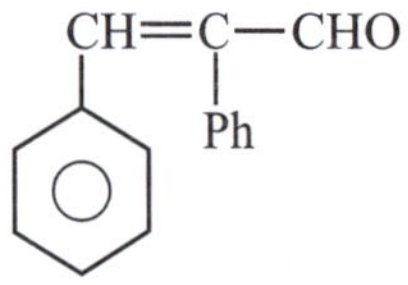

3) 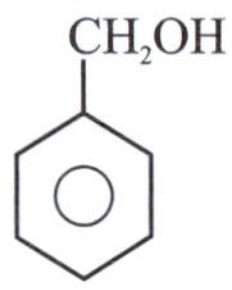

4) All of these

21. Which of the following will not give aldol condensation?
1) Phenyl acetaldehyde
2) 2 - Methyl pentanal
3) Benzaldehyde
4) 1 - Phenyl propanone

22. Which among the following is most reactive to give nucleophilic addition?
1) FCH_2CHO 2) $ClCH_2CHO$
3) $BrCH_2CHO$ 4) ICH_2CHO

23. Which of the following gives iodoform test?
1) CH_3OH 2) $CH_3COCH_2CH_3$
3) HCHO 4) CH_3COOH

24. Which compound is obtained when acetaldehyde is treated with dilute solution of caustic soda?
1) Sodium acetate 2) Butanal
3) Aldol 4) Ethyl acetate

25. The reagent with which both acetaldehyde and acetone react is:
1) Fehling's solution 2) Grignard Reagent
3) Tollen's reagent 4) Carbonic acid

26. Which of the following compounds undergoes both Cannizaro and Aldol reaction?
1) $(CH_3)_2CH.CHO$ 2) HCHO
3) C_6H_5CHO 4) CH_3CHO

27. The product formed when acetone reacts with nitro methane in the presence of a base is:

1) 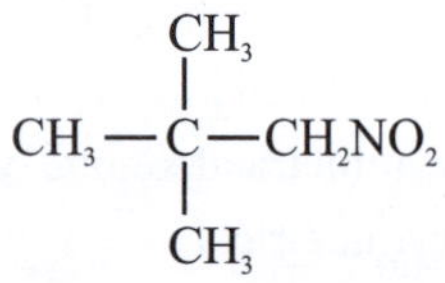

2) 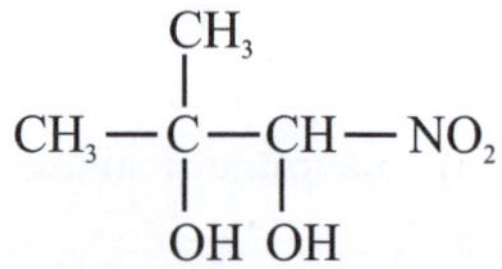

3) 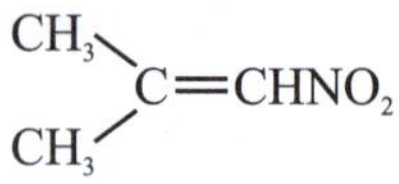

4) 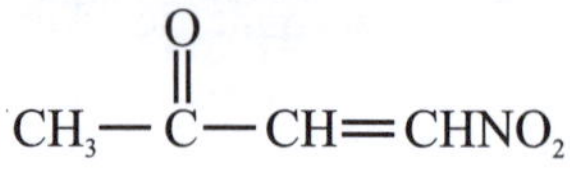

28. Predict the products in the given reaction:

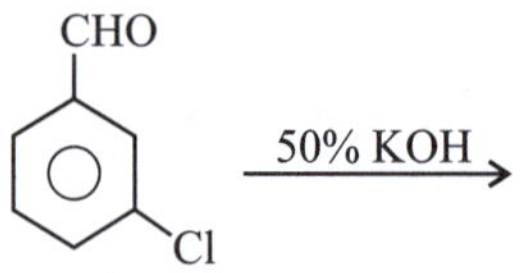

1) 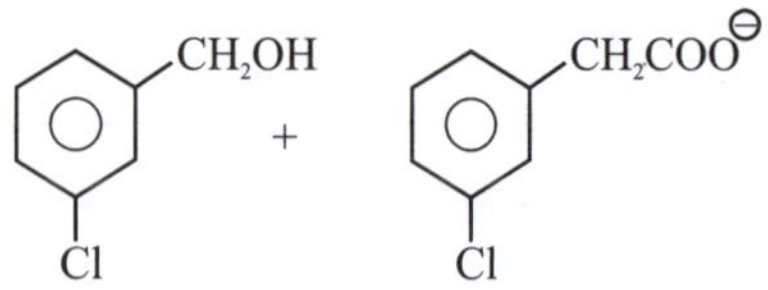

2) 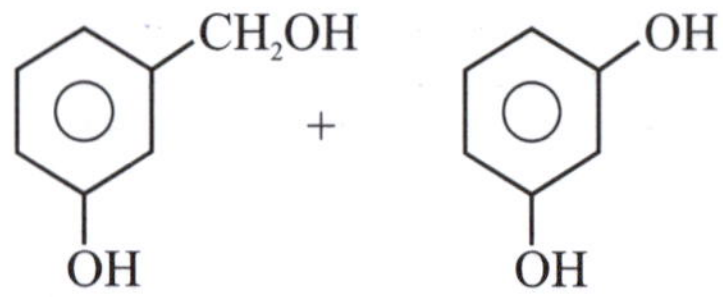

3) 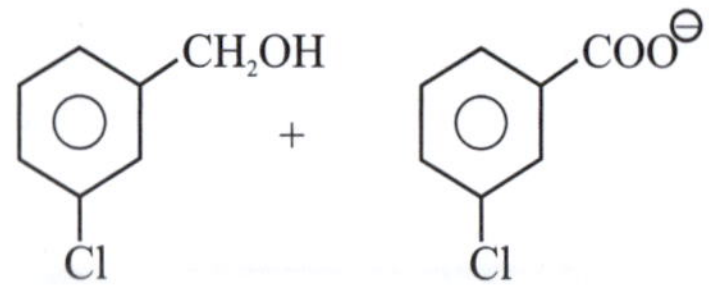

4) 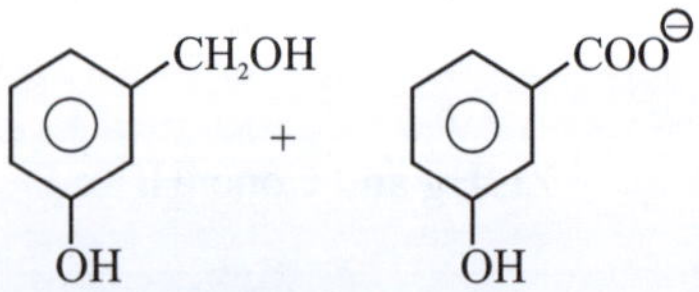

29. Which of the following reactions converts acetone into hydrocarbon having same number of carbon atoms?
    1) Wolff - Kishner reaction
    2) Hoffmann reaction
    3) Grignard reaction
    4) Reduction with $LiAlH_4$

30. The compound which does not respond to Fehling's test is/are:
    1) CH_3CHO 2) C_6H_5CHO
    3) CH_3COCH_3 4) Both (2) & (3)

31. Which of the following reaction is given by both, formaldehyde and acetaldehyde?
    1) Cannizaro's reaction 2) Aldol condensation
    3) Haloform reaction 4) None of these

32. Ethyl methyl ketone on reduction with zinc amalgam and Conc. HCl gives:
    1) Sec. alcohol 2) 1° alcohol
    3) 3° alcohol 4) n-butane

33. Which of the following is the most reactive isomer?
    1) $CH_3CH_2CH_2CH_2-C(=O)-H$
    2) $CH_3CH_2CH_2-C(=O)-CH_3$
    3) $CH_3CH_2-C(=O)-CH_2-CH_3$
    4) $CH_3-C(=O)-CH(CH_3)-CH_3$

34. When propanal reacts with 2 - methyl propanal in presence of NaOH, four different products are formed. The reaction is known as: (excluding stereoisomers)
    1) Aldol condensation
    2) Cross aldol condensation
    3) Cannizzaro reaction
    4) HVZ condensation

35. Which of the following compound will undergo self - aldol condensation in the presence of cold dilute alkali?
    1) $CH \equiv CH - CHO$
    2) $CH_2 = CHCHO$
    3) C_6H_5CHO
    4) CH_3CH_2CHO

36. The reactivity order of indicated functional groups towards a nucleophile is:

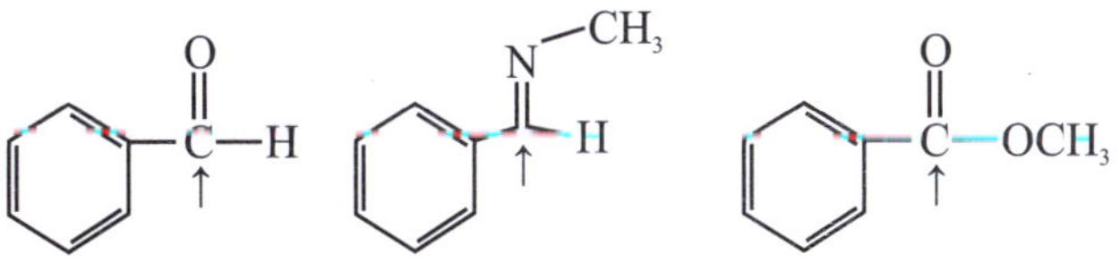

(P) (Q) (R)

    1) P > Q > R 2) Q > P > R
    3) Q > R > P 4) R > P > Q

37. The structure of addition product formed with acetone reacts with a concentrated aqueous solution of $Na^+H^-SO_3$ is:

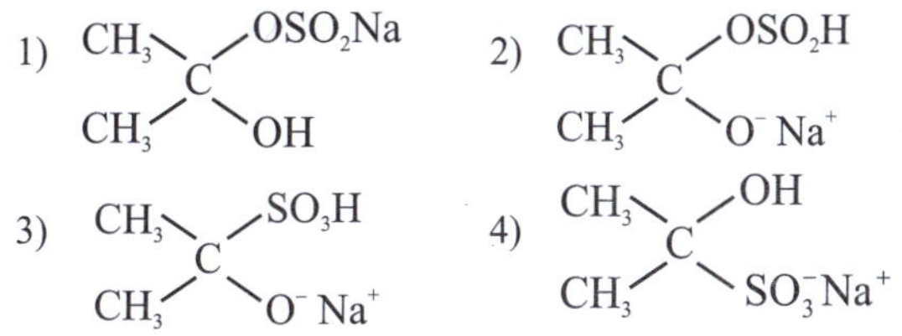

38. Which of the following will be most readily dehydrated under acidic condition ?

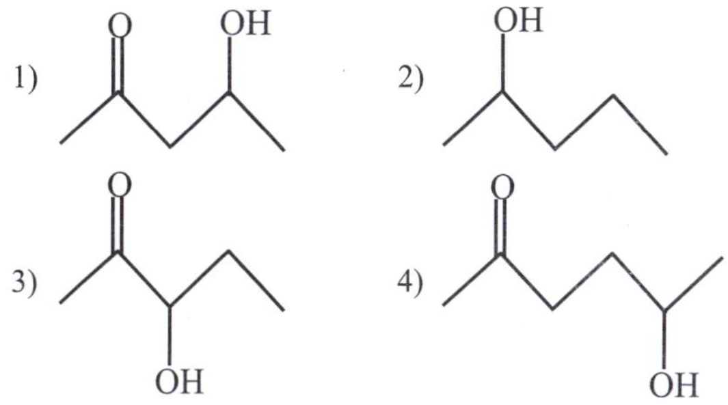

Nomenclature and Structure of Carboxyl Group

39. Write the IUPAC name of the following compound:

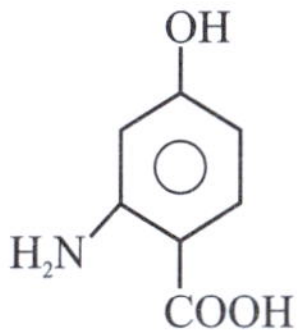

    1) 2 – amino – 4 – hydroxy benzoic acid
    2) 6 – amino – 4 hydroxy benzoic acid
    3) 3 – amino – 4 – carboxyphenol
    4) 2 – carboxy – 4 – hydroxy aniline

40. The correct structure representation of carboxylate ion is:

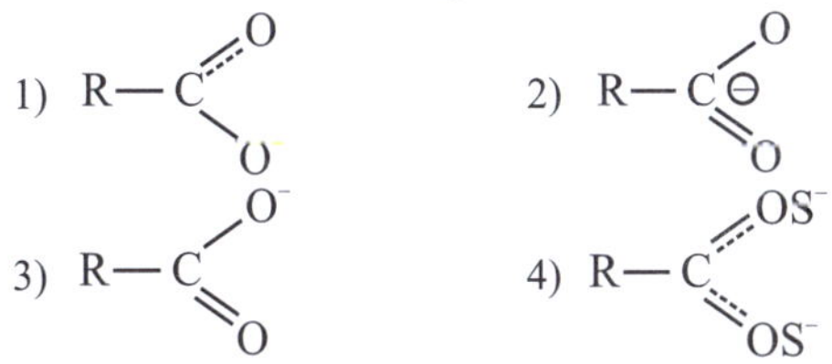

41. The hybridization of the C - atom of the carboxyl group is:
    1) sp 2) sp^2
    3) sp^3 4) None of these

Methods of Preparation of Carboxylic Acids

42. The end product (z) in the given sequence of reaction is:

$$CH_3CH = CHCHO \xrightarrow{NaBH_4} X \xrightarrow[ZnCl_2]{HCl} Y \xrightarrow[(ii)\ H_3O^+]{(i)KCN} Z$$

    1) $CH_3\ CH = CH\ CH_2COOH$
    2) $CH_3CH_2CH_2COOH$
    3) $CH_3CH = CH\ COOH$
    4) $CH_3CH(Cl)CH_2COOH$

43. An organic compound (X) with molecular formula $C_9H_{10}O$ gives positive 2, 4-DNP and Tollen's test. It undergoes Cannizarro reaction and on vigorous oxidation, it gives 1, 4-Benzenedicarboxylic acid. Compound (X) is:
   1) Benzaldehyde
   2) o - methyl benzaldehyde
   3) p - ethyl benzaldehyde
   4) 2,2 - dimethyl hexanal

44. Which one of the following cannot be formed by the reaction of dry ice with grignard's reagent?
   1) Acetic acid 2) Formic acid
   3) Propanoic acid 4) All of these

45. Esterification is a reverse of:
   1) Acidification 2) Hydration
   3) Neutralization 4) Hydrolysis

46. In the reaction, $RCN \xrightarrow[2)\ NaOH]{1)\ HOH} A + NH_3$; A is a/an:
   1) Sodium acetate 2) Acetic acid
   3) Fatty acid 4) Sodium alkanoate

47. Hydrolysis of hydrocyanic acid gives:
   1) Propanoic acid 2) Acetamide
   3) Ethanoic acid 4) Methanoic acid

48. In the reaction $\text{Ethanol} \xrightarrow{PI_3} A \xrightarrow{KCN} B \xrightarrow{H_3O^+} C$, C may be:
   1) Acetamide 2) Propionamide
   3) Propanoic acid 4) Acetic acid

Physical Properties & Chemical Properties

49. Which of the following is the correct order of acidic strength?
   1) $Cl_3CCOOH > Cl_2CHCOOH > Cl.CH_2COOH > CH_3COOH$
   2) $CH_3COOH > ClCH_2COOH > Cl_2CHCOOH > Cl_3CCOOH$
   3) $CH_3COOH > Cl_3CCOOH > Cl_2.CHCOOH > Cl\,CH_2COOH$
   4) $CH_3COOH > ClCH_2COOH > Cl_2CH\,COOH > Cl_3C\,COOH$

50. $CH_3-CH(CH_3)-COOH + AgNO_3 \longrightarrow X \xrightarrow{Br_2/CCl_4} Y$
   1) $CH_3-C(CH_3)(Br)-COOAg$
   2) $CH_3-CH(CII_3)-COOBr$
   3) $CH_3-CH(CH_3)-Br$
   4) None

51. Which acid is present in vinegar?
   1) CH_3COOH 2) H_2SO_4
   3) HCl 4) HNO_3

52. Which of the following compounds is most acidic?
   1) $BrCH_2COOH$ 2) $ClCH_2COOH$
   3) FCH_2COOH 4) ICH_2COOH

53. Pick up from the following reactions that in which formic acid differs from acetic acid?
   1) Turns red litmus blue
   2) Forms ester with alcohols
   3) Sodium replaces hydrogen from the compound
   4) Reduces ammoniacal silver nitrate

54. 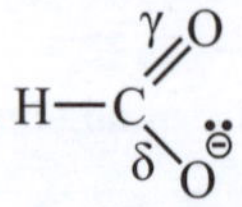

(C–O) bond length designated by γ and δ are in order-
   1) $\gamma > \delta$ 2) $\delta > \gamma$
   3) $\delta = \gamma$ 4) None

55. HVZ reaction is not given by:
   1) RCH_2COOH 2) CH_3COOH
   3) $R_3C.COOH$ 4) RCOOH

56. In HVZ reactions, the substitution takes place at:
   1) Carbon atom of the carboxyl group
   2) Carbon atom attached to carboxyl group
   3) Both (1) and (2)
   4) None of these

57. Which of the following statement is correct regarding formic acid?
   1) It is a reducing agent
   2) It is a weaker acid than acetic acid
   3) It is an oxidising agent
   4) When its calcium salt is heated, it form acetone

58. Which of the following will not undergo HVZ reaction?
   1) Propanoic acid
   2) Ethanoic acid
   3) 2 - Methylpropanoic acid
   4) 2,2 - Dimethylpropanoic acid

59. What happens when a carboxylic acid is treated with lithium aluminum hydride?
   1) Aldehyde is formed
   2) Primary alcohol is formed
   3) Ketone is formed
   4) Grignard reagent is formed

60. 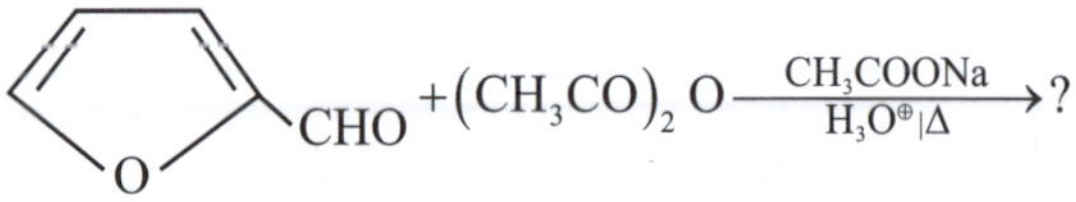

   1) 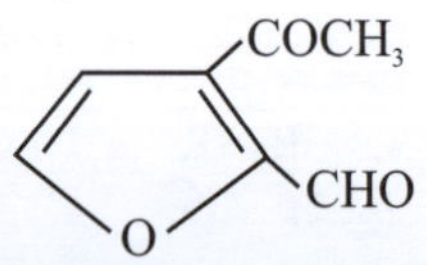

2) 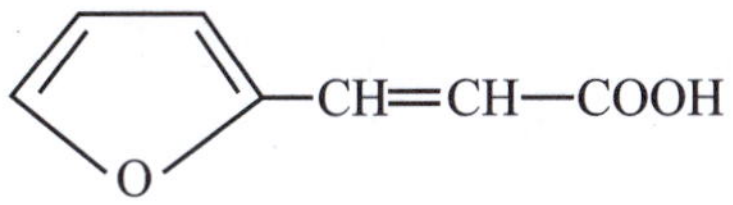

3) 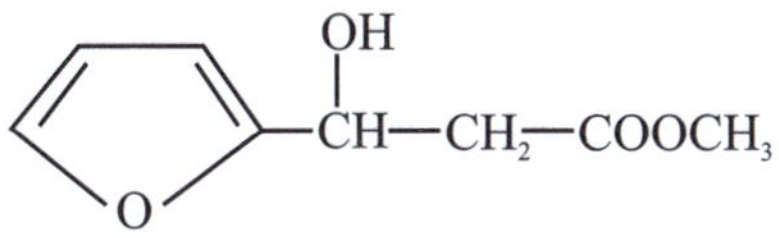

4) 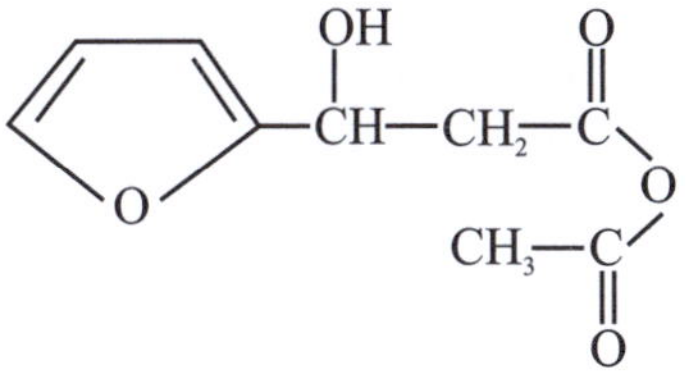

61. Identify 'Y' in;

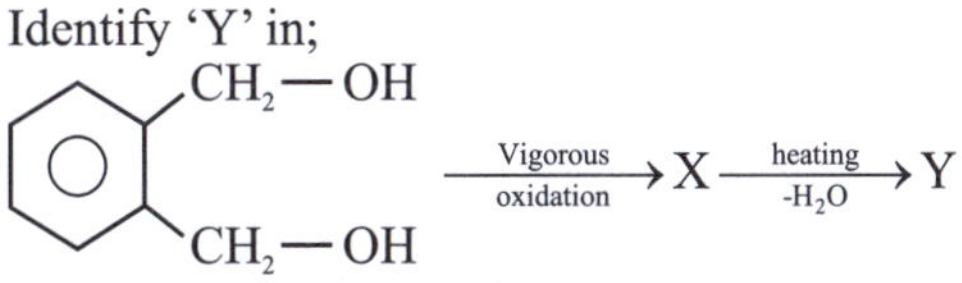

1) 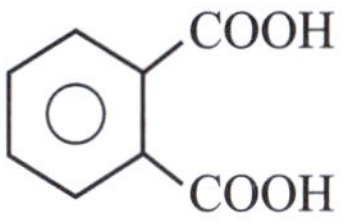
2) 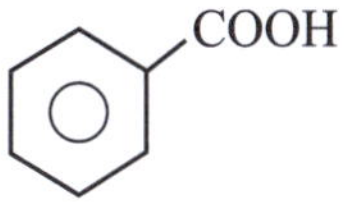
3) 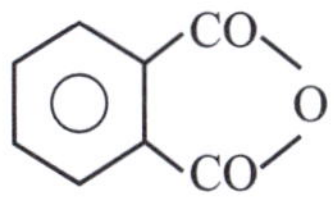
4) 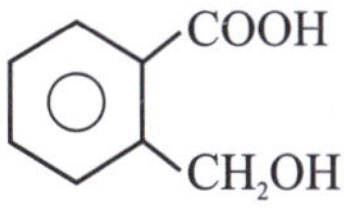

62. When $CH_2 = CH - COOH$ is reduced by $LiAlH_4$, the compound obtained is:

1) CH_3CH_2COOH
2) $CH_2 = CHCH_2OH$
3) $CH_3CH_2CH_2OH$
4) CH_3CH_2CHO

63. Main product of the reaction, $CH_3CONH_2 + HNO_2 \rightarrow$ .....is:

1) CH_3COOH
2) $CH_3CH_2NH_2$
3) CH_3NH_2
4) CH_3COONH_4

64. The reagent in friedel - craft reaction is:

1) Pyridine
2) RCOCl
3) RCOOH
4) HCl

65. What are the correct steps to convert acetaldehyde to acetone?

1) CH_3MgBr, H_2O, oxidation
2) Oxidation,$Ca(OH)_2$, Heat
3) Reduction, KCN, hydrolysis
4) Oxidation, C_2H_5O Na, Heat

66. Which of the following esters is most reactive for saponification?

1) $CH_3-\overset{O}{\overset{\|}{C}}-O-CH_3$
2) $CH_3-\overset{O}{\overset{\|}{C}}-O-CH(CH_3)-CH_3$
3) $CH_3-\overset{O}{\overset{\|}{C}}-O-C(CH_3)_2-CH_3$
4) 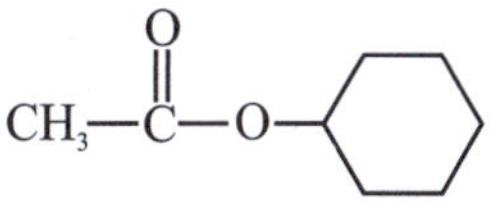

67. Carboxylic acid reacts with diazomethane to yield:

1) Amines
2) Alcohols
3) Esters
4) Amides

68. $CH_3CH_2COOH \xrightarrow[\text{red } P_4]{Cl_2} X \xrightarrow[KOH]{Alc} Y$. The compound Y is:

1) CH_3CH_2OH
2) CH_3CH_2CN
3) $CH_2 = CHCOOH$
4) CH_3CHCl

69. 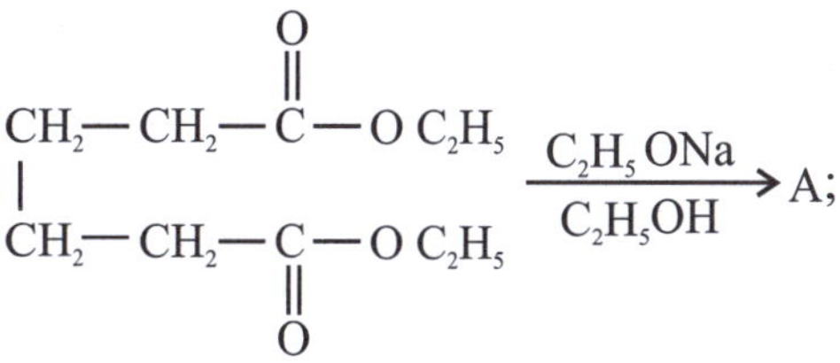

What is A?

1) 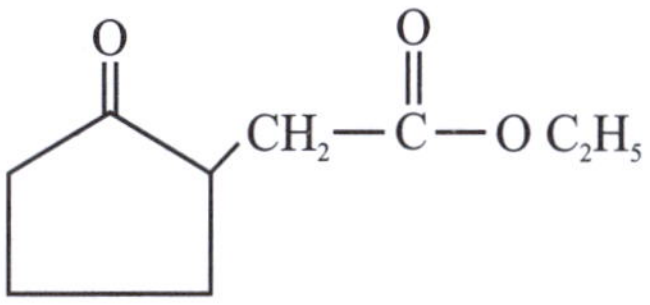
2) 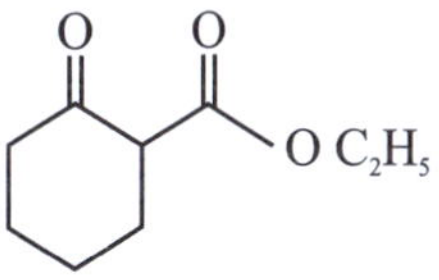
3) 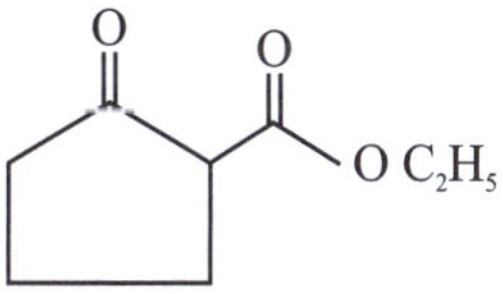
4) 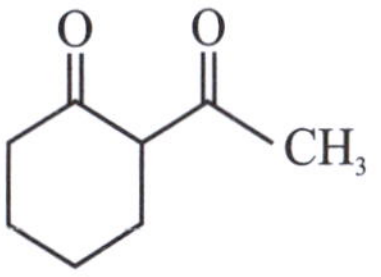

70. From given set of reactions:

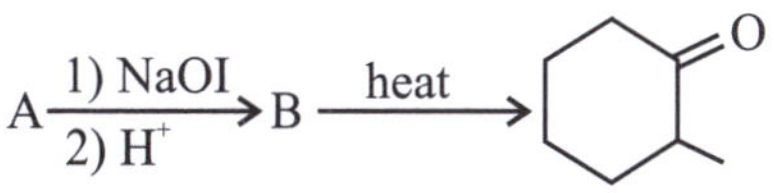

The compound A is:

1) 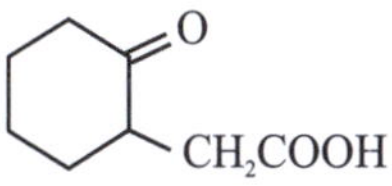
2) 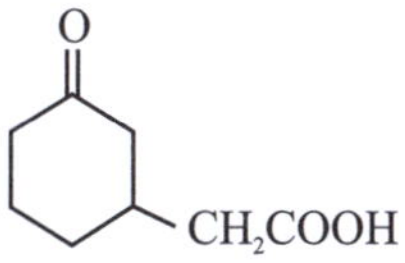
3) 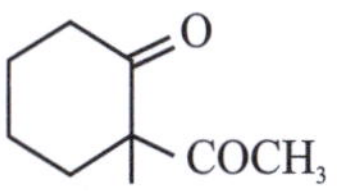
4) 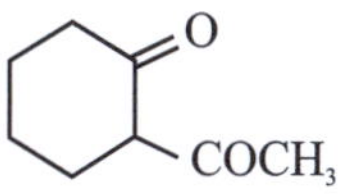

71. Compound 'X' C_4H_8O which gives 2, 4-DNP derivative and positive iodoform test is-

1) 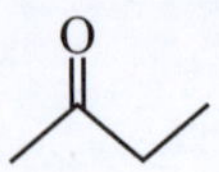

2) 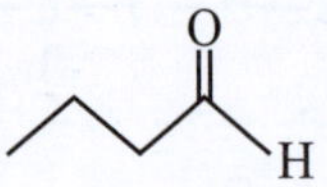

3) $CH_3-CH_2-\overset{\overset{O}{\|}}{C}-H$

4) 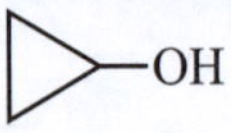

72. A mixture of benzaldehyde and formaldehyde on heating with 50% NaOH solution gives

1) Benzyl alcohol and sodium formate
2) Sodium benzoate and methyl alcohol
3) Sodium benzoate and sodium formate
4) Benzyl alcohol and methyl alcohol

73. Base catalysed aldol condensation occurs with

1) Propanal
2) Benzaldehyde
3) 2, 2-dimethyl propanal
4) (1) & (3) both correct

74. Least pK_a value is of

1) CH_3CH_2COOH
2) $(CH_3)_2$ CH-COOH
3) HCOOH
4) CH_3COOH

75. Find out structure of compound R:

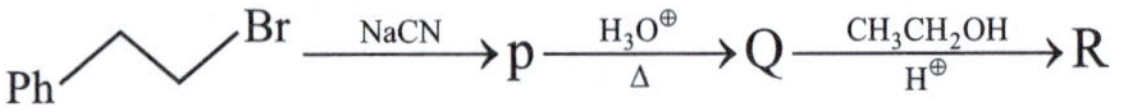

1) 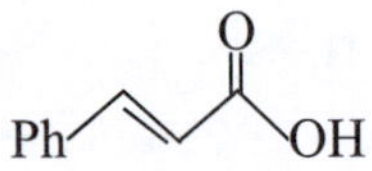

2) 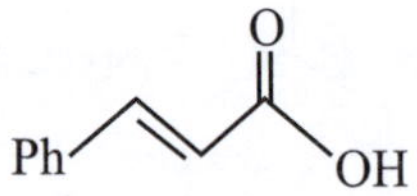

3) 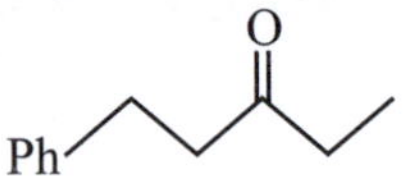

4) None of these

76. The correct order for the acidic character of the following carboxylic acids is:

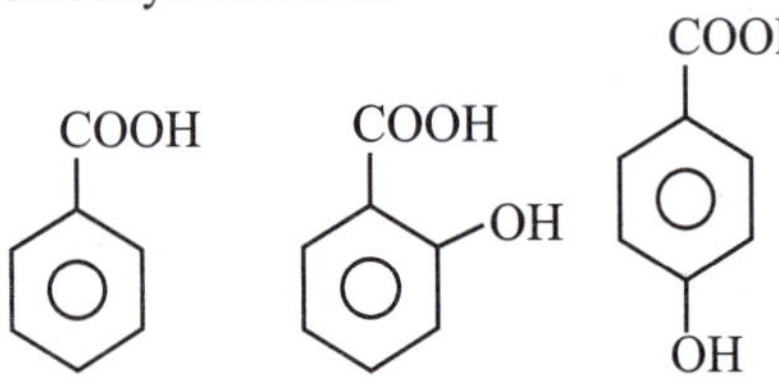

(I) (II) (III)

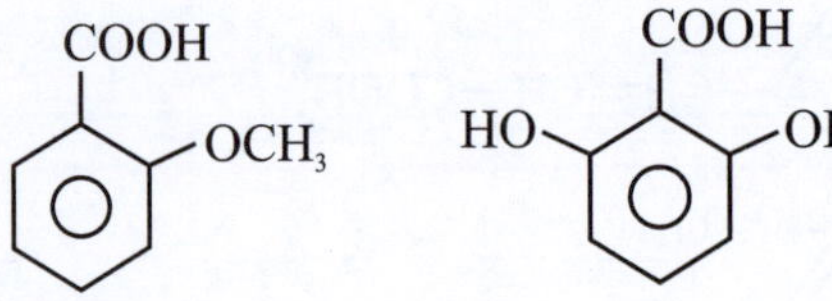

(IV) (V)

1) IV > I > II > III > V
2) V > II > IV > I > III
3) V > II > III > I > IV
4) None of these

77. 

B is:

1) 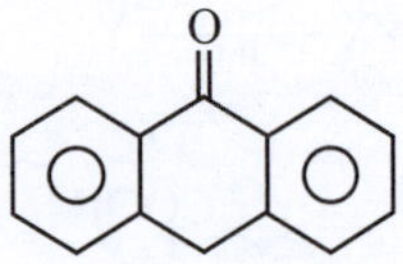

2) 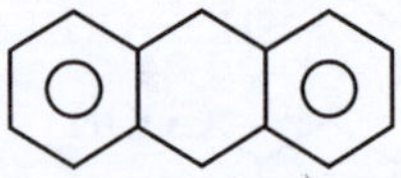

3) 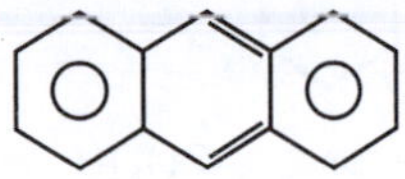

4) 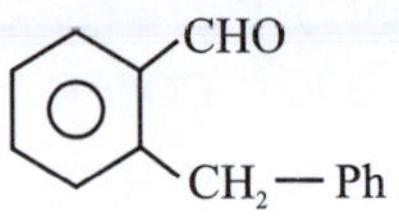

78. Which of the following will not give HVZ reaction?

1) 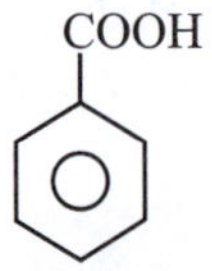

2) 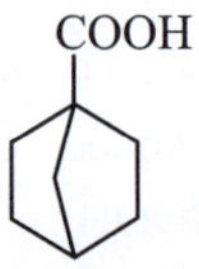

3) 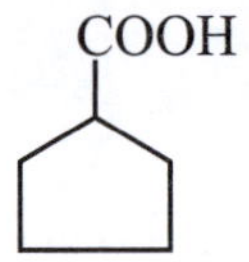

4) Both (1) & (2)

79. Maximum decarboxylation occurs in:

1) CH_3COOH
2) C_6H_5COOH
3) $C_6H_5CH_2COOH$
4) CH_3COCH_2COOH

80. Which of the following carboxylic acids is most acidic in character?

1) o-methyl benzoic acid
2) m-methyl benzoic acid
3) p-methyl benzoic acid
4) Benzoic acid

High Order Time Intensive MCQs

1. Which of the following will undergo alkaline hydrolysis most rapidly?

1) 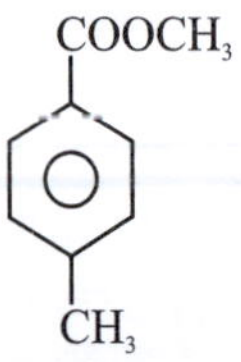

2) 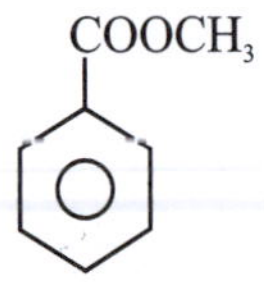

3) 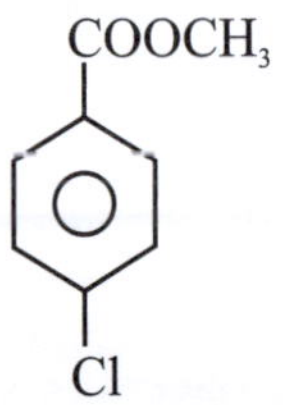

4) 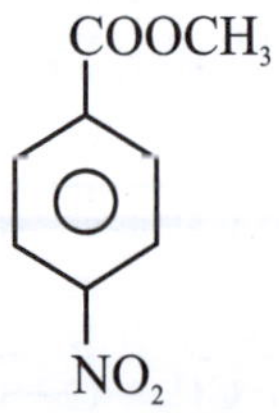

2. Predict the nature of end product in the following reaction:

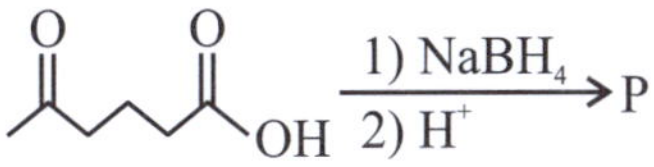

1) 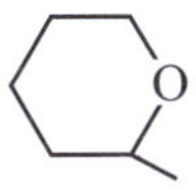
2) 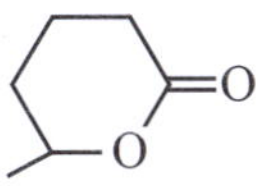
3) 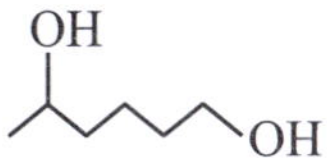
4) 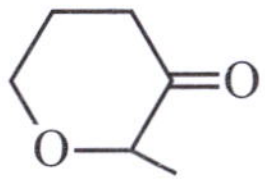

3. Identify major product 'P' formed in the following reaction.

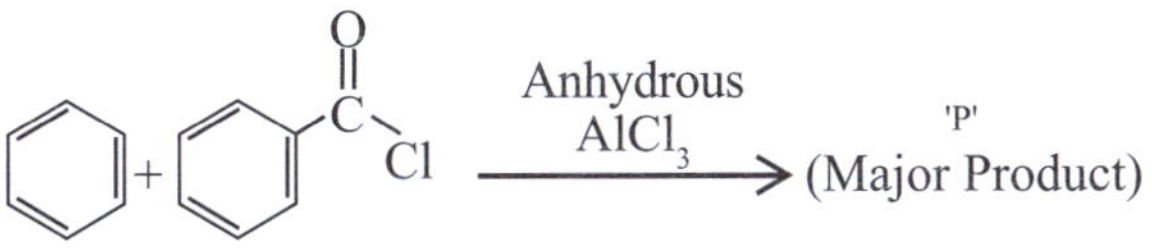

1) 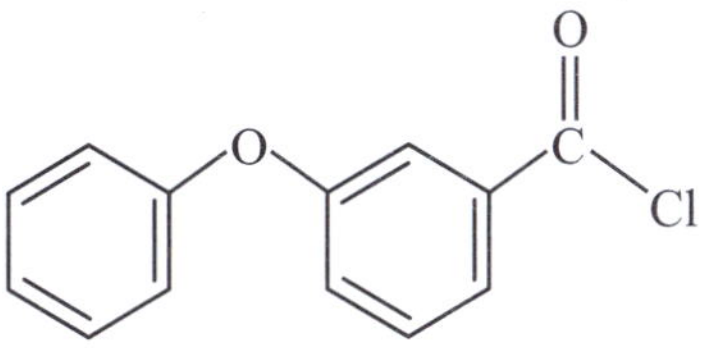
2) 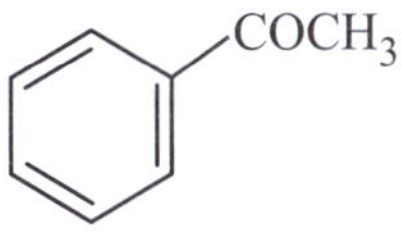
3) 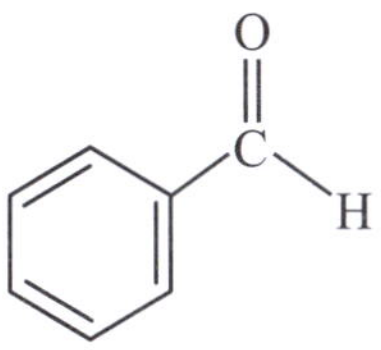
4) 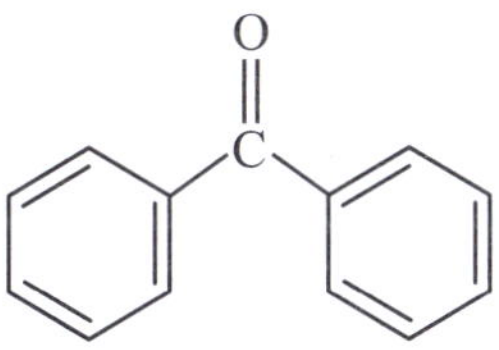

4. 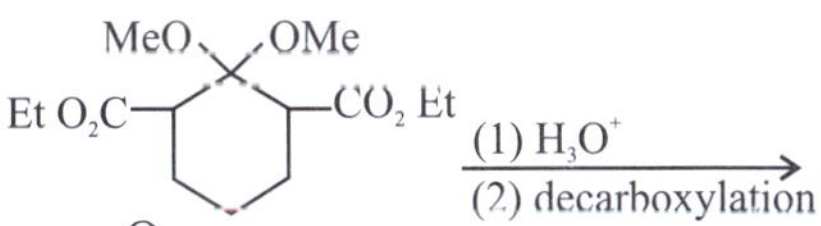 A. Product A is:

1) 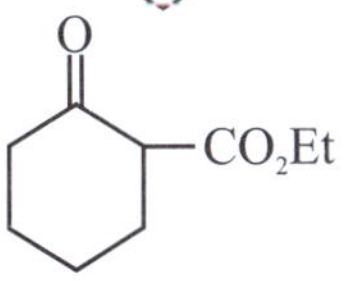
2) 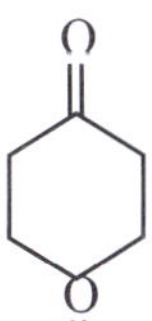
3) 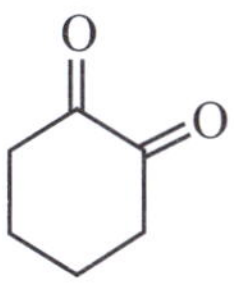
4) 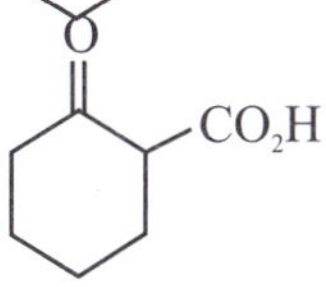

5. Identify the name reaction.

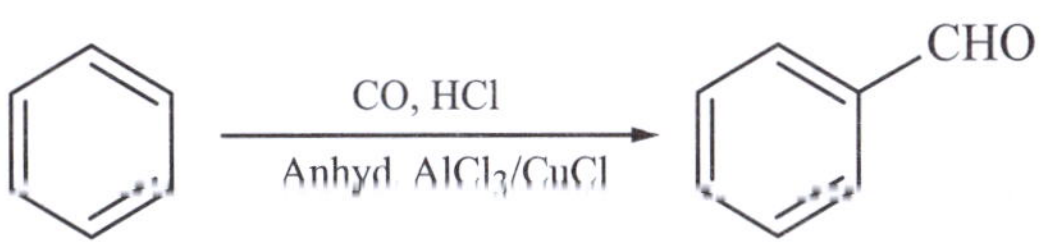

1) Stephen reaction
2) Etard reaction
3) Gatterman-koch reaction
4) Rosenmund reduction

6. The products obtained in the reaction:

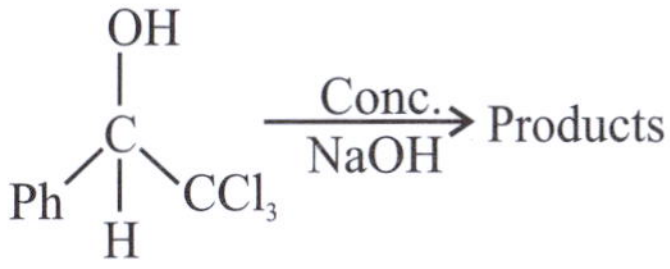

1) Ph – CHO & $CHCl_3$
2) PhCHO & $HCOO^-Na^+$
3) Ph – CH_2OH & $CHCl_3$
4) $PhCOO^- Na^+ + PhCH_2OH + HCOO^- Na^+$

7. Identify the correct reagents required for the following transformation:

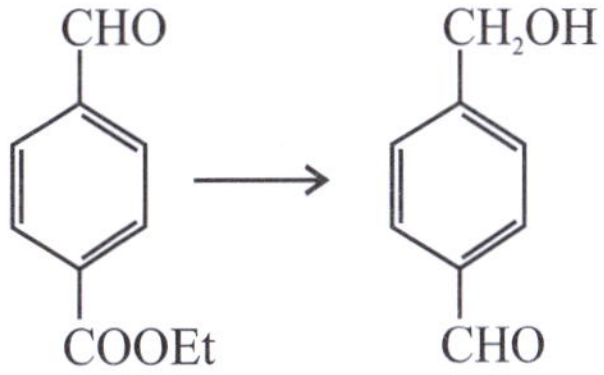

1) (i) $NaBH_4$ (ii) H_3O^+
2) (i) $LiAlH_4$ (ii) H_3O^+
3) (i) $\begin{array}{l} CH_2OH \\ | \\ CH_2OH \end{array}$; H^+ (ii) $LiAlH_4$ (iii) H_3O^+
4) (i) $\begin{array}{l} CH_2SH \\ | \\ CH_2SH \end{array}$; H^+ (ii) $LiAlH_4$ (iii) H_3O^+

8. The major product obtained in the following reaction is:

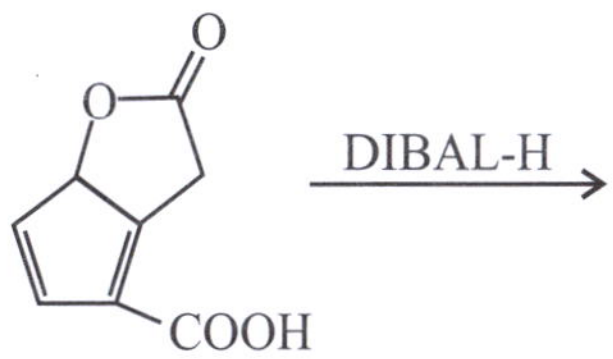

1) 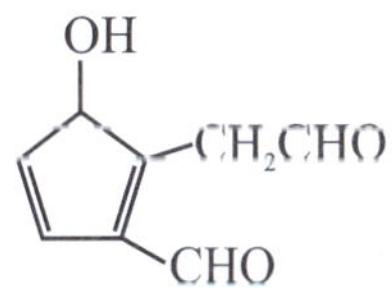
2) 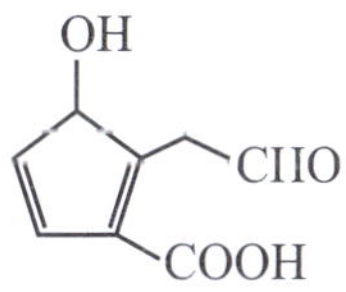
3) 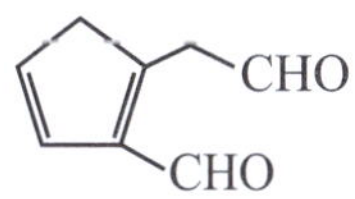
4) 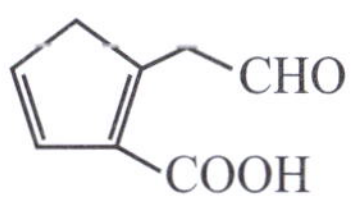

9. Oxidation of toluene to benzaldehyde can be easily carried out with which of the following reagents?

1) CrO_3/acetic acid, H_3O^+
2) CrO_3 / acetic anhydride, H_3O^+
3) $KMnO_4$ / HCl, H_3O^+
4) CO / HCl, anhydrous $AlCl_3$

10. The major product of the following chemical reaction is:

$$CH_3CH_2CN \xrightarrow[(3)\ Pd/BaSO_4,\ H_2]{(1)\ H_3O^+,\ \Delta \quad (2)\ SOCl_2} ?$$

1) $(CH_3CH_2CO)_2O$
2) CH_3CH_2CHO
3) $CH_3CH_2CH_3$
4) $CH_3CH_2CH_2OH$

11. 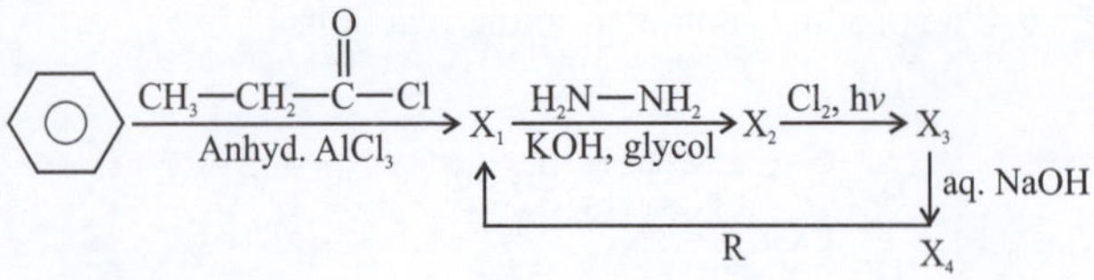

Reagent 'R' can be:

1) MnO_2 2) $Cl_2 + H_2O$
3) PCC 4) Both (1) & (3)

12. Which of the following reagent is used for the following reaction?

$CH_3CH_2CH_3 \xrightarrow{?} CH_3CH_2CHO$

1) Manganese acetate
2) Molybdenum oxide
3) Potassium permanganate
4) Copper at high temperature and pressure

13. Which among the following has highest boiling point?

1) $CH_3CH_2CH_2CH_3$ 2) $CH_3CH_2CH_2CH_2–OH$
3) $CH_3CH_2CH_2CHO$ 4) $H_5C_2–O–C_2H_5$

14. 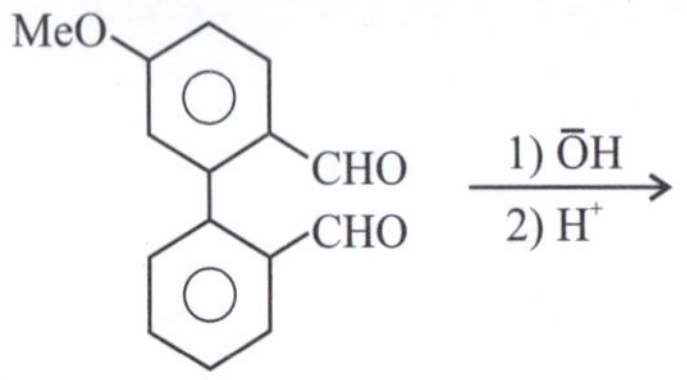

The major product which can be isolated from this reaction is:

1) 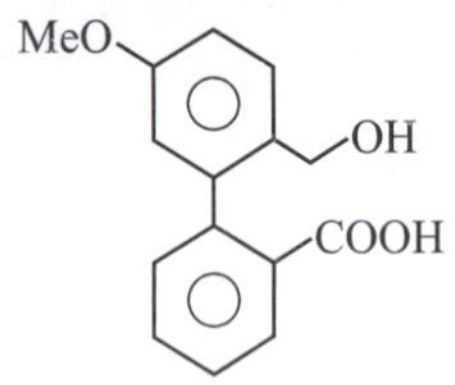 2) 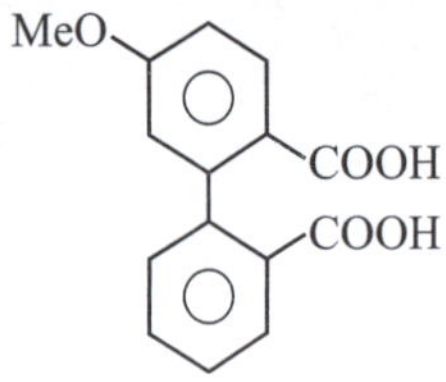

3) 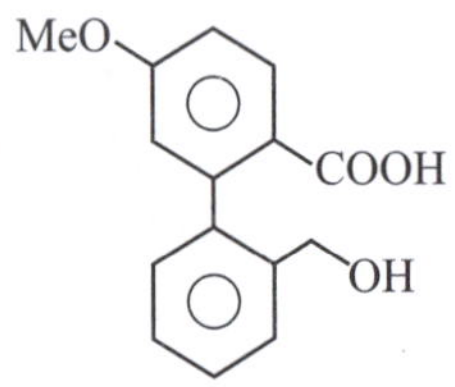 4) Both (2) & (3)

15. $CH_3–\overset{O}{\overset{||}{C}}–H + H–\overset{O}{\overset{||}{C}}–H \text{ (excess)} \xrightarrow[2)\ H^+]{1)\ ^-OH}$

The major product which can be isolated from this reaction is:

1) $CH_3COOH + CH_3OH$
2) $CH_3–CH_2OH + HCOOH$
3) $HOH_2C–C(CH_2OH)_2–COOH + CH_3OH$
4) $HOH_2C–C(CH_2OH)_2–CH_2OH + HCOOH$

16. Which of the following will give positive tollen's test?

1) $CH_3–\overset{O}{\overset{||}{C}}–H$

2) 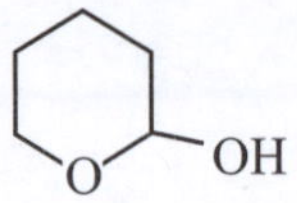

3) $CH_3–CH_2–CH(OH)(OCH_3)$

4) All of these

17. Which of following orders is correct for ease of hydride donation by these molecules?

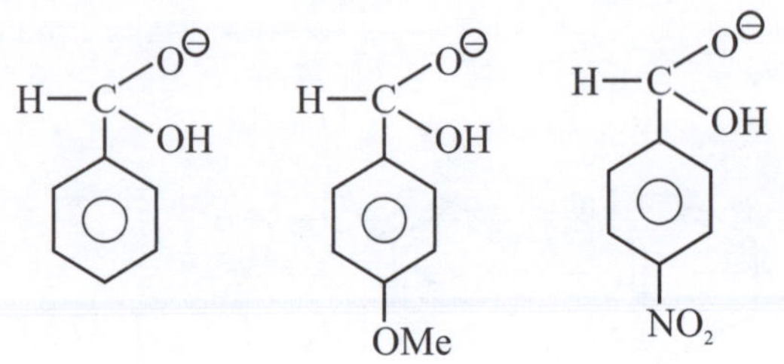

(I) (II) (III)

1) I > II > III 2) III > I > II
3) II > I > III 4) II > III > I

18. Cyclohexene on ozonolysis followed by reaction with Zn/dust & H_2O gives compound E. Compound E on further treatment with aqueous KOH yield compound F. Compound F is:

1) 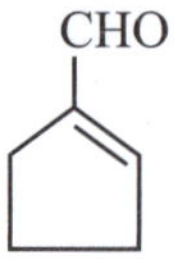 2) 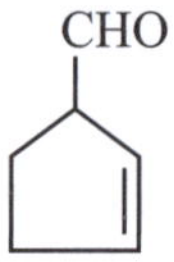

3) 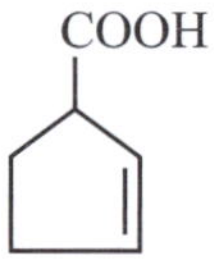 4) 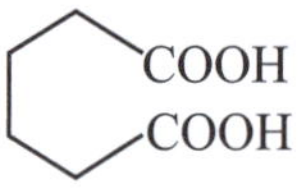

19. 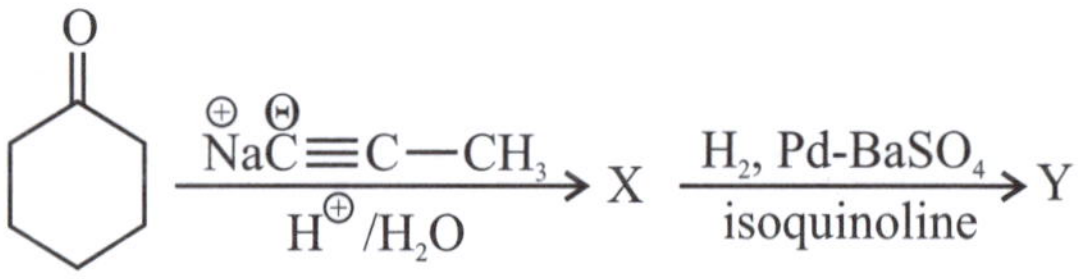

Identify structure of Y:

1) 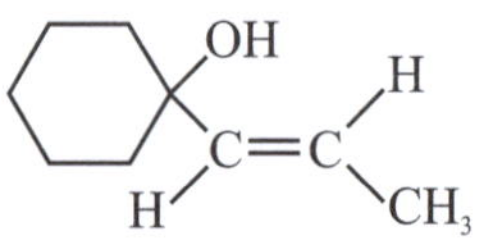

2) 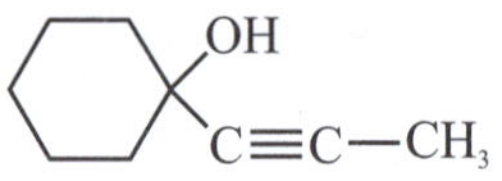

3) 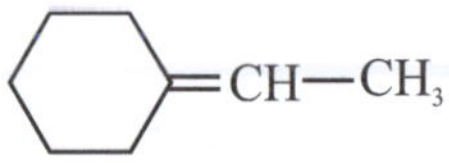

4) 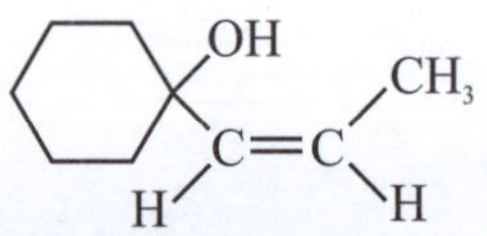

20. Find the major product of given reaction :

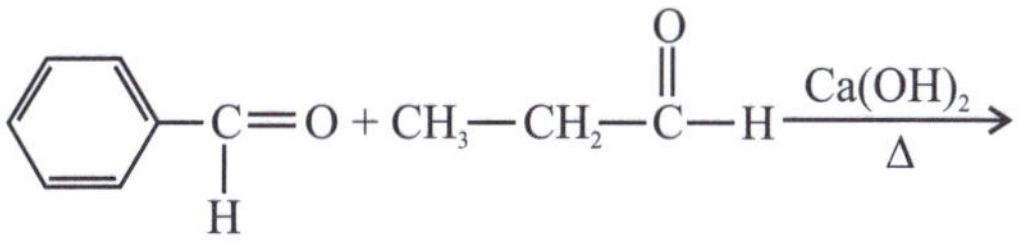

1) $Ph-CH=CH-CH_2-\overset{O}{\overset{\|}{C}}-H$

2) $CH_3CH_2-CH=\underset{CH_3}{\underset{|}{C}}-\overset{O}{\overset{\|}{C}}-H$

3) $CH_3CH_2CH=CH-CH_2-\overset{O}{\overset{\|}{C}}-H$

4) $Ph-CH=\underset{CH_3}{\underset{|}{C}}-\overset{O}{\overset{\|}{C}}-H$

21. 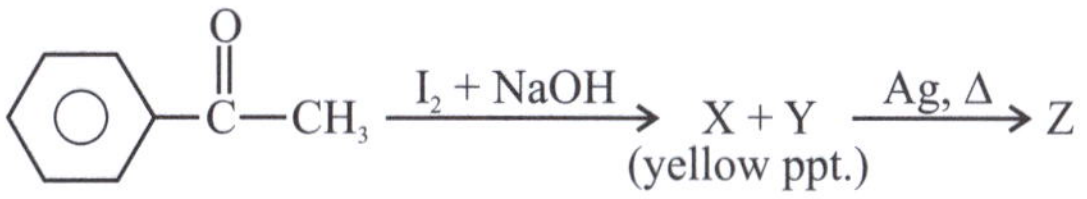

Identify final product 'Z' :

1) CHI_3
2) $HC \equiv CH$
3) $Ph-\overset{O}{\overset{\|}{C}}-\overset{\ominus}{O}\overset{\oplus}{Na}$
4) $H_2C = CH_2$

22. 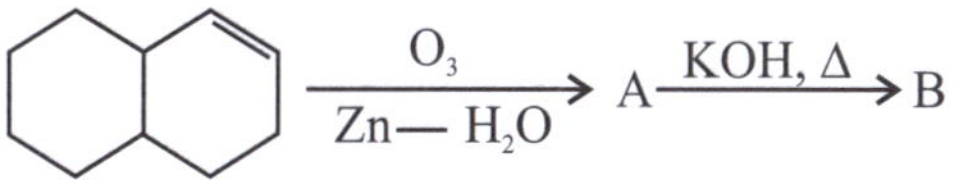

Compound B is-

1) 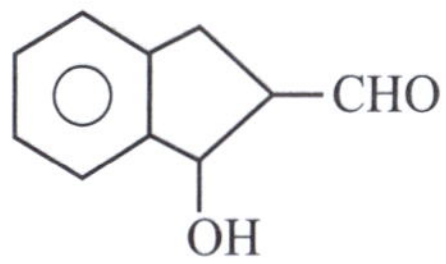
2) 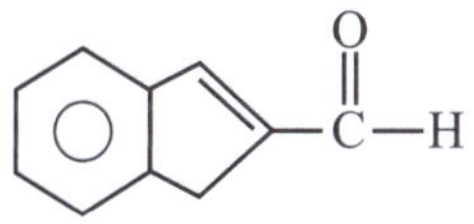
3) 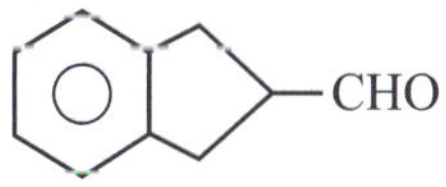
4) 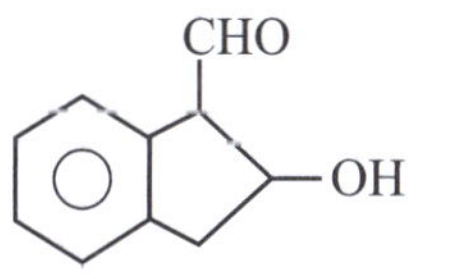

23. 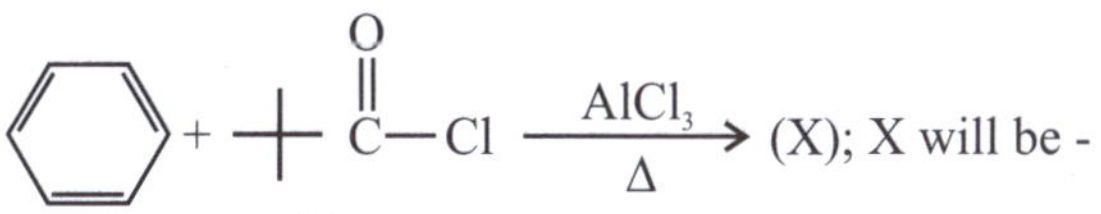

1) 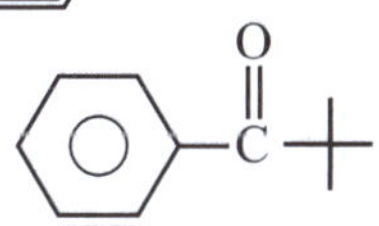
2) 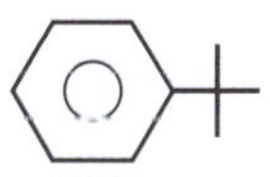
3) 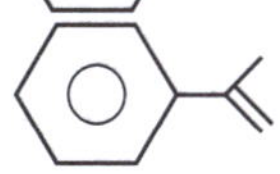
4) 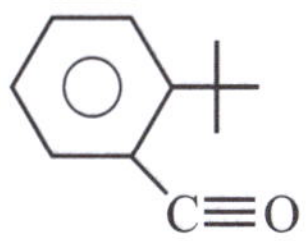

24. 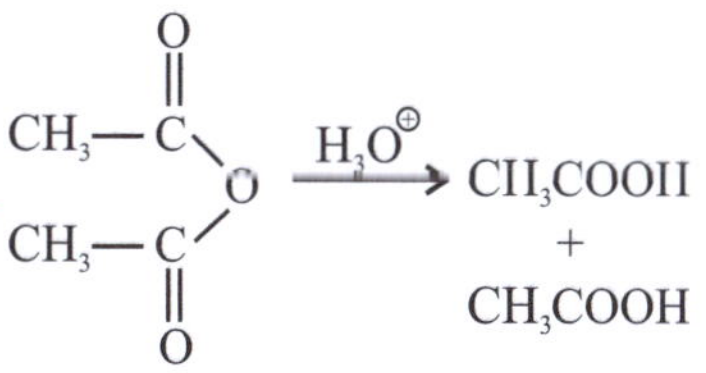

The mechanism of reaction is

1) S_N1
2) S_N2
3) SN_{AE}
4) SN_{EA}

25. Find the product of the following reaction:

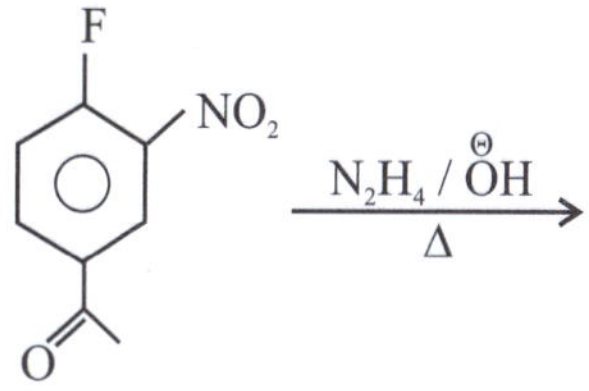

1) 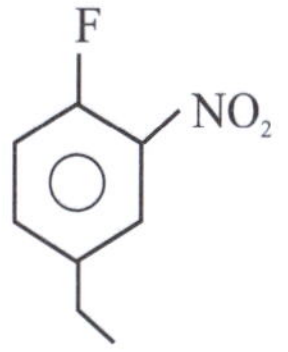
2) 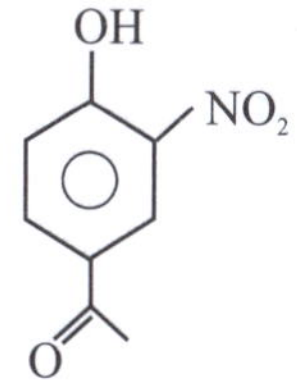
3) 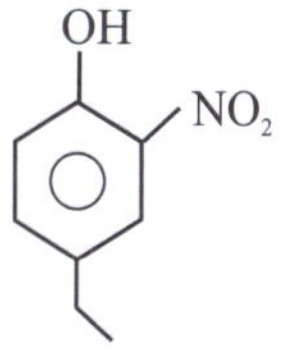
4) 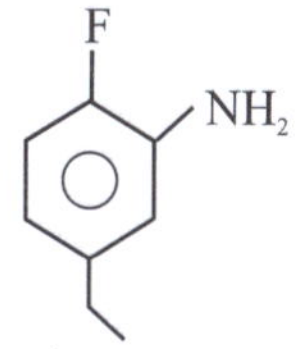

26. Identify 'A' in the following reaction:

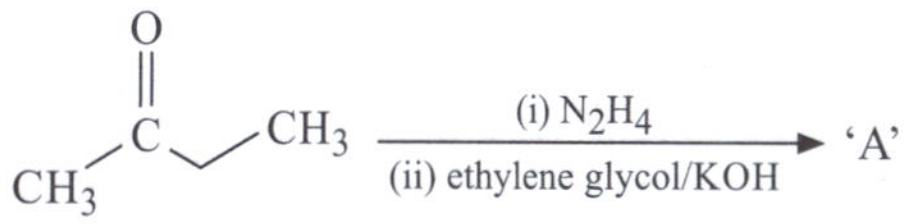

1) 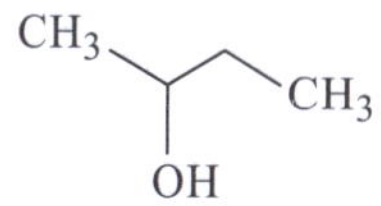
2) 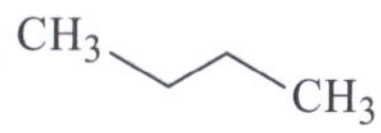
3) 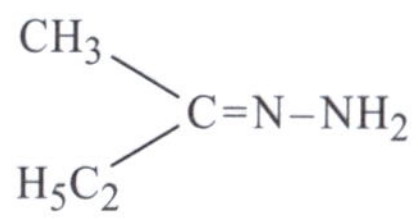
4) 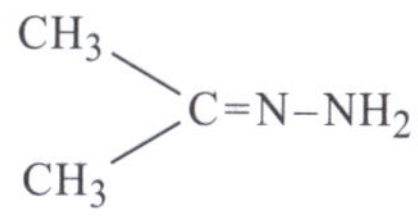

27. 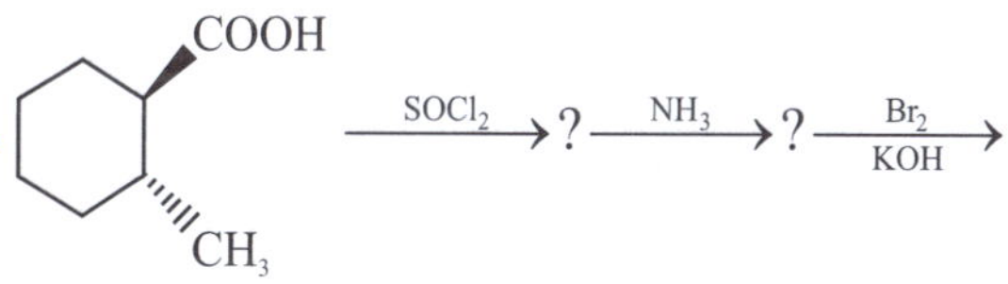

Product is-

1) 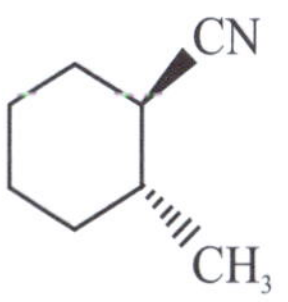
2) 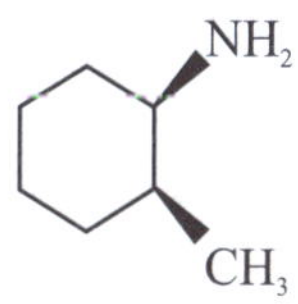
3) 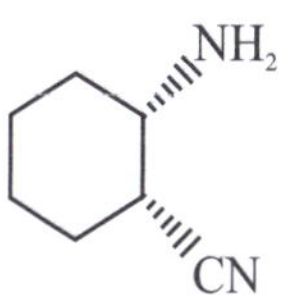
4) 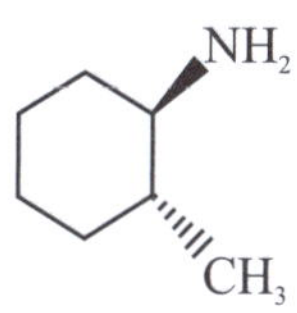

28. Which of the following reactions involve decarboxylation?

1) $CH_3COOAg + Br_2 \xrightarrow[\Delta]{CCl_4}$

2) 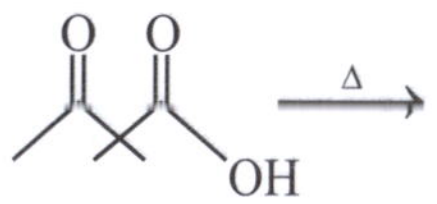

3) 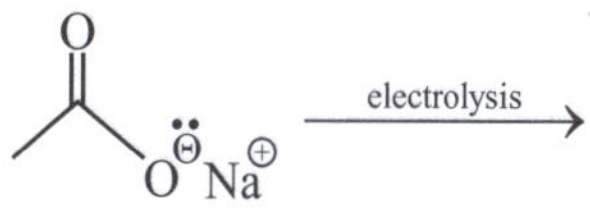

4) All of these

29. 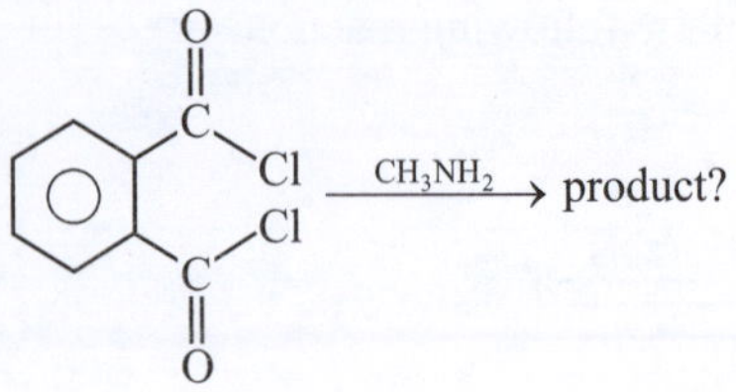

1) 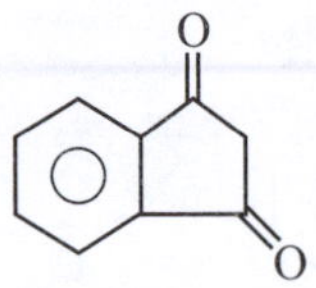
2) 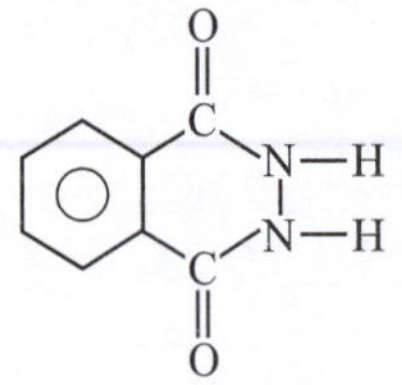
3) 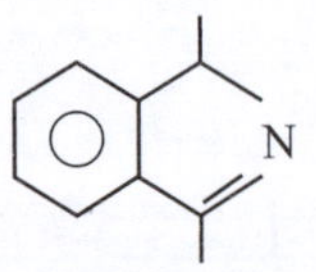
4) 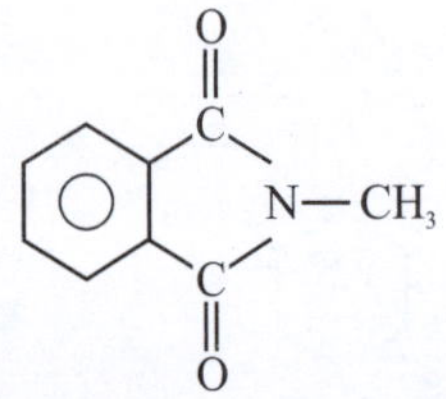

30. In the given reaction, the major product (P) will be-

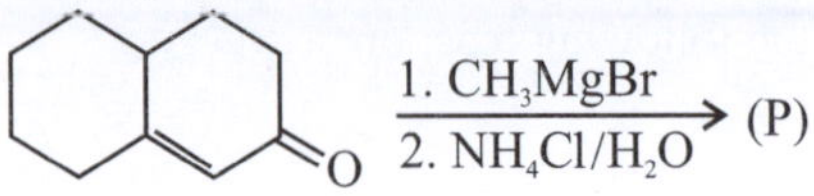

1) 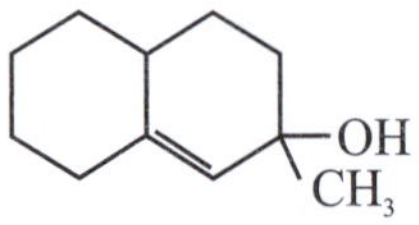
2) 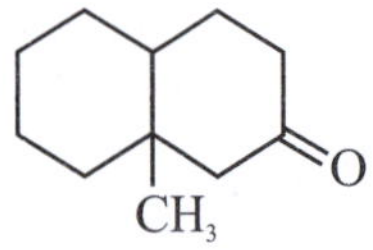
3) 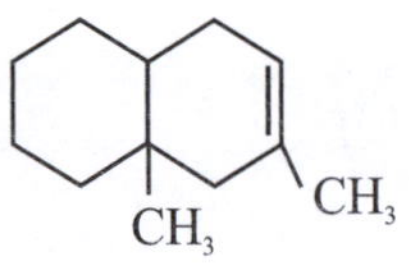
4) 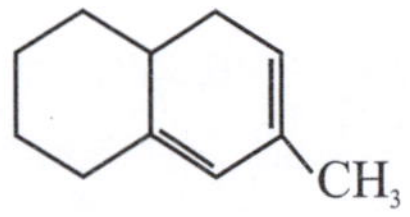

31. When CH_2=CH–COOH is reduced with $LiAlH_4$, the compound obtained will be:

1) CH_3–CH_2–COOH
2) CH_2=CH–CH_2OH
3) CH_3–CH_2–CH_2OH
4) CH_3–CH_2–CHO

32. Acid hydrolysis of which of the following compounds yields two different organic compounds?

1) CH_3COOH
2) CH_3CONH_2
3) $CH_3COOC_2H_5$
4) $(CH_3CO)_2O$

33. 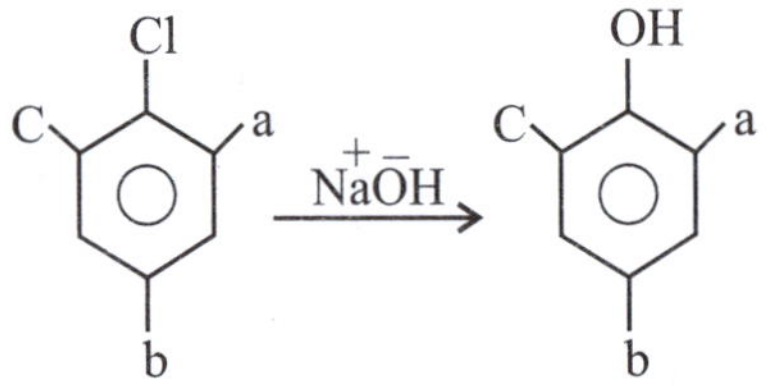

This conversion will be very fast if a, b, c respectively are-

1) $-CH_3, -NO_2, -OH$
2) $-NO_2, -NO_2, -CH_3$
3) $-NO_2, -NO_2, -NO_2$
4) $-NO_2, -CH_3, -OCH_3$

34. In which pairs, first compound is stronger acid than the second?

1) Adipic acid, succinic acid
2) Fumaric acid, maleic acid
3) Phthalic acid, terephthalic acid
4) Picric acid, o-toluic acid

35. In a set of reactions, acetic acid yielded a product D.

$$CH_3COOH \xrightarrow{SOCl_2} A \xrightarrow[\text{Anhy.}AlCl_3]{\text{Benzene}} B \xrightarrow{HCN} C \xrightarrow{H_3O^+} D$$

The structure of D would be:

1) 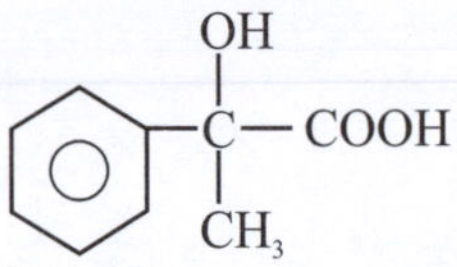
2) 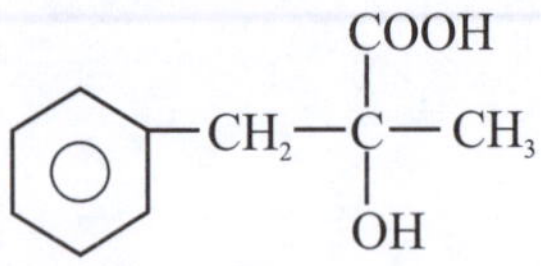
3) 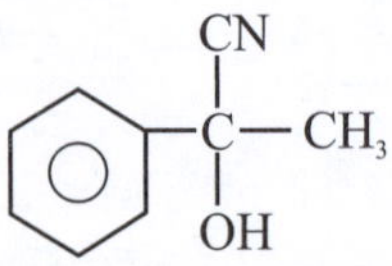
4) 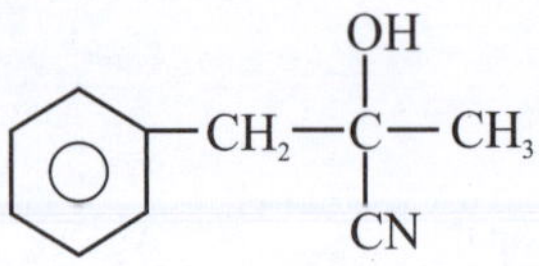

36. Which of the following compounds will undergo decarboxylation on heating?

1) 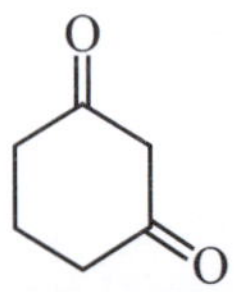
2) 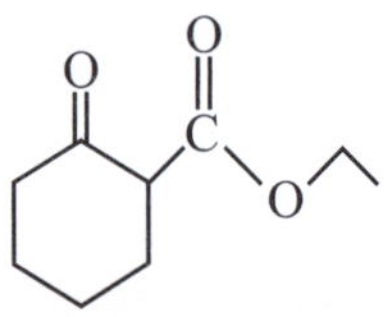
3) 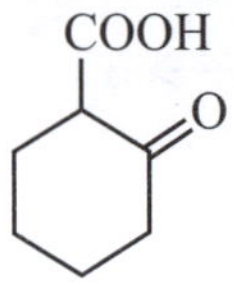
4) 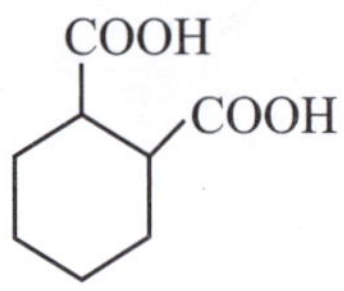

37. What product is obtained on heating Adipic acid?

1) 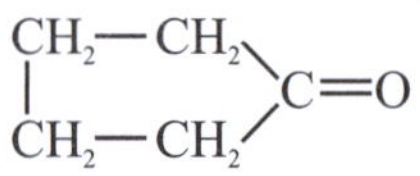
2) 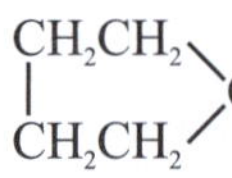
3) 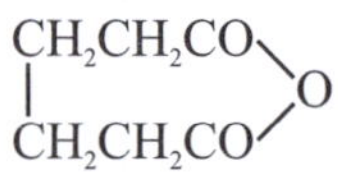
4) 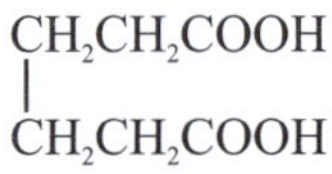

38. The reagent(s) which can be used to distinguish acetophenone from benzophenone is (are)

1) 2, 4-dinitrophenylhydrazine
2) Aqueous solution of $NaHSO_3$
3) Benedict reagent
4) I_2 and Na_2CO_3

39. Which among the following aldehydes is most reactive towards nucleophilic addition reactions?

1) H—C(=O)—H
2) C_2H_5—C(=O)—H
3) CH_3—C(=O)—H
4) C_3H_7—C(=O)—H

40. 2,4-DNP test can be used to identify :

1) halogens
2) amine
3) aldehyde
4) ether

41. Which of the following statements regarding chemical properties of acetophenone are wrong?

I. It is reduced to methyphenylcarbinol by sodium and ethanol

II. It is oxidized to benzoic acid with acidified $KMnO_4$.

III. It does not undergo aldol condensation.

IV. It does not undergo iodoform reaction with iodine and alkali

1) I and II
2) II and IV
3) III and IV
4) I and III

42. An organic compound 'A' has the molecular formula C_3H_6O. It undergoes iodoform test. When saturated with HCl it gives 'B' of molecular formula $C_9H_{14}O$. 'A' and 'B' respectively are

1) propanal and mesitylene
2) Propanone and mesityl oxide
3) Propan-2-one and 2, 6-dimethyl-2, 5-heptadien-4-one
4) Propanone and mesitylene oxide

43. 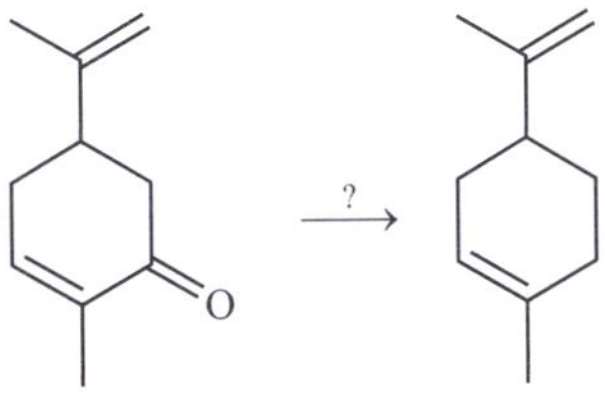

Which of the following reagent is suitable for the preparation of the product in the above reaction?

1) Red P + Cl_2
2) Ni/H_2
3) $NH_2 - NH_2/C_2H_5O^{\ominus}Na^{+}$
4) $NaBH_4$

44. 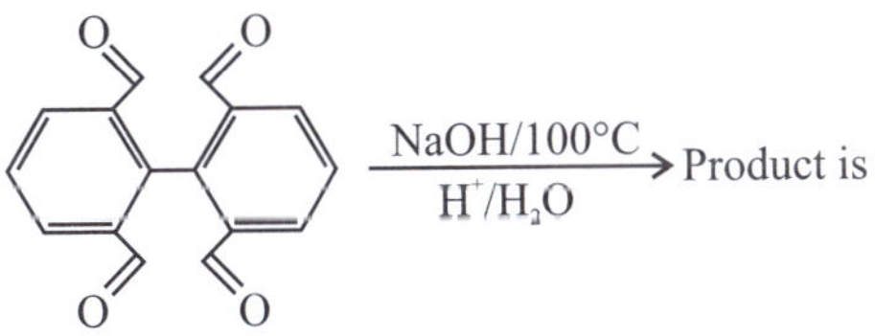

1) 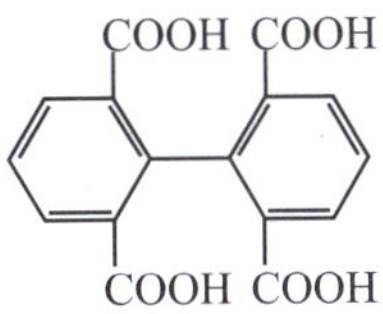
2) 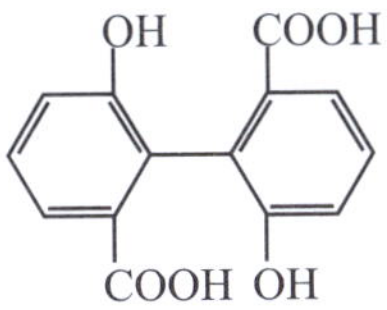
3) 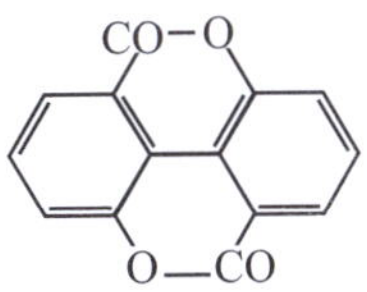
4) 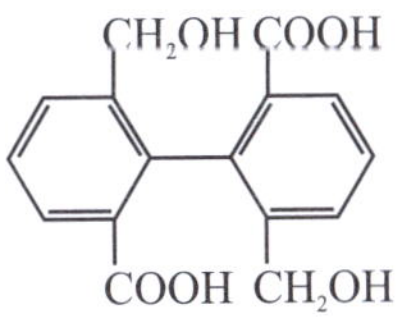

45. Suitable reagents to convert methyl acetate into ethanol is/are

1. $NaBH_4$
2. $LiAlH_4$
3. H_2/Pd

1) 1,3
2) 2,3
3) Only 3
4) Only 2

46. The products A and B in the reaction given below are:

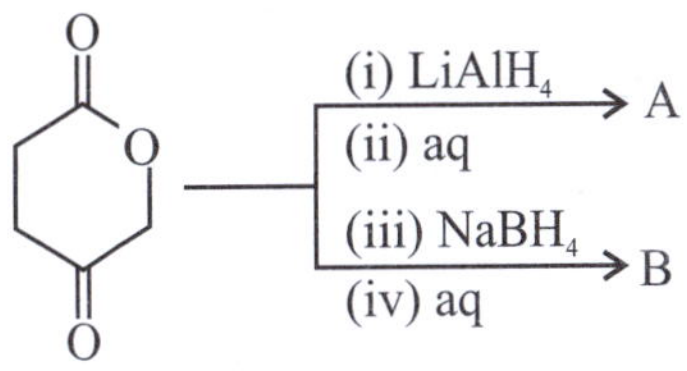

1) 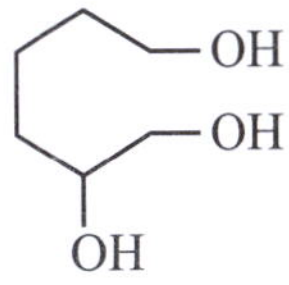 (both A and B)

2) 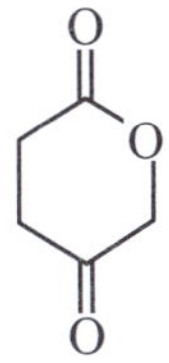 (both A and B)

3) 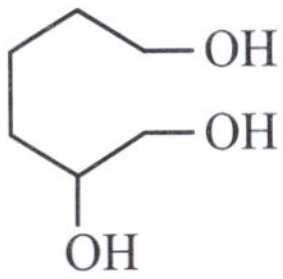 and 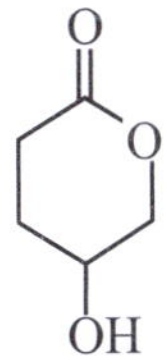

4) None of these

47. 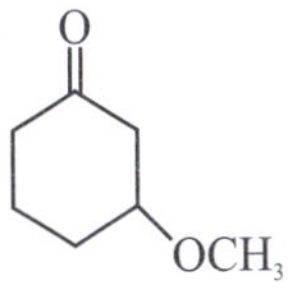
(I)
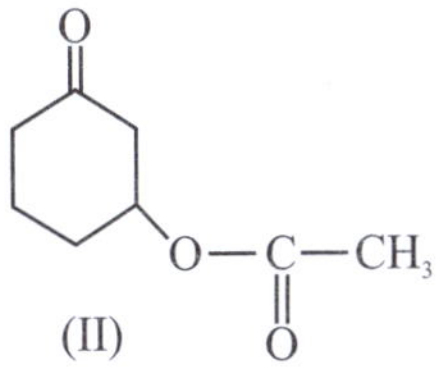
(II)

The correct statements about (I) and (II) are :

1) The acid catalysed hydrolysis of (I) and (II) yields one product identical.
2) For Wolff Kishner reduction, (I) is the better reactant.
3) Both (1) and (2).
4) Reduction by $LiAlH_4$ forms a diol in both the cases.

48. Mononitration of phenyl benzoate gives which has the major product:

1) 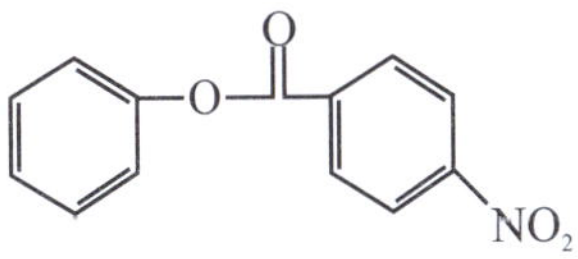
2) 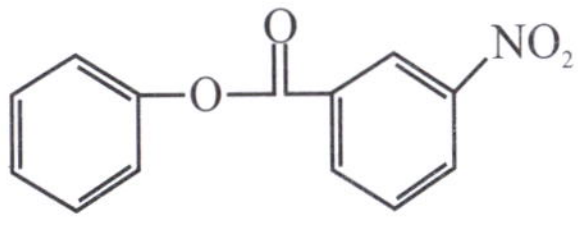
3) 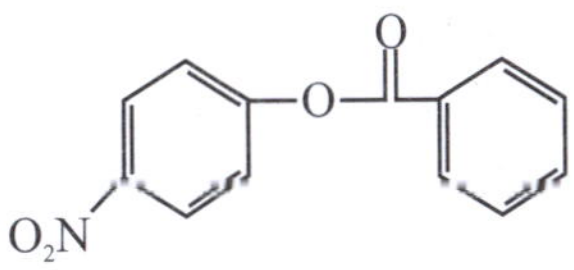
4) 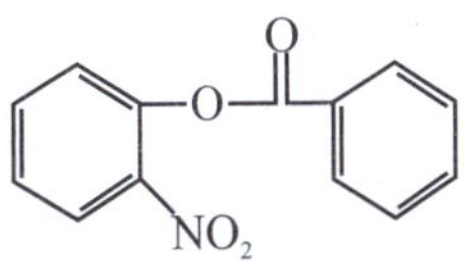

49. The increasing order of the reactivity of the following compounds in nucleophilic addition reaction is:

Propanal, Benzaldehyde, Propanone, Butanone

1) Benzaldehyde < Propanal< Propanone < Butanone
2) Butanone < Propanone < Benzaldehyde < Propanal
3) Propanal< Propanone < Butanone <Benzaldehyde
4) Benzaldehyde < Butanone < Propanone < Propanal

50. In the reaction 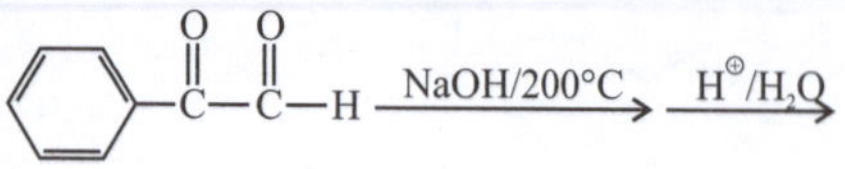

the major product is:

1) 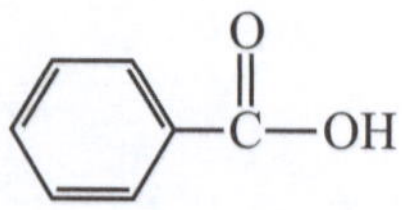 2) 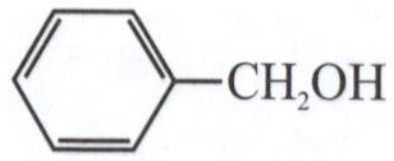

3) 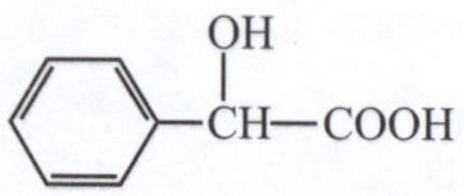 4) 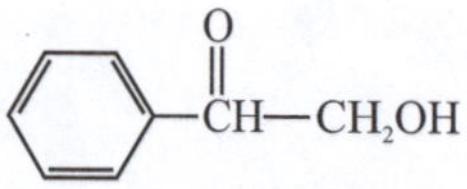

51. 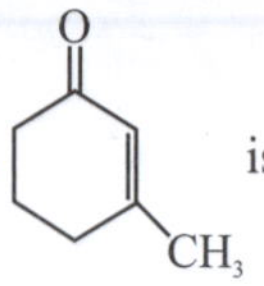 is the final product obtained when one of the following is reacted with base.

1) 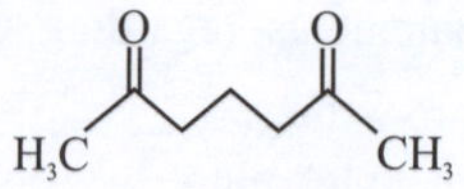

2) 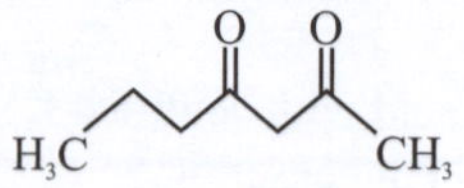

3) 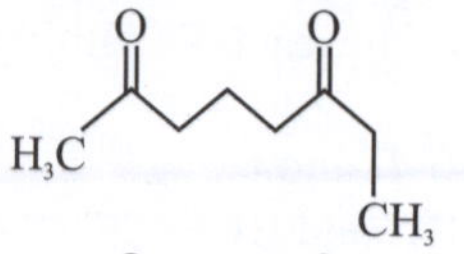

4) 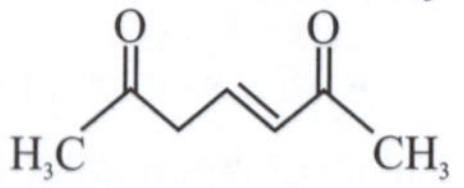

52. The compound that neither forms semicarbazone nor oxime is

1) HCHO 2) CH_3COCH_2Cl
3) CH_3CHO 4) $CH_3CONHCH_3$

53. Identify the product (P) in the following reaction:

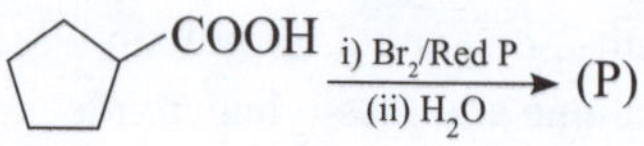

1)  2) 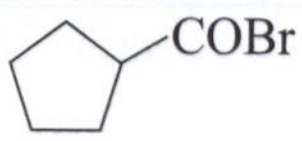

3)  4) 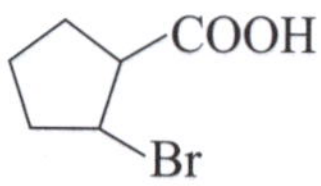

Match the Columns MCQs

1. Match the common names given in Column-I with the IUPAC names given in Column-II

Column-I (common names)		Column-II (IUPAC names)	
(a)	Cinnamaldehyde	(i)	Pentanal
(b)	Acetophenone	(ii)	Prop-2-enal
(c)	Valeraldehyde	(iii)	4-Methylpent-3-en-2-one
(d)	Acrolein	(iv)	3-Phenylprop-2-enal
(e)	Mesityl oxide	(v)	1-Phenylethanone

1) (a)-(i) (b)-(iii) (c)-(iv) (d)-(v) (e)-(ii)
2) (a)-(iv) (b)-(v) (c)-(i) (d)-(ii) (e)-(iii)
3) (a)-(iii) (b)-(i) (c)-(iv) (d)-(v) (e)-(ii)
4) (a)-(iii) (b)-(ii) (c)-(v) (d)-(i) (e)-(iv)

2. Match the acids given in Column-I with their correct IUPAC names given in Column-II.

Column-I (Acids)		Column-II (IUPAC names)	
(a)	Phthalic acid	(i)	Hexane-1,6-dioic acid
(b)	Oxalic acid	(ii)	Benzene-1,2-dicarboxylic acid
(c)	Succinic acid	(iii)	Pentane-1,5-dioic acid
(d)	Adipic acid	(iv)	Butane-1,4-dioic acid
(e)	Glutaric acid	(v)	Ethane-1,2-dioic acid

1) (a)-(ii) (b)-(i) (c)-(iii) (d)-(iv) (e)-(v)
2) (a)-(iii) (b)-(ii) (c)-(iv) (d)-(v) (e)-(i)
3) (a)-(ii) (b)-(v) (c)-(iv) (d)-(i) (e)-(iii)
4) (a)-(v) (b)-(iii) (c)-(ii) (d)-(iv) (e)-(i)

3. Match the reactions given in Column-I with the suitable reagents given in Column-II.

Column-I (Reactions)		Column-II (Reagents)	
(a)	Benzophenone → Diphenylmethane	(i)	$LiAlH_4$
(b)	Benzaldehyde → 1-Phenylethanol	(ii)	DIBAL—H
(c)	Cyclohexanone → Cyclohexanol	(iii)	Zn(Hg)/conc. HCl
(d)	Phenyl benzoate → Benzaldehyde	(iv)	CH_3MgBr

1) (a)-(iii) (b)-(iv) (c)-(i) (d)-(ii)
2) (a)-(iv) (b)-(i) (c)-(iii) (d)-(ii)
3) (a)-(ii) (b)-(i) (c)-(iii) (d)-(iv)
4) (a)-(iv) (b)-(iii) (c)-(ii) (d)-(i)

4. Match the example given in Column-I with the name of the reaction in Column-II.

	Column-I (Example)		Column-II (Reaction)
(a)	$CH_3-\overset{O}{\overset{\Vert}{C}}-Cl + H_2 \xrightarrow{Pd-C/BaSO_4} CH_3-\overset{O}{\overset{\Vert}{C}}-H$	(i)	Friedel Crafts acylation
(b)	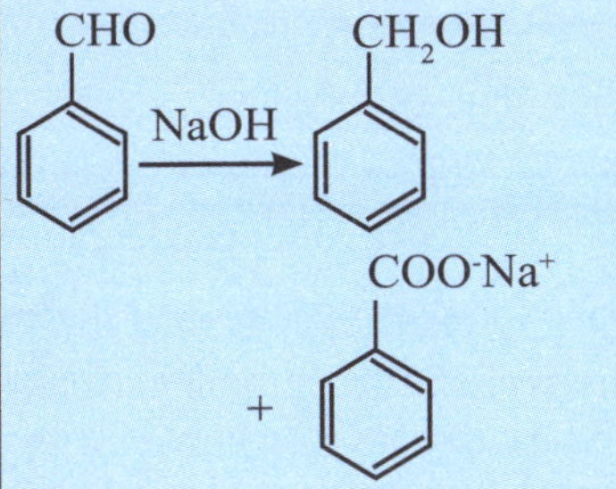	(ii)	HVZ reaction
(c)	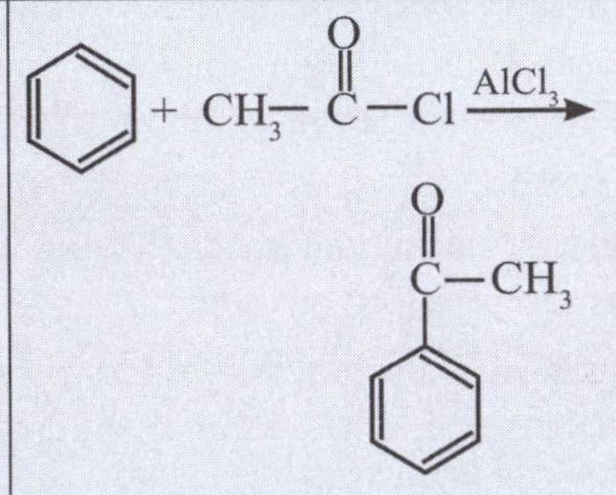	(iii)	Aldol condensation
(d)	$R-CH_2-COOH \xrightarrow{Br_2/Red\ P} R-\underset{Br}{\underset{\vert}{C}}H_2-COOH$	(iv)	Cannizaro's reaction
(e)	$CH_3-CN \xrightarrow[(ii)\ H_2O/H^+]{(i)\ SnCl_2/HCl} CH_3CHO$	(v)	Rosenmund's reduction
(f)	$2CH_3CHO \xrightarrow{NaOH} CH_3-CH=CHCHO$	(vi)	Stephen's reaction

1) (a)-(ii) (b)-(iii) (c)-(iv) (d)-(i) (e)-(vi) (f)-(v)
2) (a)-(i) (b)-(iv) (c)-(iii) (d)-(ii) (e)-(vi) (f)-(v)
3) (a)-(iii) (b)-(iv) (c)-(ii) (d)-(i) (e)-(vi) (f)-(v)
4) (a)-(v) (b)-(iv) (c)-(i) (d)-(ii) (e)-(vi) (f)-(iii)

5. Match the example given in Column-I with the name of the reaction in Column-II.

	Column-I		Column-II
(a)	$CH_3CH_2CH_2CH_2OH \xrightarrow{PCC}$	(i)	$(CH_3CH_2)_2C=O$
(b)	$C_6H_5CH_3 \xrightarrow{CrO_2Cl_2}$	(ii)	CH_3COCH_3
(c)	$CH_3COCl + (CH_3)_2Cd \xrightarrow{CrO_2Cl_2}$	(iii)	$CH_3CH_2CH_2CHO$
(d)	$(CH_3CH_2COO)_2Ca \xrightarrow{\Delta}$	(iv)	C_6H_5CHO

1) (a)-(iii) (b)-(iv) (c)-(ii) (d)-(i)
2) (a)-(iv) (b)-(iii) (c)-(ii) (d)-(i)
3) (a)-(iii) (b)-(iv) (c)-(i) (d)-(ii)
4) (a)-(i) (b)-(ii) (c)-(iii) (d)-(iv)

Correct & Incorrect MCQs

1. Which of the following statement (s) are correct?

(a) The IUPAC name of formaldehyde is Methanal.

(b) The IUPAC name of acetaldehyde is Ethanal.

(c) The IUPAC name of isobutyraldehyde is 2-Methoxypropanal.

(d) The IUPAC name of acrolcin is Prop-2 enal.

1) (a), (b) and (d) 2) Only (a) and (d)
3) (a), (c) and (d) 4) (b), (c) and (d)

2. Which of the following statement about carbonyl group is incorrect?

1) The carbonyl carbon atom is sp^2-hybridised and forms three sigma (σ) bonds.
2) The carbonyl compounds are polar than ethers.
3) The carbon-oxygen double bond is polarised due to higher electronegativity of oxygen relative to carbon.
4) The carbonyl oxygen is an electrophilic and the carbonyl carbon is a nucleophilic centre.

3. Choose the incorrect statement among the following:

1) Aldehydes and ketones are generally prepared by oxidation of primary and secondary alcohols, respectively.
2) Addition of water to ethyne in the presence of H_2SO_4 and $HgSO_4$ gives acetone.
3) In Stephen rcaction, nitriles are reduced to corresponding imine with stannous chloride in the presence of hydrochloric acid, which on hydrolysis give corresponding aldehyde.
4) Treating a nitrile with Grignard reagent followed by hydrolysis yields a ketone.

4. Select the incorrect statement:

1) Methanal is a gas at room temperature.
2) Ethanal is a volatile liquid
3) The boiling points of aldehydes and ketones are higher than hydrocarbons and ethers of comparable molecular masses.
4) The boiling points of aldehydes and ketones are more than those of alcohols of similar molecular masses.

5. Which of the following statements is incorrect ?
   1) Methanal is miscible with water in all proportions.
   2) Methanal is immiscible with water in all proportions.
   3) Ethanal is miscible with water in all proportions.
   4) Propanone is miscible with water in all proportions.

6. Correct statement(s) is/are
   (a) Chromyl chloride is used in the formation of benzaldehyde from toluene.
   (b) Aldehydes and ketones react with hydrogen cyanide (HCN) to yield cyanohydrins.
   (c) Ketones are more reactive than aldehydes in nucleophilic addition reaction.
   (d) Ketones react with ethylene glycol under similar conditions to form cyclic products known as ethylene glycol ketals.
   1) (a), (b) and (d)   2) (b), (c) and (d)
   3) (a) and (d)   4) (b) and (c)

7. Select the incorrect statement:
   1) Aldehydes and ketones are reduced to secondary and primary alcohols respectively by sodium borohydride ($NaBH_4$)or lithium aluminium hydride ($LiAlH_4$).
   2) Aldehydes are easily oxidised to carboxylic acids on treatment with common oxidising agents like nitric acid, potassium permanganate.
   3) Ketones are generally oxidised under vigorous conditions, i.e., strong oxidising agents and at elevated temperatures.
   4) Both (1) and (2)

8. Which of the following statements about Fehling's test is incorrect?
   1) Fehling solution A is aqueous copper sulphate solution.
   2) Fehling solution B is alkaline sodium potassium tartarate solution.
   3) Aldehydes are oxidised to corresponding carboxylate anion in Fehling's test
   4) Aromatic aldehydes also respond to Fehling's test

9. Select the incorrect statement:
   1) Aldehydes and ketones having at least one α-hydrogen undergo Aldol condensation reaction.
   2) Cyclohexanone undergoes Aldol condensation reaction.
   3) Formaldehyde undergoes Cannizzaro reaction.
   4) Benzaldehyde undergoes Aldol condensation reaction.

10. Which amongst the following is incorrect statement ?
   1) Aliphatic carboxylic acids upto nine carbon atoms are colourless liquids at room temperature with unpleasant odours.
   2) Carboxylic acids are lower boiling liquids than aldehydes, ketones and even alcohols of comparable molecular masses.
   3) Simple aliphatic carboxylic acids having upto four carbon atoms are miscible in water.
   4) Carboxylic acids are also soluble in less polar organic solvents (like benzene, ether, alcohol, chloroform, etc)

Statement Based MCQs

Directions: These questions consist of two statements each, printed as Statement-I and Statement-II. While answering these questions, you are required to choose any one of the following four responses.

1) Both Statement-I and Statement-II are correct.
2) Both Statement-I and Statement-II are incorrect.
3) Statement-I is correct & Statement-II is incorrect.
4) Statement-I is incorrect & Statement-II is correct.

1. **Statement-I:** Nitration of benzoic acid gives m-nitrobenzoic acid.
   Statement-II: Carboxyl group increases the electron density at meta-position.

2. **Statement-I:** In formaldehyde, all the four atoms are in the same plane.
   Statement-II: The carbon atom in formaldehyde is sp^{2-} hybridised.

3. **Statement-I:** Fluoroacetic acid is stronger than chloroacetic acid.
   Statement-II: Carboxylic acids are weak acids and turn blue litmus red.

4. **Statement-I:** CH_3CHO reacts with NH_3 to form urotropine.
   Statement-II: Urotropine is used as medicine in case of urinary troubles.

5. **Statement-I:** Formic acid reduces tollen's reagent.
   Statement-II: Tollen's reagent is Ammoniacal silver nitrate.

6. **Statement-I:** The α –hydrogen atom in carbonyl compound is basic.
   Statement-II: The basicity of α–hydrogen atoms of carbonyl compounds is due to the strong electron withdrawing effect of the carbonyl group and resonance stabilisation of the conjugate base.

7. **Statement-I:** Aldehydes and ketones can be differentiated on the reaction with tollen's reagent.
   Statement-II: Ketones react with tollen's reagent to form silver mirror but not aldehydes.

8. **Statement-I:** Aliphatic aldehyde can be differentiated from aromatic aldehyde by Fehling's test.
   Statement-II: Aromatic aldehydes reduce Fehling solution but aliphatic aldehydes do not.

9. **Statement-I:** Cannizzaro reaction is disproportionation reaction.
   Statement-II: In Cannizzaro reaction, one molecule of aldehyde which do not have α–hydrogen atom, is reduced to alcohol while another one is oxidised to carboxylic acid.

10. **Statement-I:** Etard reaction helps to stop the oxidation of toluene at the aldehyde stage.
   Statement-II: Chromyl chloride oxidises methyl group to a chromium complex, which on hydrolysis gives corresponding benzaldehyde.

Assertion & Reason MCQs

Directions: These questions consist of two statements each, printed as Assertion and Reason. While answering these questions, you are required to choose any one of the following four responses.

1) **Both Assertion and Reason are True and the Reason is a correct explanation of the Assertion.**
2) **Both Assertion and Reason are True but Reason is not a correct explanation of the Assertion.**
3) **Assertion is True but the Reason is False.**
4) **Assertion is False but Reason is True.**

1. **Assertion:** In HVZ reactions, acidic strength of carboxylic acid increases
   Reason: As -I group strength increase, acidic strength increases.
2. **Assertion :** Compounds containing —CHO group are easily oxidised to corresponding carboxylic acids.
   Reason: Carboxylic acids can be reduced to alcohols by treatment with $LiAlH_4$.
3. **Assertion:** The α-hydrogen atom in carbonyl compounds is less acidic.
   Reason: The anion formed after the loss of α-hydrogen atom is resonance stabilised.
4. **Assertion:** Aromatic aldehydes and formaldehyde undergo Cannizaro reaction.
   Reason: Aromatic aldehydes are almost as reactive as formaldehyde.
5. **Assertion:** Aldehydes and ketones, both react with Tollen's reagent to form silver mirror.
   Reason: Both, aldehydes and ketones contain a carbonyl group.
6. **Assertion:** Ethylidene chloride on treatment with aq. KOH yield ethanal.
   Reason: Ethylidene chloride is a vicinal dihalide.
7. **Assertion:** Ethyl acetate is more reactive than that of acetamide towards nucleophilic substitution reaction
   Reason: More +ve charge on carbonyl carbon in ethyl acetate than acetamide.
8. **Assertion:** Benzaldehyde is less reactive than acetaldehyde towards nucleophiles
   Reason: Benzaldehyde is heterocylic compound.
9. **Assertion:** 2-Pentanol and 3-Pentanol can be distinguished by iodoform test.
   Reason: Both 2-Pentanol and 3-Pentanol are primary alcohols.
10. **Assertion:** Formic acid gives black ppt with corrosive sublimate while acetic acid does not.
    Reason: Formic acid is stronger acid than acetic acid.

Multi Correct MCQs

1. Which of the following compounds will give silver mirror with ammonical silver nitrate?
   A. Formic acid B. Formaldehyde
   C. Benzaldehyde D. Acetone
   Choose the correct answer from the options given below:
   1) C and D only 2) A, B and C only
   3) A only 4) B and C only

2. Among the following, the number of compounds which will give positive iodoform reaction is
   A. 1-Phenylbutan-2-one B. 2-Methylbutan-2-ol
   C. 3-Methylbutan-2-ol D. 1-Phenylethanol
   E. 3,3-dimethylbutan-2-one
   F. 1-Phenylpropan-2-ol
   1) A, B, C and D 2) B, C, D and E
   3) B, D, E and F 4) C, D, E and F

3. 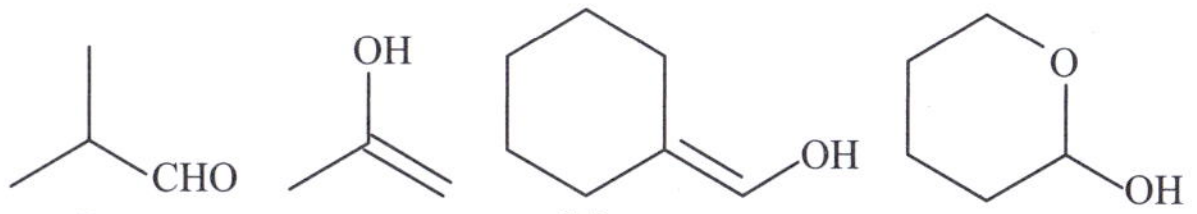

(I) (II) (III) (IV)

   Which among the above compound/s does / do not form silver mirror when treated with Tollen's reagent?
   1) Only (IV) 2) (III) and (IV) only
   3) Only (II) 4) (I), (III) and (IV) only

4. Which of the following reactions will yield benzaldehyde as a product?

A. 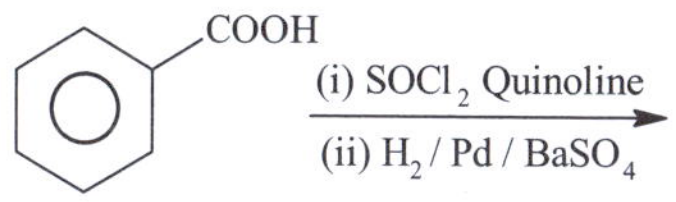

B. 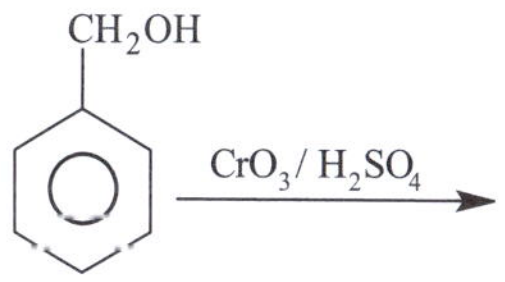

C. 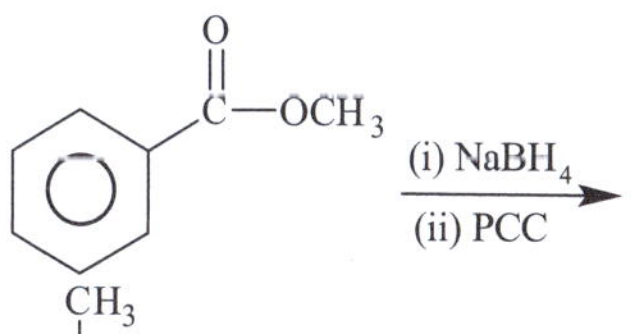

D. 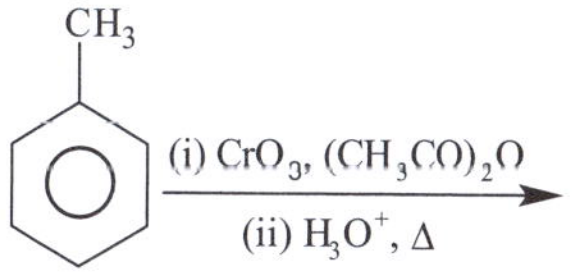

   1) (B) and (C) 2) (C) and (D)
   3) (A) and (D) 4) (A) and (C)

5. Given the following carboxylic acids:
   A. 2-Methylpropanoic acid
   B. Acetic acid
   C. 3-Methylbutanoic acid
   Arrange them in increasing order of acidity (i.e., from least acidic to most acidic):
   1) C < A < B 2) A < C < B
   3) C < B < A 4) B < A < C

NCERT Exemplar MCQs

1. Addition of water to alkynes occurs in acidic medium and in the presence of Hg^{2+} ions as a catalyst. Which of the following products will be formed on addition of water to but-1-yne under these conditions?

1) $CH_3-CH_2-CH_2-\overset{\overset{O}{\|}}{C}-H$

2) $CH_3-CH_2-\overset{\overset{O}{\|}}{C}-CH_3$

3) $CH_3-CH_2-\overset{\overset{O}{\|}}{C}-OH+CO_2\uparrow$

4) $CH_3-\overset{\overset{O}{\|}}{C}-OH+H-\overset{\overset{O}{\|}}{C}-H$

2. Which of the following compounds is most reactive towards nucleophilic addition reactions?

1) $CH_3-\overset{\overset{O}{\|}}{C}-H$

2) $CH_3-\overset{\overset{O}{\|}}{C}-CH_3$

3) 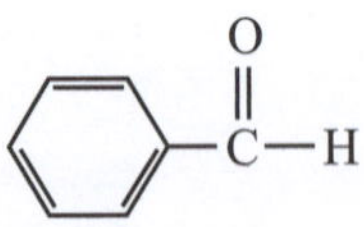

4) 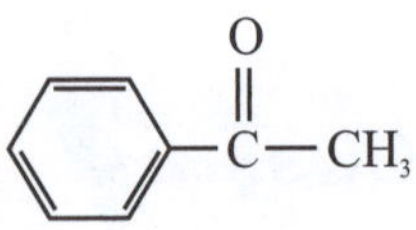

3. The correct order of increasing acidic strength is:

1) Phenol < ethanol < chloroacetic acid < acetic acid
2) Ethanol < phenol < chloroacetic acid < acetic acid
3) Ethanol < phenol < acetic acid < chloroacetic acid
4) Chloroacetic acid < acetic acid < phenol < ethanol

4. Compound $Ph-O-\overset{\overset{O}{\|}}{C}-Ph$ can be prepared by the reaction of:

1) Phenol and benzoic acid in the presence of NaOH
2) Phenol and benzoyl chloride in the presence of pyridine
3) Phenol and benzoyl chloride in the presence of $ZnCl_2$
4) Phenol and benzaldehyde in the presence of palladium

5. The reagent which does not react with both, acetone and benzaldehyde?

1) Sodium hydrogen sulphite
2) Phenyl hydrazine
3) Fehling's solution
4) Grignard reagent

6. Cannizzaro's reaction is not given by:

1) 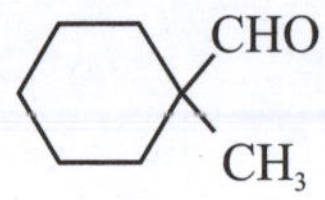

2) 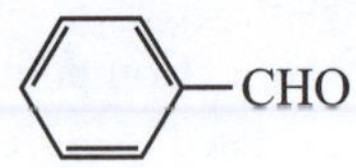

3) HCHO

4) CH_3CHO

7. Which products is formed when the compound 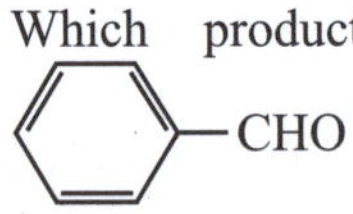 is treated with concentrated aqueous KOH solution?

1) 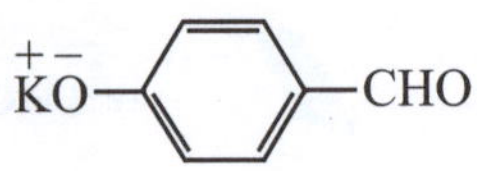

2) 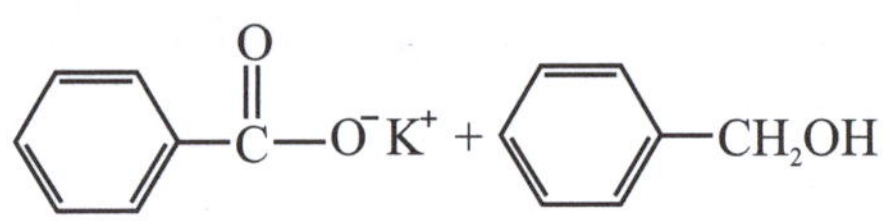

3) 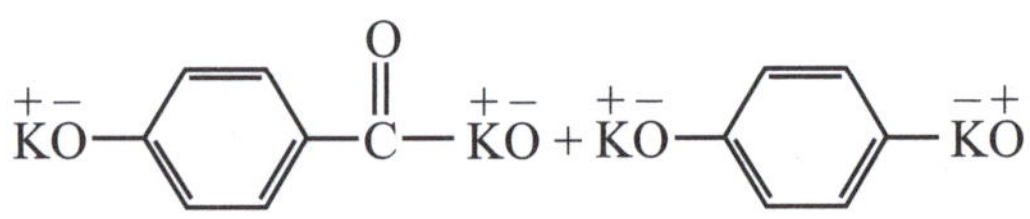

4) 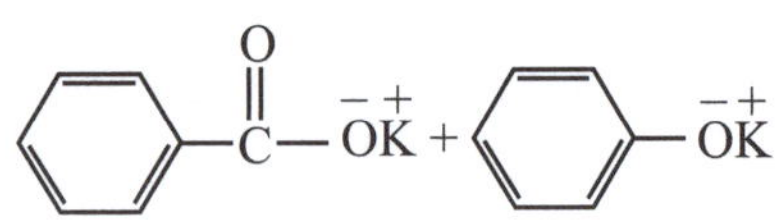

Scan Me for Detailed Explanations

Kattar NEET

1. 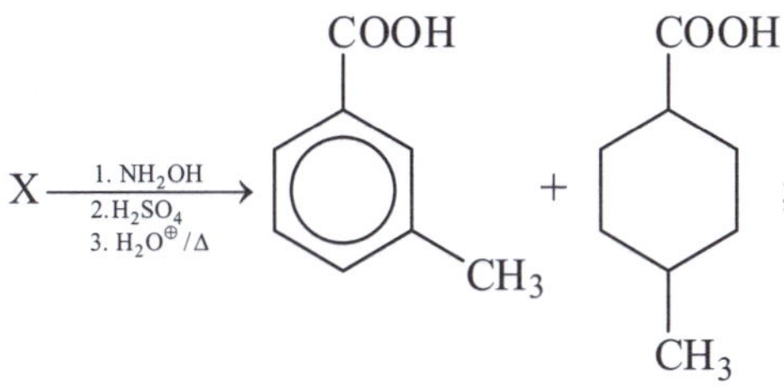 ; What is X.

1) 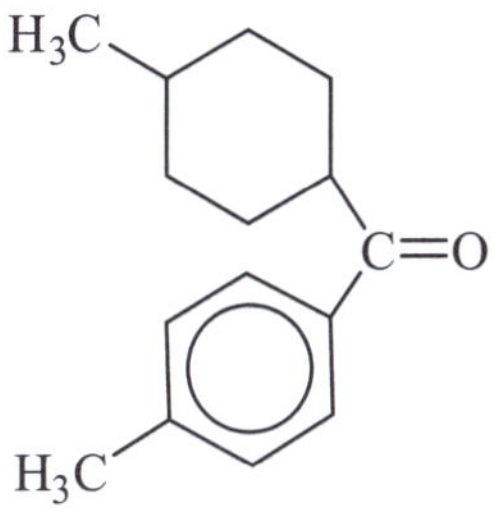

2) 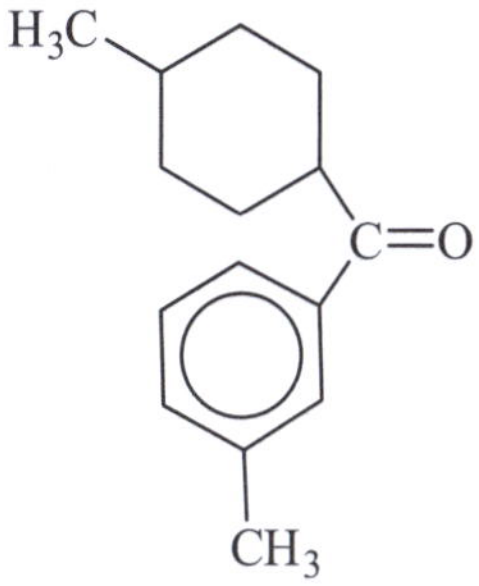

3) 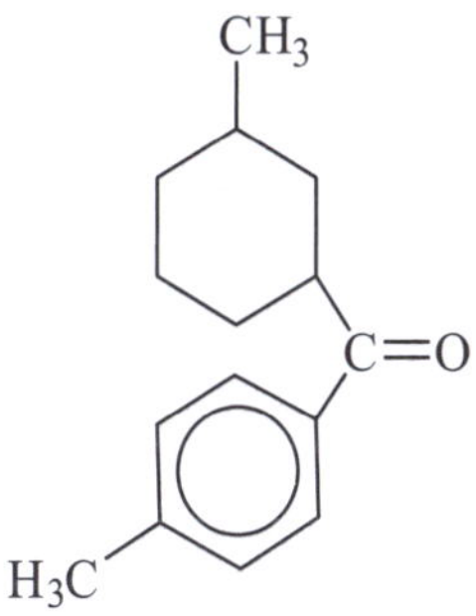

4) 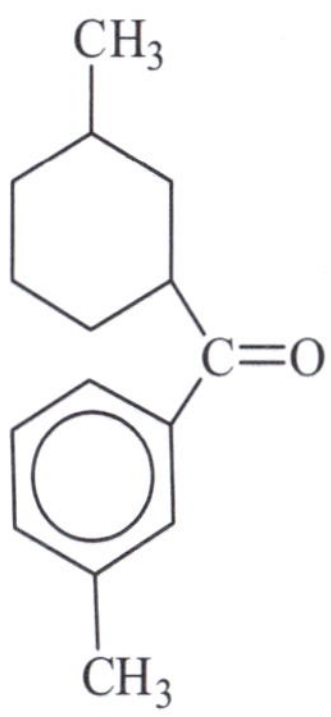

2. Which carbonyl group is protonated most readily in acidic solution, but gives nucleophilic addition by NH_2NH-Ph least readily?

1) 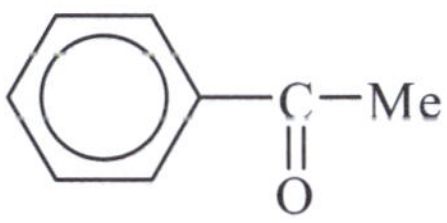

2) 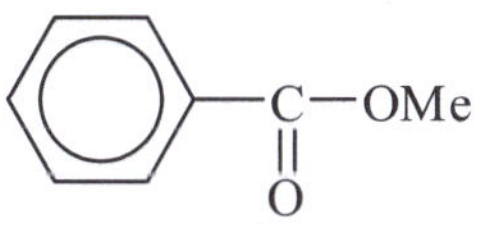

3) 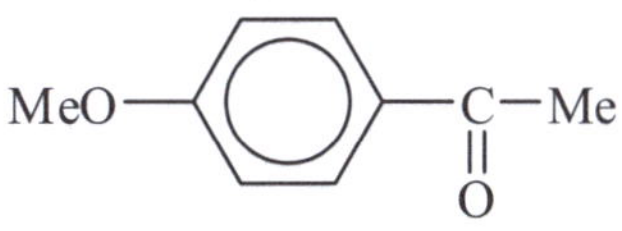

4) 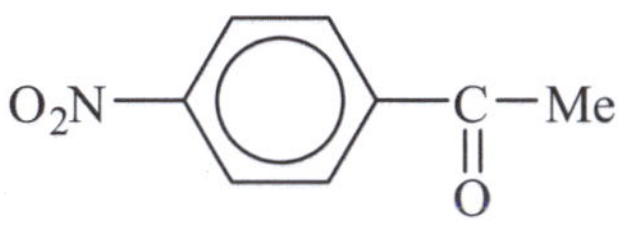

3. $PhCOCHBr_2 \xrightarrow{OH^-} A \xrightarrow{OH^-} B \xrightarrow{H^+} C$;

The compound 'C' is:

1) PhCH(OH)CHO
2) PhCH(OH)COOH
3) $PhCOCHBr_2$
4) $PhCOCH_2OH$

4. Which step is not feasible in the following loop?

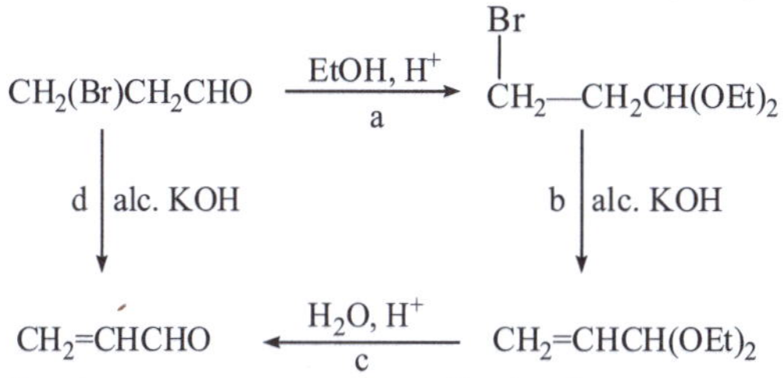

1) step-a
2) step-b
3) step-a
4) step-d

5. $CH_3-\underset{\underset{O}{\parallel}}{C}-(CH_2)_4-\underset{\underset{O}{\parallel}}{C}-CH_3 \xrightarrow[H^+]{\overset{\ominus}{O}H/\Delta} A \xrightarrow{Ph-MgBr} B \xrightarrow[\text{Conc. HCl}]{\text{Zn-Hg}} C$

1) 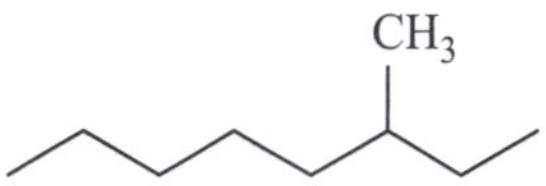

2) 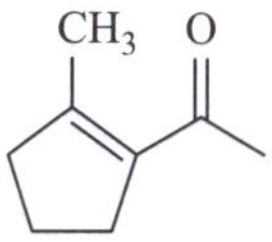

3) 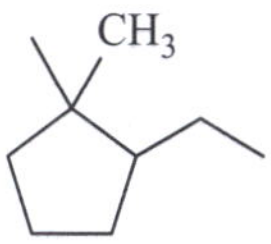

4) 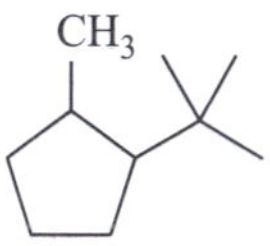

6. The final product (X) of the following reactions would be:

$C_6H_5-\underset{\underset{O}{\parallel}}{C}-C_6H_5 \xrightarrow[(1)]{NH_2OH.\ HCl} \xrightarrow[(2)]{PCl_5/\Delta} \xrightarrow[(3)]{Br_2/Fe} (X)$

1) 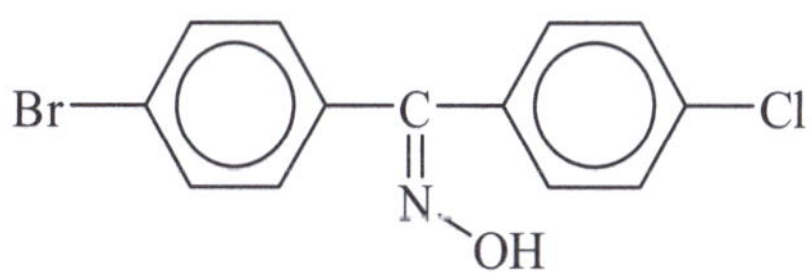

2) 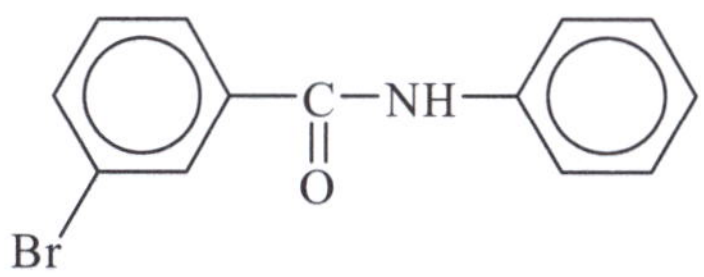

3) 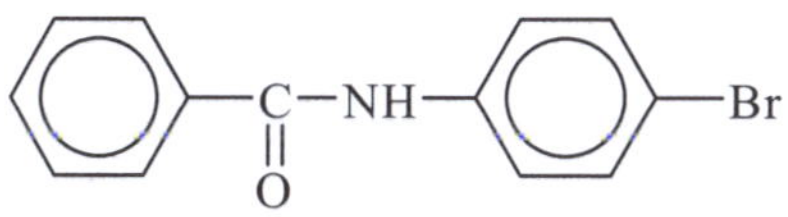

4) 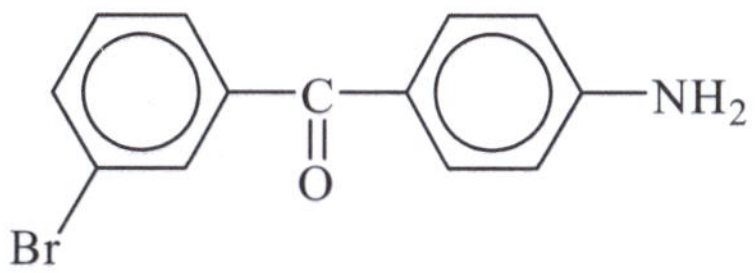

7. $CH_3COCH_3 \xrightarrow{CH_3C\equiv\overset{\oplus}{C}Na} X \xrightarrow{H_2O} Y \xrightarrow[Pd + CaCO_3]{H_2} Z \xrightarrow{Al_2O_3} W$, W is

1) $CH_2 = CH—CH = CH_2$

2) $CH_3— CH = CH — CH — CH_2$

3) 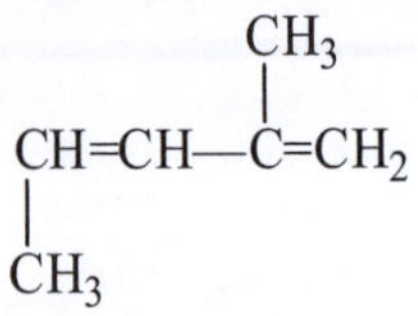

4) $CH_3— CH = CH — CH = CH — CH_3$

8. $X \xrightarrow[Zn,H_2O]{O_3} Y + Z \xrightarrow[\Delta]{OH^-}$ 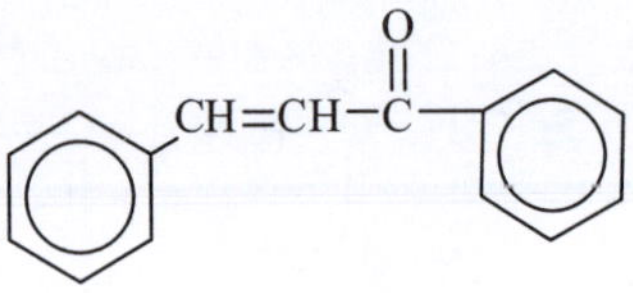

1) 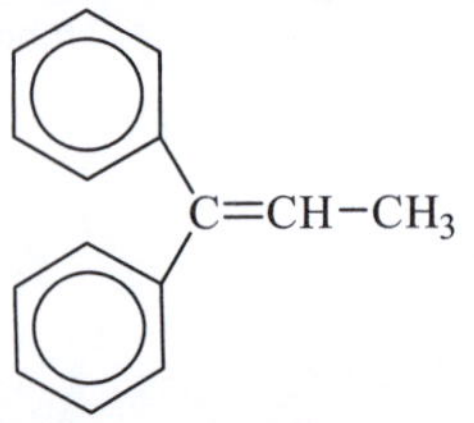

2) 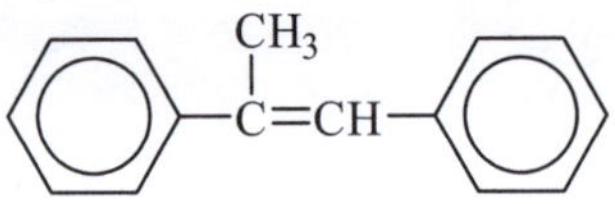

3) 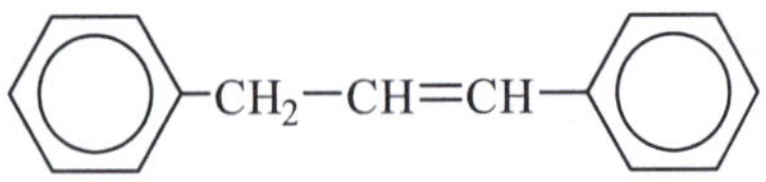

4) 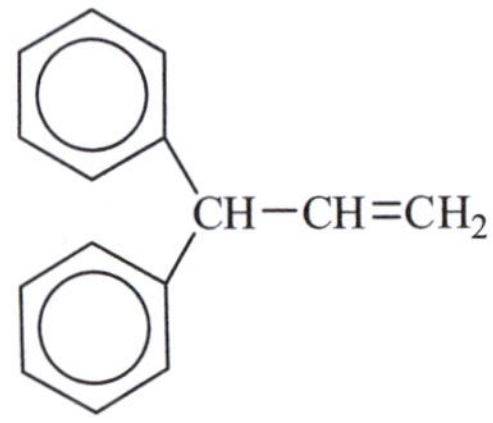

9. 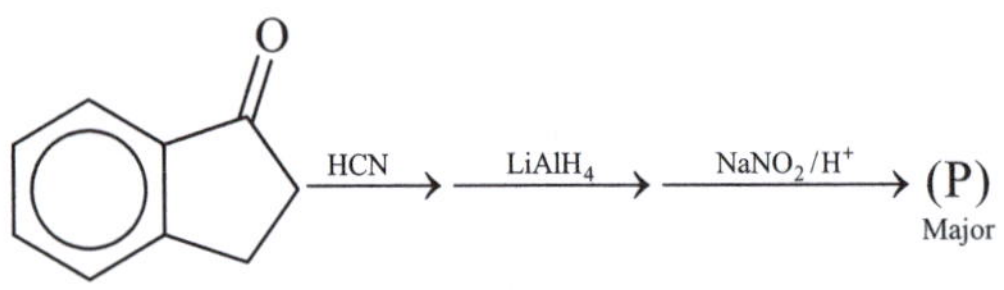

Product P will be:

1) 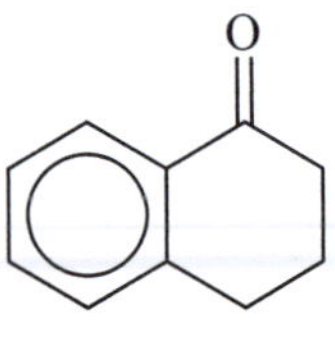
2) 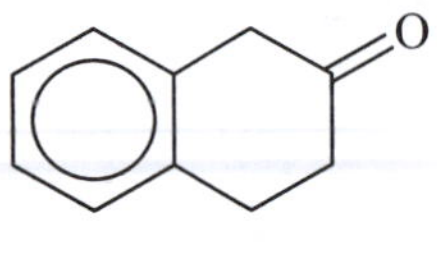

3) 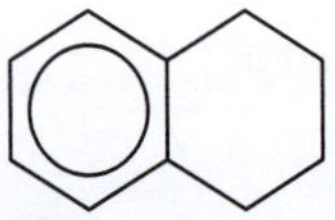
4) 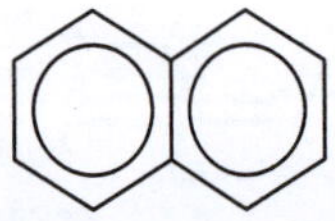

10. The product of the following reaction does not show positive test with

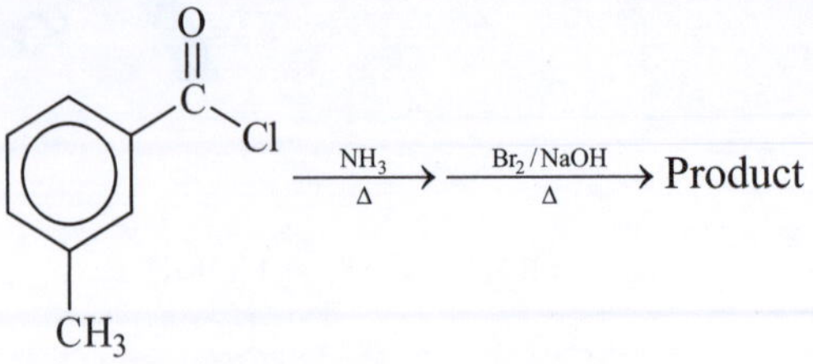

1) $(CHCl_3 + KOH)/\Delta$

2) HNO_2 followed by β-naphthol in alkali

3) $Ph—SO_2Cl/NaOH$

4) Neutral $FeCl_3$ Test

11. $H_3C-C(=O)-OH \xrightarrow{NaH\overset{14}{C}O_3} (P\ gas) \xrightarrow[2.\ H_3O^{\oplus}]{1.\ PhCH_2MgBr} (Q)$

$H_3C-S(=O)_2-OH \xrightarrow{NaH\overset{14}{C}O_3} (R\ gas) \xrightarrow[2.\ H_3O^{\oplus}]{1.\ Ph\overset{14}{C}H_2MgBr} (S)$

Product (Q) and (S) in the above reactions are respectively:

1) $PhCH_2COOH$, $Ph-\overset{14}{C}H_2-S(=O)_2-OH$

2) $PhCH_2\overset{14}{C}OOH$, $Ph\overset{14}{C}H_2COOH$

3) $PhCH_2\overset{14}{C}OOH$, $Ph\overset{14}{C}H_2\overset{14}{C}OOH$

4) $PhCH_2COOH$, $Ph\overset{14}{C}H_2\overset{14}{C}OOH$

12. 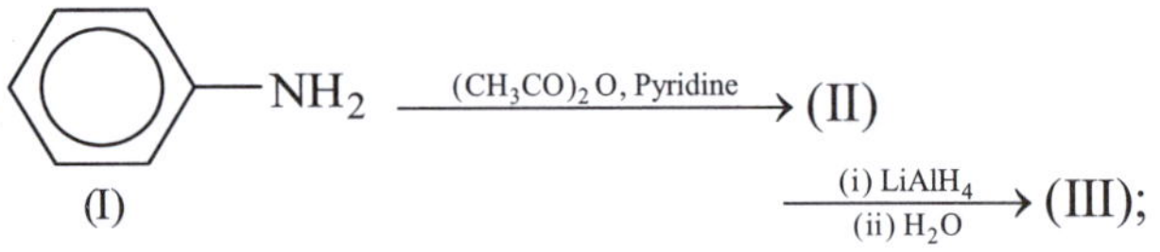

The basicity order of I, II and III is:

1) III > I > II

2) I > II > III

3) III > II > I

4) II > III > I

13. 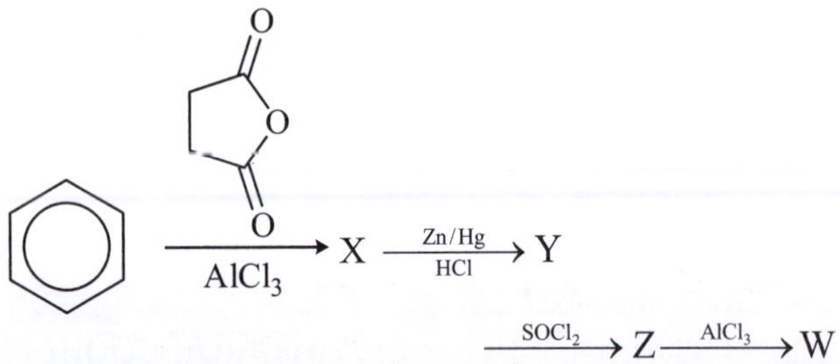

Product 'W' is:

1) 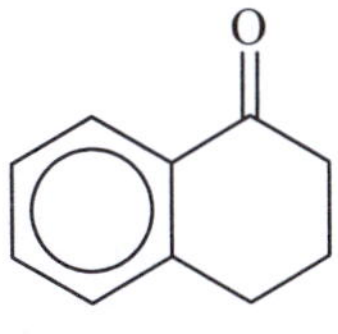
2) 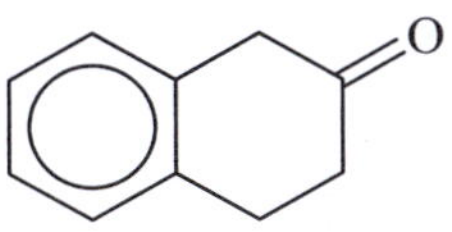
3) 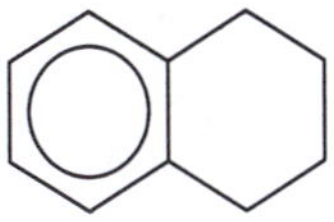
4) 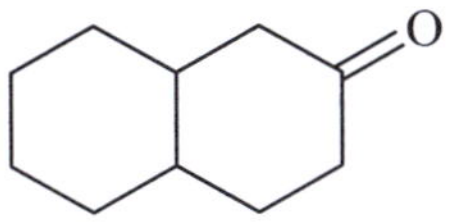

14. Identify C in the following sequence of reactions:

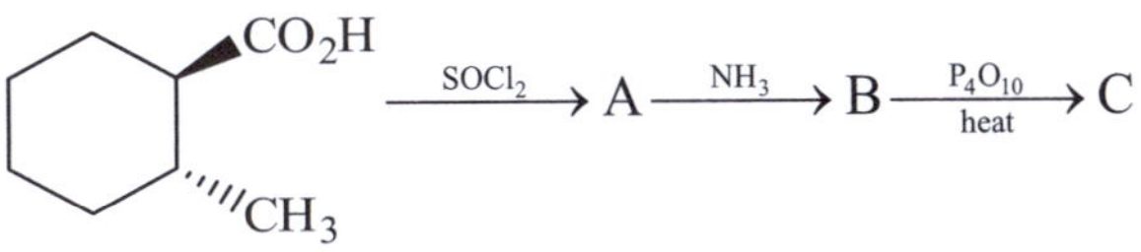

1) 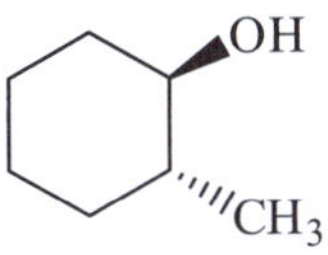
2) 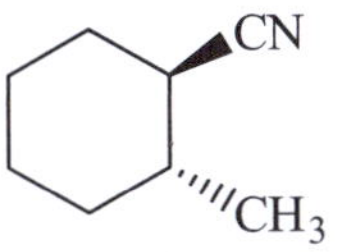
3) 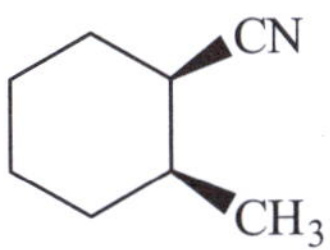
4) 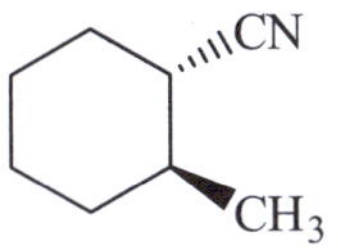

15. 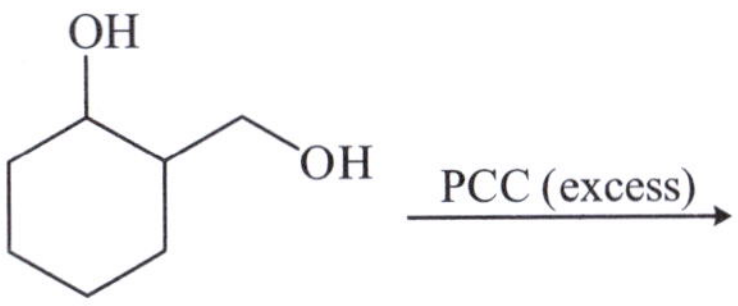

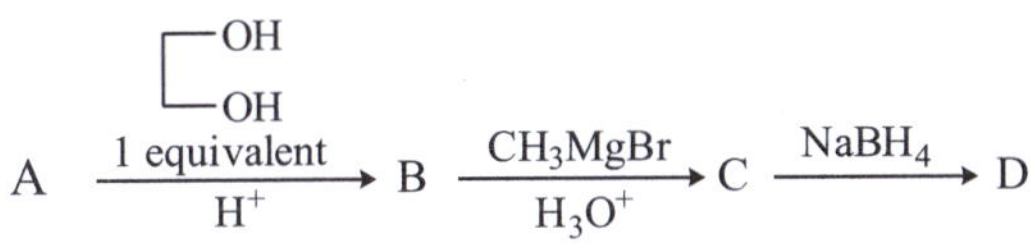

Identify product D.

1) 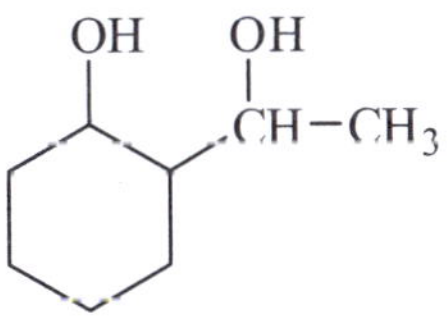
2) 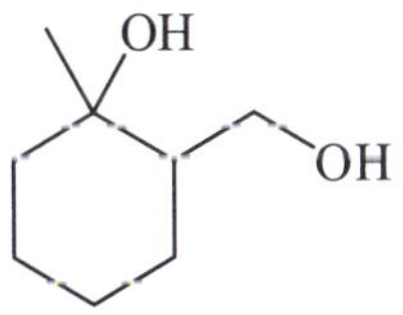
3) 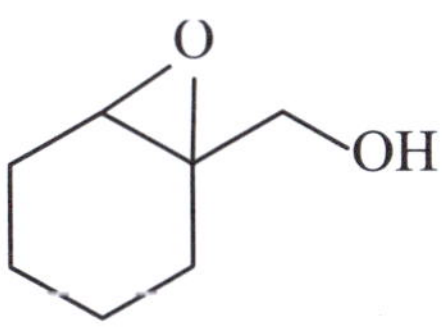
4) 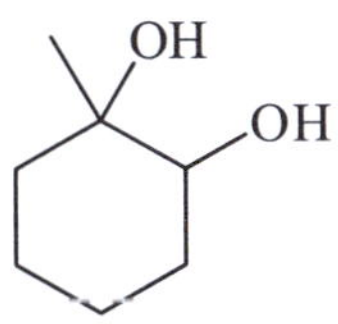

16. 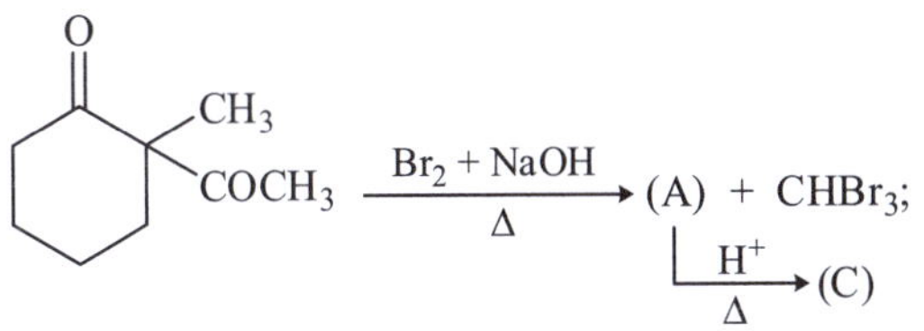

Product (C) of the reaction is

1) 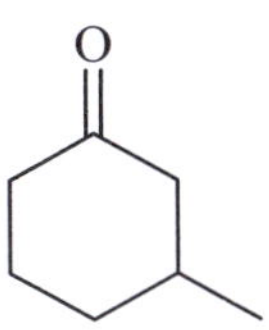
2) 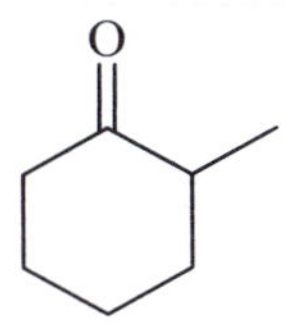
3) 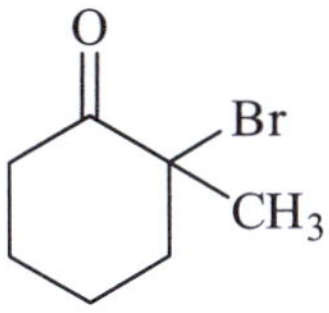
4) 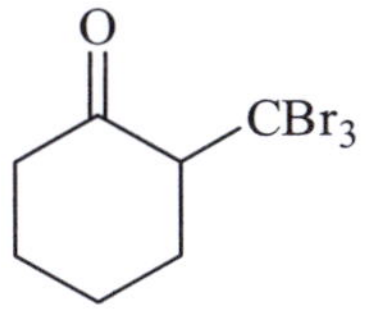

17. 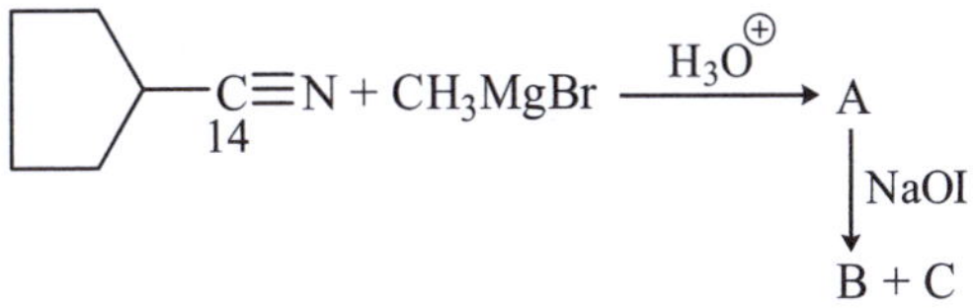

Product (A) and (C) is

1) 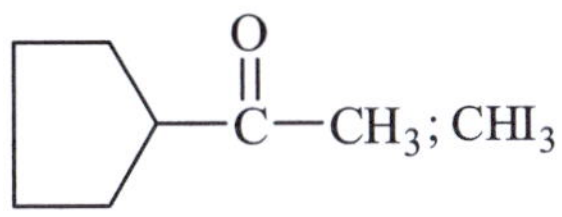
2) 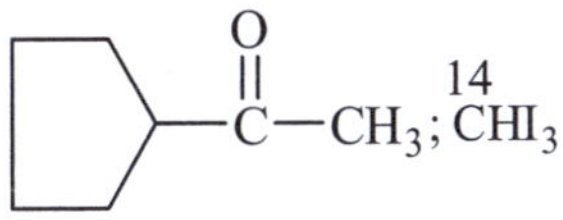
3) 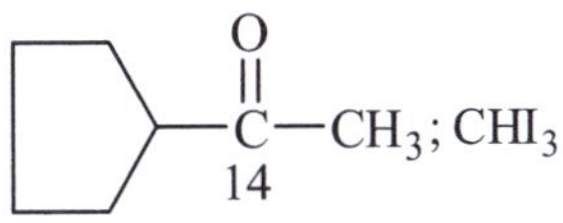
4) 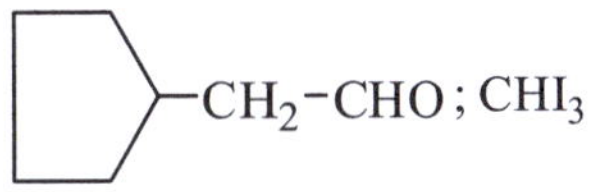

18. $A \xrightarrow[1\%\ HgSO_4]{20\%\ H_2SO_4} B \xrightarrow{I_2+NaOH} C + C_6H_5COONa$

The reactant 'A' is:

1) 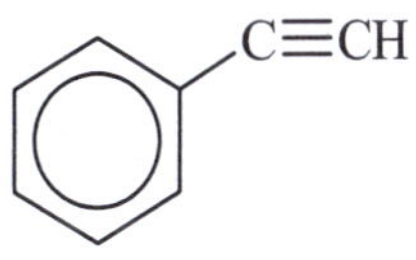
2) 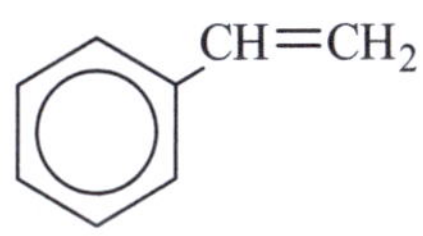
3) 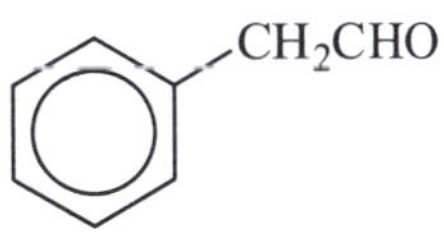
4) 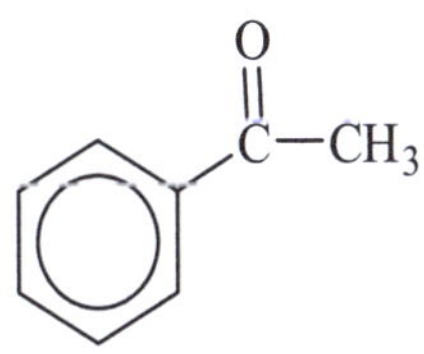

19. 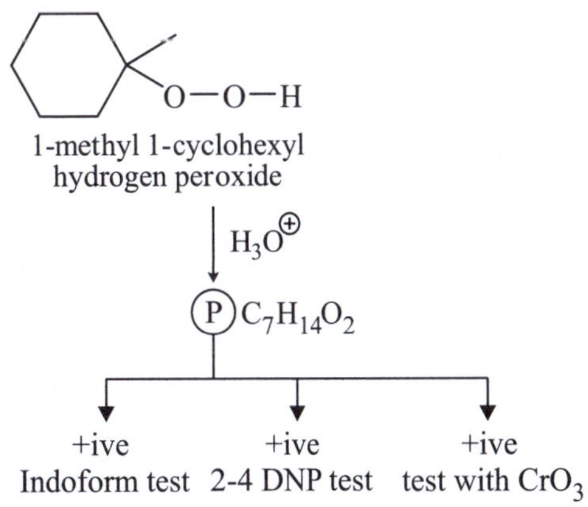

Compound (P) is:

1) 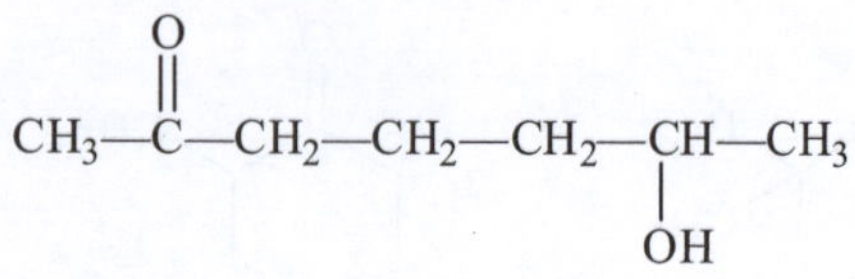

2) 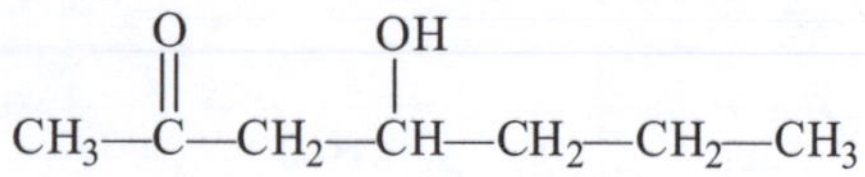

3) 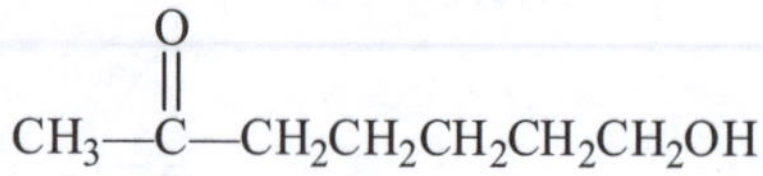

4) 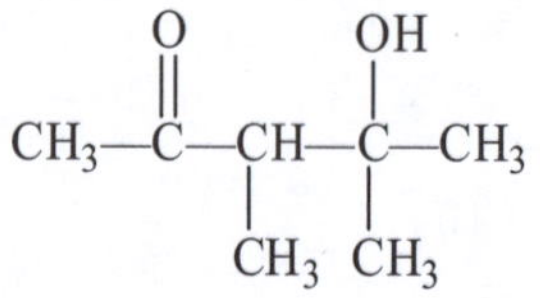

20. Which of the following compound(s) is (are) formed in the following reaction?

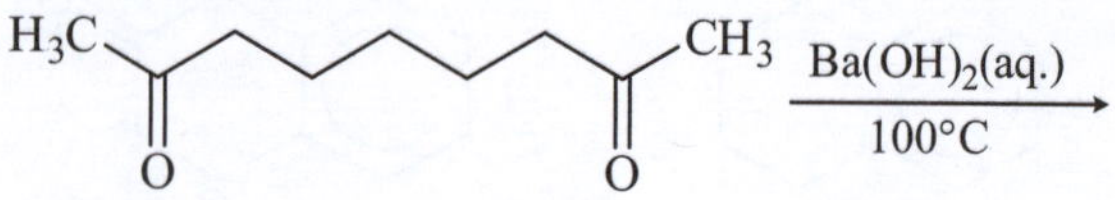

1) 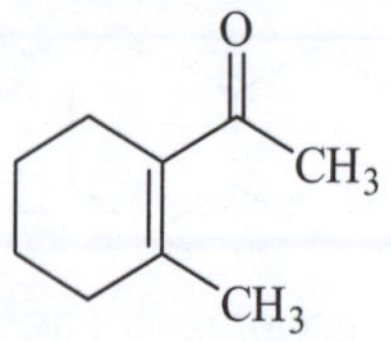

2) 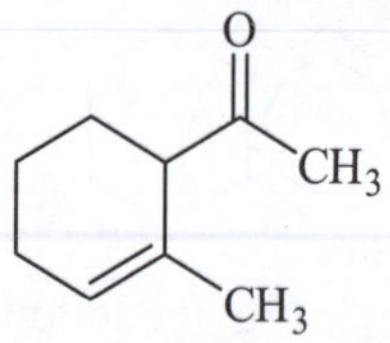

3) 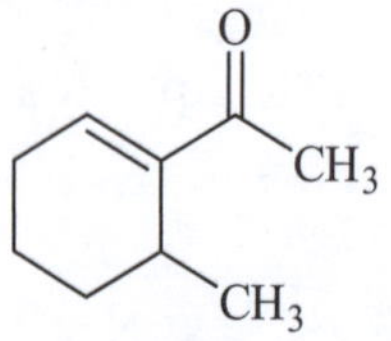

4) 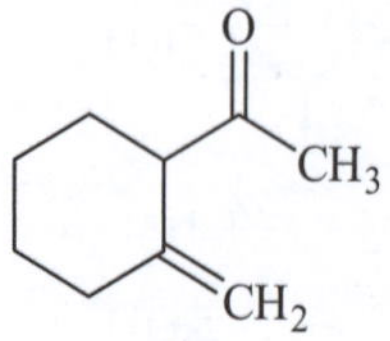

NEET & AIIMS Past Year MCQs

1. Select the incorrect reaction among the following: **(2024 Re)**

1) $CH_3COCl \xrightarrow[\Delta]{H_2O} CH_3COOH$

2) 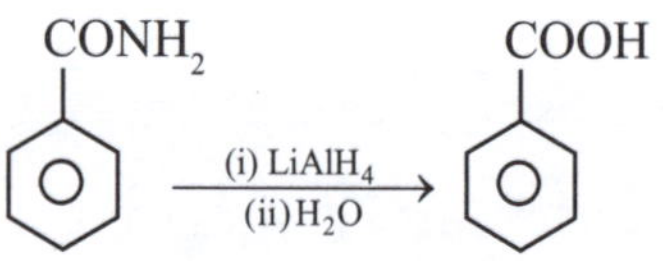

3) $CH_3CH_2OH \xrightarrow[(ii)\,H_3O^{\oplus}]{(i)\,KMnO_4/OH^-} CH_3COOH$

4) $CH_3CH_2CH_2OH \xrightarrow{CrO_3-H_2SO_4} CH_3CH_2COOH$

2. Identify D in the following sequence of reactions: **(2024 Re)**

$CH_3CH_2OH \xrightarrow{P+I_2} A \xrightarrow[\text{dry ether}]{Mg} B \xrightarrow{HCHO} C \xrightarrow{H_2O} D$ (major)

1) n-propyl alcohol 2) isopropyl alcohol
3) propanal 4) propionic acid

3. Match List-I with List-II. **(2024)**

	List-I (Reaction)		List-II (Reagents/ Condition)
A.	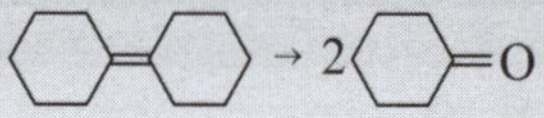	I.	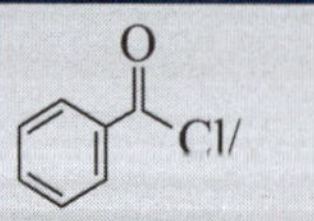 Anhyd. $AlCl_3$
B.	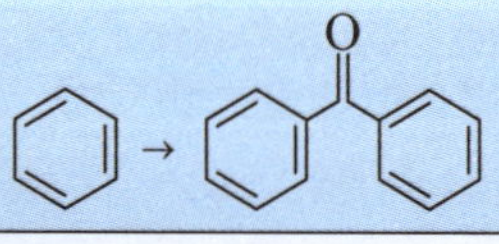	II.	CrO_3
C.	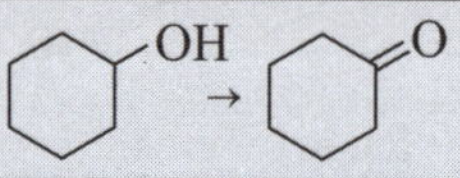	III.	$KMnO_4$/KOH, Δ
D.	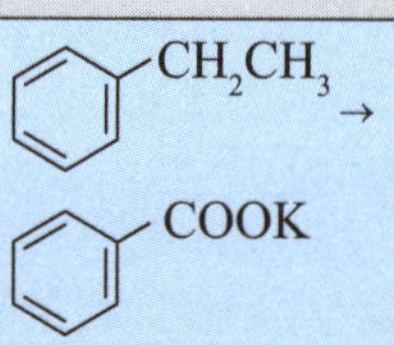	IV.	(i) O_3 (ii) Zn-H_2O

Choose the correct answer from the options given below:
1) A-IV, B-I, C-II, D-III 2) A-I, B-IV, C-II, D-III
3) A-IV, B-I, C-III, D-II 4) A-III, B-I, C-II, D-IV

4. Identify the correct reagents that would bring about the following transformation. **(2024)**

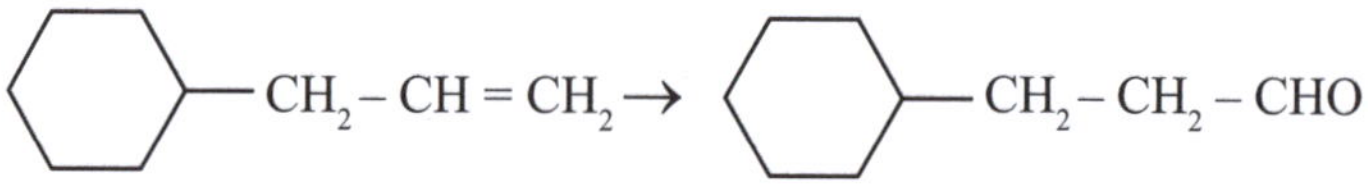

1) (i) BH_3
(ii) $H_2O_2/\overset{\ominus}{O}H$
(iii) alk. $KMnO_4$
(iv) $H_3O^{\oplus}$

2) (i) H_2O/H^+
(ii) PCC

3) (i) H_2O/H^+
(ii) CrO_3

4) (i) BH_3
(ii) $H_2O_2/\overset{\ominus}{O}H$
(iii) PCC

5. Fehling's solution 'A' is **(2024)**

1) alkaline solution of sodium potassium tartrate (Rochelle's salt)
2) aqueous sodium citrate
3) aqueous copper sulphate
4) alkaline copper sulphate

6. Identify product (A) in the following reaction: **(2023)**

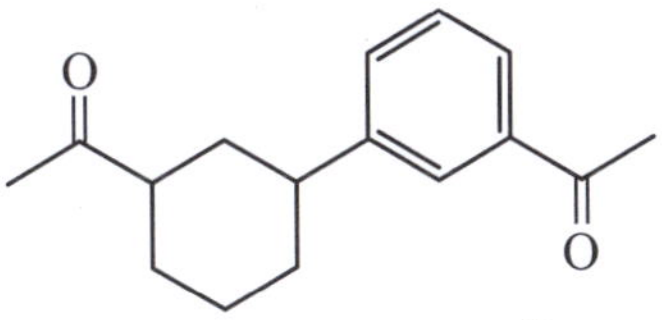 $\xrightarrow[\text{Conc.HCl}]{\text{Zn–Hg}}$ (A) + $2H_2O$

1) 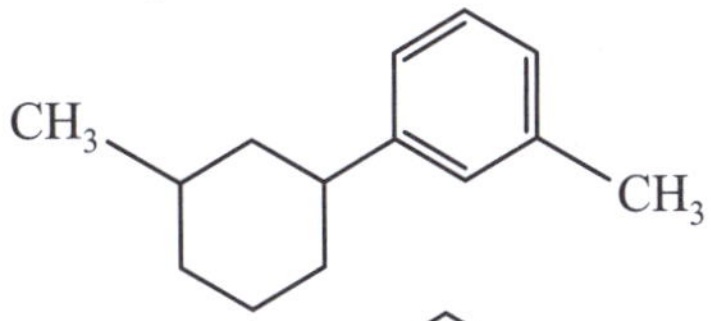

2) 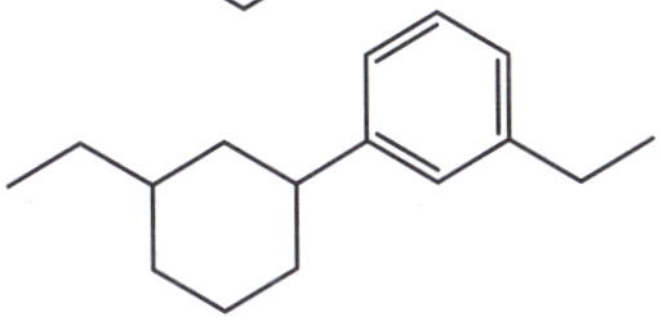

3) 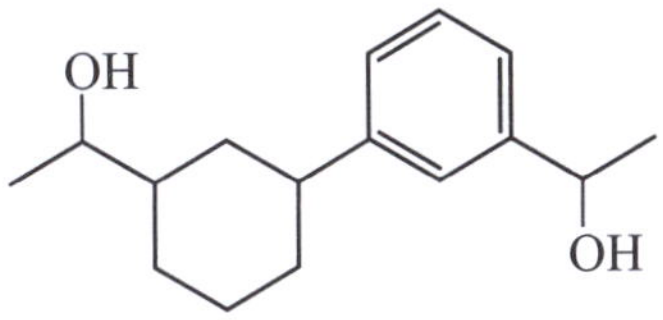

4) 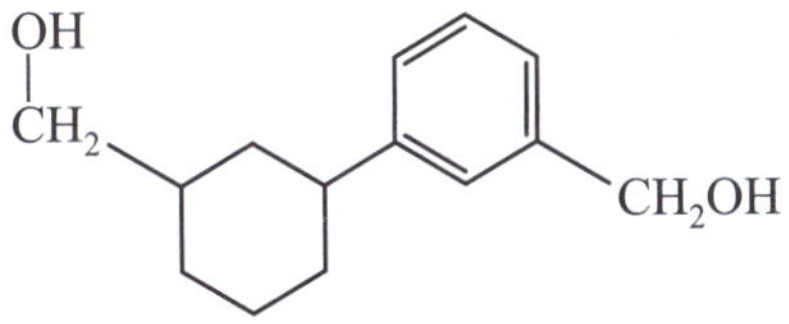

7. Complete the following reaction : **(2023)**

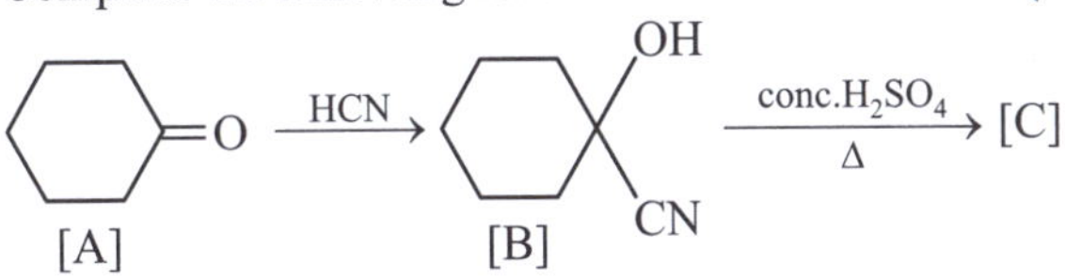

[C] is ______

1) 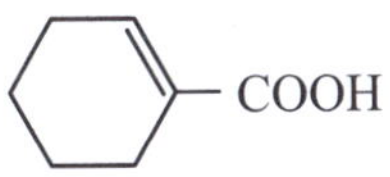
2) 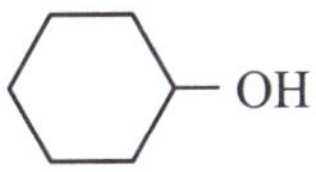
3) 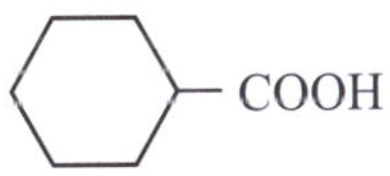
4) 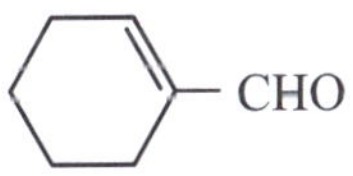

8. Identify the major product obtained in the following reaction: **(2023)**

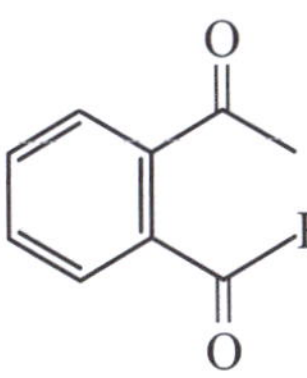 $+ 2[Ag(NH_3)_2]^+ + 3\,OH^- \xrightarrow{\Delta}$ major product

1) 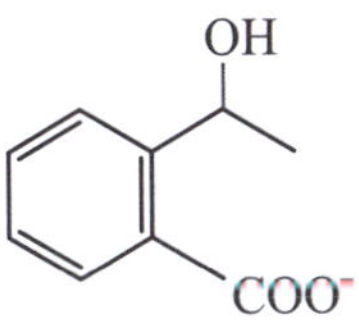
2) 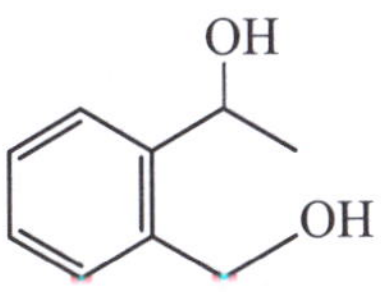
3) 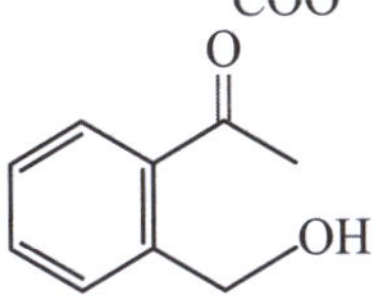
4) 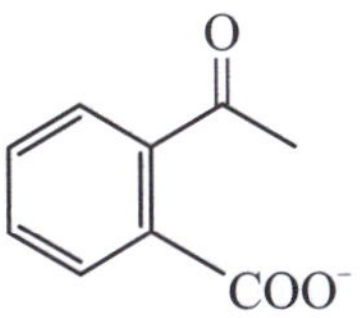

9. Identify the final product [D] obtained in the following sequence of reactions, **(2023)**

$CH_3CHO \xrightarrow[\text{(ii) } H_3O^+]{\text{(i) LiAlH}_4} [A] \xrightarrow[\Delta]{H_2SO_4} [B] \xrightarrow{\text{HBr}}$

[C] 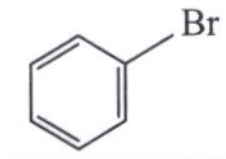 $\xrightarrow{\text{Na/dry ether}}$ [D]

1) $HC \equiv C^-Na^+$
2) 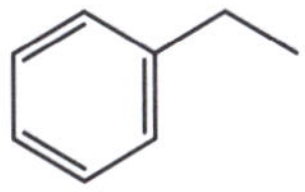
3) 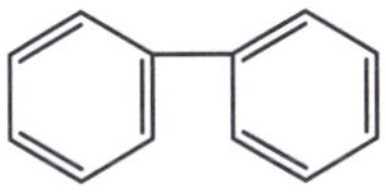
4) C_4H_{10}

10. $RMgX + CO_2 \xrightarrow[\text{ether}]{\text{dry}} Y \xrightarrow{H_3O^+} RCOOH$

What is Y in the above reaction? **(2022)**

1) $(RCOO)_2Mg$
2) $RCOO^-Mg^+X$
3) $R_3CO^-Mg^+X$
4) $RCOO^-X^+$

11. Given below are two statements: **(2022)**

Statement-I: The boiling points of aldehydes and ketones are higher than hydrocarbons of comparable molecular mass because of weak molecular association in aldehydes and ketones due to dipole-dipole interactions.

Statement-II: The boiling points of aldehydes and ketones are lower than the alcohols of similar molecular mass due to the absence of H-bonding.

In the light of the above statements, choose the most appropriate answer from the options given below

1) Statement-I is incorrect but Statement-II is correct.
2) Both Statement-I and Statement-II are correct.
3) Both Statement I and Statement-II are incorrect.
4) Statement-I is correct but Statement-II is incorrect.

12. Match Column-I with Column-II. **(2022)**

	Column-I (Products formed)		Column-II (Reaction of carbonyl compound with)
A.	Cyanohydrin	p.	NH_2OH
B.	Acetal	q.	RNH_2
C.	Schiff's base	r.	Alcohol
D.	Oxime	s.	HCN

Choose the correct answer from the options given below:

1) A-(s), B-(r), C-(q), D-(p)
2) A-(r), B-(s), C-(q), D-(p)
3) A-(q), B-(r), C-(s), D-(p)
4) A-(p), B-(r), C-(q), D-(s)

13. Which one of the following is not formed when acetone reacts with 2-pentanone in the presence of dilute NaOH followed by heating? **(2022)**

1) 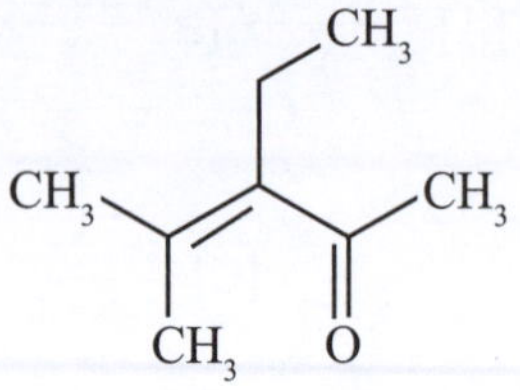

2) 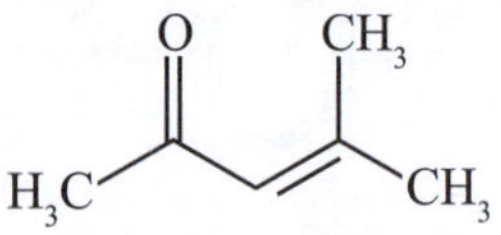

3) 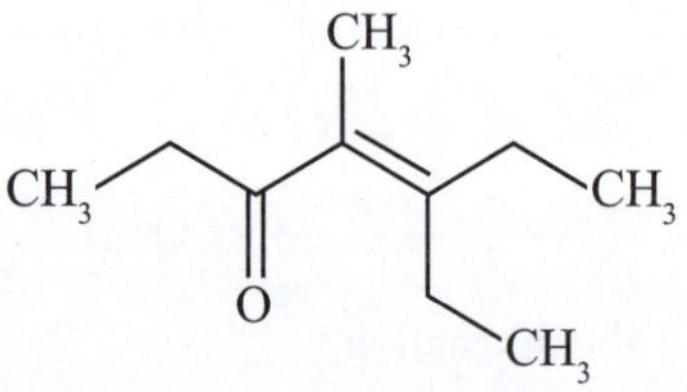

4) 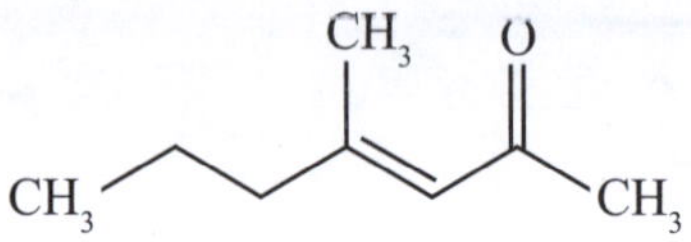

14. $CH_3CH_2COO^-Na^+ \xrightarrow[\text{Heat}]{\text{NaOH,+?}} CH_3CH_3 + Na_2CO_3$.

Consider the above reaction and identify the missing reagent/chemical. **(2021)**

1) Red Phosphorus 2) CaO
3) DIBAL-H 4) B_2H_6

15. Match Column-I with Column-II.

	Column-I		Column-II
A.	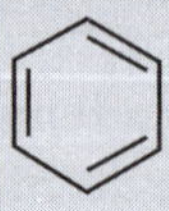 $\xrightarrow[\text{Anhyd. } AlCl_3/CuCl]{\text{CO, HCl}}$	p.	Hell-Volhard-Zelinsky reaction
B.	$R-\overset{O}{\overset{\parallel}{C}}-CH_3 + NaOX \rightarrow$	q.	Gattermann-Koch reaction
C.	$R-CH_2-OH + R'COOH \xrightarrow{\text{Conc. } H_2SO_4}$	r.	Haloform reaction
D.	$R-CH_2COOH \xrightarrow[\text{(ii) } H_2O]{\text{(i) } X_2/\text{Red P}}$	s.	Esterification

Choose the correct answer from the options given below. **(2021)**

1) A-(r); B-(q); C-(p); D-(s)
2) A-(p); B-(s); C-(r); D-(q)
3) A-(q); B-(r); C-(s); D-(p)
4) A-(s); B-(p); C-(q); D-(r)

16. 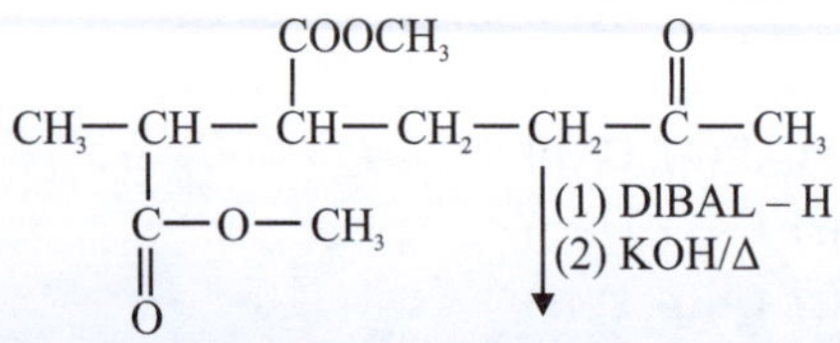 **(AIIMS 2018)**

1) 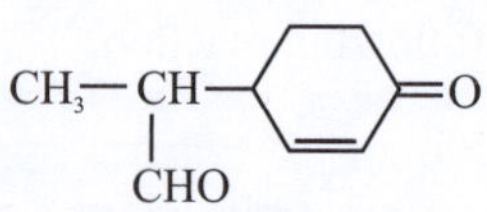

2) 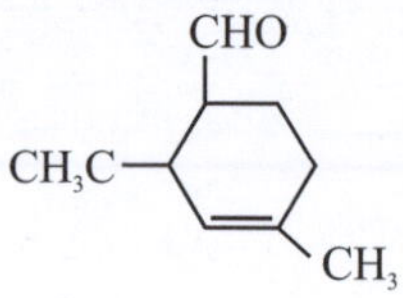

3) 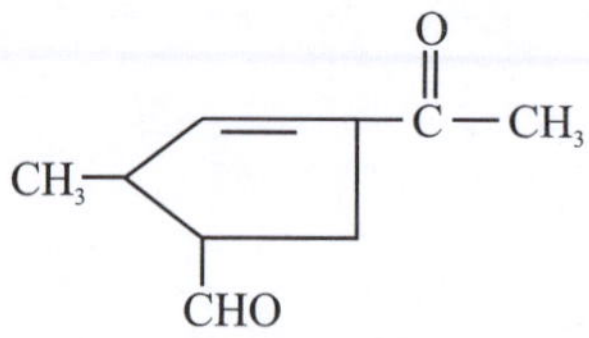

4) 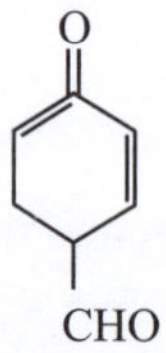

17. The correct set of reagents for the following conversion is: **(AIIMS 2017)**

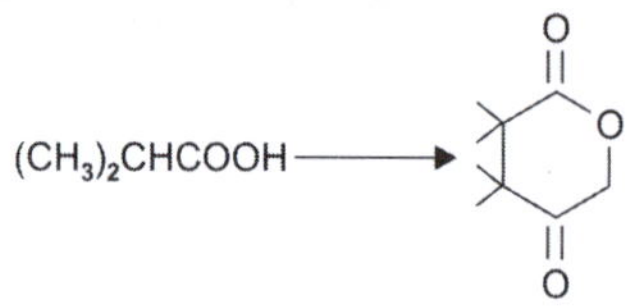

1) Pu/I_2, Na, Conc H_2SO_4
2) P_2O_5, $LiAlH_4$
3) P_2O_5/Δ, H_2O, P_4/I_2, Na
4) P_4/I_2,Na,P_2O_5/Δ

18. Consider the following set of reactions: **(AIIMS 2017)**

$$CH_3COOH \xrightarrow{SOCl_2} A \xrightarrow[\text{Anhyd, } AlCl_3]{\text{Benzene}} B \xrightarrow{HCN} C \xrightarrow{HOH} D$$

The structure of D would be

1) 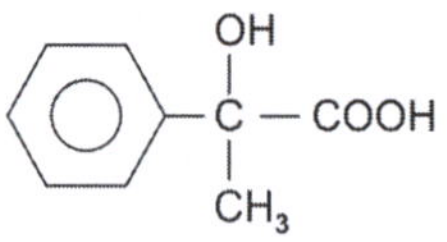 2) 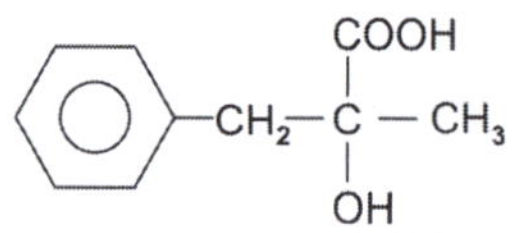

3) 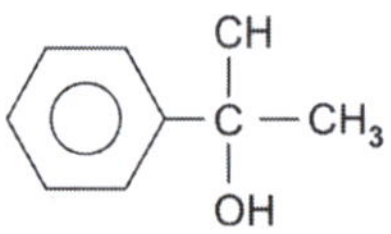 4) 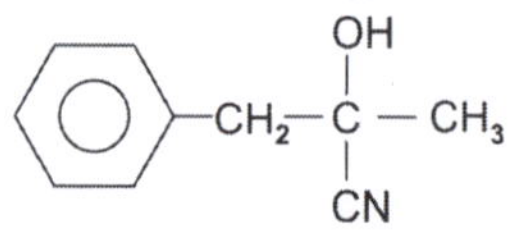

19. Order of melting point of: **(AIIMS 2016)**

(i) 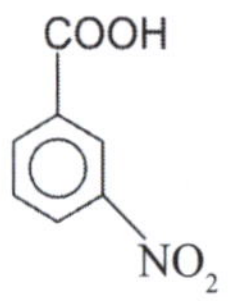 (ii) 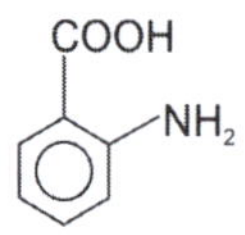

(iii) 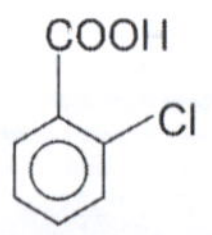

1) i > ii > iii 2) iii > ii > i
3) ii > iii > i 4) i > iii > ii

20. 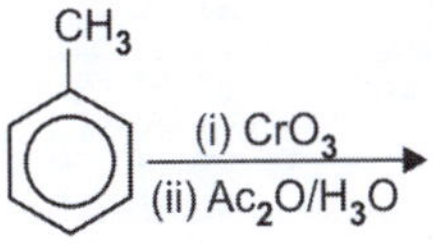 A (Major product) is (AIIMS 2013)

1) 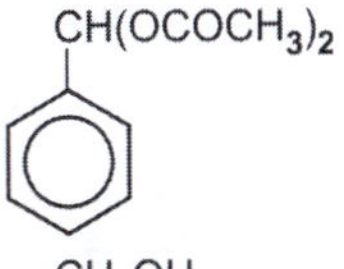

2) 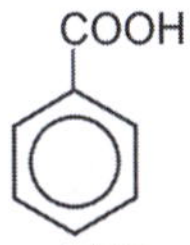

3) 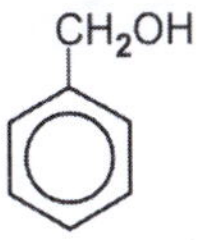

4) 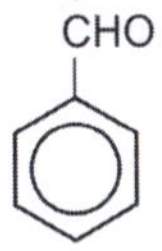

21. In one reaction: (AIIMS 2009)

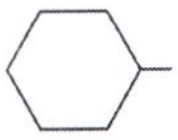 Mg—Br is treated with CH_3CN and then hydrolysed. In another reaction

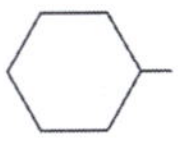 Mg-Br is treated with ethyl acetate and then hydrolysed.

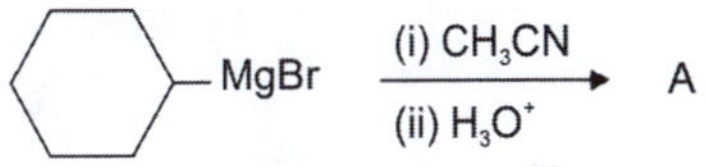

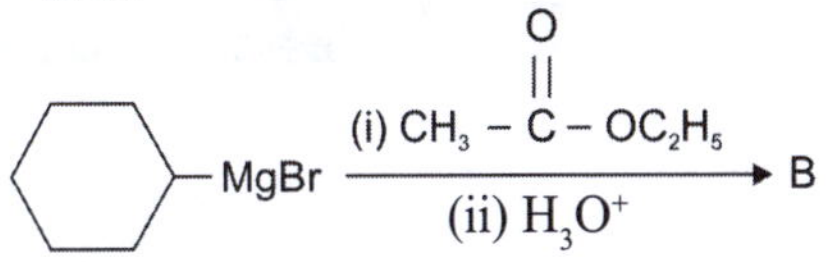

A and B are

1) 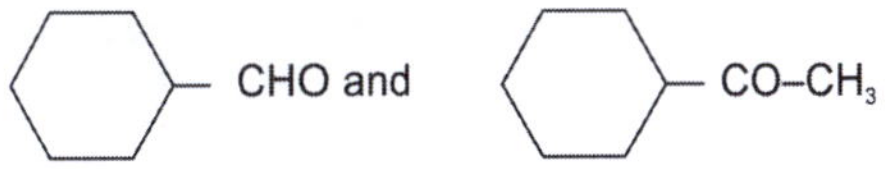

2) 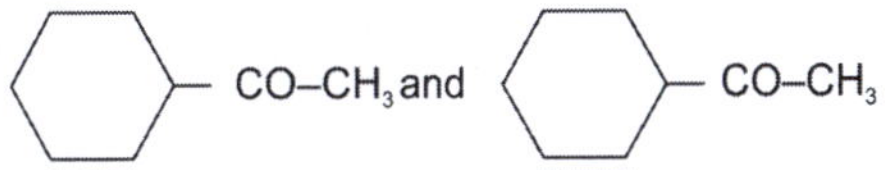

3) 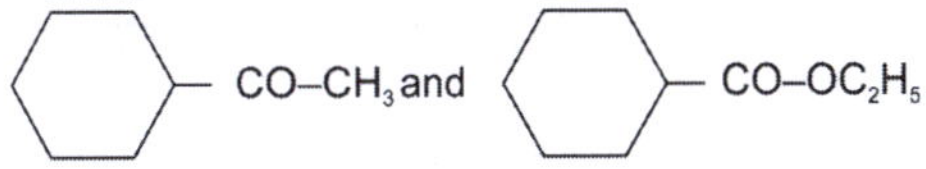

4) 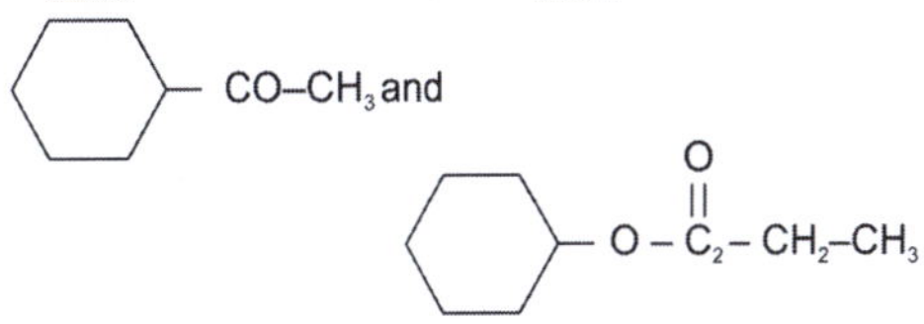

Answer Key

NCERT Topic-wise MCQs

1	2	3	4	5	6	7	8	9	10	11	12	13	14	15	16	17
3	2	4	3	4	1	2	4	1	3	1	1	1	4	1	2	1

18	19	20	21	22	23	24	25	26	27	28	29	30	31	32	33	34
4	2	4	3	1	2	3	2	1	3	3	1	4	4	4	1	2

35	36	37	38	39	40	41	42	43	44	45	46	47	48	49	50	51
4	1	4	1	1	3	2	1	3	2	4	4	4	3	1	3	1

52	53	54	55	56	57	58	59	60	61	62	63	64	65	66	67	68
3	4	3	3	2	1	4	2	2	3	2	1	2	2	1	3	3

69	70	71	72	73	74	75	76	77	78	79	80
3	3	1	1	1	3	4	3	2	4	4	1

High Order Time Intensive MCQs

1	2	3	4	5	6	7	8	9	10	11	12	13	14	15	16	17
4	2	4	2	3	2	3	1	2	2	4	2	2	1	4	4	3

18	19	20	21	22	23	24	25	26	27	28	29	30	31	32	33	34
1	4	4	2	2	2	3	3	2	4	4	4	1	2	3	3	3

35	36	37	38	39	40	41	42	43	44	45	46	47	48	49	50	51
1	3	1	4	1	3	3	3	3	4	4	3	1	3	2	3	1

52	53
4	1

Match the Columns MCQs

1	2	3	4	5
2	3	1	4	1

Correct & Incorrect MCQs

1	2	3	4	5	6	7	8	9	10
1	4	2	4	2	1	1	4	4	2

Statement Based MCQs

1	2	3	4	5	6	7	8	9	10
3	1	1	4	1	2	3	3	1	1

Assertion & Reason MCQs

1	2	3	4	5	6	7	8	9	10
1	2	4	3	4	3	1	3	3	2

Multi Correct MCQs

1	2	3	4	5
2	4	3	3	2

NCERT Exemplar MCQs

1	2	3	4	5	6	7
2	1	3	2	3	4	2

Kattar NEET

1	2	3	4	5	6	7	8	9	10	11	12	13	14	15	16	17
2	3	2	4	3	3	3	2	1	4	3	1	1	2	2	2	3

18	19	20
1	3	1

NEET & AIIMS Past Year MCQs

1	2	3	4	5	6	7	8	9	10	11	12	13	14	15	16	17
2	1	1	4	3	2	1	4	2	2	2	1	3	2	3	1	4

18	19	20	21
1	3	4	2

CHAPTER 10

Amines

Topicwise Number of Questions Analysis (2025-2016)

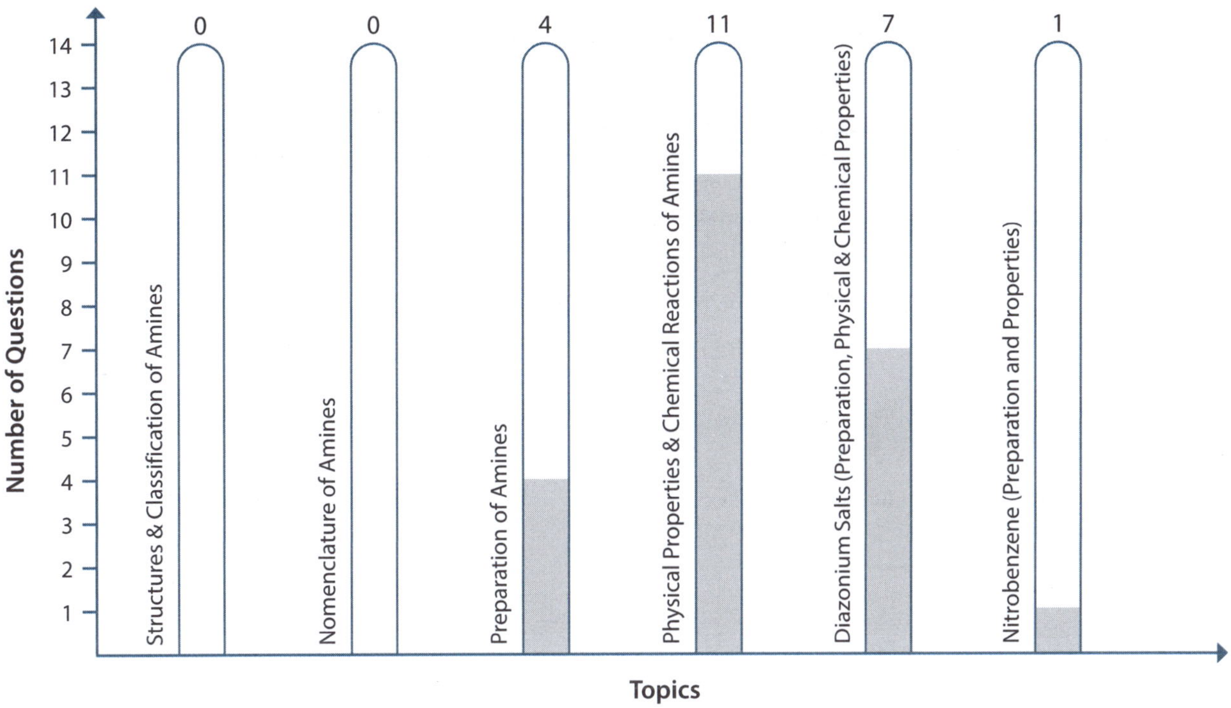

NCERT-Picks

- Amines are considered as derivatives of ammonia, obtained by replacement of one, two or all the three hydrogen atoms by alkyl and or aryl groups.
- In an amines both the C – N & N – H bonds are polar.

Classification

- Classification of amines is done on the basis of number of alkyl or aryl groups attached to nitrogen atom.

  Amines are classified as

  Primary (1°)–RNH_2

  Secondary(2°)–R_2NH

  Tertiary (3°)–R_3N

Preparation of Alkyl Amines

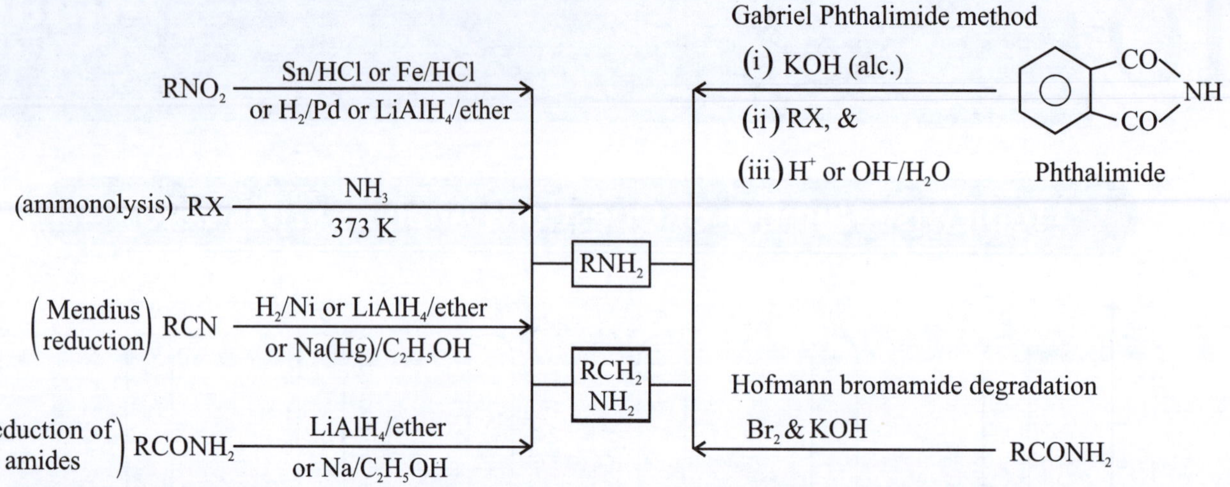

- Hofmann's bromamide reaction is given by 1°-amides.

Properties of amine

- In amines, there is the presence of unshared electron pair so, amines behave as bases and nucleophiles.
- Aliphatic amines are stronger bases than ammonia. There is the +I effect of alkyl group which leads to high electron density on the N-atom in aliphatic amines.

Chemical Reactions

Some of the reactions are given by amines are:

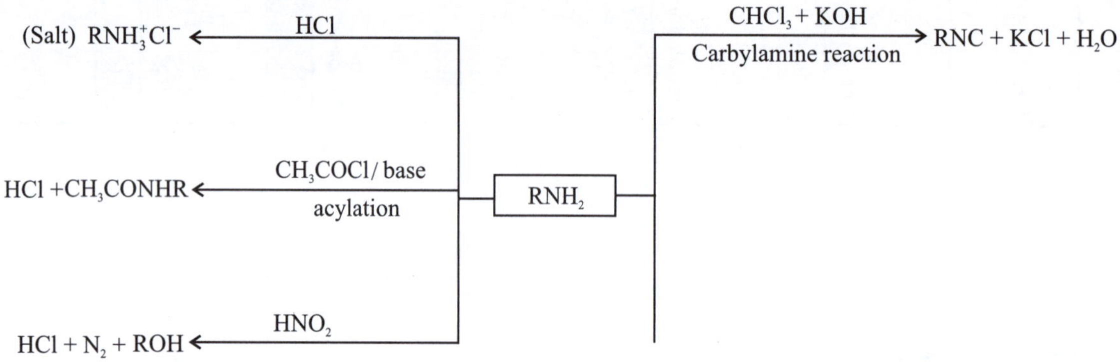

- 1° amine can be distinguished from 2° & 3° amine by carbylamines reaction. Carbylamines reaction carried out only by primary amine & gives unpleasant odours.

  Distinction between 1°, 2° and 3° amines:

  Hinsberg's test:

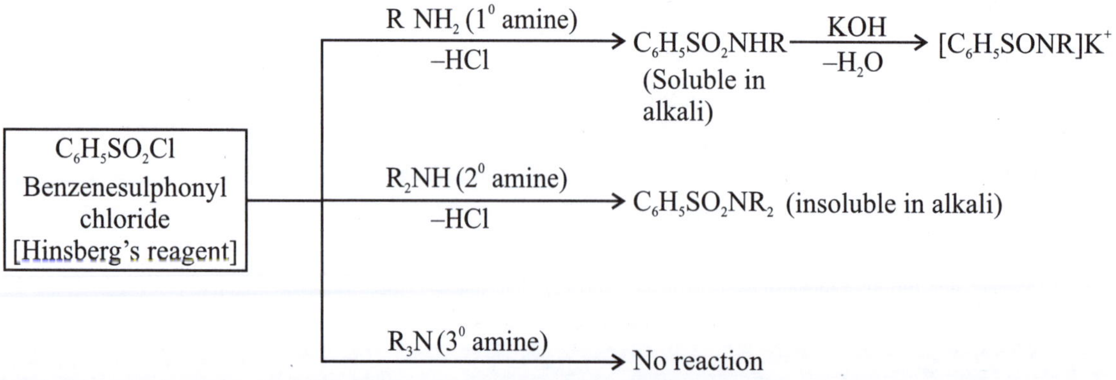

- Aliphatic tertiary amine can be distinguished from aromatic tertiary amines. Aliphatic tertiary amine react with nitrous acid to form soluble nitrites while aromatic tertiary amine give green coloured nitrosoamines.

Aniline

Preparation of aromatic amines

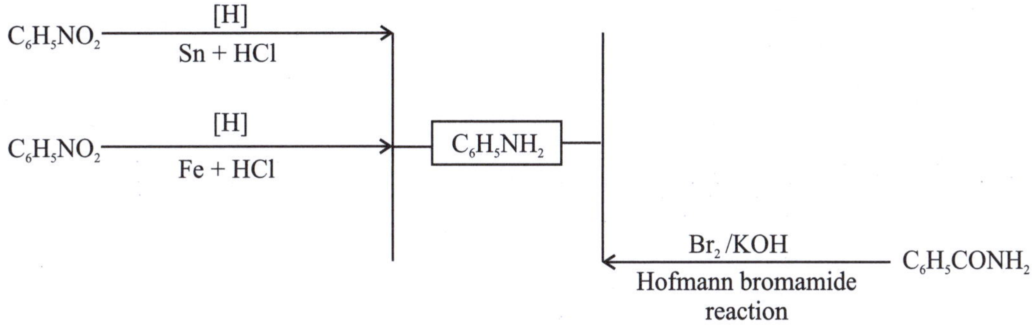

- Arylamines cannot be prepared by **Hofmann's Ammonolysis method**. Aryl halide do not undergo simple nucleophilic substitution.
- Aromatic primary amine cannot be prepared by Gabriel phthalimide synthesis. Aryl halides do not undergo nucleophilic substitution with the anion formed by phthalimide.

Basicity of aniline

- Aromatic amines are less basic than Aliphatic Amines $C_6H_5NH_2 < NH_3 < CH_3NH_2$
- pK_b value of aniline is quite high because in aniline or other arylamines, the -NH_2 group is attached directly to the benzene ring. It results in the unshared electron pair on nitrogen atom to be in conjugation with the benzene ring and thus, making it less available for protonation.
- In case of substituted aniline, electron releasing groups like OCH_3, CH_3 increase basic strength whereas electron withdrawing groups like NO_2, $-SO_3H$, $-X$ decrease the basic strength.
- Aromatic amines are weaker bases than ammonia because of the electron withdrawing nature of the aryl group.

Chemical Reactions

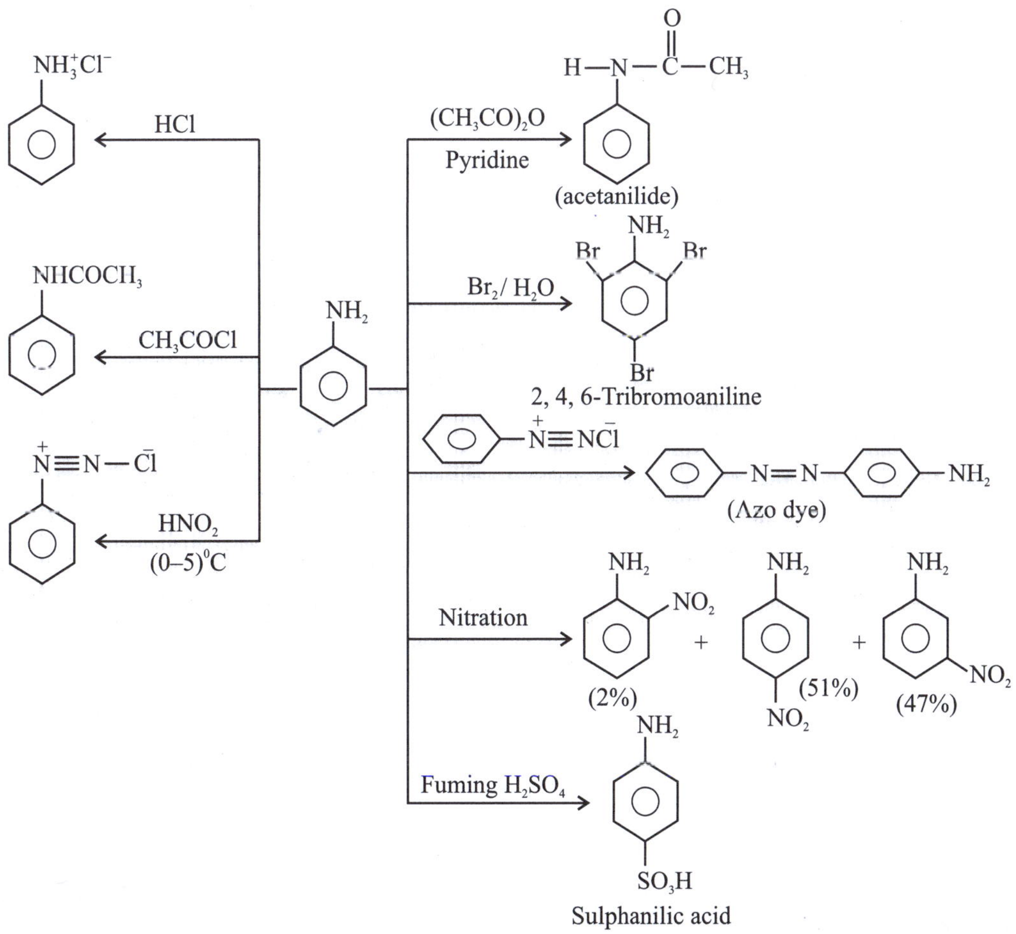

- Aniline does not undergo Friedal crafts reaction. The presence of a +ve charge on the N-atom strongly deactivates the benzene ring.
- Aromatic amines undergo electrophilic substitution reaction readily. This is due to strong activating effect of the amino group.
- Aryl amines are very reactive towards electrophilic aromatic substitution. $-NH_2$, $-NHR_2$ & $-NR_3$ are very strong activators & are ortho-para directing.
- In strongly acidic solution, aniline becomes less reactive towards electrophilic reagents. Due to protonation of amino group, the lone pair of electrons on nitrogen is not available for donation to benzene ring by the resonance.

Diazonium Salts

Preparation of diazonium salts

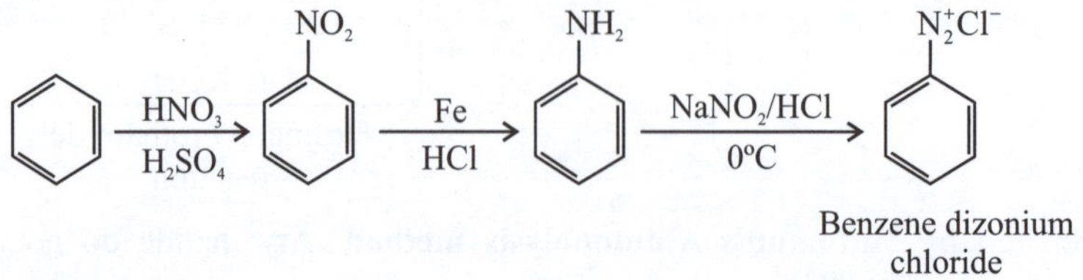

Properties

- Aromatic diazonium salts are little more stable than the aliphatic diazonium salts. This is due to the dispersal of the positive charge over the benzene ring.
- Diazonium group is substituted by Cl^-, Br^- & CN^-. Diazonium group being very good leaving group, and N_2 formed escapes from the reaction mixture.
- When aromatic amine react with nitrous acid, temperature should not be more than 278 K. At higher temp. diazonium salts will decompose to give phenols.

Reaction of diazonium compounds

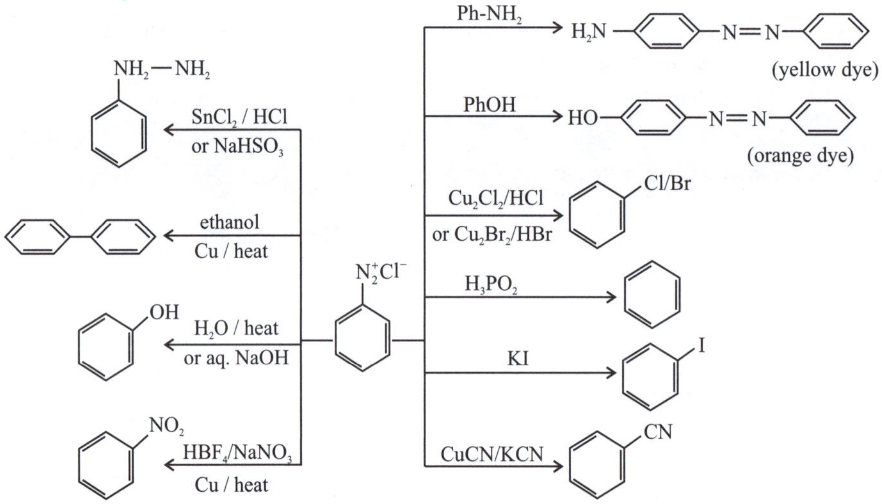

- Aromatic nitro compounds on reduction with $LiAlH_4$ give azo compounds.

Important Mechanisms

Hofmann's bromamide degradation

Deprotonation of the amide and nucleophilic attack on bromine:

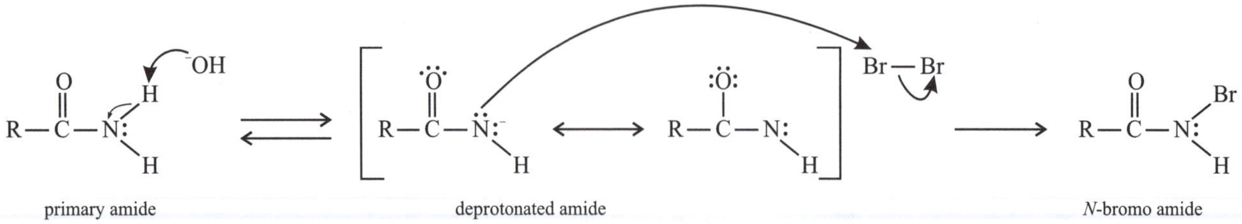

Because of the electronegative nature of bromine, the N-bromoamide is more easily deprotonated than the original primary amide. Deprotonation of the N-bromo amide gives another resonance-stabilized anion. The deprotonated amide has bromine atom present as a potential leaving group. In order for bromide to leave, the alkyl (R–) group must migrate to nitrogen. This is the actual rearrangement step, giving an isocyanate intermediate.

Second deprotonation and rearrangement with loss of bromide:

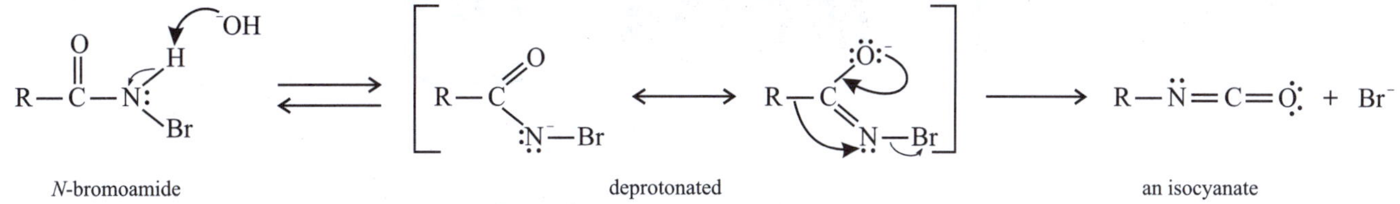

Isocyanates react rapidly with water to give carbamic acids.

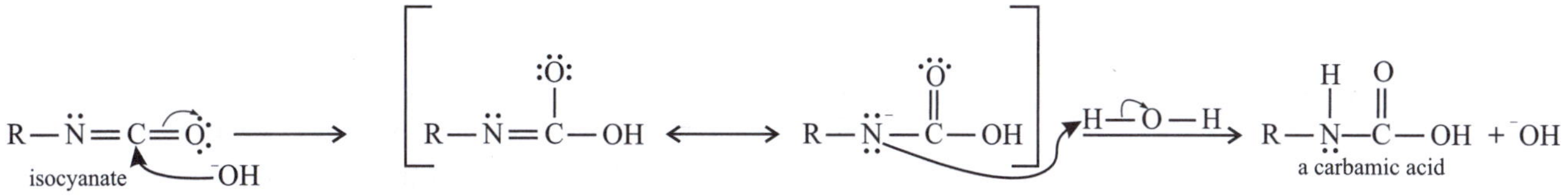

Decarboxylation of the carbamic acid gives the amine and carbon dioxide.

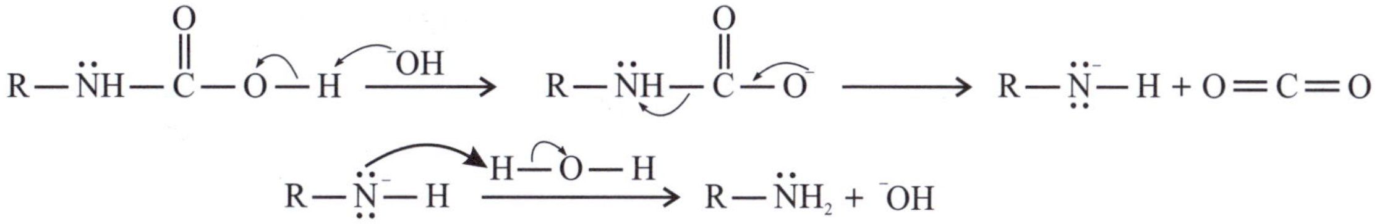

Conversion of an acid chloride to an amide

Step 1: Additon of the nucleophile.

Step 2: Elimination of the leaving group

Step 3: Loss of a proton.

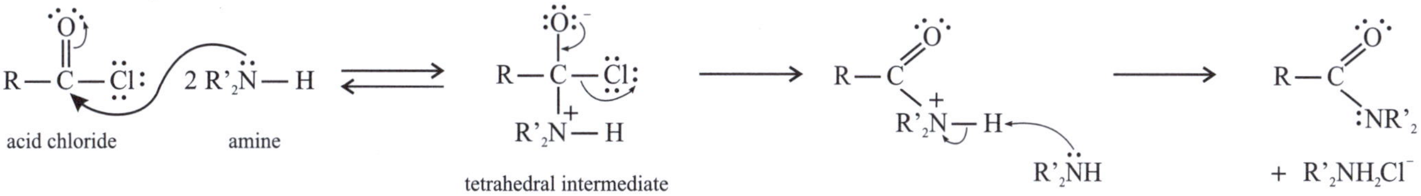

Conversion of an acid anhydride to an amide

Step 1: Additon of the nucleophile

Step 2: Elimination of the leaving group

Step 3: Loss of a proton

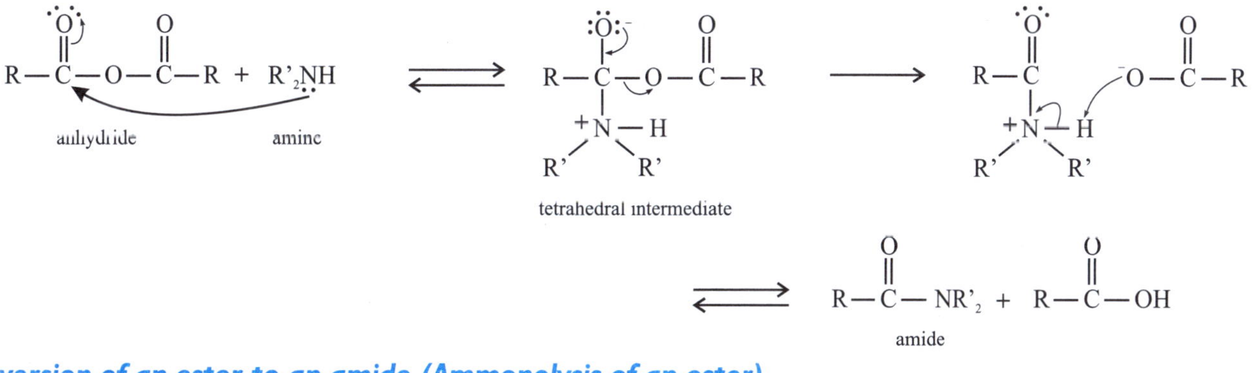

Conversion of an ester to an amide (Ammonolysis of an ester)

Step 1: Additon of the nucleophile

Step 2: Elimination of the leaving group

Step 3: Loss of a proton

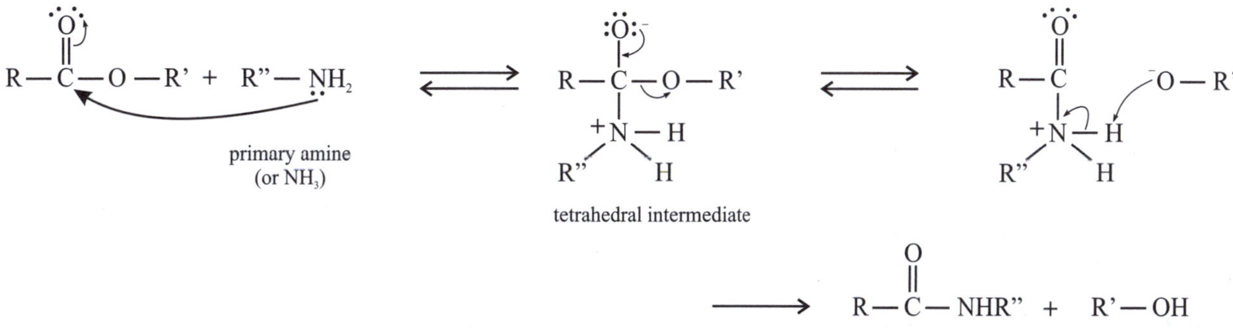

MCQs

NCERT Topic-wise MCQs

Structures of Amines and Classification

1. Hybridisation of N in amino group is:

1) sp 2) sp^2
3) sp^3 4) sp^3d

2. Which of the following is a tertiary amine?

1) 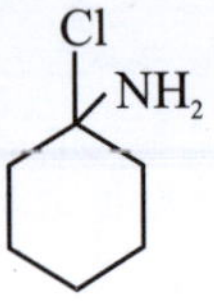

2) 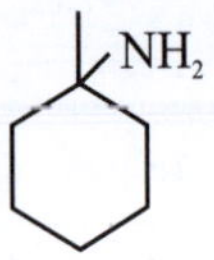

3) 

4) None

3. The structural formula of N-methyl aminomethane is

1) $(CH_3)_2CHNH_2$ 2) $(CH_3)_3N$
3) $(CH_3)_2NH$ 4) CH_3NH_2

4. Bond angle present in trimethylamine is

1) 105° 2) 109°
3) 108° 4) 107°

5. $\underset{(A)}{CH_3-CH_2-CH_2-CH_2-NH_2}$ & $H_3C-\overset{CH_3}{\underset{H}{C}}-CH_2-NH_2$ (B)

are which type of isomers

1) Positional Isomers 2) Chain isomer
3) Functional Isomers 4) None of these

Nomenclature

6. Which of the following systematic name is correct for structure given below?

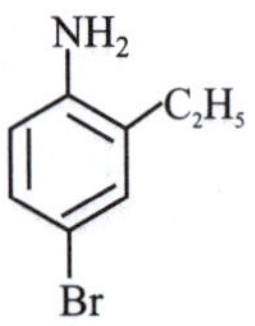

(A) 4-Bromo-2-ethyl aniline
(B) 4-Bromo-2-ethyl benzenamine
(C) 4-Bromo-2-ethyl amino benzene
(D) 3-Bromo-1-ethyl benzanamine

1) Only D 2) A and B only
3) A, B, C, only 4) A, B, C & D

7. Identify the incorrect IUPAC name:

1) $(CH_3CH_2)_2NCH_3$ = N-ethyl-N-methyl ethanamine
2) $(CH_3)_3CNH_2$ = 2-Methyl propan-2-amine
3) $CH_3NHCH(CH_3)_2$ = N-Methyl propan-2-amine
4) $(CH_3)_2CHNH_2$ = 2,2-Dimethyl-N-propanamine

Preparation of Amines

8. Amine that cannot be prepared by Gabriel-phthalimide synthesis is:

1) Aniline 2) Benzyl amine
3) Methyl amine 4) Iso-butylamine

9. Identify the compounds from the following which form primary amines under suitable reduction conditions:

(i) C_2H_5NC (ii) C_2H_6
(iii) $C_2H_5CONH_2$ (iv) $C_2H_5NO_2$

1) (i), (iv) 2) (iii), (iv)
3) (i), (iii), (iv) 4) (ii), (iii), (iv)

10. Aniline can be obtained by reduction of nitrobenzene with

1) Fe/HCl
2) Sn/HCl
3) Electrolytic reduction under weakly acidic conditions
4) All

11. $RCONH_2 + NaOH + Br_2 \rightarrow RNH_2 + 2NaBr + Na_2CO_3 + 2H_2O$. Reaction is said:

1) Hofmann bromamide reaction
2) Schmidt reaction
3) Curtius reaction
4) Beckmann reaction

12. The product formed when phthalimide is treated with a mixture of Br_2 and strong NaOH solution is:

1) Aniline
2) Phthalimide
3) Phthalic acid
4) Anthranilic acid or o-amino benzoic acid

13. Reaction $CH_3CN \xrightarrow{Na(Hg)/C_2H_5OH}$ gives:

1) CH_3Br 2) CH_4
3) CH_3OBr 4) $CH_3CH_2NH_2$

14. The most convenient method to prepare an amine containing one carbon atom less is:
 1) Gabriel phthalimide synthesis
 2) Reductive amination of aldehydes
 3) Hofmann bromamide reaction
 4) Reduction of isonitriles

15. Which of the following amides will give ethylamine on reaction with sodium hypobromide?
 1) Butanamide 2) Propanamide
 3) Acetamide 4) Benzamide

16. What is the end product in the following sequence of reaction?

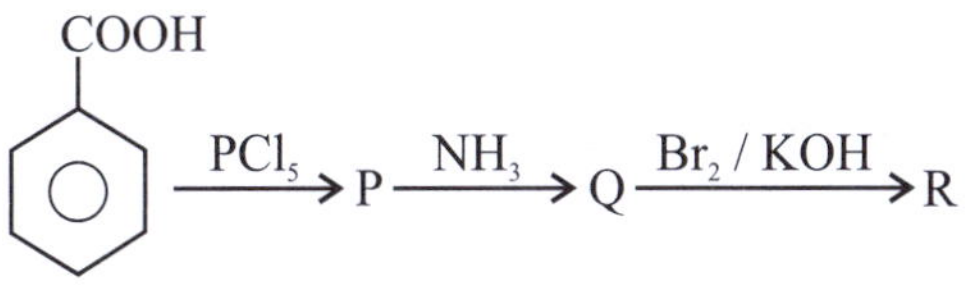

 1) Aniline
 2) Phenol
 3) Benzene
 4) Benzene diazonium chloride

17. Which of the following reactions does not yield an amine?
 1) $R{-}X + NH_3 \longrightarrow$
 2) $R-CH=NOH+[H]\xrightarrow[C_2H_5OH]{Na}$
 3) $R-CN+H_2O\xrightarrow{H^+}$
 4) $R-CONH_2+4[H]\xrightarrow{LiAlH_4}$

18. Which of the following reactions will not give a primary amine?
 1) $CH_3CONH_2\xrightarrow{LiAlH_4}$ 2) $CH_3CN\xrightarrow{LiAlH_4}$
 3) $CH_3NC\xrightarrow{LiAlH_4}$ 4) $CH_3CONH_2\xrightarrow{LiAlH_4}$

Physical Properties

19. The correct order of basic strength in benzene solution is
 1) $CH_3NH_2 > (CH_3)_2NH > (CH_3)_3N$
 2) $(CH_3)_3N > (CH_3)_2NH > CH_3NH_2$
 3) $(CH_3)_2NH > (CH_3)_3N > CH_3NH_2$
 4) $(CH_3)_3N > CH_3NH_2 > (CH_3)_2NH$

20. The correct decreasing order of Boiling points is
 1) $R-CH_2-NH_2 > RCOOH > RCH_2OH$
 2) $R-CH_2OH > RCOOH > RCH_2NH_2$
 3) $R-COOH > RCH_2OH > RCH_2NH_2$
 4) $RCH_2OH > RCH_2NH_2 > RCOOH$

21. Tertiary amines have lowest boiling points amongst isomeric amines because:
 1) They have highest molecular mass
 2) They do not form hydrogen bonds
 3) They are most basic in nature
 4) They are more polar in nature

22. Which of the following statements is Correct about ethyl amine?
 1) Ethyl amine is slightly acidic
 2) Ethyl amine is less basic than NH_3
 3) Ethyl amine is stronger base than NH_3
 4) Ethyl amine forms salts with alkali

23. Which of the following is least basic?
 1) $(CH_3CO)\ddot{N}HC_2H_5$ 2) $(C_2H_5)_3\ddot{N}$
 3) $(CH_3CO)_2\ddot{N}H$ 4) $(C_2H_5)_2\ddot{N}H$

Chemical Properties

24. The correct sequence of reaction to be performed to convert benzene into m-bromoaniline is:
 1) Nitration, reduction, bromination
 2) Bromination, nitration, reduction
 3) Nitration, bromination, reduction
 4) Reduction, nitration, bromination

25. Which of the following statements about primary amines is incorrect?
 1) Alkyl amines are stronger bases than ammonia
 2) Alkyl amines are stronger bases than aryl amines
 3) Alkyl amines react with nitrous acid to produce alcohols
 4) Aryl amines react with nitrous acid to produce phenols

26. The strongest base among the following is:
 1) $C_6H_5NH_2$ 2) $(C_6H_5)_2NH$
 3) NH_3 4) $(C_2H_5)_2NH$

27. Most basic among the following is:
 1) 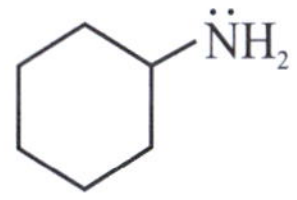 2) 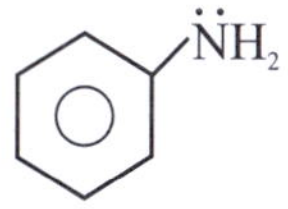
 3) 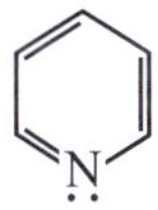 4) 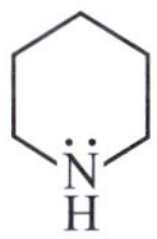

28. Which of the following is used as Hinsberg's reagent?
 1) $C_6H_5SO_2Cl$ 2) $C_6H_5SO_3H$
 3) $C_6H_5NHCH_3$ 4) $C_6H_5COCH_3$

29. Which of the following is amphoteric in nature?
 1) CH_3NH_2 2) CH_3NHCH_3
 3) CH_3CONH_2 4) $CH_3-\underset{\displaystyle CH_3}{\underset{|}{N}}-CH_3$

30. Which of the following can exist as zwitter ion?
 1) p-Amino actophenone 2) Sulphanilic acid
 3) p-Nitro aminobenzene 4) p-methoxy phenol

31. Identify 'Z' in the sequence:

$$C_6H_5NH_2\xrightarrow[273K]{NaNO_2+HCl}X\xrightarrow{CuCN}Y\xrightarrow[Boil]{H+/H_2O}Z$$

 1) C_6H_5CN 2) $C_6H_5CONH_2$
 3) C_6H_5COOH 4) $C_6H_5CH_2NH_2$

32. The strongest base among the following is:

1) $C_6H_5NH_2$ 2) p-$NH_2C_6H_4NH_2$

3) m-$NO_2C_6H_4NH_2$ 4) $C_6H_5CH_2NH_2$

33. $CH_3CH_2Cl \xrightarrow{NaCN} X \xrightarrow{Ni/H_2} Y \xrightarrow{\text{Acetic anhydride}} Z$, in this reaction, Z is:

1) $CH_3CH_2CH_2NHCOCH_3$

2) $CH_3CH_2CH_2NH_2$

3) $CH_3CH_2CH_2CONHCH_3$

4) $CH_3CH_2CH_2CONHCOCH_3$

34. Isocyanide is prepared by:

1) Friedal-Crafts reaction

2) Wurtz reaction

3) Williamson synthesis

4) Carbylamine reaction

35. Which one of the following reagent will convert acetamide to ethanamine?

1) Phosphorus pentoxide 2) Lithium aluminium hydride

3) Potassium cyanide 4) Thionyl chloride

36. $\begin{matrix} CONH_3^+ \\ | \\ CONH_3^+ \end{matrix} \xrightarrow{heat} X \xrightarrow{P_4O_{10},\ \Delta} Y$; Y is likely to be:

1) $\begin{matrix} CONH_2 \\ | \\ CONH_2 \end{matrix}$ 2) $\begin{matrix} COOH \\ | \\ COOH \end{matrix} + NH_3$

3) C_2N_2 4) $2CO_2 + 2NH_3$

37. An amide, C_3H_7NO on hydrolysis gives an amine (B) and an acid (C). B gives carbylamine reaction whereas C reduces Tollen's reagent. 'A' is:

1) $CH_3CONHCH_3$ 2) $CH_3CH_2CONO_2$

3) $HCONHC_2H_5$ 4) $HCO-\underset{\underset{CH_3}{|}}{N}-CH_3$

38. Which of the following nitro compounds does not give any reaction with HNO_2?

1) $CH_3CH_2CH_2NO_2$ 2) $CH_3\underset{\underset{NO_2}{|}}{C}HCH_2CH_3$

3) $CH_3\overset{\overset{CH_3}{|}}{C}H-CH_2-NO_2$ 4) $CH_3-\overset{\overset{CH_3}{|}}{\underset{\underset{CH_3}{|}}{C}}-NO_2$

39. Ethylamine and aniline can be distinguished by:

1) Hinsberg's test

2) Azo Dye test

3) Liebermann's nitrosoamine test

4) Carbylamine test

40. Phenol $\xrightarrow[\text{distillation}]{Zn} A \xrightarrow[\text{conc.}HNO_3(60^\circ C)]{\text{conc.}H_2SO_4} B \xrightarrow{Zn-NaOH} C$

In the above reaction, compounds A, B and C are:

1) Benzene, nitrobenzene and aniline

2) Benzene, dinitrobenzene and m-nitroaniline

3) Toluene, n-nitrobenzene and hydrazobenzene

4) Benzene, nitrobenzene, hydrazobenzene

41. $C_6H_5CONH_2 + HNO_2 \rightarrow A + B$; the products are:

1) Phenol and ammonia

2) Benzoic acid and ammonia

3) Benzoic acid and nitrogen

4) Phenol and nitrogen

42. Carbylamine test is not responded by:

1) $C_6H_5NH_2$ 2) $H_3C-CH_2-NH_2$

3) $C_6H_5CH_2NH_2$ 4) $(CH_3)_3N$

43. In the reaction sequence

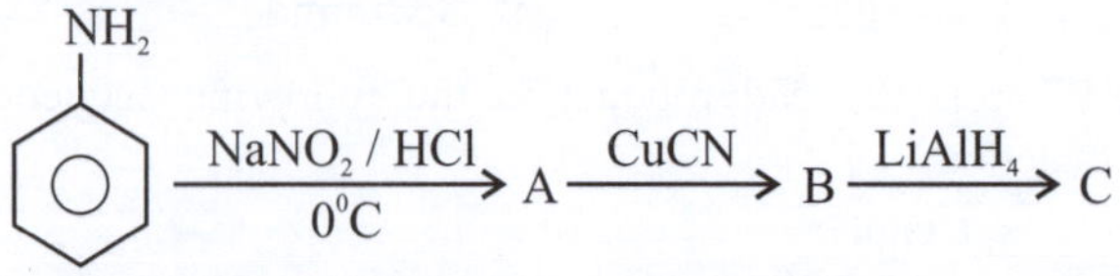

The product 'C' is:

1) Benzonitrile 2) Benzaldehyde

3) Benzoic acid 4) Benzyl amine

44. Aniline reacts with acetaldehyde to form:

1) Schiff's base 2) Carbylamine

3) Imine 4) None of these

45. The strongest base in aqueous solution among the following amines is:

1) N-N-diethylethanamine

2) N-ethylethanamine

3) N-methyl methanamine

4) Ethanamine

46. Which of the following will be obtained on acetylation of aniline?

1) N-acetylamino benzene

2) Paracetamol

3) p-amino acetophenone

4) o-amino acetophenone

47. The product of the following reaction is:

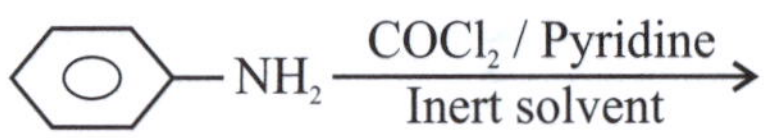

1) 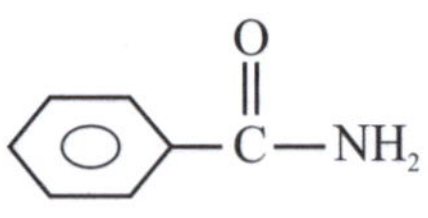

2) 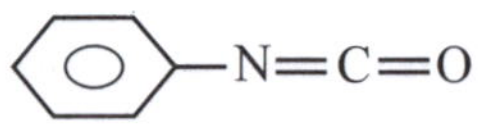

3) 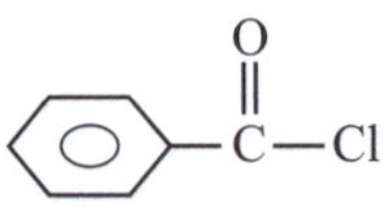

4) 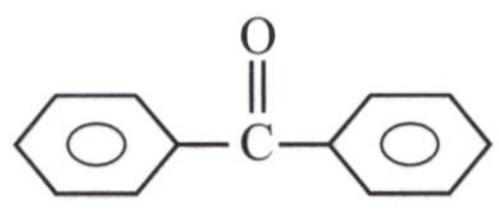

48. Which of the following species are involved in carbylamine test?

(i) RNC (ii) $CHCl_3$

(iii) $COCl_2$ (iv) $NaNO_2 + HCl$

1) (i) and (iv) 2) (i) and (ii)

3) (ii) and (iv) 4) (ii) and (iii)

49. For carbylamine reaction, we need hot alcoholic KOH and:

1) Any primary amine and chloroform

2) Chloroform and silver powder

3) A primary amine and an alkyl halide

4) A monoalkylamine and trichloromethane

50. Basic strength of different alkyl amines depends upon:

1) +I effect 2) Steric effect
3) Solvation effect 4) All of these

51. Primary, secondary and tertiary amines may be separated by using:

1) Iodoform
2) Tollen's reagent
3) Benzene sulphonyl chloride
4) Acetyl chloride

52. Which of the following compounds cannot be identified by carbylamine test?

1) $CH_3CH_2NH_2$ 2) $(CH_3)_3C-NH_2$
3) $C_6H_5NH_2$ 4) $C_6H_5NHC_6H_5$

53. Among the compounds: $C_3H_7NH_2$, CH_3NH_2, $C_2H_5NH_2$ and $C_6H_5NH_2$. Which is the least basic compound?

1) CH_3NH_2 2) $C_2H_5NH_2$
3) $C_3H_7NH_2$ 4) $C_6H_5NH_2$

54. Aniline reacts with acetic anhydride to give:

1) CH_3COOH + 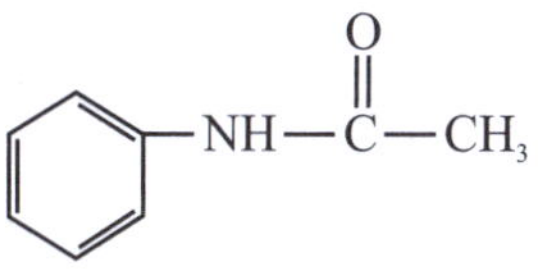

2) $HCOOH$ + 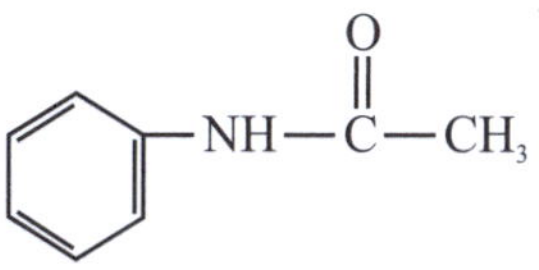

3) CH_4 + 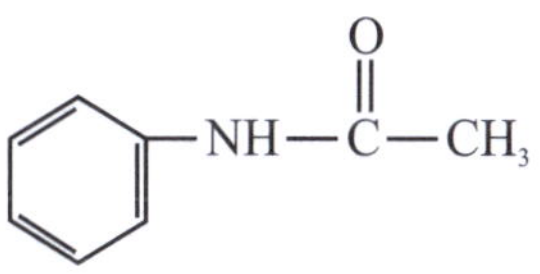

4) C_2H_6 + 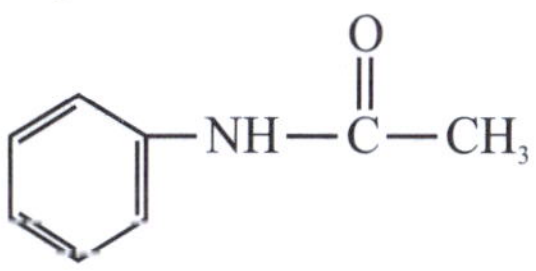

55. Reaction $CH_3CH_2NH_2 \xrightarrow{NaNO_2/HCl} X$. The product of the above reaction is :-

1) CH_3CH_2OH 2) CH_3CHO
3) CH_3-CH_3 4) $CH_3-\overset{\overset{O}{||}}{C}-CH_3$

56. A positive carbylamine test is given by

1) N,N-dimethyl aniline 2) Isopropyl amine
3) Diethyl amine 4) Trimethyl amine

57. $C_6H_5Cl \xrightarrow[Cu_2O,200°C]{NH_3} X$; $X \xrightarrow[0-5°C]{HNO_2} Z$; $X + Z \rightarrow A$; the no. of σ and π bonds in 'A' are

1) 25σ, 6π 2) 25σ, 7π
3) 27σ, 7π 4) 27σ, 6π

58. Aniline when treated with benzoyl chloride, gives benzanilide, the reaction is known as

1) Perkin 2) Hofmann
3) Schotten baumann 4) Benzoin

59. Aniline and diphenylamine may be distinguished by

1) Lassaigne's test 2) Schiff's test
3) Carbylamine reaction 4) Solubility test

60. Strongest base is

1) 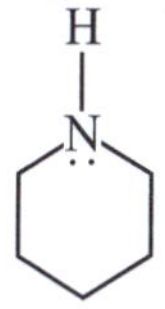 2) 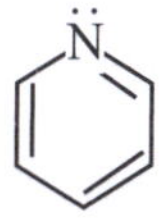

3) 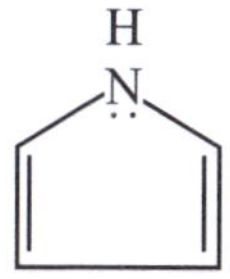 4) 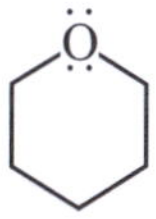

61. Increasing pK_b values of o–, m– and p–toluidines is

1) p < m < o 2) o < m < p
3) p < o < m 4) m < o < p

62. 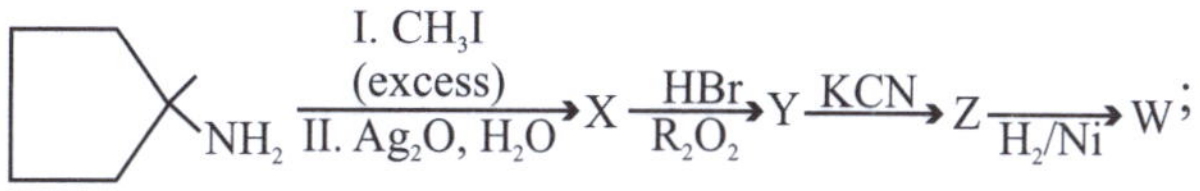

product W will be:

1) 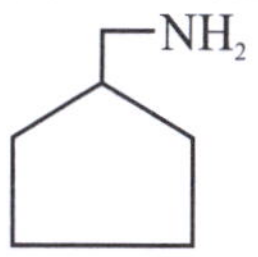 2) 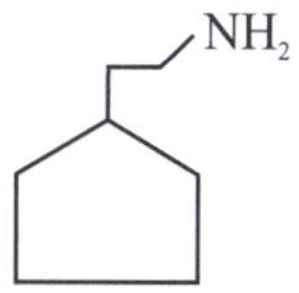

3) 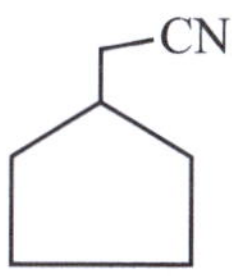 4) 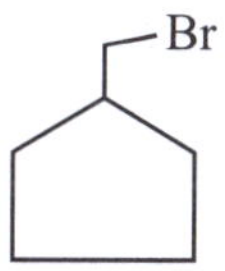

63. NH_2-group in aniline is:

1) m-directing and deactivating
2) o, p-directing and deactivating
3) o, p-directing and activating
4) m-directing and activating

64. Which of the following undergoes mustard oil reaction?

1) Primary amines 2) Secondary amines
3) Tertiary amines 4) All of these

65. An amine (X) reacts with benzenesulphonyl chloride and the product thus obtained is soluble in KOH. The amine (X) is

1) 1° Amine 2) 2° Amine
3) 3° Amine 4) Any of the three.

66. Acetoxime on reduction with Na/C_2H_5OH would give

1) Isopropylamine 2) n-Propylamine
3) Ethylmethylamine 4) Diethylamine

67. $C_2H_5Br \xrightarrow{AgCN} A \xrightarrow{H_3O^+} HCOOH + B$;

$B \xrightarrow[KOH]{CHCl_3} A \xrightarrow{Reduction} C$

A,B,C respectively in the above sequence are:

1) Ethane amine, methane nitrile and diethylamine
2) Ethane isocyanide, ethaneamine and secondary amine
3) Ethyl isocyanide, ethylamine and methyl isocyanate
4) Ethyl isocyanide, ethaneamine and ethylmethylamine

68. Ethylamine reacts with nitrosyl chloride (NOCl) to form:
1) Ethyl chloride
2) Ethyl alcohol
3) Ethyl nitrite
4) Nitroethane

69. An organic compound with M.F. C_3H_5N on hydrolysis forms an acid which reduces Fehling's solution. The compound can be
1) Ethanenitrile
2) Ethyl carbylamine
3) Ethoxyethane
4) Propanenitrile

70. 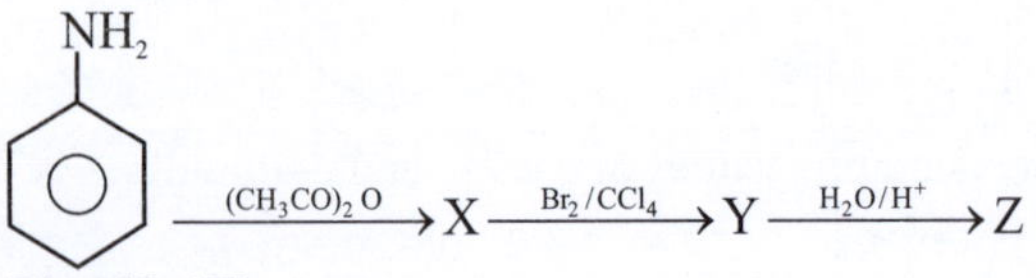

Identify 'Z'
1) p-Bromo aniline
2) p-Bromo acetophenone
3) p- Bromo acetanilide
4) o-Bromo acetophenone

71. If meta-nitroaniline is chlorinated, the major product is:

1) 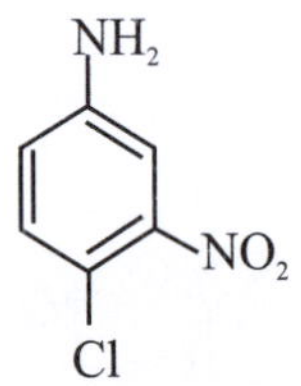

2) 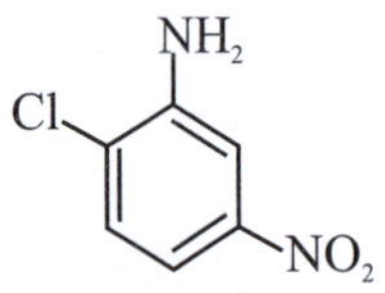

3) 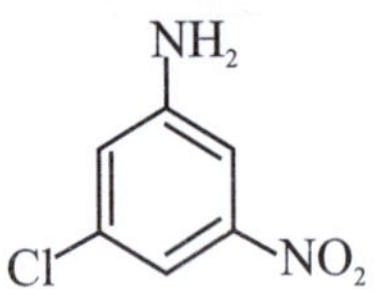

4) 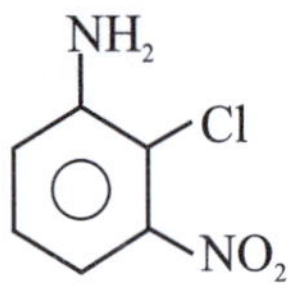

72. Hydrazobenzene on treatment with H_2SO_4 forms:
1) Azobenzene
2) Azobenzene sulphonic acid
3) Benzidine
4) None of these

73. Which of the following structures represents benzanilide?
1) $C_6H_5NHCOCH_3$
2) $C_6H_5NHCOC_6H_5$
3) $CH_3NHCOC_6H_5$
4) $CH_3NHCOCH_3$

74. Which of the following compounds will give aromatic nucleophilic substitution reaction?
1) $C_6H_5 - CH_3$
2) 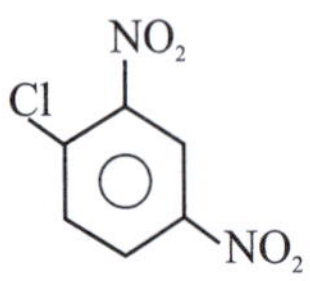
3) $C_6H_5 - OH$
4) All of these

75. Identify the compound 'X' in the following reaction:

$$CH_3NO_2 \xrightarrow[\text{Excess}]{Cl_2/NaOH} X$$

1) $ClCH_2NO_2$
2) Cl_2CHNO_2
3) Cl_3CNO_2
4) CH_3Cl

76. $CH_3CH_2{-}CH_2NH_2$ with chloroform and KOH (alc.) forms:
1) $CH_3CH_2CH_2CN$
2) $CH_3CH_2CH_2NO_2$
3) $CH_3CH_2CH_2{-}NC$
4) $CH_3CH_2CH_2{-}Cl$

77. Which one of the following substances react with nitrous acid to give a hydroxy compound?
1) $C_2H_5NH_2$
2) $(C_2H_5)_2NH$
3) $C_6H_5NH_2$
4) Both (1) and (3)

Methods of Preparation of Diazonium Salts

78. Fluorobenzene can be synthesised in the laboratory:
1) From aniline by diazotisation followed by heating the diazonium salt with HBF_4
2) By direct fluorination of benzene with F_2 gas
3) By reacting bromobenzene with NaF solution
4) By heating phenol with HF and KF

79. $CaC_2 \xrightarrow{\text{Hydrolysis}} A \xrightarrow[\text{Cu tube}]{\text{Red hot}} B \xrightarrow[50-60°C]{HNO_3} C$

$\xrightarrow{Fe+HCl} D \xrightarrow{NaNO_2 + HCl, 0°C} E$

Then, E is:
1) Aniline black
2) Benzene diazonium Chloride
3) Phenyl osazone
4) Benzoyl chloride

80. Consider the following compounds:
(i) p-methylaniline
(ii) N, N-dimethylaniline
(iii) N-ethylaniline
(iv) N-ethyl-N-methylaniline

The compounds which do not form diazonium salt with ice-cold $NaNO_2$ and HCl are:
1) (i), (ii) and (iii)
2) (ii), (iii) and (iv)
3) (i), (iii) and (iv)
4) (ii) and (iv)

Physical Properties

81. Which of the following is not a property of diazonium salts?
1) Diazonium salts are colourless crystalline solids
2) Being ionic in nature, they are soluble in water
3) Most of these salts explode when dried
4) The aqueous solutions of these salts are poor conductors of electricity

82. Which of the following diazonium salt is most stable?
1) p-Nitrobenzenediazonium chloride
2) 2, 4-Dinitrobenzenediazonium chloride
3) 2, 4, 6-Trinitrobenzendiazonium chloride
4) p-Methoxybenzenediazonium chloride

Chemical Reactions

83. Aniline in a set of reactions yields a product D:

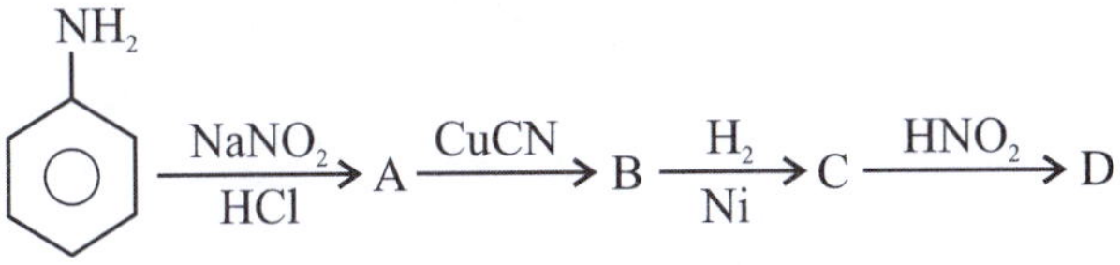

The structure of the product D would be:

1) $C_6H_5CH_2NH_2$ 2) $C_6H_5NHCH_2CH_3$

3) C_6H_5NHOH 4) $C_6H_5CH_2OH$

84. 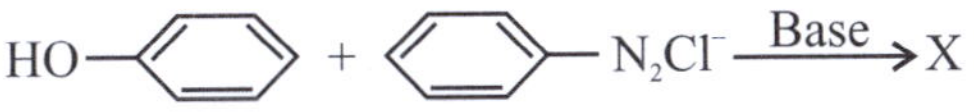 X is:

1) 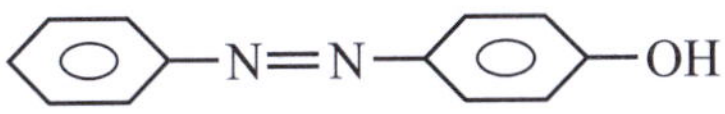

2) 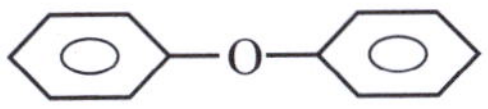

3) 

4) 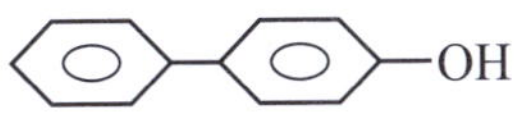

85. Benzenediazonium chloride reacts with phenol to give p-hydroxy azobenzene, an orange dye. This reaction is known as:

1) Coupling reaction

2) Carbylamine reaction

3) Gattermann reaction

4) Hinsberg's reaction

86. $C_6H_6 \xrightarrow[H_2SO_4]{HNO_3} P \xrightarrow{Sn/HCl} Q \xrightarrow[HCl]{NaNO_2} P \xrightarrow[H_2O]{H_3PO_2} S$.

The end product S in the given sequence of reaction is:

1) Benzoic acid 2) Benzene

3) Phenol 4) Chlorobenzene

87. The product 'D' in the following sequence of reactions is:

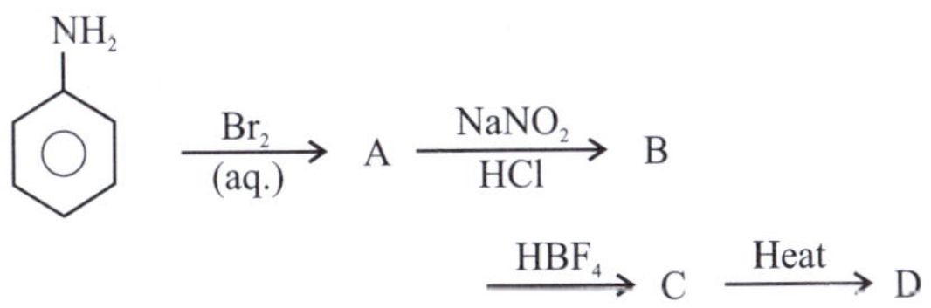

1) 2, 4, 6-tribromofluorobenzene

2) Fluorobenzene

3) p-bromofluorobenzene

4) Tribromobenzene

88. What is the observation when aniline is treated with excess bromine water at room temperature?

1) A red precipitate of monobromoaniline forms

2) A colorless solution is obtained

3) A white precipitate of 2,4,6-tribromoaniline forms

4) A yellow oily layer separates

89. Compound [X] C_7H_9N reacts with $NaNO_2$/conc. HCl at 0^oC followed by β-naphthol to give orange coloured dye. Compound [X] will be:

1) $C_6H_5-CH_2-NH_2$ 2) 

3) 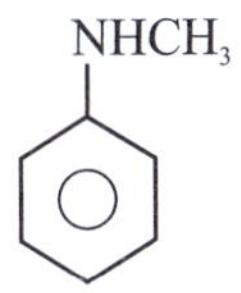 4) All of these

90. Which will not go for diazotisation?

1) $C_6H_5NH_2$ 2) $C_6H_5CH_2NH_2$

3)  4) 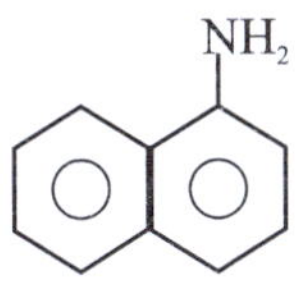

91. 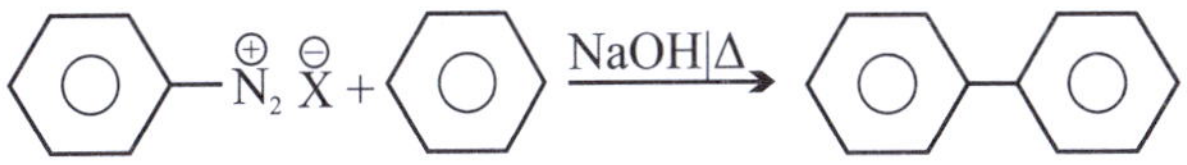

1) Swart reaction

2) Gomberg reaction

3) Coupling reaction

4) Oppenaur reaction

92. During preparation of arenediazonium salts, the excess of nitrous acid, if any, is destroyed by adding

1) aq. NaOH 2) aq. Na_2CO_3

3) aq. NH_2CONH_2 4) aq. KI

93. Benzenediazonium chloride on treatment with water gives:

1) Benzene 2) o-chlorophenol

3) Anisole 4) Phenol

High Order Time Intensive MCQs

1. Which of the following is least likely to synthesise benzylamine?

1) 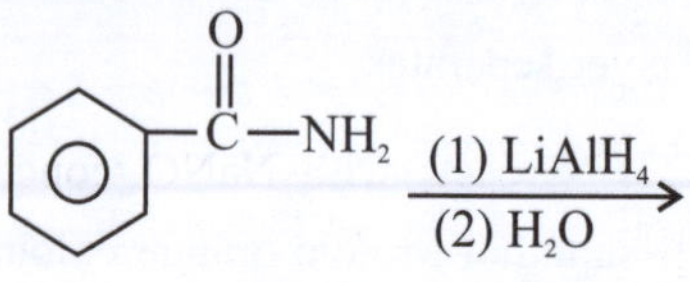

2) 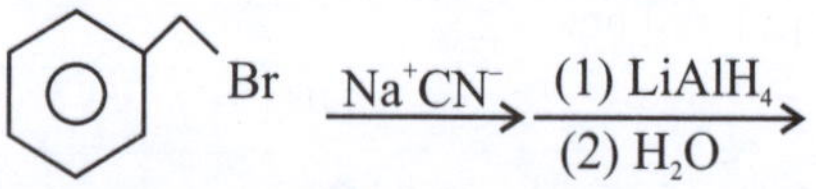

3) 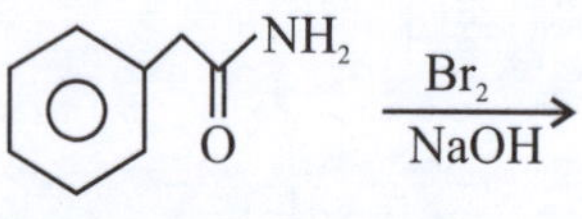

4) 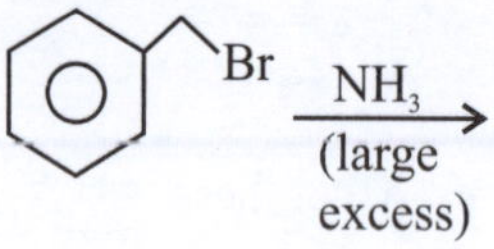

2. Which of following is incorrect?

(a) 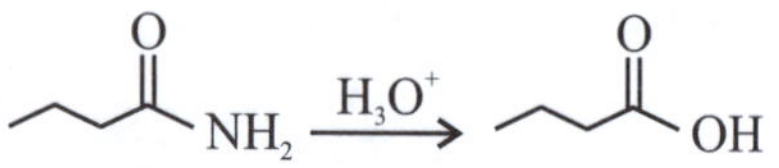

(b) 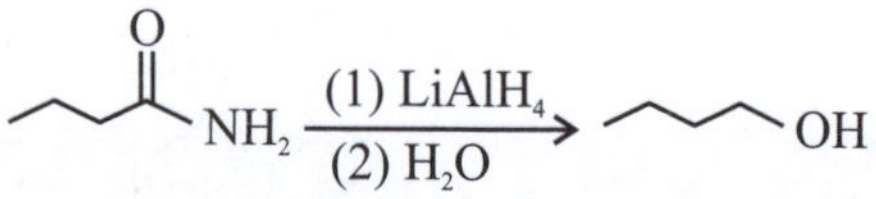

(c) 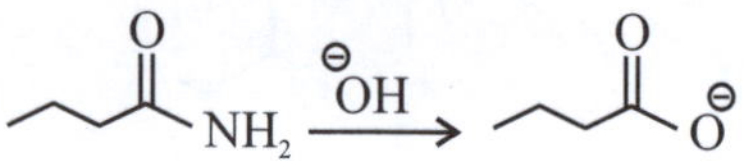

(d) 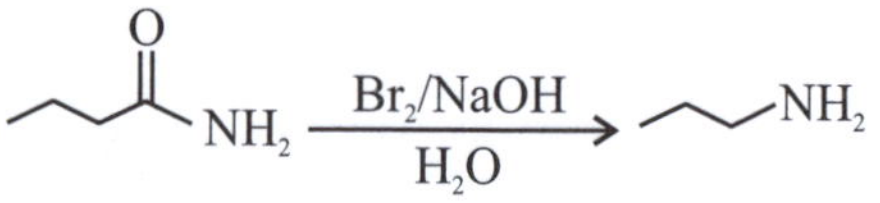

1) Only (a) 2) Only (b)
3) Both (c) & (d) 4) Only (d)

3. The among the following which can not be synthesized by Gabriel synthesis is:

1) $(CH_3)_2CH—CH_2—NH_2$

2) $CH_3CH_2NH_2$

3) 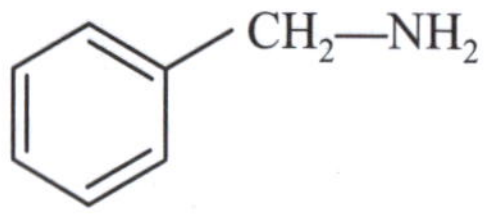

4) 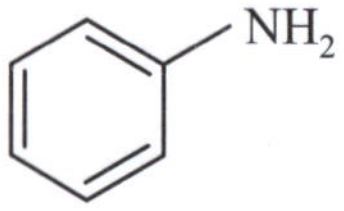

4. 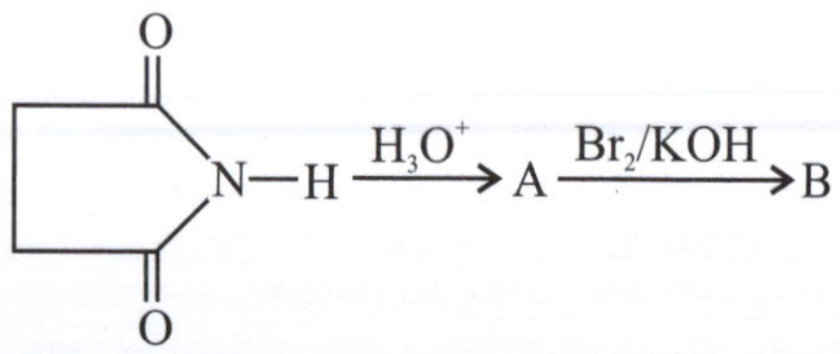

The final product B in the above reaction is:

1) 2-Amineethanoic acid 2) 1-Amineethanoic acid
3) 3-Aminobutanoic acid 4) 3-Aminopropanoic acid

5. The most appropriate reagent for conversion of C_2H_5CN into $CH_3CH_2CH_2NH_2$ is:

1) $NaBH_4$ 2) CaH_2
3) $Na(CN)BH_3$ 4) $LiAlH_4$

6. 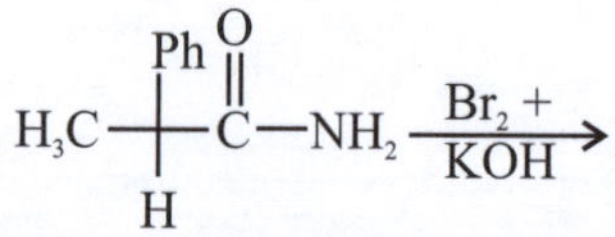 X, product 'X' is:

1) 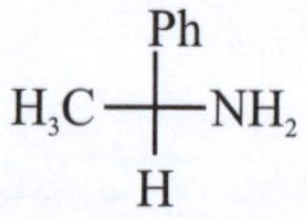 2) 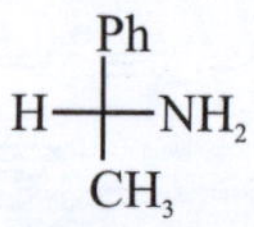

3) Both (1) & (2) 4) 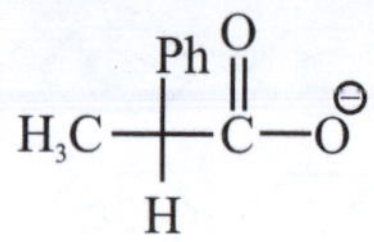

7. Identify the type of reaction in the step C → D.

P–$NO_2C_6H_5OH$

$\xrightarrow[2)EtBr]{1)\bar{O}H} A \xrightarrow[HCl]{Zn/} B \xrightarrow{NaNO_2/HCl(0\text{-}5^\circ C)} C \xrightarrow{C_6H_5OH} D$

1) It is an electrophilic substitution reaction.
2) It is coupling reaction.
3) It is nucleophillic addition reaction.
4) Both (1) and (2)

8. 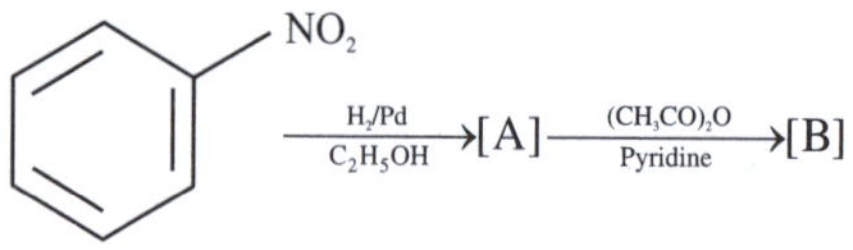

1) 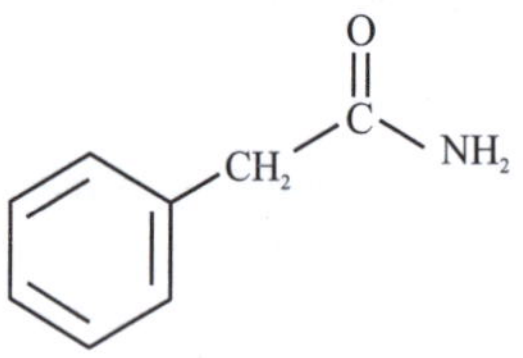

2) 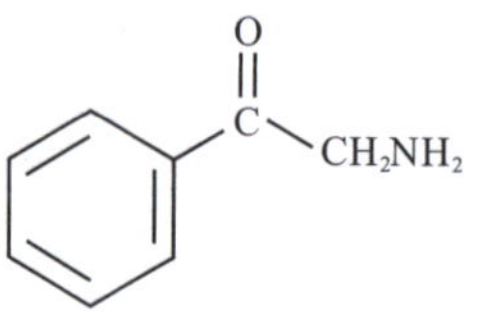

3) 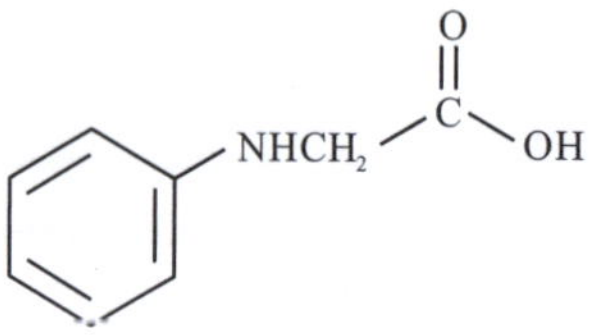

4) 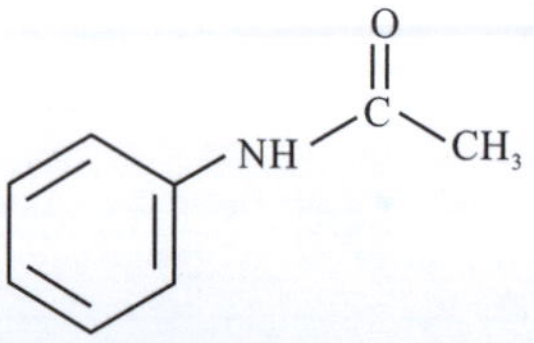

9. Which of the following is most basic?

1) 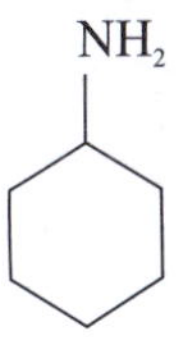

2) 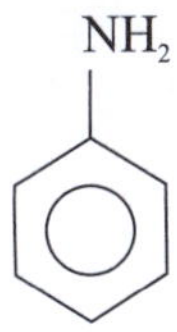

3) 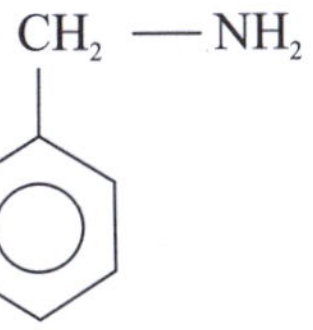

4) 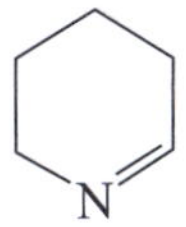

10. The correct order in aqueous medium of basic strength in case of methyl substituted amines is:

1) $Me_2NH > MeNH_2 > Me_3N > NH_3$
2) $Me_2NH > Me_3N > MeNH_2 > NH_3$
3) $NH_3 > Me_3N > MeNH_2 > Me_2NH$
4) $Me_3N > Me_2NH > MeNH_2 > NH_3$

11. $CH_3CH_2Br \xrightarrow{\text{aq.KOH}} A \xrightarrow{KMnO_4/H^+} B \xrightarrow[\Delta]{NH_3} C \xrightarrow[\text{alkali}]{Br_2} D$

D is:

1) CH_3Br
2) CH_3CONH_2
3) CH_3NH_2
4) $CHBr_3$

12. One mole of an amine (A) consumes two moles of methyl bromide to give a quaternary ammonium salt. The amine (A) is-

1) $(CH_3)_3CCH_2NH_2$
2) $(CH_3)_2NCH_2CH_3$
3) 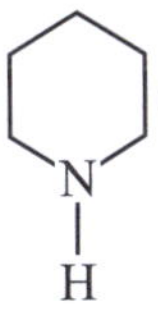 (Piperidine)
4) 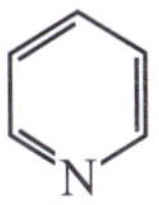 (Pyridine)

13. In the chemical reaction, the product A is?

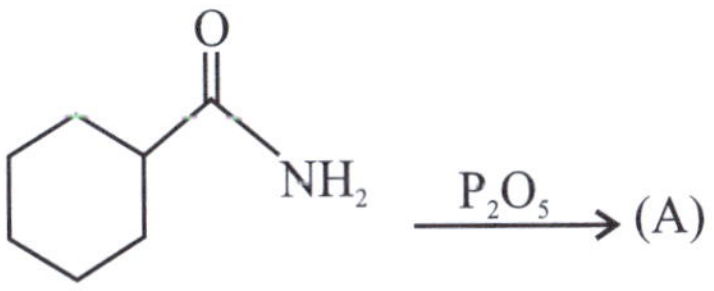

1) 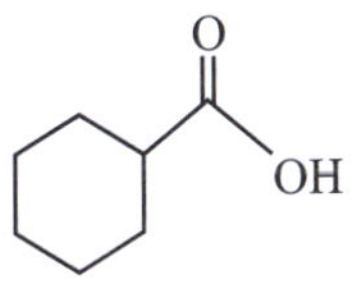

2) 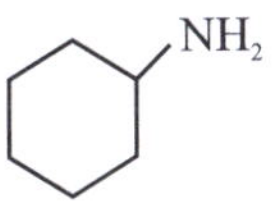

3) 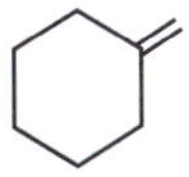

4) 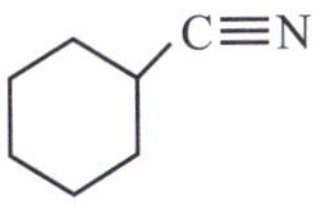

14. An organic compound 'A' having molecular formula C_2H_3N on reduction gave another compound 'B'. Upon treatment with nitrous acid, 'B' gave ethyl alcohol. On warming with chloroform and alcoholic KOH, it formed an offensive smelling compound 'C'. The compound 'C' is:

1) $CH_3CH_2NH_2$
2) $CH_3CH_2N \rightleftarrows C$
3) $CH_3C \equiv N$
4) CH_3CH_2OH

15. 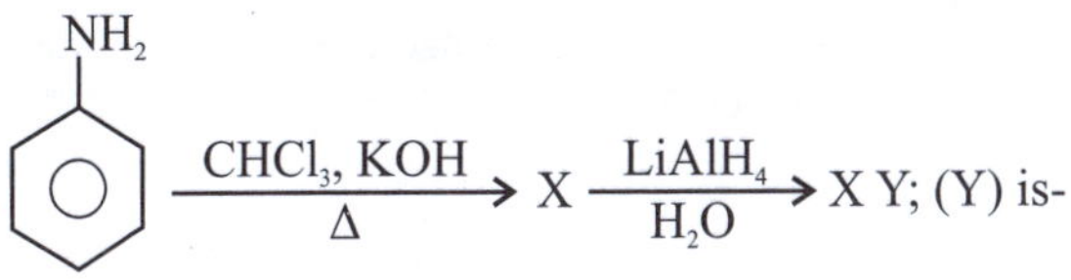

1) $Ph-NH-CH_3$
2) $Ph-NH-C(=O)-H$
3) $Ph-CH_2-CH_3$
4) $Ph-\overset{\oplus}{N} \equiv \overset{\ominus}{\ddot{C}}$

16. 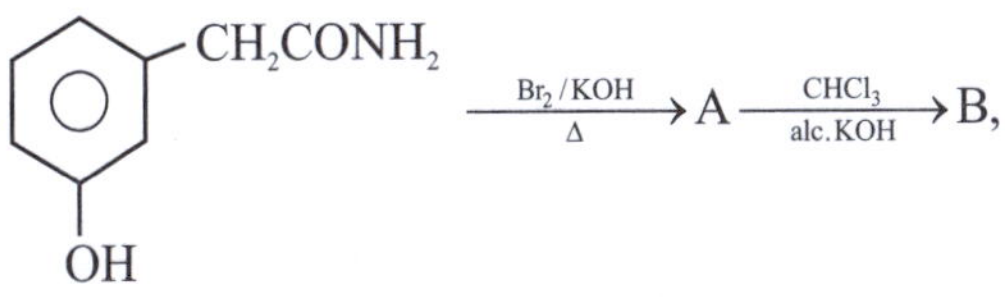

The final product B is:

1) 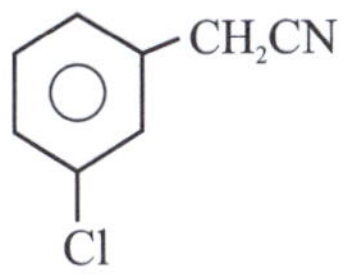

2) 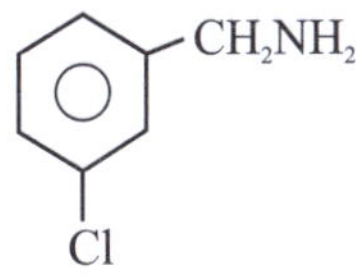

3) 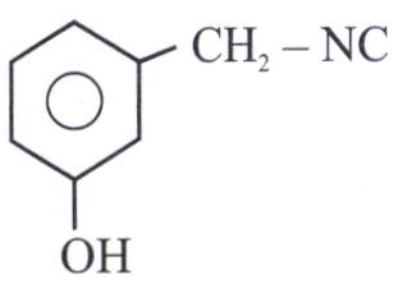

4) 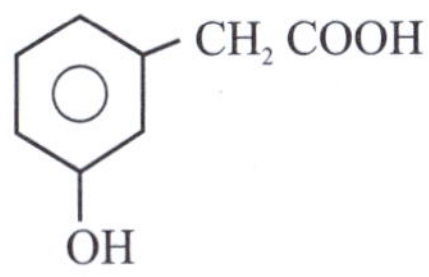

17. Following reaction sequence is given:

$$A \xrightarrow[H_2O]{Br_2/} B \xrightarrow[HCl]{NaNO_2/} C \xrightarrow[H_2O]{H_3PO_2/} D$$
$$0-5°C$$

'D' is 1, 3, 5-tibromobenzene

Then, what is A?

1) 
2) Phenol
3) Aniline
4) None of these

18. Which is the most basic in aqueous solution?

1) CH_3NH_2
2) $(CH_3)_2NH$
3) $(CH_3)_3N$
4) $Ph–NH_2$

19. Which of the following statements are incorrect for aniline?

1) Compound is heterocyclic hydrocarbon
2) Number of σ bonds are 8
3) Degree of unsaturation of the compound is 3
4) All of these

20. Which statement is NOT correct for p- toluenesulphonyl chloride?

1) It is known as Hinsberg's reagent
2) It is used to distinguish primary and secondary amines.
3) On treatment with secondary amine, it leads to a product, that is soluble in alkali.
4) It doesn't react with tertiary amines.

21. In a set of reactions m-bromobenzoic acid gave a product D. Identify the product D.

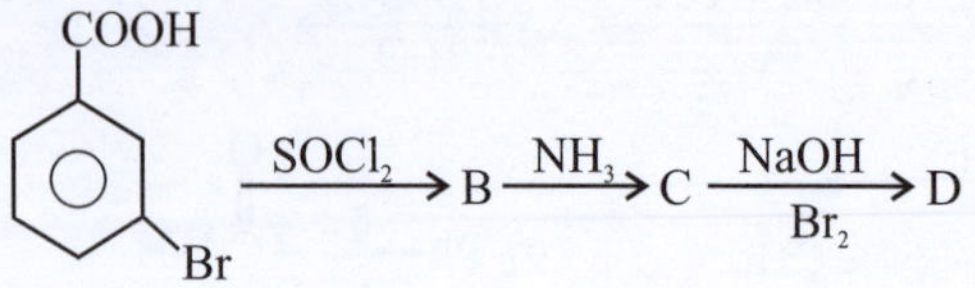

1) 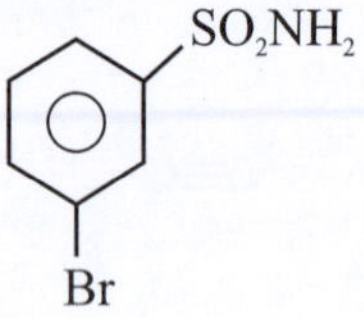
2) 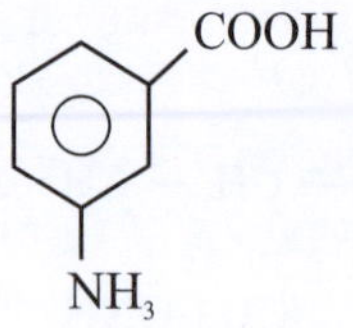
3) 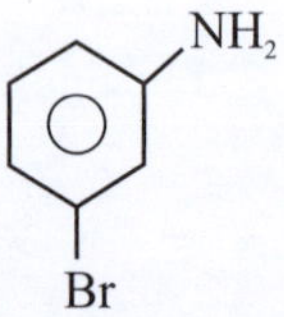
4) 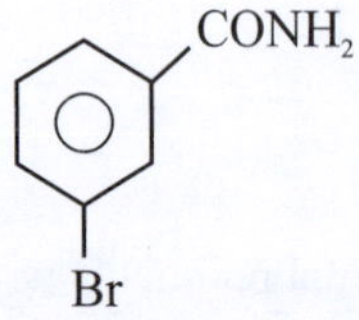

22. The product A formed in the following reaction is:

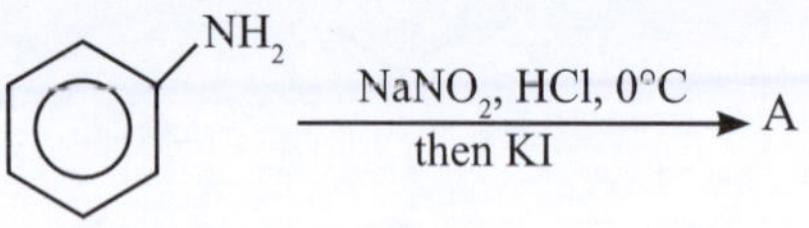

1) 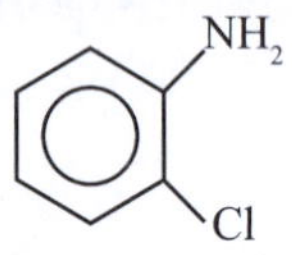
2) 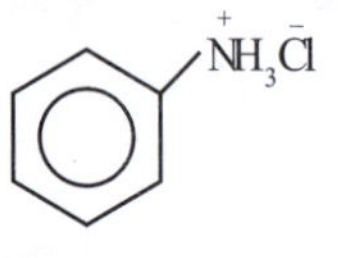
3) 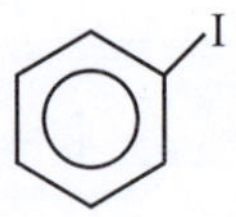
4) 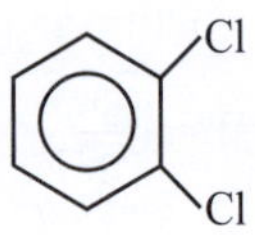

23. Which one among the following is expected to form a secondary alcohol on treatment with HNO_2?

1) 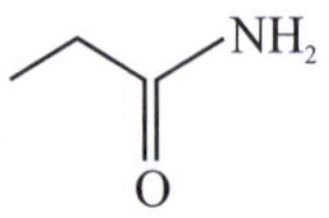
2) 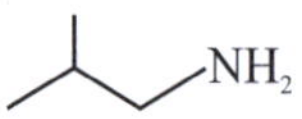
3) 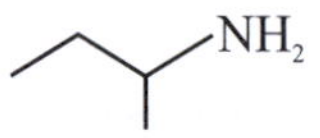
4) 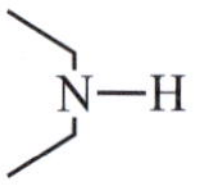

24. When aniline is treated with benzene diazonium chloride at low temperature in weakly acidic medium, the final product obtained is:

1) 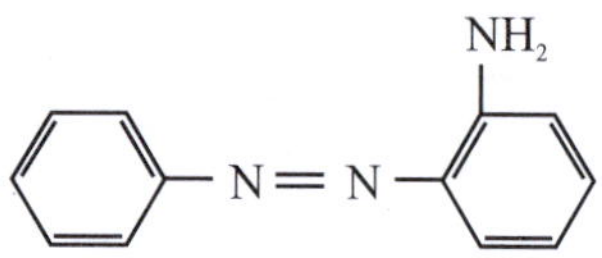
2) 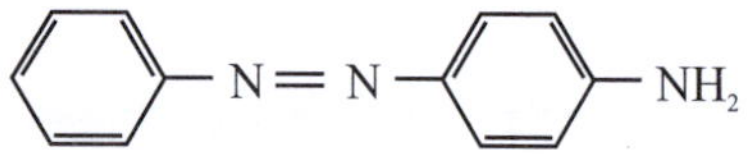
3) 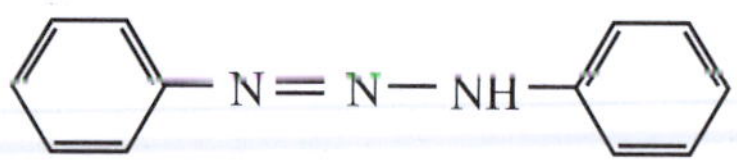
4) 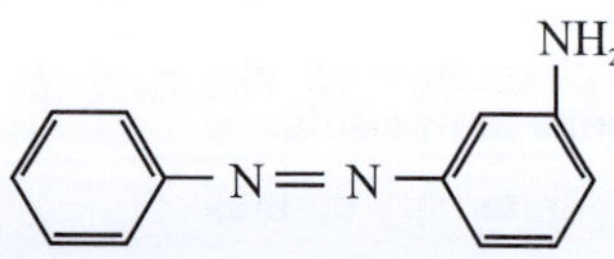

25. The end product (Z) of the following reaction is

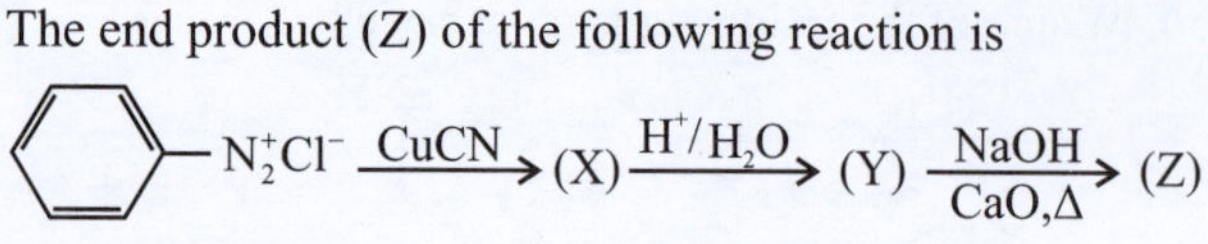

1) A cyanide 2) A carboxylic acid
3) An amine 4) An arene

26. When benzamide is heated with thionyl chloride, the main product of the reaction is:

1) C_6H_5CN 2) C_6H_5COCl
3) $C_6H_5NH_2$ 4) $C_6H_5-\overset{S}{\overset{||}{C}}-NH_2$

27. The compound $\begin{matrix} R_1 \\ R_2 \end{matrix}\!\!>\!N-R_3$ forms nitroso amines when the substituents are:

1) $R_1 = CH_3, R_2 = R_3 = H$
2) $R_1 = R_2 = H, R_3 = C_2H_5$
3) $R_1 = H$ and $R_2 = R_3 = CH_3$
4) $R_1 = CH_3, R_2 = C_2H_5, R_3 = C_3H_7$

28. Indicate which nitrogen compound amongst the following would undergo Hofmann's reaction (i.e. reaction with Br_2 and strong KOH) to furnish the primary amine $(R-NH_2)$.

1) $R-\overset{O}{\overset{||}{C}}-O-NH.CH_3$ 2) $R-\overset{O}{\overset{||}{C}}-\bar{O}.\overset{+}{N}H_4$
3) $R-\overset{O}{\overset{||}{C}}-NH_2$ 4) $R-\overset{O}{\overset{||}{C}}-NHOH$

29. 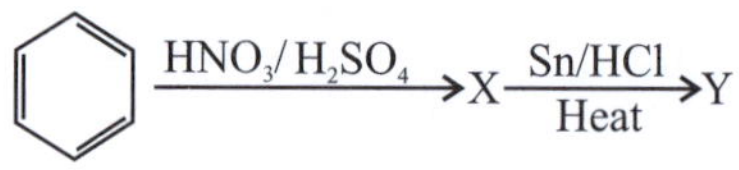

The final product 'Y' is:

1) 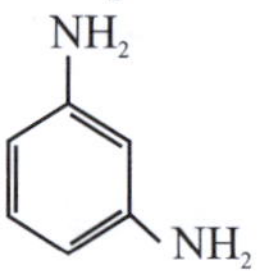
2) 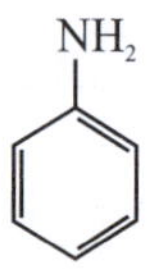
3) 
4) 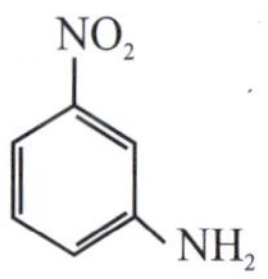

30. In the reaction

$CH_3CN + 2H \xrightarrow[SnCl_2]{HCl} X \xrightarrow{\text{Boiling } H_2O} Y,$

the term Y is:

1) Acetone 2) Ethanamine
3) Acetaldehyde 4) Dimethyl amine

31. 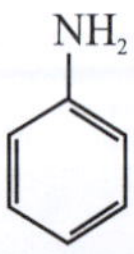 $+ CHCl_3 + KOH \longrightarrow ?$

1) Phenyl isocyanide 2) Benzyl amine
3) Benzyl chloride 4) None of these

32. RNH_2 reacts with $C_6H_5SO_2Cl$ in aqueous KOH to give a clear solution. On acidification, a precipitate is obtained which is due to the formation of:

1) $R-\overset{H}{\underset{H}{N^+}}-SO_2C_6H_5OH^-$ 2) $[R-N^-SO_2C_6H_5]K^+$

3) $R-NHSO_2C_6H_5$ 4) $C_6H_5SO_2NH_2$

33. $CH_3\ CH_2\ NH_2$ is soluble in

1) Dilute HCl 2) H_2SO_4

3) $CuSO_4$ 4) All of these

34. Which of the following reagent converts nitrobenzene to aniline?

1) Sn/HCl 2) Zn/NH_4OH

3) Zn/NaOH 4) $LiAlH_4$

35. 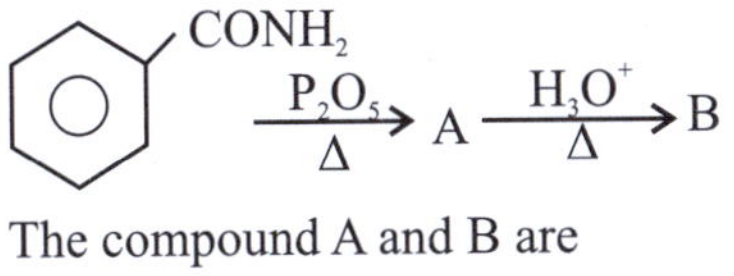

The compound A and B are

1) 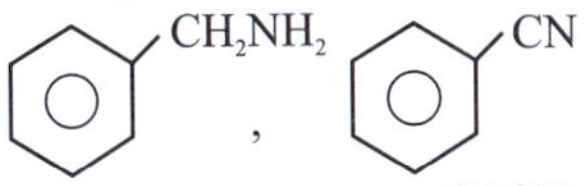

2) 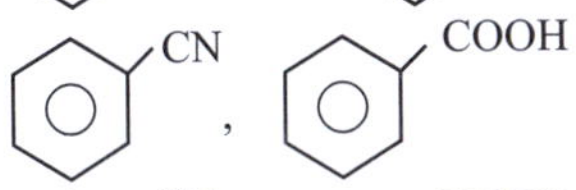

3) 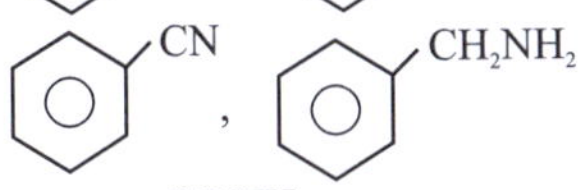

4) 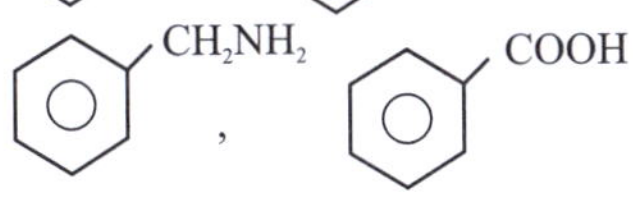

36. $CH_3CH_2NH_2 \xrightarrow[\Delta]{(CH_3CO)_2O} A + B$. (A) and (B) are :

1) $CH_3CH_2NHCOCH_3 + CH_3CH_2N \equiv C$

2) $CH_3CH_2NO_2 + CH_3COOH$

3) $CH_3CH_2NHCOCH_3 + CH_3COOH$

4) $CH_3CH_2NO_2 + CH_3CH_2NHCOCH_3$

37. Among the following compounds, which one will produce an enamine on reaction with cyclohexanone?

1) 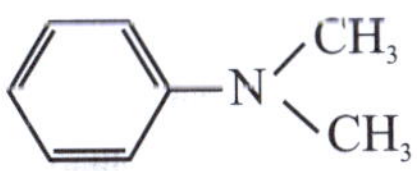 2) 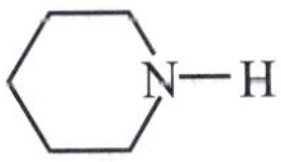

3) 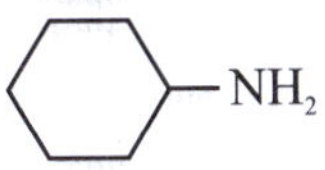 4) 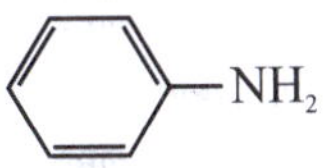

38. 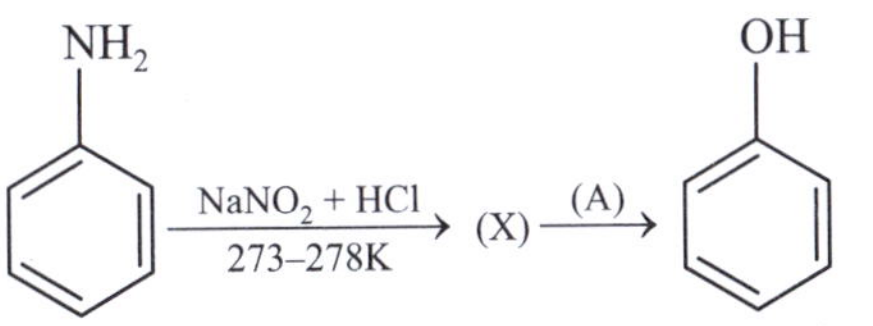

Major Product

In the above chemical reaction, intermediate (X) and reagent/condition (A) are:

1) X= 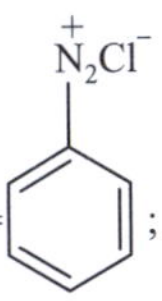 ; A=H_2O/NaOH 2) X= 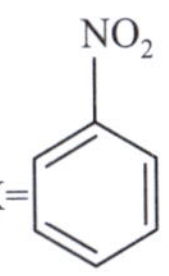 ; A=H_2O/Δ

3) X= 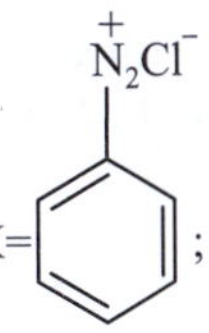 ; A=H_2O/Δ 4) X= 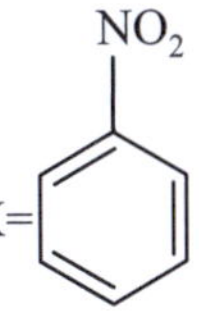 ; A=H_2O/NaOH

39. 2, 4, 6-tribromo aniline is a product of

1) Electrophilic addition on $C_6H_5NH_2$

2) Electrophilic substitution on $C_6H_5NH_2$

3) Nucleophilic addition on $C_6H_5NH_2$

4) Nucleophilic substitution on $C_6H_5NH_2$

40. Benzenediazonium chloride is reduced to benzene by

1) Water 2) Hypophosphorous acid

3) Hypophosphoric acid 4) Phosphine

41. Which of the following reactions are suitable method for the preparation of primary amines?

1) Action of ammonia on alkyl halides

2) Hofmann Bromamide reaction

3) Gabriel's phthalimide synthesis of amines

4) Both are (2) and (3) correct

42. Which of the following compound(s) do not give Friedel Crafts reaction?

(i) 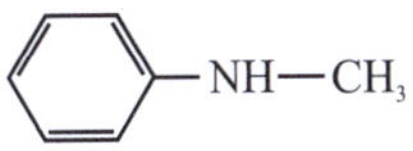 (ii) 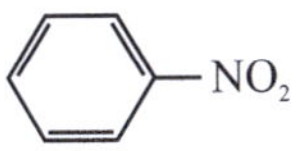

(iii) 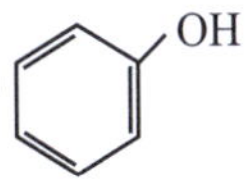 (iv) 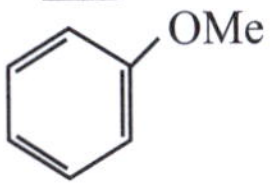

1) only (iii) 2) only (iv)

3) (i) and (iii) both 4) (i) and (ii) both

43. Maximum pK_b value is of

1) $(CH_3)_2NH$

2) $(CH_3CH_2)_2NH$

3) 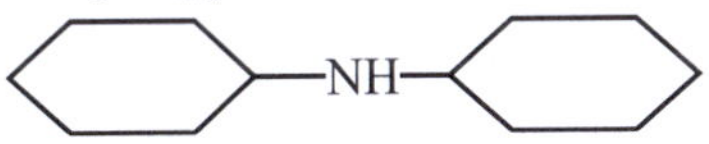

4) 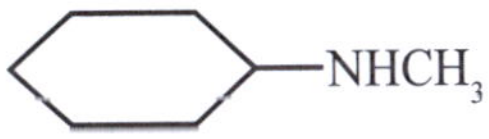

44. The reagents needed to convert is/are:

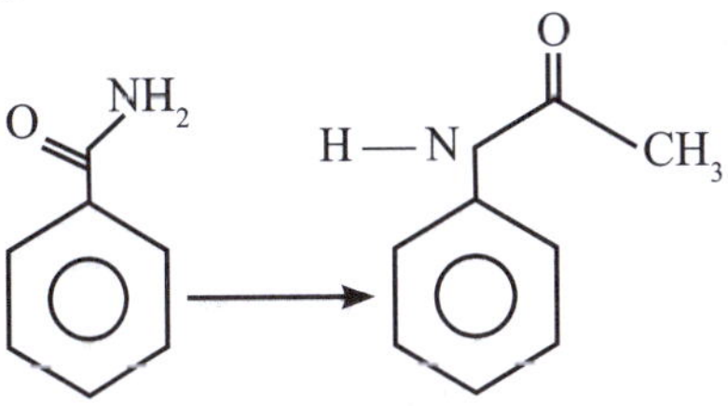

1) KOH, Br_2; $LiAlH_4$

2) KOH, Br_2; CH_3COCl

3) HONO, Cu_2Cl_2, $(CH_3CO)_2O$

4) KOH, Br_2; Ni, H_2, CH_3COCl

45. p-chloroaniline and anilinium hydrochloride cannot be distinguished by

1) Sandmeyer's reaction

2) $NaHCO_3$

3) $AgNO_3$

4) Carbylamine test

46. Aniline when diazotized in cold and then treated with dimethyl aniline give a coloured product. Its structure would be

1) 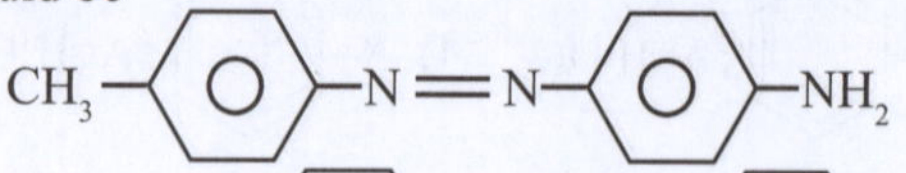

2) 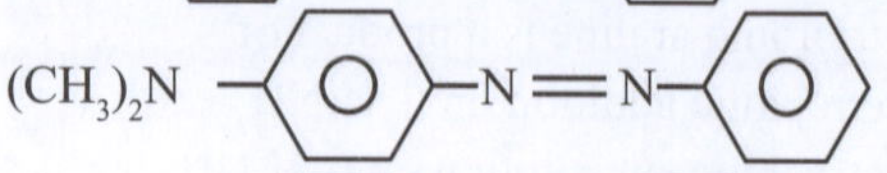

3) 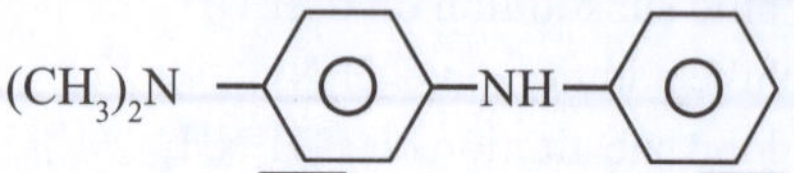

4) 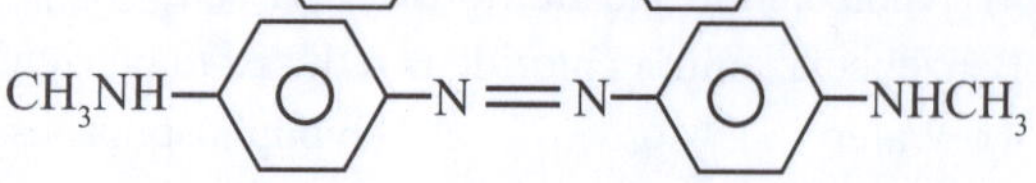

47. Amongst the following the most basic compound is

1) benzylamine
2) aniline
3) acetanilide
4) p-nitroaniline

48. Presence of a nitro group in a benzene ring

1) renders the ring basic
2) deactivates the ring towards nucleophilic substitution
3) deactivates the ring towards electrophilic substitution
4) activates the ring towards electrophilic substitution.

49. In the chemical reaction,

$CH_3CH_2NH_2 + CHCl_3 + 3KOH \longrightarrow (A) + (B) + 3H_2O$

The compounds (A) and (B) are respectively :

1) $CH_3CH_2CONH_2$ and 3KCl
2) C_2H_5NC and K_2CO_3
3) C_2H_5NC and 3KCl
4) C_2H_5CN and 3KCl

50. In the chemical reaction

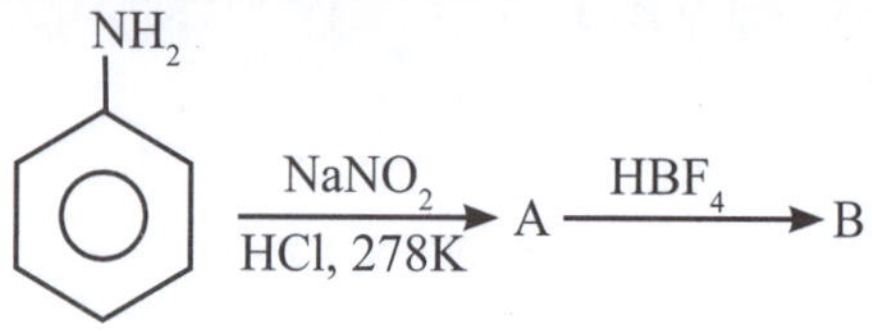

The compound 'A' and 'B' respectively are:

1) benzene diazonium chloride and fluorobenzene
2) nitrobenzene and chlorobenzene
3) nitrobenzene and fluorobenzene
4) phenol and benzene

51. What is the IUPAC name of isopropylamine?

1) Propan-2-amine
2) Ethanamine
3) 2-aminotoluene
4) Propan-1-amine

52. In the reaction

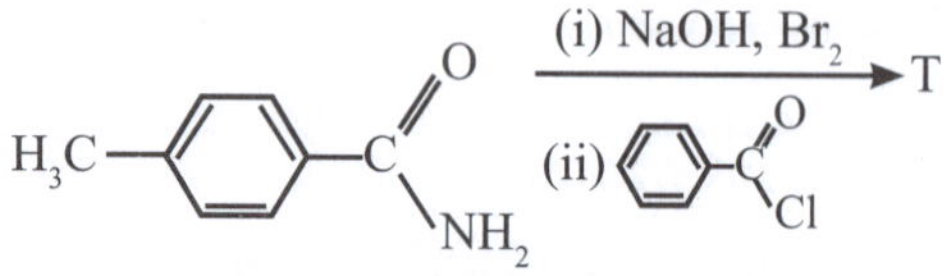

the structure of the product T is:

1) 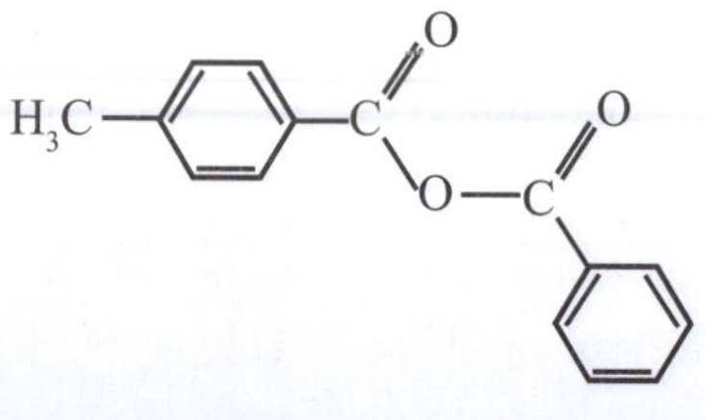

2) 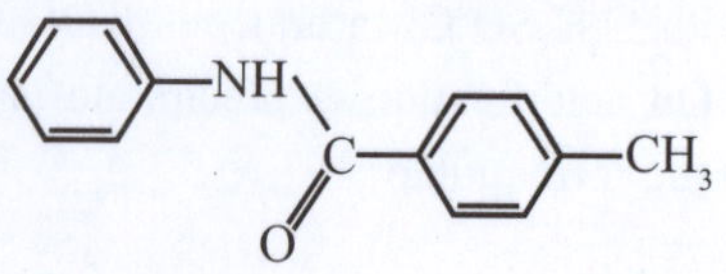

3) 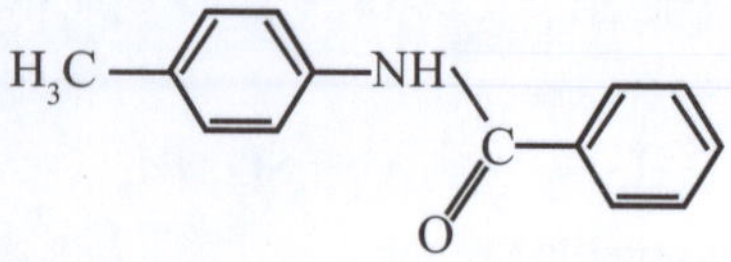

4) 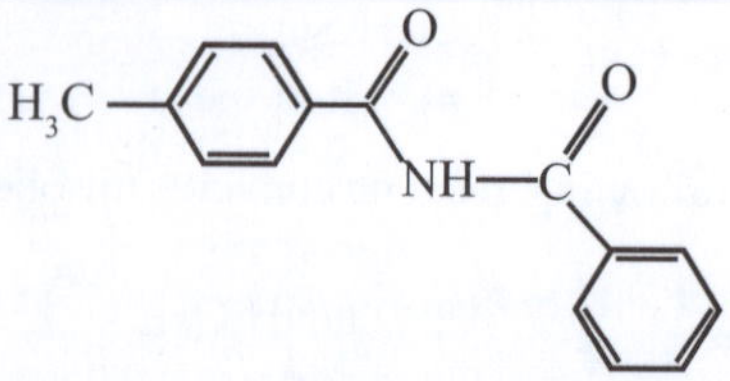

53. The correct order of basicity for the following molecules is:

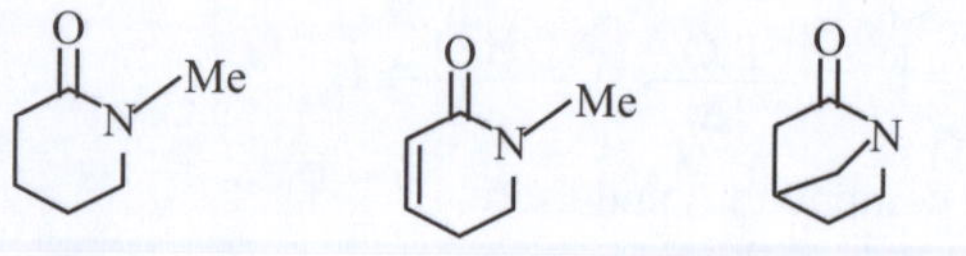

(P) (Q) (R)

1) P> Q >R
2) R > P > Q
3) Q > P >R
4) R > Q >P

54. The correct order of basic nature on aqueous solution for the bases NH_3, $H_2N–NH_2$, $CH_3CH_2NH_2$, $(CH_3CH_2)_2NH$ and $(CH_3CH_2)_3N$ is:

1) $NH_3 < H_2N–NH_2 < (CH_3CH_2)_3N < CH_3CH_2NH_2 < (CH_3CH_2)_2NH$
2) $NH_3 < H_2N–NH_2 < CH_3CH_2NH_2 < (CH_3CH_2)_2NH < (CH_3CH_2)_3N$
3) $H_2N–NH_2 < NH_3 < (CH_3CH_2)_3N < CH_3CH_2NH_2 < (CH_3CH_2)_2NH$
4) $NH_2–NH_2 < NH_3 < CH_3CH_2NH_2 < (CH_3CH_2)_3N < (CH_3CH_2)_2NH$

55. A. Phenyl methanamine

B. N, N-Dimethylaniline

C. N-Methyl aniline

D. Benzenamine

Choose the correct order of basic nature of the above amines:

1) D > C > B > A
2) D > B > C > A
3) A > C > B > D
4) A > B > C > D

56. The increasing order of pK_b values of the following compounds is:

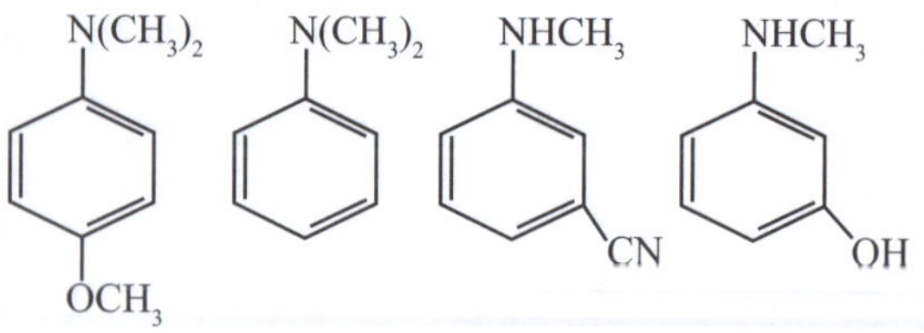

I II III IV

1) II < IV < III < I
2) I < II < IV < III
3) II < I < III < IV
4) I < II < III < IV

Match the Columns MCQs

1. Match the reactions given in Column-I with the statements given in Column-II.

Column-I		Column-II	
(a)	Ammonolysis	(i)	Amine with lesser number of carbon atoms
(b)	Gabriel phthalimide synthesis	(ii)	Detection test for primary amines
(c)	Hoffmann Bromamide reaction	(iii)	Reaction of phthalimide with KOH and R—X
(d)	Carbylamine reaction	(iv)	Reaction of alkylhalides with NH_3

1) (a)-(ii) (b)-(i) (c)-(iii) (d)-(iv)
2) (a)-(i) (b)-(iv) (c)-(ii) (d)-(iii)
3) (a)-(iii) (b)-(ii) (c)-(i) (d)-(iv)
4) (a)-(iv) (b)-(iii) (c)-(i) (d)-(ii)

2. Match the compounds given in Column-I with the items given in Column-II.

Column-I		Column-II	
(a)	Benzene sulphonyl chloride	(i)	Zwitter ion
(b)	Sulphanilic acid	(ii)	Hinsberg reagent
(c)	Alkyl diazonium salts	(iii)	Dyes
(d)	Aryl diazonium salts	(iv)	Conversion to alcohols

1) (a)-(iii) (b)-(i) (c)-(ii) (d)-(iv)
2) (a)-(ii) (b)-(i) (c)-(iv) (d)-(iii)
3) (a)-(iv) (b)-(ii) (c)-(iii) (d)-(i)
4) (a)-(iii) (b)-(i) (c)-(iv) (d)-(ii)

3. Match the Column-I with Column-II.

Column-I Amines		Column-II PK_b values	
(a)	$CH_3CH_2NH_2$	(i)	4.70
(b)	$(CH_3CH_2)_2NH$	(ii)	3.29
(c)	$C_6H_5NH_2$	(iii)	3.00
(d)	$C_6H_5CH_2NH_2$	(iv)	9.38

1) (a)-(i) (b)-(ii) (c)-(iv) (d)-(iii)
2) (a)-(iii) (b)-(iv) (c)-(i) (d)-(ii)
3) (a)-(ii) (b)-(iii) (c)-(iv) (d)-(i)
4) (a)-(iv) (b)-(i) (c)-(ii) (d)-(iii)

4. Match the Column-I with Column-II.

Column-I		Column-II	
(a)	$CH_3CH_2NH_2$	(i)	Undergoes Lieberman nitrosoamine reaction
(b)	$C_6H_5NH_2$	(ii)	Undergoes Hofmann bromamide reaction
(c)	$(CH_3)_2NH$	(iii)	Gives azo dye test
(d)	CH_3CONH_2	(iv)	With alcoholic KOH and $CHCl_3$ Produces bad smell

1) (a)-(iv) (b)-(iii) (c)-(i) (d)-(ii)
2) (a)-(iii) (b)-(i) (c)-(iv) (d)-(ii)
3) (a)-(iv) (b)-(iii) (c)-(ii) (d)-(i)
4) (a)-(iii) (b)-(ii) (c)-(i) (d)-(i)

5. Match the Column-I with Column-II.

Column-I		Column-II	
(a)	$PhNO_2 \xrightarrow{Zn/NH_4Cl}$	(i)	$Ph-N=N-Ph$
(b)	$PhNO_2 \xrightarrow[NaOH]{Zn/alc.}$	(ii)	$Ph-NH-NH-Ph$
(c)	$PhNO_2 \xrightarrow{LAH}$	(iii)	PhNHOH
(d)	$PhNO_2 \xrightarrow{Na_3AsO_3/NaOH}$	(iv)	$Ph-\overset{\overset{O}{\uparrow}}{N}=N-Ph$

1) (a)-(i) (b)-(ii) (c)-(iii) (d)-(iv)
2) (a)-(ii) (b)-(iii) (c)-(i) (d)-(iv)
3) (a)-(iii) (b)-(ii) (c)-(i) (d)-(iv)
4) (a)-(ii) (b)-(iv) (c)-(iii) (d)-(i)

Correct & Incorrect MCQs

1. Which of the following statement (s) are correct?

A. Nitrogen atom of amines is trivalent and carries an unshared pair of electrons.

B. Nitrogen orbitals in amines are sp^3 hybridised.

C. The geometry of amines is pyramidal.

D. The angle C–N–E, (where E is C or H) is more than 109.5°.

1) A, B and C   2) A and D

3) A, C and D   4) B, C and D

2. Which of the following statement is incorrect?

1) In IUPAC system, primary amines are named as alkanamines.

2) $C_6H_5NH_2$ is the simplest example of arylamine.

3) The common name of $CH_3 - CH_2 - NH_2$ is ethylamine.

4) The IUPAC name of isopropylamine is propan-1-amine.

3. Choose incorrect statement among the following:

1) The order of reactivity of halides with amines is RI > RBr >RCl.

2) Nitriles on reduction with lithium aluminium hydride ($LiAlH_4$) or catalytic hydrogenation produce primary amines.

3) Aromatic primary amines can be prepared by Gabriel phthalimide synthesis.

4) The amides on reduction with lithium aluminium hydride yield amines.

4. Select the correct statement from the following:

1) Methyl amine is less soluble than propyl amine.

2) Methyl amine is solid with fishy odour.

3) Aniline gets coloured on storage.

4) The order of boiling points of isomeric amines is : primary < secondary < tertiary.

5. Correct statement (s) is/are

A. Secondary and tertiary amines do not show carbylamine reaction.

B. Aromatic amines react with nitrous acid at low temperatures (273-278 K) to form diazonium salts.

C. Benzenesulphonyl chloride ($C_6H_5SO_2Cl$), is known as Hinsberg's reagent.

D. The reaction of benzenesulphonyl chloride with primary amine yields N-ethylbenzenesulphonyl amide which is insoluble in alkali.

1) Only B   2) A, B and C

3) A and D   4) B and C

6. Select the incorrect statement about primary amines

1) Alkyl amines are stronger bases than that of ammonia.

2) Alkyl amines are stronger bases than that of aryl amines.

3) Alkyl amines produce alcohols on reacting with nitrous acid.

4) Aryl amines react with nitrous acid to produce phenols.

7. Which of the following statements is incorrect?

1) Aniline is more acidic than ammonia.

2) 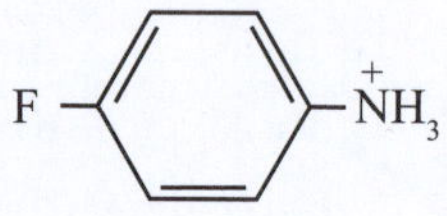 is less acidic than 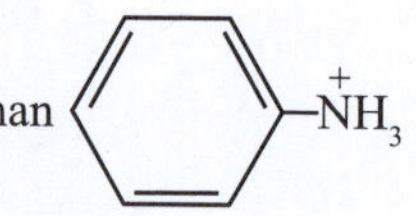

3) Amines are less acidic than alcohols of comparable molecular masses.

4) $RCONH_2$ is more acidic than RCH_2NH_2.

8. Which statement is not correct?

1) Amines form hydrogen bond.

2) Ethyl amine has higher boiling point than propane.

3) Methyl amine is more basic than ammonia.

4) Dimethyl amine is less basic than methyl amine.

9. Select the incorrect statement :

1) Benzenediazonium chloride is prepared by the reaction of aniline with nitrous acid at 273-278K.

2) Benzenediazonium chloride is a colourless crystalline solid.

3) Benzenediazonium chloride is readily soluble in water and is stable in cold but reacts with water when warmed.

4) Benzenediazonium fluoroborate is water soluble and is unstable at room temperature.

10. Which amongst the following is incorrect statement ?

1) Amines behave as Lewis acids.

2) Alkylamines are stronger bases than ammonia.

3) The order of basicity of amines in the gaseous phase follows the order: tertiary amine > secondary amine > primary amine > NH_3.

4) Both (2) and (3).

Statement Based MCQs

Directions: These questions consist of two statements each, printed as Statement-I and Statement-II. While answering these questions, you are required to choose any one of the following four responses.

1) **Both Statement-I and Statement-II are correct.**
2) **Both Statement-I and Statement-II are incorrect.**
3) **Statement-I is correct & Statement-II is incorrect.**
4) **Statement-I is incorrect & Statement-II is correct.**

1. **Statement-I:** Isocyanides are prepared by carbylamine reaction.
   Statement-II: Carbylamines on reduction always give 2° amines.
2. **Statement-I:** Nitration of aniline gives m-nitroaniline as one of a major product.
   Statement-II: Aniline gives anilinium ion in acidic medium which is strong deactivating in nature.
3. **Statement-I:** Tertiary amines give alkanes with alkyl magnesium halide.
   Statement-II: Tertiary amines contain active-H.
4. **Statement-I:** Dimethyl amine is a stronger base than that of trimethyl amine in aqueous medium.
   Statement-II: In tertiary amine, inductive effect is more than that of secondary amine.
5. **Statement-I:** The reaction between a diazo salt and an aromatic amine or a phenol, giving an amino azo or hydroxy azo compounds is called coupling reaction.
   Statement-II: Condensation of diazonium salt with phenol is carried out in weakly acidic medium.
6. **Statement-I:** In an amine, both the C – N and N – H bonds are polar.
   Statement-II: Aromatic primary amines cannot be prepared by Gabriel phthalimide synthesis.
7. **Statement-I:** Reduction of nitro compound with iron scrap and HCl is preferred over H_2 /Pd & ethanol.
   Statement-II: Nitriles on reduction with lithium aluminium hydride or catalytic hydrogenation produce tertiary amines.
8. **Statement-I:** A primary amine has higher boiling point than any tertiary amines.
   Statement-II: Pyrrole is more basic than pyridine.
9. **Statement-I:** Amines behave as electrophiles.
   Statement-II: In amines, unshared electron pair is present on nitrogen atom.
10. **Statement-I:** Acetylation of 3° amine is not possible.
    Statement-II: 3° amine contain a H – atom on N– atom.

Assertion & Reason MCQs

Directions: These questions consist of two statements each, printed as Assertion and Reason. While answering these questions, you are required to choose any one of the following four responses.

1) **Both Assertion and Reason are True and the Reason is a correct explanation of the Assertion.**
2) **Both Assertion and Reason are True but Reason is not a correct explanation of the Assertion.**
3) **Assertion is True but the Reason is False.**
4) **Assertion is False but Reason is True.**

1. **Assertion:** Acylation of amines gives a monosubstituted product whereas alkylation of amines gives polysubstituted product.
   Reason: Acyl group sterically hinders the approach of further acyl groups.
2. **Assertion:** Hoffmann's bromamide reaction is given by primary amines.
   Reason: Primary amines are more basic than secondary amines.
3. **Assertion:** N-Ethylbenzene sulphonamide is soluble in alkali.
   Reason: Hydrogen attached to nitrogen in sulphonamide is strongly acidic.
4. **Assertion:** N, N-Diethylbenzene sulphonamide is insoluble in alkali.
   Reason: Sulphonyl group attached to nitrogen atom is strong electron withdrawing group.
5. **Assertion:** Only a small amount of HCl is required in the reduction of nitro compounds with iron scrap and HCl in the presence of steam.
   Reason: $FeCl_2$ formed gets hydrolysed to release HCl during the reaction.
6. **Assertion:** Acylation of aniline decreases the reactivity of benzene ring as compared with aniline.
   Reason: Acylation of aniline results in increase in electron density at the benzene ring.
7. **Assertion:** In carbylamine test, amines are converted into isocyanides.
   Reason: Alkyl isocyanide is an isomer of 1°- amine.
8. **Assertion:** Carbylamine reaction involves the reaction between primary amine and chloroform in basic medium.
   Reason: In carbylamine reaction, $-NH_2$ group is converted into –NC group.
9. **Assertion:** Tertiary amines do not react with grignard reagent.
   Reason: Tertiary amines contain active-H
10. **Assertion:** Total possible structural amines of $C_4H_{11}N$ are eight.
    Reason: 1°, 2° and 3° amines are functional isomers to each other.

Multi Correct MCQs

1. Which of the following compounds will give reaction with Hinsberg's Reagent?

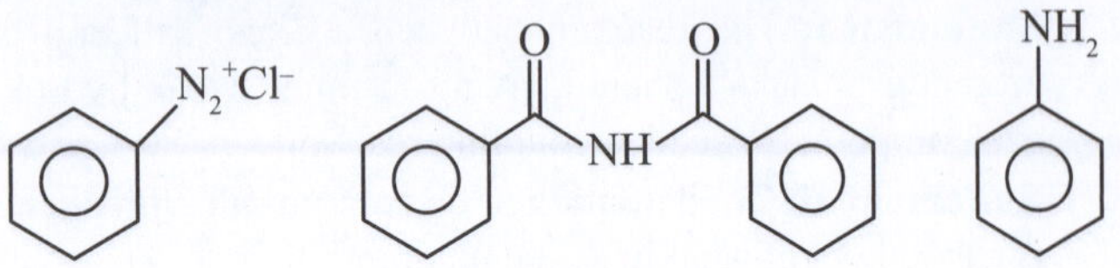

(A) (B) (C)

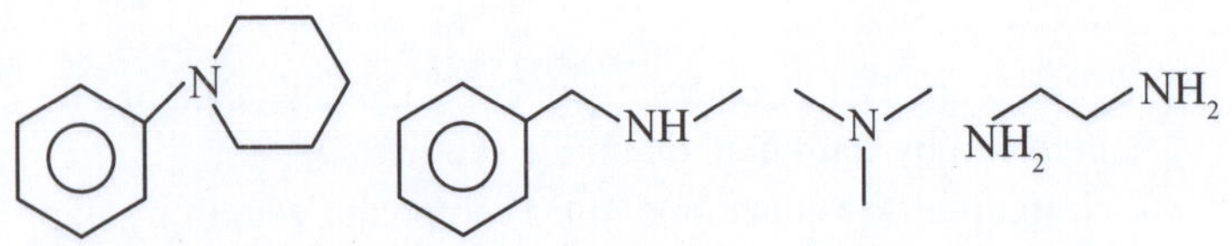

(D) (E) (F) (G)

1) A, B, C and D   2) C, E and G
3) B, C, D and E   4) C, D and F

2. In the following compound

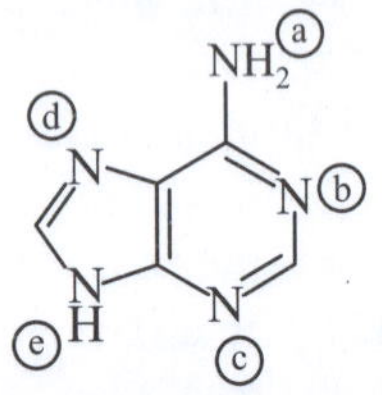

The favourable site/s for protonation is /are:
1) a and e   2) b, c and d
3) a and d   4) a

3. Benzyl isocyanide can be obtained by:

A. 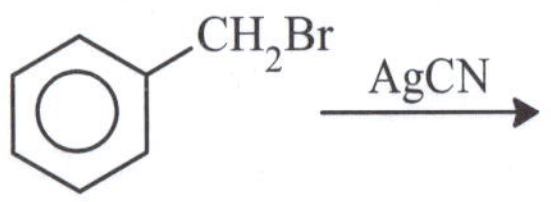

B. 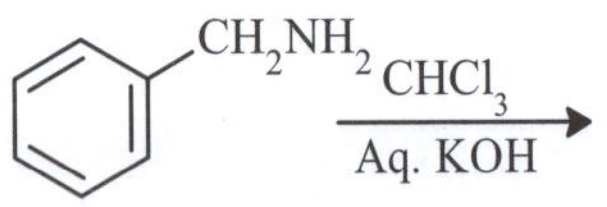

C. 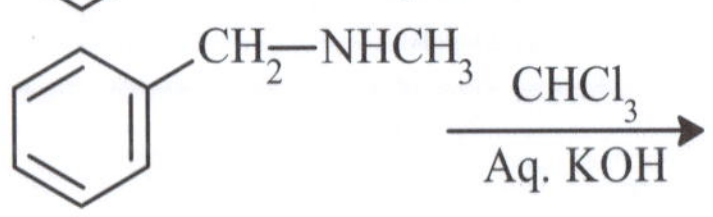

D. 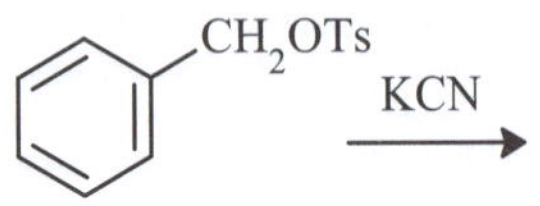

Choose the correct answer from the options given below:
1) A and D   2) Only B
3) A and B   4) B and C

4. Which of the following reaction/s will not give p-aminoazobenzene?

(I) 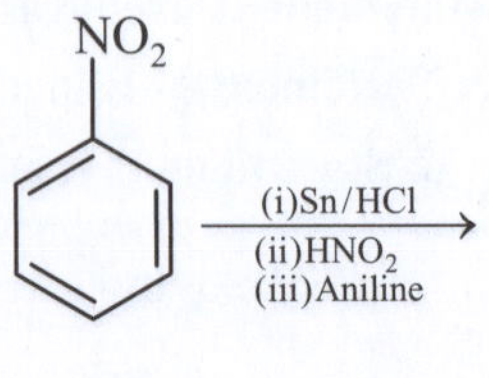

(II) 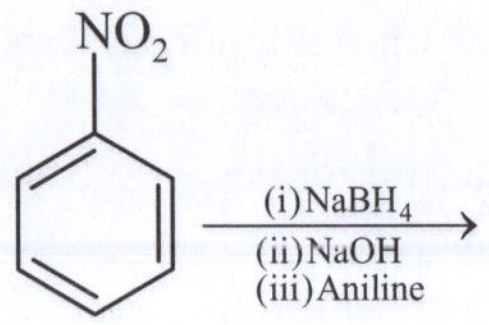

(III) 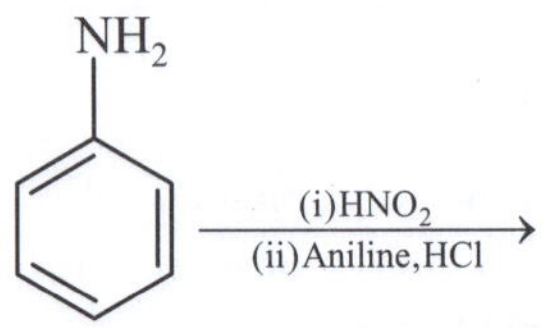

1) II only   2) I and II
3) III only   4) I and III only

5. The descending order of basicity of following amines is :

A. 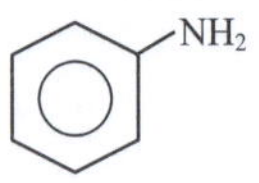
B. 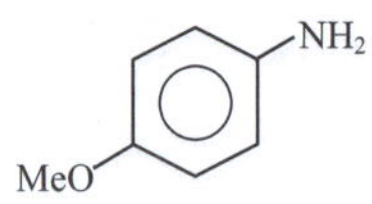
C. 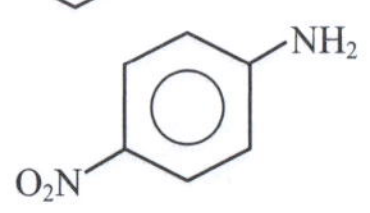
D. CH_3NH_2
E. $(CH_3)_2NH$

Choose the correct answer from the options given below :
1) B > E > D > A > C
2) E > D > B > A > C
3) E > D > A > B > C
4) E > A > D > C > B

NCERT Exemplar MCQs

1. Which of the following is a 3° amine?
   1) 1-Methylcyclohexylamine
   2) Triethylamine
   3) Tert-butylamine
   4) N-Methylaniline
2. The correct IUPAC name for $CH_2 = CHCH_2NHCH_3$ is:
   1) Alkyl methylamine 2) 2-Amino-pentene
   3) 4-Aminopent-1-ene 4) N-Methylprop-2-en1-amine
3. Amongst the following, the strongest base in aqueous medium is:
   1) CH_3NH_2 2) $NCCH_2NH_2$
   3) $(CH_3)_2NH$ 4) $C_6H_5NHCH_3$
4. Which of the following is the weakest Bronsted base?
   1) 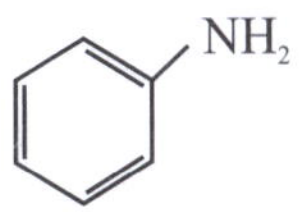 2) 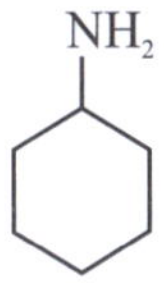
   3) 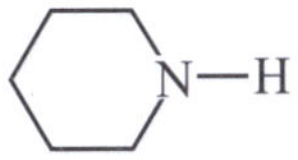 4) CH_3NH_2
5. Benzylamine may be alkylated as shown in the following equation?
   $C_6H_5CH_2NH_2 + R - X \rightarrow C_6H_5CH_2NHR$
   Which of the following alkyl halides is best suited for this reaction through S_N1 mechanism?
   1) CH_3Br 2) C_6H_5Br
   3) $C_6H_5CH_2Br$ 4) C_2H_5Br
6. Which of the following reagents would not be a good choice for reducing an aryl nitro compound to an amine?
   1) H_2(excess) / Pt 2) $LiAlH_4$ in ether
   3) Fe and HCI 4) Sn and HCl
7. In order to prepare a 1° amine from an alkyl halide with simultaneous addition of one CH_2 group in the carbon chain, the reagent used as source of nitrogen is :
   1) Sodium amide, $NaNH_2$
   2) Sodium azide, NaN_3
   3) Potassium cyanide, KCN
   4) Potassium phthalimide $C_6H_4(CO)_2 N^-K^+$
8. The source of nitrogen in Gabriel synthesis of amines is:
   1) Sodium azide, NaN_3
   2) Sodium nitrite, $NaNO_2$
   3) Potassium cyanide, KCN
   4) Potassium phthalimide $C_6H_4(CO)_2 N^-K^+$
9. Amongst the given set of reactants, the most appropriate for preparing 2° amine is:
   1) 2° R—Br + NH_3
   2) 2° R — Br + NaCN followed by H_2/ Pt
   3) 1° R—NH_2 + RCHO followed by H_2 /Pt
   4) 1° R—Br (2 mol) + potassium phthalimide followed by H_3O^+/ heat
10. The best reagent for converting 2-Phenylpropanamide into 2-Phenylpropanamine is:
    1) Excess H_2
    2) Br_2 in aqueous NaOH
    3) Iodine in the presence of red phosphorus
    4) $LiAlH_4$ in ether
11. The best reagent for converting, 2-Phenylpropanamide into 1-Phenylethanamine is:
    1) Excess H_2/Pt 2) NaOH / Br_2
    3) $NaBH_4$/ methanol 4) $LiAlH_4$/ether
12. Hoffman bromamide degradation reaction is shown by:
    1) $ArNH_2$ 2) $ArCONH_2$
    3) $ArNO_2$ 4) $ArCH_2NH_2$
13. The correct increasing order of basic strength for the following compounds is:

    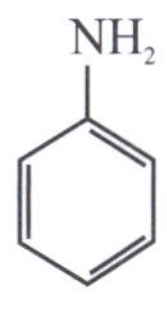 (I) 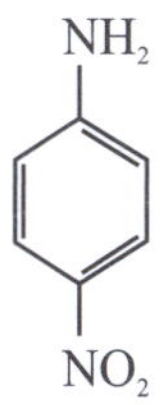 (II)  (III)

    1) II < III < I 2) III < I < II
    3) III < II < I 4) II < I < III
14. Methylamine reacts with HNO_2 to form:
    1) $CH_3 - O - N = O$ 2) $CH_3 - O - CH_3$
    3) CH_3OH 4) CH_3CHO
15. The gas evolved when methylamine reacts with nitrous acid is:
    1) NH_3 2) N_2
    3) H_2 4) C_2H_6
16. In the nitration of benzene using a mixture of conc. H_2SO_4 and conc. HNO_3, the species which initiates the reaction is:
    1) NO_2 2) NO^+
    3) NO_2^+ 4) NO_2^-
17. Reduction of aromatic nitro compounds using Fe and HCl gives:
    1) Aromatic oxime 2) Aromatic hydrocarbon
    3) Aromatic primary amine 4) Aromatic amide
18. The most reactive amine towards dilute hydrochloric acid is:
    1) $CH_3 - NH_2$ 2) 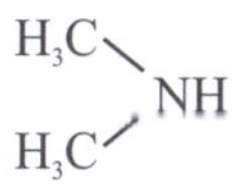
    3) 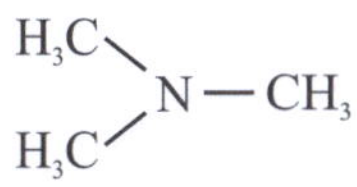 4) 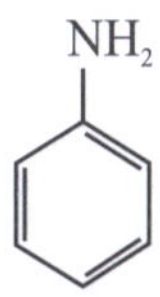

19. Acid anhydride on reaction with primary amines give:

1) Amide 2) Imide
3) Secondary amine 4) Imine

20. The reaction $Ar\overset{+}{N_2}Cl^- \xrightarrow{Cu/HCl} ArCl + N_2 + CuCl$ is named as:

1) Sandmeyer reaction
2) Gattermann reaction
3) Claisen reaction
4) Carbylamine reaction

21. Best method for preparing primary amines from alkyl halides without changing the number of carbon atoms in the chain is:

1) Hoffman bromamide reaction
2) Gabriel phthalimide synthesis
3) Sandmeyer reaction
4) Reaction with NH_3

22. Which of the following compound will not undergo azo coupling reaction with benzene diazonium chloride?

1) Aniline 2) Phenol
3) Anisole 4) Nitrobenzene

23. Which of the following compounds is the weakest Bronsted base?

1) 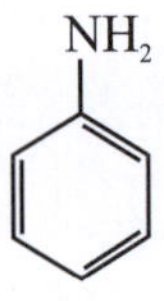

2) 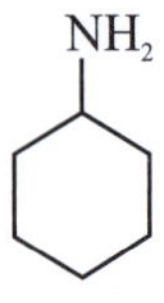

3) 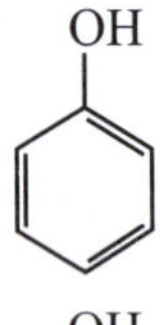

4) 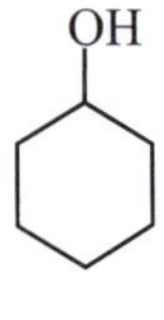

24. Among the following amines, the strongest Bronsted base is:

1) 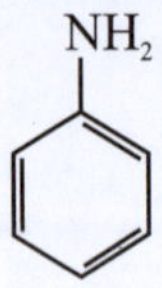
2) NH_3

3) 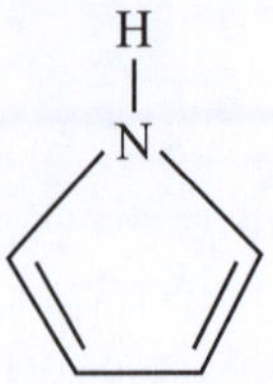
4) 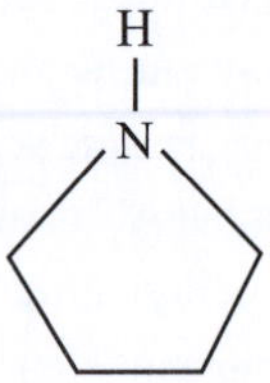

25. The correct decreasing order of basic strength of the following species is:

H_2O, NH_3, OH^-, NH_2^-

1) $NH_2^- > OH^- > NH_3 > H_2O$
2) $OH^- > NH_2^- > H_2O > NH_3$
3) $NH_3 > H_2O > NH_2^- > OH^-$
4) $H_2O > NH_3 > OH^- > NH_2^-$

26. Which of the following should be most volatile?

(i) $CH_3CH_2CH_2NH_2$ (ii) $(CH_3)_3N$
(iii) $CH_3CH_2-NH-CH_3$ (iv) $CH_3CH_2CH_3$

1) (ii) 2) (iv)
3) (i) 4) (iii)

27. Which of the following methods of preparation of amines will give same number of carbon atoms in the chain of amines as in the reactant?

1) Reaction of nitrile with $LiAlH_4$
2) Reaction of amide with $LiAlH_4$ followed by treatment with water
3) Heating alkyl halide with potassium salt of phthalimide followed by hydrolysis
4) Treatment of amide with bromine in aqueous solution of sodium hydroxide

Scan Me for Detailed Explanations

Kattar NEET

1. On hydrolysis of a compound (X), two compounds are obtained. One of which on treatment with sodium nitrite and hydrochloric acid gives a product which does not respond to iodoform test. The second one reduces Tollen's reagent and Fehling's solution. The compound (X) is :
   1) $CH_3CH_2CH_2NC$
   2) $CH_3CH_2CH_2CN$
   3) $CH_3CH_2CH_2ON{=}O$
   4) $CH_3CH_2CH_2CON(CH_3)_2$

2. Aniline in a set of reactions yielded a product D.

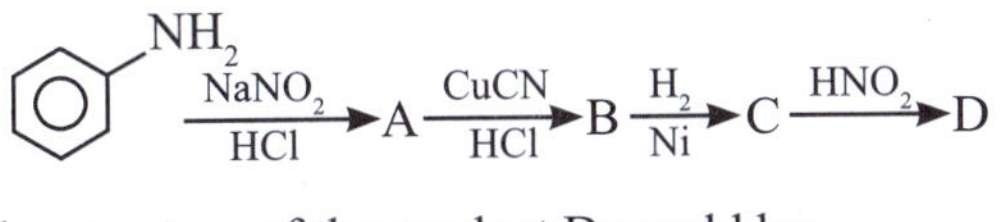

   The structure of the product D would be:
   1) C_6H_5NHOH
   2) $C_6H_5NHCH_2CH_3$
   3) $C_6H_5CH_2NH_2$
   4) $C_6H_5CH_2OH$

3. Consider the following sequence of reactions:

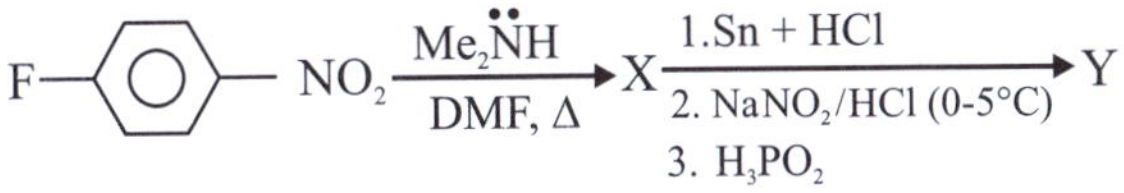

   the product Y is:

   1) 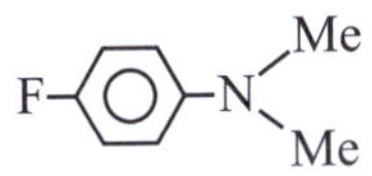
   2) 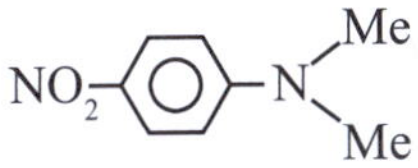
   3) 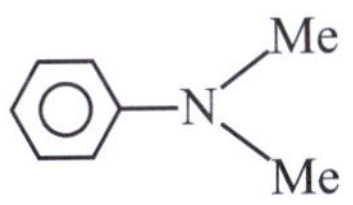
   4) 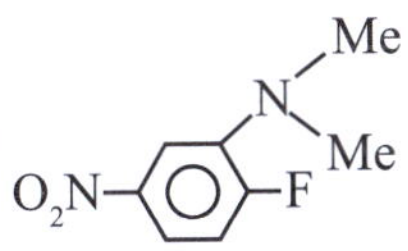

4. How many products are capable of beings formed from toluene in each of following reaction?

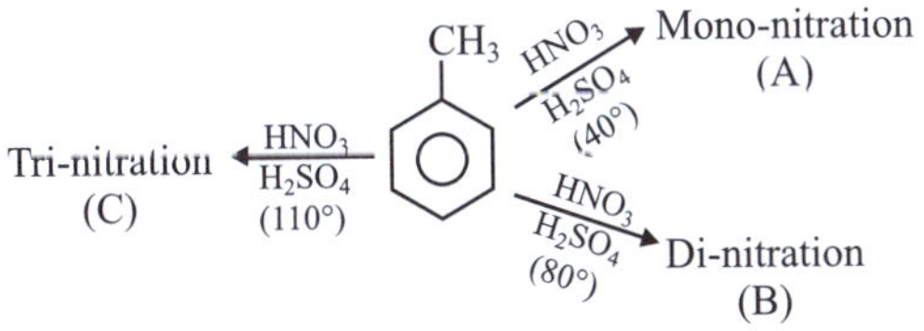

   1) A = 3, B = 6, C = 8
   2) A = 3, B = 6, C = 6
   3) A = 3, B = 6, C = 10
   4) A = 3, B = 4, C = 6

5. The best reagent for converting 2-phenyl-propanamide into 2-phenylpropanamine is.
   1) Excess H_2
   2) Br_2 in aqueous NaOH
   3) Iodine in the presence of red phosphorus
   4) $LiAlH_4$ in ether

6. An organic compound (C_3H_9N) (A), when treated with nitrous acid, gave an alcohol and N_2 gas was evolved. (A) on warming with $CHCl_3$ and caustic potash gave (C) which on reduction gave isopropylmethylamine. Predict the structure of (A).
   1) $(CH_3)_2CH{-}NH_2$
   2) $CH_3CH_2-NH-CH_3$
   3) $CH_3-N(CH_3)-CH_3$
   4) $CH_3CH_2CH_2-NH_2$

7. Aniline when diazotized in cold and then treated with dimethyl aniline gives a coloured product. Its structure would be:
   1) 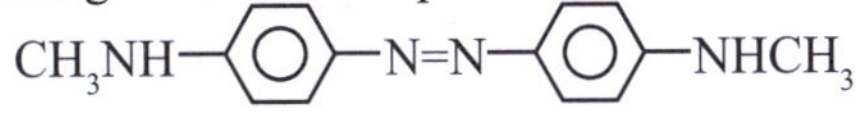
   2) 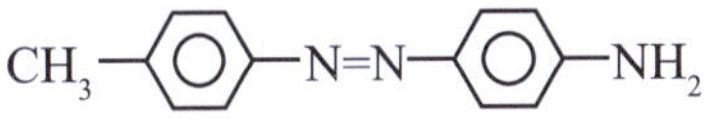
   3) 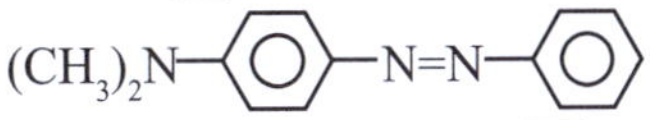
   4) 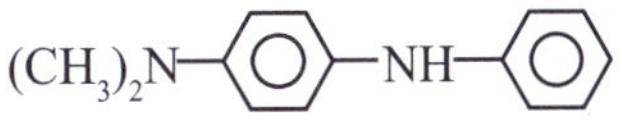

8. An organic compound A on reacting with NH_3 gives B. On heating B gives C. C in the presence of KOH reacts with Br_2 to give $CH_3CH_2NH_2$. A is:
   1) $CH_3{-}CH(CH_3){-}COOH$
   2) $CH_3CH_2CH_2CH_2COOH$
   3) CH_3CH_2COOH
   4) CH_3COOH

9. A compound with molecular mass 180 amu is acylated with CH_3COCl, to get a compound with molecular mass 390 amu . The number of amino groups present per molecule of the former compound is:
   1) 6 2) 2 3) 5 4) 4

10. Identify the final product of following sequence of reaction:

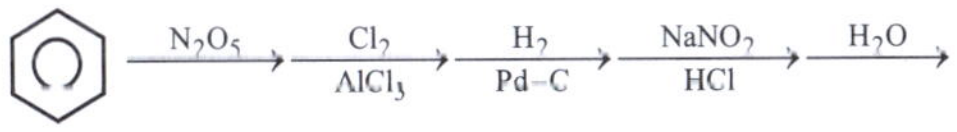

   1) 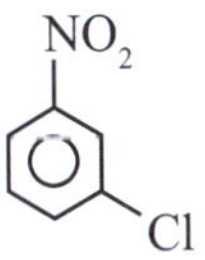
   2) 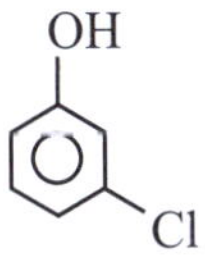
   3) 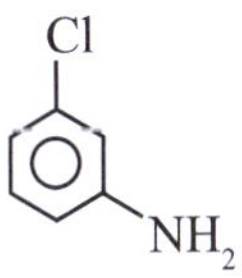
   4) 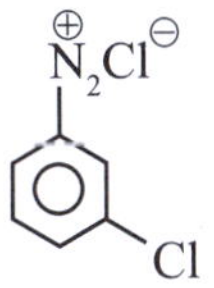

11. How many of the below listed compounds on treatment with HNO_2 can go for ring expansion?

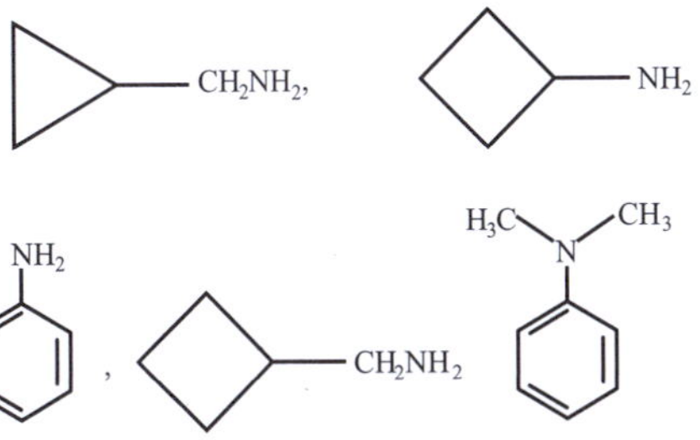

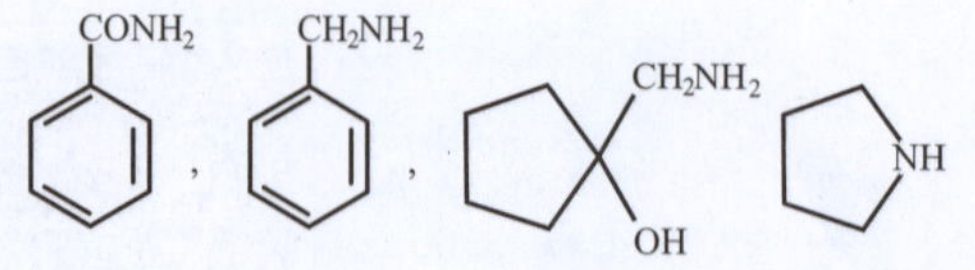

1) 2 2) 3 3) 4 4) 5

12. Gabriel-Phthalimide synthesis is suitable for preparation of

1) $CH_3CH_2NHCH_2CH_3$

2) $CH_3-C(CH_3)_2-NH_2$

3) 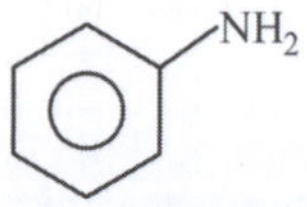

4) 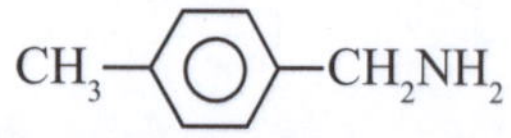

13. Match the List-I with List-II:

List-I (Group of Compounds)		List-II (Positive Test Given by)	
I.	$C_6H_5NH_2$	P.	Azo dye test
II.	$(CH_3)_2NH$	Q.	Alcohol formation on reaction with HNO_2 at 0 – 5°C
III.	$CH_3CH_2NH_2$	R.	Yellow oily liquid with HNO_2
IV.	p-Toluidine	S.	Isocyanide test
		T.	Hofmann's mustard oil test

1) I-P, S, T; II-R; III-Q, S, T; IV-P, S, T
2) I-P, S; II-Q; III-Q, T; IV-P, R
3) I-P, S, T; II-Q, S, T; III-R; IV-P, S, T
4) I-P; II-Q, S; III-Q, S, T; IV-P, T

14. Match List-I with List-II and select the correct answer using the codes give below the lists:

	List-I		List-II
I	$CH_2=CH-CHO \xrightarrow{NaBH_4}$	P	Acetal
II	$C_6H_5CHO + Ph-NH_2 \xrightarrow{H^{\oplus}}$	Q	Schiff's base
III	$C_6H_5COCH_3 + CH_3-CH_2-NH_2 \xrightarrow{H^{\oplus}}$	R	Unsaturated alcohol
IV	$RCHO + 2RCH_2OH \xrightarrow{H^{\oplus}}$	S	Enamine

1) I-R; II-P; III-Q; IV-S 2) I-P; II-R; III-S; IV-Q
3) I-R; II-Q; III-Q; IV-P 4) I-Q; II-P; III-R; IV-S

15. Match the Matrix:

List-I (Amines)		List-II (pKb Values)	
I.	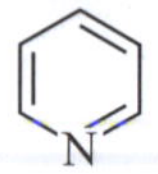	P.	3.27
II.	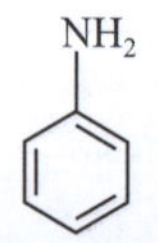	Q.	8.8
III.	CH_3NHCH_3	R.	3.35
IV.	CH_3NH_2	S.	9.4

1) I-R; II-P; III-S; IV-Q
2) I-R; II-Q; III-S; IV-P
3) I-Q; II-S; III-R; IV-P
4) I-Q; II-S; III-P; IV-R

16. $C_2H_5NH_2$ $\xrightarrow{HNO_2}$ A; $\xrightarrow{C_6H_5CHO}$ B; $\xrightarrow{NOCl}$ C; $\xrightarrow{C_6H_5SO_2Cl}$ D

Which product is a Schiff's base?

1) A 2) B
3) C 4) D

17. Which of the following statement(s) is/are correct?

(*i*) In Sandmeyer reaction, nucleophiles like Cl^-, Br^- and CN^- are introduced in benzene ring in the presence of Cu^+ ion.

(*ii*) In Gattermann reaction, nucleophiles are introduced in benzene ring in the presence of copper powder and HCl.

(*iii*) The yield in Gattermann reaction is found to be better than in Sandmeyer reaction.

1) (*i*) and (*ii*) 2) (*i*), (*ii*) and (*iii*)
3) (*ii*) and (*iii*) 4) (*i*) and (*iii*)

18. Consider the following sequence of reactions:

$$X(C_3H_9N) \xrightarrow{HNO_2} C_3H_8O \xrightarrow[H_2SO_4]{K_2Cr_2O_7} C_3H_6O_2$$

The compound (X) is:

1) $CH_3CH_2NHCH_3$ 2) $CH_3CH_2CH_2NH_2$
3) $(CH_3)_2CHNH_2$ 4) $(CH_3)_3N$

19. Aniline in a set of reactions yielded a product

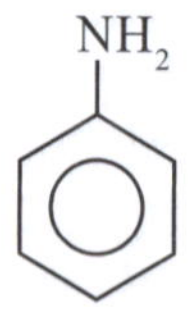 $\xrightarrow[HCl]{NaNO_2} A \xrightarrow{CuCN} B \xrightarrow[Ni]{H_2} C \xrightarrow{HNO_2} D$

The structure of the product D would be

1) C_6H_5NHOH 2) $C_6H_5NHC_2H_5$
3) $C_6H_5CH_2OH$ 4) $C_6H_5CH_2NH_2$

20. Consider the reaction, 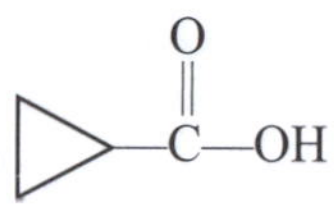 $\xrightarrow{NH_3, \Delta}$

$$A \xrightarrow[\Delta]{KOH/Br_2} B \xrightarrow[\Delta]{CHCl_3, KOH(alc.)} C \xrightarrow{H_3O^+} D + X$$

X is same as that of

1) A 2) B
3) C 4) D

NEET & AIIMS Past Year MCQs

1. The major product D formed in the following reaction sequence is: **(2024 Re)**

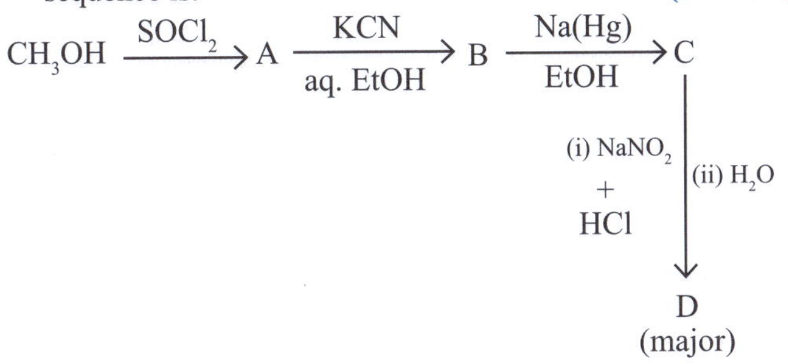

1) 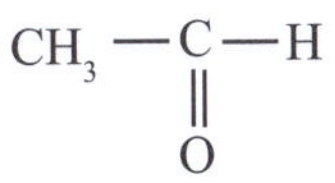
2) 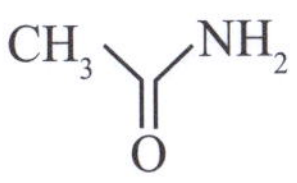
3) CH_3CH_2OH
4) CH_3CH_2Cl

2. The compound that does not undergo Friedel-Crafts alkylation reaction but gives a positive carbylamine test is: **(2024 Re)**

1) Aniline
2) Pyridine
3) N-methylaniline
4) Triethylamine

3. The major product X formed in the following reaction sequence is: **(2024 Re)**

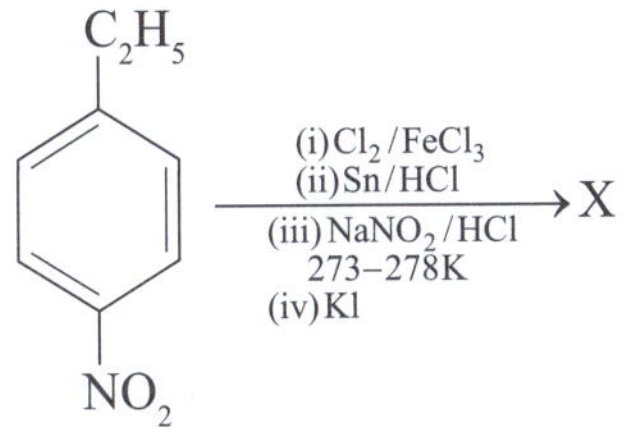

1) 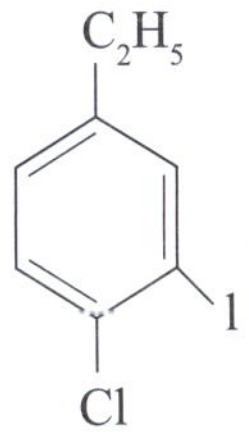
2) 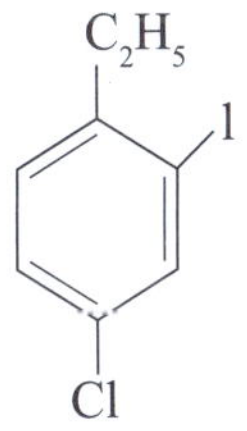
3) 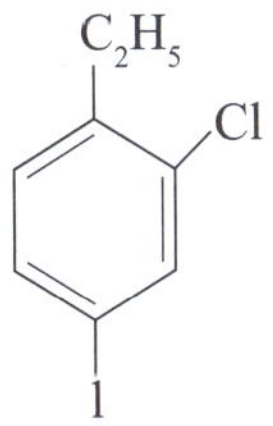
4) 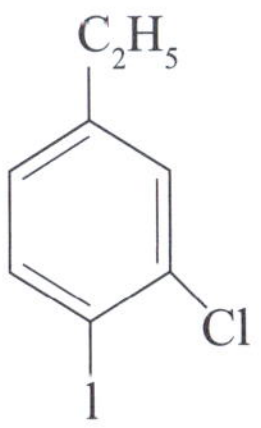

4. Given below are two statements: **(2024)**

Statement-I: Aniline does not undergo Friedel-Crafts alkylation reaction.

Statement-II: Aniline cannot be prepared through Gabriel synthesis.

In the light of the above statements, choose the correct answer from the options given below:

1) Statement-I is correct but Statement-II is false.
2) Statement-I is incorrect but Statement-II is true.
3) Both Statement-I and Statement-II are true.
4) Both Statement-I and Statement-II are false.

5. Identify the major product C formed in the following reaction sequence: **(2024)**

$$CH_3{-}CH_2{-}CH_2{-}I \xrightarrow{NaCN} A \xrightarrow[\text{Partial hydrolysis}]{OH^-} B \xrightarrow[Br_2]{NaOH} \underset{\text{major}}{C}$$

1) butanamide
2) α-bromobutanoic acid
3) propylamine
4) butylamine

6. Which of the following reactions will NOT give primary amine as the product? **(2023)**

1) $CH_3CONH_2 \xrightarrow[\text{(ii) } H_3O^{\oplus}]{\text{(i) } LiAlH_4} \text{Product}$
2) $CH_3CONH_2 \xrightarrow{Br_2/KOH} \text{Product}$
3) $CH_3CN \xrightarrow[\text{(ii) } H_3O^{\oplus}]{\text{(i) } LiAlH_4} \text{Product}$
4) $CH_3NC \xrightarrow[\text{(ii) } H_3O^{\oplus}]{\text{(i) } LiAlH_4} \text{Product}$

7. Identify the product in the following reaction: **(2023)**

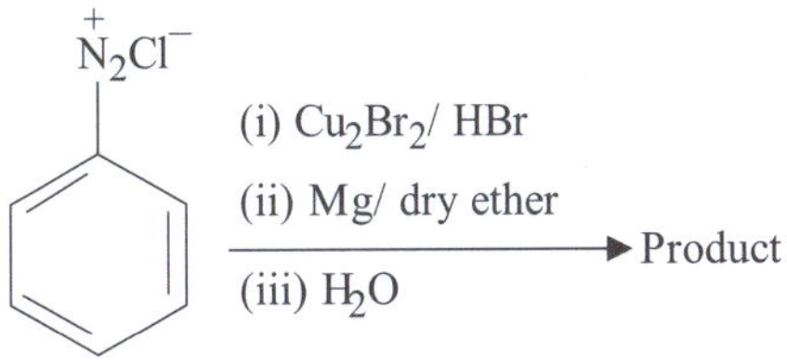

1) 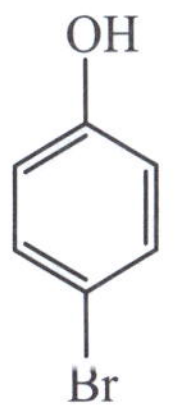
2) 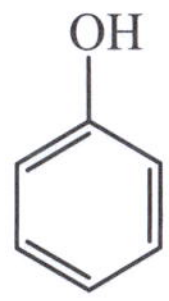
3) 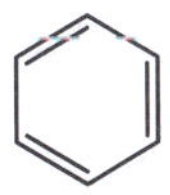
4) 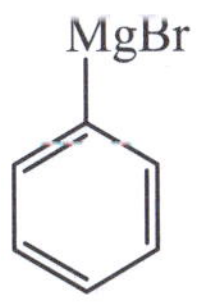

8. Given below are two statements: **(2022)**

Statement-I: Primary aliphatic amines react with HNO_2 to give unstable diazonium salts.

Statement-II: Primary aromatic amines react with HNO_2 to form diazonium salts which are stable even above 300 K.

In the light of the above Statements, choose the most appropriate answer from the options given below:

1) Statement-I is incorrect but Statement-II is correct.
2) Both Statement-I and Statement-II are correct
3) Both Statement-I and Statement-II are incorrect.
4) Statement-I is correct but Statement-II is incorrect.

9. The product formed from the following reaction sequence is: **(2022)**

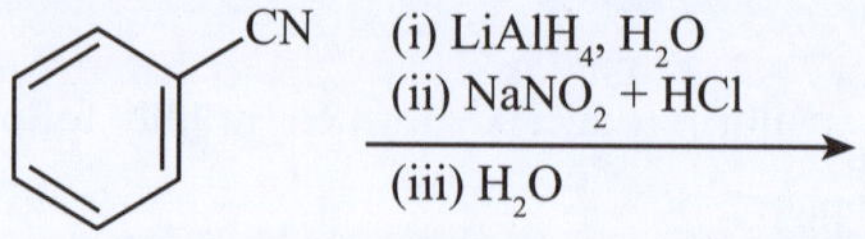

1) 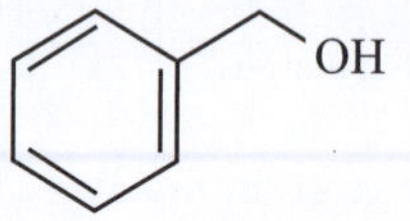

2) 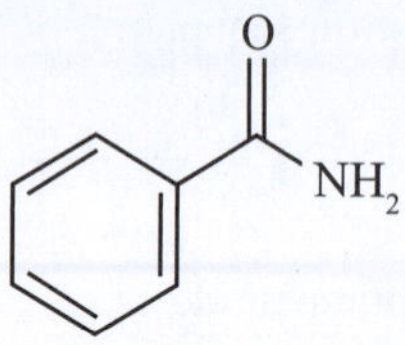

3) 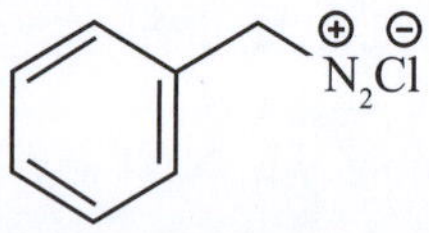

4) 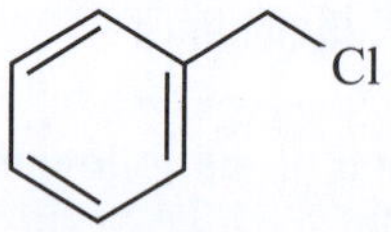

10. Identify the compound that will react with Hinsberg's reagent to give a solid which dissolves in alkali. **(2021)**

1) $CH_3-CH_2-\ddot{N}H-CH_3$
2) $CH_3-CH_2-\ddot{N}H_2$
3) $CH_3-CH_2-\ddot{N}(CH_3)-CH_2-CH_3$
4) $CH_3-CH_2-\ddot{N}O_2$

11. The reagent 'R' in the given sequence of chemical reaction is: **(2021)**

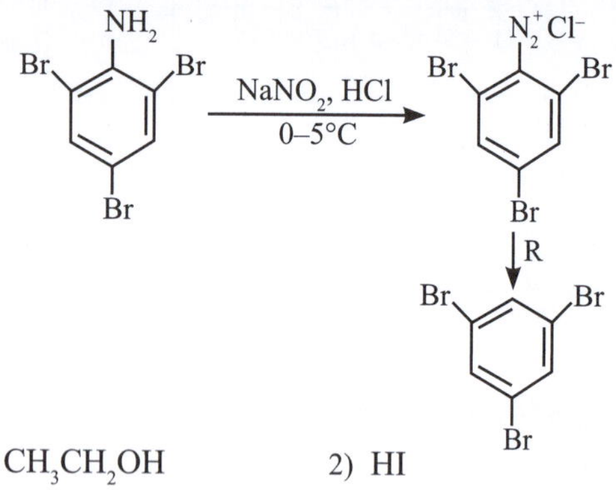

1) CH_3CH_2OH 2) HI
3) CuCN/KCN 4) H_2O

12. Which of the following does not give nitroalkane? **(AIIMS 2013)**

1) $CH_3-N(CH_3)-CH_3 \xrightarrow{KMnO_4}$
2) $C_2H_5I \xrightarrow{alc.AgNO_2}$
3) $CH_3-CH_3 \xrightarrow{\text{Fuming } HNO_3}$
4) Both (1) and (2)

13. If phthalic acid is treated with NH_3, which is first heated gently and then strongly then the product is: **(AIIMS 2012)**

1) 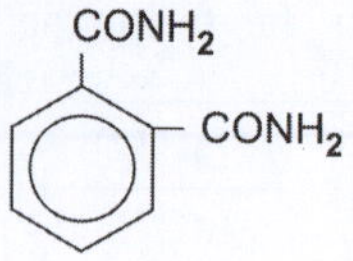

2) 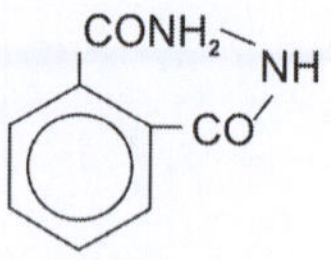

3) 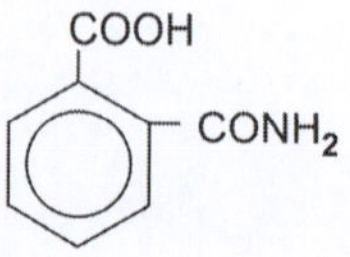

4) 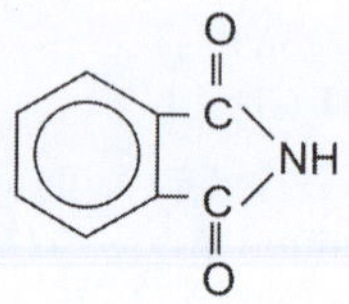

14. In the following sequence of the reactions, what is D? **(AIIMS 2007)**

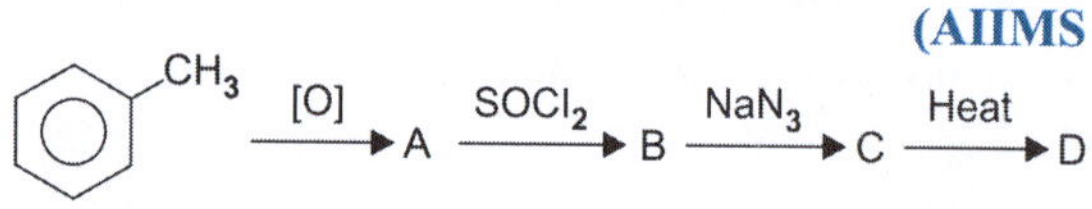

1) Primary amine
2) An amide
3) Phenyl isocyanate
4) A chain lengthed hydrocarbon

15. **Assertion:** Benzene diazonium salt on boiling with water forms phenol. **(AIIMS 2007)**
Reason: C – N bond is polar.
1) If both Assertion & Reason are True & the Reason is a correct explanation of the Assertion.
2) If both Assertion & Reason are True but Reason is not a correct explanation of the Assertion.
3) If Assertion is True but the Reason is False.
4) If both Assertion & Reason are False.

16. Among the following which one does not act as an intermediate in Hofmann rearrangement: **(AIIMS 2005)**
1) RNCO 2) RCON
3) RCONHBr 4) RNC

17. Nitrosoamines (R_2 N – N ═ O) are soluble in water. On heating them with concentrated H_2SO_4, they give secondary amines. This reaction is called: **(AIIMS 1998)**
1) Sandmeyer's reaction
2) Fittig's reaction
3) Perkin reaction
4) Liebermann nitroso reaction

Answer Key

NCERT Topic-wise MCQs

1	2	3	4	5	6	7	8	9	10	11	12	13	14	15	16	17
3	4	3	3	2	3	4	1	2	4	1	4	4	3	2	1	3
18	19	20	21	22	23	24	25	26	27	28	29	30	31	32	33	34
3	2	3	2	3	3	3	4	4	4	1	3	2	3	4	1	4
35	36	37	38	39	40	41	42	43	44	45	46	47	48	49	50	51
2	3	3	4	2	4	3	4	4	1	2	1	2	2	1	4	3
52	53	54	55	56	57	58	59	60	61	62	63	64	65	66	67	68
4	4	1	1	2	3	3	3	1	1	2	3	1	1	1	4	1
69	70	71	72	73	74	75	76	77	78	79	80	81	82	83	84	85
2	1	1	3	2	2	3	3	1	1	2	2	4	4	4	1	1
86	87	88	89	90	91	92	93									
2	1	3	2	2	2	3	4									

High Order Time Intensive MCQs

1	2	3	4	5	6	7	8	9	10	11	12	13	14	15	16	17
2	2	4	4	4	1	4	4	1	1	3	3	4	2	1	3	3
18	19	20	21	22	23	24	25	26	27	28	29	30	31	32	33	34
2	4	3	3	3	3	2	4	1	3	3	2	3	1	3	4	1
35	36	37	38	39	40	41	42	43	44	45	46	47	48	49	50	51
2	3	2	3	2	2	4	4	3	2	4	2	1	3	3	1	1
52	53	54	55	56												
3	4	4	4	2												

Match the Columns MCQs

1	2	3	4	5
4	2	3	1	3

Correct & Incorrect MCQs

1	2	3	4	5	6	7	8	9	10
1	4	3	3	2	4	2	4	4	1

Statement Based MCQs

1	2	3	4	5	6	7	8	9	10
1	1	2	1	3	1	3	3	4	3

Assertion & Reason MCQs

1	2	3	4	5	6	7	8	9	10
3	3	1	2	1	3	3	1	3	2

Multi Correct MCQs

1	2	3	4	5
2	2	3	1	2

NCERT Exemplar MCQs

1	2	3	4	5	6	7	8	9	10	11	12	13	14	15	16	17
2	4	3	1	3	2	3	4	3	4	2	2	4	3	2	3	3
18	19	20	21	22	23	24	25	26	27							
2	1	2	2	4	3	4	1	2	1,2,3							

Kattar NEET

1	2	3	4	5	6	7	8	9	10	11	12	13	14	15	16	17
1	4	3	2	4	1	3	3	3	2	2	4	1	3	4	2	1
18	19	20														
2	3	2														

NEET & AIIMS Past Year MCQs

1	2	3	4	5	6	7	8	9	10	11	12	13	14	15	16	17
3	1	3	3	3	4	3	4	1	2	1	1	4	3	2	4	4

CHAPTER

11 Biomolecules

Topicwise Number of Questions Analysis (2025-2016)

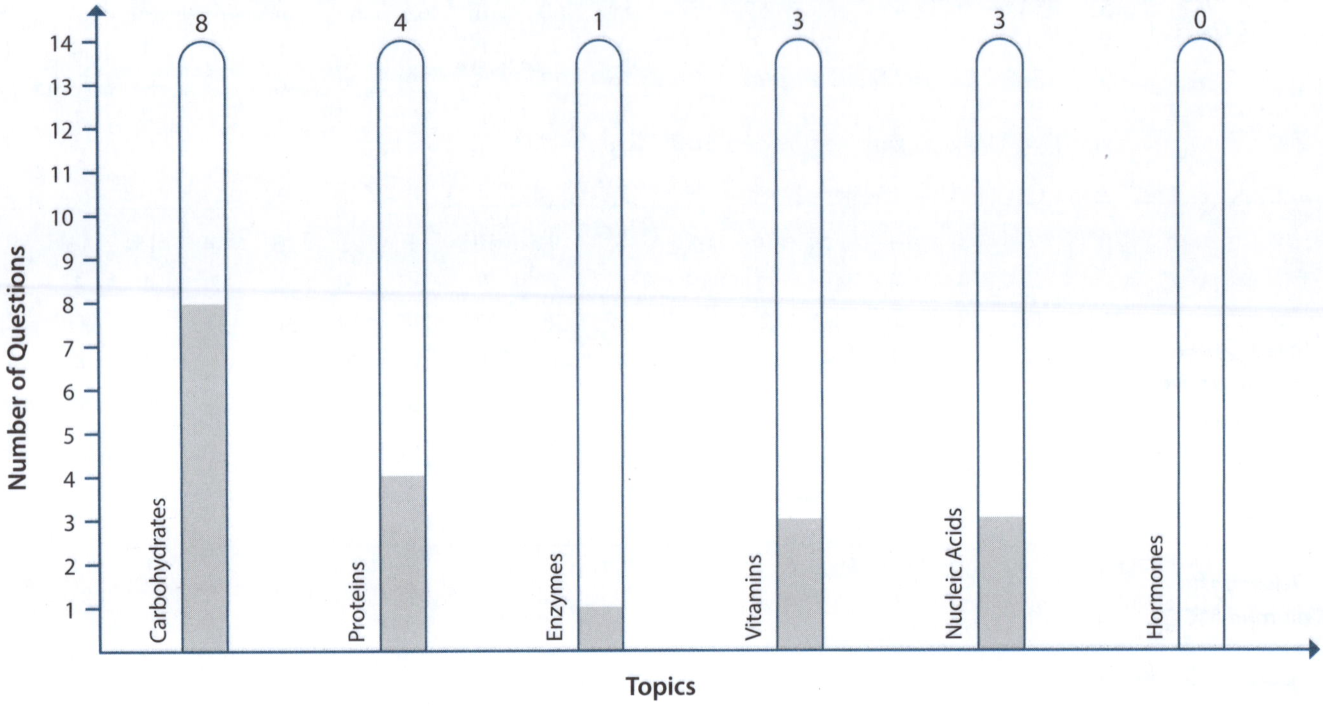

NCERT-Picks

- The branch of chemistry which deals with chemical changes occuring in living systems is known as biochemistry.

Carbohydrates

- Optically active polyhydroxy aldehydes or ketones or the compounds which produce such units on hydrolysis are called carbohydrates, eg: sugar, starches, cellulose, glycogen, dextins, gums, etc. They are generally derived from plants.
- Carbohydrates are substances having general formula $Cx(H_2O)y$

Classification of Carbohydrates

Carbohydrates

Monosaccharides	Oligosaccharides	Disaccharides	Polysaccharides
A carbohydrate that cannot be hydrolysed to simpler compounds **eg:** Glucose, fructose etc.	Carbohydrates that yield 2 to 10 monosaccharide units, on hydrolysis, **eg:** Sucrose, maltose, raffinose.	**eg:** Maltose, lactose etc. (It is a part of oligosaccharides)	A carbohydrate that can be hydrolyzed to many monosaccharide molecules **eg:** Starch, cellulose etc.

Other classification of carbohydrates:

Reducing and Non Reducing Sugar

- Reducing sugars are the carbohydrates which are reduced by Tollen's and Fehling's solution, eg: glucose and fructose
- Maltose is a reducing sugar. In maltose, the free aldehyde group can be produced at C-1 of second glucose in the solution.
- Non-reducing sugar are the carbohydrates which do not reduced by Tollen's and Fehling's solution, eg: sucrose.

Monosaccharides

- Monosaccharides are further classified on the basis of number of carbon atoms and the functional groups present in them. If a monosaccharide contains an aldehyde group, it is known as an aldose and if it contains a keto group, it is known as a ketose. Number of carbon atoms constituting the monosaccharide is also introduced in the name as is evident from the examples given in Table.

Table: Different Types of Monosaccharides

Carbon atoms	General term	Aldehyde	Ketone
3	Triose	Aldotriose	Ketotetrose
4	Tetrose	Aldotetrose	Ketoterose
5	Pentose	Aldopentose	Ketopentose
6	Hexose	Aldohexose	Ketohexose
7	Heptose	Aldoheptose	Ketoheptose

Glucose

- It is an aldohexose and is also known as dextrose. It is the monomer of many of the large carbohydrates.
- It is present in ripe grapes, honey and in most of the sweet fruits. Glucose is known as dextrose because it occurs in nature as the optically active dextrorotatory isomer.
- Glucose is sparingly soluble in alcohol but is soluble in water and insoluble in ether.
- Glucose is optically active and the ordinary naturally occuring form is (+) glucose or dextro form and it shows mutarotation.

Preparation of Glucose

1. From sucrose (Cane sugar)

$$\underset{\text{Sucrose}}{C_{12}H_{22}O_{11}} + H_2O \xrightarrow{H^+} \underset{\text{Glucose}}{C_6H_{12}O_6} + \underset{\text{Fructose}}{C_6H_{12}O_6}$$

2. From starch

$$\underset{\text{Starch or cellulose}}{(C_6H_{10}O_5)_n} + nH_2O \xrightarrow[393K;2-3atm]{H^+} \underset{\text{Glucose}}{nC_6H_{12}O_6}$$

Chemical reactions of glucose

1. Prolonged heating with HI

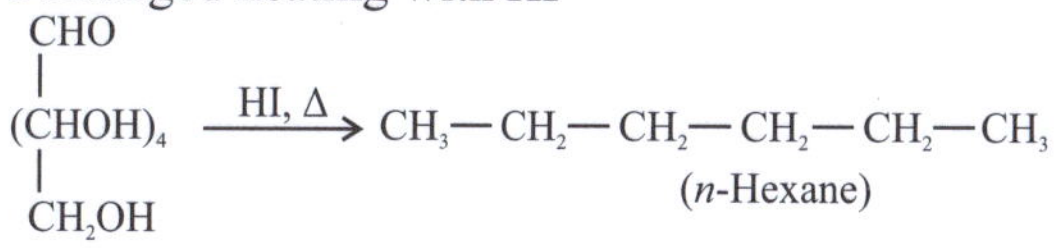

2. Glucose reacts with hydroxylamine

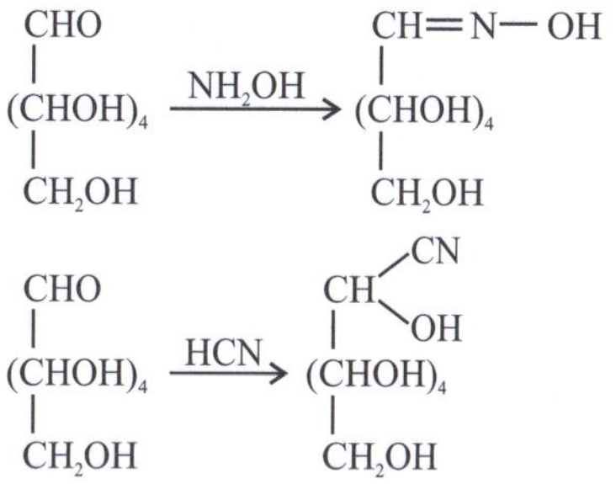

3. Glucose reacts with Br_2 water

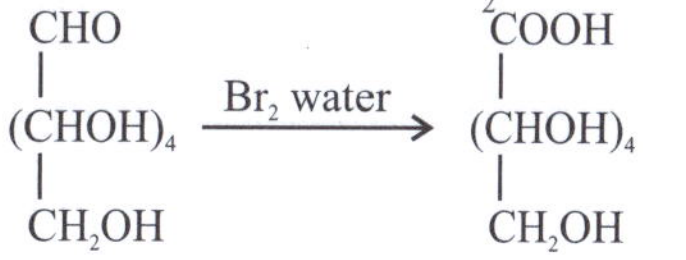

Gluconic acid

4. Acetylation of glucose

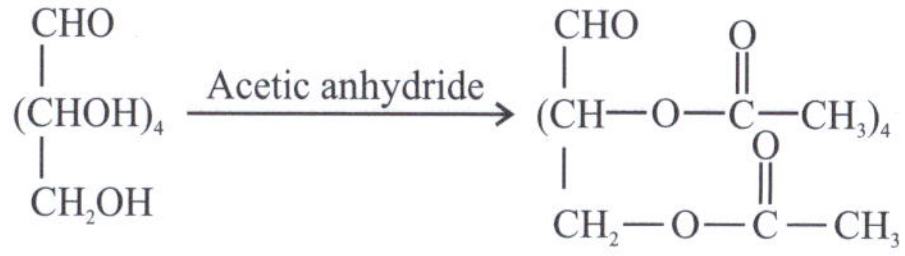

5. Oxidation with nitric acid

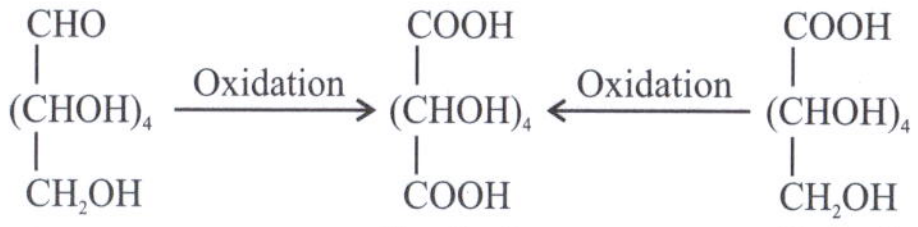

Saccharic acid　　Gluconic acid

- The pentaacetate of glucose does not react with hydroxylamine. Free aldehyde group is absent in pentaacetate.
- **Anomers** refer to a pair of stereoisomers which differ in configuration only around C_1 and C_2 carbon in hemiacetal forms of glucose is called anomeric carbon or glycosidic carbon.
- α–(D)–(+)–glucose & β–(D)–(+)– glucose are called anomers. α–(D)–(+)–glucose & β–(D)–(+)–glucose differ in the configuration at C1 [anomeric carbon].
- Cyclic structures for two anomeric forms are:

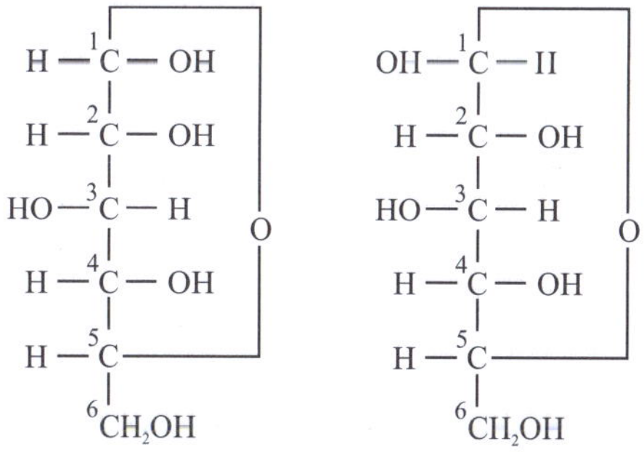

α–D–(+)–glucopyranose　　β–D–(+)–glucopyranose

Cyclic structures are also known as **hemi-acetal structures**.

- Five membered ring containing an oxygen is called furanose form.

- Six membered ring containing an oxygen is called pyranose form.

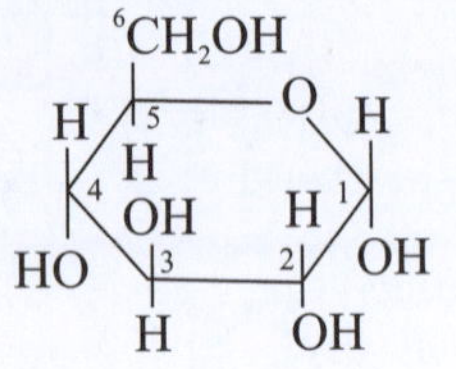

α-D-(+)-Glucopyranose

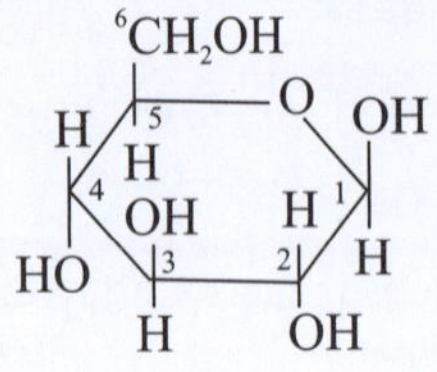

β-D-(+)–Glucopyranose

- D (+) glucose exists in two anomeric forms, i.e., α-D(+) glucopyranose and β-D(+) glucopyranose shown above. When these are separately dissolved in water, they undergo a change in specific rotation till it becomes constant after some time. The change in specific rotation of isomers in aqueous solution is called mutarotation.

Fructose

- It is considered as an important ketohexose. Sucrose on hydrolysis gives glucose along with fructose. It is present in honey, fruits, and vegetables.

Structure

- It contains a ketonic functional group at second carbon and six carbons in straight chain like in glucose.

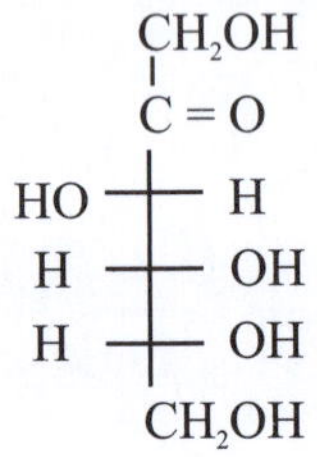

D(–) – Fructose

- It belongs to D series and laevorotatory. It also exists in two cyclic forms.

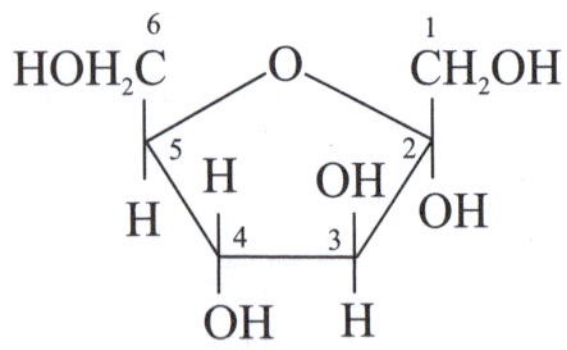

α-D-(–)-Fructofuranose

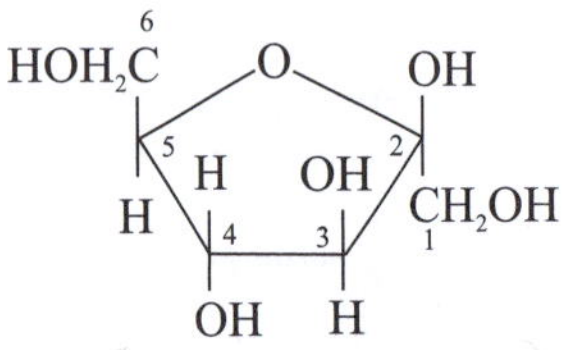

β-D-(–)-Fructofuranose

Disaccharides

Sucrose

- Sucrose is dextrorotatory but after hydrolysis gives D-(+)-glucose & D-(–)-fructose. The laevorotation of fructose (–92.4°) is more than dextrorotation of glucose (+52.5°), the mixture is laevorotatory. So, hydrolysis cause change in the sign of rotation from dextro to laevo, and the product is named as invert sugar.

$$\underset{\text{Sucrose}}{C_{12}H_{22}O_{11}} + H_2O \longrightarrow \underset{\text{D-(+)-Glucose}}{C_6H_{12}O_6} + \underset{\text{D-(–)-Fructose}}{C_6H_{12}O_6}$$

- Aqueous solution of sucrose is dextrorotatory, its specific rotation being + 66.5.
- It does not form an oxime or an osazone, and does not undergo mutarotation.

Maltose

- Maltose is composed of two α-D-glucose units in which C1 of one glucose unit is linked with C4 of another glucose unit.
- It is a reducing sugar.

Lactose

- It is more commonly known as milk sugar since this is found in milk.
- It is composed of β–D–galactose (C1) and β–D– glucose (C4).
- It is hydrolysed by dilute acids or by the enzyme lactase, to an equimolar mixture of D(+)-glucose and D(+)-galactose.
- Lactose sugar shows mutarotation. Lactose is a reducing sugar.

Polysaccharides

- Starch $(C_6H_{10}O_5)n$ consists of two compound, i.e., amylose (which is soluble in water) and amylopectin (insoluble in water).

  20% = amylose

  80% = amylopectin

- The amylose molecule is made up of D-glucose unit joined by α-glycosidic linkages between C-1 of one glucose unit and C-4 of the next glucose unit. The number of D-glucose units in amylose range from 60-300.
- Cellulose is a straight chain polysaccharide which is composed of β–D–glucose having linkage between C1 and C4.
- Glycogen is known as animal starch. Glycogen structure is similar to that of amylopectin.

Amino Acids and Proteins

- Amino acids are the building block of proteins so they are very important.

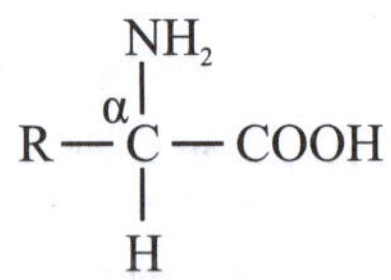

α–amino acid

Classification of amino acids

- On the basis of relative number of amine (–NH2) and carboxyl (–COOH) group,

  Neutral: eg: Glycine, Alanine, Valine

  Acidic: eg: Aspartic acid, Glutamic acid,

  Basic: eg: Lysine, Arginine, Histidine

Classification on basis of synthesis

- Essential amino acids: eg: Valine, leucine, lysine, isoleucine, arginine etc.
- Non-essential amino acids: eg: glycine, alanine, glutamic acid, aspartic acid.

Properties

- Amino acids are usually colourless, crystalline solids. These are water-soluble, high melting solids and behave like salts rather than simple amines or carboxylic acids. This behavior is due to the presence of both acidic (carboxyl group) and basic (amino group) groups in the same molecule. In aqueous solution, the carboxyl group can lose a proton and amino group can accept a proton, giving rise to a dipolar ion known as zwitter ion. This is neutral but contains both positive and negative charges.

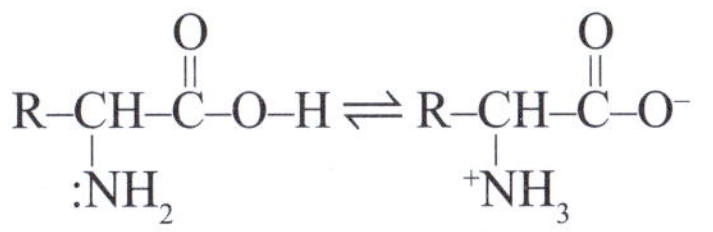

(Zwitter ion)

- In zwitter ionic form, amino acids show amphoteric behaviour as they react both with acids and bases.
- Except glycine, all other naturally occurring a-amino acids are optically active since the α-carbon atom is asymmetric. These exist both in 'D' and 'L' forms. Most naturally occurrring amino acids have L-configuration. L-Aminoacids are represented by writing the –NH2 group on left hand side.

Proteins

- Proteins are the polymers of α-amino acids and they are connected to each other by peptide bond or peptide linkage. Chemically, peptide linkage is an amide formed between –COOH group and $-NH_2$ group. The reaction between two molecules of similar or different amino acids, proceeds through the combination of the amino group of one molecule with the carboxyl group of the other. This results in the elimination of a water molecule and formation of a peptide bond –CO–NH–. The product of the reaction is called a dipeptide because it is made up of two amino acids. For example, when carboxyl group of glycine combines with amino groups of alanine, we get a dipeptide, glycylalanine.
- If a third amino acid combines to a dipeptide, the product is called a tripeptide. A tripeptide contains three amino acids linked by two peptide linkages. Similarly when four, five or six amino acids are linked, the respective products are known as tetrapeptide, pentapeptide or hexapeptide, respectively. When the number of such amino acids is more than ten, then the products are called polypeptides. A polypeptide with more than hundred amino acid residues, having molecular mass higher than 10,000u is called a protein. However, the distinction between a polypeptide and a protein is not very sharp. Polypeptides with fewer amino acids are likely to be called proteins if they ordinarily have a well defined conformation of a protein such as insulin which contains 51 amino acids.

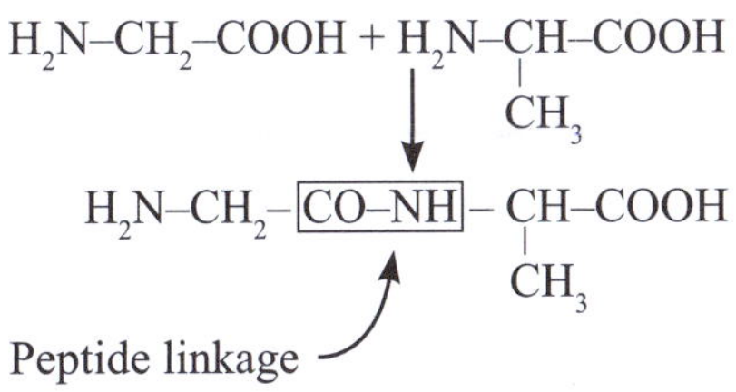

Glycylalanine (Gly-Ala)

Classification of proteins

a. **Fibrous:** In fibrous proteins, polypeptide chain are parallel and are held together by hydrogen and disulphide bonds. This is insoluble in water, eg: keratin (Present in hair wool) and myosin (present in muscles).

b. **Globular:** Polypeptide chain coil around to give three dimensional spherical shape. They are soluble in water. Eg: insulin and albumins.

- Haemoglobin is a globular protein.
- The protein conformation is of 4 types
  1. Primary
  2. Secondary
  3. Tertiary
  4. Quaternary

Denaturation of proteins

- When a protein in its native form is subjected to a physical change like change in temperature or pH, the hydrogen bonds are disturbed. As a result, globules unfold and helix get uncoiled protein loses its biological activity. This is called Denaturation of proteins. During denaturation, 2° and 3° structure are destroyed but 1° structure remains intact.

Vitamins

- Vitamins are the organic compounds (other than carbohydrates, proteins, fats or group of biomolecules) which are essential for maintaining a normal health, growth and nutrition.

Classification of vitamins

Vitamins

↓

- Water – soluble vit: B_1, B_2, B_6, C
- Fat – soluble Vit: A, D, E, K

- Vitamin 'C' must be supplied regularly in diet, because vitamin 'C' is water soluble & hence excess of it is readily excreted in urine so, it cannot be stored in body.
- Example of some important vitamins and deficiency diseases are:
- Vitamin A (Retinol): Xerophthalmia (hardening of cornea of eye) night blindness
- Vitamin B1 (Thiamine): Beriberi (paralysis of legs and general weakness) loss of appetite.
- Vitamin D (Ergocalciferol) (sun shine Vitamin): Rickets (deformation of bones), osteomalacia (soft bones and joint pain)
- Vitamin C is chemically known as ascorbic acid- Scurvy.

Nucleic Acids

- The particles in nucleus of the cell, responsible for heredity, are called chromosomes which are made up of proteins and another type of biomolecules called nuclei acids. These are mainly of two types- DNA (Deoxyribonucleic acid) and RNA (Ribonucleic acid). Nucleic acids are long chain polymers of nucleotides, so they are called polynucleotides.

The relationship can be shown as given below:

- Nucleic acid = many nucleotides
- Nucleotide = nucleosides + phosphate
- Nucleoside = sugar + nitrogenous base
- Thus, nucleotide = phosphate + sugar + nitrogenous base

Nucleic acids are of two types:

RNA	DNA
All bases are same except Thymine, Uracil is present	Complementary base pairing occur Adenine (A) Thymine (T) Guanine (G) Cytosine (C) A = T C ≡ G

Biological functions of nucleic acids

DNA is the chemical basis of heredity and may be regarded as the reserve of genetic information. DNA is exclusively responsible for maintaining the identity of different species of organisms over millions of years. A DNA molecule is capable of self duplication during cell division and identical DNA strands are transferred to daughter cells. Another important function of nucleic acids is the protein synthesis in the cell.

Hormones

Hormones are molecules that act as intercellular messengers and are produced by the endocrine glands in the body. They help to maintain the balance of biological activities in the body. Epinephrine and norepinephrine mediate responses to external stimuli. Growth and sex hormones help in growth and development. Hormones released by the adrenal cortex play very important role in the functions of the body. Hormones released by gonads are responsible to develop secondary sex characters.

MCQs

NCERT Topic-wise MCQs

Carbohydrates

1. What is the basic formula for starch?
   1) $(C_6H_{12}O_6)_n$ 2) $(C_6H_{10}O_5)_n$
   3) $C_2H_{12}O_{11}$ 4) $(C_6H_{12}O_4)_n$

2. The glycosidic linkage involved in the glucose unit in amylose part of starch is:
   1) $C_1 - C_4 - \beta$ – linkage 2) $C_1 - C_6 - \beta$ – linkage
   3) $C_1 - C_6 - \alpha$ – linkage 4) $C_1 - C_4 - \alpha$ – linkage

3. Which of the following statements is not correct for maltose?
   1) It is a disaccharide
   2) It undergoes mutarotation
   3) It is a reducing sugar
   4) It does not have hemiacetal group

4. Identify the correct statement about lactose.
   1) It consists of one galactose and one glucose unit
   2) Mutarotation is not possible
   3) Anomeric carbon of galactose is attached to 4' carbon of glucose which is β-1, 4'-glycoside bond.
   4) Lactose is used to cleave the β-1, 4'-glycoside bond

5. Which of the following is a non-reducing sugar?
   1) Glucose 2) Sucrose
   3) Maltose 4) Lactosc

6. Which of the following carbohydrates would be most abundant in the diet of strict vegetarian?
   1) Amylose 2) Glycogcn
   3) Cellulose 4) Maltose

7. Which of the following reactions establish difference between glucose and fructose?
   1) Tollen's reagent reaction
   2) Phenyl hydrazine
   3) P/HI
   4) conc. HNO_3

8. Select the incorrect statement:
   1) Starch is a homopolymer of glucose forming an α-glycosidic chain
   2) Maltose and Lactose are disaccharides
   3) Cellulose has only β-D-glucose unit which are joined by glycosidic linkage between C_1 of one glucose unit and C_4 of another glucose unit
   4) The structure of glycogen is similar to amylose

9. Which of the following pair will give positive test with tollen reagent?
   1) Glucose, sucrose 2) Sucrose, Fructose
   3) Glucose, fructose 4) Acetophenone & hexanal

10. Glucose + Tollen's reagent → silver mirror, the above process shows:
    1) Presence of –COOH group
    2) Presence of keto group
    3) Presence of –CHO group
    4) Presence of $-CONH_2$ group

11. Starch is a polymer of:
    1) Glucose 2) Fructose
    3) Both (1) & (2) 4) None of these

12. Polysaccharides have which of the following linkages?
    1) Glycosidic linkages 2) H-bond
    3) Peptide linkage 4) No-linkage

13. A mixture of amylose and amylopectin is called:
    1) Lactose 2) Starch
    3) Cellulose 4) Sucrose

14. Among the naturally occurring carbohydrates, furanose ring is found in the:
    1) Glucose unit of cane sugar
    2) Glucose unit of cellulose
    3) Fructose unit of cane sugar
    4) Galactose unit of lactose

15. Which of the following polymers is stored in the liver of animals?
    1) Amylose 2) Cellulose
    3) Amylopectin 4) Glycogen

16. The intermediate compound formed in the conversion of starch to glucose is
    1) Lactose 2) Sucrose
    3) Maltose 4) Fructose

17. Carbohydrates may be:
    1) Sugars
    2) Starch
    3) Polyhydroxy aldehyde/ketones
    4) All of the above

18. The enantiomers of α – D – (+) – glucose is:
    1) β – D – (+) – glucose 2) α – D – (–) – glucose
    3) α – L – (–) – glucose 4) β – L – (–) – glucose

19. Which of the following is an aldose?
1) Glucose
2) Both glucose & fructose
3) Fructose
4) None of these

20. Which of the following carbohydrate is not a disaccharide?
1) Sucrose 2) Lactose
3) Glucose 4) Maltose

21. Lactose on hydrolysis gives:
1) Two glucose molecules
2) Two fructose molecules
3) A galactose and a fructose molecule
4) Equimolar mixture of galactose and glucose

22. Glucose on oxidation with nitric acid as well as gluconic acid both gives:
1) Saccharic acid 2) n-Hexane
3) Fructose 4) Glucosazone

23. Which of the following is a reducing sugar?
1) Maltose 2) Glucose
3) Lactose 4) All

24. Specific rotation of equilibrium mixture of three forms of glucose is
1) 52.5° 2) 112°
3) 19.5° 4) 90°

25. Molecular formula $C_6H_{12}O_6$ is of
1) Glucose 2) Fructose
3) Both (1) and (2) 4) None of these

26. Molecular formula of pentahydroxy acid obtained when glucose is oxidised with Br_2 water is
1) $C_6H_{12}O_7$ 2) $C_6H_{12}O_8$
3) $C_6H_{18}O_4$ 4) $C_6H_{10}O_6$

27. Amylopectin is
1) Water soluble
2) Water insoluble
3) Forms colloidal solution with water
4) None of these

28. Maltose contains how many oxygen atoms?
1) 6 2) 8
3) 11 4) 22

29. A carbohydrate that cannot be hydrolysed to simpler forms is known as
1) Disaccharide 2) Monosaccharide
3) Polysaccharide 4) Pentasaccharide

30. A pyranose ring consist of skeleton of
1) 5 carbon atoms and one oxygen atoms
2) 6 carbon atoms & one oxygen atom
3) 4 carbon atoms & one oxygen atom
4) 6 carbon atoms

31. Glucose reacts with acetic anhydride to form
1) Mono-acetate 2) Tetra-acetate
3) Penta-acetate 4) Hexa-acetate

32. The disaccharide present in milk is
1) Maltose 2) Lactose
3) Sucrose 4) Cellobiose

33. To become a carbohydrate, a compound must contain at least
1) 2 carbons 2) 3 carbons
3) 4 carbons 4) 6 carbons

34. Which of the following is/are the oligosaccharides?
1) Lactose 2) Sucrose
3) Both (1) and (2) 4) None of these

35. In which of the following, high content of starch is found?
1) Roots 2) Cereals
3) Tubers 4) All of these

36. Select the incorrect statement :
1) Amylose is water soluble component
2) Amylopectin is water insoluble component
3) Amylose constitutes about 93-98% of starch
4) Amylopectin constitutes about 80- 85% of starch

Proteins

37. In aqueous solutions, amino acids mostly exist as :
1) $NH_2 - CHR - COOH$
2) $NH_2 - CHR - COO^-$
3) $NH_3^+ - CHR - COOH$
4) $H_3N^+ - CHR - COO^-$

38. Globular proteins are present in:
1) Blood 2) Egg
3) Milk 4) All of these

39. Which one of the following amino acids cannot be synthesized in the body?
1) Leucine 2) Glycine
3) Alanine 4) Glutamine

40. Acidic amino acid is:
1) Glycine 2) Valine
3) Leucine 4) Glutamic acid

41. Keratin, a structural protein is present in:
1) Hair 2) Wool
3) Silk 4) All of these

42. α-helix refers to:
1) Primary structure of protein
2) Secondary structure of protein
3) Tertiary structure of protein
4) Quaternary structure of protein

43. Most of the naturally occurring α - amino acids have:
1) D-configuration
2) L-configuration
3) Meso isomerism
4) Both (1) & (2)

44. Which of the following is more acidic?
1) H_2NCH_2COOH
2) $H_3N^+CH_2COOH$
3) CH_3COOH
4) All have equal acidic strength

45. Which of the following Statement-Is NOT true about globular proteins?
1) These are soluble in water
2) Polypeptide chains coil around to give a spherical shape.
3) Keratin is an example of globular proteins
4) Albumin in egg is an example of globular protein

46. Which one of the following is an essential amino acid?
1) Methionine 2) Tyrosine
3) Proline 4) Glycine

47. Protein is a polymer of:
1) Glucose 2) Terephthalic acid
3) Amino acid 4) Glycol

48. The helical structure of protein is stabilised by:
1) Dipeptide bonds 2) Hydrogen bonds
3) Ether bonds 4) Peptide bonds

49. Denaturation of proteins leads to loss of its biological activity by:
1) Formation of amino acids
2) Loss of primary structure
3) Loss of both primary and secondary structures
4) Loss of both secondary and tertiary structures

50. Proteins are condensation polymers of:
1) α-amino acids
2) β-amino acids
3) α-hydroxy acids
4) β-hydroxy acids

51. The number of amino acids found in proteins that a human body can synthesise:
1) 20 2) 25
3) 10 4) 100

52. The pH of the solution containing following zwitter ion species is

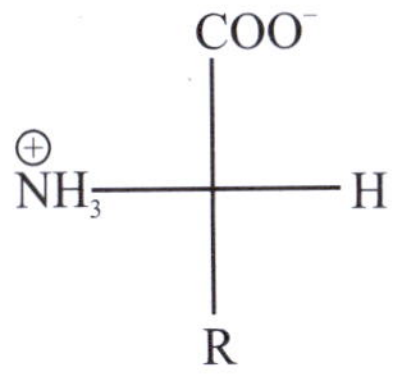

1) 4 2) 8
3) 6 4) 9

53. Peptide linkage is:
1) $-\overset{\overset{O}{||}}{C}-O-$ 2) $-\overset{\overset{O}{||}}{C}-NH_2$
3) $-\overset{\overset{O}{||}}{C}-NH-$ 4) $-\overset{\overset{O}{||}}{C}-NH-NH_2$

54. Most common types of secondary structures of protein are:
1) α-helix and β-helix structures
2) α-helix and β-pleated sheet structures
3) Right and left hand twisted structures
4) Globular and fibrous structures

55. Protein found in biological system with definite configuration and biological activity is called–
1) Amino acids 2) Native protein
3) Conformer protein 4) Inactive protein

56. A zwitter ion is:
1) A positively charged ion without a metal atom in it
2) A negatively charged ion without a metal atom in it
3) An ion with positive and negative charges at different point on it
4) A heavy ion with a small charge on it

57. In an alkaline medium, glycine predominantly exists as:
1) Cation 2) Anion
3) Zwitter ion 4) Covalent form

58. The only amino acid which is non chiral is:
1) Lysine 2) Proline
3) Glycine 4) Histidine

59. An example of a sulphur containing amino acid is:
1) Lysine 2) Serine
3) Cysteine 4) Tyrosine

60. Polypeptides having molecular weights above 10,000 are known as:
1) Amino acids 2) Hormones
3) Proteins 4) Terminal amino acids

61. Which of the following is not a function of proteins?
1) Nails formation
2) Skin formation
3) Muscle formation
4) Providing energy for metabolism

62. The proteins which are insoluble in water are:
1) Fibrous proteins 2) Globular proteins
3) Both (1) and (2) 4) None of these

63. Amino acids which cannot be synthesised in the body are known as
1) Non-essential amino acids
2) Essential amino acids
3) Laevo amino acids
4) None of these

64. On the basis of molecular structure, proteins is/are classified as:
1) fibrous protein only
2) globular proteins only
3) both fibrous and globular protein
4) none of these

65. Which of the following is obtained when carboxyl group of glycine combines with the amino group of alanine?
1) Phenylalanine 2) Glycylalanine
3) Trypsin 4) Alanine

Enzymes and Hormones

66. Estradiol is responsible for the development of
1) Primary male characters
2) Secondary female characters
3) Primary female characters
4) Secondary male characters

67. The reaction of enzymes in living system is to:
1) Supply energy to tissues
2) Create immunity
3) Circulate oxygen
4) Enhance the rate of biochemical reaction

68. Enzymes are made up of:
1) Edible proteins
2) Protein with specific structure
3) Nitrogen containing carbohydrates
4) Carbohydrates

69. Which one of the following Statement-Is incorrect about enzyme catalysts?
1) Enzymes are mostly protein in nature
2) Enzymes action is specific
3) Enzymes are denatured by ultraviolet rays and at high temperature.
4) Enzymes are least reactive at optimum temperature

70. Which of the following enzymes is not useful in the digestion of proteins ?
1) Chymotrypsin 2) Pepsin
3) Trypsin 4) Lipase

Vitamins

71. The night blindness is developed due to shortage of which vitamin?
1) Vitamin B_6 2) Vitamin C
3) Vitamin B_{12} 4) Vitamin A

72. Which of the following vitamin is water soluble?
1) Vitamin B 2) Vitamin E
3) Vitamin K 4) Vitamin A

73. Vitamin B_6 is known as:
1) Pyridoxin 2) Thiamine
3) Tocopherol 4) Riboflavin

74. Ascorbic acid is a:
1) Hormone 2) Fat soluble vitamin
3) Water soluble vitamin 4) Enzyme

75. Which of the following statements about fat soluble vitamins is incorrect ?
1) These are soluble in fat and oils
2) These are soluble in water
3) Vitamins A, D, E and K are examples of fat soluble vitamins
4) These are stored in liver and adipose tissues

76. Vitamin B_2 is known as:
1) Thiamine 2) Riboflavin
3) Pyridoxine 4) Ascorbic acid

Nucleic Acids

77. Which one of the following bases is not present in DNA?
1) Thymine 2) Quinoline
3) Adenine 4) Cytosine

78. A segment of DNA molecule which codes or specifies for one polypeptide chain is called
1) Phosphate group 2) Adenine
3) Gene 4) Amino acid

79. DNA multiplication is called:
1) Translation 2) Transduction
3) Transcription 4) Replication

80. Which of the following contributes to the double helical structure of DNA?
1) Hydrogen bond 2) Covalent bond
3) Disulphide bond 4) Van der waal's force

81. DNA and RNA contain four bases each. Which of the following bases is not present in RNA?
1) Adenine 2) Uracil
3) Thymine 4) Cytosine

82. Thymine is:
1) 5-Methyluracil 2) 4-Methyluracil
3) 3-Methyluracil 4) 1-Methyluracil

83. If one strand of DNA has the sequence ATGCTTGA, the sequence in the complimentary strand would be:
1) TCCGAACT
2) TACGTAGT
3) TACGAATC
4) TACGAACT

84. Guanine is an example of:
1) A nitrogenous base 2) A nucleoside
3) A nucleotide 4) Phosphate

85. Name the sugar moiety present in DNA molecules
1) β-D-2-deoxyribose 2) β-D-ribose
3) α-Glucose 4) α-Fructose

High Order Time Intensive MCQs

1. Which of the following forces stabilise the 2° and 3° structures of proteins?
   i. Disulphide linkages
   ii. Hydrogen bonds
   iii. Electrostatic forces of attraction
   iv. van der Waals forces
   1) Only (i) 2) Only (ii)
   3) Only (i) and (iii) 4) (i), (ii), (iii) and (iv)

2. Glycosidic linkage between C_1 of α-glucose and C_2 of α-fructose is found in
   1) Maltose 2) Sucrose
   3) Lactose 4) Amylose

3. Which one of the following does not correctly match with each other?
   1) Albumin-Globular protein
   2) Lipase-Enzyme
   3) Butter - Fat
   4) Glucagon-Enzyme

4. Characters are transferred from parents to offspring through:
   1) RNA 2) DNA 3) Lipid 4) Protein

5. Functionality of protein depends on:
   1) Its shape and structure 2) pH of medium
   3) Temperature 4) All of these

6. Vitamin necessary for blood clotting is:
   1) E 2) C 3) K 4) D

7. A peptide hormone is:
   1) Estrogen 2) Testosterone
   3) Insulin 4) Corticoid

8. Which of the following sugars occurs in the urine of diabetics?
   1) Dextrose 2) Galactose
   3) Fructose 4) Ribose

9. Fructose is an example of:
   1) Pyranose 2) Ketohexose
   3) Aldohexose 4) Heptose

10. Animal starch is the other name of:
    1) Amylose 2) Maltose
    3) Glycogen 4) Amylopectin

11. Methyl-α-D-glucoside & methyl-β-D-glucoside are:
    1) Epimers
    2) Anomers
    3) Conformations diastereomers
    4) Enantiomers

12. The following carbohydrate is:

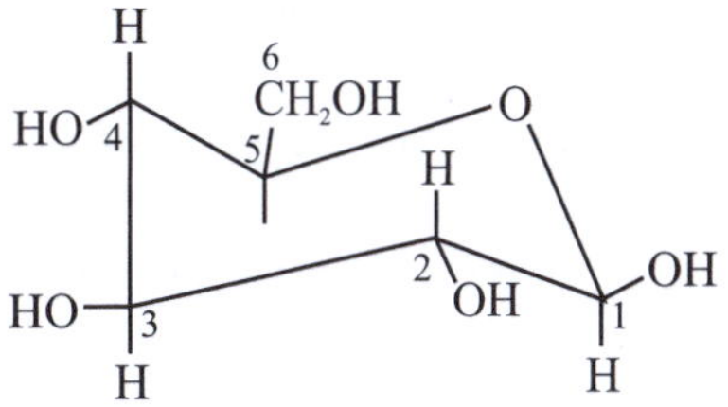

    1) A ketohexose 2) An aldohexose
    3) A α-furanose 4) An α-pyranose

13. Glucose does not react with:
    1) 2, 4-DNP 2) Schiff's reagent
    3) $NaHSO_3$ 4) All of these

14. Which of the following Statement-Is correct?
    1) All amino acids except glycine are optically active
    2) All amino acids except glutamic acid are optically active
    3) All amino acids except lysine are optically active
    4) All amino acids are optically active

15. Which of the following is/are correct?
    1) Glucose reacts with $NH_2 - OH$ to form an oxime
    2) Pentaacetate of glucose does not react with $NH_2 - OH$
    3) Red P/HI can convert glucose to n-hexane
    4) All are correct

16. Identify x, y & z where x, y & z are the number of functional groups in the final product?

    (1) Glucose $\xrightarrow{HCN}$ x (2) Glucose $\xrightarrow{HNO_3}$ y

    (3) Glucose $\xrightarrow{HI}$ z

    1) x = 2, y = 2, z = 0 2) x = 0, y = 0, z = 2
    3) x = 1, y = 1, z = 1 4) x = 2, y = 2, z = 3

17. The term anomers of glucose refers to:
    1) Isomers of glucose that differ in configuration at C - 1
    2) Enantiomers of glucose
    3) Position isomers of glucose
    4) Both (2) & (3)

18. The correct statement for the following structure:

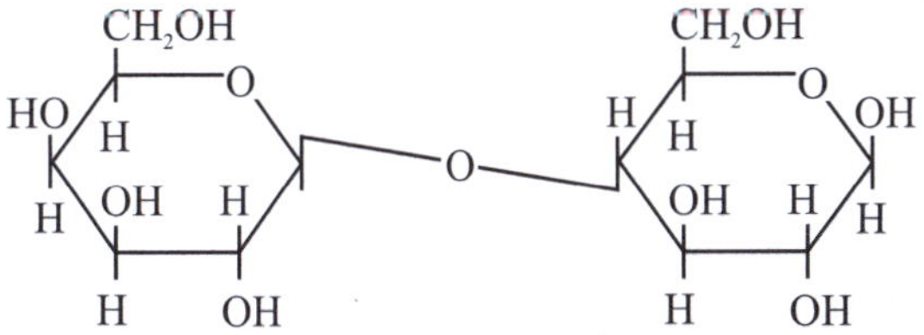

    1) On right side is galactose & on the left is glucose
    2) It is composed of β-D-glucose & β-D-fructose
    3) There is no free anomeric OH
    4) None of these

19. Which of the following statements is true regarding a carbohydrate having five carbon atoms and an aldehyde group?

1) It can have 8 stereo isomers
2) It can have 4 stereo isomers
3) It can have 2 sterero isomers
4) All the above

20. Coagulation of egg, on heating is because of:
1) Denaturation of protein occurs
2) The secondary structure of protein remains unchanged
3) Breaking of the peptide linkage in the primary
4) Biological property of protein remains unchanged

21. Amino acid 'T':
A. is the simplest amino acid with no asymmetric carbon.
B. exists as a zwitterion at physiological pH.
C. does not show optical activity.
D. A non essential amino acid

'T' is:
1) Alanine 2) Glycine
3) Serine 4) Valine

22. Hydrolysis of sucrose with dilute aqueous sulphuric acid yields
1) 1:1 D (+)-Glucose; D-(-)- fructose
2) 1:2 D -(+)-Glucose; D-(-)- fructose
3) 1:1 D -(-)-Glucose; D-(+)- fructose
4) 1:2 D -(-) Glucose; D-(+)- fructose

23. Which of the following statements about (+) (-) sucrose is not correct ?
1) It does not possess a free aldehydic (or) ketonic group
2) On hydrolysis, it produce D-(+)-glucose and D-(-)-fructose
3) It has glycosidic linkage between C_1 of α-D-glucose and C_2 of β-D-fructose
4) It is a reducing sugar.

24. A mixture of α-amino acids is obtained when proteins are hydrolysed by
1) Acids 2) Bases 3) Enzymes 4) All

25. In aqueous solutions, amino acids mostly exist as
1) $NH_2-CHR-COOH$ 2) $NH_2-CHR-COO^-$
3) $N_3H^+-CHR-COOH$ 4) $H_3N^+-CHR-COO^-$

26. An α-amino acid exists as,

$$\overset{+}{N}H_3-\underset{\underset{R}{|}}{CH}-COOH$$

at (pH = 2) and its isoelectric point is 6. The amino acid at pH 10.97 will exist as:

1) $\overset{+}{N}H_3-\underset{\underset{R}{|}}{CH}-COO^-$ 2) $NH_2-\underset{\underset{R}{|}}{CH}-COO^-$

3) $NII_2-\underset{\underset{R}{|}}{CII}-COOII$ 4) $\overset{+}{N}H_2\ \ \underset{\underset{R}{|}}{\overset{-}{C}H}\ \ COOH$

27. Molecule 'P'
A. A water soluble vitamin
B. Is stored in our body
C. Obtained from curd
D. It's deficiency can cause pernicious anaemia

'P' is:
1) Vitamin B_2 2) Vitamin B_1
3) Vitamin B_6 4) Vitamin B_{12}

Match the Columns MCQs

1. Match the vitamins given in Column-I with the deficiency disease they cause given in Column-II.

	Column-I (Vitamins)		Column-II (Diseases)
(a)	Vitamin A	(i)	Pernicious anaemia
(b)	Vitamin B_1	(ii)	Increased blood clotting time
(c)	Vitamin B_{12}	(iii)	Xerophthalmia
(d)	Vitamin C	(iv)	Rickets
(e)	Vitamin D	(v)	Muscular weakness
(f)	Vitamin E	(vi)	Night blindness
(g)	Vitamin K	(vii)	Beri Beri
		(viii)	Bleeding gums

1) (a)-(i) (b)-(iii) (c)-(v) (d)-(iv) (e)-(ii) (f)-(vi) (g)-(vii)
2) (a)-(vii) (b)-(viii) (c)-(iii) (d)-(i) (e)-(ii) (f)-(v) (g)-(vi)
3) (a)-(iii)(vi) (b)-(vii) (c)-(i) (d)-(viii) (e)-(iv) (f)-(v) (g)-(ii)
4) (a)-(i) (b)-(iii) (c)-(vii) (d)-(v) (e)-(iv) (f)-(ii) (g)-(vi)

2. Match the following enzymes given in Column-I with the reactions they catalyse given in Column-II.

	Column-I (Enzymes)		Column-II (Reactions)
(a)	Invertase	(i)	Decomposition of urea into NH_3 and CO_2
(b)	Maltase	(ii)	Conversion of glucose into ethyl alcohol
(c)	Pepsin	(iii)	Hydrolysis of maltose into glucose
(d)	Urease	(iv)	Hydrolysis of cane sugar
(e)	Zymase	(v)	Hydrolysis of proteins into peptides

1) (a)-(iv) (b)-(iii) (c)-(v) (d)-(i) (e)-(ii)
2) (a)-(ii) (b)-(i) (c)-(iii) (d)-(iv) (e)-(v)
3) (a)-(v) (b)-(iii) (c)-(ii) (d)-(i) (e)-(iv)
4) (a)-(iv) (b)-(v) (c)-(i) (d)-(iii) (e)-(ii)

3. Match the items given in Column-I with items given in Column-II.

Column-I		Column-II	
(a)	Lysine	(i)	Neutral optically active amino acid
(b)	Aspartic acid	(ii)	Basic amino acid
(c)	Valine	(iii)	Neutral optically inactive amino acid
(d)	Glycine	(iv)	Acidic amino acid

1) (a)-(iv) (b)-(iii) (c)-(ii) (d)-(i)
2) (a)-(iii) (b)-(ii) (c)-(iv) (d)-(i)
3) (a)-(i) (b)-(ii) (c)-(iii) (d)-(iv)
4) (a)-(ii) (b)-(iv) (c)-(i) (d)-(iii)

4. Match the items given in Column-I with items given in Column-II.

Column-I		Column-II	
(a)	Pepsin	(i)	Vitamin
(b)	Nucleic acid	(ii)	Male sex hormone
(c)	Ascorbic acid	(iii)	Digestive enzyme
(d)	Testosterone	(iv)	Genetic material

1) (a)-(i) (b)-(iv) (c)-(iii) (d)-(ii)
2) (a)-(ii) (b)-(iii) (c)-(iv) (d)-(i)
3) (a)-(iv) (b)-(ii) (c)-(i) (d)-(iii)
4) (a)-(iii) (b)-(iv) (c)-(i) (d)-(ii)

5. Match the items given in Column-I with items given in Column-II.

Column-I		Column-II	
(a)	Glycosidic linkage	(i)	Globular protein
(b)	Maltase	(ii)	Connects two monosaccharide units
(c)	Peptide bond	(iii)	Monomeric unit of nucleic acids
(d)	Nucleotide	(iv)	Connects two amino acid units

1) (a)-(iv) (b)-(iii) (c)-(ii) (d)-(i)
2) (a)-(iii) (b)-(iv) (c)-(i) (d)-(ii)
3) (a)-(ii) (b)-(i) (c)-(iv) (d)-(iii)
4) (a)-(i) (b)-(ii) (c)-(iii) (d)-(iv)

Correct & Incorrect MCQs

1. Which of the following statements is correct ?
   1) All monosaccharides whether aldose or ketose are reducing sugars.
   2) All monosaccharides whether aldose or ketose are non-reducing sugars.
   3) Only aldose monosaccharides are non-reducing sugars.
   4) None of these

2. Which of the following statement(s) is/are correct?
   A. Glucose is a monosaccharide.
   B. Sucrose is a disaccharide.
   C. Maltose is a monosaccharide.
   D. Starch is a polysaccharide.
   1) A, B and D 2) A and C
   3) Only A 4) B, C and D

3. Select the correct statement:
   1) Starch is the polymer of α – glucose.
   2) Amylose is a component of cellulose.
   3) Amylopectin is soluble in water and constitutes about 15-20% of starch.
   4) There are five carbons and one oxygen atom present in the ring in cyclic structure of fructose.

4. Which of the following statements is true about glucose?
   1) It is an aldohexose.
   2) Its molecular formula is $C_6H_{12}O_7$.
   3) On heating with HI, it forms n-heptane.
   4) It is present in furanose form.

5. Which of the following statements is incorrect?
   1) Glycine is an essential amino acid.
   2) Alanine is a non-essential amino acid.
   3) Histidine is an essential amino acid.
   4) Both (2) and (3)

6. Select the incorrect statement:
   1) Threonine is an essential amino acid.
   2) An amino acid may be neutral, acidic or basic.
   3) –COOH is the basic functional group in amino acid.
   4) None of these

7. Correct statements are:
   A. Fibrous proteins are generally insoluble in water
   B. Insulin and albumins are examples of fibrous proteins.
   C. Globular proteins are usually soluble in water.
   D. Keratin and myosin are the examples of globular proteins.
   1) A and C 2) B, C and D
   3) A and D 4) B and C

8. Incorrect statement/s is/are:
   A. Protein found in a biological system with a unique three-dimensional structure and biological activity is called a native protein.
   B. The primary structure of protein remains intact during denaturation of proteins.

C. The secondary and tertiary structures of proteins remain intact during denaturation of proteins.

D. The coagulation of egg white on boiling is a common example of denaturation.

1) A and B 2) B, C and D

3) Only B 4) Only C

9. Which one of the following statements is incorrect about enzyme catalysts?

1) Almost all the enzymes are globular proteins.

2) Enzymes are very specific for a particular reaction and for a particular substrate.

3) Enzyme maltase converts sucrose into glucose.

4) The enzymes which catalyse the oxidation of one substrate with simultaneous reduction of another substrate are named as oxidoreductase enzymes.

10. Which of the following statement regarding DNA and RNA is incorrect?

1) Complete hydrolysis of DNA (or RNA) yields a pentose sugar, phosphoric acid and nitrogen containing heterocyclic compounds.

2) In DNA molecules, the sugar moiety is β-D-2-deoxyribose.

3) In RNA molecule, the sugar moiety is β-D-ribose.

4) RNA molecules mainly are of five types.

Statement Based MCQs

Directions: These questions consist of two statements each, printed as Statement-I and Statement-II. While answering these questions, you are required to choose any one of the following four responses.

1) **Both Statement-I and Statement-II are correct.**
2) **Both Statement-I and Statement-II are incorrect.**
3) **Statement-I is correct & Statement-II is incorrect.**
4) **Statement-I is incorrect & Statement-II is correct.**

1. **Statement-I:** Fructose reduces Fehling's solution and Tollen's reagent.
Statement-II: Fructose does not contain any aldehyde group.

2. **Statement-I:** Sucrose is a disaccharide.
Statement-II: Sucrose is a non-reducing sugar.

3. **Statement-I:** α-amino acids exist as dipolar ions or zwitter ions.
Statement-II: α-amino acids are the building blocks of proteins.

4. **Statement-I:** Maltose is composed of equimolar mixture of D-(+)-glucose and D-(-) fructose.
Statement-II: Maltose is a reducing sugar.

5. **Statement-I:** Glucose reacts with hydroxylamine to form an oxime and adds a molecule of hydrogen cyanide to give cyanohydrin.
Statement-II: Acetylation of glucose with acetic anhydride gives n-hexane.

6. **Statement-I:** Glycogen is known as animal starch.
Statement-II: Glycogen structure is similar to that of amylopectin.

7. **Statement-I:** Complementary base pairing occur in DNA.
Statement-II: Adenine combine with thymine with triple bond and guanine combine with cytosine with double bond.

8. **Statement-I:** Deficiency of vitamin A causes rickets.
Statement-II: Deficiency of vitamin C causes scurvy.

9. **Statement-I:** Number of carbon atoms present in aldohexose are 3.
Statement-II: Glucose is an aldohexose and is also known as dextrose.

10. **Statement-I:** Fructose is obtained along with glucose by the hydrolysis of disaccharide, sucrose.
Statement-II: Fructose has the molecular formula $C_{11}H_{22}O_{18}$.

Assertion & Reason MCQs

Directions: These questions consist of two statements each, printed as Assertion and Reason. While answering these questions, you are required to choose any one of the following four responses.

1) **Both Assertion and Reason are True and the Reason is a correct explanation of the Assertion.**
2) **Both Assertion and Reason are True but Reason is not a correct explanation of the Assertion.**
3) **Assertion is True but the Reason is False.**
4) **Assertion is False but Reason is True.**

1. **Assertion:** D (+) – Glucose is dextrorotatory in nature.
Reason: 'D' represents its dextrorotatory nature.

2. **Assertion:** Vitamin D can be stored in our body.
Reason: Vitamin D is fat soluble vitamin.

3. **Assertion:** β-glycosidic linkage is present in maltose.

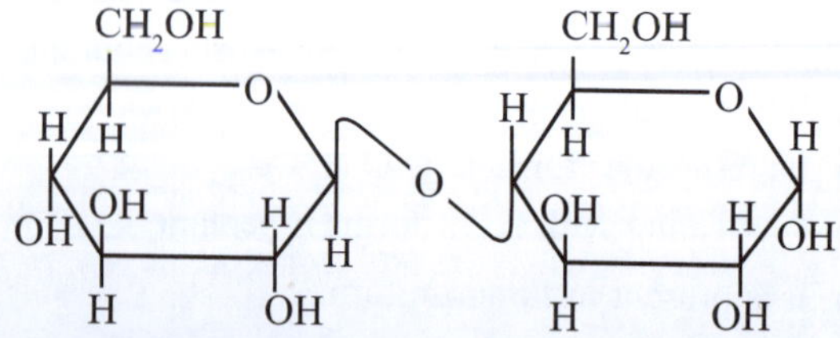

Reason: Maltose is composed of two glucose units in which C– 1 of one glucose unit is linked to C– 4 of another glucose unit.

4. **Assertion:** All naturally occurring α-aminoacids except glycine are optically active.

   Reason: Most naturally occurring amino acids have L-configuration.

5. **Assertion:** In presence of enzyme, substrate molecule can be attacked by the reagent effectively.

   Reason: Active sites of enzymes hold the substrate molecule in a suitable position.

6. **Assertion:** Sucrose is dextrorotatory, but after hydrolysis gives D(+) glucose and D(–) fructose.

   Reason: Because the leavorotation of fructose (–92.4°) is more than dextrorotation of glucose (+52.5), the mixture is leavorotatory. Hydrolysis causes change in sign of rotation from dextrorotation to laevorotation.

7. **Assertion:** Glucose and fructose give the same osazone.

   Reason: During osazone formation, stereochemistry at Cl and C2 is destroyed.

8. **Assertion:** Insulin is water soluble.

   Reason: Insulin is a globular protein.

9. **Assertion:** All Amino acids exist as Zwitter ions.

   Reason: Amino acids contain both $-NH_2$ and –COOH group.

10. **Assertion:** Glucose when treated with CH_3OH in the presence of dry HCl gas gives α and β-methyl glucosides.

    Reason: Glucose reacts with phenyl hydrazine to form crystalline osazone.

Multi Correct MCQs

1. The incorrect statements regarding enzymes are:

   A. Enzymes are biocatalysts.

   B. Enzymes are non-specific and can catalyse different kinds of reactions.

   C. Most Enzymes are globular proteins.

   D. Enzyme - oxidase catalyses the hydrolysis of maltose into glucose.

   Choose the correct answer from the option given below:

   1) (B) and (C) 2) (B), (C) and (D)
   3) (B) and (D) 4) (A), (B) and (C)

2. Which of the following are essential amino acids?

   A. Arginine, B. Phenylalanine,
   C. Aspartic acid, D. Cysteine,
   E. Histidine, F. Valine,
   G. Proline

   1) A, B, C and D 2) B , C, D and E
   3) C, D , E and F 4) A, B , E and F

3. Sulphur (S) containing amino acids from the following are:

   A. isoleucine B. cysteine
   C. lysine D. methionine
   E. glutamic acid

   1) A, D 2) B, D
   3) B, C, E 4) A, B, C

4. The Vitamins that can be stored in our body is:

   A. Vitamin A B. Vitamin B_1
   C. Vitamin B_6 D. Vitamin B_{12}
   E. Vitamin C F. Vitamin D
   G. Vitamin E

   1) Vitamin A, B, E 2) Vitamin B_1, B_6, C
   3) Vitamin C, D, E, B_{12} 4) Vitamin A, D, E and B_{12}

NCERT Exemplar MCQs

1. Glycogen is a branched chain polymer of α-D glucose units in which chain is formed by C_1—C_4 glycosidic linkage whereas branching occurs by the formation of C_1-C_6 glycosidic linkage. Structure of glycogen is similar to:

   1) Amylose 2) Amylopectin
   3) Cellulose 4) Glucose

2. A good source of vitamin A & D is

   1) Whole cereal 2) Yeast
   3) Cod liver oil 4) Watermelon

3. Sucrose (cane sugar) is a disaccharide. One molecule of sucrose on hydrolysis gives:

   1) 2 molecules of glucose
   2) 2 molecules of glucose + 1 molecule of fructose
   3) 1 molecule of glucose +1 molecule of fructose
   4) 2 molecules of fructose

4. Which of the following pairs represent anomers?

   1) 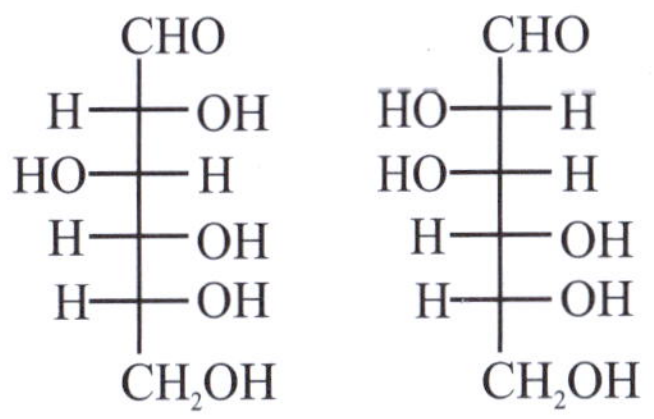

   2) 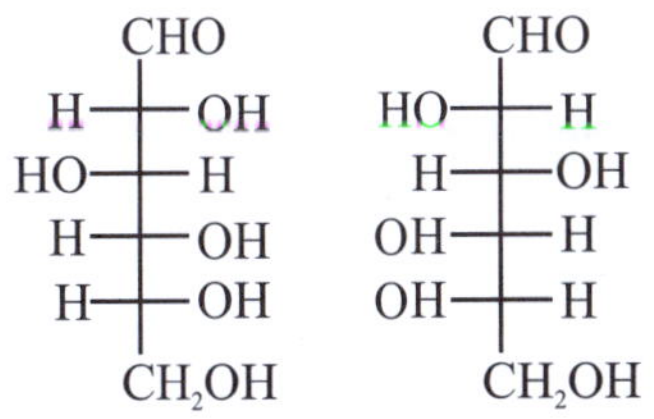

3) 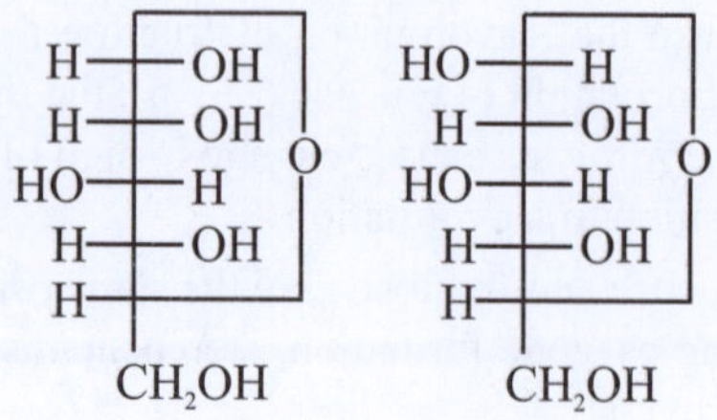

4) 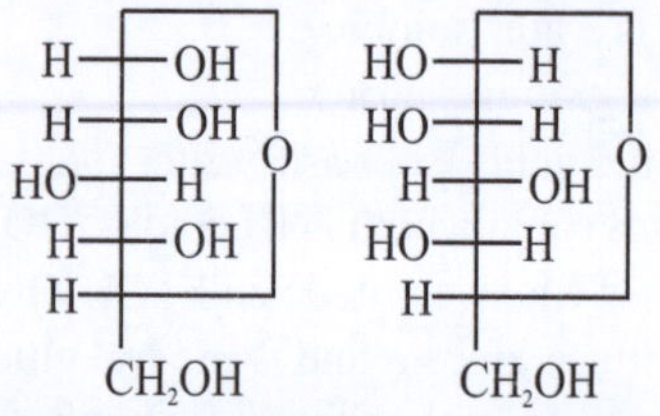

5. Proteins are found to have two different types of secondary structures viz α-helix and β-pleated sheet structure. α-helix structure of protein is stabilized by:
   1) Peptide bonds
   2) Van der Waals, forces
   3) Hydrogen bonds
   4) Dipole-dopole interactions

6. In disaccharides, if the reducing groups of monosaccharides, i.e., aldehydic or ketonic groups are bonded, these are non-reducing sugars. Which of the following disaccharide is a non-reducing sugar?

1) 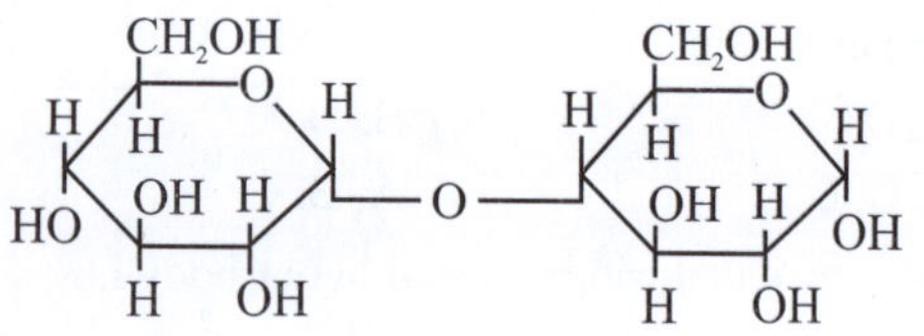

2) 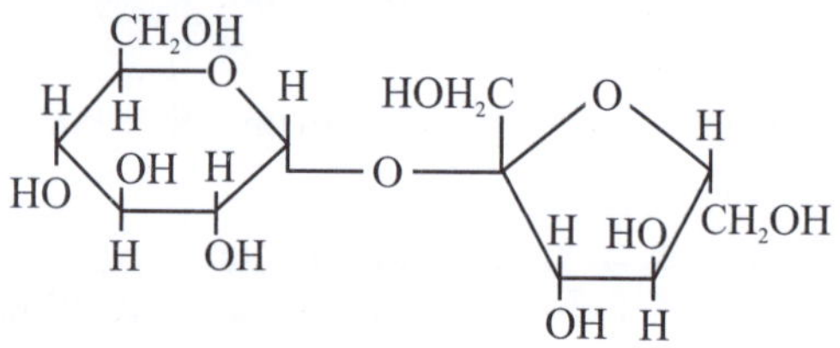

3) 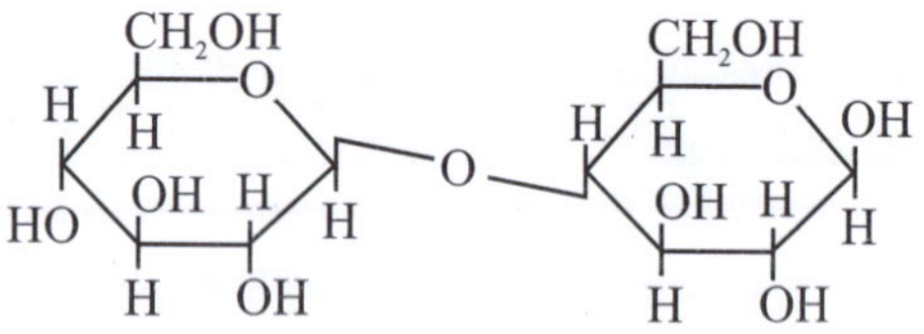

4) 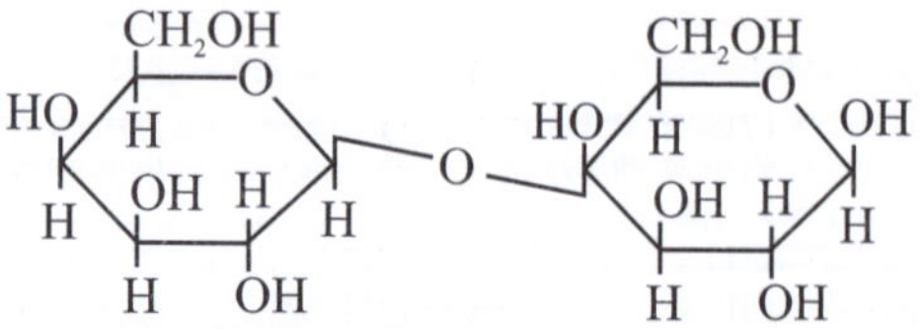

7. Which of the following acids is a vitamin?
   1) Aspartic and
   2) Ascorbic acid
   3) Adipic acid
   4) Saccharic acid

8. Dinucleotide is obtained by joining two nucleotides together by phosphodiester linkage. Between which carbon atoms of pentose sugars of nucleotides are these linkages present?
   1) 5' and 3'
   2) 1' and 5'
   3) 5' and 5'
   4) 3' and 3'

9. Nucleic acids are the polymers of:
   1) Nucleosides
   2) Nucleotides
   3) Bases
   4) Sugars

10. Which of the following statements is not true about glucose?
   1) It is an aldohexose
   2) On heating with HI, it forms n-hexane
   3) It is present in furanose form
   4) It does not give 2, 4- DNP test

11. Each polypeptide in a protein has amino acids linked with each other in a specific sequence. This sequence of amino acids is said to be:
   1) Primary structure of proteins
   2) Secondary structure of proteins
   3) Tertiary structure of proteins
   4) Quaternary structure of proteins

12. Which of the given amino acids have following unit:-

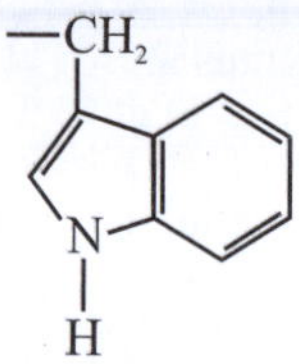

   1) Histidine
   2) Proline
   3) Tryptophan
   4) Tyrosine

13. Which of the following B group vitamins can be stored in our body?
   1) Vitamin B_1
   2) Vitamin B_2
   3) Vitamin B_6
   4) Vitamin B_{12}

14. Which of the following bases is not present in DNA?
   1) Adenine
   2) Thymine
   3) Cytosine
   4) Uracil

15. Three cyclic structures of monosaccharides are given below, which of these are anomers?

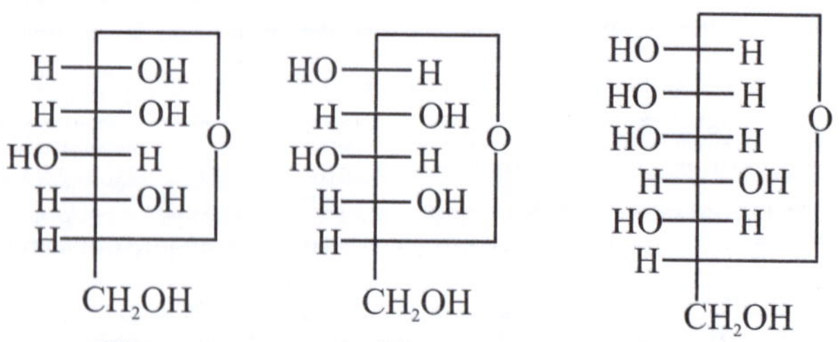

(I) (II) (III)

   1) I and II
   2) II and III
   3) I and III
   4) III is anomer of I and II

16. Which of the following reactions of glucose can be explained only by its cyclic structure?
   1) Glucose forms pentaacetate
   2) Glucose reacts with hydroxylamine to form an oxime
   3) Pentaacetate of glucose does not react with hydroxyl amine
   4) Glucose is oxidized by nitric acid to gluconic acid

17. Optical rotations of some compounds along with their structures are given below, which of them have D configuration?

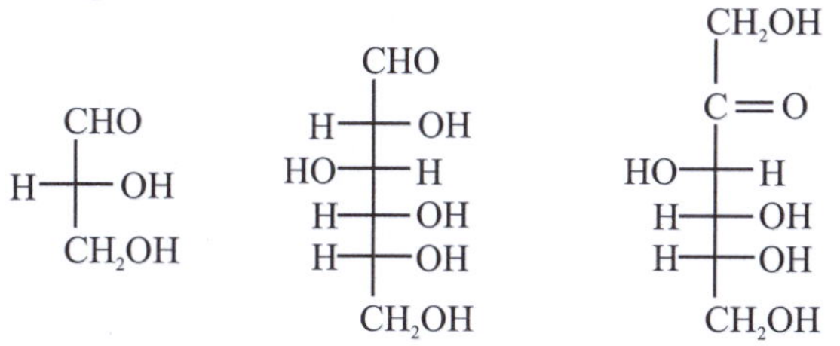

(+) rotation (I) (+) rotation (II) (–) rotation (III)

1) I, II, III
2) II, II
3) I, II
4) III

18. Structure of a disaccharide formed by glucose and fructose is given below. Identify anomeric carbon atoms in monosaccharide units.

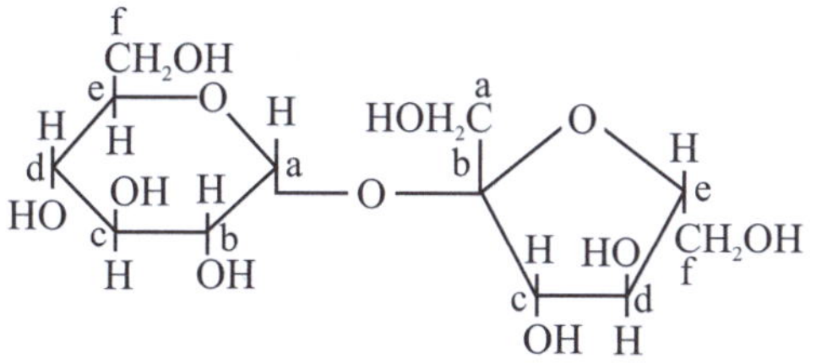

1) 'a' carbon of glucose and 'a' carbon of fructose
2) 'a' carbon of glucose and 'e' carbon of fructose
3) 'a' carbon of glucose and 'b' carbon of fructose
4) 'f' carbon of glucose and 'f' carbon of fructose

19. Three structures are given below in which two glucose units are linked. Which of these linkages between glucose units are between C_1 and C_4 and which linkages are between C_1 and C_6?

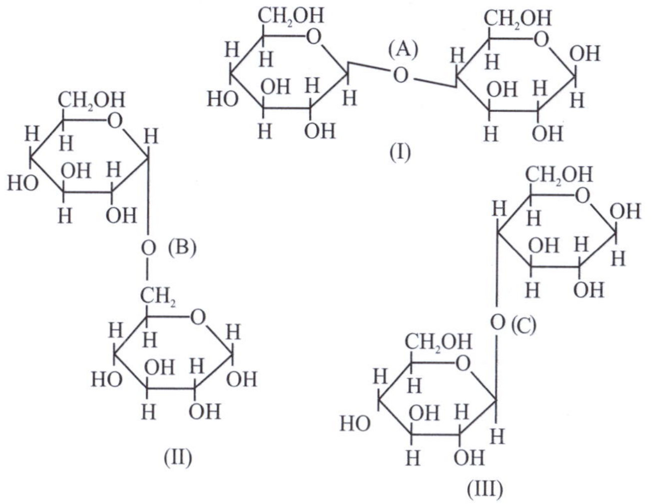

1) (A) is between C_1 and C_4, (B) and (C) are between C_1 and C_6
2) (A) and (B) are between C_1 and C_4, (C) is between C_1 and C_6
3) (A) and (C) are between C_1 and C_4, (B) is between C_1 and C_6
4) (A) and (C) are between C_1 and C_6, (B) is between C_1 and C_4

Kattar NEET

Scan Me for Detailed Explanations

1.

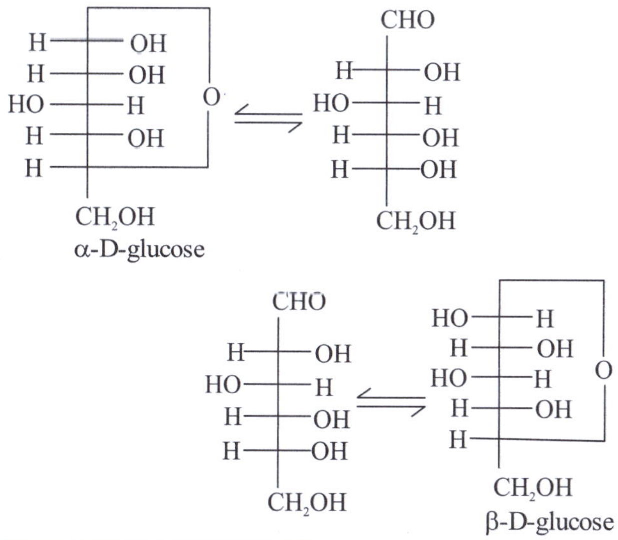

The above process in which α and β form remain in equilibrium with acyclic form and a change in optical rotation is observed which is called as-

1) Mutarotation
2) Epimerisation
3) Condensation
4) Inversion

2. Which of the following aldohexose give the same osazone derivative?

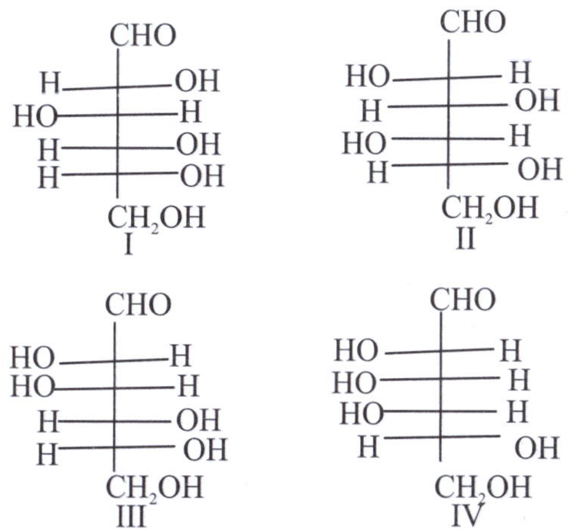

1) I and IV only
2) I and III only
3) II and III only
4) III and IV only

3. Which of the following amino acid is most basic:

1) $H_2N-C(=NH)-NH-CH_2CH_2CH_2-CH(NH_2)-COOH$

2) $H_2N-CH(CH_3)-C(=O)-NH-CH_2-COOH$

3) $HOOC-CH_2-CH(NH_2)-COOH$

4) 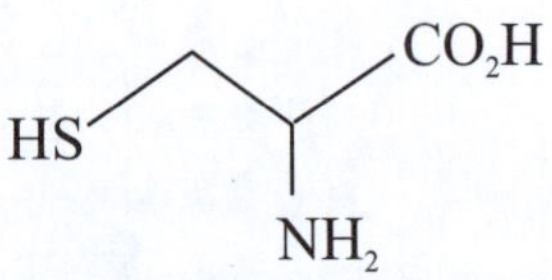

4. Identify the product 'Z' in the following series of reactions

$$C_6H_{10}O_6\,(\text{glucose}) \xrightarrow{HCN} X \xrightarrow{H_3O^+} Y \xrightarrow{(\text{Red P. HI})} Z$$

1) Hexanoic acid
2) α-methyl caproic acid
3) Heptanoic acid
4) Heptane 1-7 diol

5. Identify the correct set of stereochemical relationships amongst the following monosaccharides I - IV.

(I) 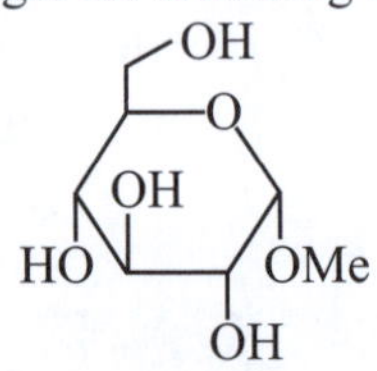
(II) 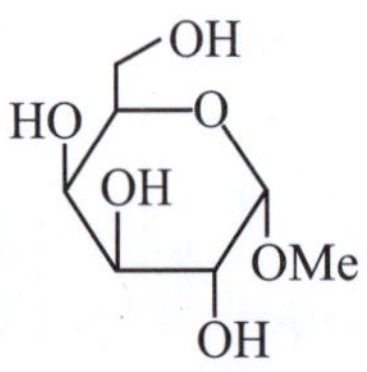
(III) 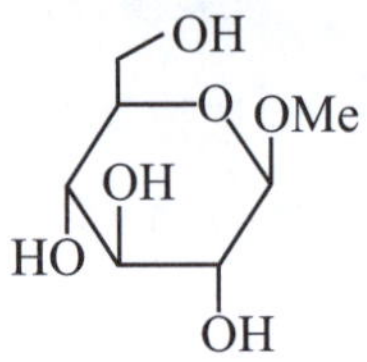
(IV) 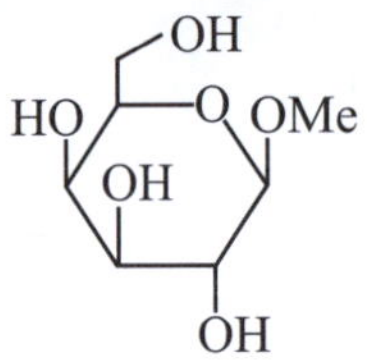

1) I, II are anomer
2) I, III are epimer
3) I, II are epimer
4) I, IV are anomer

6. The average energy of each hydrogen bond in A–T pair is x kcal mol^{-1} and that in G–C pair is y kcal mol^{-1}. Assuming that no other interaction exists between the nucleotides, the approximate energy required in kcal mol^{-1} to split the following double stranded DNA into two single strands is

A—T—A—T—G—C—A—G
: : : : : : : :
T—A—T—A—C—G—T—C

[Each dashed line may represent more than one hydrogen bond between the base pairs

1) 10x + 9y
2) 5x + 3y
3) 15x + 6y
4) 5x + 4.5y

7. A nonapeptide in rat on hydrolysis gave the following identifiable tripeptides:
Gly-Ala-Phe, Ala-Leu-Val, Gly-Ala-Leu, Phe-Glu-His and His-Gly-Ala. The sequence in the nonapetide is:

1) Gly-Ala-Leu-Val-Phe-Glu-His-His-Gly
2) Ala-Phe-Leu-Val-Gly-Leu-Phe-Glu-His
3) Gly-Ala-Phe-Glu-His-Gly-Ala-Leu-Val
4) Phe-Ala-Leu-Val-Gly-Glu-His-Gly-Ala

8. Based on the information given below, the isoelectric point (pI) of lysine is______.

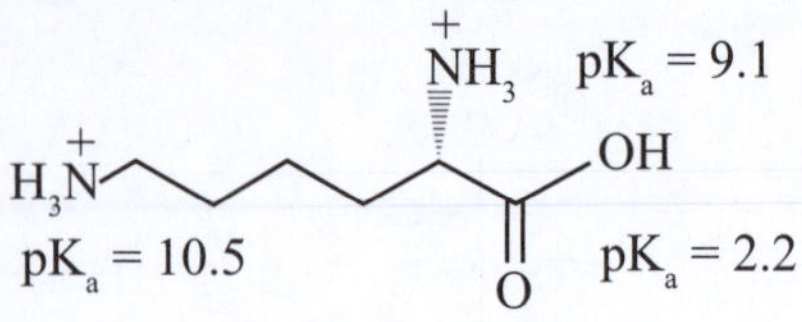

1) 5.85
2) 11.3
3) 9.80
4) 6.35

9. Match the following:

	List-I (Compounds)		List-II (Properties or Linkages Shown)
I	Glycine	P	Amide linkage
II	Protein	Q	H-bonding
III	Alanine (2-aminopropanoic acid)	R	Zwitter ion
IV	DNA	S	Deoxy ribose sugar
		T	Natural polymer

1) I-Q, R;II-P, Q, T;III-Q, R;IV-Q, S, T
2) I-Q, R, T;II-P, Q;III-Q, T;IV-Q, S
3) I-Q, R;II-P, T;III-Q, S;IV-Q, R, T
4) I-R;II-P, R, T;III-R, T;IV-Q, R, S

10. Match the following:

	List-I		List-II
I	Sucrose $\xrightarrow{H_3O^{\oplus}}$	P	Product is glucose + Fructose
II	Maltose $\xrightarrow{H_3O^{\oplus}}$	Q	Product is 2 Glucose
III	Lactose $\xrightarrow{H_3O^{\oplus}}$	R	Product is Glucose + Galactose
IV	Melibiose $\xrightarrow{H_3O^{\oplus}}$	S	Reactant is Oligosaccharide
		T	Reactant is reducing sugar

1) I-P, S; II-P, Q, T; III-Q, R; IV-Q, S, T
2) I-Q, R, T; II-P, Q, T; III-Q, T; IV-Q, S
3) I-Q, R; II-P, T; III-Q, S; IV-Q, R, T
4) I-P, S; II-Q, S, T; III-R, S, T ; IV-R, S, T

11. Find true and false from the following statements regarding carbohydrates.

S_1 All monosaccharides whether aldoses or ketoses are reducing sugars.

S_2 Bromine water can be used to differentiate between aldoses and ketoses.

S_3 A pair of diastereomeric aldoses which differ only in configuration at C-2 are anomers.

S_4 Osazone formation destroys the configuration at C-2 of an aldose, but does not affect the configuration of the rest of the molecule.

1) TTTT
2) TFTF
3) TTFT
4) FTTT

12. Match the Column-I with Column-II.

	Column-I		Column-II
A.	$RCH(\overset{+}{N}H_3)CO_2^-$	p.	Acidic amino acid
B.	Arginine	q.	Neutral amino acid
C.	Valine	r.	Zwitter ion
D.	Aspartic acid	s.	Basic amino acid

1) A → r; B → s; C → q; D → p
2) A → p; B → q; C → s; D → r
3) A → q; B → s; C → r; D → p
4) A → s; B → p; C → q; D → r

13. Which of the following gives an optically inactive dicarboxylic acid on oxidation with dilute HNO_3 acid?

1) 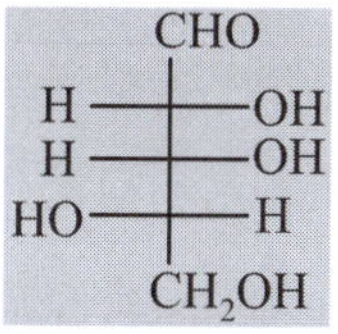
2) 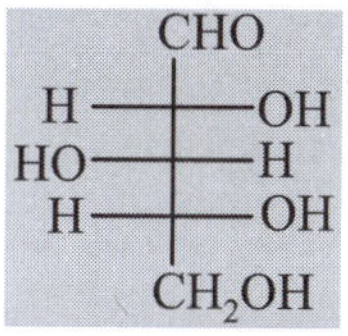
3) 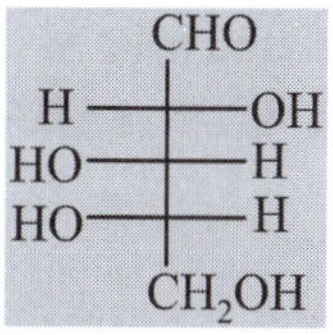
4) 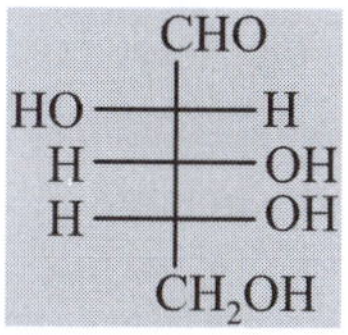

14. CHO–CH–OH–CH–OH–CH_2–OH $\xrightarrow[(3)\ H_3O^+]{(1)\ HCN \quad (2)\ H_2/Pd\text{–}BaSO_4}$ Product

The final product of the reaction, is

1) CH_2OH–$(CHOH)_3$–CH_2OH
2) CHO–$(CH\text{—}OH)_3$–$CH_2\text{—}OH$
3) COOH–$(CH\text{—}OH)_3$–$CH_2\text{—}OH$
4) COOH–$(CHOH)_3$–COOH

15. Which of the following properties of glucose cannot be explained by its open chain structure?

(*i*) Glucose does not form hydrogen sulphite with $NaHSO_3$.
(*ii*) On oxidation with HNO_3, glucose gives saccharic acid.
(*iii*) Glucose is found to exist in two different crystalline forms which are named as α and β.

1) (*ii*) only
2) (*i*) and (*iii*)
3) (*ii*) and (*iii*)
4) (*i*) and (*ii*)

16. Which of the following statement(s) is/are true?

(*i*) All amino acids contain one chiral center.
(*ii*) Some amino acids contain one, while some contain more chiral centers or even "NO" chiral center.
(*iii*) All amino acids found in proteins have L configuration.
(*iv*) All amino acids found in proteins have 1° amino group.

1) (*ii*), (*iii*) and (*iv*)
2) (*ii*) and (*iii*)
3) (*i*), (*iii*) and (*iv*)
4) (*i*) and (*iv*)

17. Assertion(A): Vitamin D cannot be stored in our body.
Reason (R): Vitamin D is a fat soluble vitamin.

1) Assertion (A) is correct, reason (R) is correct; reason (R) is a correct explanation for assertion.
2) Assertion (A) is correct, reason (R) is correct; reason (R) is not a correct explanation for assertion.
3) Assertion (A) is correct, reason (R) is incorrect.
4) Assertion (A) is incorrect, reason (R) is correct.

18. Match the Column-I with Column-II.

	Column-I		Column-II (Deficiency diseases)
A.	Vitamin A	p.	Scurvy
B.	Vitamin B_{12}	q.	Hemorrhagic condition
C.	Vitamin C	r.	Sterility
D.	Vitamin E	s.	Xeropthalmia
E.	Vitamin K	t.	Pernicious anaemia

1) A → t; B → s; C → p; D → r; E → r
2) A → s; B → t; C → p; D → q; E → r
3) A → s; B → t; C → p; D → r; E → q
4) A → t; B → s; C → p; D → r; E → q

19. Which of the following statement(s) is/are correct?

(*i*) Vitamins A, D, E and K are insoluble in water.
(*ii*) Vitamins A, D, E and K are stored in liver and adipose tissues.
(*iii*) Vitamin B and vitamin C are water soluble.
(*iv*) Water soluble vitamins should not be supplied regularly in diet.

1) (*i*), (*ii*) and (*iv*)
2) (*i*), (*ii*) and (*iii*)
3) (*i*) and (*iv*)
4) (*ii*) and (*iv*)

20. Match the Column-I with Column-II.

	Column-I (Enzymes)		Column-II (Reactions)
A.	Invertase	p.	Decomposition of urea into NH_3 and CO_2.
B.	Maltase	q.	Conversion of glucose into ethyl alcohol.
C.	Pepsin	r.	Hydrolysis of maltose into glucose.
D.	Urease	s.	Hydrolysis of cane sugar
E.	Zymase	t.	Hydrolysis of proteins into peptides.

1) A → s; B → r; C → t; D → p; E → q
2) A → r; B → q; C → s; D → p; E → t
3) A → q; B → p; C → r; D → s; E → t
4) A → s; B → p; C → t; D → q; E → r

NEET & AIIMS Past Year MCQs

1. Given below are two statements :
Statement-I: Glycogen is similar to amylose in its structure.
Statement-II: Glycogen is found in yeast and fungi also.
In the light of the above statements, choose the correct answer from the options given below: **(2024 Re)**
1) Statement-I is true but Statement-II is false.
2) Statement-I is false but Statement-II is true.
3) Both Statement-I and Statement-II are true.
4) Both Statement-I and Statement-II are false.

2. The reagents with which glucose does **not** react to give the corresponding tests/products are **(2024)**
A. Tollen's reagent B. Schiff's reagent
C. HCN D. NH_2OH
E. $NaHSO_3$
Choose the correct options from the given below:
1) B and E 2) E and D
3) B and C 4) A and D

3. Given below are two statements: **(2023)**
Statement-I: A unit formed by the attachment of a base to 1' position of sugar is known as nucleoside
Statement-II: When nucleoside is linked to phosphorous acid at 5'-position of sugar moiety, we get nucleotide.
In the light of the above statements, choose the correct answer from the options given below:
1) Statement-I is false but Statement-II is true.
2) Both Statement-I and Statement-II are true.
3) Both Statement-I and Statement-II are false.
4) Statement-I is true but Statement-II is false.

4. The incorrect statement regarding enzymes is **(2022)**
1) Enzymes are very specific for a particular reaction and substrate
2) Enzymes are biocatalysts
3) Like chemical catalysts enzymes reduce the activation energy of bio process.
4) Enzymes are polysaccharides.

5. The RBC deficiency is deficiency disease of: **(2021)**
1) Vitamin B_6 2) Vitamin B_1
3) Vitamin B_2 4) Vitamin B_{12}

6. Find the hydrolysis product when a phosphodiester bonds of nucleotide breaks: **(AIIMS 2011)**
1) 3-OH-deoxyribose-5-PO_4^{3-}
2) 5-OH-deoxyyribose-3-PO_4^{3-}
3) 2-OH-deoxyribose-2-PO_4^{3-}
4) 4-OH-deoxyribose-2-PO_4^{3-}

7. Which of the following biomolecules is insoluble in water: **(AIIMS 2005)**
1) α-Keratin
2) Haemoglobin
3) Ribonuclease
4) Adenine

8. Among the following, the achiral amino acid is: **(AIIMS 2003)**
1) 2-Ethylalanine
2) 2-Methylglycine
3) 2 Hydroxymethyl serine
4) Tryptophan

9. **Assertion:** A solution of sucrose in water is dextrorotatory. But on hydrolysis in the presence of a little hydrochloric acid, it becomes laevorotatory.
Reason: Sucrose on hydrolysis gives unequal amounts of glucose and fruc474tose. As a result of this, change in sign of rotation is observed. **(AIIMS 2001)**
1) If both Assertion & Reason are True & the Reason is a correct explanation of the Assertion.
2) If both Assertion & Reason are True but Reason is not a correct explanation of the Assertion.
3) If Assertion is True but the Reason is False.
4) If both Assertion & Reason are False.

Answer Key

NCERT Topic-wise MCQs

1	2	3	4	5	6	7	8	9	10	11	12	13	14	15	16	17
2	4	4	1	2	3	4	4	3	3	1	1	2	3	4	3	4
18	**19**	**20**	**21**	**22**	**23**	**24**	**25**	**26**	**27**	**28**	**29**	**30**	**31**	**32**	**33**	**34**
3	1	3	4	1	4	1	3	1	2	3	2	1	3	2	2	3
35	**36**	**37**	**38**	**39**	**40**	**41**	**42**	**43**	**44**	**45**	**46**	**47**	**48**	**49**	**50**	**51**
4	3	4	4	1	4	4	2	2	2	3	1	3	2	4	1	3
52	**53**	**54**	**55**	**56**	**57**	**58**	**59**	**60**	**61**	**62**	**63**	**64**	**65**	**66**	**67**	**68**
3	3	2	2	3	2	3	3	3	4	1	2	3	2	2	4	2
69	**70**	**71**	**72**	**73**	**74**	**75**	**76**	**77**	**78**	**79**	**80**	**81**	**82**	**83**	**84**	**85**
4	4	4	1	1	3	2	2	2	3	4	1	3	1	4	1	1

High Order Time Intensive MCQs

1	2	3	4	5	6	7	8	9	10	11	12	13	14	15	16	17
4	2	4	2	4	3	3	1	2	3	2	2	4	1	4	1	1
18	**19**	**20**	**21**	**22**	**23**	**24**	**25**	**26**	**27**							
4	1	1	2	1	4	4	4	2	4							

Match the Columns MCQs

1	2	3	4	5
3	1	4	4	3

Correct & Incorrect MCQs

1	2	3	4	5	6	7	8	9	10
1	1	1	1	1	3	1	4	3	4

Statement Based MCQs

1	2	3	4	5	6	7	8	9	10
1	1	1	4	3	1	3	4	4	3

Assertion & Reason MCQs

1	2	3	4	5	6	7	8	9	10
3	1	4	2	1	1	1	2	1	2

Multi Correct MCQs

1	2	3	4
3	4	2	4

NCERT Exemplar MCQs

1	2	3	4	5	6	7	8	9	10	11	12	13	14	15	16	17
2	3	3	3	3	2	2	1	2	3	1	3	4	4	1	3	1
18	**19**															
3	3															

Kattar NEET

1	2	3	4	5	6	7	8	9	10	11	12	13	14	15	16	17
1	2	1	3	3	1	3	3	1	4	3	1	2	2	2	2	4
18	**19**	**20**														
3	2	1														

NEET & AIIMS Past Year MCQs

1	2	3	4	5	6	7	8	9
2	1	4	4	4	1	1	3	3

CHAPTER

12

Principles Related to Practical Chemistry

Topicwise Number of Questions Analysis (2025-2016)

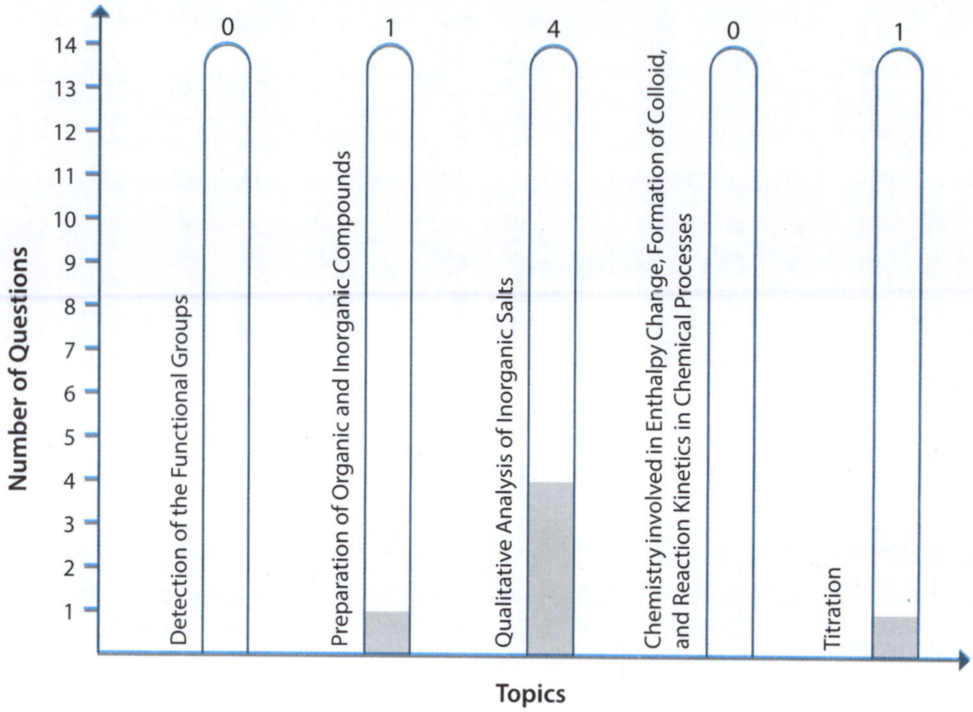

NCERT-Picks

Detection of extra elements

Nitrogen, sulphur, halogens and phosphorus present in an organic compound are detected by "Lassaigne's test". The elements present in the compound are converted from covalent form into the ionic form by fusing the compound with sodium metal. Following reaction take place:

$$Na + C + N \xrightarrow{\Delta} NaCN$$

$$2Na + S \xrightarrow{\Delta} Na_2S$$

$$Na + X \xrightarrow{\Delta} NaX \ (X = Cl, Br \text{ or } I)$$

C, N, S and X come from organic compound.

Cyanide, sulphide and halide of sodium so formed on sodium fusion are extracted from the fused mass by boiling it with distilled water. This extract is known as sodium fusion extract.

Test for Nitrogen

The sodium fusion extract is boiled with iron (II) sulphate and then acidified with concentrated sulphuric acid. The formation of Prussian blue color confirms the presence of nitrogen. Sodium cyanide first reacts with iron (II) sulphate and forms sodium hexacyanoferrate (II). On heating with concentrated sulphuric acid, some iron (II) ions are oxidised to iron (III) ions which react with sodium hexacyanoferrate (II) to produce iron (III) hexacyanoferrate (II) (ferriferrocyanide) which is Prussian blue in colour.

$$6CN^- + Fe^{2+} \longrightarrow [Fe(CN)_6]^{4-}$$

$$3[Fe(CN)_6]^{4-} + 4Fe^{3+} \xrightarrow{xH_2O} Fe_4[Fe(CN)_6]_3.xH_2O$$

(Prussian blue)

Test for Sulphur

- The sodium fusion extract is acidified with acetic acid and lead acetate is added to it. A black precipitate of lead sulphide indicates the presence of sulphur.

  $$S^{2-} + Pb^{2+} \longrightarrow \underset{\text{Black}}{PbS}$$

- On treating sodium fusion extract with sodium nitroprusside, appearance of a violet colour further indicates the presence of sulphur.

  $$S^{2-} + [Fe(CN)_5NO]^{2-} \longrightarrow \underset{\text{Violet}}{[Fe(CN)_5NOS]^{4-}}$$

In case, nitrogen and sulphur both are present in an organic compound, sodium thiocyanate is formed. It gives blood red colour and no Prussian blue since there are no free cyanide ions.

$$Na + C + N + S \longrightarrow NaSCN$$

$$Fe^{3+} + SCN^- \longrightarrow \underset{\text{Blood red}}{[Fe(SCN)]^{2+}}$$

If sodium fusion is carried out with excess of sodium, the thiocyanate decomposes to yield cyanide and sulphide. These ions give their usual tests.

$$NaSCN + 2Na \longrightarrow NaCN + Na_2S$$

Test for Halogens

- The sodium fusion extract is acidified with nitric acid and then treated with silver nitrate. A white precipitate, soluble in ammonium hydroxide shows the presence of chlorine, a yellowish precipitate; sparingly soluble in ammonium hydroxide shows the presence of bromine and a yellow precipitate, insoluble in ammonium hydroxide shows the presence of iodine.

  $$X^- + Ag^+ \longrightarrow AgX$$

  X represents a halogen: Cl, Br or I.

  If nitrogen or sulphur is also present in the compound, the sodium fusion extract is first boiled with concentrated nitric acid to decompose cyanide or sulphide of sodium formed during Lassaigne's test. Otherwise these ions would interfere with silver nitrate test for halogens.

- Beilstein's test

  Organic compound $\xrightarrow[\text{Copper wire}]{\text{Heated over}}$ Green flame, confirms the presence of halogen in the organic compound.

Detection of the functional groups present in organic compounds

Detection of Functional Groups

Detection of functional groups is an important step in the analysis of organic compounds. It largely depends on the correct determination of the constituent elements.

(1) Hydroxyl group

Alcohols may be considered as neutral compounds. They are soluble in water.

(i) Sodium metal test: Take 3 – 4 mL of the organic compound in a test-tube. Add a thin slice of sodium metal. A brisk effervescence due to the evolution of hydrogen takes place. (Organic compound should be free from water).

$$2ROH + 2Na \rightarrow 2RONa + H_2\uparrow$$

(ii) Ester test: Warm the given liquid with acetic acid in presence of sulphuric acid. A fragrant smell indicates the formation of ester.

$$ROH + CH_3COOH \xrightarrow[\text{Heat}]{H_2SO_4} CH_3COOR + H_2O$$

Alcohols give crystalline esters with 3,5-dinitrobenzoyl chloride. These esters have sharp melting points and can be used for testing a given alcohol.

(iii) Oxidation test: Take some organic compound, 5 mL of $K_2Cr_2O_7$, solution and 2–3 mL of conc. H_2SO_4 in a test-tube. Heat the test-tube after fitting a delivery tube. The distillate is passed into another test-tube containing water. Test the aqueous solution of the distilalte for aldehyde group. A positive test indicates the presence of primary alcoholic group in the given compound.

$$RCH_2OH \xrightarrow{[O]} RCHO + H_2O$$

(iv) Ceric ammonium nitrate test: Shake a few drops of ceric ammonium nitrate with 1 mL of the organic compound. A red colour is produced (Phenols give a green or brown colour).

$$2ROH + (NH_4)_2[Ce(NO_3)_6] \rightarrow [Ce(NO_3)_4(ROH)_2] + 2NH_4NO_3$$

Distinction Between Primary, Secondary and Tertiary Alcohols

Lucas test: Add 3–4 drops of alcohol to 2 mL of Lucas reagent (anhydrous $ZnCl_2$ + conc. HCl) in a test-tube. Shake the mixture and allow to stand at room temperature. If a cloudy precipitate is obtained immediately then it is a tertiary alcohol, if obtained after 2–3 minutes then it is a secondary alcohol, while no cloudy precipitate even after long standing means it is a primary alcohol.

(2) Tests for phenolic group

(i) Litmus test: On a piece of moist blue litmus paper, place a little amount of the organic compound or its aqueous solution. If the blue litmus paper turns red, this indicates the presence of a phenolic group.

(ii) Neutral $FeCl_3$ test: When phenols are treated with neutral ferric chloride solution, they form coloured complexes. The colour of the complex may be violet, red, blue or green. This is a characteristic reaction of compounds having enolic group. (=C—OH). All enols respond to this test.

With relatively milder oxidizing agents such as ferric chloride (Fe^{3+}) a resonance stabilized phenoxyl radical is formed. This radical soon undergoes dimerization which may take place by ortho-ortho, ortho-para, para-para coupling reaction.

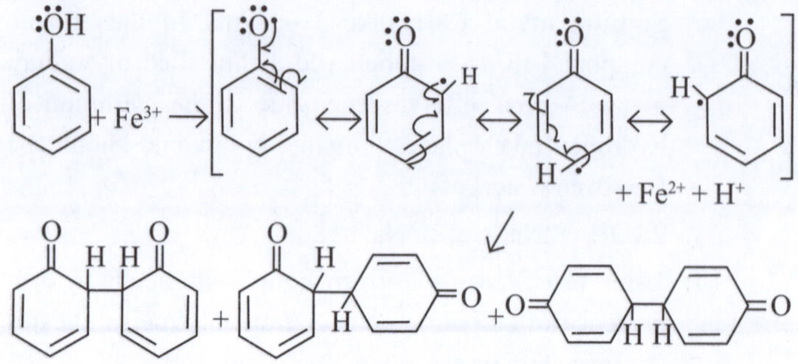

The phenoxyl radicals are highly coloured and therefore, products are also coloured.

(iii) Aniline Dye test: Dissolve a little aniline in dilute hydrochloric acid, cool in ice-cold water and add aqueous solution of sodium nitrite dropwise. To this cold solution, add organic compound already dissolved in sodium hydroxide solution. A red or orange precipitate is obtained.

$NaNO_2 + HCl \rightarrow HNO_2 + NaCl$

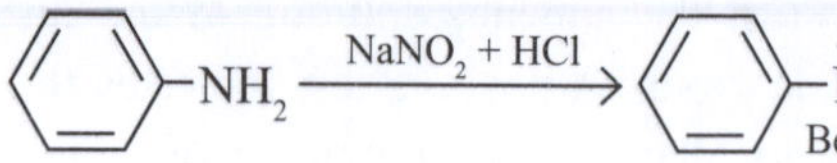

Benzene diazonium chloride

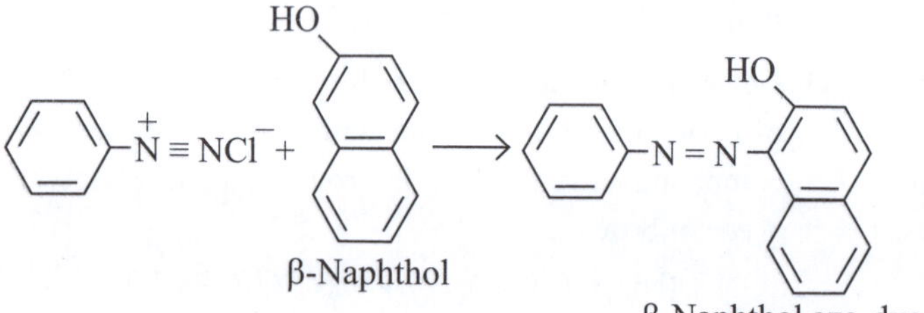

β-Naphthol

β-Naphthol azo-dye (Scarlet red)

(iv) Phthalein test (Fluorescein test): Heat gently some organic compound and phthalic anhydride with a 2-3 drops of conc. sulphuric acid in a test-tube. Pour the heated mixture to about 15 ml of water taken in a beaker and then add sodium hydroxide solution till it became alkaline. Appearance of pink, blue, green, red etc. colours indicates the presence of phenolic –OH group in the compound. However, the colour disappears on addition of large excess of sodium hydroxide solution.

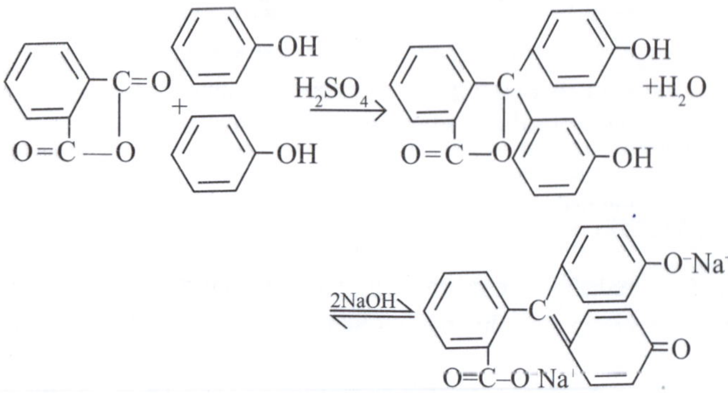

(pink or purple)

(v) Bromine water test: Take aqueous or alcoholic solution of the organic compound in a test-tube. Add excess of bromine water. A yellowish white precipitate is obtained.

(3) Tests for aldehyde group

(i) Schiff test: Take 5 ml of Schiff's reagent in a test-tube and shake with organic compound (do not heat). A pink colour is formed within two minutes.

(ii) Fehling's test: Fehling reagent comprises of two solutions, Fehling solution A and Fehling solution B. Fehling solution A is aqueous copper sulphate and Fehling solution B is alkaline sodium potassium tartarate (Rochelle salt). These two solutions are mixed in equal amounts before test.

$$R{-}CHO + 2CuO \rightarrow RCOO^- + Cu_2O\downarrow \text{ (Fehling's)}$$
(Red ppt.)

Ketones are not easily oxidised, thus they do not reduce Fehling's solution or Tollen's reagent. But α-hydroxy ketones (compounds containing the unit $-CH(OH){-}\underset{\underset{O}{\|}}{C}{-}R$) readily reduce the compounds which respond to the test.

(iii) Benedicts test: Benedict modified the original Fehling's test by using a single solution which is more convenient for the test. Benedict's solution is more stable than Fehling's reagent and can be stored for a long time. It is an alkaline solution containing a mixture of copper sulphate and sodium citrate ($2Na_3C_6H_5O_7 .11H_2O$).

$$\underset{\text{Copper sulfate pentahydrate}}{CuSO_4.5H_2O} + \underset{\text{sodium carbonate}}{Na_2CO_3} + \underset{\text{Sodium citrate}}{Na_3C_6H_5O_7} \longrightarrow \text{Benedict's Reagent}$$

In Benedict test also, Cu^{2+} ions are reduced to Cu^+ ions in the same manner as in the case of Fehling's reagent.

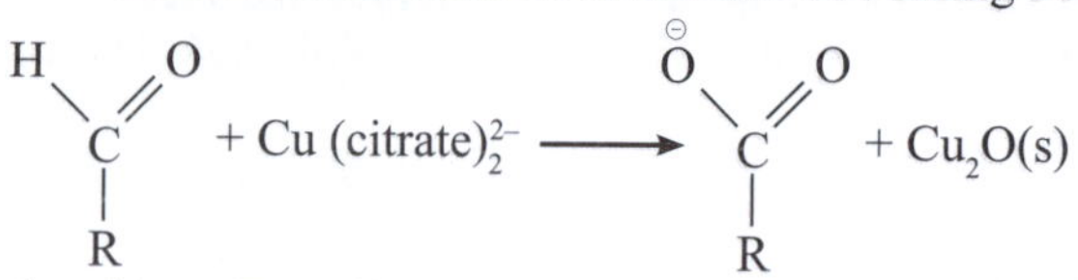

An aldose Benedict's reagent (blue solution) Carboxylate anion Brick-red precipitate

(iv) Tollen's test

Tollens' test, also known as silver-mirror test, because when silver ion is get reduce to silver it get precipitate and produce mirror like image.

Tollens' test is used as a test to distinguish between an aldehyde and a ketone. Tollens' reagent is a colorless, basic, aqueous solution containing silver ions coordinated to ammonia $[Ag(NH_3)^2]^+$. And consist of Ag^+ ions which reduce to Ag when treated with aldehyde, and aldehyde get oxidise to carboxylic acid.

Aqueous silver nitrate which consist of silver ion in +1 oxidation state is mixed with aqueous sodium hydroxide.

$$AgNO_3 + NaOH \rightarrow AgOH + NaNO_3$$
$$2AgOH \rightarrow Ag_2O + H_2O$$

Now aqueous ammonia is added drop-wise until the precipitated silver oxide completely dissolves.

$$Ag_2O + 4NH_3 + H_2O \rightarrow 2Ag(NH_3)_2^+ + 2OH^-$$

Tollens' reagent oxidizes an aldehyde into the corresponding carboxylic acid.

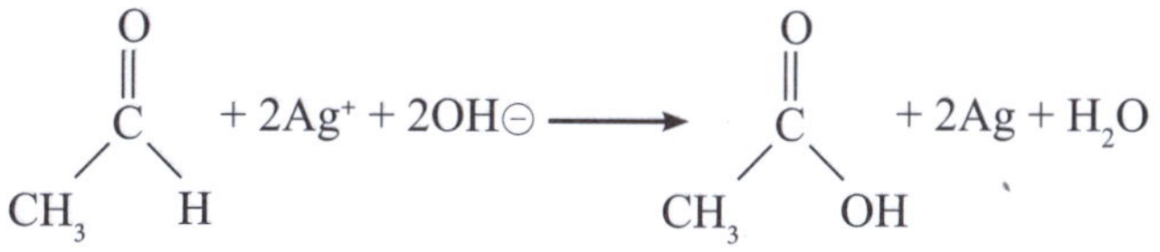

The reaction is accompanied by the reduction of silver ions in Tollens' reagent into metallic silver, which, if the test is carried out in a clean glass test tube, forms a mirror on the test tube.

(4) Tests for carboxylic group

(i) **Litmus test:** Place a small quantity of the organic compound or its aqueous solution on a piece of moist blue litmus paper. If the paper turns red, carboxylic group may be present.

(ii) **Sodium bicarbonate test:** Add a small quantity of the organic compound to about 5 mL sodium bicarbonate solution taken in a test-tube. Compound dissolves with brisk effervescences.

$RCOOH + NaHCO_3 \rightarrow RCOONa + H_2O + CO_2\uparrow$

(iii) **Ester test:** Take a small quantity of the organic compound and add 4–5 mL of alcohol in a test-tube. Add a few drops of conc. sulphuric acid and warm on a water bath. A fruity smell indicates the presence of an ester.

$$RCOOH + R'OH \xrightarrow[\Delta]{Conc.H_2SO_4} \underset{Ester}{RCOOR'} + H_2O$$

(5) Test for amino groups in organic compounds

Tests for primary amines($-NH_2$)

(i) **Carbylamine test:** Take about 0.2g of solid KOH in a dry test tube and add 2-3 ml of ethanol. Warm the contents until KOH dissolves. To this, add few drops of chloroform, and a pinch of given compound. A highly offensive smell is evolved due to formation of isocyanides.

$$RNH_2 + CHCl_3 + 3KOH \xrightarrow{Heat} R{-}NC + 3KCl + 3H_2O$$

This test is applicable to aliphatic and aromatic primary amines.

(ii) **Dye test:** This test is applicable only for aromatic primary amines. Dissolve a small amount of the organic compound in dil. HCl. Cool it in ice-cold water. Add cold solution of sodium nitrite with constant shaking and then add β-naphthol prepared in NaOH solution. An orange-red dye is obtained.

$NaNO_2 + HCl \rightarrow HNO_2 + NaCl$

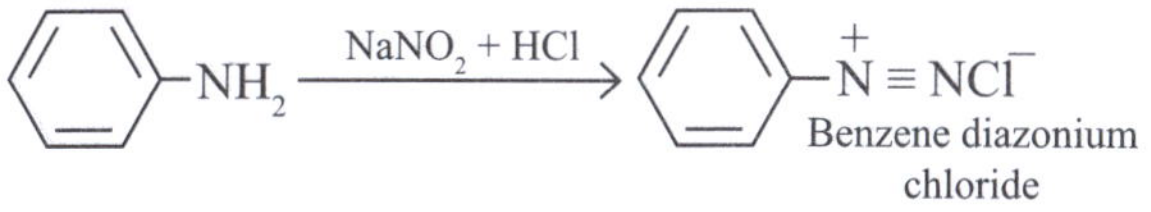

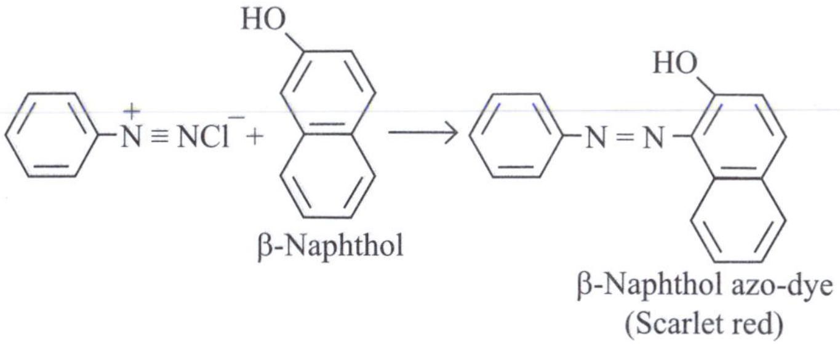

Tests for secondary amines (–NH–)

Liebermann's nitroso test: Both aliphatic and aromatic secondary amines react with HNO_2 to form N-nitrosoamines which are neutral yellow oily compounds. These on warming with a crystal of phenol and a few drops of conc. H_2SO_4 form a green solution which turns blue on adding NaOH and then red on dilution. This test is called **Liebermann's nitroso reaction and is used to distinguish secondary amines (aliphatic or aromatic) from primary and tertiary amines.**

Inorganic compounds

When a mixture containing definite proportions of potassium sulphate and aluminium sulphate or ferrous sulphate and ammonium sulphate is crystallised from its solution, a double salt is formed. The formation of double salt may be shown as follows:

$$K_2SO_4 + Al_2(SO_4)_3 + 24H_2O \longrightarrow K_2SO_4 \cdot Al_2(SO_4)_3 \cdot 24H_2O \text{ or } 2KAl(SO_4)_2 \cdot 12H_2O$$

(Potash alum)

$$FeSO_4 + (NH_4)_2SO_4 + 6H_2O \longrightarrow FeSO_4.(NH_4)_2SO_4.6H_2O$$

Ferrous ammonium sulphate (Mohr's salt)

Fe^{2+} and Al^{3+} ions undergo hydrolysis, therefore, while preparing aqueous solutions of ferrous sulphate and aluminium sulphate in water, 2–3 mL dilute sulphuric acid is added to prevent the hydrolysis of these salts.

Mohr's Salt ($FeSO_4(NH_4)_2SO_4.6H_2O$)

Preparation of Double Salt: Ferrous ammonium sulphate (Mohr's salt)

- Dissolve 3.5 g of ferrous sulphate and 1.7 g of ammonium sulphate (weighed separately), in 5 mL of distilled water contained in a 50 mL conical flask by heating. Add about 0.5 mL of dilute sulphuric acid to the flask and concentrate the solution by heating till the crystallization point is reached.
- Allow the mixture to cool to room temperature slowly.
- On cooling, light green crystals of ferrous ammonium sulphate separate out.
- Decant the mother liquor and wash the crystals by shaking with very small amounts of 1:1 cold water and alcohol mixture to remove sticking mother liquor.
- Separate the crystals by filtration wash with alcohol, dry between the folds of a filter paper and record the yield.

Potash Alum ($K_2SO_4 \cdot Al_2(SO_4)_3 \cdot 24H_2O$)

Preparation of Double Salt: Potassium Aluminium Sulphate (Potash Alum)

- Take 10 mL of distilled water in a 50 mL beaker and heat it to about 40°C. Dissolve 6.6 g of aluminium sulphate in it and add about 0.4 mL of dilute sulphuric acid.
- Weigh 2.4 g of powdered potassium sulphate and transfer it to the above solution.
- Heat the solution with constant stirring till potassium sulphate dissolves completely.
- Allow the solution to cool to room temperature slowly.

- On cooling, white crystals of potash alum separate out.
- Decant the mother liquor and wash the crystals by gently shaking with 1:1 cold water and alcohol mixture.
- Filter the crystals, dry these between the folds of a filter paper and note the yield.

Precautions

- Cool the solution slowly to get good crystals. Avoid rapid cooling.
- Do not disturb the solution while cooling.
- Avoid prolonged heating while preparing crystals of ferrous ammonium sulphate, as it may oxidise ferrous ions to ferric ions and change the stoichiometry of the crystals.

Organic Compounds

Acetanilide

Theory

The replacement of one hydrogen atom of the $-NH_2$ group of aniline by CH_3CO- group in the presence of glacial acetic acid gives acetanilide. In the laboratory, acetylation is usually carried out with acetic anhydride. Acetyl chloride may also be used for the purpose of acetylation if acetic anhydride is not available. Acetylation with CH_3COCl is usually carried out in the presence of pyridine.

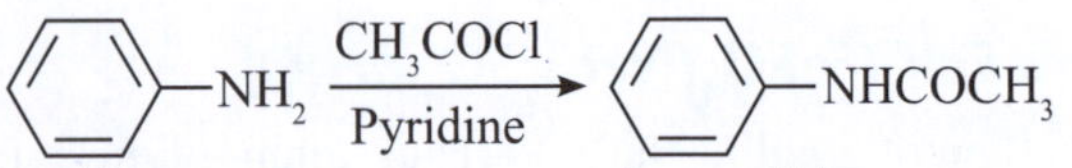

Aniline Acetanilide

Material Required

Funnel	One
Round bottomed flask (100 mL)	One
Beaker (250 mL)	One
Air condenser	One
Sand bath	One
Clamp and iron stand	One
Pumice stone	One
Melting point assembly	One
Aniline	5 mL
Acetic anhydride/Acetyl chloride	5 mL
Acetic acid/Pyridine	5 mL

Procedure

- Take 5 mL of aniline in a 100 mL round bottom flask and add acetylating mixture containing 5 mL acetic anhydride and 5 mL glacial acetic acid. Alternatively, you can use 5 mL of acetyl chloride and 5 mL of dry pyridine as the acetylating mixture.
- Fit an air condenser on the mouth of the round bottom flask after adding a few pumice stones and reflux the mixture gently for 10-15 minutes on a sand bath.
- Cool the reaction mixture and pour it slowly in 150-200 mL of ice cold water with stirring.
- Filter the solid, wash it with cold water and recrystallise a small amount of sample from hot water containing a few drops of methanol or ethanol.
- Report the yield and the melting point of the compound.

Precautions

- Handle acetic anhydride and acetyl chloride carefully as they cause irritation to the eyes and acetyl chloride also strongly fumes in air.
- Store acetylchloride under dry conditions.
- Handle pyridine with extreme caution. Dispense it in an efficient fume cupboard and wear disposable glasses while using it.
- Distil pyridine before use because it absorbs moisture and the reaction does not take place under moist conditions.
- Wash the solid 2-3 times with cold water till the filtrate is neutral to litmus.
- Determine the melting point of perfectly dried and recrystallised sample.

p-nitro acetanilide

Theory

p-Nitroacetanilide is prepared by the nitration of acetanilide by using a mixture of conc. nitric acid and conc. sulphuric acid as nitrating reagent. The mixture of the two acids releases nitronium ion (NO_2^+) which acts as an electrophile in the reaction.

$$HNO_3 + 2H_2SO_4 \longrightarrow NO_2^+ + H_3O^+ + 2HSO_4^-$$

Nitronium ion attacks the benzene ring containing anilide group, mainly at the para position to give p-nitroacetanilide as a major product. This is an example of aromatic electrophilic substitution reaction.

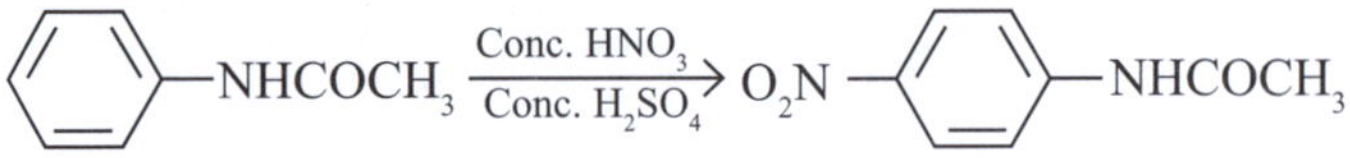

Acetanilide p-nitroacetanilide (Major product)

Material Required

Beaker (100 mL)	One
Funnel	One
Glass rod	One
Ice bath	One
Acetanilide	2 g
Glacial acetic acid	2 mL
Conc. H_2SO_4	5 mL
Conc. HNO_3	1.5 mL
Ice	As per requirement
Ethanol/methanol	As per requirement

Procedure

- Dissolve 2 g of acetanilide in 2 mL of glacial acetic acid taken in a 100 mL beaker.
- Add 4 mL of conc. H_2SO_4 gradually with stirring to the above mixture. The mixture becomes hot and clear solution is obtained. Cool the reaction mixture in an ice bath maintained at 0–5°C.

- Add a cold mixture of 1.0 mL conc. HNO_3 and 1.0 mL conc. H_2SO_4 to the viscous reaction mixture drop by drop with constant stirring, so that the temperature of the mixture does not rise above 10°C.
- Remove the beaker from the ice bath and allow the reaction mixture to attain room temperature. Let it stand at room temperature for about 30 minutes. Stir the reaction mixture continuously and pour it on 100g of crushed ice.
- Stir the mixture well and filter the compound so obtained.
- Wash the compound with cold water and dry it.
- Recrystallise a small amount of the pale yellow solid from alcohol. Colourless crystals of p-nitroacetanilide are obtained. Yellow ortho-nitroacetanilide formed in the small amount remains dissolved in the mother liquor.
- Record the yield and melting point of the pure compound.

Precautions

- Do not allow the temperature of the reaction mixture to exceed 10°C during addition of nitrating mixture.
- Add mixture of concentrated nitric acid and sulphuric acid into the solution of acetanilide slowly and carefully.

Aniline Yellow

Theory

p-aminoazobenzene can be prepared in a good yield by rearrangement reaction of diazoaminobenzene with a small quantity of aniline hydrochloride in the presence of aniline as solvent. The chemistry of this conversion is as follows:

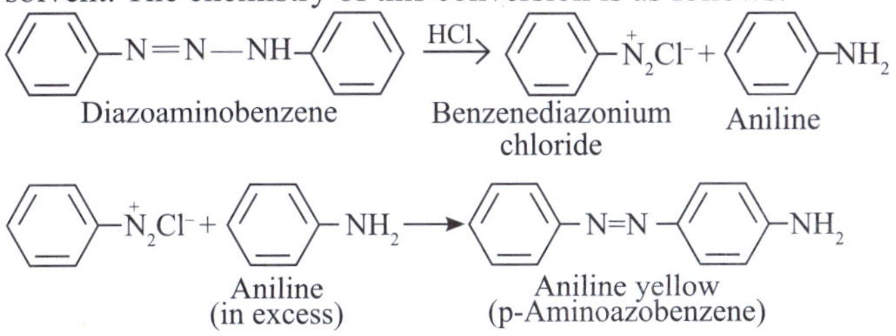

The above reaction is carried out only in weakly acidic conditions.

Material Required

- Conical Flask (100 mL) One
- Thermometer One
- Funnel One
- Melting point assembly One
- Waterbath One
- Diazoaminobenzene 3 g
- Aniline 7 mL
- Aniline hydrochloride 1.5 g
- Glacial acetic acid 9 mL
- Carbon tetrachloride 9 mL

Procedure

- Dissolve 3 g of finely powdered diazoaminobenzene in 7 mL of aniline in a 100 mL conical flask.
- Add 1.5 g of finely powdered aniline hydrochloride to the above mixture.
- Warm the mixture with occasional shaking on a water bath at about 40-45°C, for a period of about one hour.
- Remove the flask from the water bath and allow the reaction mixture to stand at room temperature for about 30 minutes.
- Add 9 mL of glacial acetic acid diluted with an equal volume of water and shake the reaction mixture thoroughly to convert excess aniline to its acetate, which is water-soluble.
- Allow the mixture to stand for 15 minutes with occasional stirring.
- Filter p-aminoazobenzene, wash with a little cold water and dry between the folds of a filter paper.
- Recrystallise a small portion of crude p-aminoazobenzene from carbon tetrachloride.
- Report the yield and melting point of p-aminoazobenzene

Precautions

- Maintain the temperature of the reaction mixture at about 40-50°C.
- Wash the crude product with a small volume of water repeatedly.
- Note the melting point of the perfectly dry sample.

Iodoform

Theory

The organic compounds which contain either $CH_3—\underset{\text{OH}}{\underset{|}{CH}}—$ group or $CH_3—\underset{\text{O}}{\underset{\|}{C}}—$ group react with potassium iodide and sodium hypochlorite to produce iodoform. Thus, iodoform can be prepared by reacting compounds like ethanol, 2-propanol or acetone with potassium iodide and sodium hypochlorite. Sodium hypochlorite oxidises potassium iodide to potassium hypoiodite. The potassium hypoiodite in turn oxidises ethanol to ethanal and converts ethanal to iodal. Sodium hydroxide present in the solution hydrolyses iodal to iodoform and sodium formate.

When acetone is used to prepare iodoform, then no initial oxidation occurs. Acetone reacts with sodium hypoiodite to form tri-iodo-acetone, which undergoes hydrolysis to yield iodoform.

The chemical reactions taking place are:

(i) Using ethanol

$$\underset{\text{Ethanol}}{CH_3CH_2OH} \xrightarrow{\text{NaOI}} \underset{\text{Ethanal}}{CH_3CHO} \xrightarrow{\text{NaOI}} \underset{\text{Iodal}}{CI_3CHO}$$

$$CI_3CHO + NaOH \longrightarrow \underset{\text{Iodoform}}{CHI_3} + HCOONa$$

(ii) Using acetone

$$\underset{\text{Acetone}}{CH_3COCH_3} \xrightarrow{\text{NaOI}} \underset{\text{Tri-iodoacetone}}{CI_3COCH_3}$$

$$CI_3COCH_3 + NaOH \longrightarrow \underset{\text{Iodoform}}{CHI_3} + CH_3COONa$$

Iodoform is obtained more readily and in a better yield from acetone than from ethanol.

Apparatus

Beaker (25 mL), Conical flask (25 mL), buchner funnel, water pump, filtration flask, water-bath, tripod stand, burner, etc.

Chemical required

Ethyl alcohol = 3 mL

Or

Acetone = 2 mL

10% Sodium hydroxide solution = 30 mL

10% Potassium iodide solution = 80 mL

5% Sodium hypochlorite solution = 80 mL

Procedure

Preparation of Crude Iodoform

1. Take 2.0 ml of acetone, 80 ml of 10% KI solution and 30 mL of 10% NaOH solution in a 500 ml conical flask.
2. To the above solution add 80 ml of 5% sodium hypochlorite solution slowly with constant shaking, Bright yellow precipitate of iodoform separates out.
3. Allow the mixture to stand at room temperature, when fine yellow coloured crystals of iodoform gradually separates out.
4. Filter the yellow solid using a Buchner funnel and a water pump and wash it with cold water.
5. Recrystallization of iodoform
   Transfer the precipitate in a clean 100 ml round bottom flask containing small quantity of rectified spirit. Fit the water condenser on the flask and heat it on a water bath. Add rectified spirit through water condenser so that the iodoform is completely dissolved. Filter the hot solution quickly from the flask into a beaker using a fluted filter paper. Cool the filtrate in ice bath. The iodoform will crystallize soon. Filter the solution. Dry the crystals by pressing between filter papers. Weight and record its yield.

Precautions

1. Add 5% sodium hypochlorite solution in small lots with constant shaking.
2. Wash iodoform precipitates with cold water.
3. Use buchner funnel for filtration to get good yield.

Titration

Titrimetric analysis involves determination of the volume of a solution of accurately known concentration, which is required to react quantitatively with the measured volume of the solution of a substance, concentration of which is to be determined. The solution of accurately known concentration is called standard solution. The mass of the substance dissolved in the solution of unknown concentration is calculated from the volume of the standard solution used, the chemical equation and the relative molecular masses of the reacting compounds. The reagent of known concentration is called titrant and the substance being titrated is termed as titrand.

To carry out titrimetric analysis, standard solution is usually added from the long graduated tube called burette. The process of adding the standard solution to the solution of unknown concentration until the reaction is just complete is called titration. The point at which reaction is completed is called equivalence point or the theoretical or stoichiometric end point. It is not possible all the time to take standard solution in the burette.

Acids and Bases

Titrimetric analysis can be performed on a variety of reactions. Titrations of bases and acids take place. In these titrations, standard solutions of acids (acidimetry) and bases (alkalimetry) are utilized. The concentration of a solution is expressed in terms of the molarity in quantitative estimate via titrimetric analysis. It is the amount of solute dissolved in one liter of solution.

Indicators in Acid Base Titration

Acid base indicators are sensitive to pH change. For most acid base titrations, it is possible to select indicators which exhibit colour change at pH close to the equivalence point. We will discuss here about only two indicators - phenolphthalein and methyl orange.

Phenolphthalein

Phenolphthalein is a weak acid, therefore it does not dissociate in the acidic medium and remains in the unionised form, which is colourless.

HPh	$\rightleftharpoons$	$H^+ + Ph^-$
Unionised		Ionised
Colourless		Pink

Ionised and unionised forms of phenophthalein are given below:

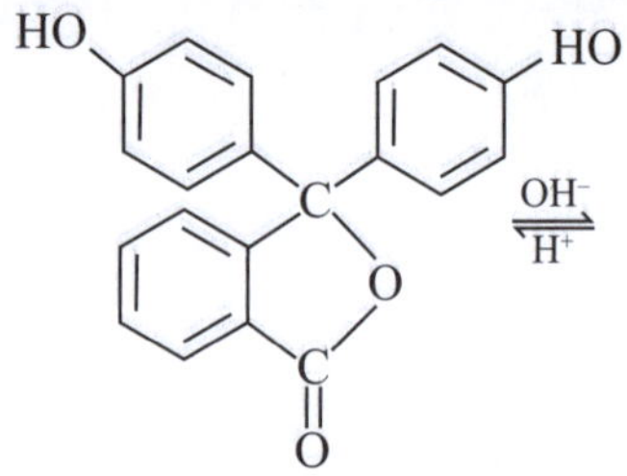

(Colourless in acid)

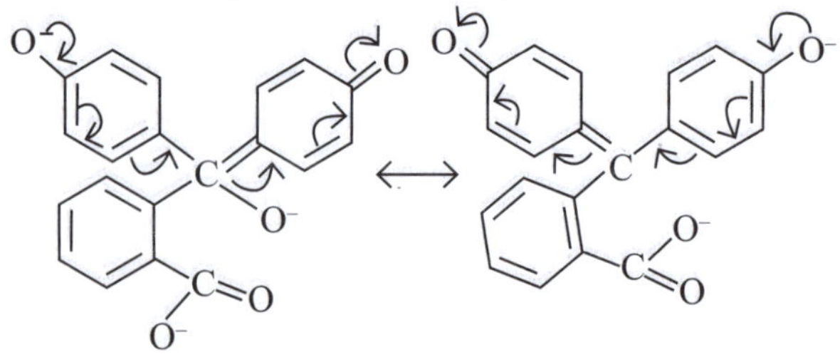

(Pink in alkali)

Phenolphthalein in acidic and basic medium

In the acidic medium, equilibrium lies to the left. In the alkaline medium, the ionisation of phenolphthalein increases considerably due to the constant removal of H^+ ions released from HPh by the OH^- ions from the alkali. So the concentration of Ph^- ion increases in the solution, which imparts pink colour to the solution.

$$HPh \rightleftharpoons H^+ + Ph^-$$
$$NaOH \rightarrow Na^+ + OH^-$$
$$H^+ + OH^- \rightarrow H_2O$$

For a weak acid vs strong alkali titration, phenolphthalein is the most suitable indicator. This is so because the last drop of added alkali brings the pH of the solution in the range in which phenolphthalein shows sharp colour change.

Methyl Orange

Methyl orange is a weak base and is yellow in colour in the unionised form. Sodium salt of methyl orange is represented as follows:

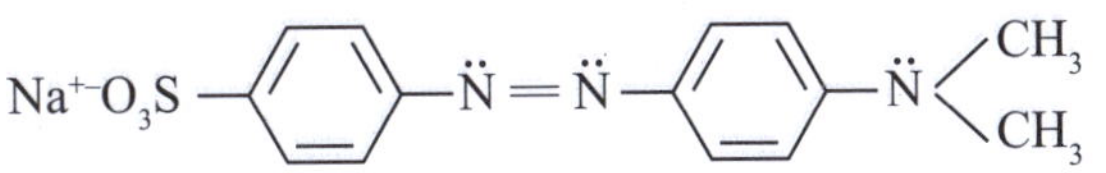

The anion formed from the indicator is an active species, which on accepting a proton (i.e acting as Bronsted Lowry base) changes from the benzenoid form to the quinonoid form. The quinonoid form is deeper in colour and thus is responsible for the colour change at the end point. This is illustrated in the following manner:

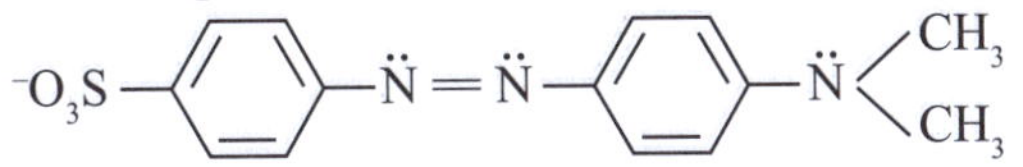

Benzenoid form of the anion
(in basic solution)
(Yellow in colour)

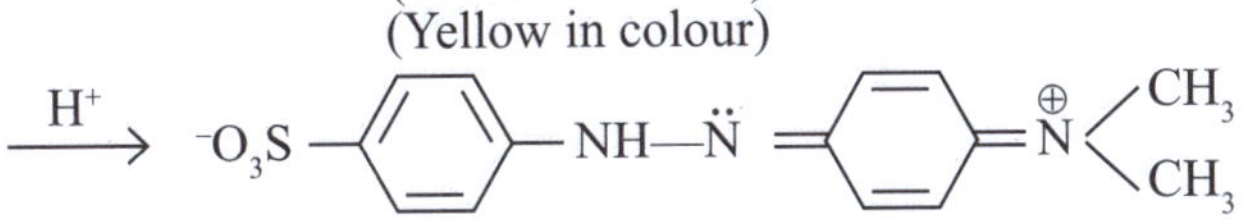

(Bronsted Lowry base)

Quinonoid form of the anion
(in acidic solution)
(Pinkish red in colour)

Choice of Indicator

Methyl orange is used as an indicator in the titration of a strong acid and a weak base. When titration between strong base and weak acid is to be performed, phenolphthalein is an excellent indicator. In this example, alkali is dispensed from the burette, and acid is taken in a tiration flask. The colour of the solution in the titration flask shifts from colourless to pink.

The human eye can easily detect this change in colour. If we put alkali in the titration flask, the colour will shift from pink to colourless, and the accuracy in noting the colour change will be reduced. In the titration of strong acid versus strong base anyone of the above indicators can be used. For the titration of weak acid vs weak base, no indicator is available.

Titrimetric Analysis (Redox Reactions)

The oxidation and reduction reactions in aqueous solutions involve the transfer of electrons from one species to another. In the oxidation of a substance, electron(s) is (are) transfered from the species and in reduction, electron(s) is (are) gained by the species. Oxidation and reduction reactions occur simultaneously. A reaction, which involves simultaneous oxidation and reduction, is called a redox reaction. The titrations involving redox reaction are called redox titrations. You know that in acid-base titrations, indicators which are sensitive to pH change are employed to note the end point. Similarly, in redox titrations there is a change in oxidation potential of the system.

The indicators used in redox reactions are sensitive to change in oxidation potential. The ideal oxidation-reduction indicators have an oxidation potential intermediate between the values for the solution being titrated and the titrant and these show sharp readily detectable colour change.

Oxalic Acid vs $KMnO_4$

Theory

In the present experiment, potassium permanganate acts as a powerful oxidising agent. Although $KMnO_4$ acts as an oxidising agent in alkaline medium also, for quantitative analysis mostly acidic medium is used. The oxidising action of $KMnO_4$ in the acidic medium can be represented by the following equation:

$$MnO_4^- + 8H^+ + 5e^- \rightarrow Mn^{2+} + 4H_2O$$

The acid used in this titration is dilute sulphuric acid. Nitric acid is not used as it is itself an oxidising agent and hydrochloric acid is usually avoided because it reacts with $KMnO_4$ according to the equation given below to produce chlorine and chlorine which is also an oxidising agent in the aqueous solution.

$$2KMnO_4 + 16HCl \rightarrow 2KCl + 2MnCl_2 + 5Cl_2 + 8H_2O$$

Since, oxalic acid acts as a reducing agent, it can be titrated against potassium permanganate in the acidic medium according to the following equation:

Reactions

A. Chemical equations

Reduction half reaction:

$$2KMnO_4 + 3H_2SO_4 \longrightarrow K_2SO_4 + 2MnSO_4 + 3H_2O + 5[O]$$

Oxidation half reaction: $H_2C_2O_4 + [O] \xrightarrow{60°C} 2CO_2 + H_2O] \times 5$

$$2KMnO_4 + 3H_2SO_4 + 5H_2C_2O_4 \longrightarrow K_2SO_4 + 2MnSO_4 + 8H_2O + 10\,CO_2$$

B. Ionic equation

Reduction half reaction:

$$MnO_4^- + 5e^- + 8H^+ \longrightarrow Mn^{2+} + 4H_2O] \times 2$$

Oxidation half reaction:

$$C_2O_4^{2-} \longrightarrow 2CO_2 + 2e^-] \times 5$$

$$2MnO_4^- + 5C_2O_4^{2-} + 16H^+ \longrightarrow 2Mn^{2+} + 10CO_2 + 8H_2O$$

In these equations, MnO_4^- is reduced to Mn^{2+} and $C_2O_4^{2-}$ is oxidised to CO_2. The oxidation number of carbon in $C_2O_4^{2-}$ changes from +3 to +4.

In these titrations, potassium permanganate acts as a self-indicator. Initially colour of potassium permanganate is discharged due to its reduction by oxalic acid. After complete consumption of oxalate ions, the end point is indicated by the appearance of a light pink colour produced by the addition of a little excess of unreacted potassium permanganate. Further, during the titration of oxalic acid against potassium permanganate, warming of oxalic acid solution (50°– 60°C) along with dilute H_2SO_4 is required.

Mohr's Salt vs $KMnO_4$

Theory

Like oxalic acid, ferrous ammonium sulphate also acts as a reducing agent in the titration against potassium permanganate. The reaction which takes place is given below:

(1) Chemical equation

Reduction half reaction:

$2KMnO_4 + 3H_2SO_4 \rightarrow K_2SO_4 + 2\,MnSO_4 + 3H_2O + 5[O]$

Oxidation half reaction:

$[2\,FeSO_4(NH_4)_2SO_4 \cdot 6H_2O + H_2SO_4 + [O] \rightarrow Fe_2(SO_4)_3 + 2(NH_4)_2SO_4 + 13H_2O] \times 5$

$2KMnO_4 + 8H_2SO_4 + 10FeSO_4\,(NH_4)_2SO_4.6H_2O \rightarrow K_2SO_4 + 2MnSO_4 + 5Fe_2(SO_4)_3 + 10(NH_4)_2SO_4 + 68H_2O$

(2) Ionic equation

Reduction half reaction: $MnO_4^- + 5e^- + 8H^+ \rightarrow Mn^{2+} + 4H_2O$

Oxidation half reaction: $[Fe^{2+} \rightarrow Fe^{3+} + e^-] \times 5$

$MnO_4^- + 5Fe^{2+} + 8H^+ \rightarrow Mn^{2+} + 5Fe^{3+} + 4H_2O$

The oxidation number of iron in Mohr's salt is +2. Iron is oxidised during the reaction and its oxidation number changes from +2 to +3. In this titration heating of ferrous ammonium sulphate solution is not required because reaction rate is very high even at room temperature. Also, at high temperatures, ferrous ions may be oxidised to ferric ions by oxygen of air and error may be introduced in the experiment.

Qualitative Analysis

Introduction

The detection of cations (basic radicals) and anions (acidic radicals) in a salt or in a mixture is known as Qualitative Analysis. The systematic procedure for qualitative analysis of an inorganic salt involves the following steps:

(1) Preliminary tests

1. Physical appearance (colour and smell).
2. Dry heating test.
3. Flame test.
4. Borax bead test.
5. Charcoal cavity test.
6. Charcoal cavity and cobalt nitrate test.

(2) Analysis of anions (acidic radicals).

(3) Analysis of cations (basic radicals).

Preliminary test

1. **Physical appearance (smell):** Take a pinch of the sample of salt between your fingers and rub with a drop of water.

Inference	Smell
NH_4^+	Ammonical smell
CH_3COO^-	Vinegar like smell
S^{2-}	Smell like that of rotten eggs

2. **Dry Heating Test:** This test is performed by heating a small amount of mixture in a dry test tube. Quite valuable information can be generated by carefully performing and noting the observations here. On heating, some salts undergo decomposition, thus evolving the gases or may undergo characteristic changes in the colour of residue. These observations are tabulated below along with the inferences that you can draw.

Gas evolved

(1) Colourless and odourless gas

CO_3^{2-} ion

CO_2 gas – Turns lime water milky.

(2) Colourless gas with odour

- **Hydrated S^{2-} or S^{2-} ion**

  H_2S gas – Smells like rotten eggs, turns lead acetate paper black.

- **SO_3^{2-} ion**

  SO_2 gas – Characteristic suffocating smell of burning sulphur turns acidified potassium dichromate solution or paper green.

- **Cl^- ion**

  HCl gas – Pungent smell, white fumes with ammonia, white precipitate with silver nitrate solution.

- **NH_4^+ ion**

  NH_3 gas – Characteristic smell, turns Nessler's solution brown.

(3) Coloured gas – Pungent smell

- **NO_2^- or NO_3^- ion**

  NO_2 gas – Reddish brown, turns ferrous sulphate solution brownish black.

- **Cl^- ion**

  Cl_2 gas – Greenish yellow, turns starch iodide paper blue.

- **Br^- ion**

  Br_2 vapours – Reddish brown, turns starch paper orange red.

- **I^- ion**

  I_2 vapours – Dark violet, turns starch paper blue.

Residue

(1) Yellow when hot, white when cold.
Eg: Zn^{2+}

(2) Brown when hot and yellow when cold.
Eg: Pb^{2+}

(3) Original salt blue in color becomes white on heating.
Eg: Hydrated $CuSO_4$ indicated

(4) Coloured salt becomes brown or black on heating.
Eg: Co^{2+}, Cu^{2+}, Mn^{2+}

3. **Flame test:** The chlorides of the metals are more volatile as compared to other salts and these are prepared in situ by mixing the compounds with a little concentrated hydrochloric acid. On heating, in a non-luminous Bunsen flame, they are volatilized and impart a characteristic colour to the flame as these absorb energy from the flame and transmit the same as characteristic colour.

Metal	Colour of flame
Li	Crimson red
Na	Golden yellow
K	Violet

Ca	Brick red
Sr	Crimson red
Ba	Apple green
Cu	Greenish blue

4. **Borax Bead test:** This test is employed only for coloured salts because borax reacts with metal salts to form metal borates or metals, which have characteristic colours. On heating, borax ($Na_2B_4O_7 \cdot 10H_2O$) forms a colourless glassy bead of $NaBO_2$ and B_2O_3.

$$Na_2B_4O_7 \cdot 10H_2O \xrightarrow{\Delta} Na_2B_4O_7 \xrightarrow{\Delta} 2NaBO_2 + B_2O_3$$

On treatment with metal salt, boric anhydride forms metaborate of the metal which gives different colours in oxidising and reducing flame. For example, in the case of copper sulphate, following reactions occurs.

$$CuSO_4 + B_2O_3 \xrightarrow{\text{Non-luminous flame}} \underset{\substack{\text{Cupric metaborate} \\ \text{Blue-green}}}{Cu(BO_2)_2} + SO_3$$

Two reactions may take place in the reducing flame:

(1) The blue Cu $(BO_2)_2$ is reduced to colourless cuprous metaborate as follows:

$$2Cu(BO_2)_2 + 2NaBO_2 + C \xrightarrow{\text{Luminous flame}} 2CuBO_2 + Na_2B_4O_7 + CO$$

(2) Cupric metaborate may be reduced to metallic copper and the bead appears red and opaque.

$$2Cu(BO_2)_2 + 4NaBO_2 + 2C \xrightarrow{\text{Luminous flame}} 2Cu + 2Na_2B_4O_7 + 2CO$$

Metal	Colour in oxidising (non-luminous) flame	
	When Hot	When Cold
Cu	Green	Blue
Fe	Yellowish brown	Pale yellow/Yellow
Cr	Green	Green
Co	Blue	Blue
Mn	Light violet/Amethyst	Light violet/Amethyst
Ni	Violet	Brown/Reddish brown

Metal	Colour in reducing (luminous) flame	
	When Hot	When Cold
Cu	Colourless	Brown red
Fe	Bottle green	Bottle green
Cr	Green	Green
Co	Blue	Blue
Mn	Colourless	Colourless
Ni	Grey	Grey

5. **Charcoal Cavity Test:** In this test metallic carbonates when heated in a charcoal cavity decompose to give corresponding oxides. The oxides appear as coloured incrustation or residue in the cavity.

In some cases, the oxides formed partially undergo reduction to the metal producing metallic beads.

For Example:

(1) $ZnSO_4 + Na_2CO_3 \xrightarrow{\text{Heat}} ZnCO_3 + Na_2SO_4$

$ZnCO_3 \xrightarrow{\text{Heat}} ZnO$ (Yellow when hot, white when cold) $+ CO_2 \uparrow$

(2) $CuSO_4 + Na_2CO_3 \xrightarrow{\text{Heat}} CuCO_3 + Na_2SO_4$

$CuCO_3 \xrightarrow{\text{Heat}} CuO + CO_2 \uparrow$

$CuO + C \xrightarrow{\text{Heat}} Cu$ (Reddish scales) $+ CO \uparrow$

Inference from the charcoal cavity test	
Observations	**Inference**
Yellow residue when hot and grey metal when cold	Pb^{2+}
White residue with the odour of garlic	As^{3+}
Brown residue	Cd^{2+}
Yellow residue when hot and white when cold	Zn^{2+}

6. **Cobalt Nitrate Test:** In case the residue is white in colour after charcoal cavity test, add a drop of cobalt nitrate $[Co(NO_3)_2]$ in the charcoal cavity. A drop of water is then added and the mass is heated in an oxidising flame. It is cooled and one or two drops of cobalt nitrate solution is added and then again heated in the oxidising flame. Different metal salts give different coloured mass.

$ZnSO_4 + Na_2CO_3 \rightarrow ZnCO_3 + Na_2SO_4$;

$ZnCO_3 \rightarrow ZnO + CO_2$

$2Co(NO_3)_2 \rightarrow 2CoO + 4NO_2 + O_2$;

$ZnO + CoO \rightarrow ZnO.CoO$ (or $CoZnO_2$)

(Rinmann's green)

S. No.	Metal	Colour of the Mass
1.	Zinc	Rinmann green
2.	Aluminium	Thenard blue
3.	Magnesium	Pink
4.	Tin	Bluish-green

Analysis of Anions (Acidic Radicals)

Methods available for the detection of anions are not as systematic as those used for the detection of cations. Futhermore, anions are classified essentially on the basis of process employed.

Class A

Includes anions that are identified by volatile products obtained on treatment with acids. It is further divided into two sub groups:

(i) Gases evolved with dil HCl/dil H_2SO_4.

Example: CO_3^{2-}, SO_3^{2-}, S^{2-}, NO_2^-

(ii) Gases or acid vapours evolved with conc. H_2SO_4.

Example: Cl^-, Br^-, I^-, NO_3^-

Preparation of Solution for the Identification of Anions

1. **Preparation of Water Extract (W.E.)**

   All common acetates, nitrites, nitrates and thiosulphates are soluble in water. Confirmatory tests for these anions can be performed with the water extract of the mixture. Water extract can be prepared by boiling 1-2 g of the mixture with 10-15 cm^3 distilled water in a boiling tube for a minute or two. Residue, if any, is filtered. The filtrate is called water extract (W.E.).

2. **Preparation of Sodium Carbonate Extract (S.E.)**

   If the mixture is found to be insoluble in water, it is boiled with saturated sodium carbonate solution. This treatment converts the anions present in mixture into soluble sodium salts as a result of double decomposition. E.g.,

   $$BaCl_2 + Na_2CO_3 \rightarrow \underset{\text{Insoluble}}{BaCO_3\downarrow} + \underset{\text{Soluble}}{2NaCl}$$

   For preparing sodium carbonate extract, take 0.5-1.0 g of powdered mixture, 1.0-2.0 g of sodium carbonate and 5-10 cm^3 of distilled water in a boiling tube or a 50 ml beaker. Heat with stirring for 5-10 minutes. Cool the contents and filter. The filtrate is called sodium carbonate extract (S.E.). This extract is used for confirming the presence of most anions. Confirmation of CO_3^{2-} is done by using aqueous solution of the salt or by using solid salt as such because sodium carbonate extract contains carbonate ions.

Individual tests of Anions

Class 'A' Radicals

Gases evolved with dil HCl/dil H_2SO_4.

1. **Carbonate (CO_3^{2-})**

   (i) **Dilute HCl:** A colourless, odourless gas is evolved with brisk effervescence due to the evolution of carbon dioxide.

   $$CaCO_3 + H_2SO_4 \rightarrow CaSO_4 + H_2O + CO_2\uparrow$$

   Lime water/Baryta water ($Ba(OH)_2$) test

   The gas gives white turbidity with lime water and baryta water.

   $$CO_2 + Ca^{2+} + 2OH^- \longrightarrow CaCO_3\downarrow + H_2O$$

   $$CO_2 + Ba^{2+} + 2OH^- \longrightarrow BaCO_3\downarrow + H_2O$$

   On prolonged passage of carbon dioxide in lime water, the turbidity slowly disappears due to the formation of soluble hydrogen carbonate.

   $$CaCO_3\downarrow + CO_2 + H_2O \longrightarrow Ca(HCO_3)_2$$

   The following tests are then performed with aqueous salt solution.

Note:

- Mercury (II) chloride does not form precipitate with hydrogen carbonate ions, while in a solution of normal carbonates a reddish–brown precipitate of basic mercury(II) carbonate ($3HgO\cdot HgCO_3 = Hg_4O_3CO_3$) is formed.

  $$CO_3^{2-} + 4Hg^{2+} + 3H_2O \rightarrow Hg_4O_3CO_3\downarrow + 6H^+$$

- Lime water test is shown by SO_2 and CO_2 both but CO_2 does not turn filter paper soaked in acidified $K_2Cr_2O_7$, green.
- Soluble bicarbonates give white precipitate with $MgSO_4$ (aq)/ $MgCl_2$(aq) only on heating.

  $$Mg^{2+} + 2HCO_3^- \rightarrow Mg(HCO_3)_2 \xrightarrow{\Delta} MgCO_3\downarrow + H_2O + CO_2$$

   (ii) **Barium chloride or Calcium chloride solution:** White ppt of barium or calcium carbonate is obtained, which is soluble in mineral acid.

   $$CO_3^{2-} + Ba^{2+} \longrightarrow BaCO_3\downarrow$$

   $$CO_3^{2-} + Ca^{2+} \longrightarrow CaCO_3\downarrow$$

   (iii) **Silver nitrate solution:** White ppt of silver carbonate is obtained.

   $$CO_3^{2-} + 2Ag^+ \longrightarrow \underset{\text{(Pale yellow)}}{Ag_2CO_3\downarrow}$$

   The ppt. so obtained is soluble in nitric acid and in ammonia, the ppt. becomes yellow or brown on addition of excess reagent and same may also happened if the mix is boiled due to the formation of silver oxide.

2. **Sulphites (SO_3^{2-})**

   (i) **Dilute HCl or Dilue H_2SO_4:** Decomposes with the evolution of sulphur dioxide.

   $$SO_3^{2-} + 2H^+ \longrightarrow SO_2 + H_2O$$

   The gas has suffocating odour of burning sulphur.

   (ii) **Acidified potassium dichromate solution:** The gas turns filter paper moistened with acidifed potassium dichromate solution, green, due to the formation of Cr^{3+} ions.

   $$3SO_2 + K_2Cr_2O_7 + H_2SO_4 \longrightarrow K_2SO_4 + \underset{\text{(Green)}}{Cr_2(SO_4)_3} + H_2O$$

   (iii) **Lime water:** On passing the gas through the lime water, a milky ppt is formed.

   $$SO_2 + Ca(OH)_2 \longrightarrow \underset{\text{(Milky)}}{CaSO_3\downarrow} + H_2O$$

   Precipitate dissolves on prolonged passage of the gas, due to the formation of soluble hydrogen sulphite ions.

   $$CaSO_3\downarrow + SO_2 + H_2O \longrightarrow Ca(HSO_3)_2$$

   (iv) **Barium chloride or Strontium chloride solution:** Salt solution gives white ppt of barium or strontium sulphite.

   $$SO_3^{2-} + Ba^{2+} \rightarrow \underset{\text{(white)}}{BaSO_3\downarrow}$$

   $$SO_3^{2-} + Sr^{2+} \rightarrow \underset{\text{(white)}}{SrSO_3\downarrow}$$

3. **Sulphide (S^{2-})**

   (i) **Dil HCl or Dil H_2SO_4:** A colourless gas with a smell of rotten eggs (H_2S) is evolved.

   $$S^{2-} + 2H^+ \longrightarrow H_2S\uparrow$$

   The gas turns lead acetate paper black.

   $$(CH_3COO)_2Pb + H_2S \longrightarrow \underset{\text{(Black)}}{PbS\downarrow} + 2CH_3COOH$$

(ii) Salt solution gives yellow ppt. with $CdCl_2$

$$Na_2S + CdCl_2 \longrightarrow \underset{(Yellow)}{CdS} \downarrow + 2NaCl$$

(iii) Silver nitrate solution: Black ppt. of silver sulphide insoluble in cold but soluble in hot dil nitric acid.

$$S^{2-} + 2Ag^+ \longrightarrow \underset{(greyish\ black)}{Ag_2S} \downarrow$$

(iv) Sodium nitroprusside solution: Turns sodium nitroprusside solution purple

$$Na_2S + Na_2[Fe(CN)_5NO] \rightarrow Na_4[Fe(CN)_5NOS]$$

4. Nitrites (NO_2^-)

(i) Dil. HCl or Dil. H_2SO_4: Adding to solid nitrite in cold yield pale blue liquid (due to the presence of free nitrous acid HNO_2 or its anhydride N_2O_3) and the evolution of brown fumes of nitrogen dioxide, the latter being largely produced by combination of nitric oxide with the oxygen of the air.

$$NO_2^- + H^+ \longrightarrow HNO_2$$

$$2HNO_2 \longrightarrow H_2O + N_2O_3$$

$$3HNO_2 \longrightarrow HNO_3 + 2NO\uparrow + H_2O$$

$$2NO\uparrow + O_2\uparrow \longrightarrow 2NO_2\uparrow$$

Following tests are performed with an aqueous salt solution.

(ii) Silver nitrate solution: White crystalline ppt is obtained.

$$NO_2^- + Ag^+ \rightarrow AgNO_2\downarrow$$

(iii) Turns acidified KI - starch paper blue.

$$2KI + 2NO_2 \longrightarrow 2KNO_2 + I_2\uparrow$$

$$Starch + I_2 \longrightarrow Blue\ complex$$

(iv) Brown ring test: When the nitrite solution is added carefully to a conc. solution of iron (II) sulphate, acidified with dil. acetic acid or with dilute sulphuric acid, a brown ring, due to the formation of $[Fe(H_2O)_5NO]SO_4$ at the junction of the two liquids is formed.

$$NO_2^- + CH_3COOH \longrightarrow HNO_2 + CH_3COO^-$$

$$3HNO_2 \longrightarrow H_2O + HNO_3 + 2NO\uparrow$$

$$Fe^{2+} + SO_4^{2-} + NO\uparrow \longrightarrow [Fe(H_2O)_5NO]SO_4$$

Gases or acid vapours evolved with conc. H_2SO_4.

1. Chloride (Cl^-)

(i) Conc. H_2SO_4: Decomposes with the evolution of HCl.

$$Cl^- + H_2SO_4 \rightarrow HCl + HSO_4^-$$

Gas so produced:

1. Turns blue litmus paper red.
2. Gives white fumes of NH_4Cl when a glass rod moistened with ammonia solution is brought to the mouth of test tube.

(ii) Manganese dioxide and conc. sulphuric acid: If a salt gives effervescence on heating with conc. H_2SO_4 and MnO_2 and a light greenish yellow pungent gas is evolved, this indicates the presence of Cl^- ions

$$MnO_2 + 2H_2SO_4 + 2Cl^- \rightarrow Mn^{2+} + \underset{(Yellowish\ green)}{Cl_2}\uparrow + 2SO_4^{2-} + 2H_2O$$

The following test are performed with the salt soluton:

(iii) Silver nitrate solution: White, curdy ppt. of AgCl is insoluble in water and dil nitric acid, but soluble in dilute ammonia solution.

$$Cl^- + Ag^+ \rightarrow AgCl\downarrow$$

$$AgCl\downarrow + 2NH_3 \rightarrow [Ag(NH_3)_2]Cl$$

White ppt. reappears on acidifying.

$$[Ag(NH_3)_2]Cl + 2H^+ \rightarrow AgCl + NH_4^+$$

(iv) Lead acetate solution: White ppt of lead chloride is formed.

$$2Cl^- + Pb^{2+} \rightarrow \underset{(white)}{PbCl_2}\downarrow$$

(v) Chromyl chloride test: When a mix containing chloride ion is heated with $K_2Cr_2O_7$ and conc. H_2SO_4 organge red fumes of chromyl chloride (CrO_2Cl_2) are formed.

$$K_2Cr_2O_7 + 4NaCl + 6H_2SO_4 \rightarrow 2KHSO_4 + 4NaHSO_4 + 2CrO_2Cl_2\uparrow + 3H_2O$$

orange - red fumes

When chromyl chloride vapours are passed into sodium hydroxide, a yellow solution of sodium chromate is formed, which when treated with lead acetate gives yellow ppt. of lead chromate.

$$CrO_2Cl_2 + 4NaOH \rightarrow \underset{(Yellow\ solution)}{Na_2CrO_4} + 2NaCl + 2H_2O$$

$$Na_2CrO_4 + (CH_3COO)_2Pb \longrightarrow 2CH_3COONa + PbCrO_4\downarrow\ (yellow\ ppt.)$$

2. Bromide (Br^-)

(i) Conc. H_2SO_4: Gives reddish brown vapours of bromine.

$$2KBr + H_2SO_4 \rightarrow K_2SO_4 + 2HBr$$

$$2HBr + H_2SO_4 \rightarrow 2H_2O + SO_2\uparrow + \underset{(Reddish\ brown)}{Br_2}\uparrow$$

(ii) Manganese dioxide and conc. sulphuric acid: When a mix of solid bromide, $MnO_2$and conc. H_2SO_4 is heated, reddish brown vapours of bromine are evolved.

$$2KBr + MnO_2 + 2H_2SO_4 \rightarrow Br_2 + K_2SO_4 + MnSO_4 + 2H_2O$$

(iii) Silver nitrate solution: Acidify the sodium carbonate extract of the salt with dil. HNO_3. Add silver nitrate ($AgNO_3$) solution and shake the test tube. A pale yellow precipitate is obtained which dissolves in ammonium hydroxide with difficulty.

$$Br^- + Ag^+ \rightarrow \underset{(pale\ yellow)}{AgBr}$$

(iv) Lead acetate solution: White crystalline ppt. of lead bromide which is soluble in boiling water.

$$2Br^- + Pb^{2+} \rightarrow \underset{(white)}{PbBr_2}\downarrow$$

(v) Chlorine water: When this solution is added to a solution of bromide and chloroform, free bromine is liberated, which colours the organic layer orange - red.

$$2KBr + Cl_2\ (water) \rightarrow 2KCl + Br_2$$

$$Br_2 + Chloroform \rightarrow Orange\ red\ colour$$

3. Iodide (I^-)

(i) Conc. H_2SO_4: Gives violet vapours of iodine

$$2I^- + 2H_2SO_4 \longrightarrow \underset{\text{violet vapours}}{I_2} + SO_4^{2-} + 2H_2O + SO_2\uparrow$$

(ii) Silver nitrate solution: Yellow, curdy ppt. of silver iodide (AgI),very slightly soluble in conc. ammonia solution and insoluble in dil nitric acid.

$$I^- + Ag^+ \rightarrow AgI$$

(iii) Lead acetate solution: Yellow, curdy ppt. of lead iodide, soluble in much hot water forming a colourless solution and yielding golden yellow plates (spangles) on cooling.

$$2I^- + Pb^{2+} \rightarrow \underset{\text{(yellow)}}{PbI_2\downarrow}$$

(iv) Potassium dichromate and conc. sulphuric acid: Iodine is liberated.

$$6I^- + Cr_2O_7^{2-} + 2H_2SO_4 \longrightarrow 3I_2\uparrow + 2Cr^{3+} + 7SO_4^{2-} + 7H_2O$$

(v) Chlorine water. Iodine is liberated by the dropwise addition of chlorine water to the salt solution containing iodide ions, and on addition of $CHCl_3$, violet coloured organic layer is obtained.

$$2I^- + Cl_2 \rightarrow I_2 + 2Cl^-$$

I_2 + Chloroform → Violet coloured layer

(vi) Copper sulphate solution: Gives brown ppt. consisting of a mixture of copper (I) iodide and iodine and on addition of hypo solution, brown ppt. changes to white ppt.

$$4I^- + 2Cu^{2+} \rightarrow 2CuI + I_2$$

$$I_2 + 2S_2O_3^{2-} \longrightarrow 2I^- + S_4O_6^{2-}$$

(vii) Mercury (II) chloride solution: Forms scarlet ppt. of HgI_2.

$$2I^- + HgCl_2 \rightarrow HgI_2\downarrow + 2Cl^-$$

This ppt. dissolves in excess of KI, forming tetra iodo mercurate (II) complex.

$$HgI_2 + 2I^- \rightarrow [HgI_4]^{2-}$$

4. Nitrate (NO_3^-)

(i) Conc. H_2SO_4: Gives reddish - brown vapours of nitrogen dioxide.

$$4NO_3^- + 2H_2SO_4 \rightarrow 4NO_2 + 2SO_4^{2-} + 2H_2O + O_2$$

(ii) Brown ring test: Take 1 mL of an aqueous solution of the salt and add 2 mL conc. H_2SO_4 slowly. Mix the solutions thoroughly and cool the test tube under the tap. Now, add freshly prepared ferrous sulphate solution along the sides of the test tube dropwise so that it forms a layer on the top of the liquid already present in the test tube. A dark brown ring is formed at the junction of the two solutions due to the formation of nitroso ferrous sulphate. Alternatively first ferrous sulphate is added and then concentrated sulphuric acid is added.

$$NaNO_3 + H_2SO_4 \rightarrow NaHSO_4 + HNO_3$$

$$6\,FeSO_4 + 3H_2SO_4 + 2HNO_3 \rightarrow 3Fe_2(SO_4)_3 + 4H_2O + 2NO$$

$$FeSO_4 + NO \rightarrow \underset{\substack{\text{Nitroso ferrous sulphate}\\\text{(Brown)}}}{[Fe(NO)]SO_4}$$

Action of heat: The result varies with the metal.

- Nitrates of sodium and potassium evolve oxygen (test with glowing splint) and leave solid nitrities (brown fumes with dilute acid).

  $2NaNO_3 \rightarrow 2NaNO_2 + O_2\uparrow$
- Ammonium nitrate yields dinitrogen oxide and steam.

  $NH_4NO_3 \rightarrow N_2O\uparrow + 2H_2O$
- Nitrates of the noble metals leave a residue of the metal and a mix of nitrogen dioxide and oxygen is evolved.

  $2AgNO_3 \rightarrow 2Ag + 2NO_2\uparrow + O_2\uparrow$
- Nitrates of other metals, such as those of lead and copper, evolve oxygen and nitrogen dioxide and leave a residue of the oxide.

  $2Pb(NO_3)_2 \rightarrow 2PbO + 4NO_2\uparrow + O_2\uparrow$

Classification of Cations

Cations are classified into six groups on the basis of their behaviour with some reagents and classification is based on whether a cation reacts with these reagents by the formation of precipitate or not (solubility difference).

Group 0:

Group Reagent : None

Ions : NH_4^+

Colour and ppt :

Group I:

Group Reagent : dil. HCl

Ions : Pb^{2+}

Colour and ppt : $PbCl_2$ white

Group II A:

Group Reagent : H_2S in presence of dil. HCl

Ions : Pb^{2+}, Cu^{2+}

Colour and ppt : PbS, CuS (Black)

Group II B:

Group Reagent : H_2S in presence of dil. HCl

Ion : As^{3+}

Colour and ppt : As_2S_3 (Yellow)

Group III:

Group Reagent : NH_4OH in presence of NH_4Cl

Ions : Fe^{3+}, Al^{3+}

Colour and ppt : $Fe(OH)_3$ - Brown; $Al(OH)_3$- White

Group IV:

Group Reagent : H_2S in presence of NH_4OH and NH_4Cl

Ions : Ni^{2+}, Co^{2+}, Zn^{2+}, Mn^{2+}

Colour and ppt : CoS, NiS - Black; ZnS - White; MnS - flesh coloured

Group V:

Group Reagent : $(NH_4)_2CO_3$ in presence of NH_4Cl and NH_4OH

Ions : Ba^{2+}, Sr^{2+}, Ca^{2+}

Colour and ppt : Carbonates – White

Group VI:

Group Reagent : None

Ion : Mg^{2+}

Group 0 Radicals

1. **Test of NH_4^+ ion**

(i) **Sodium hydroxide solution:** Ammonia gas is evolved on warming the solution containing ammonium salt and sodium hydroxide.

$NH_4Cl + NaOH \rightarrow NH_3 + H_2O + NaCl$

The gas can be identified by the following characteristics/ reactions:

- Its characteristics smell.
- The evolution of the white fumes of ammonium chloride when a glass rod dipped in dilute HCl is held in the vapour.

  $NH_3 + HCl \rightarrow NH_4Cl \uparrow$ (white fumes)
- Its ability to turn filter paper moistened with $Hg_2(NO_3)_2$ solution, black.

  $Hg_2(NO_3)_2 + 2NH_3 \rightarrow \underbrace{Hg(NH_2)NO_3 + Hg}_{\text{black}} + NH_4NO_3$

(ii) **Sodium hexanitrito–N–cobaltate (III) solution:** NH_4^+ ions give a yellow precipitate with this reagent.

$3NH_4^+ + [Co(NO_2)_6]^{3-} \rightarrow (NH_4)_3[Co(NO_2)_6]\downarrow$

(iii) **Hexachloridoplatinate (IV) solution (i.e., hexachloroplatinic acid):** NH_4^+ ions give yellow precipitate with this reagent.

$2NH_4^+ + [PtCl_6]^{2-} \rightarrow (NH_4)_2[PtCl_6]\downarrow$ (yellow)

(iv) **Saturated sodium hydrogen tartrate solution ($NaHC_4H_4O_6$):** NH_4^+ ions give white precipitate with this reagent .

$NH_4^+ + HC_4H_4O_6^- \rightarrow NH_4(HC_4H_4O_6)\downarrow$ (white)

Group I Radicals

Pb^{2+}, are precipitated as chloride because the solubility product of these chlorides ($PbCl_2$) is less than the solubility products of chlorides of all other metal ions, which remain in solution.

Lead chloride is slightly soluble in water and therefore, lead is never completely precipitated by adding dilute HCl to a sample, the rest of the lead ions are precipitated with H_2S in acidic medium together with the cations of the second group.

Test of Pb^{2+} ion

(i) **Dilute HCl:** A white ppt. in cold solution of the acid is not too dilute.

$$Pb^{2+} + 2Cl^- \rightleftharpoons \underset{\text{White}}{PbCl_2}\downarrow$$

- Soluble in hot water.
- It is also soluble in conc. HCl or conc. KCl when the tetrachloroplumbate (II) ion is formed.

  $PbCl_2\downarrow + 2Cl^- \longrightarrow [PbCl_4]^{2-}$
- If the ppt is washed and dil. NH_3 is added, no visible change occurs (difference from Hg_2^{2+} or Ag^+ ions), through a ppt. exchange reaction takes place and lead hydroxide is formed.

(ii) **NH_3 solution:** White ppt of $Pb(OH)_2$ is obtained.

$$Pb^{2+} + 2NH_3 + 2H_2O \longrightarrow \underset{\text{(white)}}{Pb(OH)_2}\downarrow + 2H^+$$

$Pb(OH)_2$ is insoluble in NH_3 solution excess.

(iii) With H_2S gas in neutral or dilute acid medium, black ppt. of lead sulphide is obtained.

$$Pb^{2+} + H_2S \longrightarrow \underset{\text{(Black)}}{PbS}\downarrow + 2H^+$$

When H_2S gas is introduced into a mixture which contains a ppt. of white lead chloride, the latter is converted into lead sulphide (black) in a precipitate exchange reaction.

$$\underset{\text{(white)}}{PbCl_2}\downarrow + H_2S \longrightarrow \underset{\text{(Black)}}{PbS}\downarrow + 2H^+ + 2Cl^-$$

If the test is carried out in the presence of larger amounts of Cl^- (KCl saturated), initially a red ppt. of lead sulphochloride is formed when $H_2S\uparrow$ is introduced.

$$2Pb^{2+} + H_2S + 2Cl^- \longrightarrow \underset{\text{(red)}}{Pb_2SCl_2}\downarrow + 2H^+$$

This, however, decomposes on dilution or on further addition of H_2S and a black ppt. of PbS is formed.

$$\underset{\text{(red)}}{Pb_2SCl_2}\downarrow \longrightarrow \underset{\text{(Black)}}{PbS}\downarrow + \underset{\text{(white)}}{PbCl_2}\downarrow$$

$$PbSCl_2 + H_2S \longrightarrow 2PbS + 2Cl^- + 2H^+$$

(iv) **NaOH:** White ppt. of $Pb(OH)_2$ is formed.

$$Pb^{2+} + 2OH^- \longrightarrow \underset{\text{(white)}}{Pb(OH)_2}\downarrow$$

The ppt. is dissolved in excess of NaOH, when tetrahydroxoplumbate (II) ions are formed.

$$Pb(OH)_2\downarrow + 2OH^- \longrightarrow [Pb(OH)_4]^{2-}$$

Thus, lead hydroxide has an amphoteric character.

(v) **H_2SO_4 (or soluble sulphate):** White ppt. of $PbSO_4$ is obtained.

$$Pb^{2+} + SO_4^{2-} \longrightarrow \underset{\text{(White)}}{PbSO_4}\downarrow$$

(vi) **K_2CrO_4:** Potassium chromate in neutral, acetic acid or ammonium solution gives yellow ppt of lead chromate.

$$Pb^{2+} + CrO_4^{2-} \longrightarrow PbCrO_4\downarrow \text{(yellow)}$$

(vii) **KI:** Yellow ppt of lead iodide is formed.

$$Pb^{2+} + 2I^- \longrightarrow \underset{\text{yellow ppt.}}{PbI_2}\downarrow$$

The ppt. is moderately soluble in boiling water to yield a colourless solution, from which it separates as golden yellow plates on cooling.

An excess of a more conc. solution of the reagent dissolves the ppt.

$PbI_2 \downarrow + 2I^- \rightleftharpoons [PbI_4]^{2-}$

The reaction is reversible, on cooling, ppt. reappears.

Group II Radicals

- Group 2nd radicals are precipitated as sulphides because of their low solubility products whereas sulphides of other metals remain in solution because of their high solubility products. HCl acts as a source of H^+ which decreases the concentration of S^{2-} due to common ion effect. Hence, the concentration of S^{2-} ion is too low that it exceeds only the solubility products of the metal sulphides of IInd group.

1 **Test of Cu^{2+} ion**

(i) Precipitation with H_2S in acidic medium: Black precipitate is formed.

$$Cu^{2+} + H_2S \xrightarrow{H^+} \underset{\text{(black)}}{CuS} \downarrow + 2H^+$$

Precipitate is insoluble in boiling dilute H_2SO_4 (distinction from cadmium), in NaOH, Na_2S and $(NH_4)_2S$. Precipitate dissolves in hot concentrated HNO_3.

$3CuS + 8HNO_3 \rightarrow 3Cu(NO_3)_2 \text{ (blue)} + 2NO + 4H_2O + 3S$

When boiled for longer, S is oxidised to H_2SO_4 and a clear solution of $Cu(NO_3)_2$ is obtained.

Note: KCN dissolves the precipitate forming a clear solution.

$2CuS \downarrow + 8\,CN^- \rightarrow 2\,[Cu(CN)_4]^{3-} + S_2^{2-}$ (disulphide ion)

(ii) Ammonia solution: When added sparingly, a blue precipitate of basic salt (basic copper sulphate) is formed with $CuSO_4$.

$2Cu^{2+} + SO_4^{2-} + 2NH_3 + 2H_2O \rightarrow Cu(OH)_2.CuSO_4 \downarrow + 2NH_4^+$

It is soluble in excess of reagent forming a deep blue colouration.

$Cu(OH)_2.CuSO_4\downarrow + 8NH_3 \rightarrow 2\,[Cu(NH_3)_4]^{2+} + SO_4^{2-} + 2OH^-$

(iii) Sodium hydroxide in cold solution: A blue precipitate is formed.

$Cu^{2+} + 2OH^- \rightarrow Cu(OH)_2 \downarrow$

$Cu(OH)_2\downarrow \xrightarrow{\text{Heat}} CuO\downarrow \text{ (black)} + H_2O$

(iv) Potassium iodide: It gives a white precipitate of Cu(I) iodide but the solution is intensely brown because of the formation of tri–iodide ions (or iodine).

$$2Cu^{2+} + 5I^- \rightarrow Cu_2I_2 \downarrow + \underset{\text{(Brown)}}{I_3^-}$$

The solution becomes colourless and a white precipitate is visible when excess of sodium thiosulphate solution is added.

$I_3^- + 2S_2O_3^{2-} \rightarrow 3I^- + S_4O_6^{2-}$

Note: These reactions are used in quantitative analysis for the iodometric determination of copper.

(v) Potassium ferrocyanide (Potassium hexacyanidoferrate (II) solution: Cu^{2+} ions give brown/chocolate brown precipitate.

$2Cu^{2+} + K_4[Fe(CN)_6] \rightarrow Cu_2[Fe(CN)_6] \downarrow + 4K^+$

Note: $2\,[Fe(CN)_6]^{3-} + 3\,Cu^{2+} \rightarrow Cu_3\,[Fe(CN)_6]_2 \downarrow$ (green)

(vi) Potassium cyanide: When added sparingly, forms first a yellow precipitate.

$Cu^{2+} + 2CN^- \rightarrow Cu(CN)_2 \downarrow$ (yellow)

Precipitate quickly decomposes into CuCN and cyanogen.

$2\,Cu(CN)_2 \downarrow \rightarrow 2CuCN \downarrow \text{ (white)} + (CN)_2 \uparrow$ (highly poisonous)

Excess reagent dissolves the precipitate forming a colourless soluble complex.

$CuCN \downarrow + 3CN^- \rightarrow [Cu(CN)_4]^{3-}$

Complex is so stable that H_2S cannot precipitate Cu(I) sulphide (distinction from cadmium).

(vii) Potassium Thiocyanate (KSCN): Black precipitate of copper (II) thiocyanate is obtained.

$$Cu^{2+} + 2SCN^- \longrightarrow \underset{\text{black}}{Cu(SCN)_2} \downarrow$$

The precipitate decomposes slowly to form white copper (I) thiocyanate and thiocyanogen is formed.

$$2Cu(SCN)_2 \downarrow \longrightarrow \underset{\text{White}}{2CuSCN} \downarrow + (SCN)_2 \uparrow$$

Thiocyanogen rapidly decomposes in aqueous solution.

2. **Test of Pb^{2+} ion**

(i) Precipitation with H_2S in acidic medium: Black precipitate is formed which is soluble in hot dilute HNO_3.

$Pb^{2+} + H_2S \rightarrow PbS \downarrow \text{ (black)} + 2H^+$

$3PbS + 8HNO_3 \rightarrow 3Pb(NO_3)_2 + 2NO + 4H_2O + 3S$

(ii) Dilute H_2SO_4: White precipitate is formed on addition of dil. H_2SO_4 and a few drops of alcohol to this solution, which is soluble in ammonium acetate.

$Pb(NO_3)_2 + H_2SO_4 \rightarrow PbSO_4 \downarrow \text{ (white)} + 2HNO_3$

$PbSO_4 + 2CH_3COONH_4 \rightarrow (CH_3COO)_2Pb + (NH_4)_2SO_4$

(iii) Potassium iodide: Yellow precipitate is formed which is soluble in excess concentrated (6M) solution of the reagent. Yellow precipitate of PbI_2 is moderately soluble in boiling water to give a colourless solution.

$(CH_3COO)_2Pb + 2KI \rightarrow 2CH_3COOK + PbI_2 \downarrow$ (yellow)

(iv) Potassium chromate: Yellow precipitate is formed.

$(CH_3COO)_2Pb + K_2CrO_4 \longrightarrow 2CH_3COOK + PbCrO_4\downarrow$ (yellow)

Group III Radicals

Group IIIrd radicals are precipitated as hydroxides and the addition of NH_4Cl suppresses the ionisation of NH_4OH so that only the group 3 cations are precipitated as hydroxides because of their low solubility products.

- Excess of NH_4Cl should not be added, as manganese will precipitate as $MnO_2.H_2O$.
- Concentrated HNO_3 is added to oxidise Fe^{2+} to Fe^{3+}, if present.

1. Test of Al^{3+} ion

(i) Precipitation with NH_4OH in presence of NH_4Cl: White gelatinous precipitate is formed which is slightly soluble in excess reagent. The solubility is decreased in the presence of ammonium salt. A small portion of the precipitate passes into the solution as colloidal $Al(OH)_3$ (Aluminium hydroxide sol), the sol is coagulated on boiling the solution or upon the addition of soluble salt yielding a precipitate of $Al(OH)_3$, known as $Al(OH)_3$ gel.

$$Al^{3+} + 3NH_4OH \xrightarrow{NH_4Cl} Al(OH)_3\downarrow + 3NH_4^+$$

(ii) Sodium hydroxide: White precipitate of $Al(OH)_3$ is obtained.

$$Al^{3+} + 3OH^- \longrightarrow \underset{\text{White}}{Al(OH)_3}\downarrow$$

White precipitate dissolves in excess sodium hydroxide according to following reaction.

$$Al(OH)_3 + OH^- \rightleftharpoons [Al(OH)_4]^-$$

The reaction is reversible and any reagent, which will reduce the hydroxyl ion concentration sufficiently should cause the reaction to proceed from right to left with the consequent precipitation of aluminium hydroxide. This may be effected with a solution of ammonium chloride (the hydroxyl ion concentration is reduced owing to the formation of the weak base ammonia, which can be readily removed as ammonia gas by heating) or by the addition of acid. In the latter case, a large excess of acid causes the precipitated hydroxide to redissolve.

$$[Al(OH)_4]^- + NH_4^+ \rightarrow Al(OH)_3\downarrow + NH_3\uparrow + H_2O$$

$$[Al(OH)_4]^- + H^+ \rightleftharpoons Al(OH)_3\downarrow + H_2O$$

$$Al(OH)_3 + 3H^+ \text{ (excess)} \rightleftharpoons Al^{3+} + 3\,H_2O$$

(iii) Dry test: Aluminium compounds when heated with sodium carbonate upon charcoal gives a white infusible solid, which glows when hot. If the residue is moistened with a little cobalt nitrate solution and again heated, a blue infusible mass is obtained.

$$2Al_2O_3 + 2Co^{2+} + 4NO_3^- \rightarrow \underset{\text{(thenard blue)}}{2CoAl_2O_4} + 4NO_2\uparrow + O_2\uparrow$$

(iv) Ammonium Sulphide Solution: A white precipitate, $Al(OH)_3$ is obtained.

$$2Al^{3+} + 3S^{2-} + 6H_2O \rightarrow 2Al(OH)_3\downarrow + 3H_2S\uparrow$$

(v) Sodium Acetate Solution: No precipitate is obtained in cold, neutral solution, but on boiling with excess reagent, a voluminous precipitate of basic aluminium acetate $Al(OH)_2(CH_3COO)$ is formed.

$$Al^{3+} + 3CH_3COO^- + 2H_2O \longrightarrow \underset{\text{White}}{Al(OH)_2(CH_3COO)}\downarrow + 2CH_3COOH$$

2. Test of Fe^{3+} ion

(i) Precipitation with NH_4OH in presence of NH_4Cl: Gelatinous reddish brown precipitate is formed which is insoluble in excess reagent but soluble in acids.

$$Fe^{3+} + 3NH_4OH \xrightarrow{NH_4Cl} Fe(OH)_3\downarrow + 3NH_4^+$$

- Precipitation of iron(II) hydroxide occurs with ammonia solution. If larger amounts of ammonium ions are present, the dissociation of NH_4OH is suppressed and the concentration of OH^- ions is lowered to such an extent that solubility product of iron (II) hydroxide, $Fe(OH)_2$ is not attained and precipitation does not occur.

(ii) Sodium hydroxide solution: Reddish brown precipitate is formed which is insoluble in excess reagent (distinction from aluminium and chromium).

$$Fe^{3+} + 3OH^- \rightarrow Fe(OH)_3\downarrow$$

(iii) H_2S gas in acidic solution: Fe^{3+} reduces to Fe^{2+}.

$$2Fe^{3+} + H_2S \rightarrow 2Fe^{2+} + 2H^+ + S\downarrow \text{ (milky-white)}$$

If a neutral solution of iron (III) chloride is added to a freshly prepared saturated solution of H_2S, bluish colouration appears first, followed by precipitation of sulphur. The blue colour is due to a colloidal solution of sulphur of extremely small particle size. This reaction can be used to test the freshness of H_2S solution.

(iv) Ammonium sulphide solution: Black precipitate consisting of Fe(II) sulphide and sulphur is formed.

$$2Fe^{3+} + 3S^{2-} \rightarrow 2FeS + S\downarrow$$

In HCl, the black precipitate of Fe(II) sulphide dissolves and white precipitate of sulphur becomes visible.

$$FeS\downarrow + 2H^+ \rightarrow H_2S\uparrow + Fe^{2+}$$

- From alkaline solutions of ammonium sulphide, black iron(III) sulphide is obtained.

$$2Fe^{3+} + 3S^{2-} \rightarrow Fe_2S_3\downarrow$$

- On acidification with hydrochloric acid, iron (III) ions are reduced to iron (II) ions and sulphur is formed.

$$Fe_2S_3\downarrow + 4H^+ \rightarrow 2Fe^{2+} + 2H_2S\uparrow + S\downarrow$$

- The damp iron (II) sulphide precipitate, when exposed to air, is slowly oxidised to brown iron(III) hydroxide.

$$4FeS\downarrow + 6H_2O + 3O_2 \rightarrow 4Fe(OH)_3\downarrow + 4S\downarrow$$

- From alkaline solutions of ammonium sulphide, black iron(III) sulphide is obtained.

$$2Fe^{3+} + 3S^{2-} \rightarrow Fe_2S_3\downarrow$$

- On acidification with hydrochloric acid, iron (III) ions are reduced to iron (II) ions and sulphur is formed.

$$Fe_2S_3\downarrow + 4H^+ \rightarrow 2Fe^{2+} + 2H_2S\uparrow + S\downarrow$$

(v) Potassium ferrocyanide (Potassium hexacyanidoferrate (II)): Intense blue precipitate (Prussian blue) of iron(III) hexacyanidoferrate(II) is formed.

$$4Fe^{3+} + 3[Fe(CN)_6]^{4-} \rightarrow Fe_4[Fe(CN)_6]_3\downarrow$$

This is insoluble in dilute acids but decomposes in concentrated HCl. A large excess of the reagent dissolves it partly or entirely, when an intense blue solution is obtained. Sodium hydroxide turns the precipitate red.

$$Fe_4[Fe(CN)_6]_3\downarrow + 12OH^- \rightarrow Fe(OH)_3\downarrow + 3[Fe(CN)_6]^{4-}$$

- Oxalic acid also dissolves Prussian blue forming a blue-solution.

- If iron(III) chloride is added to an excess of potassium hexacyanidoferrate (II), a product with the composition of $K_4Fe[Fe(CN)_6]$ is formed. This tends to form colloidal solutions ("soluble Prussian blue") and can not be filtered.

(vi) Potassium ferricyanide (Potassium hexacyanidoferrate (III)): A brown colouration is formed.

$$Fe^{3+} + [Fe(CN)_6]^{3-} \rightarrow Fe[Fe(CN)_6]$$

- Fe^{2+} gives dark blue precipitate with potassium ferricyanide. First hexacyanidoferrate (III) ions oxidise iron(II) to iron(III), when hexacyanidoferrate(II) is formed.

$$Fe^{2+} + [Fe(CN)_6]^{3-} \rightarrow Fe^{3+} + [Fe(CN)_6]^{4-}$$

- These ions combine to form a precipitate called Turnbull's blue.

$$4Fe^{3+} + 3[Fe(CN)_6]^{4-} \rightarrow Fe_4[Fe(CN)_6]_3$$

- Composition of this precipitate is identical to that of Prussian blue. Earlier the composition suggested was $Fe_3[Fe(CN)_6]_2$, hence different name.

$$3Fe^{2+} + 2K_3[Fe(CN)_6] \rightarrow Fe_3[Fe(CN)_6]_2 \text{ (ferrous ferric cyanide)} + 6K^+$$

(Turnbull's blue)

- Fe(II) in ammonical solution gives red solution with DMG, colouration fades on standing due to the oxidation of the iron(II) complex. Fe (III) does not give such complex.
- In complete absence of air, Fe(II) ions produce white precipitate with potassium hexacyanidoferrate(II).

$$Fe^{2+} + 2K^+ + [Fe(CN)_6]^{4-} \rightarrow K_2Fe\,[Fe(CN)_6] \downarrow$$

Under ordinary atmospheric conditions a pale–blue precipitate is formed.

(vii) Potassium thiocyanate (Potassium sulphocyanide): In slightly acidic medium, a deep red colouration is produced due to the formation of a non-dissociated iron(III) thiocyanate complex.

$$Fe^{3+} + SCN^- \longrightarrow [Fe(SCN)]^{2+}$$

Blood red colour

This neutral molecule can be extracted by ether or amyl alcohol.

- With pure Fe(II) ions no colouration is obtained.
- Fluorides and Hg(II) ions bleach the red colour.

$$Fe(SCN)_3 + 6F^- \rightarrow [FeF_6]^{3-} + 3SCN^-$$

$$2Fe(SCN)_3 + 3Hg^{2+} \rightarrow 2Fe^{3+} + 3Hg(SCN)_2$$

(viii) Sodium acetate solution: A deep red colouration is obtained owing to the formation of $[Fe_3(OH)_2(CH_3COO)_6]^+$. The reaction becomes complete only if the strong acid, which is formed, is removed by the addition of an excess reagent, which acts as a buffer.

$$3Fe^{3+} + 6CH_3COO^- + 2H_2O \rightleftharpoons [Fe_3(OH)_2(CH_3COO)_6]^+ + 2H^+$$

This deep red coloured solution on dilution with water and boiling gives reddish brown precipitate of basic ferric acetate.

$$[Fe_3(OH)_2(CH_3COO)_6]^+ + 4H_2O \rightarrow 3Fe(OH)_2CH_3COO \downarrow + 3CH_3COOH + H^+$$

The excess of acetate ion acts as buffer and the reaction goes to completion.

Group IV Radicals

In IVth group, ammonium hydroxide increases the ionisation of H_2S by removing H^+ from H_2S as unionised water.

$$H_2S \rightleftharpoons 2H^+ + S^{2-}$$

$$H^+ + OH^- \rightarrow H_2O$$

Now the excess of S^{2-} ions is available and hence the ionic products of group 4th group cations exceeds their solubility products and will be precipitated. In case H_2S is passed through a neutral solution, incomplete precipitation will take place due to the formation of HCl, which decreases the ionisation of H_2S.

1. Test of Zn^{2+} ion

(i) Precipitation with H_2S in presence of NH_4OH + NH_4Cl: A white precipitate is formed. In neutral solutions, precipitation is partial as H^+ ions concentration produced, depressed the ionisation of H_2S.

$$Zn^{2+} + H_2S \rightleftharpoons ZnS \downarrow + 2H^+$$

The precipitate is soluble in dilute HCl.

$$ZnS + 2H^+ \rightarrow Zn^{2+} + H_2S \uparrow$$

(ii) Sodium hydroxide solution: On addition of sodium hydroxide solution it gives a white precipitate of zinc hydroxide, which is soluble in excess of NaOH solution on heating. This confirms the presence of Zn^{2+} ions.

$$ZnCl_2 + 2NaOH \rightarrow Zn(OH)_2 + 2NaCl$$

$$Zn(OH)_2 + 2NaOH \rightarrow Na_2ZnO_2 + 2H_2O$$

(Sodium zincate)

(iii) Potassium ferrocyanide (Potassium hexacyanido-ferrate(II)) solution: When potassium ferrocyanide $K_4Fe(CN)_6$ solution is added to the solution after neutralisation by NH_4OH solution, a white or a bluish white precipitate of zinc ferrocyanide appears

$$2\,ZnCl_2 + K_4\,[Fe(CN)_6] \longrightarrow Zn_2\,[Fe(CN)_6] + 4\,KCl$$

Zinc ferrocyanide

2. Test of Co^{2+} ion:

(i) Cobalt sulphide dissolves in aqua regia in the same manner as nickel sulphide. When the aqueous solution of the residue obtained after treatment with aqua regia is treated with a strong solution of potassium nitrite after neutralisation with ammonium hydroxide and the solution is acidified with dil. acetic acid, a yellow precipitate of the complex of cobalt named potassium hexanitritocobaltate [III] is formed.

$$CoS + HNO_3 + 3HCl \rightarrow CoCl_2 + NOCl + S + 2H_2O$$

$$CoCl_2 + 7KNO_2 + 2CH_3COOH \rightarrow K_3[Co(NO_2)_6] + 2KCl + 2CH_3COOK + NO + H_2O$$

Potassium hexanitritocobaltate[III] (Yellow precipitate)

(ii) Ammonium thiocyanate solution: A neutral or acid solution of cobalt(II) gives a blue colouration in amyl alcohol or ether layer when a few crystals of ammonium thiocyanate are added.

$$Co^{2+} + 4SCN^- \rightarrow [Co(SCN)_4]^{2-}$$

- In amyl alcohol or ether, the free acid $H_2[Co(SCN)_4]$ is formed and is dissolved by the organic solvent (distinction from nickel).

(iii) Sodium bicarbonate and bromine water test: To the test solution, sodium bicarbonate is added in excess followed by the bromine water. The mixture is slightly heated, the solution turns apple green.

$CoCl_2 + 2NaHCO_3 \rightarrow Co(HCO_3)_2 + 2NaCl$

$Co(HCO_3)_2 + 4NaHCO_3 \rightarrow Na_4[Co(CO_3)_3] + 3H_2O + 3CO_2$

$Br_2 + H_2O \rightarrow 2HBr + [O]$

$2Na_4[Co(CO_3)_3] + H_2O + [O] \rightarrow \underset{\text{(green)}}{2Na_3[Co(CO_3)_3]} + 2NaOH.$

- Green solution of Sodium tris(carbonato)cobalt(III) is obtained.

(iv) Sodium hydroxide solution: Cobalt(II) nitrate in cold gives a blue basic salt.

$Co^{2+} + OH^- + NO_3^- \rightarrow \underset{\text{(Blue)}}{Co(OH)NO_3\downarrow}$

Upon warming with excess alkali, the basic salt is converted into a pink precipitate of Co(II) hydroxide.

$Co(OH)NO_3\downarrow + OH^- \rightarrow \underset{\text{(Pink)}}{Co(OH)_2\downarrow} + NO_3^-$

3. Test of Ni^{2+} ion:

(i) Precipitation with H_2S in presence of $NH_4OH + NH_4Cl$:

A black precipitate is formed which is insoluble in cold dilute HCl and CH_3COOH but dissolves in hot concentrated HNO_3 and in aquaregia.

$NiCl_2 + H_2S \xrightarrow{NH_4OH} NiS\downarrow \text{(Black)} + 2HCl$

$3NiS\downarrow + 2HNO_3 + 6H^+ \rightarrow 3Ni^{2+} + 2NO\uparrow + 3S\downarrow + 4H_2O$

$NiS + HNO_3 + 3HCl \rightarrow Ni^{2+} + S\downarrow + NOCl\uparrow + 2Cl^- + 2H_2O.$

- Solution on evaporation to dryness gives yellow residue ($NiCl_2$) which turns green on adding water.

(ii) Dimethylglyoxime reagent: A red precipitate obtained from the solution is just made alkaline or acid solutions buffered with sodium acetate.

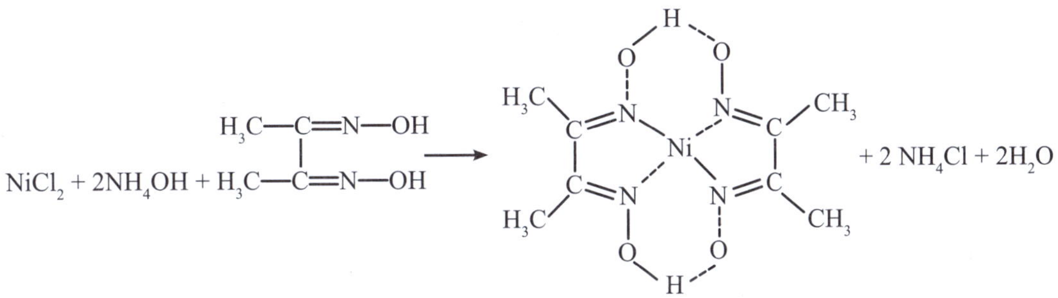

Complex of red colour
(Stable form of complex)

- Fe(II) ions give red colouration, bismuth gives yellow precipitate and cobalt gives brown colouration with DMG in ammonical solutions.
- Ni^{2+} gives black precipitate (Ni_2O_3) with sodium bicarbonate and bromine water on heating whereas Co^{2+} gives green coloured solution, this is the point of difference.

$NiCl_2 + 2NaHCO_3 \rightarrow NiCO_3 + 2NaCl + H_2O$

$2NiCO_3 + [O] \rightarrow Ni_2O_3\downarrow \text{(black)} + 2CO_2$

Group V Radicals

In V[th] group, the reagent ammonium carbonate should be added in alkaline or neutral medium.

1. Test of Ba^{2+} ion

(i) Precipitation with $(NH_4)_2CO_3$ in presence of $NH_4OH + NH_4Cl$: A white precipitate is formed which is soluble in acetic acid and dilute mineral acids.

$BaCl_2 + (NH_4)_2CO_3 \rightarrow BaCO_3\downarrow + 2NH_4Cl$

$BaCO_3 + 2CH_3COOH \rightarrow Ba(CH_3COO)_2 + H_2O + CO_2$

(ii) Potassium chromate test: A yellow precipitate is formed, practically insoluble in water.

$Ba(CH_3COO)_2 + K_2CrO_4 \rightarrow BaCrO_4\downarrow + 2CH_3COOK$

- Precipitate is insoluble in dilute acetic acid (distinction from strontium and calcium) but readily soluble in mineral acids.
- Addition of acid to K_2CrO_4 changes the yellow colour of the solution to reddish–orange due to the formation of dichromate.

$2H^+ + 2CrO_4^{2-} \rightleftharpoons Cr_2O_7^{2-} + H_2O$

(iii) Flame test: Take a platinum wire and dip it in conc. HCl. Heat it strongly until the wire does not impart any colour to the non-luminous flame. Now dip the wire in the paste of the (Group–V) precipitate in conc. HCl. Heat it in the flame. A grassy green colour of the flame confirms the presence of Ba^{2+} ions.

Group VI Radicals

Test of Mg^{2+} ion

If group–V is absent the solution may contain magnesium carbonate, which is soluble in water in the presence of ammonium salts because the equilibrium is shifted towards the right hand side.

$NH_4^+ + CO_3^{2-} \rightleftharpoons NH_3 + HCO_3^-$

The concentration of carbonate ions required to produce a precipitate is not attained. When disodium hydrogenphosphate solution is added and the inner walls of the test tube are scratched with a glass rod, a white crystalline precipitate of magnesium ammonium phosphate is formed which indicates the presence of Mg^{2+} ions.

$$Mg^{2+} + Na_2HPO_4 \rightarrow \underset{\text{Magnesium ammonium phosphate (White precipitate)}}{Mg(NH_4)PO_4} + NH_4OH + 2Na^+ + H_2O$$

- Some times precipitate of magnesium ammonium phosphate appears after some time. So warm the solution and scratch the sides of test tube after adding sodium hydrogen phosphate solution.

Chemical principles involved in the following experiments

Enthalpy of Solution of $CuSO_4$

In thermochemical measurements generally aqueous solutions are mixed therefore, water in the reaction medium and the temperature changes result due to the chemical reactions taking place in solution.

According to law of conservation of energy, the sum of enthalpy changes taking place in the calorimeter (loss and gain of energy) must be zero. Thus, we can write the following equation-

(ΔH_1) Heat gained by calorimeter, thermometer and stirrer + (ΔH_2) Enthalby change of solution/water in calorimeter + (ΔH_3) Enthalpy change of added solution/ water in calorimeter + (ΔH_4) Enthalpy change of reaction = 0

In these reactions we take the product of density and heat capacity of solutions, dC_p, to be 4.184 $J.mL^{-1}.K^{-1}$, nearly the same as that of pure water.

Solution formation often accompanies heat changes. Enthalpy of solution is the amount of heat liberated or absorbed when one mole of a solute (solid/liquid) is dissolved in such a large quantity of solvent (usually water) that further dilution does not make any heat changes.

Enthalpy of Neutralization of Strong Acid and Strong Base

A neutralisation reaction involves the combination of $H^+(aq)$ ions furnished by an acid and OH^- (aq) ions furnished by a base, evidently leading to the formation of H_2O. Since the reaction envisages bond formation, therefore, this reaction is always exothermic. Enthalpy of neutralisation is defined as the amount of heat liberated when 1 mol of H^+ ions furnished by acid combine with 1 mole of OH^- ions furnished by base to form water. Thus:

$$\underset{\text{(Acid)}}{H^+(aq)} + \underset{\text{(Base)}}{OH^-(aq)} \rightarrow H_2O\ (1), \qquad \Delta_{neut} H \text{ is negative}$$

where $\Delta_{neut} H$ is known as enthalpy of neutralisation.

If both the acid and the base are strong then for the formation of 1 mol H_2O, always a fixed amount of heat, viz, 57 kJ mol^{-1} is liberated. If any one of the acid or the base is weak or if both of these are weak, then some of the heat liberated is used for the ionisation of the acid or base or both of them (as the case may be) and the amount of heat liberated is less than 57 kJ mol^{-1}.

Preparation of Lyophilic and Lyophobic Sols

Colloidal sols are divided into two categories:

1. Lyophillic sol- Solvent attracting
2. Lyophobic sol- Solvent repelling

Lyophillic sols are more stable than the lyophobic sols because in lyophillic sols, particles of dispersed phase have an affinity for the particles of dispersion medium.

Factors responsible for the stability of sols are:

1. Charge
2. Solvation of the colloidal particles by the solvent.

- Lyophillic sols are stable due to the solvation factor.
- Lyophobic sols are stable due to the charge on the colloidal particles. Charges can be positive or negative.
- **Examples of lyophillic sols-**are Egg albumin, starch and gum.
- **Examples of lyophobic sols-**are freshly prepared ferric hydroxide, aluminium hyroxide and arsenic sulphide.
- **Positively charged sols-**hydrated ferric oxide (when $FeCl_3$ is added to excess of hot water)
- **Negatively charged sol-**starch, arsenious sulphide and hydrated ferric oxide (when $FeCl_3$ is added to NaOH solution).
- Lyophillic sols are directly formed by mixing and shaking the substance with a suitable liquid.
- Lyophobic sols cannot be prepared by direct mixing and shaking.
- Some methods for the colloids preparation are
  (1) Chemical methods
  (2) Electrical disintegration
  (3) Peptization.
- Sols are purified by dialysis.

Kinetic Study of the Reaction of Iodide Ions with Hydrogen Peroxide at Room Temperature

4. Kinetic study of the reaction of iodide ions with hydrogen peroxide at room temperature.

The reaction between iodide ions and hydrogen peroxide occurs in the acidic medium and can be represented in the following manner:

$$2I^- + H_2O_2\ (1) + 2H^+(aq) \rightarrow I_2(g) + 2H_2O(1)$$

In this reaction, hydrogen peroxide oxidises iodide ions (I^-) to molecular iodine. If calculated amount of sodium thiosulphate is added in the presence of starch solution as an indicator to the above reaction mixture, the liberated iodine reacts with thiosulphate ions as fast as it is formed and is reduced back to iodide ions till all the thiosulphate ions are oxidised to tetrathionate ions.

$$I_2(g) + 2S_2O_3^{2-}(aq) \rightarrow S_4O_6^{2-}(aq) + 2I^-(aq)$$

After the complete consumption of thiosulphate ions, the concentration of iodine liberated in the reaction of hydrogen peroxide with iodide ions increases rapidly to a point where iodine forms intense blue complex with starch. The time required to consume a fixed amount of the thiosulphate ions is reproducible. Since the time for the appearance of colour is noted, the reaction is some times called a clock reaction.

MCQs

Topic-wise MCQs

Detection of Extra Elements

1. An organic compound which produces a bluish green coloured flame on heating in presence of copper is:
   1) benzaldehyde 2) benzoic acid
   3) aniline 4) chlorobenzene

2. Lassaigne's test (with silver nitrate) is commonly used to detect halogens such as chlorine, bromine and iodine but not useful to detect fluorine because the product AgF formed is:
   1) volatile 2) reactive
   3) explosive 4) soluble in water

3. In the estimation of sulphur in an organic compound, fuming nitric acid is used to convert sulphur into:
   1) SO_2 2) H_2S
   3) H_2SO_3 4) H_2SO_4

Detection of the Functional Groups

4. Which of the following is true?
   1) Tollen's reagent gives a positive test with all aldehyde
   2) Fehling's solution gives a positive test with all aldehyde
   3) Tollen's reagent gives a positive test with all carboxylic acid
   4) Tollen's reagent gives a positive test with a-methyl ketone

5. 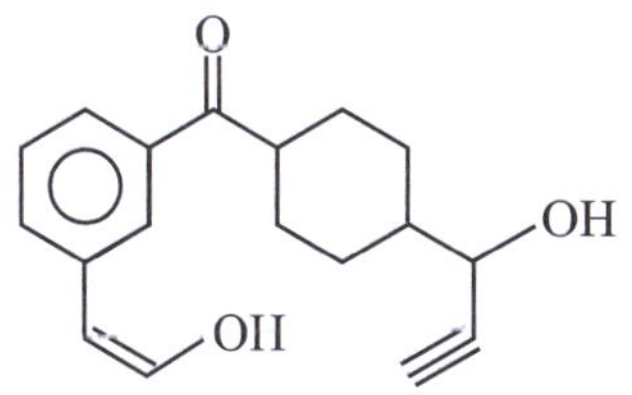

   Which of the following reagents will not react with above compound?
   1) Na metal 2) $AgNO_3 + NH_4OH$
   3) $Cu_2Cl_2 + NH_4OH$ 4) $NaHCO_3$

6. 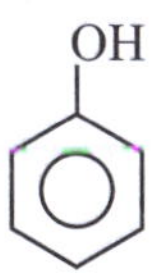 and 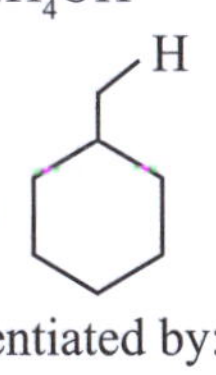

   can be differentiated by:
   1) $FeCl_3$ 2) NaOH
   3) $NaNO_2 + HCl$ 4) Fehling's solution

Preparation of Organic and Inorganic Compounds

7. What product is formed when aniline is treated with CH_3COCl?
   1) Acetanilide 2) Hydrazo benzene
   3) Phenol 4) Para acetyl aniline

8. Which of the following is the molecular formula of aniline yellow?
   1) 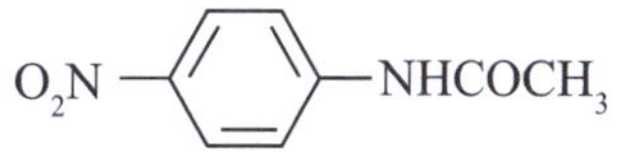
   2) 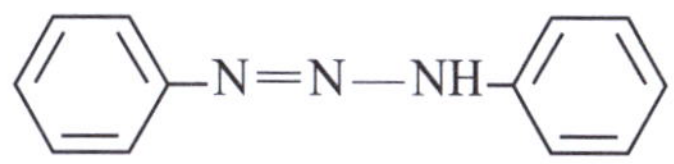
   3) 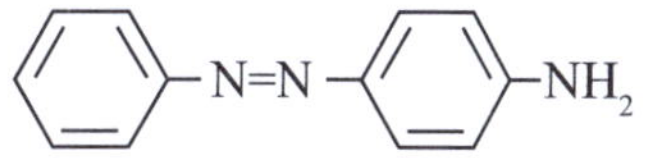
   4) None of these

9. How many water of crystallisation are in Mohr's salt?
   1) 3 2) 2
   3) 1 4) 6

Titration

10. Heat of neutralization will be maximum for:

	ACID (1 M)	Base (1M)
1.	30 ml HCl	70 ml NaOH
2.	40 ml HCl	60 ml NaOH
3.	20 ml HCl	80 ml NaOH
4.	80 ml HCl	20 ml NaOH

   1) 1 2) 2
   3) 3 4) 4

11. If 83 mL of 0.45 M NaOH solution neutralizes a 235 mL HCl solution. Calculate the molarity of the HCl solution.
   1) 0.12 2) 0.16
   3) 0.1 4) 0.2

12. Which of the following indicator should be used during titration of weak acid with a strong base?
   1) Methyl orange 2) Phenolphthalein
   3) Bromothymol blue 4) None of these

13. Order of acidic strength of the given four acids will be when they neutralize with NaOH.

	ACID	Heat of neutralization $\Delta H_{neutralization}$
1.	30 ml HCl	70 ml NaOH
2.	40 ml HCl	60 ml NaOH
3.	20 ml HCl	80 ml NaOH
4.	80 ml HCl	20 ml NaOH

1) HA > HB > HC > HD
2) HA < HB < HC < HD
3) HB > HA > HC > HD
4) HC > HA > HB > HD

Preliminary/Wet Tests

14. In which of the following reactions a reddish-brown coloured gas is evolved?
1) $KBr\,(s) + conc.H_2SO_4 \longrightarrow$
2) $NH_4NO_2 \xrightarrow{\Delta}$
3) $NaNO_3 \xrightarrow[800^\circ C]{\Delta}$
4) $Na_2S_2O_3 \longrightarrow NaI +$ _____

15. Which of the following will not react with each other when heated together?
1) $BeO + MgO$
2) $Li_2CO_3 + BeO$
3) $MgO + CaCO_3$
4) $MgCO_3 + Al_2O_3$

16. Be and Mg do not impart colour to flame:
1) Due to very high IE.
2) Due to higher ionic potential.
3) Due to their greater covalent character.
4) All of the above.

17. Which of the following can not evolve more than one gas (vapour) if heated in dry test tube?
1) $NaNO_3(s)$
2) $MgCO_3(s)$
3) $FeSO_4(s)$
4) $(NH_4)_2Cr_2O_7(s)$

18. On heating, a white amorphous inorganic compound becomes yellow and on cooling, turns white again.The salt may be:
1) $PbCO_3$
2) $MgCO_3$
3) $ZnCO_3$
4) K_2CO_3

19. The hottest part of the flame of a Bunsen burner is the
1) Zone of complete combustion.
2) Blue Zone.
3) Zone of partial combustion.
4) All parts of the flame are equally hot.

20. In the Borax bead test of Cu^{2+}, the blue colour of bead is due to the formation of:
1) B_2O_3
2) Cu_3B_2
3) $Cu(BO_2)_2$
4) CuO

Dilute HCl/dil. H_2SO_4 Group

21. If addition of conc. H_2SO_4 is made to an unknown salt, a colourless and odourless gas is produced then which of the following can be present?
1) CO_3^{2-}
2) S^{2-}
3) Cl^-
4) NO_3^-

22. 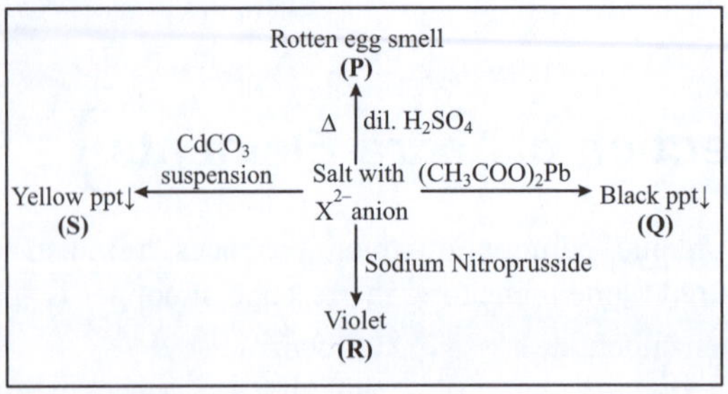

Anion (X^{2-}) is:
1) CO_3^{2-}
2) SO_3^{2-}
3) S^{2-}
4) $S_2O_3^{2-}$

23. Tests on an aqueous solution of a sodium salt having anion X^{n-} gave the following results

$X^{n-} \xrightarrow{K_2Cr_2O_7/H^+}$ green solution + gas

$X^{n-} \xrightarrow{Pb(NO_3)_2}$ black ppt

X^{n-} is:
1) I^-
2) NO_2^-
3) S^{2-}
4) SO_4^{2-}

24. Two inorganic compounds A and B were heated in a dry test tube. A evolved a colourless gas which turned lead acetate paper black and B evolved a gas which turned lime water milky. The anions in A and B respectively are:
1) SO_3^{2-}, CO_3^{2-}
2) S^{2-}, CO_3^{2-}
3) PO_4^{3-}, HCO_3^-
4) S^{2-}, NO_3^-

25. The acidic solution of a salt produces blue colour with KI starch solution. The reaction indicates the presence of:
1) Sulphite
2) Bromide
3) Nitrite
4) Chloride

26. Which of the following anions are identified by dil. HCl?
1) $NO_2^-, NO_3^-, CO_3^{2-}$
2) $NO_2^-, NO_3^-, SO_3^{2-}$
3) $S^{2-}, SO_3^{2-}, NO_2^-$
4) I^-, CO_3^{2-}

27. A mixture when rubbed with organic acid smells like vinegar. It contains:
1) Sulphate
2) Nitrate
3) Nitrite
4) Acetate

Concentrated HCl/Conc. H_2SO_4 Group

28. A solution of a salt in concentrated H_2SO_4 produced a deep blue colour with starch iodide solution. The salt may contain:
1) Chloride
2) Carbonate
3) Acetate
4) Bromide

29. Which of the following salts evolved brown fumes on treatment with conc. H_2SO_4?

1) Nitrate
2) Bromide
3) Both
4) None of these

30. A mixture upon adding conc. H_2SO_4 gives deep red fumes. Mixture may contain the anions pair:

1) $Cr_2O_7^{2-}$ and Cl^-
2) Br^- and $Cr_2O_7^{2-}$
3) NO_3^- and Cl^-
4) I^- and NO_3^{2-}

31. Which of the following pair of anions are identified by conc. H_2SO_4?

1) NO_3^-, CO_3^{2-}
2) Cl^-, NO_3^-
3) Br^-, CO_3^{2-}
4) CO_3^{2-}

Precipitation Reactions

32. Precipitate of $PbSO_4$ is soluble in:

1) Ammonium acetate (6M)
2) Dilute HCl
3) Dilute H_2SO_4
4) None of these

33. A salt which gives CO_2 with hot conc. H_2SO_4 and also decolourizes acidified $KMnO_4$ on warming is:

1) HCO_3^-
2) CO_3^{2-}
3) Oxalate
4) Acetate

34. Brown ring test is made for:

1) NO_3^-
2) Cl^-
3) I^-
4) Br^-

35. When a mixture containing phosphate is heated with conc. HNO_3 and ammonium molybdate solution, a canary yellow precipitate is formed. The formula of the yellow precipitate is:

1) $(NH_4)_3PO_4$
2) $(NH_4)_3PO_4.12MoO_4$
3) $(NH_4)_3PO_4.12MoO_3$
4) $(NH_4)_3PO_4.(NH_4)_2MO_4$

36. Which anion does not liberate any gas with dilute as well as conc. H_2SO_4?

1) NO_2^-
2) NO_3^-
3) SO_3^{2-}
4) SO_4^{2-}

Group I Radicals

37. Consider the following observation;

M^{n+} + HCl (dilute) $\longrightarrow$ white precipitate $\xrightarrow{\Delta}$ water soluble $\xrightarrow{CrO_4^{2-}}$ yellow precipitate.

The metal ion M^{n+} will be:

1) Hg^{2+}
2) Ag^+
3) Pb^{2+}
4) Sn^{2+}

38. Which of the nitrates on strong heating leaves the metal as the residue?

1) $AgNO_3$
2) $Pb(NO_3)_2$
3) $Cu(NO_3)_2$
4) $Al(NO_3)_3$

39. Three separate samples of a solution of a single salt gave these results. One formed a white precipitate with excess ammonia solution, one formed a white precipitate with dilute NaCl solution and one formed a black precipitate with H_2S. The salt could be:

1) $AgNO_3$
2) $Pb(NO_3)_2$
3) $Hg(NO_3)_2$
4) $Mn(NO_3)_2$

40. A white powder when strongly heated gives off brown fumes. A solution of this powder gives a yellow precipitate with a solution of KI. When a solution of barium chloride is added to a solution of powder, a white precipitate results. This white powder may be:

1) A soluble sulphate
2) KBr or NaBr
3) $Ba(NO_3)_2$
4) $AgNO_3$

41. A metal nitrate reacts with KI solution to give yellow precipitate which on addition of excess of more concentrated solution (6M) of KI dissolves forming a solution. The cation of metal nitrate is:

1) Hg_2^{2+}
2) Ag^+
3) Pb^{2+}
4) Cu^{2+}

Group II A Radicals

42. Which diagram correctly represents change for given nitrate solution?

1) 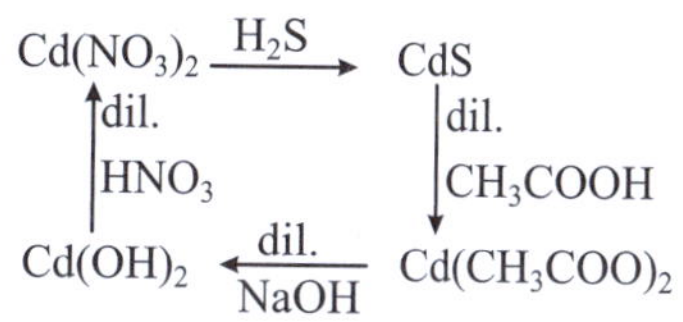

2) 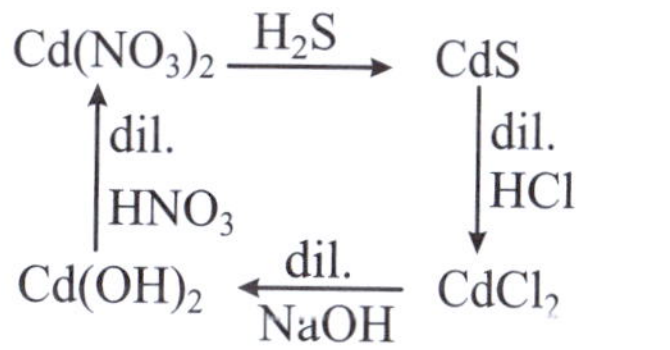

3) 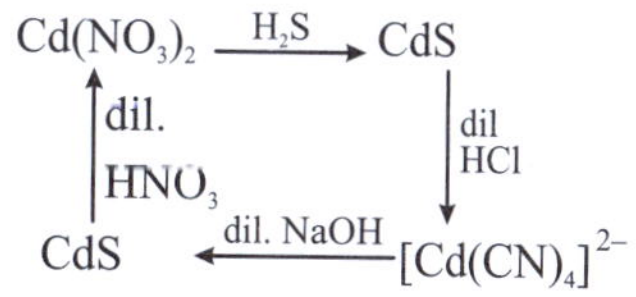

4) 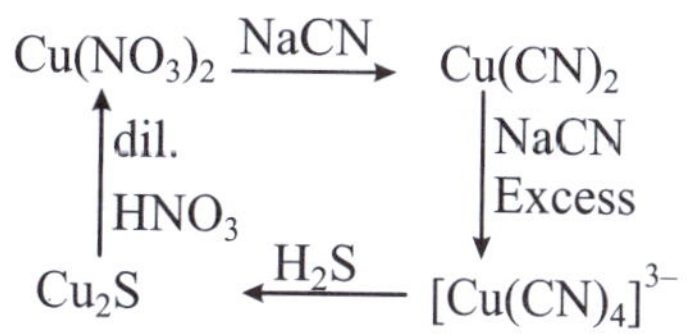

43. H_2S in the presence of HCl precipitates II group but not IV group because:

1) HCl activates H_2S.
2) HCl increases concentration of Cl^-.
3) HCl decreases concentration of S^{2-}.
4) HCl lowers the solubility of H_2S in solution.

44. Black sulphides is not formed by:

1) Cu^{2+}
2) Sb^{3+}
3) Pb^{2+}
4) Ag^+

45. A metal chloride original solution (i.e. O.S) on mixing with K_2CrO_4 solution gives a yellow precipitate soluble in aqueous sodium hydroxide. The metal may be:
1) Mercury 2) Iron
3) Silver 4) Lead

46. Sb_2S_3 is:
1) Black 2) Yellow
3) Orange 4) None

Group III Radicals

47. An original solution of an inorganic salt in dilute HCl gives a brown colouration with potassium hexacyanidoferrate (III) and reddish brown colouration with sodium acetate solution. The cation of the salt is:
1) Ni^{2+} 2) Fe^{3+}
3) Cu^{2+} 4) None of these

48. To avoid the precipitation of hydroxides of Zn^{2+}, Mn^{2+} and Ni^{2+} along with those of Fe^{3+}, Al^{3+} and Cr^{3+}, the third group solution should be:
1) Concentrated HNO_3 is added.
2) Treated with excess of NH_4Cl.
3) Concentrated H_2SO_4 is added.
4) Treated with excess of NH_4OH.

49. When NH_4Cl is added to a solution of NH_4OH:
1) The dissociation of NH_4OH increases.
2) The concentration of OH^- increases.
3) The concentrations of both OH^- an NH_4^+ increase.
4) The concentration of OH^- ion decreases.

50. $K_4[Fe(CN)_6] + Fe^{3+}(aq.) \longrightarrow$ products;
Here products is/are:
1) $K_2Fe[Fe(CN)_6]$ 2) $Fe[Fe(CN)_6]$
3) $Fe_3[Fe(CN)_6]_2$ 4) $Fe_4[Fe(CN)_6]_3$

Group IV Radicals

51. An aqueous solution of colourless metal sulphate M gives a white ppt, with NH_4OH. This is soluble in excess of NH_4OH. On passing H_2S through this solution a white ppt. is formed. The metal M in the salt is:
1) Ni 2) Ba
3) Al 4) Zn

52. A coloured solution of x metal ion salt gives following reactions:
(i) It gives white precipitate with sodium hydroxide which becomes brown on exposure to air.
(ii) It gives white precipitate with ammonia solution which insoluble in ammonium salts.
(iii) Its acidified solution containing a few drops of dilute $AgNO_3$ on boiling with ammonium peroxodisulphate gives a reddish violet solution.
The cation of the salt is:
1) Mn^{2+} 2) Zn^{2+}
3) Al^{3+} 4) Ni^{2+}

53. Co^{2+} + KCN (not in excess) → precipitate.
Select the incorrect statement with respect to the precipitate.
1) It is yellow in colour
2) It is reddish-brown in colour
3) It dissolves in excess of the reagent forming a brown solution
4) It is obtained when brown solution (option 3) is acidified with dilute HCl in the cold.

54. The ion that can not be precipited by H_2S in presence of dil. HCl:
1) Pb^{2+} 2) As^{3+}
3) Cu^{2+} 4) Co^{2+}

55. To increase significantly the concentration of free Zn^{2+} ion in a solution of the complex ion $[Zn(NH_3)_4]^{2+}$
$Zn^{2+}(aq.) + 4NH_3(aq.) \rightleftharpoons [Zn(NH_3)_4]^{2+}(aq.)$
add to the solution some:
1) H_2O 2) HCl (aq.)
3) NH_3(aq.) 4) NH_4Cl (aq.)

Group V, VI and Zero Radicals

56. An aqueous solution of salt gives white precipitate with $AgNO_3$ solution as well as with dilute $H_2SO_4$and a green flame test. It maybe:
1) $Pb(NO_3)_2$ 2) $Ba(NO_3)_2$
3) $BaCl_2$ 4) $CuCl_2$

57. A metal salt solution form a yellow ppt with potassium chromate in acetic acid, a white ppt with dilute sulphuric acid but gives no ppt with sodium chloride or iodide. The white precipitate obtained when sodium carbonate is added to the metal salt solution will consist of:
1) Lead carbonate 2) Basic lead carbonate
3) Barium carbonate 4) Strontium nitrate

58. Na_2CO_3 cannot be used in place of $(NH_4)_2CO_3$ for the precipitation of V group because:
1) NaI interferes in the detection of V group.
2) Conc. of CO_3^{2-} is very low.
3) Na will react with acid radicals.
4) Mg will be precipitated.

59. Ba ion is precipitated in fifth group as its:
1) Oxides 2) Sulphates
3) Carbonates 4) Sulphides

60. The brown precipitate formed by passing ammonia into Nessler's reagent in due to the formation of
1) HgI_4^{2-} 2) NH_2O–Hg–HgI
3) NH_2–Hg–O–Hg–I 4) NH_3–Hg–I

61. Orange coloured sodium cobaltinitrite $Na_3[Co(NO_2)_6]$ is used for the detection of K^+ ions which gives ____ ppt. due to the formation of Potassiumsodiumcobaltinitrite $K_2Na[Co(NO_2)_6]$
1) White 2) Orange
3) Yellow 4) Brown

62. The addition of K_2CO_3 (aq.) to the following solution is expected to produce a precipitate in every case but that one which does not produce precipitate is:

1) $BaCl_2$(aq.) 2) $CaBr_2$(aq.)
3) Na_2SO_4(aq.) 4) $Pb(NO_3)_2$ (aq.)

Chemistry involved in Enthalpy Change, Formation of Colloid, and Reaction Kinetics in Chemical Processes

63. The enthalpy of neutralization of a strong dibasic acid by a dilute solution of NaOH is nearly

1) –27.4 kcal/eq 2) 13.7 kcal/eq
3) –13.7 kcal/eq 4) –13.7 kcal/mol

64. Which of the following is not a lyophobic sol?

1) Ferric hydroxide 2) Egg albumin sol
3) Arsenious sulphide 4) Aluminium hydroxide

65. The enthalpy of solution of hydrated copper(II) sulfate can be determined by measuring the temperature changes when a known amount of copper(II) sulfate is added to a fixed volume of water in a calorimeter. What assumption is not made when calculating the enthalpy change?

1) There is negligible heat losses in the calorimeter
2) The specific heat capacity of the solution is the same as water
3) The density of solution is the same as water
4) The specific heat capacity of solution is different from water

66. For kinetic study of the reaction of iodide ion with H_2O_2 at room temperature:

A. Always use freshly prepared starch solution.
B. Always keep the concentration of sodium thiosulphate solution less than that of KI solution.
C. Record the time immediately after the appearance of blue colour.
D. Record the time immediately before the appearance of blue colour.
E. Always keep the concentration of sodium thiosulphate solution more than that of KI solution.

Choose the correct answer from the options given below:

1) A, B, C only 2) A, D, E only
3) D, E only 4) A, B, E only

67. What is the colour of the complex formed in kinetic study of the reaction of iodide ion with H_2O_2?

1) Red 2) White
3) Blue 4) Yellow

68. Lyophilic sols are

1) Irreversible sols
2) They are prepared from inorganic compounds
3) Coagulated by adding electrolytes
4) Self-stabilising

High Order Time Intensive MCQs

1. A substance on treatment with dilute H_2SO_4 liberates a colourless gas which produces (*i*) turbidity with baryta water and (*ii*) turns acidified dichromate solution green. The reaction indicates the presence of:

1) CO_3^{2-} 2) S^{2-}
3) SO_3^{2-} 4) NO_2^-

2. A colourless gas is dissolved in water and the resulting solution turns red litmus blue ; the gas may have been which one of the following?

1) HCl 2) H_2S
3) SO_2 4) NH_3

3. Which of the following salt will evolve sulphur dioxide gas along with formation of yellowish turbidity when treated with dilute H_2SO_4?

1) Sodium sulphide 2) Sodium sulphite
3) Sodium thiosulphate 4) Sodium sulphate

4. In the test for iodine, I_2 is treated with sodium thiosulphate ($Na_2S_2O_3$):

$Na_2S_2O_3 + I_2 \longrightarrow NaI +$ _____

1) $Na_2S_4O_6$ 2) Na_2SO_4
3) Na_2S 4) Na_3ISO_4

5. Which metal gives violet colour in oxidising flame when heated with borax?

1) Fe 2) Pb
3) Co 4) Mn

6. KBr on reaction with conc. H_2SO_4 gives reddish-brown gas:

1) Bromine 2) Mixture of bromine and HBr
3) HBr 4) NO_2

7. An inorganic salt when heated evolves coloured gas which bleaches moist litmus paper. The evolved gas is:

1) NO_2 2) SO_2
3) N_2O 4) I_2

8. Which of the following combines with Fe(II) ions to form a brown complex?

1) N_2O 2) NO
3) N_2O_3 4) N_2O_4

9. Colourless salt (A) + dil. H_2SO_4 or CH_3COOH + KI $\longrightarrow$ blue colour with starch, (A) can be:

1) K_2SO_3 2) Na_2CO_3
3) NH_4NO_2 4) NH_4Cl

10. There are four test tubes containing dilute HCl, $BaCl_2$, $CdCl_2$ and KNO_3 solutions. Which of the following reagents will help in the identification of $BaCl_2$?

1) NaOH 2) K_2CrO_4
3) $AgNO_3$ 4) Both (2) and (3)

11. Which one of the following ions does not give borax bead test?

1) Cr^{3+} 2) Cu^{2+}
3) Mn^{2+} 4) Zn^{2+}

12. A brick red colour is imparted to Bunsen flame by a:
1) Ca salt 2) Sr salt
3) Na salt 4) Co salt

13. Which one of the following metal salts produces a blue coloured bead in cobalt nitrate charcoal cavity test?
1) Zn^{2+} 2) Mg^{2+}
3) Sn^{2+} 4) Al^{3+}

14. $BaCl_2$ solution gives a white precipitate with a solution of a salt, which dissolves in dilute hydrochloric acid with the evolution of colourless, pungent smelling gas. The gas as well as the salt both are used as bleaching agent in the textile industries. The salt contains:
1) Sulphite 2) Sulphide
3) Acetate 4) Carbonate

15. Pink colour of acidified $KMnO_4$ is decolourised but there is no evolution of any gas. This may happen with the compound containing the following acid radical:
1) SO_3^{2-} 2) NO_2^-
3) S^{2-} 4) All of these

16. When KI is added to acidified solution of sodium nitrite:
1) NO gas is liberated and I_2 is set free.
2) N_2 gas is liberated and HI is produced.
3) N_2O gas is liberated and I_2 is set free.
4) N_2 gas is liberated and HOI is produced.

17. Zinc pieces are added to acidified solution of SO_3^{2-}. Gas liberated can:
1) Turn lead acetate paper black.
2) Turn lime water milky.
3) Give white precipitate with $AgNO_3$ solution.
4) None of these.

18. An inorganic salt when heated with concentrated H_2SO_4 evolves a colourless pungent smelling gas but with concentrated H_2SO_4 and MnO_2, evolves a coloured pungent smelling gas which bleaches moist litmus paper. The coloured gas is:
1) NO_2 2) Cl_2
3) Br_2 4) I_2

19. When CS_2 layer containing both Br_2 and I_2 is shaken with excess of chlorine (Cl_2) water, the violet colour due to I_2 disappears and a pale yellow colour appears in the solution. The disappearance of violet colour and appearance of pale yellow colour is due to the formation of:
1) I_3^- and Br_2 respectively.
2) HIO_3 and BrCl respectively.
3) ICl and BrCl respectively.
4) I^- and Br^- respectively.

20. Which of the following will not give positive chromyl chloride test?
1) Copper chloride, $CuCl_2$.
2) Mercuric chloride, $HgCl_2$.
3) Zinc chloride, $ZnCl_2$.
4) Anilinium chloride $C_6H_5NH_3Cl$.

21. A white sodium salt dissolves in water to give a solution which is neutral to litmus. When silver nitrate solution is added to the aforementioned solution, a white precipitate is obtained which does not dissolve in dilute HNO_3. The anion is:
1) CO_3^{2-} 2) Cl^-
3) SO_3^{2-} 4) S^{2-}

22. A one litre flask is full of reddish brown bromine fumes. The intensity of brown colour of vapour will not decrease appreciably on adding to the flask some:
1) Pieces of marble. 2) Animal charcoal powder.
3) Carbon tetrachloride. 4) Carbondisulphide.

23. Which of the following halide is soluble in water?
1) AgF 2) AgCl
3) AgBr 4) AgI

24. Which of the following radical can not be confirmed by using dil.HCl?
1) S^{2-} 2) $S_2O_3^{2-}$
3) CO_3^{2-} 4) NO_2^-

25. A white solid imparts a violet colour to a Bunsen flame. On being heated with concentrated H_2SO_4, the solid gives violet vapours that turn starch paper blue. The salt may be:
1) KI 2) NaI
3) MgI_2 4) $CaBr_2$

26. NaCl, NaBr, NaI mixture on adding conc. H_2SO_4 gives gases, respectively:
1) HCl, HBr, HI 2) HCl, Br_2, I_2
3) Cl_2, Br_2, I_2 4) None of these

27. Potassium chromate solution is added to an aqueous solution of a metal chloride. The yellow precipitate thus obtained is insoluble in acetic acid. The precipitate is subjected to flame test, the colour of the flame is:
1) Lilac 2) Apple green
3) Crimson red 4) Brick red

28. What are the products formed when an aqueous solution of magnesium bicarbonate is boiled?
1) $MgCO_3$, H_2O, CO_2 2) $Mg(HCO_3)_2$, H_2O
3) $Mg(OH)_2$, H_2O, CO_2 4) Mg, CO, H_2O

29. NaX (Sodium salt of particular anion 'X') gives brisk effervescence of Y with dilute HCl. On heating, NaX evolves gas Y which can be completely absorbed in conc. KOH solution and is colorless odourless gas. Hence X and Y respectively are:
1) HSO_3^-, SO_2 2) HS^- and H_2S
3) HCO_3^-, CO_2 4) $HC_2O_4^-$ and CO_2 + CO

30. A mixture is known to contain NO_3^- and NO_2^-. Before performing ring test for NO_3^-, the aqueous solution should be made free of NO_2^-. This is done by heating aqueous extract with:
1) Conc. HNO_3 2) dil. HNO_3
3) Urea 4) Zinc dust

31. Acetic anhydride does not react with:

1) ethanol
2) phenol
3) aniline
4) N, N-dimethyl aniline

32. Which of the following will not give positive iodoform test?

1) 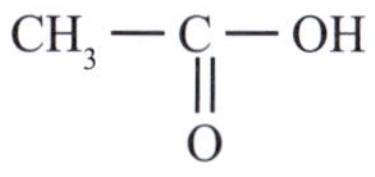

2) 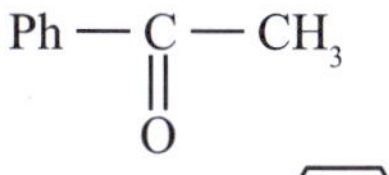

3) 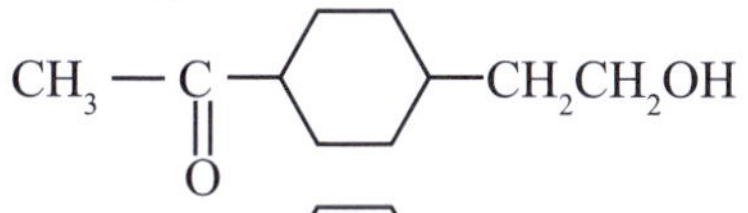

4) 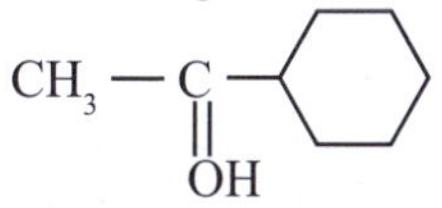

33. Cu^{2+} and Ag^{+} are both present in the same solution. To precipitate one of the ions and leaves the other in solution, add

1) $H_2S(aq)$
2) $HCl(aq)$
3) $HNO_3(aq)$
4) $NH_4NO_3(aq)$

34. Which one is soluble in dilute HCl?

1) $AgNO_3$
2) $Pb(NO_3)_2$
3) $Hg_2(NO_3)_2$
4) $Cu(NO_3)_2$

35. Which will not give colour with $FeCl_3$?

1) 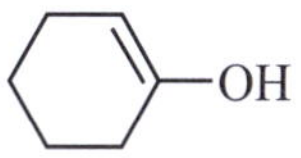

2) 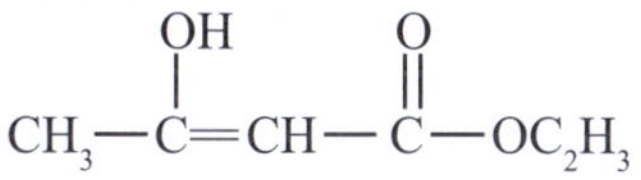

3) 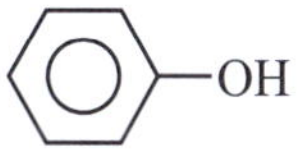

4) 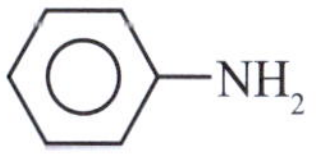

36. Which of the following compounds decolourise Br_2—H_2O and also given positive test with neutral $FeCl_3$?

1) 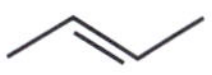
2) 
3) 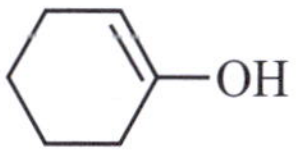
4) 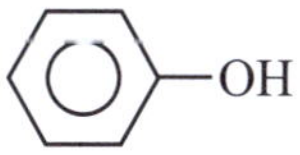

37. In the Lassaigne's test for the detection of nitrogen in an organic compound, the Prussian blue colour is due to the formation of _______________.

1) $Fe_4[Fe(CN)_6]_3$
2) $Na[Fe(CN)_5NOS]$
3) $K_3[Fe(CN)_6]$
4) $K_4[Fe(CN)_6]$

38. When a salt is heated with dilute H_2SO_4 and $KMnO_4$ solution, the pink colour of $KMnO_4$ is discharged, the salt is:

1) A sulphite
2) A carbonate
3) A nitrate
4) A bicarbonate

39. Solution of a salt in dilute H_2SO_4 or acetic acid produces deep blue colour with starch iodide solution. The salt contains:

1) Br^-
2) I^-
3) NO_3^-
4) NO_2^-

40. A test tube containing a nitrate and another containing a bromide and MnO_2 are treated with concentrated H_2SO_4. The reddish brown fumes evolved are passed through water. The water will be coloured by:

1) The nitrate
2) The bromide
3) Both (1) and 2)
4) None of these

41. A solution containing SCN^- ions can be used to test one or more out of: Fe^{3+}, Co^{2+}, Cu^{2+}.

1) Fe^{3+} and Co^{2+}
2) Co^{2+}, Cu^{2+}
3) Fe^{3+}, Cu^{2+}
4) All of these

42. A blue colouration (in solution or precipitate) is not obtained when:

1) Ammonium hydroxide dissolves in copper sulphate.
2) Cobalt chloride reacts with NH_4SCN in presence of amyl alcohol.
3) Ferric chloride reacts with sodium ferrocyanide.
4) Aqueous ammonia reacts with white precipitate of silver (I) chloride.

43. A metal nitrate solution reacts with dilute hydrochloric acid to give a white precipitate which is soluble in concentrated potassium chloride. White precipitate on passing of hydrogen sulphide gas is converted into black precipitate. The black precipitate on boiling with hydrogen peroxide (3%) is converted again to a white precipitate which is soluble in ammonium acetate. The cation of the metal nitrate is:

1) Pb^{2+}
2) Ag^{+}
3) Hg^{2+}
4) Bi^{3+}

44. A coloured solution of an inorganic salt rcacts with potassium thiocyanate to give first a black precipitate, which slowly turns white. The salt solution also gives black precipitate with H_2S gas in slightly acidic medium. The black precipitate dissolves in potassium cyanide forming a colourless solution. The basic radical present in the inorganic salt is:

1) Bi^{3+}
2) Cu^{2+}
3) Hg^{2+}
4) None of these

45. Select the incorrect statement:

1) Marsh's test involves the reduction of soluble arsenic compound to arsine by nascent hydrogen in acidic solution and its subsequent decomposition into hydrogen and metallic arsenic as a brownish-black mirror on heating in a glass tube mixed with hydrogen.
2) Ammonical silver nitrate gives metallic silver mirror with saturated solution of hydrazine sulphate.
3) Red precipitate of silver (I) chromate is soluble in dilute nitric acid and ammonia solution.
4) None of these

Match the Columns MCQs

1. Match the Column-I with Column-II.

	Column-I Cation		Column-II Group Number
(a)	Co^{2+} and Ni^{2+}	(i)	Group-II
(b)	Al^{3+}	(ii)	Group-I
(c)	Pb^{2+}	(iii)	Group-III
(d)	Cu^{2+}	(iv)	Group-V
(e)	Ba^{2+}	(v)	Group-IV

1) (a)-(v) (b)-(iii) (c)-(ii) (d)-(i) (e)-(iv)
2) (a)-(v) (b)-(iii) (c)-(ii) (d)-(iv) (e)-(i)
3) (a)-(iii) (b)-(ii) (c)-(iv) (d)-(v) (e)-(i)
4) (a)-(ii) (b)-(iv) (c)-(v) (d)-(iii) (e)-(ii)

2. $CuCO_3$ was strongly heated to obtain a residue (1) and gas (2). The residue obtained was treated with a salt of sodium (X) and oxide (Y), which produced a blue colored glassy compound (3) on heating in oxidizing flame. The same combination of (X) and (Y) gave a green colored glassy compound (4) when $Cr_2(SO_4)_3$ was heated with them in oxidizing flame. Match the following accordingly:

	Column-I Colour		Column-II Ppt.
(a)	(1)	(i)	$Cu(BO_2)_2$
(b)	(2)	(ii)	Na_2CO_3
(c)	(3)	(iii)	CuO
(d)	(X)	(iv)	CO
		(v)	Cu_2O
		(vi)	CO_2
		(vii)	$NaBO_2$
		(viii)	$Cr(BO_2)_2$

1) (a)-(vii) (b)-(vi) (c)-(v) (d)-(iii)
2) (a)-(iii) (b)-(vi) (c)-(i) (d)-(viii)
3) (a)-(iii) (b)-(vi) (c)-(i) (d)-(vii)
4) (a)-(ii) (b)-(iv) (c)-(v) (d)-(iii)

3. Match the Column-I with Column-II.

	Column-I Test		Column-II Anions
(a)	Brown ring test	(i)	NO_2^-
(b)	Chromyl chloride test	(ii)	CO_3^{2-}
(c)	Baryta water ($Ba(OH)_2$) test	(iii)	NO_3^-
(d)	Sodium nitroprusside solution	(iv)	S^{2-}
(c)	Starch iodide paper test	(v)	Cl^-

1) (a)-(i, iii) (b)-(v) (c)-(ii) (d)-(iii, iv) (e)-(i)
2) (a)-(i, iii) (b)-(v) (c)-(ii) (d)-(iv) (e)-(i)
3) (a)-(iii) (b)-(ii, iv) (c)-(iv) (d)-(iv, v) (e)-(i)
4) (a)-(ii) (b)-(iv) (c)-(v) (d)-(iii) (e)-(ii)

4. Match Column-I with Column-II and select the correct answer using the codes given below the lists:

	Column-I Colour		Column-II Ppt.
(a)	White turbidity	(i)	$IO_3^- + SO_2 + \text{Starch} \rightarrow$
(b)	Rotten egg smell	(ii)	$SO_2 + MnO_4^- \rightarrow$
(c)	Colourless solution	(iii)	$Zn + NaOH + SO_2 \rightarrow$
(d)	Blue colour	(iv)	$CO_2 + Ca(OH)_2 \rightarrow$

1) (a)-(iii) (b)-(iv) (c)-(ii) (d)-(i)
2) (a)-(i) (b)-(iii) (c)-(ii) (d)-(iv)
3) (a)-(iii) (b)-(ii) (c)-(iv) (d)-(i)
4) (a)-(iv) (b)-(iii) (c)-(ii) (d)-(i)

5. Match the reactions/reaction conditions listed in Column-I with the characteristics/precipitate colour of the reaction products listed in Column-II.

	Column-I Colour		Column-II Ppt.
(a)	$2NiS + 2HNO_3 + 6HCl \xrightarrow[\text{extract with water}]{\Delta}$	(i)	Black precipitate and its chloride imparts greenish blue flame.
(b)	$CoCl_2 + 4NH_4CNS \xrightarrow{\text{amyl alcohol}}$	(ii)	Blue colour in organic layer.
(c)	$CuCl_2 + NaOH \xrightarrow{\Delta}$	(iii)	White precipitate soluble in ammonium acetate.
(d)	$H_2SO_4 + PbCl_2 \longrightarrow$	(iv)	Green colour solution.
(e)	$HgCl_2 + KI \text{ (Not in excess)} \longrightarrow$	(v)	Scarlet / red precipitate.

1) (a)-(iv) (b)-(ii) (c)-(i) (d)-(iii) (e)-(v)
2) (a)-(iv) (b)-(ii) (c)-(i) (d)-(v) (e)-(i)
3) (a)-(iii) (b)-(v) (c)-(i) (d)-(iv) (e)-(i)
4) (a)-(ii) (b)-(v) (c)-(i) (d)-(iv) (e)-(v)

6. Match the Column-I with Column-II.

	Column-I (Colour)		Column-II (Gas)
(a)	Violet color gas	(i)	SO_2
(b)	Rotten egg smell	(ii)	NO_2
(c)	Brown color	(iii)	I_2
(d)	Suffocating smell	(iv)	H_2S
		(v)	Br_2

1) (a)-(iii) (b)-(iv) (c)-(ii, v) (d)-(i)
2) (a)-(i) (b)-(ii) (c)-(iv) (d)-(iii)
3) (a)-(iii) (b)-(ii) (c)-(i) (d)-(iv)
4) (a)-(iv) (b)-(ii) (c)-(i) (d)-(iii)

7. Match the Column-I with Column-II.

	Column-I (Reaction)		Column-II (Precipitate obtained as one of the products)
(a)	$Hg^{2+}(aq) + I^-_{(aq)} \rightarrow$	(i)	Yellow precipitate
(b)	$Cu^{2+}(aq) + [Fe(CN)_6]^{4-}(aq) \rightarrow$	(ii)	Brown precipitate
(c)	$Mg^{2+}(aq) + NH_3(aq) + HPO_4^{2-}(aq) \rightarrow$	(iii)	White precipitate
(d)	$Pb^{2+}(aq) + CrO_4^{2-}(aq) \rightarrow$	(iv)	Red precipitate

1) (a)-(i) (b)-(ii) (c)-(iii) (d)-(iv)
2) (a)-(ii) (b)-(i) (c)-(iv) (d)-(iii)
3) (a)-(iv) (b)-(ii) (c)-(iii) (d)-(i)
4) (a)-(iii) (b)-(iv) (c)-(ii) (d)-(i)

8. Match the Column-I with Column-II.

	Column-I (Colour)		Column-II (Reaction)
(a)	White ppt.	(i)	$CO_3^{2-} + BaCl_2 \rightarrow$
(b)	Brown ppt.	(ii)	$CO_3^{2-} + Pb(NO_3)_2 \rightarrow$
(c)	Black ppt.	(iii)	$Ag_2CO_3 \xrightarrow{\Delta} Ag_2O$
		(iv)	$Na_2S + (CH_3COO)_2Pb$

1) (a)-(iv) (b)-(i) (c)-(iii, ii)
2) (a)-(i, ii) (b)-(iii) (c)-(iv)
3) (a)-(iii) (b)-(ii) (c)-(iii, iv)
4) (a)-(iii, i) (b)-(ii) (c)-(iv)

9. Match the Column-I with Column-II.

	Column-I (Ppt Colour)		Column-II (Reaction)
(a)	Pale green solution	(i)	$Na_2S + Hg(NO_3)_2 \rightarrow$
(b)	Black ppt.	(ii)	$Na_2S + Na_2[Fe(CN)_5NO] \rightarrow$
(c)	Finally purple color	(iii)	$Pb^{2+} + SO_4^{2-} \rightarrow$
(d)	White ppt.	(iv)	$[Ag(NH_3)_2]Cl + 2H^+ \rightarrow$
		(v)	$Fe^{3+} + S_2O_3^{2-} \rightarrow$

1) (a)-(v) (b)-(i) (c)-(ii) (d)-(iii, iv)
2) (a)-(i) (b)-(ii) (c)-(iv) (d)-(iii, v)
3) (a)-(v, iv) (b)-(iii) (c)-(ii) (d)-(i)
4) (a)-(i) (b)-(iii) (c)-(iv, v) (d)-(ii)

10. Match the Column-I with Column-II.

	Column-I (Compound)		Column-II (Colour)
(a)	HgO	(i)	Purple solution
(b)	$BaCO_3$	(ii)	Yellow ppt
(c)	$Na_4[Fe(CN)_5NOS]$	(iii)	Dark brown
(d)	KI_3	(iv)	White ppt

1) (a)-(iii) (b)-(iv) (c)-(i) (d)-(ii)
2) (a)-(ii) (b)-(iv) (c)-(i) (d)-(iii)
3) (a)-(ii) (b)-(iv) (c)-(iii) (d)-(i)
4) (a)-(ii) (b)-(iv) (c)-(iii) (d)-(i)

11. Match the Column-I with Column-II.

	Column-I (Basic Radical)		Column-II (Group)
(a)	Fe^{+3}	(i)	II group
(b)	Ni^{+2}	(ii)	V group
(c)	Ba^{2+}	(iii)	IV group
(d)	Pb^{2+}	(iv)	III group

1) (a)-(iv) (b)-(ii) (c)-(i) (d)-(iii)
2) (a)-(ii) (b)-(iv) (c)-(i) (d)-(iii)
3) (a)-(iii) (b)-(i) (c)-(ii) (d)-(iv)
4) (a)-(iv) (b)-(iii) (c)-(ii) (d)-(i)

12. Match the following:

	Property to be determined		Method used for determination
(a)	Estimation of carbon and hydrogen in an organic compound	(i)	Lassaigne's test
(b)	Estimation of nitrogen in aniline	(ii)	Carius method
(c)	Estimation of chlorine in carbon tetrachloride	(iii)	Liebig's method
(d)	Detection of nitrogen, sulphur and halogens	(iv)	Kjeldahl's method

1) (a)-(i) (b)-(ii) (c)-(iii) (d)-(iv)
2) (a)-(iv) (b)-(iii) (c)-(i) (d)-(ii)
3) (a)-(ii) (b)-(i) (c)-(iv) (d)-(iii)
4) (a)-(iii) (b)-(iv) (c)-(ii) (d)-(i)

13. Match the column:

	Test		Detection of functional group
(a)	Schiff's Test	(i)	Carboxylic acid
(b)	Litmus Test	(ii)	Phenol
(c)	Liebermann nitroso test	(iii)	Halogen
(d)	Beilstein Test	(iv)	Aldehyde

1) (a)-(ii) (b)-(iii) (c)-(iv) (d)-(i)
2) (a)-(ii) (b)-(iv) (c)-(iii) (d)-(i)
3) (a)-(iii) (b)-(i) (c)-(iv) (d)-(ii)
4) (a)-(iv) (b)-(i) (c)-(ii) (d)-(iii)

14. Match the items given in Column-I and Column-II.

	Column-I		Column-II
(a)	Protective colloid	(i)	$FeCl_3$ + NaOH
(b)	Liquid-liquid colloid	(ii)	Lyophilic colloids
(c)	Positively charged colloid	(iii)	Emulsion
(d)	Negatively charged colloid	(iv)	$FeCl_3$ + hot water

1) (a)-(iii) (b)-(iii, iv) (c)-(ii) (d)-(i)
2) (a)-(ii, iv) (b)-(ii) (c)-(iii) (d)-(i)
3) (a)-(ii) (b)-(iii) (c)-(iv) (d)-(i)
4) (a)-(iv) (b)-(iii) (c)-(i) (d)-(ii)

15. Match the Column-I with Column-II.

	Column-I (Reaction)		Column-II (Product)
(a)	$CO_3^{2-} \xrightarrow{\text{dil } H_2SO_4}$	(i)	H_2S
(b)	$S^{2-} \xrightarrow{\text{dil } H_2SO_4}$	(ii)	SO_2
(c)	$SO_3^{2-} \xrightarrow{\text{dil } H_2SO_4}$	(iii)	CO_2
(d)	$S_2O_3^{2-} \xrightarrow{\text{dil } H_2SO_4}$	(iv)	$S + SO_2$

1) (a)-(iii) (b)-(iv) (c)-(i) (d)-(ii)
2) (a)-(ii) (b)-(i) (c)-(iv) (d)-(iii)
3) (a)-(iii) (b)-(i) (c)-(ii) (d)-(iv)
4) (a)-(ii) (b)-(iv) (c)-(i) (d)-(iii)

16. Match the Column-I with Column-II.

	Column-I (Molecule)		Column-II (Characteristic Odour)
(a)	CO_2	(i)	Rotten egg smell
(b)	SO_2	(ii)	Suffocating smell of burning sulphur
(c)	H_2S	(iii)	Vineger like smell
(d)	CH_3COOH	(iv)	Odour less

1) (a)-(iv) (b)-(ii) (c)-(i) (d)-(iii)
2) (a)-(ii) (b)-(iv) (c)-(i) (d)-(iii)
3) (a)-(iii) (b)-(i) (c)-(ii) (d)-(iv)
4) (a)-(ii) (b)-(iv) (c)-(i) (d)-(iii)

Correct & Incorrect MCQs

1. Experiment of detection of S^{-2} is given below in alkaline medium

$S^{2-} + [Fe(CN)_5NO]^{2-} \rightarrow$ product

Select the correct statement about the experiment.

1) Colour of the complex compound formed in product is purple or violet
2) Oxidation number of Fe, in complex compound formed in product is +3
3) IUPAC name of complex compound formed in product is pentacyanonitrosonium ferrate (II)
4) Colour of complex formed is green

2. Which of the following statement is incorrect?

1) Colour of lead sulphide (PbS) precipitate is yellow
2) Colour of barium sulphate ($BaSO_4$) precipitate is white.
3) Colour of nickel sulphide (NiS) precipitate is black
4) Colour of lead chromate ($PbCrO_4$) precipitate is yellow.

3. Which of the following is incorrect statement?

1) Cu^+ can undergo disproportionation in aqueous solution
2) $Cu^+(aq)$ is more stable than $Cu^{2+}(aq)$
3) Hydration energy of Cu^{2+} is more negative than Cu^+
4) All of these

4. Which of the following statement is correct?

1) Fe^{3+} gives red colour with potassium thiocyanate
2) Fe^{3+} gives blue precipitate with potassium ferricyanide
3) Fe^{2+} gives brown colour with potassium ferrocyanide
4) Fe^{3+} gives pink colour with potassium ferrocyanide

5. Select the incorrect statement out of the following.

1) Group I basic radicals precipitate as chlorides
2) Group IV basic radicals precipitate as sulphites
3) Group V basic radicals precipitate as carbonates
4) Group III basic radicals precipitate as hydroxides.

6. Select the incorrect statement about lyophilic sol.

1) Egg albumin sol is a lyophilic sol.
2) Stability of lyophilic sols is primarily due to the solvation of colloidal particles by the solvent
3) The lyophilic sols cannot be directly formed by mixing and shaking the substance with a suitable liquid.
4) Ferric hydroxide is not a lyophilic sol.

Statement Based MCQs

Directions: These questions consist of two statements each, printed as Statement-I and Statement-II. While answering these questions, you are required to choose any one of the following four responses.

1) **Both Statement-I and Statement-II are correct.**
2) **Both Statement-I and Statement-II are incorrect.**
3) **Statement-I is correct & Statement-II is incorrect.**
4) **Statement-I is incorrect & Statement-II is correct.**

1. **Statement-I:** Starch is used as an indicator in iodometric and iodimetric titrations.

   Statement-II: The complex formed by the reaction of starch and iodine is blue in colour.

2. **Statement-I:** A solution of AgCl in NH_4OH gives a white precipitate with HNO_3.

   Statement-II: $[Ag(NH_3)_2]^+$ decomposes in the presence of HNO_3.

3. **Statement-I:** Iodine is liberated when KI is added to Cu^{2+} ions but Cl_2 is not liberated when KCl is added to Cu^{2+} ions.

   Statement-II: The reducing power of I is more than Cl.

4. **Statement-I:** PbS can be separated from ZnS by electrostatic separation method.

   Statement-II: PbS is a good conductor and ZnS is a poor conductor of electricity.

5. **Statement-I:** Hg_2Cl_2 does not give chromyl chloride test.

   Statement-II: Ammonia gas evolved by the action of sodium hydroxide on ammonium salts reacts with hydrochloric acid to give ammonium chloride, which is visible as dense brown fume.

Assertion & Reason MCQs

Directions: These questions consist of two statements each, printed as Assertion and Reason. While answering these questions, you are required to choose any one of the following four responses.

1) **Both Assertion and Reason are True and the Reason is a correct explanation of the Assertion.**
2) **Both Assertion and Reason are True but Reason is not a correct explanation of the Assertion.**
3) **Assertion is True but the Reason is False.**
4) **Assertion is False but Reason is True.**

1. **Assertion (A):** During test for nitrogen with Lassaigne extract on adding $FeCl_3$ solution, sometimes a red precipitate is formed.

   Reason (R): This indicates that sulphur is also present.

2. **Assertion (A):** During digestion with conc. H_2SO_4, nitrogen of the compound is converted to $(NH_4)_2SO_4$.

   Reason (R): $(NH_4)_2SO_4$ on heating with alkali give NH_3 gas.

3. **Assertion (A):** Phenols having free para position respond positively to Liebermann's nitroso test.

   Reason (R): Secondary amines give positive Liebermann's test.

4. **Assertion (A):** The solubility of $Zn(OH)_2$ is enhanced in acidic as well as in basic medium as compared to water.

   Reason (R): $Zn(OH)_2$ is an amphoteric hydroxide.

5. **Assertion (A):** In acidic medium, Zn^{2+} is not precipitated by S^{2-} ions.

   Reason (R): Common ion effect reduces the concentration of S^{2-} ions to the minimum level.

Multi Correct MCQs

1. Which of the following metal ions characterized by flame test

   I. Sr^{2+}, II. Ba^{2+}
   III. Ca^{2+} IV. Cu^{2+}
   V. Zn^{2+} VI. Co^{2+}

   1) I, II and III 2) III, IV and V
   3) I, II, III and IV 4) III, IV, V and VI

2. The compound that are not of white in colour is

   A. ammonium sulphide
   B. lead sulphate
   C. lead iodide
   D. ammonium arsenomolybdate

   1) A and B 2) B and C
   3) Both A and D 4) A, C and D

3. Which of the following are white coloured salts?

   A. $SrSO_4$ B. $Mg(NH_4)PO_4$
   C. $BaCrO_4$ D. $Mn(OH)_2$
   E. $PbSO_4$ F. $PbCrO_4$

   1) A, B, C and D 2) B , C, D and E
   3) C, D, E and F 4) A, B , D and E

4. Which of the following nitrogen containing compounds give Lassaigne's test?

   A. Phenyl hydrazine B. Glycine
   C. Urea D. Hydrazine

   1) A and B 2) B and C
   3) Both A and D 4) A, B and C

Kattar NEET

Scan Me for Detailed Explanations

1. Lassaigne's test in qualitative analysis is used to detect:
   1) Nitrogen 2) Sulphur
   2) Chlorine 4) All of these
2. Presence of halogen in organic compounds can be detected by using:
   1) Leibig's test 2) Duma's test
   3) Kjeldahl's test 4) Beilstein test
3. In Lassaigne's test thiourea is converted into-
   1) NaCNS 2) Na_2S
   3) NaCN 4) Na_2SO_4
4. The compound that does not give a blue colour in Lassaigne's test is
   1) $C_6H_5—NH_2$ 2) CH_3CONH_2
   3) $NH_2—NH_2$ 4) $C_6H_5—NO_2$
5. Nitrogen containing organic compound when fused with sodium metal forms:
   1) $NaNO_2$ 2) NaCN
   3) $NaNH_2$ 4) NaNC
6. Select the incorrect statement regarding Lassaigne's test?
   1) Li and K are not used in this test.
   2) NH_4Cl, $NH_2–NH_2$ can't be detected by this test.
   3) Urea and Acetamide can not give this test.
   4) Benzamide and phenyl hydrazine can give this test.
7. Match the following regarding Lassaigne's test.

Column-I		Column-II	
Element		Color of ppt or solution	
A.	Nitrogen	p.	Black sulphide
B.	Sulphur	q.	Violet Complex
C.	Chloride	r.	Prussian blue solution
D.	Bromide	s.	Light yellow ppt
		t.	White ppt

   1) A → r; B → p, q; C → t; D → s
   2) A → p; B → s; C → p; D → q
   3) A → p; B → r; C → q; D → p
   4) A → q; B → p; C → r; D → s
8. If addition of dil. H_2SO_4 is made to an unknown salt, a colourless and odourless gas is produced then which of the following can be present?
   1) CO_3^{2-} 2) S^{2-}
   3) Cl^- 4) NO_3^-
9. A mixture upon adding conc. H_2SO_4 gives deep red fumes. Mixture may contain the anion pair:
   1) $Cr_2O_7^{2-}$ and Cl^- 2) Br^- and $Cr_2O_7^{2-}$
   3) NO_3^- and Cl^- 4) CrO_4^{2-} and NO_3^{2-}
10. A solution of a salt in concentrated H_2SO_4 produced a deep blue colour with starch iodide solution. The salt may contain:
    1) Chloride 2) Carbonate
    3) Acetate 4) Bromide
11. In the test for iodine, I_2 is treated with sodium thiosulphate ($Na_2S_2O_3$) to form
    1) $Na_2S_4O_6$ 2) Na_2SO_4
    3) Na_2S 4) Na_3ISO_4
12. Which anion does not liberate any gas with dilute as well as conc. H_2SO_4?
    1) NO_2^- 2) NO_3^-
    3) SO_3^{2-} 4) SO_4^{2-}
13. A white powder when strongly heated gives off brown fumes. A solution of this powder gives a yellow precipitate with a solution of KI. When a solution of barium chloride is added to a solution of powder, a white precipitate results. This white powder may be:
    1) A soluble sulphate 2) KBr or NaBr
    3) $Ba(NO_3)_2$ 4) $AgNO_3$
14. Consider the following observation;
    M^{n+} + HCl (dilute) $\longrightarrow$ white precipitate $\xrightarrow{\Delta}$ water soluble $\xrightarrow{CrO_4^{2-}}$ yellow precipitate.
    The metal ion M^{n+} will be:
    1) Hg^{2+} 2) Ag^+
    3) Pb^{2+} 4) Sn^{2+}
15. A metal chloride original solution (i.e. O.S) on mixing with K_2CrO_4 solution gives a yellow precipitate soluble in aqueous sodium hydroxide. The metal may be:
    1) Mercury 2) Iron
    3) Silver 4) Lead
16. When calomel reacts with NH_4OH solution the compound formed is:
    1) Hg_2Cl_2 2) $Hg(NH_2)Cl$
    3) $Hg(NH_3)_2Cl$ 4) $HgCl_2.NH_3$
17. Which of the sulphide is yellow?
    1) ZnS 2) CdS
    3) NiS 4) PbS
18. Match the column:

Column-I (Colour)		Column-II (Ppt.)	
A.	Co^{2+} and Ni^{2+}	p.	Group-II
B.	Al^{3+} and Cr^{3+}	q.	Group-I
C.	Ag^+ and Hg_2^{2+}	r.	Group-III
D.	Cu^{2+} and Cd^{2+}	s.	Group-V
E.	Ca^{2+} and Sr^{2+}	t.	Group-IV

1) A → t ; B → r; C → q ; D → p ; E → s
2) A → t ; B → r; C → q; D → s; E → p
3) A → r ; B → q ; C → s ; D → t ; E → p
4) A → q ; B → s ; C → t ; D → r ; E → q

19. Match column-I with column-II and select the correct answer using the codes given below the lists:

	Column-I (Colour)		Column-II (Ppt.)
A.	White turbidity	p.	$IO_3^- + SO_2$ + starch →
B.	Rotten egg smell	q.	$SO_2 + MnO_4^-$ →
C.	Colourless solution	r.	$Zn + NaOH + SO_2$ →
D.	Blue colour	s.	$CO_2 + Ca(OH)_2$ →

1) A → r ; B → s; C → q ; D → p
2) A → p ; B → r; C → q; D → s
3) A → r ; B → q ; C → s ; D → p
4) A → s; B → r; C → q; D → p

20. Match the column:

	Column-I Colour		Column-II Ppt.
A.	Brown ring test	p.	NO_2^-
B.	Chromyl chloride test	q.	BO_3^{3-}
C.	Green edged flame with ethyl alcohol	r.	NO_3^-
D.	Canary yellow precipitate	s.	PO_4^{3-}
E.	Starch iodide paper test	t.	Cl^-

1) A → p, r; B → t; C → q; D → r, s; E → p
2) A → p, r; B → t; C → q; D → s; E → p
3) A → r ; B → q, s; C → s; D → s, t; E → p
4) A → q ; B → s ; C → t ; D → r ; E → q

NEET & AIIMS Past Year MCQs

1. Which indicator is used in the titration of sodium hydroxide against oxalic acid and what is the colour change at the end point? **(2024 Re)**
   1) Phenolphthalein, pink to yellow
   2) Alkaline $KMnO_4$, colourless to pink
   3) Phenolphthalein, colourless to pink
   4) Methyl orange, yellow to pinkish red colour

2. Match List-I with List-II: **(2024 Re)**

	List-I (Test/reagent)		List-II (Radical identified)
A.	Lake Test	I.	NO_3^-
B.	Nessler's Reagent	II.	Fe^{3+}
C.	Potassium sulphocyanide	III.	Al^{3+}
D.	Brown Ring Test	IV.	NH_4^+

Choose the correct answer from the options given below :
1) A-IV, B-II, C-III, D-I 2) A-II, B-IV, C-III, D-I
3) A-II, B-III, C-IV, D-I 4) A-III, B-IV, C-II, D-I

3. Match List-I will List-II: **(2024 Re)**

	List-I Solid salt treated with dil. H_2SO_4		List-II Anion detected
A.	effervescence of colourless gas	I.	NO_2^-
B.	gas with smell of rotten egg	II.	CO_3^{2-}
C.	gas with pungent smell	III.	S^{2-}
D.	brown fumes	IV.	SO_3^{2}

Choose the correct answer from the options given below:
1) A-II, B-III, C-IV, D-I
2) A-IV, B-III, C-II, D-I
3) A-I, B-II, C-III, D-IV
4) A-II, B-III, C-I, D-IV

4. Given below are certain cations. Using inorganic qualitative analysis, arrange them in increasing group number from 0 to VI. **(2024)**
   A. Al^{3+}
   B. Cu^{2+}
   C. Ba^{2+}
   D. Co^{2+}
   E. Mg^{2+}

   Choose the correct answer from the options given below:
   1) E, C, D, B, A
   2) E, A, B, C, D
   3) B, A, D, C, E
   4) B, C, A, D, E

5. During the preparation of Mohr's salt solution (Ferrous ammonium sulphate), which of the following acid is added to prevent hydrolysis of Fe^{2+} ion? **(2024)**
   1) dilute nitric acid
   2) dilute sulphuric acid
   3) dilute hydrochloric acid
   4) concentrated sulphuric acid

Answer Key

NCERT Topic-wise MCQs

1	2	3	4	5	6	7	8	9	10	11	12	13	14	15	16	17
4	4	4	1	4	1	1	3	4	2	2	2	1	1	3	1	2
18	**19**	**20**	**21**	**22**	**23**	**24**	**25**	**26**	**27**	**28**	**29**	**30**	**31**	**32**	**33**	**34**
3	1	3	1	3	3	2	3	3	4	4	3	1	2	1	3	1
35	**36**	**37**	**38**	**39**	**40**	**41**	**42**	**43**	**44**	**45**	**46**	**47**	**48**	**49**	**50**	**51**
3	4	3	1	2	4	3	2	3	2	4	3	2	2	4	4	4
52	**53**	**54**	**55**	**56**	**57**	**58**	**59**	**60**	**61**	**62**	**63**	**64**	**65**	**66**	**67**	**68**
1	1	4	2	3	3	4	3	3	3	3	3	2	4	1	3	4

High Order Time Intensive MCQs

1	2	3	4	5	6	7	8	9	10	11	12	13	14	15	16	17
3	4	3	1	4	1	1	2	3	2	4	1	4	1	4	1	1
18	**19**	**20**	**21**	**22**	**23**	**24**	**25**	**26**	**27**	**28**	**29**	**30**	**31**	**32**	**33**	**34**
2	2	2	2	1	1	2	1	2	2	1	3	3	4	1	2	4
35	**36**	**37**	**38**	**39**	**40**	**41**	**42**	**43**	**44**	**45**						
4	3	1	1	4	2	4	4	1	2	4						

Match the Columns MCQs

1	2	3	4	5	6	7	8	9	10	11	12	13	14	15	16
1	3	2	4	1	1	3	2	1	2	4	4	4	3	3	1

Correct & Incorrect MCQs

1	2	3	4	5	6
1	1	2	1	2	3

Statement Based MCQs

1	2	3	4	5
1	1	1	1	3

Assertion & Reason MCQs

1	2	3	4	5
1	2	2	1	1

Multi Correct MCQs

1	2	3	4
3	4	4	4

Kattar NEET

1	2	3	4	5	6	7	8	9	10	11	12	13	14	15	16	17
4	4	1	3	2	3	1	1	1	4	1	4	4	3	4	2	2
18	**19**	**20**														
1	4	2														

NEET & AIIMS Past Year MCQs

1	2	3	4	5
3	4	1	3	2

Explanations

CLASS-XI-Explanations

Some Basic Concepts of Chemistry

NCERT Topic-wise MCQs

1. **(4)** Zero between '5' and '4' is also significant according to rule (2).

2. **(2)** $0.582 + 324.65 = 325.232$

   $= 325.23$ (upto second decimal place)

   As, 324.65 contains digits upto second decimal place, hence answer must be reported upto second decimal place.

3. **(3)** 7.00 is more accurate than 7.0 because the former has three significant figures while the later has two.

4. **(4)** π is irrational number (please note that $\pi \neq 22/7$ as 22/7 is a rational number) and it means it has infinite number of significant figures.

5. **(1)** $29.4406 + 3.2 + 2.25 = 34.8906$. As 3.2 has least number of decimal places, i.e., one, therefore sum should contain one decimal place only. After round off, reported sum is 34.9, which has three significant figures.

6. **(2)** Area = length × width

   $= (12.34\text{cm}) \times (1.23\text{cm}) = 15.1782\ \text{cm}^2$ (Calculated result)

   $\simeq 15.2$ (Reported answer)

   Since, three is the smallest number of significant figures used in original numbers, therefore the reported answer should contain only three significant figures.

7. **(3)** $0.2876\ g \times 9 = 2.5884\ g = 2.588\ g$

   NOTE:

   We do not round off this product to one significant figure because the number 9 is actually 9.000000 (as it is exact number and can be represented by writing infinite number of zeros after placing a decimal).

8. **(2)** According to Avogadro's law, equal volumes of all gases under similar conditions of temperature and pressure contain equal number of molecules.

   $\therefore$ x L of O_3 will contain $= 3 \times 10^{22}$ molecules

   Hence, x/2 L of O_3 will contain $= \frac{1}{2} \times 3 \times 10^{22} = 1.5 \times 10^{22}$ molecules

9. **(2)** Total mass of reactants

   $= 1.70 + (100 \times 0.1 \times 10^{-3} \times 36.5) = 2.065\ g$

   Mass of products $= 0.63 + 1.435 = 2.065\ g$

10. **(3)** 1g of X combines with Y = 35.5 g

    $\therefore$ 2 g of X combines with $Y = 2 \times 35.5\ g = 71\ g$

11. **(3) In First Oxide N_2O** : Mass of 2.24 L of nitrogen at STP = 2.8g

    $\therefore$ Mass of oxygen $= 4.4 - 2.8 = 1.6$ g

    In Second Oxide N_2O_2 : Mass of 22.4 L of nitrogen at STP = 28 g

    $\therefore$ Mass of oxygen $= 60 - 28 = 32$ g

    $\therefore$ In second oxide 2.8 g of nitrogen combines with 3.2 g of oxygen.

    Keeping the mass of nitrogen same in both the oxides, the different masses of oxygen which combines with 2.8 g of nitrogen are 1.6 g : 3.2 g or 1 : 2, This is a simple whole number ratio. This illustrates the law of multiple proportions.

12. **(2)** $BaCl_2 + H_2SO_4 \rightarrow BaSO_4 + 2HCl$

    20.8 g 9.8 g x 7.3 g

    mass of $BaSO_4$ produced,

    $x = 20.8 + 9.8 - 7.3 = 23.3$ g

13. **(2)** Law of conservation of mass not applicable to nuclear reactions.

14. **(4)**

H_2O_2	H_2O
H : O	H : O
5.93 : 94.07	11.2 : 88.8 or 5.93 : 47.0

    Ratio of different masses of O which combines with fixed mass of H is 94.07 : 47.0 or 2 : 1

15. **(3)** In B, 32 parts of X combines with Y = 84 parts

    $\therefore$ 16 parts of X combine with Y = 42 parts

    Now, number of parts of X in both B and C is equal.

    Different masses of Y which combine with same mass of X in B and C are in the ratio 3:5

    $\therefore \frac{\text{Mass of Y in B}}{\text{Mass of Y in C}} = \frac{3}{5}$

    $\therefore \text{Mass of Y in C} = \frac{5}{3} \times 42 = 70$ parts

16. **(4)** Average relative atomic weight

    $$A.W = \frac{20 \times 10 + 80 + 11}{100} = \frac{200 + 880}{100} = \frac{1080}{100} = 10.8g$$

17. **(3)** Minimum molecular mass of insulin would be the mass containing at least one atom of sulphur.

    Mass of one atom of sulphur = 32 amu

    3.4 amu S $\rightarrow$ 100 amu insulin

    $32\ \text{amu S} \rightarrow \frac{100}{3.4} \times 32 = 941$ amu

18. **(2)** Molecular mass of $C_6H_{12}O_6$
$= (6 \times 12.011) + (12 \times 1.008) + (6 \times 16.00)$
$= 72.066 + 12.096 + 96.00 = 180.162$ u
Therefore, molecular mass of glucose is 180.162 u.

19. **(3)** 1 g – atom of nitrogen $= 6.02 \times 10^{23}$ N atoms
1 mol of N = 1/2 mole of N_2 = 11.2 L of N_2 at S.T.P

20. **(2)** Molar mass of $CaCO_3 = 100$ g mol^{-1}
No. of moles in 10 g of $CaCO_3 = \frac{\text{Given mass}}{\text{Molar mass}} = \frac{10}{100}$
$= 0.1$ mol of $CaCO_3 = 0.1 \times 3$ g – atom of oxygen
$= 0.3$ g atoms of oxygen

21. **(2)** Magnesium bicarbonate is $Mg(HCO_3)_2$.
Molar mass of $Mg(HCO_3)_2 = 146$ g mol^{-1}
No. of moles of $Mg(HCO_3)_2 = \frac{\text{Given mass}}{\text{Molar mass}} = \frac{14.6}{146} = 0.1$
So, 14.6 g or 0.1 mole of $Mg(HCO_3)_2$ contains O atom $= 0.1 \times 6 \times N_A = 0.6 N_A$

22. **(2)** Molar mass of $CO_2 = 44$g mol^{-1}
$\therefore$ Mass of 4 mol of $CO_2 = 44 \times 4 = 176$ g
(3) 6.023×10^{23} atoms of hydrogen has mass = 1.008 g
$\therefore$ 12×10^{24} atoms of hydrogen has mass = 20.1 g
(4) 22.4 L of helium at N.T.P. has mass = 4 g
11.2 L of helium at N.T.P. has mass = 2 g
Thus; 4 moles of CO_2 has maximum mass

23. **(1)** Mass of 1 mole of substance remains same.

24. **(2)** Options can be verified by calculating no. of moles.

25. **(3)** Mass of 1 atom $= 1.8 \times 10^{-22}$ g
Mass of 6.022×10^{23} atoms $= 6.02 \times 10^{23} \times 1.8 \times 10^{-22} = 108.36$ g
$\therefore$ Atomic mass of element = 108.36 u

26. **(4)** 1 mole of H_2SO_4 produces ions = 3 moles
$\therefore$ 0.1 mol of H_2SO_4 produce ions = 0.3 mol
Number of ions $= 0.3 \times 6.02 \times 10^{23} = 1.8 \times 10^{23}$

27. **(3)** Mass of 1 mole of $CaCO_3 = 100$ g
$\therefore$ Mass of 0.1 mol of $CaCO_3 = 0.1 \times 100 = 10$ g
Mass of 6.023×10^{23} Ca^{2+} ions = 40
$\therefore$ Mass of 1.51×10^{23} Ca^{2+} ions $= \frac{1.51 \times 10^{23}}{6.023 \times 10^{23}} \times 40 = 10$ g
Mass of 1 mole of CO_3^{2-} ions = 60 g
Mass of 0.16 mol of CO_3^{2-} ions $= 0.16 \times 60 = 9.6$ g
Mass of 6.02×10^{23} Br atoms = 80
Mass of 7.525×10^{22} Br atoms $= \frac{7.525 \times 10^{22}}{6.02 \times 10^{23}} \times 80 = 10$ g

28. **(2)** Number of moles in 2 g of oxygen $= \frac{2}{16} = \frac{1}{8}$ mol
Number of moles in 4 g of sulphur $\frac{4}{32} = \frac{1}{8}$ mol
Hence, 2 g of oxygen and 4 g of sulphur contain same number of atoms.

29. **(1)** Moles of sugar added $= \frac{1.71}{342} = 5 \times 10^{-3}$
Carbon atoms added $= 12 \times 5 \times 10^{-3} \times 6.02 \times 10^{23}$
$= 3.61 \times 10^{22}$

30. **(2)** 0.1 mole P_4 contains $= 4 \times 0.1 \times 6.02 \times 10^{23}$ atoms
0.05 mole of S_8 contains $= 8 \times 0.05 \times 6.02 \times 10^{23}$ atoms
Hence, both contains same number of atoms.

31. **(4)** 82 g of $N_2 = \frac{82}{28}$ g-molecule = 2.92 moles

32. **(4)** Since; mass of oxygen is fixed. Therefore, the number of g-atom and also the number of atoms will be fixed.
i.e, Atoms of Oxygen $= \frac{1 \times 2}{32}$ in O_2
$= \frac{1}{16}$ in atomic Oxygen $= \frac{1 \times 3}{48}$ in O_3

33. **(4)** $25.0\text{mg} = 25.0\text{mg} \times \left(\frac{1\text{g}}{10^{-3}\text{mg}}\right) \times \left(\frac{1\text{mol}}{152\text{g}}\right) \times \left(\frac{27 \times 6.02 \times 10^{23}\text{atoms}}{1\text{ mol}}\right)$
$= \frac{25 \times 27 \times 6.02}{152} \times 10^{20}\text{atoms} = 2.67 \times 10^{21}$ atoms
Molar mass of $C_{10}H_{16}O = 120 + 16 + 16$
$= 152$ g mol^{-1}

34. **(2)** Gram molecular mass of $O_2 = 32$ g
Equivalent mass $= \frac{\text{Atomic mass of O}}{\text{valency}} = \frac{16}{2} = 8$
32 g of oxygen contain oxygen atoms $= 2 \times N_A$
Therefore, 8 g of oxygen contains oxygen atom
$= \frac{2 \times N_A}{32} \times 8 = \frac{N_A}{2}$

35. **(4)**
No of mol of triatomic gas $= \frac{224\text{ mL}}{22400\text{ mL mole}^{-1}} = 10^{-2}$ mol
No. of molecules $= 10^{-2}$mol $\times$ (6.02×10^{23} molecules mol^{-1})
$= 6.02 \times 10^{21}$ molecules
No. of atoms of gas $= 3 \times 6.02 \times 10^{21} = 18.06 \times 10^{21}$
1.806×10^{22} atoms has mass = 1 g
1 atom has mass $= \frac{1}{1.806 \times 10^{22}\text{g}} = \frac{10^{-22}\text{g}}{1.806}$
$= 0.554 \times 10^{-22}$ g $= 5.54 \times 10^{-23}$ g

36. **(3)** Mass of 1 mole of electrons
$= 9.11 \times 10^{-31}$ kg $\times 6.02 \times 10^{23} = 5.5 \times 10^{-7}$ kg mol^{-1}

37. **(3)** 12 moles of hydrogen atoms are present with 4 moles of oxygen atoms in $(NH_4)_3PO_4$.
If 3.18 mol of hydrogen atom are there, then number of moles of oxygen atoms
$= \frac{4}{12} \times 3.18 = \frac{3.18}{3}$ mol = 1.06 mol

38. (1) $\because$ 5.8 L of gas has mass = 7.5 gm

$\therefore$ 22.4 L of gas has mass = $\frac{7.5}{5.8} \times 22.4 = 28.96$

So molecular weight = 29

So, molecular formula of compound is NO

39. (3)

(1) Molar mass of CO_2 = 44 g mol^{-1}

44 g CO_2 = 3 × 6.02 × 10^{23} atoms

$\therefore$ 1 g of $CO_2 = \frac{3 \times 6.02 \times 10^{23}}{44}$ = 0.41 × 10^{23} atoms

(2) Molar mass of C_8H_8 = 104 g mol^{-1}

1g of $C_8H_8 = \frac{16 \times 6.02 \times 10^{23}}{104}$ atoms = 0.93 × 10^{23} atoms

(3) Molar mass of C_2H_6 = 30g mol^{-1}

1g of $C_2H_6 = \frac{8 \times 6.02 \times 10^{23}}{30}$ atoms

= 1.60 × 10^{23} atoms

(4) Molar mass of LiF = 7 + 19 = 26

1 g of LiF = $\frac{2 \times 6.02 \times 10^{23}}{26}$ atoms

= 0.46 × 10^{23} atoms

40. (3)

Number of atoms in 4.0 g of NaOH = $\frac{4}{40} \times N_A = \frac{N_A}{10}$ atom

Number of atoms in 0.5 M Na_2SO_4

$= \frac{0.5 \times 100 \times 2}{1000} \times N_A = \frac{N_A}{10}$ [$\because$ n = M × V (in L)]

41. (4)

$SO_2 \rightarrow$ Contains 3 atom

$n = \frac{\text{wt.}}{\text{M.wt.}} = \frac{64}{64}$

Number of atoms present in 64 gm of $SO_2 = \frac{64}{64} \times 3N_A$

= 3 × 6.02 × 10^{23}

42. (1) Total e^- + p + n = 18

Total number of protons, electrons and neutrons in 12 gm of $^{12}_{6}C$

= 18 × N_A = 1.084 × 10^{25}

43. (1) 111 g anhydrous $CaCl_2$ = 1 mole $CaCl_2$

1 mole $CaCl_2$ contain 1 mole Ca^{+2} and 2 mole Cl^-

44. (4)

1 g molecule of CO_2 = 1 mole CO_2

1 mole CO_2 occupy → 22.4 L

45. (2) 1 mole $BaCl_2.2H_2O$ contains = 2 mol H_2O

mole ($BaCl_2.2H_2O$) = $\frac{488}{244}$

[Molar mass of $BaCl_2.2H_2O$ = 244]

mole ($BaCl_2.2H_2O$) = 2

So moles of H_2O = 2 × 2 = 4 mole

46. (2) $C_2H_5OH + Na \rightarrow C_2H_5ONa + \frac{1}{2}H_2$

47. (3) 22.4 L vol. of gas at STP = 1 mole

48. (4) No. of molecules = $\frac{W}{M.w} \times N_A$

49. (4) No. of moles of A = $\frac{75}{75} = 1$

No. of moles of B = $\frac{25}{25} = 1$

$\therefore$ Molecular formula of the compound will be AB

50. (4)

	Mass in "g"	% age	$\frac{\%}{\text{At.mass}}$	Simplest ratio
C	24	40	40/12 = 3.33	1
H	4	6.66	6.66/1= 6.66	2
O	32	53.33	53.33/16 = 3.33	1

Empirical formula : CH_2O

51. (1)

Element	%	At. mass	$\frac{\%}{\text{At. mass}}$	Simplest ratio
A	25	12.5	25/12.5 = 2	1
B	75	37.5	75/37.5 = 2	1

$\therefore$ The empirical formula of the compound is AB

52. (2) As, an oxide of a metal (M) contains 40 % by mass of oxygen.

i.e., 100 g of the oxide will contain 40 g of oxygen and 60 g of metal

It is given that, atomic mass of metal (M) = 24 g

$\therefore$ 60 g of metal corresponds to = $\frac{60}{24}$ = 2.5 moles

40 g of oxygen corresponds to = $\frac{40}{16}$ = 2.5 moles

Ratio of metal and oxygen = 1 : 1

Hence, empirical formula = MO

53. (1) Mol. mass of Fe $(CHCOO)_2$ = 170 g

Fe present in 170 g of Fe $(CHCOO)_2$ = 56 g

$\therefore$ Fe present in 100 gms of Fe $(CHCOO)_2 = \frac{56 \times 100}{170}$

= 32.9 mg

Total Fe in 400mg of capsule = 32.9 mg

Percentage of Fe = $\frac{32.9 \times 100}{400}$ = 8.2%

54. (1) $\because$ 40gm NaOH contains 16 gm of oxygen

$\therefore$ 100gm of NaOH contains $\frac{16}{40} \times 100$ = 40% oxygen.

55. (2) Moles of Na = $\frac{1.15g}{23 \text{ g mol}^{-1}}$ = 0.05 mol

Moles of C = $\frac{3.01 \times 10^{22}}{6.02 \times 10^{23}}$ = 0.05 mol

Moles of O = 0.1

Moles ratio, Na : C : O = 1 : 1 : 2

∴ Empirical formula – $NaCO_2$

56. **(4)** Percentage of carbon = 75%

Percentage of hydrogen = (100 –75) = 25%

Element	% composition	Atomic mass	No. of moles of atoms	Mole ratio
C	75	12	$\frac{75}{12} = 6.25$	$\frac{6.25}{6.25} = 1$
H	25	1	$\frac{25}{6.25} = 25$	$\frac{25}{6.25} = 4$

Therefore, empirical formula of the compound = CH_4

57. **(3)** $100 \rightarrow 20$

$? \rightarrow 28$

20 gm Nitrogen → 100gm compound

$1 \text{ gm Nitrogen} = \frac{100}{20} \text{ gm compound}$

$28 \text{ gm Nitrogen} = \frac{100}{20} \times 28 = 140 \text{ gm compound}$

58. **(4)** Let us suppose, molecule of alkaloid contains x atoms of nitrogen

$\% \text{ of nitrogen} = \frac{14 \times x}{162} \times 100 = 17.28$

$\Rightarrow x = \frac{17.28 \times 162}{14 \times 100} = 2$

Hence, number of nitrogen atoms in alkaloid = 2

59. **(4)** Empirical formula of the compound = C_3H_4O

Empirical formula mass = 12 × 3+ 4 × 1 + 16 = 56

$n = \frac{170 \pm 5}{56} = 3$

∴ Molecular formula of compound = $(C_3H_4O)_3 = C_9H_{12}O_3$

60. **(1)** 44 g CO_2 = 1 mol of carbon atoms

$3.38 \text{ g } CO_2 = \frac{1}{44} \times 3.38 - 0.0768 \text{ mol carbon atoms}$

18 g H_2O = 2 mol of hydrogen atoms

$0.690 \text{ g } H_2O = \frac{2}{18} \times 0.690 = 0.0767 \text{ mol hydrogen atoms}$

Molar ratio of C : H = 1:1

∴ Empirical formula of hydrocarbon is CH

61. **(1)** Empirical formula of compound : CH_2O

Empirical formula mass of compound (CH_2O)

= 12 + 2 + 16 = 30

Given molecular mass = 90

$n = \frac{\text{Molecular mass}}{\text{Empirical formula mass}} = \frac{90}{30} = 3$

Therefore, molecular formula = $(CH_2O)_3 = C_3H_6O_3$.

Hence, molecular formula of the compound is $C_3H_6O_3$.

62. **(4)** Moles of iodine $= \frac{254}{127} = 2$

Moles of oxygen $= \frac{80}{16} = 5$

Formula : I_2O_5

63. **(3)** X_2O

X = 7

$X = 7 \times 2 = 14$ g

64. **(1)** 100 gm caffeine has 28.9 gm nitrogen

$194 \text{ gm caffeine has} = \frac{28.9}{100} \times 194 = 56.06\text{gm}$

$\therefore \text{No of atoms in caffeine} = \frac{56.06}{14} \approx 4.$

65. **(4)** The balanced equation is:

$4\,NH_3 + 5O_2 \rightarrow 4NO + 6H_2O$

4 × 17 g of NH_3 required O_2 = 5 mol

$\therefore 6.8 \text{ g of } NH_3 \text{ require } O_2 = \frac{5 \times 6.8}{4 \times 17} = 0.5 \text{ mol}$

66. **(1)** $CaCl_2 + 2AgNO_3 \rightarrow Ca(NO_3)_2 + 2\,AgCl$

111 g 2 × 143.5 g

$CaCl_2$ required to produce 2 × 143.5 g of AgCl = 111 g

∴ $CaCl_2$ required to produce 14.35 g of AgCl

$= \frac{111 \times 14.35}{2 \times 143.5} = 5.55 \text{ g}$

67. **(3)** $2\,NaOH + H_2SO_4 \rightarrow Na_2SO_4 + 2H_2O$

Pure H_2SO_4 required for 1 mole of

NaOH = 1/2 mole = 49 g

$70\% \; H_2SO_4 \text{ required for 1 mole of NaOH} = \frac{49 \times 100}{70} = 70$

68. **(1)** $2KClO_3 \rightarrow 2\,KCl + 3O_2$

2 × 122.5 g 3 × 32 g

3 × 32 g of O_2 is produced from $KClO_3$ = 245 g

$48\text{g of } O_2 \text{ is produced from } KClO_3 = \frac{245 \times 48}{3 \times 32} = 122.5 \text{ g}$

Since the $KClO_3$ is only 80% pure

$\therefore 80\% \; KClO_3 \text{ needed} = \frac{122.5 \times 100}{80} = 153.12 \text{ g}$

69. **(3)** $MCl_x + x\,AgNO_3 \rightarrow M(NO_3)_x + xAgCl$

Molar ratio: 1 x 1 x

No. of moles of $AgNO_3$ required for precipitation

$= 0.6 \times \frac{500}{1000}$ [∵ n = M × V (in L)]

– 0.3 moles

i.e., 0.1 mol of MCl_x requires $AgNO_3$ = 0.3 mol

$\therefore 1 \text{ mol of } MCl_x \text{ requires } AgNO_3 = \frac{0.3}{0.1} \text{mol} = 3 \text{ mol}$

∴ The value of x = 3

70. (2) $CaCO_3 + 2HCl \rightarrow CaCl_2 + CO_2 + H_2O$

100 g   73 g   44 g

Here, $CaCO_3$ is a limiting reagent

100 g of $CaCO_3$ produce = 44 g of CO_2

$\therefore$ 20 g of $CaCO_3$ produce = $\frac{44}{100} \times 20\text{g}$ of CO_2

= 8.8 g of CO_2

71. (1) $CaCO_3 + 2HCl \rightarrow CaCl_2 + CO_2 + H_2O$

25 ml of 0.75 molar HCl = $\frac{25}{1000} \times 0.75$

= 0.01875 moles of HCl

Moles of $CaCO_3$ required = $\frac{\text{moles of HCl}}{2} = \frac{0.01875}{2}$

= 9.375×10^{-3} moles

Mass of $CaCO_3$ required = 9.375×10^{-3} moles × 100 g/mole

= 0.94 g

72. (1) 22.4 L of SO_2 at N.T.P = 1 mole

11.2 L of SO_2 at N.T.P = 1/2 moles

$SO_2 + 2H_2S \rightarrow 3S + 2H_2O$

Here SO_2 is a limiting reactant

$\therefore$ Moles of sulphur produced = 3 × moles of SO_2 = 1.5 moles

73. (1) $I_2 + 10HNO_3 \longrightarrow 2HIO_3 + 10NO_2 + 4H_2O$

Moles of $I_2 = \frac{5}{254}$

1 mole I_2 reacts with 10 moles of HNO_3

$\frac{5}{254}$ moles of I_2 reacts with $\rightarrow 10 \times \frac{5}{254} = 0.196$

Wt. of HNO_3 = 0.196 × 63 = 12.348 g

74. (2) M = w/molar mass × 1/V in L

75. (3) $Zn + 2HCl \rightarrow ZnCl_2 + H_2$

Here: HCl is a limiting reagent

$\therefore$ Moles of H_2 produced $= \frac{\text{moles of HCl}}{2} = \frac{0.52}{2} = 0.26$

High Order Time Intensive MCQs

1. (4) $d = s \times t = (3 \times 10^8 \text{m/s})(10^{-15} \text{s})$ [$\because$ 1 femto sec = 10^{-15} sec]

$= 3 \times 10^{-7}$ m $= 3 \times 10^{-4}$ mm = 0.0003 mm

2. (4) $2C_8H_7SO_3^-Na^+ + Ca^{2+} \longrightarrow (C_8H_7SO_3)_2Ca + 2Na^+$

2 mole   1 mole

(2×206 = 412g)

412 g of resin take up Ca^{2+} ions = 1 mole

$\therefore$ 1g of resin will take up $Ca^{2+} = \frac{1}{412}$ mole

3. (1) Converting 200 mg of CO_2 into moles

Number of moles of $CO_2 = \frac{\text{Mass}}{\text{Molar mass}} = \frac{200 \times 10^{-3}}{44}$

Converting 10^{21} molecules into moles

Number of moles of CO_2 removed = $\frac{10^{21}}{6 \times 10^{23}}$

Number of moles of CO_2 left = $\frac{200 \times 10^{-3}}{44} - \frac{10^{21}}{6 \times 10^{23}}$

$= 2.88 \times 10^{-3}$

4. (3) 1 mL H_2O = 1 g of H_2O

$(\because d_{H_2O} = 1 \text{ g/mL})$

1 mL of water contains = 20 drop

20 drops of water = 1 g

1 drop of water = $\frac{1}{20}$ g = 0.05 g

0.05 g $H_2O = \frac{0.05}{18}$ mole

Number of molecules = $\frac{0.05}{18} \times 6 \times 10^{23} = 1.673 \times 10^{21}$

5. (2) For A

50 g A would contain atoms = N_A

xg A would contain atoms = $\frac{N_A}{50} \times x\text{g}$ = n atoms

For B

100 g B would contain atoms = N_A

20 xg B would contains atoms = $\frac{N_A}{100} \times 20x\text{g} = \frac{N_A}{5}x$

$\frac{N_A}{50}x$ corresponds to atoms = n

$\frac{N_A}{5}x$ corresponds to atoms = $\frac{n}{\frac{N_A x}{50}} \times \frac{N_A}{5}x$

$= \frac{n \times 50}{N_A x} \times \frac{N_A}{5}x = 10$ n

6. (2) $3NaBH_4 + 4BF_3 \longrightarrow 3NaBF_4 + 2B_2H_6$

Since BF_3 is in excess, the limiting reagent is $NaBH_4$.

To obtain 2 mole B_2H_6, $NaBH_4$ required = 3 mole

To obtain 0.200 mole B_2H_6, $NaBH_4$ required = $\frac{3}{2} \times 0.200$

= 0.300 mole

Because, the yield is 70%, Hence

$x \times \frac{70}{100} = 0.300$

$x = 0.300 \times \frac{100}{70} = \frac{3}{7} = 0.429$ moles

7. (1) $Ag_2S \rightarrow 2Ag + S$

Molar mass of Ag = 108 g mol^{-1}

1 g Ag = $\frac{1}{108}$ mol Ag

To obtain 2 mole Ag, mole of Ag_2S required = 1

To obtain $\frac{1}{108}$ mole Ag, moles of Ag_2S required

$= \frac{1}{2} \times \frac{1}{108} = \frac{1}{216}$ mole

Grams of Ag_2S required $= \frac{1}{216} \times 248$ g

[$\because$ Molar mass of $Ag_2S = 248$ g mol^{-1}]

Ag_2S required $= \frac{248}{216}$ g

1.24 g Ag_2S is obtained from ore = 100 g

$\frac{248}{216}$ g Ag_2S is obtained from ore $= \frac{100}{1.24} \times \frac{248}{216}$

$= \frac{24800}{267.84} = 92.6$ g

8. **(3)** 8 M means 8 moles of ethanol in 1 L of solution

8 moles ethanol = 46 × 8 = 368 g of ethanol

[$\because$ Molar mass of ethanol = 46 g mol^{-1}]

Let us consider one litre of this solution-

So, mass of solution = 1025 g [$\because$ $d = \frac{m}{V}$]

Mass of water = 1025 – 368 = 657 g

$\because$ Molality $= \frac{\text{No. of moles of solute}}{\text{Mass of solvent in kg}}$

$\therefore$ Molality $= \frac{8.0 \times 1000}{657} = 12.18\text{m}$

9. **(2)** $CrSO_4 \rightarrow Cr^{2+} + SO_4^{2-}$

$Cr_2(SO_4)_3 \rightarrow 2Cr^{3+} + 3SO_4^{2-}$

$Cr^{2+} \rightarrow SO_4^{2-}$

$2Cr^{3+} \rightarrow 3SO_4^{2-}$

$\frac{2}{3}Cr^{3+} \rightarrow SO_4^{2-}$

$\therefore \frac{Cr^{2+}}{Cr^{3+}} = \frac{1}{\frac{2}{3}} = 3:2$ $\because$ Sulphates are equal in number

10. **(2)** No. of moles of A = 0.1

No. of moles of B = 1

No of moles of C = 0.036

C is limiting reagent as it is present in least amount

AB_2C_3 formed $= \frac{0.036}{3} = 0.012$

So, molar mass of AB_2C_3 $= \frac{4.8}{0.012} = 400$

[$\because$ $n = \frac{\text{Mass}}{\text{Molar mass}}$]

$\Rightarrow 60 + 2x + 80 \times 3 = 400$

x = 50

11. **(3)** 100 g haemoglobin contains Fe = 0.34 g

1 mole haemoglobin contains Fe = 4 mole Fe^{2+}

= 4 × 56 = 224 g [$\because$ Mass of Fe = 56 amu]

0.34 g Fe is present in haemoglobin = 100 g

$\therefore$ 224 g Fe is present in haemoglobin $= \frac{100}{0.34} \times 224$

= 65882 g

12. **(1)** $Na_2SO_4 + BaCl_2 \rightarrow BaSO_4 + 2NaCl$

No. of moles of $BaSO_4 = \frac{10}{233} = 0.0429$

No. of moles of Na_2SO_4 required = 0.0429

Molarity of $Na_2SO_4 = \frac{\text{Moles of } Na_2SO_4}{V/1000}$

$5 = \frac{0.0429}{V/1000}$

$V = \frac{42.9}{5} = 8.58$ mL.

Hence, required volume = 8.58 mL

13. **(4)** Let mass of Mn = x gm

Moles of Mn $= \frac{x}{55} = 0.018x$

Let mass of O = x gm

Moles of O $= \frac{x}{16} = 0.0625\,x$

$\frac{0.018\,x}{0.018\,x} = 1 \Rightarrow \frac{0.0625\,x}{0.018\,x} \approx 3.5$

$Mn_1O_{3.5} \Rightarrow Mn_2O_7$

Hence, empirical formula = Mn_2O_7.

14. **(1)** In 12 g of carbon, the amount of C – 14 $= \frac{12 \times 2}{100} = 0.24$ g

C-14 atoms in 1 mole i.e., 14 g = 6.02×10^{23}

$\therefore$ C-14 atoms in 0.24 g $= \frac{0.24 \times 6.02 \times 10^{23}}{14}$

$= 1.03 \times 10^{22}$ atoms

15. **(2)** $H_2SO_4 + 2NaOH \rightarrow Na_2SO_4 + 2H_2O$

1 L of 0.1 M H_2SO_4 = 0.1 mol of H_2SO_4

1 L of 0.1 M NaOH = 0.1 mol of NaOH

As, 2 moles of NaOH react with 1 mole of H_2SO_4

$\therefore$ 0.1 mole of NaOH react with $= \frac{1}{2} \times 0.1 = 0.05$ mol of H_2SO_4

Hence, NaOH is the limiting reagent.

As, 2 moles of NaOH produce = 1mole of Na_2SO_4

$\therefore$ 0.1 mole of NaOH will produce $= \frac{1}{2} \times 0.1$

= 0.05 mol of Na_2SO_4

Therefore, mass of Na_2SO_4 = Moles × molar mass of Na_2SO_4

= 0.05 × 142 = 7.10 g

16. **(2)** 143.5 g AgCl contain 108 g Ag

$\therefore$ 2.87 g AgCl contain $= \frac{108}{143.5} \times 2.87 = 2.16\text{g Ag}$

17. **(3)** Molar mass of $KClO_3$ = 122.5g

Mass of KCl = 74.5 g

% loss $= \frac{122.5 - 74.5}{122.5} \times 100 = \frac{48}{122.5} \times 100 = 39.18$

18. **(2)** The candela is the luminous intensity of a source that emits monochromatic radiation of $\nu = 540 \times 10^{12}$ Hz, and has a radiant intensity in that direction of $\frac{1}{683}$ w/sr. It is unit of Candela.

Hence, A and B are 540 and 683 respectively.

19. **(1)** $C : H : N = 9 : 1 : 3.5$

$\therefore$ Mole ratio $= C : H : N = \frac{9}{12} : \frac{1}{1} : \frac{3.5}{14}$

Simplest ratio = 3 : 4 : 1

$\therefore$ Empirical formula = C_3H_4N

Empirical formula mass = 54

$$\therefore n = \frac{\text{Molar mass}}{\text{Empirical formula mass}} = \frac{108}{54} = 2$$

$\therefore$ Molecular formula = $(C_3H_4N)_2 = C_6H_8N_2$

20. **(2)** No. of moles of urea $= \frac{6.02 \times 10^{20}}{6.02 \times 10^{23}} = 10^{-3}$ mol

$$\therefore \text{Molarity of urea solution} = \frac{\text{No. of moles}}{\text{V(in L)}}$$

$$= \frac{10^{-3}}{0.1} = 0.01 \text{ M}$$

21. **(4)** From the equation, it is clear that, 6 mol of HCl produces 3 mol of H_2 or 1 mole of HCl produces

$$= \frac{3 \times 22.4}{6} \text{ L of } H_2 = 11.2 \text{ L of } H_2$$

22. **(2)** 4.4 g CO_2 = 0.1 mole CO_2 molecule

2.24 L H_2 at STP = 0.1 mole H_2 molecule

Total number of moles of CO_2 and H_2 molecules

= 0.2 mole molecules

Total number of molecules present in the container

$= 0.2 \times 6.023 \times 10^{23} = 1.20 \times 10^{23}$

23. **(2)** $M + H_2SO_4 \rightarrow MSO_4 + H_2$

$$\frac{\text{Mole of M}}{1} = \frac{\text{Mole of } H_2SO_4}{1}$$

$\frac{1}{a} = \frac{4.08}{98}$ [$\because$ Molecular mass of H_2SO_4 = 98 g]

$$a = \frac{98}{4.08} = 24.01$$

Thus atomic weight of M = 24.

24. **(1)** Let molar mass of compound = y g

% by wt of C = 36

So, $\frac{24}{y} \times 100 = 36$

$$\frac{2400}{36} = y$$

y = 66.6 g

So, 66.6 g has moles = 1

$\therefore$ 10 g has moles $= \frac{1}{66.6} \times 10 = 0.15$

25. **(1)** $C_8H_{18} + \frac{25}{2}O_2 \rightarrow 8CO_2 + 9H_2O$

Moles of CO_2 formed $= \frac{7.04}{44}$ moles

Now, from equation

Moles of H_2O formed when 8 moles of CO_2 produced = 9

Moles of H_2O formed when $\frac{7.04}{44}$ moles of CO_2 produced

$$= \frac{9}{8} \times \frac{7.04}{44} = \frac{63.36}{8 \times 44} = 0.18 \text{ moles}$$

Wt. of H_2O formed $= 0.18 \times 18 = 3.24$ g

26. **(3)** $2KHCO_3 \xrightarrow{\Delta} K_2CO_3 + H_2O + CO_2 \uparrow$

Loss of wt. will be because of CO_2 escaped

Total $KHCO_3$ taken = 1.50 g

Percentage purity = 80%

$\therefore$ Pure $KHCO_3 = 1.50 \times \frac{80}{100} = 1.2$ g

Molecular mass of $KHCO_3 = 39 + 1 + 12 + 3 \times 16 = 100$ g

Moles of pure $KHCO_3 = \frac{1.2}{100} = 0.012$ moles

From balanced equation-

2 moles of $KHCO_3$ yield moles of CO_2 = 1

0.012 moles of $KHCO_3$ yield moles of $CO_2 = \frac{1}{2} \times 0.012$

= 0.006 moles

Wt. of CO_2 formed $= 0.006 \times 44 = 0.264$ g

27. **(1)** Moles of N $= \frac{28}{14} = 2$

Moles of metal, M $= \frac{100 - 28}{a} = \frac{72}{a}$

Mole ratio, M:N $= \frac{72}{a} : 2 = 3:2$

$$\frac{72}{a} = 3$$

$$a = \frac{72}{3} = 24$$

28. **(1)** $MnO_2 + 4HCl \rightarrow MnCl_2 + Cl_2 + 2H_2O$

From balanced equation,

71 gm Cl_2 is produced from HCl $= 4 \times 36.5$ g

2.5 gm Cl_2 is produced from HCl $= \frac{4 \times 36.5 \times 2.5 g}{71} = 5.14$ g

Now, HCl is 36% by mass, which means-

36 g HCl is obtained from HCl solution = 100 g

$\therefore$ 5.14 g HCl is obtained from HCl solution

$$= \frac{100 \times 5.14}{36} = 14.27 \text{ g}$$

29. **(4)** Density of solution = 1.25 g/mL

29.2% (w/w) means that 29.2 g of HCl is present in 100 g of solution

So, $\rho(\text{density of solution}) = \frac{\text{Wt. of solution}}{\text{Vol. of solution}}$

$1.25 \text{ g/mL} = \frac{100}{\text{Vol. of solution}}$

$V = \frac{100}{1.25} \text{mL}$

$\text{Molarity of solution} = \frac{\text{No. of moles of solute}}{\text{Vol. of solution (in L)}}$

$M = \frac{29.2/36.5}{100/1.25} \times 1000 = 10 \text{ M}$

Apply, $M_1V_1 = M_2V_2$

$0.4 \times 200 = 10 \times V$

V = 8 mL

30. **(2)** Molality = 9 which means,

9 moles of solute (H_2SO_4) are present per kg of solvent.

So, no. of moles of solute in 1000 g solvent = 9

No. of moles of solute in 910 g solvent $= \frac{9}{1000} \times 910$

$= \frac{819}{100} = 8.19$ moles

Wt. of 8.19 moles of $H_2SO_4 = 98 \times 8.19 = 802.62$ g

% age (w/v) of $H_2SO_4 = \frac{802.62\text{g}}{1000 \text{ mL}} \times 100 = 80.26\%$

31. **(3)** $2 \text{ Mg} + O_2 \rightarrow 2 \text{ MgO}$

2×24 g Mg reacts with $O_2 = 32$ g

30 g Mg reacts with $O_2 = \frac{32}{2 \times 24} \times 30 = 20$ g

So, 10 g O_2 will be left unreacted

2×24 g Mg forms MgO = 2×40 g

$\therefore$ 30 g Mg forms MgO $= \frac{2 \times 40}{2 \times 24} \times 30 = 50$ g

32. **(2)** 8 oxygen atoms are present in molecule of $Mg_3(PO_4)_2$ = 1

$0.25 \times N_A$ oxygen atoms are present in molecule of

$Mg_3(PO_4)_2 = \frac{1}{8} \times 0.25N_A$

So, molecules of $Mg_3(PO_4)_2$ required $= \frac{0.25}{8} N_A$

Moles of $Mg_3(PO_4)_2 = \frac{\frac{0.25}{8} N_A}{N_A} = \frac{0.25}{8} N_A \times \frac{1}{N_A}$

$= 0.03125 = 3.125 \times 10^{-2}$

33. **(3)** Number of moles of $A = \frac{x}{40}$

Number of atoms of $A = \frac{x}{40} \times$ Avogadro no. = y (say)

Or $x = \frac{40y}{\text{Avogadro no.}}$

Number of moles of $B = \frac{2x}{80}$

Number of atoms of B

$= \frac{2x}{80} \times \text{Avogadro no} = \frac{2}{80} \times \frac{40y}{\text{Avogadro.no.}} \times \text{Avogadro no.} = y.$

34. **(2)** 100 g haemoglobin has Fe = 0.33 g

67200 g haemoglobin has Fe $= \frac{0.33}{100} \times 67200 = 221.76$ g

$\therefore$ No. of iron atoms per molecule of haemoglobin

$= \frac{221.76}{56} = 4$

35. **(2)** $CO_2 + 2NaOH \rightarrow Na_2CO_3 + H_2O$

$n_{NaOH} = 1$;

$\therefore$ CO_2 present in mixture = 0.5 mol and CO present = (0.8 – 0.5) mol = 0.3 mole

If these 0.3 mol of CO are also oxidised to CO_2, then 0.3 mol of CO_2 would require 0.6 mol of NaOH

36. **(1)** Let the mass of $CaCO_3$ = x g

Then, mass of $MgCO_3$ = (3.68 – x)g

Moles of $CaCO_3 = \frac{x}{100}$

Moles of $MgCO_3 = \frac{(3.68 - x)}{84}$

$\frac{x}{100} + \frac{3.68 - x}{84} = 0.04$

$\Rightarrow x = 2$

$\therefore$ $n_{CaCO_3} = \frac{2}{100} = 0.02$ and $n_{MgCO_3} = \frac{1.68}{84} = 0.02$

Mole% of $CaCO_3 = \frac{0.02}{0.04} \times 100 = 50\%$

Mole% of $MgCO_3 = \frac{0.02}{0.04} \times 100 = 50\%$

37. **(3)** Let relative abundance of heavier isotope be x %

$A = \frac{(100 - x)(A - 1) + x(A + 3)}{100}$

100 A = 100A – xA – 100 + x + 3x + xA

4x = 100

$\therefore$ x = 25.

38. **(2)** 0.5 mol of K_4 [Fe $(CN)_6$] has C $= \frac{1}{2} \times 6$ mol

= 3 mol or $3 \times 12 = 36$ g

39. **(3)** Mole $= \frac{0.224}{22.4} = 0.01$ mole H_2

Mass = $0.01 \times 2 = 0.02$ g

40. **(2)**

$\because$ Mass of 22.4 mL of hydrocarbon = 0.078 g

$\therefore$ 22.400 L or 22400 mL of gaseous hydrocarbon at STP will have mass = 78 g

$\therefore$ Molar mass of hydrocarbon = 78 g /mol

As, empirical formula of hydrocarbon = CH

$\therefore$ Empirical formula mass = 12 + 1 = 13 g

$$n = \frac{\text{Molar mass}}{\text{Empirical formula mass}}$$

$$n = \frac{78}{13} = 6$$

$\therefore$ Molecular formula = $(CH)_6 = C_6H_6$

41. (3) $N_1V_1 = N_2V_2$

Dibasic acid NaOH

Here,

$$N = \frac{w}{E} = \frac{0.16}{E}$$

$V_1 = 1000$

$$N_2 = \frac{1}{10}$$

$V_2 = 25$ mL

$$\frac{0.16}{E} \times 1000 = \frac{1}{10} \times 25$$

$$\frac{160}{E} = \frac{25}{10}$$

$$E = \frac{1600}{25} = 64$$

M = 2 × E = 2 × 64 = 128 g

42. (4) Only single solution have all these

Molar mass of NaCl = 58.5 g mol^{-1}

Molar mass of $CaCl_2$ = 111 g mol^{-1}

Molar mass of NaOH = 40 g mol^{-1}

Means 100 mL solution have 5.85 g NaCl $= \frac{5.85}{58.5} = 0.1$ mole

5.55 g $CaCl_2 = \frac{5.55}{111} = 0.05$ mole and

6 g NaOH $= \frac{6}{40} = 0.15$ mol

$$[Cl^-] = \frac{(0.1 + 0.05 \times 2) \times 1000}{100} = 2\,M$$

$$[Na^+] = \frac{(0.1 + 0.15) \times 1000}{100} = 2.5\,M$$

$$[Ca^{2+}] = \frac{0.05}{100} \times 1000 = 0.5\,M$$

$$[OH^-] = \frac{0.15}{100} \times 1000 = 1.5\,M$$

43. (1) A → q, t; B → p,r,t : C → t;D → s,t

1 mole $(NH_4)_2CO_3$ → 1 mole carbon atom
→ 3 mole O-atoms = 48 g
→ 2 mole nitrogen atom

0.5 mole NH_2CONH_2 → 0.5 mole C-atom
→ 0.5 mole O-atoms
→ 1 mole N-atom

(0.5 mole C-atom and 0.5 mole O-atoms: Both have same number of atoms)

1 mole C_6H_5N → 6 mole C-atoms
→ 1 mole N-atom

9 mole C-atoms

44. (2) Moles of Fe $= \frac{0.0056}{56} = 10^{-4}$

1 mol of alum = 2 mol of Fe

2 mol of Fe = 1 mol of alum

10^{-4} mol of Fe $= \frac{1}{2} \times 10^{-4}$ mol of alum

$= 0.5 \times 10^{-4}$ mol of alum

45. (4) According to the question,

$$M_2CO_3 + 2HCl \rightarrow 2MCl + H_2O + CO_2$$

In this equation, number of moles of M_2CO_3 is equal to that of CO_2.

i.e., $n_{M_2CO_3} = n_{CO_2}$

$$\frac{\text{Wt. of } M_2CO_3}{\text{Molar mass of } M_2CO_3} = n_{CO_2}$$

$$\frac{1g}{\text{Molar mass of } M_2CO_3} = 0.01186\,\text{mol}$$

$$\text{Molar mass of } M_2CO_3 = \frac{1}{0.01186} \approx 84.3\,\text{g mol}^{-1}$$

46. (1) Let atomic weight of element X be x and that of element Y be y.

For XY_2, $n = \frac{\text{wt.}}{\text{Mol. wt.}}$

$$0.1 = \frac{10}{x + 2y} \Rightarrow x + 2y = \frac{10}{0.1} = 100 \quad .....(i)$$

For X_3Y_2, $n = \frac{\text{wt.}}{\text{Mol. wt.}}$

$$0.05 = \frac{9}{3x + 2y} \Rightarrow 3x + 2y = \frac{9}{0.05} = 180 \quad ......(ii)$$

On solving equations (i) and (ii), we get y = 30

x + 2(30) = 100 ⇒ x = 100 – 60 = 40

47. (4) H_2SO_4 is 98% by mass.

Weight of H_2SO_4 = 98 g

Weight of solution = 100 g

$$\text{Volume of solution} = \frac{\text{Mass}}{\text{Density}} = \frac{100}{1.80}\,\text{mL}$$

= 55.55 mL = 0.0555 L

$$\text{Molarity of solution} = \frac{98}{98 \times 0.0555} \approx 18.02\,M$$

Let V mL of H_2SO_4 be used to prepare one litre of 0.1 M H_2SO_4.

V × 18.02 = 1000 × 0.1

$$V = \frac{1000 \times 0.1}{18.02} = 5.55\,\text{mL}$$

48. (4) (i) 14 g (ii) 40 g

(iii) $\frac{108}{6.022\times10^{23}}=1.79\times10^{-22}$ g

(iv) 32 g (v) 1.99 g (vi) 1 g

Hence, the correct order of increasing masses is

(iii) < (vi) < (v) < (i) < (iv) < (ii)

49. (4) $2H_2 + O_2 \rightarrow 2H_2O$

Initial mole 2 3 0

Final mole 0 3–2 = 1 2

50. (1)

$H_{2(g)} + Cl_{2(g)} \rightarrow 2HCl_{(g)}$

$112L=\frac{112}{22.4}\text{moles},\ 213g=\frac{213}{71}\text{moles}$

$=5\text{moles} \qquad =3\text{moles}$

Here, $Cl_{2(g)}$ is limiting reagent,

1 Mole $Cl_{2(g)}$ produces $HCl_{(g)} = 2 \times 6.022 \times 10^{23}$ molecules

3 Moles $Cl_{2(g)}$ produces $HCl_{(g)} = 3 \times 2 \times 6.022 \times 10^{23}$ molecules

$= 3.61 \times 10^{24}$ molecules

51. (4) $2AgNO_3 + BaCl_2 \rightarrow 2AgCl + Ba(NO_3)_2$

Moles: $\frac{20}{1000}\times0.4 \qquad \frac{15}{1000}\times0.6$

$=0.008 \qquad =0.009$

2 moles $AgNO_3$ react with 1 mole of $BaCl_2$ to give 2 moles of AgCl

$AgNO_3$ is limiting reagent

∴ 0.008 moles of $AgNO_3$ will produce = 0.008 moles of AgCl

∴ Mass of AgCl produced = 0.008 × 143.5 = 1.148 g

52. (3) $MgO + CO_2 \rightarrow MgCO_3$

40 g 44 g

40g MgO needs CO_2 = 44 g

30g MgO needs $CO_2 = \frac{44\times30}{40} = 33g$

$CaCO_3 \rightarrow CaO + CO_2$

100 g 44 g

44 g CO_2 is obtained from $CaCO_3$ = 100 g

33 g CO_2 is obtained from $CaCO_3 = \frac{100\times33}{44} = 75g$

Percentage purity of $CaCO_3$ sample

$=\frac{75}{85}\times100 = 88.24\%$

53. (3) Alternatively, $X_{CH_3COOH} = 0.1$

$X_{H_2O} = 1-0.1 = 0.9$

$\text{Molality}=\frac{1000x_B}{x_AM_A}=\frac{1000\times0.1}{0.9\times18}=6.17\ \text{mol kg}^{-1}$

54. (3) $M_{H_2SO_4}$ = 18 M = 18 mole of H_2SO_4 in 1 litre solution

density = 1.8 g/mL (given)

So, mass of solution = 1.8 × 1000 = 1800 g

Mass of solute = 18 × 98 = 1760

∴ Mass of solvent = 1800 – 1760 = 36 g

$\text{Molality} = \frac{18}{36(g)} \times 1000 = 500$ molal

55. (3) If atomic weights of X and Y are 'a' and 'b' respectively

1.5 (a + 2b) = 96 and 1.5 (a + 3b) = 120

⇒ a + 2b = 64 and a + 3b = 80

On solving, a = 32 and b = 16

Correct & Incorrect MCQs

1. **(1)** Glucose is a pure substance which contain carbon, hydrogen & oxygen. C.H.O in glucose are present in a fixed ratio, they cannot be separated by the simple physical method.
2. **(3)** The multiple of prefix nano is 10^{-9}.
3. **(4)** The mass of a substance is constant, whereas, its weight may vary from one place to another due to change in gravity.
4. **(3)** Zeros preceding to first non-zero digit are not significant, hence there is one significant figure in 0.03.
5. **(4)** According to Avogadro's law, equal volumes of all gases at the same temperature and pressure should contain equal number of molecules.
6. **(1)** Molecular mass of methane.

   (CH_4) = (12.011 u) + 4 (1.008 u) = 16.043 u
7. **(2)** One mole of a substance contains exactly 6.022×10^{23} elementary entities. This number is the fixed numerical value of the Avogadro constant, N_A.
8. **(4)** 16 g of CH_4 (g) reacts with 2×32 g of O_2 (g) to give 44 g of CO_2 (g) and 2 × 18 g of H_2O (g).
9. **(4)** Molarity is defined as the number of moles of the solute present in 1 litre of the solution. Since, volume of a solution is temperature dependent, hence molarity of a solution depends upon temperature.
10. **(3)** Given mass of CO_2 = 0.22 g

    Therefore, moles of CO_2 = 0.22/44 = 1/200 = 0.005

    Given mass of SO_2 = 320 mg = 0.32 g

    Therefore, moles of SO_2 = 0.32/64 = 1/200 = 0.005

    Given mass of NH_3 = 0.085 g

    Therefore, moles of NH_3 = 0.085/17 = 1/200 = 0.005

Statement Based MCQs

1. **(1)** Both Statement-I & Statement-II are true. Stoichiometry describes the quantitative relations among the substances as they participate in chemical reaction.
2. **(1)** The data given shows the law of multiple proportions

   Both the statement are correct.

3. **(2)** 32 g $O_2 \rightarrow$ 1 mol O_2 Similarly 48g $O_3 \rightarrow$ 1 mol O_3. Both have N_A no. of O_2 and O_3 molecules but different no. of O atoms.
4. **(3)** Fact based
5. **(2)** Here statement-I is false & Statement-II also false. As molality, mole fraction & mass fraction does not change with temperature but molarity and normality change with temperature.
6. **(1)** In a homogeneous mixture, the components completely mix with each other i.e., particles of components of the mixture are uniformly distributed throughout the bulk of the mixture and its composition is uniform throughout.

   Air is an example of homogeneous mixture.
7. **(1)** Weight of a substance may vary from one place to another due to change in gravity.
8. **(2)** Dalton's atomic theory could not explain the laws of gaseous volumes.
   According to law of conservation of mass, in all physical and chemical changes, the total mass of the reactants is equal to the mass of the products.
9. **(4)** Molality of a solution does not change with temperature, since mass remains unaffected with temperature.
10. **(3)** One mole of a substance always contain the same number of entities, no matter what the substance may be.

Assertion & Reason MCQs

1. **(3)** Volume of gas does not depend on nature of gas.
2. **(1)** 16 g $O_2 = \frac{1}{2}$ mol O_2; 16 g $O_3 = \frac{1}{3}$ mole O_3

   $2\frac{1}{2}N_A = 3\frac{1}{3}N_A$

   Therefore, the no. of oxygen atoms in 16g of oxygen and 16g of O_3 is same.
3. **(4)** In 1502, number of significant digits are 4.
4. **(4)** Gay Lussac's law deals only with gaseous reactants and products at same temperature and pressure.
5. **(1)** Molar weight of SO_2 = 64g, molar weight of SO_3 = 80g and number of moles of SO_2 & SO_3 = 1, 1
6. **(1)** Since empirical formula is simplest ratio, in C_2H_4 if we divide by whole number 2 it comes to be CH_2. That is why Empirical mass of ethene is half of its molecular mass
7. **(2)** Carbon-12 is assigned a mass of exactly 12 atomic mass unit (amu) and masses of other atoms are given relative to this standard.
8. **(2)** According to rule 4, Zeros at the end or right of a number are significant, provided they are on the right side of the decimal point, Terminal zeros are not significant if there is no decimal point.
9. **(4)** $CH_4 + 2O_2 \rightarrow CO_2 + 2H_2O$

   From reaction it can be estimated that, 1 mole of methane gives 2 moles of water, hence it can be said that 16g of Methane will give 36 g of water.
10. **(3)** One atomic mass unit is defined as a mass exactly equal to one twelfth of mass of one carbon-12 atom.

Multi Correct MCQs

1. **(3)** To determine the no. of significant figures certain rules are followed. According to the given options rules are mentioned as,

   (A) 0.00253

   Zeros to the left of first non-zero digits are not significant.

   Significant figures = 3(2, 5, 3)

   (B) 1.0003

   Zeros between non-zero digit are significant.

   Thus, 1.0003 has 5 significant figures.

   (C) 15.0

   If a number ends in zeros and if these zeros are to the right of a decimal point, then it will be significant.

   Significant figures = 3

   (D) 163

   All non zero digits are significant.

   Significant figures = 3

   Hence, number given in A, C and D have same number of significant figures.
2. **(4)** (B) Matter consists of indivisible atoms.

   (D) All the atoms of given element have same properties including mass.
3. **(1)** The empirical formula of the given compound can be calculated as:

   % of element

   $= \frac{\text{Amount of element}}{\text{Amount of substance}} \times 100$

   Hence, % of C $= \frac{48.5}{116} \times 100 = 41.8\%$

   % of H $= \frac{7.5}{116} \times 100 = 6.5\%$

   % of O $= \frac{60}{116} \times 100 = 51.7\%$

	Mass %	Moles $= \frac{w}{m}$	Simplest Ratio
C	41.8	$\frac{41.8}{12} = 3.48$	$\frac{3.48}{3.23} \approx 1$
H	6.5	$\frac{6.5}{1} = 6.5$	$\frac{6.5}{3.23} \approx 2$
O	51.7	$\frac{51.7}{16} = 3.23$	$\frac{3.23}{3.23} = 1$

Given the ratio of moles of C, H and O is 1:2:1

A. $CH_3COOH \Rightarrow C : H : O = 2 : 4 : 2 = 1 : 2 : 1$

B. $HCHO \Rightarrow C : H : O = 1 : 2 : 1$

C. $CH_3OOCH_3 \Rightarrow C : H : O = 2 : 6 : 2 = 1 : 3 : 1$

D. $CH_3CHO \Rightarrow C : H : O = 2 : 4 : 1$

So, only two compounds CH_3COOH and HCHO agrees with given data.

4. **(2)** Temperature is directly proportional to Volume, so if temperature changes, it will affect the volume and thus molarity. Mole fraction, Molality and Weight percent are not temperature dependent since, they do not contain any volume term in them.

5. **(3)** The law of definite proportions, and the law of multiple proportions directly relate to the way atoms combine to form compounds by fixed ratios.

NCERT Exemplar MCQs

1. **(2)** Average of readings of student A $= \frac{3.01+2.99}{2} = 3.00$

Average of readings of student B $= \frac{3.05+2.95}{2} = 3.00$

Correct reading = 3.00

For both the students, average value is close to the correct value.

The readings recorded by student A are both precise and accurate as they differ only by ± 0.01

2. **(3)** Since, molarity (M) is calculated by following equation

$$\text{Molarity} = \frac{\text{Weight} \times 1000}{\text{Molecular weight} \times \text{volume (mL)}}$$

$$= \frac{5.85 \times 1000}{58.5 \times 500} = 0.2 \text{ molL}^{-1}$$

3. **(2)** Given that : $M_1 = 5$ M, $V_1 = 500$ mL ,$V_2 = 1500$ mL

Let, $M_2 = M$

For dilution, a general formula is

$$\underset{\text{(Before dilution)}}{M_1V_1} = \underset{\text{(After dilution)}}{M_2V_2}$$

$500 \times 5 = 1500 \times M$

$$M = \frac{5}{3} = 1.66 \text{ M}$$

4. **(4)** For comparing number of atoms, first we calculate the moles, as all are monoatomic and hence, moles × N_A = number of atoms.

$$\text{Moles in 4 g He} = \frac{4}{4} = 1 \text{ mol}$$

$$46 \text{ g Na} = \frac{46}{23} = 2 \text{ mol}$$

$$0.40 \text{ g Ca} = \frac{0.40}{40} = 0.01 \text{ mol}$$

$$12 \text{ g He} = \frac{12}{4} = 3 \text{ mol}$$

Hence, 12 g He contains greatest number of atoms as it possesses maximum number of moles.

5. **(3)** In the given question, 0.9 g L^{-1} means that 1000 mL (or 1 L) solution contains 0.9 g of glucose.

$$\therefore \text{Number of moles} = \frac{0.9}{180} = 0.005 \text{ moles}$$

$$\therefore \text{Molarity} = \frac{0.005}{1} = 0.005 \text{ M}$$

6. **(4)** $$\text{Molality (m)} = \frac{\text{Moles of solute}}{\text{Mass of solvent (in kg)}} \quad .....(i)$$

Given that, Mass of solvent (H_2O) = 500 g = 0.5 kg

Weight of HCl = 18.25 g

Molecular weight of HCl = 36.5 g

$$\text{Moles of HCl} = \frac{18.25}{36.5} = 0.5$$

$$m = \frac{0.5}{0.5} = 1 \text{ m [from Eq. (i)]}$$

7. **(1)** One mole of any substance contains 6.022×10^{23} atoms/molecules.

Hence, Number of millimoles of H_2SO_4

= molarity × volume in mL

= 0.02 × 100 = 2 millimoles = 2×10^{-3} mol

Number of molecules = number of moles × 6.022×10^{23}

= $2 \times 10^{-3} \times 6.022 \times 10^{23} = 12.044 \times 10^{20}$ molecules

8. **(2)** Molecular mass of $CO_2 = 1 \times 12 + 2 \times 16 = 44$ g

1 g molecule of CO_2 contains 1 g atoms of carbon

∵ 44 g of CO_2 contain = 12 g of carbon

$$\therefore \quad \% \text{ of C in } CO_2 = \frac{12}{44} \times 100 = 27.27\%$$

Hence, the mass percent of carbon in CO_2 is 27.27%.

9. **(3)** Empirical formula mass = CH_2O

= 12 + 2 × 1+ 16 = 30

Molecular mass = 180

$$= \frac{180}{30} = 6$$

∴ Molecular formula = n × empirical formula

= 6 × CH_2O = $C_6H_{12}O_6$

10. **(1)** Given that density of solution = 3.12 g mL^{-1}

Volume of solution = 1.5 mL

For a solution,

Mass = Volume × density

= 1.5 mL × 3.12 gm mL^{-1} = 4.68 g

Hence, the answer is reported as 4.7 g.

11. **(3)** A compound is a pure substance containing two or more than two elements combined together in a same proportion by mass and which can be decomposed into its constituent elements by suitable chemical methods.

Further, the properties of a compound are quite different from the properties of constituent elements. e.g., water is a compound containing hydrogen and oxygen combined together in same proportion. But the properties of water are completely different from its constituents, hydrogen and oxygen.

12. **(1)** According to the law of conservation of mass,

Total mass of reactants = Total mass of products

Amount of Fe_2O_3 is decided by limiting reagent.

13. **(2)** The element, carbon, combines with oxygen to form two compounds, i.e., carbon dioxide and carbon monoxide in CO_2, 12 parts by mass of carbon combine with 32 parts by mass of oxygen while in CO, 12 parts by mass of carbon combine with 16 parts by mass of oxygen.

Therefore, the masses of oxygen combine with a fixed mass of carbon (12 parts) in CO_2 and CO are 32 and 16 respectively. These masses of oxygen have ratio of 32 : 16 or 2 : 1 to each other.

This is an example of law of multiple proportion.

NEET & AIIMS Past Year MCQs

1. (1) We know that, Number of moles = Given mass/Molar mass and number of molecules = Moles × Avogadro's number

Number of moles of $H_2 = \frac{1}{2} = 0.5$ mole

Number of molecules of $H_2 = 0.5\ N_A$. molecules

1. Number of moles of $N_2 = \frac{14}{28} = 0.5$ mole

   Number of molecules of $N_2 = 0.5\ N_A$ molecules

2. Number of moles of $H_2O = \frac{18}{18} = 1$ mole

   Number of molecules of $H_2O = 1 \times N_A = N_A$

3. Number of moles of $CO = \frac{16}{28} = 0.57$ mole

   Number of molecules of $CO = 0.57\ N_A$ molecules

4. Number of moles of $N_2 = \frac{28}{28} = 1$ mole

   Number of molecules of $N_2 = 1 \times N_A = N_A$ molecules

2. (3)

- Atoms in 4 g of Helium $= \frac{4}{4} \times N_A = N_A$
- Atoms in 2.271098 L of Helium at

STP $= \frac{2.271098}{22.4} \times N_A = 0.101\ N_A$

- Atoms in 4 mol of Helium $= 4 \times N_A = 4N_A$
- Atoms in 4u of Helium $= \frac{4}{4} = 1$

Hence, highest number of atoms are present in 4 mol of Helium.

3. (4)

Element	A	B	C
Percentage composition	32%	20%	48%
Atomic mass	64	40	32
Atomic ratio of atoms	$\frac{32}{64} = 0.5$	$\frac{20}{40} = 0.5$	$\frac{48}{32} = 1.5$
Simplest ratio of atoms	$\frac{0.5}{0.5} = 1$	$\frac{0.5}{0.5} = 1$	$\frac{1.5}{0.5} = 3$

Hence, empirical formula is ABC_3

4. (4) $NaOH + HCl \rightarrow NaCl + H_2O$

Mass of NaOH = 1 g

Moles of NaOH $= \frac{1}{40}$ moles

= 0.025 moles = 25 millimoles

Millimoles of HCl = M × V = 25 × 0.75

= 18.75 millimoles

Millimoles of NaOH unreacted

= 25 – 18.75 = 6.25 millimoles

Mass = 40 × 6.25 = 250 mg

5. (3) 20 g of 20% $CaCO_3 = 20 \times \frac{20}{100} = 4g\ CaCO_3$

$$\underset{100g}{CaCO_3} \xrightarrow{1200K} CaO + \underset{44g}{CO_2}$$

According to the reaction,

100g of $CaCO_3 \rightarrow$ 44g of CO_2

4g $CaCO_3 \rightarrow \frac{4 \times 44}{100}$ g of CO_2

= 1.76 g of CO_2

6. (3) $CaCO_{3(s)} + 2HCl_{(aq.)} \rightarrow CaCl_{2(aq.)} + CO_{2(g)} + H_2O_{(l)}$

no. of moles of $CaCO_3$(pure) $= \frac{1}{2} \times$ mole

[No. of Moles = molarity × volume of solution (in ltr.)]

$= \frac{1}{2} \times 0.5 \times \frac{50}{1000} = 0.0125$

weight of $CaCO_3$ (pure) = mole × mol. wt

= 0.0125 × 100 = 1.25 g

$$\%\text{purity} = \frac{\text{wt. of pure substance}}{\text{wt. of impure sample}} \times 100$$

wt. of impure sample $= \frac{1.25 \times 100}{95} = 1.32g$

7. (2)

Element	%	At Weight	$\frac{\%}{\text{At weight}}$	Simplest Ratio
C	78	12	6.5	1
H	22	1	22	3

Empirical formula of this compound is CH_3

8. (3) Mole ratio or atomic ratio

$= \frac{\text{Percentage}}{\text{Atomic mass}}$

For 'C' $= \frac{24}{12} = 2$

For 'H' $= \frac{4}{1} = 4$

for Cl $= \frac{72}{35.5} = 2.02 = \sim 2$

9. (2) $\frac{\text{Equivalent weight of metal}}{\text{Equivalent weight of compound}}$

$= \frac{\text{Weight of metal}}{\text{Weight of compound}}$

$\Rightarrow \frac{12}{12+35.5} = \frac{\text{Weight of metal}}{0.475}$

$\Rightarrow$ Weight wt. of metal = 0.12 g

Structure of Atom

NCERT Topic-wise MCQs

1. **(4)** It was determined by Millikan's with a help of oil drop experiment.
2. **(1)** The formula of lithium nitride is Li_3N.

$$\underset{\text{(Lithium Nitride)}}{Li_3N} \rightarrow 3Li^+ + \underset{\text{(Nitride ion)}}{N^{3-}}$$

No. of protons = 7

No. of electrons = 7 + 3 = 10

3. **(3)** Nucleus of helium is $_2He^4$ means 2 neutrons and 2 protons.
4. **(1)** Rutherford's alpha particle experiment is also known as gold foil experiment as Rutherford used gold foil in it.
5. **(2)** Because they both have 2 electrons.

Be = 4, [↑↓] $1s^2$ [↑↓] $2s^2$

Be^{2+} = 2, [↑↓] $1s^2$

Li = 3, [↑↓] $1s^2$ [↑] $2s^1$

Li^+ = 2, [↑↓] $1s^2$

6. **(2)** Number of neutrons = M – A

Number of neutrons in $_{14}Si^{30}$ = 30 – 14 = 16

Number of neutrons in $_{16}S^{32}$ = 32 – 16 = 16

Because of the presence of same number of neutrons, $_{14}Si^{30}$ and $_{16}S^{32}$ are called isotones and $_{32}Ge^{76}$ and $_{34}Se^{76}$ are isobars as they have same mass number.

Hence; $_{32}Ge^{76}$, $_{34}Se^{76}$ and $_{14}Si^{30}$ and $_{16}S^{32}$ are isobars and isotones, respectively.

7. **(3)** H does not contain neutrons, it contains only one proton and one electron.
8. **(2)** The ratio of charge and mass would be greater for electron as the mass of electron is least among the species.
9. **(1)** Atomic number = 50

No. of protons = 50

As, no. of neutrons = 68

Mass number = No. of protons + No. of neutrons

= 50 + 68 = 118

Hence, mass number is 118.

10. **(1)** Weight of $^{200}X = 0.90 \times 200 = 180.00$ u

Weight of $^{199}X = 0.08 \times 199 = 15.92$ u

Weight of $^{202}X = 0.02 \times 202 = 4.04$ u

Total weight = 199.96u ≈ 200 u

11. **(2)** Isoelectronic species have the same number of electrons.

$_{19}K^+$: Number of electrons = 19 – 1 = 18

$_{17}Cl^-$: Number of electrons = 17 + 1 = 18

$_{20}Ca^{2+}$: Number of electrons = 20 – 2 = 18

Hence, K^+, Cl^- and Ca^{2+} are isoelectronic species.

12. **(1)** Nitride ion can be represented as N^{3-}.

Number of protons = 7

Number of electrons = 7 + 3 = 10

Therefore, nitride ions contain 7-protons + 10 electrons.

13. **(2)** $^{39}K^+$ and $^{40}K^+$ both contain 18 electrons; i.e., they are isoelectronic and isotopic as they have same atomic number that is 19.
14. **(3)** Neutrons in $^{12}_{6}C$ = Mass number – Atomic number = 12 – 6 = 6

Neutrons in $^{28}_{14}Si = 14$

Ratio : 6 : 14 = 3 : 7

15. **(4)** 16 p means Z = 16, $18e^-$ means 2 unit negative charge is present.

Hence; specie is S^{2-}.

16. **(4)** In OH^-, number of protons = 8 + 1 = 9

Number of neutrons = 8 + 0 = 8

(as n_p in O-atom = 8 and n_p in H-atom = 1)

Number of electrons = 8 + 1 + 1 = 10

Hence, in OH^-, the number of protons is greater than the number of neutrons but number of protons is less than the number of electrons.

17. **(2)** Neil Bohr utilised the concept of quantisation of energy (proposed by Max Planck) to give a new model of an atom.
18. **(4)** $\frac{1}{\lambda} = R_H (Z)^2 \left(\frac{1}{(n_1)^2} - \frac{1}{(n_2)^2}\right)$

where, $R_H = R$, Z = 3 (for Li^{2+}, atomic no. 3)

$n_2 = 2, n_1 = 1$

$$\frac{1}{\lambda} = R(3)^2 \left(\frac{1}{(1)^2} - \frac{1}{(2)^2}\right)$$

$$= \frac{R \times 9 \times 3}{4} \Rightarrow \lambda = \frac{4}{27R}$$

19. **(1)** $K.E. = \frac{1}{2} m_e v^2 = h(\nu - \nu_0)$

$K.E. = (6.626 \times 10^{-34}) (1.1 \times 10^{15} - 6 \times 10^{14})$

$K.E. = (6.626 \times 10^{-34}) (5 \times 10^{14}) = 3.313 \times 10^{-19}$ J

20. **(1)** 109,677 cm^{-1}.
21. **(3)** We know that $\Delta E = hcR_H \left(\frac{1}{n_1^2} - \frac{1}{n_2^2}\right)$

For the lowest energy of spectral line in the lyman series.

$n_1 = 1, n_2 = 2$

Hence, $\Delta E = hcR_H \left(\frac{1}{(1)^2} - \frac{1}{(2)^2}\right) = hcR_H \left(\frac{1}{1} - \frac{1}{4}\right) = \frac{3hcR_H}{4}$

22. (1) Energy of emitted electrons = $hv - hv_0$ and since hv_0 is constant , therefore, the emitted electrons have energy less than a maximum value of energy depending upon the frequency of the incident radiation.

23. (2) Power of the bulb = 100 watt = 100J/s

Energy of one photon, E = hν

$$= \frac{hc}{\lambda} = \frac{6.626 \times 10^{-34}\,\text{Js} \times 3 \times 10^8\,\text{ms}^{-1}}{400 \times 10^{-9}\,\text{m}}$$

$= 4.969 \times 10^{-19}$J

Hence, number of photon emitted = $\frac{100\text{J}/\text{s}}{4.969 \times 10^{-19}\,\text{J}}$

$= 2.012 \times 10^{20}\text{s}^{-1}$

24. (1) The quantum of light energy is called photon.

25. (4) In 1885, Balmer for the first time showed that the wave numbers of spectral lines present in the visible region in hydrogen spectrum are given by:

$$\bar{v}(\text{cm}^{-1}) = 109677\left(\frac{1}{(2)^2} - \frac{1}{n_2^2}\right)$$

Here, n_2 = 3, 4, 5 ....

Hence, the Balmer spectrum of hydrogen was discovered first and it lies in visible region.

26. (4) According to Bohr, velocity of an electron (v)

$$= 2.188 \times 10^6 \frac{Z}{n} \text{m/sec}$$

For H, Z = 1, n = 2 (given)

$$\therefore V = \frac{2.188 \times 10^6 \times 1}{2} = 1.094 \times 10^6 \text{m/sec}$$

27. (3) Bohr model of an atom is applicable for those species which have only 1 electron but He^{2+} do not have any electron. Therefore, this model is not applicable to He^{2+}.

28. (3) Ionisation energy of an atom of atomic number, Z is given by

$$\text{I.E.} = \frac{(\text{I.E})_H \times Z^2}{n^2} = 13.6 \times \frac{Z^2}{n^2}\text{eV}$$

Thus, second ionisation energy for He-atom is given by

$$\text{I.E.} = \frac{13.6 \times (2)^2}{(1)^2}\text{eV} = 54.4 \text{ eV}$$

29. (1) For L-shell, n = 2 and also, for Balmer series, n = 2, hence, the lines are called Balmer series.

30. (3) The number of spectral lines = n(n – 1)/2

∴ When an electrons jump from n = 4 level to ground state,

Number of spectral line: $= \frac{(4)(4-1)}{2} \Rightarrow \frac{4 \times 3}{2} = \frac{12}{2} = 6$

31. (1) Wave number of spectral lines in emission series of hydrogen,

$$\bar{v} = R_H\left(\frac{1}{n_1^2} - \frac{1}{n_2^2}\right) \quad ...(i)$$

Given; $\bar{v} = \frac{8}{9}(R_H)$

On putting the value of $\bar{v}$ in equation (i), we get

$$\frac{8}{9}R_H = R_H\left(\frac{1}{n_1^2} - \frac{1}{n_2^2}\right)$$

$$\frac{8}{9} = \frac{1}{(1)^2} - \frac{1}{n_2^2}$$

$$1 - \frac{8}{9} = \frac{1}{n_2^2} \Rightarrow \frac{1}{9} = \frac{1}{n_2^2}$$

$n_2 = 3$

∴ Electrons jumps from $n_2 = 3$ to $n_1 = 1$

32. (2) Radius of H-like species, $r_n = \frac{0.529 \times n^2}{Z}$

For first Bohr orbit of H-atom

radius = r ⇒ 0.529 = r

So, radius of first orbit of Li^{2+} = $\frac{0.529 \times 1^2}{3} = \frac{r}{3}$

(n = 1, Z = 3)

33. (2) $\lambda_1 = 3000\,\text{Å}, \quad \lambda_2 = 6000\,\text{Å}$

$$E_1 = \frac{hc}{\lambda_1} = \frac{hc}{3000}, \quad E_2 = \frac{hc}{\lambda_2} = \frac{hc}{6000}$$

$$\frac{E_1}{E_2} = \frac{hc}{3000} \times \frac{6000}{hc} = 2:1$$

34. (3) Bohr's model is applicable only to H-atom and H-like species containing one electron system.

35. (1) Rydberg equation; $\bar{v} = R\left[\frac{1}{n_1^2} - \frac{1}{n_2^2}\right]$

For Balmer series; $n_1 = 2, n_2 = 3$

$$\bar{v} = R\left[\frac{9-4}{36}\right] = \frac{5R}{36}\text{cm}^{-1}$$

36. (3) Second Bohr orbit of hydrogen atom, i.e., n = 2.

Atomic number of hydrogen (Z) = 1

By using $r = \frac{0.529n^2}{Z} = \frac{0.529 \times (2)^2}{1}$

$= 2.116\text{A°} = 0.2116$ nm

37. (4) As $E_n = -21.8 \times 10^{-19} \times \frac{Z^2}{n^2}$ J

Thus; $E_n \propto \left(\frac{-1}{n^2}\right)$ where n is the number of orbit. Hence; as the value of "n" increases, energy of the electron also increases. Due to this reason, maximum energy is possessed by an electron when it is present at infinite distance from the nucleus.

38. (1) $\because E = \frac{hc}{\lambda}$

$\therefore E \propto \frac{1}{\lambda}$

∴ Decrease in wavelength, increases the energy.

Energy difference in $n_4 \rightarrow n_1$ transition is maximum.

∴ $n_4 \rightarrow n_1$, transition has minimum wavelength.

39. (2) $r_2Be^{3+} = \frac{0.529 \times n^2}{z} = \frac{0.529 \times 2^2}{4} = 0.529 = r_1H$

40. (1) For any H-atom like system, $r_n = \frac{a_0 n^2}{Z}$

(n = number of orbit, z = Atomic number)

Radius of the first orbit of He$^{\oplus}$ ion = $\frac{0.52 \times 10^{-8} \times 1^2}{2}$

$= 0.26 \times 10^{-8}$ cm

41. (4) Ionisation energy of H atom =13.6eV

Here, $Be^{(3+)}$ is H like atom.

$\therefore$ Ionisation energy of Be^{3+} ion = $13.6 \times Z^2 = 16 \times 13.6$eV

42. (2) As we move away from nucleus, the energy of orbit increases.

43. (1) E_1 (for H) $= -13.6 \times \left[\frac{1^2}{1^2}\right]$ ...(i)

E_2 (for Be^{+3}) $= -13.6 \times \left[\frac{4^2}{2^2}\right]$ ...(ii)

On dividing eq. (i) and (ii), we get

$$\frac{E_1}{E_2} = \frac{-13.6}{-13.6 \times 4} = \frac{1}{4}$$

$\therefore E_1 : E_2 = 1 : 4$

44. (4) When an electron is de-excited from higher to lower energy level, a quantum of energy is emitted. Thus, is true for all transitions.

45. (3) $\frac{1}{\lambda} = R_H\left[\frac{1}{2^2} - \frac{1}{n^2}\right]$

For first line n = 3,

$$\therefore \frac{1}{\lambda} = 109677\left[\frac{1}{2^2} - \frac{1}{3^2}\right]$$

$\lambda = 6569$ Å.

46. (3) For H-atom, $r_2 = 0.529 \times (2)^2$, $r_3 = 0.529 \times 3^2$

$$\therefore \frac{r_2}{r_3} = \frac{2^2}{3^2} \Rightarrow r_3 = \frac{9}{4}r_2$$

47. (1) $\frac{1}{\lambda_1} = R\left[\frac{1}{1} - \frac{1}{\infty}\right] = R$

$$\frac{1}{\lambda_2} = R\left[\frac{1}{1} - \frac{1}{4}\right] = \frac{3R}{4}$$

$$\frac{\lambda_1}{\lambda_2} = \frac{4}{3}$$

48. (1) K.E. in an orbit $= \frac{Ze^2}{8\pi\varepsilon_0 r}$ ...(i)

Magnitude of P.E. in an orbit $= \frac{Ze^2}{4\pi\varepsilon_0 r}$ ...(ii)

On comparing (i) & (ii); we get

$$K.E. = \frac{1}{2} P.E.$$

49. (2) $L = \frac{nh}{2\pi}$

n = 4

$\therefore L = \frac{4h}{2\pi} = \frac{2h}{\pi}$

50. (3) Radius of Bohr's orbit $= \frac{n^2 \times a_0}{z} = \frac{(3)^2 \times a_0}{1} = 9a_0$

51. (4) According to Heisenberg's uncertainty principle

$$m\Delta v \times \Delta x \geq \frac{h}{4\pi}$$

$m = 6.626 \times 10^{-28}$ kg

$\Delta v = 10^{-6}$ m/sec

$h = 6.626 \times 10^{-34}$ Js

$$6.626 \times 10^{-28} \times 10^{-6} \times \Delta x = \frac{6.626 \times 10^{-34}}{4\pi}$$

$$\Delta x = \frac{6.626 \times 10^{-34}}{4\pi \times 6.626 \times 10^{-34}} = \frac{1}{4\pi}$$

52. (4) According to de-Brogie, $\lambda = \frac{h}{mv}$

Mass of α–particle = 4u, Mass of deuterium = 2u

$V_D = 5V_\alpha$ (Given)

$$\lambda_D = \frac{h}{2 \times V_D}, \lambda_\alpha = \frac{h}{4V_\alpha}$$

$$\frac{\lambda_D}{\lambda_\alpha} = \frac{4V_\alpha}{2V_D} = \frac{4}{2 \times 5} = \frac{4}{10} = 0.4$$

53. (3) According to De-Broglie equation;

Wavelength $(\lambda) = \frac{h}{mv}$

Given; m = 100g, v = 100 cm/sec

$h = 6.6 \times 10^{-34}$ Js $= 6.6 \times 10^{-27}$ erg-s

On substituting values, we get

$$\lambda = \frac{6.6 \times 10^{-27}}{100 \times 100} = \frac{6.6 \times 10^{-27}}{10^4} = 6.6 \times 10^{-31} \text{cm/s}$$

($\because$ 1 erg – 1gcm²/s²)

54. (4) According to de-Broglie equation

$$\lambda = \frac{h}{mv} = \frac{6.63 \times 10^{-34}}{10^{-6} \times 10} \Rightarrow \frac{6.63 \times 10^{-34}}{10^{-5}} = 6.63 \times 10^{-29} \text{ m}$$

55. (4) According de-Broglie equation

$\lambda = \frac{h}{mv} \Rightarrow v = \frac{h}{m\lambda}$ ...(i)

K.E. = 1/2 mv² ...(ii)

On putting the value of v in equation (ii), we get

$$K.E. = \frac{1}{2}m\left(\frac{h}{m\lambda}\right)^2 = \frac{1}{2}\left(\frac{h^2}{m\lambda^2}\right)$$

Hence, K.E. ∝ 1/m (if λ is same) and order of mass is as

$m_e < m_p < m_\alpha$

Thus; the order of K.E. is:

$E_e > E_p > E_\alpha$

56. (4) As, $\lambda = \frac{h}{mv}$

$$\lambda_X = \frac{h}{m_X \times 0.1},$$

$$\lambda_Y = \frac{h}{5 \times m_X \times 0.05}$$

Therefore, ratio of wavelengths:

$$\frac{\lambda_X}{\lambda_Y} = \frac{m_X \times 0.05 \times 5}{m_X \times 0.1} = \frac{5}{2}, \therefore \lambda_X : \lambda_Y = 5 : 2$$

57. (2) Heisenberg gave the uncertainty principle in 1927.

58. (2) $h = 6.626 \times 10^{-27}$ erg-s

$m = 1mg = 10^{-3}g$

$v = 1km/s = 10^5$ cm/s

$$\lambda = \frac{h}{mv} = \frac{6.626 \times 10^{-27}}{10^{-3} \times 10^5} \Rightarrow \frac{6.626 \times 10^{-27}}{10^2} = 6.626 \times 10^{-29} cm$$

59. (3) According to Heisenberg principle

$$\Delta x = \frac{h}{4\pi m \Delta v}$$

$\Delta v = 3 \times 10^4 \times \frac{0.011}{100}$ cm/s and $m = 9.1 \times 10^{-28}$g

$$\Delta x = \frac{6.6 \times 10^{-27} \times 100}{4 \times 3.14 \times 9.1 \times 10^{-28} \times 3 \times 10^4 \times 0.011}$$

$\Delta x = 0.175$ cm

60. (1) Here; $\Delta x = \Delta v$

$$\Delta x \times \Delta p \geq \frac{h}{4\pi}$$

$$\Delta x \times m\Delta v = \frac{h}{4\pi}$$

$$(\Delta v)^2 = \frac{h}{4\pi m}$$

$(\because \Delta x = \Delta v)$

$$\Delta v = \sqrt{\frac{h}{4\pi m}}$$

$\Delta p = m. \Delta v$

$$\Delta p = m\sqrt{\frac{h}{4\pi m}} = \sqrt{\frac{mh}{4\pi}}$$

$$\Delta p = \frac{1}{2}\sqrt{\frac{mh}{\pi}}$$

61. (4) $\lambda = \frac{h}{mv}$

$$\lambda_A = \frac{h}{m_A \times 0.1}, \quad \lambda_B = \frac{h}{5 \times m_A \times 0.05}$$

$$\frac{\lambda_A}{\lambda_B} = \frac{m_A \times 0.05 \times 5}{m_A \times 0.1} = \frac{5}{2}$$

62. (2) Let, the wavelength of particle be x,

Thus, x = 100 × velocity

velocity (v) $= \frac{x}{100}, \ \lambda = \frac{h}{mv}$

$$x = \frac{h \times 100}{m \times x}$$

$$x^2 = 100 \times \frac{h}{m} \Rightarrow x = 10\sqrt{\frac{h}{m}}$$

63. (1) de-broglie equation is $\lambda = \frac{h}{mv}$

$$K.E. = \frac{1}{2}mv^2 \quad \text{or} \quad v = \sqrt{\frac{2K.E.}{m}}$$

$$\lambda = \frac{h}{\sqrt{2m K.E.}}$$

$$\frac{K.E_1}{K.E_2} = \frac{\lambda_2^2}{\lambda_1^2} = \left(\frac{5}{3}\right)^2 = \frac{25}{9}$$

64. (1) According to Heisenberg's uncertainty principle

$$\Delta x \times \Delta p \geq \frac{h}{4\pi}$$

Here, $\Delta x - \Delta p$ and $\Delta p = m.\Delta v$

$$\Delta p^2 = \frac{h}{4\pi}$$

$$m^2\Delta v^2 = \frac{h}{4\pi}$$

$$\therefore \Delta v^2 = \frac{h}{m^2 4\pi} \quad \text{or} \quad \Delta v = \frac{1}{2m}\sqrt{\frac{h}{\pi}}$$

65. (1) $\lambda = \frac{h}{\sqrt{2mE}} = \frac{6.626 \times 10^{-34}}{\sqrt{2 \times 1 \times 0.5}} = 6.626 \times 10^{-34}$ m.

66. (4) de-broglie wavelength $(\lambda) = \frac{h}{mv}$

$$\frac{\lambda_{ball}}{\lambda_e} = \frac{h}{m_{ball}v} \times \frac{m_e v}{h} = \frac{9.1 \times 10^{-31}}{0.91}$$

Hence, $\lambda_{ball} : \lambda_e = 10^{-30} : 1$

67. (3) According to Heisenberg's uncertainty principle.

$$\Delta x . \Delta p \geq \frac{h}{4\pi}$$

But, Δp (momentum) $= m\Delta v$

$$\Delta x . m\Delta v \geq \frac{h}{4\pi}$$

$$\Delta x = \frac{h}{4\pi m \Delta v}$$

68. (4) $\Delta x \times \Delta p = \frac{h}{4\pi}$

$$\Rightarrow \Delta p = \frac{h}{4\pi \Delta x} = \frac{h}{4\pi \times 0} = \infty$$

69. (1) $\lambda = \frac{h}{mv}$

$$\lambda \propto \frac{1}{mv} \ (p = mv)$$

70. (4) According to Heisenberg's uncertainty principle,

$$\Delta x \cdot \Delta p \geq \frac{h}{4\pi}$$

According to question,

$$\frac{\Delta x_e . m_e . \Delta v_{e^-}}{\Delta x_p . m_p . \Delta v_p} = 1 \qquad (\because \Delta p = m\Delta v)$$

$$\frac{\Delta v_e}{\Delta v_p} = \frac{m_p}{m_e} = \frac{1.67 \times 10^{-27}}{9.1 \times 10^{-31}} = 1836 : 1 \qquad (\because \Delta x_e = \Delta x_p)$$

71. (4) $_{20}Ca = 1s^2 2s^2 2p^6 3s^2 3p^6 4s^2$

$_{20}Ca^{2+} = 1s^2 2s^2 2p^6 3s^2 3p^6$

Hence, Ca^{2+} has 8 electrons in outermost ($3s^2$, $3p^6$) as well as in penultimate shell ($2s^2$, $2p^6$).

72. (4) The electronic configurations are:

P: $1s^2 2s^2 2p^6 3s^2 3p^3$

Fe^{2+}: [Ar] $3d^6$

Hence, P has 6 electrons in s-subshells as in d-shell of Fe^{2+}.

73. (1) If m = + 3 (maximum), then l = 3 and n can have maximum value of four. Also no. of waves in an orbit = no. of orbit.

74. (2) For any value of ℓ, the possible value of

$m = -\ell$ to $+\ell$

$\ell = 1$, $m = -1, 0, +1$

75. (4) Orbital angular momentum $= \frac{h}{2\pi}\sqrt{\ell(\ell+1)}$

For f-orbital, $\ell = 3$

$$\therefore \text{Orbital angular momentum} = \frac{h}{2\pi}\sqrt{3(3+1)} = \frac{h\sqrt{3(4)}}{2\pi}$$

$$= \frac{2h\sqrt{3}}{2\pi} = \frac{h\sqrt{3}}{\pi} = \frac{\sqrt{3}h}{\pi}$$

76. (1) Electronic configuration is

$1s^2 2s^2 2p^6 3s^2 3p^6 3d^3 4s^2$

$l = 2$ represents d-subshell.

Thus, 3 electrons are present.

77. (3) The maximum number of electrons which can be held by electron with Azimuthal quantum number 'ℓ' in an atom is given by $2(2\ell + 1)$.

78. (4) The orbital of the electron having n = 3, ℓ = 1, m = –1 is $3p_x$ and an orbital can have a maximum of two electrons with opposite spins.

Therefore, $3p_x$ orbital contain only two electrons.

79. (3) Azimuthal quantum number determines shape of the orbital, magnetic quantum number, m is associated with spatial orientation of the orbital. Principal quantum number (n) determines energy of the electron & size (volume) of the atom.

80. (1) Atomic number of element having electronic configuration $1s^2 2s^2 2p_x^1 2p_y^1 2p_z^1$ is 7. The atomic number 7 is of nitrogen atom.

81. (1) Atomic number 17, $1s^2 2s^2 2p^6 3s^2 3p^5$

Valence shell has $3s^2$ $3p^5$ orbit.

[⇅] [⇅|⇅|↑]

$3s^2$ $3p^5$

Valence shell has 3 orbitals with paired electrons.

82. (3) Electronic configuration of bromine is

$1s^2 2s^2 2p^6 3s^2 3p^6 4s^2 3d^{10} 4p^5$

Then; the total no. of p–electron in bromine is 17.

83. (1) Electronic configuration of $_{26}Fe$ is

$1s^2 2s^2 2p^6 3s^2 3p^6 3d^6 4s^2$

$Fe^{3+} = 1s^2 2s^2 2p^6 3s^2 3p^6 3d^5$

∵ Outer orbital diagram of Fe^{3+} is

[↑|↑|↑|↑|↑]

$3d^5$

∴ The no. unpaired electrons are 5.

84. (2) Orbital angular momentum $= \sqrt{\ell(\ell+1)}\frac{h}{2\pi}$

∵ For p-electron, $\ell = 1$

$$\text{Orbital angular momentum} = \sqrt{1(1+1)}\frac{h}{2\pi}$$

$$= \frac{\sqrt{2}h}{2\pi} = \frac{h}{\sqrt{2}\pi}$$

85. (1) The hydrogen atom has configuration $1s^1$.

So; 3s, 3p & 3d orbital will have the same energy with respect to 1s orbital.

86. (3) Total no. of orbital in a shell is n^2.

87. (3) $\because \ell = 0$

$\therefore m = -\ell$ to $+\ell = 0$

But m (given) = –1, so (3) is incorrect.

88. (4) n = 4

$\ell = 2$ (d-sub shell)

$m = -\ell$ to $+\ell$, i.e., – 2, – 1, 0, 1, 2 (any value)

s = –1/2

89. (3) Any orbital have maximum of two electrons with opposite spin.

90. (3) M-shell corresponds to n= 3.

$$_{19}X \rightarrow 1s^2 2s^2 2p^6 \underline{3s^2 3p^6} 4s^1$$

$$_{21}Y \rightarrow 1s^2 2s^2 2p^6 \underline{3s^2 3p^6} 4s^2 \underline{3d^1}$$

$$_{25}Z \rightarrow 1s^2 2s^2 2p^6 \underline{3s^2 3p^6} 4s^2 \underline{3d^5}$$

Hence, the correct order is $Z > Y > X$

91. (3) This electronic configuration shows atomic number '9'. So, the element is fluorine.

High Order Time Intensive MCQs

1. (4) Binding energy $= \frac{13.6 \times Z^2}{n^2}$

2nd excited state refers to 3rd shell, its means, n = 3

$$\therefore \frac{13.6 \times Z^2}{3^2} = 24$$

$$Z^2 = \frac{24 \times 9}{13.6}$$

Now, $IE = +13.6 \times \frac{Z^2}{n^2}$

$$IE = 13.6 \times \frac{\frac{24\times9}{13.6}}{1^2}$$

$$IE = 13.6 \times \frac{24\times9}{13.6} = 216 eV$$

2. (2) $2\pi r = n\lambda$

For minimum radius, n = 1

$2\pi r_{min} = \lambda$

$$r_{min} = \frac{\lambda}{2\pi}$$

3. (4) Formula for energy of photon is already mentioned,

$$\text{i.e.} = \frac{12400 eV \mathring{A}}{\lambda(\mathring{A})}$$

$$E = \frac{12400 eV \mathring{A}}{600\times10(\mathring{A})} (1nm = 10\mathring{A})$$

$$h\nu = \frac{12400}{6000} eV$$

$$h\nu = \frac{124}{60}\times1.6\times10^{-19} J \ (1 eV = 1.6 \times 10^{-19} J)$$

$$6.626\times10^{-34} Js \times \nu = \frac{124}{60}\times1.6\times10^{-19} J$$

$\nu = 5 \times 10^{14} s^{-1}$

To calculate energy of a photon,

$$E = \frac{12400 eV \mathring{A}}{6000 \mathring{A}} = 2.07 eV$$

4. (1) $E = h\nu = \frac{hc}{\lambda}$

$$E = \frac{6.626\times10^{-34}\times3\times10^8}{310\times10^{-9}} J$$

$E = 0.0641 \times 10^{-26} \times 10^9 J$

$E = 0.0641 \times 10^{-17} J$

This is the energy of one photon required to break one A_2 molecule.

To break 1 mole of A_2, Energy of photons required

$= 6.02\times10^{23}\times0.0641\times10^{-17} = 0.386\times10^6 = 386 kJ/mole$

K.E = 386 kJ/mole – 288 kJ/mole = 98 kJ/mole

Percentage of total energy that got converted into

$$KE = \frac{98}{386}\times100 = 25\%$$

5. (3) $r_n = 0.529\times\frac{n^2}{Z}$

$$\frac{4}{9} = \frac{r_{n_1}}{r_{n_2}} = \frac{0.529\times\frac{n_1^2}{Z}}{0.529\times\frac{n_2^2}{Z}}$$

$$\frac{4}{9} = \left(\frac{n_1}{n_2}\right)^2$$

$$\frac{2}{3} = \frac{n_1}{n_2}$$

n_1 could be = 2, n_1 could be = 4

n_2 could be = 3, n_2 could be = 6

Energy difference between 3rd and 2nd shell =

$$-13.6\times\frac{1^2}{3^2} - \left(-13.6\times\frac{1^2}{2^2}\right)$$

$$= \frac{-13.6}{9} + \frac{13.6}{4} = 13.6\left(\frac{1}{4}-\frac{1}{9}\right)$$

$$= 13.6\times\frac{5}{36} = 1.89 eV$$

Energy difference between 6th & 4th shell

$$= -13.6\times\frac{1^2}{6^2} - \left(-13.6\times\frac{1^2}{4^2}\right)$$

$$= \frac{-13.6}{36} + \frac{13.6}{16} = 13.6\left(\frac{1}{16}-\frac{1}{36}\right)$$

$= 0.47 eV \approx 0.49 eV$

6. (2) Last line of lyman series refers to transition $\infty \to 1$

$$\frac{1}{\lambda_1} = R_H\left(\frac{1}{1^2}-\frac{1}{\infty^2}\right)$$

$$\frac{1}{\lambda_1} = R_H \quad ...(1)$$

2nd line of Balmer series refers to transition $4 \to 2$.

$$\frac{1}{\lambda_2} = R_H\left(\frac{1}{2^2}-\frac{1}{4^2}\right)$$

$$\frac{1}{\lambda_2} = R_H\left(\frac{1}{4}-\frac{1}{16}\right)$$

$$\frac{1}{\lambda_2} = R_H\times\frac{3}{16} \quad ...(2)$$

$$\frac{\frac{1}{\lambda_1}}{\frac{1}{\lambda_2}} = \frac{R_H}{\frac{3R_H}{16}} = \frac{R_H}{1}\times\frac{16}{3R_H}$$

$$= \frac{\lambda_2}{\lambda_1} = \frac{16}{3} \Rightarrow 3\lambda_2 = 16\lambda_1 \Rightarrow \frac{16}{\lambda_2} = \frac{3}{\lambda_1}$$

7. (4) For n = 4, $\ell = 0$ to $n-1$

= 0 to 3

= 0, 1, 2, 3

$m = -\ell$ to $+\ell$

= –3, –2, –1, 0, + 1, + 2, + 3

Therefore, the value of ℓ may be 2 or 3, and s may be +1/2.

8. (3) 3rd line of Balmer series refers to transition $5 \to 2$

$$\text{Change in angular momentum} = \frac{5h}{2\pi} - \frac{2h}{2\pi} = \frac{3h}{2\pi}$$

9. (4)

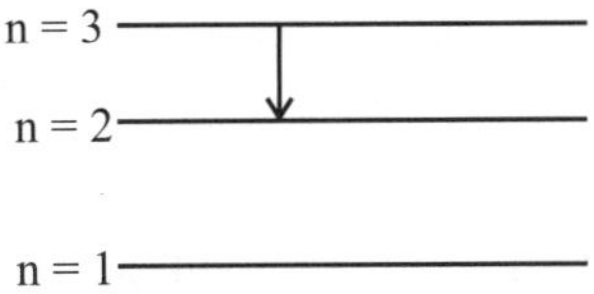

The transition $3 \rightarrow 2$ represents first line of Balmer series

Now, IE of H = 13.6 eV

It's given as x J/atom

So, T.E. of an electron $= -x \times \frac{Z^2}{n^2} J/atom$

$$\Delta E = E_3 - E_2 = -x \times \frac{2^2}{3^2} - \left(-x \times \frac{2^2}{2^2}\right)$$

$$\Delta E = -x \times \frac{4}{9} + x$$

$$\Delta E = x - \frac{4}{9}x = \frac{9x - 4x}{9} \Rightarrow \Delta E = \frac{5x}{9}$$

$\therefore$ Energy of photon emitted $= \frac{5x}{9}$

Apply $E = \frac{hc}{\lambda} \Rightarrow \frac{5x}{9} = \frac{hc}{\lambda}$

$$\lambda = \frac{9hc}{5x}$$

10. (4) $He \longrightarrow He^+ + e^-$; $IE_1 = 24.6$ eV

$He^+ \longrightarrow He^{2+} + e^-$; $IE_2 = ?$

He^+ is an unielectron specie for which we can apply,

$$IE = +13.6 \times \frac{Z^2}{n^2} eV/atom$$

$$IE = +13.6 \times \frac{2^2}{1^2} = 54.4 eV$$

So, total energy required to remove both the electrons = 24.6 + 54.4 = 79 eV

11. (4)

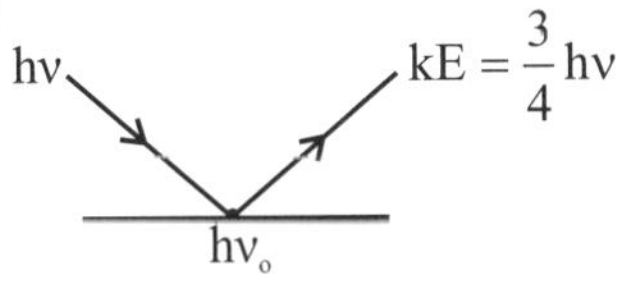

ν_0 – threshold frequency.

$h\nu - h\nu_o = kE$

Apply, $h\nu - h\nu_o = \frac{3}{4}h\nu \Rightarrow h\nu - \frac{3}{4}h\nu = h\nu_o$

$$\frac{h\nu}{4} = h\nu_o \Rightarrow \frac{\nu}{4} = \nu_o$$

$$\frac{3.2 \times 10^{16}}{4} = \nu_o = 8 \times 10^{15} Hz$$

12. (1) $T.E. = -Rhc \times \frac{Z^2}{n^2}$

For He, Z = 2

$$\frac{-Rhc}{4} = -Rhc \times \frac{4}{n^2}$$

$n^2 = 16$

$\Rightarrow n = 4$

13. (3) $H^-(2)$:- $1s^2$

It's a multielectron specie, first excited state will be 2s & 2nd excited state will be 2p. Degeneracy of 2s is 1. Degeneracy of 2p will be 3.

14. (1) $r_n = \frac{a_0 n^2}{Z}$

(A) For H, Z = 1, $r_2 = a_0 \times (2)^2 = 4\ a_0$

(B) For He^+, Z = 2, $r_1 = \frac{a_0 \times 1^2}{2} = 0.5\ a_0$

(C) For Be^{3+}, Z = 4, $r_2 = \frac{a_0 \times 2^2}{4} = a_0$

(D) In 2s orbital, the node exists at 2 a_0.

15. (1) $\Delta E(\text{in eV}) = \frac{hc}{\lambda}$

$$\Delta E \times 1.6 \times 10^{-19} J = \frac{6.626 \times 10^{-34} Js \times 3 \times 10^8 ms^{-1}}{\lambda}$$

$$\lambda = \frac{6.626 \times 10^{-34} \times 3 \times 10^8}{1.6 \times 10^{-19} \times \Delta E} m$$

$$\lambda = \frac{6.626 \times 10^{-34} \times 3 \times 10^8 \times 10^{10} \mathring{A}}{1.6 \times 10^{-19} \times \Delta E}$$

$$\lambda = \frac{12.42 \times 10^{-34} \times 10^{18} \times 10^{19} \mathring{A}}{\Delta E}$$

$$\lambda = \frac{12.42 \times 10^{-34} \times 10^{37} \mathring{A}}{\Delta E}$$

$$\lambda = \frac{12.42 \times 10^3 \mathring{A}}{\Delta E} = \frac{12420}{\Delta E}\mathring{A} \approx \frac{12375}{\Delta E}\mathring{A}$$

16. (2) $T.E = -13.6 \times \frac{Z^2}{n^2} eV/atom$

$$E = -13.6 \times \frac{1^2}{n^2} = -3.4 eV$$

$n^2 = 4$

$n = 2$ (first excited state)

17. (1) $V_n = 2.18 \times 10^6 \times \frac{Z}{n} ms^{-1}$

$$V_n = V_o \times \frac{Z}{n} ms^{-1}$$

$$V_o = \frac{1}{100} \times 3 \times 10^8 = 3 \times 10^6 ms^{-1}$$

$$V_n = 3 \times 10^6 \times \frac{Z}{n} ms^{-1}$$

Velocity in 5th Bohr orbit

$$V_{5^{th}} = 3 \times 10^6 \times \frac{1}{5} ms^{-1} = 0.6 \times 10^6 ms^{-1}$$

Velocity relative to speed of light

$$\frac{0.6 \times 10^6}{3 \times 10^8} = 0.2 \times 10^{-2} = 2 \times 10^{-3} = 0.002$$

18. (4) I.E. $= +x \times \frac{Z^2}{n^2}$ J / atom

For He^+, n = 1, Z = 2

$19.6 \times 10^{-18} = +x \times \frac{2^2}{1^2}$

$19.6 \times 10^{-18} = x \times 4$

$\frac{19.6 \times 10^{-18}}{4} = x$

$x = 4.9 \times 10^{-18}$

For Li^{2+} atom,

T. E. $= -4.9 \times 10^{-18} \times \frac{Z^2}{n^2}$ J / atom

($\because$ T.E. & I.E. have opposite signs & rest formula is the same)

$-4.9 \times 10^{-18} \times \frac{3^2}{1^2} = -4.9 \times 10^{-18} \times 9$

$= -44.1 \times 10^{-18}$ J atom^{-1} or -4.41×10^{-17} J atom^{-1}

19. (1) $h\nu_1 - h\nu_0 = kE_1$ ...(1)

$h\nu_2 - h\nu_0 = kE_2$ ...(2)

$\frac{h(\nu_1 - \nu_0)}{h(\nu_2 - \nu_0)} = \frac{2}{1}$

$\nu_1 - \nu_0 = 2\nu_2 - 2\nu_0 \Rightarrow \nu_0 = 2\nu_2 - \nu_1$

20. (2) $\frac{Area_{2^{nd}\ orbit}}{Area_{1^{st}\ orbit}} = \frac{\pi r^2_{2^{nd}\ orbit}}{\pi r^2_{1^{st}\ orbit}} = \frac{r^2_{2^{nd}\ orbit}}{r^2_{1^{st}\ orbit}}$

$= \frac{\left(0.529 \times \frac{2^2}{Z}\right)^2}{\left(0.529 \times \frac{1^2}{Z}\right)^2} = \frac{16}{1}$

21. (1) PE = – 2KE

$\therefore$ PE will change from $-2x$ to $-\frac{2x}{4}$

Change in potential energy $= \left(-\frac{2x}{4}\right) - (-2x)$

$= -\frac{x}{2} + 2x = \frac{3x}{2}$

22. (2) For H-atom, first Balmer line is $3 \rightarrow 2$ transition.

$\bar{\nu} = RZ^2\left[\frac{1}{n_1^2} - \frac{1}{n_2^2}\right] = R\,1^2\left[\frac{1}{2^2} - \frac{1}{3^2}\right] = R\left[\frac{1}{4} - \frac{1}{9}\right]$ ...(i)

For He^+, Z = 2

$\bar{\nu} = R \cdot 2^2\left[\frac{1}{n_1^2} - \frac{1}{n_2^2}\right] = R\left[\frac{4}{n_1^2} - \frac{4}{n_2^2}\right]$ ...(ii)

On comparing (i) and (ii):

$\frac{4}{n_1^2} = \frac{1}{4} \Rightarrow n_1^2 = 16$

$n_1 = 4$

$\frac{4}{n_2^2} = \frac{1}{9} \Rightarrow n_2^2 = 36$

$n_2 = 6$

$\therefore$ $6 \rightarrow 4$ transition of $He^{\oplus}$ ion shall have the same wavenumber as first line of the Balmer series in H-atom.

23. (3) According to Heisenberg uncertainty Principle,

$\Delta x. \Delta p = \frac{h}{4\pi} \Rightarrow \Delta x. m\Delta v = \frac{h}{4\pi}$

$\Delta x. \Delta v = \frac{h}{4\pi m}$

$\Delta v = \frac{0.005}{100} \times 600 = 3 \times 10^{-2}\ ms^{-1}$

$\Delta x = \frac{h}{4\pi m \Delta v} = \frac{6.626 \times 10^{-34}}{4 \times 3.14 \times 9.1 \times 10^{-31} \times 3 \times 10^{-2}}$

$\Delta x = 1.92 \times 10^{-3}$ m

24. (4) Energy absorbed - Bond energy = K.E of the molecule

$4.4 \times 10^{-19}J - 4 \times 10^{-19}J$ = K.E of molecule

$10^{-19} \times 0.4$ J = K.E of molecule

4×10^{-20} J = K.E of molecule

K.E per atom $= \frac{4 \times 10^{-20}}{2} = 2 \times 10^{-20}$ J

25. (4) 4 →

4s → 2
4p → 6
4d → 10
4f → 14

32

26. (3) $1s^2\ 2s^2\ 2p^6\ 3s^2\ 3p^6\ 4s^2\ 3d^{10}\ 4p^6$

27. (2) For IInd period,

n = 2 hence,

l	m	s
0	0	$+\frac{1}{2}, 0, -\frac{1}{2}$
1	–1	$+\frac{1}{2}, 0, -\frac{1}{2}$
	0	$+\frac{1}{2}, 0, -\frac{1}{2}$
	+1	$+\frac{1}{2}, 0, -\frac{1}{2}$

Hence, total number of elements in 2nd period having the given value of spin quantum number = 12

28. (3) The order of energy from high to low is shown as-

VIBGYOR

→ E ↓

→ λ ↑

Violet colour has highest energy therefore, transition corresponds to $4 \rightarrow 1$ will give highest energy.

Hence, $4 \rightarrow 1$ jump will given off the violet spectral line.

29. (3) Option 1 not possible because s can never have –2 value.

Option 2 not possible because n and l cannot have same values.

Option 4 not possible because s cannot have zero value.

$\therefore$ Correct answer is 3rd option, that is:

$n = 4, \ell = 3, m = -3, s = \frac{1}{2}$

30. (4) Total number of e^- = 30. Therefore, e^- configuration will be

$1s^2$	$2s^2$	$2p^6$	$3s^2$	$3p^6$	$4s^2$	$3d^{10}$
n = 1	n = 2	n = 2	n = 3	n = 3	n = 4	n = 3
$\ell = 0$	$\ell = 0$	$\ell = 1$	$\ell = 0$	$\ell = 1$	$\ell = 0$	$\ell = 2$
m = 0	m = 0	m = –1, 0, +1	m = 0	m = –1, 0, + 1	m = 0	m = –2, –1, 0, +1, +2

From the above table, it is clear that total 7 orbital have m = 0. Therefore, total number of electrons = 7 × 2 = 14

31. (1) Correct graph for probability density curve for 2s electron is:

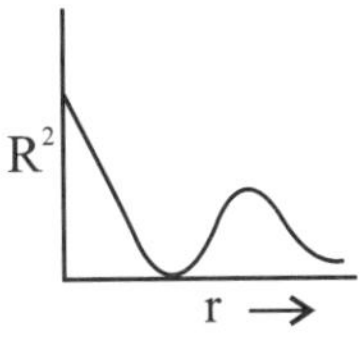

32. (3) m = represents the spatial orientation of orbital

33. (2) (ii), (iv), and (v) are not possible

(ii)	n = 2	$l = 2$	m = 1	s = +1/2	l equal to n is not possible
(iv)	n = 1	$l = 0$	m = –1	s = –1/2	Not possible (because m = –1 is not for $l = 0$)
(v)	n = 3	$l = 2$	m = 3	s = +1/2	Not possible because m = 3 is not for $l = 2$

34. (1, 3) When n = 3, ℓ cannot be 3, so (1) is incorrect and in (3), when $l = 1$, m cannot be = +2 as the value of l ranges from 0 to (n – 1) and that of m from $-l$ to $+l$.

35. (2) E = – 78.4k cal/mol

1 kcal = 4184J

1 mol = N_A = 6.022 × 10^{23}

Hence, $E = -\frac{78.4 \times 4184}{6.022 \times 10^{23}} = -5.44 \times 10^{-19}$ J

Now, $E = -\frac{2.18 \times 10^{-18}}{n^2}$ J $= -5.44 \times 10^{-19}$ J

$\Rightarrow n^2 = 4 \Rightarrow n = 2$

36. (4) $r_n \propto n^2$

But $(r_n + 1) - r_n = r_n - 1$

For n^{th} orbit, $r_n = n^2$

For $(n + 1)^{th}$ orbit, $r_{n+1} = (n + 1)^2$

For $(n - 1)^{th}$ orbit, $r_{n-1} = (n - 1)^2$

$(n + 1)^2 - n^2 = (n - 1)^2$

$\Rightarrow n^2 + 1 + 2n - n^2 = n^2 + 1 - 2n$

n = 4

37. (1) $\frac{hc}{\lambda} = \frac{430.53 \times 10^3}{6.023 \times 10^{23}} + K.E.$

$K.E. = \frac{6.626 \times 10^{-34} \times 3 \times 10^8}{253.7 \times 10^{-9}} - \frac{430.53 \times 10^3}{6.023 \times 10^{23}} = 6.9 \times 10^{-20}$

$\therefore$ Fraction $= \frac{6.9 \times 10^{-20}}{7.83 \times 10^{-19}} = 0.088 = 8.86\%$

38. (4) When intensity is doubled, number of electrons emitted per second is also doubled but average energy of photoelectrons emitted remains the same.

39. (1) $\lambda = \frac{h}{mv}$

$\because p = mv$

$p = \frac{h}{\lambda} = \frac{6.626 \times 10^{-34}}{0.33 \times 10^{-9}} = 2.0 \times 10^{-24}$ kgm/s

40. (3) In case of hydrogen like atoms, energy depends on the principal quantum number only. Hence, 2s-orbital will have energy equal to 2p-orbital.

41. (3) The orbitals described by these sets of quantum numbers are:

(i) 4s (ii) 3p (iii) 3d (iv) 3s

The energy of these orbitals follows the order:

3d > 4s > 3p > 3s

i.e., (iii) > (i) > (ii) > (iv)

42. (4) 1) Co^{3+} : $1s^2 2s^2 2p^6 3s^2 3p^6 3d^6$

$\therefore$ 4 unpaired electrons

$\therefore \mu = \sqrt{n(n+2)} = \sqrt{4(4+2)} = \sqrt{24} = 4.9$ BM

2) Number of radial nodes = $n - \ell - 1$

Number of radial nodes in 3p orbital = 3 – 1 – 1 = 1

3) Number of electrons with (m = 0) in Mn^{2+} ($1s^2 2s^2 2p^6 3s^2 3p^6 3d^5$) ion = 1s (2) + 2s(2) + 2p(2) + 3s(2) + 3p(2) + 3d(1) = 11

43. (3) According to I[st] option

No. of radial nodes = $n - l - 1$

For 3s, Radial nodes = 3 – 0 – 1 = 2

For 2s, Radial nodes = 2 – 0 – 1 = 1

For 1s, Radial nodes = 1 – 0 – 1 = 0

So, order 1 is correct.

According to 2nd option

$E_{2s(H)} > E_{2s(Li)} > E_{2s(K)}$

($\because$ K has maximum protons & exerts good attractive force on 2s)

According to 3rd option

Given Order is incorrect, because when velocities are same, then $\lambda \propto \frac{1}{mass}$ & H_2 has least mass among the given molecules. So, its de-broglie λ will be maximum.

44. (3) ${}^{76}_{32}Ge$: n = 76 – 32 = 44

i. ${}^{77}_{32}Ge$: n = 77 – 32 = 45

ii. ${}^{77}_{33}As$: n = 77 – 33 = 44

iii. ${}^{77}_{34}Se$: n = 77 – 34 = 43

iv. ${}^{78}_{34}Se$: n = 78 – 34 = 44

45. (1) Stark effect is the splitting of spectral lines under the influence of electric field.

46. (2) $E_n = \frac{-2.18\times10^{-18}}{n^2}\times Z^2 J$

: $n = 1$, $Z = 2$: For first orbit of He^+

$E_1 = \frac{-2.18\times10^{-18}\times4}{1^2} = -8.72\times10^{-18}$ Joule

Correct & Incorrect MCQs

1. (2)

	3d	4s	4p
$_{29}Cu$	↑↓ ↑↓ ↑↓ ↑↓ ↑↓	↑	□ □ □
$_{29}Cu^+$	↑↓ ↑↓ ↑↓ ↑↓ ↑↓	□	□ □ □

Cu^+ has no unpaired e^- hence diamagnetic.

2. (4) Thomson proposed that:

(i) An atom consists of a positively charged sphere and the electrons are embedded in it.

(ii) The negative and positive charges are equal in magnitude. Hence, the atom as a whole is electrically neutral.

3. (4) In Rutherford's scattering experiment, a very few α–particles (~1 in 20,000) bounced back, i.e., were deflected by nearly 180°.

4. (2) Atomic number (Z) = Number of protons in the nucleus of an atom = Number of electrons in a neutral atom.

Mass number (A) = Number of protons (Z) + Number of neutrons (n)

5. (3) Electromagnetic waves do not require medium and can move in vacuum.

6. (3) For each metal, there is a characteristic minimum frequency, ν_0 (which is also known as threshold frequency) below which photoelectric effect is not observed.

7. (4) All five 3d orbitals are equivalent in energy.

8. (4) 1) No two electrons in an atom can have the same set of four quantum numbers.

2) Only two electrons may exist in the same orbital and these electrons must have opposite spin.

3) The maximum number of electrons present in s, p and d subshell can be 2, 6 and 10.

4) The maximum number of electrons in the shell with principal quantum number n is equal to $2n^2$.

9. (2) For a given value of l, $m_l = -l$ to $+l$.

Hence, for $l = 2$, $m_l = -2, -1, 0, +1$ and $+2$.

10. (1) For 3 p orbital,

Number of nodes = n – 1 = 3 – 1 = 2

For 3p-orbital,

Number of radial nodes = $n - l - 1 = (3 - 1 - 1) = (3 - 2) = 1$

Number of angular nodes = l = 1

Number of radial and angular nodes depend on both n and l.

Statement Based MCQs

1. (1) Orbital angular momentum = $\sqrt{l(l+1)}\frac{h}{2\pi}$

For s orbital, $\ell = 0$, thus, orbital angular momentum is zero. All the s-orbitals are spherically symmetric. Thus, 1s, 2s, 3s, all have spherical shape.

2. (4) $Fe^{2+} \rightarrow [Ar]3d^6$; Cr-$[Ar]3d^5 4s^1$. Also, due to half filled configuration in Cr, it is more stable. Thus, both have different configuration.

	3d	4s	4p
Fe^{2+}:	↑↓ ↑ ↑ ↑ ↑	□	□ □ □
Cr:	↑ ↑ ↑ ↑ ↑	↑	

3. (1) The max no. of electrons in a energy level (n) is equal to $2n^2$. For n = 3, no. of e^- = 18 which is equal to the total no. of electron in the s, p & d orbitals. Hence, it does-not have any f-subshell.

For any value of n, l = 0 to (n – 1)

Hence, for n = 3, l = 0, 1, 2

4. (1) The formula for orbital angular momentum $= \sqrt{l(l+1)}\frac{h}{2\pi}$

Since all orbitals have same ℓ value i.e., 0, therefore, all orbitals have 0 angular momentum of their electrons.

5. (1) Shape of a subshell is determined by azimuthal quantum no.

For p orbital, l = 1

so, m = –1, + 0, – 1

Also p-orbitals are dumbbell shaped.

Hence, both statements are true.

6. (2) Isotopes are the species with same atomic number but different mass number. Chemical properties of atoms are controlled by the number of electrons, which are determined by number of protons in the nucleus and neutrons have very little effect on the chemical properties of an element, hence all the isotopes of a given element show similar chemical behavior.

7. (1) Rutherford's model of an atom resembles the solar system. In case of solar system, gravitational force works. Similarly in case of an atom, electron and nucleus are held together by electrostatic force of attraction

8. (3) X-rays are not deflected by the electric and magnetic fields and have a very high penetrating power through the matter and that is the reason that these rays are used to study the interior of the objects.

9. (1) Both statements are correct.

10. (2) Energy is absorbed in absorption spectrum i.e electron moves from smaller orbit to larger. Energy is emitted in the emission spectrum. Hence, in absorption spectrum, $n_f > n_i$ while in emission spectrum, $n_i > n_f$.

Assertion & Reason MCQs

1. **(4)** Electronic configuration of K(19) = $1s^2, 2s^2, 2p^6, 3s^2, 3p^6, 4s^1$. The lower the value of (n + *l*) for an orbital, the lower will be its energy.
   Since in this case, 3p and 4s have same value, therefore, the one with lower 'n' will have lower energy.
   Hence, 19th electron will enter in higher energy level 4s.
2. **(2)** At a given temperature, the intensity of radiation emitted increases with a decreases in wavelength, reaches maximum and then begin to decrease with the further decrease in wavelength.
3. **(3)** The path of an electron in an atom can never be determined or known accurately.
4. **(1)** The single electron can go to higher energy levels and on de-excitation produces different spectral lines.
5. **(2)**

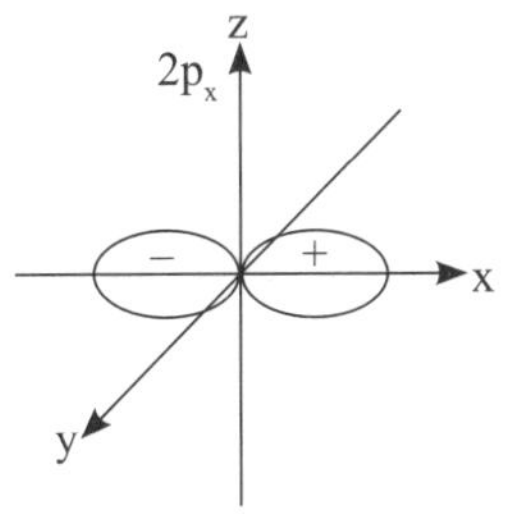

6. **(1)** Number of spherical nodes (n – 2)
   $\because n = 2$
   $\therefore n - 2 \Rightarrow 2 - 2 = 0$
   Therefore, there is no spherical node.
7. **(1)** For d-orbital, $l = 2$, Number of radial nodes = $n - l - 1 = 3 - 2 - 1 = 0$
8. **(1)** Using n + *l* rule, as d orbital has $l = 2$ and for s, $l = 0$
   Therefore, $(n + l)_{3d} > (n + l)_{4s}$.
9. **(1)** Both Assertion and Reason are True and the Reason is a correct explanation of the Assertion
10. **(4)** Most of the alpha particles passed through gold foil undeflected in Rutherford experiment

Multi Correct MCQs

1. **(2)** Degenerate orbitals must have same energy.
   $\therefore$ Orbitals with same values of n and *l* are degenerate orbitals.
2. **(3)** Rule of (n + *l*) determines the sequence of sub energy level. The one with more (n + *l*) value has more energy. And in case if (n + *l*) value is same, then the one with more 'n' value has more energy. Hence, increasing order of energy for given electrons is :
   IV < I < II < III.
3. **(4)** Radial nodes = $n - l - 1$

Orbitals	Value of *l*	Number of Radial nodes
4d	2	1
2s	0	1
3s	0	2
5f	3	1
2p	1	0

Hence, for (A), (B) and (D), the radial nodes are same.

4. **(3) Conditions:** For a valid set of quantum numbers, following conditions needs to be satisfied.
   $l = 0$ to $n - 1$, $m_l = -l$ to $+l$
   A. $n = 3$ $l = 3$ $m_l = -3$ is incorrect as l can not be equal to n.
   B. $n = 3$ $l = 2$ $m_l = -2$ is correct set.
   C. $n = 2$ $l = 1$ $m_l = +1$ is correct set
   D. $n = 2$ $l = 2$ $m_l = +2$ is incorrect set as l can not be equal to n.
   So, B and C are the correct set of quantum numbers.

NCERT Exemplar MCQs

1. **(3)** Bohr put forward the concept of electrons that move in a circular path of fixed energy called orbits. It is not derived from Rutherford's scattering experiment.
2. **(2)** Correct configuration should be $1s^2\,2s^2\,2p^6\,3s^2 3p^6\,3d^{10}\,4s^1$ for copper which has atomic number 29 ($_{29}Cu$). Due to extra stability of fully filled orbital of d-subshell, the last electron enters into d-orbital instead of s-orbital.
3. **(4)** The probability density of electrons in 2s orbital first increases then decreases and after that it begins to increase again as distance from nucleus increases.
4. **(4)** Characteristics of cathode rays do not depend upon the nature of the gas present in the cathode ray tube.
5. **(2)** The mass of electron is very small as compared to the mass of the neutron.
   Mass of electron = 9.1×10^{-31} kg
   Mass of neutron = 1.67×10^{-27} kg
6. **(1)** J J Thomson, in 1898, proposed plum pudding, (raisin pudding or watermelon) model of atom. In this model, mass of atom is assumed to be uniformly distributed over the atom. This model explained the overall neutrality of the atom.
7. **(4)** Isobars have the same mass number (i.e., sum of protons and neutrons) but different atomic number (i.e., number of protons). e.g., $_{18}Ar^{40}$ and $_{19}K^{40}$ are isobars.

$_{18}Ar^{40}$	$_{19}K^{40}$

8. **(4)** There are ($n - \ell - 1$) radial nodes, and (n – 1) total nodes.
   Number of radial nodes in 3p orbital.
   $= n - l - 1 = 3 - 1 - 1 = 1$ ($\because$ For p-orbital, $l = 1$)

9. (3) Number of angular nodes = ℓ

For 4d orbital, n = 4 and $\ell = 2$

Hence, number of angular nodes = 2

10. (2) A German physicist, Werner Heisenberg in 1927, stated uncertainty principle which states that it is not possible to determine simultaneously, the exact position and exact momentum of an electron.

Mathematically, $\Delta x \times \Delta P_x \geq \frac{h}{4\pi}$

The important implications of the Heisenberg uncertainty principle is that it rules out existence of definite paths or trajectories of electrons and other similar particles.

11. (3) Total number of orbitals associated with n^{th} shell = n^2

$\therefore$ Total number of orbitals associated with third shell $= (3)^2 = 9$

12. (1) Orbital angular momentum $= \sqrt{l(l+1)}\frac{h}{2\pi}$

Hence, it depends only on 'l', l can have values ranging from 0 to (n – 1).

(a) When $l = 0$, the subshell is s and orbital is spherical in shape.

(b) When $l = 1$, the subshell is p and orbital is dumb-bell shaped.

(c) When $l = 2$, the subshell is d and orbital is double dumb-bell shaped.

(d) When $l = 3$, the subshell is f and orbital is complicated in shape.

13. (3) The fractional atomic mass (35.5) of chlorine is due to the fact that in ordinary chlorine atom, Cl-37 and Cl-35 are present in the ratio of 1 : 3.

$\therefore$ Average atomic mass of Cl $= \frac{3\times 35 + 1\times 37}{4}$

$= \frac{105+32}{4} = 35.5$ amu

14. (2) $_{24}Cr = [Ar]3d^5, 4s^1$, $_{24}Cr^{3+} = [Ar]3d^3$

$_{26}Fe = [Ar]3d^6, 4s^2$, $_{26}Fe^{3+} = [Ar]3d^5$

$_{25}Mn = [Ar]3d^5, 4s^2$, $_{25}Mn^{2+} = [Ar]3d^5$

$_{27}Co = [Ar]3d^7, 4s^2$, $_{27}Co^{3+} = [Ar]3d^6$

$_{21}Sc = [Ar]3d^1, 4s^2$, $_{21}Sc^{3+} = [Ar]$

Thus, Fe^{3+} and Mn^{2+} have the same electronic configuration.

15. (4) (a) Electrons in 2s and 2p orbitals have different screening effect. Hence, their Z_{eff} is different. Z_{eff} of 2s orbital > Z_{eff} of 2p orbital.

Therefore, it is not correct.

(b) Energy of 2s orbital < energy of 2p orbital.

Hence, it is not correct.

(c) Z_{eff} of 1s orbital $\neq$ Z_{eff} of 2s orbital

Hence, it is incorrect.

(d) For the two electrons of 2s orbital, the value of m_s is $+\frac{1}{2}$ and $-\frac{1}{2}$. Hence, it is correct.

16. (2) From the de-Broglie equation, wavelength, $\lambda = \frac{h}{mv}$

For same speed of different particles, i.e., electron, proton, neutron and α-particle,

$\lambda \propto \frac{1}{m}$, (As h is constant).

Greater the mass of matter waves, lesser is wavelength and vice-versa. In these matter waves, alpha particle (He^{2+}) has higher mass, therefore, shortest wavelength.

NEET & AIIMS Past Year MCQs

1. (3) We know that, for balmer series; The transition occurs with $n_1 = 2$

$n_2 = 3$

Using the Rydberg formula

$$= V_v = \frac{1}{\lambda} = R_H\left[\frac{1}{n_1^2} - \frac{1}{n_2^2}\right]$$

$$\bar{v} = R_H\left[\frac{1}{4} - \frac{1}{9}\right]$$

$$\bar{v} = \frac{5R_H}{36}\ cm^{-1}$$

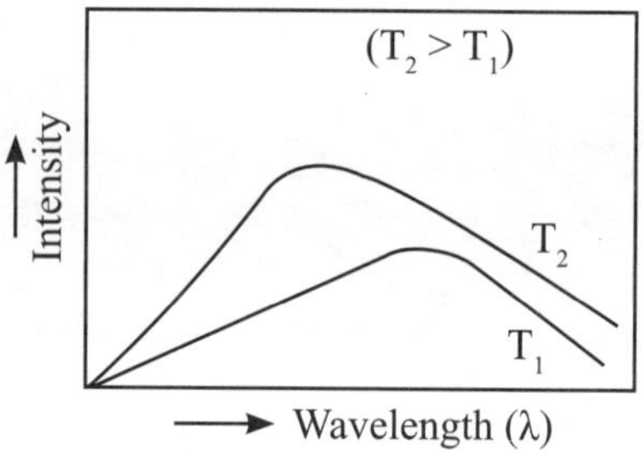

2. (2)

(I) $n = 4, l = 2, m_l = -2, s = -\frac{1}{2}$;

represents 4d (n + l = 4 + 2 = 6)

(II) $n = 3, l = 2, m_l = 1, s = +\frac{1}{2}$;

represents 3d (n + l = 3 + 2 = 5)

(III) $n = 4, l = 1, m_l = 0, s = +\frac{1}{2}$;

represents 4p (n + l = 4 + 1 = 5)

(IV) $n = 3, l = 1, m_l = -1, s = +\frac{1}{2}$;

represents 3p (n + l = 3 + 1 = 4)

The energy order is determined by the (n + l) value, with higher values corresponding to greater energy. If two orbitals have the same (n + l) value, the energy depends on the principal quantum number 'n'; higher 'n' results in higher energy.

Step-1: According to (n + l) rule

Energy = (I) > (II) = (III) > (IV)

Step-2: If n↑, then energy increases

Energy = (I) > (III) > (II) > (IV)

3. (3)

- For He^+ ion; Energy = –x J, Z = 2, n = 1

Energy $= \dfrac{-2.18\times10^{-18}Z^2}{n^2}J$

$-x = \dfrac{-2.18\times10^{-18}\times(2)^2}{(1)^2}J$

$x = 2.18 \times 10^{-18} \times 4$ J

- For Be^{3+} ion; Z = 4, n = 2

$(E_2)_{Be^{3+}} = \dfrac{-2.18\times10^{-18}Z^2}{n^2}J$

$= \dfrac{-2.18\times10^{-18}\times(4)^2}{(2)^2}J$

$= -2.18 \times 10^{-18} \times 4$ J

= –x J

4. (4)

(A) $m_l \rightarrow$ orientation of orbital.

(B) $m_s \rightarrow$ orientation of spin of electron.

(C) $l \rightarrow$ shape of orbital.

(D) n $\rightarrow$ size of orbital.

5. (1) An Atom has three fundamental particles-electron, proton and neutron.

The mass of the electron is 9.10939×10^{-31} kg.

Neutrons and protons, both are collectively known as nucleons.

All the isotopes of a given element show same chemical properties.

Dalton's atomic theory, regarded atom as an ultimate particle of matter.

6. (2) $n_m = 2l + 1$

As there are (2l + 1) number of permissible values of magnetic quantum number.

Hence, $l = \dfrac{n_m - 1}{2}$

7. (1) The shapes of all d orbitals are given as

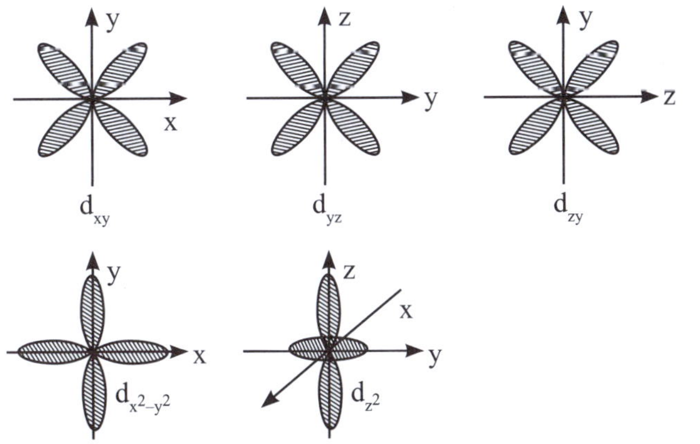

From the above figures it is clear that shape of d_{z2} is not similar to d_{x2-y2}.

8. (2) Acc. to Bohr's atomic model

$r \propto \dfrac{n^2}{Z}$

3rd orbit of Li^{+2} : $n_1 = 3, Z_1 = 3$

2nd orbit of He^+ : $n_2 = 2, Z_2 = 2$

$\dfrac{(r_3)_{Li^{+2}}}{(r_2)_{He^+}} = \dfrac{n_1^2}{n_2^2}\times\dfrac{Z_2}{Z_1}$

$\dfrac{(r_3)_{Li^{+2}}}{105.8\,pm} = \dfrac{(3)^2\times2}{(2)^2\times3}$

$(r_3)_{Li^{+2}} = 158.7$ pm

9. (4) $\lambda = \dfrac{c}{v}$

$\lambda = \dfrac{3\times10^8}{1368\times10^3} = 219.298m = 219.3m$

10. (2) $\dfrac{1}{\lambda} = R\left(\dfrac{1}{m^2}-\dfrac{1}{n^2}\right)\times Z^2$

$\lambda_{He^+} = \dfrac{400}{2^2} = \dfrac{400}{4} = 100$ nm

11. (1)

12. (3) For Lyman series,

$\dfrac{1}{\lambda} = R\left(\dfrac{1}{n_1^2}-\dfrac{1}{n_2^2}\right)Z^2$

$\dfrac{15R}{16} = R\left(1-\dfrac{1}{n_2^2}\right) \Rightarrow \dfrac{15}{16} = \left(1-\dfrac{1}{n_2^2}\right)$

$\dfrac{1}{n_2^2} = \dfrac{1}{16}$

$n_2^2 = 16$

$n_2 = 4$

13. (3) **14. (2)**

Classification of Elements and Periodicity in Properties

NCERT Topic-wise MCQs

1. (1) The third period of periodic table contains 8 elements.

2. (3) According to Dobereiner law of triads, the atomic mass of the middle element was nearly the arithmetic mean of atomic masses of other two elements.

Cl	Br	I	Arithmetic mean
35.5	80	127	$\dfrac{35.5+127}{2} = 81.2$

3. (2) The tenth element is Neon (Ne). It is a noble gas element with atomic number 10. Element with atomic number (2) is He and 54 is (Xe) and they are also noble gases. Therefore, they will resemble with each other.

4. (3) The 5th period of the periodic table contains 18 elements.

5. (2) Moseley gave the modern periodic law. He showed that atomic number is more fundamental property of an element than its atomic mass. He found that the square root of the frequency of X-rays emitted against atomic number gave a straight line.

6. (3) In the modern periodic table, elements are arranged in the increasing atomic number. Moseley found that atomic number was better fundamental property than that of atomic weight.

7. (1) Eka-silicon was the name given by Mendeleev for the unknown element Germanium in Mendeleev's classification of elements.

8. (4) The long form of periodic table consists of 7 periods and 18 groups.

9. (3) The long form of periodic table is based on atomic number of element.

10. (2) In IUPAC nomenclature, it is known as Dubnium.

Db (105) = $[Rn]^{86}\ 5f^{14}\ 6d^3\ 7s^2$

11. (1) The IUPAC official name of element with atomic number 103 is Lawrencium.

12. (3) The electronic configuration of element of group 18 can be represented by ns^2np^6.

13. (1) In transition element, the incoming electron occupies (n–1)d subshell in preference to np-level.

14. (1) They have same valence shell electronic configuration of ns^2, hence belong to same group.

15. (4) 4th transition element has atomic no. 24.

The condensed electronic configuration is [Ar] $3d^5\ 4s^1$ (exception).

The complete electronic configuration is $1s^2\ 2s^2\ 2p^6\ 3s^2\ 3p^6\ 3d^5\ 4s^1$.

16. (4) The electronic configuration of Ar is $1s^2\ 2s^2\ 2p^6\ 3s^2\ 3p^6$ which means it has 18 electrons. Except I^-, all have 18 electrons.

The number of electrons in I^- are 54.

17. (4) For elements with atomic number, 3 and 12, the electronic configuration are respectively: $1s^2\ 2s^1$ and $1s^2\ 2s^2\ 2p^6 3s^2$. Since, in both cases, e^- is present in s-orbital, hence, they belong to s-block elements.

18. (4) Helium is a p-block element with configuration of s block.

It's electronic configuration is $1s^2$.

19. (2) 28 → d-block

48 → d-block

70 → f-block

100 → f-block

55 → s-block

45 → d-block

34 → p-block

36 → p-block

20. (4) According to electronic configuration mentioned in (4), the element has atomic number 22 which corresponds to transition element, Ti.

21. (3) According to electronic configuration, the element has 29 electrons. Hence it is copper, a d-block transition element.

22. (2) The atomic number of Gd is 64. Hence, the electronic configuration of Gd is $[Xe]4f^7\ 5d^1 6s^2$.

23. (4) The first inert gas which contains d electrons is in 4th period i.e., Kr with atomic no. = 36

E.C. of Kr = $1s^2\ 2s^2\ 2p^6\ 3s^2\ 3p^6\ 4s^2\ 3d^{10}\ 4p^6$

Total number of d-electrons = 10

Total number of p-electrons = 6 + 6 + 6 = 18

Total number of s-electrons = 2 + 2 + 2 + 2 = 8

Difference in p and s electron is = 18 – 8 = 10

24. (3) An element with atomic number 29 has following electronic configuration.

Cu_{29} – [Ar] $3d^{10}4s^1$.

Since, last e^- enters in d-subshell,

∴ belongs to d-block.

25. (4) Be : $1s^2, 2s^2$

Hence, Be contains 2 electrons in its outermost orbit.

26. (2) Pd = $4d^{10}, 5s^0$ member of 4d series i.e., 5th period and 10th group.

27. (3) Silver belongs to 5th period. So, the atomic number of elements placed above and below will be 47 – 18 = 29 and 47 + 32 = 79 respectively because in 4th period we have 18 elements and that in 6th, we have 32 elements.

28. (1) The atomic number which are representative elements: 3, 33, 53, 87.

The s-block and p-block elements come under the category of representative elements.

29. (4) $\begin{bmatrix} Z = 89 = Ac = d-\text{block, } 7^{th}\text{ period} \\ Z = 115 = Uup = p-\text{block, } 7^{th}\text{ period} \end{bmatrix}$

30. (4) Radii are in the following order

$O^{2-} > F^- > O > F$

O^{2-} and F^- are isoelectronic and O,F belong to the same period.

31. (3) Isoelectronic species have same number of electrons, not same electric charge. For example N^{3-}, O^{2-}, F^- are isoelectronic but have different charges. And we know that more the –ve charge, more is the ionic size and anions have larger size than their parent atoms due to less Z_{eff}.

32. (3) On increasing effective nuclear charge, ionic size decreases.

$$r_{ionic} \propto \frac{1}{Z_{eff}}$$

33. (4) In isoelectronic species, greater the negative charge, lower the ionisation enthalpy as negative species have more size, hence less energy is required to remove an electron, S^{2-} being largest will have the lowest I.E.

34. (4) Peaks in the IE curve are occupied by the noble gases because IE of noble gases is highest in the respective periods.

35. (3) Among Rb, K and Na, the IE_1 increases as Rb < K < Na but IE_1 of Na < Mg.

36. (2) The reaction in (2) option, refers to IE_3 which has a higher value.

37. (3) Based on definition of ionisation enthalpy, an e^- is always removed from the isolated gaseous atom.

38. **(1)** Down the group; size increases with decrease in nuclear charge. Hence, Rb^+ has the largest size.

39. **(2)** Atomic radii are generally measured in pm.

40. **(4)** Electronic configuration indicates that 1 e^- is present in outermost shell. Hence, it will easily lose electrons

$\therefore$ It is metal and form basic oxide

$$2Na + \frac{1}{2}O_2 \rightarrow \underset{\text{(Sodium oxide)}}{Na_2O}$$

41. **(4)** (15)P^{3-}, (16)S^{2-}, (17)Cl^-, belong to 3rd period whereas (20) Ca^{2+} belong to 4th period.

These are isoelectronic ions and therefore size, depends on nuclear charge. More is the nuclear charge, smaller is the size. Also, size of cation is always smaller than anion. Hence, the correct order is as shown by option (4).

42. **(3)** The order of electron affinity: Cl > F > Br > I

43. **(4)** The electron affinity of Be is similar to that of Ne.

44. **(3)** Li and Mg show diagonal relationship.

45. **(1)** The electronic configuration in option (1) represents halogen i.e., Cl and the E.A of Cl is the highest.

46. **(2)** Na_2O is basic, SO_3 and P_2O_5 are acidic but Al_2O_3 can react both with acids and bases, hence amphoteric in nature.

47. **(3)** Order of electronegativity is : F > O > N > P

48. **(4)** Halogens have a high value of electron affinity. Also by gaining one electron, it acquire stable noble gas configuration. Hence, have strongest tendency to gain electrons or to form anions.

49. **(2)** Ionic size increases down the group and decreases along a period. Be^{2+} belong to 2nd period whereas Na and Mg are of 3rd period. Also, for isoelectronic species, greater the positive charge, smaller will be the radius. Hence, the correct order becomes $Be^{2+} < Mg^{2+} < Na^+$

50. **(3)** $IE_1 < IE_2 < IE_3$

Because with the removal of electron,number of protons remain same but number of electron decreases and therefore effective nuclear charge increases. Hence, it becomes difficult to remove an electron, so I.E. increases.

51. **(2)** The most electronegative element is Fluorine whose outermost electronic configuration is $2s^2\ 2p^5$.

52. **(4)** P has higher IE than S due to stable half filled configuration.

53. **(3)** Isoelectronic species have same number of electrons, and hence, same electronic configuration.

54. **(4)** Along a period, size decreases while down the group, size increases.

Group 14	Group 15	Group 16
C	N	–
	P	S

Hence, the correct order is N < C < S < P.

55. **(1)** Greater the positive charge on the ion, smaller is its size. Hence, Al^{3+} has the smallest size.

56. **(2)** The electron affinity of chlorine is greater than flourine.

57. **(4)** Smaller the size, more will be the electron affinity due to more attraction towards new electron. As size of F is less it is expected to have more electron affinity but actually F has 7 e^- in its outermost shell (and have smaller size), which causes repulsion towards new e^-.

$\therefore$ Electron affinity of Cl > Electron affinity of F (Due to smaller size of 2p - subshell)

58. **(1)** $\underbrace{Ca^{2+} < K^+}_{} < Ar < \underbrace{Cl^- < S^{2-}}_{}$

Smaller size due to more effective nuclear charge and more the positive charge smaller will be the size

Large size due to less effective nuclear charge i.e., attraction per electron decreases and more the negative charge, more will be the size.

59. **(3)** In the element with electronic configuration $1s^2\ 2s^2\ 2p^6\ 3s^1$, the removal of second electron would be from inert gas core, hence will have highest second ionisation enthalpy.

60. **(1)** I.E increases while moving from left to right in period due to decrease in atomic size and increase in Z_{eff}.

I.E. of B is less than Be because Be has fully filled stable $2s^2$ valence configuration, hence it requires more energy to remove electron from it. Now, 'N' has more I.E than 'O' due to half filled valence $2p^3$ configuration, therefore, the correct order is B < Be < C < O < N.

61. **(2)** For isoelectronic species, more is the positive charge smaller will be the size. Therefore, the correct order for the ionic size is

$\therefore Na^+ > Mg^{2+} > Si^{4+} > Cl^{7+}$

62. **(1)** For a particular atom, the successive ionisation enthalpy always increases. Since it is difficult to remove an electron from cation, thus, $E_1 < E_2 < E_3$.

63. **(3)** The tendency to lose electron is higher with elements having lower ionisation enthalpy. Hence lower I.E. is the favourable factor for cation formation.

64. **(1)** Size of cation is always smaller than parent atom due to increase in effective nuclear charge. Also, size of cation decreases as the charge on cation increases. Hence, the correct order is $Al > Al^+ > Al^{2+}$

65. **(1)** Ist ionisation energy of Mg is more than that of Na, is not an exception.

High Order Time Intensive MCQs

1. **(1)** Californium, atomic no - 98, it is f-block element in 7th period. It belongs to actinoid series.

2. **(2)** 4th period comprises of 4s, 3d and 4p.

4d elements will lie in 5th period. Hence there is no element having one or more 4d electron in 4th period.

3. **(4)** For option (1), Cl has most negative electron gain enthalpy.

For option (2), F has greater negative electron gain enthalpy than O.

For option (3), adding e^- in O^- will certainly require energy, as there will be electronic repulsions between the already existing '–ve' charge and incoming electron. So electron gain enthalpy is +ve.

4. (1) In periodic table, after certain regular interval, some properties are repeated. e.g., atomic volume, ionic and covalent radius, ionization enthalpy, electronegativity, electron affinity etc. But atomic mass is not a periodic property.

5. (4) IE_3 is always greater than IE_2 which in turn is greater than IE_1.

I.E. of O^- will be less due to its larger size than O.

6. (3) 0.02 g-atom = 0.02 mol atom = $0.02 \times N_A$ atom

$0.02 \times N_A$ atom have ionisation energy = x kJ

$$1 \text{ atom have ionisation energy} = \frac{x}{0.02 \times N_A} \text{kJ} = \frac{x \times 10^3}{0.02 \times N_A} \text{J}$$

7. (2) N, O and Ne are present in 2nd period and Na is present in the 3rd period. The I.E decreases while moving down the group, hence Na has minimum I.E among given species.

Na^+ and Ne are isoelectronic but Na^+ has more protons, thus more Z_{eff}

$\therefore$ small size. Hence, its I.E. is highest.

8. (1) Ca^{2+} , K^+ and Ar are isoelectronic species (with $18e^-$). Since, Ca^{2+} has greatest positive charge, therefore it has the smallest size.

Hence, correct order is $Ca^{2+} < K^+ < Ar$.

9. (2) Nitrogen is smaller in size and therefore has higher electron density than phosphorous, so there are additional repulsions to the incoming electron. Hence, electron affinity of nitrogen is less than that of phosphorus.

10. (4) The general outer electronic configuration of d-block elements is $(n-1)d^{1-10}\, ns^{0-2}$.

11. (2) Isoelectronic species contain same number of electrons. The number of electrons in K^+, Ca^{2+}, Sc^{3+} and Cl^- = 18

Hence, all the species are isolectronic.

12. (4) After removing 1 e^- it will get stable noble gas configuration. Hence, its second ionization energy will be more than first.

13. (1) CO, H_2O, N_2O are the examples of neutral oxides.

14. (1) $\frac{1}{2}N_A$ atoms have ionisation energy = ΔH

$$1 \text{ atom of gas X has ionisation energy} = \frac{\Delta H \times 2}{N_A}$$

15. (1) Moles of Mg $= \frac{1.2}{24} = 0.05$ mol

$Mg \rightarrow Mg^{2+} + 2e^-$

I.E. = $I.E_1 + I.E_2 = 120 + 240 - 360$ kJ/mol

$1 \text{ mol} \xrightarrow{\text{energy required}} 360 \text{ kJ/mol}$

$0.05 \text{ mol} \xrightarrow{\text{energy required}} 360 \times 0.05 = 18 \text{ kJ}$

16. (3) In isoelectronic species i.e., species containing same number of electrons.

$\begin{Bmatrix} \text{More the positive charge; smaller will be the size} \\ \text{More the negative charge; larger will be the size} \end{Bmatrix}$

17. (3) Electron gain enthalpy is negative of I.E. i.e., –5.1 eV

18. (2) Mg P Cl $\begin{Bmatrix} \text{left to right size decreases} \\ \text{down the group size increases} \end{Bmatrix}$

Ca

i.e., Ca > Mg > P > Cl

19. (3) Atomic number of element = 32. An element below it will have atomic number 32 + 18 = 50.

20. (1) Rn(86) ---------- $6p^6$

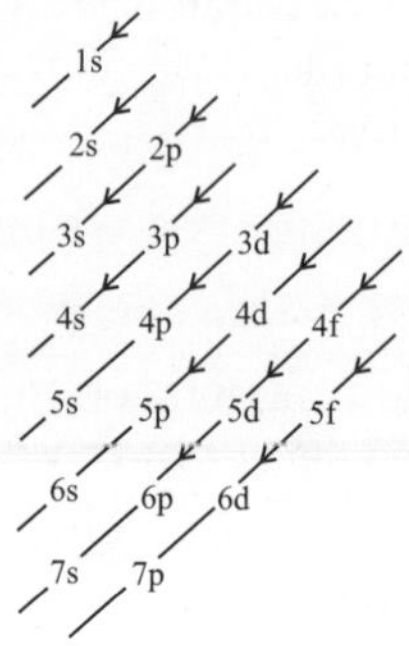

At No. 112 will have configuration $[Rn]_{86}\, 5f^{14}\, 6d^{10}\, 7s^2$ or $[Rn]_{86}\, 7s^2\ 5f^{14}\ 6d^{10}$. It's a d – block element as last electron enters the d-orbital which belongs to 7th period since, highest principal quantum number, n = 7 & 12th group.

21. (1) Order of Ist IP

F > O > C > Li, because IP increase across the period.

22. (2) The typical oxides of elements become more acidic on moving from left to right in a period since their electronegativity rises.

23. (1) Option (1) is correct

Correct order in options (2) and (3):

$C^{4-} > N^{3-} > O^{2-} > F^-$

$F^- > Na^+ > Mg^{+2} > Al^{3+}$

24. (2) On moving from right to left in a period, size increases therefore, electronegativity decreases.

25. (1) Referring to periodic table, we have,

	Cl	
Se	Br	Kr
Te	I	Xe

Hence, D is Cl which has maximum electron affinity.

26. (2) The chemical properties of an element is predicted by the position of element, atomic number and number of electrons in the outer shell.

27. (3) All these species have 18 electrons.

28. (2) [Rn] $7s^2\, 6d^2$ – f block

[Xe] $6s^2\, 4f^{14}\, 5d^1$ – f block

[Xe] $6s^2\, 4f^{14}\, 5d^{10}6p^1$ – p block

[Xe] $6s^2\, 4f^1\, 5d^1$ – f block

29. (4) IIIrd I.P. > IInd I.P. > Ist IP for any individual atom.

30. (2) I = F III = Cl
II = N IV = P
order Cl > F > P > N

31. (3) In a group, the electropositive nature increases while in a period it decreases. Hence, element X is most electropositive. V, W, Y and Z are less electropositive than X.

Atomic size decreases while moving along a period and increases while moving down the group. Hence, W < X.

32. (4) Cl^- and K^+ are isoelectronic species. Isoelectronic ions are those ions which have same number of electrons.

K = 2, 8, 8, 1 K^+ = 2, 8, 8
Cl = 2, 8, 7 Cl^- = 2, 8, 8

Potassium ion is smaller due to high nuclear charge.

33. (2) I.E. in a group decreases and reactivity increases.

34. (2) (P → (i), (ii), (iii))

Rf (Z = 104) : [Rn] $5f^{14}6d^27s^2$

Period no. 7, d-block element, group no. 4.

(Q →(i), (iii))

Rg (Z = 111) : [Rn] $5f^{14}\ 6d^{10}7s^1$

Period no. 7, group no. 11, d-block element.

(R →(i), (iv))

Th (Z = 90) : [Rn] $5f^0\ 6d^27s^2$

Period no. 7, group no. 3 , f-block element.

(S→(i), (iv))

Np (Z = 93) : [Rn] $5f^4\ 6d^17s^2$

Period no. 7, f-block element.

35. (2) With increase in nuclear charge, the ionic radius decreases for isoelectronic ions.

36. (2) N has half filled stable electronic configuration, therefore, its ionization enthalpy is greater than that of oxygen. Hence, increasing order of first ionization enthalpy is: B < C < O < N

37. (3) As the electronegativity of element increases, acidic character of oxides increases. So, in a group, basic nature increases on moving down and it decreases while moving along a period.

38. (2) IE_2 of Na is higher than that of IE_2 of Mg because in case of Na, the second electron has to be removed from the noble gas core while in case of Mg, removal of second electron gives a noble gas core.

39. (2) Elements X, Y and Z with atomic numbers 19, 37, 55 lie in group 1 (alkali metals). Within a group, IE decreases from top to bottom. Therefore, IE of Y could be between X and Z.

40. (1) The electronic configuration of the element with Z = 114 is [Rn]$5f^{14}\ 6d^{10}7s^27p^2$. Hence, it belongs to carbon family which has the same outer electronic configuration.

41. (3) (1) is incorrect because the electronic configuration of P is exactly half-filled i.e., more stable.

(2) is incorrect because s-orbital is closer to nucleus than p-orbital.

(3) is correct.

42. (1) The less electron affinity of fluorine is due to its smaller size and high charge density.

Correct & Incorrect MCQs

1. (1) In Dobereiner's Triad, the atomic weight of the middle element was arithmetic mean of the other two.

Triad:

Element	Atomic weight
Li	7
Na	23
K	39

Triad:

Element	Atomic weight
Ca	40
Sr	88
Ba	137

Triad:

Element	Atomic weight
Cl	35.5
Br	80
I	127

2. (4) Modern periodic table is based on atomic number rather than atomic mass. In long form of periodic table, elements having similar outer electronic configurations in their atoms are arranged in vertical columns, referred to as groups.

3. (2) The IUPAC name of an element with atomic number 102 is Unnilbium.

4. (3) In s-block elements, the metallic character and the reactivity increase as we go down the group, because as the size increases, the tendency to lose electrons increases, hence, the metallic character and the chemical reactivity increases.

5. (4) Since electrons are added in the same shell in a period, therefore effective nuclear charge increases while moving from left to right across a period, hence the atomic size generally decreases across a period.

6. (3) Among isoelectronic species, the cation with the greater positive charge will have a smaller radius because of the greater attraction of the electrons to the nucleus. Hence, the cation Mg^{2+} will have smaller radius than that of cation Na^+ since Mg^{2+} acquires more positive charge on it.

7. (3) The outermost electronic configuration of p-block elements varies from ns^2np^1 to ns^2np^6 in each period.

8. (4) In beryllium, the electron removed during the ionization of s-electron whereas the electron removed during ionization of boron is a p-electron. 's' electron is attracted to the nucleus more than 'p' electron. Hence, the 2p electron of boron is more shielded from the nucleus by the inner core of electrons than the 2s electrons of beryllium. Thus, it is easier to remove the 2p-electron from boron as compared to 2s-electron from beryllium. Hence, first ionization enthalpy of boron is less than that of beryllium.

9. **(2)** Noble gases have large positive electron gain enthalpies because, the electron has to enter the next higher principal quantum level leading to a very unstable electronic configuration.
10. **(1)** Generally, metallic oxides are basic in nature. Na_2O being a metallic oxide is basic in nature.

Statement Based MCQs

1. **(3)** The first ionization enthalpy of oxygen is smaller as compared to nitrogen. This is because N-atom has half-filled stable 2p-configuration. Therefore, it is difficult to remove e^- from the stable valence shell electronic configuration.
2. **(2)** Ionization enthalpy is the energy required to remove an electron from isolated gaseous atom in its ground state. Ionization enthalpy of an atom is always +ve. Not all elements have tendency to readily lose electron.
3. **(3)** The electronic configurations are: He-$1s^2$, Be-$1s^2 2s^2$.
   Be is not inert, while He is inert because it has completely filled valence shell which is stable. Therefore, He will not react with anyone and will stay inert.
4. **(1)** Al has one unpaired e^- in its highest energy orbital (3p) and in Mg, the electrons are paired in highest energy orbital (3s).
   Therefore, electrons can easily be removed from 3p orbital of Al.
   Hence, I[st] I.E of Al is lower than that of Mg.
5. **(1)** Ionization enthalpy decreases on moving down the group because the atomic size increases on moving down the group as a result force of attraction between nucleus and electrons decreases, therefore, it becomes easy to lose an e^-.
6. **(1)** Modern periodic table is based on atomic number rather than atomic mass. The horizontal rows are called periods and the vertical columns are called groups. Elements in same vertical column have similar valence shell electronic configuration, hence exhibit similar properties.
7. **(2)** On moving from left to right across a period, effective nuclear charge increases and tendency to lose valence electrons decreases, hence metallic character decreases.
   While moving down the group, the valence electron becomes farther from the nucleus and due to shielding of valence e^-, tendency to lose valence electrons increases, hence metallic character increases.
8. **(4)** Due to increased nuclear charge, atomic radius decreases across a period. Hence, Atomic radius of Mg is larger than that of S.
9. **(1)** Due to fully filled stable electronic configuration of noble gases, they have largest ionization enthalpy as they will not prefer to lose any e^-.
10. **(3)** The electronegativity values of both N and Cl is 3.0 on Pauling scale i.e. same.

Assertion & Reason MCQs

1. **(1)** Both assertion and reason are true and reason is the correct explanation of the assertion.
2. **(4)** Electron gain enthalpy not always becomes less negative as we go down the group. It can be explained from the fact that Cl has more negative electron gain enthalpy than F in Group 17.
3. **(3)** The correct order for ionisation enthalpy is F > N > O.
4. **(3)** Electronegativity is not a measurable quantity.
5. **(3)** The fluorine atom is much smaller than that of chlorine.
6. **(4)** Cl_2O_7 is an acidic oxide.
7. **(3)** The magnitude of effective nuclear charge of the outer shell electrons in Al^{3+} is greater than that in Na^+.
8. **(4)** Cations are always smaller than the neutral atom.
9. **(3)** The electronic configuration of He is $1s^2$.
   It can be seen that the last filling e^- in He goes into s-subshell. Since He resembles other noble gases in its properties e.g. it is an inert gas like other noble gases, hence it is placed in p-block elements.
10. **(3)** The first period in the long form of the periodic table contains 2 elements.

Multi Correct MCQs

1. **(2)** In general, as you move along a period from left to right in the periodic table, the size of the atoms decreases and the metallic character decreases.
   In general, as you move down a group in the periodic table, the size of the atoms increases and the metallic character increases.
   B' < A'(size) C' > A'(size)
   D' < C'(size) D' > B'(size)
   B' < A' (metallic character)
   D' < C' (metallic character)
   $B'^+ < A^+$ (size)
   $D'^+ < C^+$ (size)
   ∴ C statement is incorrect.
2. **(4)** Electron affinity is negative of electron gain enthalpy i.e.,
   E.A = $-\Delta H_{eg}$

(A)	$Be + e^-$	→	Be^-,	E.A = –ve
(B)	$N + e^-$	→	N^-	E.A = –ve
(C)	$O + e^-$	→	O^-	
	$O^- + e^-$	→	O^{-2}	E.A = –ve
(D)	$Na + e^-$	→	Na^-	E.A = +ve
(E)	$Al + e^-$	→	Al^-	E.A = + ve

   Hence, A, B and C have negative values for E.A.
3. **(4)**

	O^{2-}	F^-	Na^+	Mg^{2+}
(No. of e^-)	10	10	10	10
(Ionic radius)	O^{2-} >	F^- >	Na^+ >	Mg^{2+}
Z_{eff}	O^{2-} <	F^- <	Na^+ <	Mg^{2+}

4. **(4)** Isoelectronic species have same no. of electrons in valence shell. Among the given, isoelectronic species are O^{2-}, F^-, Mg^{2+}, Na^+, Al^{+3}

5. **(4)** Main group elements are those in groups 1, 2, and 13–18 of the periodic table. They typically have their last electron in the s- or p- orbitals.

   I. Ends in $4p^5 \rightarrow$ Group 17 (halogen), a main group element.

   II. It has last electron in d-orbital, hence, belongs to the transition metals.

   III. Ends in $4s^1 \rightarrow$ Group 1, total electrons = 19 $\rightarrow$ Potassium (also a main group element).

   IV. It has last electron in d-orbital, hence, transition metal.

   Therefore, both (I) and (III) corresponds to a main group element

NCERT Exemplar MCQs

1. **(2)** In case of isoelectronic species, greater the positive charge, smaller will be the radius. Also, greater the negative charge, more will be radius. Also, the anions have more radii than cations. Hence, the correct order of their radii is $Mg^{2+} < Na^+ < F^- < O^{2-}$
2. **(4)** Elements with atomic number, Z = 90 to 103 are called actinoids. Thus, Terbium (Z = 65) is not an actinoid. Terbium belongs to lanthanoids.
3. **(1)** For the same shell, screening effect decreases in the order s > p > d > f
4. **(1)** The electronic configurations are as follows:

   Na: $[Ne]3s^1$, Mg: $[Ne]3s^2$, Al: $[Ne]3s^23p^1$, Si: $[Ne]3s^23p^2$

   Ionization enthalpy increases as we move across the period. The I.E. of Mg will be greater than that of Na due to fully filled electronic configuration. The I.E. of Al will be smaller than that of Mg due to $1e^-$ extra than the stable electronic configuration. Hence, the correct order is Na < Mg > Al < Si.
5. **(3)** The electronic configuration of La (Z = 57) is [Xe] $5d^1 6s^2$. Therefore, further addition of electrons occurs in the lower energy 4f-orbital till it is exactly half-filled at Eu (Z = 63) Thus, the electronic configuration of Eu is [Xe] $4f^7 6s^2$. Thereafter, addition of the next electron does not occur in the more stable exactly half-filled $4f^7$ shell but occurs in the little higher energy 5d-orbital. Thus, the electronic configuration of Gd (Z = 64) is [Xe] $4f^7 5d^1 6s^2$.
6. **(3)** 3d orbital is filled when 4s orbital gets completely filled. In case of transition elements (or any elements), the order of filling of electrons in various orbital is 3p < 4s < 3d.
7. **(3)** As we move from Cl to I, the electron gain enthalpy become less and less negative due to a corresponding increase in the atomic size and decrease in z_{eff}. However, the electron gain enthalpy of F is less negative than that of Cl due to its small size and inter electronic repulsions in it.
8. **(3)** Since, each period starts with the filling of electrons in a new principal quantum number, therefore, the period number in the long form of the periodic table refers to the maximum principal quantum number of any element in the period.

   Period number = maximum 'n' of any element (where, n = principal quantum number).
9. **(3)** The elements in which electrons are progressively filled in 4f-orbital are called lanthanoids. Lanthanoids consist of elements from Z = 58 (cerium) to 71 (lutetium).
10. **(4)** Cation is formed after the loss of electron from outer shell and anion is formed after the gain of electron to the neutral atom. Also, cation has smaller size than anion and anion has bigger size than its neutral atom. Thus, the correct order is: $I^- > I > I^+$.
11. **(3)** Although O^{2-} has noble gas configuration as it is isoelectronic with Neon but its formation is unfavourable due to the strong electronic repulsion between the negatively charged O^- ion and the second electron being added.

    Hence, the electron repulsion outweighs the stability gained by achieving noble gas configuration.
12. **(3)** The element with atomic number 57 belongs to d-block element as the last electron enters the 5d-orbital against the Aufbau principle. This anomalous behaviour can be explained on the basis of greater stability of the xenon (inert gas) core.

    After barium (Z = 56) , the addition of the next electron (i.e., 57th) should occur in 4f-orbital in accordance with Aufbau principle. This will however, tend to destabilize the xenon core (Z = 54), [Kr] ($4d^{10}\ 4f^0\ 5s^2\ 5p^6\ 5d^0$) since the 4f-orbitals lie inside the core.

    Therefore, the 57th electron prefers to enter 5d-orbital which lies outside the xenon core and whose energy is only slightly higher than that of 4f-orbital. In doing so, the stability conferred on the atom due to xenon core more than compensates the slight instability caused by the addition of one electron to the higher energy 5d- orbital instead of the lower energy 4f- orbital.

    Thus, the outer electronic configuration of La(Z = 57) is $5d^1 6s^2$ rather than the expected $4f^1 6s^2$.
13. **(3)** The last element of p-block in 6th period is $_{86}Rn$. It's outer electronic configuration is $4f^{14} 5d^{10} 6s^2 6p^6$.
14. **(3)** The long form of the periodic table contain element with atomic number 1 to 118.
15. **(1)** The fifth period begins with Rb (Z = 37) and ends at Xe (Z = 54). Thus, the element with Z = 43 lies in the 5th period. Since, the 4th period has 18 elements, therefore, the atomic number of the element which lies immediately above the element with atomic number 43 is 43 – 18 = 25.

    Now, the electronic configuration of element with Z = 25 is $1s^2 2s^2 2p^6 3s^2 3p^6 4s^2 3d^5$
16. **(2)** Elements with atomic numbers 35 (36 – 1), 53 (54 – 1), and 85 (86 – 1), lie in a group before noble gases, i.e., halogens (group 17) elements because each period ends with a noble gas. The atomic number of noble gases (i.e., group 18 elements) are 2, 10, 18, 36, 54 and 86.

    Thus, the elements with atomic number 35, 53 and 85 are all belongs to halogens.
17. **(1)** Electronic configuration of elements indicate that A is a noble gas (i.e., Ne), B is oxygen (group 16), C is sodium metal (group 1) and D is fluorine (group 17).

(i) Noble gases have no tendency to gain electrons, since all their orbitals are completely filled. Thus, element A has the least electron gain enthalpy.

(ii) Since, element D has one electron less and element B has two electrons less than the corresponding noble gas configuration, hence, element D has the highest electron, gain enthalpy followed by element B.

(iii) Since, element C corresponds to sodium metal and metals lose electrons, therefore, electron gain enthalpy of element C will be less than that of element B which is oxygen, a non metal. Combining all the facts given above, the electron gain enthalpies of the four elements increase in the order A < C < B < D.

NEET & AIIMS Past Year MCQs

1. (2) In a period, as the atomic number increases, the effective nuclear charge also increases, resulting in a decrease in atomic radii across the period.

The correct order of atomic radii is: Li > Be > B > C

2. (3)

$$\left.\begin{array}{l} F = 4.0 \\ O = 3.5 \\ N = 3.0 \\ C = 2.5 \\ Si = 1.8 \end{array}\right\} \text{electronegativity}$$

Hence order will be:

Si < C < N < O < F

3. (4)

- In a period, generally on moving left to right ionization enthalpy increases. But due to more stable configuration of Be($2s^2$), its ionization enthalpy is more as compared to B($2p^1$), less stable configuration.
- So, correct order of first IE :
  Li < B < Be < C < N

4. (4) Among isoelectronic monoatomic species, size is inversely proportional to atomic number.

Hence among isoelectronic species Na^+, O^{2-}, N^{3-}, F^- (having nearest noble gas configuration);

Order of size is $Na^+ < F^- < O^{2-} < N^{3-}$

N^{3-} has least atomic number hence largest size.

5. (2) According to IUPAC nomenclature one is named as un and nine is named as enn and overall name is ended with ium. Therefore the correct name for 119 will be ununennium.

6. (3) Total no. of e^-

${}_{26}Fe \rightarrow 3d^64s^2$, $Fe^{+2} \rightarrow 3d^6$

${}_{25}Mn \rightarrow 3d^54s^2$, $Mn^{+2} \rightarrow 3d^5$

7. (4) **8. (4)** Size decreases **9. (2)**

10. (4) **11. (1)**

Chemical Bonding and Molecular Structure

NCERT Topic-wise MCQs

1. (2) Due to Fajan's rule, according to which the cation distorts the electron cloud of the anion and pulls the electron density towards itself. Thus, it introduces some covalent character in the ionic bond. Now, the smaller the cation, greater will be its polarizability.

2. (3) Ionic character increases as; $BeCl_2 < MgCl_2 < CaCl_2 < BaCl_2$ because larger the size of the cation, higher is the ionic character.

3. (1)

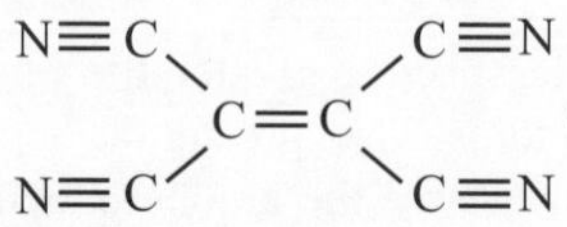

π bonds = 9

4. (2) $BeCl_2$ is an electron deficient species in which Be contains only 4 electrons.

5. (3) Formal charge (F.C) = Total number of valence electrons in free atom-total number of lone pair electrons- ½(total number of bonding electrons)

In carbonate ion,

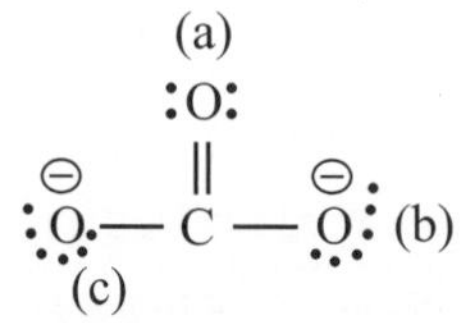

For O(a);

F.C = 6 – 6 = 0

For O(b) and O(c),

F.C = 6 – 7 = –1

6. (1) Lattice energy is directly proportional to charge and inversely proportional to size. So, according to this, NaF has the highest lattice energy.

7. (2) In HCl, H – atom and Cl – atom shares one electron pair to form covalent bond.

8. (1) Li^+ being smaller, according to Fazan rule, Li^+ has high polarising power. Hence, LiCl should have maximum covalent character.

9. (2) The conditions required for the formation of an ionic bond are :

(*i*) Ionization enthalpy [$M(g) \rightarrow M^+(g) + e^-$] of electropositive element must be low.

(*ii*) Negative value of electron gain enthalpy [$X(g) + e^- \rightarrow X^-(g)$] of electronegative element should be high.

10. (1) $CuSO_4 \cdot 5H_2O$ has all ionic, covalent, coordinate and hydrogen bonds.

$CuSO_4 . 5H_2O$ exists as $[Cu(H_2O)]^{2+} SO_4^{2-}$

Coordinate bond = $[Cu(H_2O)_4]^{2+}$ between H_2O molecules and Cu.

Ionic bond = $[Cu(H_2O)_4]^{2+}$ and SO_4^{2-}

Covalent bond = Between S and O in SO_4^{2-}

Hydrogen bond = Within the H_2O molecules

11. (3) Urea (NH_2CONH_2) has maximum covalent bonds = 8. The structure of urea is:

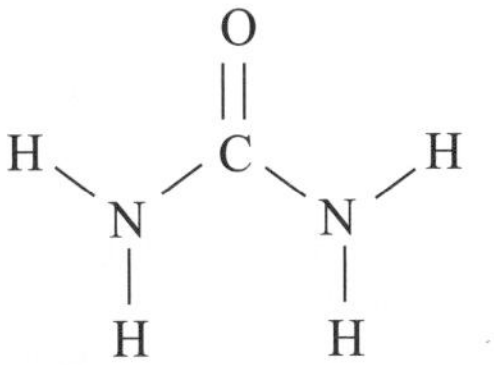

12. (4) Ethane consists of 6C–H bonds and 1C–C bond. Total number of bonds is 7. Each bond is made up of two electrons.

13. (4) HCN contains 4 bonds (1C–H single bond and 1 triple bond between C and N). Its structure is: $H-C\equiv N$

14. (1) The bond formed between them would be ionic due to large electronegativity difference.

15. (1) More electropositive and more electronegative combination gives more ionic compound.

16. (2) Smaller the size of an anion, the compound will be highly ionic in nature. Among the given options, F is the smallest in size, so, KF will be highly ionic.

17. (1) As fluorine is highly electronegative than other atoms and smallest in size, sodium fluoride is more ionic than other molecules.

18. (3) Polarity is resulting from unequal sharing of electrons. There are three sigma bonds present in hydrogen peroxide. Two of them are between hydrogen & oxygen atoms which is polar in nature. The other one is in between two oxygen atoms where there is no unequal sharing of electrons as they have same dipole moment. So, it is non-polar. So, the correct answer is H_2O_2. It's structure is given as-

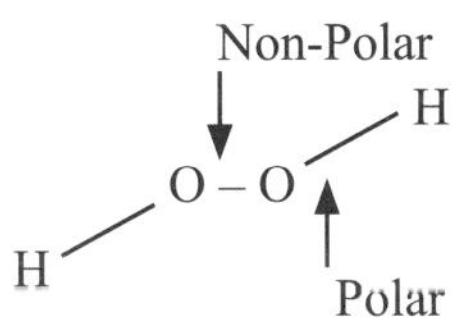

19. (3) In Cl_2 Δ.EN = 0 $\therefore$ non-polar.

20. (4) Covalent molecules are held together by covalent bonds. There are only weak van der Waal's force of attraction that held strongly the covalent bonds. Due to weak van der Waal's force of attraction, it can be broken with little energy.

21. (3) The correct Lewis dot structure of sulfite ion is:

S atom = 6 valence electrons

O atom = 6 valence electrons

Total number of electrons = 6 + 3(6) +2 (from negative charges) = 26

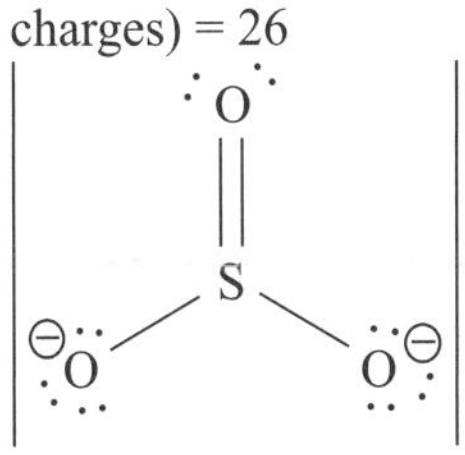

22. (2) In diamond, one carbon atom is attached to other carbon atoms by means of covalent bonds.

23. (3) The shape of NH_4^+ ion is tetrahedral having bond angle of exactly equal to 109°28'.

24. (3) Bond angle of hydrides of group 16 elements are in the order: $H_2O > H_2S > H_2Se > H_2Te$

25. (2) H_2O (104.5°) ; BeF_2 (180°)
CH_4 (109°) ; NH_3 (107°)

26. (3) In H–C≡C–H, the C–H bond is made of sp-s overlap which imparts shorter bond length.

27. (1) As the electronegativity of the central atom decreases and the size increases, the bond angle decreases.

28. (4) $:\ddot{O}=\ddot{O}\rightarrow\ddot{\underset{..}{O}}:$ is one of the resonating structure of ozone.

29. (3) Number of lone pair increase around the central, bond angle decreases.

NH_3 = 1 l.p. CH_4 = 0 l.p.
BeF_2 = 0 l.p. H_2O = 2 l.p.

30. (4) Bond angle of water is 104.5. In water, oxygen is sp^3 – hydridised with two lone pairs.

31. (4) The structure of hydrogen peroxide is an open book having bond order = 1

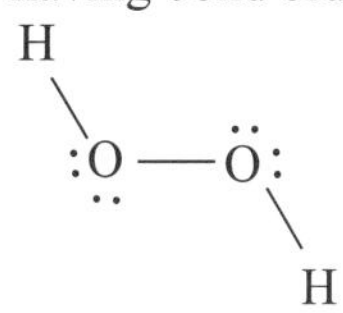

Bond order of ozone is 1.5 and its structure is:

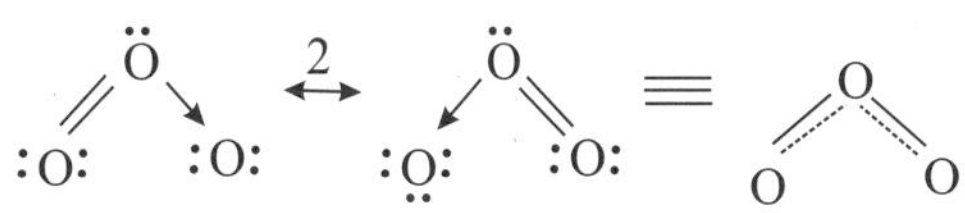

Bond order of O_2 is 2.

$\ddot{\underset{..}{O}}=\ddot{\underset{..}{O}}$

Bond order and bond length are inversely related to each other.

Hence, the O – O bond length increases in the order:

$O_2 < O_3 < H_2O_2$

32. (2)

1) 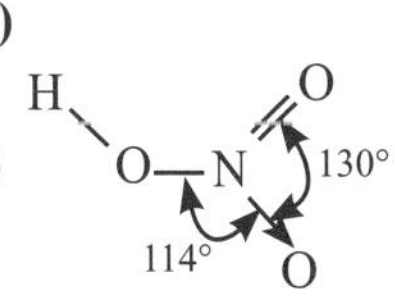

2) $O=\overset{+}{N}=O$, Bond angle is 180°.

3) 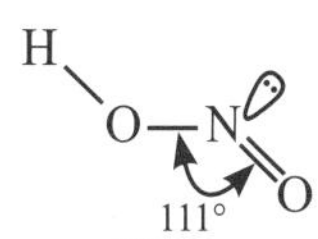

4) 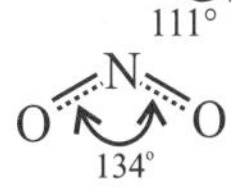

33. (2) $O=\overset{\oplus}{N}=O$ Bond order = 2

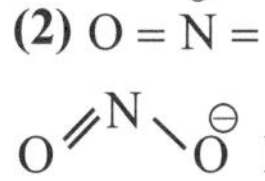 Bond order < 2

So, bond length $NO_2^{\ominus} > NO_2^{\oplus}$

Bond order and bond length are inversely related to each other.

34. (4) Higher the s-character of hybrid orbital, higher will be the bond angle.

35. (4) Among C_6H_6, CO_2, CO_3^{2-} and SiO_2, SiO_2 does not show resonance. Silicon dioxide (SiO_2) commonly known as silica or sand occurs in a number of crystalline forms. SiO_2 is a covalent three dimensional network solid in which each silicon atom is linked to four oxygen atoms through covalent bonds.

36. (3) In BeF_2, dipole moment is zero because of its linear shape while water has net dipole moment because of its angular shape.

H_2O has angular structure with bond angle 105° while BeF_2 is a linear molecule.

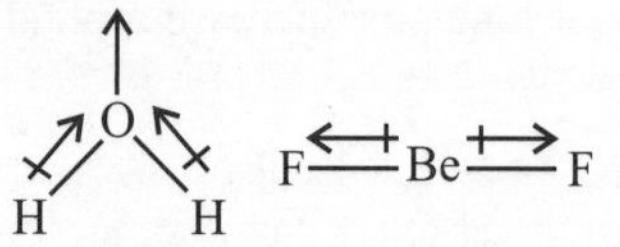

$\mu > 0$ $\mu = 0$

37. (2)

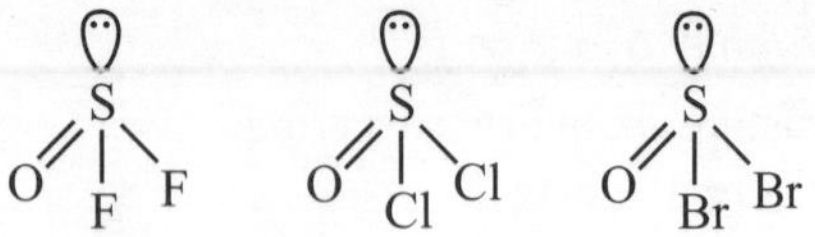

In $SOBr_2$, S–O bond has maximum bond length in comparison to S–O bond lengths in SOF_2 and $SOCl_2$, because in $SOBr_2$, S–O bond has been formed by hybrid orbital containing less s-character according to Bent Rule. Br is less electronegative than F and Cl.

38. (1) Symmetrical molecules have zero dipole moment.

Also, dipole moment $\propto \dfrac{1}{\text{Bond angle}}$

39. (2)

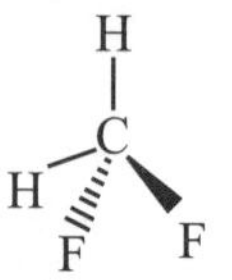

Total number of angles are six.

$H\hat{C}H, H\hat{C}F, F\hat{C}F, H\hat{C}F, H\hat{C}F, H\hat{C}F$

Maximum number of identical angles are four ($H\hat{C}F$)

40. (2) Due to absence of lone pair on N in NH_4^+, lp–lp, or lp –bp repulsion is not there. Hence, it has the highest bond angle.

41. (2) In H_2S, due to low electronegativity of sulphur, the l.p. -l.p. repulsion is more than b.p. -b.p. repulsion and hence, the bond angle is 92°.

42. (1) The order of bond angles is:

$BF_3 > SiH_4 > NH_3 > H_2S$

120° 109°28' 107° 92.5°

No. of lone pairs increases, bond angle decreases.

43. (4) The value of average bond energies is as follows-

$S=O$	523 kJ mol^{-1}
$C\equiv C$	839 kJ mol^{-1}
$C\equiv N$	891 kJ mol^{-1}
$N\equiv N$	941 kJ mol^{-1}

$\therefore$ $N\equiv N$ has the highest average bond energy

44. (3) Bond angles are:

BF_3 – 120°, NF_3 – 106°, PF_3 – 101°, ClF_3 – 90°

45. (1) Bond lengths of C–C, C–H and O–H are 154pm, 107 pm and 96 pm.

So, the order is: C–C > C–H > O–H.

46. (4) Shape of the molecule depends upon the total number of electron pairs around the central atom which include l.p as well as b.p. Therefore, AX_3 molecule has any of the given shape,

47. (2) SF_4 has see-saw shape due to. 1 1*l*p and 4bp and XeF_4 has square planar shape due to 2*l*p and 4bp. SiF_4 has 0 lone pair having tetrahedral structure.

48. (4) Sn has 2 b.p and 1 *l*.p around it. The arrangement of electron pairs is trigonal planar.

49. (3) Both structure and shape of SO_4^{2-} is tetrahedral as it contains 0 lone pairs.

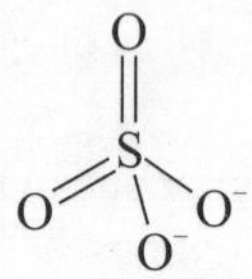

50. (4) In $AlCl_3$ molecule, central Al-atom is sp^2 hybridised.

$$H = \frac{1}{2}[3+3-0+0] = 3$$

It contains no lone pairs and its geometry will be trigonal planar.

51. (1) Because lone pair of electrons and equatorial F lies in same plane.

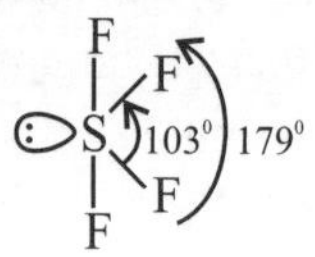

52. (2) CO_2 and BeF_2 both are linear.

53. (3) Regular geometry of CH_4 is attributed to zero dipole moment ($\mu = 0$) and absence of lone pair on carbon atom.

54. (3) Both XeF_4 and $[Ni(CN)_4]^{2-}$ are square planar. In XeF_4, Xe is sp^3d^2 hydridised with 2 *l*p and Ni in $[Ni(CN)_4]^{2-}$ is dsp^2 hybridised.

55. (3) BF_4^-, NH_4^+, $NiCl_4^{2-}$ all are tetrahedral in shape but NH_3 is pyramidal due to presence of one lone pair.

56. (2) Formaldehyde is:

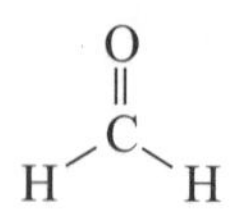

Its shape is trigonal planar as it contains no lone pairs.

57. (3) Both SO_4^{2-} and BF_4^- are tetrahedral.

58. (3) I does not have its valency 4. It has valency 1, 3, 5 & 7.

59. (3) Since the molecule possess 6 pairs of electrons, it should exhibit octahedral geometry. But as its structure is planar, it will have 2 lone pairs of electrons, and 4 sigma bonds.

60. (1) Structure of cyanogen is $N\equiv C-C\equiv N$. It has linear shape.

61. **(3)** $IF_5 = 7 + 5 \times 7 = \frac{42}{8} = 5+1$

sp^3d^2 hybridisation, hence it should be octahedral.
Due to the presence of one lone pair of electrons, its actual geometry is square pyramidal.

62. **(4)**

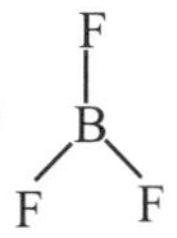

$\mu_d = 0$

Planar & non-polar

63. **(4)** Values of dipole moment can't be predicted as we don't know about the electronegativities of the atoms.

64. **(4)** 1) XeO_3 is trigonal pyramid.

2) IOF_4^+ is trigonal bipyramidal.

3) PCl_5 is trigonal bipyramidal.

4) 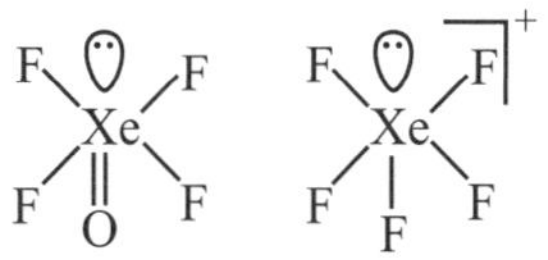

square pyramidal.

65. **(4)**

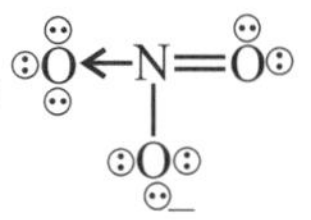

66. **(3)** According to VSEPR theory, the pairs of electrons tend to occupy such positions that place them farthest apart from each other to minimise repulsions.

67. **(4)** Lewis structure of PO_4^{3-} is:

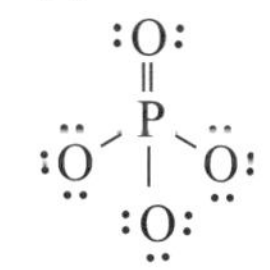

Number of bond pair = 5

Number of lone pair = 0

68. **(2)** OF_2 has 2 bond pairs and 8 lone pairs.

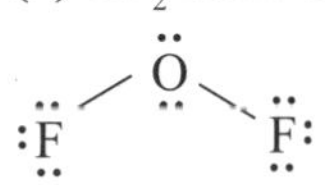

69. **(4)** In $XeOF_2$

Number of σ bonds = 3

Number of lone pairs = 2

∴ Total number of pairs of electrons
= 3 + 2 = 5

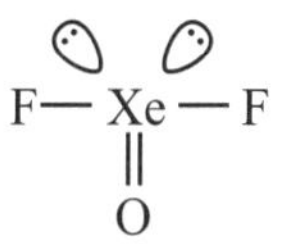

70. **(2)** H_3O^+ contains 3 B.P and 1 L.P

71. **(4)** CCl_4 is tetrahedral having no lone pairs.

72. **(1)** s-p bond involves head-on overlapping. Therefore; it is a sigma (σ) bond.

73. **(2)** S = C = S (linear) is best explained by sp hybrid state of C.

74. **(1)** Be is not sp^3–hybridized but is sp^2 hybridized in BeF_3^-.

75. **(4)** The structure of 1, 2 butadiene is $H_2C = C = CHCH_3$. C_1 is sp^2, C_2 is sp, C_3 is sp^2 and C_4 is sp^3 hybridized.

76. **(2)** As ICl_2^+ is a positive charged species, the central atom, I contains six electrons, out of which two electrons will be involved in forming 2 σ-bonds with two Cl atoms and there will two lone pairs. So, I has two bonding electron pairs and two lone pairs. Hence, the hybridization of ICl_2^+ is sp^3.

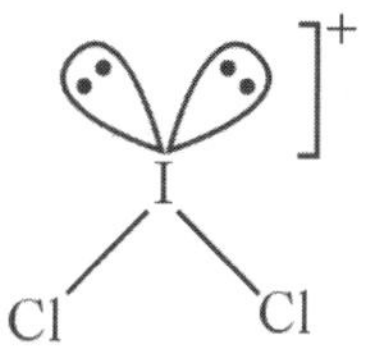

77. **(1)**

(1) I_3^- = 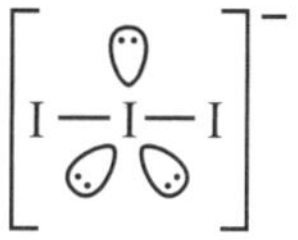 = 2σ-bonds + 3 *l.p.* = sp^3d hybridization

(2) PCl_3 = 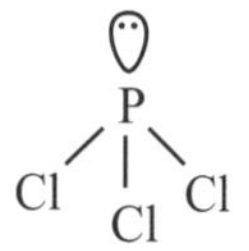 = 3σ-bonds + 1 *l.p.* = sp^3 hybridization

(3) NO_3^- = 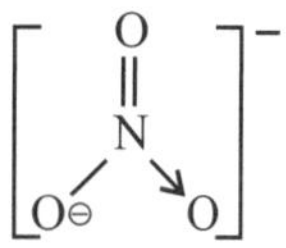 = 3σ-bonds + 0 *l.p.* = sp^2 hybridization

(4) H_2Se = 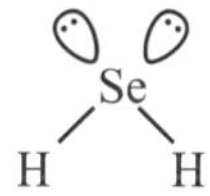 = 2σ-bonds + 2 *l.p.* = sp^3 hybridization

78. **(2)** 1. 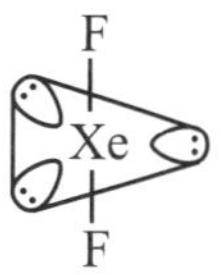

Hybridisation : sp^3d
shape : linear

O=C=O

Hybridisation : sp
shape : linear

2. 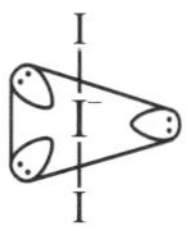

Hybridisation : sp^3d
shape : linear

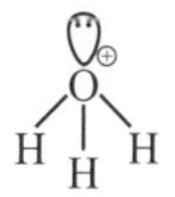

Hybridisation : sp^3
shape :pyramidal

3. 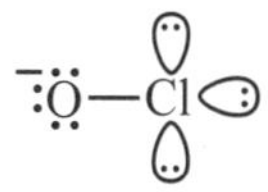

Hybridisation of Cl : sp^3
shape : linear

:C≦O:

Hybridisation of C : sp
shape : linear

4. 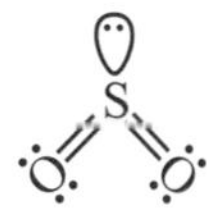

Hybridisation : sp^2
shape : V-shape/bent/angular

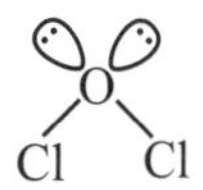

Hybridisation of O : close to sp^3
shape : V-shape/bent/angular

79. (4) In perchlorate ions,

Central Cl atom is sp^3 - hybridised.

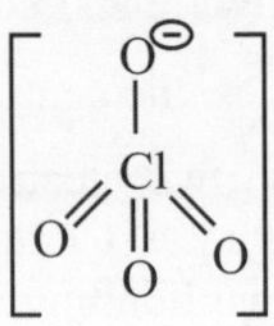

80. (3) 1) $CH_4 \rightarrow CH_3-CH_3$

4 b.p   4 b.p

Tetrahedral (sp^3)   Tetrahedral (sp^3)

2) $NH_3 \rightarrow NH^+_4$

3 b.p + 1 *l.p*   4b.p.

Pyramidal (sp^3)   Tetrahedral (sp^3)

3) $BF_3 \rightarrow BF_4^-$

Trigonal (sp^3)   Tetrahedral (sp^3)

4) $H_2O \rightarrow H_3O^+$

2b.p. + 2 *l.p*   3b.p. + 1 *l.p*

Angular (sp^3)   Pyramidal (sp^3)

Thus, conversion of BF_3 into BF_4^- involves changes in both hybridisation and shape.

81. (3) The bond angle and hybridisation in other CH_3-O-CH_3 is approximately 110° and sp^3 respectively.

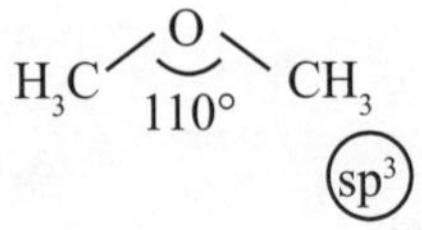

82. (2) In sp^3 hybridization, percent of p character is 75.

83. (1) When the number of hybrid orbitals is 4 then hybridization will be sp^3. Except SF_4 (sp^3d), rest all have sp^3 hybridisation.

84. (2) Hybridisation is predicted as sp^2

$X = \frac{1}{2}$ [Valence e^- + No. of monovalent – atoms around central atom charge on cation + charge on anion]

For CO_3^{2-}:

$X = \frac{1}{2}[4+0-0+2] = \frac{1}{2}(6) = 3 \Rightarrow sp^2$

85. (2) Nitrate ion (NO_3^-)

$X = \frac{1}{2}[5+0-0+1] = 3 \Rightarrow sp^2$

Hybridisation of nitrogen in nitrate ion is sp^2.

86. (1) Iodine hepta fluoride (IF_7) sp^3d^3

$X = \frac{1}{2}[7+7-0+0] = 7 \Rightarrow sp^3d^3$

87. (2) Hybridisation of Bromine in BrF_5 is sp^3d^2.

88. (4) In NO_3^-, sp^2 hybridisation is present.

89. (1) In Cl_2O, T.P. $= \dfrac{6+2}{2} = 4(sp^3)$

90. (2) A pi bond is formed by the lateral overlapping of unhybridised p-p orbitals.

91. (2) 1) σ bond is formed by axial overlapping.

2) p-orbital have both axial and side ways overlapping

92. (1)

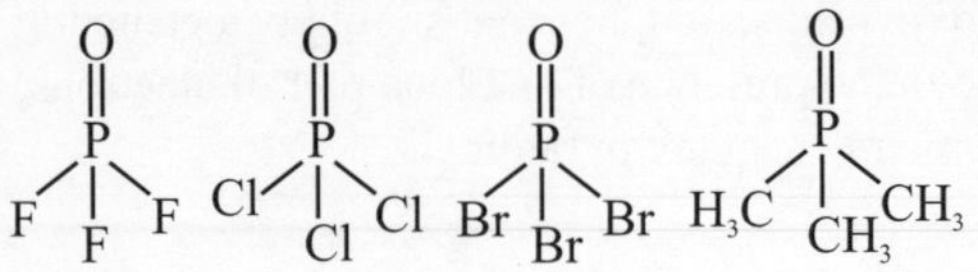

In OPF_3, molecule % s-character in O – P is maximum, hence it is strongest among all.

93. (2)

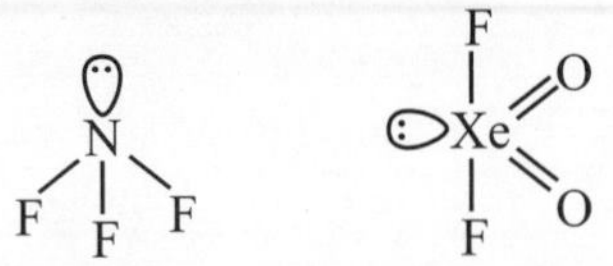

sp^3 Hybridisation sp^3d Hybridisation

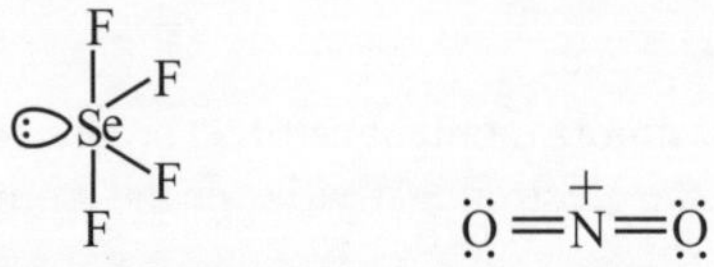

sp^3d Hybridisation   sp Hybridisation

94. (4) $TeCl_6$, 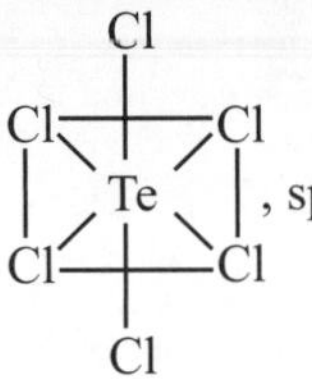 , sp^3d^2

95. (2) $\ddot{N}H_3 + H^+ \rightarrow NH_4^+$ $2NH_3 \rightarrow NH_2^- + NH_4^+$

sp^3 sp^3 sp^3 sp^3 sp^3

$H_3BO_3 + OH^- \longrightarrow [B(OH)_4]^-$ $H_2O + H^+ \rightarrow H_3O^+$

sp^2 sp^3 sp^3 sp^3 sp^3

96. (2) N_2O_5 in solid state $\rightarrow NO_2^+, NO_3^-$

$NO_3^- \longrightarrow$ hybridisation $\longrightarrow sp^2$

97. (1)

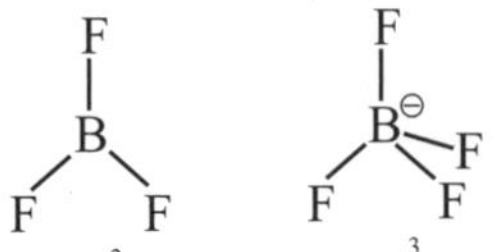

sp^2 sp^3

98. (1) 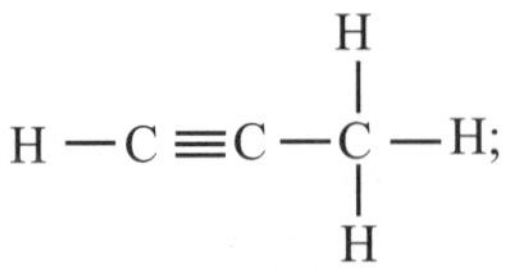

(Propyne)

Number of sigma bonds = 6 and number of pi bonds = 2

99. (2) $s + p_y + p_x \rightarrow sp^2$

$p_z + d_{z^2} \rightarrow pd$

100. (3) $2p_y$ and $2p_y$ will go sidewise overlapping which results in the formation of π bond.

101. (2) sp hybridisation is formed.

102. (4) s-orbital of hydrogen and p-orbital of fluorine overlaps.

103. (2) $NO_2^+ \rightarrow sp$

$NO_3^- \rightarrow sp^2$

$NH_4^+ \rightarrow sp^3$

104. (2) In SF_4, hybridization of S is sp^3d.

105. (1) In H_2O, hybridization is sp^3 and the shape is angular due to 2 bp & 2 *l*p.

106. (2) Due to presence of vacant 3d-orbitals, S can go into excited state with 6 unpaired electrons which are overlapped by six F electrons.

107. (1) NO^- has 16 electron system, bond order = 2

NO^-, CN^-, N_2 are isoelectronic with CO.

∴ All have same bond order = 3

108. (4) Both C_2 and He_2 do not contain unpaired electrons and are diamagnetic in nature.

109. (3) Bond order of Be_2 is zero, it does not exist.

110. (1, 3) Paramagnetic species contain unpaired electrons in their molecular orbital electronic configuration.

NO (7 + 8 = 15) = $\sigma 1s^2, \sigma^*1s^2, \sigma 2s^2, \sigma^*2s^2, \sigma 2p_z^2, \pi 2p_x^2 \approx \pi 2p_y^2, \pi^*2p_x^1 \approx \pi^*2p_y$ (paramagnetic)

NO^+ (7 + 8 – 1 = 14) = $\sigma 1s^2, \sigma^*1s^2, \sigma 2s^2, \sigma^*2s^2, \sigma 2p_z^2 \pi 2^2p_x \approx 2p_y^2$ (diamagnetic)

NO^- (7 + 8 + 1=16) = $\sigma 1s^2, \sigma^*1s^2, \sigma 2s^2, \sigma^*2s^2, \sigma 2p_z^2, \pi 2p_x^2 \approx 2p_y^2\ \pi^*2p_x^1 \approx \pi^*2p_y^1$ (Paramagnetic)

N_2(7 + 7 = 14) = same as NO^+

111. (1) Bond order ∝ bond length.

B.O. of O_2^+ = 2.5 and rest has B.O. less than 2.5.

Thus, it has the strongest O – O bond among the given.

112. (3) Total number of electrons in N_2^- = 7 + 7 + 1 = 15

Total number of electrons in N_2^{2-} 7 + 7 + 2 = 16

Total number of electrons in O_2^{2-} = 8 + 8 + 2 = 18

Total number of electrons in O_2^- = 8 + 8 + 1 = 17

Total number of electrons in O_2^+ = 8 + 8 – 1 = 15

Configuration will be same as N_2^- and O_2^+ contain the same total number of electrons. Thus, N_2^- and O_2^+ has the same bond order.

113. (2) B_2 = KK * $(\sigma 2s)^2 (\sigma^*2s)^2 (\pi 2p_x)^1 \approx (\pi 2p_y)^1$

Bond order = $\frac{1}{2}(4-2) = 1$

114. (2) Higher the bond order, smaller will be the bond length.

115. (3) No unpaired electrons are present in Li_2, therefore it is diamagnetic.

$Li_2 = 6e^- = \sigma 1s^2, \sigma^*1s^2, \sigma 2s^2$

116. (3) Mulliken gave the molecular orbital theory.

117. (1) Given electronic configuration of anion X is-

$\sigma 1s^2, \sigma^*1s^2, \sigma 2s^2, \sigma^*2s^2, \pi 2p_x^2 \approx \pi 2p_y^2, \sigma 2p_z^2, \pi^*2p_x^1$

∴ Total number of electron of anion X = 15

Hence; the anion X is N_2^-.

118. (4) (18)→$(\sigma 1s)^2$, $(\sigma^*1s)^2(\sigma 2s)^2$, $(\sigma^*2s)^2$ $(\sigma 2p_z)^2$, $(\pi 2p_x)^2$, $(\pi 2p_y)^2$, $(\pi^*2p_x)^2$, $(\pi^*2p_y)^2$

*represents antibonding molecular orbitals

Thus the no. of antibonding electrons in O_2^{2-} ion is = 8 (4 pairs)

119. (2) According to Molecular Orbital Theory, $H_2^+ = \sigma 1s^1$

Bond order $\frac{N_b - N_a}{2} = \frac{1-0}{2} = 0.5$

120. (1) Isostructural species will have same bond order.

$N_2 = 14e^-$

$O_2^{2+} = 14e^-$

121. (2) In O_2^-, the bond order is 1.5.

122. (2) N_2 molecule has bond order 3.

123. (1) $O_2^+ : B.O. = \frac{10-5}{2} = 2.5$; $NO : B.O. = \frac{10-5}{2} = 2.5$;

$NO^{2+} : B.O. = \frac{9-4}{2} = 2.5$; $CN : B.O. = \frac{9-4}{2} = 2.5$

124. (2) 'NO' has 15 electrons & it is paramagnetic in nature.

125. (2)

(1) C_2	⟶	C_2^+
B.O.= 2		B.O. = 1.5
$\mu = 0$,		$\mu \neq 0$,
i.e. diamagnetic		i.e. paramagnetic
(2) NO	⟶	NO^+
B.O.= 2.5		B.O. = 3.0
$\mu \neq 0$,		$\mu = 0$,
i.e. paramagnetic		i.e. diamagnetic
(3) O_2	⟶	O_2^+
B.O.= 2		B.O. = 2.5
$\mu \neq 0$,		$\mu \neq 0$,
i.e. paramagnetic		i.e. paramagnetic
(4) N_2	⟶	N_2^+
B.O.= 3		B.O. = 2.5
$\mu = 0$,		$\mu \neq 0$,
i.e. diamagnetic		i.e. paramagnetic

126. (3) $\sigma 1s^2\ \sigma^*1s^1$ is the molecular configuration for $H_2^{\ominus}$.

127. (1) According to MOT, when two atomic orbitals combine, formation of two molecular orbitals take place- one bonding and one antibonding

128. (4) One water molecule is joined to four water molecules - two with H - atoms and other two with O - atoms. Due to this, the maximum number of hydrogen bonds that a molecule of water can have are four.

129. (4) Hydrogen bond is formed between molecules of compounds having O, F and N with hydrogen.

130. (3) Ethyl alcohol forms stronger H - bonds than ethyl amine or ammonia due to greater electronegativity of oxygen than nitrogen atom. Diethyl ether, however, does not form H-bonds, since it does not have a H-atom attached to any highly electronegative atom.

131. (4) The higher boiling point of water is due to hydrogen bonding. O-atom of each H_2O molecule is covalently linked with two H-atoms of its own molecule and with another H-atom of adjacent H_2O molecule by H-bonding.

132. (4) Ice has H-bonding therefore; when two ice-cubes are pressed over each other, they unite to form one cube.

133. (3) In ethanol, hydrogen bonding is maximum.

134. (1) The molecule of CH_3OH are associated by hydrogen bonding.

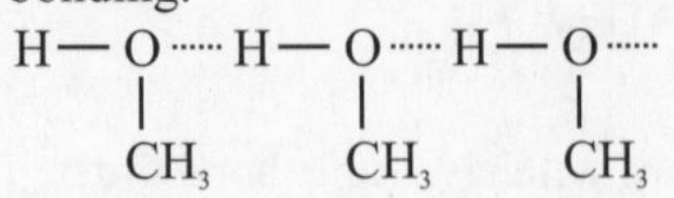

135. (1) Water (H_2O) is least volatile due to the association of H_2O molecules through H-bonding.

136. (4) Strength of hydrogen bond between H----F is the strongest and hence; it has the strongest bond.

137. (3) Intermolecular H-bonding is found in $(HF)_n$, due to higher electro-negativity of fluorine atoms. Hydrogen bonding is held in the association of HF molecule. As a result, HF form in liquid form.

138. (2)

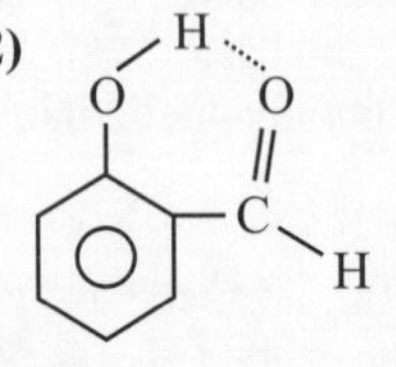

139. (2) Ortho-nitrophenol

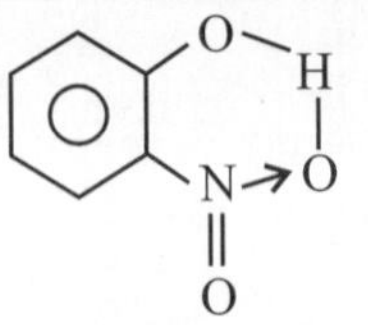

140. (2) Vander Waals' forces are present in liquid hydrogen.

141. (1) O is more electronegative than N & S. Strength of H-bonds is attributed to the electro-negativity of the atom bonded to hydrogen.

142. (2) More is the partial charges generated, more is the strength of H-bonding.

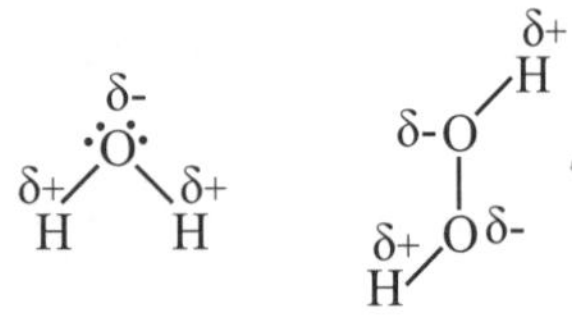

The strength of H-bonding is more in H_2O.

143. (1) Ethyl alcohol has hydrogen bonds.

144. (1) In water molecule, H-bond is formed between other molecules but not within the same molecule.

145. (2) C_2H_5OH can form intermolecular H-bonding with H_2O.

146. (4) Hydrogen bond is formed between molecules of compounds having O, F and N with hydrogen.

147. (2) Ice contains H-bonding.

148. (3) In o-nitrophenol, intramolecular H-bonding is present but in p-nitrophenol, intermolecular H-bonding is there due to which boiling point of latter is more than that of former.

High Order Time Intensive MCQs

1. (1) Bonds labelled as 1 , 2 and 3 are lying in a single plane. Bonds labelled 4 and 5 are meant to be formed along Z-axis. So, p_z and d_{z^2} orbitals hybridise to give two hybridised orbitals along the z-axis.

2. (4)

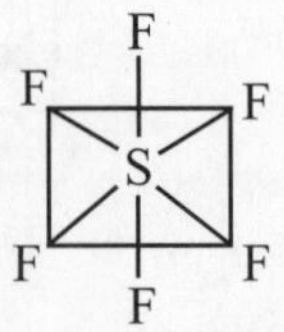

The molecule has regular octahedral geometry and a perfect symmetrical structure, b.p – b.p repulsions are uniform wrt every bond in the molecule.

So, all bond lengths are equal. In SF_6, sp^3d^2 hybridisation occurs.

3. (3) Both CO_2 and N_2O are linear in shape. CO_2 is non-polar but N_2O is polar

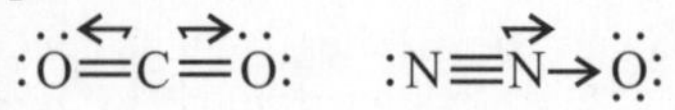

4. (2)

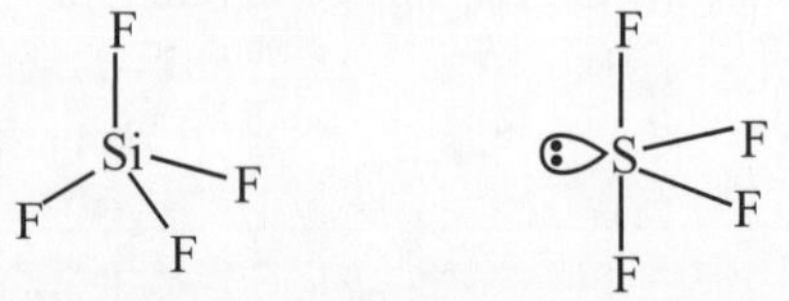

Si is sp^3 hybridised in SiF_4 and it has tetrahedral shape S is sp^3d hybridised and SF_4 has bent see-saw shape. So, not isostructural.

5. (3) If 2s–2p mixing is not operative, the increasing order of energy of M.O'S will be:

$C_2(12): \sigma 1s^2 \sigma^* 1s^2 \sigma 2s^2 \sigma^* 2s^2 \sigma 2p_z^2 \pi 2p_x^1 = \pi 2p_y^1$

6. (1) In KO_2, O_2^- (superoxide ion) has one unpaired electron.
In BaO_2, O_2^{2-} (peroxide ion) has paired electrons.
In AlO_2^-, $[Al^{3+}\ O_2^{2-}]^-$, O_2^{2-} (peroxide ion) has paired electrons

7. (4) H–C≡C–H (180°) (sp) > BF_3 (120°) (sp^2) > CH_4 (109.5°) (sp^3, tetrahedral) > NH_3 (107°) (sp^3 with one l.p) > H_2O (sp^3 with 2 large pairs) (104.5°)

$\overset{+}{N}H_4$ (109°) (sp^3 with no lp) > $\ddot{N}H_3$ (107°) (sp^3 with one lp) > $:\ddot{N}H_2$ (104°) (sp^3 with V-shape and 2 lp)

Dipole moment of $NH_3 > NF_3$.

8. (1) Bond order $\propto \dfrac{1}{\text{Bond length}}$

KO_2 has O_2^- ion, Bond order = 1.5
$O_2(AsF_6)$ has O_2^+ ion, Bond order = 2.5
O_2 has Bond order = 2
Na_2O_2 has O_2^{2-} (peroxide ion), Bond order = 1

9. (3)

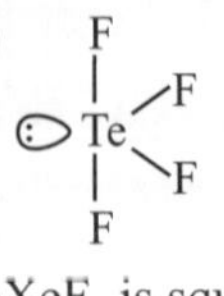

Geometry:- Trigonal bipyramidal (distorted)
Shape:- See-saw (Bent)

10. (3) XeF_4 is square planar in shape.

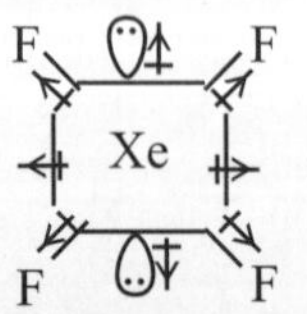

$M_R = 0$

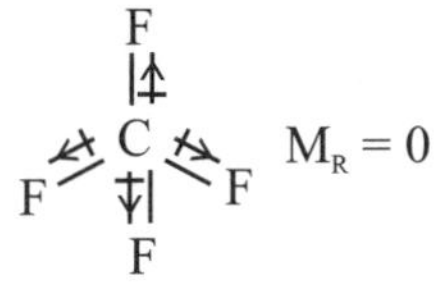

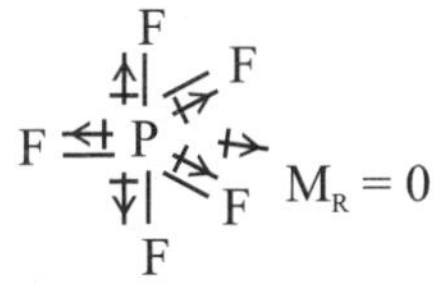

SF$_4$ has see-saw shape which is not symmetrical in nature.

11. (3)

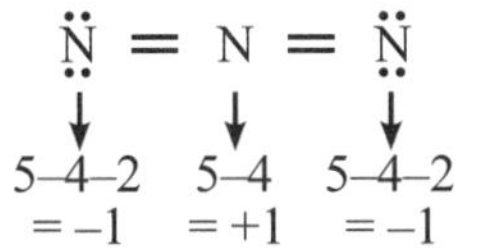

12. (3) NCl_3, N_2O_5 & Ca_3N_2 are known but NCl_5 is not due to absence of d–orbitals in valence shell.

13. (1)

BrO_3^- 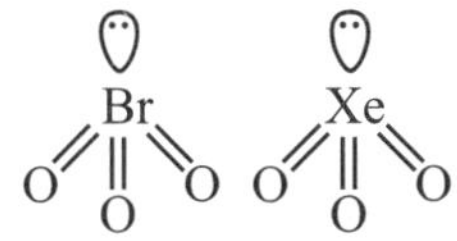 XeO_3

14. (3) The bond order of individual carbon-carbon bonds in benzene is between one and two. It is due to resonance structure of C_6H_6.

15. (3) Linear dipole is created in p-nitroaniline, dipole moment in opposite direction cancel out.

16. (1) O_2 have two unpaired electron, O_2^+ have one unpaired electron, O_2^{2-} have zero unpaired electron. As number of unpaired electrons increases, paramagnetic character increases.

17. (3) In resonance structure, there should be the same number of electron pairs.

18. (3) Only in BF_3, pπ-pπ back bonding is possible due to which B–F bond length is shortest. In other compounds, p-orbital at boron is not vacant hence, pπ-pπ back bonding is not possible.

19. (4) O_2 = Paramagnetic molecule with 2 unpaired electrons in antibonding p-orbital

O_3 = diamagnetic molecule

B_2 = Paramagnetic with 2 unpaired electron in bonding π-orbitals

C_2 = Diamagnetic molecule

$NO \rightarrow \overset{\oplus}{NO}$ (Electron removes from antibonding orbital bond length decreases)

$CO \rightarrow \overset{\oplus}{CO}$ (Electron removes from antibonding orbital bond length decreases)

20. (2) 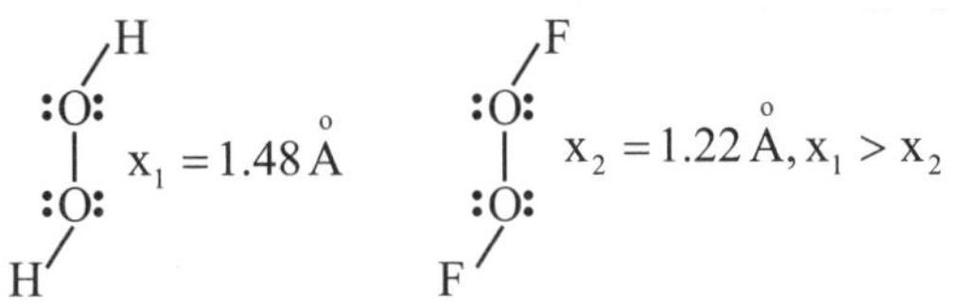

According to Bent's Rule in O_2F_2, there is more p-character in O–F bond in comparison to O–H bond in H_2O_2 ; hence s-character in O–O bond is greater in O_2F_2.

21. (4)

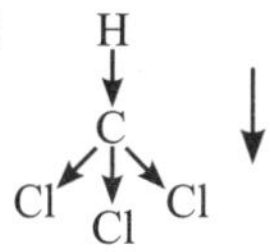

$\mu \neq 0$

$CHCl_3$ is polar molecule and rest all molecules are non-polar.

22. (4) 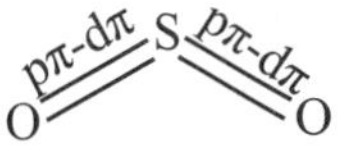

SeF_4 has sp^3d hydridisation and see-saw shape I_3^+ has sp^3 hybridisation and bent shape.

23. (4) p–p overlapping is produced by the overlapping of the p–orbitals of the atoms. In the case of chlorine molecule, it is produced by the overlapping of the $3p_z$ orbitals of two chlorine atoms.

24. (2) The boiling point of a compound is raised by intermolecular hydrogen bonding.

25. (2) The given molecules can be categorised as:

BCl_3 – electron deficient molecule

SF_6 – expanded octet molecule

NO – odd electron containing molecule

H_2SO_4 – expanded octet molecule

26. (2) Symmetrical molecule with linear structure will have zero dipole moment

27. (1) Species having odd no. of electrons, expanded octet and electron deficient do not follow octet rule.

28. (3) pi bond is formed by the pure orbitals.

29. (3) C–C bond is $sp^3 - sp^3$ and C–*Cl* is $sp^3 - p$.

30. (4) sp^3 hybridization, hence s - character is 25%.

31. (2) Resonance energy is the difference energy between the actual structure and the lowest energy resonance structure for the given compound.

32. (4) Species with odd number of electrons have fractional bond order.

33. (1) Anti bonding electrons in

$O_2^- = 7, O_2 = 6, O_2^{2-} - 8$

34. (2) Bond order $= \frac{1}{2}(N_b - N_a)$

Bond order of $Li_2 = 1, N_2 = 3, Be_2 = 0, O_2 = 2$

35. (2) All given molecules have trigonal planar geometry.

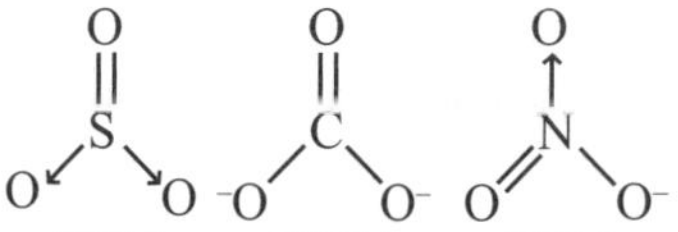

(SO_3) (CO_3^{2-}) (NO_3^-)

36. (3) A is a VII A group and C is a IA group element

∴ The compound is ionic.

37. (4) MgS has high lattice energy, so its solubility will be less.

38. (3) KCN has both ionic and covalent bonds.

39. (2) p_z orbital by convention undergoes end to end overlapping while p_x, p_y undergoes sidewise overlapping.

40. (4)

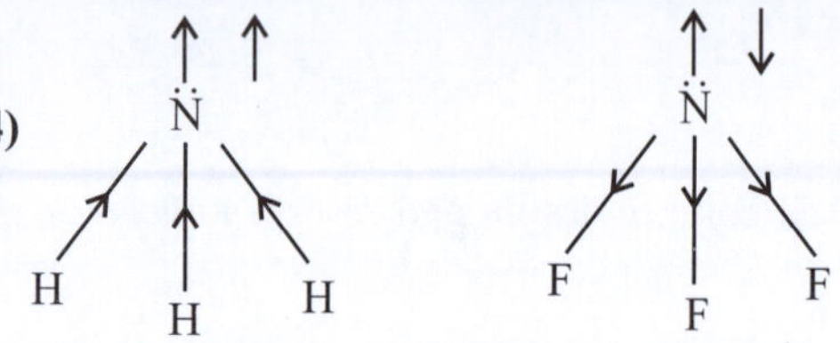

41. (3) NO_2 and O_3 have V-Shape or bent shape

∴ They have some dipole moment.

42. (4) In sp^3d^2, s-character is 16.66%.

43. (3) In $BeCl_2$, hybridization of Be is sp and the shape is linear due to 2 bp & 0 lp.

44. (2) Both O_2^+ and O_2 are paramagnetic, bond order of $O_2^+ = 2.5$ and $O_2 = 2$.

45. (3) The bonding molecular orbital has lower energy than the corresponding antibonding molecular orbital.

46. (2) Intermolecular H-bonding is formed between H_2O molecules.

47. (2) From sp^3d-hybridization, trigonal planar geometry is not possible, because in trigonal planar geometry, lone pair electrons will be placed at the axial position, which violates VSEPR theory.

48. (2) (I) $[PCl_4]^+ \rightarrow sp^3$

(II)  has 104.5° bond angle due to lp-lp repulsion.

(III) All have sp^3 hybridisation and one lone pair.

(IV) 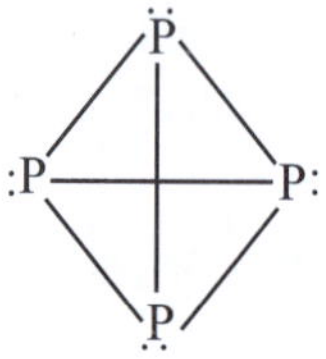

49. (1) F_2 and O_2^{2-} are isoelectronic.

$F_2 : (\sigma 1s^2) (\sigma^* 1s^2) (\sigma 2s^2) (\sigma^* 2s^2) (\sigma 2p_z^2) (\pi 2p_x^2 = \pi 2p_y^2)$
$(\pi^* 2p_x^2 = \pi^* 2p_y^2)$

$$B.O. = \frac{10-8}{2} = 1$$

$O_2^{2-} : (\sigma 1s^2) (\sigma^* 1s^2) (\sigma 2s^2) (\sigma^* 2s^2) (\sigma 2p_z^2) (\pi 2p_x^2 = \pi 2p_y^2)$
$(\pi^* 2p_x^2 = \pi^* 2p_y^2)$

$$B.O. = \frac{1}{2} \times (10-8) = 1$$

50. (1) % s-character ∝ bond angle ∝ bond strength

$$\propto \frac{1}{\text{bond length}}$$

51. (4) Larger the size of anion, greater the polarisability. Thus, $I^- > Br^- > Cl^- > F^-$ (polarisability order).

52. (3) $2s \rightarrow (\sigma 2s) (\sigma^* 2s)$ (2)

$2p \rightarrow (\sigma 2p_z) (\sigma^* 2p_z) (\pi 2p_x) (\pi^* 2p_x) (\pi 2p_y) (\pi^* 2p_y)$ (6)

Hence, 2s and 2p atomic orbitals combine to give 8 atomic orbitals.

53. (3) R forms anion readily by gaining one electron only.

54. (2) Xe in XeO_4 is sp^3 hybridised.

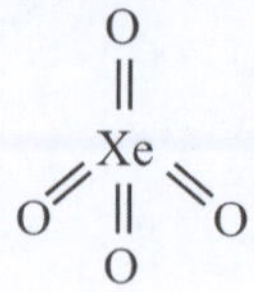

55. (3) SO_2 is a bent molecule.

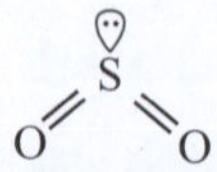

Geometry: Distorted Trigonal Planar

Shape: Bent

56. (2) BF_3 is a Non-polar molecule because its net overall dipole moment is equal to zero.

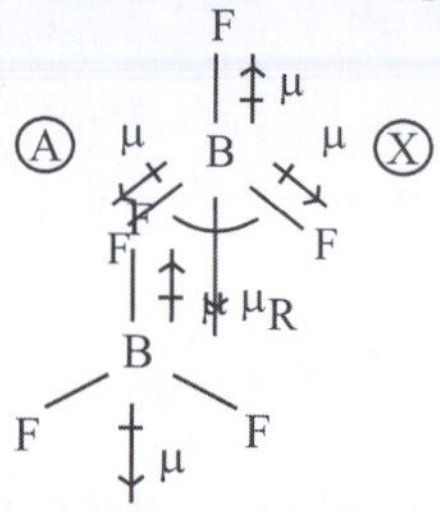

Now, both these vectors will finally cancel Out each other & net dipole moment = 0.

57. (4) 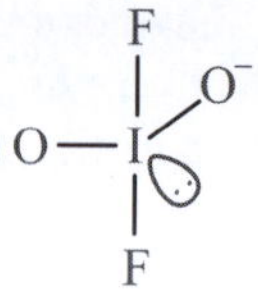 sp^3d hybridization but see-saw shape

58. (2)

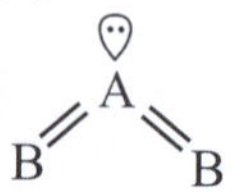

Arrangement of electron pairs → Trigonal planar hybridisation of A → sp^2

59. (2)

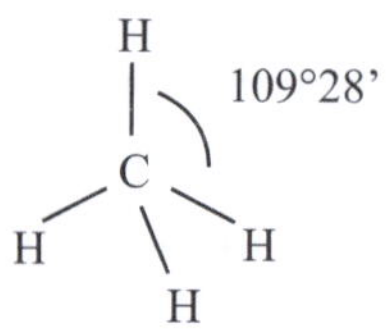

Regular tetrahedral geometry

60. (1) In C_2H_6, the C—H bond is made by sp^3–s overlap.

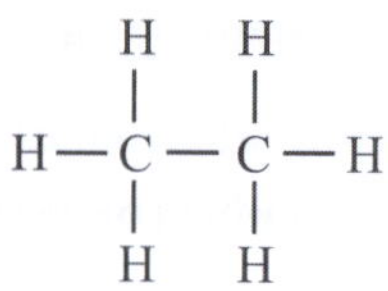

In C_2H_4, the C—H bond is made by sp^2–s overlap

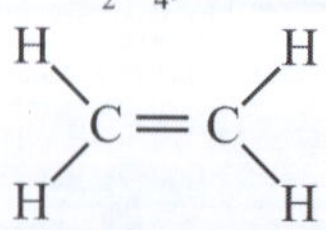

In C_2H_2, the C—H bond is made by sp-s overlap

Now, bond length follows the order:
$sp^3 - s > sp^2 - s > sp - s$

61. (4) 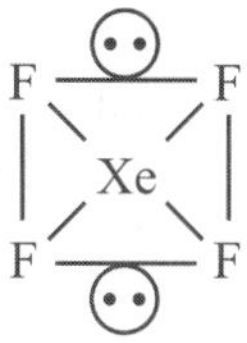 total 5 atoms lie in a plane including Xe.

62. (4) Sulphur in SF_6 is sp^3d^2 hybridised.
d-orbitals used are $d_{x^2-y^2}$ & d_{z^2} (both axial d-orbitals)
wrt option (2), phosphorous in PCl_5 is sp^3d hybridised. Here, d orbital used is d_{z^2} (axial).

wrt option (3), Xe in XeF_5^+ is sp^3d^2 hybridised. d-orbitals used are $d_{x^2-y^2}$ & d_{z^2} (axial).

63. (4) O_2^- (17 electrons):
$\sigma 1s^2, \sigma^* 1s^2, \sigma 2s^2, \sigma^* 2s^2, \sigma 2p_z^2, \pi 2p_x^2 = \pi 2p_y^2, \pi^* 2p_x^2 = \pi^* 2p_y^1, \sigma^* 2p_z^0$

$$\text{Bond Order} = \frac{N_b - N_a}{2} = \frac{10-7}{2} = \frac{3}{2}$$

So, bond order = 1.5
No. of unpaired electron = 1
O_2 (16 electrons) :-
$\sigma 1s^2, \sigma^* 1s^2, \sigma 2s^2, \sigma^* 2s^2, \sigma 2p_z^2, \pi 2p_x^2 = \pi 2p_y^2, \pi^* 2p_x^1 = \pi^* 2p_y^1$

$$\text{So, Bond Order} = \frac{N_b - N_a}{2} = \frac{10-6}{2} = \frac{4}{2} = 2$$

No. of unpaired electrons = 2

64. (4) For H_2^{2+}
Total $e^- = 0$
Bond Order = 0
For He_2,
Total $e^- = 4$
$\sigma 1s^2\ \sigma^* 1s^2$

$$\text{Bond Order} = \frac{N_b - N_a}{2} = \frac{2-2}{2} = 0$$

So, He_2 can't exist
If bond order is zero or negative, then molecule cannot exist.

65. (2) $B_2(10)$:- Violating Hund's rule

$\sigma 1s^2, \sigma^* 1s^2, \sigma 2s^2, \sigma^* 2s^2, \pi 2p_x^2 = \pi 2p_y$

B_2 is diamagnetic.

For O_2 (16 e.) violating Hund's rule

$\sigma 1s^2, \sigma^* 1s^2, \sigma 2s^2, \sigma^* 2s^2, \sigma 2p_z^2, \pi 2p_x^2 = \pi 2p_y^2, \pi^* 2p_x^2 = \pi^* 2p_y$

O_2 is also diamagnetic.

66. (3) XeF_2L_3 is a perfect example for molecule of the type AX_2L_n where bond angle is maximum when value of n = 3.

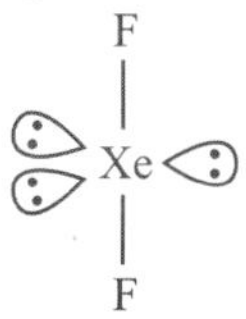

67. (4) Consider the structure of $XeF_5^{\ominus}$.

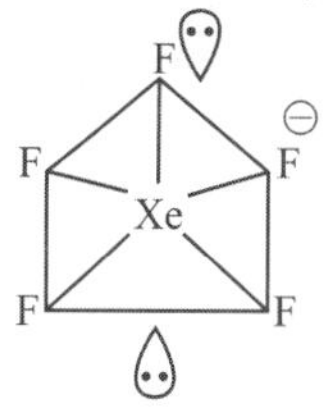

Geometry- Pentagonal bipyramidal
Shape Pentagonal planar.
wrt option (2), consider the structure of $XeOF_4$.

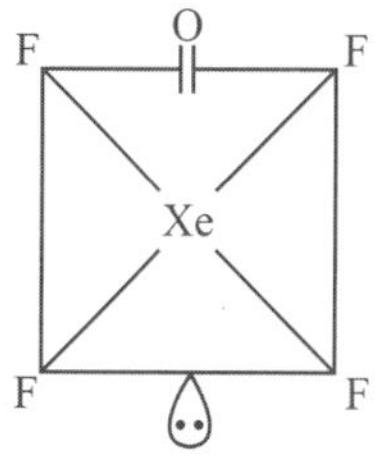

Electronic geometry– Octahedral
Shape– Square pyramidal
wrt option (3), consider the structure of H_2O.

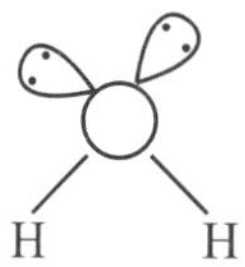

Electronic geometry :– Tetrahedral
Shape :– Bent
wrt option (4), triangular planar shape is not possible in trigonal bipyramidal electronic geometry.

68. (1) wrt CO_3^{2-} –

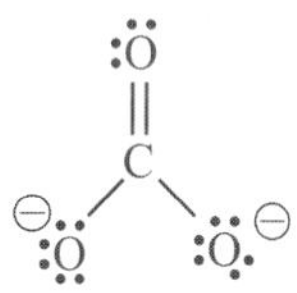

C is sp^2 hybridised

$$\text{s-character} = \frac{1}{3} \times 100 = 33.3\%$$

wrt XeF_4

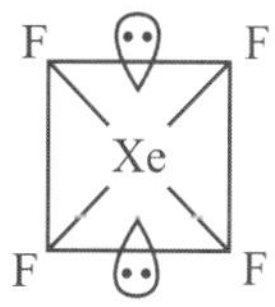

Xe is sp^3d^2 hybridised

$$\text{s-character} = \frac{1}{6} \times 100 = 16.66\%$$

wrt I_3^-

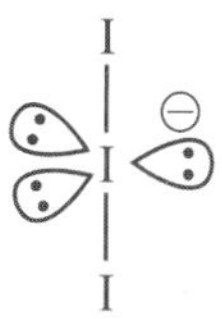

I is sp^3d hybridised

s-character = $\frac{1}{5} \times 100 = 20\%$

wrt NCl_3

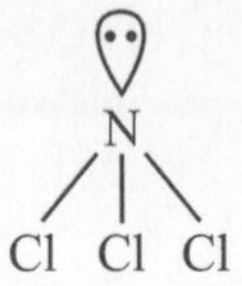

N is sp^3 hybridised s-character = $\frac{1}{4} \times 100 = 25\%$

wrt $BeCl_2$–

Cl—Be—Cl

Be is sp hybridised

s-character = $\frac{1}{2} \times 100 = 50\%$

69. (4) All the given molecules have correct hybridisation and planarity in the respective column.

70. (1) The hybrid orbitals used for forming C–H bonds contain more s–character than hybrid

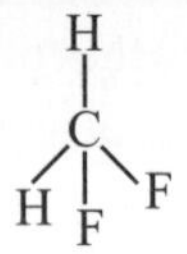

orbitals used for forming C–F bonds. As more the s–character in hybrid orbital, larger will be bond angle.

71. (3) XeF_5^- : 2 lone pairs ; pentagonal planar

BrF_3 : 2 lone pairs ; Bent-T-shape

XeF_2 : 3 lone pairs ; Linear

H_3O^+ : 1 lone pairs ; Pyramidal

72. (4) Ethylene on hydrogenation into ethane : hybridization , bond angle & C - C bond length change.

73. (2) Since anti-bonding orbitals are formed by subtraction of orbitals.

74. (2) $[ICl_2]^-$, SOF_4, ClF_3 and PCl_5 have sp^3d hybridisation.

75. (1) Complete structure of given compound is -

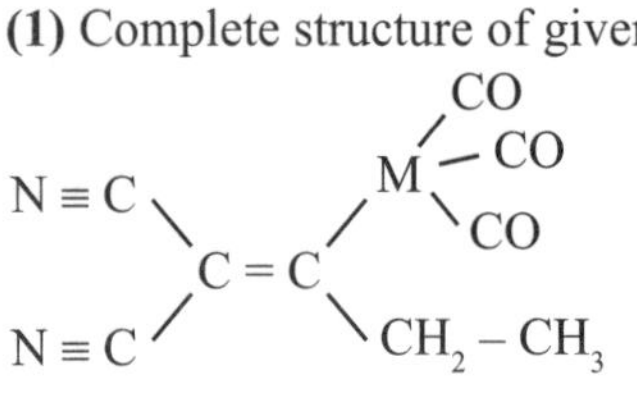

No. of σ bonds = 19

No. of π bonds = 11

76. (3) $SiF_4 \rightarrow$ Si have vacant d- orbital so can act as Lewis acid.

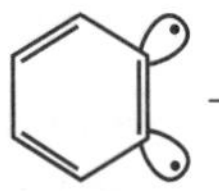

→ In benzyne all C-atoms are sp^2 hybridised.

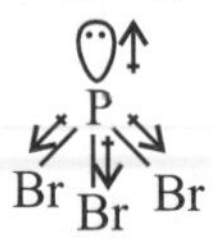

→ This is polar molecule.

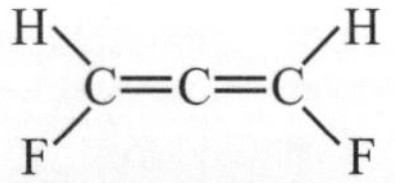

→ The nodal planes of π– bonds are perpendicular to each other.

77. (4) 1) 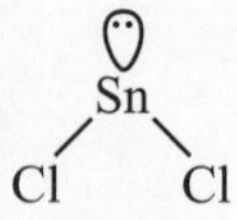

sp^2- bent

2) 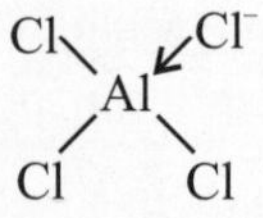

sp^3, tetrahedral

3) 

sp^3, pyramidal

4) 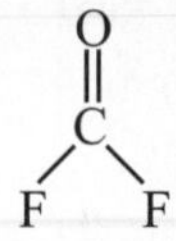

sp^2 & planar

78. (3) 1) NH_3 bond angle is 107.5°

2) 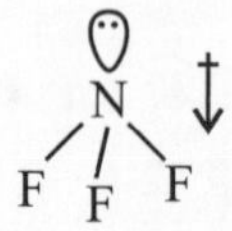

3) 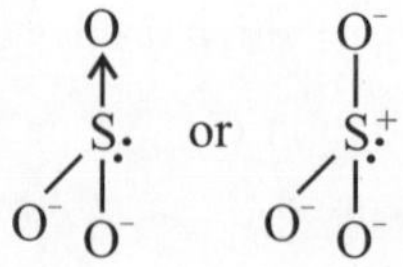

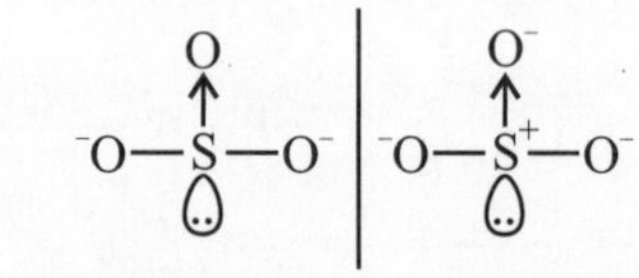

4) Percent 's' character $= \frac{1}{3} \times 100 = 33.33\%$

Percent 'p' character = 66.67 %

79. (2) In $\ddot{N}H_3$, central atom is nitrogen which is sp^3 hybridized hence it will be at the centre of tetrahedron with H–atoms at three vertices.

80. (2) $C_2: (\sigma 1s)^2 < (\sigma^* 1s)^2 < (\sigma 2s)^2 < (\sigma^* 2s)^2 < (\pi 2p_x)^2 < (\pi 2p_y)^2$

$\mu_{eff} = 0$; i.e. diamagnetic

$B_2: (\sigma 1s)^2 < (\sigma^* 1s)^2 < (\sigma 2s)^2 < (\sigma^* 2s)^2 < (\pi 2p_x)^1 = (\pi 2p_y)^1$

$\mu_{eff} \neq 0$; i.e. paramagnetic

$O_2: (\sigma 1s)^2 < (\sigma^* 1s)^2 < (\sigma 2s)^2 < (\sigma^* 2s)^2 < (\sigma 2p_z)^2 < (\pi 2p_x)^2 = (\pi 2p_y)^2 < (\pi^* 2p_x)^1 = (\pi^* 2p_y)^1$

$O_2: \mu_{eff} \neq 0$

$O_2^{2+} : \mu_{eff} = 0$

$O_2^{2-} : \mu_{eff} = 0$

$N_2: (\sigma 1s)^2 < (\sigma^* 1s)^2 < (\sigma 2s)^2 < (\sigma^* 2s)^2 < (\pi 2p_x)^2 = (\pi 2p_y)^2 < (\sigma 2p_z)^2$

N_2 : $\mu_{eff} = 0$

N_2^+ : $\mu_{eff} \neq 0$,

N_2^{2-} : $\mu_{eff} \neq 0$

N_2^{2+} : $\mu_{eff} = 0$

81. (3) Stability depends upon bond order.

Li_2 : $KK(\sigma 2s)^2$: Bond order of $Li_2 = 1$

Li_2^+ : $KK(\sigma 2s)^1$: Bond order of $Li_2^+ = \frac{1}{2}$

Li_2^- : $KK(\sigma 2s)^2(\sigma^* 2s)^1$: Bond order of $Li_2^- = \frac{1}{2}$

Li_2^+ is more stable than that of Li_2^- because it of contains less no. of antibonding electrons. Therefore,

$Li_2^- < Li_2^+ < Li_2$

Correct & Incorrect MCQs

1. **(4)** The group valence of the elements is generally either equal to the number of dots in Lewis symbols or 8 minus the number of dots or valence electrons.
2. **(1)** Ionic bonds will be formed more easily between elements with comparatively low ionization enthalpies and elements with comparatively high negative value of electron gain enthalpy.
3. **(3)** Bond length is defined as the equilibrium distance between the nuclei of two bonded atoms in a molecule.
4. **(1)** Larger the bond dissociation enthalpy, stronger will be the bond in the molecule. Hence, the increasing bond enthalpy order is $H_2 < O_2 < N_2$.
5. **(3)** Isoelectronic molecules and ions have identical bond orders. Hence, F_2 and O_2^{2-} have same bond order equal to 1.
6. **(2)** Due to large electronegativity difference, the dipole moment of H_2O is more than that of H_2S.
7. **(4)** The molecular geometry of BF_3 molecule is trigonal planar according to VSEPR theory.
8. **(2)** Larger bond dissociation enthalpy, stronger will be the bond in the molecule.
9. **(4)** O_2: $(\sigma 1s)^2\ (\sigma^*\ 1s)^2\ (\sigma 2s)^2\ (\sigma^* 2s)^2\ (\sigma 2p_z)^2$
   $(\pi 2p_x{}^2 \equiv \pi 2p_y{}^2)\ (\pi^* 2p_x{}^1 \equiv \pi^* 2p_y{}^1)$
   It contains two unpaired electrons in $\pi^*\ 2p_x$ and $\pi^*\ 2p_y$ molecular orbitals, hence, O_2 molecule is paramagnetic in nature.
10. **(3)** The hybridization of PF_5 molecule is sp^3d, hence it is trigonal bipyramidal in shape.

Statement Based MCQs

1. **(2)** $(\sigma 1s^2)\ (\sigma^* 1s^2)\ (\sigma 2s^2)\ (\sigma^* 2s^2)\ (\pi 2P_x^2 = \pi 2P_y{}^2)$ Bond order $= (8 - 4)/2 = 2$
   C_2 is diamagnetic as all e^-s are paired up.
2. **(3)** CO bond is polar that are arranged symmetrically as C and O have different electro-negatives. The electrons are not shared equally between atoms. The net dipole moment of CO_2 molecule is 0. So, it is non polar.
3. **(1)** NO_2 has a lone pair of electrons and two sigma bonds thus, it is sp^2 hybridised. By VSEPR theory, it takes trigonal planar arrangement. Both statements are correct.
4. **(2)** PCl_3 is sp^3 hybridised, PCl_5 is sp^3d hybridised (PCl_3 – Trigonal Pyramidal and PCl_5 – Trigonal bipyramidal)
5. **(3)** H_2O_2 is stronger acid than water (bronsted acid) as it protonates to large extent in aq. solution than water. H_2O_2 is non planar but H_2O is planar.
6. **(1)** In a number of compounds, there are more than eight valence electrons around the central atom. This is termed as the expanded octet. The octet rule does not apply in such cases. PF_5, SF_6, H_2SO_4 are some examples of such cases.
7. **(3)** Resonance helps in stabilizing the structure of molecules.
8. **(4)** In bonding orbital, repulsion between nuclei is very less, hence a bonding molecular orbital is more stable than antibonding molecular orbital.
9. **(1)** Due to more size difference in S-H than O-H, bond energy of S-H is lesser than that of O-H.
10. **(3)** There are four bond pairs and one lone pair (total five electron pairs of S present in SF_4); thus, expected geometry is trigonal bipyramidal but observed geometry is see-saw.

Assertion & Reason MCQs

1. **(1)** In NaCl, both Na and Cl have complete octets due to which it is a stable compound.
2. **(1)** Higher the number of lone pairs on central atom, lower will be the bond angle.
3. **(3)** In N_2, no delocalization of electrons takes place in which bond pairs are localized between two N atoms while lone pairs localized on each N atom.
4. **(1)** In the presence of lone pair, geometry of a molecule gets distorted.
5. **(2)** Both NO^+ and N_2 not contain unpaired electrons and hence are diamagnetic.
6. **(1)** The triatomic linear molecules of the type *B–A–B*, For example: CO_2 molecule have no dipole moment this is not because the individual C=O bonds are non-polar but is because the two bond moments (polarity) cancel out each other. Hence, geometry of the molecules could be predicted by the value of dipole moment.
7. **(2)** SO_2 is Bent shaped while SO_3 is trigonal planar.
8. **(3)** σ-bond is formed due to end to end overlapping of the two atomic orbitals, example: between *s* and *s*, *s* and p and p and p orbitals.
   π-bonding is formed due to side wise overlapping of the two atomic orbitals. Example: between p and p orbitals.
   Overlapping of orbitals is more effective in sigma bonding than in pi-bonding. Rotation of an atom is not possible around π-bond due to the geometry of the overlapping orbitals.
9. **(1)** The O_3 (ozone) molecule is described by an average of two Lewis structure (i.e., resonance forms). Hence, both assertion and reason are true.
10. **(1)** According to VSEPR theory NH_3 has 3 bonding pairs, 1 lone pair, hence tetrahedral and is sp^3 hybridized.

Multi Correct MCQs

1. **(3)** According to Fajan's Rule,
   A compound is said to be covalent when
   1. Size of cation should be smaller
   2. Size of anion should be larger
   3. Charge on ion should be high.
2. **(3)** Bond strength is directly proportional to Bond order.
   $N_2 \longrightarrow \sigma_{1s}{}^2 \sigma_{1s}{}^{*2} \sigma_{2s}{}^2 \sigma_{2s}{}^{*2} \pi_{2px}{}^2 = \pi_{2py}{}^2 \sigma^2{}_{2pz}$

$O_2 \longrightarrow \sigma_{1s}{}^2\sigma_{1s}{}^{*2}\sigma_{2s}{}^2\sigma_{2s}{}^{*2}\sigma_{2pz}{}^2\pi_{2px}{}^2 = \pi_{2py}{}^2\pi^*_{2px}{}^1 = \pi^*_{2py}{}^1$

$C_2 \longrightarrow \sigma_{1s}{}^2\sigma_{1s}{}^{*2}\sigma_{2s}{}^2\sigma_{2s}{}^{*2}\pi_{2px}{}^2 = \pi_{2py}{}^2$

$B_2 \longrightarrow \sigma_{1s}{}^2\sigma_{1s}{}^{*2}\sigma_{2s}{}^2\sigma_{2s}{}^{*2}\pi_{2px}{}^1 = \pi_{2py}{}^1$

NO $\longrightarrow$ 7 (from N) + 8 (from O) = $15e^-$

B.O of $N_2 = \frac{10-4}{2} = 3;$

B.O of $C_2 = \frac{8-4}{2} = 2$

B.O of $B_2 = \frac{6-4}{2} = \frac{2}{2} = 1$

B.O of NO = B.O of $N_2^- = \frac{10-5}{2} = 2.5$

When an electron is removed from antibonding molecular orbital, bond order will increase.

NO & O_2 contain valence e^- in π^* orbital.

3. (3) $NH_3 \rightarrow sp^3$

$SO_2 \rightarrow sp^2$

$SiO_2 \rightarrow sp^3$

$BeCl_2 \rightarrow sp$

$CO_2 \rightarrow sp$

$H_2O \rightarrow sp^3$

$CH_4 \rightarrow sp^3$

$BF_3 \rightarrow sp^2$

4. (4) According to VSEPR theory, the bent molecule is one which contains 2b.p +1l.p.

N_3^- : linear

NO_2^- : bent (2b.p + 1 l.p)

I_3^- : linear

O_3: bent (2b.p + 1 l.p)

SO_2 : bent

NCERT Exemplar MCQs

1. (2) (1) NF_3 is pyramidal whereas BF_3 is trigonal planar.

(2) BF_4^- and NH_4^+ ion both are tetrahedral and have sp^3 hybridisation.

(3) $BrCl_3$ is T-shaped whereas BCl_3 is trigonal planar.

(4) NO_3^- is triangular planar whereas NH_3 is pyramidal.

2. (3) CO_2 being symmetrical has zero dipole moment.

Among HI, SO_2 and H_2O, dipole moment is highest for H_2O as in it, the central atom contains 2 lone pairs.

3. (2) The type of hybrid orbitals of nitrogen can be decided by using VSEPR theory and counting b.p. and *lp* in

$NO_2^+ = 2bp + 0\ell p$ = linear = sp hybridized

$NO_3^- = 3bp + 0\ell p \Rightarrow sp^2$ hybridized

$NH_4^+ = 4bp + 0\ell p \Rightarrow sp^3$ hybridised

4. (2) Strength of H-bond is in the order H....F > H...O > H.....N.

But each H_2O molecule is linked to four other H_2O molecules through H-bonds whereas each HF molecule is linked only to two other HF molecules. Hence, b.p of H_2O > b.p of HF > b.p. of NH_3

5. (3) In PO_4^{3-} ion, formal charge on each O-atom of P–O bond

$= \frac{\text{total charge}}{\text{Number of O-atom}} = -\frac{3}{4} = -0.75$

6. (4) In N-atom, number of valence electrons = 5

Due to the presence of one negative charge, number of valence electrons = 5 + 1 = 6; one O-atom forms two bond (= bond) and two O-atom shared with two electrons of N-atom

Thus, 3 O-atoms shared with 8 electrons of N-atom.

∴ Number of bond pairs (or shared pairs) = 4

Number of lone pairs = 0

7. (1) ($BH_4^- \Rightarrow$ 4 bond pairs + 0 lone pair

$\Rightarrow sp^3$ hybridised = tetrahedral geometry)

(CO_3^{2-} = triangular planar

NH_2^- = V – shape

H_3O^+ = pyramidal)

8. (3) The given compound will have the correct structure as 19 σ-bonds are present in the given molecule i.e., there are 5 π-bonds and 8 C–H + 11 C–C σ-bonds.

9. (3) The electronic configuration of the given molecules are:

$N_2^+ = \sigma 1s^2, \sigma^* 1s^2, \sigma 2s^2, \sigma^* 2s^2, \pi p_x^2 = \pi p_y^2, \sigma 2p_z^1$

It has one unpaired electron.

$O_2 = \sigma 1s^2, \sigma^* 1s^2, \sigma 2s^2, \sigma^* 2s^2, \sigma 2p_z^2, \pi 2p_x^2 \approx \pi 2p_y^2$
$\pi^* 2p_x^1 = \pi^* 2p_y^1$

O_2 has two unpaired electrons.

$O_2^{2-} = \sigma 1s^2, \sigma^* 1s^2, \sigma 2s^2, \sigma^* 2s^2, \sigma 2p_z^2, \pi 2p_x^2 \approx$
$\pi 2p_y^2, \pi^* 2p_x^2 \approx \pi^* 2p_y^2$

Thus, O_2^{2-} has no unpaired electrons.

$B_2 = \sigma 1s^2, \sigma^* 1s^2, \sigma 2s^2, \sigma^* 2s^2 \pi 2p_x^1 \approx \pi 2p_y^1$ Thus, B_2 has two unpaired electrons.

10. (3) $XeF_4 \Rightarrow 4bp + 2\ell p$ square planar $\Rightarrow$ all bonds are equal

$BF_4^- \Rightarrow 4bp + 0\ell p \Rightarrow$ tetrahedral (all bonds are equal)

$SiF_4 \Rightarrow 4bp + 0\ell p \Rightarrow$ tetrahedral (all bonds are equal)

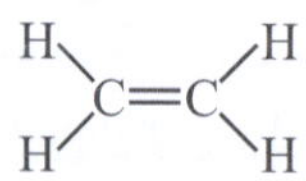

Thus, in C_2H_4 all the bonds are not equal.

11. (4) The given configuration belongs to d-block of the periodic table. In transition elements i.e., d-block elements, electrons of ns and (n – 1) d subshell take part in bond formation. Electronic configuration shows that the element is Titanium (Z = 22).

12. (2) For sp^2 hybridisation, the geometry is generally triangular planar.

Thus, bond angle is 120°.

13. (1) The given electronic configuration shows that A represents noble gas because the octet is complete. A is Neon which has 10 atomic number.

14. (2) The electronic configuration of C represents chlorine. Dichlorine is a stable form of chlorine.

15. (4) The electronic configuration shows that B represents phosphorus and C represents chlorine. The stable compound formed is PCl_3 i.e., BC_3.

16. (2) Both B and C are non-metal atoms. B represents phosphorus and C represent chlorine. Thus, the bond between B and C will be covalent.

17. (1) The correct increasing order of energies of molecular orbitals of N_2 is given below

$$\sigma 1s^2 < \sigma^* 1s < \sigma 2s < \sigma^* 2s < (\pi 2p_x = \pi 2p_y)$$
$$< \sigma 2p_z < (\pi^* 2p_x = \pi^* 2p_y) < \sigma^* 2p_z$$

18. (4) Existence of molecule, bonding nature and energy order of molecular orbitals can be explained on the basis of molecular orbital theory as follows

(i) Molecules having zero bond order never exists while molecular having non-zero bond order either exists or expected to exist.

(ii) Higher the value of bond order, higher will be its bond strength.

Electrons present in bonding molecular orbital are known as bonding electrons (N_b) and electrons present on anti-bonding molecular orbital are known as anti-bonding electrons (N_a) and half of their difference is known as bond order.

(4) is incorrect. The correct order of energies of molecular orbitals in N_2 molecule is

$$\sigma 2s < \sigma^* 2s < (\pi 2p_x \approx \pi 2py) < \sigma 2p_z < \pi^* 2p_x \approx \pi^* 2p_y < \sigma^* 2p_z$$

19. (2) Electronic configuration of O_2 (16 electrons) is:

$$= \sigma 1s^2, \sigma^* 1s^2, \sigma 2s^2, \sigma^* 2s^2, \sigma 2p_z^2, \pi 2p_x^2 \approx \pi 2p_y^2, \pi^* 2p_x^1 \approx \pi^* 2p_y^1$$

$$\text{Bond order} = \frac{1}{2}(N_b - N_a) = \frac{1}{2}(10-6) = 2$$

Electronic configuration of O_2^+ (15 electrons) is:

$$= \sigma 1s^2, \sigma^* 1s^2, \sigma 2s^2, \sigma^* 2s^2, \sigma 2p_z^2, \pi 2p_x^2 \approx \pi 2p_y^2, \pi^* 2p_x^1 \approx \pi^* 2p_y^0$$

$$\text{Bond order} = \frac{1}{2}(N_b - N_a) = \frac{1}{2}(10-5) = 2.5$$

Electronic configuration of O_2^- (17 electrons) is:

$$= \sigma 1s^2, \sigma^* 1s^2, \sigma 2s^2, \sigma^* 2s^2, \sigma 2p_z^2, \pi 2p_x^2 \approx \pi 2p_y^2, \pi^* 2p_x^2 \approx \pi^* 2p_y^1$$

$$\text{Bond order} = \frac{1}{2}(N_b - N_a) = \frac{1}{2}(10-7) = 1.5$$

Thus, the order of bond order is $O_2^- < O_2 < O_2^+$

20. (1) The electronic configuration represents

$2s^2\ 2p^5$ = fluorine = most electronegative element

$3s^2\ 3p^5$ = chlorine

$4s^2\ 4p^5$ = bromine

$5s^2\ 5p^5$ = iodine

21. (2) The electronic configuration of options (2) and (4) have exactly half-filled 3p orbitals (2) represents phosphorus and (3) represents arsenic but (2) is smaller in size than (4).

Hence, (2) has highest ionization enthalpy, ionization enthalpy increases left to right in the periodic table as the size decreases.

NEET & AIIMS Past Year MCQs

1. (2) The dipole moment of a molecule is influenced by both its shape and polarity of bond. A dipole moment depends on the charge, distance, electronegativity and shape.

Molecule	Shape	μ(Debye)
CCl_4	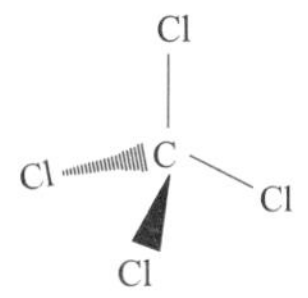	0
HI	H – I	0.38
CO_2	O = C = O	0
BF_3	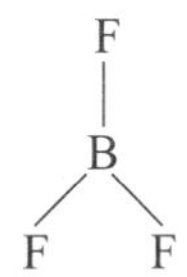	0

2. (3) Since, $CHCl_3$ is a non polar covalent compound, covalent (non-polar) compounds exhibit higher solubility in it. However, as the dipole moment of the solute increases, solubility in chloroform decreases. Therefore, the increasing order of solubility is;

$NaCl < CH_3OH < CH_3CN <$ Cyclohexane

3. (4) The correct match is; A-IV, B-III, C-I, D-II

Molecule	Bond enthalpy (kJ mol[-1])
HCl	431.0
N_2	946.0
H_2	435.8
O_2	498

4. (2)

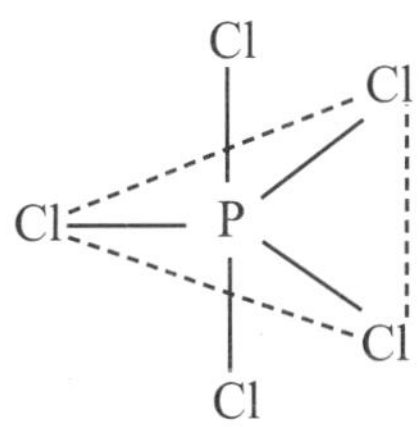

It is sp^3d hybridized with an axial to equatorial angle of 90° and equatorial bond angles of 120°, featuring five P-Cl sigma bonds. The axial bonds are longer than the equatorial bonds.

5. (2)

• Dipole Moment:

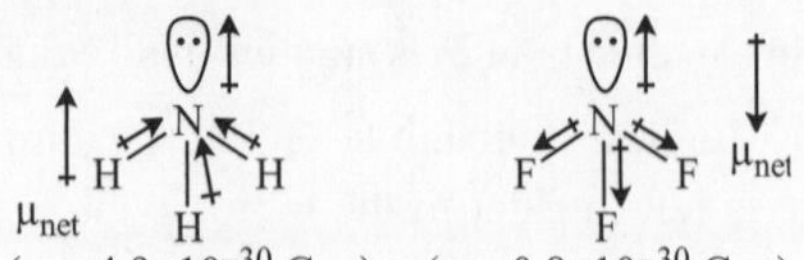

($\mu = 4.9\times10^{-30}$ C m) ($\mu = 0.8\times10^{-30}$ C m)

• CO_3^{2-} has three canonical forms.

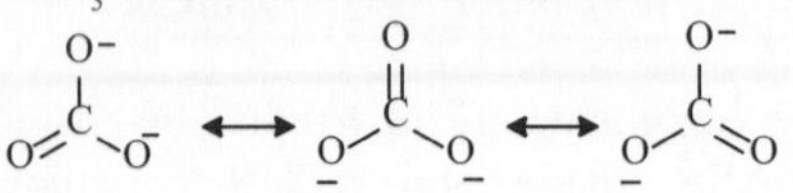

• Two resonating structures can be drawn for ozone.

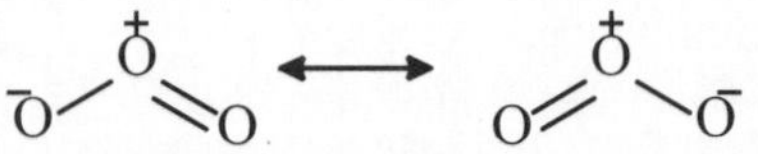

• BF_3 has trigonal planar structures. Its net dipole moment is Zero.

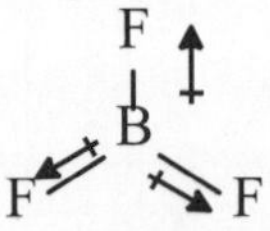

μ = Zero

6. (3)

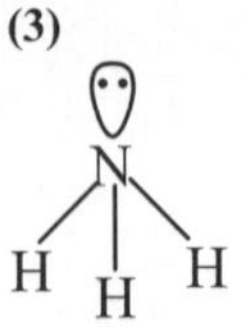

Trigonal Pyramidal

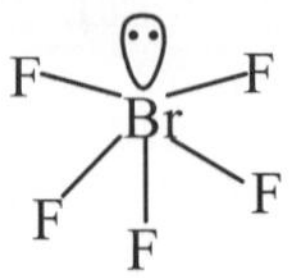

Square Pyramidal

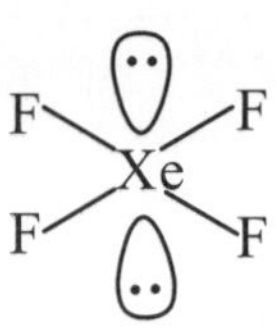

Square Planar

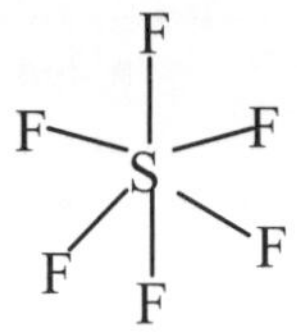

Octahedral

7. (3)

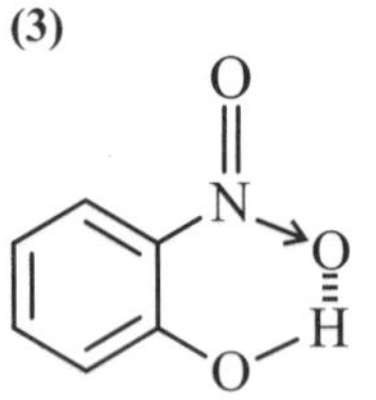

Intramolecular hydrogen bonding is observed in ortho nitrophenol.

8. (2) $AlCl_3$, $BeCl_2$ and PCl_5 does not obey octet rule.

$AlCl_3$ and $BeCl_2$ both are electron-deficient species having six electrons in valence shell of central atom whereas PCl_5 has ten electrons in valence shell of phosphorous.

The structures are :

$AlCl_3$ →

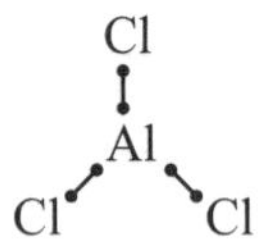

$BeCl_2$ → Cl ⟵ Be ⟵ Cl

PCl_5 →

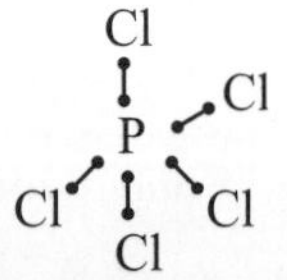

9. (2) $\sigma 1s < \sigma^* 1s < \sigma 2s < \sigma^* 2s < (\pi 2p_x = \pi 2p_y) < \sigma 2p_z < (\pi^* 2p_x = \pi^* 2p_y) < \sigma^* 2p_z$

is the correct order of energy of MO for homo nuclear diatomic species N_2.

10. (1) XeF_2

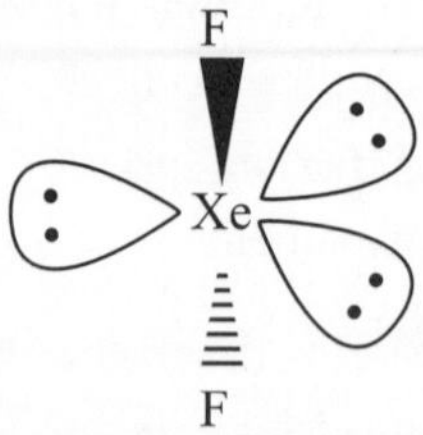

XeF_2 has maximum 3 lone-pair – lone-pair repulsions.

In ClF_3 two lone pairs, in IF_5 & SF_4, only one lone pair is present. Whereas, in XeF_2, three lone pairs are present.

11. (1) Electronic configuration of oxygen molecule is:

$(O_2) = \sigma 1s^2, \sigma^* 1s^2, \sigma 2s^2, \sigma^* 2s^2, \sigma 2P_z^2, \pi 2p_x^2 = \pi 2p_y^2, \pi^* 2p_x^1 = \pi^* 2p_y^1$

Bond order for $O_2 = \dfrac{10-6}{2} = \dfrac{4}{2} = 2$

Bond order for $O_2^+ = \dfrac{10-5}{2} = \dfrac{5}{2} = 2.5$

Bond order for $O_2^- = \dfrac{10-7}{2} = \dfrac{3}{2} = 1.5$

Bond order for $O_2^{2-} = \dfrac{10-8}{2} = \dfrac{2}{2} = 1$

O_2^+ contains unpaired e^-_s ⇒ paramagnetic

⇒ Electronic configuration of C_2 molecule is

$C_2 = \sigma 1s^2\ \sigma^* 1s^2\ \sigma 2s^2\ \sigma^* 2s^2$

$(\pi 2Px^2 = \pi 2Py^2)$

Therefore, it is clear that C_2 molecule has four electrons in its two degenerate p molecular orbitals.

⇒ No. of electrons in H_2^+ ion is = 2 – 1 = 1

12. (2)

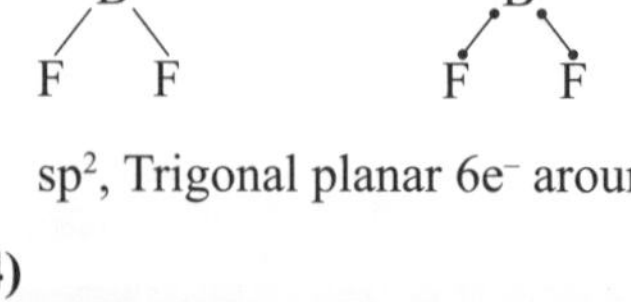

sp^2, Trigonal planar $6e^-$ around central atom.

13. (4)

	Hybridisation	L.P.	Shape
(A) PCl_5	sp^3d	0	Trigonal bipyramidal
(B) SF_6	sp^3d^2	0	Octahedral
(C) BrF_5	sp^3d^2	1	Square pyramidal
(D) BF_3	sp^2	0	Trigonal planar

14. (2) 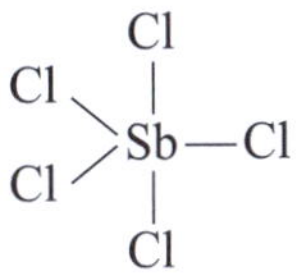

sp^3d — Dipole moment (μ) = 0

Trigonal bipyramidal — Non-polar

15. (2) In BMO, nodal plane is absent.

16. (4) 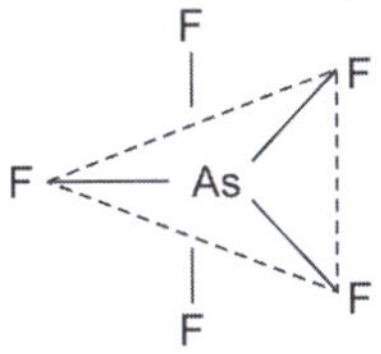

This AsF_5 molecule is trigonal bipyramidal.

It is sp^3d hybridized, So, the orbital in the bonding is

s, p_x, p_y, p_z, d_z^2

17. (4) **18. (4)** **19. (1)** **20. (4)**

21. (4) B_2H_6 is an electron deficient compound. B_2H_6 contain some unusual bonds which are called as 2-electron, 3-centre bonds.

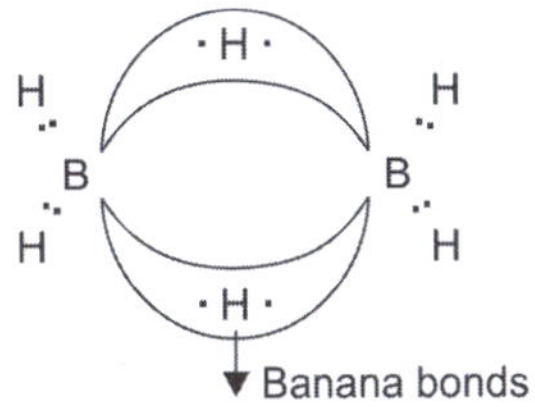

Si_2H_6 is called disilane

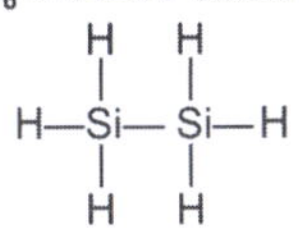

22. (4)

Thermodynamics

NCERT Topic-wise MCQs

1. **(3)** Work depends upon the path followed by the system therefore, it is a path function.
2. **(2)** Volume depends upon the amount of the substance therefore, it is an extensive property.
3. **(2)** The given process is considered as a reversible process.
4. **(2)** Thermally isolated gaseous state can exchange energy in the form of work with the surroundings.
5. **(3)** Density does not depend upon the amount of the substance therefore, it is an intensive property.
6. **(3)** For isothermal expansion of an ideal gas, the correct relation is $W_{rev} > W_{irr}$.
7. **(3)** Because there is no exchange of matter and energy.
8. **(2)** Work done during isothermal reversible expansion of a gas is maximum.
9. **(4)** Since, 1 L-atm = 101.266 J

   $\therefore$ 4 L - atm = 101.266 × 4 = 405.06 J

   $= \frac{405.06}{4.2} = 96.4 \text{ cal} \approx 96 \text{ cal} \quad [\because 1 \text{ cal} = 4.2 \text{ J}]$
10. **(1)** Heat is absorbed so it is endothermic.
11. **(1)** Internal energy of one mole of a gas is $\frac{3}{2}RT$
12. **(3)** Internal energy does not include energy arising by gravitational pull. Because it is an external energy.
13. **(3)** At A, temperature = 300 K, volume = 10 L, Pressure = P_1,

    At C, Temperature = 600K, volume = 20L, Pressure = P_2

    $\frac{P_1V_1}{T_1} = \frac{P_2V_2}{T_2} \Rightarrow \frac{P_1 \times 10}{300} = \frac{P_2 \times 20}{600}$

    $P_1 = P_2$

    Since, the pressure comes out to be same, therefore, the process is isobaric.
14. **(3)** Internal energy, enthalpy and entropy are state functions and work is a path function.
15. **(4)** Solubility increases with increase in temperature up to 60°C. Therefore, solubility curve of a hydrated salt in water shows endothermic reaction till 60°C and after 60°C solubility decreases with increase in temperature, therefore, exothermic reaction thereafter.
16. **(3)** The enthalpy of combustion (i.e., ΔH) is always negative. So, combustion of propane is an exothermic reaction.
17. **(1)** $\Delta n_g = 2 - 1 = 1$; $\Delta H = \Delta U + \Delta n_g RT$

    Thus; $\Delta H > \Delta U$
18. **(1)** A monoatomic molecule possesses only potential energy.
19. **(3)** Because ΔU is a state function.
20. **(3)** $C_6H_5COOH(s) + \frac{15}{2}O_2(g) \rightarrow 7CO_2(g) + 3H_2O(l)$

    $\Delta n_g = 7 - \frac{15}{2} = \frac{-1}{2}$

    $\Delta H = \Delta E + \Delta n_g RT$

    $= -321.30 + \left(\frac{-1}{2}\right)R \times 300 = -321.30 - 150 R$
21. **(1)** $q_v = \Delta E$, or $q_v = \Delta U$, at constant temperature and constant volume.
22. **(1)** Since $\Delta n_g = 2 - 2 = 0$

    Thus; $\Delta H = \Delta U$
23. **(2)** For a monoatomic gas, $C_P = \frac{5}{2}R$. Therefore,

    $\Delta H = -nC_p\Delta T = -2 \times \frac{5}{2}R(225 - 125)$

    $= -5 R (100) = -500 R$
24. **(4)** Heat absorbed at constant volume is equal to change in the internal energy.

25. **(1)** Work done due to change in volume against constant pressure,

$w = -P(V_2 - V_1)$

$= -1 \times 10^5\ Nm^{-2}\ (1 \times 10^{-2} - 1 \times 10^{-3})m^3$

$= -900\ Nm = -900\ J\ (1\ Nm = 1\ J)$

26. **(4)** From first law of thermodynamics,

$\Delta U = q + w$

Here; $w = +4$ kJ($\therefore$ work done on the system)

$q = -1$ kJ($\therefore$ Heat is given out by the system)

$\Delta U = +3$ kJ

27. **(2)** $w = -2.303nRT \log \frac{V_2}{V_1}$

$= -2.303 \times \frac{16}{32} \times 8.314 \times 300 \log \frac{100}{10}$

$= -2872\ J \approx -2875\ J$

28. **(1)** Work done, $w = -P_{ext}(V_2 - V_1)$

Here; $P_{ext} = 100$ kPa

$V_1 = 5m^3 = 5 \times 10^3\ dm^3$, $V_2 = 1\ dm^3$

$w = -100 \times (1 - 5000) = 499900\ kPa\ dm^3 = 499.9\ kJ$

29. **(3)** This equation is valid for a closed system at constant temperature and pressure.

30. **(4)** The released enthalpy is called enthalpy of hydration

31. **(3)** The heat change determined in bomb calorimeter is the change in internal energy. Thus $\Delta U = -870\ kcal\ mol^{-1}$

32. **(2)** Work done (w) $= -P_{ext}(V_2 - V_1)$

$= -3 \times (6 - 4) = -6$ L - atm ($\because 1dm^3 = 1L$)

$= -6 \times 101.32$ J ($\because$ 1 L - atm = 101.32 J)

$= -607.92 \approx -608$ J

33. **(4)** Work done $= -2.303nRT \log \frac{P_1}{P_2} = -2.303 \times 1 \times 8.314 \times 300 \log \frac{1}{4}$

$= 3458.3$ J or 3.458 kJ

34. **(4)** Internal energy is a function of temperature because internal energy of ideal gas consists of molecular kinetic energy which further depends upon temperature.

$\therefore dT = 0$ then $\Delta U = 0$

$\Delta U = q + w$

for an isothermal process, $\Delta U = 0$, $q = -w$

for an isochoric process; $w = 0$, $\Delta U = q$

for an adiabatic process, $q = 0$, $\Delta U = w$

for a cyclic process, $\Delta U = 0$, Therefore $q = -w$

35. **(3)** Therefore, $\Delta U = q + w$; $w = -P\Delta V$

$$P\Delta V = 0.6 \times \frac{(300 - 500)}{1000} = 0.6 \times -0.2 = -0.12\ L-atm$$

$= -0.12 \times 101.3 = -12.16$ J

$\therefore \Delta U = -10 - (-12.16) = -10 + 12.16 = 2.16$ J

36. **(4)** In the formation of CO_2 from carbon,

$C_{(s)} + O_{2(g)} \rightarrow CO_{2(g)}$

$\Delta n_g = 1 - 1 = 0$

$\therefore \Delta H = \Delta E$

37. **(3)** $CH_3CHO + \frac{5}{2}O_2 \rightarrow 2CO_2 + 2H_2O$

Xg ethanal gives heat = Y Joules

$\therefore$ 44 g ethanal gives heat $= \frac{44Y}{X}$ Joules mol^{-1}

As reaction is exothermic (due to combustion)

$\therefore \Delta U = -\frac{44Y}{X} J\ mol^{-1}$

38. **(2)** $\Delta H_{combustion} = -393.5$ kJ/mol,

i.e., 1 mole of CO_2 on combustion release $= -393.5$kJ

$\frac{35.2}{44}$ moles of CO_2 will release $= \frac{-13851.2}{44} = -314.8$ kJ

39. **(3)** For the decomposition of 9 g of water,

heat required = 142.5 kJ

We know that the molecular weight of H_2O = 18 g

Therefore; heat required for decomposition of 18 g of water.

$= \frac{18}{9} \times 142.5 = 2 \times 142.5 = 285\ kJ$

Therefore; enthalpy of formation of water $= -285$kJ

40. **(4)** I. $NH_3(g) + 3Cl_2(g) \rightleftharpoons NCl_3(g) + 3HCl(g), -\Delta H_1$ ...(i)

II. $N_2(g) + 3H_2(g) \rightleftharpoons 2NH_3(g), -\Delta H_2$ ...(ii)

III. $H_2(g) + Cl_2(g) \rightleftharpoons 2HCl(g), \Delta H_3$ ...(iii)

On multiplying eq (iii) by 3, we get

$3H_2 + 3Cl_2 \rightleftharpoons 6HCl; 3 \times \Delta H_3$ ...(iv)

On inverting eq (iv), we get

$6HCl(g) \rightleftharpoons 3H_2(g) + 3Cl_2(g); -3 \times \Delta H_3$ ...(v)

On multiplying eq (i) with 2, we get

$2NH_3 + 6Cl_2 \rightleftharpoons 2NCl_3 + 6HCl; -2 \times \Delta H_1$ ...(vi)

On adding eq (ii), (v) and (vi), we get

$N_2 + 3Cl_2 \rightleftharpoons 2NCl_3$

$\Delta H_r = -\Delta H_2 - 3 \times \Delta H_3 - 2 \times \Delta H_1$

$\Delta H_f = \frac{\Delta H_r}{2} = -\frac{\Delta H_2}{2} - \frac{3}{2}\Delta H_3 - \Delta H_1$

$\Delta H_f = -\Delta H_1 - \frac{\Delta H_2}{2} - \frac{3}{2}\Delta H_3$

41. **(1)** Graphite is burnt as, $C(s) + O_2(g) \rightarrow CO_2(g)$

$q = C \times \Delta T \times \frac{M}{m}$ (heat lost by the system is heat gained by the calorimeter)

$\therefore q = -C \times \Delta T \times \frac{M}{m}$

$= -\frac{30 \times 4 \times 12}{4} = -360\ kJ\ mol^{-1}$

42. **(4)** $2O_2 + CH_{4(g)} \rightarrow CO_{2(g)} + 2H_2O_{(l)}$

$\Delta H_c = \Delta^\circ H_{f_{CO_2}} + 2 \times \Delta^\circ H_{f_{H_2O}} - \Delta^\circ H_{f_{CH_4}}$

$= -393.5 + 2 \times -285.8 - (-74.8)$

$= -889.7\ kJ/mol \approx -890.3\ kJ\ mol^{-1}$

43. **(4)** $X_2 + Y_2 \rightarrow 2XY$

$\Delta H = (BE)_{x-x} + (BE)_{y-y} - 2(BE)_{x.y}$

If B.E of (X – Y) = a, then B.E. of (X – X)= a

and B.E of (Y – Y) = $\frac{a}{2}$

$\Delta H_f (X - Y) = 200$ kJ

$\therefore -400$ (for 2 moles of XY) $= a + \frac{a}{2} - 2a$

$-400 = \frac{-a}{2}$ or $a = +800\,kJ$

The bond dissociation energy of X_2 = 800 kJ/mol

44. **(1)** $H_2O(l) \xrightarrow{100^\circ C} H_2O(g)$

$\Delta H^0_{Vap} = \Delta_{Vap}U^0 + \Delta n_g RT$

For above reaction; $\Delta n_g = 1 - 0 = 1$

$\therefore$ 40.66 kJ mol^{-1} $= \Delta_{Vap}U + 1 \times 8.314 \times 10^{-3} \times 373$

$\Delta_{Vap}U = 40.66$ kJ mol^{-1} – 3.1 kJ mol^{-1} = +37.56 kJ/mol

45. **(2)** Some of the heat evolved during neutralization reaction is utilised to ionise the weak acid, this happen in the case of weak acid.

Thus; less heat is evolved. Hence; lower the value of heat of neutralization, weaker is the acid. Thus; B is the weakest acid.

46. **(4)**

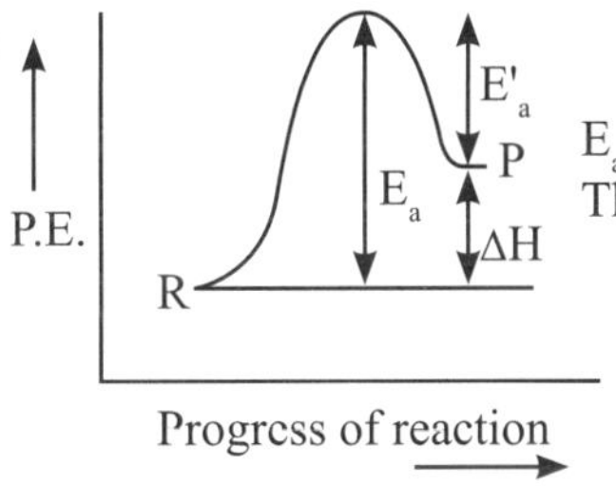

$E_a = E'_a + \Delta H$
Thus, $E_a > \Delta H$

47. **(2)** The change in Gibbs energy is given by

$\Delta G = \Delta H - T\Delta S$

where; ΔH – change in enthalpy of reaction

ΔS = change in entropy of reaction

Thus; in order to determine ΔG, the values of ΔH must be known. The value of ΔH can be calculated by the equation

$\Delta H = \Delta U + \Delta n_g RT$

where; ΔU = change in internal energy

Δn_g = (no. of moles of gaseous products)-(no. of moles of gaseous reactants)

$= 2 - 0 = 2$

$\Delta U = 2.1$ kcal $= 2.1 \times 10^3$ cal [$\therefore$ 1 kcal = 10^3 cal]

$\Delta H = (2.1 \times 10^3) + (2 \times 2 \times 300) = 3300$ cal

Hence; $\Delta G = \Delta H - T\Delta S$

$\Delta G = (3300) - (300 \times 20) = -2700$

$\Delta G = -2.7$ kcal

48. **(3)** $\Delta G = \Delta H - T\Delta S = 48000 - 400 \times 135$

$= 48000 - 54000 = -6000$ J mol^{-1}

For spontaneous reaction

$\Delta G = -ve$

For endothermic reaction

$\Delta H = +ve$

49. **(1)** $\Delta S_{vap} = \frac{\Delta H_{vap}}{T} = \frac{300\,kJ\,mol^{-1}}{300\,K} = 1000\,JK^{-1}\,mol^{-1}$

50. **(2)** The order of entropy in solid, liquid and gas is

gas > liquid > solid

Hence; in sublimation of naphthalene, maximum increase in entropy is observed.

51. **(2)** Amount of $H_2SO_4 = M \times V = 0.20 \times 350 = 70$ m mol

$\therefore$ Amount of H^+ ions $= 2 \times 70 = 140$ m mol

Amount of NaOH $= M \times V = 0.10 \times 650 = 65$ m mol

$\therefore$ Amount of OH^- ions = 65 m mol

Thus, NaOH is the limiting reagent as 65 mmol of OH^- ions will react with 65 mmol of H^+ ions to give 65 mmol of water.

1 mol of H^+ ions react with 1 mol of OH^- ions to give 1 mole of H_2O and in the process 57.1 kJ of heat is produced.

Thus, heat produced for 65mmol $= 57.1 \times 65 \times 10^{-3}$

$= 3.71$ kJ

52. **(2)** $\Delta G = \Delta H - T\Delta S$

When both ΔH and ΔS are + ve, and ΔG is –ve, then reaction is said to be spontaneous. It is only possible when,

$T\Delta S > \Delta H$

(When ΔH is +ve and ΔS is –ve, the reaction is never spontaneous).

53. **(2)** (i) $C + O_2 \rightarrow CO_2$; $\Delta H = -393.5$ kJ

(ii) $CO + 1/2\,O_2 \rightarrow CO_2$; $\Delta H = -283$ kJ

(iii) $C + 1/2\,O_2 \rightarrow CO$; $\Delta H = ?$

(iv) is formed by adding eq (i) with the reverse of eq(ii),

$\therefore \Delta H = (-393.5) + 283$ kJ $= (-393.5 + 283)$ kJ $= -110.5$ kJ

54. **(3)** $2KOH + H_2SO_4 \rightarrow K_2SO_4 + 2H_2O$

Here; H_2SO_4 in the limiting reactant

$N_1V_1 = N_2V_2$

$0.2 \times 50 = 1 \times V_2$

$V_2 = \frac{2}{10} \times 50$

$= 10$ mL

Thus; 50 ml of 0.2 N H_2SO_4 with exactly neutralise by 10 ml of 1N KOH

Heat produced on neutralisation of 1000ml of 1 N KOH – 57.3 kJ

$\therefore$ Heat produced on neutralisation of 10 mL of

1 N KOH $= \frac{57.3}{1000} \times 10 \times 1000\,J = 573\,J$

55. **(4)** $\Delta G^\circ = -RT\ln K_C$

56. (3) $NaNO_3(s) \rightleftharpoons Na^+(aq) + NO_3^-(aq)$

$NaNO_3$ is a solid, which is converted to liquid ions. As we know that, in transition from solid state to liquid state, entropy increases. Hence the value of ΔS will be greater than zero for this reaction.

57. (4) Mass/volume = density is a intensive property.

58. (1) ΔS (universe) should be greater than zero for a process to a spontaneous. $\Delta S_{universe} = \Delta S_{system} + \Delta S_{surrounding} > 0$

59. (4) $\Delta G = \Delta H - T\Delta S$

$T\Delta S = \Delta H - \Delta G$

$\Delta S = \frac{1}{T}(\Delta H - \Delta G)$

60. (4) Measure of randomness is entropy.

61. (2) For perfectly crystalline solids at 0 K, all molecular motion ceases and particles are closed packed. Thus; randomness (entropy) is minimum.

62. (4) Hydrogen being in gaseous state has highest entropy.

63. (3) For adiabatic process, $\Delta q = 0$; ΔS which is $\Delta q/T$ is also zero.

64. (2) In all natural processes the entropy of the universe is constantly increasing.

65. (1) $\Delta S = \frac{q_{rev}}{T} = \frac{J\,mol^{-1}}{K} = JK^{-1}mol^{-1}$

66. (4) Because it is a state function.

67. (2) The gas phase combustion of isobutane (C_4H_{10}) represented as

$$C_4H_{10(g)} + \frac{13}{2}O_{2(g)} \rightarrow 4CO_{2(g)} + 5H_2O_{(g)}$$

$$\Delta n_g = 9 - \left(\frac{13}{2} + 1\right) = 9 - \frac{15}{2} = +\frac{3}{2}$$

since; $\Delta n_g = +ve \therefore \Delta H = \Delta U + \Delta n_g RT \therefore \Delta H > \Delta U$

68. (3) As ΔG^o for the formation of C is + ve; ΔG for formation of A and B is –ve. Hence; A and B predominate in the reaction mixture.

69. (4) $\Delta G = \Delta H - T\Delta S$

For spontaneous process, $\Delta G = -ve$

$\therefore \Delta H < 0$ and $\Delta S > 0$

High Order Time Intensive MCQs

1. (2) For adiabatic expansion of ideal gas, the correct relation is PV^γ = constant.

2. (2) $\Delta G = \Delta H - T\Delta S$

For spontaneous process at all temperature, $\Delta G < 0$, and it is only possible when $\Delta H < 0$ and $\Delta S > 0$.

3. (4) $C_6H_6(l) \rightleftharpoons C_6H_6(g)$

$\Delta H = \Delta U + \Delta n_g RT$

$7.4 = \Delta U + 1 \times 2 \times 10^{-3} \times 353$

$7.4 = \Delta U + 0.706$

$7.4 - 0.706 = \Delta U$

6.694 kcal/mol $= \Delta U$

4. (2) Given-

$CH_3 - H \longrightarrow \dot{C}H_{3(g)} + \dot{H}_{(g)}$; $\Delta_r H = 103$ kcal/mol

$\Delta_r H = \Sigma\Delta_f H_{(P)} - \Sigma\Delta_f H_{(R)} = \Delta H_f(\dot{C}H_3, g) + \Delta H_f(\dot{H}, g) - \Delta H_f(CH_4, g)$

$103 = x + \frac{103}{2} - (-18)$

$103 = (x + 51.5) - (-18)$

$x = 33.5$ kcal/mol

5. (1) $XeF_4 \rightarrow Xe^+F^- + F_2 + F$ ... (i)

Acc. to given Eq^n

(1) Average bond enthalpy of Xe – F is 34 kcal /mol

$XeF_4 \rightarrow Xe + 4F$

$\Delta H_1 = 4 \times 34 = 136$ kcal /mol

(2) Ionisation enthalpy of Xe = 279 kcal/mol

$Xe \rightarrow Xe^+ + e^-$; $\Delta H_2 = 279$ kcal/mol

(3) Bond dissociation enthalpy of F_2 = 38 kcal/mol

$2F \rightarrow F_2$; $\Delta H_3 = -38$ kcal/mol

(4) Electron gain enthalpy of F = –85 kcal/mol

$F + e^- \rightarrow F^-$; $\Delta H_4 = -85$ kcal/mol

Now, Eq^n (i) can be written as

Eq (i) = $\Delta H_1 + \Delta H_2 + \Delta H_3 + \Delta H_4 = 136 + 279 + (-38) + (-85)$

$= 415 - 123 = 292$ kcal/mol

6. (1) Aim is:

$$\frac{1}{2}I_2 + \frac{1}{2}Cl_2(g) \longrightarrow ICl(g)$$

Given:

1. $Cl_2(g) \longrightarrow 2Cl(g)$, $\Delta H_1 = 242.3$ kJ /mol
2. $I_2(g) \longrightarrow 2I(g)$, $\Delta H_2 = 151$ kJ /mol
3. $ICl(g) \longrightarrow I(g) + Cl(g)$, $\Delta H_3 = 211.3$ kJ /mol
4. $I_2(s) \longrightarrow I_2(g)$, $\Delta H_4 = +62.8$ kJ /mol

$$\Delta H = \frac{\Delta H_1}{2} + \frac{\Delta H_2}{2} - \Delta H_3 + \frac{\Delta H_4}{2}$$

So, $\Delta H = \frac{1}{2}(242.3) + \frac{1}{2}(151) - 211.3 + \frac{1}{2} \times 62.8$

$\Delta H = 16.75$ kJ/mol

7. (4) $4H(g) \rightarrow 2H_2(g)$; $\Delta H = -869$ kJ

$2H_2(g) \rightarrow 4H(g)$; $\Delta H = +869.6$ kJ

or $H_2(g) \rightarrow 2H(g)$, $\Delta H = \frac{869.6}{2} = 434.8$ kJ

8. (1) $H_2O(\ell) \rightarrow H_2O(g)$

$\Delta H = \Delta U + \Delta n_g RT$ ($\Delta n_g = 1 - 0 = 1$)

(100°C = 273 + 100 = 373 K)

$40.66 = \Delta U + 1 \times \frac{8.314}{1000} \times 373$

$40.66 = \Delta U + 3.101$

$\Delta U = 37.56$ kJ/mol

9. (3) $C_6H_{12}O_6(s) + 6O_2(g) \rightarrow 6CO_2(g) + 6H_2O(l)$

$$\Delta_r^0H = \left(6\Delta_f^0H_{(CO_2)} + 6\Delta_f^0H_{\ H_2O}\right) - \left(\Delta_f^0H_{(C_6H_{12}O_6)}\right)$$

$$= 6\times(-400) + 6(-300) - (-1300)$$

$= -2400 - 1800 + 1300 = -2900$ kJ/mol

Hence, standard enthalpy of combustion for per gram of glucose $= \dfrac{-2900}{180}$ kJ/g $= -16.11$ kJ/g

10. (2) $\frac{1}{2}X_2 + \frac{3}{2}Y_2 \rightleftharpoons XY_3$

$$\Delta_r^0S = \Sigma\Delta_P^0S - \Sigma\Delta_R^0S$$

$$\Delta_r^0S = (50) - \left(\frac{1}{2}\times 60 + \frac{3}{2}\times 40\right)$$

$= 50 - (30 + 60) = 50 - 90 = -40$

Apply, $\Delta G = \Delta H - T\Delta S$

(At equilibrium ., $\Delta G = 0$)

$\Delta H = T\,\Delta S$; $T = \dfrac{\Delta H}{\Delta S}$

$$\frac{-30\times 1000}{-40} = T$$

$$\frac{3000}{4} = T = 750\ K$$

11. (2) $\Delta S = nR\ln\left(\dfrac{P_i}{P_f}\right)$

12. (3) For isothermal process, $\Delta T = 0$. For an ideal gas, internal energy U depend only on temperature.

So, $\Delta U = 0$

Now $\Delta U = q + w$

$q = -w = -P_{ext.}\Delta V$

For free expansion $P_{ext} = 0$

So, $q = -w = 0$

$\therefore \Delta U - 0,\ q = 0,\ w = 0$

13. (2) $CaCO_3(s) \rightleftharpoons CaO(s) + CO_2(g)$

$$K_p = (p_{CO_2})^1$$

$\Delta_fH^\circ = \Sigma\Delta_fH^\circ_{(Products)} - \Sigma\Delta_fH^\circ_{(Reactants)}$

$=(-635.1)+((-393.5))-(-1206.9)=(-635.1-393.5)+1206.9$

$\Delta_fH^\circ = +178.3$ kJ/mol

$\Delta_fS^\circ = \Sigma S_P^0 - \Sigma S_R^0 = (38.2 + 213.7) - (92.9)$

$\Delta_fS^\circ = 159$ J/mol $= 0.159$ kJ/mol

Apply, $\Delta G^\circ = \Delta H^\circ - T\Delta S^\circ = 178.3 - 1273 \times 0.159$

$= 178.3 - 202.407 = -24.107$ kJ/mol

Also, $\Delta_fG^0 = -2.303RT\log K$

$$-24.107 = -2.303\times\frac{8.314}{1000}\times 1273\log Kp$$

$$\log K_p = \frac{24.107}{24.374}$$

$\log K_p = 0.98$

$K_p = 9.75$ atm

14. (3) $C_7H_{16}(l) + 11O_2(g) \rightarrow 7CO_2(g) + 8H_2O\ (l)$

$\Delta n_g = n_p - n_r = 7 - 11 = -4$

$\because \Delta H = \Delta U + \Delta n_gRT$

$\therefore \Delta H - \Delta U = -4\ RT$

15. (4) Data given

$CuSO_4(s) + 5H_2O(g) \rightarrow CuSO_4.\ 5H_2O(s)$ is -71.50 kcal/mol ....(1)

Also, $H_2O(\ell) \rightarrow H_2O(g)$; $\Delta H = +10.5$ kcal/mol ....(2)

Multiply equation '2' by 5 & then add-

$CuSO_4(s) + 5H_2O(g) \rightarrow CuSO_4.\ 5H_2O(s)$, $\Delta H = -71.5$ kcal/mol

$5H_2O(\ell) \rightarrow 5H_2O(g)$; $\Delta H = +10.5 \times 5 = +52.5$ kcal/mol

Net equation : $CuSO_4(s) + 5H_2O(\ell) \rightarrow CuSO_4.\ 5H_2O(s)$

$\Delta H = -19$ kcal/mol

16. (2) $10C(g) + 10H_2(g) \rightarrow C_{10}H_{20}(g)$; $\Delta_rH = ?$

$\Delta_rH = \Sigma B.E_R - \Sigma B.E_P$

$\Delta_rH = (10b) - (8z + 20a - w)$

$\Delta_rH = 10b - 8z - 20a + w$

Enthalpy of atomisation of C = x kJ/mol^{-1}

$10C(g) + 10H_2(g) \rightarrow C_{10}H_{20}(g)$; $\Delta H = 10b - 8z - 20a + w$

$10C(s) \rightarrow 10C(g)$; $\Delta H = 10x$ kJ/mol (after multiplying with 10)

$10C(s) + 10H_2(g) \rightarrow C_{10}H_{20}(g)$; $\Delta H = 10b - 8z - 20a + w + 10x$

Also, $C_{10}H_{20}(\ell) \rightarrow C_{10}H_{20}(g)$; $\Delta H = y$ kJ/mol

Subtracting these two reactions

We finally get, $10C(s) + 10H_2(g) - C_{10}H_{20}(\ell) \rightarrow 0$

$10C(s)+10H_2(g)\rightarrow C_{10}H_{20}(\ell);\Delta H=10b-8z-20a+w+10x-y$

17. (1) At equilibrium. $\Delta G = 0$

Since standard conditions prevail at equilibrium,

then $\Delta G = \Delta G^\circ = 0$

Now, $\Delta G^\circ = -2.303RT\log K$

$0 = -2.303RT\log K$

$\log K = 0$, $K = 1$

18. (2) $\frac{1}{2}N_2(g) + \frac{3}{2}H_2(g) \rightarrow NH_3(g); \Delta_fH = -46$ kJ/mol

$2H(g) \rightarrow H_2(g)$; $\Delta H = -436$ kJ mol^{-1}; $B.E._{H_2} = +436$ kJ mol^{-1}

$2N(g) \rightarrow N_2(g)$; $\Delta H = -712$ kJ mol^{-1} ; $B.E._{N_2} = +712$ kJ mol^{-1}

$\Delta_rH = \Sigma B.E_R - \Sigma B.E_P$

$$-46 = \left(\frac{1}{2}\times 712 + \frac{3}{2}\times 436\right) - (3x)$$

[where x is bond energy of N–H bond]

$-46 = (356 + 654) - 3x$

$-46 - 1010 = -3x$

$$\frac{-1056}{-3} = x$$

$\therefore x = 352$ kJmol^{-1}

19. (1) $4NO_2(g) + O_2(g) \rightarrow 2N_2O_5(g)$; $\Delta H = -111$ kJ

$4NO_2(g) + O_2(g) \rightarrow 2N_2O_5(s)$; $\Delta H = -219$ kJ

(–) (–) (–)

$0 \rightarrow 2N_2O_5(g) - 2N_2O_5(s)$

$2N_2O_5(s) \rightarrow 2N_2O_5(g)$; $\Delta H = -111 - (-219) = 108$kJ

For 1 mole N_2O_5, $\Delta_{sub}H = \frac{108}{2} = 54\text{kJ} = 54{,}000\text{J}$

$$\Delta S_{sub} = \frac{\Delta_{sub}H}{T} = \frac{54000}{300} = 180 \text{ J mol}^{-1}\text{K}^{-1}$$

20. (2) $H_2 \rightarrow 2H\bullet$; $\Delta H = x_1$

$O_2 \rightarrow 2O\bullet$; $\Delta H = x_2$

$O\text{–}H \rightarrow O\bullet + H\bullet$; $\Delta H = x_3$

$H_2O(\ell) \rightarrow H_2O(g)$; $\Delta H = x_4$

$$H_2(g) + \frac{1}{2}O_2(g) \rightarrow H_2O(g)$$

$$\Delta_f H = \Sigma B.E_R - \Sigma B.E_P = \left(x_1 + \frac{x_2}{2}\right) - (2x_3)$$

$$\Delta_f H = x_1 + \frac{x_2}{2} - 2x_3$$

$$H_2(g) + \frac{1}{2}O_2(g) \rightarrow H_2O(g); \Delta H = x_1 + \frac{x_2}{2} - 2x_3$$

$H_2O(\ell) \rightarrow H_2O(g)$; $\Delta H = x_4$

(–) (–)

$$H_2(g) + \frac{1}{2}O_2(g) \rightarrow H_2O(\ell); \Delta H = x_1 + \frac{x_2}{2} - 2x_3 - x_4$$

21. (3) $\Delta S_{vap} = \frac{\Delta H_{vap}}{T_b} = \frac{540}{373}$ .....(1)

$\Delta S_{fusion} = \frac{\Delta H_{fusion}}{T_m} = \frac{80}{273}$ .....(2)

$$\frac{\Delta S_{vap}}{\Delta S_{fusion}} = \frac{\frac{540}{373}}{\frac{80}{273}} = \frac{540}{373} \times \frac{273}{80} = \frac{14742}{2984} = 4.94$$

22. (1) $\Delta H = \Delta U + \Delta(PV)$

$= \Delta U + P\Delta V + V\Delta P$

At constant pressure,

$= \Delta U + P\Delta V$

23. (1) $\Delta S_{fusion} = \frac{\Delta_f H}{T}$

$$T = \frac{30.5 \times 1000}{28.8} = 1059 \text{ K}$$

24. (1) 500 mL of 0.2M KOH (S.B.) (Use $q = mC\Delta T$)

500 of 0.2M HCl (S.A.)

The heat of neutralisation is same

$T_1 = T_2$

25. (3) $W = -P_{ext} \Delta V$

$W = -1.25 \times 32 = -40$ Lbar $= -40 \times 100$ J $= -4000$ J $= -4$ kJ

$\Delta E = q + w$

So, $q = \Delta E - w = -51 + 4 = -47$ kJ

26. (4) Condensation of steam will convert gas to liquid.

$\therefore$ Decrease in entropy

27. (1)

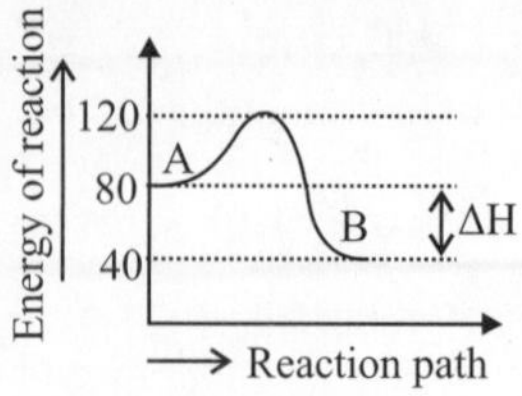

ΔH = energy of product – energy of reactant

$= 40 - 80 = -40$

28. (4) $w = -2.303RT \log\frac{V_2}{V_1}$

$= -2.303 \times 2 \times 500 \log\frac{1}{10} = -2.303 \times 10^3 \log 10^{-1}$

$= 2.303$ kcal

29. (2) $CaCO_3(s) \rightarrow CaO(s) + CO_2(g)$

$\Delta H = 176$ kJ /mole

$T = 1240$ K

$\Delta H = \Delta U + \Delta n_g RT$

$176 = \Delta U + 1 \times 8.314 \times 10^{-3} \text{ kJK}^{-1} \text{ mol}^{-1} \times 1240$ K

$\Delta U = 176 - 10.3 = 165.69$ kJ/mole

30. (2) $\Delta G = \Delta H - T\Delta S$, [given $T = 27°C = 273 + 27 = 300$]

$= 40.63 \times 10^3 - 300 \times 100 = 10630$ J/mole ; non-spontaneous.

31. (2) $\Delta G < 0$, $\Delta S > 0$

32. (4) $2N_2(g) + 6H_2(g) \rightarrow 4NH_3(g)$

$\Delta H = -46$ kJ/mole

Formation of 1 mole $NH_3 = -46$ kJ/mole

4 mole $NH_3 = -4 \times 46 = -184$ kJ

33. (4) $H_2 + Cl_2 \rightarrow 2HCl$ $\Delta H = -x$ kJ

ΔH_f of $HCl = -\frac{x}{2}$

34. (1) $H_2O(l) \rightarrow H^{\oplus}(aq) + OH^-(aq)$

$\Delta H = 57.32$ kJ

$$H_2(g) + \frac{1}{2}O_2(g) \rightarrow H_2O(l)$$

$\Delta H = -286.20$ kJ

by adding above two equations, we get,

$$H_2(g) + \frac{1}{2}O_2(g) \rightarrow H^{\oplus} + OH^-$$

$\Delta H = (57.32) + (-286.20)\text{kJ} = -228.88$ kJ

35. (3) In this graph temperature change is positive, then enthalpy change will also be positive.

$\Delta H = nC_p\Delta T$

$\Delta H = +ve$ ...(i)

As we know, $V \propto T$

i.e. if temperature increases volume of gas will also be increased.

$w = -P(V_2 - V_1) = -p\Delta V = -ve$ ...(ii)

from (i) and (ii), option 3 is correct.

36. (3) $\Delta G = \Delta H - T\Delta S$

at equilibrium $\Delta G = 0$

So, $T\Delta S = \Delta H$ ...(i)

As ΔH and ΔS should be +ve for a reaction to be spontaneous and ΔG should be –ve

$\therefore\ T\Delta S > \Delta H$ ...(ii)

From eq. (i) and (ii), we get

$T > T_e$

37. (3) $W = -P_{ext}(V_2 - V_1)$

$= -1 \text{ bar} \times (10-1)\text{lit}$

$= -9 \text{ bar lit} = -900 \text{ J} = -0.9 \text{ kJ}$

38. (1)

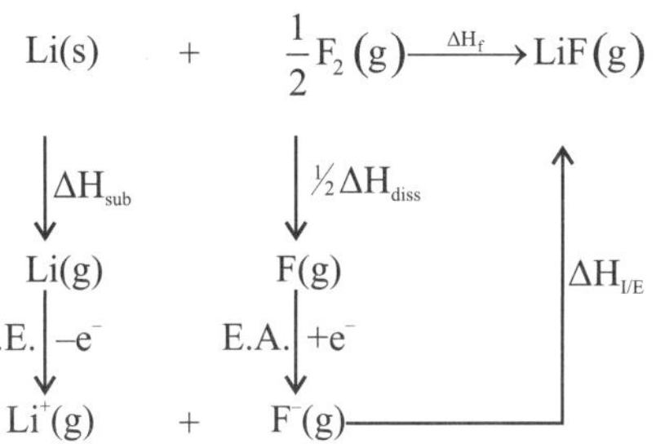

$$\therefore\ \Delta_f H = \Delta H_{sub} + \frac{1}{2}\Delta H_{diss} + \Delta H_{I.E.} + \Delta H_{E.A.} + \Delta H_{Lattice}$$

$\therefore\ \Delta H_{Lattice} = -1011.6$ kJ/mol

39. (3) As free expansion in adiabatic condition means, no transfer of heat occurs between system & its surroundings i.e., $P_{ext} = 0$,

$w = 0$,

$q = 0$

$\therefore\ \Delta U = q + w$

$= 0$

$\Delta U = nC_v\Delta T =$

and no work is done.

$w = 0$

Since, $q = 0$

$\Delta U = 0$ and initial and final state hence, are same,

hence, no exchange of temperature is expected

Hence, For free expansion of an ideal gas under adiabatic condition $q = 0$ $\Delta T = 0$ $w = 0$.

40. (4) $\dfrac{C_P}{C_V} = \dfrac{\frac{5R}{2}}{\frac{3R}{2}} = \dfrac{5}{3} = 1.67$

41. (1) $w = -2.303\ nRT \log \dfrac{V_2}{V_1}$

$$\Rightarrow w = -2.3 \times \frac{20}{40} \times 25/3 \times 300 \log \frac{10}{5} = -862.5\text{ J}$$

42. (3) All are in their standard state.

43. (2) It is free expansion; i.e., $P_{ext} = 0$, $w = 0$,

$q = 0$

$\therefore\ \Delta U = 0$

$nC_v\Delta T = 0$

Hence, $\Delta T = 0$

Therefore, $T_2 = T_1$

44. (3) $V_i = \dfrac{nRT}{P} = \dfrac{4.4}{44} \times \dfrac{0.082 \times 300}{10} = 0.246$ litre

$V_f = \dfrac{P_i V_i}{P_f} = \dfrac{10 \times 0.246}{1} = 2.46$ litres

$q = -w = P(V_f - V_i) = 1(2.46 - 0.246) \times 100 = 221.4$ J

45. (3) For isothermal reversible expansion of an ideal gas, $\Delta U = 0$

46. (1) We use the relation:

$\Delta H_f^\circ\,(Mg(OH)_2(s)] = \Delta H_f^\circ[Mg^{2+}(aq) + 2\times\ \Delta H_f^\circ[OH^-(aq)] + \Delta H_{cryst}$

Now plug in the values:

$-924 = \Delta H_f^\circ(Mg^{2+}] + 2 \times (-230) + (-8)$

$-924 = \Delta H_f^\circ[Mg^{2+}] - 460 - 8$

$\Delta H_f^\circ[Mg^{2+}] = -924 + 468 = -456$ kJ/mol

Correct & Incorrect MCQs

1. (4) Temperature measures a property of a state of a system irrespective of how it got there and hence, it is a state function.

2. (1) As, $\Delta U = q + w$

$\Delta U = q + p\Delta V$

At constant volume i.e., $\Delta V = 0$

$\Delta U = q$

Hence, heat absorbed at constant volume by a system is equal to change in its internal energy.

3. (4) Heat capacity is an extensive property. Temperature is an intensive property.

4. (3) Specific heat capacity of a gas is infinite, during isothermal process.

As $C = \dfrac{q}{m\Delta T}$

For isothermal process, $\Delta T = 0$

$\therefore\ \dfrac{q}{0} = \infty$

For adiabatic process, $q = 0$

$\therefore\ C = 0$

5. (1) Fusion is an endothermic reaction, so heat is absorbed in this reaction. Hence, all enthalpies of fusion are positive.

6. (3) The enthalpy of formation of $H_2O(l)$ is greater than that of H_2O (g).

7. (2) Since, entropy is directly proportional to disorderness, therefore when heat is added, disorderness of molecule increases. Hence, the entropy of a system increases as heat is added to it.

8. (2) ΔG is negative for spontaneous process. ΔG is positive for non-spontaneous process.

9. (2) Adiabatic system does not allow exchange of heat with surroundings.

10. (1) Enthalpy change is negative for the condensation reaction $H_2O\ (g) \rightarrow H_2O(l)$.

Statement Based MCQs

1. (1) $\Delta G = \Delta H - T\Delta S$

(When $\Delta H = -Ve$, $\Delta S = -Ve$) $\rightarrow \Delta G = -\Delta H + T\Delta S$

When T large, $\Delta G = +ve$, when T is small $\Delta G = -Ve$.

Temperature decreases randomness decreases.

2. (1) Because value of internal energy is fixed for each state of the system and in cyclic process initial and final state are same so change in internal energy is zero.

3. (3) Combustion reactions are always exothermic as they release heat so $\Delta H = -ve$

4. (3) For spontaneous reaction $\Delta G = -ve$. Spontaneous reactions may not be always exothermic.

5. (1) Hess law is also based on principles of conservation of energy so it the consequence of the first law. Enthalpy is a state function.

Both statement-I and statement-II are true.

6. (4) Strong H-bonding is present in water while weak dipole-dipole interactions are present in acetone, hence it requires less heat to vaporise 1 mol of acetone than it does to vaporize 1 mol of water.

7. (1) Since, particles are already fixed at their positions in solid state while particles are much disordered in gaseous state, hence the gaseous state of a given substance has more entropy than that in the solid state.

8. (4) Melting of a solid is endothermic process.

9. (3) In adiabatic system, no transfer of heat takes place.

10. (3) $\Delta U = Q + W$

Internal energy is state function but Q and W are path function.

Hence, addition of two path function can give state function.

Assertion & Reason MCQs

1. (2) The given assertion is true as, combustion of all organic compounds produces energy in the form of heat. Hence, it is an exothermic reaction. Reason is also true as enthalpy of all elements in their standard state is taken as zero. Therefore, Both are true but reason is not correct explanation of assertion.

2. (2) The given assertion is true because spontaneous reactions are irreversible as they can proceed on their own, but to proceed their reverse reaction it will need some external agency. For a spontaneous reaction, $\Delta G < 0$, $\Delta G = \Delta H - T\Delta S$

$\Delta H - T\Delta S < 0$, when $\Delta H < 0$ it may lead to negative value of ΔG and thus it is contributing factor for spontaneity.

Hence, both assertion & reason are true but reason is not correct explanation of assertion.

3. (1) In the crystalline process of liquid to solid, entropy decreases. This is because in solid crystals, molecules are organised in ordered manner.

Hence, both assertion & reason are true and reason is the correct explanation of assertion.

4. (1) Combustion of methane is given as,

$CH_4(g) + 2O_2(g) \rightarrow CO_2(g) + 2H_2O(\ell)$

$\Delta H = \Delta U + \Delta n_g RT$

$\Delta n_g = 1 - 3 = -2$

$\Delta H = \Delta U - 2RT$

$\Delta U = \Delta H + 2RT \quad \therefore \ \Delta U > \Delta H$

From the above explanation, it is clear that both assertion and reason are true and reason is correct explanation of assertion.

5. (3) The given assertion is true but reason is false.

6. (1) Water in liquid form has lower energy than water in gaseous form therefore exactly that much more energy is released.

7. (4) For the given reaction, $\Delta n_g = n_P - n_R$

$\Delta n_g = 1 - 1 = 0$, Therefore, $\Delta H = \Delta U$

ΔH is related to ΔU by the equation, $\Delta H = \Delta U + \Delta n_g RT$

8. (2) Formation of ammonia is an exothermic process, and process will be in equilibrium at $\Delta G = 0$

9. (3) The adiabatic process can cause the system's temperature to rise or fall. Systems that undergo such processes are thermally isolated from their surroundings.

10. (1) The energy released in making new bonds is less than energy required when bonds are broken. N–N bond is very strong due to presence of 1σ and 2π bonds.

Multi Correct MCQs

1. (1) Path function are those which depends on the initial and final state of the system.

2. (2) Only (B) and (C) are correct.

(B) $G = H - TS$

At constant T

$\Delta G = \Delta H - T\Delta S$

(A) Acc. to first law of thermodynamics

$\Delta U = Q + W$

If we apply constant P and reversible work.

$\Delta U = Q - P\Delta V$

(C) Acc. to 2nd law of thermodynamics.

$$dS = \frac{dq_{rev}}{T}$$

At constant T

$$\Delta S = \frac{q_{rev}}{T}$$

(D) $H = U + PV$

For ideal gas

$H = U + nRT$

At constant T

$\Delta H = \Delta U + \Delta nRT$

3. **(2)** Water $\xrightarrow{0°C}$ ice $\qquad \Delta S = -ve$

$H_2O(l) \xrightarrow{-10°C}$ ice

$\Delta S = -ve$

$N_2(g) + 3H_2(g) \rightarrow 2NH_3(g)$ $\Delta S = -ve$

For adsorption, $\Delta S = -ve$

4. **(4)** Extensive properties depends upon the mass/amount of matter present in a system while the intensive properties doesn't depends on the amount of matter.

Extensive properties: Mole, Volume, Gibbs free energy change.

Intensive properties: Molar mass, Molar heat capacity, Molarity, E^0 cell.

NCERT Exemplar MCQs

1. **(3)** The rate at which a reaction proceeds is not concerned with thermodynamics. Thermodynamics deals with the energy change, feasibility and the extent to which a chemical reaction proceeds.

2. **(3)** For a closed vessel made up of copper is an example of closed system. As, only energy exchange can occur through its walls, matter can't exchange between the system and the surroundings.

3. **(4)** The state of a gas can be described by quoting the relationship between pressure, volume, temperature and amount of substance. The ideal gas equation is :

$pV = nRT$

Thus, p, V, T and n describe the state of the system.

4. **(3)** The specific heat will remain constant even when the volume of gas is reduced to half from its original volume. Specific heat is an intensive property depending only on the nature of the gas.

Hence, when such a gas expands under adiabatic conditions into a vacuum, no external work is done to separate the molecules as no heat is absorbed or evolved.

5. **(3)** The complete combustion of one mole of butane is represented by thermochemical reaction as

$C_4H_{10}(g) + \frac{13}{2}O_2(g) \rightarrow 4CO_2(g) + 5H_2O(l)$ We have to take the combustion of one mole of C_4H_{10} and $\Delta_{comb}H$ should be negative and have a value of 2658 kJ mol^{-1}

6. **(2)** Reaction for formation of CH_4

$C_{graphite} + 2H_2(g) \rightarrow CH_4(g)$

$\Delta H = \Delta U + \Delta n_g RT$

Δn_g = No. of moles of gaseous products – No. of moles of gaseous reactants

$\Delta n_g = 1 - 2 = -1$

$\Delta H = \Delta U - RT$

$\Delta H = 393 - RT$

$\therefore \Delta H < \Delta U$

7. **(3)** Free expansion, $W = 0$

Adiabatic process, $q = 0$

$\Delta U = q + W = 0$, this means that internal energy remains constant. Therefore, $\Delta T = 0$.

8. **(2)** The correct option is W (reversible) < W (irreversible). This is because area under the curve is always more in irreversible compression as can be seen from given figure.

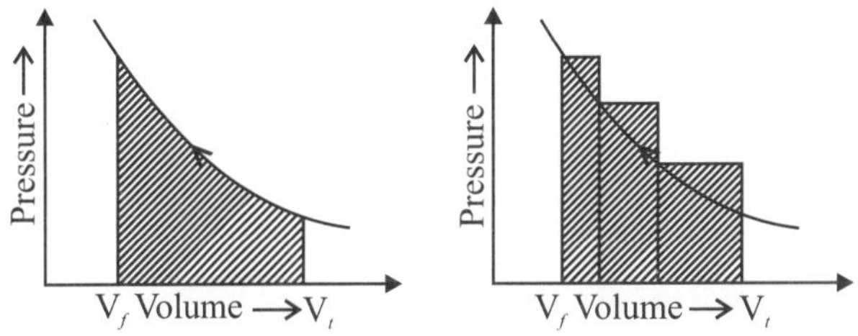

(a) Reversible compression (b) Irreversible compression

9. **(3)** By using the following expression, the entropy change can be calculated.

$\Delta S = \frac{q_{rev}}{T}$.

Entropy of the surrounding increases when water freezes in a glass beaker, ΔS (system) decreases because molecules in solid ice are less random than in liquid water. However, when water freezes to ice, heat is released which is absorbed by the surroundings.

10. **(3)** The algebraic relationships of the given reaction is

equation (a) = equation (b) + equation (c)

(a) $C(\text{graphite}) + O_2(g) \rightarrow CO_2(g); \Delta_r H = x \text{ kJ mol}^{-1}$

(b) $C(\text{graphite}) + \frac{1}{2}O_2(g) \rightarrow CO(g); \Delta_r H = y \text{ kJ mol}^{-1}$

(c) $CO(g) + \frac{1}{2}O_2(g) \rightarrow CO_2(g); \Delta_r H = z \text{ kJ mol}^{-1}$

Hence, $x = y + z$

11. **(3)** Same bonds are formed in reaction (a) and (b) but no bond is broken in reaction (a) whereas bonds in the reactant molecules are broken in reaction (b). As energy is absorbed when bonds are broken, energy released in reaction (a) is greater than that in reaction (b) hence, $x > y$.

12. **(3)** Enthalpy of formation can be positive or negative as combustion of elements to form a compound can be exothermic or endothermic. *e.g.*, $C + O_2 \rightarrow CO_2$ is exothermic.

whereas, $C + 2S \rightarrow CS_2$ is endothermic

13. **(1)** Enthalpy of sublimation of a substance is equal to enthalpy of fusion + enthalpy of vaporisation.

Sublimation is, direct conversion of solid to vapour

Writing in two steps, we have solid → liquid → vapour,

solid → liquid requires enthalpy of fusion

liquid → vapour requires enthalpy of vaporisation

14. (2) ΔG gives a criteria for spontaneity at constant pressure and temperature.

(i) If ΔG is negative (< 0), the process is spontaneous.

(ii) If ΔG is positive (> 0), the process is non-spontaneous.

(iii) If ΔG is zero then reaction is equilibrium.

NEET & AIIMS Past Year MCQs

1. (3) Given that, it is isothermal, $\Delta T = 0$ .....(i)

$\Delta U = nC_v\Delta T = 0$ .....(ii)

Since expansion is taking place against vacuum

$P_{ext} = 0$

$W = -P_{ext}\Delta V = 0$ .....(iii)

From first law of thermodynamics,

$\Delta U = q + W$

$0 = q + 0$ .....From (ii) and (iii)

$q = 0$

Hence, work done will be zero and heat will also be zero.

2. (1) For endothermic reactions: $\Delta_r H$ is positive, as heat is absorbed during the reaction.

$\Delta H = E_{a(f)} - E_{a(b)}$

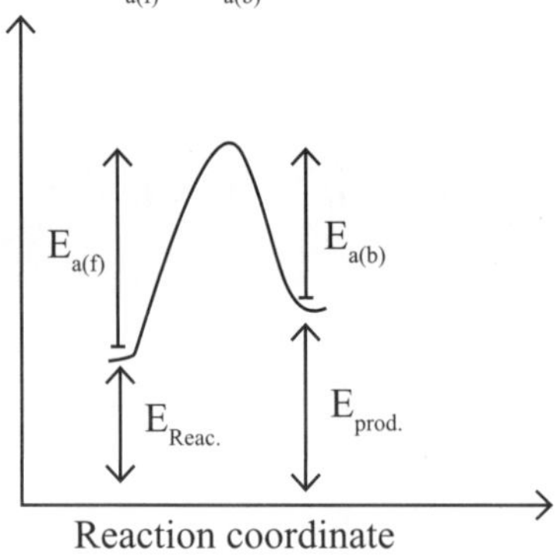

$q_p = +ve$

3. (1) From the given reaction:

$A_2(g) + 3B_2(g) \rightarrow 2AB_3(g)$

The change in the number of moles of gas (Δn_g) is calculated as:

$\Delta n_g = n_P - n_R$

$= 2 - 3 - 1 = -2$

The relationship between ΔH and ΔU is:

$\Delta H = \Delta U + \Delta n_g RT$

$15\ J \times 1000 = \Delta U - 2 \times 8.314\ J/K\ mol^{-1} \times 300\ K$

$\Delta U = 15000 + 600 \times 8.314$

$\Delta U = 19988.4\ J$

4. (4) $W = -2.303nRT \log\left(\frac{P_1}{P_2}\right)$

$= -2.303 \times 1 \times 2 \times 298 \times \log\left(\frac{20}{10}\right)$

$= -2.303 \times 2 \times 298 \times 0.301$

$= -413.14$ Calories

5. (2) Isothermal process $\rightarrow$ at constant temperature

Isochoric process $\rightarrow$ at constant volume

Isobaric process $\rightarrow$ at constant pressure

Adiabatic process $\rightarrow$ No heat exchange

6. (1)

(A) When liquid evaporates to vapour, entropy increases due to increase in randomness.

(B) When temperature of crystalline solid lowered from 130 K to 0 K, the random collisions will decrease, resulting in decrease in entropy.

(C) In the given reaction,

$2NaHCO_3(s) \rightarrow Na_2CO_3(s) + CO_2(g) + H_2O(g)$

Solid is converting to gas, due to which randomness increases and hence, entropy increase.

(D) In the given conversion, $Cl_2(g) \rightarrow 2Cl(g)$ number of gaseous atom increases, due to which entropy increases.

Hence, Entropy will increase in case of A, C, D.

7. (3) $\Delta H = \Delta U + \Delta n_g RT$

8. (3) Work done is basically area under the curve in the PV diagram. In the given figures, third figure has maximum area.

9. (1) For one mole of an ideal gas $C_p - C_v = R$

10. (2) For Irreversible expansion of an ideal gas under Isothermal condition

$\Delta U = 0$

$\Delta S_{Total} \neq 0$

11. (1) Process is adiabatic $\therefore\ q = 0$

$\therefore\ \Delta U = w = -P_{ext}\Delta V$

$= -1\ (10 - 2)$ atm L

$= -8$ atm L

12. (1) For adiabatic conditions, $PV^{\gamma} =$ Constant

$P_1V_1^{\gamma} = P_2V_2^{\gamma};\quad V_2 = \frac{1}{2}V_1$

$P_2 = P_1\left(\frac{V_1}{V_2}\right)^{\gamma}$ [For diatomic gas, $\gamma = 1.4$]

$P_2 = P_1\left(\frac{V_1 \times 2}{V_1}\right)^{1.4}$

$P_2 = P_1(2)^{1.4} = P(2)^{1.4}$

13. (3) In adiabatic process no exchange of heat take place b/w system and surrounding i.e. $q = 0$.

14. (2)

Equilibrium

NCERT Topic-wise MCQs

1. (3) Vapours of Iodine are in violet colour.

2. (2) $\Delta V = -ve$ for $H_2O(s) \rightarrow H_2O(l)$

Increase of pressure would cause the decrease in volume. This is achieved by melting of ice.

3. **(3)** A reaction reaches a state of equilibrium only when the products react together at the same rate at which they are formed.

4. **(1)** Either reactants or products, one of them must be coloured.

5. **(3)** The continuous changes observed in the measurable properties of the system, which is not general characteristic of the equilibrium for physical processes.

6. **(3)** Others are not characteristic of reversible reactions.

7. **(1)** A reaction is said to be in equilibrium when the rate of transformation of reactants to products is equal to the rate of transformation of products to the reactants.

8. **(4)** KCl and O_2 do not react to form $KClO_3$.

9. **(3)** Number of moles of reactants and products are not always equal. Ex. $N_2 + 3H_2 \rightleftharpoons 2NH_3$

10. **(3)** In closed system, CaO and CO_2 can combine to form $CaCO_3$

11. **(4)** Initial : $PCl_5 \rightleftharpoons PCl_3 + Cl_2$

at equilibrium; $1-\alpha \quad \alpha \quad \alpha$

Total moles; $1-\alpha+\alpha+\alpha = 1+\alpha$

Total pressure = P

$$\therefore\ P_{PCl_5} = \frac{(1-\alpha)P}{1+\alpha}, P_{PCl_3} = \frac{\alpha P}{1+\alpha}, P_{Cl_2} = \frac{\alpha P}{1+\alpha}$$

$$K_p = \frac{P_{PCl_3} \times P_{Cl_2}}{P_{PCl_5}} = \frac{\alpha^2 P^2}{(1+\alpha)(1-\alpha)(P)} = \frac{\alpha^2 P}{1-\alpha^2}$$

12. **(2)** On adding all three reactions, final reaction can be obtained. So,

$K = K_1.K_2.K_3$

13. **(3)** $K_p = K_c RT^{\Delta ng}$ $\quad [\Delta n_g = -1]$

$K_p = K_c\, RT^{-1}$

$$\therefore\ K_p = \frac{K_c}{RT}$$

14. **(2)** From the equation, it is clear that 2 moles of A combine with one mole of B. Let at equilibrium "a" moles of B have reacted.

This means that 2a moles of A have reacted

∴ Equilibrium concentration of A = 2 – 2a

Equilibrium concentration of B = 2 – a

Hence ; [A] < [B]

15. **(4)** $N_2(g) + O_2(g) \rightleftharpoons 2NO(g)$, K_1

$2NO(g) + O_2(g) \rightleftharpoons 2NO_2(g)$, K_2

$N_2(g) + 2O_2(g) \rightleftharpoons 2NO_2(g)$, $K = K_1 \times K_2$

∴ for $NO_2(g) \rightleftharpoons 1/2\, N_2(g) + O_2(g)$

$$K = \left[\frac{1}{K_1K_2}\right]^{1/2}$$

16. **(3)** $K_p = K_c\,(RT)^{\Delta ng}$

For equation ; $CaCO_{3(s)} \rightleftharpoons CaO_{(s)} + CO_{2(g)}$, $\Delta n_g = 1$

T = 800 + 273 = 1073 K

$$K_c = \frac{K_p}{(RT)^{\Delta n_g}} = \frac{167}{(0.0821 \times 1073)^1} = 1.89$$

17. **(3)** $A_{(g)} + B_{(g)} \rightleftharpoons C_{(g)} + D_{(g)}$

	A	B	C	D
T = 0;	2	2	0	0
At equilibrium ;	(2 – x)	(2 – x)	x	x

At equilibrium given that 3[B] = [C]

$3(2-x) = x \Rightarrow x = 3/2$

$$K_c = \frac{[C][D]}{[A][B]} = \frac{x^2}{(2-x)^2} = \frac{\left(\frac{3}{2}\right)^2}{\left(2-\frac{3}{2}\right)^2} = 9$$

18. **(4)** $AB(g) \rightleftharpoons A(g) + B(g)$

	AB(g)	A(g)	B(g)
t = 0	1	0	0
33%	1-0.33	0.33	0.33

Total pressure at equilibrium

$= 1 - 0.33 + 0.33 + 0.33 \Rightarrow 1 + 0.33 = 1.33$

$$P_A = \frac{0.33}{1.33}P \Rightarrow P_B = \frac{0.33P}{1.33} \Rightarrow P_{AB} = \frac{0.67}{1.33}P$$

(Where. P_A, P_B and P_{AB} are partial pressure of A, B and AB)

$$K_p = \frac{P_A.P_B}{P_{AB}} = \frac{\left(\frac{0.33P}{1.33}\right)\left(\frac{0.33P}{1.33}\right)}{\left(\frac{0.67P}{1.33}\right)} = \frac{(0.33)^2 P}{1.33 \times 0.67} = \frac{0.1089P}{0.8911}$$

$$K_p = 0.122P \Rightarrow P = \frac{K_p}{0.122} = 8K_p$$

19. **(4)** The value of equilibrium constant is temperature dependent. Its value is independent of the concentration of reactants. Thus, when concentrations of H_2 and I_2 are increased, the equilibrium constant will remain same as given, i.e., 6.4.

20. **(2)** $2AB_2(g) \rightleftharpoons 2AB(g) + B_2(g)$

	$2AB_2(g)$	$2AB(g)$	$B_2(g)$
t = 0	2	0	0
equilibrium	2 – 2x	2x	x

Total no. of moles at equilibrium = 2 – 2x + 2x + x = 2 + x

$$\text{So; } P_{AB_2} \frac{2(1-x)P}{2+x}, P_{AB} = \frac{2xP}{2+x},\ P_{B_2} = \frac{xP}{2+x}$$

$$K_p = \frac{(P_{AB})^2(P_{B_2})}{(P_{AB_2})^2} = \frac{\left(\frac{2xP}{2+x}\right)^2\left(\frac{xP}{2+x}\right)}{\left(\frac{2(1-x)}{2+x}P\right)^2} = \frac{4x^3P}{4(2+x)(1-x)^2}$$

$$=\frac{x^3P}{(2+x)(1-x)^2}$$

$[\therefore x <<< 1 \text{ and } 2, \text{So} : (1-x) \approx 1, (2+x) \approx 2]]$

$$K_p = \frac{x^3P}{2} \quad x = \left(\frac{2K_p}{P}\right)^{\frac{1}{3}}$$

21. (1)

	X(g)	⇌ Y(g) +	Z(g)
t = 0	1	0	0
Equilibrium	0.5	0.5	0.5
Partial pressure;	$\frac{0.5P}{1.5}$	$\frac{0.5P}{1.5}$	$\frac{0.5P}{1.5}$

$$K_p = \frac{P_y.P_z}{P_x}$$

$$1 = \frac{\frac{P}{3}\times\frac{P}{3}}{\frac{P}{3}} \Rightarrow p = 3\,\text{atm}$$

Partial pressure of X = $\frac{P}{3} = 1\,\text{atm}$

22. (4) Lime kiln is open vessel therefore CO_2 formed is escape out.

23. (3) Rate ∝ product of partial pressures of reactants

24. (2) $K_{c_2} = \frac{1}{K_{c_1}} = \frac{1}{0.6} = 1.66$

25. (1) K_c depends only on temperature as for endothermic reaction $K_c \propto T$ for exothermic reaction $K_c \propto \frac{1}{T}$

26. (4) Position of equilibrium can be changed by changing the concentration of reactants or products.

27. (3) If the reaction having equilibrium constant K is reversed then resulting equation has eq. constant = 1/K

28. (1) $\Delta n_g = 0$

$\Delta n_g = 2\text{-}2 = 0$

$K_p = K_c\,(RT)^{\Delta n_g} = K_c(RT)^0$

$K_p = K_c$

29. (3) For solids, partial pressure values will be 1.

$$\therefore K_p = \frac{P_{CO_2}.1}{1}$$

$K_p = P_{CO_2}$

30. (3) For $A \rightleftharpoons B, K = \frac{[B]}{[A]}$ and for $mA \rightleftharpoons mB$

$$K' = \frac{[B]^m}{[A]^m}$$

$\therefore K' = (K)^m$

31. (2) $4\,NH_3(g) + 5O_2\,(g) \rightleftharpoons 4\,NO\,(g) + 6\,H_2\,O(g)$

$\Delta n_g = 10 - 9 = 1$

$$K_c = \frac{[NO]^4[H_2O]^6}{[NH^3]^4\cdot[O_2]^5} = \frac{(\text{conc.})^{10}}{(\text{conc.})^9} = (\text{conc})^1 \quad ...(1)$$

as we know $K_c = (\text{conc})^{\Delta n_g}$ ...(2)

Comparing eq. (1) and (2) we get,

$\Delta n_g = 1$

32. (2) $K_p \longrightarrow (\text{atm})^{\Delta n_g} = (\text{atm})^{-2}$

33. (3) If, $K_c = \frac{[N_2]^{1/2}[H_2]^{3/2}}{[NH_3]^1} = 4$

then $\frac{[NH_3]^2}{[N_2][H_2]^3} = (4)^{-2}$

T = 527 + 273 = 800 K

also $\Delta n_g = 2 - 4 = -2$

$$K_p = K_c.(RT)^{\Delta n_g} = (4)^{-2}\,(800\,R)^{-2} = \left(\frac{1}{4\times800R}\right)^2$$

34. (2) Dissociation of PCl_5 involves increase of moles

35. (3) $K_p = K_c\,(RT)^{\Delta n_g}, \Delta n_g = -1, \therefore K_p = \frac{K_c}{RT}$

for (a) $\Delta n_g = 1$ (c) $\Delta n_g = -1$

(b) $\Delta n_g = 0$ (d) $\Delta n_g = 0$

36. (3) $K_c = \left(\frac{\text{Mole}}{\text{Lit}}\right)^{\Delta n_g}$, Δn_g = (total no. of moles of gaseous products - total no of moles of gaseous reactants)

37. (1) $K_p = K_c\,(RT)^{\Delta n}$ or $K_p = K_c(RT)^{\Delta ng}$

$\Delta n = 1\text{–}2 = -1$

$$\therefore \frac{K_P}{K_C} = \frac{1}{RT}$$

38. (3) $K_c = \frac{[NO_2]^2}{[N_2O_4]} = \frac{1.2\times1.2\times10^{-2}\times10^{-2}}{4.8\times10^{-2}} = 3.3\times10^{-3}\,\text{molL}^{-1}$

39. (4)

	N_2O_4	⇌	$2NO_2$
Initial	1		0
Final	1–α		2α

$$K_p = \frac{(n_{NO_2})^2}{n_{N_2O_4}} \times \left[\frac{P}{\Sigma n}\right]^1$$

For 33% dissociation : $K_p = \frac{(2\times0.33)^2}{0.67}\times\left[\frac{P}{1.33}\right]$

For 40% dissociation : $K_p = \frac{(2\times0.40)^2}{0.60}\times\left[\frac{P}{1.40}\right]$

$$\therefore \frac{P_1}{P_2} = 1.56 \approx 1.60 = \frac{8}{5}$$

40. (2)

	A	+ B	⇌ AB;
Initial:	1	1	0
Final:	(1–x)	(1–x)	x

Given, x = 0.4

$\therefore$ Percentage of A changing to AB $= \frac{0.4\times100}{1} = 40\%$

41. (4)

	$2AB_3\,(g)$	⇌ $A_2\,(g)$	$+ 3B_2\,(g)$
t=0	8	0	0
At. equilibrium:	(8–a)	a/2	$\frac{3a}{2}$

Thus, $K_c = \frac{[A_2][B_2]^3}{[AB_3]^2}$; Also, $\frac{a}{2} = 2 \therefore a = 4$

$\therefore [AB_3] = \frac{4}{1}; [A_2] = \frac{2}{1}; [B_2] = \frac{6}{1}$

Thus, $K_c = \frac{2\times 6^3}{4^2} = 27\ mol^2L^{-2}$

42. (1) $NH_2COONH_4(s) \rightleftharpoons 2NH_3(g) + CO_2(g)$

Suppose at equilibrium,

$p_{CO_2} = P$

$p_{NH_3} = 2P$

Total pressure = 3atm

$P + 2P = 3$

$3P = 3 \Rightarrow P = 1$

$\therefore p_{CO_2} = 1$

$P_{NH_3} = 2$

$K_p = p^2_{NH_3} \cdot p_{CO_2} = 2^2.\ 1 = 4.1 = 4$

43. (2) $NH_4HS_{(s)} = NH_{3(g)} + H_2S_{(g)}$

at equilibrium concentration (interms of P) of both NH_3 and H_2S will be P.

$\therefore p_{NH_3} = P, p_{H_2S} = P$

$P + P = 30$ atm

$P = 15$ atm

$2x = 30$ atm

$x = 15$ atm

$K_p = p_{NH_3} \cdot p_{H_2S} = 15\times 15 = 225\ atm^2$

44. (1) On increasing pressure, volume decrease. Hence; when there are less number of gaseous moles, the reaction will move in the same direction (according to Le-Chatelier's principle).

$2SO_2 + O_2 \rightleftharpoons 2SO_3$

2 + 1   2 mol

Hence; the reaction will move towards right, to overcome change.

45. (3) Here; water is reactant so increase in concentration of reactant shifts the equilibrium in the forward direction. Hence; on diluting the solution; the equilibrium shifts in the $COCl_4^{2-}$

46. (3) K_p will remain same on adding inert gas at constant volume. The total pressure will be increased but the partial pressures will remain same of individual components.

47. (2) if $\Delta H > 0$ endothermic reaction $\rightarrow$ increase T favours reaction in forward direction.

48. (1) Pressure change does not exist in this case, as the number of moles in both the reactant and product are same.

49. (3) $BaO_2(s) \rightleftharpoons BaO(s) + O_2(g)$; $\Delta H = +ve$

According to law of mass action,

The rate of forward reaction = R

$R_1 \propto [BaO_2]$ or $R_1 = K_1 [BaO_2]$

But concentration of solid = 1

Then $R_1 = K_1$

similarly ; the rate of backward reaction = R

$R_2 \propto [BaO][O_2]$

$R_2 = K_2 [BaO] [O_2]$ $\therefore$ Conc. of [BaO] = 1

$R_2 = K_2 [O_2]$

At equilibrium; $R_1 = R_2$

$K_1 = K_2 [O_2]$ or $K_1 = K_2 . P_{O_2}$

where; P_{O_2} = partial partial pressure of O_2

$$\frac{K_1}{K_2} = P_{O_2} \quad (\text{equilibrium constant})$$

$$\frac{K_1}{K_2} = K \quad \text{or} \quad K = P_{O_2}$$

So; from above, it is clear that pressure of O_2 does not depend upon concentrations of reactants. If the temperature of such reaction is increased, then dissociation of BaO_2 would increase; and more O_2 is produced. The given equation is an endothermic reaction.

50. (2) At equilibrium, low pressure favours those reaction which are accompanied by increase in total number of moles and high pressure favours those reaction which takes place with decrease in total number of moles. However; pressure is not dependent on an equilibrium reaction which proceeds with no change in total number of moles. Hence; the reaction

$PCl_{5(g)} \rightleftharpoons PCl_3(g) + Cl_{2(g)}\ \Delta_{ng} - n_p - n_R = 2 - 1 = 1$

is favored by low pressure.

51. (3) Equilibrium constant do not change with concentration.

52. (2) If pressure is increased, equilibrium shifts in a direction where the number of moles decrease.

53. (1) If pressure is increased, equilibrium shifts in a direction where the number of moles decrease. Therefore, formation of NH_3 takes place in a direction where the number of moles decrease.

54. (1) Increase of products concentration favours the backward reaction.

55. (2) $K_c = 7.4\times 10^{-26} = \frac{[N_2H_4]}{[N_2][H_2]^2}$

The value of K_c shows that the equilibrium will be in backward direction, hence the equilibrium mixture contains mostly reactants.

56. (3) Equilibrium constant depends upon temperature for endothermic reaction, $K_c \propto T$

For exothermic reaction, $K_c \propto \frac{1}{T}$

57. (2) Increase in temperature, K_c value is increased, means reaction is endothermic.

58. (2) K_c is high, reaction goes to more completion.

59. (1) At constant temperature and pressure, $\Delta G = 0$

60. (2) Greater the K_p, more is the tendency to proceed to completion.

61. (3) Transition of Rhombic sulphur to Monoclinic sulphur (physical equilibrium).

62. (4) For the reaction

$N_2O_4(g) \rightleftharpoons 2NO_2(g)$

$$K_c = \frac{4a^2\alpha^2}{(a - a\alpha).V} \quad ....(1)$$

where 'a' is initial mole of N_2O_4 present in V litre and 'α' is its degree of dissociation,

Also, $K_p = K_c(RT)^{\Delta ng}$

On reducing the volume of container to $\frac{V}{2}$L, initial concentration of N_2O_4 becomes $\frac{2a}{V}$. An increase in concentration leads to more dissociation of N_2O_4 in order to have K_c constant, a characteristic constant for a given reaction at a temperature.

63. **(4)** K_p is a characteristic constant for a given reaction and changes only with temperature.

64. **(4)** For $A(s) \rightleftharpoons 2B(g) + 3C(g)$

$\therefore K_c = [C]^3[B]^2$; if [C] becomes twice,

Then let conc. of B becomes B', then

$K_c = [2C]^3[B']^2$ or $[C]^3.[B]^2 = [2C]^3.[B']^2$

$$\therefore \frac{[B']}{[B]} = \frac{1}{8} = \frac{1}{2\sqrt{2}}$$

65. **(4)** Catalyst does not disturb the equilibrium state. In the presence of catalyst, equilibrium attains quickly.

66. **(2)** Equilibrium constant doesn't depends on increasing volume.

67. **(3)** The strength of oxyacids can decide oxidation number of central atom. Higher the oxidation number of central atom, more acidic is the oxyacid.

$$\overset{+6}{H_2SO_4}, \overset{+7}{HClO_4}, \overset{+4}{H_2SO_3}, \overset{+5}{HClO_3}$$

Since; in $HClO_4$, oxidation number of Cl is highest, So; $HClO_4$ is the strongest acid among the given.

68. **(3)** Strong acid have a weak conjugate base and vice-versa.

$$\xrightarrow[\text{Increasing order of basic strength}]{H_2O < CH_3OH < OH^- < CH_3O^-}$$

69. **(1)** The bronsted base is NO_3^- because a bronsted base is the one that receives a proton. By accepting proton (H^+), it becomes HNO_3 (a strong acid).

70. **(1)** $AlCl_3$ is sp^2 hybridised when it combine with six H_2O molecules, one of the vacant p-orbital and five d-orbitals are filled by electron pairs, so; will not act as electron acceptor or lewis acid. As a result of which it cannot accept more electron.

71. **(2)** Transfer of proton from acid to base is called Bronsted neutralisation.

72. **(2)** $H_3PO_3 + H_2O \rightarrow H_2PO_3^- + H_3O^+$

$H_2PO_3^- + H_2O \rightarrow HPO_3^{2-} + H_3O^+$

conjugate acid-base pairs are

H_3PO_3 and $H_2PO_3^-$, $H_2PO_3^-$ and HPO_3^{2-}, H_2O and H_3O^+.

73. **(4)** Acid – $H^+ \rightarrow$ conjugate base

$\therefore H_2PO_4^- \rightarrow HPO_4^{2-}$

So, HPO_4^{2-} is the conjugate base of $H_2PO_4^-$

74. **(1)** Every protonic acid has conjugate base, but oxide (O^{-2}) ion has no protons.

75. **(2)** Conjugate base of weak acid is strong.

76. **(4)** Conjugate acid base pair is differ by a single proton.

77. **(3)** $N_3H \xrightarrow{-H^+} N_3^-$

78. **(3)** $[Al(H_2O)_6]^{+3} \xrightarrow{-H^+} [Al(H_2O)_5(OH)]^{+2}$

79. **(1)** Weakest conjugate base has strongest conjugate acid.

80. **(4)** All the given species are e^- deficient, therefore, a pair of electron can be given.

81. **(3)** $BaCl_2$ is a salt of strong base and strong acid, hence, acts as a strong electrolyte.

82. **(2)** $C\alpha = [H^+]$

$$\alpha = \frac{10^{-5}}{5 \times 10^{-3}} = \frac{1}{5} \times 10^{-2}$$

$$\% \text{ ionisation} = \frac{1}{5} \times 10^{-2} \times 100 = 0.2\%$$

83. **(4)** $\alpha = \sqrt{\frac{K_a}{c}} = \sqrt{\frac{1 \times 10^{-5}}{0.1}}$

$\sqrt{1 \times 10^{-4}} = 1 \times 10^{-2}$

84. **(2)** The strongest acid shows highest dissociation constant value, i.e., 2×10^{-2}, because it is highest value among all the given values.

85. **(3)** Ionization depends upon concentration, temperature, nature of solute and nature of solvent. Ionization increase on increasing dilution.

86. **(3)** pK_b for $CN^- = 4.7$

pK_a for conjugate acid HCN = 14 – 4.7 [$P_{ka} + P_{kb} = P_{kw}$]

pK_a for HCN = 9.3

NaCN is made up with strong Base NaOH and weak acid HCN. [It is similar to CH_3COONa]

Now salt hydrolysis of NaCN is

$$pH = \frac{1}{2}[pK_w + pK_a + \log C]$$

$$= \frac{1}{2}[14 + 9.3 + \log(0.5)]$$

$$= \frac{1}{2}[14 + 9.3 - 0.30]$$

$$= \frac{1}{2}[23] \Rightarrow P_H = 11.5$$

87. **(2)** Given self ionisation constant (SIC)

$K_{SIC} = 10^{-30}$

$N_A = 6.0 \times 10^{23}$

	$2NH_3$ (l) $\rightleftharpoons$	NH_4^+ +	NH_2^-
at t = 0	n	0	0
at equilibrium	n – 2x	x	x

$K_{SIC} = 10^{-30} = [NH_4^+][NH_2^-]$

$10^{-30} = x.x \Rightarrow 10^{-30} = x^2 \Rightarrow x = 10^{-15}$ M

$x = 10^{-15}$ moles/lit – $10^{-15} \times 6.0 \times 10^{23}$ no. of molecules/lit

No. of amide ions $[NH_2^-]$ in 1L = $6 \times 10^{23} \times 10^{-15}$

i.e. 1L = 6×10^8 molecules $\Rightarrow 10^3$ cm^3 = 6×10^8 molecules

1 cm^3 = $6 \times 10^8 \times 10^{-3}$ molecules = 6×10^5 molecules

Hence, no. of amide ions per cm^3 = 6×10^5 molecules

88. (1) pH = 1
pH = – log[H^+] ⇒ [H^+] = 10^{-pH}
10^{-1} M = 10^{-1} moles/lit
The mass of 1 mole of a substance in grams is called its molar mass.
Hence, (molar mass of HCl is 36.5) 1 mole HCl = 36.5 g
1 mole HCl = 36.5 ×.1 = 3.65 g

89. (2) pH = 12 ⇒ [H^+] = 10^{-12} i.e. [OH^-] = 10^{-2} M
molar mass of NaOH = 40 g/L
1 mole NaOH = 40 g
10^{-2} mole NaOH = 40 × 10^{-2} = 0.40 g

90. (2) pH = 3, ∴ [H^+] = 10^{-3} M
1 lit solution contains [H^+] = 10^{-3} moles
i.e., (1000 ml) solution contains [H^+] = 10^{-3} moles
10 ml will contain [H^+] = $\frac{10^{-3}}{1000}\times 10 = 10^{-5}$ moles
No. of [H^+] ions in 1 mole = 6.022 × 10^{23}
[H^+] ions 10^{-5} 1 mole = 10^{-5} × 6.022 × 10^{23} = 6.022 × 10^{18}

91. (3) In all the given options, only option (3) is not correct.
For acidic solution, pH should be less
than 7 means [H^+] > 10^{-7} M or [H^+] > [OH^-] or [OH^-] < 10^{-7} M

92. (3) Acetic acid (CH_3(COOH) is a weak acid it's dissociation equilibrium is shown as,
$CH_3COOH \rightleftharpoons CH_3COO^- + H^+$
When, NaOH is added into the solution, the hydroxide ion react H^+ ions to form water.
$H^+ + OH^- \rightarrow H_2O$
Thus, the concentration of H^+ ions decreases and pH of the solution increases.
Hence, the equilibrium for the dissociation of acetic acid will shift towards right, this will increase the acetate ion concentration.
Therefore, option 3 is correct.

93. (1) Methyl orange is an indicator, its colour below pH 3 is red and above pH 4.5 (approx) is yellow

94. (3) For weak acid and strong base;
$$pH = 7+\frac{1}{2}pK_a+\frac{1}{2}\log c = 7+\frac{4.74}{2}+\frac{1}{2}\log 10^{-1} = 8.87$$

95. (2) CH_3COONa being strong electrolyte ionizes into CH_3COO^- and Na^+ ions and it suppresses the ionization of CH_3COOH
(CHCOOH ⇌ $CH_3COO^- + H^+$)
thereby decreasing the H^+ ion concentration. This could increase pH to beyond 4.3.

96. (3) For anionic hydrolysis, pH is given by-
$$\frac{1}{2}pK_w+\frac{1}{2}pK_a+\frac{1}{2}\log c$$

97. (1) [H_3O^+] = $\sqrt{K_a \times c}$
Let conc. of acetic acid is C_1 and that of formic acid is C_2
$$\sqrt{1.8\times10^{-5}\times C_1} = \sqrt{1.8\times10^{-4}\times0.001}$$
or $$C_1 = \frac{1.8\times10^{-4}\times0.001}{1.8\times10^{-5}} = 0.01$$

98. (1) pH remains unchanged because the salt of strong acid and strong base does not undergo hydrolysis.

99. (1) H_2O is a weak electrolyte.
$H_2O \rightleftharpoons H^+ + OH^-$
as Temperature increases more and more H^+ come to solution, and same for OH^- K_w = [H^+] [OH^-]
Both increases, hence K_w increase.

100. (3) Ammonium acetate is a salt of weak acid and weak base. When solution are of weak acid and weak base we can calculate the pH of solution, from the following relationship $pH = 7+\frac{1}{2}pK_a-\frac{1}{2}pK_b$ [$pK_a \approx pK_b$ [given].]
∴ pH = 7

101. (1) The relative strength of weak acids
$$= \sqrt{\left(\frac{K_{a_1}}{K_{a_2}}\times\frac{C_1}{C_2}\right)}$$
∴ Relative strength = $\sqrt{\left(\frac{K_{a_1}}{K_{a_2}}\right)}$ (∵ $C_1 = C_2$)
$$= \sqrt{\left(\frac{2\times10^{-4}}{2\times10^{-5}}\right)}$$
Relative strength for HCOOH to CH_3COOH = $\sqrt{10}:1$

102. (4) Aqueous solution of HCOONa, $C_6H_5NH_3^+Cl^-$, KCN are:
HCOONa + H_2O ⇌ HCOOH (weak acid) + NaOH (strong base)
(basic solution)
$C_6H_5NH_3Cl_{(aq)}$ ⇌ $C_6H_5NH_2$ (weak base) + HCl (strong acid)
KCN + H_2O ⇌ KOH (strong base) + HCN (weak acid)
(basic solution)

103. (1) For hydrolysis of salt,
Hydrolysis constant,
$$K_h = \frac{K_w}{K_b} = \frac{1\times10^{-14}}{1.77\times10^{-15}} = 5.65\times10^{-10}$$

104. (3) [H^+] < 10^{-7} therefore, basic.

105. (2) Due to common ion effect, degree of ionisation of NH_4OH is suppressed by NH_4Cl.

106. (3) NH_4 OH + NH_4 Cl pair will show common ion effect.
Since; NH_4OH is a weak electrolyte whereas NH_4Cl is a strong electrolyte.

107. (1) Total milliequivalents of H^+ = $20\times\frac{1}{2}+30\times\frac{1}{3} = 20$
Total milliequivalents of OH^- = $40\times\frac{1}{4} = 10$

Thus; milliequivalents of H^+ left = 20 – 10 = 10

Total volume of solution = 1 dm^3 = 1000mL

$\therefore [H^+] = \frac{10}{1000} = 10^{-2}$

pH = –log $[H^+]$= –log $[10^{-2}]$ = 2

108. **(4)** Sodium chloride is a salt of strong base (NaOH) and strong acid, (HCl) In its aqueous solution, following equilibrium is observed:

$NaCl + H_2O \rightleftharpoons Na^+ + Cl^- + H_2O$

Hence; no hydrolysis takes place in this case, and the solution remains neutral. Thus; here the product of molar concentration of hydroxide and hydrogen ions is ionic product of water.

109. **(4)** pH = – log $[H^+]$

pH of solution = 9, $[H^+] = 10^{-9}$

New Hydrogen ion concentration is,

$[H^+] = 10^{-9} \times 10^3 = 10^{-6} \Rightarrow pH = -\log [H^+] = -\log [10^{-6}] = 6$

110. **(3)** H_2SO_4 is strong acid & it dissociates completely as $H_2SO_4 \rightarrow 2H^+ + SO_4^{2-}$

Concentration of H_2SO_4 = 0.05 × 2 = 0.1 M

pH = – log $[H^+]$ = – log $[10^{-1}]$ = 1

111. **(1)** An example of acidic buffer solution is a pair of HNO_2 and $NaNO_2$. This pair constitute a buffer because HNO_2 is weak acid and $NaNO_2$ is a salt of weak acid (HNO_2) with strong base (NaOH).

112. **(3)** $pOH = pK_b + \log\frac{[salt]}{[base]} = 4.74 + \log\frac{0.20}{0.30}$

$= 4.74 + (0.301 - 0.477) = 4.74 - 0.176 = 4.56$

pH = 10 – 4.56 = 9.44

113. **(4)** $pOH = pK_b + \log\frac{[salt]}{[base]}$

$pOH = 5 + \log\frac{0.01}{0.1} = 5 + \log(10^{-1}) = 5 - 1 = 4$

$pH = 14 - pOH \Rightarrow pH = 14 - 4 = 10$

114. **(3)** It is a mixture of weak acid and salt of a strong base.

115. **(2)** Buffer is a equimolar mixture of weak acid and its salt with strong base.

116. **(2)** The removal of PO_4^{3-} ions is done by Buffer solution.

117. **(2)** $pH = pK_a + \log\frac{[salt]}{[Acid]}$

$= -\log(1\times10^{-8}) + \log\frac{HX}{HX}$ ($[HX] = [X^-]$ given)

$= -\log 10^{-8} + 0 = 8$

118. **(4)** In mixture (before dilution)

$[CH_3COOH] = \frac{0.01\times100}{300} = \frac{0.01}{3}$ M

$[CH_3COONa] = \frac{0.02\times200}{300} = \frac{0.04}{3}$ M

$pH = pK_a + \log\frac{[salt]}{[acid]} = 4.74 + \log\frac{0.04}{3}\times\frac{3}{0.01}$

$= 4.74 + \log 4 = 4.74 + 0.60 = 5.34$

After further dilution

$[CH_3COOH] = \frac{0.01}{10}$

$[CH_3COONa] = \frac{0.04}{10}$

After dilution, pH still Remains same.

119. **(3)** There are two kinds of buffer solutions :

(1) Acidic buffers : A weak acid and the salt of its conjugate base.

Ex. 1.0 M CH_3COOH & 1.0 M CH_3 WO NA

(2) Basic buffer : A weak base and the salt of its conjugate acid.

ex. 1.0 M CH_3 & 1.0 M NH_4Cl.

Here, option (1) is not correct because, HCl is a strong acid. option (2) is also incorrect because, H_2CO_3 is a salt of weak acid and CH_3COOH is also a weak acid.

option (4) is again incorrect. because, NaCl is a salt of strong acid and NaOH is a strong base.

Hence, option (3) is correct as NaCN is a salt of weak acid & strong base and HCN is a weak acid. Therefore, these can act as a buffer.

120. **(2)** It is a mixture of weak acid and its salt with strong base.

121. **(3)** $NH_4OH + HCl \rightarrow NH_4Cl + H_2O$. (Here, NH_4OH is present in an excess amount).

So, the mixture contains a mixture of weak base and salt of strong acid ($NH_4OH + NH_4Cl$) which acts as a basic buffer solution.

122. **(1)** $pH = pKa + \log\frac{[Salt]}{[Acid]}$

This equation shows that on increasing the concentration of salt, pH will increase.

123. **(2)** The salt with highest K_{sp} is most soluble. Thus; FeS will be most soluble.

124. **(3)** $Ag_2CrO_4 \rightleftharpoons 2Ag^+ + CrO_4^{2-}$

$K_{sp} = [Ag^+]^2 [CrO_4^{2-}] = (2S)^2 (S) = 4S^3$

125. **(4)** $MX \rightarrow M^+ + X^-$ $\qquad K_{sp} = S^2$

S S $\qquad S = \sqrt{K_{sp}}$

$S = \sqrt{4\times10^{-8}} = 2\times10^{-4}$ M

$MX_2 \rightleftharpoons M^{+2} + 2X^-$

S 2S

$K_{SP} = (S)(2S)^2 = 4S^3$

$\therefore S = \left(\frac{K_{sp}}{4}\right)^{1/3} = \left(\frac{3.2\times10^{-14}}{4}\right)^{1/3} = 2\times10^{-5}$ M

$M_3X \rightleftharpoons 3M^+ + X^{3-}$ $\qquad K_{SP} = (3S)^3 (S) = 27S^4$

3S S

$\therefore S = \left(\frac{K_{sp}}{27}\right)^{1/4} = \left(\frac{2.7\times10^{-15}}{27}\right)^{1/4} = 10^{-4}$

$\therefore MX > M_3X > MX_2$

$2\times10^{-4} > 10^{-4} > 2\times10^{-5}$

126. (3) $AX_2 \rightleftharpoons A^{2+} + 2X^-$

S 2S

$K_{sp} = [A^{2+}][X^-]^2 = S \times (2S)^2 = 4S^3$

$$S^3 = \frac{K_{sp}}{4} = \frac{3.2\times10^{-11}}{4} = 8 \times 10^{-12}$$

$S = 2 \times 10^{-4}$ mol/L

127. (3) $A_3B_{2'} \rightleftharpoons 3A^{2+} + 2B^{-3}$

$K_{sp} = (3S)^3.(2S)^2 = 27 \times 4 \times S^5 = 108\ S^5 = 108\ x^5$

128. (2) $A_2B \rightleftharpoons 2A^+ + B^{-2}$

2S S

$$K_{sp} = (2S)^2 \times S \Rightarrow S = \sqrt[3]{\frac{K_{sp}}{4}}$$

129. (3) CaF_2 is a AB_2 Type of salt

$CaF_2 \rightarrow Ca^{+2} + 2F^-$

S 2S

$K_{sp} = [Ca^{+2}][F^-]^2$

S 2S

$K_{sp} = 4S^3$; $S = 2\times10^{-4}$

$K_{sp} = 4(2 \times 10^{-4})^3 = 4\times8\times10^{-12}$

130. (4) $AgCl \rightleftharpoons Ag^+ + Cl^-$

x x

$\because$ NaCl is strong electrolyte and it completely ionised it shall provide Cl^- ion conc^n 0.1 M.

$\therefore$ Total conc^n of Cl^- $[Cl^-] = x + 0.1$

$K_{sp} = x.(x + 0.1) = x^2 + 0.1\ x$

$[x <<< 1]$

$\therefore x^2$ is neglected

$0.1x = 1.2\times10^{-10}$

$$x = \frac{1.2\times10^{-10}}{0.1} = 1.2 \times 10^{-10} = 1.2 \times 10^{-9}$$

High Order Time Intensive MCQs

1. (1) $A \underset{K_2}{\overset{K_1}{\rightleftharpoons}} B \qquad K_C = \frac{K_1}{K_2}$

$r_f = K_1[A] = K_1(a - x)$

$r_b = K_2[B] = K_2(b + x)$

At equilibrium (Rate of forward Reaction = Rate of backward Reaction)

$r_f = r_b$

$K_1(a - x) = K_2(b + x)$

$$\frac{b+x}{a-x} = \frac{K_1}{K_2}$$

$K_2b + K_2x = K_1a - K_1x$

$K_2x + K_1x = K_1a - K_2b$

$x(K_2 + K_1) = K_1a - K_2b$

$$x = \frac{K_1a - K_2b}{K_2 + K_1}$$

2. (1) $N_2(g) + 3H_2(g) \rightleftharpoons 2NH_3(g)$

At t = 0	0.2	0.6	0
At t = teq	0.2–x	0.6–3x	2x

N_2 & H_2 mixture has 0.8 moles. 40% of its reacts So, mixture reacted $= 0.8\times\frac{40}{100} = 0.32$ moles

At equilibrium, n moles N_2 and 3n moles H_2 have reacted

$\therefore 4n = 0.32$

$$n = \frac{32}{4}\times\frac{1}{100} = 0.08$$

Total number of final gaseous moles

$= 0.2 - x + 0.6 - 3x + 2x$ (at eq)

$= 0.8 - 4x + 2x = 0.8 - 2x = 0.8 - 2 \times 0.08$

$= 0.8 - 0.16 = 0.64$ moles

Total number of initial gaseous moles = 0.8

Apply PV = nRT,

$P_fV_f = n_fRT$

$P \times V_f = 0.64 \times R \times T$ ... (1)

$P \times V_i = 0.8 \times R \times T$ ...(2)

Divide 1 & 2

$$\frac{V_f}{V_i} = \frac{0.64}{0.8} = \frac{64}{8}\times\frac{10}{100} = \frac{8}{10} = \frac{4}{5}$$

3. (2) $2NOBr(g) \rightleftharpoons 2NO(g) + Br_2(g)$

t = 0	p	0	0
$t = t_{eq}$	p – 2x	2x	x

$P = p - 2x + 2x + x$

$P = p + x$ $\left[\text{and } x = \frac{P}{9}\right]$

$$K_p = \frac{\left(2\times\frac{P}{9}\right)^2\left(\frac{P}{9}\right)}{\left(\frac{2P}{3}\right)^2}$$

$$K_p = \frac{\frac{4\times P^3}{9\times9\times9}}{\frac{4P^2}{9}} \Rightarrow \frac{4\times P^3\times9}{9\times9\times9\times4\times P^2}$$

$$K_p = \frac{P}{81}$$

4. (1) $N_2(g) + 3H_2(g) \rightleftharpoons 2NH_3(g)$

At t = 0	1	3	0
At t = equilibrium	1 – x	3 – 3x	2x

$$K_c = \frac{[NH_3]^2}{[N_2][H_2]^3} = \frac{(2x)^2}{(1-x)(3-3x)^3}$$

In terms of pressure, total no. of moles at equilibrium

$= 1 - x + 3 - 3x + 2x = 4 - 2x$

If P is the total pressure of the reaction mixture,

$$p_{N_2} = \frac{1-x}{4-2x} \times P$$

$$p_{H_2} = \frac{3-3x}{4-2x} \times P$$

$$p_{NH_3} = \frac{2x}{4-2x} \times P$$

Substitute in the expression

$$K_p = \frac{p^2_{NH_3}}{p_{N_2} \cdot p^3_{H_2}}$$

$$K_p = \frac{\left(\frac{2x}{4-2x}P\right)^2}{\left(\frac{1-x}{4-2x}P\right)\left(\frac{3-3x}{4-2x}P\right)^3}$$

x is very small

w. r. to 1 (x <<< 1)

$$K_p = \frac{4x^2.P^2}{(4-2x)^2} \times \frac{(4-2x)^4}{(1-x)(3-3x)^3 P^4}$$

$$K_p = \frac{4x^2P^2}{P^4(1-x)(3-3x)^3} \times (4-2x)^2$$

$$K_p = \frac{4x^2P^2}{P^4 \times 1 \times 27} \times 16$$

$$K_p P^2 \times \frac{27}{64} = x^2$$

$$\frac{3P\sqrt{3K_p}}{8} = x$$

5. (2) $A(g) + B(g) \rightleftharpoons C(g) + D(g)$

At t = 0	x	x	0	0
At t = eq	x–y	x–y	y	y

According to the given conditions, at equilibrium (x–y) = 2y for both A and B

$$K_C = \frac{y \times y}{(x-y)(x-y)} = \frac{y^2}{2y \times 2y} = \frac{y^2}{4y^2}$$

$$K_C = \frac{1}{4}$$

$$\& \; K_C = \frac{K_f}{K_b} \qquad [K_f = 2 \times 10^{-3} \text{ is given}]$$

$$\frac{1}{4} = \frac{2 \times 10^{-3}}{K_b}$$

$K_b = 8 \times 10^{-3} mol^{-1}\, L\, sec^{-1}$

6. (1) Helium is an inert gas. It will not react with Neither reactant nor products.

Therefore, there will be no change on equilibrium.

7. (4) $AB(g) \rightleftharpoons A(g) + B(g)$ [x = degree of dissociation]

t = 0	P	0	0
$t = t_{eq}$	P – Px	Px	Px

Total equilibrium pressure = P + Px

$$K_P = \frac{Px \times Px}{P - Px}$$

$$\frac{P+Px}{3} = \frac{P^2x^2}{P-Px}$$

$$P^2 - P^2x^2 = 3P^2x^2$$

$$P^2 = 4P^2x^2$$

$$x^2 = \frac{1}{4} \quad \left[x = \frac{1}{2}\right]$$

x = 50%

8. (3) $N_2 + 3H_2 \rightleftharpoons 2NH_3$

At t = 0	1	3	0
At t = teq	(1 – α)	(3 – 3α)	2α

Initially, = 4

At equilibrium, Degree of dissociation α = 0.2

Moles of N_2 = 1 – α = 1 – 0.2 = 0.8

Moles of H_2 = 3 – 3α = 3 – 3 (0.2) = 3 – 0.6 = 2.4

Moles of NH_3 = 2α = 2 × 0.2 = 0.4

Total No. of moles at equilibrium = 0.8 + 2.4 + 0.4 = 3.6

Using ideal gas equation

PV = nRT

Initial, P_1 = 100 atm, n_1 = 4 moles

$P_1V = n_1RT$ ...(1)

Final P_2 = ? , n_2 = 3.6 moles $P_2V = n_2RT$ ... (2)

from (1) & (2)

$$\frac{P_1V}{P_2V} = \frac{n_1 RT}{n_2 RT} \Rightarrow \frac{100}{P_2} = \frac{4}{3.6} \Rightarrow P_2 = \frac{100 \times 3.6}{4} = 90 \text{ atm.}$$

9. (2) The equation given in question is,

$$\log\left(\frac{K_P}{K_C} + RT\right) = 0$$

$$\log\left(\frac{K_P}{K_c} + RT\right) = 0$$

$$\log\frac{K_P}{K_c} = -\log RT$$

$$\frac{K_P}{K_c} = -RT$$

$$\frac{K_P}{K_c} = \frac{1}{RT}$$

$K_p = K_c (RT)^{-1}$

∴ $\Delta n_g = -1$ it is valid for option 2 only.

10. (3) Variation of equilibrium constant with temperature is given by Van't Hoff equation.

$$\log\frac{K_2}{K_1} = \frac{\Delta H}{2.303R}\frac{(T_2 - T_1)}{T_1T_2}$$

11. (1) $A_2(g) + 3B_2(g) \rightleftharpoons 2AB_3(g)$

Initial conc.	n	3n	0
At eqm.	n(1-x)	3n(1-x)	2nx

Mole fraction at eqm.	$\frac{1-x}{4-2x}$	$\frac{3(1-x)}{4-2x}$	$\frac{2x}{4-2x}$

($\because$ total no. of moles $= n(1-x)+3n(1-x)+2nx = n(4-2x)$)

Also, $\chi_{AB_3} = \frac{2x}{4-2x} = 0.6$

$x = 0.75$

So, $\chi_{A_2} = 0.1$ & $\chi_{B_2} = 0.3$

Use the expression, $K_p = \frac{(pAB_3)^2}{(pA_2)(pB_2)^3} = 3.33\times10^{-3}$

12. (3) $K_p = \frac{P_{CH_3OH}}{P_{CO}\times(p_{H_2})^2} = \frac{2}{1\times(0.1)^2} = 200$

K_p for reverse reaction will be $= \frac{1}{K_p} = \frac{1}{200} = 5\times10^{-3}\,atm^2$

13. (2) $CH_3COCH_3(g) \rightleftharpoons CH_3-CH_3(g)+CO(g)$

At t = 0	100	0	0
at eq	100 – P	P	P

Since, rest all parameters are constant (V, T), we can say that

$P_{CO_{(at\ eq)}} = \frac{1}{3}P_{Total_{(at\ eq)}}$

$P = \frac{1}{3}(100-P+P+P) \Rightarrow P = \frac{1}{3}(100+P)$

$3P = 100 + P \Rightarrow 2P = 100$

P = 50 mm of Hg

$K_p = \frac{(PCH_3-CH_3)(PCO)}{(PCH_3COCH_3)} = \frac{50\times50}{(100-50)} = \frac{50\times50}{50} = 50$ mm

14. (1) $CaCO_3 \rightleftharpoons CaO + CO_2$

(s) (s) (g)

$K_p = p_{CO_2} = 1.16$ atm

Apply, $PV = nRT \Rightarrow n = \frac{PV}{RT}$

$\frac{1.16\times10}{0.0821\times1073} = 0.132$ moles

Initial moles of $CaCO_3 = \frac{20}{100} = 0.2$

Mole of $CaCO_3$ reacted = 0.132

Mole of $CaCO_3$ unreacted = 0.2 – 0.132 = 0.068

as weight of $CaCO_3$ is calculated $= \frac{W}{M} = n$

W = n × M (moles)

(weight) mass

So, weight of Ca CO_3 unreacted = 0.068 × 100 = 6.8

% age of $CaCO_3$ unreacted by weight $= \frac{6.8}{20}\times100 \approx 35\%$

15. (3) $2HI \rightleftharpoons H_2 + I_2$

Initial	2 moles	0	0
At eqm.	$2-\frac{20}{100}\times2$	0.2	0.2

$= 2 - 0.4 = 1.6$

$K = \frac{[H_2][I_2]}{[HI]} = \frac{0.2\times0.2}{(1.6)^2} = \frac{1}{64}$

16. (4) $\log\frac{K_{p_2}}{K_{p_1}} = \frac{\Delta H}{2.303R}\left(\frac{1}{T_1}-\frac{1}{T_2}\right)$

$\log\frac{K_{p_2}}{1.6\times10^{-4}} = \frac{-25.14}{2.303\times2\times10^{-3}}\;\frac{(773-673)}{773\times673}$

$K_{p2} = 1.46\times10^{-5}$ atm

$K_{p2} = P_{O2} = 1.46\times10^{-5}$ atm

Apply, PV = n RT

$1.46\times10^{-5}\times2 = n\times0.0821\times773$

$n = \frac{1.46\times10^{-5}\times2}{0.0821\times773} = \frac{2.92\times10^{-5}}{63.46} = 0.046\times10^{-5}$

17. (4) Since inert gas addition has no effect at const. volume.

18. (4) (A) Equilibrium constant depends on temperature.

(B) Equilibrium constant is a constant.

(C) p_{NH_3} and p_{H_2S} depend on K.

$K = p_{NH_3}\times p_{H_2S}$

so if at equilibrium p_{NH_3} is increased, then automatically p_{H_2S} will decrease.

19. (2) $b(MV)_A = 2\times\frac{1}{5}\times200 = 80$ millimoles

$a(MV)_B = 1\times\frac{1}{2}\times150 = 75$ millimoles

$\Rightarrow$ Acid dominates, pH < 7

20. (2) Since slope is negative hence reaction is endothermic. So high temperature favours forward reaction similarly high pressure favours forward reaction.

21. (2) $\Delta G° = -RT\ln K_C$

$K_C = \frac{\Delta G°}{-RT}$

$\Delta G° = 13.8$ kJ/mol $= 13.8\times10^3$ J mol^{-1}

T = 298 K

R = 8.314 JK^{-1} mol^{-1}

$\ln K_C = -\frac{13.8\times10^3\ J\,mol^{-1}}{298K\times8.314\ JK^{-1}\,mol^{-1}}$

$= -\frac{13.8\times10^3}{298\times8.314} = -5.56997$

$K_C = e^{-5.56997}$

22. (1) $2NH_3(g) \rightleftharpoons N_2(g) + 3H_2(g)$

$NH_3(g) \rightleftharpoons \frac{1}{2}N_2(g) + \frac{3}{2}H_2(g)$

Initially : a mole 0 0

At equi. (a–x) mole $\frac{x}{2}$ mole $\frac{3x}{2}$ mole

At equilibrium, total moles = $a - x + \frac{x}{2} + \frac{3x}{2} = (a + x)$

Partial pressure $\frac{(a-x)}{(a+x)} \times P, \quad \frac{x}{2(a+x)} \times P, \quad \frac{3x}{2(a+x)} \times P$

$$\therefore K_P = \frac{(p_{N_2})^1 \times (p_{H_2})^3}{(p_{NH_3})^3} = \frac{27p^2x^4}{16(a-x)^2(a+x)^2}$$

23. (1) The salt $Ca(RCOO)_2$ is a salt of weak acid and strong base. Such salts undergo anionic hydrolysis i.e., it will be $RCOO^-$ that undergoes hydrolysis.

Now, concentration of the salt = $\frac{0.6}{2} = 0.3$ mol / ltr.

If it is dissociated to the extent of 90%-

$$(RCOO)_2Ca \rightarrow 2RCOO^- + Ca^{2+}$$

at, t = 0 0.3 0 0

at equilibrium $2 \times \frac{90}{100} \times 0.3 = 0.54$ mol ltr^{-1}

$pH = \frac{1}{2}(pK_w + pK_a + \log[C])$ (pH + pOH = 14)

$8.4 = \frac{1}{2}(14 + pK_a + \log 0.54)$ (pH + 5.60 = 14, pH = 8.4)

$pK_a = 2.8 - \log 0.54$

24. (1) For weak acid,

$$HA \rightleftharpoons H^+ + A^-$$

(aq) (aq) (aq)

At t = 0 C(0.1) 0 0

At eq C–Cα Cα Cα

$$K_a = \frac{(H^+)(A^-)}{(HA)}$$

for strong Base complete dissociation will occur.

$HB \rightarrow H^+ + B^-$

0.1 0 0

0 0.1 0.1

(H^+) in the solution = 0.1M ($\because$ of strong acid)

K_a will remain same

$$\therefore 10^{-6} = \frac{(0.1)(C\alpha)}{(C - C\alpha)}$$

$$10^{-6} = \frac{0.1(C\alpha)}{C(1-\alpha)}$$

$$10^{-6} = \frac{0.1 \times \alpha}{1-\alpha}$$

$1 - \alpha \approx 1$

$10^{-6} = 0.1 \times \alpha \Rightarrow 10^{-6} = 10^{-1} \times \alpha \Rightarrow \alpha = 10^{-5}$

25. (1) Case I : 1 mole CH_3COONa + 1 mole HCl.

Acid reacts with salt as-

$$CH_3COONa + HCl \rightarrow CH_3COOH + NaCl$$

1 mole 1 mole

X X 1 mole

The resultant solution will be acidic due to acetic acid.

Apply, $pH = \frac{1}{2}(pK_a - \log c)$ (Formula to calculate pH of weak monoprotic acid)

$$pH = \frac{1}{2}(pK_a - \log\frac{1}{1})$$

$pH_1 = \frac{1}{2}pK_a$ ......(1)

Case II : 1 mole CH_3COONa + 1 mole CH_3COOH

This mixture will act as Acidic Buffer

So, $pH = pK_a + \log\frac{[Salt]}{[Acid]}$

Since, (Salt) = (Acid)

$pH_2 = pK_a$.....(2)

The ratio $\frac{pH_1}{pH_2} = \frac{\frac{pK_a}{2}}{pK_a} = 1:2$

26. (4) Apply, $pOH = pK_b + \log\frac{[Salt]}{[Base]}$

$$pOH = pK_b + \log\frac{0.1}{0.1}$$

$pOH = pK_b = 4.74$

$pH = 14 - 4.74 = 9.26$

After 0.05 moles of NaOH are added, it reacts with salt

$$NH_4Cl + NaOH \rightarrow NaCl + NH_4OH$$

	NH_4Cl	NaOH	NaCl	NH_4OH
At t = 0	0.1	0.05	0	0
At t = t	0.1-0.05	0	0.05	0.05

Concentration of NH_4OH now = 0.1 + 0.05 = 0.15

Concentration of NH_4Cl now = 0.05

Apply Handerson's equation again,

$$pOH = pK_b + \log\frac{[Salt]}{[Base]}$$

$$pOH = pK_b + \log\frac{0.05}{0.15}$$

pOH = 4.27

pH = 9.73

$\Delta pH = 9.73 - 9.26 = 0.47$

27. (1) $Ba(OH)_2 \rightleftharpoons Ba^{2+} + 2OH^-$

pH of solution = 12

pOH = 2

$(OH) = 10^{-2}$M

$(Ba^{2+}) = \frac{1}{2} \times 10^{-2}$M

$$K_{sp} = [Ba^{2+}][OH^-]^2 = \left(\frac{10^{-2}}{2}\right)(10^{-2})^2$$

$$= \frac{10^{-2}}{2} \times 10^{-4} = 0.5 \times 10^{-6}$$

$= 5 \times 10^{-7}$ M^3

28. (1) $pH = 1 \Rightarrow (H^+) = 10^{-1}$ M

Which means 10^{-1} moles of H^+ are present in 1 litre of solution

pH = 2 is desired. So $(H^+) = 10^{-2}$ M

$$10^{-2}\ (M) = \frac{10^{-1}}{V(L)}(\text{no. of mole of } H^+)$$

$V = 10^{-1} \times 10^{2} = 10$ L

i.e., now the volume of solution = 10 L

Initially, it was 1L, so 9L of water is required to be added.

29. (3) $HClO_4$ is a very strong acid

So, $HClO_4 \rightarrow H^+ + ClO_4^-$

$[H^+] = 2 \times 10^{-3}$ M

These 2×10^{-3} moles of H^+ are present in 1 litre of solution initially. Now, when $KClO_4$ is mixed along (which is a salt of strong base & strong acid, the resultant solution of which is neutral). The contribution of $KClO_4$ will just be in terms of volume.

$$\text{So, now new } [H^+] = \frac{2\times10^{-3}}{2} = 10^{-3}$$

$[H^+] = 10^{-3}$ M

pH = 3

30. (4) $Pb(OH)_2 \rightleftharpoons Pb^{2+} + 2OH^-$

s 2s

$$K_{sp} = [Pb^{2+}][OH^-]^2 = (6\times10^{-6})(2\times6\times10^{-6})^2$$
$$= 6\times10^{-6}\times144\times10^{-12}$$
$$= 864\times10^{-18} M$$

Now, buffer has pH = 8

pOH = 6

$[OH^-] = 10^{-6}$ M

So, buffer will maintain $[OH^-] = 10^{-6}$ M

K_{sp} of $Pb(OH)_2$ will remain constant

$K_{sp} = [Pb^{2+}][OH^-]^2$

$864 \times 10^{-18} = [Pb^{2+}][10^{-6}]^2$

$$\frac{864\times10^{-18}}{10^{-12}} = (Pb^{2+})$$

$864 \times 10^{-6} = (Pb^{2+})$

31. (4) $A_3B_2 \rightleftharpoons 3A^{2+} + 2B^{3-}$

3s 2s

$K_{sp} = [A^{2+}]^3 [B^{3-}]^2 = (3s)^3 (2s)^2 = 27s^3 \times 4s^2 = 108\ s^5$

$$\text{Ratio of } \frac{(B^{3-})}{K_{sp}} = \frac{2s}{108s^5} = \frac{1}{54s^4}$$

Now, solubility is given = X g/L

$$S = \frac{X}{M}\text{ mol / ltr}$$

Where M = Molar mass of A_3B_2

$$\text{Ratio} = \frac{1}{54s^4} = \frac{1}{54\left(\frac{X}{M}\right)^4} = \frac{1}{54X^4}\times M^4 = \frac{M^4}{54X^4}$$

32. (3) Na_2CO_3 is a salt of weak Acid and strong Base.

33. (4) $Ag_2S \rightleftharpoons 2Ag^+ + S^{2-}$

$K_{sp} = [Ag^+]^2 [S^{2-}]$

$$[Ag^+]^2 = \frac{K_{sp}}{[S^{2-}]}$$

$$[Ag^+] = \left(\frac{K_{sp}}{S^{2-}}\right)^{1/2} = \left(\frac{10^{-51}}{0.1}\right)^{1/2} = 10^{-25} M$$

34. (4) $[H^+] = y \times 10^{-x}$

$\log [H^+] = \log y - x$

$pH = x - \log y$

(B) $[H^+] = \frac{1}{y}\times10^{-x}$

$\log [H^+] = -x - \log y$

$pH = x + \log y$

(C) pH = 14 – pOH

or

pH = 14 + log OH

All are correct.

35. (2) $AgIO_3(s) \rightleftharpoons Ag^+(aq.) + IO_3^-\ (aq.)$

Let the solubility of $AgIO_3$ be S

$K_{sp} = [Ag^+] [IO_3^-]$

$S = (K_{sp})^{1/2}$

$= (1 \times 10^{-8})^{1/2}$

$= 1 \times 10^{-4}$

In 1000 mL of $AgIO_3$ dissolved = 1×10^{-4} mol

In 100 mL of mole of $AgIO_3$ dissolved = 1×10^{-5} mol

Mass of $AgIO_3$ in 100 mL = $1 \times 10^{-5} \times 283 = 2.83 \times 10^{-3}$ g

36. (2) Higher the value of K_{sp}, higher will be its solubility.

37. (1) $PCl_5 \rightleftharpoons PCl_3 + Cl_2$

α = 0.2, initially,

$$K_p = \frac{\alpha^2}{1-\alpha^2}P = \frac{(0.2)^2}{1-(0.2)^2}\times1 = \frac{0.04}{0.96} = 0.042$$

$$\text{If } \alpha = 0.5, \text{ thus, } \frac{(0.5)^2}{1-(0.5)^2}\times P = 0.042,\ P = 0.126$$

38. (3) $[Ag^+] [CO_3^{2-}] - K_{sp}$

$(2s)^2 \times 0.1 = 4 \times 10^{-13}$

$(2s)^2 = 4 \times 10^{-12}$

$2s = 2 \times 10^{-6}$

$s = 1 \times 10^{-6}$ M

39. (1) $$[Ba^{2+}] = \frac{K_{sp}}{[CO_3^{2-}]} = \frac{5\times10^{-9}}{5.1\times10^{-4}}$$

$[Ba^{2+}] = 5.1 \times 10^{-5}$ M

40. (3) $Mg(OH)_2 \rightleftharpoons Mg^{2+} + 2OH^-$

$K_{sp} = [Mg^{2+}] [OH^-]^2$

$$[OH^-] = \sqrt{\frac{K_{sp}}{[Mg^{2+}]}} = 10^{-4}$$

pOH = 4, and pH = 10

41. (2) 10^{-2} M Ca^{2+} and 10^{-3} M F^-

42. (1) $K_C = \frac{[\text{Products}]}{[\text{Reactants}]}$

$$K_C = \frac{[(FeSCN)^{2+}]}{[Fe^{3+}][SCN^-]}$$

43. (1) $pH = pK_a + \log\frac{[\text{salt}]}{[\text{Acid}]}$

$$4 = 4 + \log\frac{[\text{salt}]}{(0.5)} \Rightarrow \frac{[\text{salt}]}{(0.5)} = 1$$

[salt] = 0.5 M

$K_a = 10^{-4} \Rightarrow pK_a = -\log 10^{-4} = -(-4)\log 10 = 4$

$$pH = pK_a + \log\frac{[\text{salt}]}{[\text{Acid}]}$$

$$6 = 4 + \log\frac{[\text{salt}]}{(0.5)} \Rightarrow \frac{[\text{salt}]}{(0.5)} = 100 \Rightarrow [\text{salt}] = 50\,M$$

After mixing,

$$pH = pK_a + \log\frac{[\text{salt}]}{[\text{acid}]}$$

$$pH = 4 + \log\frac{\left(\frac{50.5}{2}\right)}{\left(\frac{1}{2}\right)}$$

pH = 4 + log 50.5

pH = 5.703

44. (3) $pOH = 7 - 0.5\,pK_a + 0.5\,pK_b$

or

$pH = 7 + 0.5\,pK_a - 0.5\,pK_b$ is valid for salt of wA and wB.

45. (3) ${}^{n}CH_3COOH = 0.1 \times 0.1 = 0.01$ mol.

${}^{n}HCl = 0.01$ mol

$$[H^+]_f = \frac{0.01}{0.1} = 0.1\,M$$

pH = 1

$$\begin{array}{lcl} CH_3COOH & \rightleftharpoons & CH_3COO^- + H^+ \\ 0.1 - x & & x \quad (x + 0.1) \end{array}$$

$$2\times10^{-5} = \frac{x \times 0.1}{0.1}$$

$x = 2 \times 10^{-5}$

$$\alpha\, CH_3COOH = \frac{2\times10^{-5}}{0.1} = 2\times10^{-4}$$

$H_2O \rightleftharpoons H^+ + OH^-$, $10^{-14} = 0.1 \times [OH^-]$

$[OH^-] = 10^{-13}$ M

$$\alpha(H_2O) = \frac{10^{-13}}{\frac{1000}{18}} = 18 \times 10^{-16} = 1.8 \times 10^{-15}$$

46. (3) $K_{H_2O} = \frac{[H^+][OH^-]}{[H_2O]} = \frac{10^{-12}}{55.55}$

$\therefore K_{H_2O} = 1.8\times10^{-14}$

47. (3) Milk of magnesia will have highest pH i.e. 10.5.

48. (3) If total pressure is decreased, equilibrium will shift in the direction in which no. of moles of gas will increase.

$N_2O_2 \rightleftharpoons 2NO_2$

49. (3) Formation of NH_3 is exothermic. When the temperature is increased, the reaction will proceed in the backward direction so as to absorb heat and nullify the effect of increasing temperature. In other words, the equilibrium will shift to left.

50. (1) $M_1V_1 = M_2V_2$ for Ca^{2+} ion

$50 \times 0.04 = M_2 \times 200$ (Total volume 50 + 150 = 200)

$$M_2 = \frac{50\times0.04}{200} = 0.01$$

Therefore, $[Ca^{2+}] = 0.01$

$[SO_4^{-2}] = M_1V_1 = M_2V_2$

$150 \times 0.0008 = M_2 \times 200$

$$M_2 = \frac{150\times0.0008}{200} = 0.0006$$

$[SO_4^{-2}] = 0.0006$

$K_{IP} = [Ca^{2+}] \times [SO_4^{-2}] = (0.01) \times (0.0006) = 6 \times 10^{-6}$

Therefore, $K_{sp} > K_{IP}$

51. (4) $CH_3COOH + NaOH \rightarrow CH_3COONa + H_2O$

1	0.5	0	0
0.5	0	0.5	0.5

This solution contains a weak acid and salt with a strong base and thus, acts as a buffer.

52. (3) I and II takes place with the absorption of heat energy.

53. (4) For reaction (1) $K_1 = \frac{[HA^-][H^+]}{[H_2A]}$

For reaction (2) $K_2 = \frac{[A^{2-}][H^+]}{[HA^-]}$

For reaction (3) $K = \frac{[H^+]^2[A^{2-}]}{[H_2A]}$

Multiplying K_1 & K_2, we get,

$$K_1.K_2 = \frac{[HA^-][H^+]}{[H_2A]} \times \frac{[A^{2-}][H^+]}{[HA^-]} = \frac{[H^+]^2[A^{2-}]}{[H_2A]} = K$$

54. (4) Due to common ion effect, ionisation decreases.

55. (2) pH of water decreases with increase in temperature and with addition of acid.

56. (4) (A) K_w is temperature dependent & it increases with increase in temperature.

(B) pH of $[H^+]$ of 1M is zero.

(C) pOH = 14 – pH is always correct at 25°C

57. (2) Common ion effect

58. (1) Decomposition of $CaCO_3$ is endothermic reaction.

The forward reaction is favored by increasing temperature.

Thus, when the temperature is increased, more carbon dioxide is formed which increases the pressure of carbon dioxide until it reaches a constant value.

59. (2) [solid] = 1

60. (2) Equilibrium constant of forward reaction is the inverse of backward reaction.

First reaction is reversed and added to the second reaction to obtain the third reaction.

Hence, the equilibrium constant for the third reaction is the ratio of the equilibrium constants of second and first reaction.

61. (3) $2AB \rightleftharpoons A_2 + B_2; K_1 = 49$

$$AB \rightleftharpoons \frac{1}{2}A_2 + \frac{1}{2}B; K_2 = ?$$

$$K_2 = \sqrt{K_1}$$

62. (2) Compound having low K_{sp} precipitated first.

63. (1) The solubility decreases by common ion effect.

A= in water solubility of AgCl is very less

$B[Cl^-] = 0.1$

$C[Cl^-] = 0.2$

D = Soluble complex compound is formed

$\because$ $[Ag(NH_3)_2]Cl$

$\therefore$ Solubility of AgCl is D > A > B > C

64. (2) Buffer capacity= No of moles of an acid per litre of solution/ Change in pH

pH change = (8/40)/0.2

= 0.2/0.2

=1

65. (1) K_2SO_4 is a salt and it forms from a strong base KOH and strong acid H_2SO_4. Therefore, it will be neutral salt, hence pH will not be affected on addition of water.

66. (2) Sulphuric acid is a diprotic acid, meaning it can donate two protons in a stepwise manner:

$H_2SO_4 \rightarrow H^+ + HSO_4^-$ ...(K_{a_1})

$HSO_4^- \rightleftharpoons H^+ + SO_4^{2-}$ ...(K_{a_2})

A. H_2SO_4 is a **stronger acid** than HSO_4^-, since its ionization is complete (strong acid), whereas HSO_4^- is only partially ionized (weak acid).

B. The overall ionization constant K is the product of the two steps:

$K = \log K_{a_1} \cdot K_{a_2} + \Rightarrow \log K = \log K_{a_1} + \log K_{a_2}$

C. $K_{a_1} = K_{a_2}/2$

There is no such direct mathematical relationship between K_{a_1} and $K_{a_2} \cdot K_{a_1}$ is significantly larger.

D. $K_{a_1} > K_{a_2}$

67. (4) The equilibrium constant for this reaction is: $K = \frac{K_f}{K_b}$

Given $K_b = 5000 \times K_f$

So, $K_c = K_f/5000\ K_f$

$K_c = 0.0002$

For the given reaction, $\Delta n_g = 2$

$K_p = K_c(RT)^{\Delta n_g}$

$K_p = 0.0002(0.0831 \times 1200)^2 = 1.98$

Correct & Incorrect MCQs

1. (1) Volatile liquids have higher vapour pressure than other liquid at the same temperature. More volatile liquid has lower boiling point and hence can be vaporised easily.

2. (2) Since, there is a difference in the solubility of carbon dioxide at different pressure, hence when a soda water bottle is opened, some of the carbon dioxide gas dissolved in it fizzes out rapidly.

3. (1) Equilibrium in a system having more than one phase is known as heterogeneous equilibrium. The equilibrium between water vapour and liquid water in a closed container is an example of heterogeneous equilibrium.

4. (4) Equilibrium constant is temperature dependent having one unique value for a particular reaction represented by a balanced equation at a given temperature.

5. (3) If $K_c > 10^3$, products predominate over reactants. If $K_c < 10^{-3}$, reactants predominate over products. If K_c is in the range of 10^{-3} to 10^3, appreciable concentrations of both reactants and products are present.

6. (2) if $Q_c < K_c$, net reaction goes from left to right and if $Q_c > K_c$, net reaction goes from right to left. If $Q_c = K_c$, no net reaction occurs.

7. (3) The conjugate base of H_2O is OH^-.

8. (3) The equilibrium constant for an exothermic reaction (negative ΔH) decreases as the temperature increases.

9. (4) NH_3 can donate a pair of electrons, hence acts as a Lewis base.

10. (2) Since, $K_p = K_c(RT)^{\Delta n}$ depends on the number of moles of gaseous reactant and products, hence K_p can be equal to or greater than or less than K_c.

Equilibrium constant changes on changing the temperature.

Statement Based MCQs

1. (3) K_C & K_P are equilibrium constants of gaseous mixture, defined by molar concentration and partial pressures of gases inside a closed system, which are further following a linear relation i.e. partial pressure is directly proportional to molar concentration.

2. (4) $Q_c < K_c$ then reaction move in direction of products.

3. (3) HCl being strongest acid undergoes disssociation as compared to acetic acid. In equimolar solution, no. of tilrable proton in HCl is greater than present in acetic acid.

4. (2) Since NaCl completely dissociates in aqueous solution and a neutral solution is obtained. But Na_2CO_3 is alkaline in nature, it is a salt of weak acid and strong base i.e. carbonic acid and NaOH and the salt undergoes hydrolysis in water to produce OH^- ions. NaCl does not undergo hydrolysis.

5. (1) Gaseous moles of reactant = gaseous moles of product and since $K_p = K_c(RT)^{\Delta n}$ & $\Delta n = 0$, So, $K_p = K_c$

6. (4) Since, molar concentration of pure solid or liquid is constant, hence the concentration of partial pressure of pure solid or liquid do not appear in the expression of the equilibrium constant.

7. (3) Production of ammonia is an exothermic reaction. While production of ammonia, if temperature is raised then the equilibrium shift towards left.

8. (3) As size of X(X = F, Cl, Br, I) increases, H – X bond strength decreases, hence the acidic strength increases as HF < HCl < HBr < HI.

9. (1) Both statements are individually correct.

10. (4) As, NH_4Cl is a salt of weak base NH_4OH and strong acid HCl whereas CH_3COONa is salt of a weak acid CH_3COOH and strong base NaOH, hence the pH of NH_4Cl solution in water is less than 7 and pH of CH_3COONa solution is more than 7.

Assertion & Reason MCQs

1. (1) As size increases, the strength of H – X Bond decreases. Hence, the acidity increases.

2. (2) Both are correct but reason is not correct explanation because H_2S already a weak acid and its dissociation is further suppressed by strong acid HCl.

3. (3) Equilibrium constant is dependent on temperature and given by Van' Hoff Equation.

4. (1) The relation for pH is given by

$$pH = \frac{1}{2}[pK_w + pK_a + pK_b]$$

5. (1) As no. of moles of gaseous product & reactants are same, there will be no effect of pressure change.

6. (4) K_C & K_P are equilibrium constants of gaseous mixture, defined by molar concentration and partial pressures of gases inside a closed system, which are further following a linear relation i.e. partial pressure is directly proportional to molar concentration.

7. (4) Acidic character increase with increase in oxidation no. $HClO_4$ is strongest acid among all mineral acids

8. (1) pH is equal to $-\log_{10}$ (concentration) or $-\log(H^+)$

9. (1) Both Assertion and Reason are True and the Reason is a correct explanation of the Assertion.

10. (1) Anions of strong acid does not hydrolyse nor does the cation of strong base. But the anions and cations of weak bases and acids do hydrolyze. Since Br^- does not hydrolyze since it is the anion of strong acid but NH_4^+ will react with water $NH_4^+ + H_2O \rightleftharpoons NH_3(aq) + H_3O^+$

Multi Correct MCQs

1. (2) According to Le-Chatelier's Principle, adding more O_2 and SO_2 will increase the yield of SO_3. Also, by increasing pressure, the reaction will move forward, hence SO_2 yield will get increase.

2. (2) As per Le-Chatelier's principle, when solid, Fe_2O_3 is added, there is no effect on equilibrium.

3. (1) pOH $\propto$ Acidic strength.

4. (1) PCl_5 cannot act as Lewis base, due to non-availability of l.p on P, while rest can.

NCERT Exemplar MCQs

1. (4) The relationship between K_p and K_c is

$K_p = K_c (RT)^{\Delta n_g}$

where, Δn_g = (number of moles of gaseous products) — (number of moles of gaseous reactants)

For the reaction,

$$NH_4Cl(s) \rightleftharpoons NH_3(g) + HI(g)$$

$\Delta n = 2 - 0 = 2$

2. (4) $\Delta G^\ominus$ and K are related as

$\Delta G^\ominus = -RT \ln K_C$

when $\Delta G^\ominus > 0$ means ΔG° is positive. This can be so only if $\ln K_C$ is negative i.e., $K_C < 1$.

3. (3) At the stage of equilibrium involving physical processes like melting of ice and freezing of water etc., process does not stop but the opposite processes i.e., forward and reverse process occur with the same rate.

4. (2) For the reaction,

$$PCl_5 \rightleftharpoons PCl_3 + Cl_2$$

At 500K in a closed container, $[PCl_5] = 0.8 \times 10^{-3}$ mol L^{-1}

$[PCl_3] = 1.2 \times 10^{-3}$ mol L^{-1}

$[Cl_2] = 1.2 \times 10^{-3}$ mol L^{-1}

$$K_c = \frac{[PCl_3][Cl_2]}{[PCl_5]} = \frac{(1.2\times10^{-3})\times(1.2\times10^{-3})}{(0.8\times10^{-3})} = 1.8\times10^{-3}$$

5. (2) In the reaction,

$$Fe^{3+} + SCN^- \rightleftharpoons \underset{(Red)}{[Fe(SCN)]^{2+}}$$

Equilibrium shifts towards backward direction and intensity of red colour decreases.

6. (1) In the reaction,

$$\underset{(pink)}{[Co(H_2O_6)]^{3+}}(aq) + 4Cl^-(aq) \rightleftharpoons \underset{(Blue)}{[CoCl_4]^{2-}}(aq) + 6H_2O(l)$$

On heating, the equilibrium shifts in forward direction. or on cooling, the equilibrium shifts in backward direction. Hence, reaction is endothermic. i.e., $\Delta H > 0$.

7. (3) As temperature increases, pH decreases. So, correct option is (3).

8. (2) As the acidity or K_a value increases, pH decreases thus, the order of pH value of the acids is

$$\underset{(3.8\times10^{-8})}{\text{Hypochlorous acid}} \rangle \underset{(1.74\times10^{-5})}{\text{Acetic acid}} \rangle \underset{(18\times10^{-4})}{\text{Formic acid}}$$

9. (1) For the reaction.

$$H_2S \rightleftharpoons H^+ + HS^-$$

$$K_{a_1} = \frac{[H^+][HS^-]}{[H_2S]}$$

For the reaction, $HS^- \rightleftharpoons H^+ + HS^{2-}$

$$K_{a_2} = \frac{[H^+][S^{2-}]}{[HS^-]}$$

When, the above two reactions are added, their equilibrium constants are multiplied, thus

$$K_{a_3} = \frac{[H^+]^2[S^{2-}]}{[H_2S]} = K_{a_1} \times K_{a_2}$$

Hence, $K_{a_3} = K_{a_1} \times K_{a_2}$

10. (3) An acid is a species which accepts an electron pair and base which donates an electron pair. As BF_3 is an electron deficient compound. Hence, it is a Lewis acid.

11. (3) A mixture of weak base and its conjugate acid is obtained when the concentration of NH_4OH (weak base) is higher than the strong acid (HCl), which acts as basic buffer.

	NH_4OH +	HCl	$\rightarrow NH_4Cl + H_2O$
Initially	0.1 M	0.05 M	0
After reaction	0 05 M	0	0.05 M

12. (4) Among the given solvent, AgCl is most soluble in aqueous ammonia solution. It react with aqueous ammonia to form a complex $[Ag(NH_3)_2]^+ Cl^-$.

13. (1) Given that, $K = 1.74 \times 10^{-5}$

Concentration of $CH_3COOH = 0.01 \text{ mol dm}^{-3}$

$$[H^+] = \sqrt{K_a.C} = \sqrt{1.74 \times 10^{-6} \times 0.001} = 4.17 \times 10^{-4}$$

$pH = -\log [H^+] = -\log(4.17 \times 10^{-4}) - 3.4$

14. (3) Given that,

K_a for $CH_3COOH = 1.8 \times 10^{-5}$

K_b for $NH_4OH = 1.8 \times 10^{-5}$

Ammonium acetate is a salt of weak acid and weak base. For such salts

$$pH = 7 + \frac{pK_a - pK_b}{2} = 7 + \frac{[\log 1.8 \times 10^{-5}] - [-\log 1.8 \times 10^{-5}]}{2}$$

$$= 7 + \frac{4.74 - 4.74}{2} = 7.00$$

15. (1) As we know that

$\Delta G^{\ominus} = -RT \ln K$

At the stage of half completion of the reaction,

$$A \rightleftharpoons B, [A] = [B]$$

Therefore, $K = 1$

Thus, $\Delta G^{\ominus} = 0$

16. (1) In the reaction,

$$N_2(g) + 3H_2(g) \rightleftharpoons 2NH_3(g)$$

If the total pressure at which the equilibrium is established, is increased without changing the temperature, K will remain same, K is temperature dependent.

17. (2) The given compounds are

$\underset{\text{(Maximum b.p)}}{\text{Water}}$, acetone, $\underset{\text{(minimum b.p)}}{\text{ether}}$

Greater the boiling point, lower is the vapour pressure of the solvent. Hence, the correct order of vapour pressure will be

Water < acetone < ether.

18. (1) For the reaction,

$$\frac{1}{2}H_2(g) + \frac{1}{2}I_2(g) \rightleftharpoons HI(g)$$

$$K_c = \frac{[HI]}{[H_2]^{1/2}[I_2]^{1/2}} = 5$$

Thus, for the reaction,

$$2HI(g) \rightleftharpoons H_2(g) + I_2(g)$$

$$K_{c_1} = \frac{[H_2][I_2]}{[HI]^2} = \left(\frac{1}{K_c}\right)^2 = \left(\frac{1}{5}\right)^2 = \frac{1}{25} = 0.04$$

19. (4) In these reactions, at constant volume

$$H_2(g) + I_2(g) \rightleftharpoons 2HI(g)$$

$$PCl_5(g) \rightleftharpoons PCl_3(g) + Cl_2(g)$$

$$N_2(g) + 3H_2(g) \rightleftharpoons 2NH_3(g)$$

The equilibrium remains unaffected on addition of inert gas in all the three cases.

NEET & AIIMS Past Year MCQs

1. (4) $3A_2 + B_2 \rightleftharpoons 2A_3B; K_1$ ...(i)

$2A_3B \rightleftharpoons 3A_2 + B_2; K' = \frac{1}{K_1}$ ...(ii)

Equation (ii) is divided by 2. Then, we have

$$A_3B \rightleftharpoons \frac{3}{2}A_2 + \frac{1}{2}B_2; K_2 = \sqrt{K'}$$

Now, $K_2 = \sqrt{K'} = \frac{1}{\sqrt{K_1}}$ ....From (ii)

2. (4) $N_2(g) + 3H_2(g) \rightleftharpoons 2NH_3(g)$, $\Delta H = -Q$

According to Le Chatelier's principle.

- This reaction is exothermic reaction, as indicated by the negative enthalpy change ($\Delta H = -Q$) and low temperatures favour exothermic reactions.
- An increase in pressure shifts the reaction towards the direction with fewer moles. Therefore, the given reaction shifts forward upon increasing pressure.
- Increasing the concentration of reactants drives the reaction forward. Therefore, a high concentration of H_2 shifts the reaction in the forward direction.

3. (1) Solubility of AgCl in 0.1 M KCl

The dissociation of AgCl in 0.1 M KCl is represented as:

$$\underset{}{AgCl(aq)} \rightleftharpoons \underset{s}{Ag^+(aq)} + \underset{(s+0.1)}{Cl^-(aq)}$$

The solubility product is given by:

$K_{sp} = (s)(s + 0.1)$

since s << 0.1 ∴ (s + 0.1) 0.1 M

∴ $10^{-10} = s \times 0.1$

$$\therefore s = \frac{10^{-10}}{0.1} = 10^{-9}\ M$$

Solubility of AgCl in water

The dissociation of AgCl in water is represented as:

$$AgCl(aq) \rightleftharpoons \underset{s}{Ag^+(aq)} + \underset{s}{Cl^-(aq)}$$

Now, $K_{sp} = S \times S$

∴ $10^{-10} = s^2$

$$\therefore s = \sqrt{10^{-10}} = 10^{-5}\ M$$

Hence, the ratio of solubility of AgCl in 0.1 M KCl solution to the solubility of AgCl in water = $10^{-9} : 10^{-5} = 10^{-4} : 1$

4. (1)

$2A \rightleftharpoons B + C$

$$Q_C = \frac{[B][C]}{[A]^2} = \frac{[2\times10^{-3}][2\times10^{-3}]}{[2\times10^{-3}]^2}$$

$Q_C = 1$

$Q_C > K_C$

Reaction has a tendency to go in backward direction.

$Q_C < K_C$ Reaction has a tendency to go in forward direction.

5. (3)

(1) $CO(g) + H_2O(g) \rightleftharpoons CO_2(g) + H_2(g)$

$$K_C = \frac{[CO_2][H_2]}{[CO][H_2O]}$$

$\Delta n_g = 2 - 2 = 0$

$K_p = K_C(RT)^{\Delta n_g}$

$K_p = K_C(RT)^0$

$K_p = K_C$

(2) $2BrCl(g) \rightleftharpoons Br_2(g) + Cl_2(g)$

$\Delta n_g = 2 - 2 = 0$

$$K_C = \frac{[Br_2][Cl_2]}{[BrCl]^2}$$

$K_p = K_C(RT)^{\Delta n_g}$

$K_p = K_C(RT)^0$

$K_p = K_C$

(3) $PCl_5(g) \rightleftharpoons PCl_3(g) + Cl_2(g)$

$\Delta n_g = 2 - 1 = 1$

$$K_C = \frac{[PCl_3][Cl_2]}{[PCl_5]}$$

$K_p = K_C(RT)^{\Delta n_g}$

$K_p = K_C(RT)^1$

Here, $K_p \neq K_C$

(4) $H_2(g) + I_2(g) \rightleftharpoons 2HI(g)$

$\Delta n_g = 2 - 2 = 0$

$$K_C = \frac{[HI]^2}{[H_2][I_2]}$$

$K_p = K_C(RT)^{\Delta n_g}$

$K_p = K_C(RT)^0$

Here, $K_p = K_C$

6. (2) $2NO(g) \rightleftharpoons N_2(g) + O_2(g)$

At t = 0 0.1 M – –

At time = t $(0.1 - 0.1\,\alpha)M$ $\left(\frac{0.1\alpha}{2}\right)M$ $\left(\frac{0.1\alpha}{2}\right)M$

$$\therefore K_c = \frac{3\times10^{-3}\times4.2\times10^{-3}}{(2.8\times10^{-3})^2} = \frac{\frac{0.1\alpha}{2}\times\frac{0.1\alpha}{2}}{(0.1-0.1\alpha)^2}$$

$$1.6 = \frac{\left(\frac{0.1\alpha}{2}\right)^2}{(0.1-0.1\alpha)^2}$$

$$1.27 = \frac{\left(\frac{0.1\alpha}{2}\right)}{0.1-0.1\alpha}$$

$\alpha = 0.717$

7. (4) $K_{eq.} = \frac{[C][D]}{[A][B]}$

$K_{eq} = (10 \times 6)/(2 \times 3) = 10$

$\Delta G^0 = -2.303\ RT\log K_{eq.} = -2.303(2)(300)(\log 10)$

$= -1381.8$ cal (∵ R = 2 cal/mol k)

8. (2) According to Henderson equation

pH of acidic buffer is-

$$pH = pKa + \log\frac{[salt]}{[acid]}$$

$$= 4.57 + \log\frac{0.10}{0.01}$$

$= 4.57 + [\log 10 - \log 1]$

$= 4.57 + (1 - 0)$

$= 4.57 + (1)$

$= 5.57$

9. (2) $3O_2(g) \rightleftharpoons 2O_3(g)$

$$K_C = \frac{[O_3]^2}{[O_2]^3}$$

$$3 \times 10^{-59} = \frac{[O_3]^2}{\left[4\times10^{-2}\right]^3}$$

$[O_3]^2 = 3 \times 10^{-59} \times 64 \times 10^{-6}$

$[O_3]^2 = 19.2 \times 10^{-64}$

$[O_3] = 4.38 \times 10^{-32}$ M

10. (2) $pH = 7 + \frac{1}{2}(pK_a - pK_b)$

$= 7 + \frac{1}{2}(4.77\text{-}3.27)$

$= 7 + 0.75$

$= 7.75$

11. (2) $2ICl \rightleftharpoons I_2 + Cl_2$

0.6

$0.6 - 2x \quad x \quad x$

$$K_C = 0.14 = \frac{x^2}{(0.6-2x)^2}$$

$$0.37 = \frac{x}{0.6-2x}$$

$0.222\text{-}0.74\,x = x$

$1.74x = 0.222$

$x = 0.128$

12. (3) $K_a = 5\times10^{-10}$, $pH = 9$

$$pH = pK_a + \log\frac{[Salt]}{[Aid]}$$

$$pH - pK_a + \log\frac{[KCN]}{[HCN]}$$

$$9 = -\log(5 \times 10^{-10}) + \log\left[\frac{5V}{V+10} \times \frac{V+10}{10\times2}\right]$$

$$9 = -\log(5 \times 10^{-10}) + \log\left[\frac{5V}{20}\right]$$

$V - 1.99 \approx 2$ ml

13. (2) The given reaction is

$$\frac{1}{2}H_2(g) + \frac{1}{2}I_2(g) \Leftrightarrow HI(g) \quad \text{(i)}$$

$$\text{Hence, } K_c = \frac{[HI]}{[H_2]^{1/2}[I_2]^{1/2}} \quad \text{(ii)}$$

Now, reverse the eqn. (i) and multiply by 2, we get

$2HI \Leftrightarrow H_{2(g)} + I_{2(g)}$

$$\text{Hence, } K_c = \frac{[H_2][I_2]}{[HI]^2} \quad \text{(iii)}$$

Equating equations (ii) and (iii), we get

$$K'_c = \frac{1}{(K_c)^2}$$

14. (4) **15. (3)**

16. (3) **17. (2)** **18. (1)**

Redox Reactions

NCERT Topic-wise MCQs

1. (3) Oxidation is the addition of electronegative element or removal of electropositive element.

$Cl_2 + \underline{2KI} \rightarrow 2KCl + I_2$

In this reaction, the underlined species is oxidised as it involve the removal of electropositive element, i.e., potassium.

2. (1) In the given reaction, sodium is oxidised and hydrogen is reduced.

3. (1) Element Mg in the given reaction gets oxidized on account of addition of oxygen to it.

4. (4) All the given statements are correct regarding reduction process.

5. (1) $\overset{+4\ -2}{MnO_2} + 4\overset{+1\ -1}{HCl} \longrightarrow \overset{+2}{Mn}Cl_2 + \overset{0}{Cl_2} + 2\overset{+1\ -2}{H_2O}$

In this reaction, same element does not get reduced and oxidised, hence it does not represent disproportionation reaction.

6. (2) Oxidation number increases and number of electrons decreases in oxidation process.

7. (1) In the reaction,

$$HOCl + H_2\overset{-1}{O_2} \xrightarrow{H^+} H_3O^+ + Cl + \overset{0}{O_2}$$

↑ in O.N.

Oxidation number of oxygen increases from -1 to 0. Hence, H_2O_2 acts as reducing agent.

8. (1)

↓ in O.N. (reduction)

$$\overset{+1}{Ag_2O} + H_2\overset{-1}{O_2} \longrightarrow 2\overset{0}{Ag} + H_2O + \overset{0}{O_2}$$

↑ in O.N. (oxidation)

Hence, H_2O_2 acts as a reducing agent as it is undergoing oxidation.

9. (2)

Reduction

$$\overset{2+}{Cu} + \overset{0}{Ag} \longrightarrow \overset{0}{Cu} + \overset{+1}{Ag^+}$$

Oxidation

So, oxidation-half reaction is $Ag \rightarrow Ag^+$

10. **(4)** (1) $I_2 + 10H\overset{+5}{N}O_3 \rightarrow 2HIO_3 + 10\overset{+4}{N}O_2 + 4H_2O$

In this reaction, HNO_3 is acting as an oxidising agent because oxidising state of N decrease from + 5 to +4

(2) $3Cu + 8H\overset{+5}{N}O_3 \rightarrow 3Cu(NO_3)_2 + 2\overset{+2}{N}O + 4H_2O$

It is a reaction in which HNO_3 is acting as an oxidising agent because oxidation number of nitrogen is decreasing.

(3) $4\overset{0}{Zn} + 10H\overset{+5}{N}O_3 \rightarrow 4\overset{+2}{Zn}(NO_3)_2 + \overset{-3}{N}H_4\overset{+5}{N}O_3 + 3H_2O$

∴ In this reaction also, HNO_3 act as oxidising agent as oxidation state of N changes from to +5 to –3 in NH_4NO_3, that is, decreasing.

(4) $2H\overset{+5}{N}O_3 + \overset{+5}{P_2}O_5 \rightarrow 2H\overset{+5}{P}O_3 + \overset{+5}{N_2}O_5$

In this reaction, there is no change in oxidation number of N, therefore HNO_3 does not act as an oxidising agent.

11. **(1)** The reaction which involves oxidation, requires an oxidising agent. $S_2O_3^{2-} \rightarrow S_4O_6^{2-}$

S is undergoing oxidation. Oxidation state of S changes from 2 to 2.5 and hence, this reaction requires an oxidising agent.

12. **(3)** The species in which an element has oxidation number lying in between its minimum and maximum values can act as oxidant and reductant both.

Oxidation state of O in species H_2O_2	Minimum oxidation state of O	Maximum oxidation state of O
–1	–2	0

Thus; hydrogen peroxide (H_2O_2) can act both as oxidising as well as a reducing agent.

H_2O_2 as a reducing agent,

$Ag_2O + H_2O_2 \rightarrow 2Ag + H_2O + O_2$

H_2O_2 as an oxidising agent,

$PbS + 4H_2O_2 \rightarrow PbSO_4 + 4H_2O$

13. **(1)**

Reduction

$2\overset{+2}{Cu}SO_4 + 4K\overset{-1}{I} \longrightarrow \overset{+1}{Cu_2}I_2 + 2K_2SO_4 + \overset{0}{I_2}$

Oxidation

14. **(2)** Oxidation takes place with loss of electrons.

15. **(3)** The oxidation number of hydrogen in CaH_2 is –1.

16. **(1)** The oxidation number of sulphur in S_8 molecule is zero. In S_8 molecule, each sulphur atom is connected to two other sulphur atoms, therefore covalency is 2.

17. **(4)** Let the oxidation state of N in HNO_3 is x.

$\therefore +1 + x + (-2 \times 3) = 0$

$x + 1 - 6 = 0$

$x = +5$

18. **(3)** NO_3^- has uni-negative charge.

∴ In $Fe(NO_3)_2$, Fe has oxidation number = + 2

$C_2O_4^{2-}$ has bi-negative charge.

∴ In FeC_2O_4, Fe has oxidation number = + 2

In $[Fe(H_2O)_6]Cl_3$, the cation $[Fe(H_2O)_6]^{3+}$ is present and H_2O is neutral, therefore oxidation number of Fe is +3.

19. **(1)** Let the oxidation state of P in the given compounds be x.

$H_4P_2O_5 : 2x + 4 - 10 = 0$, $H_4P_2O_6 : 2x + 4 - 12 = 0$

$2x = 6$ $\qquad 2x = 8$

$x = +3$ $\qquad x = +4$

$H_4P_2O_7 : 2x + 4 - 14 = 0$

$2x = 10$

$x = +5$

20. **(2)** Let the oxidation number of carbonyl carbon in methanal (HCHO) and methanoic acid (HCOOH) be x and y respectively.

In HCHO, $2(+1) + x + (-2) = 0$

$2 + x - 2 = 0$

$x = 0$

In HCOOH, $2(+1) + y + 2(-2) = 0$

$2 + y - 4 = 0$

$y = +2$

21. **(2)** As we know electron releasing tendency of the copper metal is greater than silver, therefore copper is more easily oxidised and silver is more easily reduced.

22. **(3)** Oxidation state of I in IO_3^- is:

$x + (-2)3 = -1 \Rightarrow x = +5$

Oxidation state of I in IO_4^- is:

$x + (-2)4 = -1 \Rightarrow x = +7$

Oxidation state of I in KI is:

$+1 + x = 0 \Rightarrow x = -1$

Oxidation state of I in I_2, x = 0

Hence; the oxidation number of iodine in IO_3^-, IO_4^-, KI and I_2 are + 5, + 7, –1 and 0 respectively.

23. **(3)** Acidified $K_2Cr_2O_7$ solution oxidises SO_2 into $Cr_2(SO_4)_3$.

$$3\overset{+4}{S}O_2 + K_2Cr_2O_7 + H_2SO_4 \rightarrow K_2SO_4 + Cr_2\left(\overset{+6}{S}O_4\right)_3 + H_2O$$

Hence; the oxidation state of sulphur changes from +4 to +6.

24. **(1)** HNC

suppose O.S. of C is x.

$(+1) + 1(-3) + x = 0$

$-2 + x = 0$

$x = +2$

25. **(1)** In $Ba(H_2PO_2)_2$, let the oxidation state of P be x.

$2(+1) + 2[2(+1) + x + 2(-2)] = 0$

$2 + 2(2 + x - 4) = 0$

$2 + 4 + 2x - 8 = 0$

$2 + 2x - 4 = 0$

$2x = 2 \Rightarrow x = +1$

26. (3) $Mn^{2+} + NO_3^- + OH^- \rightarrow \overset{(+6)}{Mn}O_4^{2-} + H_2O + NO$

The oxidation number of Mn changes from + 2 to + 6.

27. (3) When H_2O_2 is added to an acidified solution of a dichromate $Cr_2O_7^{2-}$, a deep blue coloured complex, chromic peroxide, CrO_5 [or $CrO(O_2)_2$] is formed.

$Cr_2O_7^{2-} + 2H^+ + 4H_2O_2 \rightarrow 2CrO(O_2)_2 + 5H_2O$

This deep blue coloured complex has the following structure :

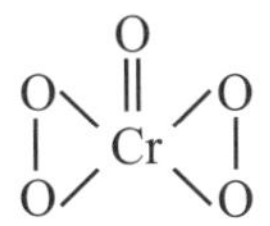

Oxidation state of Cr is +6 due to the presence of two peroxide linkages, which can be calculated as:

In $CrO(O_2)_2$, let the oxidation state of Cr be x

$x + (-1) \times 4 + (-2) = 0$

$x - 6 = 0$

$x = +6$

28. (3) The following reaction occurs when a mixture of potassium chlorate, oxalic acid and sulphuric acid is heated:

$$\overset{+1}{K}\overset{+5}{Cl}\overset{-2}{O_3} + \overset{+1}{H_2}\overset{+3}{C_2}\overset{-2}{O_4} + \overset{+1}{H_2}\overset{+6}{S}\overset{-2}{O_4} \rightarrow \overset{+1}{K_2}\overset{+6}{S}\overset{-2}{O_4} + \overset{+1}{K}\overset{-1}{Cl} + \overset{+4}{C}\overset{-2}{O_2} + \overset{+1}{H_2}\overset{-2}{O}$$

Thus, the element which undergoes maximum change in the oxidation state is Cl.

29. (4) $Na_2\overset{2.5}{S_4}O_6 : H_2\overset{+6}{S_2}O_7 : Na_2\overset{+2}{S_2}O_3 : \overset{0}{S_8}$

$H_2\overset{+6}{S}O_4 : \overset{+4}{S}O_2 : H_2\overset{-2}{S} : H_2\overset{+7}{S_2}O_8$

$\overset{+6}{S}O_2^{2+} ; \overset{+6}{S}O_4^{2-} ; \overset{+4}{S}O_3^{2-} : H\overset{+6}{S}O_4^-$

$H_2\overset{+6}{S}O_4 ; H_2\overset{+4}{S}O_3 ; \overset{+2}{S}Cl_2 ; H_2\overset{-2}{S}$

30. (3) The tetrathionate anion, $S_4O_6^{2-}$ is a sulphur oxoanion derived from the compound tetrathionic acid, $H_4S_4O_6$. Anion $S_4O_6^{2-}$ can be viewed as the product formed from reaction between lewis base, S_2^{2-} and lewis acid, SO_3

$S_2^{2-} + 2SO_3 \rightarrow S_4O_6^{2-}$

Two of the sulphur atoms are present in the oxidation state of zero and two are in oxidation state +5.

The structure of $S_4O_6^{2-}$ is:

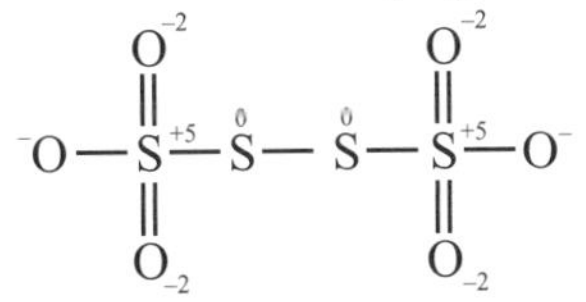

Since; each of the terminal sulphur atom is connected to two oxygen atom by a double bond and to one oxygen atom by a single bond, the oxidation state of these terminal sulphur atoms is +5. Since; two central sulphur atoms are linked to each other by a single bond, and the sulphur is further attached to similar species on either side, the electron pair forming S-S bond remains in the centre and hence; each of the two central sulphur atoms has an oxidation state of zero. Thus; the average oxidation state of the four sulphur atom is:

$(2 \times 5 + 2 \times 0)/4 = 2.5$

31. (1) Bromine has zero oxidation state because it is in the free state.

$Br_2^0 \rightarrow BrO_3^-$

Let oxidation state of Br in BrO_3^- be x.

$x + (-2 \times 3) = -1$

$x + (-6) = -1$

$x = +6 = +5$

So: oxidation number changes from 0 to + 5.

32. (3) H_2SO_5 (Peroxomonosulphuric acid)

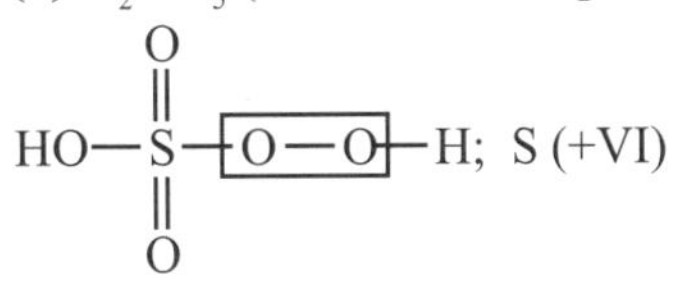

$2 \times (+1) + x + 3 \times (-2) \times 2 \times (-1) = 0$

(for H) (for S) (for O) (for peroxide)

$2 + x - 6 - 2 = 0 \Rightarrow x = +6$

$H_2S_2O_8$ (peroxodisulphuric acid)

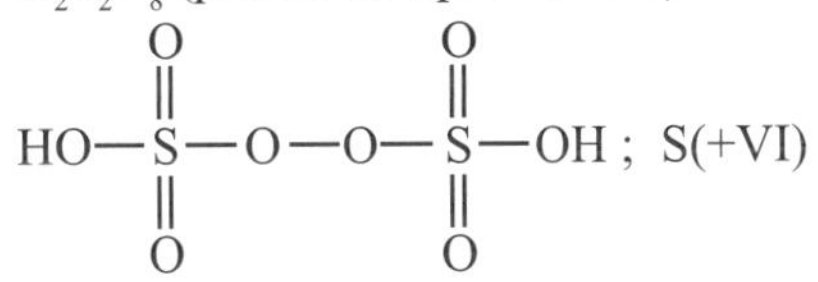

The oxidation number of these two oxygen atoms will be –1 each because it contains peroxide linkage.

$2 \times (+1) + 2 \times x + 6 \times (-2) + 2 \times (-1)$

(for H) (for S) (for O) (for peroxide)

$2 + 2x - 12 - 2 = 0$

$2x = 12$

$x = +6$

33. (3) In Fe_3O_4, let the oxidation number of Fe = x

$3x + 4(-2) = 0$

$x = \frac{8}{3}$

34. (2) The reaction of aqueous $KMnO_4$ with H_2O_2 in acidic medium is:

$3H_2SO_4 + 2KMnO_4 + 5H_2O_2 \rightarrow 5O_2 + 2MnSO_4 + 8H_2O + K_2SO_4$

In the above reaction, $KMnO_4$ oxidises H_2O_2 to O_2 and itself (MnO_4^-) gets reduced to Mn^{2+} ion in acidic conditions.

35. (4) 1. Oxidation state of C in CO_2

$x + 2(-2) = 0 \Rightarrow x = +4$

2. Oxidation state of C in CF_4

$x + 4(-1) = 0 \Rightarrow x = +4$

3. Oxidation state of C in CCl_4

$x + 4(-1) = 0 \Rightarrow x = +4$

4. Oxidation state of C in CH_4

$x + 4(+1) = 0 \Rightarrow x = -4$

36. (1) $H_4P_2O_7$, let the oxidation number of P be x.

$2x + 4 - 14 = 0 \Rightarrow 2x - 10 = 0 \Rightarrow x = +5$

37. (1) The oxidation number of nitrogen in different compounds are:

1) Mg_3N_2: – 3 2) NH_2OH : – 1

3) $(N_2H_5)_2SO_4$: – 2 4) $[Co(NH_3)_5Cl]Cl_2$: +3

38. (1) In N_3H (hydrazoic acid), let the oxidation state of N be x

$3x + 1 = 0 \Rightarrow 3x = -1 \Rightarrow x = -1/3$

39. (3) Let the oxidation number of Co be x.

In $K[Co(CO)_4]$, $1 + x + 0 = 0$ [as CO is neutral]

$x = -1$

40. (1) 1. Oxidation state of P in H_3PO_3

$3(+1) + x + 3(-2) = 0 \Rightarrow 3 + x - 6 = 0$

$x - 3 = 0 \Rightarrow x = +3$

2. Oxidation state of P in H_3PO_4

$3(+1) + x + 4(-2) = 0$

$3 + x - 8 = 0 \Rightarrow x - 5 = 0 \Rightarrow x = +5$

3. Oxidation State of P in HPO_3^-

$+1 + x + 3(-2) = -1$

$+1 + x - 6 = -1 \Rightarrow x - 5 = -1 \Rightarrow x = +4$

4. Oxidation state of P in $H_4P_2O_7$

$4(+1) + 2x + 7(-2) = 0$

$4 + 2x - 14 = 0 \Rightarrow 2x - 10 = 0 \Rightarrow x = +5$

41. (2) In the given reaction, oxidation number of Br change from 0 to –1 in $Br^{\ominus}$ and +5 in $BrO_3^{\ominus}$ hence, bromine is reduced as well as oxidised.

42. (1) $\overset{+2\ +4}{CaCO_3} \longrightarrow \overset{+2}{CaO} + \overset{+4}{CO_2}$

As, the oxidation states of Ca and C are +2 and +4 in both reactant and product, hence it is not a redox reaction.

43. (3) 1. $2H_2\overset{-2}{S} + \overset{+4}{SO_2} \longrightarrow 2H_2O + 3\overset{0}{S}$ (Oxidation: S from –2 to 0; Reduction: S from +4 to 0)

2. $2\,\overset{+2\ -2}{BaO} + \overset{0}{O_2} \longrightarrow 2\,\overset{+4\ -2}{BaO_2}$ (Reduction, Oxidation)

3. $Ba\overset{-2}{O_2} + H_2\overset{+6}{SO_4} \longrightarrow Ba\overset{+6}{SO_4} + H_2\overset{-1}{O_2}$ (Oxidation)

Reduction does not occur in this reaction.

4. $2\,\overset{+1\ +5\ -2}{KClO_3} \longrightarrow 2\,\overset{+1\ -1}{KCl} + 3\,\overset{0}{O_2}$ (Reduction, Oxidation)

44. (2) F shows only (–1) oxidation state.

45. (4) In reaction (4), oxidation number changes from + 4 in NO_2 to + 3 in HNO_2 and + 5 in HNO_3.

46. (2) $\overset{+1\ -2}{H_2O} + \overset{0}{C} \longrightarrow \overset{+2\ -2}{CO} + \overset{0}{H_2}$ (Oxidation: C; Reduction: H)

Hence, H_2O acts as the oxidising agent.

47. (3) HOCl

Let oxidation state of Cl be x

$+1 + (-2) + x = 0 \Rightarrow x - 1 = 0 \Rightarrow x = +1$

48. (4) All the given species act as both oxidising and reducing agents.

49. (4) NH_3 is a strong reducing agent.

50. (3) Oxidation number of S in H_2SO_5 is 6. This is observed due to the presence of peroxide linkage in the molecule.

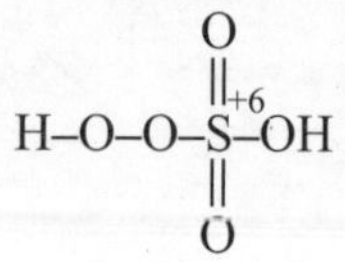

51. (3) ClF_3

Let oxidation state of Cl be x

$x + (-1 \times 3) = 0 \Rightarrow x - 3 = 0 \Rightarrow x = +3$

52. (2) Reduction is gain of electrons by any species.

53. (1) An oxidising agent is acceptor of electrons.

54. (4) In this reaction, 5 electrons are required for the reduction.

$MnO_4^- + 5e^- \rightarrow Mn^{2+}$

55. (3) $Cu^{2+} + \overset{0}{Ag} \longrightarrow \overset{0}{Cu} + Ag^+$ (Reduction: Cu^{2+} to Cu; Oxidation: Ag to Ag^+)

Therefore, reduction-half reaction in the given reaction is $Cu^{2+} \rightarrow Cu$.

56. (1) Oxidation No. of Au in the given compound is +3. Hence, the correct representation is $HAu(III)Cl_4$.

57. (1) Cl_2O_7

$2x + 7(-2) = 0$

$x = +7$

58. (3) The balanced chemical reaction is:

$14H^+ + Cr_2O_7^{2-} + 6I^- \rightarrow 2Cr^{3+} + 3I_2 + 7H_2O$

Thus; when KI reacts with 1 mole of potassium dichromate, 3 moles of I_2 are evolved.

59. (4) The balanced reaction is as follows :

$8KMnO_4 + 3NH_3 \rightarrow 8MnO_2 + 3KNO_3 + 5KOH + 2H_2O$

60. (4) Given, $NO_3^- \rightarrow N_2H_4$

The balanced reaction is as follows:

$2NO_3^- + 16H^+ + 14e^- \rightarrow N_2H_4 + 6H_2O$

∴ Number of electrons involved in the reduction of two nitrate ions = $14e^-$

∴ One nitrate ion will involve $= \frac{14}{2}e^- = 7e^-$

61. (2) Balanced equation is:

$BiO_3^- + 6H^+ + 2e^- \rightarrow Bi^{+3} + 3H_2O$

$\therefore$ The value of x is 2.

62. (1) The balanced chemical reaction is as follows :

$3UO^{2+} + Cr_2O_7^{2-} + 8H^+ \rightarrow 3UO_2^{2+} + 2Cr^{3+} + 4H_2O$

63. (3) Balanced equation is:

$2IO_3^- + 5SO_2 + 4H_2O \rightarrow I_2 + 5SO_4^{2-} + 8H^+$

The coefficient of SO_2 is 5.

64. (1) $I^- + \overset{+5}{I}O_3^- + H^+ \rightarrow \overset{0}{I_2} + H_2O$

$2I^- \rightarrow I_2 + 2e^-$ (i)

$10e^- + 2(\overset{+5}{I}O_3^-) \rightarrow \overset{0}{I_2}$ (ii)

Multiplying eq (i) by 5 and then adding to eq (ii), in order to cancel the electrons, we get

$10I^- + 2IO_3^- \rightarrow 6I_2$

To balance O-atom, add $6H_2O$ molecules on RHS and $12H^+$ on L.H.S, then

$10I^- + 2IO_3^- + 12H^+ \rightarrow 6I_2 + 6H_2O$

or $5I^- + IO_3^- + 6H^+ \rightarrow 3I_2 + 3H_2O$

65. (3) $ClO_2 \rightarrow Cl^-$

$ClO_2 + 2H_2O + 5e^- \rightarrow Cl^- + 4OH^-$

$H_2O_2 \rightarrow O_2$

$H_2O_2 + 2OH^- \rightarrow O_2 + 2H_2O + 2e^-$

$[ClO_2 + 2H_2O + 5e^- \rightarrow Cl^- + 4OH^-] \times 2$

$[H_2O_2 + 2OH^- \rightarrow O_2 + 2H_2O + 2e^-] \times 5$

$2ClO_2 + 5H_2O_2 + 2OH^- \rightarrow 2Cl^- + 5O_2 + 6H_2O$

According to the reaction:

2 mol of ClO_2 oxidises 5 moles of H_2O_2.

Hence, one mole of ClO_2 oxidises 2.5 moles of H_2O_2

66. (3) $4Zn + NO_3^- + 7H_2O \rightarrow 4Zn^{2+} + NH_4^+ + 10\,OH^-$

67. (1) Balanced chemical reaction :

$MnO_4^- + 5Fe^{2+} + 8H^+ \longrightarrow Mn^{2+} + 5Fe^{3+} + 4H_2O$

68. (1) F_2, being most electronegative, is the most powerful oxidising agent with standard oxidation potential 2.87 V.

69. (4) Magnesium is more electropositive than zinc.

70. (1) A negative E^0 means that the redox couple is a stronger reducing agent than the H^+/H_2 couple.

High Order Time Intensive MCQs

1. (1) In $MnSO_4$, oxidation state of Mn = +2. In Mn_2O_3, oxidation state of Mn = +3. There is an increase of oxidation state by 1 when $MnSO_4$ changes to Mn_2O_3. Hence; its equivalent mass is equal to its molar mass.

$$\text{Equivalent weight} = \frac{\text{Molecular weight}}{\text{Oxidation state}} = \frac{\text{Molecular wt.}}{1}$$

2. (4) Here; equivalent weight of salt =

$$\frac{\text{Molecular weight}}{\text{Number of electron lost or gained}}$$

Number of electron lost or gained = 5

$$\therefore \text{Equivalent mass} = \frac{M}{5}$$

3. (2) $E.wt = \dfrac{M}{\text{change in O.S}}$

4. (2) $3\overset{0}{Fe} \longrightarrow \overset{+8/3}{Fe_3O_4}$ ($8e^-$)

$$\therefore E = \frac{M}{8/3} = \frac{56 \times 3}{8} = 21$$

5. (4) In alkaline medium, following reaction occurs,

$2K\overset{+7}{Mn}O_4 + 2KOH \rightarrow 2K_2\overset{+6}{Mn}O_4 + H_2O + [O]$

Change in oxidation state of Mn = +1

$\therefore$ Equivalent mass of $KMnO_4$

$$= \frac{\text{Molar mass}}{\text{Change in oxidation state}} = \frac{\text{Molar mass}}{1} = \text{Molar mass}$$

6. (1) $Fe\overset{+2}{}C_2\overset{+3}{}O_4 \longrightarrow Fe^{3+} + \overset{+4}{C}O_2$ ($3e^-$)

$$\therefore \text{Equivalent weight} = \frac{M}{3}$$

7. (1) Equivalent mass of $NH_3 = \dfrac{17}{3}$

Equivalent mass of $N_2 = \dfrac{28}{6}$

$$E_1 - E_2 = \frac{17}{3} - \frac{28}{6} = \frac{34 - 28}{6} = \frac{6}{6} = 1$$

8. (2) $3\,Cu + 8\,H\overset{+5}{N}O_3 \longrightarrow 3Cu(NO_3)_2 + 2\overset{+2}{N}O + 4H_2O$

$\therefore$ Change in oxidation state = 3

Hence, equivalent weight $= \dfrac{63}{3} = 21$

$$N = \frac{\text{Weight}}{\text{Equivalent weight} \times V\,(L)}$$

$$2 = \frac{W}{\frac{63}{3} \times 1}$$

W = 21 × 2

W = 42g

9. (2) 1) $8H^+ + MnO_4^- + 5Fe^{2+} \rightarrow 5Fe^{3+} + Mn^{2+} + 4H_2O$

2) $14H^+ + Cr_2O_7^{2-} + 6Fe^{2+} \rightarrow 6Fe^{3+} + 2Cr^{3+} + 7H_2O$

From 2 balanced equations, it is evident that Fe^{2+} will be oxidised more by $K_2Cr_2O_7$.

10. (1) $KHC_2O_4 \rightleftharpoons K^+ + HC_2O_4^-$

$HC_2O_4^- \rightleftharpoons H^+ + C_2O_4^{2-}$

$KOH \rightleftharpoons K^+ + OH^-$

There is one replaceable H in KHC_2O_4, so its n - factor is 1.

Its equivalent weight $= \dfrac{128}{1} = 128.$

11. (2) $48H^+ + 6MnO_4^- + 5Fe_2(C_2O_4)_3 \rightarrow 6Mn^{2+} + 30CO_2 + 10Fe^{3+} + 24H_2O$

5 moles of ferric oxalate require $MnO_4^- = 6$ moles

1 mole of ferric oxalate require $MnO_4^- = \frac{6}{5} = 1.2$

12. (4) There is peroxy linkage in $S_2O_8^{2-}$ & H_2SO_5. As a result, they have +6 oxidation state.

13. (3) $8\overset{+6}{Cr}O_4^{2-} + 3\overset{+2}{S_2}O_3^{2-} + 17H_2O \rightarrow 6\overset{+6}{S}O_4^{2-} + 8\overset{+3}{Cr}(OH)_4^- + 2OH^-$

Normality of 0.154 M $CrO_4^{2-} = 0.154 \times 3N$

(Normality = Molarity × n)

Normality of 0.246 M $S_2O_3^{2-} = 0.246 \times 8N$

$\because N_1V_1 = N_2V_2$

$0.154 \times 3 \times V_1 = 0.246 \times 8 \times 40$

$V_1 = 170.4$ mL

14. (4) All the given reactions are redox reactions.

15. (2) $2Cr(s) + 2OH^-(aq) + 6H_2O \rightarrow 2Cr(OH)_4^-(aq) + 3H_2(g)$

$2 + 2 + 6 + 2 + 3 = 15$

16. (3) $5As_2O_3 + 4MnO_4^- + 9H_2O + 12H^+ \rightarrow 10H_3AsO_4 + 4Mn^2$

According to reaction,

5 moles of $As_2O_3 \rightarrow 4$ moles of $MnO_4^{\ominus}$

Also, 0.1097 g of $As_2O_3 = 0.000554$ mol of As_2O_3

Thus, 0.000554 mol of $As_2O_3 \rightarrow \frac{4 \times 0.000554}{5}$ moles of $MnO_4^{\ominus}$

Molarity of $KMnO_4 = \frac{n_{KMnO_4}}{V(l)} = \frac{4 \times 0.000554 \times 1000}{5 \times 36.10}$

$= 0.0122$ M

17. (3) The redox couple with maximum reduction potential will be best oxidising agent and with minimum reduction potential will be best reducing agent.

18. (2) $\overset{+5}{Br}O_3^- + 6H^+ + 6e^- \longrightarrow Br^- + 3H_2O$

$\because$ The oxidation number of Br changes from +5 to –1, therefore n factor = 6

Normality = n × Molarity

$\therefore$ Molarity $= \frac{0.672}{6} = 0.112$ M

$$\text{Molarity} = \frac{\text{Weight of NaBrO}_3\ (W_{NaBrO_3})}{\text{Molar mass of NaBrO}_3\ (M_{NaBrO_3})} \times \frac{1}{V(\text{in L})}$$

$\therefore W_{NaBrO_3} = \text{Molarity} \times M_{NaBrO_3} \times V_{mL} \times 10^{-3}$
$= 0.112 \times 151 \times 55.5 \times 10^{-3} = 0.9386$ g

19. (1) $\overset{+7}{Mn}O_4^- + Fe\overset{+2}{}\overset{+3}{C_2}O_4 + H^+ \longrightarrow \overset{2+}{Mn} + \overset{3+}{Fe} + 2\overset{+4}{C}O_2 + H_2O$

(n factor = 5 for Mn; n factor = 3 for FeC_2O_4)

Equivalents of $KMnO_4$ = Equivalents of FeC_2O_4

Moles × n factor of $KMnO_4$ = Moles × n factor of FeC_2O_4

$\Rightarrow 0.1 \times V \times 5 = \frac{\text{Mass} \times \text{n factor}}{\text{Molar mass}}$

$\Rightarrow 0.1 \times V \times 5 = \frac{100 \times 3}{1000 \times 143.91}$

($\because$ Molar mass of $FeC_2O_4 = 143.91$ g)

$\Rightarrow V = 4.1 \times 10^{-3}\ L \Rightarrow V = 4.1$ mL

20. (1) $NO_3^- + 2H^+ + 1e^- \rightarrow NO_2 + H_2O$ $\quad x = 1$

$NO_3^- + 6H^+ + 5e^- \rightarrow \frac{1}{2}N_2 + 3H_2O$ $\quad y = 5$

$NO_3^- + 5H^+ + 4e^- \rightarrow \frac{1}{2}N_2O + \frac{5}{2}H_2O$ $\quad z = 4$

21. (3) As the molecule is neutral, therefore the sum of oxidation numbers will be zero

$A_3(BC_4)_2$

$3(+2) + 2[1(+5) + 4(-2)] = 6 + 2[5 - 8]$

$= 6 + 2(-3) = 6 - 6 = 0$

22. (4) The two sulphur atoms in $Na_2S_4O_6$ which are linked to each other have zero oxidation number. The oxidation number of other two sulphur atoms can be calculated as follows:

$2x + (2 \times 0) + (6 \times -2) = -2$

$2x - 12 = -2$

$x = +5$

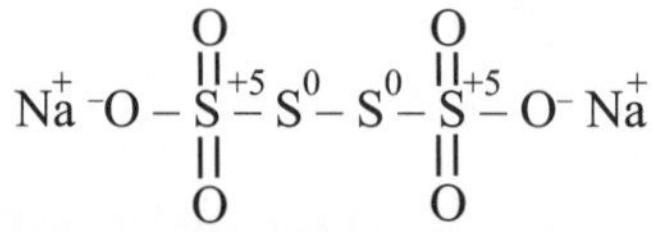

23. (4) 1. H_2S

$2(+1) + x = 0$

$x = -2$

2. H_2SO_5

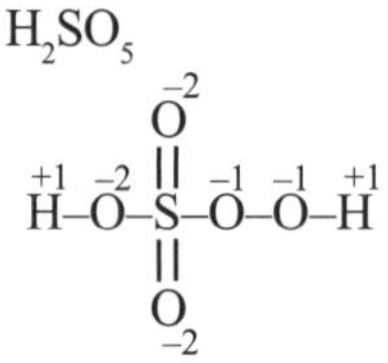

$2(+1) + x + 3(-2) + 2(-1) = 0$

$2 + x - 6 - 2 = 0$

$x = +6$

3. H_2SO_4

$2(+1) + x + 4(-2) = 0$

$2 + x - 8 = 0$

$x = +6$

4. H_2SO_3

$2(+1) + x + 3(-2) = 0$

$2 + x - 6 = 0$

$x = +4$

24. (4) $\overset{0}{O_3} + 6I^- + 6H^+ \longrightarrow 3I_2 + 3H_2\overset{-2}{O}$

i.e., 2 e⁻ change for one oxygen

$\therefore$ 6 e⁻ change for 3 oxygen

Hence, n factor = 6

$\therefore$ Equivalent weight of $O_3 = \frac{M}{6}$

25. (3) X_2H_4

$\because$ E· N of X > H

$\therefore$ H is in +1 oxidation state and X is in $\frac{-1}{2}$ oxidation state.

$\therefore -\frac{1}{2}\times 2+4(+1)=-1+4=3$

Hence, 3 electrons must be liberated.

26. (2) The structure of O_2PtF_6 is known and there is an O_2 unit. The O_2 is best formulated as O_2^+ (dioxygenyl) and oxidation number of each O atom is $+\frac{1}{2}$.

27. (1) $\overset{-3}{(NH_4)_2}\overset{+6}{Cr_2}O_7 \xrightarrow{heat} \overset{+3}{Cr_2}O_3 + \overset{0}{N_2} + 4H_2O$

NH_4^+	$Cr_2O_7^{2-}$	Cr_2O_3
$x+4(+1)=+1$	$2x+7(-2)=-2$	$2x-6=0$
$x=-3$	$2x-14=-2$	$2x=6$
	$x=+6$	$x=+3$

28. (3) Beryllium carbide reacts with water to give methane

$Be_2C + 4H_2O \longrightarrow \underset{(X)}{CH_4} + 2Be(OH)_2$

CH_4

$x+4(+1)=0$

$x=-4$

Hence, oxidation state of heaviest constituent element of gas X is –4.

29. (2) $2S_2O_3^{2-} + I_2 \longrightarrow S_4O_6^{2-} + 2I^-$; only I_2, being a weaker oxidising agent oxidises $S_2O_3^{2-}$ ion.

Hence, X in the given reaction will be I.

30. (4) All the given reactions are disproportionation reactions.

a. $\overset{+1\ -1}{H_2O_2} \longrightarrow 2\,H_2\overset{-2}{O} + \overset{0}{O_2}$ (Oxidation: $-1 \to 0$; Reduction: $-1 \to -2$)

b. $\overset{0}{S_8} + 12\,OH^- \longrightarrow 4\,\overset{2-}{S} + 2\overset{+2}{S_2}O_3^{2-}$ (Oxidation: $0 \to +2$; Reduction: $0 \to -2$)

c. $\overset{0}{Cl_2} + 2\,OH^- \longrightarrow \overset{+1}{Cl}O^- + Cl^- + H_2O$ (Oxidation: $0 \to +1$; Reduction: $0 \to -1$)

31. (1) $N_2H_4 \longrightarrow Y + 10\,e^-$

Total oxidation number of two N atoms in N_2H_4 is

$2x+4(+1)=0$

$2x+4=0$

$2x=-4$

As, it loses 10 moles of electrons, therefore total oxidation number of two nitrogen atoms increase by 10.

i.e., Total oxidation number of two nitrogen atoms in Y is: $-4+10=+6$

$\therefore$ Oxidation number of each nitrogen atom in Y is = +3.

32. (4) $2\overset{0}{H_2} + \overset{0}{O_2} \longrightarrow 2\overset{+1\ -2}{H_2O}$

$\therefore O_2$ acts as oxidizing agent

2 e^- change for 1 oxygen atom

$\therefore$ 4 e^- change for 2 oxygen atoms

Hence, equivalent weight = $\frac{\text{Molar mass}}{\text{Change in oxidation number}}$

$=\frac{32}{4}=8$

33. (3) $\overset{+3\ -2}{As_2S_3} + H^+ + NO_3^- \longrightarrow NO + H_2O + \overset{+5}{As}O_4^{3-} + \overset{+6}{S}O_4^{2-}$

Valence factor = 2 × (+2) + 3× (8) = 28

Equivalent weight = $\frac{\text{Molecular weight}}{\text{Valence factor}} = \frac{M}{28}$

34. (1) F_2 has, the maximum E° value (+2.87 V) among the halogens, hence it is the best oxidant among halogens.

35. (2) F does not show disproportionation tendency as it is most electronegative atom.

36. (2) $Cr_2O_7^{2-} + 6e^{\ominus} \rightarrow Cr^{3+}$

Meq. of ABD = Meq. of $K_2Cr_2O_7$

$3.26\times10^{-3}\times x = 1.68\times10^{-3}\times 6$ or $x=3$

New oxidation number of A^{-n} will be $=-n+3$ or $3-n$

37. (1) $CH_2 = CH_2$ gets reduced to $CH_3 - CH_3$ while H_2 gets oxidised as its oxidation state changes to +1.

38. (3) Equivalent weight in disproportionation reaction

$=\frac{M}{n\ factor_1}+\frac{M}{n\ factor_2}$

$96=\frac{160}{2}+\frac{160}{n\ factor_2} \Rightarrow 16=\frac{160}{n\ factor_2}$

$\therefore$ n $factor_2 = 10$

which is only possible if the oxidised product is BrO_3^-.

39. (3) 2 HgO (s) → 2 Hg (l) + O_2 (g)

This reaction represents reduction process on account of removal of oxygen from mercuric oxide.

40. (3) Oxidation state of nitrogen in NH_3 is $x + 3\times(+1) = 0$ or $x = -3$

Oxidation state of nitrogen in HNO_3 is $1 + x + 3\times(-2) = 0$ or $x = +5$

Oxidation state of N in NaN_3 is $+1 + 3x = 0$ or $x = -1/3$

Oxidation state of N in Mg_3N_2 is $3\times2 + 2x = 0$ or $x = -3$

Hence, 3 molecules (i.e. NH_3, NaN_3 and Mg_3N_2) among the given have nitrogen in negative oxidation state.

41. (4) Among the given species, ClO_4^- does not disproportionate because in this oxoanion, chlorine is present in its highest oxidation state i.e., +7.

Rest three oxoanions of chlorine show disproportionation reaction:

$3\overset{+1}{Cl}O^- \longrightarrow 2\overset{-1}{Cl^-} + \overset{+5}{Cl}O_3^-$

$6\overset{+3}{Cl}O_2^- \xrightarrow{hv} 4\overset{+5}{Cl}O_3^- + 2\overset{-1}{Cl^-}$

$4\overset{+5}{Cl}O_3^- \longrightarrow \overset{-1}{Cl^-} + 3\overset{+7}{Cl}O_4^-$

42. (2) Oxidation number of O in OF_2, O_3, KO_2 and BaO_2 are + 2, 0, – 1/2 and – 1 respectively.

43. (2) The balanced chemical reaction is:

$Cr_2O_7^{2-}$ (aq) + 3 SO_3^{2-} (aq) + 8 H^+ (aq) → 2 Cr^{3+} (aq) + 3 SO_4^{2-} (aq) + 4 H_2O (l)

Hence, a, b, x, y and z in the reaction correspond to 3,8,2,3,4.

44. (4) The oxidation numbers of P in PO_4^{3-}, of S in SO_4^{2-} and that of Cr in $Cr_2O_7^{2-}$ are + 5, + 6 and + 6 respectively.

Correct & Incorrect MCQs

1. (4) Each oxygen atom is assigned an oxidation number of –1 in peroxides like H_2O_2 and Na_2O_2.

2. (3) The algebraic sum of the oxidation number of all the atoms in a compound must be zero.

3. (1) Fluorine is the strongest oxidising agent.

4. (2) ClO_4^- does not disproportionate because in this oxoanion, chlorine is present in its highest oxidation state i.e., +7.

5. (3) In redox reactions, electrons are transferred from the reductant to the oxidant.

release of $2e^-$ (Cu → Cu^{2+})

$$Cu(s) + 2Ag^+(aq) \longrightarrow Cu^{2+}(aq) + 2Ag(s)$$

gain of $2e^-$ (Ag^+ → Ag)

Hence, this reaction is a redox reaction.

6. (3) $Cr_2O_7^{2-}$ is not a self indicator but is used in the titration. It oxidises the indicator substance diphenylamine just after the equivalence point to produce an intense blue colour, thus signaling the end point.

7. (3) The oxidation state of N in HNO_2 is +3 which can change to –3 and +5, hence HNO_2 can act both as an oxidising as well as a reducing agent.

8. (4) In $AgCN + KCN \rightarrow K[Ag(CN)_2]$, no oxidation or reduction occur as oxidation number remains same.

$\overset{+4}{Mn}O_2 + 4\overset{+1}{H}Cl \rightarrow \overset{+2}{Mn}Cl_2 + \overset{0}{Cl_2} + 2H_2O$, MnO_2 acts as an oxidising agent and HCl acts as reducing agent.

9. (3) In a redox reaction, reductant gets oxidised and oxidant gets reduced.

10. (4) Aluminium has most common +3 oxidation number in all its compounds.

Statement Based MCQs

1. (2) Cu + HCl → No reaction (in accordance with electrochemical series, H is more electropositive than Cu). Position of Cu is below H_2 in series.

2. (1) $\overset{0}{Sn}(s) + 8H^+(aq.) + 4NO_3^- + 6Cl^- \rightarrow \overset{+4}{Sn}Cl_6^{2-} + 4NO_2 + 4H_2O$

Oxidation : Reducing Agent.

$4e^-$ are added in the balancing to make:

$S_2O_3^{2-} + 6H^+ + 4e^- \rightarrow 2S + 3H_2O$

3. (1) CrO_5 :

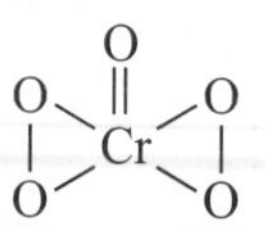

$x + 4(-1) + 1(-2) = 0$

$\boxed{x = +6}$

Here, 4 oxygen are involved in peroxy linkage.

4. (1) Aluminum is more reactive than Fe. Thus, displacement occurs; $Fe_2O_3 + 2Al \rightarrow Al_2O_3 + 2Fe$.

5. (1) $\overset{0}{N_2} \rightarrow 2\overset{-3}{N}H_3$

Equivalent weight of $NH_3 = \frac{14+3}{3} = \frac{17}{3}$

(Molecular weight of NH_3 = 14 + 3 = 17)

$N_2 + 6e^{\ominus} \rightarrow 2N^{3-}$

Equivalent weight of $N_2 = \frac{14 \times 2}{6} = \frac{28}{6}$.

6. (3) Since, fluorine is the most electronegative element, hence it has an oxidation number of -1 in all its compounds.

7. (4) In a disproportionation reaction, an element in one oxidation state is simultaneously oxidised and reduced.

8. (4) Oxidation number can be calculated using some rules. H is assigned +1 oxidation state and O has oxidation number –2.

Let O. N of C in CH_2O be x

O. N of C = x + 2 – 2 = 0 $\Rightarrow$ x = 0

$\therefore$ O. N of C = 0

9. (2) E° of Cu^{2+}/Cu is + 0.34 V and positive E° means that the redox couple is a weaker reducing agent than H^+/H_2 couple.

10. (1) Number of valence shell electrons increases in same shell across the period, hence oxidation number exhibited by an atom generally increases across the period in periodic table.

In ClO_4^- oxoanion, chlorine is present in its highest oxidation state +7, hence ClO_4^- does not disproportionate.

Assertion & Reason MCQs

1. (2) F always shows an oxidation number of –1 as being the most electronegative element, hence, F can only be reduced. Thus, it is the best oxidant.

2. (3) $2K\overset{+7}{Mn}O_4 + 2KI \rightarrow 2\overset{+4}{Mn}O_2 + I_2 + 4KOH$

Reduction

Hence, manganese changes its oxidation state from +7 to +4 during the reaction.

3. (1) Any reaction in which same element undergoes reduction as well as oxidation is considered as a disproportionation reaction.

4. (2) Reason is not the correct explanation.

5. (2) Zn is more reactive than Ag and hence precipitates Ag from $AgNO_3$.

6. (3) As, nitrogen is more electronegative, it will pull the electron of hydrogen towards itself, hydrogen has +1 oxidation state, therefore N will have (–1/3) O.S.

7. (1) Since oxidation number of Br changes from –1 to 5, thus undergoing oxidation, therefore, acts as a reducing agent.

8. (1) Species in highest oxidation state behave as oxidant only.

9. (4) As oxidation number of Ag remains same, therefore no oxidation or reduction occur in this reaction.

10. (4) HNO_3 can only be reduced so it is an oxidant as N in it has maximum oxidation state of +5.

Multi Correct MCQs

1. **(4)** $2Fe^{3+} + 2I^- \longrightarrow 2Fe^{2+} + I_2$

$2Fe^{2+} + S_2O_8^{2-} \longrightarrow 2Fe^{3+} + 2SO_4^{2-}$

(A) Fe^{+3} oxidises I^- to I_2 and convert itself into Fe^{+2}.

(D) This Fe^{+2} reduces $S_2O_8^{2-}$ to SO_4^{2-} and converts itself into Fe^{+3}.

2. **(1)** $2\left[\overset{+1}{Au}(CN)_2\right]^- + \overset{0}{Zn}(s) \longrightarrow 2\overset{0}{Au} + \left[\overset{+2}{Zn}(CN)_4\right]^{-2}$

It is a redox reaction in which Zn displaced Au^+

Reduction and Oxidation both are taking place simultaneously.

3. **(4)** H_2SO_3, $SOCl_2$, SF_4

Compounds	O.S. of Sulphur
SO_3	+6
H_2SO_3	+4
$SOCl_2$	+4
SF_4	+4
$BaSO_4$	+6
$H_2S_2O_7$	+6

4. **(2)** Elements having intermediate oxidation state can under go disproportionation reaction.

H_2O_2, ClO_3^-, P_4, Cl_2, Cu^{+1}, NO_2

NCERT Exemplar MCQs

1. **(4)** Following are the examples of redox reaction

(1) $\overset{+2}{Cu}\overset{-2}{O} + \overset{0}{H_2} \rightarrow \overset{0}{Cu} + \overset{+1}{H_2}\overset{-2}{O}$

(2) $\overset{+3}{Fe_2}O_3 + 3\overset{+2}{C}O \rightarrow 2\overset{0}{Fe} + 3\overset{+4}{C}O_2$

(3) $2\overset{0}{K} + \overset{0}{F_2} \rightarrow 2\overset{+1}{K}\overset{-1}{F}$

(4) $\overset{+2}{Ba}\overset{-1}{Cl_2} + \overset{+1}{H_2}\overset{+6}{S}\overset{-2}{O_4} \rightarrow \overset{+2}{Ba}\overset{+6}{S}\overset{-2}{O_4} + 2\overset{+1}{H}\overset{-1}{Cl}$

Option (4) is not an example of redox reaction as there is no change in the oxidation number of any element.

2. **(4)** Given E° values of

$Br_2/Br^- = +1.90$

$Ag^+/Ag(s) = +0.80$

$Cu^{2+}/Cu(s) = +0.34$

$I_{2(s)}|I^- = +0.54$

Copper will reduce Br_2, if the E° of the following redox reaction is positive.

$2Cu + Br_2 \rightarrow CuBr_2 + 2Br^-$

Now, $Cu \rightarrow Cu^{2+} + 2e^-$; E° = –0.34 V

$Br_2 + 2e^- \rightarrow 2Br^-; E = +1.90\,V$

$Cu + Br_2 \rightarrow CuBr_2 + 2Br^{\ominus}; \Delta E = +1.56\,V$

Since, E° of this reaction is positive, therefore, Cu can reduce Br_2. While other reaction will give negative value.

3. **(4)** $2Fe^{3+} + 2e^- \longrightarrow 2Fe^{2+}$; E° = +0.77 V

$2I^- \rightarrow I_2 + 2e^-$; E° = –0.54 V (sign of E° is reversed)

$Fe^{+3} + 2I^- \rightarrow 2Fe^{2+} + I_2$; $E^\circ_{cell} = +0.23$ V

This reaction is feasible since E°_{cell} is positive.

(2) $Cu \rightarrow Cu^{2+} + 2e^-$; $E^\circ_{cell} = -0.34$ V (sign of E° has been reversed)

$2Ag^+ + 2e^- \rightarrow 2Ag$; E° = 0.80 V

$Cu + 2Ag^+ \rightarrow Cu^{2+} + 2Ag$; E° = +0.46 V

This reaction is feasible since E°cell is positive.

(3) $2Fe^{3+} + 2e^- \rightarrow 2Fe^{2+}$; E° = +0.77 V

$Cu \rightarrow Cu^{2+} + 2e^-$; E° = –0.34 V (sign of E° is reversed)

$2Fe^{3+} + Cu \rightarrow 2Fe^{2+} + Cu^{2+}$; E° = +0.43 V

This reaction is feasible since E°_{cell} is positive.

(4) $Ag \rightarrow Ag^+ + e^-$; E° = –0.80 V (sign of E° is reversed)

$Fe^{3+} + e^- \rightarrow Fe^{2+}$; E° = +0.77V

$Ag + Fe^{3+} \rightarrow Ag^+ + Fe^{2+}$; $E^0 = -0.03$V

This reaction is not feasible since E°cell is negative.

4. **(1)** $2\overset{+2}{S_2}\overset{-2}{O_3^{2-}}(aq) + I_2^0(s) \rightarrow \overset{2.5}{S_4}O_6^{2-}(aq) + 2I^-(aq)$

$\overset{+2}{S_2}\overset{-2}{O_3^{2-}}(aq) + 2\overset{0}{Br_2}(l) + 5H_2O(l) \rightarrow +2\overset{+6}{S}O_4^{2-}(aq) + 4Br^-(aq) + 10H^+(aq)$

Bromine being stronger oxidising agent than I_2, oxidises S of $S_2O_3^{2-}$ to SO_4^{2-} whereas I_2 oxidises it only into $S_4O_6^{2-}$ ion.

5. **(1)** Oxidation number of hydrogen is always +1 is a wrong rule since, it is +1 in hydrogen halides, – 1 in hydrides and zero in H_2 molecule.

All the other three statements (2), (3) and (4) are correct.

6. **(2)** NH_4NO_3 is actually NH_4^+ and NO_3^-. It is an ionic compound.

Nitrogen in the two species have different oxidation no. as shown below:

Let oxidation number of N in NH_4^+ is x.

$\Rightarrow x + (4 \times 1) = +1$

or $x + 4 = +1$ or $x = -3$

Let oxidation number of N in NO_3^- is x

$\Rightarrow x + (3 \times -2) = -1$

or $x - 6 = -1$ or $x = +5$

7. **(1)** Writing the oxidation number (O.N.) of Cr, Cl and Mn on each species in the four set of ions, then,

(1) $\overset{+3}{Cr}O_2^-, \overset{+5}{Cl}O_3^-, \overset{+6}{Cr}O_4^{2-}, \overset{+7}{Mn}O_4^-$

(2) $\overset{+5}{Cl}O_3^-, \overset{+6}{Cr}O_4^{2-}, \overset{+7}{Mn}O_4^-, \overset{+3}{Cr}O_2^-$

(3) $\overset{+3}{Cr}O_2^-, \overset{+5}{Cl}O_3^-, \overset{+7}{Mn}O_4^-, \overset{+6}{Cr}O_4^{2-}$

(4) $\overset{+6}{Cr}O_4^{2-}, \overset{+7}{Mn}O_4^-, \overset{+3}{Cr}O_2^-, \overset{+5}{Cl}O_3^-$

Only in the arrangement (1), the O. N. of central atom increases from left to right. Therefore, option (1) is correct.

8. (4) Highest oxidation number of any transition element = (n – 1) d electrons + ns electrons. Therefore, large the number of electrons in the 3d-orbitals, higher is the maximum oxidation number.

(1) $3d^1 4s^2 = 3$

(2) $3d^3 4s^2 = 3 + 2 = 5$

(3) $3d^5 4s^1 = 5 + 1 = 6$, and

(4) $3d^5 4s^2 = 5 + 2 = 7$

Thus, option (4) is correct

9. (4) Disproportionation reactions are reactions in which the same substance is oxidised as well as reduced. Writing the oxidation number of each element above its symbol in the given reactions.

(1) $\overset{-4+1}{CH_4} + \overset{0}{2O_2} \rightarrow \overset{+4-2}{CO_2} + \overset{+1\ -2}{2H_2O}$

(2) $\overset{-4+1}{CH_4} + \overset{0}{4Cl_2} \rightarrow \overset{+4-1}{CCl_4} + \overset{+1-1}{4HCl}$

(3) $\overset{0}{2F_2} + \overset{-2+1}{2OH^-} \rightarrow \overset{-1}{2F^-} + \overset{+2\ -1}{OF_2} + \overset{+1\ -2}{H_2O}$

(4) $\overset{+4-2}{2NO_2} + \overset{-2+1}{2OH^-} \rightarrow \overset{+3-2}{NO_2^-} + \overset{+5-2}{NO_3^-} + \overset{+1\ -2}{H_2O}$

Thus, in reaction (4), N is both oxidised as well as reduced since the O.N. of N increases from + 4 in NO_2 to + 5 in NO_3^- and decreases from + 4 in NO_2 to + 3 in NO_2^-.

10. (3) F always shows an oxidation number of –1 as being the most electronegative element, hence, F can only be reduced. Further, due to the absence of d-orbitals, it cannot be oxidised and hence it does not show positive oxidation numbers.

In other words, F cannot be oxidised as well as reduced simultaneously and hence does not show disproportionation reactions.

NEET & AIIMS Past Year MCQs

1. (4) In the given reaction:

$$3MnO_4^{2-} + 4H^+ \longrightarrow 2MnO_4^- + MnO_2 + 2H_2O$$

Oxidation state of Mn	Species
+6	MnO_4^{2-}
+7	MnO_4^-
+4	MnO_2

Thus, manganese exhibits oxidation states of +6, +7, and +4 in this reaction. However, it does not exhibit oxidation states of +2 or +3.

2. (2) Redox reactions involve change in oxidation number of the interacting species.

$\overset{0}{H_2} + \overset{0}{Cl_2} \rightarrow \overset{+1\ -1}{2HCl}$ (Redox)

$\overset{+2\ -1}{BaCl_2} + \overset{+1\ -2}{Na_2SO_4} \rightarrow \overset{+2\ -2}{BaSO_4} + \overset{+1\ -1}{2NaCl}$

(No change in oxidation state, hence not redox reaction)

$\overset{0}{Zn} + \overset{+2}{CuSO_4} \rightarrow \overset{+2}{ZnSO_4} + \overset{0}{Cu}$ (Redox reaction)

$\overset{+5}{2KClO_3} + \overset{0}{I_2} \rightarrow \overset{+5}{2KIO_3} + \overset{0}{Cl_2}$ (Redox reaction)

3. (2) Reduction half reaction.

$Cr_2O_7^{2-} + 14H^+ + 6e^- \rightarrow 2Cr^{+3} + 7H_2O$ ... (i)

Oxidation half reaction

$SO_3^{2-} + H_2O \rightarrow SO_4^{2-} + 2e^-] \times 3$ ... (ii)

Oxygen is balanced by adding water and hydrogen is balanced by adding H^+ and the charge is balanced by electrons.

Add(eq. (i)) + (3 × eq. (ii))

$Cr_2O_7^{2-} + 3SO_3^{2-} + 8H^+ \rightarrow 2Cr^{+3} + 3SO_4^{2-} + 4H_2O$

a = 1 b = 3 c = 8

4. (1) SRP ⇒ Standard Reduction Potential values:

Therefore SRP order is-

$E^o_{Zn^{2+}/Zn} < E^o_{Fe^{2+}/Fe} < E^o_{Cu^{2+}/Cu} < E^o_{Ag^+/Ag}$

Reactivity order : Zn > Fe > Cu > Ag

In case of displacement reaction, more reactive metals (lower SRP) can displace less reactive metals (higher SRP) from their salt solution.

$CuSO_{4(aq.)} + 2Ag_{(s)} \rightarrow Cu_{(s)} + Ag_2SO_{4(aq.)}$

Reaction is not possible as Ag is less reactive metal as compared to Cu.

5. (1) Aluminium is more electropositive than Cr, so it displaced chromium from Cr_2O_3.

$Cr_2O_3 + Al \xrightarrow{\Delta} Al_2O_3 + Cr$

6. (1) The reaction of MnO_4^{2-} in aqueous medium takes place as

$3MnO_4^{2-} + 2H_2O \longrightarrow MnO_2 + 2MnO_4^- + 4OH^-$

If one mole of MnO_4^{2-} is taken then

$MnO_4^{2-} + 2/3\ H_2O \longrightarrow 1/3\ MnO_2 + 2/3\ MnO_4^- + 4/3\ OH^-$

Therefore, $\frac{2}{3}$ mol of MnO_4^- and 1/3 mole of MnO_2

7. (1) **8. (4)** **9. (4)** **10. (4)**

The p-Block Elements

NCERT Topic-wise MCQs

1. (4) Due to inert pair effect, thallium trihalides are less stable.

2. (1) Al and Ga have nearly same atomic radii due to inert pair-effect.

3. (3) With the increase in the size of the halogen atom, the acidic strength of boron halides increase.

4. (1) The strongest Lewis acid is BI_3.

5. (3) A mineral ore of aluminum: $Al_2O_3 . 2H_2O$ is bauxite.

6. **(2)** As we move down the group from Al to Tl, inert pair effect becomes more and more pronounced. In other words, stability of +1 oxidation state increases down the group, i.e.,

Al < Ga < In < Tl

7. **(3)** Tl^{3+} acts as oxidising agent.

8. **(3)** Al is rendered passive by formation of an impervious layer of Al_2O_3 on its surface when brought in contact with conc. HNO_3.

9. **(2)** Greater the charge and smaller the size of the cation (i.e., Al^{3+}), greater the charge density, greater the polarisation and hence, greater is the covalent character i.e., $AlCl_3$.

10. **(2,3)** The maximum oxidation state of Tl is +3, therefore, Tl in $TlCl_3$ cannot increase its oxidation state beyond +3. Further, since +1 oxidation state of Tl is much more stable than +3, therefore, Tl in TlCl is not likely to increase its oxidation state to +3. Thus; both $TlCl_3$ and TlCl does not undergo disproportionation.

11. **(1)** BF_3 is electron deficient compound, hence, it is known as lewis acid.

12. **(2)** Aluminium has very high metallic character.

13. **(3)** Boron does not exhibit inert pair effect.

14. **(3)** The general outer electronic configuration of group-13 elements is $ns^2\ np^1$.

15. **(1)** The general valence shell electronic configuration of p-block elements is $ns^2\ np^{1-6}$. Therefore, the maximum oxidation state shown by a p-block element is equal to the total number of valence electrons i.e., the sum of the s and p-electrons.

16. **(1)** The electronegativity of group-13 elements first decreases from B to Al and then increases marginally.

17. **(1)** Boron is non-metallic in nature. Hence, option (1) is incorrect.

18. **(1)** Due to high electronegativity

19. **(1)** Aluminium chloride in acidified aqueous solution forms octahedral $[Al(H_2O)_6]^{3+}$ion.

20. **(4)** Amorphous boron and aluminium metal on heating in air form B_2O_3 and Al_2O_3 respectively.

21. **(3)** Aluminium and gallium oxides are amphoteric in nature.

22. **(2)** The maximum covalence of B is 4 due to the absence of d orbitals.

23. **(1)** In B_2H_6, each B - atom uses sp^3 - hybrid orbitals for bonding. Out of 4 sp^3 - hybrid orbitals, one orbital is empty. Here, one half filled orbital of B - atoms overlaps with one empty orbitals of other B - atom and half - filled orbital of hydrogen atom resulting in 3 centered two e^- bridge bonds. Thus, these bonds are electron deficient.

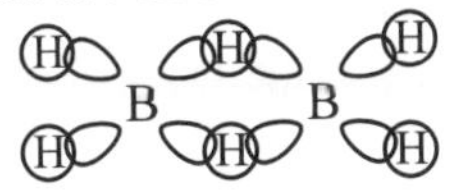

24. **(3)** Borazine, $B_3N_3H_6$ is known as inorganic benzene due to its resemblance in structure and some physical and chemical properties with benzene.

25. **(2)** Four isomers are possible. These are:-

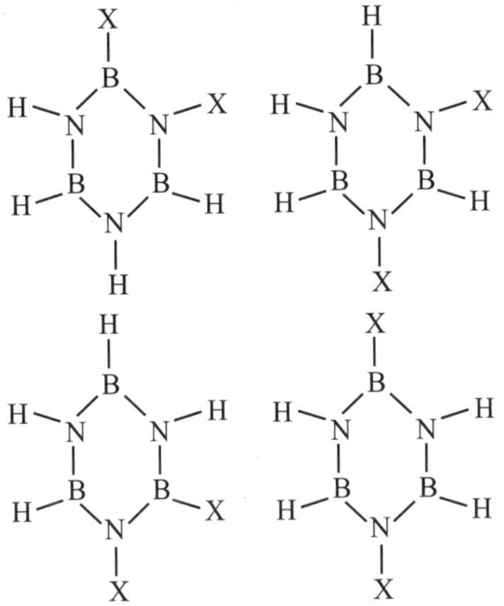

26. **(3)** B_2H_6 has two 3c-2e⁻ bonds.

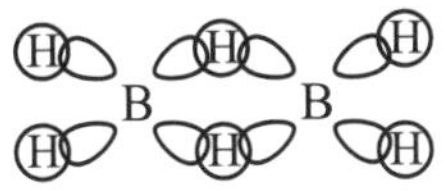

27. **(3)** B_2H_6 - Tetrahedral, sp^3

28. **(2)** B-B bond does not exist in diborane.

29. **(2)** 2 Boron atoms and four terminal hydrogen atoms are present in same plane.

30. **(1)** A blue coloured $Co(BO_2)_2$ bead is formed when borax is heated in a Bunsen burner flame with CoO on a loop of platinum wire.

31. **(2)** Metaboric acid is formed when orthoboric acid is heated above 370K.

$H_3BO_3 \xrightarrow{\Delta} HBO_2$

32. **(3)** HBO_2 (obtained on heating orthoboric acid above 370 K) on further heating yields boric oxide i.e., B_2O_3.

$HBO_2 \xrightarrow{\Delta} B_2O_3$

33. **(1)** Diborane (B_2H_6) is produced on an industrial scale by the reaction of BF_3 with sodium hydride.

$2BF_3 + 6NaH \xrightarrow{450\ K} B_2H_6 + 6NaF$

34. **(2)** B_3H_6 is not a borane.

35. **(2)** Boron fibre is used in making bullet-proof vest.

36. **(4)** Borax and boric acid are used in the manufacture of heat resistant glasses, glass-wool and fiberglass.

37. **(2)** Borax is used as a flux for soldering metals.

38. **(3)** An aqueous solution of orthoboric acid is generally used as a mild antiseptic.

39. **(4)** Aluminium is a bright silvery-white metal. It has high tensile strength and it has a high electrical and thermal conductivity. Hence, all the given statements are correct.

40. **(3)** Si is a semi-conductor and used in transistors.

41. **(4)** The stability of dihalides increases down the group because inert pair effect becomes more predominant down the group.

42. **(1)** BeO is amphoteric, CaO is basic while B_2O_3 and SiO_2 are acidic. Therefore; CaO will not react with alkali.

43. (3) Due to inert pair effect, +2 oxidation states of Pb is more stable while +4 oxidation state of tin is more stable.

44. (4) Carbon due to its small size and more bond strength.

45. (4) Due to inert pair effect, the reducing character (M^{2+} species) in dihalides decreases in order:-

$GeCl_2 > SnCl_2 > PbCl_2$

46. (1) SiO_2 is acidic among given options.

47. (1) Due to the absence of d-orbitals, CCl_4 cannot accommodate the electron pair supplied by H_2O in the first step of hydrolysis reaction.

48. (1) +4 oxidation state of Ge is more stable than that of Sn therefore; strongest reductant is $GeCl_2$.

49. (2) The outer shell electronic configuration of group 14 elements is ns^2np^2.

50. (1) The common oxidation states exhibited by group 14 elements are +4 and +2.

51. (2) Lead compounds in +4 oxidation states tend to obtain lower oxidation states by getting reduced and hence, behave as strong oxidizing agents.

52. (4) Among monoxides of group 14 elements, CO is neutral, GeO is acidic, SnO and PbO are amphoteric in nature.

53. (4) Among the given group-14 elements, tin decomposes steam to form dioxide and dihydrogen gas.

$$Sn + 2H_2O \xrightarrow{\Delta} SnO_2 + 2H_2$$

54. (1) SnF_4 and PbF_4 are ionic in nature.

55. (4) All the given options are correct.

56. (2) The geometry of carbon is tetrahedral in diamond, each carbon atom is sp^3 - hybridised.

57. (2) The hybridization state of carbon in fullerene is sp^2.

58. (3) Since; C-C bond in graphite has double bond character, therefore, its bond order is higher than that in diamond.

59. (2) Graphite has hexagonal sheet type structure.

60. (2) The structural unit of silicone:

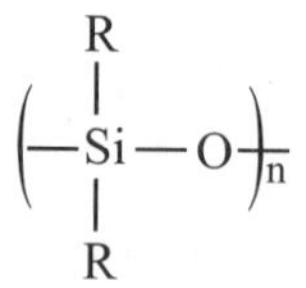

61. (3) Me_3SiCl is not a monomer for a high molecular mass silicone polymer because it generates Me_3SiOH when subjected to hydrolysis which contains only one reacting site.

Hence; the polymerisation reaction stop just after first step.

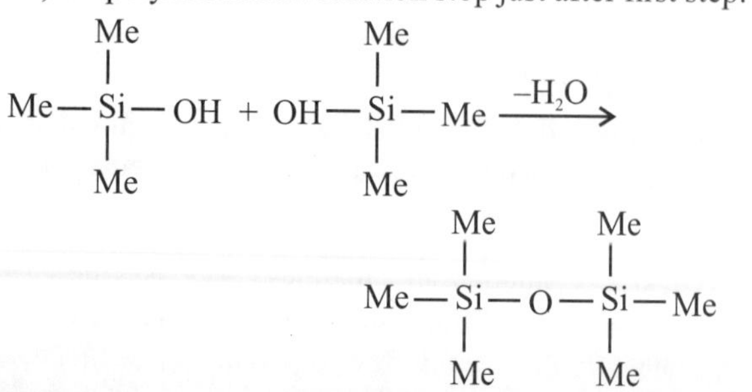

62. (1) Silica is SiO_2. So, ratio is 1:2.

63. (2) $SiO_2 + 2NaOH \rightarrow Na_2SiO_3 + H_2O$

64. (1) Si–O–Si

65. (2) A mixture of CO and N_2 is known as producer gas.

66. (3) Carbon monoxide gas is highly poisonous in nature because it forms a complex with haemoglobin which is about 300 times more stable than the oxygen-haemoglobin complex. This prevents haemoglobin in the red blood corpuscles from carrying oxygen round the body and ultimately resulting in death.

67. (4) CO_2 is non supporter of combustion, hence, it is used as fire extinguisher.

68. (4) Silicones are widely used as sealant, greases, electrical insulators and for water proofing of fabrics.

69. (1) ZSM-5 is a type of zeolite which is used to convert alcohols directly into gasoline.

High Order Time Intensive MCQs

1. (4) The vacant orbital of B participates in a 3c-2e- bond in B_2H_6 and Cl^- donates a pair of electrons to the vacant orbital of Al in Al_2Cl_6.

2. (4) Borazine ($B_3N_3H_6$) is polar, so, it is more reactive than organic benzene.

3. (2) The chemical formula of Borazine is $B_3N_3H_6$ and its chemical structure is

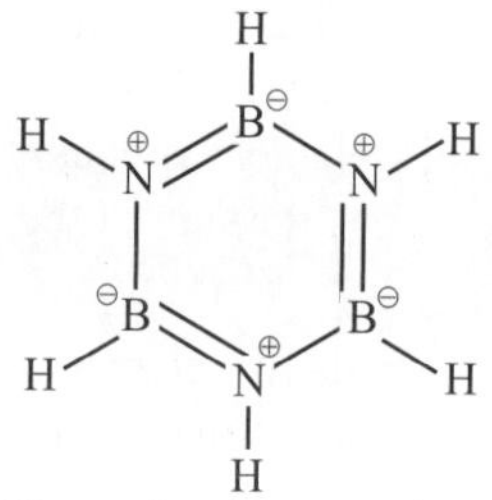

From the structure, it can be concluded that it has electronic delocalization and is a cyclic compound.

The reaction with water takes place as:

$B_3N_3H_6 + 9H_2O \rightarrow 3NH_3 + 3H_3BO_3 + 3H_2$

4. (1) Due to inert pair effect, in boron family, lower O.S. (+1) becomes more stable down the group.

5. (2) C–C = 83 kcal / mol; Si–Si = 54 kcal / mol; Ge–Ge = 40 kcal / mol; Sn–Sn = 37 kcal / mol.

6. (3) $\underset{}{Na_2B_4O_7 . 10H_2O} \xrightarrow{\Delta} \underset{(X)}{B_2O_3} + NaBO_2 + H_2O$

$B_2O_3 + Cr_2O_3 \xrightarrow{\Delta} \underset{(Y)\ Green}{Cr(BO_2)_3}$

7. (3) 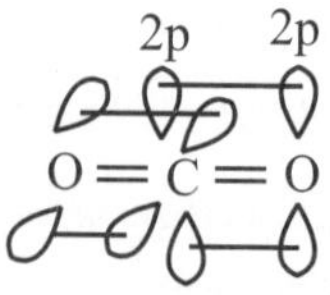 2pπ – 2pπ bonding in CO_2

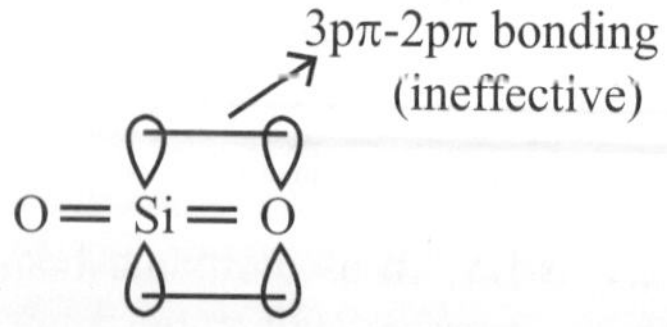

8. (3) Atomic weight is common in diamond & Graphite.

9. (4) $SiO_2 + 4HF \rightarrow SiF_4 + 2H_2O$

10. (2) $4BCl_3 + 3LiAlH_4 \rightarrow \underset{(A)}{2B_2H_6} + 3LiCl + 3AlCl_3$

$B_2H_6 + 6H_2O \rightarrow \underset{(B)}{2B(OH)_3} + 6H_2$

$2B(OH)_3 \xrightarrow{\Delta} \underset{(C)}{B_2O_3} + 3H_2O$

11. (2) Hydride of super halogen = HF

$SiO_2 + 4HF \rightarrow \underset{(X)}{SiF_4} + 2H_2O$

$3SiF_4 + 4H_2O \rightarrow \underset{(Y)}{H_4SiO_4} + 2H_2SiF_6$

$H_4SiO_4 \xrightarrow{1000°C} \underset{(Z)}{SiO_2} + 2H_2O$

$X = SiF_4$, $Y = H_4SiO_4$, $Z = SiO_2$

$Si + O_2 \rightarrow SiO_2$

12. (2) SiO_2 reacts with molten NaOH.
$2NaOH + SiO_2 \rightarrow Na_2SiO_3 + H_2O$

13. (1) Order of I.P. = C > Si > Ge > Pb > Sn
or C > Si > Ge > Sn < Pb

14. (3) An aqueous solution of borax is alkaline in nature.

15. (3) Borax : $Na_2B_4O_7 . 10H_2O$
Tetra Boric acid : $H_2B_4O_7$

16. (4) Thallium is more stable in +1 oxidation state due to inert pair effect.

17. (3) Lead does not give coloured bead.

18. (4) Borax: $Na_2B_4O_7 . 10H_2O$ or $Na_2[B_4O_5(OH)_4].8H_2O$

19. (1) Orthoboric acid contains triangular BO_3^{3-} units

20. (1) H_3BO_3 is a Monobasic acid and weak lewis acid as it accepts electron from OH ions from water.

21. (2) Due to small size and high ionization enthalpy, B forms covalent trihalides.

22. (1) Boron and silicon reacts with metals M to form boranes and silanes respectively.

X-B and Y is Si

23. (2) Aluminate ion in aqueous solution exists as:

$[Al(OH)_4(H_2O)_2]^-$

24. (4) Al is passive towards conc. HNO_3 due to the formation of Al_2O_3 layer.

25. (3) On moving down the group, shielding decreases, therefore Al and Ga have same covalent radius.

26. (1) The maximum covalency of boron is 4 due to the absence of d-orbitals. Hence, $[BF_6]^{3-}$ does not exist.

27. (4) Pb does not show catenation property.

28. (1) Boron halides are electron deficient compounds because the central atom i.e. B has incomplete octet.

29. (2) $2NaH + B_2H_6 \xrightarrow[(Y)]{(C_2H_5)_2O} \underset{\text{(tetrahedral)}}{\underset{Z}{2NaBH_4}}$

$NaBH_4$ is used as reducing agent in organic synthesis.

Y - $C_2H_5OC_2H_5$; Z - $NaBH_4$

30. (1) The maximum valency of beryllium is +4 (due to absence of d-orbitals) while that of aluminium is +6 (due to presence to d-orbitals).

31. (2) $NaNO_3 + Al + \underset{conc}{NaOH} + H_2O \rightarrow 8\ Na[Al(OH)_4] + 3NH_3$

32. (2) Inert pair effect explain stability of +2 oxidation state in lower elements of carbon family.

33. (1) Only option (1), is correct i.e., strength of Lewis acids decreases in the order $BCl_3 > AlCl_3 > GaCl_3$ while all other options are incorrect as explained below:

Al does not show inert pair effect

Tl^{3+} has the highest oxidising power

Ionization enthalpy of Tl is higher than that of Al.

B > Tl > Al

34. (4) There are four electrons in two bridges.

35. (4) $BeCl_2$ and $AlCl_3$ have bridged chlorine structures.

36. (2)

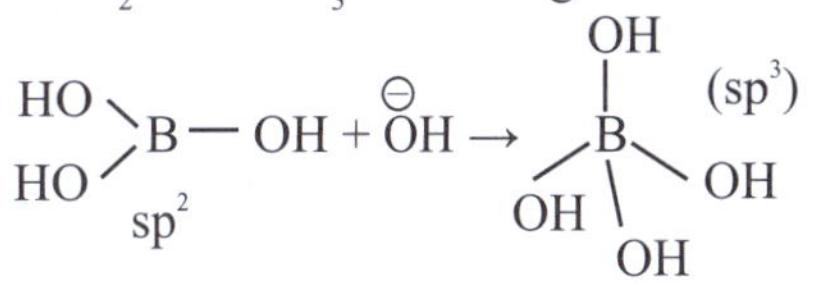

37. (3) Graphite contains free & mobile e^- that is responsible to conduct electricity.

38. (1) Due to absence of d-orbitals, B cannot extend its co-valency beyond 4 and hence, it forms only BF_4^- and not BF_6^{3-}.

39. (2) By reducing silica (rocks), amorphous silicon is prepared. Extra pure silicon is obtained by the removal of SiO_2 by:

$SiO_2 + 4HF \rightarrow SiF_4 + 2H_2O$

40. (4) CO

41. (1) $Al_2(SO_4)_3$ is used in making fire proof clothes.

42. (2) Al (Aluminum) is a chief component of silvery paints.

43. (2) Si-O-Si

44. (2) The basic structural unit of all silicates is the tetrahedral SiO_4^{4-}. It is represented as:

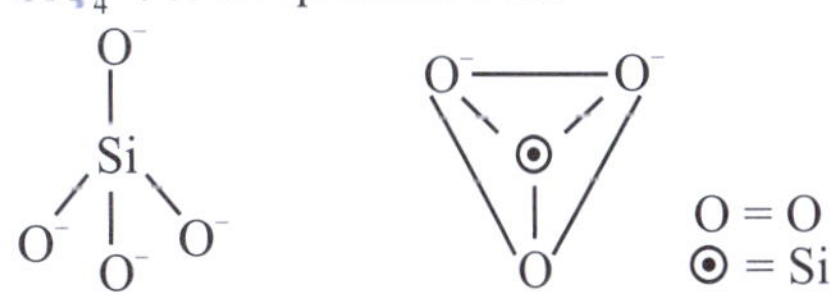

Structure of SiO_4^{4-} ion

45. (3) $2BCl_3 + 6H_2 \rightarrow B_2H_6 + 6HCl$

$3LiAlH_4 + 4BF_3 \rightarrow 3LiF + 3AlF_3 + 2B_2H_6$

46. (3) Pyrosilicate contains two units of SiO_4^{4-} joined together through one corner oxygen atom giving $(Si_2O_7)^{6-}$ anion.

47. (3) $RSiCl_3$ gives cross-linked polymer.

48. (4) Chain structure of silicates contain $(SiO_3^{2-})_n$ anions.

49. (3) There are six groups of p–block elements in the periodic table numbering from group 13 to group 18.

50. (2) In boron and carbon families, the group oxidation state is the most stable state for the lighter elements in the group.

51. (1) The non-metals and metalloids exist only in the p-block elements of the periodic table.

52. **(3)** The 3d orbitals of Al are involved in this complex ion $[Al(H_2O)_6]^{3+}$, and the hybridisation state of Al is sp^3d^2.

53. **(2)** Lithium and sodium tetra-hydridoborates, also known as borohydrides are prepared by the reaction of metal hydrides with B_2H_6(diborane) in diethyl ether.

$2MH + B_2H_6 \rightarrow 2M^+[BH_4]^-$ (M = Li or Na)

54. **(2)** Gallium has an unusually low melting point, allowing it to exist in a liquid state during high summer temperatures.

Statement (1) is incorrect because boron is non-metallic, extremely hard, and has a very high melting point, contrary to what is stated.

Statement (3) is incorrect because the rest of the members of Group 13 (except boron) are soft metals with low melting point and high electrical conductivity.

Statement (4) is incorrect. The density of Group 13 elements increases as one moves down the group from boron to thallium. Gallium has a higher density (5.90 g/cm^3) compared to Aluminium (2.70 g/cm^3).

Correct & Incorrect MCQs

1. **(1)** Since, the total number of valence electrons increases towards right across a period, hence the number of possible oxidation states of p-block elements increases towards right across a period.

2. **(4)** The presence of f-electrons in Tl have low screening effect to compensate the increase in the nuclear charge, hence the ionization enthalpy of Tl is larger than that of In.

3. **(1)** Boron is non-metallic in nature.

4. **(4)** The sum of first three ionization enthalpies is very high for B while it considerably decreases for Al. Hence, B does not form B^{3+} ions while Al is able to form Al^{3+} ions.

Aluminium oxides are amphoteric in nature.

5. **(3)** Due to poor shielding effect of intervening d and f orbitals, the ionization enthalpy of Pb is slightly greater than that of Sn.

Carbon cannot exceed its covalence more than 4.

6. **(2)** Down the group, the size increases and electronegativity decreases, and, hence, tendency to show catenation decreases. Therefore, the order of catenation is $C >> Si > Ge \approx Sn$.

7. **(4)** Fullerenes contain twenty six- membered rings and twelve five-membered rings.

8. **(3)** The four terminal hydrogen atoms and the two boron atoms in diborane lie in one plane.

9. **(3)** The mixture of CO and H_2 is known as water gas or synthesis gas.

In CO molecule, there are one sigma and two π bonds between carbon and oxygen, $:C \equiv O:$.

In CO_2 molecule, carbon atom undergoes sp hybridisation.

10. **(4)** The basic structural unit of silicates is SiO_4^{4-}.

Statement Based MCQs

1. **(1)** Pb^{4+} will take up electrons from I^- to oxidise it to I_2, and itself will reduce to Pb^{2+} to form PbI_2.

2. **(3)** $2Al + 2NaOH + 2H_2O \rightarrow 2NaAlO_2 + 3H_2$

$Zn + 2NaOH + 2H_2O \rightarrow Na_2[Zn(OH)_4] + H_2$

H_2 produces in the above reaction and it corrodes the metal. Therefore, NaOH is not stored in Al or Zn containers.

3. **(1)** BF_3 is an electron deficient molecule and have tendency to accept a pair of electrons to achieve stable electronic configuration and hence, behaves as Lewis acids. Carbon cannot exceed its covalence more than 4 due to the absence of d orbital.

4. **(3)** In carbon dioxide, the carbon is sp hybridized.

5. **(1)** Both statements are individually correct.

6. **(4)** Boron-10 isotope has high ability to absorb neutrons, hence metal borides are used in nuclear industry as protective shields and control rods.

7. **(4)** $[SiF_6]^{2-}$ is known whereas $[SiCl_6]^{2-}$ is not known. Si^{4+} has smaller size, therefore six large chloride ions cannot be accommodated around Si^{4+} and also interaction between lone pair of chloride ion and Si^{4+} is not very strong, hence $[SiCl_6]^{2-}$ is not known.

8. **(1)** Since, diamond is a rigid three dimensional network of carbon atoms, hence, it has high melting point.

9. **(3)** In graphite, electrons are mobile, hence, graphite conducts electricity along the sheet.

10. **(4)** Carbon monoxide is highly poisonous gas because it forms a complex with haemoglobin and prevents it from carrying oxygen around the body.

Assertion & Reason MCQs

1. **(1)** Aluminum is trivalent while silicon is tetravalent. By replacing few aluminum atoms by silicon, the overall structure acquires a negative charge.

2. **(2)** Silicones are water repelling in nature because these are surrounded by non-polar alkyl groups.

3. **(2)** The compound called inorganic benzene is borazine because its structure is like benzene.

4. **(3)** In diamond, all C-atoms are sp^3 hybridized.

5. **(2)** Boron is non-metallic in nature because its radius is very small and has relatively high nuclear charge.

6. **(1)** Both Assertion and Reason are True and the Reason is the correct explanation of the Assertion

7. **(3)** Both aluminium and gallium oxides are amphoteric in nature.

8. **(3)** Silicon form single bond with itself.

9. **(1)** Both Assertion and Reason are True and Reason is the correct explanation of the Assertion

10. **(1)** Pb^{2+} compounds are more stable than Pb^{4+} compounds due to inert pair effect.

Multi Correct MCQs

1. **(3)** Both SnO_2 (tin dioxide) and PbO_2 (lead dioxide) exhibit amphoteric properties. This means that they can act as both an acid and a base, reacting with both acidic and basic substances.

2. **(3)** Among the all given materials, Graphite is conductor and rest are semiconductors.

3. **(4)** Pb^{+4} and Tl^{3+} can behave as oxidising agents, due to inert pair effect.

4. **(2)** Electron deficient species have less than 8 electrons (or two electrons for H) in their valence (incomplete octet) B_2H_6, BCl_3, have incomplete octet.

NCERT Exemplar MCQs

1. **(3)** Ga exists as liquid from 30°C upto 2000°C and hence, it is used in high temperature thermometer. In gallium, the crystal structure is different, suggesting that Ga consists of almost discrete Ga_2 molecule, that's why its melting point is lowest.

2. **(1)** $AlCl_3$ is an electron deficient compound.
   Despite of sharing electrons with chlorine, the octet of aluminium is incomplete. To complete the octet, it needs electrons and thus, acts as a Lewis acid.
   $_{13}Al = 2, 8, 3$ and $_{17}Cl = 2, 8, 7$

3. **(1)** Structure of $B(OH)_4^-$ is:

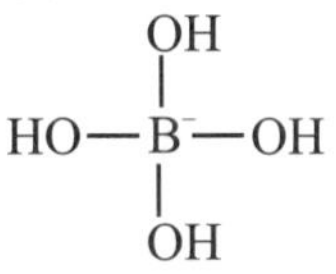

   Hybridisation — sp^3
   Geometry — Tetrahedral

4. **(1)** On moving down the group the acidic nature of oxides changes from acidic to basic through amphoteric.
   $$\underbrace{B_2O_3}_{\text{More acidic less basic}}, \underbrace{Al_2O_3 \text{ and } Ga_2O_3}_{\text{Amphoteric}}, \underbrace{In_2O_3 \text{ and } Tl_2O_3}_{\text{Basic}}$$

5. **(1)** Boron has atomic number 5 and do not have vacant d-orbital. The element M in the complex ion MF_6^{3-} has a coordination number of six. The lowest atomic number among the given is of boron. Boron can show maximum coordination number of 4. Thus, B can not form complex of the type MF_6^{3-}.

6. **(3)** Lewis acids are substances whose octet is incomplete and accept electrons.
   Boric acid is a monobasic weak acid. It does not liberate H^+ ion but accepts electrons from OH^- ion i.e., behaves as Lewis acid.
   $$H_3BO_3 + H_2O \rightarrow \left[B(OH)\right]_4^- + H^+$$
   or $$B(OH)_3 + 2H_2O \rightarrow \left[B(OH)_4\right]^- + H_3O^+$$
   The structure of H_3BO_3 is as shown below, where the octet of boron in H_3BO_3 is incomplete.

7. **(2)** The tendency to form long open or closed atom chains by combination of same atoms is known as catenation. The catenation is maximum in carbon and decreases down the group.
   $C >> Si > Ge \approx Sn > Pb$
   Down the group, size increases and electronegativity decreases, thereby, tendency to show catenation decreases.

8. **(3)** The chain length of silicon polymer can be controlled by adding Me_3SiCl which blocks the ends.

9. **(4)** The decrease in ionization enthalpy from B to Al is associated with increase in size. On moving from Al to Ga, the ionisation enthalpy increases slightly, because on moving from Al to Ga, both nuclear charge and shielding effect increase but due to poor shielding of d-electron in Ga, effective nuclear charge on valence electron increases resulting in *d*-block contraction, that's why ionisation enthalpies increase.
   On moving from Ga to In, again there is slight decrease in ionisation enthalpies due to increased shielding effect by additional ten 4*d* electrons, which outweighs the effect of increased nuclear charge.
   On moving from In to Tl, ionisation enthalpies show the increase again because fourteen 4*f* electrons shield valence electron poorly (order of shielding effect $s > p > d > f$) and so, effective nuclear charge increases, consequently ionisation enthalpies increase.

10. **(2)** Boron is trivalent, we would expect a simple hydride BH_3. However, BH_3 is not stable. Boron possess incomplete octet and BH_3 dimerises to form B_2H_6 molecule and forms three centre 2-electron bond. The simplest boron hydride is diborane B_2H_6.
    As seen from the structure drawn, 6 electrons are required for the formation of conventional covalent bond structure by B-atom, whereas in diborane, there are 12 valence electrons, three from each boron atoms and six from the six hydrogen atoms. The geometry of B_2H_6 can be represented as:

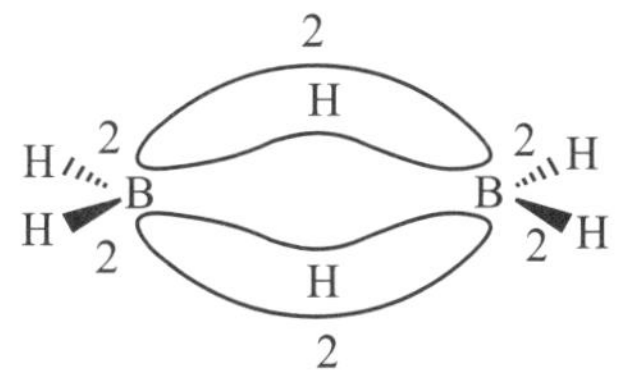

    The four terminal hydrogen atoms and two boron atoms lie in one plane. Above and below the plane, there are two bridging hydrogen atoms. Each boron atom forms four bonds even though it has only three electrons. The terminal B—H bonds are regular bonds but the bridge B—H bonds are different.
    Each bridge hydrogen is bonded to the two boron atoms only by sharing of two electrons. Such covalent bond is called 3-centre-2-electron bond or a multi centre bond or banana bond.

11. **(1)** (i) $B_2H_6.2NH_3$ which is formulated as $[BH_2(NH_3)_2]^+[BH_4]^-$ is formed initially by reaction of ammonia with diborane, further heating gives borazine, $B_3N_3H_6$.
    $$B_2H_6 + 6NH_3 \rightarrow 3[BH_2(NH_3)_2]^+[BH_4]^-$$
    $$\downarrow \Delta$$
    $$2B_3N_3H_6 + 12H_2$$
    Inorganic benzene
    Borazole has cyclic structure and is isoelectronic and isosteric with benzene and thus called inorganic benzene.
    (ii) By the reduction of BF_3 with lithium aluminium hydride in diethyl ether, diborane can be prepared.
    $$4BF_3 + 3LiAlH_4 \rightarrow 2B_2H_6 + 3AlF_3 + 3LiF$$

12. **(2)** Quartz, cristobalite and tridymite are some of the crystalline forms of silica and they are interconvertable at suitable temperature. Quartz is extensively used as a piezoelectric material.

13. **(4)** Those substances which reduces other substances and itself gets oxidised are reducing agents.
    In $SnCl_2$, Sn exists in +2 oxidation state, thus, acts as a strong reducing agent. i.e.,
    $$SnCl_2 + 2FeCl_3 \rightarrow 2FeCl_2 + SnCl_4$$
    $$SnCl_2 + 2CuCl_2 \rightarrow 2CuCl + SnCl_4$$

14. (3) Carbon dioxide can be obtained as a solid in the form of dry ice allowing the liquified CO_2 to expand rapidly.

Note : Dry ice is also called cardice. It is obtained when CO_2 is cooled under pressure (50-60 atm). Dry ice is used for making cold baths in the laboratory by mixing it with volatile organic solvents. It is also used as a coolant for preserving perishable articles in food industry, curing local burns and in hospitals for surgical operation of sores.

15. (2) Cement is a product obtained by combining a material rich in lime CaO with other material like clay which contains silica SiO_2 along with the oxides of aluminium, iron and magnesium. The average composition of Portland cement is:

CaO (50 – 60%)

SiO_2 (20 – 25%)

Al_2O_3 (5 – 10%)

Fe_2O_3 (1 – 2%)

SO_3 (1 – 2%)

MgO (2 – 3%)

Thus, it contains elements of group 2 (Ca), group 13 (Al) and group 14 (Si).

NEET & AIIMS Past Year MCQs

1. (1) Tl^{+3} is less stable than Tl^{+1} (inert pair effect)

Going down the group 13, stability of lower oxidation state increases.

In case of B, Al, Ga and In, higher O.S. +3 remains more stable than lower O.S. +1.

But, in last stable element, thallium (Tl), lower O.S. +1 become more stable than higher O.S. +3.

2. (4) Coke- reducing agent

Diamond- sp^3 carbons

Fullerenes- cage like structure

Graphite- used as lubricant

3. (4) Only graphite is used as a dry lubricant. Both diamond and graphite are covalent in nature.

In diamond, each carbon is bonded with four other carbon atoms. So, hybridisation of carbon atom is sp^3.

In graphite, each carbon is bonded with three other carbon atoms. So, hybridisation of carbon atom is sp^2.

4. (1)

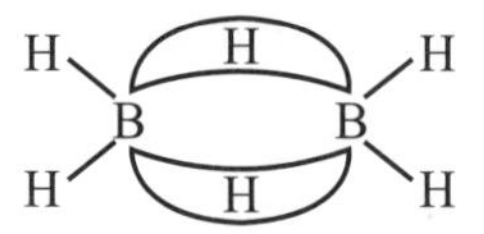

B_2H_6

B is sp^3 hybridized.

3 centre 2 electron bond which is called bridge bonding or banana bond.

There are two different bonds that is 3 centre 2 electron and 2 centre 2 electron. So, bond lengths are different.

Molecule is non-planar as 4 terminal hydrogen atoms and two boron atoms lie in one plane & rest two hydrogen lies in another plane.

5. (4) Because size of the hydrogen is very small as compare to the size of the chloride [factual question]

6. (3) **7. (1)** **8. (4)**

9. (1) **10. (2)** **11. (3)** **12. (3)**

Organic Chemistry-Some Basic Principles and Techniques

NCERT Topic-wise MCQs

1. (1)

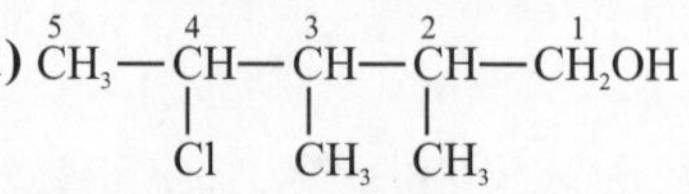

Longest chain-5 carbon

IUPAC name 4-Chloro-2, 3-dimethyl pentan-1-ol

2. (4) Select the longer possible chain, count no. of carbon atom present and write the name of compound according to IUPAC rules.

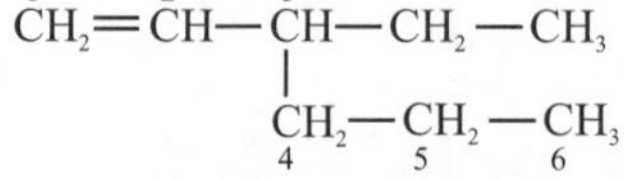

3-Ethyl hex-1-ene

3. (2)

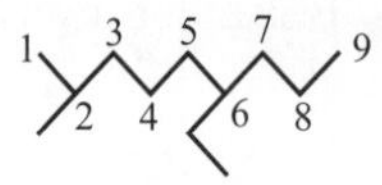

The correct name 6-ethyl-2-methyl nonane.

4. (2) The 1UPAC name is 3, 4, 4-trimethyl octane.

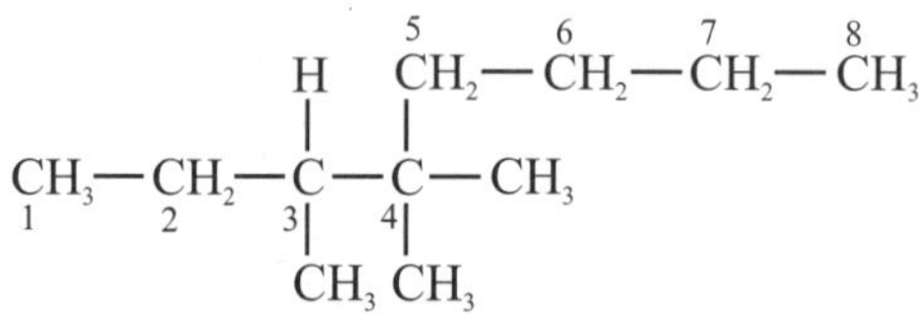

5. (2) $CH_2 = C = CH - CH_3$

Buta-1, 2 diene or 1, 2-Butadiene

6. (3)

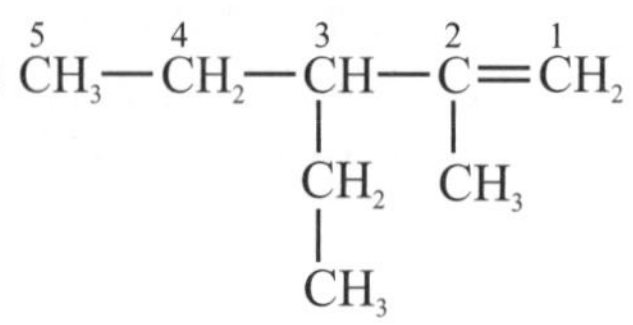

3-ethyl-2-methyl-pentene

7. (2)

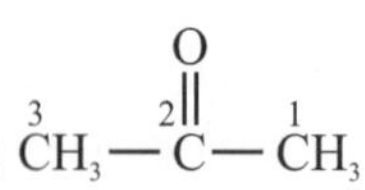

2-propanone

8. (2)

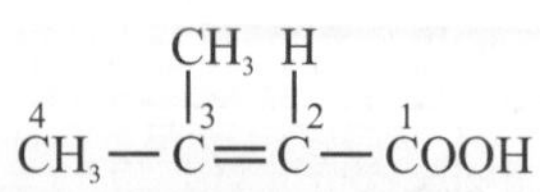

3-methyl but-2-enoic acid

∴ Order or preference COOH > double bond > alkyl group.

9. (2) 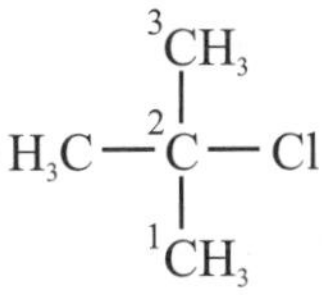

(IUPAC name: 2-Chloro-2-methyl propane)

10. (2) 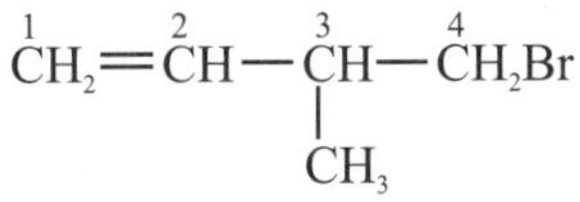

(4-Bromo-3-methyl but-1-ene).

11. (1) 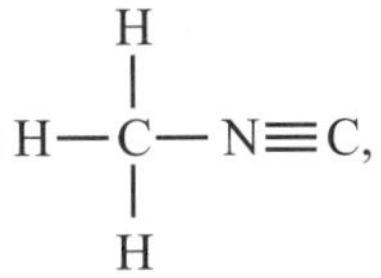

CH_3NC may be called methyl isocyanide, methyl carbylamine or methyl isonitrile.

12. (4) 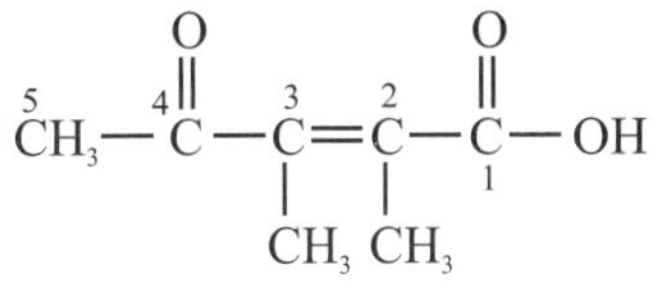

IUPAC name:

2, 3-Dimethyl-4-oxo-pent-2-en-1-oic acid.

13. (1) 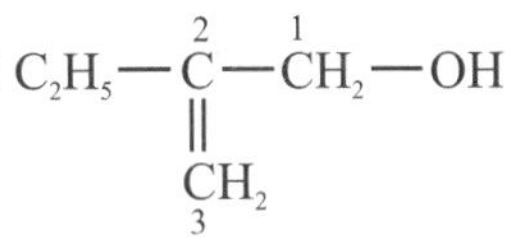

(2-Ethylprop-2-en-1-ol)

14. (1) 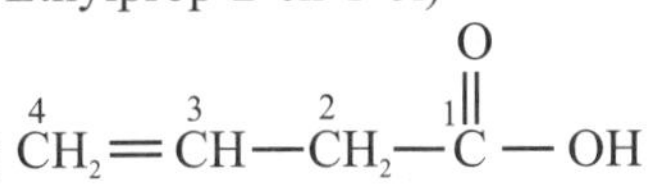

(But-3-enoic acid)

15. (4) 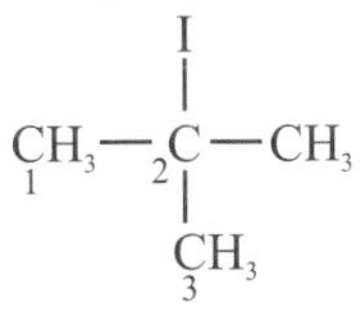

2-Iodo-2-methyl propane

16. (1) 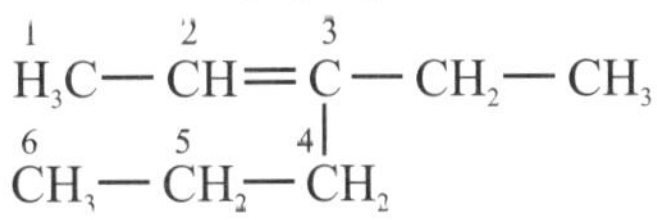

(3-Ethyl hex-2-ene)

17. (4) 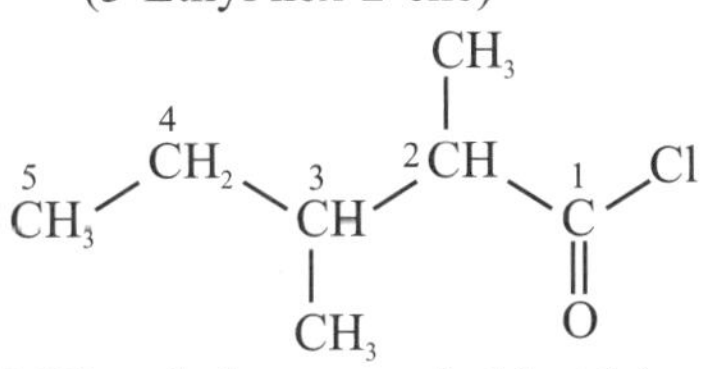

(2,3-Dimethyl pentanoyl chloride)

18. (4) 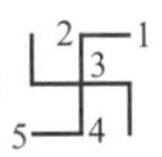

The IUPAC name for the hydrocarbons is 3, 3-Diethylpentane.

19. (4) 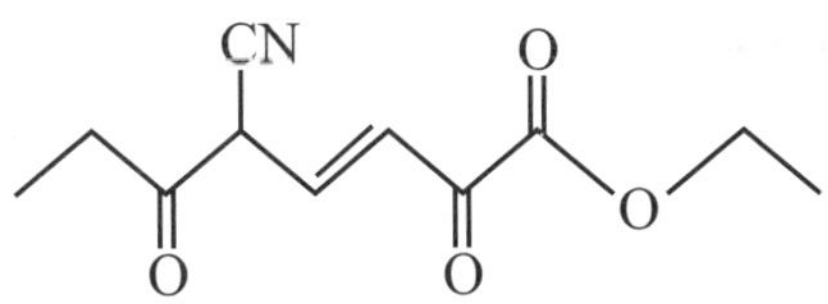

Ethyl 5-cyano 2, 6 dioxo oct-3en – 1 – oate

20. (4) Aromatic compounds contain (4n + 2) πe^-

21. (4) All the given compounds are aromatic. All are represented as-

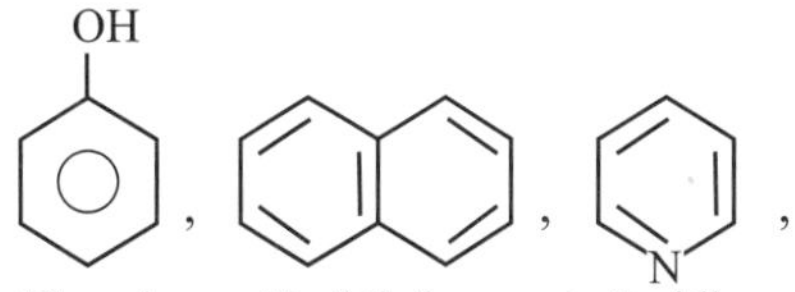

Phenol, Naphthalene, Pyridine

All these compounds have (4n + 2) e^- and are planar. All double bonds are in conjugation. Hence, all follows Huckel's rule & therefore aromatic in nature.

22. (4) All the given functional groups follow the order,

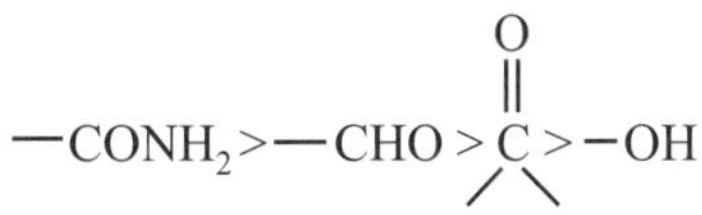

Hence, $-CONH_2$ will have highest preference.

23. (1) 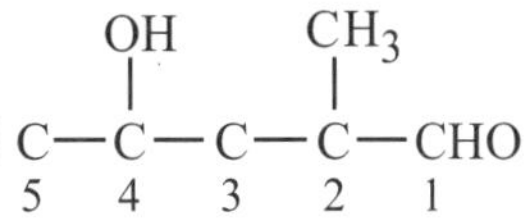

24. (2) $C_6H_5CH_2OH$ is phenyl carbinol

CCl_3CHO is chloral

25. (3) The carbon atoms which are directly attached with sp hybridised carbon atoms arrange linearly.

$HC-C\equiv C-CH_2$

① ② ③ ④

26. (2) As the number of bonds increase between carbon-carbon, the bond strength also increases.

27. (3) Ylidene suffix is used for gem bivalent. Therefore 3 is the correct option

28. (3) No. of $\sigma\uparrow$ bonds = 7

No. of $\pi\uparrow$ = 2

No. of lp = 2

Total sum: 11

29. (3) Triphenylmethyl radical (often shorted to trityl radical)

30. (4) For the given column correct option is 4.

31. (1) Anthranilic acid = 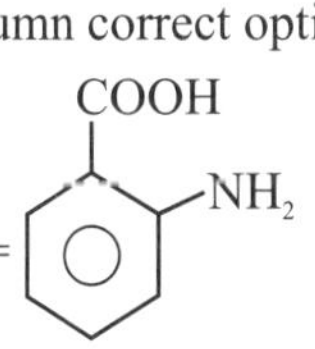

Sulfanilic acid = 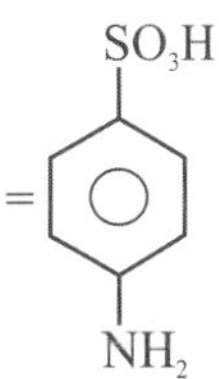

32. (2) Isobutyl isonitrile

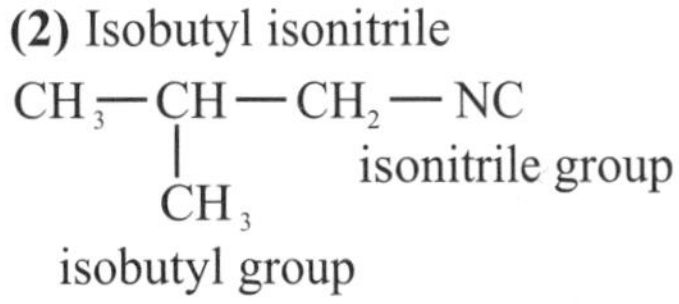

33. (1) Groups attached at one carbon are same. Hence can not show G.I.

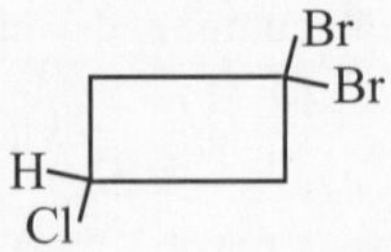

34. (3) CH_3-O-CH_3 — Methoxy methane (ether) (C_2H_6O)

C_2H_5OH — ethanol — alcohol functional group (C_2H_6O) molecular formula

In methoxy methane and ethanol both have same molecular formula, but functional groups are different. So, they are functional isomers.

35. (4) $CH_3COCH_3 \xrightarrow{NaBH_4} CH_3-CH(OH)-CH_3$, achiral

$CH_3COCl \xrightarrow{\text{Rosenmund's reaction}} CH_3CHO + HCl$, achiral

$CH_3CH_2COCH_2CH_3 \xrightarrow{Sn,\ HCl} CH_3CH_2CH_2OH + CH_3CH_2OH$, achiral

$CH_3CH_2COCH_3 \xrightarrow{LiAlH_4} CH_3CH_2CH(OH)-CH_3$, chiral

36. (2) Geometrical isomers are the compound which have atleast one double bond and the groups attached with double bonded carbon atoms are different.

$CH_3CH = CHCH_2C_6H_5$ 1-phenyl-2-butene exhibits geometrical isomerism.

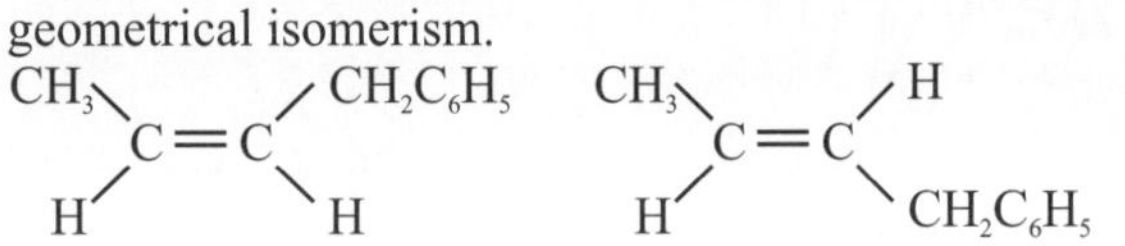

(cis-form) (trans-form)

37. (2) Conformations or conformational isomers are the different arrangement of atoms in space that results from the carbon - carbon single bond free rotation by 360°, the phenomenon is known as conformational isomerism.

38. (4) But-2-ene exhibits geometrical isomensim. It is shown below-

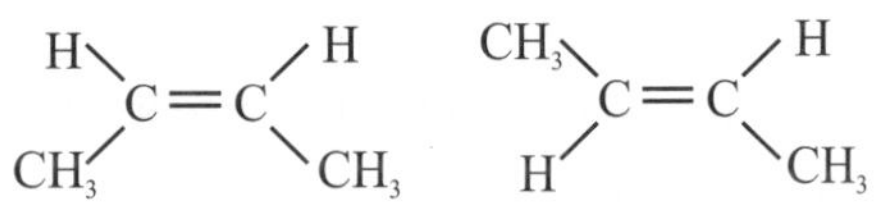

cis trans

39. (4) Chiral molecule is the one having asymmetric carbon atom and is not superimposable on its mirror image. On the other hand, molecule which is superimposable on its mirror image are called achiral hence, the molecule in option (4) is achiral.

40. (2) (1) $CH_3-CH(OH)-CH_3$ — 2-propanol

(2) $CH_3-CH_2-CH(OH)-CH_3$ — 2-Butanol

(3) CH_3-CH_2-OH — ethanol

(4) CH_3OH

Therefore; butanol-2 shows optical isomerism.

41. (4) Tautomerism is observed in urea.

$H_2N-C(=O)-NH_2 \rightleftharpoons HN=C(OH)-NH_2$

keto-form enol-form

42. (3) A compound could be optically active only when it contains atleast one asymmetric carbon atom or a chiral centre.

43. (3) Isomers of $C_4H_{10}O$ are as follows

(i) $CH_3-CH_2-CH_2-CH_2OH$ — butan-1-ol (alcohol)

(ii) $CH_3-CH(OH)-CH_2-CH_3$ — butan-2-ol (alcohol)

(iii) $CH_3-CH(CH_3)-CH_2-OH$ — 2-methyl propan-1-ol (alcohol)

(iv) $CH_3-C(CH_3)_2-OH$ — 2-methyl propan-2-ol (alcohol)

(v) $CH_3-CH_2-O-CH_2-CH_3$ — diethyl ether (ether)

(vi) $CH_3-O-CH(CH_3)_2$ — methyl isopropyl ether (ether)

(vii) $CH_3-O-CH_2-CH_2-CH_3$ — methyl propyl ether (ether)

Hence; three isomeric ethers are possible.

44. (2) For optical isomerism, there must be a chiral C-atom.

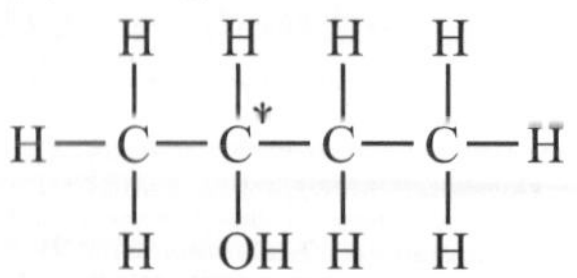

(C* = chiral C– atom)

45. (2) The main conditions for exhibiting geometrical isomers are

(i) Presence of different groups on same double bonded carbon.

(ii) Restricted rotation

$\therefore$ C_3H_6 ($\overset{3}{C}H_3-\overset{2}{C}H=\overset{1}{C}H_2$) does not exhibit geometric isomerism due to presence of same group on double bonded carbon atom (C_1).

46. (3) $CH_3-\overset{*}{C}H(OH)-\overset{*}{C}H(OH)-CHO$

2, 3 dihydroxy butanal

Number of optical isomers = $(2)^n = (2)^2 = 4$

47. (4) A molecule is said to be chiral or asymmetric, if it does not possess any element of symmetry and non-superimposable on its mirror images and this property of molecule to show non-superimposability is called chirality.

$(H)(Cl)C=C=C(CH_2-CH_3)(H)$

(allene)

Allenes exhibit optical isomerism (chiral), provided that the two groups attached to each terminal carbon atoms are different.

48. (2)

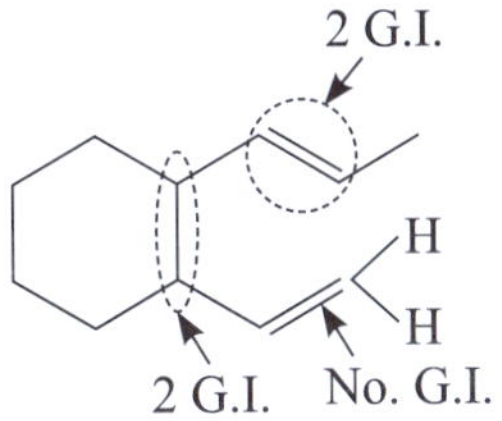

Total = 4

49. (4) Asymmetric carbon atom is the carbon, four valencies of which are satisfied by four different groups. The structure of α-hydroxypropanoic acid is

$CH_3-\overset{*}{C}H(OH)-COOH$

(C* = asymmetric carbon atom)

50. (1) In vicinal-dihalildes, the two – Cl atoms are present at adjacent positions. In case of $C_3H_6Cl_2$ only one such structure is possible.

Hence; only one vicinal-dihalides is possible for $C_3H_6Cl_3$

$CH_3-CH(Cl)-CH_2(Cl)$

51. (3)

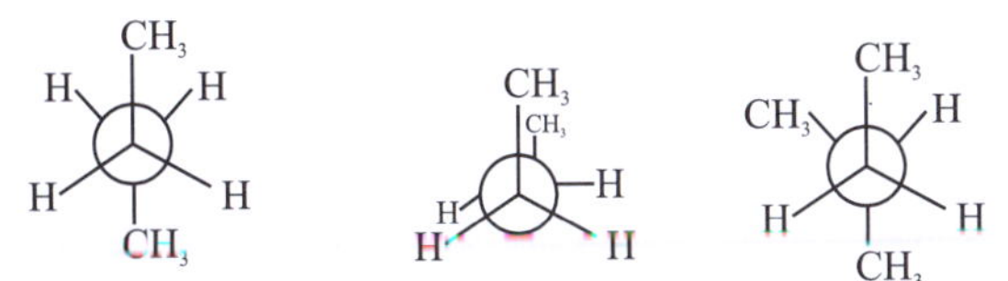

Staggered or anti form (most stable) | eclipsed (least stable) | skew form

Since; in staggered conformation, the groups are present as far apart as possible, it is the most stable conformation of n-butane.

52. (2) Structures are given as-

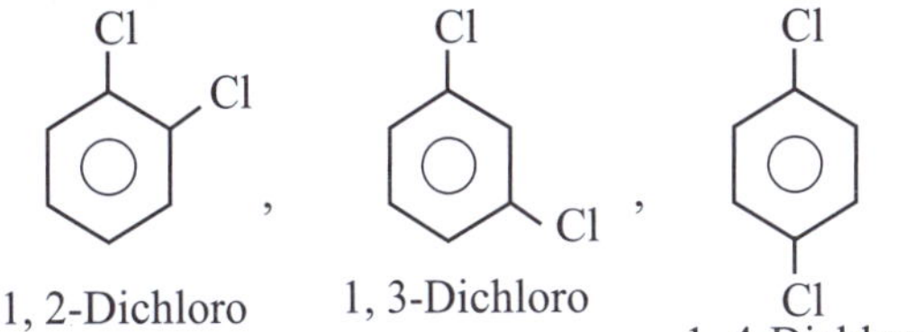

1, 2-Dichloro benzene, 1, 3-Dichloro benzene, 1, 4-Dichloro benzene

53. (1) Alkanols are alcohols and alkoxyalkanes are ethers. Alcohols and ethers are functional group isomers.

54. (1) Due to difference in the position of – OH group n-propyl alcohol and isopropyl alcohol are examples of position isomers. The stroctures of both are given as,

n-propyl alcohol (Propan-1-ol) | Isopropyl alcohol (Propan-2-ol)

55. (3) Structural isomers of $C_4H_{10}O$ are given as,

C—C—C—C—OH, C—C(OH)—C—C, C—C(C)—C—OH

& C—C(C)(C)—OH

56. (3)

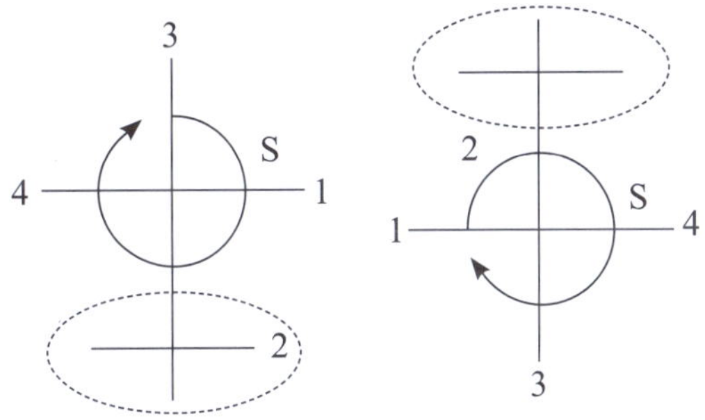

57. (1) Only one form is possible. Structure is

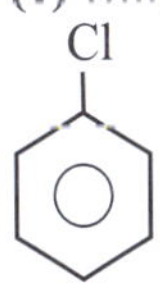

58. (1) $C_4H_{10}O$ compound shows two type of structural isomers i.e. alcohol and ether. Total structures of alcohols and ether are given as,-

$H_3C-CH_2-CH_2-OH$, $H_3C-CH(OH)-CH_2-CH_3$,

$CH_3-CH(CH_3)-CH_2-OH$, $H_3C-C(CH_3)_2-OH$,

$H_3C-CH_2-O-CH_2-CH_3$, $H_3C-O-CH(CH_3)-CH_3$

$CH_3-O-CH_2-CH_2-CH_3$

Total isomers = 4 + 3 = 7

59. (4) Structure of diethyl ether is $H_5C_2-O-C_2H_5$ It's molecular formula is $C_4H_{10}O$

Structure if the given compounds are-

n-propyl methyl ether

$H_3C-O-CH_2-CH_2-CH_3$

Butan-1-of $CH_3-CH_2-CH_2-CH_2-OH$

2-Methyl propan-2-ol $CH_3-\underset{OH}{\overset{CH_3}{C}}-CH_3$

Butanone- $H_3C-CH_2-\overset{O}{\overset{||}{C}}-CH_3$

Butanone is not isomeric with diethyl ether.

60. (4) These are not isomers because molecular formula is different in both structures.

61. (3) Positional isomers differ in position of C = C or C ≡ C or substituent.

$\boxed{C_4H_6}$

$CH_3-CH_2-C\equiv CH$ (1 - butyne) $CH_3-C\equiv C-CH_3$ (but - 2 - yne)

62. (3) 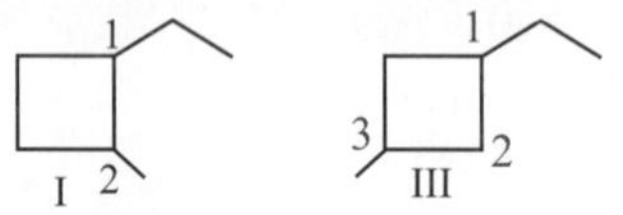

1-ethyl-2-methyl cyclobutane   1-ethyl-3-methyl cyclobutane

In both str. position of substituent is different so, they are called positional isomer.

63. (4) $C_5H_{12}O$

(A,B & C are Positional isomers)

$CH_3-CH_2-CH_2-CH_2-CH_2-OH$ (A) 1-pentanol

$CH_3-C-C-\underset{OH}{C}-C$ (B) 2 - pentanol

$C-C-\underset{OH}{C}-C-C$ (C) 3 - pentanol

(A, D, E are chain isomers)

$\overset{4}{C}H_3-\overset{3}{C}H_2-\overset{2}{C}H(CH_3)-\overset{1}{C}H_2-OH$ (D) 2-methyl - butanol

$CH_3-\underset{CH_3}{\overset{CH_3}{C}}-CH_2-OH$ (E) 2, 2-dimethyl-propanol

Fuctional group isomers are ⟶ [A & F, B & F, C & F, D & F, E & F]

$CH_3-O-CH_2-CH_2-CH_2-CH_3$ (F)

$CH_3-O-CH(CH_3)(CH_2-CH_3)$ (G)

F & H are metamers $[CH_3-CH_2-O-CH_2\,CH_2CH_3]$ (H)

$CH_3-O-\underset{CH_3}{\overset{CH_3}{C}}-CH_3$ (I)

F,G,I are chain isomers.

64. (2) $C_3H_6Br_2$

1. $CH_3-CH_2-CHBr_2$
2. $CH_3-\underset{Br}{\overset{Br}{C}}-CH_3$
3. $CH_3-\underset{Br}{CH}-CH_2-Br$
4. $\underset{Br}{CH_2}-CH_2-\underset{Br}{CH_2}$

65. (3) 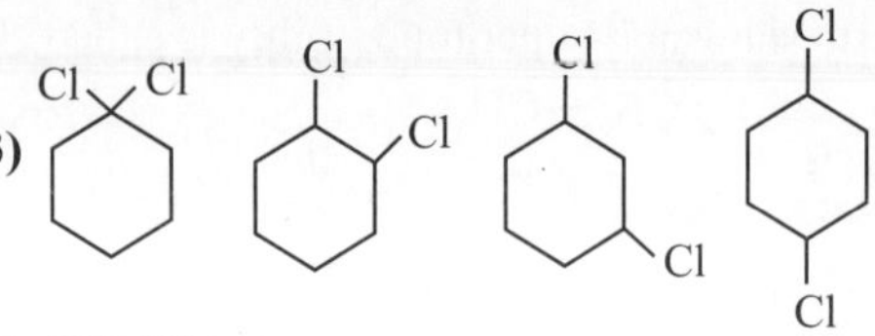

66. (4) $C_3H_6ClBr \rightarrow$

1. $H_3C-\underset{Br}{\overset{Cl}{C}}-CH_3$
2. $\underset{Br}{\overset{Cl}{HC}}-CH_2-CH_3$
3. $\underset{Cl}{H_2C}-CH_2-\underset{Br}{CH_2}$
4. $CH_3-\underset{Cl}{CH}-\underset{Br}{CH_2}$
5. $\overset{3}{H_3C}-\underset{Br}{\overset{2}{CH}}-\underset{Cl}{\overset{1}{CH_2}}$

67. (4) Compounds having same molecular formula but possessing open chain and cyclic structures are called ring chain isomers

68. (2) In functional group isomerism, isomers have different functional group.

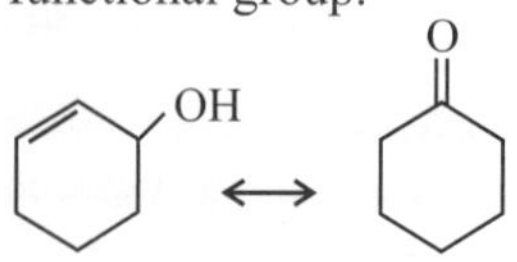

Both are functional group isomers, one has keto group and other has alcoholic group.

69. (4) B.P. ∝ Polarity

m.p. ∝ Packing efficiency

stability $\propto \dfrac{1}{\text{steric hinderance}}$

70. (3) $H_3C-CH_2-CH_3 \rightarrow \overset{\ominus}{C}H_3 + \overset{\oplus}{C}H_2-CH_3$

71. (1) The given reaction is an example of free radical substitution reaction.

72. (3) $\overset{\oplus}{N}O_2$ is an electrophilic reagent.

73. (4) NH_3^+ is most stronger electron withdrawing since it is electron deficient and carrying a positive charge. Nitro group is also an electron withdrawing group. CN – is more withdrawing group than F-because fluorine being smaller in size, is already carrying a negative charge and cannot withdraw more electrons.

74. (4) The reaction intermediate carbenes are produced form diazomethane, ketene and $CHCl_3/C_2H_5ONa$.

75. (3) %S character ∝ Electronegativity /– I effect

	$HC\equiv CH$	> $CH_2=CH_2$	> CH_3-CH_3
	sp	sp^2	sp^3
%S ⇒	50%	33%	25%

Strongest –I ⇒ 50% ⇒ sp Hybridization

76. (3)

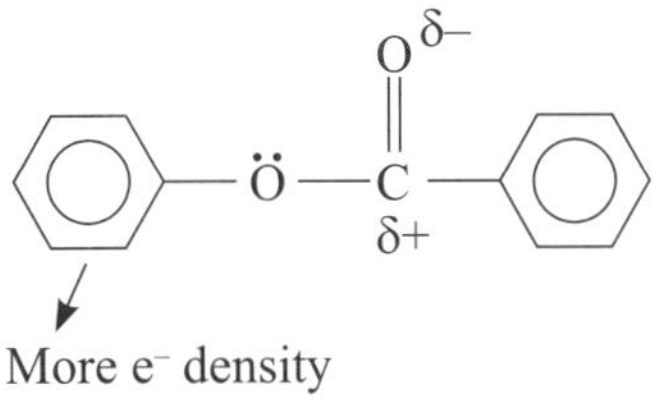

77. (1) Acidity ∝ Acidic strength $\propto K_a \propto \dfrac{1}{pK_a} \propto \dfrac{1}{K_b} \propto pK_b$

Acidity $\propto pK_b$

(A) $F-CH_2-COOH$

(B) $Cl-\underset{\displaystyle Cl}{\underset{|}{C}H}-CH_2-COOH$

(C) $F-CH_2-CH_2-COOH$

(D) $Br-CH-CH_2-COOH$

Electronegativity of halogen are in the order → F > Cl > Br > I

As electronegativity increases, acidic nature also increase,

As the distance of halogen increases from acid, acidity decreases.

In option (A) F is at least distance (in all) from acidic group, therefore it will be most acidic

In option (A) 2-Cl group is present at 2nd carbon from the acidic group, therefore, it will be less acidic than (A)

In option (C) F is present of 2nd carbon from acidic groups, therefore, it is little les acidic than (B).

In option (D) Br is present at 2nd carbon from acidic group, therefore, it is least acidic in all four.

Hence, the correct order of acidity is A > B > C > D

pK_b order is A > B > C > D

The stability of carbocation follows the order -3° > 2° > 1°

78. (3) e⁻ withdrawing group (EWG) decrease stability of carbocation.

All the given carbocations are primary. But in II electron withdrawing Cl group is present nearest the +ve charge, which decreases its stability. Hence it will be least stable. In I EWG is present on 2nd carbon and in III EWG is present on 3rd carbon therefore, I will be less stable than IIIrd. In IVth No EWG is present therefore, it will be most stable.

Hence, the correct order of stability of following carbocation is-

IV > III > I > II

79. (4) upcoming group is decided by +M

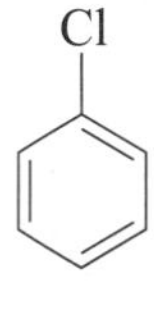

80. (2) Acidic strength $\propto K_a$ value $\propto \dfrac{1}{pK_a \text{ value}}$

$CH_3SO_3H > CH_3COOH > C_6H_5OH > CH_3OH$

81. (2) Carboxylic acids are always more acidic than phenols because carboxylate ion is more stable than phenoxide ion.

82. (2) Stability of Alkene ∝ No. of α H atoms. No. of α–H ∝ Hyperconjugation structure

stability of alkane ∝ Hyperconjugation structure

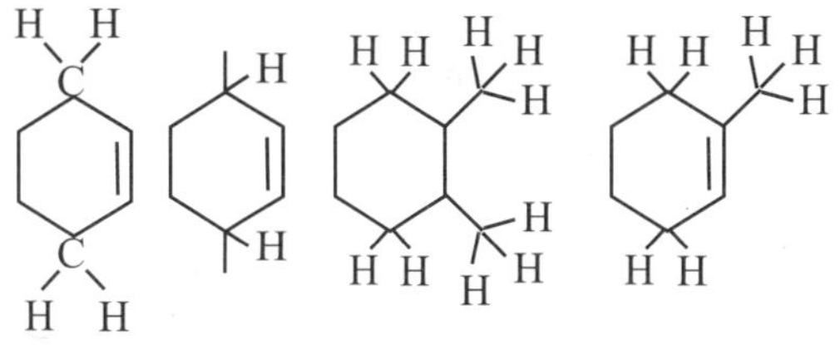

4 α - H 2α - H 10α - H 7α - H

Hence the stability order is-

(C > D > A > B)

83. (1) Heat of hydrogenation of alkenes is a measure of the stability of Carbon-Carbon double bond.

The smaller the numerical value of heat of hydrogenation of an alkene, the more stable the double bond there in.

Stability of alkane (or double bond of alkene) ∝ Heat of hydrogenation

Stability of alkene ∝ No. of α-H atoms

Heat of Hydrogenation $\propto \dfrac{1}{\text{stability of Alkane}}$

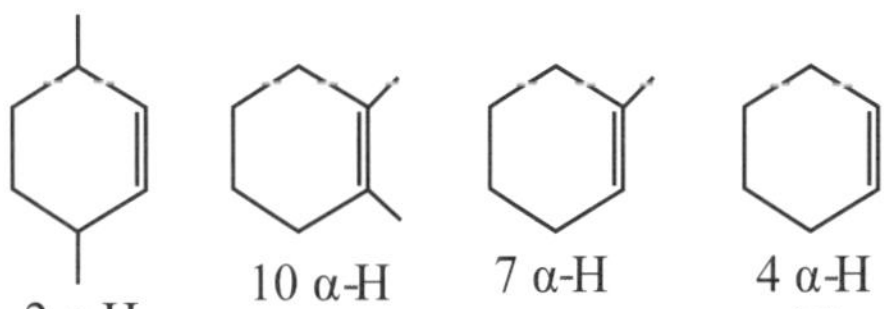

2 α-H (a) 10 α-H (b) 7 α-H (c) 4 α-H (d)

(a) Alkene is least stable so, highest heat of hydrogenation.

84. (1) The direction of dipole shifts where there is more electron density.

In pyrrole, the lone pair is involved in resonance & hence N assumes +ve charge.

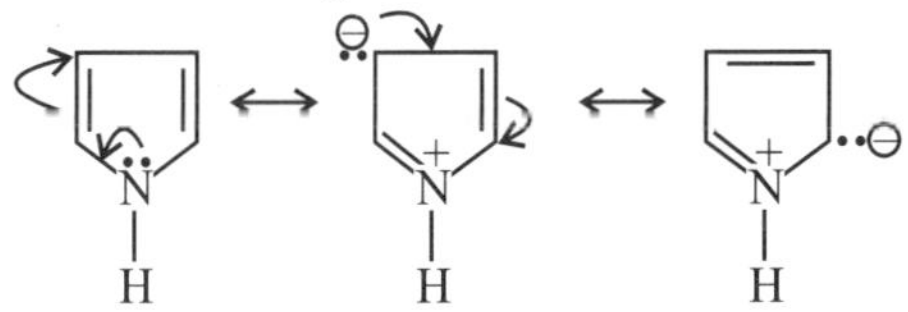

On the other hand, the lone pair of pyridine is not involved in resonance. So, electron density remains towards nitrogen

85. (4) The electron density in the aromatic ring will increase with the presence of electron donating group as a substituent.

The order of electron donating power of the substituent is-

$-NHCOCH_3 > -OCOEt > -CH_3 > -COOEt$

This order is based upon resonance effect

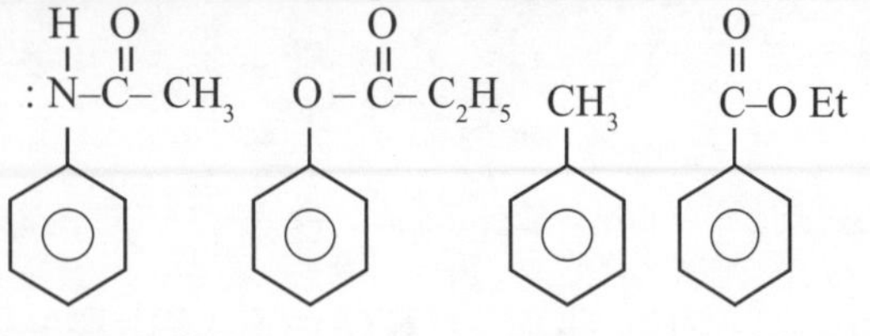

86. (3) The method used to separate a mixture of two miscible liquids having different boiling points is fractional distillation. The more volatile liquid is distilled over first and is collected in a receiver.

The two fractions are re-distilled separately a number of times to ensure complete purity of the liquids.

87. (4) Ortho and pare –nitrophenols have significant differences in their boiling points because ortho-nitrophenol forms intramolecular hydrogen bonds & has lower boiling point than para nitrophenol which has intermolecular hydrogen bonding. Hence, steam distillation method is used to separate them.

88. (4) Sublimation is the process in which solid substance converts into vapours directly. Iodine is found in solid state while F_2 and Cl_2 are found in gaseous state and Br_2 is found in liquid state. So; iodine can be purified by sublimation. It is represented as, I_2 (solid) $\rightleftharpoons I_2$ (g)

89. (1) Aniline is purified with the help of steam distillation as aniline boils at 98.5°C in presence of steam and hence; can be purified by steam distillation.

90. (1) Sublimation is a process in which a solid changes into the vapour form on heating without passing through the intermediate liquid state and vice-versa on cooling.

91. (1) Camphor is a sublime solid, hence mixture of camphor & benzoic acid can be separated by sublimation. Benzoic acid remains in the loot where as camphor Vapourise and collected separately.

92. (4) The best method for the separation of naphthalene and benzoic acid from their mixture is sublimation.

Naphthalene is a volatile compound whereas benzoic acid is non-volatile compound. Therefore, naphthalene sublimes easily. Basically naphthalene is non-polar molecule.

93. (1) Modern technique, used for the purification / separation of mixtures into its components, purification of compounds and also to test the purity of compounds is known as chromatography.

94. (3) To separate a mixture of nitrobenzene and aniline, aqueous HCl is used.

Aniline + aq . HCl → salt, which is water soluble

Nitrobenzene + aq . HCl → No reaction.

95. (2) A liquid which decomposes at its normal boiling point can be purified by vacuum distillation i.e., distillation under reduced pressure.

96. (1) Steam distillation method is used for those liquids which are not volatile and are immisible with water. The boiling point of turpentine oil is between 90-115°C.

97. (2) Partition chromatography

98. (4) Substances which decompose below their boiling points are purified by vacuum distillation.

99. (4) Naphthalene is purified by sublimation.

100. (2) Camphor is volatile & KCl is non-volatile. Therefore, these are separated by sublimation.

101. (1) In steam distillation the mixture begins to boil when the sum of the vapour pressure of the organic liquid and that of water becomes equal to that of atmospheric pressure.

102. (4) Lassaigne's test detected presence of extra elements (such as N,S and halogens).

103. (1) The sodium fusion extract is boiled with iron (II) sulphate and then acidified with Conc^n H_2SO_4. The formation of prussian blue colour confirms the presence of nitrogen.

Sodium cyanide first reacts with iron (II)sulphate & forms sodium hexacyanidoferrate (II). on heating with Conc^n H_2SO_4 some iron (II) irons are oxidised to iron (III) iorns which react with sodium hexacyanidoferrate (II) (ferriferrocyanide) which is Prussian blue is colour.

$$6CN^- + Fe^{2+} \rightarrow [Fe(CN)_6]^{4-}$$

$$3[Fe(CN)_6]^{4-} + 4Fe^{3+} \xrightarrow{x\,H_2O} \underset{\text{Prussian blue}}{Fe_4[Fe(CN)_6]_3 \cdot x\,H_2O}$$

$$3Na_4[Fe(CN)_6] + 4Fe^{3+} \rightarrow Fe_4[Fe(CN)_6]_3 + 12Na$$

Iron (III) hexacyanoferrate (II) (Prussian blue)

104. (1) For the detection of halogens, nitrogen and sulphur Lassaigne's test is used.

105. (2) If both N and S are present, it may result in the formation of NaCNS during fusion with sodium metal

$$Na + C + N + S \rightarrow NaCNS$$

If sodium metal is present in excess, the sulphocyanide formed decomposes into sodium cyanide and sulphide.

$$NaCNS + 2Na \rightarrow NaCN + Na_2S$$

106. (4) By Beilstein test presence of halogens in organic compounds can be detected.

107. (2) In Lassaigne's test for nitrogen in sodium extract, add fresh and saturated solution of $FeSO_4$ followed by NaOH. Now boil and cool the content. Then, add dil H_2SO_4 and $FeCl_3$ solution. Formation of Prussian blue precipitate shows the presence of nitrogen.

$$FeSO_4 + 2NaOH \rightarrow Fe(OH)_2 \downarrow + Na_2SO_4$$

$$Fe(OH)_2 + 6NaCN \rightarrow Na_4[Fe(CN)_6] + 2NaOH$$

From sodium extract

$$2Na_4[Fe(CN)_6] + 4FeCl_3 \rightarrow Fe_4[Fe(CN)_6]_3 + 12NaCl$$

Iron (III) hexa cyano ferrate (II)

108. (1) In carius method, the weighed compound is heated with concentrated HNO_3 followed by $MgCl_2$ in the presence NH_4Cl and NH_4OH. By this magnesium ammonium phosphate is formed which on ignition gives magnesium pyrophoshate ($Mg_2P_2O_7$). From the amount of magnesium pyrophosphate, the % of phosphorus is calculated as

$$\% \text{ of P} = \frac{62}{222} \times \frac{\text{weight of } Mg_2P_2O_7}{\text{weight of substance taken}} \times 100$$

109. (3) When organic compound is fused with sodium metal, nitrogen of the compound is converted into sodium cyanide as

$Na + C + N \rightarrow NaCN$

110. (1) Hydrazine ($NH_2 - NH_2$) does not contain carbon, on fusion with Na it can't form NaCN.

111. (3) Sulphur of organic compound is detected by using sodium nitroprusside. It forms a violet colour complex called sodium thionitro prusside with sulphur.

$Na_2S + Na_2[Fe(CN)_5NO] \rightarrow Na_4[Fe(CN)_5NOS]$

Sodium nitroprusside   Sodium thionitroprusside

112. (1) In lassaigne method, Organic compound is fused with Na-metal.

113. (2) Sodium extract is basic in nature due to the formation of NaCN or Na_2S etc.

114. (4) $2Na + S \rightarrow Na_2S$

115. (3) Camphor is a sublime solid, hence mixture of camphor & Calcium sulphate can be separated by sublimation

116. (2) A compound which does not give a positive result in the lassaigne's test for nitrogen is Hydroxyl ammonium chloride.

Because it doesn't not contain carbon.

117. (4) $Na + C + N + S \rightarrow NaSCN$

$Fe^{+3} + SCN^- \rightarrow Fe(SCN)_3$ Blood red colour

118. (3) The compound must contain both N and C.

119. (2) Volume of nitrogen collected at 300 K and 715 mm pressure is 50 mL

Actual pressure = 715 – 15 = 700 mm

$$\text{Volume of nitrogen at STP} = \frac{273 \times 700 \times 50}{300 \times 760} = 41.9\text{mL}$$

22, 400 ml of N at STP weights – 28 g

$$41.9 \text{ ml of nitrogen weights} = \frac{28 \times 41.9}{22400} \text{g}$$

$$\% \text{ of nitrogen} = \frac{28 \times 41.9 \times 100}{22400 \times 0.3} = 17.46\%$$

120. (1) Molar mass of AgBr = 108 + 80 = 188 g/mL

188 g AgBr contains 80g bromide

$$0.12\text{g AgBr contains} = \frac{80 \times 0.12}{188} \text{g bromide} = 0.05\text{g}$$

$$\% \text{ of bromine} = \frac{0.05}{0.15} \times 100 = 34.04$$

121. (2) For estimation of nitro and azo compounds kjeldahl's method is not used.

122. (4) Per cent of carbon in compound

$$\frac{12}{44} \times \frac{\text{weight of } CO_2}{\text{weight of compound}} \times 100$$

$$\frac{12}{44} \times \frac{12}{4} \times 100 = 81.81\%$$

% of H = 100 – 81.81 = 18.19 %

Element	%	Atomic weight	Relative no. of atoms	Simplest ratio of atoms	Simple ratio
C	81.81	12	$\frac{81.18}{12} = 6.81$	$\frac{6.81}{6.81} = 1$	$1 \times 3 = 3$
H	18.91	1	$\frac{81.19}{1} = 18.19$	$\frac{18.19}{6.81} = 2.67$	$2.67 \times 3 = 8$

Hence; empirical formula of the compound = C_3H_8

123. (1) From Kjeldahl's method

$$\% \text{ of nitrogen} = \frac{1.4 \times N \times V}{w}$$

$$\frac{1.4 \times 0.1 \times 30}{5} = 0.84\%$$

124. (4) Empirical formula = CH_2O

∴ Empirical formula weight = 30

$$n = \frac{240}{30} = 8$$

Molecular formula = (Empirical formula)$_n$ = $(CH_2O)_8$ = $C_8H_{16}O_8$

125. (3) $\% \text{ of carbon} = \frac{12 \times 0.198 \times 100}{44 \times 0.246} = 21.95\%$

$$\% \text{ of hydrogen} = \frac{2 \times 0.1014 \times 100}{18 \times 0.246} = 4.58\%$$

126. (2) Volume of acid taken = 50 mL of 0.5 M H_2SO_4

= 25 mL of 1 M H_2SO_4

Volume of alkali used for neutralization of excess acid

= 80 mL of 0.5 M NaOH

= 40 mL of 1 M NaOH

Now; 1 mole of H_2SO_4 neutralizes 2 moles of NaOH i.e.,

$H_2SO_4 + 2NaOH \rightarrow Na_2SO_4 + 2H_2O$

∴ 40 ml of 1M NaOH – 20 mL of 1 M H_2SO_4

∴ Volume of acid used by NH_3 = 25 – 20 = 5 mL of 1 N H_2SO_4

$$\text{Now; \% N} \ \frac{1.4 \times 1 \times 2 \times 5}{0.5} = 28$$

127. (1) $\% \text{ S} = \frac{32}{233} \times \frac{\text{wt. of } BaSO_4}{\text{wt. of oganic compound}} \times 100$

$$\% \text{ S} = \frac{32}{233} \times \frac{0.699}{0.480} \times 100 = 20\%$$

128. (1) (i) Carbon and hydrogen are estimated by Leibig's method. In this method carbon and hydrogen are quantitatively converted into CO_2 and H_2O. By knowing the weight of CO_2 and H_2O formed, percentage of carbon and hydrogen in the compound can be calculated.

(ii) Chlorine and bromine are estimated by Carius method.

(iii) Nitrogen is estimated by Duma's and Kjeldahl's method

129. (4) In Carius method halogens are estimated as AgX.

130. (1) In Duma's method nitrogen in organic compound is estimated in the form of N_2.

131. (1) In organic compounds, sulphur is estimated as $BaSO_4$.

132. (4) Kjeldahl's method cannot be used for azo compounds, nitro compounds and if N is present in the ring.

133. (2) Organic compound is first converted into $(NH_4)_2SO_4$

134. (3) Pure and dry O_2

135. (2) If molecular formula is different then molecular weight is also different

136. (1) $\%N = \frac{28}{22400} \times \frac{\text{Vol. of } N_2 \text{ at STP(ml)}}{\text{wt. of org. comp}} \times 100$

$\%N = \frac{28g}{22400\,ml} \times \frac{0.112\,ml}{0.5g} \times 100 = 28\%$

137. (1) $\%C = \frac{12}{44} \times \frac{\text{wt. of } CO_2}{\text{wt. of org. comp.}} \times 100$

$\%C = \frac{12}{44} \times \frac{1.32g}{0.73g} \times 100 = 49.32\%$

138. (1) $\%S = \frac{32}{233} \times \frac{\text{wt. of } BaSO_4}{\text{wt. of org. comp.}} \times 100$

$\%S = \frac{32}{233} \times \frac{0.5825g}{0.2175g} \times 100$

$\frac{18.64}{50.6775} \times 100 = 36.78$

High Order Time Intensive MCQs

1. (3) The carbocation 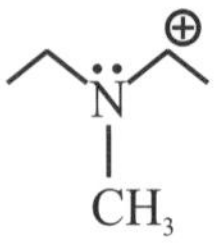 is maximum stabilised by Mesomeric effect.

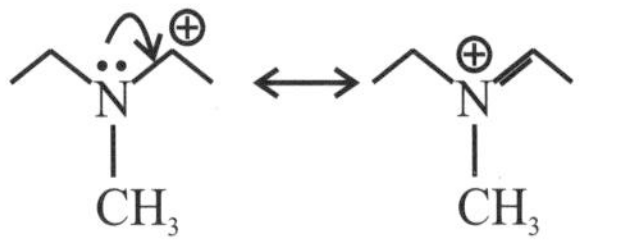

Effective (2p–2p) overlap occurs between vacant 2p orbital of carbocation & 2p orbital of N containing the lone pair and in all cases +I effect and Hyperconjugation is same.

In option a, mesomeric effect does occur but positive charge rests more on electronegative atom i.e., O which decreases the extent of stabilisation.

In option (3), there is no resonance. Rather, O exerts - I effect and destablilises the carbocation.

2. (1) $\text{Basicity} \propto \frac{1}{\text{Electronegativity}}$

Therefore, order would be:

i > iv > ii > iii > v.

3. (3) In option I, the indicated C–Cl bond has a partial double bond character, so this bond is toughest to break. Hence, it has maximum bond energy .

In option II, 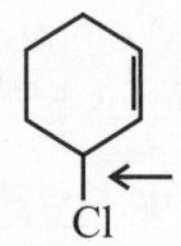 the C-Cl bond energy will be least.

The indicated C-Cl bond once broken leads to the formation of a carbocation which is highly resonance stabilised.

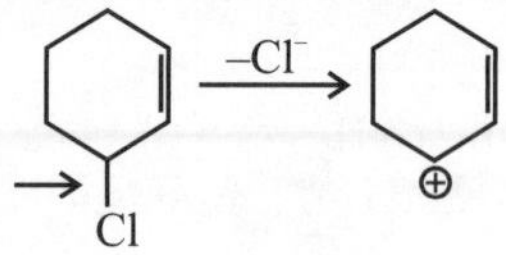

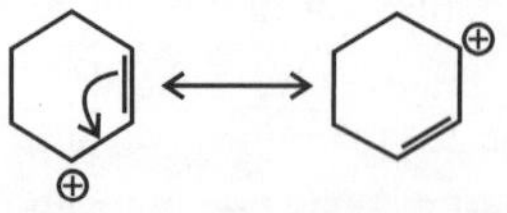

Hence the correct energy order for the heterolytic cleavage is I > III > II

4. (4) 'S' atom has vacant d-orbital where the negative charge of a carbanion can be delocalised.

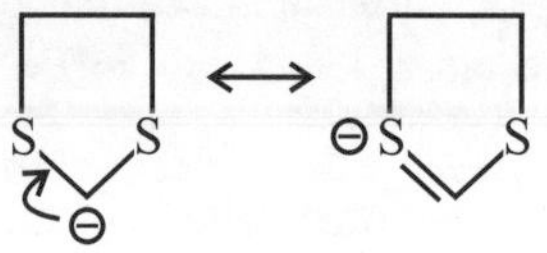

5. (3)

(1) $H_2C{=}CH{-}\overset{\oplus}{C}H_2 \longleftrightarrow H_2\overset{\oplus}{C}{-}CH{=}CH_2$
(I) (II)

No extra bond formed

(2) 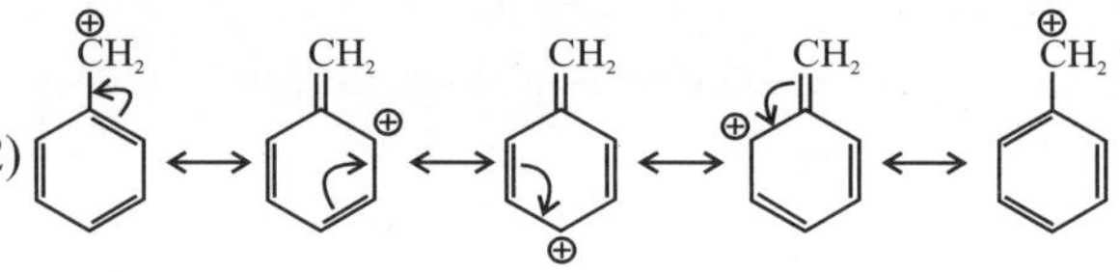

No extra bond formed

(3) $H_3C{-}\overset{\oplus}{C}H{-}\ddot{O}CH_3 \longleftrightarrow H_3C{-}CH{=}\overset{\oplus}{O}CH_3$
(II)

Extra bond formed here

(4) 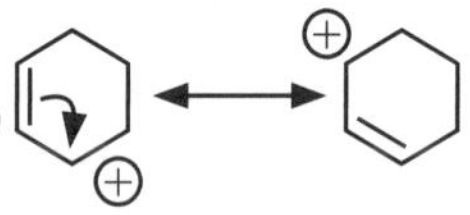

No extra bond formed here.

6. (4) Due to two bulky methyl group carbon of carbonation will be in different plane as that of benzene ring.

So resonance will not be possible.

7. (4) Option '4' is maximum stable carbanion. $-NO_2$ group exerts –M & –I effects in order to stabilise the carbanion

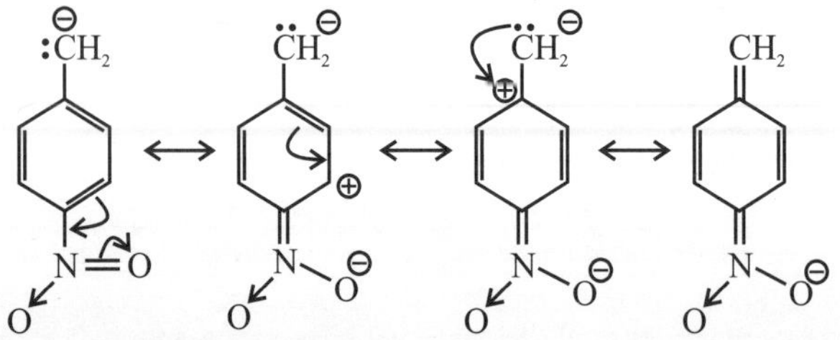

8. (1) Option a can also be re-drawn as-

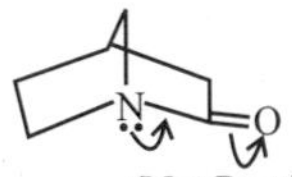

(Not Possible)

N is at bridge head position & here π-bond can't be formed as it violates BREDT'S RULE.

9. (3) Apply DNP rule: Distance > Number > Power

II > I > IV > III

Acidity or Acidic strength $\propto K_a \propto \dfrac{1}{pK_a}$

10. (1) In the given compounds, I[st] compound i.e. 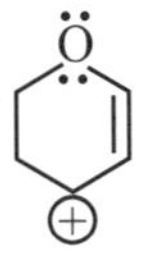 is resonance stabilised. Therefore, it is the most stable compound.

A resonance structure in which positive charge is present on the most electro - negative atom is highly unstable.

Since structure (III) has a positive charge on most Electronegative atom (i.e.O), Therefore, it is most unstable.

Hence the correct order of stability is I > II > III

11. (1) Here, lone pair is not involved in conjugation.

12. (4) $CH_3-\underset{\underset{CH_3}{|}}{C}=CH_2 \longleftrightarrow \overset{H^{\oplus}}{H_2C}=\underset{\underset{CH_3}{|}}{C}-\overset{\ominus}{C}H_2$

Due to Hyperconjugation single bond character arises in double bond so bond length increase

B.L. of double bond ∝ no. of ∝ H atoms ∝ hyperconjugation

13. (4) Stability order of Carbon free radical is-

3° > 2° > 1°

Bond dissociation energy

Carbon free radical $\propto \dfrac{1}{\text{stability of Carbon free radical}}$

In the given structure, the differentiation of carbon free radical is -

C_1–H = 1° Carbon free radical

C_2–H = 3° Carbon free radical

C_3–H = 2° Carbon free radical

C_4–H = 1° Carbon free radical

Bond dissociation energy order = C_1–H < C_4–H < C_3–H < C_2–H

14. (2) As we know stability of carbanion is directly proportional to –M and –I effect i.e. E.W. groups

Stability of anions $\propto -M \propto -I \propto \dfrac{1}{+M} \propto \dfrac{1}{+I} \propto \dfrac{1}{+H}$

In the given compounds structure (P) will be least stable because $-CH_3$ (+I) group is preset at pare position as it is an electron releasing group.

In rest compounds I effect observed and follows the order CCl_3>$-CCl_2H$>$-CClH_2$ ∴ Stability order of the compound is S > R > Q > P

15. (4) Here the stability of given compounds (anions) depend upon the group attached to the para position. All groups are showing +I effect & therefore destabilising the ring.

Groups present at para position following the order -

$H_3C-\underset{CH_3}{\overset{CH_3}{C}}- \;>\; -\underset{CH_3}{\overset{CH_3}{C}}- \;>\; H-\underset{H}{\overset{CH_3}{C}} \;>\; CH_3$

∴ the stability order of the given anions is - P > Q > R > S

16. (4) Nucleophilicity ∝ Electron density $\propto \dfrac{1}{\text{Electronegativity}}$

17. (4) Hyperconjugation is directly proportion to the No. of α-hydrogens. It is represented below-

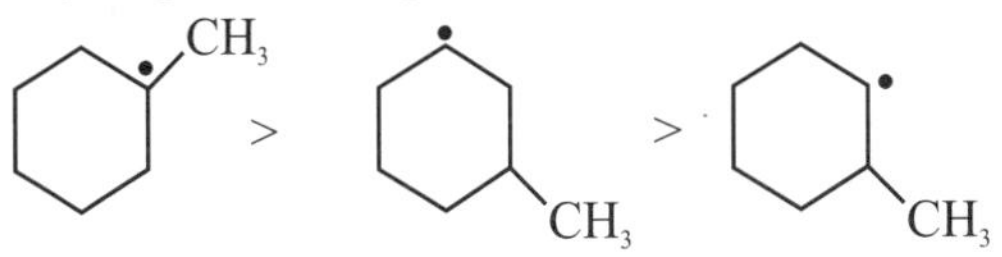

7 α – H   4 α – H   3 α – H

18. (3) As hyper conjunction is directly proportional to the No. of α-hydrogens.

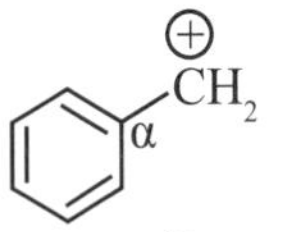

α Hydrogen = zero

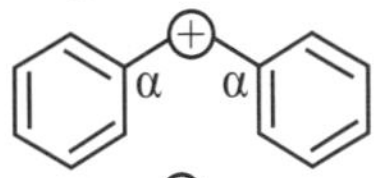

α Hydrogen = zero

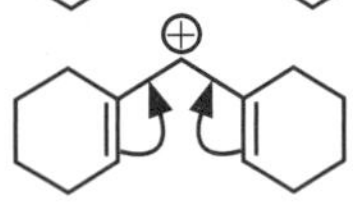 with the help of resonating structure, this structure is represented in two ways

Hence, option (3) is correct

19. (4) % s Character ∝ Electronegativity in $HC\equiv C^-$, negative change is present on most electronegative sp hybridised (% s = 50%) carbon atom, therefore it will be most electronegative as well as most stable carbanion.

20. (2) A compound is said to be aromatic it obey Huckel's rule i.e. [4n + 2]π rule.

21. (3) 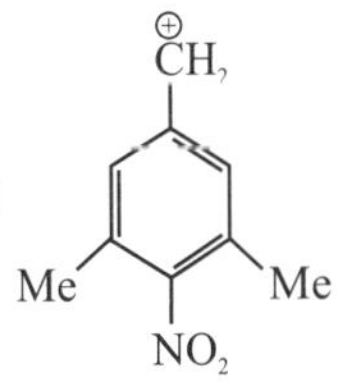 more stable then 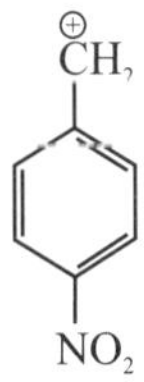

due to SIR effect

22. (3) The compound 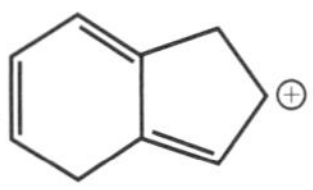 is stable due to extended (continuous) Conjugation. It is shown as,

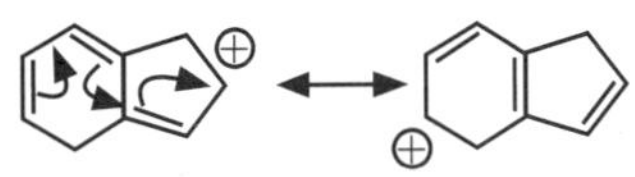

In this compound longest conjugation is present in comparison to rest compounds.

23. (2) Alcohols undergo dehydration to form alkene on treating with a protic acid (Ex. Conc[n] H_2SO_4 or H_3PO_4 or Catalyst such as Anhy. Zn Cl_2) .

2° and 3° alcohols are dehydrated under milder conditions. During this process carbocation intermediate is formed and we know that stability of carbocation follows the order-3° > 2° > 1°

Therefore, tertiary butyl alcohol forms stable carbocation or carbonium ion on dehydration.

24. (2) $CH_3-\overset{\oplus}{C}(CH_3)-CH_3$ > $CH_3-\overset{\oplus}{C}H-CH_3$ > 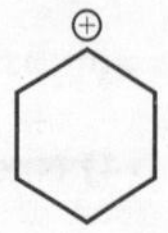

9α – H; 6α – H; 4α – H

$CH_3-CH(CH_3)-CH_2-\overset{\oplus}{C}H_2$ > $CH_3-CH_2-\overset{\oplus}{C}H_2$

2α – H (+I more); 2α – H

25. (4) Decreasing order of stability carbocation is-

IV > III > II > I

3° carbocation is more stable which is option (IV)

Stability ∝ No. of α-H.

26. (4) 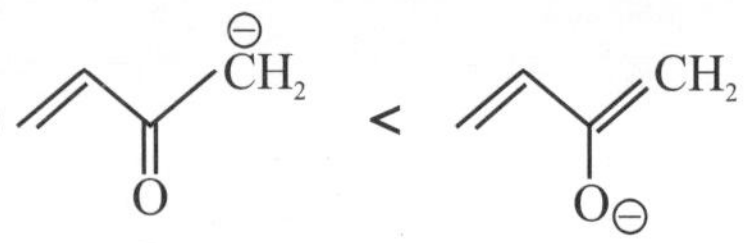

I < II

In I[st] structure less EN atom is negatively charged but in IInd structure more EN atom is negatively charged so, IInd structure is more stable.

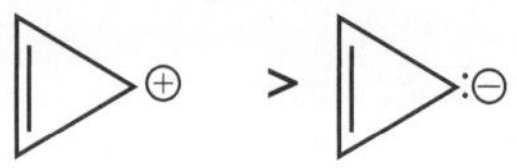

Aromatic I; Antiaromatic II

So I[st] structure is more stable than IInd.

in option (2) stability order is incorrect.

27. (4) Stability ∝ resonance ∝ Hyperconjugation

length of conjugation is equal in all I, II, III & IV.

In structure I, No of ∝–H = 3

In structure II, No of ∝–H = 4

In structure III, No of ∝–H = 7

In structure IV, No of ∝–H = 13

∴ Stability order = IV > III > II > I

28. (4) More the charge separation, more will be dipole moment. More the electronegativity difference, more will be dipole moment. In the given compounds, Decreasing order of dipole moment for given compounds

I > IV > III > II

29. (4) $-\overset{O}{\overset{\|}{C}}-NH-CH_3$ do not contain lone pair or negative charge on first atom.

30. (4) $H_3C-\overset{②}{C}(\underset{①}{CH_3})=\overset{③}{C}(\underset{④}{CH_2}-\underset{⑤}{CH_3})-CH_3$

2, 3-dimethyl pent-2-ene

31. (2) $\underset{4}{CH}\equiv\underset{3}{C}-\underset{2}{C}=\underset{1}{C}$

32. (2) 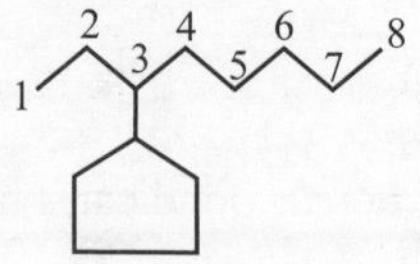

33. (4) The given compound has the following structure:

$\overset{5}{C}H_3-\overset{4}{C}H(CH_3)-\underset{3}{C}H=\underset{2}{C}H-\underset{1}{\overset{O}{\overset{\|}{C}}}-H$

4-Methylpent-2-enal

34. (1) 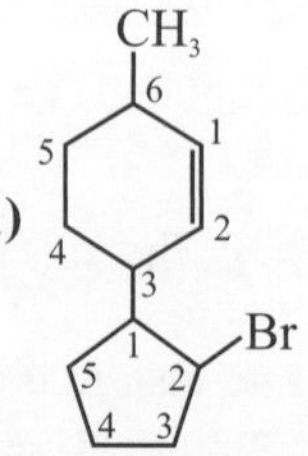

35. (2) 1, 3 and 4 have different functional group, So, cannot be homologous.

36. (3) 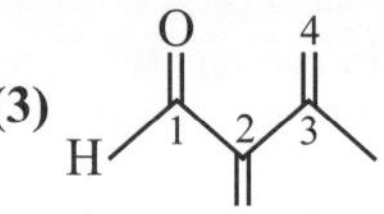 3-Methyl-2-methylenebut-3-enal

37. (4) in option (4) both the given compounds are identical.

38. (3) Chain same

39. (1) Compounds given in '1' are not metamers. Metamers are the isomers having same molecular formula but different alkyl groups on two sides of functional groups

ex. $CH_3OC_3H_7$, $C_2H_5OC_2H_5$

40. (1) (a) → (iii); (b) → (ii); (c) → (iv); (d) → (i)

$CH_3-\overset{O}{\overset{\|}{C}}-O-C_3H_7$ and $C_2H_5-\overset{O}{\overset{\|}{C}}-O-C_2H_5$

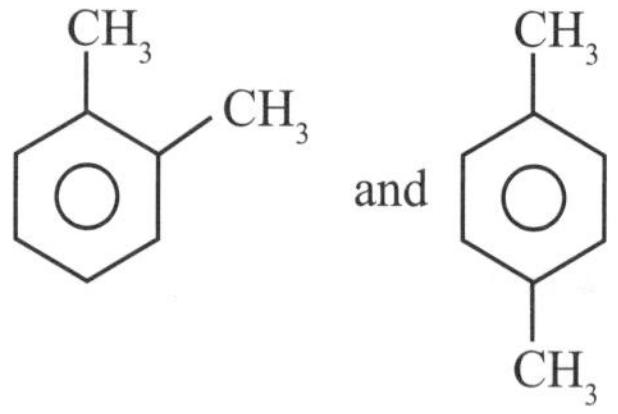

are position isomers.

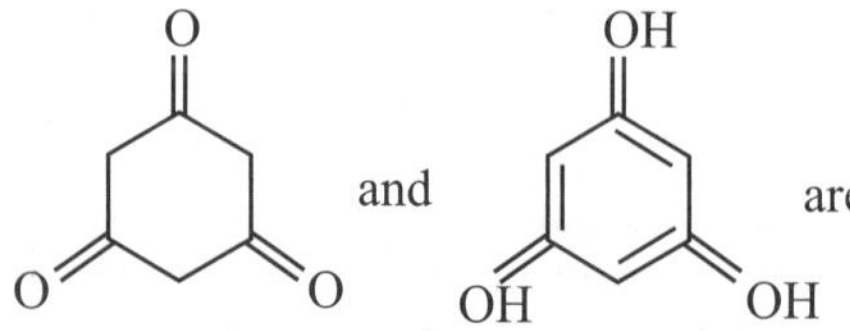

are Tautomers.

$H_3C-CH(CH_3)-OH$ and $H_3C-CH_2-CH_2-OH$ chain isomer

41. (3) –O–, –m–, –p–, chlorotoluene & benzylchloride

Molecular formula C_7H_7Cl four Benzene derivatives are possible-

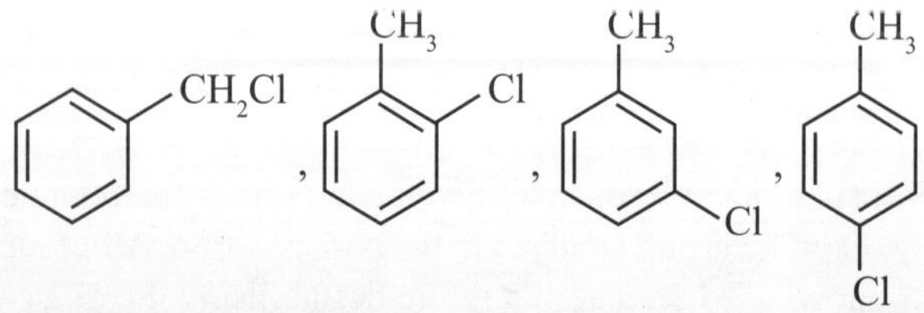

42. **(4)** The correct arrangement of the given functional group is-
–COOH > –CHO > –R–CO–R > –OH

43. **(2)** The decreasing order of priority of functional group is,
$-SO_3H > -COCl > -CONH_2 > -CN$

44. **(1)** In the given compounds, structure I and II are identical.

Structure I

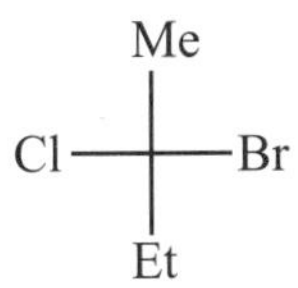

Priority order Br > Cl > Et > Me

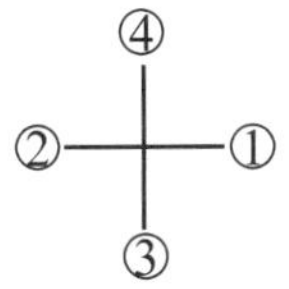

Structure II

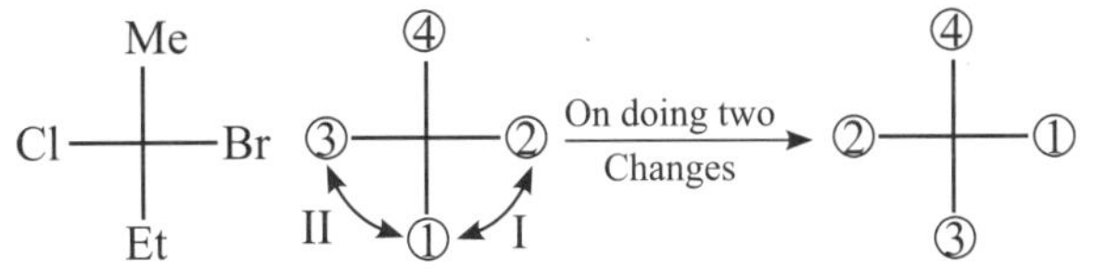

New structure obtained is similar to structure I.

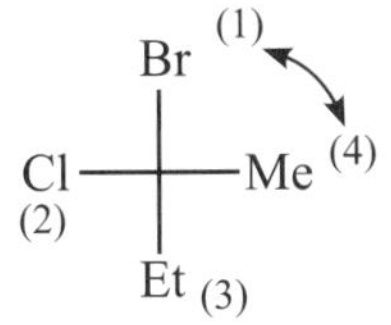

Here on doing only one change similar structure is obtained. Which is not permissible condition for the structure to be identical.

45. **(3)** 1. $CH_3-CH_2-CH_2-CH_2-CHO$

2. $CH_3-CH_2-CH(CH_3)-CHO$ 3. $CH_3-CH(CH_3)-CH_2-CHO$

5. $CH_3-C(=O)-CH_2-CH_2-CH_3$ 4. $CH_3-C(CH_3)_2-CHO$

6. $CH_3-C(=O)-CH(CH_3)-CH_3$ 7. $CH_3-CH_2-C(=O)-CH_2-CH_3$

46. **(4)** Conditions for the geometrical isomers-

(1) Compond must contains at least one carbon-carbon double bond

(2) The atoms or groups attached to the double bond must be different.

(3) For any ring, cis-trans isomers also results in geometrical isomerism.

In the given compounds, all compound will shown geometrical isomerism.

47. **(4)** Correct name of the given compounds are correctly matched according to option 4.

48. **(1)** 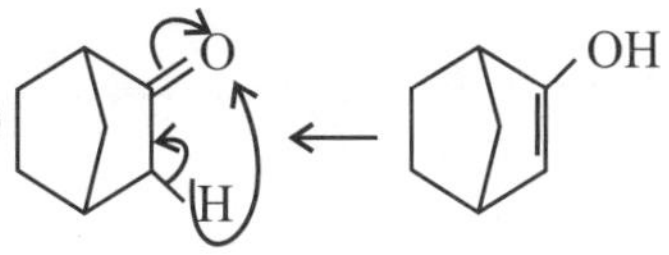

49. **(3)** Carbocation is a six electron species.

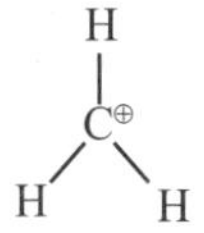

50. **(2)** The most stable resonance structure is one which is neutral and has maximum number of double bonds.

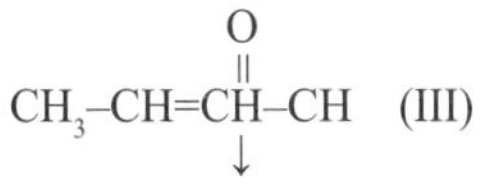

Non Polar R.S.

More number of double bond

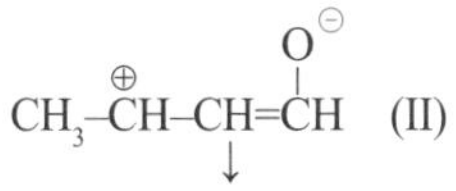

Has –ve charge on more electronegative atom

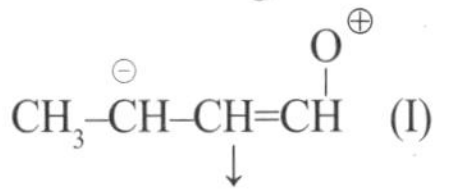

Has –ve charge on less electronegative atom

Thus, stability order: III>II>I

51. **(4)** In the given compounds, first three compounds will show cis-trans isomers. In compound (4) two symmetrical groups are present are side of the double bond. Therefore, compound (4) will not show geometrical isomerism.

52. **(3)** Three same substituent at 120°
∴ net dipole moment is zero

53. **(3)** 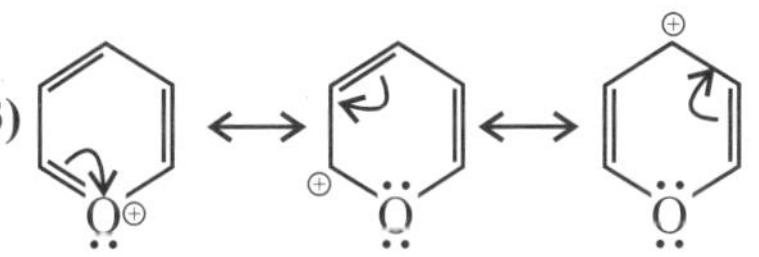

54. **(2)** Conditions for the resonance energy
No. of π bond ∝ resonance energy ∝ stability
Contributing structures should be that negative charges reside on an electronegative element, and positive energy resides on an electropositive element.
In contributing structures, like charges should not reside on atoms close to each other and unlike charges should be widely separated.
Electronegativity order ⇒ O > N > S > C

55. **(3)** (i) 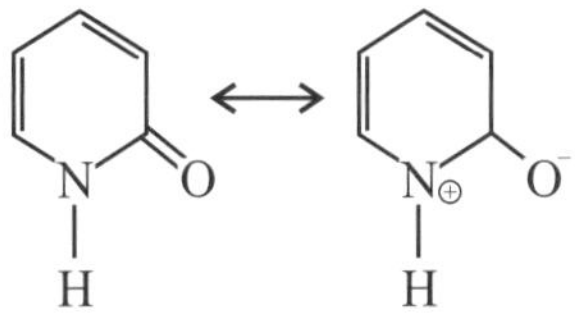 Aromatic

(ii) 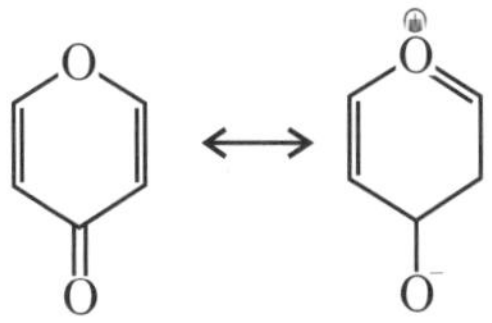 Aromatic

(iii) 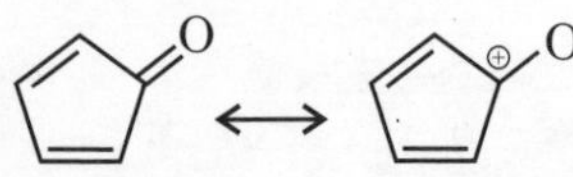 Anti aromatic

(iv) 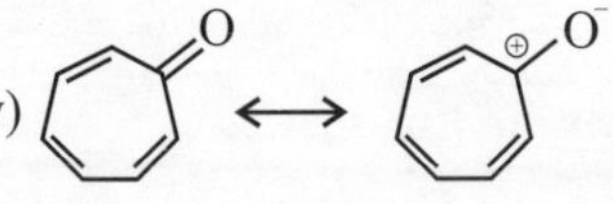 Aromatic

(v) 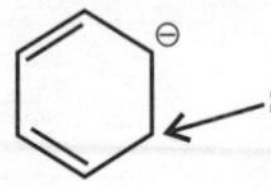 sp^3 non-aromatic (non planar)

56. (4) Structure I and II are tautomers

Tautomers are structural isomers which are readily intervertebral.

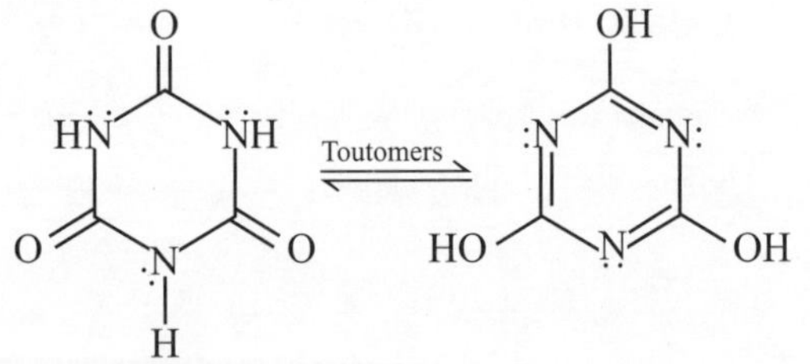

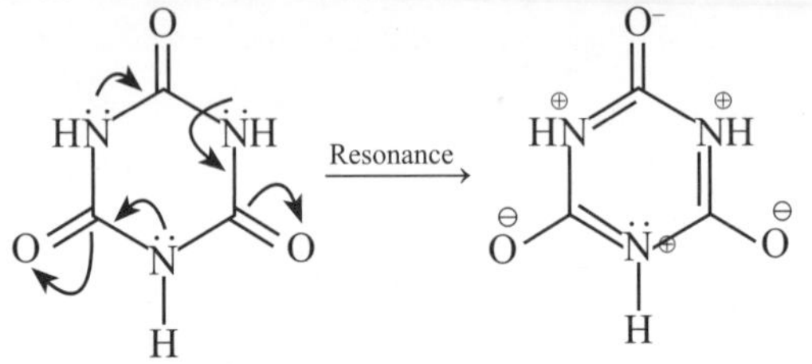

Structure I and II are resonating structure.

57. (2) In K_{a_1}, Acidic strength depends upon –I effect of –COOH group

In K_{a_2} Acidic strength depends upon +I effect of $-COO^-$ group after removal of H^+

58. (3)

decreasing stability of alkoxide (↓)

$CH \equiv C - CH_2O^{\ominus}$ (sp)

\>

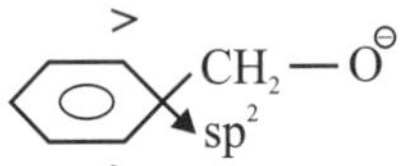

\>

$CH_2 = CH - CH_2 - O^{\ominus}$ (sp^2)

\>

$CH_3CH_2CH_2 - O^{\ominus}$ (sp^3)

Acidity of Alcohol ∝ stability of Alkoxide ion.

59. (1) +I or –I effect depends on EN. as well as distance. In I[st] structure only +I group present so it increase stability of carbocation.

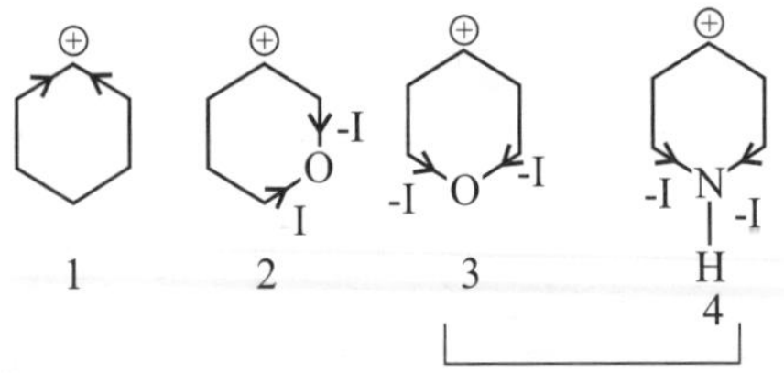

1 2 3 4 1 > 4 > 3 > 2

–Id 3 > –I of 4

distance of –I (↑)

60. (2) The name of given structures is correctly matched as- (a)-(ii), (b)-(i), (c)-(iv), (d)-(iii)

61. (1) 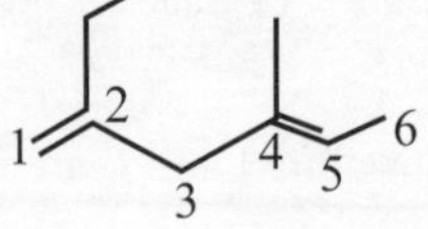

2-Ethyl-4-methyl hexa 1, 4 diene

62. (1) Compound "a" will not show tautomerism due to unavailability of α-hydrogens.

Tautomerised structure of (2), (3) and (4) are given as-

2) $H_3C-N=C(OH)-CH_3$

4) 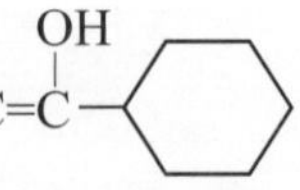

3) $H_3C-C(CH_3)=\overset{+}{N}(OH)(O^-)$

63. (2) 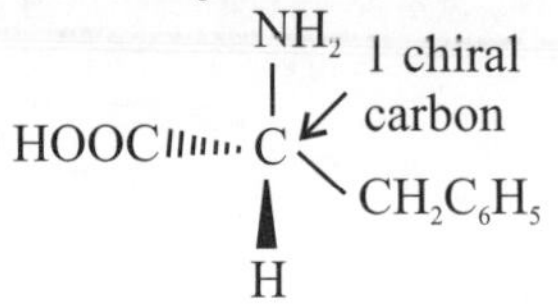

Molecule is chiral, so it shows optical isomerism.

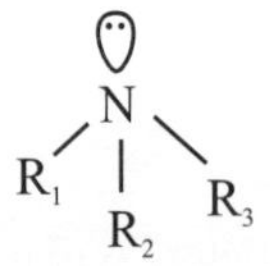

Chiral but optically inactive i.e., it does not show optical isomerism. (due to pyramidal inversion).

Compound and d will not show optical isomerism due to presence of planarity.

64. (1) Compounds 1, 2, 3, 4 will show aromaticity due to following Huckel Rule. i.e., Planarity, (4n + 2) π e⁻ and conjugation rule. Compound (5) is not aromatic because it does not follow (4n + 2) π e⁻ rule. It has only 4n π e⁻. Therefore, It is anti-aromatic.

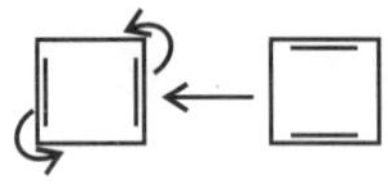 $4n\pi e^-$ delocalized

∴ Antiaromatic.

65. (3) In option 1, 2nd compound is tropylium cation which is more stable due to aromaticity.

In option "4" 2nd compound is more stable than I[st] because I[st] compound has ring strain

$$CH_3-O-\overset{\oplus}{C}H-CH_3 > CH_3-\overset{\oplus}{C}(CH_3)-CH_2-CH_3$$

(I) ↕ (II)

$$CH_3-\overset{\oplus}{O}=CH-CH_3$$

I[st] is more stable than II[nt] because in resonating structure of I[st] compound, octet of carbon is complete. So it is highly stable.

66. (3) The correct order of ortho para directing nature of the given compounds follow the order

$O^{\ominus} > -\overset{\ominus}{N}H > -NHCOCH_3 > NO_2$

67. (3) 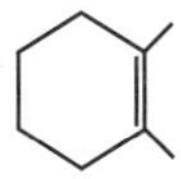

Due to maximum number of $\propto$ Hydrogens the compound shows highest hyperconjugating structures & due to which double bond have highest single bond character so, bond length is highest.

68. (3) Correct matching of the given compounds is-
(a)-(iii), (b)-(iv), (c)-(i), (d)-(ii)

69. (2) Compound 1 and 4 are trans in nature therefore, they have zero dipole moment.
Compound b is also trans but trans groups are not exactly same

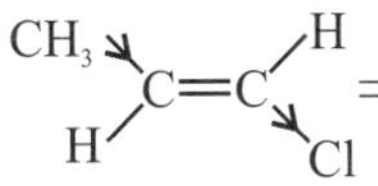 $\Rightarrow$ Highest dipole moment

70. (4) –I power $-F > -Cl$
+M power $-F > -Cl$
So statement 4 is correct

71. (1) Descending order of conjugate base is

$CH \equiv C^{\ominus} > CH_2 = H\overset{\ominus}{C} > H_3C - \overset{\ominus}{C}H_2 > (CH_3)_3C^{\ominus}$

So, Stability of conjugate base $\propto$ acidic strength is
$C < D < B < A$

72. (3) Based on the organic compounds nature, and impurity present in it, they are purified.

73. (3) 1^0, 2^0, 3^0 amines are functional isomers

74. (2) Acidity order: $CHCl_3 > CHF_3 > CHBr_3 > CHI_3$
due to $d\pi$-$p\pi$ back bonding. $CHCl_3$ is more acidic than CHF_3.

75. (2) Acidic strength order
$[CH_3 - CH_3 < CH_2 = CH_2 < NH_3 < CH \equiv CH < H_2O]$

Correct & Incorrect MCQs

1. **(1)** A carbon atom having an sp hybrid orbital with 50% s character is more electronegative than that possessing sp^2 or sp^3 hybridised orbitals.
2. **(1)** In a π (pi) bond formation, parallel orientation of the two p orbitals on adjacent atoms is necessary for a proper sideways overlap.
3. **(3)** A heterocyclic compound has at least two different elements as a member of its ring.
4. **(3)** Acetaldehyde is an example of acyclic compound. Cyclopropane is an example of homocyclic compound. Tetrahydrofuran is an example of heterocyclic compound.
5. **(2)** In Lassaigne's test ,the elements present in the compound are converted from covalent form into the ionic form by fusing the compound with sodium metal.
6. **(1)** A species having a carbon atom possessing sextet of electrons and a positive charge is called a carbocation.
7. **(3)** Carbon in carbanion is generally sp^3 hybridised and its structure is distorted tetrahedron.
8. **(2)** Negatively charged ions with lone pair of electrons such as hydroxide (HO^-), cyanide (NC^-)
ions acts as nucleophile. Neutral molecules such as $H_2\ddot{O}$; R_3N:, $R_2\ddot{N}H$ etc., also act as nucleophiles due to the presence of lone pair of electrons.
Carbocations $\overset{+}{C}H_3$ act as an electrophile.
9. **(3)** Mixture of camphor and benzoic acid cannot be separated by the method of sublimation because both of them sublime on heating.
10. **(1)** Methyl ($-CH_3$) and ethyl ($-CH_2 - CH_3$) are usually considered as electron donating groups. Electromeric effect is a temporary effect. Hyperconjugation is a permanent effect.

Statement Based MCQs

1. **(1)** F show stronger +m effect because of resonance b/w 2 p orbitals of F and 2p orbitals of carbon attached to it
2. **(2)**
Pyrrole Pyridine
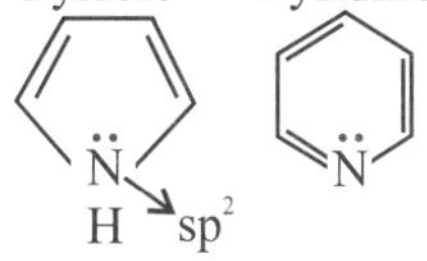
lone pair of N in pyrrole is in resonance so lone pair can't be donated. Pyridine more basic than pyrrole.
3. **(1)** Since glycerol decomposes at temp of 598 K at a reduced pressure, liquid boils at a low temperature than its normal boiling point and will therefore not decompose.
Both statements are true.
4. **(1)** Inductive effect arises because of electronegativity difference between atoms. This cause shift in electron density resulting in polar covalent bond which in turn polarises adjacent bond.
Both statements are true.
5. **(1)** As there is difference in arrangement of atoms so it will have different property depending on stability, polarity etc.
6. **(1)** Two or more compounds having the same molecular formula but different functional groups are called functional isomers. Propanone and propanal have molecular formula C_3H_6O, hence C_3H_6O shows functional group isomerism.
7. **(4)** Since, electrophiles have a strong tendency to attract electrons from other sources, hence electrophiles behave as lewis acid.
8. **(4)** The energy of actual structure of the molecule (the resonance hybrid) is lower than that of any of the canonical structures.
9. **(3)** Alkyl radical stability increases as we proceed from primary to tertiary:
$\overset{+}{C}H_3 < \overset{+}{C}H_2CH_3 <. (\overset{+}{C}H_3)_2 < \overset{+}{C}(CH_3)_3$
10. **(1)** The change in hybridisation affects the electronegativity of carbon. The greater the s character of the hybrid orbitals, the greater is the electronegativity.

Assertion & Reason MCQs

1. **(1)** Simple distillation is used where Boiling point difference is adequate otherwise both substances will vapourise at same temperature and separation does not occur.
2. **(1)** Position isomers only differ in position of functional group.
3. **(3)** In carius method sulphur is Oxidised to H_2SO_4 and $BaSO_4$ gives the light yellow solid.
4. **(1)** In Chromatographic technique liquid phase is allowed to move on stationary phase, movement depends upon solubility of components in pure solvent.
5. **(2)** Glycerol decomposes below it's Boiling Point. So, Reduced pressure is used
6. **(3)** All C atoms in alkanes are sp^3 hybridized
7. **(2)** OR (OCH_3) is ortho-para directing (cause + R effect)
8. **(2)** Heterolytic fission will form carbocations or carbanions
9. **(2)** It is so because of resonance as carbonate ion has 3 canonical forms, so average of them i.e. resonance hybrid will have bond length of all 3 bonds equal.
10. **(1)** Propyne has sp hybridisation of its middle carbon whereas in propene it is sp^2 hybridised so C atom in propyne will have more s character and hence attracts more electron density from sp^3– hybridised carbon of methyl group i.e. it is more electronegative.

Multi Correct MCQs

1. **(3)** The most common polar and acidic adsorbent used in adsorption chromatography is silica. Alumina is the example of polar and basic adsorbent that is used in adsorption chromatography.
2. **(3)** Heterolytic cleavage of a bond leads to the formation of ions.
3. **(4)** Only (A) and (B) structures show resonance effect. In (A), benzyllic resonance is present while in (B), allylic resonance of carbocation takes place.
4. **(4)** Compounds III and IV will show geometrical isomerism.

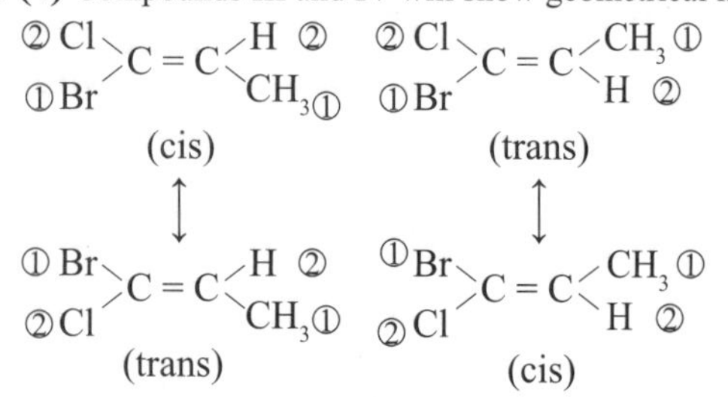

NCERT Exemplar MCQs

1. **(1)** The structure

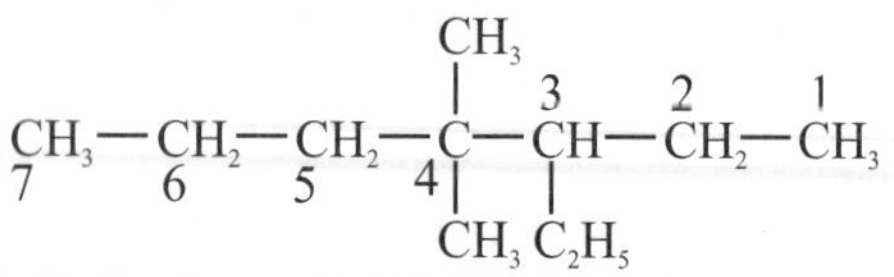

while writing IUPAC name, alkyl groups are written in alphabetical priority, thus lower locant 3 is assigned to ethyl.

2. **(4)** When more than one functional group lie in the main chain, nomenclature is done according to that functional group which has higher priority.

Carboxylic acid (—COOH) has more priority than ketone ($>C=O$). Hence the name is written as,-

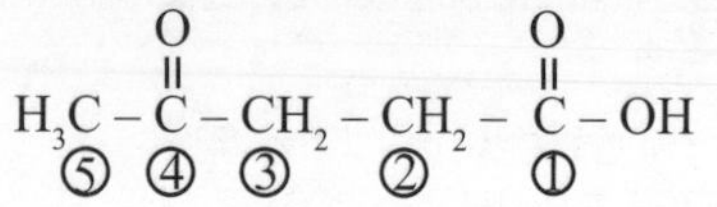

4 - Oxopentanoic acid

3. **(2)** For tri or higher substituted benzene derivatives, the compounds are named by identifying substituent positions on the ring followed by the lowest locant rule. Substituent of the base compound is assigned number 1 and then the direction of numbering is chosen such that the next substituent gets the lowest number. The substituents appear in the name in alphabetical order.

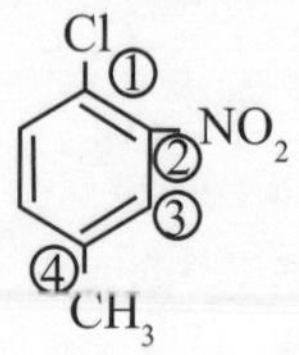

1-Chloro-4-methyl-2-nitrobenzenc

(not 1-chloro-2-nitro-4-methyl benzene)

4. **(3)** Electronegativity of carbon is state of hybridization dependent. % s-character ∝ Electronegativity

sp^3 < sp^2 < s

25% s 33% s 50% s

Hence in option (3), sp-hydridised carbon will be more electronegative.

5. **(3)** Functional isomers are two or more compounds having the same molecular formula but different functional groups.

Functional isomer of alcohol is ether.

Functional isomer of aldehyde is ketone.

Functional isomer of cyanide is isocyanide

However, alkyl halides do not show functional isomerism. Hence, option (3) is correct.

6. **(4)** As we know, essential oils are insoluble in water and have high vapour-pressure at 373K but are miscible with water-vapour in vapor phase, it means these are steam volatile. For the extraction of essential oils steam distillation technique is used.
7. **(4)** TLC stands for Thin Layer Chromatography which is an another type of adsorption which involves separation of substances of a mixture over a thin layer of an adsorbent coated on a glass plate.
8. **(2)** Partition chromatography is based on continuous differential partition of components of a mixture between stationary and mobile phases. Paper chromatography is a type of partition chromatography.
9. **(1)** Stability of the given cations can be understood by the following structures

I. $CH_3 \rightarrow \overset{\oplus}{C}H \leftarrow CH_3$

Weak + I-effect of the two methyl groups stabilises carbocation (I)

II. $CH_3 - \overset{\oplus}{C}H \leftarrow \ddot{O} - CH_3$

Strong +M-effect of $-OCH_3$ group stabilises carbocation (II)

III. $CH_3—\overset{\oplus}{C}H—CH_2 \rightarrow OCH_3$

-I-effect of $-OCH_3$ group destabilises the carbocation (III)

Hence, the stability of carbocation decreases in the following order.

II > I > III

10. (2) IUPAC name of the given compound is given as, -

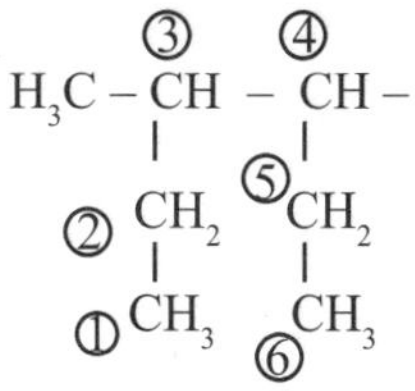

3,4 – Dimethylhexame

11. (1) Electronegativity of *Cl, Br, C* and *Mg* follows the order *Cl > Br > C > Mg*

Due to presence of highly -ve Cl^- atom compound (i) will have highest positive charge.

$\overset{*}{C}H_3 - CH_2 - Cl \quad (-I - \text{effect})$

$\overset{*}{C}H_3 - CH_2 - Mg^+Cl^-$

NEET & AIIMS Past Year MCQs

1. (1)

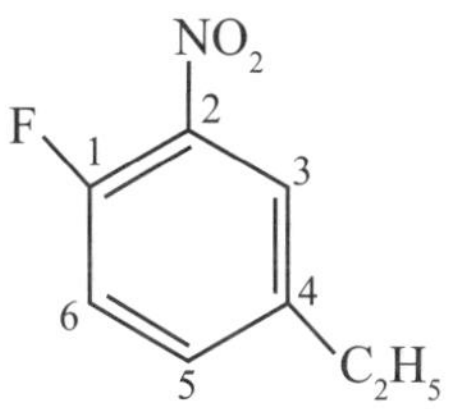

4-ethyl-1-fluoro-2-nitrobenzene

2. (3) The presence of a methyl group connected to a positively charged carbon atom enhances the stability of the carbocation through hyperconjugation and +I effect.

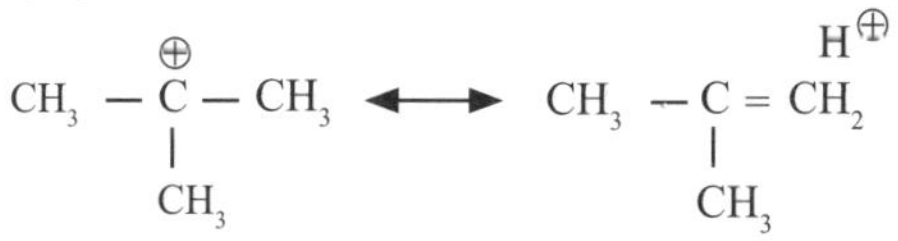

3. (4) If one of the substances in the mixture is water and the other, a water insoluble substance, then the mixture will boil close to but below, 373 k (100°C).

4. (2) Percentage of carbon in the organic compound:

$$\text{Percentage of carbon} = \frac{12 \times m_2 \times 100}{44 \times m}$$

Where, m = mass of organic compound = 0.3 g

m_2 = mass of carbon dioxide = 0.2 g

$$\therefore \%C = \frac{12 \times 0.2 \times 100}{44 \times 0.3} = 18.18\%$$

Percentage of hydrogen in the organic compound:

$$\text{Percentage of hydrogen} = \frac{2 \times m_1 \times 100}{18 \times m}$$

where, m_1 = mass of water = 0.1 g

$$\therefore \%H = \frac{2 \times 0.1 \times 100}{18 \times 0.3} = 3.70\%$$

5. (4) On heating, solid substances change from solid to vapour state without passing through liquid state. This purification technique is known as sublimation and is used to separate sublimable compound from non-sublimable impurities.

6. (1)

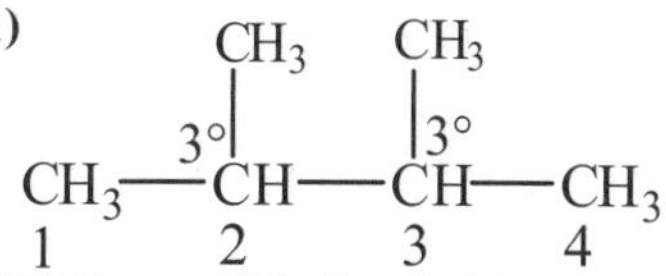

IUPAC name: 2,3-dimethyl butane

7. (1)

(A) 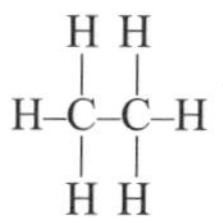 $\rightarrow$ One σ bond

(B) $H_2C{=}CH_2 \rightarrow$ One σ-bond and one π-bond

(C) $C_2 \rightarrow$ Two π-bonds

(D) $HC{\equiv}CH \rightarrow$ One σ-bond and two π-bonds

8. (2)

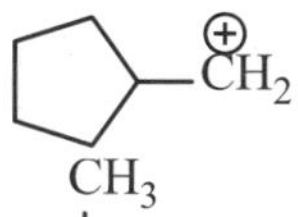

: 1° carbocation

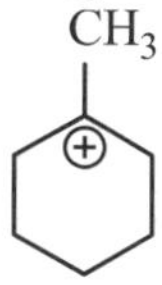

: 3° carbocation

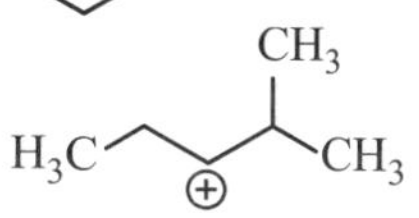

: 2° carbocation

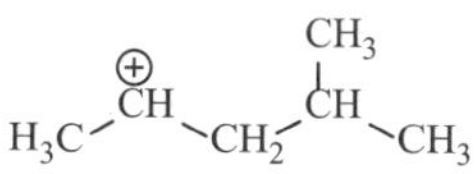

: 2° carbocation

- Stability order of carbocation : 3° > 2° > 1°.
- Hence, most stable carbocation is present in option (2).

9. (1) In case, nitrogen and sulphur both are present in an organic compound, sodium thiocyanate is formed.

$Na + C + N + S \rightarrow NaSCN$

Which further reacts as:

$Fe^{3+} + SCN^- \rightarrow [Fe(SCN)]^{2+}$
(Blood red color)

10. (4) The structure of pyridine is:

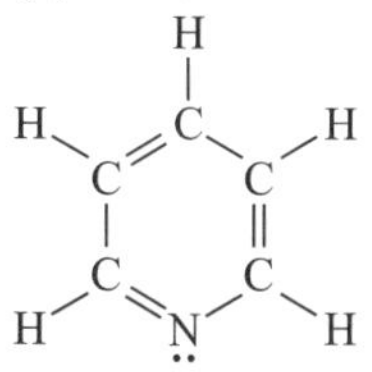

Hence, it has 11 sigma, 3 π bonds and one lone pair.

11. (4) 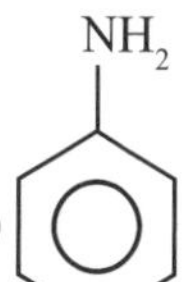

Kjeldahl's method is not applicable to the compounds containing nitrogen having nitro and azo group and nitrogen present in the ring (pyridine), as nitrogen of these compounds does not change to ammonium sulphate under these conditions.

12. (2)

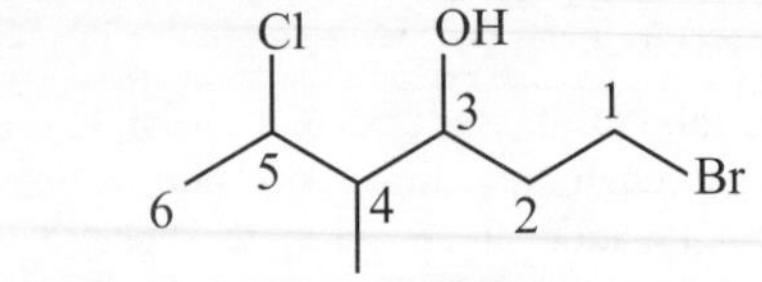

1-Bromo-5-chloro-4-methylhexan-3-ol

13. (3) $CH_3-CH_2-O-CH_2-CH_3$, $CH_3-O-\underset{\displaystyle CH_3}{\underset{|}{CH}}-CH_3$ and

$CH_3-O-CH_2-CH_2-CH_3$

14. (4) The correct structure is:

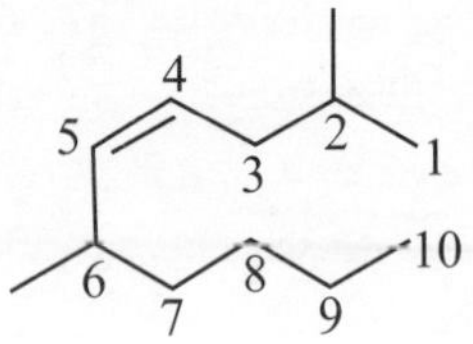

15. (3)

16. (2)

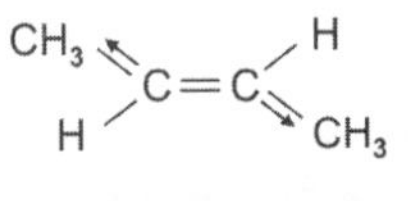

trans–but–2–ene

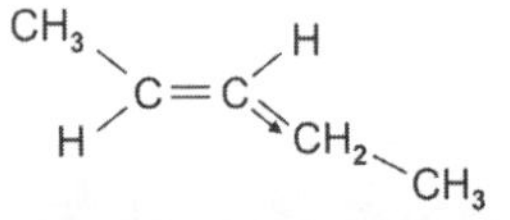

trans–pent–2–ene

In trans–but–2–ene the direction of CH_3 are opposite to each other and cancel the vector direction of each other but not in trans-pent-2-ene

17. (1)

18. (3) It views a carbon-carbon bond from an oblique angle. This helps in visualize the molecule as a whole in 3-D.

19. (1)

20. (1) The Beilstein test is used for the detection of halogens, specifically chlorine (Cl^-).

Hydrocarbons

NCERT Topic-wise MCQs

1. (4) This is wurtz reaction and proceeds as follows:

$$CH_3CH_2CH_2-\underset{\displaystyle C_2H_5}{\underset{|}{CH}}-Br + 2Na \longrightarrow$$

$$Br-\underset{\displaystyle C_2H_5}{\underset{|}{CH}}-CH_2-CH_2-CH_3$$

$$\xrightarrow{\text{dry ether}} \overset{8}{C}H_3-\overset{7}{C}H_2-\overset{6}{C}H_2-\underset{\displaystyle C_2H_5}{\underset{|}{\overset{5}{C}H}}-\underset{\displaystyle C_2H_5}{\underset{|}{\overset{4}{C}H}}-\overset{3}{C}H_2-\overset{2}{C}H_2-\overset{1}{C}H_3$$

4, 5-Diethyloctane

2. (3) Due to hyper conjugation and resonance the stability order of free radicals is $3° > 2° > 1°$.

$CH_3\dot{C}HC_6H_5$ is most stable.

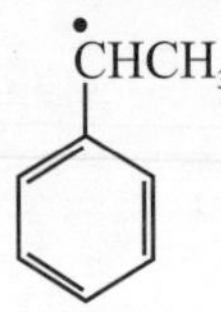

shows both resonance and hyper conjugation

3. (1) At room temperature lower alkanes (C_1 to C_4) are in gaseous state.

4. (1) The major constituent of natural gas is methane.

5. (2) The process of aromatisation occurs.

$$\underset{\text{(Aromatisation)}}{CH_3CH_2CH_2CH_2CH_2CH_2CH_3} \xrightarrow[Cr_2O_3,\ 773\ K]{Al_2O_3}$$

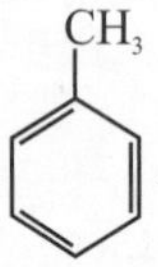

$+ 3H_2$

6. (4) Compounds containing two or more carbon atoms are prepared by Wurtz reaction. Thus; methane, CH_4 cannot be prepared by wurtz reaction.

7. (3) Sodium salt of any carboxylic acid with hot soda lime undergo decarboxylation to form alkane which has one carbon atom less than the parent molecule.

$$\underset{\text{propanoic acid}}{C_2H_5COOH} \xrightarrow[\Delta]{NaOH / CaO} \underset{\text{ethane}}{C_2H_6}\uparrow + Na_2CO_3$$

8. (4) Methane is called marsh gas, it is found in swamps or marshy places and can be obtained by bacterial decomposition of fossil of plants and animals.

9. (4) As number of C atoms increases in alkanes, the Van der waal's forces also increase. However, branching results in decrease in boiling point due to decreased surface area which results in weaker Van der waal forces, Thus; order of boiling point of given alkanes is

Hexane > 2-Methyl pentane> 3 - methyl pentane > 2 ; 2 - dimethyl butane.

10. (1) Boiling point $\propto$ molecular mass $\propto \dfrac{1}{\text{branching}}$

($\therefore$ surface area decreases)

$\therefore$ n-hexane has the highest boiling point among the given alkanes.

11. (4) $\underset{\text{Propyl bromide}}{CH_3CH_2CH_2Br} \xrightarrow{Mg/ether} \underset{\text{Grignard Reagent}}{CH_3CH_2CH_2MgBr} \xrightarrow[H_2O]{\Delta} CH_3CH_2CH_3$

12. (4) $\underset{\text{Sodium ethanoate}}{2CH_3COO^-Na^+} \longrightarrow CH_3-CH_3 + 2CO_2$

13. (2) 2, 3 - Dimethylbutane has only two types of hydrogens and hence forms only two mono chlorinated compounds.

14. (2) $LiAlH_4$ reduces 1° and 2° alkyl halides but brings about elimination in 3° alkyl halides. In contrast, $NaBH_4$ reduces both 2° and 3° alkyl halides to alkanes.

15. (4) (1), (2) and (3) are all isomer but (4) is not.

16. (1) 3-methyl hexane has all the seven carbon atoms different and hence forms seven different mono-chlorination products.

$$\overset{1}{C}H_3-\overset{2}{C}H_2-\overset{\displaystyle CH_3}{\overset{|}{\overset{3}{C}H}}-\overset{4}{C}H_2-\overset{5}{C}H_2-\overset{6}{C}H_3$$

17. (1) $CH_4 + HNO_3 \xrightarrow[400°c]{\Delta} CH_3 - NO_2$

Nitromethane

18. (2) The dihedral angle between the H-atoms of 2-methyl groups in staggered conformation of ethane is 60°.

19. (3) The most stable conformation of ethane is staggered. In staggered conformer of ethane the substituents on two carbon atoms are at the maximum distance from each others thus, have having less tortional strain. This makes the staggered conformer of ethane more stable than its eclipsed and skew conformers.

Staggered form of ethane is represented as,

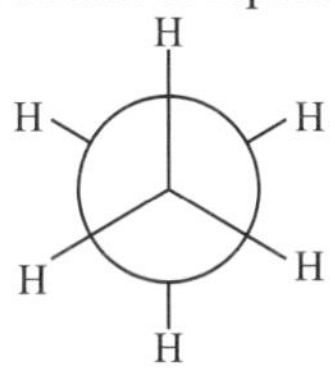

20. (1) In Kolbe's electrolysis, the alkane is formed when an aqueous solution of Na or K salt of carboxylic acid is electrolysed. The alkane formed has, two or more carbon atoms. This method can be used to prepare alkanes with even number of carbon atoms.

21. (2) IUPAC name is - 2, 2-Dimethyl Propane

22. (3) Wurtz reaction

23. (2) $CH_4 + O_2 \xrightarrow[\Delta]{Mo_2O_3} HCHO$

24. (3) Pyrolysis is endothermic reactions

25. (4) Alkanes are soluble in organic, non-polar solvents

26. (3) Cracking involves decomposition of higher alkanes to lower alkanes

27. (3) Homolysis.

28. (4) $C_nH_{2n+2} + \left(\frac{3n+1}{2}\right)O_2 \rightarrow nCO_2 + (n+1)H_2O$

29. (3) Anh.$AlCl_3$/HCl

30. (3) In ethane the number of σ bond formed by the overlapping of $sp^3 - sp^3$ hybridisation carbon is 1.

$$\underset{sp^3}{H_3C} - \underset{sp^3}{CH_3}$$

31. (2) $\underset{\text{ethylene}}{H_2C \overset{(sp^2)}{=} CH_2} \rightarrow \underset{\text{ethane}}{\overset{(sp^3)}{CH_3}-CH_3}$

In ethylene hybridisation is sp^2 and in ethane sp^3

32. (4) Relative stability order of conformers of butane anti staggered > Gauche > Partial Eclipsed > Fully Eclipsed

$$\text{Stability} \propto \frac{1}{\text{energy}}$$

So Anti confirmation has lowest energy

33. (1) $CH_2{=}\underset{|}{\overset{CH_3}{C}}{-}CH{=}CH_2$ (Isoprene)

2 methyl 1,3 Butadiene

$CH_2{=}\overset{Cl}{C}{-}CH{=}CH_2$ 2 chloro 1,3 Butadiene

(Chloroprene)

34. (4) Cyclohexane adopts a chair conformation is which all the bond angles between carbon atoms are close to tetrahedral angle, i.e, 109.5. This conformation is free from angular strain as well as torsional strain.

chain conformation

35. (4) The angle strain is maximum in cyclopropane. Hence, it is highly strained molecule and consequently most unstable. The cyclobutane is more stable. Hence; angle strain in cyclobutane is less than cyclopropane.

This stability increase up to 6 membered rings, then decreases from 7 to 11 membered rings and from the 12 membered rings onwards attains the satbility of 6 membered rings. Heat of combustion is a method of measuring chemical stability.

Hence; cyclohexane has the lowest heat of combustion.

36. (3) $Br_2 + h\nu \longrightarrow Br^\bullet + Br^\bullet$

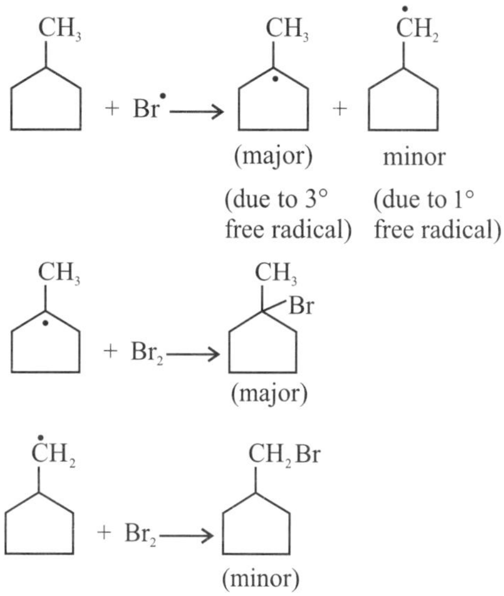

37. (2) Cycloalkanes are isomeric with alkenes because they have same general formula C_nH_{2n} (i.e., same molecular formula) but possessing different structures. They show ring chain isomerism.

38. (2) Addition of HBr gives anti-Markonikoff product in presence of peroxide. Anti-markonikoff rule is only pronounced in case of HBr. Therefore in case of HCl the reaction is given as-

$$H_3C - CH{=}CH_2 \xrightarrow[\text{Peroxide}]{HCl} H_3C{-}\underset{Cl}{CH}{-}CH_3$$

(markonikoff product)

39. (3)

trans-2-Chlorobut-2-ene cis-2-Chlorobut-2-ene

40. (1) The group, iodide is bigger as down the group size of the halogen increases. Thus, it is better leaving group.

∴ Iodide > Bromide > Chloride.

41. (3)

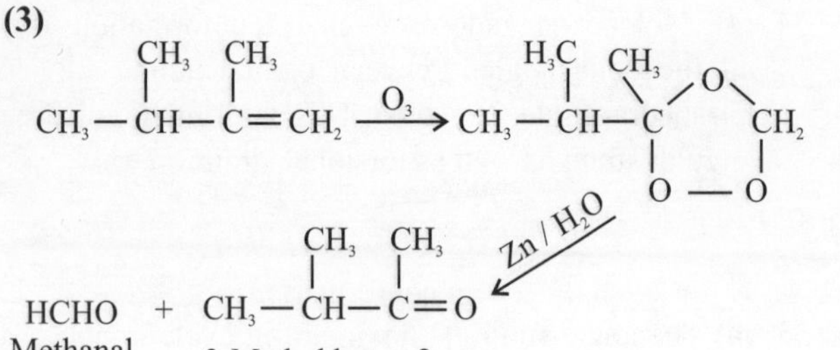

42. (4) More stable alkene will be formed by β - elimination, only when there are more number of β-substituents. Since (i) has two, (iii) has one β - methyl substituent while (ii) has no β - methyl substitutive therefore; reactivity β - elimination decrease in order:

(i) > (iii) > (ii)

43. (1)

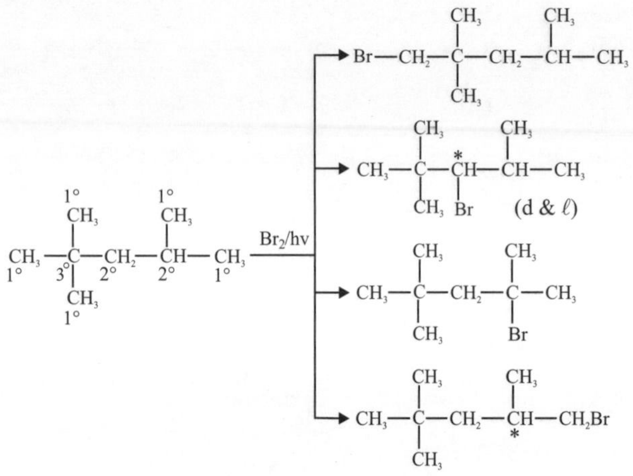

44. (4) $CH_3CH{=}CH_2 + \overset{\delta-\ \delta+}{HOBr} \longrightarrow CH_3{-}CH(OH){-}CH_2{-}Br$

1-Bromo-2-propanol

45. (3) $CH_2{=}CH{-}CH_2Cl + CH_3MgI \longrightarrow$
Allyl chloride

$CH_2{=}CH{-}CH_2CH_3 + Mg(I)Cl$
1-butene

46. (2) Since, the alkene on oxidation gives only acetic acid, therefore; the alkene must be symmetrical containing two carbon - atoms on either side of the double bond, i.e., 2 - butene.

$CH_3CH = CHCH_3 \xrightarrow{KMnO_4} 2CH_3COOH$

47. (2) Propene give HCHO and CH_3CHO on ozonolysis.

48. (3) Due to the presence of acidic hydrogen ammonical Cu_2Cl_2 will give red copper acetylide ppt. with Cu_2Cl_2, whereas C_2H_4 will be neutral to ammonical Cu_2Cl_2.

49. (2) $\begin{matrix} CH_2-COONa \\ | \\ CH_2-COONa \end{matrix} + 2H_2O \xrightarrow{\text{electrolysis}} \underbrace{\begin{matrix} CH_2 \\ \| \\ CH_2 \end{matrix} + 2CO_2}_{\text{At Anode}} + \underbrace{2NaOH + H_2}_{\text{At Cathode}}$

50. (2) $CH_2 = CH - CH = CH - CH = CH - CH = CH_2$

No. of σ bonds = 17, No. of π-bonds = 4

51. (3) The compound does not exhibit geometrical isomerism when two groups attached to a double bonded carbon atoms are same. Only 2- butene exhibits Cis-trans isomerism.

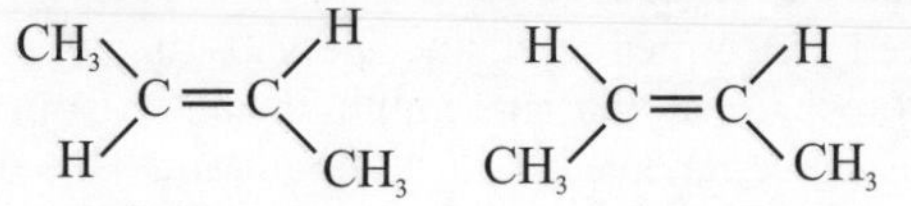

trans-2-Butene cis-2-Butene

52. (1) $\begin{matrix} CH_2Br \\ | \\ CH_2Br \end{matrix} + Zn \xrightarrow{\text{alcohol, heat}} \begin{matrix} CH_2 \\ \| \\ CH_2 \end{matrix} + ZnBr_2$

53. (1)

54. (2) This reaction is known as oxymercuration demercuration. It involve addition of water without rearrangement.

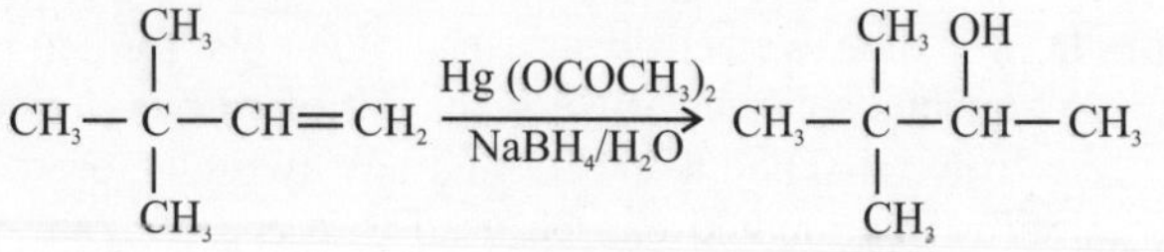

55. (4) $CH_2 = CH -$ is vinyl group

56. (1) Addition of halogens in non polar solvent occurs in anti manner.

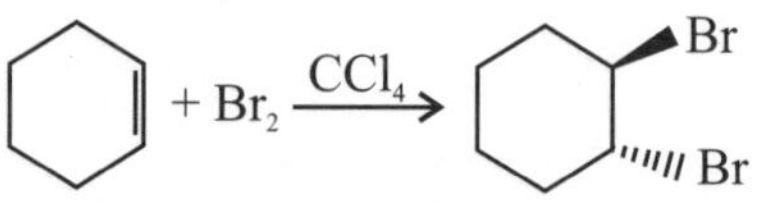

57. (1) The IUPAC name of given compound is-
2, 8-Dimethyldeca-3, 6-diene

58. (2) The IUPAC name of the given compound is 4-methylocta-1, 3, 5, 7-tetraene

59. (2)

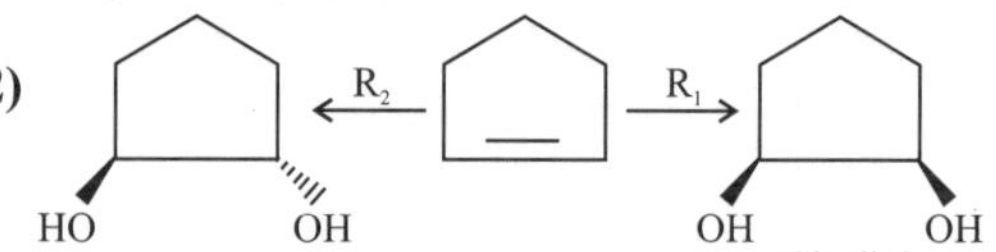

Trans diol Cis diol

R_1 = cold alkaline $KMnO_4$ (Baeyer's Reagent) is used for Cis hydroxylation

R_2 = Peracid/ $H_3O^{\oplus}$ is used for formation of transdiol.

60. (2) Pd/$BaSO_4$ + Quinoline used for partial hydrogenation

61. (2) Unsaturated hydrocarbons undergo addition reactions

62. (1) Ozonolysis gives carbonyl compounds

63. (2) Alumina acts as a dehydrating agent. Therefore, ethene is formed.

64. (2) Unsymmetrical alkenes with HBr only shows anti-Markonikov's rule.

65. (4) 2-Pentene gives C_2H_5CHO and CH_3CHO on ozonolysis.

66. (2) Ethylene oxide

67. (3) $C_2H_4 + H_2SO_4 \xrightarrow{80^0 C} C_2H_5HSO_4 \xrightarrow[\Delta]{H_2O} C_2H_5OH$

68. (3) $CH_3 - CH = CH_2 \xrightarrow{HBr} CH_3 - CH(Br) - CH_3$

$\xrightarrow{alc\,KOH} CH_3 - CH = CH_2 \xrightarrow{O_3,\ Zn/H_2O} CH_3CHO + HCHO$

69. (1) Markovnikov's rule and antiMarkovnikov's rule

70. (1) x = Na/NH_3(liq)

y = $Pd/BaSO_4 + H_2$

71. (4) $KMnO_4$ is very powerful oxidising agent in acidic medium.

72. (3) Heat of hydrogenation decreases as the conjugation increases,. Thus, the alkene (3) with two isolated double bonds has the highest heat of hydrogenation.

73. (2) $H-C\equiv C-H + HCl \xrightarrow[60°C]{HgCl_2} CH_2=CH-Cl$

Vinyl chloride

74. (3)

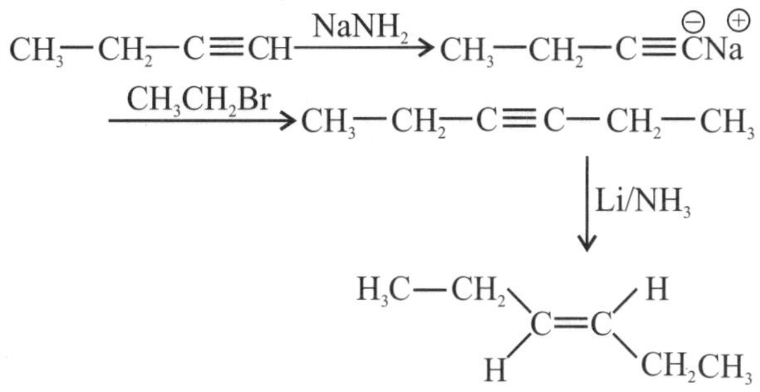

Trans alkene

75. (4) Due to 50% s-character $CH\equiv CH$ is more acidic.

76. (3) Ethane will form when CaC_2 reacts with water, it gives

$CaC_2 + H_2O \rightarrow CH\equiv CH + Ca(OH)_2$

77. (1) $CH_3-C\equiv C-CH_2-CH_3 + O_3 \longrightarrow$

2-Pentyne

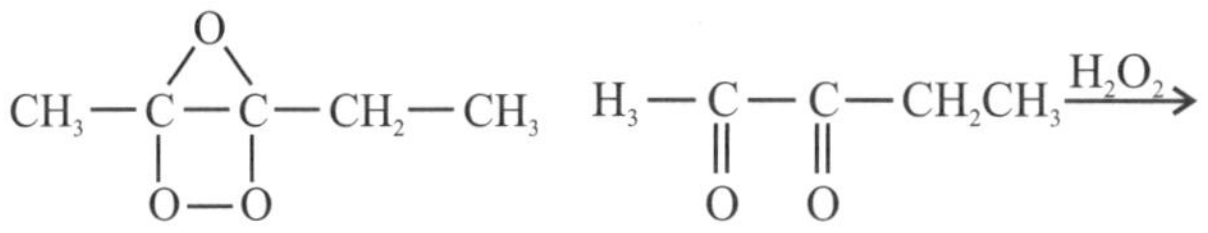

$CH_3COOH + CH_3CH_2COOH$

78. (2) $CH_3C\equiv CH \xrightarrow{HBr} CH_3-C(Br)=CH_2 \xrightarrow{HBr} CH_3-CBr_2-CH_3$

2,2-Dibromopropane

79. (2)

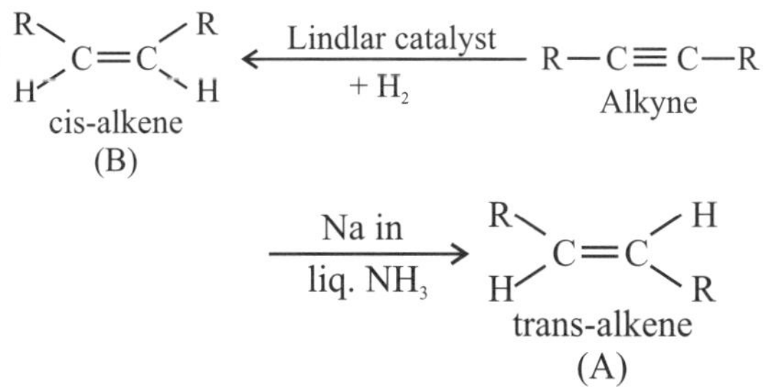

80. (4) Alkynes are converted into alkenes by using lindlar's catalyst (H_2/pd)

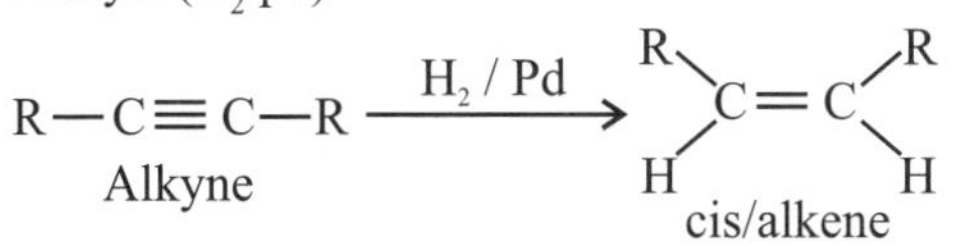

81. (3)

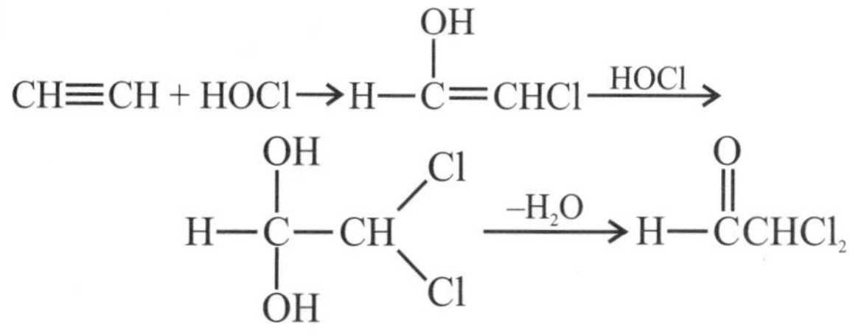

82. (2) $CH\equiv CH + HCl \longrightarrow CH_2=CH-Cl \xrightarrow{HCl} CH_3CHCl_2$

Acetylene Vinyl chloride

(Alkene due to presence of only 1° of unsaturation, form only monochloro derivative).

83. (1)

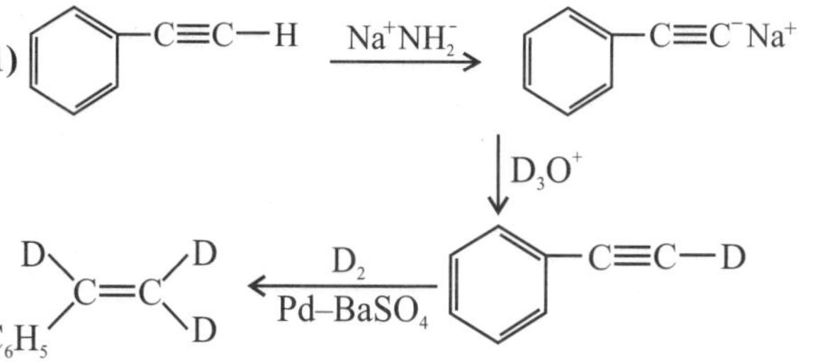

84. (4)

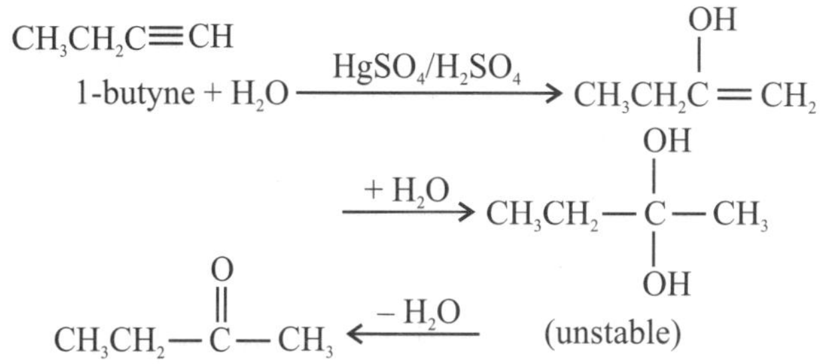

Butane-2-one

85. (2) Hybridisation of C in O=C=O is sp. In, H–C≡C–H, C is sp hybridised.

86. (3) Only alkynes react with $AgNO_3$ to give white precipitate while alkene and alkynes both react with $KMnO_4$ and decolorise it.

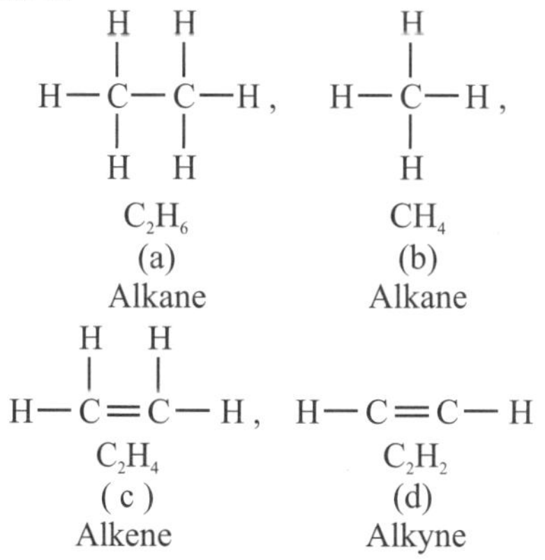

∴ C_2H_4 (an alkene) reacts with $KMnO_4$ and decolorises it but does not react with $AgNO_3$. Whereas C_2H_6 and CH_4 are alkanes they do not react with both $KMnO_4$ and $AgNO_3$

87. (2) Propyne is formed when magnesium carbide undergoes hydrolysis.

$Mg_2C_3 + 4H_2O \rightarrow CH_3C\equiv CH + 2Mg(OH)_2$

88. (3) Due to the acidic nature of hydrogen atoms attached to a triple bonded carbon atoms, acetylene and terminals alkynes form metal acetylides.

$CH_3CH_2C \equiv CH + AgNO_3 + NH_4OH \rightarrow CH_3CH_2C \equiv CAg + NH_4NO_3 + H_2O$

Silver butylide

89. (2) C_5H_8 can have the following structure with one triple bond.

$CH_3-CH_2-CH_2-C\equiv CH$, $CH_3CH_2-C\equiv C-CH_3$,

Pent-1-yne, Pent-2-yne

$CH_3-CH(CH_3)-C\equiv CH$

3-Methylbut-1-yne

90. (3) As the s - character increase, acidity of hydrogen atom increases.

Ethane:	CH_3-CH_3	sp^3	25%	s - Character
Ethane:	$CH_2=CH_2$	sp^2	33%	s - Character
Ethane:	$CH\equiv CH$	sp	50%	s - Character
Ethane:	C_6H_5	sp^2	33%	s - Character

91. (1) $H-C\equiv C-H + 2CH_3COOH \rightarrow CH_3-CH(OCOCH_3)_2$

Both the protons go to same carbon atom.

92. (2) $CH_3-CH_2-C\equiv CH \xrightarrow{NaNH_2} CH_3-CH_2-C\equiv \overset{\ominus\oplus}{C}Na \xrightarrow{CH_3CH_2Br} CH_3-CH_2-C\equiv C-CH_2CH_3$

$CH_3-CH_2-C\equiv C-CH_2CH_3 \xrightarrow[BasO_4]{Pd} CH_3-CH_2-CH{=}CH-CH_2-CH_3 \xrightarrow[Zn/H_2O]{O_3} CH_3CH_2CHO + CH_3CH_2CHO$

Propanal

93. (4) Alkynes exhibits all the given type of isomerism

94. (4) Double bonds have more priority over triple bond when they are equidistant

95. (1) Kolbe's electrolysis

96. (3) Dehydrohalogenation in two steps

97. (2) $CHI_3 + 6Ag + CHI_3 \xrightarrow{\Delta} \underset{\text{Acetylene}}{H-C\equiv C-H} + 6AgI$

98. (4) Dehydrohalogenating agent Zn dust is used in preparation of acetylene from tetra halides.

99. (2) Due to presence of PH_3 and H_2S

100. (2) 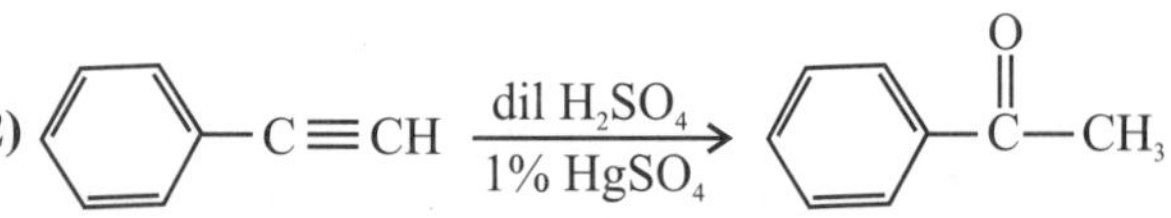

101. (1) Birch reduction

$CH_3-C\equiv C-CH_3 \xrightarrow{Na/Liq\ NH_3} CH_3(H)C{=}C(H)CH_3$

Trans alkene

102. (1) $CH_3-CHBr_2 \xrightarrow{KOH(AlC)} \underset{\text{Acetylene}}{H-C\equiv C-H}$

103. (3) 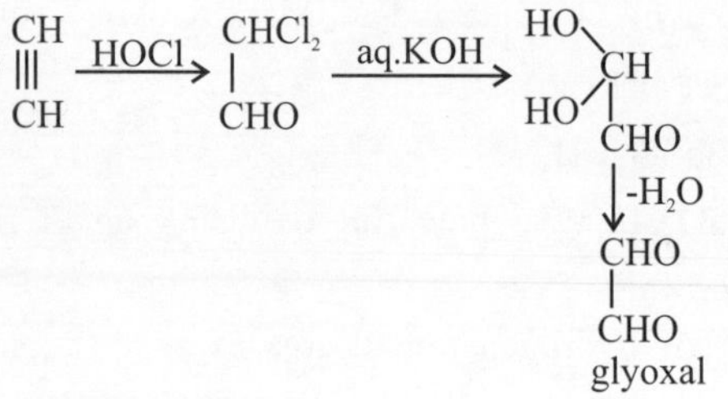

104. (2) It gives trans-2-Butene

105. (4) Isobutane gives minimum boiling point.

106. (2) 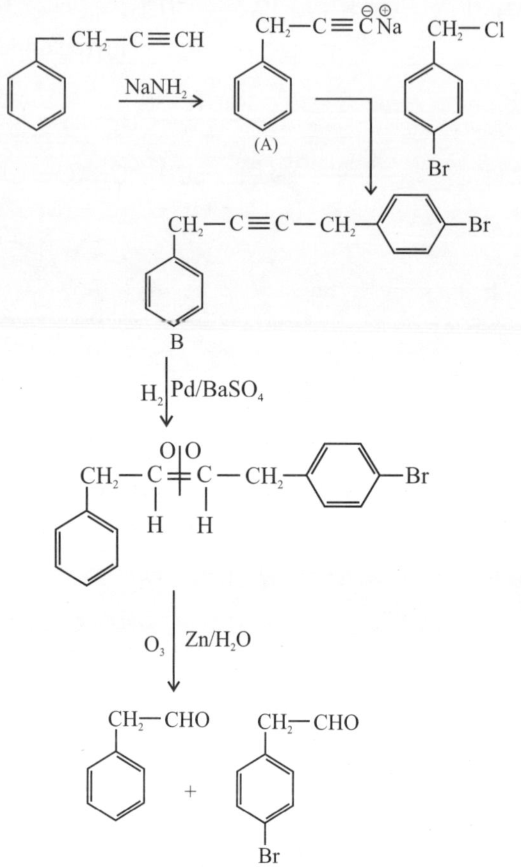

107. (2) $C_7H_{12} \xrightarrow[H_2O]{O_3} CH_3-CH(CH_3)-COOH + CH_3-CH_2COOH$

$CH_3-CH(CH_3)-C\equiv C-CH_2-CH_3$

2-methyl hex-3-yne

108. (1) Acetylene is a very weak acid. It cannot react with strong base NaOH

109. (3) $CH_3-C\equiv CH \xrightarrow[NaOH|H_2O_2]{B_2H_6} \underset{(A)}{CH_3-CH_2-CHO}$

Acetaldehyde

$CH_3-C\equiv CH \xrightarrow[HgSO_4]{dil\ H_2SO_4} \underset{(B)}{CH_3-\overset{O}{\overset{\|}{C}}-CH_3}$ Ketone

A & B are functional group isomers.

110. (2) $CH\equiv CH + HCN \rightarrow CH_2=CH-CN$

111. (3) X = C_2H_2 and y = C_2H_4

112. **(1)** 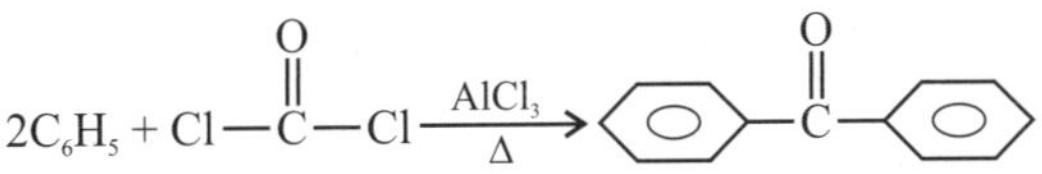

113. **(2)** t - butyl benzene is produced when benzene reacts with iso - butene in the presence of H_2SO_4. The reaction takes place as follows

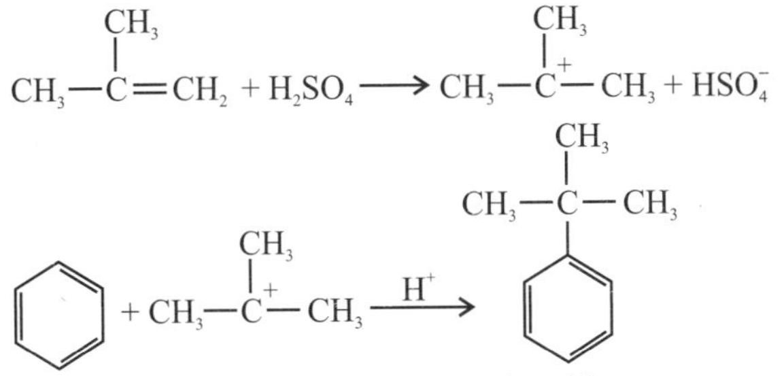

t-butyl benzene

114. **(2)** Due to greater +M effect of –OH group.

115. **(3)** Phenol undergoes bromination easily due to presence of –OH group the ring becomes much more active in substitution reaction.

116. **(3)** While 2,4,6-trinitrophenol, benzoic acid and benzene sulphonic acid are soluble in $NaHCO_3$. In fact,

Acid + $NaHCO_3 \rightarrow$ salt + H_2CO_3

This is possible if acid is more acidic than H_2CO_3. o-nitrophenol is less acidic than H_2CO_3 reaction. Hence, It is not soluble in sodium hydrogen carbonate.

117. **(3)** Nitration is carried out by the help of electrophile NO_2^+

$HNO_3 + 2H_2SO_4 \rightarrow NO^+_2 + 2HSO^-_4 + H_3O^+$

Halogenation is carried out by the help of electrophile Cl^+ (X^+)

$Cl_2 + FeCl_3 \rightarrow Cl^+ + FeCl^-_4$

118. **(1)** 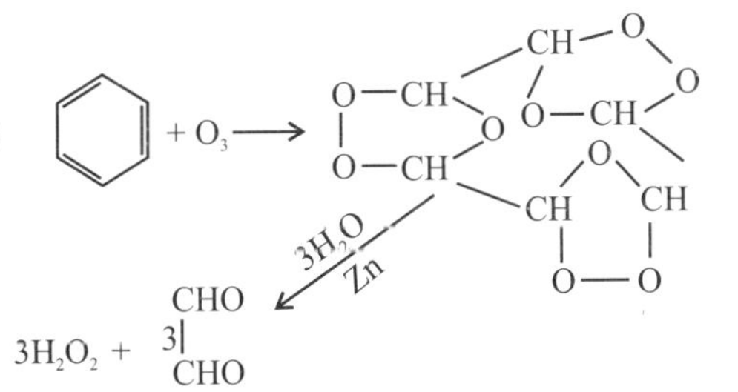

119. **(2)** Alkylation and acylation of benzene in the presence of anhydrous aluminum chloride are known as friedel - crafts reaction.

120. **(2)** Due to delocalisation of π-electrons benzene does not show addition reaction like alkenes.

121. **(3)** Electrophilic substitution reactions are the most common reaction shown by benzene.

122. **(4)** –COOH and $-NO_2$ deactivates the ring. $-CH_3$ group activates the benzene ring, undergoes nitration easily.

123. **(3)** $-\ddot{O}H$ is +M group

so attack of Br^+ take place on para position of –OH group.

124. **(3)**

125. **(4)** Formation of all the three compounds are the result of addition reactions of benzene which confirms the unsaturation character of benzene.

126. **(1)** Due to strong electron withdrawing (–M effect) effect of NO_2 group, $C_6H_5NO_2$ shows least reactivity towards electrophilic substitution

127. **(2)** Carbon percentage is more

128. **(2)** All bond length are in between single and double bonds

129. **(2)** 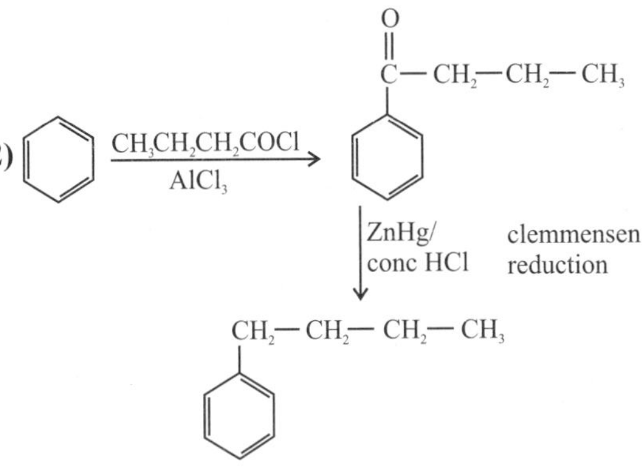

130. **(4)** 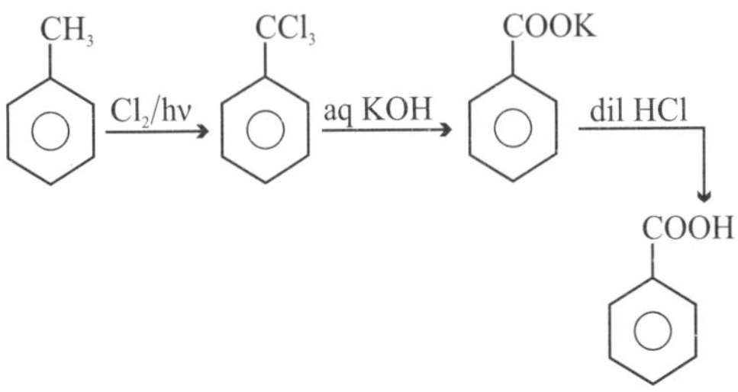

131. **(2)** Benzene hexachloride

132. **(1)** $C_6H_5OH \xrightarrow[\Delta]{Zn} ZnO + C_6H_6$

133. **(3)** Resonance

134. **(3)** Cyclic but not following Huckel's Rule because it have 4πe's.

135. **(2)** Formation of carbocation is the rate determining step.

136. **(4)** Alkane cannot undergo oxidation with O_3

137. **(3)** In presence of sunlight addition occurs

138. **(3)** $X = C_6H_6$

139. **(3)** In presence of catalyst electrophilic substitution occurs

140. **(1)** More the number of α- hydrogen present in the alkyl group attached to the benzene ring more pronounced will be the hyperconjugation and the benzene ring will be more electron rich and easily be attacked by an electrophile.

α - hydrogen in $-CH_3$, $-CH_2-CH_3$, $-CH(CH_3)_2$ and $-C(CH_3)_3$ Respectively are three, two, one and zero respectively.

High Order Time Intensive MCQs

1. **(2)** The reaction follow free radical mechanism. (Kolbe's electrolysis)

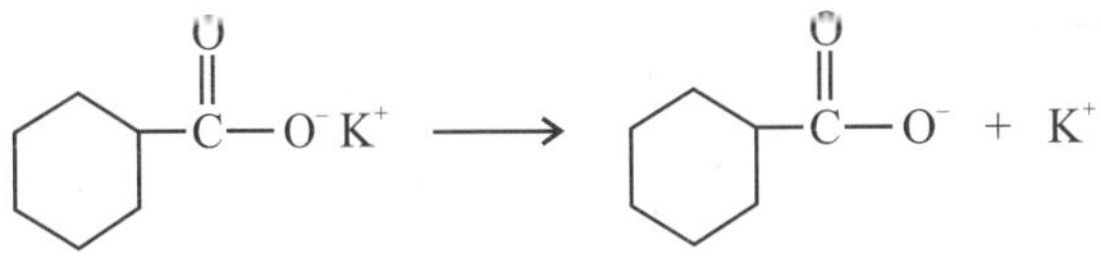

At anode

(oxidation occurs)

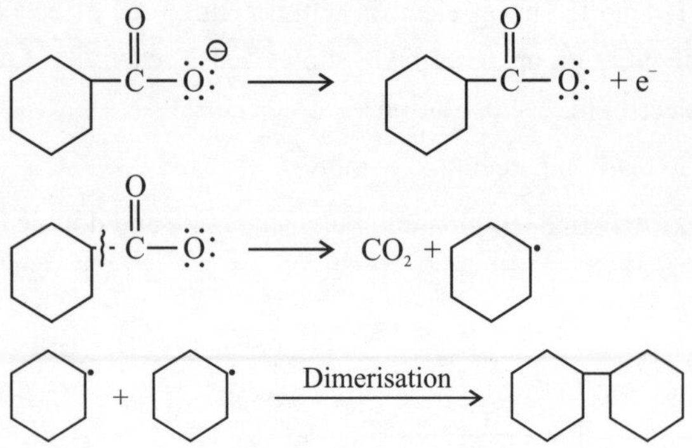

2. (1)

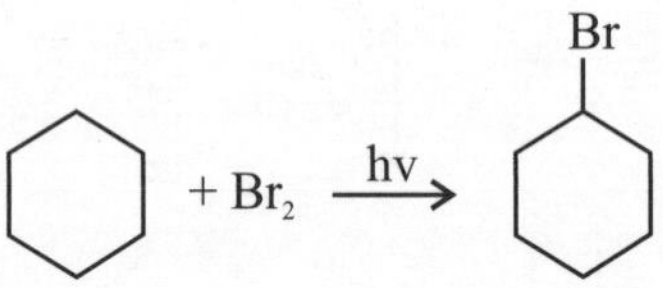

(Free radical halogenation occurs)

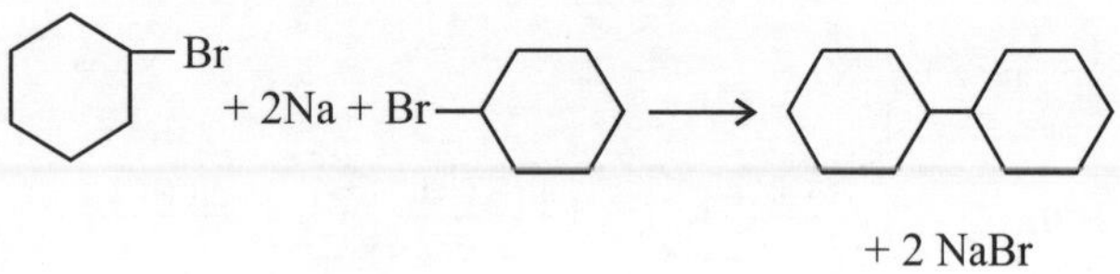

3. (3)

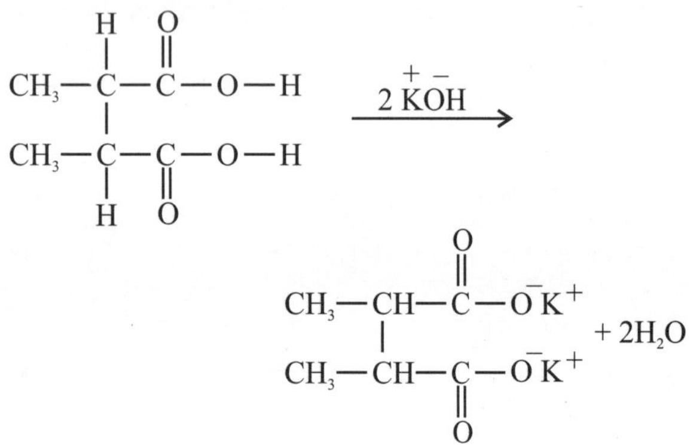

Kolbe's electrolysis follows free radical mechanism.

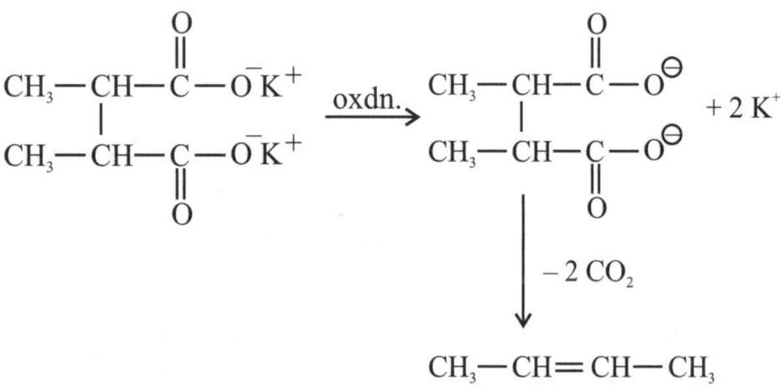

(But - 2 - ene can exist as geometical isomers)

4. (4)

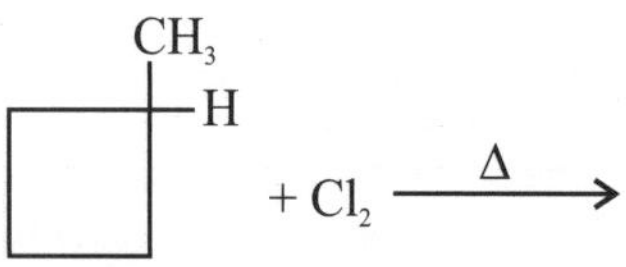

Free radical halogenation occurs. Possible products are-

(1)

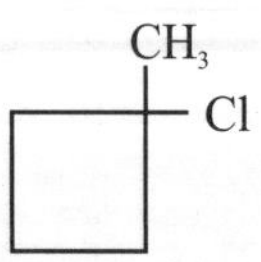

(2)

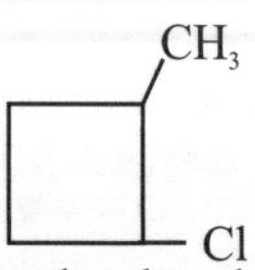

This molecule exhibits geometrical isomerism.

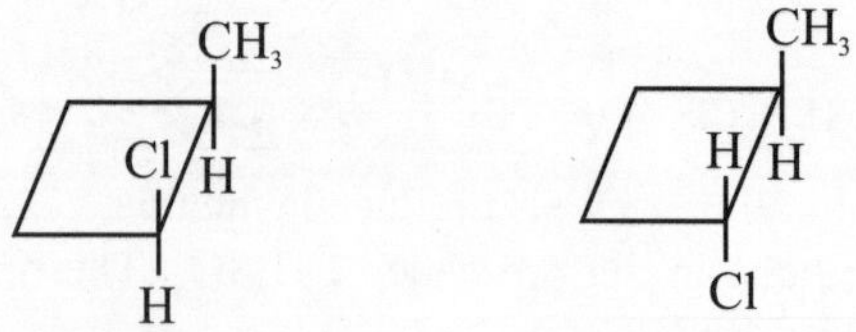

Cis - 1 - chloro - 2 -methyl cyclobutane

Trans - 1 - chloro - 2 - methyl-cyclobutane

Further

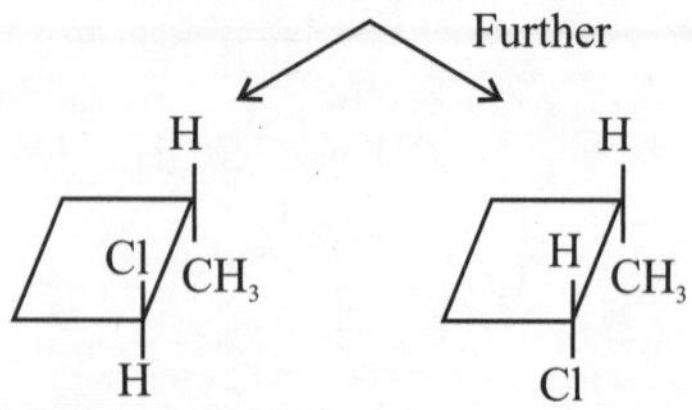

(3)

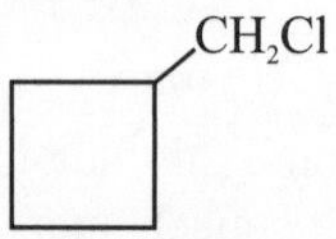

(4)

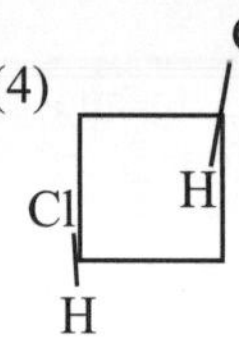

Cis - 1 - chloro - 3 - methyl cyclobutane (optically inactive, meso)

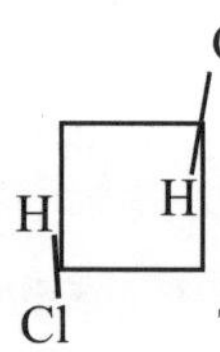

Trans - 1 - chloro - 3 - methyl cyclobutane (optically inactive, meso)

Total = 8

5. (2) The reaction is an example of electrophilic addition reaction & the intermediate here is a carbocation.

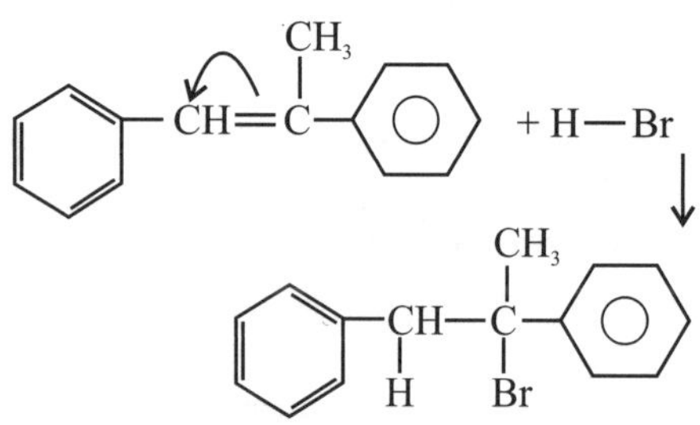

6. (1)

$$CH_3-CH_2-C\equiv C-H \xrightarrow[\text{(Amide ion acts as base)}]{Na^+ NH_2^{\ominus}} CH_3-CH_2-C\equiv C^- Na^+$$

Sodium butoxide + NH_3

$$CH_3-CH_2-C\equiv C^- Na^+ + CH_3-CH_2-Br \xrightarrow{S_N2}$$

$$CH_3-CH_2-C\equiv C-CH_2-CH_3$$

7. (4) All these are electrophilic substitution reactions.

In option (1),

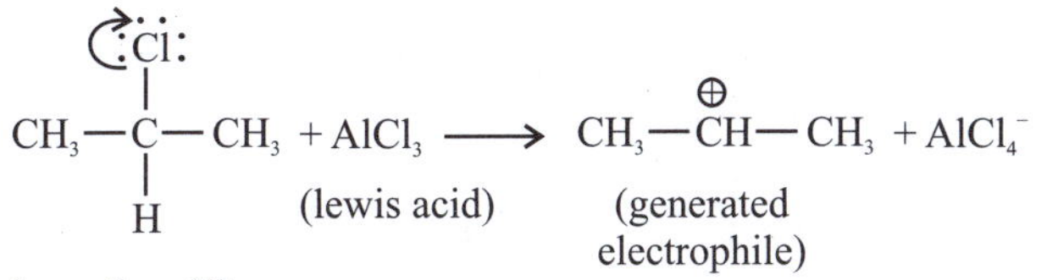

In option (2),

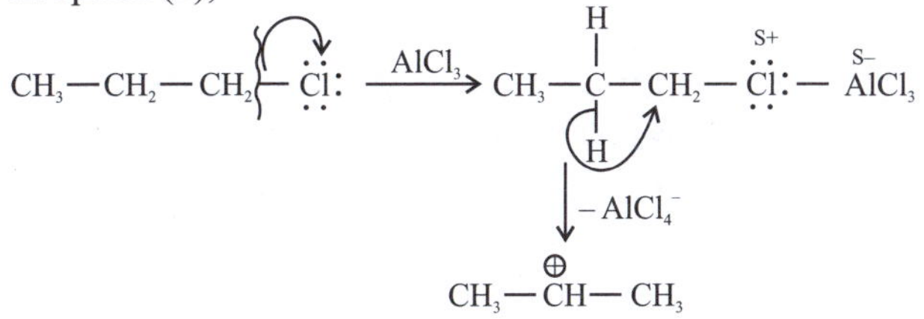

In option (3). $CH_3-CH=CH_2 + H^+ \longrightarrow CH_3-\overset{\oplus}{C}H-CH_3$

All generated electrophiles react with benzene, formation of cumene occurs-

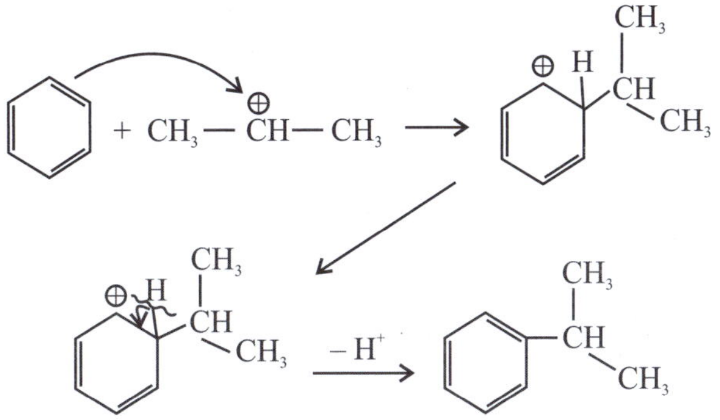

8. (4)

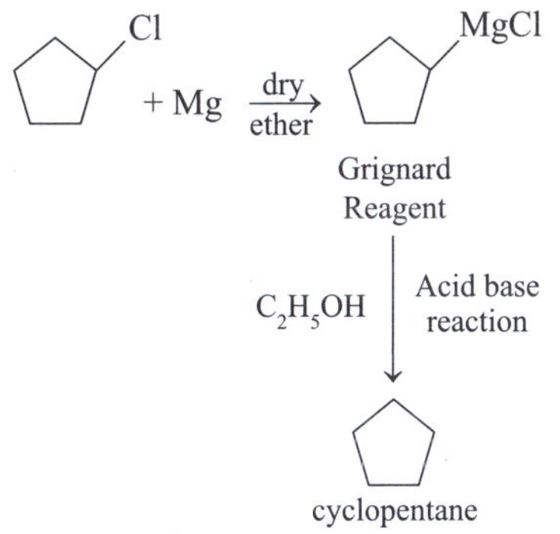

9. (1) Mo_2O_3 is an aromatisation reagent.

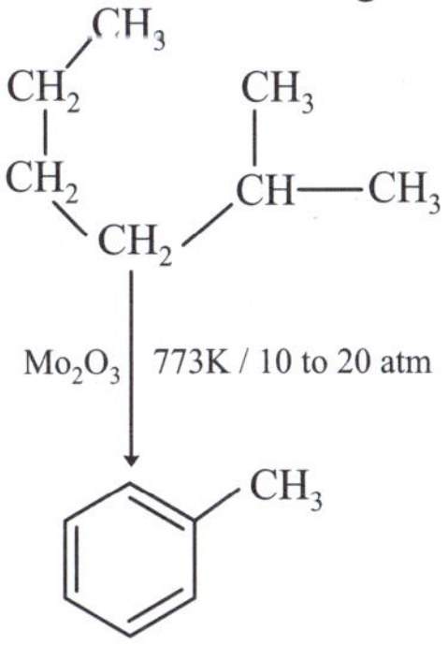

10. (4) This reaction occurs via cyclic halonium ion formation.

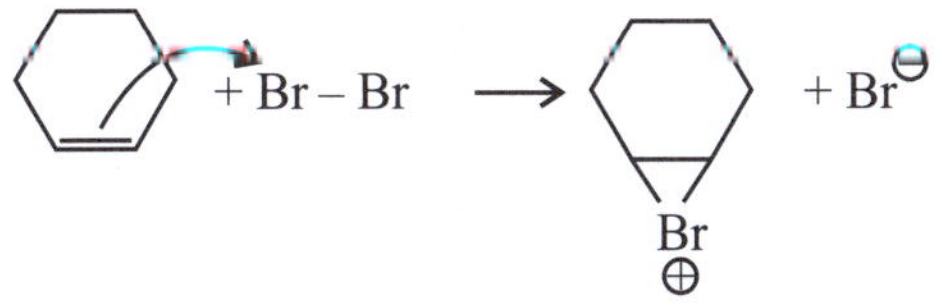

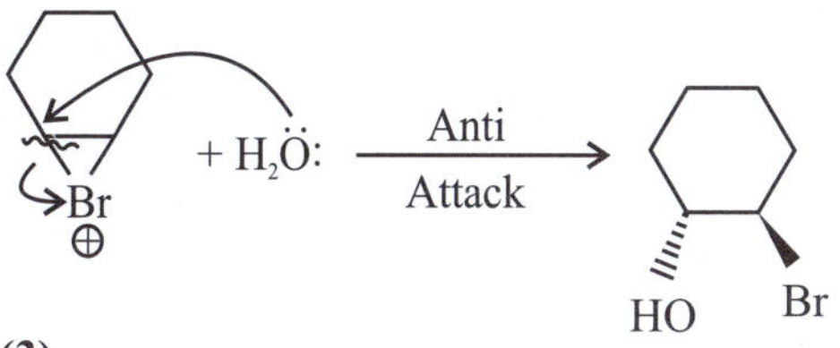

11. (3)

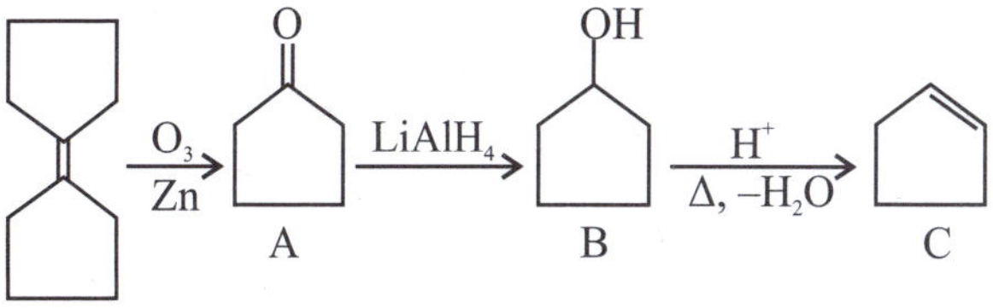

12. (3) The molecules

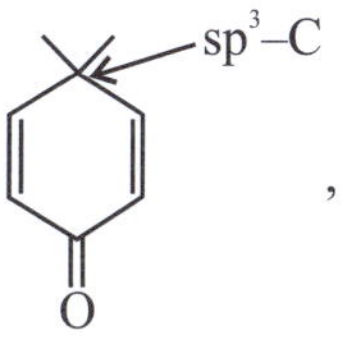

, is not aromatic because it is non-aromatic. It is planar due the presence of sp^3 hybridised carbon. (conjugation also broken)

13. (4) $CH_2=CH_2 + Br_2 \longrightarrow$

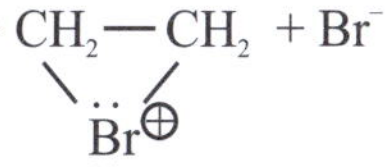

This 3 membered ring can be opened as follows:

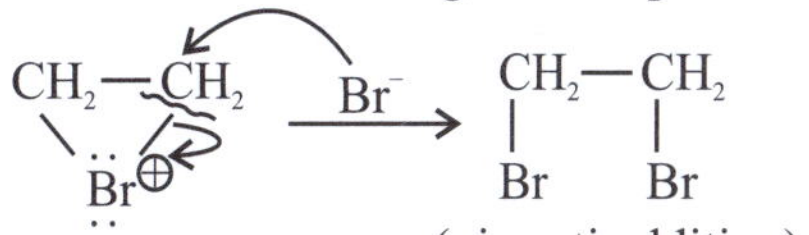

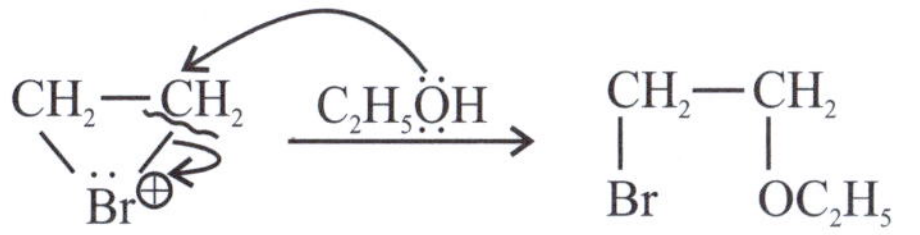

14. (3) $CH_3-CH=CH-CH_3 + HCl \rightarrow$

(a) $CH_3-CH=CH-CH_3 + H^+ \xrightarrow{\text{slow}} CH_3-\overset{\oplus}{C}H-CH_2-CH_3$

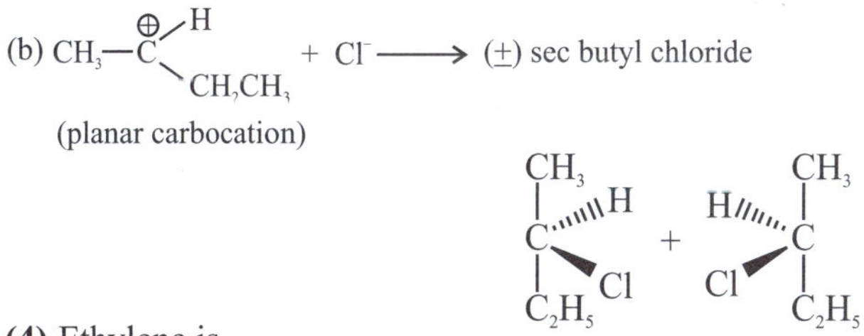

15. (4) Ethylene is

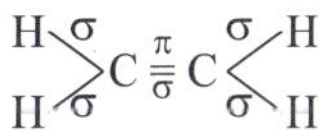

No. of σ bonds = 5

No. of π bonds = 1

16. (4) Electrophilic addition reaction involves formation of a carbocation.

More stable carbocation is formed in propene than ethene.

Catalytic hydrogenation is not an electrophilic addition reaction. Rather, stability of alkene is to be kept in mind.

Since, propene is more stable than ethene (on the basis of hyper conjugation), it is less reactive towards hydrogenation.

Carbocation in electrophilic addition on ethene involves 1^0 carbocation which is relatively more stable than vinyl carbocation in case of ethyne. So, ethene is more reactive than ethyne towards electrophilic addition.

(a) $CH_2{=}CH_2 + E^+ \longrightarrow \overset{}{CH_2}(E)-\overset{\oplus}{C}H_2$

(relatively stable)

(b) $H-C{\equiv}C-H + E^+ \longrightarrow CH(E){=}\overset{\oplus}{C}H$

(relatively unstable)

17. (3)

(i) $R-CH{=}CH_2 \xrightarrow{BH_3} (R-CH_2-CH_2)_3B \xrightarrow{H-OOCCH_3} R-CH_2-CH_3$

(ii)

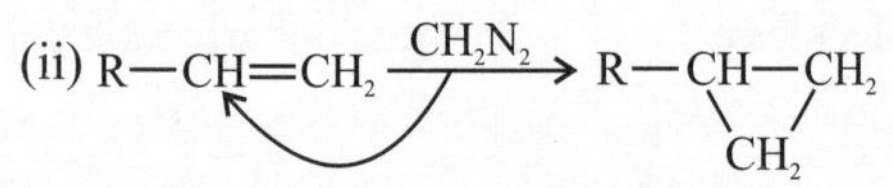

(iii)

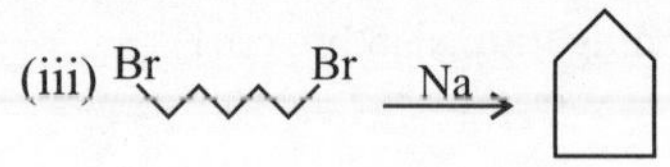

(iv)

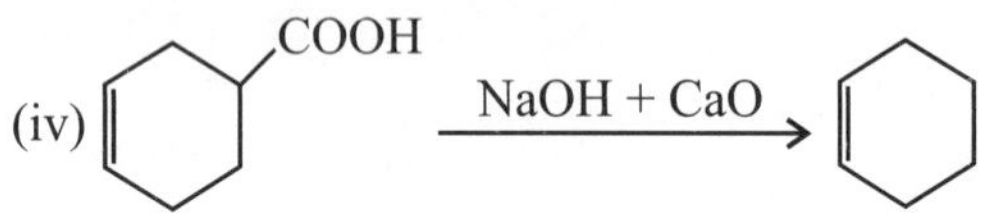

unsaturated

18. (1) Wurtz reaction is a coupling reaction, which is used to prepare symmetrical alkanes. It is basically a reaction in which higher alkanes are synthesised by a reaction in between alkyl halides and metallic sodium. This reaction takes place in presence of dry ether.

As we know that less stable radical is more reactive towards wurtz reaction. Therefore, the order of reactivity towards wurtz reaction is 1°- carbocation (least stable) > 2°-carbocation (less stable) > 3°-carbocation (most stable) in option 2 and 3 most stable 2° and 3° carbocations are formed which are stablized by phenyl ring. In option 4, 3° stable carbocation is formed. Hence, $H_3C-C(Br)(H)-CH_3$ will be most reactive.

19. (2)

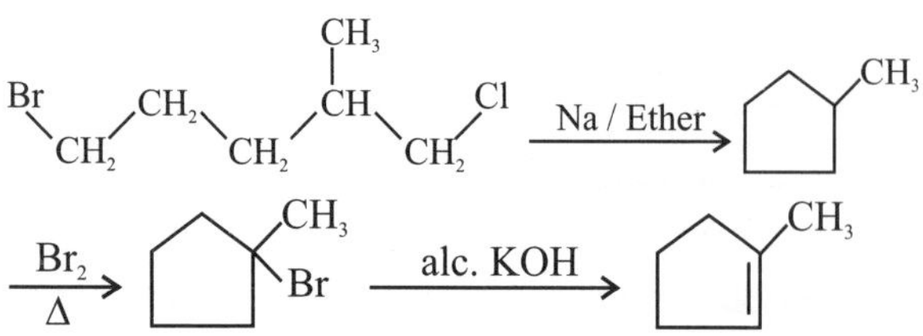

20. (4)

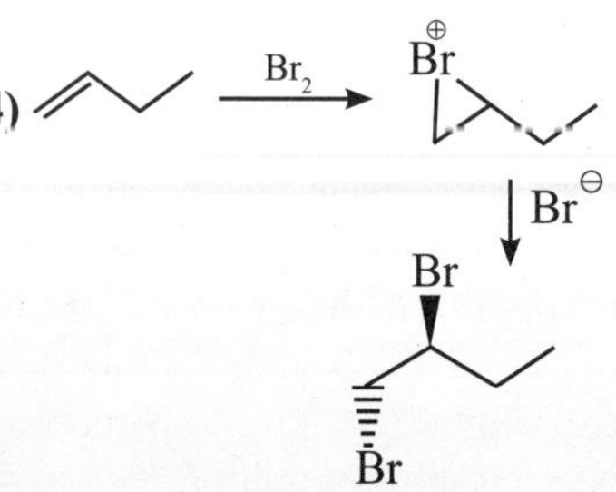

21. (3) Reagent is alcoholic alkali and $\overset{\ominus}{O}R$ is a strong base so elimination dominates.

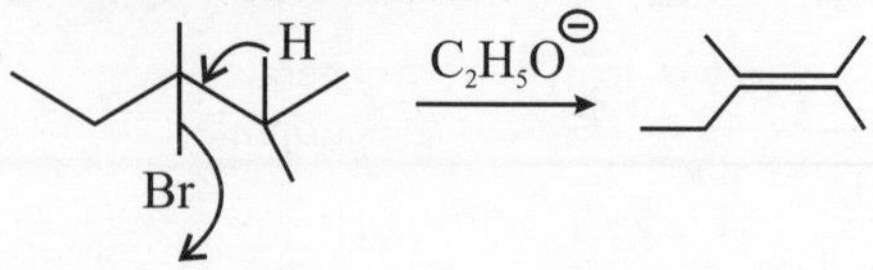

22. (2) Option (2) is correct i.e., Acetylene is less reactive than ethylene towards electrophilic attack.

23. (1) Reverse reaction of ozonolysis is given as-

$(C_2H_4O)\ CH_3-CH{=}O + O{=}C(CH_3)-CH_2-CH_3 \rightarrow$ (2-Butanone)

$CH_3-CH{=}C(CH_3)-CH_2-CH_3$

3-Methyl pent-2-ene

24. (4)

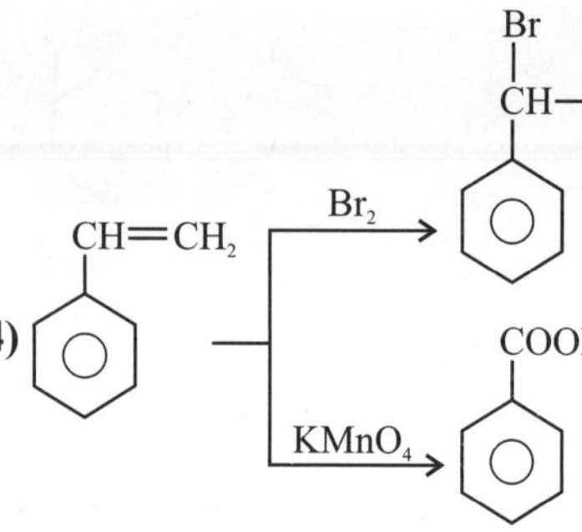

25. (2)

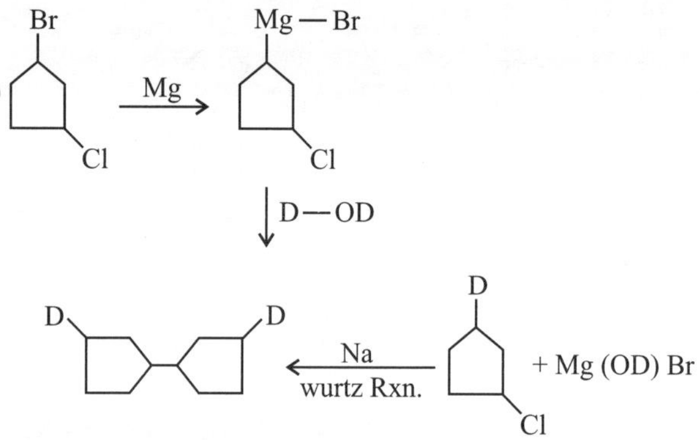

26. (1) Reaction of alkene with H_2SO_4 at high temp. $-SO_3H$ group is attached on alkene. At low temp. it will from higher alkene.

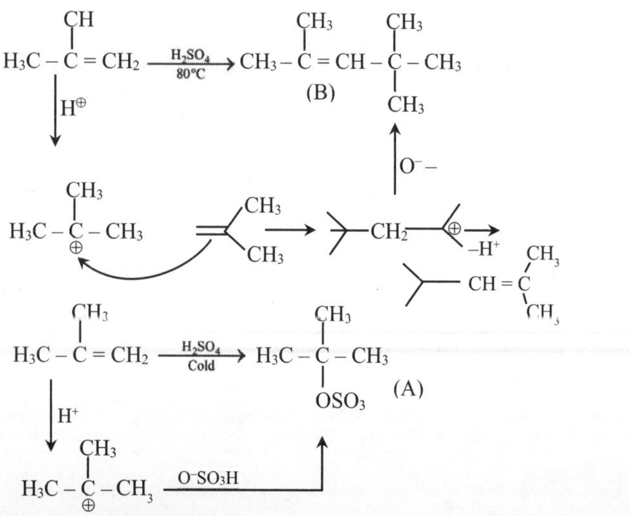

27. (4)

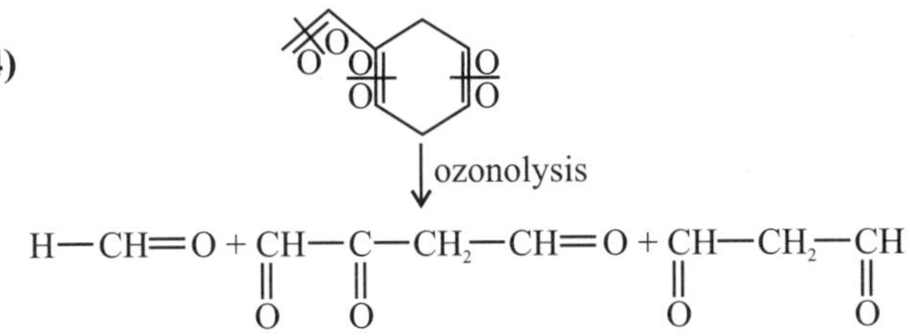

28. (3)

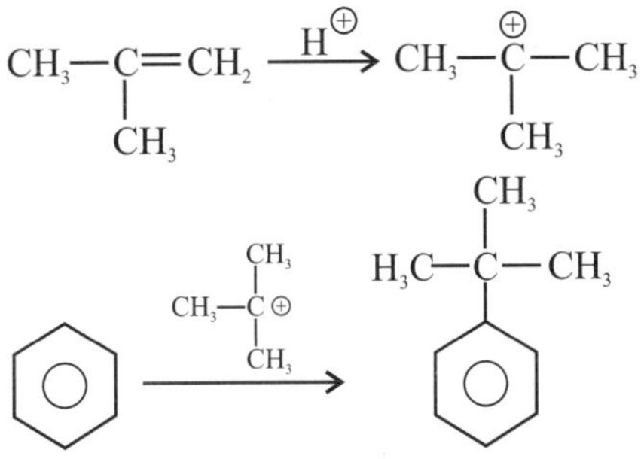

29. (3)

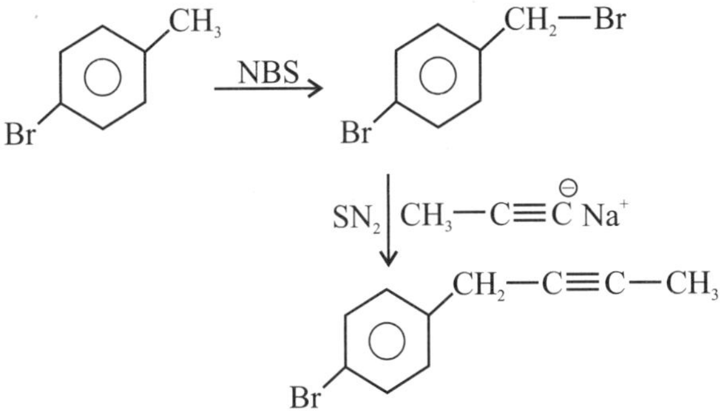

30. (1)

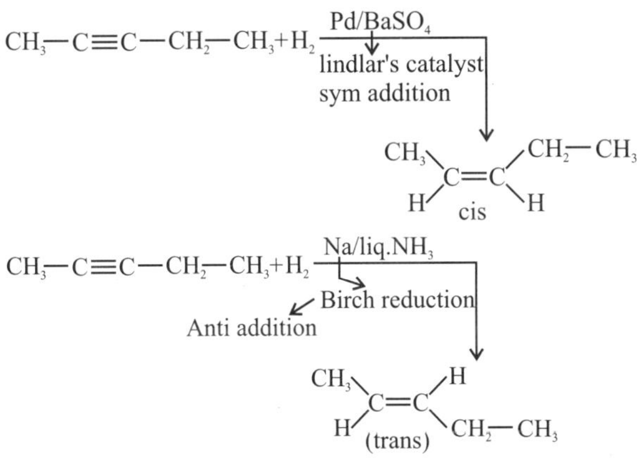

31. (3) Electophillic addition of halogen is an anti addition reaction so racemic mixture is obtained with cis-alkene.

32. (4)

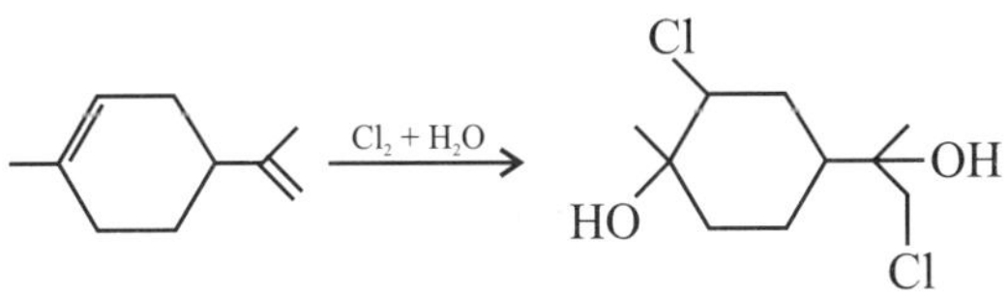

33. (4)

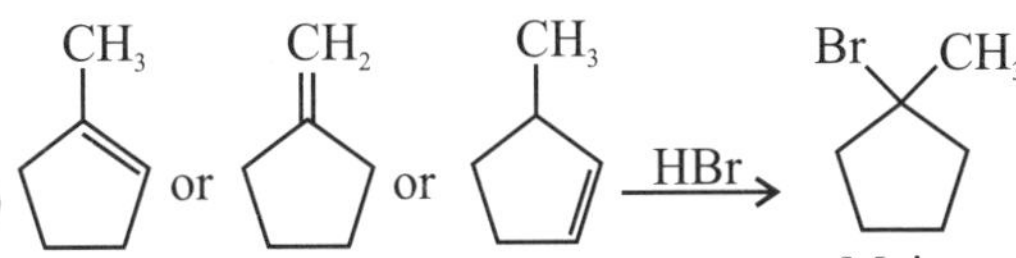

Major

In (3) major product is formed by rearrangement of carbocation.

34. (2) At anode alkane and CO_2.

35. (2) Grignard reagent on hydrolysis give alkane

36. (1) 2-hexene on ozonolysis in the presence of water give acetic acid and butanoic acid.

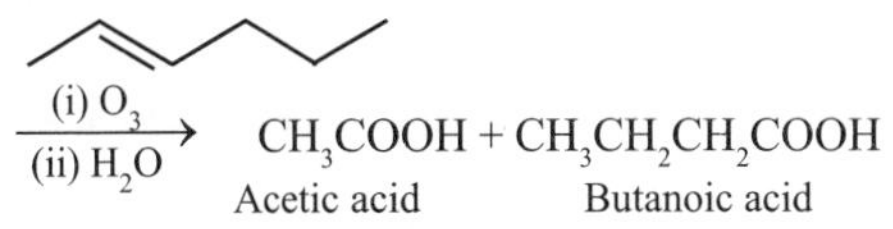

it is oxidative ozonolysis.

37. (2) Wurtz reaction is used to prepare symmetrical alkanes having even number of carbon atoms.

38. (1) $C_4H_{10} + \frac{13}{2}O_2 \rightarrow 4CO_2 + 5H_2O + \text{Heat}$

For 58 g of butane 208 g oxygen is required for complete combustion.

For 29 g of butane 104 g oxygen will be consumed.

39. (1) $C_2H_4 \xrightarrow{HBr} C_2H_5Br \xrightarrow{\text{Alc. KOH}} A\ C_2H_4$

40. (1) In $H_2C = CH_2$, sp^2 hybridised carbons are present which shows planar geometry.

41. (2) Vinyl ethers are highly reactive towards electrophilic addition due to +M effect of alkoxy group

42. (1)

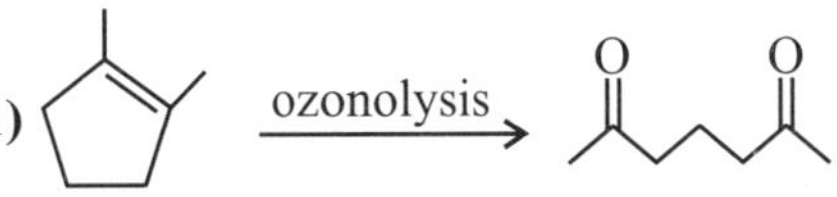

43. (3)

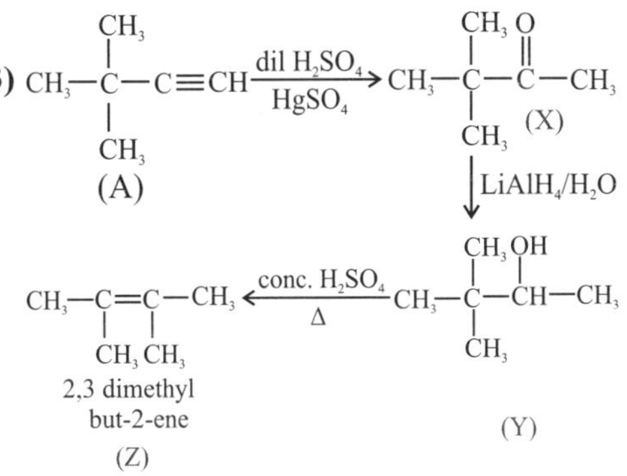

44. (2) Free radical substitution takes place at more stable alkyl radical at 3° benzyl carbon.

45. (2) $CH_3-C\equiv CH \xrightarrow{HgSO_4, H_2SO_4} CH_3-\overset{O}{\overset{\|}{C}}-CH_3 \xrightarrow{NaBH_4} CH_3-\overset{OH}{\overset{|}{C}H}-CH_3$

Step-1 is Markovnikov addition of water and step-2 is reduction of ketone.

46. (3) The given columns are based on Ozonolysis process.

47. (1)

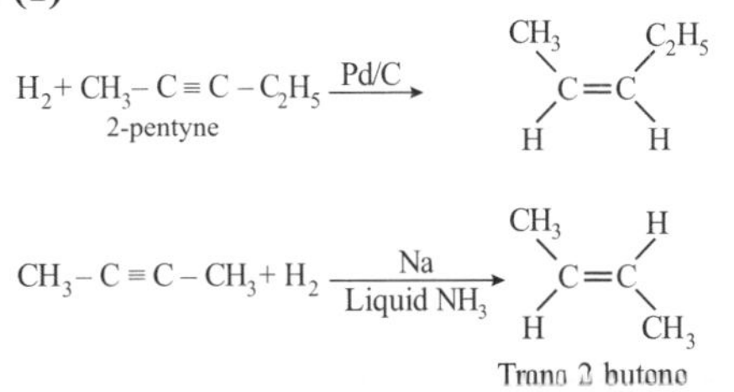

48. (3) $\begin{matrix}CH\\|||\\CH\end{matrix} + 2[O] \xrightarrow[\text{dil alk}]{KMnO_4} \left[\begin{matrix}CH=O\\|\\CH=O\end{matrix}\right] \xrightarrow{[O]} \begin{matrix}COOH\\|\\COOH\end{matrix}$

oxalic acid

49. (2) Due to presence of acidic H

50. (2) $H-C\equiv C-Na + CH_3Cl \rightarrow H-C\equiv C-CH_3 + NaCl$

51. (1) Both can be distinguished by silver mirror test.

52. (2) With Lindlar's catalyst, cis–alkene is formed which undergoes anti addition with Br_2 to give Racemic mixture.

53. (1) Nitration is electrophilic substitution reaction and electrophile is nitronium ion ($-NO_2^+$) . It is produced by transfer of proton from sulphuric acid to nitric acid. Hence HNO_3 acts as a base

54. (4) Ethane is saturated compound

55. (1) $-O > -OH > -OCH_3 > -OCOCH_3$

56. (4) NH_2 is ortho & para orienting group

57. (2) Conc. H_2SO_4 is a dehydrating agent

58. (3) For a compound to be aromatic, it should be planar, must follow Huckel's rule and should be cyclic.

The compound 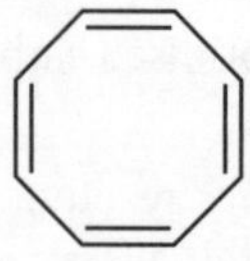 has non-planar shape and it exist in tub-form 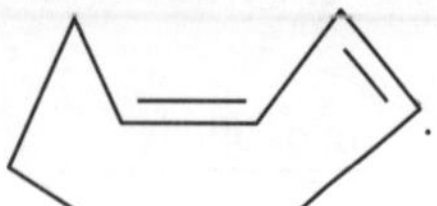. Thus it non-aromatic.

59. (2) [Decarboxylation ∝ stability of anion]

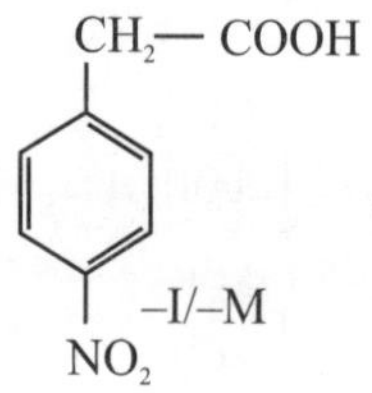

-I/-M group increase stability

60. (2) 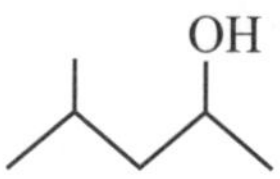 $\xrightarrow[\Delta]{\text{conc. } H_2SO_4}$ major

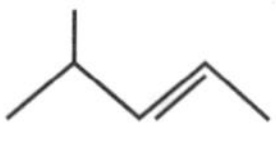

2 methyl pent - 2-ene

61. (3)

$$R-Br \xrightarrow[\text{dryether}]{Mg} RMgBr \xrightarrow{D_2O} CH_3-\underset{D}{\overset{H}{C}}-CH_3 + Mg(OD)Br$$

$$R-Br = CH_3-\overset{CH_3}{CH}-Br \qquad RMgBr = CH_3-\underset{CH_3}{CH}MgBr$$

$$CH_3-\overset{CH_3}{CH}-Br + 2Na + Br-\overset{CH_3}{CH}-CH_3 \longrightarrow CH_3-\overset{CH_3}{CH}-\overset{CH_3}{CH}-CH_3$$

2,3 dimethyl butane

62. (2)

$$\underset{(A)}{CH_3-C\equiv CH} + C_6H_5MgBr \longrightarrow \overset{(X)}{C_6H_6} + \underset{(W)}{CH_3-C\equiv \overset{\ominus\oplus}{C}MgBr}$$

$$\underset{(W)}{CH_3-C\equiv \overset{\ominus\oplus}{C}MgBr} \xrightarrow{D_2O} \underset{(Y)}{CH_3-C\equiv C-D} + Mg(OD)Br$$

A & Y both show acidic behaviour

63. (3) $CH_3-C\equiv C-CH_3 \xrightarrow{Na/NH_3(\ell)}$ $\overset{CH_3}{\underset{H}{>}}C=C\overset{H}{\underset{CH_3}{<}}$

Trans But -2ene

$\downarrow Br_2/CCl_4$

y

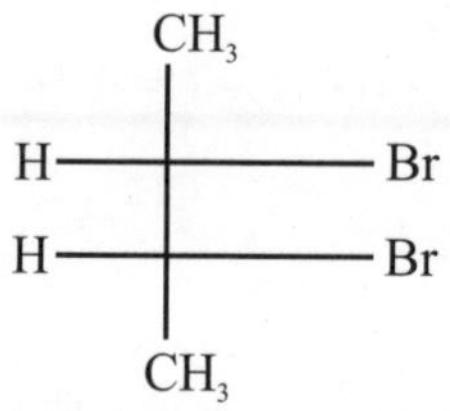

(Meso)2, 3-dibromo butane

64. (2) $CH_3-CH_2-CH_2-CH_2-OH \xrightarrow[H_2SO_4/\Delta]{\text{conc.}} CH_3-CH=CH-CH_3$

But-2-ene

$CH_3-CH_2-CH_2-CH_2-Cl \xrightarrow[\Delta]{\text{alc KOH}} CH_3-CH_2-CH=CH_2$

But-1-ene

Both are position isomers

65. (2) 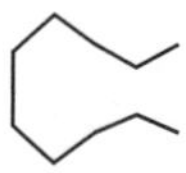 $\xrightarrow[500-600°C]{Al_2O_3-Cr_2O_3}$ 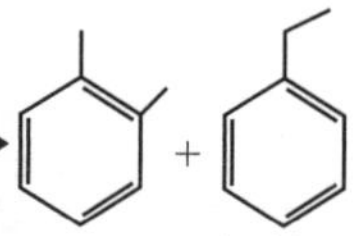

n - octane → o/m/p xylene + ethyl benzene

66. (2) The above given columns are based on different reaction processes

67. (2) $C_3H_8 + 5O_2 \rightarrow 3CO_2 + 4H_2O$

1 lit 5 lit

5 lit 25 lit

20 lit O_2 is present in 100 lit air

25 lit O_2 is present is $= \frac{100}{20} \times 25 = 125\,\text{lit}$

68. (1) $H_3C-\underset{CH_3}{C}\begin{smallmatrix}Br\\Br\end{smallmatrix} + \text{Zn dust} \longrightarrow CH_3-\underset{CH_3}{C}=\underset{CH_3}{C}-CH_3$

$\downarrow HI$

$$CH_3-\underset{CH_3}{\overset{I}{C}}-\underset{CH_3}{CH}-CH_3$$

69. (1) $\underset{(a)}{CH_3CH=CHCH_2C_6H_5}$ $\quad \underset{(b)}{CH_3-\underset{Ph}{CH_2}-CH=CH_2}$

$\underset{(c)}{CH_3CH_2\underset{C_6H_5}{C}=CH_2}$ $\quad \underset{(d)}{CH_3-CH=\underset{C_6H_5}{C}-C_6H_5}$

Only (a) can show geometrical isomerism.

70. (3)

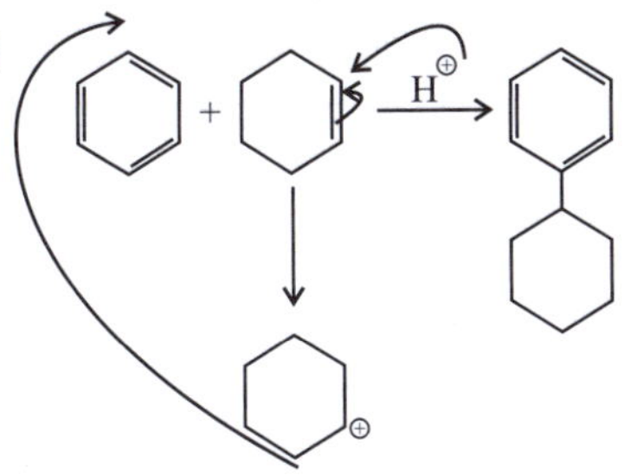

71. (2) It has symmetrical structure.

72. (2) $\underset{\substack{| \\ CH_3}}{H_3C-\overset{3^\circ}{C}H-CH_2-CH_3} \xrightarrow{Br_2,\ h\nu} CH_3-\underset{\substack{| \\ CH_3}}{\overset{\substack{Br \\ |}}{C}}-CH_2-CH_3$

This is because reactivity order of abstraction of H atom from alkanes is 3° > 2° > 1°.

73. (1)

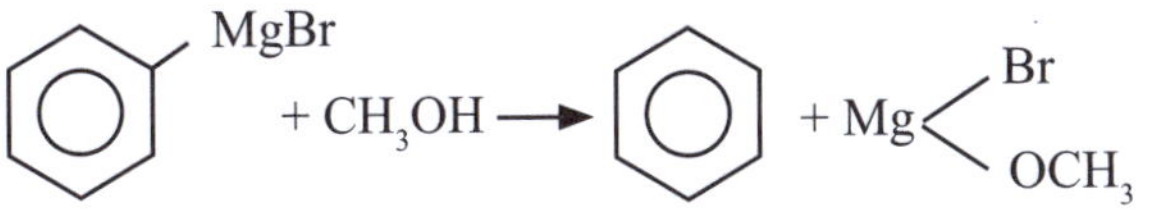

74. (3) IUPAC name of the given compound is -

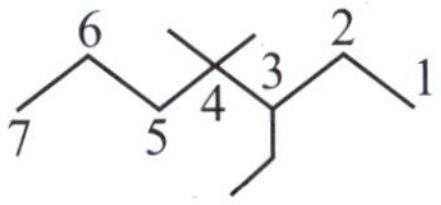

3-Ethyl-4, 4-dimethyl-Heptane

75. (4) 2,2-Dimethylbutane will give two monochlorinated compounds whereas 2-methylpentane will give 5-monochlorinated derivatives.

76. (4) $CH_3-\underset{\substack{| \\ CH_3}}{C}=\underset{\substack{| \\ CH_3}}{C}-CH_3 \xrightarrow{O_3}$

2,3-Dimethyl but-2-ene

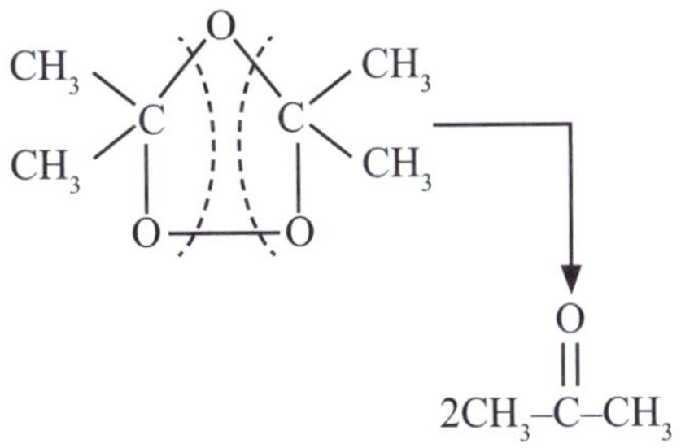

Acetone

Correct & Incorrect MCQs

1. **(1)** Wurtz reaction is used for the preparation of higher alkanes containing even number of carbon atoms.
2. **(4)** The intermolecular van der Waals forces increases with increase in molecular size or surface area of the molecule, hence there is a steady increase in boiling point of alkanes with increase in molecular mass.
3. **(1)** As branching increases, surface area in contact decreases so, boiling point decreases.
4. **(3)** The staggered form of ethane has the least torsional strain and the eclipsed form has the maximum torsional strain. Therefore, staggered conformation of ethane is more stable than the eclipsed conformation.
5. **(2)** The presence of π-bond makes alkenes behave as a source of loosely held electrons, hence alkenes are easily attacked by electrophilic reagents.
6. **(2)** Cis form of alkene is found to be more polar than the trans form.
7. **(3)** Partially deactivated palladised charcoal is known as Lindlar's catalyst. Alkynes on reduction with sodium in liquid ammonia form trans alkenes.
8. **(3)** Peroxide effect is observed in addition of HBr but not in addition of HCl because H — Cl bond is stronger than H — Br bond and is not cleaved by the free radical.
9. **(4)** Alkynes are lighter than water and immiscible with water but soluble in organic solvents like ethers, carbon tetrachloride and benzene.
10. **(3)** Chlorobenzene is less reactive than benzene towards the electrophilic substitution reaction due to –I effect.

Statement Based MCQs

1. **(3)** $CH \equiv CH + NaNH_2 \longrightarrow CH \equiv C^- Na^+ + NH_3$

   $CH \equiv CH$ is stronger acid because conjugate base $CH \equiv C^-$ is stable due to sp carbon.
2. **(1)** Benzene is electron rich molecule. Thus it is readily attacked by electrophiles

   Due to presence of pi bonds benzene is unsaturated compound.
3. **(3)** Since nitro group is meta directing & strongly deactivating group, it lowers the density at ortho & para positions of the benzene ring
4. **(4)**
5. **(1)**

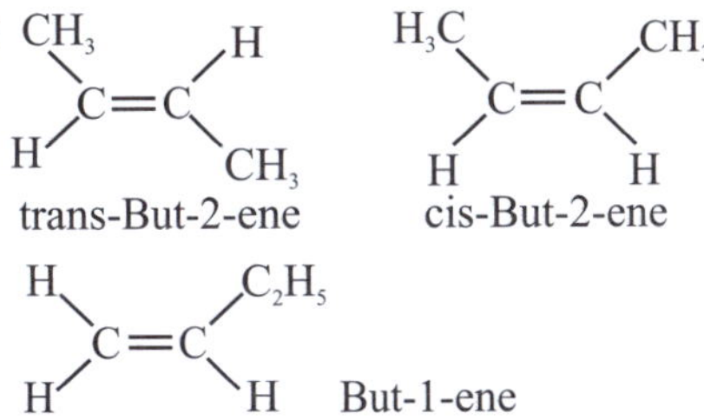

But-1-ene does not show geometrical isomerism.

6. **(4)** Since, iodination of alkanes is very slow and a reversible process, hence iodination of alkanes is carried out in the presence of an oxidizing agent like HIO_3 or HNO_3.
7. **(1)** Due to the evolution of large amount of heat during combustion, alkanes are used as fuels.
8. **(3)** In staggered form, the electron clouds of C – H bonds are as far as possible. Therefore, there are minimum repulsive forces, minimum energy and maximum stability of the molecule. In eclipsed form, the electron clouds of the carbon – hydrogen bonds come closer to each other resulting in increase in electron cloud repulsions. Hence, staggered form of ethane is the most stable and eclipsed form is the least stable.
9. **(1)** Tertiary (3°) carbon atom is attached to three carbon atoms. 2,2-Dimethylbutane does not contain any tertiary carbon atom. It contain quaternary carbon.
10. **(3)** The position of the incoming group is determined by the nature of the group already present in monosubstituted benzene ring and it does not depend on the nature of the entering group.

Assertion & Reason MCQs

1. **(1)** The system although has conjugated π-bonds but, it does not follow 4n + 2 rule. Compound contains more than 8 carbons and loses planarity.

2. **(1)** 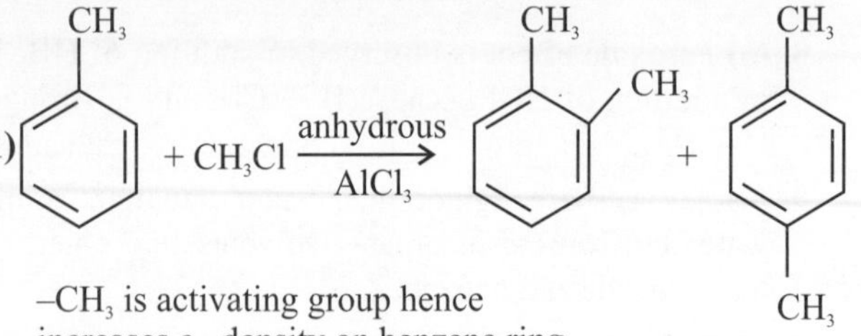

$-CH_3$ is activating group hence increases e– density on benzene ring.

3. **(1)** HNO_3 acts as base and H_2SO_4 acts as an acid

$$HNO_3 + H_2SO_4 \xrightarrow[\text{(can(.)}]{} NO_2^+ + HSO_4^- + H_2O$$

$\overset{+}{N}O_2$ acts as an electrophile.

4. **(3)** Tropylium is planar, having conjugated double bonds and follow 4n + 2 rule.

5. **(3)** 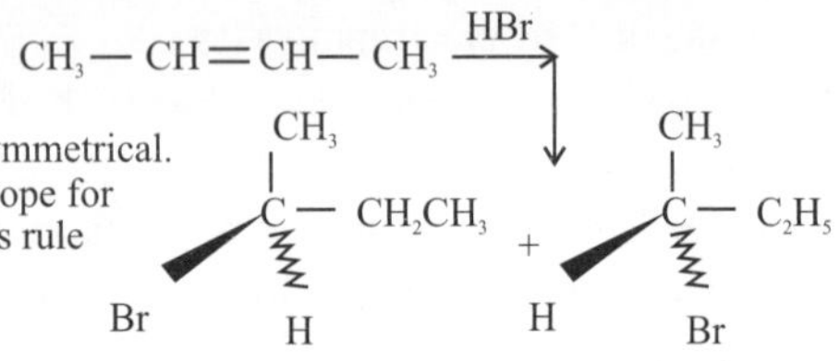

2 Butene is symmetrical. There is no scope for markovnikov's rule

6. **(2)** 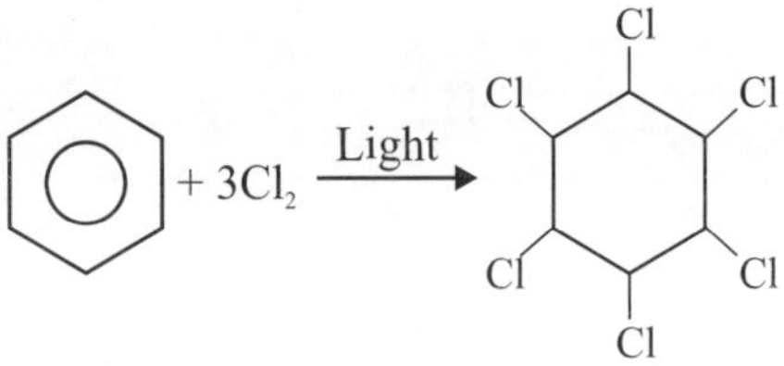

7. **(1)** H_2SO_4 here acts as catalyst and is responsible for formation of electrophile i.e. NO_2^+ (nitronium ion). H_2SO_4 is a stronger acid than HNO_3, it gets protonated easily than nitric acid that causes loss of water molecule and form electrophile.

8. **(4)** Propene reacts with HBr in the presence of Benzoyl peroxide to form 1-bromopropane and the mechanism is free radical mechanism.

9. **(4)** With increase in branching, intermolecular forces becomes weaker and the boiling point decreases.

10. **(2)** Propene being unsaturated hydrocarbon undergoes addition reactions. On the other hand, cyclopropane being strained readily undergoes ring cleavage to give addition products.

Multi Correct MCQs

1. **(1)** Reactions A and C are not possible because carbon-chlorine bond cleavage is not possible.

2. **(4)** As we know that compound (d) is aromatic and the compound (a) is anti-aromatic. Hence compound (d) is most stable and compound (a) is least stable. Among these in compound (c) is more stable than (b) due to its bond angle that provide minimum strain.

Therefore, correct order of stability of given carbanions is: (d) > (c) > (b) > (a)

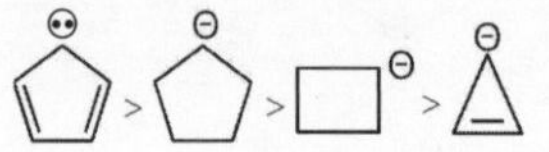

3. **(1)**

A. 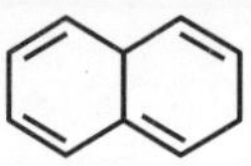 ⟶ Non aromatic

B. 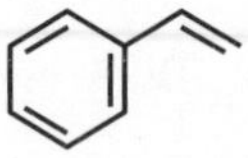 ⟶ Aromatic

C. 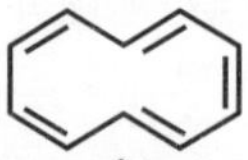 ⟶ Non aromatic

D. 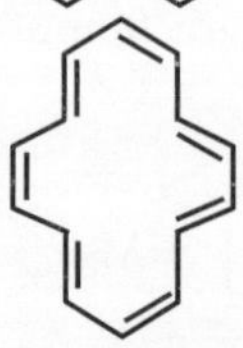 ⟶ Aromatic

Hence, B & D are aromatic.

4. **(4)** 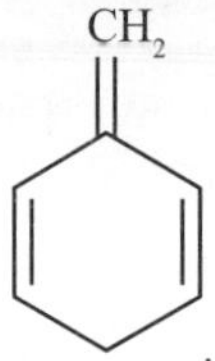

It is non-aromatic due to the presence of one sp^3 hybrid carbon.

It is aromatic as it has 6π electrons.	It is aromatic as it has 6π electrons.	It is anti-aromatic as it has 4π electrons.

5. **(4)** Bromine water is commonly used as a qualitative test to detect unsaturation in organic compounds, particularly alkenes and alkynes.

Cyclohexane, being a saturated hydrocarbon (alkane), does not react with bromine water and therefore does not cause any decolourisation.

Benzene also does not decolourise bromine water. Although benzene contains pi electrons, it is an aromatic compound with a stable delocalized electron system, which makes it resistant to addition reactions such as those with bromine water.

Hence, Only compounds (B) and (C) can decolourise bromine water.

NCERT Exemplar MCQs

1. **(4)** Boiling point increases as the number of carbon atom increases, . Boiling point decreases with branching

2, 2-dimethyl propane,

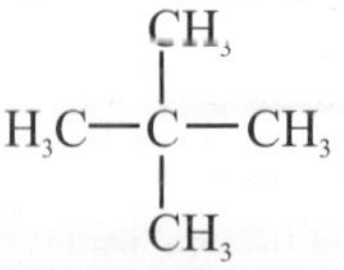

b.pt = 282.5K

n-pentane,

$H_3C—CH_2—CH_2—CH_2—CH_3$

b.pt = 309.1 K

2-methyl butane

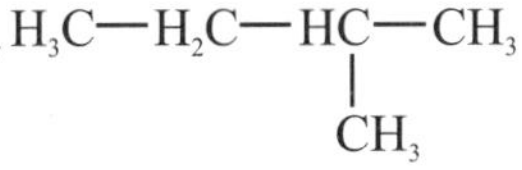

b.pt = 301 K

n-butane

$H_3C—H_2C—CH_2—CH_3$

b.pt = 273 K

(4 carbon atoms with no branching)

In the given compounds, n-pentane will have highest b.p. then 2-methyl butane and than n-butane. 2, 2-dimethyl propane will have least b.p.

2. **(1)** Rate of reaction of alkanes with halogens is $F_2 > Cl_2 > Br_2 > I_2$
Reactivity decreases with decrease in electronegativity and electronegativity decrease down the group. Alkane react with F_2 vigorously and with I_2 the reaction is too slow that it requires a catalyst.

3. **(2)** The reactivity of halogens with alkane is $F_2 > Cl_2 > Br_2 > I_2$. Hence, reduction of alkyl halide with Zn and dilute HCl follows reverse order i.e., R – I > R – Br > R – Cl. Further, the reactivity of this reduction increases as the strength of C – X bond decreases.

4. **(1)** The correct IUPAC name is 3,6-diethyl-2-methyl-octane. numbering is shown as-

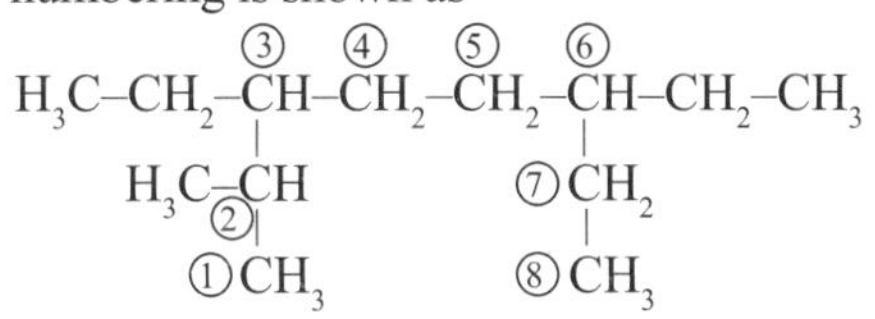

5. **(1)** The alkene is unsymmetrical, hence will follow Markownikoff's rule to give major product.
$CH_3—CH_2—CH{=}CH_2 + H—Br \longrightarrow$
Butene-1
$C_2H_5—CHBr—CH_3 + CH_3—CH_2—CH_2—CH_2Br$
(A) (major product) (B) (minor product)

Product A exists in two forms and these are shown as-

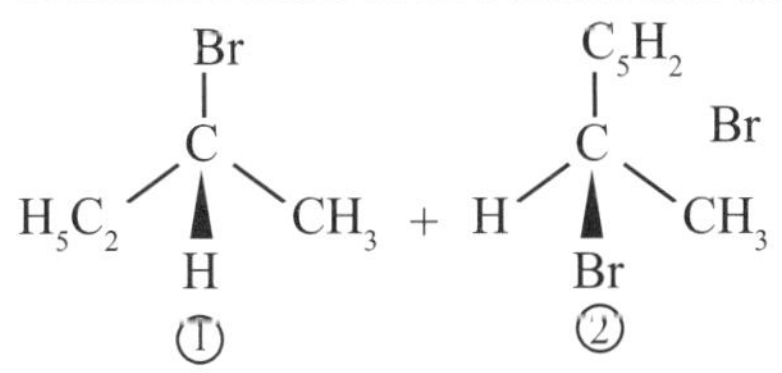

6. **(4)** In option (4), in carbon with double bond has two same functional groups (CH_3) are attached. The rotation around carbon will not produce a new compound. Hence, geometrical isomerism is not possible.

7. **(3)** Bond energy of HI is 296.8 kJ/mol, HBr is 36.7 kJ/mol and HCl is 430.5 kJ/mol. Hence, HI > HBr > HCl is the order of reactivity with propene.

8. **(2)** +I effect decreases the stability of carbon anion. Since, (CH_3) group has +I effect, therefore, it intensifies the negative charge and hence destabilizes (A) relative to (B), sp hybridized carbanion is more stabilized than sp^3

$$\underset{(B)^{sp}}{CH{\equiv}C^-} > \underset{(A)^{sp}}{CH_3—C{\equiv}C^-} > \underset{(C)\ \ sp^3}{CH_3—\overset{\ominus}{C}H_2}$$

Hence, $B > A > C$

9. **(4)** Alkyl halides on heating with alcoholic potash eliminates one atoms of halogen and are atom of hydrogen acid to form alkene. Hydrogen is eliminated from β-carbon atom.

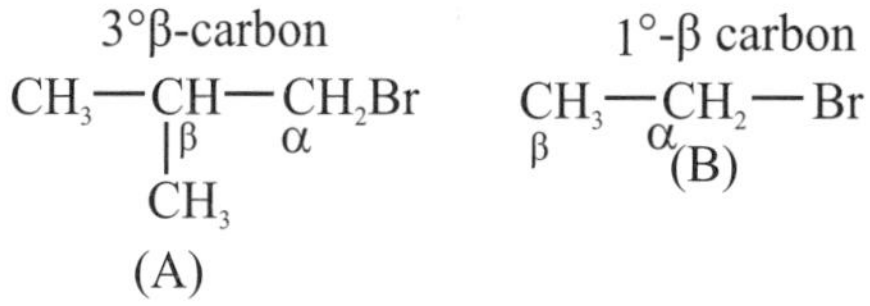

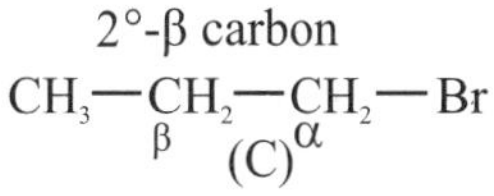

more the number of β-substituent (alkyl groups) more stable alkene swill be formed and more will be the reaction thus the decreasing order of β eliminate is A > C > B.

10. **(3)** During incomplete combustion of alkanes with insufficient amount of air or dioxygen, carbon black is formed which is used in the manufacture of ink, printer ink, black pigments and as filters. Thus,

$$CH_4(g) + O_2(g) \xrightarrow[\text{combustion}]{\text{Incomplete}} C(s) + 2H_2O(l)$$

NEET & AIIMS Past Year MCQs

1. **(2)** In general, alkanes exhibit resistance to oxidation; however, alkanes containing tertiary hydrogen atom(s) can be oxidized to form the corresponding alcohols through the use of $KMnO_4$.

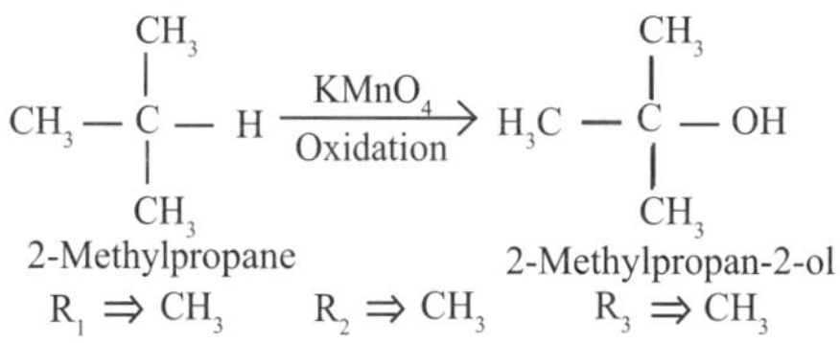

2. **(3)** Baeyer's reagent is a cold, dilute, and aqueous solution of potassium permanganate ($KMnO_4$).
It is commonly used as a test for unsaturation.

3. **(3)** Boiling point decreases with increase in branching as the intermolecular forces decreases
Boiling point: $CH_3—CH_2—CH_2—CH_2—CH_3 >$
(n-pentane)

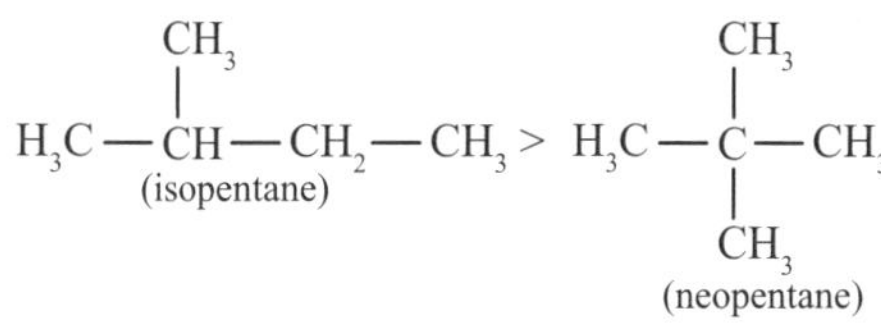

4. **(4)**

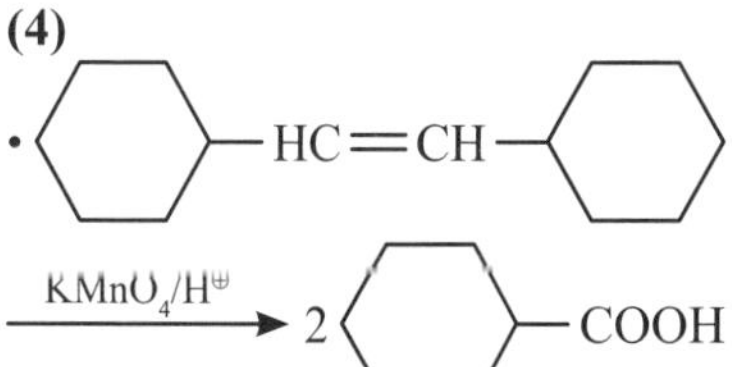

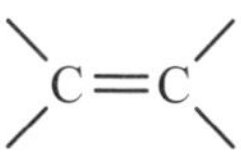 Will be completely oxidized in presence of acidified $KMnO_4$ into carboxylic acid.

5. (3) Sodium ethanoate is CH_3COONa and given process is soda-lime decarboxylation.

$$CH_3COONa + NaOH \xrightarrow{CaO} CH_4 + Na_2CO_3$$

Methane is obtained having molar mass 16. Two moles would be 32 g.

6. (2) Acc. to Huckel rule, A compound/molecule is said to be aromatic if it have cyclic conjugate and $[4n+2]\pi$ electron system

$\Rightarrow$ Molecule must be sp^2 hybridised

$\Rightarrow$ Molecule must be planar.

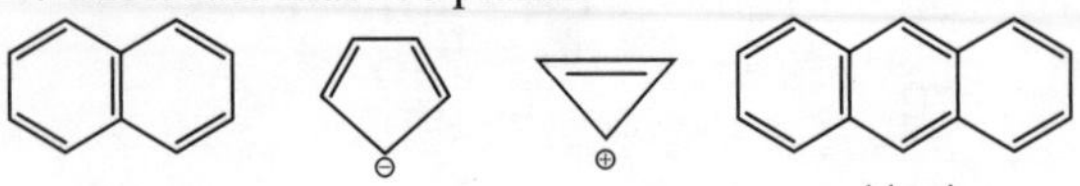

10π e's   6π e's   2π e's   14π e's

All these are aromatic species acc. to Huckel rule.

7. (1) Conditions for the compound to be aromatic-

(i) Planarity

(ii) Complete delocalisation of π electrons in the ring.

(iii) Presence of $(4n + 2)\pi$ electrons i.e., Huckel rule.

In the given compounds only compound "a" is not aromatic because it contains sp^3 hybridised carbon which is not planar & does not follow delocalisation.

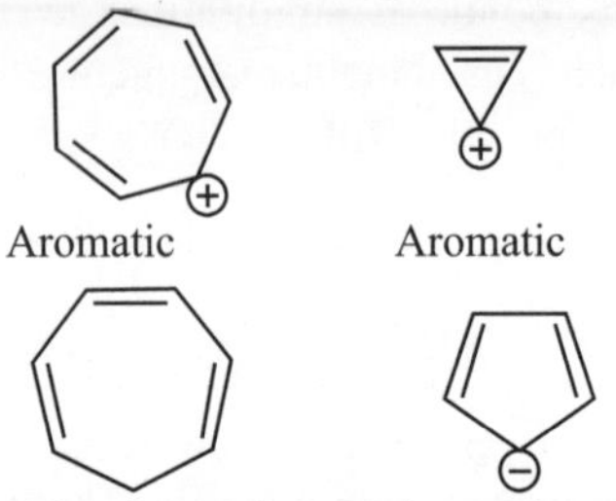

8. (2) The suitable method for the synthesis of chlorobenzene is given as-

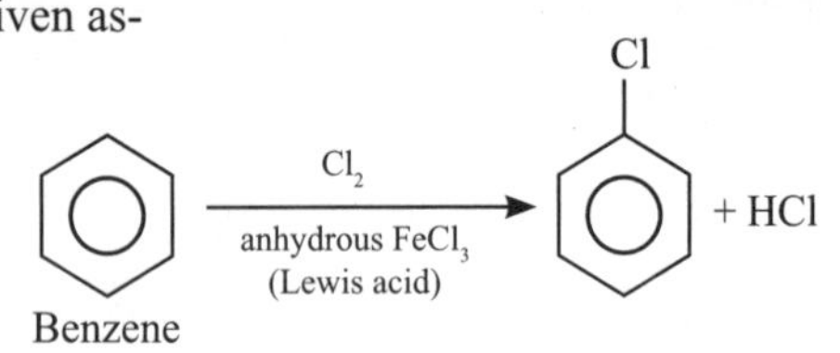

9. (2) $CH_3\text{-}CH(CH_3)\text{-}CH{=}CH_2 \xrightarrow[\text{(i) } Zn + H_2O]{\text{(i) } O_3} H_3C\text{-}CH(CH_3)\text{-}CHO + H\text{-}CHO$

3-Methylbut-1-ene → 2-Methypropanal + Formaldehyde

10. (1)

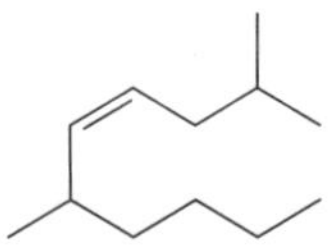

2,6-Dimethyl-dec-4-ene

11. (4) Dihedral angle (D.A.) of least stable conformer of ethane = 0°

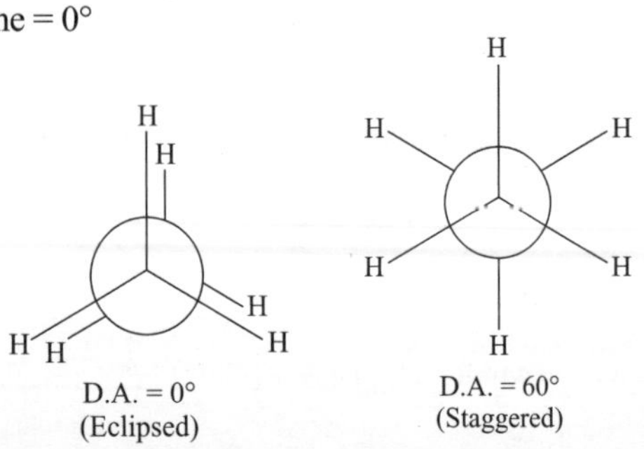

12. (1)

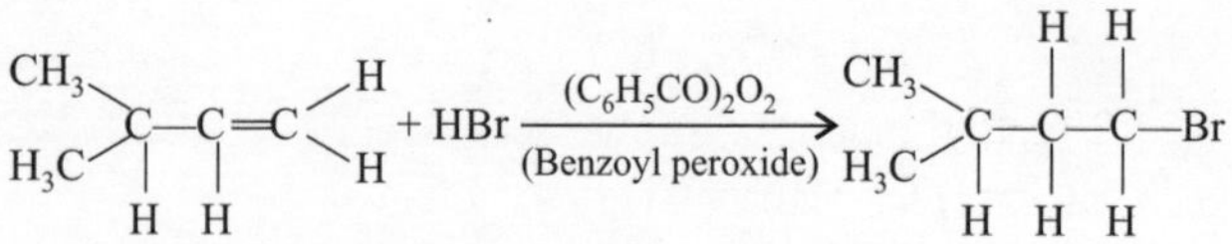

In the presence of peroxide, addition of HBr to unsymmetrical alkenes take place by antimarkonikov's rule/peroxide effect/kharash effect.

13. (3)

$$CH_3\text{—}CH_2\text{—}\overset{O}{\overset{\|}{C}}\text{—}\overset{-}{O}\overset{+}{Na} \xrightarrow[\text{Heat}]{NaOH + ?} CH_3\text{—}CH_3 + Na_2CO_3$$

Decarboxylation takes place by soda-lime (NaOH + CaO).

14. (1) Bond angles are different due to steric crowding of CH_3 group.

15. (1) $HC{\equiv}C^-$ has 50% s-character and $H_2C{=}CH^-$ has 33% s-character. Stability of carbanions increases with an increase in the s-character at the carbanion. So $HC \equiv C^-$ is more stable than $H_2C = CH^-$.

16. (4)

17. (4) $AlCl_3$ acts as a Lewis acid and helps in generating the electrophile by coordinating to the halogen in the alkyl halide, making it more electrophilic and ready for the substitution reaction.

18. (1) The reaction involves the nitration of chlorobenzene, followed by chlorination. Based on the sequence, the nitro group ($-NO_2$) would likely direct the new substitution to the meta position, which is the 3rd position relative to the existing chlorine (Cl) group.

19. (3) This structure satisfies the conditions of aromaticity (planarity, conjugated π-electrons, and following Hückel's rule for $4n + 2$ π-electrons).

20. (3)

21. (2) $CH_2{=}C(CH_3)\text{-}CH_3 + H^+ \longrightarrow CH_3\text{-}C^+(CH_3)\text{-}CH_3$

Highly stable

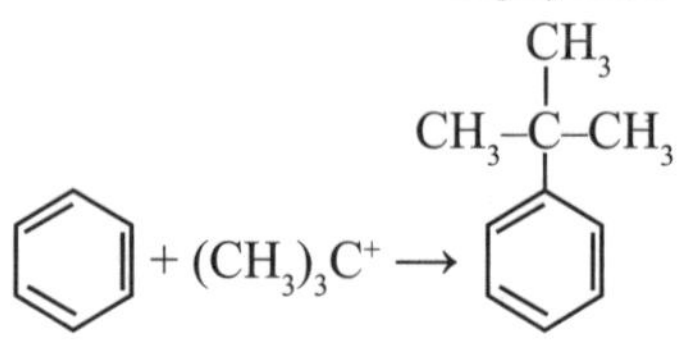

22. (2) Nomenclature in benzene ring:

For only one substituent, no numbering is required. According to latest IUPAC system of nomenclature, the lowest set of locants is preferred even if it violates the lowest sum rule.

CLASS-XII-Explanations

Solutions

NCERT Topic-wise MCQs

1. (3) $m = \frac{(125 \times 0.8) \times 1000}{32 \times 375} = 8.33$

[$\because$ Mass = density × volume]

2. (3) $M \propto \frac{1}{V}$

3. (1) No. of moles of glucose $= \frac{10}{180} = 0.0555$

No. of moles of water $= \frac{90}{18} = 5$

No. of moles of solution = 5.0555

Mole fraction of glucose $= \frac{\text{No. of moles of glucose}}{\text{No. of moles of solution}}$

$= \frac{0.0555}{5.0555} = 0.01$

4. (1) Molecular mass of ethyl alcohol = 46 g/mol

Molecular mass of water = 18 g/mol

Number of moles of $C_2H_5OH = \frac{138}{46} = 3$

Number of moles of $H_2O = \frac{72}{18} = 4$

Mole fraction of ethyl alcohol and water:

$\chi_{C_2H_5OH} = \frac{3}{3+4} = \frac{3}{7}$

$\chi_{H_2O} = \frac{4}{3+4} = \frac{4}{7}$

$\therefore$ Ratio, $\frac{\chi_{C_2H_5OH}}{\chi_{H_2O}} = \frac{\frac{3}{7}}{\frac{4}{7}} = \frac{3}{4} \Rightarrow 3:4$

5. (3) $V_1 = 2.5\ cm^3 = 2.5 \times 10^{-3}\ dm^3$

$M_1 = 0.2$ M

$N_1 = 2 \times M_1 = 2 \times 0.2 = 0.4$ N

$V_2 = 0.5\ dm^3$

$N_1V_1 = N_2V_2$

$0.4 \times 2.5 \times 10^{-3} = N_2 \times 0.5$

$N_2 = 2 \times 10^{-3} = 0.002$ N

6. (1) Normality = Molarity × n_f

$\Rightarrow$ Molarity $= \frac{\text{Normality}}{n_f}$ $\left(\because n_f = 2 \text{ for } Na_2CO_3\right)$

$\therefore$ Molarity $= \frac{0.2}{2} = 0.1M$

7. (3) 70% by weight means 70 g of orthophosphoric acid is present in 100 g of acid.

$N = \frac{w}{\text{eq.wt.}} \times \frac{1000}{V_{(cc)}}$

w = 70 g

$\text{Eq. wt.} = \frac{\text{Molecular mass}}{\text{number of replaceable H} - \text{atom}} = \frac{98}{3}$

$V = \frac{\text{Mass}}{\text{density}} = \frac{100}{1.54}$

$N = \frac{70 \times 1000 \times 1.54 \times 3}{98 \times 100} = 33$ N

8. (1) Equivalents of acid = Equivalents of base

$\frac{0.126}{\text{Eq. wt. of acid}} = (N \times V)_{\text{Base}}$

$\Rightarrow \frac{0.126}{\text{Eq. wt.}} = 0.1 \times \frac{20}{1000}$

$\Rightarrow \text{Eq. wt.} = \frac{0.126 \times 1000}{0.1 \times 20} = 63$ g/eq

$\therefore$ Equivalent weight of acid = 63 g/eq.

9. (1) Let the solution be x molal, then moles of benzene present in 1000g of benzene $= \frac{1000}{78} = 12.82$ moles

$\therefore$ Mole fraction of solute $= \frac{x}{x + 12.82}$

$0.2 = \frac{x}{x + 12.82}$

$0.2\,(x + 12.82) = x$

$0.2x + 2.564 = x$

$2.564 = x - 0.2x = 0.8x$

$x = \frac{2.564}{0.8} = 3.2$

10. (1) For H_2SO_4

N = M × basicity

N = 1 × 2 = 2

For NaOH

N = M × acidity

N = 1 × 1 = 1

Now from normality equation

$N_1V_1 = N_2V_2$

$2 \times V_1 = 1 \times 20$

$V_1 = \frac{20}{2} = 10$ mL of H_2SO_4

11. (1) A solution which is described as 10 % glucose in water by mass means that 10 g of glucose is dissolved in 90 g of water resulting in a 100 g solution.

12. (1) Given,

Mass of benzene = 7.8 g

Mass of toluene = 46.0 g

$$\text{Mole fraction of } C_6H_6 = \frac{\frac{7.8}{78}}{\frac{7.8}{78}+\frac{46}{92}} = \frac{1}{6}$$

13. (4) Molarity is a function of temperature.

14. (3) $\text{Molarity} = \frac{w}{\text{m.wt.} \times \text{volume in litre}} = \frac{171}{342 \times 1} = 0.5\text{M}$.

15. (4) No. of moles of $H_2S = 0.195$

$$\text{No. of moles of } H_2O = \frac{1000}{18} = 55.55 \text{ mol}$$

$$\text{Mole fraction of } H_2S = \frac{0.195}{0.195+55.55} = 0.0035$$

Pressure at STP = 0.987 bar

According to Henry's law, $p = K_H x$

$$\text{or } K_H = \frac{p_{H_2S}}{x_{H_2S}} = \frac{0.987}{0.0035} = 282 \text{ bar}$$

16. (2) Henry's law states that the solubility of a gas in a liquid is directly proportional to the partial pressure of the gas.

$p = K_H x$

17. (3) The value of Henry's law constant decreases with increases in solubility. Among the given set of gases, CO_2 is the most soluble in water.

18. (1) According to Henry's law

$p = K_H x$

Where, K_H is the Henry's law constant

19. (1) Solubility of gases in liquids increases with decrease in temperature.

20. (1) Deep sea divers use mixture of helium-oxygen gas because of its very low solubility in blood.

21. (4) Given, $\frac{p_A^o}{p_B^o} = \frac{1}{2}, \frac{x_A}{x_B} = \frac{1}{2}$

Suppose, $p_A^o = p$, then $p_B^o = 2p$

In the similar way,

Suppose, $x_A = x$, then $x_B = 2x$

Now, mole fraction of A in vapour phase

$$y_A = \frac{p_A}{p_{total}}$$

$$y_A = \frac{p_A}{p_A + p_B} \quad ...(i)$$

Where, $p_A = p_A^o x_A = px$

$p_B = p_B^o x_B = 2p \times 2x = 4px$

On putting values of p_A and p_B in eq. (i), we get

$$\therefore y_A = \frac{px}{px+4px}$$

$$= \frac{px}{5px} = \frac{1}{5} = 0.2$$

22. (4) Number of moles of $CHCl_3$,

$$n_A = \frac{W}{M} = \frac{25.5}{119.5} = 0.213$$

Number of moles of CH_2Cl_2,

$$n_B = \frac{40}{85} = 0.47$$

Mole fraction of $CHCl_3$,

$$x_A = \frac{n_A}{n_A + n_B} = \frac{0.213}{0.684} = 0.31$$

Mole fraction of CH_2Cl_2

$x_B = 1 - x_A$

$= 1 - 0.31 = 0.69$

$p_{Total} = p_A^0 x_A + p_B^0 x_B$

$= 200 \times 0.31 + 41.5 \times 0.69$

$= 62 + 28.63 = 90.63$ mm Hg.

23. (4) Mercury is non-volatile and non-volatile substances have low vapour pressure.

24. (2) According to Raoult's law

$p_A \propto x_A$ and $p_B \propto x_B$ for a binary solution of A and B

25. (3) $\text{Mole fraction of P} = \frac{3}{3+2} = 0.6$

$$\text{Mole fraction of Q} = \frac{2}{3+2} = 0.4$$

$p_{Total} = p_p + p_Q$

$= p_P^0 x_p + p_Q^0 x_Q$

$= 80 \times 0.6 + 60 \times 0.4 = 72$ torr

26. (1) If the total number of particles of solute changes in the solution due to association or dissociation, Raoult's law is not applicable. Among the given compounds, NaCl undergoes dissociation and forms Na^+ and Cl^- ions. Thus, Raoult's law is not applicable to NaCl.

27. (4) Molar ratio of 1 : 1 means $x_X = x_Y = 0.5$

$p_{total} = x_X p_X^\circ + x_Y p_Y^\circ$

$\therefore \; 400 = 0.5 \, p_X^\circ + 0.5 \, p_Y^\circ$

or $p_X^\circ + p_Y^\circ = 800$ ...(i)

Molar ratio of 1 : 2 means, $x_X = \frac{1}{3}, x_Y = \frac{2}{3}$

$$\therefore \quad 350 = \frac{1}{3} p_X^\circ + \frac{2}{3} p_Y^\circ$$

or $p_X^\circ + 2\,p_Y^\circ = 1050$ ...(ii)

Eqn (ii) - Eqn. (i) gives

$p_Y^\circ = 250$ mm

On putting value of p_y^o in eq. (i), we get

$p_X^\circ = 550$ mm

28. (3) $x_X = \frac{1}{4} = 0.25,\ x_Y = \frac{3}{4} = 0.75$

$P_{total} = x_X \times p_X{}^\circ + x_Y \times p_Y{}^\circ$

i.e., $550 = 0.25\ p_X{}^\circ + 0.75\ p_Y{}^\circ$

or $2200 = p_X{}^\circ + 3\ p_Y{}^\circ$ ...(i)

After adding 1 mol of Y,

$x_X = \frac{1}{5} = 0.20,\ x_Y = \frac{4}{5} = 0.80$

$\therefore\ 560 = 0.20\ p_X{}^\circ + 0.80\ p_Y{}^\circ$

or $2800 = p_X{}^\circ + 4\ p_Y{}^\circ$ ...(ii)

Solving the two equations, we get

$p_Y{}^\circ = 600$ mm, $p_X{}^\circ = 400$ mm

29. (4) $P_T = P_P^0 x_P + P_Q^0 x_Q\ ; P_T = 80 \times \frac{3}{5} + 60 \times \frac{2}{5} = 72$ torr

30. (3) Maximum boiling azeotrope solution show negative deviation from Raoult's law. A mixture of acetone and chloroform forms a solution with negative deviation from Raoult's law and hence forms maximum boiling azeotrope.

31. (2) A solution of acetone in ethanol shows a positive deviation from Raoult's law.

32. (4) $CH_3-\overset{\overset{O}{||}}{C}-CH_3$ and $CHCl_3$

Acetone Chloroform

Acetone and chloroform will show a negative deviation from Raoult's law due to their association after mixing.

33. (3) Raoult's law is applicable only for ideal solutions. These two components A and B follows the condition of Raoult's law, if the forces of attraction between A and B is similar to the forces of attraction between A-A & B-B.

34. (1) For an ideal solution, ΔH and ΔV for mixing should be zero. $P_{Total} = p_A + p_B$ and A-A, B-B and A-B interactions are nearly same.

35. (2) $p_{Total} = 2.619 + 4.556 = 7.175$ kPa

$X_{MeOH} = \frac{p_{MeOH}}{p_{Total}} = \frac{2.619}{7.175} = 0.365$

$X_{EtOH} = 1 - 0.365 = 0.635$

36. (2) H_2O (43%) and HI (57%) forms maximum boiling azeotropic mixture on boiling at 400K.

37. (3) Acetone + chloroform shows negative deviation while acetone + ethyl alcohol solution shows positive deviation from Raoult's law.

Other examples:

Positive deviations: Acetone + ethyl alcohol, acetone + benzene, water + ethyl alcohol.

Negative deviations: Nitric acid + water, benzene + chloroform.

38. (1) Since, the solution boils at relatively higher temperature, vapour pressure of solution would be less. Therefore, it shows –ve deviation from Raoult's law.

39. (1) In (A) para-nitrophenol, intermolecular (between two molecules) H-bonding exists, while in (B) ortho-nitrophenol, intramolecular H-bonding exists. The boiling point of (B) is lower than (A) and, thus, (B) is more volatile, i.e., (B) has higher vapour pressure as compared to (A).

40. (1) $\frac{p^0 - p_s}{p^0} = x_{solute} \Rightarrow \frac{0.8-0.6}{0.8} \Rightarrow x_{solute} = 0.25$

41. (3) $p^0 = 760$ torr; 1 molal aqueous solution contains 1 mole of solute in 1000 g (55.5 mol) of water.

$x_B = \frac{1}{1+55.5} = \frac{1}{56.5}$

$\frac{\Delta p}{p^0} = x_B,\ \Delta p = p^0 x_B \Rightarrow \frac{760 \times 1}{56.5} = 13.45$ torr

42. (2) If $p^0 = 100$ mm, $p_s = 100 - 10 = 90$ mm

$\frac{p^0 - p_s}{p^0} = \frac{w_2/M_2}{w_1/M_1} = \frac{w_2}{M_2} \times \frac{M_1}{w_1}$

$\frac{100-90}{100} = \frac{w_2}{60} \times \frac{18}{180}$

or $w_2 = \frac{10}{100} \times 60 \times 10 = 60$ g

43. (4) $M = \frac{K_b \times w_B}{\Delta T_b \times w_A} = \frac{0.52 \times 12.5}{0.80 \times 0.185}$ ($\because$ 185 g = 0.185 kg)

$= \frac{52 \times 125}{0.80 \times 185} = 43.92\ g\ mol^{-1}$

44. (3) Isotonic solutions have same osmotic pressure at a given temperature.

$\therefore\ C_1RT = C_2RT$

i.e., $C_1 = C_2$

$\frac{n_1}{V} = \frac{n_2}{V} \Rightarrow n_1 = n_2$

No. of moles of urea $= \frac{10}{60} = \frac{1}{6}$

Weight of solute, X = 6 g

No. of moles of solute $X = \frac{6}{M}$

For isotonic solution, $n_1 = n_2$

or $\frac{1}{6} = \frac{6}{M}$ or $M = 36\ g\ mol^{-1}$

45. (1) $\Pi = \frac{n_B RT}{V} : n_B = \frac{w_B}{M_B}$

$\Pi = \frac{w_B}{M_B} \times \frac{RT}{V}$

$\Rightarrow M_B = \frac{w_B}{V} \times \frac{RT}{\Pi} = \frac{2 \times 0.0821 \times 300 \times 760}{0.3 \times 20}$

($\because$ 1 atm = 760 mm Hg)

$= 0.0821 \times 100 \times 760 = 6239.6\ g\ mol^{-1}$

46. (1) $M = \dfrac{w_B RT}{\Pi V} = \dfrac{20\times0.0821\times288\times1000}{(600/760)\times500} = 1198\ g\ mol^{-1}$

47. (2) Vapour pressure of pure water at boiling point = 1 atm = 1.013 bar

Vapour pressure of solution (p_s) = 1.004 bar

Let the mass of solution = 100 g

Mass of solvent = 100 – 2 = 98 g

$$\frac{p^0 - p_s}{p^0} = \frac{w_2/M_2}{w_1/M_1}$$

$$\frac{1.013-1.004}{1.013} = \frac{2\times18}{M_2\times98}$$

$$\Rightarrow M_2 = \frac{2\times18}{98}\times\frac{1.013}{0.009} = 41.35 g\ mol^{-1}$$

48. (3) Solubility is temperature dependent. A plant cell shrinks in hypertonic solution.

49. (1) For isotonic solution,

$$\Pi_{Glycerine} = \Pi_{Glucose}$$

$$\frac{n_1RT}{V_1} = \frac{n_2RT}{V_2}$$

$$\frac{10.2}{M}\times\frac{1}{1} = \frac{2}{180}\times\frac{1000}{100} \Rightarrow M = \frac{10.2\times18}{2} = 91.8 g$$

50. (2) $\Delta T_f = \dfrac{K_f\times w_2\times1000}{M_2\times w_1} \Rightarrow w_2 = \dfrac{\Delta T_f\times M_2\times w_1}{K_f\times1000}$

$$w_2 = \frac{10\times92\times600}{1.86\times1000} = 296.77 g$$

51. (3) $\Delta T_b = K_b \times m$

$$K_b = \frac{\Delta T_b}{m} = \frac{1}{\frac{1}{50}\times\frac{1000}{50}} = \frac{5}{2} = 2.5$$

52. (3) Properties of dilute solutions containing non-volatile solutes whose values depend upon the concentration and number of solute particles in the solution but not on the individuals identity of the solute, are known as colligative properties. Various colligative properties are

(i) Relative lowering of vapour pressure

(ii) Elevation in boiling point

(iii) Depression in freezing point

(iv) Osmotic pressure

So; optical activity is not a colligative property.

53. (3) $\Delta T_b = \dfrac{K_b\times1000\times w_2}{M_2\times w_1}$

$$\Delta T_b = \frac{K_b\times1000\times10}{100\times100} \therefore \Delta T_b = K_b$$

54. (4) The osmotic pressure is expressed in the unit of atmosphere.

55. (4) Depression in freezing point,

$\Delta T_f = K_f \times m$

Where, m = molality =

$$\frac{\text{Weight of solute}\times1000}{\text{Molecular weight of solute}\times\text{Weight of solvent (in g)}}$$

$$= \frac{1.5\times1000}{60\times250} = 0.1$$

$\Delta T_f = K_f \times 0.1 \Rightarrow 0.01 = K_f \times 0.1 \Rightarrow K_f = 0.1$

56. (2) The solution which has higher osmotic pressure than its standard solution is known as hypertonic solution.

57. (2) $\Delta T_f = 5.48 - 4.47 = 1.01$, $\Delta T_f = K_f \times m$

$$1.01 = 1.93 \times \frac{20\times1000}{M\times500} \Rightarrow M = 76.4$$

58. (1) Given that, w = 0.5143 g, W = 35 g

$K_b' = 3.9\ K\ mol^{-1}\ kg$, $\Delta T_b = 0.323\ K$

$$\because \Delta T_b = \frac{1000K_b'w}{MW}$$

$$\therefore 0.323 = \frac{1000\times3.9\times0.5143}{M\times35}$$

M = 177.42 g/mol

59. (3) In order to produce same lowering in vapour pressure for same solvent, the mole fraction of solute should be same.

Thus, $\dfrac{w_B' M_A}{M_B' w_A'} = \dfrac{w_B M_A}{M_B w_A}$

or $w_B' = \dfrac{w_B M_B' w_A'}{M_B w_A} = \dfrac{1\times180\times100}{60\times50} = 6 g$

60. (1) $\Delta T_f = K_f \times m$

$$\Delta T_f = K_f \times \frac{w_{solute}}{M_{solute}} \times \frac{1000}{w_{solvent}}$$

Lesser the molar mass of the solute, greater will be the freezing point depression.

61. (1) $p^0 - p_s = 10$

Apply, $\dfrac{p^0 - p_s}{p^0} = x_B$

$$\frac{10}{p^0} = 0.2 \Rightarrow p^0 = 50$$

Then, $p^0 - p_s = 20$

$$\frac{p^0 - p_s}{p^0} = x_B \Rightarrow \frac{20}{50} = x_B = 0.4$$

$\because x_A + x_B = 1$

$\therefore x_A = 0.6$

62. (2) The unit of molal depression constant (K_f) is $K\ kg\ mol^{-1}$.

63. (2) Given, mass of sucrose added = 100 g and mass of water = 1000 g

Moles of sucrose (n) $= \dfrac{100}{342} = 0.292$ mole

Moles of water (N) $= \frac{1000}{18} = 55.5$ mole

Vapour pressure of pure water, $p^0 = 23.8$ mm Hg

According to Raoult's law,

$$\frac{\Delta p}{p^0} = \frac{n}{n+N} \Rightarrow \frac{\Delta p}{23.8} = \frac{0.292}{0.292+55.5}$$

$$\Delta p = \frac{23.8 \times 0.292}{55.792} = 0.125 \text{ mm Hg.}$$

64. (3) $K_b = \frac{\Delta T_b}{m} = \frac{0.1 \times 100}{\frac{1.8}{180} \times 1000} = 1\,K/m$.

65. (1) By using, $m = \frac{K_f \times 1000 \times w}{\Delta T_f \times w_{Solvent}(gm)} = \frac{5.12 \times 1000 \times 1}{0.40 \times 50}$

= 256 gm/mol

Hence, molecular mass of the solute = 256 gm mol^{-1}

66. (1) $i = \frac{\text{Normal molar mass}}{\text{Observed molar mass}}$

$= \frac{164}{65.6} = 2.5$

$\alpha = \frac{i-1}{n-1}$, n = 3 (number of ions)

$= \frac{2.5-1}{3-1} = \frac{1.5}{2} = 0.75$

∴ Percentage ionization of MX_2 will be 75%

67. (3) Depression in freezing point is a colligative property. Thus, the compound which produces maximum ions has the least freezing point.

∵ Concentration is same

Sucrose → No ions

$NaCl \rightarrow \underbrace{Na^+ + Cl^-}_{\text{Two ions}}$

$CaCl_2 \rightarrow \underbrace{Ca^{2+} + 2Cl^-}_{\text{Three ions}}$

Glucose → No ions.

Thus; 1% $CaCl_2$ has the least freezing point.

68. (2) For NaCl, i = 2

$\Delta T_f = iK_f m$

$= 2 \times 1.86 \times 1 = 3.72$

$T_f = T^0_f - \Delta T_f = 0 - 3.72 = -3.72°C$

69. (4) i =

$$\frac{\text{Total number of particles after association / dissociation}}{\text{Total number of particles before association / dissociation}}$$

70. (1) $K_4[Fe(CN)_6] \rightarrow 4K^+ + [Fe(CN)_6]^{4-}$

$Al_2(SO_4)_3 \rightarrow 2Al^{3+} + 3SO_4^{2-}$

71. (2) $Mg(NO_3)_2 \rightarrow Mg^{2+} + 2NO_3^-$

$1-\alpha \quad \alpha \quad 2\alpha$

$i = 1 - \alpha + \alpha + 2\alpha$; $2.74 = 1 + 2\alpha$

$\alpha = 0.87 = 87\%$

72. (4) $C_6H_{12}O_6$ is a non-electrolyte, hence furnishes minimum number of particles and will have maximum freezing point.

$\Delta T_f = ik_f m$ or $\Delta T_f \propto i$

73. (3) $Na_2SO_4 \rightarrow 2Na^+ + SO_4^{2-}$

$1 - \alpha \quad 2\alpha \quad \alpha$

Van't hoff factor (i) $= 1 - \alpha + 2\alpha + \alpha = 1 + 2\alpha$

74. (1) $KCl \rightleftharpoons K^+ + Cl^-$ (n = 2)

$$\alpha = \frac{i-1}{n-1} = \frac{1.95-1}{2-1} = 0.95$$

75. (1) Since, HA is 20% ionised, van't hoff factor, $i = 1 + (y-1)\alpha$

(∵ y = 2, as two particles are produced)

$i = 1 + (2-1)\,0.2 = 1.2$

$\Delta T_f = iK_f m = 1.2 \times 1.86 \times 0.2 = 0.45°$ C

$\Delta T_f = T_1 - T_2$

$T_2 = 0°\,C - 0.45°\,C = -0.45°$ C

76. (1) Van't Hoff factor (i)

i = 1, for non-electrolytes

i > 1, for dissociation

i < 1, for association

77. (1) $\Delta T_f = i \times K_f \times m$

Here; degree of dissociation = 80%

$KBr \rightleftharpoons K^+ + Br^-$

$\because i = 1 - \alpha + n\alpha$

$\therefore i = 1 - 0.8 + 2 \times 0.8 = 1.8$

$\therefore \Delta T_f = 1.8 \times 1.86 \times 0.5 \Rightarrow \Delta T_f = 1.674$ K

$\Delta T_f = T^0_f - T_f \Rightarrow 1.674 = 273 - T_f$

$T_f = 273 - 1.674 \Rightarrow T_f = 271.326$ K

78. (3) $Ca(NO_3)_2 \rightleftharpoons Ca^{2+} + 2NO_3^-$

Its van't Hoff factor is 3 because it furnishes three ions per formula unit.

79. (2) When benzoic acid dissolves in benzene it will be dimerised due to the inter molecular hydrogen bonding

$$2C_6H_5COOH \rightleftharpoons (C_6H_5COOH)_2$$

Von't Hoff factor =

$$\frac{\text{Number of particles after dissociation / association}}{\text{Number of particles before dissociation / association}} = \frac{1}{2}$$

= 0.5

80. (3) Using; $\Delta T_f = i \times K_f \times m$

$$i = \frac{\Delta T_f \times w_A}{K_f \times n_B \times 1000}$$

$$i = \frac{3.82 \times 45}{1.86 \times \frac{5}{142} \times 1000} = 2.63$$

81. (4) In case of dissociation, Van't Hoff factor i > 1.

In case of association, Van't Hoff factor i < 1.

82. (2) $\Delta T_f = i \times K_f \times m$ $\quad HBr \rightarrow H^+ + Br^-$

$\qquad\qquad\qquad\qquad\qquad 1-\alpha \quad \alpha \quad \alpha$

Total ions = $1 - \alpha + \alpha + \alpha = 1 + \alpha$

$\therefore\ i = 1 + \alpha$

Given; $K_f = 1.86$ K kg mol^{-1}

Mass of HBr = 8.1 g

Mass of H_2O = 100 g

(α) = degree of ionisation = 90%

$$m(\text{molality}) = \frac{\text{mass of solute / mol wt. of solute}}{\text{mass of solvent in kg}}$$

$$= \frac{8.1/81}{100/1000}$$

$$i = 1 + \alpha = 1 + \frac{90}{100} = 1.9$$

$$\Delta T_f = i \times K_f \times m$$

$$= 1.9 \times 1.86 \times \frac{8.1/81}{100/1000} = 3.534°C$$

ΔT_f (Depression in freezing point) = freezing point of water- Freezing point of solution.

3.534 = 0 – Freezing point of solution

$\therefore$ Freezing point of solution = –3.534 °C

83. (2) Lowering of vapour pressure is a colligative property i.e., it depends only upon the number of particles of solute and not on the nature of solute.

0.1 M glucose $\rightarrow$ remains undissociated

0.1 M $BaCl_2 \rightarrow Ba^{2+} + 2Cl^- \Rightarrow$ 3 ions

0.1 M $MgSO_4 \rightarrow Mg^{2+} + SO_4^{2-} \Rightarrow$ 2 ions

0.1 M $NaCl \rightarrow Na^+ + Cl^- \Rightarrow$ 2 ions

$\because$ 0.1 M $BaCl_2$ gives maximum number of particles, hence it exhibits maximum lowering of vapour pressure.

84. (2) $Hg_2Cl_2 \underset{n=3}{\rightleftharpoons} Hg_2^{2+} + 2Cl^-$

$$\alpha = \frac{i-1}{n-1};\quad 0.8 = \frac{i-1}{3-1};\quad i = 2.6$$

High Order Time Intensive MCQs

1. (2) Chloroform and acetone form a non-ideal solution in which A–B type interactions are more than A–A and B–B type interaction due to H-bonding. Hence, the solution shows negative deviation from Raoult's law

i.e., $\Delta V_{mix} = -ve$; $\Delta H_{mix} = -ve$

$\therefore$ Total volume of mixture = < (30 + 50) mL

2. (1) Solution in 'A' solvent $\qquad$ Solution in 'B' solvent

$\qquad w_{solute} \qquad\qquad\qquad w_{solute}$

$\qquad M_{solute} \qquad\qquad\qquad M_{solute}$

$$\left(\frac{p^0 - p_s}{p^0}\right)_A = X_{solute} \text{ in solvent A}$$

$$\left[\frac{p^0 - p_s}{p^0}\right]_B = X_{solute} \text{ in solvent B}$$

According to question,

$$\left(\frac{p^0 - p_s}{p^0}\right)_B = 2\left(\frac{p^0 - p_s}{p^0}\right)_A$$

$$\left(\frac{n_{solute}}{n_{solute} + n_{solvent}}\right)_{in\,B} = 2\left(\frac{n_{solute}}{n_{solute} + n_{solvent}}\right)_{in\,A}$$

X_{solute} in solvent B = 2 X_{solute} in solvent A

Since solution is very dilute-

$$n_{solute} + n_{solvent} \approx n_{solvent}$$

$$\therefore \left(\frac{n_{solute}}{n_{solvent}}\right)_{in\,B} = 2\left(\frac{n_{solute}}{n_{solvent}}\right)_{in\,A}$$

Now, no. of moles of solute in solvent A & B are same,

$$\therefore \frac{1}{n_{solvent\ in\ B}} = 2\left[\frac{1}{n_{solvent\ in\ A}}\right]$$

$$\frac{1}{w_{solvent}} \times M_{solvent_B} = 2 \times \frac{1}{w_{solvent}} \times M_{solvent_A}$$

(Given, weights of solvents are also equal)

$\therefore M_B = 2M_A$

3. (1) $K_3[Fe(CN)_6]$ ionises as

$K_3[Fe(CN)_6] \rightarrow 3K^+ + [Fe(CN)_6]^{3-}$

For dissociation,

$$\alpha = \frac{i-1}{n-1}$$

$\because i = (3\alpha + 1)$

$$\text{Now, } i = \frac{\text{Normal molar mass of solute}}{\text{Abnormal molar mass of solute}}$$

$$1 + 3\alpha = \frac{M_{normal}}{\text{Abnormal molar mass of solute}}$$

$\therefore$ Abnormal molar mass of solute = $M_{normal}\,(1 + 3\alpha)^{-1}$

4. (1)

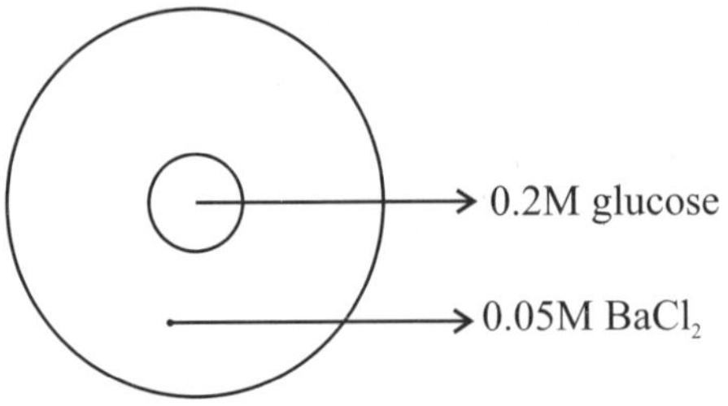

$\pi = (i_1C_1 - i_2C_2)RT$

$= (1 \times 0.2 - 3 \times 0.05) \times 0.0821 \times 300$

$= (0.2 - 0.15) \times 0.0821 \times 300$

$= 0.05 \times 0.0821 \times 300$

$= 1.23$ atm

5. (3) Concentration in case I

10 gms of solute 'A' + 20g of solute 'B' dissolved in 500 mL H_2O.

$$M = \frac{\frac{10}{M_A} + \frac{20}{M_B}}{\frac{500}{1000}} = \left(\frac{10}{M_A} + \frac{20}{M_B}\right) \times \frac{1000}{500}$$

$$M = \frac{20}{M_A} + \frac{40}{M_B} = \frac{20 M_B + 40 M_A}{M_A M_B}$$

Concentration in Case II

6.67 g of A + 30 g of B in 500 mL H_2O

$$M = \frac{\frac{6.67}{M_A} + \frac{30}{M_B}}{\frac{500}{1000}} = \left(\frac{6.67}{M_A} + \frac{30}{M_B}\right) \times \frac{1000}{500}$$

$$M = \frac{13.34}{M_A} + \frac{60}{M_B} = \frac{13.34 M_B + 60 M_A}{M_A M_B}$$

Since solutions are isotonic, therefore concentrations must be same:

$$\therefore \frac{20 M_B + 40 M_A}{M_A M_B} = \frac{13.34 M_B + 60 M_A}{M_A M_B}$$

$20\, M_B - 13.34\, M_B = 60\, M_A - 40\, M_A$

$6.66\, M_B = 20\, M_A$

$$\frac{M_A}{M_B} = \frac{6.66}{20} = \frac{1}{3}$$

6. (2) $\frac{p^0 - p_S}{p^0} = \frac{n_2}{n_1 + n_2}$

1 molal solution means 1 mole of solute is dissolved in 1 kg of solvent (water here)

$$\frac{12.3 - p_S}{12.3} = \frac{1}{1 + \frac{1000}{18}} = \frac{1}{1 + 55.56}$$

$$\frac{12.3 - p_S}{12.3} = 0.0177$$

$12.3 - p_S = 0.2177$

$p_S = 12.08$ kPa

7. (2) $\Delta T_b = iK_b \times m$

$T_b - T_b^0 = iK_b m$

$T_b = iK_b m + T_b^0$

In option 1:

For glucose, i = 1

$T_b = 1 \times K_b \times 0.015 + T_b^0$

$T_b = 0.015\, K_b + T_b^0$

In option 2:

For Na_2SO_4, i = 3

$T_b = 3K_b \times 0.10 + T_b^0$

$T_b = 0.3K_b + T_b^0$

In option 3:

For KNO_3, i = 2

$T_b = 2 \times K_b \times 0.01 + T_b^0$

$T_b = 0.02K_b + T_b^0$

In option 4:

For urea, i = 1

$T_b = 1 \times K_b \times 0.015 + T_b^0$

$T_b = 0.015\, K_b + T_b^0$

8. (3) $p_A = p_A^0 X_A$ (Raoult's law)

$p_A = p_{total}\, Y_A$ (Dalton's law)

$\therefore p_A^0 X_A = p_{total} Y_A$

$$p_A^0 = \frac{p_{total} Y_A}{X_A} \quad ...(1)$$

$$\text{Similarly, } p_B^0 = \frac{p_{total} Y_B}{X_B} \quad ...(2)$$

Now, $p_A^0 > p_B^0$

$$\therefore p_{total}\frac{Y_A}{X_A} > p_{total}\frac{Y_B}{X_B}$$

$$\frac{Y_A}{X_A} > \frac{Y_B}{X_B} \Rightarrow \frac{Y_A}{Y_B} > \frac{X_A}{X_B}$$

9. (2) $n_{C_6H_{12}O_6} = \frac{18}{180} = 0.1$

$$n_{H_2O} = \frac{178.2}{18} = 9.9$$

$$\frac{p^0 - p}{p^0} = X_B \Rightarrow \frac{p^0 - p}{p^0} = \frac{0.1}{0.1 + 9.9} = \frac{0.1}{10}$$

$$p^0 - p = \frac{0.1}{10} \times 760 = 7.6$$

$p = p^0 - 7.6$

$= 760 - 7.6$

$= 752.4$ torr

10. (2) $T_b - T_f = 105°C$ (given)

$\Delta T_b = K_b \times m$

$T_b^s - T_b^0 = K_b \times m$ ......(1)

$\Delta T_f = K_f \times m$

$T_f^0 - T_f^s = K_f \times m$ ....(2)

Add 1 & 2

$$(T_b^s - T_b^0) + (T_f^0 - T_f^s) = K_b \times m + K_f \times m$$

$$(T_b^s - 100) + (0 - T_f^s) = m(K_b + K_f)$$

$$105 - 100 = \frac{\frac{w}{342}}{\frac{100}{1000}} \times 2.37 \Rightarrow w = 72 \text{ gm}$$

11. (2) Apply $\Delta T_f = i\, K_f\, m$

$$\Delta T_f = i \times 1.86 \times \frac{\frac{3}{60}}{\frac{500}{1000}} \ ...(1) \left[\begin{array}{l} \because \rho \text{ of } H_2O = 1\, g/cm^3 \\ \therefore 500\, cm^3\ H_2O = 500\, g\ H_2O \end{array}\right]$$

Also, $\alpha = \frac{i-1}{n-1} \Rightarrow 0.23 = \frac{i-1}{2-1}$

$\therefore i = 1.23$ ...(2)

Placing the value of (2) in equation 1, we get

$$\Delta T_f = 1.23 \times 1.86 \times \frac{3}{60} \times \frac{1000}{500}$$

$\Delta T_f = 0.228$ K

12. (4) Apply, $\Delta T_f = K_f m$

$0.48 = K_f \times m$ .....(1)

$\Delta T_b = i K_b m$

$0.26 = i K_b m$ ......(2)

On dividing eq. (1) and (2), we get

$$\frac{0.48}{0.26} = \frac{1}{i} \times 3.7 \quad \left(\because \frac{K_f}{K_b} = 3.7\right)$$

$$i = \frac{3.7}{1.84} = 2$$

Since, van't hoff factor, $i > 1$, its a case of dissociation.

13. (1) Apply, $\Delta T_f = K_f m$

$0 - (-34) = 1.86 \times m$

$$\frac{34}{1.86} = m = 18.27$$

$$\text{Molality} = \frac{\text{no. of moles of solute}}{\text{wt of solvent (in kg)}}$$

$$18.27 = \frac{\frac{w}{62}}{\frac{50}{1000}}$$

$$18.27 = \frac{w}{62} \times \frac{1000}{50}$$

$$\Rightarrow w = \frac{18.27 \times 62 \times 50}{1000}$$

$\Rightarrow w = 56.6$ g

$$\rho \text{ of ethylene glycol} = \frac{56.6 \text{ g}}{50 \text{ cm}^3} = 1.13 \text{ g/cm}^3$$

$$\left(\because \text{density} = \frac{\text{mass}}{\text{volume}}\right)$$

14. (3) $\pi = CRT$

T = osmotic press.

Given: (mass of urea = 0.6g)

(mass of glucose = 1.8 g)

(Volume of H_2O = 10 mL)

(T = 27°c)

$$= \left(\frac{6}{60} + \frac{18}{180}\right) \times 0.821 \times 300$$

$= 0.2 \times 0.82 \times 300 = 4.926$ atm.

15. (1) $\text{Molarity} = \frac{\text{Moles of solute}}{\text{Volume of solution}}$

Molarity will change upon change in temperature due to change in volume with change in temperature

16. (1) (*i*) 10 moles urea in 1000 gm of water

$$X_{urea} = \frac{10}{10 + \frac{1000}{18}} = \frac{10}{65.55} = 0.152$$

(*ii*) % *w*/*w* of urea $= \frac{10 \times 60}{10 \times 60 + 1000} \times 100$

$= 37.5\%$

17. (2) $\frac{p_A^0 - p_A}{p_A^0} = X_{solute} = \frac{n_B}{n_A + n_B}$

$$\frac{59.2 - 51.3}{59.2} = X_{solute} = \frac{7.9}{59.2} = 0.1334$$

$X_{solvent} = 0.8666$

$$\text{Molality (m)} = \frac{X_{solute} \times 1000}{X_{solvent} \times M_{solvent}}$$

$$= \frac{0.1334 \times 1000}{0.8666 \times 46} = 3.34 \text{ m}$$

18. (1) For 1 molal aq. urea solution,

$\Delta T_f = K_f m$ ($\because$ i for urea solution = 1)

$T^0 - T_{f(s)} = K_f m$

$0 - (-1.86) = K_f \times 1 \Rightarrow 1.86 = K_f$

Since solvent is same for both, i.e. acetic acid solution & urea solution, the value for K_f is same.

$\Delta T_f = iK_f m$

$0.02046 = i \times 1.86 \times m$

$0.02046 = i \times 1.86 \times 0.01$

$i = 1.1$

CH_3COOH	$\rightleftharpoons$ CH_3COO^-	$+ H^+$
(aq.)	(aq.)	(aq.)
c	0	0
c – cα	cα	cα

Now,

$$\alpha = \frac{i-1}{n-1} \Rightarrow \alpha = \frac{1.1-1}{2-1}$$

$\alpha = 0.1$

$[H^+] = c\alpha = 0.01 \times 0.1 = 10^{-3}$m

$pH = -\log [H^+] = -\log[10^{-3}] = 3$

19. (1)

	HA	$\rightleftharpoons$	H^+	$+ A^-$
	(aq.)		(aq.)	(aq.)
At t = 0	c		0	0
At t = t	c – cα		cα	cα

pH = 2

$\therefore [H^+] = 10^{-2}$M

$\therefore c\alpha = 10^{-2} \Rightarrow 1 \times \alpha = 10^{-2}$

$\alpha = 0.01$

$\alpha = \frac{i-1}{n-1} \Rightarrow 0.01 = \frac{i-1}{2-1}$

$i = 1.01$

20. (4) 1 M aqueous glucose solution means 1 mole of glucose is present per litre of solution.

So, no. of moles of solute = 1

$\therefore$ Weight of solute = 180 g

ρ of solution = 1.2 g/mL

So, 1 mL of solution weighs = 1.2 g

1000 mL of solution weighs = 1.2 × 1000 = 1200 g

$\therefore$ Wt. of solvent = 1200 – 180 = 1020 g

$\Delta T_b = K_b m$

$= K_b \times \frac{1}{1020/1000} \Rightarrow K_b \times \frac{1000}{1020}$

$\Delta T_b = 0.98\, K_b$

21. (2) Mass of MeOH = 100 × 0.791 = 79.1 g

Mass of EtOH = 100 × 0.998 = 99.8 g

$M_1 = \frac{79.1 \times 1000}{32 \times 200} = 12.35$

$M_2 = \frac{99.8 \times 1000}{46 \times 200} = 10.84$, Thus $M_1 > M_2$.

22. (2) $p_A = p_A^o.x_A = 240 \times \frac{2}{3} = 160$ torr;

$p_B = p_B^o.x_B = 150 \times \frac{1}{3} = 50$ torr;

$p_{Total} = 160 + 50 = 210$ torr

Since observed vapour pressure is less than expected, this means that solution exhibits negative deviation.

23. (2) 3 M means 3 moles are present in 1 L solution

Molality, $m = \frac{n_{solute}}{W_{solvent(kg)}}$

$m = \frac{3 \times 1000}{1.25 \times 1000 - [3 \times 58.5]} = 2.79$ m

($\because$ Density $= \frac{\text{Mass of solution}}{\text{Volume of solution}}$,

thus, mass of solution = (1.25 × 1000)g

Mass of solute = $n_{solute} \times M.W_{(solute)}$

= 3 × 58.5 g)

24. (4) The vapour pressure of a mixture of two immiscible liquids is the sum of their vapour pressures in the pure states, independent of their relative amounts. Hence, boiling point of the mixture will be less than that of either of the liquids, remaining constant throughout.

25. (3) Given, $\Delta T_f = 3.72$ °C

$\Delta T_f = i\, K_f m$

3.72 = i × 1.86 × 1

$\therefore$ i = 2

This means that the salt ionizes to give two particles. The formula pertaining this observation is $[Pt(H_2O)_3Cl_3]Cl \cdot 3H_2O$ which ionises in solution as $[Pt(H_2O)_3Cl_3]^+$ and Cl^- ions.

26. (2) $p_A = 0.01 \times 0.082 \times 300 = 0.2463$ atm;

$p_B = 0.001 \times 0.082 \times 300 = 0.02463$ atm

The movement of solvent particles will occur from dilute solution (B) to (A). Hence, the pressure of (0.2463 – 0.02463 = 0.22167 atm.) should be applied on concentrated solution i.e., solution A.

27. (3) $P_S = P_A^0.x_A + P_B^0.x_B$

$= P_A^0(1 - x_B) + P_B^0.x_B$

$P_S = P_A^0 + x_B(P_B^0 - P_A^0)$ ...(1)

Given equation,

$P_S = 120 - 75\, x_B$ ...(2)

On comparing eq. (1) & (2), we get

$P_A^0 = 120$, $P_B^0 - P_A^0 = -75$

$P_B^0 - 120 = -75$

$P_B^0 = 45$

28. (1) $p_A = p_A^0 . x_2$; $p_B = p_B^0.(1 - x_2)$

$p_{Total} = p_A^0 . x_2 + p_B^0\ (1 - x_2)$ ...(i)

Also $x_1 = \frac{p_A}{p_A + p_B} = \frac{p_A}{p_A^0 x_2 + p_B^0(1 - x_2)}$

or $p_A^0.x_2 + p_B^0(1 - x_2) = \frac{p_A}{x_1}$ ...(ii)

On putting value of eq. (ii) in eq. (i), we get

$p_{Total} = \frac{p_A}{x_1} = \frac{p_A^0 x_2}{x_1}$

29. (3) $\Delta T_b = K_b \times m$ ...(1)

If solution is diluted with equal volume of water, then concentration of solution is $\frac{m}{2}$

So, $\Delta T_f = K_f \times \frac{m}{2}$ ...(2)

On dividing eq. (1) and (2), we get

$\frac{\Delta T_b}{\Delta T_f} = \frac{K_b \times m}{K_f \times \frac{m}{2}}$

$\frac{0.15}{\Delta T_f} = \frac{0.512 \times 2}{1.86}$

$\Delta T_f = 0.272$

$\Delta T_f = (F.P.)_{solvent} - (F.P.)_{solution}$

$0.272 = 0°C - (F.P.)_{solution}$

$(F.P.)_{solution} = -0.272°C$

30. (4) Using, $\pi_1 V_1 + \pi_2 V_2 = \pi_R(V_1 + V_2)$

$$\pi_R = \frac{2.4V_1 + 4.6V_2}{V_1 + V_2} = 2.95 \text{ atm}$$

(assume temperature constant)

$$\frac{V_1}{V_2} = 3$$

31. (2,4) Those which have same no of ions after dissociation will have same Van't Hoff factor.

$K_4[Fe(CN)_6]$ & $FeSO_4 . (NH_4)_2SO_4 . 6H_2O$ have 5 ions each

$FeSO_4 (NH_4)_2SO_4 . 6H_2O$ & $KCl. MgCl_2 . 6H_2O$ have 5 ions each

32. (2) $P_{Total} = 0.4 \times 80 + 0.6 \times 120 = 104$

As $p_A + p_B < p^0_A \; x_A + p^0_B \; x_B$

Thus, solution shows negative deviation and for negative deviation, $\Delta H_{mix} < 0$.

33. (3) $M = \dfrac{n}{V(L)}$

$$n = \frac{120}{60} = 2$$

$\because$ Mass of solution = (1000 + 120)g = 1120g

$$V = \frac{\text{Mass}}{\text{density}} = \frac{1120 \text{ g}}{1.15 \text{ g/mL}} = 974 \text{ mL} = 974 \times 10^{-3} \text{ L}$$

$$M = \frac{2 \times 10^3}{974} = 2.05 \text{ M}$$

Correct & Incorrect MCQs

1. **(2)** Polar solutes dissolve in polar solvents and non-polar solutes dissolve in non-polar solvents. Hence, sodium chloride and sugar dissolve readily in water while naphthalene and anthracene dissolve readily in benzene.
2. **(4)** With increase in intermolecular forces, escaping tendency decreases, therefore rate of evaporation decreases.

   Order of intermolecular forces is A < B < C

   Therefore, order of rate of evaporation is C < B < A

   Hence, A evaporates more readily than C.
3. **(1)** There will be no net movement of the solvent through the semipermeable membrane between two solutions of equal concentration.
4. **(1)**

   (1) Mole fraction being a ratio has no units.

   (2) Unit of molarity is mol L^{-1}.

   (3) Since, mass does not depend on temperature, hence mass percentage is independent of temperature.

   (4) Since, volume depends on temperature, hence molarity is dependent on the temperature.
5. **(3)** The solubility of gases in liquid increases with decrease of temperature, that's why the aquatic species are more comfortable in cold waters rather than in warm waters.
6. **(2)** C_2H_5OH + acetone shows positive deviation from Raoult's law. A mixture of chloroform and acetone forms a solution with negative deviation from Raoult's law.
7. **(1)** The flow of solvent from dilute solution to the concentrated solution across a semipermeable membrane is due to osmosis.
8. **(4)** Since, isotonic solutions have same osmotic pressure, hence osmosis does not take place in two isotonic solutions separated by semipermeable membrane.
9. **(3)** In case of association, value of i is less than unity.
10. **(2)** At a given pressure, higher the value of K_H, the lower is the solubility of the gas in the liquid.

Statement Based MCQs

1. **(2)** Relative lowering of vapour pressure is equal to mole fraction of solute.

   Relative lowering of vapour pressure is a colligative property.
2. **(1)** In case of positive deviation from Raoult's law, A-B interactions are weaker than those between A-A or B-B while in case of negative deviations from Raoult's law, the intermolecular attractive forces between A-A and B-B are weaker than those between A-B.
3. **(1)** Molality depends upon mass and not volume. Mass does not change with temperature and therefore molality does not change with change in temperature.

$$\text{Molality}(m) = \frac{\text{Moles of solute (mole)}}{\text{Mass of solvent (kg)}}$$

4. **(1)** Polar compounds tend to dissociate in polar solvent like NaCl in water and acids tends to associate in benzene solvents.

   A molar mass that is either lower or higher than the expected or normal value is called as abnormal molar mass and it is obtained when the substance in the solution undergoes dissociation or association.
5. **(3)** In all cases both solute and solvent form vapours.
6. **(1)** Both statements are correct.
7. **(3)** In case of negative deviations from Raoult's law, the intermolecular attractive forces between A-A and B-B are weaker than those between A-B.
8. **(4)** Since, KCl dissociates to produce more number of particles in solution, hence 1 M solution of KCl has greater osmotic pressure than 1 M solution of glucose at the same temperature.
9. **(3)** The solutions that show large negative deviation from Raoult's law form maximum boiling azeotrope at a specific composition. Example: Nitric acid and water.
10. **(4)** The solutions which obey Raoult's law over the entire range of concentration are known as ideal solutions.

Assertion & Reason MCQs

1. **(1)** Molarity (M) $= \dfrac{\text{Moles of solute}}{\text{Volume of solution in litre}}$

   Molarity is a function of temperature. This is because volume depends on temperature. Hence, molarity changes with change in temperature.

2. **(3)** When methyl alcohol is being added to the water, the boiling point of water decreases since on the addition of a volatile solute to a volatile solvent, depression in boiling point is observed.
3. **(1)** A depression in freezing point is observed when NaCl is added to water. This is due to the lowering of vapour pressure of a solution. The lowering of vapour pressure occurs due to intermolecular interaction of solvent-solute particles.
4. **(2)** Both Assertion and Reason are true but Reason is not the correct explanation of the Assertion.
5. **(2)** Mole fraction being a ratio has no units.
6. **(3)** Solubility depends upon the nature of solute and solvent as well as temperature and pressure.
7. **(2)** The value of 'i' is same for both urea and sucrose, therefore both solutions will freeze at the same temperature.
   $\Delta T_f = K_f m$
   $(T_f^0 - T_f) = K_f m$
   On increasing m, ΔT_f will increase and hence freezing point will decrease.
8. **(3)** Solution having same composition throughout is homogeneous
9. **(4)** Freezing point of substance may be defined as the temperature at which vapour pressure of substance in liquid phase is equal to vapour pressure in solid phase
10. **(2)** Osmotic pressure is a colligative property as it depends on the number of solute molecules and not on their identity.

Multi Correct MCQs

1. **(1)** As V.P. is same for all, High temperature is required.
   $\because$ attraction between the particle is strong for the given liquids.
2. **(1)** Mass percent, mole fraction, molarity, ppm, molality are used to measure concentration terms.
3. **(4)** $\pi = iCRT$
   $\pi \propto iC$
   where, i is Van't Hoff factor and C is the concentration.
   Hence, A, B, D and E have same value of osmotic pressure.
4. **(1)** For C_2H_5OH, i = 1
   The comp. which have i > 1 shows, F.P. lower than that of 0.1 M C_2H_5OH.
   $Ba_3(PO_4)_2 \rightarrow 3Ba^{2+} + 2PO_4^{3-}$, i = 5
   $Na_2SO_4 \rightarrow 2Na^+ + SO_4^{2-}$, i = 3
   KCl, i = 2; Li_3PO_4, i = 4

NCERT Exemplar MCQs

1. **(1)** According to Henry's law, partial pressure of gas in the solution is proportional to the mole fraction of gas in the solution.
   $p = K_H x$
   Where, K_H = Henry's constant
   i.e., The vapour pressure of a solution is related to its concentration in terms of mole fraction.
2. **(4)** Dissolution of sugar in water will be most rapid when it is dissolved in hot water in powdered form because powdered form can easily insert in the vacancies of liquid particles.
   Further dissolution of sugar in water is an endothermic process. Hence, temperature will favour the dissolution of sugar in water.
3. **(3)** At equilibrium, the rate of dissolution of solid in a volatile liquid solvent is equal to the rate of crystallization.
4. **(2)** When solute is added to the solution three cases may arise
   (i) It dissolves into solution then solution is unsaturated.
   (ii) It does not dissolve in the solution then solution is known as saturated.
   (iii) When solute gets precipitated, solution is known as supersaturated solution.
5. **(3)** Maximum amount of solid that can be dissolved in a given amount of solvent does not depend upon pressure because solid and liquid are highly incompressible and practically do not get affected by change in pressure.
6. **(2)** Low concentration of oxygen in the blood and tissues of people living at high altitude is due to low atmospheric pressure.
7. **(1)** In pure methanol, molecules are hydrogen bonded. On adding acetone, its molecules get in between the host molecules and break some of the hydrogen bonds between them.
   Therefore, the intermolecular attractive forces between the solute-solvent molecules become weaker.
8. **(2)** Colligative properties depend upon the number of solute particles in solution irrespective of their nature. Colligative property is used to determine the molecular mass of particle.
9. **(1)** As we know from elevation in boiling point that,
   $\Delta T_b = K_b m$
   $$\text{Unit of } K_b - \frac{\text{Unit of } \Delta T_b}{\text{Unit of m}}$$
   $$= \frac{K}{\text{mol kg}^{-1}} = \text{K mol}^{-1}\text{ kg}$$
10. **(3)** Depression in freezing point is directly related to van't Hoff factor (i) according to which greater the value of i greater will be the depression in freezing point.

Solution	*i*
0.01 M glucose	1
0.01 M $MgCl_2$	3

   Hence, depression in freezing point of $MgCl_2$ is about 3 times of glucose.
11. **(4)** When an unripe mango is placed in a concentrated salt solution to prepare pickle, then mango loses water due to osmosis and get shriveled.
12. **(1)** According to definition of osmotic pressure, π = CRT. For concentrated solution, C has higher value than dilute solution.

13. (1) According to definition of depression in freezing point,

$$\Delta T_f = K_f m$$

where, K_f = freezing point depression constant, value of K_f depends upon nature of solvent. Two different solutions of sucrose of same molality prepared in different solvents will have different depression in freezing point.

14. (2)

Substances	van't Hoff factor (*i*)
For KCl	2
For $NaCl$	2
For K_2SO_4	3

15. (2) In reverse osmosis, solvent molecules move through a semipermeable membrane from a region of higher concentration of solute to lower concentration.

16. (1) Value of Henry's constant (K_H) increases with increase in temperature representing the decrease in solubility.

17. (2) According to Henry's law

$\Rightarrow p = K_H x$

As value of K_H rises, solubility of gases decreases.

18. (2) We know that, if a pressure higher than the osmotic pressure is applied on the solution, the solvent will flow from the solution into the pure solvent through the semi-permeable membrane. This process is called reverse osmosis.

19. (3) Since, van't hoff factor depends on the number of dissociated ions. Hence, the order will be: $i_A = i_B = i_c$.

20. (1) For an ideal solution, the A–A or B–B type intermolecular interaction is nearly equal to A–B type interaction. Here, a mixture of bromoethane and chloroethane is an example of ideal solution.

On the other hand, chloroform and acetone mixture is an example of non-ideal solution having negative deviation. So, (A–A) or (B–B) interaction must be weaker than A-B type interaction. While ethanol-acetone mixture shows positive deviation due to weaker A–B interaction in comparison to A–A or A–B interaction.

21. (1) When salt is added to water to make the solution, the vapour pressure of solution decreases. This is due to decrease in surface covered by solvent molecule which lead to decrease in number of solvent molecule escaping from the surface corresponding to pure solvent.

22. (2, 3) The two solutions having same osmotic pressure are known as isotonic solutions. The solute and solvent particles may or may not be same but osmotic pressure must be same as it is a colligative property.

23. (1, 4) Depending on the vapour pressures of the pure components 1 and 2, total vapour pressure over the solution decreases or increases with the increase of the mole fraction of component 1.

24. (1) At specific composition, methanol- acetone mixture will form minimum boiling azeotrope and will show positive deviation. This is due to weaker A – B interaction than A – A or B – B interaction. *i.e.*, A – B <A – A and B – B.

25. (3) Value of K_H depends upon nature of gases dissolved in water

Gas	Temperature (K)	K_H/k bar
Ar	298 K	40.3
CO_2	298 K	1.67
CH_4	298 K	0.413
HCHO	298 K	1.83×10^{-5}

26. (1, 2) Solubility of gaseous solute in the fixed volume of liquid solvent always depends upon nature of solute but it depends upon pressure at constant temperature and depends upon temperature at constant pressure.

27. (3, 4) The solution which follows Raoult's law is known as ideal solution.

$\Delta V_{mix} = 0$ and $\Delta H_{mix} = 0$

Thus, the mixture of benzene and toluene is an example of ideal solution.

Minimum boiling azeotropes are formed by non-ideal solutions.

28 (1, 2) Relative lowering of vapour pressure is a colligative property because

(i) It does not depend upon nature of solute.

(ii) It depends upon number of solute particles.

(iii) It depends upon concentration of non-electrolyte solution.

29. (1, 3) Van't Hoff factor (i) is a measure of extent of association or dissociation of solute particles which can be calculated as:

$$i = \frac{\text{normal molar mass}}{\text{abnormal molar mass}}$$

$$i = \frac{\text{observed colligative property}}{\text{calculated colligative property}}$$

30. (3, 4) Isotonic solutions have same osmotic pressure and same concentration. Elevation in boiling point and depression in freezing point are the colligative properties. These two colligative properties depend upon concentration.

As the molar concentration is same for isotonic solutions, so, elevation in boiling point and depression in freezing point of isotonic solutions must be same.

31. (2, 3) Mixtures having same composition in liquid and vapour phase are known as azeotropes. Azeotropes boils at same temperature.

Here, water-nitric acid and water-ethanol mixtures are non-ideal solution. Hence, water-nitric acid and water-ethanol are examples of azeotropes.

While benzene-toluene and n-hexane—*n*-heptane are examples of ideal solution.

NEET & AIIMS Past Year MCQs

1. (1) We know that, Molarity,

$$M = \frac{w_2 \times 1000}{M_2 \times (V)}$$

w_2 = Amount of glucose; M_2 = Molar mass of glucose

And, given molarity (M) $= \frac{1}{20}$

$$\frac{1}{20} = \frac{w_2 \times 1000}{180 \times 250}$$

$$w_2 = \frac{180 \times 250}{20 \times 1000} = 2.25 \text{ g}$$

The amount of glucose required is 2.25 g

2. (4) For isotonic solutions, the osmotic pressure (π) must be equal. This is expressed as:

Where:

C_1 and C_2 are the molar concentrations of the two solutions.

R and T are constants (cancel out because they are the same).

$\pi_1 = \pi_2$

Thus: $C_1RT = C_2RT$

$C_1 = C_2$

$$\frac{m}{180 \times 1} = \frac{15}{60 \times 1} \qquad \text{(Where, m is the mass of glucose)}$$

$$m = \frac{180 \times 15}{60}$$

$m = 180 \times 0.25 = 45$ g

3. (4) Lesser the value of K_H, more will be the solubility of gas in water.

So, solubility order:

$B(K_H = 2 \times 10^{-5}) > C(K_H = 35) > A(K_H = 145)$

4. (3) $\pi - CRT$

$\therefore$ slope = RT = 25.73 L bar mol^{-1}

$$\therefore T = \frac{25.73}{0.083}$$

$= 310$K

$= 37$°C

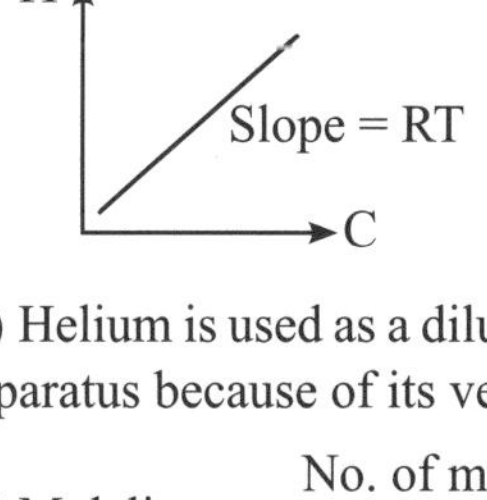

5. (4) Helium is used as a diluent for oxygen in modern diving apparatus because of its very low solubility in blood.

6. (3) Molality $= \frac{\text{No. of moles of solute}}{\text{Weight of solvent (kg)}}$

$$1 = \frac{0.5}{\text{Weight of solvent (kg)}}$$

Weight of solvent (kg) = 0.5 kg = 500 g

7. (1) According to Dalton's law of partial pressure:

Partial pressure of gas = mole fraction of gas in gaseous mixture × Total pressure of gaseous mixture.

$p_1 = x_1p$

$p_2 = x_2p$

$p_3 = x_3p$

Total pressure,

$p = p_1 + p_2 + p_3$

8. (4) As,

$$\pi = CRT \Rightarrow \pi = \frac{n_2RT}{V} = \frac{w_2RT}{MV}$$

So, osmotic pressure is directly proportional to the molar concentration of the solution and inversely proportional to molar mass of solute.

Molar mass of the given solutes decreases in the order:

Sucrose > Glucose > Urea

Hence, decreasing order of osmotic pressure of these solutions is :

$P_3 < P_1 < P_2$

9. (2) Given,

Vapour pressure of benzene (P_A) = 280 mm Hg

Vapour pressure of octane (P_B) = 420 mm Hg

And,

Total molar ratio = 3 + 2 = 5

$$x_{C_6H_6} = \frac{3}{5}, x_{C_8H_{18}} = \frac{2}{5}$$

Now,

$P_T = P_Ax_A + P_Bx_B$

$$= \left(280 \times \frac{3}{5}\right) + \left(420 \times \frac{2}{5}\right) = \frac{840 + 840}{5} = 336 \text{ mm of Hg}$$

10. (1) Vapour pressure of liquid and solid are equal at freezing point. Reduction in V.P occurs when solute is added.

11. (4) Elevation in boiling point (ΔT_b) = K_b × m. Depression in freezing point (ΔT_f) = K_f × m. Elevation in boiling point and depression in freezing point are colligative properties *i.e.*, they depend only on the number of particles of the solute. Value of K_b and K_f are different, so ΔT_b and ΔT_f are also different.

12. (2) **13. (1)**

14. (1) **15. (2)** **16. (1)**

Electrochemistry

NCERT Topic-wise MCQs

1. (1) The conversion of the chemical energy liberated during redox reaction to electrical energy is done by Daniel cell and it has an electrode potential of 1.1 V when concentration of Zn^{2+} and Cu^{2+} ions is unity.

$$Zn(s) + Cu^{2+}(aq) \rightarrow Zn^{2+}(aq) + Cu(s)$$

2. **(2)** The electrochemical cell stops working when the electrode potential of both the electrodes become equal.

3. **(3)** Electrical energy is converted into chemical energy in the electrolytic cell.

4. **(3)** $E_{cell} = E^0_{Cathode} - E^0_{Anode}$

$= -0.40 - (-0.74)$

$= -0.40 + 0.74 = 0.34$ V

5. **(2)** $E_{cell} = E^0_{Cathode} - E^0_{Anode}$

$= 0.34 - (-0.76) = 1.1$ V

6. **(1)** $E_{cell} = E^0_{Cathode} - E^0_{Anode}$

$= +1.23 - (-0.44)$

$= +1.67$ V

7. **(1)** Cell representation is done as follows:

Anode | Anode electrolyte || Cathode electrolyte | Cathode

Oxidation takes place at anode and it involves loss of electron.

Reduction is gain of electron and it takes place at cathode.

∴ For cell reaction:

$Zn + Cu^{2+} \rightarrow Zn^{2+} + Cu$

Zn is anode and Cu is cathode

∴ Cell representation is:

$Zn/Zn^{2+}||Cu^{2+}/Cu$

8. **(3)** At anode, oxidation takes place while at cathode, reduction takes place.

9. **(4)** At cathode, reduction takes place.

10. **(4)** Higher $E^\circ_{reduction}$ means good oxidizing agent.

11. **(1)** Anode has negative polarity.

12. **(1)** A potential difference developed between the electrode and the electrolyte is known as electrode potential.

13. **(3)** The direction of current flow is opposite to that of the electron flow.

14. **(2)** The relation between standard reduction potential of a cell and equilibrium constant is shown by

$$E^\circ_{cell} = \frac{0.059}{n}\log K_c$$

15. **(3)** $\Delta G^\circ = -nFE^\circ$ ...(i)

$\Delta G^\circ = -2.303\ RT \log K_{eq}$ ...(ii)

From eq (i), if E° is positive

then ΔG° is negative

In equation (ii), ΔG° is negative

So; $\log K > 0 \Rightarrow K > 1$

16. **(3)** $E^\circ_{cell} = E^\circ_{Cu^{2+}/Cu} - E^\circ_{Fe^{2+}/Fe}$

$= 0.34 - (-0.44) = 0.78$

$\Delta G^\circ = -n FE^\circ_{cell}$

$= -2 \times 96500 \times 0.78$

$= -150540$ J

$= -150.5$ kJ

17. **(1)** $2H_2O \rightarrow H_3O^+ + OH^-$

$$E^\circ_{cell} = \frac{0.0591}{n}\log K$$

$$\log K = \frac{E^\circ_{cell} \times n}{0.0591} = \frac{-0.8277 \times 1}{0.0591} = -14$$

$K = 10^{-14}$

18. **(2)** $E^\circ_{cell} = E^\circ_{Ag_2O/Ag} - E^\circ_{Zn^{2+}/Zn}$

$= 0.344 - (-0.76)$

$= 1.104$ V

$\Delta G^\circ = -nFE^\circ_{cell}$

$= -2 \times 96500 \times 1.104 = -2.13 \times 10^5\ J\ mol^{-1}$

19. **(3)** At anode: $Fe \rightarrow Fe^{2+}_{(0.001M)} + 2e^-$

At cathode: $2H^+(1M) + 2e^- \rightarrow H_2(1atm)$

Net reaction: $Fe + 2H^+ \rightarrow Fe^{2+} + H_2$

Nernst equation for the given cell is:

$$E_{cell} = E^0_{cell} - \frac{0.0591}{2}\log\frac{[Fe^{2+}][H_2]}{[Fe][H^+]^2}$$

20. **(4)** $\frac{1}{2}H_2 \rightarrow H^+ + e^-$

$$E_{cell} = -\frac{0.0591}{1}\log\frac{[H^+]}{(P_{H_2})^{\frac{1}{2}}}$$

$$\text{if } \log\frac{[H^+]}{(P_{H_2})^{\frac{1}{2}}} > 0 \text{ then } E_{cell} < 0$$

21. **(1)** The standard free energy (ΔG°) is related to standard electrode potential (E°) and equilibrium constant as

$\Delta G^\circ = -nFE^\circ_{cell}$

$\Delta G = -RT \log K_c$

Thus ; If E°_{cell} is –ve, then ΔG° = +ve thus; ΔG° > 0

Hence ; K_{eq} would be negative and less than 1, $K_{eq} < 1$

22. **(1)** $Cl_2 \rightarrow 2Cl^- + 2e^-$

$$E = E^\circ - \frac{0.0591}{n}\log_{10}\frac{[Cl^-]^2}{P_{Cl_2}}$$

$$E = 1.36 - (0.059/2)\log\frac{(0.1)^2}{10} = 1.36 - 0.0295 \times -3 = 1.45\ V$$

23. **(4)** In a cell reaction, electrical energy is produced. The electrical work is done by the system and it results in decrease in the free energy of the system.

24. **(3)** Due to lower value of reduction potential, alkali metals get oxidised easily. Hence, they behave as good reducing agent.

25. **(4)** In electrochemical series, Zn is present above H_2. So; it liberates hydrogen gas from dilute HCl/ H_2SO_4 but HNO_3 is oxidising agent. So, it oxidises H_2 and NO^-_3 is itself reduced.

26. (1) $E = E^\circ - \frac{2.303RT}{nF}\log\frac{1}{[Ni^{2+}]}$

$E = -0.25 - \frac{0.0591}{2}\log\left[\frac{1}{0.1}\right]$

$E = -0.25 - 0.03 \times 1$

$E = -0.28$ V

27. (4) $2H^+ + 2e^- \rightarrow H_2$

According to nernst equation,

$E_{H^+/H_2} = E^0_{H^+/H_2} - \frac{0.0591}{n}\log\frac{1}{[H^+]^2}$

$= 0 - \frac{0.0591}{2}(-\log[H^+]^2)$

$= -0.0591(-\log[H^+])$

$= -0.0591$ pH[$\therefore -\log[H^+] = pH$]

28. (1) $E_{cell} = E^0_{cell} - \frac{0.059}{n}\log\frac{[Cr^{3+}]^2[H_2O]^7}{[Cr_2O_7^{2-}][H^+]^{14}}$

$1.06 = 1.33 - \frac{0.059}{6}\log\frac{(15\times10^{-3})^2(1)^7}{(4.5\times10^{-3})[H^+]^{14}}$

$0.27 = \frac{0.059}{6}\log\left[\frac{225\times10^{-6}}{(4.5\times10^{-3})[H^+]^{14}}\right]$

$27.45 = \log\left[\frac{50\times10^{-3}}{[H^+]^{14}}\right]$

$27.45 = \log[50\times10^{-3}] - \log[H^+]^{14}$

$27.45 = -1.3010 - 14\log[H^+]$

$27.45 + 1.3010 = -14\log[H^+]$

$28.751 = 14 \times -\log[H^+]$

$28.041 = 14$ pH

pH = 2

29. (2) $E = \frac{-RT}{F}\ln\frac{(P_{H_2})^{1/2}}{[H^+]}$ when E = 0, then

$(P_{H_2})^{1/2} = [H^+] = 10^{-7}$; $P_{H_2} = 10^{-14}$ atm.

30. (3) $\Delta G^\circ = -nFE^\circ = -2 \times 96500 \times [1.23 - (-0.44)]$ J

$= -322310$ J $= -322.31$ kJ

31. (2) Given, $E^\circ_{cell} = 1.10$ V

As, $E^\circ_{cell} = \frac{0.059}{n}\log K$

$\log K = \frac{1.10\times2}{0.059} = 37.2881$

or $K = 10^{+37}$

32. (1) A positive E° means that the redox couple is a weaker reducing agent than the H^+/H_2 couple and a negative E° means that the redox couple is a stronger reducing agent than the H^+/H_2 couple.

33. (4) Since, Cu^{+2}/Cu has higher standard reduction potential than Ni^{+2}/Ni & Zn^{+2}/Zn,

$\therefore$ Cu^{+2} gets reduced faster than Ni^{+2} & Zn^{+2}.

34. (2) Higher the reduction potential, stronger is the oxidising agent.

35. (3) Higher the reduction potential, stronger is the oxidising agent.

$2I^- \rightarrow I_2 + 2e^-$ [oxidation]

$Br_2 + 2e^- \rightarrow 2Br^-$ [Reduction]

$2I^- + Br_2 \rightarrow I_2 + 2Br^-$ [Net reaction]

36. (3) Copper lies below the Hydrogen in the electrochemical series. So, it can not reduce water.

37. (4) Mg > Ca > Ba > K; This is the order of SRP.

38. (2) Higher SRP will undergo reduction.

39. (1) $\Lambda_{eq.[H_2O]} = \lambda_{eq.[H^+]} + \lambda_{eq.[OH^-]} = 349.8 + 198.5 = 548.3$

40. (2) $\kappa = \text{Conductance} \times \frac{l}{A}$

$\frac{l}{A} = \kappa \times \frac{1}{\text{Conductance}} = \kappa \times R = 0.0212 \times 55 = 1.166\ cm^{-1}$

41. (1) Depending on the charges on the cation and anion produced on dissociation of the electrolyte in solution, the type of electrolytes are known as 1 – 1, 2 –1 and 2 –2 electrolytes respectively.

$NaCl \rightarrow Na^+ + Cl^-$ (1 – 1)

$MgCl_2 \rightarrow Mg^{2+} + 2Cl^-$ (2 – 1)

$CaSO_4 \rightarrow Ca^{2+} + SO_4^{2-}$ (2 – 2)

42. (3) $\Lambda_m = \frac{\kappa\times1000}{M} = \frac{0.0152\times1000}{0.15} = 101\Omega^{-1}cm^2mol^{-1}$

43. (1) $\Lambda_m = \frac{\kappa\times1000}{M}$

$= \frac{1.01\times10^{-2}\times1000}{0.1} = 1.01\times10^2\ \Omega^{-1}cm^2mol^{-1}$

44. (2) $\Lambda^\circ_m NaBr = \Lambda^\circ_m NaCl + \Lambda^\circ_m KBr - \Lambda^\circ_m KCl$

45. (1) $\Lambda^\circ_{m\ HCOOH} = \lambda^\circ_{H^+} + \lambda^\circ_{HCOO^-}$

$= 349.6 + 54.6 = 404.2\ Scm^2\ mol^{-1}$

$\alpha = \frac{\Lambda_m}{\Lambda^0_m} = \frac{46.1}{404.2} = 0.114$

$K_a = \frac{c\alpha^2}{1-\alpha} = \frac{0.025\times(0.114)^2}{1-0.114} = \frac{0.025\times0.114\times0.114}{0.886}$

$K_a = 3.67\times10^{-4}\ mol\ L^{-1}$

46. (3) $\Lambda^\circ_{Ba(OH)_2} = \lambda^\circ_{Ba^{2+}} + 2\lambda^\circ_{OH^-}$

$\Lambda^\circ_{BaCl_2} = \lambda^\circ_{Ba^{2+}} + 2\lambda^\circ_{Cl^-}$

$\Lambda^\circ_{NH_4Cl} = \lambda^\circ_{NH_4^+} + \lambda^\circ_{Cl^-}$

$\Lambda^\circ_{NH_4OH} = \frac{\Lambda^\circ_{Ba(OH)_2} + 2\Lambda^\circ_{NH_4Cl} - \Lambda^\circ_{BaCl_2}}{2}$

$\Lambda^\circ_{NH_4OH} = \lambda^\circ_{NH_4^+} + \lambda^\circ_{OH^-}$

47. (3) Specific conductance : Sm^{-1}

48. (2) Number of ions per unit volume decreases results in decrease of conductivity.

49. (2) $R = \rho\frac{l}{a} \Rightarrow R = \frac{l}{\kappa a} \Rightarrow \kappa = \frac{l}{aR}$

$$\kappa = \frac{2}{40\times 5}$$

$$\Lambda_{eq} = \frac{\kappa\times 1000}{N} = \frac{2\times 1000}{40\times 5\times 0.5} = 20\Omega^{-1}cm^2eq^{-1}$$

50. (1)

$BaCl_2 \rightarrow Ba^{+2} + 2Cl^-$

$$\Lambda^o_{eq(BaCl_2)} = \frac{\lambda^o_{eq(Ba^{+2})}}{2} + \lambda^o_{eq(Cl^-)}$$

$$\Lambda^o_{eq(BaCl_2)} = \frac{127}{2} + 76\left(\Omega^{-1}cm^2eq^{-1}\right) = 139.5\ \Omega^{-1}\ cm^2eq^{-1}$$

51. (4) The electronic conductance is dependent on all the factors.

The electronic conductance depends on:

i. The nature and structure of the metal

ii. The number of valence electrons per atom

iii. Temperature (it decreases with increase of temperature)

52. (1) $Al_2(SO_4)_3 \rightarrow 2Al^{3+} + 3SO_4^{2-}$

$858 = 2x + 3 \times 160$

$858 = 2x + 480$

$378 = 2x$

$x = 189\ Scm^2\ mol^{-1}$

53. (4) For strong electrolytes, the plot between Λ_m and $C^{1/2}$ is linear.

54. (2) A → (i), (B) → (ii), (C) → (iii), (D) → (iv)

55. (4) $\Lambda_m = \frac{\kappa\times 1000}{M} \Rightarrow M = \frac{\kappa\times 1000}{\Lambda_m}$

$$M = \frac{1.85\times 10^{-7}\times 1000}{140}$$

$= 0.01321 \times 10^{-4} = 1.321 \times 10^{-6}$

$AB \rightleftharpoons A^+ + B^-$

$K_{sp} = [A^+][B^-] = (s)^2 = (1.32 \times 10^{-6})^2 = 1.74 \times 10^{-12}$

56. (1) $\Lambda^o_{CH_3COOH} = \lambda^o_{CH_3COO^-} + \lambda^o_{H^+}$

$= 35 + 315 = 350\ mho\ cm^2\ eq^{-1}$

57. (4) $\Lambda = \frac{\kappa\times 1000}{N} = \frac{\kappa\times 1000}{0.1} = \kappa\times 10000 \quad (N = n_f \times M)$

58. (1) Λ v/s $\sqrt{c}$ curve is linear for strong electrolytes.

59. (3) Pure water does not conduct electricity because it does not have ions.

60. (1) In aqueous solutions, strong electrolytes will undergo complete dissociation into ions.

61. (2) C_2H_5OH does not dissociate much into its ions as C_2H_5OH is a very weak acid.

62. (4) All the given statements are correct.

63. (4) Graphite, zinc wire and copper wire are electronic conductors, so, composition does not change by passing current through them.

64. (2) Conductivity of a solution is directly proportional to the number of ions present in the solution.

65. (2) The degree of dissociation does not depend on catalyst.

66. (1) Conductivity is represented by symbol κ.

67. (2) The SI unit of conductance is siemens which is represented by the symbol 'S'.

68. (4) The dimension of cell constant is $length^{-1}$.

69. (2) The cell constant is given by the equation: Cell constant = G* = Conductivity × resistance = 1.29 S/m × 100 Ω = 129 m^{-1}
Conductivity of 0.02 mol L^{-1} KCl solution = cell constant / resistance

$$= \frac{G^*}{R} = \frac{129m^{-1}}{520\Omega} = 0.248\ S\ m^{-1}$$

Concentration = 0.02 mol L^{-1}

= 1000 × 0.02 mol m^{-3} = 20 mol m^{-3}

$$\text{Molar conductivity} = \Lambda_m = \frac{\kappa}{C}$$

$$= \frac{248\times 10^{-3}\ Sm^{-1}}{20\ mol\ m^{-3}} = 124\times 10^{-4}\ Sm^2mol^{-1}$$

70. (2) NaCl in solid state, does not dissociate into ions. Hence, it does not conduct electricity.

71. (4) When concentration approaches zero, the molar conductivity is known as limiting molar conductivity.

72. (3) Among the given electrolytes, $CaCl_2$ is known as 2 –1 type electrolyte.

73. (4)

$$\Lambda^\circ_{m(HAc)} = \Lambda^\circ_{H^+} + \Lambda^\circ_{AC^-} - \lambda^\circ_{H^+} + \lambda^\circ_{Cl^-} + \lambda^\circ_{AC^-} + \lambda^\circ_{Na^+} - \lambda^\circ_{Cl^-} - \lambda^\circ_{Na^+}$$

$$= \Lambda^\circ_{m(HCl)} + \Lambda^\circ_{m(NaAc)} - \Lambda^\circ_{m(NaCl)}$$

$= (425.9 + 91.0 – 126.4)\ S\ cm^2\ mol^{-1} = 390.5\ S\ cm^2mol^{-1}$

74. (3) Apply, $\kappa = \frac{1}{R}\frac{\ell}{a}$

$$\Rightarrow 50\Omega = \frac{1}{\kappa}\times\frac{2.2\times 10^{-2}m}{4.4\times\left(10^{-2}m\right)^2}$$

$\kappa = 1\Omega^{-1}\ m^{-1}$ or $0.01\Omega^{-1}\ cm^{-1}$

$$\Lambda_m = \frac{\kappa}{C}\times 1000 = \frac{0.01\times 1000}{0.5}$$

$= 20\ S\ cm^2\ mol^{-1}$ or $0.002\ S\ m^2\ mol^{-1}$

75. (1) With increase in temperature, vibration of Kernel (Cation) increases and therefore, conduction decreases and hence, resistance of the metallic conductor increases.

76. (2) 1 F deposits 108 g of Ag ($Ag^+ + e^- \rightarrow Ag$)

54 g of Ag will be deposited by $\frac{1}{108} \times 54 = \frac{1}{2}F$

3 F deposited 27 g of Al ($Al^{3+} + 3e^- \rightarrow Al$)

$\frac{1}{2}F$ will deposit $\frac{27}{3} \times \frac{1}{2} = 4.5$ g of Al.

77. (1) In $ZnSO_4$, $Zn^{2+} + 2e^- \rightarrow Zn$
(1 mol) (2 mole)

or

(2 faraday).

In $AlCl_3$, $Al^{3+} + 3e^- \rightarrow Al$
1 mol 3F

In $AgNO_3$; $Ag^+ + e^- \rightarrow Ag$
1 mol 1F

Thus; the ratio of the amounts of electricity required is 2 : 3 : 1.

78. (2) 1000 mL of $AgNO_3$ solution contains 108 g Ag

∴ 125 mL of $AgNO_3$ solution will contain

$$= \frac{108 \times 125}{1000} = 13.5\text{g Ag}$$

$Ag^+ + e^- \rightarrow Ag$

(1F = 96500) 108 g

108 g of Ag is deposited by 96500 C

∴ 13.5 of Ag is deposited by $= \frac{96500}{108} \times 13.5 = 12062.5$ C

Q = It

$$t = \frac{Q}{I} = \frac{12062.5}{241.25} = 50 \text{ sec}$$

79. (3) Electricity passed = 9.65 coulomb

The amount of silver deposited = ?

We know that 96500 coulomb electricity deposits Ag = 108g mol^{-1}

∴ 9.65 coulomb electricity deposits Ag

$$= \frac{9.65}{96500} \times 108 = 108 \times 10^{-4} \text{g} = 10.8 \text{ mg}$$

80. (3) $MnO_4^{2-} \xrightarrow[\text{[-1F]}]{\text{1 mole } e^-} MnO_4^-$

(+6) (+7)

As per the equation, for 1 mole of MnO_4^{2-}, 1F of electricity is required. Thus, for 0.1 mole of MnO_4^{2-}, 0.1 F of electricity is required.

Since, 1 F = 96500 C

∴ 0.1 F = 0.1 × 96500 C = 9650 C

Hence; 9650 C of electricity is required to completely oxidise MnO_4^{2-} to MnO_4^-

81. (3) Products of electrolysis depend on both the factors i.e., the nature of material being electrolyzed and the type of electrodes being used.

82. (4) The overall reaction for lead storage battery during discharging is:

$Pb(s) + PbO_2(s) + 2H_2SO_4(aq) \rightarrow 2PbSO_4(s) + 2H_2O(l)$

Hence; H_2SO_4 is consumed when lead storage battery is discharges.

83. (4) All the given statements about fuel cells are correct.

84. (1) At anode : $2H_2(g) + 4OH^-(aq) \rightarrow 4H_2O(\ell) + 4e^-$

At cathode : $O_2(g) + 2H_2O(\ell) + 4e^- \rightarrow 4OH^-(aq)$

Overall reaction : $2H_2(g) + O_2(g) \rightarrow 2H_2O(l)$

85. (2) Among the given options, Leclanche cell is an example of primary battery.

86. (4) Hydrogen–oxygen fuel cells are used to supply power and drinking water to astronauts in space programme.

87. (3) When iron is rusted, it is oxidised to hydrated ferric oxide.

$4Fe + 3O_2 + xH_2O \rightarrow 2Fe_2O_3 . xH_2O$

Hydrated ferric oxide (Rust)

88. (3) The chemical formula of rust is $Fe_2O_3 . xH_2O$.

89. (4) Corrosion is usually defined as the degradation of metals due to an electrochemical process. The rusting of iron, tarnishing of silver, development of green coating on copper and bronze are some of the examples of corrosion.

High Order Time Intensive MCQs

1. (1) $Cu^{2+} + 2e^- \rightarrow Cu$

$$E_{red} = E^o_{red} - \frac{0.0591}{2}\log\frac{1}{(Cu^{2+})}$$

$$E_{red} = E^o_{red} + \frac{0.0591}{2}\log(Cu^{2+})$$

↓ ↓ ↓ ↓

y c + m x

$E^o_{red} = 0.34V \Rightarrow E^o_{oxdn} = -0.34V$

$Cu \rightarrow Cu^{2+} + 2e$

$$E_{ox} = -0.34 - \frac{0.0591}{2}\log(0.1) \Rightarrow E_{ox} = -0.34 + \frac{0.0591}{2}$$

2. (4) $\alpha = \frac{\Lambda^c_m}{\Lambda^\infty_m}$; $\Lambda^\infty_m(H_2O) = \lambda^\infty_m(H^+) + \lambda^\infty_m(OH^-)$

$= 200 + 350 = 550$

$$\Rightarrow 1.9 \times 10^{-9} = \frac{\Lambda^c_m}{550 \text{ Scm}^2\text{mol}^{-1}}$$

$\Lambda^c_m = 550 \times 1.9 \times 10^{-9}$ Scm^2mol^{-1}

$= 1.045 \times 10^{-6} Scm^2mol^{-1}$

3. (4) In electrochemical series, Mg is present above Cu and hence, Cu electrode acts as cathode.

$E^o_{cell} = E^o_{Cu^{++}/Cu} - E^o_{Mg^{++}/Mg}$

$2.70 \text{ V} = 0.34 - E^o_{Mg^{++}/Mg}$

$E^o_{Mg^{++}/Mg} = -2.36$ V

4. (2) $E_{cell} = E^0_{cell} - \frac{0.059}{2}\log\frac{[Zn^{2+}]}{[Cu^{2+}]}$

$$0 = 1.10 - \frac{0.059}{2}\log\frac{[Zn^{+2}]}{[Cu^{+2}]}$$

$$-1.10 = \frac{-0.059}{2}\log\frac{[Zn^{+2}]}{[Cu^{+2}]}$$

$$\frac{-2.20}{-0.059} = \log\frac{[Zn^{+2}]}{[Cu^{+2}]} \Rightarrow 37.25 = \log\frac{[Zn^{+2}]}{[Cu^{+2}]}$$

$$\log\frac{[Zn^{+2}]}{[Cu^{+2}]} = 10^{37.2}$$

5. (2) $M \rightarrow M^{n+} + ne^-$

$$nH^+ + ne^- \rightarrow \frac{n}{2}H_2$$

$$M + nH^+ \rightarrow M^{n+} + \frac{n}{2}H_2; E^o = 0.76\ V$$

$$E_{cell} = E^o - \frac{0.0591}{n}\log\frac{0.02}{1}$$

$$0.81 = 0.76 - \frac{0.0591}{n}\log\frac{0.02}{1} \Rightarrow n = 2$$

6. (1) Since pH = 14

$[H^+] = 10^{-14}$

$K_w = [H^+][OH^-]$

$$[OH^-] = \frac{10^{-14}}{10^{-14}}$$

= 1M

$$Cu(OH)_2 \rightleftharpoons Cu^{2+} + 2OH^-$$

$K_{sp} = [Cu^{2+}][OH^-]^2$

$10^{-19} = [Cu^{2+}]1^2$

$[Cu^{2+}] = 10^{-19}$ M

$Cu^{2+} + 2e^- \rightarrow Cu$

$$E_{red} = 0.34 - \frac{0.0591}{2}\log\frac{1}{[Cu^{2+}]}$$

$$= 0.34 - \frac{0.0591}{2}\log\frac{1}{10^{-19}} \Rightarrow 0.34 - \frac{0.0591}{2}\log 10^{19}$$

$$= 0.34 - \frac{0.0591}{2}\times 19 \Rightarrow E_{red} = -0.22\ V$$

7. (3) $6 \times 10^{22}\ e^-$ deposit metal = 1.9 gm

$6 \times 10^{23}\ e^-$ deposit metal $= \frac{1.9}{6\times10^{22}} \times 6\times10^{23} = 19$ gm

(1 Faraday)

19 gm is the equivalent weight of metal.

Equivalent wt. $= \frac{\text{Atomic wt.}}{\text{n factor}}$

$$19 = \frac{57}{\text{valency}} \Rightarrow \text{valency} = 3$$

8. (3) $Cu \rightarrow Cu^{2+} + 2e^-$

63.5 g → 2F

$$16\times10^{-3}\,g \rightarrow \frac{2}{63.5}\times16\times10^{-3}\,F \Rightarrow 0.0005F$$

$2H^+ \rightarrow H_2 + 2e^-$

2F → 22.4 L H_2 liberated

$$0.0005F \rightarrow \frac{22.4}{2}\times 0.0005\,\text{mol } H_2 \text{liberated}$$

$= 5.6 \times 10^{-3} \Rightarrow 5.6$ mL at STP

9. (1) At cathode: $2H_2O + 2e^- \rightarrow H_2 + 2OH^-$

2F charge generate moles of $OH^- = 2$

0.04F charge generate moles of $OH^- = 0.04$

Moles of $OH^- = 0.04$

Volume of solution = 400mL

$$[OH^-] = \frac{0.04}{400}\times 1000 = 10^{-1}M \left[\because M = \frac{\text{No. of moles}}{\text{V(in L)}}\right]$$

pOH = –log [OH⁻] = –log[10⁻¹] = 1

pH = 13

10. (1, 3) For electrolysis of aqueous solution of $CuSO_4$:

$CuSO_4(aq) \rightarrow Cu^{2+} + SO_4^{2-}$

$H_2O \rightarrow 2H^+ + O^{2-}$

At anode $2O^{2-} \rightarrow O_2 + 2e^-$

At cathode : $Cu^{2+} + 2e^- \rightarrow Cu(s)$

11. (4) $\alpha = \frac{\Lambda_{eq.}}{\Lambda^{\infty}_{eq.}}$ where $\Lambda_{eq.} = \frac{\kappa\times1000}{N}$

$$\text{Resistivity} = \frac{1}{\kappa}$$

$$\kappa = \frac{1}{18.5}\text{ohm}^{-1}\text{cm}^{-1}$$

$$\Lambda_{eq} = \frac{\kappa\times1000}{N} = \frac{1\times1000\times49}{18.5\times15} \quad \left(N = \frac{n}{V}\right)$$

$$N = \frac{15}{49\times1} = \frac{15}{49}$$

$$\alpha = \frac{1\times1000\times49}{18.5\times15\times348} = 0.507$$

12. (4) $Zn^{2+} + 2e^- \rightarrow Zn$

$$E_{RP} = E^{\circ}_{RP} - \frac{0.0591}{2}\log\frac{1}{[Zn^{2+}]}$$

$$E_{RP} = -0.76 - \frac{0.0591}{2}\log\frac{1}{[Zn^{2+}]}$$

$$-0.82 = -0.76 - \frac{0.06}{2}\log\frac{1}{[Zn^{2+}]}$$

$$-0.06 = \frac{-0.06}{2}\log\frac{1}{[Zn^{2+}]}$$

$$2 = \log \frac{1}{[Zn^{2+}]}$$

$$\frac{1}{[Zn^{2+}]} = 10^2$$

$$[Zn^{2+}] = \frac{1}{100} = 0.01\,M$$

So, concentration of $ZnSO_4 = 0.01$ M

13. (2) At cathode

$Cu^{2+} + 2e^- \rightarrow Cu$

2F charge deposit, Cu = 1 mole

Since 2 moles (1 M, 2L of $CuSO_4$ were present initially & 1 mole Cu is deposited. So, moles left (after electrolysis) = 1

Molarity of solution left now $= \frac{1}{2} = 0.5$ M

14. (1) 'A' Conductivity = Conductance × Cell constant

$$\kappa_1 = \frac{1}{40} \times x$$

$$\text{'B' Conductivity, } \kappa_2 = \frac{1}{70} \times x$$

Conductivity of the mixture will be $\frac{1}{2}(\kappa_1 + \kappa_2)$

$\therefore$ For the mixture: $\frac{1}{2}(\kappa_1 + \kappa_2) = \frac{1}{R} \times x$

[R = resistance of the mixture]

R = 50.9 ohm

15. (2) $Zn \rightarrow Zn^{2+} + 2e^-$

$Cu^{2+} + 2e^- \rightarrow Cu$

$Zn + Cu^{2+} \rightarrow Zn^{2+} + Cu$

$$E = E^\circ - \frac{2.303RT}{nF} \log \frac{[Zn^{2+}]}{[Cu^{2+}]}$$

$$E = 1.1 - \frac{2.303RT}{2F} \log 10 \; [\therefore \log 10 = 1]$$

$$E = 1.1 - \frac{2.303RT}{2F}$$

$\Delta G = -nF\,Ecell$

$$= -2 \times F \times \left(1.1 - \frac{2.303RT}{2F}\right)$$

$$= -2 \times F \times 1.1 + 2F \times \frac{2.303RT}{2F}$$

$\Delta G = -2.2F + 2.303\,RT$

16. (3) $\alpha = \frac{\Lambda_M}{\Lambda_M^\infty} = \frac{5.2}{390.7} = 1.33\%$

17. (4) There will be no reaction as copper is less reactive than aluminum.

18. (3) According to standard reduction potential the Electrochemical series is

$Mg^{+2} / Mg = -2.37$

$H^+ / H = 0$

$Cu^{+2} / Cu = 0.34$

$Hg_2^{+2} \mid 2\,Hg = 0.79$

$Ag^+ / Ag = +0.8$

So, correct order of deposition on cathode is:

Ag, Hg, Cu, H_2, Mg

So, correct order is:

Ag, Hg, Cu

19. (2) The cell reaction is:

$H_2(g) + I_2(s) \rightarrow 2H^+(aq) + 2I^-(aq)$

$$0.7714 = 0.535 - \frac{0.0591}{2} \log \frac{[H^+]^2 [I^-]^2}{P_{H_2}}$$

pH = 3

20. (3) $16OH^- + Mn_3O_4 \rightarrow 3MnO_4^{2-} + 8H_2O + 10e^-$

For 1 mol, electrons required = 10

2 mol = 2 × 10 = 20

Charge required = 20 × F = 20 F

21. (2) The oxidation potential of Zn is higher as compared to oxidation potential of iron.

22. (3) According to Nernst equation,

$$E = E^\circ - \frac{0.059}{n} \log Q \text{ at } 25°C$$

At equilibrium, E = 0, Q = K

$$\therefore 0 = E^\circ - \frac{0.059}{n} \log K \Rightarrow E^\circ = \frac{0.059}{n} \log K$$

$$\log K = \frac{E^\circ \times n}{0.059} \Rightarrow \log K - \frac{0.295 \times 2}{0.059}$$

$$K = \text{antilog} \frac{0.295 \times 2}{0.059} \Rightarrow K = 1 \times 10^{10}$$

23. (3) For this concentration cell-

$$Cu(s) + \underset{(0.04M)}{Cu^{+2}} \rightarrow \underset{(0.01M)}{Cu^{+2}} + Cu(s)$$

$$E_{cell} = 0 - \frac{RT}{2F} \ln \frac{0.01}{0.04} = -\frac{RT}{2F} \ln (0.5)^2$$

$$E_{cell} = -\frac{RT}{F} \ln (0.5)$$

24. (4) Ammonia produced due to reaction

$MnO_2 + NH_4^+ + 1e^- \rightarrow MnO(OH) + NH_3$

Combines with Zn^{+2} ions to form $[Zn(NH_3)_4]^{2+}$

25. (3)

$$Cu^{+2} + e^- \rightarrow Cu^+, \Delta G_1$$
$$Cu^+ + e^- \rightarrow Cu, \Delta G_2$$
$$\text{On add}^n, Cu^{+2} + 2e^- \rightarrow Cu, \Delta G_3$$

$\Delta G_3 = \Delta G_1 + \Delta G_2$

$-2FE^0 = -1\ F \times 0.15 + (-1\ F \times 0.5)$

$-2FE^0 = -0.15F - 0.50\ F$

$-2FE^0 = -F\ (0.15 + 0.50)$

$E^0 = \dfrac{0.65}{2} = 0.325\ V$

26. (4) $Cu(s) + 2Ag^+(aq) \rightarrow Cu^{2+}(aq) + 2Ag(s)$

$\Delta G = -nFE°$

$\Delta G = -2.303\ RT \log K$

$\dfrac{+nFE°}{2.303RT} = \log K = \dfrac{2 \times 96500 \times 0.46}{2.303 \times 8.31 \times 298}$

$\log K_{eq} = 15.56$

$K_c = \text{antilog } 15.56 = 3.89 \times 10^{15} \approx 4 \times 10^{15}$

27. (3) Ionic mobility is fast in aqueous solution of the salts.

28. (2) $W_{metal} = \dfrac{EIt}{96500} = \dfrac{E \times 3 \times 50 \times 60}{96500}$

$E = \dfrac{96500 \times W}{60 \times 3 \times 50} = \dfrac{96500 \times 1.8}{3 \times 50 \times 60} = 19.3$

29. (1) Dilute sulphuric acid contains H^+, SO_4^{2-} and OH^- ions.

At cathode;

H^+ ion undergoes reduction by taking up electrons to form H_2

$2H^+(aq) + 2e^- \rightarrow H_2(g)$

At anode;

For maintaining a reaction, sulphate ions are not oxidised but remain in the solution at the end. The concentration of OH^- ions is too low. Thus, OH^- ions are discharged in preference over SO_4^{2-} giving off O_2 gas.

$4OH^-(aq) \rightarrow O_2(g) + 2H_2O(l) + 4e^-$

30. (3) In Daniel cell, following reaction occurs:

$Zn + Cu^{2+} \rightarrow Zn^{2+} + Cu$

At cathode ; $Cu^{2+} + 2e^- \rightarrow Cu$

At anode ; $Zn \rightarrow Zn^{2+} + 2e^-$

Electrons move from anode (Zn) to cathode (Cu) and electricity flows from Cu to Zn. Here, cations moves towards cathode copper electrode.

31. (1) Nickel - Cadmium battery

Anode - Cd ; cathode - NiO_2 , electrolyte – KOH

At anode : $Cd\ (s) + 2OH^-(aq) \rightarrow Cd(OH)_2 + 2e^-$

At cathode: $NiO_2(s) + 2H_2O(l) + 2e^- \rightarrow Ni(OH)_2(s) + 2OH^-(aq)$

$Cd(s) + NiO_2(s) + 2H_2O(l) \rightarrow Cd\ (OH)_2(s) + Ni(OH)_2(s)$

32. (2) Fe^{2+} ions does not react with dry oxygen.

33. (1) $\Lambda_m = \dfrac{\kappa}{c} = \dfrac{4.95 \times 10^{-5}\ S\ cm^{-1}}{0.001028\ mol\ L^{-1}} \times \dfrac{1000\ cm^3}{L}$

$= 48.15\ S\ cm^3 mol^{-1}$

$\alpha = \dfrac{\Lambda_m}{\Lambda_m^\circ} = \dfrac{48.15\ S\ cm^2 mol^{-1}}{390.5\ S\ cm^2 mol^{-1}} = 0.1233$

$k = \dfrac{c\alpha^2}{(1-\alpha)} = \dfrac{0.001028\ mol\ L^{-1} \times (0.1233)^2}{1 - 0.1233} = 1.78 \times 10^{-5} mol\ L^{-1}$

34. (4) $\Delta G° = -nE°F$

$Fe^{2+} + 2e^- \rightarrow Fe$ ...(i)

$\Delta G° = -2 \times F \times (-0.440\ V) = 0.880\ F$

$Fe^{3+} + 3e^- \rightarrow Fe$ ...(ii)

$\Delta G° = -3 \times F \times (-0.036) = 0.108\ F$

On subtracting equation (i) from (ii)

$Fe^{3+} + e^- \rightarrow Fe^{2+}$

$\Delta G° = 0.108\ F - 0.880\ F = -0.772\ F$

$E°$ for the reaction $= -\dfrac{\Delta G°}{nF}$

$= -\dfrac{(-0.772\ F)}{1 \times F} = +0.772V$

35. (3) $\kappa = C \times \text{Cell constant}$

$\therefore \text{Cell constant} = \dfrac{\kappa}{C} = \dfrac{0.2}{0.04} = 5\ cm^{-1}$

Correct & Incorrect MCQs

1. **(4)** In Daniell cell, when $E_{ext} = 1.1$ V, electricity is not produced because no flow of electrons takes places and no chemical reaction occur.
2. **(3)** Fluoride ion is the weakest reducing agent.
3. **(4)** Lithium has the lowest electrode potential.
4. **(1)** The electrical resistance of any object is directly proportional to its length and inversely proportional to its area of cross section. Hence, if length of any object is increased, its electrical resistance also increases.
5. **(2)** The electronic conductance depends on the nature and structure of the metal, the number of valence electrons per atom and temperature.
6. **(2)** Molar conductivity increases with decrease in concentration.
7. **(3)** When electrolytes are dissolved in water, they furnish their own ions in the solution and therefore its conductivity also increases. The conductance of electricity by ions present in the solutions is called electrolytic or ionic conductance.
8. **(4)** Metals like platinum or gold are used as inert electrode, they do not participate in the chemical reaction and acts only as source or sink for electrons.
9. **(2)** Secondary cells are rechargeable and can be used again and again.
10. **(3)** The chemical formula of rust is $Fe_2O_3 . x\ H_2O$.

Statement Based MCQs

1. **(3)** Fe is more reactive than Cu therefore, a reaction is takes place between Fe and Cu Hence, $CuSO_4$ can't be stored in iron vessel.
2. **(1)** Absolute value of E°_{red} cannot be determined because a half cell reaction cannot take place independently. Moreover we have to use a reference electrode.
3. **(1)** Salt bridge is used to connect the oxidation and reduction half cells of a galvanic cell. It also maintains electroneutrality within the internal circuit.
4. **(4)** $\frac{\ell}{A}$ = Cell constant

   The cell constant depends upon distance and area of its electrode.
5. **(1)** Moles $= \frac{It}{nF}$.

   The n factor is different for silver and copper
6. **(4)** In standard hydrogen electrode, the concentration of both the reduced & oxidised forms of hydrogen is maintained at unity which implies that the pressure of hydrogen gas is 1 bar and the concentration of hydrogen ion in the solution is 1 molar.
7. **(1)** KCl is a solution whose conductivity is known accurately at various concentrations, hence KCl solutions are generally used to determine cell constant.
8. **(3)** Inert electrodes do not participate while reactive electrode participate in the electrode reaction.
9. **(3)** Standard electrode potential of copper is 0.34 V.
10. **(1)** Both statements are correct.

Assertion & Reason MCQs

1. **(3)** Standard reduction potential of Cu^{2+}/Cu is +0.34 V.
2. **(3)** $E_{cell} = E_{cathode} - E_{anode}$

   For a positive value of E_{cell}:

   $E_{cathode} > E_{anode}$
3. **(1)** For all types of electrolytes, conductivity tends to decrease with dilution because of decrease in number of ions per unit volume.
4. **(1)** For weak electrolytes, molar conductivity increases steeply with dilution because of increase in number of ions.
5. **(4)** Mercury cell gives steady potential.
6. **(1)** Both Assertion and Reason are True and the Reason is the correct explanation of the Assertion.
7. **(1)** At $E_{cell} = 0$, reaction reaches the equilibrium. Hence, current stops flowing when $E_{cell} = 0$.
8. **(1)** Lower the reduction potential, higher is the reducing power.
9. **(3)** Na^+ ions migrate towards negative electrode, where there are two substances that can be reduced, i.e., Na^+ ions and water molecules. It is much easier to reduce water molecules than Na^+ ions, hence only H_2 gas is formed.

   Cathode reaction:

   $H_2O + 2e^- \rightarrow H_2(g) + 2OH^-$

   Anode reaction:

   $Cl^- \rightarrow \frac{1}{2}Cl_2(g) + 1e^-$
10. **(3)** Here, A is true but R is false as cations & anions have different mobilities so, current due to both will be different.

Multi Correct MCQs

1. **(1)** Manganese (Mn), Nickel (Ni), and Cadmium (Cd) are metals commonly used in battery industries.
2. **(1)** As these are acids and having equal concentration, so, now electrical conductivity depends upon their ions. Since ions are same, ultimately now, it depends on acidic strength. i.e., I, III, II

   Acidic strength increases, degree of ionisation increases and hence electrical conductivity: I > III > II.
3. **(4)** Fuel cell is used in spaceship. It is a type of galvanic cell.
4. **(4)** The conductivity of an electrolytic (ionic) solutions is influenced by several factors, including the concentration and nature of the electrolyte, as well as the nature of the solvent.

NCERT Exemplar MCQs

1. **(3)** Standard electrode potential of copper electrode can be calculated by constructing a concentration cell composed of two half cell reactions in which concentration of species on left hand side and right hand side are unity. In such case, cell potential is equal to standard electrode potential.

   $$\underbrace{Pt(s)\,|\,H_2(g,1\,bar)\,||\,H^+(aq,1M)}_{\text{Oxidation half cell reaction}}|\underbrace{Cu^{+2}(aq,1M)\,|\,Cu}_{\text{Reduction half cell reaction}}$$
2. **(3)** E_{cell} is an intensive property as it does not depend upon mass of species but Δ_rG of the cell reaction is an extensive property because this depends upon mass of species.
3. **(2)** Cell emf: The difference between the electrode potential of two electrodes when no current is drawn through the cell is called cell emf. It takes place in open type system.
4. **(4)** An inert electrode in a cell provide surface for either oxidation or for reduction reaction by conduction of electrons through its surface but does not participate in the cell reaction.
5. **(3)** If an external opposite potential is applied on the galvanic cell and increased reaction continues to take place till the opposing voltage reaches the value 1.1V.

   At this stage no current will flow through the cell because if external potential further increase then reaction will occur in opposite direction or reversible reaction.

   Hence, this works as an electrolytic cell.
6. **(3)** Solution consists of electrolytes is known as electrolytic solution and conductivity of electrolytic solution depends upon the following factors

   (i) Size of ions As size of ion size increases, the mobility of ion decreases and conductivity decreases.

$$\text{Size of ion} \propto \frac{1}{\text{ion mobility}} \propto \frac{1}{\text{conductivity}}$$

(ii) Solvation of ions : Greater the solvation of ions of an electrolyte, lesser will be the conductivity of the solution.

$$\text{Solvation : solvation} \propto \frac{1}{\text{conductivity}}$$

(iii) Temperature of medium: Conductivity of solution increases with increase in temperature.

7. **(2)** Standard reduction potential of chromium has higher negative value hence, most powerful reducing agent is chromium.

8. **(3)** Higher the positive value of standard reduction potential of metal ion means higher will be its oxidising capacity.

   Since, $E^\circ MnO_4^- / Mn^{2+}$ has value equal to 1.51 V hence, it is the strongest oxidising agent.

9. **(2)**

Species (ions)	SRP values
Mn^{2+}	1.51 V
Cl^-	1.36 V
Cr^{3+}	1.33 V
Cr	– 0.74 V

On moving top to bottom, SRP value decreases from positive to negative value which will increase the reducing capacity because top to bottom, metallic character increases.

10. **(4)** $E^\circ_{MnO_4^-/Mn^{2+}}$ has most + ve value equal to 1.51 V which is highest among given four choices. So, Mn^{2+} is most stable ion in its reduced form.

11. **(1)** $E^\circ_{Cr^{3+}/Cr}$ has most – ve value equal to – 0.74 among given four choices. So, Cr^{3+} is the most stable oxidized species.

12. **(3)** The quantity of charge to obtain one mole of aluminium from Al_2O_3 is equal to number of electron required to reduce Al_2O_3 to Al.

$$Al^{3+}(aq) \xrightarrow{+3e} Al(s)$$

Hence, total 3 F is required.

13. **(4)** Cell constant is defined as the ratio of length of object and area of cross section.

$$\text{Cell constant} = \frac{l}{A}$$

Since, l and A remain constant for any particular object hence, value of cell constant always remains constant.

14. **(1)** The electrode reactions are as follows:

At cathode

$$PbSO_4(s) + 2e^- \rightarrow Pb^{+2}(s) + SO_4^{2-}(aq) \text{ (Reduction)}$$

At anode

$$PbSO_4(s) + 2H_2O \rightarrow PbO_2(s) + SO_4^{2-} + 4H^+ + 2e^-(aq)$$

(Oxidation)

Overall reaction

$$2PbSO_4(s) + 2H_2O \rightarrow Pb(s) + PbO_2(s) + 4H^+(aq.) + 2SO_4^{2-}(aq.)$$

15. **(4)** In case of electrolysis of aqueous NaCl, oxidation reaction occurs at anode as follows:

$$Cl^-(aq) \rightarrow \frac{1}{2}Cl_2(g) + e^- \quad E^\circ = 1.36\,V$$

$$2H_2O(l) \rightarrow O_2(g) + 4H^+(aq) + 4e^- \quad E^\circ_{cell} = 1.23\,V$$

But due to lower E°_{cell} value, water should get oxidised in preference of Cl^- (aq).

However, the actual reaction taking place in the concentrated solution of NaCl is (4) and not (2) *i.e.*, Cl_2 is produced and not O_2.

This unexpected result is explained on the basis of the concept of 'over voltage', i.e., water needs greater reduction voltage for oxidation to O_2 (as it is kinetically slow process) than that needed for oxidation of Cl^- ions to Cl_2. Thus, the correct option is (4) not (2).

16. **(2, 4)** "Lesser the E° value of redox couple, higher the reducing power"

$$Cu^{2+}/Cu \quad E^\circ = 0.34\,V$$

$$2H^+ + 2e^- \rightarrow H_2 \quad E^\circ = 0.00\,V$$

Since, $2H^+/H_2$ has lesser SRP than Cu^{2+}/Cu redox couple. Therefore,

(i) This redox couple is a stronger oxidising agent than H^+/H_2.

(ii) Cu can't displace H_2 from acid.

Hence, (2) and (4) are correct.

17. **(2, 3)** Left side of cell reaction represents oxidation half cell *i.e.*, oxidation of Mg and right side of cell represents reduction half cell reactions i.e., reduction of copper.

(i) Reduction occurs at cathode means Cu gets reduced.

(ii) Oxidation occurs at anode means that Mg gets oxidised.

(iii) Whole cell reaction can be written as:

$$\overbrace{Mg + Cu^{2+} \rightarrow Mg^{2+}}^{\text{Oxidation}} + Cu \quad (\underbrace{Cu^{2+} \rightarrow Cu}_{\text{Reduction}})$$

Hence, options (2) and (3) both are correct choices.

18. **(1, 2)** Conductivity of electrolytic solution is due to mobility of ions in the solution. This type of conductance is known as ionic conductance. Conductivity of these type of solutions depend upon

(i) the nature of electrolyte added

(ii) size of the ion produced and their solvation

(iii) electrolyte concentration

(iv) nature of solvent and its viscosity

(v) temperature

While power of source or distance between electrodes has no effect on conductivity of electrolyte solution.

Hence, options (1) and (2) are the correct choices.

19. **(1, 3)** For electrolysis of aqueous solution of $CuSO_4$:

$$CuSO_4(aq) \rightarrow Cu^{2+} + SO_4^{2-}$$

$$H_2O \rightarrow 2H^+ + O^{2-}$$

At anode $2O^{2-} \rightarrow O_2 + 2e^-$

At cathode $Cu^{2+} + 2e^- \rightarrow Cu(s)$

20. (2) If coating is broken, iron will corrode faster if coated with tin. As iron lose electron more easily than tin due to low reactivity of tin.

This is because their standard oxidation potentials are:

$E^o_{Zn, Zn^{2+}} = +0.76$ V, $E^o_{Fe, Fe^{2+}} = +0.44$ V, $E^o_{Sn, Sn^{2+}} = +0.14$ V

21. (1, 2) As we know that, conductance is reciprocal of resistance. Also, conductivity is reciprocal of resistivity.

$$\kappa = \frac{1}{\rho}$$

$$R = \rho\frac{l}{A}$$

$$\rho = \frac{R.A}{l} \Rightarrow \kappa = \frac{1}{\left(\frac{R.A}{l}\right)}$$

$$\kappa = \frac{1}{R}.\frac{l}{A} = \frac{1}{R}\times G^* = \frac{G^*}{R}$$

Hence, options (1) and (2) are the correct choices.

22. (1, 3) Molar conductivity is the conductivity due to ions furnished by one mole of electrolyte in solution. Molar conductivity of ionic solution also depends upon.

(i) Temperature: Molar conductivity of electrolyte solution increases with increase in temperature.

(ii) Concentration of electrolytes in solution: when concentration of electrolyte increases, molar conductivity decreases.

$$\therefore \Lambda_m = \frac{\kappa}{c}$$

23. (4) $Fe^{3+}/Fe^{2+} = +0.77$

$I_2(s)/I^- = +0.54$;

$Cu^{2+}/Cu = +0.34$; $Ag^+/Ag = 0.80$ V

Ag^+ is the strongest oxidising agent as E° of the redox couple Ag^+/Ag is the most positive, i.e., 0.80 V.

NEET & AIIMS Past Year MCQs

1. (1) The Gibbs free energy change is calculated using the formula:

$$\Delta_r G^{\ominus} = -nFE^{\ominus}_{cell}$$

For the given reaction, n = 2

$\therefore \Delta_r G° = -2 \times 96487 \times 0.32$

$= -61751.68$ J mol^{-1}

$= -61.751$ kJ mol^{-1}

2. (2) Corrosion gradually covers the surfaces of metal objects with oxides or salts of the metal. Examples include the rusting of iron, tarnishing of silver, and the formation of a green coating on copper and bronze.

The production of H_2 through the electrolysis of water is not an example of corrosion.

3. (3)

A) $H_2O \rightarrow \frac{1}{2}O_2 + 2e^- + 2H^+$ (2F)

B) $MnO_4^- + 5e^- + 8H^+ \rightarrow Mn^{2+} + 4H_2O$ (5F)

C) Molten $CaCl_2 \rightarrow Ca^{2+} + 2e^-$ for 1.5 mole Ca^{2+} (3F)

D) $Fe^{2+} \rightarrow Fe^{3+} + e^-$ (1F)

4. (4) $W = Z \times i \times t$

$$= \frac{63}{96487 \times 2} \times 9.6487 \times 100 = 0.315\text{g}$$

5. (4) Conductivity = Conductance × Cell constant

Conductivity = (1/resistance)

Cell constant = Conductivity × Resistance

Cell constant = 0.0210 × 60 = 1.26 cm^{-1}

6. (2) Free energy is extensive property, and E^o_{Cell} is an intensive property $\Delta_r G$ is depends on the 'n' which is number of electron transferred in the reaction $\Delta_r G = -nFE$

7. (2) The first half cell reaction is given as,

(+7 → +2) Reduction occurs

$$MnO_4^- + 8H^+ + 5e^- \rightarrow Mn^{2+} + 2H_2O$$

Here oxidation state of Mn in reactant is +7 and in product, is +2. It means reduction is occurring here. Therefore, it will take place at cathode.

Second half cell reaction is given as:

$$\frac{1}{2}O_2 + 2H^+ + 2e^- \rightarrow 2H_2O$$

(0 → −2) Oxidation occurs

Therefore, it will take place at anode. From both the half cell reactions, the complete balanced reaction will be:

$$2MnO_4^- + 6H^+ \rightarrow 2Mn^{+2} + 3H_2O + \frac{5}{2}O_2$$

$E°_{cell} - E°_{cathode} - E°_{anode}$

$= 1.510 - 1.223 = 0.287$ V

Positive $E°_{cell}$ value suggests that the given hall cell reactions will occur.

8. (None) $Ni(s) + 2Ag^+ (0.001\text{ M}) \rightarrow Ni^{+2} (0.001\text{ M}) + 2Ag(s)$

$$E_{cell} = E^o_{cell} - \frac{0.059}{n}\log\frac{[Ni^{+2}]^1}{[Ag^+]^2}$$

$$E_{cell} = 10.5 - \frac{0.059}{2}\log\frac{10^{-3}}{(10^{-3})^2}$$

$$= 10.5 - \frac{0.059}{2}\log 10^{+3}$$

$$= 10.5 - \frac{0.059}{2} \times 3\log 10$$

$$= 10.5 - \frac{0.059}{2} \times 3$$

= 10.4115 V

(Calculated answer is not given in options)

9. (1) Given,

$\Lambda^\circ_m(NaCl) = 126.45$ S cm² mol⁻¹

$\Lambda^\circ_m(HCl) = 426.16$ S cm² mol⁻¹

$\Lambda^\circ_m(CH_3COONa) = 91.0$ S cm² mol⁻¹

According to Kohlrausch law of independent migration of ions:

$\Lambda^\circ_m(CH_3COONa) =$

$\Lambda^\circ_m(CH_3COONa) + \Lambda^\circ_m(HCl) - \Lambda^\circ_m(NaCl)$

$91.0 \text{ S cm}^2 \text{ mol}^{-1} + 426.16 \text{ S cm}^2 \text{ mol}^{-1} - 126.45 \text{ S cm}^2 \text{ mol}^{-1}$

$= 390.71 \text{ S cm}^2 \text{ mol}^{-1}$

Hence, the molar conductance of CH_3COOH at infinite dilution = 390.71 S cm² mol⁻¹

10. (2) $\Lambda_m = 20 \text{ S cm}^2 \text{ mol}^{-1}$

$\Lambda^\circ_{m\, CH_3COOH} = \lambda^\circ_{CH_3COO^-} + \lambda^\circ_{H^+}$

$= 50 + 350 = 400 \text{ S cm}^2 \text{ mol}^{-1}$

$$\alpha = \frac{\Lambda_m}{\Lambda^\circ_m} = \frac{20}{400} = \frac{1}{20}$$

$$K_a = \frac{C\alpha^2}{1-\alpha} \simeq C\alpha^2 = 7\times10^{-3}\times\left(\frac{1}{20}\right)^2 = 7\times10^{-3}\times\frac{1}{4}\times10^{-2}$$

$K_a = 1.75\times10^{-5} \text{ mol L}^{-1}$

Dissociation constant = 1.75×10^{-5} mol L⁻¹

11. (1) $E^0_{cell} = \frac{0.059}{2}\log K$

$$E^0_{cell} - E_{A^{2+}/A} = \frac{0.059}{2}\log K$$

$$E^0_{B^+/B} - 0.34 = \frac{0.059}{2}\times 15.6 \Rightarrow E^0_{B^+/B} = 0.80 \text{ V}$$

12. (1) Sulphate is consumed from electrolyte due to formation of lead sulphate thus, density decreases.

13. (2)

Chemical Kinetics

NCERT Topic-wise MCQs

1. (3) Rate $= \frac{\Delta x}{\Delta t} \Rightarrow$ Rate $\propto \frac{1}{\text{time}}$

∴ As the reaction proceeds with time, rate decreases.

2. (3) $2NH_3 \rightarrow N_2 + 3H_2$

$$-\frac{1}{2}\frac{d[NH_3]}{dt} = \frac{d[N_2]}{dt} = \frac{1}{3}\frac{d[H_2]}{dt}$$

$$\frac{1}{2}\times K_1 = K_2 = \frac{K_3}{3}$$ [–ve sign shows decrement in reactant concentration]

$$\frac{3}{2}K_1 = 3K_2 = K_3 \Rightarrow 1.5K_1 = 3K_2 = K_3$$

3. (1) $\text{Rate} = \frac{-1}{2}\frac{d[x]}{dt} = \frac{-1}{2}\frac{[2-3]}{5} = 0.1 \text{ mol L}^{-1}\text{ min}^{-1}$

4. (2) $\text{Rate} = \frac{1}{4}\frac{d[NO_2]}{dt} = \frac{0.0125}{4} = 0.0031 \text{ mol L}^{-1}\text{s}^{-1}$

5. (4) $2H_2S + O_2 \rightarrow$ products

$\text{Rate} = [p_{H_2S}]^2[p_{O_2}]$ sssss

$= [3p_{H_2S}]^2[3p_{O_2}] = 27[p_{H_2S}]^2[p_{O_2}] = 27$ (initial rate)

6. (4) $\text{Rate} = \frac{\text{Conc}}{t}\times\frac{1}{(\text{Conc})^2} = \frac{\text{mol L}^{-1}}{s}\times\frac{1}{\text{mol}^2\text{L}^{-2}}$

$= \text{mol}^{-1}\text{L s}^{-1}$

7. (2) Rate = k[A]

1.8×10^{-3} mol L⁻¹ min⁻¹ = k × 0.3 mol L⁻¹

$$\therefore k = \frac{1.8\times10^{-3}}{0.3\times60 \text{ sec}} = 1\times10^{-4}\text{s}^{-1}$$

8. (1) $\text{Rate} = \frac{k[I^-][OCl^-]}{[OH^-]}$

order: 1 + 1 – 1 = 1

9. (4) The order of reaction is decided by mechanism of reaction as well as relative concentration of reactants.

10. (1) For a zero order reaction, rate = k $[A]^0$ = k
Unit = mol L⁻¹ time⁻¹

11. (2) Molecularity of a reaction is the number of molecules of the reactants taking part in a single step of a reaction.

12. (3) Sum of power of concentration in terms of the rate law is the order of reaction.

Rate = k $[A]^c[B]^d \Rightarrow$ order = c + d

13. (3) $\text{Rate} = \frac{\text{Conc}}{\text{time}}\times\frac{1}{(\text{Conc})^3} = \frac{\text{mol L}^{-1}}{s}\times\frac{1}{\text{mol}^3\text{L}^{-3}}$

Rate = $\text{mol}^{-2}\text{L}^2\text{s}^{-1}$

The given unit of rate constant indicates that reaction is of third order.

14. (3) In actual, unit of first order rate constant is independent of unit of concentration. The unit of k is s⁻¹

15. (1) Order of reaction can be 0, 1, 2, 3, or even a fractional number.

16. (1) Rate = k $[N_2O_5]$
$= 6.2 \times 10^{-4} \times 1.25 = 7.75 \times 10^{-4}$ mol L⁻¹ s⁻¹

17. (4) $$k = \frac{2.303}{20}\log\frac{a}{\frac{a}{4}} = \frac{2.303}{20}\log\frac{4a}{a}$$

$$k = \frac{2.303}{20}[\log 4] = \frac{2.303}{20}[0.60]$$

$$k = \frac{2.303\times0.60}{20}$$

If Initial concentration is reduced by a factor of 1/16. Then

$$k = \frac{2.303}{t}\log\frac{a}{a/16}$$

$$t = \frac{2.303}{k}\log\frac{16a}{a}$$

$$t = \frac{2.303\times 20}{2.303\times 0.60}\log\frac{a}{\frac{a}{16}} = \frac{20}{0.60}\log\frac{16a}{a}$$

$$t = \frac{20}{0.60}[\log 16] = \frac{20}{0.60}[1.20]$$

$$t = 40\,\text{min}$$

18. (4) According to the rate equation given, the reaction is of zero order reaction,

∴ General eq^n for zero order reaction,

$a = a_0 - kt$ ...(1)

When $t = t_{1/2}$ & $a = \frac{a_0}{2}$

eq^n (1) becomes

$$\frac{a_0}{2} = a_0 - kt_{1/2} \Rightarrow t_{1/2} = \frac{a_0}{2k}$$

In all options, initial concentration is represented by a.

19. (1) It is independent of initial concentration as $t_{1/2} = \frac{0.0693}{k}$ for first order reaction.

20. (1) $t = \frac{2.303}{k}\log\frac{a}{a-x}$

$$t = \frac{2.303}{2.303\times 10^{-2}}\log\frac{x}{\frac{x}{10}} = 100\log\frac{10x}{x}$$

$t = 100 \times [\log 10] = 100[1]$

$t = 100 \times 1 = 100$ sec

21. (2) $K = \frac{2.303}{t}\log\frac{a}{a-x}$

$t = t_{1/2}$, 50% reaction completed

$$K = \frac{2.303}{t_{1/2}}\log\frac{a}{a-\frac{a}{2}} = \frac{2.303}{t_{1/2}}\log\frac{2a}{a}$$

$$K = \frac{2.303\times 0.30}{t_{1/2}} = \frac{0.693}{t_{1/2}}$$

∴ $t_{1/2}$ is independent of initial concentration in first order reaction.

22. (1) Integrated rate law for a zero order reaction is

$$k = \frac{R_o - R}{t}$$

For half-life, $t = t_{1/2}$, $R = \frac{R_o}{2}$

The rate constant at $t_{1/2}$ becomes,

$$k = \frac{R_o - \frac{R_o}{2}}{t_{\frac{1}{2}}}$$

$$\boxed{t_{\frac{1}{2}} = \frac{R_o}{2k}}$$

Therefore, it is directly proportional to initial concentration.

23. (4) Let $R_o = 100$

given x = 75

$$k = \frac{2.303}{t}\log\frac{R_o}{R_o - x}$$

$$k = \frac{2.303}{30}\log\frac{100}{25}$$

$$k = \frac{2.303}{30}\times 0.602$$

For 93.75% reaction, t is given by

$$t = \frac{2.303}{k}\log\frac{a}{a-x}\quad\begin{bmatrix}a = 100\\ x = 93.75\end{bmatrix}$$

$$t = \frac{2.303\times 30}{2.303\times 0.602}\log\frac{100}{100-93.75}$$

$$t = \frac{30}{0.602}\log 16 = \frac{30\times 1.20}{0.602} = 59.8 \approx 60\,\text{min}$$

24. (1) $k = \frac{2.303}{t}\log\frac{a}{a-x}$

$$k = \frac{2.303}{60}\log\frac{100}{100-60} = \frac{2.303}{60}\log\frac{100}{40}$$

$$k = \frac{2.303}{60}\log 2.5 = \frac{2.303\times 0.397}{60}$$

$$t - \frac{2.303}{k}\log\frac{a}{a-x}$$

$$t = \frac{2.303\times 60}{2.303\times 0.39}\log\frac{100}{50}$$

$46.15 \approx 45\,\text{min}$

25. (3) Rate = $k[A]^2 = x$

If conc.is tripled i.e., A' = [3A]

Rate' = $k\,[3A]^2 = k.9[A]^2 = x'$

∴ $\frac{x'}{x} = 9$ i.e., becomes nine times (9x).

26. (1) For zero order reaction,

$$[R] = [R]_0 - kt$$

For completion of reaction,

$$[R] = 0,\ t = \frac{[R_0]}{k}$$

27. (3) $k = \frac{2.303}{10}\log\frac{100}{100-20}$

$k = \frac{2.303}{10}\log\frac{100}{80}$

$k = \frac{2.303}{10}\log\frac{10}{8} = \frac{2.303}{10}[\log 10 - \log 8]$

$k = \frac{2.303}{10}[1-0.90]$

$k = \frac{2.303}{10}\times 0.1 = 0.0233\,\text{min}^{-1}$

28. (3) $k = \frac{2.303}{t}\log\frac{R_o}{R}$

$k = \frac{2.303}{200}\log\frac{400}{25}$

$= 0.01386\ s_1^{-1} = 1.386\times10^{-2}s^{-1}$

29. (3) $k = \frac{2.303}{t}\log\frac{R_o}{R}$

$t = \frac{2.303}{6.909}\log\frac{100}{25} = \frac{1}{3}\log 4$

$t = \frac{1}{3}\log 2^2 \Rightarrow \frac{2}{3}\log 2$

30. (1) $t_{1/2} = \frac{0.693}{k_1}$

31. (3) $R \to P \quad \text{Rate} = \frac{-d[R]}{dt} = k[R]$

or $\frac{-d[R]}{[R]} = kdt$ ...(i)

$\int\frac{d[R]}{[R]} = -k\int dt$

$ln[R] = -kt + C$ ...(ii)

At $t = 0$, $[R] = [R]_0$

$ln[R_0] = C$

Put the value of C in eq. (ii)

$ln[R] = -kt + ln[R_0]$

$kt = ln[R_o] - ln[R]$

$k = \frac{1}{t}ln\left[\frac{R_o}{R}\right]$

$k = \frac{1}{t}ln\frac{a}{a-x}$

32. (3) $t_{1/2} = 14$ sec , $N = \frac{N_o}{8} \to N_o = 8N$

The expression for finding the order of reaction from initial concentration is given as,

$\frac{N}{N_o} = \left(\frac{1}{2}\right)^n$

where, n = order of reaction

$\frac{N}{8N} = \left(\frac{1}{2}\right)^n \Rightarrow \left(\frac{1}{2}\right)^3 = \left(\frac{1}{2}\right)^n \Rightarrow n = 3$

$t = n\,t_{1/2} = 3\times 14 = 42$ sec

33. (1) $k = \frac{2.303}{t}\log\frac{R_0}{R} \quad [\because R = [A]-[B]\ \&\ R_o = [A]]$

$k = \frac{2.303}{10}\log\frac{0.2}{0.2-0.18}$

$k = \frac{2.303}{10}\log\frac{0.2}{0.02} = \frac{2.303}{10}\times 1 = 0.2303$

34. (1) $t_{7/8} = \frac{2.303}{k}\log\frac{a}{a-\frac{7}{8}a} = \frac{2.303}{k}\log\frac{a}{\frac{a}{8}} = \frac{2.303}{k}\log 2^3$

$t_{1/2} = \frac{2.303}{k}\log\frac{a}{\left(a-\frac{a}{2}\right)} = \frac{2.303}{k}\times\log 2$

$\therefore \quad \frac{t_{7/8}}{t_{1/2}} = \frac{\frac{2.303}{k}\log 2^3}{\frac{2.303}{k}\log 2} = 3$

35. (3) $t_{1/2} = \frac{a}{2k}$ ...(i)

$T = 2\times t_{1/2}$ ...(ii)

From (i) $t_{1/2} = \frac{1}{2\times 0.001} = 500$ sec, from (ii) $T = 1000$ sec

36. (2) Hydrolysis of an ester is pseudo first order reaction.

37. (4) Combination of H_2 and Br_2 is a second order reaction

38. (2) A bimolecular reaction becomes kinetically of first order and known as pseudo first order reaction. Pseudo first order reaction is when one of the reacting species in bimolecular reaction is in large excess, then the reaction becomes first order.

39. (3)

$$\underset{\text{Carboxylic acid}}{R-\overset{\overset{O}{\|}}{C}-OH} + \underset{\text{Alcohol}}{HO-R^1} \xrightarrow{H^+} \underset{\text{Ester}}{R-\overset{\overset{O}{\|}}{C}-OR^1} + H_2O$$

Rate of reaction = k [carboxylic acid] [alcohol]

So; order of reaction = 2

40. (4) $H_2(g) + Br_2(g) \to 2HBr(g)$

The molecularity of the reaction is 1 + 1 = 2

Rate of reaction = $k[H_2][Br_2]^{1/2}$

Order of reaction = $1 + \frac{1}{2} = \frac{3}{2}$

Unit of k = $mol^{-1/2}L^{1/2}sec^{-1}$

Hence, correct option is (4).

41. (4) The difference in energy of intermediate complex and the average energy of reactants and products is called activation energy.

42. (1) $k = Ae^{-E_a/RT}$ gives Arrhenius equation.

43. (4) The equilibrium is attained faster in case of reaction in which it lowers the activation energy by the use of catalyst.

44. (4) $k = Ae^{-E_a/RT}$

$\ln k = \ln A - E_a/RT$

$2.303 \log k = 2.303 \log A - \frac{E_a}{RT}$

$\log k = \log A - (35000/2.303 \times 2 \times 573)$

$\left[\because R = 2 \text{cal mol}^{-1}\text{K}^{-1}\right]$

$\log k = 11.16 - 13.26 = -2.1$

$k = 7.94 \times 10^{-3}\ s^{-1}$

45. (3) A catalyst accelerate rate of reaction by bringing down the activation energy.

46. (2) For 10° rise in temperature n = 1

So; rate $= (2)^n = (2)^1 = 2$

When temperature is increased from 10°C to 100°C, change is = 100 – 10 = 90°C

n = 9

rate $= (2)^9 = 512$ times

47. (4) $k_1 = 10^{16}.e^{-2000/T}$ $\quad k_2 = 10^{15}.e^{-1000/T}$

$k_1 = k_2$

$10^{16}.e^{-2000/T} = 10^{15}.e^{-1000/T} \Rightarrow \frac{10^{16}}{10^{15}} \cdot e^{\frac{-2000}{T}} = e^{\frac{-1000}{T}}$

$\ln 10 + \ln\left(\frac{-2000}{T}\right) = \ln\left(\frac{-1000}{T}\right)$

$\log 10 - \frac{2000}{2.303T} = \frac{-1000}{2.303T}$

$1 = \frac{-1000}{2.303T} + \frac{2000}{2.303T}$

$1 = \frac{1000}{2.303T} \Rightarrow T = \frac{1000}{2.303} K$

48. (3) Activation energy of forward reaction = a

Threshold energy = b

$\therefore$ Heat of reaction, $\Delta H_r = c = \Delta H°$

$\therefore$ Reaction is exothermic

49. (1) Arrhenius equation is $k = Ae^{-E_a/RT}$

On taking log on both sides,

we get, $\log k = \log A - \frac{E_a}{2.303RT}$

When log k is plotted with 1/T, the slope of the reaction will be $\frac{-E_a}{2.303R}$ and intercept given by log A

50. (1) According to Arrhenius, with rise in temperature the rate of reaction also increases because by the rise in temperature, the number of activated reactant molecules increases.

51. (2) According to Arrhenius equation, activation energy (E_a) and rate constants (k_1 and k_2) are related as in-

$$\log \frac{k_2}{k_1} = \frac{E_a}{R}\left(\frac{1}{T_1} - \frac{1}{T_2}\right) = \frac{-E_a}{R}\left(\frac{1}{T_2} - \frac{1}{T_1}\right)$$

52. (3)

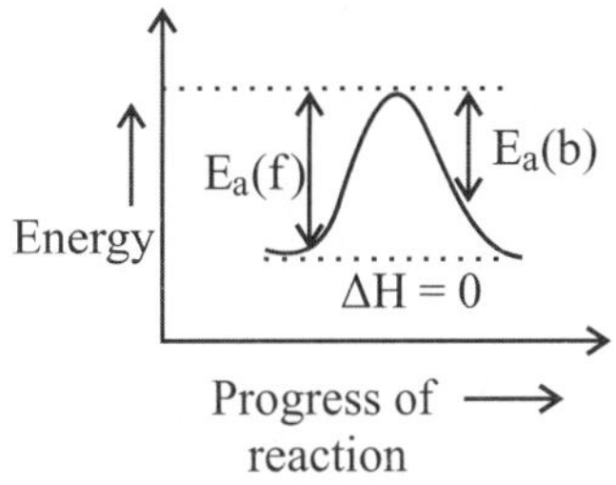

Energy profile diagram for a reaction. From the figure, it is clear that

$(E_a)_f = (E_a)_b$

$\therefore \Delta H = 0$

53. (2) The effect of temperature on reaction rate is shown by Arrhenius equation $k = Ae^{-E_a/RT}$

54. (2) Temperature coefficient $= \frac{k_t + 10}{k_t}$

$2 = \frac{10^{-3}}{k_t} \Rightarrow k_t = \frac{10^{-3}}{2} = \frac{10^{-4} \times 10}{2}$

$k_t = 5 \times 10^{-4}$

55. (4) $N_2(g) + 3H_2(g) \rightleftharpoons 2NH_3(g) + 22\,\text{kcal}$

$\therefore$ The activation energy for the forward reaction = 50 kcal

$\therefore$ The activation energy for the backward reaction = 50 + 22 = 72 kcal

56. (2) Reaction having high E_a are slow reactions and reactions having low E_a are fast reactions.

If E_a is high, then the number of effective collisions will be small and the reaction will be slow.

57. (3) $k_1 = A_1 e^{-E_{\alpha_1}/RT}$

$k_2 = A_2 e^{-E_{\alpha_2}/RT}$

$\frac{k_1}{k_2} = \left(\frac{A_1}{A_2}\right) e^{\frac{-E_{\alpha_1} + E_{\alpha_2}}{RT}}$

Now, $\frac{A_1}{A_2}$ = Constant and $E_{\alpha_2} = 2E_{\alpha_1}$

$\frac{k_1}{k_2} = Ae^{\frac{-E_{\alpha_1} + 2E_{\alpha_1}}{RT}}$

or $\frac{k_1}{k_2} = Ae^{E_{\alpha_1}/RT}$ or $k_1 = Ak_2 e^{E\alpha_1/RT}$

58. (2) $k = Ae^{-E_a/RT}$

$\log k = \log A - \frac{E_a}{RT \times 2.303}$

$\log k - \log A = \frac{-E_a}{RT}$

$$\log\frac{k}{A}=\frac{-E_a}{2.303RT}=\frac{-2.303RT}{2.303RT}$$

$$\frac{k}{A}=\text{anti}\log -1=1\times10^{-1}=0.1$$

59. (3) Large E_a and low T has smallest rate constant.

60. (2) Lesser the activation energy, more will be the rate of the reaction.

61. (3) Activation energy is inversely proportional to rate of the reaction.

62. (2) A positive catalyst increases the rate of the reaction by decreasing the activation energy.

63. (2) Reaction which have high E_a values are slow reactions because number of effective collisions will be small and reaction will become slow.

64. (4) In Arrhenius equation, the fraction

$$\text{Fraction of molecules}=e^{\frac{-E_a}{RT}}=\frac{-E_a}{10^{2.303RT}}=10^{\frac{-E_a}{2.303\times R\times 400}}$$

$$=10^{\frac{-E_a}{921.2R}}$$

65. (2) Arrhenius equation, $k=Ae^{-E_a/RT}$, Given equation is $k=(4.5\times10^{11}s^{-1})\,e^{-28000\,K/T}$

Comparing both the equations, we get

$$-\frac{E_a}{RT}=-\frac{28000\,K}{T}$$

$E_a=28000K\times R=28000\,K\times 8.314\,JK^{-1}mol^{-1}$

$=232.79\ kJ\ mol^{-1}$

66. (2) $\because k=Ae^{-E_a/RT}$ ...(1)

as $T\to\infty$, eq^n (1) becomes

$k=Ae^{-Ea/R\times\infty}\Rightarrow k=Ae^{-0}$

$\Rightarrow k=A$ as $T\to\infty$

$k=6.0\times10^{14}\ s^{-1}$

67. (1) Threshold and activation energies depend on the nature of the reactants.

68. (2) Collision frequency increases as the concentration of reactants increase, the probability of collisions increase.

69. (4) If $E_a=0$

$\because k=Ae^{-E_a/RT}$

and $k=Ae^0=A$

Hence, k becomes independent of T.

70. (3) According to the collision theory, only a small fraction of collisions is effective in bringing about the chemical reaction and the rest of the collisions are ineffective.

For effective collision (to yield product) the colliding molecules must have more than or equal to certain minimum amount of energy called threshold energy.

Threshold energy = Activation energy + average K.E. of molecules.

71. (3) Lower the activation energy, faster is the reaction

72. (2) Apart from energy considerations, the colliding molecules should have proper orientation for effective collisions.

73. (4) Only those collisions in which molecules collide with sufficient energy and proper orientation are effective collisions. Rest of the molecule collides and bounce back.

High Order Time Intensive MCQs

1. (3)

	$A_{(g)}$	$+\ 2B_{(g)}$	$\rightarrow C_{(g)}$	$+\ D_{(g)}$
Initial pressure at t = 0	0.60	0.80	0	0
Pressure at t = t	0.60-P'	0.80-2P'	P'	P'
Final pressure	0.40	0.40	0.20	0.20

Since it is an elementary reaction. Therefore,

Rate of reaction = $k[A]\,[B]^2$

Rate of reaction (initial) = $k[0.60]\,[0.80]^2$ ......(1)

Rate of reaction (final) = $k\,[0.40]\,[0.40]^2$ ....(2)

$$\frac{\text{Rate}_{\text{final}}}{\text{Rate}_{\text{initial}}}=\frac{(0.40)(0.40)^2}{(0.60)(0.80)^2}=\frac{1}{6}$$

2. (4) Rate of formation of B is

$$\frac{d[B]}{dt}=k_1[A]-k_2[B]$$

Given, $\frac{d[B]}{dt}=0$

$0=k_1[A]-k_2[B]$

$k_2[B]=k_1[A]$

$\left(\frac{k_1}{k_2}\right)[A]=[B]$

3. (2) The graph is of second order. For 2nd order reaction:

$$kt=\frac{1}{a-x}-\frac{1}{(a)}$$

$kt=(a-x)^{-1}-(a)^{-1}$

$(a-x)^{-1}=kt+(a)^{-1}$

↓

$y=mx+c$

Slope = k = 0.5 (given)

$c=OA=\frac{1}{a}=2\Rightarrow a=0.5$

Initial rate of reaction = $k\,(a)^2=0.5\times(0.5)^2$

$=0.125\ mol\ ltr^{-1}\ min^{-1}$

4. (3) For 1st order, $t=\frac{2.303}{k}\log\frac{[A_0]}{[A]}$

$$t_{90\%}-\frac{2.303}{k}\log\frac{100}{10}$$

$$t_{50\%}=\frac{2.303}{k}\log\frac{100}{50}$$

$$\frac{t_{90\%}}{t_{50\%}}=\frac{\log 10}{\log 2}=\frac{1}{0.301}=3.32$$

5. (2) $R \rightarrow P$

$Rate_1 = k[R]^1$, $0.04 = k[R]^1_{initial}$

$Rate_2 = k[R]^1$, $0.03 = k[R]^1_{final}$

$$\frac{4}{3} = \frac{(R)_{initial}}{(R)_{final}}$$

Now, apply integrated equation for I[st] law for the time gap of 10 second (i.e., between 10 seconds and 20 seconds)

$$kt = 2.303 \log \frac{[R]_0}{[R]}$$

$$k \times 10 = 2.303 \log \frac{4}{3}$$

$$k = \frac{2.303}{10} \times 0.1250$$

Half life for first order reaction is given as,

$$t_{1/2} = \frac{0.693}{k} = \frac{0.693 \times 10}{2.303 \times 0.1250} = 24.1 \sec$$

6. (4) $H_2O_2 \rightarrow H_2O + \frac{1}{2}O_2 \uparrow$

$$2 \times t_{\frac{1}{2}} = 50 \min$$

$$t_{\frac{1}{2}} = 25 \min$$

$$k = \frac{0.693}{t_{\frac{1}{2}}}$$

$$k = \frac{0.693}{25}$$

Rate of decomposition of H_2O_2

$= k[H_2O_2]$

$= \frac{0.693}{25} \times 0.05$

$= 1.39 \times 10^{-3}$ mol/min

Rate of formation of $O_2 = \frac{1}{2} \times 1.39 \times 10^{-3}$

$= 6.93 \times 10^{-4}$ mol / min

7. (3) $k_1 = Ae^{-Ea_1/RT}$; $k_2 = Ae^{-Ea_2/RT}$

As it is given that activation energy of R_1 then, exceeds R_2 it will be $= E_{a_1} - E_{a_2} = 10$ kJ mol^{-1}

The new expression becomes,

$k_1 = Ae^{-Ea_1/RT}$; $k_2 = Ae^{-(Ea_1-10)/RT}$

On solving the above equations, we get

$$\ell n\left(\frac{k_2}{k_1}\right) = \frac{10}{RT} = \frac{10}{8.314 \times 10^{-3} \times 300} = 4$$

8. (2) According to the given conditions, the expression for activation energies is given as,

$$\Delta H = E_{a_{(f)}} - E_{a_{(b)}}$$

$-40 = 2x - 3x \Rightarrow -x = -40$, $x = 40$

$E_{a_f} = 80$kJ/mol

$E_{a_b} = 120$kJ/mol

9. (4) $k = Ae^{-E_a/RT}$

if $T \rightarrow \infty$

$k = Ae^0 \Rightarrow k = A$

(2) & (3) options are facts.

10. (4) Zero order reaction means rate is independent of concentration.

100 g $\rightarrow$ 50 g

t = 10 min

means $t_{1/2} = 10$ min

for 200 g $\rightarrow$ 100 g

$t_{1/2} = 10$ min

$\therefore$ 200 g $\rightarrow$ 50 g

$t = t_{1/2} + t_{1/2} = 20$ min

11. (2) $E_a = E_P - E_R$

12. (4) $\Delta H = E_a(f) - E_a(b)$

13. (4) $2.303 \log \frac{k_2}{k_1} = \frac{E_a}{R}\left[\frac{T_2 - T_1}{T_1T_2}\right]$

$$\therefore \quad 2.303 \log \frac{k_2}{k_1} = \frac{9}{2 \times 10^{-3}}\left[\frac{10}{298 \times 308}\right]$$

$$\therefore \quad \frac{k_2}{k_1} = 1.63,$$

% $\uparrow$ in rate constant,

$$= \left(\frac{k_2 - k_1}{k_1}\right) \times 100$$

$$= \left(\frac{k_2}{k_1} - 1\right) \times 100$$

$= (1.63 - 1) \times 100 = 63\%$

14. (4) In option 1, for first order reaction, if degree of dissociation is α, then

$$k = \frac{2.303}{t} \log_{10} \frac{1}{1-\alpha}$$

$$\text{or } kt = \log_e \frac{1}{1-\alpha} = -\log_e(1-\alpha)$$

$e^{-kt} = 1 - \alpha$

$\alpha = 1 - e^{-kt}$

In option, 3 according to Arrhenius equation:

$k = A.\, e^{-E_a/RT}$

A has Time$^{(-1)}$ dimension because, $e^{-E_a/RT}$ is a dimensionless quantity.

15. **(4)** Consider a reaction; A → Product

$\therefore\ r_1 = k[A]^1$ for I order ...(i)

$\therefore\ r_2 = k[A]^2$ for II order ...(ii)

$\therefore\ r_3 = k[A]^3$ for III order ...(iii)

$\because$ k is same in all.

If $[A] = 1, r_1 = r_2 = r_3$

If $[A] < 1, r_1 > r_2 > r_3$

If $[A] > 1, r_3 > r_2 > r_1$

16. **(2)** Since $t_{1/2}$ does not depend upon the sugar concentration means it is first order with respect to sugar concentration.

$t_{1/2} \propto [\text{sugar}]^1$, For $[H^+]$

$t_{1/2} \times a^{n-1} = k$

$$\frac{(t_{1/2})_1}{(t_{1/2})_2} = \frac{[H^+]_1^{1-n}}{[H^+]_2^{1-n}}$$

$$\frac{500}{50} = \left(\frac{10^{-5}}{10^{-6}}\right)^{1-n}$$

$10 = (10)^{1-n}$

Hence, n = 0

17. **(4)** $2A + 3B + \frac{3}{2}C \rightarrow 3P$

$$\text{Rate} = -\frac{1}{2}\frac{dn_A}{dt} = -\frac{1}{3}\frac{dn_B}{dt} = -\frac{1}{3/2}\frac{dn_C}{dt}$$

$$\frac{dn_A}{dt} = \frac{2}{3}\frac{dn_B}{dt} = \frac{4}{3}\frac{dn_C}{dt}$$

18. **(3)** $k_1 = Ae^{-E_{a_1}/RT}, k_2 = Ae^{-E_{a_2}/RT}$

$$\frac{k_1}{k_2} = e^{(-E_{a_1}+E_{a_2})/RT} = e^{-[(2\times10^3)/(2\times400)]}$$

$$\frac{k_1}{k_2} = e^{-2.5} = \frac{r_1}{r_2}$$

$$\therefore \frac{r_2}{r_1} = e^{2.5}$$

19. **(2)** 2A + B → product

Let, rate law be as follows-

Rate of reaction $= k[A]^x[B]^y$

when concentration of B was doubled, the half life did not change which means reaction is first order with respect to B. If the concentration of A is doubled the rate is also getting doubled which means reaction is first order with respect to A also.

$\therefore$ Rate of reaction $= k[A]^1[B]^1$

$\text{mol L}^{-1}\text{s}^{-1} = k\ (\text{mol L}^{-1})(\text{mol L}^{-1})$

$\text{mol L}^{-1}\text{s}^{-1} = k \times \text{mol}^2\text{L}^{-2}$

$$k = \frac{\text{mol L}^{-1}\text{s}^{-1}}{\text{mol}^2\text{L}^{-2}} = \text{mol}^{-1}\text{L s}^{-1}$$

20. **(1)** Apply, $kt = 2.303 \log\frac{(A)_o}{(A)}$

$$k_1 \times t = 2.303 \log\frac{(A)_o}{90(A)_o}\times 100 \quad ......(1)$$

$$k_2 \times t = 2.303 \log\frac{(A)_o}{75(A)_o}\times 100 \quad .....(2)$$

$$\frac{k_2}{k_1} = 2.728$$

Now,

$$\log\frac{k_2}{k_1} = \frac{E_a}{2.303R}\frac{(T_2-T_1)}{T_1T_2}$$

$$\log(2.728) = \frac{E_a}{2.303\times8.314\times10^{-3}}\left(\frac{308-298}{308\times298}\right)$$

$E_a = 76.623$ kJ/mol

21. **(2)** $a \xrightarrow{30\,min} \frac{a}{2} \xrightarrow{60\,min} \frac{a}{4} \xrightarrow{90\,min} \frac{a}{8}$

So, order is 1.

22. **(1)** 1. $k = Ae^{-Ea/RT}$

2. rate $= k_n[C]^n \Rightarrow \ln \text{rate} = \ln k_n + n \ln C$

3. A → product

$a \xrightarrow{\frac{1}{2}hr} \frac{a}{2} \xrightarrow{\frac{1}{2}hr} \frac{a}{4} \xrightarrow{\frac{1}{2}hr} \frac{a}{8} \Rightarrow 1^{st}$ order reaction.

23. **(3)** $\frac{K_2}{K_1} = 16 = x^{\frac{T_2-T_1}{10}}$

$16 = x^{(70-30)/10}$

$16 = x^4$

$x = 2$

24. **(4)** $$\frac{r_{new}}{r} = \frac{k[2A]^n\left[\frac{1}{2}B\right]^m}{k[A]^n[B]^m} = 2^n \times \frac{1}{2^m} = 2^{(n-m)}$$

25. **(3)** $k_1 = A_1e^{-E_{a_1}/RT}; k_2 = A_2e^{-E_{a_2}/RT}$

$$\frac{k_1}{k_2} = \left(\frac{A_1}{A_2}\right)e^{\frac{-E_{a_1}+E_{a_2}}{RT}}$$

Now, $\frac{A_1}{A_2}$ = constant and $E_{a_2} = 2E_{a_1}$

$$\frac{k_1}{k_2} = e^{\frac{-E_{a_1}+2E_{a_1}}{RT}} \Rightarrow \frac{k_1}{k_2} = e^{E_{a_1}/RT}$$

or $k_1 = k_2e^{E_{a_1}/RT}$

26. **(4)** $K_A = Ae^{-E_A/RT}$ $K_B = Ae^{-E_B/RT}$

Also given $K_A > K_B$ $E_A < E_B$

(1) So at high temp. $K_A > K_B$

(2) At lower temp. K_A & K_B will differ more and $K_A > K_B$

27. (3) $t_{\frac{1}{2}} = 10\,\text{min} \quad R = ?$

$$k = \frac{0.693}{t_{\frac{1}{2}}} = \frac{0.693}{10}$$

$$k = \frac{2.303}{t}\log\frac{R_o}{R}$$

$$\frac{0.693}{10} = \frac{2.303}{100}\log\frac{R_o}{R_o - x}$$

$$\log\frac{R_o}{R_o - x} = \frac{0.693 \times 10}{2.303}$$

$$\log\frac{R_o}{R_o - x} = 3.00$$

Say $R_0 - x = R$

$$\Rightarrow \log\frac{R_0}{R} = 3 \Rightarrow R_0 = 10^3 R \qquad ...(1)$$

Now, % of reactant left after 100 min $= \frac{R}{R_0} \times 100$

$$= \frac{R}{10^3 R} \times 100 \ \ (\text{from}(1))$$

$= 0.1\ \%$

$\therefore$ % of reaction completed

$= 100 - 0.1$

$= 99.9\ \%$

28. (3) $t_{1/2} = 10$ min, $t = 20$ min, $N_o = 10$ M

$$t = nt_{\frac{1}{2}} \Rightarrow n = \frac{t}{t_{\frac{1}{2}}} = \frac{20}{10} = 2$$

$$\frac{N}{N_o} = \left(\frac{1}{2}\right)^n \Rightarrow \frac{N}{10} = \left(\frac{1}{2}\right)^2 \Rightarrow \frac{N}{10} = \frac{1}{4} \Rightarrow N = \frac{10}{4}$$

Therefore, the concentration of reactant is 2.5 mol/L

Rate = k[reactant]

$$= \frac{0.693}{t_{\frac{1}{2}}} \times 2.5 = \frac{0.693}{10} \times 2.5 = 0.0693 \times 2.5\ \text{M min}^{-1}$$

29. (2) As rate of reaction is doubled then pseudo first order rate constant is doubled as well. It may be noted that in presence of acid, hydrolysis of ethyl acetate is called as a pseudo-unimolecular reaction but the actual value of k depends upon the concentration of H^+ ions, otherwise rate constant of a reaction is constant at constant temperature.

30. (2) Arrhenius equation is

$$k = Ae^{-E_a/RT} \qquad ...(i)$$

In the presence of catalyst,

$$\text{Rate constant } k' = Ae^{-E_{a'}/RT} \qquad ...(ii)$$

On dividing eq (ii) by eq (i), we get

$$\frac{k'}{k} = e^{(E_a - E_{a'})} / RT$$

$$\frac{k'}{k} = e^{\left(\frac{2000}{2\times300}\right)} = e^{3.33}$$

$\because$ on taking log both sides,

$$\log\frac{k'}{k} = \frac{3.33}{2.303}$$

$$\frac{k'}{k} = \text{anti}\log 1.446 = 27.92 \approx 28$$

we know that, rate (R) $\propto$ k

$$\therefore \frac{R'}{R} = \frac{k'}{k} = 28$$

$$\therefore R' = 28R$$

31. (2) Rate = $[CO]^2$. Rate will become 4 times on doubling the concentration of CO.

32. (4) $-\frac{2d[A]}{dt} = \frac{1}{2}\frac{d[B]}{dt}$

$$\therefore -\frac{d[A]}{dt} = \frac{1}{4}\frac{d[B]}{dt}$$

33. (1) For a first order reaction: $k = \frac{2.303}{t}\log\frac{[A]_0}{[A]}$

At t = 40 min, $[A]_0 = 0.1$ M, $[A] = 0.025$ M

$$\therefore k = \frac{2.303}{40}\log\frac{0.1}{0.025}$$

$$\therefore = \frac{2.303}{40}\log 4 = 0.0347\ \text{min}^{-1}$$

Rate = k [A] = 0.0347 × 0.01 = 3.47×10^{-4} M/min.

34. (1) $p_1 = 80$ k/Pa, $p_2 = 40$ k/Pa

$(t_{1/2})_1 = 350$ sec $(t_{1/2})_2 = 175$ sec

As we know the relation between half life and concentration (or pressure) is given as, $t_{1/2} \propto p^{1-n}$

where, n is the order of the reaction.

Therefore,

$$\frac{\left(t_{\frac{1}{2}}\right)_1}{\left(t_{\frac{1}{2}}\right)_2} = \left(\frac{p_2}{p_1}\right)^{n-1}$$

$$\left(\frac{350}{175}\right) = \left(\frac{40}{80}\right)^{n-1}$$

$$2 = \left(\frac{1}{2}\right)^{n-1}$$

$$\left(\frac{1}{2}\right)^{-1} = \left(\frac{1}{2}\right)^{n-1}$$

$n - 1 = -1$

$n = -1 + 1 = 0$

n = zero i.e., zero order reaction

35. (1) The logarithmic form of Arrhenius reaction is given as,

$$2.303 \log k = 2.303\log A - \frac{E_a}{RT} \qquad ...(i)$$

$$E_a = 2.303\ RT \qquad ...(ii)$$

From equation (i) and (ii),

$$2.303\log k = 2.303\log A - \frac{2.303RT}{RT}$$

$$\log k = \log A - 1$$

$$\log\left(\frac{k}{A}\right) = -1 \text{ or } \frac{k}{A} = 10^{-1}$$

36. (2) According to rate Law

$$\text{rate} = K[A]^x[B]^y$$

$$0.045 = K(0.05)^x(0.05)^y \quad \ldots (i)$$

$$0.090 = K(0.10)^x(0.05)^y \quad \ldots (ii)$$

$$0.72 = K(0.20)^x(0.10)^y \quad \ldots (iii)$$

eq. (*i*) ÷ (*ii*)

$$\frac{0.45}{0.090} = \left(\frac{0.05}{0.10}\right)^x \Rightarrow x = 1$$

eq. (*ii*) ÷ (*iii*)

$$\frac{0.090}{0.720} = \left(\frac{0.10}{0.20}\right)^x\left(\frac{0.05}{0.10}\right)^y \Rightarrow y = 2$$

Hence, $r = K[A][B]^2$

37. (1) $k = Ae^{-E_a/RT}$

∴ As T increases, k increases exponentially.

38. (1) Rate = $K[A]^n$

$$K = \frac{\text{Rate}}{[A]^n} = \frac{\text{moleL}^{-1}\text{ s}^{-1}}{(\text{moleL}^{-1})^n}$$

$= \text{mole}^{1-n}\text{ L}^{n-1}\text{ s}^{-1}$

Correct & Incorrect MCQs

1. **(1)** Rate of reaction depends upon the experimental conditions such as concentration of reactants (pressure in case of gases), temperature and catalyst.
2. **(3)** For complex reaction, order is given by the slowest step.
3. **(4)** Molecularity is applicable only for elementary reactions.
4. **(4)** The unit of rate constant of a zero order reaction is mol $L^{-1}s^{-1}$.
5. **(2)** For a zero order reaction, $t_{1/2}$ is directly proportional to the initial concentration of the reactants and inversely proportional to the rate constant. Hence, if concentration of the reactant is doubled, then half life period is also doubled in the zero order reaction.
6. **(2)** Activation energy is inversely proportional to the rate of the reaction. Lesser the activation energy, more will be the rate of the reaction.
7. **(3)** A catalyst does not alter Gibbs energy, ΔG of a reaction.
8. **(1)** Threshold energy is the minimum amount of energy required by the colliding molecules to yield the products.
9. **(1)** For a chemical reaction with rise in temperature by 10°, the rate constant is nearly doubled.
10. **(3)** $\ln k = \frac{E_a}{RT} + \ln A$

    Hence, the plot of ln k vs 1/T gives a straight line according to the above equation.

Statement Based MCQs

1. **(1)** Since rate is independent of the concentration of reactants, hence it is constant for zero order reactions.
2. **(3)** Half life period for first order reaction

    $$t_{1/2} = \frac{0.693}{k} \text{ or, } t_{1/2} = \frac{2.303}{k}\log_{10} 2$$
3. **(3)** Effective collisions are those that result in product formation due to an increase in the rate of a chemical reaction that occurs when the two reactant molecules are correctly oriented and have attained the threshold value at the time of the collision.
4. **(1)** Both the given statements are correct.
5. **(3)** Molecularity of slowest step is order of overall complex reaction.
6. **(3)** Half – life period of a zero order reaction is directly proportional to initial concentration, $t_{1/2} \propto [A]_0$.
7. **(4)** The order of a reaction can have a fractional value.
8. **(1)** At a given temp, reaction having lesser activation energy will be executed faster than other.
9. **(3)** A catalyst does not change the equilibrium constant of a reaction rather, it helps in attaining the equilibrium faster.
10. **(1)** Both statements are correct as on increasing concentration of reactants, number of effective collisions increases and as a result, rate increases.

Assertion & Reason MCQs

1. **(2)** Order of reaction can be zero or fractional as order of reaction is directly related to sum of power of reactants. Reason is a correct statement.

    The order of reaction does not depend on the stoichiometric coefficients corresponding to each species in the balanced reaction. Therefore, it cannot be determined from balanced chemical equation.

    Hence, both assertion and reason are true but reason is not a correct explanation of the assertion.
2. **(4)** Order and molecularity can be same only for elementary reaction and it is different for complex reaction. Hence, assertion is false.

    Order is determined experimentally and molecularity is the sum of the stoichiometric coefficient of rate-determining elementary step. Hence, reason is true.
3. **(1)** Enthalpy of reaction remains constant throughout the reaction process in the presence of catalyst and catalyst lowers down the activation energy.

    Hence, both assertion and reason are correct and reason is a correct explanation of the assertion.
4. **(4)** Collisions without proper orientation or without sufficient kinetic energy do not lead to the formation of a compound.

    Whereas, only those collisions in which molecules have correct orientation and sufficient kinetic energy lead to compound formation.

    Therefore, assertion is false and reason is true.

5. **(3)** Rate constant determined from Arrhenius equation are fairly accurate for simple and complex molecules because only those molecules which have proper orientation during collision results in effective collision. Hence, assertion is true but reason is false.

6. **(2)** Both Assertion and Reason are True but Reason is not the correct explanation of the Assertion.

7. **(3)** Assertion is True but the Reason is False.

8. **(3)** Conc. of water in hydrolysis of ethyl acetate reaction is taken as a constant throughout reaction hence, rate constant will depend only on conc. of ethyl acetate hence, it is called pseudo 1st order reaction.

9. **(3)** Activation energy is fixed for a particular reaction. It is the K.E of molecules that changes with temperature.

10. **(1)** Ratio of rate constant at two different temperatures are separated by 10° C is called temperature coefficient hence, R & A are true & R gives correct explanation of A.

Multi Correct MCQs

1. **(3)** $A: k = Ae^{\frac{-E_a}{RT}}$

As E_a increases k decreases

B: Temperature coefficient $= \frac{k_{T+10}}{k_T}$

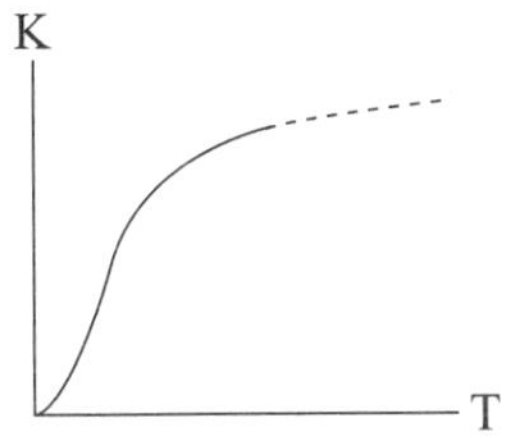

Statement (c) is wrong. k may be greater or lesser depending on temperature.

$D: \ln k = \ln A - \frac{E_a}{RT}$

2. **(3)**

(A)

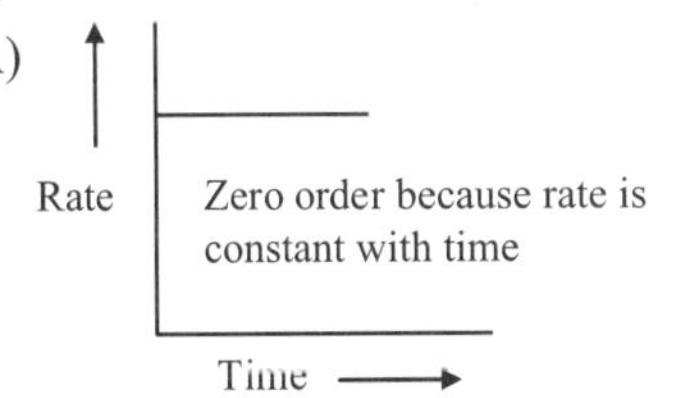

(B)

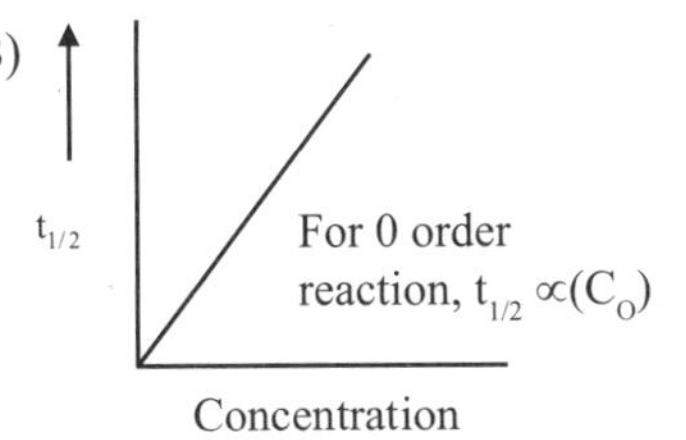

(C)

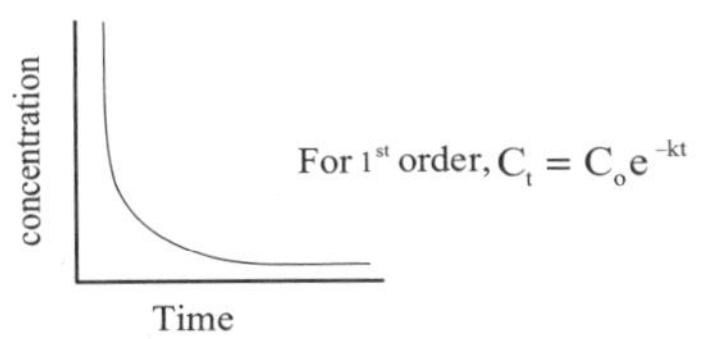

(D)

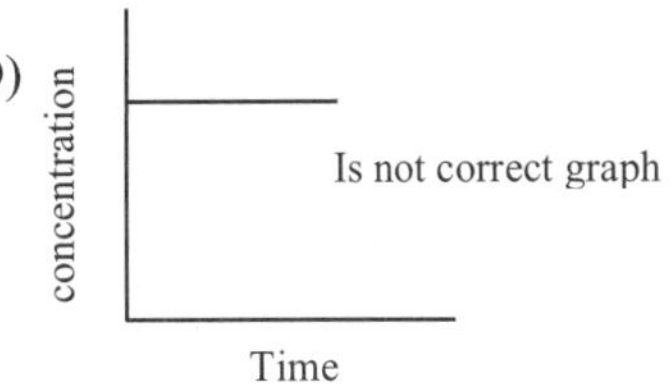

(E)

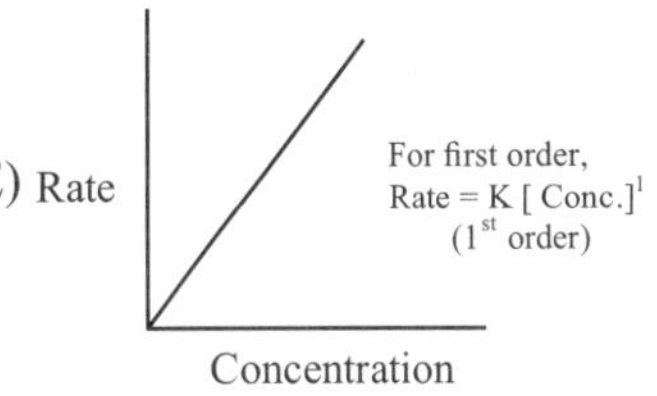

3. **(2)** Catalyst provides an alternate pathway or reaction mechanism by reducing the activation energy between reactants and products and hence lowering the potential energy barrier.

4. **(1)**

Units of rate constant	
Reaction	Units of rate constant
Zero order reaction	$molL^{-1}s^{-1}$
First order reaction	S^{-1}/Min^{-1}
Second order reaction	$mol^{-1}Ls^{-1}/mol^{-1}Lmin^{-1}$

NCERT Exemplar MCQs

1. **(3)** The role of a catalyst is to change the activation energy of reaction. This is done by either increasing or decreasing activation energy of molecule.

Note : Catalyst are of two types one is positive catalyst which increases rate of reaction by decreasing activation energy and another is negative catalyst which decreases rate of reaction by increasing energy of activation.

2. **(3)** In the presence of catalyst, the heat absorbed, or evolved during the reaction remains unchanged as there is no change in stability of reactant and product.

3. **(2)** Activation energy of a chemical reaction is related to rate constant of a reaction at different temperatures *i.e.*, k_1 and k_2 respectively.

$$\ln\left(\frac{k_1}{k_2}\right) = \frac{E_a}{R}\left[\frac{1}{T_1} - \frac{1}{T_2}\right]$$

where, E_a = activation energy

T_2 = higher temperature

T_1 =lower temperature

k_1= rate constant at temperature T_1

k_2 = rate constant at temperature T_2

This equation is known as Arrhenius equation.

4. **(1)** Activation energy is the minimum energy required to convert reactant molecules to product molecules. Here, the energy gap between reactants and activated complex is sum of E_1 and E_2.

$\therefore$ Activation energy = $E_1 + E_2$

Product is less stable than reactant as energy of product is greater than the reactant.

5. (2) $A(g) \rightarrow B(g)+C(g)$

	A(g)	B(g)	C(g)
Initially	p_i	0	0
At time t	$p_i - x$	x	x

$P_t = p_i - x + x + x = p_i + x$

For first order reaction

$x = p_t - p_i$

$$k = \frac{2.303}{t}\log\frac{p_i}{pi - x}$$

$$k = \frac{2.303}{t}\log\frac{p_i}{p_i - (p_t - p_i)}$$

$$k = \frac{2.303}{t}\log\frac{p_i}{2pi - p_t}$$

6. (1) According to Arrhenius equation, $k = A\,e^{-Ea/RT}$

Taking log on both side $\ln k = \ln\left(A.e^{\frac{-E_a}{RT}}\right)$

$$\ln k = \ln A - \frac{E_a}{RT}$$

$$\ln k = -\frac{E_a}{R}\times\frac{1}{T} + \ln A$$

This equation can be related to equation of straight line y = mx + c as shown above.

From the graph (a), it is very clear that slope of the plot (m) $= -\frac{E_a}{R}$ and intercept = ln A.

7. (4) According to Arrhenius equation $k = Ae^{-E_a/RT}$

Here $k \propto e^{-E_a}$

$k \propto e^{-\frac{1}{T}}$

$\propto e^{T}$

Which indicates that as activation energy decreases rate constant increases and as temperature increases rate of reaction increases.

8. (3) $Zn + \text{Dil. } HCl \rightarrow ZnCl_2 + H_2\uparrow$

$$\text{Average rate of reaction} = \frac{\text{Change in concentration of } H_2}{\text{Change in time}}$$

$$= \frac{V_3 - 0}{40 - 0} = \frac{V_3}{40}$$

9. (3) Out of the given four statements, option (3) is not correct.

Order of reaction is equal to the sum of powers of concentration of the reactants in rate law expression.

For any chemical reaction:

$xA + yB \rightarrow \text{Product}$

Rate = $k\,[A]^x[B]^y$

Order = $x + y$

Order of reaction is not always equal to sum of the stoichiometric coefficients of reactants in the balanced chemical equation. For of reaction it may or may not be equal to sum of stoichiometric coefficient of reactants. Order of reaction can be a fraction also.

10. (1) Rate of reaction is defined as rate of decrease of concentration of any one of reactions with passage of time

$$\text{Rate of reaction} = \frac{\text{Rate of disappearance of reactant}}{\text{Time taken}}$$

$$r = \frac{-dx}{dt}$$

Thus, as the concentration of reactant decreases with passage of time, rate of reaction decreases.

11. (1) The chemical reaction in which energy is evolved during the reaction is known as exothermic reaction i.e., activation energy of product is greater than activation energy of reactants.

Here, only (I) denotes correct picture of exothermic reaction.

12. (2) Rate law can be written as

Rate = k[A][B]

Rate of reaction *w.r.t* B is of first order.

$R_1 = k\,[A]\,[B]$

when concentration of reactant `B' is doubled, then rate (R_2)

$R_2 = k\,[A]\,[2B]$

$R_2 = 2k\,[A]\,[B]$

$R_2 = 2\,R_1$

Therefore; as concentration of B is doubled keeping the concentration of A constant, rate of reaction doubles.

13. (3) According the postulates of collision theory, there are following necessary conditions for any reaction to be occur:

(i) Molecule should collide with sufficient threshold energy.

(ii) Their orientation must be proper.

(iii) The collision must be effective.

14. (4) The time the concentration of a reactant is reduced to half of its original value is called half-life period of the reaction.

But it is impossible to perform 100% of the reaction. Whole of the substance never reacts because in every half-life, 50% of the substance reacts. Hence, time taken for 100% completion of a reaction is infinite.

15. (2) *Characteristics of catalyst*

(1) It is a substance catalyses the forward and backward reaction to the same extent as it decreases energy of activation hence, increases the rate of both the reactions.

(2) Since, reaction quotient is the relation between concentration of reactants and products. Hence, catalyst does not alter reaction quotient. Thus, Gibbs free energy does not change during the reaction when catalyst is added to it.

$\Delta G -\;\; RT \ln Q$

where, Q = reaction quotient

(3) It doesn't alter equilibrium of reaction as equilibrium constant is also concentration dependent term.

(4) It provides an alternate mechanism by reducing activation energy between reactants and products.

16. (2) Pseudo first order reaction is a chemical reaction in which rate of reaction depends upon concentration of only one reactant (present in small amount) while concentration of another reactant (present in excess) has no effect on rate of reaction.

e.g., hydrolysis of ethyl acetate in presence of excess of water.

$$CH_3COOC_2H_5 + \underset{\text{excess}}{H_2O} \xrightarrow{H^+} CH_3COOH + C_2H_5OH$$

$r = k[CH_3COOC_2H_5]$

where $k = k'[H_2O]$

$\therefore$ Rate constant of pseudo 1st order reaction depends on conc. of reactant present in excess.

17. (2) $A \to B$

Concentration of reactants and products varies exponentially w.r.t time.

(i) Concentration of reactant (here. A) decreases exponentially w.r.t time.

(ii) Concentration of product (here. B) increases exponentially w.r.t time. Correct graph representing the above reaction is (2).

18. (1, 3, 4) Rate law can be determined from balanced chemical equation if it is an elementary reaction.

19. (1, 4) For a balanced chemical equation of an elementary reaction order is same as molecularity and molecularity can never be zero. If molecularity of a reaction is considered to be zero it mean that no reactant is going to transform into product. Consider a chemical reaction:

$2NO(g) + O_2(g) \to 2NO_2(g)$

Differential rate law expression can be written as:

$$\frac{dR}{dt} = k[NO]^2[O_2]$$

Here, molecularity = 3, order = 3

20. (1, 2) Since, the reaction is an unimolecular reaction. Hence, in the slowest step i.e., in the rate determining step the only one reacting species is involved. Therefore, order of reaction and molecularity of reaction is equal to one.

21. (1, 4) (1) For a complex reaction, order of overall reaction = molecularity of slowest step

As rate of overall reaction depends upon total number of molecules involved in slowest step of the reaction. Hence, molecularity of the slowest step is equal to order of overall reaction reactants.

(4) Since, the completion of any chemical reaction is not possible in the absence of reactants hence, slowest step of any chemical reaction must contain at least one reactant. Thus, molecularity of the slowest step is never zero or non-integer.

22. (1, 3, 4) Given, chemical reaction is

$$2NH_3(g) \xrightarrow[\text{Platinum catalyst}]{1130\text{ K}} N_2(g) + 3H_2(g)$$

At very high pressure, reaction become independent of concentration of ammonia *i.e.*, zero order reaction.

Hence, $\text{Rate} = k\left[p_{NH_3}\right]^0$

$\text{Rate} = k$

(a) Rate of reaction = Rate constant

(b) Rate of decomposition of ammonia will remain constant until ammonia disappears completely.

(c) Since, formation of ammonia is a reversible process further increase in pressure will change the rate of reaction. According to Le-Chatelier principle increase in pressure will favour in backward reaction.

23. (1, 4) When the reactant molecules collide each other they lead to formation of an activated complex. It has highest energy among reactants, products and activated complex. When it decomposes to give product, energy is released and stability of product increases.

Since, the entire concentration of activated complex do not convert into products while, some activated complex may give reactants also.

24. (1, 3) Distribution of kinetic energy may be described by plotting a graph of fraction of molecules versus kinetic energy.

Kinetic energy of maximum fraction of molecule is known as most probable kinetic energy. It is important to note that with increase of temperature, peak shifts forward but downward. This means that with increase of temperature,

(i) most probable kinetic energy increases.

(ii) the fractions of molecules possessing most probable kinetic energy decreases.

25. (1, 4) According to Maxwell Boltzmann distribution curve, area under the curve must not change with increase in temperature. But with increase in temperature curve broadens and shift towards right hand side due to decrease in fraction of molecules having most probable kinetic energy.

26. (1, 2) Arrhenius equation can be written as $k = A.e^{\frac{-Ea}{RT}}$

$k \propto e^{-E_a}$ *i.e.*, rate of reaction increases with decrease in activation energy.

$k \propto e^{-\frac{1}{T}}$

$k \propto e^{T}$ *i.e.*, rate of reaction increases with increase in temperature.

27. (1, 4) Function of Catalyst as the catalyst is added to the reaction medium rate of reaction increases or decreases. By decreasing or increasing activation energy of molecule respectively. As there are +ve & –ve catalysts Hence, it follows an alternative pathway.

Catalyst does not change the enthalpy change of reaction. Energy of reactant and product remain same in both catalysed and uncatalysed reaction.

Hence, (2) (4) are incorrect statements.

28. (1, 4) For a zero order reaction

$$[R] = (-k)t + [R]_0 \quad \text{...(i)}$$

$$\uparrow \quad \uparrow \quad \uparrow$$

$$y = m \times x + c$$

On comparing with Equation of straight line

$y = [R]$ concentration

$x = t$ time

Slope $(m) = -k$ rate constant

Intercept $(c) = [R]_0$ initial concentration

On rearranging Eq. (i)

$$\frac{[R]-[R]_0}{t} = -k$$

$$\frac{[R]-[R]_0}{t} = -kt^0$$

Rate $\propto t^0$

NEET & AIIMS Past Year MCQs

1. **(3)** Let's consider the rate equation as:

Rate = $k[A]^x [B]^y$

Where x is the order of the reaction with respect to A, and y is the order with respect to B.

Thus, we can write

$2 \times 10^{-3} = k[0.1]^x [0.1]^y$ ...(i)

$4 \times 10^{-3} = k[0.2]^x [0.1]^y$ ...(ii)

$1.6 \times 10^{-2} = k[0.2]^x [0.2]^y$ ...(iii)

(ii) divided by (i);

$$\frac{4\times10^{-3}}{2\times10^{-3}} = \frac{k[0.2]^x[0.1]^y}{k[0.1]^x[0.1]^y}$$

$$\Rightarrow \frac{2}{1} = \frac{(0.2)^x}{(0.1)^x} = \left(\frac{2}{1}\right)^x$$

$\therefore$ x = 1

(ii) divided by (iii);

$$\frac{4\times10^{-3}}{1.6\times10^{-2}} = \frac{k[0.2]^x[0.1]^y}{k[0.2]^x[0.2]^y} \Rightarrow \frac{1}{4} = \frac{(0.1)^y}{(0.2)^y} = \left(\frac{1}{2}\right)^y$$

$\therefore$ y = 2

$\therefore$ Rate = $k[A]^1[B]^2$

First order with respect to A and second order with respect to B.

2. **(3)** Using Arrhenius equation,

$k = Ae^{-E_a/RT}$ ....(i)

Taking ln of both sides then we have,

$$\ln k = \ln A - \frac{E_a}{RT}$$

$\ln k = -E_a/RT + \ln A$

And compare with y = mx + c, $\text{slope}(m) = -\frac{E_a}{R}$

Intercept = lnA

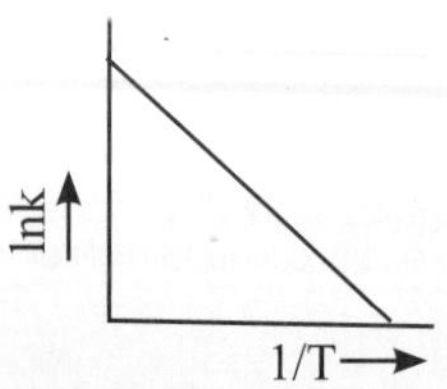

3. **(3)** According to Arrhenius equation;

$K = Ae^{-Ea/RT}$

After taking in both side then we have;

$$\ln K = \ln A - \frac{E_a}{RT}$$

For two different temperatures T_1 and T_2, the rate constants k_1 and k_2 are:

$$\ln K_1 = \ln A - \frac{E_a}{RT_1} \text{ at temp. } T_1 \quad ......(i)$$

$$\ln K_2 = \ln A - \frac{E_a}{RT_2} \text{ at temp. } T_2 \quad ......(ii)$$

Now; (ii) - (i)

$$\ln K_2 - \ln K_1 = \frac{E_a}{R}\left[\frac{1}{T_1} - \frac{1}{T_2}\right]$$

$$\ln\frac{K_2}{K_1} = \frac{E_a}{R}\left[\frac{1}{500} - \frac{1}{700}\right]$$

$$\ln\frac{0.14}{0.04} = \frac{E_a}{R}\left[\frac{700-500}{500\times700}\right]$$

$$\ln\frac{14}{4} = \frac{E_a}{R}\left[\frac{200}{500\times700}\right]$$

$$\log 3.5 = \frac{E_a}{2.303\times R}\left[\frac{1}{250\times7}\right]$$

$$0.5441 = \frac{E_a}{2.303\times8.31}\left[\frac{1}{250\times7}\right]$$

$E_a = 0.5441 \times 8.31 \times 250 \times 7 \times 2.303$

= 18222.65

$\approx$ 18219 J

4. **(2)** Activation energy of any chemical reaction can be calculated if one knows the value of rate constant at two different temperature by given formula:

$$\log\frac{k_1}{k_2} = \frac{E_a}{2.303R}\left[\frac{1}{T_2} - \frac{1}{T_1}\right]$$

5. **(2)** Arrhenius equation is;

$$\ln k = -\frac{E_a}{R} \times \frac{1}{T} + \log A$$

Graph between ln k vs $\frac{1}{T}$ with intercept = log A and slope = $-\frac{E_a}{R}$

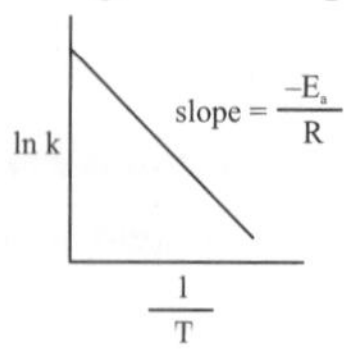

6. **(3)** $\log\left(\frac{k_{330}}{k_{300}}\right) = \frac{E_a}{2.303R} \times \left[\frac{330-300}{300\times330}\right]$

Given: $\frac{k_{330}}{k_{300}} = 4$

$$\therefore \log 4 = \frac{E_a}{2.303R} \times \frac{30}{300 \times 330}$$

$$\therefore E_a = \frac{0.602 \times 2.303 \times 8.314 \times 300 \times 330}{30} J / mol$$

$= 38037.7 J / mol$

$= 38.0377 kJ / mol$

$\simeq 38.04 kJ / mol$

7. (4) Let the new rate be, R'.

A' = [3A], B' = [B]

$R' = K[A']^2 [B']$

$= K[3A]^2 [B]$

$= 9K[A]^2 [B]]$

$= 9 \times R$

Hence, rate would become nine times.

8. (3) A reaction can have zero activation energy, for example, $^{\cdot}CH_3 + {}^{\cdot}CH_3 \rightarrow CH_3–CH_3$ has zero activation energy.

9. (4)

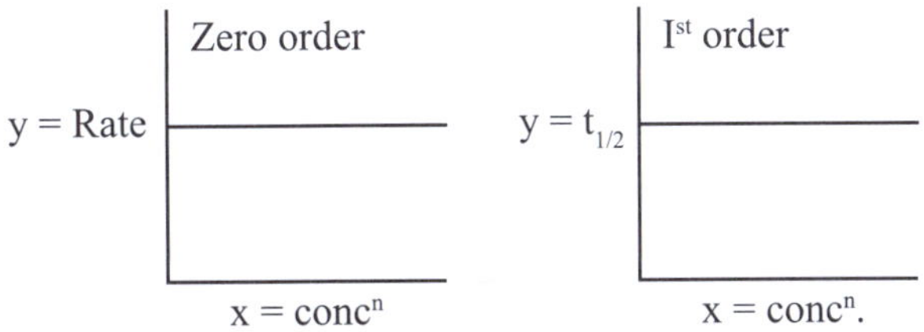

Figure (I) is suitable for zero order if y = rate and x = concentration because in case of zero order reaction, rate is constant and does not depend on conc^n.

Figure (II) is suitable for first order if y = $t_{1/2}$ and x = conc^n because in case of first order, $t_{1/2}$ does not depend on conc^n.

10. (3) A → Products

Initial conc, $[A_o] = 0.1$ M

Concentration after 5 min $[A_t] = 0.001$ M

t = 5 min.

For first order reaction

$$k = \frac{2.303}{t}\log\left(\frac{A_0}{A_t}\right) = \frac{2.303}{5}\log\left(\frac{0.1}{0.001}\right)$$

$k = 0.9212 \text{ min}^{-1}$

11. (1) $\Delta H_{rxn} = E_{a(forward)} - E_{a(backward)}$

$-4.2 = (E_{a(forward)} - E_{a(backward)})$

$-4.2 = 9.6 - (E_a)_{backward}$

$E_{a(backward)} = 9.6 + 4.2 = 13.8 \text{ kJ mol}^{-1}$

Thus, reaction is exothermic.

Activation energy of backward reaction is more than that of forward reaction.

12. (4) According to Arrhenius equation:

$k = Ae^{-E_a/RT}$

$$\ln k = \ln A - \frac{E_a}{RT}$$

Slope of (ln k v/s 1/T) curve,

$$m = -\frac{E_a}{R} \quad -5 \times 10^3 = -\frac{E_a}{R}$$

$E_a = 5 \times 10^3 \times 8.314$ J/mol

$= 41.57 \times 103$ J/mol

$\simeq 41.5$ kJ/mol

13. (3) For zero order

$A = A_0 . kt$

$$t_{\frac{1}{2}} = \frac{A_0}{2k}$$

$$t_{\frac{1}{4}} = \frac{A_0}{4k} \qquad \frac{t_{\frac{1}{2}}}{t_{\frac{1}{4}}} = \frac{2}{1}$$

14. (1) $$\frac{0.8 \times 10^{-5}}{131} \times 6 \times 10^{23} \times \frac{2.3 \times 0.3}{8.1 \times 24 \times 60 \times 60}$$

$= 3.6 \times 10^9$

15. (4) For a nth order reaction,

$$t_{1/2} \propto \frac{1}{[a]^{n-1}}$$

Where n-order of reaction

$$t_{1/2} \propto \frac{1}{[a]^{3-1}} \quad [\because n = 3]$$

$$\boxed{t_{1/2} \propto \frac{1}{[a]^2}}$$

16. (2) **17. (1)** **18. (3)** **19. (1)** **20. (3)**

The p-Block Elements (Group 15 to 18)

NCERT Topic-wise MCQs

1. (3) The order of boiling point will be:

$H_2O > SbH_3 > NH_3 > AsH_3 > PH_3$

2. (2)

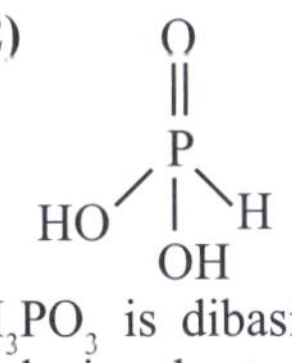

H_3PO_3 is dibasic because it has two P - OH bond and reducing due to the presence of one P-H bond.

3. (2) $(NH_4)_2 Cr_2O_7 \xrightarrow{\Delta} N_2 + Cr_2O_3 + 4H_2O$

4. (3) A blue ppt of $Cu(OH)_2$ is first obtained which further dissolves in excess of NH_4OH to give a deep blue solution of complex i.e., tetraammine copper (II) sulphate, $[Cu(NH_3)_4]SO_4$.

5. (1) $NH_3 = x + 3 = 0$, $x = -3$ $\qquad$ $HN_3 = +1 + 3x = 0$, $3x = -1$, $x = -1/3$

$N_2H_4 = 2x + 4 = 0$, $2x = -4$, $x = -2$ $\qquad$ $NO_2 = x + 2(-2) = 0$, $x = 4$, $x = +4$

6. (3) $\underset{\substack{\text{(Calcium phosphide)}\\ \text{1 mol}}}{Ca_3P_2} + 6H_2O \rightarrow 3Ca(OH)_2 + \underset{\substack{\text{(Phoshine)}\\ \text{2 mol}}}{2PH_3}$

7. (1) NH_3 has greater dipole moment among all.

8. (4) $\overset{0}{P_4} + 3NaOH + 3H_2O \rightarrow \overset{-3}{P}H_3 + 3NaH_2\overset{+1}{P}O_2$

Since, P is undergoing both oxidation and reduction, therefore, shows here disproportionation reaction.

9. (1) $H_3PO_4 \rightleftharpoons H^+ + H_2PO_4^- \rightleftharpoons H^+ + HPO_4^{2-} \rightleftharpoons H^+ + PO_4^{3-}$

10. (1) $H_3PO_3 \Rightarrow$ 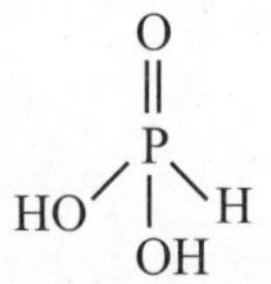

$\Rightarrow$ It is dibasic in nature, hence have two replaceable H-atoms. Thus, can form two series of salts.

11. (1) Phosphorus is kept in water because it burnt at 30ºC and readily catches fire.

12. (4) $NH'_4 = x + 4 = +1$, $\quad NO_3^- = x + 3(-2) = -1$

$x = +1 - 4 \quad x - 6 = -1$

$x = -3 \quad x = +6 - 1$

$x = +5$

13. (1) $N_2O_3 + H_2O \rightarrow 2HNO_2$ (Nitrous acid)

14. (4) 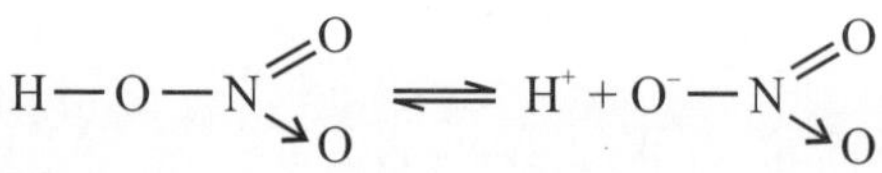 [Nitrate ion]

15. (2) Nitrogen atom has two stable isotopes which are ^{14}N and ^{15}N.

16. (2) Oxidation states of nitrogen are

$HNO_3 = +5, \quad NO = +2, \quad N_2 = 0, \quad NH_4Cl = -3$

17. (1) Due to unavailability of d-orbitals in nitrogen, it can form only NCl_3 and not NCl_5.

18. (2) $NH_4NO_3 \xrightarrow{\Delta} N_2O + 2H_2O$

19. (3) The structure of the white phosphorus is tetrahedral.

20. (1) $PCl_3 + 3H_2O \rightarrow H_3PO_3 + 3HCl$

Thus in this reaction, ortho phosphorous acid (H_3PO_3) is produced.

21. (3) Pyrophosphoric acid ($H_4P_2O_7$) is tetrabasic acid.

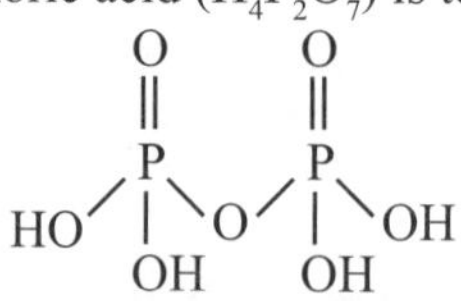

22. (4) Nitrous oxide is neutral gas.

23. (1) $NO_3^- + 3Fe^{2+} + 4H^+ \rightarrow NO + 3Fe^{3+} + 2H_2O$

$[Fe(H_2O)_6]^{2+} + NO \rightarrow \underset{\text{brown}}{[Fe(H_2O)_5(NO)]^{2+}} + H_2O$

24. (4) 'Bi' $Fe(H_2O)_5(NO)]^{2+}$ doesnot show allotropy.

25. (3) Nitrogen exhibits allotropy in solid state

26. (1) Dinitrogen is inert at room temperature because it has high bond enthalpy of N≡N bond.

27. (2) From Bi to N bond length decreases

28. (2) As electronegativity decreases down the group, the electron density on central atom of hydride decreases and thus repulsion between bond pair and lone pair of electrons decreases which ultimately decreases H – M – H bond angle.

Therefore NH_3 will be having highest bond angle.

29. (4) Hydrides of nitrogen are NH_3, N_2H_4, HN_3 etc.

30. (1) $N_2 + O_2 \underset{arc}{\overset{electric}{\rightarrow}} 2NO$

31. (2) Three bond pairs and one lone pair of electrons are present in ammonia molecule.

32. (3) As, lone pair density on nitrogen in NH_3 is the largest, therefore it has maximum complex forming ability.

33. (1) Due to the presence of of a lone pair of electrons on the nitrogen atom of the ammonia molecule, it acts as a Lewis base.

34. (3) BaO_2 is barium peroxide, containing O_2^{2-} peroxide ion.

35. (2) Transition temperature is the temperature at which both the allotropes of sulphur coexist.

$\alpha - \text{sulphur} \underset{< 369\,K}{\overset{> 369\,K}{\rightleftharpoons}} \beta - \text{sulphur}$

36. (2) α-Sulphur and β-sulphur are allotropic forms of sulphur. α-Sulphur is stable below 369 K and transforms into β-sulphur above this.

37. (4) Ozone O_3 is an allotropic form of oxygen.

38. (2) SF_4 is a good fluorinating agent

39. (3) S in excited state $\Rightarrow$ 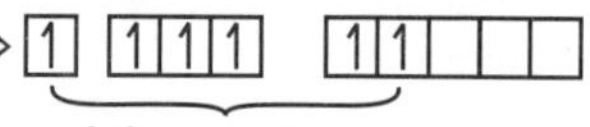

40. (4) 1. $\overset{+5}{PCl_5} \rightarrow \overset{+5}{POCl_3}$

2. $\overset{-1}{HI} \rightarrow \overset{0}{I_2}$ (Oxidation)

3. $\overset{0}{S} \rightarrow \overset{+4}{SO_2}$ (Oxidation)

Since, H_2SO_4 cause oxidation of HI and S in the reaction, 2 and 3, therefore, acts as an oxidising agent.

41. (3) SO_3^{2-} is 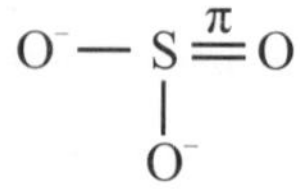

π – bond involves d - orbitals of S and p - orbital of O. N, B and C does not have d - orbitals.

42. (4) H_2SO_4 is 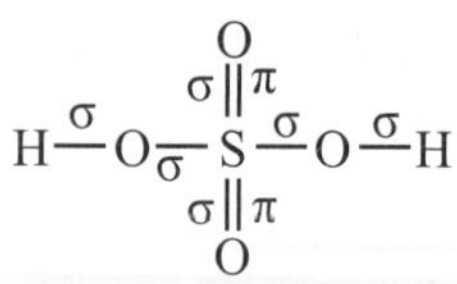

43. (2) Further oxidation is not possible in Mn because Mn in $KMnO_4$ has the highest oxidation state, i.e., +7.

44. (1) A given acid decomposes the salt of that acid which is more volatile than that.

45. (2) Sulphuric anhydride = SO_3

'S' is sp^2 hybridised, there are three σ bonds formed between the three sp^2 hybridised orbitals of sulfur and 3 pure p-orbitals of three O-atoms.

Sulfur has one pure p-orbital and two pure d-orbitals. These three orbitals are involved in one pπ-pπ bond and two pπ-dπ bonds.

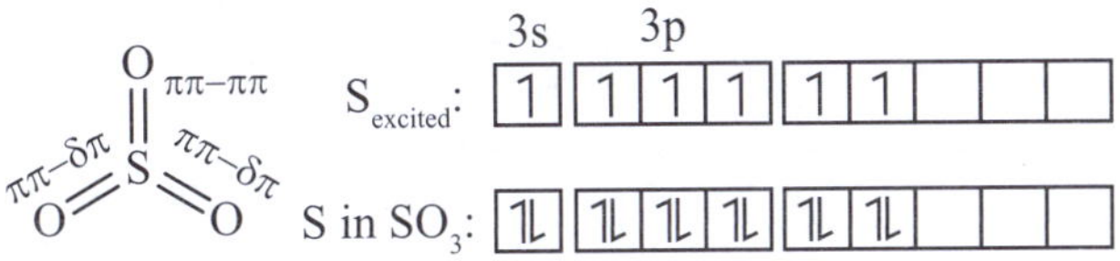

46. (1) Ozone is diamagnetic gas.

47. (2) Sulphuric acid dissolved in water is accompanied by the release of a large amount of energy, hence it is an exothermic process.

48. (1) Peroxodisulphuric acid is the name of the compound $H_2S_2O_8$.

49. (1) O_3 is a disinfectant.

50. (1) The rate of reaction is very slow even at (400-500)°C. Therefore, to increase the reaction velocity, a suitable catalyst is used. Nowadays most of the sulphuric acid plants use V_2O_5 as a catalyst.

51. (3) $BaO_2 + O_3 \rightarrow BaO + 2O_2$

52. (4) Sulphur exists as octa-atomic molecule.

53. (1) Dithionic acid – $H_2S_2O_6$ = +5

Sulphurous acid – H_2SO_3 = +4

Sulphuric acid – H_2SO_4 = +6

Pyrosulphuric acid – $H_2S_2O_7$ = +6

54. (1) $2Na + S \rightarrow Na_2S$ (Sulphide S^{-2})

55. (2) O, S, Se, Te are called chalcogens

56. (1) H_2SO_5 – Caro's acid.

57. (4) $H_2S_2O_7$ does not contain S—S bond.

58. (2) O – $1s^2$ $2s^2$ $2p^4$

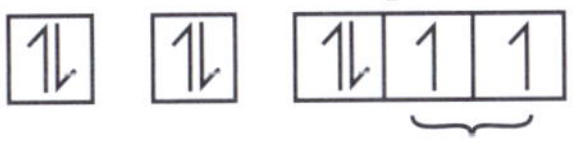

2 unpaired electrons

59. (4) Reducing power of hydrides

$H_2O < H_2S < H_2Se < H_2Te$

60. (1) Due to absence of d orbital, oxygen shows maximum covalency of 4.

61. (4) Anhydride is formed by eliminating water from another compound.

$H_2SO_4 \rightarrow SO_3 + H_2O$

62. (3) Peroxide ion is given by this symbol.

63. (4) $H_2SO_4 + SO_3 \rightarrow H_2S_2O_7$(Oleum)

64. (3) $H_2SO_4 + CaF_2 \rightarrow CaSO_4 + 2HF$

65. (2) $2KClO_3 \xrightarrow[MnO_2]{Heat} 2KCl + 3O_2$

Hence, A and B in the given reaction are KCl and O_2 respectively.

66. (3) Since, halogens have little tendency to lose electron, hence they have very high ionisation enthalpy.

67. (3) Since, one corner of the octahedron is occupied by lone pair. Hence; the molecule is square pyramidal.

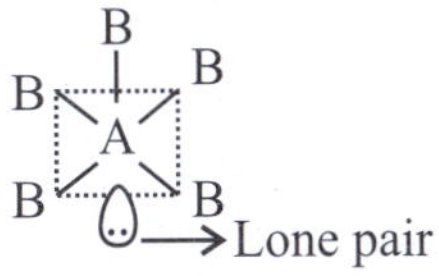

68. (1)

KI_3	KIO_3
$+1 + 3x = 0$	$+1 + x + 3(-2) = 0$
$3x = -1$	$+1 + x - 6 = 0$
$x = -1/3$	$x - 5 = 0$
	$x = +5$
KI	IF_5
$+1 + x = 0$	$x + 5(-1) = 0$
$x = -1$	$x - 5 = 0$
	$x = +5$

69. (1) $Cl_2O + H_2O \rightarrow 2HOCl$ [Hypochlorous acid]

70. (4) $Cl_2O_6 + H_2O \rightarrow \underset{\text{Chloric acid}}{HClO_3} + \underset{\text{Perchloric acid}}{HClO_4}$

71. (1) It is IF_7 that has maximum number of halogen atoms.

72. (3) $IF_5 + F_2 \rightarrow IF_7$

73. (2) The half filled d-orbital of Cl overlaps with p-orbital of O to form dπ–pπ bond.

74. (1) It can be only reduced and not oxidised because Cl in ClO_4^- has highest oxidation state, i.e., +7.

75. (3) $2F_2 + 2H_2O \rightarrow 4HF + O_2$

$3F_2 + 3H_2O \rightarrow 6HF + O_3$

76. (1) All hydrogen halides are gases, except HF, which is a liquid with boiling point 293K. This behaviour is due to association of HF molecules through hydrogen bonding.

H — F -----H — F ---- H — F ---- H — F

77. (1) The bond dissociation enthalpy of F_2 is smaller than that of chlorine due to large electron-electron repulsion among the lone pairs in F_2 molecule.

78. (1) $2CaOCl_2 + H_2SO_4 \rightarrow CaCl_2 + CaSO_4 + 2HClO$

[limited]

$HClO \rightarrow HCl + [O]$

On account of evolution of nascent oxygen, it acts an oxidising and bleaching agent.

79. (1) Size varies inversely with hydration energy. Among the given oxyacids, ClO_4^- has the smallest size because of the presence of Cl^{7+} followed by ClO_3^- (with Cl^{+5}), ClO_2^- (with Cl^{3+}). ClO^- has the largest size with Cl^+ ion. Hence, the order of increasing hydration energy is

$ClO^- < ClO_2^- < ClO_3^- < ClO_4^-$

80. (3) The acidic strength of these acids varies in the order:

HF < HCl < HBr < HI

81. (2) Both BrO_2^- (35 + 2 × 8 + 1 = 52) has 52 electrons and BrF_2^+ (35 + 2 × 9 – 1 = 52) has also 52 electrons.

82. (1) Iodine is solid at room temperature.

83. (2) I_2 forms soluble KI_3 compound with 'KI.

84. (3) I_2 is a solid at room temperature, hence can sublime.

85. (4) I^-, I^+ can exist.

86. (2) Fluorine exhibits only -1 oxidation state. Fluorine being the most electronegative element, cannot exhibit any positive oxidation state.

87. (3) The oxidation numbers of iodine in HIO_4, ICl_5, I_2 and HI. are +7, +5, 0, –1 respectively.

88. (2) F_2 bond dissociation energy is low because of repulsion between non-bonding electrons.

89. (2) The correct order of reducing power: HI > HBr > HCl > HF

90. (1) Acidic nature of oxyacids increases with increase of oxidation number of central atom.

91. (4) ClO^- has 3 lone pairs, ClO_2^- has 2 lone pairs, ClO_3^- has 1 lone pair, ClO_4^- has zero lone pair of electrons.

92. (3) $HClO_4$ is having weak oxidising nature, hence is weak reducing agent.

93. (4) The geometry of XeF_4 is square planar as the two corners of the octahedral are occupied by lone pairs. Hence, hybridisation is sp^3d^2 as per VSEPR theory.

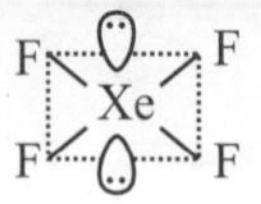

94. (4) Xe because of its low ionisation energy forms many compounds however, no true compounds of He, Ne and Ar are known. Xe forms most of the compounds with F and O, like XeF_2, XeF_4, XeF_6, and XeO_3 etc.

95. (3) Argon is the most abundant noble gas in the atmosphere.

96. (2) Bartlett reacted xenon and platinum hexafluoride in gas phase and a solid of $XePtF_6$ was obtained which is orange yellow in colour.

$Xe_{(g)} + PtF_{6(g)} \rightarrow Xe^+ [PtF_6]^-_{(s)}$
orange yellow

97. (4) Since the only type of interatomic interaction in noble gas elements is weak dispersion forces, hence their melting and boiling points are very low.

98. (2) Helium (He) is a noble gas which exist as monoatomic.

99. (4)

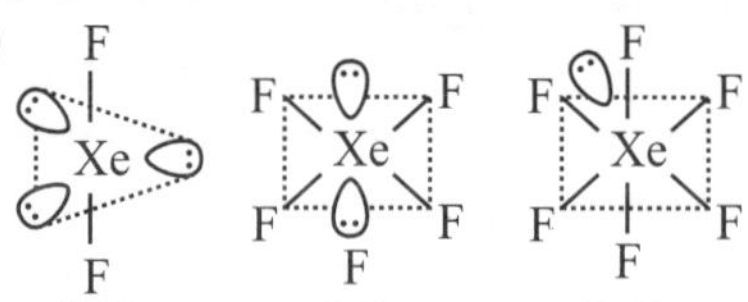

XeF_2 XeF_4 XeF_6

100. (1) $XeF_4 + O_2F_2 \rightarrow XeF_6 + O_2$
Hence, 'A' in the given reaction is XeF_6.

101. (3) $XeF_6 \rightleftharpoons XeF_5^+ + F^-$
$\downarrow$
$sp^3d^2\,(5\sigma + 1LP)$ square pyramidal

102. (3) $1s^2\,2s^2\,2p^6\,3s^2\,3p^6$ is the electronic configuration of argon.

103. (1) Rn is radioactive in nature, hence is not found in atmosphere.

104. (1) Inert gases do not contain unpaired electrons

105. (4) Noble gases form compounds very easily with fluorine.

106. (2) He is non-inflammable and lighter gas

107. (3) XeF_2 molecule is linear in shape.

108. (4) The bond order and stabilization energy of He_2 molecule are zero.
Hence, the He_2 molecule does not exist.

109. (3) $XeF_6 + 3H_2O \rightarrow XeO_3 + 6HF$

110. (3) $XeOF_4$ molecule contains one lone pair of electrons.

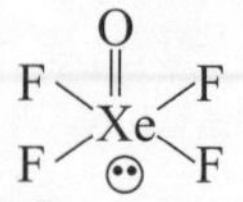

111. (3) Radon (Rn) is radioactive element so, it can be used in cancer treatment.

High Order Time Intensive MCQs

1. (4) $(NH_4)_2Cr_2O_7 \xrightarrow{\Delta} N_2 + Cr_2O_3 + 4H_2O$
$Zn(ClO_3)_2 \xrightarrow{\Delta} ZnCl_2 + 3O_2$
$KClO_3 \xrightarrow{\Delta} KCl + 3/2O_2$
$2K_2Cr_2O_7 \xrightarrow{\Delta} 2K_2CrO_4 + Cr_2O_3 + 3/2O_2$

2. (4) The most abundant rare gas in the atmosphere is Argon.

3. (4) Thermal and electrical conductivity is highest in Bi.

4. (4) Bi_2O_5 is most acidic in nature as it in the highest oxidation state among all of +5.

5. (3) Maximum covalency of Sb will be 5.

6. (2) $(NH_4)_2Cr_2O_7 \xrightarrow{\Delta} Cr_2O_3 + N_2 + 4H_2O$

7. (4) $2NaNO_3 \xrightarrow{500^0C} 2NaNO_2 + O_2$
$2NaNO_3 \xrightarrow{800^0C} Na_2O + 5/2O_2 + N_2$

8. (3) Barium azide on thermal decomposition can produce very pure nitrogen.
$Ba(N_3)_2 \rightarrow Ba + 3N_2$

9. (2, 4)

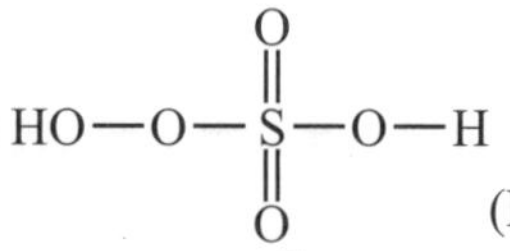

(H_2SO_5)

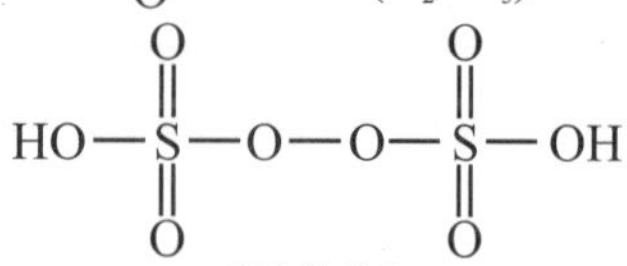

($H_2S_2O_8$)

10. (4)

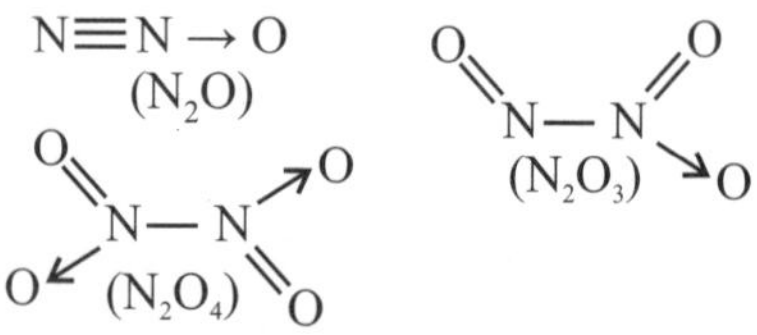

11. (1, 3) SO_2 acts as a bleaching agent in moist condition as it released nascent hydrogen
$SO_2 + 2H_2O \rightarrow H_2SO_4 + 2[H]$
(Nascent)
Coloured matter + H $\rightarrow$ Colourless
The dilute solution of SO_2 is used as a disinfectant for killing disease causing germs, fungi and certain moulds.

12. (4) Fluorine is the most electronegative element of the periodic table. Therefore; it can oxidise water to oxygen.
$2F_2 + 2H_2O \rightarrow 4HF + O_2$.

13. (2) $\underset{(A)}{PCl_5} + H_2O \rightarrow \underset{(B)}{POCl_3} + 2HCl$

$\underset{(B)}{POCl_3} + 3H_2O \rightarrow \underset{(C)}{H_3PO_4} + 3HCl$

14. (1) Aqua regia is a mixture of conc. HCl and HNO_3 in a molar ratio of 3 : 1

15. (3) $SO_2 + 2H_2O \rightarrow H_2SO_4 + 2(H)$

16. (4) N_2O is relatively unreactive

17. (3) As the ratio between radii of X and X′ increases, the number of atoms per molecule also increases. Thus, IF_7 have maximum number of atoms as the ratio of radii between I and F is maximum.

18. (1) The reaction of Iodine with concentrated nitric acid is given by

$I_2 + 10HNO_3 \rightarrow \underset{(Y)}{2HIO_3} + 10NO_2 + 4H_2O$

Let oxidation state of I in HIO_3 be x.

$\Rightarrow +1 + x - 6 = 0$

$\Rightarrow x = 5$

19. (3) From NH_3 to BiH_3, the electron pair donating nature gradually decreases, as down the group, size increase, therefore lone pair occupies more volume.

20. (2) $FeSO_4$ forms brown ring with NO.

21. (3) Both rhombic sulphur (α-sulphur) and monoclinic sulphur (β-sulphur) are stable at 369 K, this temperature is known as transition temperature.

22. (2) NH_3 is more stable due to inter molecular H-bond.

23. (2) In hydrazoic acid (HN_3), $1 + 3x = 0$

$$3x = -1;\ x = \frac{-1}{3}$$

24. (1) $Au + 4H^+ + NO_3^- + 4Cl^- \rightarrow AuCl_4^- + NO + 2H_2O$

$3Pt + 16H^+ + 4NO_3^- + 18\ Cl^- \rightarrow 3PtCl_6^{-2} + 4NO + 8H_2O$

25. (1) Because NF_3 is a stable halide.

26. (1) NCl_3 undergoes hydrolysis easily as it is quite unstable.

27. (3) $3O_2 \underset{\text{electric discharge}}{\overset{\text{silent}}{\rightleftharpoons}} 2O_3$

28. (1) Amongst H_2O, H_2S, H_2Se and H_2Te, the one with the highest boiling point is H_2O because of hydrogen bonding.

29. (4) SO_2 has one pπ-dπ & SO_3 has two dπ-pπ bonds.

30. (3) Among the given options, (H_2Te) give H^+ ions most easily because the bond between (H_2Te) is weakest hence it gives H^+ ion easily.

31. (4) SF_6 is inert due to hindrance of S atom.

32. (4) Peroxymonosulphuric acid is also known as Caro's acid, H_2SO_5.

33. (2) $O_3 \rightarrow O_2 + (O)$

34. (2) All SO_2 is oxidized to SO_3 by atmospheric oxygen in the presence of V_2O_5 as a catalyst in contact process of manufacturing H_2SO_4.

35. (1) Fluorine is a stronger oxidizing agent than chlorine due to low enthalpy of dissociation of F – F bond and high hydration enthalpy of F^-.

36. (1) As the size of halogen increases bond energy decreases

37. (4) All the given statements are correct.

38. (4) Anomalous behavior of nitrogen is due to

1. Small size
2. Absence of 'd' orbitals
3. Ease of multiple bond formation

39. (2) Concentrated H_2SO_4 is diluted by adding the concentrated H_2SO_4 solution in water dropwise with constant stirring because it is an exothermic reaction and by doing so, heat is generated slowly and gets dissipated in the atmosphere.

40. (4) ClO_2 is powerful oxidising agent, also strong chlorinating agent. Its bleaching powder is almost 30 times stronger than Cl_2. In alkaline solution, it undergoes disproportionation.
$2ClO_2 + 2NaOH \rightarrow NaClO + NaClO_3 + H_2O$

41. (2) 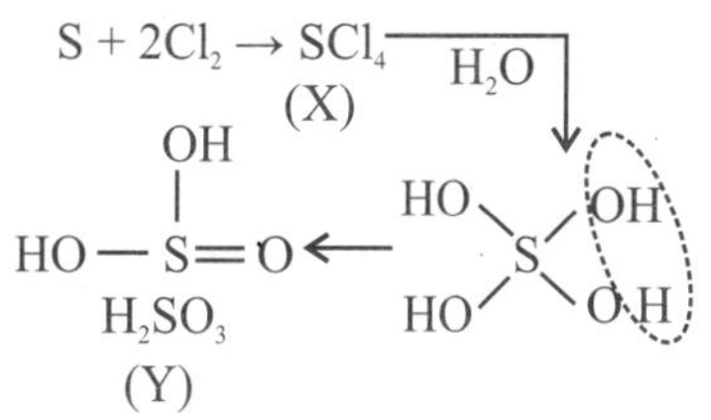

Hybridisation of S in H_2SO_3 is sp^3. (3σ bond + 1*l*.p)

42. (4) In red, black, white and yellow allotropic forms of phosphorus, black phosphorus is the most thermodynamically stable form because its ignition temperature is highest.

43. (3) Radon is not present in the atmosphere.

44. (1) Ozone acts as a powerful oxidising agent and it oxidises lead sulphide to lead sulphate.

45. (2) $SO_3 + D_2O \rightarrow D_2SO_4$
In D_2SO_4, S-undergoes sp^3 hybridization

46. (4) Argon is the most abundant rare gas found in the atmosphere.

47. (4) In halogens, oxidizing ability decreases while moving down the group.

48. (3) $\underset{\text{Slaked lime}}{Ca(OH)_2} + Cl_2 \rightarrow Ca(OCl)_2 + 2H_2O + CaCl_2$
Hence, the main reaction product is $CaOCl_2$.

49. (1) $2NaClO_4 + H_2SO_4 \rightarrow Na_2SO_4 + 2HClO_4$

50. (2) Shape of $XeOF_4$ is square pyramidal.

51. (1) The reaction between hydrogen and fluorine is the most violent in nature because of the large electro negativity difference between two atoms.

52. (4) $2F_2 + 2NaOH \rightarrow 2NaF + OF_2 + H_2O$

53. (1) $Ca(OCl)Cl \rightarrow Ca^{+2} + Cl^- + ClO^-$

54. (2) Bleaching powder with hot water gives:
$3CaOCl_2 \rightarrow 3Ca^{2+} + 5Cl^- + ClO_3^-$

55. (1) $2NO + N_2O_4 \xrightarrow{250\ K} 2N_2O_3$

56. **(2)** Na_2O = Basic
As_2O_3 = Amphoteric
N_2O = Neutral
NO = Neutral
Cl_2O_7 = Acidic

57. **(2)** N cannot go beyond +5 oxidation state.

58. **(4)** In the reaction, $\overset{+4}{XeF_4} + \overset{+1}{O_2}F_2 \longrightarrow \overset{+6}{XeF_6} + \overset{0}{O_2}$ (Oxidation: Xe +4 → +6; Reduction: O +1 → 0)
Hence, an example of redox reaction.

59. **(3)** The structure of various given oxides of N are:
N_2O
$\overset{+}{N}\equiv N—\overset{-}{O} \longleftrightarrow \overset{-}{N}=\overset{+}{N}=O$
N_2O_3
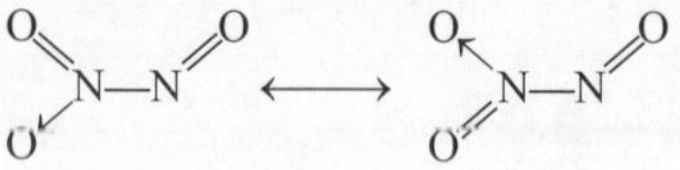
N_2O_4
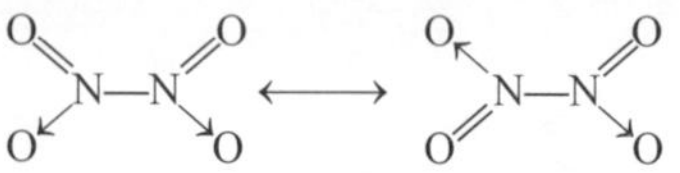
N_2O_5
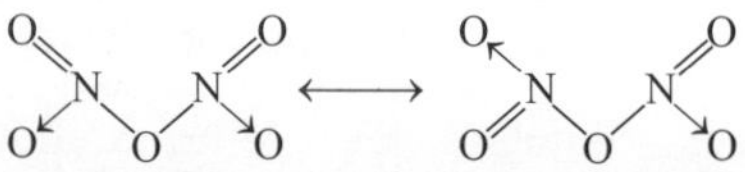
Hence, N_2O, N_2O_3 and N_2O_4 has N–N bond.

60. **(4)** Bismuth forms **unstable hydrides** and rarely exhibits the +5 oxidation state due to the **inert pair effect**.
Hence, option (4) is **incorrect.**

Correct & Incorrect MCQs

1. **(3)** Ionisation enthalpy of group 15 element is much greater than that of 14 elements, in corresponding period.
2. **(3)** Bond enthalpy of nitrogen (N≡N) is very high.
3. **(4)** NH_3 exhibits hydrogen bonding in solid as well as liquid state due to high electronegativity and small size of nitrogen. Hence, NH_3 has higher melting and boiling points than that of PH_3.
4. **(4)** Due to non-availability of the d orbitals in its valence shell, nitrogen does not form pentahalide.
5. **(2)** The acids containing P–H bond have strong reducing properties. Hypophosphorous acid contains two P–H bonds, hence it acts as a good reducing agent.
6. **(3)** Due to the compact nature of oxygen atom, it has less negative electron gain enthalpy than sulphur. The thermal stability of hydrides decreases from H_2O to H_2Po due to the decrease in enthalpy for the dissociation of H–E bond down the group.
7. **(3)** In group 16 elements, reducing property of dioxide decreases from SO_2 to TeO_2 thus SO_2 is reducing while TeO_2 is an oxidising agent.
8. **(3)** Due to small size of fluorine atom, there is more electronic repulsions, as a result the incoming electron is unstable. Hence, the negative electron gain enthalpy of fluorine is less than that of chlorine.
9. **(3)** Chlorine on reaction with cold and dilute alkalis produces a mixture of chloride and hypochlorite but on reaction with hot and concentrated alkalies, it gives chloride and chlorate.
10. **(3)** The only type of interatomic interaction in noble gas elements is weak dispersion forces, hence they have very low melting and boiling points.

Statement Based MCQs

1. **(1)** Liquid NH_3 is an effective cooling agent because it has a high latent heat of vaporization.
And, dielectric constant of NH_3 is lower than that of H_2O.
2. **(3)** Bleaching action of SO_2 is temporary and it is due to reduction, while bleaching action by Cl_2 is permanent and it is due to oxidation.
3. **(2)** Basic character decrease from top to bottom due to increase in size of atom and decrease in electron density.
Electronegativity of N is greater than that of P.
4. **(2)** Oxidation state of Cl in $HClO_4$ is +7. The structure is
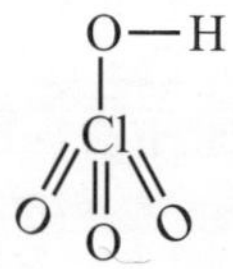
5. **(1)** In case of NI_3, the lone pair dipole moment adds on the resultant of NI dipole moment but in case of NF_3 the lone pair dipole moment on N partly cancels the resultant N-F dipole moment.
NX_3 (nitrogen halides) have trigonal pyramidal structure.
6. **(3)** In group 15 elements, electronegativity decreases down the group with increase in atomic size. Hence, Nitrogen has higher electronegativity than that of Bismuth.
7. **(1)** Due to the ease with which O_3 liberates atoms of nascent oxygen [$O_3 \rightarrow O_2 + [O]$], it acts as a powerful oxidising agent.
8. **(3)** The halogens have the smallest atomic radii in their respective periods due to maximum effective nuclear charge.
9. **(2)** Since, X–X′ bond in interhalogens is weaker than X–X bond in halogens except F–F bond, hence interhalogen compounds in general are more reactive than halogens (except fluorine).
10. **(2)** Hydrolysis of XeF_4 and XeF_6 with water gives XeO_3.
$6XeF_4 + 12\,H_2O \rightarrow 4Xe + 2XeO_3 + 24\,HF + 3O_2$
$XeF_6 + 3H_2O \rightarrow XeO_3 + 6\,HF$
Partial hydrolysis of XeF_6 gives oxyfluorides, $XeOF_4$ and XeO_2F_2.
$XeF_6 + H_2O \rightarrow XeOF_4 + 2\,HF$
$XeF_6 + 2H_2O \rightarrow XeO_2F_2 + 4HF$

Assertion & Reason MCQs

1. **(3)** Nitrogen has less electron gain enthalpy than phosphorous.
2. **(3)** HNO_3 makes iron passive and its passivity is attained by formation of a thin film of oxide on iron. Fe does not dissolve in conc. HNO_3 solution.
3. **(2)** HI being a powerful reducing agent, when produced during the process is oxidised to I_2.
4. **(1)** Oxygen forms $p\pi - p\pi$ multiple bond due to small size and small bond length but $p\pi - p\pi$ bonding is not possible in sulphur. Due to this, oxygen exists as O_2 and rhombic and monoclinic sulphur exist as S_8.
5. **(1)** When NaCl reacts with conc. H_2SO_4, the following reaction takes place:
   $2NaCl + H_2SO_4 \rightarrow Na_2SO_4 + 2HCl$
   Thus when NaCl reacts with concentrated H_2SO_4, hydrochloric gas (HCl) is produced as a by-product.
   On adding MnO_2, the reaction will take place as:
   $MnO_2 + 4HCI \rightarrow MnCl_2 + 2H_2O + Cl_2$
   Thus, MnO_2 oxidizes HCI to chlorine gas (Cl_2).
   The greenish-yellow fumes are of chlorine gas
6. **(1)** Six F atoms in SF_6 prevent the attack of H_2O on sulphur atom of SF_6, hence SF_6 cannot be hydrolysed but SF_4 can be hydrolyzed.
7. **(1)** Formation of pentahalides requires sp^3d hybridization of central atom. This is obtained by excitation of electrons of central atom to empty d-orbitals. Since, nitrogen does not contain d-orbitals, therefore it does not form pentahalides.
8. **(3)** F does not contains d orbitals.
9. **(2)** When green leaf is introduced in Cl_2, colors disappears due production of nascent oxygen which combines with coloured substance & makes it colourless.
10. **(3)** H_2O has stronger H bonding as compared to H_2S because of small size and high electronegativity of O, hence water exists as liquid but H_2S exists as a gas. Molecular weight of H_2S is higher than that of H_2O.

Multi Correct MCQs

1. **(2)** $BrF_5 \rightarrow$ Square pyramidal (sp^3d^2)
   $IF_5 \rightarrow$ Square Pyramidal (sp^3d^2)
   $ClF_5 \rightarrow$ Square Pyramidal (sp^3d^2)
2. **(2)** The order of the E-H bond's bond dissociation energy in group 16 hydrides is:
   $H_2O > H_2S > H_2Se > H_2Te$
3. **(1)** Among all, only $H_2S_2O_8$ contains peroxo bond as shown below:

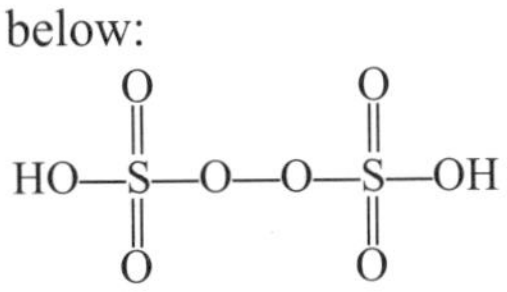

4. **(3)** The geometry of all the given species is shown as,

XeF_5^+

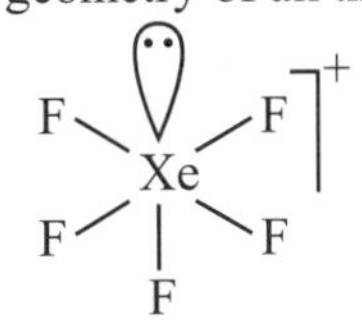

XeO_3

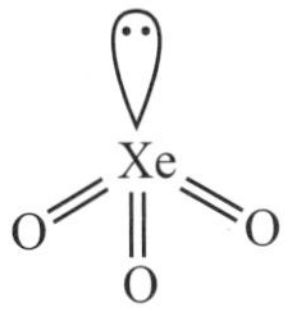

XeO_2F_2

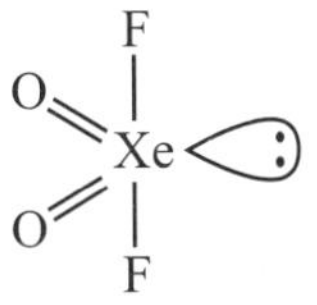

XeF_5^-

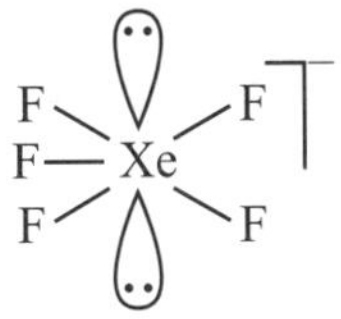

XeO_3F_2

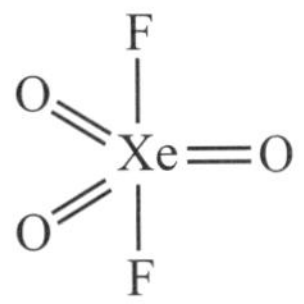

$XeOF_4$

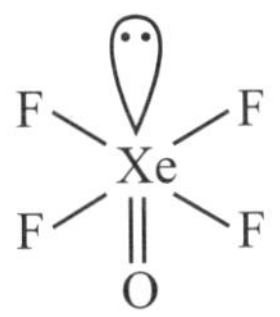

XeF_4

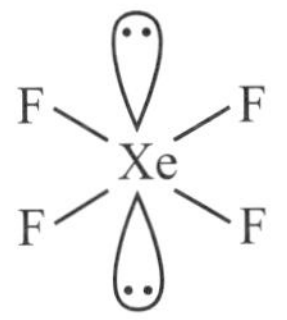

From all the above given structures, it is clear that the species XeF_5^+, XeO_3, $XeOF_4$, and XeO_2F_2 carries single lone pair of electrons.

NCERT Exemplar MCQs

1. **(3)** Hydrogen iodide (HI) is more stronger oxidising agent than H_2SO_4. So, it reduces H_2SO_4 to SO_2 and itself gets oxidised to I_2. Colour of I_2 is violet, hence on adding conc. H_2SO_4 to HI, it gets oxidised to I_2.

$$H_2SO_4 + 2HI \rightarrow SO_2 + \underset{\text{Violet colour}}{I_2} + 2H_2O$$

2. **(2)** In qualitative analysis, when H_2S is passed through an aqueous solution of salt acidified with dil. HCl, a black ppt. of CuS is obtained.

$$CuSO_4 + H_2S \xrightarrow{\text{dil. HCl}} \underset{\text{Black ppt}}{CuS} + H_2SO_4$$

On boiling CuS with dil. HNO_3, it forms a blue coloured solution.

$$CuS + 2HNO_3 \rightarrow \underset{\text{(Blue solution)}}{Cu(NO_3)_2} + H_2S$$

Addition of excess of aqueous solution of ammonia to this solution gives deep blue solution of $[Cu(NH_3)_4]^{2+}$

$Cu(NO_3) + 4NH_3 \rightarrow [Cu(NH_3)_4]^{2+} + 2NO_3^-$
(deep blue solution)

3. **(3)** Cyclotrimetaphosphoric acid contains three double bond and 9 single bonds as shown below a, b, c are three double bonds and numerics 1 to 12 are single bonds.

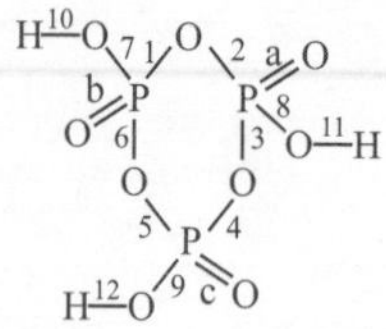

4. **(3)** Among given four elements i.e., carbon, nitrogen, phosphorus and boron. Only phosphorus has vacant d-orbital, so only phosphorus has ability to form $p\pi - d\pi$ bonding.

5. **(1)** Compounds having same value of total number of electrons are known as isoelectronic.

For CO_3^{2-}	For NO_3^-
Total number of electrons	Total number of electrons
$= 6 + 8 \times 3 + 2 = 32$	$= 7 + 8 \times 3 + 1 = 32$

Hence, CO_3^{2-} and NO_3^- are isoelectronic. These two ions have similar structure so they are isostructural.

Both have triangular planar structure as in both the species carbon and nitrogen are sp^2 hybridised. Hence, (1) is the correct choice.

6. **(1)** On moving top to bottom, size of halogen atom increases, hence bond length increases. Therefore, bond dissociation enthalpy decreases.

7. **(4)** On moving top to bottom, size of central atom increases. Bond length of X–H bond increases and bond dissociation energy decreases. Hence, reducing nature increases.

8. **(3)** White phosphorous on reaction with NaOH solution in the presence of inert atmosphere of CO_2 produces phosphine gas which is less basic than NH_3.

$$P_4 + 3NaOH + 3H_2O \longrightarrow PH_3 + \underset{\text{(sodium hypophosphite)}}{3NaH_2PO_2}$$

9. **(3)** Structure of H_3PO_4 is

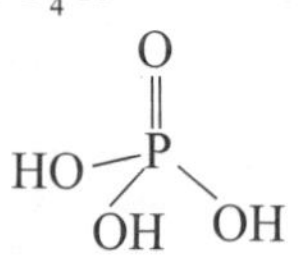

H_3PO_4 has 3–OH groups *i.e.*, has three ionisable H-atoms and hence forms three series of salts. These three possible series of salts for H_3PO_4 are as follows

NaH_2PO_4, Na_2HPO_4 and Na_3PO_4

10. **(2)** On heating lead nitrate, it produces brown coloured nitrogen dioxide (NO_2) and lead (II) oxide

$$2Pb(NO_3)_2 \xrightarrow{\Delta} 4NO_2 + 2PbO + O_2$$

11. **(1)** Nitrogen does not show allotropy due to its weak N–N single bond. Therefore, ability of nitrogen to form polymeric structure or more than one structure or form become less. Hence, nitrogen does not show allotropy.

12. **(3)** Maximum covalency of nitrogen is 4 since only four (one s and three p) orbitals are available for bonding.

13. **(1)** True statement is that single N–N bond is weaker than the single P–P bond. This is why phosphorous show allotropy but nitrogen does not.

(2) PH_3 acts as a ligand in the formation of coordination compound due to presence of lone pair of electrons.

(3) NO_2 is paramagnetic in nature due to presence of the unpaired electron.

(4) Covalency of nitrogen in N_2O_5 is 4.

14. **(2)** Stability of + 5 oxidation state decreases from top to bottom and + 3 oxidation state increases from top to bottom due to inert pair effect. Meanwhile compound having + 5 oxidation state of Bi is BiF_5. It is due to smaller size and high electronegativity of fluorine.

15. **(1)** On heating ammonium dichromate and barium azide it produces N_2 gas separately.

$$(NH_4)_2Cr_2O_7 \xrightarrow{\Delta} N_2 + 4H_2O + Cr_2O_3$$

$$Ba(N_3)_2 \longrightarrow Ba + 3N_2$$

16. **(1)** Catalytic oxidation of ammonia in preparation of nitric acid is shown as.

$$4NH_3 + 5O_2 \xrightarrow[\text{Pt/Rh gauge catalyst}]{\Delta} 4NO(g) + 6H_2O(l)$$

According to above reaction, 4 moles of $NH_3 \rightarrow$ 4 moles of NO

Hence, 2 moles of $NH_3 \rightarrow$ 2 moles of NO

17. **(3)** Let oxidation state of P in $H_2PO_2^-$ be *x*.

$(2 \times 1) + x + (2 \times -2) = -1$

$2 + x - 4 = -1$

$x - 2 = -1$

$x = +1$

18. **(3)** SF_4 has see-saw shaped as shown below

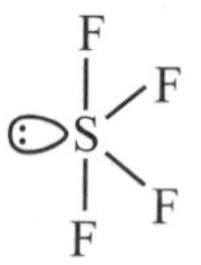

It has trigonal bipyramidal geometry having sp^3d hybridisation.

19. **(1)** H_2SO_5 and $H_2S_2O_8$ contains one peroxide linkage. Hence, both are peroxoacids of sulphur. Their structures are:

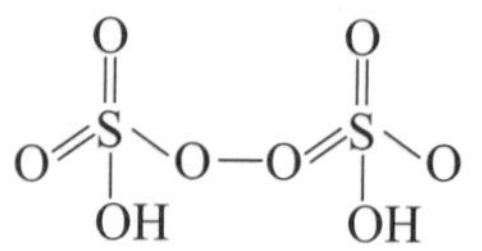

($H_2S_2O_8$)

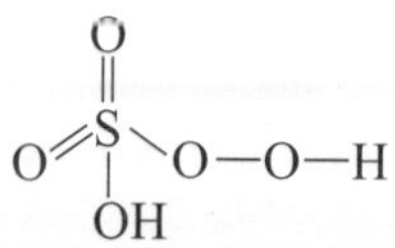

(H_2SO_5)

20. **(3)** H_2SO_4 is a moderately strong oxidising agent which oxidises both metals and non-metals as shown below

$$Cu + 2H_2SO_4(conc) \longrightarrow CuSO_4 + SO_2 + 2H_2O$$

$$S + 2H_2SO_4(conc) \longrightarrow 3SO_2 + 2H_2O$$

While carbon on oxidation with H_2SO_4 produces two types of oxides CO_2 and SO_2.

$$C + 2H_2SO_4(conc) \longrightarrow CO_2 + 2SO_2 + 2H_2O$$

21. **(1)** Black coloured compound MnO_2 reacts with HCl to produce greenish yellow coloured gas of Cl_2

$$\underset{\text{(Black)}}{MnO_2} + 4HCl \longrightarrow MnCl_2 + 2H_2O + \underset{\text{(greenish yellow gas)}}{Cl_2}$$

Excess Cl_2 on further treatment with NH_3 produces NCl_3.

$$\overset{-3}{N}H_3 + 3Cl_2 \longrightarrow \overset{+3}{N}Cl_3 + 3HCl$$

NH_3 (– 3) changes to NCl_3 (+ 3) in the above reaction. Hence, (1) is the correct choice.

22. **(3)** Bartlett had taken $O_2^+PtF_6^-$ as a base compound because O_2 and Xe both have almost same ionisation enthalpy. The ionisation enthalpies of noble gases are the highest in their respective periods due to their stable electronic configurations.

23. **(4)** In solid state, PCl_5 exists as an ionic solid with $[PCl_4]^+$ tetrahedral and $[PCl_6]^-$ octahedral.

24. **(3)** Greater the standard reduction potential (SRP) value of species, higher will be its oxidizing power.

Species	E^0_{cell}
ClO_4^-	1.19 V
IO_4^-	1.65 V
BrO_4^-	1.74 V

∵ SRP of species increases, oxidizing power increase.

25. **(1, 3)** When chlorine gas is passed through hot NaOH solution it produces NaCl and $NaClO_3$.

Oxidation state varies from 0 to – 1 and 0 to + 5.

Hence, (1) and (3) are correct choices.

26. **(2, 3)** 1. $F_2 > Cl_2 > Br_2 > I_2$ As ability to gain electron increases oxidising property increases. Here, F is the most electronegative element having highest value of SRP hence it has highest oxidising power.

2. MI > MBr > MCl > MF

This is the incorrect order of ionic character of metal halide. Correct order can be written as

MI < MBr < MCl < MF

As electronegativity difference between metal and halogen increases, ionic character increases.

3. $F_2 > Cl_2 > Br_2 > I_2$

This is incorrect order of bond dissociation energy. Correct order is $Cl_2 > Br_2 > F_2 > I_2$ due to electronic repulsion among lone pairs in F_2 molecule.

4. HI < HBr < HCl < HF

As the size of the halogen atom increases, the halogen-hydrogen bond strength decreases.

27. **(2, 4)** Structure of P_4 molecule can be represented as below

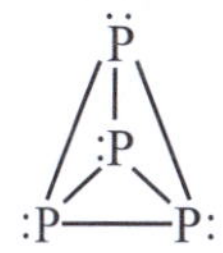

It has total four lone pairs of electrons situated at each P-atom.

It has six P – P single bond.

28. **(1, 3, 4)** (1) Among halogens, radius ratio between iodine and fluorine is maximum because iodine has maximum radius and fluorine has minimum radius.

(2) It can be correctly stated as in general interhalogen compounds are more reactive than halogens. This is because X–X' bond in interhalogen is weaker than that of X–X bond in halogens except F–F bond.

(3) Among interhalogen compounds maximum number of atom are present in iodine fluoride because radius ratio of iodine and fluorine has maximum value.

(4) Interhalogen compounds are more reactive than halogen due to weaker X – X' bond as compared to X – X of halogen compounds.

29. **(3, 4)** (1) All the three N – O bond lengths in HNO_3 are not equal.

(2) All P – Cl bond length in PCl_5 molecule in gaseous state are not equal. Axial bond is longer than equatorial bond.

(3) P_4 molecule in white phosphorous has angular strain therefore white phosphorus is very reactive.

(4) PCl_5 is ionic in solid state in which cation is tetrahedral and anion is octahedral.

Cation – $[PCl_4]^+$

Anion – $[PCl_6]^-$

30. **(1, 4)** (1) $\xrightarrow[\text{acidic strength increases}]{As_2O_3 < SiO_2 < P_2O_3 < SO_2}$

(2) Correct order is $NH_3 > AsH_3 > PH_3$

(3) $O < S < F < Cl$ More negative electron gain enthalpy

(4) $H_2O > H_2S > H_2Se > H_2Te$ Thermal stability decreases on moving top to bottom due to increase in its bond length.

31. **(1, 2)** (1) Structure of $H_2S_2O_6$ is as shown below

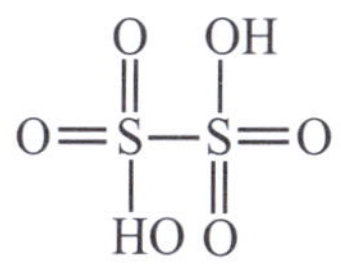

It contains one S–S bond.

(2) In peroxosulphuric acid (H_2SO_5) sulphur is in + 6 oxidation state.

Let oxidation state of S = x

$$2 \times (+1) + x + 3 \times (-2) + 2 \times (-1) = 0$$

$$x - 6 = 0$$

$$x = 6$$

(3) During preparation of ammonia, iron oxide with small amount of K_2O and Al_2O_3 is used as a catalyst to increase the rate of attainment of equilibrium.

(4) Change in enthalpy is negative for preparation of SO_3 by catalytic oxidation of SO_2.

32. (2, 3) In the given four reactions, (2) and (3) represent oxidising behaviour of H_2SO_4. As we know that oxidising agent reduces itself,

Here,

$$2H\overset{-1}{I} + H_2\overset{+6}{S}O_4 \longrightarrow \overset{0}{I_2} + \overset{+4}{S}O_2 + 2H_2O$$

$$\overset{0}{Cu} + 2H_2\overset{+6}{S}O_4 \longrightarrow \overset{+2}{Cu}SO_4 + \overset{+4}{S}O_2 + 2H_2O$$

Since, S in H_2SO_4 is reducing in both the reactions, therefore, it acts as the oxidising reagent in both of these reactions.

33. (1, 2) (1) The only type of interactions between particles of noble gases are due to weak dispersion forces.

(2) Ionization enthalpy of molecular oxygen is very close to that of xenon.

(3) Hydrolysis of

$XeF_6 (\overset{+6}{Xe}\overset{-1}{F_6} + 3\overset{+1}{H_2}\overset{-2}{O} \longrightarrow \overset{+6}{Xe}\overset{-2}{O_3} + 6\overset{+1}{H}\overset{-1}{F})$ is not a redox reaction.

(4) Xenon fluorides are highly reactive and readily undergoes hydrolysis even by traces of water.

34. (3) The reducing character of H_3PO_2 acid is due to two P–H bonds and presence of one –OH group

35. (1) In the ClF_3, Cl atom is sp^3d hybridised, having trigonal bipyramidal geometry, in which axial bonds are longer than equatorial bonds.

The structure is:

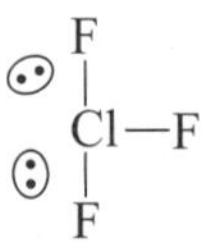

Hence, two Cl–F (axial bonds) are not equal to Cl–F (equatorial bond).

36. (1) Basicity of an acid is the number of replaceable H^+ ion the acid can furnish in its aqueous state.

Phosphinic acid is H_3PO_2 which is monobasic acid, hence basicity is 1.

37. (1) (A) Phosphoric acid $-H_3PO_4$. It contains P in +5 oxidation state.

(B) Marshall's acid $-H_2S_2O_8$. It contains sulphur in its +6 oxidation state and a peroxide group.

(C) Perchloric acid-$HClO_4$. It contains Cl in +7 oxidation state.

(D) Sulphurous acid-H_2SO_3. It contains S in + 4 oxidation state.

Thus, A-2, B-1, C-4, D-3

NEET & AIIMS Past Year MCQs

1. (4)

(Atom/Molecule)	(Property)
Nitrogen atom	Element with highest I.E. in group 15
Fluorine molecule	Strongest oxidising agent
Oxygen molecule	Paramagnetic in nature
Xenon atom	Most reactive element in group 18

The correct match is A-III, B-IV, C-I, D-II

2. (2)

- PEt_3 and $AsPh_3$ can act as ligands that are capable of creating $d\pi$-$d\pi$ bonds with transition metals.
- The N – N single bond is weaker than the P – P single bond due to the increased inter-electronic repulsion of the non-bonding electrons.
- Nitrogen possesses a distinct capability to establish $p\pi$-$p\pi$ multiple bonds with itself, carbon, and oxygen.
- Unlike other heavier elements in its group, nitrogen is unable to form $d\pi$-$p\pi$ bonds.

3. (1) Higher the bond dissociation energy, lower is the degree of ionization and hence weaker is the acid. Since the bond dissociation energies of the halogen acids increase in the order:

HF > HCl > HBr > HI.

Therefore, strength of the acids increases in the reverse direction, i.e.,

HF < HCl < HBr < HI

4. (3) The boiling point of hydrides of group 16 are in the order $H_2O > H_2Te > H_2Se > H_2S$.

It is due to the presence of extensive H-bonding in H_2O.

5. (2) Po does not show –2 oxidation state because on going down the group, electropositive nature increases.

6. (4) BF_3 behaves as Lewis acid due to incomplete octet in valence shell of Boron. Hence it can accept a lone pair of electrons.

7. (3)

$H_2S_2O_8$, peroxodisulphuric acid

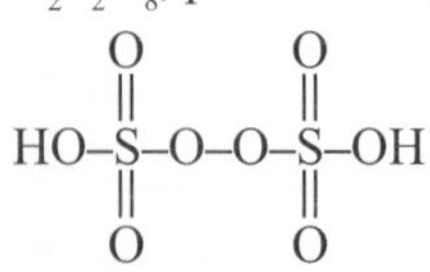

(Two S-OH, Four S=O,One S-O-O-S)

H_2SO_3, sulphurous acid

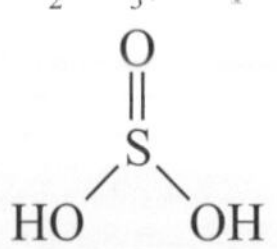

(Two S-OH, One S=O bond)

$H_2S_2O_7$, pyrosulphuric acid

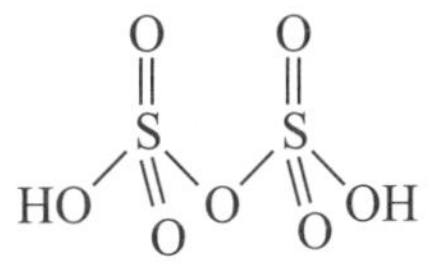

(Two S-OH, Four S = O, One S-O-S bond)

H_2SO_4, sulphuric acid

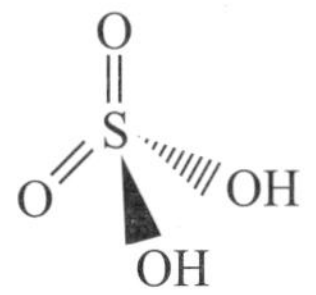

(Two S-OH, Two S=O)

8. (2) X–X' bond in interhalogens is weaker than X–X bond in halogens except F–F bond. Therefore, I–Cl bond is weaker than I–I bond. So, the reactivity of I–Cl compound is greater than I–I compound.

9. (3) The correct order of boiling point of given hydrides is

$H_2O > H_2Se > H_2S > H_2Te$

Molar mass of H_2Se is greater than H_2O but boiling point of H_2O is greater than H_2Se.

Therefore, statements I & II both are incorrect.

10. (1) Noble gases have low melting and boiling points because of presence of weak intermolecular forces between them.

11. (1) The acidic character is explained on the basis of bond dissociation enthalpy. On moving down the group, bond dissociation enthalpy of hydrides of group 16 elements decreases.

Hence, acidic character increase from H_2O To H_2Te.

Thus the pK_a value decreases because stronger is the acid, lower is the value of pK_a.

Correct order of pK_a value will be

$H_2O > H_2S > H_2Se > H_2Te$

12. (2) N_2 from azide is also produced by NaN_3. Hence mass of Ba is irrelevant.

13. (4) HOCl bond angle is greater than HOF bond angle, this is due to very small size of fluorine on due to bp-bp repulsion.

Oxygen is more electronegative than all halogen except florine (most electronegative).

14. (3) S^{2-}

15. (1) Both assertion and Reason are true and reason is correct explantion of asseration given.

16. (1) $2NaOH + Cl_2 \rightarrow NaCl + NaOCl + H_2O$; hence Cl^- and ClO^-

17. (4) $PH_4I + KOH \rightarrow PH_3 + KI + H_2O$

The d-and f-Block Elements

NCERT Topic-wise MCQs

1. (3) The correct statement is that iron belongs to first transition series of elements. It is called 3d series. The members of this series are

Sc Ti V Cr Mn Fe Co Ni Cu Zn

The electronic configuration of Fe^{26} is $3d^6 4s^2$

2. (1) $Ti^{4+} = (22 - 4 = 18) = [Ar]\ 3d^0$
$V^{4+} = (23 - 4 = 19) = [\ Ar\ ]\ 3d^1$
$Mn^{2+} = (25 - 2 = 23) = [Ar]\ 3d^5$
$Fe^{3+} = (26 - 3 = 23) = [Ar]\ 3d^5$
$\because$ Ti is not having any unpaired electron. It is colourless in nature.

3. (2) Zn (30) : $[Ar]\ 4s^2 3d^{10}$

4. (2) $Zn^{2+} = 3d^{10}$ = No unpaired electrons
$Ni^{2+} = 3d^8$ = 2 unpaired electrons
$Fe^{2+} = 3d^6$ = 4 unpaired electrons
$Cu^+ = 3d^{10}$ = no unpaired electrons

5. (4) Fe = 26 = $[Ar]\ 3d^6\ 4s^2$
$Fe^{3+} = [Ar]\ 3d^5$
$\therefore$ n =5
$\mu = \sqrt{n(n+2)} = \sqrt{5(5+2)} = \sqrt{5(7)} = \sqrt{35}$
$= \sqrt{35} = 5.91 \approx 6$ BM

6. (2) +8 is the highest oxidation state exhibited by transition metal.

7. (4) $\mu = \sqrt{n(n+2)}$ BM

8. (4) Paramagnetic behaviour of transition metals is due to the presence of unpaired electrons.

9. (2) Fe = 26 = $[Ar]\ 3d^6\ 4s^2$
$Fe^{3+} = [Ar]\ 3d^5$
Mn = 25 = $[Ar]\ 3d^5\ 4s^2$
$Mn^{2+} = [Ar]\ 3d^5$

10. (1) $\mu = \sqrt{n(n+2)}$
$\because$ Atomic number is 25, therefore metal is Mn.
$Mn^{+2} = [Ar]\ 3d^5$ = [1|1|1|1|1]
$\therefore$ n = 5
$\mu = \sqrt{5(5+2)} = \sqrt{5(7)} = \sqrt{35} = 5.9$BM

11. (3) $Fe^{3+} = 3d^5$ = [1|1|1|1|1] = 5
$Cr^{3+} = 3d^3$ = [1|1|1| |] = 3
$Ni^{2+} = 3d^8$ = [1l|1l|1l|1|1] = 2
$Cu^{2+} = 3d^9$ = [1l|1l|1l|1l|1] = 1

12. (4) Sc → $[Ar]\ 3d^1\ 4s^2$ [1| | | |]
Cu → $[Ar]\ 3d^{10}\ 4s^1$ [1l|1l|1l|1l|1l] [1]

13. (3) As energy difference between 3d and 4s is small, both these electrons are involved in bonding. Hence, the oxidation state exhibited by the elements of 3d-series are maximum.

14. (2) Zn has completely filled 3d-subshell hence, its oxidation state is decided only by 4s electrons, i.e., + 2

15. (1) Outer configurations are : Cr = $3d^5\ 4s^1$, Cu = $3d^{10}\ 4s^1$

16. (4) Ti^{3+} = [Ar] $3d^1\ 4s^0$ $Ti^{4+} = 3d^0\ 4s^0$

Mn^{3+} = [Ar]$3d^4\ 4s^0$ Mn^{2+} = [Ar] $3d^5$

Na = $1s^2\ 2s^2\ 2p^6\ 3s^1$ $Na^+ = 1s^2\ 2s^2\ 2p^6$

Fe^{3+} = [Ar]$3d^5$ (half filled) Fe^{2+} = [Ar]$3d^6$

17. (4) Ni_{28} = [Ar] $3d^8\ 4s^2$

Ni^{2+} = [Ar] $3d^8$ [⇅|⇅|⇅|↑|↑]

∴ n =2

$\mu = \sqrt{n(n+2)} = \sqrt{2(2+2)} = \sqrt{2(4)} = \sqrt{8} = 2.84\ BM$

18. (1) Cu^+ = [Ar] $3d^{10}\ 4s^0$

∴ There is no unpaired electron. Hence, the transition metal Cu^+ is colourless.

19. (1) $Ti^{3+} = 3d^1$ = [↑| | | |] = 1.75 BM.

$Ni^{2+} = 3d^8$ = [⇅|⇅|⇅|↑|↑] = 2.76 BM.

$Co^{2+} = 3d^7$ – [⇅|⇅|↑|↑|↑] – 3.86 BM.

$Mn^{2+} = 3d^5$ = [↑|↑|↑|↑|↑] = 5.96 BM.

20. (2) Homogeneous mixture of two or more metals is alloy. Mercury forms amalgams (alloy) with gold, silver, tin, Na. But it does not react with iron and platinum.

21. (3) An alloy of metals and mercury is amalgam.

∴ Mercury (Hg) is the essential component of amalgam.

22. (2) $CuCl_2 = Cu^{2+}$ [Ar] $3d^9\ 4s^0$ = [⇅|⇅|⇅|⇅|↑] = 1

$TiCl_3 = Ti^{3+}$ = [Ar] $3d^1$, $4s^0$ = [↑| | | |] = 1

∴ $CuCl_2$ and $TiCl_3$ has same number of unpaired electrons.

23. (2) Fe = [Ar] $3d^64s^2$ Fe^{3+} = [Ar] $3d^5$

Thus; ferric ion is quite stable due to half-filled d-orbitals.

24. (2) $Na_2\ [CuCl_4] \Rightarrow$ Cu = +2 = $Cu^{2+} = 3d^9$ [⇅|⇅|⇅|⇅|↑]

$Na_2\ [CdCl_4] \Rightarrow$ Cd = + 2 = $Cd^{2+} = 4d^{10}$ [⇅|⇅|⇅|⇅|⇅]

$K_4\ [Fe(CN)_6] \Rightarrow$ Fe = + 2 = $Fe^{2+} = 3d^6$ [⇅|↑|↑|↑|↑]

$K_3\ [Fe(CN)_6] \Rightarrow$ Fe = +3 = $Fe^{3+} = 3d^5$ [↑|↑|↑|↑|↑]

25. (2) The electronic configuration of X^{3+} is [Ar] $3d^5$

∴ Atomic number of X = 18 + 5 + 3 = 26

26. (3) Among the given, Fe^{3+} is not an amphoteric ion.

27. (1) Both elements belong to 3d series.

Cu → [Ar] $3d^{10}\ 4s^1$

Zn → [Ar] $3d^{10}\ 4s^2$

Same configuration of (n –1) d subshell.

28. (2) Cr(24) → [Ar] $3d^5\ 4s^1$

Cr has higher second ionisation enthalpy as after removing one electron from Cr, the resulting configuration, i.e., [Ar] $3d^5$ becomes more stable.

29. (4) Fe ($3d^6\ 4s^2$), n = 4 ; Fe^{2+} ($3d^6$), n = 4

Fe^{4+} ($3d^4$), n = 4 ; Fe^{3+} ($3d^5$), n = 5

30. (4) $Fe^{2+} \rightarrow 3d^6$, $Sc^{4+} \rightarrow 3p^5$, $Mn^{3+} \rightarrow 3d^4$

$Mn^{2+} \rightarrow 3d^5$ (Exactly half-filled, hence stable)

31. (1) No. of unpaired electrons are:

Ti^{3+} = 1 , Cu^+ = 0, Zn^{2+} = 0, Sc^{3+} = 0

32. (2) Configuration of Cu is [Ar] $3d^{10}\ 4s^1$. Hence; the electron is lost from 4s subshell.

33. (3) Isoelectronic species are those species which contains same number of electrons

34. (1) $Cr^{3+} \rightarrow 3d^3$, contains three unpaired electrons.

35. (1) Ti (22) and Ni (28) both contains 2 unpaired electrons in 3d and 2 electrons in 4s.

36. (1) V^{3+} contains 2 unpaired electrons, but others do not.

37. (4) The number of unpaired electrons are

Co^{2+} = 3, Ni^{2+} = 2, Fe^{2+} = 4, Mn^{2+} = 5

38. (4) As they do not contain any unpaired electrons, hence; this ion is colourless in nature.

39. (4) The elements of 3d-series will give coloured compounds, if its ions contain vacant or unpaired electron in 3d orbitals.

40. (3) Both Ti^{4+} and Cu^+ do not contain any unpaired electron.

41. (3) Density of transition elements increase along the period.

42. (3) The d-block elements include only metals.

43. (1) Cr = 24 = [Ar] $3d^5\ 4s^1$

44. (2) Transition metal in lower oxidation state form cations. These cations will act as lewis acids.

e.g: $Fe^{+2} + 2e^- \rightarrow Fe$

(Lewis acid)

45. (2) Cr^{+2} and Fe^{+2}

$Cr^{+2} – 3d^4$ (4 unpaired electrons)

$Fe^{+2} – 3d^6$ (4 unpaired electrons)

46. (4) p-electrons in Cl^- (atomic no. of Cl = 17)

$Cl^- \rightarrow 1s^2\ 2s^2\ 2p^6\ 3s^2\ 3p^6$

$Fe^{2+} \rightarrow 1s^2\ 2s^2\ 2p^6\ 3s^2\ 3p^6\ 3d^6$

In Fe^{2+}, total number of d–e^- = 6 which is not equal to p–e^- in Cl^- = 12.

47. (2) V = $1s^2\ 2s^2\ 2p^6\ 3s^2\ 3p^6\ 3d^3\ 4s^2$

Cr = $1s^2\ 2s^2\ 2p^6\ 3s^2\ 3p^6\ 3d^5\ 4s^1$

Mn = $1s^2\ 2s^2\ 2p^6\ 3s^2\ 3p^6\ 3d^5\ 4s^2$

Fe = $1s^2\ 2s^2\ 2p^6\ 3s^2\ 3p^6\ 3d^6\ 4s^2$

In second ionization enthalpy, Cr^+ has exact half filled d-sub shell.

48. (3) Magnetic moment is depends on the no. of unpaired electrons.

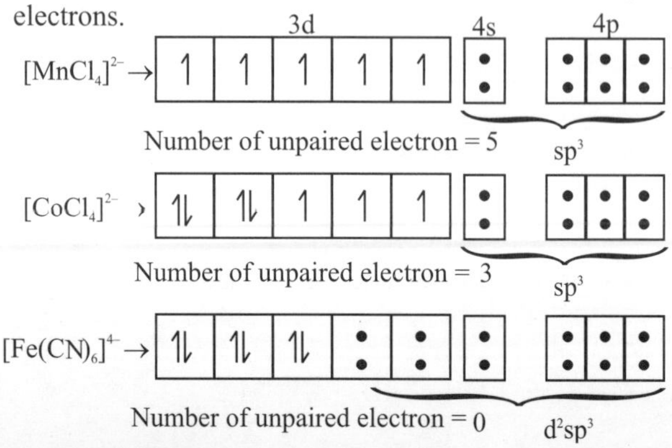

49. (1) Fe^{3+} ion has $[Ar]3d^5$ configuration, hence number of unpaired electron is 5.

Magnetic moment = $\sqrt{n(n+2)}$

$$= \sqrt{5(5+2)} = \sqrt{35} = 5.92\,\text{B.M.}$$

50. (4) Ni^{2+} in presence of CN^-

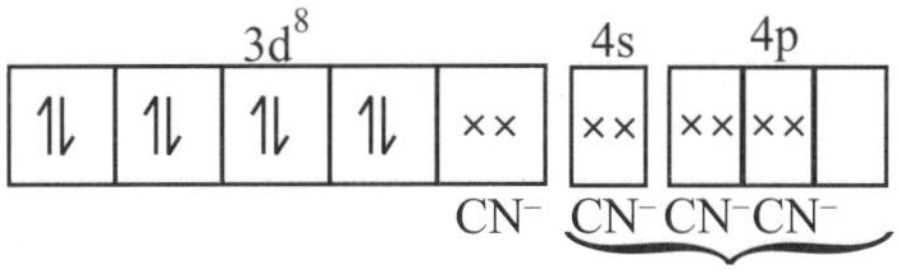

dsp^2 Hybridization

As $[Ni(CN)_4]^{2-}$ has no unpaired electron, hence it is diamagnetic.

51. (2) Mn is used to make alloy steel for armour plates, safes and helmets.

52. (2) Solder – Sn 67% and Pb 33%.

53. (2) Half filled orbital

54. (3) Among the given elements, Pd belongs to second (4d) transition series, rest other elements belong to first transition series.

55. (1) Cr has highest melting point and boiling point due to maximum number of unpaired electrons.

56. (3) The increasing order of density is as follows:

Zn < Cr < Fe < Co < Cu

	Zn	Cr	Fe	Co	Cu
Density/g cm^{-3}	7.1	7.19	7.8	8.7	8.9

57. (2) $[Ar]\ 3d^5\ 4s^1$, $[Ar]\ 3d^{10}\ 4s^1$ are the correct electronic configurations of Cr and Cu respectively.

58. (2) When d-block elements form ions, electrons from ns orbitals are lost first then electrons lost from (n – 1) d orbitals.

59. (2) Mn = +2, + 3, + 4, + 5, + 6, + 7

60. (2) Electronic configuration of Co is $1s^2\ 2s^2\ 2p^63\ s^23p^6\ 3d^74s^2$

Therefore, no. of unpaired electrons = 3

61. (4) Highest oxidation state would be

no. of s-e$^-$ + no. of d-e$^-$

2 + 2 = 4

62. (1) Magnetic moment = $\sqrt{24} = \sqrt{4(4+2)}$

Therefore, number of unpaired e$^-$ = 4

Fe^{2+} : $3d^6\ 4s^0$ (n = 4)

As, n = 4 ∴ Fe^{2+}

Hence, Fe must have +2 charge.

63. (3) The ability of d-block elements to form complexes is due to the small and highly charged ions & vacant low energy orbital to accept lone pair of electrons from ligands.

64. (1) The most oxidizing state in aqueous solution of manganese is +7 ($KMnO_4$) .

65. (3)

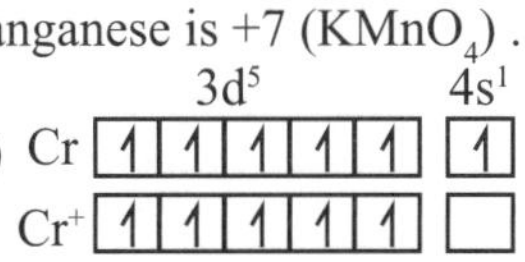

Hence, 5 unpaired electrons would be present in Cr^+.

66. (4) There are 6 electrons present in the outermost shell of Cr, hence the highest oxidation state exhibited by Cr would be 6.

67. (4) Transition elements form co-ordinate compounds due to

(i) High nuclear charge

(ii) Small size

(iii) Vacant d-orbital

68. (1) Ni^{2+} and Cr^{3+} are coloured due to presence of unpaired electrons, but Zn^{2+} is colourless due to absence of unpaired electrons.

Zn^{2+} : $3d^{10}\ 4s^0$

Ni^{2+} : $3d^8\ 4s^0$

Cr^{3+} : $3d^3\ 4s^0$

69. (2) Zinc and mercury have completely filled orbitals, hence they do not show any variable valency.

70. (2) Diamagnetic substances get repelled by the applied field while the paramagnetic substances are attracted.

71. (1) Ti^{4+} metal ion is colourless due to no unpaired electrons in d-orbitals.

72. (1) Colourless compounds are those which have no unpaired electrons and paramagnetic substance contain unpaired electrons. Therefore, paramagnetic substance possess colour.

73. (2) $2MnO_2 + 4KOH + O_2 \rightarrow 2K_2MnO_4 + 2H_2O$

74. (4) Increase in atomic number will generally decrease basic character.

75. (2) $Cr_2O_7^{-2} + 6I^- + 14H^+ \rightarrow 2Cr^{+3} + 3I_2 + 7H_2O$

76. (1) MnO_2, MnO_2, Mn^{2+}

In neutral medium :

$2KMnO_4 + 3MnSO_4 + 2H_2O \rightarrow K_2SO_4 + 2H_2SO_4 + 5MnO_2$

In alkaline medium :

$2KMnO_4 + H_2O \rightarrow 2MnO_2 + 2KOH + 3\,[O]$

In acidic medium :

$$2KMnO_4 + 3H_2SO_4 \rightarrow K_2SO_4 + 2\overset{+2}{Mn}SO_4 + 5H_2O + 5O$$

77. (3)

K_2CrO_4	$K_2Cr_2O_7$
$2(+1) + x + 4(-2) = 0$	$2(+1) + 2x + 7(-2) = 0$
$2 + x - 8 = 0$	$2 + 2x - 14 = 0$
$x - 6 = 0$	$2x - 12 = 0$
$x = +6$	$x = +6$

78. (2) K_2MnO_4

$2(+1) + x + 4(-2) = 0$

$2 + x - 8 = 0$

$x - 6 = 0 \Rightarrow x = +6$

79. (1)

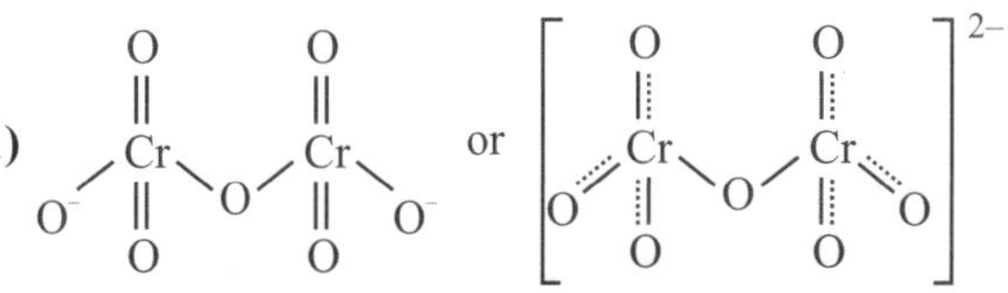

80. (1) Oxidation number of Cr in options 1, 2, 3 and 4 are +6, +4, + 3, + 2 respectively.

In given options (1) has highest oxidation number, therefore its radii will be small. Atomic radii decreases with increase in oxidation number.

81. (2) $KMnO_4$ is first reduced to manganate and then to insoluble manganese dioxide. Colour changes first from purple to green and finally becomes colourless.

82. (1) CrO_3 and Mn_2O_7 are acidic oxide since they react with water to form acid.

$$CrO_3 + H_2O \rightarrow \underset{\text{Chromic acid}}{H_2CrO_4}\ ;\ Mn_2O_7 + H_2O \rightarrow \underset{\text{Permanganic acid}}{2HMnO_4}$$

83. (3) In acidic medium, CrO_4^{2-} changes to $Cr_2O_7^{2-}$

84. (1) The chromate ion (yellow) is the predominant species in alkaline solutions, dichromate ion (orange) is the predominant species in acidic solutions. When an alkali is added to an orange red solution of dichromate, a yellow solution is obtained due to the formation of chromate ions.

85. (4) Oxidation state of Mn in $KMnO_4$ is +7.

$KMnO_4$

$1 + x - 8 = 0 \Rightarrow x = 7$

86. (1) $KMnO_4$ forms dark purple (almost black) crystals which are isostructural with those of $KClO_4$.

87. (2) Oxidising action of potassium dichromate in acidic solution can be represented as:

$$Cr_2O_7^{2-} + 14H^+ + 6e^- \longrightarrow 2Cr^{3+} + 7N_2O$$

88. (1) $K_2Cr_2O_7$ is not hygroscopic in nature and hence is used as a primary standard in volumetric analysis.

89. (1) The green manganate ion is paramagnetic because of one unpaired electron.

90. (2) The common oxidation state of lanthanoids is + 3

91. (2) The atomic and ionic radii of Zr and Hf are equal in size due to lanthanoid contraction.

92. (1) The general electronic configuration of f-block elements is $(n-2)f^{1-14}\,(n-1)d^{0-1}\,ns^2$

93. (3) In lanthanoids, electrons enter 4f - subshell

$\therefore n - 2 = 4 \quad \text{or} \quad n = 6$

94. (1) The accepted electronic configuration of Yb (70) is $4f^{14}\,5d^0\,6s^2$. In 4f subshell, all electrons are paired.

95. (2) $Gd \rightarrow 4f^7\,5d^1\,6s^2$

Number of unpaired electrons : 7 in 4f and 1 in 5d.

96. (3) In +3 oxidation state, its configuration is $4f^7\,5d^0\,6s^0$, which is stable as per the rule of extra stability.

97. (1) Lanthanum is a d-block element which resembles lanthanides.

98. (4) Different amount of shielding is experienced by 4f and 5f as they belong to different shell and hence, 4f and 5f orbitals are not equally shielded.

99. (2) The atomic number of 4f-series range from 58 to 71

100. (4) La :– [Xe] $5d^1 6s^2$, La^{3+} :– [Xe] $5d^0\,6s^0$

Lu:– $4f^{14}\,5d^1\,6s^2$, Lu^{3+} :– $4f^{14}\,5d^0\,6s^0$

Yb:– $4f^{14}\,5d^1\,6s^2$, Yb^{2+}:– $4f^{14}\,5d^0\,6s^0$

These are diamagnetic, but

Sm:– $4f^6\,5d^0\,6s^2$; Sm^{3+}:– $4f^5\,5d^0\,6s^0$ (Paramagnetic)

101. (4) Total number of inner transition elements in periodic table is 28 i.e., 14 each in lanthanoid and actinoid series.

102. (2) The electronic configuration of Ce : [Xe] $4f^1\,5d^1\,6s^2$

103. (2) Due to the smaller energy gap between 5f, 6d and 7s orbitals, actinoids show variable oxidation states but the actinoids show in general +3 oxidation state.

104. (4) Since the properties of Lanthanum are very similar to the elements of 4f block, hence it is grouped with f-block elements.

105. (4) Among the given options, Terbium belongs to 4f- series with electronic configuration $[Xe]4f^9\,6s^2$, hence Terbium is lanthanide. However, rest other members belong to 5f-series (actinides).

106. (1) + 4 oxidation state of Ce is also known in the solution.

107. (2) The ground state electronic configuration of Th is [Rn] $6d^2\,7s^2$.

108. (4) In the Wacker process, the oxidation of ethyne to ethanal is catalysed by $PdCl_2$.

109. (1) TiO is used in the pigment industry.

110. (4) MnO_2 is used in dry battery cells.

111. (2) Wacker process - $PdCl_2$

Ziegler catalyst - $TiCl_4 + Al(CH_3)_3$

Haber process - Fe

High Order Time Intensive MCQs

1. (3) In reaction 3, CrO_2Cl_2 will be formed rather than CrO_3. It is chromyl chloride test.

2. (1) Os shows + 8 oxidation state in OsO_4.

3. (1) Cr^{2+} acts as reducing agent because it itself gets oxidised to Cr^{3+} giving it d^3 configuration, whereas Mn^{3+} acts as oxidising agent giving it $3d^5$ configuration.

4. (4) $H_2SO_4 + KMnO_4 \rightarrow K_2SO_4 + Mn_2O_7 + H_2O$

5. (2) Resonance occurs

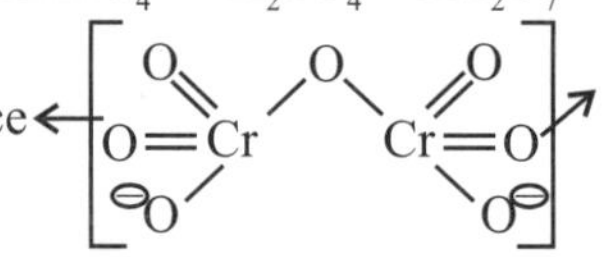

Resonance occurs

All 6 - bonds are equivalent due to resonance.

6. (4) $M(s) \rightarrow M^{n+}(aq) + ne^-$

Above change involves sublimation, ionization and hydration.

7. (1) Poor shielding of one of the 4f - electrons by another in the sub - shell causes lanthanide contraction.

8. (3) MnO_2 Fuses with KOH in presence of air or oxidising agents like KNO_3, $KClO_3$ etc.

$$2MnO_2 + 4KOH + O_2 \xrightarrow{\text{fuse}} 2K_2MnO_4 + 2H_2O$$

$$MnO_4^{2-} \xrightarrow{\text{electrolytic oxidation}} MnO_4^- + e^-$$

9. (4) All the given metals react with chlorine and form their respective chlorides.

10. (4) Lanthanide contraction occurs because the poor shielding effect of 4f electrons is coupled with increased attraction between the nucleus and the added electrons.

11. (4) The electronic configuration of Lu is [Xe] $4f^{14}\,5d^1\,6s^2$. Lu is a lanthanide element.

12. **(2)** Transition metal oxides in higher oxidation states are acidic in nature. Mn_2O_7 is acidic as it has highest O.S.

13. **(4)** La^{3+} - $4f^0$

Lu^{3+} - $4f^{14}$

Yb^{2+} - $4f^{14}$

Sm^{3+} contains unpaired electrons.

$Sm^{3+} = 4f^5$

14. **(3)** [Xe] $4f^{14}$ $5d^{10}$ $6s^1$ among the given electronic configurations is not the configuration of lanthanoid.

15. **(3)** Zr and Hf have almost identical radii due to lanthanoid contraction.

16. **(2)** Due to lanthanide contraction as the size of lanthanide ions decreases from La^{3+} to Lu^{3+}, the covalent character of the hydroxides increases & hence the basic strength decreases. Hence, $La(OH)_3$ is most basic.

17. **(4)** These are p block elements.

18. **(2)** Cr_2O_3 is amphoteric in nature.

CrO_3 is acidic

CrO is basic.

19. **(2)** The inner transition elements are the elements in which the added electrons go to (n – 2) f - orbital.

20. **(3)** Configuration of Cr is supposed to be $3d^44s^2$ but actual configuration is $3d^54s^1$.

21. **(4)** All the given factors make the separation of lanthanides a formidable task.

22. **(2)** Lanthanides are characterized by the filling of the antipenultimate 4f energy level.

23. **(1)** HgO is in +2 O.S, hence is weakly amphoteric in nature.

24. **(4)** In Gun metal alloy, transition metal is Cu.

25. **(4)** Non-typical transition Metals are those having completely filled d-orbital.

Non typical transition elements are Zn, Cd, and Hg.

26. **(4)** Ag belongs to the second transition series.

27. **(3)** $10\ FeSO_4 + 2KMnO_4 + 8H_2SO_4 \rightarrow 5Fe_2(SO_4)_3 + K_2SO_4 + 2\ MnSO_4 + 8H_2O$

$FeSO_4$ is oxidised to $Fe_2(SO_4)_3$

$KMnO_4$ is reduced to $MnSO_4$

28. **(4)** The d -block of the periodic table contains the elements of the groups 3-12.

29. **(3)** The general electronic configuration of d - block elements is $(n-1)d^{1-10}ns^{1-2}$.

30. **(4)** Steel is an alloy which contains iron (a metal) and carbon (a non metal).

31. **(3)** Cu - [Ar] $3d^{10}4s^1$

Pd - [Kr] $4d^{10}5s^0$

Pt - [Xe] $4f^{14}5d^96s^1$

Because in these elements the electrons are filled in higher energy level first then in lower energy level.

Aufbau rule of filling electrons follow the sequence:

$1s > 2s > 2p > 3s > 3p > 4s > 3d > 4p > 5s > 4s$

32. **(3)** $Mn \rightarrow [Ar]\ 3d^54s^2$

In MnO_4, Mn shows + 8 oxidation state

Mn cannot show +8 oxidation state.

33. **(1)** Among the given elements, Mn exhibits maximum oxidation state

34. **(3)** Zinc belongs to d-block but it is not regarded as transition element because its d-orbital is completely filled.

35. **(1)**

Metal	Chloride in + 1	+ 2 oxidation states
Cu	CuCl	Cu_2Cl_2
Ag	AgCl	–
Na	NaCl	–

36. **(2)** Transition metals form interstitial compounds.

37. **(4)** When $KMnO_4$ acts as oxidising agent, the oxidation state of Mn decreases from +7 to +2.

$Mn^{+7} \rightarrow Mn^{2+}$

Thus, the oxidation state decreases by 5.

38. **(2)** $CrO_2Cl_2 + 4NaOH \rightarrow \underset{\text{yellow}}{Na_2CrO_4} + 2NaCl + 2H_2O$

39. **(2)** $K_2Cr_2O_7 \xrightarrow{\Delta} K_2CrO_4 + \underset{\text{green}}{Cr_2O_3} + O_2\uparrow$

40. **(4)** (1) Ce^{4+} has a noble gas configuration. So, Ce exhibits +4 oxidation state.

(2) La, Gd and Lu are stable in +3 oxidation state.

(3) As the size of lanthanoid ions decreases from Lu^{3+} to La^{3+}, the basic strength decreases from $La(OH)_3$ to $Lu(OH)_3$.

41. **(1)**

	Ions	Number of unpaired electrons
1.	V^{2+}	3
2.	Cr^{2+}	4
3.	Mn^{2+}	5
4.	Fe^{2+}	4

So, order of paramagnetic behaviour

$Mn^{2+} > Cr^{2+} = Fe^{2+} > V^{2+}$

42. **(2)** $K_2\overset{+6}{Cr_2}O_7 + 4H_2SO_4 + 3H_2S \longrightarrow$

$K_2SO_4 + \overset{+3}{Cr_2}(SO_4)_3 + 7H_2O + 3S$

Therefore, the oxidation number of chromium decreases from + 6 to +3

43. **(3)** Potassium dichromate, on heating gives oxygen and chromic oxide (Cr_2O_3).

$4K_2Cr_2O_7$

$\downarrow$

$4K_2Cr_2O_4 + 2Cr_2O_3 + 3O_2\uparrow$

44. **(1)** Due to Lanthanoid contraction, order of ionic radii will be:

$Yb^{+3} < Pm^{3+} < Ce^{+3} < La^{3+}$

45. **(3)** The colour changes due to the conversion of dichromate to chromate.

$K_2Cr_2O_7 + 2KOH \rightarrow \underbrace{2K_2CrO_4}_{\text{yellow}} + H_2O$

46. **(3)** Stainless steel contains Fe + Cr + Ni.

47. **(2)** Stainless steel does not rust because chromium forms an oxide layer and protect iron from rusting.

48. **(3)** The transition series are those which have incomplete inner orbitals.

49. (3) Strong oxidising agents such as PbO_2 oxidise Mn^{2+} to MnO_4^- or Mn^{7+}.

50. (2) $K_2Cr_2O_7 + 3Na_2SO_3 + 4H_2SO_4 \rightarrow 3Na_2SO_4 + K_2SO_4 + 4H_2O + Cr_2(SO_4)_3$

51. (1) Electronic configuration of Neodymium (Z = 60) is; [Xe] $4f^4 6s^2$

52. (1) With an increase in oxygen percentage, an element's oxide becomes more acidic and less basic.

Hence, the correct order is given by option (1).

53. (3) $2HgO \xrightarrow{\Delta} 2Hg + O_2$

54. (1) Ag_2O is mild oxidising agent as greater the oxidation number of metal, stronger is the oxidising agent.

55. (3) Amalgams are alloys which contain mercury as one of the contents.

56. (3) The passivity of iron is due to the formation of a thin insoluble and invisible iron film on surface which prevents its further reactions. The film is due to the formation of Fe_3O_4.

57. (3) Silver; $AgNO_3 + NaCl \rightarrow \underset{\text{White ppt.}}{AgCl} + NaNO_3$

58. (4) $Zn + 2NaOH \rightarrow \underset{\text{(Sodium Zincate)}}{Na_2ZnO_2} + H_2$

59. (3) Very dil HNO_3: $\rightarrow$ Ammonium nitrate is formed

$4\,Zn + 10\,HNO_3 \rightarrow 4\,Zn(NO_3)_2 + NH_4NO_3 + 3H_2O$

60. (3) Tc & Re belong to same group. The atomic radius of elements of second transition series is similar to that of third transition elements, due to poor shielding of f-orbitals.

61. (2) Across lanthanide series, due to lanthanide contraction, size of cation decreases, covalent character increase and basic character decrease. So; basicity of hydroxides decreases.

62. (4) The 4f-electrons do not affect the properties of lanthanoids because they lie well in the interior of the atom and hence are not exposed of surrounding.

63. (2) A: Number of unpaired electrons in Sc^{3+} is zero.

Number of electrons in La^{2+} is one.

Hence, Sc^{3+} has less number of unpaired electrons than La^{2+}.

B: Ferromagnetism is considered as an extreme form of paramagnetism.

Correct & Incorrect MCQs

1. (1) Since, Zn, Cd and Hg of group 12 have full d^{10} configuration in their ground state as well as in their common oxidation states and hence, they are not regarded as transition metals.

2. (3) Since half and completely filled sets of orbitals are relatively more stable, hence the electronic configuration of Chromium is $3d^5 4s^1$ instead of $3d^4 4s^2$.

3. (2) The melting and boiling points of transition metals are high because of the involvement of greater number of electrons from (n-1)d in addition to the ns electrons in the interatomic metallic bonding.

4. (3) Titanium (IV) is more stable than Ti(III) or Ti(II). In group 6, Mo(VI) and W(VI) are found to be more stable than Cr(VI).

5. (4) The $E^\ominus$ value for copper is positive. The unique behaviour of Cu, having a positive $E^\ominus$, accounts for its inability to liberate H_2 from acids.

6. (4) MnO_2

Let oxidation state of Mn be x.

x + 2(-2) = 0

x = +4

Hence, the oxidation state of Mn in MnO_2 is +4.

Mn_2O_3

Let oxidation state of Mn be x.

2x + 3(-2) = 0

2x -6 = 0

x = +3

Hence, the oxidation state of Mn in Mn_2O_3 is +3.

7. (4) Since, number of unpaired electron = 1

$\mu = \sqrt{n(n+2)} = \sqrt{1(1+2)} = \sqrt{3} = 1.73$ B.M

8. (4) Decomposition of $KMnO_4$ is carried in presence of sunlight.

9. (3) Sodium dichromate is more soluble than potassium dichromate.

10 (1) The earlier members of the lanthanoid series are quite reactive similar to calcium.

Statement Based MCQs

1. (4) Mn^{2+} is a coloured ion because it contains unpaired electrons.

Mn^{2+} : $3d^5\ 4s^0$

2. (1) Color of a cation depends on no. of unpaired electrons. Half filled or fully filled lanthanides are colourless.

(xf) and (14 – x)f will have same no. of unpaired electron, so they will show same colour.

3. (1) MnO_4^- has +7 oxidation state, therefore there is no electron in d orbital. So, it is a strong oxidising agent.

Mn: $3d^5\ 4s^2$

Mn^{+7}: $3d^0\ 4s^0$

4. (3) $K_2Cr_2O_7$ can be found in pure state and can be accurately weighed, not hygroscopic in nature so it is used as primary standard solution.

Sodium dichromate is more soluble than potassium dichromate.

5. (1) Fe^{3+} : $3d^5$ (Half filled)

Fe^{2+} : $3d^6$

6. (2) Sc^{3+} : $3d^0$ (No unpaired e-)

Ti^{4+} : $3d^0$ (No unpaired e-)

Hence, Sc^{3+} and Ti^{4+} are colourless ions.

Ti^{3+} : $3d^1$ (1 unpaired e-)

V^{4+} : $3d^1$ (1 unpaired e-)

Hence, Ti^{3+} and V^{4+} are coloured ions.

7. (1) Fe^{2+} : $3d^6\ 4s^0$

Since, Fe^{2+} has 4 unpaired electrons, hence Fe^{2+} is paramagnetic in nature.

8. (1) The value for Ce^{4+} / Ce^{3+} is + 1.74 V which suggests that it can oxidise water. However, the reaction rate is very slow and hence Ce(IV) is a good analytical reagent.

Tb^{IV} has half-filled f-orbitals and hence, acts as an oxidant.

9. (2) Cr_2O_3 is amphoteric in nature and CrO is basic in nature.

10. (1) The transition metals and their compounds are known for their catalytic activity. This activity is due to their ability to adopt multiple oxidation states and to form complexes.

Assertion & Reason MCQs

1. (1) CuI_2 is not known as Cu oxidises I^- to I_2.

2. (2) Separation of Zr and Hf is difficult due to lanthanoid contraction both of them have similar radii.

3. (4) Actinoids form more stable complexes as compare to lanthanoids due to presence of weakly bound 5f and 6d orbitals. 5f, 6d have same energy and both of these orbitals uses for bonding.

4. (1) Metals having negative reduction potential liberate H_2 gas.

5. (2) Highest oxidation state of osmium is +8 because it can expand its octet.

6. (2) Both Assertion and Reason are True but Reason is not the correct explanation of the Assertion.

7. (4) The electronic configuration of Mn is $3d^5\,4s^2$.

8. (2) Color is due to LMCT.

9. (3) Cuprous ion is colourless because it has no unpaired electrons in 3d subshell as it has $3d^{10}$configuration. The cupric ion has one unpaired electron as it has configuration $3d^9$.

10. (1) ZnO reacts with HCl forming $ZnCl_2$ & water. With NaOH,it forms sodium zincate and water

Multi Correct MCQs

1. (1) The electronic configurations of metals are given below:

A $_{62}Sm$: $4f^6 6s^2$
B $_{63}Eu$: $4f^7 6s^2$
C $_{65}Tb$: $4f^9 6s^2$
D $_{64}Gd$: $4f^7 5d^1 6s^2$
E $_{61}Pm$: $4f^5 6s^2$

2. (4) Oxidizing agents are those which itself get reduce and oxidizes others. Among the given, Ce^{+4} and Tb^{+4} act as oxidising agent.

3. (1) Disproportionation is a redox reaction in which intermediate oxidation state of compound is converted into two compounds one is of lower oxidation state and other is of higher oxidation state.

These two reactions given below are disproportionation reaction.

$Cu^+ \rightarrow Cu^{2+} + Cu$

$3MnO_4^{2-} + 4H^+ \rightarrow 2MnO_4^- + MnO_2 + 2H_2O$

4. (1) Noble gas configuration = ns^2np^6

Electronic configuration of ions are given below:

$[Sr^{2+}] = [Kr]$
$[Cs^+] = [Xe]$
$[Yb^{2+}] = [Xe]\,4f^{14}$
$[La^{2+}] = [Xe]\,5d^1$
$[Pb^{2+}] = [Xe]\,4f^{14}5d^{10}6s^2$
$[Fe^{2+}] = [Ar]\,3d^6$

5. (4)

Element	Electronic Configuration	Unpaired electron
Sc	$[Ar]\,4s^2\,3d^1$	1
Cr	$[Ar]\,4s^1\,3d^5$	6
V	$[Ar]\,4s^2\,3d^3$	3
Ti	$[Ar]\,4s^2\,3d^2$	2
Mn	$[Ar]\,4s^2\,3d^5$	5

NCERT Exemplar MCQs

1. (4) (1) Potassium permanganate ($KMnO_4$) forms dark purple colored crystals because it absorbs light in the visible range of the electromagnetic spectrum.

(2) $5SO_2 + 2KMnO_4 + 2H_2O \rightarrow 2H_2SO_4 + 2MnSO_4 + K_2SO_4$

(3) As a ligand-to-metal charge transfer takes place between the oxygen's p- orbitals and the empty d-orbitals on the metal. This charge transfer takes place when a photon of light is absorbed, which leads to the purple color of the compound/ion

2. (1) Cu(II) is more stable than Cu(I). As it is known that, Cu(I) has $3d^{10}$ stable configuration while Cu(II) has $3d^9$ configuration. But Cu(II) is more stable due to greater effective nuclear charge of Cu(II)i.e., it hold 17 electrons instead of 18 in Cu(I).

3. (4) On moving left to right along period, mass increases while metallic radius decreases. Decrease in metallic radius coupled with increase in atomic mass results in increase in density of metal.

Hence, among the given four choices, Cu belongs to right side of Periodic Table in transition metal, and it has highest density (89 g/cm^3).

4. (2) Transition elements form coloured salt due to the presence of unpaired electrons. In CuF_2, Cu(II) contain one unpaired electron hence, CuF_2 is coloured in solid state.

5. (1) Green oily compound Mn_2O_7 is obtained on addition of $KMnO_4$ to concentrated H_2SO_4, which is highly explosive in nature.

$2KMnO_4 + 2H_2SO_4\,(conc.) \rightarrow Mn_2O_7 + 2KHSO_4 + H_2O$

6. (2) Greater the number of unpaired electrons, higher will be its value of magnetic moment. Since, $3d^5$ has 5 unpaired electrons hence, highest magnetic moment.

$\mu = \sqrt{5(5+2)} = \sqrt{35} = 5.95\,BM$

7. (2) Lanthanoids show common oxidation state of + 3. Some of which also show + 2 and + 4 stable oxidation state along with + 3 oxidation state. These are shown by those elements which by losing 2 or 4 electrons acquire a stable configuration of f^0, f^7 or f^{14}. e.g.,

Eu^{2+} is $[Xe]4f^7$, Yb^{2+} is $[Xe]4f^{14}$, Ce^{4+} is $[Xe]4f^0$ and Tb^{4f} is $[Xe]4f^7$.

8. (1) Disproportionation reaction is the reaction in which oxidation as well as reduction, occur upon same atom simultaneously.

9. (4) When $KMnO_4$ solution is added to oxalic acid solution, the decolourisation is slow in the beginning but becomes instantaneous after some time because Mn^{2+} acts as autocatalyst.

Reduction half:
$MnO_4^- + 8H^+ + 5e^- \rightarrow Mn^{2+} + 4H_2O] \times 2$
Oxidation half:
$C_2O_4^{2-} \rightarrow 2CO_2 + 2e^-] \times 2$
Overall equation is:
$2MnO_4^- + 16H^+ + 5C_2O_4^{2-} \rightarrow 2Mn^{2+} + 10CO_2 + 8H_2O$
End point of this reaction → Colourless to light pink

10. (3) Tm (Z = 69) does not belong to actinoid series. The actinoid series is with atomic numbers 90 to 103. Thulium (Tm) has atomic number 69 belongs to lanthanoids (4f series).

11. (1) The reaction of $KMnO_4$ in which it acts as an oxidising agent in acidic medium is:

$$2KMnO_4 + 3H_2SO_4 \rightarrow K_2SO_4 + 3H_2O + 5[O]$$
$$H_2S + [O] \rightarrow H_2O + S] \times 5$$
$$2KMnO_4 + 3H_2SO_4 + 5H_2S \rightarrow K_2SO_4 + 2MnSO_4 + 8H_2O + 5S$$

5 moles of S^{2-} ions react with 2 moles of $KMnO_4$. So, 1 mole of S^{2-} ion will react with $\frac{2}{5}$ moles of $KMnO_4$.

12. (1) V_2O_5 and Cr_2O_3 both react with alkalies as well as acids are amphoteric oxides.

13. (1) Gadolinium belongs to 4f series and has atomic number 64. The correct electronic configuration of gadolinium is:
$_{64}Gd = {}_{54}[Xe]4f^7 5d^1 6s^2$
It has extra stability due to half-filled 4f subshell.

14. (4) When small atoms are trapped inside the crystal lattice of metals, interstitial compounds are formed. Some of their important characteristics are as follows:
(i) They are very hard and rigid.
(ii) They have high melting point which are higher than those of the pure metals.
(iii) They show conductivity like that of the pure metal.
(iv) They acquire chemical inertness.

15. (2) The magnetic moment is associated with its spin angular momentum and orbitals angular momentum.
Spin only magnetic moment value of Cr^{3+} ion is $3d^3$.
Hence, magnetic moment $(\mu) = \sqrt{n(n+2)}BM$
$= \sqrt{3(3+2)} = \sqrt{15} = 3.87$ BM

16. (3) $KMnO_4$ acts as an oxidising agent in alkaline medium. When alkaline $KMnO_4$ is treated with KI, iodide ion is oxidised to IO_3^-
Reaction
$$2KMnO_4 + H_2O + KI \rightarrow 2MnO_4 + 2KOH + KIO_3$$
or, $$I^- + 6OH^- \rightarrow IO_3^- + 3H_2O + 6e^-$$

17. (1) Copper does not liberate H_2 from acids as Copper lies below hydrogen in the electrochemical series. Therefore, option (1) is not correct. Other three options (2, 3, 4) are correct.

18. (3) When acidified $K_2Cr_2O_7$ solution is added to Sn^{2+} salts then Sn^{2+} changes to Sn^{4+}.

19. (4) Highest oxidation state of manganese in fluoride is +4(MnF_4) but highest oxidation state in oxides is +7(Mn_2O_7). Because in covalent compounds, fluorine can form single bond while oxygen forms double bond.

20. (3) Due to lanthanoid contraction, Zr and Hf possess nearly same atomic and ionic radii *i.e.,* Zr = 160 pm and Hf = 159 pm, Zr^{4+} = 79 pm and Hf^{4+} = 78 pm. Therefore, these two elements show similar properties (physical and chemical properties).

21. (2) HCl is not used to make the medium acidic in oxidation reactions of $KMnO_4$ in acidic medium because if HCl is used, the oxygen produced from $KMnO_4$ + HCl is partially utilised in oxidising HCl to Cl_2 which itself acts as an oxidising agent and partially oxidises the reducing agent.

22. (1, 2) Not because of the presence of unpaired electrons $KMnO_4$ is coloured due to the charge transfer. Similarly, oxidation state of *Ce* in $Ce(SO_4)_2$ is + 4 with $4f^0$ electronic configuration. It is also coloured (yellow) due to charge transfer and not due to $f-f$ transition.

23. (1, 4) Electronic configuration Co^{2+} = [Ar] $3d^7$; Number of unpaired electrons = 3
Electronic configuration Cr^{2+} = [Ar] $3d^4$; Number of unpaired electrons = 4
Electronic configuration Mn^{2+} = [Ar] $3d^5$; Number of unpaired electrons = 5
Electronic configuration Cr^{3+} = [Ar] $3d^3$; Number of unpaired electrons = 3
Hence, it is clearly seen that both Co^{2+} and Cr^{3+} have same number of unpaired electrons *i.e.,* 3.

24. (2, 3) In *d*-block elements, for heavier elements. the higher oxidation states are more stable. Hence, Mo(VI) and W(VI) are more stable than Cr (VI). That why, Cr (VI) in the form of dichromate is a stronger oxidising agent in acidic medium whereas MoO_3 and WO_3 are not.

25. (2, 4) The oxidation states of the following actinoids are
(a) Americium (Z = 95); Electronic configuration = [Rn] $5f^7 6d^0 7s^2$
Oxidation states shown by Am = + 3, + 4, + 5, + 6.
(b) Plutonium (Z = 94); Electronic configuration = [Rn] $5f^6 6d^0 7s^2$
Oxidation states shown by Pu = + 3, + 4, + 5, + 6. + 7.
(c) Uranium (Z = 92); Electronic configuration = [Rn] $5f^3 6d^1 7s^2$
Oxidation states shown by U = + 3, + 4, + 5, + 6.
(d) Neptunium (Z = 93): Electronic configuration = [Rn] $5f^4 6d^1 7s^2$
Oxidation states shown by Np = + 3, + 4, + 5, + 6, + 7.

26. (1, 2) General electronic configuration of actinoids is $(n-1)f^{1-14}(n-1)d^{0-2}ns^2$. U and Np each have one electron in 6d orbital.

27. (2, 3) (1) Cerium (Z = 57) ⇒ Electronic configuration = $[Xe]4f^1 5d^1 6s^2$
Oxidation state of Ce = + 3, + 4
(2) Europium (Z = 63) ⇒ Electronic configuration = $[Xe]4f^7 5d^0 6s^2$
Oxidation state of Eu = +2, + 3
(3) Ytterbium (Z = 70) ⇒ Electronic configuration = $[Xe]4f^{14} 5d^0 6s^2$
Oxidation state of Yb = + 2, + 3
(4) Holmium (Z = 67) ⇒ Electronic configuration = $[Xe]4f^{11} 5d^0 6s^2$
Oxidation state of Ho = + 3

28. (2, 3) As, $Ti^{3+} = [Ar]3d^1$

$Mn^{2+} = [Ar]3d^5, (t_{2g}^3 e_g^2)$

$Fe^{3+} = [Ar]3d^5, (t_{2g}^3 e_{2g}^2)$

$Co^{3+} = [Ar]3d^6, (t_{2g}^4 e_g^2)$

Magnetic moment depends on the no. of unpaired electrons.

29. (1, 2) Transition elements such as Cr and Co form binary compounds with halogens, *i.e.*, CrF_3 and CoF_3 whereas Cu and Ni do not form CuF_3 and NiF_3.

30. (2, 3) A species can act as oxidising agent only when metal is present in high oxidation state but lower oxidation state show stability. As higher oxidations states of W and Mo are more stable, therefore they will not act as oxidising agents.

31. (2,3) Electronic configuration of $_{58}Ce = 54[Xe]4f^2 5d^0 6s^2$.

Therefore, electronic configuration of $Ce^{4+} = {}_{54}[Xe]4f^0$.

Thus, it has a tendency to attain noble gas configuration and attain f^0 configuration.

NEET & AIIMS Past Year MCQs

1. (1) The most stable oxidation state of lanthanoids is +3. Ce^{4+} and Tb^{4+} will readily undergo reduction and act as strong oxidizing agents.

2. (2) The UV-visible absorption bands in the spectra of lanthanoid ions arise due to electronic transitions within the f orbitals (4f). These 4f orbitals are shielded by the 5s and 5p orbitals, so the transitions are less affected by the environment or ligand field. Because of the shielding, the absorption bands are narrow.

These narrow bands are characteristic of f–f transitions in lanthanoid ions.

3. (3) Element Block/group in periodic table

Ce (Z = 58) → Lanthanoid

As (Z = 33) → p-block element

Cs (Z = 55) → s-block element

Mn (Z = 25) → d-block element

The correct match is; A-I, B-IV, C-II, D-III

4. (4) Magnetic moment is calculated by; $\mu = \sqrt{n(n+2)}$ BM

Where, n ⇒ Number of unpaired electrons

Ion	n	μ(BM)
Zn^{2+}	0	0
Ti^{2+}	2	$\sqrt{8}$
Cr^{2+}	4	$\sqrt{24}$
Fe^{2+}	4	$\sqrt{24}$
Ti^{3+}	1	$\sqrt{3}$
Cu^{2+}	1	$\sqrt{3}$
V^{2+}	3	$\sqrt{15}$
Cu^{+}	0	0

Therefore, (Cr^{2+}, Fe^{2+}) and (Ti^{3+}, Cu^{2+}) exhibit the same magnetic moment.

5. (3) $\underset{(Z=64)}{Gd^{3+}}$: $[Xe]4f^7$ (7 unpaired e^-)

∴ $\underset{(Z=63)}{Eu^{3+}}$: $[Xe]4f^6$ (6 unpaired e^-)

∴ $\underset{(Z=61)}{Pm^{3+}}$: $[Xe]4f^4$ (4 unpaired e^-)

∴ $\underset{(Z=62)}{Sm^{3+}}$: $[Xe]4f^5$ (5 unpaired e^-)

∴ $\underset{(Z=58)}{Ce^{4+}}$: $[Xe]4f^0$ (No unpaired e^-)

∴ $\underset{(Z=70)}{Yb^{2+}}$: $[Xe]4f^{14}$ (No unpaired e^-)

∴ $\underset{(Z=58)}{Ce^{3+}}$: $[Xe]4f^1$ (1 unpaired e^-)

∴ $\underset{(Z=63)}{Eu^{2+}}$: $[Xe]4f^7$ (7 unpaired e^-)

Species having no unpaired e^- [Ce^{4+} and Yb^{2+}] are diamagnetic in nature.

6. (3) Much larger third ionisation energy of Mn (where the required change is d^5 to d^4) is mainly responsible for this. This also explains why the +3 state of Mn is of little importance.

7. (3) Spin only magnetic moment

$= \sqrt{n(n+2)}$

n = number of unpaired electrons

A	Ti^{3+}	d^1	1.73
B	Cr^{2+}	d^4	4.90
C	Mn^{2+}	d^5	5.92
D	Fe^{2+}	d^6	4.90
E	Sc^{3+}	d^0	0

$B = Cr^{2+} = 4.90 \quad D = Fe^{2+} = 4.90$

8. (4) The greater stability of Cu^{2+} (aq) rather than Cu^+ (aq) is due to the much more negative $\Delta_{hyd}H$ of Cu^{2+}(aq) than Cu^+, which more than compensates for the second ionisation enthalpy of Cu.

9. (4) The highest oxidation number corresponding to the group number in transition metal oxides is attained in Sc_2O_3 to Mn_2O_7.

CrO is basic but Cr_2O_3 is amphoteric.

Note:

All the transition metals except scandium form MO oxides which are ionic (Only for 3d series, this statement is true)

Hence, (A), (C) and (D) are incorrect. But not given in the options.

10. (3) The electronic configuration of f-block element Gd is

$_{64}Gd = [Xe]\, 6s^2\, 4f^7\, 5d^1$

$Gd^{+2} = [Xe]\, 4f^7\, 5d^1$

After losing 5d electron, 4f has maximum exchange energy so, Gd has low value of third ionisation energy.

11. (2) $K\overset{+7}{Mn}O_4 + I^- \xrightarrow[\text{or weakly alkaline medium}]{\text{Neutral}} \overset{+4}{Mn}O_2 + IO_3^-$

Therefore, the change in oxidation state of Mn is from +7 to +4.

12. (2) The almost identical radii of Zr (160 pm) and Hf (159 pm) is a consequence of the lanthanoid contraction, causes the radii of third transition series member to very similar to corresponding member of second series.

13. (1) Many trivalent lanthanoid ions are coloured both in the solid state and in aqueous solutions. Colour of these ions may be attributed to the presence of f electrons.

14. (3) Na_3VO_4

$3(+1) + x + 4(-2) = 0$

$3 + x - 8 = 0$

$x - 5 = 0$

$x = +5$

$V = 23 = [Ar]\ 3d^3\ 4s^2$

$V = (23 - 5) = [Ar]$

∴ There is no unpaired electron. Hence, it is colourless in nature.

15. (2) Cu and Ni

16. (3) $H_2N - Hg - Cl + Hg$

17. (1) According to Le Chatelier's principle, on adding conc. HCl, the equilibrium shifts in forward direction giving blue colour. And when this blue colour is diluted, equilibrium shifts in backward direction leading to pink colour.

18. (1) In acidic solution Mn(VI) become unstable and undergo disproportionation to form Mn (VII) and Mn (IV)

$5MnO_3 + 2H^+ \rightarrow 3MnO_2 + 2MnO_4^- + H_2O$

19. (1)

20. (1) When mercury is exposed to ozone it gets superficially oxidised and loses its meniscus and sticks to the glass.

Coordination Compounds

NCERT Topic-wise MCQs

1. (3) According to Werner's theory of coordination compounds, the ionisation of given complex is,

$[Pt(NH_3)_6]Cl_4 \rightarrow [Pt(NH_3)_6]^+ + 4Cl^-$

1 + 4 = 5

2. (1) Primary valency is the valency which are present outside the square bracket, and this valency is ionisable.

3. (2) According to Werner's theory, the sterochemistry of a complex is determined by secondary valencies.

4. (4) Total number of coordinate bonds through which central metal atom/ion is attached with ligand is called coordination number. It is represented by the structures of the given complexes.

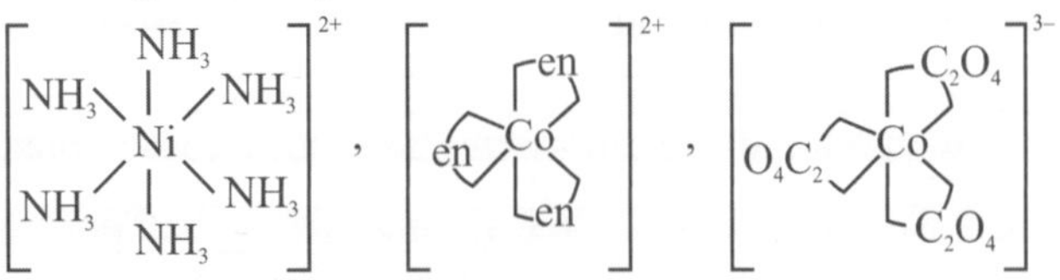

5. (2) $K_4[Fe(CN)_6] \rightarrow [Fe(CN)_6]^{4-} = x + 6(-1) = -4$

$= x - 6 = -4 = x = -4 + 6 = +2$

$K_3[Fe(CN)_6] \rightarrow [Fe(CN)_6]^{3-} = x + 6(-1) = -3$

$= x - 6 = -3 \Rightarrow x = -3 + 6 \Rightarrow x = +3$

Fe^{2+} = Ferous ion, Fe^{3+} = Ferric ion

6. (3) Based on Werner's co-ordination theory.

7. (1) $[Co(NH_3)_3(NO_2)_3] \Rightarrow x + 3(0) + 3(-1) = 0$

$x - 3 = 0 \Rightarrow x = +3$

8. (2) Coordination Number

9. (3) Primary valencies of the above complexes are 3, 2, 1 respectively.

$[Cr(NH_3)_6]Cl_3$, $[Cr(NH_3)_5Cl]Cl_2$, $[Cr(NH_3)_4Cl_2]Cl$

10. (1) Four ions are produced.

The ionisation is shown as,

$[Co(NH_3)_6]Cl_3 \rightleftharpoons [Co(NH_3)_6]^{3+} + 3Cl^-$

11. (3) The metal carbonyl compound given in option '3' has zero oxidation state because CO is a neutral ligand.

12. (2) $K_3[Fe(CN)_6] \rightarrow 3K^+ + [Fe(CN)_6]^{3-} = 4$ moles

13. (2) $AgCl + 2NH_3 \rightarrow [Ag(NH_3)_2]Cl$

14. (2) $CuSO_4 + 4NH_3 \rightarrow [Cu(NH_3)_4]SO_4 \equiv [Cu(NH_3)_4]^{2+}\ SO_4^{2-}$

15. (1) BF_3 is a Lewis acid. It can't donate the lone pair of electrons, as it is electron deficient species.

16. (2) $[Co(NH_3)_3Cl_3]$ is a neutral complex. So it cannot be ionised.

17. (1) $[Co(NH_3)_4Cl_2]Cl \rightarrow [Co(NH_3)_4Cl_2]^+ + Cl^-$

Total ions = 2

18. (2) $[Co(NH_3)_3Cl_3]$ cannot give a precipitate as it is a neutral complex.

19. (1) $[Cr(NH_3)_4Cl_2]Cl$ ionises to give one Cl^- ion. So, it gives 1 mole of AgCl, on the addition of $AgNO_3$.

20. (4) A double salt is a compound prepared by the combination of two different salt compounds & they completely dissociates into its ions in water. Potash alum is the example of double salt. Its formula is-

$K_2SO_4.Al_2(SO_4)_3.24H_2O$

21. (4) In the given salts, 4 $KCN.Fe(CN)_2$ is a complex salt.

Double salts are those addition compounds which are stable in the solid state but break up into its constituent in dissolved state. But in such compounds the individual properties of the constituent are not lost.

Ex- $KClMgCl_2.6H_2O$, $FeSO_4(NH_4)_2SO_4.6H_2O$, etc.

$Fe(CN)_2 + 4KCN \rightarrow Fe(CN)_2.4KCN$

Potassium ferrocyanide

In this reaction, a complex compound is formed which cannot further dissociates into the simpler ions. Therefore, individual properties are lost. structure of this complex salt is given as,

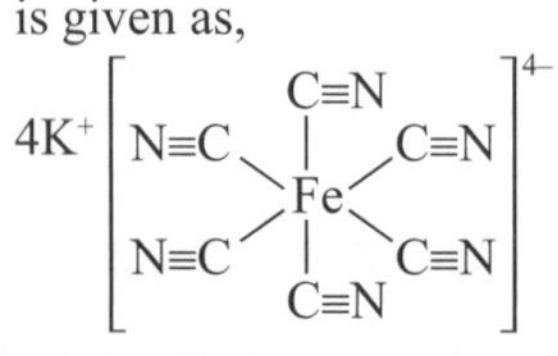

22. (4) Ligands are those entities whose number is known as secondary valencies.

23. (3) Negative ions.

24. (2) The complex is represented as $[Co(NH_3)_5.Cl]Cl_2$

ionisable $Cl^- = 2$

Non-ionisable $Cl^- = 1$

25. (2) Chelate ligand are the ligand which can form a ring structure with central metal ion.

Oxalate forms chelate complex as given in figure.

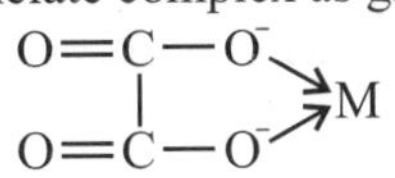

26. (1) $[MnO_4]^- = x + 4(-2) = -1$

$x - 8 = -1$

$x = -1 + 8$

$x = +7$

Mn = 25, its electronic configenatic is- $1s^2\ 2s^2\ 2p^6\ 3s^2\ 3p^6\ 4s^2\ 3d^5$

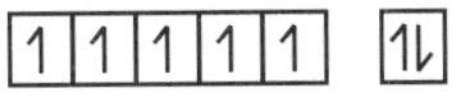

Hence, Mn^{7+} has no d-orbital electrons.

27. (3) NO_2 is present as NO_2^- = anionic ligand

H_2O, NH_3, CO = neutral ligand

28. (3) It is a tetradentate ligand as number of donor atoms in $N(CH_2CH_2NH_2)_3$ is four.

It is represented as-

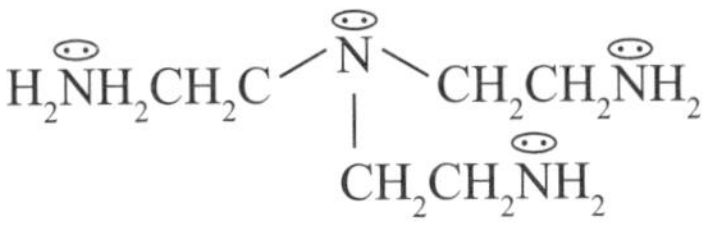

4 lone pairs are present on 4 N-atoms.

29. (4) Coordination number of the central atom or ion is determined by the number of sigma bonds formed by the ligand with the central atom or ion.

EDTA (ethylene diamminetertaactetate) has co-ordination number = 6

It is shown in the given figure.

Structure of EDTA ion

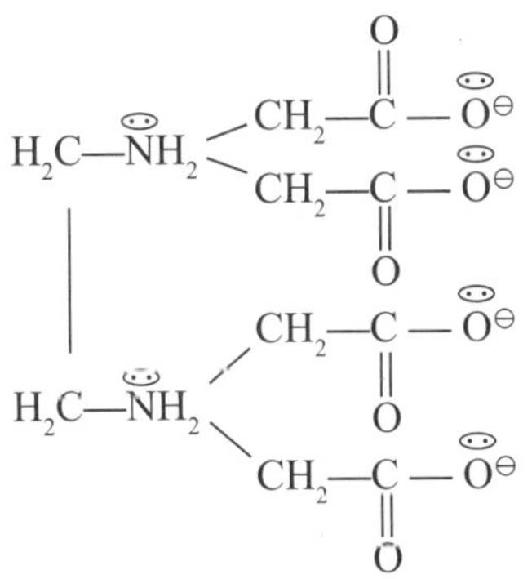

Total 6 donar sites.

30. (4) Ambidenate ligands are those ligands which have more than one donar atoms. NO_2 group can be attached through nitrogen or oxygen and ^-NC group can bc attached through nitrogen or carbon.

31. (4) Ligand can act as Lewis bases. Lewis base is a base which can donate their lone pair of electrons.

32. (3) The sulphite ligand is SO_3^{2-}.

33. (4) Both Ethane-1, 2-diamine and dimethyl glyoximato are bidentate ligand.

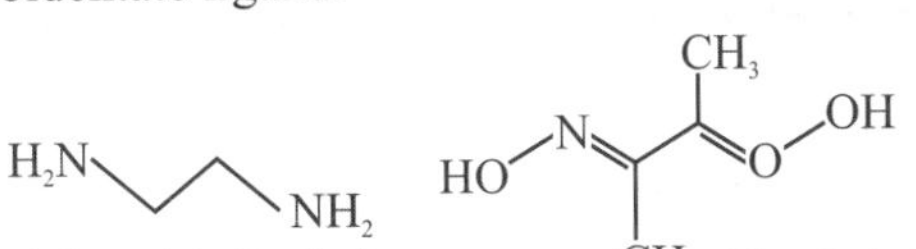

(ethane-1,2-diamine) (dimethyl glyoximato)

34. (3) EDTA is a hexadentate ligand.

35. (4) Complexes in which more than one kind of donor atoms are present, are called heteroleptic complexes. Both 2 & 3 complexes are heteroleptic. As in 2) chloride & ammine two types of ligands are present. In 3) also amine and nitrito-O ligands are present.

36. (4) SCN^- is a monodendate ligand. Hence; it does not show chelation.

37. (4) Due to small size of cation, large ionic charge and vacant d orbitals, transition elements form complexes readily.

38. (4) It is chelating ligand.

Diethylenetriamine is a tridentate neutral molecule with three donor nitrogen atoms. It's structure is given as-

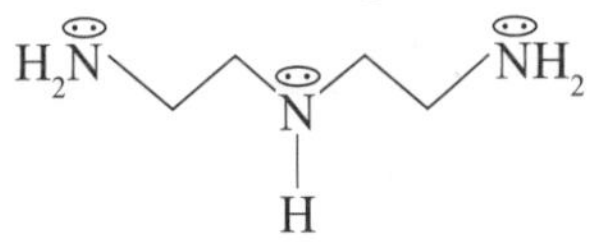

(Three donar atoms)

(Diethylenetriamine)

39. (2) Acetylacetonate is a bidentate ligand carrying unit negative charge. It coordinates through two oxygen atoms. It's structure is given as-

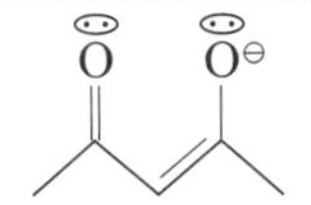

(Two donar) atoms with one –ve charge

Acetylacetonate

Oxalato structure is given as-

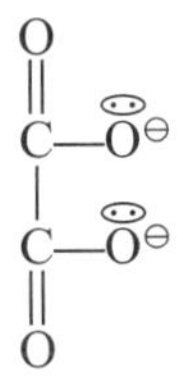

Two donar atoms with two –ve charges

Dimethylglyoximato structure is given as-

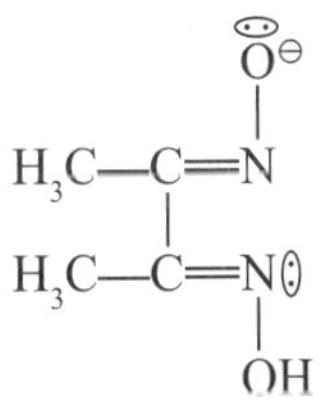

Two donar atoms with one –ve charge

Dimethylglyoximato is a bidentate ligand. It is a monoanionligand.

Oxalato ligand is a bidentate ligand and has two negative charges.

40. (1) The structure of respective ligands are given as-

en → $H_2\ddot{N}-CH_2-CH_2-\ddot{N}H_2$

Ethylene diamine

dmg →

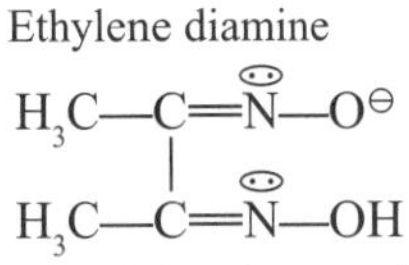

Dimethylglyoximato anion

EDTA →

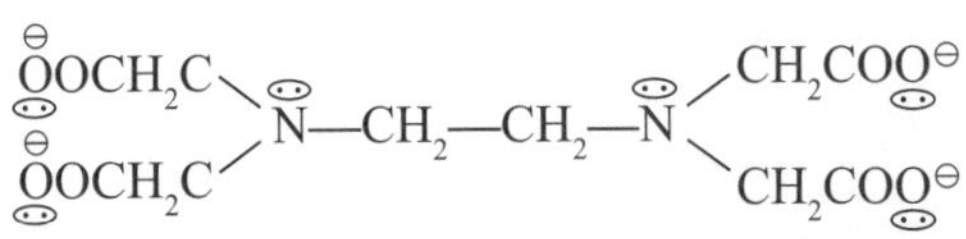

Ethylene diaminetetraacetate ion

gly → $\ddot{N}H_2—CH_2—CO\ddot{O}^{\ominus}$
glycine

ox →

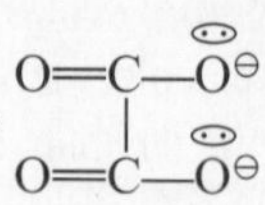

oxalate ion

41. (4) $–NO_2^-$ and $–SCN^-$ are ambidentate ligands which have two donor atoms.

42. (4) Ligands having three donor atoms which can attach to a single metal cation or atom are called the tridentate ligands. Here, dien is the tridentate ligand. It's structure is given as,

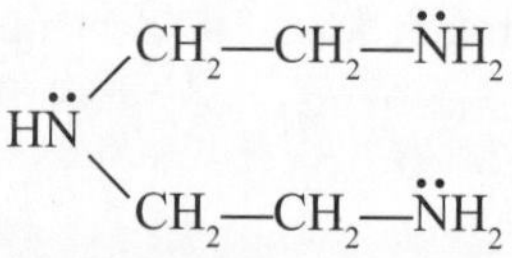

Diethylenetriamine (dien)

43. (2) Oxidation state of Ni in $K_2[Ni(CN)_4]$ is calculated as-
$K_2[Ni(CN)_4] = 2K^+ + [Ni(CN)_4]^{2-}$
$[Ni(CN)_4]^{2-} \Rightarrow x + 4(-1) = -2$
$x = -2 + 4$
$x = +2$

44. (4) $[Cu(NH_3)_4]SO_4$ is a cationic complex. It dissociates as-
$[Cu(NH_3)_4]SO_4 \rightleftharpoons [Cu(NH_3)_4]^{2+}SO_4^{2-}$

45. (2) Chelate ligand are the ligand which can form a ring structure with central metal ion. Ex: oxalate

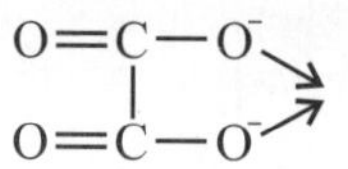

Rest ligands are the simple ligands as they can't form ring structure with central metal.
Acetate ⇒ $CH_3CO\ddot{O}^{\ominus}$
Cyanide ⇒ $\ddot{C}N^{\ominus}$
Ammonia ⇒ $\ddot{N}H_3$

46. (2) 6 CN^- ions are the ligands in the complex $K_4[Fe(CN)_6]$

47. (2) In metal carbonyl complex, carbonyl ligand (CO) is present. Ex: $[Ni(CO)_4]$, $[Fe(CO)_5]$.

48. (3) Potassium ferrocyanide is a co-ordination compound.

49. (1) The given complex is $[Co(NH_3)_5SO_4]Cl$.
It's structure is-

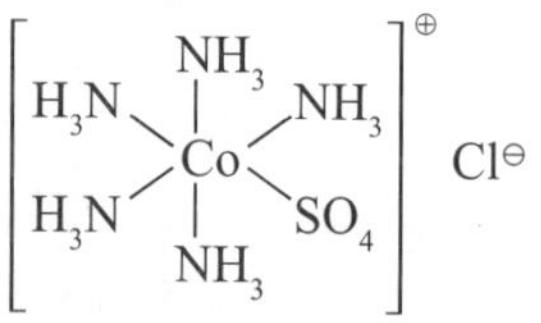

Co-ordination no = 6
Oxidation no. of Co is calculated as-
$[Co(NH_3)SO_4]^+$
$x + 0 + (-2) = +1$
$x = +3$
oxidation state = +3

50. (2) 2,4-Dinitrophenylhydrazine is a monodentate ligand . The structure of given ligand is-

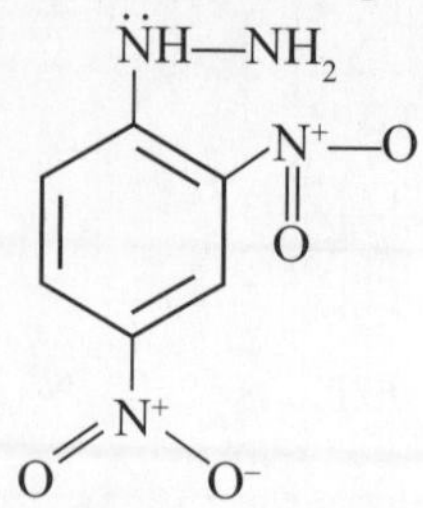

51. (1) The formula of dibromidobis (ethylene diamine) chromium (III) bromide is $[Cr(en)_2Br_2]Br$.
"Bis" is used for the bidentate ligand. Here ethylene diamine or "en" is a bidentate ligand. 2 Br-ligand inside the ring are named as dibromido and outside 1 Br-ligand is named as bromide.

52. (1) ∵ Zn is stabilized in +2 O.S
∴ Charge on $[ZnCl_4]$ complex is –2.
∴ $[Cr(NH_3)_5(NCS)]\ [ZnCl_4] \rightarrow [Cr(NH_3)_5(NCS)]^{+2} + [ZnCl_4]^{-2}$
Therefore the correct IUPAC name is Pentaammineisothiocyanato chromium (III) tetrachlorozincate (II).

53. (3) The IUPAC name of this compound is
Triamminebromochloronitroplatinum (IV) chloride

54. (1) IUPAC name of $K_2[Ni(CN)_4]$ is potassium tetracyanonickelate (II)

55. (1) Solution containing conc. HCl and conc . HNO_3 in the ratio of 3:1 is Aqua - regia.
It dissolutes noble metals like Au, Pt etc.
$HNO_3 + 3HCl \rightarrow 2H_2O + 2Cl + NOCl$
$Au + 3Cl \rightarrow AuCl_3$
$AuCl_3 + HCl \rightarrow AuCl_4^- + H^+$
$AuCl_4^-$ = Tetra chloridoaurate (III)

56. (2) According to the IUPAC nomenclature the name of $[Cr(H_2O)_4Cl_2]Cl$ is Tetraaquadichlorochromium (III) chloride.

57. (2) According to IUPAC, name of the compound $[Cu(NH_3)_4(NO_3)_2]$ is tetraaminedinitrato copper (II)

58. (3) Potassium ferrocyanide : $K_4[Fe(CN)_6]$: Fe is present as Fe^{2+}
The IUPAC name of potassium ferrocyanide is Potassium hexacyanoferrate (II).

59. (3) According to IUPAC nomenclature the formula of Tetraamminechloronitroplatinum (IV) sulphate is $[Pt(NH_3)_4(NO_2)Cl]SO_4$.

60. (1) The IUPAC name of $[Co(NCS)(NH_3)_5]Cl_2$ is pentaamminethiocyanato-N-cobalt (III) chloride.

61. (2) The IUPAC name of the complex $[Pt(NH_3)_2Cl_2]$ is Diamminedichloridoplatinum (II).

62. (3) IUPAC name of the complex $[Cu(NH_3)_4]SO_4$ is Tetraamminecopper(II) sulphate.

63. (1) Hexaaquatitanium(III) chloride is represented as $[Ti(H_2O)_6]Cl_3$.

64. (2) The correct name of $KAl(SO_4)_2.12H_2O$ is- Potassium aluminium (III) sulphate.$12H_2O$ or (with 12 water of crystallisation).

65. (2) The IUPAC name of $Ba[BrF_4]_2$ is Bariumtetrafluorobromate (III).

66. (3) The IUPAC name of the given compound is Tetraamminenickel(II)-tetrachloronickelete(II).

67. (1) The IUPAC name of the given compound is pentaaminenitritrocobalt(III)ion

68. (1) IUPAC name of $Na_3[CoCl(NO_2)_5]$ is Sodium chloropentanitrocobaltate (III).

69. (1) The IUPAC name of Wilkinson's catalyst $[RhCl(PPh_3)_3]$ is Chlorotris (triphenylposphine) rhodium (I).

70. (2) The ligands are interchanged in both the anionic and cationic entities of different metal ions present in a complex to form a coordination isomers.

71. (1)

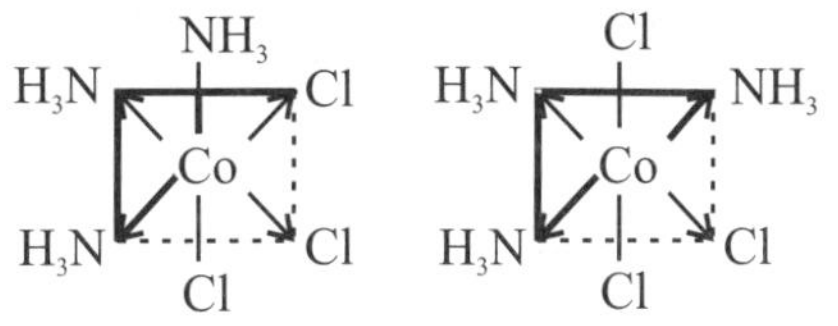

fac mer

72. (4) In the given complexes $[Cr(SCN)_2(NH_3)_4]^+$ has the maximum number of isomers. It shows geometrical (cis-trans) isomerism as well as linkage isomerism, due to SCN^- ligand. SCN^- is an ambidentate ligand and it can coordinate through either S or N atom.

Rest $[Ni(en)(NH_3)_4]^{2+}$, $[Co(NH_3)_4Cl_2]$ and $[Ni(C_2O_4)(en)_2]^{2-}$ have 2 geometrical isomers each.

73. (2) Both the given compound shows ionisation isomerism, because they furnish different ion in solution.

$[PtCl_2 (NH_3)_4]Br_2 \rightleftharpoons [Pt\ Cl_2(NH_3)_4]^{2+} + 2Br^-$

$[PtBr_2(NH_3)_4]Cl_2 \rightleftharpoons [Pt\ Br_2\ (NH_3)_4]^{2+} + 2Cl^-$

74. (4) The ligands are interchanged in both the anionic and cationic entities of different metal ions present in a complex to form a coordination isomers

75. (3) The existence of two different coloured complexes is due to geometrical isomerism as cis and trans forms are present.

76. (4) Ionisation isomerism, because it will give different ions in solution. It's another isomer is $[Co(NH_3)_5SO_4]Br$.

77. (3) The ligand which gets attached to the central atom by any of the two atom of ambidentate ligand.

78. (3) The octahedral complex $[Ma_3b_3]$ gives two types of isomers. Which are known as fac or facial isomer and mer or meridional isomer. If three identical ligands occupy positions at the corners of the same face of octahedron, it will call as fac-isomer. It is represented as-

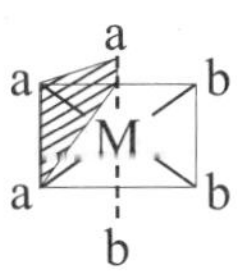

fac-isomer

79. (1) The given complex exhibits ionization isomerism, geometrical isomerism as well as linkage isomerism.

80. (4) Complexes having same molecular formula but furnish different ions in solution are called ionisation isomers. Here NO_2 & Cl ligands are responsible to show ionisation isomerism.

81. (2) II & III are Non super mirror image
I & II are geometrical isomers
I is cis wrt Cl
II is trans wrt Cl

82. (4) $[Co(NH_3)_3Cl_3]$ will not exhibit optical isomerism due to presence of plane of symmetry. It exhibit fac-mer isomerism.

83. (4) $CrCl_3.6H_2O$ show solvate isomerism due to the presence of water molecules. Water molecules are exchanged inside and outside of coordination sphere. These isomers are also called hydrate isomers. Examples of such isomers are:

(i) $[Cr(H_2O)_6]Cl_3$ – Violet in colour
(ii) $[Cr(H_2O)_5Cl]Cl_2.H_2O$ – Blue-green in colour
(iii) $[Cr(H_2O)_4Cl]Cl.2H_2O$ – Green colour

84. (2) A square planar complex does not exhibit optical activity because it has a horizontal plane of symmetry.

85. (2)

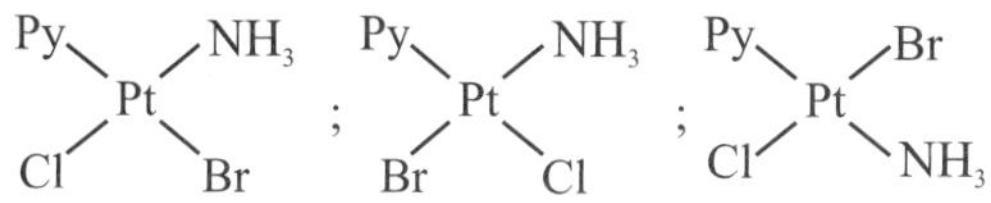

$[Pt(Py)(NH_3)(BrCl)]$ have 3 geometrical isomers.

86. (1) C_2O_4 is a bidentate ligand therefore, no geometrical isomer is possible.

87. (3) Coordination isomerism

88. (1) $[Pt(NH_3)_2Cl_2]$ will show geometrical isomerism as it has square planar geometry. It is represented as-

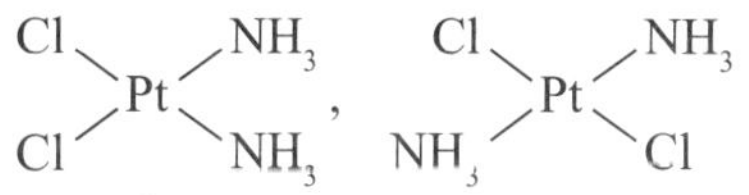

cis trans

89. (4) trans-$[CoCl_2(en)_2]^+$ is not optically active because it has plane of symmetry.

90. (3) Cis-trans isomerism is found in square planar complexes of the molecular formula Ma_2b_2 and it is represented as-

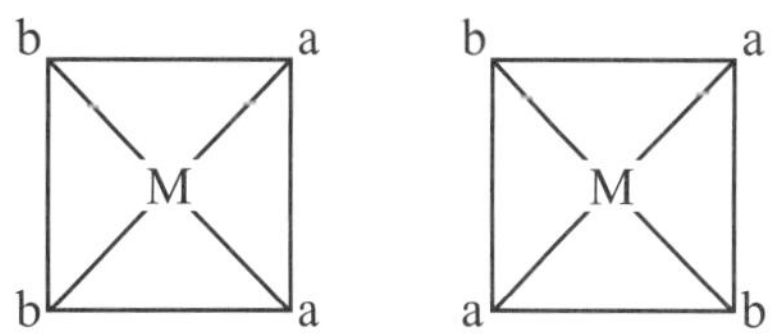

cis-isomer trans-isomer

91. (1) $[Co(en)(SCN)_4]^-$ will not show geometrical isomerism.

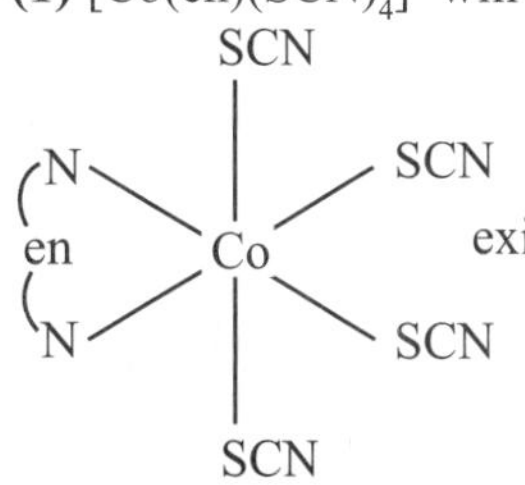

exists only in one form

92. (1) $[NiCl_4]^{2-} \Rightarrow x - 4 = -2$

$x = -2 + 4$

$x = +2$

$Ni^{2+} = [Ar]\ 3d^8$

Ni^{2+} = [1↓|1↓|1↓|1|1] ∴ n = 2 (n = no. of unpaired)

Magnetic moment $(\mu) = \sqrt{n(n+2)} = \sqrt{2(2+2)}$

$= \sqrt{2(4)} = \sqrt{8} = 2.82BM$

93. (3) $[FeF_6]^{3-} \Rightarrow x - 6 = -3$ Fe^{3+} = [1|1|1|1|1]

$x = -3 + 6$

$x = +3$

Fluorine is a weak field ligand, ∴ pairing does not occur.

$3d^5$ [1|1|1|1|1] → [1|1] e_g ; → [1|1|1] t_{2g}

94. (2) Due to the presence of H_2O molecule, d-d transition occur in Cu atom, which result in blue colour of $CuSO_4.5H_2O$. Whereas, in $CuSO_4$, due to absence of any ligand crystal field splitting is not possible. Hence, No colour is observed.

95. (3) $[Fe(CN)_6]^{3-} \Rightarrow x - 6 = -3$

$x = -3 + 6$

$x = +3$

$Fe = 1s^2\ 2s^2\ 2p^6\ 3s^2\ 3p^6\ 4s^2\ 3d^6$

$Fe^{3+} = 1s^2\ 2s^2\ 2p^6\ 3s^2\ 3p^6\ 3d^5$

Fe^{3+} = 23 = [1|1|1|1|1]

CN is strong ligand, it creates pairing

[1↓|1↓|1|×|×] [×] [×|×|×]

∴ The hybridisation used here is d^2sp^3

96. (1) In the presence of strong field ligand, low spin complex will form means 4 electrons present in d orbitals will be paired up and hence only two unpaired electrons will be left.

97. (2) The complex $[Fe(CO)_5]$ follows EAN rule as it has total (26 + 10) 36 electrons in its surroundings.

98. (4)

$[Co(NH_3)_6]^{3+}$	$[MnO_4]^-$	$[Fe(CN)_6]^{3-}$	$[Cr(CN)_6]^{3-}$
x = + 3	x + 4 (–2) = –1	x – 6 = –3	x – 6 = –3
Co = [1↓|1|1|1|1]	x – 8 = –1	x = –3 + 6	x = –3 + 6
d electrons = 6	x = –1 + 8 = +7	x = + 3	x = + 3
	d electrons = 0	Fe^{3+} = [1|1|1|1|1]	Cr^{3+} = [1|1|1|	]
		d electrons = 5	d electrons = 3

99. (3) According to CFT, Δ_o = Crystal field splitting energy.

100. (1) [| | | |] → [|] e_g ; → [| |] t_{2g} $e_g > t_{2g}$

101. (2) $[Ni(CN)_4]^{2-}$, $x - 4 = -2$

$x = -2 + 4$

$x = +2$

Ni^{2+} = [1↓|1↓|1↓|1|1]

CN^- is a strong ligand

∴ pairing takes place [1↓|1↓|1↓|1↓|××] [××] [××|××|]

∴ hybridisation is dsp^2

Thus, the shape is square planar.

102. (3)

$[Sc(H_2O)_3(NH_3)_3]^{3+}$	$[Ti(en)_2(NH_3)_2]^{4+}$	$[Cr(NH_3)_6]^{3+}$	$[Zn(NH_3)_6]^{+2}$
x = + 3	x = + 4	x = + 3	x = +2
d-electrons = 0	d-electrons = 0	d-electrons = 3	d-electrons = 10
∴ No absorption of visible light	∴ No absorption of visible light	∴ complex ion is expected to absorb visible light.	∴ No absorption of visible light

103. (3) Fe^{3+} = [1|1|1|1|1]

CN^- is a strong ligand. Therefore pairing of electrons will take place.

∴ [1↓|1↓|1| |]

one unpaired electron is left

$\mu = \sqrt{n(n+2)}$

$= \sqrt{(1)(1+2)} = \sqrt{3} = 1.738$ BM

104. (3) Increasing order of ligand field strength

$I^- < Br^- < SCN^- < Cl^- < S^{2-} < F^- < OH^- < C_2O_4^{2-} < H_2O < NCS^- < EDTA^{4-} < NH_3 < en < NO_2^- < CN^- < CO$

105. (4) $Cr(CO)_6 \Rightarrow x + 0 = 0$

Cr = [1|1|1|1|1] [1]

CO is a strong ligand

∴ Pairing takes place

Cr = [1↓|1↓|1↓| |]

no unpaired electron is present ∴ μ = 0

106. (4) $[Co(C_2O_4)_3]^{3-} \Rightarrow x + 3(-2) = -3$

$x - 6 = -3$

$x = -3 + 6$

$x = +3$

Co^{3+} = [1↓|1|1|1|1]

$C_2O_4^{2-}$ is a strong ligand, pairing occurs

[1↓|1↓|1↓|××|××] [××] [××|××|××]

∴ The hybridization is d^2sp^3

107. (2) $3d^4$ = [1|1|1|1|], $3d^5$ = [1|1|1|1|1]

$3d^6$ = [1↓|1|1|1|1], $3d^7$ = [1↓|1↓|1|1|1]

Paramagnetism ∝ Number of unpair electrons
Therefore, $3d^5$ shows maximum paramagnetism.

108. (4) $[Fe(H_2O)_6]^{3+}$ complex shows zero crystal field stabilisation energy. It is shown as, $Fe^{+3} \Rightarrow 3d^5\ 4s^o$ H_2O is a weak field ligand so it will form high spin complex.

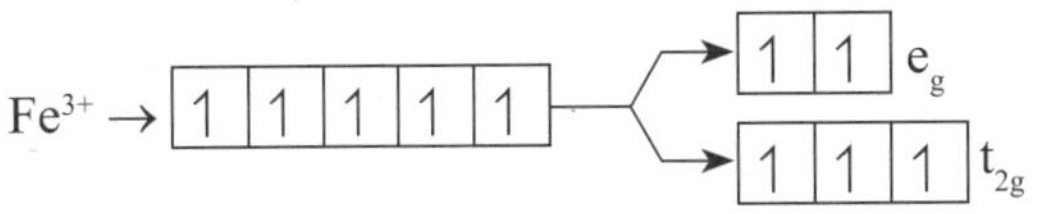

CFSE = –3 × 0.4 + 2 × 0.6 = –1.2 + 1.2 = 0

109. (4) Ionic bond between K^+ and $[Fe(CN)_6]^{4-}$; covalent bond between C and N [C ≡ N], dative bond (coordinate bond) between Fe^{+2} and CN^-. Therefore, all three types of bonds are present in $K_4[Fe(CN)_6]$.

110. (1) It is outer orbital complex. Coordination number is 4. Shape is tetrahedral.

111. (2) CFSE depends on the relative magnitude of crystal field splitting, Δ_O and pairing energy and in turns Δ_O depends upon the field produced by ligand and charge on the metal ion. Here, metal and charge present on it is constant i.e. Co^{3+}. The order of increasing crystal field strength is $F^- < H_2O < NH_3 < CN^-$. Therefore, CFSE follows the order: Q > R > S > P.

112. (4) The shape of the complex species depends upon coordination number. If coordination number = 4, then shape either tetrahedral or square planar.

113. (4) In both $[CoF_6]^{3-}$ & $[Cu(H_2O)_6]^{2+}$ complexes sp^3d^2 hybridisation is present.

114. (2) $[Ni(NH_3)_6]^{2+}$ and $[Zn(NH_3)_6]^{2+}$ Both have sp^3d^2 hybridization. It contains two and zero unpaired electrons respectively.

115. (4) $[Cu(NH_3)_4]^{2+}$ $Cu^{2+} \Rightarrow 1s^2\ 2s^2\ 2p^6\ 3s^2\ 3p^6\ 3d^9\ 4s^0$

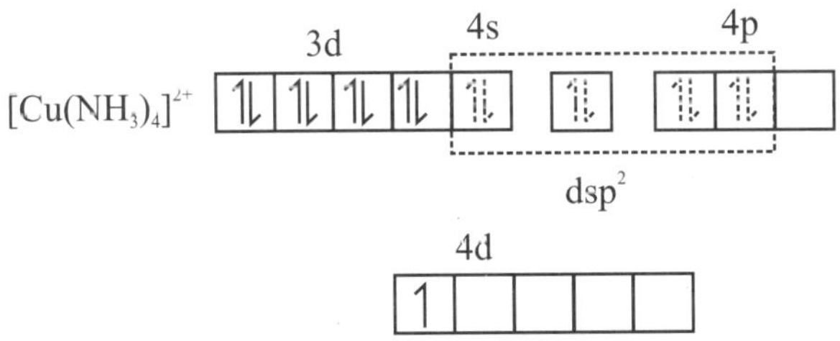

Unpaired electron = 1

$\mu = \sqrt{1(1+2)} = 1.73$ B.M

116. (3) In $VOSO_4$ O.S. of V = +4
Electronic configuration of V is V(23) = $1s^2\ 2s^2\ 2p^6\ 3s^2\ 3p^6\ 4s^2\ 3d^3$
$V^{4+}(19) = 1s^2\ 2s^2\ 2p^6\ 3s^2\ 3p^6\ 3d^1$
Here in d orbital one unpaired e^- is present. Therefore, it is paramagnetic & coloured in nature.

117. (2) Let the oxidation state of Fe in Fe $(CO)_5$ be x
Fe $(CO)_5$
x + 0 × 5 = 0
x = 0

118. (3) $[Ni(CO)_4]$
x + 4(0) = 0
x = 0

119. (4) $Na[Co(CO)_4]$: 1 + x + 4 × 0, ⇒ x = –1

120. (2) Zeigler - Natta catalyst is $Et_3Al + TiCl_4$

121. (1) EDTA is used in the treatment of lead poisoning.

122. (1) Compounds containing at least one metal - carbon bond are called organometallic compounds. Thus, CH_3MgBr is an organometallic compound.

123. (1) In heme, the central metal ion is Fe^{+2}

124. (1) In oxyhaemoglobin, the coordination number of Fe(II) is 6.

High Order Time Intensive MCQs

1. (1) The complex is $[CoCl_3(NH_3)_3]$ is non - ionic. Therefore it is non - conducting .

2. (3) $[Fe(H_2O)_6]^{3+}$ is an outer orbital complex involving sp^3d^2 hybridisation. Because H_2O is a weak field ligand

3. (2) In all the given complexes Co is present in +3 oxidation in state. Therefore electronic configuration of all the d - orbital will be same. All the given complexes have octahedral geometry therefore splitting pattern in all the complexes depends upon the nature of ligands attached. The order of ligand's increasing field strength is given as

$C_2O_4^{2-} < H_2O < NH_3 < CN^-$

As CN^- is the strongest ligand in the given complexes. Therefore $[CO(CN)_6]^{3-}$ will have highest Δ_0 value

4. (2) Linkage isomerism arises in a complex compound containing ambidentate ligand. In the given complex compounds, NCS (thiocyanate) is an ambidentate ligand. Therefore, option (2) is correct.

5. (4) $[Co(en)_3]^{3+}$ exists as enantiomeric pair and shows optical activity.

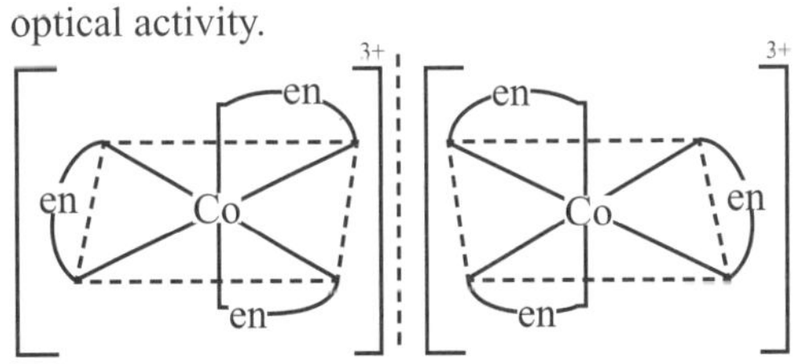

6. (3) The complex involves d^2sp^3 hybridisation and is an inner orbital complex.

7. (1) Brown ring test is used to detect the presence of nitrates in the solution. This test is usually carried out by adding dilute ferrous sulphate solution to it. The reaction is shown as,

$$NO_3^- + 3Fe^{2+} + 4H^+ \rightarrow NO + 3Fe^{3+} + 2H_2O$$

$$\overset{+2}{[Fe}(H_2O)_6]^{2+} + NO \longrightarrow \underset{\text{(Brown)}}{\overset{+3}{[Fe}(H_2O)_5(\overset{+1}{NO})]^{2+}}$$

8. (1) In the given ligands, a negatively charged bidendate ligand is - dimethyl glyoximato. It is represented as

$H_3C - C = N - \ddot{O}^-$
$\quad\quad\;|$
$H_3C - C = N - OH$

9. (4) Effective atomic number for $[Rh(H_2O)_6]^{3+}$ is calculated as -

EAN = z – x + y

where, z = atomic number

x = charge on the metal

y = total electrons donated by ligand

At. No. of Rh = 45

EAN = 45 – 3 + 2 × 6 = 54

10. (1) $[NiCl_4]^{2-}$

$Ni^{+2} \Rightarrow 3d^8$; Cl^- weak field ligand so no pairing

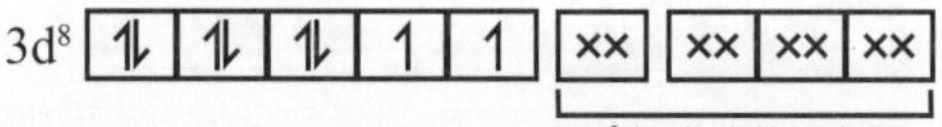

sp^3 hybridisation

Paramagnetic due to two unpaired electrons

$[Ni(H_2O)_6]^{+2}$

$Ni^{+2} \Rightarrow 3d^8$; H_2O weak field ligand, so no pairing.

Therefore, it is paramagnetic.

$[Ni(PPh_3)_2Cl_2]$

$Ni^{+2} \Rightarrow PPh_3$, Cl weak field ligand $\Rightarrow$ No pairing

Therefore, paramagnetic.

$[Ni(CO)_4]$

$Ni^0 \Rightarrow 3d^{10}$; all paired so diamagnetic.

$[Ni(CN)_4]^{2-}$ is also diamagnetic in nature.

11. (2) In $[CoF_6]^{3-}$ cobalt has $3d^6$ configuration and the complex is outer orbital complex. It has 4 unpaired electrons. $[Co(NH_3)_6]^{3+}$ is diamagnetic (inner orbital complex),$[NiCl_4]^{2-}$ is tetrahedral having 2 unpaired electrons; $[Ni(CN)_4]^{2-}$ is square planar and diamagnetic , $[Ni(CO)_4]$ is tetrahedral and diamagnetic.

12. (2) Cl^- is replaced by NO_2 in ionisation sphere and is an ionisation isomer.

13. (1) P: $[FeF_6]^{3-}$; Fe^{3+} : $3d^5$

$[FeF_6]^{3-}$;

sp^3d^2 hybridisation

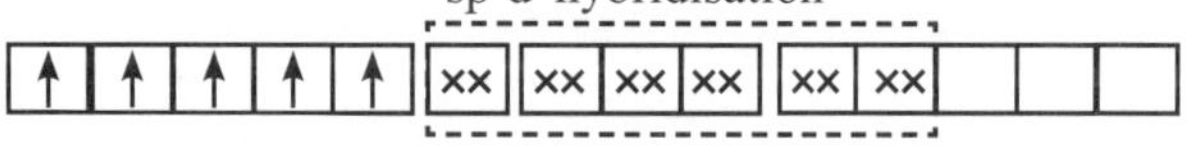

Number of unpaired electrons = 5

$\mu = \sqrt{5(5+2)} = \sqrt{35} = 5.92$ B.M

$Q:\left[V(H_2O)_6\right]^{2+}$; V^{2+} : $3d^3$

$[V(H_2O)_6]^{2+}$:

d^2sp^3 hybridisation

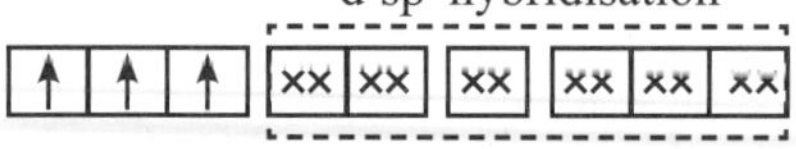

No. of unpaired electrons = 3

$\mu = \sqrt{3(3+2)} = \sqrt{15} = 3.87$ B.M.

$R:\left[Ni(H_2O_6)\right]^{2+}$; Ni^{2+} : $3d^8$

$[Ni(H_2O)_6]^{2+}$:

sp^3d^2 hybridisation

No. of unpaired electrons = 2

$\mu = \sqrt{2(2+2)} = \sqrt{8} = 2.9$ B.M.

The correct order of spin- only magnetic moment R < Q < P.

14. (3) Metal carbonyls follow EAN rule by loosing 1 e^- from $[Mn(CO)_6]$ & by gaining 1 e^- from $[V(CO)_6]$. Therefore both acts reducing agent and oxidising agent respectively.

$$\left[Mn(CO)_6\right] \xrightarrow{-e^-} \left[Mn(CO)_6\right]^+$$

EAN = 37 EAN = 36

$$\left[V(CO)_6\right] \xrightarrow{+e^-} \left[V(CO)_6\right]^-$$

EAN = 35 EAN = 36

15. (3) Only cis Pt (NH_3) Cl_2 reacts with oxalic acid $(H_2C_2O_4)$ to form $[Pt\ Cl_2(ox)]^{2-}$ and not the trans isomer because the oxalate ion is a bidentate ligand which occupies adjacent positions only

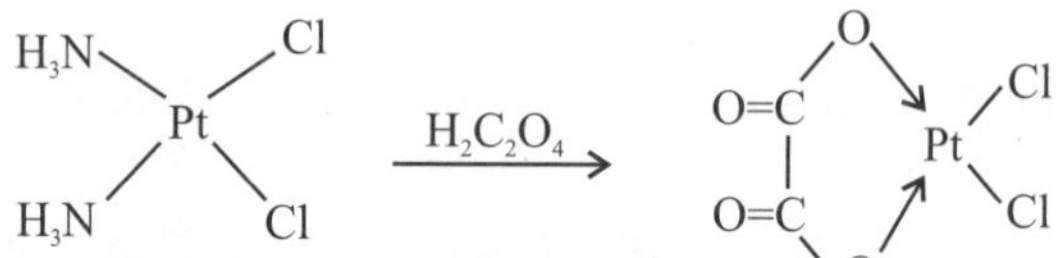

cis-form

16. (4) After splitting of d-orbital in octahedral field, electrons are distributed according to Hund's rule.

- "NO" is a $3e^-$ donar ligand.
- Dithioxalate is a bidentate as well as ambidentate ligand.

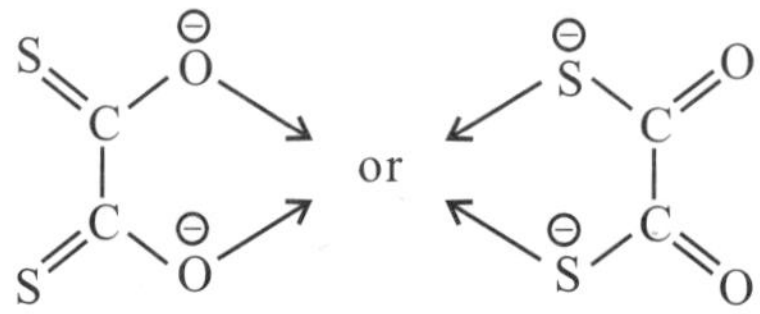

17. (2) Mabcd type compound having tetrahedral geometry will show optical isomerism. Because it has not plane of symmetry which is necessary condition of this type of compound having optical isomerism.

18. (4) An octahedral complexes having CFSE –0.8 Δ_O must have two unpaired e^-. According to given weak field ligand condition, pairing of electron is not possible. Therefore, two type of electronic configuration is possible-

CFSE = –0.8 → $t_{2g}^5\ e_g^2$; → $t_{2g}^2\ e_g^0$

For $t_{2g}^5\ e_g^2$

t_{2g}^5 e_g^2

Total number of unpaired e^- = 3

Magnetic moment μ

$= \sqrt{n(n+2)} = \sqrt{3(3+2)} = \sqrt{15}$ B.M.

For $t_{2g}^2\ e_g^0$

[1][1][ ] [ ][ ]

Total number of unpaired $e^- = 2$

$\mu = \sqrt{2(2+2)} = \sqrt{8}$ B.M

19. (4) All of these possess plane of symmetry. Therefore, no one will show optical isomerism.

20. (3) For ionisation isomerism, ligands are ionisable (exchangeable) in co-ordination sphere & outside co-ordination sphere. $[Cr(NH_3)_6]Cl_3$ & $[Cr(en)_3]Cl_3$ have not inside ionisable ligand. Whereas $[Co(NH_3)_5Br]SO_4$ will show ionisation isomerism as-

$[Co(NH_3)_5Br]SO_4 \rightleftharpoons [Co(NH_3)_5.SO_4]Br$

21. (4) All the given compounds will show d^2sp^3 hybridisation.

$Ni^{4+} = (Ar)\ 3d^6$ 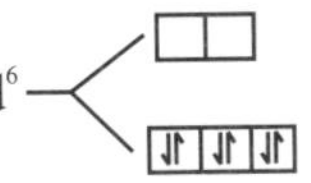 $\Rightarrow$ inner orbital complex

$Ir^{3+} : 5d^6$ 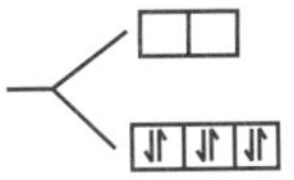 $\Rightarrow$ inner orbital complex

H_2O behaves as a strong field ligand for $Co^{3+} \Rightarrow$ inner orbital complex

22. (3) $Co(NO_2)_3 + 3KNO_2 \rightarrow K_3[Co(NO_2)_6] \rightleftharpoons 3K^+[Co(NO_2)_6]^{3-}$
potassium cobalt nitrite.

Total charge = 6 [3 positive + 3 negative]

Total particle = 4

23. (1) In the given options, option "1" is the correct match of the given columns.

24. (1) Complexes of the type $MA_3\ B_3$ cannot show optical isomerism because of plane of symmetry.

25. (1)

(i) $[Co(CN)_6]^{-4}$
EAN = 25 + 12 = 37 contain one more e^- than 36 so it can easily loose it and act as reducing agent.

(ii) $[Co(CN)_6]^{3-}$
EAN 24 + 12 = 36

(iii) $[Ag(CN)_2]^{-1}$
Stable (Experimental) and contain lesser than 54 electron

(iv) $[Co(CN)_4]^{3-}$
EAN = 27 + 8 − 3 = 32

26. (1) $[Fe(SCN)_6]^{3-}$

SCN is a weak field ligand with one negative charge.

The oxidation state of Fe is calculated as,

$x + (-6) = -3$

$x - 6 = -3$

$x = +3$

Fe(26) : (Ar) $4s^2\ 3d^6$

Fe^{3+} : (Ar) $3d^5$

[1][1][1][1][1]

Number of unpaired $e^- = 5$

$M = \sqrt{5(5+2)} = \sqrt{35}$ B.M

$[Fe(CN)_6]^{3-}$

CN^- is a strong field ligand

Fe = +3

Fe^{3+} : (Ar) $3d^5$

Before pairing [1][1][1][1][1]

After pairing [1↓][1↓][1][ ][ ]

$M = \sqrt{1(1+2)} = \sqrt{3}$ B.M

$= \sqrt{35} - \sqrt{3} = 4$

27. (1) In the given options, option "1" is correct match of the given columns.

28. (3) $[Ni(CN)_4]^{2-}$ has square planar geometry and other have tetrahedral geometry

29. (4) $[Fe(en)_3]^{3+}$ is most stable due to chelate formation among the given complexes.

30. (1) The molecule, cis $[Co(en)_2Cl_2]Cl$ lacks plane of symmetry. So, it is optically active.

31. (4) The given complex is present in tetrahedral geometry as-

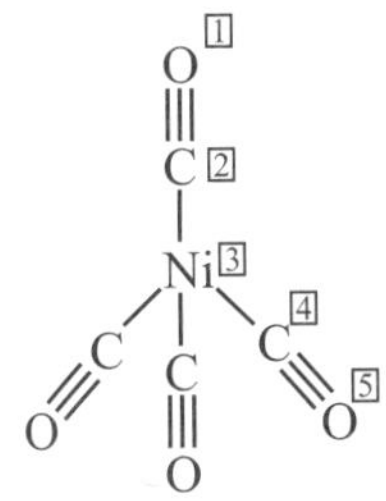

The atoms labelled are in same plane.

32. (3) The given complexes will show co-ordination isomerism.

33. (3) $[Cr(ox)_3]^{3-}$ will show only optical isomerism not geometrical isomerism as it has no plane of symmetry.

34. (4) Zeise's salt has the formula $[Pt(C_2H_4)Cl_3]$. It is a square planar complex having central atom Pt in +2 oxidation state. Ligand C_2H_4 is present as $H_2C{=}CH_2$ in a perpendicular to $PtCl_3$ plane. Therefore, all statements are correct.

35. (3) Ligands which can ligate through either of two different atoms present in it are called ambidentate ligands. Examples of such ligands are the CN^-, NO_2^- and SCN^- ions. NH_3 is not an ambidentate ligand, it is a monodentate ligand.

36. (4) (1) $[Pt(en)_3]Cl_4$: 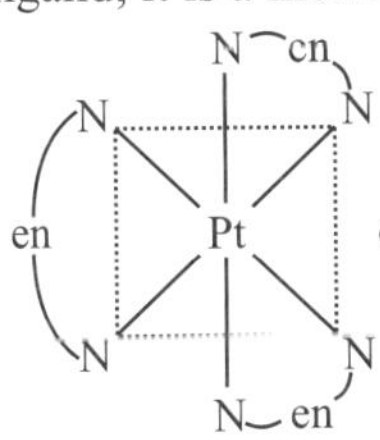 optically active

(2) $[Pt(NH_3)_2(CN)_2]$: Pt is in +2 oxidation state having $5d^8$ configuration. Hence the hybridisation of complex is dsp^2 and geometry is square planar. It has geometrical isomerism.

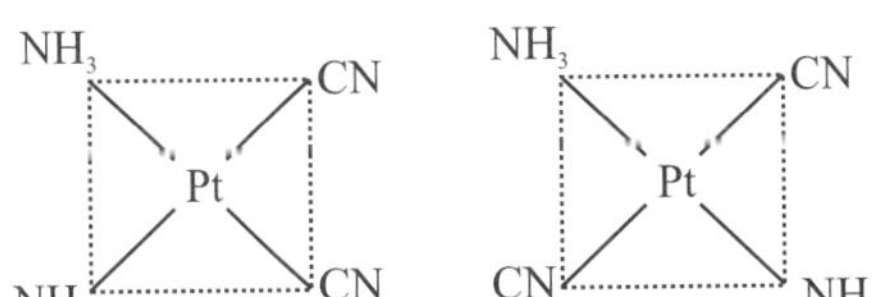

cis trans

(3) $[NiCl_2Br_2]^{2-}$ = tetrahedral, neither geometrical isomerism nor optical isomerism

37. (4) Splitting of d orbital in octahedral complexes is shown as-

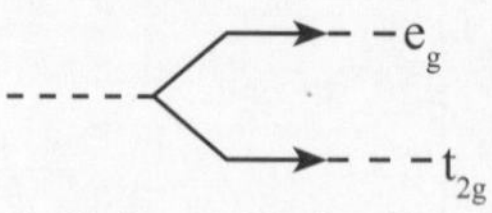

Splitting of e^- in d^3 orbital is represent as (whether strong or weak field)

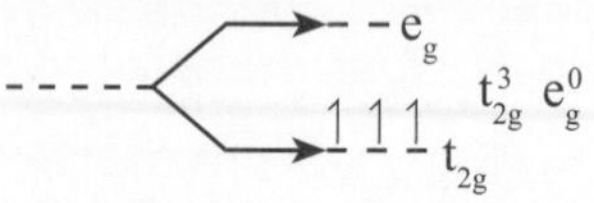

Splitting of e^- in d^8 orbital in both strong & weak field-

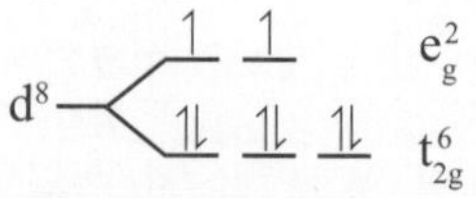

due to unpaired e_s^- coloured compounds may obtained. Therefore option "1" is incorrect.

No. of e_s^- in t_{2g} orbitals are higher than in e_g orbitals and it can have either d^3 or d^8 configuration. Hence, both statements given in option "2" & "3" are correct.

Therefore, option "4" is correct.

38. (3) The given complex is actually

$$\left[\underset{+2}{Fe}\left(\underset{-1}{Cl}\right)\left(\underset{-4}{CN}\right)_4\left(\underset{-1}{O_2}\right)\right]^{4-}$$

hence Fe(II) is $t_{2g}{}^6$, $e_g{}^0$ due to effect of strong ligands but it is paramagnetic due to $O_2{}^{-1}$ ligand.

39. (2) If $\Delta_o < P$, high spin complex is formed by association of weak field ligand with metal ion. So fourth e^- enter one of e_g orbital and electronic configuration becomes $t_{2g}^3 e_g^1$.

40. (2) $Mn(CO)_5$ does not obey EAN rule but it exist as a dimer & it becomes $Mn_2(CO)_{10}$.

41. (3) Mn_2O_7 is acidic, V_2O_5 amphoteric & CrO is basic.

42. (1) NO_2^- is a monodentate as well as an ambidentate ligand.

43. (3) $[Cu(NH_3)_4]SO_4\ (aq) \rightleftharpoons [Cu(NH_3)_4]^2\ (aq) + SO_4^{2-}\ (aq)$

$SO_4^{2-}\ (aq) + Ba^{2+} \rightarrow BaSO_4 \downarrow$ (white)

SO_4^{2-} is present in ionisation sphere.

44. (4) In both (1) and (3) dipole moment is zero. $Ni(CN)_4]^{2-}$ have square planar geometry so due to symmetrical geometry it will show zero dipole moment. Similarly due to trans geometry $Pt(NH_3)_2Cl_2$ show zero dipole moment.

45. (4) Every solution contains three moles of Cl^- ions.

46. (4) $$2K^+\left[Cr(\overset{-2}{CN})_2\overset{-2}{O_2}\overset{-4}{(O)_2}\overset{0}{(NH_3)}\right]^{2-}$$

Oxidation state of chromium = x + 2(–1) + (–2) + 2(–2) + (0) = –2.

∴ x = + 6.

As per IUPAC name.

Complex is anionic, so metal ion will be named as chromate with +6 oxidation state. So, potassium amminedicyanodioxoperoxochromate(VI).

47. (1) Four pairs of electrons from four Cl^-

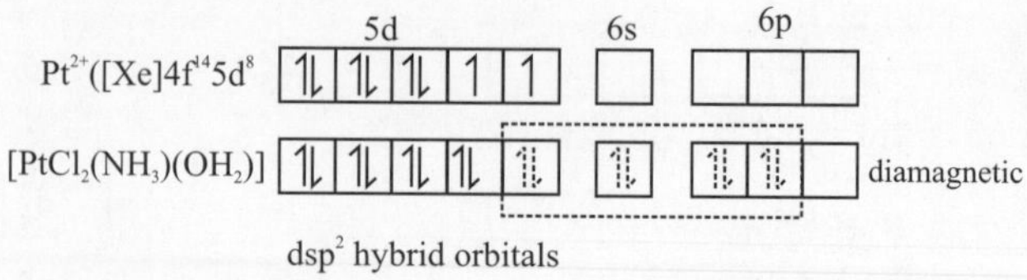

Ma_2bc have cis- and trans isomerism not hydrate isomerism. It has square planar geometry not Td geometry.

48. (1) It exhibits linkage isomerism, ionization isomerism and geometrical isomerism.

49. (4)

(a) Coordination isomerism arises due to exchange of ligands between cationic and anionic complexes.

(b) SCN^- is an ambidentate ligand and thus can attach either through nitrogen or sulphur to central metal ion causing linkage isomerism.

(c) Exchange of NO_2^- (in coordination sphere) and SO_4^{2-} (in ionization sphere) produces ionization isomerism.

50. (2) Ammonia is a stronger field ligand than water. So $[Ni(NH_3)_6]^{2+}$ will have higher CFSE than $[Ni(H_2O)_6]^{2+}$. Both complexes $[Ni(H_2O)_6]^{2+}$ and $[Ni(NH_3)_6]^{2+}$ have sp^3d^2 hybridisation with two unpaired electrons. So, they possess same magnetic moment ('spin only')

51. (3) In complex nickel is in 0 oxidation state. The hybridisation scheme is as shown below in figure.

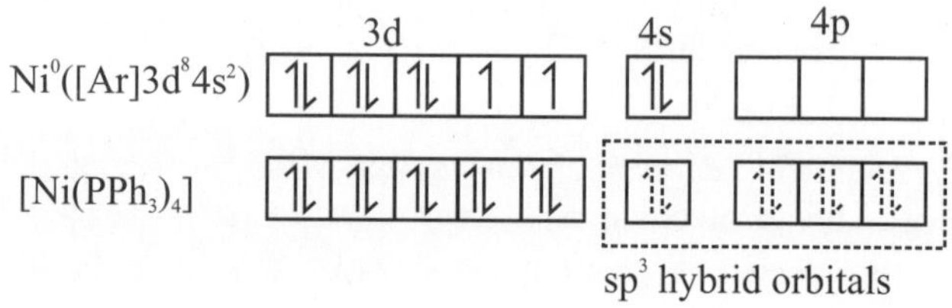

Four pairs of electrons from four CO.

It is tetrahedral and as all electrons are paired so diamagnetic.

52. (2) For $K_2[Ni(CN)_4]$

EAN = 28 – 2 + 2 × 4 = 34

53. (1) (i) $[Fe(H_2O)_6]^{2+}$, $3d^6$ 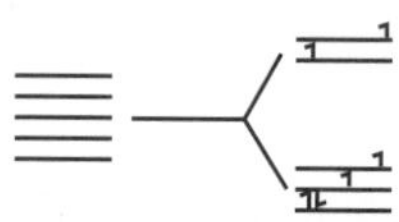

n = 4 (weak field ligand)

(ii) $[Fe(CN)_6]^{3-}$, $3d^5$ 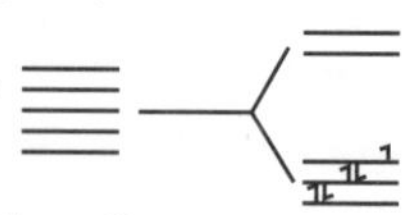

n = 1 (Strong field ligand)

(iii) $[Fe(CN)_6]^{4-}$, $3d^6$ 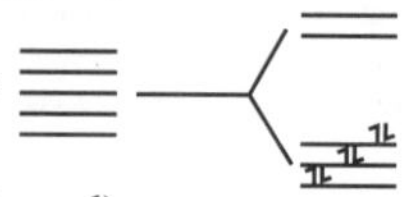

n = 0 (strong field ligand)

(iv) $[Fe(H_2O)_6]^{3-}$, $3d^5$ 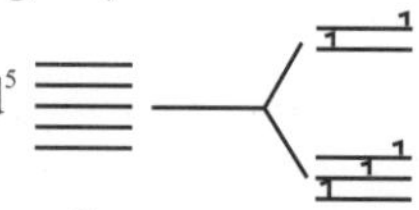

n = 5 (weak field ligand)

54. (1) $AgCl + 2NH_3 \rightarrow [Ag(NH_3)_2]Cl \rightleftharpoons [Ag(NH_3)_2]^+Cl^-$

$AgCl + 2KCN \rightarrow K[Ag(CN)_2] \rightleftharpoons K^+[Ag(CN)_2]^- + KCl$

$AgCl + 2Na_2S_2O_3 \rightarrow Na_3[Ag(S_2O_3)_2]$

$\rightleftharpoons 3Na^+[Ag(S_2O_3)_2]^{3-} + NaCl$

55. (2) cis-$[PtCl_2(NH_3)_2]$ is used in chemotherapy to inhibits the growth of tumors.

56. (1) Chlorophylls are green pigments in plants and contain magnesium

57. (1) Co^{3+}

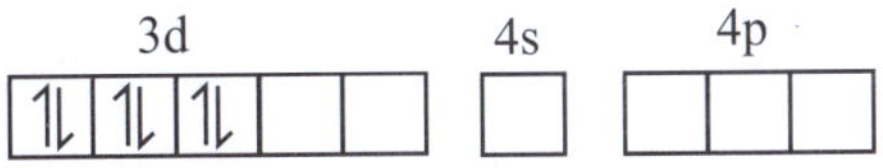

58. (4)

(1) Cr^{3+}, $3d^3$; d^2sp^3 hybridisation with three unpaired electrons (weak field ligand).

Fe^+, $3d^7$; d^2sp^3 hybridisation with three unpaired electron (one 4s electron jumps to 3d orbital for the pairing)

(2) Mn^{2+}, $3d^5$; d^2sp^3 hybridisation with one unpaired electron after pairing (CN is strong field ligand).

Fe^{3+}, $3d^5$; d^2sp^3 hybridisation with one unpaired electron after pairing (CN is strong field ligand).

(3) Ni, $3d^8$; sp^3 hybridisation and all electrons are paired because CO is strong field ligand (4s-electrons jumps to 3d-orbitals for the pairing)

Zn^{2+}, $3d^{10}$; sp^3 hybridisation and all electrons are paired.

59. (4) Effective atomic number (EAN). in a complex is defined as the total number of electrons present around the central metal ion. EAN rule is often referred to as 18 e^- rule. This rule states that the total electrons donated by the ligands to the metal should be equal to the nearest noble gas present in that period.

It is calculated by the given formula, (EAN) = z – x + y

Where, z = Atomic number of central metal

x = No. of e^- loosed by central metal ion (O.S.)

y = No. of e^- shared by ligands

EAN of $[Ni(CO)_4]$ = 28 – 0 + 2 × 4 = 36, So it is stable. All other compounds are also stable as they have EAN=36

60. (3) $[Cr(NH_3)_6]Cl_3 \rightleftharpoons [Cr(NH_3)_6]^{3+}\ 3Cl^-$ O.S. of Cr = +3

E.A.N of Cr = 24 – 3 + 2 × 6 = 24 – 3 + 12 = 36 – 3 = 33

61. (2) The most stable complex is $[Fe(CO)_5]$ because it follows EAN rule. It is calculated as

EAN = z – x + y = 26 – 0 + 2 × 5 = 26 + 10 = 36

Correct & Incorrect MCQs

1. **(2)** The primary valences are normally ionisable and the secondary valences are non ionisable.
2. **(2)** $CoCl_3.5NH_3$ i.e., $[CoCl(NH_3)_5]^{2+}2Cl^-$
   Hence, 1 mol $CoCl_3.5NH_3$ will give 2 mol AgCl on adding excess of silver nitrate solution.
3. **(3)** Carnallite i.e., $KCl.MgCl_2.6H_2O$, is an example of double salt. Potassium ferrocyanide i.e., $K_4[Fe(CN)_6]$ is an example of complex salt.
4. **(3)** Ethylenediaminetetraacetate ion can bind through two nitrogen and four oxygen atoms to a central metal ion. Hence, it is a hexadentate ligand.
5. **(4)** Complexes in which a metal is bound to more than one kind of donor groups are known as heteroleptic complex. Example : $[Co(NH_3)_4Cl_2]^+$
6. **(2)** Geometrical isomerism is not possible for a tetrahedral geometry. This is because change in the position of ligands in space in tetrahedral complex does not change with respect to each other.
7. **(2)** Only the cis-isomer of the coordination entity of the type $[PtCl_2(en)_2]^{2+}$ shows optical activity, due to lack of symmetry.
8. **(4)** In the absence of ligand, crystal field splitting does not occur and hence the substance is colourless. Therefore, anhydrous $CuSO_4$ is white, but $CuSO_4.5H_2O$ is blue in colour.
9. **(1)**
   1. Tetrahedral complexes are sp^3 hybridised.
   2. Square planar complexes are dsp^2 hybridised.
   3. Trigonal bipyramidal complexes are sp^3d hybridised.
   4. Octahedral complexes are d^2sp^3 or sp^3d^2 hybridised.

10 **(3)** The complex $[Co(NH_3)_6]^{3+}$ is an inner orbital complex.

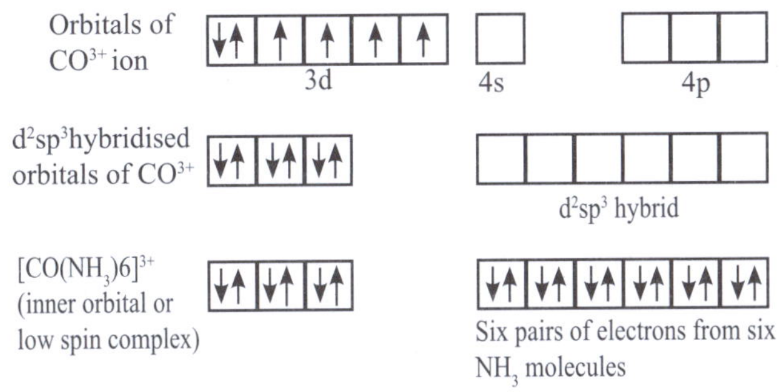

Statement Based MCQs

1. **(1)** Optical isomerism is not shown by square planar complexes (Ma_2b_2, Mabcd) as they have plane of symmetry
2. **(2)** $[Fe(CN)_6]^{4-}$ hence charge on Fe → +2
   $[Fe(CN)_6]^{3-}$ hence charge of Fe → +3

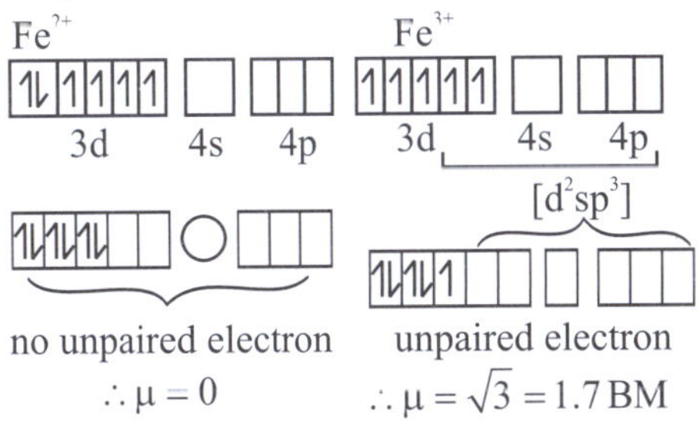

Hence, $K_3[Fe(CN)_6]$ and $K_4[Fe(CN)_6]$ do not have same magnetic moment, it depends on unpaired e^-_s, not paired e^-_s.

3. **(1)** Chelates are more stable than non chelated complex as polydentate ligand use two or more donor atoms to bind single metal ion thus forming a large close stable structure.
   Those complexes in which ligands can easily be replaced by other ligands are known as labile complexes.
4. **(3)** In ferrocyanide, Fe is in oxidation state +2
   In ferricyanide, Fe is in oxidation state +3
   Higher the oxidation state, greater is the crystal field splitting.

5. **(2)** $[Ni(en)_3]Cl_2$ is a chelating compound. Chelating complexes are more stable than similar complexes with unidentate ligands, as dissociation of the complex involves breaking of two bonds rather than one. In $[Ni(en)_3]Cl_2$, Ni with d^8 configuration shows octahedral geometry, six electron will occupy the t_{2g} orbitals and the two electrons will occupy the e_g orbitals.

6. **(2)** In metal carbonyls, the metal-carbon bond possess both σ and π character. The M – C σ bond is formed by the donation of lone pair of electrons on the carbonyl carbon into a vacant orbital of the metal. The M–C π bond is formed by the donation of a pair of electrons from a filled d orbital of metal into the vacant antibonding π^* orbital of carbon monoxide.

7. **(3)** NO_2^- and SCN^- ion are ambidentate ligand. SCN^- and NO_2^- ion can coordinate either through 'N' or 'O' to a central atom or ion.

8. **(2)** In homoleptic complex, metal is bound to only one kind of donor groups and in heteroleptic complex, a metal is bound to more than one kind of donor groups.

   Hence, $[CO(NH_3)_6]^{3+}$ is a homoleptic complex and $[Co(NH_3)_4Cl_2]^+$ is a heteroleptic complex.

9. **(2)** Inner orbital complexes are low spin complexes while outer orbital complexes are high spin complexes.

10. **(1)** Linkage isomerism arises in a coordination compound containing ambidentate ligand and ambidentate ligand contain two different donor atoms.

Assertion & Reason MCQs

1. **(1)** Chelating ligands have two or more donar sites therefore, they form cyclic ring with the metal. Which gives stability to the complex called chelating complex. In this way toxic metal ions are removed by this.

2. **(2)** Both the given complexes are reducing in nature, because both Cr and Fe are present in +2 oxidation state. On removing one electron both get stabilised by forming stable d-electron configurations.

   In both the complexes unpaired d-electrons are present.

3. **(1)** Ligand which has two different donor atoms and either of the two ligetes in the complex is called **ambidentate ligand**. Examples of such ligands are the NO_2^- and SCN^- ions.

   Linkage isomerism arises in a coordination compound containing ambidentate ligand. A simple example is provided by complexes containing the thiocyanate ligand, NCS^-, which may bind through the nitrogen to give M-NCS or through sulphur to give M-SCN.

4. **(2)** Due to presence of same type of ligands, MX_6 and MX_5 do not show geometrical isomerism. Geometrical isomerism can be shown by the complexes of co-ordination number-6. For ex: MA_3B_3, MA_4B_2 etc will show fac & mer geometrical isomers.

5. **(4)** $[Fe(CN)_6]^{3-}$

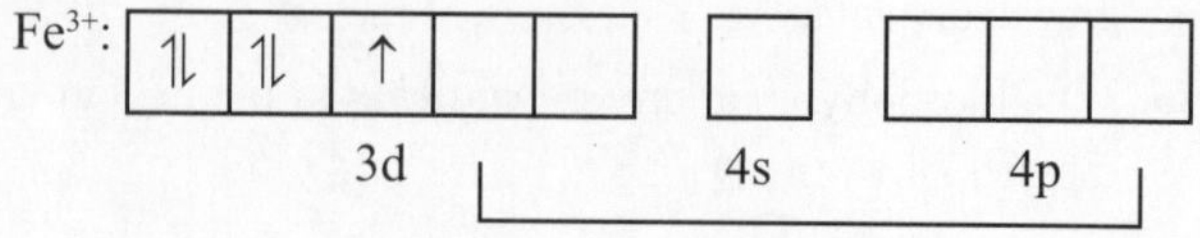

d²sp³ hybridisation

So the assertion is false as it contain only one unpaired electron.

Due to d^2sp^3 hybridisation $[Fe(CN)_6]^{3-}$ ion shows magnetic moment corresponding to one unpaired electron present in it. So the reason is true.

6. **(1)** Since, mohr's salt is a double salt i.e. it is formed by 2 or more stable compounds which exists only in solid state but dissociates into ions when dissolved in water

7. **(2)** Ligands which can ligate through two different atoms are called ambidentate ligands and these show linkage isomerism.

8. **(3)** The optical isomerism is present in the given compound due to absence of plane of symmetry.

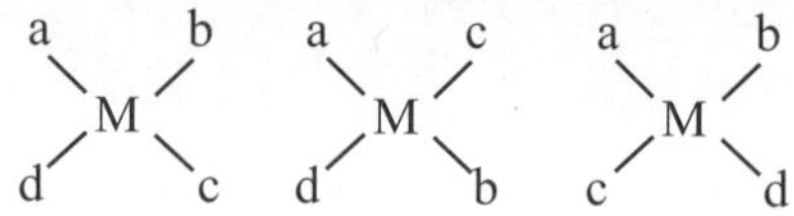

9. **(2)** Ni^{2+}: $[Ar]3d^8$, now even if we consider pairing up of electrons, then d^2sp^3 hybridisation will not be possible. Hence the hybridisation will have to be sp^3d^2 i.e they must be outer orbital complexes.

10. **(1)** The given complex gives white ppt with $AgNO_3$ means one chloride ion is present outside the sphere and the chloride ion is ionizable i.e. free Cl^-.

Multi Correct MCQs

1. **(2)** $[NiCl_2Br_2]^{2-}$ is a tetrahedral complex due to presence of WFL. As Ni is replaced by Pt, Pt belongs to 5d-series and it will form square planar complex no matter whether the ligand is SFL and WFL. and Square planar complexes show geometrical isomerism while tetrahedral show optical isomerism.

2. **(3)** $[FeF_6]^{3-}$: Fe^{+3} : $[Ar]\ 3d^5$

   F^- : Weak field Ligand

   No. of unpaired electron's = 5

   $\mu = \sqrt{5(5+2)}$

   $\mu = \sqrt{35}$ BM

   $[V(H_2O)_6]^{+2}$: V^{+2} : $3d^3$

1	1	1		

   No. of unpaired electron's = 3

   $\mu = \sqrt{3(3+2)}$

   $\mu = \sqrt{15}$ BM

   $[Fe(H_2O)_6]^{+2}$: Fe^{+2} : $3d^6$

   H_2O : Weak field Ligand

1⇂	1	1	1	1

No. of unpaired electron's = 4

$\mu = \sqrt{4(4+2)}$

$\mu = \sqrt{24}$ BM

3. (3) $[V(H_2O)_6]^{3+}$

V^{+3}:- [Ar] $3d^2$, n = 2 (even number of unpaired e^-)

$[Cr(H_2O)_6]^{2+}$

E.C of Cr^{+2}:- [Ar] $3d^4$, n = 4 (even number of unpaired e^-)

$[Fe(H_2O)_6]^{3+}$

E.C of Fe^{3+}:- [Ar] $3d^5 4s^0$

Hence, n = 5 (odd number of unpared e^-)

$[Ni(H_2O)_6]^{2+}$

E.C of Ni^{2+}: [Ar] $3d^8$, hence, n = 2 (even number of unpaired e^-)

$[Cu(H_2O)_6]^{2+}$

E.C of Cu^{2+}: [Ar] $3d^9 4s^0$

Hence, n = 1 (odd number of unpared e^-)

Thus, three complexes have even number of unpaired 'd' electrons.

4. (2)

Unpaired electrons	Oxidation	Electronic Configuration	Unpaired electrons
$[Fe(CN)_6]^{3-}$	Fe^{3+}	$[Ar]3d^5$	1
$[FeF_6]^{3-}$	Fe^{3+}	$[Ar]3d^5$	5
$[CoF_6]^{3-}$	Co^{3+}	$[Ar]3d^6$	4
$[Cr(oxalate)_3]^{3-}$	Cr^{3+}	$[Ar]3d^3$	3
$[Ni(CO)_4]$	Ni^0	$[Ar]3d^8 4d^2$	0

NCERT Exemplar MCQs

1. (2) Greater the value of log K, greater will be the stability of complex compound formed.

For reaction, B:-

$Cu^{2+} + 4\,CN^- \rightarrow [Cu(CN)_4]^{2-}$

$$K = \frac{[(Cu(CN)_4)^{2-}]}{[Cu^{2+}][CN^-]^4} \text{ and } \log K = 27.3$$

For this reaction, log K has highest value among the given four reactions means K value will be highest. Hence, stability of the complexes will be highest.

2. (3) As we know that, strong field ligand split the five degenerate energy levels with more energy separation than weak field ligand, i.e., as strength of ligand increases crystal field splitting energy increases. The order of ligand field strength is $CN^- > NH_3 > H_2O$.

Hence, $\Delta E = \frac{hc}{\lambda}$

$$\Rightarrow \quad \Delta E \propto \frac{1}{\lambda} \Rightarrow \lambda \propto \frac{1}{\Delta E}$$

As energy separation increases, the wavelength decreases. Thus, the correct order is

$[Co(H_2O)_6]^{3+} > [Co(NH_3)_6]^{3+} > [Co(CN)_6]^{3-}$

Here, strength of ligand increases, ΔE increases, CFSE increases and λ absorbed decreases.

3. (2) One mole of $AgNO_3$ precipitates one mole of chloride ion. In the above reaction, when 0.1 mole $CoCl_3(NH_3)_5$ is treated with excess of $AgNO_3$, 0.2 mole of AgCl are obtaine thus, there must be two free chloride ions in the solution of electrolyte.

So, molecular formula of complex will be $[Co(NH_3)_5Cl]Cl_2$ and electrolytic solution must contain $[Co(NH_3)_5Cl]^{2+}$ and two Cl^- as constituent ions. Thus, it is 1 : 2 electrolyte.

4. (4) 1 mole of $AgNO_3$ precipitates one free chloride ion (Cl^-).

Here, 3 moles of AgCl are precipitated by excess of $AgNO_3$. Hence, there must be three free Cl^- ions.

So, the formula of the complex is: $[Cr(H_2O)_6]\,Cl_3$ and correct choice is (4).

5. (1) $[Pt(NH_3)_2Cl_2]$.

The ligands present in the compound are

(i) NH_3 – neutral ligand represented as amine.

(ii) Cl – anion ligand (ending with-o-) represented as chlorido, di prefix is used to represent two ligands.

The oxidation number of platinum in the compound is 2. Hence, correct IUPAC name of $[Pt(NH_3)_2Cl_2]$ is Diamminedichloridoplatinum (II)

So, (1) option is correct.

6. (3) Here, only $[Fe(C_2O_4)_3]^{3-}$ is a coordination compound which contains oxalate ion as a chelating ligand. Hence it stabilises coordination compound by chelating Fe^{3+} ion.

7. (1) $[Cr(H_2O)_4Cl_2]^+$ shows geometrical isomerism because it is a MA_4B_2 type coordination compound which contains two set of equivalent ligands, four H_2O and 2Cl. Octahedral complexes of the type MA_4B_2 exists as cis and trans isomers.

8. (3) CFSE for octahedral and tetrahedral complexes are closely related to each other by the formula:-

$$\Delta_t = \frac{4}{9}\Delta_0.$$

where, Δ_0 = CFSE for octahedral, Δ_t = CFSE for tetrahedral complex

According to question, $\Delta_0 = 18,000\ cm^{-1}$

$$\therefore \quad \Delta_t = \frac{4}{9}\Delta_0 = \frac{4}{9} \times 18,000\ cm^{-1}$$

$4 \times 2,000\ cm^{-1} = 8,000\ cm^{-1}$

Hence, correct choice is (3).

9. (1) The ligand (s) which has two different bonding sites are known as ambidentate ligands e.g., NCS, NO_2 etc.

Here, NCS has two binding sites at N and S.

Hence, NCS (thiocyanate) can bind to the metal ion in two ways

$M \leftarrow NCS$ or $M \rightarrow SNC$

Here, (←) denotes binding site.

Thus, coordination compounds containing NCS as a ligand show linkage isomerism *i.e.*, $[Pd(C_6H_5)_2(SCN)_2]$ and $[Pd(C_6H_5)_2(NCS)_2]$ are linkage isomers. Hence, correct choice is (1).

10. **(4)** Compounds having same molecular formula but different structural formula are known as isomers. $[Co(SO_4)(NH_3)_5]Br$ and $[Co(SO_4)(NH_3)_5]Cl$ have not same molecular formula. Hence, they are not isomers.

11. **(1)** A chelating ligand has two or more binding donor atoms to a single metal ion e.g., oxalato, glycinato etc.

Thiosulphato ($S_2O_3^{2-}$) is not a chelating ligand because geometrically it is not favorable to chelate a metal ion.

12. **(2)** Ligand must donate a pair of electron or loosely held electron pair to metal and form a M–L bond.

Among these, NH_4^+ does not have any pair of electron. Hence NH_4^+ is not a ligand.

13. **(2)** Solvate isomerism is shown when two compounds having same molecular formula differ by whether the solvent molecule is directly bonded to metal ion or is present as free solvent molecules in the crystal lattice.

When water is present as solvent and show this type of isomerism then it is known as hydrate isomerism.

Coordination compound $[Cr(H_2O)_6]Cl_3$ and $[Cr(H_2O)_5Cl]Cl_2.H_2O$ are solvate isomers, because water is exchanged by chloride ion. This is why both of them show different colour on exposure to sunlight.

14. **(3)** The ligands present in the given coordination compound are

(i) (NH_3) represented as amine

(ii) Cl^- represented as chlorido

(iii) NO_2^- represented as nitrito-

According to IUPAC rule, ligands are named in an alphabetical order before atom. Prefex di will be used to indicate the number of NH_3 ligands present.

Oxidation state of metal is indicated by Roman numeral in parenthesis.

So. IUPAC name will be diamminechloronitrito-N-platinum (II)

Hence, option (3) is correct.

NEET & AIIMS Past Year MCQs

1. **(1)** Ethylene diaminetetraacetate ion

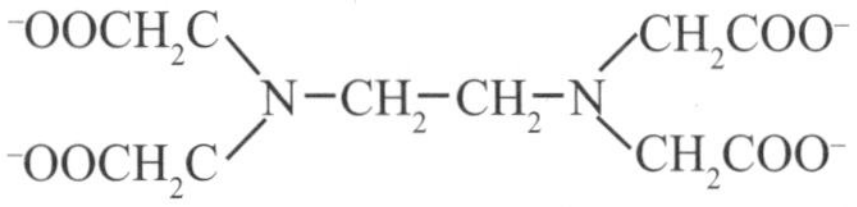

It is hexadentate since it can bind to a central metal ion through two nitrogen and four oxygen atoms.

2. **(1)** Ambidentate ligands are characterized by having two different donor atoms, with one of the donor atoms being attached to the metal during complex formation.

(1) 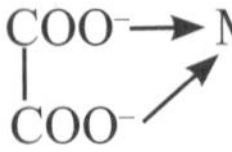

It is bidentate ligand because it can bind through two donor atoms.

(2) $M \leftarrow SCN$ (thiocyanato-S)

$M \leftarrow NCS$ (thiocyanato-N)

(Ambidentate)

(3) $M \leftarrow N=O$ (nitrito-N)

$\quad\quad\ |$

$\quad\quad O$

$M \leftarrow O-N=O$ (nitrito-O)

(Ambidentate)

(4) $M \leftarrow CN$ (cyanide)

$M \leftarrow NC$ (isocyanide)

(Ambidentate)

3. **(4)**

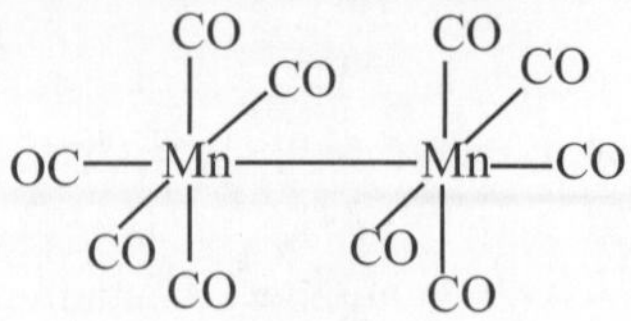

$[Mn_2(CO)_{10}]$

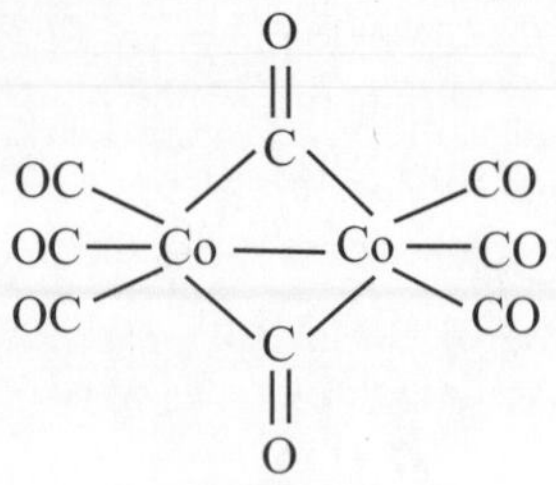

$[Co_2(CO)_8]$

A. Both complexes contain metal-metal bonds.

B. Both complexes contain terminal CO groups.

C. $[Mn_2(CO)_{10}]$ does not have any bridging CO groups, as all CO ligands are terminal.

D. In both complexes, metals are in the zero oxidation state.

4. **(3)**

(A) $[Co(NH_3)_5(NO_2)]Cl_2$ shows linkage isomerism. Its linkage Isomer is $[Co(NH_3)_5(ONO)]Cl_2$

(B) $[Co(NH_3)_5(SO_4)]Br$ shows ionization isomerism. Its ionization isomer is $[Co(NH_3)_5(Br)]SO_4$

(C) $[Co(NH_3)_6][Cr(CN)_6]$ shows coordination isomerism. Its coordination isomer is $[Cr(NH_3)_6][Co(CN)_6]$

(D) $[Co(H_2O)_6]Cl_3$ shows solvate isomerism. Its one of the solvate isomer is $[Co(H_2O)_5Cl]Cl_2.H_2O$

So correct option is:

$A \rightarrow II, B \rightarrow III, C \rightarrow IV, D \rightarrow I$

5. **(3)** Both $[Co(NH_3)_6]^{3+}$ and $[CoF_6]^{3-}$ are octahedral complex because coordination number are six.

Electronic configuration of Co^{3+} in $[Co(NH_3)_6]^{3+}$ is

Co^{3+}: $[Ar]3d^6 4s^0 4p^0$

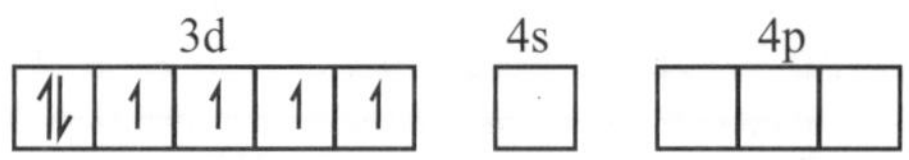

As NH_3 is a strong ligand, so pairing takes place

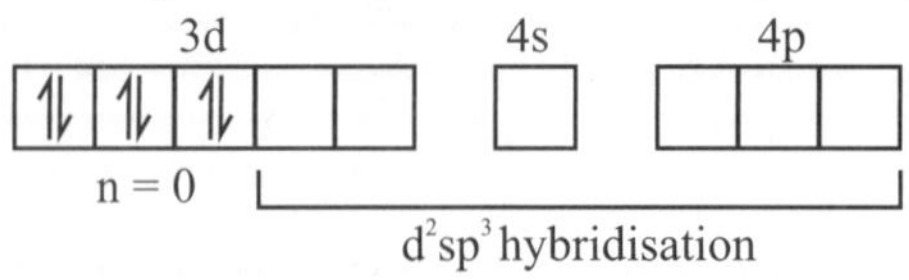

No unpaired electron, so it is diamagnetic in nature

Electronic configuration of Co^{3+} in $[CoF_6]^{3-}$ is

Co^{3+} : $[Ar]3d^6 4s^0 4p^0$

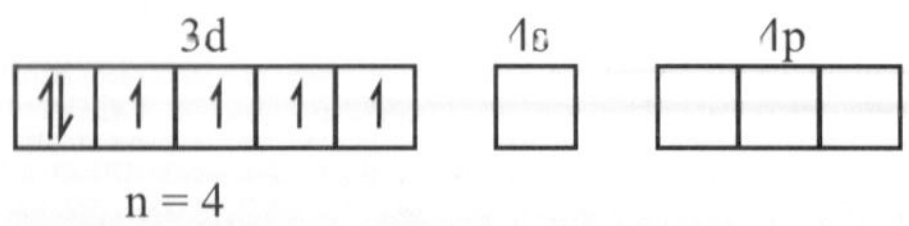

As F^- is a weak ligand, therefore no pairing takes place

So, it is paramagnetic in nature.

6. (1)

- Homoleptic complex is one which contains only one kind of ligands. Heteroleptic complex is one which contains two or more kinds of ligand.

$[Co(NH_3)_6]^{3+}$ has only one kind of ligands but

$[Co(NH_3)_4Cl_2]^+$ has more than one kind of ligands.

7. (2) Homoleptic complexes have all ligands identical.

Potassium trioxalatoaluminate (III) is $K_3[Al(C_2O_4)_3]$ which has only oxalate ion as ligand.

All others have more than one type of ligands.

8. (4) Ethylene diamine, en is bidentate, chelating ligand. Chelating ligands increase stability due to higher entropy factor.

9. (1) IUPAC name of the given co-ordination complex is given as:

$[Ag(H_2O)_2]\,[Ag(CN)_2]$

Oxidation state of Ag = +1

Therefore, the correct name of the compound is Diaquasilver (I) dicyanidoargentate (I)

10. (4) Higher the number of en(strong ligands), larger will be the splitting.

11. (4) Ethylene diaminetetraacetate (EDTA) ion is hexadentate ligand with four "O" and two "N" donor atoms.

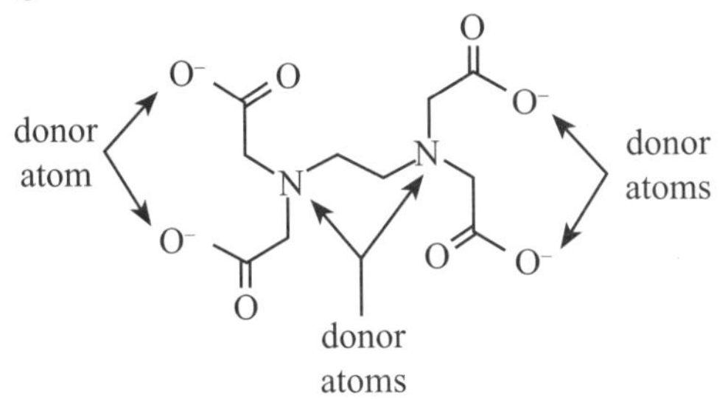

12. (3)

(A) $[Fe(CN)_6]^{3-}$

$Fe^{3+} - 3d^5$

↑	↑	↑	↑	↑

In presence of strong ligand pairing occur

↑↓	↑↓	↑		

n = 1

$\mu = \sqrt{n(n+2)} = \sqrt{3} = 1.73$ BM

(B) $[Fe(H_2O)_6]^{3+}$

Fe^{3+} in presence of weak ligand.

↑	↑	↑	↑	↑

n = 5

$\mu = \sqrt{35} = 5.92$ BM

(C) $[Fe(CN)_6]^{4-}$

$Fe^{2+} - 3d^6$

CN is strong ligand so, all electron are paired so

n = 0

$\mu = 0$

(D) $[Fe(H_2O)_6]^{2+}$

n = 4

$\mu = \sqrt{24} = 4.90$ BM

13. (1) Diamagnetic complex is are $[CO(ox)_3]^{3-}$, $[Fe(CN)_6]^{4-}$

14. (1) As in complex $[Co(NH_3)_5Cl]^{2+}$ unpaired electrons are present, so it requires less energy for excitation. Thus, the wavelength of light absorbed will be highest (*i.e.* $E \propto 1/\lambda$).

15. (2) **16. (1)** **17. (1)** **18. (3)** **19. (3)**

20. (2) **21. (1)**

Haloalkanes and Haloarenes

NCERT Topic-wise MCQs

1. (4) Neo - hexylchloride is a primary halide

Structures of all the given halides are given as,

1) Iso - propyliodide -

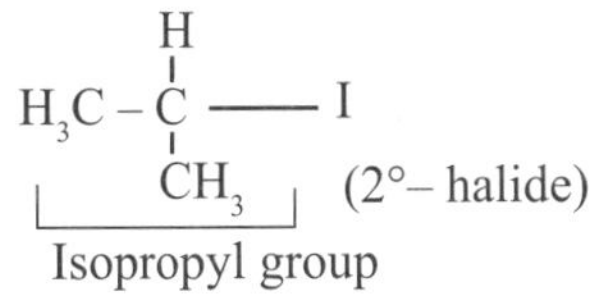

2) Sec - Butylliodide

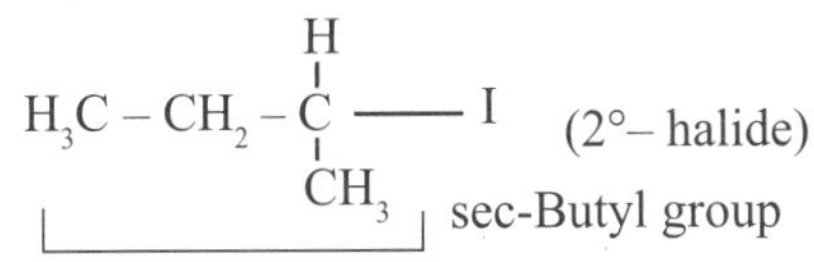

3) tert - Bulylbromide

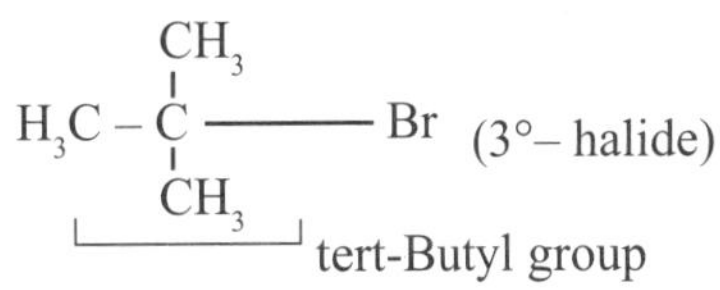

4) Neo - hexylchloride -

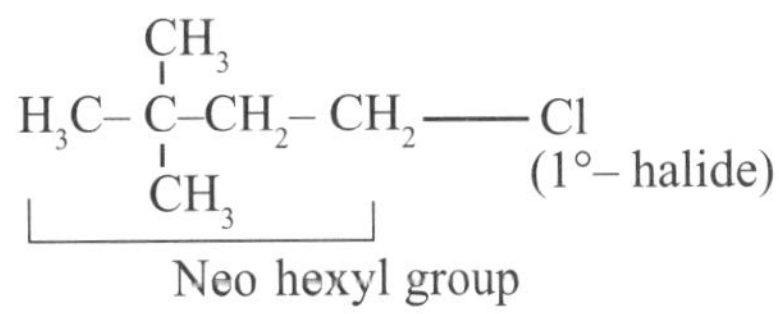

2. (4) 4-Bromobut-1-ene is not an allylic halide.

3. (2) When two halogen atoms are attached to same carbon atom then it is gem-dihalide.

It's structure is given below

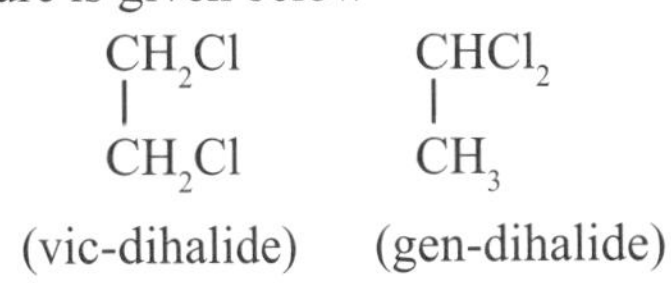

4. (4) C_4H_9X consists of all type (1°, 2° 3°) of alkyl halides. All structures are given below.

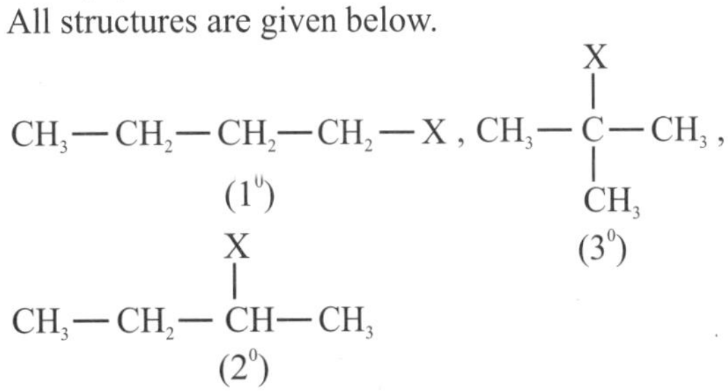

5. (1) $CH_3-\underset{CH_3}{\overset{Cl}{C}}-CH_3$ 2 - Chloro - 2 - methyl propane.

6. (3) Common name for the molecule $CH_2 = CH - CH_2 - Cl$ is Allyl chloride.

7. (3) The structure of 2-Bromopentane is-

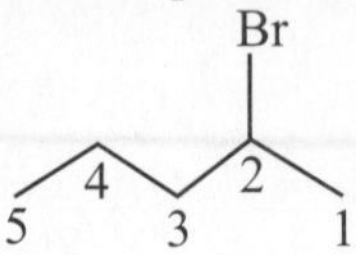

8. (3) The structure of 2-Bromo-2-methyl propane is-

$H_3C - \underset{CH_3}{\overset{Br}{C}} - CH_3$ (carbons numbered 1, 2, 3)

9. (4) All the given IUPAC names are correct.

10. (2) Product A is shown below-

$R-OH \xrightarrow{Red\ P + I_2} RI + H_3PO_3$ (A)

11. (1) $CH_2{=}CH-CH_2-CH_3 + HBr \longrightarrow CH_3-\overset{Br}{CH}-CH_2-CH_3$

(sec-Butyl bromide)
or
(2-Bromobutane)

12. (4) $CH_3-C{\equiv}CH \xrightarrow{Na} CH_3-C{\equiv}C^-Na^+ \xrightarrow[S_N2]{CH_3-CH_2-I} CH_3-C{\equiv}C-CH_2-CH_3 + NaI$

13. (3) $CH_3CH{=}CH_2 + HBr \xrightarrow[Addition]{M.K.} CH_3-\overset{Br}{CH}-CH_3$ (major) (M.K. product) $+ CH_3-CH_2-CH_2-Br$ (minor)

14. (1)

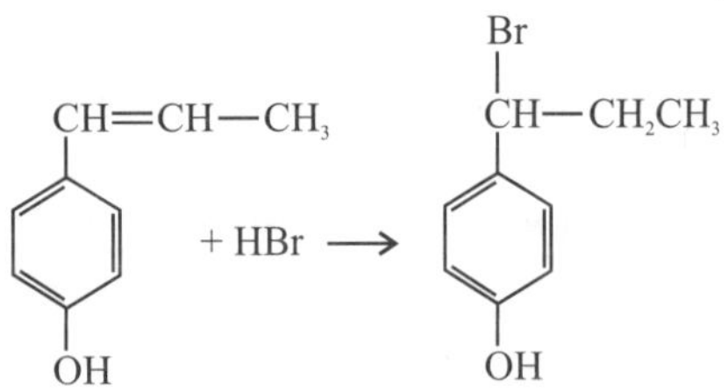

15. (3) Alkene on reaction with $SOCl_2$ doesn't give alkyl halide

16. (2)

$CH_3CH_2CH_2OH \xrightarrow{PCl_5} \underset{A}{CH_3CH_2CH_2Cl} \xrightarrow{alc.KOH} \underset{B}{CH_3CH = CH_2}$

17. (4)

$CH_3-\overset{CH_3}{CH}-\overset{OH}{CH}-CH_3 \xrightarrow{H^+/\Delta} CH_3-\overset{CH_3}{C}{=}CH-CH_3$ (A) major $+ H_3C-\underset{H}{\overset{CH_3}{C}}-\underset{H}{C}=CH_2$ (B) minor

$CH_3-\overset{CH_3}{C}{=}CH-CH_3 \xrightarrow{HBr/dark} CH_3-\underset{Br}{\overset{CH_3}{C}}-CH_2CH_3$ (C) major $+ H_3C-\underset{H}{\overset{CH_3}{C}}-\underset{Br}{CH}-CH_3$ (D) minor

18. (1) All the hydrogen atoms in 2,2- Dimethyl propane are equivalent. Hence, it can form only one chlorinated product. The reaction is given as,

$CH_3-\underset{CH_3}{\overset{CH_3}{C}}-CH_3 + Cl_2 \xrightarrow{hv} CH_3-\underset{CH_3}{\overset{CH_3}{C}}-CH_2Cl$

19. (2) In the given reaction Grignard reagent reacts with ethylene oxide. It gives addition product which on hydrolysis gives a primary alcohol as a final product. Reaction is given as-

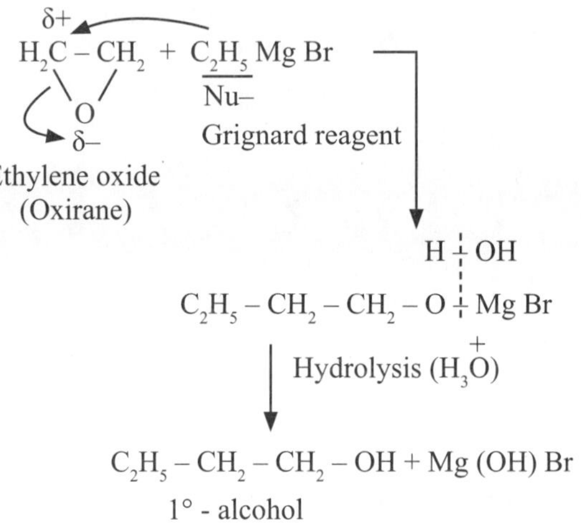

$C_2H_5 - CH_2 - CH_2 - O - MgBr \xrightarrow{Hydrolysis\ (H_3\overset{+}{O})} C_2H_5 - CH_2 - CH_2 - OH + Mg(OH)Br$

1° - alcohol

20. (4)

$CH_3-CH_2-CH_2-Cl \xrightarrow[dehydrohalogenation]{alc.\ KOH} \underset{(B)}{CH_3-CH{=}CH_2} \xrightarrow{HX} CH_3-\underset{X}{CH}-CH_3$

(C)-sec alkyl halide

21. (1)

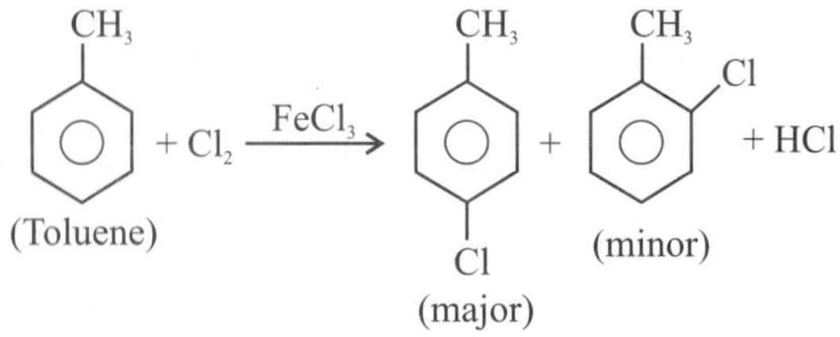

22. (3)

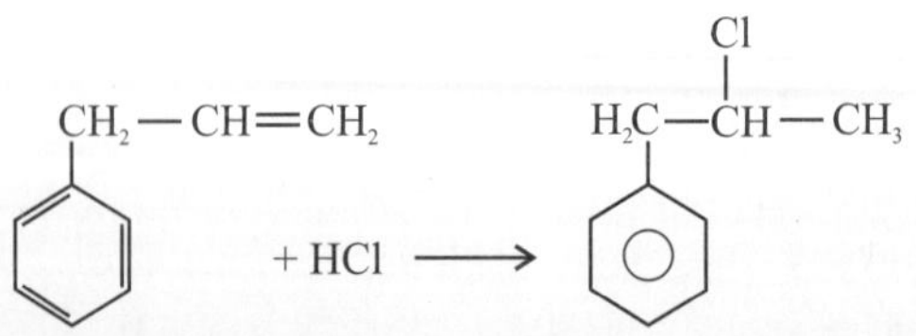

23. (1)

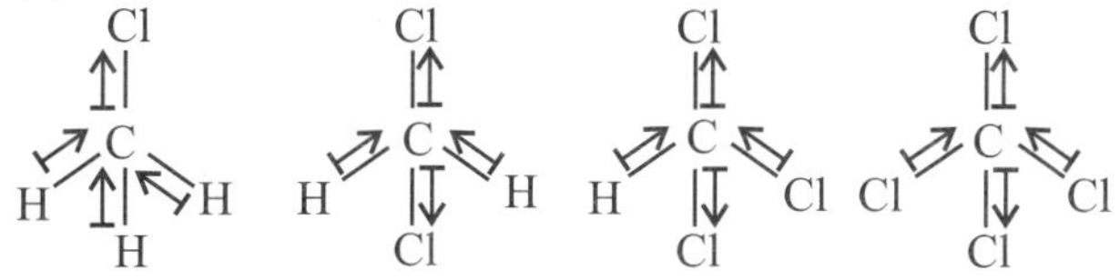

D = 1.94 D = 1.6 D = 1.03 D = 0

24. (2) As we as go down the group, leaving group property increases the size of halogen increases.

25. (2) CH_2OH | $CHOH$ | CH_2OH glycerol $CH_3CH_2CH_2OH$ propanol

The density of glycerol is higher than propanol due to extensive inter molecular hydrogen bonding. Glycerol containg three –OH groups while propanol contains only one –OH group.

26. (1) Boiling point is a property which depends upon the interaction of a particle with other particle. 1° molecule has was greater surface area to interact than the 3° molecule. Therefore, the correct order of b.p. is –1° > 2° > 3°.

27. (1) For the same type of alkyl halides, boiling point increases as the size of the halogen increases. Corresponding alkane will have least b.p.

28. (2) Primary alkyl halide prefers to undergoes S_N2 reaction. Since S_N2 reaction requires the approach of the nucleophile to the carbon bearing the leaving group, the presence of bulky substituents on or near the carbon atom have a inhibiting effect. Of the simple alkyl halides, methyl halides react most rapidly in S_N2 reactions because there are only three hydrogen atoms. Tertiary halides are the least reactive because bulky groups hinder the approaching nucleophiles. Thus, the order of reactivity followed is -1° > 2° > 3° - halide

29. (3) When halogen gets attached to benzene ring, it will deactivate the ring, as halogen is electron withdrawing in nature. Because it attached to benzene ring through σ bond and due to resonance, this effect is mainly observed at ortho and para position.

30. (2) In S_N2 reaction inversion of configuration takes place. This can be represented as-

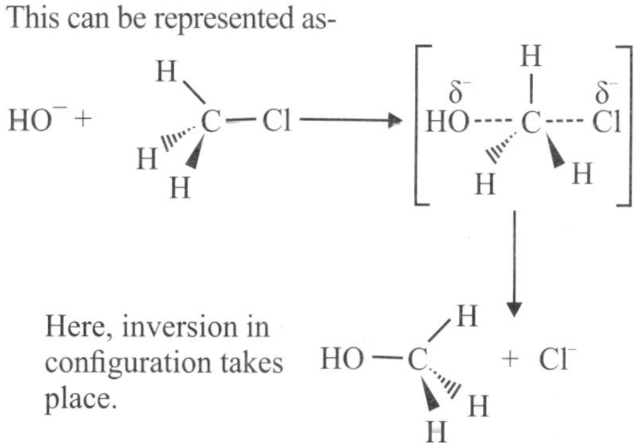

Here, inversion in configuration takes place.

31. (1) If reaction proceed via S_N1 mechanism, rearrangement of the carbocation is expected in the slow step. Since no such rearrangement is actually noticed in the products in both the reaction, this means that they both follows S_N2 mechanism.

32. (3) A = RNC, B = RCN, as both C and N can donate electron pairs. Since KCN is predominantly ionic and C - C bond is more stable than C - N bond, Therefore, in case of KCN, attack takes place through carbon atom. AgCN is predominantly covalent hence only nitrogen pair is available for bonding resulting in the formation of isocyanide.

33. (4) R Br + KOH → ROH + K Br
Alkyl halide (aqueous) Alcohol
Here, substitution reaction takes place.

34. (3) $RX + Mg \xrightarrow[\text{ether}]{\text{dry}} RMgX$

35. (4) 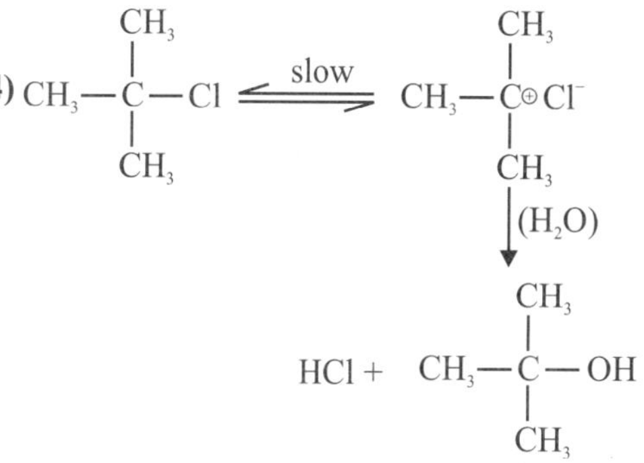

In the above reaction, the reactant is tertiary chloride and reaction occurs in polar protic solvent. Therefore, S_N1 mechanism is followed.

36. (2) Reactivity of alkyl halides in S_N2 mechanism:
1° > 2° > 3°

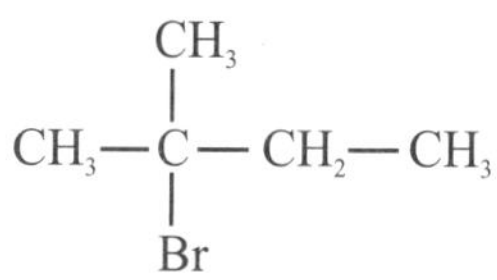

In this compound bromine is attached to 3° carbon.

Therefore, It will be least reactive.

37. (3) 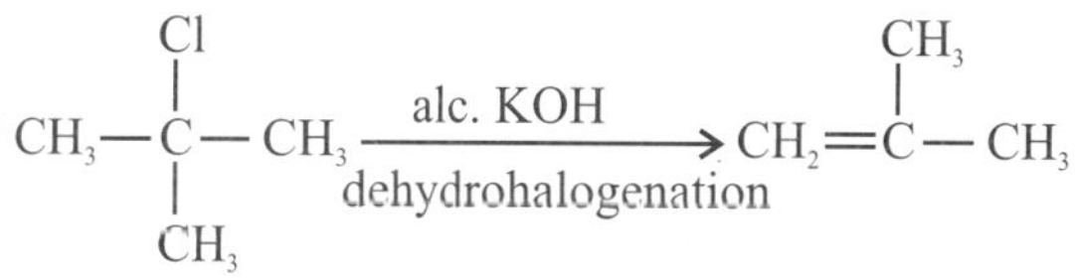

38. (3) For S_N1 reaction, the reactivity order of haloalkane is 3° > 2° > 1°

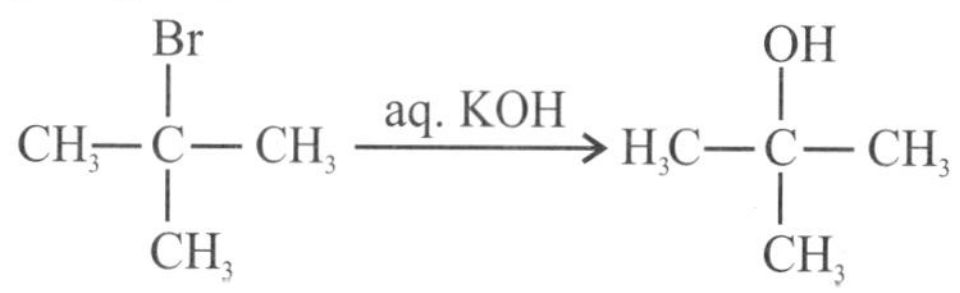

Tertiary carbocation formed as an intermediate in the reaction is most stable.
Hence, 2-Bromo-2-methyl propane reacts most easily with aq. KOH.

39. (2) The order of reactivity of S_N1 mechanism: 3° > 2° > 1°

40. (4) The alkyl halides are highly reactive as the order of reactivity is as follows-

Iodide > Bromide > Chloride as I^- is good leaving group.

Tertiary carbocation formed in the reaction is most stable.

Hence 3° > 2° > 1°(order of reactivity of S_N1 mechanism)

Thus, 2 - bromopropane is the most reactive.

41. (2) Williamson synthesis is an important laboratory method for the preparation of symmetrical and unsymmetrical ethers. The reaction is given as,.

$$C_2H_5Cl + RO^- Na^+ \rightarrow C_2H_5 - O - R + NaCl$$

42. (4)

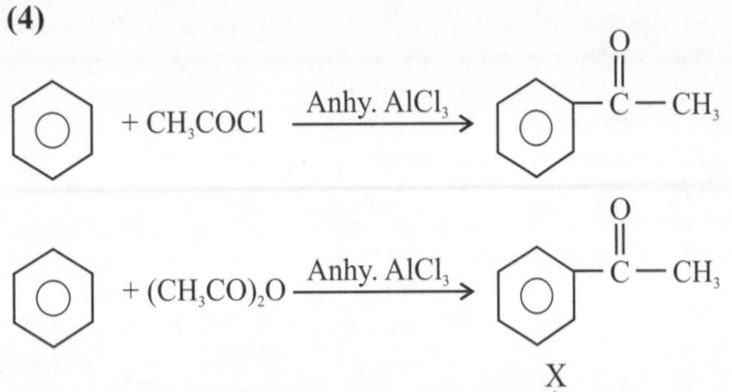

43. (4) $C_2H_5{-}CH{=}CH_2 \xrightarrow{HX} C_2H_5{-}CH(X){-}CH_3$ (B)

$\xrightarrow[AgOH]{(moist\ Ag_2O)} C_2H_5{-}CHOH{-}CH_3 + AgCl$ (C)

sec. alcohol

(in case of dry Ag_2O ether will form) in that case the reaction is given as,

$$C_2H_5{-}CH(X){-}CH_3 \xrightarrow[Ag_2O]{dry} C_2H_5{-}CH(CH_3){-}O{-}CH(CH_3){-}C_2H_5 + 2AgX$$

44. (2) Alc . KOH is used for dehydrohalogenation.

When a haloalkane with β-Hydrogen atom is heated with alcoholic solution of KOH, there is elimination of hydrogen atom from β-carbon & a halogen atom from α - carbon atom. As a result, an alkene is formed as a product. Since β - hydrogen atom is involved in elimination it is often called β - elimination. It is also known as dehydrohalogenation.

If there is possibility of formation of more than one alkene due to availability of more than one β - hydrogen atoms, usually one alkene is formed as a major product. This is observed by Russian chemist Alexander Zaitsev (or Saytzeff), who formulated a rule, which is summerised as,- In dehydrohalogenation reactions, the preferred product is that alkene which has the greater number of alkyl groups attached to doubly bonded carbon atoms. It is shown by the following example-

$H_3C{-}CH(H②){-}CH(Br){-}CH_2(H①)$ (β-hydrogen ②, β-hydrogen ①)

$\xrightarrow{alc.\ KOH(1)} H_3C{-}CH(H){-}CH{=}CH_2$ Pent–1–ene

$\xrightarrow{alc.\ KOH(2)} H_3C{-}CH{=}CH{-}CH_3$ Pent–2– ene

In both the reaction products the major product will be pent-2-ene because according to Satzeff rule, the preferred product should be most substituted

Hence, pent-2-ene will be 81% and

pent-1-ene will be 19%

45. (3) $C_2H_5I \xrightarrow[dehydrohalo-genation]{alc.\ KOH} C_2H_4$ (X) $\xrightarrow{Br_2} BrCH_2{-}CH_2Br$ (Y) $\xrightarrow{KCN} CH_2NC{-}CH_2CN$ (Z)

46. (2) Tertiary alkyl halides do not undergo Wurtz reaction.

47. (4) $CH_3{-}CH(Br){-}CH_2{-}CH_3 \xrightarrow{alc.\ KOH} CH_3{-}CH{=}CH{-}CH_3$ (major) $+ CH_2{=}CHCH_2CH_3$ (minor)

48. (2) Alkyl iodides are often prepared by the reaction of alkyl chlorides/bromides with NaI in dry acetone. This reaction is known as Finkelstein reaction.

$$R - X + NaI \rightarrow R - I + NaX$$

$$X = Cl, Br$$

49. (1)

$CH_3CH_2CH_2OH \xrightarrow{HBr} CH_3CH_2CH_2Br$ (A) $\xrightarrow[dehydrohalo-genation]{alc.\ KOH} CH_3CH{=}CH_2$ (B)

$\xrightarrow[Peroxide]{HBr}$ (anti-markov-nikov's rule) $CH_3CH_2{-}CH_2Br$ (C) $\xrightarrow[(Substitution\ rea^n)]{aq.\ KOH} CH_3{-}CH_2{-}CH_2{-}OH$ (D)

50. (4)

$CH_3 - Br \xrightarrow{KCN} CH_3CN$ (A) $\xrightarrow[(hydrolysis)]{H_3O^+} CH_3COOH$ (B) $\xrightarrow[ether\ reduction]{LiAlH_4} CH_3CH_2OH$ (C)

51. (1) $C_2H_5Cl + KCN \rightarrow C_2H_5CN + KCl$
(propanenitrile)

52. (3) 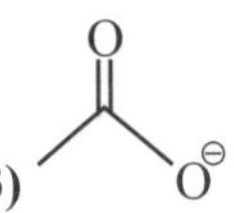 is resonance stabilised.

53. (1) Aryl S_N2 (addition elimination) will occur. More EWG will stabilise the carbanion and hence accelerate the rate of reaction.

$$Ar - Cl + Me - O - Na \rightarrow Ar - OMe + NaCl$$

54. (2) In the given reaction cyclicalkyl halide reacts with Mg in the presence of dry ether as a result corresponding reagent will form, which is highly reactive & reacts with any source of proton to give corresponding hydro carbon. The reaction is given as,

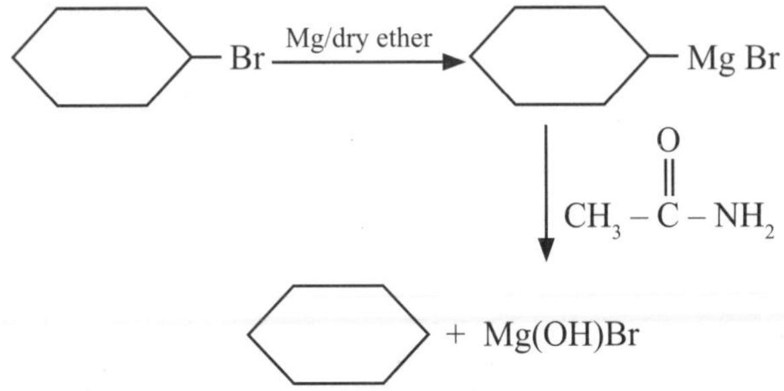

Cyclohexane

55. (3) The given reaction is Finkelstein reaction, due to halogen exchange of I with Cl.

56. (2) The given reaction is an example of elimination reaction in the presence of alcoholic KOH. One molecule of HBr is eliminated from alkyl halide to give alkene. It is also known as dehydrohalogenation reaction

57. (1) Reactivity series for S_N2 is $1^\circ > 2^\circ > 3^\circ$

In (2), it is also 1° alkyl halide but due to steric hindrance it is less reactive than (1).

58. (3) 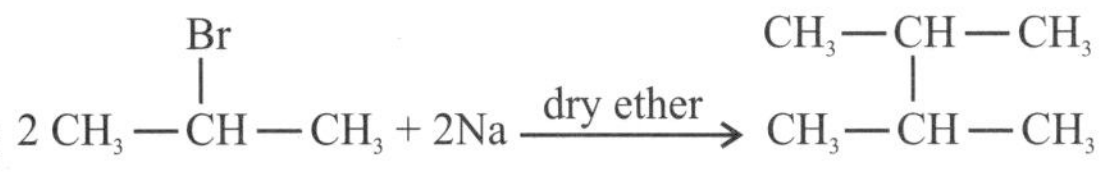

iso-Propyl bromide

59. (3) Refrigerant used in refrigerators and air conditioners are chlorofluorocarbon (CFC's) or freons.

60. (1) Gammexane is chemically known as benzene hexachloride (BHC)

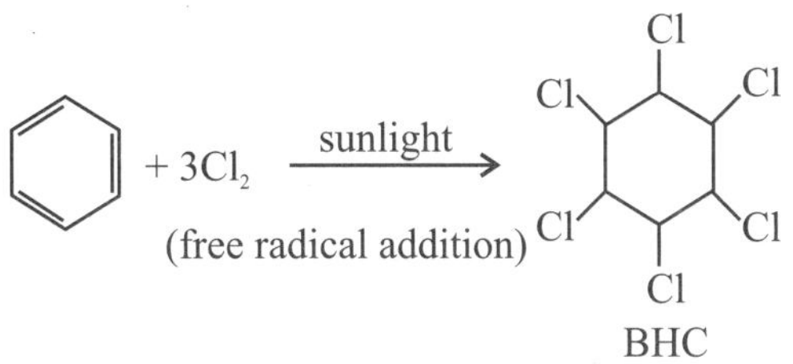

61. (3) $CCl_2F_2 + 2HCl$ is freons; used as refrigerant.

$$CCl_4 + 2HF \xrightarrow{Sb_2F_5} CF_2Cl_2 + 2HCl$$

freon

(dichlorodifluoromethane)

62. (1) Pyrene has highest chlorine content. Its formula is CCl_4.

High Order Time Intensive MCQs

1. (3) OH^- is a poor leaving group. So, it can not be substituted directly.

2. (2) NH_2^- is a strong base, it will initiate E_2 elimination here.

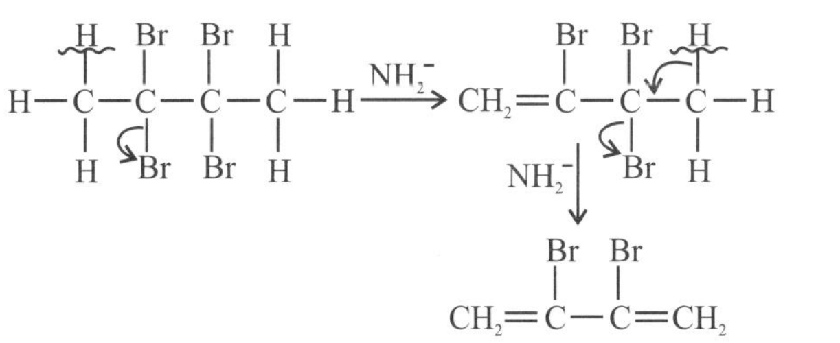

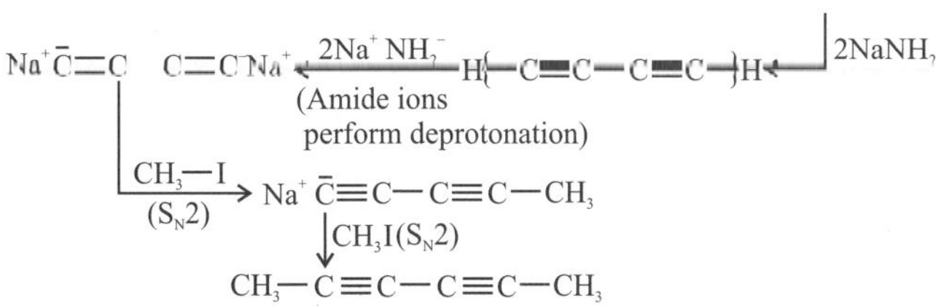

3. (2) Stability of carbocastion is the major concern for S_N1 reaction.

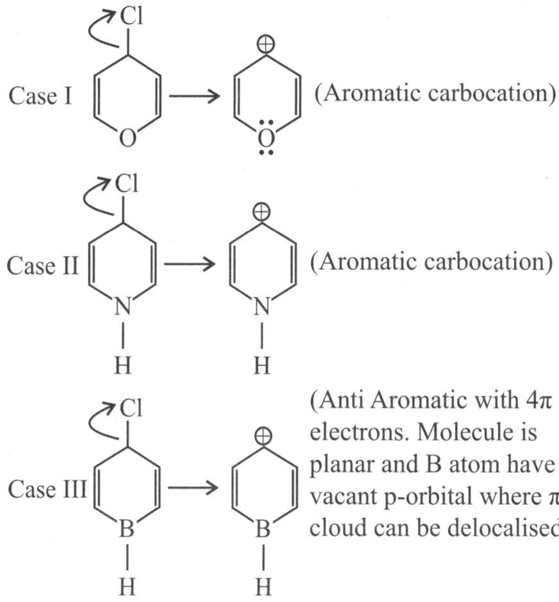

Out of case I & case II, case II, carbocation is more stable because N shows Heigher +M (being less electro –ve) than O. IIIrd is least stable due to antiaromatic nature.

4. (1) The reaction occurs via formation of a carbanion. Presence of $-NO_2$ groups at ortho & para positions will stabilize the carbanion maximum via – M & – I.

5. (4) α-carbonyl compounds are most reactive towards S_N2 because of stability of transition state by delocalisation of π–electron cloud.

6. (1) 2nd option is neopentyl chloride which is least reactive towards S_N2 & prefers to underges S_N1. 3rd option i.e., methyl chloride prefers S_N2 owing to least steric hindrance. Bridge head halides neither undergo S_N2 (3° halide) nor S_N1 (carbocation not stable).

7. (3) $R\ \ Cl \xrightarrow{(i)\,KCN} R\ \ C\equiv N \xrightarrow{LiAlH_4} R\ \ CH_2\ \ NH_2$
$R-Cl \xrightarrow{AgCN} R-N\equiv C \xrightarrow{LiAlH_4} R-NH-CH_3$ →

Functional group isomer

8. (1)

$$CH_2=CH-CH_2-OH \xrightarrow[excess]{HBr} CH_3-\underset{\displaystyle Br}{\underset{|}{CH}}-CH_2-Br$$

9. (1) 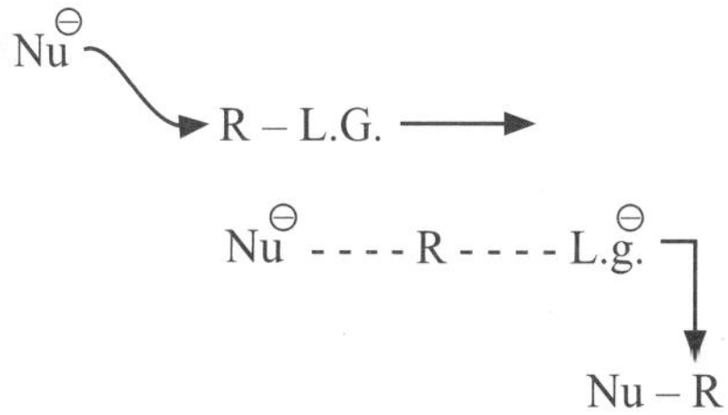

Rate ∝ [Substrate] [Nucleophile]

∵ Substrate is constant here,

∴ Rate ∝ [Nu] Hence, rate will be increased by a factor of 10.

10. (2)

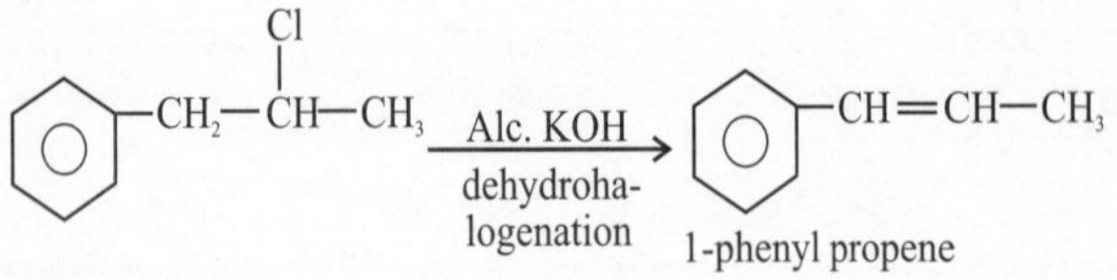

11. (2)

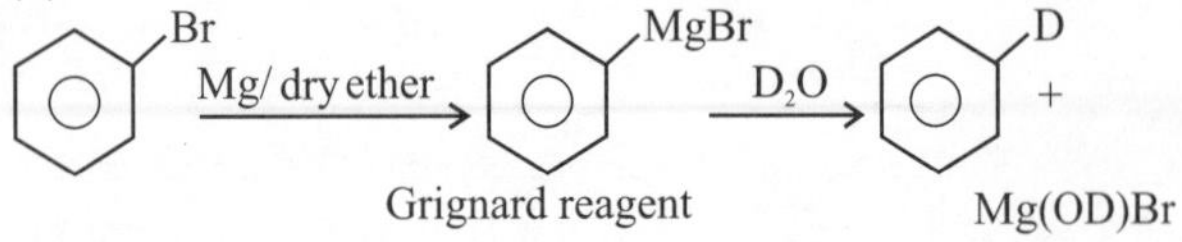

12. (1) S_N1 reactions generally carried out in polar protic solvents (like water, alcohol, acetic acid etc.) S_N1 reactions follows the following order - 3° > 2° > 1°.
Hence, option (1) will follow S_N1 mechanism.

13. (3)

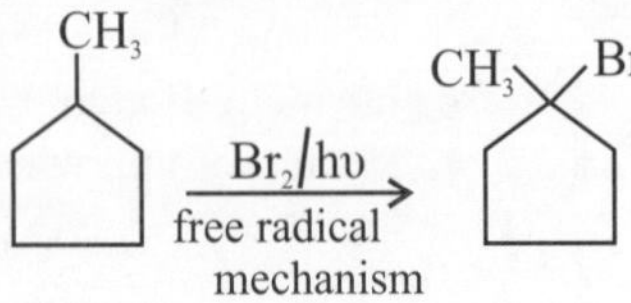

removal of 3°H is easier.

14. (1)

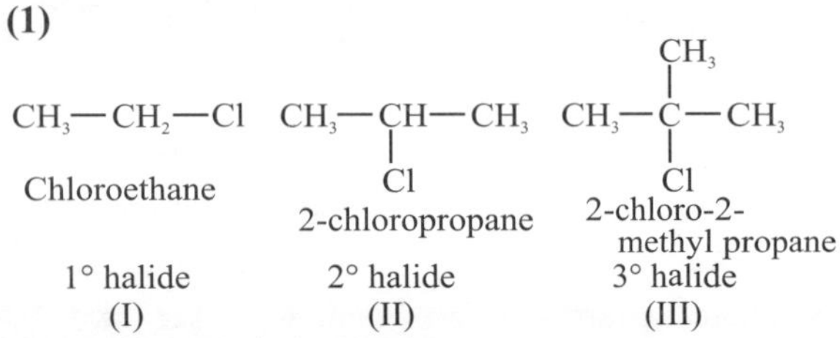

Rate of dehydrohalogenation

3° (RX) > 2° (RX) > 1° (RX), So (III) > (II) > (I) is the correct option

15. (1) $(A) \xrightarrow{\text{Alc. KOH}} C_4H_8 \xrightarrow{\text{ozonolysis}}$

$CH_3CH_2CHO + HCHO$

Propionaldehyde formaldehyde

$C_4H_8 \rightarrow$ Alkene $\rightarrow CH_3 - CH_2 - CH = CH_2$

Reactant $\rightarrow (A) \rightarrow CH_3 - CH_2 - CH_2 - CH_2Br$

16. (1) Vinyl carbon is sp^2 hybridized aliphatic carbon

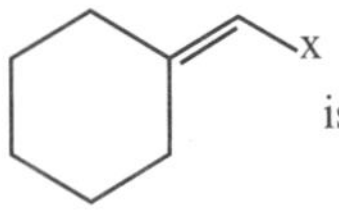

is vinly halide

While 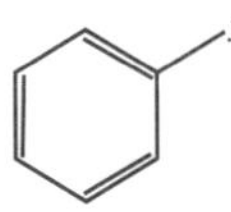

is aryl halide

and 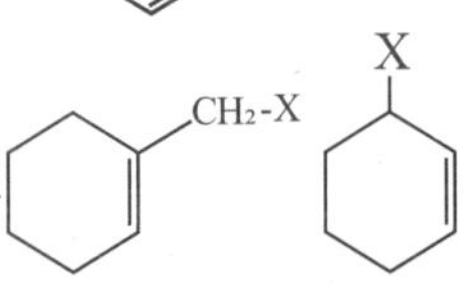

are allyl halide

17. (2)

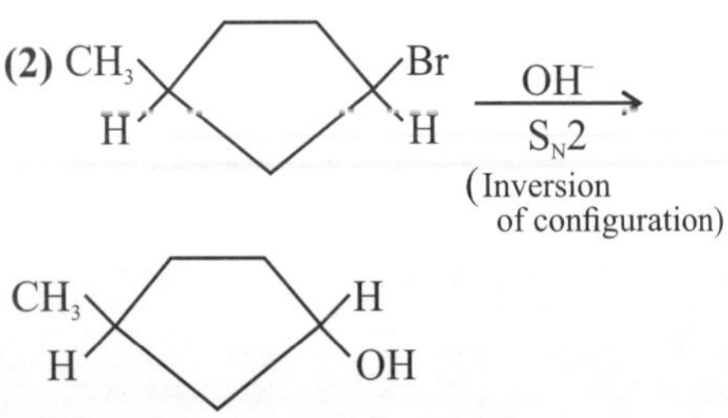

18. (1) Vinyl chloride cannot be solvolysed easily because $CH_2{=}\overset{\oplus}{C}H$ vinyl carbocation is unstable.

19. (1) –I and –M group increases rate of ArS_N2 reaction.

Nucleophilic aromatic substitution reaction is also known as S_NAr reaction. To perform S_NAr reaction benzene ring should have Positive charge on it self. As benzene ring has π e^- cloud therefore it self acts as a nucleophile & in normal circumstances it cannot undergo S_NAr reaction. The Essential condition for S_NAr is "a strong electron withdrawing group must be present at directly attached position of the ring.

in this case NO_2 group shows –M & –I effect and generates +ve charge on the benzene ring. In this way Benzene ring behaves as an electrophile and a Nu^- can easily attack on it to perform S_N Ar reaction.

Mechanism.

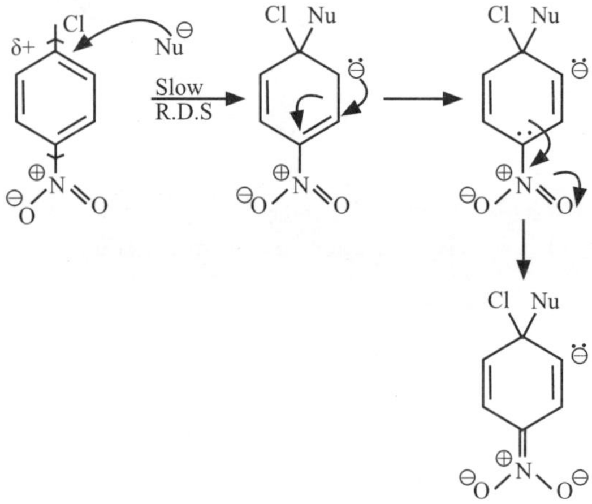

Note- $-NO_2$ group should present at o - p positions not at meta position to perform S_NAr reaction.

The order of –M effect is

$-NO_2 > Cl > -OCH_3 > CH_3$

∴ 

will undergo at fastest rate.

20. (2) The given reaction undergoes ArS_N2 reaction. The $-NO_2$ group present at para - position facilitates this reaction. The Nucleophilic attack of $\overset{\ominus}{S}CH_3$ is shown below -

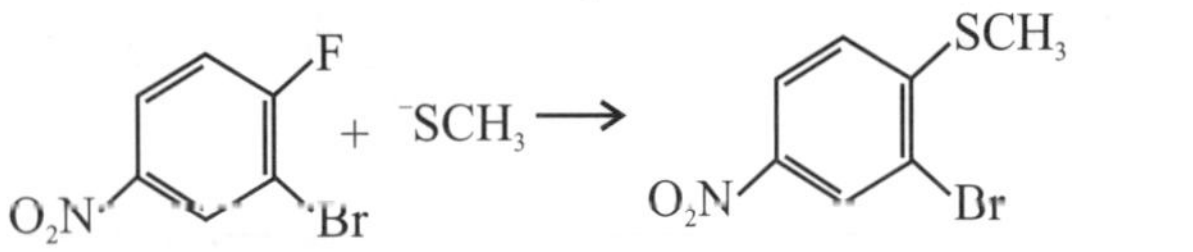

21. (4) The given reaction is an example of ArS_N2 reaction. Here, $-NO_2$ group present at para position shows both –M and –I effects facilitates the reaction towards ArS_N2. Here, the halogen group show the reaction rate in the following manner-

–F > –Cl > –Br > –I.

This can be understood by the following mechanism.

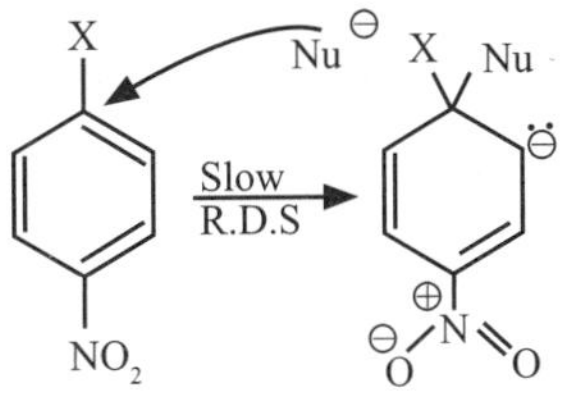

To create a nucleophilic centre at the benzene ring, x must be more electronegative. Therefore halogen will follow the following order, –F > –Cl > –Br > – I

Here rate of ArS_N2 reaction

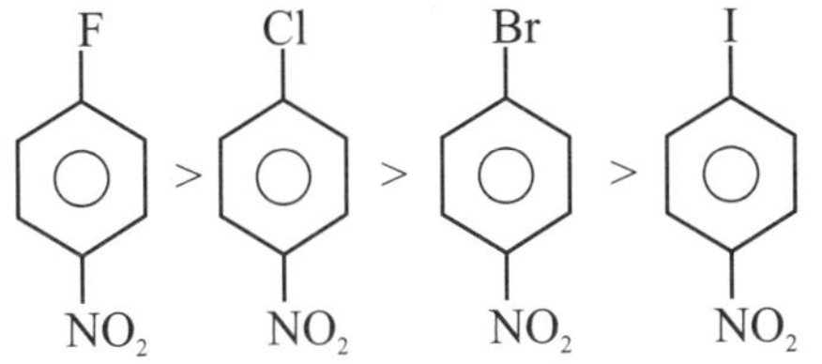

22. (4) Chlorination of cyclohexene is shown as:

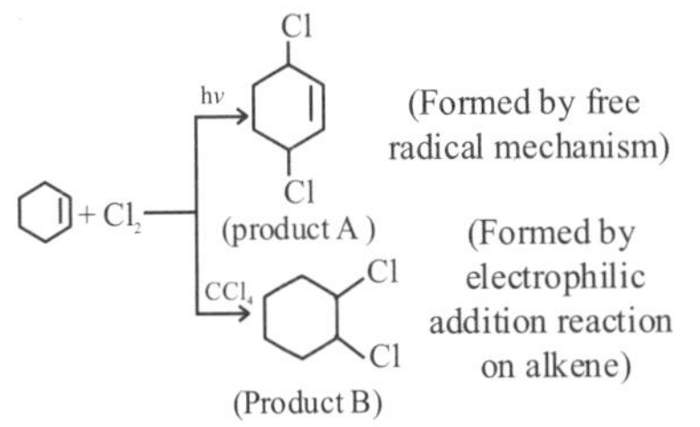

23. (3) From the given rate law it is clear that reaction proceeds via both the routes S_N1 as well as S_N2. Because in the given rate law concentration of both the reactants is involved.

24. (4) AgCN is a covalent compound and its reaction with RX gives isocyanide

25. (4) On going top to bottom, n group nucleophilicity increases.

26. (3) Cyanide ion (CN^-) is an ambident nucleophile because it has two lone pair donor atom (C and N).

27. (1) Sulphur belongs to IIIrd period so it has maximum nucleophilicity.

28. (4)

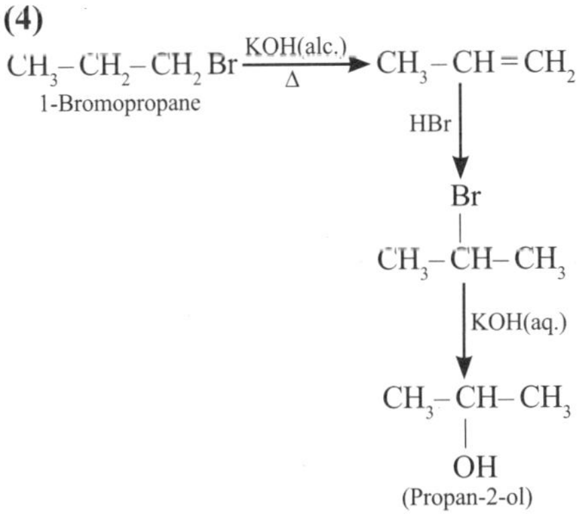

29. (2) Inversion takes place in S_N2 reaction while configuration of nucleophile remains same.

30. (3)

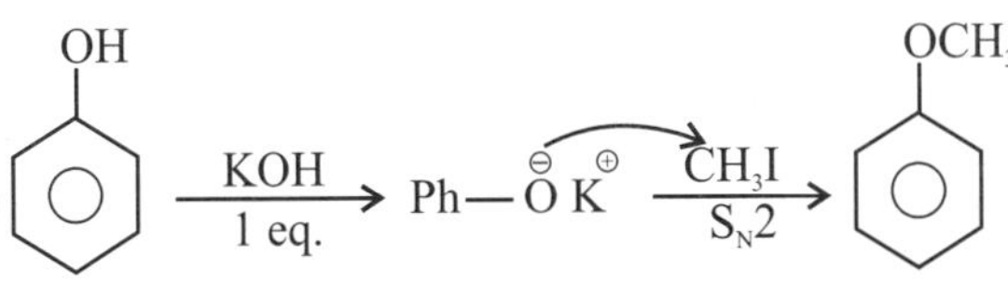

31. (2) $CH_3-CH_2-C(Br)(CH_3)-C_6H_5$ produce most stable carbocation.

32. (4) t-BuO$^\ominus$K is a bulky base, elimination takes place.

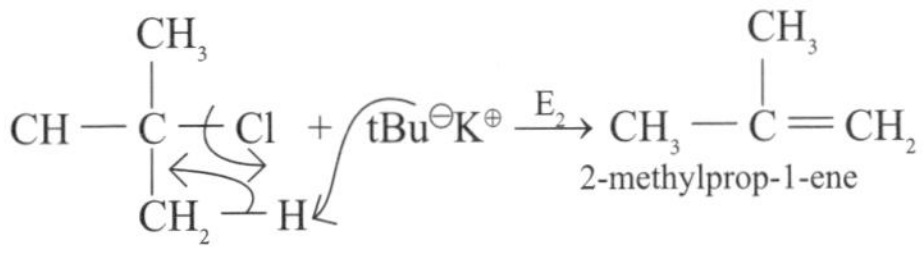

Base is tert-butoxide thus favours elimination with 3o alkyl halide and gives alkene.

33. (1) Rate of S_N2 reaction : 1° > 2° > 3°, as branching increases, steric crowding increases, rate decreases.

34. (4) Side chain reaction occurs forming benzotrichloride.

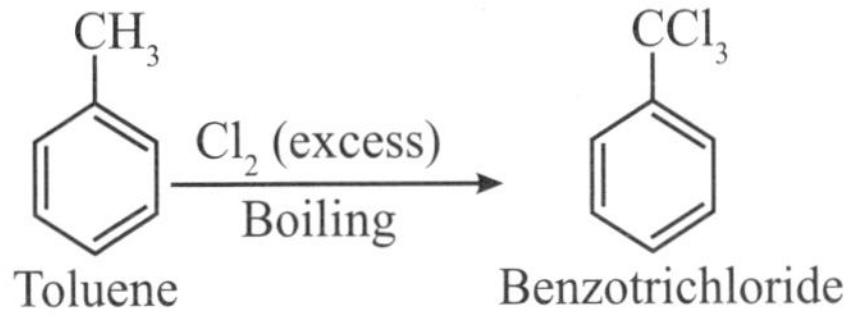

35. (4)

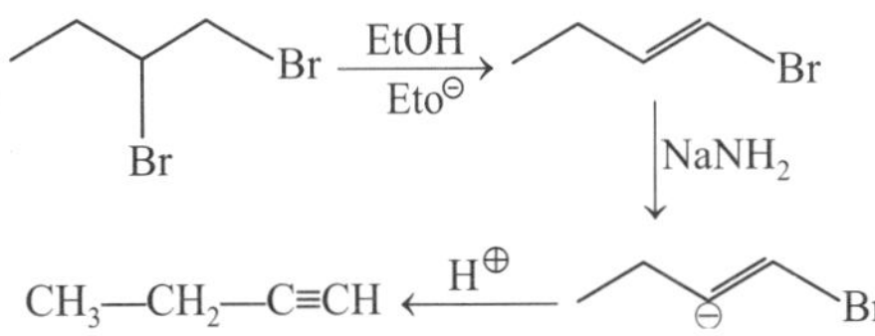

36. (2) Thionyl chloride method is preferred over phosphorous pentachloride method for the preparation of alkyl chloride because, the by products being gases escape into the atmosphere leaving behind almost pure alkyl chloride.

37. (1) Ter-butyl carbocation is most stable among these therefore, it will give fastest S_N1.

38. (1) On going left to right in period nucleophilicity decreases.

39. (3) $CH_3-C(=O)-CH_2-Cl$ is most reactive due to –I effect of C=O group.

40. (3) $CH_3-CH_3 \xrightarrow{Cl_2/h\nu} CH_3-CH_2-Cl \xrightarrow{KCN} CH_3-CH_2-CN$

41. (4) Rate of solvolysis ∝ stability of carbocation

Carbocation formed in option R is least stable whereas, formed in option P is most stable.

42. (4) H_2SO_4, Al_2O_3 and H_3PO_4 all are dehydrating agent.

43. (3)

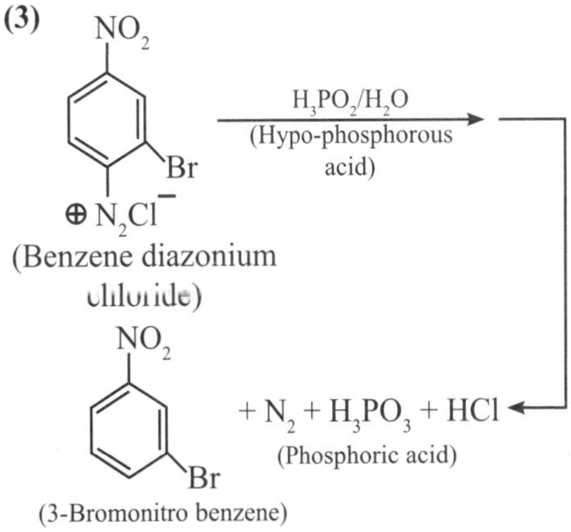

Hypophosphorous acid (H_2PO_3) is a mild reducing agent. Therefore, it reduces diazonium salts a arenes and themselves get oxidized to phosphoric acid.

44. (3) CF_3COO^- (Triflate ion) is the best leaving group amongst given options.

45. (4) Addition elimination occurs in reaction 1. Reaction 2 & 3 cannot occur because of partial double bond character in C – X (halogen) bond.

46. (4) All the mechanisms are correctly placed.

47. (3)

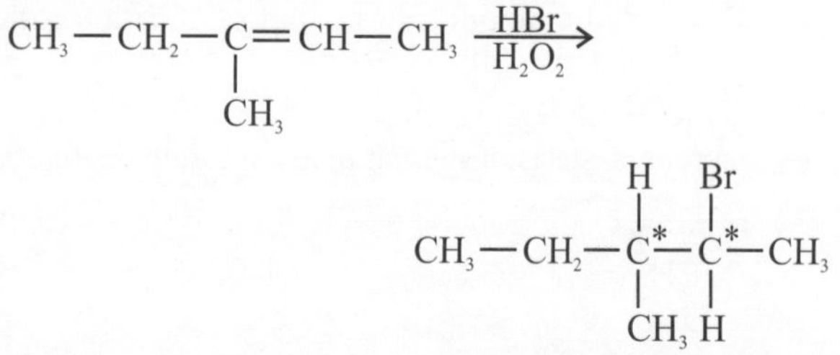

Total stereocentres = 2
Total stereoisomers = 4

48. (4) In all these reactions, the reaction conditions are most suitable for S_N2.

49. (2) Benzyl halide are more reactive than vinyl & aryl halide are less reactive than alkyl halide. Hence, option (2) is correct

50. (2) The given reaction is shown below-

$$H_3C-CH_2-CH_2Br \xrightarrow[\substack{20°C \\ \text{dehydrohalogenation} \\ (-HBr)}]{\text{dil.aq. NaOH}} H_3C-CH=CH_2$$

$$\xrightarrow[\substack{\text{Acetic acid} \\ \text{at } 20°C \\ \text{(Markovnikov Rule)}}]{HBr} H_3C-\underset{\underset{Br}{|}}{CH}=CH_3$$

51. (4) It is a Wurtz type reaction, therefore, only choice is

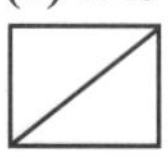

52. (3) In polar aprotic solvent, ionic compound NaCl precipitates out as product so that substitution of Cl^- takes place in preference to I^-.

53. (2) This is the only tertiary substrate.

54. (2) $-CCl_3$ is an electron withdrawing group, hence is meta directing.

55. (4) Since the reaction follows S_N1 mechanism. The alkyl halide giving most stable carbocation will be most reactive.

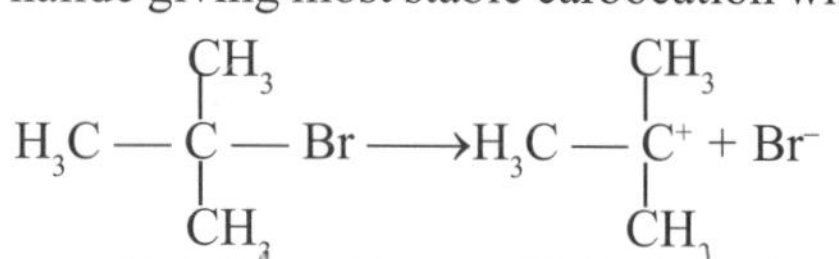

t-Butylbromide 3° Carbocation

Ethyl bromide and n - propyl bromide give primary carbocation, isopropyl bromide gives secondary carbocation, t-butyl bromide gives 3° carbocation. Order of stability of carbocation is 3° > 2° > 1°. Hence t-butyl bromide is most stable.

Correct & Incorrect MCQs

1. (2) Haloalkanes contain halogen atom(s) attached to the sp^3 hybridised carbon atom of an alkyl group and haloarenes contain halogen atom(s) attached to sp^2 hybridised carbon atom(s) of an aryl group.

2. (3) In vinylic halides, the halogen atom is bonded to a sp^2-hybridised carbon atom of a carbon-carbon double bond.

3. (2) The IUPAC name of isopropyl chloride is 2-Chloropropane.

4 (3) In gem-dihalides, both the halogen atoms are present on the same carbon atom and in vic-dihalides, both the halogen atoms are present on adjacent carbon atoms. Hence, 1,1-Dichloroethane is an example of gem-dihalide.

5 (3) The size of halogen atom increases as we go down the group in the periodic table. Hence, the carbon-halogen bond length also increases from C— F to C— I.

6 (3) In Finkelstein reaction, alkyl iodides are often prepared by the reaction of alkyl chlorides/ bromides with NaI in dry acetone.

7 (2) Methyl chloride, methyl bromide, ethyl chloride and some chlorofluoromethanes are gases at room temperature while higher members are liquids or solids.

8 (3) Treatment of chloroethane with a saturated solution of AgCN gives ethyl isocyanide as the major product. AgCN is mainly covalent in nature and acts as a nucleophile through lone pair of electrons present on N to give ethyl isocyanide.

9 (2) Tertiary alkyl halides would undergo S_N1 reaction very fast.

10 (4) When carbon tetrachloride (CCl_4) is released into the air, it rises to the atmosphere and depletes the ozone layer.

Statement Based MCQs

1. (3) Chiral carbon are asymmetric carbon, have all 4 different groups attached to it, hence it will produce non-superimposable mirror image. A molecule without having chiral center can be a chiral molecule if it does not have any plane of symmetry. Chiral molecules are dissymmetric in nature.

2. (4) Aryl halides are very less reactive towards nucleophilic substitution reaction under normal condition. Due to the presence of lone pairs on halogen atom, they have partial double bond character. Due to partial double bond character in carbon-halogen bond, it becomes shorter and stronger so cannot be easily replaced by nucleophile so nucleophilic reaction is not possible.

3. (1) Stereo chemical inversion of a molecule is known as Walden inversion i.e., chiral center inversion in a molecule in chemical reaction.

4. (3) Silver nitrite is a covalent compound. Attack occurs through 'N' of NO_2^- & hence nitro compounds are formed.

5. **(1)** Due to steric hindrance in the molecules and presence of bulkier group, 3° alkyls gives S_N1 most easily & hardly gives S_N2.

6. **(4)** Since, C – I bond is the weakest and I– is a better leaving group than Br^-, Cl^- and F^- ions, hence C_2H_5I is the most reactive alkyl halide.

7. **(4)** With the increase in size and mass of halogen atom, the magnitude of van der Waal forces increases. Hence, for the same alkyl group, the boiling points of alkyl halides decreases in the order: RI > RBr > RCl > RF.

8. **(1)** Due to greater s-character, the sp^2 hybrid of C – Cl bond in chlorobenzene has less tendency to release electrons to Cl than a sp^3 hybrid carbon of cyclohexyl chloride. Hence, the dipole moment of chlorobenzene is lower than that of cyclohexyl chloride.

9. **(4)** In S_N2 reactions, the attack of nucleophile occurs from a side opposite to the halogen atom resulting in inversion of the configuration.

10. **(1)** Both statements are correct. As the C–Cl bond acquires partial double bond character due to resonance.

Assertion & Reason MCQs

1. **(2)** For the same hydrocarbon part boiling point depends upon the atomic mass of halogen atom. Higher the mass of the halogen atom, higher will be the bioling point. Therefore, b.p. decreases with decrease in atomic mass of halogen atoms.

   Halides are polar molecules, hence, they have greater boiling points than hydrocarbon of comparable molecular mass.

2. **(4)** KCN reacts with methyl chloride to give methyl cyanide and CN^- is an ambidentate nucleophile. Therefore, Assertion is false but reason is true.

3. **(1)** Both Assertion and Reason are true and the reason is a correct explanation of the Assertion

   The reaction is given as

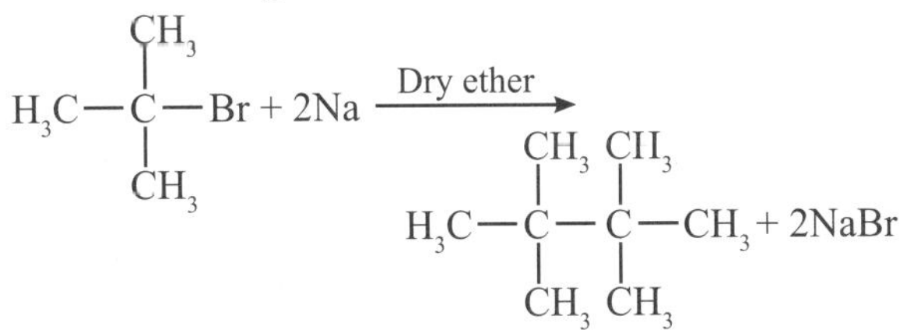

4. **(1)** Both assertion and reason are true and the reason is a correct explanation of the assertion.

5. **(2)** In monohaloarenes, halogen atom increases the electron density at ortho and para position. So, further electrophilic substitution occurs at ortho and para positions.

   Halogen atoms are highly electronegative therefore, highly deactivating.

6. **(3)** Assertion is true, the reaction is given as.

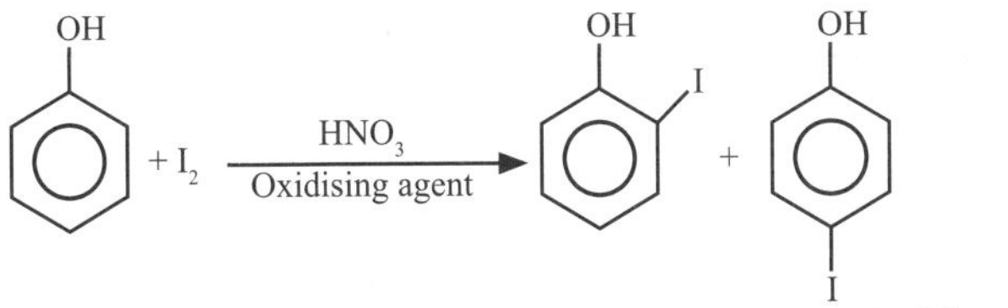

   Oxidising agent is HNO_3 not I_2, therefore reason is false.

7. **(1)** Both assertion and reason are true and reason is the correct explanation of assertion because it is difficult to replace chlorine by –OH in chlorobenzene due to partial double bond character.

8. **(3)** Assertion is true but reason is false.

   The reaction of hydrolysis of 2-bromooctane is,

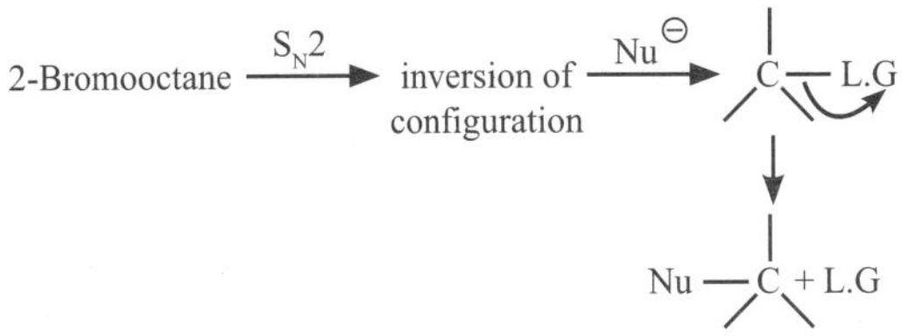

   The above reaction follows S_N2 path therefore, formation of carbocation is not possible here. But that will be possible in S_N1 reaction. Therefore, reason is false.

9. **(4)** Assertion is false but the reason is true.

   As not the nitration of chlorobenzene leads to the formation of m-nitrochlorobenze, but chlorination of nitrobenzene gives the same product, and $-NO_2$ is a m-directing group so desired product is formed.

10. **(3)** Alkyl halides form alkenes when they are heated above 300°C in presence of alc. KOH. CH_3CH_2I reacts more rapidly with strong base in comparsion to CD_3CH_2I.

Multi Correct MCQs

1. **(4)** Since $CH_3–CH = \overset{+}{C}H$ is highly unstable therefore, $CH_3–CH = CH – Cl$ cannot give S_N1 reaction.

2. **(2)** Chiral carbons in a compound are marked by:

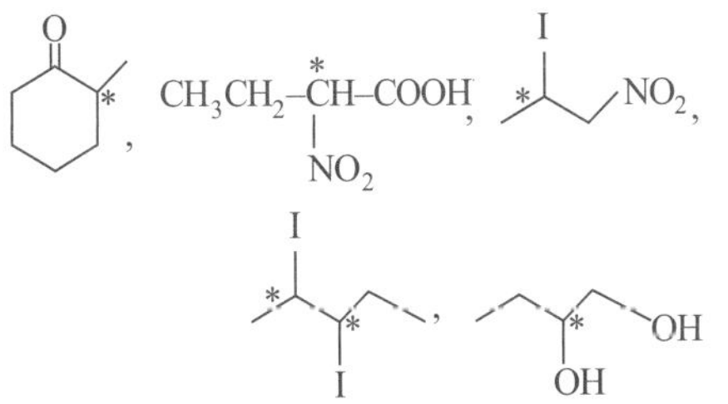

3. **(2)** Bromobenzene and Chlorobenzene are prepared by Sandmeyer's reaction.

4. **(2)** S_N1 reaction mechanism is favoured by polar protic solvents, weak nucleophiles, bulky substituents and the reaction is accompanied by racemization.

NCERT Exemplar MCQs

1. **(2)** Reaction between alcohols and halogen acid follows S_N1 mechanism. In S_N1 Mechanism carbocations are formed as intermediates.

   Let us consider the formation of carbocations with the given three alcohols.

$$CH_3-CH_2-CH_2-OH \rightarrow CH_3-CH_2-\overset{+}{C}H_2 + {}^{-}OH$$

   In this case, 1° carbocation is formed. It is least stable. So, here S_N2 mechanism is followed.

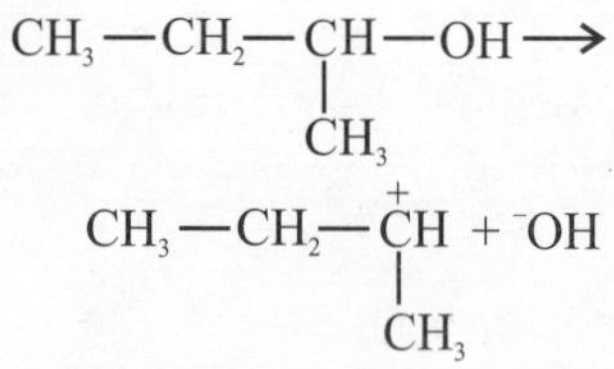

2° carbocation (more stable than 1° carbocation)

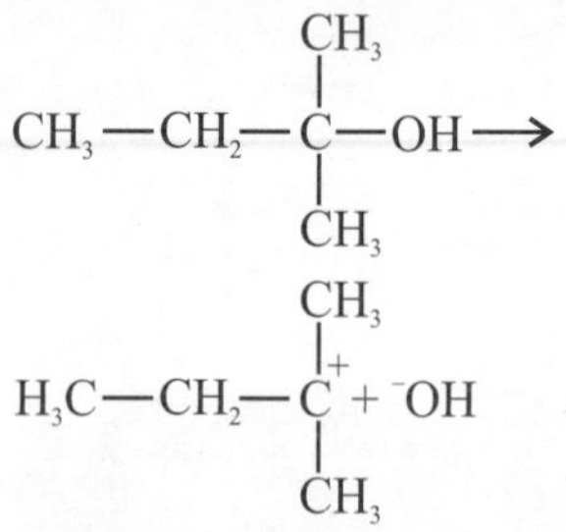

3° carbocation (most stable)

Note : Higher the stability of intermediate, higher will be the reactivity of compound and higher will be the yield of the desired product.

Hence, the reactivity order of the given alcohol is - c > b > a

2. **(4)** When alcohol gets treated with conc. HCl at room temperature then alkyl chloride is formed. This reaction follows S_N1 mechanism. S_N1 mechanism completes in two steps. In first step, a carbocation is formed and this carbocation is attacked by nucleophile in second step.

The attack of nucleophile to the carbocation is possible only if the carbocation is stable. Compound present in option (4) will give tertiary carbocation in step I. Tertiary carbocation is most stable that's why it is further attacked by Cl^- nucleophile as follows

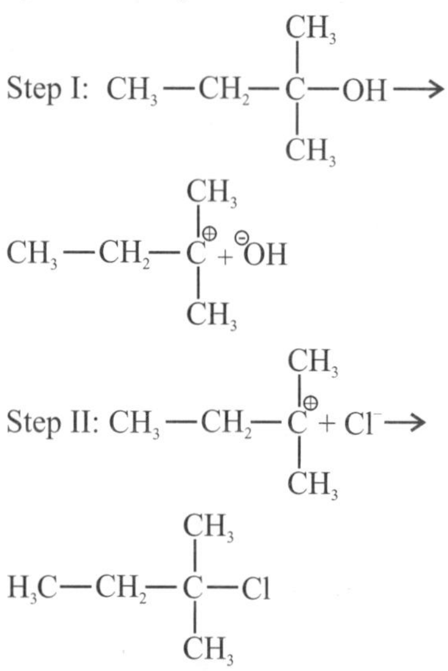

3. **(1)** A diazonium salt is formed when a primary aromatic amine, dissolved or suspended in cold aqueous mineral acid and treated with sodium nitrite. When this freshly prepared diazonium salt is mixed with cuprous chloride, then diazonium group is replaced by –Cl. Therefore, in this reaction chlorobenzene is formed.

4. **(2)** In the presence of iron (III) chloride toluene reacts with a halogen giving ortho and para halo compounds. The reaction will be electrophilic substitution reaction.

It has the following mechanism:

In this mechanism, electrophile Cl^+ attacks to electron rich benzene ring and replaces hydrogen. So, the reaction is electrophilic substitution reaction.

5. **(1)** Those reactions in which one halide replaces another (as in option (1), halogen (–X) is replaced by iodine) are halogen exchange reactions. This reaction is named as Finkelstein reaction.

In option (2), there is the addition of hydrogen halide on alkene.

In option (3), halogen replaces alcoholic group.

While in option (4) halogen replaces the hydrogen of benzene ring.

6. **(1)** The given reaction is a substitution reaction. It include the replacement of 1° and 2° hydrogen of alkanes by chlorine. It occurs in the presence of ultraviolet light or at high temperature the reaction.

The chlorination does not take place at room temperature or in absence of light. In reaction, light is absorbed by the chlorine molecule and activated chlorine initiates.

7. **(1)** Density is directly related to molecular mass. Higher the molecular mass of the compound, higher will be the density. Among the four given compounds, the order of molecular mass is

benzene < chlorobenzene < dichlorobenzene < bromochlorobenzene

Therefore, the increasing order of their densities are same as above.

Hence, option (1) is correct.

8. **(3)** Boiling point of a compound depends on the surface area. Higher the surface area, then higher will be the boiling point of a compound, with increase in branching surface area decreases. If the compound has maximum branching, its boiling point will be minimum.

9. **(2)** Asymmetric/chiral carbon atom is that in which all of its four valencies are satistied with four different groups or atoms. In compound (iv), carbon has satisfied two of its valencies with two hydrogen atoms i.e., similar atom.

So, it is not an asymmetric carbon atom while rest of the three molecules have asymmetric carbon as each carbon has satisfied by four valencies with four different groups or atoms.

So, the correct option is (2).

10. **(1)** Enantiomers are the stereoisomers related to each other as non-superimposable mirror images. Enantiomers possess identical physical properties. They only differ with respect to the rotation of plane polarised light. If one of the enantiomer dextrorotatory, then the other will be levorotatory.

Hence, the enantiomer of molecule (A) is (1).

11. **(2)** Those halides in which two halogen atoms are present on the two adjacent carbon atoms are vic-dihalides.

12. **(1)** Allyl halides are those compounds in which the halogen atom is bonded to sp^3 hybridised carbon atom next to carbon-carbon-double bond e.g:-

$CH_2 = CH - CH_2X$ and $CH_3CH = CHC(Br)(CH_3)_2$.

The compounds in which the halogen atom is bonded to the sp^2 hybridised carbon atom of an aromatic ring are aryl halides.

The compounds in which the halogen atom is bonded to the sp^2 hybridised carbon atom of a carbon-carbon double bond are vinyl halides.

e.g., $CH_2 = CH - X$

The compounds in which the halogen atom is bonded to the sp^3 hybridised carbon atom which is further bonded with two alkyl groups and one hydrogen atom are secondary alkyl halides.

13. **(2)** In this reaction, $AlCl_3$ is a catalyst which activate the chlorine molecule to show heterolytic cleavage. $AlCl_3$ is electron deficient molecule and form $AlCl_4^-$ and Cl^+ when reacts with Cl_2. This Cl^+ electrophile attacks on electron rich benzene ring.

$$AlCl_3 + Cl_2 \rightarrow [AlCl_4]^- + Cl^+$$

14. **(2)** In vic-dihalides, halogen atoms are present on the adjacent carbon atoms.

In gem-dihalides, halogen atoms are present on the same carbon atom. They are known as alkylidene halides.

In allylic halides, halogen atom is bonded to sp^3 hybridised carbon atom next to carbon-carbon double bond

In vinylic halides, halogen atom is bonded to sp^2 hybridised carbon atom of a carbon-carbon double bond.

15. **(3)** Accoring to Markovnikov's rule In this reaction, addition of HCl takes place on double bonded carbons i.e., addition of negative addendum will take place on that carbon which has lesser number of hydrogen.

16. **(2)** A primary alkyl halide would prefer to undergo S_N2 reaction.

(1) S_N1 reactions occur only if the intermediate carbocation is stable i.e., 3° carbocation.

(2) S_N2 reactions occur if there is less steric hinderance on the α carbon of alkyl halide. In case of primary alkyl halides, carbocation is highly unstable and steric hinderance is very less. So, primary alkyl halide would prefer to undergo S_N2 reaction.

(3) In α-elimination, proton and the leaving group are present on same atom.

(4) The process of conversion of enantiomer into a racemic mixture is racemisation.

17. **(4)** All the given compounds are tertiary alkyl halides but the bond formed between carbon and iodine (C–I) is the weakest bond due to large difference in the size of carbon and iodine. So, $(CH_3)_3$ C–I gives S_N1 reaction most readily. In other words, iodine is a better leaving group.

18. **(3)** The correct IUPAC name of the given compound will be 1-Bromo-2-methylbutane.

19. **(2)** Structure of the diethylbromomethane is given below

$$H_3\underset{5}{C}-H_2\underset{4}{C}-H\overset{Br|}{\underset{3}{C}}-\underset{2}{C}H_2-\underset{1}{C}H_3$$

So, the IUPAC name is 3-Bromopentane.

20. **(4)** The reaction of toluene with chlorine is carried out in the presence of iron and in the absence of light, so the substitution occurs on the benzene ring. The $-CH_3$ group of toluene is o and p-directing so, the product is the mixture of (2) and (3) o-chlorotoluene and p-chlorotoluene.

21. **(3)** The bond formed between C of benzene ring and halogen is more stable because of resonance. It has partial double bond character. So, rate of reaction towards nucleophilic substitution gets slow. This substitution is facilitated by the presence of electron withdrawing group at ortho and para position because electron density is high at this positions.

Compound (ii) and (iii) both have one electron withdrawing group but in compound (ii) electron withdrawing ($-NO_2$) group is present at ortho position, so rate of reaction in compound (ii) is more than that of (iii) while (i) has no electron withdrawing group.

Hence, the correct option is (3).

22. **(4)** The rate of nucleophilic substitution reaction decreases in presence of electron releasing group at ortho or para position. In compound (iii), electron releasing group is present at meta position w.r.t. chlorine, so the impact is less but in compound (ii) it will be present at ortho position.

Thus, the rate of reaction towards nucleophilic substitution is least in compound (ii) and highest in compound (i) as there is no electron releasing group in this compound.

23. **(4)** The nucleophilic substitution reaction is facilitated by the presence of electron withdrawing group at ortho and para position. Hence, it enhances the rate of reaction.

Compound (iii) has three electron withdrawing groups at ortho and para positions w.r.t. chlorine while compound (ii) has only one electron withdrawing group and there is no electron withdrawing group in compound (i). So, the increasing order of rate of reaction towards nucleophilic substitution is: (i) < (ii) < (iii).

24. **(3)** The rate of nucleophilic substitution reaction decreases in the presence of electron releasing group at ortho and para position w.r.t. to chlorine. Compound (iii) has two electron releasing groups and compound (ii) has one electron releasing group w.r.t chlorine while compound (i) has no electron releasing group.

So, the rate of nucleophilic substitution reaction is highest in compound (i) and order is (iii) < (ii) < (i).

25. **(4)** There is no chiral carbon in 1-Chloropentane. Therefore it is not chiral. The structure of compound 1-Chloropentane is,

$$CH_3-CH_2-CH_2-CH_2-\underset{\underset{Cl}{|}}{CH_2}$$

26. **(4)** With the increase in boiling point, the size of hydrocarbon part for the same haloalkanes also increases. All the given haloalkanes contain same halogen atom i.e., bromine but the number of carbon atoms in hydrocarbon part of the molecule is increasing. So, the boiling point is minimum for 1-Bromoethane and maximum for 1-bromobenzene.

NEET & AIIMS Past Year MCQs

1. (2)

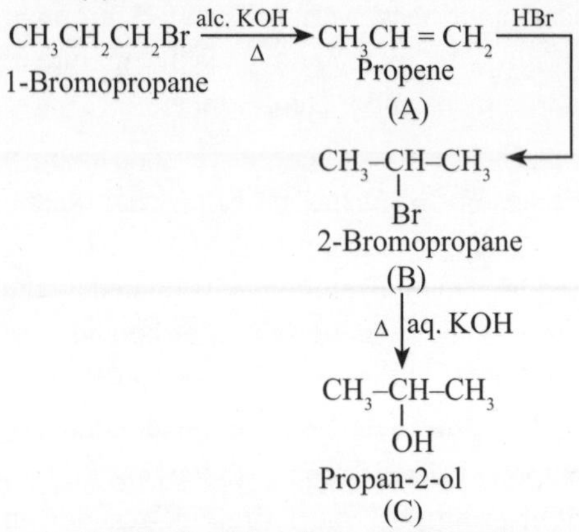

2. (1) Aryl chlorides and bromides can be conveniently synthesized through electrophilic substitution of arenes (such as toluene) using Cl_2 and Br_2 respectively, in the presence of a Lewis acid catalyst (Fe in the dark). The reaction with I_2 is reversible and necessitates the inclusion of an oxidizing agent.

The formation of the corresponding fluoroarene is not feasible through this process because of the high reactivity of fluorine. Therefore, the identity of 'X' is determined as F.

3. (3)

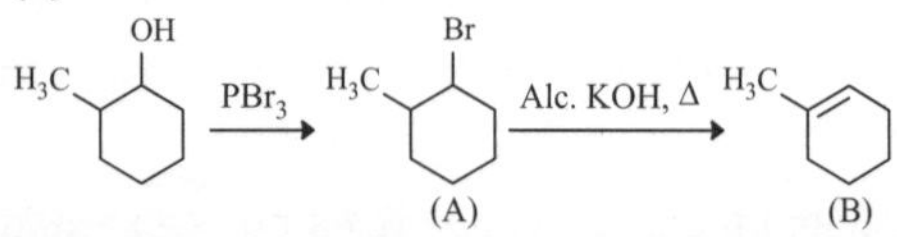

4. (2)

- S_N1 reaction are fastest in those compound, where the intermediate carbocation is resonance stablised.
- 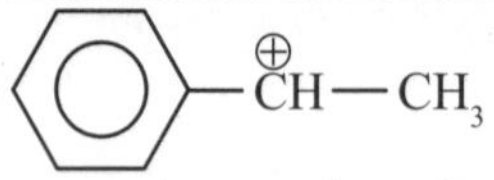 is resonance stablised carbocation

among given options. So, rate of S_N1 reaction will be fastest in 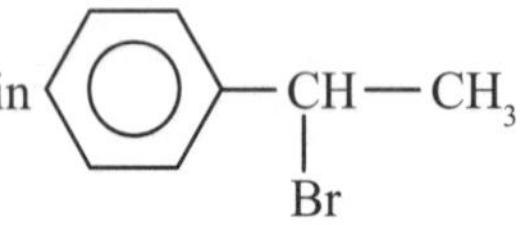

5. (2) Initially, after protonation followed by loss of water, secondary carbocation is formed, further Hydride shift leads to 3° carbocation.

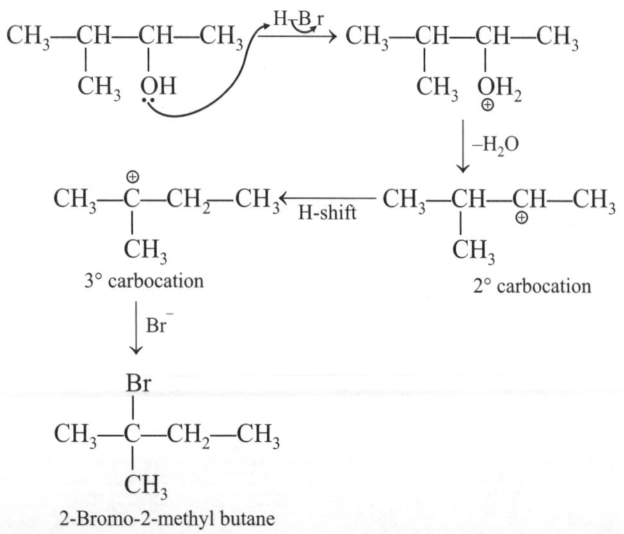

6. (4) Allylic halide have halogen bonded to sp^3 carbon which is adjacent to >C = C<

7. (4) Enantiomers are non-superimposable mirror images of each other. Therefore, statement 4 is incorrect, rest all the given statements are correct.

8. (1) Bond length from C-F to C-I increases as we move from F to I (due to increase in size of halogen atoms). Hence, bond enthalpy decreases from CH_3–F to CH_3–I.

	CH_3-F	CH_3-Cl	CH_3-Br	CH_3-I
B.E (kJ/mol)	452	351	293	234

9. (4) According to Saytzeff's rule in dehydrohalogenation reactions, alkene which contains the greater number of alkyl groups attached to the doubly bonded carbon atoms is the preffered product.

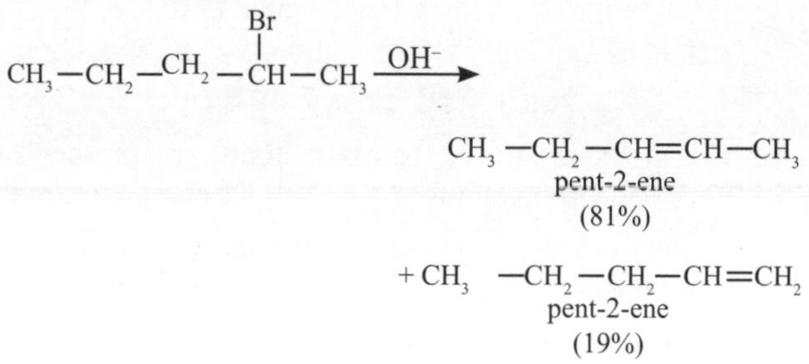

As, dehydrohologenation of 2-Bromo pentane follows thermodynamically more stable Product (i.e., more substituted alkene will be the major product). Hence, this is known as saytzeff's rule.

10. (1) **11. (3)** **12. (1)** **13. (2)** **14. (3)**
15. (1) **16. (3)**

Alcohols, Phenols and Ethers

NCERT Topic-wise MCQs

1. (2) Ether

2. (1) Ethanol

3. (3) $\underset{①}{CH_3}-\underset{②}{\overset{OH}{\overset{|}{C}H}}-\underset{③}{CH_2}-\underset{④}{CH_3}$

As (–OH) group is attached to two alkyl group so, it is secondary alcohol.

4. (3) $CH_2=CH-\underset{OH}{\underset{|}{CH_2}}$ is an example of allylic alcohol because —OH group is present on allyl position.

5. (1) Monohydric, due to presence of one —OH group. Structure of o-cresol is:

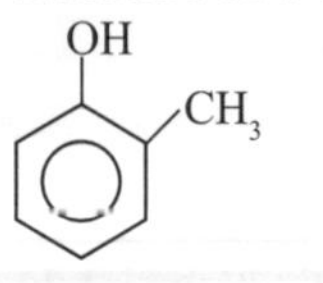

6. (3) The formula C_3H_8O represents both alcohol and ether.

$CH_3-CH_2-O-CH_3$, $CH_3CH_2CH_2-OH$
ether alcohol

as ether and alcohols are F.G. Isomers

7. **(2)** Ethers are anhydrides of alcohols, they can be obtained by elimination of a water molecule from two alcohol molecules.

$$R-OH + HO-R' \rightarrow \underset{\text{ether}}{R-O-R'} + H_2O$$

In simple ether, R and R' are same. eg: CH_3OCH_3, $C_2H_5OC_2H_5$ etc. While in mixed ether R and R' are different. eg; $CH_3OC_2H_5$.

8. **(2)** The structure of 3 - methylbutan - 2 - ol is -

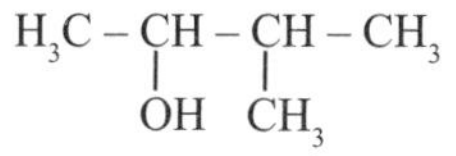

9. **(2)** 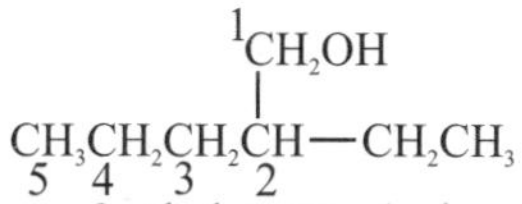

2-ethyl pentan-1-ol

10. **(3)** Glycerol

11. **(1)**

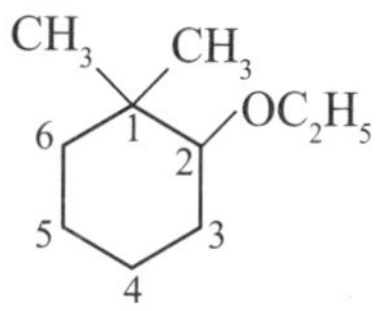

2-ethoxy-1, 1,-dimethylcyclohexane

12. **(4)** Due to resonance, lone-pair of oxygen involved in conjugation with benzene ring and the partial double bond character will appear which makes the bond length of phenol lesser than alcohol.

13. **(3)** C – O – C angle in ethers is about 110^0. Due to presence of two bulky groups, repulsion will takes place between them and increases the bond angle.

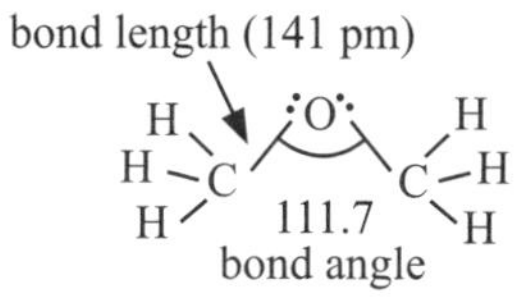

14. **(4)** 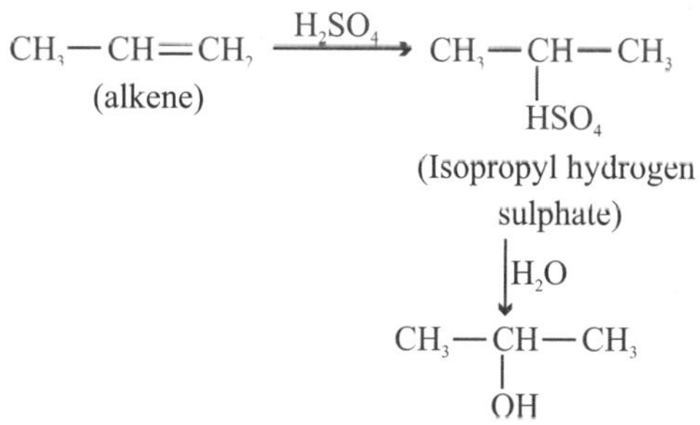

(Propan-2-ol)

The mechanism of above reaction is given as:

Step-1: protonation of alkene to form carbocation by electrophilic attack of H_3O^+

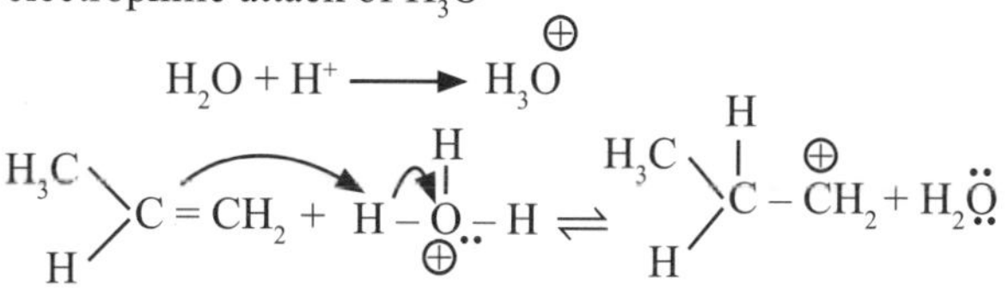

$CH_3-CH_2-\overset{\oplus}{C}H_2$ is 1°-carbocation which is least stable. Here 1, 2 - hydride shift will take place to create more stable 2° - carbocation.

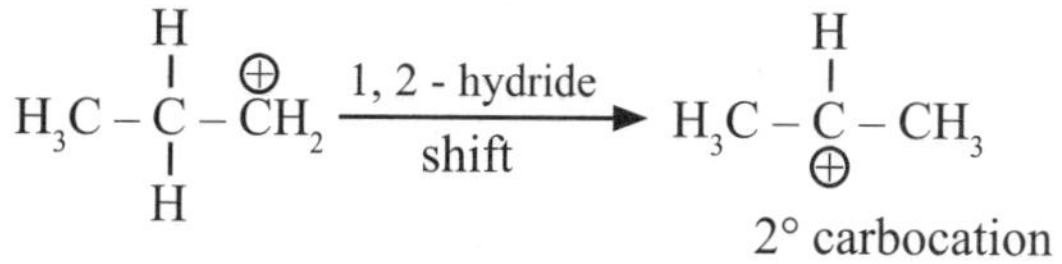

2° carbocation

Step-2: Nucleophilic attack of water on carbocation

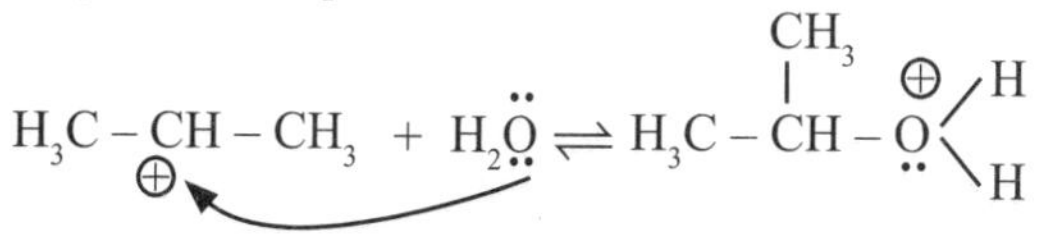

Step-3: Deprotonation to form an alcohol

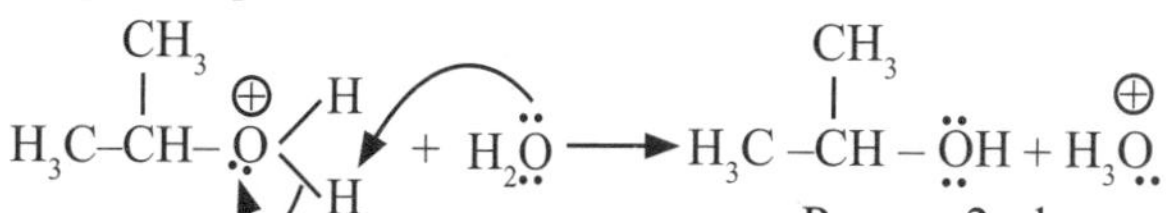

Propan-2-ol

15. **(1)**

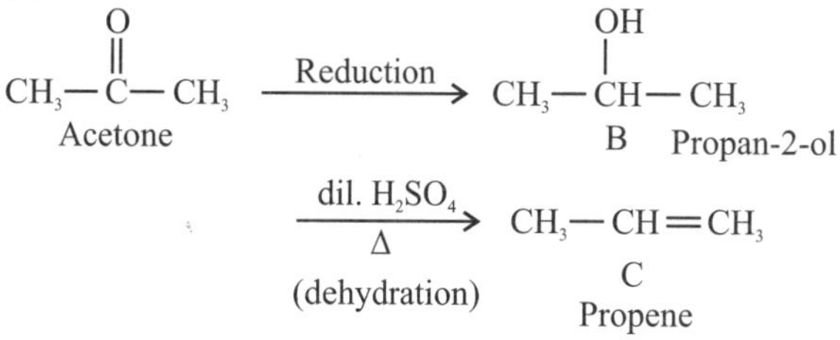

16. **(1)** Lucas reagent is Conc. HCl and anhydrous $ZnCl_2$.

17. **(2)** 

The reagent must be B_2H_6, H_2O_2/OH^-. It is known as Hydroboration - oxidation.

18. **(1)** 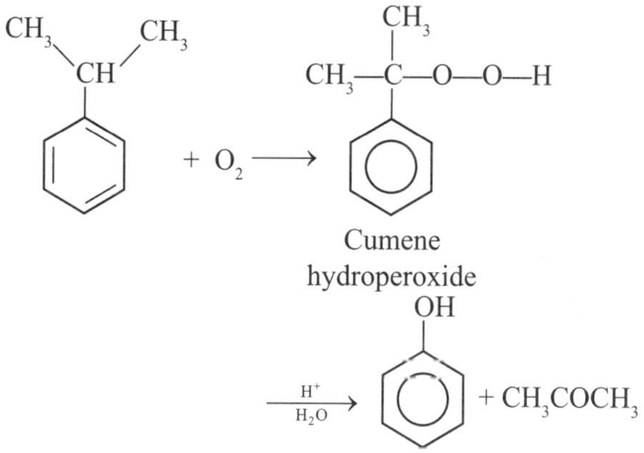

19. **(1)** 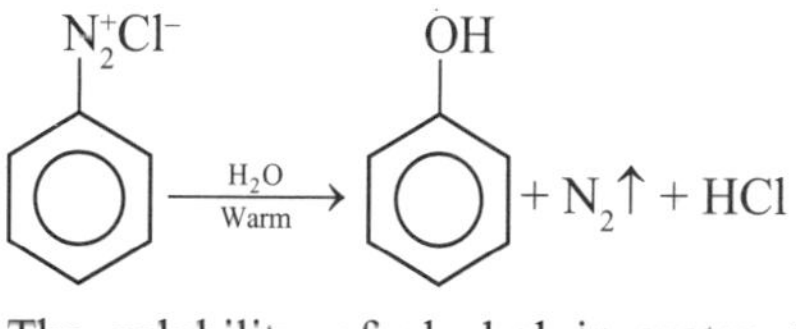

20. **(1)** The solubility of alcohol in water decreases as the number of carbon atoms increases. Hence, lower alcohols are soluble in water.

21. **(2)** Due to the presence of hydrogen bonding; methanol and ethanol are miscible in water.

22. **(3)** The compound $CH_3-\underset{|}{\overset{OH}{C}}H-CH_2-CH_2-OH$ have higher boiling point, as it forms maximum hydrogen bonds.

23. **(4)** The given conversion is represented as -

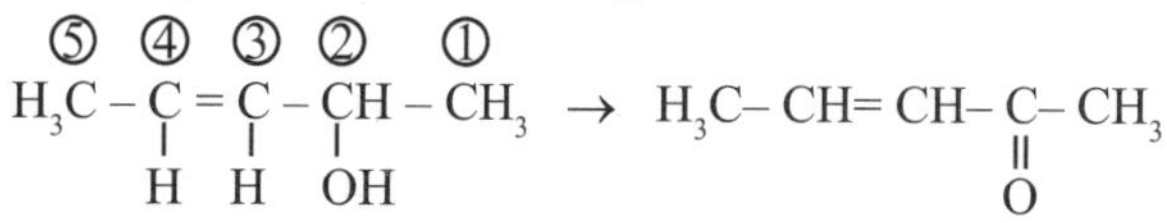

Pent-3 en-2-ol → (Pent-3-en-2-one)

PCC converts secondary alcohols to ketones. In the presence of carbon - carbon double bond pyridinium chlorochromate oxidises an alcoholic group selectively.

24. (1)

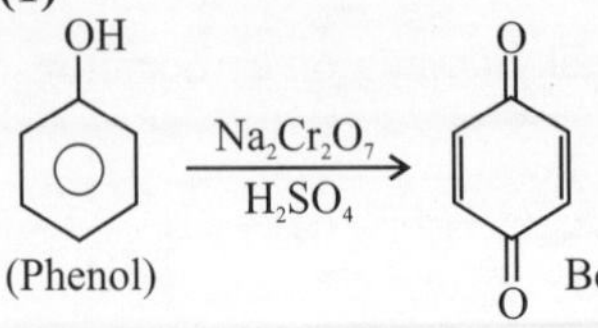

(Phenol) Benzoquinone

Oxidation of phenol with chromic acid produces a conjugated diketone known as benzoquinone.

Chromic acid is also called tetrachromic acid or chromic (IV) acid.

It is usually a mixture made by adding concentrated H_2SO_4 to a dichromate.

25. (3)

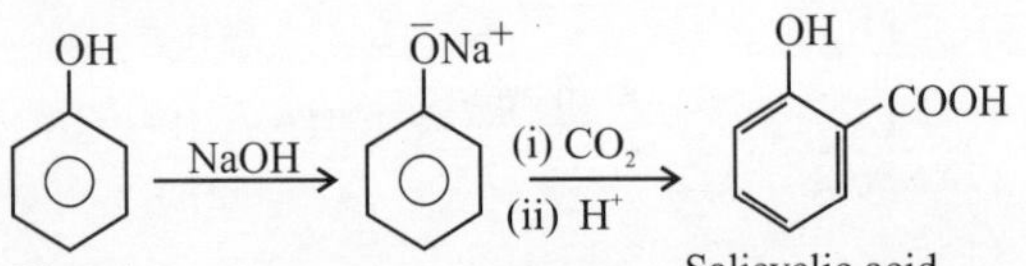

Phenol Salicyclic acid

This reaction is known as Kolbe's reaction. It's mechanism is given below,

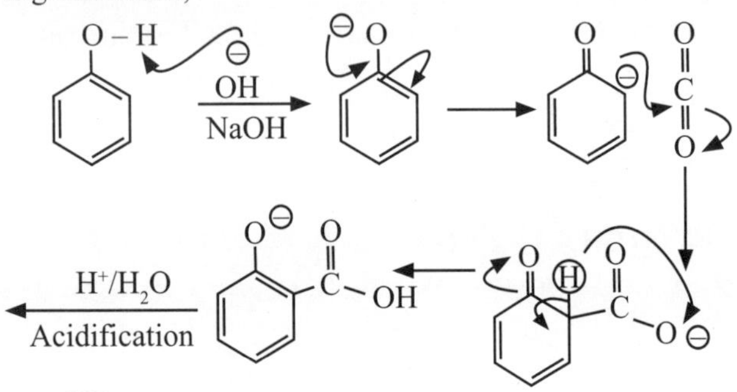

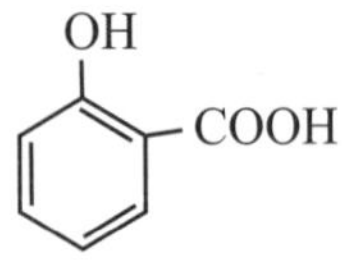

Salicylic acid

(2 - Hydroxy benzoic acid)

26. (1) $CH_3-\underset{CH_3}{\overset{CH_3}{C}}-OH \xrightarrow[\text{anhy. }ZnCl_2]{\text{conc. HCl/}} CH_3-\underset{CH_3}{\overset{CH_3}{C}}-Cl$

3° - Butyl alcohol gives 3° - butyl chloride on treatment with Lucas reagent (anhydrous $ZnCl_2$ + Conc. HCl).

Lucas reagent is used to differentiate between 1°, 2° and 3° alcohols. 3° - alcohol reacts immediately with Lucas reagent as evidenced by turbidity. 2° - alcohols react with in 5 - 10 min (depending solubility). 1° - alcohol do not react with Lucas reagent at room temp.

27. (1) Carbocation is formed in this reaction, as there is cleavage of C – O bond takes place. The stability of carbocation is as: tertiary > secondary > Primary

28. (4) Lucas reagent reacts fastest with tertiary alcohol.

Lucas reagent: Solution of anhy. $ZnCl_2$ and conc. HCl

$CH_3-\underset{CH_3}{\overset{OH}{C}}-CH_3$

29. (1) $C_2H_5OH \xrightarrow[\text{Lucas reagent}]{HX,\ ZnCl_2} \underset{B}{C_2H_5X} \xrightarrow[\text{Wurtz Reaction}]{\text{Na/ether}} \underset{C}{C_4H_{10}}$

30. (1)

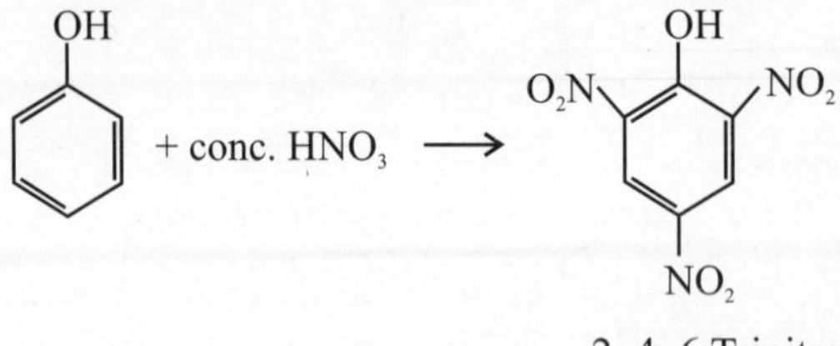

2, 4, 6 Trinitrophenol
(Picric acid)

31. (2) $\underset{4}{CH_3}-\underset{3}{CH_2}-\underset{2}{CH_2}-\underset{1}{CH_2}-OH$

$\downarrow Al_2O_3, \Delta$ (dehydration)

$CH_3-CH_2-CH=CH_2$
(But-1-ene)

32. (4) $\underset{CH_2OH}{\overset{CH_2OH}{|}} + \underset{\text{periodic acid}}{HIO_4} \rightarrow HCHO + HCHO + HIO_3 + H_2O$

33. (2)

$CH_3CH_2CH_2OH \xrightarrow{PCl_5} \underset{A}{CH_3CH_2CH_2Cl} \xrightarrow[\text{dehydrohalogenetion}]{\text{alc.KOH}} \underset{B}{CH_3CH=CH_2}$

34. (2) $C_2H_5OH + 4I_2 + Na_2CO_3 \rightarrow CHI_3$
Iodoform
(Yellow ppt.)

$CH_3OH + I_2 + Na_2CO_3 \rightarrow$ No reaction

35. (3)

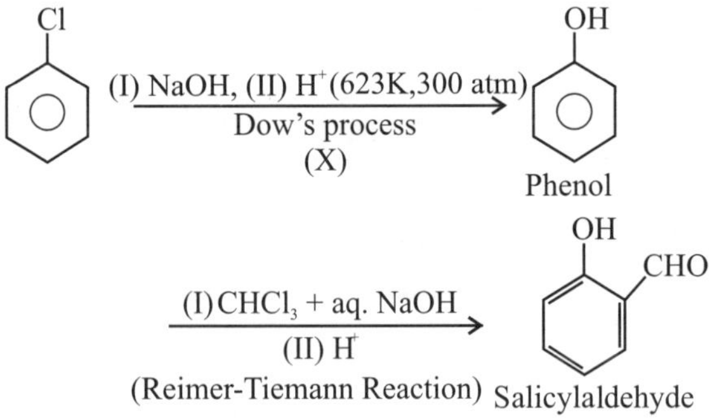

36. (2)

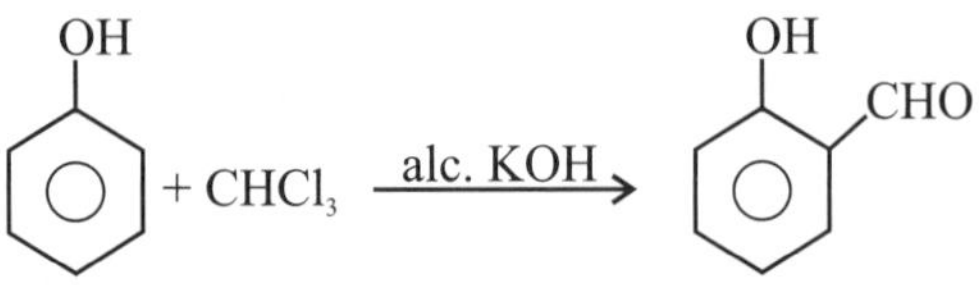

Salicylaldehyde

37. (2)

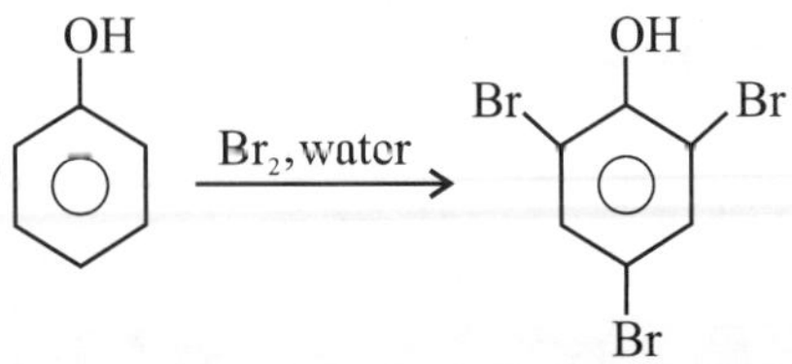

2, 4, 6-tribromophenol
(white ppt.)

38. (3)

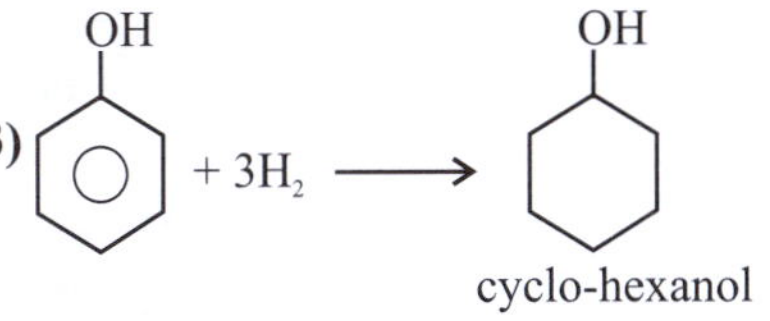

39. (2)

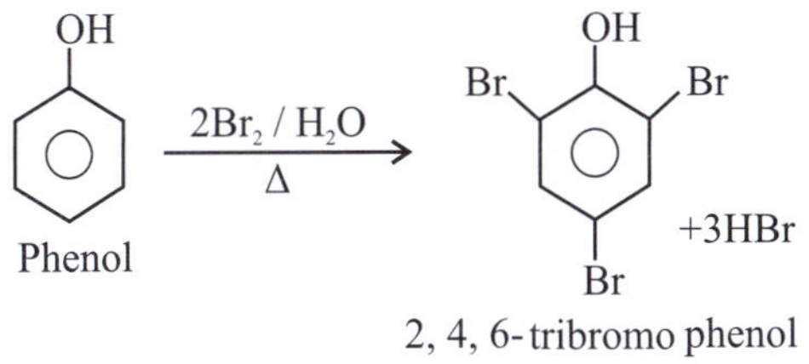

40. (2) When the reaction is carrried out in solvents of low polarity such as $CHCl_3$ or CS_2 and at low temp, mono -bromophenols are formed.

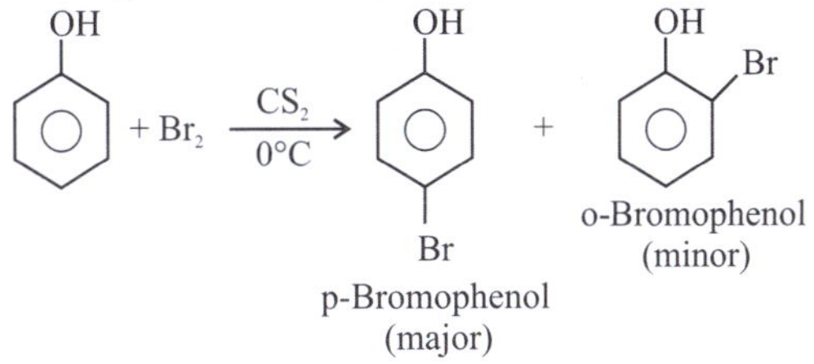

41. (2) Aspirin is an acetylation product of o-Hydroxy benzoic acid. The reaction is given as,

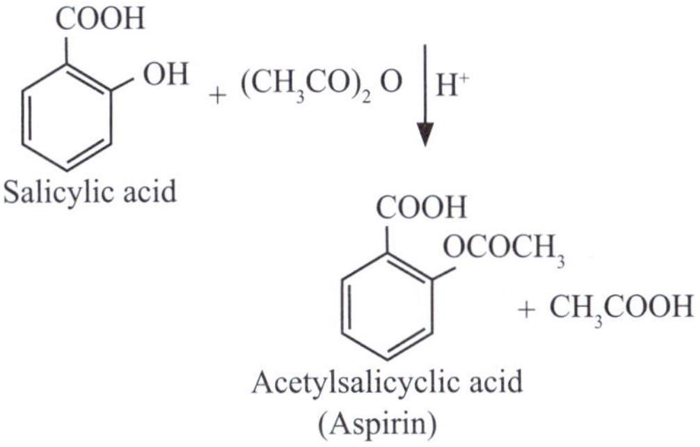

42. (2) $CH_3-\underset{|}{\overset{OH}{CH}}-CH_3 \xrightarrow{[O]} CH_3-\overset{O}{\overset{\|}{C}}-CH_3$

Secondary alcohol   Ketone

43. (2) $-CH_3$ is the electron donating in nature.

$-NO_2$ is the electron withdrawing in nature. So, it will increase acidity due to –M effect.

∴ The acidic order is:

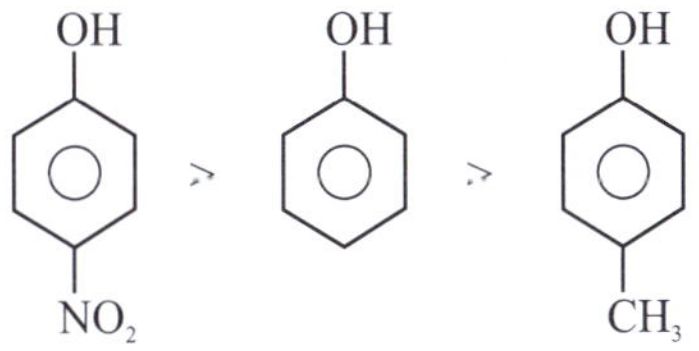

44. (3)

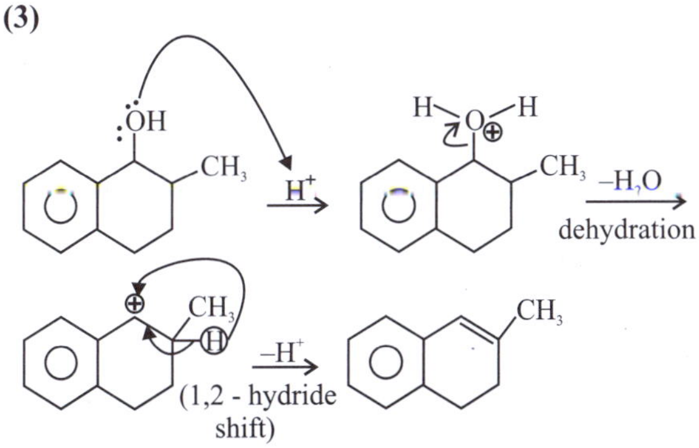

45. (1)

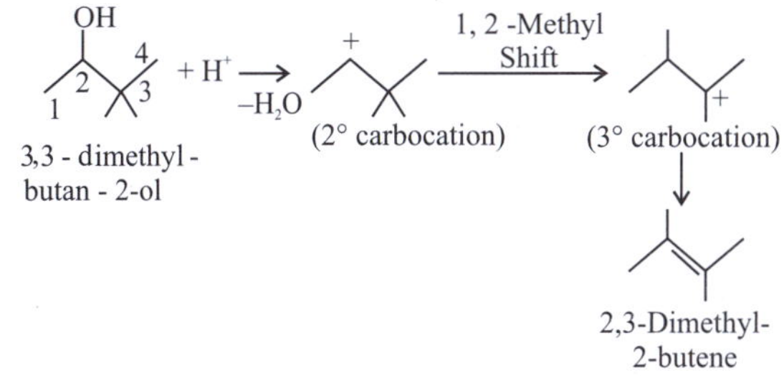

46. (3) Electron releasing group decreases the acidity of phenol whereas electron withdrawing group increases the acidity of phenol.

$-OCH_3$ is an electron releasing group, whereas rest all are electron withdrawing group.

47. (2) Nitro group is strongly electron withdrawing group, at ortho position. Intra-molecular hydrogen bond formation occurs & it reduces the acidic strength.

Whereas, $-NO_2$ at para position stabilize the electron density through benzene ring.

∴ (2) is the correct answer

48. (4)

$$\underset{(A)}{CH_3CH_2\underset{OH}{\underset{|}{CH}}-CH_3} \xrightarrow[\Delta \ (-H_2O)]{\text{conc. } H_2SO_4} CH_3CH=CHCH_3 \xrightarrow{O_3/Zn} \underset{\text{Acetaldehyde}}{2CH_3CHO}$$

49. (3) For the manufacture of a drug, aspirin, phenol is used as a starting material.

50. (4) $pk_a \propto \frac{1}{\text{acidity}}$

Phenol has least acidic strength than the o-nitrophenol, p-nitrophenol and m-nitrophenol.

Ortho and para form shows intramolecular and intermolecular H-bonding while in case of meta form –I effect favours. These two factors to decrease the acidity of phenolic group.

51. (3) CH_3 group is an electron releasing group which decreases the acidic strength.

o-cresol and p-cresol show stronger +I effect due to which they have less acidic strength.

52. (2) $\underbrace{CO + H_2}_{\text{Synthesis gas}} \xrightarrow[Cr_2O_3/ZnO]{300^\circ C/300 \text{ atm}} CH_3OH$

53. (1) Drinking ethanol mixed with methanol is called denaturated alcohol.

54. (2) $\underset{\text{Glucose}}{C_6H_{12}O_6} \xrightarrow{\text{zymase}} \underset{\text{Ethanol}}{2C_2H_5OH} + 2CO_2$

55. (2) Commonly known name of denaturated alcohol is methylated spirit.

56. (2) The manufacture of ethyl alcohol from starch consists of the following steps:

(i) Hydrolysis of starch:

$$\underset{\text{Starch}}{2(C_6H_{10}O_5)_n} + nH_2O \xrightarrow[(50-60)^\circ C]{\text{Diastase}} \underset{\text{Maltose}}{nC_{12}H_{22}O_{11}}$$

(ii) Fermentation:

$$\underset{\text{Maltose}}{C_{12}H_{22}O_{11}} + H_2O \xrightarrow[\text{yeast}]{\text{Maltase}} \underset{\text{Glucose}}{2C_6H_{12}O_6} \xrightarrow[\text{yeast}]{\text{zymase}} \underset{\text{Ethanol}}{2C_2H_5OH} + 2CO_2$$

57. (4) 6-10% of ethyl alcohol.

58. (3) Another name of methanol is wood spirit.

59. (4) Drinking of methyl alcohol mainly cause blindness.

60. (3) Primary alcohol in excess readily form ether when heated with conc. H_2SO_4

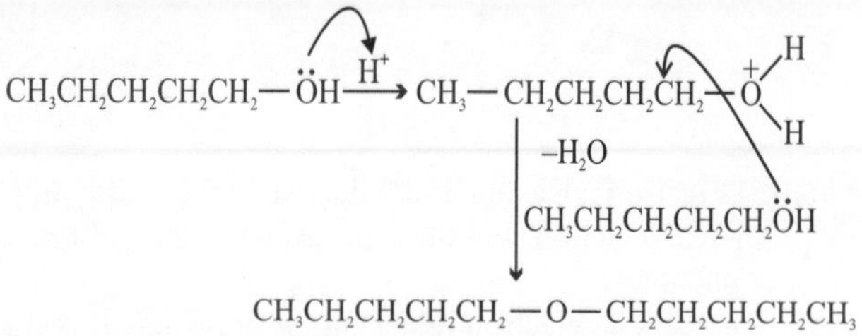

61. (3) $R - X + RO^-Na^+ \rightarrow R - O - R$ This reaction

Represents a Williamson ether synthesis. It involves nucleophilic substitution of the halide ion from the alkyl halide by the alkoxide ion via S_N2 mechanism

62. (2)

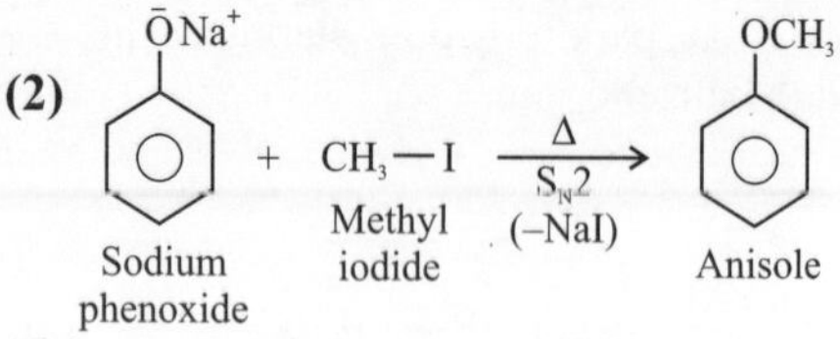

This reaction is known as Williamson's ether synthesis.

63. (3) $\underset{\text{(Excess)}}{CH_3CH_2OH} \xrightarrow[413K]{H_2SO_4} C_2H_5OC_2H_5$

64. (4) $C_2H_5—OH + H—O—SO_3H \longrightarrow \underset{\text{(ester of sulphuric acid)}}{C_2H_5—O—SO_3H} + H_2O$

65. (1) Sodium ethoxide causes dehydro halogenation of 3° alkyl and give alkene as sodium ethoxide is a strong base.

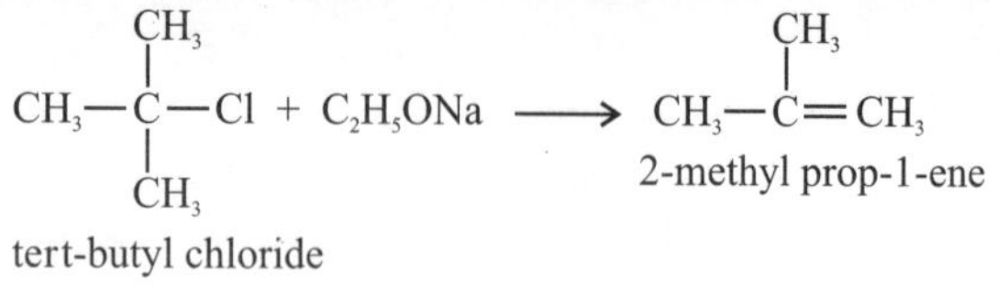

This is the limitations of williamson's synthesis.

66. (3)

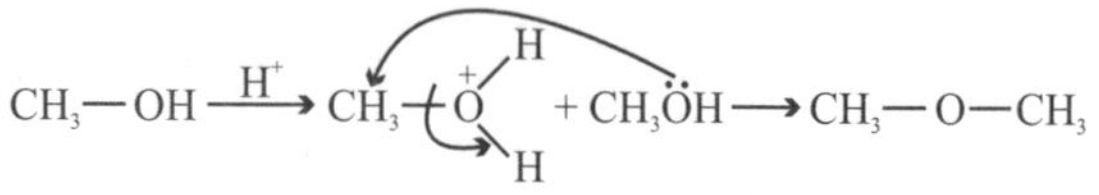

67. (3) Phenyl - N - propyl ether will form as-

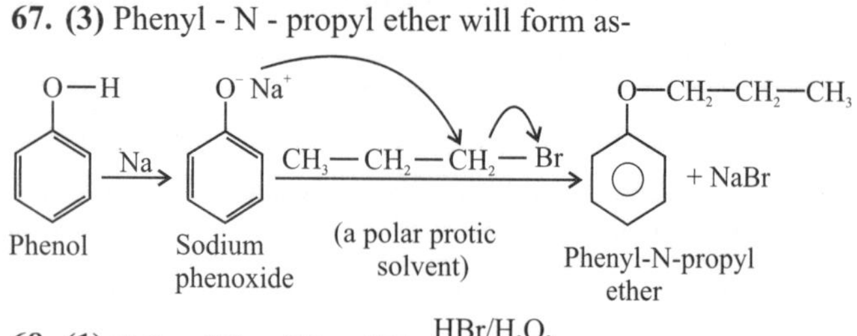

68. (1) $H_3C—CH_2—CH=CH_2 \xrightarrow{HBr/H_2O_2}$

anti markovnikov rule

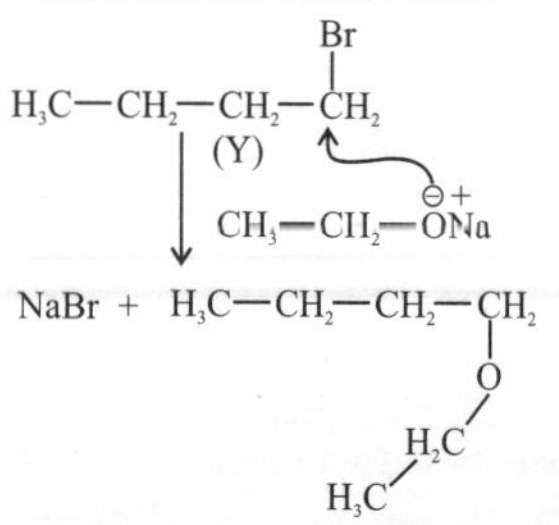

$\therefore$ Z is $H_3C - (CH_2)_3 - O - CH_2 - CH_3$ (Ethoxy butane)

69. (3) As alcohol form hydrogen bonding and boiling point increases with increase of carbon number, therefore alcohol have higher boiling point.

70. (4) Alkyl ethyl ether on reaction with HI produces ethanal and ethyl iodide by means of nucleophilic substitution reaction as shown

$$CH_2 = CH - O - CH_2 - CH_3 + H - I \rightarrow CH_3CH_2I + CH_3CHO$$

71. (4) $C_2H_5OCH_3 + HI \rightarrow C_2H_5OH + CH_3I$

Halogen goes with smaller group on cleavage of ether.

72. (2) Alkyl aryl ethers undergo Electrophilic substitution reactions.

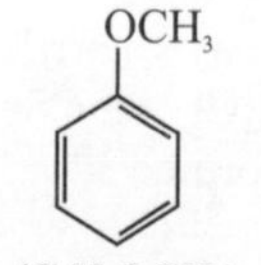

$(C_6H_5OCH_3)$

73. (4) $C_6H_5 - O - CH_3 + H^+ \rightarrow C_6H_5-\overset{+}{O}(H)-CH_3 \xrightarrow{Br^-} C_6H_5OH + CH_3Br$

74. (4) The hydrolysis gives alcohols which are successive members of a homologous series, hence the alkyl radicals in these two alcohols must differ by one $– CH_2 –$ group i.e, $CH_3 –$ and $CH_3CH_2–$

High Order Time Intensive MCQs

1. (1) The given compounds are phenol and its derivatives (also known as isomeric dihydroxy benzenes).

As the distance between the –OH group increases, the repulsion between them decreases. In case of quinol intermolecular H-bonding takes place while in case of catechol intra molecular H-bonding takes place. Intermolecular H-bonding are more stronger. So, Boiling point of quinol is higher than catechol.

Therefore, decreasing order of boiling point is;

IV > III > II > I

The name of the given compounds with their boiling point is -

I - Phenol (181.7 °C)

II - Catechol (245.5 °C)

III - Resorcinol (277 °C)

IV - Quinol (285 °C)

2. (2)

$$CH_3-CH_2-OH \xrightarrow{P, I_2} CH_3CH_2I \xrightarrow[Mg]{ether} CH_3CH_2MgI \xrightarrow{HCHO} CH_3CH_2CH_2\overset{-}{O}\overset{+}{Mg}I \xrightarrow{H_2O} CH_3CH_2CH_2OH + Mg(OH)I$$

3. (3)

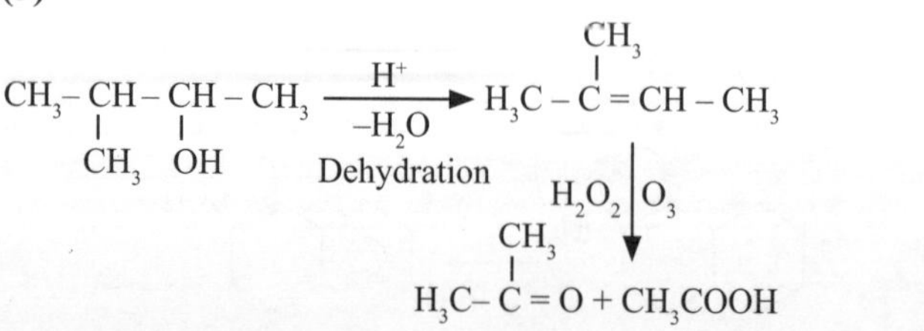

4. (3) The order of reactivity depends upon the stability of carbocations formed. The relative stability of the carbocations follows the order -

$$\mathrm{Ph-\overset{\oplus}{C}H_2 > H_3C-\overset{\oplus}{C}H-CH_3 > F-CH_2-CH_2-\overset{\oplus}{C}H-CH_3}$$
$$\mathrm{> F-CH_2-\overset{\oplus}{C}H-CH_3}$$

Therefore, the order of reactivity of corresponding alcohols follows the order -

IV > III > II > I

5. (3) 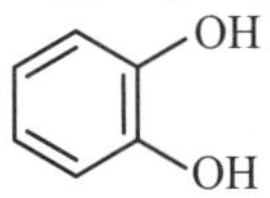

IUPAC name is Benzene-1, 2-diol

Common name is Catechol

6. (4)

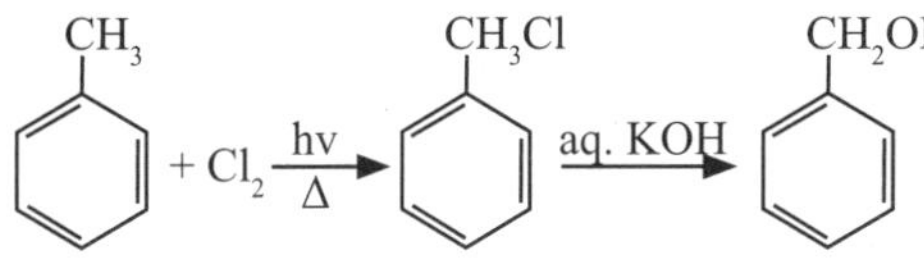

7. (4)

$$\mathrm{{}_1CH_3 - {}_2\overset{*}{C}H(OH) - {}_3CH = {}_4CH - {}_5CH_3}$$

The carbon at position 2 is a chiral center, hence, contributes to 2 stereoisomers (R/S configurations).

The C = C double bond at position 3 allows for cis-trans isomerism.

Therefore, it has total of 4 stereoisomers

8. (1) 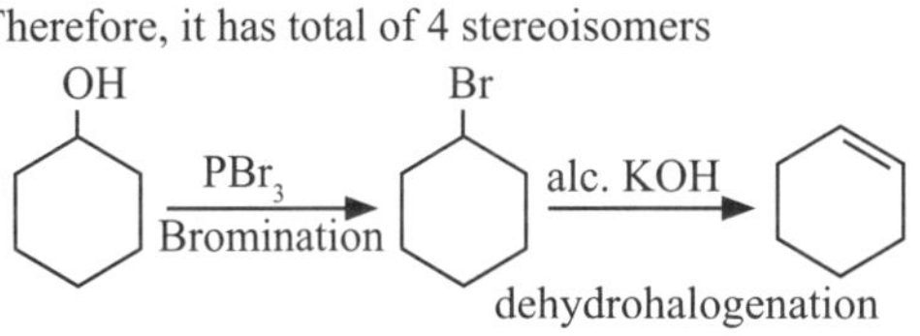

9. (4)

P is 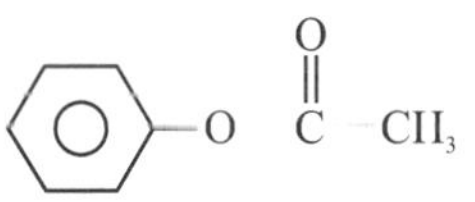

Q is 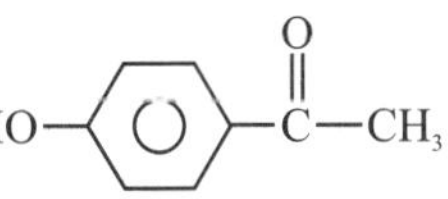

10. (4) To find acidic nature of the compound, first make their conjugated base as -

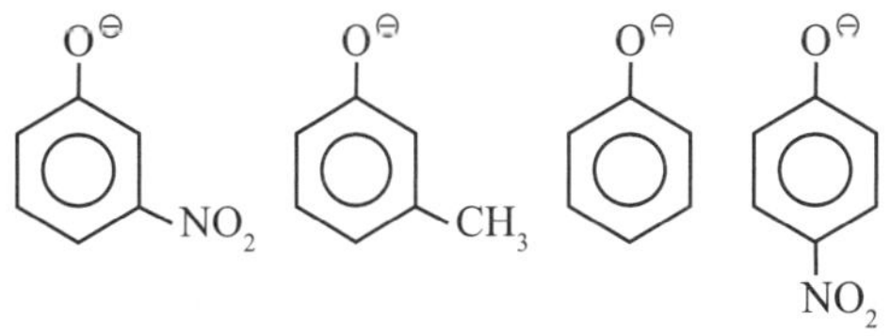

$-NO_2$ is an electron withdrawing group and $-CH_3$ is an electron releasing group.

Electron withdrawing group stabilizes the aromatic ring. NO_2 group present at para position stabilizing the ring most therefore, it will be most acidic.

Due to Electron releasing ($-CH_3$) group compound II will be least acidic. Hence the correct order of acidity will be-

IV > I > III > II

11. (1) PCC is Pyridinium chlorochromate & it is better reducing agent for the oxidation of primary alcohol.

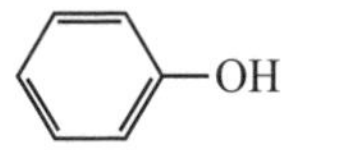

Phenol cannot be oxidised by PCC.

12. (1) 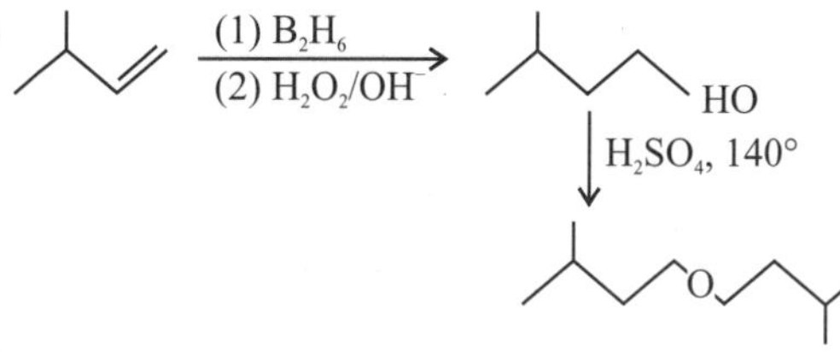

13. (3)

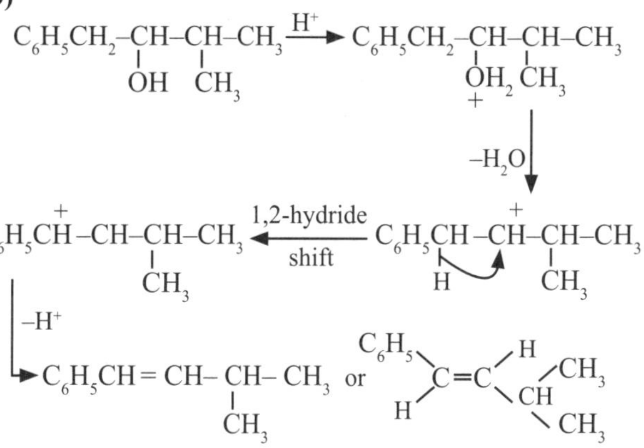

14. (1) 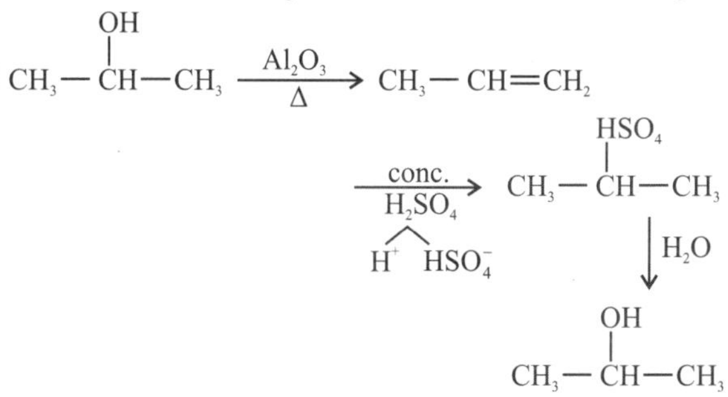

15. (2) Phthalein Dye Test is used for identification of phenols.

Carbylamine Test is used for identification of primary amines.

Lucas Test is used for differentiation between 1°, 2° and 3° alcohols.

Tollen's Test is used for identification of Aldehydes.

16. (2) cis-cylopenta-1, 2-diol can form cyclic ketal whereas trans-cyclopenta-1, 2-diol can't form cyclic ketal.

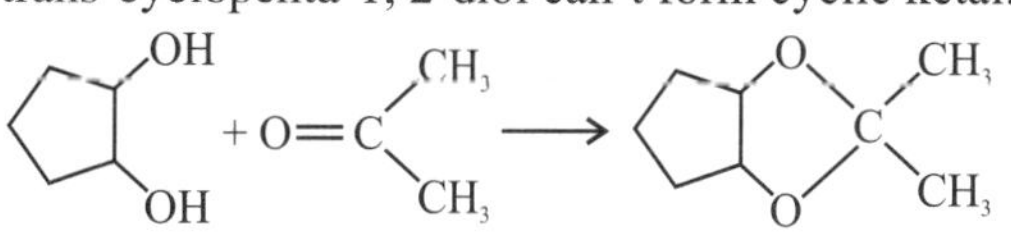

Hence, C is -cyclopenta- 1, 2 - diol & its trans form can be distinguished by acetone.

17. (1) The correct acidic order is,

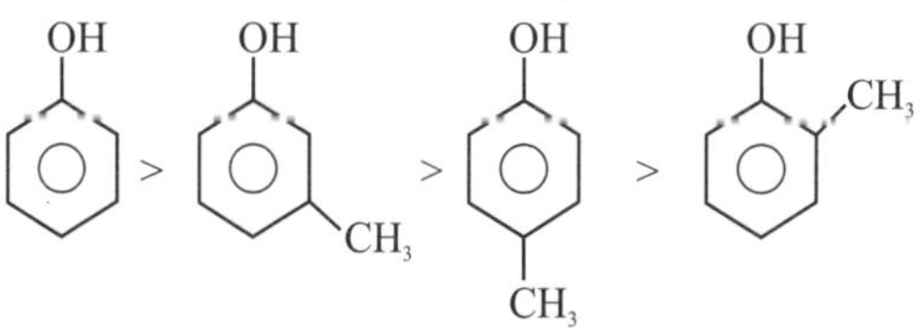

Here, CH_3 is EDG & it reduces phenolic acidity, its impact at meta position will be lesser than ortho & para position.

18. (4) More stable carbocation can't be rearranged to a less stable carbocation.

Acid catalysed dehydration of the given compound is shown as, -

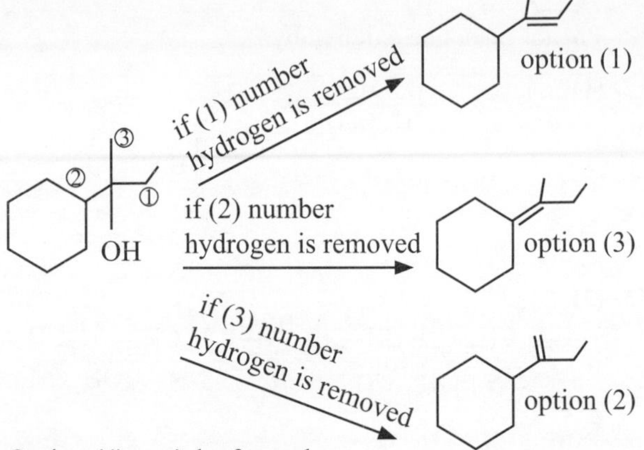

Option (4) can't be formed.

19. (1)

Clearly, B is 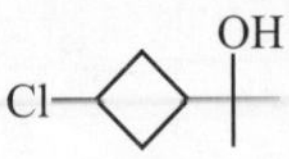

∵ Br has more tendency to form RMgX, than 'Cl' because of greater size. So, reaction will take place from 'Br' side. i.e.,

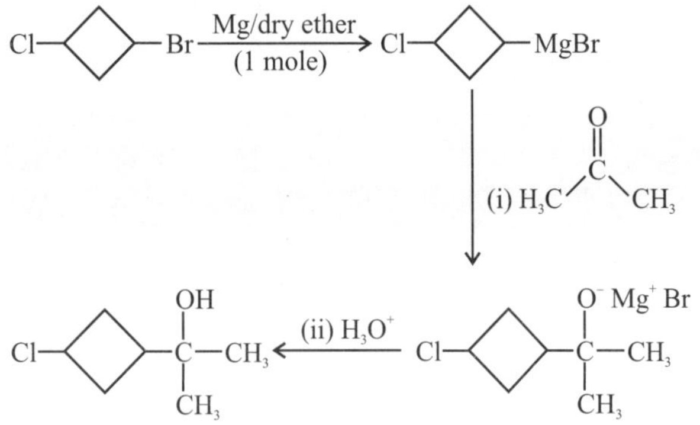

20. (1)

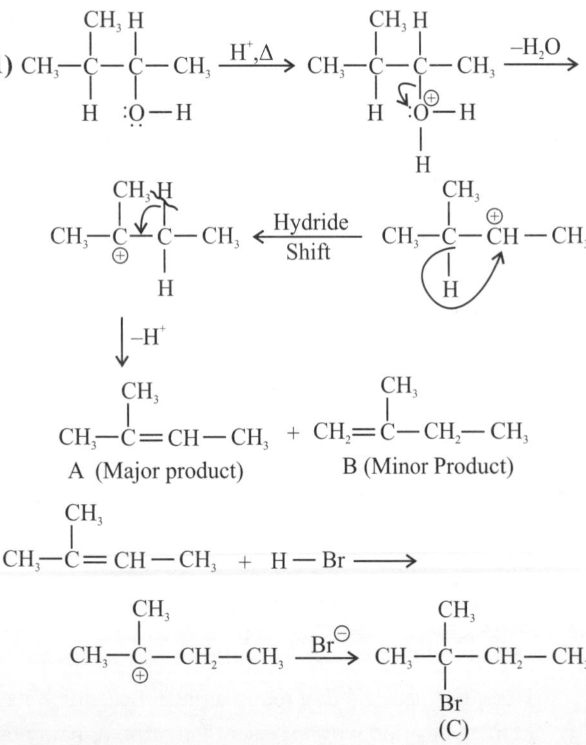

21. (3) As carbocation is the reaction intermediate, more the stability of carbocation, faster the rate of dehydration.

22. (4)

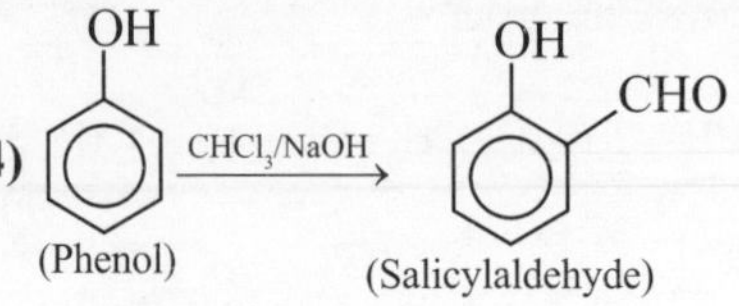

23. (3) The given compound is prepared from methanal and a Grignard reagent, which is given as -

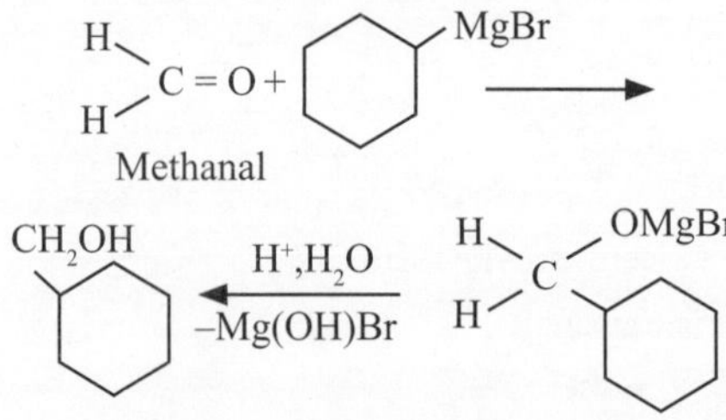

24. (3) Esters on reduction with $LiAlH_4$ will give alcohol.

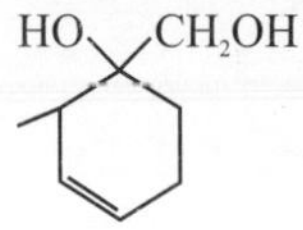

25. (4)

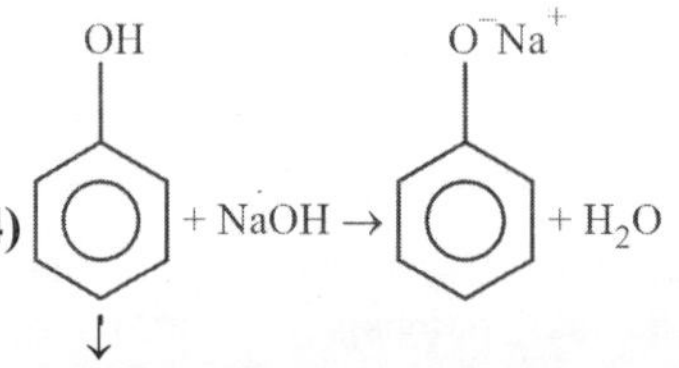

This will readily react with dilute NaOH.

26. (2) This reaction is an example of base catalysed hydrolysis of an ester.

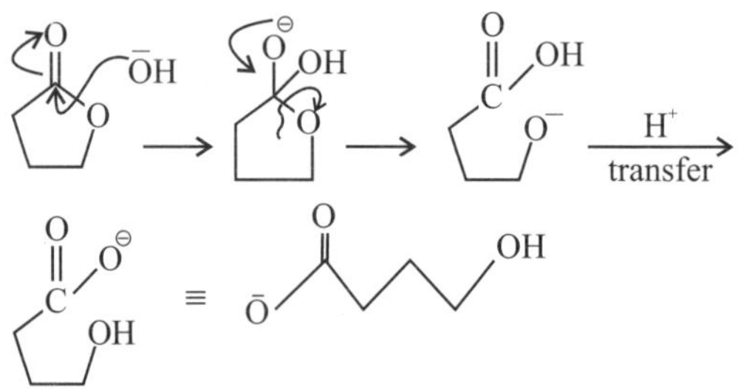

27. (4)

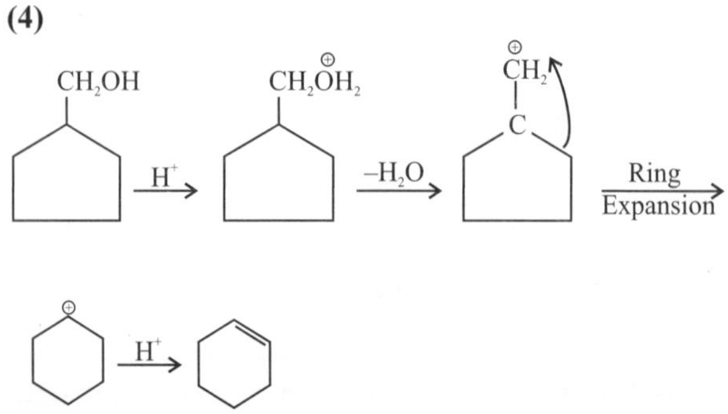

28. (3) It is a S_N1 reaction, so rearrangement takes place in carbocation

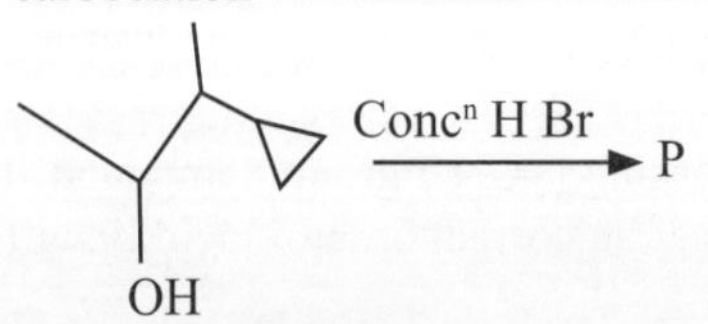

Mechanism:- The given alcohol is a secondary alcohol, therefore HBr will react here without catalyst. The reaction will proceed via S_N1 mechanism.

$HBr \longrightarrow H^+ + Br^\ominus$

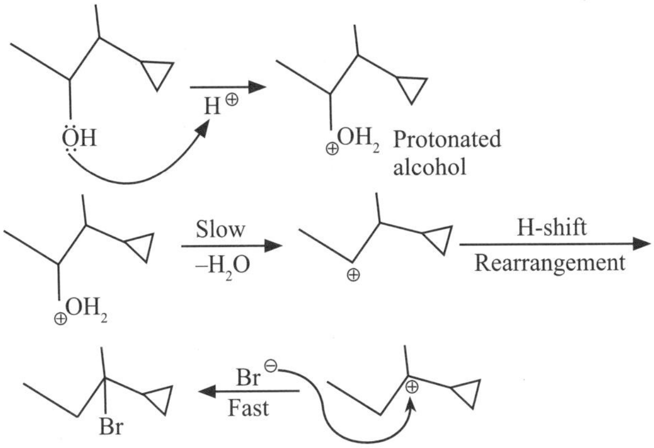

29. (2) The solvent CS_2 is non-ionisable and non-polar which produce only p-substituted compound due to the release of less Br^+ ion, while, the solvent H_2O is polar and release sufficient concentration of Br^+ to attack on –o and –p position.

30. (1) $NaBH_4$ is a H^- ion donor (Nucleophile). Most reactive towards nucleophilic addition is q > p > r

31. (3) The most common use of NBS is allylic Bromination.

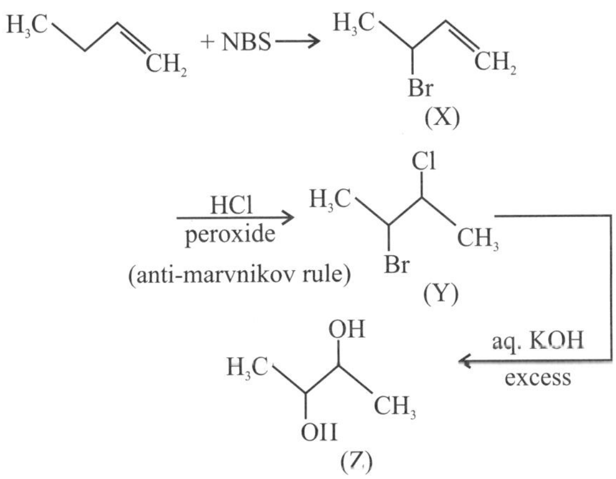

32. (2) $NaBH_4$ does not reduce double bond, it only reduces carbonyl group whereas H_2/Pt reduces both double bond as well as aldehydic group.

33. (1) Electron with drawing group ($-NO_2$) increases the acidity of phenol. Its effect is more dominant when it is present on o-& p-position rather than on m-position. The acidic order is Para-nitrophenol > Meta-nitrophenol > Phenol > Methyl Phenol.

34. (4) Electron with drawing group ($-NO_2$) increases acidity of phenol whereas, electron releasing group ($-OCH_3$, $-CH_3$) decreases the acidity of phenol.

Therefore acidic strength follows the order-

p - Nitrophenol > p - Methylphenol > p - Methoxyphenol

35. (1)

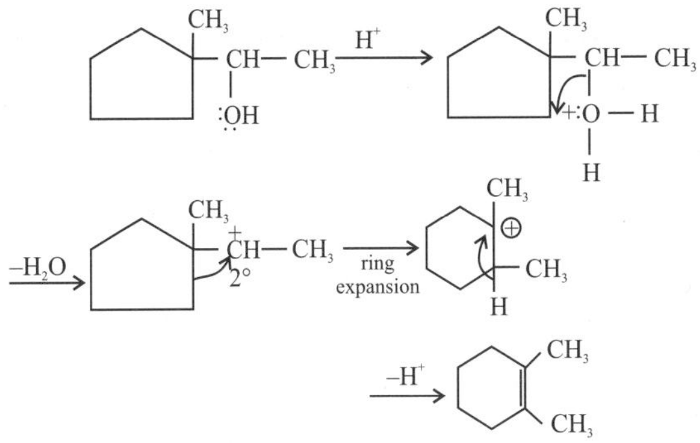

36. (3)

$$H_3C-CH_2-\underset{}{\overset{CH_3}{\overset{|}{C}H}}-CH_2-O-CH_2-CH_3 \xrightarrow[\text{Heat}]{HI}$$

$$\underset{[A]}{CH_3-CH_2-\overset{CH_3}{\overset{|}{C}H}-CH_2-OH} + CH_3CH_2-I$$

$$\downarrow H_2SO_4/\Delta$$

$$CH_3-CH=\overset{CH_3}{\overset{|}{C}}-CH_3$$

37. (4) The given reaction is reimer - Tiemann reaction.

In the first step $CHCl_3$ reacts with NaOH and generate dichloro carbene. It is given as-

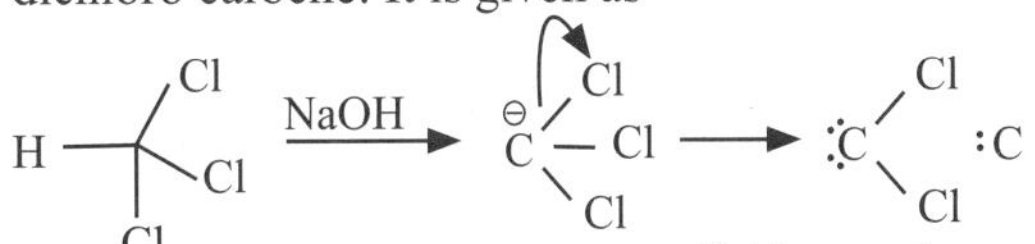

dichloro carbene

(This generated dichloro carbene is act as an electrophile for the further reaction)

Further reaction is given as -

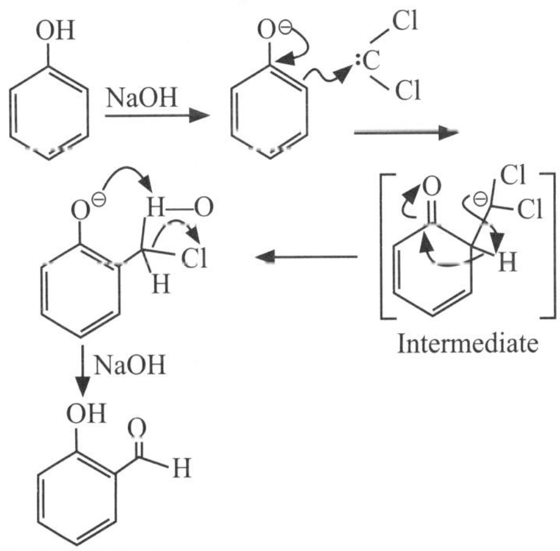

Salicyladehyde

38. (2) Tertiary alkyl bromide gives alkene as a major product in presence of sodium ethoxide because tertiary carbocation readily gives elimination reaction and is converted into most stable alkene.

$$CH_3-\underset{CH_3}{\underset{|}{\overset{CH_3}{\overset{|}{C}}}}-Br + CH_3CH_2ONa \longrightarrow CH_3-\underset{CH_2}{\underset{||}{\overset{CH_3}{\overset{|}{C}}}} + NaBr + CH_3CH_2OH$$

39. (4)

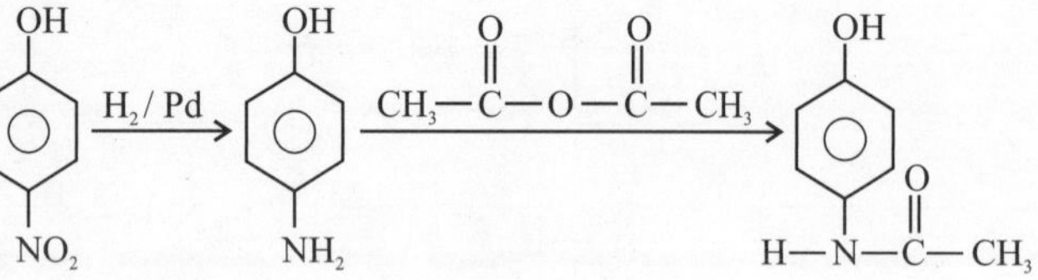

(NH_2 is more $Nu^{\ominus}$ than –OH) (paracetamol)

40. (4) The reaction between ether and HI is

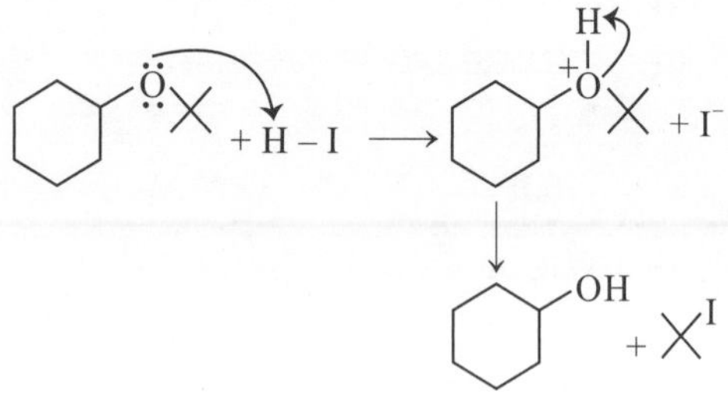

Correct & Incorrect MCQs

1. **(2)** In symmetrical ether, the alkyl or aryl groups attached to the oxygen atom are the same and in unsymmetrical ether, the alkyl or aryl groups attached to the oxygen atom are different.
2. **(3)** The IUPAC name of isobutyl alcohol is 2-methylpropan-1-ol.
3. **(3)**
   1) The IUPAC name of dimethyl ether is Methoxymethane.
   2) The IUPAC name of diethyl ether is Ethoxyethane.
   3) The IUPAC name of methyl n-propyl ether is 1-Methoxypropane.
   4) The IUPAC name of ethyl phenyl ether is Ethoxybenzene.
4. **(1)** The carbon– oxygen bond length (136 pm) in phenol is slightly less than that in methanol. This is due to (i) partial double bond character on account of the conjugation of unshared electron pair of oxygen with the aromatic ring and (ii) sp^2 hybridised state of carbon to which oxygen is attached.
5. **(1)** The solubility of alcohols in water decreases with increase in size of alkyl groups because higher the hydrocarbon part, lower will be the extent of hydrogen bonding and hence lower will be the solubility.
6. **(2)** In alcohols, the boiling points decrease with increase of branching in carbon chain due to decrease in van der Waals forces with decrease in surface area.
7. **(4)** In phenoxide ion, the charge is delocalised, hence phenoxide ion is more stable than alkoxide ion.
8. **(2)** Presence of electron donating group i.e., methyl group in cresol makes it less acidic than phenol.
9. **(1)** In Williamson ether synthesis, better results are obtained when primary alkyl halides are used.
10. **(3)** The alkoxy group in ethers is ortho, para directing in nature and activates the aromatic ring towards electrophilic substitution reactions.

Statement Based MCQs

1. **(1)** No intermolecular H-bonding takes place in ether so it has lower boiling point than alcohols.
2. **(1)** Since benzene pulls the electron density of oxygen towards –OH itself so lone pair of oxygen in OH in phenol is not available for the protonation. Attachment of proton will give a +ve charge on electronegative atom and destabilize the molecule.
3. **(4)** p-hydroxybenzoic acid has a higher boiling point than ortho isomer due to intermolecular H bonding.
4. **(3)** Statement I is correct. Dissociation constant of alcohols are much less than the dissociation constant of water. This clearly indicates that alcohols are weaker acid than even water.
5. **(1)** Acidic strength order:

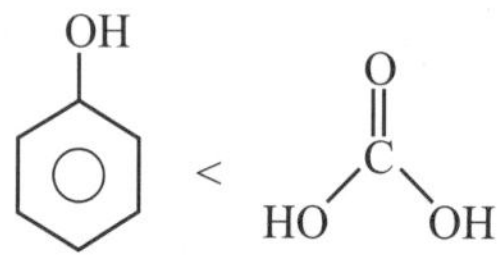

So, 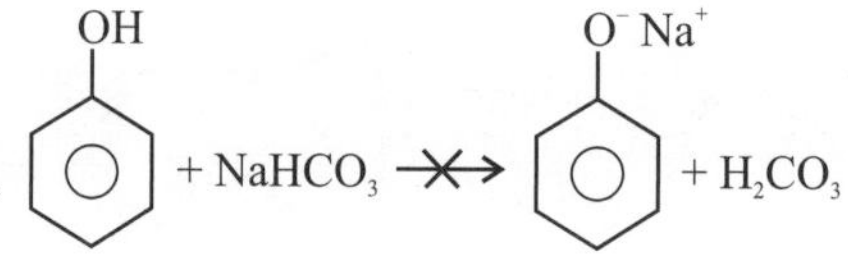

Reaction is not possible.

6. **(4)** The bond angle of ether is slightly greater than the tetrahedral angle due to the repulsive interaction between the two bulky (–R) groups.
7. **(3)** An electron – releasing group increases electron density on oxygen tending to decrease the polarity of O – H bond.

   The acid strength of alcohols decrease in the following order:

   $$R-CH_2OH > R_2CHOH >> R_3C-OH$$
8. **(4)** In Lucas test, turbidity is produced immediately with tertiary alcohols as they form the halides easily.
9. **(4)** In substituted phenols, the presence of electron withdrawing groups such as nitro group, enhances the acidic strength of phenol.
10. **(1)** The commercial alcohol is made unfit for drinking by mixing in it some $CuSO_4$ (to give it a colour) and pyridine (a foul smelling liquid), this is known as denaturation of alcohol.

Assertion & Reason MCQs

1. **(1)** p-nitro phenol is more acidic due to –R effect of nitro group it helps in stabilisation of phenoxide ion.

2. (4) 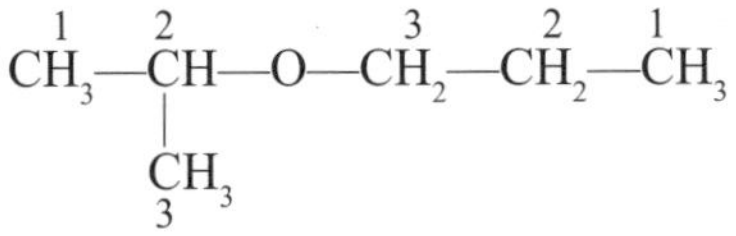

2-propoxy-propane.

Ether group is the derivative of hydrocarbon. It can be derived by replacement of hydrogen in hydrocarbon by –OR/–OAr group.

3. **(4)** Bond angle of ether is slightly greater than tetrahedral bond angle due to two bulky alkyl group which creates repulsion between these group and increases the bond angle.
4. **(4)** Bromation of phenol will takes place in the presence of CS_2 or water. Lewis acid like BF_3 polarise the Br_2 molecule.
5. **(2)** Due to presence of intramolecular H-bonding in o-nitrophenol it can't make bond with solvent molecule. So, it is least soluble in H_2O. While m-nitrophenol and p-nitrophenol have intermolecular H-bonding. So they exist in associated form.
6. **(3)** Ethanol is a weaker acid than phenol because phenoxide ion is resonance stabilised.

   Sodium ethoxide may be prepared by reaction of Na with ethanol.
7. **(4)** Phenol give 2,4,6-tri-nitrophenol when it will react with conc. HNO_3 in the presence of H_2SO_4.

   As, –OH group is ortho-para directing. So, NO_2 occupies all the ortho and para position.
8. **(2)** No intermolecular hydrogen bonding takes place in ether, therefore it has lower boiling point than alcohols.
9. **(2)** Relative ease of dehydration of alcohols will follow the order: tertiary > secondary > primary

   Tert butyl alcohol will undergo dehydration readily than that of propanol, due to the formation of more stable carbocation.
10. **(3)** Good yields of ethers are obtained when primary alkyl halides are treated with alkoxides derived from any 1°, 2° or 3° alcohol.

Multi Correct MCQs

1. **(4)** Order of acidic strength

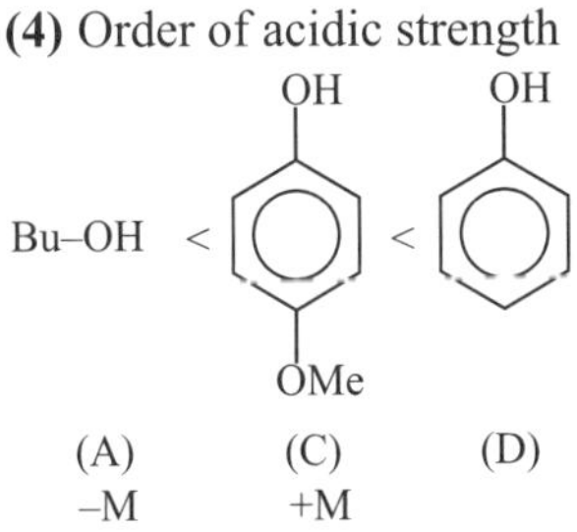

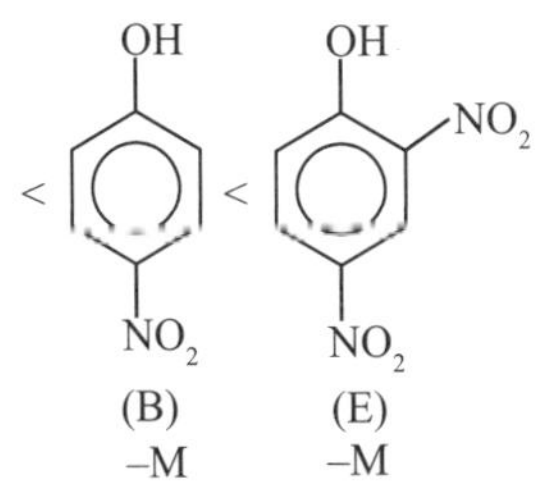

Hence acidic strength of compound (B) > (C) and (E) > (D)

2. **(2)** Intra molecular H- bonding is present in (A)

   Inter molecular H- bonding is present in (B)

   (C) cannot show intermolecular H- bonding due to high steric hindrance at o-position of benzene ring.
3. **(2)** The below reaction is not possible.

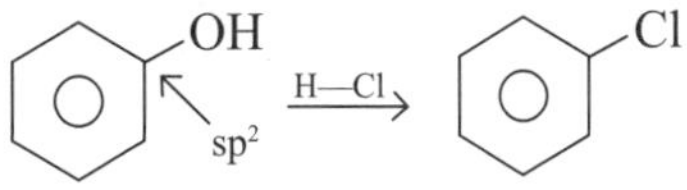

4. **(2)** 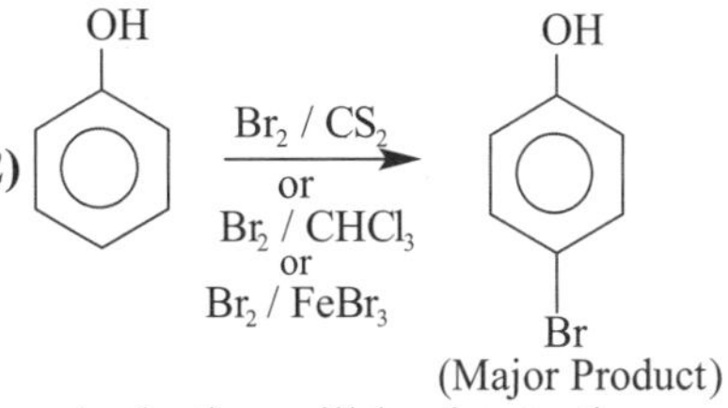

(less ionisation will be due to the presence of non- polar solvent)

(1) Phenol with Br_2/H_2O gives

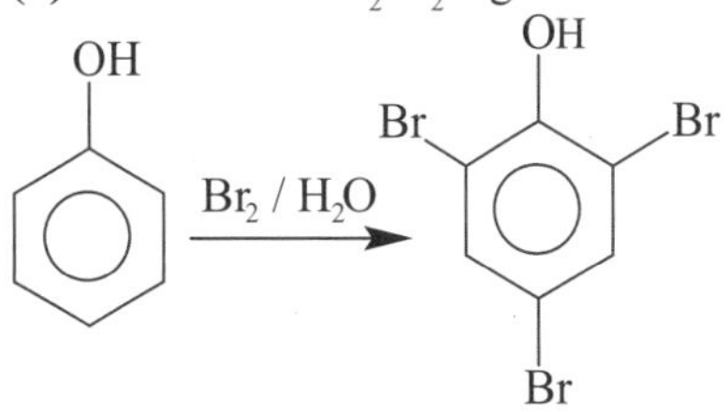

(High ionisation due to presence of polar solvent and high activation of ring)

NCERT Exemplar MCQs

1. **(4)** Mono-chlorination of toluene in sunlight gives benzyl chloride. Benzyl chloride shows nucleophilic substitution reaction to give benzyl alcohol on hydrolysis with aq. NaOH.

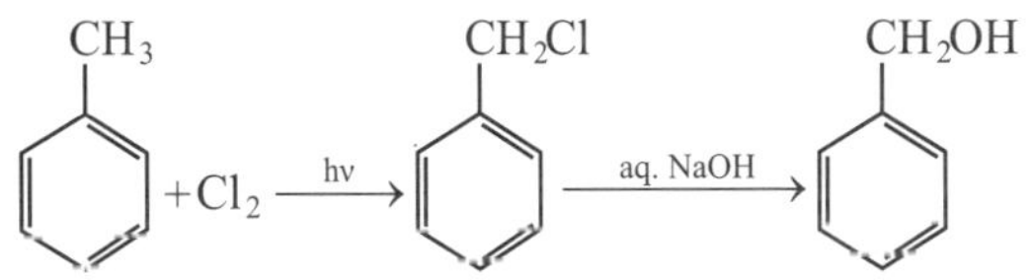

(Toluene) (Benzyl chloride) (Benzyl alcohol)

2. **(1)** The three isomers of butanol are possible. Structural formula of these isomers are given below

   (i) $CH_3-CH_2-CH_2-CH_2-OH$

   (Butan-1-ol)

   No carbon is chiral in this compound as none of the four carbon is bonded to four different substituents.

   (ii) 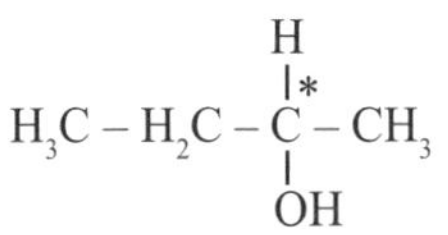

   (Butan-2-ol)

   In this compound, asterisk marked carbon is chiral carbon as all four substituents, attached to it are different and these are $-CH_3$, $-H$, $-CH_2-CH_3$, $-OH$.

   (iii) 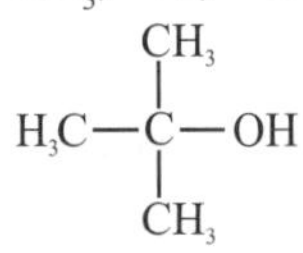

   (2-methylpropan-2-ol)

Here, again carbon is not chiral in nature.

So, only one alcohol is chiral in nature and the correct option is (1).

3. **(3)** The given reaction is nucleophilic substitution (S_N1) reaction in which –OH group is replaced by –Cl. Tertiary alcohols, when react with HCl in presence of $ZnCl_2$, (Lucas regent) form tertiary carbocation.

This intermediate 3° carbocation is more stable than 2° carbocation as well as 1° carbocation. Higher the stability of intermediate, higher will be the reactivity of reactant molecule.

So, the order of reactivity of alcohols in the given reaction is 3° > 2° > 1° and correct option is (3).

4. **(3)** PCC oxidises primary alcohols to aldehydes. Strong oxidising agents such as $KMnO_4$ are used for getting carboxylic acid from alcohols. PCC is pyridinium chlochromate.

The oxidation process can be stopped at the aldehyde stage if Cr(VI) reagents such as Collins reagent($CrO_3 \cdot 2C_5H_5N$), Corey's reagent or pyridinium chlorochromate and pyridinium dichromate $\left[(C_5H_5NH_2)^{2+}(Cr_2O_7^{2-})\right]$ in anhydrous medium are used as the oxidising agent. So, the correct option is (3).

$$\underset{\text{Ethanol}}{CH_3CH_2OH} \xrightarrow{PCC} \underset{\text{Ethanal}}{CH_3CHO}$$

5. **(2)** A reaction where two or more molecules combine to form a larger one is an addition reaction. These reactions occur to change the unsaturated compound to saturated compound.

In dehydrohalogenation reaction, alkyl halides give alkenes. Rearrangement gives the structural isomers of the reactant, while in substitution reaction one of the group or an atom is replaced by other group. Therefore, the process of converting alkyl halides into alcohols involves substitution reaction.

$$\underset{\text{Alkyl halide}}{R-X} \xrightarrow{OH^-} \underset{\text{Alcohol}}{R-OH}$$

Reaction of alkyl halides:

Primary alkyl halides – S_N2

Secondary alkyl halides – S_N1

Tertiary alkyl halides – S_N1

6. **(3)** Another name of phenol is "carbolic acid" and it cannot be considered as aromatic alcohol. It is quite separate branch of compound called phenols. So, compound (1) i.e., phenol and compound (4) i.e., a derivative of phenol cannot be considered as aromatic alcohol.

On the other hand, in compound (2) and (3), –OH group is bonded to sp^3 hybridised carbon which in turn is bonded to benzene ring. Hence, the correct option is (3).

7. **(3)**

$$\overset{6}{CH_3}-\underset{Cl}{\overset{5}{\underset{|}{CH}}}-\overset{4}{CH_2}-\overset{3}{CH_2}-\underset{OH}{\overset{2}{\underset{|}{CH}}}-\overset{1}{CH_3}$$

The correct IUPAC name of the compound is 5-chlorohexan-2-ol.

Hence, option (c) is the correct answer.

8. **(1)** The structure of m-cresol is:

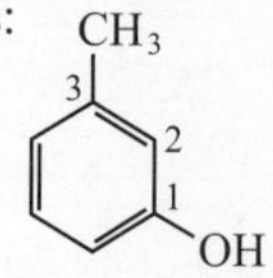

IUPAC name will be 3-methylphenol because –OH is the functional group and the methyl is substituent.

9. **(3)** $\overset{3}{CH_3}-\underset{\underset{CH_3}{|1}}{\overset{2}{CH}}-OCH_3$: 2-methoxypropane

IUPAC name of the above compound is 2-methoxypropane and correct option is (3).

10. **(2)** The strongest conjugate base is obtained from a weakest acid · ROH is the acid of RO^- conjugate base, HOH is the acid of ^-OH, C_6H_5OH is the acid of $C_6H_5O^-$ and 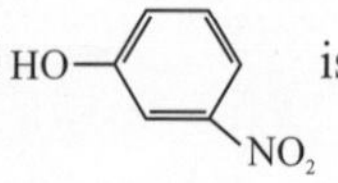 is the acid of 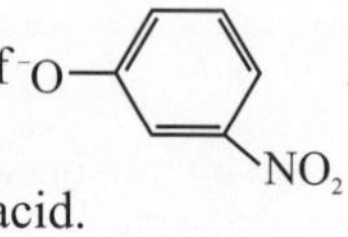. Among all these acids, ROH is the weakest acid.

Therefore, the strongest base is RO^- and the correct option is (2).

11. **(1)** The loss of one proton phenol is more acidic in nature, it gives phenoxide ion. This phenoxide ion is resonance stabilized. As phenoxide ion is a stable intermediate so, the tendency to give proton is more in phenol than the others. Phenols are more acidic than alcohols, therefore get dissolved in NaOH.

12. **(2)** In o-nitrophenol, nitro group is present at ortho position. Presence of electron withdrawing group at ortho position increases the acidic strength. On the other hand in o-methylphenol and in o-methoxyphenol, electron releasing group (–CH_3, –OCH_3) are present.

Presence of these groups at ortho or para positions of phenol decreases the acidic strength of phenols. Therefore, phenol is less acidic than o–nitrophenol.

13. **(4)** Alcohols are less acidic than phenols. Further, electron with drawing group (like-Cl) increases the acidity of phenol, therefore m-chlorophenol is most acidic among all of these.

14. **(2)** Presence of electron withdrawing group on phenols, increases its acidic strength. So, both compounds i.e., p-nitrophenol (II) and m-nitrophenol (IV) are stronger acid than (I). If –NO_2 group is present at m-position, then it exerts – I effect mly. Therefore, p-nitrophenol is much stronger acid than m-nitrophenol.

On the other hand, presence of electron releasing group on phenol, decreases its acidic strength. If –OCH_3 group is present at meta position, it will not exert + R effect but exert – I effect.

But, if it is present at para position, then it will exert + R effect. Therefore, m-methoxyphenol is more acidic than p methoxy phenol.

Hence, the correct option is (2).

15. **(3)** Reaction of the given compounds with HBr/HCl is a nucleophilic substitution reaction. It follows S_N1 mechanism. S_N1 mechanism depends upon the stability of carbocation.

The stability of carbocation decreases by presence of electron withdrawing group. In compound (II) and (III) EWG is present at para position.

Since, $-NO_2$ group is a stronger electron withdrawing group than –Cl. So, $NO_2-C_6H_5-\overset{+}{C}H_2$ carbocation will be less stable than $Cl-C_6H_5-\overset{+}{C}H_2$ carbocation. Thus, the order of stability of carbocation is (II < III < I)

Therefore, compound (II) is least reactive and correct option is (3).

16. **(1)** Boiling point increases with increase in the number of carbon atoms due to the increase in molecular mass. Therefore, the boiling point of pentan-1-ol is more than that of all other given compounds. Further, among isomeric alcohols, 1° alcohols have higher boiling points than 2° alcohols because of higher surface area in 1° alcohols.

Therefore, boiling point increase in the order:

Propan-1-ol < butan-2-ol < butan-1-ol < pentan-1-ol.

Hence, the correct option is (1).

NEET & AIIMS Past Year MCQs

1. **(2)** $CH_3CH=CH_2+(H-BH_2)_2 \longrightarrow CH_3-CH_2-CH_2-BH_2$

$\xrightarrow{CH_3CH=CH_2} (CH_3-CH_2-CH_2)_2BH$

$\xrightarrow{CH_3CH=CH_2} (CH_3-CH_2-CH_2)_3B$

$\xrightarrow{H_2O_2/\overline{O}H} CH_3-CH_2-CH_2OH$ Propanol (Major)

$((CH_3)_2CH)_3B \xrightarrow{H_2O_2/\overline{O}H} CH_3-\underset{OH}{\underset{|}{C}H}-CH_3$ Propan-2-ol

Statement I is incorrect but Statement II is correct.

2. **(2)** 3° alcohols react with Lucas reagent instantaneously.

2° alcohols react with Lucas reagent after 5 minute.

1° alcohols do not react with Lucas reagent at ordinary temperature.

$CH_3-\underset{CH_3}{\underset{|}{\overset{CH_3}{\overset{|}{C}}}}-OH$ it is tertiary alcohol which give Lucas test instantaneously.

3. **(2)**

$3ROH + PCl_3 \rightarrow 3RCl + \underset{(A)}{H_3PO_3}$

$ROH + PCl_5 \rightarrow RCl + HCl + \underset{(B)}{POCl_3}$

Hence, A : H_3PO_3 and B : $POCl_3$

4. **(4)** Reaction follows S_N1 pathway involving Benzylic carbocation.

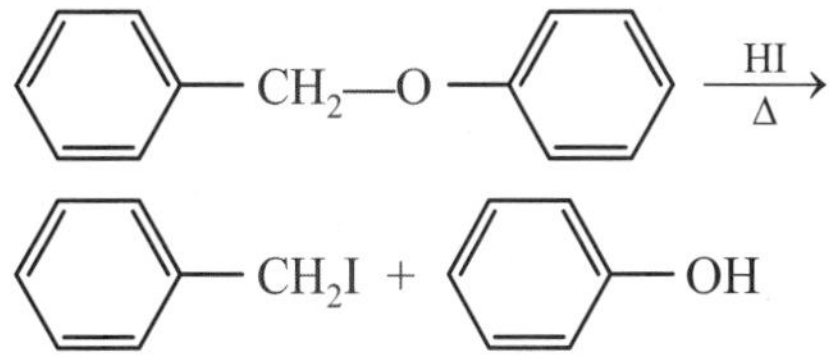

5. **(3)** 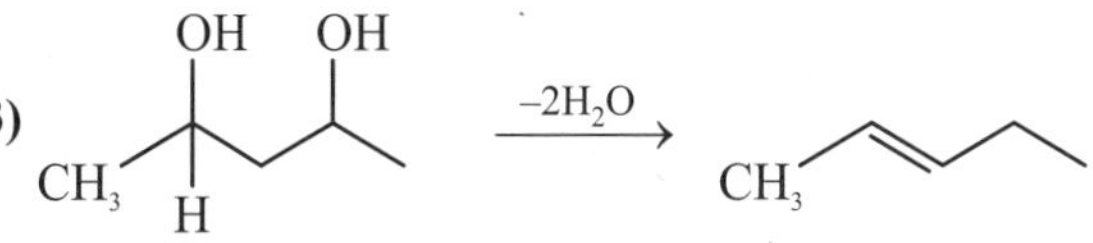

Formation of Conjugated diene in option 3 make the given reactant most reactive towards dehydration in acidic conditions.

6. **(4)** Nitro group reduces the electron density in the benzene ring due to its strong-I effect. In this way the –ve charge present on phenolic oxygen stabilises the phenoxide ring. Hence statement I is correct. The position of $-NO_2$ group present on the ring shows different activity, therefore statement II is not correct.

Therefore, order of acidic strength of the phenolic compounds is,

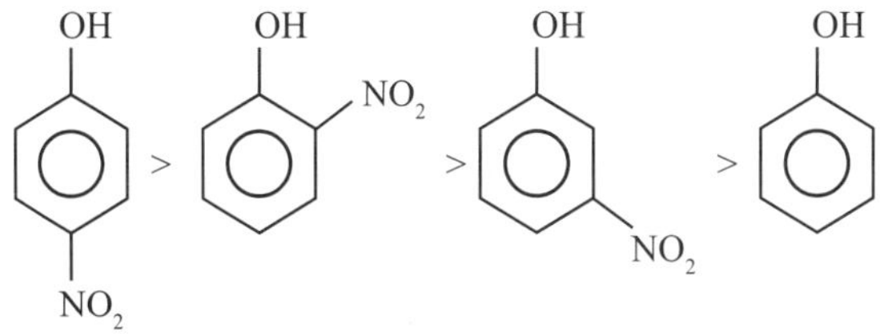

7. **(4)** Primary, secondary and tertiary alcohols are distinguished by Lucas test on the basis of the time taken for turbidity to appear

$R-CH_2-OH \xrightarrow{\text{Conc. HCl + Anhy.ZnCl}_2} R-CH_2-Cl$
1° alcohol (Turbidity in 30 min.)

$R_2CH_2-OH \xrightarrow{\text{Conc. HCl + Anhy.ZnCl}_2} R_2CH_2-Cl$
2° alcohol (Turbidity in 5 min.)

$R_3CH_2-OH \xrightarrow{\text{Conc. HCl + Anhy.ZnCl}_2} R_3CH_2-OH$
3° alcohol (Immediate turbidity appears)

Reactivity of alcohols toward Lucas regent:

3° > 2° > 1°

8. **(3)** $CH_3-\overset{O}{\overset{||}{C}}-CH_3 \xrightarrow{C_2H_5MgBr,\ \text{dry ether}} H_3C-\underset{C_2H_5}{\underset{|}{\overset{O-MgBr}{\overset{|}{C}}}}-CH_3$

$\downarrow H_3O^+$ (Hydrolysis)

$H_3C-\underset{CH_2-CH_3}{\underset{|}{\overset{O-H}{\overset{|}{C}}}}-CH_3$

2 – Methylbutan–2–ol

9. (3)

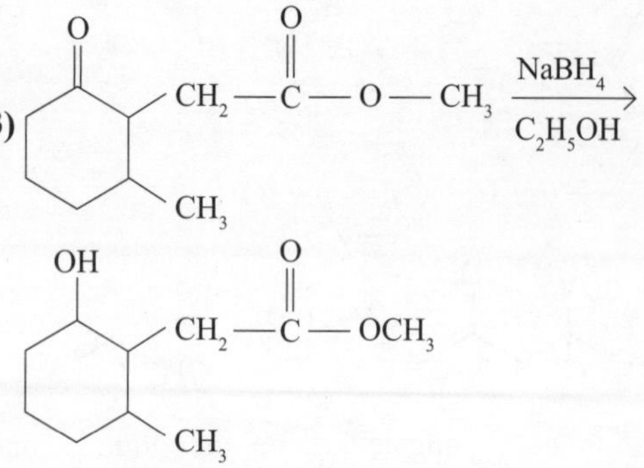

NaBH$_4$ reduces aldehyde/ketone but does not reduce ester.

10. (1) $(CH_3)_3C-\overset{-}{O}\overset{\oplus}{Na} + Ph-CH_2-Cl \rightarrow (CH_3)_3C-O-CH_2-Ph$

This is a Williamson ether synthesis in which alkyl halide should be either primary or secondary.

11. **(1)** The reaction of phenol and chloroform in presence of alkali at (333-343) K to give salicylaldehyde is Reimer-Tiemann reaction

12. **(1)** Phenol is a o, p-directing group but it can polarises to form 2, 4, 6-trinitrophenol even in the absence of lewis acid.

13. **(2)** Ethers are readily cleaved by action of HI to form alcohol and alkyl halide.

$R-O-R + HX \rightarrow RX + R-OH$

If excess of halogen acid is used, then alcohol formed reacts further with halogen acid to produce alkyl halide.

Aldehydes, Ketones and Carboxylic Acids

NCERT Topic-wise MCQs

1. **(3)** $CH_3-CH_2-\underset{|}{\overset{OH}{CH}}-CHO$

2-hydroxybutanal (IUPAC)

2. **(2)** $-\overset{O}{\overset{||}{C}}-$ is a bivalent group, whereas the remaining – CHO, – COOH and – OH are monovalent groups.

3. **(4)** $CH_3CH_2CH_2CH_2CHO$

an aldehyde

$CH_3CH_2-\overset{O}{\overset{||}{C}}-CH_2CH_3$, (Symmetrical ketone) $\quad CH_3-\overset{O}{\overset{||}{C}}-CH_2CH_2CH_3$ (unsymmetrical ketone)

4. **(3)** $\underset{\text{ethane nitrile A}}{CH_3C\equiv N} \xrightarrow{CH_3-MgBr} (CH_3)_2C=NMgBr \xrightarrow{H_3O^+} \underset{R}{(CH_3)_2C=O}$

$CH_3-\overset{O}{\overset{||}{C}}-CH_3 + CH_3-\overset{O}{\overset{||}{C}}-CH_3 \xrightarrow[\text{(self-aldol)}]{\text{dil. NaOH, aldol condensation}} CH_3-\underset{CH_3}{\underset{|}{\overset{OH}{\overset{|}{C}}}}-CH_2-\overset{O}{\overset{||}{C}}-CH_3$

$\xrightarrow[\text{dehydration}]{\Delta\ (-H_2O)} CH_3-\underset{CH_3}{\underset{|}{C}}=CH-\overset{O}{\overset{||}{C}}-CH_3$

5. **(4)**

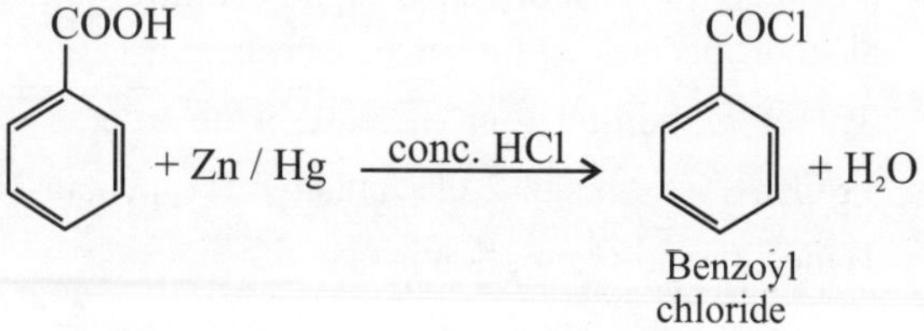

6. **(1)**

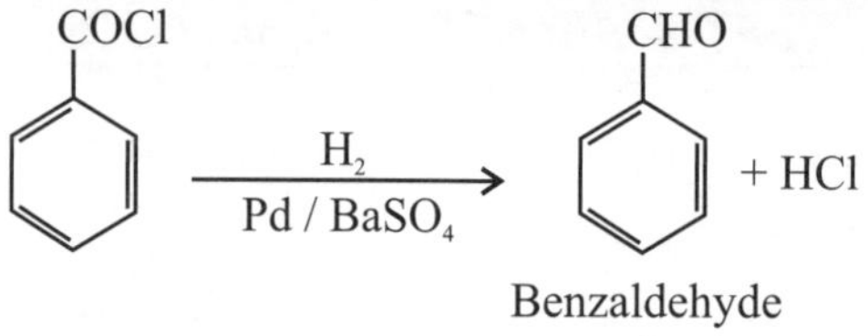

7. **(2)** $HCHO + CH_3MgBr \xrightarrow[\text{Catalyst}]{\text{Dry ether}} H-\underset{CH_3}{\underset{|}{\overset{H}{\overset{|}{C}}}}-OMgBr$

$\xrightarrow{H^+, H_2O} \underset{\text{(Primary alcohol)}}{CH_3CH_2OH} + MgBr(OH)$

8. **(4)** In Gattemann Koch synthesis, carbon monoxide & hydrogen chloride in the presence of anhydrous aluminium chloride or cuprous chemical is used as a reagent.

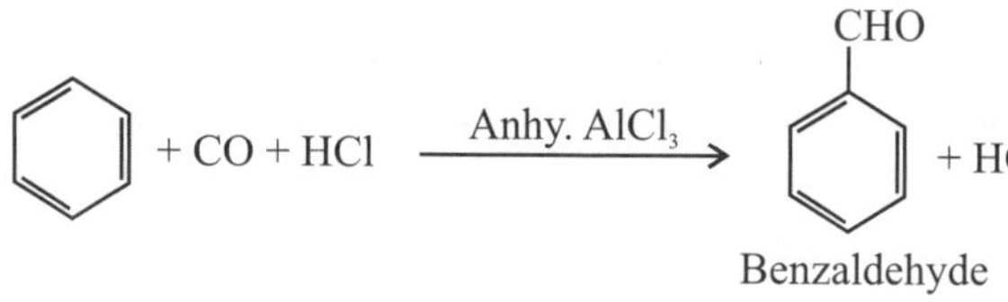

9. **(1)**

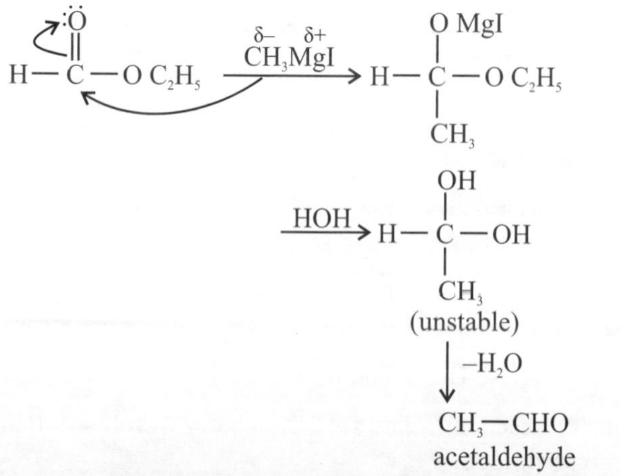

10. (3) $CH_3-C\equiv N \xrightarrow{C_2H_5MgI} CH_3-\underset{\underset{C_2H_5}{|}}{C}=NMgI$ (A) $\xrightarrow{H_3O^+} CH_3-\underset{\underset{C_2H_5}{|}}{C}=O$

Butan-2-one

11. (1) 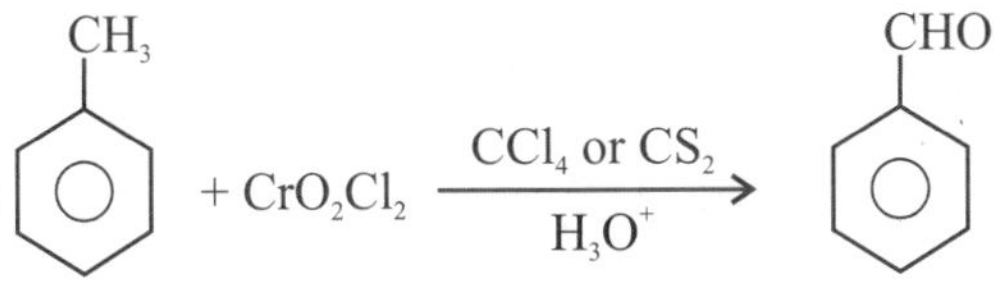

This reaction is known as Etard reaction.

12. (1) $RCOCl \xrightarrow[Pd/BaSO_4]{H_2} RCHO + HCl$

13. (1)

$$2CH\equiv CH + 3H_2O \xrightarrow[\text{Mg or zinc}]{420^\circ C} CH_3COCH_3 + CO_2 + H_2O$$

14. (4) Ketone is much less soluble than aldehyde because in ketone, both the sides bulky groups are attached whereas, in aldehyde only one side bulky group is present except $H-\overset{\overset{O}{||}}{C}-H$ (methanal).

Therefore, aldehyde possess more strong hydrogen bonding than ketone with water.

15. (1) In butan - 1 - ol, H_3C - CH_2 - CH_2 - CH_2 - OH intermolecular hydrogen bonding is present due to this, there is a large difference in boiling point.

16. (2) Acetaldehyde and Propyne can be distinguished by Iodoform test. Acetaldehyde reacts with iodine in presence of NaOH to form iodoform whereas propyne does not react with iodine.

17. (1) Nitriles are reduced to corresponding imine with stannous chloride in the presence of hydrochloric acid which on hydrolysis give corresponding aldehyde. It is Stephen's reaction.

18. (4) $H_3C-CH_2-OH \xrightarrow{LiAlH_4} CH_3CH_2CH_2OH$

$$H_3C-CH_2-\underset{\underset{O}{||}}{C}-O-CH_3 \xrightarrow{LiAlH_4} CH_3-CH_2-CH_2-OH + CH_3OH$$

$$H_3C-CH_2-O-CH_2-CH_2-CH_3 \xrightarrow{LiAlH_4} \text{No reduction possible}$$

$$H_3C-\underset{\underset{O}{||}}{C}-CH_3 \xrightarrow{LiAlH_4} H_3C-\underset{\underset{OH}{|}}{CH}-CH_3$$

Only $H_3C-\underset{\underset{O}{||}}{C}-R$ and $H_3C-\underset{\underset{OH}{|}}{C}-R$ gives iodoform test.

Hence, the last compound gives positive iodoform test.

19. (2) Due to electron withdrawing nature of $-NO_2$ group the partial positive charge on the carbon atom of the C=O group in p-nitrobenzaldehyde increases and becomes more susceptible to nucleophilic attack by the $CN^\ominus$ ion.

20. (4) Due to Aldol condensation as well as cross cannizaro's reaction, all the given products will form.

21. (3) Benzaldehyde 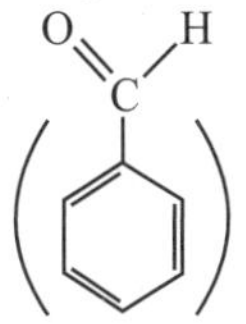 do not have α- hydrogen.

Thus; this will not show aldol condensation reaction.

22. (1) In FCH_2CHO group, strong nucleophilic addition reaction takes place because fluorine is a strong electron withdrawing in nature. This makes - CO group more polar.

23. (2) The compounds containing CH_3CO – OR $CH_3CH(OH)$– group give iodoform test.

24. (3) $CH_3CHO + CH_3CHO \rightarrow CH_3-\overset{\overset{OH}{|}}{\underset{\underset{CH_2-CHO}{|}}{C}}-H$

β–hydroxybutanal

25. (2) Acetaldehyde and acetone both react with a grignard reagent to give secondary & tertiary alcohols respectively. Whereas, fehling's Reagent, Tollen's reagent & schiff's reagent are reduced by only aldehydes.

$$\underset{\text{Acetaldehyde}}{H_3C-\underset{\underset{H}{|}}{C}=O} + CH_3MgX \longrightarrow H_3C-\overset{\overset{CH_3}{|}}{\underset{\underset{H}{|}}{C}}-OMgX \xrightarrow{H_2O} \underset{\text{Secondary alcohol}}{H_3C-\overset{\overset{CH_3}{|}}{\underset{\underset{H}{|}}{C}}-CH_3}$$

$$(H_3C)_2C=O + CH_3MgX \longrightarrow (H_3C)_2C(OMgX)(CH_3) \xrightarrow{H_2O} \underset{\text{Tertiary alcohol}}{(CH_3)_2C(OH)(H_3C)}$$

26. (1) $(CH_3)_2CH.CHO$ will undergo aldol condesation as-

$$CH_3-\overset{\overset{H}{|}}{\underset{\underset{CH_3}{|}}{C}}-CHO + H-\overset{\overset{CH_3}{|}}{\underset{\underset{CH_3}{|}}{C}}-CHO \xrightarrow{\text{dil. NaOH}}$$

$$\underset{\text{(Aldol)}}{CH_3-CH-\overset{\overset{OH}{|}}{CH}-\overset{\overset{CH_3}{|}}{\underset{\underset{CH_3}{|}}{C}}-CHO}$$

$(CH_3)_2CHCHO$ will undergo cannizzaro reaction as:

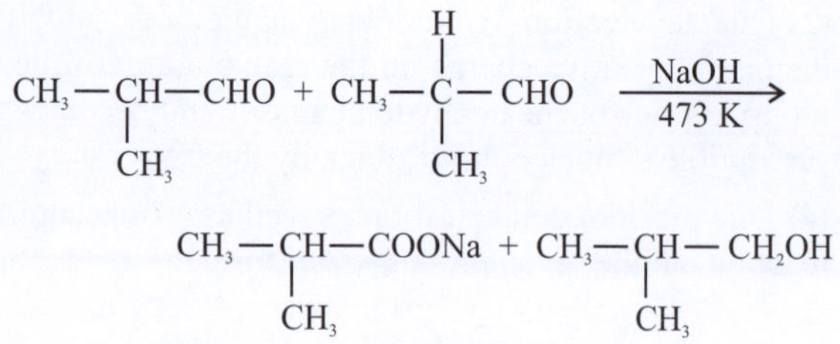

27. (3)

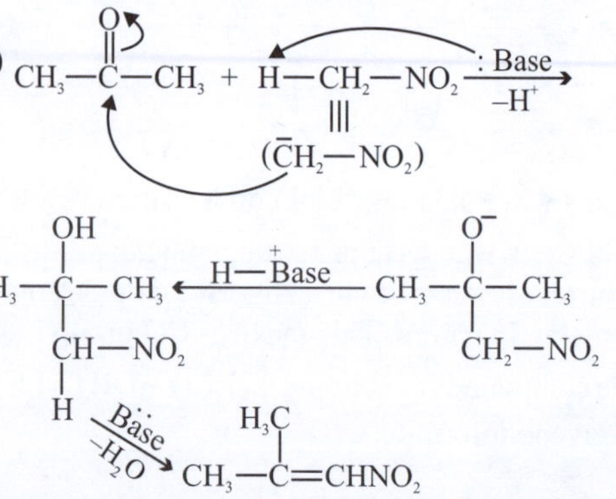

28. (3) The given reaction is a cannizzaro reaction. Aldehydes which do not have an α-hydrogen atom undergo self oxidation and reduction (disproportionation) reaction on heating with concⁿ alkali.

Mechanism:

Step-I:

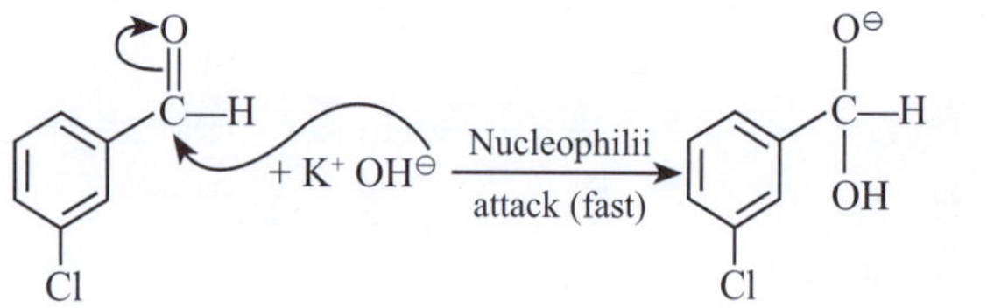

Step-II:

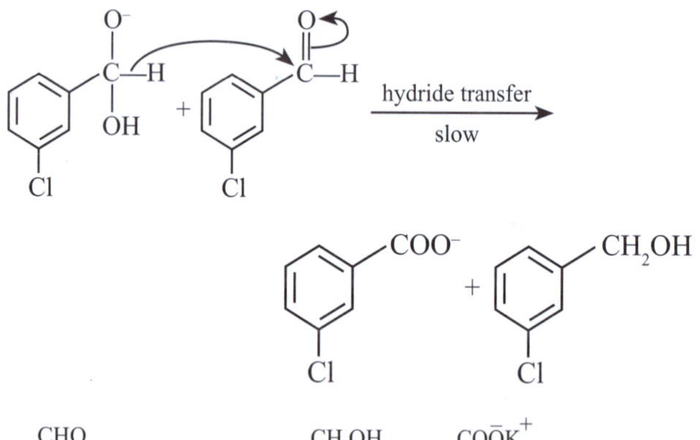

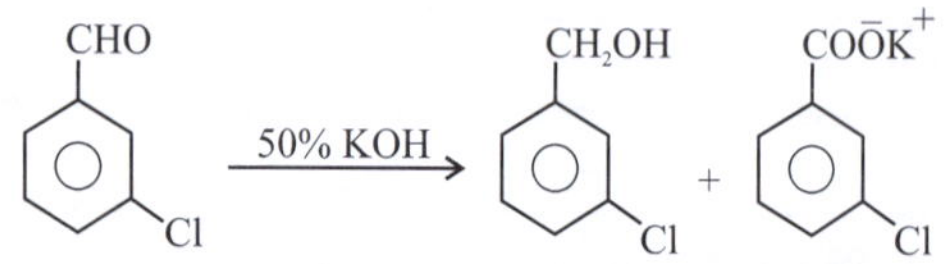

29. (1)

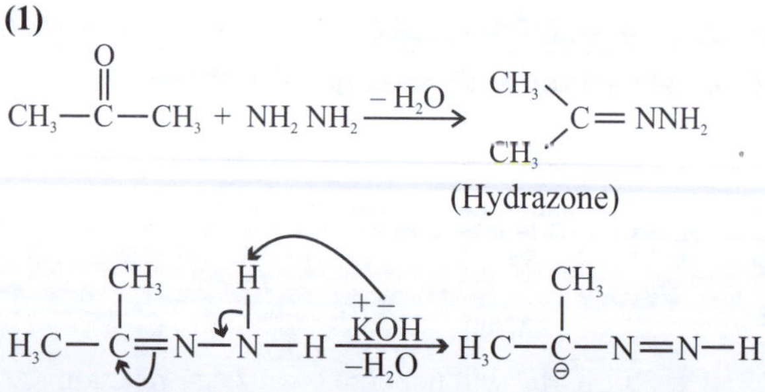

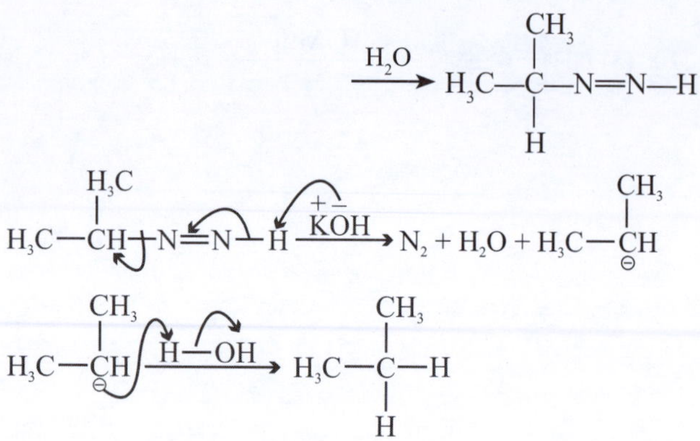

30. (4) Both compounds does not respond to Fehling's test Compound (B) do not have α - hydrogen & Ketone doesn't respond to Fehling test.

31. (4) Acetaldehyde has 3 α-hydrogens whereas formaldehyde has no α-hydrogen

32. (4) $C_2H_5-\overset{O}{\overset{\|}{C}}-CH_3 \xrightarrow[\text{Conc. HCl}]{\text{Zn – Hg}} \underset{\text{n-Butane}}{C_2H_5CH_2CH_3}$

33. (1) Aldehydes are more reactive than ketones.

34. (2) Both the given aldehydes are different therefore, they will undergo cross aldol reaction.

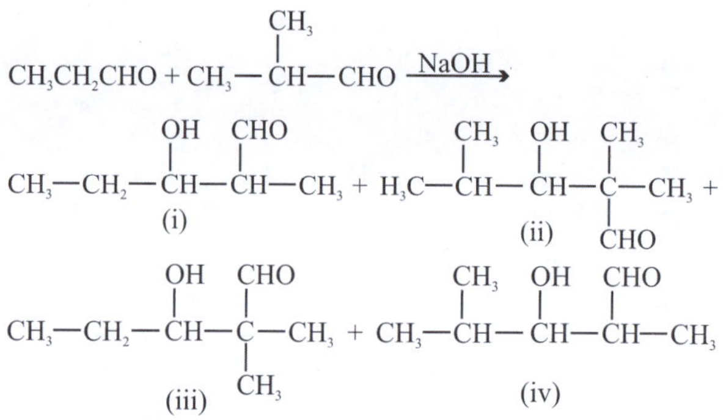

35. (4) α-hydrogen present on sp^3-hybridised carbon. In (1) & (2), the α-carbon is sp & sp^2 hybridised respectively. And the (3) have no α-Hydrogen. Therefore, 1 & 2 will not give aldol condensation.

36. (1) Amongst P & Q, O is more electronegative than N, so carbonyl C is more prone to attack by a nucleophile. R will be least reactive due to the presence of $-OCH_3$ group.

37. (4) When aldehyde/ketone reacts with $Na^+HSO_3^-$, 'S' atom acts as nucleophillic atom i.e.,

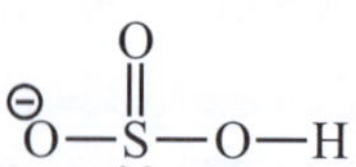

Nucleophile

The reaction is given as:

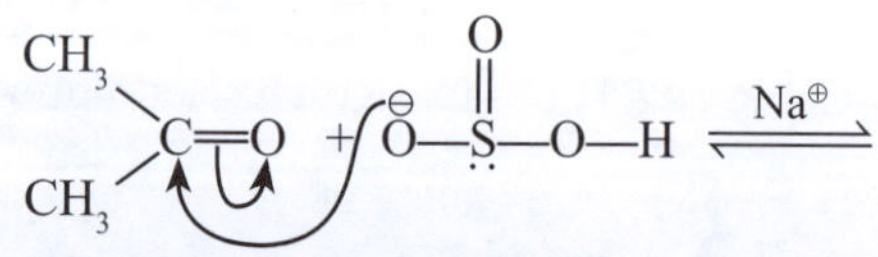

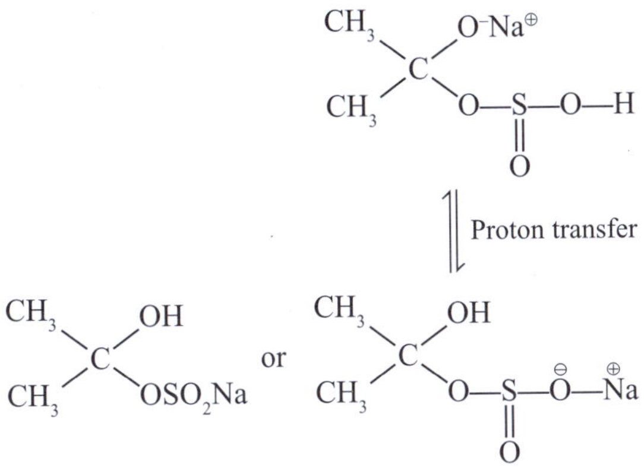

38. (1) β hydroxy aldehydes or β - hydroxy ketones readily undergo dehydration to give α,β - unsaturated aldehyde or ketones.

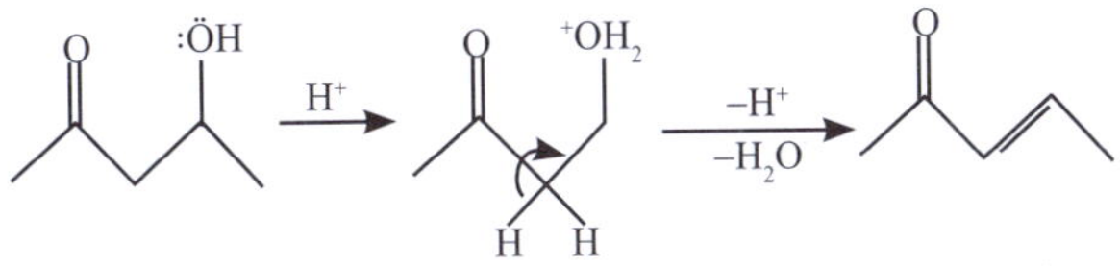

39. (1) Priority order of functional group is

$-COOH > -OH > -NH_2$

but according to lowest locant rule,

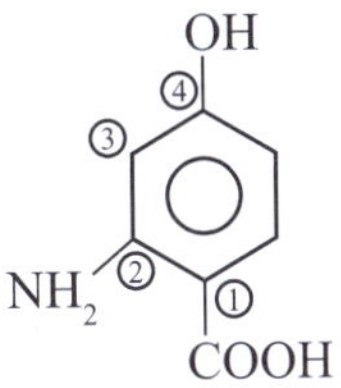

∴ IUPAC name of the compound is:

2 - amino - 4 - hydroxy benzoic acid.

40. (3) 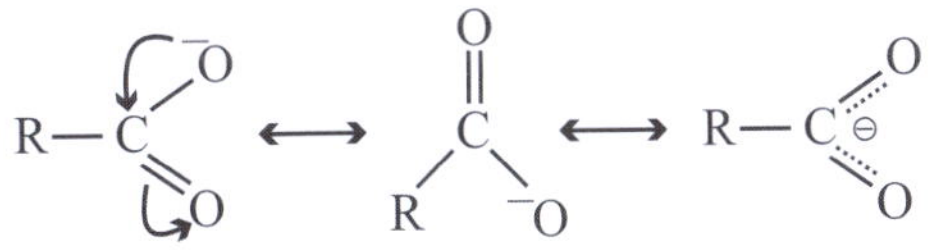

41. (2) $-\overset{\overset{O}{\|}}{C}-$ ⇒ sp^2 hybridization

42. (1)

$$CH_3CH{=}CHCHO \xrightarrow{NaBH_4} \underset{(X)}{CH_3CH{=}CHCH_2OH}$$

$$\xrightarrow[\text{ZnCl}_2 \text{ (Lucas Reagent)}]{HCl} \underset{(Y)}{CH_3CH{=}CHCH_2Cl}$$

$$\xrightarrow{(1)\ KCN,\ (2)\ H^+} CH_3CH{=}CHCH_2COOH$$

43. (3) (i) 2, 4-DNP test → carbonyl group present

(ii) Tollen's test → aldehyde group present

(iii) Cannizzaro reaction → No α-hydrogen present

(iv) Given 1, 4-Benzenedicarboxylic acid means benzene with 1,4 substitution is present in which one group is aldehyde.

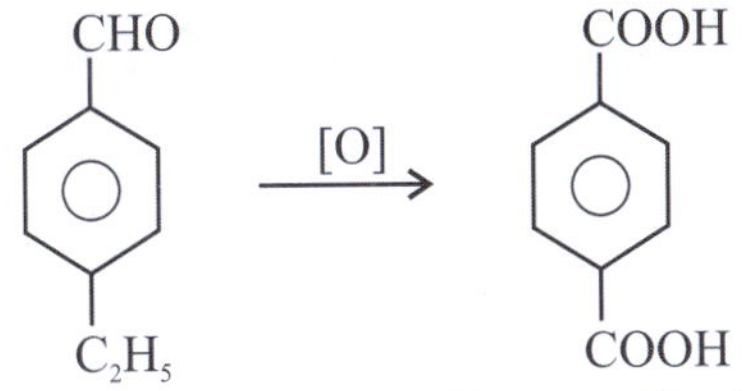

(X) 1, 4-Benzenedicarboxylic acid

44. (2) The reaction is:

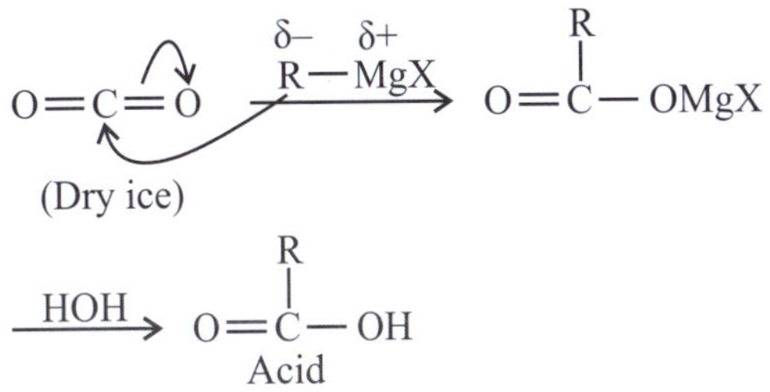

∴ R ≠ H means R will be any alkyl group always therefore formic acid cannot be obtained in this way.

45. (4) Esterification is a reverse of hydrolysis.

46. (4) $RCN \xrightarrow[NaOH]{HOH} R\,COONa + NH_3$

47. (4) $HCN \xrightarrow{\text{hydrolysis}} HCOOH$

48. (3)

$$CH_3CH_2OH \xrightarrow{PI_3} \underset{(A)}{CH_3CH_2I} \xrightarrow{KCN} \underset{(B)}{CH_3CH_2CN} \xrightarrow[H_3O^+]{HOH} \underset{(C)}{CH_3CH_2COOH}$$

49. (1) –Cl increases the polarity of $-\overset{\overset{O}{\|}}{C}-$ group because –Cl is electron withdrawing in nature.

Therefore; the acidic strength of the compounds are: $Cl_3C.COOH > Cl_2CHCOOH > ClCH_2COOH > CH_3COOH$

50. (3) 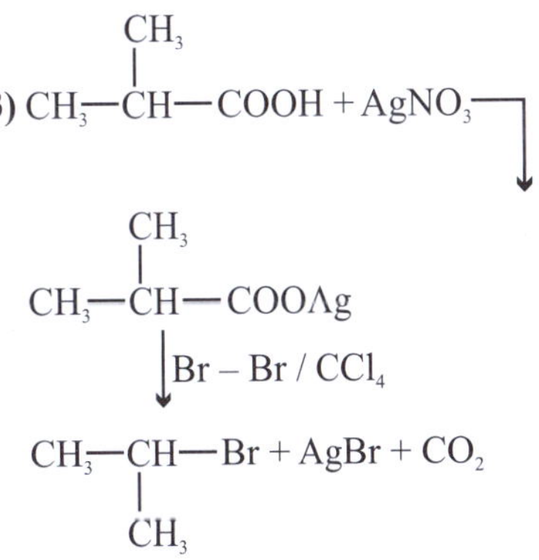

51. (1) CH_3COOH (acetic acid) is present in vinegar.

52. (3) Fluorine is stronger electron withdrawing among all given. It will increase the polarity and acidic strength of the carboxylic acid.

53. (4) Formic acid contains H – C = O or (–CHO) group and hence reduces ammonical silver nitrate. While other acids do not contain this group, therefore Tollen's test is not given by other carboxylic acids.

54. (3) Bond length order ⇒

Double bond < partial double bond < single bond

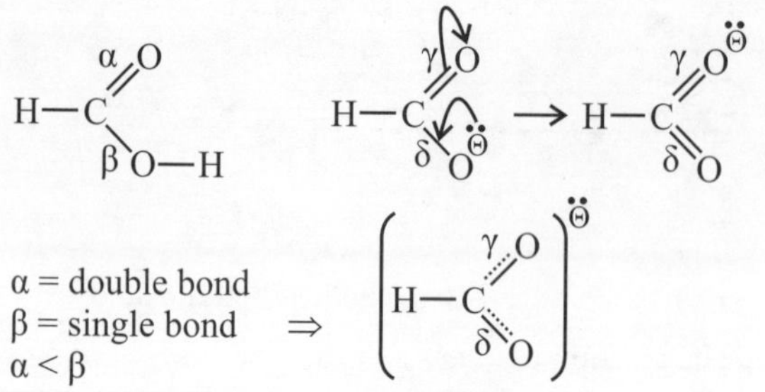

α = double bond
β = single bond
α < β

γ = δ partial double bond character

55. (3) HVZ reaction is given only by those acids which contain α - hydrogen atoms which are replaced by halogen atoms.

56. (2) In HVZ reactions; α - carbon which is attached to carboxyl group must have α - hydrogen.

$$RCHCOOH \xrightarrow[(ii)\,H_2O]{(i)\,X_2/Red\,P} R-\underset{X}{\underset{|}{CH}}-COOH\ (X = Cl, Br)$$

57. (1) Formic acid can reduce fehling's and tollen's reagent. Formic acid acts a reducing agent.

58. (4)

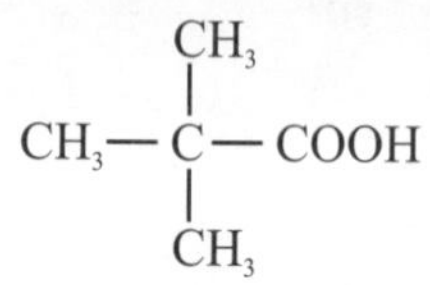

2, 2 - Dimethyl propanoic acid will not undergoe HVZ reaction due to the absence of α - H atom.

59. (2) $RCOOH + 4H \xrightarrow{LiAlH_4} RCH_2OH + H_2O$

60. (2)

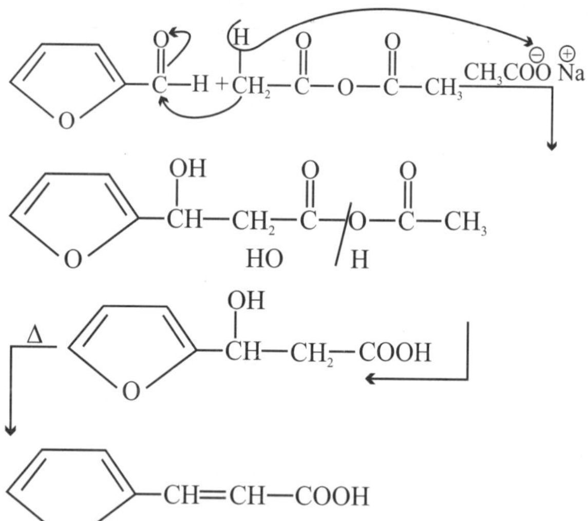

61. (3)

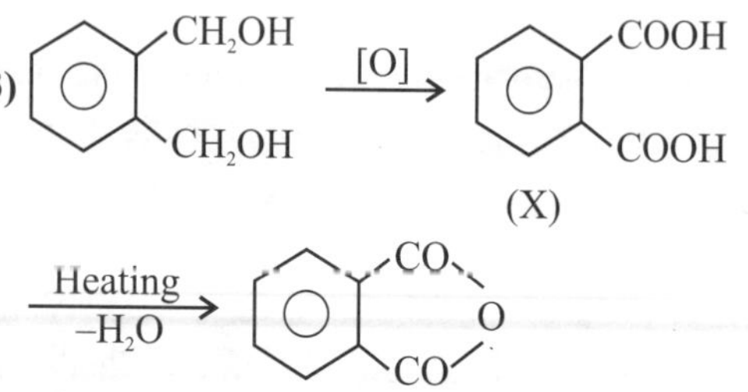

62. (2) $LiAlH_4$ does not affect the C = C or C ≡ C bond. Therefore; the product of reduction of $CH_2 = CH - COOH$ with $LiAlH_4$ is $CH_2 = CH - CH_2OH$.

63. (1) $CH_3-\overset{O}{\overset{||}{C}}-\boxed{NH_2} \rightarrow CH_3COOH + N_2 + H_2O$
$\boxed{NOOH}$

64. (2) Friedal - crafts alkylation: In presence of anhydrous $AlCl_3$, RX reacts with benzene and in friedal - craft acylation, RCOCl (acid chloride) or acid anlydride reacts with benzene in presence of anhydrous $AlCl_3$

65. (2)

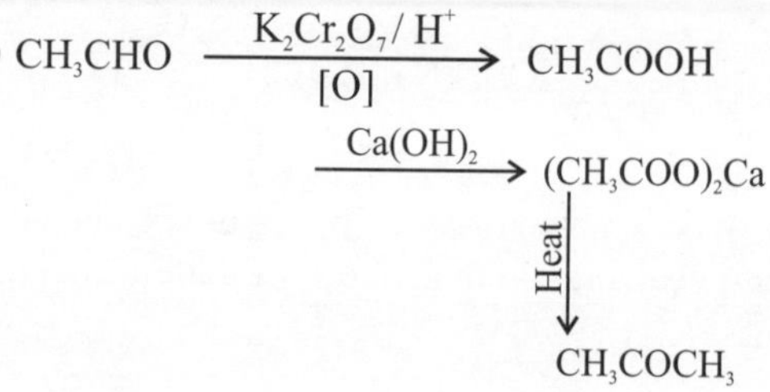

66. (1) $CH_3-\overset{O}{\overset{||}{C}}-O-CH_3$ is the most reactive towards saponification.

67. (3) Diazomethane is $^{-}CH_2 - N_2^{+}$, it is a source of methylene group, $-CH_2$.

$$R-\overset{O}{\overset{||}{C}}-OH + CH_2N_2 \longrightarrow \underset{\text{ester}}{R-\overset{O}{\overset{||}{C}}-O-CH_3} + N_2$$

68. (3) $CH_3-CH_2-COOH \xrightarrow[\text{red P}]{Cl_2} \underset{(X)}{CH_3-\underset{Cl}{\underset{|}{CH}}-COOH}$

$$\xrightarrow[\text{dehydro-halogenation}]{\text{alc. KOH}} \underset{(Y)}{CH_2{=}CH-COOH}$$

69. (3) The mechanism is:

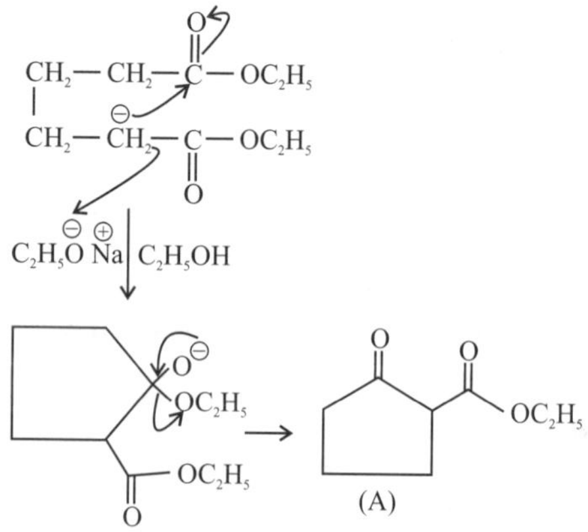

70. (3)

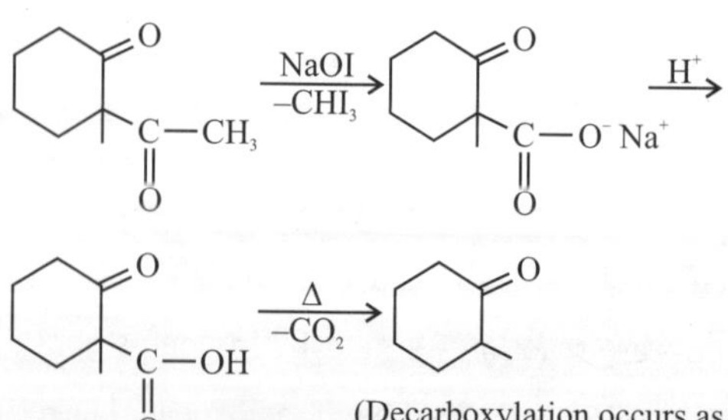

(Decarboxylation occurs as it is α, β– keto acid)

71. (1) Aldehyde & ketone gives 2DNP test $\left(CH_3-\overset{O}{\overset{\|}{C}}-\right)$ group gives positive iodoform test.

Option (1) fullfilling both the conditions.

72. (1) $Ph-CHO + HCHO \xrightarrow[\text{Cannizaro}]{\text{NaOH}}$

$Ph-CH_2OH + HCOONa$

Always HCHO undergoes Oxidation.

73. (1) Ph-CHO & 2,2 – dimethyl propanal have no α -H.

∴ Therefore, no aldol condensation reaction will occur.

74. (3) $pK_a \propto \frac{1}{K_a}$

$K_a \propto$ acidity, as no. of alkyl group increases, acidity decreases.

HCOOH (formic acid) has least pK_a value.

75. (4)

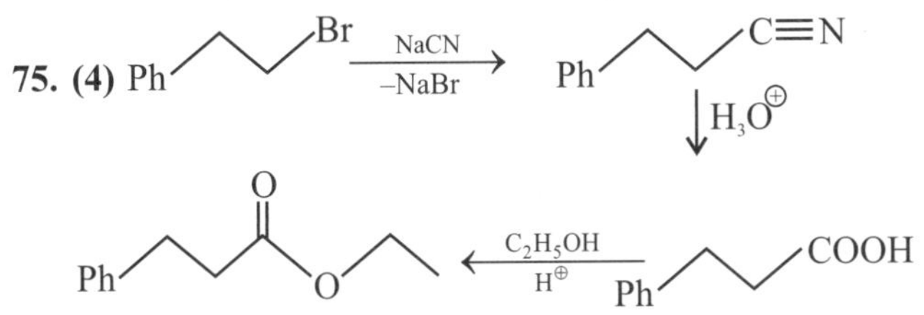

76. (3) In option (V), the conjugate base is stabilized by intramolecular H-bonding from 2 ends.

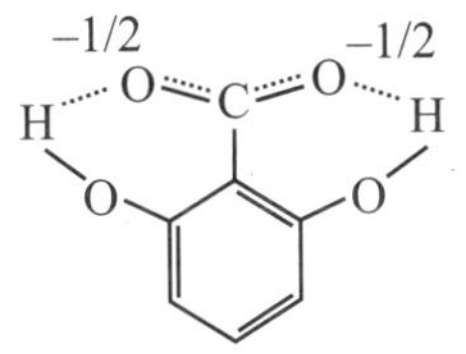

77. (2)

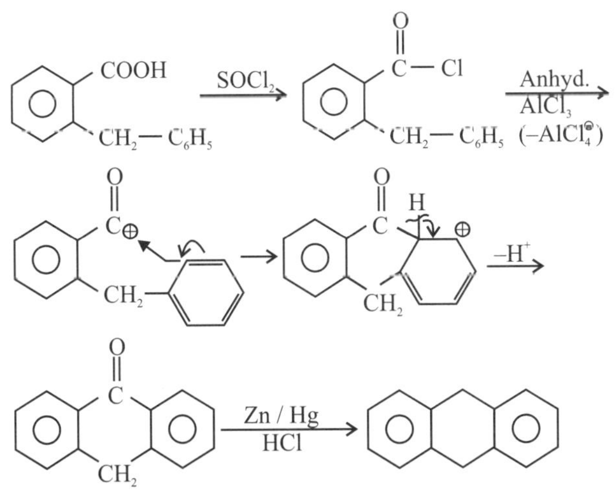

78. (4) Both 1 & 2 will not gives HVZ reaction because there is no α–H in (1) & (2).

79. (4) CH_3COCH_2COOH is a β-keto acid. Thus decarboxylation is maximum in a carboxylic acid containing an electron withdrawing group such as >CO ∝ –COOH at the β-carbon atom with respect to the –COOH group.

80. (1) Due to ortho effect.

Ortho substituted benzoic acid is stronger than benzoic acid and its meta-para substituted isomers regardless of the electronic influence of the substitutent. This effect is also known as ortho effect.

High Order Time Intensive MCQs

1. (4) Electron withdrawing group increases reactivity towards nucleophilic attack $-NO_2$ is an EWG.

2. (2) Reduction of $>C=O$ to –OH group followed by intramolecular condensation (removal of H_2O molecule).

3. (4)

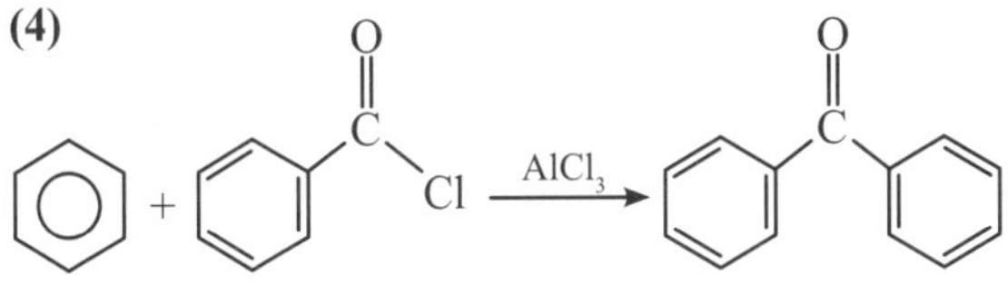

4. (2) Acidic hydrolysis of ether & ester takes place

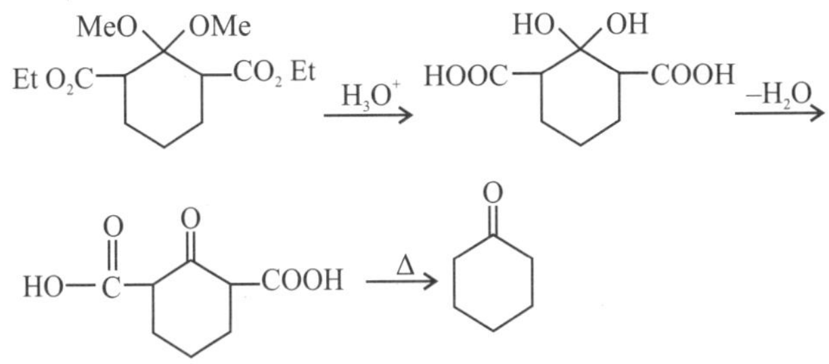

(β-keto acid)

5. (3)

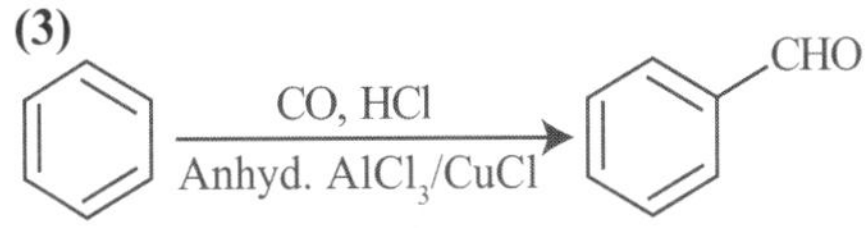

The Gattermann-Koch reaction is a chemical reaction that involves the synthesis of aldehydes from aromatic compounds using carbon monoxide (CO) and hydrogen chloride (HCl) in the presence of a catalyst such as copper(I) chloride (CuCl) or aluminum chloride ($AlCl_3$).

6. (2)

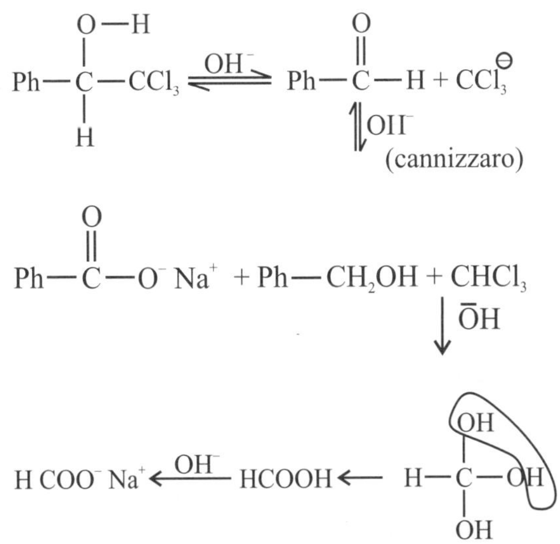

7. (3)

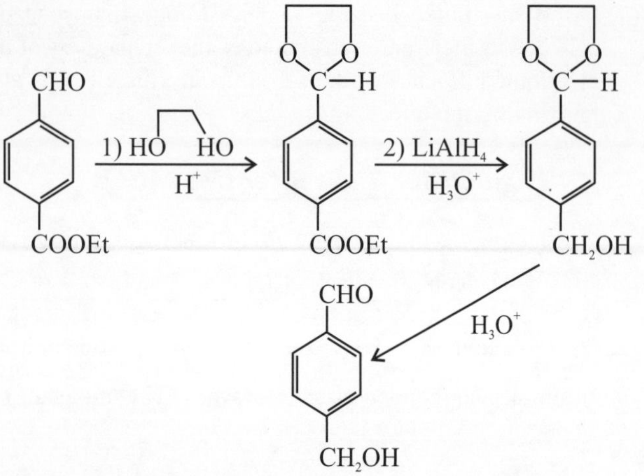

8. (1) DIBAL–H reduces ester & carboxylic acid into aldehyde at low temperature.

9. (2)

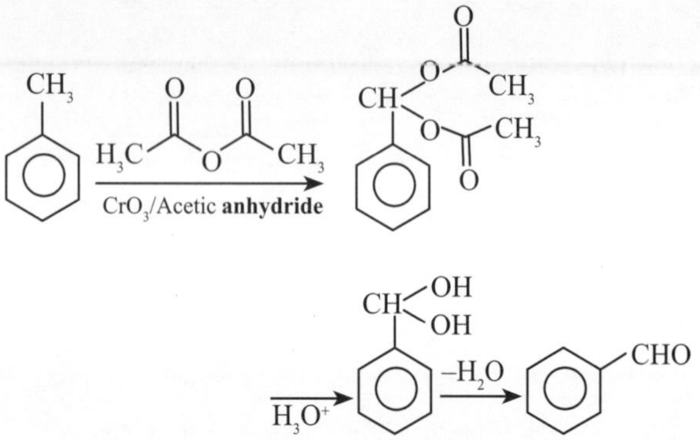

10. (2)

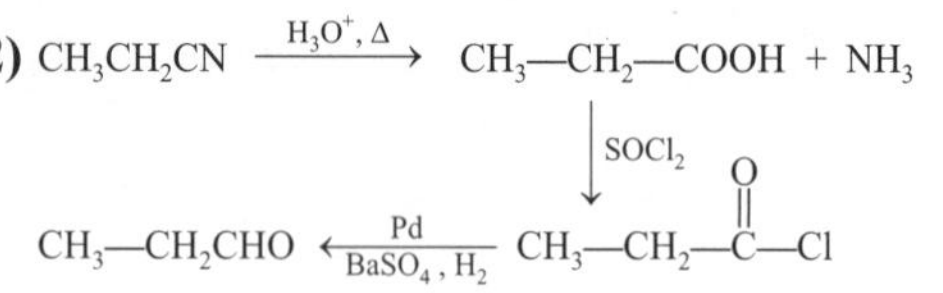

11. (4)

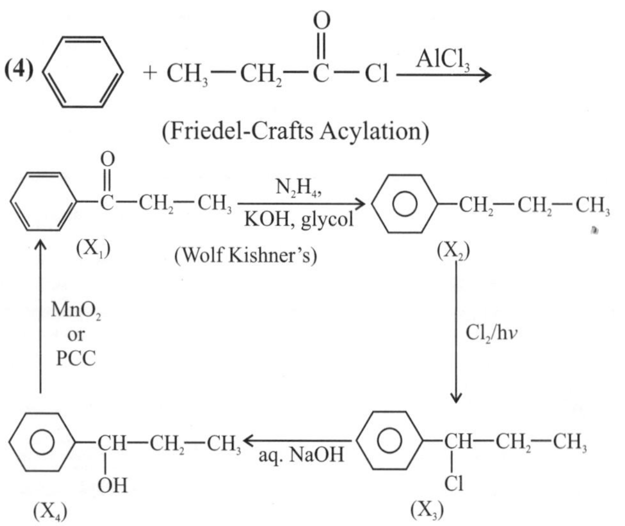

12. (2) Molybdenum oxide converts alkane to aldehyde i.e propane to propanal.

13. (2) The high boiling point of alcohols is attributed to the presence of hydrogen bonding between alcohol molecules.

14. (1)

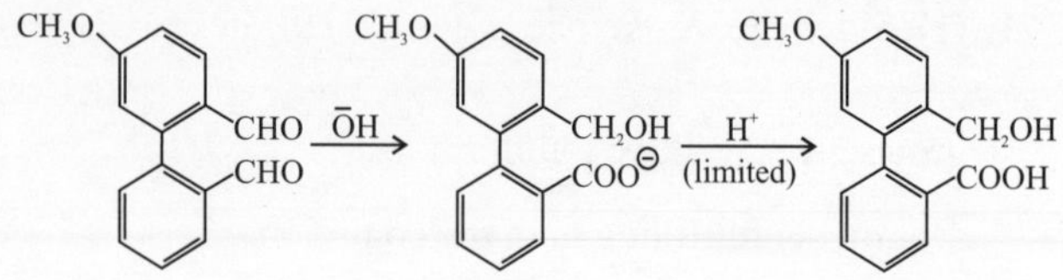

This reaction is an example of internal (intramolecular) cannizaro's reaction followed by acidification. The ring that lacks $-OCH_3$ group involves addition of nucleophile (–OH) which result in –CHO group getting oxidised there.

15. (4)

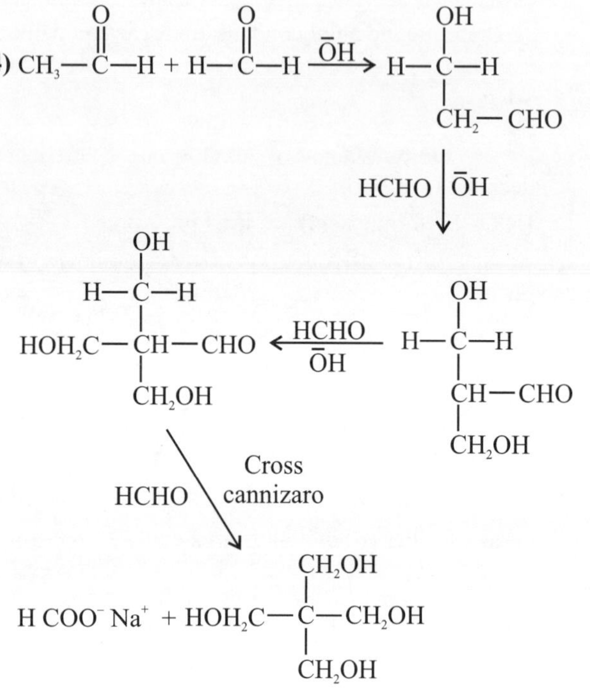

16. (4) Aldehydes give positive tollens test, Hemi-acetals also give positive tollen's test.

17. (3) Electron rich rings/groups increase the ease of hydride ion donation.

18. (1)

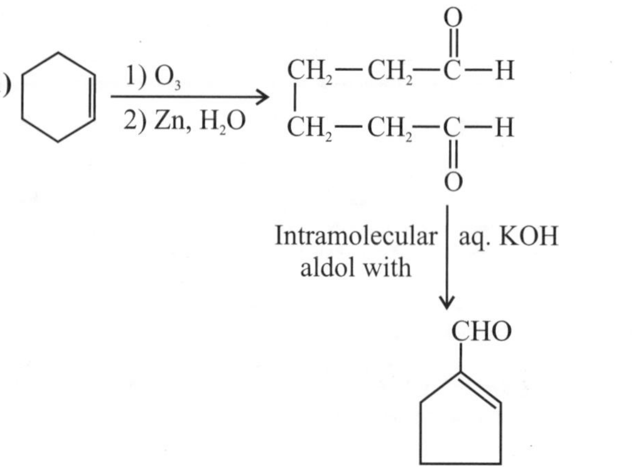

19. (4)

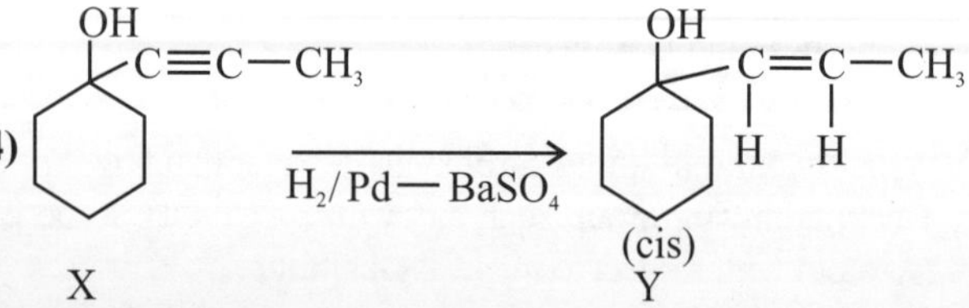

20. (4) α-H atoms involved in condensation (claisen- schmidt condensation).

21. (2) Haloform reaction than formation of acetylene.

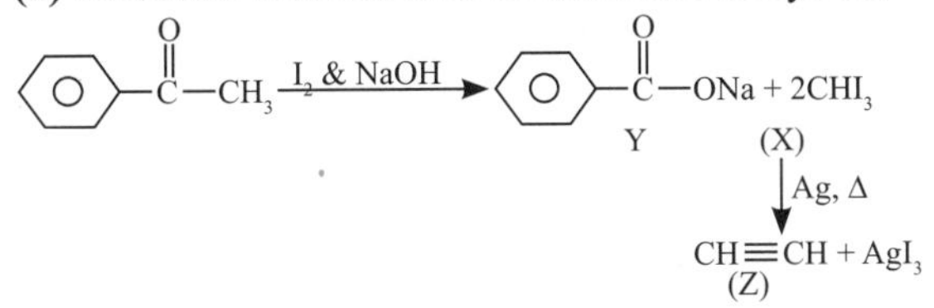

22. (2) Ozonolysis (A)

Aldol condensation reaction (B)

23. (2) Freidel craft acetylation reaction

24. (3)

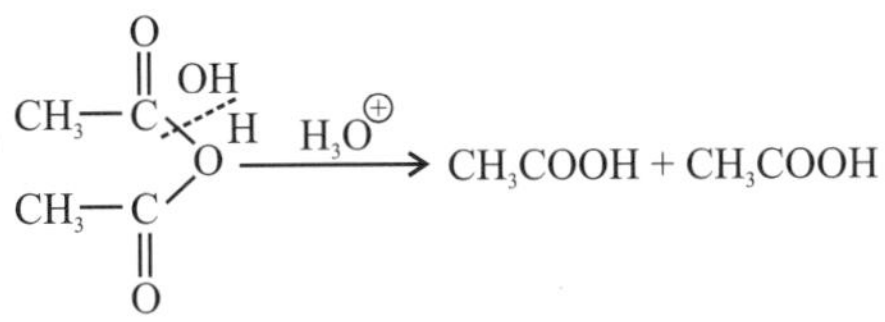

The mechanism of reaction is SN_{AE}.

25. (3) Wolf kishner reduction

Nucleophillic substitution reaction

26. (2)

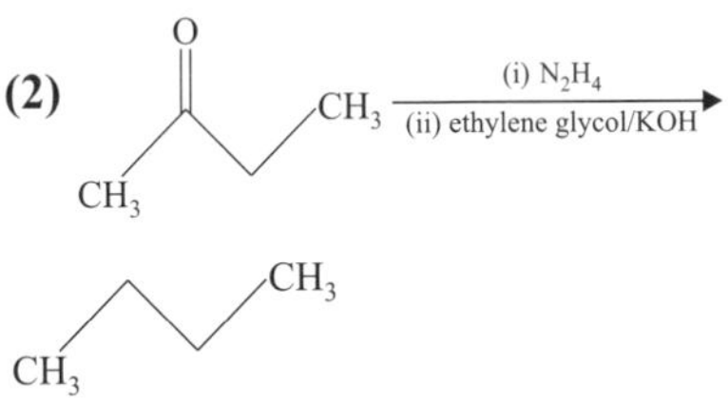

This is Wolf-Kishner reduction.

27. (4)

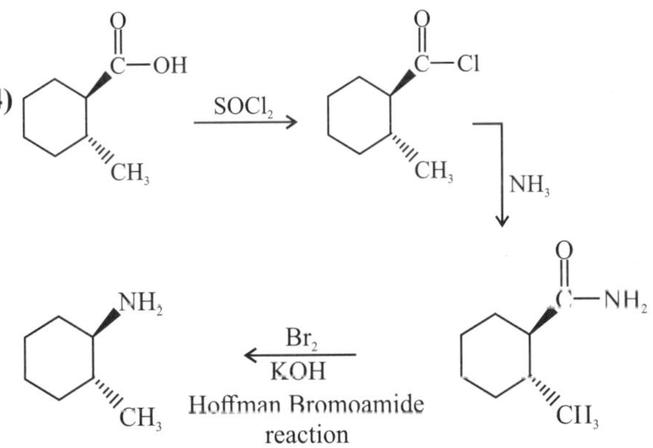

28. (4) In all the given reactions, decarboxylation process will involve.

29. (4)

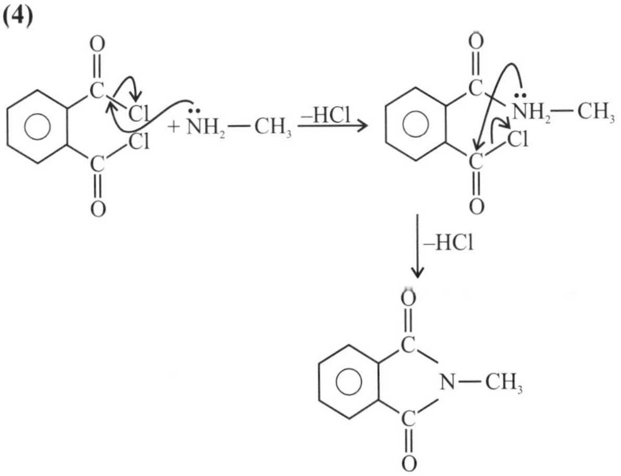

30. (1) (1,2 - addition occurs with hard nucleophiles)

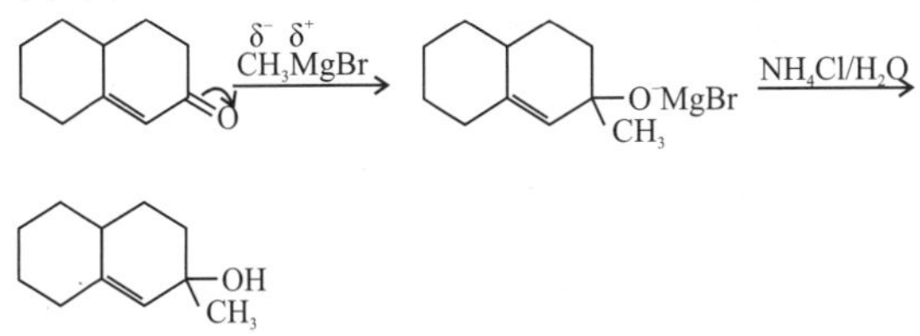

31. (2) $LiAlH_4$ reduces –COOH to $–CH_2OH$ without effecting C=C bond.

32. (3) Acid hydrolysis of an ester gives two different organic compounds.

$$CH_3COOC_2H_5 + H_3O^+ \rightarrow CH_3COOH + C_2H_5OH$$

While the acid hydrolysis of an amide gives one organic and one inorganic product as:

$$CH_3CONH_2 + H_3O^+ \rightarrow \underset{\text{organic compound}}{CH_3COOH} + \underset{\text{Inorganic compound}}{NH_3}$$

33. (3) It is an example of addition-elimination reaction via formation a carbanion. So, presence of EWG will stabilise the carbanion & accelerate the rate of reaction.

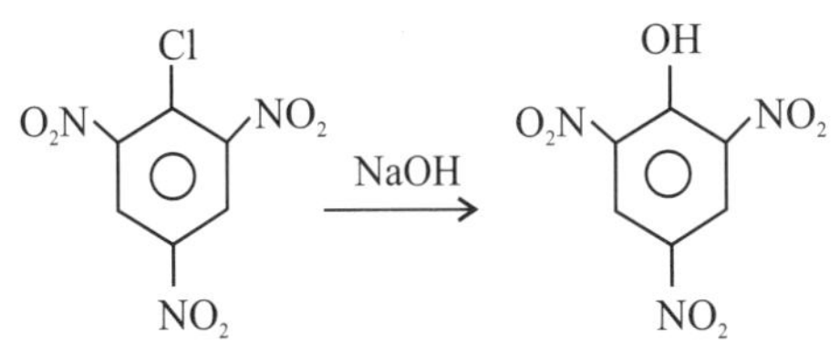

34. (3) Phthalic acid is stronger acid due to intramolecular hydrogen bonding.

35. (1) $CH_3COOH \xrightarrow{SOCl_2} CH_3COCl \xrightarrow[\text{Anhy. AlCl}_3]{\text{Benzene}}$

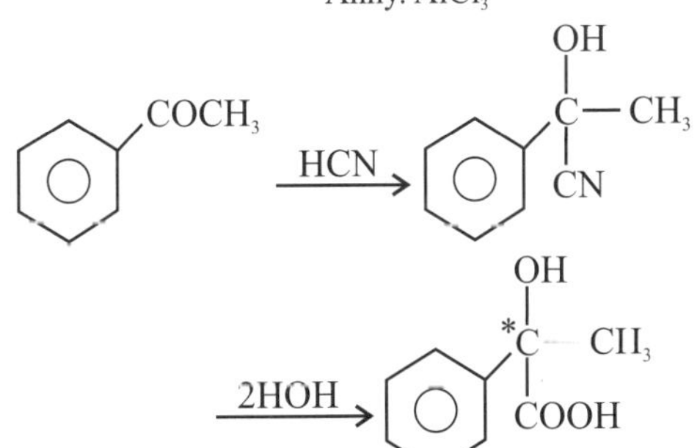

36. (3)

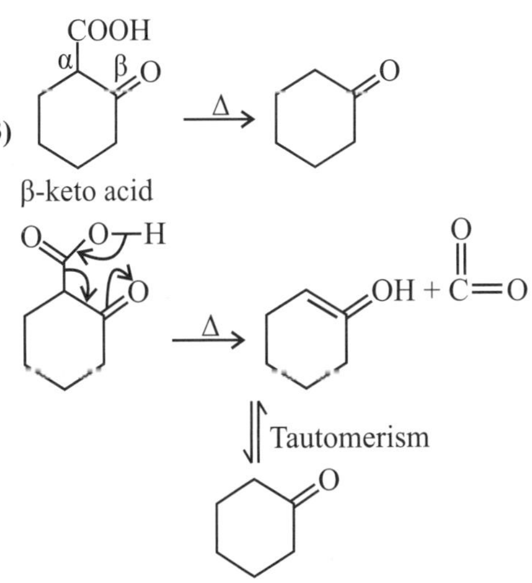

37. (1)

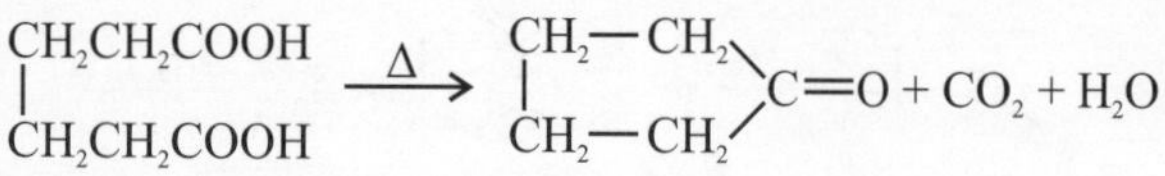

38. (4) $Ph-\overset{O}{\overset{\|}{C}}-CH_3$ gives +ve Iodoform test. But $Ph-\overset{O}{\overset{\|}{C}}-Ph$ cannot give iodoform test.

39. (1) $H-\overset{O}{\overset{\|}{C}}-H$ has the low steric hindrance at carbony1 carbon (–C=O) and high partial positive charge (+ve) at carbony1 carbon (–C=O).

40. (3) 2,4- DNP (Brady's Reagent) is used to identity carbonyl compounds.

41. (3) Acetophenone is $CH_3-\overset{O}{\overset{\|}{C}}-Ph$. It has α – H so undergoes Aldol. It has $CH_3-\overset{O}{\overset{\|}{C}}-$ group so, undergoes Iodoform.

42. (3) 3 Molecules of propanone undergoes condensation.

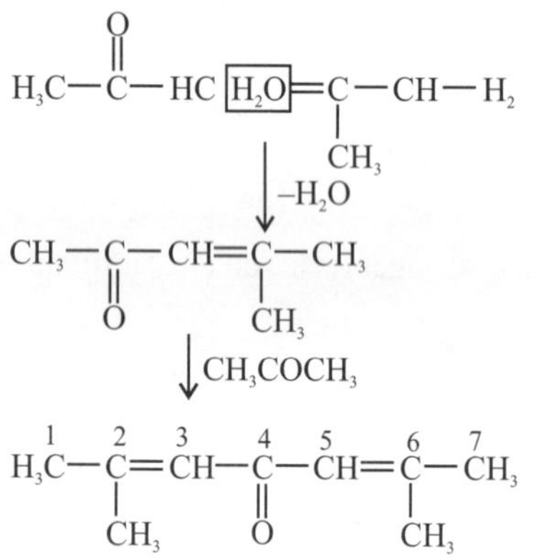

2,6-dimethyl -2,5 –heptadien -4 one

43. (3) This is Wolff-Kishner reduction.

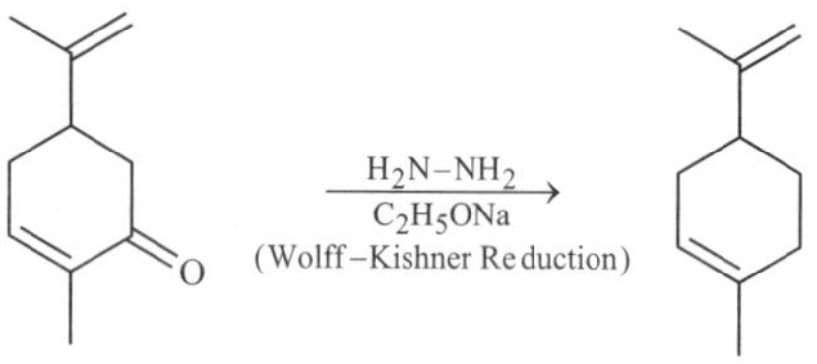

44. (4) Cannizaro reaction

45. (4) $CH_3-\overset{O}{\overset{\|}{C}}-OCH_3 \xrightarrow{LiAlH_4} CH_3CH_2OH + CH_3OH$

$LiAlH_4$ is powerful reducing agent. It reduces ester but other are weak reducing agent.

46. (3) $LiAlH_4$ is a strong reducing agent. It will reduce ester ketone and aldehydes to the corresponding alcohols.

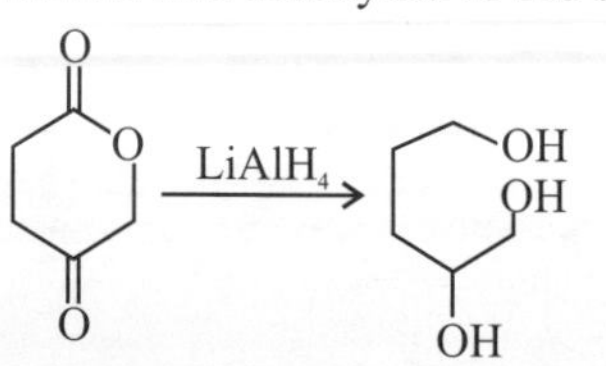

$NaBH_4$ is a mild reducing agent therefore, it will reduce only aldehydes and ketones to the corresponding alcohols.

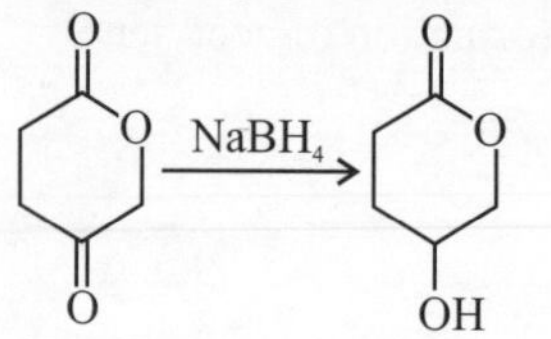

$NaBH_4$ cannot reduce esters

47. (1)

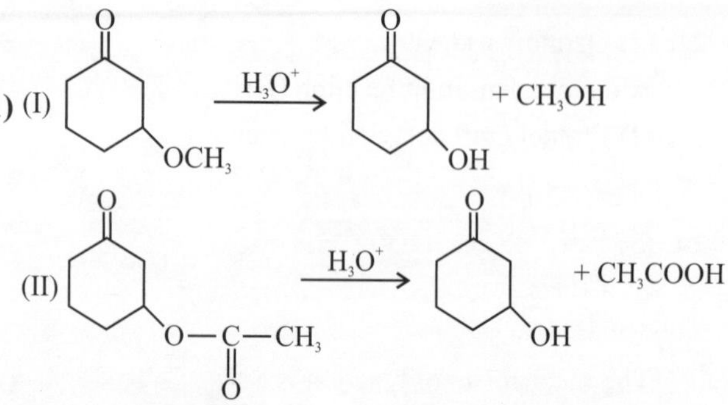

Wolf Kishner reduction is not used in presence of base sensitive group.

48. (3) Mono Nitration (Any Electrophilic substitution) reaction determine by +M group present in Benzene nucleus.

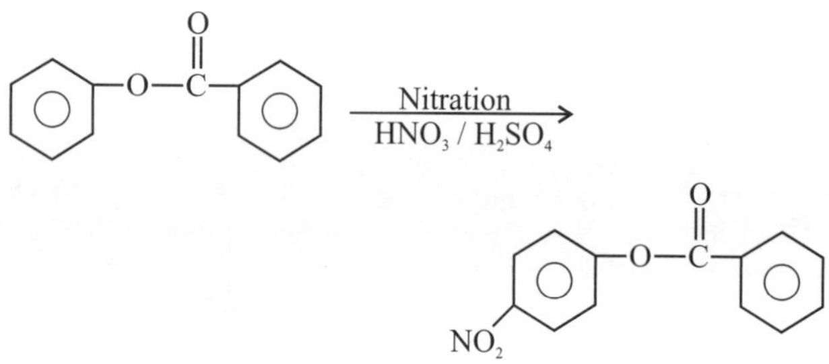

49. (2) Increasing hindrence and electron releasing group (ERG) attached on carboxyl group decrease the reactivity. Correct order of nucleophilic addition reaction is:

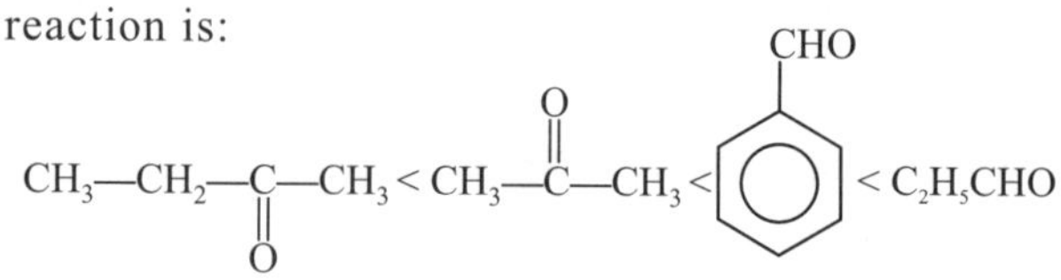

50. (3) Intramolecular Cannizzarro's reaction

51. (1)

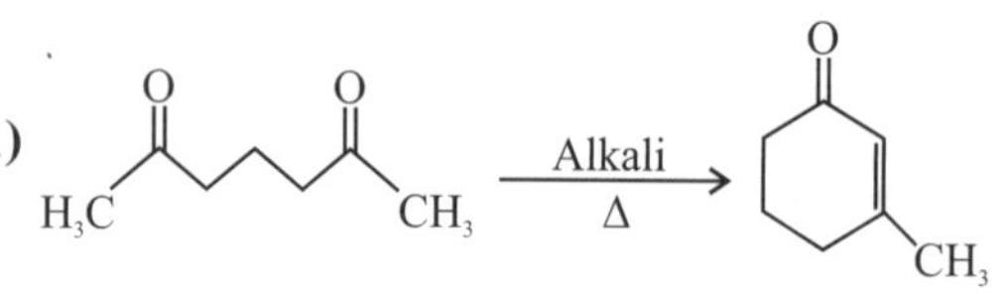

52. (4) A semicarbazone is a derivative of imines formed by a condensation reaction between a ketone or aldehyde and semicarbazide.

Oximes can be synthesized by condensation of an aldehyde or a ketone with hydroxylamine.

53. (1) As per, Hell Volhard Zelinsky reaction,

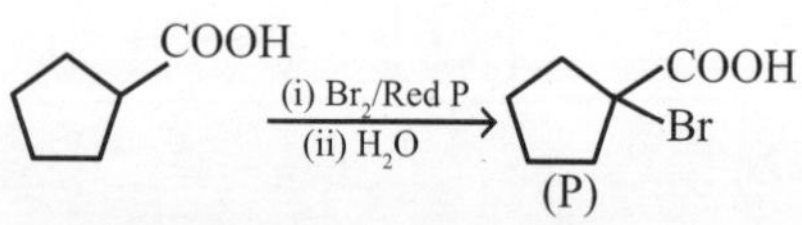

Correct & Incorrect MCQs

1. **(1)** The Iupac name of isobutyraldehyde 2-methylpropanal.
2. **(4)** In carbonyl compounds, the carbonyl carbon is an electrophilic (Lewis acid), and carbonyl oxygen, a nucleophilic (Lewis base) centre.
3. **(2)** Addition of water to ethyne in the presence of H_2SO_4 and $HgSO_4$ gives acetaldehyde. All other alkynes give ketones in this reaction.
4. **(4)** The boiling points of aldehydes and ketones are lower than those of alcohols of similar molecular masses due to absence of intermolecular hydrogen bonding.
5. **(2)** The lower members of aldehydes and ketones like methanal, ethanal and propanone are miscible with water in all proportions. Hence, option (2) is incorrect.
6. **(1)** Due to steric and electronic reasons, aldehydes are generally more reactive than ketones in nucleophilic addition reactions.
7. **(1)** Aldehydes and ketones are reduced to primary and secondary alcohols respectively by sodium borohydride ($NaBH_4$) or lithium aluminium hydride ($LiAlH_4$) as well as by catalytic hydrogenation.
8. **(4)** Aromatic aldehydes do not respond to Fehling's test.
9. **(4)** Benzaldehyde do not contain α - hydrogen atom, hence it would not undergo Aldol condensation reaction rather it would undergo Cannizzaro reaction.
10. **(2)** Carboxylic acids are higher boiling liquids than those of aldehydes, ketones and even alcohols of comparable molecular masses. This is due to more extensive association of carboxylic acid molecules through intermolecular hydrogen bonding.

Statement Based MCQs

1. **(3)** –COOH is a meta directing group, as it is an electron withdrawing group and it withdraw the electron from benzene ring through resonance effect, reducing electron density at the benzene ring. And it makes benzene ring less susceptible to electrophilic attack.
2. **(1)** Since it is sp^2 hybridised it has trigonal planar shape.
3. **(1)** F is strong electron withdrawing group than Cl, it lowers the electron density on OH bond which makes it easier to give up an H^+or proton i.e. it becomes more acidic.
   As, carboxylic acids are weak acids, hence turns blue litmus red.
4. **(4)** Instead of ethanal, it is synthesized using methanal or formaldehyde & ammonia.
5. **(1)** Formic acid has aldehyde group. Hence, formic acid reduces Tollen's reagent.
   Tollen's reagent is an alkaline solution of ammonical silver nitrate.
6. **(2)** The α–hydrogen atom in carbonyl compound is acidic in nature. The acidity of α– hydrogen atoms of carbonyl compounds is due to the strong electron withdrawing effect of the carbonyl group and resonance stabilisation of the conjugate base.
7. **(3)** Aldehyde reacts with tollen's reagent to form silver mirror but not ketones.
8. **(3)** Aliphatic aldehydes reduce Fehling solution but aromatic aldehydes do not reduce fehling solution, hence aliphatic aldehyde can be differentiated from aromatic aldehyde by Fehling's test.
9. **(1)** In Cannizzaro reaction, one molecule of aldehyde which do not have α–hydrogen atom, is reduced to alcohol while another one is oxidised to carboxylic acid. Hence, Cannizzaro reaction is disproportionation reaction.
10. **(1)** In Etard reaction, chromyl chloride oxidises methyl group to a chromium complex, which on hydrolysis gives corresponding benzaldehyde. It helps to stop the oxidation of toluene at the aldehyde stage.

Assertion & Reason MCQs

1. **(1)** In Hvz reaction, halogen group is attached on th α – position of the carboxylic group. Due to electronegativity of halogen group, it will increase the acidity of carboxylic acid.
2. **(2)** Compound containing –CHO groups are easily oxidised in the presence of mild oxidising agent and carboxylic acid reduced in to alcohol in the presence of strong reducing agent i.e., $LiAlH_4$ both statements are not related to each other.
3. **(4)** α – H in carbonyl carbon is more acidic and the anion formed after loss of α – H is resonance stabilised.
4. **(3)** Aromatic aldehydes and formaldehydes both undergoes cannizzaro reaction due to absence of α-H. But formaldehyde is more reactive than benzaldehyde.
5. **(4)** Only aldehyde reacts with Tollen's reagent while both of them have carbonyl group.
6. **(3)**

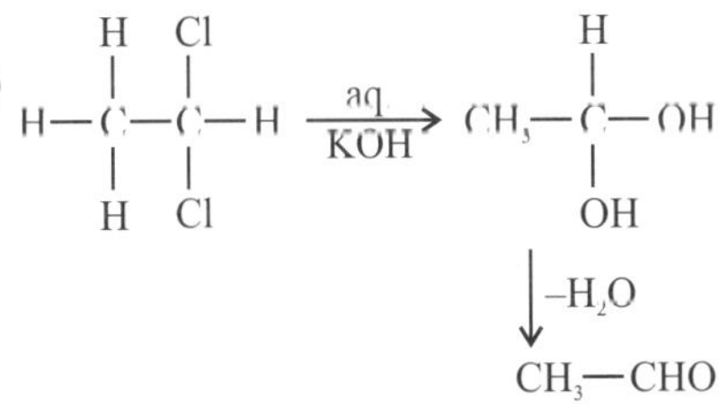

(ethanal)

Ethylidene chloride is a gem dihalide.

7. **(1)** As, (–OC_2H_5) group is more electron with drawing than –NH_2 group due to which there is more +ve charge on the carbonyl carbon in ethyl acetate than that of acetamide, therefore, making it more reactive towards nucleophilic substitution reaction (NSR) than acetamide. Further, -OMe group is a better leaving group than NH_2.
8. **(3)** Due to conjugation, benzaldehyde is less reactive than acetaldehyde.
9. **(3)** 2-Pentanol can give iodoform test. 2-pentanol and 3-pentanol are secondary alcohols.
10. **(2)** Formic acid is stronger acid than acetic acid because acetic have +I effect of alkyl group.

Multi Correct MCQs

1. **(2)** Formic acid, H–C(=O)–OH also gives silver (Ag) mirror test with ammonical silver nitrate, apart from aldehydes.

2. **(4)** Compounds having $CH_3–\overset{O}{\overset{\|}{C}}–$ or $CH_3–\underset{OH}{\underset{|}{CH}}$ group can give positive iodoform test

A. 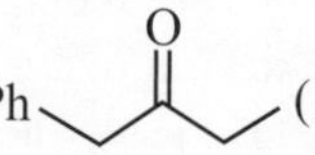 (1-Phenylbutan-2-one)

B. 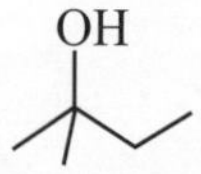 (2-Methylbutan-2-ol)

C. 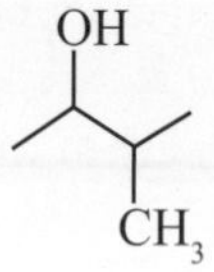 (3-Methylbutan-2-ol)

(positive for iodoform)

D. $Ph–\underset{OH}{\underset{|}{CH}}–CH_3$ (1-Phenylethanol)

(positive for iodoform)

E. 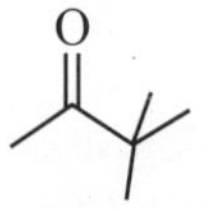 (3,3-dimethylbutan-2-one)

(positive for iodoform)

F. 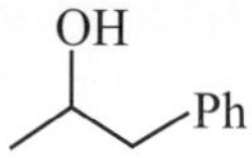 (1-Phenylpropan-2-ol)

(positive for iodoform)

3. **(3)** Except structure (II), all can produce –CHO functional group which gives +ve Tollen's test. In structure (II), tautomerisation will take place.

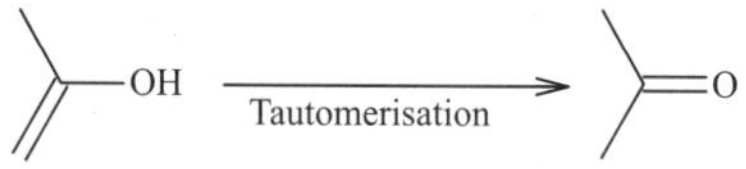

4. **(3)** The following reactions takes place as:

(*a*)

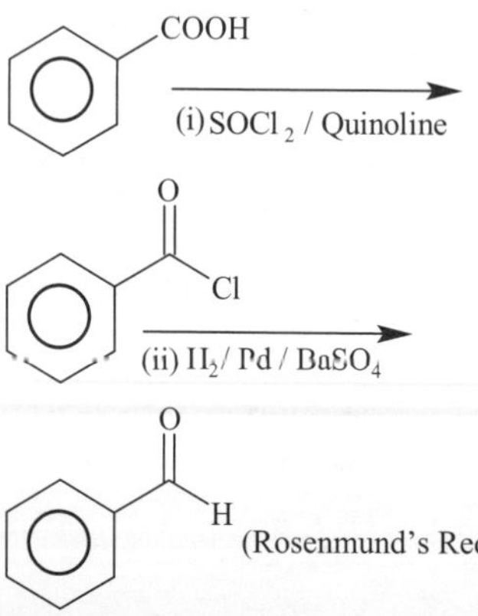

(Rosenmund's Reduction)

(*b*)

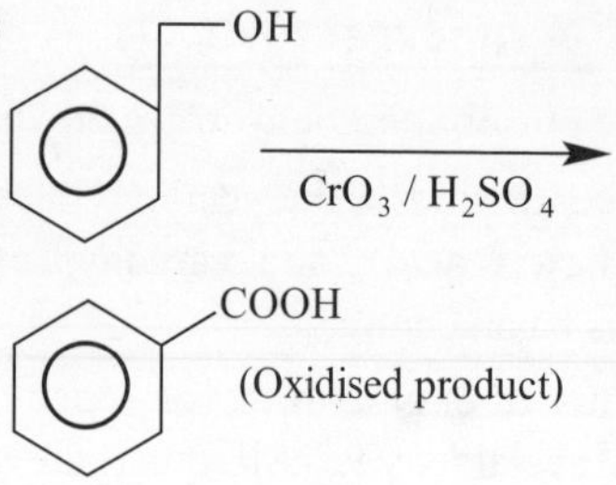

(*c*)

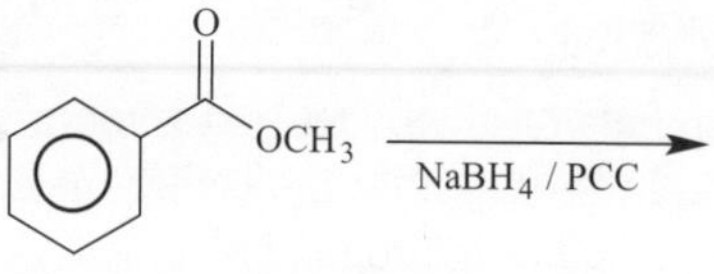

No reaction as $NaBH_4$ cannot reduce esters.

(*d*)

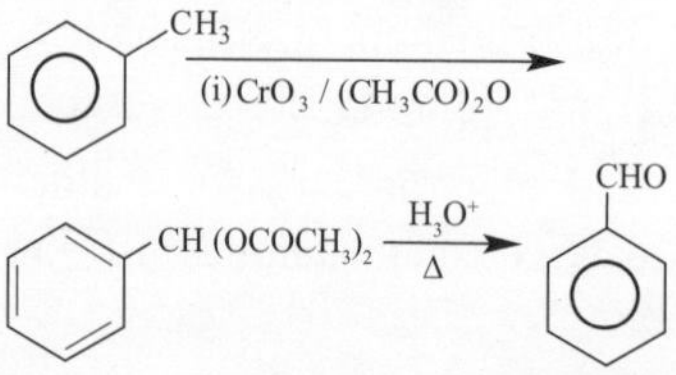

Benzaldehyde

5. **(2)** Alkyl groups are electron-donating, reducing acidity by destabilizing the conjugate base. Since 2-methylpropanoic acid (A) has more alkyl groups near the COOH than 3-methylbutanoic acid (C), and acetic acid (B) has the least, the acidity order is A < C < B.

NCERT Exemplar MCQs

1. **(2)** But-1-yne on reaction with water in presence of Hg^{2+} ions as catalyst produces butan-2-one. The reaction is given as-

$$H_3C—CH_2—C{\equiv}CH + H_2O \xrightarrow[HgSO_4]{dil.\ H_2SO_4} H_3C—CH_2—\overset{O}{\overset{\|}{C}}—CH_3$$

2. **(1)** Reactivity of carbonyl compounds can be, decided by two factors:

(i) Electronic factor:- greater the number of alkyl groups, lesser will be its electrophilic affect.

(ii) Steric factor:- Lesser the steric factor, greater will be its reactivity.

CH_3CHO is most reactive towards nucleophilic addition reactions. Carbonyl compounds are polar with positive charge on carbon atom which is attacked by nucleophiles. 2-ERG in ketones make carbon less e^- deficient than aldehydes. Benzene ring exhibits +R effect which thereby decreases the ease of nucleophilic addition reaction in benzaldehyde & acetophenone. Hence, the reactivity order is-

$$\underset{H}{\overset{H}{>}}\overset{\delta+}{C}=\overset{\delta-}{O} \;>\; \underset{R}{\overset{H}{>}}\overset{\delta+}{C}=\overset{\delta-}{O} \;>\; \underset{R}{\overset{R}{>}}\overset{\delta+}{C}=\overset{\delta-}{O}$$

3. **(3)** Due to formation of more stable conjugate base after removal of H from phenol, therefore phenol is more stable than alcohol. On the other hand, carboxylic acid is more acidic than phenol due to formation of more stable conjugate base after removal of H as compared to phenol. Chloroacetic acid is more acidic than acetic acid due to the presence of electron withdrawing chlorine group attached to a carbon of carboxylic acid.

Hence, correct choice is (3).

4. **(2)** Compound Ph—COO—Ph can be prepared by the reaction phenol and benzoyl chloride in the presence of pyridine. The reaction is given below-

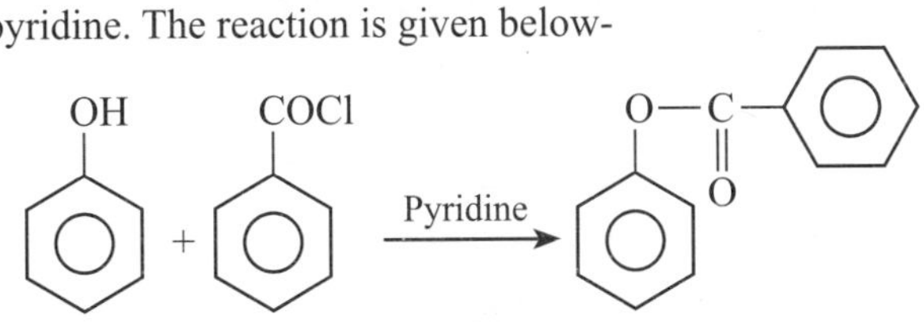

This is an example of Schotten-Baumann reaction.

5. **(3)** Fehling's solution does not react with ketones and aromatic aldehydes. Hence, both acetone and benzaldehyde do not react with fehling's solution, as acetone is a ketone and benzaldehyde is an aromatic aldehyde having absence of α-hydrogen.

6. **(4)** Necessary condition for Cannizzaro reaction is absence of alpha - hydrogen. So, CH_3CHO cannot give Cannizaro reaction, while other three compounds have no α-hydrogen. Hence, they will give Cannizaro reaction.

7. **(2)** On reaction of aqueous KOH solution with benzaldehyde, it undergo Cannizzaro reaction to produce benzyl alcohol and potassium benzoate because benzaldehyde is an aromatic aldehyde having no α-hydrogen.

NEET & AIIMS Past Year MCQs

1. **(2)**

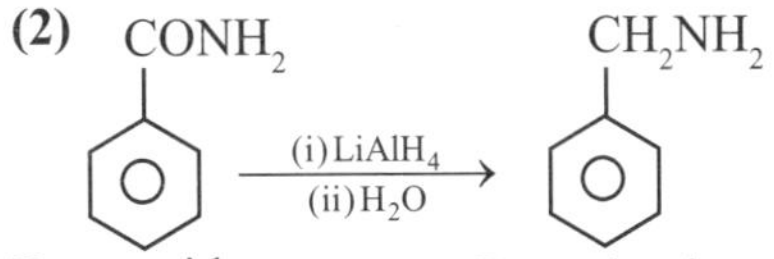

Benzamide Benzylamine

2. **(1)**

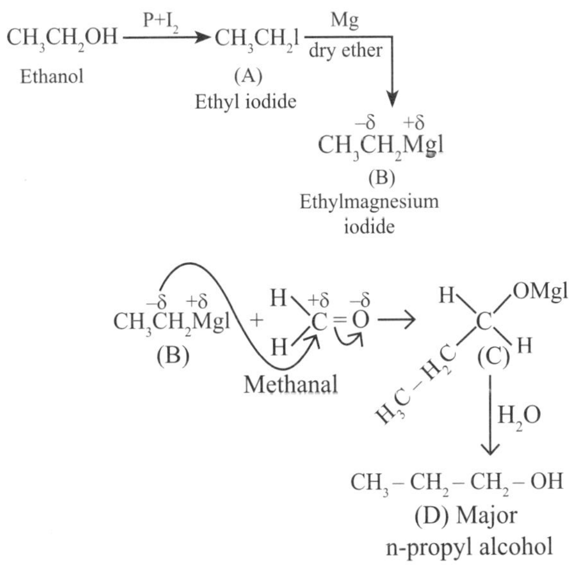

3. **(1)**

(1) 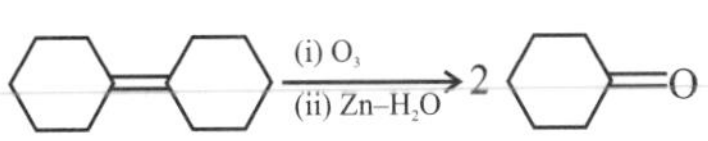

(2) 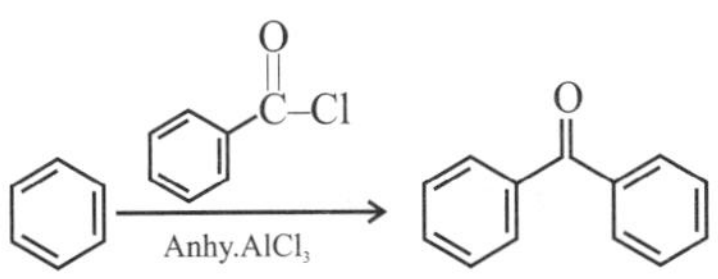

(3) 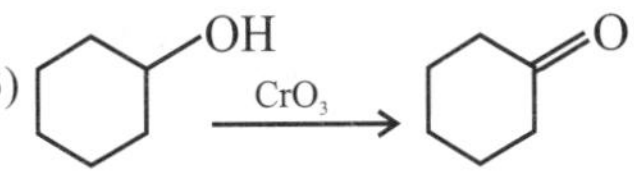

(4) 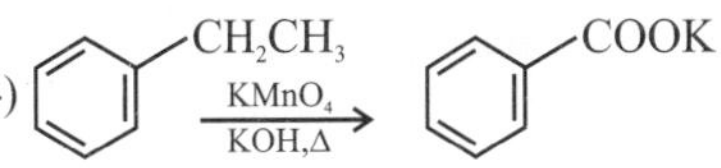

4. **(4)**

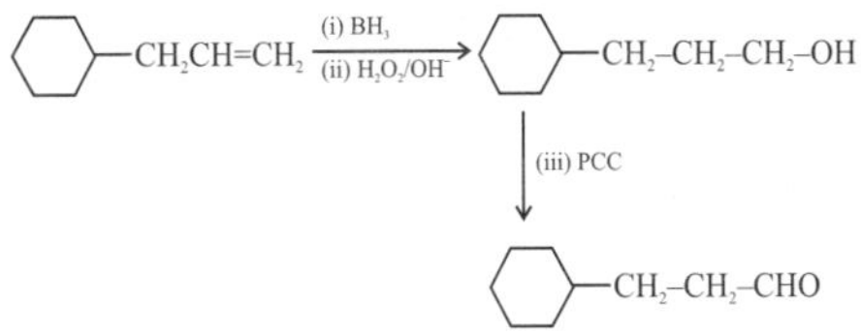

5. **(3)** Fehling's solution 'A' is aqueous copper sulphate.

6. **(2)** Ketones are reduced into hydrocarbons using Zn-Hg/HCl (Clemmenson reduction).

$>C{=}O + Zn\text{-}Hg/HCl \rightarrow >CH_2$

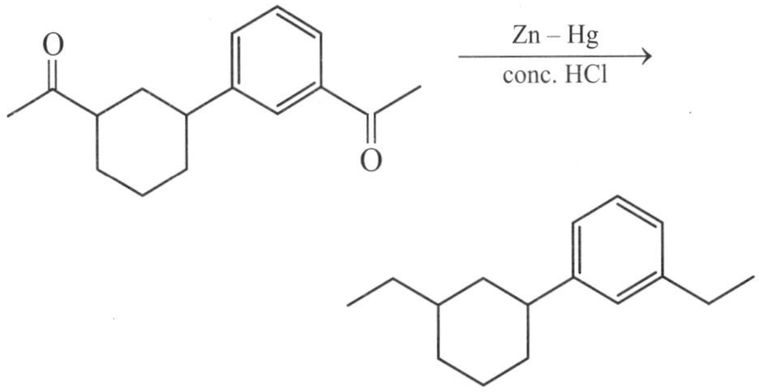

7. **(1)**

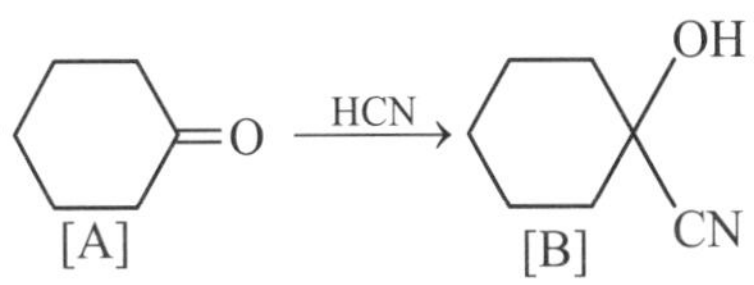

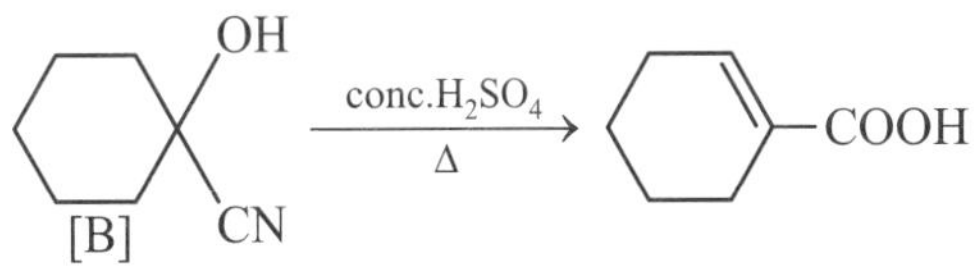

8. **(4)** Tollen's reagent oxidises aldehydes into carboxylate ion whereas ketone is not oxidised by tollen's reagent.

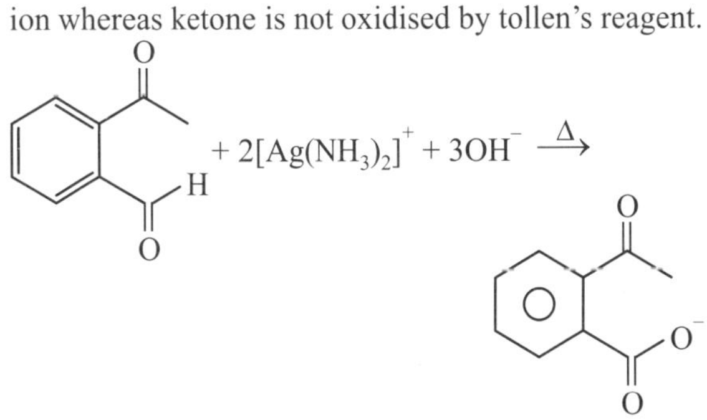

9. **(2)**

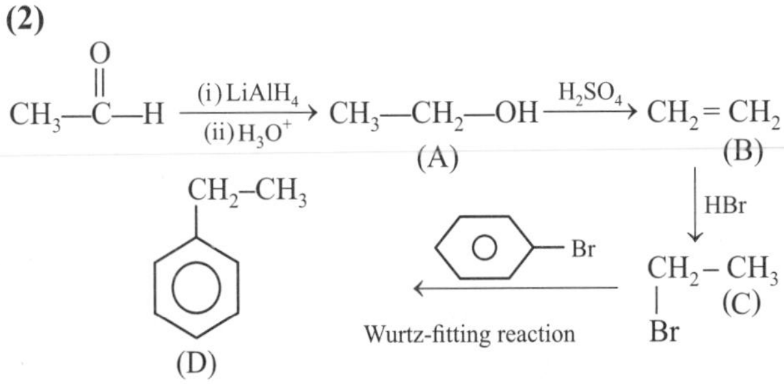

$LiAlH_4$ used for reduction of carbonyl group into alcohol and H_2SO_4 used for dehydration of alcohol.

10. (2) RMgX is a Grignard reagent and it is used to prepare carboxylic acid. The suitable reaction is given as:

$$\overset{-\delta\;+\delta}{RMgX} + O{=}C{=}O \rightarrow R{-}\overset{O}{\overset{\|}{C}}{-}\bar{O}\overset{+}{Mg}X\ (Y) \xrightarrow{H_3O^{\oplus}} R{-}\underset{O}{\underset{\|}{C}}{-}OH$$

11. (2) Boiling point of comparable molecular mass compounds is given as,

R–OH	>	Aldehyde – Ketone	>	Alkane
H–bonding		Dipole-dipole interaction		Non-polar
(strong molecular association)		(weak molecular association)		

From the above given comparison, it is clear that the boiling point of aldehydes and ketones are higher than hydrocarbons and lower than alcohols of the comparable molecular mass.

12. (1) $>C{=}C + HCN \xrightarrow{OH^{\ominus}} >C<^{OH}_{CN}$ Cyanohydrin

$C{=}O + 2ROH \xrightarrow{H^+} \,^{R}_{H}>C<^{OR}_{OR}$ Acetal
(Alcohol)

$>C{=}O + R{-}NH_2 \xrightarrow{H^+} >C{=}N{-}R$ Schiff's base

$>C{=}O + NH_2OH \xrightarrow{H^+} >C{=}N{-}OH$ Oxime

13. (3) Option (3) is neither a cross aldol nor an aldol product of acetone and 2-pentanone.

14. (2) $CH_3{-}CH_2CO\bar{O}\overset{+}{Na} \xrightarrow[Heat]{NaOH + CaO} CH_3{-}CH_3 + Na_2CO_3$

$$CH_3 - CH_2 - COONa \xrightarrow[Heat]{NaOH + CaO} CH_3 - CH_3 + Na_2CO_3$$

Decarboxylation takes place by Soda-lime (NaOH + CaO)

15. (3)

1.  $+ \xrightarrow[\text{Anhyd } AlCl_3/CuCl]{CO,\ HCl}$ Gattermann - Koch reaction

2. $R - \overset{O}{\overset{\|}{C}} - CH_3 + NaOX \rightarrow$ Haloform reaction

3. $R - CH_2 - OH + R'COOH \xrightarrow{Conc\ H_2SO_4}$ Esterification

4. $R - CH_2COOH \xrightarrow[(ii)\ H_2O]{(i)\ X_2/RedP}$ Hell-Volhard Zelinsky reaction

16. (1) $CH_3{-}\underset{\underset{O}{\overset{\|}{C}}{-}O{-}CH_3}{CH}{-}\overset{COOCH_3}{CH}{-}CH_2{-}CH_2{-}\overset{O}{\overset{\|}{C}}{-}CH_3$

$\downarrow$ DIBAL-H (Reduction of ester)

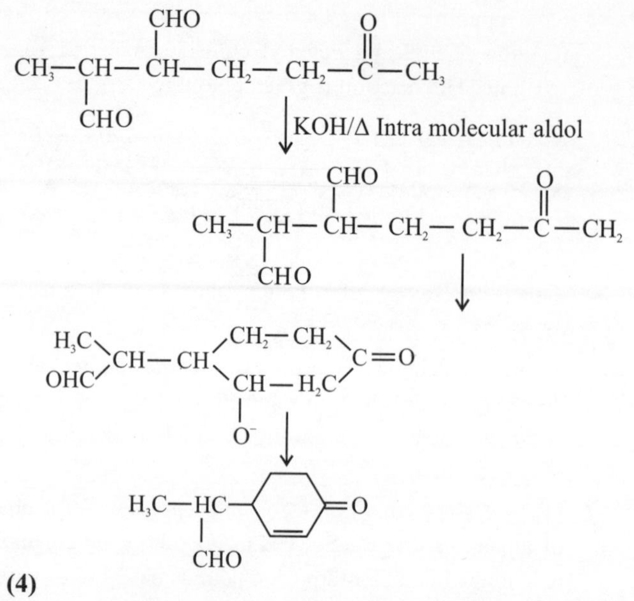

17. (4)

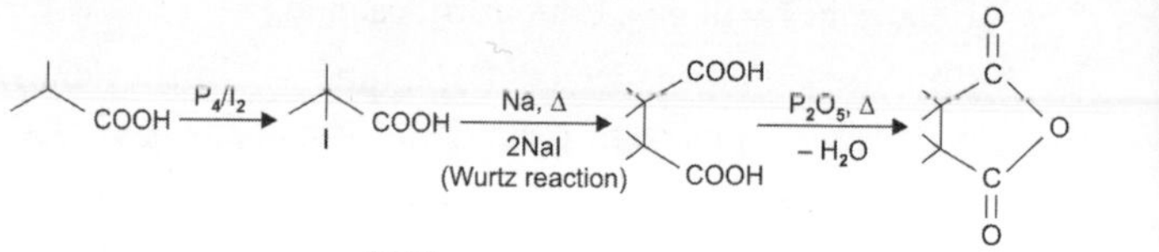

18. (1) $CH_3COOH \xrightarrow{SOCl_2} CH_3\ COCl$

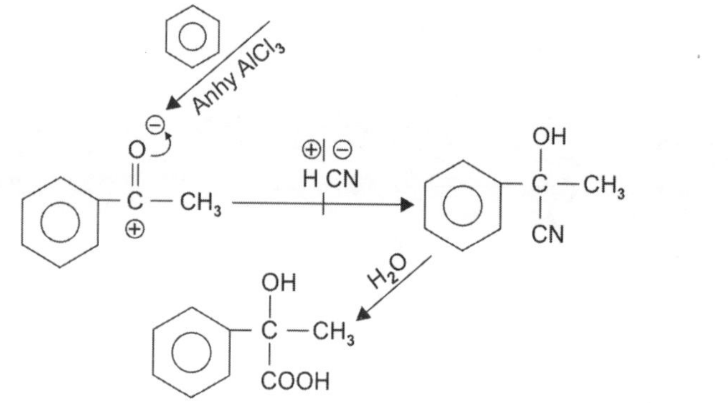

19. (3) On the basis of +I effect & –I effect

It effect shows by amines it has higher melting point than hydrocarbon, alcohol & halogen but $-NO_2$, –Cl are E.W.G.

20. (4) Toluene is converted to benzylidene diacetate on treating with chromic oxide in acetic anhydride. The benzylidene diacetate can be hydrolysed to corresponding. benzaldehyde with aq. acid.

21. (2)

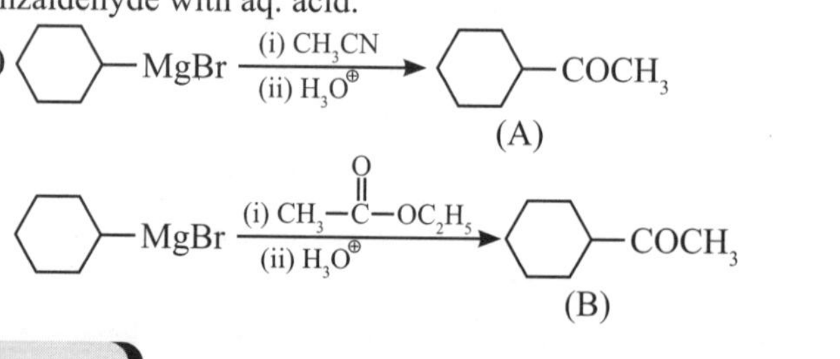

Amines

NCERT Topic-wise MCQs

1. (3) N atom of amino group has sp^3 hybridization. All the three hybridised orbitals get included in sigma bond formation while fourth hybridised orbital contains a lone pair of electrons.

2. **(4)** Degree of amine can not be defined by degree of carbon. All are primary amines.

3. **(3)** The structural formula of N-Methyl amino methane is -

$H_3C-\ddot{N}(CH_3)-H$

4. **(3)** Chemical formula of trinmethyl amine is $(CH_3)_3\ddot{N}$. Its hybridization = sp^3

Bond angle = 108°

5. **(2)** IUPAC names of the given compounds A and B are –

$CH_3-CH_2-CH_2-CH_2-NH_2$ $H_3C-CH(CH_3)-CH_2-NH_2$

1-Butanamine (A)   (B) 2-Methyl-1– propanamine

Both the given compound are chain isomers.

6. **(3)** Lowest locant rule

If benzene ring is disubstituted or higher sustituted, the position of substituents is defined by numbering the carbon atoms of the ring such that the substituents are located at the lowest numbers possible.

The name of the given structure is:

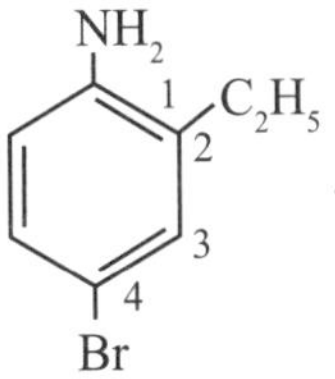

4-Bromo – 2-ethylaniline

7. **(4)** The IUPAC name of the given compounds are-

1) $(H_3C-CH_2)_2NCH_3 \Rightarrow H_3\overset{2}{C}-H_2\overset{1}{C}-N(CH_2-CH_3)-CH_3$

N-Ethyl - N-methyl ethanamine

2) $(H_3C)_3-C\,NH_2 \Rightarrow H_3C-\overset{2}{C}(\overset{1}{C}H_3)(\overset{3}{C}H_3)-NH_2$

2-methylpropan-2-amine

3) $(H_3C\,NHCH\,(CH_3)_2 \Rightarrow H_3C-N(H)-\overset{2}{C}(H)(\overset{3}{C}H_3)-\overset{1}{C}H_3$

$\Rightarrow$ N-Methyl propan-2-amine

4) $(H_3C)_2\,CHNH_2 \Rightarrow H-\overset{2}{C}(\overset{1}{C}H_3)(\overset{3}{C}H_3)-NH_2$

$\Rightarrow$ Propan-2-amine not 2, 2-Dimethyl-N-Propanamine

8. **(1)** By Gabriel-phthalimide synthesis, only pure 1° amines can be prepared. Aniline is not formed as aryl halide is not formed by nucleophilic substitution with potassium phthalimide.

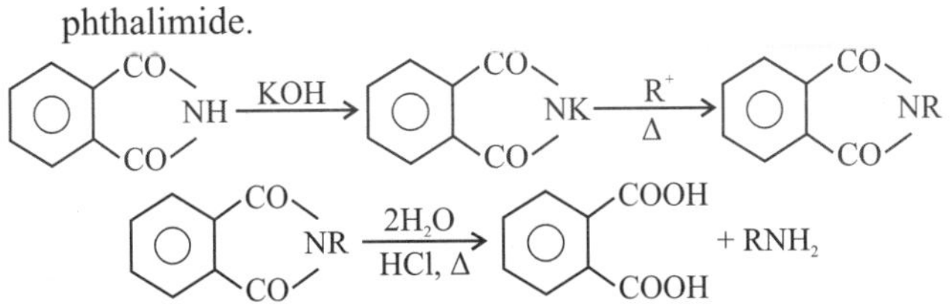

9. **(2)** Hofmann bromamide degradation reaction may form primary amines.

$RCONH_2 + Br_2 + 4KOH \xrightarrow{\Delta} RNH_2 + 2KBr + K_2CO_3 + 2H_2O$

$C_2H_5CONH_2 + Br_2 + 4KOH \rightarrow C_2H_5NH_2 + 2KBr + K_2CO_3 + 2H_2O$

Primary amines may also be formed by reduction of nitroalkanes, nitrobenzene etc.

$C_2H_5NO_2 + 3H_2 \xrightarrow[\text{or LAH or } H_2/Pt \text{ or Metal acid}]{Ni/250^0C} C_2H_5NH_2 + 2H_2O$

$C_6H_5NO_2 + 2H_2 \xrightarrow[LAH]{Ni/250^0C} C_6H_5NH_2 + 2H_2O$

10. **(4)** Nitrobenzene reduced into aniline by metal in the presence of acid as well as by the electrolytic reduction under weakly acidic condition.

11. **(1)**

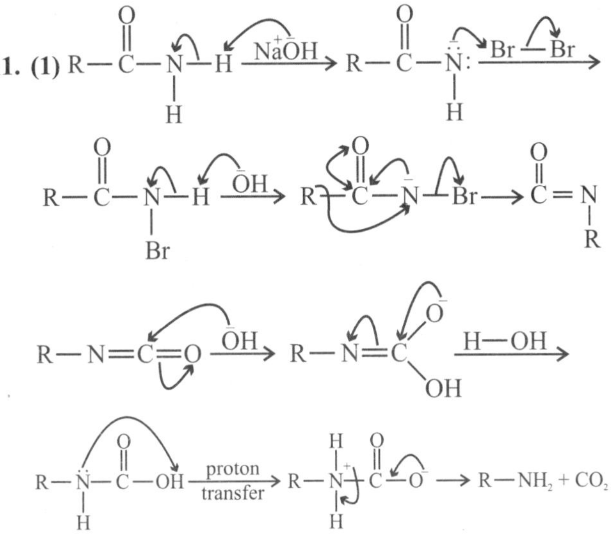

The above reaction is Hoffmann bromamide reaction.

12. **(4)**

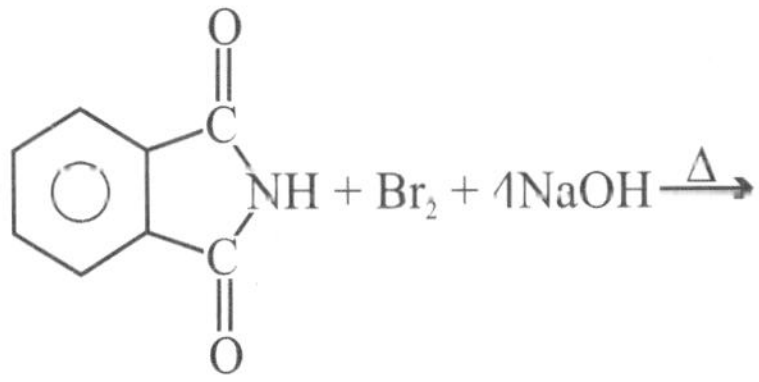

Phthalimide

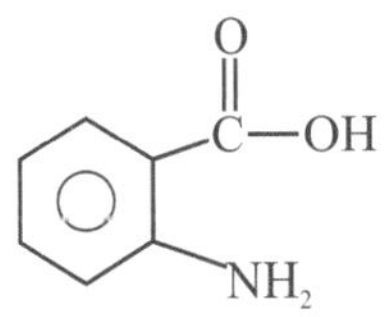

$+ 2NaBr + Na_2CO_3 + H_2O$

o-Amino benzoic acid

13. **(4)** This is a Mendius reduction.

$CH_3CN \xrightarrow{Na(Hg)/C_2H_5OH} CH_3CH_2NH_2$

14. **(3)** The best method to reduce one carbon atom in the amine from amide is Hofmann bromamide reaction.

$CH_3CH_2CONH_2 + Br_2 + 4KOH \xrightarrow{\Delta} CH_3CH_2NH_2 + 2KBr + K_2CO_3 + 2H_2O$

15. **(2)** $CH_3CH_2CONH_2 \xrightarrow[(NaOBr)]{Br_2/NaOH} CH_3CH_2NH_2 + NaBr + Na_2CO_3 + H_2O$

16. (1)

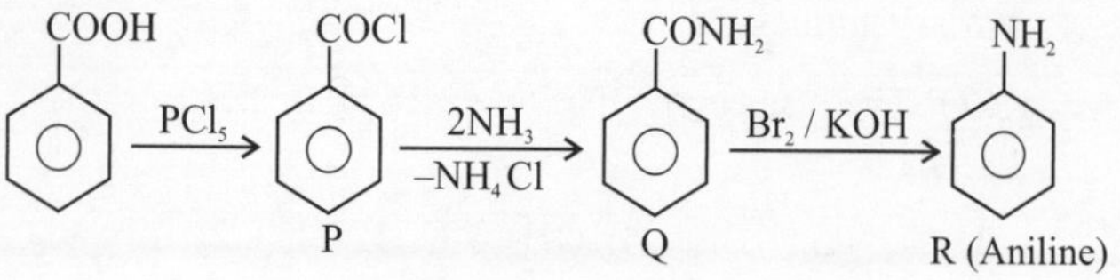

17. (3) $R-C\equiv N+H_2O \xrightarrow{H^+} R-COOH+NH_3$ (Hydrolysis reaction).

18. (3) Reduction of isocyanides gives 2° amines while all other reactions give 1° amines.

19. (2) On increasing positive inductive groups, basic strength increases. That's why in non-polar solvent or gaseous medium, order of basic strength is.

$\xrightarrow{3°\text{amine} > 2°\text{amine} > 1°\text{amine}}$

20. (3) Boiling point of carboxylic acid is highest due to strong intermolecular H-bonding.

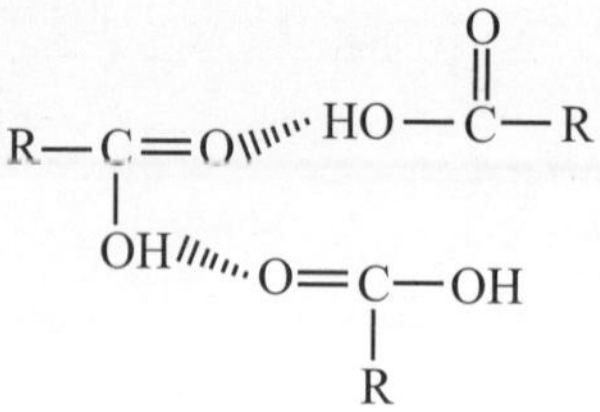

Boiling point of alcohol is higher than amines due to high electronegativity of oxygen than nitrogen so, alcohols form more stronger H-bonding than amines

$\xrightarrow[\text{order of Boiling point}]{RCOOH>R-OH>RNH_2}$

21. (2) $CH_3-N(CH_3)-CH_3$ (3° amine) do not undergo hydrogen bonding.

22. (3) As the alkyl group increases, basicity increases due to +I effect.

23. (3) $(CH_3CO)_2\ddot{N}H \rightarrow$ least basic.

24. (3)

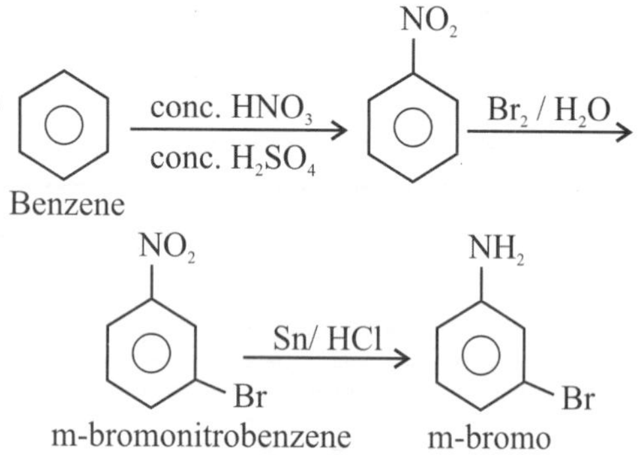

m-bromonitrobenzene → m-bromo aniline

25. (4) Aryl amines react with nitrous acid to form diazonium salts not phenols.

26. (4) The correct order of basicily in amines is- 2° > 1° > 3° >. Here, $(C_6H_5)_2$ NH and $(C_2H_5)_2$ NH, both are 2° amines but $(C_2H_5)_2$ NH, is more basic because it has no conjugation whereas 1° $(C_6H_5)_2$ NH, conjugation is present.

27. (4) The lone pair on nitrogen in option (4) and option (1) is free and not included in any conjugation so,it is easily available for donation. 'N' in piperidine is 2° amine which is more basic.

28. (1)

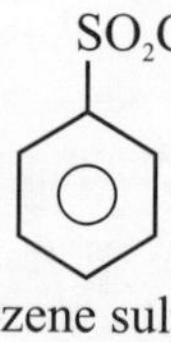

acts as Hinsberg's reagent.

Benzene sulphonyl chloride

29. (3) Amides are amphoteric in nature and amines are basic in nature.

30. (2)

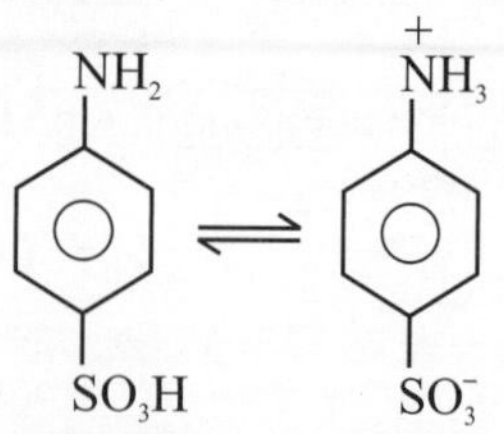

Sulphanilic acid ⇌ Zwitter ion

31. (3) $C_6H_5NH_2 \xrightarrow[273K]{NaNO_2+HCl} \underset{X}{C_6H_5N_2^+Cl^-} \xrightarrow{CuCN}$

$C_6H_5CN \xrightarrow[\text{Boil}]{H^+/H_2O} C_6H_5COOH$

32. (4) $C_6H_5CH_2NH_2$ is the strongest base since the lone pair on nitrogen is not included in any conjugation.

33. (1)

$CH_3CH_2Cl \xrightarrow{NaCN} \underset{X}{CH_3CH_2CN} \xrightarrow{Ni/H_2} \underset{Y}{CH_3CH_2CH_2NH_2}$

$\xrightarrow{\text{Acetic anhydride}} \underset{Z}{CH_3CH_2CH_2NHCOCH_3}$

34. (4) By carbylamine reaction, isocyanides are prepared. In this, reaction primary amine is treated with chloroform and NaOH or KOH to give isocyanide (unpleasant smelling compound).

$CHCl_3 + 3KOH + \underset{\text{Primary amine}}{R-NH_2} \rightarrow \underset{\text{isocyanide}}{R-NC} + 3H_2O + 3KCl$

35. (2) $LiAlH_4$ convert C = O group into CH_2 group thus, it can be used to convert acetamide into ethanamine.

$\underset{\text{Acetamide}}{CH_3CONH_2} + 4[H] \xrightarrow{LiAlH_4} \underset{\text{Ethanamine}}{CH_3CH_2NH_2} + H_2O$

36. (3) $\begin{matrix}CONH_3^{\oplus}\\|\\CO\overset{\ominus}{N}H_3\end{matrix} \xrightarrow{\text{Heat}} \begin{matrix}CONH_2\\|\\CONH_2\end{matrix} \xrightarrow[\Delta]{P_4O_{10}} \begin{matrix}CN\\|\\CN\end{matrix}$

37. (3) As an acid (C) reduces Tollen's reagent, (C) is a formic acid. B gives carbylamine reaction. So, B is a primary amine with two carbon atoms. B is ethyl amine. Therefore, compound A has the following structure:

$H-\overset{O}{\overset{\|}{C}}-NH-CH_2CH_3$

38. (4) Tertiary nitro compounds do not react with nitrous acid as it does not have alpha-hydrogen,

39. (2) Ethylamine cannot react with dye but aniline gives reaction with dye.

40. (4)

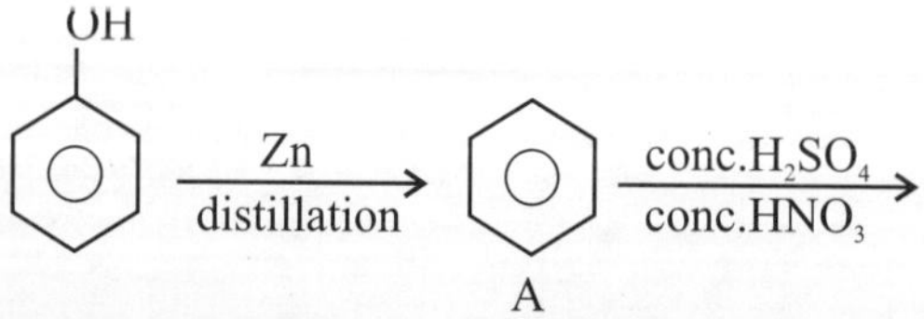

(Benzene)

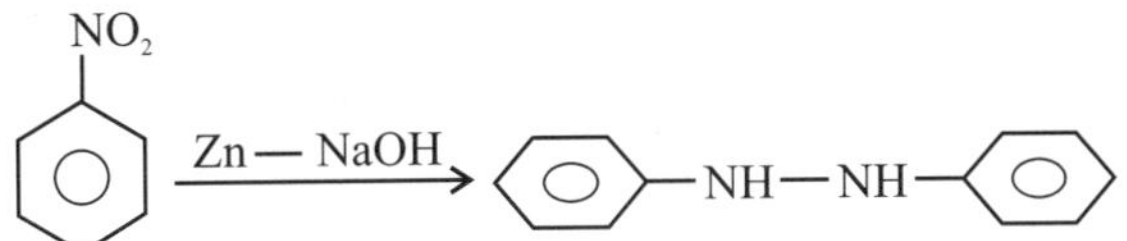

B (Nitrobenzene) C (Hydrazobenzene)

41. (3) 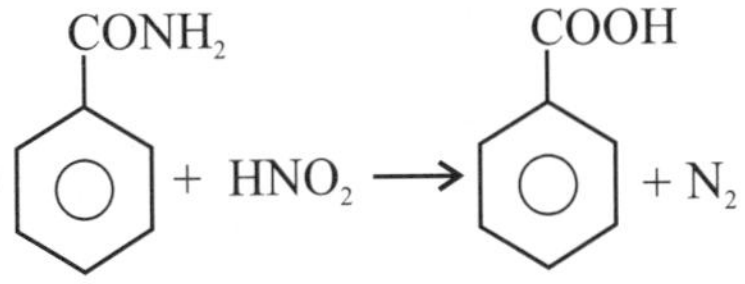

42. (4) Carbylamine test can be given by only primary amines (aromatic as well as aliphatic), i.e., the compounds containing $-NH_2$ group, and form a foul smelling poisonous compound carbylamine on reaction with chloroform in alkaline solution.

$$R-NH_2+CHCl_3+3KOH \xrightarrow{\Delta} R-N\equiv C+3KCl+3H_2O$$

43. (4) 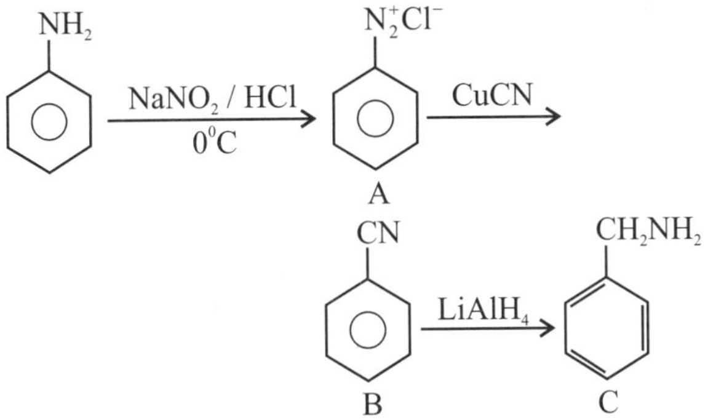

(Benzylamine)

44. (1) Aniline react with acetaldehyde to form Anil (Schiff's base).

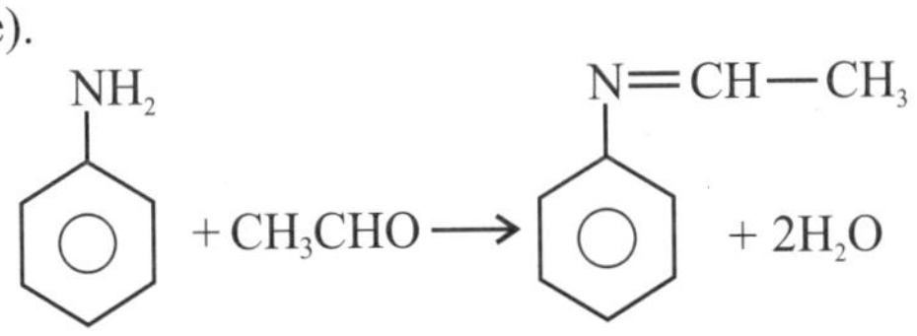

Schiff's base

45. (2) $(C_2H_5)_2NH$, i.e., N-ethylethanamine is the most basic amine among the given compounds.

46. (1) Aniline gives N-acetyl amino benzene on acetylation.

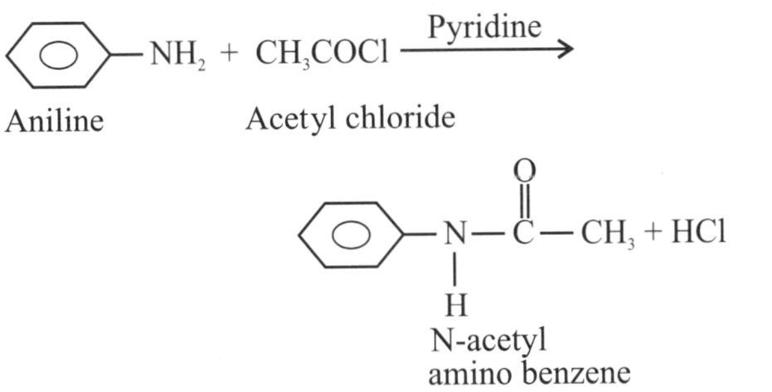

47. (2) This reaction illustrates the electrophilic character of $COCl_2$ and nucleophilic character of Ar-NH_2.

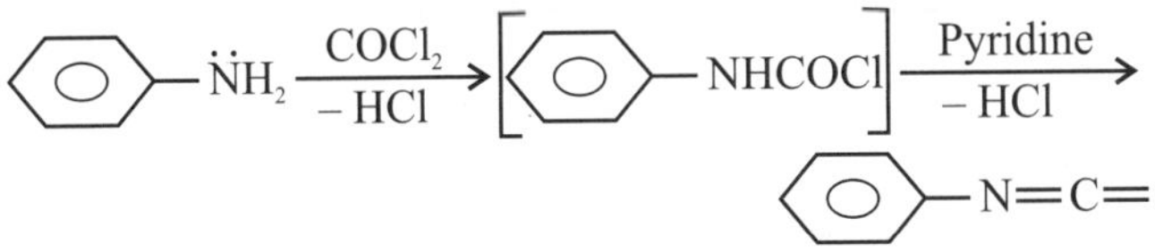

Phenyl isocyanate

48. (2) Only RNC and $CHCl_3$ are involved in carbylamine reaction, $RNH_2 + CHCl_3 + 3KOH \xrightarrow{\Delta} RNC + 3HCl + 3H_2O$.

49. (1) In carbylamines reaction, primary amines on heating with chloroform in presence of alcoholic KOH form isocyanides (or carbylamines). It is used to distinguish 1° amines from 2° and 3° amines. 1° amines react and give products where as 2° & 3° amines do not react.

$$\underset{\text{(alc.)}}{R\text{–}NH_2 + CHCl_3 + 3KOH} \rightarrow RNC + 3KCl + 3H_2O$$

50. (4) Basic strength of different alkyl amines depends upon all the given effects i.e. +I effect, solvation effect, and steric effect.

51. (3) By Hinsberg reagent, benzene sulphonyl chloride, primary, secondary and tertiary amines can be separated.

52. (4) Carbylamine test cannot be given by 2° amines.

53. (4) Aryl amines are less basic than alkyl amines because phenyl group exerts –I effect resulting in decrease in electron density in nitrogen atom for protonation.

54. (1)

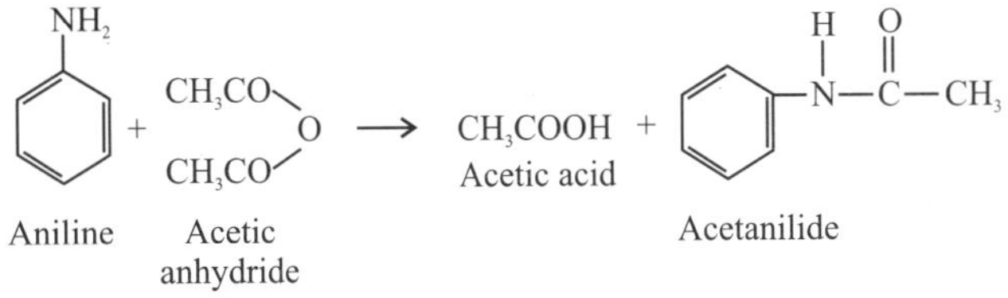

55. (1) $CH_3-CH_2-NH_2 \xrightarrow{NaNO_2/HCl}$

$$\underset{\text{(Major)}}{CH_3-CH_2-OH}+CH_3-CH_2-O-N=O+ CH_3-CH_2-Cl+CH_2=CH_2$$

56. (2) Isopropyl amine is a primary amine. It can give positive carbylamine test.

57. (3) $X = C_6H_5NH_2$, $Z = C_6H_5N_2Cl$, $A = C_6H_5-N=N-C_6H_4NH_2$

58. (3) Schotten baumann reaction.

59. (3) Diphenyl amine is 2° amine and the carbylamine test is given by 1° amine i.e., aniline in this case.

60. (1) Amines are more basic than ethers, therefore, (4) is least basic. Out of (1), (2), (3), (3) is least basic due to donation of its lone pair of electrons to maintain aromaticity. It we compare options (1) and (2), in case, I[st] 'N' atom is sp^3-hybridised and as compared to in case of IInd 'N' is sp^2-hypridised having lesser s-character I[st]. (1) is more basic than (2).

61. (1) The basicity of o-, m- and p-toluidines decreases in the order p > m > o due to steric inhibition of resonance basicity of o-toluidines is less as compared to p-toluidines and m-toluidines. In other words, their pK_b values will increase in the reverse oder, *i.e.*, option (1) is correct, i.e., p < m < o.

62. (2) 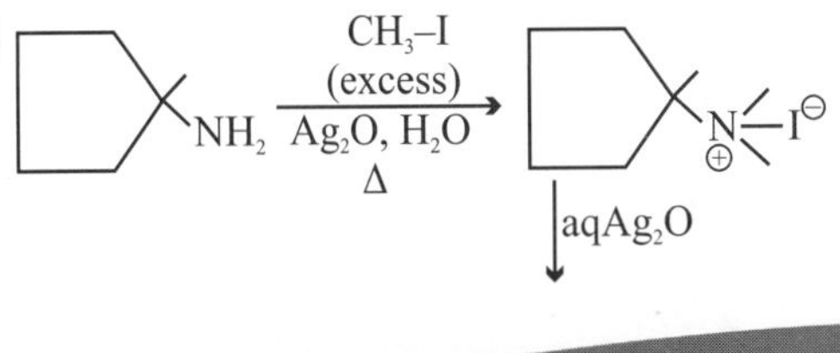

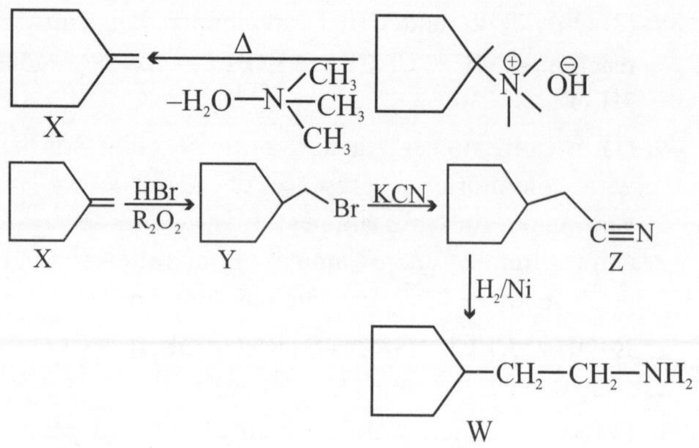

Product W will be primary amine

63. (3) Due to the lone pair of e⁻ N-atom, the group is activating & o, p directing.

64. (1) Mustard oil reaction is given by 1° amines because it has 2 active -H atoms.

65. (1) $PhSO_2Cl + \underset{(1^\circ amine)}{RNH_2} \longrightarrow Ph-SO_2-NH-R \xrightarrow{KOH} Ph-SO_2-\overset{\ominus\oplus}{NK}$ (with R on N) (soluble)

66. (1) $(CH_3)_2C=N-OH \xrightarrow{Na/C_2H_5OH} (CH_3)_2CH-NH_2$

67. (4) $C_2H_5Br \xrightarrow{AgCN} C_2H_5NC \xrightarrow{H_3O^+} HCOOH + C_2H_5NH_2$

$C_2H_5NH_2 \xrightarrow[KOH]{CHCl_3} C_2H_5NC \xrightarrow{Reduction} C_2H_5-NH-CH_3$

68. (1) $C_2H_5NH_2 \xrightarrow{NOCl} C_2H_5Cl + N_2 + H_2O$ (Cl|N|O exchanged with C_2H_5|N|H_2)

69. (2) Carbylamines on hydrolysis always give HCOOH as one of the products which reduces Fehling's solution.

$\underset{\text{Ethyl carbylamine}}{C_2H_5N \equiv C} + 2H_2O \xrightarrow{H^+} \underset{\text{Ethylamine}}{C_2H_5NH_2} + HCOOH$

$C_2H_5NH_2$ reduces Fehling's solutions.

70. (1)

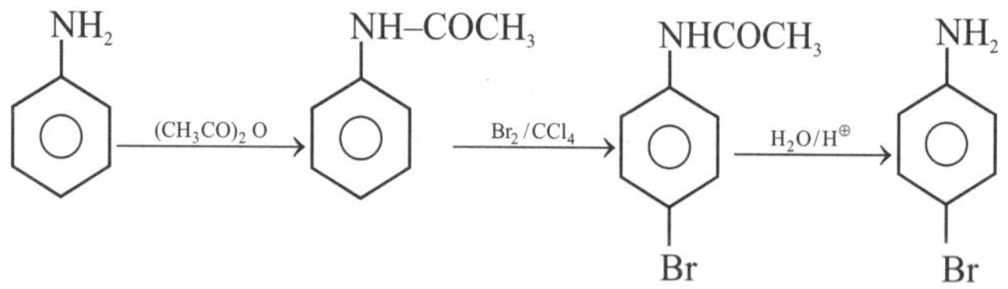

71. (1)

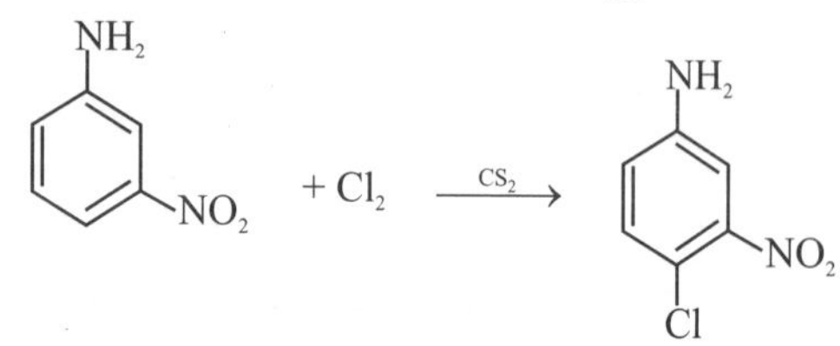

$-NH_2$ group increases e⁻ density at o- & p- positions & $-NO_2$ is meta-directing group.

∵ Chlorination is electrophilic substitution reaction, position of attack is directed by $-NH_2$ group. So, Cl will be attached to o- or p- position of $-NH_2$ group.

In option (1) & (3), –I effect of $-NO_2$ group is more pronounced at o-position wrt $-NO_2$ group than at p-position which decreases e⁻ density.

∴ (1) is correct.

72. (3) Hydrazobenzene rearranges to benzidine in acidic medium.

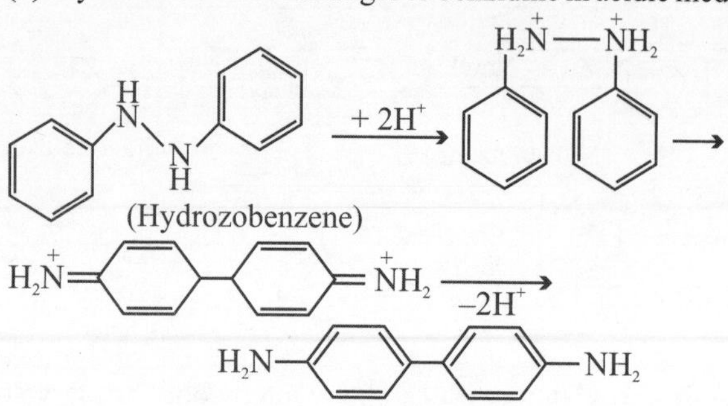

(Benzidine)

73. (2) Aniline → N-phenyl derivative of an amide.
Benzanilide is a N-phenyl derivative of benzamide, $C_6H_5CONH_2$
Benzanilide structure is $C_6H_5CONHC_6H_5$:

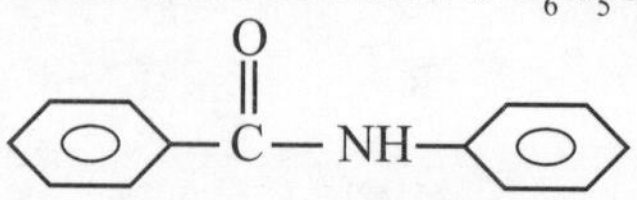

74. (2) The electron density in benzene ring in the presence of $-NO_2$ group decreases. Thus, decreasing the reactivity towards electrophilic substitution and increasing reactivity towards nucleophilic substitution reactions.

75. (3) There are three acidic α-hydrogens.

$CH_3NO_2 \xrightarrow{NaOH} H_2O + Na^+\,\bar{C}H_2-NO_2 \xrightarrow{Cl_2}$

$ClCH_2-NO_2 \xrightarrow[(ii)\ Cl_2]{(i)\ NaOH} \underset{\text{nitrochloroform}}{Cl_3CNO_2}$

76. (3) It forms $CH_3CH_2CH_2$–NC

77. (1) To distinguish 1°, 2° and 3° amine we use HNO_2 reaction.

(1°) $R-NH_2 + HNO_2 \rightarrow R-OH + N_2 + H_2O$

(2°) $R_2NH + HNO_2 \rightarrow \underset{\text{nitrosoamines}}{R_2NNO} + H_2O$

(3°) $R_3N + HNO_2 \rightarrow \underset{\text{trialkyl ammonium nitrite}}{R_3NH^+ONO^-}$

78. (1) Balz Schiemann reaction is when diazonium salt is treated with fluoroboric acid (HBF_4), the fluorobenzene is formed.

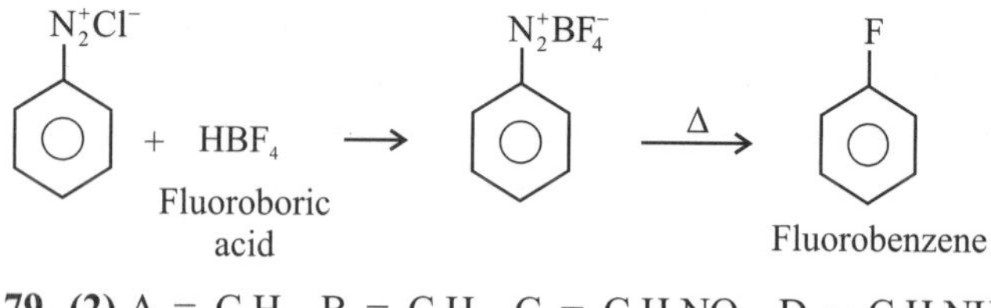

79. (2) A = C_2H_2, B = C_6H_6, C = $C_6H_5NO_2$, D = $C_6H_5NH_2$, E = $C_6H_5N_2Cl$

80. (2) Only primary aromatic amine can give diazotisation reaction.

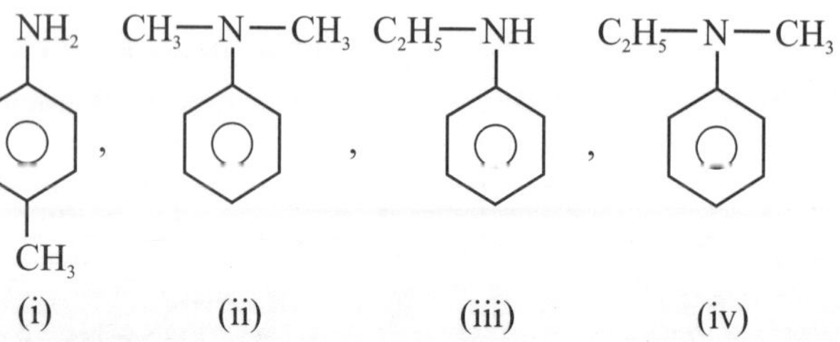

As a result, compounds (ii), (iii) and (iv) do not form diazonium salt.

81. (4) The aqueous solutions of diazonium salts are good conductors of electricity.

82. (4) Electron-donating groups increase the stability of diazonium cations by dispersing the +ve charge on the N-atom.

83. (4)

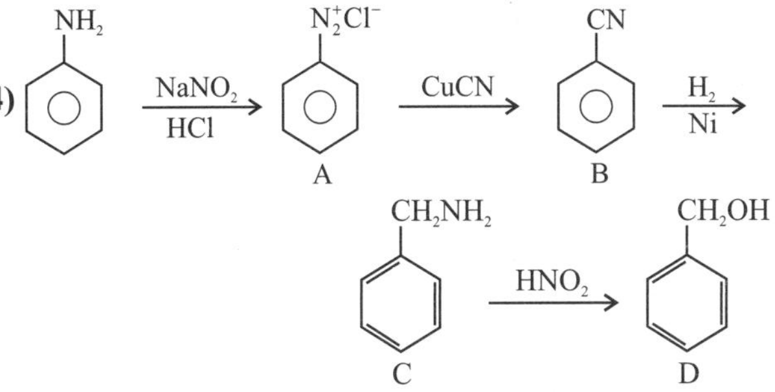

84. (1)

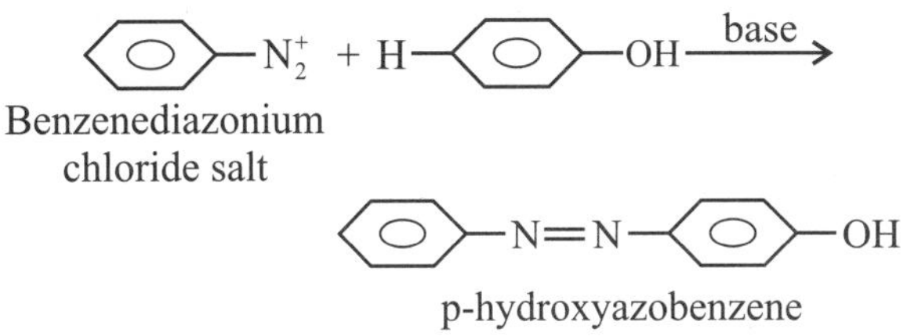

85. (1)

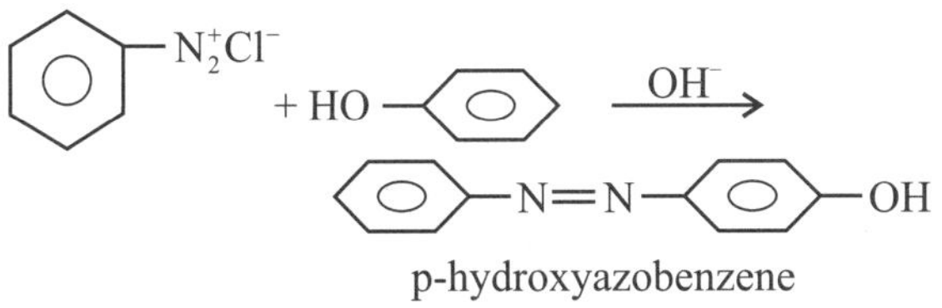

This is known as coupling reaction

86. (2)

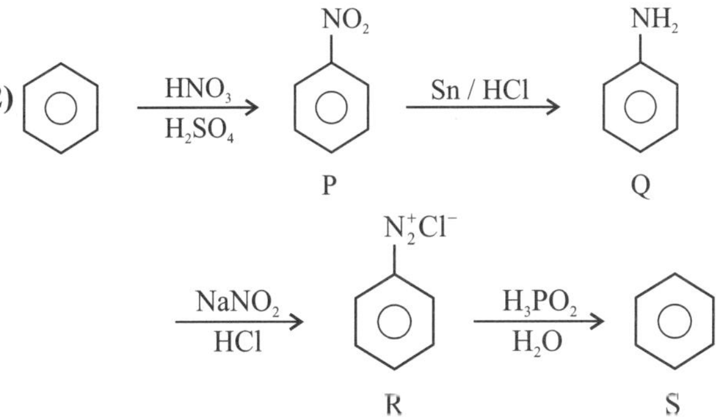

87. (1)

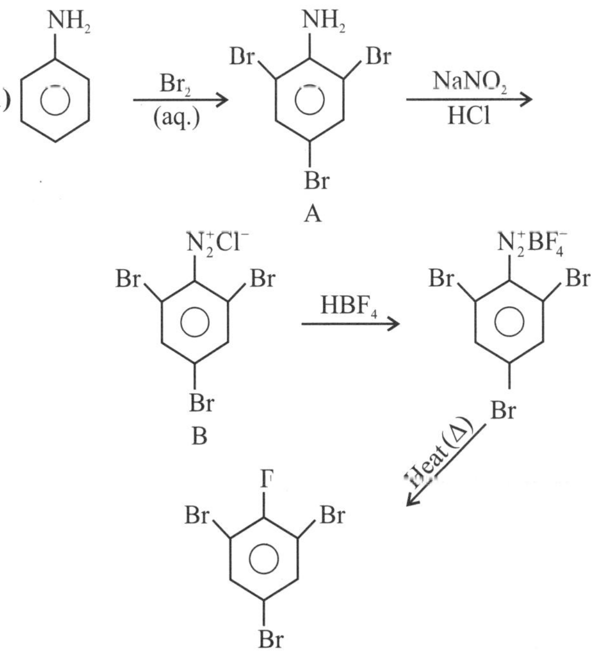

88. (3)

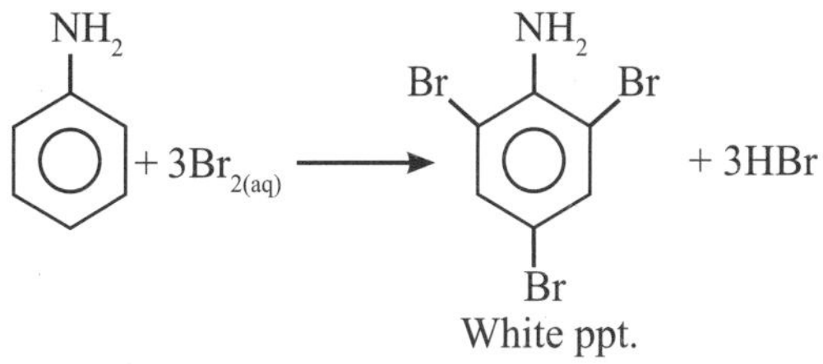

89. (2) Only primary aromatic amines give diazotisation reaction. The compound (X) is p-toluidine.

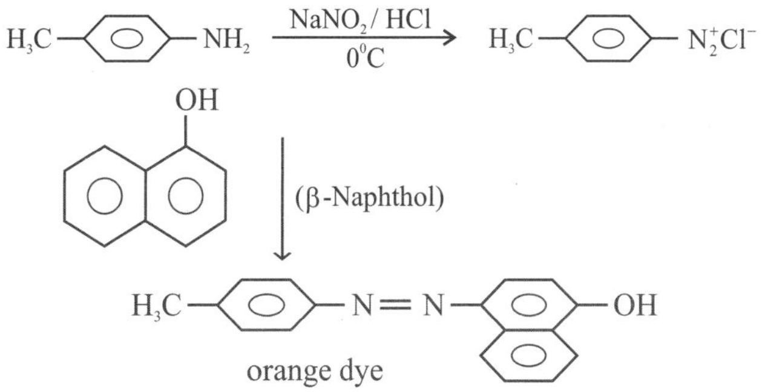

90. (2) Diazotisation is given by an aromatic primary amine ($-NH_2$ attached to benzene).

91. (2) Gomberg reaction

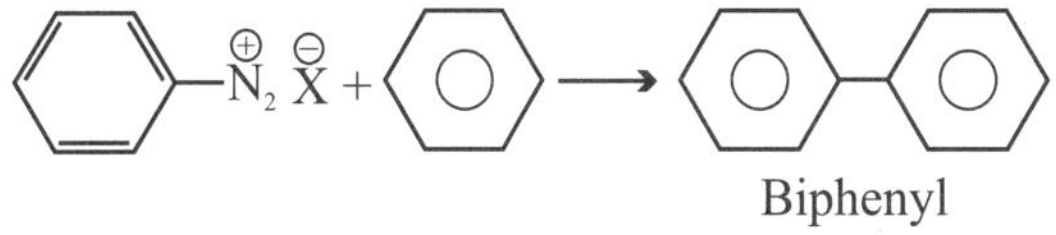

92. (3) Urea decomposes HNO_2 to give N_2, CO_2 and H_2O

$$NH_2CONH_2 + 2\ HONO \longrightarrow 2\ N_2\uparrow + CO_2\uparrow + 3\ H_2O$$

93. (4) When benzene diazonium chloride is treated with water, the diazonium group is replaced by –OH and phenol is obtained.

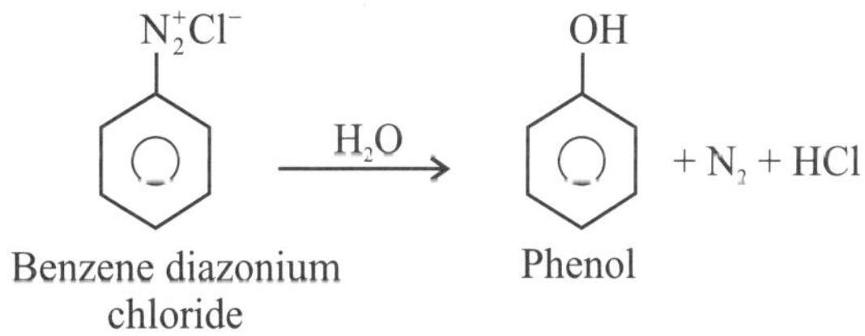

High Order Time Intensive MCQs

1. (2)

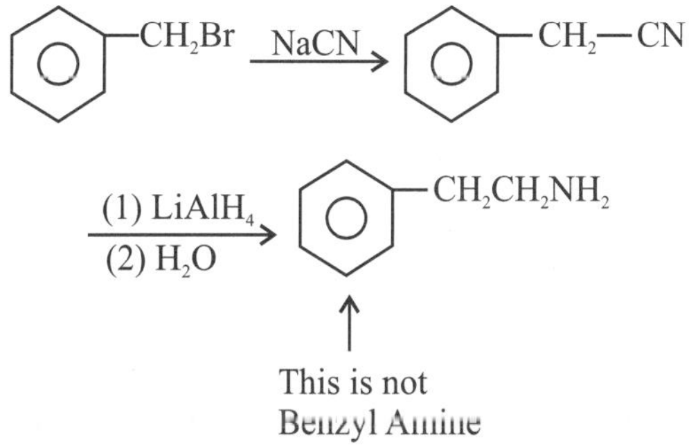

2. (2) On reduction, amine should be formed.

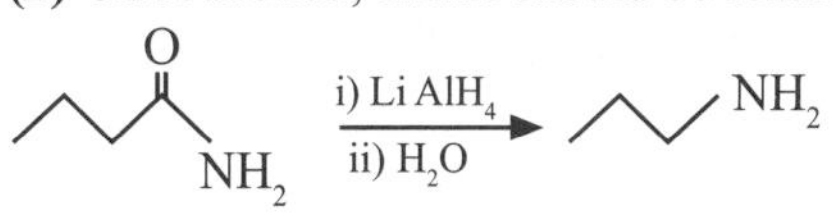

3. (4) Primary amines can be synthesized by Gabriel synthesis

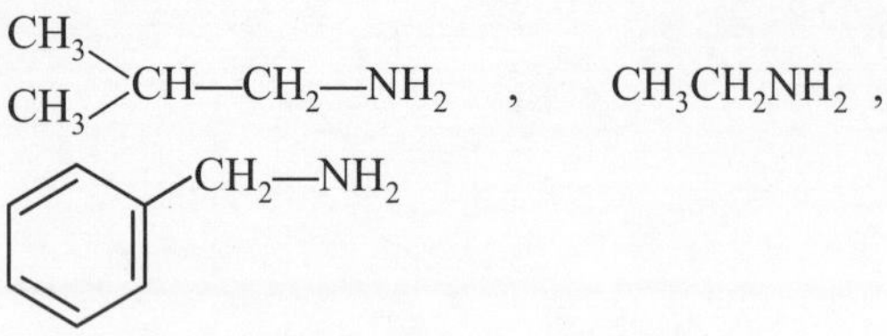

Aromatic amines can not be zynthesises by Gaberial synthesis.

4. (4)

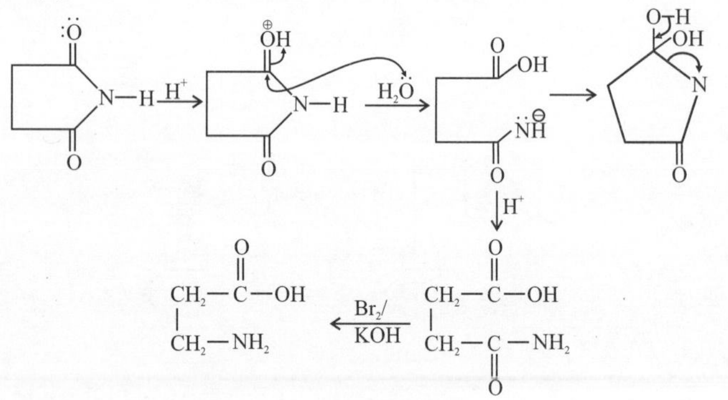

5. (4) $C_2H_5CN \xrightarrow{LiAlH_4} C_2H_5CH_2NH_2$

6. (1) Hoffmann's Bromamide proceeds with retention of configuration.

7. (4)

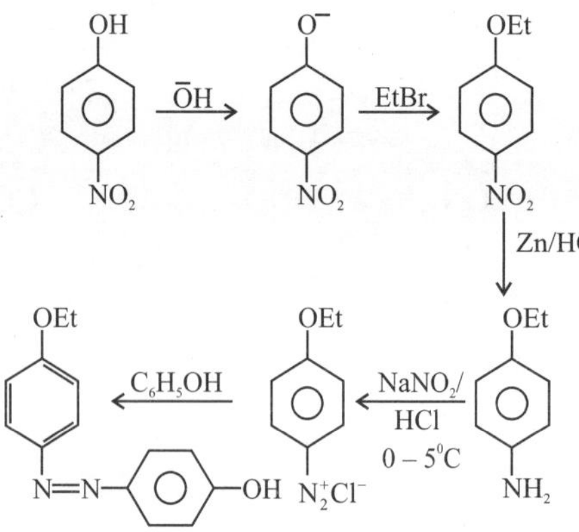

8. (4) H_2/Pd will reduce NO_2 to NH_2 and acetic anhydride will convert NH_2 to $NHCOCH_3$.

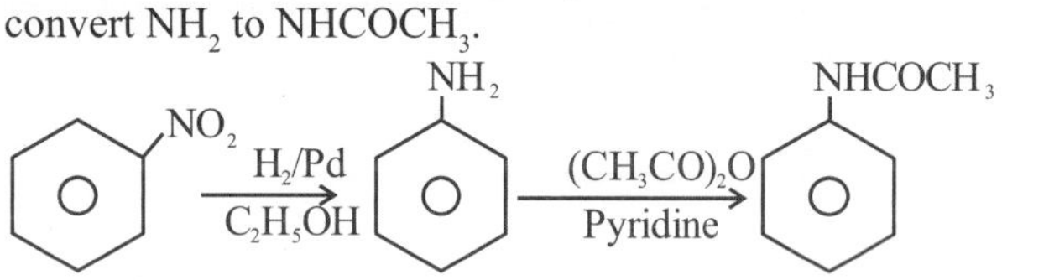

9. (1) 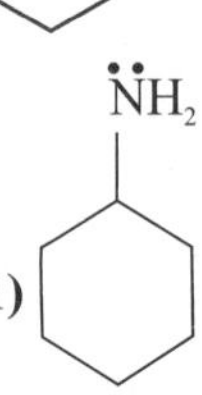 most basic because lone pair is localized on

Electronegative 'N' and also +I of cyclohexyl is present.

10. (1) In aqueous medium basic strength is dependent on electron density on nitrogen as well as solvation of cation formed after accepting H^+. After considering all these factors overall basic strength order is

$Me_2NH > MeNH_2 > Me_3N > NH_3$

$2° > 1° > 3° > NH_3$

11. (3)

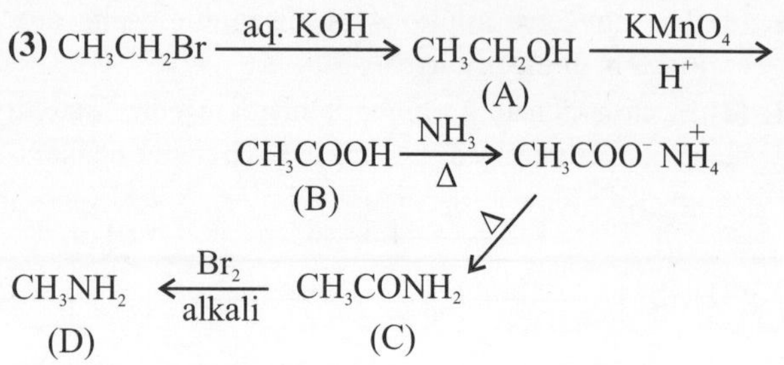

12. (3)

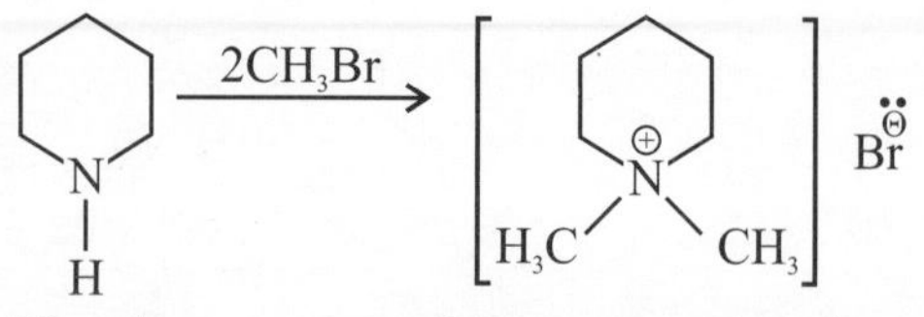

Due to SIR effect, the lone pair of N atom is not in resonance with benzene ring.

13. (4)

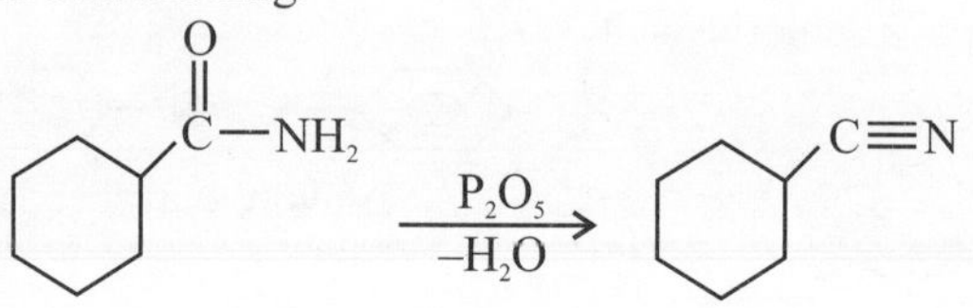

14. (2)

$$\underset{A(M.F.C_2H_3N)}{CH_3C\equiv N} \xrightarrow{Redn.} \underset{Ethylamine\ (B)}{CH_3CH_2NH_2} \xrightarrow{HNO_2} \underset{Ethylalcohol}{CH_3CH_2OH}$$

$$\underset{Ethylamine\ (B)}{CH_3CH_2NH_2} \xrightarrow{CHCl_3/Alc.KOH} \underset{Ethylisocyanide\ (C)}{CH_3CH_2NC}$$

15. (1) $Ph—NH_2 \xrightarrow[\ominus OH]{CHCl_3} Ph—\overset{\oplus}{N}\equiv \overset{\ominus}{C}$

$\xrightarrow{LiAlH_4} Ph—NH—CH_3$

16. (3) It is Hoffmann bromamide degradation reaction followed by carbylamine reaction

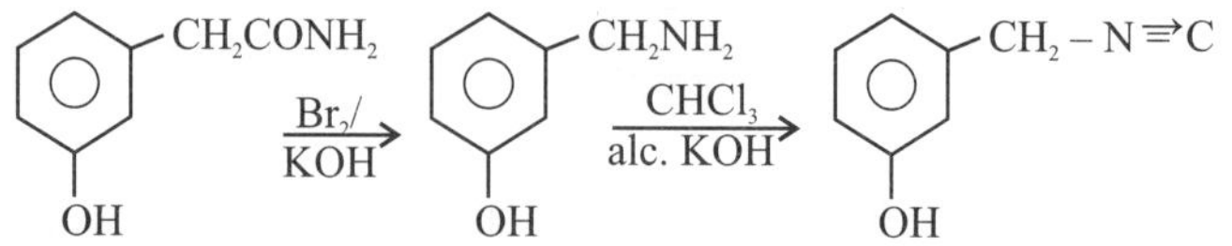

17. (3)

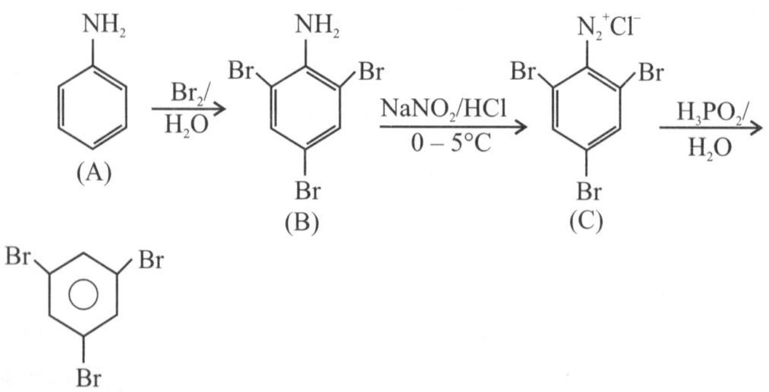

18. (2) Secondary amine is most basic in aqueous solution among aliphatic amines.

19. (4) The number of σ bonds are 14 and Degree of unsaturation = 4.

20. (3) Option (3) is correct.

Product formed from secondary amine is insoluble in alkali.

21. (3)

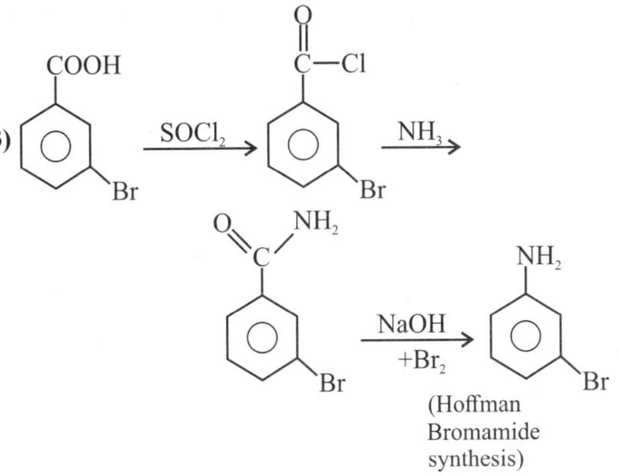

22. (3) Sandmeyer Reaction is

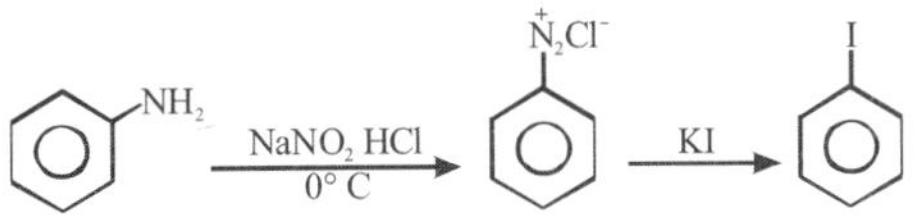

23. (3)

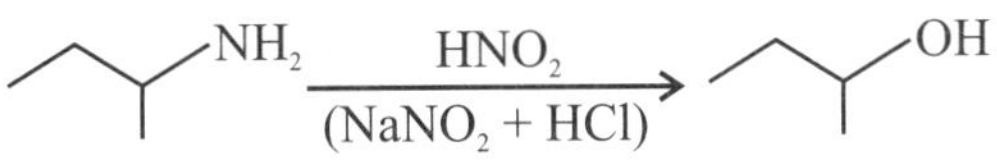

Due to formation of 2° carbocation.

24. (2) Diazonium cation reacts with aniline in weakly acidic medium resulting in N, N-coupling rather than C-coupling (at low temperature).

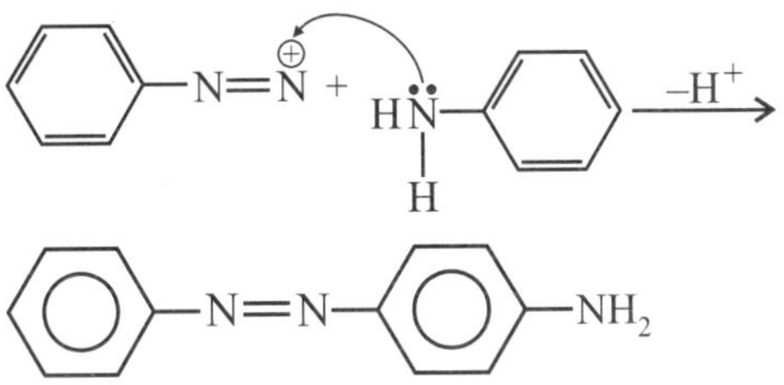

p-amino azo benzene

25. (4)

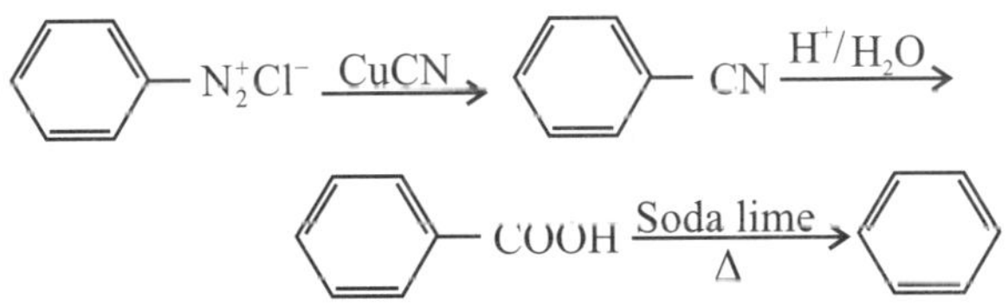

26. (1) $SOCl_2$ here can only act as a dehydrating agent.

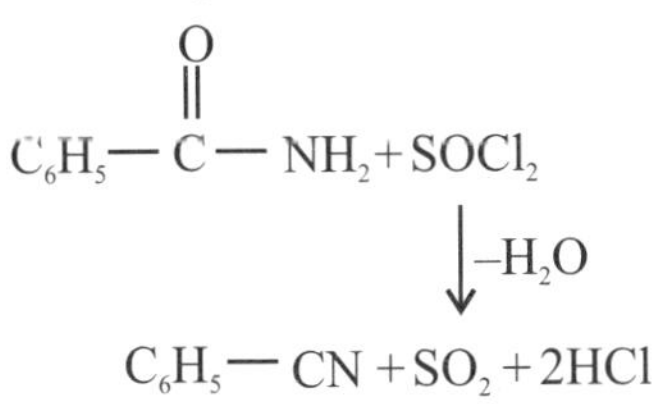

27. (3)

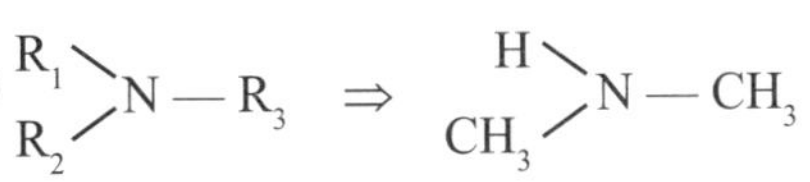

2°-Amine reacts with nitrous acid to form nitroso amine (yellow liquid).

28. (3) Hoffmann degradation of amide

$$R\text{–}CONH_2 + Br_2 + 4KOH \rightarrow R\text{–}NH_2 + 2KBr + K_2CO_3 + 2H_2O$$

Primary amine

29. (2)

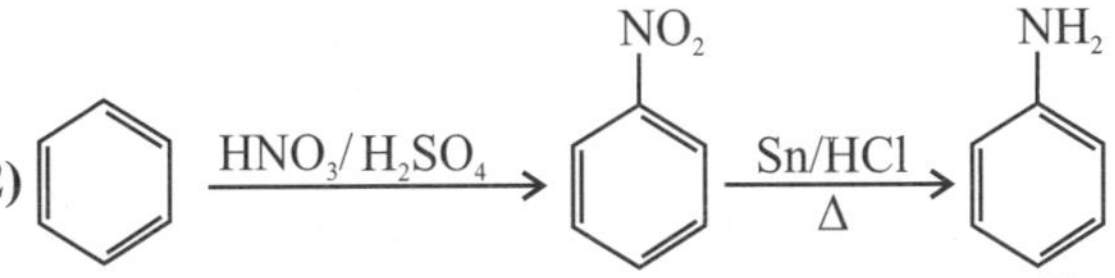

Nitrobenzene Aniline

30. (3) $CH_3CN + 2H \xrightarrow[SnCl_2]{HCl} CH_3 - CH = NH$

$\downarrow H_2O$ boil

$CH_3 - CH = O + NH_3$

Acetaldehyde

31. (1) $C_6H_5 - NH_2 + CHCl_3 + 3KOH \rightarrow C_6H_5NC + 3KCl + 3H_2O$

32. (3) $C_6H_5SO_2Cl + RNH_2 \rightarrow RNHSO_2C_6H_5 \xrightarrow{KOH}$

$\left[R - N^{-}SO_2C_6H_5\right]K^{+} \xrightarrow{HCl} R - NHSO_2C_6H_5$

33. (4) Amines being basic in nature dissolve in dilute HCl and dil. H_2SO_4. They can also coordinate with Cu^{2+} and Ag^{+} ions to form soluble complexes as they can act as good ligands.

34. (1)

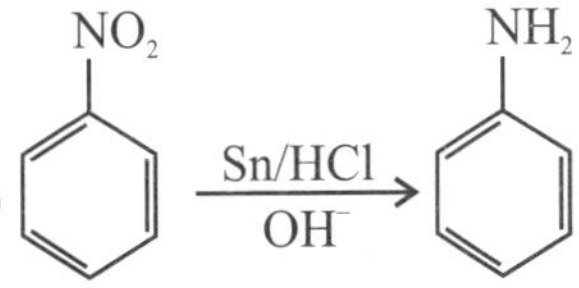

Aniline

35. (2)

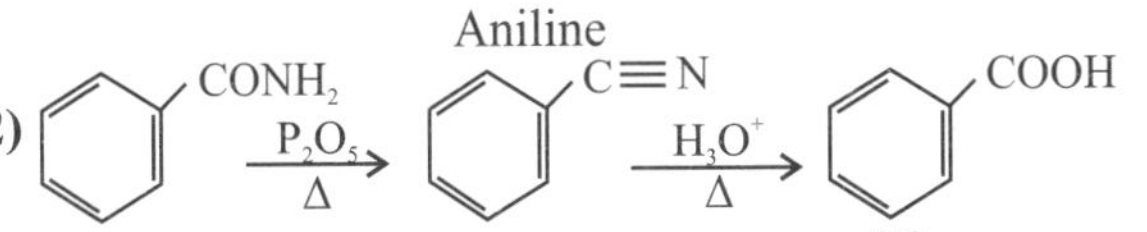

(A) (B)

36. (3) $CH_3CH_2NH_2 \xrightarrow[\Delta]{(CH_3CO)_2O} CH_3CH_2NHCOCH_3 +$ CH_3COOH. This is the acetylation of ethyl amine.

37. (2) Secondary amine reacts with cyclohexanone to give enamine.

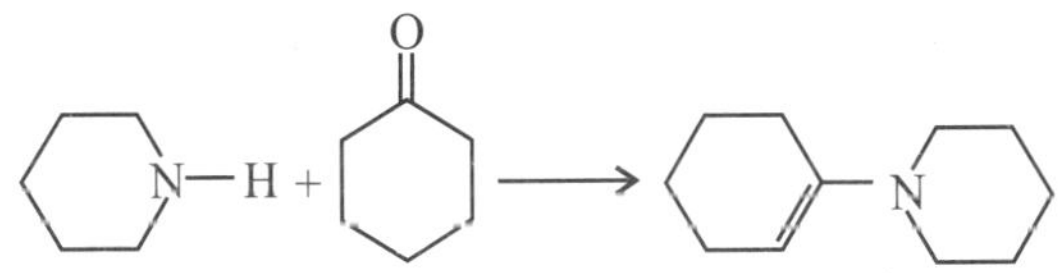

Enamine

38. (3)

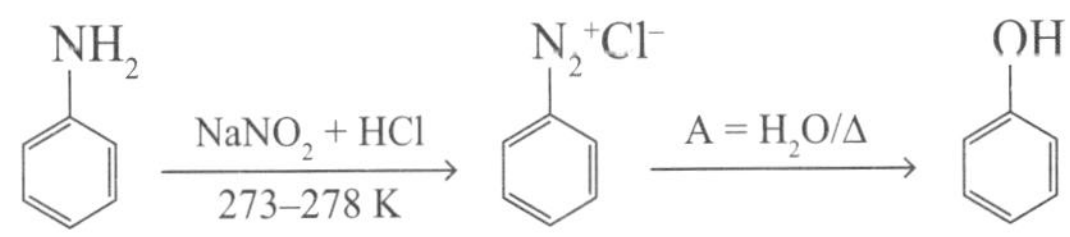

(X)

39. (2)

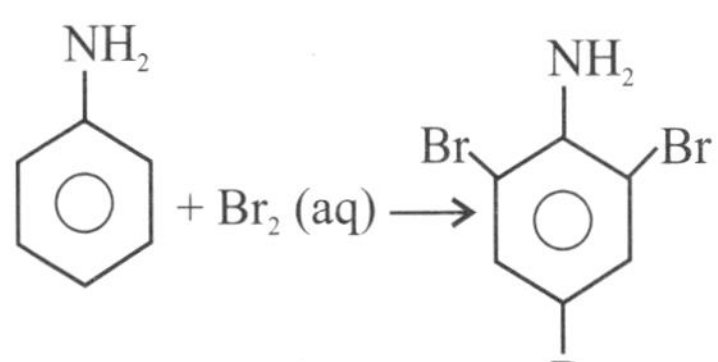

(NH_2 group is highly ring activating and o-, p-directing group for further electrophilic substitution.)

40. (2)

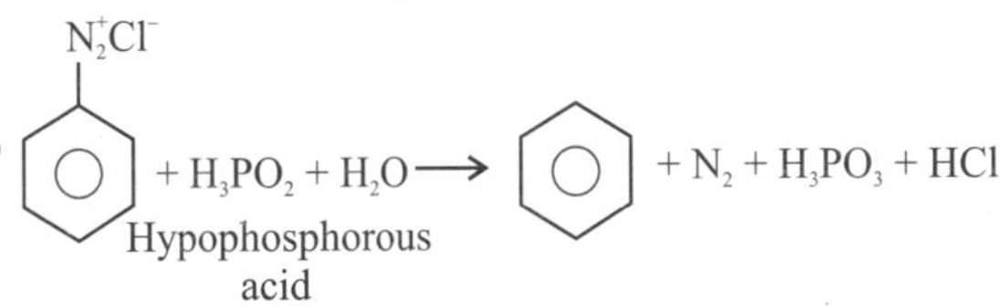

41. (4) (Hoffmann bromamide reaction)

$$RCONH_2 \xrightarrow{Br_2/KOH} \overset{RNH_2}{(1^\circ - \text{Amine})}$$

Gabriel's pthalimide synthesis is given as also used for the preparation of primary amines.

42. (4) 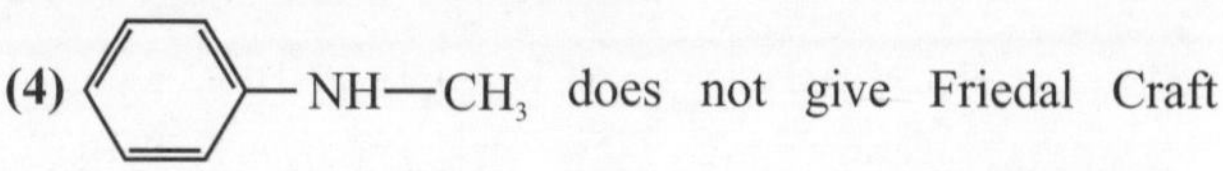 does not give Friedal Craft reaction because it forms additive compound with $AlCl_3$.

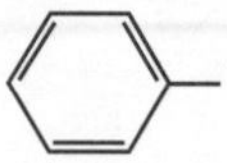 does not give Friedal Craft reaction due to the deactivation of benzene ring.

43. (3) $pK_b \propto \frac{1}{K_b}$ & K_b α basicity

$$pK_b \propto \frac{1}{\text{basicity}}$$

In the given compounds, basicity is shown by the lone pair present on Nitrogen.

In "1", 2 methyl group inhance the donar ability, In "2", 2 ethyl groups are doing the same. In "4" and "3" , only "4" will have high K_b value as it has methyl group.

Therefore, "3" will have least basic and highest pK_b value.

44. (2)

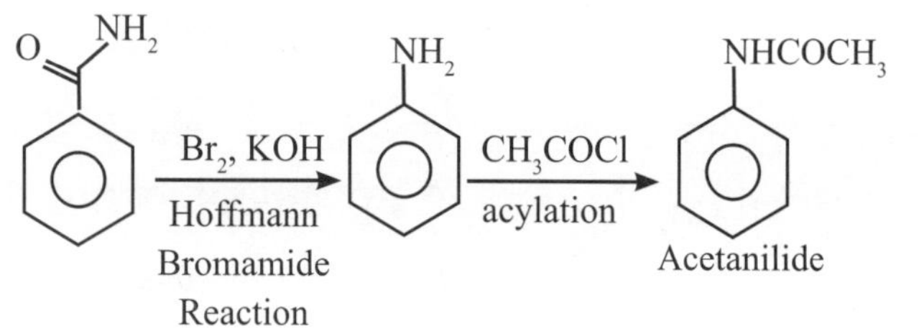

45. (4) Sandmeyer's reaction with chloroaniline gives p-dichlorobenzene which is solid (m.p. 325 K) while that with anilinium chloride give chlorobenzene, which is liquid (m.p. 405 K). Anilinium hydrochloride is an acidic salt and therefore, liberates CO_2 with $NaHCO_3$ and gives white ppt. of AgCl with $AgNO_3$. However, both are primary amines and give carbylamine test and therefore, cannot be distinguished.

46. (2)

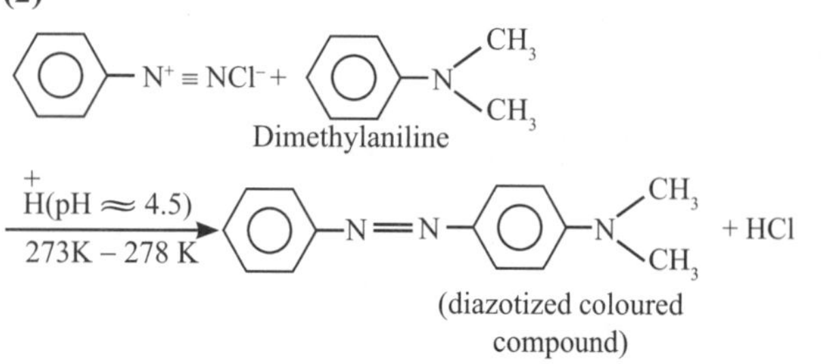

The product is used as a dye.

47. (1)

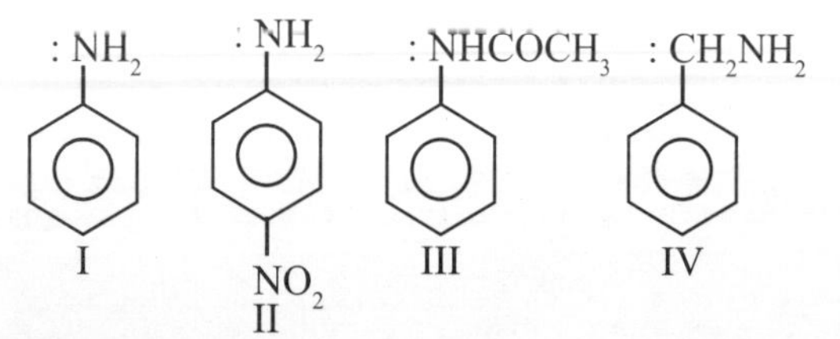

Due to resonance of electron pair in aniline, the basic strength decreases. In benzylamine, electron pair on N is free and is not involved in resonance. Therefore, benzylamine is most basic.

48. (3) $-NO_2$ group in benzene ring deactivates the ring towards electrophilic substitution and activates towards nucleophilic substitution.

49. (3) $CH_3CH_2NH_2 + CHCl_3 + 3KOH \longrightarrow$

$$\underset{\text{Ethyl isocyanides}}{CH_3CH_2NC} + 3KCl + 3H_2O$$

This is Carbylamine reaction.

50. (1)

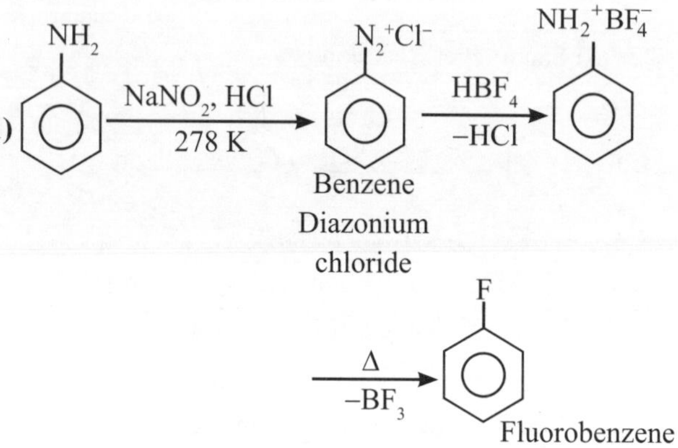

51. (1) The IUPAC name of isoproylamine is Propan-2-amine, The structure of the given compound is:

$$H_3C - \underset{\underset{NH_2}{|}}{CH} - CH_3$$

propan-2-amine

52. (3)

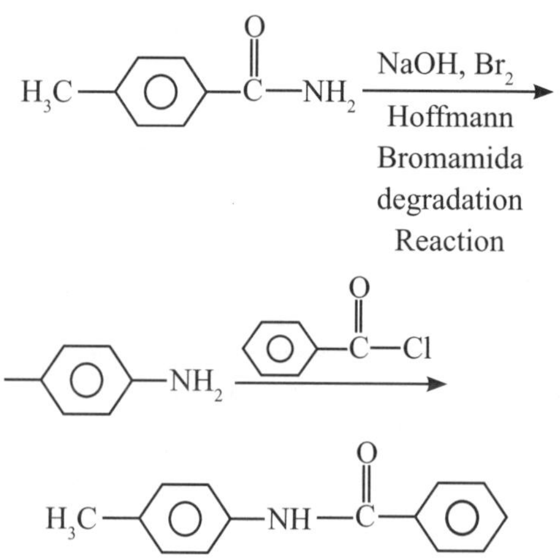

53. (4) The molecule (R) has a localized lone pair, which can participate in conjugation. This enhances its basicity compared to (Q) and (P), which have less favorable electron donation characteristics.

According to Bredt's rule, the lone pairs in (R) are more effectively delocalized, increasing its basicity.

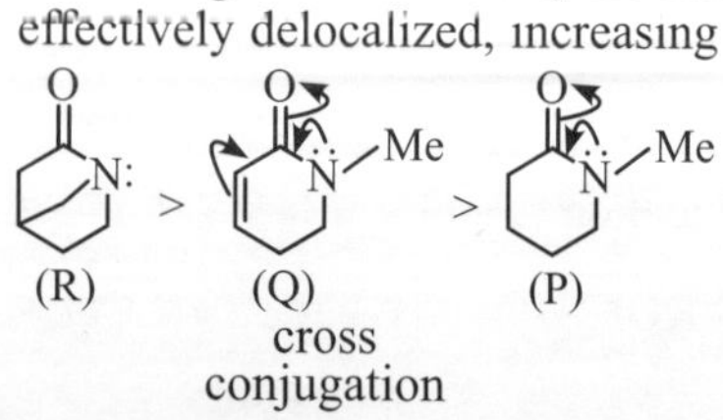

54. **(4)** The order of basic strength in case of ethyl substituted amines in aqueous solution is as follows:

$(C_2H_5)_2NH > (C_2H_5)_3N > C_2H_5NH_2 > NH_3$

Also, NH_3 is more basic than $NH_2\text{-}NH_2$.

Hence, the correct order is:

$NH_2\text{–}NH_2 < NH_3 < CH_3CH_2NH_2 < (CH_3CH_2)_3N < (CH_3CH_2)_2NH$

55. **(4)**

(A)  > (B) 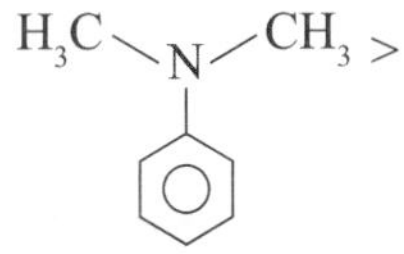 >

(C) 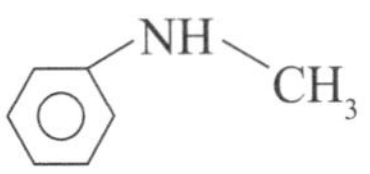 > (D) 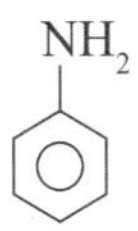

56. **(2)** pK_b α Acidic strength

Electron withdrawing group → increases acidic strength

Electron releasing group → decreases acidic strength

Correct & Incorrect MCQs

1. **(1)** Due to the presence of unshared pair of electrons, the angle C–N–E, (where E is C or H) is less than 109.5°.
2. **(4)** The IUPAC name of isopropylamine is named as propan-2-amine.
3. **(3)** Since, aryl halides do not undergo nucleophilic substitution with the anion formed by phthalimide, hence aromatic primary amines cannot be prepared by this method.
4. **(3)** (1) Methyl amine is more soluble than propyl amine because extent of hydrogen bonding is more in methyl amine.

   (2)The lower aliphatic amines are gases with fishy odour. Hence, methyl amine is a gas with fishy odour.

   (3)Due to atmospheric oxidation, aniline gets coloured on storage.

   (4)The order of boiling points of isomeric amines is: primary > secondary > tertiary.
5. **(2)** N-ethylbenzenesulphonyl amide is obtained by the reaction of benzenesulphonyl chloride with primary amine.

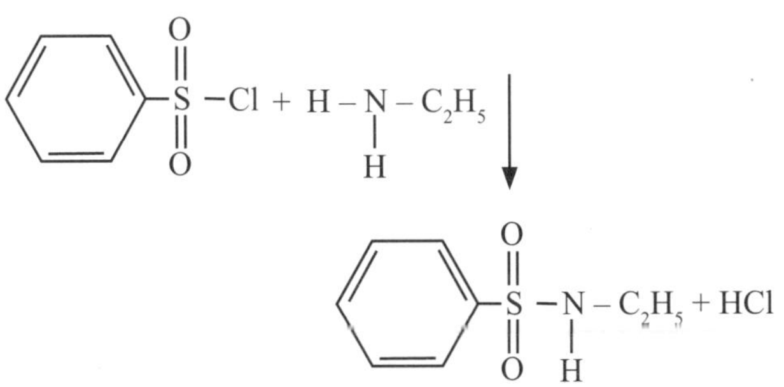

N-Ethylbenzenesulphonamide

The hydrogen attached to nitrogen in sulphonamide is strongly acidic due to the presence of strong electron withdrawing sulphonyl group. Hence, it is soluble in alkali.

6. **(4)** Aryl amines react with nitrous acid to form diazonium salts.
7. **(2)** Electron density in the N–H decreases due to –I effect of F, and therefore release of a proton becomes easier from p-fluoroanilinium ion than from anilinium ion. Rest other statements are correct.
8. **(4)** Dimethyl amine is more basic than methylamine.

   $(CH_3)_2 \rightarrow NH$ and $CH_3 \rightarrow NH_2$
9. **(4)** Benzenediazonium fluoroborate is water insoluble and is stable at room temperature.
10. **(1)** Amines have an unshared pair of electrons on nitrogen atom due to which they behave as Lewis base.

Statement Based MCQs

1. **(1)** Aliphatic amines on heating with chloroform and ethanolic potassium hydroxide forms foul smelling isocyanides or carbylamines. Reduction of isocyanides gives 2° amine.
2. **(1)** Meta product is formed because protonation of aniline by HNO_3 occur to give anilinium ion which is deactivating and meta directing.
3. **(2)** Tertiary amines lacks a reactive hydrogen since it is bonded to 3 alkyl groups & a nitrogen, hence it does not react with Grignard reagent at all.
4. **(1)** Because in case of trimethylamine, lone pair of nitrogen are sterically hindered by presence of 3 bulkier methyl group. So, it can not easily donate its lone pair of electrons hence it is weaker base than dimethylamine.

   Inductive effect is more in case of tertiary amines than that of secondary amine.
5. **(3)** Condensation of diazonium salt with phenol is carried out in weakly alkaline medium (pH = 9). This is due to the fact that in strongly acidic medium, the –OH of a phenol remains unionised, and an amine forms a salt. Phenol exists as phenoxide ion and the latter is readily substituted by electrophiles than phenol itself. Thus, in phenol, coupling is carried out in alkaline medium.
6. **(1)** In an amine, both the C – N and N – H bonds are polar due to the electronegativity of the N – atom.

   Since, aryl halides do not undergo nucleophilic substitution with the anion formed by phthalimide, hence aromatic primary amines cannot be prepared by Gabriel phthalimide synthesis.
7. **(3)** Nitriles on reduction with lithium aluminium hydride ($LiAlH_4$) or catalytic hydrogenation produce primary amines.
8. **(3)** In pyrrole, nitrogen is sp^2 hybridized and the lone pair of electrons on N is involved in aromatic sextet formation, hence, pyrrole is less basic than pyridine.
9. **(4)** In amines, unshared electron pair is present on nitrogen atom, hence amines behave as nucleophiles.

10. (3) 3° amine do not contain a H – atom on N– atom.

Tertiary (3°) amine

Assertion & Reason MCQs

1. **(3)** Conceptual
2. **(3)** Hoffmann's bromamide reaction is used to convert amide in to amine. It is used for preparation of 1° amine. 2° amine are more basic than 1° amine.
3. **(1)** N–Ethylbenzene sulphonamide is soluble in alkali because it has acidic Hydrogen.
4. **(2)** N–diethylbenzene sulphonamide is insoluble in alhali because of absence of acidic hydrogen.
5. **(1)** Conceptual
6. **(3)** The action will decrease the reactivity of ring as it is meta directing group and also a bulkier group too
7. **(3)** In isocyanide test or carbylamine test, amines get converted into foul smelling isocyanides in the presence of chloroform and alcoholic KOH.
8. **(1)** Secondary and tertiary amines do not show this reaction and this is why this reaction is used to differentiate between amines
9. **(3)** Tertiary amines lacks a reactive hydrogen, as it is bonded to 3 alkyl groups & a nitrogen, therefore, it does not react with Grignard reagent at all.
10. **(2)** Both Assertion and Reason are True but Reason is not the correct explanation of the Assertion.

Multi Correct MCQs

1. **(2)**

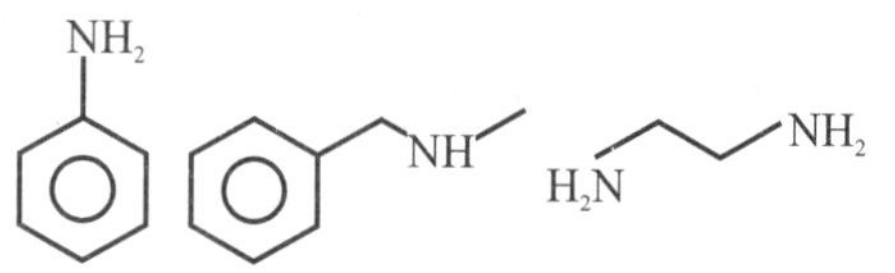

Only 1° and 2° amines give reaction with Hinsberg's Reagent.

2. **(2)** Since, b, c and d have localised l.p on nitrogen, so most favaourable sites for protonation.

3. **(3)**

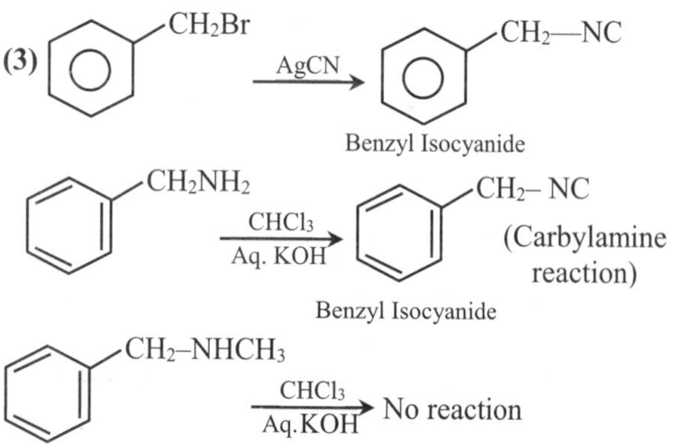

Only Primary amine can give carbylamine reaction

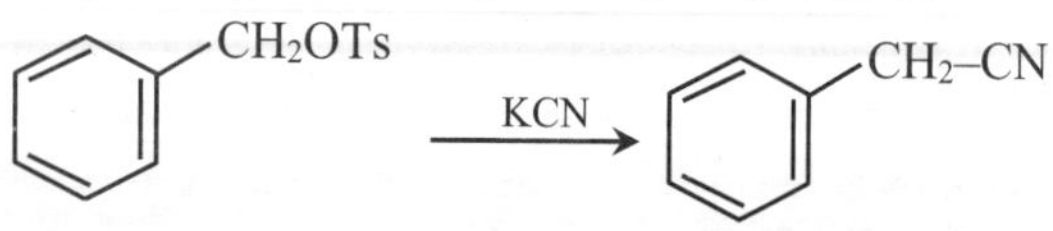

4. **(1)** (I) and (III) will give p-aminoazobenzene But (II) will not give p- aminoazobenzene

5. (2)

1. Aromatic amines are less basic than aliphatic amines because the nitrogen's lone pair is delocalized in aromatic amines, making it less available for protonation.

2. Secondary amines (E > D) are more basic than primary amines due to electron-donating effects from alkyl groups.

3. In aromatic amines, electron-withdrawing groups like NO_2 decrease basicity, so C (with NO_2) is less basic than B.

Order of basicity: E > D > B > A > C.

NCERT Exemplar MCQs

1. **(2)** is a trimethyl amine because three groups are attached to nitrogen instead of hydrogen.

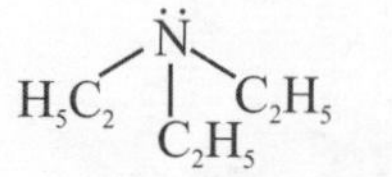

2. **(4)** IUPAC name of $\overset{3}{C}H_2 = \overset{2}{C}H\overset{1}{C}H_2\ NHCH_3$ is N-Methylprop-2-en-1-amine

   Hence, option (4) is correct.

3. **(3)** $(CH_3)_2\ NH$ is most basic in the given compounds as it is 2°-amine with 2 EDG.

	Compound	Factors responsible for basic character are
(a)	$CH_3 - NH_2$	Inductive effect (+ I)
(b)	$NC - CH_2 - NH_2$	Inductive effect (– I)
(c)	$(C_6H_5)NH\ CH_3$	– I effect and resonance

   Since, + I effect and solvation increases basic character but – I effect and resonance decreases basic character. Hence, correct choice is (3).

4. **(1)** Aniline is weakest Bronsted base among the given four compounds due to resonance present in case of aniline.

   Hence, lone pair of nitrogen are less available for donation to the acid.

5. **(3)** S_N1 reaction proceeds through formation of carbocation, Hence, more stable is the carbocation, more reactivity towards S_N1 mechanism.

   Hence, the reaction will proceed through S_N1 mechanism when $C_6H_5CH_2Br$ is the substrate because in ionisation, it gives a resonance stabilized carbocation $(C_6H_5\ CH_2)^+$.

6. **(2)** Aryl nitro compound can not converted into amine using $LiAlH_4$ in ether.

7. **(3)** In order to prepare 1° amine from an alkyl halide with simultaneous addition of one CH_2 group in the carbon chain, the reagent used as a source of nitrogen is KCN. Chemical transformation can be shown as:

   $CH_3Cl + KCN \rightarrow CH_3CN + KCl$

8. **(4)** Source of nitrogen in Gabriel phthalimide synthesis is potassium phthalimide.

9. (3) Chemical transformation for the preparation of 2°-amine can be shown as:

$$R{-}\ddot{N}H_2 + R{-}\overset{O}{\overset{\|}{C}}{-}H \longrightarrow R{-}N{=}CH{-}R \xrightarrow{H_2/Pt} R{-}NH{-}CH_2{-}R$$

While other given set of reactants gives primary amine only.

10. (4) The best reagent for converting 2-phenylpropanamide into 2-Phenylpropanamine is $LiAlH_4$ in ether.

11. (2) The best reagent for converting 2-Phenylpropanamide into 1-Phenylethanamine is $NaOH/Br_2$. The chemical transformation can be shown as:

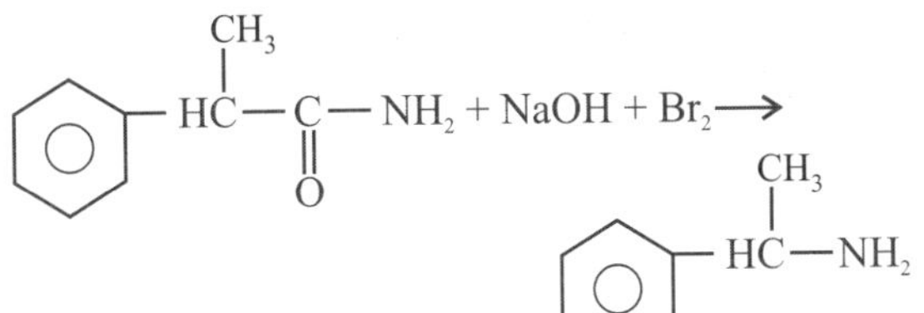

This occurs due to intramolecular migration of alkyl group. It is an example of Hoffman bromamide reaction.

12. (2) Hoffman bromamide degradation is shown by $Ar{-}\overset{O}{\overset{\|}{C}}{-}NH_2$ by which amide is converted into amine via intramolecular migration of phenyl group.

13. (4) The correct increasing order of basic strength is as follows,

Greater the electron density towards ring, greater will be its basic strength.

Electron withdrawing group decreases basic strength but electron donating group increases basic strength.

14. (3) Methylamine reacts with HNO_2 (nitrous acid) to form methanol

15. (2) Chemical reaction takes place during reaction of methylamine with nitrous acid is as follows:

$$CH_3{-}NH_2 + HNO_2 \rightarrow CH_3OH + N_2 + H_2O$$

16. (3) Nitration of benzene using a mixture of conc. H_2SO_4 and conc. HNO_3 proceeds as via NO_2^+.

This reaction is known as electrophilic substitution reaction.

17. (3) Aromatic nitro compound on reaction with Fe and HCl gives aromatic primary amine as shown below:

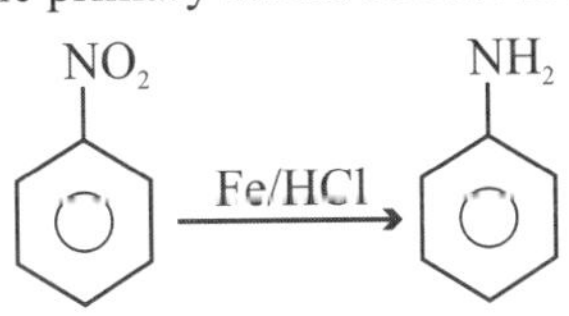

18. (2) Greater the strength of base, greater will be its reactivity towards dilute HCl. Hence, $(CH_3)_2NH$ has highest basic strength, thus it has highest reactivity.

19. (1) Acid anhydride on reaction with primary amines produces amide as:

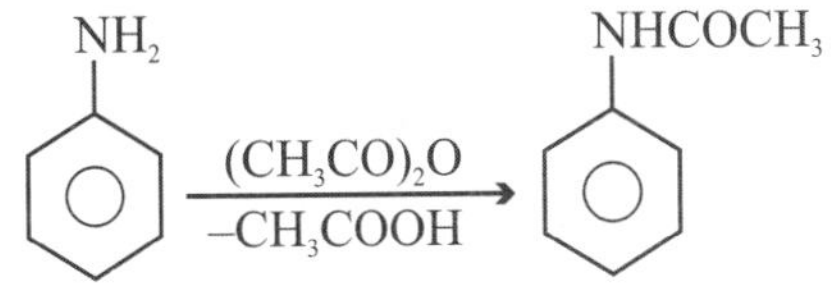

20. (2) This reaction is called Gatterman reaction. In this reaction, Cl, Br and CN can be introduced into the benzene ring by simply treating diazonium salts with HCl, HBr, KCN, respectively in presence of copper powder instead of using Cu (I) salts.

21. (2) Best method for preparing primary amines from alkyl halides without changing the number of carbon atoms in the chain is Gabriel phthalimide synthesis

22. (4) Nitrobenzene will not undergo azo coupling reaction with benzene diazonium chloride but other three undergo diazo coupling reaction very easily. Diazonium cation is a weak E^+ and hence reacts with electron rich compounds containing electron donating group i.e., – OH–NH_2 and –OCH_3 groups but not with compounds containing electron withdrawing group i.e., NO_2 etc.

23. (3) Phenol is weakest Bronsted base as phenol after losing a lone pair produces least stable conjugate acid among the compounds.

Phenol has the least tendency to accept a proton and hence it is weak Bronsted base. Hence, phenol is least basic among given four choices.

24. (4) Aniline is a weaker base than NH_3 due to delocalization of lone pair of electrons of the N-atom over the benzene ring. Pyrrole is not more basic because the lone pair electrons on the N-atom is donated towards aromatic sextet formation.

Therefore, pyrrolidine is strongest base as lone pair of nitrogen does not included in resonance and also due to presence of alkyl ring residue, basic strength becomes high among given four compounds.

25. (1) Basic strength of the above species can be explained on the basis electronegativity of central atom and its proton accepting tendency. Hence, amide ion is most basic among given compounds due to presence of negative charge and two pair of electrons on nitrogen atom.

26. (2) 1° and 2° amines has higher boiling points due to intermolecular H-bonding but less volatile than 3° amines and hydrocarbons of comparable molecular mass. Further, because of polar C-N bonds, 3° amines are more polar than hydrocarbons which are almost non-polar. Hence, due to weak dipole-dipole interactions. 3° amines have higher boiling point (i.e., less volatile) than hydrocarbons.

In other words, hydrocarbons are more volatile among given compounds as amines are less volatile than hydrocarbon.

27. (1, 2, 3) Aliphatic and arylalkyl primary amines can be easily prepared by the reduction of the corresponding nitriles with $LiAlH_4$.

$$\underset{1^\circ\text{amide}}{R-C\equiv N} \text{ or }$$

$$\underset{\text{Arynitrile}}{Ar-C\equiv N} \rightarrow \underset{1^\circ\text{ amine}}{RCH_2NH_2 \text{ or } ArCH_2NH_2}$$

Heating alkyl halide with Primary, secondary and tertiary amine can be prepared by reduction of $LiAlH_4$ followed by treating with water.

$$\underset{1^\circ \text{ amide}}{R-CONH_2} \xrightarrow[\text{(ii) } H_2O]{\text{(i) } LiAlH_4/\text{ether}} R-CH_2-NH_2$$

NEET & AIIMS Past Year MCQs

1. (3)

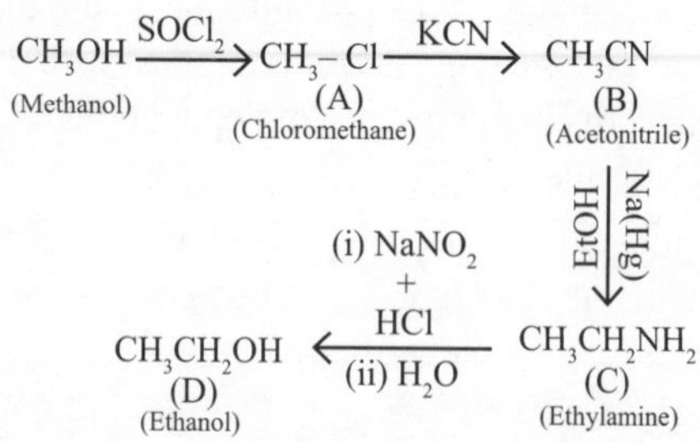

2. (1) The positive carbylamine test indicates the presence of a primary amine.

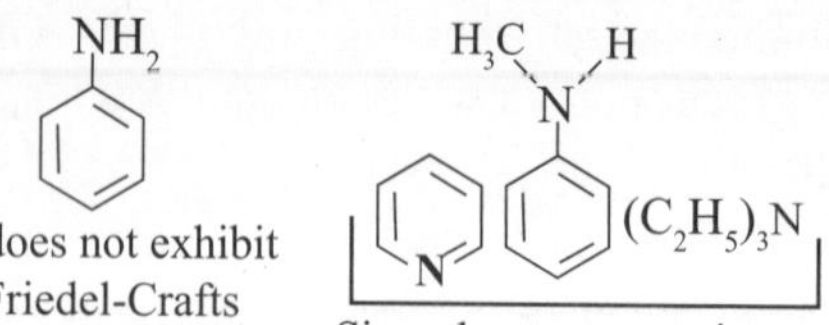

It does not exhibit Friedel-Crafts alkylation but yields a positive carbylamine test.

Since these are not primary amines, they do not give the carbylamine test.

3. (3)

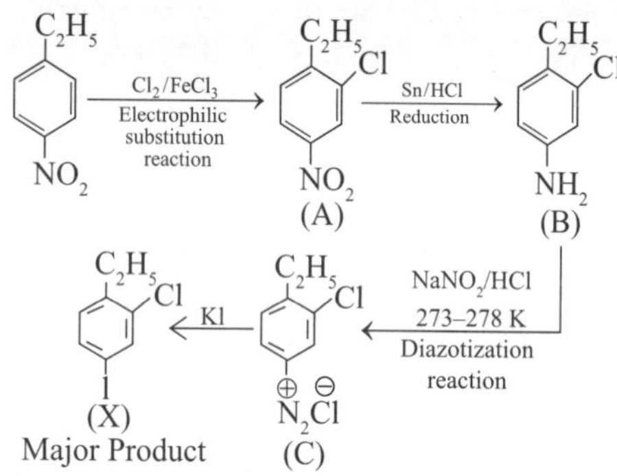

4. (3)

- Aniline does not undergo Friedel-Craft reaction due to salt formation with anhy. aluminium chloride (the lewis acid) which is used as a catalyst.
- Gabriel synthesis is used for the preparation of aliphatic primary amines only. Aromatic primary amines cannot be prepared by this method because aryl halide do not undergo nucleophilic substitution reactions.

5. (3)

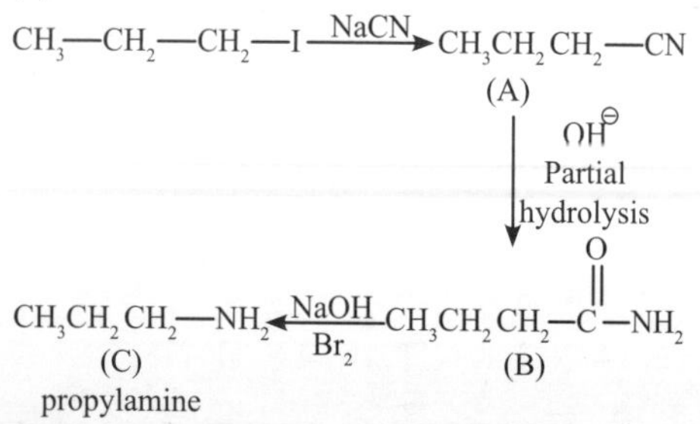

6. (4) $CH_3CONH_2 \xrightarrow[\text{(ii) } H_3O^{\oplus}]{\text{(i) } LiAlH_4} CH_3CH_2NH_2$

$CH_3CONH_2 \xrightarrow{Br_2/KOH} CH_3NH_2$

$CH_3CN \xrightarrow[\text{(ii) } H_3O^{\oplus}]{\text{(i) } LiAlH_4} CH_3CH_2NH_2$

$CH_3NC \xrightarrow[\text{(ii) } H_3O^{\oplus}]{\text{(i) } LiAlH_4} CH_3NHCH_3$

Methyl isocyanide gives a secondary amine, CH_3NHCH_3 upon reduction.

7. (3) Bromobenzene is formed in first step (Sandmeyer reaction), which further gives phenyl magnesium bromide. Phenyl magnesium bromide further gives benzene with water.

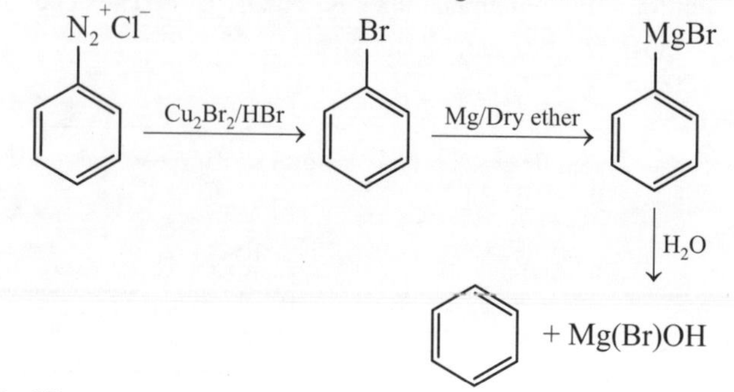

8. (4)

$R-NH_2 \xrightarrow{NaNO_2+HCl} [R-N_2^{\oplus}Cl] \xrightarrow{H_2O} ROH + N_2 + HCl$

Alkyl diazonium ion (unstable)

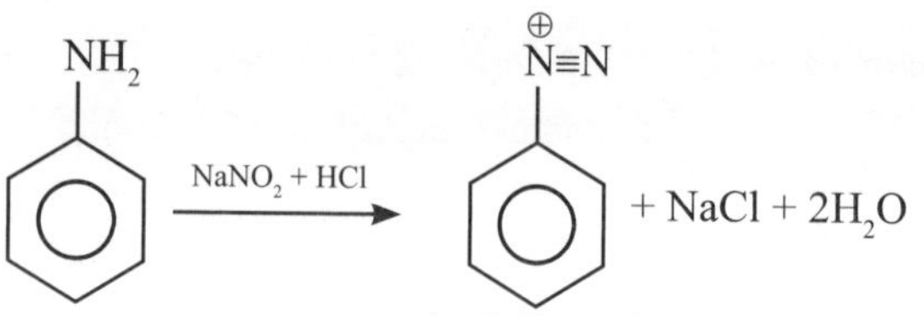

Aryl diazonium ion stable at low temperature (0–5°C)

9. (1)

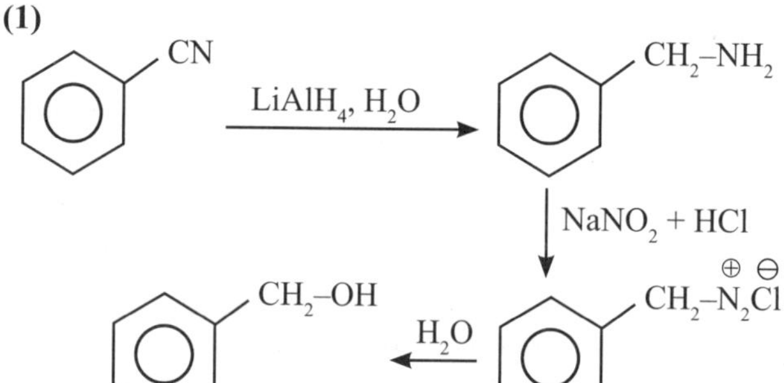

10. (2) 1° amines react with Hinsberg's reagent to give a solid, which dissolves in alkali.

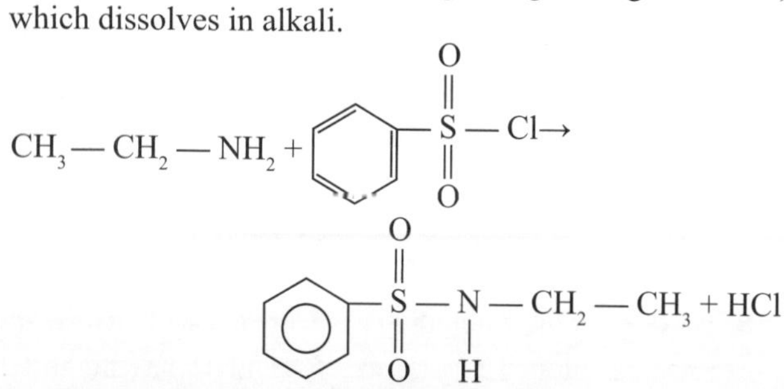

Soluble in alkali

11. (1) 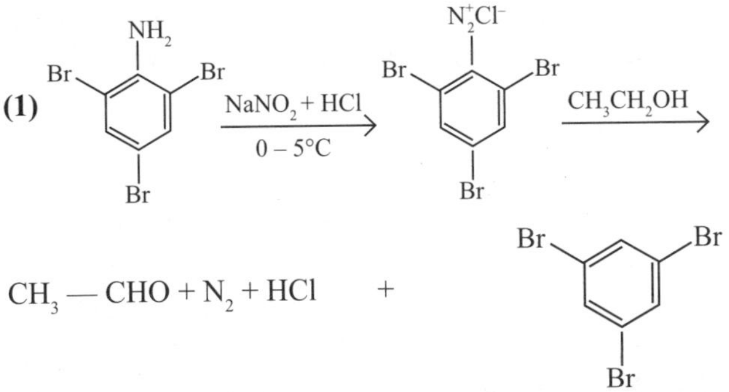

Certain mild reducing agents like hypophosphorus acid or ethanol reduce diazonium salts to arene and themselves get oxidised to phosphorous acid and ethanal respectively.

12. **(1)** Tertiary amines are not oxidised by $KMnO_4$.

13. **(4)**

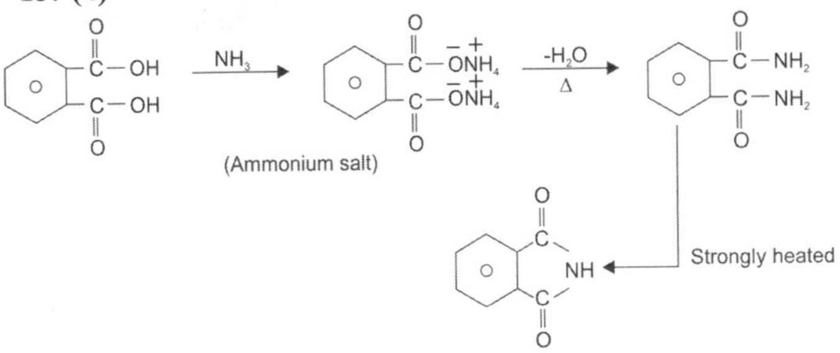

14. **(3)** 15. **(2)** 16. **(4)** 17. **(4)**

Biomolecules

NCERT Topic-wise MCQs

1. **(2)** The formula of starch is $(C_6H_{10}O_5)_n$
2. **(4)** $C_1 - C_4 - \alpha$–linkage
3. **(4)** It contains hemiacetal group.
4. **(1)** Lactose consists of one galactose and one glucose unit.
5. **(2)** Non-reducing sugars are the carbohydrates which are not reduced by Tollen's and Fehling's test.
6. **(3)** Cellulose is the carbohydrate which would be most abundant in the diet of strict vegetarian.
7. **(4)** Glucose $\xrightarrow{\text{conc.}HNO_3} COOH-(CHOH)_4\ COOH$

   Fructose $\xrightarrow{\text{conc.}HNO_3} HOOC-(CHOH)_2-COOH+HOCH_2-COOH$
8. **(4)** The structure of glycogen is similar to amylopectin.
9. **(3)** All monosaccharides & disaccharides except sucrose are reducing sugar & reduce tollen's reagent.
10. **(3)** This test indicates the presence of –CHO group.
11. **(1)** Starch $(C_6H_{10}O_5)_n$ is a polymer of glucose $(C_6H_{12}O_6)$.
12. **(1)** Polysaccharides, contain a large number of monosaccharide units joined together by glycosidic linkages.
13. **(2)** A mixture of amylose and amylopectin is called starch. Amylopectin is water insoluble component while amylose is water soluble fraction component.
14. **(3)** Furanose ring is found in the fructose unit of cane sugar. In cane sugar, the fructose molecule have a five membered ring structure (furanose ring).
15. **(4)** Glycogen is stored in liver of animals.
16. **(3)** Starch $\xrightarrow{\text{Diastase}}$ Maltose $\xrightarrow{\text{Maltase}}$ glucose
17. **(4)** Carbohydrates are polyhydroxy aldehydes or ketones. Examples are cane sugar, glucose, starch.
18. **(3)** The enantiomers of α – D – (+) – glucose is α – L – (–) – glucose.
19. **(1)** Monasaccharide containing aldehydic group is known as an aldose. Among the given, glucose is an aldose.
20. **(3)** Glucose is a monosaccharide not disaccharide.
21. **(4)** Lactose = glucose + galactose
22. **(1)** Saccharic acid – $C_6H_{10}O_8$
23. **(4)** All the given are reducing sugars.
24. **(1)**
25. **(3)** Molecular formula $C_6H_{12}O_6$ is of both glucose and fructose.
26. **(1)** Glucose gets oxidised to gluconic acid ($C_6H_{12}O_7$) on reaction with a mild oxidising agent like bromine water.
27. **(2)** Amylopectin is not soluble in water.
28. **(3)** $C_{12}H_{22}O_{11}$ – 11 O atoms
29. **(2)** A carbohydrate that cannot be hydrolysed further to give simpler unit of polyhydroxy aldehyde or ketone is known as a monosaccharide.
30. **(1)** Pyranose ring - 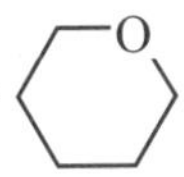
31. **(3)** Glucose reacts with acetic anhydride to form penta-acetate since a molecule of glucose contains five -OH groups.
32. **(2)** The disaccharide lactose is present in milk. It is composed of β-D-galactose and β-D-glucose.
33. **(2)** To become a carbohydrate, a compound must contain at least 3 carbons.
34. **(3)** Oligosaccharides are the compounds which give 2 to 10 monosaccharides on hydrolysis. For example-sucrose, lactose, maltose etc.
35. **(4)** High content of starch is found in cereals, roots, tubers and some vegetables.
36. **(3)** Amylose constitutes about 15-20% of starch.
37. **(4)** In aqueous solutions, the α-amino acids exist as: $H_3N^+ - CHR - COO^-$
38. **(4)** Globular proteins are usually soluble in water and are present in blood, milk, egg etc. All of these are globular protein and soluble in water.
39. **(1)** Leucine is an essential amino acid.
40. **(4)** 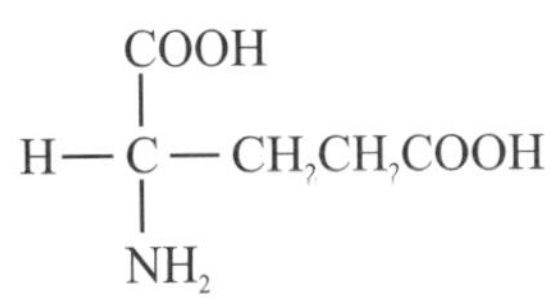

   (Glutamic acid)

   It is acidic in nature because in this group, there are two carboxylic acid and one amino acid.

41. **(4)** Keratin is a fibrous protein, it is present in hair, wool, silk, muscles. etc.
42. **(2)** α-helix refers to secondary structure of protein.
43. **(2)** Except glycine, all amino acids have a chiral centre and almost all naturally occuring amino acids have L-configuration at the α-carbon.

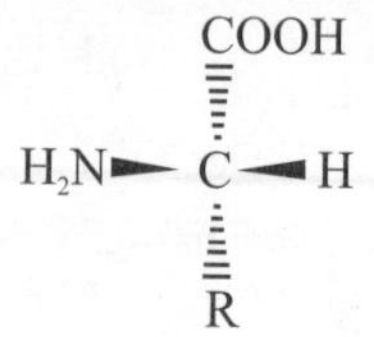

44. **(2)** As this group has positive charge on amino group and carboxylic acid is also present.
45. **(3)** Globular proteins results when the polypeptide chains coil around to give 3-D spherical shape, they are soluble in water. Insulin and albumins are the common examples of globular proteins.
46. **(1)** Among the given amino acids, methionine is an essential amino acid, as it is not synthesised in the body.
47. **(3)** Protein is a polymer of amino acid.
48. **(2)** The helical structure of protein is stabilised by hydrogen bonds.
49. **(4)** Denaturation of proteins lead to loss of biological activity of secondary and tertiary structures but does not affect the primary structure of protein.
50. **(1)** Proteins are condensation polymers of α-amino acids.
51. **(3)** In human body, there are 10 amino acids that are synthesised.
52. **(3)** The pH of the solution would be approximately equal to 6.
53. **(3)** —CO—NH—linkage is a peptide linkage.
54. **(2)** α Helix and β-pleated sheet structures are the most common types of secondary structures of protein.
55. **(2)** Native Protein
56. **(3)** A zwitterion is a molecule that contains both +vely and –vely charged functional groups, net charge of the entire molecule is zero. Amino acids are the best-known examples of zwitterions.
57. **(2)** In an alkaline medium, glycine predominantly exists as an anion.
58. **(3)** Glycine is the only amino acid which is not chiral.
59. **(3)** Cysteine is an amino acid which contain sulphur.

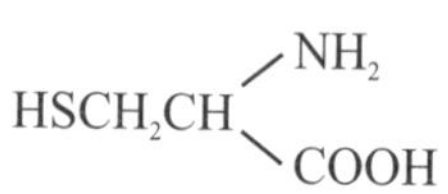

60. **(3)** Proteins are polypeptides with molecular weights above 10,000.
61. **(4)** Providing energy for metabolism is not a function of proteins. Because protein is a body building substance not energy giving substance.
62. **(1)** Fibrous proteins are insoluble in water.
63. **(2)** Amino acids which can't be synthesised in the body and must be taken through diet, are known as essential amino acids.
64. **(3)** On the basis of molecular structure, proteins are classified into two types that is fibrous protein and globular proteins.
65. **(2)** A dipeptide, glycylalanine is obtained when carboxyl group of glycine combines with the amino group of alanine.
66. **(2)** Estradiol is responsible for developing secondary female characters.
67. **(4)** The reaction of enzymes in living system is to enhance the rate of biochemical reaction.
68. **(2)** Enzymes are made up of protein with specific structure.
69. **(4)** Enzymes works at a high rate at optimum temperature.
70. **(4)** Lipase is not useful in the digestion of proteins. It is used in lipid metabolism.
71. **(4)** Deficiency of Vitamin A causes night blindness.
72. **(1)** Vitamins A, D E and K are fat soluble vitamins and group B vitamins and vitamin C are water soluble vitamins.
73. **(1)** Scientific name of vitamin B_6 is pyridoxin.
74. **(3)** Ascorbic acid is the chemical name of vitamin C which is a water soluble vitamin.
75. **(2)** Fat soluble vitamins are insoluble in water.
76. **(2)** Vitamin B_2 is known as Riboflavin.
77. **(2)** DNA contains adenine (A), thymine (T), guanine (G) and cytosine (C) bases.
78. **(3)** A segment of DNA molecule which codes or specifies for one polypeptide chain is called Gene.
79. **(4)** DNA multiplication is called replication.
80. **(1)** In DNA, double helical structure is stabilised by hydrogen bonding.
81. **(3)** Uracil is present in RNA while thymine is present in DNA.
82. **(1)** Thymine is 5-methyluracil.
83. **(4)** In a DNA molecule, A = T, C ≡ G.
    So, the complimentary sequence of ATGCTTGA is TACGAACT.
84. **(1)** Guanine is a heterocyclic nitrogenous purine base present in nucleic acids.
85. **(1)** The sugar moiety present in DNA molecules is β-D-2-deoxyribose.

High Order Time Intensive MCQs

1. **(4)** The secondary and tertiary structures of proteins are stabilized by hydrogen bonds, disulphide linkages, van der Waals and electrostatic forces of attraction.
2. **(2)** Glycosidic linkage is between C_1 of α-glucose and C_2 of β-fructose in sucrose.
3. **(4)** Glucagon-Hormone
4. **(2)** DNA transfers characters from parents to offspring.
5. **(4)** Functionality of protein depends on all of these factors, i.e., shape, pH, temperature, etc.
6. **(3)** Vitamin K helps in the clotting of blood.
7. **(3)** Insulin contains two polypeptide chains with 21 and 30 amino acids which are joined together by sulphur bridges. Hence, it is a peptide hormone.

8. (1) Dextrose occurs in the urine of diabetics.

9. (2) Fructose is a ketohexose.

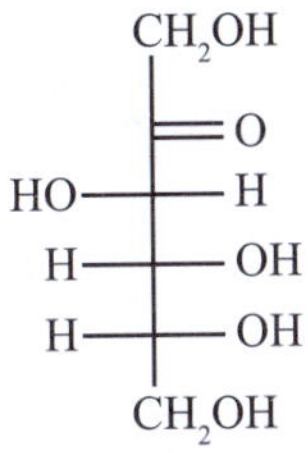

10. (3) Glycogen

11. (2)

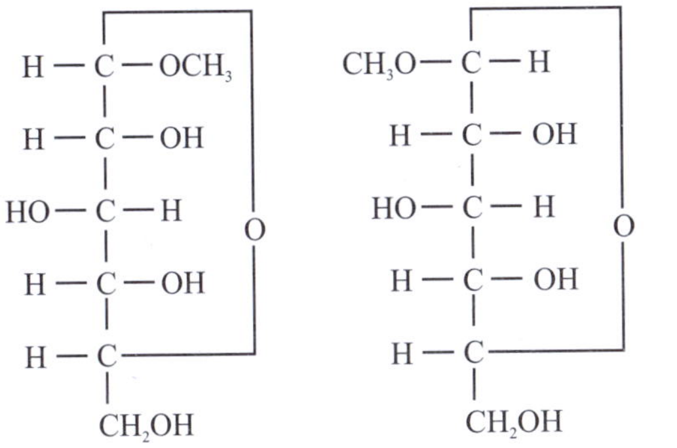

Methyl–α–D–glucoside   Methyl–β–D–glucoside

Both are anomers, differ in configuration at carbon -1.

12. (2) It is a six membered ring (Pyranose) but not α-pyranose. It has β-orientation at C-1. Since there is one H & one OH at C-1, it is an aldohexose.

13. (4) The reactions of glucose with Schiff's reagent, sodium bisulphite, 2, 4-DNP & ammonia are reversible & hence open chain form always remains in equilibrium with the two cyclic hemi-acetal forms. As a result, these reagents cannot disturb the equilibrium to generate more of open chain form from the cyclic hemi acetal and hence, do not react with glucose.

14. (1) Glycine has the following structure:

All amino acids except glycine are optically active in nature.

15. (4) Glucose react with NH_2OH to from oxime.

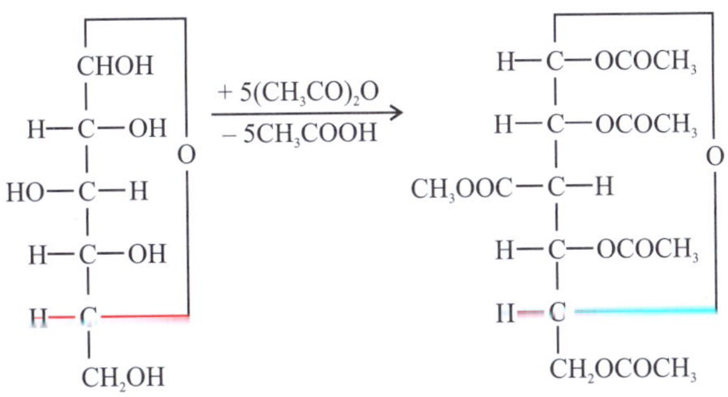

The pentacetate does not have free –OH group at C_1 & are not hydrolysed to produce open chain & hence do not react with NH_2OH to form oxime.

16. (1) Glucose $\xrightarrow{RedP/HI}$ n-hexane.

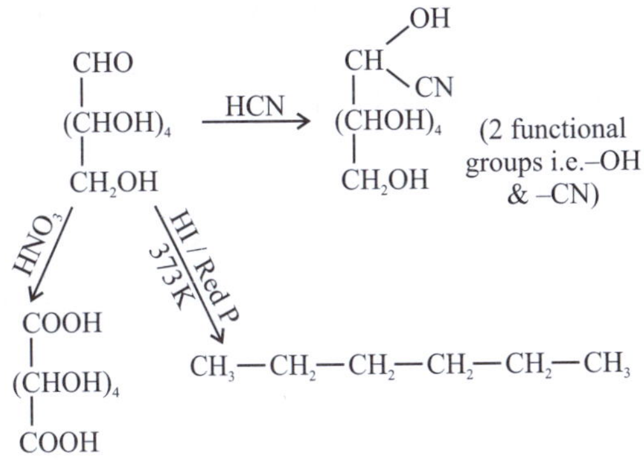

Saccharic acid
(2 functional groups i.e., –COOH & –OH)

17. (1) Anomers of glucose refers to the isomers of glucose that differ in configuration at C-1.

18. (4) It is actually lactose. It is made up of β-D-galactose & β-D-glucose. On the left hand side is galactose & on the right hand side is glucose. There is free anomeric –OH.

19. (1) It has three chiral carbons and may have symmetry. (2^3) = 8 stereoisomers.

20. (1) Coagulation of egg give 1° structure of protein, which is known as denaturation of protein.

21. (2) Glycine is the simplest amino acid (–H as side chain), lacks a chiral center, hence is not optically active, and a non-essential amino acid.

22. (1) Sucrose on hydrolysis give 1:1 ratio of D-(+)–glucose and D-(-)- fructose.

23. (4) (+)(-) Sucrose is a non-reducing sugar.

24. (4) Proteins are hydrolysed by acids, bases and enzymes to give a mixture of α-amino acids.

25. (4) In aqueous solutions, the α-amino acids exist as:

$H_3N^+ - CHR - COO^-$

26. (2) At higher pH, an amino acid exists as an anion.

27. (4) Vitamin B_{12} is stored in our body despite of being water soluble and is obtained from egg, curd etc and it's deficiency causes pernicious anaemia.

Correct & Incorrect MCQs

1. (1) All monosaccharides whether aldose or ketose are reducing sugars because they either have an aldehyde group or can tautomerize in solution to form an aldehyde group.

2. (1) Maltose is a disaccharide which is composed of two α-D-glucose units.

3. (1) 1. Starch is a polymer of α - glucose.

2. Amylose and amylopectin are components of starch.

3. Amylopectin is insoluble in water and constitutes about 80- 85% of starch.

4. There are four carbons and one oxygen atom present in the ring in cyclic structure of fructose.

4. (1) It is an aldohexose.

5. **(1)** Glycine is a non-essential amino acid.

6. **(3)** –COOH is the acidic functional group in amino acid.

7. **(1)** Keratin and myosin are examples of fibrous proteins while insulin and albumins are the examples of globular proteins.

   Fibrous proteins are insoluble in water while Globular proteins are double in water.

8. **(4)** The secondary and tertiary structures are destroyed while primary structure remains intact during denaturation.

9. **(3)** Enzyme maltase converts maltose into glucose.

10. **(4)** RNA molecules mainly are of three types i.e., messenger RNA (m-RNA), ribosomal RNA (r-RNA) and transfer RNA (t-RNA).

Statement Based MCQs

1. **(1)** When fructose is warmed with dilute alkali, it gives rise to an equilibrium mixture of glucose, fructose and mannose. Fructose does not contain any aldehyde group. The ability of fructose to reduce Fehling solution and Tollen's reagent is probably due to the isomerisation of fructose to glucose and mannose.

2. **(1)** Disaccharides are the carbohydrates which upon hydrolysis yield two molecules of the same or different monosaccharides. For example, sucrose on hydrolysis give each molecule of glucose and fructose. Sucrose is a non-reducing sugar.

3. **(1)** In acidic medium, $-NH_2$ group accepts a proton and acts as the base. Hence, α-amino acids exist as cations in acidic medium and migrate towards cathode under the influence of an electric field. In alkaline medium, –COOH group acts as the acid and loses a proton. Therefore, α-amino acids exist as anion and migrate towards anode under the influence of an electric field.

   Also, α-amino acids are the building blocks of proteins.

4. **(4)** Maltose is composed of two α-D-glucose units in which C_1 of one glucose is linked to C_4 of another glucose unit.

5. **(3)** Acetylation of glucose with acetic anhydride gives glucose pentaacetate.

6. **(1)** The structure of glycogen is similar to that of amylopectin and is known as animal starch.

7. **(3)** Adenine combine with thymine with double bond and guanine combine with cytosine with triple bond.

8. **(4)** Deficiency of vitamin A causes night blindness.

9. **(4)** Number of carbon atoms present in aldohexose are 6.

10. **(3)** Fructose has the molecular formula $C_6H_{12}O_6$.

Assertion & Reason MCQs

1. **(3)** (+) represents the dextrorotatory nature of glucose.

2. **(1)** Vitamin D is soluble in fat and oil and can be stored in liver and adipose tissues.

3. **(4)** α-glycosidic linkage is present in maltose.

4. **(2)** All other naturally occurring α-amino acids are optically active in nature except glycine since α-carbon atom is asymmetric in nature.

5. **(1)** Both assertion and reason are true and reason is the correct explanation of assertion.

6. **(1)** Sucrose is dextrorotatory, but when sucrose undergoes hydrolysis it gives dextrotatory glucose and laevorotatory fructose. Its the laevorotation of fructose (–92.4°) is more than dextrorotation of glucose (+52.5°), the overall mixture is laevorotatory.

7. **(1)** Glucose and fructose are differ from each other only at C_1 and C_2 positions (anomeric carbons). According to the mechanism, the first formed phenyl hydrazone undergoes rearrangement through a cyclic intermediate in which the secondary hydroxyl group at C_2 becomes a ketone group and this ketone group then condenses with phenyl hydrazine to form the osazone.

8. **(2)** Insulin is a globular protein. This protein has three-dimensional folded structure. These are stabilised by internal hydrogen bonding. Hence, they are water soluble.

9. **(1)** All amino acids possess both amino and carboxylic group. $-NH_2$ group is basic in nature while –COOH group is acidic in nature. Therefore, they behave as zwitter ion (dipolar ion).

10. **(2)** Due to ring structure, glucose forms two methyl glucosides, i.e., α-and β-methyl glucosides.

Multi Correct MCQs

1. **(3)** Enzymes are reactions specific and one catalyst cannot catalyse more than one reaction.

   Enzyme maltase causes hydrolyses of maltose into glucose.

2. **(4)** In the given list, the essential amino acids are :-

   Arginine, Phenylalanine, Histidine, Valine

   Hence, there are four essential amino acids, and rest are non-essential.

3. **(2)** Cysteine and methionine both are sulphur containing amino acid.

   (B) cysteine 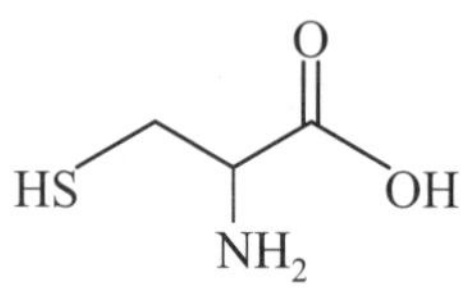 and

   (D) methionine

   $$CH_3-S-CH_2-CH_2-\underset{\underset{NH_2}{|}}{CH}-\overset{\overset{O}{||}}{C}-OH$$

4. **(4)** Vitamins A, D, E and B_{12} are stored in adipose tissue.

NCERT Exemplar MCQs

1. **(2)** Glycogen is a branched chain polymer of α–D glucose units in which chain is formed by C_1 — C_4 glycosidic linkage whereas branching occurs by the formation of C_1 — C_6 glycosidic linkage. Structure of glycogen is similar to the structure of amylopectin.

   Glycogen is also known as animal starch and is present in liver, muscles and brain.

2. **(3)** Cod liver oil is a good source of Vitamin A & D.

3. **(3)** Sucrose (cane sugar) is a disaccharide. One molecule of sucrose on hydrolysis gives one molecule of glucose and one molecule of fructose.

$$\underset{\text{Cane sugar}}{C_{12}H_{22}O_{11}} \xrightarrow[H^+]{H_2O} \underset{\text{glucose}}{C_6H_{12}O_6} + \underset{\text{fructose}}{C_6H_{12}O_6}$$

4. **(3)** Cyclic hemiacetal forms of Monosaccharide which differ only in the configuration of the hydroxyl group at C1 are anomers.

5. **(3)** Secondary structure of protein denotes the shape in which a long polypeptide chain exists. The secondary structure exist in two type of structures i.e., α-helix and β-pleated structure.

   In α-helix structure, a polypeptide chain forms all possible hydrogen bonds by twisting into a right handed screw with —NH group of each amino acid rest hydrogen bonded to C = O of adjacent amino acid, which form a helix.

6. **(2)** This structure represents sucrose in which α-D glucose and β-D-fructose is attached to each other by $C_1 - C_2$ glycosidic linkage.

   Since, reducing groups of glucose and fructose are involved in glycosidic bond formation, this is considered as non-reducing sugar.

7. **(2)** The chemical name of vitamin C is Ascorbic acid. While others are not vitamins, aspartic acid is an amino acid, adipic acid is a dicarboxylic acid having 8 carbon chain.

   Saccharic acid is a dicarboxylic acid obtained by oxidation of glucose using HNO_3.

8. **(1)** Dinucleotides are formed by phosphodiester linkage between 5' and 3' carbon atoms of the pentose sugar.

9. **(2)** Nucleic acids are long chain polymers nucleotides. These are mainly of two types ie., DNA and RNA.

10. **(3)** Glucose is a aldohexose.

    Glucose on heating with HI produces n-hexane.

$$C_6H_{12}O_6 \xrightarrow{HI} C_6H_{14}$$

    Glucose does not give 2, 4-DNP test duc to its existence as cyclic structure as shown below;

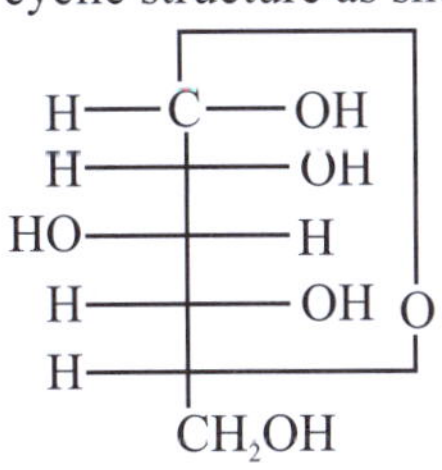

    It is present in pyranose form, as shown below:

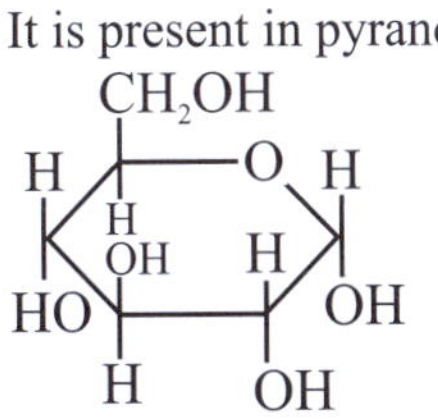

    Pyranose means pyran (six membered ring containing oxygen) like structure.

11. **(1)** Each polypeptide in a protein has amino acids linked with each other in a specific sequence. This sequence of amino acids is said to be the primary structure of proteins.

12. **(3)** Tryptophan

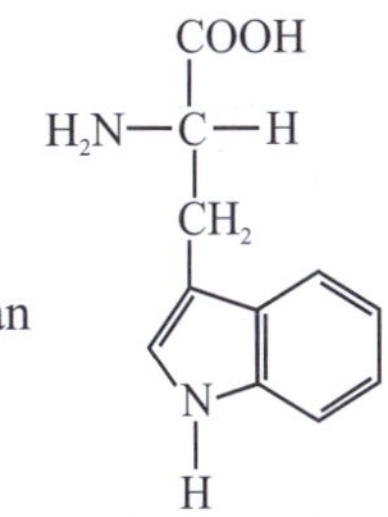

13. **(4)** Vitamin B_{12} can be stored in our body.

14. **(4)** DNA contains following four bases

    (1) Adenine (A) (2) Thymine (T)
    (3) Guanine (G) (4) Cytosine (C)

15. **(1)** Anomer cyclic structures of monosaccharides which differ in structure at carbon-1 are known as anomers. Here, I and II are anomers because they differ from each other at carbon-1 only.

16. **(3)** "Pentaacetate of glucose does not react with hydroxylamine" shows absence of free CHO group. This can not be explained by open structure of glucose. While all other properties are easily explained by open structure of glucose.

17. **(1)** D and L configuration are relative configuration decided by relating structure of given saccharide with D or L glyceraldehyde.

    D configuration is represented when OH on lowest asymmetric carbon is written at right hand side and when OH is written on left hand side, it is represented as L configuration.

18. **(3)** Anomeric carbon is carbon adjacent to oxygen atom in the cyclic structure of glucose or fructose. As shown in the structure, 'a' and 'b' are present at adjacent to oxygen atom. Both carbons differ in configurations of the hydroxyl group.

19. **(3)** Numbering of glucose starts from adjacent carbon to O-atom to the other carbon atom ending at last CH_2OH group.

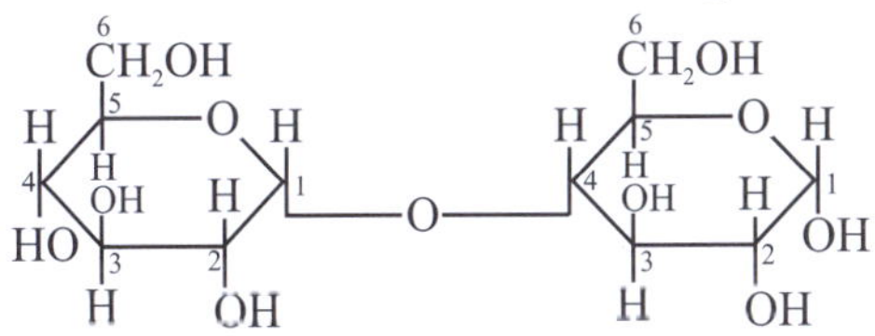

    In this way, numbering for the disaccharides can be done.

NEET & AIIMS Past Year MCQs

1. **(2)** Glycogen is also called animal starch. It is a storage polysaccharide found in animals as well as in yeast and fungi Its structure is similar to that of amylopectin.

2. **(1)** Despite having the aldehyde group, glucose does not give Schiff's test and it does not form the hydrogensulphite addition product with $NaHSO_3$.

3. **(4)** When nucleoside is linked to **phosphoric acid** at 5'-position of sugar moiety, we get nucleotide.

4. **(4)** Except option 4, rest are correct statements. As enzymes are also known as biocatalyst and are specific in nature. Like chemical catalysts, enzymes reduce the activation energy of bio process.

5. **(4)** Vitamin-B_{12}

6. **(1)** In phosphodiester hydrolysis, water is disassociated into H^+ and OH^- The OH^- act as nucleophile in the reaction and the reaction is catalysed by phoshodiesterase.

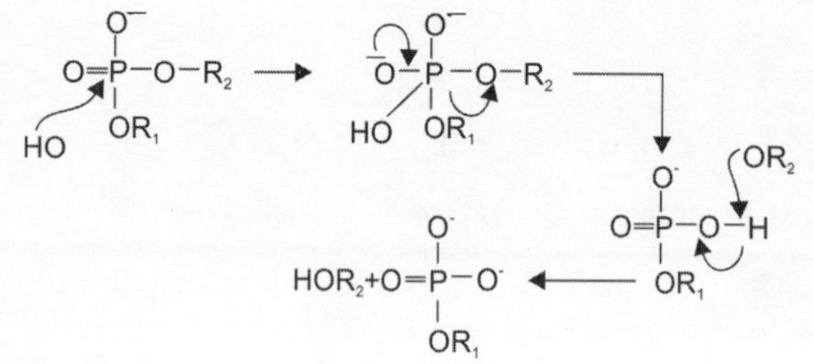

7. **(1)**

8. **(3)**

9. **(3)** The hydrolysis of sucrose by boiling with a mineral acid (HCl), produces a mixture of equal molecules of D-glucose and D-fructose. Sucrose solution is dextrorotatory having specific rotation = + 66.5°. But on hydrolysis, it becomes laevorotatory. The specific rotation of D-glucose is + 52° and of D-fructose is –92°. Therefore, the net specific rotation of an equimolar mixture of both is:

$$\frac{+52° - 92°}{2} = -20°(\text{laevorotatory})$$

Principles Related to Practical Chemistry

Topic-wise MCQs

1. **(4)** An organic compound which produces a bluish green flame when heated on a copper wire is chlorobenzene. This is called Beilstein's test and this test is used to detect the presence of halogen in an organic compound. Green color of the flame is due to the formation of copper halide.

2. **(4)** AgF is highly soluble in H_2O therefore, NaF formed in the Lassaigne's extract does not produce a ppt. with $AgNO_3$.

3. **(4)** Estimation of Sulphur: In this method, a known quantity of organic compounds is heated with either fuming nitric acid or sodium peroxide in a hard glass tube called the Carius tube. Sulphur present in the compound is oxidized to form sulphuric acid.

4. **(1)** Tollen's reagent gives a positive test with all aldehydes. Tollen's reagent oxidizes an aldehyde into the corresponding carboxylic acid. Fehling's test is used to distinguish aldehydes from ketones and aliphatic aldehydes from aromatic aldehydes.

5. **(4)** $NaHCO_3$ will not react with the given compound, because $NaHCO_3$ reacts with strong acidic groups.

6. **(1)** Phenol can be distinguished from any other functional group by reacting it with a neutral $FeCl_3$ solution.

7. **(1)**

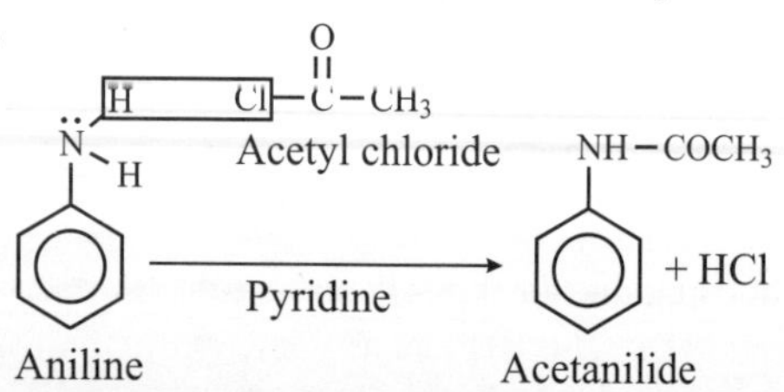

8. **(3)** Aniline Yellow structure

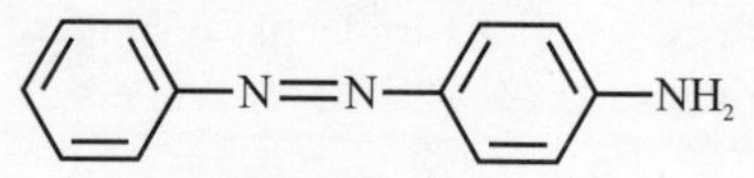

Aniline yellow
(p-Aminoazobenzene)

9. **(4)** Mohr's salt contains 6 molecules of water. Ferrous ammonium sulphate is called Mohr's salt.

Molecular formula of Mohr's salt is $FeSO_4.(NH_4)_2SO_4.6H_2O$. It is a double salt of ammonium sulphate and iron sulphate.

10. **(2)** First, 30 mL HCl is neutralized by 70 mL NaOH. But HCl is the limiting factor here so, overall 30 mL neutralization takes place.

In the 2nd, 40 mL HCl is neutralized by 60 mL NaOH. But HCl is the limiting factor here, so overall 40 mL neutralization takes place.

In the 3rd, 20 mL HCl is neutralized by 80 mL NaOH. But HCl is the limiting factor here so, overall 20 mL neutralization takes place.

In 4th, 80 mL HCl is neutralized by 20 mL NaOH. NaOH is the limiting factor here so, overall 20 mL neutralization takes place.

Heat of neutralization will be more for reaction of 40 mL HCl and 60 mL NaOH.

11. **(2)** HCl(aq) + NaOH(aq) → NaCl(aq) + H2O(l)

From the molarity equation: M_AV_A/n_A (acid) = M_BV_B/n_B (base)

Hence, $M_A = M_BV_Bn_A/n_BV_A$ = 0.45M×83 mL × 1/235 mL × 1 = 0.16M

The molarity of the HCl solution is 0.16M.

12. **(2)** Phenolphthalein is an indicator used for titrations of a weak acid and strong base, and itself is a weak acid. The weak acid is colorless and its ion is bright pink. It distinguishes the pH range from 8 to 9.6. Adding extra hydrogen ions shifts the position of equilibrium to the left, and turns the indicator colorless. Adding hydroxide ions removes the hydrogen ions from the equilibrium which shifts to the right to replace them - turning the indicator pink.

Methyl orange, Methyl red are used for the titration of strong acid and strong base or strong acid and weak base while Bromothymol blue is a dye used as an indicator in determining pH. Bromothymol blue is a weak acid. It can be in acid or base form, depending on the pH of the solution.

13. **(1)** Greater the heat of neutralization, stronger is the acid or more is the acidic strength. Acidic strength of acid proportional to magnitude of $\Delta_{neut}H$.

14. **(1)** KBr, on reaction with conc. H_2SO_4, gives reddish-brown bromine gas.

$KBr + H_2SO_4 \rightarrow KHSO_4 + HBr$,

$2HBr + H_2SO_4 \rightarrow Br_2 + 2H_2O + SO_2$

15. **(3)** $BeO + MgO \xrightarrow{\Delta} MgBeO_2$

$Li_2CO_3 + BeO \xrightarrow{\Delta} Li_2BeO_2 + CO_2\uparrow$

$MgCO_3 + Al_2O_3 \xrightarrow{\Delta} Mg(AlO_2) + CO_2\uparrow$

16. **(1)** Be and Mg have small atomic size and high ionization energies when compared to other alkaline earth metals. The energy of flame is not sufficient to excite the electrons of Be and Mg to higher energy levels. Hence, Be and Mg do not give any color in bunsen flame.

17. (2) (1) $2NaNO_3(s) \xrightarrow[800°C]{\Delta} 2NaNO_2(s) + O_2(g)\uparrow$

$$2NaNO_2(s) \xrightarrow{\Delta} Na_2O(s) + \frac{3}{2}O_2(g) + N_2(g)$$

(2) $MgCO_3(s) \xrightarrow{\Delta} MgO(s) + CO_2(g)$

(3) $2FeSO_4(s) \xrightarrow{\Delta} Fe_2O_3 + SO_2\uparrow + SO_3\uparrow$

(4) $(NH_4)_2Cr_2O_7(s) \xrightarrow{\Delta} Cr_2O_3(s) + N_2(g) + 4H_2O(g)$

18. (3) Pure $ZnCO_3$ is a white powder crystalline. Zinc oxide is thermochromic, changing from white to yellow when heated in air and reverting to white on cooling.

$$\underset{\text{White}}{ZnCO_3} \xrightarrow{\Delta} \underset{\text{Yellow}}{ZnO} + CO_2\uparrow$$

19. (1) Zone of complete combustion is the hottest part of the flame of a bunsen burner.

20. (3) The compound formed in the borax-bead test of Cu^{2+} ion in oxidizing flame is $Cu(BO_2)_2$. It gives green colour in hot oxidizing flame and blue colour in cold oxidizing flame.

21. (1) Confirmation test for CO_3^{2-}

$Na_2CO_3 + H_2SO_4 \rightarrow Na_2SO_4 + H_2O + CO_2\uparrow$

The evolution of colourless and odourless gas (CO_2) indicates the presence of CO_3^{2-}

For S^{2-}:

H_2S is liberated which has a rotten egg like smell.

For NO_3^-:

When a freshly prepared saturated solution of iron (II) sulphate is added to nitrate solution and then concentrated H_2SO_4 is added slowly from side of the test tube, a brown ring is obtained at the junction of two layers.

For Cl^-

On treatment with warm conc. H_2SO_4, colourless pungent smelling gas HCl is evolved.

22. (3) Metallic sulphides react with dil. sulphuric acid to form respective salt and hydrogen sulphide gas is evolved. The gas is colourless having a foul smell of rotten eggs and it turns moist litmus paper red.

$$Na_2S + \underset{\text{Sodium nitroprusside}}{Na_2[Fe(CN)_5NO]} \rightarrow Na_4^{+1}[Fe(CN)_5NOS]^{1}$$

$$\text{or } Na^+ \text{ and } \underset{\text{(Violet color)}}{[Fe(CN)_5NOS]^{4-}}$$

When H_2S gas i.e. hydrogen sulphide gas is passed through cadmium chloride solution, yellow colored precipitate of cadmium sulphide (CdS) is formed and aqueous hydrochloric acid (HCl) is formed.

$H_2S(g) + CdCl_2(aq) \longrightarrow CdS(s) + 2HCl(aq)$

$$\underset{\text{(Hydrogen sulphide)}}{H_2S(g)} + \underset{\text{(Lead acetate)}}{Pb(CH_3COO)_2(aq)} \rightarrow$$

$$\underset{\text{(Lead sulphide)}}{PbS(\downarrow)(black)(s)} + \underset{\text{(Acetic acid)}}{2CH_3COOH(aq)}$$

23. (3) $3Na_2S + K_2Cr_2O_7 + 7H_2SO_4 \rightarrow K_2SO_4 + \underset{\text{Green solution}}{Cr_2(SO_4)_3} + 3S$

$$+ 3Na_2SO_4 + 7H_2O$$

$$S + Pb(NO_3)_2 \rightarrow \underset{\text{Black PPt.}}{PbS} + 2NO_3$$

24. (2) A contains S^{2-} and B contains CO_3^{2-}

$$\underset{(A)}{S^{2-}} \xrightarrow{\Delta} H_2S \xrightarrow{Pb(CH_3COO)_2} \underset{\text{Black PPt.}}{PbS\downarrow}$$

$$\underset{(B)}{CO_3^{2-}} \xrightarrow{\Delta} CO_2 \xrightarrow{Ca(OH)_2} \underset{\text{White turbidity}}{CaCO_3}$$

25. (3) Confirmatory test: Nitrites react with potassium iodide in the presence of dilute sulphuric acid to liberate iodine. Iodine forms a blue-black complex with starch.

$$2KI + 2H_2SO_4 + 2HNO_2 \rightarrow 2KHSO_4 + \underset{\text{Iodine}}{I_2} + 2NO + 2H_2O$$

I_2 + starch → Blue-black complex.

26. (3) Anions that produce a positive test with dilute sulphuric acid and produce a gas during the chemical reaction include carbonate, bicarbonate, thiosulphate, sulphide, nitrite, and sulphite ions.

27. (4) A mixture when rubbed with organic acid smells like vinegar. It contains acetate.

28. (4) The salt is a bromide salt. When it reacts with concentrated sulphuric acid, bromine gas is released which turns starch containing potassium iodide blue.

$Br^- + 2H_2SO_4 \rightarrow Br_2\downarrow + SO_2 + SO_4^{2-} + H_2O$

$2I^- + Br_2 \rightarrow 2Br^- + I_2\downarrow$

I_2 + starch → blue colour solution

29. (3) Nitrate and bromide both give brown fumes when added H_2SO_4.

30. (1) $Cr_2O_7^{2-}$ and Cl^- mixture upon adding concentrated sulphuric acid gives deep red fumes. The reaction is as follows,

$4Cl^- + Cr_2O_7^{2-} + 6H^+ \rightarrow 2CrO_2Cl_2\uparrow$ (deep red) + $3H_2O$

Then, the mixture is added conc. H_2SO_4 gives deep red fumes.

31. (2) Concentrated acid group: NO_3^-, CH_3COO^-, Cl^-, Br^-, I^-.

32. (1) Solubility of $PbSO_4$ in Ammonium acetate:

When $PbSO_4$ is added to a solution of ammonium acetate, the acetate ions form soluble complexes with Pb^{2+} ions.

The complexation reaction is:

$PbSO_4 + 2NH_4C_2H_3O_2 \rightarrow Pb(C_2H_3O_2)_2 + (NH_4)_2SO_4$.

The resulting product, lead acetate, is soluble in water and does not precipitate.

Therefore, $PbSO_4$ is soluble in ammonium acetate solution.

33. (3) Oxalates are strong reducing agents and decolourize acidified $KMnO_4$.

$2KMnO_4 + 3H_2SO_4 \rightarrow K_2SO_4 + 2MnSO_4 + 3H_2O + 5[O]$

$H_2C_2O_4 + O \rightarrow 2CO_2 + H_2O$

34. (1) Brown ring test is a chemical test carried out to detect the presence of nitrate ions in a given sample. It is also known as the nitrate test. It is a common qualitative test performed on any solution which results in the formation of the brown-coloured ring in order to confirm the presence of nitrate ions.

35. (3) When a mixture containing phosphate is heated with conc. HNO_3 and Ammonium Molybdate solution, a canary yellow precipate of Ammonium Phosphomolybdate is formed.

$Na_3PO_4 + 3HNO_3 \rightarrow H_3PO_4 + 3NaNO_3$

$$H_3PO_4 + \underset{\text{Ammonium molybdate}}{12(NH_4)_2MoO_4} + 21HNO_3 \rightarrow$$

$$\underset{\text{Ammonium phosphomolybdate}}{(NH_4)_3PO_4.12MoO_3} + 21NH_4NO_3 + 12H_2O$$

36. (4) H_2SO_4 acts as an oxidising agent. Sulphur in SO_4^{2-} is in its highest oxidising state i.e. +6, so it cannot be further oxidised. Also it has a common ion i.e. SO_4^{2-}, so it will not react and liberate any gas.

37. (3) $Pb^{2+} + HCl \rightarrow PbCl_2 \xrightarrow{\Delta}$ water soluble.

So, the metal ion M^{n+} will be Pb^{+2}

38. (1) $AgNO_3$ is often called Lunar caustic since it creates a burning sensation such as caustic and leaves stain similar to moon luna (black stain while decomposition to metallic silver). This is the reason it is stored in coloured bottles. Nitrate of silver on strong heating leaves the metal as the residue.

39. (2) The salt could be $Pb(NO_3)_2$.

It formed a white precipitate with excess ammonia solution:

$Pb(NO_3)_2 + 2NH_4OH \rightarrow Pb(OH)_2\downarrow$ (white ppt) $+ 2NH_4NO_3$

It formed a white precipitate with dil. NaCl solution:

$Pb(NO_3)_2 + 2NaCl \rightarrow PbCl_2\downarrow$ (white ppt) $+ 2NaNO_3$

It formed a black precipitate with H_2S:

$Pb(NO_3)_2 + H_2S \rightarrow PbS\downarrow$ (black ppt) $+ 2HNO_3$

40. (4) As the compound gives brown fumes, the anion is nitrate and compound reacts with KI and barium chloride then the cation will be Ag^+.

$AgNO_3 \rightarrow Ag + NO_2 + \frac{1}{2}O_2$

$AgNO_3 + KI \rightarrow AgI + KNO_3$;

$BaCl_2 + 2AgNO_3 \rightarrow 2AgCl + Ba(NO_3)_2$

41. (3)

$Pb(NO_3)_2 \xrightarrow{KI} \underset{\text{(Yellow)}}{PbI_2} \xrightarrow[\text{Excess}]{KI(>6M)} \underset{\text{(Soluble)}}{[PbI_4]^{2-}}$

$[PbI_4]^{2-} \xrightarrow{\text{Dilution}} PbI_2$

42. (2) (1)

$Cd(CH_3COO)_2 + 2NaOH \rightarrow 2CH_3COONa + CdO + H_2O$

(2) $Cd(NO_3)_2 + H_2S \rightarrow CdS + 2HNO_3 \xrightarrow{\text{dil HCl}} CdCl_2 + H_2S$

$CdCl_2 + 2NaOH \rightarrow Cd(OH)_2 + 2NaCl$

$Cd(OH)_2 + 2HNO_3 \rightarrow Cd(NO_3)_2 + 2H_2O$

(3) No reaction

(4) $Cu(CN)_2 + NaCN \rightarrow [Cu(CN)_4]^{2-}$

Cu_2S does not form when $[Cu(CN)_4]^{2-}$ reacts with H_2S.

43. (3) In presence of HCl, H_2S results the precipitation of group-2 elements but not group-4 elements due to lower concentration of S^{2-} ion.

44. (2) Sb^{+3} does not form black color sulphide.

45. (4) A lead chloride original solution (i.e., O.S) on mixing with K_2CrO_4 solution gives a yellow precipitate soluble in aqeuous sodium hydroxide.

$Pb^{2+} + CrO_4^{2-} \rightarrow PbCrO_4\downarrow$ (yellow),

$PbCrO_4\downarrow + 4OH^- \rightarrow Pb(OH)_4^{2-} + CrO_4^{2-}$

46. (3) Sb_2S_3 is orange precipitate.

47. (2) $Fe^{3+} + [Fe(CN)_6]^{3-} \rightarrow Fe[Fe(CN)_6]$ brown colouration

$3Fe^{+3} + 6CH_3COO^- + 2H_2O \rightarrow [Fe_3(OH)_2(CH_3COO)_6]^+ + 2H^+$

48. (2) The third group solution should be treated with excess of NH_4Cl. The presence of NH_4Cl in excess reduces degree of dissociation of NH_4OH to a considerable amount, and thus, $[OH^-]$ is just sufficient to cross over the ionic concentration product of hydroxides of III group elements to their K_{sp}.

49. (4) When NH_4Cl is added to NH_4OH solution, concentration of NH_4^+ ions increases so the equilibrium shift towards left. So the dissociation of ammonium hydroxide is reduced.

Concentration of OH^- decreases, due to the common ion effect.

50. (4) $4FeCl_3 + 3K_4(Fe(CN)_6) \rightarrow Fe_4(Fe(CN)_6)_3 + 12KCl$

51. (4) All the given metals form white ppt. with NH_4OH but only ppt. of Zn is soluble in excess of NH_4OH and on passing H_2S it gives white ppt. of ZnS, so the metal is Zn and reactions takes place as follows

$Zn^{2+} + 2NH_4OH \rightarrow \underset{\text{White ppt.}}{Zn(OH)_2\downarrow} + 2NH_4^+$

$Zn(OH)_2 + 2NH_4OH \rightarrow \underset{\text{Soluble}}{(NH_4)_2ZnO_2} + 2H_2O$

$(NH_2)_2ZnO_2 + H_2S \rightarrow \underset{\text{White ppt}}{ZnS} + 2NH_4OH$

52. (1)

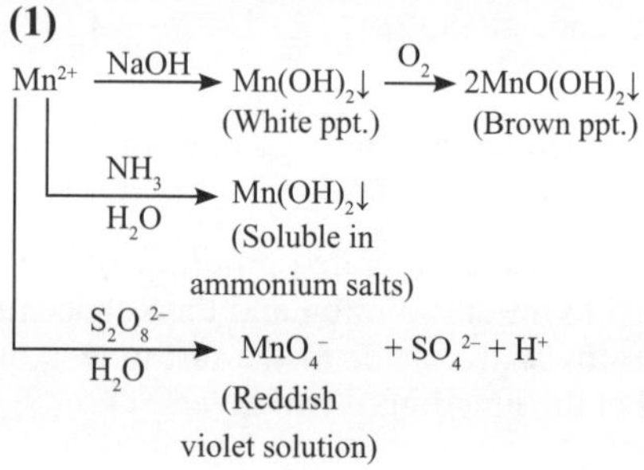

53. (1) $Co^{2+} + 2CN^- \rightarrow Co(CN)_2\downarrow$ (reddish-brown)

$Co(CN)_2\downarrow + 4CN^-$ (excess) $\rightarrow [Co(CN)_6]^{4-}$

(brown solution)

$[Co(CN)_6]^{4-} + 4H^+ \rightarrow Co(CN)_2\downarrow + 4HCN$

54. (4) Co^{2+} will not be precipitated by H_2S in the presence of dil HCl as dil HCl will suppress the ionization of H_2S.

To affect the precipitation, H_2S should be used in the ammoniacal solution.

55. (2) In equilibrium

$Zn^{2+}(aq) + 4NH_3 \rightleftharpoons [Zn(NH_3)_4]^{2+}(aq)$

We have to increase significantly the concentration of free Zn^{2+} ion in solution.

So, we have to make equilibrium shift in reverse direction this can only happen if concentration of reactants are decreased. So, when we add some HCl,

$NH_3 + HCl \rightarrow NH_4Cl$

(Ammonium chloride salt)

Due to salt formation $[NH_3]$ will decrease and more Zn^{2+} ion will start producing in solution.

56. (3) Flame colour of Ba salt is green.

$2AgNO_3 + BaCl_2 \rightarrow \underset{\text{White ppt.}}{2AgCl} + Ba(NO_3)_2$

$H_2SO_4 + BaCl_2 \rightarrow \underset{\text{White ppt.}}{BaSO_4} + 2HCl$

57. **(3)** Pb and Ba both give yellow ppt with K_2CrO_4 in acetic and white precipitate with dil. H_2SO_4. But lead salts give a precipitate with sodium chloride or iodide also whereas barium does not. It means given salt is barium salts and with sodium carbonate, it gives a white precipitate of barium carbonate.

58. **(4)** Na_2CO_3 furnishes appreciable CO_3^{2-} to precipitate Mg.

59. **(3)** Ca, Ba, and Sr ions arc precipitated in fifth group as their carbonates. Group reagent for V group is ammonium carbonate in the presence of NH_4Cl and NH_4OH.

60. **(3)** Nessler's reagent is an alkaline solution of Potassium tetraiodomercurate(II)- $K_2[HgI_4]$.

 On being passed through a solution of ammonia, it gives a dirty brown precipitate. The reaction involved is:

 $2K_2[HgI_4] + 3KOH + NH_3 \rightarrow H_2N.HgO.HgI + 7KI + 2H_2O$

 H_2N-Hg-O-Hg-I is called Iodide of Million's base which is an insoluble dirty brown precipitate. It is also used in test for ammonium ion.

61. **(3)** $Na_3[Co(NO_2)_6] \rightarrow$ Yellow colour

 $2K^+ + Na_3[Co(NO_2)_6] \rightarrow K_2Na[Co(NO_2)_6] + 2Na^+$

 ⇓ Orange colour ⇓ Yellow ppt

62. **(3)** The carbonates of barium, calcium and lead are insoluble in water as they have very low values of solubility products. Sodium carbonate on the other hand is soluble in water. Thus, when potassium carbonate is added to sodium sulphate, sodium carbonate is obtained which is soluble in water. $K_2CO_3 + Na_2SO_4 \rightarrow Na_2CO_3 + K_2SO_4$

63. **(3)** The enthalpy of neutralisation of strong acid and strong base is constant, i.e., -13.7 kcal eq^{-1}

64. **(2)** Egg albumin is lyophilic sol.

65. **(4)** The specific heat capacity of water is used to calculate the heat absorbed or released by the solution. If the specific heat capacity of the solution were different, the calculated enthalpy change would be inaccurate.

66. **(1)** The correct options for the kinetic study of the reaction of the iodide ion with H_2O_2 at room temperature are (A), (B), and (C). The time is recorded immediately after the blue colour appears, and sodium thiosulfate is kept in a limited amount.

67. **(3)** Starch complex formed is blue in colour.

68. **(4)** Lyophilic sols are self stabilising because these sols are reversible and are highly hydrated in the solution.

High Order Time Intensive MCQs

1. **(3)** Dilute H_2SO_4 or dilute HCI are used to identify Group I acidic radicals. Sulphite produces colourless sulphur dioxide gas when it reacts with diluted H_2SO_4.

 When baryta water $Ba(OH)_2$ is treated with this gas, it gives turbidity due to the formation of barium sulphite.

 $Ba(OH)_2 + SO_2 \rightarrow BaSO_3\downarrow + H_2O$

 The solution turns green as chromic sulphate forms when sulphur dioxide is treated with acidified potassium dichromate.

 $K_2Cr_2O_7 + H_2SO_4 + 3SO_2 \rightarrow K_2SO_4 + Cr_2(SO_4)_3 + H_2O$

 The sulphite (SO_3^{2-}) ion is therefore present, as shown by the reaction.

2. **(4)** NH_3 is a colourless gas and a lewis base. It's aqueous solution is alkaline in nature. Hence, when NH_3 is dissolved in water, the resulting solution turns red litmus blue. On the other hand, HCl, H_2S and SO_2 gases are lewis acids. Their aqueous solutions are acidic in nature. When these gases are dissolved in water, the resulting solution turns blue litmus red.

3. **(3)** Sodium thiosulphate will evolve sulphur dioxide gas along with formation of yellowish turbidity when treated with dilute H_2SO_4.

 Adding dilute sulphuric acid to a solution with thiosulphate ions yields a solution with precipitate, elemental sulfur, and release a burnt rubber smell (gaseous sulphur dioxide)

 $2H^+ + S_2O_3^{2-} \rightarrow S + SO_2 + H_2O$

4. **(1)** $2Na_2S_2O_3 + I_2 \rightarrow 2NaI + Na_2S_4O_6$

5. **(4)**

Oxidising Flame		Reducing Flame		Metal
Hot	Cold	Hot	Cold	
Green	Blue	Colourless	Opaque red-brown	Copper
Yellow wish-brown	Yellow	Green	Green	Iron
Yellow	Green	Green	Green	Chromium
Violet (amethyst)	Amethyst	Colourless	Colourless	Mangenese
Blue	Blue	Blue	Blue	Cobalt
Yellow	Colourless	Brown	Brown	Molybdenum
Violet	Reddish -brown	Grey	Grey	Nickel

6. **(1)** KBr, on reaction with conc. H_2SO_4, gives reddish-brown bromine gas.

 $KBr + H_2SO_4 \rightarrow KHSO_4 + HBr,$

 $2HBr + H_2SO_4 \rightarrow Br_2 + 2H_2O + SO_2$

7. **(1)** An inorganic salt, when heated, evolves colored gas which bleaches moist litmus paper. The evolved gas is NO_2.

 Some nitrates on heating give NO_2 which bleaches most litmus paper due to its oxidizing nature.

8. **(2)** NO reacts with $FeSO_4$ to form a brown ring complex, $[Fe(H_2O)_5NO]SO_4$.

9. **(3)** The colourless salt (A) can be ammonium nitrite (NH_4NO_2).

 It reacts with dil sulphuric acid to form nitrous acid which converts KI to iodine which is violet in colour.

 $NH_4NO_2 + H_2SO_4 \rightarrow NH_4SO_4 + 2HNO_2$

 $3HNO_2 \rightarrow HNO_3 + 2NO + H_2O$

 $2NO + O_2 \rightarrow 2NO_2$

 $2KI + 2NO_2 \rightarrow KNO_2 + I_2$

10. **(2)** Ba(II) ion react with potassium chromate to give yellow precipitate of barium chromate.

 $Ba^{2+} + K_2CrO_4 \rightarrow \underset{\text{yellow ppt.}}{BaCrO_4\downarrow} + 2K^+$

 Hence, potassium chromate will identify barium chromate.

11. **(4)** Borax bead test is performed by coloured cations only and as Zn^{2+} is colourless due to the absence of unpaired electron, so it cannot be detected by the borax bead test.

12. **(1)** The brick's red colour is produced by calcium metal in flame. It is due to the excitation of electrons by the thermal energy which is then followed by de-excitation of an electron to ground state along with the liberation of light in the visible range.

13. **(4)** Aluminium metal salts produce a blue coloured bead in the cobalt nitrate charcoal cavity test.

$Al_2O_3.CoO$ formed in the test is blue. It is called thenard's blue.

This precipitate is coloured blue by the absorbing of litmus solution. In this test, the reaction between the aluminium oxide generated in the charcoal cavity test and the cobalt nitrate test results in the creation of Al_2O_3 and a blue mass.

14. **(1)** The reaction is,

$BaCl_2 + H_2SO_3 \rightarrow BaSO_3 + 2HCl$

White ppt makes up this barium sulphite. The acid radical in the equation could SO_3^{2-}.

On dissolving in HCl solution

$BaSO_3 + 2HCl \rightarrow BaCl_2 + SO_2 + H_2O$

SO_2 gas emits a strong odour.

Here, the salt contains is sulphite.

15. **(4)** $5SO_3^{2-} + 2MnO_4^- + 6H^+ \rightarrow 2Mn^{2+} + 5SO_4^{2-} + 3H_2O$

$2MnO_4^- + 5NO_2^- + 6H^+ \rightarrow 2Mn^{2+} + 5NO_3^- + 3H_2O$

$2MnO_4^- + H_2S + 6H^+ \rightarrow 2Mn^{2+} + 5S^{2-} + 8H_2O$

16. **(1)** When KI is added to acidified solution of sodium nitrite NO gas is liberated and I_2 is set free.

$2I^- + 4H^+ + 2NO_2^- \rightarrow 2NO + I_2 + 2H_2O$

17. **(1)** Zinc reacts with acidified solution of sulphite to liberate hydrogen sulphide gas.

Hydrogen sulphide gas turns lead acetate paper black.

$SO_3^{2-} + Zn + 8H^+ \rightarrow H_2S\uparrow + 3Zn^{2+} + 3H_2O$

$Pb^{2+} + S^{2-} \rightarrow PbS\downarrow$ (black)

18. **(2)**

$Cl^- + H_2SO_4 \rightarrow HCl\uparrow$ (colourless)HSO_4^-

$MnO(OH)_2 + 2H_2SO_4 + 2Cl^- \rightarrow Mn^{2+} + Cl_2\uparrow + 2SO_4^{2-} + 3H_2O$

$Cl_2 + H_2O \rightarrow 2HCl + [O]$

Litmus $+[O] \rightarrow$ colourless oxidised form

Cl_2 is a yellowish green gas which bleaches litmus paper by oxidation.

19. **(2)** In chlorine water test for iodine and bromine $\rightarrow$

$2Br^- + Cl_2 \rightarrow 2Cl^- + Br_2\uparrow$

$Br_2 + CS_2, CHCl_3 \rightarrow Br_2$ (in org layer, reddish-brown)

$I_2 + CHCl_3 \rightarrow I_2$ (in org layer violet)

When excess Cl_2 added$\rightarrow$

$Br_2\uparrow + Cl_2\uparrow \rightarrow 2BrCl$(yellow)

$I_3^- + 8Cl_2\uparrow + 9H_2O \rightarrow 3IO_3^- + 16Cl^- + 18H^+$

20. **(2)** Mercuric chloride, $HgCl_2$ will not give positive chromyl chloride test. Chromyl chloride test fails in ease of chlorides of Hg, Sn, Ag, Pb, and Sb.

21. **(2)** A white sodium salt dissolves readily in water to give a solution which is neutral to litmus. When silver nitrate solution is added to the aforementioned solution, a white precipitate is obtained which does not dissolve in dil. nitric acid. The anion is Cl^-.

22. **(1)** Marble ($CaCO_3$) do not react, absorb or dissolve Br_2. As such there is no change in colour of Br_2. Remaining dissolves or absorb bromine dissolve partially in water.

23. **(1)** Unlike the other silver halides, AgF is highly soluble in water.

24. **(2)** CO_3^{2-} radical cannot be confirmed by using dil. HCl.

Carbonates can be easily tested by sulphuric acid. With dil HCl both carbonates and bicarbonates gives same test.

$$S^{2-} + 2HCl \xrightarrow{\Delta} \underset{\text{Rotten egg smell (Specific smell)}}{H_2S\downarrow} + 2Cl^-$$

$$S_2O_3^{2-} + 2HCl \xrightarrow{\Delta} \underset{\text{Suffocating smell of burning sulphur}}{SO_2\uparrow} + 2Cl^- + \underset{\text{Yellow turbidity or white turbidity}}{S\downarrow} + H_2O$$

$$CO_3^{2-} + 2HCl \xrightarrow[\text{or cold}]{\Delta} \underset{\text{Odorless gas evolves with brisk effervescences; } HCO_3^- \text{ also behave similarly}}{CO_2\uparrow} + H_2O + 2Cl^-$$

$$NO_2^- + HCl \xrightarrow{\Delta} HNO_2 + Cl^-$$

$\downarrow$

$NO\uparrow$(colourless)

$\downarrow$ atmospheric air

$NO_2\uparrow$(Brown color gas)

25. **(1)** A white solid imparts a violet color to a bunsen flame. On being heated with concentrated H_2SO_4, the solid gives violet vapors that turn starch paper blue. The salt may be KI as potassium cation imparts a violet color to borax bead test.

26. **(2)** NaCl, NaBr, NaI mixture on adding conc. H_2SO_4 gives HCl, Br_2, I_2 gases, respectively.

HBr and HI are strong reducing agents and are oxidized by H_2SO_4 to Br_2 and I_2, respectively.

$NaCl + \text{conc. } H_2SO_4 \rightarrow Na_2SO_4 + HCl$

$NaBr + \text{conc. } H_2SO_4 \rightarrow Na_2SO_4 + Br_2 + H_2O + SO_2$

$NaI + \text{conc. } H_2SO_4 \rightarrow Na_2SO_4 + I_2 + H_2O + H_2S$

27. **(2)** Potassium chromate solution is added to an aqueous solution of a metal chloride. The precipitates thus obtained are insoluble in acetic acid. These are subjected to flame test; the color of the flame is apple green.

Chromium formed in the borax-bead test in oxidizing flame is $Cu(BO_2)_3$, it gives green color in hot and cold oxidizing flame.

28. **(1)** Aqueous solution of magnesium bicarbonate on boiling gives $MgCO_3 + CO_2 + H_2O$

29. **(3)**

$$\underset{(NaX)}{NaHCO_3} + HCl \rightarrow NaCl + H_2O + \underset{(Y)}{CO_2\uparrow}$$

$$2NaHCO_3 \xrightarrow{\Delta} Na_2CO_3 + H_2O + CO_2\uparrow$$

$$CO_2 + 2KOH \rightarrow K_2CO_3 + H_2O + CO_2\uparrow$$

$$CO_2 + 2KOH \rightarrow K_2CO_3 + H_2O$$

CO_2 is colourless and odourless gas.

30. **(3)** In the mixture of nitrite and nitrate ions, the nitrite ions can be removed by heating with urea and sulphuric acid.

$2NaNO_2 + H_2SO_4 \rightarrow Na_2SO_4 + 2HNO_2$

$2HNO_2 + CO(NH_2)_2 \rightarrow CO_2 + N_2 + 3H_2O$

31. **(4)** Acetic anhydride does not react with N, N-dimethyl aniline.

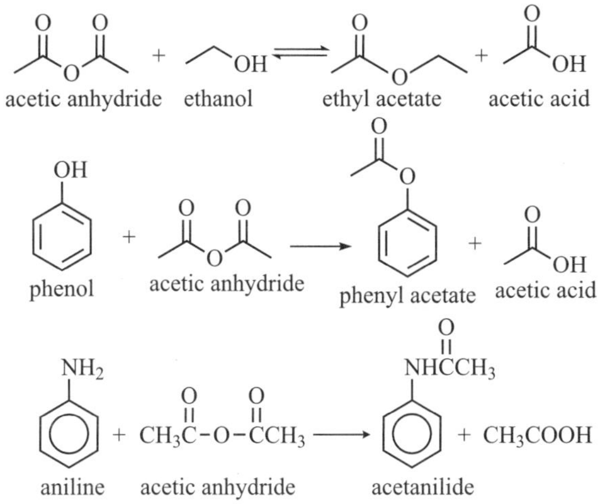

32. **(1)** Acetic acid, CH_3COOH would not respond to iodoform test. In the mechanism of haloform reaction, in the first step, base deprotonates most acidic alpha H atom (of methyl keto group).

In acetic acid, the most acidic proton is attached to O atom. So deprotonation of α-hydrogen does not occur and hence no haloform reaction.

33. **(2)** Cu^{2+} and Ag^+ are both present in the same solution. To precipitate one of the ions and leave the other in solution, add HCl(aq). Ag^+ belongs to I group and get precipitated first by dil. $HCl_{(aq)}$ and then when the filtrate is warmed up and H_2S gas is passed on it leads to the precipitation of Cu^{2+} ions.

$CuCl_2$ is soluble while AgCl is insoluble.

34. **(4)** Among these options, only copper (II) nitrate is soluble in hydrochloric acid. It is an inorganic compound which is soluble in HCl, water, ethanol, ammonia etc. But other three solutions are reacted with hydrochloric acid and there is a formation of respective chloride as precipitate. But when $Cu(NO_3)_2$ adds in HCl, there is no reaction.

35. **(4)** Aniline does not undergo in reaction because the reagent $FeCl_3$ (the lewis acid which is used as a catalyst), being electron deficient acts as a lewis base and attacks on the lone pair of nitrogen present in aniline to form an insoluble complex which precipitates out and the reaction does not proceed.

36. **(3)** Enolic group, shows positive ferric chloride test, and unsaturation will decolourise the bromine water.

37. **(1)** In the Lassaigne's test for nitrogen in an organic compound, the sodium fusion extract is boiled with iron (II) sulphate and then acidified with sulphuric acid.

In the process, sodium cyanide first reacts with iron (II) sulphate and forms sodium hexacyanoferrate (II). Then, on heating with sulphuric acid, some iron (II) gets oxidised to form iron (III) hexacyanoferrate (II), which is Prussian blue in colour. The chemical equations involved in the reaction can be represented as

$6CN + Fe^{2+} \rightarrow [Fe(CN)_6]^{4-}$

$3[Fe(CN)_6]^{4-} + 4Fe^{3+} \xrightarrow{xH_2O} \underset{\text{Prussian blue}}{Fe_4[Fe(CN)_6]_3.xH_2O}$

Hence, the Prussian blue colour is due to the formation of $Fe_4[Fe(CN)_6]_3$.

38. **(1)** $KMnO_4$ is an oxidising agent, Hence salt must contain anion that can be oxidise to higher oxidation state. Hence salt must contain sulphide (SO_3^{2-}) which on oxidation with $KMnO_4$ give (SO_4^{2-}) sulphate and $KMnO_4$ himself reduces to MnO_2 i.e. pink colour of $KMnO_4$ discharge.

$2H^+ + 2MnO_4^- + 3SO_3^{2-} \rightarrow 2MnO_2 + H_2O + 3SO_4^{2-}$

(pink)

39. **(4)** The acidic solution of a salt produced a deep blue colour with starch iodide solution. The salt may be nitrite. Acidic solution of nitrite first forms nitrous acid which decomposes to NO, which is then oxidised to nitrogen dioxide.

$2NaNO_2 + H_2SO_4 \rightarrow Na_2SO_4 + 2HNO_2$

$3HNO_2 \rightarrow HNO_3 + 2NO\uparrow + H_2O$

$2NO + O_2 \rightarrow 2NO_2$

Nitrogen dioxide gas decomposes iodide to free iodine which gives blue colour with starch solution.

$2KI + 2NO_2 \rightarrow 2KNO_2 + I_2$

I_2 + starch $\rightarrow$ starch iodide

40. **(2)** The water will be colored by vapours evolved from the test tube containing bromide.

$Br_2 + H_2O \rightarrow$ Brown color

$NO_2 + H_2O \rightarrow$ Colorless.

Br_2 impart a brown color to the aqueous solution while NO_2 does not.

41. **(4)**

$Co^{2+}(aq) + 4SCN^-(aq) \rightarrow Co(SCN)_4^{2-}(aq)$ppt.

$Fe^{3+} + 3SCN^- \longrightarrow Fe(SCN)_3 \xrightarrow[\text{or } HgCl_2]{NaF}$ colour becomes faded

With potassium thiocyanate, a black precipitate of copper(II) thiocynate

$Cu^{2+} + 2SCN^- \longrightarrow Cu(SCN)_2\downarrow$

The precipitate decomposes slowly to form white copper (I) thiocyanate

$2Cu(SCN)_2\downarrow \longrightarrow \underset{\text{White ppt}}{2CuSCN\downarrow} + (SCN)_2\uparrow$

42. **(4)** Silver (I) chloride is created as a white precipitate when the silver cation (Ag^+) and chlorine anion (Cl^-) combine in the presence of soluble chlorides such hydrochloric acid. In contrast to acids like HNO_3, this precipitate dissolves in aqueous ammonia and forms a complex ion as

$AgCl + NH_3 \rightarrow [Ag(NH_3)_2]^+ + Cl^-$

Here, blue colouration (in solution or precipitate) is not obtained.

43. **(1)** $Pb^{2+} + 2Cl^- \rightarrow PbCl_2\downarrow$ (white),

$PbCl_2 + 2Cl^- \rightarrow [PbCl_4]^{2-}$

(colourless soluble complex)

$$PbCl_2 \downarrow + H_2S \rightarrow PbS \downarrow (black) + 2HCl$$
$$PbS + 4H_2O_2 \rightarrow PbSO_4(white) + 4H_2O$$
$$PbSO_4 \downarrow + 2CH_3COONH_4 \rightarrow (NH_4)_2SO_4 + (CH_3COO)_2Pb$$

44. (2)

$$Cu^{2+} \xrightarrow{SCN^-} Cu(SCN)_2 \downarrow \longrightarrow CuSCN \downarrow + (SCN)_2 \uparrow$$
$$\downarrow H_2S, H^+$$
$$\underset{(Black)}{CuS \downarrow} \xrightarrow[excess]{KCN} [Cu(CN)_4]_3$$

45. (4)

$$4[Ag(NH_3)_2]^+ + H_2N-NH_2 \cdot H_2SO_4 \rightarrow \underset{(Silver\ mirror)}{4Ag \downarrow} + N_2 \uparrow + 6NH_4^+ + 2NH_3 + SO_4^{2-}$$

$$As_2O_3 + 4Zn + 11H^+ \rightarrow AsH_3 \uparrow + 4Zn^{2+}$$
$$4AsH_3 \xrightarrow{\Delta} 4As \downarrow + 6H_2 \uparrow$$

Correct & Incorrect MCQs

1. (1) $S^{-2} + [Fe(CN)_5NO]^{2-} \rightarrow [Fe(CN)_5(NOS)]^{4-}$
↓
Violet colour

2. (1) The correct colour of lead sulphide (PbS) precipitate is actually black, not yellow.

$(CH_3COO)_2Pb + H_2S \rightarrow PbS + 2CH_3COOH$
Lead sulphide
Black precipitate

3. (2) Cu^{2+} ions are more stable in aqueous solutions than Cu^+ ions due to their significantly more negative hydration energy. This higher negative hydration energy for Cu^{2+} more than offsets the energy required for copper's second ionization, making Cu^{2+} the more stable state in an aqueous environment.

4. (1) • Fe^{3+} gives a red colour with potassium thiocyanate

$Fe^{3+} + SCN^- \rightarrow [Fe(SCN)]^{2+}$
Blood red colour

• Fe^{3+} gives blue precipitate with potassium ferrocyanide and Fe^{2+} gives blue precipitate with potassium ferricyanide.

$4FeCl_3 + 3K_4[Fe(CN)_6] \rightarrow Fe_4[Fe(CN)_6]_3 + 12KCl$
Potassium ferrocyanide — Prussian blue precipitate

$Fe^{2+} + K_3Fe(CN)_6 \rightarrow KFe(Fe(CN)_6]$
Turnbull's Blue

5. (2) • Group I (Pb^{2+}) precipitate as chlorides.
• Group II (Pb^{2+}, Cu^{2+}, As^{3+}) precipitate as sulphides.
• Group III (Fe^{3+}, Al^{3+}) precipitate as hydroxides.
• Group IV (Co^{2+}, Ni^{2+}, Mn^{2+}, Zn^{2+}) precipitate as sulphides.
•Group V (Ba^{2+}, Sr^{2+}, Ca^{2+}) precipitate as carbonates.

6. (3) They can be prepared directly.

Statement Based MCQs

1. (1) Starch reacts with Iodine to form an intensely colored blue complex, which is visible at very low concentrations of Iodine, making it a very good indicator in both iodometric and iodimetric titrations.

2. (1)

$AgCl + NH_4OH \rightarrow [Ag(NH_3)_2]^+Cl^-$ ... (*i*)

Now acidifying the product with HNO_3

$[Ag(NH_3)_2]^+Cl^- + HNO_3 \rightarrow \underset{White\ ppt.}{AgCl} + NH_4NO_3$ ... (*ii*)

It is clear from 2nd equation that,
$[Ag(NH_3)_2]^+$ decomposes in presence of HNO_3.
So statement-1 is true, statement 2 is true and statement-2 is correct explanation for statement-1.

3. (1) I^- ion is a strong reducing agent. It reduces Cu^{2+} ion to Cu^+ ion while Cl^- ion is weak reducing agent. Thus it does not reduce Cu^{2+} ion.

$$2Cu^{2+} + 4KI \rightarrow Cu_2I_2 + I_2 + 4K^+$$

4. (1) In electrostatic method, one component should be good conductor and one should be poor or non conductor of electricity. Here PbS is a good conductor and ZnS is a poor conductor of elecricity.

5. (3) Statement I is correct. Since the chlorides of antimony, silver, lead, and mercury are covalent in nature, they do not produce Cl^- ions and do not give chromyl chloride test.

Statement II is incorrect. Ammonia gas evolved by the action of sodium hydroxide on ammonium salts reacts with hydrochloric acid to give ammonium chloride, which is visible as dense white fume.

$(NH_4)_2SO_4 + 2NaOH \longrightarrow Na_2SO_4 + 2NH_3 + 2H_2O$
$NH_3 + HCl \longrightarrow NH_4Cl$

Assertion & Reason MCQs

1. (1) On adding $FeCl_3$ solution to sodium extract during testing for nitrogen a red precipitate is obtained. It is due to the presence of sulphur also.

$Na + C + N + S \longrightarrow NaSCN$
$Fe^{3+} + SCN^- \longrightarrow [Fe(SCN)]^{2+}$
Blood red

2. (2) 3. (2) 4. (1)

5. (1) In acidic solution, the concentration of actual sulphide ion is very low. This means that a quite basic cation is required to react with the available hydrogen sulphide (due to common ion effect).

Copper ions are more basic than zinc ions, so the copper can react with hydrogen sulphide, strip away the hydrogen, and form the insoluble CuS. Zinc is not as basic and cannot strip away the H atoms to form the sulphide.

Multi Correct MCQs

1. (3) The following metal ions will give positive result to flame test.
Sr^{2+}, Ba^{2+}, Ca^{2+}, Cu^{2+}

2. (4) Lead sulphate is a white substance, while ammonium sulphide is generally yellow and soluble in water. Lead iodide is known for its bright yellow colour, and ammonium arsenomolybdate appears yellow as well.

3. (4) The colour of the following compounds is as follows:

$SrSO_4$ – white
$Mg(NH_4)PO_4$ – white
$BaCrO_4$ – yellow
$Mn(OH)_2$ – white
$PbSO_4$ – white
$PbCrO_4$ – yellow

4. (4) Lassaigne's test will not be shown by Hydrazine ($NH_2–NH_2$), since it has no carbon atom. While all the other compounds give Lassaigne's test.

NEET & AIIMS Past Year MCQs

1. (3) In the titration of a weak acid with a strong base, phenolphthalein is employed as an indicator. It remains colorless in an acidic environment but turns pink in an alkaline medium.

2. (4)

The correct match is; A-III, B-IV, C-II, D-I

Lake test	-	Al^{3+}
Nessler's reagent	-	NH_4^+
Potassium sulphocyanide	-	Fe^{3+}
Brown ring test	-	NO_3^-

3. (1)

	Anion	Observation on treatment with dil. H_2SO_4
(I)	Carbonate (CO_3^{2-})	Brisk effervescence of colourless and odourless gas (CO_2)
(II)	Sulphide (S^{2-})	Gas with smell of rotten egg (H_2S)
(III)	Sulphite (SO_3^{2-})	Gas with pungent smell (SO_2)
(IV)	Nitrite (NO_2^-)	Brown fumes (NO_2)

The correct match is; A - II, B-III, C-IV, D-I

4. (3) Below given the ions and the group they belong as per qualitative analysis.

Cation	Group
Cu^{2+}	2
Al^{3+}	3
Co^{2+}	4
Ba^{2+}	5
Mg^{2+}	6

5. (2) Fe^{2+} ion undergoes hydrolysis, therefore, while preparing aqueous solution of ferrous sulphate in water, 2-3 ml of dilute sulphuric acid is added to prevent the hydrolysis of Mohr's salt.

NEET Solved Paper

Year

2025 NEET Solved Paper

1. The ratio of the wavelengths of the light absorbed by a Hydrogen atom when it undergoes n = 2 → n = 3 and n = 4 → n = 6 transitions, respectively, is

1) $\frac{1}{36}$ 2) $\frac{1}{16}$

3) $\frac{1}{9}$ 4) $\frac{1}{4}$

2. Which of the following statements are true?

A. Unlike Ga that has a very high melting point, Cs has a very low melting point.

B. On Pauling scale, the electronegativity values of N and Cl are not the same.

C. Ar, K^+ , Cl^-, Ca^{2+} and S^{2-} are all isoelectronic species.

D. The correct order of the first ionization enthalpies of Na, Mg, Al, and Si is Si > Al Mg > Na.

E. The atomic radius of Cs is greater than that of Li and Rb.

Choose the **correct** answer from the options given below :

1) A, B, and E only 2) C and E only

3) C and D only 4) A, C, and E only

3. Match **List-I** with, **List-II**

List-I (Ion)	**List-II** (Group Number in Cation Analysis)
A. Co^{2+}	I. Group-I
B. Mg^{2+}	II. Group-III
C. Pb^{2+}	III. Group-IV
D. Al^{3+}	IV. Group-VI

Choose the **correct** answer from the options given below :

1) A-III, B-IV, C-II, D-I 2) A-III, B-IV, C-I, D-II

3) A-III, B-II, C-IV, D-I 4) A-III, B-II, C-I, D-IV

4. Predict the major product 'P' in the following sequaence of reations-

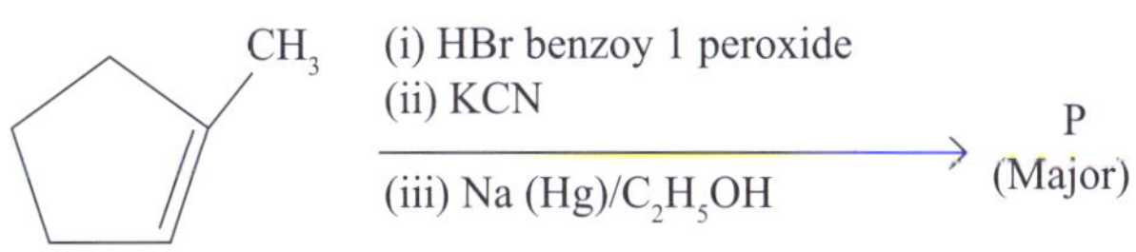

1) 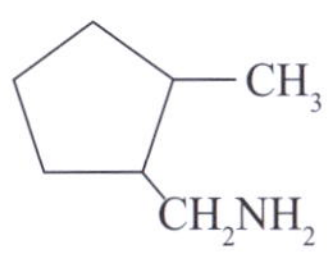

2) 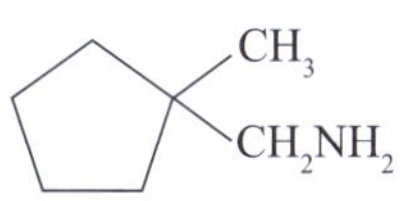

3) 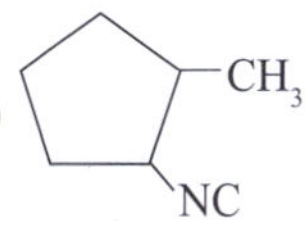

4) 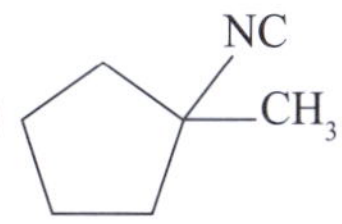

5. Energy and radius of first Bohr orbit of He^+ and Li^{2+} are

[Given $R_H = 2.18 \times 10^{-18}$ J, $a_0 = 52.9$ pm]

1) $E_n(Li^{2+}) = -19.62 \times 10^{-18}$J;
$r_n(Li^{2+}) = 17.6$ pm
$E_n(He^+) = -8.72 \times 10^{-18}$ J;
$r_n(He^+) = 26.4$ pm

2) $E_n(Li^{2+}) = -8.72 \times 10^{-18}$J;
$r_n(Li^{2+}) = 26.4$ pm
$E_n(He^+) = -19.62 \times 10^{-18}$J;
$r_n(He^+) = 17.6$ pm

3) $E_n(Li^{2+}) = -19.62 \times 10^{-16}$ J;
$r_n(Li^{2+}) = 17.6$ pm
$E_n(He^+) = -8.72 \times 10^{-16}$ J;
$r_n(He^+) = 26.4$ pm

4) $E_n(Li^{2+}) = -8.72 \times 10^{-16}$J;
$r_n(Li^{2+}) = 17.6$ pm
$E_n(He^+) = -19.62 \times 10^{-16}$J;
$r_n(He^+) = 17.6$ pm

6. Which of the following are paramagnetic?

A. $[NiCl_4]^{2-}$

B. $Ni(CO)_4$

C. $[Ni CN)_4]^2$

D. $[Ni(H_2O)_6]^{2+}$

E. $Ni(PPh_3)^4$

Choose the **correct** answer from the options given below:

1) A and C only

2) B and E only

3) A and D only

4) A, D and E only

7. Given below are two statements:

Statement I: Like nitrogen that can form ammonia, arsenic can form arsine.

Statement II: Antimony cannot form antimony pentoxide

In the light of the above statements, choose the **most appropriate** answer from the options given below:

1) Both Statement I and Statement II are correct
2) Both Statement I and Statement II are incorrect
3) Statement I is correct but Statement II is incorrect
4) Statement I is incorrect but Statement II is correct

8. Which among the following electronic configurations belong to main group elements?

A. $[Ne]3s^1$ B. $[Ar]3d^34s^2$
C. $[Kr]4d^{10}5s^25p^5$ D. $[Ar]3d^{10}4s^1$
E. $[Rn]5f^06d^27s^2$

Choose the **correct** answer from the option given below:

1) B and E only 2) A and C only
3) D and E only 4) A, C and D only

9. Dalton's Atomic theory could not explain which of the following?

1) Law of conservation of mass
2) Law of constant proportion
3) Law of multiple proportion
4) Law of gaseous volume

10. Consider the following compounds:

$\underline{K}O_2$, $H_2\underline{O}_2$ and $H_2\underline{S}O_4$.

The oxidation states of the underlined elements in them are, respectively,

1) +1, –1, and +6 2) +2, –2, and +6
3) +1, –2, and +4 4) +4, –4, and +6

11. If the half-life ($t_{1/2}$) for a first order reaction is 1 minute, then the time required for 99.9% completion of the reaction is closest to:

1) 2 minutes 2) 4 minutes
3) 5 minutes 4) 10 minutes

12. The correct order of the wavelength of light absorbed by the following complexes is,

A. $[Co(NH_3)_6]^{3+}$ B. $[Co(CN)_6]^{3-}$
C. $[Cu(H_2O)_4]^{2+}$ D. $[Ti(H_2O)_6]^{3+}$

Choose the **correct** answer from the options given below:

1) B < D < A < C 2) B < A < D < C
3) C < D < A < B 4) C < A < D < B

13. Which one of the following compounds can exist as cis-trans isomers?

1) Pent-1-ene
2) 2-Methylhex-2-ene
3) 1, 1-Dimethylcyclopropane
4) 1, 2-Dimethylcyclohexane

14. Phosphoric acid ionizes in three steps with their ionization constant values K_{a_1} K_{a_2} and K_{a_3}, respectively, while K is the overall ionization constant.

Which of the following statements are true?

A. $\log(K) = \log K_{a_1} + \log K_{a_2} + \log K_{a_3}$
B. H_3PO_4 is a stronger acid than $H_2PO_4^-$ and $H_2PO_4^{2-}$
C. $K_{a_1} > K_{a_2} > K_{a_3}$
D. $K_{a_1} = \dfrac{K_{a_3} + K_{a_2}}{2}$

Choose the **correct** answer from the options given below:

1) A and B only
2) A and C only
3) B, C and D only
4) A, B and C only

15. Which one of the following reactions does **NOT** give benzene as the product?

1) 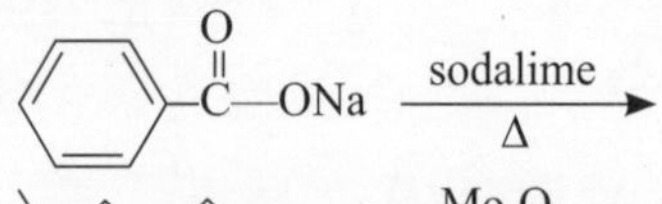

2) 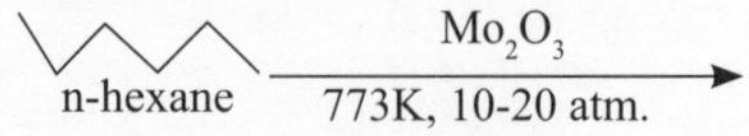

3) $H{-}C{\equiv}C{-}H \xrightarrow[\text{at 873 K}]{\text{red hot Iron Tube}}$

4) 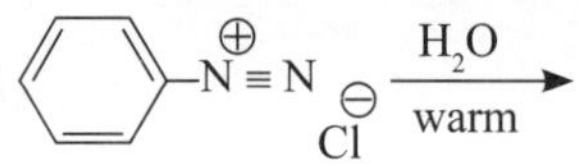

16. If the molar conductivity (Λ_m) of a 0.050 mol L^{-1} solution of a monobasic weak acid is 90 S cm^2 mol^{-1}, its extent (degree) of dissociation will be

[Assume $\Lambda^\circ_+ = 349.6$ S cm^2 mol^{-1} and $\Lambda^\circ_- = 50.4$ S cm^2 mol^{-1}.]

1) 0.115 2) 0.125
3) 0.225 4) 0.215

17. Given below are two statements:

Statement I: A hypothetical diatomic molecule with bond order zero is quite stable.

Statement II: As bond order increases, the bond length increases.

In the light of the above statements, choose the **most appropriate** answer from the options given below:

1) Both Statement I and Statement II are true
2) Both Statement I and Statement II are false
3) Statement I is true but Statement II is false
4) Statement I is false but Statement II is true

18. Out of the following complex compounds, which of the compoundwillbehavingtheminimumconductanceinsolution?

1) $[Co(NH_3)_3Cl_3]$
2) $[Co(NH_3)_4Cl_2]$
3) $[Co(NH_3)_6]Cl_3$
4) $[Co(NH_3)_5Cl]Cl$

19. Match **List-I** with **List-II**

List-I	List-II
A. XeO_3	I. sp^3d; linear
B. XeF_2	II. sp^3; pyramidal
C. $XeOF_4$	III. sp^3d^3; distorted octahedral
D. XeF_6	IV. sp^3d^2; square pyramidal

Choose the **correct** answer from the options given below:

1) A-II, B-I, C-IV, D-III
2) A-II, B-I, C-III, D-IV
3) A-IV, B-II, C-III, D-I
4) A-IV, B-II, C-I, D-III

20. $C(s) + 2H_2(g) \rightarrow CH_4(g)$; $\Delta H = -74.8 kJ\ mol^{-1}$ Which of the following diagrams gives an accurate representation of the above reaction?

[R → reactants; P → products]

1) 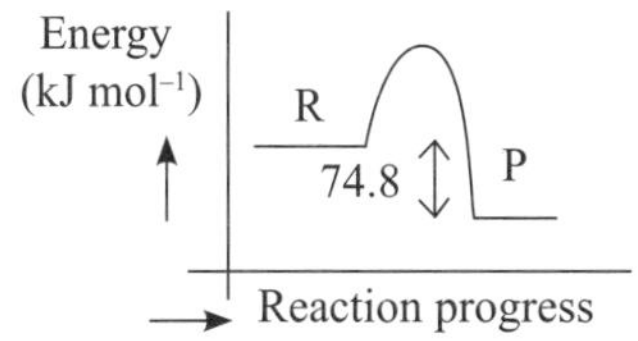

2) 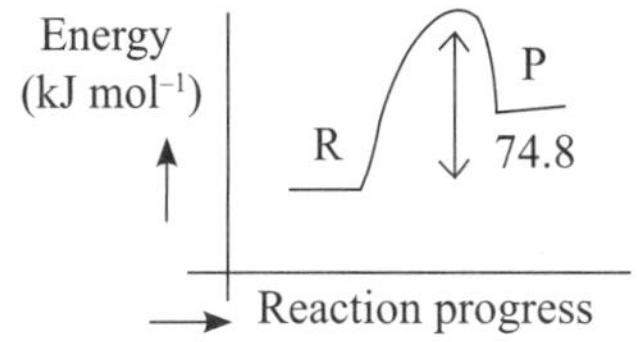

3) 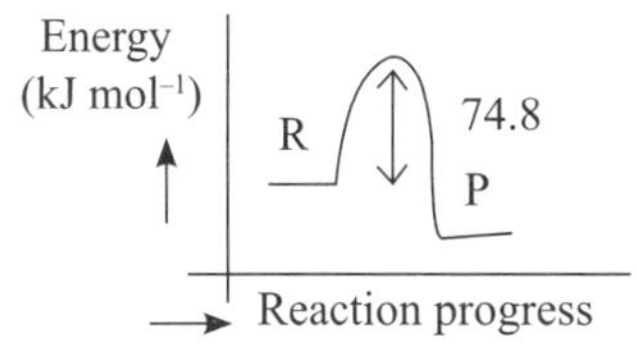

4) 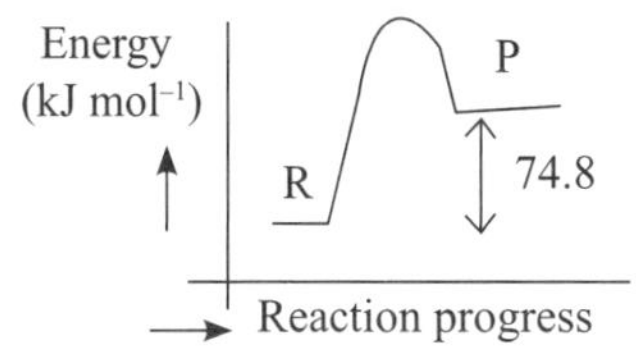

21. Match **List-I** with **List-II**

List-I (Example)	**List-II (Type of Solution)**
A. Humidity	I. Solid in solid
B. Alloys	II. Liquid in gas
C. Amalgams	III. Solid in gas
D. Smoke	IV. Liquid in solid

Choose the **correct** answer from the options given below:

1) A-II, B-IV, C-I, D-III
2) A-II, B-I, C-IV, D-III
3) A-III, B-I, C-IV, D-II
4) A-III, B-II, C-I, D-IV

22. The correct order of deereasing basic strength of the given amines is:

1) N-methylaniline > benzenamine > ethanamine > N-ethylethanamine
2) N-ethylethanamine > ethanamine > benzenamine > N methylaniline
3) N-ethylethanamine > ethanamine > N-methylaniline > benzenamine
4) benzenamine > ethanamine > N-methylaniline > N-ethylethanamine

23. Among the following, choose the ones with equal number of atoms.

A. 212 g of Na_2CO_3 (s) [molar mass = 106g]
B. 248 g of Na_2O (s) [molar mass = 62g]
C. 240 g of NaOH(s) [molar mass = 40g]
D. 12 g of H_2 (g) [molar mass = 2g]
E. 220 g of CO_2 (g) [molar mass = 44g]

Choose the **correct** answer from the options given below:

1) A, B, and C only
2) A, B, and D only
3) B, C and D only
4) B, D, and E only

24. Match **List-I** with **List-II**.

List-I **(Name of Vitamin)**	**List-II** **(Deficiency disease)**
A. Vitamin B_{12}	I. Cheilosis
B. Vitamin D	II. Convulsions.
C. Vitamin B_2	III. Rickets
D. Vitamin B_6	IV. Pernicious anaemia

Choose the correct answer from the options given below:

1) A-I, B-III, C-II, D-IV
2) A-IV, B-III, C-I, D-II
3) A-II, B-III, C-I, D-IV
4) A-IV, B-III, C-II, D-I

25. The correct order of decreasing acidity of the following aliphatic acids is:

1) $(CH_3)_3CCOOH > (CH_3)_2CHCOOH > CH_3COOH > HCOOH$
2) $CH_3COOH > (CH_3)_2CHCOOH > (CH_3)_3CCOOH > HCOOH$
3) $HCOOH > CH_3COOH > (CH_3)_2CHCOOH > (CH_3)_3CCOOH$
4) $HCOOH > (CH_3)_3CCOOH > (CH_3)_2CHCOOH > CH_3COOH$

26. Given below are two statements :

Statement I : Ferromagnetism is considered as an extreme form of paramagnetism.

Statement II : The number of unpaired electrons in a Cr^{2+} ion (Z = 24) is the same as that of a Nd^{3+} ion (Z = 60).

In the light of the above statements, choose the **correct** answer from the options given below:

1) Both Statement I and Statement II are true
2) Both Statement I and Statement II are false
3) Statement I is true but Statement II is false
4) Statement I is false but Statement II is true

27. Match **List-I** with **List-II**

List-I **(Mixture)**	**List-II** **(Method of Separation)**
A. $CHCl_3 + C_6H_5NH_2$	I. Distillation under reduced pressure
B. Crude oil in petroleum industry	II. Steam distillation
C. Glycerol from spent-lye	III. Fractional distillation
D. Aniline water	IV. Simple distillation

Choose the **correct** answer from the options given below:

1) A-IV, B-III, C-I, D-II 2) A-IV, B-III, C-II, D-I
3) A-III, B-IV, C-I, D-II 4) A-III, B-IV, C-II, D-I

28. For the reaction $A(g) \rightleftharpoons 2B(g)$, the backward reaction rate constant is higher than the forward reaction rate constant by a factor of 2500, at 1000 K.

[Given: $R = 0.0831$ L atm mol^{-1} K^{-1}] Kp for the reaction at 1000 K is

1) 83.1 2) 2.077×10^5
3) 0.033 4) 0.021

29. Given below are two statements:

Statement I: Benzenediazonium salt is prepared by the reaction of aniline with nitrous acid at 273-278 K. It decomposes easily in the dry state.

Statement II: Insertion of iodine into the benzene ring is difficult and hence iodobenzene is prepared through the reaction of benzenediazonium salt with KI.

In the light of the above statements, choose the **most appropriate** answer from the options given below:

1) Both Statement I and Statement II are correct
2) Both Statement I and Statement II are incorrect
3) Statement I is correct but Statement II is incorrect
4) Statement I is incorrect but Statement II is correct

30. How many products (including stereoisomers) are expected from monochlorination of the following compound?

$(H_3C)_2CH\text{-}CH_2\text{-}CH_3$

1) 2 2) 3
3) 5 4) 6

31. Among the given compounds I-III, the correct order of bond dissociation energy of C-H bond marked with * is:

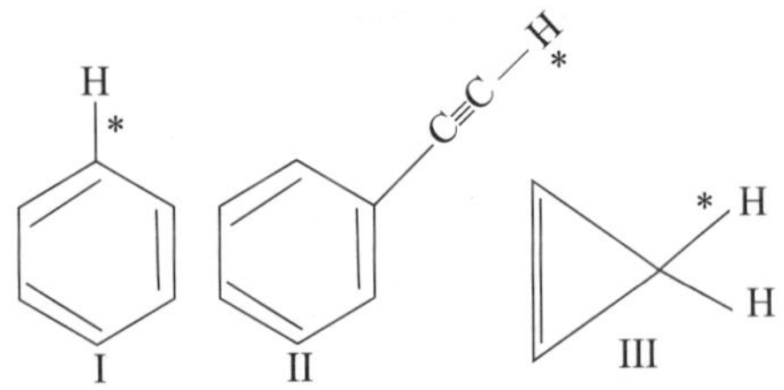

1) II > I > III 2) I > II > III
3) III > II > I 4) II > III > I

32. Which one of the following compounds **does not** decolourize bromine water?

1) 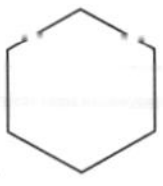 2) 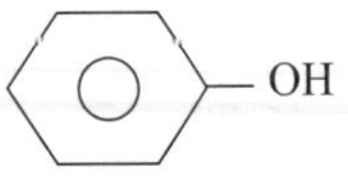

3) 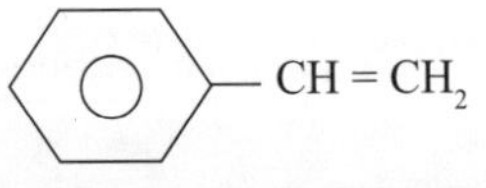 4) 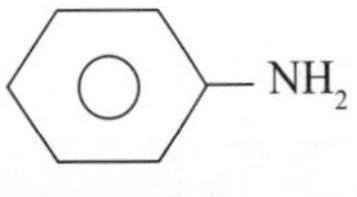

33. The major product of the following reaction is:

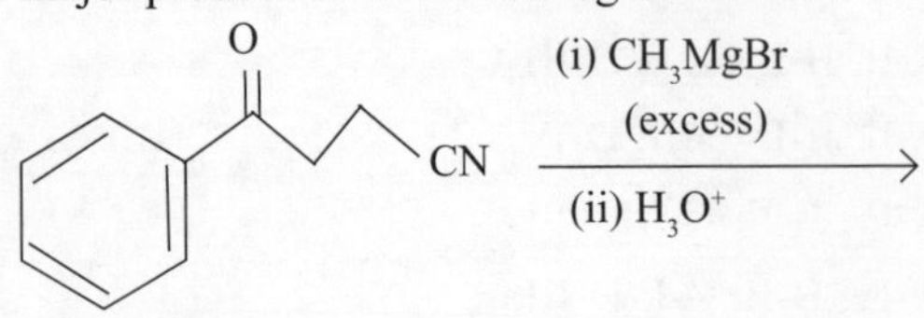

1) 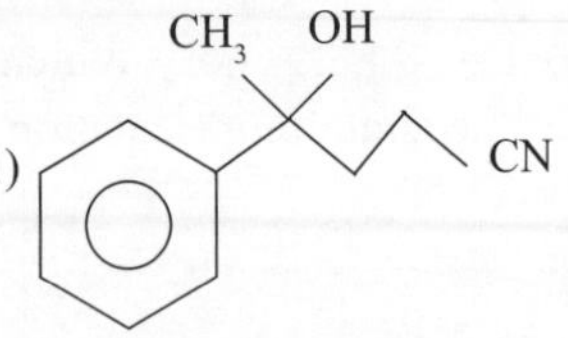

2) 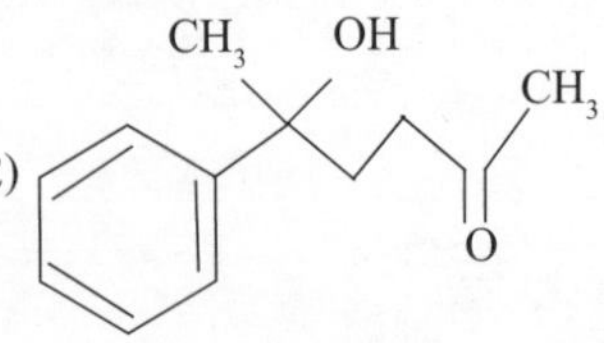

3) 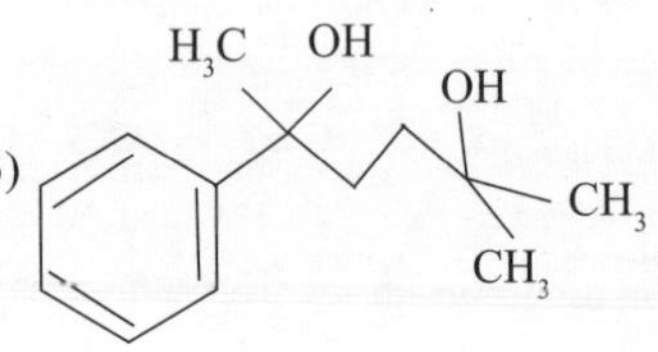

4) 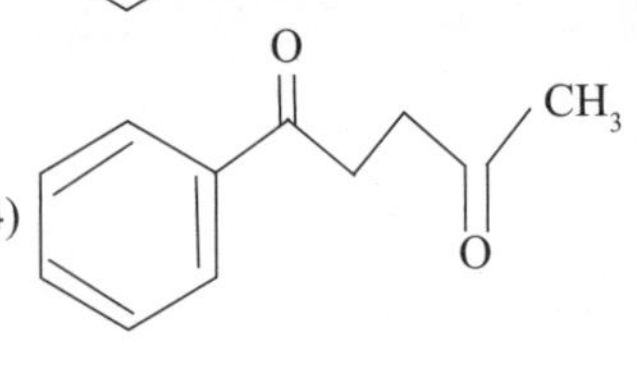

34. Which of the following aqueous solution will exhibit highest boiling point?

1) 0.01 M Urea 2) 0.01 M KNO_3
3) 0.01 M Na_2SO_4 4) 0.015 M $C_6H_{12}O_6$

35. Match **List-I** with **List-II**

List-I	List-II
A. Haber process	I. Fe catalyst
B. Wacker oxidation	II. $PdCl_2$
C. Wilkinson catalyst	III. $[(PPh_3)_3RhCl]$
D. Ziegler catalyst	IV. $TiCl_4$ with $Al(CH_3)_3$

Choose the **correct** answer from the options given below:

1) A-I, B-II, C-IV, D-III
2) A-II, B-III, C-I, D-IV
3) A-I, B-II, C-III, D-IV
4) A-I, B-IV, C-III, D-II

36. 5 moles of liquid X and 10 moles of liquid Y make a solution having a vapour pressure of 70 torr. The vapour pressures of pure X and Y are 63 torr and 78 torr respectively. Which of the following is true regarding the described solution?

1) The solution shows positive deviation.
2) The solution shows negative deviation.
3) The solution is ideal.
4) The solution has volume greater than the sum of individual volumes.

37. Sugar 'X'

A. is found in honey.

B. is a keto sugar.

C. exists in α and β-anomeric forms.

D. is laevorotatory.

'X' is:

1) D-Glucose 2) D-Fructose

3) Maltose 4) Sucrose

38. Identify the suitable reagent for the following conversion.

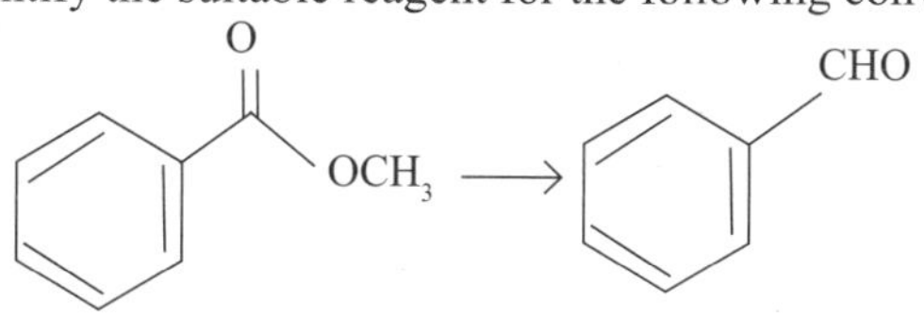

1) (i) $LiAlH_4$, (ii) H^+/H_2O

2) (i) $AlH(iBu)_2$ (ii) H_2O

3) (i) $NaBH_4$, (ii) H^+/H_2O

4) H_2/Pd-$BaSO_4$

39. Given below are two statements : one is labelled as Assertion (A) and the other is labelled as Reason (R).

Assertion (A): 

undergoes S_N2 reaction faster than 

Reason (R): Iodine is a better leaving group because of its large size.

In the light of the above statements, choose the correct answer from the options given below:

1) Both A and R are true and R is the correct explanation of A

2) Both A and R are true but R is not the correct explanation of A

3) A is true but R is false

4) A is false but R is true.

40. The standard heat of formation, in kcal/mol of Ba^{2+} is :

[Given : standard heat of formation of so_4^{2-} ion (aq) = –216 kcal/mol, standard heat of crystallisation of $BaSO_4$(s) = –4.5 kcal/mol, standard heat of formation of $BaSO_4$(s) = –349 kcal/mol]

1) – 128.5 2) – 133.0

3) + 133.0 4) + 220.5

41. Total number of possible isomers (both structural as well as stereoisomers) of cyclic ethers of molecular C_4H_8O is:

1) 6 2) 8

3) 10 4) 11

42. Identify the correct orders against the property mentioned

A. $H_2O > NH_3 > CHCl_3$ – dipole moment

B. $XeF_4 > XeO_3 > XeF_2$ – number of lone pairs on central atom

C. O – H > C – H > N – O – bond length

D. $N_2 > O_2 > H_2$ – bond enthalpy

Choose the **correct** answer from the options given below:

1) A, D only

2) B, D only

3) A, C only

4) B, C only

43. Higher yield of NO in

$N_2(g) + O_2(g) \rightleftharpoons 2NO(g)$ can be obtained at

[ΔH of the reaction = + 180.7 kJ mol^{-1}]

A. higher temperature

B. lower temperature

C. higher concentration of N_2

D. higher concentration of O_2

Choose the **corret** answer from the options given below:

1) A, D Only 2) B,C Only

3) B, C, D only 4) A, C, D, only

44. If the rate constant of a reaction is 0.03 s^{-1}, how much time does it take for 7.2 mo L^{-1} concentration of the reactant to get reduced to 0.9 mol L^{-1}?

(Given: log 2 = 0.301)

1) 69.3s 2) 23.1s

3) 210 s 4) 21.0s

45. Which one of the following reactions does **NOT** belongs to Lassaigne's test?

1) $Na + C + N \xrightarrow{\Delta} NaCN$

2) $2Na + S \xrightarrow{\Delta} Na_2S$

3) $Na + X \xrightarrow{\Delta} + NaX$

4) $2CuO + C \xrightarrow{\Delta} 2Cu + CO_2$

Answer Key

1. (4)	**2.** (2)	**3.** (2)	**4.** (1)	**5.** (1)	**6.** (3)	**7.** (3)	**8.** (2)	**9.** (4)	**10.** (1)
11. (4)	**12.** (2)	**13.** (4)	**14.** (4)	**15.** (4)	**16.** (3)	**17.** (2)	**18.** (1, 2)	**19.** (1)	**20.** (1)
21. (2)	**22.** (3)	**23.** (2)	**24.** (2)	**25.** (3)	**26.** (3)	**27.** (1)	**28.** (3)	**29.** (1)	**30.** (4)
31. (1)	**32.** (1)	**33.** (2)	**34.** (3)	**35.** (3)	**36.** (2)	**37.** (2)	**38.** (2)	**39.** (1)	**40.** (1)
41. (3)	**42.** (1)	**43.** (4)	**44.** (1)	**45.** (4)					

Explanations

1. (4) (NCERT XI, Part-1, Ch-Structure of atom, Page-47)

To find the ratio of wavelengths, we can use the Rydberg formula:

$$1/\lambda = R \times \left(1/n_1{}^2 - 1/n_2{}^2\right)$$

where λ is the wavelength, R is the Rydberg constant, and n_1 and n_2 are the principal quantum numbers.

For the n = 2 to n = 3 transition:

$$\frac{1}{\lambda_1} = R\times\left(\frac{1}{2^2}-\frac{1}{3^2}\right) = R\times\left(\frac{1}{4}-\frac{1}{9}\right)$$

$$= R\times\left(\frac{5}{36}\right)$$

For the n = 4 to n = 6 transition:

$$\frac{1}{\lambda_2} = R\times\left(\frac{1}{4^2}-\frac{1}{6^2}\right) = R\times\left(\frac{1}{16}-\frac{1}{36}\right)$$

$$= R\times\left(\frac{5}{144}\right)$$

Now, let's find the ratio of the wavelengths:

$$\frac{\lambda_1}{\lambda_2} = \left(\frac{\frac{1}{\lambda_2}}{\frac{1}{\lambda_1}}\right) = \left(\frac{\frac{R\times 5}{144}}{\frac{R\times 5}{36}}\right) = \left(\frac{\frac{5}{144}}{\frac{5}{36}}\right) = \frac{36}{144} = \frac{1}{4}$$

So, the ratio of the wavelengths is $\frac{\lambda_1}{\lambda_2} = \frac{1}{4}$

2. (2) (NCERT XI, Part-1, Ch-Classification of Elements and Periodicity in Properties, page-85)

Isoelectronic species contain the same number of electrons. All the elements given in option (3) contain 18 electrons. Down the group, atomic radius increases which means atomic radius of Cs will be greater than both Li and Rb.

3. (2)

A. Co^{2+} = Group IV

B. Mg^{2+} = Group VI

C. Pb^{2+} = Group I

D. Al^{3+} = Group III

4. (1) (NCERT XII, Part-2, Ch-Amines, page-263)

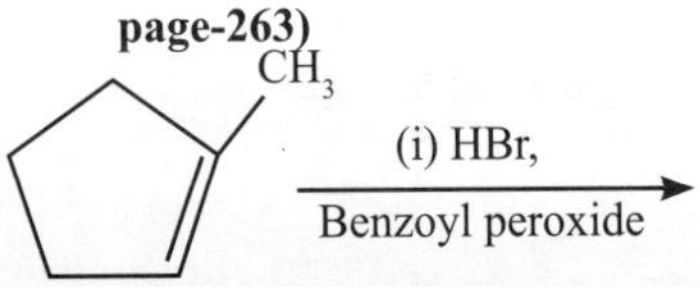

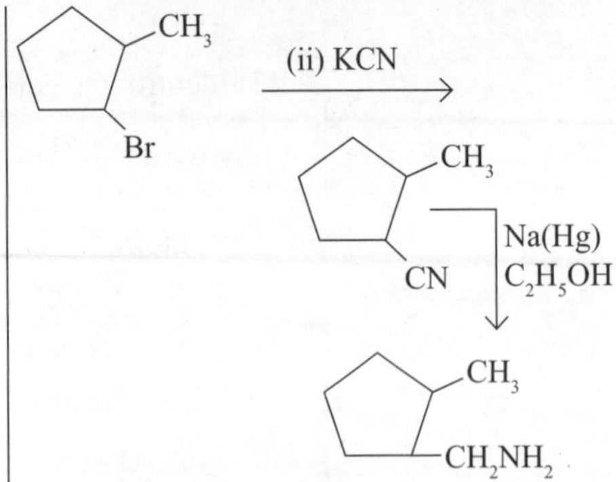

5. (1) (NCERT XI, Part-1, Ch-Structure of atom, Page-47)

Calculations for He^+ (Z = 2)

Radius: $r = \left(\frac{0.529\text{Å}}{2}\right)\times\left(\frac{1\text{Å}}{100\text{pm}}\right) = 26.45\text{pm}$

Energy: E =–54.4 eV × (1.602 × 10^{-19} J/eV) = –8.71=10^{-18}J

Calculations for Li^{2+} (Z = 3)

Radius: $r = \left(\frac{0.529\text{Å}}{3}\right)\times\left(\frac{1\text{Å}}{100\text{pm}}\right) = 17.63\,\text{pm}$

Energy: E = –122.4eV × (1.602 × 10^{-19} J/eV) = –1.96 × 10^{-17}J

6. (3) (NCERT XII, Part-1, Ch-Coordination compounds, page-130)

In $[NiCl_4]^{2-}$, Ni is in +2 oxidation state having electronic configuration $3d^8$. Since Cl^- is a weak field ligand, it will not cause pairing, hence, this complex is paramagnetic having 2 unpaired electrons. In $[Ni(H_2O)_6]^{2+}$, Ni is in +2 oxidation state having electronic configuration $3d^8$. Since H_2O is a weak field ligand, it will not cause pairing, hence, this complex is paramagnetic having 2 unpaired electrons.

7. (3) (NCERT XI, Part-1, Ch-Classification of Elements and Periodicity in Properties, Page-81)

Statement II is incorrect as antimony can form antimony pentoxide, Sb_2O_5.

8. (2) (NCERT XI, Part-1, Ch-Classification of Elements and Periodicity in Properties, Page-81)

Main group elements lie in group 1, 2 and from 13 to 18 respectively.

A. Given electronic configuration represents Na which is a main group element.

B. Given electronic configuration represents V which is a transition element.

C. Given electronic configuration represents I which is a main group element.

D. Given electronic configuration represents Cu which is a transition element.

E. Given electronic configuration represents Th which is an inner-transition element.

9. (4) (NCERT XI, Part-1, Ch-Some basic concepts of chemistry, Page-16)

Dalton's atomic theory could not explain the law of gaseous volume, which relates to the volumes of gases reacting or produced in chemical reactions. This law was later explained by Avogadro's hypothesis, which states that equal volumes of gases at the same temperature and pressure contain an equal number of molecules.

10. (1) (NCERT XI, Part-2, Ch-Redox reactions, page-240)

In KO_2 (potassium superoxide), oxygen has an oxidation state of $\frac{-1}{2}$ (since it's a superoxide, O_2^-). Let's denote the oxidation state of K as x.

The sum of the oxidation states in KO_2 should be zero:

$$x + 2\times\left(\frac{-1}{2}\right) = 0$$

$$x - 1 = 0$$

$$x = +1$$

Let x be the oxidation state of O in H_2O_2.

$$2(+1) + 2x = 0$$

$$2x = -2$$

$$x = -1$$

Let x be the oxidation state of S in H_2SO_4.

$$2(+1) + x + 4(-2) = 0$$

$$2 + x - 8 = 0$$

$$x = +6$$

11. (4) (NCERT XII, Part-1, Ch-Chemical Kinetics, page -76)

$$t_{1/2} = 1\text{min}$$

$$k = \frac{0.693}{\frac{t_1}{2}} = 0.693$$

$$Now,\ t = \frac{2.203}{k}\log\left[\frac{A_o}{A}\right]$$

$$= \frac{2.303}{0.693}\log\left[\frac{100}{0.1}\right] = 10\text{ min}$$

12. (2) (NCERT XII, Part-1, Ch-Coordination compounds, Page-133)

The wavelength of light absorbed by a complex depends on crystal field splitting energy (Δ), which is influenced by:

Ligand strength (from the spectrochemical series: $H_2O < NH_3 < CN^-$)

Oxidation state and size of the metal ion

Geometry of the complex

Tetrahedral complexes have lower Δ than octahedral ones (CFSE is $\sim 4/9$ less).

Compare Δ based on ligand strength and metal properties.

Δ Order:

$[Co(CN)_6]^{3-} > [Co(NH_3)_6]^{3+} > [Ti(H_2O)_6]^{3+} > [Cu(H_2O)_4]^{2+}$

Since, $\lambda \propto 1/\Delta$ the order of wavelength absorbed is: C > D > A > B

13. (4) (NCERT XI, Part-2, Ch-Hydrocarbons, Page- 308)

1. Pent-1-ene: No cis-trans isomerism.

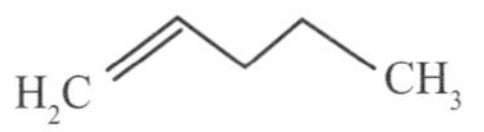

2. 2-Methylhex-2-ene: Can't have cis-trans isomerism.

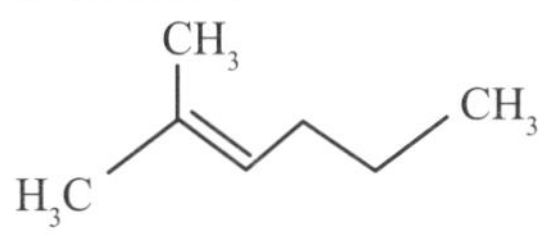

3. 1,1-Dimethylcyclopropane: No cis-trans isomerism because both methyl groups are on the same carbon.

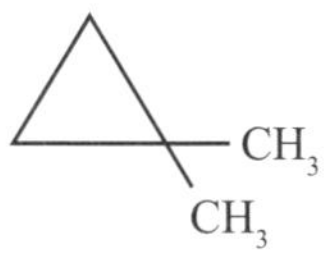

4. 1,2-Dimethylcyclohexane: cis-trans isomerism in cyclohexane is based on axial and equatorial positions.

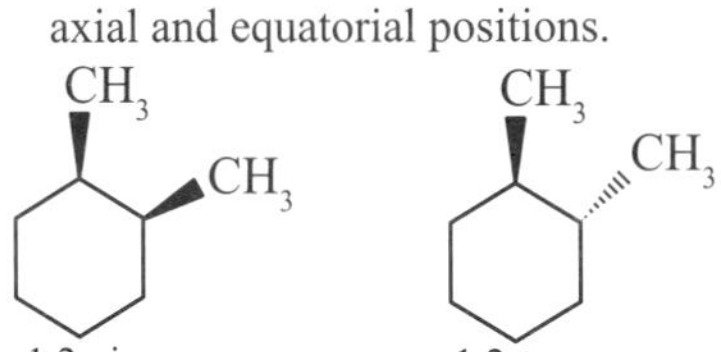

1,2-cis 1,2-trans

14. (4) (NCERT XI, Part-1, Ch-Equilibrium, Page- 200)

Phosphoric acid is a triprotic acid, meaning it can donate three protons in a stepwise manner:

1. $H_3PO_4 \rightleftharpoons H^+ + H_2PO_4^-$

(Ionization constant: K_{a_1})

2. $H_2PO_4^- \rightleftharpoons H^+ + HPO_4^{2-}$

(Ionization constant: K_{a_2})

3. $HPO_4^{2-} \rightleftharpoons H^+ + PO_4^{3-}$

(Ionization constant: K_{a_3})

The ionization constants follow the order:

$K_{a_1} > K_{a_2} > K_{a_3}$,

because it becomes progressively harder to remove a proton from an increasingly negative species.

The overall ionization constant (K) for complete ionization of H_3PO_4 is the product of all three steps:

$K = K_{a_1} \times K_{a_2} \times K_{a_3}$

Taking logarithm on both sides gives:

$\log K = \log K_{a_1} + \log K_{a_2} + \log K_{a_3}$

Among the species H_3PO_4, $H_2PO_4^-$, and HPO_4^{2-}

H_3PO_4 is the strongest acid, as it donates its first proton most readily.

15. (4) (NCERT XII, Part-2, Ch-Amines, Page- 276)

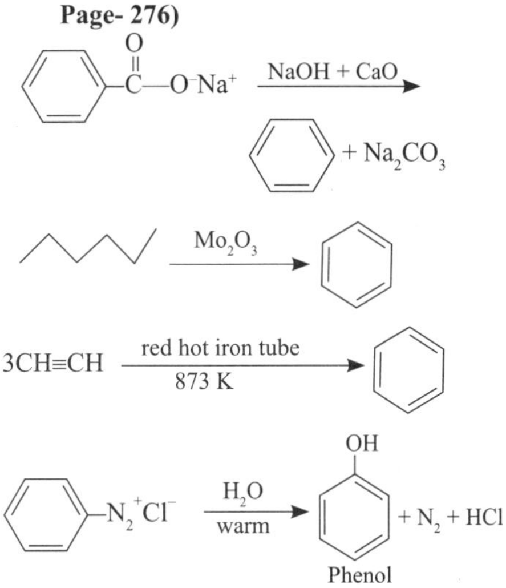

16. (3) (NCERT XII, Part-1, Ch-Electrochemistry, Page- 50)

$$\alpha = \frac{\Lambda_m}{\Lambda_m^0}$$

Where:

Λ_m = molar conductivity at given concentration = 90 S cm² mol⁻¹

Λ_m^0 molar conductivity at infinite dilution

$= \Lambda_{H^+}^0 + \Lambda_A^0$

$= 349.6 + 50.4 = 400$ S cm² mol⁻¹

$$\alpha = \frac{90}{400} = 0.225$$

17. (2) (NCERT XI, Part-1, Ch-Chemical bonding and molecular structures, Page- 129)

Statement I is incorrect;Bond order 0 means no bond exists, so the molecule is unstable and doesn't exist under normal conditions.

Statement II is also incorrect. Higher is the bond order and shorter will be the bond length. Hence, Both Statement I and Statement II are false

18. (1, 2) (NCERT XII, Part-2, Ch-Coordination compounds, Page- 119)

Conductance in solution depends on the number of ions produced upon dissociation.

Option	Complex	Dissociation in Solution	No. of Particles Formed
(a)	$[Co(NH_3)Cl_3]$	No dissociation (neutral complex)	1molecule
(b)	$[Co(NH_3)_4Cl_2]$	No dissociation (neutral complex)	1molecule
(c)	$[Co(NH_3)_6]Cl_3$	$[Co(NH_3)_6]^{3+} + 3Cl^-$	4 ions
(d)	$[Co(NH_3)_5Cl]Cl$	$[Co(NH_3)_5Cl]^+ + Cl^-$	2 ions

The higher the number of molecules, the higher the conductance.

19. (1) A. $XeO_3 \rightarrow sp^3$, pyramidal → II

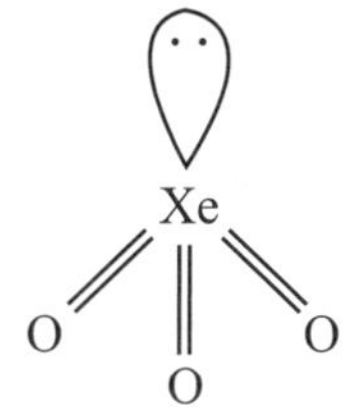

B. $XeF_2 \rightarrow sp^3d$, linear → I

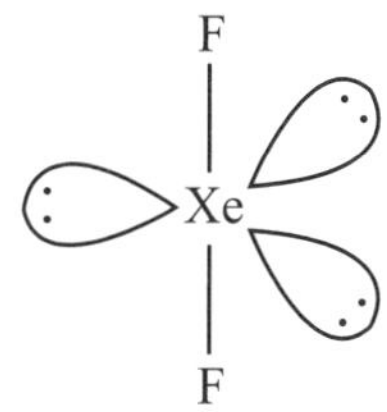

C. $XeOF_4 \rightarrow sp^3d^2$, square pyramidal → IV

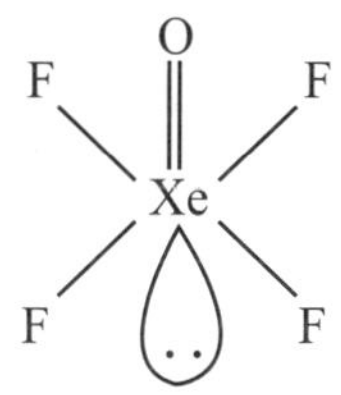

D. $XeF_6 \rightarrow sp^3d^3$, distorted octahedral → III

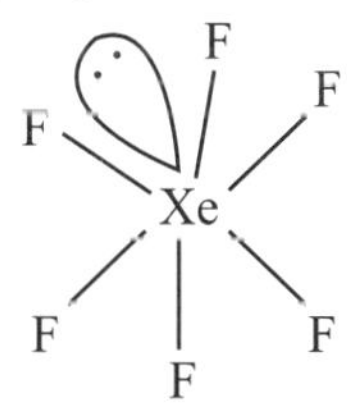

20. (1) (NCERT XII, Part-1, Ch- Chemical kinetics, Page-79)

$C(s) + 2H_2(g) \rightarrow CH_4(g)$

$\Delta H = -74.8$ kJ/mol (Negative sign means the reaction is exothermic)

In an exothermic reaction:

Products (*P*) have lower energy than reactants (*R*)

And the graph for the exothermic reaction is shown as:

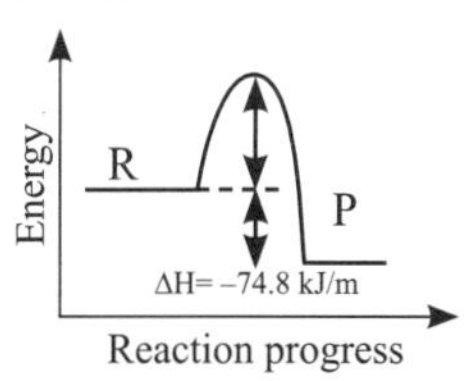

21. (2) (NCERT XII, Part-1, Ch- Solutions, Page- 2)

List-I (Example)	Type of Solution	Explanation
A. Humidity	Liquid in gas (II)	Water vapour in air
B. Alloys	Solid in solid (I)	Metals mixed together (e.g., brass)
C. Amalgams	Liquid in solid (IV)	Mercury with another metal
D. Smoke	Solid in gas (III)	Solid particles dispersed in air

22. (3) [NCERT XII, Part-2, Ch-Amines, Page- 267]

Secondary amines (like N-ethyl ethanamine) are the most basic because two alkyl groups donate electrons, increasing the availability of the nitrogen lone pair.

Primary aliphatic amines (like ethylamine) are also quite basic due to one electron-donating alkyl group.

Aromatic amines (like N-methylaniline and benzenamine) are less basic because the nitrogen lone pair is delocalized into the benzene ring by resonance, making it less available for bonding with H^+.

So, The order of basic strength are:

N-ethylethanamine > ethanamine > N-methylaniline > benzenamine

23. (2) [NCERT XI, Part-1, Ch-Some Basic concepts of chemistry, Page- 18]

A. 212 g of Na_2CO_3

Molar mass = 106 g

Moles $= \frac{212}{106} = 2$mol

Atoms per formula unit

= 2 Na + 1 C + 3 O = 6 atoms

→ Total atoms = 2 × 6 = 12

B. 248 g of Na_2O

Molar mass = 62 g

Moles $= \frac{248}{62} = 4$ mol

Atoms = 2 Na + 1 O = 3 atoms

→ Total atoms = 4 × 3 = 12

C. 240 g of NaOH

Molar mass = 40 g

Moles $= \frac{240}{40} = 6$ mol

Atoms = 1 Na + 1 O + 1 H = 3 atoms

→ Total atoms = 6 × 3 = 18

D. 12 g of H_2

Molar mass = 2g

Moles $= \frac{12}{2} = 6$ mol

Atoms per H_2 = 2 → Total atoms = 6 × 2 = 12

E. 220 g of CO_2

Molar mass = 44g

Moles $= \frac{220}{44} = 5$ mol

Atoms 1 C + 2 O = 3

→ Total atoms = 5 × 3 = 15

Equal total atoms in A, B, D

24. (2) (NCERT XII, Part-2, Ch-Biomolecules, Page- 296)

Vitamin (List I)		Deficiency Disease (List II)	
A.	Vitamin B_{12}	IV.	Pernicious anaemia
B.	Vitamin D	III.	Rickets
C.	Vitamin B_2	I.	Cheilosis
D.	Vitamin B_6	II.	Convulsions

25. (3) (NCERT XII, Part-2, Ch- Aldehydes, Ketones and Carboxylic Acids, Page- 250)

Acidity of acids is directly proportional to the the stability of the carboxylate anion. More the alkyl groups, less stable the carboxylate anion. Hence, more alkyl groups usually make the acid weaker due to electron-donating effects. Based on this, option (3) is most likely correct:

$HCOOH > CH_3COOH > (CH_3)_2CHCOOH > (CH_3)_3CCOOH$.

Formic acid (HCOOH) is the most acidic because it has no alkyl groups, and as alkyl groups increase, the acidity decreases.

26. (3) (NCERT XII, Part-1, Ch- The d- and f-Block Elements, Page- 102)

Statement I is true. Ferromagnetism is indeed a stronger form of paramagnetism, where the magnetic moments of atoms align in the same direction, leading to a stronger and permanent magnetization.

Statement II is false.

Cr^{2+} (Z = 24) has the electron configuration [Ar]$3d^4\ 4s^0$. This means it has 4 unpaired electrons.

Nd^{3+} (Z = 60) has the electron configuration [Xe] $4f^3$. This means it has 3 unpaired electrons.

27. (1) (NCERT XI, Part-2, Ch- Organic Chemistry - Some Basic Principles and Techniques, Page 279)

A-IV, B-III, C-I, D-II

A. $CHCl_3 + C_6H_5NH_2$ (Mixture): Since chloroform ($CHCl_3$) and aniline are quite different in their boiling points, simple distillation would be the appropriate method of separation.

B. Crude Oil: Crude oil is separated into its components using fractional distillation due to its multiple components with different boiling points.

C. Glycerol from spent-lye: Glycerol is typically separated from spent lye in the soap industry using distillation under reduced pressure. This technique is used because glycerol has a high boiling point and decomposes at temperatures close to its boiling point.

D. Aniline + water: Aniline and water can be separated by steam distillation since both are volatile. Steam distillation is a method of purifying or separating substances that are volatile in steam and immiscible with water.

28. (3) (NCERT XI, Part-1, Ch- Equilibrium, Page 178)

The equilibrium constant for this reaction is: $K_c = \frac{k_f}{k_b}$

Given $k_b = 2500k_f$, so:

$$K_c = \frac{k_f}{2500k_f} = \frac{1}{2500} = 0.0004$$

For $A(g) \rightleftharpoons 2\,B(g)$, the change in moles of gas, $\Delta n = 2 - 1 = 1$.

$$K_p = K_c(RT)^{\Delta n}$$

$$K_p = 0.0004 \times 83.1 = 0.03324$$

29. (1) (NCERT XII, Part-2, Ch- Amines, Page 274 and 275)

Both Statement I and Statement II are correct.

30. (4) (NCERT XII, Part-2, Ch- Haloalkanes and Haloarenes, Page 164)

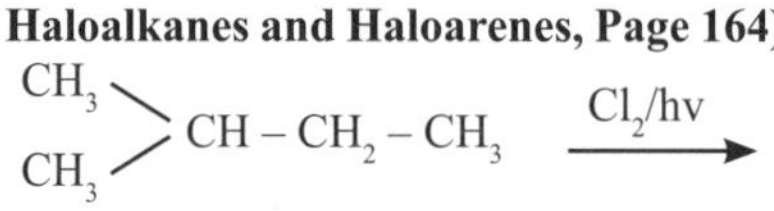

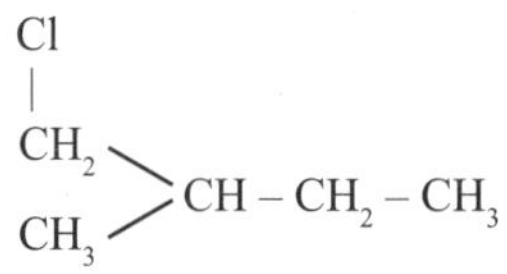

[d + l]

(2 stereoisomers)

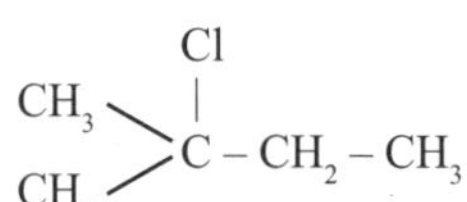

[1 stereoisomers]

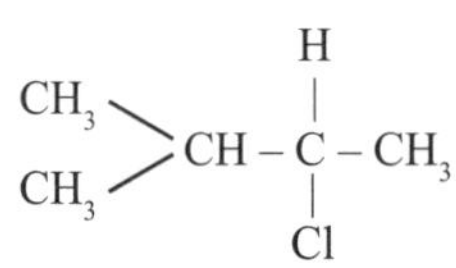

[d + l]

(2- stereoisomers)

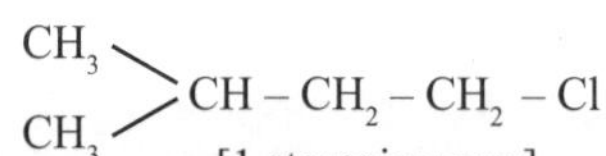

[1 stereoisomers]

Hence, there are total [2 + 1 + 2 + 1 =6] stereoisomers.

31. (1) (NCERT XI, Part-2, Ch- Organic Chemistry - Some Basic Principles and Techniques, Page 272)

Bond dissociation energy (BDE) increases with decrease stability of free radical formed after dissociation of C-H bond fission homolytically.

Stability order: III > I > II

I is more stable as it is resonance stabilised

Hence, Bond Dissociation energy order: II > I > III

32. (1) (NCERT XI, Part-2, Ch- Hydrocarbons, Page 317)

Bromine water is a common qualitative test for determining the presence of unsaturation in organic compounds, particularly alkenes and alkynes. It also reacts with certain functional groups like phenols and anilines, causing a color change that indicates their presence. Cyclohexane, being a saturated hydrocarbon (alkane), does not decolorize bromine water.

33. (2) (NCERT XII, Part-2, Ch- Aldehydes, Ketones and Carboxylic Acids, Page 234)

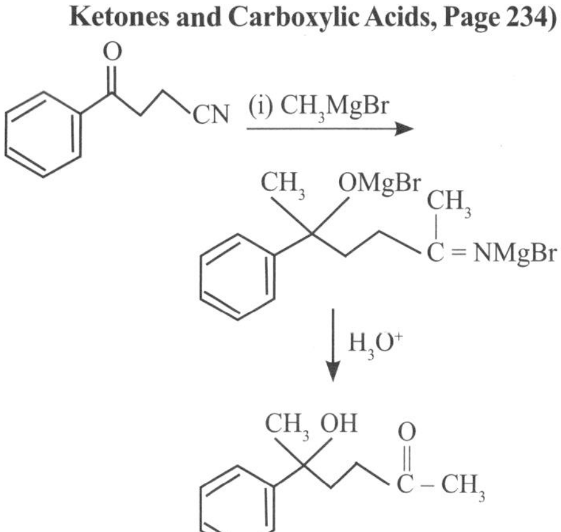

34. (3) (NCERT XII, Part-1, Ch- Solutions, Page 24)

Boiling Point Elevation is a colligative property and depends on the number of particles (ions or molecules) in solution.

More the number of particles, more the boiling point.

Urea: $i \cdot m = 1 \times 0.01 = 0.01$

$KNO_3 : i \cdot m = 2 \times 0.01 = 0.02$

$Na_2SO_4 : i \cdot m = 3 \times 0.01 = 0.03$

Glucose: $i \cdot m = 1 \times 0.015 = 0.015$

The Na_2SO_4 solution has the highest i.m product (0.03), so it will exhibit the highest boiling point.

35. (3) (NCERT XII, Part-1, Ch- The d-and f-Block Elements, Page 104)

A. The catalyst used in the Haber process, which synthesizes ammonia from nitrogen and hydrogen, is iron (Fe).

B. The Wacker oxidation reaction uses a palladium (Pd) catalyst, typically in the form of palladium (II) chloride ($PdCl_2$).

C. Wilkinson's catalyst, formally named chlorotris(triphenylphosphine) rhodium(I), is a coordination complex of rhodium with the formula $[RhCl(PPh_3)_3]$.

D. A common Ziegler-Natta catalyst formula is $TiCl_4 + Al(CH_3)_3$

36. (2) (NCERT XII, Part-1, Ch- Solutions, Page- 13)

According to Raoult's law:-

$P_{solution} = P_x + P_y$

$P_X^o x_X + P_Y^o x_Y$

Given, $n_X = 5$ and $n_Y = 10$

(where, n_X and n_Y represents number of moles of X and Y respectively)

$$\therefore x_X = \frac{n_X}{n_X + n_Y} = \frac{5}{5+10} = \frac{5}{15}$$

$$x_Y = \frac{n_Y}{n_X + n_Y} = \frac{10}{5+10} = \frac{10}{15}$$

Therefore,

$$p_{solution} = 63 \times \frac{5}{15} + 78 \times \frac{10}{15}$$

$$= 21 + 52 = 73 \text{ torr}$$

(Given, $P_X^o = 63$ torr and $P_Y^o = 78$ torr)

However, the observed vapor pressure is 70 torr, which is lower than the calculated value of 73 torr (by Raoult's law). This indicates that the actual vapor pressure is lower than expected, which suggests negative deviation.

37. (2) (NCERT XII, Part-2, Ch- Biomolecules, Page-286)

1. D-Glucose is an aldohexose, not a ketose, and is dextrorotatory. While found in honey, it doesn't fit all the criteria.
2. D-Fructose is a keto sugar, found in honey, exists in alpha and beta anomeric forms, and is laevorotatory.
3. Maltose is a disaccharide made up of two glucose molecules, not a keto sugar.
4. Sucrose is a disaccharide composed of glucose and fructose. It is not a keto sugar (since it is a non-reducing sugar).

38. (2) (NCERT XII, Part-2, Ch- Aldehydes, Ketones and Carboxylic Acids, Page-232)

1. $LiAlH_4$ reduces esters to alcohols.
2. DIBAL-H (diisobutylaluminium hydride) reduces esters to aldehydes.

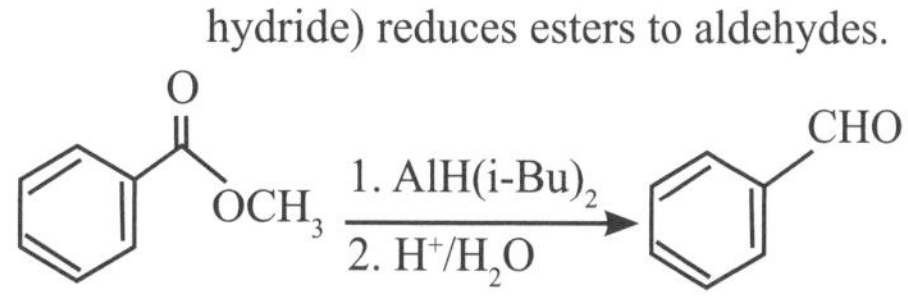

3. $NaBH_4$ does not reduce esters.
4. H_2/Pd-$BaSO_4$ converts acyl chloride to aldehydes.

39. (1) (NCERT XII, Part-2, Ch- Haloalkanes and Haloarenes, Page-174)

 (Iodobutane) undergoes S_N2 reaction faster than 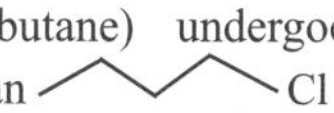 (chlorobutane) because iodine is a better leaving group than Cl due to its large size.

40. (1) (NCERT XI, Part-1, Ch- Thermodynamics, Page-151)

(1) Given:

Standard heat of formation of

SO_4^{2-} ion $(aq) = -216$ kcal / mol

Standard heat of crystallization of

$BaSO_4(s) = -4.5$ kcal / mol

Standard heat of formation of

$BaSO_4(s) = -349$ kcal / mol

The standard heat of formation of Ba^{2+} (aq) is related to the heat of formation of $BaSO_4$(s) by the following process:

$$Ba^{2+}(aq) + SO_4^{2-}(aq) \rightarrow BaSO_4(s)$$

Using Hess's Law:

$\Delta H_{\text{formation of } Ba^{2+}}$

$= \Delta H_{\text{formation of } BaSO_4(s)}$

$-\Delta H_{\text{formation of } SO_4^{2-}(aq)}$

$-\Delta H_{\text{crystallization of } BaSO_4(s)}$

Now, plug in the values:

$$\Delta H_{\text{formation of } Ba^{2+}} = (-349 \text{kcal / mol}) - (-216 \text{kcal / mol}) - (-4.5 \text{kcal / mol})$$

$$\Delta H_{\text{formation of } Ba^{2+}} = -349 + 216 + 4.5 = -128.5 \text{kcal / mol}$$

41. (3) (NCERT XI, Part-2, Ch- Organic Chemistry - Some Basic Principles and Techniques , Page-270)

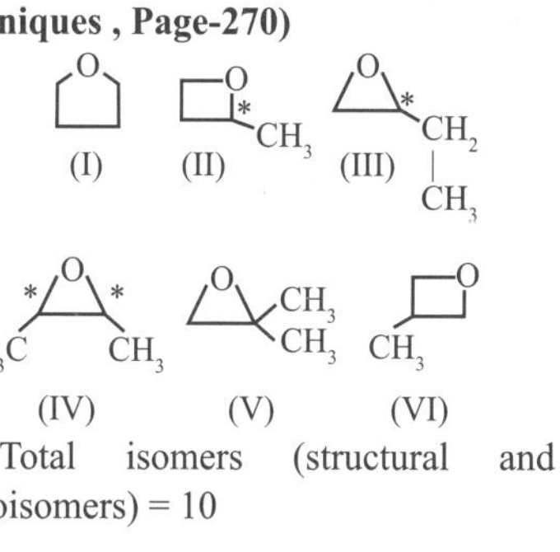

∴ Total isomers (structural and stereoisomers) = 10

42. (1) (NCERT XI, Part-1, Ch-Chemical Bonding and Molecular Structure, Page-108, 112 & 115)

A. Correct

The correct order of dipole moment is: H_2O (1.85 D) > NH_3 (1.47 D) > $CHCl_3$ (1.04 D)

B. Incorrect

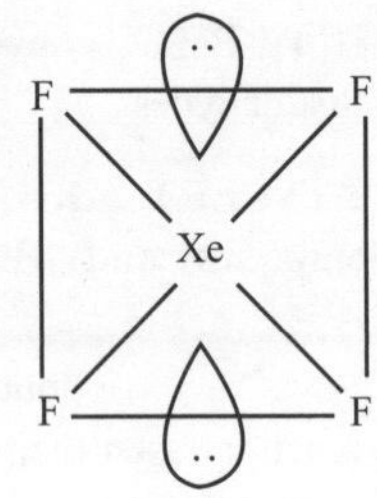

(XeF_4)

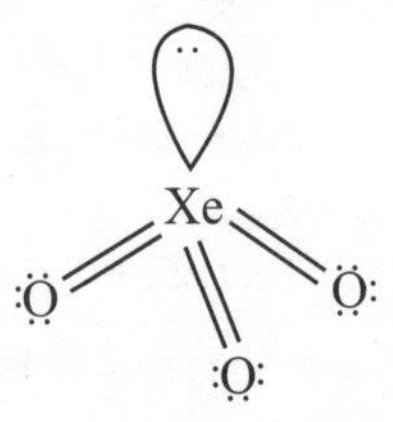

(XeO_3)

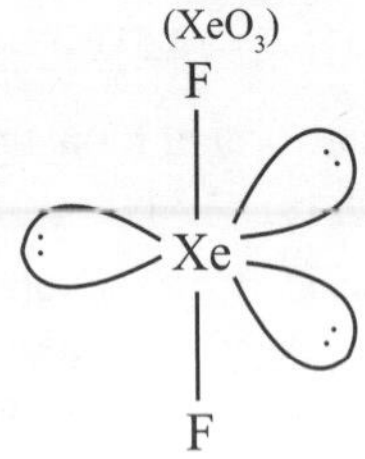

(XeF_2)

Therefore, the correct order with number of lone pairs on central atom is: XeF_2 (3 *lp*) > XeF_4 (2 *lp*) > XeO_3 (1 *lp*)

C. Incorrect

The correct order of bond length is :

N–O (136 pm) > C-H (107 pm) > O-H (96 pm)

D. Correct

N_2 has triple bond, O_2 has double bond and H_2 has single bond The correct order of bond enthalpy is:

$N_2 > O_2 > H_2$

43. (4) (NCERT XI, Part-1, Ch- Equilibrium, Page-185, 187)

Since the reaction is endothermic (as the value of ΔH is positive), therefore, increasing the temperature will shift the equilibrium towards the formation of more NO (the forward direction). Hence, a higher temperature will result in a higher yield of NO. According to Le Chatelier's Principle, increasing the concentration of reactants (i.e., N_2 and O_2) will shift the equilibrium to the right, producing more NO.

44. (1) (NCERT XII, Part-1, Ch- Chemical Kinetics, Page-73)

For first order reaction,

$$kt = 2.303 \log \frac{[A]_0}{[A]_t}$$

$$\Rightarrow t = \frac{2.303}{k} \log \frac{[A]_0}{[A]_t}$$

$$= \frac{2.303}{0.03} \log \frac{[7.2]}{[0.9]}$$

$$= \frac{2.303}{0.03} \log 8$$

$$= \frac{2.303}{0.03} \log (2)^3$$

$$= \frac{2.303}{0.03} \times 3 \log 2$$

$$= \frac{2.303}{0.03} \times 3 \times 0.301$$

$$= 69.3 \text{ sec}$$

45. (4) (NCERT XII, Part-2, Ch- Organic Chemistry – Some Basic Principles and Techniques, Page-284)

Nitrogen, sulphur, halogens and phosphorus present in an organic compound are detected by Lassaigne's test. Hence, the given below reactions belong to the Lassaigne's test:

$$Na + C + N \xrightarrow{\Delta} NaCN$$

$$2Na + S \xrightarrow{\Delta} Na_2S$$

$$Na + X \xrightarrow{\Delta} Na\,X$$

(X = Cl, Br or I)